The Ultimate Digital Study Tool

Self-Test

A comprehensive quizzing section lets you test your ability to identify anatomical structures in a simulated practical exam. Additional quizzes provide review of structure functions and features.

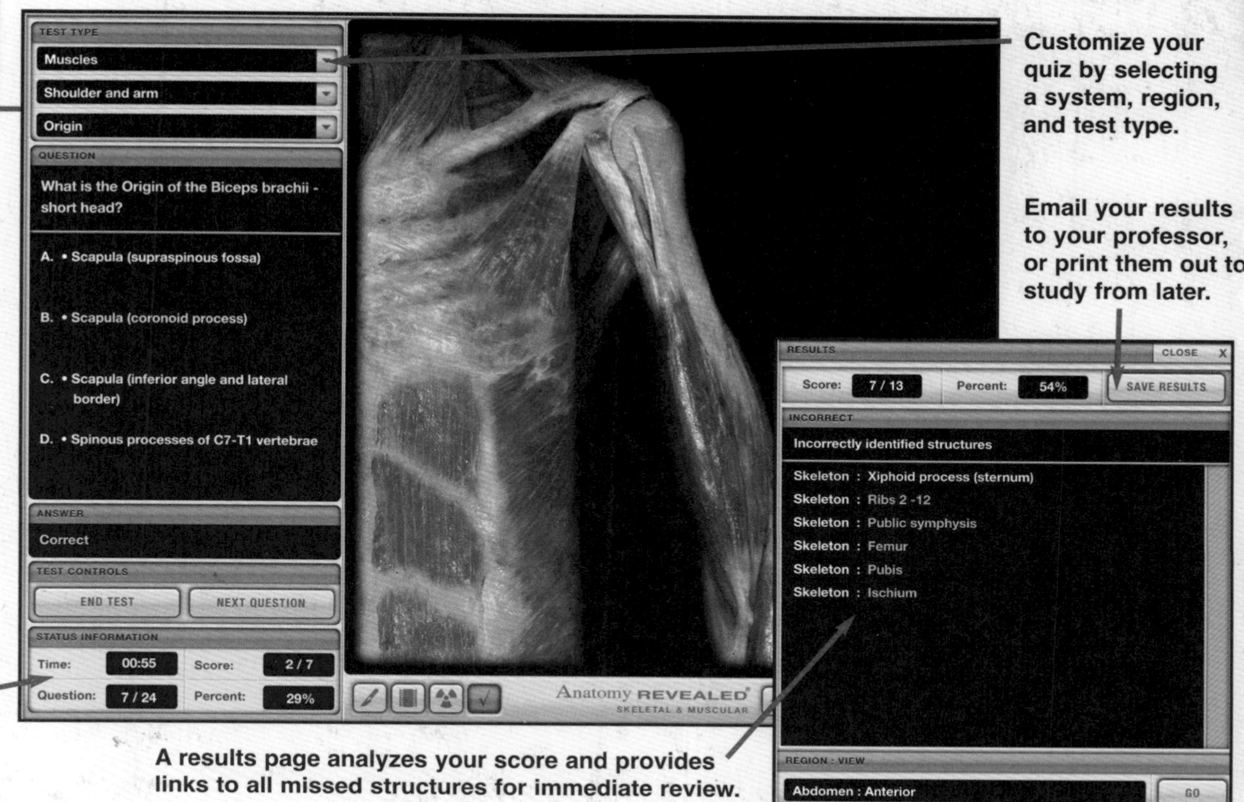

Customize your quiz by selecting a system, region, and test type.

Email your results to your professor, or print them out to study from later.

Diagnostics displayed throughout the test track your progress.

A results page analyzes your score and provides links to all missed structures for immediate review.

Animation

Compelling animations help clarify anatomical relationships or explain difficult physiological concepts.

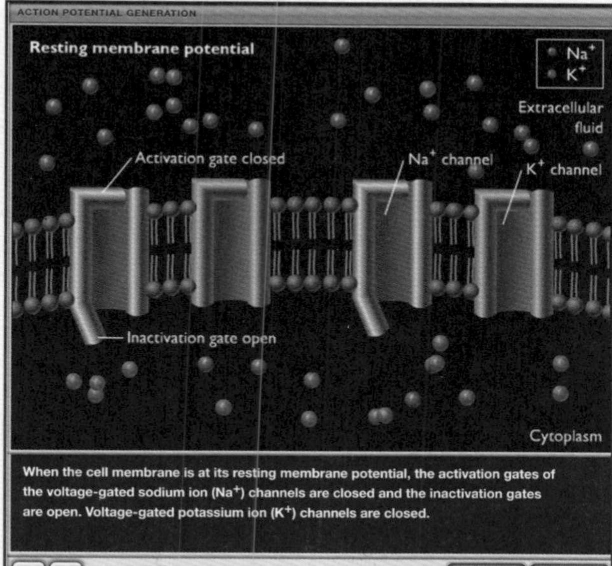

Imaging

Labeled X-ray, MRI, and CT images help you learn to recognize key anatomical structures as seen through various medical imaging techniques.

Pin and label important structures.

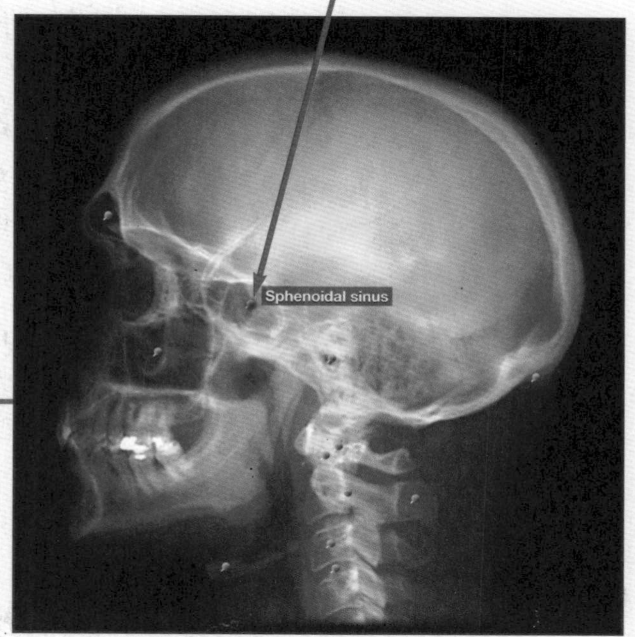

ANATOMY & PHYSIOLOGY

EIGHTH EDITION

ROD R. SEELEY
IDAHO STATE UNIVERSITY

TRENT D. STEPHENS
IDAHO STATE UNIVERSITY

PHILIP TATE
PHOENIX COLLEGE

CONTRIBUTIONS BY:

Shylaja R. Akkaraju
Bronx Community College

Christine M. Eckel
Salt Lake Community College

Jennifer L. Regan
University of Southern Mississippi

Andrew F. Russo
University of Iowa

Cinnamon L. VanPutte
Southwestern Illinois College

Boston Burr Ridge, IL Dubuque, IA New York San Francisco St. Louis
Bangkok Bogotá Caracas Kuala Lumpur Lisbon London Madrid Mexico City
Milan Montreal New Delhi Santiago Seoul Singapore Sydney Taipei Toronto

The McGraw-Hill Companies

ANATOMY & PHYSIOLOGY, EIGHTH EDITION

Published by McGraw-Hill, a business unit of The McGraw-Hill Companies, Inc., 1221 Avenue of the Americas, New York, NY 10020.

This book is printed on recycled, acid-free paper containing 10% postconsumer waste.

3 4 5 6 7 8 9 0 QPD/QPD 0 9 8

ISBN 978–0–07–296557–5
MHID 0–07–296557–6

Publisher: *Michelle Watnick*
Senior Sponsoring Editor: *James F. Connely*
Director of Development: *Kristine Tibbetts*
Senior Developmental Editor: *Kathleen R. Loewenberg*
Marketing Manager: *Lynn M. Breithaupt*
Lead Project Manager: *Mary E. Powers*
Senior Production Supervisor: *Laura Fuller*
Senior Media Project Manager: *Tammy Juran*
Lead Media Producer: *John J. Theobald*
Designer: *Rick D. Noel*
Cover Designer: *Terry Julien*
Interior Designer: *Elise Lansdon*
Cover Illustration: *Scott Holladay*
Senior Photo Research Coordinator: *John C. Leland*
Photo Research: *Jerry Marshall*
Compositor: *Techbooks*
Typeface: *10/12 Minion*
Printer: *Quebecor World Dubuque, IA*

The credits section for this book begins on page C-1 and is considered an extension of the copyright page.

Library of Congress Cataloging-in-Publication Data

Seeley, Rod R.
Anatomy & physiology / Rod R. Seeley, Philip Tate, Trent D. Stephens. – 8th ed.
p. cm.
Includes index.
ISBN 978–0–07–296557–5 — ISBN 0–07–296557–6 (hard copy : alk. paper)
1. Human anatomy. I. Tate, Philip. II. Stephens, Trent D. III. Title. IV. Title: Anatomy and physiology.
QP34.5.S4 2008
612--dc22

2006102703

www.mhhe.com

DEDICATION

This text is dedicated to the students of human anatomy and physiology. Helping students develop a working knowledge of anatomy and physiology is a satisfying challenge, and we have a great appreciation for the effort and enthusiasm of so many who want to know more. It is difficult to imagine anything more exciting, or more important, than being involved in the process of helping people learn about the subject we love.

ABOUT THE AUTHORS

Rod Seeley, Trent Stephens, and Phil Tate in Dubuque, Ia, where they met to discuss the plan for the eighth edition. The bluffs of the Mississippi River can be seen in the background.

ROD R. SEELEY

Professor of Physiology at Idaho State University

Rod has extensive experience teaching introductory biology, anatomy and physiology, pathobiology, endocrinology, and more advanced physiology courses. He has won numerous teaching awards and is actively involved in the supervision of doctoral students in biological education. With a B.S. in zoology from Idaho State University and an M.S. and Ph.D. in zoology from Utah State University, Rod has built a solid reputation as an author of journal and other professionally related articles, as well as a public lecturer.

TRENT D. STEPHENS

Professor of Anatomy and Embryology at Idaho State University

An award-winning educator and researcher, Trent Stephens teaches human anatomy, human head and neck anatomy, and human embryology. He also has many years of experience teaching neurobiology. His skill as a biological illustrator has greatly influenced the illustrations in this textbook. He has a B.S. in microbiology and a B.S. in zoology, as well as an M.S. in zoology from Brigham Young University. His Ph.D. in anatomy is from the University of Pennsylvania. Trent is actively involved in research on limb development and birth defects caused by thalidomide. He has authored numerous papers in these fields.

PHILIP TATE

Instructor of Anatomy and Physiology at Phoenix College

Phil Tate earned a B.S. in zoology, a B.S. in mathematics, and an M.S. in ecology at San Diego State University and a Doctor of Arts (D.A.) in biological education from Idaho State University. He is an award-winning instructor who has taught a wide spectrum of students at the four-year and community college levels. Phil has served as the annual conference coordinator, president-elect, president, and past president of the Human Anatomy and Physiology Society (HAPS).

Special Contributions By:

Shylaja R. Akkaraju
Bronx Community College

Christine M. Eckel
Salt Lake Community College

Jennifer L. Regan
University of Southern Mississippi

Andrew F. Russo
University of Iowa

Cinnamon L. VanPutte
Southwestern Illinois College

BRIEF CONTENTS

PART 1
ORGANIZATION OF THE HUMAN BODY

PART 2
SUPPORT AND MOVEMENT

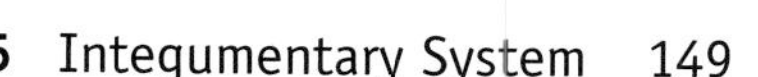

PART 3
INTEGRATION AND CONTROL SYSTEMS

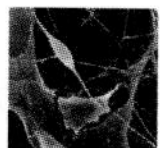

PART 4
REGULATIONS AND MAINTENANCE

PART 5
REPRODUCTION AND DEVELOPMENT

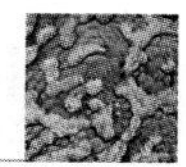

APPENDICES

CONTENTS

PART 3
INTEGRATION AND CONTROL SYSTEMS

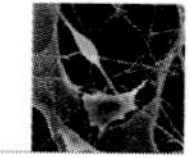

PART 5 REPRODUCTION AND DEVELOPMENT

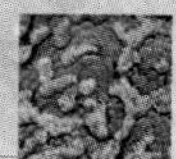

APPENDICES

PREFACE

Anatomy and Physiology is designed to help students develop a solid understanding of the concepts of anatomy and physiology and to use this knowledge to solve problems. This is accomplished via a carefully planned learning system that serves as the framework for virtually all the elements in the text. From the opening chapter previews to the end-of-chapter multi-level review questions, *Anatomy and Physiology* is the textbook that will motivate and teach your students who are going into health-related careers.

AUDIENCE

Anatomy and Physiology is written for the two-semester anatomy and physiology course. The writing is comprehensive enough to provide the depth necessary for those courses not requiring prerequisites, and yet presented with such clarity that it nicely balances the thorough coverage. Clear descriptions and exceptional illustrations combine to help students develop a solid understanding of the concepts of anatomy and physiology and to also teach them how to use that information.

WHAT SETS THIS BOOK APART?

Seeley Learning System—Connecting Students to Their Future

It begins with the micrograph on the chapter-opening page, carefully chosen to pique interest and bring into focus a close-up view of the subject at hand. Accompanying text previews the material to come and even includes a mini-review, complete with relevant page numbers. Once into the core of the chapter, students will benefit from an abundance of clinical content, step-by-step Process Figures, in-chapter Review and Predict Questions, macro-to-micro art, unique Homeostasis Summary Figures, cadaver images, and more. A Chapter Summary provided in outline form covers all the main points of the chapter and serves as an excellent study guide. Finally, the Review and Comprehension Questions, and the Critical Thinking Questions, based on Bloom's Taxonomy, allow students to test their understanding in stages of comprehension. Within this learning system are three major emphases:

1. ***Clinical Coverage****—New Case Studies Build on Rich Clinical Foundation*

Examples of diseases, responses to exercise, clinical case studies, aging, and environmental conditions are all used to explain how our bodies function and to describe the consequences when systems do not operate normally. These conditions are also used to enhance comprehension of the relationship between structure and function. *Anatomy and Physiology* has always had a strong emphasis on clinical material. In the eighth edition, that emphasis is now strengthened with the addition of *Case Studies*. These brief, real-life scenarios, combined with the popular *Clinical Asides,* the more in-depth *Clinical Focus Readings,* and the *Systems Pathology* spreads, provide a thorough clinical education that fully supports surrounding textual material. Also new, the *Clinical Genetics Essays* emphasize the connection between genetics and certain diseases and frequently tie in with *Clinical Focus Readings.*

2. ***Critical Thinking****—Recall Isn't Enough—Learning Needs to Be Developed and Applied*

A critical thinking approach is integrated throughout this textbook. It can be found in the way the narrative and the figures are designed and coordinated; in the way Process Figures explain step-by-step how mechanisms respond to a variety of stimuli; in the way Homeostasis Summary Figures explain the means by which homeostasis is maintained; and in the way Clinical Aside Boxes, Clinical Focus Readings, and new Case Studies encourage students to apply information they have learned to practical "real-life" scenarios. Finally, critical thinking is especially integrated into the way the unique Predict Questions and Critical Thinking Questions encourage students to go beyond rote memorization.

3. ***Exceptional Art****—Accuracy, Consistency, Logic Underscore Visuals*

The illustrations in *Anatomy and Physiology* are also an integral part of the Seeley Learning System. Accurate, attractive, and clearly presented, the visual program enhances comprehension in a number of ways: Tables are often combined with illustrations, relevant photos are side-by-side with drawings, cadaver photos are included where appropriate, step-by-step Process Figures explain physiologic processes, and the distinctive Homeostasis Summary Figures include explanations that are necessary to understand mechanisms and their roles in the maintenance of homeostasis. The images reflect a contemporary style and are coordinated so that colors and styles of structures in multiple figures are consistent with one another throughout the book. See the "Guided Tour" following this Preface for more details on the unique Seeley Learning System in *Anatomy and Physiology.*

Clarity and Comprehensiveness—The Right Amount of Information Presented Clearly

Not everything that is known about human anatomy and physiology can be included in a single book, and new information is accumulating at a rapid pace. For example, molecular techniques continue to identify the mechanisms that control gene expression, reveal how genes determine the structural and functional characteristics of humans, and demonstrate how alterations of genes can be responsible for abnormalities and diseases in humans. A major challenge in writing any textbook is to clearly present vital concepts that are consistent with the massive body of contemporary knowledge in a way that encourages readers to grasp these key concepts and think critically by applying them to realistic situations. *Anatomy and Physiology* is written in succinct, understandable language. We continue to improve this aspect of the text because we believe that content must be presented and explained clearly and in sufficient detail to support critical thinking. All of us make a concerted effort to maintain congruity between the explanations and the problems presented in each chapter. Whether or not critical thinking is a major emphasis in your course, this text is a valuable asset for students because of its depth and understandable language.

EIGHTH EDITION CHANGES—WHAT'S NEW?

The eighth edition of *Anatomy and Physiology* is the result of extensive analysis of the text and evaluation of input from contributing authors and instructors who have thoroughly reviewed chapters. We are grateful to these professionals and have used their constructive comments in our continuing efforts to enhance the strengths of our textbook.

Contributing Authors

Five contributing authors have extensively examined and, where appropriate, revised material in 10 of the 29 chapters. Beyond what they have contributed to their specific chapters, these talented professors brought a fresh perspective to the entire book. They have worked very closely with us to produce up-to-date and clear presentations that are consistent with the objectives of this textbook.

Clinical Case Studies

Reviewers of the seventh edition asked for more real-life scenarios, such as the kind their students may encounter. New Case Studies now appear in nearly all the chapters. They are brief examples of how alterations in anatomy and physiology result in diseases and include suggestions on how they can be treated. The Case Studies often illustrate how multiple systems are affected and how they respond in an attempt to maintain homeostasis. Each of the Case Studies is followed by a Predict Question, which helps students think critically about the application of anatomical and physiologic concepts to the situation and predict the consequences of additional changes.

Genetics Coverage

Modern genetics has made it possible to understand the connection between the structure of genes on chromosomes and many diseases. Some of these diseases that have a genetic basis are highlighted in new Clinical Genetics essays. To provide an early overview of genetic concepts, essential material has been taken from chapter 29, updated, and moved to chapter 3. These include conditions that result from inheritance or mutations in single genes that are dominant, recessive, or X-linked, such as neurofibromatosis, cystic fibrosis, and Duchenne muscular dystrophy, respectively. Conditions that involve alteration of multiple genes that are inherited, such as Type 2 diabetes mellitus and celiac disease, and those that involve mutations in multiple genes, such as cancer, are also described. These text revisions and new content will better prepare students to understand the relationship between genetics and many of the cases they may encounter in health-related careers.

Anatomy & Physiology | REVEALED® Integration

This is the first edition of *Anatomy and Physiology* to feature chapter correlations to the popular AP | REVEALED® student tutorial. Students across the country are improving their grades using this unique multimedia study aid that offers "melt-away" layers of dissection, animations, imaging, and self-testing to study cadaver specimens. The appropriate section, or body system, within the tutorial is listed on all applicable chapter opening pages. Even more specific connections between AP | REVEALED® and the text can be found on a correlation guide on the ARIS website that accompanies this textbook.

Instructor Resource Guide

McGraw-Hill Higher Education has developed several resources to assist professors teaching anatomy and physiology. To take advantage of this content and to make creating your lectures easier, this edition features a bound-in Resource Guide with listings of available case studies, animations, exercises, images, questions, and so on all in one handy chart and arranged by chapters.

Improved Art

Substantial changes have been made to improve the clarity of the art in the eighth edition. We have created 34 new figures, and two-thirds of the remaining art program has been revised to improve the quality of the illustrations. Additionally, over 40 new photographs have been added to this edition. Some of the enhancements include

- New photomicrographs of connective tissues show low and high power magnifications.
- Homeostasis Summary Figures were revised to provide a more concise and easy-to-read review of the mechanisms that maintain homeostasis. These figures have also been improved by adding a "Start" icon, making it easier to follow the color-coded directional arrows when the value of a variable increases or decreases.
- More Process Figures have been added to the text and several have been improved.

Refined and Updated Narrative

The eighth edition has undergone a complete examination and revision. Reports of new discoveries have been researched and evaluated. We have listened to suggestions from instructors who teach anatomy and physiology, as well as to our contributing authors, and have consequently scrutinized the text carefully. Explanations have been made clearer, terminology made more consistent, content reorganized to enhance clarity, facts corrected or updated, questions revised or added, and figure captions modified.

ACKNOWLEDGMENTS

A great deal of effort is required to produce a heavily illustrated textbook such as *Anatomy and Physiology.* Many hours of work are required to organize and develop the components of the textbook while creating and designing illustrations, but no text is solely the work of the authors. It is not possible to adequately acknowledge the support and encouragement provided by our loved ones. They have had the patience and understanding to tolerate our absences and our frustrations. They have also been willing to provide assistance and unwavering support.

Many hands besides our own have touched this text, guiding it through various stages of development and production. We wish to express our gratitude to the staff of McGraw-Hill for their help and encouragement. We sincerely appreciate Publisher Michelle Watnick, Sponsoring Editor James Connely, and Developmental Editor Kathy Loewenberg for their many hours of work, suggestions, and tremendous patience and encouragement. Thanks are gratefully offered to Copy Editor Debra DeBord for carefully polishing our words. We also thank Project Manager Mary Powers, Photo Editor John Leland, Production Supervisor Laura Fuller, and Designer Rick Noel for their time spent turning manuscript into a book; Media Producer Jake Theobald, Project Coordinator Melissa Leick, and Media Project Manager Tammy Juran for their assistance in building the various products that support our text; and Marketing Manager Lynn Kalb-Breithaupt for her enthusiasm in promoting this book. The McGraw-Hill employees with whom we have worked are excellent professionals. They have been consistently helpful and their efforts are truly appreciated. Their commitment to this project has clearly been more than a job to them.

We are especially grateful to contributing authors Shylaja Akkaraju, Christine Eckel, Jennifer Regan, Andrew Russo, and Cinnamon VanPutte for their involvement in this edition. Discussions with these professionals were delightful, insightful, and valuable. Their input and contributions have made this textbook substantially better.

We also extend our appreciation to the many illustrators who worked on the development and execution of the illustration program, and to those who provided photographs and photomicrographs for the eighth edition of *Anatomy and Physiology.* The art program for this textbook represents a monumental effort, and we are grateful for their contribution to the overall appearance and pedagogical value of the photos and illustrations.

Finally, we sincerely thank the reviewers and the teachers who have provided us with exceptional constructive criticism. The remuneration they received represents only a token payment for their efforts. To review a textbook conscientiously requires a true commitment and dedication to excellence in teaching. Their helpful criticisms and suggestions for improvement were significant in revising the seventh edition. We gratefully acknowledge them by name in the next section.

Rod Seeley
Trent Stephens
Phil Tate

REVIEWERS

Terry A. Austin
Temple College

Gail Baker
LaGuardia Community College/CUNY

David M. Bastedo
San Bernardino Valley College

Alease S. Bruce
University of Massachusetts–Lowell

Nishi S. Bryska
University of North Carolina–Charlotte

Patrick D. Burns
University of Northern Colorado

Brad Caldwell
Greenville Technical College

Ana Christensen
Lamar University

Nathan L. Collie
Texas Tech University

David T. Corey
Midlands Techical College

Ethel R. Cornforth
San Jacinto College–South

Cara L. Davies
Ohio Northern University

Richard Doolin
Daytona Beach Community College

Kathryn A. Durham
Lorain County Community College

Adam Eiler
San Jacinto College–South

Lee F. Famiano
Cuyahoga Community College

Kathy E. Ferrell
Greenville Technical College

Edward R. Fliss
St. Louis Community College at Florissant Valley

Paul Florence
Jefferson Community College

Cliff Fontenot
Southeastern Louisiana University

Allan Forsman
East Tennessee State University

Ralph F. Fregosi
The University of Arizona

Paul Garcia
Houston Community College–Southwest

Chaya Gopalan
St. Louis Community College at Florissant Valley

Jean C. Jackson
Lexington Community College

Amy E. Jetton
Middle Tennessee State University

Jody E. Johnson
Arapahoe Community College

Sally Johnston
Community College of Southern Nevada

Beverly P. Kirk
Northeast Mississippi Community College

Michael Kopenits
Amarillo College

Karen K. McLellan
Indiana University–Purdue University Fort Wayne

Mara L. Manis
Hillsborough Community College

Glenn Merrick
Lake Superior College

Ralph R. Meyer
University of Cincinnati

Amy Griffin Ouchley
University of Louisiana–Monroe

Mark Paternostro
Pennsylvania College of Technology

Jennifer L. Regan
University of Southern Mississippi

Jackie Reynolds
Richland College

Tim Roye
San Jacinto College–South

Gerald Schafer
Mission College–Thailand

Marilyn Shannon
Indiana University–Purdue University Fort Wayne

Jeff S. Simpson
Metropolitan State College of Denver

Claudia Stanescu
The University of Arizona

William Stewart
Middle Tennesee State University

Janis Thompson
Lorain County Community College

Katherine M. Van de Wal
Community College of Baltimore County, Essex

Jyoti R. Wagle
Houston Community College System–Central

Mark D. Womble
Youngstown State University

Linda S. Wooten
Bishop State Community College

Michelle Zurawski
Moraine Valley Community College

FOCUS GROUP PARTICIPANTS

Pegge Alciatore
University of Louisiana–Lafayette

Sara Brenizer
Shelton State Community College

Kathy Bruce
Parkland College

Jorge Cortese
Durham Technical Community College

Juville Dario-Becker
Central Virginia Community College

Smruti Desai
Cy-Fair College

Don Steve Dutton
Amarillo College

Glen Early
Jefferson Community and Technical College

Clair Eckersell
Brigham Young University–Idaho

Jim Ezell
J. Sargeant Reynolds Community College

Deanna Ferguson
Gloucester County College

Pamela Fouche
Walters State Community College

Mary Fox
University of Cincinnati

Bagie George
Georgia Perimeter College–Lawrenceville Campus

Peter Germroth
Hillsborough Community College–Dale Mabry Campus

Richard Griner
Augusta State University

Mark Hubley
Prince George's Community College

Sister Carol Makravitz
Luzerne County Community College

Ronald Markle
Northern Arizona University

Carl McAllister
Georgia Perimeter College

Alfredo Munoz
University of Texas at Brownsville

Necia Nicholas
Calhoun Community College

Robyn O'Kane
LaGuardia Community College/CUNY

Karen Payne
Chattanooga State Technical Community College

Joseph Schiller
Austin Peay State University

Colleen Sinclair
Towson University

Mary Elizabeth Torrano
American River College

Cinnamon VanPutte
Southwestern Illinois College

Chuck Venglarik
Jefferson State Community College

Mark Wygoda
McNeese State University

GUIDED TOUR

THE SEELEY LEARNING SYSTEM—CONNECTING STUDENTS TO THEIR FUTURE

The Seeley Learning System in *Anatomy and Physiology* is designed to help you learn in a systematic fashion. The textual material builds from simple facts to explanations of more complex concepts and is presented within a supporting framework of features that help you review what you have read, evaluate your comprehension of the content, and use what you have learned. Here is how your book can help you learn and improve your grade:

Respiratory System

From our first breath at birth, the rate and depth of our breathing is unconsciously matched to our activities, whether studying, sleeping, talking, eating, or exercising. We can voluntarily stop breathing, but within a few seconds we must breathe again. Breathing is so characteristic of life that, along with the pulse, it is one of the first things we check for to determine if an unconscious person is alive.

Breathing is necessary because all living cells of the body require oxygen and produce carbon dioxide. The respiratory system allows the exchange of these gases between the air and the blood, and the cardiovascular system transports them between the lungs and the cells of the body. The capacity to carry out normal activity is reduced without healthy respiratory and cardiovascular systems.

Respiration includes (1) ventilation, the movement of air into and out of the lungs; (2) gas exchange between the air in the lungs and the blood, sometimes called external respiration; (3) the transport of oxygen and carbon dioxide in the blood; and (4) gas exchange between the blood and the tissues, sometimes called internal respiration. The term *respiration* is also used in reference to cell metabolism, which is discussed in chapter 25.

This chapter explains the *functions of the respiratory system* (p. •••), the *anatomy and histology of the respiratory system* (p. •••), *ventilation* (p. •••), *measurement of lung function* (p. •••), *physical principles of gas exchange* (p. •••), *oxygen and carbon dioxide transport in the blood* (p. •••), *regulation of ventilation* (p. •••), and *respiratory adaptations to exercise* (p. •••). The chapter concludes by looking at the *effects of aging on the respiratory system* (p. •••).

Colorized scanning electron micrograph of the lung, showing alveoli, which are small chambers where gas place between the air and the blood.

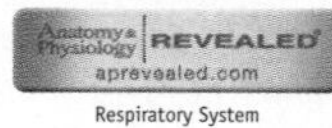

Respiratory System

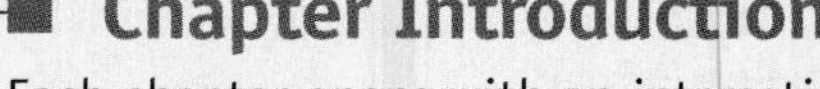

■ Chapter Introduction

Each chapter opens with an interesting photomicrograph, which ties in with the topic. The paragraphs that follow introduce the topic and include a brief overview of the key points of the chapter. At the bottom of this page, if applicable, is the correlating system in Anatomy & Physiology | REVEALED®, a multimedia study aid that allows you to "melt" away layers of dissection on cadaver specimens, view animations, examine different types of imaging, and take practice quizzes. Just pop in the correct CD, or visit the program online, and go to the system listed in the text for assistance in understanding the chapter material.

4. The pseudostratified ciliated columnar epithelium lining the larynx produces mucus, which traps debris in air. The cilia move the mucus and debris into the pharynx.

5 ***Name and describe the three unpaired cartilages of the larynx. What are their functions?***

6 ***Distinguish between the vestibular and vocal folds. How are sounds of different loudness and pitch produced by the vocal folds?***

7 ***How does the position of the arytenoid cartilages change when a person is simply breathing versus making low-pitched and high-pitched sounds?***

■ In-Chapter Section Reviews

Review questions at the end of each section *within* the chapter prompt you to test your understanding of key concepts. Use them as a self-test to determine whether you have a sufficient grasp of the information before proceeding to the next section.

PREDICT 2

Explain what happens to the shape of the trachea when a person swallows a large mouthful of food. Why is this change of shape advantageous?

■ Predict Questions

These innovative critical thinking questions encourage you to become an active learner as you read. Predict Questions challenge you to use your understanding of new concepts to solve a problem. Answers to the questions are provided at the end of the book, allowing you to evaluate your responses and to understand the logic used to arrive at the correct answer.

INSTRUCTIVE ARTWORK MAKES THE DIFFERENCE

A picture is worth a thousand words—especially when you are learning anatomy and physiology. Because words alone cannot convey the nuances of anatomy or the intricacies of physiology, *Anatomy and Physiology* uses a dynamic program of full-color illustrations and photographs to support and further clarify the textual explanations. Brilliantly rendered and carefully reviewed for accuracy and consistency, the precisely labeled illustrations and photos provide concrete, visual reinforcement of important topics discussed throughout the text.

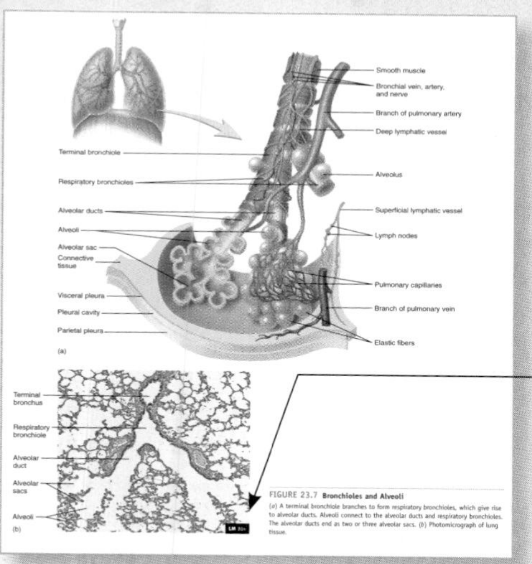

■ Combination Art

Drawings are often paired with photographs to enhance visualization and comprehension of structures.

■ Histology Micrographs

Light micrographs, as well as scanning and transmission electron micrographs, are used in conjunction with illustrations to present a true picture of anatomy and physiology from the cellular level.

Magnifications are indicated to help you estimate the size of structures shown in the photomicrographs.

Reference diagrams orient you to the view or plane an illustration represents.

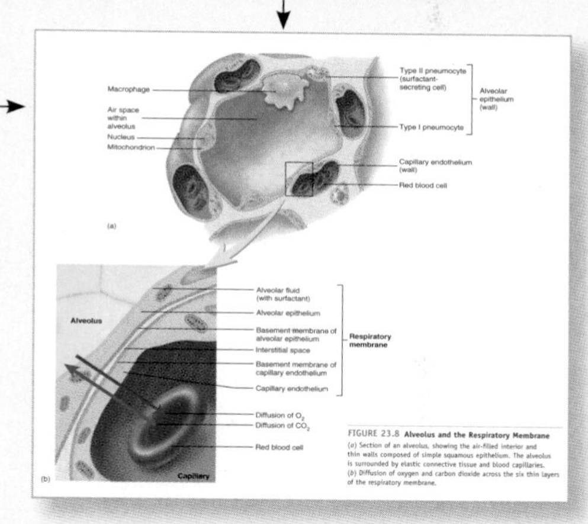

■ Macro-to-Micro Art

Illustrations depicting complex structures or processes combine macroscopic and microscopic views to help you see the relationships between increasingly detailed drawings.

■ Atlas-Quality Cadaver Images

Clearly labeled photos of dissected human cadavers provide detailed views of anatomical structures, capturing the intangible characteristics of actual human anatomy that can be appreciated only when viewed in human specimens.

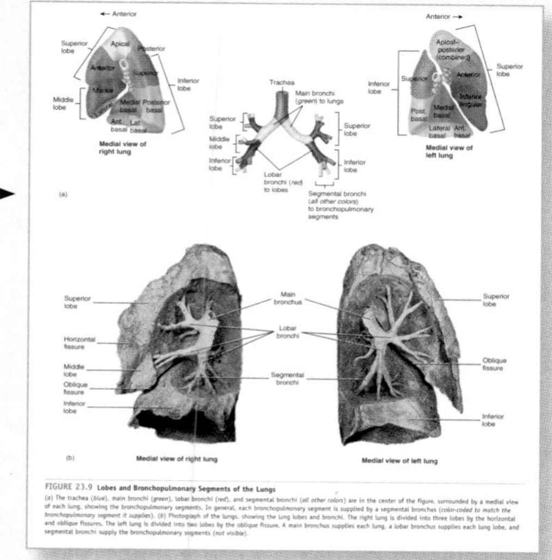

■ Realistic Anatomical Art

The anatomical figures in *Anatomy and Physiology* have been carefully drawn to convey realistic, three-dimensional detail. Richly textured bones and artfully shaded muscles, organs, and vessels lend a sense of realism to the figures that helps you envision the appearance of actual structures within the body. The colors used to represent different anatomical structures have been applied consistently throughout the book to help you easily identify structures in every figure.

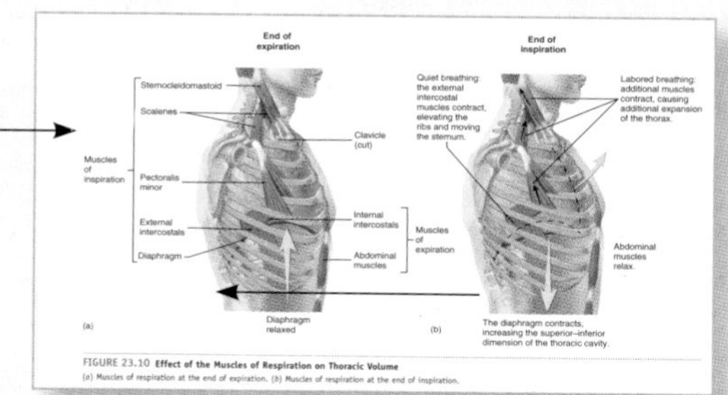

CLINICAL CONTENT PUTS KNOWLEDGE INTO PRACTICE

Anatomy and Physiology provides clinical examples to demonstrate the application of basic knowledge in interesting and relevant clinical context. Exposure to clinical information is especially beneficial if you are planning on using your knowledge of anatomy and physiology in a health-related career.

New! Clinical Genetics

Today's anatomy and physiology student knows that a basic understanding of genetics is critical to learning about various diseases and their impact on the human body. This information takes on more importance almost daily as genetic research continues to contribute to possible cures. New to this edition, Clinical Genetics boxes define diseases, describe symptoms and genetic components, and discuss possible treatments.

CLINICAL GENETICS

Alpha-1 Antitrypsin Deficiency

Emphysema (em-fi-zē′mă) is a condition in which lung alveoli become progressively destroyed and enlarged. Individuals suffering from emphysema experience shortness of breath and coughing. Chemicals in cigarette smoke damage lung tissues and stimulate inflammation. As part of the inflammatory response, neutrophils and macrophages release **proteases,** which are enzymes that break down proteins. Proteases in the lungs provide protection against some bacteria and foreign substances. Too much protease activity, however, can be harmful because it results in the breakdown of lung tissue proteins, especially elastin in elastic fibers. **Alpha-1 antitrypsin (AAT),** which is synthesized in the liver, is a **protease inhibitor (Pi).** Normally, AAT inhibits protease activity, preventing the destruction of lung tissue. Excess protease production stimulated by cigarette smoke, however, can cause lung damage, leading to emphysema.

Although cigarette smoking is the major risk factor for emphysema, approximately 1%–2% of emphysema cases are due to a deficiency of AAT caused by defects of the AAT gene located on chromosome 14. Multiple alleles for AAT have been identified. The normal allele is designated M. Individuals who are homozygous for the normal allele are designated PiMM, and they produce normal levels of AAT. That is, each M gene is responsible for 50% of the AAT produced. The most common abnormal allele is designated Z. Individuals with only one copy of Z (PiMZ) have about 60% of normal levels of AAT, which is sufficient to prevent protease damage. Individuals with two copies of the Z allele (PiZZ) produce only about 15%–20% of normal AAT levels. Smoking by these individuals accelerates the development of emphysema by 10–15 years. Other variant alleles cause different levels of AAT. The most severe form results in no AAT AAT production in the liver. In addition, individuals may receive intravenous infusions of AAT, a process called **alpha-1 antitrypsin augmentation.**

Treatment of AAT deficiency follows the normal course of treatment for emphysema. Stopping smoking reduces the destruction of lung tissue by removing the stimulus for excess protease activity. Drugs, such as danazol and tamoxifen, can stimulate increased AAT production in the liver. In addition, individuals may receive intravenous infusions of AAT, a process called **alpha-1 antitrypsin augmentation.**

Effect of Spinal Cord Injury on Ventilation

The diaphragm is supplied by the phrenic nerves, which arise from spinal nerves C3–C5 (see figure 12.16), descend along each side of the neck to enter the thorax, and pass to the diaphragm. The intercostal muscles are supplied by the intercostal nerves (see figure 12.15), which arise from spinal nerves T1–T11 and extend along the spaces between the ribs. Spinal cord injury superior to the origin of the phrenic nerves causes paralysis of the diaphragm and intercostal muscles and results in death unless artificial respiration is provided. A high spinal cord injury below the origin of the phrenic nerves causes paralysis of the intercostal muscles. Even though the diaphragm can function maximally, ventilation is drastically reduced because the intercostal muscles no longer prevent the thoracic wall from collapsing inward. Vital capacity is reduced to about 300 mL. With low spinal cord injury, below the origin of the intercostal nerves, both the diaphragm and the intercostal muscles function normally.

Clinical Topics

Interesting clinical sidebars reinforce or expand on the facts and concepts discussed within the narrative. Once you have learned a concept, applying that information in a clinical context shows you how your new knowledge can be put into practice.

Clinical Focus

These in-depth boxed essays explore relevant topics of clinical interest. The subjects covered include pathologies, current research, sports medicine, exercise physiology, and pharmacology.

CLINICAL FOCUS

Cough and Sneeze Reflexes

The function of both the cough reflex and the sneeze reflex is to dislodge foreign matter or irritating material from the respiratory passages. The bronchi and trachea contain sensory receptors that are sensitive to foreign particles and irritating substances. The cough reflex is initiated when the sensory receptors detect such substances and initiate action potentials that pass along the vagus nerves to the medulla oblongata, where the cough reflex is triggered.

The movements resulting in a cough occur as follows: Approximately 2.5 L of air are inspired; the vestibular and vocal folds close tightly to trap the inspired air in the lungs; the abdominal muscles contract to force the abdominal contents up against the diaphragm; and the muscles of expiration contract forcefully. As a consequence, the pressure in the lungs increases to 100 mm Hg or more. Then the vestibular and vocal folds open suddenly, the soft palate is elevated, and the air rushes from the lungs and out the oral cavity at a high velocity, carrying foreign particles with it.

The sneeze reflex is similar to the cough reflex, but it differs in several ways. The source of irritation that initiates the sneeze reflex is in the nasal passages instead of in the trachea and bronchi, and the action potentials are conducted along the trigeminal nerves to the medulla oblongata, where the reflex is triggered. During the sneeze reflex, the soft palate is depressed so that air is directed primarily through the nasal passages, although a considerable amount passes through the oral cavity. The rapidly flowing air dislodges particulate matter from the nasal passages and can propel it a considerable distance from the nose. About 17%–25% of people have a photic sneeze reflex, in which exposure to bright light, such as the sun, can stimulate a sneeze reflex. The pupillary reflex causes the pupils to constrict in response to bright light. It is speculated that the complicated "wiring" of the pupillary and sneeze reflexes are intermixed in some people so that, when bright light activates a pupillary reflex, it also activates a sneeze reflex. Sometimes the photic sneeze reflex is fancifully called ACHOO, which stands for autosomal dominant compelling *h*elio-*o*phthalmic outburst. As the name suggests, the reflex is inherited as an autosomal-dominant trait. A person needs to inherit only one copy of the gene to have a photic sneeze reflex.

New! Case Studies

New to this edition, these specific yet brief examples of how alternations of structure and/or function result in diseases help you better understand the practical application of anatomy and physiology. These boxed summaries are placed strategically in the text, so that you can immediately start to see connections between learned concepts and real events.

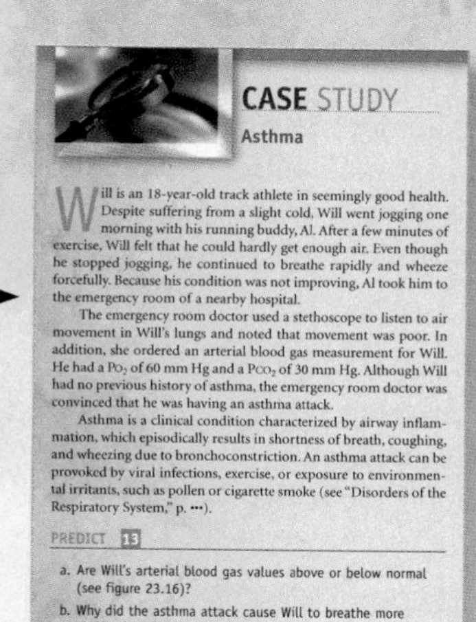

CASE STUDY

Asthma

Will is an 18-year-old track athlete in seemingly good health. Despite suffering from a slight cold, Will went jogging one morning with his running buddy, Al. After a few minutes of exercise, Will felt that he could hardly get enough air. Even though he stopped jogging, he continued to breathe rapidly and wheeze forcefully. Because his condition was not improving, Al took him to the emergency room of a nearby hospital.

The emergency room doctor used a stethoscope to listen to air movement in Will's lungs and noted that movement was poor. In addition, she ordered an arterial blood gas measurement for Will. He had a P_{O_2} of 60 mm Hg and a P_{CO_2} of 30 mm Hg. Although Will had no previous history of asthma, the emergency room doctor was convinced that he was having an asthma attack.

Asthma is a clinical condition characterized by airway inflammation, which episodically results in shortness of breath, coughing, and wheezing due to bronchoconstriction. An asthma attack can be provoked by viral infections, exercise, or exposure to environmental irritants, such as pollen or cigarette smoke (see "Disorders of the Respiratory System," p. •••).

PREDICT 13

a. Are Will's arterial blood gas values above or below normal (see figure 23.16)?
b. Why did the asthma attack cause Will to breathe more rapidly (see figure 23.22)?
c. Why did the asthma attack cause Will to wheeze forcefully?
d. Did Will's rapid, forceful wheezing restore homeostasis? Explain.
e. Explain Will's blood P_{O_2} and P_{CO_2} values.
f. Is Will's blood pH lower or higher than normal? What effect does this blood pH normally have on respiration rate? Why didn't that happen?
g. Explain how β-adrenergic agents (see "The influence of Drugs on the Autonomic Nervous System," chapter 16) or inhaled glucocorticoids (see chapter 18) can help Will.

Systems Pathology

These spreads explore a specific disorder or condition related to a particular body system. Presented in a simplified case study format, each Systems Pathology box begins with a patient history followed by background information about the featured topic.

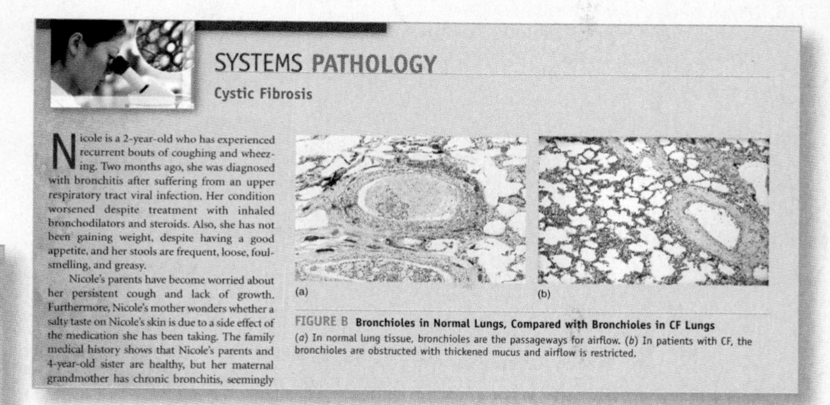

SYSTEMS PATHOLOGY

Cystic Fibrosis

Nicole is a 2-year-old who has experienced recurrent bouts of coughing and wheezing. Two months ago, she was diagnosed with bronchitis after suffering from an upper respiratory tract viral infection. Her condition worsened despite treatment with inhaled bronchodilators and steroids. Also, she has not been gaining weight, despite having a good appetite, and her stools are frequent, loose, foul-smelling, and greasy.

Nicole's parents have become worried about her persistent cough and lack of growth. Furthermore, Nicole's mother wonders whether a salty taste on Nicole's skin is due to a side effect of the medication she has been taking. The family medical history shows that Nicole's parents and 4-year-old sister are healthy, but her maternal grandmother has chronic bronchitis, seemingly

(a) (b)

FIGURE B **Bronchioles in Normal Lungs, Compared with Bronchioles in CF Lungs**
(*a*) In normal lung tissue, bronchioles are the passageways for airflow. (*b*) In patients with CF, the bronchioles are obstructed with thickened mucus and airflow is restricted.

because of the depletion of the PCL and a reduction in the water content of mucus, which causes the mucus to be thicker than normal (figure B).

A standard test for CF diagnosis is the **sweat-chloride test,** in which the chemical **pilocarpine** (pi-lō-kar′pēn) is swabbed onto the skin and a mild electric current is applied. Pilocarpine is a muscarinic agent that stimulates receptors in the sweat glands (see "The Influence of Drugs on the Autonomic Nervous System," chapter 16). The mild electric current drives the medication into the skin, producing localized sweating and avoiding systemic drug effects. The sweat that is produced is collected and tested for abnormally high levels of salt (NaCl). Normally, sweat glands produce a very dilute liquid, which cools the body without depleting salt from it. In CF, the malfunctioning CFTR results in a failure to absorb the normal amount of NaCl from sweat, resulting in high NaCl content in sweat.

Although CF tends to be primarily associated with respiratory malfunctions, the production of thickened mucus also has profound digestive tract effects. In fact, the original name of the disease was cystic fibrosis of the pancreas because, in 90% of CF patients, the pancreas is gradually destroyed and infiltrated by fibrous cysts. The pancreatic ducts of CF patients can become obstructed with sticky mucus, which prevents the secretion of adequate amounts of digestive enzymes, particularly fat-digesting enzymes. Children with CF can have severe nutritional deficiencies because of the decreased absorption of proteins and fat-soluble vitamins, such as vitamins A, D, E, and K. To aid food digestion and promote growth, children with CF may be given powdered digestive enzymes. Supplemental overnight feeding through a gastrostomy (gas-tros′tō-mē) tube (stomach tube, T-tube) may also be beneficial.

The main goal of CF treatment is to reduce lung infections, clear the lungs of mucus, improve airflow, and maintain sufficient calories and nutrition. People with CF must undergo **chest physical therapy,** also called **chest clapping** or **chest percussion.** This involves manually pounding the back and chest for 30 to 40 minutes three or four times daily to dislodge mucus trapped in the chest. Automated chest clappers are preferred by some CF patients. Antibiotics may be prescribed to help control lung infections. Mucus-thinning drugs, such as Pulmozyme, and bronchodilators can be inhaled to improve mucus clearance and open airways. Eventually, if breathing problems become too severe or the patient becomes resistant to antibiotics, a lung transplant may be necessary. The downside of a lung transplant is the need to take immunosuppressive drugs for life to prevent rejection of the transplanted lungs. These drugs produce side effects, such as increased susceptibility to infections, diabetes, tumors, and osteoporosis. The upside of lung transplantation is that it is a partial "cure" because the transplanted lung cells do not have the genetic defect. However, cells with the defective CFTR gene are still present elsewhere in the body. Scientists are also investigating the use of gene therapy, wherein a copy of the normal CFTR gene is inserted into epithelial cells by a harmless virus. So far, the effects of gene therapy have lasted for only a few days. With treatment, the current life expectancy for persons with cystic fibrosis is into the mid-30s. In 95% of CF cases, the patient dies due to complications from lung infections.

PREDICT 16

As cystic fibrosis becomes advanced, what happens to forced expiratory volume in 1 second (FEV1), residual volume, and physiologic dead space?

SYSTEM INTERACTIONS Effect of Cystic Fibrosis on Other Systems

SYSTEM	INTERACTIONS
Integumentary	Two to five times the normal amount of salt is secreted in sweat, which can cause rapid dehydration in hot conditions. Clubbing is an enlargement of the fingertips and toes due to a proliferation of connective tissue; the mechanism that produces clubbing is unclear, but it may be related to insufficient oxygen delivery, which stimulates an inflammatory response.
Skeletal	Low bone density is common because insufficient vitamin D is absorbed from the diet when the pancreatic ducts become blocked.
Cardiovascular	Lung disease may eventually cause the right ventricle of the heart to fail due to the increased force necessary to pump blood into damaged lungs.
Digestive	Mucus blockage of pancreatic ducts and liver bile ducts decreases fat digestion capabilities, resulting in bowel blockage; foul-smelling, greasy stools; and chronic diarrhea. Autodigestion of the pancreas by enzymes trapped in the pancreas can occur. Liver duct blockage may eventually lead to cirrhosis of the liver and gallstones.
Respiratory	Mucus buildup causes coughing, wheezing, and recurrent chest infections because bacteria are not effectively removed. Eventually, lung bleeding (hemoptysis) or collapsed lung (atelactasis) may result. There may also be polyps in the nasal cavity and paranasal sinuses due to thickening of the mucosa. Frequent instances of sinusitis are common.
Reproductive	Ninety-eight percent of men with CF are infertile because of a failure of the ductus deferens to develop. Up to 20% of women with CF may experience infertility related to mucus blockage of the uterine tubes or depression of the menstrual cycle because of malnutrition.
Immune	A decrease in innate immunity occurs because the thickened mucus in the respiratory tract impairs cilia movement. The beating of cilia in the respiratory tract is one of the important mechanical mechanisms that prevents the entry of microorganisms into the body.

An Interactions Table at the end of every Systems Pathology reading summarizes how the condition impacts each body system.

SPECIALIZED FIGURES CLARIFY TOUGH CONCEPTS

Studying physiology does not have to be an intimidating task mired in memorization. *Anatomy and Physiology* uses two special types of illustrations to help you not only learn the steps involved in specific processes but also apply this knowledge as you predict outcomes in similar situations. Process Figures organize the key occurrences of physiologic processes in an easy-to-follow format. Homeostasis Summary Figures detail the mechanisms of homeostasis by illustrating the means by which a system regulates a parameter within a narrow range of values.

■ Process Figures

Process Figures break down physiologic processes into a series of smaller steps, allowing you to build your understanding by learning each important phase.

1. A secretion introduced into the digestive tract or food within the tract begins in one location.
2. Segments of the digestive tract alternate between contraction and relaxation.
3. Material (*brown*) in the intestine is spread out in both directions from the site of introduction.
4. The secretion or food is spread out in the digestive tract and becomes more diffuse (*lighter color*) through time.

Secretion or food

Contraction waves

Contraction waves

PROCESS FIGURE 24.3 Segmental Contractions

Circled numbers indicate the sequence within the artwork and correspond to numbered explanations. The numbers are placed carefully, allowing you to zero right in to where the action described in each step takes place.

Process Figures and Homeostasis Summary Figures are identified next to the figure number. The accompanying caption provides additional explanation.

■ Homeostasis Summary Figures

These specialized flowcharts illustrate the mechanisms that body systems use to maintain homeostasis.

Changes caused by an increase of a variable outside its normal range are shown in the green boxes across the top.

The normal range for a given value is represented by the graphs in the center of each figure. Begin at the new yellow "Start" oval and follow the green arrows to learn about the chain of events triggered by an increase in the variable, or follow the red arrows for events resulting from a decrease in the variable.

Changes caused by a decrease of a variable outside its normal range are shown in the red boxes across the bottom of the figure.

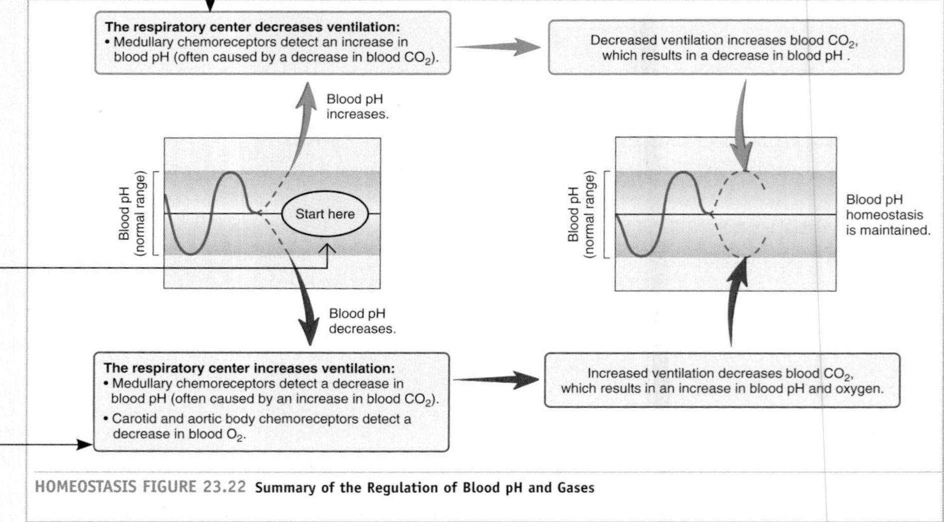

HOMEOSTASIS FIGURE 23.22 **Summary of the Regulation of Blood pH and Gases**

STUDY FEATURES ENSURE SUCCESS

Learning anatomy and physiology is, in many ways, like learning a new language. Mastering the terminology is critical to building your knowledge base. Once you understand many of the word roots, the task becomes easier. This textbook includes a variety of vocabulary aids, helpful chapter summaries, and multi-level review questions to offer you the very best learning system available.

Other respiratory system infections include the bacterial infections **diphtheria** (dif-thē′rē-ă), **whooping cough** (pertussis; per-tŭs′is), and **tuberculosis** (tū-ber′kū-lō′sis) and the fungal infections **histoplasmosis** (his′tō-plaz-mō′sis) and **coccidioidomycosis** (kok-sid-ē-oy′dō-mī-kō′sis). Vaccines against diphtheria and whooping cough are part of the normal vaccination procedure for children in the United States.

Key terms are set in boldface where they are defined in the chapter, and most terms are included in the glossary at the end of the book. Pronunciation guides are provided for difficult words. Because knowing the original meaning of a term can enhance understanding and retention, derivations of key words are given when they are relevant. Additionally, a handy list of prefixes, suffixes, and combining forms is printed on the inside back cover as a quick reference to help you identify commonly used word roots. A list of abbreviations used throughout the text is also provided.

Chapter Summary

The summary outline briefly states the important facts and concepts covered in each chapter to provide a convenient "big picture" of the chapter content.

SUMMARY

Respiration includes the movement of air into and out of the lungs, the exchange of gases between the air and the blood, the transport of gases in the blood, and the exchange of gases between the blood and tissues.

Functions of the Respiratory System (p. •••)

The major functions of the respiratory system are gas exchange, regulation of blood pH, voice production, olfaction, and protection against some microorganisms.

Anatomy and Histology of the Respiratory System (p. •••)

Nose

1. The nose consists of the external nose and the nasal cavity.
2. The bridge of the nose is bone, and most of the external nose is

Trachea

1. The trachea connects the larynx to the main bronchi.
2. The trachealis muscle regulates the diameter of the trachea.

Tracheobronchial Tree

1. The trachea divides to form two main bronchi, which go to the lungs. The main bronchi divide to form lobar bronchi, which divide to form segmental bronchi, which divide to form bronchioles, which divide to form terminal bronchioles.
2. The trachea to the terminal bronchioles is a passageway for air movement.
   - The area from the trachea to the terminal bronchioles is ciliated to facilitate the removal of inhaled debris.
   - Cartilage helps hold the tube system open (from the trachea to the bronchioles).
   - Smooth muscle controls the diameter of the tubes (terminal bronchioles).
3. Terminal bronchioles divide to form respiratory bronchioles, which

Review and Comprehension

These multiple-choice practice questions cover the main points of the chapter. Completing this self-test helps you gauge your mastery of the material. Answers are provided in Appendix E.

REVIEW AND COMPREHENSION

1. The nasal cavity
   a. has openings for the paranasal sinuses.
   b. has a vestibule, which contains the olfactory epithelium.
   c. is connected to the pharynx by the nares.
   d. has passageways called conchae.
   e. is lined with squamous epithelium, except for the vestibule.
2. The nasopharynx
   a. is lined with moist stratified squamous epithelium.
   b. contains the pharyngeal tonsil.
   c. opens into the oral cavity through the fauces.
   d. extends to the tip of the epiglottis.
   e. is an area through which food, drink, and air pass.
3. The larynx
   a. connects the oropharynx to the trachea.
   b. has three unpaired and six paired cartilages.
   c. contains the vocal folds.
   d. contains the vestibular folds.
   e. all of the above.
4. The trachea contains
   a. skeletal muscle.
   b. pleural fluid glands.
   c. C-shaped pieces of cartilage.
   d. all of the above.

Critical Thinking

These innovative exercises encourage you to apply chapter concepts to solve problems. Answering these questions helps build your working knowledge of anatomy and physiology while developing reasoning and critical thinking skills. Answers are provided in Appendix F.

CRITICAL THINKING

1. What effect does rapid (respiratory rate equals 24 breaths per minute), shallow (tidal volume equals 250 mL per breath) breathing have on minute ventilation, alveolar ventilation, and alveolar P_{O_2} and P_{CO_2}?
2. A person's vital capacity is measured while standing and while lying down. What difference, if any, in the measurement do you predict and why?
3. Ima Diver wanted to do some underwater exploration. She did not want to buy expensive SCUBA equipment, however. Instead, she bought a long hose and an inner tube. She attached one end of the hose to the inner tube so that the end was always out of the water, and she inserted the other end of the hose in her mouth and went diving. What happened to her alveolar ventilation and why? How can she compensate for this change? How does diving affect lung compliance and the work of ventilation?
4. The bacteria that cause gangrene (*Clostridium perfringens*) are anaerobic microorganisms that do not thrive in the presence of oxygen. Hyperbaric oxygenation (HBO) treatment places a person in a chamber that contains oxygen at three to four times normal atmospheric pressure. Explain how HBO helps in the treatment of gangrene.

During inspiration, does the left side of the diaphragm move superiorly, move inferiorly, or stay in place?

8. Suppose that the thoracic wall is punctured at the end of a normal expiration, producing a pneumothorax. Does the thoracic wall move inward, outward, or not move?
9. During normal, quiet respiration, when does the maximum rate of diffusion of oxygen in the pulmonary capillaries occur? The maximum rate of diffusion of carbon dioxide?
10. There is experimental evidence that the overuse of erythropoietin (EPO; see chapter 19) reduces athletic performance. What side effects of EPO abuse reduce exercise stamina?
11. Predict what would happen to tidal volume if the vagus nerves were cut, the phrenic nerves were cut, or the intercostal nerves were cut.
12. You and your physiology instructor are trapped in an overturned ship. To escape, you must swim under water a long distance. You tell your instructor it would be a good idea to hyperventilate before making the escape attempt. Your instructor calmly replies, "What good would that do, since your pulmonary capillaries are already 100% saturated with oxygen?" What would you do and why?

TEACHING AND LEARNING SUPPLEMENTS

McGraw-Hill offers various tools and technology products to support the eighth edition of *Anatomy and Physiology.* Students can order supplemental study materials by contacting their campus bookstore. Instructors can obtain teaching aids by calling the McGraw-Hill Customer Service Department at 1-800-338-3987, by visiting our A&P catalog at www.mhhe.com/ap, or by contacting their local McGraw-Hill sales representative.

ARIS TEXT WEBSITE

The ARIS website that accompanies this textbook includes tutorials, animations, practice quizzing, case studies, lab exercises, and more for students. Instructors will find a complete electronic homework and course management system where they can create and share course materials and assignments with colleagues in just a few clicks of the mouse. Instructors can also edit questions, import their own content, and create announcements and/or due dates for assignments. ARIS offers automatic grading and reporting of easy-to-assign homework, quizzing, and testing. Also included on the ARIS website is a detailed correlation guide for this text and Anatomy & Physiology | REVEALED®.

Check out www.aris.mhhe.com, select your subject and textbook, and start benefiting today!

NEW ONLINE PRESENTATION CENTER FOR INSTRUCTORS

Build instructional materials whereever, whenever, and however you want!

Part of the ARIS website, the Presentation Center is a digital library containing assets, such as photos, artwork, animations, PowerPoints, and other media resources, that can be used to create customized lectures, to visually enhance tests and quizzes, and to design compelling course websites or attractive printed support materials.

Nothing could be easier!

Accessed from your textbook's ARIS website, the Presentation Center's dynamic search engine allows instructors to explore by discipline, course, textbook chapter, asset type, or keyword. Simply browse, select, and download the files needed to build engaging course presentations. All assets are copyrighted by McGraw-Hill Higher Education but can be used by instructors for classroom purposes. The visual resources included in this collection are located easily within each chapter:

- **Art.** Full-color digital files of all illustrations in the book can be readily incorporated into lecture presentations, exams, or custom-made classroom materials. In addition, all files are pre-inserted into blank PowerPoint slides for ease of lecture preparation.
- **Photos.** The photos collection contains digital files of all the photographs from the text, which can be reproduced for multiple classroom uses.
- **Tables.** Every table that appears in the text has been saved in electronic form for use in classroom presentations and/or quizzes.

In addition to the content found within each chapter, the Presentation Center contains the following multimedia instructional materials:

- **Active Art.** Active Art consists of art files that have been converted to a format that allows the artwork to be edited inside PowerPoint. Each piece can be broken down to its core elements, grouped or ungrouped, and edited to create customized illustrations.
- **Animations.** Numerous full-color animations illustrating physiologic processes are also provided. Harness the visual impact of processes in motion by importing these files into classroom presentations or online course materials.
- **Lecture Outlines.** Specially prepared custom outlines for each chapter are offered in easy-to-use PowerPoint slides.

INSTRUCTOR'S TESTING AND RESOURCE CD

This cross-platform CD-ROM provides a wealth of resources for the instructor. One of the supplements featured on this CD is EZ Test, a flexible and easy-to-use electronic testing program. This program allows instructors to create tests from book-specific items and accommodates a wide range of question types, including the option for instructors to add their own questions. Multiple versions of the test can be created, and any test can be exported for use with course management systems, such as WebCT, BlackBoard, or PageOut. The instructor's manual is also included with this CD.

AP | REVEALED® STUDENT TUTORIAL

AP | REVEALED® is a unique multimedia study aid designed to help you learn and review human anatomy using digital cadaver specimens. Dissections, animations, imaging, and self-tests all work together as an exceptional tool for the study of structure and function.

The AP | REVEALED® CD series includes

Volume 1—Skeletal and Muscular Systems
Volume 2—Nervous System
Volume 3—Cardiovascular, Respiratory, and Lymphatic Systems
Volume 4–Digestive, Urinary, Reproductive, and Endocrine Systems

A new online version of AP | REVEALED® includes the Integumentary System and expanded physiology and histology content.

Visit www.mhhe.com/aprevealed for more information.

VIRTUAL ANATOMY DISSECTION REVIEW (AVAILABLE ONLINE OR AS A CD-ROM)

This multimedia program contains high-quality cat dissection photographs correlated to illustrations and photos of human structures. The format makes it easy to identify and review cat anatomy and to relate the cat specimen to corresponding human structures.

PHYSIOLOGY INTERACTIVE LAB SIMULATIONS (PH.I.L.S)

This unique study tool contains 26 lab simulations that allow students to perform experiments without using expensive lab equipment or live animals. The easy-to-use interface offers students the flexibility to change the parameters of every lab experiment, with no limit to the amount of times they can repeat experiments or modify variables. The power to manipulate each experiment reinforces key physiology concepts by helping students view outcomes, make predictions, and draw conclusions.

MEDIAPHYS TUTORIAL

This physiology study aid offers detailed explanations, high-quality illustrations, and animations to provide a thorough introduction to the world of physiology. MediaPhys is filled with interactive activities and quizzes to help reinforce physiology concepts that are often difficult to understand.

e-INSTRUCTION WITH CPS

The Classroom Performance System (CPS) is an interactive system that allows the instructor to administer in-class questions electronically. Students answer questions via hand-held remote control keypads (clickers), and their individual responses are logged into a gradebook. Aggregated responses can be displayed in graphical form. Using this immediate feedback, the instructor can quickly determine if students understand the lecture topic, or if more clarification is needed. CPS promotes student participation, class productivity, and individual student confidence and accountability. Specially designed questions for e-Instruction to accompany *Anatomy and Physiology* are provided through the book's ARIS website.

TRANSPARENCIES

The set of transparency acetates that accompanies this text includes 1200 full-color images identified by the authors as the most useful figures to incorporate into lecture presentations.

COURSE DELIVERY SYSTEMS

In addition to McGraw-Hill's ARIS course management options, instructors can also design and control their course content with help from our partners WebCT, Blackboard, Top-Class, and eCollege. Course cartridges containing website content, online testing, and powerful student tracking features are readily available for use within these platforms.

PART 1

ORGANIZATION OF THE HUMAN BODY

The Human Organism 1

What lies ahead is an astounding adventure—learning about the structure and function of the human body and how they are regulated by intricate systems of checks and balances. For example, tiny collections of cells embedded in the pancreas affect the uptake and use of blood sugar in the body. Eating a candy bar results in an increase in blood sugar, which acts as a stimulus. Pancreatic cells respond to the stimulus by secreting insulin. Insulin moves into blood vessels and is transported to cells throughout the body, where it increases the movement of sugar from the blood into cells, thereby providing the cells with a source of energy and causing blood sugar levels to decrease.

Knowledge of the structure and function of the human body is the basis for understanding disease. In one type of diabetes mellitus, cells of the pancreas do not secrete adequate amounts of insulin. Not enough sugar moves into cells, which deprives them of a needed source of energy, and they malfunction.

Knowledge of the structure and function of the human body is essential for those planning a career in the health sciences. It is also beneficial to nonprofessionals because it helps with understanding overall health and disease, with evaluating recommended treatments, and with critically reviewing advertisements and articles.

This chapter defines *anatomy and physiology* (p. 2). It also explains the body's *structural and functional organization* (p. 2) and provides an overview of the *characteristics of life* (p. 6), *biomedical research* (p. 9), and *homeostasis* (p. 9). Finally, the chapter presents *terminology and the body plan* (p. 12).

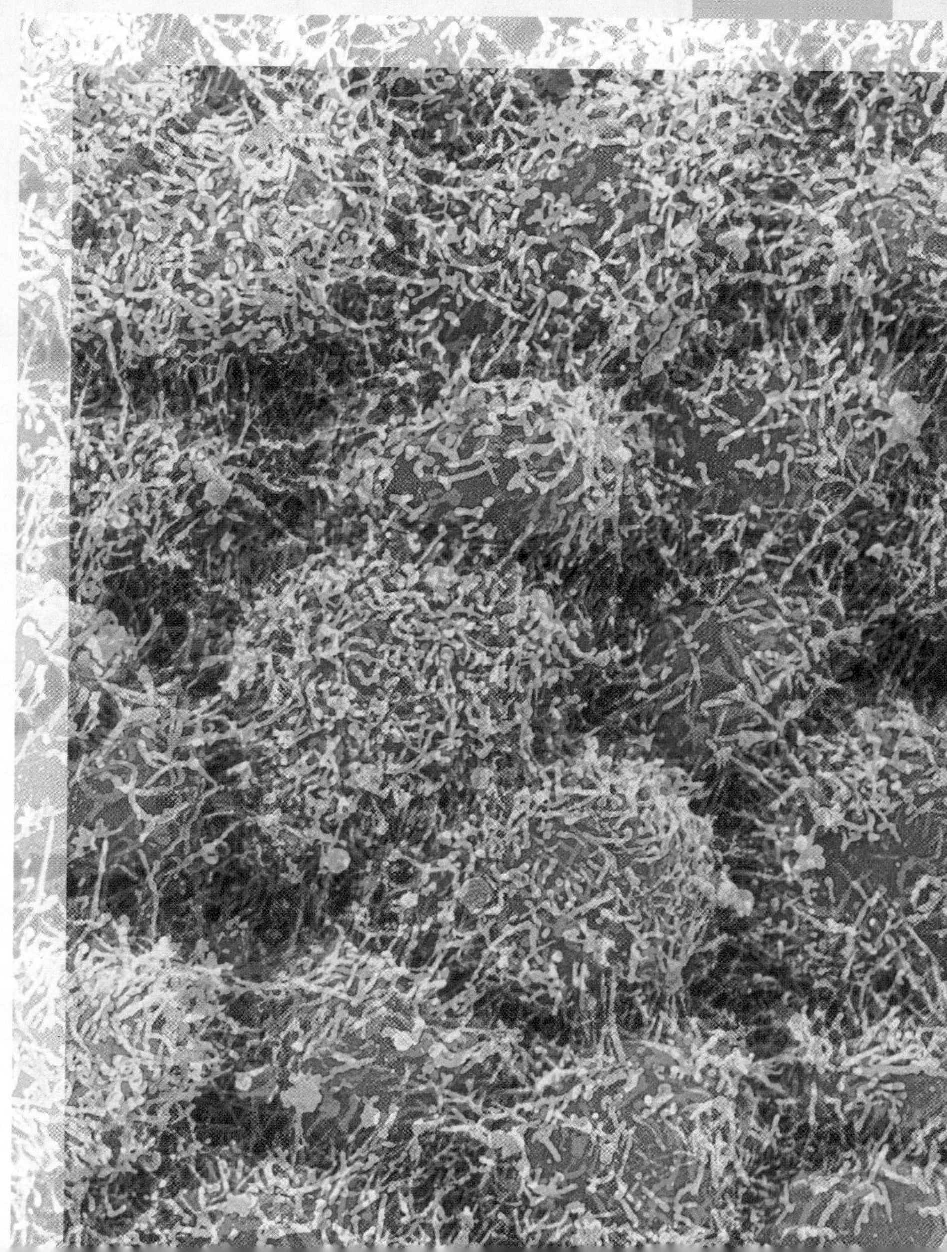

The human organism has many membranes that enclose and protect underlying structures. The colorized scanning electron micrograph shows the cells forming the peritoneum, a membrane covering abdominopelvic organs and the inside wall of the abdominopelvic cavity. These cells have many short, hair-like projections called microvilli. The microvilli increase the surface area of the cells, enabling them to secrete a slippery lubricating fluid that protects organs from friction as they rub against one another or the inside of the abdominopelvic wall.

ANATOMY AND PHYSIOLOGY

Anatomy is the scientific discipline that investigates the body's structure. For example, anatomy describes the shape and size of bones. In addition, anatomy examines the relationship between the structure of a body part and its function. Just as the structure of a hammer makes it well suited for pounding nails, the structure of a specific body part allows it to perform a particular function effectively. For example, bones can provide strength and support because bone cells surround themselves with a hard, mineralized substance. Understanding the relationship between structure and function makes it easier to understand and appreciate anatomy.

Anatomy can be considered at many different levels. **Developmental anatomy** is the study of the structural changes that occur between conception and adulthood. **Embryology** (em-brē-ol′ō-jē), a subspeciality of developmental anatomy, considers changes from conception to the end of the eighth week of development. Most birth defects occur during embryologic development.

Some structures, such as cells, are so small that they are best studied using a microscope. **Cytology** (sī-tol′ōjē) examines the structural features of cells, and **histology** (his-tol′ōjē) examines tissues, which are cells and the materials surrounding them.

Gross anatomy, the study of structures that can be examined without the aid of a microscope, can be approached from either a systemic or a regional perspective. In **systemic anatomy,** the body is studied system by system, which is the approach taken in this and most other introductory textbooks. A system is a group of structures that have one or more common functions. Examples are the circulatory, nervous, respiratory, skeletal, and muscular systems. In **regional anatomy,** the body is studied area by area, which is the approach taken in most graduate programs at medical and dental schools. Within each region, such as the head, abdomen, or arm, all systems are studied simultaneously.

Surface anatomy is the study of the external form of the body and its relation to deeper structures. For example, the sternum (breastbone) and parts of the ribs can be seen and palpated (felt) on the front of the chest. These structures can be used as anatomical landmarks to identify regions of the heart and points on the chest where certain heart sounds can best be heard. **Anatomical imaging** uses radiographs (x-rays), ultrasound, magnetic resonance imaging (MRI), and other technologies to create pictures of internal structures (see Clinical Focus "Anatomical Imaging," p. 4). Both surface anatomy and anatomic imaging provide important information about the body for diagnosing disease.

Physiology is the scientific investigation of the processes or functions of living things. Although it may not be obvious at times, living things are ever-changing, not static. The major goals of physiology are to understand and predict the body's responses to stimuli and to understand how the body maintains conditions within a narrow range of values in a constantly changing environment.

Like anatomy, physiology can be considered at many different levels. **Cell physiology** examines the processes occurring in cells and **systemic physiology** considers the functions of organ systems. **Neurophysiology** focuses on the nervous system and **cardiovascular physiology** deals with the heart and blood vessels. Physiology often examines systems rather than regions because portions of a system in more than one region can be involved in a given function.

The study of the human body must encompass both anatomy and physiology because structures, functions, and processes are interwoven. **Pathology** (pa-thol′ō-jē) is the medical science dealing with all aspects of disease, with an emphasis on the cause and development of abnormal conditions, as well as the structural and functional changes resulting from disease. **Exercise physiology** focuses on changes in function, and in structure, caused by exercise.

1 ***Define anatomy and physiology. Describe different levels at which each can be considered.***

2 ***Define pathology and exercise physiology.***

Anatomical Anomalies

No two humans are structurally identical. For instance, one person may have longer fingers than another person. Despite this variability, most humans have the same basic pattern. Normally, we each have 10 fingers. **Anatomical anomalies** are structures that are unusual and different from the normal pattern. For example, some individuals have 12 fingers.

Anatomical anomalies can vary in severity from the relatively harmless to the life-threatening. For example, each kidney is normally supplied by one blood vessel, but in some individuals a kidney is supplied by two blood vessels. Either way, the kidney receives adequate blood. On the other hand, in the condition called "blue baby" syndrome certain blood vessels arising from the heart of an infant are not attached in their correct locations; blood is not effectively pumped to the lungs, resulting in tissues not receiving adequate oxygen.

STRUCTURAL AND FUNCTIONAL ORGANIZATION

The body can be considered to have six levels of organization: the chemical, cell, tissue, organ, organ system, and complete organism levels (figure 1.1).

1. *Chemical level.* The chemical level involves interactions between atoms, which are tiny building blocks of matter. Atoms can combine to form molecules, such as water, sugar, fats, and proteins. The function of a molecule is related intimately to its structure. For example, collagen molecules are ropelike protein fibers that give skin structural strength and flexibility. With old age, the structure of collagen changes, and the skin becomes fragile and is torn more easily. A brief overview of chemistry is presented in chapter 2.

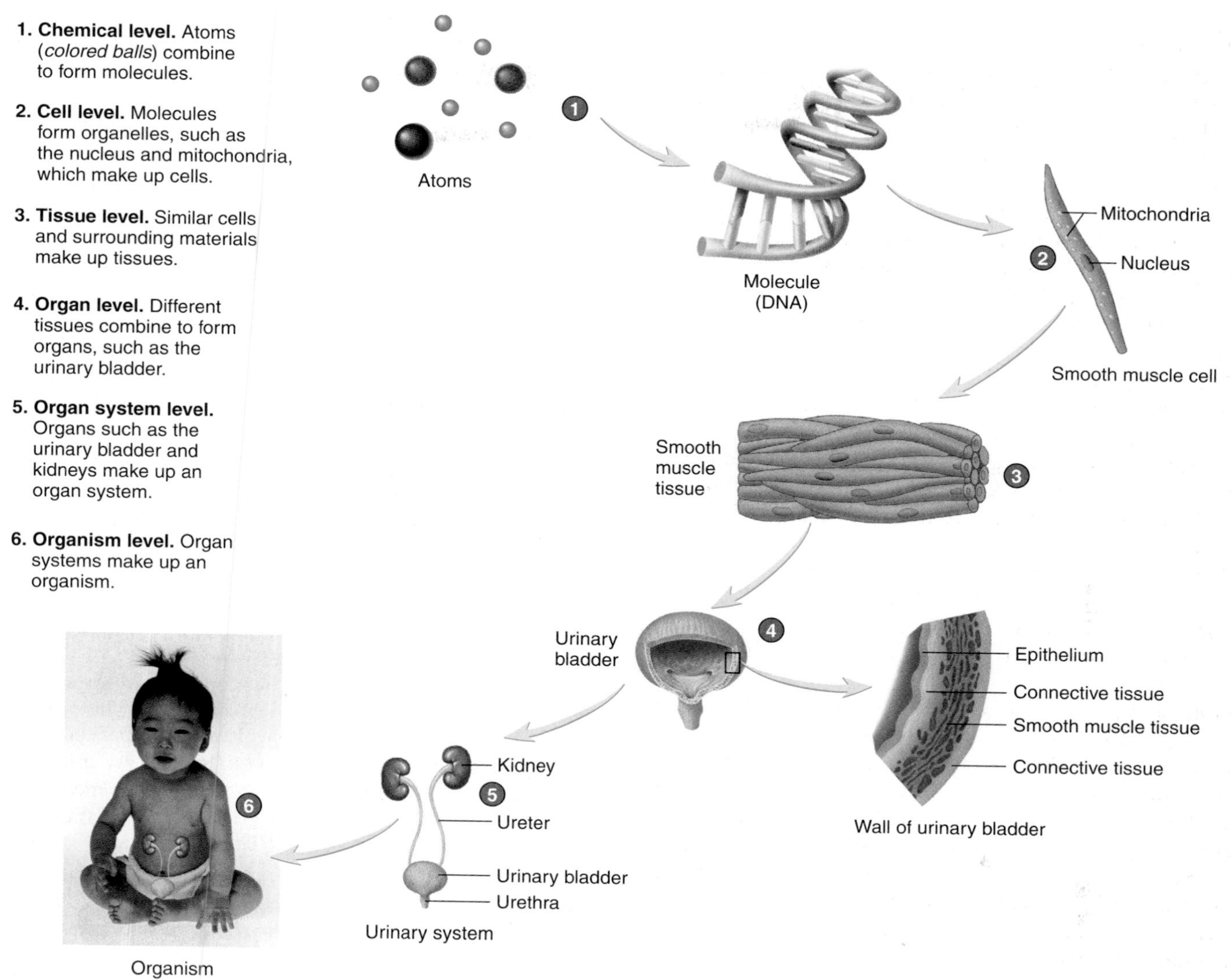

FIGURE 1.1 Levels of Organization
Six levels of organization for the human body are the chemical, cell, tissue, organ, organ system, and organism levels.

2. *Cell level.* **Cells** are the basic structural and functional units of organisms, such as plants and animals. Molecules can combine to form **organelles** (or′gă-nelz), which are the small structures that make up cells. For example, the nucleus contains the cell's hereditary information, and mitochondria manufacture adenosine triphosphate (ATP), which is a molecule used by cells for a source of energy. Although cell types differ in their structure and function, they have many characteristics in common. Knowledge of these characteristics and their variations is essential to a basic understanding of anatomy and physiology. The cell is discussed in chapter 3.
3. *Tissue level.* A **tissue** is a group of similar cells and the materials surrounding them. The characteristics of the cells and surrounding materials determine the functions of the tissue. The numerous different tissues that make up the body are classified into four basic types: epithelial, connective, muscle, and nervous. Tissues are discussed in chapter 4.
4. *Organ level.* An **organ** is composed of two or more tissue types that perform one or more common functions. The urinary bladder, heart, skin, and eye are examples of organs (figure 1.2).
5. *Organ system level.* An **organ system** is a group of organs that have a common function or set of functions and are therefore viewed as a unit. For example, the urinary system consists of the kidneys, ureter, urinary bladder, and urethra. The kidneys produce urine, which is transported by the ureters to the urinary bladder, where it is stored until eliminated from the body by passing through the urethra. In this book, the body is considered to have 11 major organ systems:

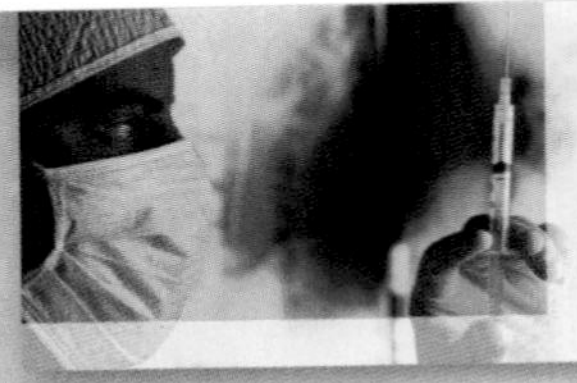

CLINICAL FOCUS

Anatomical Imaging

Anatomical imaging has revolutionized medical science. Some estimate that during the past 20 years as much progress has been made in clinical medicine as in all its previous history combined, and anatomical imaging has made a major contribution to that progress. Anatomical imaging allows medical personnel to look inside the body with amazing accuracy and without the trauma and risk of exploratory surgery. Although most of the technology of anatomical imaging is very new, the concept and earliest technology are quite old.

Wilhelm Roentgen (1845–1923) was the first to use **x-rays** in medicine in 1895 to see inside the body. The rays were called x-rays because no one knew what they were. This extremely shortwave electromagnetic radiation (see chapter 2) moves through the body, exposing a photographic plate to form a **radiograph** (rā′dē-ō-graf). Bones and radiopaque dyes absorb the rays and create underexposed areas that appear white on the photographic film (figure A). X-rays have been in common use for many years and have numerous applications. Almost everyone has had a radiograph taken, either to visualize a broken bone or to check for a cavity in a tooth. A major limitation of radiographs, however, is that they give only a flat, two-dimensional (2-D) image of the body, which is a three-dimensional (3-D) structure.

Ultrasound is the second oldest imaging technique. It was first developed in the early 1950s as an extension of World War II sonar technology and uses high-frequency sound waves. The sound waves are emitted from a transmitter–receiver placed on the skin over the area to be scanned. The sound waves strike internal organs and bounce back to the receiver on the skin. Even though the basic technology is fairly old, the most important advances in the field occurred only after it became possible to analyze the reflected sound waves by computer. Once the computer analyzes the pattern of sound waves, the information is transferred to a monitor, where the result is visualized as an ultrasound image called a **sonogram** (son′ō-gram) (figure B). One of the more recent advances in ultrasound technology is the ability of more advanced computers to analyze changes in position through time and to display those changes as "real-time" movements. Among other medical uses, ultrasound is commonly used to evaluate the condition of the fetus during pregnancy.

Computer analysis is also the basis of another major medical breakthrough in imaging. **Computed tomographic** (tō′mō-graf′ik) **(CT) scans,** developed in 1972 and originally called **computerized axial tomographic (CAT) scans,** are computer-analyzed x-ray images. A low-intensity x-ray tube is rotated through a 360-degree arc around the patient, and the images are fed into a computer. The computer then constructs the image of a "slice" through the body at the point where the x-ray beam was focused and rotated (figure C). It is also possible with some computers to take several scans short distances apart and stack the slices to produce a 3-D image of a part of the body (figure D).

Dynamic spatial reconstruction (DSR) takes CT one step further. Instead of using a single rotating x-ray machine to take single slices and add them together, DSR uses about 30 x-ray tubes. The images from all the tubes are compiled simultaneously to rapidly produce a 3-D image. Because of the speed of the process, multiple images can be compiled to

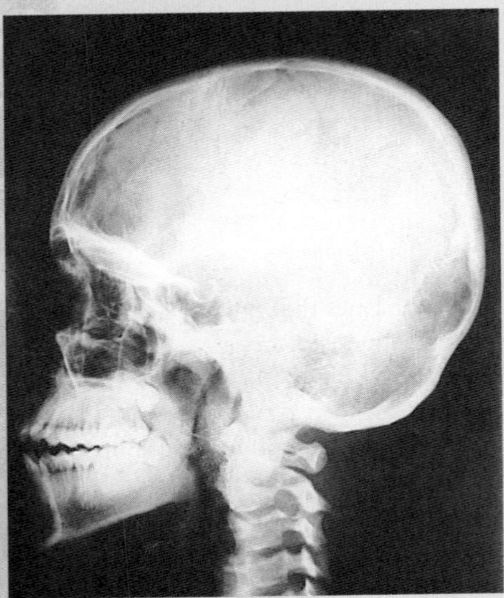

FIGURE A X-ray
Radiograph produced by x-rays shows a lateral view of the head and neck.

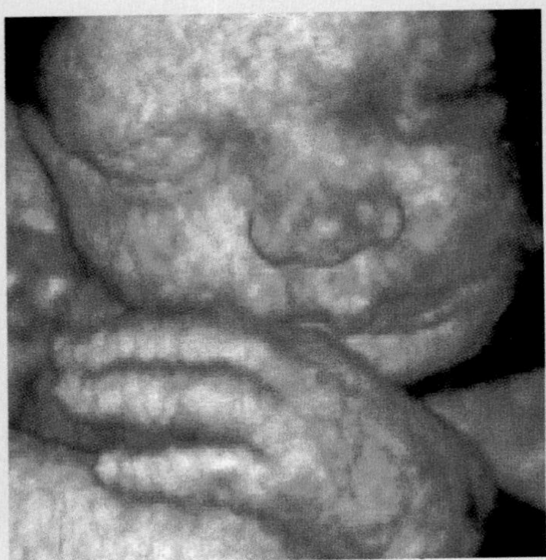

FIGURE B Ultrasound
Sonogram produced with ultrasound shows the face and hand of a fetus within the uterus.

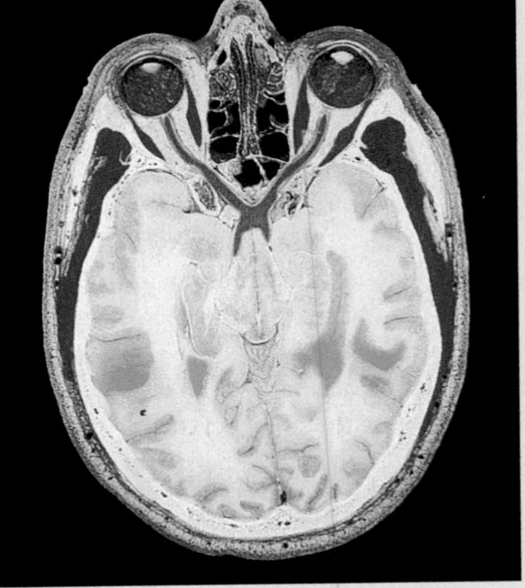

FIGURE C Computed Tomography
Transverse section through the skull at the level of the eyes.

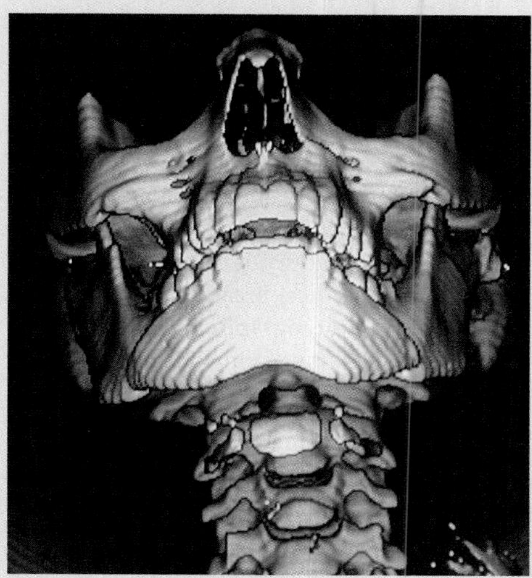

FIGURE D Computed Tomography (CT)
Stacking of images acquired using CT technology.

show changes through time, thereby giving the system a dynamic quality. This system allows us to move away from seeing only static structure and toward seeing dynamic structure and function.

Digital subtraction angiography (an-jē-og′ră-fē) (**DSA**) is also one step beyond CT scans. A 3-D radiographic image of an organ, such as the brain, is made and stored in a computer. A radiopaque dye is injected into the circulation, and a second radiographic computer image is made. The first image is subtracted from the second one, greatly enhancing the differences, with the primary difference being the presence of the injected dye (figure E).

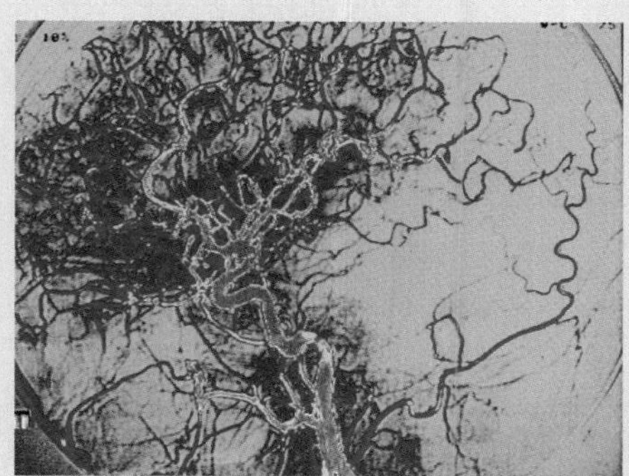

FIGURE E Digital Subtraction Angiography (DSA)
Lateral view of the head reveals the major blood vessels of the brain.

These computer images can be dynamic and used, for example, to guide a catheter into a carotid artery during angioplasty, which is the insertion of a tiny balloon into a carotid artery to compress material clogging the artery.

Magnetic resonance imaging (**MRI**) directs radio waves at a person lying inside a large electromagnetic field. The magnetic field causes the protons of various atoms to align (see chapter 2). Because of the large amounts of water in the body, the alignment of hydrogen atom protons is at present most important in this imaging system. Radio waves of certain frequencies, which change the alignment of the hydrogen atoms, then are directed at the patient. When the radio waves are turned off, the hydrogen atoms realign in accordance with the magnetic field. The time it takes the hydrogen atoms to realign is different for various tissues of the body. These differences can be analyzed by computer to produce very clear sections through the body (figure F). The technique is also very sensitive in detecting some forms of cancer and can detect a tumor far more readily than can a CT scan.

Positron emission tomographic (**PET**) **scans** can identify the metabolic states of various tissues. This technique is particularly useful in analyzing the brain. When cells are active, they are using energy. The energy they need is supplied by the breakdown of glucose (blood sugar). If radioactively treated, or "labeled," glucose is given to a patient, the active cells take up the labeled glucose. As the radioactivity in the glucose decays, positively charged subatomic particles called positrons are emitted. When the positrons collide with electrons, the two particles annihilate each other, and gamma rays are given off. The gamma rays can be detected, pinpointing the cells that are metabolically active (figure G).

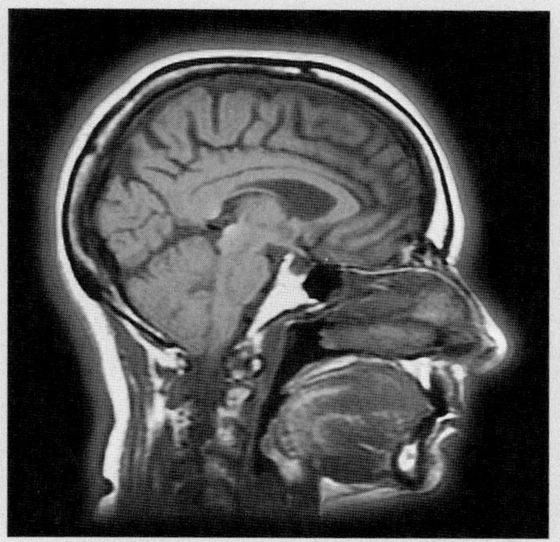

FIGURE F Magnetic Resonance Imaging (MRI)
Sagittal section of the head and neck.

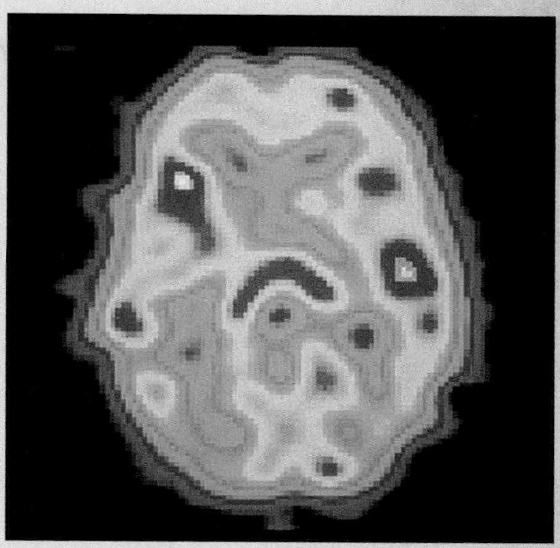

FIGURE G Positron Emission Tomography (PET)
Transverse section through the skull. The highest level of brain activity is indicated in red, with successively lower levels represented by yellow, green, and blue.

Whenever the human body is exposed to x-rays, ultrasound, electromagnetic fields, or radioactively labeled substances, a potential risk exists. In the medical application of anatomical imaging, the risk must be weighed against the benefit. Numerous studies have been conducted and are still being done to determine the outcomes of diagnostic and therapeutic exposures to x-rays.

The risk of anatomical imaging is minimized by using the lowest possible doses that provide the necessary information. For example, it is well known that x-rays can cause cell damage, particularly to the reproductive cells. Thus, the number of x-rays and the level of exposure are kept to a minimum; the x-ray beam is focused as closely as possible to avoid scattering of the rays; areas of the body not being x-rayed are shielded; and personnel administering x-rays are shielded. No known risks exist from ultrasound or electromagnetic fields at the levels used for diagnosis.

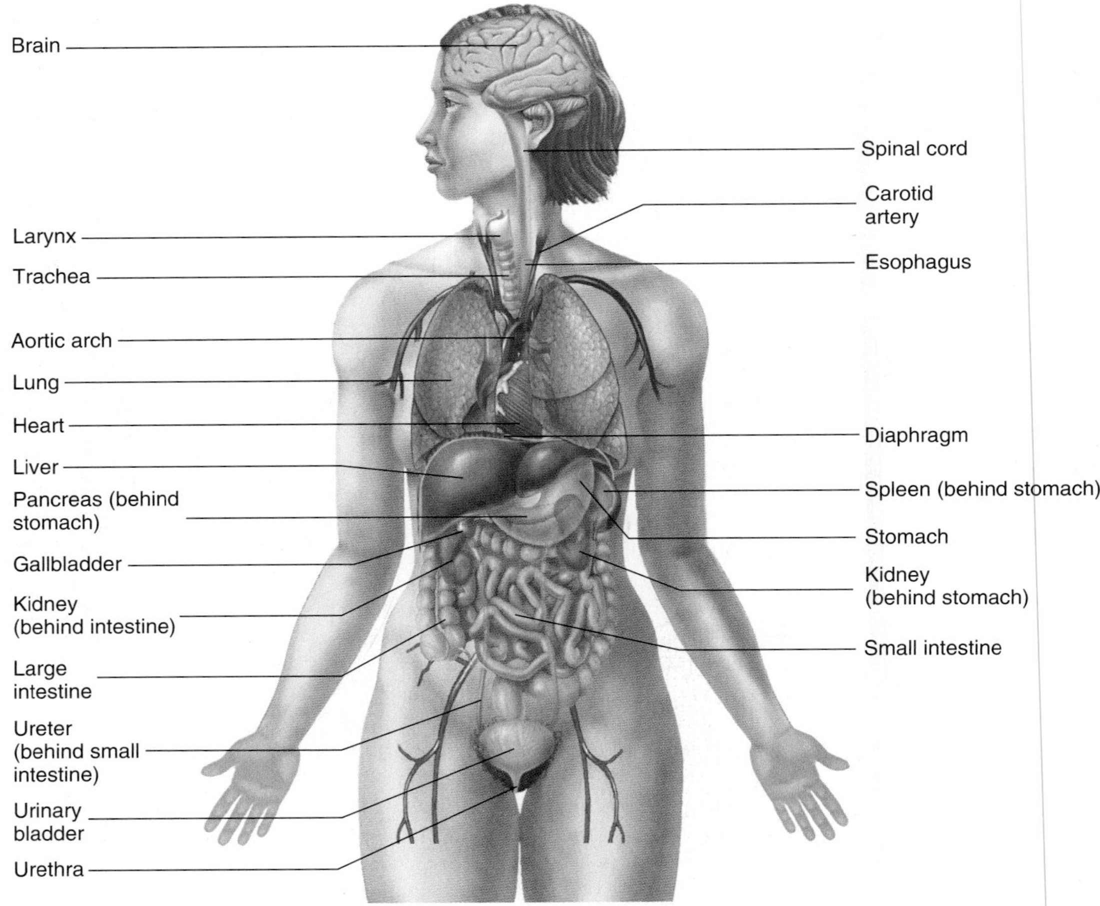

FIGURE 1.2 **Organs of the Body**

the integumentary, skeletal, muscular, nervous, endocrine, cardiovascular, lymphatic, respiratory, digestive, urinary, and reproductive systems. Figure 1.3 presents a brief summary of the organ systems and their functions.

6. *Organism level.* An **organism** is any living thing considered as a whole—whether composed of one cell, such as a bacterium, or of trillions of cells, such as a human. The human organism is a complex of organ systems, all mutually dependent on one another.

3 ***From smallest to largest, list and define the body's six levels of organization.***

4 ***What are the four primary tissue types?***

5 ***Which two organ systems are responsible for regulating the other organ systems (see figure 1.3)? Which two are responsible for support and movement?***

6 ***What are the functions of the integumentary, cardiovascular, lymphatic, respiratory, digestive, urinary, and reproductive systems (see figure 1.3)?***

PREDICT 1

In one type of diabetes, the pancreas (an organ) fails to produce insulin, which is a chemical normally made by pancreatic cells and released into the circulation. List as many levels of organization as you can in which this disorder could be corrected.

CHARACTERISTICS OF LIFE

Humans are organisms, sharing characteristics with other organisms. The most important common feature of all organisms is life. Organization, metabolism, responsiveness, growth, development, and reproduction are life's essential characteristics.

Organization is the condition in which the parts of an organism have specific relationships to each other and the parts interact to perform specific functions. Living things are highly organized. All organisms are composed of one or more cells. Cells in turn are composed of highly specialized organelles, which depend on the precise organization of large molecules. Disruption of this organized state can result in loss of functions, even death.

Metabolism (mĕ-tab′ō-lizm) is all of the chemical reactions taking place in an organism. It includes an organism's ability to break down food molecules, which it uses as a source of energy and raw materials to synthesize its own molecules. Energy is also used when one part of a molecule moves relative to another part, resulting in a change in shape of the molecule. Changes in molecular shape in turn can change the shape of cells, which can produce movements of the organism. Metabolism is necessary for vital functions, such as responsiveness, growth, development, and reproduction.

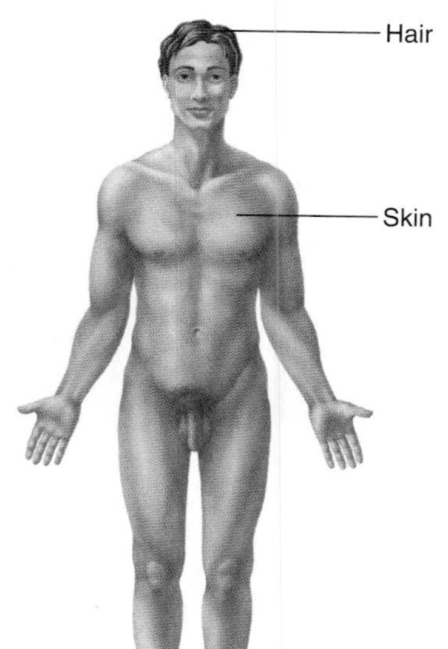

Integumentary System

Provides protection, regulates temperature, reduces water loss, and produces vitamin D precursors. Consists of skin, hair, nails, and sweat glands.

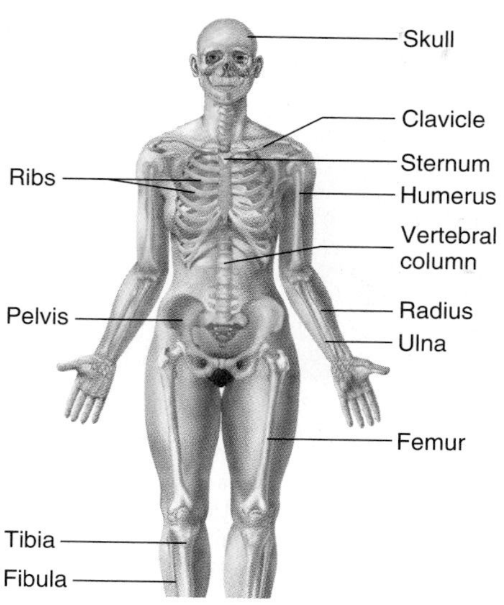

Skeletal System

Provides protection and support, allows body movements, produces blood cells, and stores minerals and fat. Consists of bones, associated cartilages, ligaments, and joints.

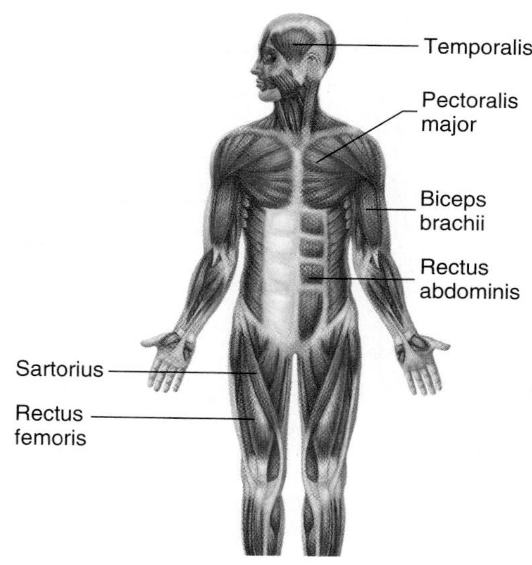

Muscular System

Produces body movements, maintains posture, and produces body heat. Consists of muscles attached to the skeleton by tendons.

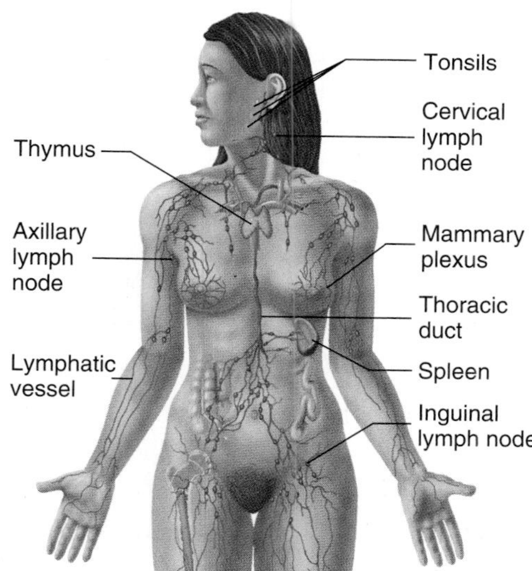

Lymphatic System

Removes foreign substances from the blood and lymph, combats disease, maintains tissue fluid balance, and transports fats from the digestive tract. Consists of the lymphatic vessels, lymph nodes, and other lymphatic organs.

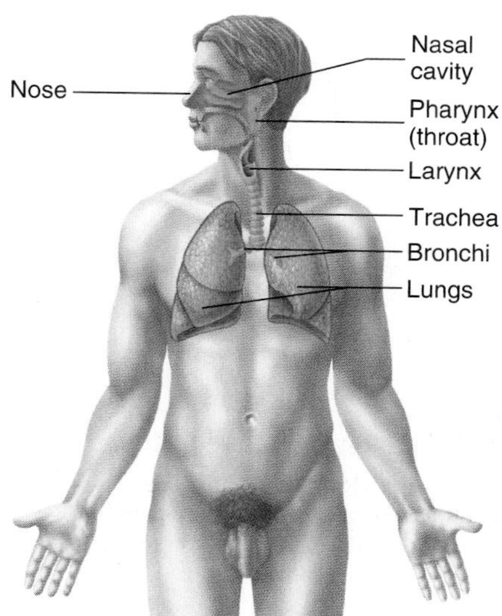

Respiratory System

Exchanges oxygen and carbon dioxide between the blood and air and regulates blood pH. Consists of the lungs and respiratory passages.

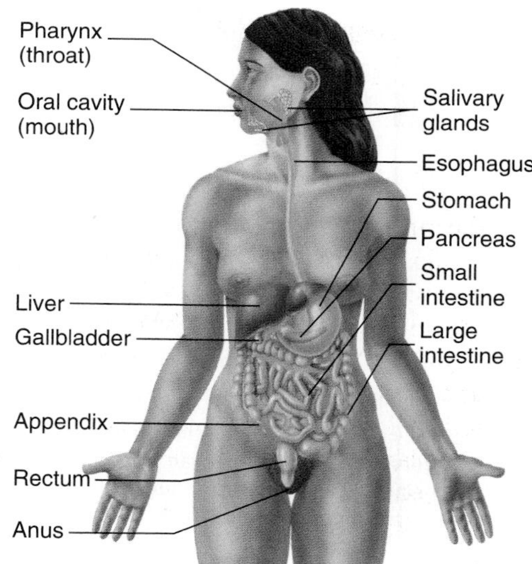

Digestive System

Performs the mechanical and chemical processes of digestion, absorption of nutrients, and elimination of wastes. Consists of the mouth, esophagus, stomach, intestines, and accessory organs.

FIGURE 1.3 **Organ Systems of the Body**

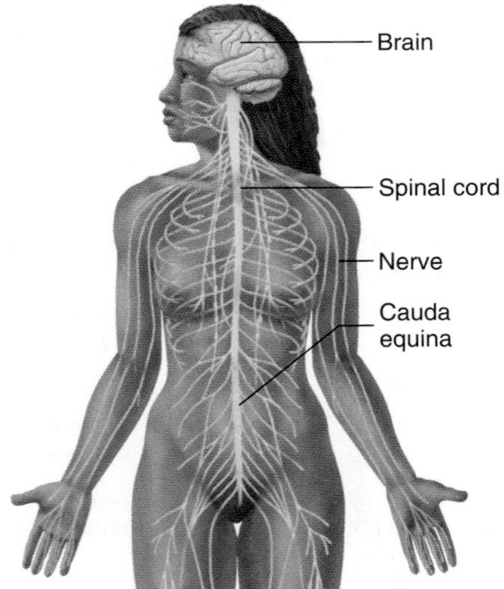

Nervous System

A major regulatory system that detects sensations and controls movements, physiologic processes, and intellectual functions. Consists of the brain, spinal cord, nerves, and sensory receptors.

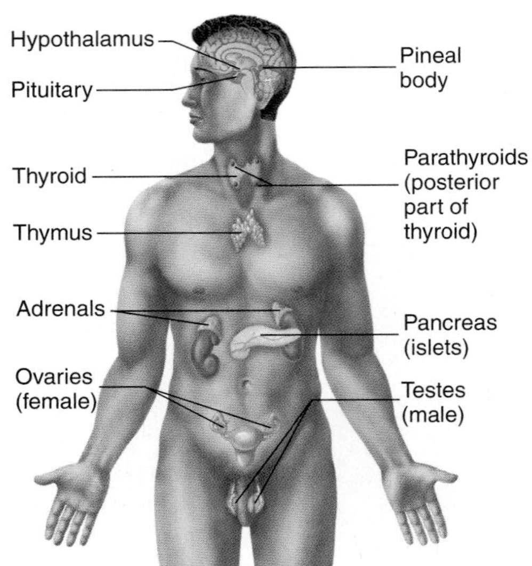

Endocrine System

A major regulatory system that influences metabolism, growth, reproduction, and many other functions. Consists of glands, such as the pituitary, that secrete hormones.

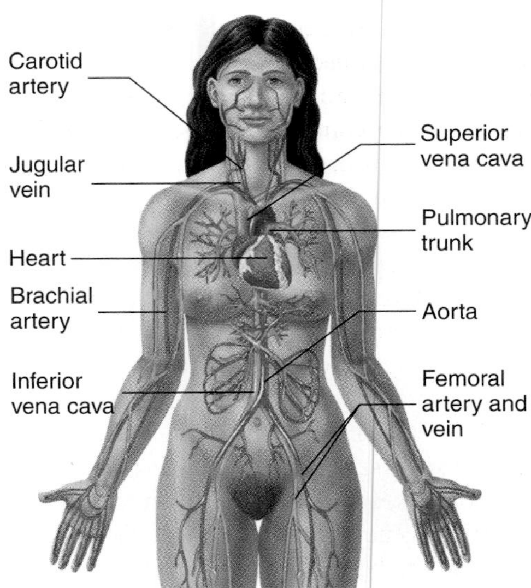

Cardiovascular System

Transports nutrients, waste products, gases, and hormones throughout the body; plays a role in the immune response and the regulation of body temperature. Consists of the heart, blood vessels, and blood.

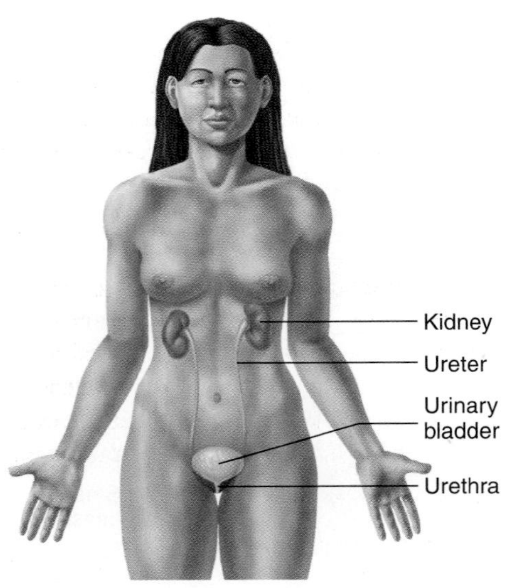

Urinary System

Removes waste products from the blood and regulates blood pH, ion balance, and water balance. Consists of the kidneys, urinary bladder, and ducts that carry urine.

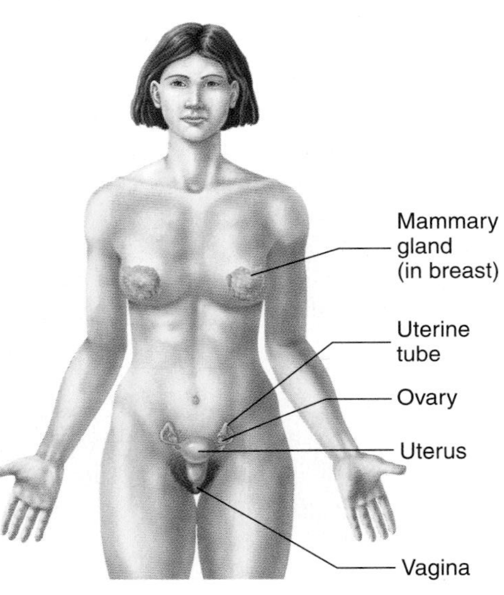

Female Reproductive System

Produces oocytes and is the site of fertilization and fetal development; produces milk for the newborn; produces hormones that influence sexual functions and behaviors. Consists of the ovaries, vagina, uterus, mammary glands, and associated structures.

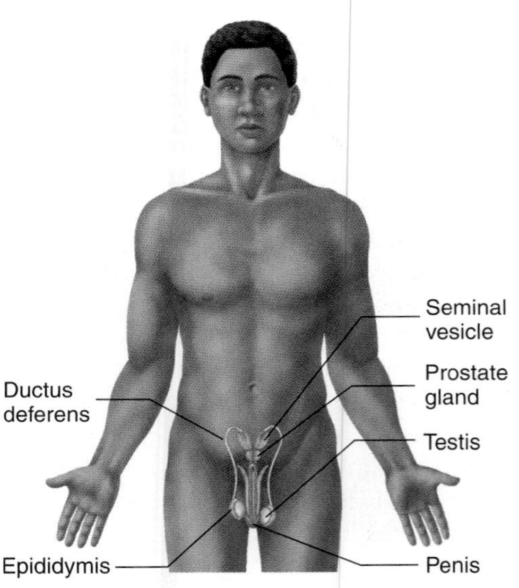

Male Reproductive System

Produces and transfers sperm cells to the female and produces hormones that influence sexual functions and behaviors. Consists of the testes, accessory structures, ducts, and penis.

FIGURE 1.3 ***(continued)***

Responsiveness is an organism's ability to sense changes in its external or internal environment and adjust to those changes. Responses include such things as moving toward food or water and away from danger or poor environmental conditions. Organisms can also make adjustments that maintain their internal environment. For example, if body temperature increases in a hot environment, sweat glands produce sweat, which can lower body temperature back toward normal levels.

Growth happens when cells increase in size or number, which produces an overall enlargement of all or part of an organism. For

example, a muscle enlarged by exercise has larger muscle cells than an untrained muscle, and the skin of an adult has more cells than the skin of an infant. An increase in the materials surrounding cells can also contribute to growth. For instance, the growth of bone results from an increase in cell number and the deposition of mineralized materials around the cells.

Development includes the changes an organism undergoes through time; it begins with fertilization and ends at death. The greatest developmental changes occur before birth, but many changes continue after birth, and some continue throughout life. Development usually involves growth, but it also involves differentiation and morphogenesis. **Differentiation** is change in cell structure and function from generalized to specialized, and **morphogenesis** (mōr-fō-jen′ĕ-sis) is change in the shape of tissues, organs, and the entire organism. For example, following fertilization, generalized cells specialize to become specific cell types, such as skin, bone, muscle, or nerve cells. These differentiated cells form the tissues and organs.

Reproduction is the formation of new cells or new organisms. Without reproduction, growth and development are not possible. Without reproduction of the organism, species become extinct.

BIOMEDICAL RESEARCH

Studying other organisms has increased our knowledge about humans because humans share many characteristics with other organisms. For example, studying single-celled bacteria provides much information about human cells. Some biomedical research, however, cannot be accomplished using single-celled organisms or isolated cells. Sometimes other mammals must be studied. For example, great progress in open-heart surgery and kidney transplantation was made possible by perfecting surgical techniques on other mammals before attempting them on humans. Strict laws govern the use of animals in biomedical research—laws designed to ensure minimum suffering on the part of the animal and to discourage unnecessary experimentation.

Although much can be learned from studying other organisms, the ultimate answers to questions about humans can be obtained only from humans because other organisms are different from humans in significant ways.

Human Versus Animal-Based Knowledge

Failure to appreciate the differences between humans and other animals led to many misconceptions by early scientists. One of the first great anatomists was a Greek physician, Claudius Galen (ca. 130–201). Galen described a large number of anatomical structures supposedly present in humans but observed only in other animals. For example, he described the liver as having five lobes. This is true for rats, but not for humans, who have four-lobed livers. The errors introduced by Galen persisted for more than 1300 years until a Flemish anatomist, Andreas Vesalius (1514–1564), who is considered the first modern anatomist, carefully examined human cadavers and began to correct the textbooks. This example should serve as a word of caution: Some current knowledge in molecular biology and physiology has not been confirmed in humans.

7 ***Describe six characteristics of life.***

8 ***Why is it important to realize that humans share many, but not all, characteristics with other animals?***

HOMEOSTASIS

Homeostasis (hō′mē-ō-stā′sis) is the existence and maintenance of a relatively constant environment within the body. A small amount of fluid surrounds each cell of the body. For cells to function normally, the volume, temperature, and chemical content—conditions known as **variables** because their values can change—of this fluid must remain within a narrow range. Body temperature is a variable that can increase in a hot environment or decrease in a cold one.

Homeostatic mechanisms, such as sweating or shivering, normally maintain body temperature near an ideal normal value, or **set point** (figure 1.4). Note that these mechanisms are not able to maintain body temperature precisely at the set point. Instead, body temperature increases and decreases slightly around the set point to produce a **normal range** of values. As long as body temperature remains within this normal range, homeostasis is maintained.

The organ systems help control the body's internal environment so that it remains relatively constant. For example, the digestive, respiratory, circulatory, and urinary systems function together so that each cell in the body receives adequate oxygen and nutrients and so that waste products do not accumulate to a toxic level. If the fluid surrounding cells deviates from homeostasis, the cells do not function normally and can even die. Disease disrupts homeostasis and sometimes results in death.

Negative Feedback

Most systems of the body are regulated by **negative-feedback** mechanisms, which maintain homeostasis. *Negative* means that any deviation from the set point is made smaller or is resisted. For example, maintaining normal blood pressure is necessary for homeostasis because pressure is required to move blood from the heart through the tissues. The blood supplies the tissues with oxygen and nutrients

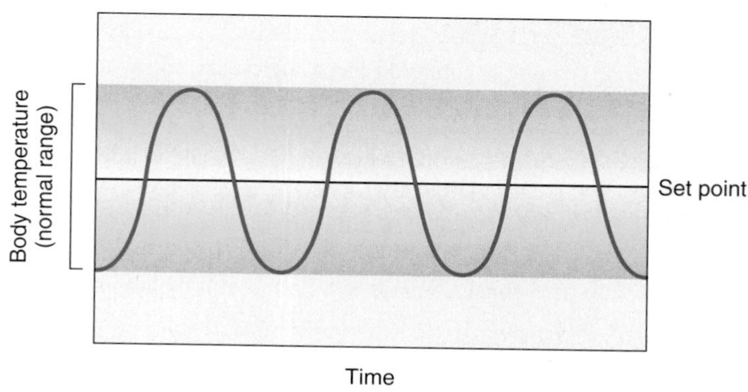

FIGURE 1.4 Homeostasis

Homeostasis is the maintenance of a variable around an ideal normal value, or set point. The value of the variable fluctuates around the set point to establish a normal range of values.

and removes waste products, thus maintaining tissue homeostasis. If blood pressure deviates from its set point, negative-feedback mechanisms return blood pressure toward the set point.

Many negative-feedback mechanisms have three components: a **receptor,** which monitors the value of a variable; a **control center,** which receives information about the variable from the receptor, establishes the set point, and controls the effector; and an **effector,** which produces responses that change the value of the variable. Several negative-feedback mechanisms regulate blood pressure, and they are described more fully in chapters 20 and 21. One negative-feedback mechanism regulating blood pressure is described here. Receptors that monitor blood pressure are located within large blood vessels near the heart and head. A control center located in the brain receives signals sent through nerves from the receptors. The control center evaluates the information and sends signals through nerves to the heart. The heart is the effector, and the heart rate increases or decreases in response to signals from the brain (figure 1.5).

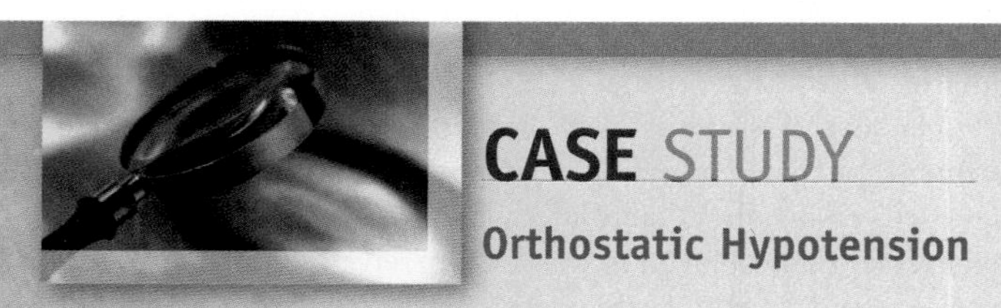

CASE STUDY

Orthostatic Hypotension

Molly is a 75-year-old widow who lives alone. For 2 days, she had a fever and chills and stayed mostly in bed. On rising to go to the bathroom, she felt dizzy, fainted, and fell to the floor. Molly quickly regained consciousness and managed to call her son, who took her to the emergency room, where a diagnosis of orthostatic hypotension was made.

Orthostasis literally means to stand and *hypotension* refers to low blood pressure. **Orthostatic hypotension** is a significant drop in blood pressure on standing. When a person changes position from lying down to standing, blood "pools" within the veins below the heart because of gravity, and less blood returns to the heart. Consequently, blood pressure decreases because the heart has less blood to pump.

PREDICT 2

Although orthostatic hypotension has many causes, in the elderly it can be due to age-related decreased neural and cardiovascular responses. Dehydration can result from decreased fluid intake while feeling ill and from sweating as a result of a fever. Dehydration can decrease blood volume and lower blood pressure, increasing the likelihood of orthostatic hypotension. Use figure 1.6 to answer the following:

a. Describe the normal response to a decrease in blood pressure on standing.
b. What happened to Molly's heart rate just before she fainted? Why did Molly faint?
c. How did Molly's fainting and falling to the floor assist in establishing homeostasis (assuming she was not injured)?

If blood pressure increases slightly, the receptors detect the increased blood pressure and send that information to the control center in the brain. The control center causes heart rate to decrease, resulting in a decrease in blood pressure. If blood pressure decreases slightly, the receptors inform the control center, which increases heart rate, thereby producing an increase in blood pressure (figure 1.6). As a result, blood pressure constantly rises and falls within a normal range of values.

Although homeostasis is the maintenance of a normal range of values, this does not mean that all variables are maintained within the same narrow range of values at all times. Sometimes a deviation from the usual range of values can be beneficial. For example, during exercise the normal range for blood pressure differs from the range under resting conditions, and the blood pressure is significantly elevated (figure 1.7). Muscle cells require increased oxygen and nutrients and increased removal of waste products to support their increased level of activity during exercise. Increased blood pressure increases blood delivery to muscles, which maintains muscle cell homeostasis during exercise by increasing the delivery of oxygen and nutrients and the removal of waste products.

9 ***Define homeostasis, variable, and set point. If a deviation from homeostasis occurs, what mechanism restores it?***

10 ***What are the three components of many negative-feedback mechanisms? How do they maintain homeostasis?***

PREDICT 3

Explain how negative-feedback mechanisms control respiratory rates when a person is at rest and when a person is exercising.

Positive Feedback

Positive-feedback responses are not homeostatic and are rare in healthy individuals. *Positive* implies that, when a deviation from a normal value occurs, the system's response is to make the deviation even greater (figure 1.8). Positive feedback therefore usually creates a cycle that leads away from homeostasis and, in some cases, results in death.

A cardiac (heart) muscle receiving an inadequate amount of blood is an example of positive feedback. Contraction of cardiac muscle generates blood pressure and moves blood through blood vessels to tissues. A system of blood vessels on the outside of the heart provides cardiac muscle with a blood supply sufficient to allow normal contractions to occur. In effect, the heart pumps blood to itself. Just as with other tissues, blood pressure must be maintained to ensure adequate delivery of blood to cardiac muscle. Following extreme blood loss, blood pressure decreases to the point that delivery of blood to cardiac muscle is inadequate. As a result, cardiac muscle homeostasis is disrupted, and cardiac muscle does not function normally. The heart pumps less blood, which causes the blood pressure to drop even further. This additional decrease in blood pressure means that even less blood is delivered to cardiac muscle, and the heart pumps even less blood, which again decreases the blood pressure (figure 1.9). If the process continues until the blood pressure is too low to sustain the cardiac muscle, the heart stops beating, and death results.

1. Receptors monitor the value of a variable—in this case, receptors in the wall of a blood vessel monitor blood pressure.
2. Information about the value of the variable is sent to a control center. In this case, information is sent by nerves to the part of the brain responsible for regulating blood pressure.
3. The control center compares the value of the variable against the set point.
4. If a response is necessary to maintain homestasis, the control center causes an effector to respond. In this case, information is sent by nerves to the heart.
5. An effector produces a response that maintains homeostasis. In this case, changing heart rate changes blood pressure.

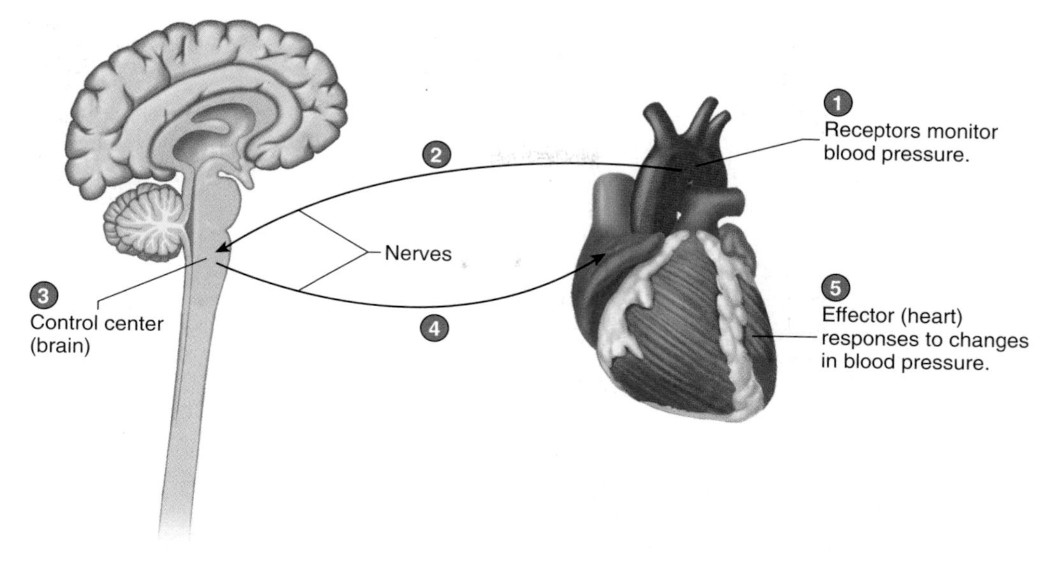

PROCESS FIGURE 1.5 Negative-Feedback Mechanism: Blood Pressure

Following a moderate amount of blood loss (e.g., after a person donates a pint of blood), negative-feedback mechanisms produce an increase in heart rate and other responses that restore blood pressure. If blood loss is severe, however, negative-feedback mechanisms may not be able to maintain homeostasis, and the positive-feedback effect of an ever-decreasing blood pressure can develop. Circumstances in which negative-feedback mechanisms are not adequate to maintain homeostasis illustrate a basic principle. Many disease states result from failure of negative-feedback mechanisms to maintain homeostasis. Medical therapy seeks to

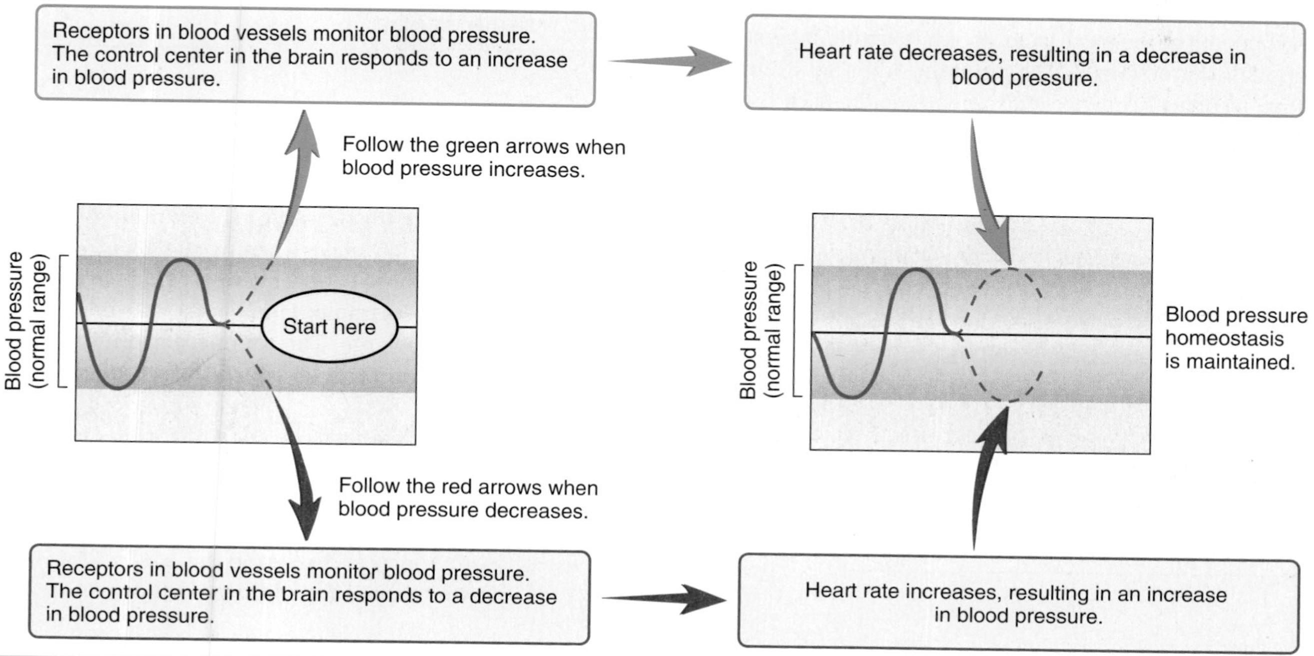

HOMEOSTASIS FIGURE 1.6 Summary of Negative Feedback Mechanism: Blood Pressure

Blood pressure is maintained within a normal range by negative-feedback mechanisms.

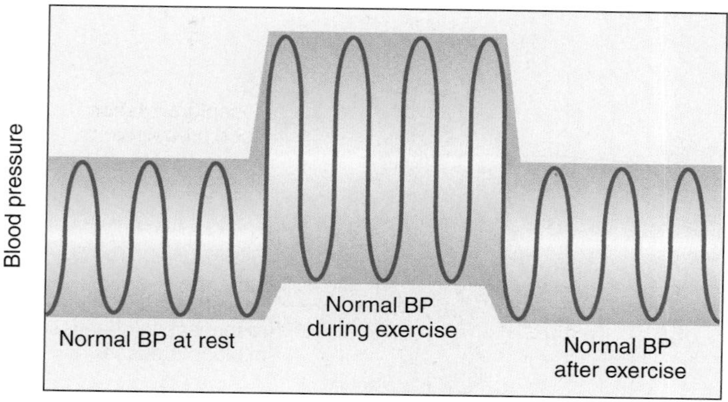

FIGURE 1.7 Changes in Blood Pressure During Exercise
During exercise, the demand for oxygen by muscle tissue increases. An increase in blood pressure (BP) results in an increase in blood flow to the tissues. The increased blood pressure is not an abnormal or a nonhomeostatic condition but is a resetting of the normal homeostatic range to meet the increased demand. The reset range is higher and broader than the resting range. After exercise ceases, the range returns to that of the resting condition.

overcome illness by aiding negative-feedback mechanisms (e.g., a transfusion reverses a constantly decreasing blood pressure and restores homeostasis).

A few positive-feedback mechanisms do operate in the body under normal conditions, but in all cases they are eventually limited in some way. Birth is an example of a normally occurring positive-feedback mechanism. Near the end of pregnancy, the baby's larger size stretches the uterus. This stretching, especially around the opening of the uterus, stimulates contractions of the uterine muscles. The uterine contractions push the baby against the opening of the uterus and stretch it further. This stimulates additional contractions, which result in additional stretching. This positive-feedback sequence ends only when the baby is delivered from the uterus and the stretching stimulus is eliminated.

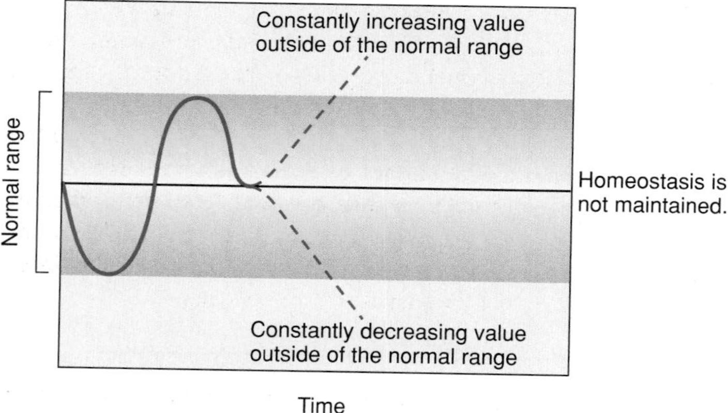

FIGURE 1.8 Positive Feedback
Deviations from the normal set point value cause an additional deviation away from that value in either a positive or negative direction.

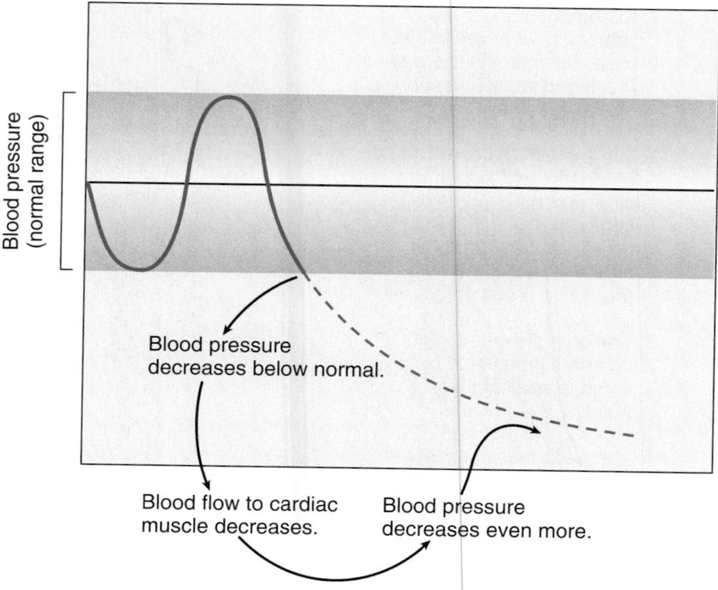

FIGURE 1.9 Example of Harmful Positive Feedback
A decrease in blood pressure below the normal range causes decreased blood flow to the heart. The heart is unable to pump enough blood to maintain blood pressure, and blood flow to the cardiac muscle decreases. Thus, the heart's ability to pump decreases further, and blood pressure decreases even more.

11 ***Define positive feedback. Why are positive-feedback mechanisms often harmful?***

PREDICT 4

Is the sensation of thirst associated with a negative- or a positive-feedback mechanism? Explain.

TERMINOLOGY AND THE BODY PLAN

You will be learning many new words as you study anatomy and physiology. Knowing the derivation, or **etymology** (et'uh-mol'ŏ-jē), of these words can make learning them easy and fun. Most words are derived from Latin or Greek, which are very descriptive languages. For example, *foramen* is a Latin word for hole, and *magnum* means large. The foramen magnum is therefore a large hole in the skull through which the spinal cord attaches to the brain.

Prefixes and suffixes can be added to words to expand their meaning. The suffix *-itis* means an inflammation, so appendicitis is an inflammation of the appendix. As new terms are introduced in this book, their meanings are often explained. The glossary and the list of word roots, prefixes, and suffixes on the inside back cover of the book provide additional information about the new terms.

It is very important to learn these new words so that when you speak to colleagues or write reports your message is clear and correct.

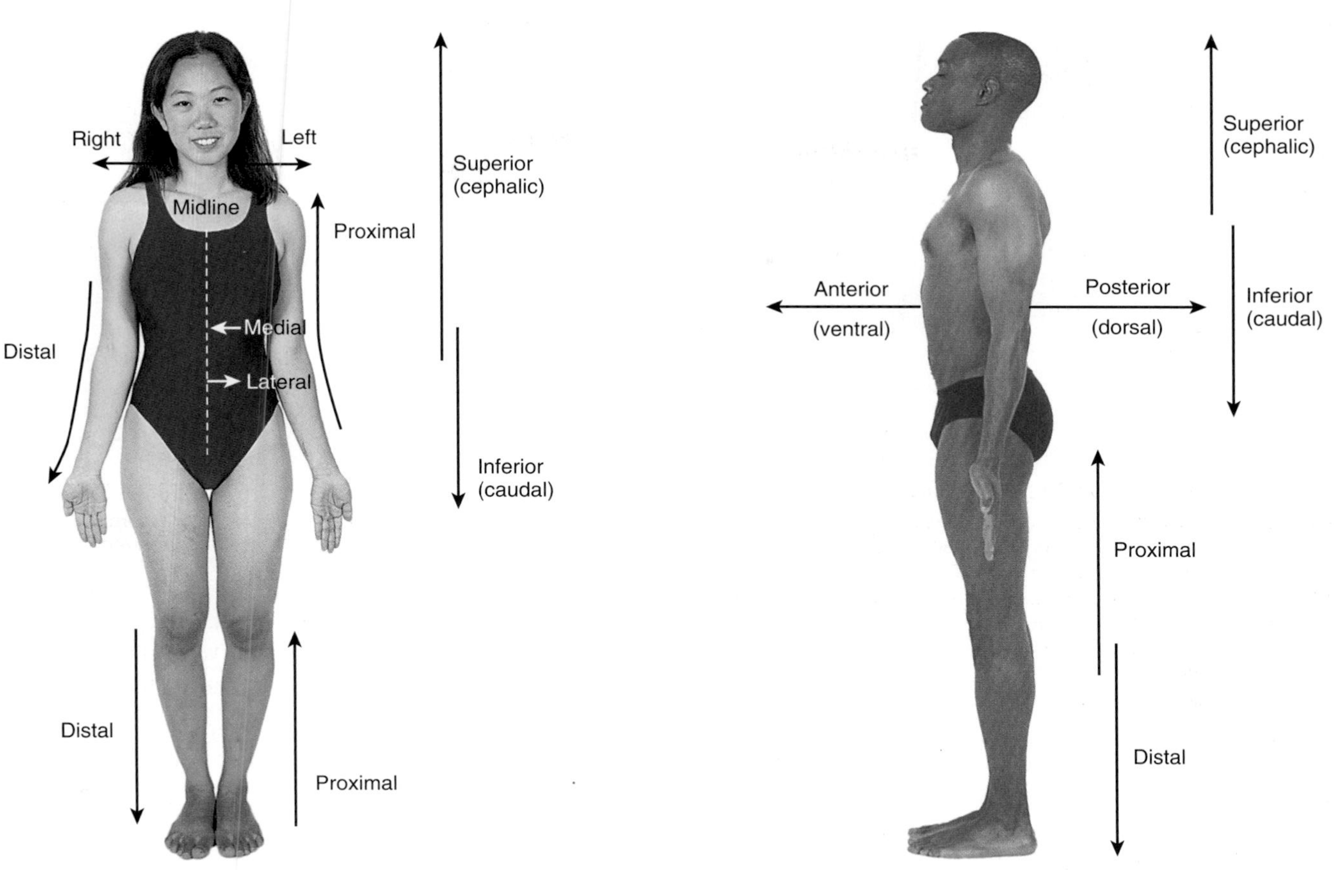

FIGURE 1.10 Directional Terms

All directional terms are in relation to a person in the anatomical position: a person standing erect with the face directed forward, the arms hanging to the sides, and the palms of the hands facing forward.

Body Positions

The **anatomical position** refers to a person standing erect with the face directed forward, the upper limbs hanging to the sides, and the palms of the hands facing forward (figure 1.10). A person is **supine** when lying face upward and **prone** when lying face downward.

The position of the body can affect the description of body parts relative to each other. In the anatomical position, the elbow is above the hand, but, in the supine or prone position, the elbow and hand are at the same level. To avoid confusion, relational descriptions are always based on the anatomical position, no matter the actual position of the body. Thus, the elbow is always described as being above the wrist, whether the person is lying down or is even upside down.

12 ***What is the anatomical position in humans? Why is it important?***

13 ***Define supine and prone.***

Directional Terms

Directional terms describe parts of the body relative to each other. Important directional terms are illustrated in figure 1.10 and summarized in table 1.1. It is important to become familiar with these directional terms as soon as possible because you will see them repeatedly throughout the book. *Right* and *left* are retained as directional terms in anatomical terminology. *Up* is replaced by **superior,** *down* by **inferior,** *front* by **anterior,** and *back* by **posterior.**

In humans, *superior* is synonymous with **cephalic** (se-fal′ik), which means toward the head, because, when we are in the anatomical position, the head is the highest point. In humans, the term *inferior* is synonymous with **caudal** (kaw′dăl), which means toward the tail, which would be located at the end of the vertebral column if humans had tails. The terms *cephalic* and *caudal* can be used to describe directional movements on the trunk, but they are not used to describe directional movements on the limbs.

The word *anterior* means that which goes before, and **ventral** means belly. The anterior surface of the human body is therefore the ventral surface, or belly, because the belly "goes first" when we are walking. The word *posterior* means that which follows, and **dorsal** means back. The posterior surface of the body is the dorsal surface, or back, which follows as we are walking.

PREDICT 5

The anatomical position of a cat refers to the animal standing erect on all four limbs and facing forward. On the basis of the etymology of the directional terms, what two terms indicate movement toward the head? What two terms mean movement toward the back? Compare these terms with those referring to a human in the anatomical position.

TABLE 1.1 Directional Terms for Humans

Terms	Etymology*	Definition	Example
Right		Toward the right side of the body	The right ear
Left		Toward the left side of the body	The left eye
Superior	L., higher	A structure above another	The chin is superior to the navel.
Inferior	L., lower	A structure below another	The navel is inferior to the chin.
Cephalic	G. *kephale,* head	Closer to the head than another structure (usually synonymous with *superior*)	The chin is cephalic to the navel.
Caudal	L. *cauda,* a tail	Closer to the tail than another structure (usually synonymous with *inferior*)	The navel is caudal to the chin.
Anterior	L., before	The front of the body	The navel is anterior to the spine.
Posterior	L. *posterus,* following	The back of the body	The spine is posterior to the breastbone.
Ventral	L. *ventr-,* belly	Toward the belly (synonymous with *anterior*)	The navel is ventral to the spine.
Dorsal	L. *dorsum,* back	Toward the back (synonymous with *posterior*)	The spine is dorsal to the breastbone.
Proximal	L. *proximus,* nearest	Closer to the point of attachment to the body than another structure	The elbow is proximal to the wrist.
Distal	L. *di-* plus *sto,* to stand apart or be distant	Farther from the point of attachment to the body than another structure	The wrist is distal to the elbow.
Lateral	L. *latus,* side	Away from the midline of the body	The nipple is lateral to the breastbone.
Medial	L. *medialis,* middle	Toward the midline of the body	The bridge of the nose is medial to the eye.
Superficial	L. *superficialis,* toward the surface	Toward or on the surface (not shown in figure 1.10)	The skin is superficial to muscle.
Deep	O.E. *deop,* deep	Away from the surface, internal (not shown in figure 1.10)	The lungs are deep to the ribs.

*Origin and meaning of the word: L., Latin; G., Greek; O.E., Old English.

Proximal means nearest, whereas **distal** means distant. These terms are used to refer to linear structures, such as the limbs, in which one end is near another structure and the other end is farther away. Each limb is attached at its proximal end to the body, and the distal end, such as the hand, is farther away.

Medial means toward the midline, and **lateral** means away from the midline. The nose is located in a medial position in the face, and the eyes are lateral to the nose. The term **superficial** refers to a structure close to the surface of the body, and **deep** is toward the interior of the body. The skin is superficial to muscle and bone.

14 ***List two terms that in humans indicate* toward the head. *Name two terms that mean the opposite.***

15 ***List two terms that indicate* the back *in humans. What two terms mean* the front?**

16 ***Define the following terms, and give the word that means the opposite:* proximal, lateral, *and* superficial.**

PREDICT 6

Use as many directional terms as you can to describe the relationship between your kneecap and your heel.

Body Parts and Regions

A number of terms are used when referring to different parts or regions of the body (figure 1.11). The upper limb is divided into the arm, forearm, wrist, and hand. The **arm** extends from the shoulder to the elbow, and the **forearm** extends from the elbow to the wrist. The lower limb is divided into the thigh, leg, ankle, and foot. The **thigh** extends from the hip to the knee, and the **leg** extends from the knee to the ankle. Note that, contrary to popular usage, the terms *arm* and *leg* refer to only a part of the respective limb.

The central region of the body consists of the **head, neck,** and **trunk.** The trunk can be divided into the **thorax** (chest), **abdomen** (region between the thorax and pelvis), and **pelvis** (the inferior end of the trunk associated with the hips).

The abdomen is often subdivided superficially into **quadrants** by two imaginary lines—one horizontal and one vertical—which intersect at the navel (figure 1.12*a*). The quadrants formed are the right-upper, left-upper, right-lower, and left-lower quadrants. In addition to these quadrants, the abdomen is sometimes subdivided into nine **regions** by four imaginary lines: two horizontal and two vertical. These four lines create an imaginary tic-tac-toe figure on the abdomen, resulting in nine regions: epigastric, right and left hypochondriac, umbilical, right and left lumbar, hypogastric, and right and left iliac (figure 1.12*b*). Clinicians use the quadrants and regions as reference points for locating underlying organs. For example, the appendix is located in the right-lower quadrant, and the pain of an acute appendicitis is usually felt there.

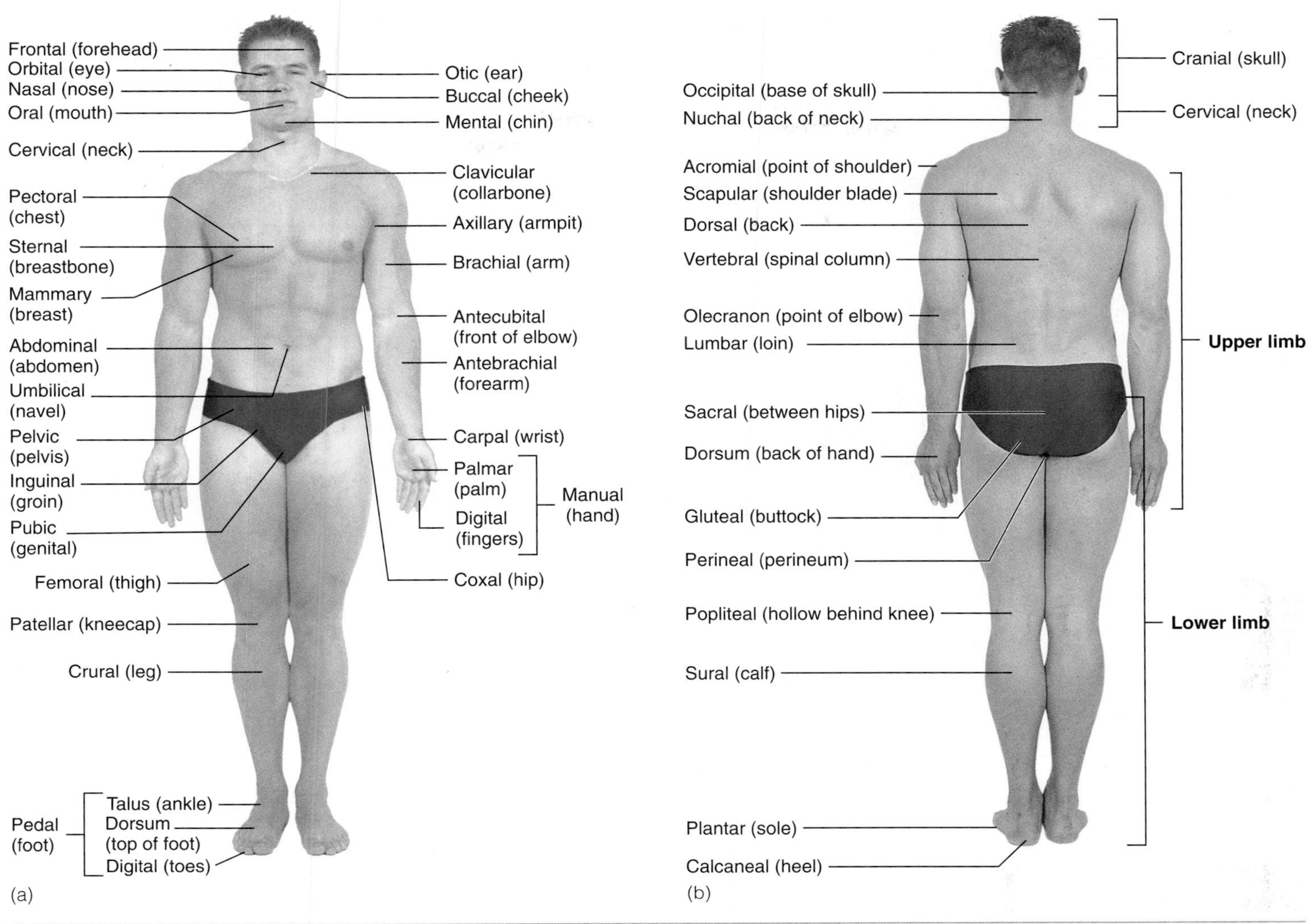

FIGURE 1.11 Body Parts and Regions

The anatomical and common (*in parentheses*) names are indicated for some parts and regions of the body. (*a*) Anterior view. (*b*) Posterior view.

17 ***What is the difference between the arm and the upper limb and the difference between the leg and the lower limb?***

18 ***Describe the quadrant and the nine-region methods of subdividing the abdominal region. What is the purpose of these subdivisions?***

PREDICT 7

If a bullet passed through the left upper quadrant in an anterior to posterior direction, which of the following organs could be damaged (see figures 1.2 and 1.12)? gallbladder, heart, kidney, liver, pancreas, spleen, stomach, urinary bladder

Planes

At times, it is conceptually useful to describe the body as having imaginary flat surfaces called **planes** passing through it (figure 1.13). A plane divides, or sections, the body, making it possible to "look inside" and observe the body's structures. A **sagittal** (saj′i-tăl) plane runs vertically through the body, separating it into right and left portions. The word *sagittal* literally means "the flight of an arrow" and refers to the way the body would be split by an arrow passing anteriorly to posteriorly. A **median** plane is a sagittal plane that passes through the midline of the body, dividing it into equal right and left halves. A **transverse,** or **horizontal,** plane runs parallel to the ground, dividing the body into superior and inferior portions. A **frontal,** or **coronal** (kōr′ō-năl, kō-rō′năl; crown), plane runs vertically from right to left and divides the body into anterior and posterior parts.

Organs are often sectioned to reveal their internal structure (figure 1.14). A cut through the long axis of the organ is a **longitudinal** section, and a cut at right angles to the long axis is a **cross,** or **transverse,** section. If a cut is made across the long axis at other than a right angle, it is called an **oblique** section.

19 ***Define the three planes of the body.***

20 ***In what three ways can an organ be cut?***

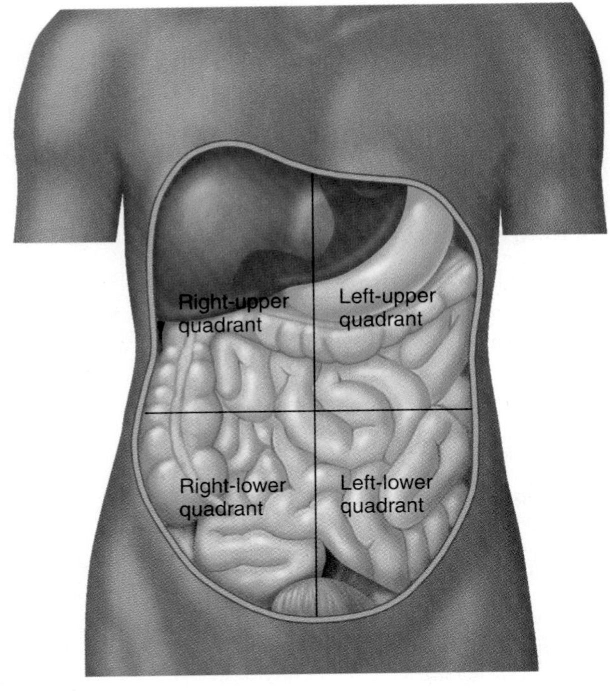

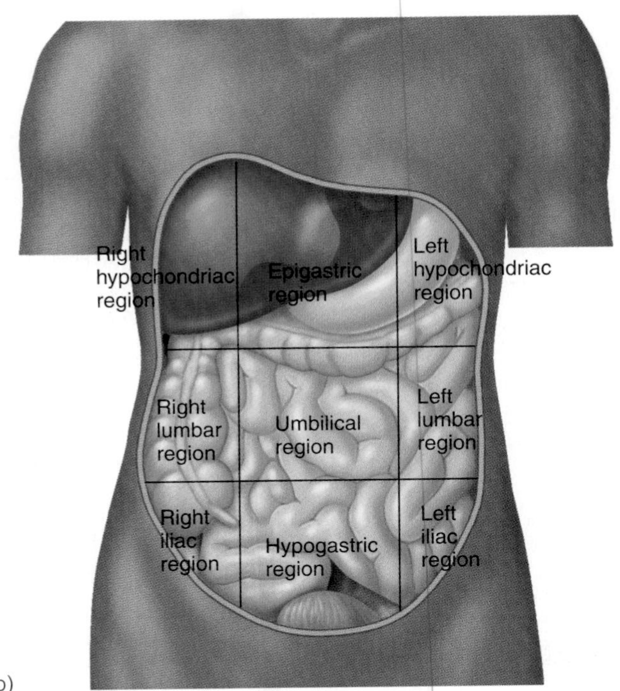

FIGURE 1.12 Subdivisions of the Abdomen

Lines are superimposed over internal organs to demonstrate the relationship of the organs to the subdivisions. (*a*) Abdominal quadrants consist of four subdivisions. (*b*) Abdominal regions consist of nine subdivisions.

Body Cavities

The body contains many cavities. Some of these, such as the nasal cavity, open to the outside of the body, and some do not. The trunk contains three large cavities that do not open to the outside of the body: the thoracic, the abdominal, and the pelvic (figure 1.15). The rib cage surrounds the **thoracic cavity,** and the muscular diaphragm separates it from the abdominal cavity. The thoracic cavity is divided into right and left parts by a median partition called the **mediastinum** (mē′dē-as-tī′nŭm; middle wall). The mediastinum contains the heart, the thymus, the trachea, the esophagus, and other structures, such as blood vessels and nerves. The two lungs are located on each side of the mediastinum.

Abdominal muscles primarily enclose the **abdominal cavity,** which contains the stomach, intestines, liver, spleen, pancreas, and kidneys. Pelvic bones encase the small space known as the **pelvic cavity,** where the urinary bladder, part of the large intestine, and the internal reproductive organs are housed. The abdominal and pelvic cavities are not physically separated and sometimes are called the **abdominopelvic cavity.**

21 ***What structure separates the thoracic cavity from the abdominal cavity? The abdominal cavity from the pelvic cavity?***

22 ***What structure divides the thoracic cavity into right and left parts?***

Serous Membranes

Serous (sēr′ŭs) **membranes** cover the organs of the trunk cavities and line the trunk cavities. Imagine an inflated balloon into which a fist has been pushed (figure 1.16). The fist represents an organ; the inner balloon wall in contact with the fist represents the **visceral** (vis′er-ăl; organ) **serous membrane** covering the organ; and the outer part of the balloon wall represents the **parietal** (pă-rī′ĕ-tăl; wall) **serous membrane.** The cavity, or space, between the visceral and parietal serous membranes is normally filled with a thin, lubricating film of serous fluid produced by the membranes. As organs rub against the body wall or against another organ, the combination of serous fluid and smooth serous membranes reduces friction.

The thoracic cavity contains three serous membrane–lined cavities: a pericardial cavity and two pleural cavities. The **pericardial** (per-i-kar′dē-ăl; around the heart) **cavity** surrounds the heart (figure 1.17*a*). The visceral pericardium covers the heart, which is contained within a connective tissue sac lined with the parietal pericardium. The pericardial cavity, which contains pericardial fluid, is located between the visceral pericardium and the parietal pericardium.

A **pleural** (ploor′ăl; associated with the ribs) **cavity** surrounds each lung, which is covered by visceral pleura (figure 1.17*b*). Parietal pleura line the inner surface of the thoracic wall, the lateral surfaces of the mediastinum, and the superior surface of the diaphragm. The pleural cavity lies between the visceral pleura and the parietal pleura and contains pleural fluid.

The abdominopelvic cavity contains a serous membrane–lined cavity called the **peritoneal** (per′i-tō-nē′ăl; to stretch over) **cavity** (figure 1.17*c*). Visceral peritoneum covers many of the organs of the abdominopelvic cavity. Parietal peritoneum lines the wall of the abdominopelvic cavity and the inferior surface of the diaphragm. The peritoneal cavity is located between the visceral peritoneum and the parietal peritoneum and contains peritoneal fluid.

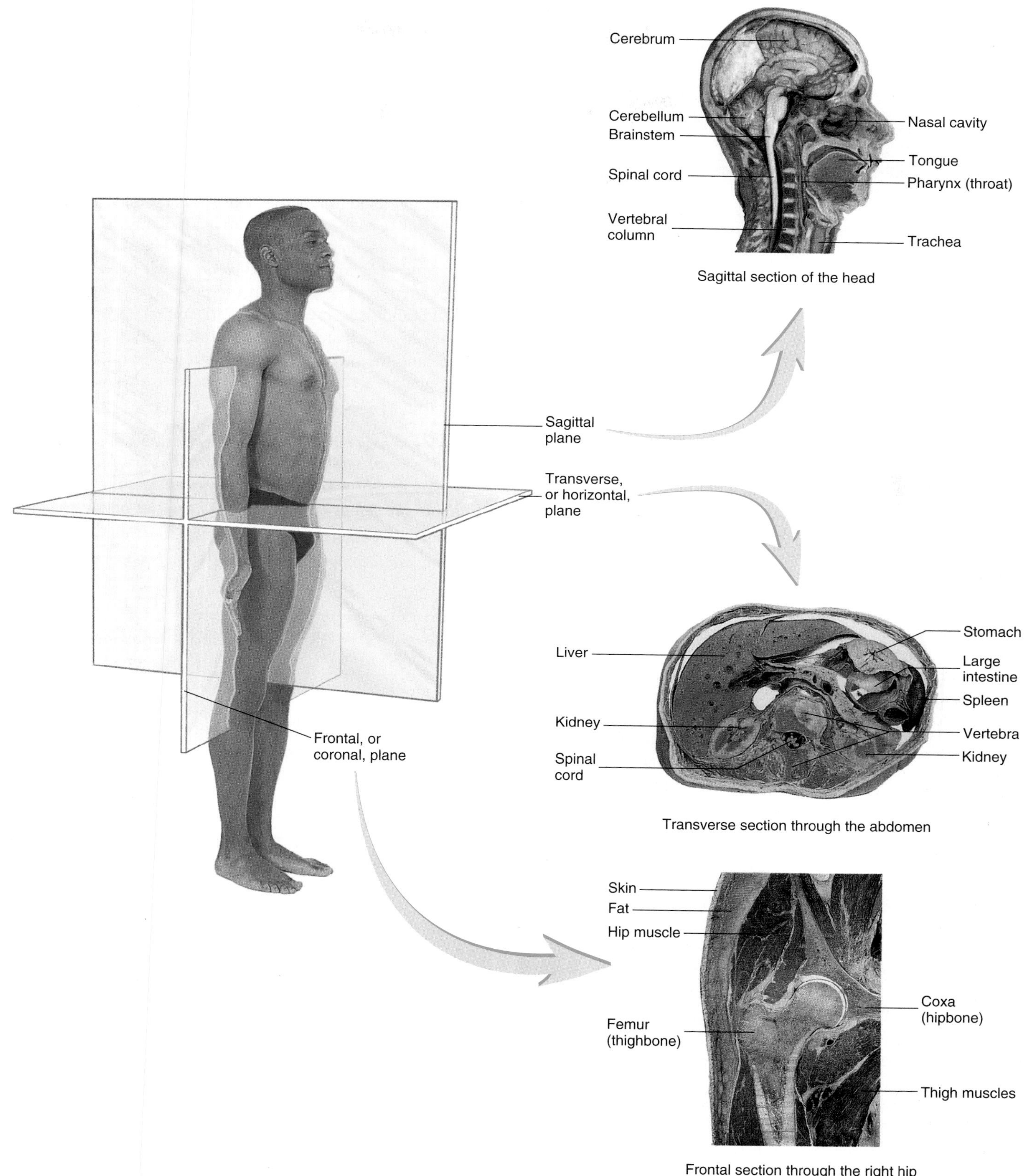

FIGURE 1.13 Planes of the Body

Planes through the whole body are indicated by "glass" sheets. Actual sections through the head (viewed from the right), abdomen (*inferior view*), and hip (*anterior view*) are also shown.

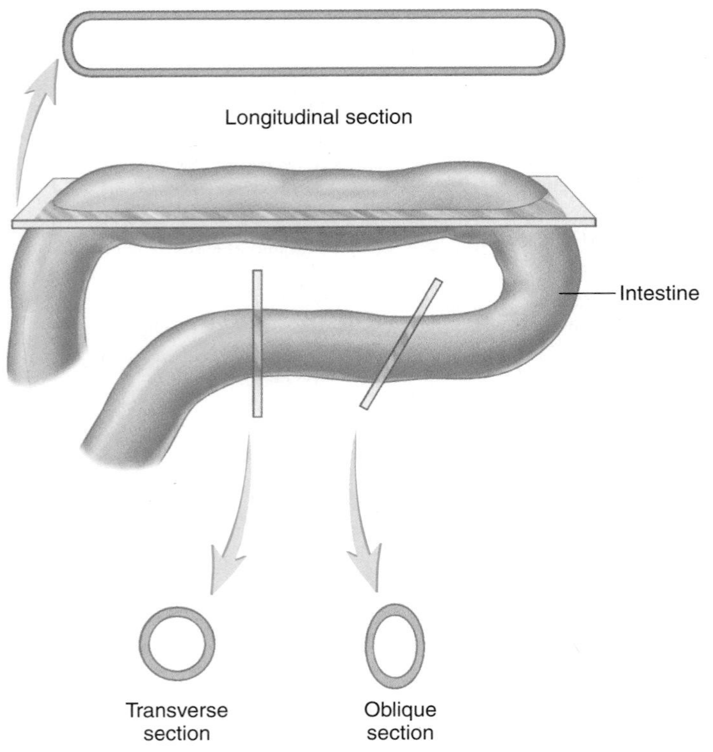

FIGURE 1.14 Planes Through an Organ

Planes through the small intestine are indicated by "glass" sheets. The views of the small intestine after sectioning are also shown. Although the small intestine is basically a tube, the sections appear quite different in shape.

Inflammation of Serous Membranes

The serous membranes can become inflamed, usually as a result of an infection. **Pericarditis** (per'i-kar-dī'tis) is inflammation of the pericardium, **pleurisy** (ploor'i-sē) is inflammation of the pleura, and **peritonitis** (per'i-tō-nī'tis) is inflammation of the peritoneum.

Mesenteries (mes'en-ter-ēz), which consist of two layers of peritoneum fused together (see figure 1.17*c*), connect the visceral peritoneum of some abdominopelvic organs to the parietal peritoneum on the body wall or to the visceral peritoneum of other abdominopelvic organs. The mesenteries anchor the organs to the body wall and provide a pathway for nerves and blood vessels to reach the organs. Other abdominopelvic organs are more closely attached to the body wall and do not have mesenteries. Parietal peritoneum covers these other organs, which are said to be **retroperitoneal** (re'trō-per'i-tō-nē'ăl; behind the peritoneum). The retroperitoneal organs include the kidneys, the adrenal glands, the pancreas, parts of the intestines, and the urinary bladder (see figure 1.17*c*).

23 ***Define serous membranes. Differentiate between parietal and visceral serous membranes. What is the function of the serous membranes?***

24 ***Name the serous membrane–lined cavities of the trunk.***

25 ***What are mesenteries? Explain their function.***

26 ***What are retroperitoneal organs? List five examples.***

PREDICT 8

Explain how an organ can be located within the abdominopelvic cavity but not be within the peritoneal cavity.

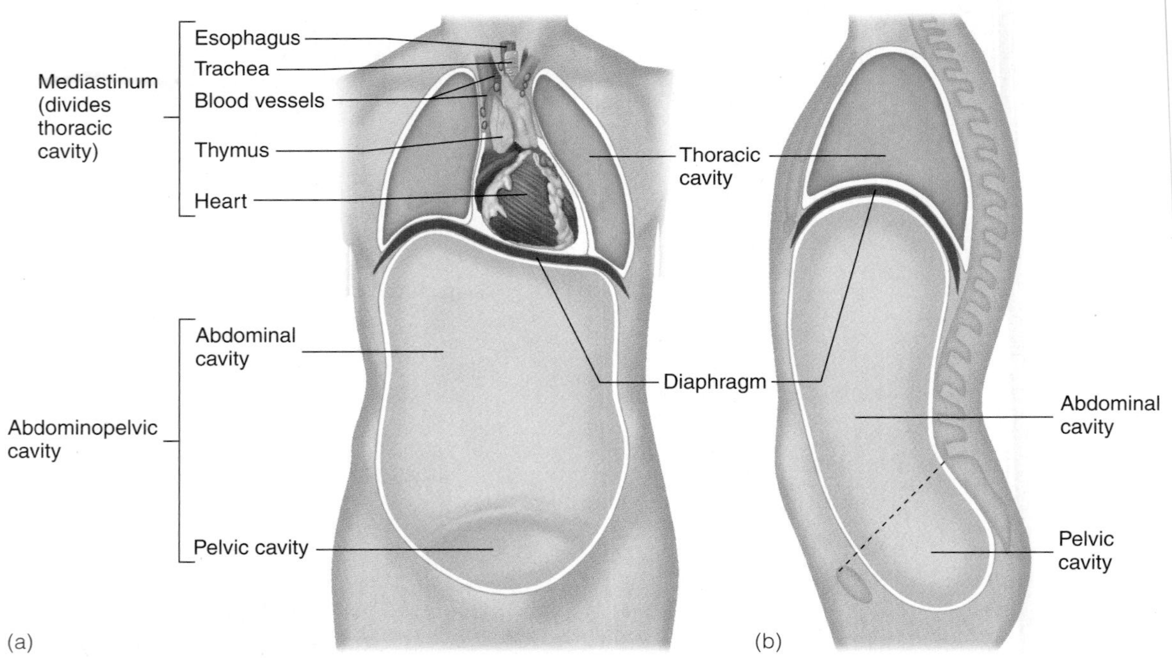

FIGURE 1.15 Trunk Cavities

(*a*) Anterior view showing the major trunk cavities. The diaphragm separates the thoracic cavity from the abdominal cavity. The mediastinum, which includes the heart, is a partition of organs dividing the thoracic cavity. (*b*) Sagittal section of the trunk cavities viewed from the left. The *dashed line* shows the division between the abdominal and pelvic cavities. The mediastinum has been removed to show the thoracic cavity.

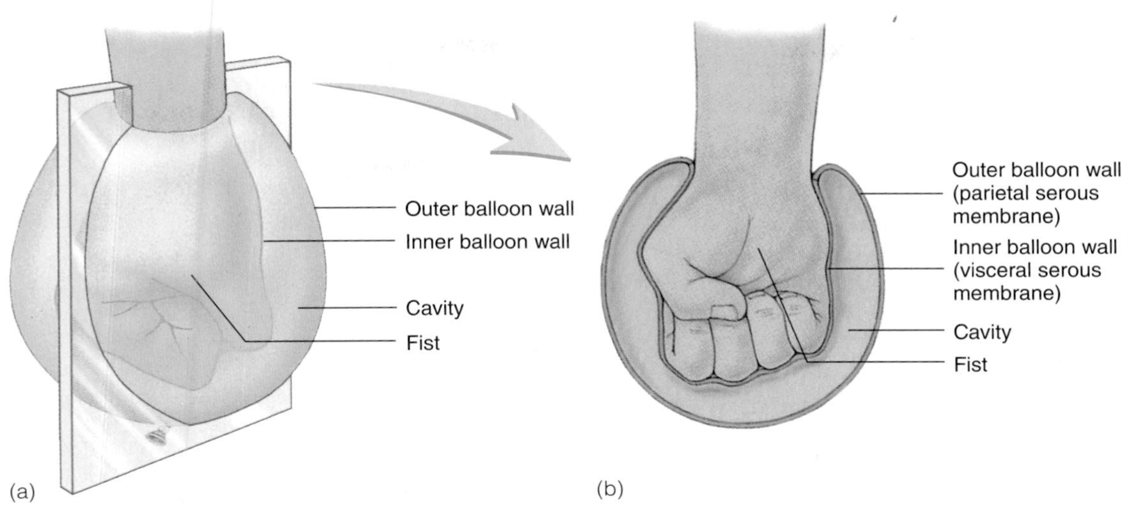

FIGURE 1.16 Serous Membranes

(*a*) Fist pushing into a balloon. A "glass" sheet indicates the location of a cross section through the balloon. (*b*) Interior view produced by the section in (*a*). The fist represents an organ, and the walls of the balloon the serous membranes. The inner wall of the balloon represents a visceral serous membrane in contact with the fist (organ). The outer wall of the balloon represents a parietal serous membrane.

Parietal pericardium
Visceral pericardium
Pericardial cavity containing pericardial fluid
Heart
(a)

Parietal pleura
Visceral pleura
Pleural cavity containing pleural fluid
Lung
Diaphragm
(b)

Parietal peritoneum
Visceral peritoneum
Organs surrounded by visceral peritoneum
Peritoneal cavity containing peritoneal fluid
Retroperitoneal organs
Mesenteries
Mesentery
Retroperitoneal organs
(c)

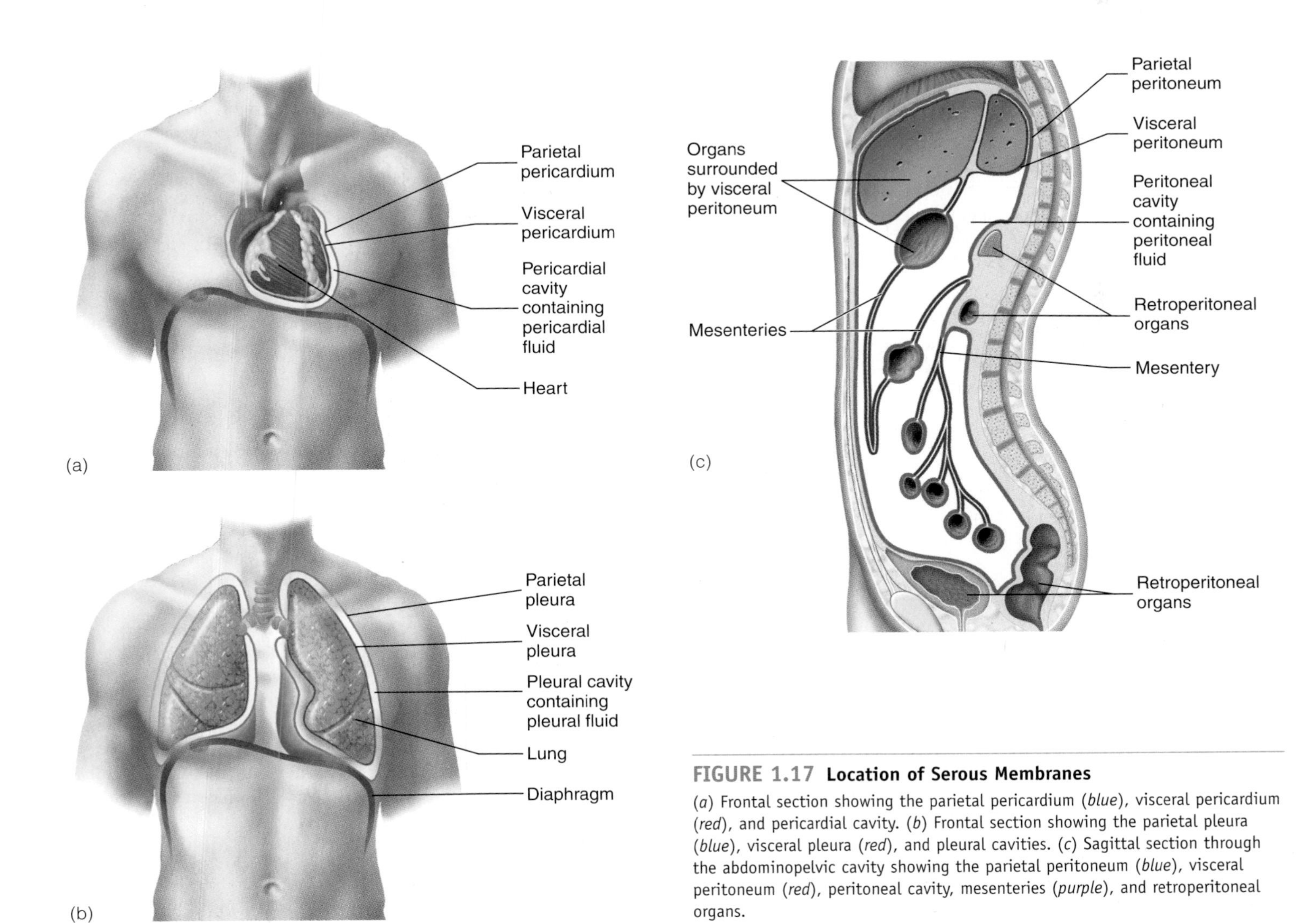

FIGURE 1.17 Location of Serous Membranes

(*a*) Frontal section showing the parietal pericardium (*blue*), visceral pericardium (*red*), and pericardial cavity. (*b*) Frontal section showing the parietal pleura (*blue*), visceral pleura (*red*), and pleural cavities. (*c*) Sagittal section through the abdominopelvic cavity showing the parietal peritoneum (*blue*), visceral peritoneum (*red*), peritoneal cavity, mesenteries (*purple*), and retroperitoneal organs.

SUMMARY

A functional knowledge of anatomy and physiology can be used to solve problems concerning the body when healthy or diseased.

Anatomy and Physiology (p. 2)

1. Anatomy is the study of the body's structures.
   - Developmental anatomy considers anatomical changes from conception to adulthood. Embryology focuses on the first 8 weeks of development.
   - Cytology examines cells, and histology examines tissues.
   - Gross anatomy emphasizes organs from a systemic or regional perspective.
2. Surface anatomy uses superficial structures to locate deeper structures, and anatomical imaging is a noninvasive technique for identifying deep structures.
3. Physiology is the study of the body's functions. It can be approached from a cellular or systems point of view.
4. Pathology deals with all aspects of disease. Exercise physiology examines changes caused by exercise.

Structural and Functional Organization (p. 2)

1. Basic chemical characteristics are responsible for the structure and functions of life.
2. Cells are the basic structural and functional units of organisms, such as plants and animals. Organelles are small structures within cells that perform specific functions.
3. Tissues are groups of cells of similar structure and function and the materials surrounding them. The four primary tissue types are epithelial, connective, muscle, and nervous tissues.
4. Organs are structures composed of two or more tissues that perform specific functions.
5. Organs are arranged into the 11 organ systems of the human body (see figure 1.2).
6. Organ systems interact to form a whole, functioning organism.

Characteristics of Life (p. 6)

Humans have many characteristics, such as organization, metabolism, responsiveness, growth, development, and reproduction, in common with other organisms.

Biomedical Research (p. 9)

Much of what is known about humans is derived from research on other organisms.

Homeostasis (p. 9)

Homeostasis is the condition in which body functions, body fluids, and other factors of the internal environment are maintained at levels suitable to support life.

Negative Feedback

1. Negative-feedback mechanisms operate to maintain homeostasis.
2. Many negative-feedback mechanisms consist of a receptor, control center, and effector.

Positive Feedback

1. Positive-feedback mechanisms usually increase deviations from normal.
2. Although a few positive-feedback mechanisms normally exist in the body, most positive-feedback mechanisms are harmful.

Terminology and the Body Plan (p. 12)

Body Positions

1. A human standing erect with the face directed forward, the arms hanging to the side, and the palms facing forward is in the anatomical position.
2. A person lying face upward is supine; a person lying face downward is prone.

Directional Terms

Directional terms always refer to the anatomical position, no matter what the actual position of the body (see table 1.1).

Body Parts and Regions

1. The body can be divided into the limbs, upper and lower, and a central region consisting of the head, neck, and trunk regions.
2. Superficially, the abdomen can be divided into quadrants or nine regions. These divisions are useful for locating internal organs or describing the location of a pain or tumor.

Planes

1. Planes of the body
   - A sagittal plane divides the body into right and left parts. A median plane divides the body into equal right and left halves.
   - A transverse (horizontal) plane divides the body into superior and inferior portions.
   - A frontal (coronal) plane divides the body into anterior and posterior parts.
2. Sections of an organ
   - A longitudinal section of an organ divides it along the long axis.
   - A cross (transverse) section cuts at a right angle to the long axis of an organ.
   - An oblique section cuts across the long axis of an organ at an angle other than a right angle.

Body Cavities

1. The mediastinum subdivides the thoracic cavity.
2. The diaphragm separates the thoracic and abdominal cavities.
3. Pelvic bones surround the pelvic cavity.

Serous Membranes

1. Serous membranes line the trunk cavities. The parietal portion of a serous membrane lines the wall of the cavity, and the visceral portion is in contact with the internal organs.
   - The serous membranes secrete fluid, which fills the space between the visceral and parietal membranes. The serous membranes protect organs from friction.
   - The pericardial cavity surrounds the heart, the pleural cavities surround the lungs, and the peritoneal cavity surrounds certain abdominal and pelvic organs.
2. Mesenteries are parts of the peritoneum that hold the abdominal organs in place and provide a passageway for blood vessels and nerves to the organs.
3. Retroperitoneal organs are located "behind" the parietal peritoneum.

REVIEW AND COMPREHENSION

1. Physiology
   a. deals with the processes or functions of living things.
   b. is the scientific discipline that investigates the body's structures.
   c. is concerned with organisms and does not deal with different levels of organization, such as cells and systems.
   d. recognizes the static (as opposed to the dynamic) nature of living things.
   e. can be used to study the human body without considering anatomy.
2. The following are conceptual levels for considering the body.
   1. cell
   2. chemical
   3. organ
   4. organ system
   5. organism
   6. tissue

   Choose the correct order for these conceptual levels, from smallest to largest.
   a. 1,2,3,6,4,5 c. 3,1,6,4,5,2 e. 1,6,5,3,4,2
   b. 2,1,6,3,4,5 d. 4,6,1,3,5,2

For questions 3–8, match each organ system with its correct function.
   a. regulates other organ systems
   b. removes waste products from the blood; maintains water balance
   c. regulates temperature; reduces water loss; provides protection
   d. removes foreign substances from the blood; combats disease; maintains tissue fluid balance
   e. produces movement; maintains posture; produces body heat

3. endocrine system
4. integumentary system
5. lymphatic system
6. muscular system
7. nervous system
8. urinary system
9. The characteristic of life that is defined as "all the chemical reactions taking place in an organism" is
   a. development. c. metabolism. e. responsiveness.
   b. growth. d. organization.
10. Negative-feedback mechanisms
    a. make deviations from the set point smaller.
    b. maintain homeostasis.
    c. are associated with an increased sense of hunger the longer a person goes without eating.
    d. all of the above.
11. The following events are part of a negative-feedback mechanism.
    1. Blood pressure increases.
    2. Control center compares actual blood pressure to the blood pressure set point.
    3. The heart beats faster.
    4. Receptors detect a decrease in blood pressure.

    Choose the arrangement that lists the events in the order they occur.
    a. 1,2,3,4 b. 1,3,2,4 c. 3,1,4,2 d. 4,2,3,1 e. 4,3,2,1
12. Which of these statements concerning positive feedback is correct?
    a. Positive-feedback responses maintain homeostasis.
    b. Positive-feedback responses occur continuously in healthy individuals.
    c. Birth is an example of a normally occurring positive-feedback mechanism.
    d. When the cardiac muscle receives an inadequate supply of blood, positive-feedback mechanisms increase blood flow to the heart.
    e. Medical therapy seeks to overcome illness by aiding positive-feedback mechanisms.
13. The clavicle (collarbone) is ______ to the nipple of the breast.
    a. anterior c. superficial e. ventral
    b. distal d. superior
14. A term that means nearer the attached end of a limb is
    a. distal. c. medial. e. superficial.
    b. lateral. d. proximal.
15. Which of these directional terms are paired most appropriately as opposites?
    a. superficial and deep d. superior and posterior
    b. medial and proximal e. anterior and inferior
    c. distal and lateral
16. The part of the upper limb between the elbow and the wrist is called the
    a. arm. c. hand. e. lower arm.
    b. forearm. d. inferior arm.
17. A patient with appendicitis usually has pain in the ______ quadrant of the abdomen.
    a. left-lower c. left-upper
    b. right-lower d. right-upper
18. A plane that divides the body into anterior and posterior parts is a
    a. frontal (coronal) plane.
    b. sagittal plane.
    c. transverse plane.
19. The pelvic cavity contains the
    a. kidneys. c. spleen. e. urinary bladder.
    b. liver. d. stomach.
20. The lungs are
    a. part of the mediastinum.
    b. surrounded by the pericardial cavity.
    c. found within the thoracic cavity.
    d. separated from each other by the diaphragm.
    e. surrounded by mucous membranes.
21. Given these characteristics:
    1. reduce friction between organs
    2. line fluid-filled cavities
    3. line trunk cavities that open to the exterior of the body

    Which of the characteristics describe serous membranes?
    a. 1,2 b. 1,3 c. 2,3 d. 1,2,3

22. Given these organ and cavity combinations:
    1. heart and pericardial cavity
    2. lungs and pleural cavity
    3. stomach and peritoneal cavity
    4. kidney and peritoneal cavity

    Which of the organs is correctly paired with a space that surrounds that organ?
    a. 1,2 b. 1,2,3 c. 1,2,4 d. 2,3,4 e. 1,2,3,4
23. Which of these membrane combinations are found on the surface of the diaphragm?
    a. parietal pleura—parietal peritoneum
    b. parietal pleura—visceral peritoneum
    c. visceral pleura—parietal peritoneum
    d. visceral pleura—visceral peritoneum
24. Mesenteries
    a. are found in the pleural, pericardial, and abdominopelvic cavities.
    b. consist of two layers of peritoneum fused together.
    c. anchor organs, such as the kidneys and urinary bladder, to the body wall.
    d. are found primarily in body cavities that open to the outside.
    e. all of the above.
25. Which of the following organs is *not* retroperitoneal?
    a. adrenal glands c. kidneys e. stomach
    b. urinary bladder d. pancreas

Answers in Appendix E

CRITICAL THINKING

1. Exposure to a hot environment causes the body to sweat. The hotter the environment, the greater the sweating. Two anatomy and physiology students are arguing about the mechanisms involved: Student A claims that they are positive feedback, and student B claims they are negative feedback. Do you agree with student A or student B and why?
2. A male has lost blood as a result of a gunshot wound. Even though bleeding has been stopped, his blood pressure is low and dropping and his heart rate is elevated. Following a blood transfusion, his blood pressure increases and his heart rate decreases. Which of the following statement(s) is (are) consistent with these observations?
   a. Negative-feedback mechanisms can be inadequate without medical intervention.
   b. The transfusion interrupted a positive-feedback mechanism.
   c. The increased heart rate after the gunshot wound and before the transfusion is a result of a positive-feedback mechanism.
   d. a and b
   e. a, b, and c
3. Provide the correct directional term for the following statement: When a boy is standing on his head, his nose is ______________ to his mouth.
4. Complete the following statements, using the correct directional terms for a human being. Note that more than one term can apply.
   a. The navel is ______________ to the nose.
   b. The nipple is ______________ to the lung.
   c. The arm is ______________ to the forearm.
   d. The little finger is ______________ to the index finger.
5. In which quadrants and regions are the pancreas and the urinary bladder located?
6. Given the following procedures:
   1. Make an opening into the mediastinum.
   2. Lay the patient supine.
   3. Lay the patient prone.
   4. Make an incision through the pericardial serous membranes.
   5. Make an opening into the abdomen.

   Which of the procedures should be accomplished to expose the anterior surface of a patient's heart?
   a. 2,1,4
   b. 2,5,4
   c. 3,1,4
   d. 3,5,4
7. During pregnancy, which of the mother's body cavities increases most in size?
8. A bullet enters the left side of a man, passes through the left lung, and lodges in the heart. Name, in order, the serous membranes and their cavities through which the bullet passes.
9. A woman falls while skiing and accidentally is impaled by her ski pole. The pole passes through the abdominal body wall and into and through the stomach, pierces the diaphragm, and finally stops in the left lung. List, in order, the serous membranes the pole pierces.

Answers in Appendix F

Visit this book's website at www.mhhe.com/seeley8 for chapter quizzes, interactive learning exercises, and other study tools.

The Chemical Basis of Life 2

Chemicals compose the structures of the body, and the interactions of chemicals with one another are responsible for the functions of the body. Nerve impulse generation, digestion, muscle contraction, and metabolism can be described in chemical terms. Many abnormal conditions and illnesses, as well as their treatments, can also be explained in chemical terms. For example, Parkinson disease, which causes uncontrollable shaking movements, results from a shortage of a chemical called dopamine in certain nerve cells in the brain. It can be treated by giving patients another chemical, which brain cells convert to dopamine.

To understand anatomy and physiology, it is essential to have a basic knowledge of chemistry—the scientific discipline concerned with the atomic composition and structure of substances and the reactions they undergo. This chapter outlines *basic chemistry* (p. 24), *chemical reactions and energy* (p. 32), *inorganic chemistry* (p. 36), and *organic chemistry* (p. 39). It is not a comprehensive review of chemistry, but it does review some of the basic concepts. When necessary, refer back to this chapter when chemical phenomena are discussed later in the book.

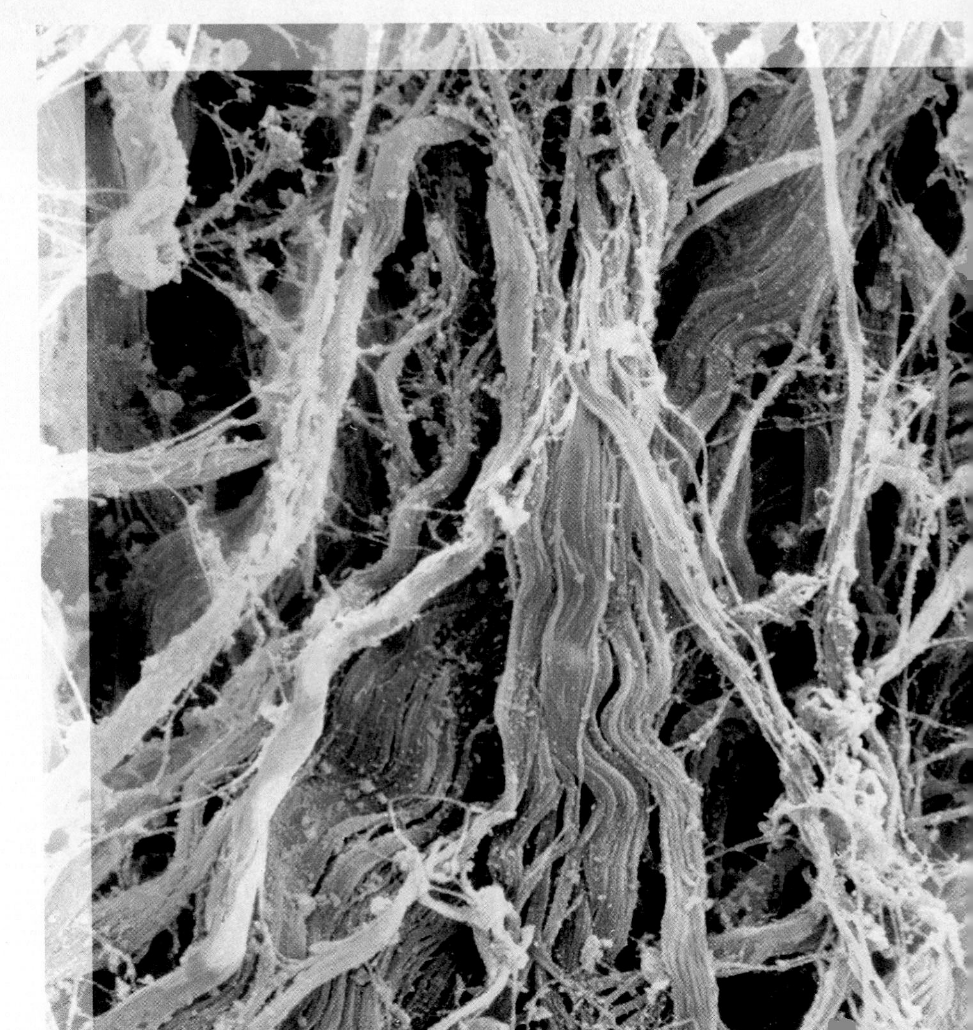

The chemical composition of the body's structures determines their function. This colorized scanning electron micrograph shows bundles of collagen fibers (*brown*) and elastic fibers (*blue*). The molecules forming collagen fibers are like tiny ropes, and collagen fibers make tissues strong yet flexible. The molecules forming elastic fibers resemble microscopic coiled springs. The elastic fibers allow tissues to be stretched and then recoil back to their original shape.

BASIC CHEMISTRY

Matter, Mass, and Weight

All living and nonliving things are composed of **matter,** which is anything that occupies space and has mass. **Mass** is the amount of matter in an object, and **weight** is the gravitational force acting on an object of a given mass. For example, the weight of an apple results from the force of gravity "pulling" on the apple's mass.

PREDICT 1

The difference between mass and weight can be illustrated by considering an astronaut. How does an astronaut's mass and weight in outer space compare with the astronaut's mass and weight on the earth's surface?

The **kilogram (kg),** which is the mass of a platinum–iridium cylinder kept at the International Bureau of Weights and Measurements in France, is the international unit for mass. The mass of all other objects is compared with this cylinder. For example, a 2.2-pound lead weight or 1 liter (L) (1.06 qt) of water each has a mass of approximately 1 kg. An object with 1/1000 the mass of a kilogram is defined as having a mass of 1 **gram (g).**

Chemists use a balance to determine the mass of objects. Although we commonly refer to "weighing" an object on a balance, we are actually "massing" the object because the balance compares objects of unknown mass with objects of known mass. When the unknown and known masses are exactly balanced, the gravitational pull of the earth on both of them is the same. Thus, the effect of gravity on the unknown mass is counteracted by the effect of gravity on the known mass. A balance produces the same results on a mountaintop as at sea level because it does not matter that gravitational pull at sea level is stronger than on a mountaintop. It only matters that the effect of gravity on both the unknown and known masses is the same.

Elements and Atoms

An **element** is the simplest type of matter with unique chemical properties. A list of the elements commonly found in the human body is given in table 2.1. About 96% of the weight of the body results from the elements oxygen, carbon, hydrogen, and nitrogen.

An **atom** (*atomos,* indivisible) is the smallest particle of an element that has the chemical characteristics of that element. An element is composed of atoms of only one kind. For example,

TABLE 2.1 Common Elements

Element	Symbol	Atomic Number	Mass Number	Atomic Mass	Percent in Human Body by Weight (%)	Percent in Human Body by Number of Atoms (%)
Hydrogen	H	1	1	1.008	9.5	63.0
Carbon	C	6	12	12.01	18.5	9.5
Nitrogen	N	7	14	14.01	3.3	1.4
Oxygen	O	8	16	16.00	65.0	25.5
Fluorine	F	9	19	19.00	Trace	Trace
Sodium	Na	11	23	22.99	0.2	0.3
Magnesium	Mg	12	24	24.31	0.1	0.1
Phosphorus	P	15	31	30.97	1.0	0.22
Sulfur	S	16	32	32.07	0.3	0.05
Chlorine	Cl	17	35	35.45	0.2	0.03
Potassium	K	19	39	39.10	0.4	0.06
Calcium	Ca	20	40	40.08	1.5	0.31
Chromium	Cr	24	52	52.00	Trace	Trace
Manganese	Mn	25	55	54.94	Trace	Trace
Iron	Fe	26	56	55.85	Trace	Trace
Cobalt	Co	27	59	58.93	Trace	Trace
Copper	Cu	29	63	63.55	Trace	Trace
Zinc	Zn	30	64	65.39	Trace	Trace
Selenium	Se	34	80	78.96	Trace	Trace
Molybdenum	Mo	42	98	95.94	Trace	Trace
Iodine	I	53	127	126.9	Trace	Trace

the element carbon is composed of only carbon atoms, and the element oxygen is composed of only oxygen atoms.

An element, or an atom of that element, often is represented by a symbol. Usually, the first letter or letters of the element's name are used—for example, C for carbon, H for hydrogen, Ca for calcium, and Cl for chlorine. Occasionally, the symbol is taken from the Latin, Greek, or Arabic name for the element—for example, Na from the Latin word *natrium* is the symbol for sodium.

Atomic Structure

The characteristics of matter result from the structure, organization, and behavior of atoms (figure 2.1). Atoms are composed of subatomic particles, some of which have an electric charge. The three major types of subatomic particles are neutrons, protons, and electrons. A **neutron** has no electric charge, a **proton** has one positive charge, and an **electron** has one negative charge. The positive charge of a proton is equal in magnitude to the negative charge of an electron. The number of protons and electrons in each atom is equal, and the individual charges cancel each other. Therefore, each atom is electrically neutral.

Protons and neutrons form the **nucleus** at the center of an atom, and electrons are moving around the nucleus (see figure 2.1). The nucleus accounts for 99.97% of an atom's mass but only 1 ten-trillionth of its volume. Most of the volume of an atom is occupied by the electrons. Although it is impossible to know precisely where any given electron is located at any particular moment, the region where it is most likely to be found can be represented by an **electron cloud.** The likelihood of locating an electron at a specific point in a region correlates with the darkness of that region in the diagram. The darker the color, the greater the likelihood of finding the electron there at any given moment.

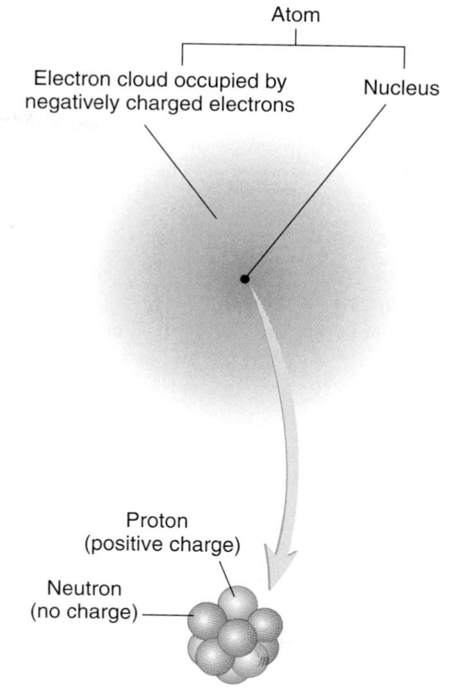

FIGURE 2.1 Model of an Atom

The tiny, dense nucleus consists of positively charged protons and uncharged neutrons. Most of the volume of an atom is occupied by rapidly moving, negatively charged electrons, which can be represented as an electron cloud. The probable location of an electron is indicated by the color of the electron cloud. The darker the color in each small part of the electron cloud, the more likely the electron is located there.

Atomic Number and Mass Number

The **atomic number** of an element is equal to the number of protons in each atom, and, because the number of electrons and protons is equal, the atomic number is also the number of electrons. Each element is uniquely defined by the number of protons in the atoms of that element. For example, only hydrogen atoms have one proton, only carbon atoms have six protons, and only oxygen atoms have eight protons (figure 2.2; see table 2.1). There are 90 naturally occurring elements, but additional elements have been synthesized by altering atomic nuclei. See the periodic table in appendix A for additional information about the elements.

Protons and neutrons have about the same mass, and they are responsible for most of the mass of atoms. Electrons, on the other hand, have very little mass. The **mass number** of an element is the

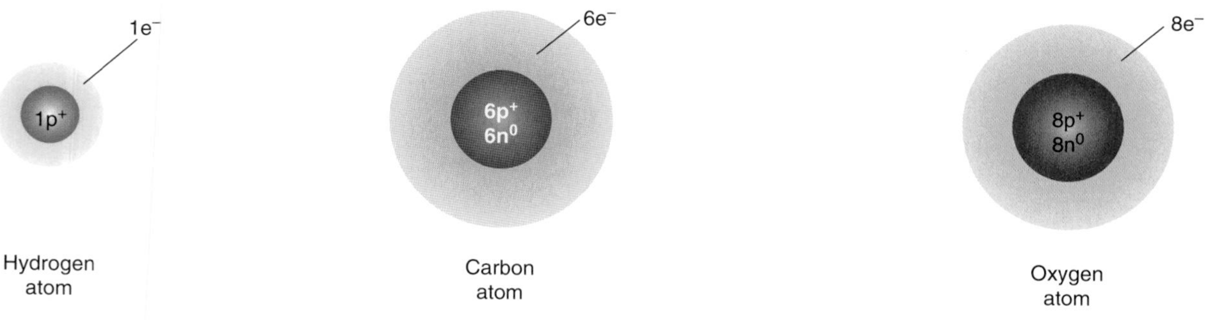

FIGURE 2.2 Hydrogen, Carbon, and Oxygen Atoms

Within the nucleus, the number of positively charged protons (p^+) and uncharged neutrons (n^0) is indicated. The negatively charged electrons (e^-) are around the nucleus. Atoms are electrically neutral because the number of protons and the number of electrons within an atom are equal.

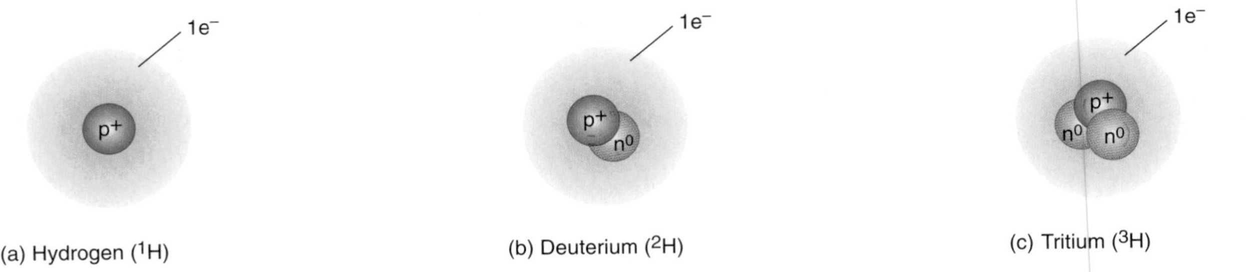

FIGURE 2.3 Isotopes of Hydrogen

(*a*) Hydrogen has one proton and no neutrons in its nucleus. (*b*) Deuterium has one proton and one neutron in its nucleus. (*c*) Tritium has one proton and two neutrons in its nucleus.

number of protons plus the number of neutrons in each atom. For example, the mass number for carbon is 12 because it has six protons and six neutrons.

PREDICT 2

The atomic number of potassium is 19, and the mass number is 39. What is the number of protons, neutrons, and electrons in an atom of potassium?

Isotopes and Atomic Mass

Isotopes (ī′sō-tōpz) are two or more forms of the same element that have the same number of protons and electrons but a different number of neutrons. Thus, isotopes have the *same* atomic number but *different* mass numbers. There are three isotopes of hydrogen: hydrogen, deuterium, and tritium. All three isotopes have one proton and one electron, but hydrogen has no neutrons in its nucleus, deuterium has one neutron, and tritium has two neutrons (figure 2.3). Isotopes can be denoted using the symbol of the element preceded by the mass number (number of protons and neutrons) of the isotope. Thus, hydrogen is ^{1}H, deuterium is ^{2}H, and tritium is ^{3}H.

Individual atoms have very little mass. A hydrogen atom has a mass of 1.67×10^{-24} g (see appendix B for an explanation of the scientific notation of numbers). To avoid using such small numbers, a system of relative atomic mass is used. In this system, a **unified atomic mass unit (u),** or **dalton (Da),** is 1/12 the mass of ^{12}C, a carbon atom with six protons and six neutrons. Thus, ^{12}C has an atomic mass of exactly 12 u. A naturally occurring sample of carbon, however, contains mostly ^{12}C but also a small quantity of other carbon isotopes, such as ^{13}C, which has six protons and seven neutrons. The **atomic mass** of an element is the *average* mass of its naturally occurring isotopes, taking into account the relative abundance of each isotope. For example, the atomic mass of the element carbon is 12.01 u (see table 2.1), which is slightly more than 12 u because of the additional mass of the small amount of other carbon isotopes. Because the atomic mass is an average, a sample of carbon can be treated as if all the carbon atoms had an atomic mass of 12.01 u.

The Mole and Molar Mass

Avogadro's number is the number of atoms in exactly 12 g of ^{12}C. This enormous number is 6.022×10^{23}. A **mole (mol)** of a substance contains Avogadro's number of entities, such as atoms, ions, or molecules. The **molar mass** of a substance is the mass of one mole of the substance expressed in grams. Because 12 g of ^{12}C is used as the standard, the atomic mass of an entity expressed in unified atomic mass units is the same as the molar mass expressed in grams. Thus, carbon atoms have an atomic mass of 12.01 u, and 12.01 g of carbon have Avogadro's number (1 mol) of carbon atoms.

Just as a grocer sells eggs in lots of 12 (a dozen), a chemist groups atoms in lots of 6.022×10^{23} (Avogadro's number, 1 mol). The molar mass is used as a convenient way to determine the number of atoms in a sample of an element. For example, 1.008 g of hydrogen (1 mol) has the same number of atoms as 12.01 g of carbon (1 mol).

PREDICT 3

Are there more or fewer atoms in 12.01 g of carbon than in 12.01 g of magnesium? (*Hint:* Use table 2.1.)

1. ***Define matter. How is the mass and the weight of an object different?***
2. ***Define element and atom. What four elements are found in the greatest abundance in humans?***
3. ***For each subatomic particle of an atom, state its charge and location. Which subatomic particles are most responsible for the mass and volume of an atom? Which subatomic particles determine atomic number and mass number?***
4. ***Define isotopes and give an example. Define atomic mass. Why is the atomic mass of most elements not exactly equal to the mass number?***
5. ***What is Avogadro's number? Define a mole and molar mass.***

Electrons and Chemical Bonding

The outermost electrons of an atom determine its chemical behavior. When these outermost electrons are transferred, or shared, between atoms, **chemical bonding** occurs. Two major types of chemical bonding are ionic and covalent bonding.

Ionic Bonding

An atom is electrically neutral because it has an equal number of protons and electrons. If an atom loses or gains electrons, the number of protons and electrons are no longer equal, and

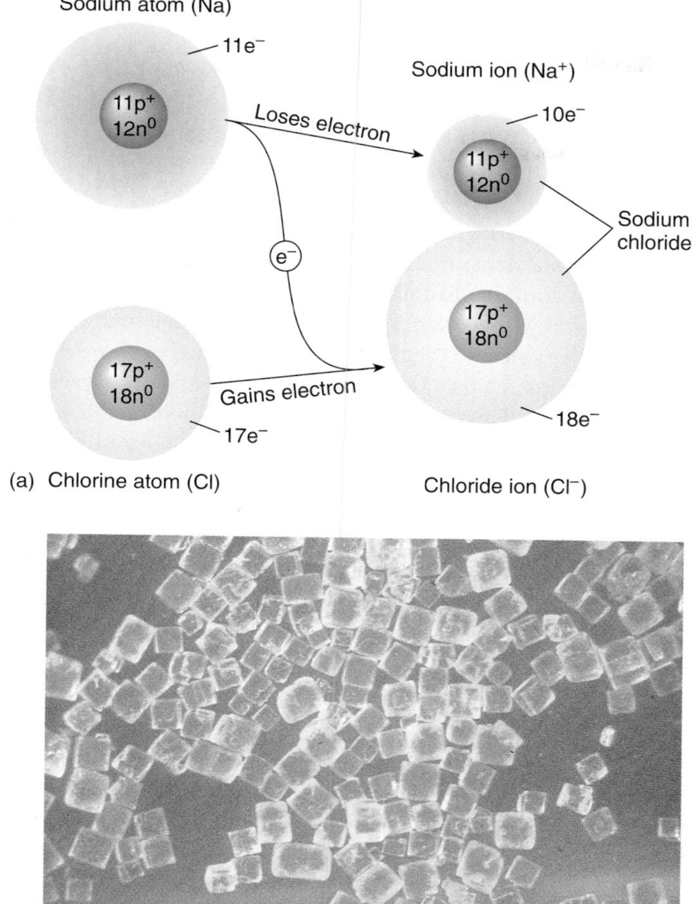

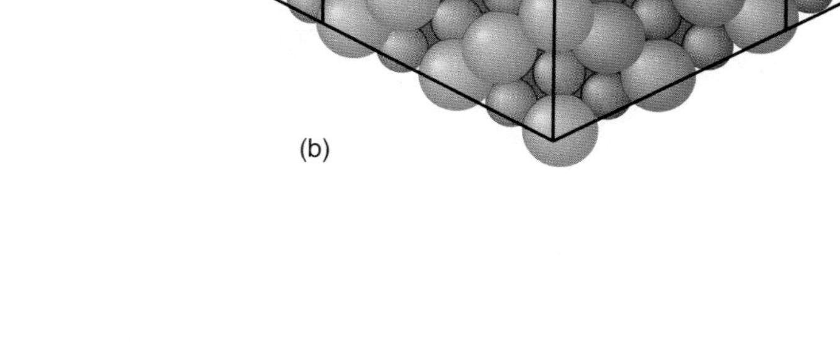

FIGURE 2.4 Ionic Bonding

(*a*) A sodium atom loses an electron to become a smaller positively charged ion, and a chlorine atom gains an electron to become a larger negatively charged ion. The attraction between the oppositely charged ions results in ionic bonding and the formation of sodium chloride. (*b*) The sodium and chlorine ions are organized to form a cube-shaped array. (*c*) Microphotograph of salt crystals reflects the cubic arrangement of the ions.

a charged particle called an **ion** (ī′on) is formed. After an atom loses an electron, it has one more proton than it has electrons and is positively charged. A sodium atom (Na) can lose an electron to become a positively charged sodium ion (Na^+) (figure 2.4*a*). After an atom gains an electron, it has one more electron than it has protons and is negatively charged. A chlorine atom (Cl) can accept an electron to become a negatively charged chloride ion (Cl^-).

Positively charged ions are called **cations** (kat′ī-onz), and negatively charged ions are called **anions** (an′ī-onz). Because oppositely charged ions are attracted to each other, cations and anions tend to remain close together, which is called **ionic** (ī-on′ik) **bonding.** For example, Na^+ and Cl^- are held together by ionic bonding to form an array of ions called sodium chloride (NaCl), or table salt (see figure 2.4*b* and *c*). Some ions commonly found in the body are listed in table 2.2.

Covalent Bonding

Covalent bonding results when atoms share one or more pairs of electrons. The resulting combination of atoms is called a molecule. An example is the covalent bond between two hydrogen atoms

TABLE 2.2 Important Ions

Common Ions	Symbols	Significance*
Calcium	Ca^{2+}	Part of bones and teeth, blood clotting, muscle contraction, release of neurotransmitters
Sodium	Na^+	Membrane potentials, water balance
Potassium	K^+	Membrane potentials
Hydrogen	H^+	Acid-base balance
Hydroxide	OH^-	Acid-base balance
Chloride	Cl^-	Water balance
Bicarbonate	HCO_3^-	Acid-base balance
Ammonium	NH_4^+	Acid-base balance
Phosphate	PO_4^{3-}	Part of bones and teeth, energy exchange, acid-base balance
Iron	Fe^{2+}	Red blood cell formation
Magnesium	Mg^{2+}	Necessary for enzymes
Iodide	I^-	Present in thyroid hormones

*The ions are part of the structures or play important roles in the processes listed.

There is no interaction between the two hydrogen atoms because they are too far apart.

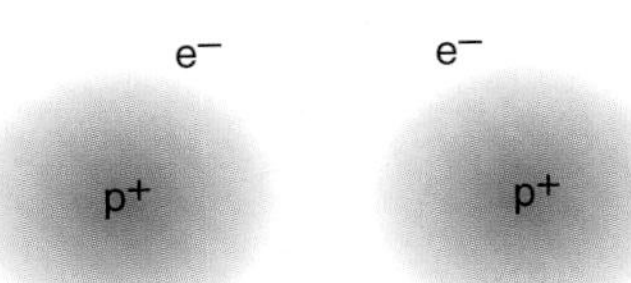

The positively charged nucleus of each hydrogen atom begins to attract the electron of the other.

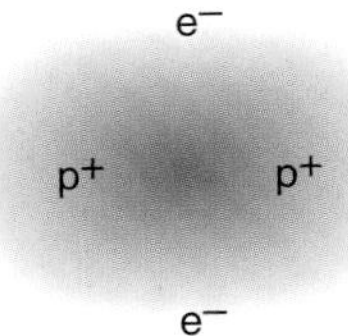

A covalent bond is formed when the electrons are shared between the nuclei because the electrons are equally attracted to each nucleus.

FIGURE 2.5 Covalent Bonding

to form a hydrogen molecule (figure 2.5). Each hydrogen atom has one electron. As the two hydrogen atoms get closer together, the positively charged nucleus of each atom begins to attract the electron of the other atom. At an optimal distance, the two nuclei mutually attract the two electrons, and each electron is shared by both nuclei. The two hydrogen atoms are now held together by a covalent bond.

When an electron pair is shared between two atoms, a **single covalent bond** results. A single covalent bond is represented by a single line between the symbols of the atoms involved (e.g., H—H). A **double covalent bond** results when two atoms share four electrons, two from each atom. When a carbon atom combines with two oxygen atoms to form carbon dioxide, two double covalent bonds are formed. Double covalent bonds are indicated by a double line between the atoms (O=C=O).

When electrons are shared equally between atoms, as in a hydrogen molecule, the bonds are called **nonpolar covalent bonds.** Atoms bound to one another by a covalent bond do not always share their electrons equally, however, because the nucleus of one atom attracts the electrons more strongly than does the nucleus of the other atom. Bonds of this type are called **polar covalent bonds** and are common in both living and nonliving matter.

Polar covalent bonds can result in polar molecules, which are electrically asymmetric. For example, oxygen atoms attract electrons more strongly than do hydrogen atoms. When covalent bonding between an oxygen atom and two hydrogen atoms forms a water molecule, the electrons are located in the vicinity of the oxygen nucleus more than in the vicinity of the hydrogen nuclei. Because electrons have a negative charge, the oxygen side of the molecule is slightly more negative than the hydrogen side (figure 2.6).

Molecules and Compounds

A **molecule** is formed when two or more atoms chemically combine to form a structure that behaves as an independent unit. The atoms that combine to form a molecule can be of the same type, such as two hydrogen atoms combining to form a hydrogen molecule. More typically, a molecule consists of two or more different types of atoms, such as two hydrogen atoms and an oxygen atom forming water. Thus, a glass of water consists of a collection of individual water molecules positioned next to one another.

A **compound** is a substance composed of two or more *different* types of atoms that are chemically combined. Not all molecules are compounds. For example, a hydrogen molecule is not a compound because it does not consist of different types of atoms. Some compounds are molecules and some are not. Covalent compounds,

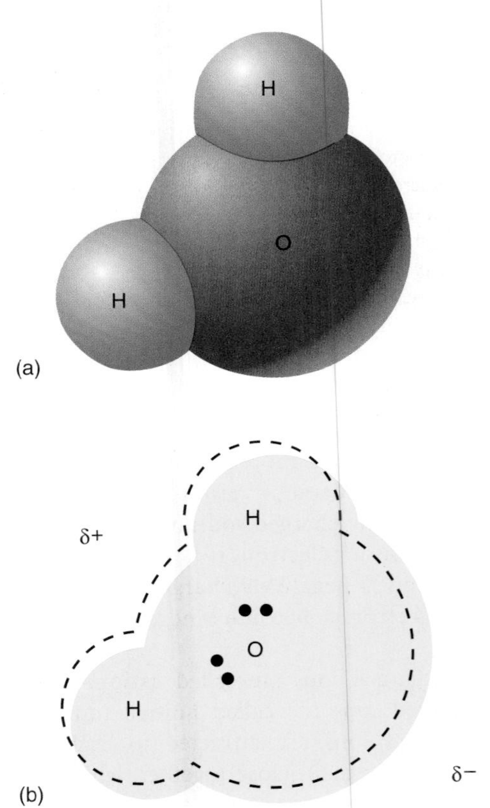

FIGURE 2.6 Polar Covalent Bonds

(*a*) A water molecule forms when two hydrogen atoms form covalent bonds with an oxygen atom. (*b*) Electron pairs (*indicated by dots*) are shared between the hydrogen atoms and oxygen. The dashed outline shows the expected location of the electron cloud if the electrons are shared equally. The electrons are shared unequally, as shown by the electron cloud (*yellow*) not coinciding with the dashed outline. Consequently, the oxygen side of the molecule has a slight negative charge (*indicated by* δ^-) and the hydrogen side of the molecule has a slight positive charge (*indicated by* δ^+).

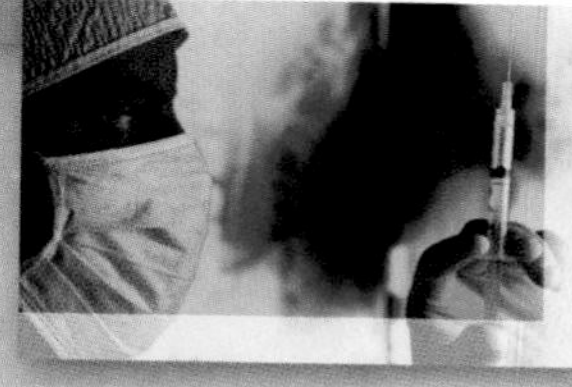

CLINICAL FOCUS

Radioactive Isotopes and X-Rays

Protons, neutrons, and electrons are responsible for the chemical properties of atoms. They also have other properties that can be useful in a clinical setting. For example, they have been used to develop methods for examining the inside of the body. **Radioactive isotopes** are commonly used by clinicians and researchers because sensitive measuring devices can detect their radioactivity, even when they are present in very small amounts.

Radioactive isotopes have unstable nuclei that spontaneously change to form more stable nuclei. As a result, either new isotopes or new elements are produced. In this process of nuclear change, alpha particles, beta particles, and gamma rays are emitted from the nuclei of radioactive isotopes. Alpha (α) particles are positively charged helium ions (He^{2+}), which consist of two protons and two neutrons. Beta (β) particles are electrons formed as neutrons change into protons. An electron is ejected from the neutron, and the proton that is produced remains in the nucleus. Gamma (γ) rays are a form of electromagnetic radiation (high-energy photons) released from nuclei as they lose energy.

All isotopes of an element have the same atomic number, and their chemical behavior is very similar. For example, ^{3}H (tritium) can substitute for ^{1}H (hydrogen), and either ^{125}iodine or ^{131}iodine can substitute for ^{126}iodine in chemical reactions.

Several procedures that are used to determine the concentration of substances such as hormones depend on the incorporation of small amounts of radioactive isotopes, such as ^{125}iodine, into the substances being measured. Clinicians using these procedures can more accurately diagnose disorders of the thyroid gland, the adrenal gland, and the reproductive organs.

Radioactive isotopes are also used to treat cancer. Some of the particles released from isotopes have a very high energy content and can penetrate and destroy tissues. Thus, radioactive isotopes can be used to destroy tumors because rapidly growing tissues, such as tumors, are more sensitive to radiation than are healthy cells. Radiation can also be used to sterilize materials that cannot be exposed to high temperatures (e.g., some fabric and plastic items used during surgical procedures). In addition, radioactive emissions can be used to sterilize food and other items.

X-rays are electromagnetic radiations with a much shorter wavelength than visible light. When electric current is used to heat a filament to very high temperatures, energy of the electrons becomes so great that some electrons are emitted from the hot filament. When these electrons strike a positive electrode at high speeds, they release some of their energy in the form of x-rays.

X-rays do not penetrate dense material as readily as they penetrate less dense material, and x-rays can expose photographic film. Consequently, an x-ray beam can pass through a person and onto photographic film. Dense tissues of the body absorb the x-rays, and in these areas the film is underexposed and so appears white or light in color on the developed film. On the other hand, the x-rays readily pass through less dense tissue, and the film in these areas is overexposed and appears black or dark in color. In an x-ray film of the skeletal system, the dense bones are white, and the less dense soft tissues are dark, often so dark that no details can be seen. Because the dense bone material is clearly visible, x-rays can be used to determine whether bones are broken or have other abnormalities.

Soft tissues can be photographed by using low-energy x-rays. Mammograms are low-energy x-rays of the breast that can be used to detect tumors because tumors are slightly denser than normal tissue.

Radiopaque substances are dense materials that absorb x-rays. If a radiopaque liquid is given to a patient, the liquid assumes the shape of the organ into which it is placed. For example, if a barium solution is swallowed, the outline of the upper digestive tract can be photographed using x-rays to detect such abnormalities as ulcers.

in which different types of atoms are held together by covalent bonds, are molecules because the sharing of electrons results in the formation of distinct, independent units. Water is an example of a substance that is a compound and a molecule.

On the other hand, ionic compounds are not molecules because the ions are held together by the force of attraction between opposite charges. A piece of NaCl does not consist of NaCl molecules positioned next to each other. Instead, table salt is an organized array of Na^{+} and Cl^{-} in which each charged ion is surrounded by several ions of the opposite charge (see figure 2.4*b*). Sodium chloride is an example of a substance that is a compound but is not a molecule.

The kinds and numbers of atoms (or ions) in a molecule or compound can be represented by a formula consisting of the symbols of the atoms (or ions) plus subscripts denoting the number of each type of atom (or ion). The formula for glucose (a sugar) is $C_6H_{12}O_6$, indicating that glucose has 6 carbon, 12 hydrogen, and 6 oxygen atoms (table 2.3).

The **molecular mass** of a molecule or compound can be determined by adding up the atomic masses of its atoms (or ions). The term *molecular mass* is used for convenience for ionic compounds, even though they are not molecules. For example, the atomic mass of sodium is 22.99 and chloride is 35.45. The molecular mass of NaCl is therefore 58.44 (22.99 + 35.45).

6 ***Describe how ionic bonding occurs. What is a cation and an anion?***

7 ***Describe how covalent bonding occurs. What is the difference between polar and nonpolar covalent bonds?***

8 ***Distinguish between a molecule and a compound. Are all molecules compounds? Are all compounds molecules?***

9 ***Define molecular mass.***

TABLE 2.3 Picturing Molecules

Representation	Hydrogen	Carbon Dioxide	Glucose
Chemical Formula The formula shows the kind and number of atoms present.	H_2	CO_2	$C_6H_{12}O_6$
Electron-Dot Formula The bonding electrons are shown as dots between the symbols of the atoms.	H:H Single covalent bond	O::C::O Double covalent bond	Not used for complex molecules
Bond-Line Formula The bonding electrons are shown as lines between the symbols of the atoms.	H—H Single covalent bond	O=C=O Double covalent bond	CH_2OH, O, OH, HO, OH, OH
Models Atoms are shown as different-sized and different-colored spheres.	Hydrogen atom	Oxygen atom, Carbon atom	

PREDICT 4

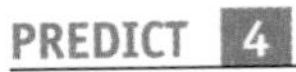

What is the molecular mass of a molecule of glucose? (Use table 2.1.)

Intermolecular Forces

Intermolecular forces result from the weak electrostatic attractions between the oppositely charged parts of molecules, or between ions and molecules. Intermolecular forces are much weaker than the forces producing chemical bonding.

Hydrogen Bonds

Molecules with polar covalent bonds have positive and negative "ends." Intermolecular force results from the attraction of the positive end of one polar molecule to the negative end of another polar molecule. When hydrogen forms a covalent bond with oxygen, nitrogen, or fluorine, the resulting molecule becomes very polarized. If the positively charged hydrogen of one molecule is attracted to the negatively charged oxygen, nitrogen, or fluorine of another molecule, a **hydrogen bond** is formed. For example, the positively charged hydrogen atoms of a water molecule form hydrogen bonds with the negatively charged oxygen atoms of other water molecules (figure 2.7).

Hydrogen bonds play an important role in determining the shape of complex molecules because the hydrogen bonds between different polar parts of the molecule hold the molecule in its normal three-dimensional shape (see "Proteins," p. 44; and "Nucleic Acids: DNA and RNA," p. 47).

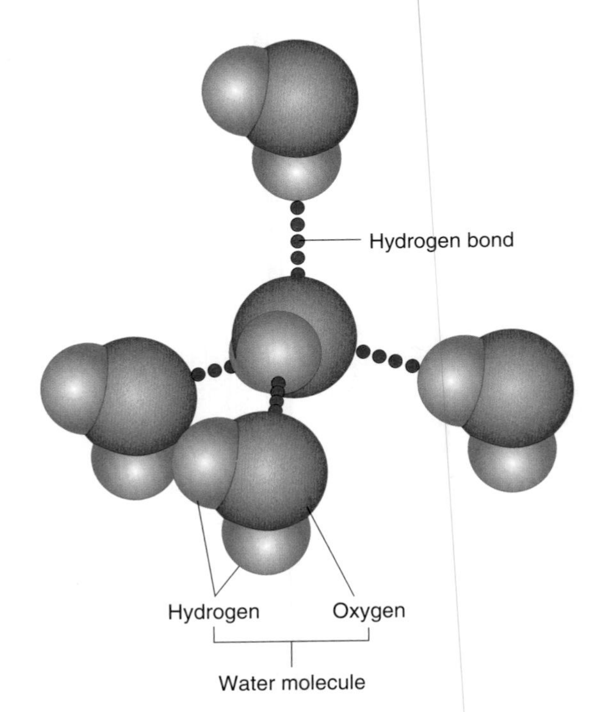

FIGURE 2.7 Hydrogen Bonds

The positive hydrogen part of one water molecule forms a hydrogen bond (*red dotted line*) with the negative oxygen part of another water molecule. As a result, hydrogen bonds hold the water molecules together.

Table 2.4 summarizes the important characteristics of chemical bonding (ionic and covalent) and intermolecular forces (hydrogen bonds).

Solubility and Dissociation

Solubility is the ability of one substance to dissolve in another—for example, when sugar dissolves in water. Charged substances, such as sodium chloride, and polar substances, such as glucose, dissolve in water readily, whereas nonpolar substances, such as oils, do not. We all have seen how oil floats on water. Substances dissolve in water when they become surrounded by water molecules. If the positive and negative ends of the water molecules are attracted more to the charged ends of other molecules than they are to each other, the hydrogen bonds between the ends of the water molecules are broken, and the water molecules surround the other molecules, which become dissolved in the water.

When ionic compounds dissolve in water, their ions **dissociate,** or separate, from one another because the cations are attracted to the negative ends of the water molecules, and the anions are attracted to the positive ends of the water molecules. When sodium chloride dissociates in water, the sodium and chloride ions separate, and water molecules surround and isolate the ions, thereby keeping them in solution (figure 2.8).

TABLE 2.4 Comparison of Bonds

Bond	Example
Ionic Bond A complete transfer of electrons between two atoms results in separate positively charged and negatively charged ions.	Na^+Cl^- Sodium chloride
Polar Covalent Bond An unequal sharing of electrons between two atoms results in a slight positive charge (δ^+) on one side of the molecule and slight negative charge (δ^-) on the other side of the molecule.	H δ^+ O δ^- H Water
Nonpolar Covalent Bond An equal sharing of electrons between two atoms results in an even charge distribution among the atoms of the molecule.	H H—C—H H Methane
Hydrogen Bond The attraction of oppositely charged ends of one polar molecule to another polar molecule holds molecules or parts of molecules together.	H O···H—O H H Water molecules

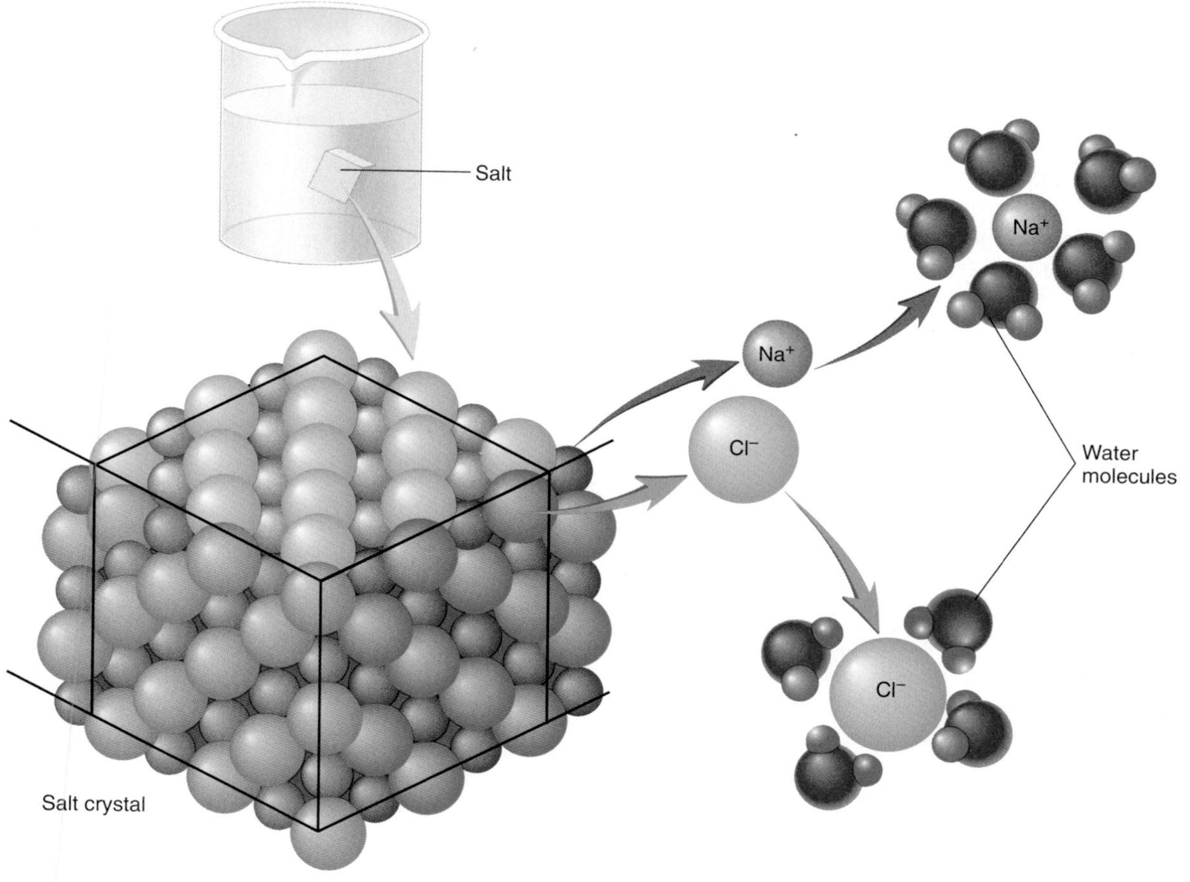

FIGURE 2.8 Dissociation

Sodium chloride (table salt) dissociating in water. The positively charged Na^+ are attracted to the negatively charged oxygen (*red*) end of the water molecule, and the negatively charged Cl^- are attracted to the positively charged hydrogen (*blue*) end of the water molecule.

When molecules (covalent compounds) dissolve in water, they usually remain intact, even though they are surrounded by water molecules. Thus, in a glucose solution, glucose molecules are surrounded by water molecules.

Cations and anions that dissociate in water are sometimes called **electrolytes** (ē-lek′trō-lītz) because they have the capacity to conduct an electric current, which is the flow of charged particles. An electrocardiogram (ECG) is a recording of electric currents produced by the heart. These currents can be detected by electrodes on the surface of the body because the ions in the body fluids conduct electric currents. Molecules that do not dissociate form solutions that do not conduct electricity and are called **nonelectrolytes.** Pure water is a nonelectrolyte.

10 ***Define hydrogen bond, and explain how hydrogen bonds hold polar molecules, such as water, together. How do hydrogen bonds affect the shape of a molecule?***

11 ***Define solubility. How do ionic and covalent compounds typically dissolve in water?***

12 ***Distinguish between electrolytes and nonelectrolytes.***

CHEMICAL REACTIONS AND ENERGY

In a **chemical reaction,** atoms, ions, molecules, or compounds interact either to form or to break chemical bonds. The substances that enter into a chemical reaction are called **reactants,** and the substances that result from the chemical reaction are called **products.**

For our purposes, three important points can be made about chemical reactions. First, in some reactions, less complex reactants are combined to form a larger, more complex product. An example is the synthesis of the complex molecules of the human body from basic "building blocks" obtained from food (figure 2.9*a*). Second, in other reactions, a reactant can be broken down, or decomposed, into simpler, less complex products. An example is the breakdown of food molecules into basic building blocks (figure 2.9*b*). Third, atoms are generally associated with other atoms through chemical bonding or intermolecular forces; therefore, to synthesize new products or break down reactants, it is necessary to change the relationship between atoms.

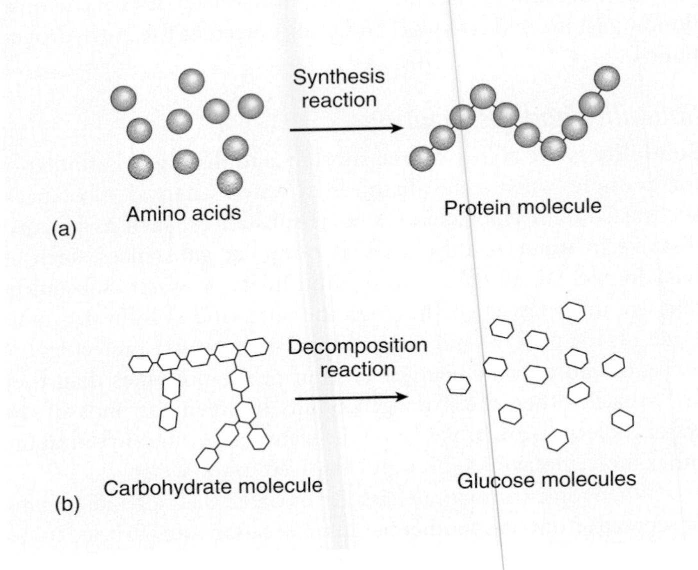

FIGURE 2.9 Synthesis and Decomposition Reactions

(*a*) Synthesis reaction in which amino acids, the basic "building blocks" of proteins, combine to form a protein molecule. (*b*) Decomposition reaction in which a complex carbohydrate breaks down into smaller glucose molecules, which are the "building blocks" of carbohydrates.

Synthesis Reactions

When two or more reactants chemically combine to form a new and larger product, the process is called a **synthesis reaction.** An example of a synthesis reaction is the combination of two amino acids to form a dipeptide (figure 2.10*a*). In this synthesis reaction, water is formed as the amino acids are bound together. Synthesis reactions in which water is a product are called **dehydration** (water out) **reactions.** Old chemical bonds are broken and new chemical bonds are formed as the atoms rearrange as a result of a synthesis reaction.

Another example of a synthesis reaction in the body is the formation of adenosine triphosphate (ATP). In ATP, *A* stands for adenosine; *T* stands for tri-, or three; and *P* stands for phosphate group (PO_4^{3-}). Thus, ATP consists of adenosine and three phosphate groups (see p. 50 for the details of the structure of ATP). ATP is synthesized from adenosine diphosphate (ADP), which has two phosphate groups, and an inorganic phosphate ($H_2PO_4^-$), which is often symbolized as P_i.

A-P-P (ADP)	+	P_i (Inorganic phosphate)	→	A-P-P-P (ATP)

Synthesis reactions produce the molecules characteristic of life, such as ATP, proteins, carbohydrates, lipids, and nucleic acids. All of the synthesis reactions that occur within the body are referred to collectively as **anabolism** (ă-nab′ō-lizm). The growth, maintenance, and repair of the body could not take place without anabolic reactions.

Decomposition Reactions

A **decomposition reaction** is the reverse of a synthesis reaction—a larger reactant is chemically broken down into two or more smaller products. The breakdown of a disaccharide (a type of carbohydrate) into glucose molecules (figure 2.10*b*) is an example. Note that this reaction requires that water be split into two parts and that each part be contributed to one of the new glucose molecules. Reactions that use water in this manner are called **hydrolysis** (hī-drol′i-sis; water dissolution) **reactions.**

The breakdown of ATP to ADP and an inorganic phosphate is another example of a decomposition reaction.

A-P-P-P (ATP)	→	A-P-P (ADP)	+	P_i (Inorganic phosphate)

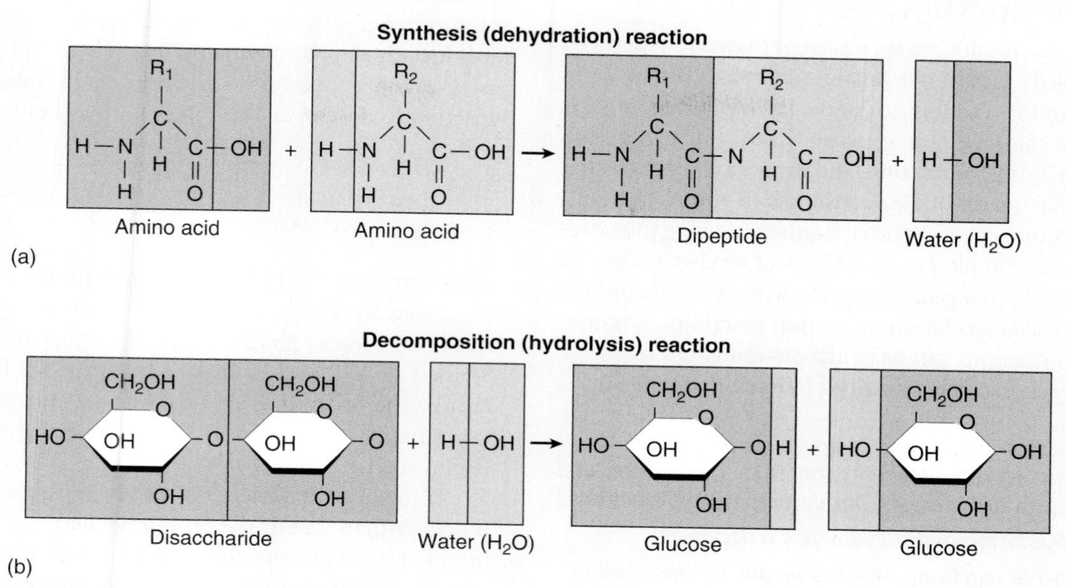

FIGURE 2.10 Synthesis (Dehydration) and Decomposition (Hydrolysis) Reactions
(*a*) Synthesis reaction in which two amino acids combine to form a dipeptide. This reaction is also a dehydration reaction because it results in the removal of a water molecule from the amino acids. (*b*) Decomposition reaction in which a disaccharide breaks apart to form glucose molecules. This reaction is also a hydrolysis reaction because it involves the splitting of a water molecule.

The decomposition reactions that occur in the body are collectively called **catabolism** (kă-tab′-ō-lizm). They include the digestion of food molecules in the intestine and within cells, the breakdown of fat stores, and the breakdown of foreign matter and microorganisms in certain blood cells that function to protect the body. All of the anabolic and catabolic reactions in the body are collectively defined as **metabolism.**

Reversible Reactions

A **reversible reaction** is a chemical reaction in which the reaction can proceed from reactants to products or from products to reactants. When the rate of product formation is equal to the rate of the reverse reaction, the reaction system is said to be at **equilibrium.** At equilibrium, the amount of reactants relative to the amount of products remains constant.

The following analogy may help clarify the concept of reversible reactions and equilibrium. Imagine a trough containing water. The trough is divided into two compartments by a partition, but the partition contains holes that allow water to move freely between the compartments. Because water can move in either direction, this is like a reversible reaction. Let the water in the left compartment be the reactant and the water in the right compartment be the product. At equilibrium, the amount of reactant relative to the amount of product in each compartment is always the same because the partition allows water to pass between the two compartments until the level of water is the same in both compartments. If additional water is added to the reactant compartment, water flows from it through the partition to the product compartment until the level of water is the same in both compartments. Likewise, if additional reactants are added to a reaction system, some will form product until equilibrium is reestablished. Unlike this analogy, however, the amount of the reactants compared with the amount of products of most reversible reactions is not one to one. Depending on the specific reversible reaction, one part reactant to two parts product, two parts reactant to one part product, or many other possibilities can occur.

An important reversible reaction in the human body involves carbon dioxide and hydrogen ions. The reaction between carbon dioxide (CO_2) and water (H_2O) to form carbonic acid (H_2CO_3) is reversible. Carbonic acid then separates by a reversible reaction to form hydrogen ions (H^+) and bicarbonate ions (HCO_3^-).

$$CO_2 + H_2O \rightleftarrows H_2CO_3 \rightleftarrows H^+ + HCO_3^-$$

If CO_2 is added to H_2O, additional H_2CO_3 forms, which in turn causes more H^+ and HCO_3^- to form. The amount of H^+ and HCO_3^- relative to CO_2 therefore remains constant. Maintaining a constant level of H^+ is necessary for proper functioning of the nervous system. This can be achieved, in part, by regulating blood CO_2 levels. For example, slowing down the respiration rate causes blood carbon dioxide levels to increase.

PREDICT 5

If the respiration rate increases, CO_2 is eliminated from the blood. What effect does this change have on blood H^+ ion levels?

Oxidation–Reduction Reactions

Chemical reactions that result from the exchange of electrons between the reactants are called oxidation–reduction reactions. When sodium and chlorine react to form sodium chloride, the sodium atom loses an electron, and the chlorine atom gains an electron. The loss of an electron by an atom is called **oxidation,** and the gain of an electron is called **reduction.** The transfer of the electron can be complete, resulting in an ionic bond, or it can be a partial transfer, resulting in a covalent bond. Because the complete or partial loss of an electron by one atom is accompanied by the gain of that electron by another atom, these reactions are called **oxidation–reduction reactions.** Synthesis and decomposition reactions can be oxidation–reduction reactions. Thus, a chemical reaction can be described in more than one way.

13 ***Define a chemical reaction and compare synthesis and decomposition reactions. How do anabolism, catabolism, and metabolism relate to synthesis and decomposition reactions?***

14 ***Describe a dehydration and a hydrolysis reaction.***

15 ***Describe reversible reactions. What is meant by the equilibrium condition in reversible reactions?***

16 ***What is an oxidation–reduction reaction?***

PREDICT 6

When hydrogen gas combines with oxygen gas to form water, is the hydrogen reduced or oxidized? Explain.

Energy

Energy, unlike matter, does not occupy space and has no mass. **Energy** is the capacity to do **work**—that is, to move matter. Energy can be subdivided into potential energy and kinetic energy. **Potential energy** is stored energy that could do work but is not doing so. **Kinetic** (ki-net′ik) **energy** is the form of energy that actually does work and moves matter. A ball held at arm's length above the floor has potential energy. No energy is expended as long as the ball does not move. If the ball is released and falls toward the floor, however, it has kinetic energy.

According to the conservation of energy principle, the total energy of the universe is constant. Therefore, energy is neither created nor destroyed. Potential energy, however, can be converted into kinetic energy, and kinetic energy can be converted into potential energy. For example, the potential energy in the ball is converted into kinetic energy as the ball falls toward the floor. Conversely, the kinetic energy required to raise the ball from the floor is converted into potential energy.

Potential and kinetic energy can be found in many different forms. **Mechanical energy** is energy resulting from the position or movement of objects. Many of the activities of the human body, such as moving a limb, breathing, and circulating blood, involve mechanical energy. Other forms of energy are chemical energy, heat energy, electric energy, and electromagnetic (radiant) energy.

Chemical Energy

The **chemical energy** of a substance is a form of stored (potential) energy within its chemical bonds. In any chemical reaction, the potential energy contained in the chemical bonds of the reactants can be compared with the potential energy in the chemical bonds of the products. If the potential energy in the chemical bonds of the reactants is less than that of the products, then energy must be supplied for the reaction to occur. An example is the synthesis of ATP from ADP.

$ADP + H_2PO_4^-$	+	Energy	→	$ATP + H_2O$
(Less potential energy in reactants)				(More potential energy in products)

For simplicity, the H_2O is often not shown in this reaction, and P_i is used to represent inorganic phosphate ($H_2PO_4^-$). For this reaction to occur, bonds in $H_2PO_4^-$ are broken and bonds are formed in ATP and H_2O. As a result of the breaking of existing bonds, the formation of new bonds, and the input of energy, these products have more potential energy than the reactants (figure 2.11*a*).

If the potential energy in the chemical bonds of the reactants is greater than that of the products, energy is released by the reaction. For example, the chemical bonds of food molecules contain more potential energy than the waste products that are produced when food molecules are decomposed. The energy released from the chemical bonds of food molecules is used by living systems to synthesize ATP. Once ATP is produced, the breakdown of ATP to ADP results in the release of energy.

$ATP + H_2O$	→	$ADP + H_2PO_4^-$	+	Energy
(More potential energy in reactants)		(Less potential energy in products)		

For this reaction to occur, the bonds in ATP and H_2O are broken and bonds in $H_2PO_4^-$ are formed. As a result of breaking the existing bonds and forming new bonds, these products have less potential energy than the reactants, and energy is released (figure 2.11*b*). Note that the energy released does not come from breaking the phosphate bond of ATP because breaking a chemical bond requires the input of energy. It is commonly stated, however, that the breakdown of ATP results in the release of energy, which is true when the overall reaction is considered. The energy released when ATP is broken down can be used to synthesize other molecules; to do work, such as muscle contraction; or to produce heat.

Heat Energy

Heat energy is the energy that flows between objects that are at different temperatures. Temperature is a measure of how hot or cold a substance is relative to another substance. Heat is always transferred from a hotter object to a cooler object, such as from a hot stove top to a finger.

All other forms of energy can be converted into heat energy. For example, when a moving object comes to rest, its kinetic energy is converted into heat energy by friction. Some of the potential energy of chemical bonds is released as heat energy during chemical reactions. Human body temperature is maintained by heat produced as a by-product of chemical reactions.

17 ***How is energy different from matter? How are potential and kinetic energy different from each other?***

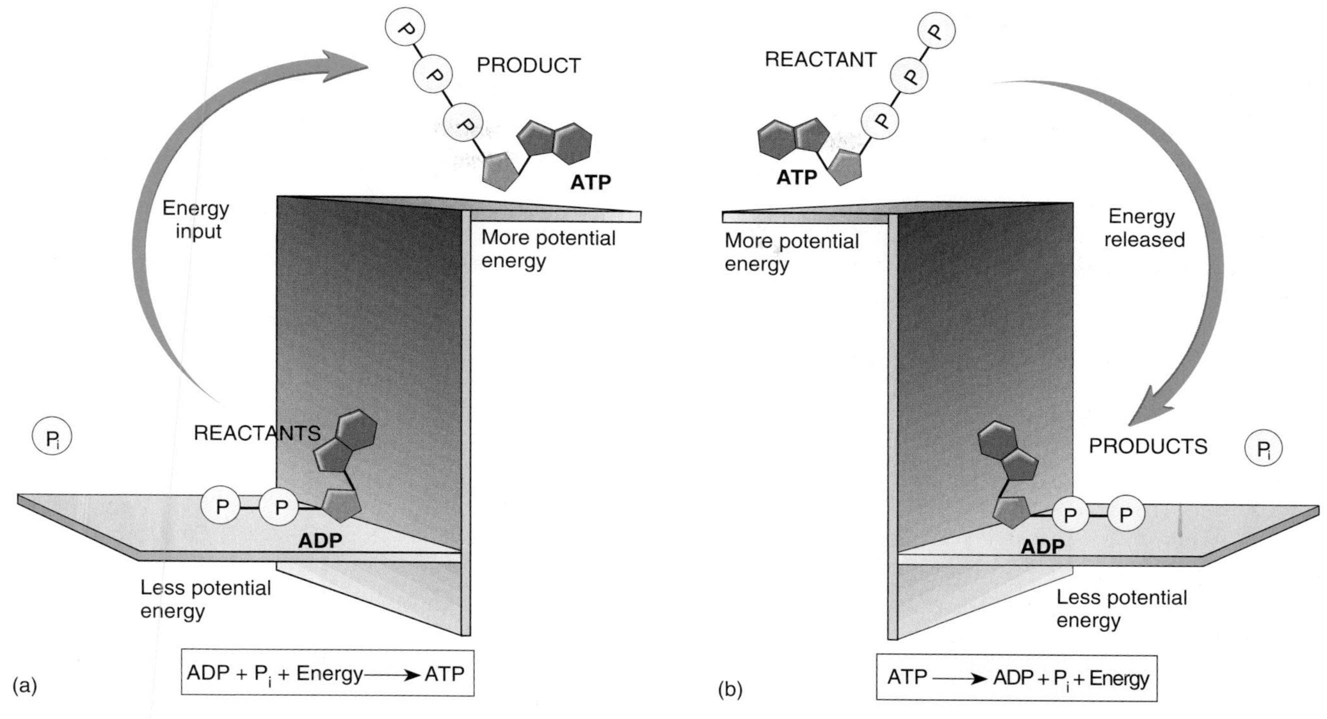

FIGURE 2.11 Energy and Chemical Reactions

In each figure, the upper shelf represents a higher energy level, and the lower shelf represents a lower energy level. (*a*) Reaction in which the input of energy is required for the synthesis of ATP. (*b*) Reaction in which energy is released as a result of the breakdown of ATP.

18 ***Define mechanical energy, chemical energy, and heat energy. How is chemical energy converted to mechanical energy and heat energy in the body?***

19 ***Use ATP and ADP to illustrate the release or input of energy in chemical reactions.***

PREDICT 7

Energy from the breakdown of ATP provides the kinetic energy for muscle movement. Why does body temperature increase during exercise?

Speed of Chemical Reactions

Molecules are constantly in motion and therefore have kinetic energy. A chemical reaction occurs only when molecules with sufficient kinetic energy collide with each other. As two molecules move closer together, the negatively charged electron cloud of one molecule repels the negatively charged electron cloud of the other molecule. If the molecules have sufficient kinetic energy, they overcome this repulsion and come together. The nuclei in some atoms attract the electrons of other atoms, resulting in the breaking and formation of new chemical bonds. The **activation energy** is the minimum energy that the reactants must have to start a chemical reaction (figure 2.12*a*). Even reactions that result in a release of energy must overcome the activation energy barrier for the reaction to proceed. For example, heat in the form of a spark is required to start the reaction between oxygen and gasoline vapor. Once some oxygen molecules react with gasoline, the energy released can start additional reactions.

Given any population of molecules, some of them have more kinetic energy and move about faster than others. Even so, at normal body temperatures, most of the chemical reactions necessary for life proceed too slowly to support life because few molecules have enough energy to start a chemical reaction. **Catalysts** (kat′ă-listz) are substances that increase the rate of chemical reactions without being permanently changed or depleted. **Enzymes** (en′zīmz), which are discussed in greater detail on p. 47, are protein catalysts. Enzymes increase the rate of chemical reactions by lowering the activation energy necessary for the reaction to begin (figure 2.12*b*). As a result, more molecules have sufficient energy to undergo chemical reactions. With an enzyme, the rate of a chemical reaction can take place more than a million times faster than without the enzyme.

Temperature can also affect the speed of chemical reactions. As temperature increases, reactants have more kinetic energy, move at faster speeds, and collide with one another more frequently and with greater force, thereby increasing the likelihood of a chemical reaction. When a person has a fever of only a few degrees, reactions occur throughout the body at an accelerated rate, resulting in increased activity in the organ systems, such as increased heart and respiratory rates. When body temperature drops, various metabolic processes slow. In cold weather, the fingers are less agile largely because of the reduced rate of chemical reactions in cold muscle tissue.

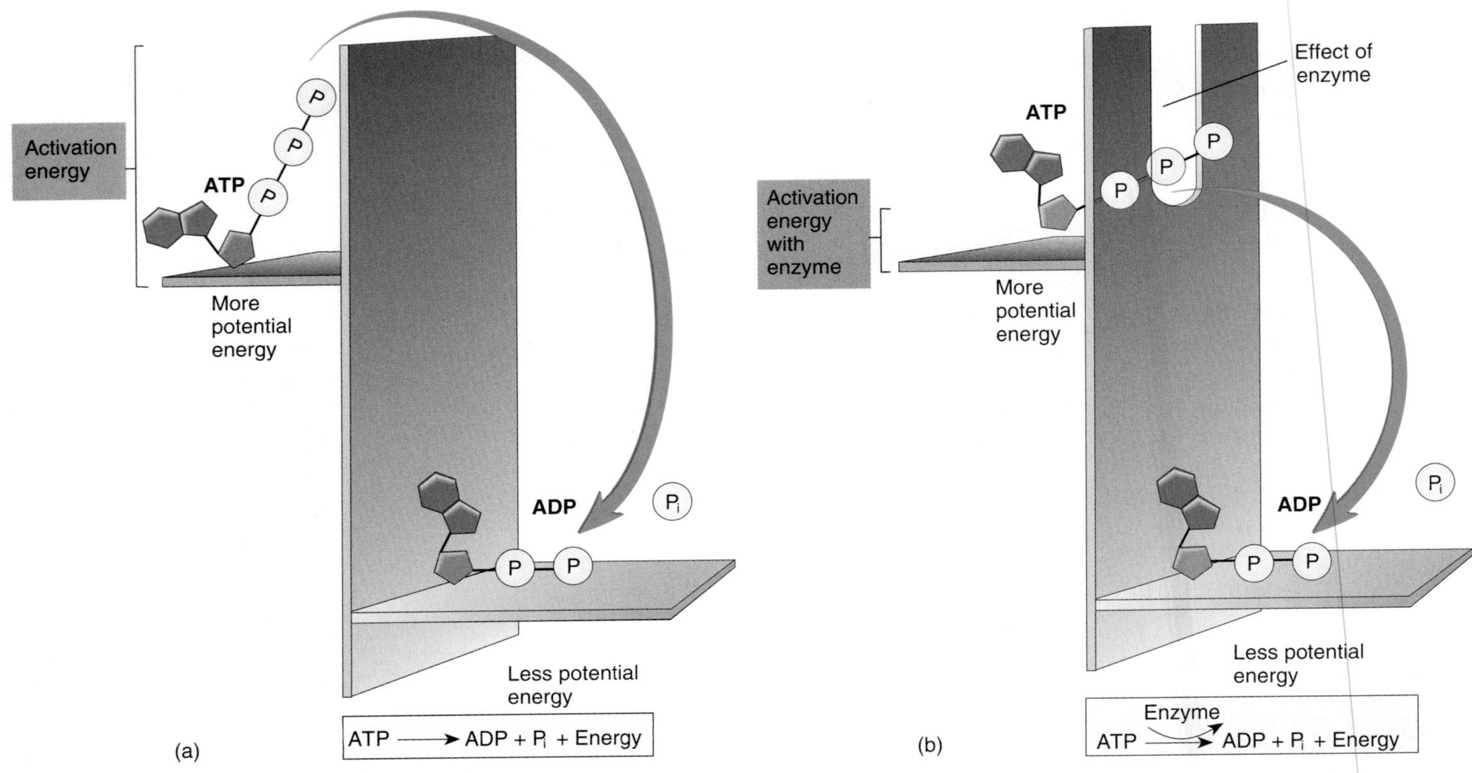

FIGURE 2.12 Activation Energy and Enzymes
(*a*) Activation energy is needed to change ATP to ADP. The upper shelf represents a higher energy level, and the lower shelf represents a lower energy level. The "wall" extending above the upper shelf represents the activation energy. Even though energy is given up moving from the upper to the lower shelf, the activation energy "wall" must be overcome before the reaction can proceed. (*b*) The enzyme lowers the activation energy, making it easier for the reaction to proceed.

Within limits, the greater the concentration of the reactants, the greater the rate at which a given chemical reaction proceeds. This occurs because, as the concentration of reactants increases, they are more likely to come into contact with one another. For example, the normal concentration of oxygen inside cells enables oxygen to come into contact with other molecules and produce the chemical reactions necessary for life. If the oxygen concentration decreases, the rate of chemical reactions decreases. A decrease in oxygen in cells can impair cell function and even result in death.

20 ***Define activation energy, catalysts, and enzymes. How do enzymes increase the rate of chemical reactions?***

21 ***What effect does increasing temperature or increasing concentration of the reactants have on the rate of a chemical reaction?***

INORGANIC CHEMISTRY

It was once believed that inorganic substances were those that came from nonliving sources and organic substances were those extracted from living organisms. As the science of chemistry developed, however, it became apparent that organic substances can be manufactured in the laboratory. As defined currently, **inorganic chemistry** generally deals with those substances that do not contain carbon, whereas **organic chemistry** is the study of carbon-containing substances. These definitions have a few exceptions. For example, carbon monoxide (CO), carbon dioxide (CO_2), and bicarbonate ion (HCO_3^-) are classified as inorganic molecules.

Water

A molecule of **water** is composed of one atom of oxygen joined to two atoms of hydrogen by covalent bonds. Water molecules are polar, with a partial positive charge associated with the hydrogen atoms and a partial negative charge associated with the oxygen atom. Hydrogen bonds form between the positively charged hydrogen atoms of one water molecule and the negatively charged oxygen atoms of another water molecule. These hydrogen bonds organize the water molecules into a lattice, which holds the water molecules together (see figures 2.6 and 2.7).

Water accounts for approximately 50% of the weight of a young adult female and 60% of a young adult male. Females have a lower percentage of water than males because they typically have more body fat, which is relatively free of water. Plasma, the liquid portion of blood, is 92% water. Water has physical and chemical properties well suited for its many functions in living organisms. These properties are outlined in the following sections.

Stabilizing Body Temperature

Water has a high **specific heat,** meaning that a relatively large amount of heat is required to raise its temperature; therefore, it tends to resist large temperature fluctuations. When water evaporates, it changes from a liquid to a gas, and, because heat is required for that process, the evaporation of water from the surface of the body rids the body of excess heat.

Protection

Water is an effective lubricant that provides protection against damage resulting from friction. For example, tears protect the surface of the eye from the rubbing of the eyelids. Water also forms a fluid cushion around organs that helps protect them from trauma. The cerebrospinal fluid that surrounds the brain is an example.

Chemical Reactions

Many of the chemical reactions necessary for life do not take place unless the reacting molecules are dissolved in water. For example, sodium chloride must dissociate in water into Na^+ and Cl^- before they can react with other ions. Water also directly participates in many chemical reactions. As previously mentioned, a dehydration reaction is a synthesis reaction in which water is produced, and a hydrolysis reaction is a decomposition reaction that requires a water molecule (see figure 2.10).

Mixing Medium

A **mixture** is a combination of two or more substances physically blended together, but not chemically combined. A **solution** is any mixture of liquids, gases, or solids in which the substances are uniformly distributed with no clear boundary between the substances. For example, a salt solution consists of salt dissolved in water, air is a solution containing a variety of gases, and wax is a solid solution of several fatty substances. Solutions are often described in terms of one substance dissolving in another: The **solute** (sol′ūt) dissolves in the **solvent.** In a salt solution, water is the solvent and the dissolved salt is the solute. Sweat is a salt solution in which sodium chloride and other solutes are dissolved in water.

A **suspension** is a mixture containing materials that separate from each other unless they are continually, physically blended together. Blood is a suspension; red blood cells are suspended in a liquid called plasma. As long as the red blood cells and plasma are mixed together as they pass through blood vessels, the red blood cells remain suspended in the plasma. If the blood is allowed to sit in a container, however, the red blood cells and plasma separate from each other.

A **colloid** (kol′oyd) is a mixture in which a dispersed (solutelike) substance is distributed throughout a dispersing (solventlike) substance. The dispersed particles are larger than a simple molecule but small enough that they remain dispersed and do not settle out. Proteins, which are large molecules, and water form colloids. For instance, the plasma portion of blood and the liquid interior of cells are colloids containing many important proteins.

In living organisms, the complex fluids inside and outside cells consist of solutions, suspensions, and colloids. Blood is an example of all of these mixtures. It is a solution containing dissolved nutrients, such as sugar; a suspension holding red blood cells; and a colloid containing proteins.

Water's ability to mix with other substances enables it to act as a medium for transport, moving substances from one part of the body to another. Body fluids, such as plasma, transport nutrients, gases, waste products, and a variety of molecules involved in regulating body functions.

Solution Concentrations

The concentration of solute particles dissolved in solvents can be expressed in several ways. One common way is to indicate the percent of solute by weight per volume of solution. A 10% solution of sodium chloride can be made by dissolving 10 g of sodium chloride into enough water to make 100 mL of solution.

Physiologists often determine concentrations in **osmoles** (os′mōlz), which express the number of particles in a solution. A particle can be an atom, an ion, or a molecule. An osmole (Osm) is Avogadro's number of particles of a substance in 1 kilogram (kg) of water. The **osmolality** (os-mō-lal′i-tē) of a solution is a reflection of the number, not the type, of particles in a solution. For example, a 1 Osm glucose solution and a 1 Osm NaCl solution both contain Avogadro's number of particles per kg water. The glucose solution has 1.0 Osm of glucose molecules, whereas the NaCl solution has 0.5 Osm of Na^+ and 0.5 Osm of Cl^- because NaCl dissociates into Na^+ and Cl^- in water.

Because the concentration of particles in body fluids is so low, the measurement **milliosmole** (mOsm), 1/1000 of an osmole, is used. Most body fluids have a concentration of about 300 mOsm and contain many different ions and molecules. The concentration of body fluids is important because it influences the movement of water into or out of cells (see chapter 3). Appendix C contains more information on calculating concentrations.

22 ***Define inorganic and organic chemistry.***

23 ***List four functions that water performs in living organisms and give an example of each.***

24 ***Describe solutions, suspensions, and colloids, and give an example of each. Define solvent and solute.***

25 ***How is the osmolality of a solution determined? What is a milliosmole?***

Acids and Bases

Many molecules and compounds are classified as acids or bases. For most purposes, an **acid** is defined as a proton donor. A hydrogen ion (H^+) is a proton because it results from the loss of an electron from a hydrogen atom, which consists of a proton and an electron. Therefore, a molecule or compound that releases H^+ is an acid. Hydrochloric acid (HCl) forms hydrogen ions (H^+) and chloride ions (Cl^-) in solution and therefore is an acid.

$$HCl \rightarrow H^+ + Cl^-$$

A **base** is defined as a proton acceptor, and any substance that binds to (accepts) H^+ ions is a base. Many bases function as proton acceptors by releasing hydroxide ions (OH^-) when they dissociate. The base sodium hydroxide (NaOH) dissociates to form Na^+ and OH^-.

$$NaOH \rightarrow Na^+ + OH^-$$

The OH^- are proton acceptors that combine with H^+ to form water.

$$OH^- + H^+ \rightarrow H_2O$$

Acids and bases are classified as strong or weak. Strong acids or bases dissociate almost completely when dissolved in water. Consequently, they release almost all of their H^+ or OH^-. The more completely the acid or base dissociates, the stronger it is. For example, HCl is a strong acid because it completely dissociates in water.

$$HCl \rightarrow H^+ + Cl^-$$
Not freely reversible

Weak acids or bases only partially dissociate in water. Consequently, they release only some of their H^+ or OH^-. For example, when acetic acid (CH_3COOH) is dissolved in water, some of it dissociates, but some of it remains in the undissociated form. An equilibrium is established between the ions and the undissociated weak acid.

$$CH_3COOH \rightleftarrows CH_3COO^- + H^+$$
Freely reversible

For a given weak acid or base, the amount of the dissociated ions relative to the weak acid or base is a constant.

The pH Scale

The **pH scale** is a means of referring to the H^+ concentration in a solution (figure 2.13). Pure water is defined as a **neutral solution** and has a pH of 7. A neutral solution has equal concentrations of H^+ and OH^-. Solutions with a pH less than 7 are **acidic** and have a greater concentration of H^+ than OH^-. **Alkaline** (al′kă-līn), or **basic,** solutions have a pH greater than 7 and have fewer H^+ than OH^-.

A change in the pH of a solution by 1 pH unit represents a 10-fold change in the H^+ concentration. For example, a solution of pH 6 has a H^+ concentration 10 times greater than a solution of pH 7 and 100 times greater than a solution of pH 8. As the pH value becomes smaller, the solution has more H^+ and is more acidic; as the pH value becomes larger, the solution has fewer H^+ and is more basic. Appendix D considers pH in greater detail.

Acidosis and Alkalosis

The normal pH range for human blood is 7.35 to 7.45. **Acidosis** results if blood pH drops below 7.35, in which case the nervous system becomes depressed and the individual can become disoriented and possibly comatose. **Alkalosis** results if blood pH rises above 7.45. Then the nervous system becomes overexcitable, and the individual can be extremely nervous or have convulsions. Both acidosis and alkalosis can be fatal.

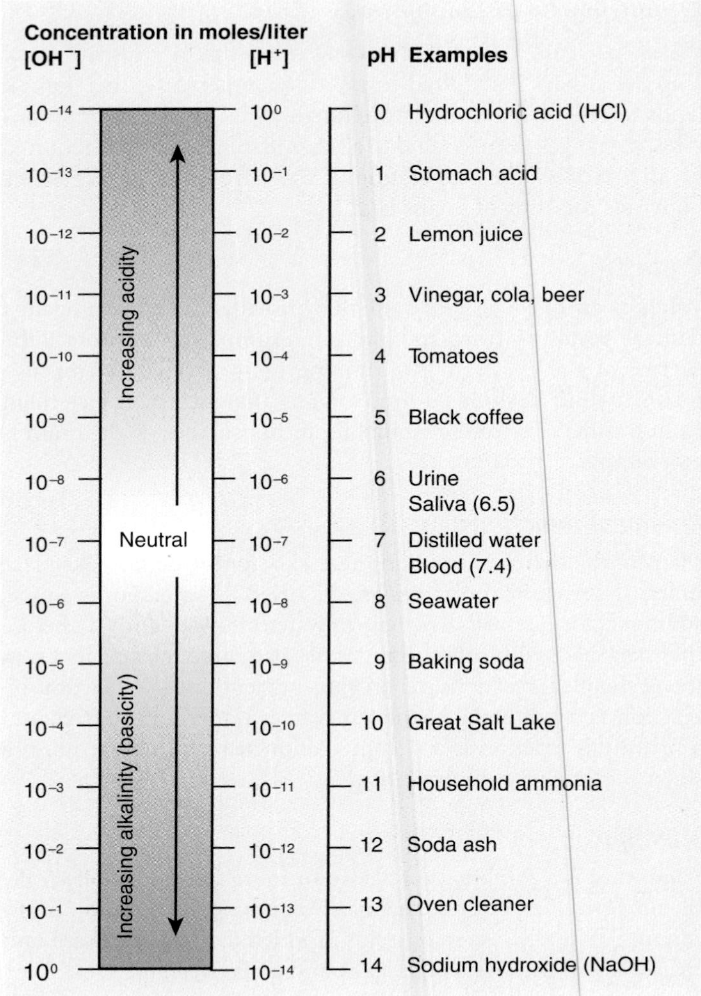

FIGURE 2.13 The pH Scale
A pH of 7 is considered neutral. Values less than 7 are acidic (the lower the number, the more acidic). Values greater than 7 are basic (the higher the number, the more basic). Representative fluids and their approximate pH values are listed.

Salts

A **salt** is a compound consisting of a cation other than a H^+ and an anion other than a OH^-. Salts are formed by the interaction of an acid and a base in which the H^+ of the acid are replaced by the positive ions of the base. For example, in a solution when hydrochloric acid (HCl) reacts with the base sodium hydroxide (NaOH), the salt sodium chloride (NaCl) is formed.

$$\underset{\text{(Acid)}}{HCl} + \underset{\text{(Base)}}{NaOH} \rightarrow \underset{\text{(Salt)}}{NaCl} + \underset{\text{(Water)}}{H_2O}$$

Typically, when salts such as sodium chloride dissociate in water, they form positively and negatively charged ions (see figure 2.8).

Buffers

The chemical behavior of many molecules changes as the pH of the solution in which they are dissolved changes. For example, many enzymes work best within narrow ranges of pH. The survival of an organism depends on its ability to regulate body fluid pH within a narrow range. Deviations from the normal pH range for human blood are life-threatening.

One way body fluid pH is regulated involves the action of buffers, which resist changes in solution pH when either acids or bases are added. A **buffer** is a solution of a conjugate acid–base pair in which the acid component and the base component occur in similar concentrations. A conjugate base is everything that remains of an acid after the H^+ (proton) is lost. A conjugate acid is formed when an H^+ is transferred to the conjugate base. Two substances related in this way are a **conjugate acid–base pair.** For example, carbonic acid (H_2CO_3) and bicarbonate ion (HCO_3^-), formed by the dissociation of H_2CO_3, are a conjugate acid–base pair.

$$H_2CO_3 \rightleftarrows H^+ + HCO_3^-$$

In the forward reaction, H_2CO_3 loses a H^+ to produce HCO_3^-, which is a conjugate base. In the reverse reaction, a H^+ is transferred to the HCO_3^- (conjugate base) to produce H_2CO_3, which is a conjugate acid.

For a given condition, this reversible reaction results in an equilibrium, in which the amounts of H_2CO_3 relative to the amounts of H^+ and HCO_3^- remain constant. The conjugate acid–base pair can resist changes in pH because of this equilibrium. If an acid is added to a buffer, the H^+ from the added acid can combine with the base component of the conjugate acid–base pair. As a result, the concentration of H^+ does not increase as much as it would without this reaction. If H^+ are added to a H_2CO_3 solution, many of the H^+ combine with HCO_3^- to form H_2CO_3.

On the other hand, if a base is added to a buffered solution, the conjugate acid can release H^+ to counteract the effects of the added base. For example, if OH^- are added to a H_2CO_3 solution, the OH^- combine with H^+ to form water. As the H^+ are incorporated into water, H_2CO_3 dissociates to form H^+ and HCO_3^-, thereby maintaining the H^+ concentration (pH) within a normal range.

The greater the buffer concentration, the more effective it is in resisting a change in pH, but buffers cannot entirely prevent some change in the pH of a solution. For example, when an acid is added to a buffered solution, the pH decreases, but not to the extent it would have without the buffer. Important buffers found in living systems have bicarbonate, phosphates, amino acids, and proteins as components.

26 ***Define acid and base, and describe the pH scale. What is the difference between a strong acid or base and a weak acid or base?***

27 ***Define acidosis and alkalosis, and describe the symptoms of each.***

28 ***What is a salt? What is a buffer, and why are buffers important to organisms?***

PREDICT 8

Dihydrogen phosphate ion ($H_2PO_4^-$) and monohydrogen phosphate ion (HPO_4^{2-}) form the phosphate buffer system.

$$H_2PO_4^- \rightleftarrows H^+ + HPO_4^{2-}$$

Identify the conjugate acid and conjugate base in the phosphate buffer system. Explain how they function as a buffer when either H^+ or OH^- are added to the solution.

Oxygen

Oxygen (O_2) is an inorganic molecule consisting of two oxygen atoms bound together by a double covalent bond. About 21% of the gas in the atmosphere is oxygen, and it is essential for most animals. Oxygen is required by humans in the final step of a series of reactions in which energy is extracted from food molecules (see chapter 25).

Carbon Dioxide

Carbon dioxide (CO_2) consists of one carbon atom bound to two oxygen atoms. Each oxygen atom is bound to the carbon atom by a double covalent bond. Carbon dioxide is produced when organic molecules, such as glucose, are metabolized within the cells of the body (see chapter 25). Much of the energy stored in the covalent bonds of glucose is transferred to other organic molecules when glucose is broken down, and carbon dioxide is released. Once carbon dioxide is produced, it is eliminated from the cell as a metabolic by-product, transferred to the lungs by blood, and exhaled during respiration. If carbon dioxide is allowed to accumulate within cells, it becomes toxic.

29 ***What are the functions of oxygen and carbon dioxide in living systems?***

ORGANIC CHEMISTRY

The ability of carbon to form covalent bonds with other atoms makes possible the formation of the large, diverse, complicated molecules necessary for life. A series of carbon atoms bound together by covalent bonds constitutes the "backbone" of many large molecules. Variation in the length of the carbon chains and the combination of atoms bound to the carbon backbone allows for the formation of a wide variety of molecules. For example, some protein molecules have thousands of carbon atoms bound by covalent bonds to one another or to other atoms, such as nitrogen, sulfur, hydrogen, and oxygen.

The four major groups of organic molecules essential to living organisms are carbohydrates, lipids, proteins, and nucleic acids. Each of these groups has specific structural and functional characteristics.

Carbohydrates

Carbohydrates are composed primarily of carbon, hydrogen, and oxygen atoms and range in size from small to very large. In most carbohydrates, for each carbon atom there are approximately two hydrogen atoms and one oxygen atom. Note that this ratio of hydrogen atoms to oxygen atoms is two to one, the same as in water (H_2O). They are called carbohydrates because carbon (*carbo*) atoms are combined with the same atoms that form water (*hydrated*). The large number of oxygen atoms in carbohydrates makes them relatively polar molecules. Consequently, they are soluble in polar solvents, such as water.

Carbohydrates are important parts of other organic molecules, and they can be broken down to provide the energy necessary for life. Undigested carbohydrates also provide bulk in feces, which helps maintain the normal function and health

TABLE 2.5 Role of Carbohydrates in the Body

Role	Example
Structure	Ribose forms part of RNA and ATP molecules, and deoxyribose forms part of DNA.
Energy	Monosaccharides (glucose, fructose, galactose) can be used as energy sources. Disaccharides (sucrose, lactose, maltose) and polysaccharides (starch, glycogen) must be broken down to monosaccharides before they can be used for energy. Glycogen is an important energy-storage molecule in muscles and in the liver.
Bulk	Cellulose forms bulk in the feces.

of the digestive tract. Table 2.5 summarizes the role of carbohydrates in the body.

Monosaccharides

Large carbohydrates are composed of numerous, relatively simple building blocks called **monosaccharides** (mon-ō-sak′ă-rīdz; the prefix *mono-* means one; the term *saccharide* means sugar), or simple sugars. Monosaccharides commonly contain three carbons (trioses), four carbons (tetroses), five carbons (pentoses), or six carbons (hexoses).

The monosaccharides most important to humans include both five- and six-carbon sugars. Common six-carbon sugars, such as glucose, fructose, and galactose, are **isomers** (ī′sō-merz), which are molecules that have the same number and types of atoms but differ in their three-dimensional arrangement (figure 2.14). Glucose, or blood sugar, is the major carbohydrate found in the blood and is a major nutrient for most cells of the body. Fructose and galactose are also important dietary nutrients. Important five-carbon sugars include ribose and deoxyribose (see figure 2.24), which are components of ribonucleic acid (RNA) and deoxyribonucleic acid (DNA), respectively.

Disaccharides

Disaccharides (dī-sak′ă-rīdz; *di-* means two) are composed of two simple sugars bound together through a dehydration reaction. Glucose and fructose, for example, combine to form a disaccharide called **sucrose** (table sugar) plus a molecule of water (figure 2.15*a*). Several disaccharides are important to humans, including sucrose, lactose, and maltose. Lactose, or milk sugar, is glucose combined with galactose; and maltose, or malt sugar, is two glucose molecules joined together.

Polysaccharides

Polysaccharides (pol-ē-sak′ă-rīdz; *poly-* means many) consist of many monosaccharides bound together to form long chains that are either straight or branched. **Glycogen,** or animal starch, is a

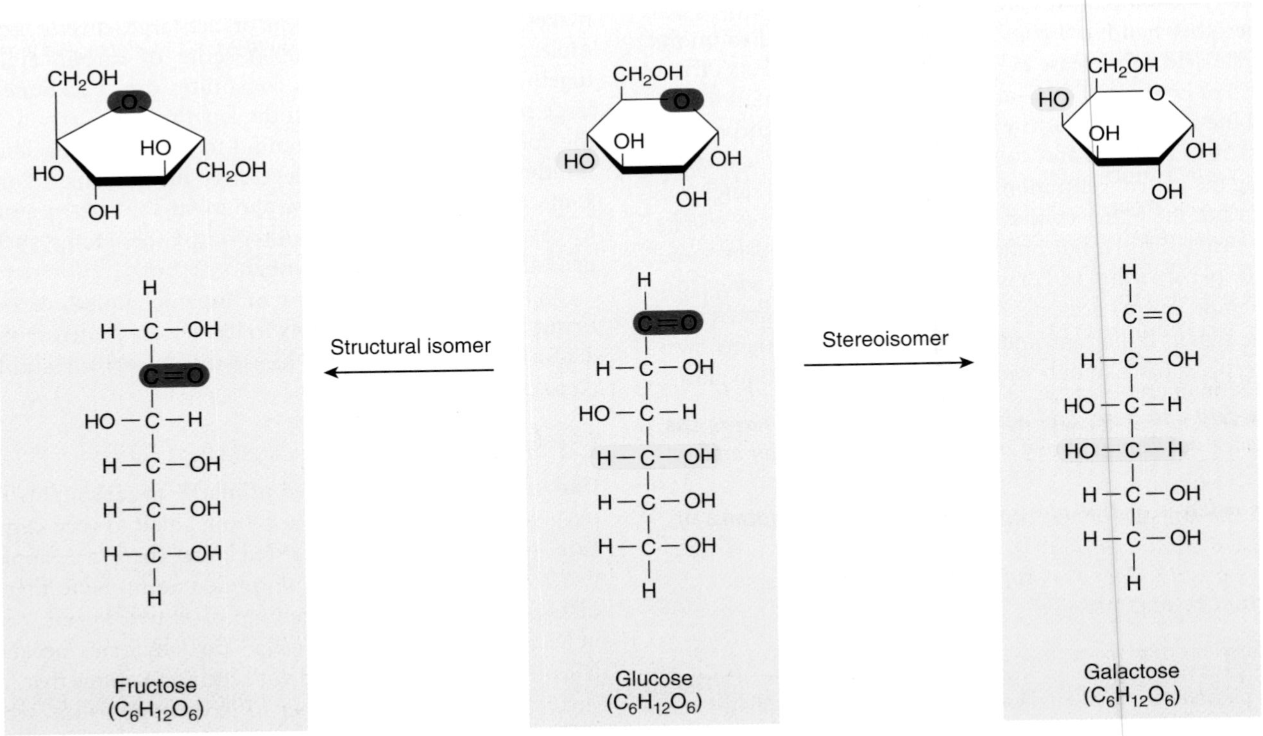

FIGURE 2.14 Monosaccharides

These monosaccharides almost always form a ring-shaped molecule. They are represented as linear models to more readily illustrate the relationships between the atoms of the molecules. Fructose is a structural isomer of glucose because it has identical chemical groups bonded in a different arrangement in the molecule (*red shading*). Galactose is a stereoisomer of glucose because it has exactly the same groups bonded to each carbon atom but located in a different three-dimensional orientation (*yellow shading*).

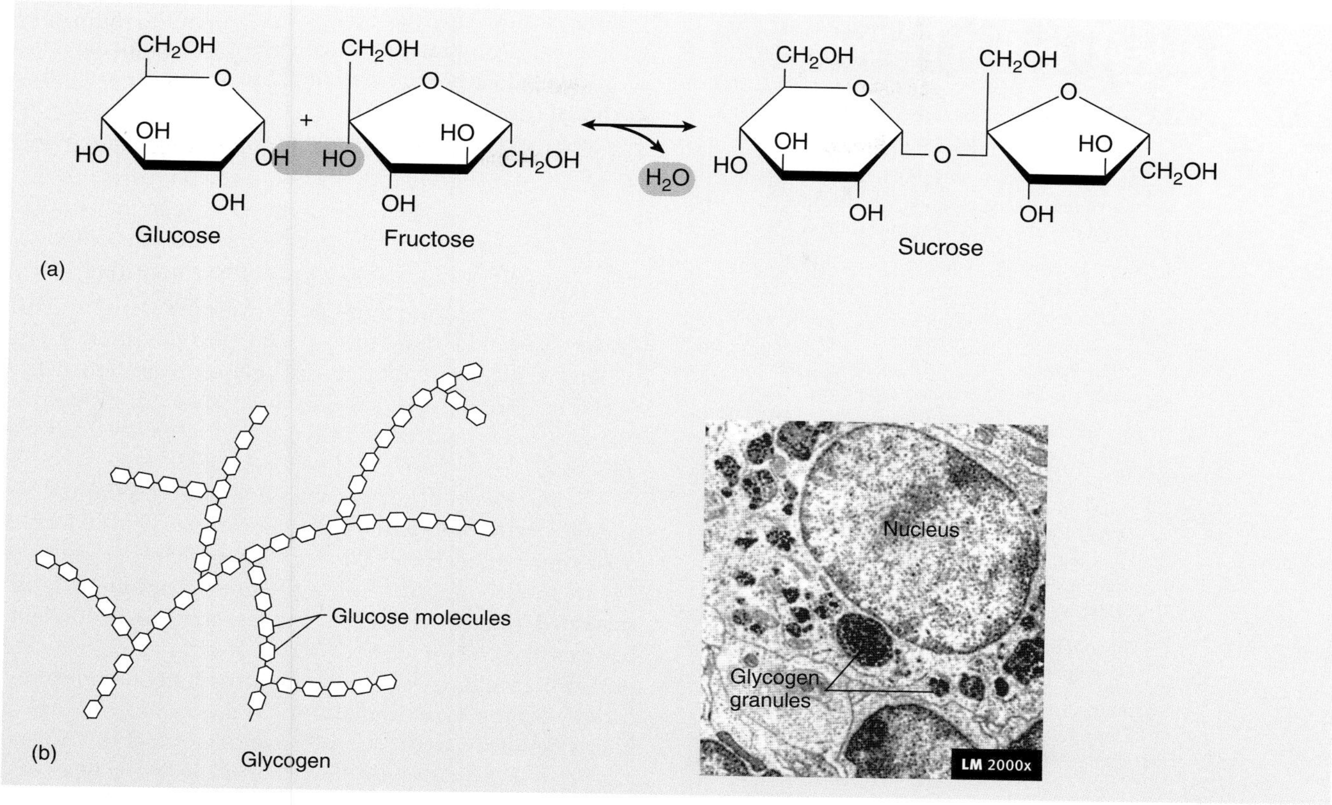

FIGURE 2.15 Disaccharide and Polysaccharide
(*a*) Formation of sucrose, a disaccharide, by a dehydration reaction involving glucose and fructose (monosaccharides). (*b*) Glycogen is a polysaccharide formed by combining many glucose molecules. The photomicrograph shows glycogen granules in a liver cell.

polysaccharide composed of many glucose molecules (figure 2.15*b*). Because glucose can be metabolized rapidly and the resulting energy can be used by cells, glycogen is an important energy-storage molecule. A substantial amount of the glucose that is metabolized to produce energy for muscle contraction during exercise is stored in the form of glycogen in the cells of the liver and skeletal muscles.

Starch and **cellulose** are two important polysaccharides found in plants, and both are composed of long chains of glucose. Plants use starch as an energy-storage molecule in the same way that animals use glycogen, and cellulose is an important structural component of plant cell walls. When humans ingest plants, the starch can be broken down and used as an energy source. Humans, however, do not have the digestive enzymes necessary to break down cellulose. The cellulose is eliminated in the feces, where it provides bulk.

30 ***Name the basic building blocks of carbohydrates.***

31 ***List the five- and six-carbon sugars that are important for humans.***

32 ***What is an isomer?***

33 ***What are disaccharides and polysaccharides and how are they formed?***

34 ***Which carbohydrates are used for energy? What is the function of starch and cellulose in plants? The function of glycogen and cellulose in animals?***

Lipids

Lipids are a second major group of organic molecules common to living systems. Like carbohydrates, they are composed principally of carbon, hydrogen, and oxygen, but other elements, such as phosphorus and nitrogen, are minor components of some lipids. Lipids contain a lower ratio of oxygen to carbon than do carbohydrates, which makes them less polar. Consequently, lipids can be dissolved in nonpolar organic solvents, such as alcohol or acetone, but they are relatively insoluble in water.

Lipids have many important functions in the body. They provide protection and insulation, help regulate many physiologic processes, and form plasma membranes. In addition, lipids are a major energy-storage molecule and can be broken down and used as a source of energy. Table 2.6 summarizes the many roles of lipids in the body. Several kinds of molecules, such as fats, phospholipids, steroids, and prostaglandins, are classified as lipids.

Fats are a major type of lipid. Like carbohydrates, fats are ingested and broken down by hydrolysis reactions in cells to release energy for use by those cells. Conversely, if intake exceeds need, excess chemical energy from any source can be stored in the body as fat for later use as energy is needed. Fats also provide protection by surrounding and padding organs, and under-the-skin fats act as an insulator to prevent heat loss.

TABLE 2.6 Role of Lipids in the Body

Role	Example
Protection	Fat surrounds and pads organs.
Insulation	Fat under the skin prevents heat loss. Myelin surrounds nerve cells and electrically insulates the cells from one another.
Regulation	Steroid hormones regulate many physiologic processes. For example, estrogen and testosterone are sex hormones responsible for many of the differences between males and females. Prostaglandins help regulate tissue inflammation and repair.
Vitamins	Fat-soluble vitamins perform a variety of functions. Vitamin A forms retinol, which is necessary for seeing in the dark; active vitamin D promotes calcium uptake by the small intestine; vitamin E promotes wound healing; and vitamin K is necessary for the synthesis of proteins responsible for blood clotting.
Structure	Phospholipids and cholesterol are important components of the membranes of cells.
Energy	Lipids can be stored and broken down later for energy; per unit of weight, they yield more energy than carbohydrates or proteins.

Triglycerides (trī-glis′er-īdz) constitute 95% of the fats in the human body. Triglycerides, which are sometimes called **triacylglycerols** (tri-as′il-glis′er-olz), consist of two different types of building blocks: one glycerol and three fatty acids. **Glycerol** is a three-carbon molecule with a hydroxyl group attached to each carbon atom, and **fatty acids** consist of a straight chain of carbon atoms with a carboxyl group attached at one end (figure 2.16). A **carboxyl** (kar-bok′sil) **group** (—COOH) consists of both an oxygen atom and a hydroxyl group attached to a carbon atom.

$$-\overset{\overset{\displaystyle O}{\|}}{C}-OH \quad \text{or} \quad HO-\overset{\overset{\displaystyle O}{\|}}{C}-$$

The carboxyl group is responsible for the acidic nature of the molecule because it releases hydrogen ions into solution. Glycerides can be described according to the number and kinds of fatty acids that combine with glycerol through dehydration reactions. Monoglycerides have one fatty acid, diglycerides have two fatty acids, and triglycerides have three fatty acids bound to glycerol.

Fatty acids differ from one another according to the length and the degree of saturation of their carbon chains. Most naturally occurring fatty acids contain an even number of carbon atoms, with 14- to 18-carbon chains being the most common. A fatty acid is **saturated** (figure 2.17) if it contains only single covalent bonds between the carbon atoms. Sources of saturated fats include beef, pork, whole milk, cheese, butter, eggs, coconut oil, and palm oil. The carbon chain is **unsaturated** if it has one or more double covalent bonds between carbon atoms. Because the double covalent bonds can occur anywhere along the carbon chain, many types of unsaturated fatty acids with an equal degree of unsaturation are possible. **Monounsaturated fats,** such as olive and peanut oils, have one double covalent bond between carbon atoms. **Polyunsaturated fats,** such as safflower, sunflower, corn, and fish oils, have two or more double covalent bonds between carbon atoms. Unsaturated fats are

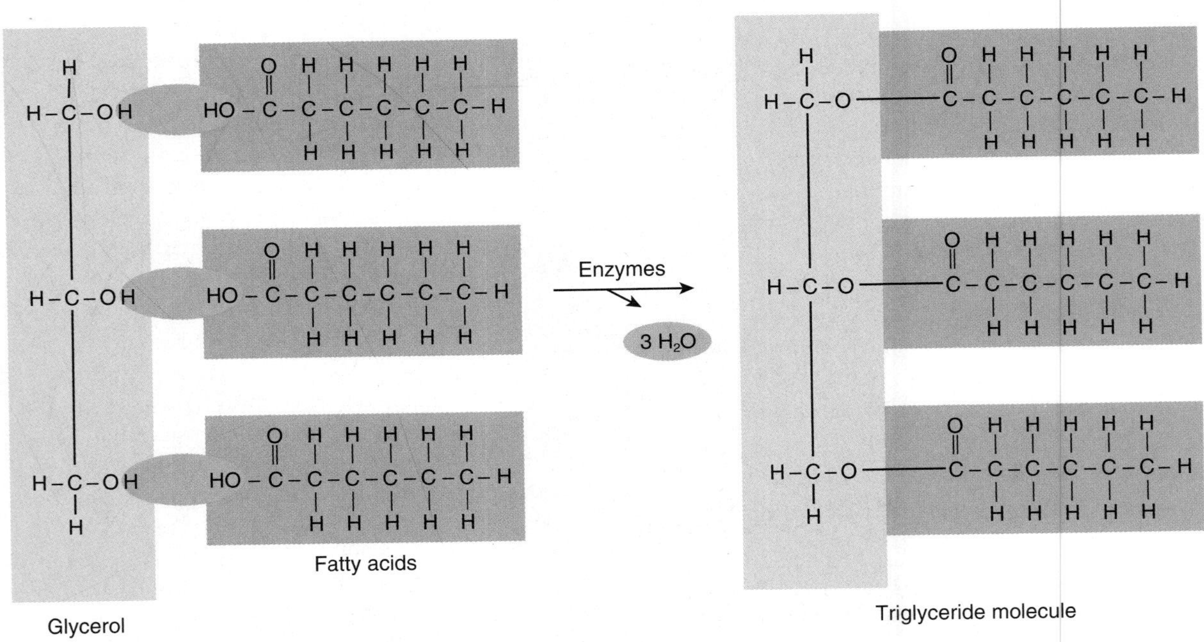

FIGURE 2.16 Triglyceride
Production of a triglyceride from one glycerol molecule and three fatty acid molecules. One water molecule (H_2O) is given off for each covalent bond formed between a fatty acid molecule and glycerol.

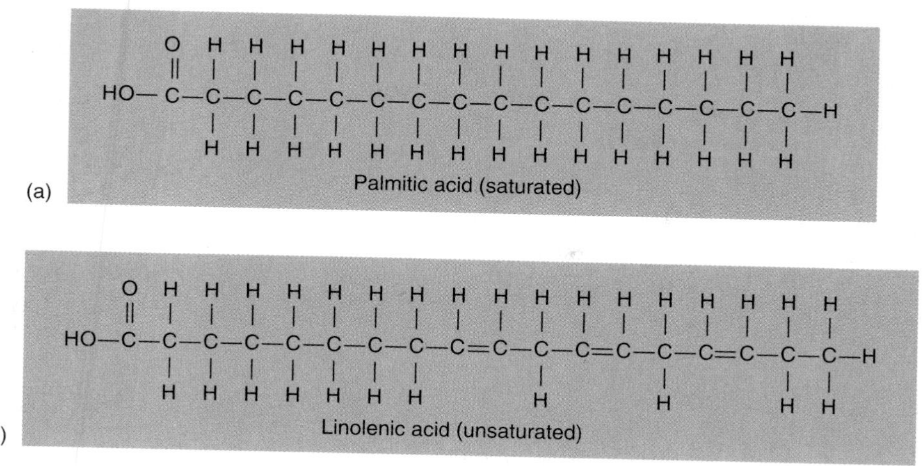

FIGURE 2.17 Fatty Acids
(*a*) Palmitic acid (saturated with no double bonds between the carbons). (*b*) Linolenic acid (unsaturated with three double bonds between the carbons).

the best type of fats in the diet because, unlike saturated fats, they do not contribute to the development of cardiovascular disease.

Phospholipids are similar to triglycerides, except that one of the fatty acids bound to the glycerol is replaced by a molecule containing phosphate and, usually, nitrogen (figure 2.18). They are polar at the end of the molecule to which the phosphate is bound and nonpolar at the other end. The polar end of the molecule is attracted to water and is said to be **hydrophilic** (water-loving). The nonpolar end is repelled by water and is said to be **hydrophobic** (water-fearing). Phospholipids are important structural components of the membranes of cells (see figure 3.2).

The **eicosanoids** (ī′kō-să-noydz) are a group of important chemicals derived from fatty acids. They include **prostaglandins** (pros′tā-glan′dinz), **thromboxanes** (throm′bok-zānz), and **leukotrienes** (loo-kō-trī′ēnz). Eicosanoids are made in most cells and are important regulatory molecules. Among their numerous effects is their role in the response of tissues to injuries. Prostaglandins have been implicated in regulating the secretion of some hormones, blood clotting, some reproductive functions, and many other processes. Many of the therapeutic effects of aspirin and other anti-inflammatory drugs result from their ability to inhibit prostaglandin synthesis.

Steroids differ in chemical structure from other lipid molecules, but their solubility characteristics are similar. All steroid molecules are composed of carbon atoms bound together into four ringlike structures (figure 2.19). Important steroid molecules include cholesterol, bile salts, estrogen, progesterone, and testosterone. Cholesterol is an important steroid because other molecules are synthesized from it. For example, bile salts, which increase fat absorption in the intestines, are derived from

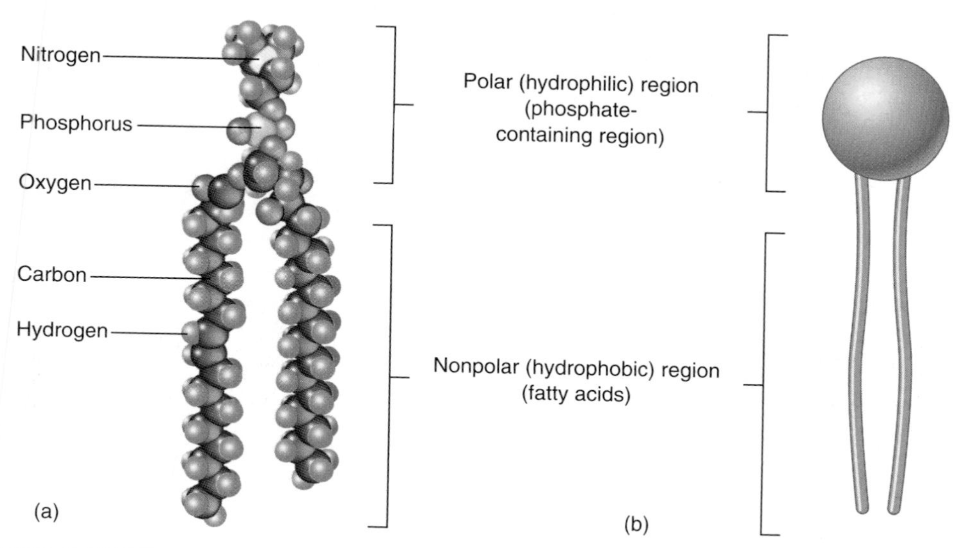

FIGURE 2.18 Phospholipids
(*a*) Molecular model of a phospholipid. (*b*) Simplified way in which phospholipids are often depicted.

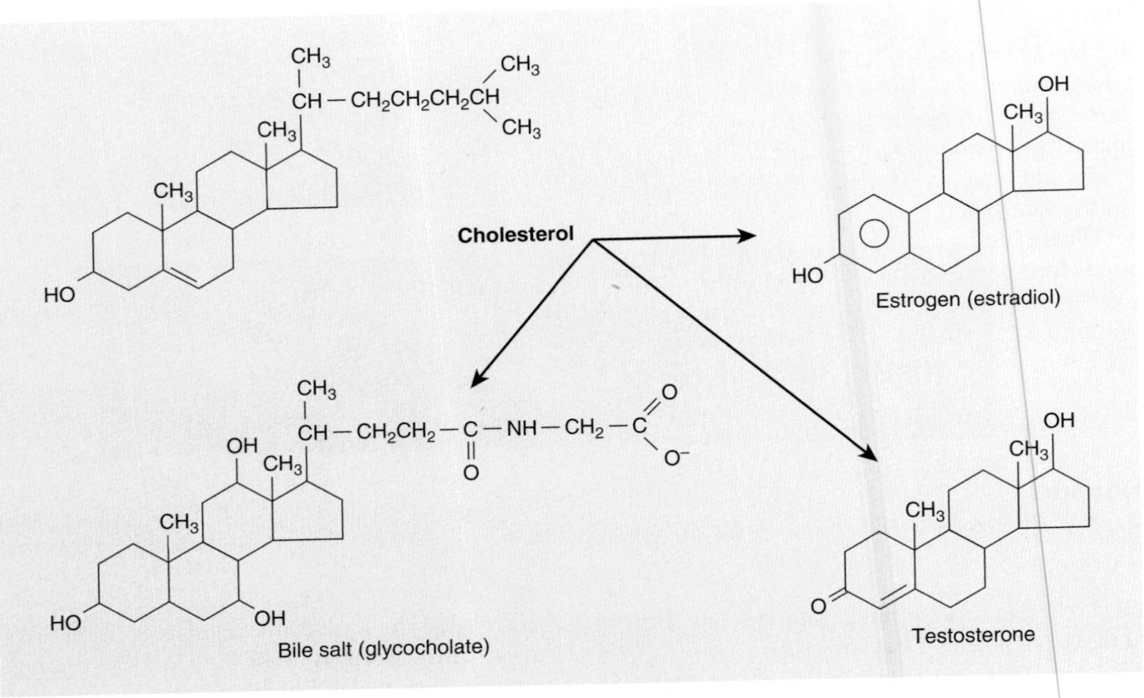

FIGURE 2.19 **Steroids**
Steroids are four-ringed molecules that differ from one another according to the groups attached to the rings. Cholesterol, the most common steroid, can be modified to produce other steroids.

cholesterol, as are the reproductive hormones estrogen, progesterone, and testosterone. In addition, cholesterol is an important component of plasma membranes. Although high levels of cholesterol in the blood increase the risk of cardiovascular disease, a certain amount of cholesterol is vital for normal function.

Another class of lipids is the **fat-soluble vitamins.** Their structures are not closely related to one another, but they are nonpolar molecules essential for many normal functions of the body.

35 ***Give four functions of fats. Name the most common type of fat in the human body. What are the basic building blocks of this type of fat?***

36 ***What is the difference between a saturated and an unsaturated fat? Between monounsaturated and polyunsaturated fats?***

37 ***Describe the structure of phospholipids. Which end of the molecule is hydrophilic?***

38 ***What are the functions of eicosanoids? Give three examples of eicosanoids.***

39 ***Why is cholesterol an important steriod?***

Proteins

All **proteins** contain carbon, hydrogen, oxygen, and nitrogen bound together by covalent bonds, and most proteins contain some sulfur. In addition, some proteins contain small amounts of phosphorus, iron, and iodine. The molecular mass of proteins can be very large. For the purpose of comparison, the molecular mass of water is approximately 18, sodium chloride 58, and glucose 180, but the molecular mass of proteins ranges from approximately 1000 to several million.

Proteins regulate bodily processes, act as a transportation system in the body, provide protection, help muscles contract, and provide structure and energy. Table 2.7 summarizes the role of proteins in the body.

TABLE 2.7 **Role of Proteins in the Body**

Role	Example
Regulation	Enzymes control chemical reactions. Hormones regulate many physiologic processes; for example, insulin affects glucose transport into cells.
Transport	Hemoglobin transports oxygen and carbon dioxide in the blood. Plasma proteins transport many substances in the blood. Proteins in plasma membranes control the movement of materials into and out of the cell.
Protection	Antibodies and complement protect against microorganisms and other foreign substances.
Contraction	Actin and myosin in muscle are responsible for muscle contraction.
Structure	Collagen fibers form a structural framework in many parts of the body. Keratin adds strength to skin, hair, and nails.
Energy	Proteins can be broken down for energy; per unit of weight, they yield as much energy as carbohydrates.

Protein Structure

The basic building blocks for proteins are the 20 **amino** (ă-mē′nō) **acid** molecules. Each amino acid has an amine (ă-mēn′) group (—NH_2), a carboxyl group (—COOH), a hydrogen atom, and a side chain designated by the symbol *R* attached to the same carbon atom. The side chain can be a variety of chemical structures, and the differences in the side chains make the amino acids different from one another (figure 2.20).

Covalent bonds formed between amino acid molecules during protein synthesis are called **peptide bonds** (figure 2.21). A dipeptide is two amino acids bound together by a peptide bond, a tripeptide is three amino acids bound together by peptide bonds,

The general structure of an amino acid showing the amine group (—NH_2), carboxyl group (—COOH), and hydrogen atom highlighted in yellow. The R side chain is the part of an amino acid that makes it different from other amino acids.

Amine group

Carboxyl group

Glycine is the simplest amino acid. The side chain is a hydrogen atom.

Glycine

Tyrosine, which has a more complicated side chain, is an important component of thyroid hormones.

Tyrosine

Improper metabolism of phenylalanine in the genetic disease phenylketonuria (PKU) can cause mental retardation.

Phenylalanine

Aspartic acid combined with phenylalanine forms the artificial sweetener aspartame (Nutrasweet™ and Equal™).

Aspartic acid

FIGURE 2.20 Amino Acids

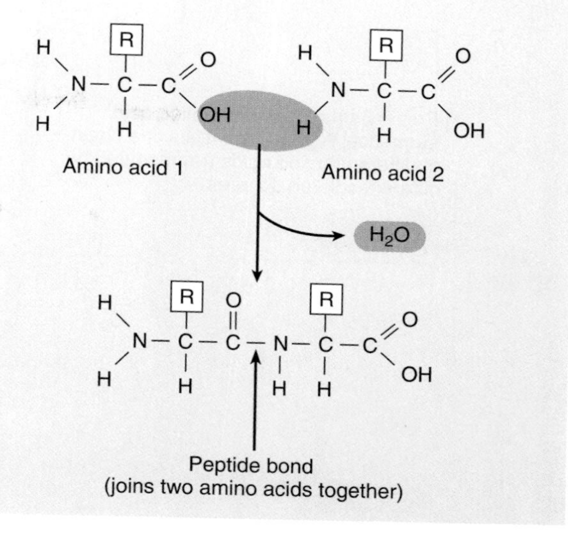

FIGURE 2.21 Peptide Bond

A dehydration reaction between two amino acids forms a dipeptide and a water molecule. The covalent bond between the amino acids is called a peptide bond.

and a polypeptide is many amino acids bound together by peptide bonds. Proteins are polypeptides composed of hundreds of amino acids.

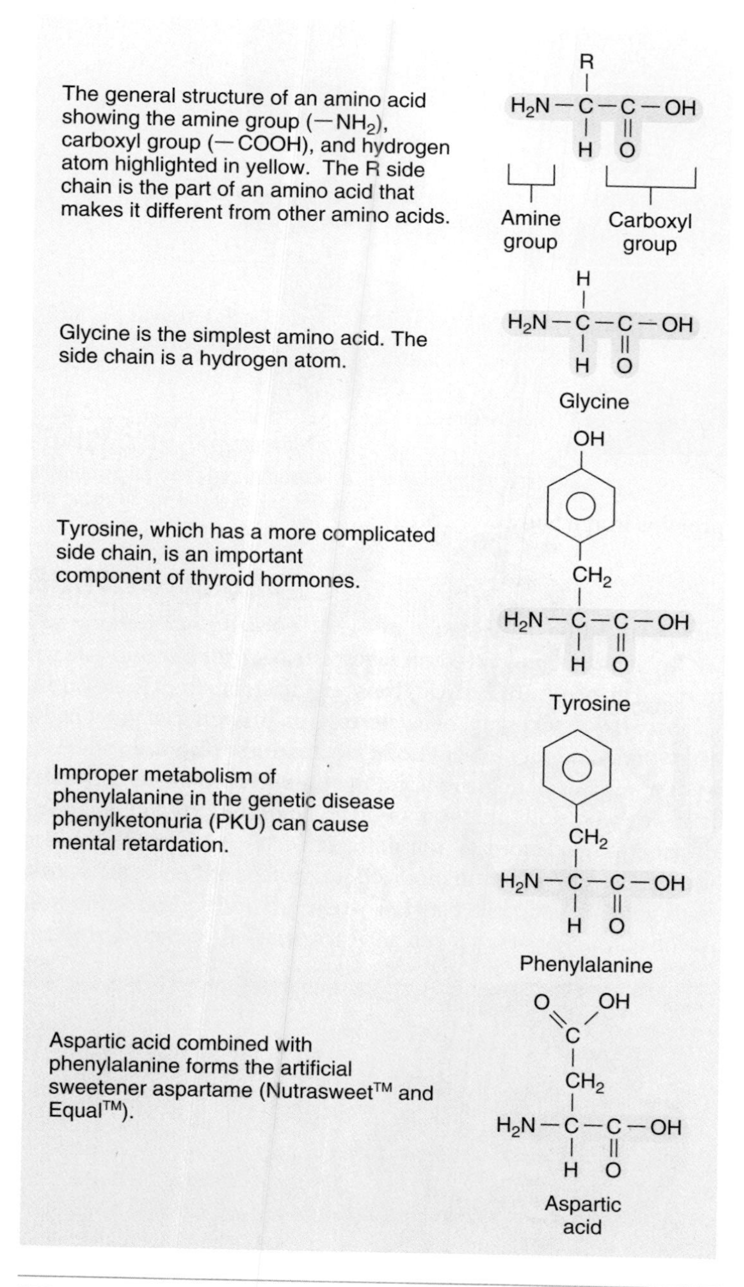

The **primary structure** of a protein is determined by the sequence of the amino acids bound by peptide bonds (figure 2.22*a*). The potential number of different protein molecules is enormous because 20 different amino acids exist, and each amino acid can be located at any position along a polypeptide chain. The characteristics of the amino acids in a protein ultimately determine the three-dimensional shape of the protein, and the shape of the protein determines its function. A change in one, or a few, amino acids in the primary structure can alter protein function, usually making the protein less or even nonfunctional.

The **secondary structure** results from the folding or bending of the polypeptide chain caused by the hydrogen bonds between amino acids (figure 2.22*b*). Two common shapes that result are helices (coils) and pleated (folded) sheets. If the hydrogen bonds that maintain the shape of the protein are broken, the protein becomes nonfunctional. This change in shape is called **denaturation,** and it can be caused by abnormally high temperatures or changes in the pH of body fluids. An everyday example of denaturation is the change in the proteins of egg whites when they are cooked.

The **tertiary structure** results from the folding of the helices or pleated sheets (figure 2.22*c*). Some amino acids are quite polar and therefore form hydrogen bonds with water. The polar portions of proteins tend to remain unfolded, maximizing their contact with water, whereas the less polar regions tend to fold into a globular shape, minimizing their contact with water. The formation of covalent bonds between sulfur atoms of one amino acid and sulfur atoms of another amino acid located at a different place in the sequence of amino acids can also contribute to the tertiary structure of proteins. The tertiary structure determines the shape of a **domain,** which is a folded sequence of 100–200 amino acids within

Amino acids

Peptide bond

(a) Primary structure—the amino acid sequence. A protein consists of a chain of different amino acids (represented by different colored spheres).

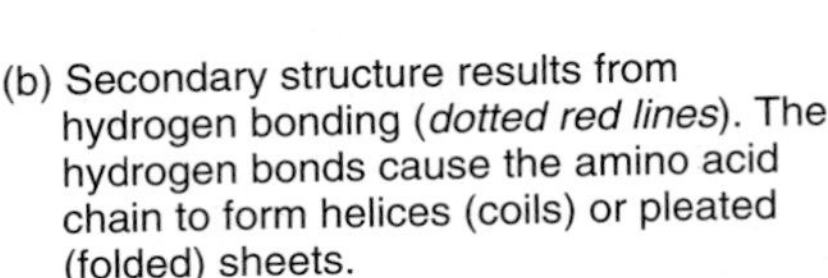

(b) Secondary structure results from hydrogen bonding (*dotted red lines*). The hydrogen bonds cause the amino acid chain to form helices (coils) or pleated (folded) sheets.

Pleated sheet

Helix

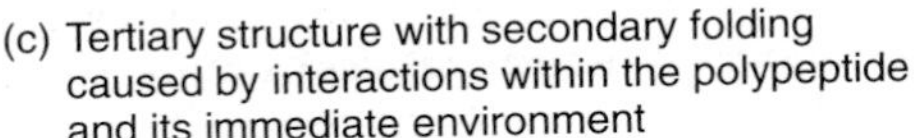

(c) Tertiary structure with secondary folding caused by interactions within the polypeptide and its immediate environment

(d) Quaternary structure—the relationships between individual subunits

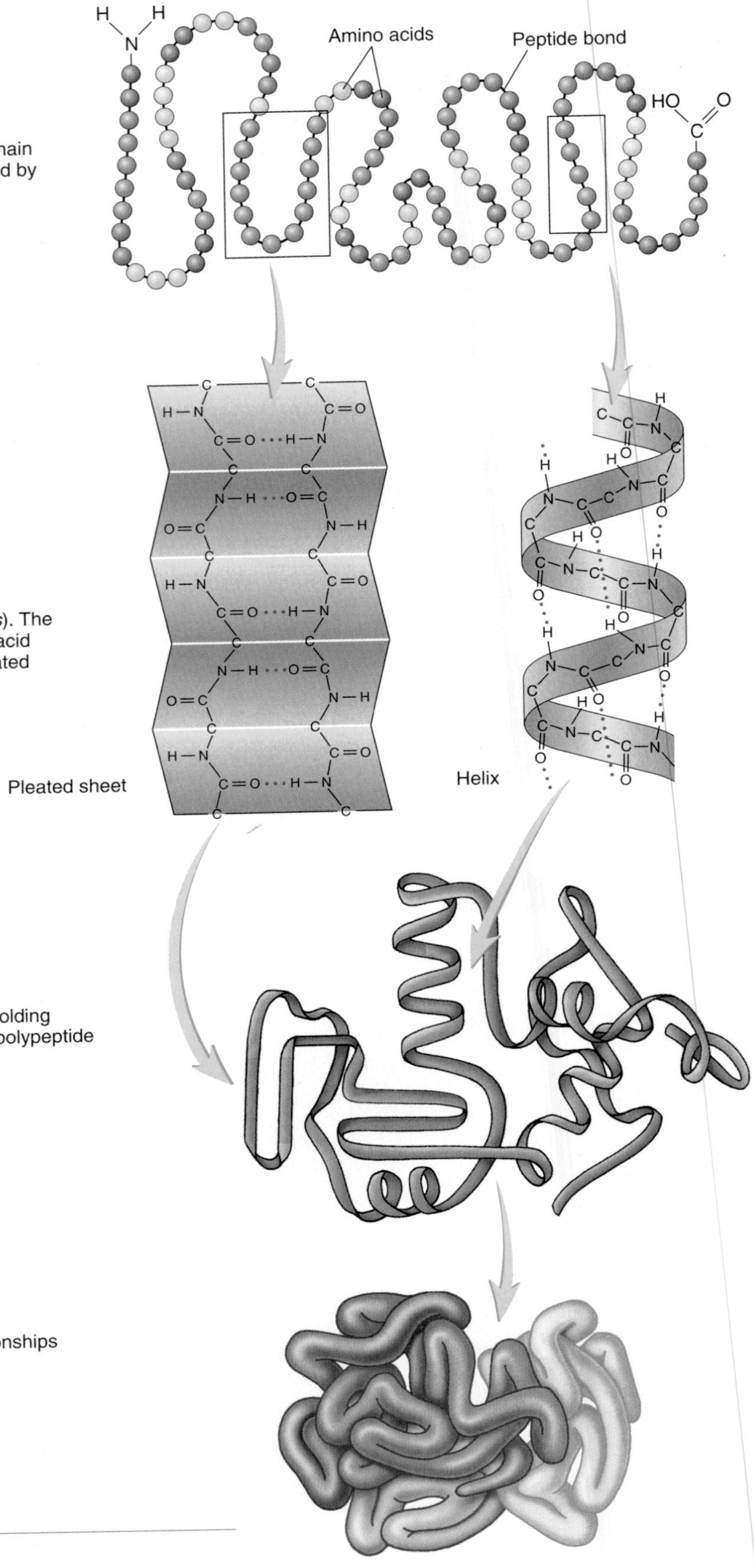

FIGURE 2.22 Protein Structure

a protein. The functions of proteins occur at one or more domains. Therefore, changes in the primary or secondary structure that affect the shape of the domain can change protein function.

If two or more proteins associate to form a functional unit, the individual proteins are called subunits. The **quaternary structure** is the spatial relationships between the individual subunits (figure 2.22*d*).

Enzymes

Proteins perform many roles in the body, including acting as enzymes. An **enzyme** is a protein catalyst that increases the rate at which a chemical reaction proceeds without the enzyme being permanently changed. The three-dimensional shape of enzymes is critical for their normal function because it determines the structure of the enzyme's **active site.** According to the **lock-and-key model** of enzyme action, a reaction occurs when the reactants (key) bind to the active site (lock) on the enzyme. This view of enzymes and reactants as rigid structures fitting together has been modified by the **induced fit model,** in which the enzyme is able to slightly change shape and better fit the reactants. The enzyme is like a glove that does not achieve its functional shape until the hand (reactants) moves into place.

At the active site, reactants are brought into close proximity (figure 2.23). After the reactants combine, they are released from the active site, and the enzyme is capable of catalyzing additional reactions. The activation energy required for a chemical reaction to occur is lowered by enzymes (see figure 2.12) because they orient the reactants toward each other in such a way that it is more likely a chemical reaction will occur.

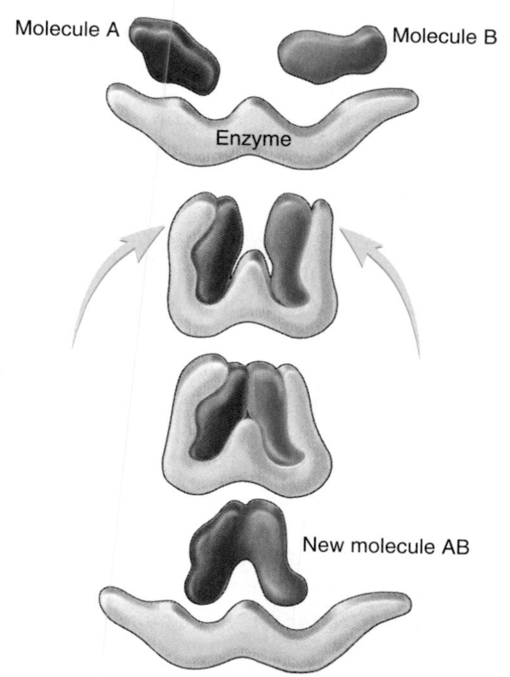

FIGURE 2.23 Enzyme Action
The enzyme brings the two reacting molecules together. After the reaction, the unaltered enzyme can be used again.

Slight changes in the structure of an enzyme can destroy the ability of the active site to function. Enzymes are very sensitive to changes in temperature or pH, which can break the hydrogen bonds within them. As a result, the relationship between amino acids changes, thereby producing a change in shape that prevents the enzyme from functioning normally.

To be functional, some enzymes require additional, nonprotein substances called **cofactors.** A cofactor can be an ion, such as magnesium or zinc, or an organic molecule. Cofactors that are organic molecules, such as certain vitamins, may be referred to as **coenzymes.** Cofactors normally form part of the enzyme's active site and are required to make the enzyme functional.

Enzymes are highly specific because their active site can bind only to certain reactants. Each enzyme catalyzes a specific chemical reaction and no others. Many different enzymes are therefore needed to catalyze the many chemical reactions of the body. Enzymes often are named by adding the suffix *-ase* to the name of the molecules on which they act. For example, an enzyme that catalyzes the breakdown of lipids is a **lipase** (lip′ās, lī′pās), and an enzyme that breaks down proteins is called a **protease** (prō′tē-ās).

Enzymes control the rate at which most chemical reactions proceed in living systems. Consequently, they control essentially all cellular activities. At the same time, the activity of enzymes themselves is regulated by several mechanisms within the cells. Some mechanisms control the enzyme concentration by influencing the rate at which the enzymes are synthesized; others alter the activity of existing enzymes. Much of what is known about the regulation of cellular activity involves knowledge of how enzyme activity is controlled.

40 ***What are the building blocks of proteins? Define a peptide bond.***

41 ***What determines the primary, secondary, tertiary, and quaternary structures of proteins? Define denaturation and name two things that can cause it to occur.***

42 ***Compare the lock-and-key model and the induced fit model of enzyme activity. Define cofactor and coenzyme.***

Nucleic Acids: DNA and RNA

Deoxyribonucleic (dē-oks′ē-rī′bō-noo-klē′ik) **acid (DNA)** is the genetic material of cells, and copies of DNA are transferred from one generation of cells to the next generation. DNA contains the information that determines the structure of proteins. **Ribonucleic** (rī′bō-noo-klē′ik) **acid (RNA)** is structurally related to DNA, and three types of RNA also play important roles in protein synthesis. In chapter 3, the means by which DNA and RNA direct the functions of the cell are described.

The **nucleic** (noo-klē′ik, noo-klā′ik) **acids** are large molecules composed of carbon, hydrogen, oxygen, nitrogen, and phosphorus. Both DNA and RNA consist of basic building blocks called **nucleotides** (noo′klē-ō-tīdz). Each nucleotide is composed of a monosaccharide to which a nitrogenous organic base and a phosphate group

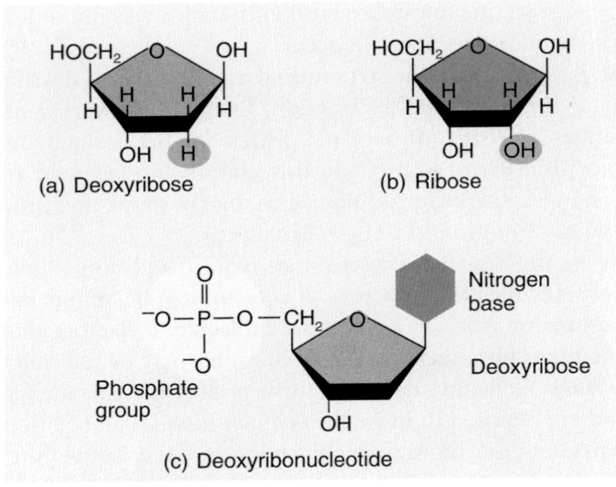

FIGURE 2.24 Components of Nucleotides

(*a*) Deoxyribose sugar, which forms nucleotides used in DNA production.
(*b*) Ribose sugar, which forms nucleotides used in RNA production. Note that deoxyribose is ribose minus an oxygen atom. (*c*) Deoxyribonucleotide consisting of deoxyribose, a nitrogen base, and a phosphate group.

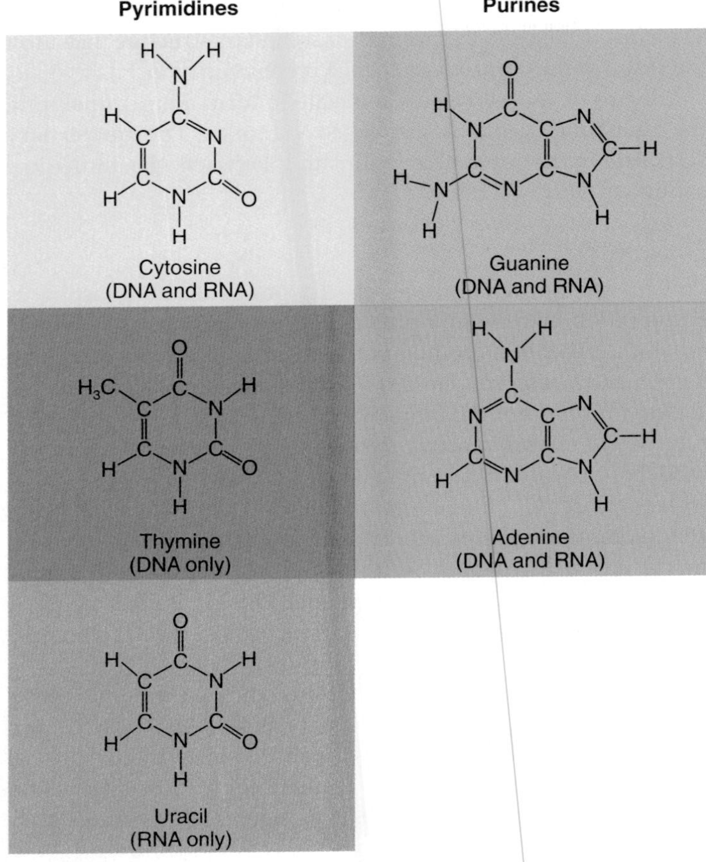

FIGURE 2.25 Nitrogenous Organic Bases

The organic bases found in nucleic acids are separated into two groups. Purines are double-ringed molecules, and pyrimidines are single-ringed molecules.

are attached (figure 2.24). The five-carbon monosaccharide is deoxyribose for DNA; it is ribose for RNA. The nitrogenous organic bases consist of carbon and nitrogen atoms organized into rings (figure 2.25). They are bases because the nitrogen atoms tend to take up H^+ from solution. The nitrogenous organic bases are thymine (thī′mēn, thī′min), cytosine (sī′tō-sēn), and uracil (ūr′ă-sil), which are single-ringed pyrimidines (pī-rim′i-dēnz), and adenine (ad′ĕ-nēn) and guanine (gwahn′ēn), which are double-ringed purines (pūr′ēnz).

DNA has two strands of nucleotides joined together to form a twisted, ladderlike structure called a double helix (figure 2.26). The uprights of the ladder are formed by covalent bonds between the deoxyribose molecules and phosphate groups of adjacent nucleotides. The rungs of the ladder are formed by the bases of the nucleotides of one upright connected to the bases of the other upright by hydrogen bonds. Each nucleotide of DNA contains one of the organic bases: adenine, thymine, cytosine, or guanine. **Complementary base pairs** are organic bases held together by hydrogen bonds. Adenine and thymine are complementary base pairs because the structure of these organic bases allows two hydrogen bonds to form between them. Cytosine and guanine are complementary base pairs because the structure of these organic bases allows three hydrogen bonds to form between them. The two strands of a DNA molecule are said to be complementary. If the sequence of bases in one DNA strand is known, the sequence of bases in the other strand can be predicted because of complementary base pairing.

The two nucleotide strands of a DNA molecule are **antiparallel,** meaning that the two strands lie side by side, but their sugar-phosphate "backbones" extend in opposite directions because of the orientation of their nucleotides (see figure 2.26). A nucleotide has a 5′ end and a 3′ end. The prime sign is used to indicate the carbon atoms of the deoxyribose sugar, which are numbered 1′ to 5′.

The sequence of organic bases in DNA is a "code" that stores information used to determine the structures and functions of cells. A sequence of DNA bases that directs the synthesis of proteins or RNA molecules is called a **gene** (see chapter 3 for more information on genes). Genes determine the type and sequence of amino acids found in protein molecules. Because enzymes are proteins, DNA structure determines the rate and type of chemical reactions that occur in cells by controlling enzyme structure. The information contained in DNA therefore ultimately defines all cellular activities. Other proteins, such as collagen, that are coded by DNA determine many of the structural features of humans.

RNA has a structure similar to a single strand of DNA. Like DNA, four different nucleotides make up the RNA molecule, and the organic bases are the same, except that thymine is replaced with uracil (see figure 2.25). Uracil can bind only to adenine.

43 ***Name two types of nucleic acids. What are their functions?***

44 ***What are the basic building blocks of nucleic acids? What kinds of sugar and bases are found in DNA? In RNA?***

45 ***DNA is like a twisted ladder. What forms the uprights and rungs of the ladder?***

46 ***Name the complementary base pairs in DNA and RNA.***

47 ***How are DNA strands complementary and antiparallel?***

48 ***Define gene and explain how genes determine the structures and functions of cells.***

1. The building blocks of nucleic acids are nucleotides, which consist of a phosphate group, a sugar, and a nitrogen base.
2. The phosphate groups connect the sugars to form two strands of nucleotides (*purple columns*).
3. Hydrogen bonds (*dotted red lines*) between the nucleotides join the two nucleotide strands together. Adenine binds to thymine and cytosine binds to guanine.
4. Deoxyribose carbon atoms are numbered. One end of a DNA strand has a 3′end because of the orientation of its nucleotides.
5. The other end of a DNA strand has a 5′end.
6. The complementary strands are antiparallel in that the 5′→ 3′ direction of one strand runs counter to the 5′→ 3′direction of the other strand.
7. The nucleotide strands coil to form a double-stranded helix.

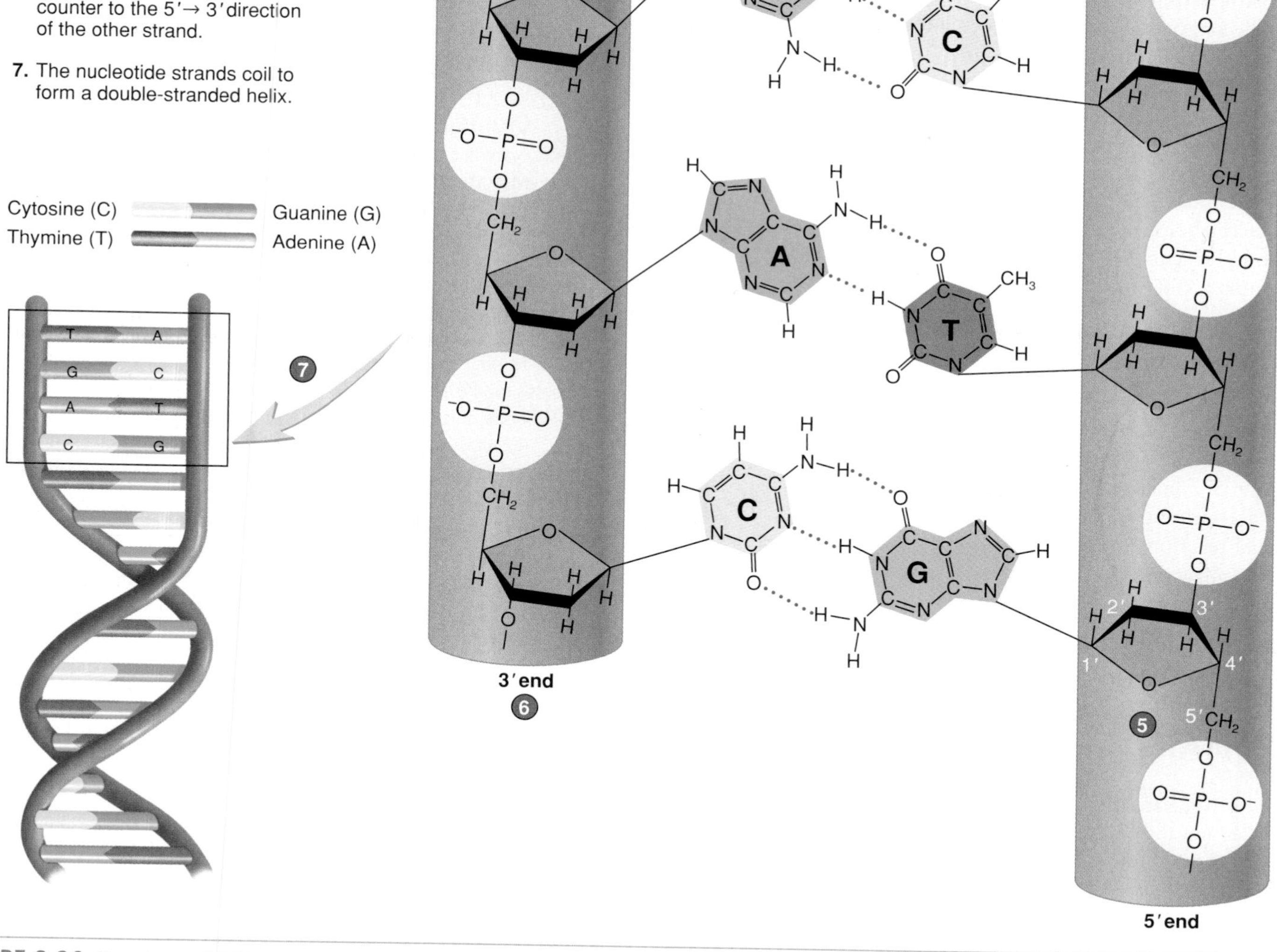

FIGURE 2.26 Structure of DNA

1. Adenosine is adenine, which is one of the nitrogenous bases in DNA, combined with the sugar ribose.
2. Adenosine diphosphate (ADP) is adenosine with two phosphate groups.
3. Adenosine triphosphate (ATP) is adenosine with three phosphates.

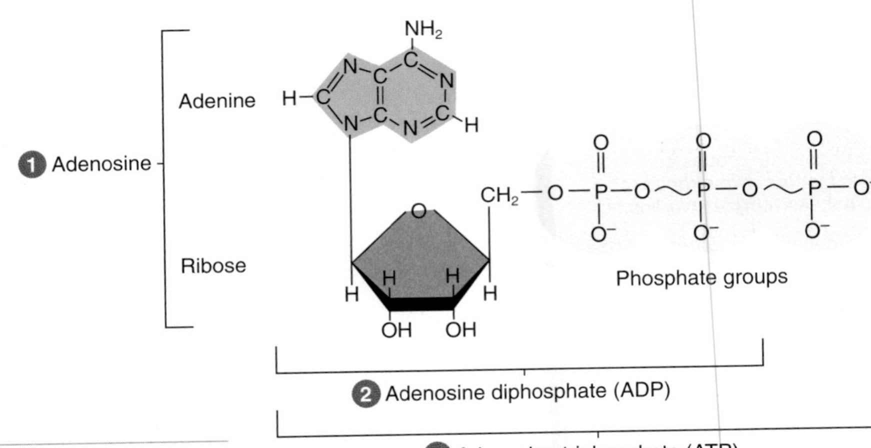

FIGURE 2.27 Adenosine Triphosphate (ATP) Molecule

Adenosine Triphosphate

Adenosine triphosphate (ă-den′ō-sēn trī-fos′fāt) **(ATP)** is an especially important organic molecule found in all living organisms. It consists of adenosine and three phosphate groups (figure 2.27). Adenosine is the sugar ribose with the organic base adenine. The potential energy stored in the covalent bond between the second and third phosphate groups is important to living organisms because it provides the energy used in nearly all of the chemical reactions within cells.

The catabolism of glucose and other nutrient molecules results in chemical reactions that release energy. Some of that energy is used to synthesize ATP from ADP and an inorganic phosphate group (P_i):

$$\text{ADP} + P_i + \text{Energy (from catabolism)} \rightarrow \text{ATP}$$

The transfer of energy from nutrient molecules to ATP involves a series of oxidation–reduction reactions in which a high-energy electron is transferred from one molecule to the next molecule in the series. In chapter 25, the oxidation–reduction reactions of metabolism are considered in greater detail.

Once produced, ATP is used to provide energy for other chemical reactions (anabolism) or to drive cell processes, such as muscle contraction. In the process, ATP is converted back to ADP and an inorganic phosphate group.

$$\text{ATP} \rightarrow \text{ADP} + P_i + \text{Energy (for anabolism and other cell processes)}$$

ATP is often called the energy currency of cells because it is capable of both storing and providing energy. The concentration of ATP is maintained within a narrow range of values, and essentially all energy-requiring chemical reactions stop when there is an inadequate quantity of ATP.

Cyanide Poisoning

Cyanide compounds can be lethal to humans because they interfere with the production of ATP in mitochondria (see chapter 25). Without adequate ATP, cells die because there is inadequate energy for anabolic chemical reactions, active transport, and other energy-requiring cell processes. The heart and brain are especially susceptible to cyanide poisoning. Inhalation of smoke released by the burning of rubber and plastic by household fires is the most common cause of cyanide poisoning. Cyanide poisoning by inhalation or absorption through the skin can also occur in certain manufacturing processes, and cyanide gas was used during the Holocaust to kill people. Deliberate suicide by ingesting cyanide is rare but was made famous by the suicide capsules in spy movies. In 1982, seven people in the Chicago area died after taking Tylenol that some unknown person had laced with cyanide. Subsequent copycat tamperings occurred and led to the widespread use of tamper-proof capsules and packaging.

49 ***Describe the structure of ATP. Where does the energy to synthesize ATP come from? What is the energy stored in ATP used for?***

SUMMARY

Chemistry is the study of the composition, structure, and properties of substances and the reactions they undergo. Much of the structure and function of healthy or diseased organisms can be understood at the chemical level.

Basic Chemistry (p. 24)

Matter, Mass, and Weight

1. Matter is anything that occupies space.
2. Mass is the amount of matter in an object.
3. Weight results from the force exerted by earth's gravity on matter.

Elements and Atoms

1. An element is the simplest type of matter with unique chemical and physical properties.
2. An atom is the smallest particle of an element that has the chemical characteristics of that element. An element is composed of only one kind of atom.
3. Atoms consist of protons, neutrons, and electrons.
   - Protons are positively charged, electrons are negatively charged, and neutrons have no charge.
   - Protons and neutrons are found in the nucleus; electrons, which are located around the nucleus, can be represented by an electron cloud.

4. The atomic number is the unique number of protons in an atom. The mass number is the sum of the protons and the neutrons.
5. Isotopes are atoms that have the same atomic number but different mass numbers.
6. The atomic mass of an element is the average mass of its naturally occurring isotopes weighted according to their abundance.
7. A mole of a substance contains Avogadro's number (6.022×10^{23}) of atoms, ions, or molecules. The molar mass of a substance is the mass of 1 mole of the substance expressed in grams.

Electrons and Chemical Bonding

1. The chemical behavior of atoms is determined mainly by their outermost electrons. A chemical bond occurs when atoms share or transfer electrons.
2. Ions are atoms that have gained or lost electrons.
   - An atom that loses one or more electrons becomes positively charged and is called a cation. An anion is an atom that becomes negatively charged after accepting one or more electrons.
   - Ionic bonding is the attraction of the oppositely charged cation and anion to each other.
3. A covalent bond is the sharing of electron pairs between atoms. A polar covalent bond results when the sharing of electrons is unequal and can produce a polar molecule that is electrically asymmetric.

Molecules and Compounds

1. A molecule is two or more atoms chemically combined to form a structure that behaves as an independent unit. A compound is two or more *different* types of atoms chemically combined.
2. The kinds and numbers of atoms (or ions) in a molecule or compound can be represented by a formula consisting of the symbols of the atoms (or ions) plus subscripts denoting the number of each type of atom (or ion).
3. The molecular mass of a molecule or compound can be determined by adding up the atomic masses of its atoms (or ions).

Intermolecular Forces

1. A hydrogen bond is the weak attraction that occurs between the oppositely charged regions of polar molecules. Hydrogen bonds are important in determining the three-dimensional structure of large molecules.
2. Solubility is the ability of one substance to dissolve in another. Ionic substances that dissolve in water by dissociation are electrolytes. Molecules that do not dissociate are nonelectrolytes.

Chemical Reactions and Energy (p. 32)

Synthesis Reactions

1. Synthesis reactions are the chemical combination of two or more substances to form a new or larger substance.
2. Dehydration reactions are synthesis reactions in which water is produced.
3. Anabolism is the sum of all the synthesis reactions in the body.

Decomposition Reactions

1. Decomposition reactions are the chemical breakdown of a larger substance to two or more different smaller substances.
2. Hydrolysis reactions are decomposition reactions in which water is depleted.
3. All of the decomposition reactions in the body are called catabolism.

Reversible Reactions

Reversible reactions produce an equilibrium condition in which the amount of reactants relative to the amount of products remains constant.

Oxidation–Reduction Reactions

Oxidation–reduction reactions involve the complete or partial transfer of electrons between atoms.

Energy

1. Energy is the ability to do work. Potential energy is stored energy, and kinetic energy is energy resulting from movement of an object.
2. Chemical energy
   - Chemical bonds are a form of potential energy.
   - Chemical reactions in which the products contain more potential energy than the reactants require the input of energy.
   - Chemical reactions in which the products have less potential energy than the reactants release energy.
3. Heat energy
   - Heat energy is energy that flows between objects that are at different temperatures.
   - Heat energy is released in chemical reactions and is responsible for body temperature.

Speed of Chemical Reactions

1. Activation energy is the minimum energy that the reactants must have to start a chemical reaction.
2. Enzymes are specialized protein catalysts that lower the activation energy for chemical reactions. Enzymes speed up chemical reactions but are not consumed or altered in the process.
3. Increased temperature and concentration of reactants can increase the rate of chemical reactions.

Inorganic Chemistry (p. 36)

Inorganic chemistry is mostly concerned with noncarbon-containing substances but does include some carbon-containing substances, such as carbon dioxide and carbon monoxide.

Water

1. Water is a polar molecule composed of one atom of oxygen and two atoms of hydrogen.
2. Water stabilizes body temperature, protects against friction and trauma, makes chemical reactions possible, directly participates in chemical reactions (e.g., dehydration and hydrolysis reactions), and is a mixing medium (e.g., solutions, suspensions, and colloids).
3. A mixture is a combination of two or more substances physically blended together, but not chemically combined.
4. A solution is any liquid, gas, or solid in which the substances are uniformly distributed with no clear boundary between the substances.
5. A solute dissolves in the solvent.
6. A suspension is a mixture containing materials that separate from each other unless they are continually, physically blended together.
7. A colloid is a mixture in which a dispersed (solutelike) substance is distributed throughout a dispersing (solventlike) substance. Particles do not settle out of a colloid.

Solution Concentrations

1. One way to describe solution concentration is an osmole, which contains Avogadro's number (6.022×10^{23}) of particles (i.e., atoms, ions, or molecules) in 1 kilogram water.
2. A milliosmole is 1/1000 of an osmole.

Acids and Bases

1. Acids are proton (i.e., H^+) donors, and bases (e.g., OH^-) are proton acceptors.
2. A strong acid or base almost completely dissociates in water. A weak acid or base partially dissociates.
3. The pH scale is the H^+ concentration in a solution.
   - A neutral solution has an equal number of H^+ and OH^- and is assigned a pH of 7.
   - Acid solutions, in which the number of H^+ is greater than the number of OH^-, have pH values less than 7.
   - Basic, or alkaline, solutions have more OH^- than H^+ and a pH greater than 7.
4. A salt is a molecule consisting of a cation other than H^+ and an anion other than OH^-. Salts are formed when acids react with bases.
5. A buffer is a solution of a conjugate acid–base pair that resists changes in pH when acids or bases are added to the solution.

Oxygen

Oxygen is necessary in the reactions that extract energy from food molecules in living organisms.

Carbon Dioxide

During metabolism when the organic molecules are broken down, carbon dioxide and energy are released.

Organic Chemistry (p. 39)

Organic molecules contain carbon atoms bound together by covalent bonds.

Carbohydrates

1. Monosaccharides are the basic building blocks of other carbohydrates. They, especially glucose, are important sources of energy. Examples are ribose, deoxyribose, glucose, fructose, and galactose.
2. Disaccharide molecules are formed by dehydration reactions between two monosaccharides. They are broken apart into monosaccharides by hydrolysis reactions. Examples of disaccharides are sucrose, lactose, and maltose.
3. Polysaccharides are many monosaccharides bound together to form long chains. Examples include cellulose, starch, and glycogen.

Lipids

1. Triglycerides are composed of glycerol and fatty acids. One, two, or three fatty acids can attach to the glycerol molecule.
   - Fatty acids are straight chains of carbon molecules with a carboxyl group. Fatty acids can be saturated (only single covalent bonds between carbon atoms) or unsaturated (one or more double covalent bonds between carbon atoms).
   - Energy is stored in fats.
2. Phospholipids are lipids in which a fatty acid is replaced by a phosphate-containing molecule. Phospholipids are a major structural component of plasma membranes.
3. Steroids are lipids composed of four interconnected ring molecules. Examples include cholesterol, bile salts, and sex hormones.
4. Other lipids include fat-soluble vitamins, prostaglandins, thromboxanes, and leukotrienes.

Proteins

1. The building blocks of protein are amino acids, which are joined by peptide bonds.
2. The number, kind, and arrangement of amino acids determine the primary structure of a protein. Hydrogen bonds between amino acids determine secondary structure, and hydrogen bonds between amino acids and water determine tertiary structure. Interactions between different protein subunits determine quaternary structure.
3. Enzymes are protein catalysts that speed up chemical reactions by lowering their activation energy.
4. The active sites of enzymes bind only to specific reactants.
5. Cofactors are ions or organic molecules, such as vitamins, that are required for some enzymes to function.

Nucleic Acids: DNA and RNA

1. The basic unit of nucleic acids is the nucleotide, which is a monosaccharide with an attached phosphate and organic base.
2. DNA nucleotides contain the monosaccharide deoxyribose and the organic base adenine, thymine, guanine, or cytosine. DNA occurs as a double strand of joined nucleotides. Each strand is complementary and antiparallel to the other strand.
3. A gene is a sequence of DNA nucleotides that determines the structure of a protein or RNA.
4. RNA nucleotides are composed of the monosaccharide ribose. The organic bases are the same as for DNA, except that thymine is replaced with uracil.

Adenosine Triphosphate

ATP stores energy derived from catabolism. The energy is released from ATP and is used in anabolism and other cell processes.

REVIEW AND COMPREHENSION

1. The smallest particle of an element that still has the chemical characteristics of that element is a (an)
   a. electron.
   b. molecule.
   c. neutron
   d. proton.
   e. atom.
2. The number of electrons in an atom is equal to the
   a. atomic number.
   b. mass number.
   c. number of neutrons.
   d. isotope number.
   e. molecular mass.
3. ^{12}C and ^{14}C are
   a. atoms of different elements.
   b. isotopes.
   c. atoms with different atomic numbers.
   d. atoms with different numbers of protons.
   e. compounds.

4. A cation is a (an)
   a. uncharged atom.
   b. positively charged atom.
   c. negatively charged atom.
   d. atom that has gained an electron.
   e. both c and d.
5. A polar covalent bond between two atoms occurs when
   a. one atom attracts shared electrons more strongly than another atom.
   b. atoms attract electrons equally.
   c. an electron from one atom is completely transferred to another atom.
   d. the molecule becomes ionized.
   e. a hydrogen atom is shared between two different atoms.
6. Table salt (NaCl) is
   a. an atom. c. a molecule. e. a cation.
   b. organic. d. a compound.
7. The weak attractive force between two water molecules forms a (an)
   a. covalent bond. c. ionic bond. e. isotope.
   b. hydrogen bond. d. compound.
8. Electrolytes are
   a. nonpolar molecules.
   b. covalent compounds.
   c. substances that usually don't dissolve in water.
   d. found in solutions that do not conduct electricity.
   e. cations and anions that dissociate in water.
9. In a decomposition reaction,
   a. anabolism occurs.
   b. proteins are formed from amino acids.
   c. large molecules are broken down to form small molecules.
   d. a dehydration reaction may occur.
   e. all of the above.
10. Oxidation–reduction reactions
    a. can be synthesis or decomposition reactions.
    b. have one reactant gaining electrons.
    c. have one reactant losing electrons.
    d. can create ionic or covalent bonds.
    e. all of the above.
11. Potential energy
    a. is energy caused by movement of an object.
    b. is the form of energy that is actually doing work.
    c. includes energy within chemical bonds.
    d. can never be converted to kinetic energy.
    e. all of the above.
12. Which of these descriptions of heat energy is *not* correct?
    a. Heat energy flows between objects that are at different temperatures.
    b. Heat energy can be produced from all other forms of energy.
    c. Heat energy can be released during chemical reactions.
    d. Heat energy must be added to break apart ATP molecules.
    e. Heat energy is always transferred from a hotter object to a cooler object.
13. A *decrease* in the speed of a chemical reaction occurs if
    a. the activation energy requirement is increased.
    b. catalysts are increased.
    c. temperature increases.
    d. the concentration of the reactants increases.
    e. all of the above.
14. Which of these statements concerning enzymes is correct?
    a. Enzymes increase the rate of reactions but are permanently changed as a result.
    b. Enzymes are proteins that function as catalysts.
    c. Enzymes increase the activation energy requirement for a reaction to occur.
    d. Enzymes usually can only double the rate of a chemical reaction.
    e. Enzymes increase the kinetic energy of the reactants.
15. Water
    a. is composed of two oxygen atoms and one hydrogen atom.
    b. has a low specific heat.
    c. is composed of polar molecules into which ionic substances dissociate.
    d. is produced in a hydrolysis reaction.
    e. is a very small organic molecule.
16. When sugar is dissolved in water, the water is called the
    a. solute. b. solution. c. solvent.
17. Which of these is an example of a suspension?
    a. sweat
    b. water and proteins inside cells
    c. sugar dissolved in water
    d. red blood cells in plasma
18. A solution with a pH of 5 is ______________ and contains ______________ H^+ than a neutral solution.
    a. a base, more d. an acid, less
    b. a base, less e. neutral, the same number of
    c. an acid, more
19. A buffer
    a. slows down chemical reactions.
    b. speeds up chemical reactions.
    c. increases the pH of a solution.
    d. maintains a relatively constant pH.
    e. works by forming salts.
20. A conjugate acid–base pair
    a. acts as a buffer.
    b. can combine with H^+ in a solution.
    c. can release H^+ to combine with OH^-.
    d. describes carbonic acid (H_2CO_3) and bicarbonate ion (HCO_3^-).
    e. all of the above.
21. Carbon dioxide
    a. consists of two oxygen atoms ionically bonded to carbon.
    b. becomes toxic if allowed to accumulate within cells.
    c. is mostly eliminated by the kidneys.
    d. is combined with fats to produce glucose during metabolism within cells.
    e. is taken into cells during metabolism.
22. Which of these is a carbohydrate?
    a. glycogen c. steroid e. triglyceride
    b. prostaglandin d. DNA
23. The polysaccharide used for energy storage in the human body is
    a. cellulose. c. lactose. e. starch.
    b. glycogen. d. sucrose.
24. The basic units or building blocks of triglycerides are
    a. simple sugars (monosaccharides).
    b. double sugars (disaccharides).
    c. amino acids.
    d. glycerol and fatty acids.
    e. nucleotides.

25. A ______________ fatty acid has one double covalent bond between carbon atoms.
    a. cholesterol
    b. monounsaturated
    c. phospholipid
    d. polyunsaturated
    e. saturated
26. A peptide bond joins together
    a. amino acids.
    b. fatty acids and glycerol.
    c. monosaccharides.
    d. disaccharides.
    e. nucleotides.
27. The ______________ structure of a protein results from the folding of the helices or pleated sheets.
    a. primary
    b. secondary
    c. tertiary
    d. quaternary
28. According to the lock-and-key model of enzyme action,
    a. reactants must first be heated.
    b. enzyme shape is not important.
    c. each enzyme can catalyze many types of reactions.
    d. reactants must bind to an active site on the enzyme.
    e. enzymes control only a small number of reactions in the cell.
29. DNA molecules
    a. contain genes.
    b. contain a single strand of nucleotides.
    c. contain the nucleotide uracil.
    d. have three different types that have roles in protein synthesis.
    e. contain up to 100 organic bases.
30. ATP
    a. is formed by the addition of a phosphate group to ADP.
    b. is formed with energy released during catabolism reactions.
    c. provides the energy for anabolism reactions.
    d. contains three phosphate groups.
    e. all of the above.

Answers in Appendix E

CRITICAL THINKING

1. Iron has an atomic number of 26 and a mass number of 56. How many protons, neutrons, and electrons are in an atom of iron? If an atom of iron lost three electrons, what would the charge of the resulting ion be? Write the correct symbol for this ion.
2. Which of the following pairs of terms applies to the reaction that results in the formation of fatty acids and glycerol from a triglyceride molecule?
    a. Decomposition or synthesis reaction
    b. Anabolism or catabolism
    c. Dehydration or hydrolysis reaction
3. A mixture of chemicals is warmed slightly. As a consequence, although no more heat is added, the solution becomes very hot. Explain what has occurred to make the solution so hot.
4. Two solutions, when mixed together at room temperature, produce a chemical reaction. When the solutions are boiled and allowed to cool to room temperature before mixing, however, no chemical reaction takes place. Explain.
5. In terms of the potential energy in food, explain why eating food is necessary for increasing muscle mass.
6. Solution A has a pH of 2, and solution B has a pH of 8. If equal amounts of solutions A and B are mixed, is the resulting solution acidic or basic?
7. Given a buffered solution that is based on the following equilibrium,

$$CO_2 + H_2O \rightleftarrows H_2CO_3 \rightleftarrows H^+ + HCO_3^-$$

   what happens to the pH of the solution if $NaHCO_3$ is added?
8. An enzyme, E, catalyzes the following reaction:

$$A + B \xrightarrow{E} C$$

   The product, C, however, binds to the active site of the enzyme in a reversible fashion and keeps the enzyme from functioning. What happens if A and B are continually added to a solution that contains a fixed amount of the enzyme?
9. Given the materials commonly found in a kitchen, explain how one can distinguish between a protein and a lipid.
10. A student is given two unlabeled substances: one a typical phospholipid and one a typical protein. She is asked to determine which substance is the protein and which is the phospholipid. The available techniques allow her to determine the elements in each sample. How can she identify each substance?

Answers in Appendix F

Cell Biology and Genetics 3

The cell is the basic structural and functional unit of all living organisms. The characteristic functions of cells include cell metabolism and energy use; synthesis of molecules, such as proteins and nucleic acids; communication between cells; reproduction; and inheritance. Cells are like very complex but minute factories that are always active, carrying out the functions of life. These microscopic factories are so small that an average-sized cell is only one-fifth the size of the smallest dot you can make on a sheet of paper with a sharp pencil. Each human body is made up of trillions of cells. If each cell were the size of a standard brick, the colossal human statue made from those bricks would be 6 miles high!

All the cells of an individual originate from a single fertilized cell. During development, cell division and specialization give rise to a wide variety of cell types, such as nerve, muscle, bone, fat, and blood cells. Each cell type has important characteristics that are critical to the normal function of the body as a whole. One of the important reasons for maintaining homeostasis is to keep the trillions of cells that form the body functioning normally.

This chapter outlines *functions of the cell* (p. 56), *how we see cells* (p. 58), the composition of the *plasma membrane* (p. 58), *membrane lipids* (p. 58), and *membrane proteins* (p. 58). It addresses *movement through the plasma membrane* (p. 64) and *endocytosis and exocytosis* (p. 71). The chapter then addresses the *cytoplasm* (p. 76) and *organelles* (p. 77). It presents *genes and gene expression* (p. 86), the *cell life cycle* (p. 91), and *genetics* (p. 93).

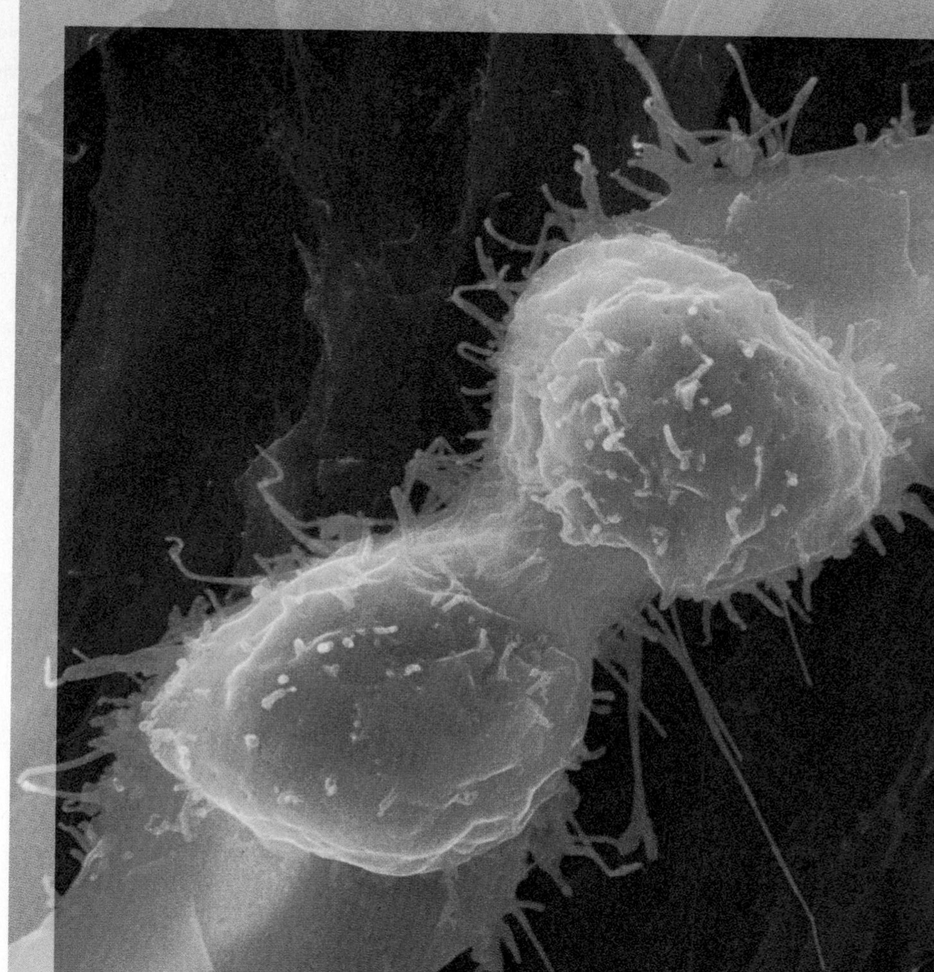

Colorized scanning electron micrograph (SEM) of a dividing cell. Cell division is one of the major features of organisms and is the basis of growth and repair of tissues and organs.

FUNCTIONS OF THE CELL

Because cells are the basic units of all living things, they are the smallest parts of an organism, such as a human, that have the characteristics of life. Although cells may have quite different structures and functions, they share several characteristics (figure 3.1; table 3.1). The **plasma** (plaz′mă), or **cell, membrane** forms the outer boundary of the cell, through which the cell interacts with its external environment. The **nucleus** (noo′klē-ŭs) is usually located centrally; it directs cell activities, most of which take place in the **cytoplasm** (sī′tō-plazm), located between the plasma membrane and the nucleus. Within cells, specialized structures called **organelles** (or′gă-nelz) perform specific functions.

The characteristic functions of the cell include the following:

1. *Cell metabolism and energy use.* The chemical reactions that occur within cells are referred to as metabolic reactions and collectively as cell metabolism. The energy released from some metabolic reactions provides the energy necessary for cellular activities, such as the synthesis of molecules and muscle contraction. During some metabolic reactions, energy is released as heat, which helps maintain body temperature.
2. *Synthesis of molecules.* Cells synthesize various types of molecules, including proteins, nucleic acids, and lipids. The molecules synthesized by the various cells of the body differ. Therefore, the structural and functional characteristics of cells are determined by the types of molecules they produce.
3. *Communication.* Cells produce and respond to chemical and electrical signals that allow them to communicate with one another. For example, nerve cells produce chemical signals by which they communicate with muscle cells, and muscle cells respond by contracting or relaxing.
4. *Reproduction and inheritance.* Most cells contain a complete copy of all the genetic information of the individual. This genetic information ultimately determines the structural and functional characteristics of the cell. During the growth of an individual, cells divide to produce new cells, each containing the same genetic information. Specialized cells of the body, called gametes, are responsible for transmitting genetic information to the next generation.

1 ***What are the characteristic functions of the cell?***

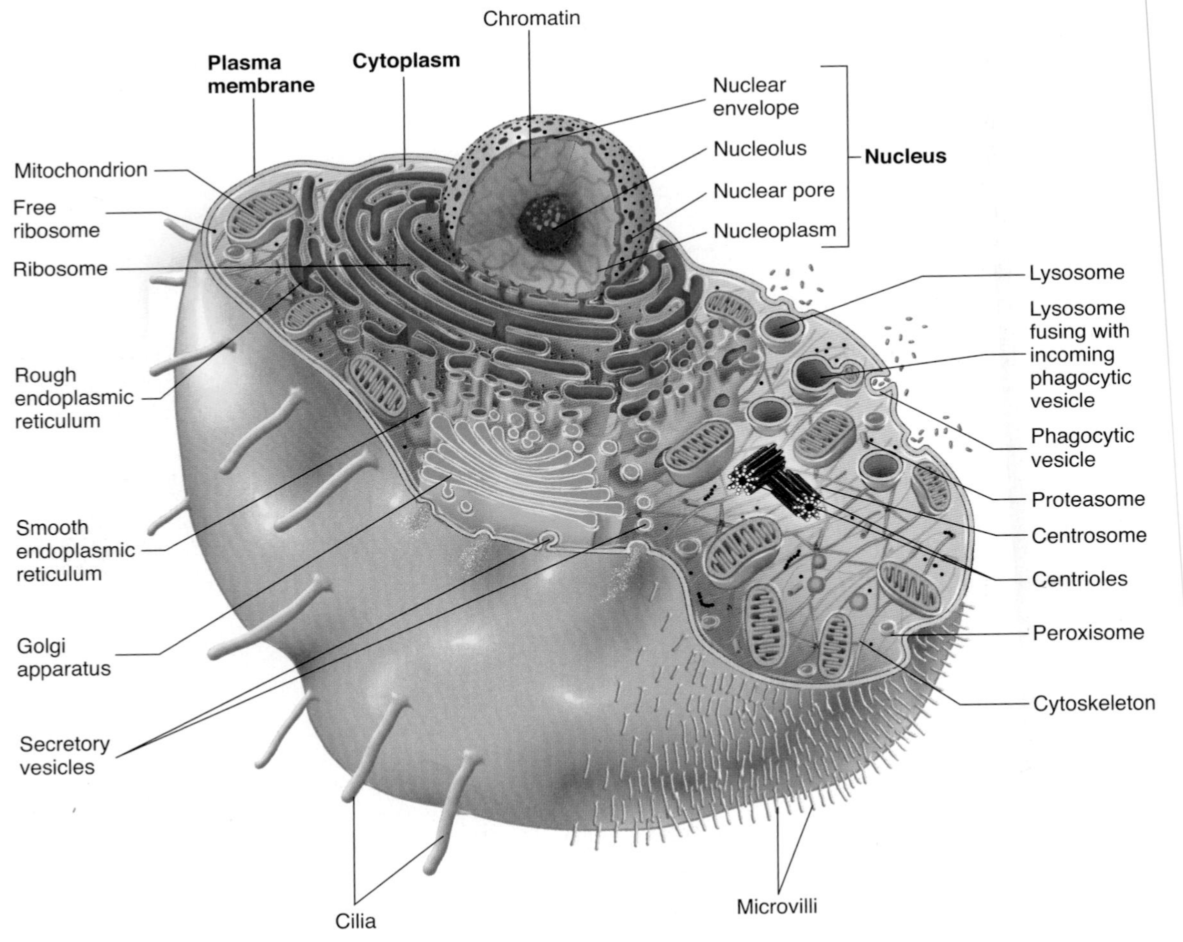

FIGURE 3.1 Cell

A generalized human cell showing the plasma membrane, nucleus, and cytoplasm with its organelles. Although no single cell contains all these organelles, many cells contain a large number of them.

TABLE 3.1 Summary of Cell Parts and Functions

Cell Parts	Structure	Function
Plasma Membrane	Lipid bilayer composed of phospholipids and cholesterol with proteins that extend across or are embedded in either surface of the lipid bilayer	Outer boundary of cells that controls entry and exit of substances; receptor molecules function in intercellular communication; marker molecules enable cells to recognize one another
Cytoplasm: Cytosol		
Fluid part	Water with dissolved ions and molecules; colloid with suspended proteins	Contains enzymes that catalyze decomposition and synthesis reactions; ATP is produced in glycolysis reactions
Cytoskeleton		
Microtubules	Hollow cylinders composed of the protein tubulin; 25 nm in diameter	Support the cytoplasm and form centrioles, spindle fibers, cilia, and flagella; responsible for movement of structures in the cell
Actin filaments	Small fibrils of the protein actin; 8 nm in diameter	Provide structural support to cells, support microvilli, responsible for cell movements
Intermediate filaments	Protein fibers; 10 nm in diameter	Provide structural support to cells
Cytoplasmic inclusions	Aggregates of molecules manufactured or ingested by the cell; may be membrane-bound	Function depends on the molecules: energy storage (lipids, glycogen), oxygen transport (hemoglobin), skin color (melanin), and others
Organelles		
Nucleus		
Nuclear envelope	Double membrane enclosing the nucleus; the outer membrane is continuous with the endoplasmic reticulum; nuclear pores extend through the nuclear envelope	Separates nucleus from cytoplasm and regulates movement of materials into and out of the nucleus
Chromatin	Dispersed, thin strands of DNA, histones, and other proteins; condenses to form chromosomes during cell division	DNA regulates protein (e.g., enzyme) synthesis and therefore the chemical reactions of the cell; DNA is the genetic, or hereditary, material
Nucleolus	One or more dense bodies consisting of ribosomal RNA and proteins	Assembly site of large and small ribosomal subunits
Cytoplasmic Organelles		
Ribosome	Ribosomal RNA and proteins form large and small subunits; attached to endoplasmic reticulum or free ribosomes are distributed throughout the cytoplasm	Site of protein synthesis
Rough endoplasmic reticulum	Membranous tubules and flattened sacs with attached ribosomes	Protein synthesis and transport to Golgi apparatus
Smooth endoplasmic reticulum	Membranous tubules and flattened sacs with no attached ribosomes	Manufactures lipids and carbohydrates; detoxifies harmful chemicals; stores calcium
Golgi apparatus	Flattened membrane sacs stacked on each other	Modifies, packages, and distributes proteins and lipids for secretion or internal use
Secretory vesicle	Membrane-bound sac pinched off Golgi apparatus	Carries proteins and lipids to cell surface for secretion
Lysosome	Membrane-bound vesicle pinched off Golgi apparatus	Contains digestive enzymes
Peroxisome	Membrane-bound vesicle	One site of lipid and amino acid degradation; breaks down hydrogen peroxide
Proteasomes	Tubelike protein complexes in the cytoplasm	Break down proteins in the cytoplasm
Mitochondria	Spherical, rod-shaped, or threadlike structures; enclosed by double membrane; inner membrane forms projections called cristae	Major site of ATP synthesis when oxygen is available
Centrioles	Pair of cylindrical organelles in the centrosome, consisting of triplets of parallel microtubules	Centers for microtubule formation; determine cell polarity during cell division; form the basal bodies of cilia and flagella
Spindle fibers	Microtubules extending from the centrosome to chromosomes and other parts of the cell (i.e., aster fibers)	Assist in the separation of chromosomes during cell division
Cilia	Extensions of the plasma membrane containing doublets of parallel microtubules; 10 μm in length	Move materials over the surface of cells
Flagellum	Extension of the plasma membrane containing doublets of parallel microtubules; 55 μm in length	In humans, responsible for movement of spermatozoa
Microvilli	Extension of the plasma membrane containing microfilaments	Increase surface area of the plasma membrane for absorption and secretion; modified to form sensory receptors

HOW WE SEE CELLS

Most cells are too small to be seen with the unaided eye. As a result, it is necessary to use microscopes to study them. **Light microscopes** allow us to visualize general features of cells. **Electron microscopes,** however, must be used to study the fine structure of cells. A **scanning electron microscope (SEM)** allows us to see features of the cell surface and the surfaces of internal structures. A **transmission electron microscope (TEM)** allows us to see "through" parts of the cell and thus to discover other aspects of cell structure. A more detailed description of microscopes and their use can be found on p. 112.

2 ***Which cell features can be seen with light and electron microscopes?***

PLASMA MEMBRANE

The plasma membrane is the outermost component of a cell. It functions as a boundary separating the substances inside the cell, which are **intracellular,** from substances outside the cell, which are **extracellular.** Sometimes extracellular substances are referred to as **intercellular,** meaning between cells. The plasma membrane encloses and supports the cell contents. It attaches cells to the extracellular environment or to other cells. The cells' ability to recognize and communicate with each other takes place through the plasma membrane. In addition, the plasma membrane determines what moves into and out of cells. As a result, the intracellular contents of cells is different from the extracellular environment.

The regulation of ion movement by cells results in a charge difference across the plasma membrane called the **membrane potential.** The outside of the plasma membrane is positively charged, compared with the inside, because there are more positively charged ions immediately on the outside of the plasma membrane and more negatively charged ions and proteins inside. The membrane potential, an important feature of a living cell's normal function, will be considered in greater detail in chapters 9 and 11.

The plasma membrane consists of 45%–50% lipids, 45%–50% proteins, and 4%–8% carbohydrates (figure 3.2). The carbohydrates combine with lipids to form glycolipids and with proteins to form glycoproteins. The **glycocalyx** (glī-kō-kā′liks) is the collection of glycolipids, glycoproteins, and carbohydrates on the outer surface of the plasma membrane. The glycocalyx also contains molecules absorbed from the extracellular environment, so there is often no precise boundary where the plasma membrane ends and the extracellular environment begins.

3 ***Define intracellular, extracellular, and intercellular.***

4 ***What is the membrane potential? Is the outside of the plasma membrane positively or negatively charged, compared with the inside?***

5 ***What is the glycocalyx? What types of molecules are found in the glycocalyx?***

MEMBRANE LIPIDS

The predominant lipids of the plasma membrane are phospholipids and cholesterol. **Phospholipids** readily assemble to form a **lipid bilayer,** a double layer of phospholipid molecules, because they have a polar (charged) head and a nonpolar (uncharged) tail (see chapter 2). The polar **hydrophilic** (water-loving) heads are exposed to water inside and outside the cell, whereas the nonpolar **hydrophobic** (water-fearing) tails face one another in the interior of the plasma membrane (see figure 3.2). The modern concept of the plasma membrane, the **fluid-mosaic model,** suggests that the plasma membrane is neither rigid nor static in structure but is highly flexible and can change its shape and composition through time. The lipid bilayer functions as a liquid in which other molecules, such as proteins, are suspended. The fluid nature of the lipid bilayer has several important consequences. It provides an important means of distributing molecules within the plasma membrane. In addition, slight damage to the membrane can be repaired because the phospholipids tend to reassemble around damaged sites and close them. The fluid nature of the lipid bilayer also enables membranes to fuse with one another.

Cholesterol is the other major lipid in the plasma membrane (see chapter 2). It is interspersed among the phospholipids and accounts for about one-third of the total lipids in the plasma membrane. The hydrophilic OH group of cholesterol extends between the phospholipid heads to the hydrophilic surface of the membrane, whereas the hydrophobic part of the cholesterol molecule lies within the hydrophobic region of the phospholipids. The amount of cholesterol in a particular plasma membrane is a major factor in determining the fluid nature of the membrane. Cholesterol limits the movement of phospholipids, providing stability to the plasma membrane.

6 ***How do hydrophilic heads and hydrophobic tails of phospholipid molecules result in a plasma membrane?***

7 ***What is the function of cholesterol in plasma membranes?***

MEMBRANE PROTEINS

The basic structure of the plasma membrane and some of its functions are determined by its lipids, but many functions of the plasma membrane are determined by its proteins. Some protein molecules, called **integral,** or **intrinsic, proteins,** penetrate deeply into the lipid bilayer, in many cases extending from one surface to the other (figure 3.3), whereas other proteins, called **peripheral,** or **extrinsic, proteins,** are attached to either the inner or outer surfaces of the lipid bilayer. Integral proteins consist of regions made up of amino acids with hydrophobic R groups and other regions of amino acids with hydrophilic R groups (see chapter 2). The hydrophobic regions are located within the hydrophobic part of the membrane, and the hydrophilic regions are located at the inner or outer surface of the membrane or line channels through the membrane. Some

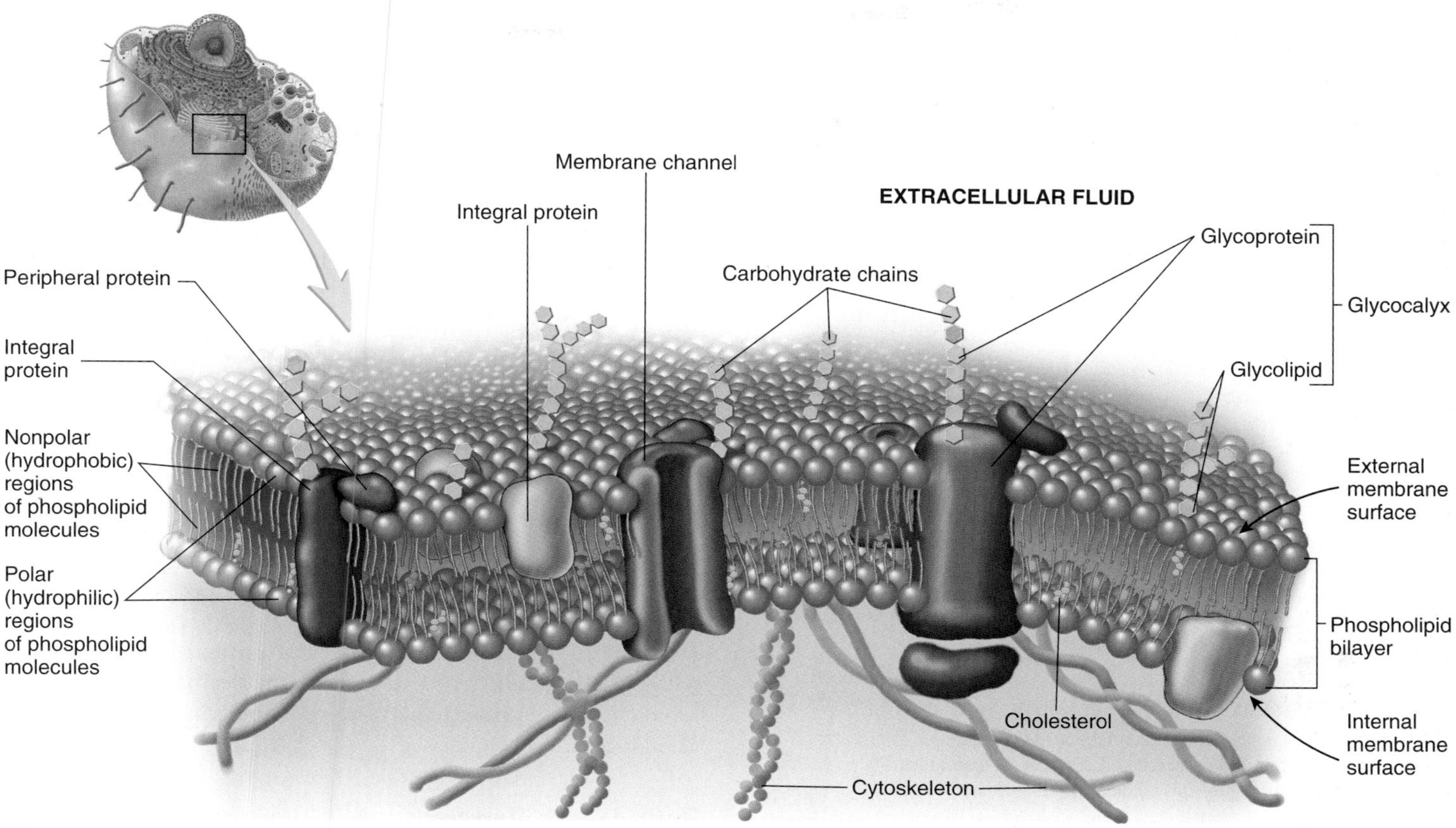

Membrane channel
Integral protein
EXTRACELLULAR FLUID
Peripheral protein
Carbohydrate chains
Glycoprotein
Glycocalyx
Integral protein
Glycolipid
Nonpolar (hydrophobic) regions of phospholipid molecules
External membrane surface
Polar (hydrophilic) regions of phospholipid molecules
Phospholipid bilayer
Cholesterol
Internal membrane surface
Cytoskeleton
CYTOPLASM (Intracellular Fluid)

(a)

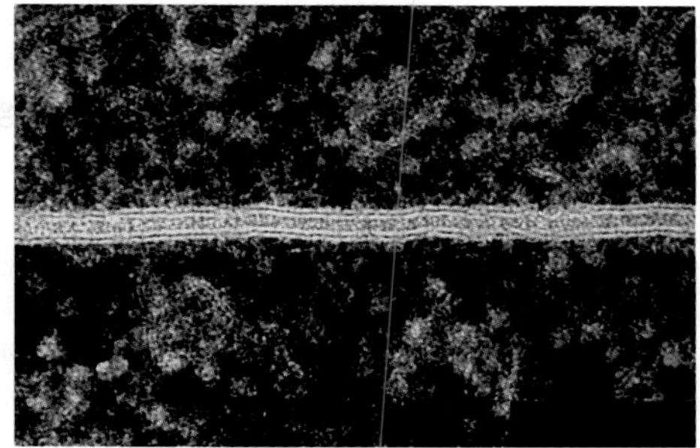

(b)

FIGURE 3.2 Plasma Membrane

(*a*) Fluid-mosaic model of the plasma membrane. The membrane is composed of a bilayer of phospholipids and cholesterol with proteins "floating" in the membrane. The nonpolar hydrophobic region of each phospholipid molecule is directed toward the center of the membrane and the polar hydrophilic region is directed toward the water environment either outside or inside the cell. (*b*) Colorized transmission electron micrograph of plasma membranes of two adjacent cells. Proteins at either surface of the lipid bilayer stain more readily than the lipid bilayer does and give each membrane the appearance of consisting of three parts: the two yellow, outer parts are proteins and the phospholipid heads, and the red, central part is the phospholipid tails and cholesterol.

peripheral proteins may be bound to integral proteins, whereas others are bound to the polar heads of the phospholipid molecules. Membrane proteins can function as marker molecules, attachment proteins, transport proteins, receptor proteins, or enzymes (table 3.2). The ability of membrane proteins to function depends on their three-dimensional shapes and their chemical characteristics.

Marker Molecules

Marker molecules are cell surface molecules that allow cells to identify one another or other molecules. They are mostly **glycoproteins** (proteins with attached carbohydrates) or **glycolipids** (lipids with attached carbohydrates). The protein portions of glycoproteins may be either integral or peripheral proteins (figure 3.4). Examples include recognition of an oocyte by a sperm cell and the ability of the immune system to distinguish between self-cells and foreign cells, such as bacteria or donor cells in an organ transplant. Intercellular communication and recognition are important because cells are not isolated entities; they must work together to ensure normal body functions.

Attachment Proteins

Integral proteins may function as **attachment proteins,** which allow cells to attach to other cells or to extracellular molecules

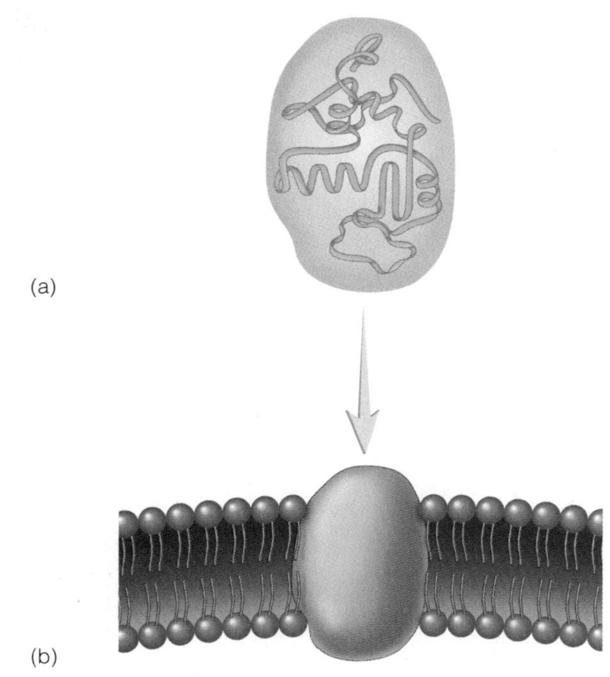

FIGURE 3.3 Proteins in the Plasma Membrane
(*a*) Proteins are commonly depicted as ribbons (see chapter 2). The space occupied by the protein ribbon can be enclosed by a three-dimensional, shaded region. (*b*) The shaded region can be depicted as a three-dimensional integral protein inserted into the plasma membrane.

TABLE 3.2 Functions of Membrane Proteins

Membrane Proteins	Protein Function
Marker molecules	Allow cells to identify one another or other molecules
Attachment proteins	Anchor cells to other cells (cadherins) or to extracellular molecules (integrins)
Transport proteins	
Channel proteins	Form passageways through the plasma membrane, allowing specific ions or molecules to enter or exit the cell; may be gated or nongated
Carrier proteins (transporters)	Move ions or molecules across the membrane; binding of specific chemical to carrier proteins causes changes in the shape of the carrier proteins; the carrier proteins then move the specific chemical across the membrane
ATP-powered pumps	Move specific ions or molecules across the membrane; require ATP molecules to function
Receptor proteins	Function as binding sites for chemical signals in the extracellular fluid; binding of chemical signals to receptors triggers cellular responses
Enzymes	Catalyze chemical reactions either inside or outside cells

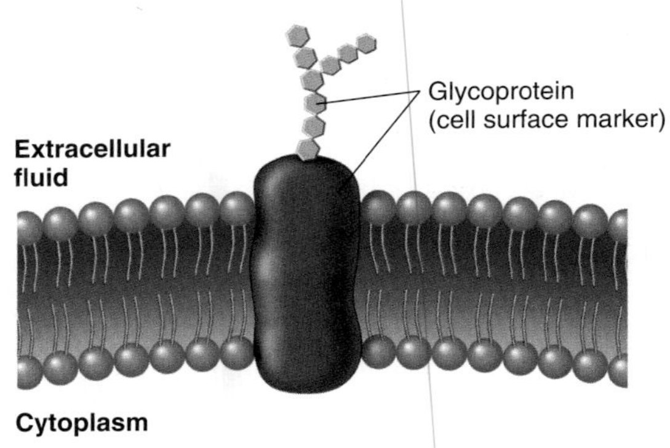

FIGURE 3.4 Cell Surface Marker
Glycoproteins that extend to the cell surface allow cells to identify one another or other molecules.

(figure 3.5). Many attachment proteins also attach to intracellular molecules. **Cadherins** are proteins that attach cells to other cells. **Integrins** are proteins that attach cells to extracellular molecules. Integrins function in pairs of integral proteins, which interact with both intracellular and extracellular molecules. Because of the interaction with intracellular molecules, integrins also function in cellular communication.

Transport Proteins

Transport proteins are integral proteins that allow ions or molecules to move from one side of the plasma membrane to the other. Transport proteins include channel proteins, carrier proteins, and ATP-powered pumps.

Channel Proteins

Channel proteins are one or more integral proteins arranged so that they form a tiny channel through the plasma membrane (figure 3.6). The hydrophobic regions of the proteins face outward

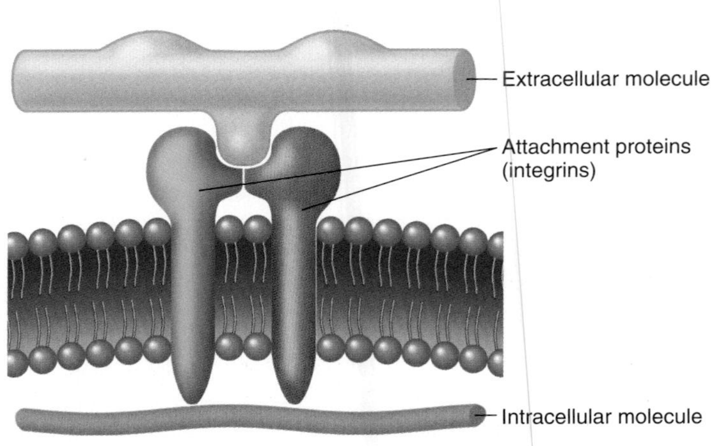

FIGURE 3.5 Attachment Proteins
Proteins (integrins) in the plasma membrane attach to extracellular molecules.

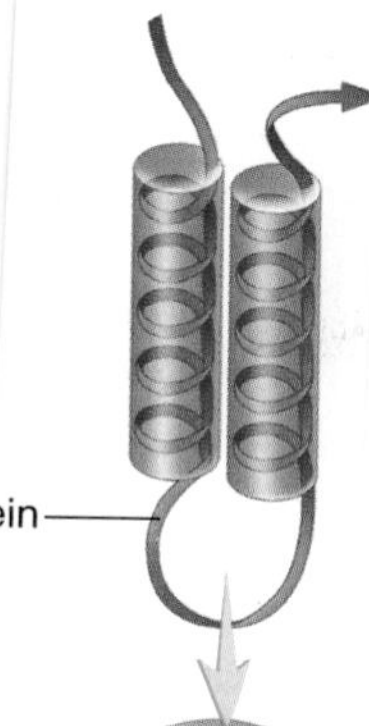

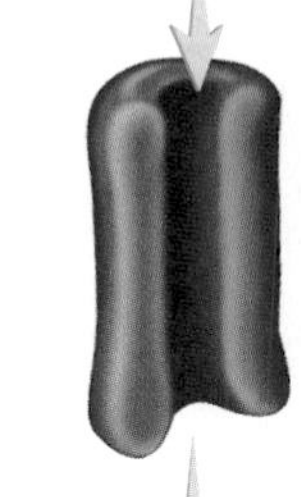

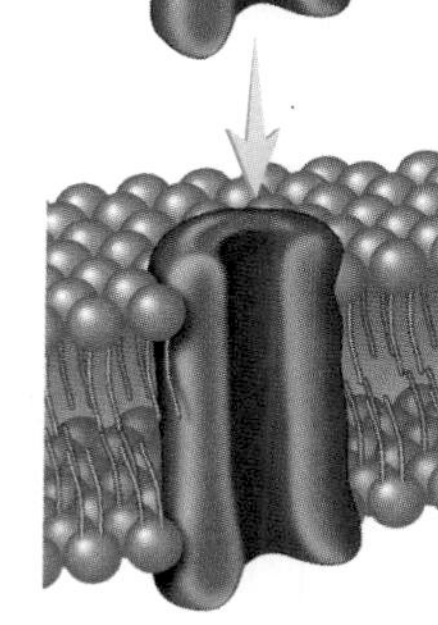

FIGURE 3.6 Channel Protein

toward the hydrophobic part of the plasma membrane, and the hydrophilic regions of the protein face inward and line the channel. Ions or small molecules of the right size, charge, and shape can pass through the channel. The charges in the hydrophilic part of the channel proteins determine which types of ions can pass through the channel.

Some channel proteins, called **nongated ion channels,** are always open and are responsible for the permeability of the plasma membrane to ions when the plasma membrane is at rest. Other channels, called **gated ion channels,** can be open or closed. Some gated ion channels open or close in response to **chemical signals,** or **ligands** (lig′andz, lī′gandz), which are small molecules that bind to the proteins or glycoproteins. These are called **ligand-gated ion channels.** Other gated ion channels open or close when there is a change in charge across the plasma membrane. These are called **voltage-gated ion channels.**

Carrier Proteins

Carrier proteins, or **transporters,** are integral membrane proteins that move ions or molecules from one side of the plasma membrane to the other. The carrier proteins have specific binding sites to which ions or molecules attach on one side of the plasma membrane. The carrier proteins change shape to move the bound ions or molecules to the other side of the plasma membrane, where they are released (figure 3.7). The carrier protein then resumes its original shape and is available to transport more molecules.

Movement of ions or molecules by carrier proteins can be classified as uniport, symport, or antiport. **Uniport** is the movement of one specific ion or molecule across the membrane. **Symport** is the movement of two different ions or molecules in the same direction across the plasma membrane, whereas **antiport** is the movement of two different ions or molecules in opposite directions across the plasma membrane. Transport proteins involved in these types of movement are called **uniporters, symporters,** and **antiporters,** respectively.

ATP-Powered Pumps

ATP-powered pumps are transport proteins that move specific ions and molecules from one side of the plasma membrane to the other. Unlike carrier proteins, however, the movement of ions or molecules across the membrane is fueled by the breakdown of adenosine triphosphate (ATP). Recall from chapter 2 that energy stored in ATP molecules is used to power many cellular activities. ATP-powered pumps have binding sites, to which specific ions or molecules can bind, as well as a binding site for ATP. The breakdown of ATP to adenosine diphosphate (ADP) releases energy, changing the shape of the protein, which moves the ion or molecule across the membrane (figure 3.8).

Receptor Proteins

Receptor proteins (figure 3.9) are proteins or glycoproteins in the plasma membrane with an exposed **receptor site** on the outer cell surface, which can attach to specific chemical signals. Many receptors and the chemical signals they bind are part of an intercellular communication system that coordinates cell activities. One cell can release a chemical signal that diffuses to another cell and binds to its receptor. The binding acts as a signal that triggers a response. The same chemical signal would have no effect on other cells that lacked the specific receptor molecule.

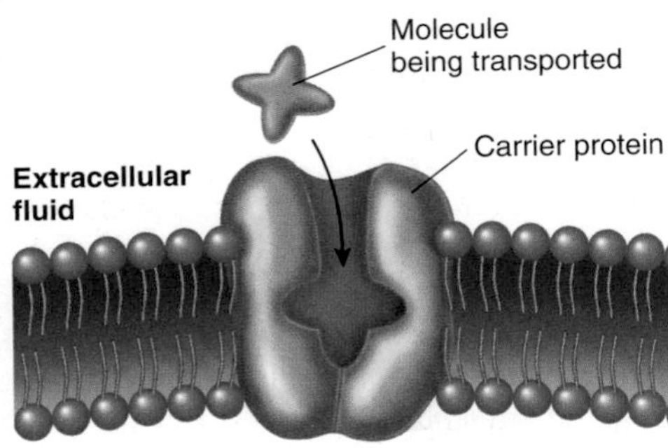

1. A molecule enters the carrier protein from one side of the plasma membrane.

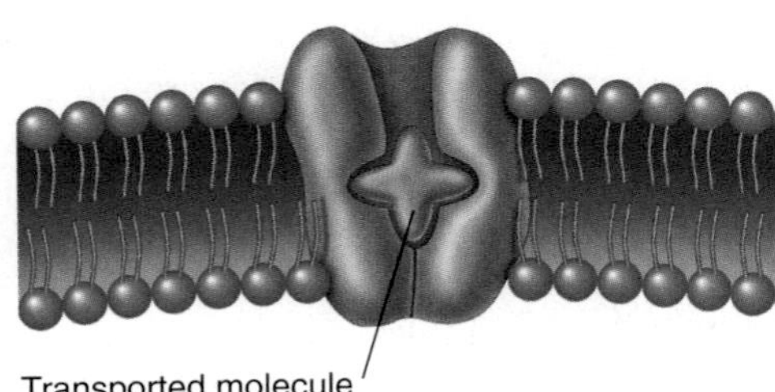

2. The carrier protein briefly binds the transported molecule.

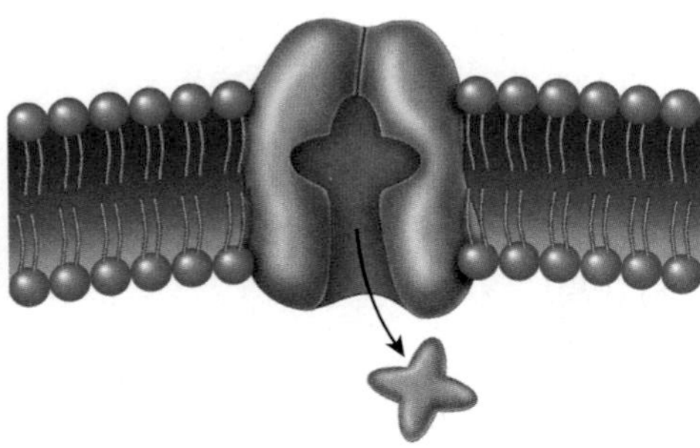

3. The carrier protein changes shape and releases the transported molecule on the other side of the plasma membrane. The carrier protein then changes back to its original shape (go to step 1).

PROCESS FIGURE 3.7 Carrier Protein

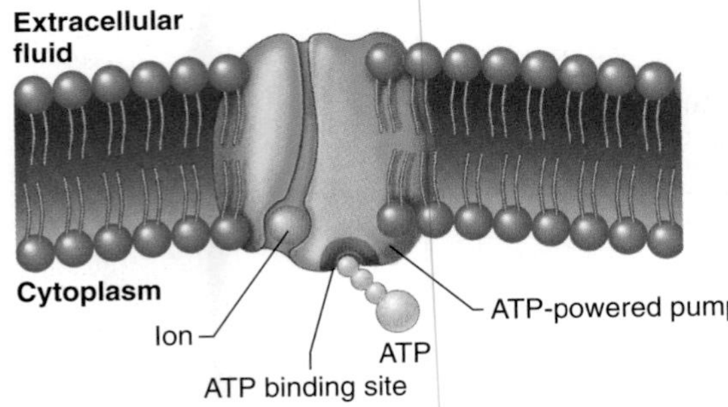

1. ATP and ion bind to the ATP-powered pump.

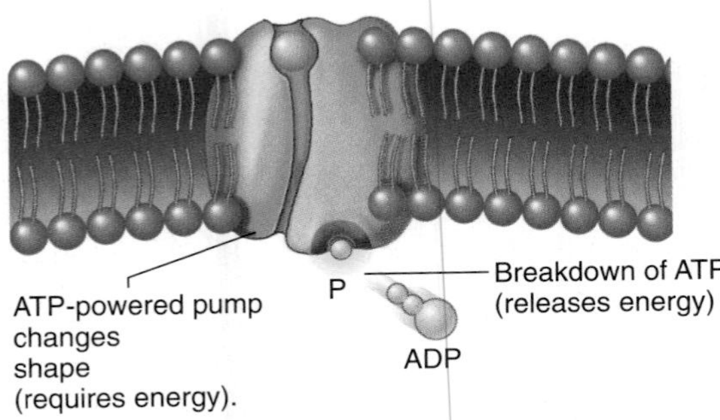

2. The ATP breaks down to adenosine diphosphate (ADP) and a phosphate (P) and releases energy, which powers the shape change in the ATP-powered pump. As a result, the ion moves across the membrane.

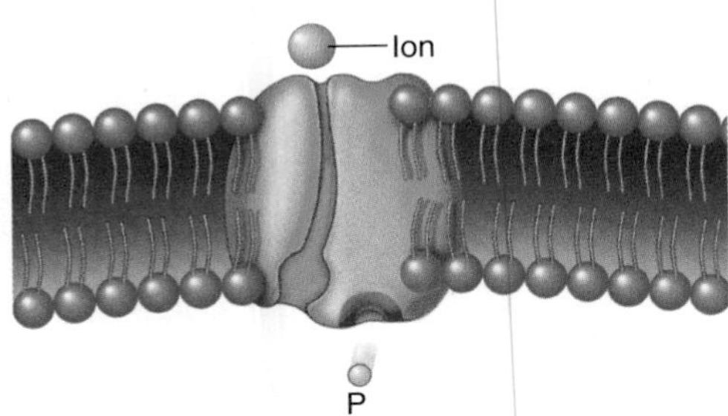

3. The ion and phosphate are released from the ATP-powered pump. The pump resumes its original shape (go to step 1).

PROCESS FIGURE 3.8 ATP-Powered Pump

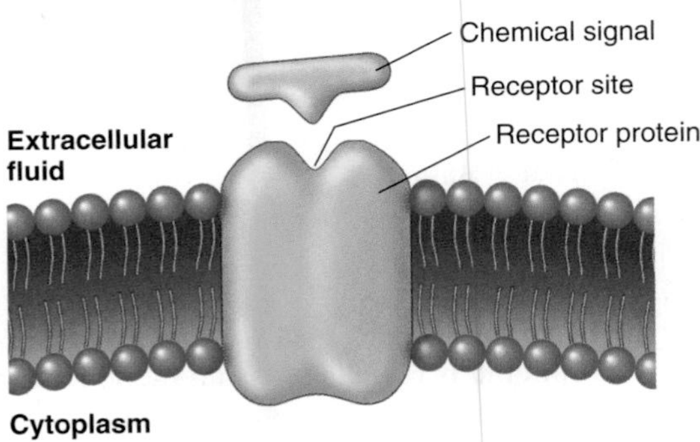

FIGURE 3.9 Receptor Protein

A protein in the plasma membrane with an exposed receptor site, which can attach to specific chemical signals.

Receptors Linked to Channel Proteins

Some membrane-bound receptors also help form ligand-gated ion channels. The ion channels are composed of proteins that span the plasma membrane. Parts of one or more of the channel proteins form receptors on the cell surface. When chemical signals, or ligands, bind to these receptors, the combination alters the three-dimensional structure of the proteins of the ion channels, causing the channels either to open or to close. The result is a change in the permeability of the plasma membrane to the specific ions passing through the ion channels (figure 3.10). For example, acetylcholine released from nerve cells is a chemical signal that combines with membrane-bound receptors of skeletal muscle cells. The combination of acetylcholine molecules with the receptor sites opens Na^+ channels in the plasma membrane. Consequently, the Na^+ diffuse into the skeletal muscle cells and trigger events that cause them to contract.

Cystic Fibrosis

Cystic fibrosis is a genetic disorder that affects chloride ion channels. There are three types of cystic fibrosis. In about 70% of cases, a defective channel protein is produced that fails to reach the plasma membrane from its site of production inside the cell. In the remaining cases, the channel protein is incorporated into the plasma membrane but does not function normally. In some cases, the channel protein fails to bind ATP. In others, ATP binds to the channel protein, but the channel does not open. Failure of these ion channels to function results in the affected cells producing thick, viscous secretions. Although cystic fibrosis affects many cell types, its most profound effects are in the pancreas and in the lungs. In the pancreas, the thick secretions block the release of digestive enzymes, resulting in an inability to digest certain types of food and sometimes leading to serious cases of pancreatitis (inflammation of the pancreas). In the lungs, the thick secretions block airways and make breathing difficult. A more detailed description of cystic fibrosis and its consequences can be found in chapter 23.

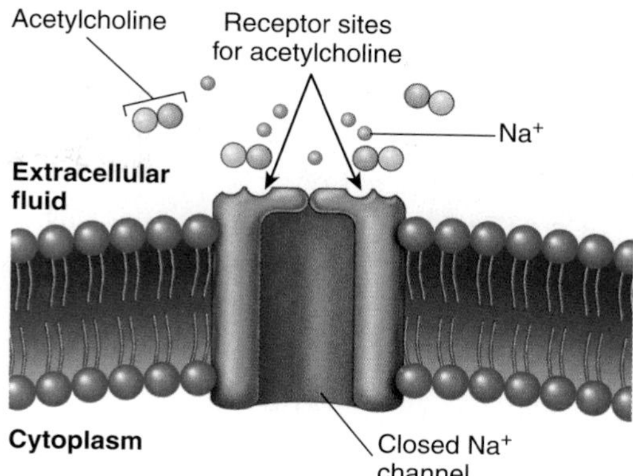

1. The Na^+ channel has receptor sites for the chemical signal, acetylcholine. When the receptor sites are not occupied by acetylcholine, the Na^+ channel remains closed.

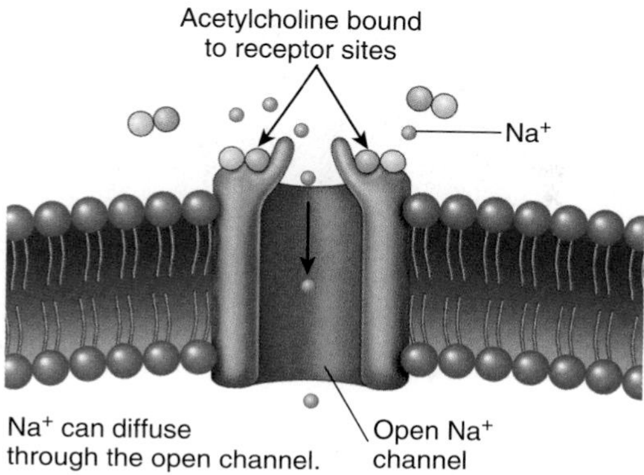

2. When two acetylcholine molecules bind to their receptor sites on the Na^+ channel, the channel opens to allow Na^+ to diffuse through the channel into the cell.

PROCESS FIGURE 3.10 **Receptors Linked to a Channel Protein**

Receptors Linked to G Protein Complexes

Some membrane-bound receptor molecules function by altering the activity of a **G protein complex** located on the inner surface of the plasma membrane (figure 3.11), which acts as an intermediate between a receptor and other cellular proteins. The G protein complex consists of three proteins, called alpha (α), beta (β), and gamma (γ) proteins. The G protein complex will only associate with a receptor that has a chemical signal bound to it. In its unassociated state, the α subunit of the G protein complex has guanosine diphosphate (GDP) attached to it (figure 3.11, step 1). When a chemical signal binds to the receptor, the receptor becomes associated with the G protein complex. The α subunit releases the GDP and attaches to guanosine triphosphate (GTP) (figure 3.11, step 2). The G protein complex separates from the receptor and the α subunit separates from the β and γ subunits (figure 3.11, step 3). The activated α subunit can stimulate a cell response in at least three ways: (1) by means of intracellular chemical signals, (2) by the opening of ion channels in the plasma membrane, and (3) by the activation of enzymes associated with the plasma membrane.

Drugs and Receptors

Drugs with structures similar to specific chemical signals may compete with those chemical signals for their receptor sites. Depending on the exact characteristics of a drug, it may either bind to a receptor site and activate the receptor or bind to a receptor site and inhibit the action of the receptor. For example, some drugs compete with the chemical signal epinephrine for its receptor sites. Some of these drugs activate epinephrine receptors; others inhibit them.

Enzymes

Some membrane proteins function as **enzymes,** which can catalyze chemical reactions on either the inner or the outer surface of the plasma membrane. For example, some enzymes on the surface

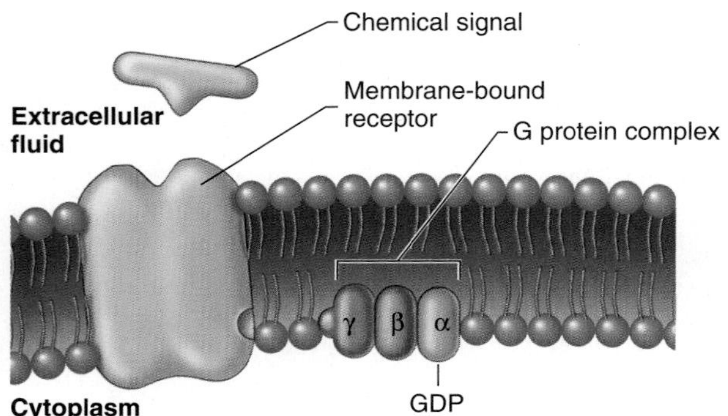

1. A G protein complex will only associate with a receptor that has a chemical signal bound to it. In its unassociated state, the α subunit of the G protein complex has guanosine diphosphate (GDP) attached to it.

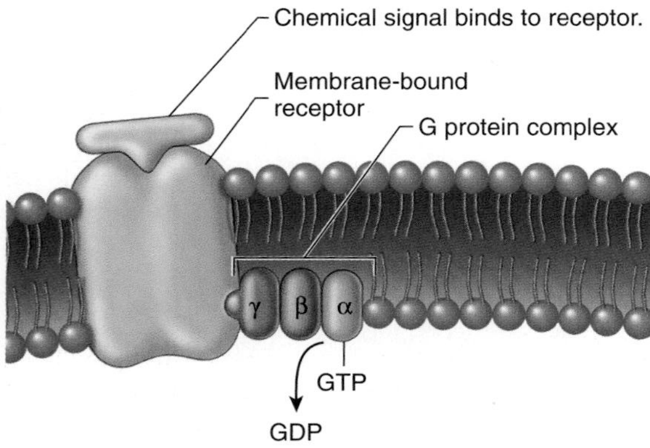

2. When a chemical signal binds to the receptor, the receptor becomes associated with the G protein complex. GDP is released from the α subunit and a guanosine triphosphate (GTP) is attached to it.

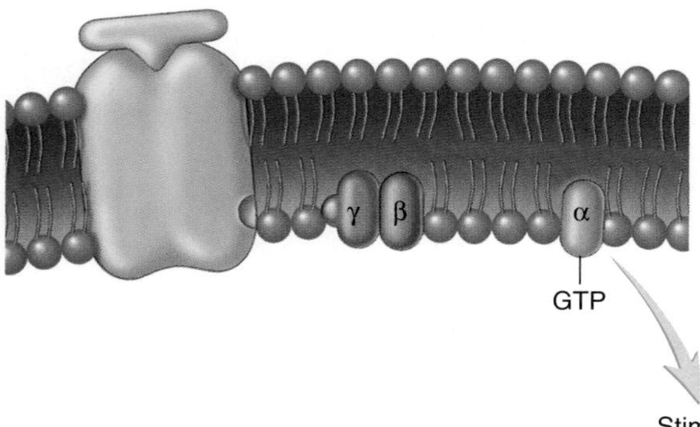

3. The G protein complex separates from the receptor and the α subunit separates from the other subunits. The α subunit stimulates a cell response.

PROCESS FIGURE 3.11 Receptor Linked to a G Protein Complex

of cells in the small intestine break the peptide bonds of dipeptides to form two single amino acids (figure 3.12). Some membrane-associated enzymes are always active. Others are activated by membrane-bound receptors or G protein complexes.

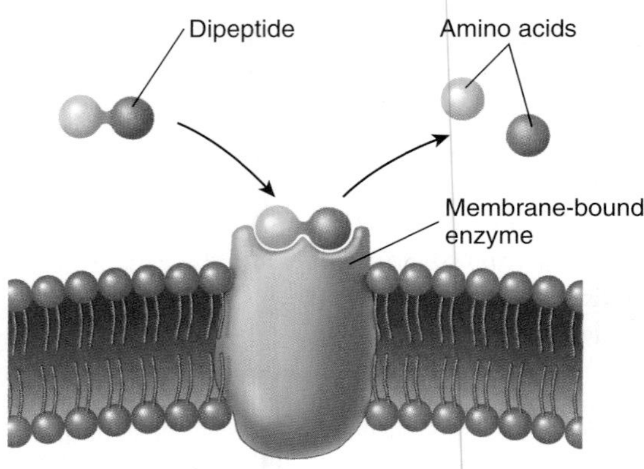

FIGURE 3.12 Enzyme in the Plasma Membrane
This enzyme in the plasma membrane breaks the peptide bond of a dipeptide to produce two amino acids.

8 ***Describe the difference between integral and peripheral proteins in the plasma membrane. Define glycolipid and glycoprotein.***
9 ***List two functions of marker molecules.***
10 ***Describe and give the function of cadherins and integrins.***
11 ***What are the three classes of transport proteins?***
12 ***Define nongated ion channels, ligand-gated ion channels, and voltage-gated ion channels.***
13 ***Describe how carrier proteins and ATP-powered pumps move ions or molecules across the plasma membrane.***
14 ***Define uniport, symport, and antiport.***
15 ***To what part of a receptor molecule does a chemical signal attach? Explain how a chemical signal can bind to a receptor in the plasma membrane and cause a change in membrane permeability.***
16 ***Describe how receptors alter the activity of G protein complexes. List three ways in which activated α subunits can stimulate a cell response.***
17 ***Give an example of the action of an enzyme in the plasma membrane.***

MOVEMENT THROUGH THE PLASMA MEMBRANE

The plasma membrane separates extracellular material from intracellular material and is **selectively permeable**—that is, it allows only certain substances to pass through it. The intracellular material has a different composition than the extracellular material, and the cell's survival depends on the maintenance of these differences. Enzymes, other proteins, glycogen, and potassium ions are found in higher concentrations intracellularly; sodium, calcium, and chloride ions are found in greater concentrations extracellularly.

In addition, nutrients must continually enter the cell, and waste products must exit, but the volume of the cell remains unchanged. Because of the plasma membrane's permeability characteristics and its ability to transport molecules selectively, the cell is able to maintain homeostasis. Rupture of the membrane, alteration of its permeability characteristics, or inhibition of transport processes can disrupt the normal concentration differences across the plasma membrane and lead to cell death.

Molecules and ions can pass through the plasma membrane in four ways:

1. *Directly through the phospholipid membrane.* Molecules that are soluble in lipids, such as oxygen, carbon dioxide, and steroids, pass through the plasma membrane readily by dissolving in the lipid bilayer. The phospholipid bilayer acts as a barrier to most substances that are not lipid-soluble, but some small, nonlipid-soluble molecules, such as carbon dioxide and urea, can diffuse between the phospholipid molecules of the plasma membrane.
2. *Membrane channels.* There are several types of protein channels through the plasma membrane. Each channel type allows only certain molecules to pass through it. The size, shape, and charge of molecules determine whether they can pass through a given channel. For example, sodium ions pass through sodium channels, and potassium and chloride ions pass through potassium and chloride channels, respectively. Rapid movement of water across the cell membrane also occurs through membrane channels.
3. *Transport proteins.* Large polar molecules that are not lipid-soluble, such as glucose and amino acids, cannot pass through the plasma membrane in significant amounts unless they are moved across by transport proteins. Substances that are moved across the plasma membrane by transport proteins are said to be transported by **mediated processes.** Specific molecules bind to specific transport proteins that carry them across the plasma membrane. Transport proteins that move glucose across the plasma membrane do not move amino acids, and transport proteins that move amino acids across the plasma membrane do not move glucose.
4. *Vesicles.* Large nonlipid-soluble molecules, small pieces of matter, and even whole cells can be transported across the plasma membrane in a **vesicle,** which is a small sac surrounded by a membrane. Because of the fluid nature of membranes, the vesicle and the plasma membrane can fuse, allowing the contents of the vesicle to cross the plasma membrane.

18 ***List four ways that substances move across the plasma membrane.***

Diffusion

A solution consists of one or more substances called **solutes** dissolved in the predominant liquid or gas, which is called the **solvent. Diffusion** is the movement of solutes from an area of higher solute concentration to an area of lower solute concentration (figure 3.13). Diffusion is a product of the constant random motion of all atoms, molecules, or ions in a solution. Because there are more solute particles in an area of higher concentration than in an area of lower concentration and because the particles move randomly, the chances are greater that solute particles will move from the higher to the lower concentration than in the opposite direction. Thus, the overall, or net, movement is from the area of higher solute concentration to that of lower solute concentration. At equilibrium, the net movement of solutes stops, although the random molecular motion continues, and the movement of solutes in any one direction is balanced by an equal movement in the opposite direction. The movement and distribution of smoke or perfume throughout a room without air currents and of a dye throughout a beaker of still water are examples of diffusion.

A concentration difference occurs when the concentration of a solute is greater at one point than at another point in a solvent. The concentration difference between two points, divided by the distance

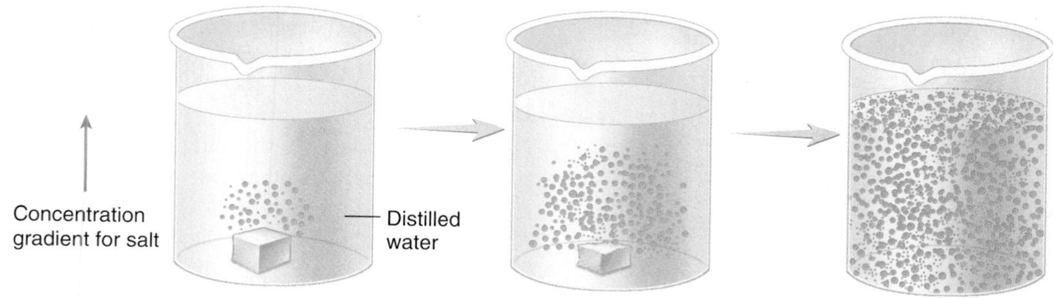

PROCESS FIGURE 3.13 Diffusion

between the two points, is called the **concentration,** or **density, gradient.** Solutes diffuse down their concentration gradients (from a higher to a lower solute concentration) until an equilibrium is achieved. The greater the concentration gradient, the greater the rate of diffusion of a solute down that gradient. Increasing the concentration difference or decreasing the distance between the two points increases the concentration gradient, whereas decreasing the concentration difference or increasing the distance between the two points decreases the concentration gradient.

The rate of diffusion is influenced by the magnitude of the concentration gradient, the temperature of the solution, the size of the diffusing molecules, and the viscosity of the solvent. The greater the concentration gradient, the greater the number of solute particles moving from a higher to a lower solute concentration. As the temperature of a solution increases, the speed at which all molecules move increases, resulting in a greater diffusion rate. Small molecules diffuse through a solution more readily than do large ones. **Viscosity** is a measure of how easily a liquid flows; thick solutions, such as syrup, are more viscous than water. Diffusion occurs more slowly in viscous solvents than in thin, watery solvents.

Diffusion of molecules is an important means by which substances move between the extracellular and intracellular fluids in the body. Substances that can diffuse through either the lipid bilayer or the membrane channels can pass through the plasma membrane (figure 3.14). Some nutrients enter and some waste products leave the cell by diffusion, and maintenance of the appropriate intracellular concentration of these substances depends to a large degree on diffusion. For example, if the extracellular concentration of oxygen is reduced, inadequate oxygen diffuses into the cell, and normal cell function cannot occur. Some lipid-soluble chemical signals can diffuse through the plasma membrane and attach to receptors inside the cell (figure 3.15).

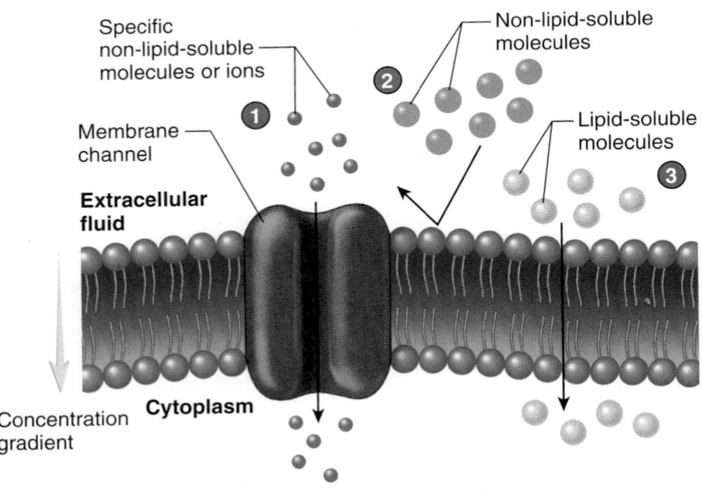

PROCESS FIGURE 3.14 Diffusion Through the Plasma Membrane

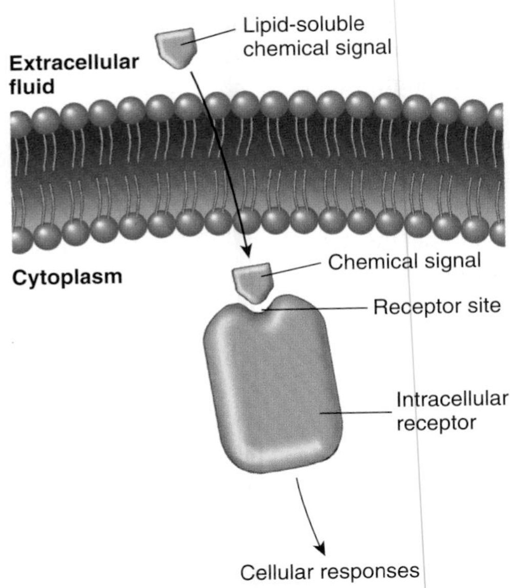

FIGURE 3.15 Intracellular Receptor
A small, lipid-soluble chemical signal diffuses through the plasma membrane and combines with the receptor site of an intracellular receptor.

19 ***Define solute, solvent, and concentration gradient. Do solutes diffuse with (down) or against their concentration gradient?***

20 ***How is the rate of diffusion affected by an increased concentration gradient? By increased temperature of a solution? By increased viscosity of the solvent?***

PREDICT 1

Urea is a toxic waste produced inside cells. It diffuses from the cells into the blood and is eliminated from the body by the kidneys. What would happen to the intracellular and extracellular concentration of urea if the kidneys stopped functioning?

Osmosis

Osmosis (os-mō′sis) is the diffusion of water (solvent) across a selectively permeable membrane, such as a plasma membrane. A selectively permeable membrane is a membrane that allows water but not all the solutes dissolved in the water to diffuse through the membrane. **Aquaporins,** or water channel proteins, increase membrane permeability to water in some cell types, such as kidney cells. Water diffuses from a solution with proportionately more water, across a selectively permeable membrane, and into a solution with proportionately less water. Because solution concentrations are defined in terms of solute concentrations, not in terms of water content (see chapter 2), water diffuses from the less concentrated solution (fewer solutes, more water) into the more concentrated solution (more solutes, less water). Osmosis is important to cells because large volume changes caused by water movement disrupt normal cell function.

Osmotic pressure is the force required to prevent the movement of water by osmosis across a selectively permeable membrane.

The osmotic pressure of a solution can be determined by placing the solution into a tube that is closed at one end by a selectively permeable membrane (figure 3.16). The tube is then immersed in distilled water. Water molecules move by osmosis through the membrane into the tube, forcing the solution to move up the tube. As the solution rises into the tube, its weight produces hydrostatic pressure, which moves water out of the tube back into the distilled water surrounding the tube. At equilibrium, net movement of water stops, which means the movement of water into the tube by osmosis is equal to the movement of water out of the tube caused by hydrostatic pressure. The osmotic pressure of the solution in the tube is equal to the hydrostatic pressure that prevents net movement of water into the tube.

The osmotic pressure of a solution provides information about the tendency for water to move by osmosis across a selectively permeable membrane. Because water moves from less concentrated solutions (fewer solutes, more water) into more concentrated solutions (more solutes, less water), the

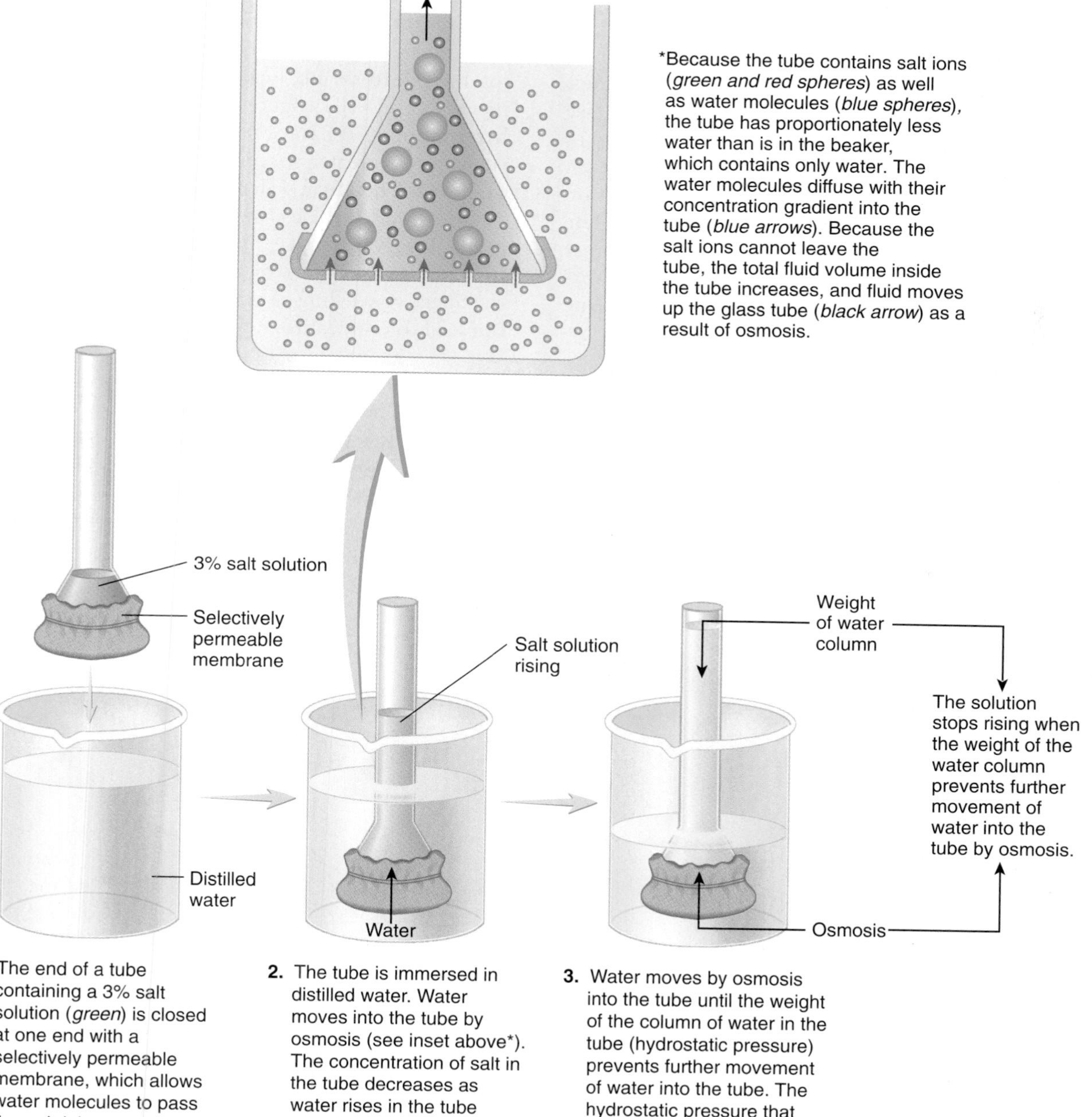

PROCESS FIGURE 3.16 Osmosis

greater the concentration of a solution (the less water it has), the greater the tendency for water to move into the solution, and the greater the osmotic pressure to prevent that movement. Thus, the greater the concentration of a solution, the greater the osmotic pressure of the solution, and the greater the tendency for water to move into the solution.

PREDICT 2

Given the demonstration in figure 3.16, what would happen to osmotic pressure if the membrane were not selectively permeable but instead allowed all solutes and water to pass through it?

Three terms describe the osmotic pressure of solutions. Solutions with the same concentration of solute particles (see chapter 2) have the same osmotic pressure and are referred to as **isosmotic** (ī'sos-mot'ik). The solutions are still isosmotic even if the types of solute particles in the two solutions differ from each other. If one solution has a greater concentration of solute particles, and therefore a greater osmotic pressure than another solution, the first solution is said to be **hyperosmotic** (hī'per-oz-mot'ik) compared with the more dilute solution. The more dilute solution, with the lower osmotic pressure, is **hyposmotic** (hī-pos-mot'ik) compared with the more concentrated solution.

Three additional terms describe the tendency of cells to shrink or swell when placed into a solution. If a cell is placed into a solution in which it neither shrinks nor swells, the solution is said to be **isotonic** (ī-sō-ton'ik). In an isotonic solution, the shape of the cell remains constant, maintaining its internal tension or tone, a condition called **tonicity** (tō-nis'i-tē). If a cell is placed into a solution and water moves out of the cell by osmosis, causing the cell to shrink, the solution is called **hypertonic** (hī-per-ton'ik). If a cell is placed into a solution and water moves into the cell by osmosis, causing the cell to swell, the solution is called **hypotonic** (hī-pō-ton'ik; figure 3.17*a*).

An isotonic solution may be isosmotic to the cytoplasm. Because isosmotic solutions have the same concentration of solutes and water as the cytoplasm of the cell, no net movement of water occurs, and the cell neither swells nor shrinks (figure 3.17*b*). Hypertonic solutions can be hyperosmotic and have a greater concentration of solute molecules and a lower concentration of water than the cytoplasm of the cell. Therefore, water moves by osmosis from the cell into the hypertonic solution, causing the cell to shrink, a process called **crenation** (krē-nā'shŭn) in red blood cells (figure 3.17*c*). Hypotonic solutions can be hyposmotic and have a smaller concentration of solute molecules and a greater concentration of water than the cytoplasm of the cell. Therefore, water moves by osmosis into the cell, causing it to swell. If the cell swells enough, it can rupture, a process called **lysis** (lī'sis; see figure 3.17*a*). Solutions injected into the circulatory system or the tissues must be isotonic because shrinkage or swelling of cells disrupts their normal function and can lead to cell death.

The *-osmotic* terms refer to the concentration of the solutions, and the *-tonic* terms refer to the tendency of cells to swell or shrink. These terms should not be used interchangeably. Not all isosmotic solutions are isotonic. For example, it is possible to prepare a solution of glycerol and a solution of mannitol that are isosmotic to the cytoplasm of the cell. Because the solutions are isosmotic, they have the same concentration of solutes and

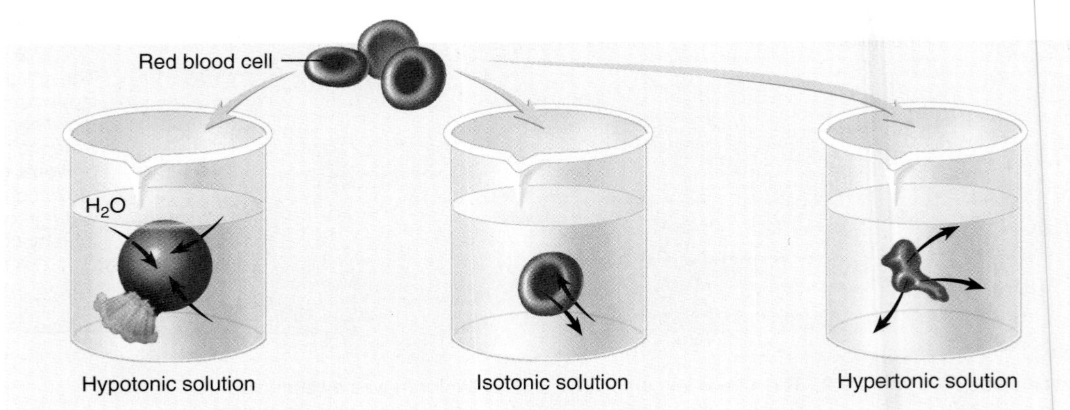

(a) A hypotonic solution with a low solute concentration results in swelling of the red blood cell placed into the solution. Water enters the cell by osmosis (*black arrows*), and the red blood cell lyses (bursts; *puff of red in the lower part of the cell*).

(b) An isotonic solution with a concentration of solutes equal to that inside the cell results in a normally shaped red blood cell. Water moves into and out of the cell at the same rate (*black arrows*), but there is no net water movement.

(c) A hypertonic solution, with a high solute concentration, causes shrinkage (crenation) of the red blood cell as water moves by osmosis out of the cell and into the hypertonic solution (*black arrows*).

FIGURE 3.17 Effects of Hypotonic, Isotonic, and Hypertonic Solutions on Red Blood Cells

water as the cytoplasm. Glycerol, however, can diffuse across the plasma membrane, but mannitol cannot. When glycerol diffuses into the cell, the solute concentration of the cytoplasm increases, and its water concentration decreases. Therefore, water moves by osmosis into the cell, causing it to swell, and the glycerol solution is both isosmotic and hypotonic. In contrast, mannitol cannot enter the cell, and the isosmotic mannitol solution is also isotonic.

21 ***Define osmosis and osmotic pressure. As the concentration of a solution increases, what happens to its osmotic pressure and to the tendency for water to move into it?***

22 ***Compare isosmotic, hyperosmotic, and hyposmotic solutions with isotonic, hypertonic, and hypotonic solutions. What type of solution causes crenation of a cell? What type of solution causes lysis of a cell?***

Filtration

Filtration results when a partition containing small holes is placed in a stream of moving liquid. The partition works as a minute sieve. Particles small enough to pass through the holes move through the partition with the liquid, but particles larger than the holes are prevented from moving beyond the partition. In contrast to diffusion, filtration depends on a pressure difference on either side of the partition. The liquid moves from the side of the partition with the greater pressure to the side with the lower pressure.

Filtration occurs in the kidneys as a step in urine formation. Blood pressure moves fluid from the blood through a partition, or filtration membrane. Water, ions, and small molecules pass through the partition, whereas most proteins and blood cells remain in the blood.

23 ***Define filtration and give an example of where it occurs in the body.***

Mediated Transport

Many essential molecules, such as amino acids and glucose, cannot enter the cell by diffusion, and many products, such as proteins, cannot exit the cell by diffusion. **Mediated transport** is the process by which transport proteins mediate, or assist in, the movement of large, water-soluble molecules or electrically charged molecules or ions across the plasma membrane. All three types of transport proteins—carrier proteins (transporters), ATP-powered pumps, and channel proteins (ion channels)—are involved in mediated transport.

Mediated transport has three characteristics: specificity, competition, and saturation. **Specificity** means that each transport protein binds to and transports only a single type of molecule or ion (figure 3.18*a*). For example, the transport protein that moves glucose does not move amino acids or ions. The chemical structure of the binding site determines the specificity of the transport protein (see figure 3.7). **Competition** is the result of similar molecules binding to the transport protein. Although

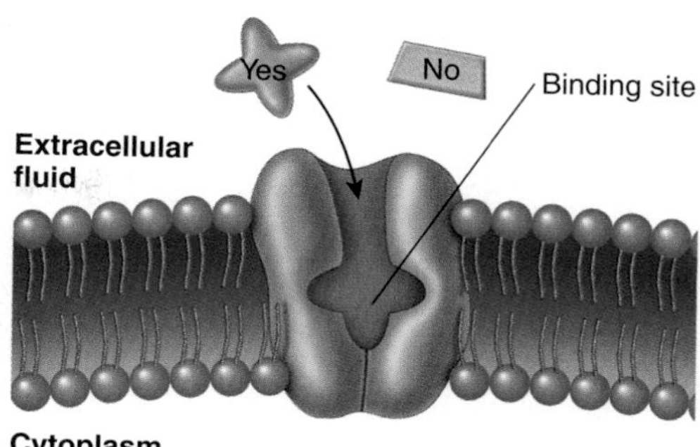

(a) Specificity. Only molecules that are the right shape to bind to the binding site are transported.

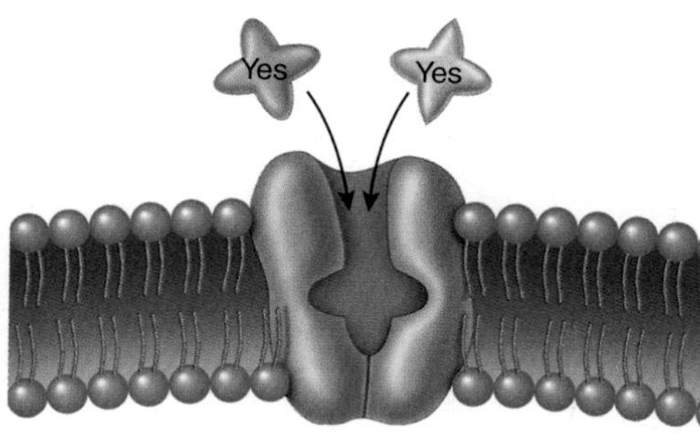

(b) Competition. Similarly shaped molecules can compete for the same binding site.

FIGURE 3.18 Mediated Transport: Specificity and Competition

the binding sites of transport proteins exhibit specificity, closely related substances, in which regions of two different molecules have the same shape, may bind to the same binding site. The substance in the greater concentration or the substance that binds to the binding site more readily is moved across the plasma membrane at the greater rate (figure 3.18*b*). **Saturation** means that the rate of movement of molecules across the membrane is limited by the number of available transport proteins. As the concentration of a transported substance increases, more transport proteins have their binding sites occupied. The rate at which the substance is moved across the plasma membrane increases; however, once the concentration of the substance is increased so that all the binding sites are occupied, the rate of movement remains constant, even though the concentration of the substance increases further (figure 3.19).

Ion channels often are thought of as simple tubes, with or without gates, through which ions pass. Many ion channels, however, appear to be more complex than once thought. It now appears that ions briefly bind to specific sites inside channels, and that there is a change in the shape of those channels as ions are transported through them. The size and charge within a channel determine the channel's specificity. For example, Na^+ channels do not transport K^+ and vice versa. In addition, similar ions moving into and binding within an ion channel are in competition with each other. Furthermore, the number of ions moving into an ion

The rate of transport of molecules into a cell is plotted against the concentration of those molecules outside the cell minus the concentration of those molecules inside the cell. As the concentration difference increases, the rate of transport increases and then levels off.

Rate of molecule transport

Molecule concentration difference across the plasma membrane

Extracellular fluid

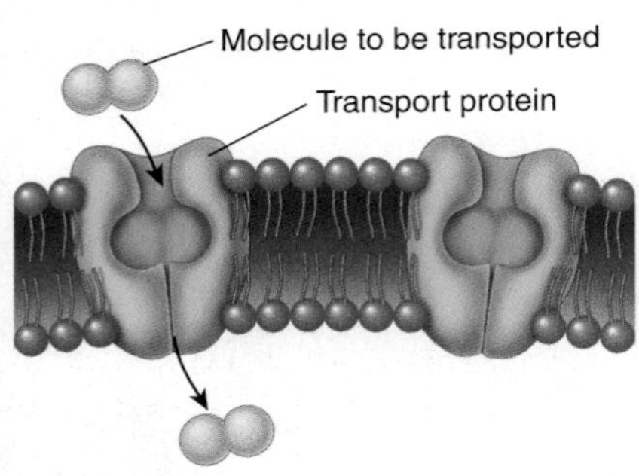

Cytoplasm

1. When the concentration of molecules outside the cell is low, the transport rate is low because it is limited by the number of molecules available to be transported.

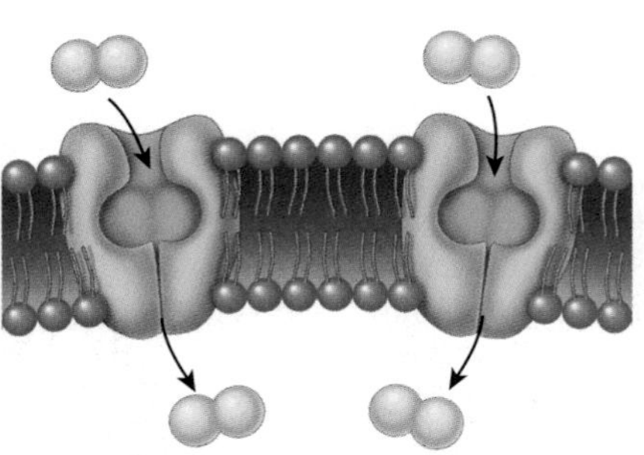

2. When more molecules are present outside the cell, as long as enough transport proteins are available, more molecules can be transported, and therefore the transport rate increases.

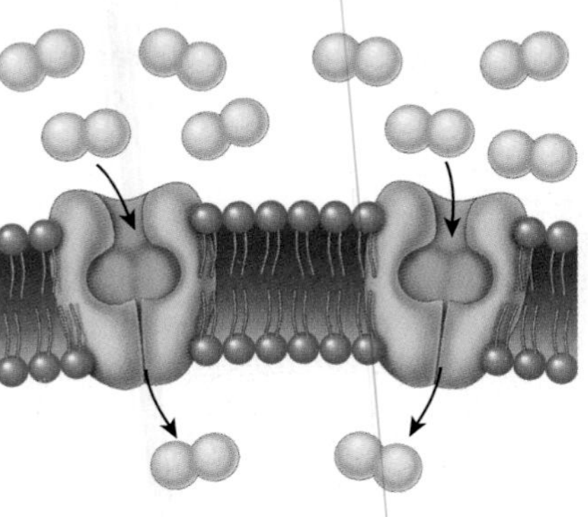

3. The transport rate is limited by the number of transport proteins and the rate at which each transport protein can transport solutes. When the number of molecules outside the cell is so large that the transport proteins are all occupied, the system is saturated and the transport rate cannot increase.

PROCESS FIGURE 3.19 Saturation of a Transport Protein

channel can exceed the capacity of the channel, thus saturating the channel. Therefore, ion channels exhibit specificity, competition, and saturation.

Three kinds of mediated transport exist: facilitated diffusion, active transport, and secondary active transport.

Facilitated Diffusion

Facilitated diffusion is a carrier-mediated or channel-mediated process that moves substances into or out of cells from a higher to a lower concentration (figure 3.20). Facilitated diffusion does not require metabolic energy to transport substances across the plasma membrane. The rate at which molecules or ions are transported is directly proportional to their concentration gradient up to the point of saturation, when all the carrier proteins or channels are occupied. Then the rate of transport remains constant at its maximum rate.

PREDICT 3

The transport of glucose into and out of most cells, such as muscle and fat cells, occurs by facilitated diffusion. Once glucose enters a cell, it is rapidly converted to other molecules, such as glucose-6-phosphate or glycogen. What effect does this conversion have on the cell's ability to acquire glucose? Explain.

Active Transport

Active transport is a mediated transport process that requires energy provided by ATP (figure 3.21). Movement of the transported substance to the opposite side of the membrane and its subsequent release from the ATP-powered pump are fueled by the breakdown of ATP. The maximum rate at which active transport proceeds depends on the number of ATP-powered pumps in the plasma membrane and the availability of adequate ATP. Active-transport processes are important because they can move

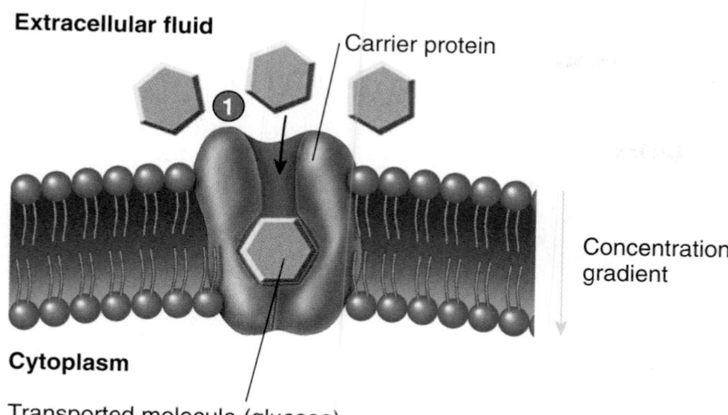

1. The carrier protein binds with a molecule, such as glucose, on the outside of the plasma membrane.

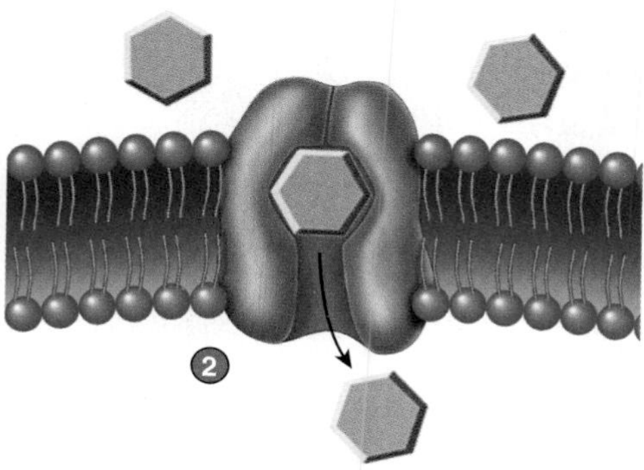

2. The carrier protein changes shape and releases the molecule on the inside of the plasma membrane.

PROCESS FIGURE 3.20 Facilitated Diffusion

substances against their concentration gradients—that is, from lower concentrations to higher concentrations. Consequently, they can accumulate substances on one side of the plasma membrane at concentrations many times greater than those on the other side. Active transport can also move substances from higher to lower concentrations.

Some active-transport mechanisms exchange one substance for another. For example, the **sodium–potassium (Na^+–K^+) pump** moves Na^+ out of cells and K^+ into cells (see figure 3.21). The result is a higher concentration of Na^+ outside the cell and a higher concentration of K^+ inside the cell. Because ATP is broken down during the transport of Na^+ and K^+, the pump is also called **sodium-potassium ATP-ase.** The Na^+–K^+ pump is very important to a number of cell functions. These are discussed in chapters 9 and 11.

Secondary Active Transport

Secondary active transport involves the active transport of an ion, such as sodium, out of a cell, establishing a concentration gradient, with a higher concentration of the ions outside the cell. The tendency for the ions to move back into the cell, down their concentration gradient, provides the energy necessary to move a different ion or some other molecule into the cell. For example, glucose is moved from the lumen of the intestine into epithelial cells by secondary active transport (figure 3.22). This process requires two transport proteins: (1) a Na^+–K^+ pump actively moves Na^+ out of the cell, and (2) a carrier protein facilitates the movement of Na^+ and glucose into the cell. Both Na^+ and glucose are necessary for the carrier protein to function.

The movement of Na^+ down their concentration gradient provides the energy to move glucose molecules into the cell against their concentration gradient. Thus, glucose can accumulate at concentrations higher inside the cell than outside. Because the movement of glucose molecules against their concentration gradient results from the formation of a concentration gradient of Na^+ by an active-transport mechanism, the process is called secondary active transport.

24 ***What is mediated transport? What types of molecules are moved through the plasma membrane by mediated transport?***

25 ***Describe specificity, competition, and saturation as characteristics of mediated transport mechanisms.***

26 ***Contrast facilitated diffusion and active transport in relation to energy expenditure and movement of substances with or against their concentration gradients.***

27 ***What is secondary active transport?***

PREDICT 4

In cardiac (heart) muscle cells, the force of contraction increases as the intracellular Ca^{2+} concentration increases. Intracellular Ca^{2+} concentration is regulated in part by secondary active transport involving a Na^+–Ca^{2+} antiporter. The movement of Na^+ down their concentration gradient into the cell provides the energy to transport Ca^{2+} out of the cell against their concentration gradient. Digitalis, a drug often used to treat congestive heart failure, slows the active transport of Na^+ out of the cell by the Na^+–K^+ pump, thereby increasing intracellular Na^+ concentration. Should the heart beat more or less forcefully when exposed to this drug? Explain.

ENDOCYTOSIS AND EXOCYTOSIS

Endocytosis (en′dō-sī-tō′sis), or the internalization of substances, includes both phagocytosis and pinocytosis and refers to the uptake of material through the plasma membrane by the formation of a vesicle. A **vesicle** is a membrane-bound sac found within the cytoplasm of a cell. A portion of the plasma membrane wraps around a particle or droplet and fuses so that the particle or droplet is surrounded by a membrane. That portion of the membrane then "pinches off" so that the particle or droplet, surrounded by a membrane, is within the cytoplasm of the cell, and the plasma membrane is left intact.

Phagocytosis (fāg-ō-sī-tō′sis) literally means cell-eating (figure 3.23) and applies to endocytosis when solid particles are ingested and phagocytic vesicles are formed. White blood cells and some other cell types phagocytize bacteria, cell debris, and foreign

1. Three sodium ions (Na^+) and adenosine triphosphate (ATP) bind to the Na^+–K^+ pump .
2. The ATP breaks down to adenosine diphosphate (ADP) and a phosphate (P) and releases energy. That energy is used to power a shape change in the Na^+–K^+ pump. Phosphate remains bound to the Na^+–K^+–ATP binding site.
3. The Na^+–K^+ pump changes shape, and the Na^+ are transported across the membrane.
4. The Na^+ diffuse away from the Na^+–K^+ pump.
5. Two potassium ions (K^+) bind to the Na^+–K^+ pump.
6. The phosphate is released from the Na^+–K^+ pump binding site.
7. The Na^+–K^+ pump resumes its original shape, transporting K^+ across the membrane, and the K^+ diffuse away from the pump. The Na^+–K^+ pump can again bind to Na^+ and ATP.

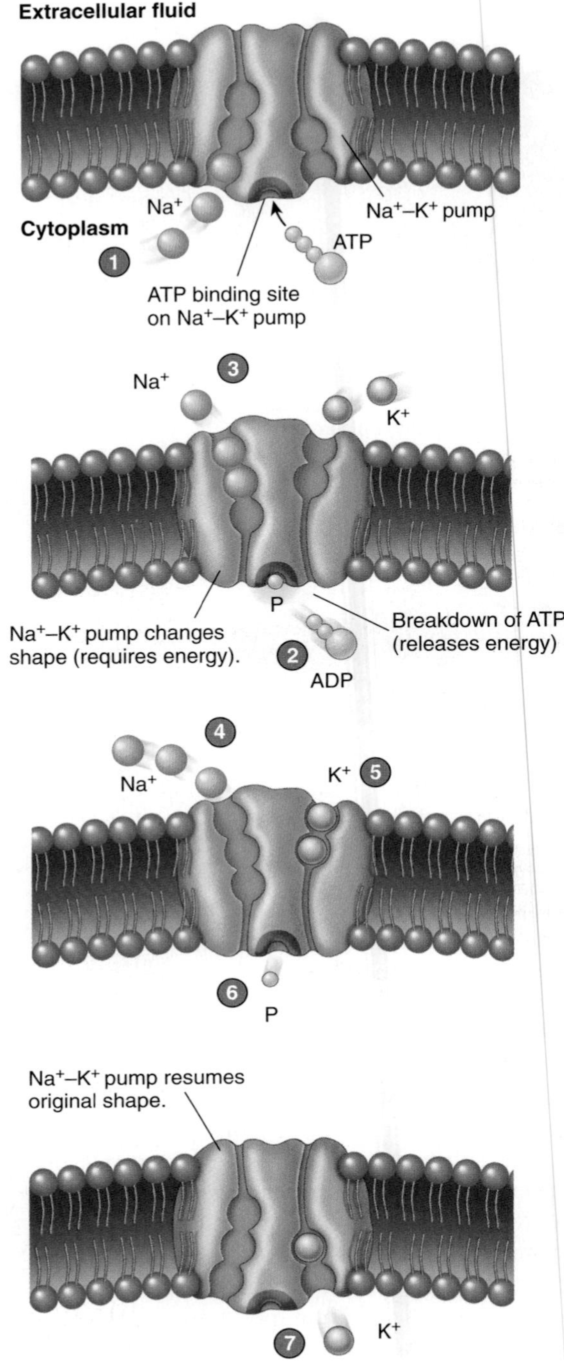

PROCESS FIGURE 3.21 Sodium–Potassium Pump

particles. Phagocytosis is therefore important in the elimination of harmful substances from the body.

Pinocytosis (pin′ō-sī-tō′sis) means cell-drinking and is distinguished from phagocytosis in that smaller vesicles are formed, and they contain molecules dissolved in liquid rather than particles (figure 3.24). Pinocytosis often forms vesicles near the tips of deep invaginations of the plasma membrane. It is a common transport phenomenon in a variety of cell types and occurs in certain cells of the kidneys, epithelial cells of the intestines, cells of the liver, and cells that line capillaries.

Endocytosis can exhibit specificity. For example, cells that phagocytize bacteria and necrotic tissue do not phagocytize healthy cells. The plasma membrane may contain specific receptor molecules that recognize certain substances and allow them to be

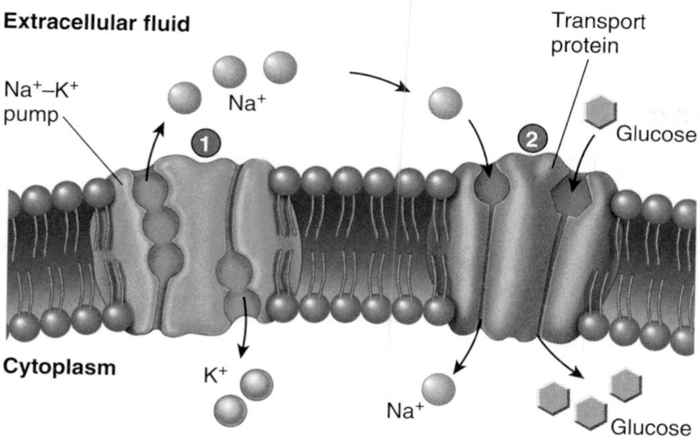

1. A Na^+–K^+ pump maintains a concentration of Na^+ that is higher outside the cell than inside.
2. Sodium ions move back into the cell through a transport protein that also moves glucose. The concentration gradient for Na^+ provides energy required to move glucose against its concentration gradient.

PROCESS FIGURE 3.22 Secondary Active Transport (Symport) of Na^+ and Glucose

transported into the cell by phagocytosis or pinocytosis. This is called **receptor-mediated endocytosis,** and the receptor sites combine only with certain molecules (figure 3.25). This mechanism increases the rate at which specific substances are taken up by the cells. Cholesterol and growth factors are examples of molecules that can be taken into a cell by receptor-mediated endocytosis. Both phagocytosis and pinocytosis require energy in the form of ATP and therefore are active processes. Because they involve the bulk movement of material into the cell, however, phagocytosis and pinocytosis do not exhibit the degree of specificity or saturation that active transport exhibits.

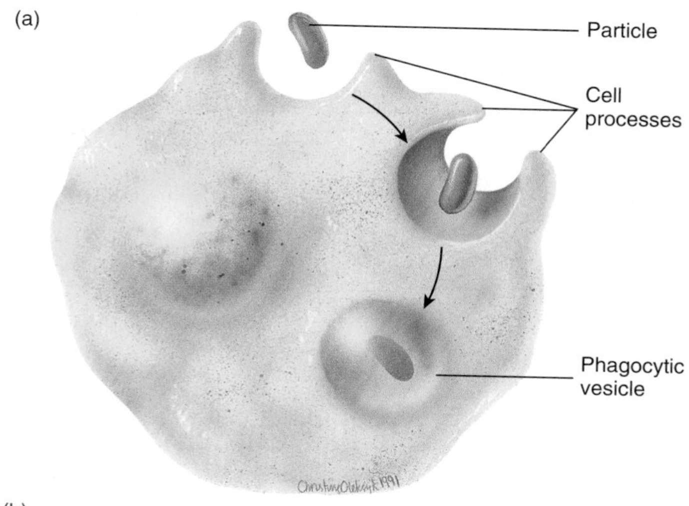

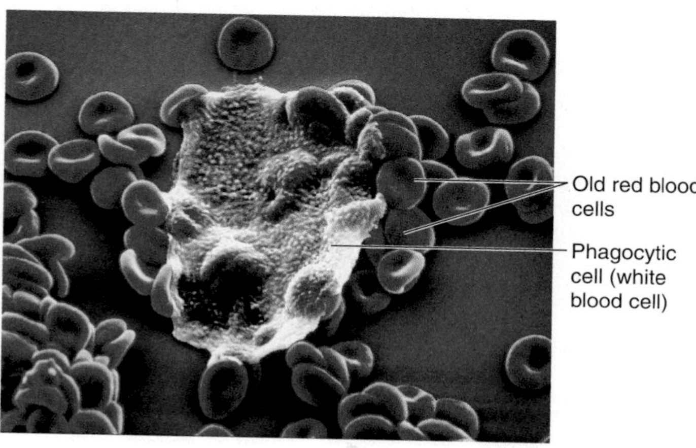

FIGURE 3.23 Endocytosis

(*a*) Phagocytosis. (*b*) Scanning electron micrograph of phagocytosis of red blood cells.

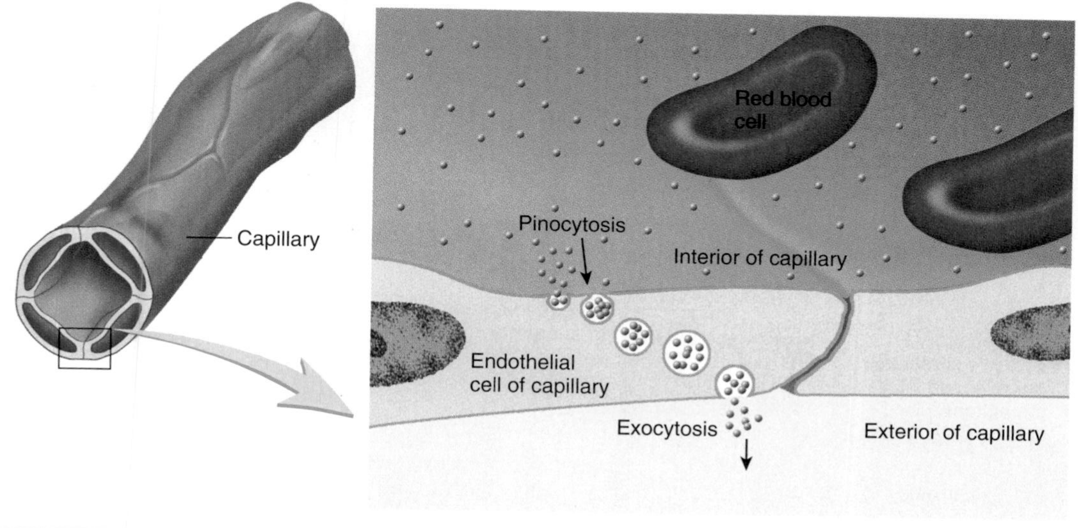

FIGURE 3.24 Pinocytosis

Pinocytosis is much like phagocytosis, except that the cell processes and therefore the vesicles formed are much smaller and the material inside the vesicle is liquid rather than particulate. Pinocytotic vesicles form on the internal side of a capillary, are transported across the cell, and open by exocytosis outside the capillary.

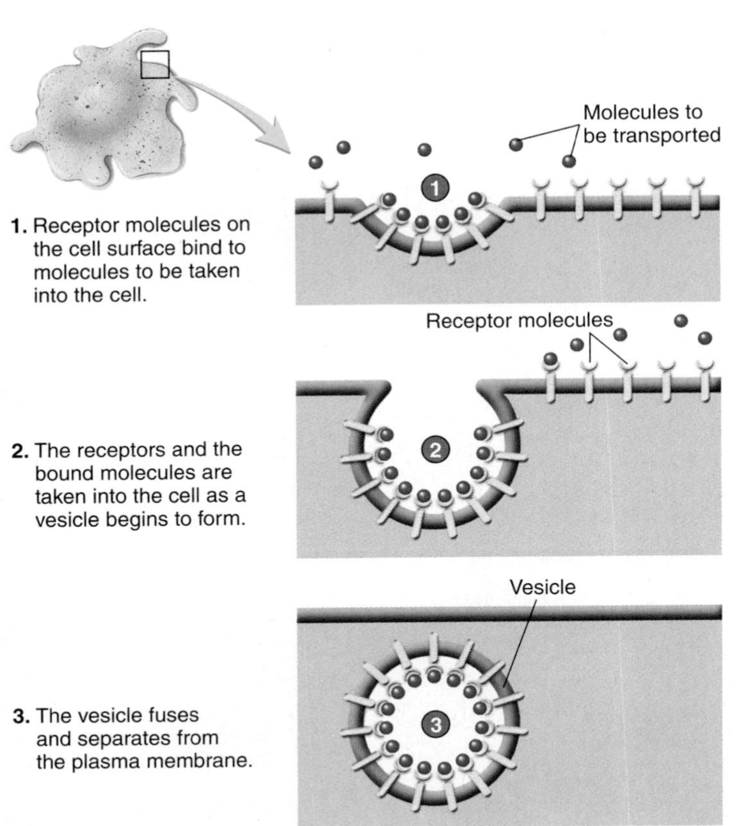

PROCESS FIGURE 3.25 Receptor-Mediated Endocytosis

Hypercholesterolemia

Hypercholesterolemia is a common genetic disorder affecting 1 in every 500 adults in the United States. It consists of a reduction in or absence of low-density lipoprotein (LDL) receptors on cell surfaces. This interferes with receptor-mediated endocytosis of LDL cholesterol. As a result of inadequate cholesterol uptake, cholesterol synthesis within these cells is not regulated, and too much cholesterol is produced. The excess cholesterol accumulates in blood vessels, resulting in atherosclerosis. Atherosclerosis can result in heart attacks or strokes. A more detailed description of hypercholesterolemia can be found in chapter 24.

In some cells, secretions accumulate within vesicles. These secretory vesicles then move to the plasma membrane, where the membrane of the vesicle fuses with the plasma membrane and the content of the vesicle is expelled from the cell. This process is called **exocytosis** (ek′sō-sī-tō′sis; figure 3.26). Secretion of digestive enzymes by the pancreas and secretion of mucus by the salivary glands are examples of exocytosis.

Table 3.3 summarizes and compares the mechanisms by which different kinds of molecules are transported across the plasma membrane.

28 ***Define endocytosis and vesicle. How do phagocytosis and pinocytosis differ from each other?***

29 ***What is receptor-mediated endocytosis?***

30 ***Describe and give examples of exocytosis.***

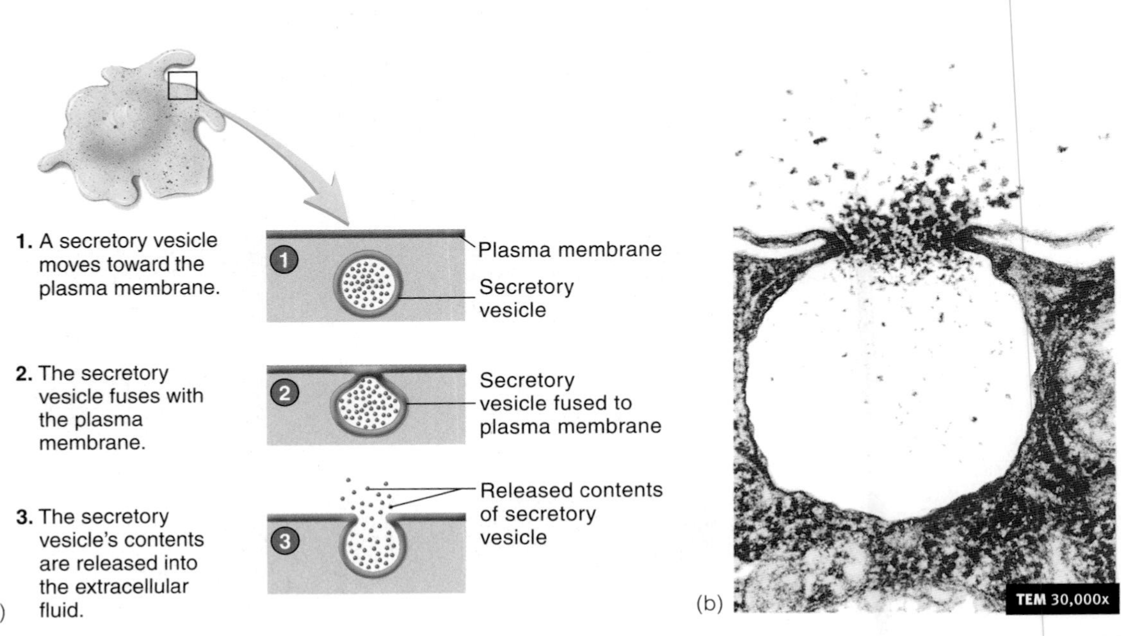

PROCESS FIGURE 3.26 Exocytosis

(*a*) Diagram of exocytosis. (*b*) Transmission electron micrograph of exocytosis.

TABLE 3.3 Comparison of Membrane Transport Mechanisms

Transport Mechanism	Description	Substances Transported	Example
Diffusion	Random movement of molecules results in net movement from areas of higher to lower concentration.	Lipid-soluble molecules dissolve in the lipid bilayer and diffuse through it; ions and small molecules diffuse through membrane channels.	Oxygen, carbon dioxide, and lipids, such as steroid hormones, dissolve in the lipid bilayer; Cl^- and urea move through membrane channels.
Osmosis	Water diffuses across a selectively permeable membrane.	Water diffuses through the lipid bilayer.	Water moves from the intestines into the blood.
Filtration	Liquid moves through a partition that allows some, but not all, of the substances in the liquid to pass through it; movement is due to a pressure difference across the partition.	Liquid and substances pass through holes in the partition.	Filtration in the kidneys allows removal of everything from the blood except proteins and blood cells.
Facilitated diffusion	Carrier proteins combine with substances and move them across the plasma membrane; no ATP is used; substances are always moved from areas of higher to lower concentration; it exhibits the characteristics of specificity, saturation, and competition.	Some substances too large to pass through membrane channels and too polar to dissolve in the lipid bilayer are transported.	Glucose moves by facilitated diffusion into muscle cells and fat cells.
Active transport	ATP-powered pumps combine with substances and move them across the plasma membrane; ATP is used; substances can be moved from areas of lower to higher concentration; it exhibits the characteristics of specificity, saturation, and competition.	Substances too large to pass through channels and too polar to dissolve in the lipid bilayer are transported; substances that are accumulated in concentrations higher on one side of the membrane than on the other are transported.	Ions, such as Na^+, K^+, and Ca^{2+}, are actively transported.
Secondary active transport	Ions are moved across the plasma membrane by active transport, which establishes an ion concentration gradient; ATP is required; ions then move back down their concentration gradient by facilitated diffusion, and another ion or molecule moves with the diffusion ion (symport) or in the opposite direction (antiport).	Some sugars, amino acids, and ions are transported.	There is a concentration gradient for Na^+ into intestinal epithelial cells. This gradient provides the energy for the symport of glucose. As Na^+ enter the cell, down their concentration gradient, glucose also enters the cell. In many cells, H^+ are moved in the opposite direction of Na^+ (antiport).
Endocytosis	The plasma membrane forms a vesicle around the substances to be transported, and the vesicle is taken into the cell; this requires ATP; in receptor-mediated endocytosis, specific substances are ingested.	Phagocytosis takes in cells and solid particles; pinocytosis takes in molecules dissolved in liquid.	Immune system cells called phagocytes ingest bacteria and cellular debris; most cells take in substances through pinocytosis.
Exocytosis	Materials manufactured by the cell are packaged in secretory vesicles that fuse with the plasma membrane and release their contents to the outside of the cell; this requires ATP.	Proteins and other water-soluble molecules are transported out of cells.	Digestive enzymes, hormones, neurotransmitters, and glandular secretions are transported, and cell waste products are eliminated.

CYTOPLASM

Cytoplasm, the cellular material outside the nucleus but inside the plasma membrane, is about half cytosol and half organelles.

Cytosol

Cytosol (sī'tō-sol) consists of a fluid portion, a cytoskeleton, and cytoplasmic inclusions. The fluid portion of cytosol is a solution with dissolved ions and molecules and a colloid with suspended molecules, especially proteins. Many of these proteins are enzymes that catalyze the breakdown of molecules for energy or the synthesis of sugars, fatty acids, nucleotides, amino acids, and other molecules.

Cytoskeleton

The **cytoskeleton** supports the cell and holds the nucleus and other organelles in place. It is also responsible for cell movements, such as changes in cell shape and the movement of cell organelles. The cytoskeleton consists of three groups of proteins: microtubules, actin filaments, and intermediate filaments (figure 3.27).

Microtubules are hollow tubules composed primarily of protein units called **tubulin.** The microtubules are about 25 nanometers (nm) in diameter, with walls about 5 nm thick. Microtubules vary in length but are normally several micrometers (μm) long. Microtubules play a variety of roles within cells. They help provide support and structure to the cytoplasm of the cell, much like an internal scaffolding. They are involved in the process of cell division and in the transport of intracellular materials, and they form essential components of certain cell organelles, such as centrioles, spindle fibers, cilia, and flagella.

Actin filaments, or **microfilaments,** are small fibrils about 8 nm in diameter that form bundles, sheets, or networks in the cytoplasm of cells. These filaments have a spiderweb-like appearance within the cell. Actin filaments provide structure to the cytoplasm and mechanical support for microvilli. Actin filaments support the plasma membrane and define the shape of the cell. Changes in cell shape involve the breakdown and reconstruction of actin filaments. These changes in shape allow some cells to move about. Muscle cells contain a large number of highly organized actin filaments, which are responsible for the muscle's contractile capabilities (see chapter 9).

Intermediate filaments are protein fibers about 10 nm in diameter. They provide mechanical strength to cells. For example, intermediate filaments support the extensions of nerve cells, which have a very small diameter but can be a meter in length.

Cytoplasmic Inclusions

The cytosol also contains **cytoplasmic inclusions,** which are aggregates of chemicals either produced by the cell or taken in by the cell.

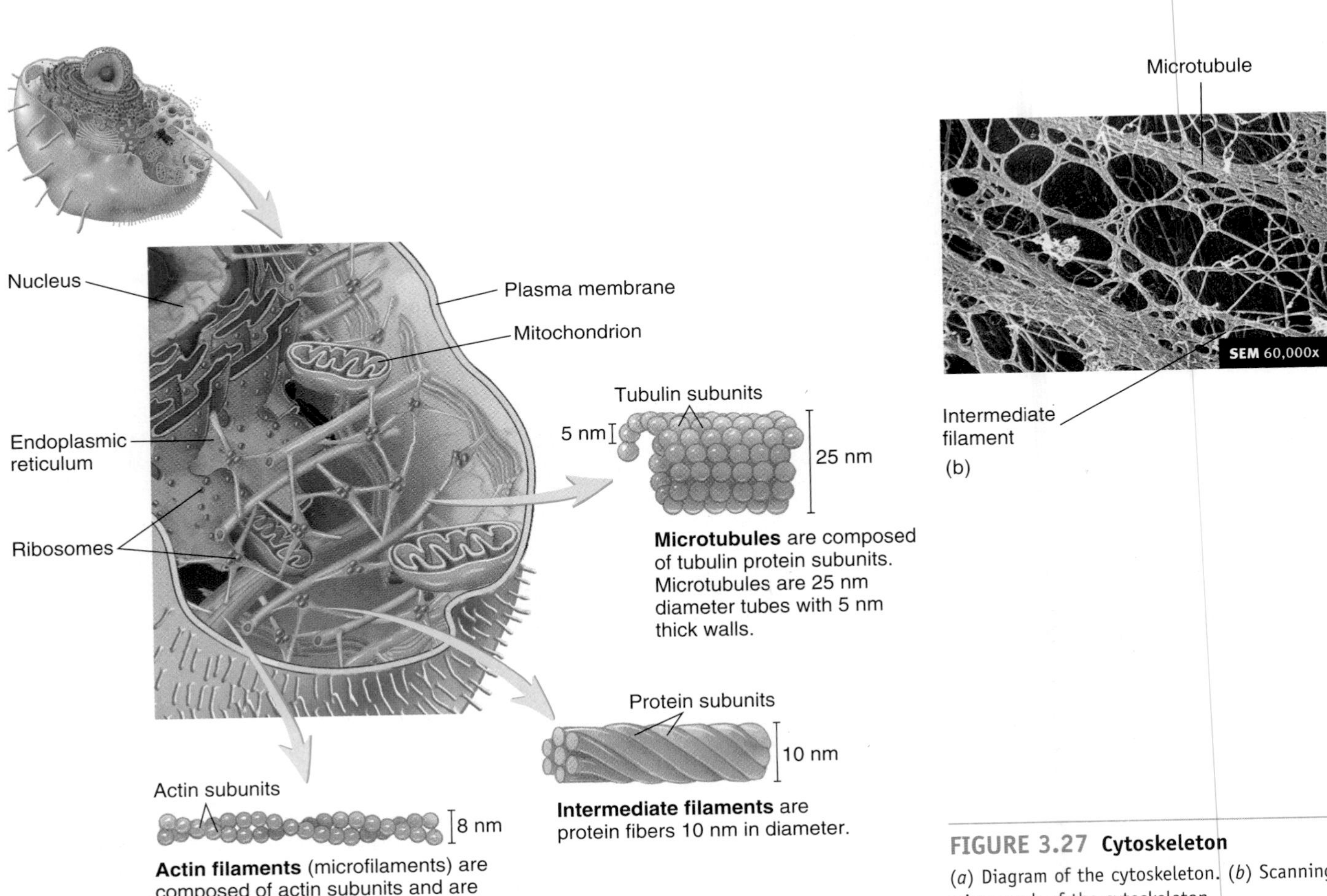

FIGURE 3.27 Cytoskeleton
(*a*) Diagram of the cytoskeleton. (*b*) Scanning electron micrograph of the cytoskeleton.

For example, lipid droplets or glycogen granules store energy-rich molecules; hemoglobin in red blood cells transports oxygen; melanin is a pigment that colors the skin, hair, and eyes; and **lipochromes** (lip′ō-krōmz) are pigments that increase in amount with age. Dust, minerals, and dyes can also accumulate in the cytoplasm.

31 ***Define cytoplasm and cytosol.***

32 ***What are the two general functions of the cytoskeleton?***

33 ***Describe and list the functions of microtubules, actin filaments, and intermediate filaments.***

34 ***Define and give examples of cytoplasmic inclusions. What are lipochromes?***

THE NUCLEUS AND CYTOPLASMIC ORGANELLES

Organelles are structures within cells that are specialized for particular functions, such as manufacturing proteins or producing ATP. Organelles can be thought of as individual workstations within the cell, each responsible for performing specific tasks. Most, but not all, organelles have membranes that are similar to the plasma membrane. The membranes separate the interior of the organelles from the cytoplasm, creating subcellular compartments with their own enzymes that are capable of carrying out unique chemical reactions. The nucleus is the largest organelle of the cell. The remaining organelles are referred to as cytoplasmic organelles (see table 3.1).

The number and type of cytoplasmic organelles within each cell are related to the specific structure and function of the cell. Cells secreting large amounts of protein contain well-developed organelles that synthesize and secrete protein, whereas cells actively transporting substances, such as sodium ions, across their plasma membrane contain highly developed organelles that produce ATP. The following sections describe the structure and main functions of the nucleus and major cytoplasmic organelles found in cells.

Nucleus

The **nucleus** is a large, membrane-bound structure usually located near the center of the cell. It may be spherical, elongated, or lobed, depending on the cell type. All cells of the body have a nucleus at some point in their life cycle (see p. 91), although some cells, such as red blood cells, lose their nuclei as they develop. Other cells, such as skeletal muscle cells and certain bone cells, called osteoclasts, contain more than one nucleus.

The nucleus consists of **nucleoplasm** surrounded by a **nuclear envelope** (figure 3.28) composed of two membranes separated by a space. At many points on the surface of the nuclear envelope, the inner and outer membranes fuse to form porelike structures

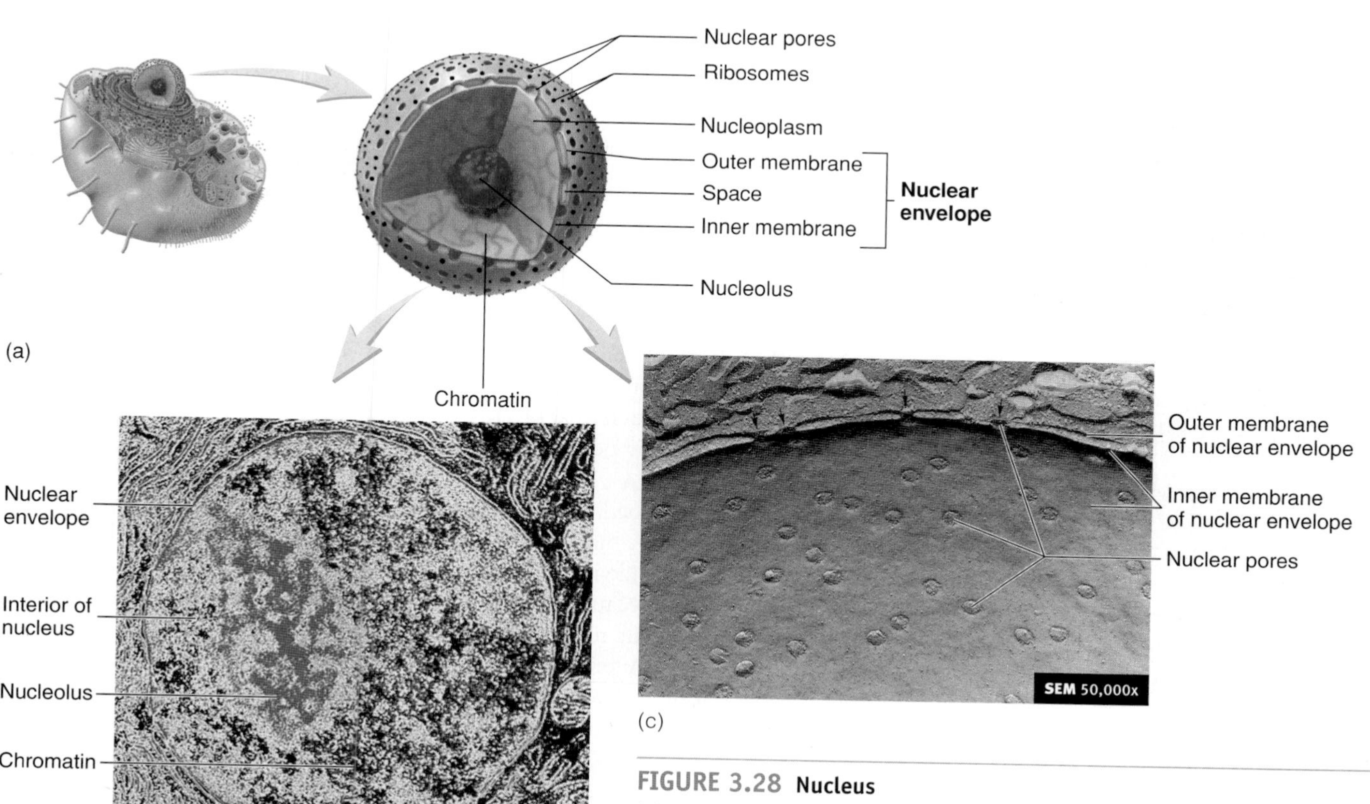

FIGURE 3.28 Nucleus

(*a*) The nuclear envelope consists of inner and outer membranes that become fused at the nuclear pores. The nucleolus is a condensed region of the nucleus not bound by a membrane and consisting mostly of RNA and protein. (*b*) Transmission electron micrograph of the nucleus. (*c*) Scanning electron micrograph showing the inner surface of the nuclear envelope and the nuclear pores.

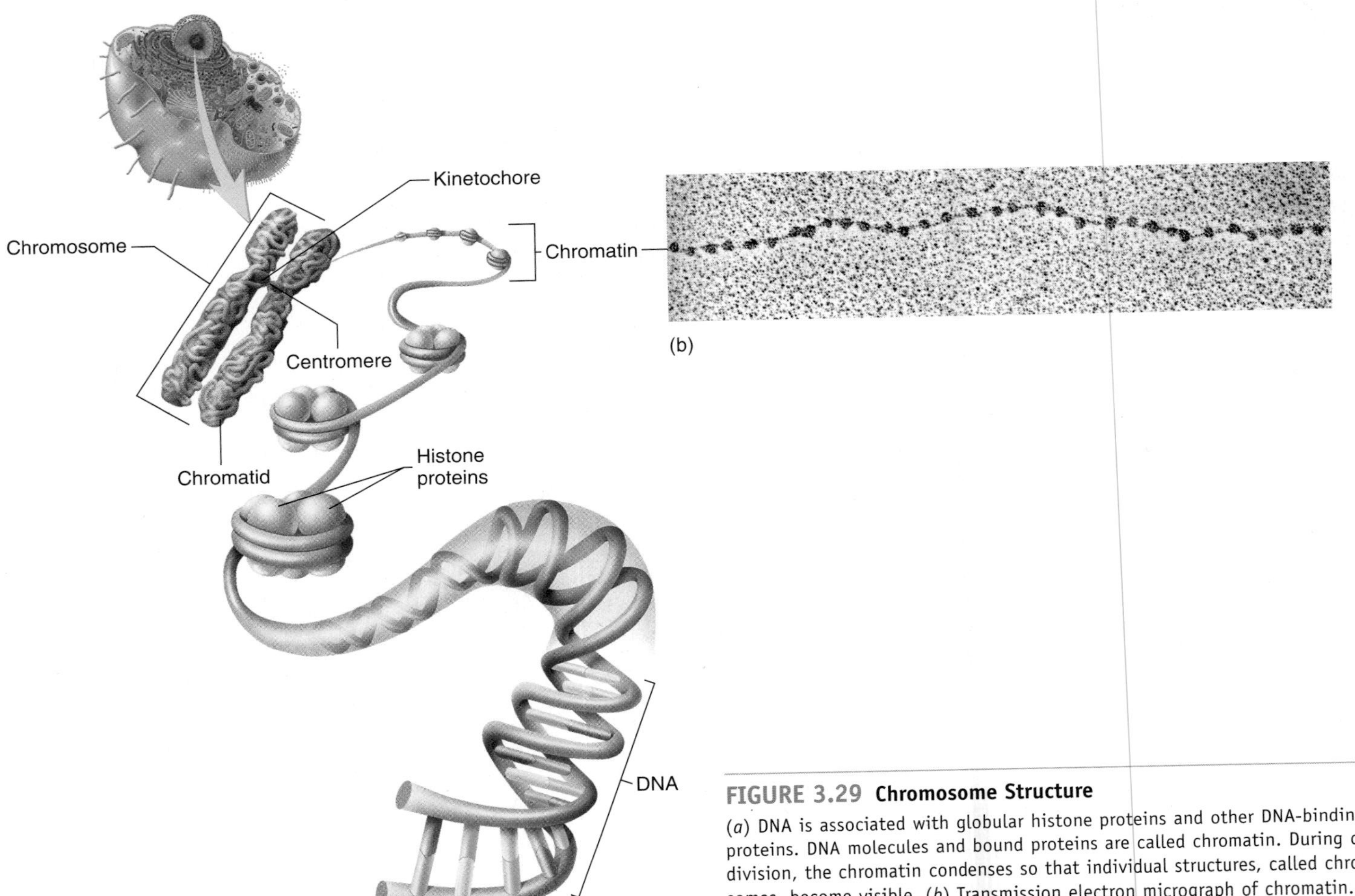

FIGURE 3.29 Chromosome Structure

(*a*) DNA is associated with globular histone proteins and other DNA-binding proteins. DNA molecules and bound proteins are called chromatin. During cell division, the chromatin condenses so that individual structures, called chromosomes, become visible. (*b*) Transmission electron micrograph of chromatin.

called **nuclear pores.** Molecules move between the nucleus and the cytoplasm through these nuclear pores.

Deoxyribonucleic acid (DNA) is mostly found within the nucleus (see figure 2.26), although small amounts of DNA are also found within mitochondria (see p. 83). Nuclear DNA and associated proteins are organized into discrete structures called **chromosomes** (krō′mō-sōmz; colored bodies; figure 3.29). The proteins include **histones** (his′tōnz), which are important for the structural organization of DNA, and other proteins that regulate DNA function. During most of the life cycle of a cell, the chromosomes are dispersed throughout the nucleus as delicate filaments referred to as **chromatin** (krō′ma-tin; colored material; see figures 3.28 and 3.29). During cell division (see p. •••), the dispersed chromatin becomes densely coiled, forming compact chromosomes. At the beginning of cell division, each chromosome consists of two **chromatids** (krō′ma-tids), which are attached at a single point called the **centromere** (sen′trō-mēr). The **kinetochore** (ki-nē′tō-kōr, ki-net′ō-kōr), a protein structure within the centromere, provides a point of attachment for microtubules during cell division.

DNA determines the structural and functional characteristics of the cell by specifying the structure of proteins. Proteins form many structural components of the cell and all the enzymes, which regulate most chemical reactions in the cell. DNA establishes the structure of proteins by specifying the sequence of their amino acids (see figure 2.22). DNA is a large molecule that does not leave the nucleus but functions by means of an intermediate, **ribonucleic acid (RNA),** which can leave the nucleus through nuclear pores. DNA determines the structure of messenger RNA (mRNA), ribosomal RNA (rRNA), and transfer RNA (tRNA) (all described in more detail on p. 86). A sequence of nucleotides in a DNA molecule that specifies the structure of a protein or RNA molecule is called a **gene.**

Because mRNA synthesis occurs within the nucleus, cells without nuclei accomplish protein synthesis only as long as the mRNA produced before the nucleus degenerates remains functional. The nuclei of developing red blood cells are expelled from the cells before the red blood cells enter the blood, where they survive without a nucleus for about 120 days. In comparison, many cells with nuclei, such as nerve and skeletal muscle cells, survive as long as the person survives.

A **nucleolus** (noo-klē′ō-lŭs) is a somewhat rounded, dense region within the nucleus that lacks a surrounding membrane (figure 3.30). Usually, one nucleolus exists per nucleus, but

1. Ribosomal proteins, produced in the cytoplasm, are transported through nuclear pores into the nucleolus.
2. rRNA, most of which is produced in the nucleolus, is assembled with ribosomal proteins to form small and large ribosomal subunits.
3. The small and large ribosomal subunits leave the nucleolus and the nucleus through nuclear pores.
4. The small and large subunits, now in the cytoplasm, combine with each other and with mRNA during protein synthesis.

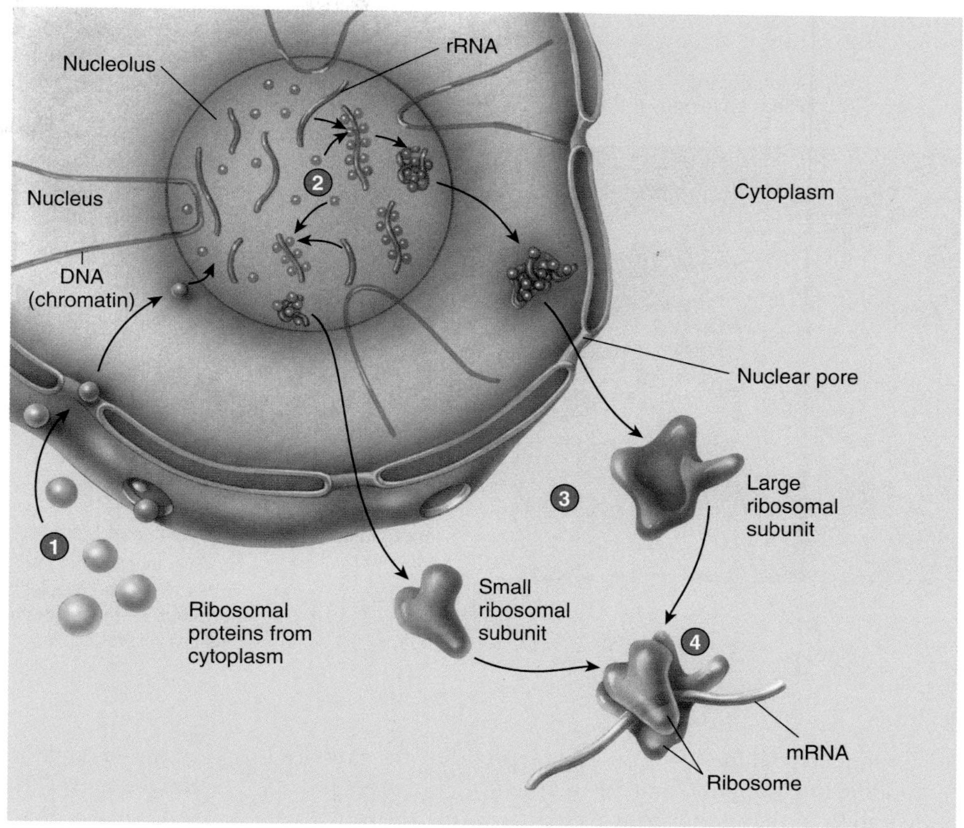

PROCESS FIGURE 3.30 Production of Ribosomes

several nucleoli may also be seen in the nuclei of rapidly dividing cells. The nucleolus incorporates portions of 10 chromosomes (5 pairs), called **nucleolar organizer regions.** These regions contain DNA from which rRNA is produced. Within the nucleolus, the subunits of ribosomes are manufactured (see the section "Ribosomes" below).

35 ***Define organelles.***

36 ***Describe the structure of the nucleus and nuclear envelope. What is the function of the nuclear pores?***

37 ***What molecules are found in chromatin? When does chromatin become a chromosome?***

38 ***List the types of RNA whose structure is determined by DNA. How can DNA control the structural and functional characteristics of the cell without leaving the nucleus?***

39 ***Describe the nucleolus. Define and give the function of the nucleolar organizer regions.***

Ribosomes

Ribosomes (rī′bō-sōms) are the sites of protein synthesis. Each ribosome is composed of a large subunit and a smaller one. The ribosomal subunits, which consist of **ribosomal RNA (rRNA)** and proteins, are produced separately in the nucleolus of the nucleus. The ribosomal subunits then move through the nuclear pores into the cytoplasm, where the ribosomal subunits assemble with mRNA to form the functional ribosome during protein synthesis (see figure 3.30). Ribosomes can be found free in the cytoplasm or associated with an intracellular membrane complex called the endoplasmic reticulum. **Free ribosomes** primarily synthesize proteins used inside the cell, whereas ribosomes attached to the endoplasmic reticulum usually produce proteins that are secreted from the cell.

40 ***What kinds of molecules combine to form ribosomes? Where are ribosomal subunits formed and assembled?***

41 ***Compare the functions of free ribosomes and ribosomes attached to the endoplasmic reticulum.***

Endoplasmic Reticulum

The outer membrane of the nuclear envelope is continuous with a series of membranes distributed throughout the cytoplasm of the cell (see figure 3.1), collectively referred to as the **endoplasmic reticulum** (en′dō-plas′mik re-tik′ū-lŭm; network inside the cytoplasm; figure 3.31). The endoplasmic reticulum consists of broad, flattened, interconnecting sacs and tubules. The interior spaces of those sacs and tubules are called **cisternae** (sis-ter′nē) and are isolated from the rest of the cytoplasm.

The **rough endoplasmic reticulum** is called "rough" because it has ribosomes attached to it. The ribosomes of the rough

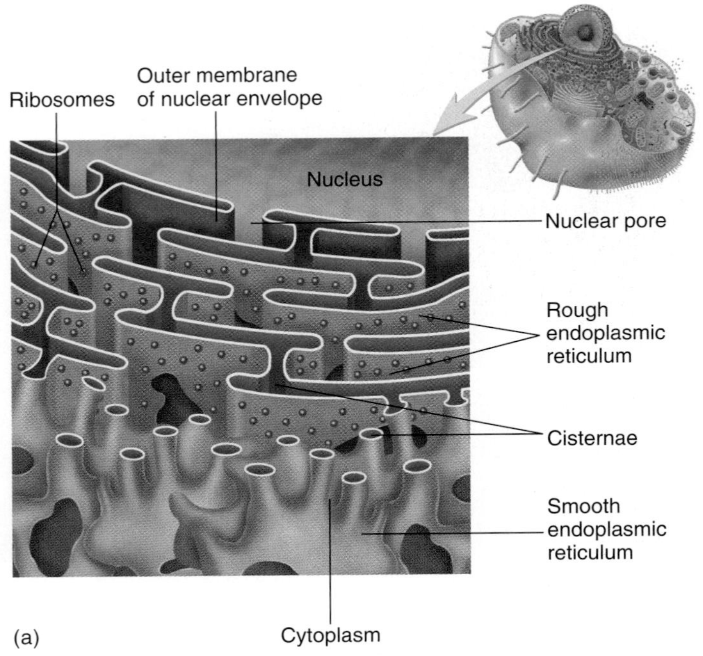

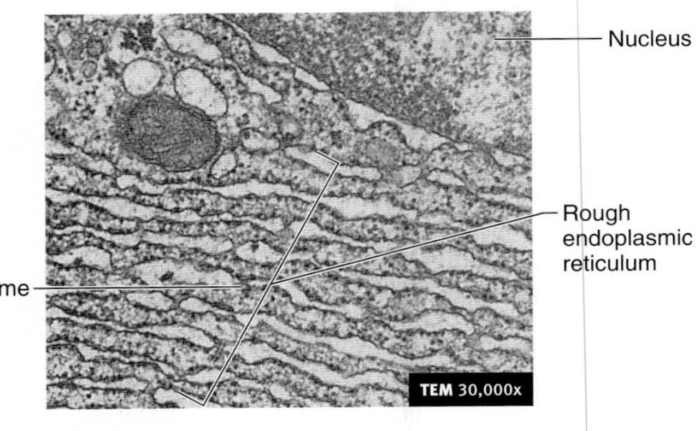

FIGURE 3.31 Endoplasmic Reticulum

(*a*) The endoplasmic reticulum is continuous with the nuclear envelope and occurs either as rough endoplasmic reticulum (with ribosomes) or as smooth endoplasmic reticulum (without ribosomes). (*b*) Transmission electron micrograph of the rough endoplasmic reticulum.

endoplasmic reticulum are sites where proteins are produced and modified for secretion and for internal use. The amount and configuration of the endoplasmic reticulum within the cytoplasm depend on the cell type and function. Cells with abundant rough endoplasmic reticulum synthesize large amounts of protein that are secreted for use outside the cell.

Smooth endoplasmic reticulum, which is endoplasmic reticulum without attached ribosomes, manufactures lipids, such as phospholipids, cholesterol, steroid hormones, and carbohydrates. Many phospholipids produced in the smooth endoplasmic reticulum help form vesicles within the cell and contribute to the plasma membrane. Cells that synthesize large amounts of lipid contain dense accumulations of smooth endoplasmic reticulum. Enzymes required for lipid synthesis are associated with the membranes of the smooth endoplasmic reticulum. Smooth endoplasmic reticulum also participates in the detoxification processes by which enzymes act on chemicals and drugs to change their structure and reduce their toxicity. The smooth endoplasmic reticulum of skeletal muscle stores calcium ions that function in muscle contraction.

Golgi Apparatus

The **Golgi** (gōl′jē) **apparatus** (figure 3.32) is composed of flattened, membranous sacs, containing cisternae, that are stacked on each other, like dinner plates. The Golgi apparatus can be thought of as a packaging and distribution center because it modifies, packages, and distributes proteins and lipids manufactured by the rough and smooth endoplasmic reticula (figure 3.33). Proteins produced at the ribosomes of the rough endoplasmic reticulum enter the endoplasmic reticulum and then are surrounded by a **vesicle** (ves′i-kl), or little sac, that forms from the membrane of the endoplasmic reticulum. This vesicle, called a **transport vesicle,** moves to the Golgi apparatus, fuses with its membrane, and releases the protein into its cisterna. The Golgi apparatus concentrates and, in some cases, chemically modifies the proteins by synthesizing and attaching carbohydrate molecules to the proteins to form glycoproteins or by attaching lipids to proteins to form lipoproteins. The proteins are then packaged into vesicles that pinch off from the margins of the Golgi apparatus and are distributed to various locations. Some vesicles carry proteins to the plasma membrane, where the proteins are secreted from the cell by exocytosis; other vesicles contain proteins that become part of the plasma membrane; and still other vesicles contain enzymes that are used within the cell.

The Golgi apparatus is most highly developed in cells that secrete large amounts of protein or glycoproteins, such as cells in the salivary glands and the pancreas.

42 ***How is the endoplasmic reticulum related to the nuclear envelope? How are the cisternae of the endoplasmic reticulum related to the rest of the cytoplasm?***

43 ***What are the functions of smooth endoplasmic reticulum?***

44 ***Describe the structure and function of the Golgi apparatus.***

45 ***Describe the production of a protein at the endoplasmic reticulum and its distribution to the Golgi apparatus. Name three ways in which proteins are distributed from the Golgi apparatus.***

Secretory Vesicles

The membrane-bound **secretory vesicles** (see figure 3.32) that pinch off from the Golgi apparatus move to the surface of the cell, their membranes fuse with the plasma membrane, and the

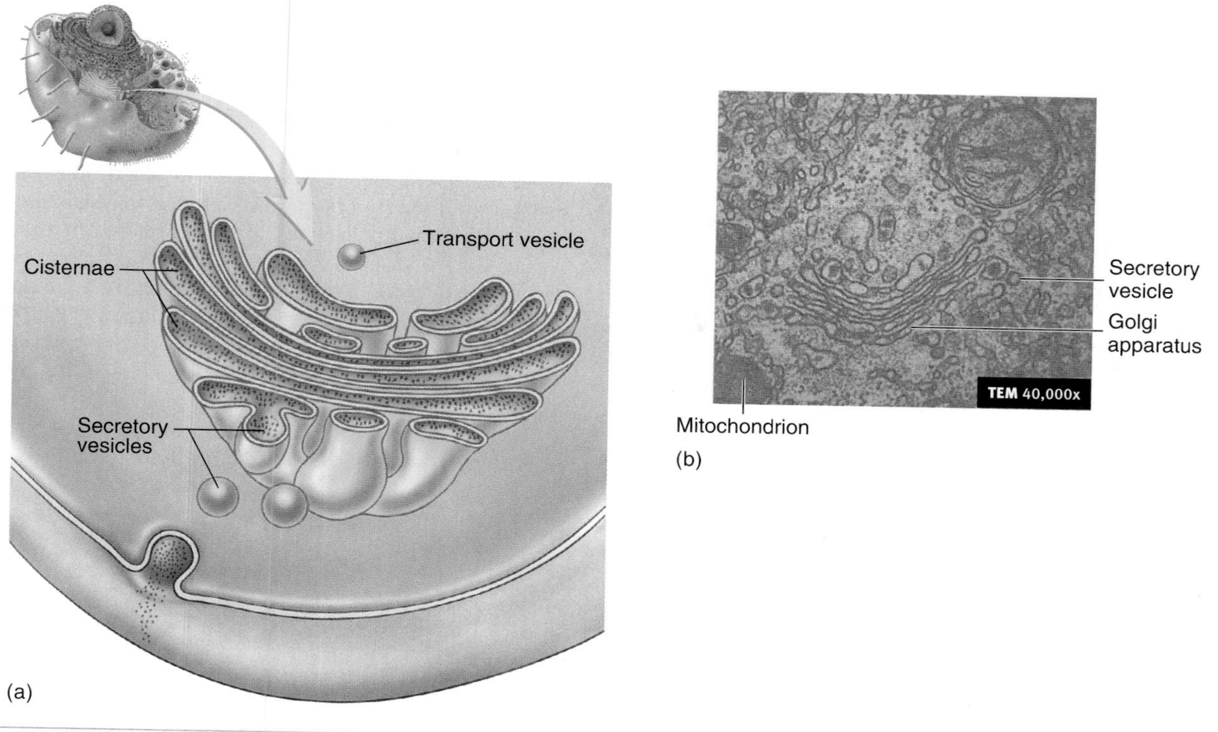

FIGURE 3.32 Golgi Apparatus

(*a*) The Golgi apparatus is composed of flattened, membranous sacs, containing cisternae, and resembles a stack of dinner plates or pancakes. (*b*) Transmission electron micrograph of the Golgi apparatus.

1. Some proteins are produced at ribosomes on the surface of the rough endoplasmic reticulum and are transferred into the cisterna as they are produced.
2. The proteins are surrounded by a vesicle that forms from the membrane of the endoplasmic reticulum.
3. This transport vesicle moves from the endoplasmic reticulum to the Golgi apparatus, fuses with its membrane, and releases the proteins into its cisterna.
4. The Golgi apparatus concentrates and, in some cases, modifies the proteins into glycoproteins or lipoproteins.
5. The proteins are packaged into vesicles that form from the membrane of the Golgi apparatus.
6. Some vesicles, such as lysosomes, contain enzymes that are used within the cell.
7. Secretory vesicles carry proteins to the plasma membrane, where the proteins are secreted from the cell by exocytosis.
8. Some vesicles contain proteins that become part of the plasma membrane.

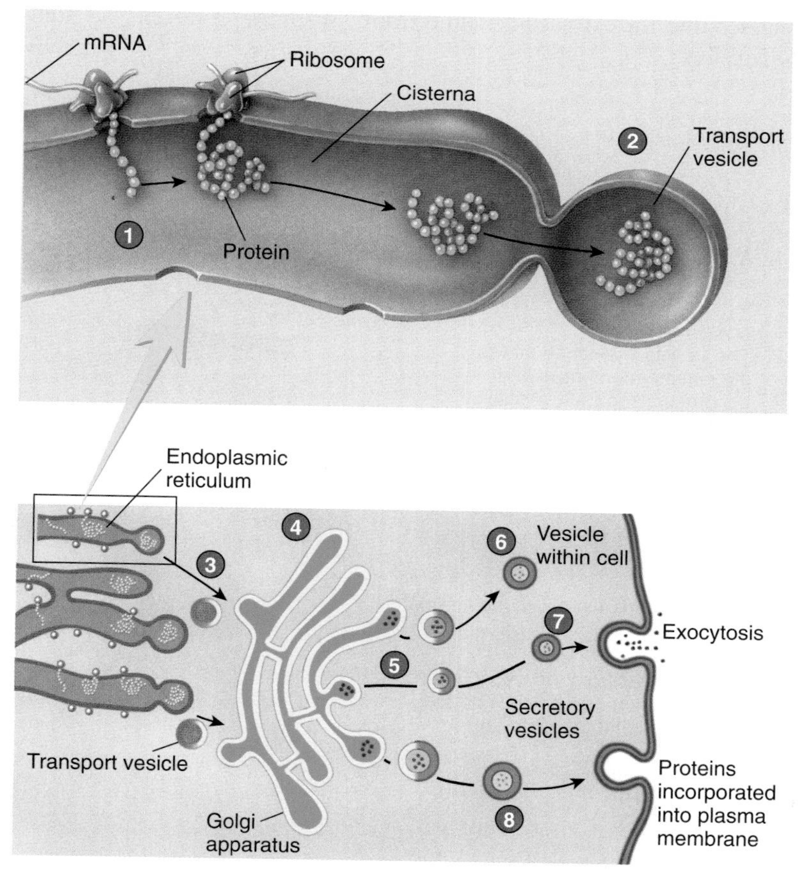

PROCESS FIGURE 3.33 Function of the Golgi Apparatus

contents of the vesicle are released to the exterior by exocytosis. The membranes of the vesicles are then incorporated into the plasma membrane.

Secretory vesicles accumulate in some cells, but their contents frequently are not released to the exterior until the cell receives a signal. For example, secretory vesicles that contain the hormone insulin do not release it until the concentration of glucose in the blood increases and acts as a signal for the secretion of insulin from the cells.

Lysosomes

Lysosomes (lī′sō-sōmz) are membrane-bound vesicles that pinch off from the Golgi apparatus (see figure 3.33). They contain a variety of hydrolytic enzymes that function as intracellular digestion systems. Vesicles taken into the cell fuse with the lysosomes to form one vesicle and to expose the endocytized materials to hydrolytic enzymes (figure 3.34). Various enzymes within lysosomes digest nucleic acids, proteins, polysaccharides, and lipids. Certain white blood cells have large numbers of lysosomes that contain enzymes to digest phagocytized bacteria. Lysosomes also digest the organelles of the cell that are no longer functional in a process called **autophagia** (aw-tō-fā′jē-ă; self-eating). In other cells, the lysosomes move to the plasma membrane, and the enzymes are secreted by exocytosis. For example, the normal process of bone remodeling involves the breakdown of bone tissue by specialized bone cells. Enzymes responsible for that degradation are released into the extracellular fluid from lysosomes produced by those cells.

Diseases of Lysosomal Enzymes

Some diseases result from nonfunctional lysosomal enzymes. For example, **Pompe disease** is a rare genetic disorder that results from the inability of lysosomal enzymes to break down glycogen. The glycogen accumulates in large amounts in the heart, liver, and skeletal muscles, an accumulation that often leads to heart failure. **Familial hyperlipoproteinemia** is a group of genetic disorders characterized by the accumulation of large amounts of lipids in phagocytic cells that lack the normal enzymes required to break down the lipid droplets. Symptoms include abdominal pain, enlargement of the spleen and liver, and eruption of yellow nodules in the skin filled with the affected phagocytic cells. **Mucopolysaccharidoses,** such as **Hurler syndrome,** are genetic diseases in which lysosomal enzymes are unable to break down mucopolysaccharides (glycosaminoglycans), so these molecules accumulate in the lysosomes of connective tissue cells and nerve cells. People affected by these diseases suffer mental retardation and skeletal deformities.

Peroxisomes

Peroxisomes (per-ok′si-sōmz) are membrane-bound vesicles that are smaller than lysosomes. Peroxisomes contain enzymes that break down fatty acids and amino acids. Hydrogen peroxide (H_2O_2), which can be toxic to the cell, is a by-product of that breakdown. Peroxisomes also contain the enzyme **catalase,** which breaks down hydrogen peroxide to water and oxygen. Cells that are active in detoxification, such as liver and kidney cells, have many peroxisomes.

1. A vesicle forms around material outside the cell.
2. The vesicle is pinched off from the plasma membrane and becomes a separate vesicle inside the cell.
3. A lysosome is pinched off the Golgi apparatus.
4. The lysosome fuses with the vesicle.
5. The enzymes from the lysosome mix with the material in the vesicle, and the enzymes digest the material.

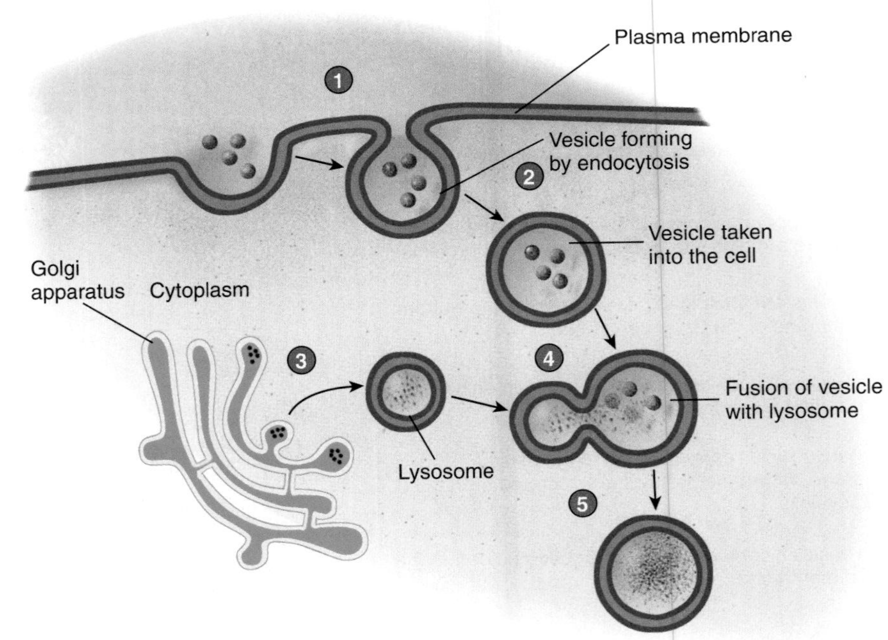

PROCESS FIGURE 3.34 Action of Lysosomes

Proteasomes

Proteasomes (prō′tē-ă-sōmz) consist of large protein complexes, including several enzymes that break down and recycle proteins within the cell. Proteasomes are not surrounded by membranes. They are tunnel-like structures, similar to channel protein complexes; the inner surfaces of the tunnel have enzymatic regions that break down proteins. Smaller protein subunits close the ends of the tunnel and regulate which proteins are taken into it for digestion.

Mitochondria

Mitochondria (mī-tō-kon′drē-ă) provide energy for the cell. Consequently, they are often called the cell's power plants. Mitochondria are usually illustrated as small, rod-shaped structures (figure 3.35). In living cells, time-lapse photomicrography reveals that mitochondria constantly change shape from spherical to rod-shaped or even to long, threadlike structures. Mitochondria are the major sites of ATP production, which is the major energy source for most energy-requiring chemical reactions within the cell. Each mitochondrion has an inner and an outer membrane, separated by an intermembranous space. The outer membrane has a smooth contour, but the inner membrane has numerous infoldings called **cristae** (kris′tē; sing. *crista*) that project like shelves into the interior of the mitochondria.

A complex series of mitochondrial enzymes forms two major enzyme systems that are responsible for oxidative metabolism and most ATP synthesis (see chapter 25). The enzymes of the citric acid (or Krebs) cycle are found in the **matrix,** which is the substance located in the space formed by the inner membrane. The enzymes of the electron-transport chain are embedded within the inner membrane. Cells with a greater energy requirement have more mitochondria with more cristae than do cells with lower energy requirements. Within the cytoplasm of a cell, the mitochondria are more numerous in areas in which ATP is used. For example, mitochondria are numerous in cells that perform active transport and are packed near the membrane where active transport occurs.

Increases in the number of mitochondria result from the division of preexisting mitochondria. When muscles enlarge as a result of exercise, the number of mitochondria within the muscle cells increases to provide the additional ATP required for muscle contraction.

The information for making some mitochondrial proteins is stored in DNA contained within the mitochondria themselves, and those proteins are synthesized on ribosomes within the mitochondria. The structure of many other mitochondrial proteins is determined by nuclear DNA, however, and these proteins are synthesized on ribosomes within the cytoplasm and then transported into the mitochondria. Both the mitochondrial DNA and

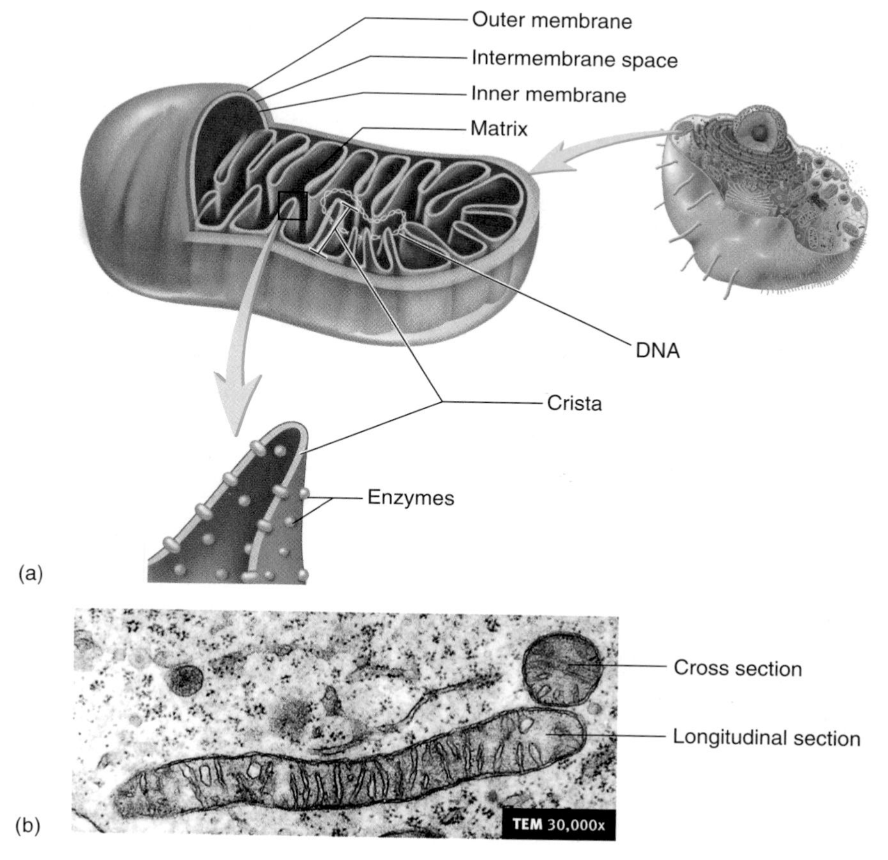

FIGURE 3.35 Mitochondrion

(*a*) Typical mitochondrion structure. (*b*) Transmission electron micrograph of mitochondria in longitudinal and cross section.

mitochondrial ribosomes are very different from those within the nucleus and cytoplasm of the cell, respectively. Mitochondrial DNA is a closed circle of about 16,500 base pairs (bp) coding for 37 genes, compared with the open strands of nuclear DNA, which is composed of 3 billion bp coding for 30,000 genes. In addition, unlike nuclear DNA, mitochondrial DNA does not have associated histone proteins.

Mitochondrial Diseases

Mitochondria play a major role in the synthesis of ATP. Each mitochondrion contains a single DNA molecule with at least 37 genes, 13 of which code for proteins that are important for ATP synthesis. The other 24 genes are important for the expression of mitochondrial genes. Therefore, mutations, or changes, in mitochondrial genes can lead to disruptions in normal ATP synthesis, reducing the amount of ATP produced by the cells. Disorders that result from such mutations are collectively called **mitochondrial diseases.** The effects of these diseases are most obvious in tissues that require large amounts of ATP, particularly nervous and muscle tissues. As a consequence, the common symptoms of mitochondrial diseases are loss of neurological function and defects in muscular activity. For example, **Leber hereditary optic neuropathy** results in sudden vision loss due to optic nerve degeneration. Mutations associated with this disorder have been found in the genes that function in ATP synthesis, especially in the cells of the optic nerve. Because of their high energy demands, the cells of the optic nerve are damaged or die because of the lack of ATP.

In humans, mitochondria are passed only from the mother to her offspring because the mitochondria of sperm cells do not enter the oocyte during fertilization (see chapter 29). Therefore, mitochondrial diseases, involving the mitochondrial genes, show a pattern of maternal inheritance in which a mother afflicted with a mitochondrial disease passes it to all of her offspring, but a father suffering from the same disorder passes it to none of his offspring.

46 ***Define secretory vesicles.***

47 ***Describe the process by which lysosomal enzymes digest phagocytized materials. Define autophagia.***

48 ***What is the function of peroxisomes? How does catalase protect cells?***

49 ***Describe the structure and function of proteasomes.***

50 ***What is the function of mitochondria? What enzymes are found on the cristae and in the matrix? How can the number of mitochondria in a cell increase?***

PREDICT 5

Describe the structural characteristics of cells that are highly specialized to do the following: (a) synthesize and secrete proteins, (b) actively transport substances into the cell, (c) synthesize lipids, and (d) phagocytize foreign substances.

Centrioles and Spindle Fibers

The **centrosome** (sen′trō-sōm), a specialized zone of cytoplasm close to the nucleus, is the center of microtubule formation. It contains two **centrioles** (sen′trē-ōlz). Each centriole is a small, cylindrical organelle about 0.3–0.5 μm in length and 0.15 μm in diameter, and the two centrioles are normally oriented perpendicular to each other within the centrosome (see figure 3.1). The wall of the centriole is composed of nine evenly spaced, longitudinally oriented, parallel units, or triplets. Each unit consists of three parallel microtubules joined together (figure 3.36).

Microtubules appear to influence the distribution of actin and intermediate filaments. Through its control of microtubule formation, the centrosome is closely involved in determining cell shape and movement. The microtubules extending from the centrosomes are very dynamic—constantly growing and shrinking.

Before cell division, the two centrioles double in number, the centrosome divides into two, and one centrosome, containing two

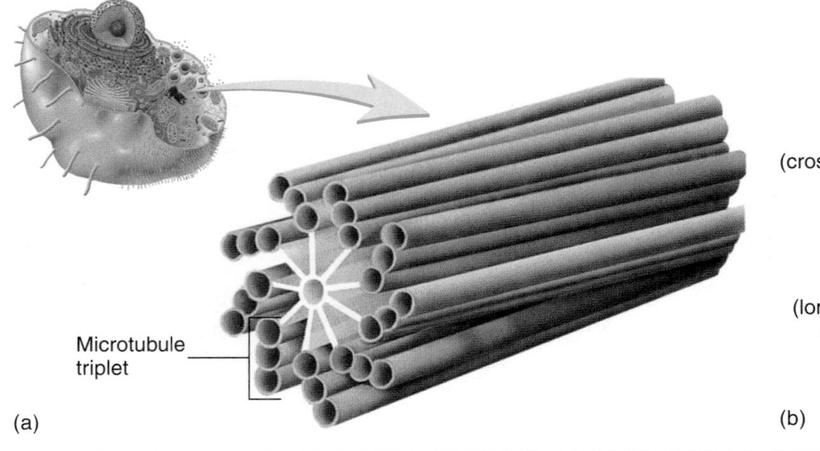

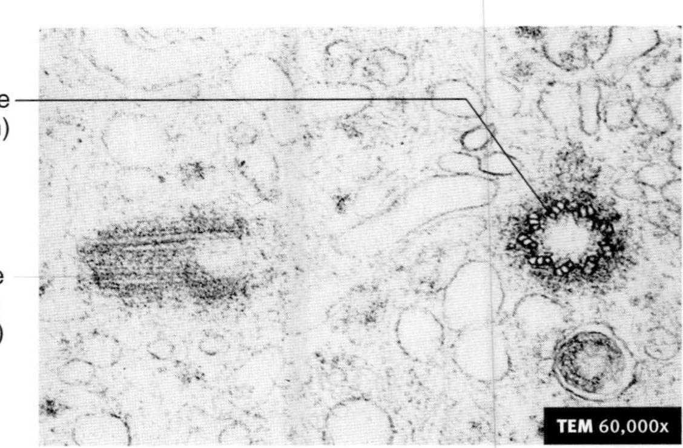

FIGURE 3.36 Centriole

(*a*) Structure of a centriole, which comprises nine triplets of microtubules. Each triplet contains one complete microtubule fused to two incomplete microtubules. (*b*) Transmission electron micrograph of a pair of centrioles, which are normally located together near the nucleus. One is shown in cross section and one in longitudinal section.

centrioles, moves to each end of the cell. Microtubules called **spindle fibers** extend out in all directions from the centrosome. These microtubules grow and shrink even more rapidly than those of nondividing cells. If the extended end of a spindle fiber comes in contact with a kinetochore, a specialized region in the centromere of each chromosome, the spindle fiber attaches to the kinetochore and stops growing or shrinking. Eventually, spindle fibers from each centrosome bind to the kinetochores of all the chromosomes. During cell division, the microtubules facilitate the movement of chromosomes toward the two centrosomes (see "Cell Division," p. 92).

Cilia and Flagella

Cilia (sil′ē-ă) are structures that project from the surface of cells and are capable of movement. They vary in number from one to thousands per cell. Cilia are cylindrical in shape, about 10 μm in length and 0.2 μm in diameter, and the shaft of each cilium is enclosed by the plasma membrane. Two centrally located microtubules and nine peripheral pairs of fused microtubules, the so-called 9 + 2 arrangement, extend from the base to the tip of each cilium (figure 3.37). Movement of the microtubules past each other, a process that requires energy from ATP, is responsible for movement of the cilia. **Dynein arms,** proteins connecting adjacent

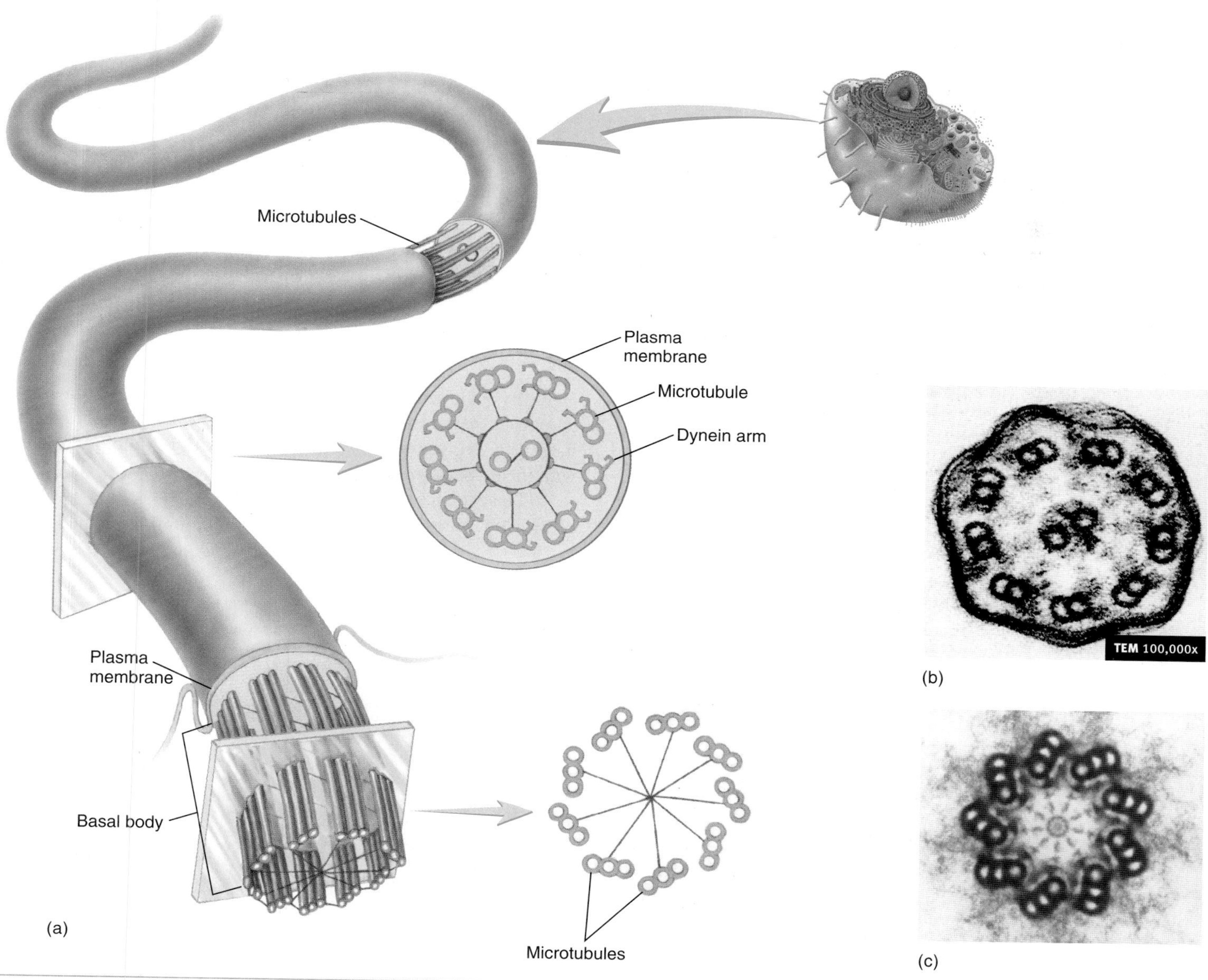

FIGURE 3.37 Cilia and Flagella

(*a*) Ciliary and flagellar structures. The shaft is composed of nine microtubule doublets around its periphery and two in the center. Dynein arms are proteins that connect one pair of microtubules to another pair. Dynein arm movement, which requires ATP, causes the microtubules to slide past each other, resulting in bending or movement of the cilium or flagellum. A basal body attaches the cilium or flagellum to the plasma membrane. (*b*) Transmission electron micrograph through a cilium. (*c*) Transmission electron micrograph through the basal body of a cilium.

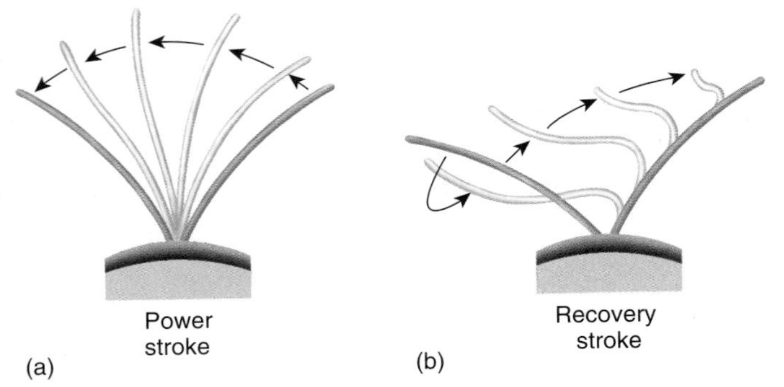

FIGURE 3.38 Ciliary Movement
(*a*) Power and (*b*) recovery strokes.

pairs of microtubules, push the microtubules past each other. A **basal body** (a modified centriole) is located in the cytoplasm at the base of the cilium. Cilia are numerous on surface cells that line the respiratory tract and the female reproductive tract. In these regions, cilia move in a coordinated fashion, with a power stroke in one direction and a recovery stroke in the opposite direction (figure 3.38). Their motion moves materials over the surface of the cells. For example, cilia in the trachea move mucus embedded with dust particles upward and away from the lungs. This action helps keep the lungs clear of debris.

Flagella (flă-jel′ă) have a structure similar to cilia but are longer (45 μm). Sperm cells are the only human cells to possess flagella and usually only one flagellum exists per cell. Furthermore, whereas cilia move small particles across the cell surface, flagella move the entire cell. For example, each sperm cell is propelled by a single flagellum. In contrast to cilia, which have a power stroke and a recovery stroke, flagella move in a wavelike fashion.

Microvilli

Microvilli (mī-krō-vil′ī; figure 3.39) are cylindrically shaped extensions of the plasma membrane about 0.5–1.0 μm in length and 90 nm in diameter. Normally, many microvilli are on each cell, and they function to increase the cell surface area. A student looking at photographs may confuse microvilli with cilia. Microvilli, however, are only one-tenth to one-twentieth the size of cilia. Individual microvilli can usually be seen only with an electron microscope, whereas cilia can be seen with a light microscope. Microvilli do not move, and they are supported with actin filaments, not microtubules. Microvilli are found in the intestine, kidney, and other areas in which absorption is an important function. In certain locations of the body, microvilli are highly modified to function as sensory receptors. For example, elongated microvilli in hair cells of the inner ear respond to sound.

51 ***List and describe the functions of centrosomes. Explain the structure of centrioles.***

52 ***What are spindle fibers? Explain the relationship among centrosomes, spindle fibers, and the kinetochores of chromosomes during cell division.***

53 ***Contrast the structure and function of cilia and flagella.***

54 ***Describe the structure and function of microvilli. How are microvilli different from cilia?***

GENES AND GENE EXPRESSION

Heredity is the genetic transmission of characteristics, or traits, from parents to their offspring. **Genes** are the functional units of heredity. Each gene is a segment of a DNA molecule. DNA molecules, along with their associated proteins, form the chromosomes.

The sequence of nucleotides in a DNA molecule is a method of storing information that is based on a triplet code. Three

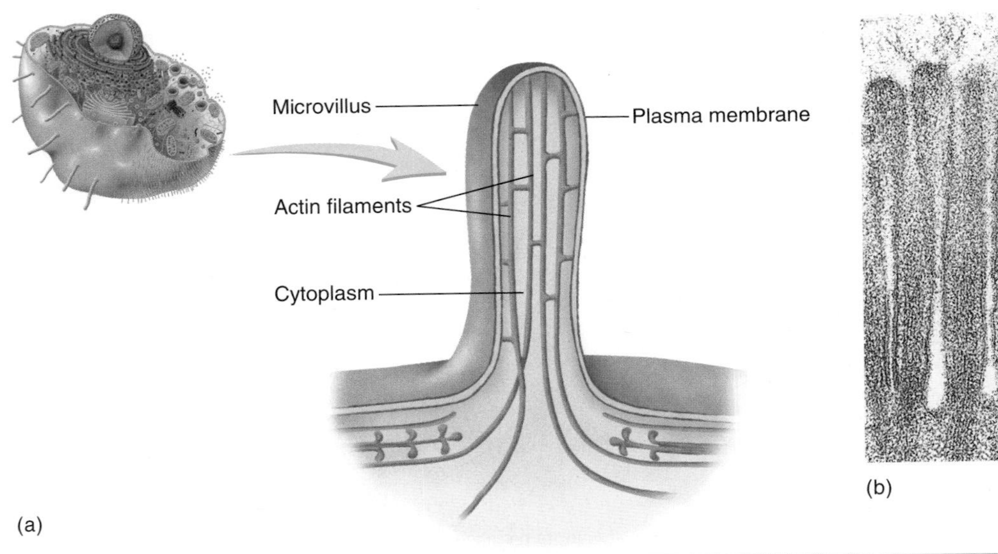

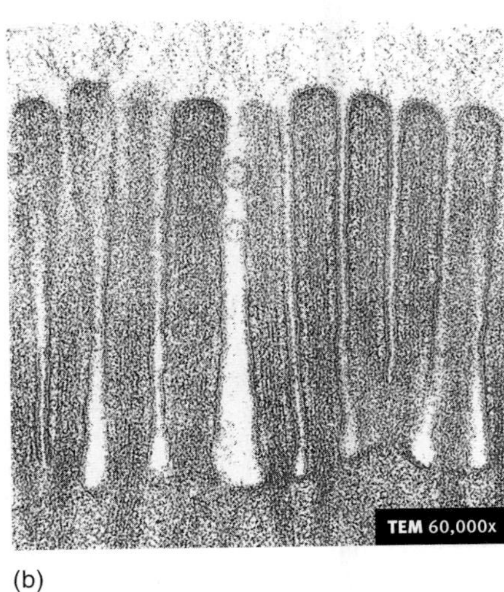

FIGURE 3.39 Microvillus
(*a*) A microvillus is a tiny, tubular extension of the cell; it contains cytoplasm and some actin filaments (microfilaments). (*b*) Transmission electron micrograph of microvilli.

consecutive nucleotides, called **triplets,** form the words of the triplet code. Just as the sequence of letters "seedogrun" can be deciphered to mean "see dog run," a sequence of bases, such as "CATGAGTAG," has meaning, which is used to construct other DNA molecules, RNA molecules, or proteins. A molecular definition of a gene is all the triplets necessary to make a functional RNA molecule or protein.

Most chromosomes contain thousands of genes. There are two major types of genes: structural and regulatory. **Structural genes** are those DNA sequences that determine specific amino acid sequences in proteins, such as enzymes; hormones; or structural proteins, such as collagen. **Regulatory genes** are segments of DNA involved in controlling which structural genes are expressed in a given tissue. By determining the structure of proteins and by regulating which proteins are produced by cells, genes are responsible for the characteristics of cells, and therefore the inherited characteristics, or traits, of the entire organism.

The production of proteins from the information stored in DNA is called **gene expression** (figure 3.40). Gene expression involves two steps, transcription and translation, which can be illustrated with an analogy. Suppose a cook wants a cake recipe that is found only in a reference book in the library. Because the book cannot be checked out, the cook makes a copy, or **transcription,** of the recipe. Later in the kitchen, the information contained in the copied recipe is used to make the cake. The changing of something from one form to another (from recipe to cake) is called **translation.** In this analogy, DNA is the reference book that contains many recipes (genes) for making different proteins. DNA, however, is too large a molecule to pass through the nuclear envelope to go to the cytoplasm (the kitchen), where the proteins are synthesized. Just as the reference book stays in the library, DNA remains in the nucleus. Therefore, through transcription, the cell makes a copy of the gene (the recipe) necessary to make a particular protein (the cake). The copy, which is called mRNA, travels from the nucleus to ribosomes (the kitchen) in the cytoplasm, where the information in the copy is used to construct a protein (i.e., translation). Of course, to turn a recipe into a cake, ingredients are needed. The ingredients necessary to synthesize a

1. DNA contains the information necessary to produce proteins.
2. Transcription of one DNA strand results in mRNA, which is a complementary copy of the information in the DNA strand needed to make a protein.
3. The mRNA leaves the nucleus and goes to a ribosome.
4. Amino acids, the building blocks of proteins, are carried to the ribosome by tRNAs.
5. In the process of translation, the information contained in mRNA is used to determine the number, kinds, and arrangement of amino acids in the polypeptide chain.

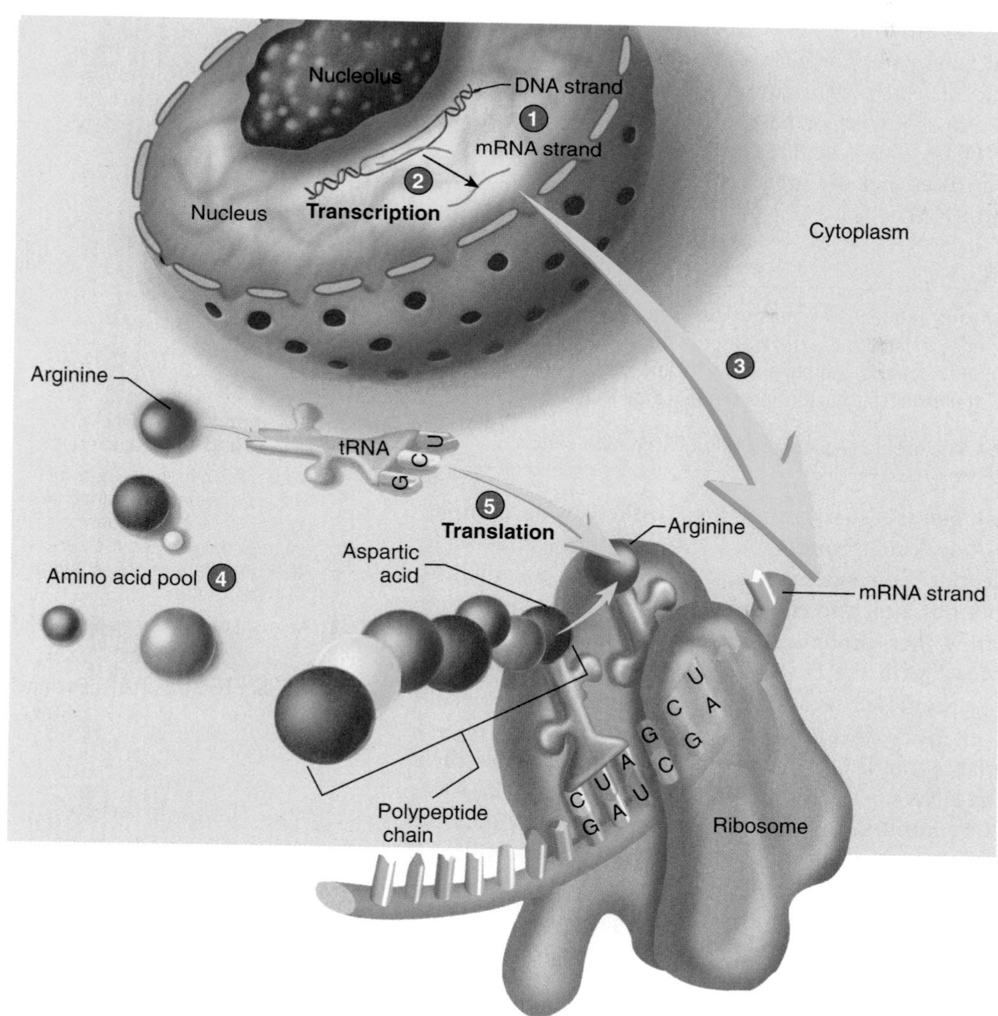

PROCESS FIGURE 3.40 Overview of Gene Expression

protein are amino acids. Specialized transport molecules, called **transfer RNA (tRNA),** carry the amino acids to the ribosomes (see figure 3.40).

In summary, gene expression involves transcription, making a copy of a small part of the stored information in DNA, and translation, converting that copied information into a protein. The details of transcription and translation are considered next.

55 ***What is a gene, and how are genes responsible for the structure and function of cells?***

56 ***What type of molecule is produced as a result of transcription? Of translation? Where do these events take place?***

Transcription

Transcription is the synthesis of mRNA, tRNA, and rRNA based on the nucleotide sequence in DNA (figure 3.41). Transcription occurs when a section of a DNA molecule unwinds and its complementary strands separate. One of the DNA strands serves as the template strand for the process of transcription. Nucleotides that form RNA align with the DNA nucleotides in the template strand by complementary base pairing. An adenine aligns with a thymine of DNA, cytosine aligns with a guanine of DNA, and guanine aligns with a cytosine of DNA. Instead of thymine, uracil of RNA (see figure 2.25), aligns with adenine of DNA. Thus, the sequence of bases that aligns with the TCGA sequence of DNA is AGCU. This pairing relationship between nucleotides ensures that the information in DNA is transcribed correctly into RNA.

PREDICT 6

Given the following sequence of nucleotides of a DNA molecule, write the sequence of mRNA that is transcribed from it. What is the nucleotide sequence of the complementary strand of the DNA molecule? How does it differ from the nucleotide sequence of RNA?

DNA nucleotide sequence: CGTACGCCGAGACGTCAAC

RNA polymerase is an enzyme that synthesizes the complementary RNA molecule from DNA. RNA polymerase attaches to a DNA nucleotide sequence called a **promoter.** RNA polymerase, however, does not attach to the promoter by itself. It must first associate with other proteins called **transcription factors** in order to interact with the DNA. The attachment of RNA polymerase to the promoter causes a portion of the DNA molecule to unwind, exposing the nucleotide sequence for that region of the template strand. Complementary RNA nucleotides then align with the DNA nucleotides of the template strand. The RNA nucleotides are combined by dehydration reactions, catalyzed by RNA polymerase, to form mRNA. Only a small portion of the DNA molecule unwinds at any one time. As complementary nucleotides are added to the mRNA, RNA polymerase moves along the DNA unwinding the next portion, while the previously unwound section of DNA strands winds back together. When RNA polymerase encounters a DNA nucleotide sequence called the **terminator,** it detaches from the DNA releasing the newly formed mRNA.

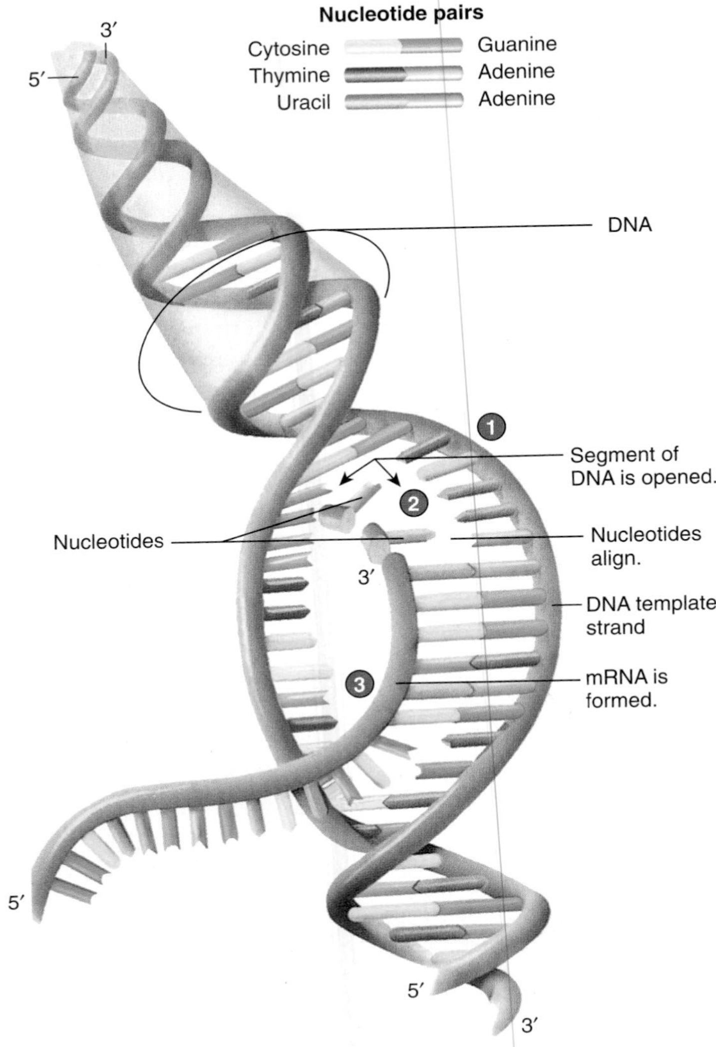

1. The strands of the DNA molecule separate from each other. One DNA strand serves as a template for mRNA synthesis.
2. Nucleotides that will form mRNA pair with DNA nucleotides according to the base-pair combinations shown in the key at the top of the figure. Thus, the sequence of nucleotides in the template DNA strand (*purple*) determines the sequence of nucleotides in mRNA (*gray*). RNA polymerase (the enzyme is not shown) joins the nucleotides of mRNA together.
3. As nucleotides are added, an mRNA molecule is formed.

PROCESS FIGURE 3.41 Formation of mRNA by Transcription of DNA

The region of a DNA molecule between the promoter and terminator is a gene. The structure of a gene is more complex than just the nucleotides that code for a protein; some regions that are transcribed to form mRNA do not code for parts of a protein. The DNA sequence of a gene contains a protein-coding region and flanking untranslated regions. During transcription, the sequence of nucleotides in DNA between the promoter and the terminator serve as a template to make an RNA molecule called **pre-mRNA** (figure 3.42). The untranslated regions are

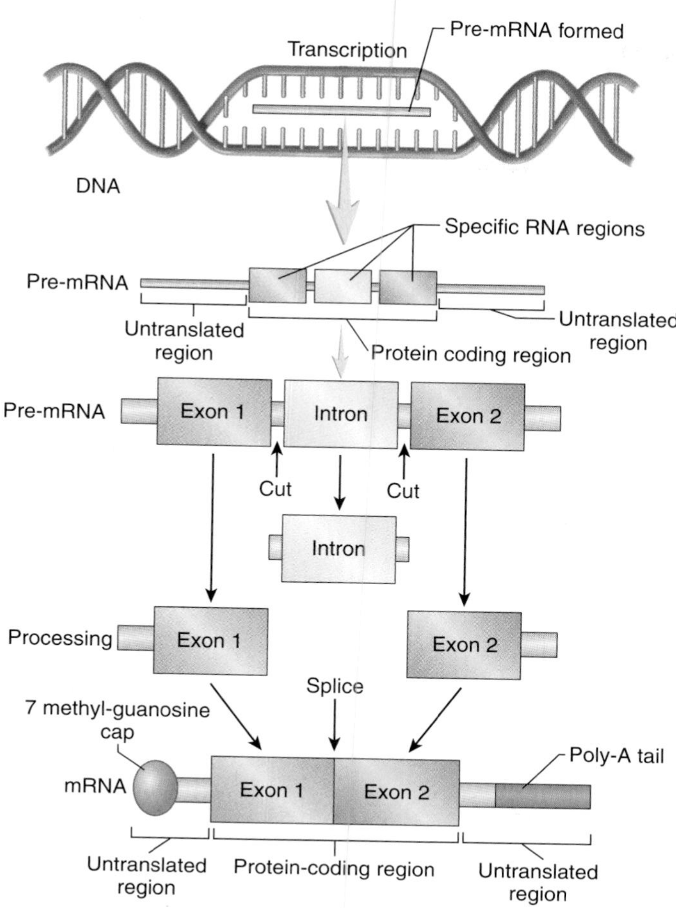

FIGURE 3.42 Posttranscriptional Change in mRNA

The two DNA strands separate, and one strand is transcribed to produce a pre-mRNA strand. An intron is cleaved from between two exons and is discarded. The exons are spliced together by spliceosomes (not shown) to make the functional mRNA. Also, a 7-methyl guanosine cap and a poly-A tail are added to the mRNA.

transcribed into pre-mRNA but do not code for parts of a protein. Instead, these regions function in the regulation of translation. The protein-coding region of RNA contains sections, called **exons,** that code for parts of a protein and sections, called **introns,** that do not code for parts of a protein.

Before pre-mRNA leaves the nucleus, it undergoes several modifications called **posttranscriptional processing,** which produces the functional mRNA that is used in translation to produce a protein or part of a protein (see figure 3.42). A **7-methyl guanosine cap** is added to one end of mRNA and a series of adenine nucleotides, called a **poly-A tail,** is added to the other end. These modifications to the ends of the mRNA ensure that mRNA travels from the nucleus to the cytoplasm and interacts with ribosomes during translation. The introns are removed from the pre-mRNA and the exons are spliced together by enzymes called **spliceosomes.** The functional mRNA consists only of exons.

In a process called **alternative splicing,** various combinations of exons are incorporated into mRNA. Which exons, and how many exons, are used to make mRNA can vary between cells of different tissues, resulting in different mRNAs transcribed from the same gene. Alternative splicing allows a single gene to produce more than one specific protein; however, the various proteins usually have similar functions in different tissues.

Thalassemia

Hemoglobin is an oxygen-carrying protein molecule composed of four polypeptides. **Thalassemia** is a group of genetic disorders in which one or more of the polypeptides of hemoglobin is produced in decreased amounts as the result of defective post-transcriptional processing. The decreased amount of hemoglobin in the blood causes anemia, which reduces the oxygen-carrying capacity of the blood.

Genetic Code

The information contained in mRNA, called the **genetic code,** is carried in sets of three nucleotide units called **codons.** A codon specifies an amino acid during translation. For example, the codon GAU specifies the amino acid aspartic acid, and the codon CGA specifies arginine. Although there are only 20 different amino acids commonly found in proteins, 64 possible codons exist. Therefore, an amino acid can have more than one codon. The codons for arginine include CGA, CGG, CGU, and CGC. Furthermore, some codons act as signals during translation. AUG, which specifies methionine, also acts as a **start codon,** which signals the beginning of translation. UAA, UGA, and UAG act as **stop codons,** which signal the end of translation. Unlike the start codon, those codons do not specify amino acids. Therefore, the protein-coding region of an mRNA begins at the start codon and ends at a stop codon.

Translation

The synthesis of a protein at the ribosome in response to the codons of mRNA is called translation (figure 3.43). In addition to mRNA, translation requires ribosomes and tRNA. Ribosomes consist of **ribosomal RNA (rRNA)** and proteins. Like mRNA, tRNA and rRNA are produced in the nucleus by transcription.

The function of tRNA is to match a specific amino acid to a specific codon of mRNA. To do this, one end of each kind of tRNA combines with a specific amino acid. Another part of the tRNA called the **anticodon** consists of three nucleotides and is complementary to a particular codon of mRNA. On the basis of the pairing relationships between nucleotides, the anticodon can combine only with its matched codon. For example, the tRNA that binds to aspartic acid has the anticodon CUA, which combines with the codon GAU of mRNA. Therefore the codon GAU codes for aspartic acid.

Ribosomes align the codons of the mRNA with the anticodons of tRNA and then enzymatically join the amino acids of adjacent tRNA molecules. The mRNA moves through the ribosome one codon at a time. With each move, a new tRNA enters the ribosome and the amino acid is linked to the growing chain, forming a polypeptide. The step-by-step process of gene expression at the ribosome is described in detail in figure 3.43.

1. To start protein synthesis, a ribosome binds to mRNA. The ribosome also has two binding sites for tRNA, one of which is occupied by a tRNA with its amino acid. Note that the first codon to associate with a tRNA is AUG, the start codon,which codes for methionine. The codon of mRNA and the anticodon of tRNA are aligned and joined. The other tRNA binding site is open.

2. By occupying the open tRNA binding site, the next tRNA is properly aligned with mRNA and with the other tRNA.

3. An enzyme within the ribosome catalyzes a synthesis reaction to form a peptide bond between the amino acids. Note that the amino acids are now associated with only one of the tRNAs.

4. The ribosome shifts position by three nucleotides. The tRNA without the amino acid is released from the ribosome, and the tRNA with the amino acids takes its position. A tRNA binding site is left open by the shift. Additional amino acids can be added by repeating steps 2 through 4.

5. Eventually, a stop codon in the mRNA, such as UAA, ends the process of translation. At this point, the mRNA and polypeptide chain are released from the ribosome.

6. Multiple ribosomes attach to a single mRNA to form a polyribosome. As the ribosomes move down the mRNA, proteins attached to the ribosomes lengthen and eventually detach from the mRNA.

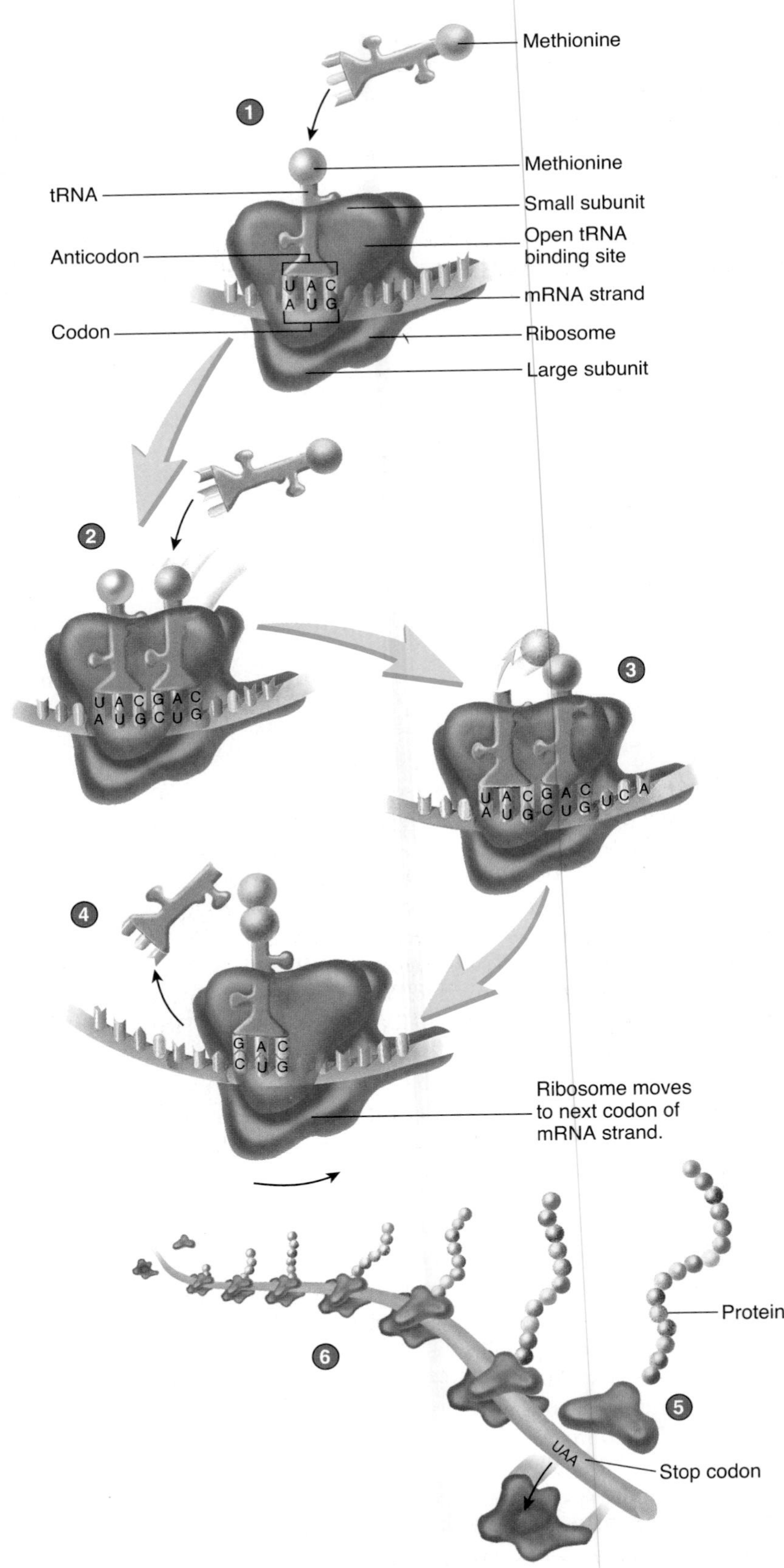

PROCESS FIGURE 3.43 Translation of mRNA to Produce a Protein

Many proteins are longer when first made than they are in their final, functional state. These proteins are called **proproteins,** and the extra piece of the molecule is cleaved off by enzymes to make the proprotein into a functional protein. Many proteins are enzymes, and the proproteins of those enzymes are called **proenzymes.** If many proenzymes were made within cells as functional enzymes, they could digest the cell that made them. Instead, they are made as proenzymes and are not converted to active enzymes until they reach a protected region of the body, such as inside the small intestine, where they are functional. Many proteins have side chains, such as polysaccharides, added to them following translation. Some proteins are composed of two or more amino acid chains that are joined after each chain is produced on separate ribosomes. These various modifications to proteins are referred to as **posttranslational processing.**

After the initial part of mRNA is used by a ribosome, another ribosome can attach to the mRNA and begin to make a protein. The resulting cluster of ribosomes attached to the mRNA is called a **polyribosome** (see figure 3.43). Each ribosome in a polyribosome produces an identical protein, and polyribosomes are an efficient way to use a single mRNA molecule to produce many copies of the same protein.

Regulation of Gene Expression

Most of the cells in the body have the same DNA. The transcription of mRNA in cells is regulated, however, so that all portions of all DNA molecules are not continually transcribed. The proteins associated with DNA in the nucleus play a role in regulating the transcription. As cells differentiate and become specialized by function during development, part of the DNA becomes nonfunctional and is not transcribed, whereas other segments of DNA remain very active. For example, in most cells the DNA coding for hemoglobin is nonfunctional, and little if any hemoglobin is synthesized. In developing red blood cells, however, the DNA coding for hemoglobin is functional, and hemoglobin synthesis occurs rapidly.

Gene expression in a single cell is not normally constant, but it occurs more rapidly at some times than others. Regulatory molecules that interact with the nuclear proteins can either increase or decrease the transcription rate of specific DNA segments. For example, triiodothyronine (T3), a hormone released by cells of the thyroid gland, enters cells, such as skeletal muscle cells; interacts with specific nuclear proteins; and increases specific types of mRNA transcription. Consequently, the production of certain proteins increases. As a result, an increase in the number of mitochondria and an increase in metabolism occur in these cells.

57 ***How are triplets and genes related?***

58 ***In what molecules are triplets, codons, and anticodons found? What is the genetic code?***

59 ***Describe the role of mRNA, rRNA, and tRNA in the production of a protein at a ribosome. What is a polyribosome?***

60 ***What are exons and introns? How are they related to pre-mRNA and posttranscriptional processing?***

61 ***What are start and stop codons? How are they different from promoters and terminators?***

62 ***Define proprotein, proenzyme, and posttranslational processing.***

63 ***State two ways the cell controls what DNA is transcribed.***

CELL LIFE CYCLE

The **cell life cycle** includes the changes a cell undergoes from the time it is formed until it divides to produce two new cells. The life cycle of a cell has two stages, an interphase and a cell division stage (figure 3.44).

Interphase

Interphase is the phase between cell divisions; 90% or more of the life cycle of a typical cell is spent in interphase. During this time, the cell carries out the metabolic activities necessary for life and performs its specialized functions—for example, secreting digestive enzymes. In addition, the cell prepares to divide. This preparation includes an increase in cell size, because many cell

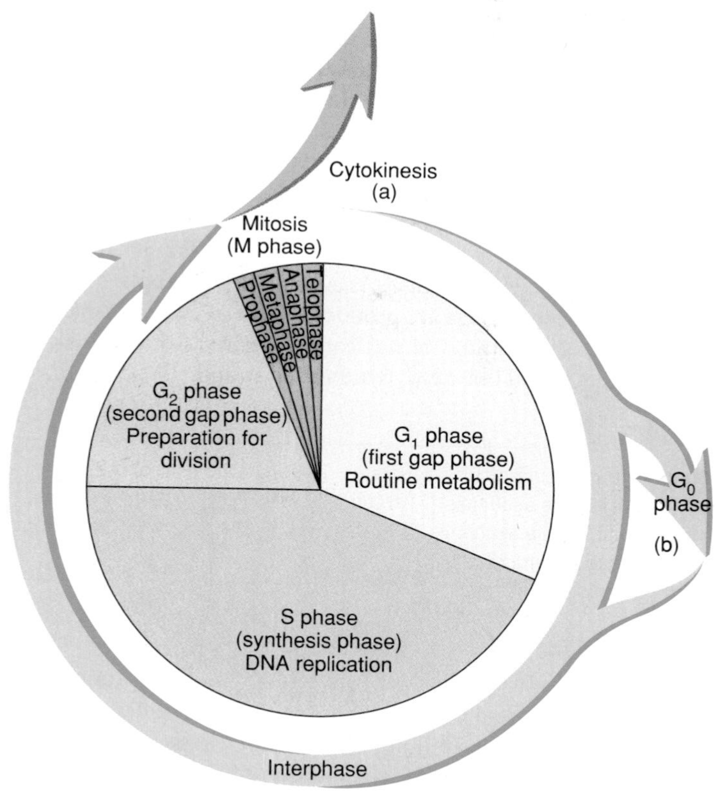

FIGURE 3.44 Cell Cycle

The cell cycle is divided into interphase and cell division. Interphase is divided into G_1, S, and G_2 subphases. During G_1, the cell carries out routine metabolic activities. During the S phase, DNA is replicated. During the G_2 phase, the cell prepares for division. Cell division includes mitosis and cytokinesis. (*a*) Following mitosis, two cells are formed by the process of cytokinesis. Each new cell begins a new cell cycle. (*b*) Many cells exit the cell cycle and enter the G_0 phase, where they remain until stimulated to divide, at which point they reenter the cell cycle.

components double in quantity, and a replication of the cell's DNA. The centrioles within the centrosome are also duplicated. Consequently, when the cell divides, each new cell receives the organelles and DNA necessary for continued functioning.

Interphase can be divided into three subphases, called G_1, S, and G_2. During G_1 (the first *gap* phase), the cell carries out routine metabolic activities. During the S phase (the *synthesis* phase), the DNA is replicated (new DNA is synthesized). During the G_2 phase (the second *gap* phase), the cell prepares for cell division. Many cells in the body do not divide for days, months, or even years. These "resting" cells exit the cell cycle and enter what is called the G_0 phase, in which they remain unless stimulated to divide.

DNA Replication

DNA **replication** is the process by which two new strands of DNA are made, using the two existing strands as templates. During interphase, DNA and its associated proteins appear as dispersed chromatin threads within the nucleus. When DNA replication begins, the two strands of each DNA molecule separate from each other for some distance (figure 3.45). Each strand then functions as a template, or pattern, for the production of a new, complementary strand of DNA, which is formed as complementary nucleotides pair with the existing nucleotides of each strand of the separated DNA molecule. The production of the new nucleotide strands is catalyzed by DNA **polymerase,** an enzyme that adds new nucleotides to the 3′ end of the growing strands. Because of the antiparallel orientation of the two DNA strands, the strands are formed differently. One strand, called the **leading strand,** is formed as a continuous strand, whereas the other strand, called the **lagging strand,** is formed in short segments called *Okazaki fragments.* The Okazaki fragments are then spliced by **DNA ligase.** As a result of DNA replication, two identical DNA molecules are produced. Each of the two new DNA molecules has one strand of nucleotides derived from the original DNA molecule and one newly synthesized strand.

PREDICT 7

Suppose a molecule of DNA separates, forming strands 1 and 2. Part of the nucleotide sequence in strand 1 is ATGCTA. From this template, what would be the sequence of nucleotides in the DNA replicated from strand 1 and strand 2?

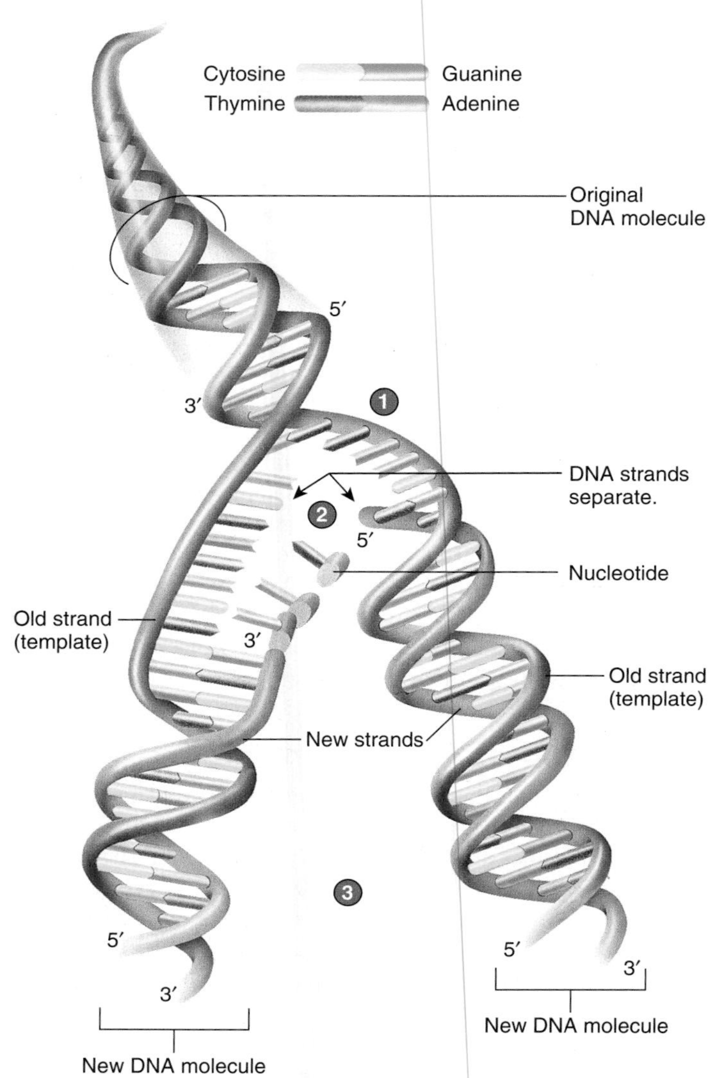

1. The strands of the DNA molecule separate from each other.
2. Each old strand (*dark purple*) functions as a template on which a new, complementary strand (*light purple*) is formed. The base-pairing relationship between nucleotides determines the sequence of nucleotides in the newly formed strands.
3. Two identical DNA molecules are produced.

PROCESS FIGURE 3.45 Replication of DNA

Replication of DNA during the S phase of interphase produces two identical molecules of DNA.

Cell Division

New cells necessary for growth and tissue repair are produced by cell division. A parent cell divides to form two daughter cells, each of which has the same amount and type of DNA as the parent cell. Because DNA determines cell structure and function, the daughter cells tend to have the same structure and perform the same functions as the parent cell. During development and cell differentiation, however, the functions of daughter cells may differ from each other and from that of the parent cell.

Cell division involves two major events: the division of the nucleus to form two new nuclei and the division of the cytoplasm to form two new cells. Each of the new cells contains one of the newly formed nuclei. The division of the nucleus occurs by mitosis, and the division of the cytoplasm is called cytokinesis.

Mitosis

Mitosis (mī-tō′sis) is the division of the nucleus into two nuclei, each of which has the same amount and type of DNA as the original nucleus. The DNA, which was replicated during interphase, is dispersed as chromatin. During mitosis, the chromatin becomes very densely coiled to form compact chromosomes. These chromosomes are discrete bodies easily seen with a light microscope after staining the nuclei of dividing cells. Because the DNA has been replicated, each chromosome consists of two chromatids, which are attached at a single point called the centromere (figure 3.46; see figure 3.29). Each chromatid contains a DNA molecule. As two

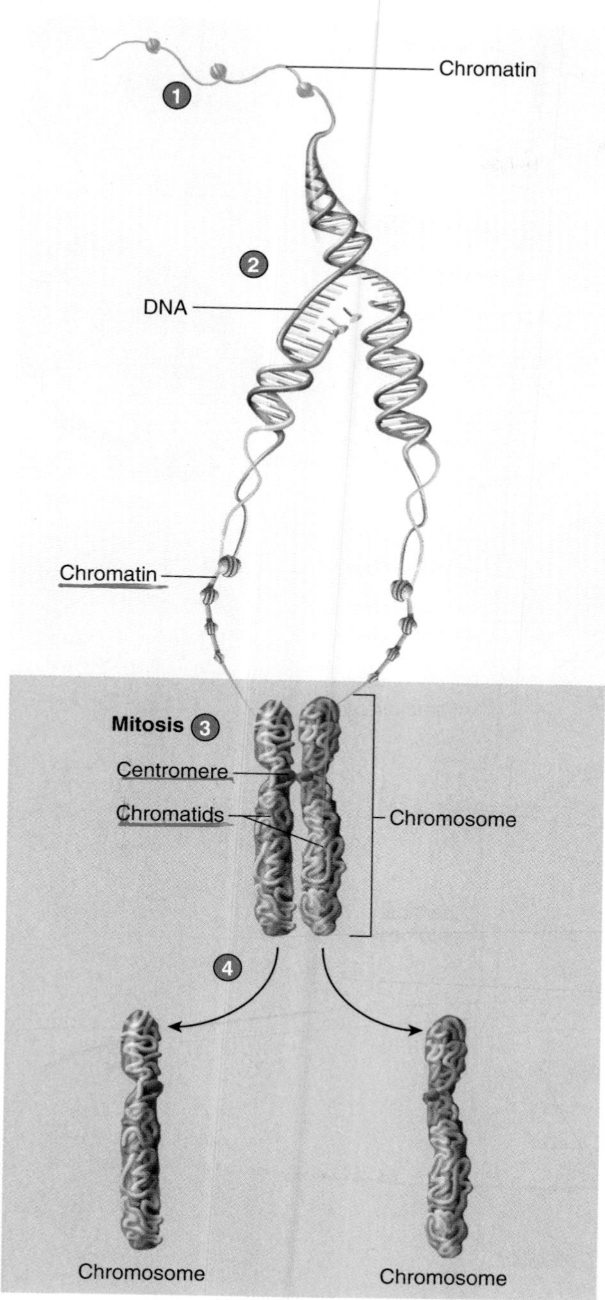

1. The DNA of a chromosome is dispersed as chromatin.
2. The DNA molecule unwinds and each strand of the molecule is replicated.
3. During mitosis the chromatin from each replicated DNA strand condenses to form a chromatid. The chromatids are joined at the centromere to form a single chromosome.
4. The chromatids separate to form two new, identical chromosomes. The chromosomes will unwind to form chromatin in the nuclei of the two daughter cells.

PROCESS FIGURE 3.46 **Replication of a Chromosome**

daughter cells are formed, the chromatids separate, and each is now called a chromosome. Each daughter cell receives one of the chromosomes. Thus, the daughter cells receive the same complement of chromosomes and DNA and are genetically identical.

For convenience of discussion, mitosis is divided into four phases: **prophase, metaphase, anaphase,** and **telophase** (tel′ō-fāz). Although each phase represents major events, mitosis is a continuous process, and no discrete jumps occur from one phase to another. Learning the characteristics associated with each phase is helpful, but a more important concept is that each daughter cell obtains the same number and type of chromosomes as the parent cell. The major events of mitosis are summarized in figure 3.47.

Cytokinesis

Cytokinesis (sī′tō-ki-nē′sis) is the division of the cytoplasm of the cell to produce two new cells. Cytokinesis begins in anaphase, continues through telophase, and ends in the following interphase (see figure 3.47). The first sign of cytokinesis is the formation of a **cleavage furrow,** or an indentation of the plasma membrane, which forms midway between the centrioles. A contractile ring composed primarily of actin filaments pulls the plasma membrane inward, dividing the cell into halves. Cytokinesis is complete when the membranes of the halves separate at the cleavage furrow to form two separate cells.

64 ***Define interphase. What percent of the cell life cycle is typically spent in interphase?***

65 ***Describe the cell's activities during G_1, S, and G_2 phases of the cell life cycle.***

66 ***Describe the process of DNA replication. What are the functions of DNA polymerase and DNA ligase?***

67 ***Define mitosis. How do the two nuclei that are produced in mitosis compare with the original nucleus?***

68 ***Describe the differences among chromatin, chromatids, and chromosomes.***

69 ***List the events that occur during interphase, prophase, metaphase, anaphase, and telophase of mitosis.***

70 ***Describe cytokinesis.***

GENETICS

Genetics is the study of heredity—that is, those characteristics inherited by children from their parents. Although the environment can influence gene expression, people's physical characteristics and abilities are largely determined by their genetic makeup. Many of a person's abilities, susceptibility to disease, and even life span are influenced by the genes they inherit from their parents. Because many of the diseases caused by microorganisms now are preventable or treatable, diseases that have a genetic basis are receiving more attention.

An understanding of genetics is an important tool for medical professionals. Patients are asked questions about their family medical history to help diagnose many diseases. The family medical history also allows doctors to determine the probability that patients will develop certain diseases, such as heart disease or

1. **Interphase** is the time between cell divisions. DNA is found as thin threads of chromatin in the nucleus. DNA replication occurs during the S phase of interphase. Organelles, other than the nucleus, and centrioles duplicate during interphase.

2. In **prophase,** the chromatin condenses into chromosomes. Each chromosome consists of two chromatids joined at the centromere. The centrioles move to the opposite ends of the cell, and the nucleolus and the nuclear envelope disappear. Microtubules form near the centrioles and project in all directions. Some of the microtubules end blindly and are called astral fibers. Others, known as spindle fibers, project toward an invisible line called the equator and overlap with fibers from opposite centrioles.

3. In **metaphase,** the chromosomes align in the center of the cell in association with the spindle fibers. Some spindle fibers are attached to kinetochores in the centromere of each chromosome.

4. In **anaphase,** the chromatids separate, and each chromatid is then referred to as a chromosome. Thus, when the centromeres divide, the chromosome number is double, and there are two identical sets of chromosomes. The chromosomes, assisted by the spindle fibers, move toward the centrioles at each end of the cell. Separation of the chromatids signals the beginning of anaphase, and, by the time anaphase has ended, the chromosomes have reached the poles of the cell. Cytokinesis begins during anaphase as a cleavage furrow forms around the cell.

5. In **telophase,** migration of each set of chromosomes is complete. The chromosomes unravel to become less distinct chromatin threads. The nuclear envelope forms from the endoplasmic reticulum. The nucleoli form, and cytokinesis continues to form two cells.

6. Mitosis is complete, and a new interphase begins. The chromosomes have unraveled to become chromatin. Cell division has produced two daughter cells, each with DNA that is identical to the DNA of the parent cell.

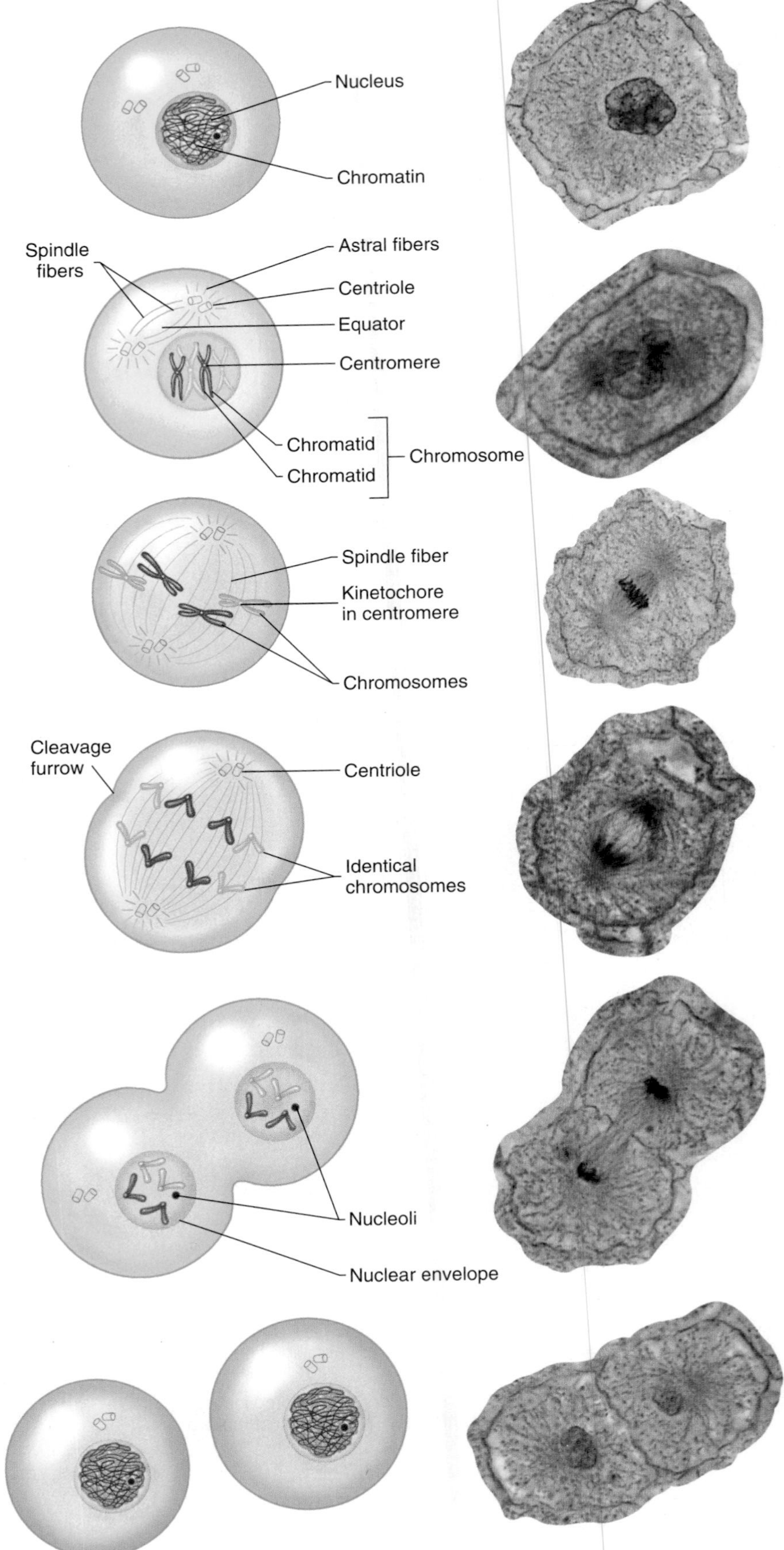

PROCESS FIGURE 3.47 Mitosis and Cytokinesis

cancer, and to suggest preventive measures. Recent advances in the field of genetics have shown how genes influence health and have provided new methods for treating certain diseases.

Some genetic diseases, such as hemophilia, have a simple pattern of inheritance, usually involving one or a few genes, and it is easy to predict whether a person will develop the disease. However, other genetic diseases have a more complex pattern of inheritance involving many more genes. **Mendelian genetics,** the study of how certain genetic traits are passed from parent to offspring, has been used to determine an individual's risk of developing certain genetic diseases.

Over the past 10 years, the human genome, all of the genes in human chromosomes, has been sequenced. The results of this project have provided a wealth of information useful to the medical field. This information will allow researchers to better identify genes associated with particular diseases, to understand the biochemical relationship between genes and diseases, and to create new treatments for disease. The new, genetic approach to the diagnosis and management of diseases is known as **genomic medicine.**

Mendelian Genetics

Gregor Mendel (1822–1884) was an Austrian monk who made revolutionary discoveries about inheritance patterns in pea plants. For his discoveries, Mendel is known as the "father of genetics." Mendel found that the transmission of traits, such as purple versus white flowers, from one generation of pea plants to another can be explained by discrete units he called "heritable factors." He concluded that each pea plant has two heritable factors for a characteristic, such as flower color. He proposed that, during the production of **gametes,** or sex cells, each gamete receives one of these factors. Then two gametes combine to produce the next generation so that each member of the next generation of pea plants has two heritable factors. Mendel's "heritable factors" are now called genes. The genes an organism has for a given trait is called the **genotype** (jēn′ō-tīp). The expression of the genes as traits is called the **phenotype** (fē′nō-tīp).

Environmental effects can interact with gene expression to determine the phenotype of many traits. Even genetically identical twins can have different phenotypes because of environmental effects. For example, in addition to genes, height is affected by nutrition, and skin color by exposure to the sun. Knowledge of environmental effects can be used to improve our genetic potential and to prevent harmful effects. For example, a healthy diet can promote growth or help prevent diabetes, and not smoking can reduce the risk of developing cancer.

Mendel proposed that genes can occur in dominant and recessive forms. Alternate forms of genes are now called **alleles** (ă-lēlz′). By definition, the effects of a **dominant allele** for a trait mask the effects of the **recessive allele** for that trait. For example, in pea plants, the allele for purple flowers is dominant over the allele for white flowers. By convention, dominant alleles are indicated by uppercase letters, and recessive alleles are indicated by lowercase letters. For example, the letter *P* designates the dominant allele for purple flower color, and the letter *p* represents the recessive allele for white flower color. A plant with the *Pp* genotype has purple flowers because purple is dominant over white.

An organism is **homozygous** (hō-mō-zī′gŭs) for a trait if the two alleles for the trait are identical, whereas an organism is **heterozygous** (het′er-ō-zī′gŭs) for a trait if the two alleles for the trait are different. The possible genotypes and phenotypes for purple flower color are

Alleles	Genotype	Phenotype
PP	Homozygous dominant	Purple
Pp	Heterozygous	Purple
pp	Homozygous recessive	White

Note that the recessive trait is expressed when it is not masked by the dominant trait.

The relationship between genotype and phenotype can be understood on the molecular level. For example, there is a recessive trait in humans called **albinism** (al′bĭ-nĭzm), which results in a lack of normal coloring of the skin, hair, and eyes. Several human genes produce the enzymes that are necessary for the synthesis of melanin, the pigment responsible for skin, hair, and eye color (see chapter 5). An **albino,** a person with albinism, has light blonde or white hair and light colored eyes and skin, with shades of pink, blue, and yellow (figure 3.48). The pink and blue colors result from blood seen in the eyes or through the skin. The yellow is from the natural accumulation of ingested yellow plant pigments in the skin (see "Skin Color," chapter 5).

The normal alleles for the melanin-synthesizing enzymes produce normal, functional enzymes capable of catalyzing the steps in melanin synthesis. When normal enzymes are produced, a person has normal pigmentation. An abnormal allele for any one of the enzymes in the melanin pathway will produce a defective enzyme. The defective enzyme will block the pathway and stop the production of melanin. Type 1 albinism results from a defective enzyme that fails to initiate the synthesis of melanin from tyrosine. The normal allele of the gene for this enzyme,

FIGURE 3.48 Albinism
Photograph of an albino male with his normally pigmented father.

designated *A*, is dominant over the abnormal allele, designated *a*. The possible genotypes and phenotypes for albinism are

Alleles	Genotype	Phenotype
AA	Homozygous dominant	Normal pigmentation
Aa	Heterozygous	Normal pigmentation
aa	Homozygous recessive	Albino

A person with the genotype *AA* has the phenotype of normal pigmentation because the functional enzyme for melanin synthesis is produced by both alleles. A person with the genotype *Aa* has the phenotype of normal pigmentation. Even though the *a* allele produces an abnormal, nonfunctional enzyme for melanin synthesis, the *A* allele produces sufficient amounts of the normal enzyme so that pigmentation is normal. A person with the genotype *aa* has the phenotype of albinism because only the abnormal, nonfunctional enzyme for melanin synthesis is produced.

Not all dominant traits are the normal condition, and not all recessive traits are abnormal. In many cases, the dominant trait is abnormal. For example, a person with **polydactyly** (pol-ē-dak′ti-lē) has extra fingers or toes (figure 3.49). One allele for polydactyly, which results in extra fingers or toes, is dominant over the recessive, normal allele which results in the normal number of fingers or toes.

71 ***What is genetics?***
72 ***What is the difference between genotype and phenotype?***
73 ***Define homozygous dominant, heterozygous, and homozygous recessive.***

PREDICT 8

List all the possible genotypes and phenotypes for polydactyly. Use the letters *D* and *d* for the genotypes.

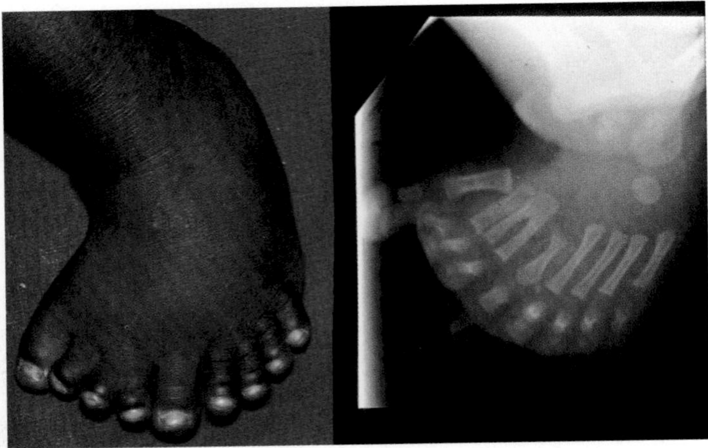

FIGURE 3.49 Polydactyly
Polydactyly, having extra fingers or toes, is determined by a dominant gene.

Modern Concepts of Genetics

Although the existence of chromosomes was known in Mendel's time, the relationship between genes and chromosomes was not confirmed until T. H. Morgan was able to induce mutations (gene changes) in fruit flies in 1910. DNA was first isolated as the genetic material in 1944, and James Watson and Francis Crick resolved the structure of DNA in 1953, which led to an understanding of the genetic code and protein synthesis. The sequence of nucleotides of a human gene, specifically the gene that causes cystic fibrosis, was determined for the first time in 1989. The sequencing of 99% of the human genome was accomplished in 2003, with the completion of the Human Genome Project (see p. 102).

Chromosomes

In modern terminology, Mendel hypothesized that organisms have genes that control the expression of traits. Most genes, which are segments of DNA, occur in pairs of alleles. Chromosomes are made up of DNA and associated proteins found in the nuclei of somatic cells and gametes. **Somatic** (sō-mat′ik) **cells** are all the cells of the body except for the gametes. Examples of somatic cells are epithelial cells, muscle cells, neurons, fibroblasts, lymphocytes, and macrophages. In males, the gametes are **sperm cells;** in females, the gametes are **oocytes** (egg cells; see chapter 28).

The normal number of chromosomes in a somatic cell is called the **diploid** (dip′loyd; twofold) **number.** The normal number of chromosomes in a gamete is the **haploid** (hap′loyd; single) **number.** In humans, the diploid number of chromosomes is 46. Chromosomes in diploid cells are paired, so humans have 23 pairs of chromosomes. The haploid number of chromosomes is 23.

PREDICT 9

Why does it make sense that the number of chromosomes in a gamete is half the number in a somatic cell?

The 23 pairs of chromosomes are divided into autosomal and sex chromosomes. Humans have 22 pairs of **autosomal** (aw-tō-sō′măl) **chromosomes,** which are all the chromosomes but the sex chromosomes, and one pair of **sex chromosomes,** which determines the sex of the individual. Sex chromosomes are denoted as **X** or **Y chromosomes.** A normal female has two X chromosomes (XX) in each somatic cell. One X chromosome of a female is derived from her mother; the other is derived from her father. A normal male has one X chromosome and one Y chromosome (XY) in each somatic cell. The X chromosome of a male is derived from his mother; the Y chromosome is derived from his father.

Genetic traits can be classified by the type of chromosome their alleles are located on and by whether their alleles are dominant or recessive. Thus, traits can be autosomal dominant or recessive, or they can be sex-linked dominant or recessive. For example, albinism is an autosomal recessive trait.

A **karyotype** (kar′ē-ō-tīp) is a display of the chromosomes of a somatic cell during metaphase of mitosis. It is produced

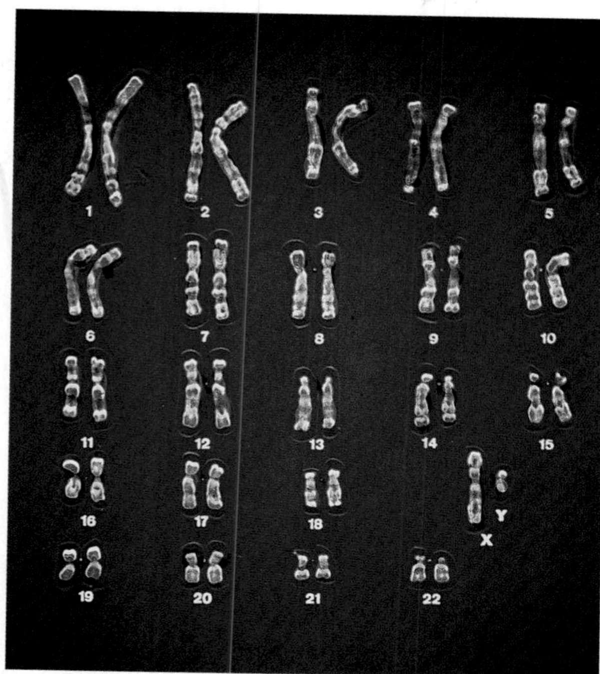

FIGURE 3.50 Human Karyotype

The 23 pairs of chromosomes in humans consist of 22 pairs of autosomal chromosomes (numbered 1–22) and 1 pair of sex chromosomes. This karyotype is of a male and has an X and a Y sex chromosome. A female karyotype would have two X chromosomes.

by photographing the cell's stained chromosomes through a microscope and arranging the photographed chromosomes in pairs (figure 3.50). For convenience, the autosomal chromosomes are numbered, from largest to smallest, 1 through 22. The sex chromosomes are denoted with an X or a Y. Note that a chromosome in a karyotype is a replicated chromosome—that is, it has two chromatids (see figure 3.50). The chromatids of each chromosome are attached at the centromere so that each replicated chromosome appears as a single structure.

Chromosome pairs are called **homologous** (hŏ-mŏl′ō-gŭs) **pairs,** and each member of the pair is called a **homolog** (hŏm′ō-lŏg). One homolog is derived from a person's father, and the other is derived from a person's mother. A **genome** (jē′nōm, jĕ′nōm) consists of all the genes found in the haploid number of chromosomes from one parent. The combined genomes from both parents are responsible for all of a person's genetic traits. Each gene of the genome occupies a specific **locus,** or location, on a chromosome (figure 3.51). The locus on one chromosome may have one allele type; whereas the locus on its homolog may have another allele type.

74 ***How do the chromosomes in somatic cells and gametes differ from each other?***

75 ***What are the number and type of chromosomes in the karyotype of a human somatic cell? How do the chromosomes of a male and female differ from each other?***

76 ***What are homologous chromosomes?***

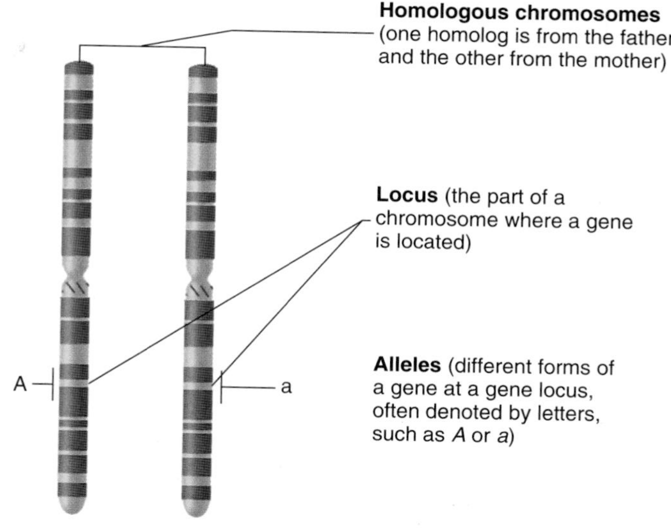

FIGURE 3.51 Homologous Chromosomes

A drawing of a pair of unduplicated homologous chromosomes. For simplicity, the chromosomes are shown as compact bodies. In the unduplicated state, chromosomes are dispersed within the nucleus as chromatin.

Multiple Alleles

Mendel believed that alleles came in two forms, dominant and recessive. It is now known that alleles can exist in many forms, called **multiple alleles.** An individual has only two alleles for a given gene, one on each homologous chromosome. At the population level, however, there can be many forms of an allele.

Differences in alleles arise by mutation (see p. 100), in which the DNA nucleotide sequence is altered. Many alleles are possible in a population because any change in the nucleotide sequence of DNA, even of one base pair, potentially produces a different allele. A different form of an allele is called an **allelic variant,** a **mutated allele (gene),** or a **polymorphism** (many forms).

Allelic variants can result in either no effect on the phenotype or minor to major phenotypic changes. Allelic variants at a locus may encode for different sequences of amino acids in proteins. Recall from chapter 2 that the sequence of amino acids in a protein affects the shape of the protein, including the protein's domain, which is the functional part of the protein. A mutated allele that causes an amino acid change in a protein that does not significantly change protein shape may not affect the phenotype. The greater the effect of an amino acid change on a protein's shape, however, the greater the effect on the protein's function, and the greater the effect on the phenotype. Millions of allelic variants have been identified. The significance of most of them is unknown.

The allelic variant present at a locus can affect a person's phenotype. For example, **phenylketonuria (PKU)** (fen′il-kē′tō-noo′rē-ă) is an autosomal recessive trait with multiple alleles. On chromosome 12 is a gene that encodes for an enzyme that converts the amino acid phenylalanine to the amino acid tyrosine. If phenylalanine is not converted to tyrosine, high levels of phenylalanine can accumulate, leading to brain cell damage. The more defective the enzyme, the greater the accumulation of phenylalanine and the greater the damage to the brain, in most cases. Over 400 disease-causing allelic variants of the

PKU gene are known. The severity of PKU depends in part on which two allelic variants are present. The less severe of the two alleles determines the severity of the disorder. Symptoms of untreated PKU can vary from profound mental retardation to nearly normal mental abilities. Thus, even though PKU is an autosomal recessive trait, the expression of the recessive condition exhibits considerable variability because of multiple alleles. Furthermore, reducing phenylalanine in the diet can prevent the development of mental retardation.

Gene Dominance

In Mendel's experiments with pea plants, each dominant allele produced enough of its protein product to cause the maximum phenotypic response in the heterozygote. Thus, the homozygous dominant and heterozygote had the same phenotype—that is, purple flowers. This is called **complete dominance.** It is now known that many other types of gene expression exist.

Complete dominance is at one end of a continuum of genetic expression. At the other end is codominance, and between them is incomplete dominance. In **codominance,** two alleles at the same locus are expressed so that separate, distinguishable phenotypes occur at the same time. ABO blood types are an example of codominance. A gene with three alleles on chromosome 9 encodes for enzymes that add sugar molecules to certain carbohydrates found on the surface of red blood cells. The carbohydrates are part of glycoproteins (see figure 3.2), called A and B antigens (see chapter 19). These carbohydrates consist of different sugars joined together. Traditionally, in designating alleles for blood type, the A and B alleles are superscripted to a capital letter "I" and the O allele is designated with the lower case letter "i." The I^A allele encodes for an enzyme that adds a particular sugar to the ends of the carbohydrate, producing the A antigen. The I^B allele encodes for a different enzyme that adds a different sugar to the ends of the carbohydrate, producing the B antigen. The i allele encodes for no functional enzyme and therefore is recessive to A and B. The possible genotypes and phenotypes for the ABO blood group are

Genotype	Phenotype
I^AI^A or I^Ai	Type A blood (A antigen only)
I^BI^B or I^Bi	Type B blood (B antigen only)
I^AI^B	Type AB blood (A and B antigens)
ii	Type O blood (neither A nor B antigen)

Type A blood results from the expression of only the I^A allele, type B blood from the expression of only the I^B allele, and type O blood from the expression of neither the I^A allele nor the I^B allele. Type AB blood illustrates codominance and results from the expression of both the I^A allele and the I^B allele at the same time.

In **incomplete dominance,** the dominant allele does not completely mask the effects of the recessive allele in the heterozygote. The heterozygote produces less of the protein product than the homozygous dominant and has phenotypic characteristics intermediate between the homozygous dominant and the homozygous recessive. For example, **beta thalassemia** (thăl-ă-sē′mē-ă) is a disorder of a gene on chromosome 11. It affects the synthesis of β-globulin polypeptide chains, which are part of the hemoglobin in red blood cells. Hemoglobin is a protein that transports oxygen. If normal amounts of the β-globulin polypeptide are produced, the β-globulin polypeptides join with other proteins to form hemoglobin. If lower than normal amounts of β-globulin polypeptide are synthesized, lower than normal amounts of hemoglobin are produced. In the homozygous dominant state, two normal alleles for β-globulin synthesis are present, normal amounts of hemoglobin are produced, and the phenotype is normal, meaning that there is normal transport of oxygen. In the homozygous recessive condition, called **major thalassemia,** two abnormal alleles are present, and much lower than normal amounts of β-globulin polypeptide chains are synthesized. The result is severe **anemia** (ă-nē′mē-ă), which is a deficiency of hemoglobin in the blood. Symptoms include pallor, weakness, fatigue, and spleen enlargement. Blood transfusions are necessary to maintain hemoglobin levels. In the heterozygous condition, called **minor thalassemia,** one normal allele and one abnormal allele are present. The production of hemoglobin is intermediate between the normal phenotype and major thalassemia, and mild anemia results.

Polygenic Traits

In some cases, one gene (usually with two alleles) determines one phenotype. In other cases, many genes determine a phenotype. These **polygenic traits** result from the interactions of many genes. Although all the genes contribute to the phenotype without being dominant or recessive to each other, each gene has its own characteristics, such as multiple alleles or incomplete dominance. Some of the genes may be more important than others for the expression of the phenotype, and the genes can be on different chromosomes. The result of all the gene interactions is a phenotype with a great amount of variability. Examples of polygenic traits are height, intelligence, eye color, and skin color.

Even though a complex combination of genes determines a polygenic trait, a defect in one of them can sometimes have a dramatic effect on phenotype. For example, even though many genes contribute to skin color, one defective gene can eliminate skin color completely, resulting in albinism.

Sex-Linked Traits

Sex-linked traits are traits affected by genes on the sex chromosomes. **X-linked traits** are on the X chromosome, and **Y-linked traits** are on the Y chromosome. Most sex-linked traits are X-linked, whereas only a few Y-linked traits exist, mainly because the Y chromosome is very small. An example of an X-linked recessive trait is **hemophilia A** (classic hemophilia), in which the ability to produce certain blood clotting factors is not present (see chapter 19). Consequently, clotting is impaired and persistent bleeding occurs either spontaneously or as a result of an injury. Traditionally, X-linked genes are designated by letters superscripted to the letter "X." The possible genotypes and phenotypes for hemophilia A are

Genotype	Phenotype
X^HX^H or X^HX^h	Normal female
X^hX^h	Hemophiliac female
X^HY	Normal male
X^hY	Hemophiliac male

Note that a female must have both recessive alleles (X^hX^h) to exhibit hemophilia, whereas a male, because he has only one X chromosome, has hemophilia if he has only one of the recessive alleles. For this reason, X-linked recessive traits are seen more frequently in males than in females.

77 ***Explain multiple allelism.***

78 ***Distinguish among complete dominance, incomplete dominance, and codominance. Give examples of each.***

79 ***What is the difference between multiple allelism and polygenic traits?***

80 ***How are sex-linked traits inherited? Give an example.***

Meiosis and the Transmission of Genes

Gametes are haploid cells that are derived from diploid cells. In **meiosis** (mī-ō′sĭs) gametes are produced that have one homolog from each of the homologous pairs of chromosomes. Therefore gametes have one half the number of chromosomes and one half the alleles as the original diploid cells.

Before meiosis begins, DNA replication occurs so that replicated chromosomes are formed, each consisting of two chromatids (figure 3.52). As a result of the first meiotic division, each daughter cell receives one replicated member of a homologous pair. No replication of DNA takes place between the first and second meiotic divisions. During the second meiotic division, the chromatids of each chromosome separate. As a result of the second meiotic division, each cell receives one chromatid from each chromosome. That single chromatid is now referred to as a chromosome. Each gamete now has a haploid number of chromosomes. See chapter 28 for a more detailed discussion of gamete production.

In figure 3.52, the alleles for albinism are shown on the homologous chromosomes. On one homolog, the normal, dominant allele is indicated with an uppercase *A;* on the other homolog, the abnormal allele is indicated with a lowercase *a.* Figure 3.52 illustrates how the alleles on homologous chromosomes are distributed to gametes. In this case, gametes with either an *A* or an *a* allele are produced.

1. A pair of unduplicated homologous chromosomes is shown in the nucleus of a somatic cell. For simplicity, the chromosomes are shown as compact bodies. In the unduplicated state, chromosomes are dispersed within the nucleus as chromatin.
2. The dominant (*A*) and recessive (*a*) alleles for albinism are shown on the homologous chromosomes.
3. The DNA has replicated, and each replicated chromosome consists of two chromatids (see figure 3.46).
4. Each replicated homologous chromosome has two alleles because the DNA has replicated.
5. As a result of the first meiotic division, each new nucleus has a replicated homologous chromosome.
6. DNA is not replicated between the first and second meiotic divisions. As a result of the second meiotic division, the chromatids separate and are now called chromosomes.
7. Each homologous chromosome has an allele for albinism. Two types of gametes have been produced, those carrying the *A* allele and those carrying the *a* allele.

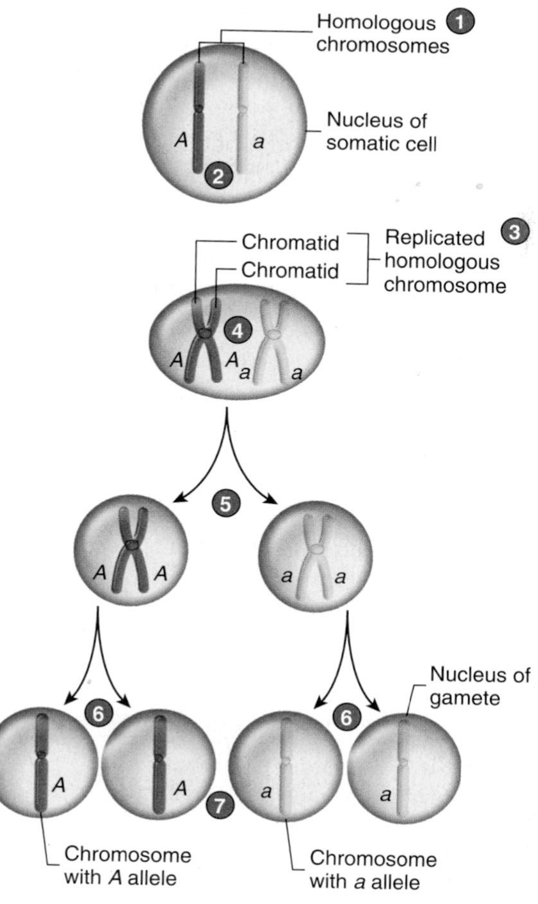

PROCESS FIGURE 3.52 Meiosis and the Distribution of Alleles

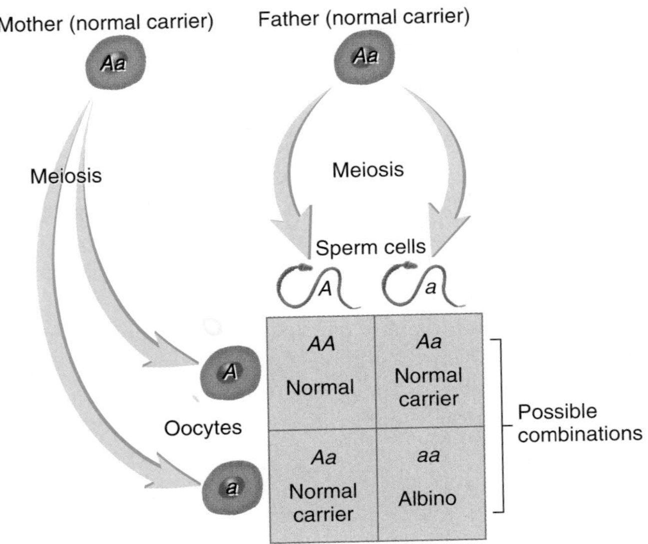

FIGURE 3.53 Inheritance of a Recessive Trait: Albinism

A represents the normal, pigmented condition, and *a* represents the recessive, unpigmented condition. The figure shows a Punnett square of a mating between two normal carriers.

The probability of the transmission of alleles to the next generation can be determined using a **Punnett square** if the genotypes of the parents are known (figure 3.53). The possible alleles found in the gametes of one parent form the rows of the square, and the possible alleles found in the gametes of the other parent form the columns. The body of the Punnett square shows all possible combinations of the parents' gametes. For example, suppose both parents are heterozygous (*Aa*) for albinism. The heterozygous parents can produce two types of gametes. One gamete has the *A* allele, and the other has the *a* allele. The probability that the child will be homozygous dominant (*AA*) is one out of four; heterozygous (*Aa*), two out of four; and homozygous recessive (*Aa*), one out of four.

Albinism is an autosomal recessive trait. The homozygous dominant and heterozygous conditions result in normal pigmentation, whereas the homozygous recessive condition results in albinism. Both heterozygous parents have normal pigmentation, but there is a one out of four chance that any one of their children will be an albino. A **carrier** for a recessive trait is a person who is heterozygous for that trait, with one normal allele and one disorder-causing allele. The carrier does not exhibit the disorder but can pass it on to his or her children.

PREDICT 10

If a carrier for albinism mates with a homozygous normal person, what is the likelihood that any of the children will be albinos? Explain.

81 ***What is meiosis? How does it differ from mitosis?***

82 ***What is a carrier?***

Genetic Disorders

A **genetic disorder** is a failure of structure, function, or both as a result of abnormalities in a person's genetic makeup—that is, his or her DNA. Humans suffer from a variety of genetic disorders, some of which are highlighted throughout the remainder of this book. Genetic disorders determined by one gene are described by Mendelian characteristics—for example, as autosomal recessive. In some cases, this type of description is accurate, but, in other cases, the reality is more complex. For example, PKU is said to be autosomal recessive, but PKU has multiple alleles, which result in a spectrum of defects. As you study the genetic disorders in later chapters, keep in mind the concepts discussed in this chapter.

Genetic disorders involve either a single gene or an entire chromosome, as in the case of aneuploidy (table 3.4). Recall that a mutation is a change in a gene that usually involves a change in the number or kinds of nucleotides composing the DNA (see chapter 2). Mutations are known to occur by chance (randomly without known cause), but they can also be caused by chemicals, radiation, and viruses. Agents that cause mutations are called **mutagens** (mū'tă-jenz). In most cases, a specific cause of a mutation cannot be determined. Once a mutation has occurred, however, the abnormal trait can be passed from one generation to the next.

A mutation involving a single nucleotide change is called a point mutation. Mutations that change the sequence or number of nucleotides involve changes in chromosome structure, and therefore are called structural mutations. Structural mutations can involve deletions, the loss of part of a chromosome; duplications, the addition of a copied section of a chromosome; translocations, in which part of one chromosome becomes attached to another chromosome; and inversions, in which the DNA sequence of a section of a chromosome is reversed.

Furthermore, segregation errors that occur during meiosis result in changes in the chromosome number in the gametes. As the chromosomes separate during meiosis, the two members of a homologous pair may become "sticky" and not segregate as they normally do. As a result, one of the daughter cells receives both chromosomes and the other daughter cell receives none. This event is called **nondisjunction.** When the gametes are fertilized, the resulting zygote has either 47 chromosomes or 45 chromosomes rather than the normal 46, a condition called **aneuploidy** (an'ū-ploy-dē). Aneuploidies include monosomies, in which a chromosome is missing, or trisomies, in which an extra chromosome is present. Aneuploidies are usually lethal and is one reason for a high rate of early embryo loss. **Down syndrome,** or **trisomy 21,** is a type of aneuploidy in which three chromosomes 21 are present. A **syndrome** (sĭn'drōm) is a set of signs and symptoms occurring together as the result of a single cause, such as a single mutation or one extra chromosome (a trisomy). Down syndrome is an example of an aneuploidy that is not always lethal. However, individuals with this genetic disorder exhibit physical and mental developmental problems, as well as an increased probability of developing certain cancers. These individuals also have a shortened life span.

TABLE 3.4 Genetic Disorders

Disorder	Description
Dominant Traits	
Achondroplasia	Dwarfism characterized by shortening of the upper and lower limbs
Huntington chorea	Severe degeneration of the basal nuclei and frontal cerebral cortex; characterized by purposeless movements and mental deterioration; onset is usually between 40 and 50 years of age
Hypercholesterolemia	Elevated blood cholesterol levels that contribute to atherosclerosis and cardiovascular disease
Marfan syndrome	Abnormal connective tissue, resulting in increased height, elongated digits, and weakness in the aortic wall
Neurofibromatosis	Small, pigmented lesions (café-au-lait spots) in the skin and disfiguring tumors (noncancerous) caused by the proliferation of Schwann cells along nerves
Osteogenesis imperfecta	Abnormal collagen synthesis, resulting in brittle bones that break repeatedly
Recessive Traits	
Albinism	Lack of an enzyme necessary to produce the pigment melanin; characterized by lack of skin, hair, and eye coloration
Cystic fibrosis	Impaired transport of chloride ions across plasma membranes; resulting in excessive production of thick mucus, which blocks the respiratory and gastrointestinal tract; the most common fatal genetic disorder
Phenylketonuria	Lack of the enzyme necessary to convert the amino acid phenylalanine to the amino acid tyrosine; an accumulation of phenylalanine leads to mental retardation
Severe combined immune deficiency	Inability to form the white blood cells (B cells, T cells, and phagocytes) necessary for an immune system response
Sickle-cell disease	Inability to produce normal hemoglobin; resulting in abnormally shaped red blood cells that clog capillaries or rupture
Tay-Sachs disease	Lack of the enzyme necessary to break down certain fatty substances; an accumulation of fatty substances impairs action potential propagation, resulting in deterioration of mental and physical functions and death by 3–4 years of age
Thalassemia	Decreased rate of hemoglobin synthesis; resulting in anemia, enlargement of the spleen, increased cell numbers in red bone marrow, and congestive heart failure
Sex-Linked Traits	
Duchenne muscular dystrophy	Deletion or alteration of part of the X chromosome; resulting in progressive weakness and wasting of muscles
Hemophilia	Most commonly, failure to produce blood clotting factors, caused by a recessive gene; resulting in prolonged bleeding
Red-green color blindness	Most commonly, deficiency in functional green-sensitive cones, caused by a recessive gene; inability to distinguish between red and green colors
Chromosomal Disorders	
Down syndrome	Caused by having three chromosomes 21; resulting in mental retardation, short stature, and poor muscle tone
Klinefelter syndrome	Caused by two or more X chromosomes in a male (XXY); resulting in small testes, sterility, and development of femalelike breasts
Turner syndrome	Caused by having only one X chromosome; resulting in immature uterus, lack of ovaries, and short stature

Sex Chromosome Abnormalities

A wide range of sex chromosome abnormalities exist. The presence of a Y chromosome makes a person genetically a male, and the absence of a Y chromosome makes a person genetically a female, regardless of the number of X chromosomes. Individuals with XO (Turner syndrome), XX, XXX, or XXXX karyotypes are therefore females, and individuals with XY, XXY, XXXY, or XYY karyotypes are males. A YO condition is lethal because the genes on the X chromosome are necessary for survival. Secondary sexual characteristics are usually underdeveloped in both the XO female and the XXY male (called Klinefelter syndrome), and additional X chromosomes (XXXX or XXXY) are often associated with some degree of mental retardation. As a result of hormonal imbalances, the morphological features (phenotype) may be reversed from the genetic constitution (genotype).

Cancer usually results from a series of somatic mutations that cause uncontrolled cell division and changes in the normal functions of cells. The genetics of cancer is discussed in the "Clinical Genetics" essay in chapter 4.

83 ***What is a genetic disorder?***

84 ***What is a mutagen?***

85 ***What is the cause of the genetic disorder Down Syndrome?***

86 ***Why is cancer considered a genetic disease?***

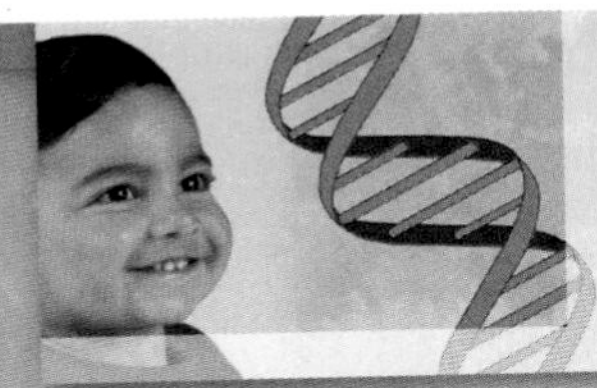

CLINICAL GENETICS

The Human Genome Project

The human genome consists of all the genes found in one homologous set of human chromosomes. It is estimated that humans have 25,000–30,000 genes. A **genomic map** is a description of the DNA nucleotide sequences of the genes and their locations on the chromosomes (figure A). Celera, a private corporation working on the **Human Genome Project,** announced on February 12, 2001, that it had completed sequencing 99% of the genome. Online Mendelian Inheritance in Man (OMIM) is a database that contains information about many human genes and genetic disorders (www.ncbi.nlm.nih.gov/omim).

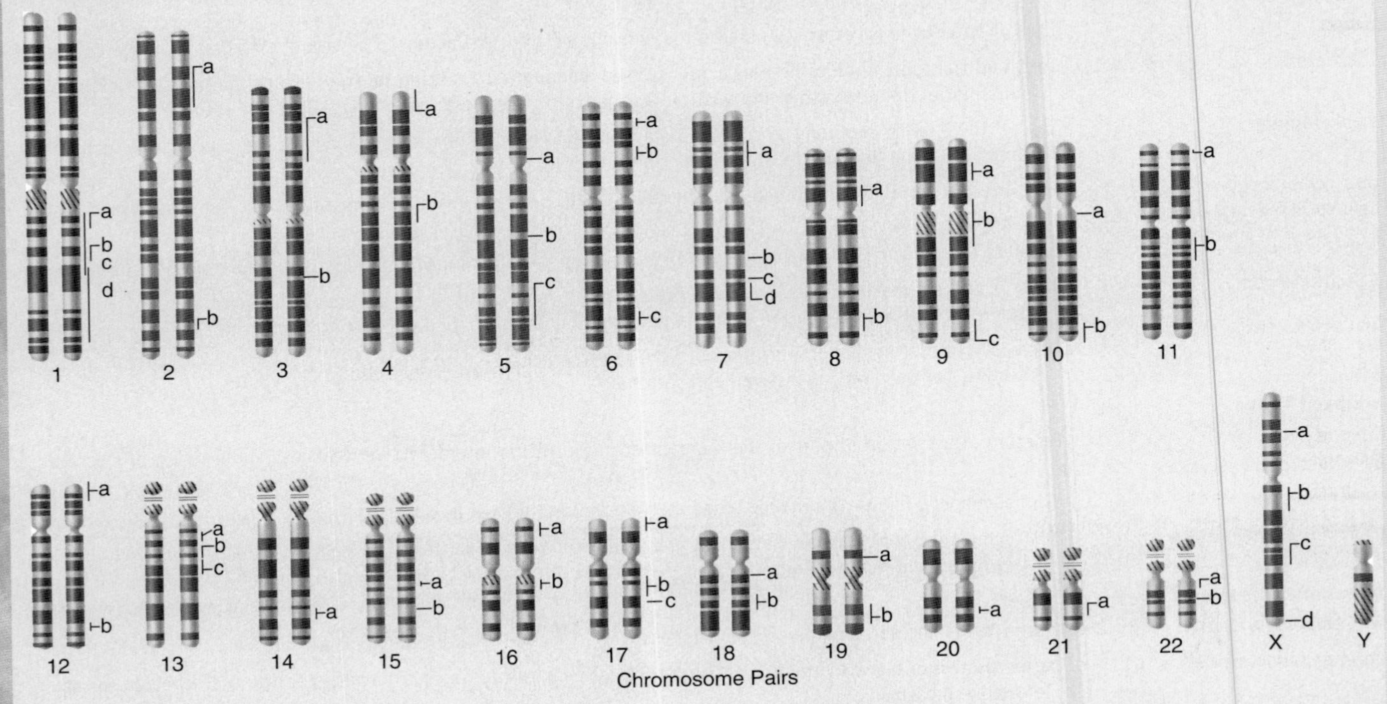

1. a. Gaucher disease
   b. Prostate cancer
   c. Glaucoma
   d. Alzheimer disease*
2. a. Familial colon cancer*
   b. Waardenburg syndrome
3. a. Lung cancer
   b. Retinitis pigmentosa*
4. a. Huntington chorea
   b. Parkinson disease
5. a. Cockayne syndrome
   b. Familial polyposis of the colon
   c. Asthma
6. a. Spinocerebellar ataxia
   b. Diabetes*
   c. Epilepsy*
7. a. Diabetes*
   b. Osteogenesis imperfecta
   c. Cystic fibrosis
   d. Obesity*
8. a. Werner syndrome
   b. Burkitt lymphoma
9. a. Malignant melanoma
   b. Friedreich ataxia
   c. Tuberous sclerosis
10. a. Multiple endocrine neoplasia, type 2
    b. Gyrate atrophy
11. a. Sickle-cell disease
    b. Multiple endocrine neoplasia
12. a. Zellweger syndrome
    b. Phenylketonuria (PKU)
13. a. Breast cancer*
    b. Retinoblastoma
    c. Wilson disease
14. a. Alzheimer disease*
15. a. Marfan syndrome
    b. Tay-Sachs disease
16. a. Polycystic kidney disease
    b. Crohn disease*
17. a. Tumor suppressor protein
    b. Breast cancer*
    c. Osteogenesis imperfecta
18. a. Amyloidosis
    b. Pancreatic cancer*
19. a. Familial hypercholesterolemia
    b. Myotonic dystrophy
20. a. Severe combined immunodeficiency
21. a. Amyotrophic lateral sclerosis*
22. a. DiGeorge syndrome
    b. Neurofibromatosis, type 2

X a. Duchenne muscular dystrophy
  b. Menkes syndrome
  c. X-linked severe combined immunodeficiency
  d. Factor VIII deficiency (hemophilia A)

*Gene responsible for only some cases.

FIGURE A **Human Genomic Map**
Representative genetic defects mapped to date. The bars and lines indicate the location of the genes listed for each chromosome.

Armed with a knowledge of the human genome and its effects on a person's physical, mental, and behavioral abilities, medicine and society will be transformed in many ways. Medicine, for example, will shift emphasis from the curative to the preventive. The disorders or diseases a person is likely to develop can be prevented or their severity lessened. When prevention is not possible, knowledge of the enzymes or other molecules involved in a disorder may result in new drugs and techniques that can compensate for the genetic disorder. Knowledge of the genes involved in a disorder may result in **gene therapy,** or **genetic engineering,** that repairs or replaces defective genes, resulting in cures of or treatments for genetic disorders.

Despite the great promise of benefits from the Human Genome Project, the knowledge produced has raised a number of ethical and legal questions for society. Should a person's genomic information be public knowledge? Should persons with a genome that predisposes them to cancer or behavioral disorders be barred from certain types of employment or be refused medical insurance because they are a high risk? Can a person demand to know a prospective mate's genome? Should parents know the genome of their fetus and be allowed to make decisions regarding abortion based on this knowledge? Should the same genetic-engineering techniques that provide alteration of the genome to cure genetic disorders be used to create genomes that are deemed to be superior? Such questions raise the specter of genetic discrimination.

SUMMARY

1. The plasma membrane forms the outer boundary of the cell.
2. The nucleus directs the activities of the cell.
3. Cytoplasm, between the nucleus and plasma membrane, is where most cell activities take place.

Functions of the Cell (p. 56)

1. Cells metabolize and release energy.
2. Cells synthesize molecules.
3. Cells provide a means of communication.
4. Cells reproduce and provide for inheritance.

How We See Cells (p. 58)

1. Light microscopes allow us to visualize the general features of cells.
2. Electron microscopes allow us to visualize the fine structure of cells.

Plasma Membrane (p. 58)

1. The plasma membrane passively or actively regulates what enters or leaves the cell.
2. The plasma membrane is composed of a phospholipid bilayer, in which proteins are suspended (fluid-mosaic model).

Membrane Lipids (p. 58)

Lipids give the plasma membrane most of its structure and some of its function.

Membrane Proteins (p. 58)

1. Membrane proteins function as marker molecules, attachment proteins, transport proteins, receptor proteins, and enzymes.
2. Transport proteins include channel proteins, carrier proteins (transporters), and ATP-powered pumps.
3. Some receptor molecules are linked to and control channel proteins.
4. Some receptor molcules are linked to G protein complexes, which in turn control numerous cellular activities.

Movement Through the Plasma Membrane (p. 64)

1. Lipid-soluble molecules pass through the plasma membrane readily by dissolving in the lipid bilayer.
2. Small molecules pass through membrane channels. Most channels are positively charged, allowing negatively charged ions and neutral molecules to pass through more readily than positively charged ions.
3. Large polar substances (e.g., glucose and amino acids) are transported through the membrane by carrier molecules.
4. Larger pieces of material enter cells in vesicles.

Diffusion

1. Diffusion is the movement of a substance from an area of higher solute concentration to one of lower solute concentration (down a concentration gradient).
2. The concentration gradient is the difference in solute concentration between two points divided by the distance separating the points.
3. The rate of diffusion increases with an increase in the concentration gradient, an increase in temperature, a decrease in molecular size, and a decrease in viscosity.
4. The end result of diffusion is a uniform distribution of molecules.
5. Diffusion requires no expenditure of energy.

Osmosis

1. Osmosis is the diffusion of water (solvent) across a selectively permeable membrane.
2. Osmotic pressure is the force required to prevent the movement of water across a selectively permeable membrane.

3. Isosmotic solutions have the same concentration of solute particles, hyperosmotic solutions have a greater concentration, and hyposmotic solutions have a lesser concentration of solute particles than a reference solution.
4. Cells placed in an isotonic solution neither swell nor shrink. In a hypertonic solution, they shrink (crenate); in a hypotonic solution, they swell and may burst (lyse).

Filtration

1. Filtration is the movement of a liquid through a partition with holes that allow the liquid, but not everything in the liquid, to pass through them.
2. Liquid movement results from a pressure difference across the partition.

Mediated Transport

1. Mediated transport is the movement of a substance across a membrane by means of a transport protein. The substances transported tend to be large, water-soluble molecules.
   - The transport protein has binding sites that bind with either a single molecule or a group of similar molecules. This selectiveness is called specificity.
   - Similar molecules can compete for transport proteins, with each reducing the rate of transport of the other.
   - Once all the transport proteins are in use, the rate of transport cannot increase further (saturation).
2. Three kinds of mediated transport can be identified.
   - Facilitated diffusion moves substances down their concentration gradient and does not require energy (ATP) expenditure.
   - Active transport can move substances against their concentration gradient and requires ATP. An exchange pump is an active-transport mechanism that simultaneously moves two substances in opposite directions across the plasma membrane.
   - In secondary active transport, an ion is moved across the plasma membrane by active transport, and the energy produced by the ion diffusing back down its concentration gradient can transport another molecule, such as glucose, against its concentration gradient.

Endocytosis and Exocytosis (p. 71)

1. Endocytosis is the bulk movement of materials into cells.
   - Phagocytosis is the bulk movement of solid material into cells by the formation of a vesicle.
   - Pinocytosis is similar to phagocytosis, except that the ingested material is much smaller or is in solution.
2. Exocytosis is the secretion of materials from cells by vesicle formation.
3. Endocytosis and exocytosis use vesicles, can be specific (receptor-mediated endocytosis) for the substance transported, and require energy.

Cytoplasm (p. 76)

The cytoplasm is the material outside the nucleus and inside the plasma membrane.

Cytosol

1. Cytosol consists of a fluid part (the site of chemical reactions), the cytoskeleton, and cytoplasmic inclusions.
2. The cytoskeleton supports the cell and enables cell movements. It consists of protein fibers.
   - Microtubules are hollow tubes composed of the protein tubulin. They form spindle fibers and are components of centrioles, cilia, and flagella.
   - Actin filaments are small protein fibrils that provide structure to the cytoplasm or cause cell movements.
   - Intermediate filaments are protein fibers that provide structural strength to cells.
3. Cytoplasmic inclusions, such as lipochromes, are not surrounded by membranes.

The Nucleus and Cytoplasmic Organelles (p. 77)

Organelles are subcellular structures specialized for specific functions.

Nucleus

1. The nuclear envelope consists of a double membrane with nuclear pores.
2. DNA and associated proteins are found inside the nucleus as chromatin.
3. DNA is the hereditary material of the cell. It controls the activities of the cell by producing proteins through RNA.
4. A gene is a portion of a DNA molecule. Genes determine the proteins in a cell.
5. Nucleoli consist of RNA and proteins and are the sites of ribosomal subunit assembly.

Ribosomes

1. Ribosomes consist of small and large subunits manufactured in the nucleolus and assembled in the cytoplasm.
2. Ribosomes are the sites of protein synthesis.
3. Ribosomes can be free or associated with the endoplasmic reticulum.

Endoplasmic Reticulum

1. The endoplasmic reticulum is an extension of the outer membrane of the nuclear envelope; it forms tubules or sacs (cisternae) throughout the cell.
2. The rough endoplasmic reticulum has ribosomes and is a site of protein synthesis and modification.
3. The smooth endoplasmic reticulum lacks ribosomes and is involved in lipid production, detoxification, and calcium storage.

Golgi Apparatus

The Golgi apparatus is a series of closely packed, modified cisternae that function to modify, package, and distribute lipids and proteins produced by the endoplasmic reticulum.

Secretory Vesicles

Secretory vesicles are membrane-bound sacs surrounded by membranes that carry substances from the Golgi apparatus to the plasma membrane, where the contents of the vesicle are released by exocytosis.

Lysosomes

1. Lysosomes are membrane-bound sacs containing hydrolytic enzymes. Within the cell, the enzymes break down phagocytized material and nonfunctional organelles (autophagia).
2. Enzymes released from the cell by lysis or enzymes secreted from the cell can digest extracellular material.

Peroxisomes

Peroxisomes are membrane-bound sacs containing enzymes that digest fatty acids and amino acids, as well as enzymes that catalyze the breakdown of hydrogen peroxide.

Proteasomes

Proteasomes are large, multienzyme complexes, not bound by membranes, which digest selected proteins within the cell.

Mitochondria

1. Mitochondria are the major sites of the production of ATP, which is used as an energy source by cells.
2. The mitochondria have a smooth outer membrane and an inner membrane that is infolded to produce cristae.
3. Mitochondria contain their own DNA, can produce some of their own proteins, and can replicate independently of the cell.

Centrioles and Spindle Fibers

1. Centrioles are cylindrical organelles located in the centrosome, a specialized zone of the cytoplasm. The centrosome is the site of microtubule formation.
2. Spindle fibers are involved in the separation of chromosomes during cell division.

Cilia and Flagella

1. Movement of materials over the surface of the cell is facilitated by cilia.
2. Flagella, much longer than cilia, propel sperm cells.

Microvilli

Microvilli increase the surface area of the plasma membrane for absorption or secretion.

Genes and Gene Expression (p. 86)

1. During transcription, information stored in DNA is copied to form mRNA.
2. During translation, the mRNA goes to ribosomes, where it directs the synthesis of proteins.

Transcription

1. DNA unwinds and, through nucleotide pairing, produces pre-mRNA (transcription).
2. Introns are removed and exons are spliced by spliceosomes during post-transcriptional processing.
3. Modifications to the ends of mRNA also occur during post-transcriptional processing.

Genetic Code

The genetic code, which specifies amino acids, consists of codons, which are sequences of three nucleotides in mRNA.

Translation

1. The mRNA moves through the nuclear pores to ribosomes.
2. Transfer RNA (tRNA), which carries amino acids, interacts at the ribosome with mRNA. The anticodons of tRNA bind to the codons of mRNA, and the amino acids are joined to form a protein (translation).
3. Proproteins, some of which are proenzymes, are modified into proteins, some of which are enzymes, during posttranslational processing.

Regulation of Genetic Expression

1. Cells become specialized because of inactivation of certain parts of the DNA molecule and activation of other parts.
2. The level of DNA activity and thus protein production can be controlled internally or can be affected by regulatory substances secreted by other cells.

Cell Life Cycle (p. 91)

The cell life cycle has two stages: interphase and cell division.

Interphase

- Interphase is the period between cell divisions. This is the time of DNA replication.

DNA Replication

- DNA unwinds, and each strand produces a new DNA molecule during replication.

Cell Division

1. Cell division includes nuclear division and cytoplasmic division.
2. Mitosis is the replication of the nucleus of the cell, and cytokinesis is division of the cytoplasm of the cell.
3. Mitosis is a continuous process divided into four phases.
   - *Prophase.* Chromatin condenses to become visible as chromosomes. Each chromosome consists of two chromatids joined at the centromere. Centrioles move to opposite poles of the cell, and astral fibers and spindle fibers form. Nucleoli disappear, and the nuclear envelope degenerates.
   - *Metaphase.* Chromosomes align at the equatorial plane.
   - *Anaphase.* The chromatids of each chromosome separate at the centromere. Each chromatid then is called a chromosome. The chromosomes migrate to opposite poles.
   - *Telophase.* Chromosomes unravel to become chromatin. The nuclear envelope and nucleoli reappear.
4. Cytokinesis begins with the formation of the cleavage furrow during anaphase. It is complete when the plasma membrane comes together at the equator, producing two new daughter cells.

Genetics (p. 93)

1. Genetics is the study of heredity, that is, characteristics inherited by children from their parents.
2. Genomic medicine uses an understanding of the biochemical relationship between genes and disease to diagnosis and manage disease.

Mendelian Genetics

1. The genes an organism has for a given trait is called the genotype. The expression of the genes is called the phenotype.
2. Alleles are alternate forms of genes. A dominant allele masks the effects of a recessive allele for the same trait.
3. An organism homozygous for a trait has two identical alleles for the trait, whereas an organism heterozygous for a trait has two different alleles for the trait.

Modern Concepts of Genetics

1. Chromosomes.
   - Somatic cells have a diploid number of chromosomes, whereas gametes have a haploid number. In humans, the diploid number is 46 and the haploid number is 23.
   - Humans have 22 pairs of autosomal chromosomes and one pair of sex chromosomes. Females have the sex chromosomes XX and males have XY.
   - A karyotype is a display of the chromosomes of a somatic cell during metaphase of mitosis.
   - Chromosome pairs are called homologous chromosomes.
2. The genome consists of all the genes found in the haploid number of chromosomes from one parent.

3. Each gene occupies a specific locus, or location on a chromosome.
4. Alleles exist in many different forms, called multiple alleles.
5. Gene dominance.
   - In complete dominance, the dominant allele masks the effects of the recessive allele.
   - In codominance, two alleles at the same locus are expressed so that separate, distinguishable phenotypes occur at the same time.
   - In incomplete dominance, the dominant allele does not completely mask the effects of the recessive allele.
6. Polygenic traits result from the interaction of many genes.
7. Sex linked traits.
   - Sex-linked traits are traits affected by genes on the sex chromosomes.
   - X-linked traits are affected by genes on the X chromosome, and Y-linked traits are traits affected by genes on the Y chromosome.
   - X-linked traits are seen more frequently in males than in females because males have only one X chromosome.

Meiosis and the Transmission of Genes

1. Meiosis results in the production of gametes.
2. A Punnett square can determine the probability of the transmission of alleles to the next generation.
3. A carrier for a recessive trait is a person who is heterozygous for the trait, with one normal allele and one disorder-causing allele.

Genetic Disorders

1. A mutation is a change in the number or kinds of nucleotides in DNA.
2. Some genetic disorders result from an abnormal distribution of chromosomes during gamete formation.
3. Cancer results from a series of somatic mutations.

REVIEW AND COMPREHENSION

1. In the plasma membrane, ____________ form(s) the lipid bilayer, ____________ determine(s) the fluid nature of the membrane, and ____________ mainly determine(s) the function of the membrane.
   a. phospholipids, cholesterol, proteins
   b. phospholipids, proteins, cholesterol
   c. proteins, cholesterol, phospholipids
   d. cholesterol, phospholipids, proteins
   e. cholesterol, proteins, phospholipids
2. Which of the following are functions of the proteins found in the plasma membrane?
   a. channel proteins
   b. marker molecules
   c. receptor molecules
   d. enzymes
   e. all of the above
3. Integrins in the plasma membrane function as
   a. channel proteins.
   b. marker molecules.
   c. attachment proteins.
   d. enzymes.
   e. receptor molecules.
4. In general, lipid-soluble molecules diffuse through the ____________; small, water-soluble molecules diffuse through the ____________.
   a. membrane channels, membrane channels
   b. membrane channels, lipid bilayer
   c. lipid bilayer, carrier molecules
   d. lipid bilayer, membrane channels
   e. carrier proteins, membrane channels
5. Small pieces of matter, and even whole cells, can be transported across the plasma membrane in
   a. membrane channels.
   b. carrier molecules.
   c. receptor molecules.
   d. marker molecules.
   e. vesicles.
6. The rate of diffusion increases if the
   a. concentration gradient decreases.
   b. temperature of a solution decreases.
   c. viscosity of a solution decreases.
   d. all of the above.
7. Concerning the process of diffusion, at equilibrium
   a. the net movement of solutes stops.
   b. random molecular motion continues.
   c. there is an equal movement of solute in opposite directions.
   d. the concentration of solute is equal throughout the solution.
   e. all of the above.
8. Which of these statements about osmosis is true?
   a. Osmosis always involves a membrane that allows water and all solutes to diffuse through it.
   b. The greater the solute concentration, the smaller the osmotic pressure of a solution.
   c. Osmosis moves water from a greater solute concentration to a lesser solute concentration.
   d. The greater the osmotic pressure of a solution, the greater the tendency for water to move into the solution.
   e. Osmosis occurs because of hydrostatic pressure outside the cell.
9. If a cell is placed in a ____________ solution, lysis of the cell may occur.
   a. hypertonic
   b. hypotonic
   c. isotonic
   d. isosmotic
10. Container A contains a 10% salt solution, and container B contains a 20% salt solution. If the two solutions are connected, the net movement of water by diffusion is from ____________ to ____________, and the net movement of salt by diffusion is from ____________ to ____________.
    a. A, B; A, B
    b. A, B; B, A
    c. B, A; A, B
    d. B, A; B, A
11. Suppose that a woman ran a long-distance race in the summer. During the race, she lost a large amount of hyposmotic sweat. You would expect her cells to
    a. shrink.
    b. swell.
    c. stay the same.
12. Suppose that a man is doing heavy exercise in the hot summer sun. He sweats profusely. He then drinks a large amount of distilled water. After he drank the water, you would expect his tissue cells to
    a. shrink.
    b. swell.
    c. remain the same.

13. Unlike diffusion and osmosis, filtration depends on a __________ on the two sides of the partition.
    a. concentration gradient
    b. pressure difference
    c. difference in electric charge
    d. difference in osmotic pressure
    e. hyposmotic solution
14. Which of these statements about facilitated diffusion is true?
    a. In facilitated diffusion, net movement is down the concentration gradient.
    b. Facilitated diffusion requires the expenditure of energy.
    c. Facilitated diffusion does not require a carrier protein.
    d. Facilitated diffusion moves materials through membrane channels.
    e. Facilitated diffusion moves materials in vesicles.
15. Which of these statements concerning the symport of glucose into cells is true?
    a. The sodium–potassium exchange pump moves Na^+ into cells.
    b. The concentration of Na^+ outside cells is less than inside cells.
    c. A carrier protein moves Na^+ into cells and glucose out of cells.
    d. The concentration of glucose can be greater inside cells than outside cells.
    e. As Na^+ is actively transported into the cell, glucose is carried along.
16. A white blood cell ingests solid particles by forming vesicles. This describes the process of
    a. exocytosis.
    b. facilitated diffusion.
    c. secondary active transport.
    d. phagocytosis.
    e. pinocytosis.
17. Given these characteristics:
    1. requires energy
    2. requires carrier proteins
    3. requires membrane channels
    4. requires vesicles

    Choose the characteristics that apply to exocytosis.
    a. 1, 2
    b. 1, 4
    c. 1, 3, 4
    d. 1, 2, 3
    e. 1, 2, 3, 4
18. Cytoplasm is found
    a. in the nucleus.
    b. outside the nucleus and inside the plasma membrane.
    c. outside the plasma membrane.
    d. inside mitochondria.
    e. everywhere in the cell.
19. Which of these elements of the cytoskeleton is composed of tubulin and forms essential components of centrioles, spindle fibers, cilia, and flagella?
    a. actin filaments b. intermediate filaments c. microtubules
20. Mature red blood cells cannot
    a. synthesize ATP.
    b. transport oxygen.
    c. synthesize new protein.
    d. use glucose as a nutrient.
21. A large structure, normally visible in the nucleus of a cell, where ribosomal subunits are produced is called a (an)
    a. endoplasmic reticulum.
    b. mitochondria.
    c. nucleolus.
    d. lysosome.
22. A cell that synthesizes large amounts of protein for use outside the cell has a large
    a. number of cytoplasmic inclusions.
    b. number of mitochondria.
    c. amount of rough endoplasmic reticulum.
    d. amount of smooth endoplasmic reticulum.
    e. number of lysosomes.
23. Which of these organelles produces large amounts of ATP?
    a. nucleus
    b. mitochondria
    c. ribosomes
    d. endoplasmic reticulum
    e. lysosomes
24. Cylindrically shaped extensions of the plasma membrane that do not move, are supported with actin filaments, and may function in absorption or as sensory receptors are
    a. centrioles.
    b. spindle fibers.
    c. cilia.
    d. flagella.
    e. microvilli.
25. A portion of an mRNA molecule that determines one amino acid in a polypeptide chain is called a (an)
    a. nucleotide.
    b. gene.
    c. codon.
    d. exon.
    e. intron.
26. In which of these organelles is mRNA synthesized?
    a. nucleus
    b. ribosome
    c. endoplasmic reticulum
    d. nuclear envelope
    e. peroxisome
27. During the cell life cycle, DNA replication occurs during the
    a. G_1 phase.
    b. G_2 phase.
    c. M phase.
    d. S phase.
28. Given the following activities:
    1. repair
    2. growth
    3. gamete production
    4. differentiation

    Which of the activities is (are) the result of mitosis?
    a. 2
    b. 3
    c. 1, 2
    d. 3, 4
    e. 1, 2, 4
29. A gene is
    a. the functional unit of heredity.
    b. a certain portion of a DNA molecule.
    c. a part of a chromosome.
    d. all of the above.
30. Which of these terms is correctly matched with its definition?
    a. autosome—an X or a Y chromosome
    b. phenotype—the genetic makeup of an individual
    c. allele—genes occupying the same locus on homologous chromosomes
    d. heterozygous—having two identical genes for a trait
    e. recessive—a trait expressed when the genes are heterozygous
31. Which of these genotypes is heterozygous?
    a. *DD*
    b. *Dd*
    c. *dd*
    d. both a and c
32. The ABO blood group is an example of
    a. dominant versus recessive alleles.
    b. incomplete dominance.
    c. codominance.
    d. a polygenic trait.
    e. sex-linked inheritance.
33. Assume that a trait is determined by an X-linked dominant gene. If the mother exhibits the trait, but the father does not, then their
    a. sons are more likely than their daughters to exhibit the trait.
    b. daughters are more likely than their sons to exhibit the trait.
    c. sons and daughters are equally likely to exhibit the trait.

Answers in Appendix E

CRITICAL THINKING

1. Why does a surgeon irrigate a surgical wound from which a tumor has been removed with sterile distilled water rather than with sterile isotonic saline?
2. Solution A is hyperosmotic to solution B. If solution A is separated from solution B by a selectively permeable membrane, does water move from solution A into solution B or vice versa? Explain.
3. A researcher wants to determine the nature of the transport mechanism that moved substance X into a cell. She could measure the concentration of substance X in the extracellular fluid and within the cell, as well as the rate of movement of substance X into the cell. She does a series of experiments and gathers the data shown in the following graph. Choose the transport process that is consistent with the data.
   a. diffusion
   b. active transport
   c. facilitated diffusion
   d. not enough information to make a judgment

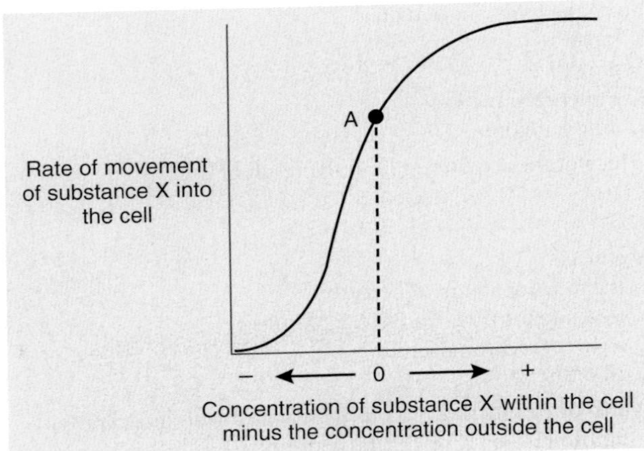

Graph depicting the rate of movement of substance X from a fluid into a cell (*y* axis) versus the concentration of substance X within the cell minus the concentration outside the cell (*x* axis). At point *A*, the extracellular concentration of substance X is equal to the intracellular concentration of substance X (designated *0* on the *x* axis).

4. A dialysis membrane is selectively permeable, and substances smaller than proteins are able to pass through it. If you wanted to use a dialysis machine to remove only urea (a small molecule) from blood, what could you use for the dialysis fluid?
   a. a solution that is isotonic and contains only large molecules, such as protein
   b. a solution that is isotonic and contains the same concentration of all substances except that it has no urea
   c. distilled water, which contains no ions or dissolved molecules
   d. blood, which is isotonic and contains the same concentration of all substances, including urea
5. Predict the consequence of a reduced intracellular K^+ concentration on the resting membrane potential.
6. If you had the ability to inhibit mRNA synthesis with a drug, explain how you could distinguish between proteins released from secretory vesicles in which they had been stored and proteins released from cells in which they had been newly synthesized.
7. Given the following data from electron micrographs of a cell, predict the major function of the cell:
   - moderate number of mitochondria
   - well-developed rough endoplasmic reticulum
   - moderate number of lysosomes
   - well-developed Golgi apparatus
   - dense nuclear chromatin
   - numerous vesicles
8. Nondisjunction during the first meiotic division is called primary nondisjunction, whereas nondisjunction during the second meiotic division is called secondary nondisjunction. What type of nondisjunction produces the higher number of abnormal gametes? Explain.
9. Dimpled cheeks are inherited as a dominant trait. Is it possible for two parents, each of whom has dimpled cheeks, to have a child without dimpled cheeks? Explain.
10. The ability to roll the tongue to form a "tube" results from a dominant gene. Suppose that a woman and her son can roll their tongues, but her husband cannot. Is it possible to determine if the husband is the father of her son?
11. The ABO antigens are a group of molecules found on the surface of red blood cells. Can an individual who has blood type AB be a parent of a child with blood type O? Why or why not?

Answers in Appendix F

Histology

The Study of Tissues

4

In some ways, the human body is like a complex machine, such as a car. Both consist of many parts, which are made of materials consistent with their specialized functions. For example, car tires are made of synthetic rubber reinforced with a variety of fibers, the engine is made of a variety of metal parts, and the windows are made of transparent glass. All parts of a car cannot be made of a single type of material. Metal capable of withstanding the heat of the engine cannot be used for windows or tires. Similarly, the many parts of the human body are made of collections of specialized cells and the materials surrounding them. Muscle cells, which contract to produce body movements, are structurally different from and have different functions than those of epithelial cells, which protect, secrete, or absorb. Also, cells in the retina of the eye, specialized to detect light and allow us to see, do not contract as muscle cells do or exhibit the functions of epithelial cells.

The structure and function of tissues are so closely related that you should be able to predict the function of a tissue when given its structure, and vice versa. Knowledge of tissue structure and function is important in understanding the structure and function of organs, organ systems, and the complete organism. This chapter begins with brief discussions of *tissues and histology* (p. 110) and the development of *embryonic tissue* (p. 110) and then describes the structural and functional characteristics of the major tissue types: *epithelial tissue* (p. 110), *connective tissue* (p. 120), *muscle tissue* (p. 134), and *nervous tissue* (p. 136). In addition, the chapter provides an explanation of *membranes* (p. 137), *inflammation* (p. 138), *tissue repair* (p. 140), and *tissues and aging* (p. 142).

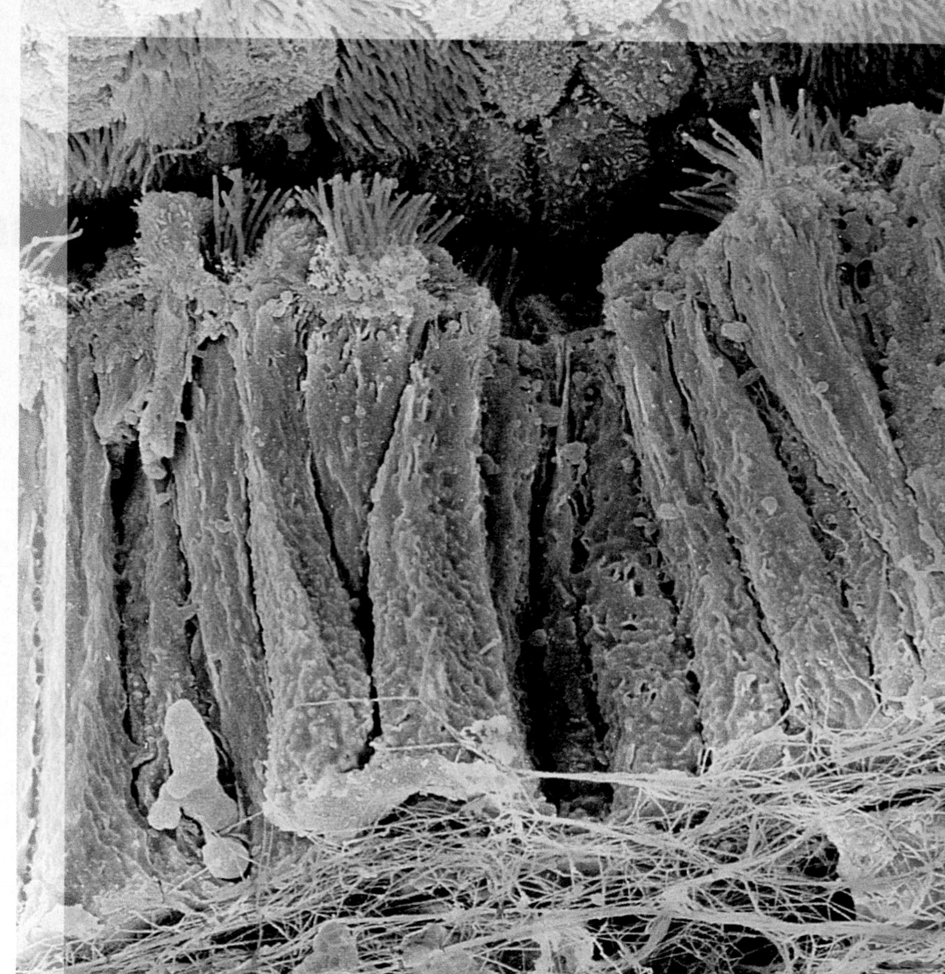

Colored SEM of simple columnar epithelial cells, with cilia, of the uterine tube.

TISSUES AND HISTOLOGY

Tissues (tish'ūz) are collections of similar cells and the substances surrounding them. Specialized cells and the extracellular matrix surrounding them form all the tissue types found at the **tissue level of organization.** The classification of tissue types is based on the structure of the cells; the composition of the noncellular substances surrounding cells, called the **extracellular matrix;** and the functions of the cells. The four primary tissue types, which include all tissues, and from which all organs of the body are formed, are

1. epithelial tissue
2. connective tissue
3. muscle tissue
4. nervous tissue

Epithelial and connective tissues are the most diverse in form. The different types of epithelial and connective tissues are classified by structure, including cell shapes, relationships of cells to one another, and the materials making up the extracellular matrix. In contrast, muscle and nervous tissues are classified by functional and structural characteristics.

The tissues of the body are interdependent. For example, muscle tissue cannot produce movement unless it receives oxygen carried by red blood cells, and new bone tissue cannot be formed unless epithelial tissue absorbs calcium and other nutrients from the digestive tract. Also, all tissues in the body die if cancer or another disease destroys the tissues of vital organs, such as the liver or kidneys.

Histology (his-tol'ō-jē) is the microscopic study of tissues. Much information about a person's health can be gained by examining tissues. A **biopsy** (bī'op-sē) is the process of removing tissue samples from patients surgically or with a needle for diagnostic purposes. Examining tissue samples from individuals with various disorders can distinguish the specific disease. For example, some red blood cells have an abnormal shape in people suffering from sickle-cell disease, and red blood cells are smaller than normal in people with iron-deficiency anemia. Cancer is identified and classified based on characteristic changes that occur in tissues. Changes in the structure of epithelial cells are used to identify cancer of the uterine cervix, and changes in white blood cells are used to identify people who have leukemia. Also, the number of white blood cells can be greatly increased in people who have infections. Epithelial cells from respiratory passages have an abnormal structure in people with chronic bronchitis and in people with lung cancer.

Tissue samples can be sent to a laboratory and are prepared for examination. Results are obtained from the examination of the prepared tissues. In some cases, tissues are removed surgically, they are prepared quickly, and the results are reported while the patient is still anesthetized. The appropriate therapy is based to a large degree on the results. For example, the amount of tissue removed as part of breast or other types of cancer treatment can be determined by the results.

An **autopsy** (aw'top-sē) is an examination of the organs of a dead body to determine the cause of death or to study the changes caused by a disease. Microscopic examination of tissue is often part of an autopsy.

1 ***Name the four primary tissue types, and list three characteristics used to classify them. How does the classification of epithelial and connective tissue differ from the classification of muscle and nervous tissue?***

2 ***Define histology. Explain how microscopic examination of tissues taken by biopsy or autopsy can be used to diagnose some diseases.***

EMBRYONIC TISSUE

Approximately 13 or 14 days after fertilization, the cells that give rise to a new individual, called embryonic stem cells, form a slightly elongated disk consisting of two layers called the ectoderm and the endoderm. Cells of the ectoderm then migrate between the two layers to form a third layer called the mesoderm. The ectoderm, mesoderm, and endoderm are called germ layers because the beginning of all adult structures can be traced back to one of them and they give rise to all the tissues of the body (see chapter 29).

The **endoderm** (en'dō-derm), the inner layer, forms the lining of the digestive tract and its derivatives. The **mesoderm** (mez'ō-derm), the middle layer, forms tissues such as muscle, bone, and blood vessels. The **ectoderm** (ek'tō-derm), the outer layer, forms the skin, and a portion of the ectoderm, called **neuroectoderm** (noor-ō-ek'tō-derm), becomes the nervous system (see chapter 29). Groups of cells that break away from the neuroectoderm during development, called **neural crest cells,** give rise to parts of the peripheral nerves (see chapter 29), skin pigment (see chapter 5), the medulla of the adrenal gland (see chapter 18), and many tissues of the face.

3 ***Name the three embryonic germ layers.***

4 ***What adult structures are derived from the endoderm, mesoderm, ectoderm, and neuroectoderm and neural crest cells?***

EPITHELIAL TISSUE

Epithelium (ep-i-thē'lē-ŭm; pl. *epithelia,* ep-i-thē'lē-ă), or **epithelial tissue,** can be thought of as a protective covering of surfaces, both outside and inside the body. Characteristics common to most types of epithelium are (figure 4.1)

1. Epithelium consists almost entirely of cells, with very little extracellular matrix between them.
2. Epithelium covers surfaces of the body and forms glands that are derived developmentally from body surfaces. The body surfaces include the outside surface of the body, the lining of the digestive and respiratory tracts, the heart and blood vessels, and the linings of many body cavities.
3. Most epithelial tissues have one **free,** or **apical** (ap'i-kăl), **surface** not attached to other cells; a **lateral surface** attached to other epithelial cells; and a **basal surface.** The free surface often lines the lumen of ducts, vessels, or cavities. The basal surface of most epithelial tissues is attached to a **basement membrane.** The basement membrane is a specialized type of extracellular material secreted by the epithelial cells and by connective tissue cells. It is like the adhesive on Scotch™ tape. It helps attach the epithelial cells to the underlying

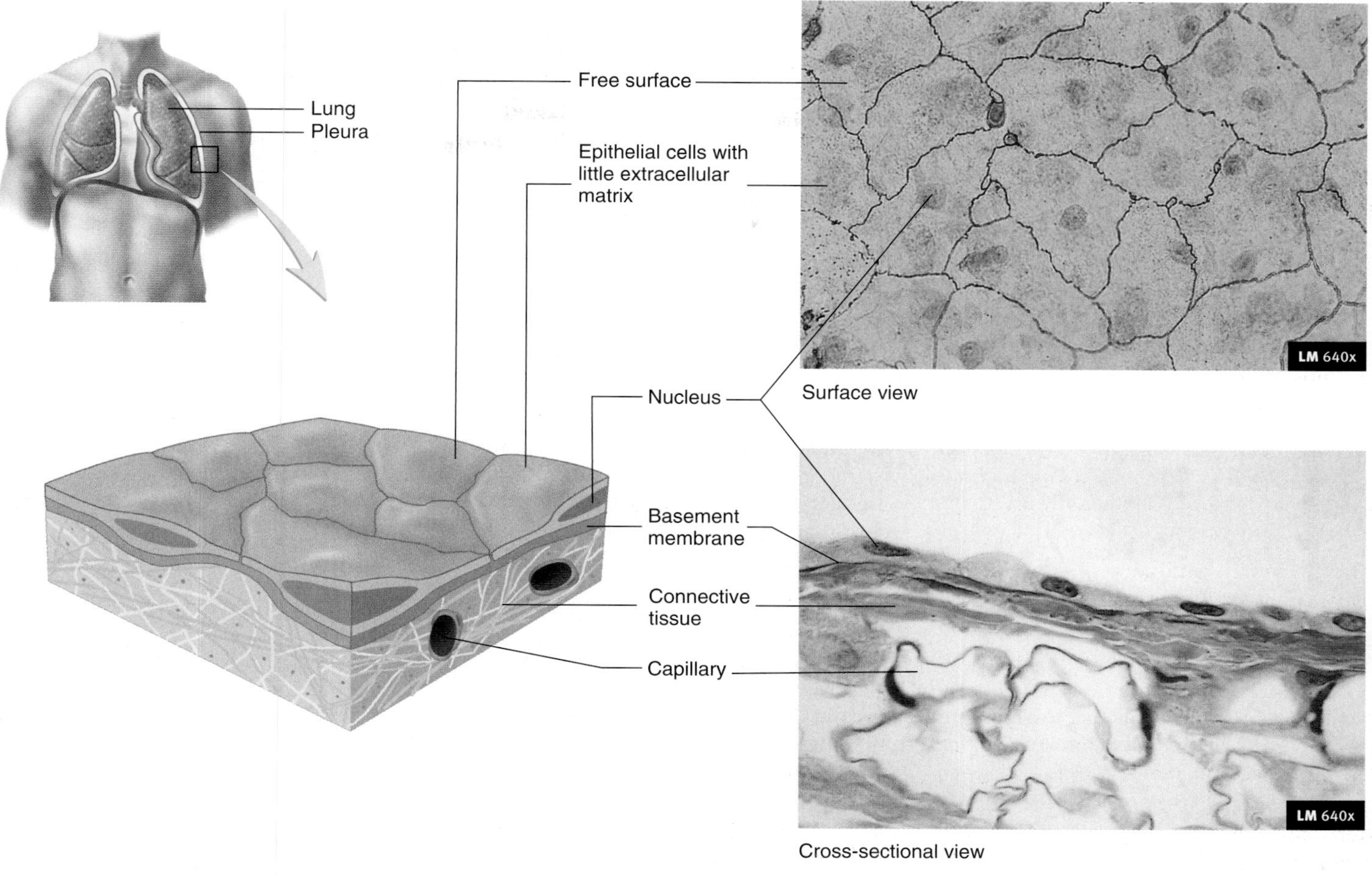

FIGURE 4.1 Characteristics of Epithelium

Surface and cross-sectional views of epithelium illustrate the following characteristics: little extracellular material between cells, a free surface, and a basement membrane attaching epithelial cells to underlying tissues. Capillaries in connective tissue do not penetrate the basement membrane. Nutrients, oxygen, and waste products must diffuse across the basement membrane between the capillaries and the epithelial cells.

tissues, and it plays an important role in supporting and guiding cell migration during tissue repair. A few epithelial tissues, such as in lymphatic capillaries and liver sinusoids, do not have basement membranes, and some epithelial tissues, such as in some endocrine glands, do not have a free surface or a basal surface with a basement membrane.

4. Specialized cell contacts bind adjacent epithelial cells together.
5. Blood vessels in the underlying connective tissue do not penetrate the basement membrane to reach the epithelium; thus, all gases and nutrients carried in the blood must reach the epithelium by diffusing from blood vessels across the basement membrane. In epithelial tissues with many layers of cells, the most metabolically active cells are close to the basement membrane.
6. Epithelial cells retain the ability to undergo mitosis and therefore are able to replace damaged cells with new epithelial cells. Undifferentiated cells (stem cells) continuously divide and produce new cells. In some types of epithelial tissues, such as in the skin and in the digestive tract, cells that are lost or die are continuously replaced by new cells.

5 ***List six characteristics common to most types of epithelium. Define free (apical), lateral, and basal surfaces of epithelial cells.***

6 ***What is the basement membrane and what are its functions? Why must metabolically active epithelial cells be close to the basement membrane?***

Functions of Epithelial Tissue

Major functions of epithelial tissue include

1. *Protecting underlying structures.* Examples are the outer layer of the skin and the epithelium of the oral cavity, which protect the underlying structures from abrasion.
2. *Acting as barriers.* Epithelium prevents the movement of many substances through the epithelial layer. For example, the skin acts as a barrier to water and reduces water loss from the body. The skin is also a barrier that prevents the entry of many toxic molecules and microorganisms into the body.
3. *Permitting the passage of substances.* Epithelium allows the movement of many substances through the epithelial layer. For

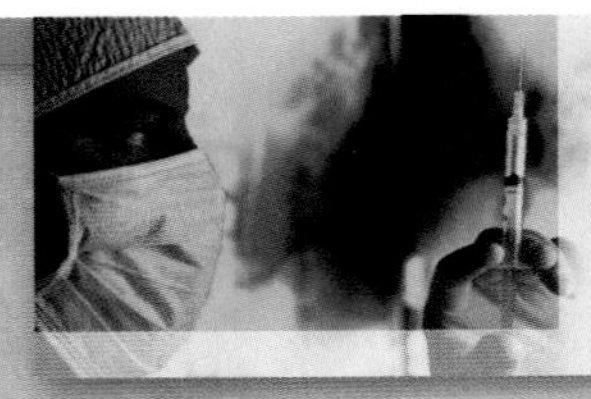

CLINICAL FOCUS
Microscopic Imaging

We see objects because light either passes through them or is reflected off them and enters our eyes (see chapter 15). We are limited, however, in what we can see with the unaided eye. Without the aid of magnifying lenses, the smallest objects we can resolve, or identify as separate objects, are approximately 100 μm, or 0.1 mm, in diameter, which is approximately the size of a fine pencil dot. The details of cells and tissues, which are much smaller than 100 μm, cannot be examined without the aid of a microscope.

Two basic types of microscopes have been developed: light microscopes and electron microscopes. As their names imply, light microscopes use light to produce an image, and electron microscopes use beams of electrons. **Light microscopes** usually use transmitted light, which is light that passes through the object being examined, but some light microscopes are equipped to use reflected light. Glass lenses are used in light microscopes to magnify images. The images can be observed directly by looking into the microscope, or the light from the images can be used to expose photographic film to make a photomicrograph of the images. Video or digital cameras are also used to record images, which can be stored or analyzed and enhanced electronically. The resolution of light microscopes is limited by the wavelength of light, the lower limit of which is approximately 0.1 μm—about the size of a small bacterium.

Recall that a biopsy is the process of removing tissue from living patients for diagnostic examination. For example, changes in tissue structure allow pathologists to identify tumors and to distinguish between noncancerous (benign) and cancerous (malignant) tumors. Light microscopy is used on a regular basis to examine biopsy specimens. Light microscopy is used instead of electron microscopy because less time and effort are required to prepare materials for examination, and the resolution is adequate to diagnose most conditions that cause changes in tissue structure.

Because images are usually produced using transmitted light, tissues to be examined must be cut very thinly to allow the light to pass through them. Sections are routinely cut between 1 and 20 μm thick to make them thin enough for light microscopy. To cut such thin sections, the tissue must be fixed, or frozen, which is a process that preserves the tissue and makes it more rigid. Fixed tissues are then embedded in a material, such as wax or plastic, that makes the tissue rigid enough for cutting into sections. Frozen sections, which can be prepared rapidly, are rigid enough for sectioning, but tissue embedded in wax or plastic can be cut much thinner, which makes the image seen through the microscope clearer. Because most tissues are colorless and transparent when thinly sectioned, the tissue must be colored with a stain or dye so that the structural details can be seen. As a result, the colors seen in color photomicrographs are not the true colors of the tissue but instead are the colors of the stains used. The color of the stain can also provide specific information about the tissue, because special stains color only certain structures.

To see objects much smaller than a cell, such as cell organelles, an **electron microscope,** which has a limit of resolution of approximately 0.1 nm, must be used; 0.1 nm is about the size of some molecules. In objects viewed through an electron microscope, a beam of electrons either is passed through objects using a transmission electron microscope (TEM) or is reflected off the surface of objects using a scanning electron microscope (SEM). The electron beam is focused with electromagnets. For both processes, the specimen must be fixed, and for TEM the specimen must be embedded in plastic and thinly sectioned (0.01–0.15 μm thick). Care must be taken when examining specimens in an electron microscope because a focused electron beam can cause most tissues to disintegrate quickly. Furthermore, the electron beam is not visible to the human eye; thus, it must be directed onto a fluorescent or photographic plate on which the electron beam is converted into a visible image. Because the electron beam does not transmit color information, electron micrographs are black and white unless color enhancement has been added using computer technology.

The magnification ability of SEM is not as great as that of TEM; however, depth of focus of SEM is much greater and allows for the production of a clearer three-dimensional image of tissue structure.

example, oxygen and carbon dioxide are exchanged between the air and blood by diffusion through the epithelium in the lungs. Epithelium acts as a filter in the kidney, allowing many substances to pass from the blood into the urine but retaining other substances, such as blood cells and proteins, in the blood.

4. *Secreting substances.* Mucous glands, sweat glands, and the enzyme-secreting portions of the pancreas secrete their products onto epithelial surfaces or into ducts that carry them to other areas of the body.
5. *Absorbing substances.* The plasma membranes of certain epithelial tissues contain carrier proteins (see chapter 3), which regulate the absorption of materials.

7 ***List the major functions of epithelial tissue.***

Classification of Epithelium

Epithelium is classified primarily according to the number of cell layers and the shape of the superficial cells. There are three major types of epithelium based on the number of cell layers in each type.

1. **Simple epithelium** consists of a single layer of cells, with each cell extending from the basement membrane to the free surface.
2. **Stratified epithelium** consists of more than one layer of cells, but only the basal layer of cells attaches the deepest layer to the basement membrane.
3. **Pseudostratified columnar epithelium** is a special type of simple epithelium. The prefix *pseudo-* means false, so this type of epithelium appears to be stratified but is not.

It consists of one layer of cells, with all the cells attached to the basement membrane. There is an appearance of two or more layers of cells because some of the cells are tall and extend to the free surface, whereas others are shorter and do not extend to the free surface.

There are three types of epithelium based on the shape of the epithelial cells.

1. **Squamous** (skwā′mŭs; flat) cells are flat or scalelike.
2. **Cuboidal** (cubelike) cells are cube-shaped; they are about as wide as they are tall.
3. **Columnar** (tall and thin, similar to a column) cells are taller than they are wide.

In most cases, an epithelium is given two names, such as simple squamous, stratified squamous, simple columnar, or pseudostratified columnar. The first name indicates the number of layers, and the second indicates the shape of the cells (table 4.1) at the free surface. Tables 4.2–4.4 provide an overview of the major types of epithelial tissues and their distributions.

Simple squamous epithelium consists of one layer of flat, or scalelike, cells that rest on a basement membrane (see table 4.2*a*). Stratified squamous epithelium consists of several layers of cells. Near the basement membrane, the cells are more cuboidal, but at the free surface the cells are flat or scalelike (see table 4.3*a*). Pseudostratified columnar epithelial cells are columnar in shape in that they are taller than they are wide and, although they appear to consist of more than one layer, all of the cells rest on the basement membrane (see table 4.4*a*).

Stratified squamous epithelium can be classified further as either nonkeratinized (moist) or keratinized, according to the condition of the outermost layer of cells. **Nonkeratinized**

TABLE 4.1 Classification of Epithelium

Number of Layers or Category	Shape of Cells
Simple (single layer of cells)	Squamous Cuboidal Columnar
Stratified (more than one layer of cells)	Squamous Nonkeratinized (moist) Keratinized Cuboidal (very rare) Columnar (very rare)
Pseudostratified (modification of simple epithelium)	Columnar
Transitional (modification of stratified epithelium)	Roughly cuboidal to columnar when not stretched and squamouslike when stretched

TABLE 4.2 Simple Epithelium

(a) Simple Squamous Epithelium

Structure: Single layer of flat, often hexagonal cells; the nuclei appear as bumps when viewed as a cross section because the cells are so flat

Function: Diffusion, filtration, some secretion and some protection against friction

Location: Lining of blood vessels and the heart, lymphatic vessels (endothelium) and small ducts, alveoli of the lungs, portions of the kidney tubules, lining of serous membranes (mesothelium) of body cavities (pleural, pericardial, peritoneal), and inner surface of the eardrums

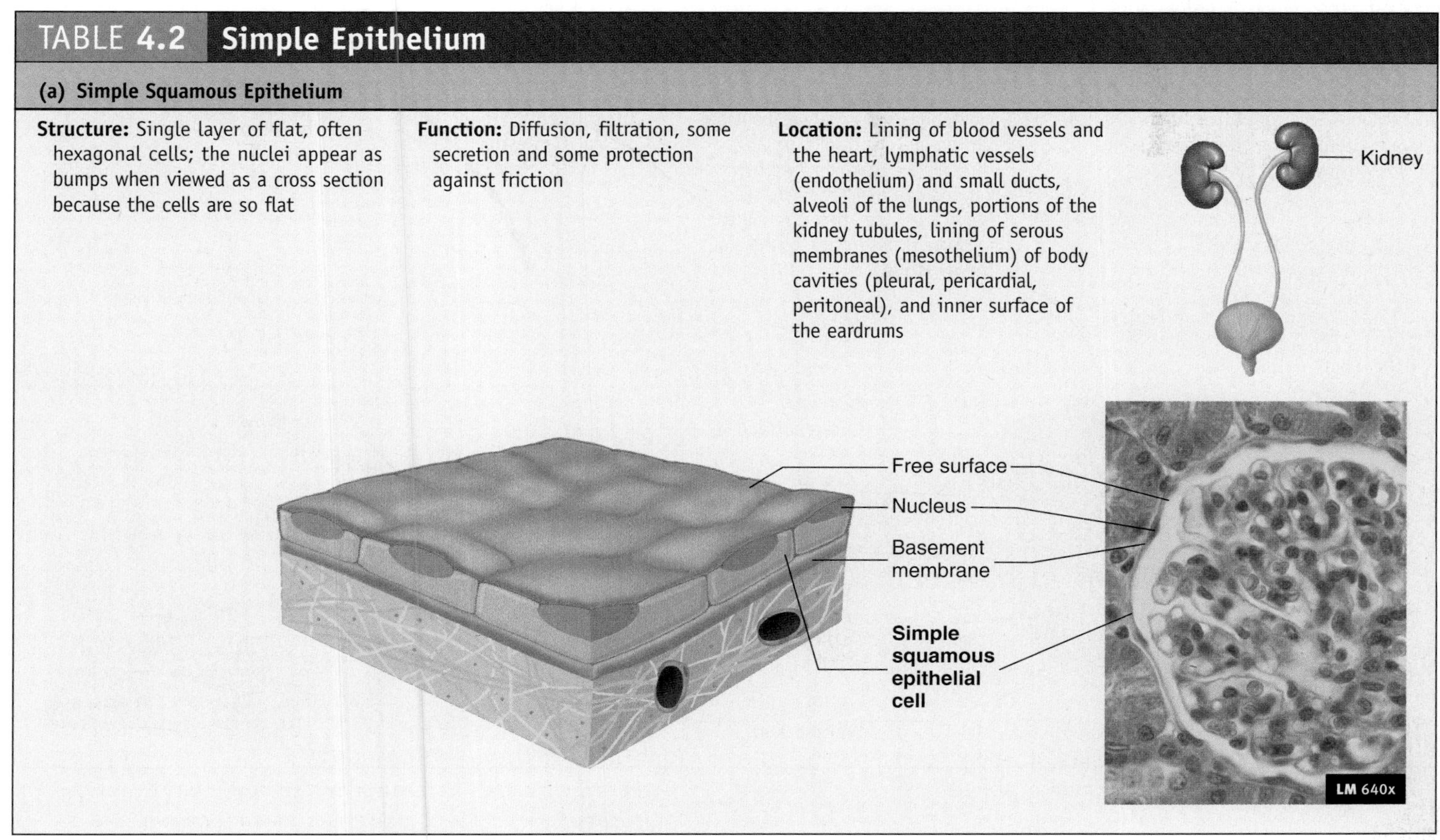

TABLE 4.2 Simple Epithelium—Continued

(b) Simple Cuboidal Epithelium

Structure: Single layer of cube-shaped cells; some cells have microvilli (kidney tubules) or cilia (terminal bronchioles of the lungs)

Function: Active transport and facilitated diffusion result in secretion and absorption by cells of the kidney tubules; secretion by cells of glands and choroid plexuses; movement of particles embedded in mucus out of the terminal bronchioles by ciliated cells

Location: Kidney tubules, glands and their ducts, choroid plexuses of the brain, lining of terminal bronchioles of the lungs, surfaces of the ovaries

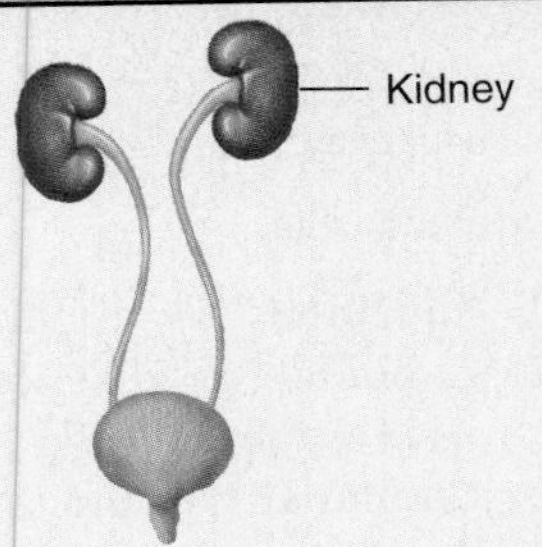

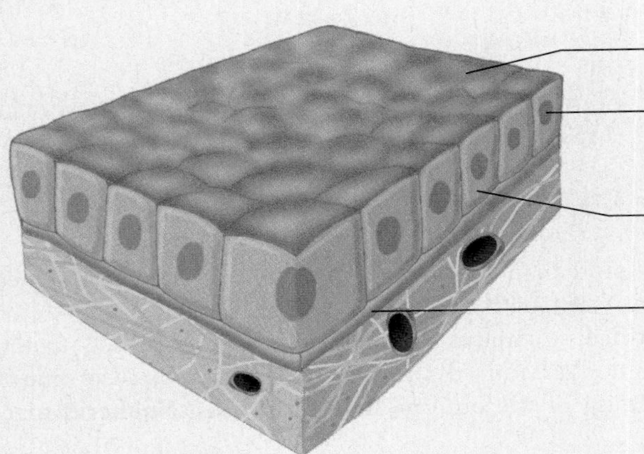

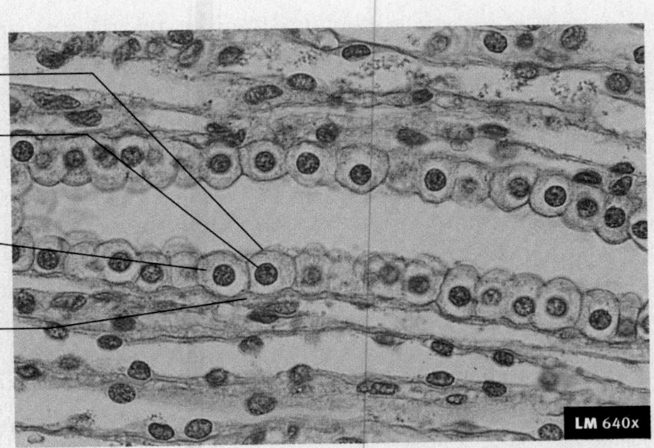

(c) Simple Columnar Epithelium

Structure: Single layer of tall, narrow cells; some cells have cilia (bronchioles of lungs, auditory tubes, uterine tubes, and uterus) or microvilli (intestines)

Function: Movement of particles out of the bronchioles of the lungs by ciliated cells; partially responsible for the movement of oocytes through the uterine tubes by ciliated cells; secretion by cells of the glands, the stomach, and the intestines; absorption by cells of the intestine

Location: Glands and some ducts, bronchioles of lungs, auditory tubes, uterus, uterine tubes, stomach, intestines, gallbladder, bile ducts, ventricles of the brain

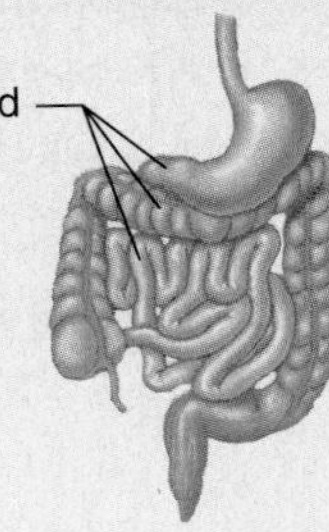

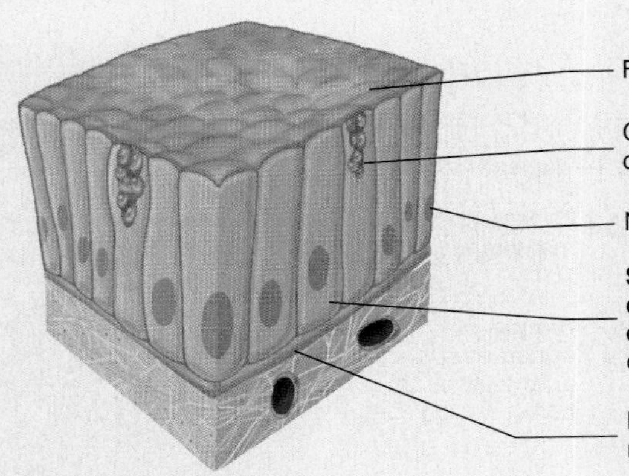

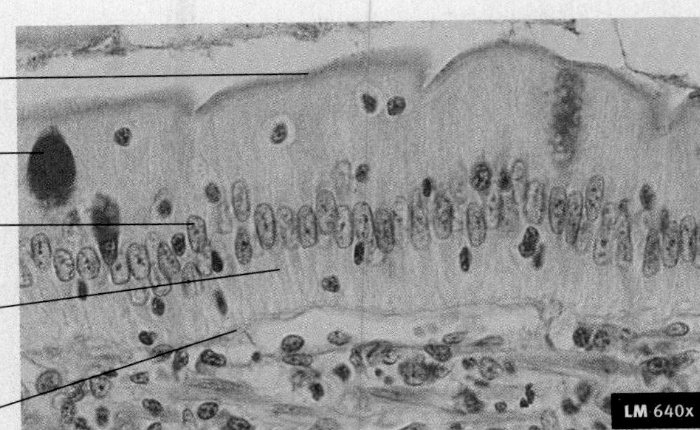

TABLE 4.3 Stratified Epithelium

(a) Stratified Squamous Epithelium

Structure: Multiple layers of cells that are cuboidal in the basal layer and progressively flattened toward the surface; the epithelium can be nonkeratinized (moist) or keratinized; in nonkeratinized stratified squamous epithelium, the surface cells retain a nucleus and cytoplasm, in keratinized stratified epithelium; the cytoplasm of cells at the surface is replaced by a protein called keratin, and the cells are dead

Function: Protection against abrasion, barrier against infection, reduction of water loss from the body

Location: Keratinized—skin; nonkeratinized (moist)—mouth, throat, larynx, esophagus, anus, vagina, inferior urethra, cornea

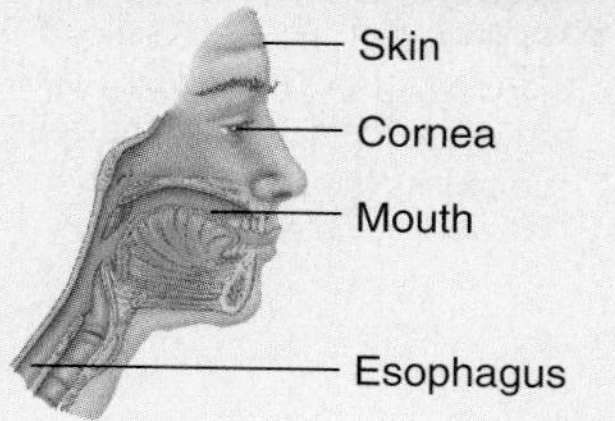

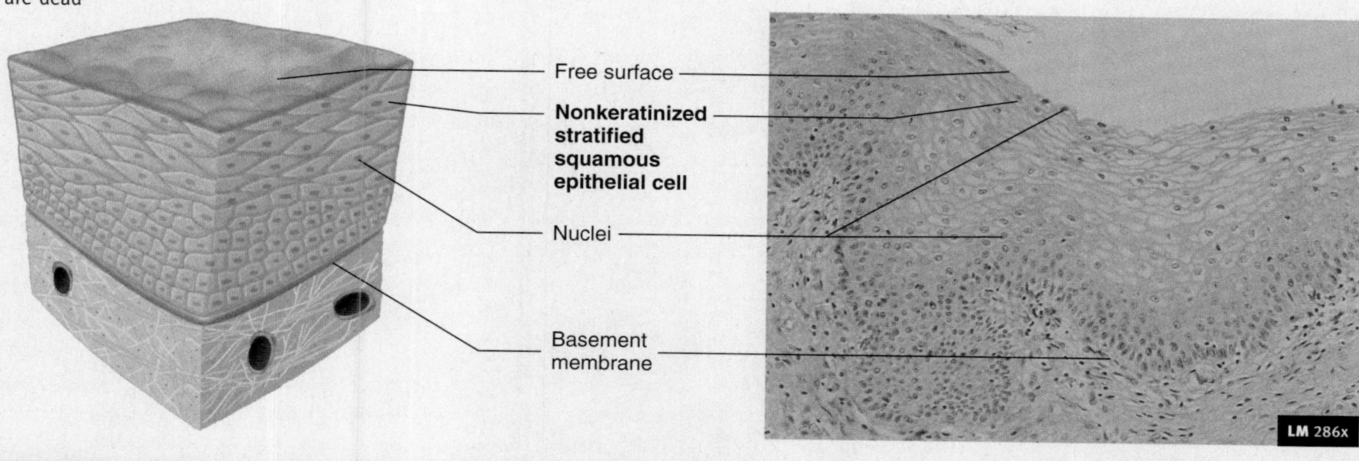

(b) Stratified Cuboidal Epithelium

Structure: Multiple layers of somewhat cube-shaped cells

Function: Secretion, absorption, protection against infection

Location: Sweat gland ducts, ovarian follicular cells, salivary gland ducts

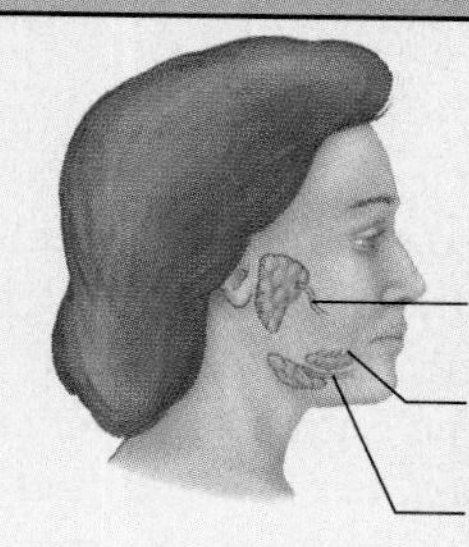

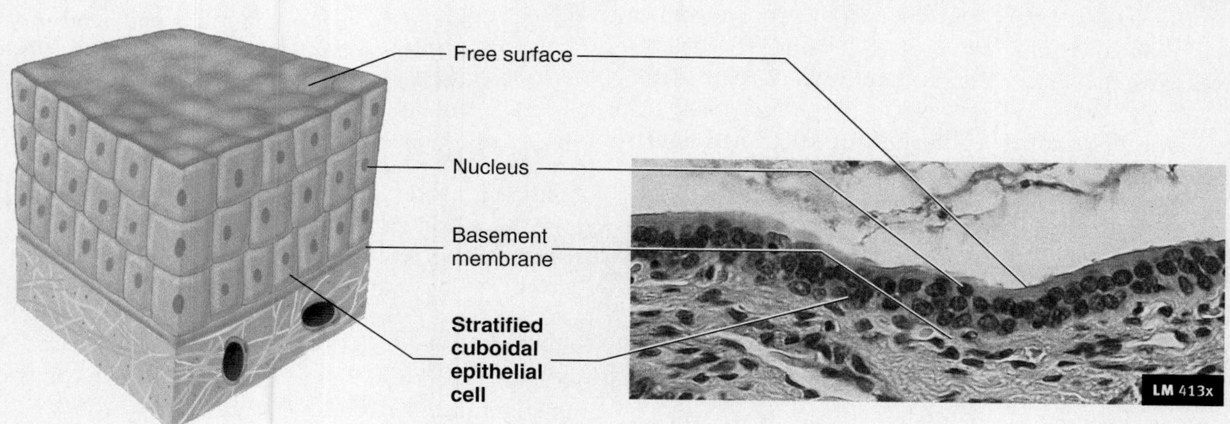

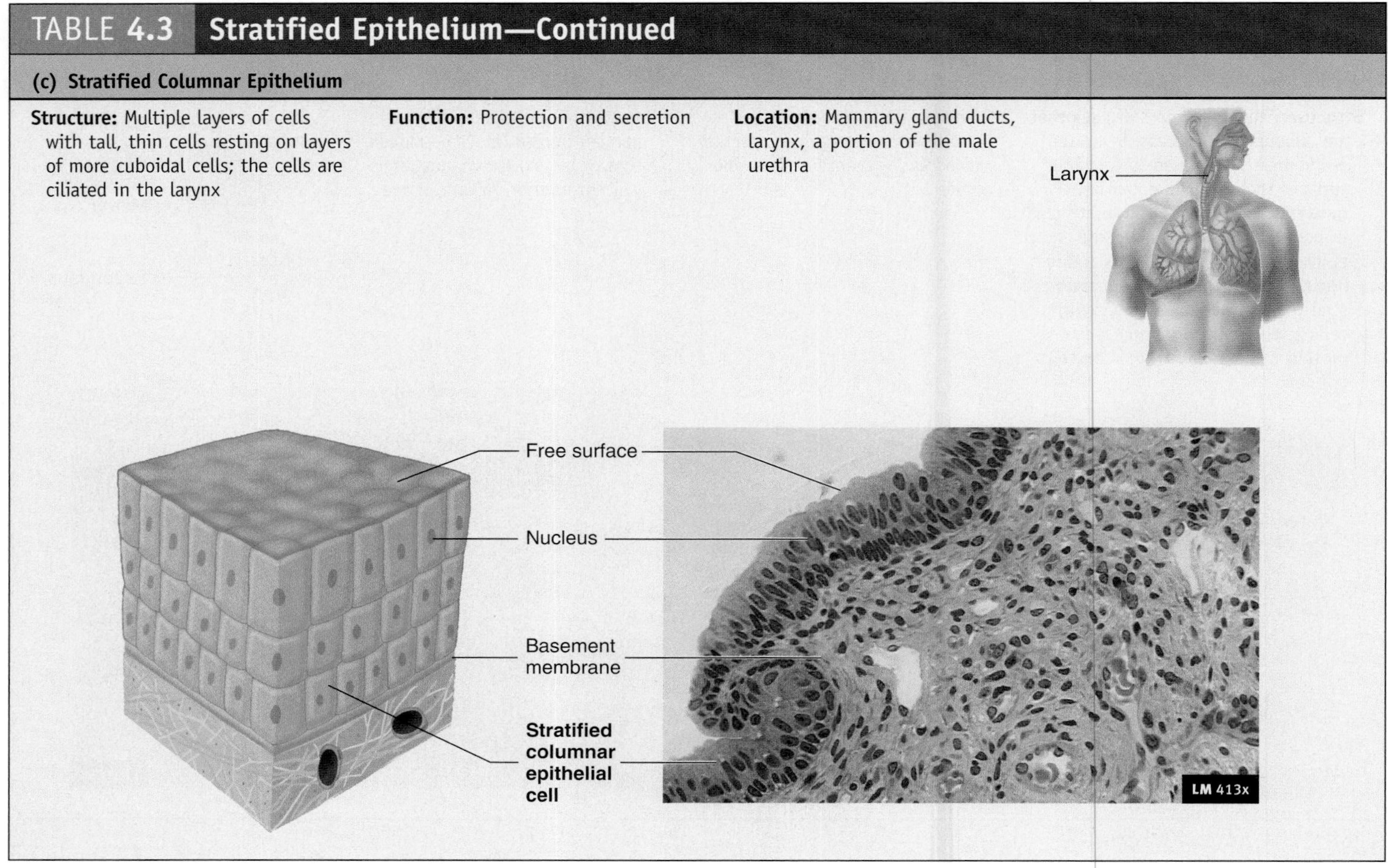

TABLE 4.3 Stratified Epithelium—Continued

(c) Stratified Columnar Epithelium

Structure	Function	Location
Multiple layers of cells with tall, thin cells resting on layers of more cuboidal cells; the cells are ciliated in the larynx	Protection and secretion	Mammary gland ducts, larynx, a portion of the male urethra

(moist) **stratified squamous epithelium** (see table 4.3*a*), found in areas such as the mouth, esophagus, rectum, and vagina, consists of living cells in the deepest and outermost layers. A layer of fluid covers the outermost layers of cells, which makes them moist. In contrast, **keratinized** (ker′ă-ti-nizd) **stratified squamous epithelium,** found in the skin (see chapter 5), consists of living cells in the deepest layers, and the outer layers are composed of dead cells containing the protein keratin. The dead, keratinized cells give the tissue a durable, moisture-resistant, dry character.

A unique type of stratified epithelium called **transitional epithelium** (see table 4.4*b*) lines the urinary bladder; ureters; pelvis of the kidney, including the major and minor calyces (kal′-i-sēz); and superior part of the urethra (see chapter 26). These are structures where considerable expansion can occur. The shape of the cells and the number of cell layers vary, depending on the degree to which transitional epithelium is stretched. The surface cells and the underlying cells are roughly cuboidal or columnar when the epithelium is not stretched, and they become more flattened or squamouslike as the epithelium is stretched. Also, the number of layers of epithelial cells decreases in response to stretch. As the epithelium is stretched, the epithelial cells can shift on one another so that the number of layers decreases from five or six to two or three.

8 ***Describe simple, stratified, and pseudostratified epithelial tissues. Distinguish among squamous, cuboidal, and columnar epithelial cells.***

9 ***How do nonkeratinized (moist) stratified squamous epithelium and keratinized stratified squamous epithelium differ? Where is each type found?***

10 ***Describe the changes in cell shape and number of cell layers in transitional epithelium as it is stretched. Where is transitional epithelium found?***

Functional Characteristics

Epithelial tissues have many functions (table 4.5), including forming a barrier between a free surface and the underlying tissues and secreting, transporting, and absorbing selected molecules. The type and arrangement of organelles within each cell (see chapter 3), the shape of cells, and the organization of cells within each epithelial type reflect these functions. Accordingly, structural specializations of epithelial cells are consistent with the functions they perform.

Cell Layers and Cell Shapes

Simple epithelium, with its single layer of cells, covers surfaces and allows diffusion of gases (lungs), filters blood (kidneys), secretes

TABLE 4.4 Pseudostratified Columnar Epithelium and Transitional Epithelium

(a) Pseudostratified Columnar Epithelium

Structure: Single layer of cells; some cells are tall and thin and reach the free surface, and others do not; the nuclei of these cells are at different levels and appear stratified; the cells are almost always ciliated and are associated with goblet cells that secrete mucus onto the free surface

Function: Synthesize and secrete mucus onto the free surface and move mucus (or fluid) that contains foreign particles over the surface of the free surface and from passages

Location: Lining of nasal cavity, nasal sinuses, auditory tubes, pharynx, trachea, bronchi of lungs

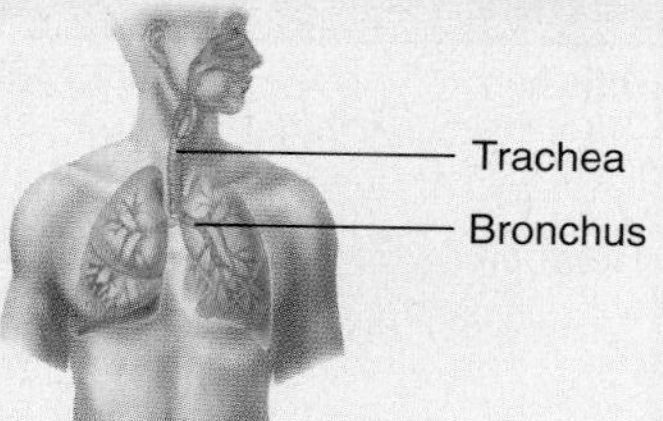

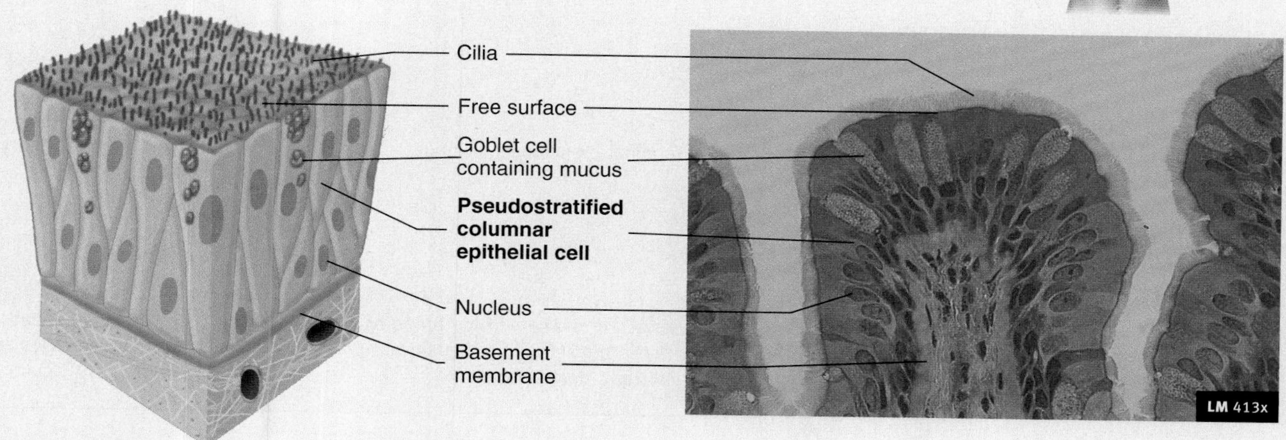

(b) Transitional Epithelium

Structure: Stratified cells that appear cuboidal when the organ or tube is not stretched and squamous when the organ or tube is stretched by fluid

Function: Accommodates fluctuations in the volume of fluid in organs or tubes; protects against the caustic effects of urine

Location: Lining of urinary bladder, ureters, superior urethra

Ureter

Urinary bladder

Urethra

Free surface

Transitional epithelial cell

Nucleus

Basement membrane

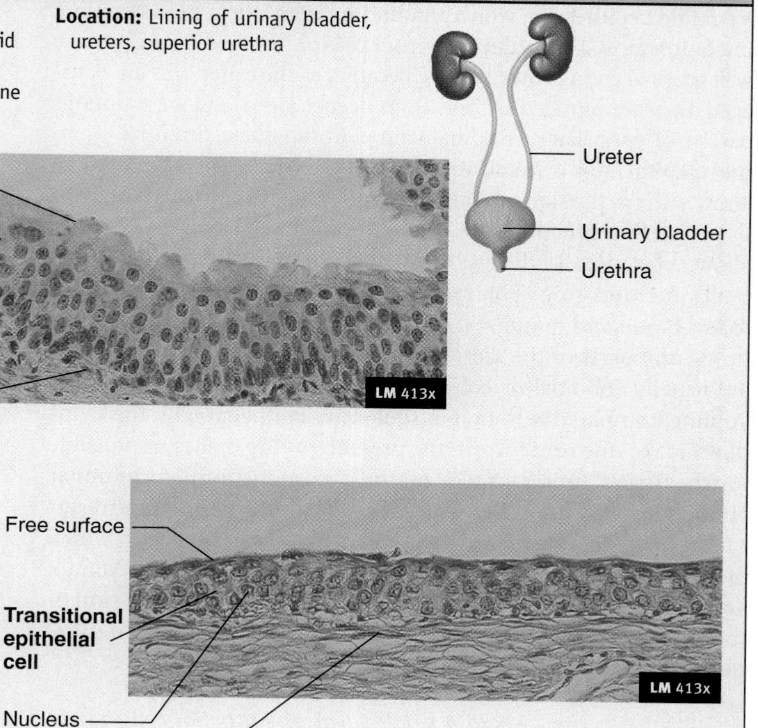

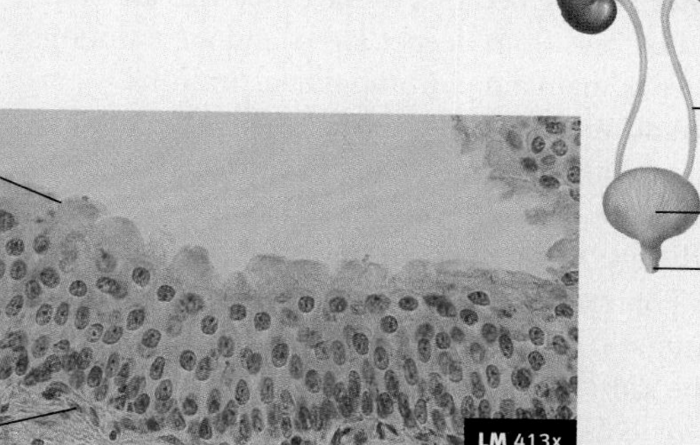

Tissue not stretched

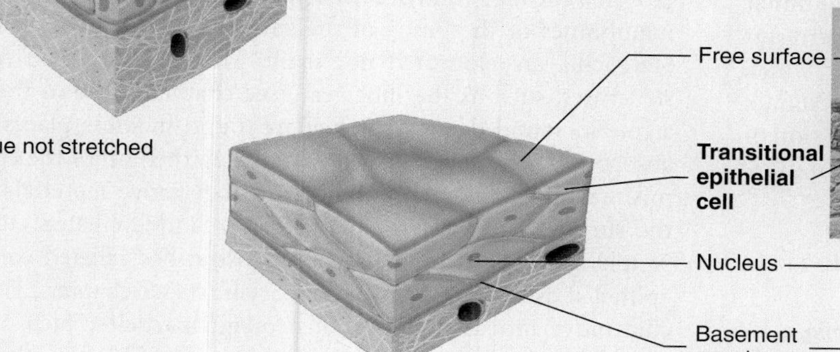

Tissue stretched

TABLE 4.5 Function and Location of Epithelial Tissue

Function	Simple Squamous Epithelium	Simple Cuboidal Epithelium	Simple Columnar Epithelium
Diffusion	Blood and lymph capillaries, alveoli of lungs, thin segments of loops of Henle		
Filtration	Bowman's capsules of kidneys		
Secretion or absorption	Mesothelium (serous fluid)	Choroid plexus (produces cerebrospinal fluid), part of kidney tubules, many glands and their ducts	Stomach, small intestine, large intestine, uterus, many glands
Protection (against friction and abrasion)	Endothelium (e.g., epithelium of blood vessels) Mesothelium (e.g., epithelium of body cavities)		
Movement of mucus (ciliated)		Terminal bronchioles of lungs	Bronchioles of lungs, auditory tubes, uterine tubes, uterus
Capable of great stretching			
Miscellaneous	Inner part of the eardrums, smallest ducts of glands	Surface of ovaries, inside lining of eyes (pigmented epithelium of retina), ducts of glands	Bile duct, gallbladder, ependyma (lining of brain ventricles and central canal of spinal cord), ducts of glands

cellular products (glands), and absorbs nutrients (intestines). The selective movement of materials through epithelium is hindered by a stratified epithelium, which is found in areas where protection is a major function. The multiple layers of cells in stratified epithelium are well adapted for a protective role because, as the outer cells are damaged, they are replaced by cells from deeper layers and a continuous barrier of epithelial cells is maintained in the tissue. Stratified squamous epithelium is found in areas of the body where abrasion can occur, such as the skin, mouth, throat, esophagus, anus, and vagina.

Differing functions are also reflected in cell shape. Cells that allow substances to diffuse through them and that filter are normally flat and thin. For example, simple squamous epithelium forms blood and lymphatic capillaries, the alveoli (air sacs) of the lungs, and parts of the kidney tubules. Cells that secrete or absorb are usually cuboidal or columnar. They have greater cytoplasmic volume, compared with that of squamous epithelial cells; this cytoplasmic volume results from the presence of organelles responsible for the tissues' functions. For example, pseudostratified columnar epithelium, which secretes large amounts of mucus, lines the respiratory tract (see chapter 23) and contains large **goblet cells,** which are specialized columnar epithelial cells. The goblet cells contain abundant organelles responsible for the synthesis and secretion of mucus, such as ribosomes, endoplasmic reticulum, Golgi apparatuses, and secretory vesicles filled with mucus.

11 ***What functions would a single layer of epithelial cells be expected to perform? A stratified layer?***

12 ***In locations in which diffusion or filtration is occurring, what shape would you expect the epithelial cells to be?***

13 ***Why are cuboidal or columnar cells found where secretion or absorption is occurring?***

PREDICT 1

Explain the consequences of having (a) nonkeratinized stratified epithelium rather than simple columnar epithelium lining the digestive tract, (b) nonkeratinized stratified squamous epithelium rather than keratinized stratified squamous epithelium in the skin, and (c) simple columnar epithelium rather than nonkeratinized stratified squamous epithelium lining the mouth.

Cell Surfaces

The free surfaces of epithelial tissues can be smooth, contain microvilli, be ciliated, or be folded. Smooth surfaces reduce friction. The lining of blood vessels is a simple squamous epithelium that reduces friction as blood flows through the vessels (see chapter 21).

Microvilli and cilia were described in chapter 3. Microvilli are nonmotile and contain microfilaments. They greatly increase surface area and are found in cells that absorb or secrete, such as serous membranes or the lining of the small intestine (see chapter 24). Stereocilia are elongated microvilli. They are found in sensory structures, such as the inner ear (see chapter 15), and they play a role in sound detection. They are found in some places where absorption is important, such as in the epithelium of the epididymis. Cilia contain microtubules and they move materials across the surface of the cell (see chapter 3). Simple ciliated cuboidal, simple ciliated columnar, and pseudostratified ciliated columnar epithelial tissues are in the respiratory tract (see chapter 23), where cilia move mucus that contains foreign particles, such as dust, out of the respiratory passages. Cilia are also found on the apical surface of the simple columnar epithelial cells of the uterus and uterine tubes. The cilia help move mucus and oocytes.

Stratified Squamous Epithelium	Stratified Cuboidal Epithelium	Stratified Columnar Epithelium	Pseudostratified Columnar Epithelium	Transitional Epithelium
Skin (epidermis), corneas, mouth and throat, epiglottis, larynx, esophagus, anus, vagina				
			Larynx, nasal cavity, paranasal sinuses, nasopharynx, auditory tubes, trachea, bronchi of lungs	
				Urinary bladder, ureter, superior part of urethra
Lower part of urethra, sebaceous gland ducts	Sweat gland ducts	Part of male urethra, epididymides, ductus deferens, mammary gland ducts	Part of male urethra, salivary gland ducts	

Transitional epithelium has a rather unusual plasma membrane specialization: More rigid sections of membrane are separated by very flexible regions in which the plasma membrane is folded. When transitional epithelium is stretched, the folded regions of the plasma membrane can unfold. Transitional epithelium is specialized to expand. It is found in the urinary bladder, ureters, kidney pelvis, calyces of the kidney, and superior part of the urethra.

14 ***What is the function of an epithelial free surface that is smooth, has cilia, has microvilli, or is folded? Give an example of epithelium in which each surface type is found.***

Cell Connections

Lateral and basilar surfaces have structures that hold cells to one another or to the basement membrane (figure 4.2). These structures do three things: (1) mechanically bind the cells together, (2) help form a permeability barrier, and (3) provide a mechanism for intercellular communication. Epithelial cells secrete glycoproteins that attach the cells to the basement membrane and to one another. This relatively weak binding between cells is reinforced by **desmosomes** (dez′mō-sōmz), disk-shaped structures with especially adhesive glycoproteins that bind cells to one another and intermediate filaments that extend into the cytoplasm of the cells. Many desmosomes are found in epithelial tissues that are subjected to stress, such as the stratified squamous epithelium of the skin. **Hemidesmosomes,** similar to one-half of a desmosome, attach epithelial cells to the basement membrane.

Tight junctions hold cells together and form a permeability barrier (see figure 4.2). They consist of a zonula adherens and a zonula occludens, which are found in close association with each other. The **zonula adherens** (zō′nū-lă, zon′ū-lă ad-hēr′enz) is located between

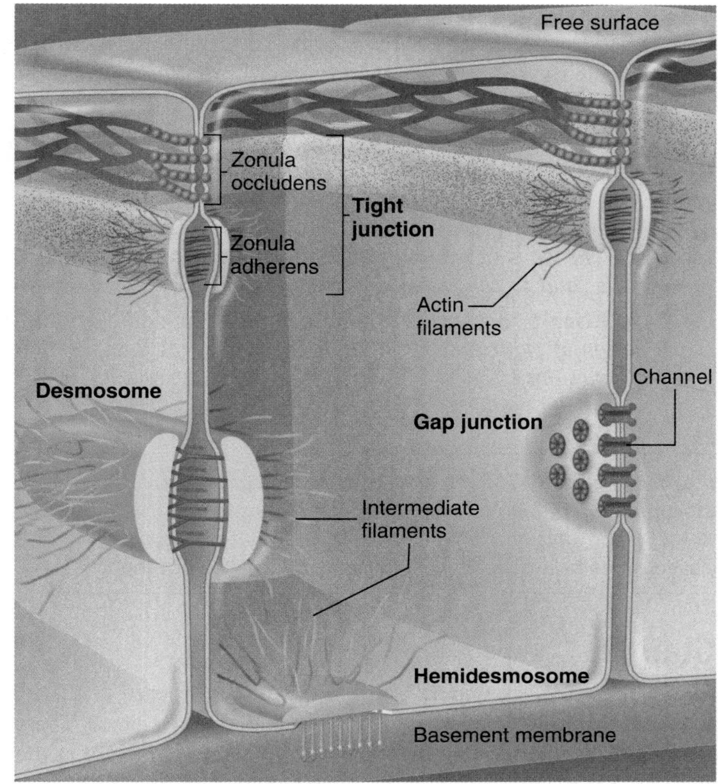

FIGURE 4.2 Cell Connections

Desmosomes anchor cells to one another and hemidesmosomes anchor cells to the basement membrane. Tight junctions consist of a zonula occludens and zonula adherens. Gap junctions allow adjacent cells to communicate with each other. Few cells have all of these different connections.

the plasma membranes of adjacent cells and acts as a weak glue that holds cells together. The zonulae adherens are best developed in simple epithelial tissues; they form a girdle of adhesive glycoprotein around the lateral surface of each cell, binding adjacent cells together. These connections are not as strong as that of desmosomes.

The **zonula occludens** (ō-klood′enz) forms a permeability barrier. It is formed by plasma membranes of adjacent cells that join one another in a jigsaw fashion to form a tight seal (see figure 4.2). Near the free surface of simple epithelial cells, the zonulae occludens form a ring that completely surrounds each cell and binds adjacent cells together. The zonulae occludens prevent the passage of materials between cells. For example, in the stomach and in the urinary bladder, chemicals cannot pass between cells. Thus, water and other substances must pass through the epithelial cells, which can actively regulate what is absorbed or secreted. Zonulae occludens are found in areas where a layer of simple epithelium forms a permeability barrier. For example, water can diffuse through epithelial cells, and active transport, symport, and facilitated diffusion move most nutrients through the epithelial cells of the intestine.

A **gap junction** is a small, specialized contact region between cells containing protein channels that aid intercellular communication by allowing ions and small molecules to pass from one cell to another (see figure 4.2). The function of gap junctions in epithelium is not entirely clear. Gap junctions between ciliated epithelial cells may coordinate the movements of cilia. Gap junctions are important in coordinating the function of cardiac and smooth muscle tissues. Because ions can pass through the gap junctions from one cell to the next, electric signals can pass from cell to cell to coordinate the contraction of cardiac and smooth muscle cells. Thus, electric signals that originate in one cell of the heart can spread from cell to cell and cause the entire heart to contract. The gap junctions between cardiac muscle cells are found in specialized cell-to-cell connections called **intercalated disks** (see chapter 20).

15 ***Name the ways in which epithelial cells are bound to one another and to the basement membrane.***

16 ***In addition to holding cells together, name an additional function of tight junctions. What is the general function of gap junctions?***

PREDICT 2

If a simple epithelial type has well-developed zonula occludens, explain how NaCl can be moved from one side of the epithelial layer to the other, what type of epithelium it is likely to be, and how the movement of NaCl causes water to move in the same direction.

Glands

Glands are secretory organs. Many glands are composed primarily of epithelium, with a supporting network of connective tissue. These glands develop from an infolding or outfolding of epithelium in the embryo. If the gland maintains an open contact with the epithelium from which it developed, a duct is present. Glands with ducts are called **exocrine** (ek′sō-krin) **glands,** and their ducts are lined with epithelium. Alternatively, some glands become separated from the epithelium of their origin. Glands that have no ducts are called **endocrine** (en′dō-krin) **glands.** Endocrine glands have extensive blood vessels in the connective tissue of the glands. The cellular products of endocrine glands, which are called **hormones** (hōr′mōnz), are secreted into the bloodstream and are carried throughout the body. Some of the endocrine glands form from nonepithelial tissue. For example, the medulla of the adrenal gland is formed from neural crest cells, and the cortex of the adrenal gland is formed from mesoderm. In addition, cardiac muscle cells can function as endocrine glands.

Most exocrine glands are composed of many cells and are called **multicellular glands,** but some exocrine glands are composed of a single cell and are called **unicellular glands** (figure 4.3*a*). **Goblet cells** (see table 4.2*c*) of the respiratory system are unicellular glands that secrete mucus. Multicellular glands can be classified further according to the structure of their ducts (figure 4.3*b*–*i*). Glands that have ducts with few branches are called **simple,** and glands with ducts that branch repeatedly are called **compound.** Further classification is based on whether the ducts end in **tubules** (small tubes), in saclike structures called **acini** (as′i-nī; grapes, suggesting a cluster of grapes or small sacs), or in **alveoli** (al-vē′ō-lī; hollow sacs). Tubular glands can be classified as straight or coiled. Most tubular glands are simple and straight, simple and coiled, or compound and coiled. Acinar glands can be simple or compound.

Exocrine glands can also be classified according to how products leave the cell. **Merocrine** (mer′ō-krin) **glands,** such as water-producing sweat glands and the exocrine portion of the pancreas, secrete products with no loss of actual cellular material (figure 4.4*a*). Secretions are either actively transported or packaged in vesicles and then released by the process of exocytosis at the free surface of the cell. **Apocrine** (ap′ō-krin) **glands,** such as the milk-producing mammary glands, discharge fragments of the gland cells in the secretion (figure 4.4*b*). Products are retained within the cell, and portions of the cell are pinched off to become part of the secretion. Apocrine glands, however, also produce merocrine secretions. Most of the volume secreted by apocrine glands are merocrine in origin. **Holocrine** (hol′ō-krin) **glands,** such as sebaceous (oil) glands of the skin, shed entire cells (figure 4.4*c*). Products accumulate in the cytoplasm of each epithelial cell, the cell ruptures and dies, and the entire cell becomes part of the secretion.

Endocrine glands are so variable in their structure that they are not classified easily. They are described in chapters 17 and 18.

17 ***Define the term*** **gland.** ***Explain how epithelial tissues give rise to exocrine and some endocrine glands, and give examples of endocrine glands formed by nonepithelial tissues.***

18 ***Distinguish between exocrine and endocrine glands. Describe the classification scheme for multicellular exocrine glands on the basis of their duct systems.***

19 ***Describe three ways in which exocrine glands release their secretions. Give an example of each method.***

CONNECTIVE TISSUE

Connective tissue is abundant, and it makes up part of every organ in the body. The major structural characteristic that distinguishes connective tissue from the other three tissue types is that it consists of cells separated from each other by abundant extracellular matrix. Connective tissue structure is diverse, and it performs a variety of important functions.

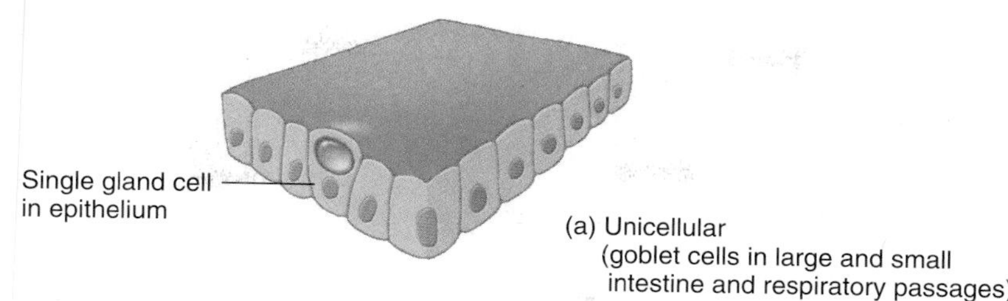

(a) Unicellular (goblet cells in large and small intestine and respiratory passages)

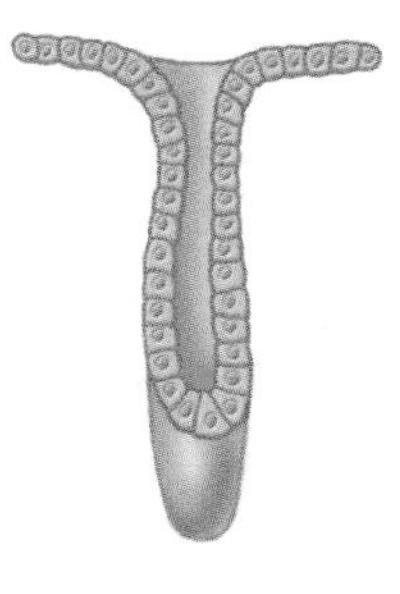

(b) Simple straight tubular (glands in stomach and colon)

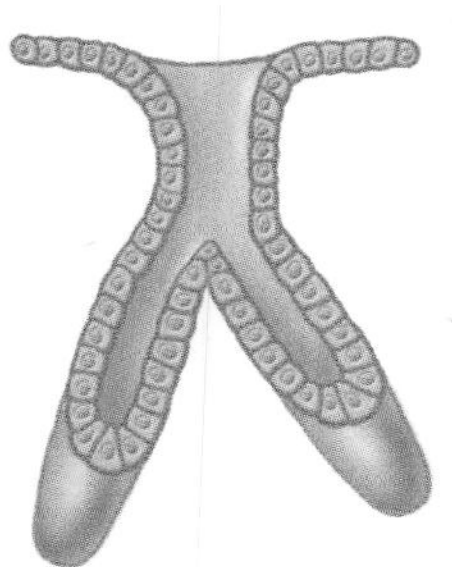

(c) Simple branched tubular (glands in lower portion of stomach)

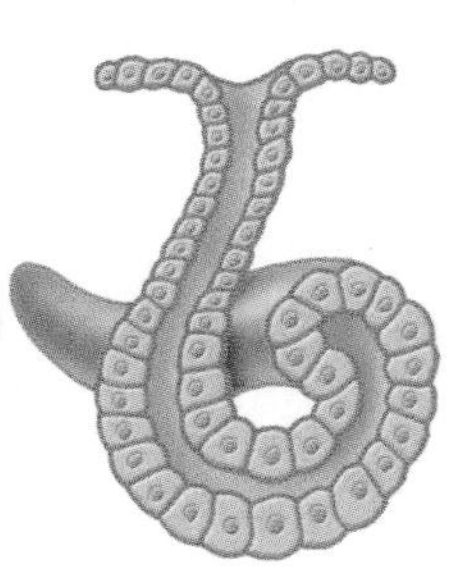

(d) Simple coiled tubular (lower portion of stomach and small intestine)

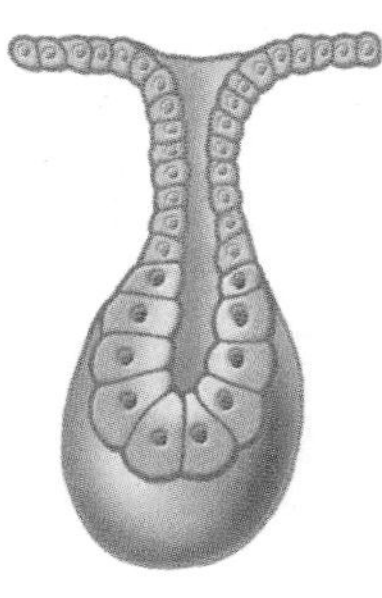

(e) Simple acinar (sebaceous glands of skin)

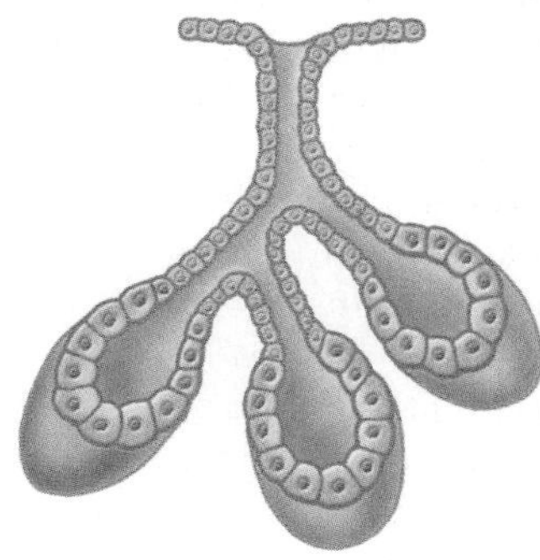

(f) Simple branched acinar (sebaceous glands of skin)

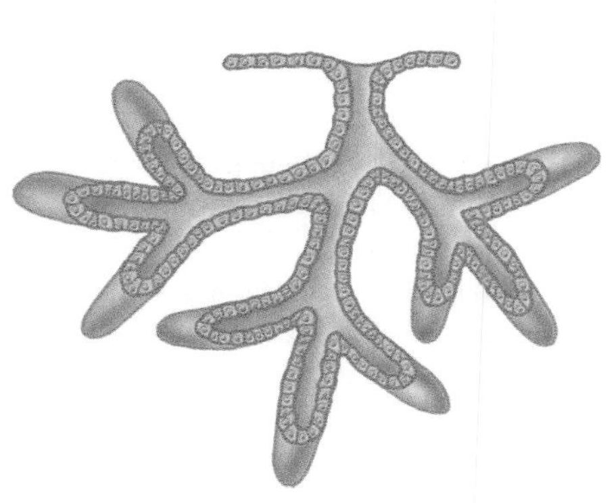

(g) Compound tubular (mucous glands of duodenum)

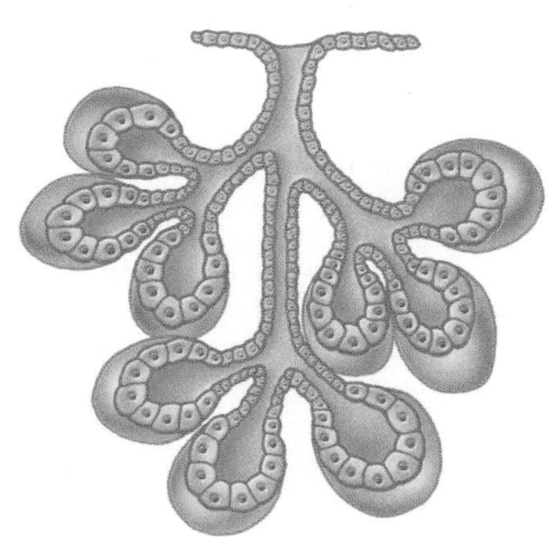

(h) Compound acinar (mammary glands)

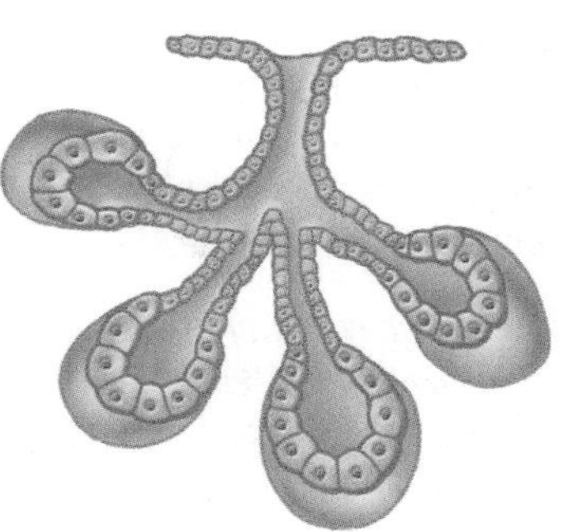

(i) Compound tubuloacinar (pancreas)

FIGURE 4.3 Structure of Exocrine Glands
The names of exocrine glands are based on the shapes of their secretory units and their ducts.

Functions of Connective Tissue

Connective tissues perform the following major categories of functions:

1. *Enclosing and separating.* Sheets of connective tissues form capsules around organs, such as the liver and kidneys. Connective tissue also forms layers that separate tissues and organs. For example, connective tissues separate muscles, arteries, veins, and nerves from one another.
2. *Connecting tissues to one another.* For example, tendons are strong cables, or bands, of connective tissue that attach muscles to bone, and ligaments are connective tissue bands that hold bones together.
3. *Supporting and moving.* Bones of the skeletal system provide rigid support for the body, and the semirigid cartilage supports structures such as the nose, ears, and surfaces of joints. Joints between bones allow one part of the body to move relative to other parts.
4. *Storing.* Adipose tissue (fat) stores high-energy molecules, and bones store minerals, such as calcium and phosphate.
5. *Cushioning and insulating.* Adipose tissue cushions and protects the tissue it surrounds and provides an insulating layer beneath the skin that helps conserve heat.

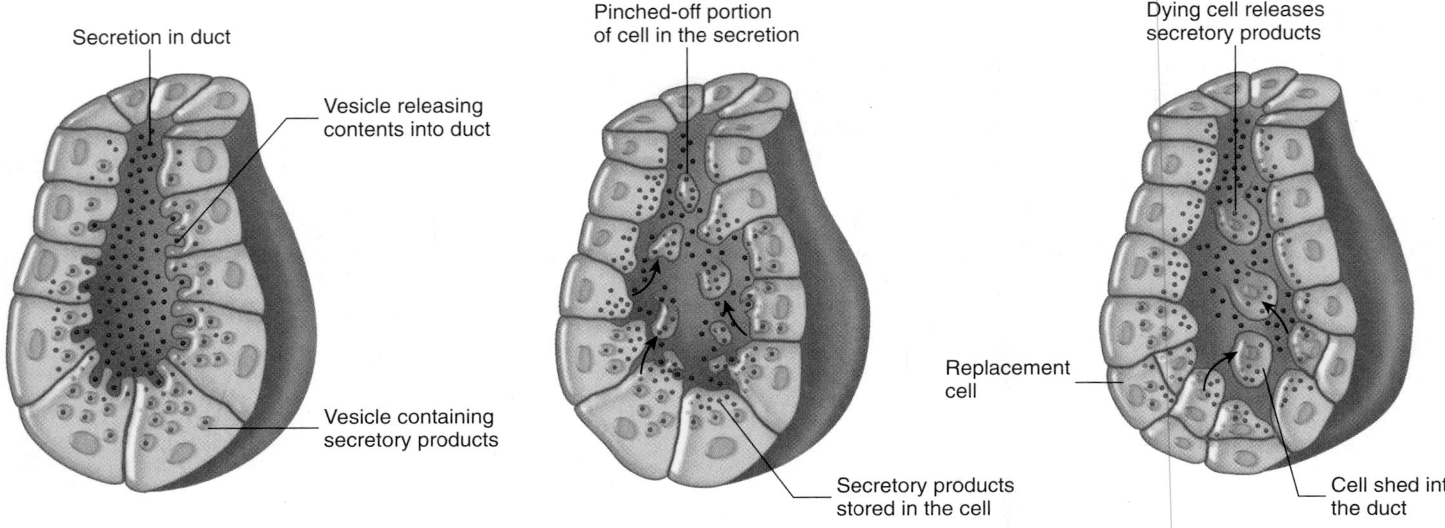
Secretion in duct
Vesicle releasing contents into duct
Vesicle containing secretory products
Pinched-off portion of cell in the secretion
Secretory products stored in the cell
Dying cell releases secretory products
Replacement cell
Cell shed into the duct
(a) Merocrine gland
Cells of the gland produce secretions by active transport or produce vesicles that contain secretory products, and the vesicles empty their contents into the duct through exocytosis.
(b) Apocrine gland
Secretory products are stored in the cell near the lumen of the duct. A portion of the cell near the lumen containing secretory products is pinched off the cell and joins secretions produced by a merocine process.
(c) Holocrine gland
Secretory products are stored in the cells of the gland. Entire cells are shed by the gland and become part of the secretion. The lost cells are replaced by other cells deeper in the gland.

FIGURE 4.4 Exocrine Glands and Secretion Types
Exocrine glands are classified according to the type of secretion.

6. *Transporting.* Blood transports substances throughout the body, such as gases, nutrients, enzymes, hormones, and cells of the immune system.
7. *Protecting.* Cells of the immune system and blood provide protection against toxins and tissue injury, as well as from microorganisms. Bones protect underlying structures from injury.

20 ***What is the major characteristic that distinguishes connective tissue from other tissues?***

21 ***List the major functions of connective tissues, and give an example of a connective tissue that performs each function.***

Cells of Connective Tissue

The specialized cells of the various connective tissues produce the extracellular matrix. The names of the cells end with suffixes that identify the cell functions as blasts, cytes, or clasts. **Blasts** create the matrix, **cytes** maintain it, and **clasts** break it down for remodeling. For example, **fibroblasts** are cells that form fibrous connective tissue and **fibrocytes** maintain it. **Chondroblasts** form cartilage (*chondro-* refers to cartilage) and **chondrocytes** maintain it. **Osteoblasts** form bone (*osteo-* means bone), **osteocytes** maintain it, and **osteoclasts** break it down (see chapter 6).

Adipose (ad′i-pōs; fat), or **fat, cells,** also called **adipocytes** (ad′i-pō-sītz), contain large amounts of lipid. The lipid pushes the rest of the cell contents to the periphery so that each cell appears to contain a large, centrally located lipid droplet with a thin layer of cytoplasm around it. Adipose cells are rare in some connective tissue types, such as cartilage; are abundant in others, such as loose connective tissue; or are predominant, such as in adipose tissue.

Mast cells are commonly found beneath membranes in loose connective tissue and along small blood vessels of organs. They contain chemicals such as heparin, histamine, and proteolytic enzymes. These substances are released in response to injury, such as trauma and infection, and play important roles in inflammation.

White blood cells, or **leukocytes** (see chapter 18) continuously move from blood vessels into connective tissues. The rate of movement increases dramatically in response to injury or infection. In addition, accumulations of lymphocytes, a type of white blood cell, are common in some connective tissues, such as in the connective tissue beneath the epithelial lining of certain parts of the digestive system.

Macrophages are found in some connective tissue types. They are derived from monocytes, a white blood cell type. Macrophages are either **fixed** and do not move through the connective tissue in which they are found or are **wandering macrophages** and move by ameboid movement through the connective tissue. Macrophages phagocytize foreign and injured cells, and they play a major role in providing protection against infections.

Undifferentiated mesenchymal cells, sometimes called **stem cells,** are embryonic cells that persist in adult connective tissue. They have the potential to differentiate to form adult cell types, such as fibroblasts or smooth muscle cells, in response to injury.

22 ***Explain the difference among connective tissue cells that are termed blast, cyte, or clast cells.***

23 ***Describe and give the functions of the cells of connective tissue.***

Extracellular Matrix

The extracellular matrix of connective tissue has three major components: (1) protein fibers, (2) ground substance consisting of nonfibrous protein and other molecules, and (3) fluid. The structure of the matrix gives connective tissue types most of their functional characteristics, such as the ability of bones and cartilage to bear weight, of tendons and ligaments to withstand tension, and of the skin's dermis to withstand punctures, abrasions, and other abuses.

Protein Fibers of the Matrix

Three types of protein fibers—collagen, reticular, and elastic fibers—help form connective tissue.

Collagen (kol′a-jen) **fibers** consist of collagen, which accounts for one-fourth to one-third of total body protein, or 6% of total body weight. Within fibroblasts, chains of amino acids called **collagen α-chains** are synthesized, and three collagen α-chains wind around each other to form a microscopic, ropelike helix called a **collagen molecule.** After collagen molecules are secreted, they are modified. Some amino acids are cleaved off of the collagen molecules to produce **tropocollagen** (trō-pō-kol′a-jen) molecules. The tropocollagen molecules are then linked together to make long **collagen fibrils.** The collagen fibrils are then joined together in bundles to form collagen fibers (figure 4.5*a*). Approximately 25 types of collagen α-chains, each encoded by a separate gene, are known, and they combine to form approximately 20 types of collagen. Collagen fibers are very strong and flexible but quite inelastic. There are at least 20 types of collagen fibers, many of which are specific to certain tissues. Collagen fibers differ in the types of amino acids that make up the polypeptide chains. Of the 20 types of collagen fibers, 6 types are most common. Tendons, ligaments, skin, and bone contain mainly type I collagen, cartilage is mainly type II collagen, and reticular fibers are mainly type III collagen.

Reticular (re-tik′ū-lār; netlike) **fibers** are very fine collagen fibers and therefore are not a chemically distinct category of fibers. They are very short, thin fibers that branch to form a network and appear different microscopically from other collagen fibers. Reticular fibers are not as strong as most collagen fibers, but networks of reticular fibers fill spaces between tissues and organs.

Elastic fibers consist of proteins called **elastin** (e-las′tin). As the name suggests, this protein has the ability to return to its original shape after being stretched or compressed. Elastin gives the tissue in which it is found an elastic quality. Fibroblasts secrete polypeptide chains, which are then linked together to form a network. The polypeptide chains recoil when they are stretched (figure 4.5*b*).

Other Matrix Molecules

Two types of large, nonfibrous molecules, called hyaluronic acid and proteoglycans, are part of the extracellular matrix. These molecules constitute most of the **ground substance** of the matrix, the "shapeless" background against which the collagen fibers are seen through the microscope. The molecules themselves, however, are not shapeless but are highly structured. **Hyaluronic** (hī′ă-loo-ron′ik; glassy appearance) **acid** is a long, unbranched polysaccharide chain

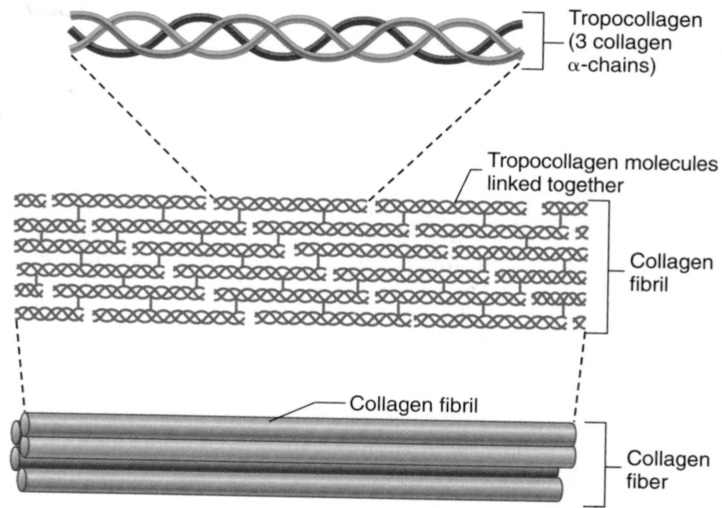

(a) Collagen fibers. Collagen molecules are secreted by fibroblasts and enzymes cleave off amino acids and the 3 collegen α-chains form tropocollagen. Each tropocollagen molecule consists of 3 collagen α-chains coiled together. Tropocollagen molecules are then linked to form collagen fibrils. The collagen fibrils are joined together to form collagen fibers.

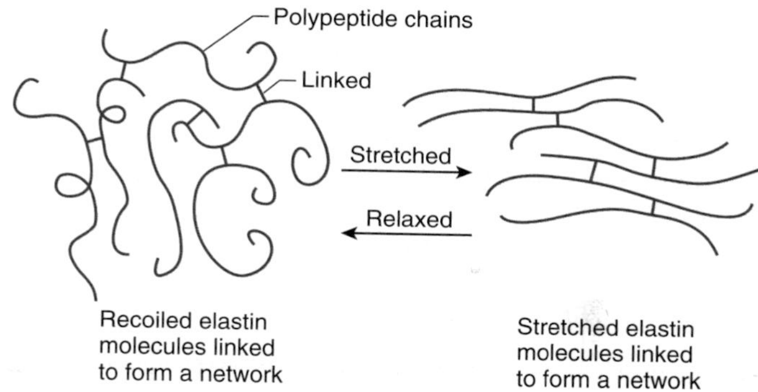

(b) Elastic fibers. Fibroblasts secrete polypeptide chains, which are then linked together to form a network of polypeptide chains. The network stretches in response to a force. When the force is removed, the network recoils.

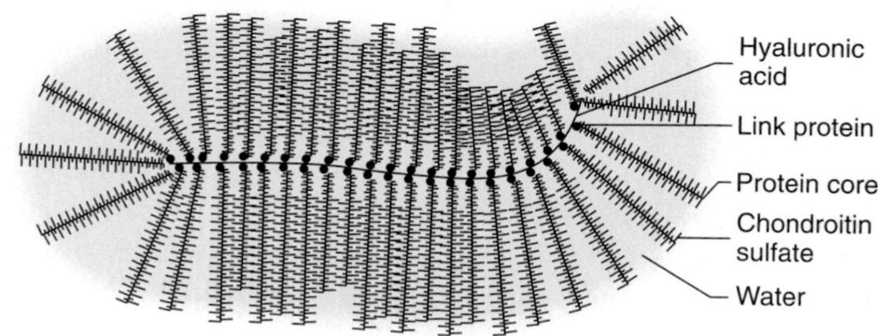

(c) Proteoglycan aggregates. Proteoglycan monomers are formed from 80 to 100 glycosaminoglycans, such as chondroitin sulfate, attached by one end to a protein core. The protein cores of many proteoglycan monomers can attach through link proteins to a long molecule of hyaluronic acid. Water molecules are trapped by the glycosaminoglycans.

FIGURE 4.5 Molecules of Connective Tissue Matrix

composed of repeating disaccharide units. It gives a very slippery quality to the fluids that contain it; for that reason, it is a good lubricant for joint cavities (see chapter 8). Hyaluronic acid is also found in large quantities in connective tissue and is the major component of the vitreous humor of the eye (see chapter 15).

A **proteoglycan** (prō′tē-ō-glī′kan; formed from proteins and polysaccharides) **monomer** is a large molecule that consists of 80 to 100 polysaccharides, called **glycosaminoglycans** (glīkōs-am-i-nō-glī′kanz), such as **chondroitin** (kon-drō′i-tin) **sulfate,** each attached by one end to a protein core. The protein cores of many proteoglycan monomers can attach through link proteins to a long molecule of hyaluronic acid to form a **proteoglycan aggregate** (figure 4.5*c*). Proteoglycan aggregates trap large quantities of water, which allows them to return to their original shape when compressed or deformed. There are several types of glycosaminoglycans, and their abundance varies with each connective tissue type.

Several **adhesive molecules** are found in ground substance. These adhesive molecules hold the proteoglycan aggregates together and to structures such as the plasma membranes. A specific adhesive molecule type predominates in certain types of ground substance. For example, **chondronectin** is in the ground substance of cartilage, **osteonectin** is in the ground substance of bone, and **fibronectin** is in the ground substance of fibrous connective tissues.

24 ***What three components are found in the extracellular matrix of connective tissue?***

25 ***Contrast the structure and characteristics of collagen fibers, reticular fibers, and elastin fibers.***

26 ***Describe the structure and function of hyaluronic acid and proteoglycan aggregates. What is the function of adhesive molecules?***

Classification of Connective Tissue

Connective tissue types blend into one another, and the transition points cannot be defined precisely. As a result, the classification scheme for connective tissues is somewhat arbitrary. Classification schemes for connective tissue are influenced by (1) protein fibers and the arrangement of protein fibers in the extracellular matrix, (2) protein fibers and ground substance in the extracellular matrix, and (3) a fluid extracellular matrix. The classification of connective tissues used in this book is presented in table 4.6.

The two major categories of connective tissue are embryonic and adult connective tissue.

TABLE 4.6 Classification of Connective Tissue

A. Embryonic connective tissue
   1. Mesenchyme
   2. Mucous
B. Adult connective tissue
   1. Loose (areolar)
   2. Dense
      a. Dense, regular collagenous
      b. Dense, regular elastic
      c. Dense, irregular collagenous
      d. Dense, irregular elastic
   3. Special properties
      a. Adipose
      b. Reticular
   4. Cartilage
      a. Hyaline
      b. Fibrocartilage
      c. Elastic
   5. Bone
      a. Cancellous
      b. Compact
   6. Blood and hemopoietic tissue

Embryonic Connective Tissue

Embryonic connective tissue is called **mesenchyme** (mez′en-kīm). It is made up of irregularly shaped fibroblasts surrounded by abundant, semifluid extracellular matrix in which delicate collagen fibers are distributed (table 4.7*a*). It forms in the embryo during the third and fourth weeks of development from mesoderm and neural crest cells (see chapter 29), and all adult connective tissue types develop from it. By 8 weeks of development, most of the mesenchyme has become specialized to form the types of connective tissue seen in adults, as well as muscle, blood vessels, and other tissues. The major source of remaining embryonic connective tissue in the newborn is found in the umbilical cord, where it is called **mucous connective tissue,** or **Wharton's jelly** (table 4.7*b*). The structure of mucous connective tissue is similar to that of mesenchyme.

Adult Connective Tissue

Adult connective tissue consists of six types: loose, dense, connective tissue with special properties, cartilage, bone, and blood and hemopoietic tissue.

Loose Connective Tissue

Loose connective tissue (table 4.8), which is sometimes referred to as **areolar** (ă-rē′ō-lăr; area) **tissue,** consists of protein fibers that form a lacy network, with numerous fluid-filled spaces. Areolar tissue is the "loose packing" material of most organs and other tissues; it attaches the skin to underlying tissues. It contains collagen, reticular, and elastic fibers and a variety of cells. For example, fibroblasts produce the fibrous matrix; macrophages move through the tissue, engulfing bacteria and cell debris; mast cells contain chemicals that help mediate inflammation; and lymphocytes are involved in immunity. The loose packing of areolar tissue is often associated with other connective tissue types, such as reticular tissue and fat (adipose tissue).

Dense Connective Tissue

Protein fibers of **dense connective tissue** form thick bundles and fill nearly all of the extracellular space. Most of the cells of developing

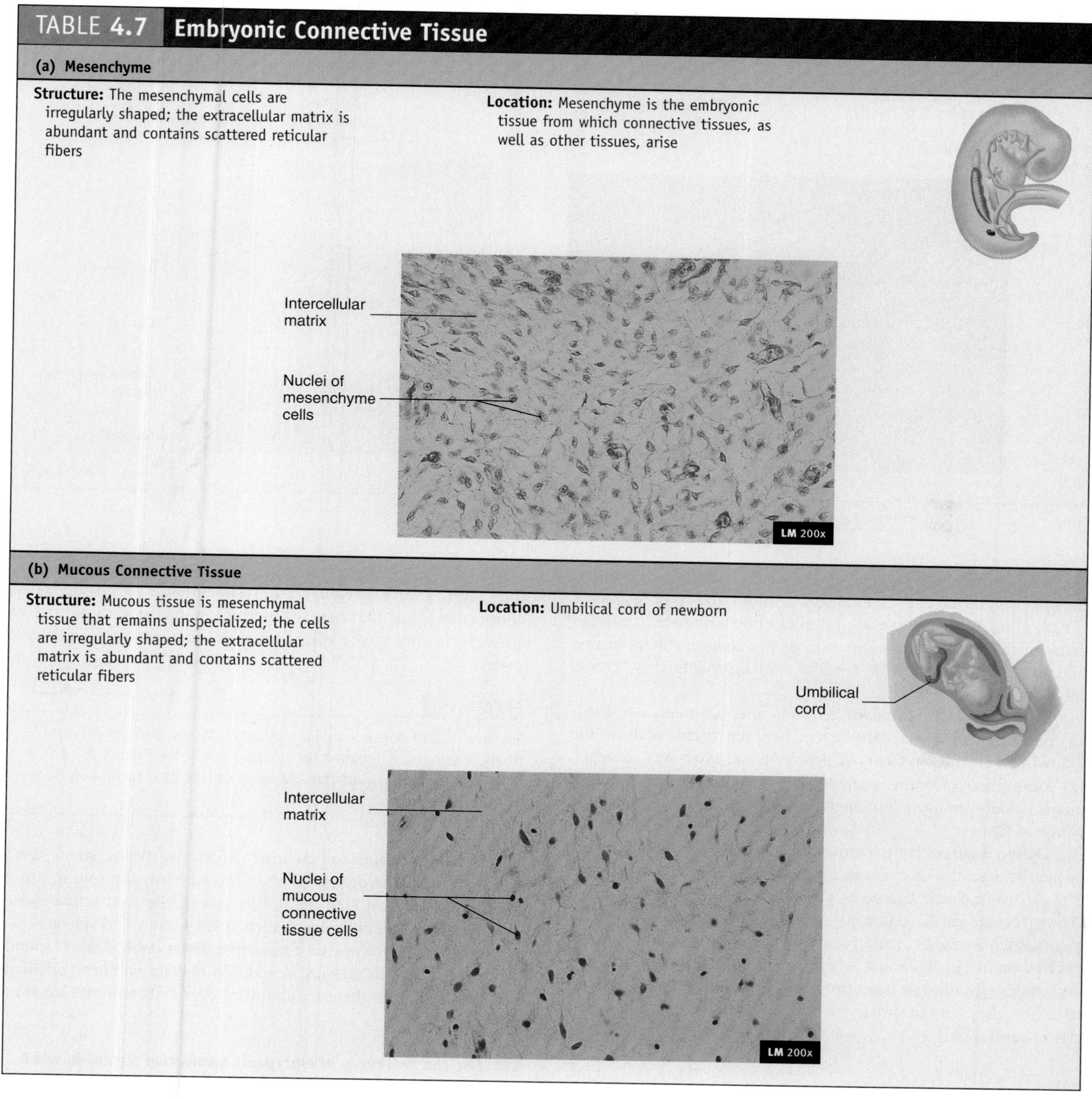

dense connective tissue are spindle-shaped fibroblasts. Once the fibroblasts become completely surrounded by matrix, they are fibrocytes. Dense connective tissue can be subdivided into two major groups: regular and irregular.

Dense regular connective tissue has protein fibers in the extracellular matrix that are oriented predominantly in one direction. **Dense regular collagenous connective tissue** (table 4.9*a*) has abundant collagen fibers. The collagen fibers give this tissue a white appearance. Dense regular collagenous connective tissue forms structures such as tendons, which connect muscles to bones (see chapter 11), and most ligaments, which connect bones to bones (see chapter 8). The collagen fibers of dense connective

TABLE 4.8 Loose Connective Tissue

Structure: Cells (e.g., fibroblasts, macrophages, and lymphocytes) within a fine network of mostly collagen fibers; often merges with denser connective tissue

Function: Loose packing, support, and nourishment for the structures with which it is associated

Location: Widely distributed throughout the body; substance on which epithelial basement membranes rest; packing between glands, muscles, and nerves; attaches the skin to underlying tissues

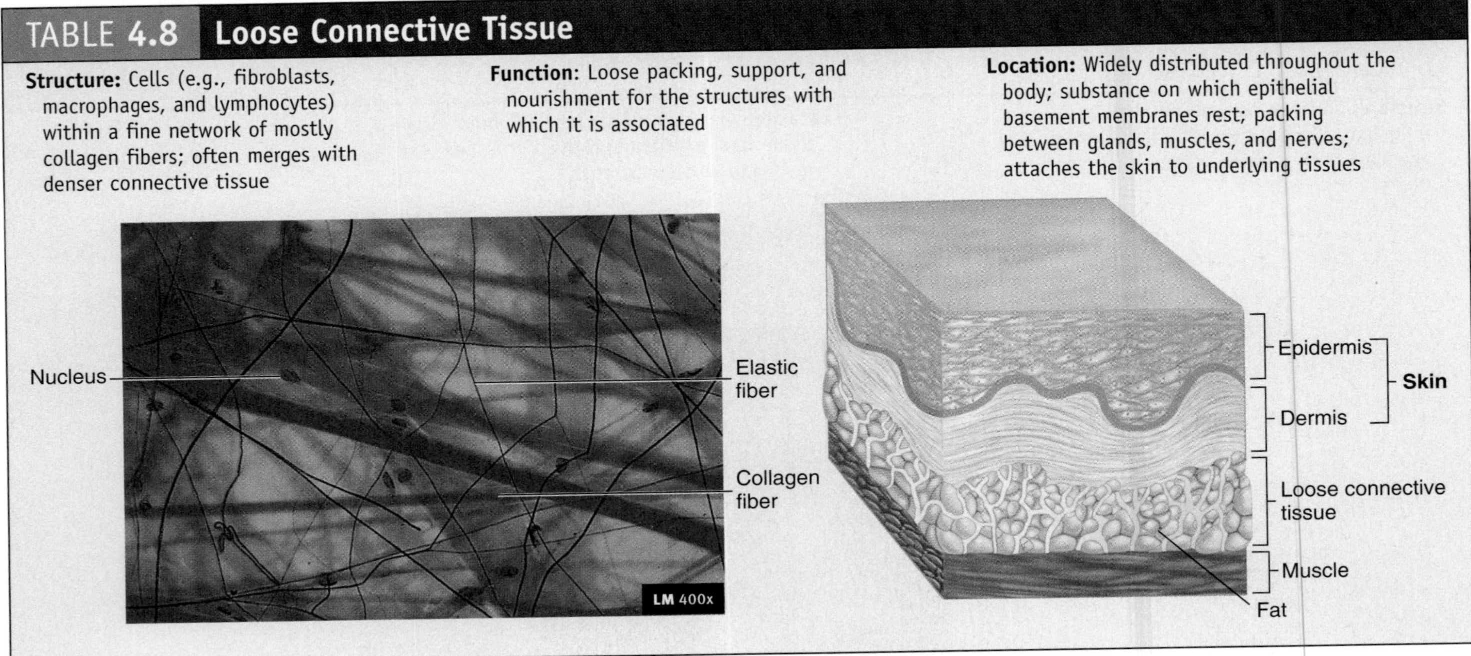

tissue resist stretching and give the tissue considerable strength in the direction of the fiber orientation. Tendons and most ligaments consist almost entirely of thick bundles of densely packed parallel collagen fibers with the orientation of the collagen fibers in one direction, which makes the tendons and ligaments very strong, cablelike structures.

The general structures of tendons and ligaments are similar, but there are major differences between them, such as the following: (1) collagen fibers of ligaments are often less compact, (2) some fibers of many ligaments are not parallel, and (3) ligaments usually are more flattened than tendons and form sheets or bands of tissues.

Dense regular elastic connective tissue (table 4.9*b*) consists of parallel bundles of collagen fibers and abundant elastic fibers. The elastin in elastic ligaments gives them a slightly yellow color. Dense regular elastic connective tissue forms some elastic ligaments, such as those in the vocal folds and the **nuchal** (noo′kăl; back of the neck) **ligament,** which lies along the posterior of the neck, helping hold the head upright. When elastic ligaments are stretched, they tend to shorten to their original length, much as an elastic band does.

PREDICT 3

Explain the advantages of having elastic ligaments that extend from vertebra to vertebra in the vertebral column and why it would be a disadvantage if tendons, which connect skeletal muscles to bone, were elastic.

Dense irregular connective tissue contains protein fibers arranged as a meshwork of randomly oriented fibers. Alternatively, the fibers within a given layer of dense irregular connective tissue can be oriented in one direction, whereas the fibers of adjacent layers are oriented at nearly right angles to that layer. Dense irregular connective tissue forms sheets of connective tissue that have strength in many directions, but less strength in any single direction than does regular connective tissue.

PREDICT 4

Scars consist of dense irregular connective tissue made of collagen fibers. Vitamin C is required for collagen synthesis. Predict the effect of scurvy, which is a nutritional disease caused by vitamin C deficiency, on wound healing.

Dense irregular collagenous connective tissue (table 4.9*c*) forms most of the dermis of the skin, which is the tough, inner portion of the skin (see chapter 5) and of the connective tissue capsules that surround organs such as the kidney and spleen.

Dense irregular elastic connective tissue (table 4.9*d*) is found in the wall of elastic arteries. In addition to collagen fibers, oriented in many directions, there are abundant elastic fibers in the layers of this tissue.

27 ***List the two types of embryonic connective tissue. To what does mesenchyme give rise in the adult?***

28 ***Describe the fiber arrangement in loose (areolar) connective tissue. What are the functions of this tissue type?***

29 ***Structurally and functionally, what is the difference between dense regular connective tissue and dense irregular connective tissue?***

30 ***Name the two kinds of dense regular connective tissue, and give an example of each. Do the same for dense irregular connective tissue.***

TABLE 4.9 Dense Connective Tissue

(a) Dense Regular Collagenous Connective Tissue

Structure: Matrix composed of collagen fibers running in somewhat the same direction

Function: Ability to withstand great pulling forces exerted in the direction of fiber orientation, great tensile strength and stretch resistance

Location: Tendons (attach muscle to bone) and ligaments (attach bones to each other)

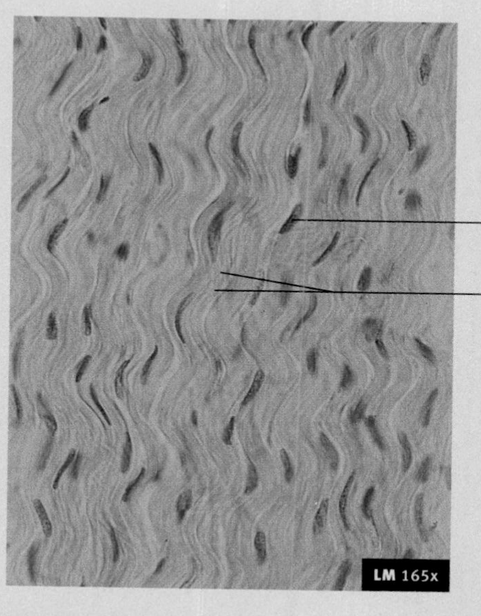

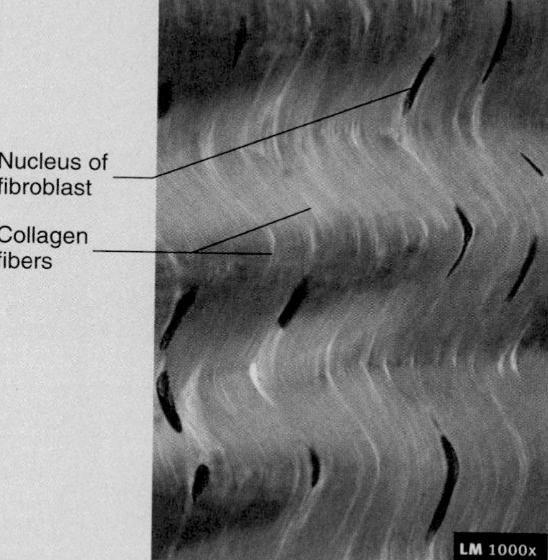

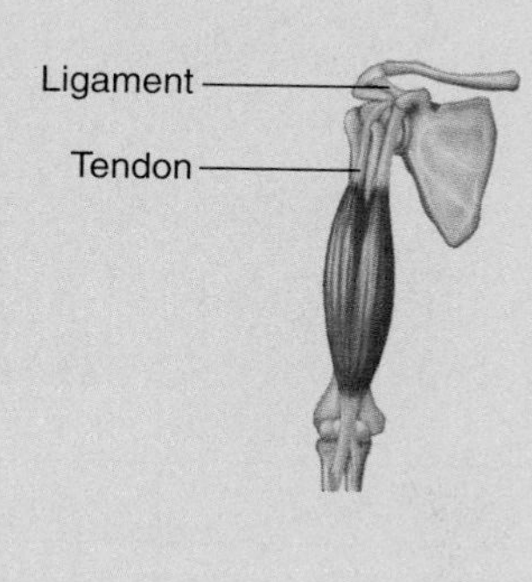

(b) Dense Regular Elastic Connective Tissue

Structure: Matrix composed of regularly arranged collagen fibers and elastin fibers

Function: Capable of stretching and recoiling like a rubber band, with strength in the direction of fiber orientation

Location: Ligaments between the vertebrae and along the dorsal aspect of the neck (nucha) and in the vocal cords

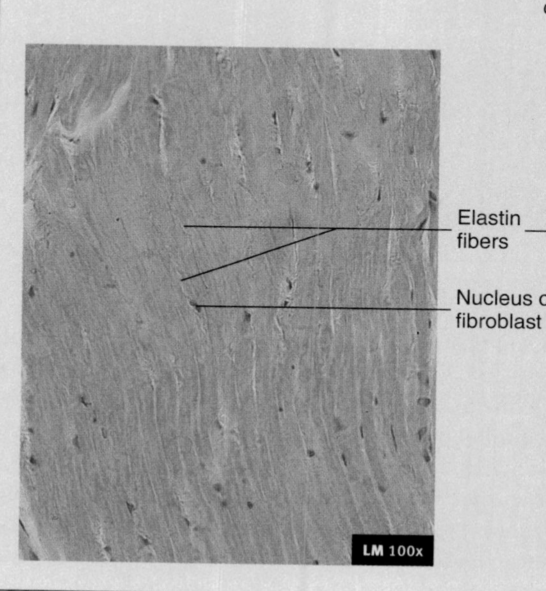

LM 100x

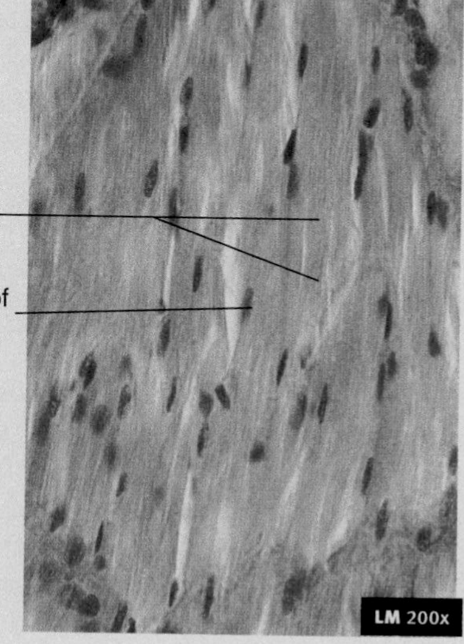

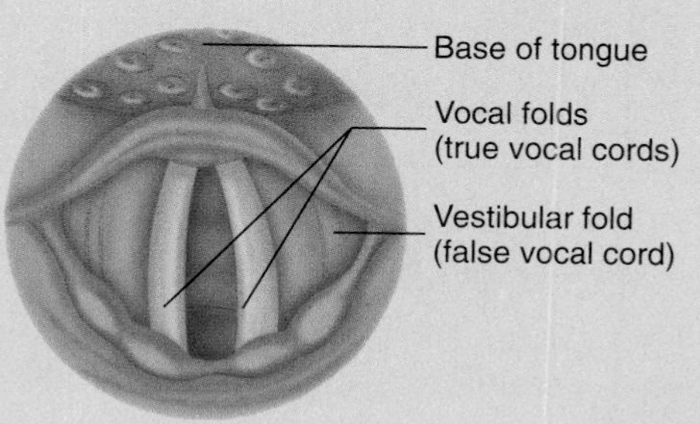

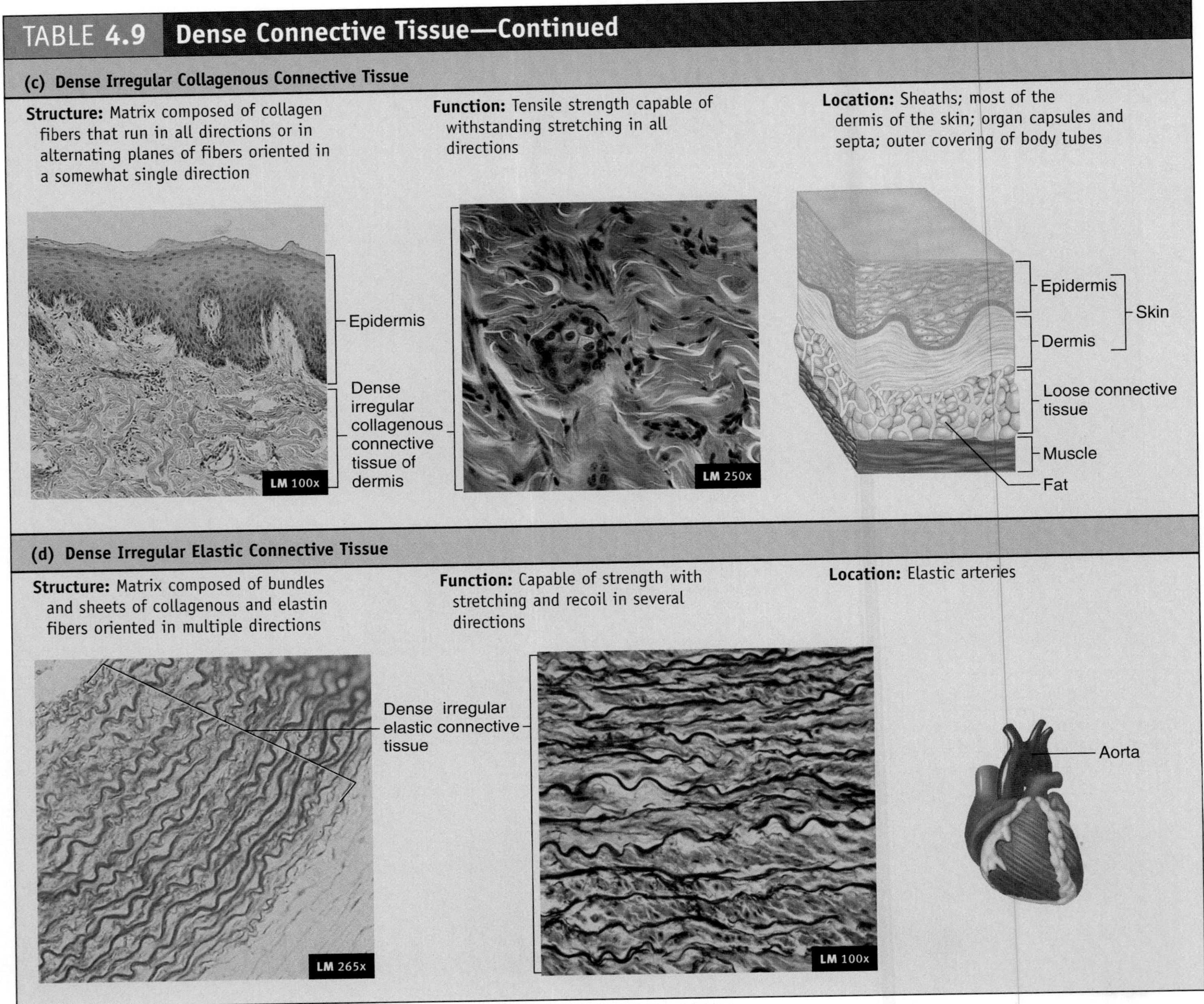

Connective Tissue with Special Properties

Adipose tissue and reticular tissue are connective tissues with special properties. **Adipose tissue** (table 4.10*a*) consists of adipocytes, or fat cells, which contain large amounts of lipid. Unlike other connective tissue types, adipose tissue is composed of large cells and a small amount of extracellular matrix, which consists of loosely arranged collagen and reticular fibers with some scattered elastic fibers. Blood vessels form a network in the extracellular matrix. The fat cells are usually arranged in clusters or lobules separated from one another by loose connective tissue. Adipose tissue functions as an insulator, a protective tissue, and a site of energy storage. Lipids take up less space per calorie than either carbohydrates or proteins and therefore are well adapted for energy storage.

Adipose tissue exists in both yellow (white) and brown forms. **Yellow adipose** tissue is by far the most abundant. Yellow adipose tissue appears white at birth, but it turns yellow with age because of the accumulation of pigments, such as carotene, a plant pigment that humans can metabolize as a source of vitamin A. Storage, insulation, and protection are the functions of yellow adipose tissue. Brown adipose tissue is found only in specific areas of the body, such as the axillae (armpits), the neck, and near the kidneys. The brown color results from the cytochrome pigments in its numerous mitochondria and its abundant blood supply. Although brown fat is much more prevalent in babies than in adults, it is difficult to distinguish brown fat from yellow fat in babies because the color difference between them is not great. Brown fat is specialized to generate heat as a result of oxidative metabolism of lipid

TABLE 4.10 Connective Tissue with Special Properties

(a) Adipose Tissue

Structure: Little extracellular matrix surrounding cells; the adipocytes, or fat cells, are so full of lipid that the cytoplasm is pushed to the periphery of the cell

Function: Packing material, thermal insulator, energy storage, and protection of organs against injury from being bumped or jarred

Location: Predominantly in subcutaneous areas, in mesenteries, in renal pelvis, around kidneys, attached to the surface of the colon, in mammary glands, in loose connective tissue that penetrates into spaces and crevices

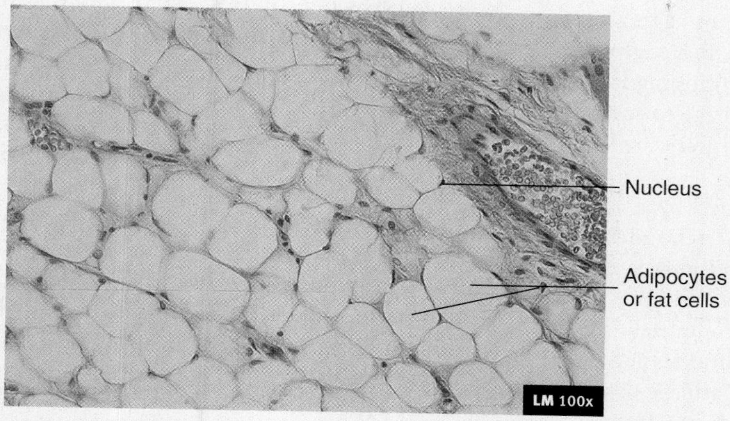

Nucleus

Adipocytes or fat cells

LM 100x

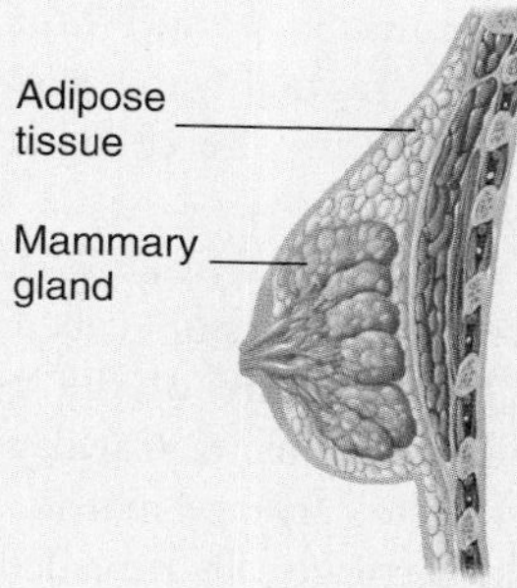

Adipose tissue

Mammary gland

(b) Reticular Tissue

Structure: Fine network of reticular fibers irregularly arranged

Function: Provides a superstructure for the lymphatic and hemopoietic

Location: Within the lymph nodes, spleen, and bone marrow

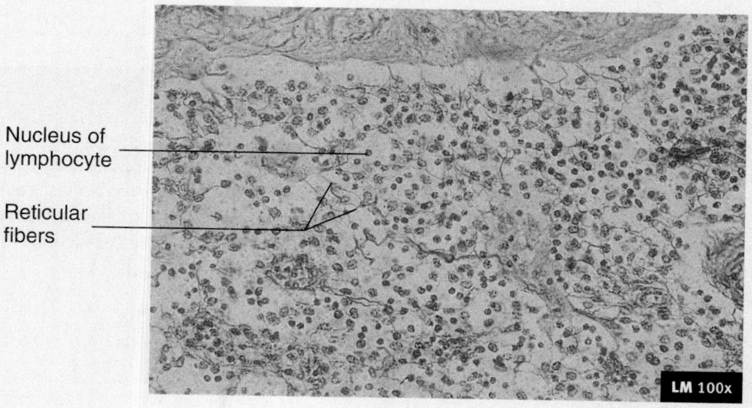

Nucleus of lymphocyte

Reticular fibers

LM 100x

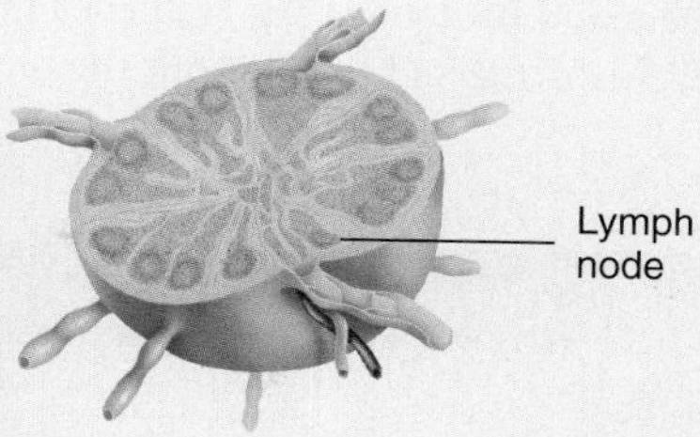

Lymph node

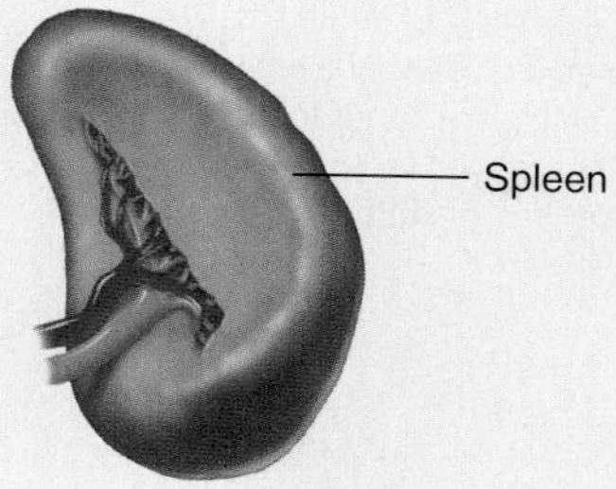

Spleen

molecules in mitochondria and can play a significant role in body temperature regulation in newborn babies.

Reticular tissue forms the framework of lymphatic tissue (table 4.10*b*), such as in the spleen and lymph nodes, as well as in bone marrow and the liver. It is characterized by a network of reticular fibers and reticular cells. **Reticular cells** produce the reticular fibers and remain closely attached to them. The spaces between the reticular fibers can contain a wide variety of other cells, such as dendritic cells, which look very much like reticular cells but are cells of the immune system, macrophages, and blood cells (see chapter 22).

31 ***What feature of the extracellular matrix distinguishes adipose tissue from other connective tissue types? What is an adipocyte?***

32 ***List the functions of adipose tissue. Name the two types of adipose tissue. Which one is important in generating heat?***

33 ***What is the function of reticular tissue? Where is it found?***

Cartilage

Cartilage (kar′ti-lij) is composed of cartilage cells, or **chondrocytes** (kon′drō-sītz), located in spaces called **lacunae** (lă-koo′nē) within an extensive and relatively rigid matrix. Next to bone, cartilage is the firmest structure in the body. The matrix contains protein fibers, ground substance, and fluid. The protein fibers are collagen fibers or collagen and elastic fibers. The ground substance consists of proteoglycans and other organic molecules. Most of the proteoglycans in the matrix form aggregates with hyaluronic acid. Within the cartilage matrix, proteoglycan aggregates function as minute sponges capable of trapping large quantities of water. This trapped water allows cartilage to spring back after being compressed. The collagen fibers give cartilage considerable strength.

The surface of nearly all cartilage is surrounded by a layer of dense irregular connective tissue called the **perichondrium** (per-i-kon′drē-ŭm). The structure of the perichondrium is described in more detail in chapter 6. Cartilage cells arise from the perichondrium and secrete cartilage matrix. Once completely surrounded by matrix, the cartilage cells are called chondrocytes and the spaces in which they are located are called lacunae. Cartilage has no blood vessels or nerves, except those of the perichondrium; it therefore heals very slowly after an injury because the cells and nutrients necessary for tissue repair cannot reach the damaged area easily.

There are three types of cartilage.

1. **Hyaline** (hī′ă-lin) **cartilage** has large amounts of both collagen fibers and proteoglycans (table 4.11*a*). Collagen fibers are evenly dispersed throughout the ground substance, and in joints hyaline cartilage has a very smooth surface. Specimens appear to have a glassy, translucent matrix when viewed through a microscope. Hyaline cartilage is found in areas in which strong support and some flexibility are needed, such as in the rib cage and the cartilage within the trachea and bronchi (see chapter 23). It also covers the surfaces of bones that move smoothly against each other in joints. Hyaline cartilage forms most of the skeleton before it is replaced by bone in the embryo, and it is involved in growth that increases the length of bones (see chapter 6).
2. **Fibrocartilage** has more collagen fibers than proteoglycans (table 4.11*b*). Compared with hyaline cartilage, fibrocartilage has much thicker bundles of collagen fibers dispersed through its matrix. Fibrocartilage is slightly compressible and very tough. It is found in areas of the body where a great deal of pressure is applied to joints, such as in the knee, in the jaw, and between vertebrae.
3. **Elastic cartilage** has elastic fibers in addition to collagen and proteoglycans (table 4.11*c*). The numerous elastic fibers are dispersed throughout the matrix of elastic cartilage. It is found in areas, such as the external ears, that have rigid but elastic properties.

PREDICT 5

One of several changes caused by rheumatoid arthritis in joints is the replacement of hyaline cartilage with dense irregular collagenous connective tissue. Predict the effect of replacing hyaline cartilage with fibrous connective tissue.

TABLE 4.11 Connective Tissue: Cartilage

(a) Hyaline Cartilage

Structure: Collagen fibers are small and evenly dispersed in the matrix, making the matrix appear transparent; the cartilage cells, or chondrocytes, are found in spaces, or lacunae, within the firm but flexible matrix

Function: Allows growth of long bones; provides rigidity with some flexibility in the trachea, bronchi, ribs, and nose; forms rugged, smooth, yet somewhat flexible articulating surfaces; forms the embryonic skeleton

Location: Growing long bones, cartilage rings of the respiratory system, costal cartilage of ribs, nasal cartilages, articulating surface of bones, embryonic skeleton

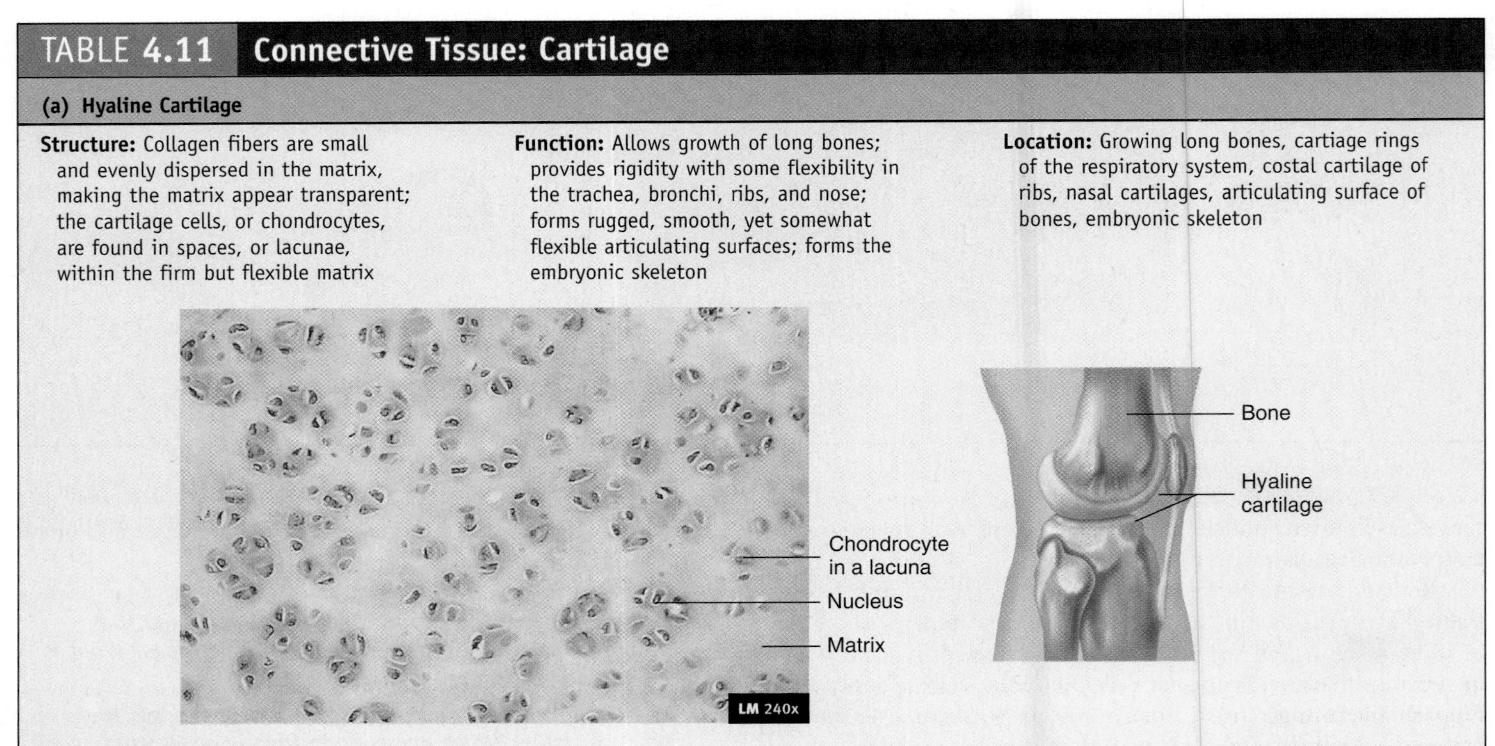

TABLE 4.11 Connective Tissue: Cartilage—Continued

(b) Fibrocartilage

Structure: Collagenous fibers similar to those in hyaline cartilage; the fibers are more numerous than in other cartilages and are arranged in thick bundles

Function: Somewhat flexible and capable of withstanding considerable pressure; connects structures subjected to great pressure

Location: Intervertebral disks, symphysis pubis, articular disks (e.g., knee and temporomandibular [jaw] joints)

(c) Elastic Cartilage

Structure: Similar to hyaline cartilage, but matrix also contains elastin fibers

Function: Provides rigidity with even more flexibility than hyaline cartilage because elastic fibers return to their original shape after being stretched

Location: External ears, epiglottis, auditory tubes

Bone

Bone is a hard connective tissue that consists of living cells and mineralized matrix. Bone matrix has an organic and an inorganic portion. The organic portion consists of protein fibers, primarily collagen, and other organic molecules. The mineral, or inorganic, portion consists of specialized crystals called **hydroxyapatite** (hī-drok′sē-ap-ă-tīt), which contain calcium and phosphate. The strength and rigidity of the mineralized matrix allow bones to support and protect other tissues and organs. Bone cells, or **osteocytes** (os′tē-ō-sītz), are located within holes in the matrix, which are called lacunae and are similar to the lacunae of cartilage.

Two types of bone exist.

1. **Cancellous** (kan′sē-lŭs), or **spongy, bone** has spaces between **trabeculae** (tră-bek′ū-lē; beams), or plates, of bone and therefore resembles a sponge (table 4.12*a*).
2. **Compact bone** is more solid with almost no space between many thin layers, or **lamellae** (lă-mel′ē sing. lă-mel′ă) of bone (table 4.12*b*).

TABLE 4.12 Connective Tissue: Bones

(a) Cancellous Bone

Structure: Latticelike network of scaffolding characterized by trabeculae with large spaces between them filled with hemopoietic tissue; the osteocytes, or bone cells, are located within lacunae in the trabeculae

Function: Acts as a scaffolding to provide strength and support without the greater weight of compact bone

Location: In the interior of the bones of the skull, vertebrae, sternum, and pelvis, and in the ends of the long bones

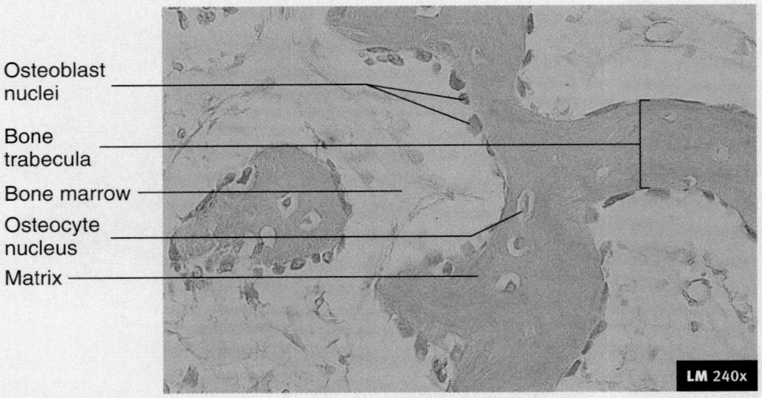

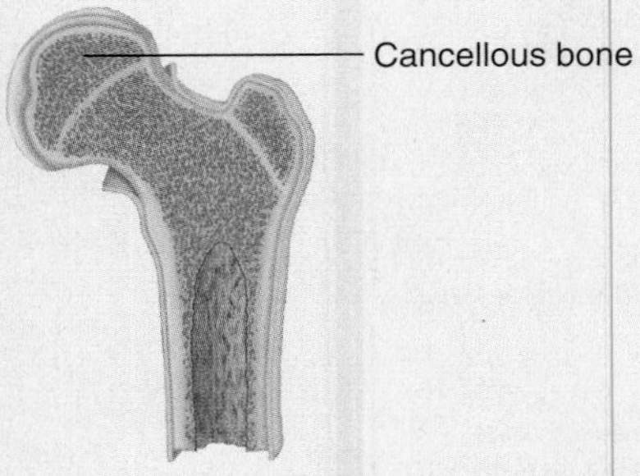

(b) Compact Bone

Structure: Hard, bony matrix predominates; many osteocytes (not seen in this bone preparation) are located within lacunae that are distributed in a circular fashion around the central canals; small passageways connect adjacent lacunae

Function: Provides great strength and support; forms a solid outer shell on bones that keeps them from being easily broken or punctured

Location: Outer portions of all bones, the shafts of long bones

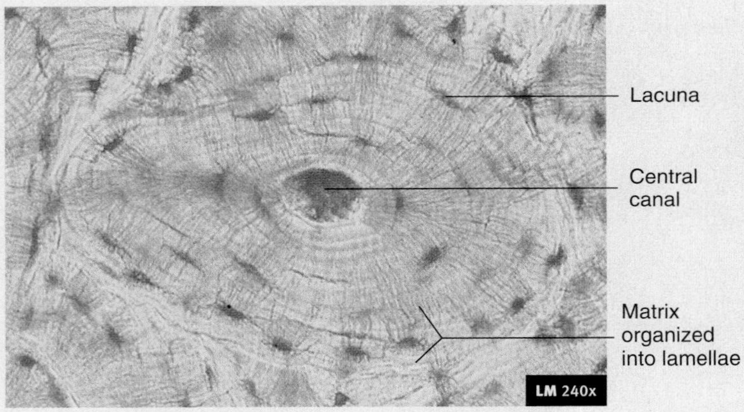

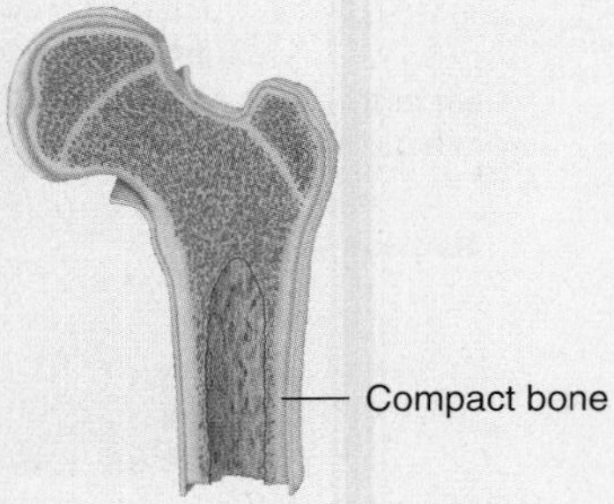

Bone, unlike cartilage, has a rich blood supply. For this reason, bone can repair itself much more readily than can cartilage. Bone is described more fully in chapter 6.

Blood and Hemopoietic Tissue

Blood is unusual among the connective tissues because the matrix between the cells is liquid (table 4.13*a*). Like many other connective tissues, blood has abundant extracellular matrix. The cells of most other connective tissues are more or less stationary within a relatively rigid matrix, but blood cells are free to move within a fluid matrix. Some blood cells leave the bloodstream and wander through other tissues. The liquid matrix of blood allows it to flow rapidly through the body, carrying food, oxygen, waste products, and other materials. The matrix of blood is also unusual in that most of it is produced by cells contained in other tissues, rather than by blood cells. Blood is discussed more fully in chapter 19.

TABLE 4.13 Connective Tissue: Blood and Hemopoietic Tissue

(a) Blood

Structure: Blood cells and a fluid matrix

Function: Transports oxygen, carbon dioxide, hormones, nutrients, waste products, and other substances; protects the body from infections and is involved in temperature regulation

Location: Within the blood vessels; produced by the hemopoietic tissues; white blood cells frequently leave the blood vessels and enter the interstitial spaces

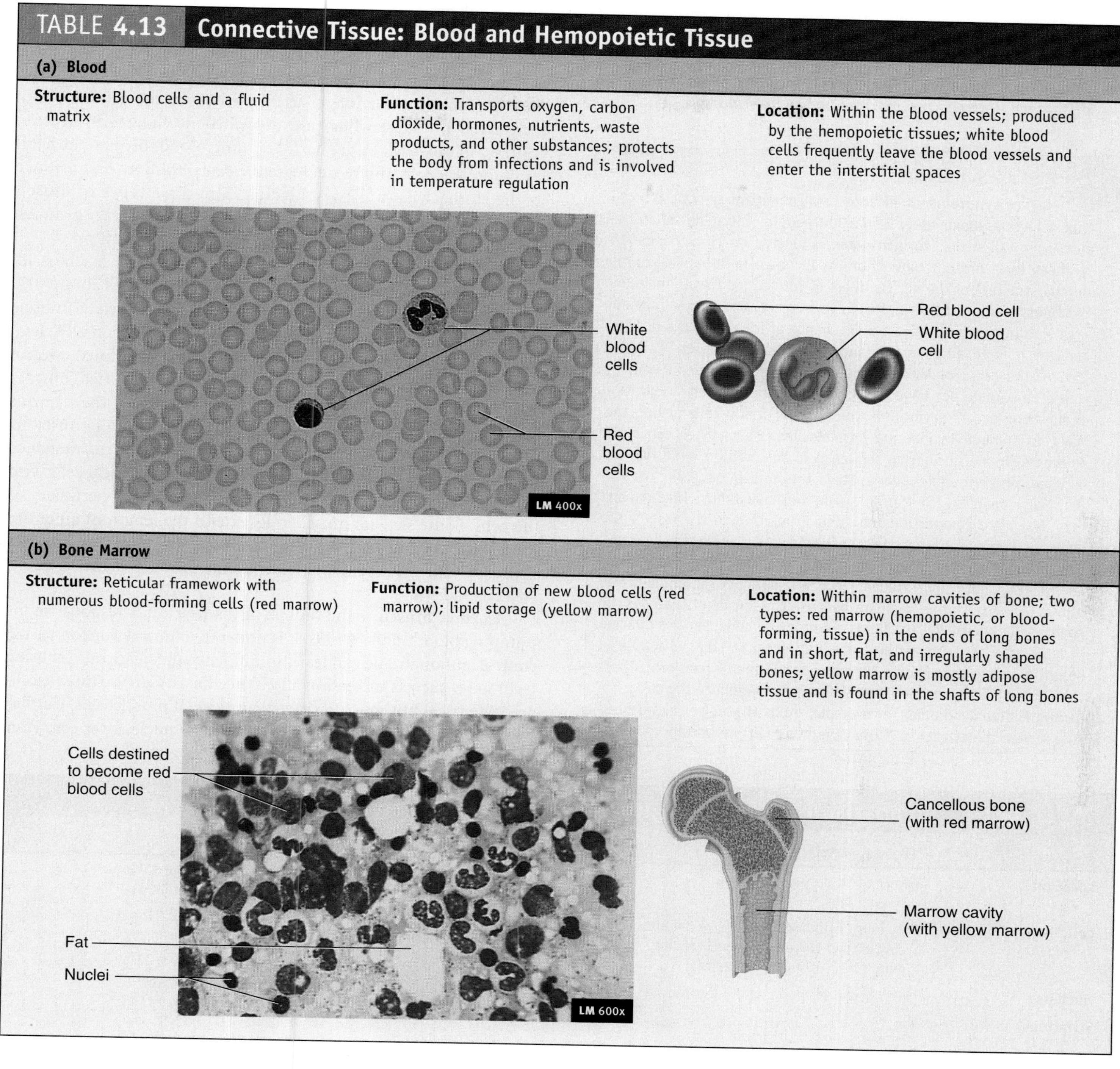

(b) Bone Marrow

Structure: Reticular framework with numerous blood-forming cells (red marrow)

Function: Production of new blood cells (red marrow); lipid storage (yellow marrow)

Location: Within marrow cavities of bone; two types: red marrow (hemopoietic, or blood-forming, tissue) in the ends of long bones and in short, flat, and irregularly shaped bones; yellow marrow is mostly adipose tissue and is found in the shafts of long bones

Hemopoietic (hē′mō-poy-et′ik) **tissue** forms blood cells. In adults, hemopoietic tissue is found in **bone marrow** (mar′ō; table 4.13*b*), which is the soft connective tissue in the cavities of bones. There are two types of bone marrow: **red marrow** and **yellow marrow** (see chapter 6). Red marrow is hemopoietic tissue surrounded by a framework of reticular fibers. Hemopoietic tissue produces red and white blood cells; it is described in detail in chapter 19. In children, the marrow of most bones is red marrow. Yellow marrow consists of yellow adipose tissue and does not produce blood cells. As children grow, yellow marrow replaces much of the red marrow in bones (see chapter 6).

34 ***Describe the cells and matrix of cartilage. What are lacunae? What is the perichondrium? Why does cartilage heal slowly?***

35 ***How do hyaline cartilage, fibrocartilage, and elastic cartilage differ in structure and function? Give an example of each.***

36 ***Describe the cells and matrix of bone. Differentiate between cancellous bone and compact bone.***

37 ***What characteristic separates blood from the other connective tissues?***

38 ***Describe the function of hemopoietic tissue. Explain the difference between red marrow and yellow marrow.***

Marfan Syndrome

Marfan syndrome is an autosomal dominant genetic disorder that affects approximately 1 in 5000 people. The gene for Marfan syndrome, called the **fibrillin gene,** is located on chromosome 15. It codes for a protein called fibrillin-1, which is necessary for the normal structure of the elastic fibers of connective tissue, including the fibers holding the lens of the eye in place. Children of a person with Marfan syndrome have a 50% chance of inheriting the disorder because it is an autosomal dominant trait (see chapter 3). About 25% of the cases of Marfan syndrome, however, occur in children whose parents do not have the disorder. In these cases, a mutation of the gene occurs during the formation of sperm cells or oocytes. Many different mutations are possible because changes can occur anywhere in the nucleotide sequence of the fibrillin gene. Several hundred different alleles (variants) of the fibrillin gene are known.

Many people with Marfan syndrome have limbs, fingers, and toes that are disproportionately long in relation to the rest of the body. Connective tissues are weakened. As a consequence, the heart valves, which are comprised largely of connective tissue, do not function normally, resulting in heart murmurs (abnormal heart sounds). Poor vision is common because the lenses of the eyes are positioned abnormally. The lungs are prone to collapse, and dilation of large arteries, such as the aorta, can occur. A common cause of death in people with Marfan syndrome is rupture of the aorta. There is no cure for the condition, but treatments can reduce the danger of having Marfan syndrome. For example, drugs that lower blood pressure reduce the effects of Marfan syndrome on the aorta.

MUSCLE TISSUE

The main characteristic of **muscle tissue** is that it contracts, or shortens, with a force and therefore is responsible for movement. Muscle contraction is accomplished by the interaction of contractile proteins, which are described in chapter 9. Muscles contract to move the entire body, to pump blood through the heart and blood vessels, and to decrease the size of hollow organs, such as the stomach and urinary bladder. The three types of muscle tissue—skeletal, cardiac, and smooth muscle—are grouped according to both structure and function (table 4.14).

Skeletal muscle is what normally is thought of as "muscle" (table 4.15*a*). It is the meat of animals and constitutes about 40% of a person's body weight. As the name implies, skeletal muscle attaches to the skeleton and enables body movement. Skeletal muscle is described as being under voluntary (conscious) control because one can purposefully cause skeletal muscle contraction to achieve specific body movements. However, the nervous system can cause skeletal muscles to contract without conscious involvement, such as during reflex movements and maintenance of muscle tone. Skeletal muscle cells are long, cylindrical cells with several nuclei per cell. The nuclei are located near the periphery of the cells. Some skeletal muscle cells extend the length of an entire muscle. Skeletal muscle cells are **striated** (strī′āt-ed), or banded, because of the arrangement of contractile proteins within the cells (see chapter 9).

Cardiac muscle is the muscle of the heart; it is responsible for pumping blood (table 4.15*b*). It is under involuntary (unconscious) control, although one can learn to influence the heart rate by using techniques such as mediation and biofeedback. Cardiac muscle cells are cylindrical but much shorter than skeletal muscle cells. Cardiac muscle cells are striated and usually have one nucleus per cell. They

TABLE 4.14 Comparison of Muscle Types

	Skeletal Muscle	Cardiac Muscle	Smooth Muscle
Location	Attached to bones	Heart	Walls of hollow organs, blood vessels, eyes, glands, skin
Cell Shape	Very long, cylindrical cells (1–4 cm and may extend the entire length of the muscle; 10–100 μm in diameter)	Cylindrical cells that branch (100–500 μm in length; 12–20 μm in diameter)	Spindle-shaped cells (15–200 μm in length; 5–8 μm in diameter)
Nucleus	Multinucleated, peripherally located	Single, centrally located	Single, centrally located
Striations	Yes	Yes	No
Control	Voluntary (conscious)	Involuntary (unconscious)	Involuntary (unconscious)
Ability to Contract Spontaneously	No	Yes	Yes
Function	Body movement	Contraction provides the major force for moving blood through the blood vessels.	Movement of food through the digestive tract, emptying of the urinary bladder, regulation of blood vessel diameter, change in pupil size, contraction of many gland ducts, movement of hair, and many more functions
Special Features		Branching fibers, intercalated disks containing gap junctions joining the cells to each other	Gap junctions

TABLE 4.15 Muscle Tissue

(a) Skeletal Muscle

Structure: Skeletal muscle cells or fibers appear striated (banded); cells are large, long, and cylindrical, with many nuclei located at the periphery

Function: Movement of the body; under voluntary (conscious) control

Location: Attaches to bone or other connective tissue

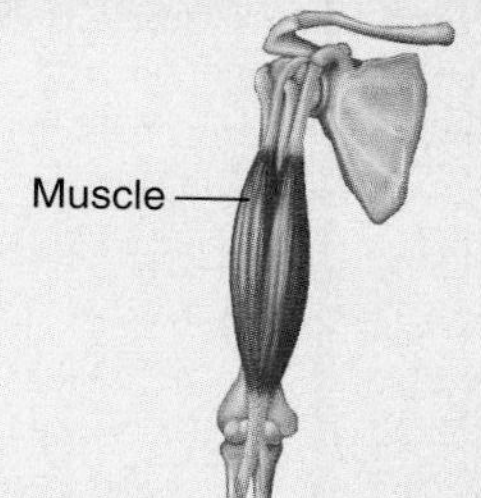

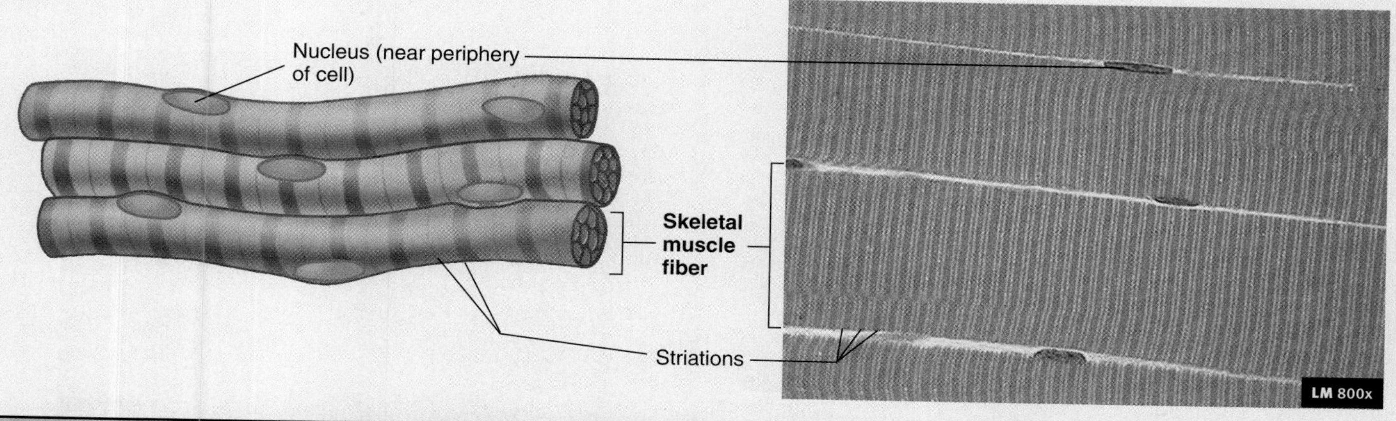

(b) Cardiac Muscle

Structure: Cardiac muscle cells are cylindrical and striated and have a single, centrally located nucleus; they are branched and connected to one another by intercalated disks, which contain gap junctions

Function: Pumps the blood; under involuntary (unconscious) control

Location: In the heart

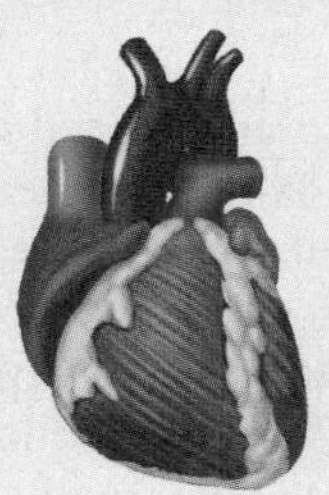

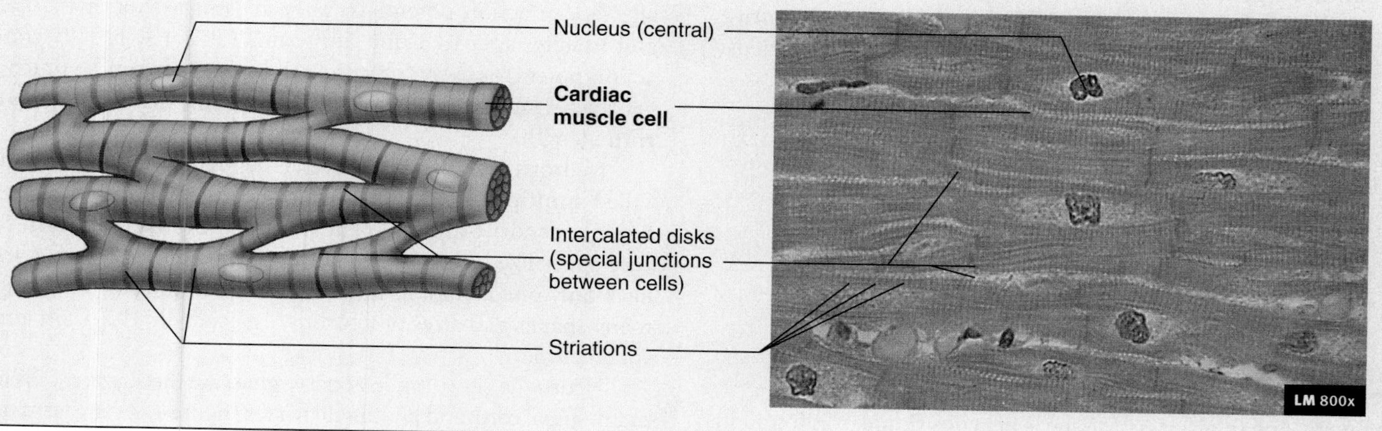

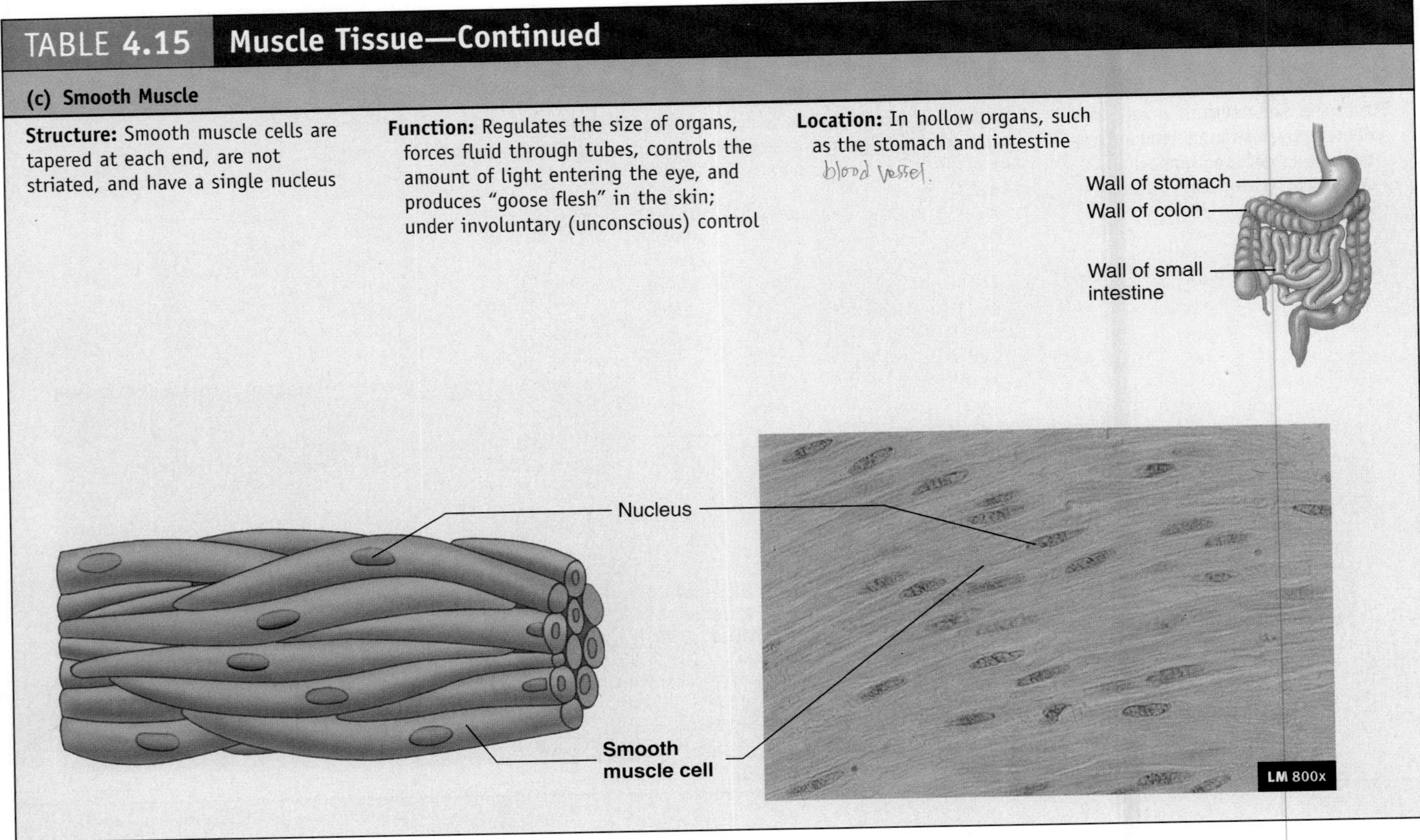

often are branched and connected to one another by **intercalated** (in-ter′kă-lā-ted, inserted between) **disks.** The intercalated disks, which contain specialized gap junctions, are important in coordinating the contractions of the cardiac muscle cells (see chapter 20).

Smooth muscle forms the walls of hollow organs (except the heart); it is found in the skin and the eyes (table 4.15*c*). It is responsible for a number of functions, such as movement of food through the digestive tract and emptying of the urinary bladder. Like cardiac muscle, smooth muscle is controlled involuntarily. Smooth muscle cells are tapered at each end, have a single nucleus, and are not striated.

39 ***Functionally, what is unique about muscle tissue?***

40 ***Compare the structure of skeletal, cardiac, and smooth muscle cells.***

41 ***Which of the muscle types is under voluntary control?***

42 ***What tasks does each muscle type perform?***

NERVOUS TISSUE

The fourth, and final, class of tissue is **nervous tissue.** It is found in the brain, spinal cord, and nerves and is characterized by the ability to conduct electric signals called **action potentials.** It consists of neurons, which are responsible for this conductive ability, and support cells called neuroglia.

Neurons, or **nerve cells** (table 4.16), are the conducting cells of nervous tissue. Just as an electrical wiring system transports electricity throughout a house, neurons transport electric signals throughout the body. They are composed of three major parts: cell body, dendrites, and axon. The **cell body** contains the nucleus and is the site of general cell functions. Dendrites and axons are two types of nerve cell processes, both consisting of projections of cytoplasm surrounded by membrane. **Dendrites** (den′drītz) usually receive action potentials. They are much shorter than axons and usually taper to a fine tip. **Axons** (ak′sonz) usually conduct action potentials away from the cell body. They can be much longer than dendrites, and they have a constant diameter along their entire length.

Neurons that possess several dendrites and one axon are called **multipolar neurons** (table 4.16*a*). Neurons that possess a single dendrite and an axon are called **bipolar neurons.** Some very specialized neurons, called **unipolar neurons** (table 4.16*b*), have only one axon and no dendrites. Within each subgroup are many shapes and sizes of neurons, especially in the brain and the spinal cord.

Neuroglia (noo-rog′lē-ă; nerve glue) are the support cells of the brain, spinal cord, and peripheral nerves (figure 4.6). The term *neuroglia* originally referred only to the support cells of the central nervous system, but it is now applied to cells in the peripheral nervous system as well. Neuroglia nourish, protect, and insulate neurons. Neurons and neuroglial cells are described in greater detail in chapter 11.

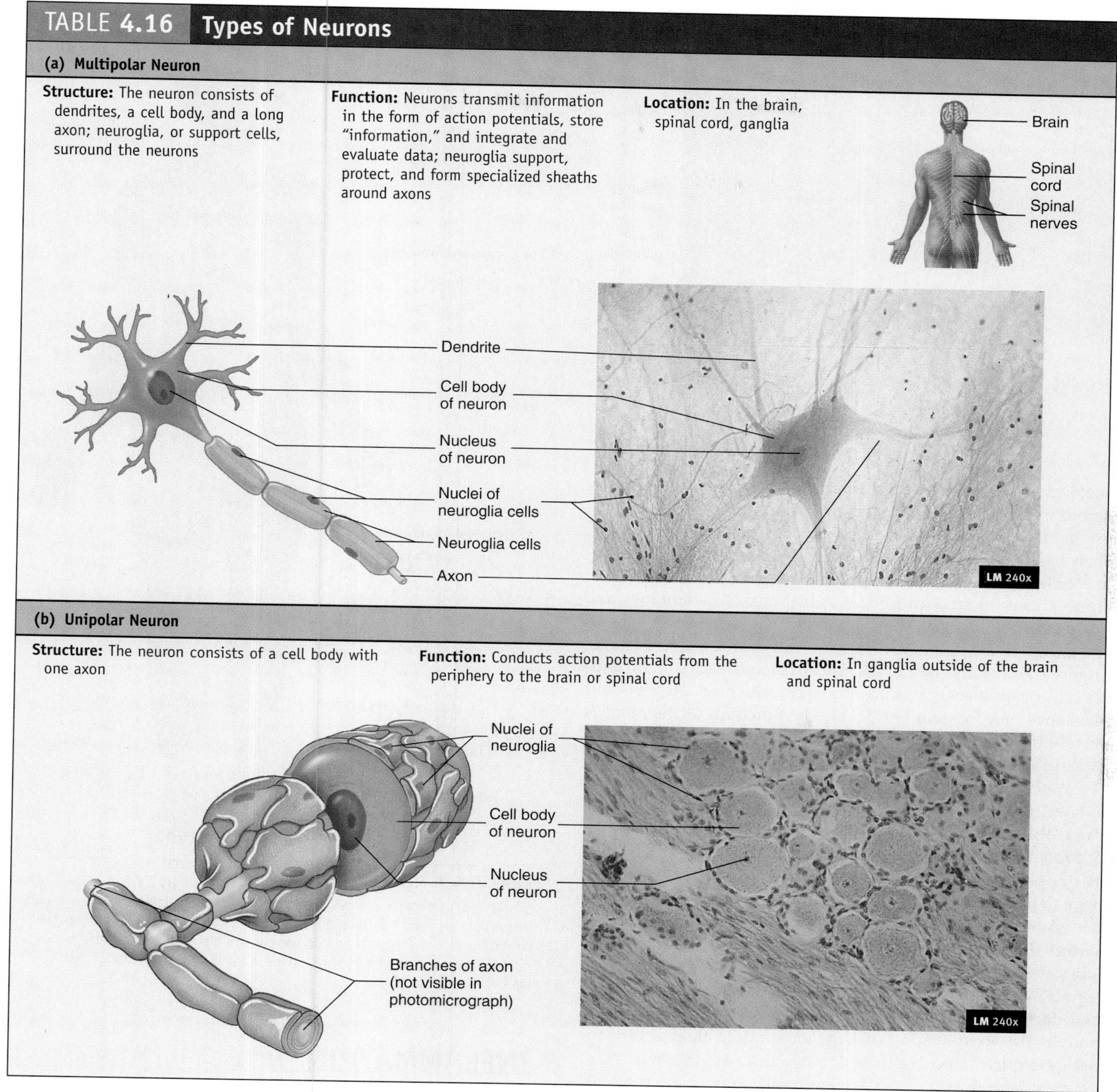

TABLE 4.16 Types of Neurons

(a) Multipolar Neuron

Structure: The neuron consists of dendrites, a cell body, and a long axon; neuroglia, or support cells, surround the neurons

Function: Neurons transmit information in the form of action potentials, store "information," and integrate and evaluate data; neuroglia support, protect, and form specialized sheaths around axons

Location: In the brain, spinal cord, ganglia

(b) Unipolar Neuron

Structure: The neuron consists of a cell body with one axon

Function: Conducts action potentials from the periphery to the brain or spinal cord

Location: In ganglia outside of the brain and spinal cord

43 ***Functionally, what is unique about nervous tissue?***

44 ***Define and list the functions of the cell body, dendrites, and axons of a neuron.***

45 ***Differentiate among multipolar, bipolar, and unipolar neurons.***

46 ***What is the general function of neuroglia?***

MEMBRANES

A membrane is a thin sheet or layer of tissue that covers a structure or lines a cavity. Most membranes are formed from epithelium and the connective tissue on which it rests. The skin, or cutaneous, membrane (see chapter 5) is the external membrane. The three

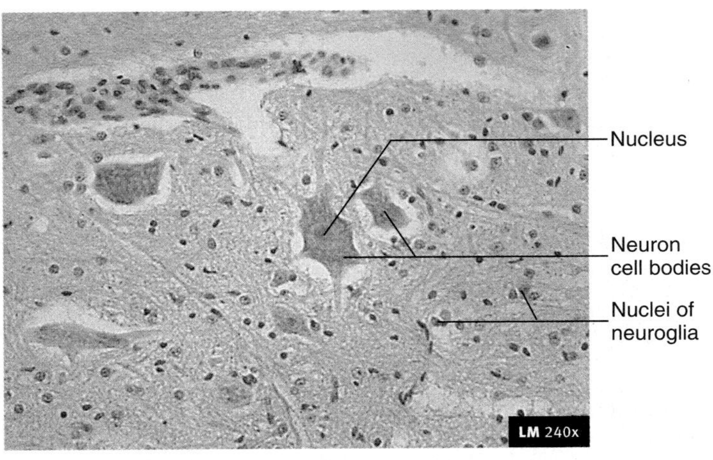

FIGURE 4.6 **Neuroglia**

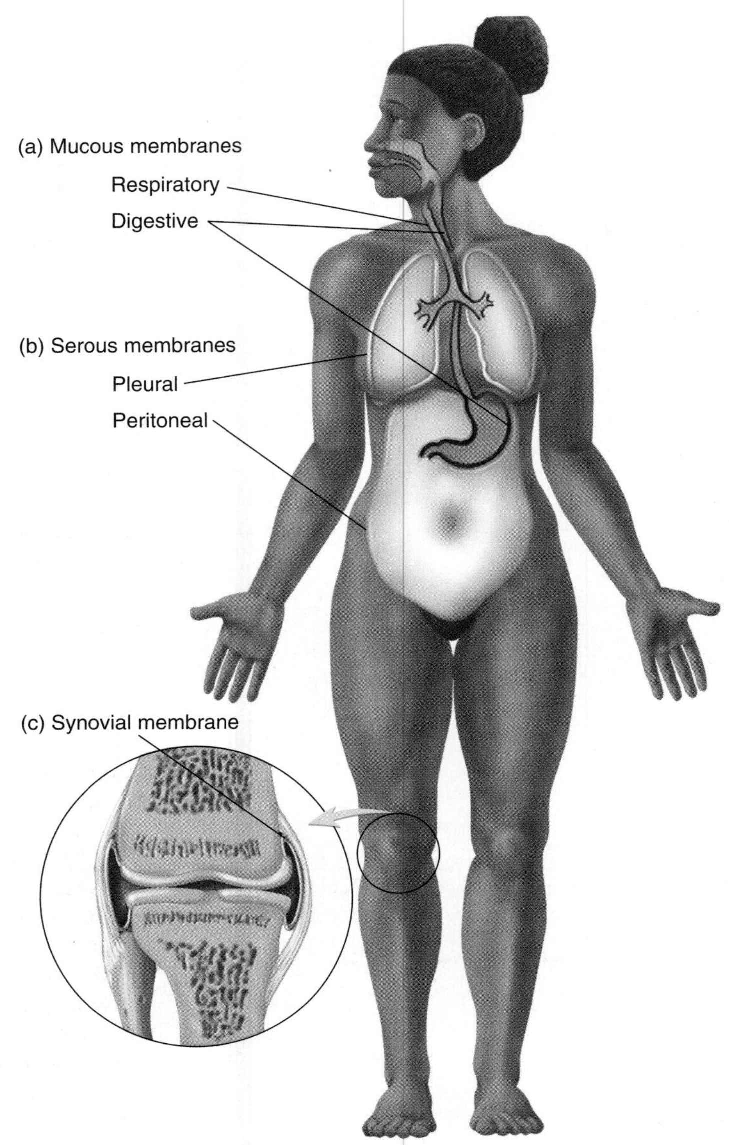

FIGURE 4.7 **Membranes**

(*a*) Mucous membranes line cavities that open to the outside and often contain mucous glands, which secrete mucus. (*b*) Serous membranes line cavities that do not open to the exterior and do not contain glands but do secrete serous fluid. (*c*) Synovial membranes line cavities that surround synovial joints.

major categories of internal membranes are mucous membranes, serous membranes, and synovial membranes.

A **mucous** (mū′kŭs) **membrane** consists of epithelial cells, their basement membrane, a thick layer of loose connective tissue called the **lamina propria** (lam′i-nă prō′prē-ă), and sometimes a layer of smooth muscle cells. Mucous membranes line cavities and canals that open to the outside of the body, such as the digestive, respiratory, excretory, and reproductive passages (figure 4.7*a*). Many, but not all, mucous membranes contain goblet cells or multicellular mucous glands, which secrete a viscous substance called **mucus** (mū′kŭs). The functions of the mucous membranes vary, depending on their location, and include protection, absorption, and secretion.

A **serous** (ser′ŭs) **membrane** consists of three components: a layer of simple squamous epithelium called **mesothelium** (mez-ō-thē′lē-ŭm), its basement membrane, and a delicate layer of loose connective tissue. Serous membranes line cavities, such as the pericardial, pleural, and peritoneal cavities, that do not open to the exterior (figure 4.7*b*). Serous membranes do not contain glands but are moistened by a small amount of fluid called **serous fluid,** produced by the serous membranes. Serous fluid lubricates the serous membranes, making their surfaces slippery. Serous membranes protect the internal organs from friction, help hold them in place, and act as selectively permeable barriers to prevent the accumulation of large amounts of fluid within the serous cavities.

A **synovial** (si-nō′vē-ăl) **membrane** consists of modified connective tissue cells, either intermixed with part of the dense connective tissue of the joint capsule or separated from the capsule by areolar or adipose tissue. Synovial membranes line freely movable joints (see chapter 8; figure 4.7*c*). They produce a fluid rich in hyaluronic acid, which makes the joint fluid very slippery, thereby facilitating smooth movement within the joint.

47 ***Compare mucous, serous, and synovial membranes according to the type of cavities they line and their secretions.***

INFLAMMATION

The inflammatory response occurs when tissues are damaged (figure 4.8) or when associated with an immune response. Although many agents cause injury, such as microorganisms, cold, heat, radiant energy, chemicals, electricity, and mechanical trauma, the inflammatory response to all causes is similar. The inflammatory response mobilizes the body's defenses, isolates and destroys microorganisms and other injurious agents, and removes foreign materials and damaged cells so that tissue repair can proceed (see chapter 22).

1. A splinter in the skin causes damage and introduces bacteria. Mediators of inflammation are released or activated in injured tissues and adjacent blood vessels. Some blood vessels are ruptured, causing bleeding.

2. Mediators of inflammation cause capillaries to dilate, causing the skin to become red. Mediators of inflammation also increase capillary permeability, and fluid leaves the capillaries, producing swelling (*arrows*).

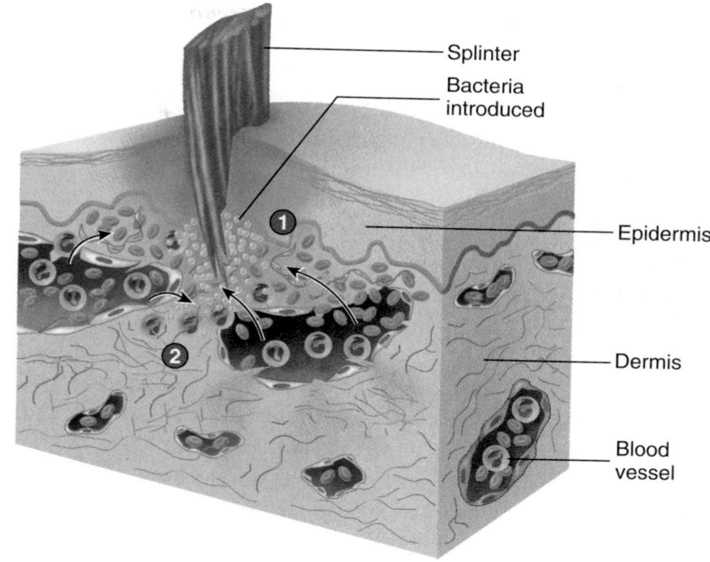

3. White blood cells (e.g., neutrophils and macrophages) leave the dilated blood vessels and move to the site of bacterial infection, where they begin to phagocytize bacteria and other debris.

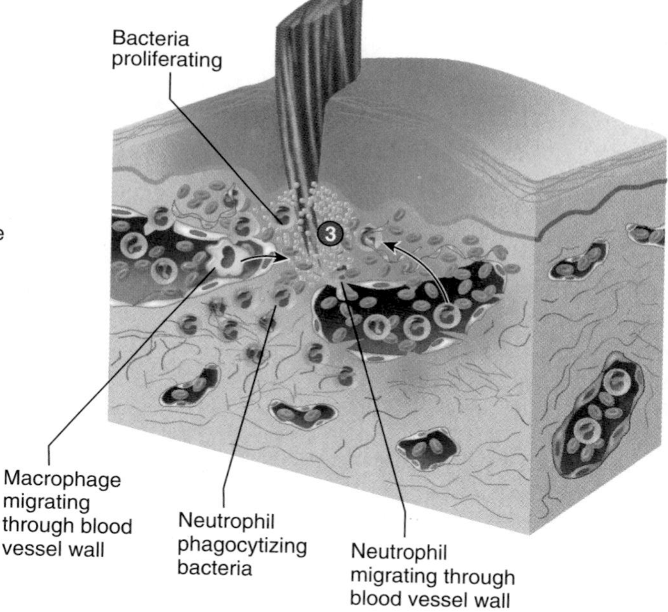

PROCESS FIGURE 4.8 Inflammation

Inflammation has five major manifestations: redness, heat, swelling, pain, and disturbance of function. Although unpleasant, these processes usually benefit recovery, and each of the symptoms can be understood in terms of events that occur during the inflammatory response.

After a person is injured, chemical substances called **mediators of inflammation** are released or activated in the tissues and the adjacent blood vessels. The mediators include histamine, kinins, prostaglandins, leukotrienes, and others. Some mediators induce dilation of blood vessels and produce redness and heat. Dilation of blood vessels is beneficial because it increases the speed with which white blood cells and other substances important for fighting infections and repairing the injury arrive at the site of injury.

Mediators of inflammation also stimulate pain receptors and increase the permeability of blood vessels. The increased permeability allows the movement of materials, such as clotting proteins and white blood cells, out of the blood vessels and into the tissue, where they can deal directly with the injury. As proteins from the blood move into the tissue, they change the osmotic relationship between the blood and the tissue. Water follows the proteins by osmosis, and the tissue swells, producing **edema** (e-dē′mă). Edema increases the pressure in the tissue, which can also stimulate neurons and cause pain.

Clotting proteins found in blood diffuse into the interstitial spaces and form a clot. Clotting also occurs in the more severely injured blood vessels. Clotting isolates the injurious agent and separates it from the rest of the body. Foreign particles and microorganisms at the site of injury are "walled off" from tissues by the clotting process. Pain, limitation of movement resulting from edema, and tissue destruction all contribute to the disturbance of function. This disturbance can be valuable because it warns the person to protect the injured structure from further damage. Sometimes the inflammatory response lasts longer or is more intense than is desirable, and drugs are used to suppress the symptoms. Antihistamines block the effects of histamine, aspirin prevents the synthesis of prostaglandins, and cortisone reduces the release of several mediators of inflammation. On the other hand, the inflammatory response by itself may not be enough to combat the effects of injury or fight off an infection. Medical intervention, such as administering antibiotics, may be required.

48 ***What is the function of the inflammatory response?***

49 ***Name five manifestations of the inflammatory response, and explain how each is produced.***

PREDICT 6

In some injuries, tissues are so severely damaged that some cells are killed and blood vessels are destroyed. For such injuries, where do the signs of inflammation, such as redness, heat, edema, and pain, occur?

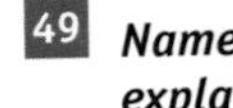

Chronic Inflammation

When the agent responsible for an injury is not removed or if interference occurs with the process of healing, the inflammatory response persists and is called **chronic inflammation.** For example, a lung infection can result in a brief period of inflammation followed by repair, but a prolonged infection causes chronic inflammation, which results in tissue destruction and permanent damage to the lung. Chronic inflammation of the stomach or small intestine may result in an ulcer. Prolonged infections; prolonged exposure to irritants, such as silica, in the lung; or abnormal immune responses can result in chronic inflammation. White blood cells invade areas of chronic inflammation, and ultimately healthy tissues are destroyed and replaced by a fibrous connective tissue, which is an important cause of the loss of organ function. Chronic inflammation of the lungs, the liver, the kidney, or other vital organs can lead to death.

TISSUE REPAIR

Tissue repair is the substitution of viable cells for dead cells, and it can occur by regeneration or replacement. In **regeneration** (rē′jen-er-ā′shŭn), the new cells are the same type as those that were destroyed, and normal function is usually restored. In **replacement,** a new type of tissue develops that eventually causes scar production and the loss of some tissue function. Most wounds heal through regeneration and replacement; which process dominates depends on the tissues involved and the nature and extent of the wound.

Cells are classified into three groups—labile, stable, or permanent cells—according to their ability to regenerate. **Labile cells,** including cells of the skin, mucous membranes, and hemopoietic and lymphatic tissues, continue to divide throughout life. Damage to these cells can be repaired completely by regeneration. **Stable cells,** such as the cells of connective tissues and glands, including the liver, pancreas, and endocrine glands, do not divide after growth ceases, but they retain the ability to divide and are capable of regeneration in response to injury. **Permanent cells** have very limited ability to replicate and, if killed, are usually replaced by a different type of cell. Neurons fit into this category, although neurons are able to recover from damage. If the cell body of a neuron is not destroyed, most neurons can replace a damaged axon or dendrite, but if the neuron cell body is destroyed, the remainder of the neuron dies. Some undifferentiated cells of the central nervous system can undergo mitosis and form functional neurons, although the degree to which mitosis occurs and its functional significance is not clear. Undifferentiated cells of skeletal and cardiac muscle also have very limited ability to regenerate in response to injury, although individual skeletal and cardiac muscle cells can repair themselves. In contrast, smooth muscle readily regenerates following injury.

Skin repair is a good example of tissue repair (figure 4.9). The basic pattern of the repair is the same as for other tissues, especially ones covered by epithelium. If the edges of the wound are close together, such as in a surgical incision, the wound heals by a process called **primary union,** or **primary intention.** If the edges are not close together, or if extensive loss of tissue has occurred, the process is called **secondary union,** or **secondary intention.**

In primary union, the wound fills with blood, and a clot forms (see chapter 19). The clot contains a threadlike protein, **fibrin** (fī′brin), which binds the edges of the wound together. The surface of the clot dries to form a **scab,** which seals the wound and helps prevent infection. An inflammatory response induces vasodilation and brings more blood cells and other substances to the area. Blood vessel permeability increases, resulting in edema. Fibrin and blood cells move into the wounded tissues because of the increased vascular permeability. Fibrin isolates and walls off microorganisms and other foreign matter. Some of the white blood cells that move into the tissue are phagocytic cells called **neutrophils** (noo′trō-filz; see figure 4.9). They ingest bacteria, thus helping fight infection, and they ingest tissue debris and clear the area for repair. Neutrophils are killed in this process and can accumulate as a mixture of dead cells and fluid, called **pus** (pŭs).

Fibroblasts from surrounding connective tissue migrate into the clot and produce collagen and other extracellular matrix components. Capillaries grow from blood vessels at the edge of the wound and revascularize the area, and fibrin in the clot is broken down and removed. The result is the replacement of the clot by a delicate connective tissue, called **granulation tissue,** which consists of fibroblasts, collagen, and capillaries. A large amount of granulation tissue is converted to a **scar,** which consists of dense irregular collagenous connective tissue. At first, a scar is bright red because

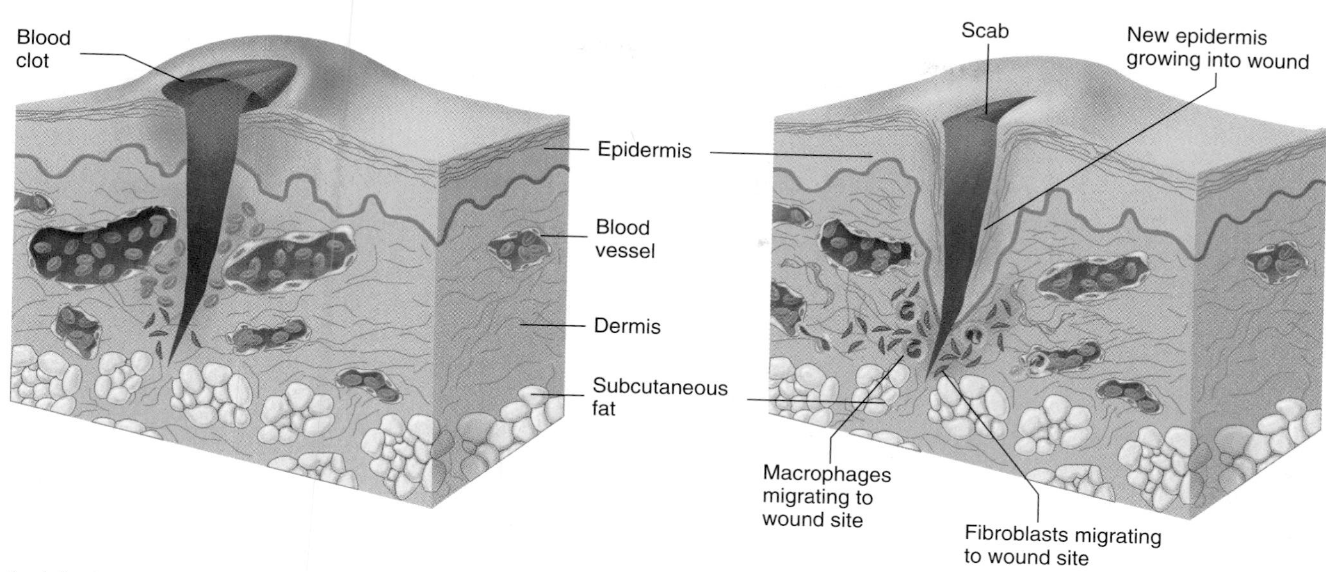

1. A fresh wound cuts through the epithelium (epidermis) and underlying connective tissue (dermis), and a clot forms.

2. Approximately 1 week after the injury, a scab is present, and epithelium (new epidermis) is growing into the wound.

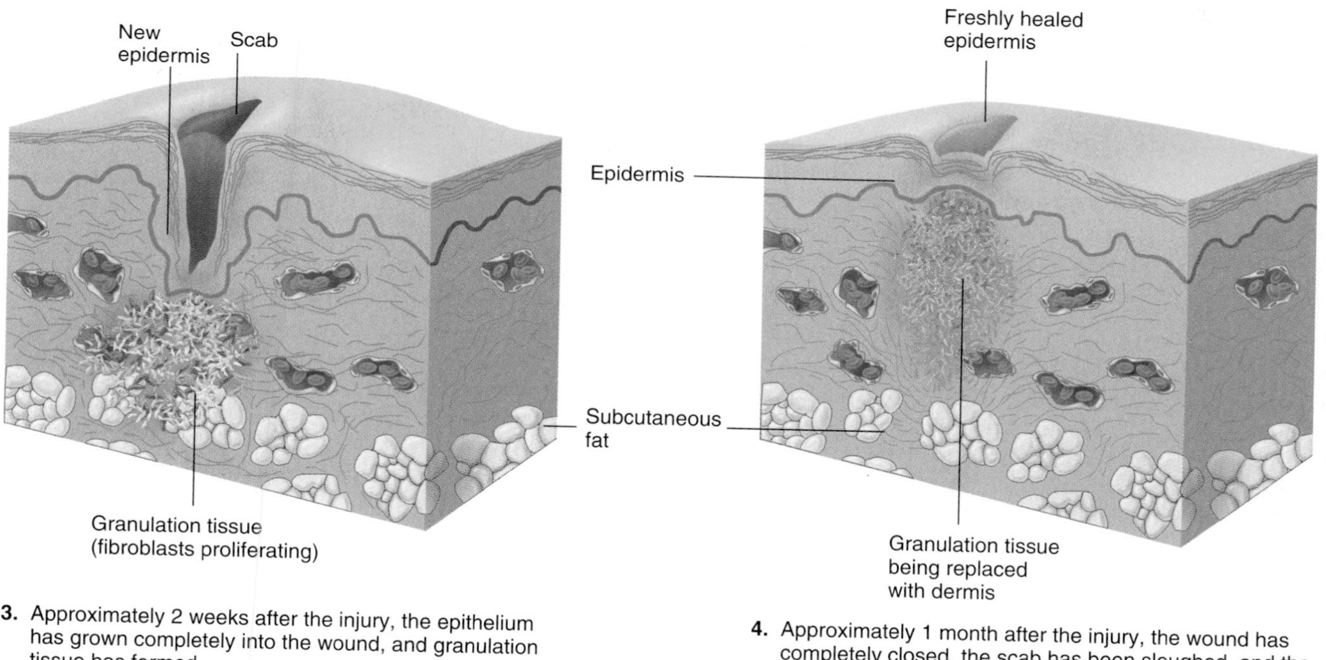

3. Approximately 2 weeks after the injury, the epithelium has grown completely into the wound, and granulation tissue has formed.

4. Approximately 1 month after the injury, the wound has completely closed, the scab has been sloughed, and the granulation tissue is being replaced with dermis.

PROCESS FIGURE 4.9 **Tissue Repair**

of vascularization of the tissue. Later, the scar becomes white as collagen accumulates and the vascular channels are compressed.

Repair by secondary union proceeds in a similar fashion, but some differences exist. Because the wound edges are far apart, the clot may not close the gap completely, and it takes the epithelial cells much longer to regenerate and cover the wound. With increased tissue damage, the degree of inflammation is greater, there is more cell debris for the phagocytes to remove, and the risk of infection is greater. Much more granulation tissue forms, and **wound contraction** occurs as a result of the contraction of fibroblasts in the granulation tissue. Wound contraction leads to disfiguring and debilitating scars. Thus, it is advisable to suture a large wound so that it can heal by primary rather than secondary union. Healing is faster, the risk of infection is lowered, and the degree of scarring is reduced.

CLINICAL FOCUS

Cancer Tissue

Cancer (kan′ser) is a malignant, spreading tumor, as well as the illness that results from it. A **tumor** (too′mŏr) is any swelling, although modern usage has limited the term to swellings that involve neoplastic tissue. **Neoplasm** (nē′ō-plazm) means new growth and refers to abnormal tissue growth, resulting in rapid cellular proliferation, which continues after the growth of normal tissue has stopped or slowed considerably. **Oncology** (ong-kol′ō-jē) is the study of tumors and the problems they cause. A neoplasm can be either **benign** (bē-nīn′; L. kind), not inclined to spread and not likely to become worse, or **malignant** (ma-lig′nănt; with malice or intent to cause harm), able to spread and become worse. Although benign tumors are usually less dangerous than malignant tumors, they can cause problems. As a benign tumor enlarges, it can compress surrounding tissues and impair their functions. In some cases, such as in some benign brain tumors, the result can be death.

Cells of malignant neoplasms, or cancer cells, differ from cells of normal tissues in several ways; the greater the degree to which they differ, the more dangerous they are. Cancer cells are more spherical because they do not adhere tightly to surrounding normal cells. They appear to be more embryonic, or less mature, than the normal tissue from which they arise. For example, a skin cancer cell is more spherical and softer than the stratified squamous epithelial cells of the skin. Cancer cells are also invasive. That is, they have the ability to squeeze into spaces and enter surrounding tissues. They secrete enzymes that cut paths through healthy tissue, so they are able to grow irregularly, sending processes in all directions. Cancer cells can dislodge; enter blood vessels, lymphatic vessels, or body cavities; and travel to distant sites, where they attach and invade tissues. The process by which cancer spreads to distant sites is called **metastasis** (me-tas′ta-sis). Cancer cells secrete substances that cause blood vessels to grow into the tumor and supply oxygen and nutrients. Cancer cells also produce a number of substances that can be found on their plasma membranes or are secreted. For example, prostate specific antigen (PSA) is an enzyme, produced by prostate gland cells, that is involved with the liquefaction of semen. Normal blood levels of PSA are low. Prostate cancer cells secrete PSA in increasing amounts, and these proteins are released into the blood. Therefore, blood levels of PSA can be monitored to determine if a person is likely to have prostate cancer.

There are many types of cancer and special terms to name them. For example, a **carcinoma** (kar-si-nō′ma) is a cancer derived from epithelial tissue. **Basal cell** and **squamous cell carcinomas** are types of skin cancer derived from epithelial tissue. **Adenocarcinomas** (ad′ĕ-nō-kar-si-nō′maz) are derived from glandular epithelium. Most breast cancers are adenocarcinomas. A **sarcoma** (sar-kō′mă) is cancer derived from connective tissue. For example, an **osteosarcoma** (os′tē-ō-sar-kō′mă) is cancer of bone and a **chondrosarcoma** (kon′drō-sar-kō′mă) is cancer of cartilage.

Cancer therapy concentrates primarily on trying to confine and then kill the malignant cells. This goal is accomplished currently by killing the tissue with x-rays or lasers, by removing the tumor surgically, or by treating the patient with drugs that kill rapidly dividing cells or reduce the blood supply to the tumor. The major problem with current therapy is that some cancers cannot be removed completely by surgery or killed completely by x-rays or laser therapy. These treatments can also kill normal tissue adjacent to the tumor. Many drugs used in cancer therapy kill not only cancer tissue but also other rapidly growing tissues, such as bone marrow, where new blood cells are produced, and the lining of the intestinal tract. Loss of these tissues can result in anemia, caused by the lack of red blood cells, and nausea, caused by the loss of the intestinal lining.

A newer class of drugs eliminates these unwanted side effects. These drugs prevent blood vessel development, thus depriving the cancer tissue of a blood supply, rather than attacking dividing cells. Other normal tissues, in which cells divide rapidly, have well-established blood vessels and are therefore not affected by these drugs.

Promising anticancer therapies are also being developed in which the cells responsible for immune responses can be stimulated to recognize tumor cells and destroy them. A major advantage in such anticancer treatments is that the cells of the immune system can specifically attack the tumor cells and not other, healthy tissues.

50 ***Define tissue repair. Differentiate between tissue repair that occurs by regeneration and by replacement.***

51 ***Compare labile cells, stable cells, and permanent cells. Give examples of each type. What is the significance of these cell types to tissue repair?***

52 ***Describe the process of wound repair. Contrast healing by primary union and healing by secondary union.***

53 ***What is pus? Describe granulation tissue. How does granulation tissue contribute to scars and wound contraction?***

TISSUES AND AGING

Age-related changes—for example, reduced visual acuity and reduced smell, taste, and touch sensation—are well documented. A clear decline in many types of athletic performance can be measured after approximately age 30–35. Ultimately, there is a substantial decrease in the number of neurons and muscle cells. The functional capacity of systems, such as the respiratory and cardiovascular systems, declines. The rate of healing and scarring are very different in the elderly than in the very young, and

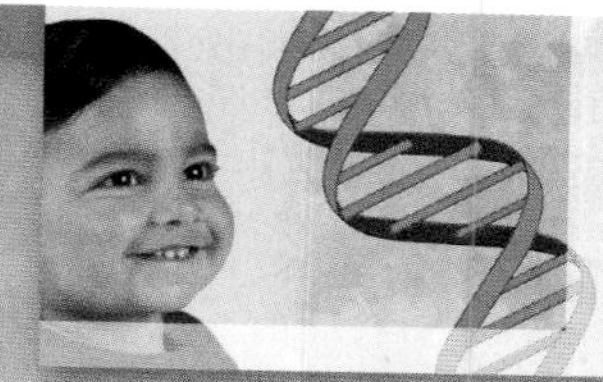

CLINICAL GENETICS

Genetic Changes in Cancer Cells

Most cancers are caused by mutations of genes within somatic cells. When DNA is replicated prior to cell division, a small number of replication errors occur in which the nucleotide sequence in the replicated DNA is different from the original DNA. A DNA sequence, with replication errors in it, is a mutation (see chapter 3). Factors such as radiation, certain chemicals and toxins, and some viruses also cause mutations because they damage or alter DNA. Because mutations are most likely to occur during DNA replication, cancer usually develops in tissues that are undergoing frequent cell divisions.

If mutations affect genes that regulate cell divisions, and if the result is uncontrolled cell divisions, a neoplasm can be produced. A neoplasm can become cancer if additional mutations change the structure and function of its cells. For example, some mutations increase the ability of cancer cells to invade and destroy surrounding tissues, some mutations allow cancer cells to metastasize, and some mutations make cancer cells resistant to drug treatments, such as the drugs used in chemotherapy.

Two major mechanisms help prevent the development of cancer in cells: (1) DNA repair enzymes detect and correct errors that occur during replication and (2) a self-destruction mechanism destroys cells with abnormal DNA. **Apoptosis** (ap′op-to′sis) is a process by which cells self destruct. Many cells, such as epithelial cells, have a limited life span, whereas others, such as neurons and skeletal muscle cells last a lifetime. Apoptosis is a normal process involved in the self-destruction of cells that have a limited life span. Apoptosis can also cause self destruction in cells with damaged DNA. Therefore, apoptosis can cause self-destruction in cells with mutations, and apoptosis can, therefore, remove cells with mutations before cancer develops. The likelihood that cancer will develop is increased if the genes controlling DNA repair enzymes undergo mutations, resulting in defective DNA repair enzymes, so that mutated genes persist in cells. Mutation of genes responsible for apoptosis can also result in the persistence of cells with mutations, and these cells can continue to divide.

Because mutations are most likely to occur when DNA replicates, mutations leading to cancer are most likely to occur in cells undergoing cell division, such as rapidly dividing epithelial cells, rather than in nondividing cells, such as in skeletal muscle cells and neurons.

Cancer develops in somatic cells because of mutations that occur during cell divisions. Therefore, few cancers are inherited. Approximately 10% of human cancers, however, are inherited because of an increased genetic susceptibility to cancer. For example, a person can inherit a normal allele and a mutated allele for a regulatory gene that is involved in the development of cancer. As long as the normal allele functions, the effect of the mutated allele is masked. However, if the normal allele mutates during cell division of a somatic tissue, expression of the allele involved in the development of cancer can be expressed. Compared with a person with two normal alleles for a regulatory gene involved in the development of cancer, a person with a genetic susceptibility for cancer is much more likely to develop cancer because only one allele has to mutate, instead of two. Certain kinds of colon cancers and breast cancers are examples of genetic susceptibility to cancer.

The accumulation of mutations resulting in cancer occurs over many generations of cells and may require several years to develop. This is one reason that cancer becomes more common in people as they become older. Once a mutation alters a gene in a cell, that altered gene can be passed to the daughter cells when the parent cell undergoes cell division. For cells that survive and undergo cell division, altered genes are passed to all of the offspring of the original cell in which the mutation occurred. For example, a single mutation may cause a cell to undergo cell division at an increased rate. This mutation is passed to the offspring of that cell. Another mutation may occur in one of these cells and it, along with the original mutation, is passed to its offspring. In this fashion, mutations responsible for the development of cancer accumulate in cells. Therefore, cancer is polygenic (see chapter 3).

Some genes promote cell division, whereas others suppress it. Genes that promote cell division are called proto-oncogenes. Mutations in proto-oncogenes can give rise to abnormal regulatory genes, called **oncogenes** (ong′ko-jenz), which increase the rate of cell division. Oncogenes can code for growth factors, growth factor receptors, or chemical signals that control cell divisions. Many types of oncogenes have been identified in human cancers. **Tumor suppressor genes** are normal genes that slow or stop cell division. Mutations that delete or inactivate tumor suppressor genes can also increase the rate of cell division by taking off the brakes, so to speak, of the processes that promote cell division. Many altered tumor suppressor genes have been identified in human cancer cells.

Additional mutations cause the structure and functions of the cancer cells to differ from those of normal cells. For example, these mutations increase the ability of cancer cells to invade and destroy surrounding tissues and to metastasize. The continued accumulation of mutations in cancer cells is also responsible for changes in the characteristics of the cancer cells over time. These changes can result in cancer cells in a tumor becoming less sensitive to treatment designed to kill the cancer cells. Some mutations help make cancer cells less sensitive to chemotherapeutic drugs.

major changes in the structural characteristics of the skin develop. Characteristic alterations in brain function also develop in the elderly. All these changes result in the differences among young, middle-age, and older people.

At the tissue level, age-related changes affect cells and the extracellular materials they produce. In general, cells divide more slowly in older than in younger people. Collagen fibers become more irregular in structure, even though they may increase in number. As a consequence, connective tissues with abundant collagen, such as tendons and ligaments, become less flexible and more fragile. Elastic fibers fragment, bind to calcium ions, and become less elastic. Consequently, elastic connective tissues become less elastic.

Changes in the structure of elastic and collagen fibers of arterial walls cause them to become less elastic. Atherosclerosis results as plaques form in the walls of blood vessels, which contain collagen fibers, lipids, and calcium deposits (see chapter 21). These changes result in reduced blood supply to tissues and increased susceptibility to blockage and rupture. The rate of red blood cell synthesis declines in the elderly as well. Reduced flexibility and elasticity of connective tissue is responsible for increased wrinkling of skin, as well as the increased tendency for bones to break in older people.

Injuries in the very young heal more rapidly and more completely than in older people. A fracture in an infant's femur is likely to heal quickly and eventually leave no evidence of the fracture in the bone. A similar fracture in an adult heals more slowly and a scar, seen in x-rays of the bone, is likely to persist throughout life.

54 ***Describe the age-related changes that occur in cells such as nerve cells, muscle cells, and cells of hemopoietic tissues.***

55 ***Describe the age-related changes in tissues with abundant collagen and elastic fibers.***

SUMMARY

Tissues and Histology (p. 110)

1. Tissues are collections of similar cells and the substances surrounding them.
2. The four primary tissue types are epithelial, connective, muscle, and nervous tissues.
3. Histology is the microscopic study of tissues.

Embryonic Tissue (p. 110)

All four of the primary tissue types are derived from each of the three germ layers (mesoderm, ectoderm, and endoderm).

Epithelial Tissue (p. 110)

1. Epithelium consists of cells with little extracellular matrix, it covers surfaces, it has a basement membrane, and it does not have blood vessels.
2. The basement membrane is secreted by the epithelial cells and attaches the epithelium to the underlying tissues.

Functions of Epithelial Tissues

Epithelial tissues (1) protect underlying structures, act as barriers, permit the passage of some substances through epithelial layers, secrete substances, and absorb substances.

Classification of Epithelium

1. Simple epithelium has a single layer of cells, stratified epithelium has two or more layers, and pseudostratified epithelium has a single layer that appears stratified.
2. Cells can be squamous (flat), cuboidal, or columnar.
3. Stratified squamous epithelium can be nonkeratinized or keratinized.
4. Transitional epithelium is stratified, with cells that can change shape from cuboidal to flattened.

Functional Characteristics

1. Simple epithelium is usually involved in diffusion, filtration, secretion, or absorption. Stratified epithelium serves a protective role. Squamous cells function in diffusion and filtration. Cuboidal or columnar cells, with a larger cell volume that contains many organelles, secrete or absorb.
2. A smooth free surface reduces friction (mesothelium and endothelium), microvilli increase absorption (intestines), and cilia move materials across the free surface (respiratory tract and uterine tubes). Transitional epithelium has a folded surface that allows the cell to change shape, and the number of cells making up the epithelial layers changes.
3. Cells are bound together mechanically by glycoproteins, desmosomes, and the zonulae adherens and to the basement membrane by hemidesmosomes. The zonulae occludens and zonulae adherens form a permeability barrier or tight junction, and gap junctions allow intercellular communication.

Glands

1. Glands are organs that secrete. Exocrine glands secrete through ducts, and endocrine glands release hormones that are absorbed directly into the blood.
2. Glands are classified as unicellular or multicellular. Goblet cells are unicellular glands. Multicellular exocrine glands have ducts, which are simple or compound (branched). The ducts can be tubular or end in small sacs (acini or alveoli). Tubular glands can be straight or coiled.
3. Glands are classified according to their mode of secretion. Merocrine glands (pancreas) secrete substances as they are produced, apocrine glands (mammary glands) accumulate secretions that are released when a portion of the cell pinches off, and holocrine glands (sebaceous glands) accumulate secretions that are released when the cell ruptures and dies.

Connective Tissue (p. 120)

Connective tissue is distinguished by its extracellular matrix.

Functions of Connective Tissue

Connective tissues enclose and separate organs and tissue; connect tissues to one another; play a role in support for movement; store high-energy molecules; cushion; insulate; transport; and protect.

Cells of Connective Tissue

1. The extracellular matrix results from the activity of specialized connective tissue cells; in general, blast cells form the matrix, cyte cells maintain it, and clast cells break it down. Fibroblasts form protein fibers of many connective tissues, osteoblasts form bone, and chondroblasts form cartilage.
2. Adipose (fat) cells, mast cells, white blood cells, macrophages, and mesenchymal cells (stem cells) are commonly found in connective tissue.

Extracellular Matrix

1. The extracellular matrix of connective tissue has protein fibers, ground substance, and fluid as major components.
2. Protein fibers of the matrix have the following characteristics
   - Tropocollagens are linked together to form collagen fibrils, which are joined to form collagen fibers. The collagen fibers resemble ropes. They are strong and flexible but resist stretching.
   - Reticular fibers are fine collagen fibers that form a branching network that supports other cells and tissues.
   - Elastin fibers have a structure similar to that of a spring. After being stretched, they tend to return to their original shape.
3. Ground substance has the following as major components:
   - Hyaluronic acid makes fluids slippery.
   - Proteoglycan aggregates trap water, which gives tissues the capacity to return to their original shape when compressed or deformed.
   - Adhesive molecules hold proteoglycans together and to plasma membranes.

Classification of Connective Tissue

Connective tissue is classified according to the type of protein and the proportions of protein, ground substance, and fluid in the matrix.

Embryonic Connective Tissue

Mesenchyme arises early, consists of irregularly shaped cells and abundant matrix, and gives rise to adult connective tissue.

Adult Connective Tissue

1. *Loose connective tissue*
   - Loose (areolar) connective tissue has many different cell types and a random arrangement of protein fibers with space between the fibers. This tissue fills spaces around the organs and attaches the skin to underlying tissues.
2. *Dense connective tissue*
   - Dense regular connective tissue is composed of fibers arranged in one direction, which provides strength in a direction parallel to the fiber orientation. Two types of dense regular connective tissue exist: collagenous (tendons and most ligaments) and elastic (ligaments of vertebrae).
   - Dense irregular connective tissue has fibers organized in many directions, which produces strength in different directions.

     Two types of dense irregular connective tissue exist: collagenous (capsules of organs and dermis of skin) and elastic (large arteries).
3. *Connective tissue with special properties*
   - Adipose tissue has fat cells (adipocytes) filled with lipid and very little extracellular matrix (a few reticular fibers).

     Adipose tissue functions as energy storage, insulation, and protection.

     Adipose tissue can be yellow (white) or brown. Brown fat is specialized for generating heat.
   - Reticular tissue is a network of reticular fibers and forms the framework of lymphoid tissue, bone marrow, and the liver.
   - Hemopoietic tissue, or red bone marrow, is the site of blood cell formation, and yellow bone marrow is a site of fat storage.
4. *Cartilage*
   - Cartilage has a relatively rigid matrix composed of protein fibers and proteoglycan aggregates. The major cell type is the chondrocyte, which is located within lacunae.

     Hyaline cartilage has evenly dispersed collagen fibers that provide rigidity with some flexibility. Examples include the costal cartilage, the covering over the ends of bones in joints, the growing portion of long bones, and the embryonic skeleton.

     Fibrocartilage has collagen fibers arranged in thick bundles; it can withstand great pressure, and it is found between vertebrae, in the jaw, and in the knee.

     Elastic cartilage is similar to hyaline cartilage, but it has elastin fibers. It is more flexible than hyaline cartilage. It is found in the external ear.
5. *Bone*

   Bone cells, or osteocytes, are located in lacunae that are surrounded by a mineralized matrix (hydroxyapatite) that makes bone very hard. Cancellous bone has spaces between bony trabeculae, and compact bone is more solid.
6. *Blood and hemopoietic tissue*
   - Blood cells are suspended in a fluid matrix.
   - Hemopoietic tissue forms blood cells.

Muscle Tissue (p. 134)

1. Muscle tissue has the ability to contract.
2. Skeletal (striated voluntary) muscle attaches to bone and is responsible for body movement. Skeletal muscle cells are long, cylindrically shaped cells with many peripherally located nuclei.
3. Cardiac (striated involuntary) muscle cells are cylindrical, branching cells with a single, central nucleus. Cardiac muscle is found in the heart and is responsible for pumping blood through the circulatory system.
4. Smooth (nonstriated involuntary) muscle forms the walls of hollow organs, the iris of the eye, and other structures. Its cells are spindle-shaped with a single, central nucleus.

Nervous Tissue (p. 136)

1. Nervous tissue is able to conduct electric impulses and is composed of neurons (conductive cells) and neuroglia (support cells).
2. Neurons have cell processes called dendrites and axons. The dendrites can receive electric impulses, and the axons can conduct them.

Neurons can be multipolar (several dendrites and an axon), bipolar (one dendrite and one axon), or unipolar (one axon).

Membranes (p 137)

1. Mucous membranes consist of epithelial cells, their basement membrane, the lamina propria, and sometimes smooth muscle cells; they line cavities that open to the outside and often contain mucous glands, which secrete mucus.
2. Serous membranes line cavities that do not open to the exterior and do not contain glands but do secrete serous fluid.
3. Synovial membranes are formed by connective tissue and line joint cavities.

Inflammation (p. 138)

1. The function of the inflammatory response is to isolate injurious agents from the rest of the body and to attack and destroy the injurious agent.
2. The inflammatory response produces five symptoms: redness, heat, swelling, pain, and disturbance of function.

Tissue Repair (p. 140)

1. Tissue repair is the substitution of viable cells for dead ones. Tissue repair occurs by regeneration or replacement.
   - Labile cells divide throughout life and can undergo regeneration.
   - Stable cells do not ordinarily divide after growth is complete but can regenerate if necessary.
   - Permanent cells cannot replicate. If killed, permanent tissue is repaired by replacement.
2. Tissue repair by primary union occurs when the edges of the wound are close together. Secondary union occurs when the edges are far apart.

Tissues and Aging (p. 142)

1. Age-related changes in tissues result from reduced rates of cell division and changes in the extracellular fibers.
2. Collagen fibers become less flexible and have reduced strength.
3. Elastic fibers become fragmented and less elastic.

REVIEW AND COMPREHENSION

1. Given these characteristics:
   1. capable of contraction
   2. covers free body surfaces
   3. lacks blood vessels
   4. composes various glands
   5. anchored to connective tissue by a basement membrane

   Which of these are characteristics of epithelial tissue?
   a. 1,2,3
   b. 2,3,5
   c. 3,4,5
   d. 1,2,3,4
   e. 2,3,4,5
2. Which of these embryonic germ layers gives rise to muscle, bone, and blood vessels?
   a. ectoderm
   b. endoderm
   c. mesoderm
3. A tissue that covers a surface, is one cell layer thick, and is composed of flat cells is
   a. simple squamous epithelium.
   b. simple cuboidal epithelium.
   c. simple columnar epithelium.
   d. stratified squamous epithelium.
   e. transitional epithelium.
4. Epithelium composed of two or more layers of cells with only the deepest layer in contact with the basement membrane is
   a. stratified epithelium.
   b. simple epithelium.
   c. pseudostratified epithelium.
   d. columnar epithelium.
   e. cuboidal epithelium.
5. Stratified epithelium is usually found in areas of the body where the principal activity is
   a. filtration.
   b. protection.
   c. absorption.
   d. diffusion.
   e. secretion.
6. Which of these characteristics do *not* describe nonkeratinized stratified squamous epithelium?
   a. many layers of cells
   b. surface cells are flat
   c. surface cells are living
   d. found in the skin
   e. outer layers covered by fluid
7. In parts of the body, such as the urinary bladder, where considerable expansion occurs, one can expect to find which type of epithelium?
   a. cuboidal
   b. pseudostratified
   c. transitional
   d. squamous
   e. columnar
8. A tissue that contains cells with these characteristics:
   1. covers a surface
   2. one layer of cells
   3. cells are flat

   Performs which of the following functions?
   a. phagocytosis
   b. active transport
   c. secretion of many complex lipids and proteins
   d. allow certain substances to diffuse across it
   e. protection from abrasion
9. Epithelial cells with microvilli are most likely found
   a. lining blood vessels.
   b. lining the lungs.
   c. lining the uterine tube.
   d. lining the small intestine.
   e. in the skin.
10. Pseudostratified ciliated columnar epithelium can be found lining the
    a. digestive tract.
    b. trachea.
    c. thyroid gland.
    d. kidney tubules.
    e. urinary bladder.
11. A type of cell connection whose *only* function is to prevent the cells from coming apart is the
    a. desmosome.
    b. gap junction.
    c. tight junction.
12. The glands that lose their connection with epithelium during embryonic development and secrete their cellular products into the bloodstream are called ______________ glands.
    a. apocrine
    b. endocrine
    c. exocrine
    d. holocrine
    e. merocrine
13. Glands that accumulate secretions and release them only when the individual secretory cells rupture and die are called ______________ glands.
    a. apocrine
    b. holocrine
    c. merocrine

14. A ______________ gland has a duct that branches repeatedly, and the ducts end in saclike structures.
    a. simple tubular
    b. compound tubular
    c. simple coiled tubular
    d. simple acinar
    e. compound acinar
15. The fibers in dense connective tissue are produced by
    a. fibroblasts.
    b. adipocytes.
    c. osteoblasts.
    d. osteoclasts.
    e. macrophages.
16. Mesenchymal cells
    a. form embryonic connective tissue.
    b. give rise to all adult connective tissues.
    c. in adults produce new connective tissue cells in response to injury.
    d. all of the above.
17. A tissue with a large number of collagen fibers organized parallel to each other would most likely be found in
    a. a muscle.
    b. a tendon.
    c. adipose tissue.
    d. a bone.
    e. cartilage.
18. Extremely delicate fibers that make up the framework for organs such as the liver, spleen, and lymph nodes are
    a. elastic fibers.
    b. reticular fibers.
    c. microvilli.
    d. cilia.
    e. collagen fibers.
19. In which of these locations are dense irregular elastic connective tissue found?
    a. ligaments
    b. nuchal ligament
    c. dermis of the skin
    d. large arteries
    e. adipose tissue
20. Which of these is *not* true of adipose tissue?
    a. It is the site of energy storage.
    b. It is a type of connective tissue.
    c. It acts as a protective cushion.
    d. Brown adipose is found only in babies.
    e. It functions as a heat insulator.
21. Which of these types of connective tissue has the smallest amount of extracellular matrix?
    a. adipose
    b. bone
    c. cartilage
    d. loose connective tissue
    e. blood
22. Given these characteristics:
    1. cells located in lacunae
    2. proteoglycans in ground substance
    3. no collagen fibers present
    4. perichondrium on surface
    5. heals rapidly after injury

    Which of these characteristics apply to cartilage?
    a. 1,2,3
    b. 1,2,4
    c. 2,4,5
    d. 1,2,4,5
    e. 2,3,4,5
23. Fibrocartilage is found
    a. in the cartilage of the trachea.
    b. in the rib cage.
    c. in the external ear.
    d. on the surface of bones in moveable joints.
    e. between vertebrae.
24. A tissue in which cells are located in lacunae surrounded by a hard matrix of hydroxyapatite is
    a. hyaline cartilage.
    b. bone.
    c. nervous tissue.
    d. dense regular collagenous connective tissue.
    e. fibrocartilage.
25. Which of these characteristics apply to smooth muscle?
    a. striated, involuntary
    b. striated, voluntary
    c. unstriated, involuntary
    d. unstriated, voluntary
26. Which of these statements about nervous tissue is *not* true?
    a. Neurons have cytoplasmic extensions called axons.
    b. Electric signals (action potentials) are conducted along axons.
    c. Bipolar neurons have two axons.
    d. Neurons are nourished and protected by neuroglia.
    e. Dendrites receive electric signals and conduct them toward the cell body.
27. The linings of the digestive, respiratory, excretory, and reproductive passages are composed of
    a. serous membranes.
    b. mucous membranes.
    c. mesothelium.
    d. synovial membranes.
    e. endothelium.
28. Chemical mediators of inflammation
    a. cause blood vessels to constrict.
    b. decrease the permeability of blood vessels.
    c. initiate processes that lead to edema.
    d. help prevent clotting.
    e. decrease pain.
29. Which of these types of cells are labile?
    a. neurons
    b. skin
    c. liver
    d. pancreas
30. Permanent cells
    a. divide and replace damaged cells in replacement tissue repair.
    b. form granulation tissue.
    c. are responsible for removing scar tissue.
    d. are usually replaced by a different cell type if they are destroyed.
    e. are replaced during regeneration tissue repair.

Answers in Appendix E

CRITICAL THINKING

1. Given the observation that a tissue has more than one layer of cells lining a free surface, (1) list the possible tissue types that exhibit those characteristics, and (2) explain what additional observations need to be made to identify the tissue type.
2. A patient suffered from kidney failure a few days after he was exposed to a toxic chemical. A biopsy of his kidney indicated that many of the thousands of epithelium-lined tubules that make up the kidney had lost many of the simple cuboidal epithelial cells that normally line them, although the basement membranes appeared to be mostly intact. Predict how likely this person is to recover fully.
3. Compare the cell shapes and surface specializations of an epithelium that resists abrasion with those of an epithelium that carries out the absorption of materials.

4. Tell how to distinguish between a gland that produces a merocrine secretion and a gland that produces a holocrine secretion. Assume that you have the ability to chemically analyze the composition of secretions.
5. Name a tissue that has the following characteristics: abundant extra cellular matrix consisting almost entirely of collagen fibers that are parallel to each other. Injuries of which of the following types result from damage to this kind of tissue: dislocated neck vertebrae, a torn tendon, a ruptured intervertebral disk?
6. Indicate whether the following statement is correct or not: "If a tissue is capable of contracting, is under involuntary control, and has mononucleated cells, it is smooth muscle." Explain your answer.
7. Antihistamines block the effect of a chemical mediator of inflammation called histamine, which is released during the inflammatory response. What effect does administering antihistamines have on the inflammatory response, and is the use of an antihistamine beneficial?
8. A gland produces a watery secretion containing solutes and is close to being isotonic. Which of the processes responsible for producing secretions functions in this gland?
9. A tissue has the following characteristics: a free surface; a single layer of cells, cells; narrow, tall cells; microvilli, many mitochondria, and globlet cells. Describe the tissue type and as many functions of the cell as possible.
10. Most people who develop breast cancer do not have a known family history of breast cancer. However, approximately 7% of breast cancers are associated with the dominant breast cancer susceptibility genes called BRCA1, located on chromosome 17, and BRCA2, located on chromosome 13. The BRCA1 gene is also associated with an increased probability of ovarian cancer. However, not all people who have these genes develop cancer. Explain how that might occur.

Answers in Appendix F

5 Integumentary System

The **integumentary** (in-teg-ū-men′tă-rē) **system** consists of the skin and accessory structures, such as hair, nails, and glands. *Integument* means covering, and the integumentary system is familiar to most people because it covers the outside of the body and is easily observed. In addition, humans are concerned with the appearance of the integumentary system. Skin without blemishes is considered attractive, whereas acne is a source of embarrassment for many people. The development of wrinkles and the graying or loss of hair are signs of aging that some people find unattractive. Because of these feelings, much time, effort, and money are spent on changing the appearance of the integumentary system. For example, people apply lotion to their skin, color their hair, and trim their nails. They also try to prevent sweating with antiperspirants and body odor with washing, deodorants, and perfumes.

The appearance of the integumentary system can indicate physiological imbalances in the body. Some disorders, such as acne or warts, affect just the integumentary system. Disorders of other parts of the body can be reflected there, and thus the integumentary system is useful for diagnosis. For example, reduced blood flow through the skin during a heart attack can cause a pale appearance, whereas increased blood flow as a result of fever can cause a flushed appearance. Also, the rashes of some diseases are very characteristic, such as the rashes of measles, chicken pox, and allergic reactions.

This chapter provides an *overview of the integumentary system* (p. 150) and an explanation of the *skin* (p. 150), the *hypodermis* (p. 157), and the *accessory skin structures* (p. 158). A *summary of integumentary system functions* (p. 163) and the *effects of aging on the integumentary system* (p. 165) are also presented.

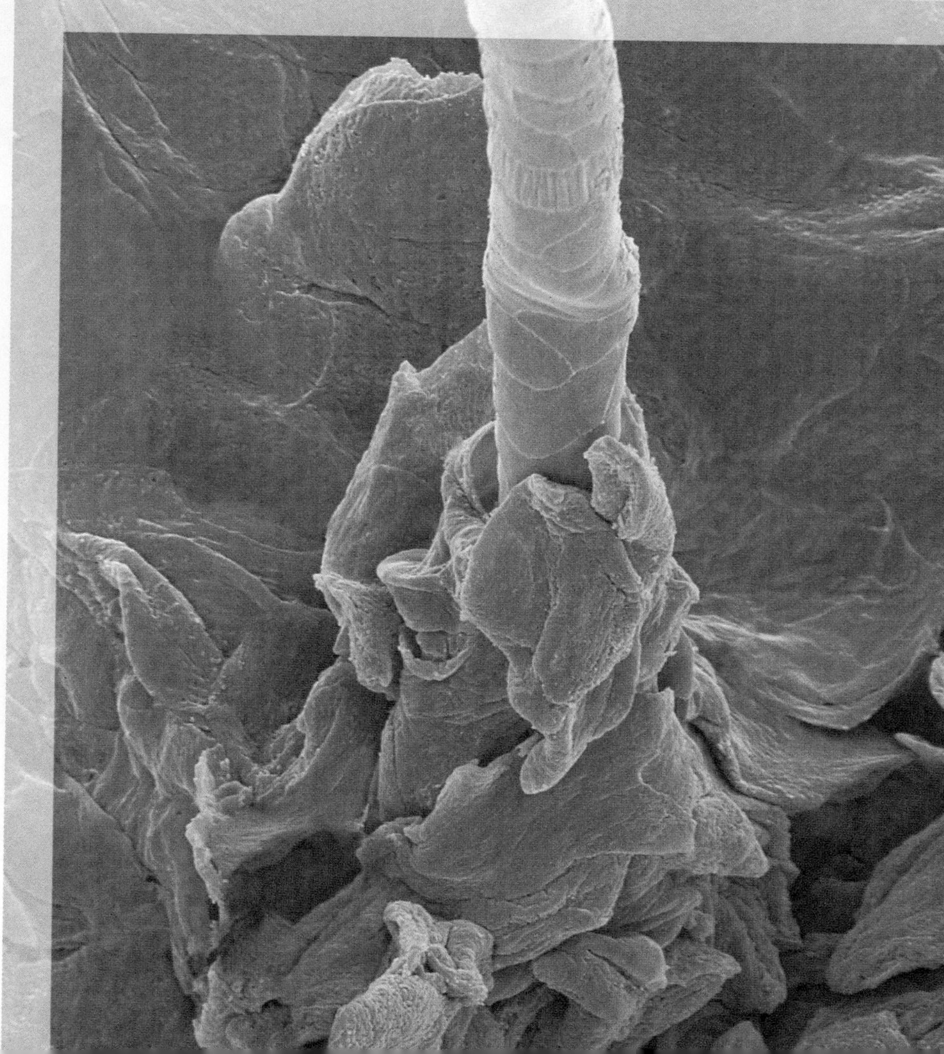

Skin, hair, nails, and glands form the integumentary system. This colorized scanning electron micrograph shows the shaft of a hair (*yellow*) protruding through the surface of the skin. Tightly bound epithelial cells form the hair shaft, and flat, scalelike epithelial cells form the skin's surface. Like the shingles on the roof of a house, the skin's tough cells protect underlying structures. Dandruff is excessive flaking of these epithelial cells from the scalp.

Anatomy & Physiology | REVEALED®
aprevealed.com

Integumentary System (online only)

OVERVIEW OF THE INTEGUMENTARY SYSTEM

Although we are often concerned with how the integumentary system looks, it has many important functions that go beyond appearance. The integumentary system forms the boundary between the body and the external environment, thereby separating us from the external environment while allowing us to interact with it. The following are the major functions of the integumentary system:

1. *Protection.* The skin protects against abrasion and ultraviolet light. It also prevents the entry of microorganisms and prevents dehydration by reducing water loss from the body.
2. *Sensation.* The integumentary system has sensory receptors that can detect heat, cold, touch, pressure, and pain.
3. *Temperature regulation.* Body temperature is regulated through the control of blood flow through the skin and the activity of sweat glands.
4. *Vitamin D production.* When exposed to ultraviolet light, the skin produces a molecule that can be transformed into vitamin D.
5. *Excretion.* Small amounts of waste products are lost through the skin and in gland secretions.

1 ***Provide an example for each function of the integumentary system.***

SKIN

The skin is made up of two major tissue layers. The **epidermis** (ep-i-derm′is; on the dermis) is the most superficial layer of the skin; it consists of epithelial tissue (figure 5.1). The epidermis resists abrasion on the skin's surface and reduces water loss through the skin. The epidermis rests on the **dermis** (derm′is; skin), which is a layer of connective tissue. The dermis is responsible for most of the structural strength of the skin. The strength of the dermis is seen in leather, which is produced from the hide (skin) of an animal. The epidermis is removed, and the dermis is preserved by tanning.

The skin rests on the **hypodermis** (hi-pō-der′mis; below the skin), which is a layer of loose connective tissue (see figure 5.1). The hypodermis is not part of the skin or the integumentary system, but it does connect the skin to underlying muscle or bone. Table 5.1 summarizes the structures and functions of the skin and hypodermis.

Epidermis

The epidermis is stratified squamous epithelium, and it is separated from the dermis by a basement membrane. The epidermis is not as thick as the dermis, contains no blood vessels, and is

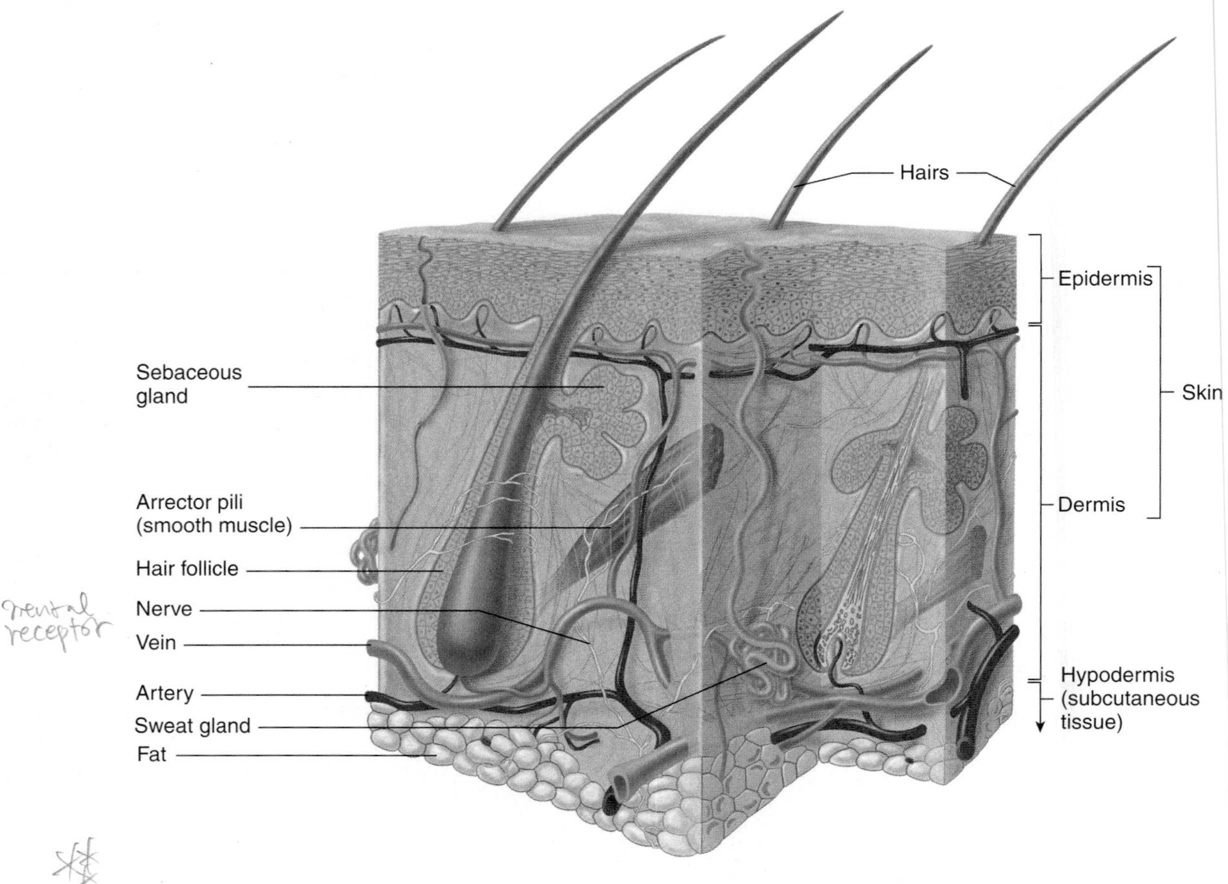

FIGURE 5.1 Skin and Hypodermis

The skin, consisting of the epidermis and the dermis, is connected by the hypodermis to underlying structures. Note the accessory structures (hairs, glands, and arrector pili), some of which project into the hypodermis, and the large amount of fat in the hypodermis.

TABLE 5.1 Comparison of the Skin (Epidermis and Dermis) and Hypodermis

Part	Structure	Function
Epidermis	Superficial part of skin; stratified squamous epithelium; composed of four or five strata	Barrier that prevents water loss and the entry of chemicals and microorganisms; protects against abrasion and ultraviolet light; produces vitamin D; gives rise to hair, nails, and glands
Stratum corneum	Most superficial strata of the epidermis; 25 or more layers of dead squamous cells	Provision of structural strength by keratin within cells; prevention of water loss by lipids surrounding cells; sloughing off of most superficial cells resists abrasion
Stratum lucidum	Three to five layers of dead cells; appears transparent; present in thick skin, absent in most thin skin	Dispersion of keratohyalin around keratin fibers
Stratum granulosum	Two to five layers of flattened, diamond-shaped cells	Production of keratohyalin granules; lamellar bodies release lipids from cells; cells die
Stratum spinosum	A total of 8–10 layers of many-sided cells	Production of keratin fibers; formation of lamellar bodies
Stratum basale	Deepest strata of the epidermis; single layer of cuboidal or columnar cells; basement membrane of the epidermis attaches to the dermis	Production of cells of the most superficial strata; melanocytes produce and contribute melanin, which protects against ultraviolet light
Dermis	Deep part of skin; connective tissue composed of two layers	Responsible for the structural strength and flexibility of the skin; the epidermis exchanges gases, nutrients, and waste products with blood vessels in the dermis
Papillary layer	Papillae project toward the epidermis; loose connective tissue	Brings blood vessels close to the epidermis; dermal papillae form fingerprints and footprints
Reticular layer	Mat of collagen and elastin fibers; dense irregular connective tissue	Main fibrous layer of the dermis; strong in many directions; forms cleavage lines
Hypodermis	Not part of the skin; loose connective tissue with abundant fat deposits	Attaches the dermis to underlying structures; fat tissue provides energy storage, insulation, and padding; blood vessels and nerves from the hypodermis supply the dermis

nourished by diffusion from capillaries of the dermis (figure 5.2). Most cells of the epidermis are called **keratinocytes** (ke-rat′i-nō-sītz) because they produce a protein mixture called **keratin** (ker′ă-tin), which makes the cells hard. Keratinocytes are responsible for the ability of the epidermis to resist abrasion and reduce water loss. Other cells of the epidermis include **melanocytes** (mel′ă-nō-sītz), which contribute to skin color, **Langerhans cells,** which are part of the immune system (see chapter 22), and **Merkel cells,** which are specialized epidermal cells associated with nerve endings responsible for detecting light touch and superficial pressure (see chapter 14).

Cells are produced by mitosis in the deepest layers of the epidermis. As new cells are formed, they push older cells to the surface, where they slough off, or **desquamate** (des′kwă-māt). The outermost cells in this stratified arrangement protect the cells underneath, and the deeper replicating cells replace cells lost from the surface. As they move from the deeper epidermal layers to the surface, the cells change shape and chemical composition. This process is called **keratinization** (ker′ă-tin-i-zā′shŭn) because the cells become filled with keratin. During keratinization, these cells eventually die and produce an outer layer of dead, hard cells that resists abrasion and forms a permeability barrier.

Keratinization and Disease

The study of keratinization is important because many skin diseases result from malfunctions in this process. For example, large scales of epidermal tissue are sloughed off in **psoriasis** (sō-rī′ă-sis; see "Clinical Focus: Clinical Disorders of the Integumentary System," p. 167). By comparing normal and abnormal keratinization, scientists may be able to develop effective therapies.

Although keratinization is a continual process, distinct transitional stages can be recognized as the cells change. On the basis of these stages, the many layers of cells in the epidermis are divided into regions, or **strata** (sing. *stratum;* see figure 5.2; figure 5.3). From the deepest to the most superficial, the five strata are the stratum basale, stratum spinosum, stratum granulosum, stratum lucidum, and stratum corneum. The number of cell layers in each stratum and even the number of strata in the skin vary, depending on their location in the body.

Stratum Basale

The deepest portion of the epidermis is a single layer of cuboidal or columnar cells called the **stratum basale** (bā′să-lē), or **stratum germinativum** (jer′mi-nă-tīv′um; see figures 5.2 and 5.3, step 1).

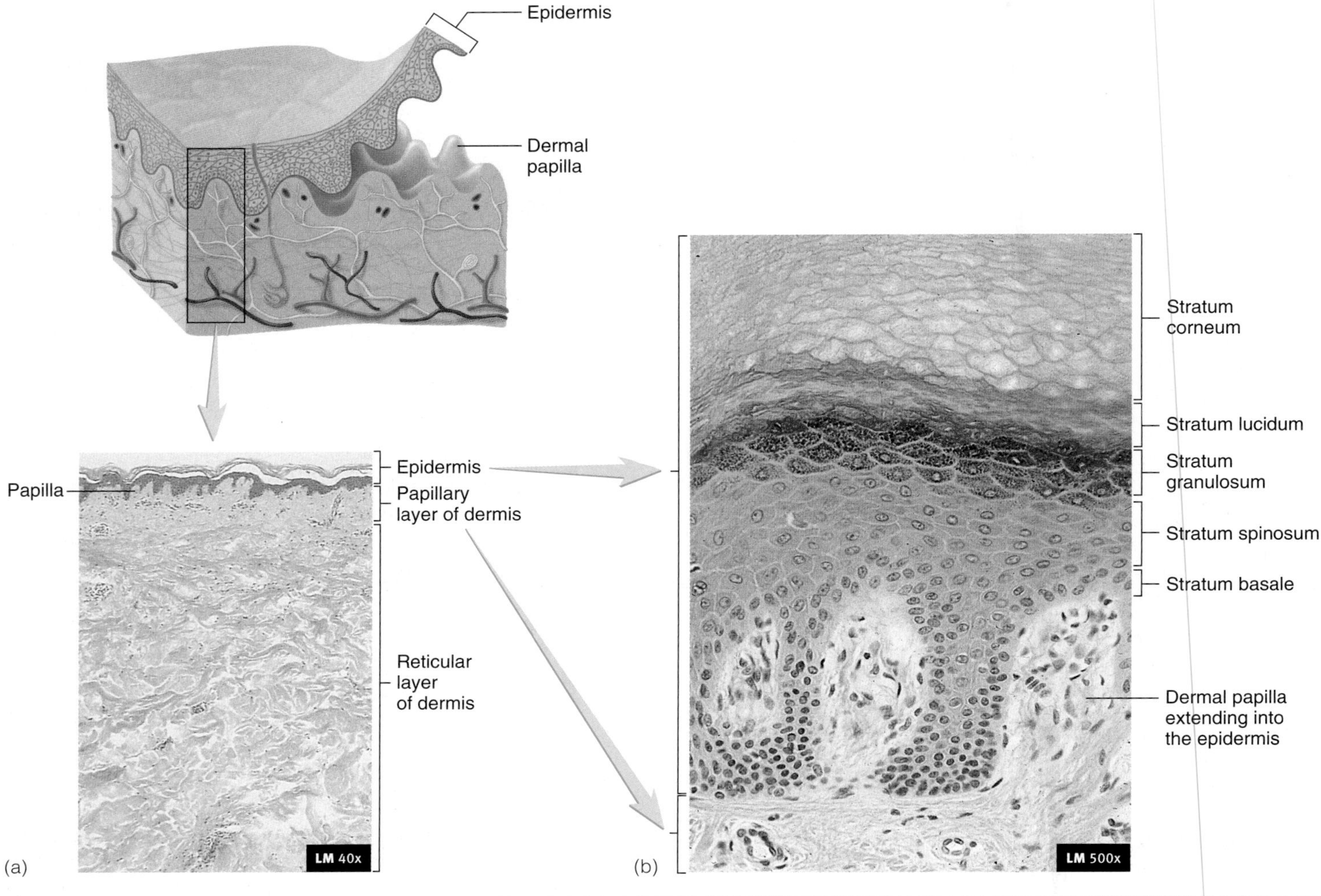

FIGURE 5.2 Dermis and Epidermis

(*a*) Photomicrograph of dermis covered by the epidermis. The dermis consists of the papillary and reticular layers. The papillary layer has projections, called papillae, that extend into the epidermis. (*b*) Higher-magnification photomicrograph of the epidermis resting on the papillary layer of the dermis. Note the strata of the epidermis.

Structural strength is provided by hemidesmosomes, which anchor the epidermis to the basement membrane, and by desmosomes, which hold the keratinocytes together (see chapter 4). Keratinocytes are strengthened internally by keratin fibers (intermediate filaments) that insert into the desmosomes. Keratinocytes undergo mitotic divisions approximately every 19 days. One daughter cell becomes a new stratum basale cell and divides again, but the other daughter cell is pushed toward the surface and becomes **keratinized** (ker′ă-ti-nīzd). It takes approximately 40–56 days for the cell to reach the epidermal surface and desquamate.

Stratum Spinosum

Superficial to the stratum basale is the **stratum spinosum** (spī-nō′sŭm), consisting of 8–10 layers of many-sided cells (see figures 5.2 and 5.3, step 2). As the cells in this stratum are pushed to the surface, they flatten; desmosomes are broken apart, and new desmosomes are formed. During preparation for microscopic observation, the cells usually shrink from one another, except where they are attached by desmosomes, causing the cells to appear spiny—hence the name stratum spinosum. Additional keratin fibers and lipid-filled, membrane-bound organelles called **lamellar** (lam′ĕ-lăr, lă-mel′ar) **bodies** are formed inside the keratinocytes. A limited amount of cell division takes place in the stratum spinosum. Mitosis does not occur in the more superficial strata.

Stratum Granulosum

The **stratum granulosum** (gran-ū-lō′sŭm) consists of two to five layers of somewhat flattened, diamond-shaped cells with long axes that are oriented parallel to the surface of the skin (see figures 5.2 and 5.3, step 3). This stratum derives its name from the nonmembrane-bound protein granules of **keratohyalin** (ker′ă-tō-hī′ă-lin), which accumulate in the cytoplasm of the cell. The lamellar bodies of these cells move to the plasma membrane and

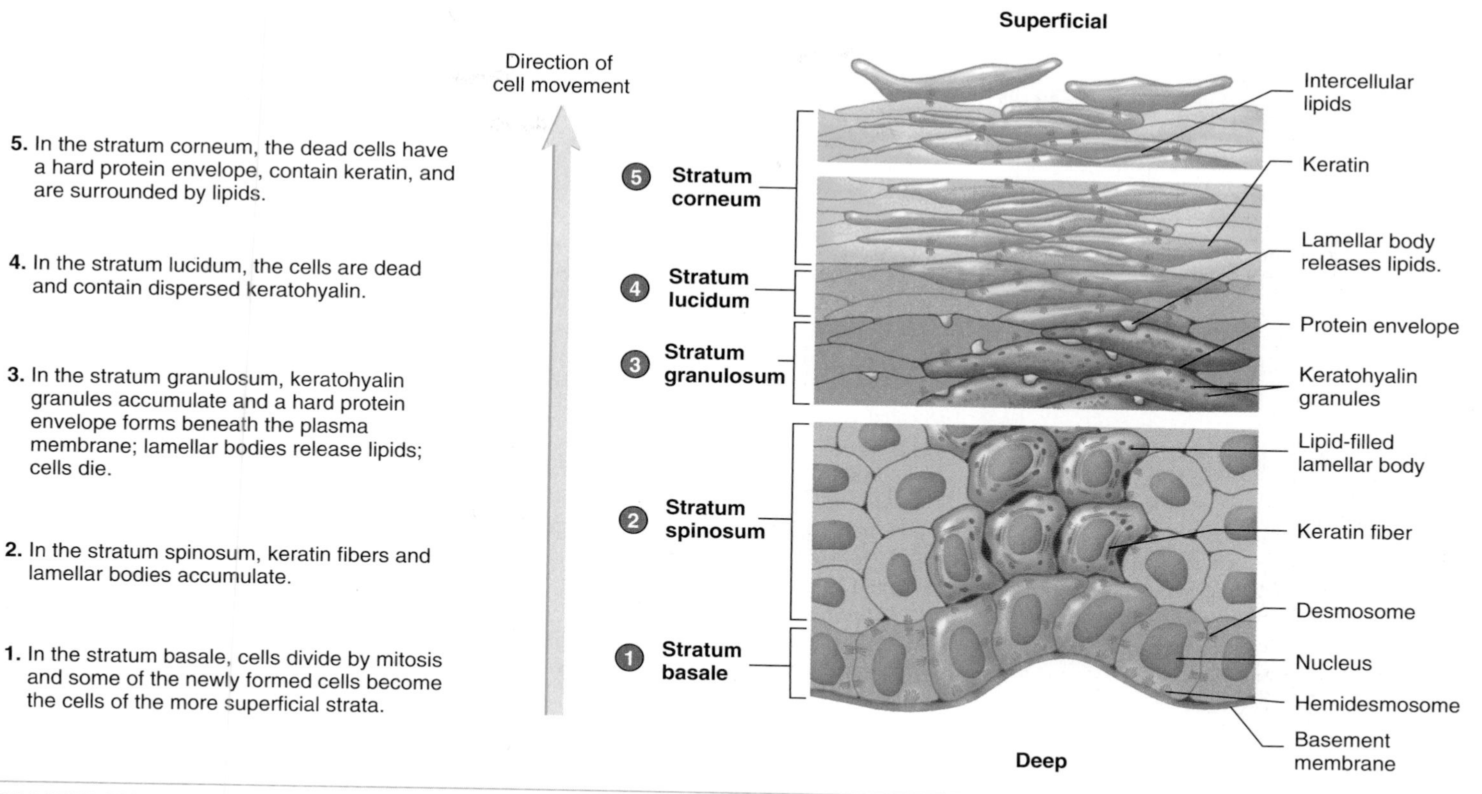

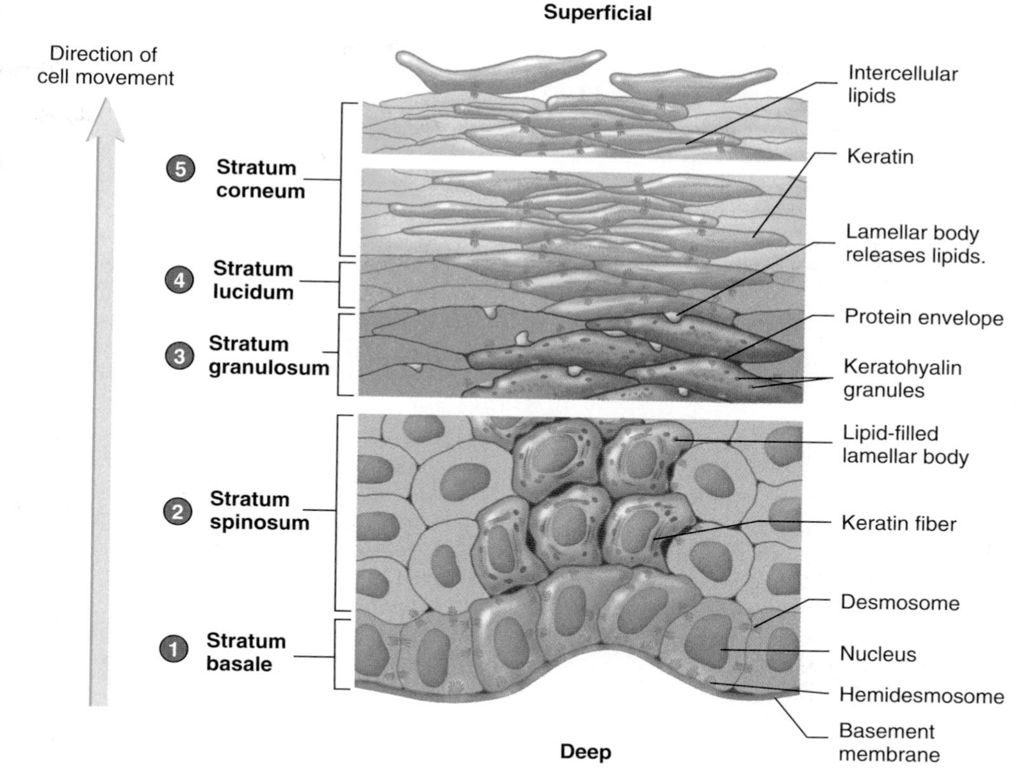

PROCESS FIGURE 5.3 **Epidermal Layers and Keratinization**

release their lipid contents into the intercellular space. Inside the cell, a protein envelope forms beneath the plasma membrane. In the most superficial layers of the stratum granulosum, the nucleus and other organelles degenerate, and the cell dies. Unlike the other organelles, however, the keratin fibers and keratohyalin granules do not degenerate.

Stratum Lucidum

The **stratum lucidum** (loo′si-dŭm) appears as a thin, clear zone above the stratum granulosum (see figures 5.2 and 5.3, step 4); it consists of several layers of dead cells with indistinct boundaries. Keratin fibers are present, but the keratohyalin, which was evident as granules in the stratum granulosum, has dispersed around the keratin fibers, and the cells appear somewhat transparent. The stratum lucidum is present in only a few areas of the body (see "Thick and Thin Skin" on p. 154).

Stratum Corneum

The last, and most superficial, stratum of the epidermis is the **stratum corneum** (kōr′nē-ŭm; see figures 5.2 and 5.3, step 5). This stratum is composed of 25 or more layers of dead squamous cells joined by desmosomes. Eventually, the desmosomes break apart, and the cells are desquamated from the surface of the skin. Excessive desquamation of the stratum corneum of the scalp is called dandruff. Less noticeably, cells are continually shed as clothes rub against the body or as the skin is washed.

The stratum corneum consists of **cornified cells,** which are dead cells, with a hard protein envelope, filled with the protein keratin. **Keratin** is a mixture of keratin fibers and keratohyalin. The envelope and the keratin are responsible for the structural strength of the stratum corneum. The type of keratin found in the skin is soft keratin. Another type of keratin, hard keratin, is found in nails and the external parts of hair. Cells containing hard keratin are more durable than cells with soft keratin, and they do not desquamate.

Surrounding the cells are the lipids released from lamellar bodies. The lipids are responsible for many of the permeability characteristics of the skin.

2 ***From deepest to most superficial, name and describe the five strata of the epidermis. In which strata are new cells formed by mitosis? Which strata have live cells, and which have dead cells?***

3 ***Describe the structural features resulting from keratinization that make the epidermis structurally strong and resistant to water loss.***

PREDICT 1

Some drugs are administered by applying them to the skin (e.g., a nicotine skin patch to help a person stop smoking). The drug diffuses through the epidermis to blood vessels in the dermis. What kind of substances can pass easily through the skin by diffusion? What kind have difficulty?

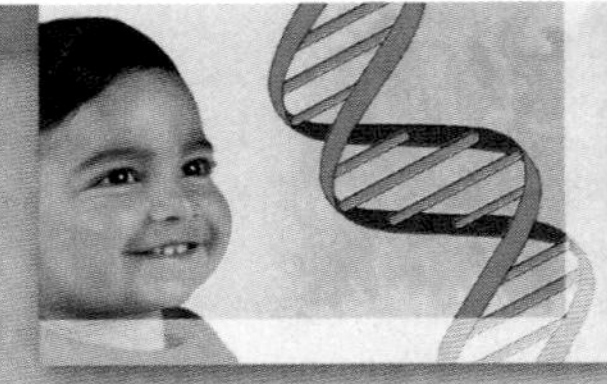

CLINICAL GENETICS

Skin Cancer

Skin cancer is the most common type of cancer. Most skin cancers result from damage caused by the ultraviolet (UV) radiation in sunlight. Some skin cancers are induced by chemicals, x-rays, depression of the immune system, and inflammation, and some are inherited.

UV radiation damages the genes (DNA) in epidermal cells, producing mutations. If a mutation is not repaired, when a cell divides by mitosis the mutation is passed to one of the two daughter cells. An accumulation of mutations affecting oncogenes and tumor suppressing genes in epidermal cells can lead to uncontrolled cell division and skin cancer (see "Cancer," chapter 4).

The amount of protective melanin in the skin affects the likelihood of developing skin cancer. Fair-skinned individuals, with less melanin, are genetically predisposed to develop skin cancer, compared with dark-skinned individuals, who have more melanin. Long-term or intense exposure to UV radiation also increases the risk of developing skin cancer. Thus, individuals who are older than 50, who have engaged in repeated recreational or occupational exposure to the sun, or who have experienced sunburn are at increased risk. The sites of development of most skin cancers are the parts of the body most exposed to sunlight, such as the face, neck, ears, and dorsum of the forearm and hand.

There are three types of skin cancer: basal cell carcinoma, squamous cell carcinoma, and melanoma (figure A). **Basal cell carcinoma,** the most common type of skin cancer, arises from cells in the stratum basale. Basal cell carcinomas have a varied appearance. Some are open sores that bleed, ooze, or crust for several weeks. Others are reddish patches; shiny, pearly, or translucent bumps; or scarlike areas of shiny, taut skin. A physician should be consulted for the diagnosis and treatment of all skin cancers. Removal or destruction of the tumor cures most cases of basal cell carcinoma.

An **actinic keratosis** (ak-tin′ik ker-ă-tō′sis) is a small, scaly, crusty bump that arises on the surface of the skin. If untreated, about 2%–5% of actinic keratoses can progress to **squamous cell carcinoma,** the second most common type of skin cancer. Squamous cell carcinoma arises from cells in the stratum spinosum and can appear as a wartlike growth; a persistent, scaly red patch; an open sore; or an elevated growth with a central depression. Bleeding from these lesions can occur. Removal or destruction of the tumor cures most cases of squamous cell carcinoma.

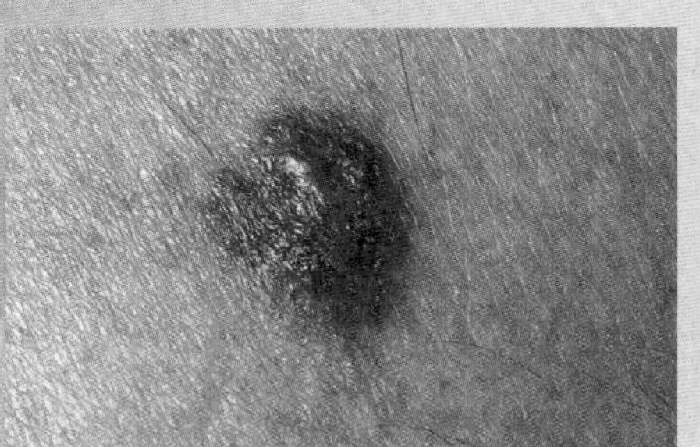

(a) Basal cell carcinoma

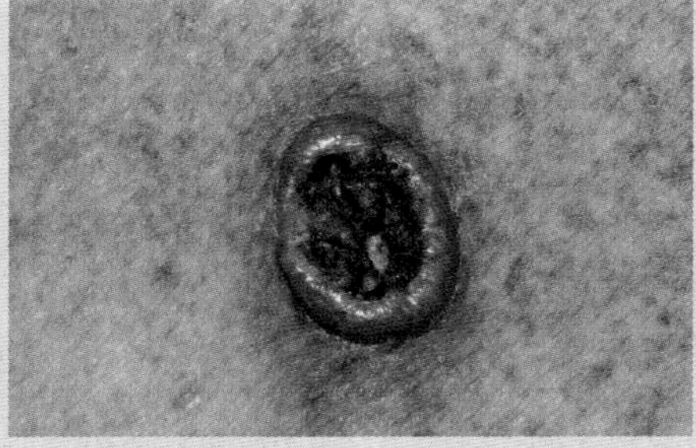

(b) Squamous cell carcinoma

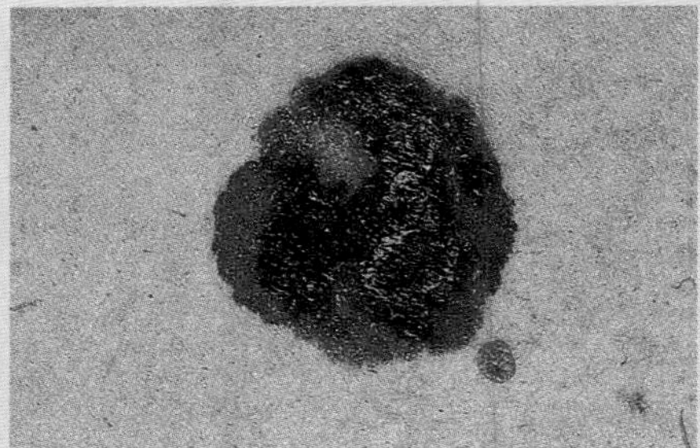

(c) Melanoma

FIGURE A Cancer of the Skin

Thick and Thin Skin

When we say a person has thick or thin skin, we are usually referring metaphorically to the person's ability to take criticism. However, all of us in a literal sense have both thick and thin skin. Skin is classified as thick or thin on the basis of the structure of the epidermis. **Thick skin** has all five epithelial strata, and the stratum corneum has many layers of cells. Thick skin is found in areas subject to pressure or friction, such as the palms of the hands, the soles of the feet, and the fingertips.

Thin skin covers the rest of the body and is more flexible than thick skin. Each stratum contains fewer layers of cells than are found in thick skin; the stratum granulosum frequently consists of only one or two layers of cells, and the stratum lucidum generally is absent. Hair is found only in thin skin.

The entire skin, including both the epidermis and the dermis, varies in thickness from 0.5 mm in the eyelids to 5.0 mm on the back and shoulders. The terms *thin* and *thick,* which refer to the epidermis only, should not be used when total skin thickness is considered. Most of the difference in total skin thickness results from variation in the thickness of the dermis. For example, the skin of the back is thin skin, whereas that of the palm is thick skin; however, the total skin thickness of the back is greater than that of the palm because more dermis exists in the skin of the back.

In skin subjected to friction or pressure, the number of layers in the stratum corneum greatly increases to produce a thickened area called a **callus** (kal′ŭs). The skin over bony prominences may develop a cone-shaped structure called a **corn.** The base of the

Melanoma (mel′ă-nō′mă) is the least common, but most deadly, type of skin cancer, accounting for over 77% of the skin cancer deaths in the United States. Because they arise from melanocytes, most melanomas are black or brown, but occasionally a melanoma stops producing melanin and appears skin-colored, pink, red, or purple. About 40% of melanomas develop in preexisting moles. Treatment of melanomas when they are confined to the epidermis is almost always successful. If a melanoma invades the dermis and metastasizes to other parts of the body, it is difficult to treat and can be deadly.

Early detection and treatment of melanoma before it metastasizes can prevent death. Detection of melanoma can be accomplished by routine examination of the skin and the application of the **ABCDE rule,** in which *A* stands for asymmetry (one side of the lesion does not match the other side), *B* is for border irregularity (the edges are ragged, notched, or blurred), *C* is for color (pigmentation is not uniform), *D* is for diameter (greater than 6 mm), and *E* is for evolving (lesion changes over time). Evolving lesions change size, shape, elevation, or color; they may bleed, crust, itch, or become tender.

In order for cancer cells to metastasize, they must leave their site of origin, enter the circulation, and become established in a new location. For example, melanoma cells first spread within the epidermis. Melanoma cells that break through the basement membrane and invade the dermis may enter lymphatic or blood vessels and spread to other parts of the body. The ability to metastasize requires an accumulation of mutations that enables cells to detach from similar cells, recognize and digest the basement membrane, and become established elsewhere when surrounded by different cell types.

Basal cell carcinomas very rarely metastasize and only 2%–6% of squamous cell carcinomas metastasize. Compared with keratinocytes, melanocytes may be more likely to give rise to tumors that metastasize because, in their developmental past, they had the ability to migrate and become established in new locations. In the embryo, melanocytes are derived from a population of cells, called neural crest cells, that originate along the midline of the back (see "Development of the CNS," chapter 13). Whereas many neural crest cells give rise to parts of the nervous system, others migrate to the skin and become melanocytes. A gene called Slug on chromosome 8 regulates neural crest cell migration. In normal melanocytes, the Slug gene is inactive, but in metastasizing melanoma cells it is reactivated. The reactivation of other genes regulating the migration of embryonic cells may play a role in other metastasizing cancers.

Most skin cancers result from a series of genetic changes in somatic cells. In some cases, however, there is a genetic susceptibility to developing skin cancer. For example, the risk of developing melanoma is two to three times greater than for the general population if a person's parent, sibling, or child has melanoma. Defects in tumor suppressor genes on chromosome 9 have been found in some cases of familial melanoma. Individuals who inherit these defective genes have already started down the cancer-developing pathway at birth. They are at increased risk of developing melanoma because fewer mutations later in life are required to develop melanoma.

Xeroderma pigmentosum (zēr′ō-der′mă pig′men-tō′sŭm) is a rare, autosomal recessive trait in which a DNA repair gene is defective. The severity of the disorder varies, depending on which of seven different genes is affected and on how the gene is affected—that is, what alleles are present. The most common defective gene is located on chromosome 9. Because damage to genes by UV radiation is not repaired, exposure to UV radiation results in the development of fatal skin cancers in childhood. Minimizing UV radiation exposure to a level as close as possible to zero can delay the development of the disorder.

Limiting exposure to the sun and using sunscreens can reduce the likelihood of developing skin cancer in everyone. Two types of UV radiation play a role. Ultraviolet-B (UVB; 290–320 nm) radiation is the most potent for causing sunburn, is the main cause of basal and squamous cell carcinomas, and is a significant cause of melanoma. Ultraviolet-A (UVA; 320–400 nm) also contributes to skin cancer development, especially melanoma. It also penetrates into the dermis, causing wrinkling and leathering of the skin. A broad-spectrum sunscreen, which protects against both UVB and UVA, with a sun protection factor (SPF) of at least 15, is recommended by the Skin Cancer Foundation.

Someday, increased protection against UV radiation may be achieved by stimulating tanning. Melanotan I is being tested. It is a synthetic version of melanocyte-stimulating hormone, which stimulates increased melanin production by melanocytes.

Sunless tanning stains the dead surface cells of the epidermis with dihydroxyacetone (DHA). At best, DHA provides a SPF of only 2 to 4. Although it does not provide adequate protection against UV radiation, it is beneficial in the sense that a tanned appearance is achieved without exposing the skin to the damaging effects of UV radiation.

cone is at the surface, but the apex extends deep into the epidermis, and pressure on the corn may be quite painful. Calluses and corns can develop in both thin and thick skin.

4 ***Compare the structure and location of thick skin and thin skin. Is hair found in thick or thin skin?***

Skin Color

Pigments in the skin, blood circulating through the skin, and the thickness of the stratum corneum together determine skin color. **Melanin** (mel′ă-nin) is the group of pigments responsible for skin, hair, and eye color. Melanin provides protection against ultraviolet light from the sun. Large amounts of melanin are found in certain regions of the skin, such as freckles, moles, nipples, areolae of the breasts, axillae, and genitalia. Other areas of the body, such as the lips, palms of the hands, and soles of the feet, contain less melanin.

In the production of melanin, the enzyme tyrosinase (tī′rō-si-nās, tir′ō-si-nās) converts the amino acid tyrosine to dopaquinone (dō′pă-kwin′ōn, dō′pă-kwī-nōn). Dopaquinone can be converted to a variety of related molecules, most of which are brown to black pigments, but some of which are yellowish or reddish.

Melanin is produced by **melanocytes** (mel′ă-nō-sītz), irregularly shaped cells with many long processes that extend between the keratinocytes of the stratum basale and the stratum spinosum (figure 5.4). The Golgi apparatuses of the melanocytes package melanin into vesicles called **melanosomes** (mel′ă-nō-sōmz), which

1. Melanosomes are produced by the Golgi apparatus of the melanocyte.
2. Melanosomes move into melanocyte cell processes.
3. Epithelial cells phagocytize the tips of the melanocyte cell processes.
4. The melanosomes, which were produced inside the melanocytes, have been transferred to epithelial cells and are now inside them.

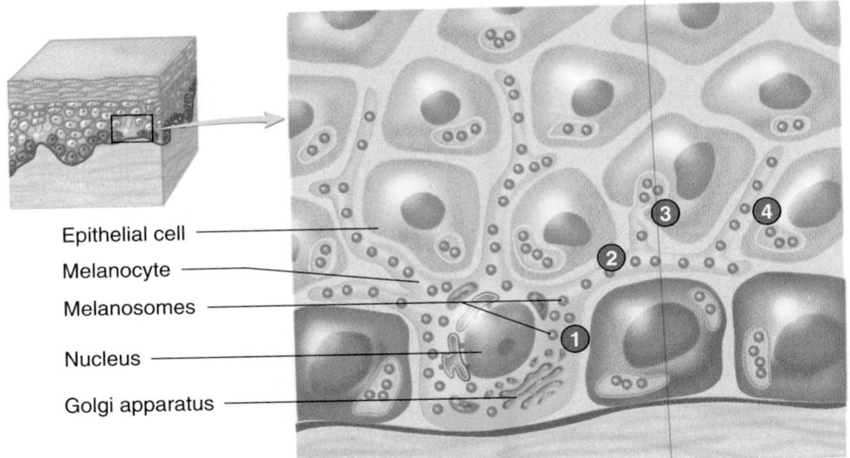

PROCESS FIGURE 5.4 Melanin Transfer from Melanocyte to Keratinocytes
Melanocytes make melanin, which is packaged into melanosomes and transferred to many keratinocytes.

move into the cell processes of the melanocytes. Keratinocytes phagocytize (see chapter 3) the tips of the melanocyte cell processes, thereby acquiring melanosomes. Although all keratinocytes can contain melanin, only the melanocytes produce it.

Melanin production is determined by genetic factors, exposure to light, and hormones. Genetic factors are primarily responsible for the variations in skin color among different races and among people of the same race. The amount and types of melanin produced by the melanocytes, as well as the size, number, and distribution of the melanosomes, are genetically determined. Skin colors are not determined by the number of melanocytes because all races have essentially the same number. Although many genes are responsible for skin color, a single mutation (see chapter 3) can prevent the manufacture of melanin. **Albinism** (al′bi-nizm) usually is a recessive genetic trait causing an inability to produce tyrosinase. The result is a deficiency or absence of pigment in the skin, hair, and eyes.

Exposure to ultraviolet light darkens melanin already present and stimulates melanin production, resulting in tanning of the skin.

During pregnancy, certain hormones, such as estrogen and melanocyte-stimulating hormone, cause an increase in melanin production in the mother, which in turn causes darkening of the nipples, areolae, and genitalia. The cheekbones, forehead, and chest also may darken, resulting in the "mask of pregnancy," and a dark line of pigmentation may appear on the midline of the abdomen. Diseases, such as Addison disease, that cause an increased secretion of adrenocorticotropic hormone and melanocyte-stimulating hormone also cause increased pigmentation.

Blood flowing through the skin imparts a reddish hue. **Erythema** (er-ĭ-thē′mă) is increased redness of the skin resulting from increased blood flow through the skin. An inflammatory response (see chapter 4) stimulated by infections, sunburn, allergic reactions, insect bites, or other causes can produce erythema, as can exposure to the cold, blushing, or flushing when angry or hot. A decrease in blood flow, such as occurs in shock, can make the skin appear pale, and a decrease in the blood oxygen content produces **cyanosis** (sī-ă-nō′sis), a bluish skin color.

Carotene (kar′ō-tēn) is a yellow pigment found in plants, such as carrots and corn. Humans normally ingest carotene and use it as a source of vitamin A. Carotene is lipid-soluble, and, when large amounts of carotene are consumed, the excess accumulates in the stratum corneum and in the adipose cells of the dermis and hypodermis, causing the skin to develop a yellowish tint, which slowly disappears once carotene intake is reduced.

The location of pigments and other substances in the skin affects the color produced. If a dark pigment is located in the dermis or hypodermis, light reflected off the dark pigment can be scattered by collagen fibers of the dermis to produce a blue color. The same effect produces the blue color of the sky as light is reflected from dust particles in the air. The deeper within the dermis or hypodermis any dark pigment is located, the bluer the pigment appears because of the light-scattering effect of the overlying tissue. This effect causes the blue color of tattoos, bruises, and some superficial blood vessels.

5 ***Which cells of the epidermis produce melanin? What happens to the melanin once it is produced?***

6 ***How do genetic factors, exposure to light, and hormones determine the amount of melanin in the skin?***

7 ***How do melanin, carotene, and blood affect skin color?***

PREDICT 2

Explain the differences in skin color between (a) the palms of the hands and the lips, (b) the palms of the hands of a person who does heavy manual labor and one who does not, (c) the anterior and posterior surfaces of the forearm, and (d) the genitals and the soles of the feet.

Dermis

The dermis is connective tissue with fibroblasts, a few adipose cells, and macrophages. Collagen is the main connective tissue fiber, but elastin and reticular fibers are also present. Adipose cells and blood vessels are scarce in the dermis, compared with the hypodermis. Nerve endings, hair follicles, smooth muscles, glands, and lymphatic vessels are also in the dermis (see figure 5.1). The nerve endings are varied in structure and function: free nerve endings for pain, itch, tickle, and temperature sensations; hair follicle receptors for light touch; Pacinian corpuscles for deep pressure; Meissner corpuscles for the ability to detect simultaneous stimulation at two points on the skin; and Ruffini end organs for continuous touch or pressure (see figure 14.1).

The dermis is divided into two layers (see figures 5.1 and 5.2): the deeper **reticular** (re-tik′ū-lăr) **layer** and the superficial **papillary** (pap′i-lār-ē) **layer.** The papillary layer derives its name from projections called **dermal papillae** (pă-pil′ē) that extend toward the epidermis (see figure 5.2). The papillary layer is loose connective tissue with thin fibers that are somewhat loosely arranged. The papillary layer also contains blood vessels that supply the overlying epidermis with nutrients, remove waste products, and aid in regulating body temperature.

The dermal papillae underlying thick skin are in parallel, curving ridges that shape the overlying epidermis into fingerprints and footprints. The ridges increase friction and improve the grip of the hands and feet.

Fingerprints and Criminal Investigations

Fingerprints were first used in criminal investigation in 1880 by Henry Faulds, a Scottish medical missionary. Faulds used a greasy fingerprint left on a bottle to identify a thief who had been drinking purified alcohol from a dispensary. Everyone has unique fingerprints and footprints, even identical twins.

The reticular layer, which is composed of dense irregular connective tissue, is the main layer of the dermis. It is continuous with the hypodermis and forms a mat of irregularly arranged fibers that are resistant to stretching in many directions. The elastin and collagen fibers are oriented more in some directions than in others and produce **cleavage,** or **tension, lines** in the skin (figure 5.5). Knowledge of cleavage line directions is important because an incision made parallel to the cleavage lines is less likely to gap than is an incision made across them. The closer together the edges of a wound, the less likely is the development of infections and the formation of considerable scar tissue.

If the skin is overstretched, the dermis may rupture and leave lines that are visible through the epidermis. These lines of scar tissue, called **striae** (strī′ē), or **stretch marks,** can develop on the abdomen and breasts of a woman during pregnancy.

8 ***Name and compare the two layers of the dermis. Which layer is responsible for most of the structural strength of the skin?***

9 ***What are cleavage lines and striae?***

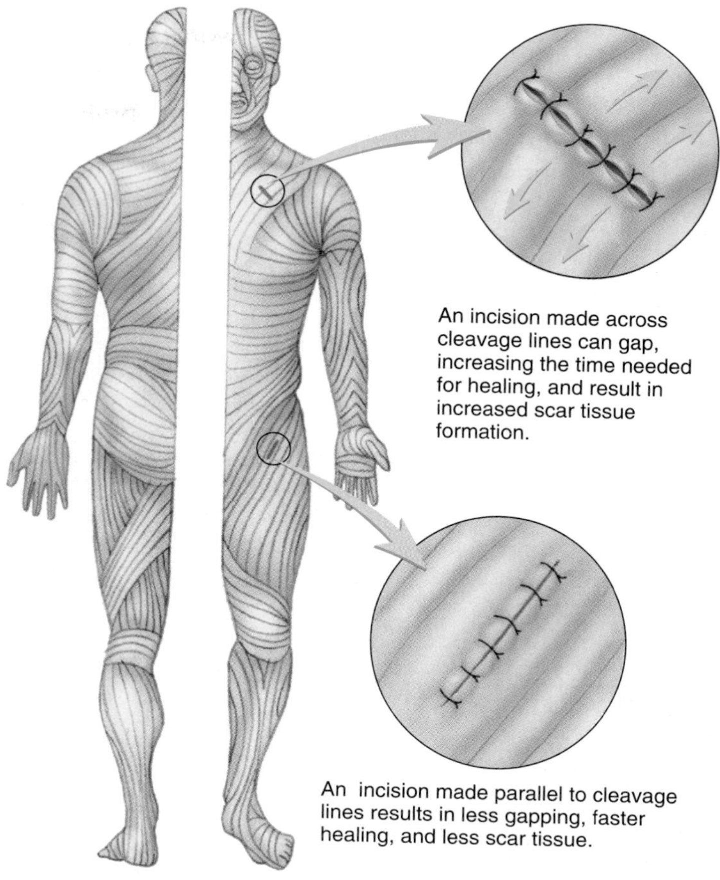

FIGURE 5.5 Cleavage Lines
The orientation of collagen fibers produces cleavage, or tension, lines in the skin.

HYPODERMIS

Just as a house rests on a foundation, the skin rests on the **hypodermis** (hī-pō-der′mis), which attaches it to underlying bone and muscle and supplies it with blood vessels and nerves (see figure 5.1). The hypodermis consists of loose connective tissue with collagen and elastin fibers. The main types of cells within the hypodermis are fibroblasts, adipose cells, and macrophages. The hypodermis, which is not part of the skin, is sometimes called **subcutaneous** (sŭb-koo-tā′nē-ŭs) **tissue,** or **superficial fascia** (fash′ē-ă).

Approximately half the body's stored fat is in the hypodermis, where it functions as a source of energy, insulation, and padding. The hypodermis can be used to estimate total body fat. The skin is pinched at selected locations, and the thickness of the fold of skin and underlying hypodermis is measured. The thicker the fold, the greater the amount of total body fat.

The amount of fat in the hypodermis varies with age, sex, and diet. Most babies have a chubby appearance because they have proportionately more fat than adults. Women have proportionately more fat than men, especially over the thighs, buttocks, and breasts, which accounts for some of the differences in body shape between women and men.

The amount of fat in the hypodermis is also responsible for some of the differences in body shape between individuals of the same sex.

10 ***Name the types of tissue forming the hypodermis.***

11 ***How is the hypodermis related to the skin?***

12 ***List the functions of the fat within the hypodermis.***

Injections

There are three types of injections. An **intradermal injection,** such as for the tuberculin skin test, is an injection into the dermis. It is administered by drawing the skin taut and inserting a small needle at a shallow angle into the skin. A **subcutaneous injection** is an injection into the fatty tissue of the hypodermis, such as for an insulin injection. This injection is achieved by pinching the skin to form a "tent" into which a short needle is inserted. An **intramuscular injection** is an injection into a muscle deep to the hypodermis. It is accomplished by inserting a long needle at a 90-degree angle to the skin. Intramuscular injections are used for injecting most vaccines and certain antibiotics.

ACCESSORY SKIN STRUCTURES

Accessory skin structures include hair and hair follicles, smooth muscles called the arrector pili, sweat and sebaceous glands, and nails.

Hair

The presence of **hair** is one of the characteristics common to all mammals; if the hair is dense and covers most of the body surface, it is called fur. In humans, hair is found everywhere on the skin except the palms, the soles, the lips, the nipples, parts of the external genitalia, and the distal segments of the fingers and toes.

By the fifth or sixth month of fetal development, delicate, unpigmented hair called **lanugo** (lă-noo′gō) has developed and covered the fetus. Near the time of birth, **terminal hairs,** which are long, coarse, and pigmented, replace the lanugo of the scalp, eyelids, and eyebrows. **Vellus** (vel′ŭs) **hairs,** which are short, fine, and usually unpigmented, replace the lanugo on the rest of the body. At puberty, terminal hair, especially in the pubic and axillary regions, replaces much of the vellus hair. The hair of the chest, legs, and arms is approximately 90% terminal hair in males, compared with approximately 35% in females. In males, terminal hairs replace the vellus hairs of the face to form the beard. The beard, pubic, and axillary hair are signs of sexual maturity. In addition, pubic and axillary hair may function as wicks for dispersing odors produced by secretions from specialized glands in the pubic and axillary regions. It also has been suggested that pubic hair protects against abrasion during intercourse and axillary hair reduces friction when the arms move.

Hair Structure

A hair is divided into the **shaft** and the **root** (figure 5.6*a*). The shaft protrudes above the surface of the skin, and the root is located below the surface. The base of the root is expanded to form the **hair bulb** (figure 5.6*b*). Most of the root and the shaft of the hair are composed of columns of dead keratinized epithelial cells arranged in three concentric layers: the medulla, the cortex, and the cuticle (figure 5.6*c*). The **medulla** (me-dool′ă) is the central axis of the hair; it consists of two or three layers of cells containing soft keratin. The **cortex** forms the bulk of the hair; it consists of cells containing hard keratin. The **cuticle** (kū′ti-kl) is a single layer of cells that forms the hair surface. The cuticle cells contain hard keratin, and the edges of the cuticle cells overlap like shingles on a roof.

Hard keratin contains more sulfur than does soft keratin. When hair burns, the sulfur combines with hydrogen to form hydrogen sulfide, which produces the unpleasant odor of rotten eggs. In some animals, such as sheep, the cuticle edges of the hair are raised; during textile manufacture, they catch each other and hold together to form threads.

The **hair follicle** consists of a **dermal root sheath** and an **epithelial root sheath.** The dermal root sheath is the portion of the dermis that surrounds the epithelial root sheath. The epithelial root sheath is divided into an external and an internal part (see figure 5.6*b*). At the opening of the follicle, the external epithelial root sheath has all the strata found in thin skin. Deeper in the hair follicle, the number of cells decreases until at the hair bulb only the stratum basale is present. This has important consequences for the repair of the skin. If the epidermis and the superficial part of the dermis are damaged, the undamaged part of the hair follicle that lies deep in the dermis can be a source of new epithelium. The internal epithelial root sheath has raised edges that mesh closely with the raised edges of the hair cuticle and hold the hair in place. When a hair is pulled out, the internal epithelial root sheath usually comes out as well and is plainly visible as whitish tissue around the root of the hair.

The hair bulb is an expanded knob at the base of the hair root (see figure 5.6*a* and *b*). Inside the hair bulb is a mass of undifferentiated epithelial cells, the **matrix,** which produces the hair and the internal epithelial root sheath. The dermis of the skin projects into the hair bulb as a **hair papilla;** it contains blood vessels that provide nourishment to the cells of the matrix.

Hair Growth

Hair is produced in cycles that involve a **growth stage** and a **resting stage.** During the growth stage, hair is formed by matrix cells that differentiate, become keratinized, and die. The hair grows longer as cells are added at the base of the hair root. Eventually, hair growth stops; the hair follicle shortens and holds the hair in place. A resting period follows, after which a new cycle begins and a new hair replaces the old hair, which falls out of the hair follicle. Thus, loss of hair normally means that the hair is being replaced. The length of each stage depends on the hair—eyelashes grow for approximately 30 days and rest for 105 days, whereas scalp hairs grow for 3 years and rest for 1–2 years. At any given time, an estimated 90% of the scalp hairs are in the growing stage, and loss of approximately 100 scalp hairs per day is normal.

The most common kind of permanent hair loss is "pattern baldness." Hair follicles are lost, and the remaining hair follicles revert to producing vellus hair, which is very short, transparent,

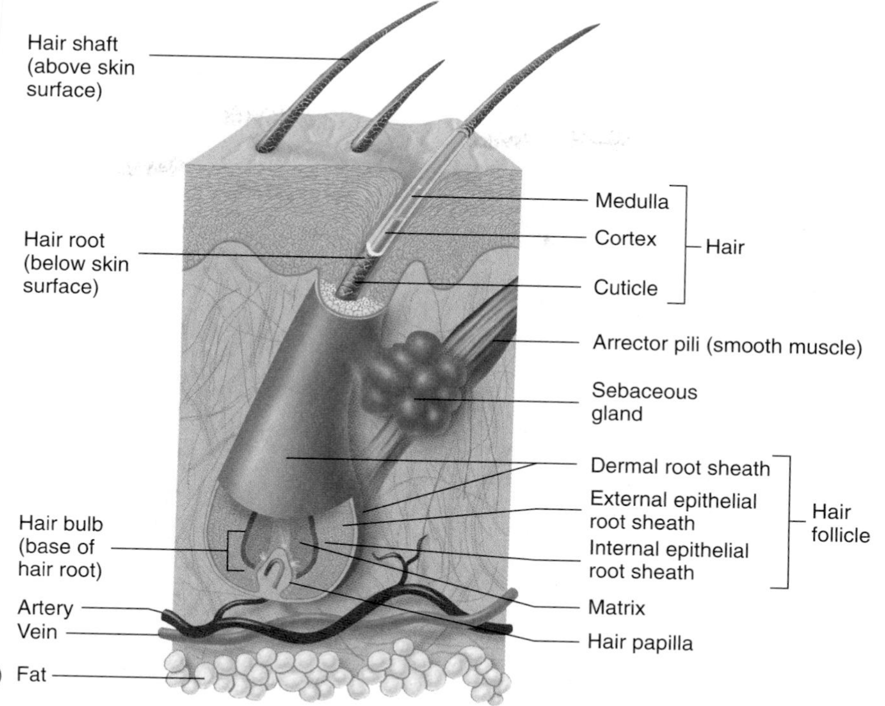

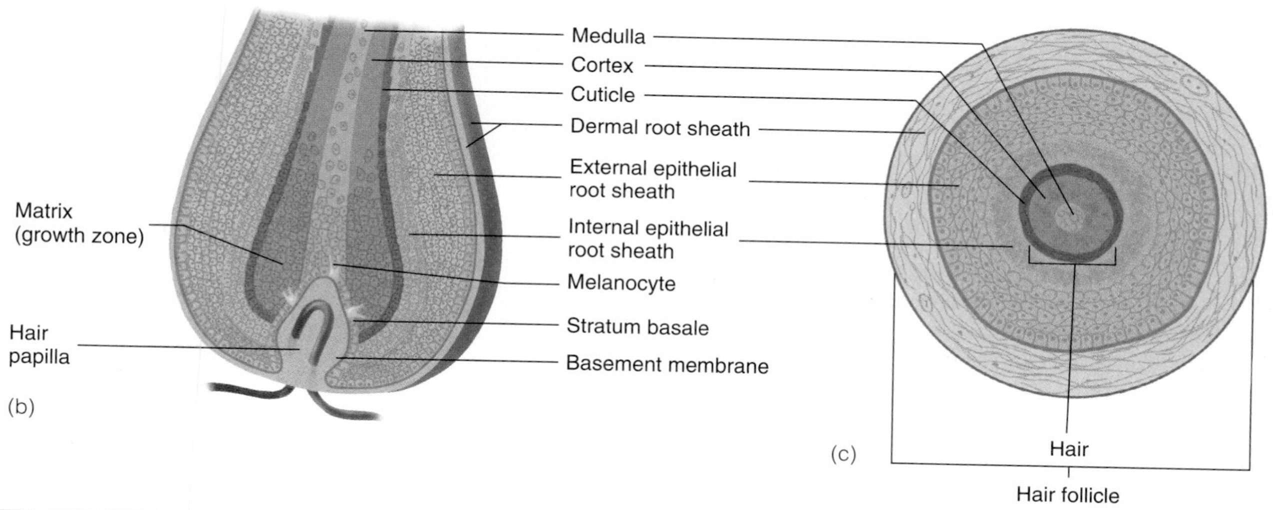

FIGURE 5.6 Hair Follicle

(*a*) The hair follicle contains the hair and consists of a dermal and an epithelial root sheath. (*b*) Enlargement of the hair follicle wall and hair bulb. (*c*) Cross section of a hair within a hair follicle.

and for practical purposes invisible. Although more common and more pronounced in certain men, baldness can also occur in women. Genetic factors and the hormone testosterone are involved in causing pattern baldness.

The average rate of hair growth is approximately 0.3 mm per day, although hairs grow at different rates even in the same approximate location. Cutting, shaving, or plucking hair does not alter the growth rate or the character of the hair, but hair can feel coarse and bristly shortly after shaving because the short hairs are less flexible. Maximum hair length is determined by the rate of hair growth and the length of the growing phase. For example, scalp hair can become very long, but eyelashes are short.

Hair Color

Melanin is produced by melanocytes within the hair bulb matrix and is passed to keratinocytes in the hair cortex and medulla. As with the skin, varying amounts and types of melanin cause different shades of hair color. Blonde hair has little black-brown melanin, whereas jet black hair has the most. Intermediate

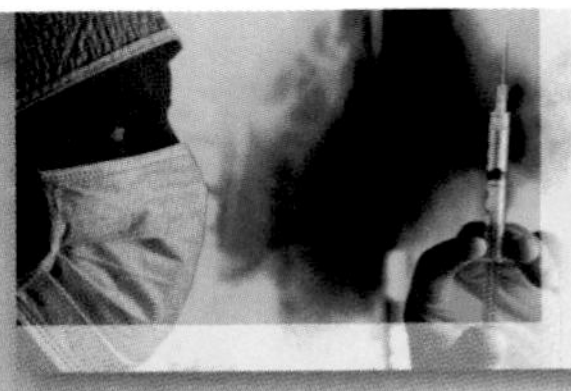

CLINICAL FOCUS

Burns

A **burn** is injury to a tissue caused by heat, cold, friction, chemicals, electricity, or radiation. Burns are classified according to the extent of surface area involved and the depth of the burn. For an adult, the surface area that is burned can be conveniently estimated by "the rule of nines," in which the body is divided into areas that are approximately 9%, or multiples of 9%, of the body surface area (BSA) (figure B). For younger patients, surface area relationships are different. For example, in an infant, the head and neck are 21% of BSA, whereas in an adult they are 9%. For burn victims younger than age 15, a table specifically developed for them should be consulted.

On the basis of depth, burns are classified as either partial-thickness or full-thickness burns (figure C). **Partial-thickness burns** are subdivided into first- and second-degree burns. **First-degree burns** involve only the epidermis and are red and painful, and slight edema (swelling) may occur. They can be caused by sunburn or brief exposure to hot or cold objects, and they heal in a week or so without scarring.

Second-degree burns damage the epidermis and the dermis. Minimal dermal damage causes redness, pain, edema, and blisters. Healing takes approximately 2 weeks, and no scarring results. If the burn goes deep into the dermis, however, the wound appears red, tan, or white; may take several months to heal; and might scar. In all second-degree burns, the epidermis regenerates from epithelial tissue in hair follicles and sweat glands, as well as from the edges of the wound.

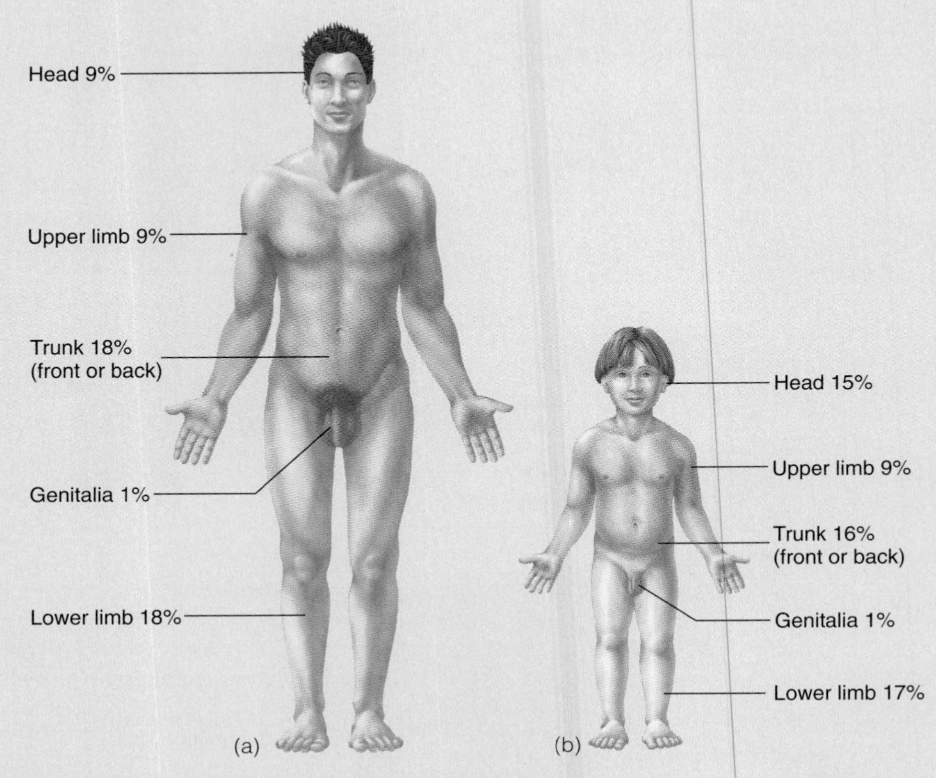

FIGURE B The Rule of Nines

(*a*) In an adult, surface areas can be estimated using the rule of nines: Each major area of the body is 9%, or a multiple of 9%, of the total body surface area. (*b*) In infants and children, the head represents a larger proportion of surface area. The rule of nines is not as accurate for children, as can be seen in this 5-year-old child.

amounts of melanin account for different shades of brown. Red hair is caused by varying amounts of a red type of melanin. Hair sometimes contains both black-brown and red melanin. With age, the amount of melanin in hair can decrease, causing hair color to fade or become white (i.e., no melanin). Gray hair is usually a mixture of unfaded, faded, and white hairs. Hair color is controlled by several genes, and dark hair color is not necessarily dominant over light.

PREDICT 3

Marie Antoinette's hair supposedly turned white overnight after hearing she would be sent to the guillotine. Explain why you believe or disbelieve this story.

Muscles

Associated with each hair follicle are smooth muscle cells, the **arrector pili** (ă-rek′tōr pī′lī), which extend from the dermal root sheath of the hair follicle to the papillary layer of the dermis (see figure 5.6*a*). Normally, the hair follicle and the hair inside it are at an oblique angle to the surface of the skin. When the arrector pili muscles contract, however, they pull the follicle into a position more perpendicular to the surface of the skin, causing the hair to "stand on end." Movement of the hair follicles produces raised areas called "gooseflesh," or "goose bumps."

Contraction of the arrector pili muscles occurs in response to cold or to frightening situations, and in animals with fur the

Full-thickness burns are also called **third-degree burns.** The epidermis and dermis are completely destroyed, and deeper tissue may be involved. Third-degree burns are often surrounded by first- and second-degree burns. Although the areas that have first- and second-degree burns are painful, the region of third-degree burn is usually painless because of destruction of sensory receptors. Third-degree burns appear white, tan, brown, black, or deep cherry red in color. Skin can regenerate in a third-degree burn only from the edges, and skin grafts are often necessary.

The depth of burns and the percent of BSA affected by the burns can be combined with other criteria to classify the seriousness of burns. A **major burn** is a third-degree burn over 10% or more of the BSA; a second-degree burn over 25% or more of the BSA; or a second- or third-degree burn of the hands, feet, face, genitals, or anal region. Facial burns are often associated with damage to the respiratory tract, and burns of joints often heal with scar tissue formation that limits movement. A **moderate burn** is a third-degree burn of 2%–10% of the BSA or a second-degree burn of 15%–25% of the BSA. A **minor burn** is a third-degree burn of less than 2% or a second-degree burn of less than 15% of the BSA.

Deep partial-thickness and full-thickness burns take a long time to heal and form scar tissue with disfiguring and debilitating wound contracture. Skin grafts are performed to prevent these complications and to speed healing. In a split skin graft, the epidermis and part of the dermis are removed from another part of the body and are placed over the burn. Interstitial fluid from the burned area nourishes the graft until its dermis becomes vascularized. At the graft donation site, part of the dermis is still present. The deep parts of hair follicles and sweat gland ducts are in this remaining dermis. These hair follicles and sweat gland ducts are a source of epithelial cells, which form a new epidermis that covers the dermis. This is the same process of epidermis formation that occurs in superficial second-degree burns.

When it is not possible or practical to move skin from one part of the body to a burn site, artificial skin or grafts from human cadavers or pigs are used. These techniques are often unsatisfactory because the body's immune system recognizes the graft as a foreign substance and rejects it. A solution to this problem is laboratory-grown skin. A piece of healthy skin from the burn victim is removed and placed into a flask with nutrients and hormones that stimulate rapid growth. The skin that is produced consists only of epidermis and does not contain glands or hair.

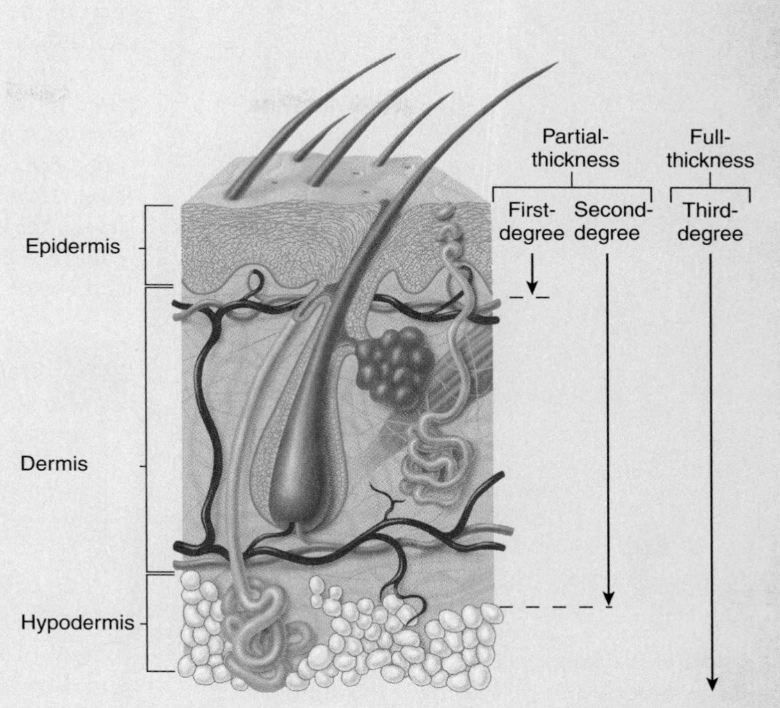

FIGURE C Burns

Parts of the skin damaged by burns of different degrees. Partial thickness burns are subdivided into first-degree burns (damage to only the epidermis) and second-degree burns (damage to the epidermis and part of the dermis). Full-thickness, or third-degree, burns destroy the epidermis, dermis, and sometimes deeper tissues.

response increases the thickness of the fur. When the response results from cold temperatures, it is beneficial because the fur traps more air and thus becomes a better insulator. In a frightening situation, the animal appears larger and more ferocious, which might deter an attacker. It is unlikely that humans, with their sparse amount of hair, derive any important benefit from either response and probably retain this trait as an evolutionary holdover.

13 ***When and where are lanugo, vellus, and terminal hairs found in the skin?***

14 ***Define the root, shaft, and hair bulb of a hair. Describe the three parts of a hair seen in cross section.***

15 ***Describe the parts of a hair follicle. How is the epithelial root sheath important in the repair of the skin?***

16 ***In what part of a hair does growth take place? What are the stages of hair growth?***

17 ***What determines the different shades of hair color?***

18 ***Explain the location and action of arrector pili muscles.***

Glands

The major glands of the skin are the sebaceous glands and the sweat glands (figure 5.7).

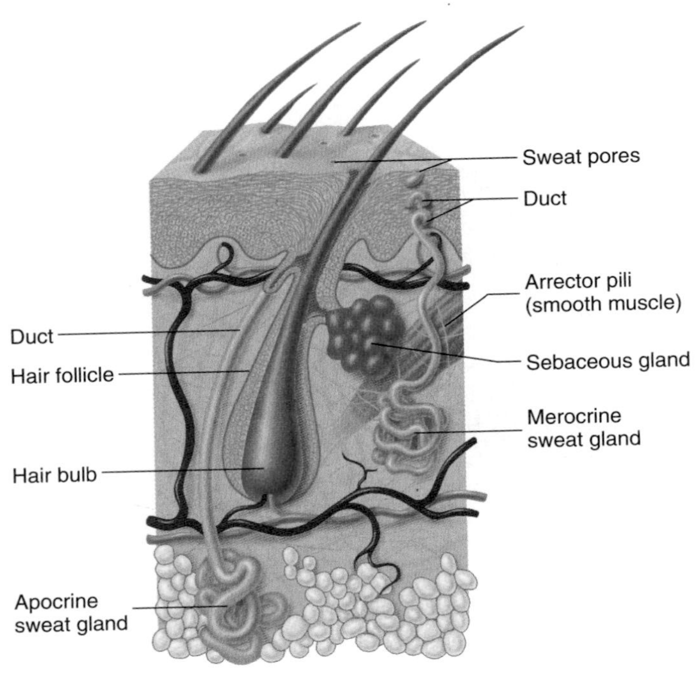

FIGURE 5.7 Glands of the Skin

Sebaceous and apocrine sweat glands empty into the hair follicle. Merocrine sweat glands empty onto the surface of the skin.

Sebaceous Glands

Sebaceous (sē-bā'shŭs) **glands,** located in the dermis, are simple or compound alveolar glands that produce **sebum** (sē'bŭm), an oily, white substance rich in lipids. Because sebum is released by the lysis and death of secretory cells, sebaceous glands are classified as holocrine glands (see chapter 4). Most sebaceous glands are connected by a duct to the upper part of the hair follicles, from which the sebum oils the hair and the skin surface. This prevents drying and protects against some bacteria. A few sebaceous glands located in the lips, in the eyelids (meibomian glands), and in the genitalia are not associated with hairs but open directly onto the skin surface.

Sweat Glands

There are two types of **sweat,** or **sudoriferous** (soo-dō-rif'er-ŭs), **glands,** which are called merocrine and apocrine glands. At one time, it was believed that secretions were released in a merocrine fashion from merocrine glands and in an apocrine fashion from apocrine glands (see chapter 4). Accordingly, they were called merocrine and apocrine sweat glands. It is now known that apocrine sweat glands also release some of their secretions in a merocrine fashion, and possibly some in a holocrine fashion. Traditionally, they are still referred to as apocrine sweat glands.

Merocrine (mer'ō-krin, mer'ō-krīn, mer'ō-krēn), or **eccrine** (ek'rin), **sweat glands** are the most common type of sweat gland. They are simple, coiled, tubular glands that open directly onto the surface of the skin through sweat pores (see figure 5.7). Merocrine sweat glands can be divided into two parts: the deep coiled portion, which is located mostly in the dermis, and the duct, which passes to the surface of the skin. The coiled part of the gland produces an isotonic fluid that is mostly water but also contains some salts (mainly sodium chloride) and small amounts of ammonia, urea, uric acid, and lactic acid. As this fluid moves through the duct, sodium chloride moves by active transport from the duct back into the body, thereby conserving salts. The resulting hyposmotic fluid that leaves the duct is called **sweat.** When the body temperature starts to rise above normal levels, the sweat glands produce sweat, which evaporates and cools the body. Sweat also can be released in the palms, soles, and axillae as a result of emotional stress.

> **Detecting Lies**
>
> Emotional sweating is used in lie detector (polygraph) tests because sweat gland activity can increase when a person tells a lie. The sweat produced, even in small amounts, can be detected because the salt solution conducts electricity and lowers the electric resistance of the skin.

Merocrine sweat glands are most numerous in the palms of the hands and the soles of the feet but are absent from the margin of the lips, the labia minora, and the tips of the penis and clitoris. Only a few mammals, such as humans and horses, have merocrine sweat glands in hairy skin. Dogs, on the other hand, keep cool by water lost through panting instead of sweating.

Apocrine (ap'ō-krin) **sweat glands** are simple, coiled, tubular glands that usually open into hair follicles superficial to the opening of the sebaceous glands (see figure 5.7). In other mammals, these glands are widely distributed throughout the skin; they help regulate temperature. In humans, apocrine sweat glands are found in the axillae and genitalia (scrotum and labia majora) and around the anus, and they do not help regulate temperature. In humans, apocrine sweat glands become active at puberty as a result of the influence of sex hormones. Their secretions contain organic substances, such as 3-methyl-2-hexenoic acid, that are essentially odorless when first released but are quickly metabolized by bacteria to cause what commonly is known as body odor. Many mammals use scent as a means of communication, and it has been suggested that the activity of apocrine sweat glands may be a sign of sexual maturity.

Other Glands

Other skin glands include the ceruminous glands and the mammary glands. The **ceruminous** (sĕ-roo'mi-nŭs) **glands** are modified merocrine sweat glands located in the ear canal (external auditory meatus). **Cerumen,** or earwax, is the combined secretions of ceruminous glands and sebaceous glands. Cerumen and hairs in the ear canal protect the eardrum by preventing the entry of dirt and small insects. An accumulation of cerumen, however, can block the ear canal and make hearing more difficult.

The **mammary glands** are modified apocrine sweat glands located in the breasts. They produce milk. The structure and regulation of mammary glands are discussed in chapters 28 and 29.

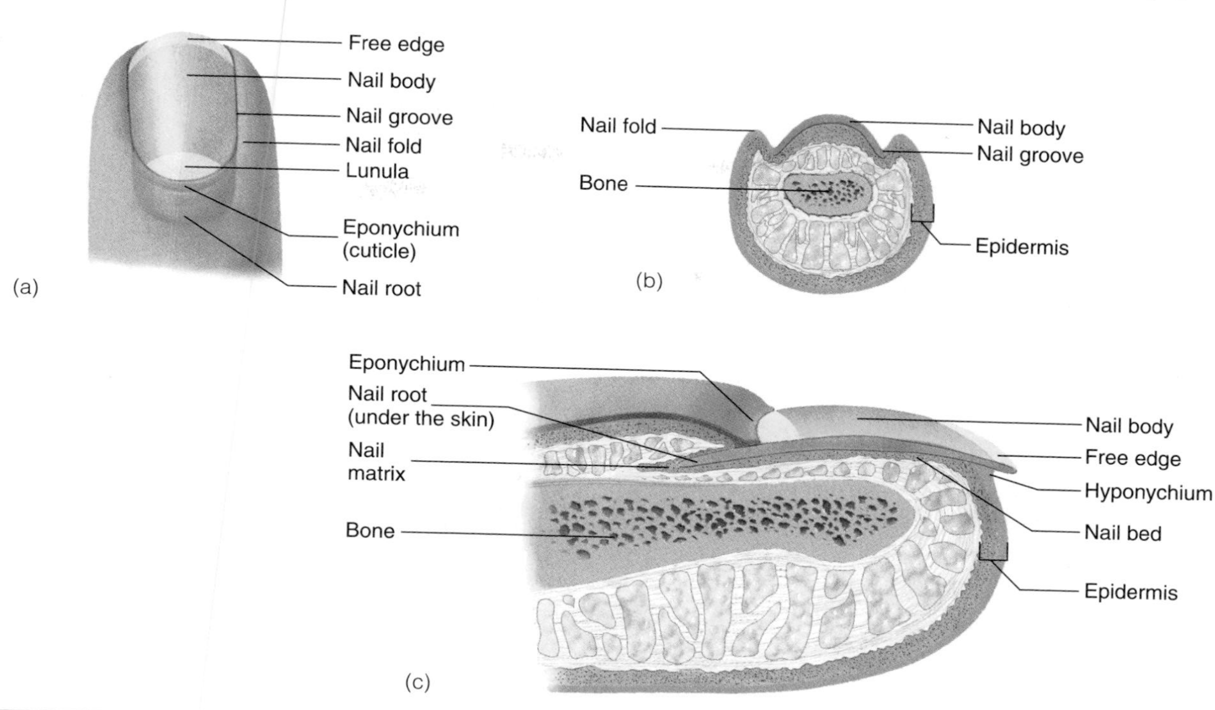

FIGURE 5.8 Nail
(*a*) Dorsal view. (*b*) Cross section. (*c*) Longitudinal section.

19 ***What secretion is produced by the sebaceous glands? What is the function of the secretion?***

20 ***Which glands of the skin are responsible for cooling the body? Which glands are involved with the production of body odor?***

21 ***Where are ceruminous and mammary glands located, and what secretions do they produce?***

Nails

The distal ends of primate digits have nails, whereas most other mammals have claws or hooves. Nails protect the ends of the digits, aid in manipulation and grasping of small objects, and are used for scratching.

A **nail** consists of the proximal **nail root** and the distal **nail body** (figure 5.8*a*). The nail root is covered by skin, and the nail body is the visible portion of the nail. The lateral and proximal edges of the nail are covered by skin called the **nail fold,** and the edges are held in place by the **nail groove** (figure 5.8*b*). The stratum corneum of the nail fold grows onto the nail body as the **eponychium** (ep-ō-nik′ē-ŭm), or cuticle. Beneath the free edge of the nail body is the **hyponychium** (hī′pō-nik′ē-ŭm), a thickened region of the stratum corneum (figure 5.8*c*).

The nail root extends distally from the **nail matrix.** The nail also attaches to the underlying **nail bed,** which is located between the nail matrix and the hyponychium. The nail matrix and bed are epithelial tissue, with a stratum basale that gives rise to the cells that form the nail. The nail matrix is thicker than the nail bed and produces nearly all of the nail. The nail bed is visible through the clear nail and appears pink because of blood vessels in the dermis.

A small part of the nail matrix, the **lunula** (loo′noo-lă), is seen through the nail body as a whitish, crescent-shaped area at the base of the nail. The lunula, seen best on the thumb, appears white because the blood vessels cannot be seen through the thicker nail matrix.

The nail is stratum corneum. It contains a hard keratin, which makes the nail hard. As the nail is formed in the nail matrix and bed, it slides over the nail bed toward the distal end of the digit. Nails grow at an average rate of 0.5–1.2 mm per day, and fingernails grow more rapidly than toenails. Unlike hair, they grow continuously throughout life and do not have a resting phase.

22 ***Name the parts of a nail. Which part produces most of the nail? What is the lunula?***

23 ***What makes a nail hard? Do nails have growth stages?***

SUMMARY OF INTEGUMENTARY SYSTEM FUNCTIONS

Protection

The integumentary system is the body's fortress, defending it from harm. It performs many protective functions.

1. The skin protects underlying structures from mechanical damage. The dermis provides structural strength, preventing tearing of the skin. The stratified epithelium of the epidermis protects against abrasion. As the outer cells of the stratum

corneum are desquamated, they are replaced by cells from the stratum basale. Calluses develop in areas subject to heavy friction or pressure.

2. The skin prevents the entry of microorganisms and other foreign substances into the body. Secretions from skin glands produce an environment unsuitable for some microorganisms. The skin contains components of the immune system that act against microorganisms (see chapter 22).
3. Melanin absorbs ultraviolet light and protects underlying structures from its damaging effects.
4. Hair provides protection in several ways. The hair on the head acts as a heat insulator and protects against ultraviolet light and abrasion. The eyebrows keep sweat out of the eyes, eyelashes protect the eyes from foreign objects, and hair in the nose and ears prevents the entry of dust and other materials. Axillary and pubic hair are a sign of sexual maturity and protect against abrasion.
5. Nails protect the ends of the digits from damage and can be used in defense.
6. The intact skin plays an important role in reducing water loss because its lipids act as a barrier to the diffusion of water.

Administering Medications Through the Skin

Some lipid-soluble substances readily pass through the epidermis. Lipid-soluble medications can be administered by applying them to the skin, after which the medication slowly diffuses through the skin into the blood. For example, nicotine patches are used to help reduce withdrawal symptoms in those attempting to quit smoking.

CASE STUDY

Frostbite

Billy was hiking in the mountains during the fall. Unexpectedly, a cold front moved in and the temperature dropped to well below freezing. Billy was unprepared for the temperature change, and he did not have a hat or ear muffs. As the temperature dropped, his ears and nose became pale in color. After a continued decrease in temperature and continued exposure to the cold, every 15–20 minutes his ears and nose turned red for 5–10 minutes, then became pale again. After several hours, Billy managed to hike back to the trail head. By then, he was very chilled and had no sensation in his ears or nose. As he looked in the rearview mirror of his car, he could see that the skin of his ears and nose had turned white. It took Billy 2 hours to drive to the nearest emergency room, where he was informed that the white-colored skin meant he had frostbite of his ears and nose. About 2 weeks later, the frostbitten skin peeled and, despite treatment with an antibiotic, Billy developed an infection of his right ear. Eventually, he recovered, but he lost part of his right ear.

PREDICT 4

Frostbite is the most common type of freezing injury. When skin temperature drops below 0°C (32°F), the skin freezes and ice crystal formation damages tissues.

a. Using figure 5.9, describe the mechanism responsible for Billy's ears and nose becoming pale. How is this beneficial when the ambient temperature is decreasing?
b. Explain what happened when Billy's ears and nose periodically turned red. How is this beneficial when the ambient temperature is very cold?
c. What is the significance of Billy's ears and nose turning and staying white?
d. Why is a person with frostbite likely to develop an infection of the affected part of the body?

Sensation

The body feels pain, heat, and cold because the integumentary system has sensory receptors in all its layers. For example, the epidermis and dermal papillae are well supplied with touch receptors. The dermis and deeper tissues contain pain, heat, cold, touch, and pressure receptors. Hair follicles (but not the hair) are well innervated, and the movement of hair can be detected by sensory receptors surrounding the base of hair follicles. Sensory receptors are discussed in more detail in chapter 14.

Temperature Regulation

Body temperature is affected by blood flow through the skin. When blood vessels (arterioles) in the dermis dilate, there is increased flow of warm blood from deeper structures to the skin and heat loss increases (figure 5.9, steps 1 and 2). Body temperature tends to increase as a result of exercise, fever, or an increase in environmental temperature. Homeostasis is maintained by the loss of excess heat. To counteract environmental heat gain or to get rid of excess heat produced by the body, sweat is produced. The sweat spreads over the surface of the skin; as it evaporates, heat is lost from the body.

When blood vessels in the dermis constrict, there is decreased flow of warm blood from deeper structures to the skin and heat loss decreases (figure 5.9, steps 3 and 4). If body temperature begins to drop below normal, heat can be conserved by a decrease in the diameter of dermal blood vessels.

Contraction of the arrector pili muscles causes hair to stand on end, but, with the sparse amount of hair covering the body, this does not significantly reduce heat loss in humans. Hair on the head, however, is an effective insulator. General temperature regulation is considered in chapter 25.

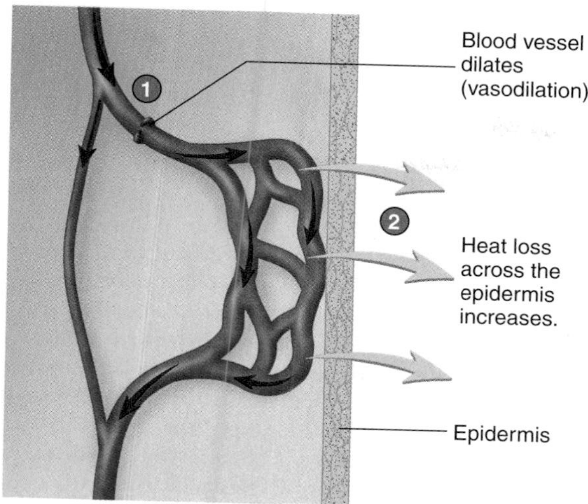

1. Blood vessel dilation results in increased blood flow toward the surface of the skin.
2. Increased blood flow beneath the epidermis results in increased heat loss (*gold arrows*).

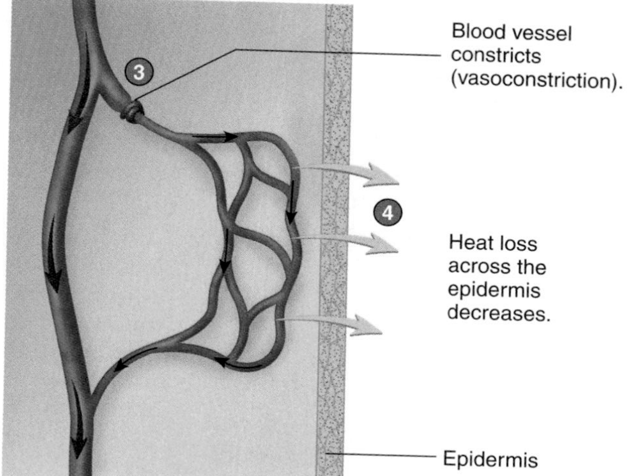

3. Blood vessel constriction results in decreased blood flow toward the surface of the skin.
4. Decreased blood flow beneath the epidermis results in decreased heat loss.

PROCESS FIGURE 5.9 Heat Exchange in the Skin

Vitamin D Production

Vitamin D functions as a hormone to stimulate the uptake of calcium and phosphate from the intestines, to promote their release from bones, and to reduce calcium loss from the kidneys, resulting in increased blood calcium and phosphate levels. Adequate levels of these minerals are necessary for normal bone metabolism (see chapter 6), and calcium is required for normal nerve and muscle function (see chapter 9).

Vitamin D synthesis begins in skin exposed to ultraviolet light, and humans can produce all the vitamin D they require by this process if enough ultraviolet light is available. Because humans live indoors and wear clothing, however, their exposure to ultraviolet light may not be adequate for the manufacture of sufficient vitamin D. This is especially likely for people living in cold climates because they remain indoors or are covered by warm clothing when outdoors. Fortunately, vitamin D can also be ingested and absorbed in the intestine. Natural sources of vitamin D are liver (especially fish liver), egg yolks, and dairy products (e.g., butter, cheese, and milk). In addition, the diet can be supplemented with vitamin D in fortified milk or vitamin pills.

Vitamin D synthesis begins when the precursor molecule 7-dehydrocholesterol (7-dē-hī′drō-kō-les′ter-ol) is exposed to ultraviolet light and is converted into cholecalciferol (kō′lē-kal-sif′er-ol). Cholecalciferol is released into the blood and modified by hydroxylation (hydroxide ions are added) in the liver and kidneys to form active vitamin D (calcitriol; kal-si-trī′ol).

Excretion

Excretion is the removal of waste products from the body. In addition to water and salts, sweat contains a small amount of waste products, such as urea, uric acid, and ammonia. Compared with the kidneys, however, the quantity of waste products eliminated in the sweat is insignificant, even when large amounts of sweat are lost.

24 ***In what ways does the skin provide protection?***

25 ***What kind of sensory receptors are found in the skin, and why are they important?***

26 ***How does the skin help regulate body temperature?***

27 ***Name the locations where cholecalciferol is produced and then modified into vitamin D. What are the functions of vitamin D?***

28 ***What substances are excreted in sweat? Is the skin an important site of excretion?***

EFFECTS OF AGING ON THE INTEGUMENTARY SYSTEM

As the body ages, the skin is more easily damaged because the epidermis thins and the amount of collagen in the dermis decreases. Skin infections are more likely, and repair of the skin occurs more slowly. A decrease in the number of elastic fibers in the dermis and loss of fat from the hypodermis cause the skin to sag and wrinkle.

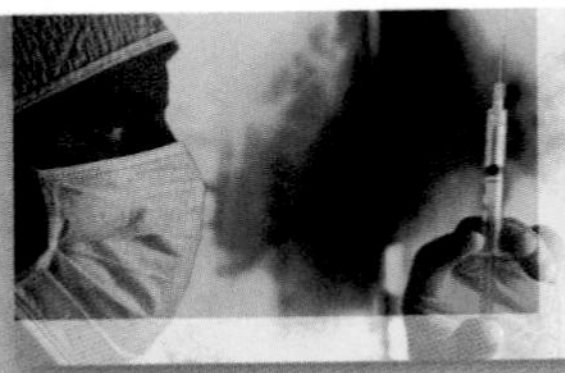

CLINICAL FOCUS

Clinical Disorders of the Integumentary System

THE INTEGUMENTARY SYSTEM AS A DIAGNOSTIC AID

The integumentary system can be used in diagnosis because it is easily observed and often reflects events occurring in other parts of the body. For example, **cyanosis** (sī-ă-nō′sis), a bluish color to the skin that results from decreased blood oxygen content, is an indication of impaired circulatory or respiratory function. When red blood cells wear out, they are broken down, and part of their contents is excreted by the liver as bile pigments into the intestine. **Jaundice** (jawn′dis), a yellowish skin color, occurs when excess bile pigments accumulate in the blood. If a disease, such as viral hepatitis, damages the liver, bile pigments are not excreted and accumulate in the blood.

Rashes and lesions in the skin can be symptomatic of problems elsewhere in the body. For example, scarlet fever results from a bacterial infection in the throat. The bacteria release a toxin into the blood that causes the pink-red rash for which this disease was named. In allergic reactions (see chapter 22), a release of histamine into the tissues produces swelling and reddening. The development of a rash (hives) in the skin can indicate an allergy to foods or drugs, such as penicillin.

The condition of the skin, hair, and nails is affected by nutritional status. In vitamin A deficiency, the skin produces excess keratin and assumes a characteristic sandpaper texture, whereas, in iron-deficiency anemia, the nails lose their normal contour and become flat or concave (spoon-shaped).

Hair concentrates many substances, which can be detected by laboratory analysis, and a comparison of a patient's hair with "normal" hair can be useful in certain diagnoses. For example, lead poisoning results in high levels of lead in the hair. The use of hair analysis as a screening test to determine the general health or nutritional status of an individual is unreliable, however.

BACTERIAL INFECTIONS

Staphylococcus aureus is commonly found in pimples, boils, and carbuncles and causes **impetigo** (im-pe-tī′gō), a disease of the skin that usually affects children. It is characterized by small blisters containing pus that easily rupture and form a thick, yellowish crust. *Streptococcus pyogenes* causes **erysipelas** (er-i-sip′ĕ-las), swollen red patches in the skin. Burns are often infected by *Pseudomonas aeruginosa,* which produces a characteristic blue-green pus caused by bacterial pigment.

Acne is a disorder of the hair follicles and sebaceous glands that affects almost everyone. Although the exact cause of acne is unknown, four factors are believed to be involved: hormones, sebum, abnormal keratinization within hair follicles, and the bacterium *Propionibacterium acnes.* Acne lesions apparently begin with a hyperproliferation of the hair follicle epidermis, and many cells are desquamated. These cells are abnormally sticky and adhere to one another to form a mass of cells mixed with sebum, which blocks the hair follicle. During puberty, hormones, especially testosterone, stimulate the sebaceous glands to increase sebum production. Because both the testes and the ovaries produce testosterone, the effect is seen in both males and females. An accumulation of sebum behind the blockage produces a whitehead, which may continue to develop into a blackhead or a pimple. A blackhead results if the opening of the hair follicle is pushed open by the accumulating cornified cells and sebum. Although it is generally agreed that dirt is not responsible for the black color of blackheads, its exact cause is disputed. Once the wall of the follicle ruptures, *P. acnes* and other microorganisms stimulate an inflammatory response, which results in the formation of a red pimple filled with pus. If tissue damage is extensive, scarring occurs.

The skin becomes drier with age as sebaceous gland activity decreases. A decrease in the activity of sweat glands and a decrease in the blood supply to the dermis result in a poor ability to regulate body temperature. Death from heat prostration can occur in elderly individuals who do not take proper precautions.

The number of functioning melanocytes generally decreases, but, in some localized areas, especially on the hands and the face, melanocytes increase in number to produce age spots. (Age spots are different from freckles, which are caused by an increase in melanin production, not an increase in melanocyte numbers.) White or gray hairs also occur because of a decrease in or lack of melanin production.

Skin that is exposed to sunlight appears to age more rapidly than nonexposed skin. This effect is observed on areas of the body, such as the face and hands, that receive sun exposure (figure 5.10). The effects of chronic sun exposure on the skin, however, are different from the effects of normal aging. In skin exposed to sunlight, normal elastic fibers are replaced by an interwoven mat of thick, elasticlike material, the number of collagen fibers decreases, and the ability of keratinocytes to divide is impaired.

29 ***Compared with young skin, why is aged skin more likely to be damaged, wrinkled, and dry?***

30 ***Why is heat potentially dangerous to the elderly?***

31 ***Explain age spots and white hair.***

32 ***What effect does exposure to sunlight have on skin?***

Treatment of Skin Wrinkles

Retin-A (tretinoin; tret′i-nō-in) is a vitamin A derivative that is being used to treat skin wrinkles. It appears to be effective in treating fine wrinkles on the face, such as those caused by long-term exposure to the sun, but is not effective in treating deep lines. One ironic side effect of Retin-A use is increased sensitivity to the sun's ultraviolet rays. Doctors prescribing this cream caution their patients to always use a sunblock when they are going to be outdoors.

VIRAL INFECTIONS

Some of the well-known viral infections of the skin include **chicken pox** (varicella-zoster), **measles, German measles** (rubella), and **cold sores** (herpes simplex). **Warts,** which are caused by a viral infection of the epidermis, are generally harmless and usually disappear without treatment.

FUNGAL INFECTIONS

Ringworm is a fungal infection that affects the keratinized portion of the skin, hair, and nails and produces patchy scaling and an inflammatory response. The lesions are often circular with a raised edge, and in ancient times they were thought to be caused by worms. Several species of fungus cause ringworm in humans and are usually described by their location on the body; in the scalp, the condition is ringworm; in the groin, it is jock itch; in the feet, it is athlete's foot.

DECUBITUS ULCERS

Decubitus (dē-kū'bi-tŭs) **ulcers,** also known as bedsores, or pressure sores, develop in patients who are immobile (e.g., bedridden or confined to a wheelchair). The weight of the body, especially in areas over bony projections, such as the hipbones and heels, compresses tissues and causes **ischemia** (is-kē'mē-ă), or reduced circulation. The consequence is destruction, or **necrosis** (nĕ-krō'sis), of the hypodermis and deeper tissues, which is followed by necrosis of the skin. Once skin necrosis occurs, microorganisms gain entry to produce an infected ulcer.

BULLAE

Bullae (bul'ē), or blisters, are fluid-filled areas in the skin that develop when tissues are damaged and the resultant inflammatory response produces edema. Infections or physical injuries can cause bullae or lesions in different layers of the skin.

PSORIASIS

Psoriasis (sō-rī'ă-sis) is characterized by a thicker than normal stratum corneum that sloughs to produce large, silvery scales. If the scales are scraped away, bleeding occurs from the blood vessels at the top of the dermal papillae. These changes result from increased cell division in the stratum basale, abnormal keratin production, and elongation of the dermal papillae toward the skin surface. Evidence suggests that the disease has a genetic component and that the immune system stimulates the increased cell divisions. Psoriasis is a chronic disease that can be controlled with drugs and phototherapy (ultraviolet light) but as yet has no cure.

ECZEMA AND DERMATITIS

Eczema (ek'zĕ-mă, eg'zĕ-mă, eg-zē'mă) and **dermatitis** (der-mă-tī'tis) are general terms used for inflammatory conditions of the skin. The cause of the inflammation can be an allergy, an infection, poor circulation, or exposure to physical factors, such as chemicals, heat, cold, or sunlight.

BIRTHMARKS

Birthmarks are congenital (present at birth) disorders of the capillaries in the dermis of the skin. Usually, they are of concern only for cosmetic reasons. A **strawberry birthmark** is a mass of soft, elevated tissue that appears bright red to deep purple. In 70% of patients, strawberry birthmarks disappear spontaneously by age 7. **Portwine stains** appear as flat, dull red or bluish patches that persist throughout life.

VITILIGO

Vitiligo (vit-i-lī'gō) is the development of patches of white skin because the melanocytes in the affected area are destroyed, apparently by an autoimmune response (see chapter 22).

MOLES

A **mole** is an elevation of the skin that is variable in size and is often pigmented and hairy. Histologically, a mole is an aggregation, or "nest," of melanocytes in the epidermis or dermis. They are a normal occurrence, and most people have 10–20 moles, which appear in childhood and enlarge until puberty.

(a)

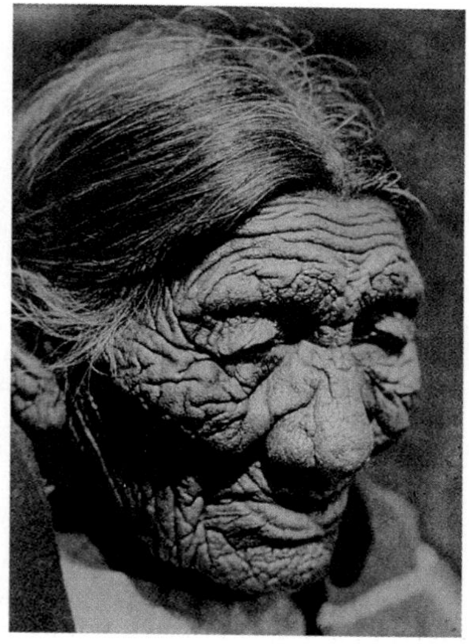

(b)

FIGURE 5.10 Effects of Sunlight on Skin
(*a*) A 91-year-old Japanese monk who has spent most of his life indoors. (*b*) A 62-year-old Native American woman who has spent most of her life outdoors.

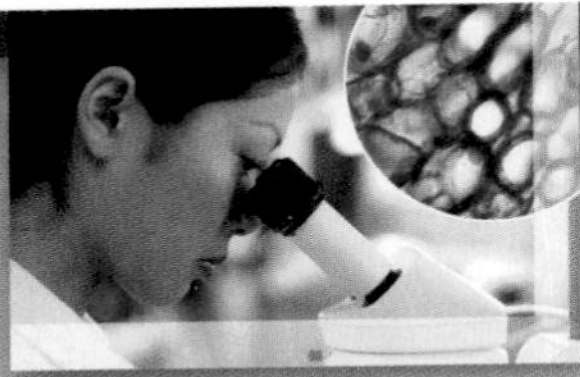

SYSTEMS PATHOLOGY

Burns

Sam is a 23-year-old man who had difficulty falling asleep at night. He often stayed up late, watching television or reading, until he fell asleep. Sam was also a chain smoker. One night, he took several sleeping pills. Unfortunately, he fell asleep before putting out his cigarette, which started a fire. As a result, Sam was severely burned and received full-thickness and partial-thickness burns (figure D*a*). He was rushed to the emergency room and was eventually transferred to a burn unit.

For the first day after his accident, his condition was critical because he went into shock. Administration of large volumes of intravenous fluid stabilized his condition. As part of his treatment, Sam was also given a high-protein, high-calorie diet.

A week later, dead tissue was removed from the most serious burns (figure D*b*), and a skin graft was performed. Despite the use of topical antimicrobial drugs and sterile bandages, some of the burns became infected. An additional complication was the development of a venous thrombosis in his leg.

Although the burns were painful and the treatment was prolonged, Sam made a full recovery. He no longer smokes.

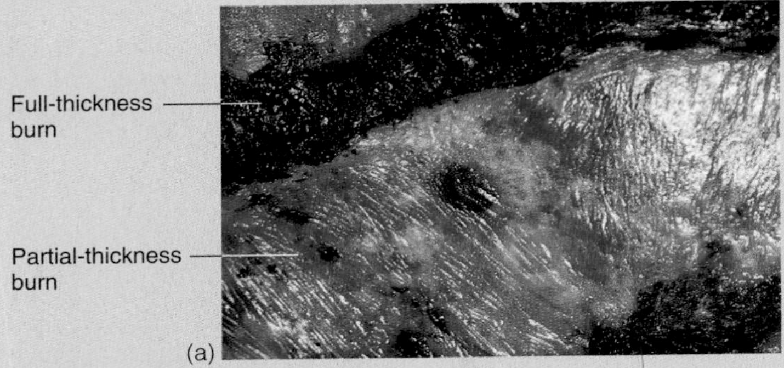

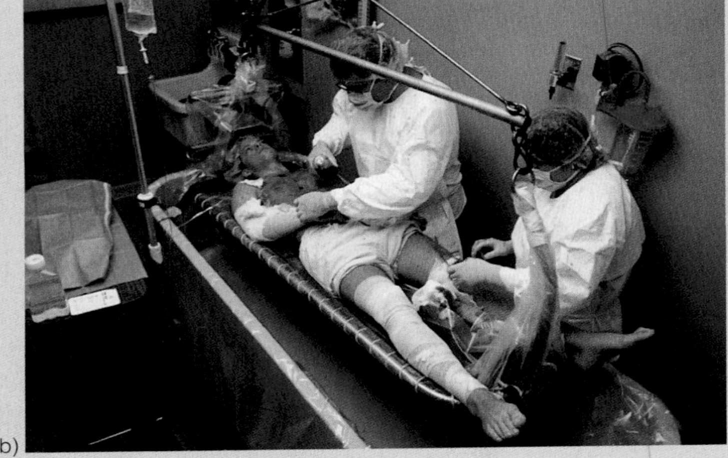

FIGURE D Burn Victim
(*a*) Partial and full-thickness burns. (*b*) Patient in a burn unit.

BACKGROUND INFORMATION

Major burns can cause systemic effects that can be life-threatening. One effect is on capillaries, which are the small blood vessels in which fluid, gases, nutrients, and waste products are normally exchanged between the blood and tissues. Within minutes of a major burn injury, capillaries become more permeable at the burn site and throughout the body. As a result, fluid and electrolytes (see chapter 2) are lost from the burn wound and into tissue spaces. The loss of fluid decreases blood volume, which decreases the heart's ability to pump blood. The resulting decrease in blood delivery to tissues can cause tissue damage, shock, and even death. Treatment consists of administering intravenous fluid at a faster rate than it leaks out of the capillaries. Although this can reverse the shock and prevent death, fluid continues to leak into tissue spaces, causing pronounced **edema,** a swelling of the tissues.

Typically, after 24 hours, capillary permeability returns to normal, and the amount of intravenous fluid administered can be greatly decreased. How burns result in capillary permeability changes is not well understood. It is clear, however, that, following a burn, immunologic and metabolic changes occur that affect not only capillaries but the rest of the body as well. For example, mediators of inflammation (see chapter 4), which are released in response to the tissue damage, contribute to changes in capillary permeability throughout the body.

Substances released from the burn may also play a role in causing cells to function abnormally. Burn injuries result in an almost immediate hypermetabolic state, which persists until wound closure. Also contributing to the increased metabolism is a resetting of the temperature control center in the brain to a higher temperature and an increase in the hormones released by the endocrine system. For example, epinephrine and norepinephrine from the adrenal glands increase cell metabolism. Compared with a normal body temperature of approximately 37°C (98.6°F), a body temperature of 38.5°C (101.3°F) is typical in burn patients, despite the higher loss of water by evaporation from the burn.

In severe burns, the increased metabolic rate can result in weight loss as great as 30%–40% of the patient's preburn weight. To help compensate, caloric intake may double or even triple. In addition, the need for protein, which is necessary for tissue repair, is greater.

The skin normally maintains homeostasis by preventing the entry of microorganisms. Because burns damage and even completely destroy the skin, microorganisms can cause infections. For this reason, burn patients are maintained in an aseptic environment, which attempts to prevent the entry of microorganisms into the wound. They are also given antimicrobial drugs, which kill microorganisms or suppress their growth. **Debridement** (dā-brēd-mon′), the removal of dead tissue from the burn, helps prevent infections by cleaning the wound and removing tissue in which infections could develop. Skin grafts, performed within a week of the injury, also prevent infections by closing the wound and preventing the entry of microorganisms.

Despite these efforts, however, infections still are the major cause of death of burn victims. Depression of the immune system during the first or second week after the injury contributes to the high infection rate. The thermally altered tissue is recognized as a foreign substance, stimulating the immune system. As a result, the immune system is overwhelmed as its cells become less effective and the production of the chemicals that normally provide resistance to infections decreases (see chapter 22). The greater the magnitude of the burn, the greater the depression of the immune system, and the greater the risk of infection.

Venous thrombosis, the development of a clot in a vein, is also a complication of burns. Blood normally forms a clot when exposed to damaged tissue, such as at a burn site, but clotting can also occur elsewhere, such as in veins. Clots can block blood flow, resulting in tissue destruction. The concentration of chemicals in the blood that cause clotting increases for two reasons: an increased release of clotting factors from the liver and loss of fluid from the burn.

PREDICT 5

When Sam was first admitted to the burn unit, the nurses carefully monitored his urine output. Why does that make sense in light of his injuries?

SYSTEM INTERACTIONS — Effect of Burns on Other Systems

SYSTEM	INTERACTIONS
Skeletal	Red bone marrow replaces red blood cells destroyed in the burned skin.
Muscular	Loss of muscle mass results from the hypermetabolic state caused by the burn.
Nervous	Pain is sensed in the partial-thickness burns. The temperature-regulatory center in the brain is set to a higher temperature, which contributes to increased body temperature. Abnormal K^+ concentrations disturb normal nervous system activity: elevated levels are caused by the release of K^+ from damaged tissues; low levels can be caused by rapid loss of K^+ in fluid from the burn.
Endocrine	Increased secretion of epinephrine and norepinephrine from the adrenal gland in response to the injury contributes to increased body temperature by increasing cell metabolism.
Cardiovascular	Increased capillary permeability causes decreased blood volume, resulting in decreased blood delivery to tissues, edema, and shock. The heart's pumping effectiveness is impaired by electrolyte imbalance and substances released from the burn. Increased blood clotting causes venous thrombosis. Preferential delivery of blood to the injury promotes healing.
Lymphatic and immune	Inflammation increases in response to tissue damage. Later, depression of the immune system can result in infection.
Respiratory	Airway obstruction is caused by edema. Increased respiration rate is caused by increased metabolism and lactic acid buildup.
Digestive	Decreased blood delivery as a result of the burn causes degeneration of the intestinal lining and liver. Bacteria from the intestine can cause systemic infections. The liver releases blood clotting factors in response to the injury. Increased nutrients necessary to support increased metabolism and for repair of the integumentary system are absorbed.
Urinary	The kidneys compensate for the increased fluid loss caused by the burn by greatly reducing or even stopping urine production. Decreased blood volume causes decreased blood flow to the kidneys, which reduces urine output but can cause kidney tissue damage. Hemoglobin, released from red blood cells damaged in the burnt skin, can decrease urine production by blocking fluid movement from the blood into the urine.

SUMMARY

The integumentary system consists of the skin, hair, nails, and a variety of glands.

Overview of the Integumentary System (p. 150)

The integumentary system separates and protects us from the external environment. Other functions include sensation, temperature regulation, vitamin D production, and excretion of small amounts of waste products.

Skin (p. 150)

Epidermis

1. The epidermis is stratified squamous epithelium divided into five strata.
2. The stratum basale consists of keratinocytes, which produce the cells of the more superficial strata.
3. The stratum spinosum consists of several layers of cells held together by many desmosomes.
4. The stratum granulosum consists of cells filled with granules of keratohyalin. Cell death occurs in this stratum.
5. The stratum lucidum consists of a layer of dead, transparent cells.
6. The stratum corneum consists of many layers of dead squamous cells. The most superficial cells are desquamated.
7. Keratinization is the transformation of the living cells of the stratum basale into the dead squamous cells of the stratum corneum.
   - Keratinized cells are filled with keratin and have a protein envelope, both of which contribute to structural strength. The cells are also held together by many desmosomes.
   - Intercellular spaces are filled with lipids from the lamellae that contribute to the impermeability of the epidermis to water.
8. Soft keratin is found in skin and the inside of hairs, whereas hard keratin occurs in nails and the outside of hairs. Hard keratin makes cells more durable, and these cells do not desquamate.

Thick and Thin Skin

1. Thick skin has all five epithelial strata.
2. Thin skin contains fewer cell layers per stratum, and the stratum lucidum is usually absent. Hair is found only in thin skin.

Skin Color

1. Melanocytes produce melanin inside melanosomes and then transfer the melanin to keratinocytes. The size and distribution of melanosomes determine skin color. Melanin production is determined genetically but can be influenced by ultraviolet light (tanning) and hormones.
2. Carotene, an ingested plant pigment, can cause the skin to appear yellowish.
3. Increased blood flow produces a red skin color, whereas a decreased blood flow causes a pale skin. Decreased oxygen content in the blood results in a bluish color called cyanosis.

Dermis

1. The dermis is connective tissue divided into two layers.
2. The papillary layer has projections called dermal papillae and is loose connective tissue that is well supplied with capillaries.
3. The reticular layer is the main layer. It is dense irregular connective tissue consisting mostly of collagen.

Hypodermis (p. 157)

1. Located beneath the dermis, the hypodermis is loose connective tissue that contains collagen and elastin fibers.
2. The hypodermis attaches the skin to underlying structures and is a site of fat storage.

Accessory Skin Structures (p. 158)

Hair

1. Lanugo (fetal hair) is replaced near the time of birth by terminal hairs (scalp, eyelids, and eyebrows) and vellus hairs. At puberty, vellus hairs can be replaced with terminal hairs.
2. Hair is dead keratinized epithelial cells consisting of a central axis of cells with soft keratin, known as the medulla, which is surrounded by a cortex of cells with hard keratin. The cortex is covered by the cuticle, a single layer of cells filled with hard keratin.
3. A hair has three parts: shaft, root, and hair bulb.
4. The hair bulb produces the hair in cycles involving a growth stage and a resting stage.
5. Hair color is determined by the amount and kind of melanin present.
6. Contraction of the arrector pili muscles, which are smooth muscles, causes hair to "stand on end" and produces "gooseflesh."

Glands

1. Sebaceous glands produce sebum, which oils the hair and the surface of the skin.
2. Merocrine sweat glands produce sweat, which cools the body. Apocrine sweat glands produce an organic secretion that can be broken down by bacteria to cause body odor.
3. Other skin glands include ceruminous glands, which make cerumen (earwax), and mammary glands, which produce milk.

Nails

1. The nail root is covered by skin, and the nail body is the visible part of the nail.
2. Nearly all of the nail is formed by the nail matrix, but the nail bed contributes.
3. The lunula is the part of the nail matrix visible through the nail body.
4. The nail is stratum corneum containing hard keratin.

Summary of Integumentary System Functions (p. 163)

Protection

1. The skin protects against abrasion and ultraviolet light, prevents the entry of microorganisms, helps regulate body temperature, and prevents water loss.
2. Hair protects against abrasion and ultraviolet light and is a heat insulator.
3. Nails protect the ends of the digits.

Sensation

The skin contains sensory receptors for pain, touch, hot, cold, and pressure that allow proper response to the environment.

Temperature Regulation

1. Through dilation and constriction of blood vessels, the skin controls heat loss from the body.
2. Sweat glands produce sweat, which evaporates and lowers body temperature.

Vitamin D Production

1. Skin exposed to ultraviolet light produces cholecalciferol, which is modified in the liver and then in the kidneys to form active vitamin D.
2. Vitamin D increases blood calcium levels by promoting calcium uptake from the intestine, the release of calcium from bone, and the reduction of calcium loss from the kidneys.

Excretion

Skin glands remove small amounts of waste products (e.g., urea, uric acid, and ammonia) but are not important in excretion.

Effects of Aging on the Integumentary System (p. 165)

1. As the body ages, blood flow to the skin declines, the skin becomes thinner, and elasticity is lost.
2. Sweat and sebaceous glands are less active, and the number of melanocytes decreases.

REVIEW AND COMPREHENSION

1. A layer of skin (where mitosis occurs) that replaces cells lost from the outer layer of the epidermis is the
   a. stratum corneum.
   b. stratum basale.
   c. stratum lucidum.
   d. reticular layer.
   e. hypodermis.
2. If a splinter penetrates the skin of the palm of the hand to the second epidermal layer from the surface, the last layer damaged is the
   a. stratum granulosum.
   b. stratum basale.
   c. stratum corneum.
   d. stratum lucidum.
   e. stratum spinosum.

For questions 3–7, match the layer of the epidermis with the correct description or function:

   a. stratum basale
   b. stratum corneum
   c. stratum granulosum
   d. stratum lucidum
   e. stratum spinosum

3. Production of keratin fibers; formation of lamellar bodies; limited amount of cell division
4. Desquamation occurs; 25 or more layers of dead squamous cells
5. Production of cells; melanocytes produce and contribute melanin; hemidesmosomes present
6. Production of keratohyalin granules; lamellar bodies release lipids; cells die
7. Dispersion of keratohyalin around keratin fibers; layer appears transparent; cells dead
8. In which of these areas of the body is thick skin found?
   a. back of the hand
   b. abdomen
   c. over the shin
   d. bridge of the nose
   e. heel of the foot
9. The function of melanin in the skin is to
   a. lubricate the skin.
   b. prevent skin infections.
   c. protect the skin from ultraviolet light.
   d. reduce water loss.
   e. help regulate body temperature.
10. Concerning skin color, which of these statements is *not* correctly matched?
   a. skin appears yellow—carotene present
   b. no skin pigmentation (albinism)—genetic disorder
   c. skin tans—increased melanin production
   d. skin appears blue (cyanosis)—oxygenated blood
   e. African-Americans darker than Caucasians—more melanin in African-American skin

For questions 11–14, match the layer of the dermis with the correct description or function:

   a. papillary layer
   b. reticular layer

11. Layer of dermis closest to the epidermis
12. Layer of dermis responsible for most of the structural strength of the skin
13. Layer of dermis responsible for fingerprints and footprints
14. Layer of dermis responsible for cleavage lines and striae
15. After birth, the type of hair on the scalp, eyelids, and eyebrows is
   a. lanugo.
   b. terminal hair.
   c. vellus hair.
16. Hair
   a. is produced by the dermal root sheath.
   b. consists of living keratinized epithelial cells.
   c. is colored by melanin.
   d. contains mostly soft keratin.
   e. grows from the tip.
17. Given these parts of a hair and hair follicle:
   1. cortex
   2. cuticle
   3. dermal root sheath
   4. epithelial root sheath
   5. medulla

   Arrange the structures in the correct order from the outside of the hair follicle to the center of the hair.
   a. 1,4,3,5,2
   b. 2,1,5,3,4
   c. 3,4,2,1,5
   d. 4,3,1,2,5
   e. 5,4,3,2,1
18. Concerning hair growth,
   a. hair falls out of the hair follicle at the end of the growth stage.
   b. most of the hair on the body grows continuously.
   c. cutting or plucking the hair increases its growth rate and thickness.
   d. genetic factors and the hormone testosterone are involved in "pattern baldness."
   e. eyebrows have a longer growth stage and resting stage than scalp hair.
19. Smooth muscles that produce "goose bumps" when they contract and are attached to hair follicles are called
   a. external root sheaths.
   b. arrector pili.
   c. dermal papillae.
   d. internal root sheaths.
   e. hair bulbs.

For questions 20–22, match the type of gland with the correct description or function:

a. apocrine sweat gland
b. merocrine sweat gland
c. sebaceous gland

20. Alveolar glands that produce a white, oily substance; usually open into hair follicles
21. Coiled, tubular glands that secrete a hyposmotic fluid that cools the body; most numerous in the palms of the hands and soles of the feet
22. Secretions from these coiled, tubular glands are broken down by bacteria to produce body odor; found in the axillae, in the genitalia, and around the anus
23. The lunula of the nail appears white because
    a. it lacks melanin.
    b. blood vessels cannot be seen through the thick nail matrix.
    c. the eponychium decreases blood flow to the area.
    d. the nail root is much thicker than the nail body.
    e. the hyponychium is thicker than the eponychium.
24. The stratum corneum of the nail fold grows onto the nail body as the
    a. eponychium.
    b. hyponychium.
    c. lunula.
    d. nail bed.
    e. nail matrix.
25. Most of the nail is produced by the
    a. eponychium.
    b. hyponychium.
    c. nail bed.
    d. nail matrix.
    e. dermis.
26. The skin aids in maintaining the calcium and phosphate levels of the body at optimum levels by participating in the production of
    a. vitamin A.
    b. vitamin B.
    c. vitamin D.
    d. melanin.
    e. keratin.
27. Which of these processes increase(s) heat loss from the body?
    a. dilation of dermal arterioles
    b. constriction of dermal arterioles
    c. increased sweating
    d. both a and c
    e. both b and c
28. In third-degree (full-thickness) burns, both the epidermis and dermis of the skin are destroyed. Which of the following conditions does *not* occur as a result of a third-degree burn?
    a. dehydration (increased water loss)
    b. increased likelihood of infection
    c. increased sweating
    d. loss of sensation in the burned area
    e. poor temperature regulation in the burned area

Answers in Appendix E

CRITICAL THINKING

1. The skin of infants is more easily penetrated and injured by abrasion than that of adults. Based on this fact, which stratum of the epidermis is probably much thinner in infants than that in adults?
2. Melanocytes are found primarily in the stratum basale of the epidermis. In reference to their function, why does this location make sense?
3. The rate of water loss from the skin of a hand was measured. Following the measurement, the hand was soaked in alcohol for 15 minutes. After all the alcohol had been removed from the hand, the rate of water loss was again measured. Compared with the rate of water loss before soaking the hand in alcohol, what difference, if any, would you expect in the rate of water loss after soaking the hand in alcohol?
4. It has been several weeks since Goodboy Player has competed in a tennis match. After the match, he discovers that a blister has formed beneath an old callus on his foot, and the callus has fallen off. When he examines the callus, he discovers that it appears yellow. Can you explain why?
5. Harry Fastfeet, a Caucasian man, jogs on a cold day. What color would you expect his skin to be (a) just before starting to run, (b) during the run, and (c) 5 minutes after the run?
6. A woman has stretch marks on her abdomen, yet she states that she has never been pregnant. Is this possible?
7. The lips are muscular folds forming the anterior boundary of the oral cavity. A mucous membrane covers the lips internally and the skin of the face covers them externally. The vermillion border, which is the red part of the lips, is covered by keratinized epithelium that is the transition between the epithelium of the mucous membrane and the facial skin. The vermillion border can become chapped (dry and cracked), whereas the mucous membrane and the facial skin do not. Propose as many reasons as you can to explain why the vermillion border is more prone to drying than the mucous membrane or facial skin.
8. Why are your eyelashes not a foot long? Your fingernails?
9. Pulling on hair can be quite painful, yet cutting hair is not painful. Explain.
10. Given what you know about the cause of acne, propose some ways to prevent or treat the disorder.
11. A patient has an ingrown toenail, a condition in which the nail grows into the nail fold. Would cutting the nail away from the nail fold permanently correct this condition? Why or why not?
12. Consider the following statement: Dark-skinned children are more susceptible to rickets (insufficient calcium in the bones) than fair-skinned children. Defend or refute this statement.

Answers in Appendix F

Visit this book's website at www.mhhe.com/seeley8 for chapter quizzes, interactive learning exercises, and other study tools.

Skeletal System 6

Bones and Bone Tissue

Sitting, standing, walking, picking up a pencil, and taking a breath all involve the skeletal system. It is the structural framework that gives the body its shape and provides protection for internal organs and soft tissues. The term *skeleton* is derived from a Greek word meaning dried, indicating that the skeleton is the dried, hard parts left after the softer parts are removed. Even with the flesh and organs removed, the skeleton is easily recognized as human. Despite its association with death, however, the skeletal system actually consists of dynamic, living tissues that are capable of growth, adapt to stress, and undergo repair after injury.

This chapter describes the *functions of the skeletal system* (p. 174), provides an explanation of *cartilage* (p. 174), and examines *bone histology* (p. 175), *bone anatomy* (p. 180), *bone development* (p. 183), *bone growth* (p. 185), *bone remodeling* (p. 191), *bone repair* (p. 192), *calcium homeostasis* (p. 194), and the *effects of aging on the skeletal system* (p. 198).

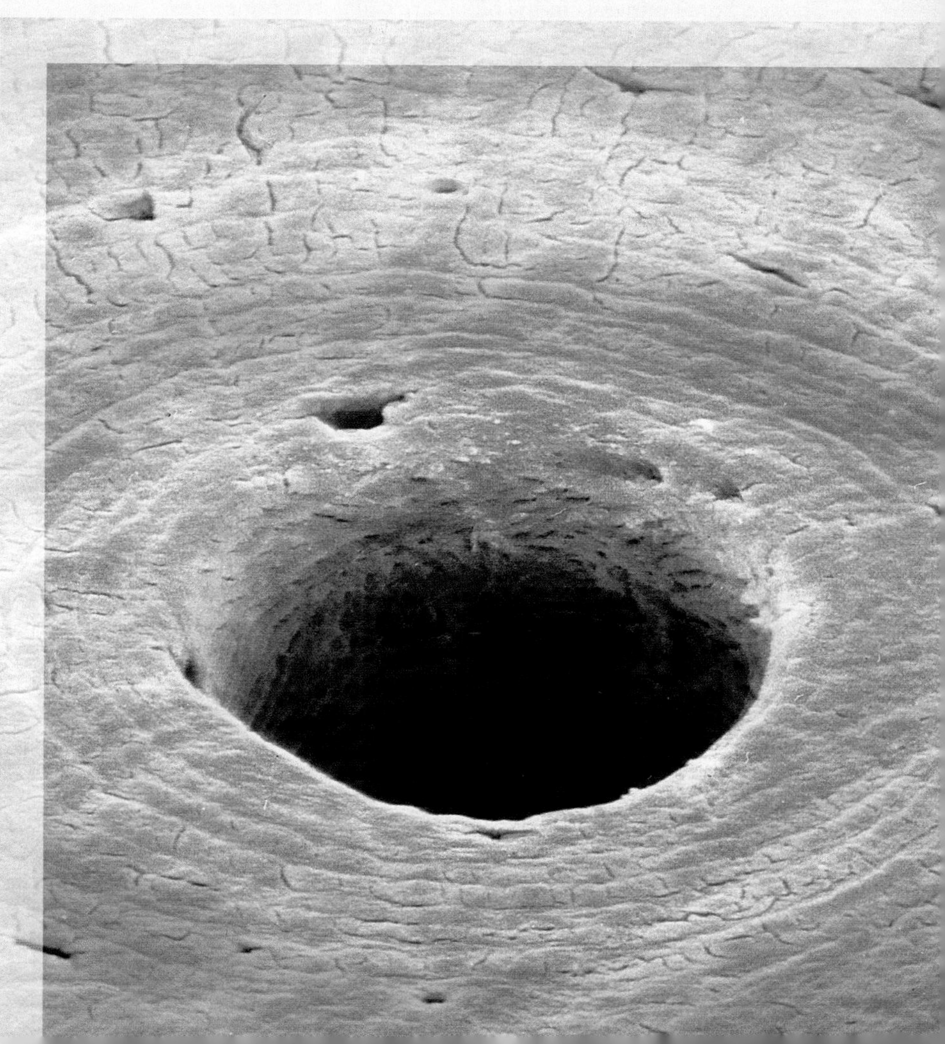

Colorized scanning electron micrograph of an osteon. The large opening is the space through which blood vessels bring blood to the bone. The surrounding bone matrix is organized into circular layers.

Skeletal System

FUNCTIONS OF THE SKELETAL SYSTEM

The skeletal system has four components: bones, cartilage, tendons, and ligaments. The skeletal system provides support and protection, allows body movements, stores minerals and fats, and is the site of blood cell production.

1. *Support.* Rigid, strong bone is well suited for bearing weight and is the major supporting tissue of the body. Cartilage provides a firm yet flexible support within certain structures, such as the nose, external ear, thoracic cage, and trachea. Ligaments are strong bands of fibrous connective tissue that attach to bones and hold them together.
2. *Protection.* Bone is hard and protects the organs it surrounds. For example, the skull encloses and protects the brain, and the vertebrae surround the spinal cord. The rib cage protects the heart, lungs, and other organs of the thorax.
3. *Movement.* Skeletal muscles attach to bones by tendons, which are strong bands of connective tissue. Contraction of the skeletal muscles moves the bones, producing body movements. Joints, which are formed where two or more bones come together, allow movement between bones. Smooth cartilage covers the ends of bones within some joints, allowing the bones to move freely. Ligaments allow some movement between bones but prevent excessive movements.
4. *Storage.* Some minerals in the blood are taken into bone and stored. Should blood levels of these minerals decrease, the minerals are released from bone into the blood. The principal minerals stored are calcium and phosphorus. Fat (adipose tissue) is also stored within bone cavities. If needed, the fats are released into the blood and used by other tissues as a source of energy.
5. *Blood cell production.* Many bones contain cavities filled with red bone marrow, which gives rise to blood cells and platelets (see chapter 19).

1 ***Name the four components of the skeletal system. List the five functions of the skeletal system.***

CARTILAGE

Cartilage comes in three types: hyaline cartilage, fibrocartilage, and elastic cartilage (see chapter 4). Although each type of cartilage can provide support, hyaline cartilage is most intimately associated with bone. An understanding of hyaline cartilage is important because most of the bones in the body develop from it. In addition, the growth in length of bones and bone repair often involve the production of hyaline cartilage, followed by its replacement with bone.

Hyaline cartilage consists of specialized cells that produce a matrix surrounding the cells (figure 6.1). The cells that produce new cartilage matrix are **chondroblasts** (kon′drō-blastz; *chondro* is from the Greek word *chondrion,* cartilage). When matrix surrounds a chondroblast, it becomes a **chondrocyte** (kon′drō-sīt), which is a rounded cell that occupies a space called a **lacuna** (lă-koo′nă) within the matrix. The matrix contains collagen, which provides strength, and proteoglycans, which make cartilage resilient by trapping water (see chapter 4).

The **perichondrium** (per-i-kon′drē-ŭm) is a double-layered connective tissue sheath covering most cartilage (see figure 6.1). The outer layer of the perichondrium is dense irregular connective tissue containing fibroblasts. The inner, more delicate layer has fewer fibers and contains chondroblasts. Blood vessels and nerves penetrate the outer layer of the perichondrium but do not enter the cartilage matrix so that nutrients must diffuse through the cartilage matrix to reach the chondrocytes. **Articular** (ar-tik′ū-lăr) **cartilage,** which is the cartilage covering the ends of bones where they come together to form joints, has no perichondrium, blood vessels, or nerves.

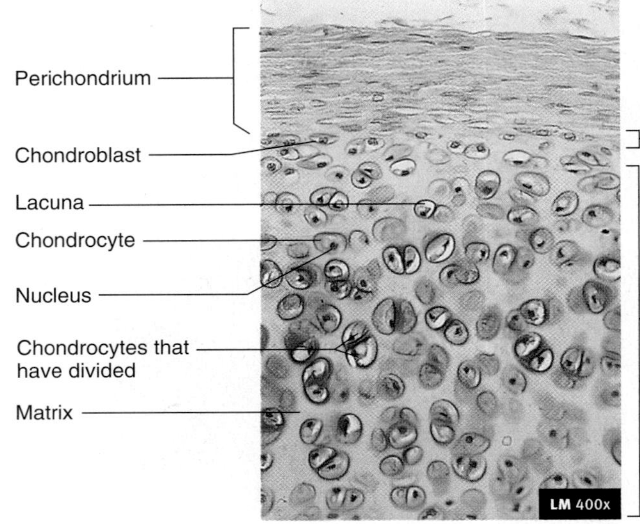

FIGURE 6.1 Hyaline Cartilage

Photomicrograph of hyaline cartilage covered by perichondrium. Chondrocytes within lacunae are surrounded by cartilage matrix.

PREDICT 1

Explain why damaged cartilage takes a long time to heal. What are the advantages of articular cartilage having no perichondrium, blood vessels, or nerves?

Cartilage grows in two ways. Through **appositional growth,** chondroblasts in the perichondrium lay down new matrix and add new chondrocytes to the outside of the tissue; through **interstitial growth,** chondrocytes within the tissue divide and add more matrix between the cells (see figure 6.1).

2 ***Describe the structure of hyaline cartilage. Name two types of cartilage cells. What is a lacuna?***

3 ***Describe the connective tissue and cells found in both layers of the perichondrium. How do nutrients from blood vessels in the perichondrium reach the chondrocytes?***

4 ***Describe appositional and interstitial growth of cartilage.***

BONE HISTOLOGY

Bone consists of extracellular bone matrix and bone cells. The composition of the bone matrix is responsible for the characteristics of bone. The bone cells produce the bone matrix, become entrapped within it, and break it down so that new matrix can replace the old matrix.

Bone Matrix

By weight, mature bone matrix normally is approximately 35% organic and 65% inorganic material. The organic material consists primarily of collagen and proteoglycans. The inorganic material consists primarily of a calcium phosphate crystal called **hydroxyapatite** (hī-drok′sē-ap-ă-tīt), which has the molecular formula $Ca_{10}(PO_4)_6(OH)_2$.

The collagen and mineral components are responsible for the major functional characteristics of bone. Bone matrix might be said to resemble reinforced concrete. Collagen, like reinforcing steel bars, lends flexible strength to the matrix, whereas the mineral components, like concrete, give the matrix compression (weight-bearing) strength.

If all the mineral is removed from a long bone, collagen becomes the primary constituent, and the bone becomes overly flexible. On the other hand, if the collagen is removed from the bone, the mineral component becomes the primary constituent, and the bone is very brittle (figure 6.2).

PREDICT 2

In general, the bones of elderly people break more easily than the bones of younger people. Give as many possible explanations as you can for this observation.

Bone Cells

Bone cells are categorized as osteoblasts, osteocytes, and osteoclasts, which have different functions and origins.

Osteoblasts

Osteoblasts (os′tē-ō-blastz) have an extensive endoplasmic reticulum and numerous ribosomes. They produce collagen and proteoglycans, which are packaged into vesicles by the Golgi apparatus and released from the cell by exocytosis. Osteoblasts also release **matrix vesicles,** which are membrane-bound sacs formed when the plasma membrane buds, or protrudes outward, and pinches off. The matrix vesicles concentrate Ca^{2+} and PO_4^{3-} and

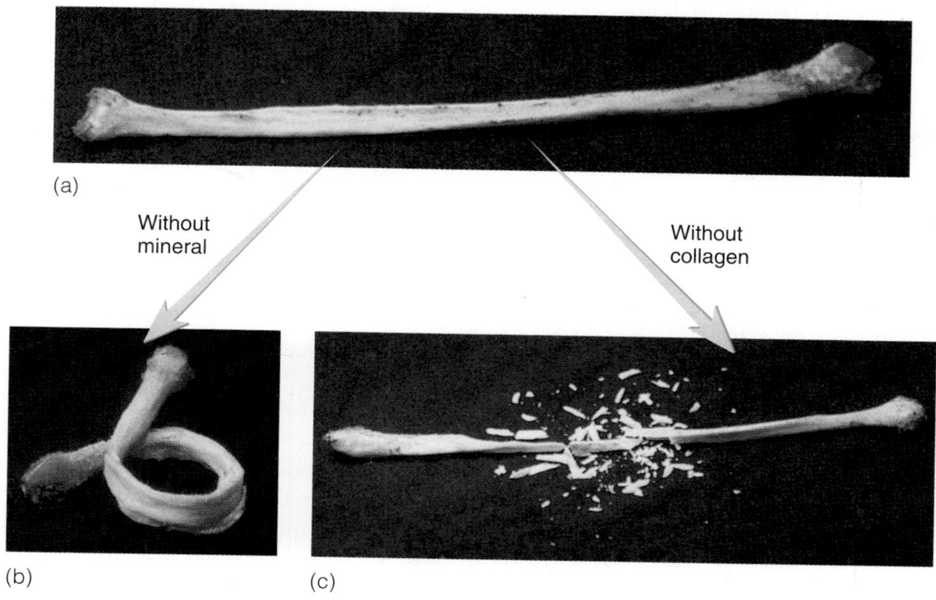

FIGURE 6.2 Effects of Changing the Bone Matrix

(*a*) Normal bone. (*b*) Demineralized bone, in which collagen is the primary remaining component, can be bent without breaking. (*c*) When collagen is removed, mineral is the primary remaining component, thus making the bone so brittle that it is easily shattered.

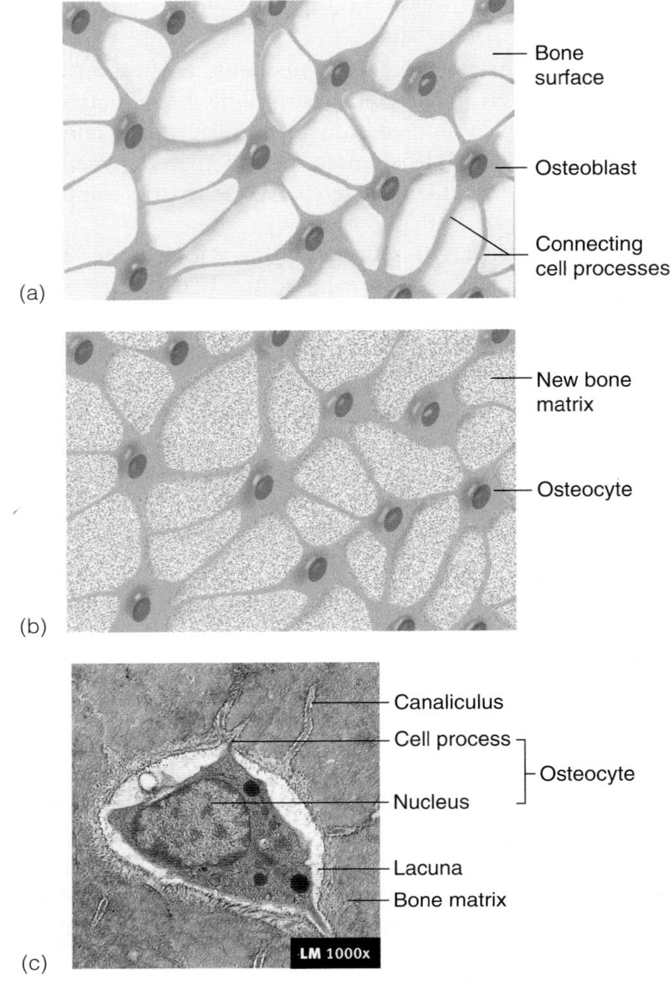

FIGURE 6.3 Ossification

(*a*) Osteoblasts on a preexisting surface, such as cartilage or bone. The cell processes of different osteoblasts join together. (*b*) Osteoblasts have produced bone matrix. The osteoblasts are now osteocytes. (*c*) Photomicrograph of an osteocyte in a lacuna with cell processes in the canaliculi.

form needlelike hydroxyapatite crystals. When the hydroxyapatite crystals are released from the matrix vesicles, they act as templates, or "seeds," which stimulate further hydroxapatite formation and mineralization of the matrix.

Ossification (os′i-fi-kā′shŭn), or **osteogenesis** (os′tē-ō-jen′ĕ-sis), is the formation of bone by osteoblasts. Ossification occurs by appositional growth on the surface of previously existing bone or cartilage. For example, osteoblasts beneath the periosteum cover the surface of preexisting bone (figure 6.3). Elongated cell processes from osteoblasts connect to cell processes of other osteoblasts through gap junctions (see chapter 4). Bone matrix produced by the osteoblasts covers the older bone surface and surrounds the osteoblast cell bodies and processes. The result is a new layer of bone.

Osteocytes

Once an osteoblast becomes surrounded by bone matrix, it is a mature bone cell called an **osteocyte** (os′tē-ō-sīt). Osteocytes become relatively inactive, compared with most osteoblasts, but it is possible for them to produce the components needed to maintain the bone matrix.

The spaces occupied by the osteocyte cell bodies are called **lacunae** (lă-koo′nē), and the spaces occupied by the osteocyte cell processes are called **canaliculi** (kan-ă-lik′ū-lī; little canals; see figure 6.3). In a sense, the cells and their processes form a "mold" around which the matrix is formed. Bone differs from cartilage in that the processes of bone cells are in contact with one another through the canaliculi. Instead of diffusing through the mineralized matrix, nutrients and gases can pass through the small amount of fluid surrounding the cells in the canaliculi and lacunae or pass from cell to cell through the gap junctions connecting the cell processes.

Osteoclasts

Osteoclasts (os′tē-ō-klastz) are large cells with several nuclei; they are responsible for the **resorption,** or breakdown, of bone. Where the plasma membrane of osteoclasts contacts bone matrix, it forms many projections called a **ruffled border.** Hydrogen ions are pumped across the ruffled border and produce an acid environment that causes decalcification of the bone matrix. The osteoclasts also release enzymes that digest the protein components of the matrix. Through the process of endocytosis, some of the breakdown products of bone resorption are taken into the osteoclast.

Osteoclasts break down bone best when they are in direct contact with mineralized bone matrix. Osteoblasts assist in the resorption of bone by osteoclasts by producing enzymes that break down the thin layer of unmineralized organic matrix normally covering bone. Removal of this layer by osteoblasts enables the osteoclasts to come into contact with the mineralized bone.

Origin of Bone Cells

Connective tissue develops embryologically from mesenchymal cells (see chapter 4). Some of the mesenchymal cells become **stem cells,** which can replicate and give rise to more specialized cell types. **Osteochondral progenitor cells** are stem cells that can become osteoblasts or chondroblasts. Osteochondral progenitor cells are located in the inner layer of the perichondrium, the inner layer of the periosteum, and the endosteum. From these locations, they are a potential source of new osteoblasts or chondroblasts.

Osteoblasts are derived from osteochondral progenitor cells, and osteocytes are derived from osteoblasts. Whether or not osteocytes freed from their surrounding bone matrix by resorption can revert to active osteoblasts is a debated issue. Osteoclasts are not derived from osteochondral progenitor cells but instead are derived from stem cells in red bone marrow (see chapter 19). The bone marrow stem cells that give rise to a type of white blood cell, called a monocyte, also are the source of osteoclasts. The multinucleated osteoclasts probably result from the fusion of many stem cell descendants.

5 ***Name the components of bone matrix, and explain their contribution to bone flexibility and the ability of bones to bear weight.***

6 ***What are the functions of osteoblasts, osteocytes, and osteoclasts?***

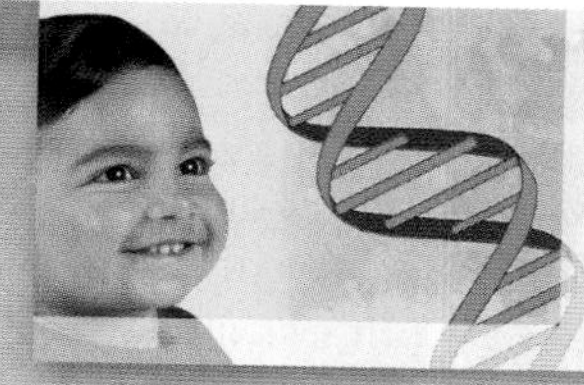

CLINICAL GENETICS

Osteogenesis Imperfecta

Collagen molecules consist of three polypeptide chains, called **collagen α-chains,** which wind around each other to form a ropelike helix. Collagen molecules combine to form collagen fibers (see chapter 4). Approximately 25 different collagen α-chains, each encoded by a separate gene, are known, and they combine to form approximately 20 types of collagen molecules. Type I collagen is the major collagen of bone, tendon, and skin. It consists of two collagen α-chains encoded by a gene on chromosome 17 called the COL1A1 gene (figure A) and a different collagen α-chain encoded by a gene on chromosome 7 called the COL1A2 gene.

Osteogenesis imperfecta (OI; os′tē-ō-jen′ĕ-sis im-per-fek′tă; imperfect bone formation) is a connective tissue disorder caused by mutations in either of these collagen genes. Consequently, less collagen than normal is produced. Depending on the type of osteogenesis imperfecta, the collagen can have a normal or an abnormal structure, resulting in bones with decreased flexibility and a tendency to break more easily than normal. OI is also known as brittle bone disorder.

In the United States, 20,000 to 50,000 people may have OI. Seven types of OI are recognized, and there is great variation in the appearance and severity of OI both between and within the types. Type I OI, the mildest form, is caused by too little formation of normal type I collagen. It may be characterized by any of the following: bones predisposed to fracture, especially before puberty; normal or near-normal stature; minimal or absent bone deformities; a tendency to develop spinal curvature; loose joints; brittle teeth; hearing loss; and a blue, purple, or gray tint to the whites of the eyes. The number of fractures over a lifetime can vary from a few to more than 100. The fractures usually heal normally without deformity. Children with mild type I OI may have few obvious clinical features, except for a history of broken bones. It is important for children with OI to be properly diagnosed because broken bones can be associated with child abuse.

The other types of OI are more severe than type I, and they result from a reduced amount of abnormal type I collagen. Type II is the most severe OI and most often is lethal within the first week of life. Death usually is caused by an inability to breathe because of a small thorax and rib fractures. Type III OI is characterized by bones that fracture very easily, often with fractures occurring before and during birth. Fractures of the extremities occurring before birth often heal in poor alignment, causing the limbs to be bent and short. Other characteristics of type III OI include short stature; severe bone deformities; barrel-shaped rib cage; spinal curvature; loose joints; brittle teeth; hearing loss; and a blue, purple, or gray tint to the whites of the eyes. Types IV to VII OI are between types I and III in severity.

Almost all cases of OI are caused by autosomal-dominant mutations. The vast majority of type I OI cases result from mutations in the COL1A1 gene in which the normal nucleotide sequence is altered to produce a stop codon. Messenger RNA that codes for the collagen α-chain is not produced because transcription of the gene is prematurely stopped. Recall from chapter 3 that genes (alleles) occur in pairs on homologous chromosomes. In type I OI, one of the COL1A1 genes is mutated and nonfunctional, whereas the other is normal. A person with type I OI produces approximately half as much normal type I collagen as usual. Thus, type I OI is a collagen-deficiency disorder.

Other types of OI occur when mutated COL1A1 or COL1A2 genes are transcribed and the resulting mRNA codes for defective collagen α-chains. The more defective the collagen α-chains, the weaker the collagen fibers and the greater the severity of the disorder. In addition, bone cells produce less bone matrix because they are inefficient at making new bone matrix containing defective collagen α-chains. Thus, types II–VII OI are defective collagen *and* collagen-deficiency disorders.

There is a 50% chance that a child will inherit OI from an affected parent because it is an autosomal-dominant trait. However, because of the disorder's great variability, the child may be affected in different ways than the parent. For example, their tendency for fractures, bone deformity, and so on may be different. Approximately 25% of children with OI have parents who do not have the disorder. In these cases, a new mutation, which occurred during the formation of sperm cells or oocytes, is responsible.

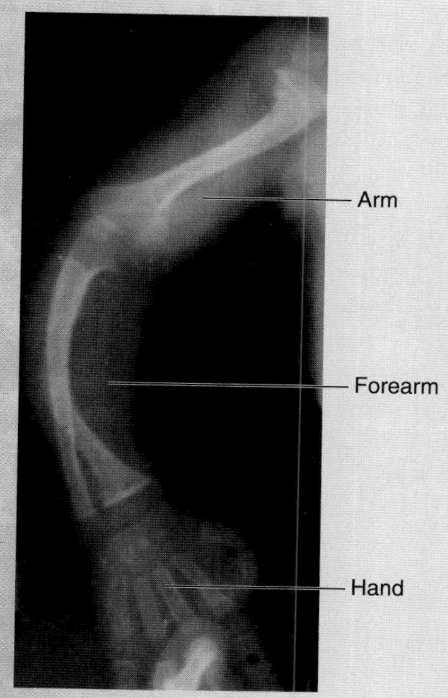

FIGURE A **Osteogenesis imperfecta**

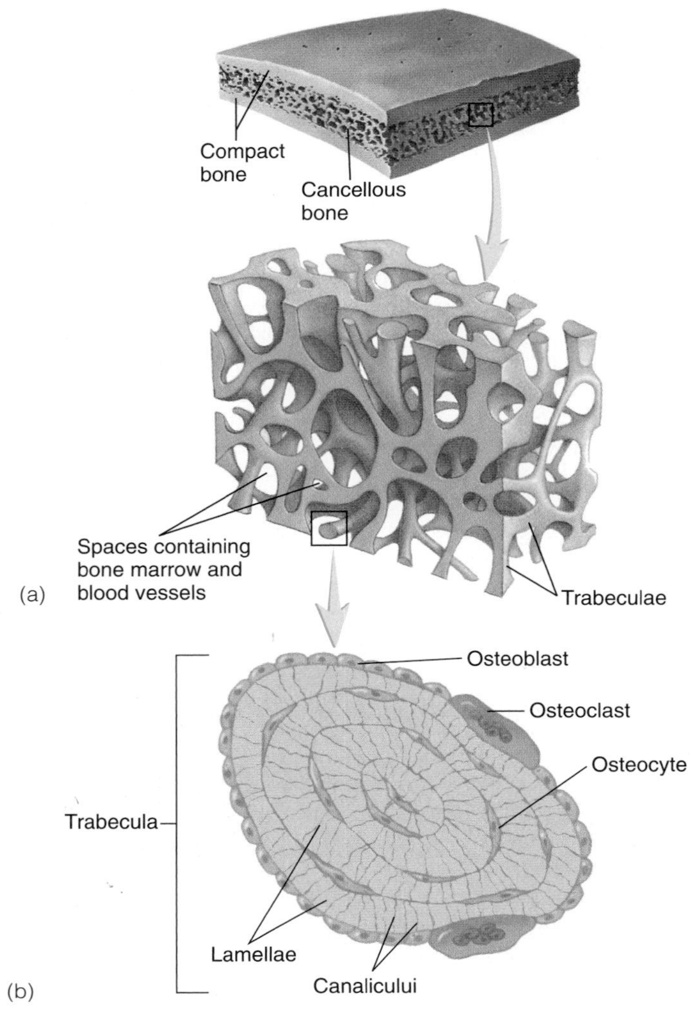

FIGURE 6.4 Cancellous Bone

(*a*) Beams of bone, the trabeculae, surround spaces in the bone. In life, the spaces are filled with red or yellow bone marrow and with blood vessels. (*b*) Transverse section of a trabecula.

7 ***Describe the formation of new bone by appositional growth. Name the spaces that are occupied by osteocyte cell bodies and cell processes.***

8 ***What cells give rise to osteochondral progenitor cells? What kinds of cells are derived from osteochondral progenitor cells? What types of cells give rise to osteoclasts?***

Woven and Lamellar Bone

Bone tissue is classified as either woven or lamellar bone, according to the organization of collagen fibers within the bone matrix. In **woven bone,** the collagen fibers are randomly oriented in many directions. Woven bone is first formed during fetal development or during the repair of a fracture. After its formation, osteoclasts break down the woven bone and osteoblasts build new matrix. The process of removing old bone and adding new bone is called **remodeling.** It is an important process discussed later in this chapter (see "Bone Remodeling," p. 191). Woven bone is remodeled to form lamellar bone.

Lamellar bone is mature bone that is organized into thin sheets or layers approximately 3–7 micrometers (μm) thick called **lamellae** (lă-mel′ē). In general, the collagen fibers of one lamella lie parallel to one another, but at an angle to the collagen fibers in the adjacent lamellae. Osteocytes, within their lacunae, are arranged in layers sandwiched between lamellae.

Cancellous and Compact Bone

Bone, whether woven or lamellar, can be classified according to the amount of bone matrix relative to the amount of space present within the bone. Cancellous bone has less bone matrix and more space than compact bone, which has more bone matrix and less space than cancellous bone.

Cancellous bone (figure 6.4*a*) consists of interconnecting rods or plates of bone called **trabeculae** (tră-bek′ū-lē; beam). Between the trabeculae are spaces that, in life, are filled with bone marrow and blood vessels. Cancellous bone is sometimes called spongy bone because of its porous appearance.

Most trabeculae are thin (50–400 μm) and consist of several lamellae with osteocytes located in lacunae between the lamellae (figure 6.4*b*). Each osteocyte is associated with other osteocytes through canaliculi. Usually, no blood vessels penetrate the trabeculae, so osteocytes must obtain nutrients through their canaliculi. The surfaces of trabeculae are covered with a single layer of cells consisting mostly of osteoblasts with a few osteoclasts.

Trabeculae are oriented along the lines of stress within a bone (figure 6.5). If the stress on a bone is changed slightly (e.g., because

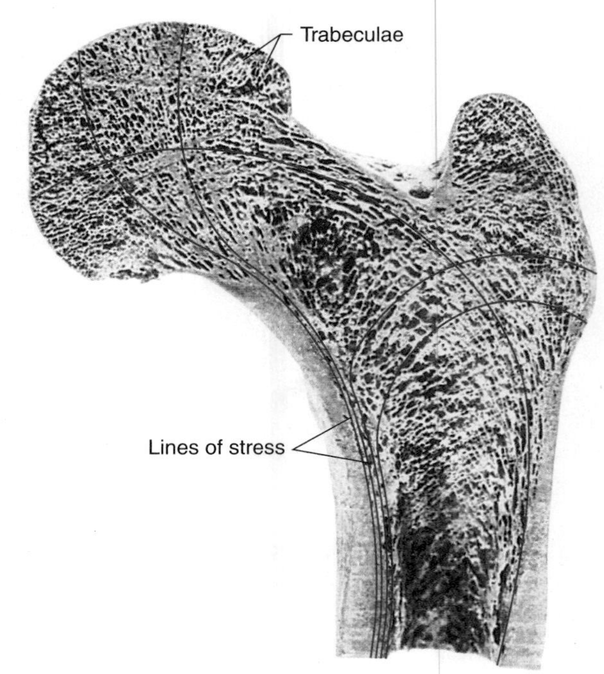

FIGURE 6.5 Trabeculae Oriented Along Lines of Stress

The proximal end of a long bone (femur) showing trabeculae oriented along lines of stress (red lines). The trabeculae bear weight and help bones resist bending and twisting.

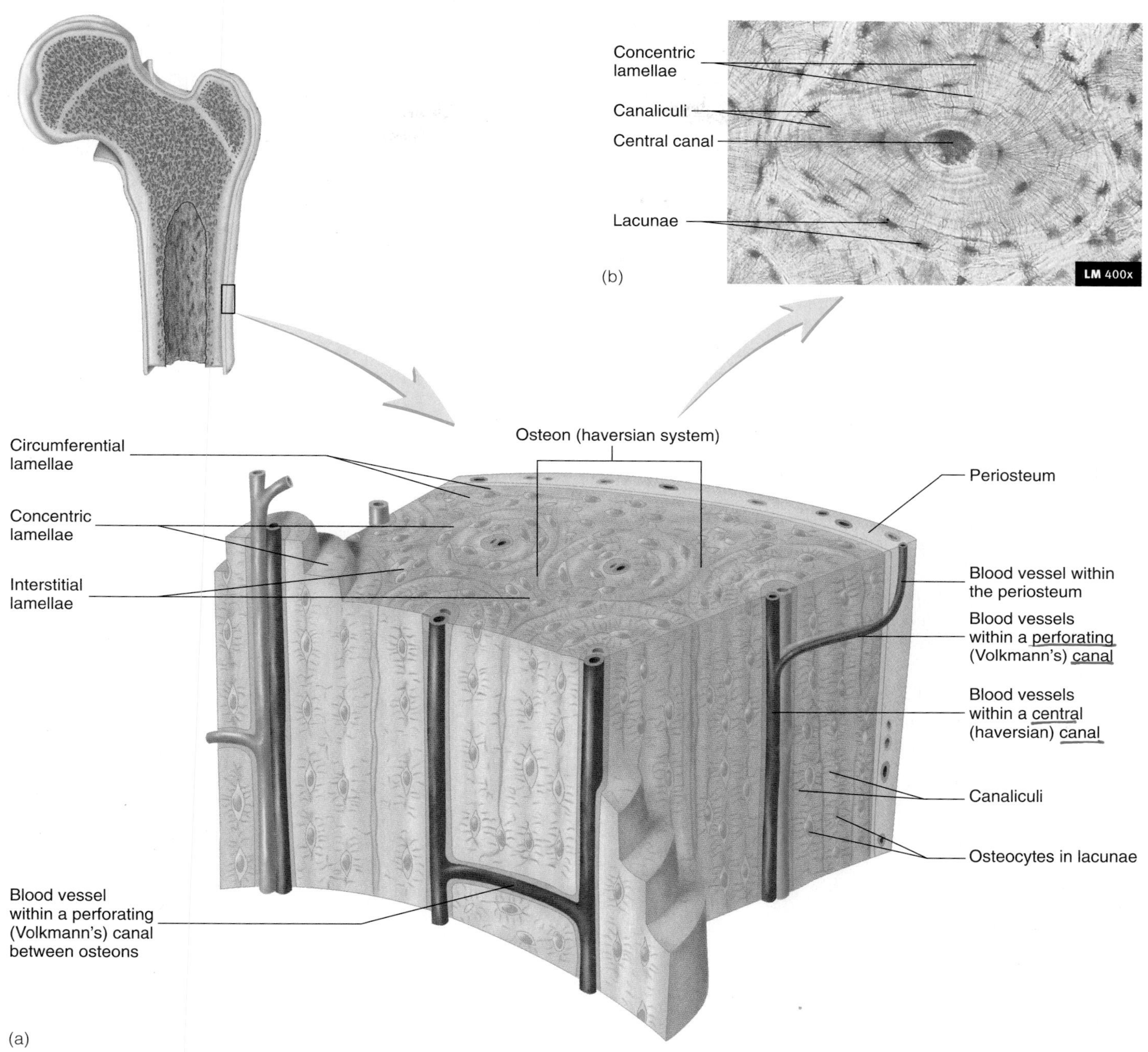

FIGURE 6.6 Compact Bone

(*a*) Compact bone consists mainly of osteons, which are concentric lamellae surrounding blood vessels within central canals. The outer surface of the bone is formed by circumferential lamellae, and bone between the osteons consists of interstitial lamellae. (*b*) Photomicrograph of an osteon.

of a fracture that heals improperly), the trabecular pattern realigns with the new lines of stress.

Compact bone (figure 6.6) is denser and has fewer spaces than cancellous bone. Blood vessels enter the substance of the bone itself, and the lamellae of compact bone are primarily oriented around those blood vessels. Vessels that run parallel to the long axis of the bone are contained within **central,** or **haversian** (ha-ver′shan), **canals.** Central canals are lined with endosteum and contain blood vessels, nerves, and loose connective tissue.

Concentric lamellae are circular layers of bone matrix that surround a common center, the central canal. An **osteon** (os′tē-on), or **haversian system,** consists of a single central canal, its contents, and associated concentric lamellae and osteocytes. In cross section, an osteon resembles a circular target; the "bull's-eye" of the target is the central canal, and 4–20 concentric lamellae form the rings. Osteocytes are located in lacunae between the lamellar rings, and canaliculi radiate between lacunae across the lamellae, producing the appearance of minute cracks across the rings of the target.

The outer surfaces of compact bone are formed by **circumferential lamellae,** which are thin plates that extend around the bone (see figure 6.6). In some bones, such as certain bones of the face, the layer of compact bone can be so thin that no osteons exist, and the compact bone is composed of only circumferential lamellae. In between the osteons are **interstitial lamellae,** which are remnants of concentric or circumferential lamellae that were partially removed during bone remodeling.

Osteocytes receive nutrients and eliminate waste products through the canal system within compact bone. Blood vessels from the periosteum or medullary cavity enter the bone through **perforating,** or **Volkmann's, canals,** which run perpendicular to the long axis of the bone (see figure 6.6). Perforating canals are not surrounded by concentric lamellae but pass through the concentric lamellae of osteons. The central canals receive blood vessels from perforating canals. Nutrients in the blood vessels enter the central canals, pass into the canaliculi, and move through the cytoplasm of the osteocytes that occupy the canaliculi and lacunae to the most peripheral cells within each osteon. Waste products are removed in the reverse direction.

9 ***Distinguish between woven bone and lamellar bone. Where is woven bone found?***

10 ***Describe the structure of cancellous bone. What are trabeculae, and what is their function? How do osteocytes within trabeculae obtain nutrients?***

11 ***Describe the structure of compact bone. What is an osteon? Name three types of lamellae found in compact bone.***

12 ***Trace the pathway nutrients must follow to go from blood vessels in the periosteum to osteocytes within osteons.***

PREDICT 3

Compact bone has perforating and central canals. Why isn't such a canal system necessary in cancellous bone?

BONE ANATOMY

Bone Shapes

Individual bones are classified according to their shape as long, short, flat, or irregular (figure 6.7). **Long bones** are longer than they are wide. Most of the bones of the upper and lower limbs are long bones. **Short bones** are about as broad as they are long. They are nearly cube-shaped or round and are exemplified by the bones of the wrist (carpals) and ankle (tarsals). **Flat bones** have a relatively thin, flattened shape and are usually curved. Examples of flat bones are certain skull bones, the ribs, the breastbone (sternum), and the shoulder blades (scapulae). **Irregular bones,** such as the vertebrae and facial bones, have shapes that do not fit readily into the other three categories.

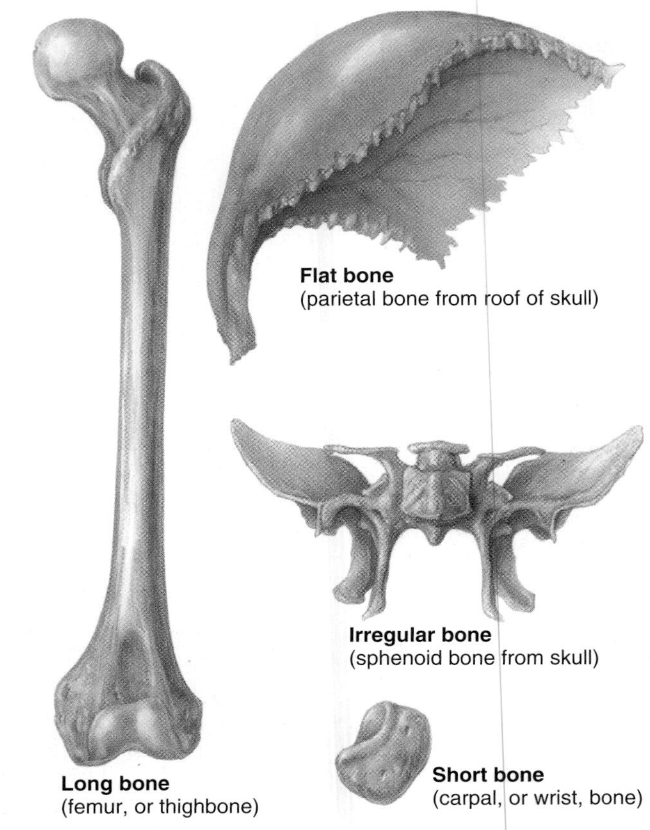

FIGURE 6.7 Bone Shapes

Structure of a Long Bone

The **diaphysis** (dī-af′i-sis), or shaft, of a long bone is composed primarily of compact bone, but it can also contain some cancellous bone (figure 6.8*c*; table 6.1). The end of a long bone is mostly cancellous bone, with an outer layer of compact bone. Within joints, the end of a long bone is covered with hyaline cartilage called **articular cartilage** (figure 6.8*a* and *b*).

During bone formation and growth, bones develop from centers of ossification (see "Bone Development," p. 183). The primary ossification center is in the diaphysis. An **epiphysis** (e-pif′i-sis) is a part of a long bone that develops from a center of ossification distinct from that of the diaphysis. Each long bone of the arm, forearm, thigh (see figure 6.8*a* and *b*), and leg has one or more epiphyses (e-pif′i-sēz) on each end of the bone. Each long bone of the hand and foot has one epiphysis, which is located on the proximal or distal end of the bone.

The **epiphyseal** (ep-i-fiz′ē-ăl), or **growth, plate** separates the epiphysis from the diaphysis (see figure 6.8*a*). Growth in bone length (see p. 186) occurs at the epiphyseal plate. Consequently, growth in length of the long bones of the arm, forearm, thigh, and leg occurs at both ends of the diaphysis, whereas growth in length of the hand and foot bones occurs at one end of the diaphysis. When bone stops growing in length, the epiphyseal plate becomes ossified and is called the **epiphyseal line** (see figure 6.8*b*).

In addition to the small spaces within cancellous bone and compact bone, the diaphysis of a long bone can have a large internal space called the **medullary cavity.** The cavities of cancellous bone and the medullary cavity are filled with marrow. **Red marrow** is

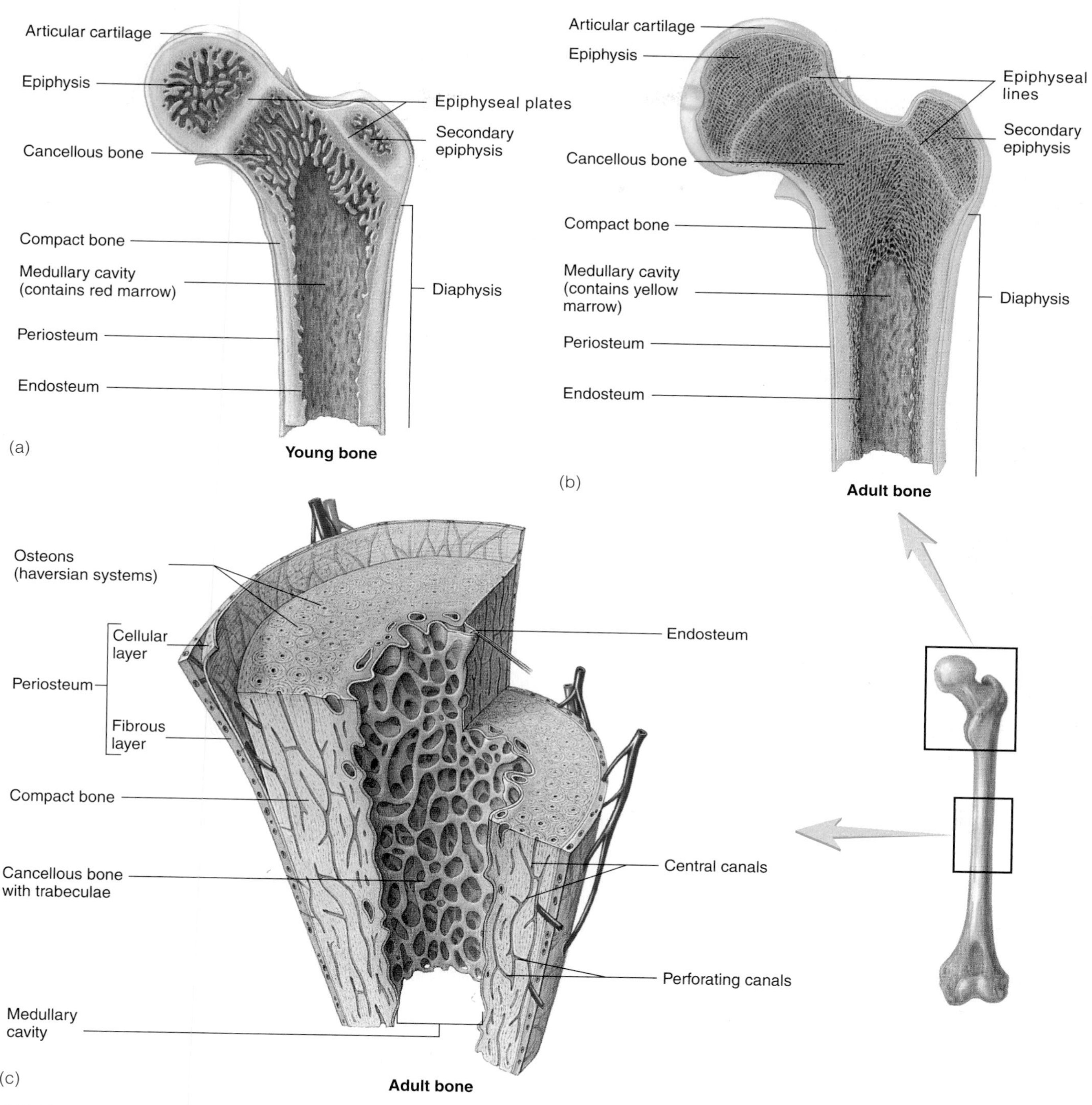

FIGURE 6.8 Long Bone

(*a*) Young long bone (the femur) showing epiphyses, epiphyseal plates, and diaphysis. The femur is unusual in that there are two epiphyses at the proximal end of the bone. (*b*) Adult long bone with epiphyseal lines. (*c*) Internal features of a portion of the diaphysis in (*a*).

the site of blood cell formation, and **yellow marrow** is mostly adipose tissue. In the fetus, the spaces within bones are filled with red marrow. The conversion of red marrow to yellow marrow begins just before birth and continues well into adulthood. Yellow marrow completely replaces the red marrow in the long bones of the limbs, except for some red marrow in the proximal part of the arm and thigh bones. Elsewhere, varying proportions of yellow and red marrow are found. In some locations, red marrow is completely replaced by yellow marrow; in others, there is a mixture of red and yellow marrow. For example, part of the hip bone (ilium) may contain 50% red marrow and 50% yellow marrow. Marrow from the hip bone is used as a source for donating red bone marrow

TABLE 6.1 Gross Anatomy of a Long Bone

Part	Description
Diaphysis	Shaft of the bone
Epiphysis	End of a long bone that develops from a center of ossification distinct from the diaphysis
Periosteum	Double-layered cnnective tissue membrane covering the outer surface of bone except where articular cartilage exists; ligaments and tendons attach to bone through the periosteum; blood vessels and nerves from the periosteum supply the bone; the periosteum is the site of bone growth in diameter
Endosteum	Thin connective tissue membrane lining the inner cavities of bone
Articular cartilage	Thin layer of hyaline cartilage covering a bone where it forms a joint (articulation) with another bone
Epiphyseal plate	Area of hyaline cartilage between the diaphysis and epiphysis; cartilage growth followed by endochondral ossification results in bone growth in length
Cancellous (spongy) bone	Bone having many small spaces; found mainly in the epiphysis; arranged into trabeculae
Compact bone	Dense bone with few internal spaces organized into osteons; forms the diaphysis and covers the cancellous bone of the epiphyses
Medullary cavity	Large cavity within the diaphysis
Red marrow	Connective tissue in the spaces of cancellous bone or in the medullary cavity; the site of blood cell production
Yellow marrow	Fat stored within the medullary cavity or in the spaces of cancellous bone

because it is a large bone with more marrow than smaller bones and it is accessed relatively easily.

The **periosteum** (per-ē-os′tē-ŭm) is a connective tissue membrane that covers the outer surface of a bone (figure 6.8*c*). The outer fibrous layer is dense irregular collagenous connective tissue that contains blood vessels and nerves. The inner layer is a single layer of bone cells, which includes osteoblasts, osteoclasts, and osteochondral progenitor cells (see "Bone Cells," p. 175). Where tendons and ligaments attach to bone, the collagen fibers of the tendon or ligament become continuous with those of the periosteum. In addition, some of the collagen fibers of the tendons or ligaments penetrate the periosteum into the outer part of the bone. These bundles of collagen fibers are called **perforating,** or **Sharpey's, fibers,** and they strengthen the attachment of the tendons or ligaments to the bone.

The **endosteum** (en-dos′tē-ŭm) is a single layer of cells that lines the internal surfaces of all cavities within bones, such as the medullary cavity of the diaphysis and the smaller cavities in cancellous and compact bone (see figure 6.8*c*). The endosteum includes osteoblasts, osteoclasts, and osteochondral progenitor cells.

Structure of Flat, Short, and Irregular Bones

Flat bones contain an interior framework of cancellous bone sandwiched between two layers of compact bone (figure 6.9). Short and irregular bones have a composition similar to the epiphyses of long bones. They have compact bone surfaces that surround a cancellous bone center with small spaces that usually are filled with marrow. Short and irregular bones are not elongated and have no diaphyses. Certain regions of these bones, however, such as the processes (projections) of irregular bones, possess epiphyseal growth plates and therefore have small epiphyses.

Some of the flat and irregular bones of the skull have air-filled spaces called **sinuses** (sī′nŭs-ĕz; see chapter 7), which are lined by mucous membranes.

13 ***List the four basic shapes of individual bones, and give an example of each.***

14 ***Define the diaphysis, epiphysis, epiphyseal plate, and epiphyseal line of a long bone.***

15 ***What are red marrow and yellow marrow? Where are they located in a child and in an adult?***

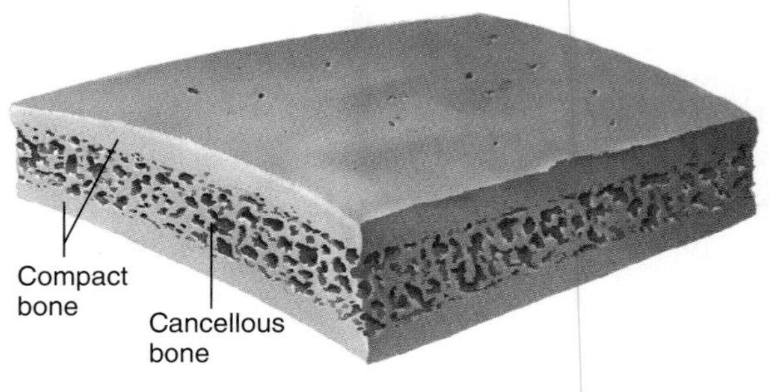

FIGURE 6.9 Structure of a Flat Bone
Outer layers of compact bone surround cancellous bone.

16 ***Where are the periosteum and endosteum located? What types of cells are found in the periosteum and endosteum? What is the function of perforating (Sharpey's) fibers?***

17 ***Compare the structure of long bones with the structure of flat, short, and irregular bones. How are compact bone and cancellous bone arranged in each?***

BONE DEVELOPMENT

During fetal development, bone formation occurs in two patterns—**intramembranous** and **endochondral ossification.** The terms describe the tissues in which bone formation takes place: intramembranous ossification in connective tissue membranes and endochondral ossification in cartilage. Both methods initially produce woven bone, which is then remodeled. After remodeling, bone formed by intramembranous ossification cannot be distinguished from bone formed by endochondral ossification. Table 6.2 compares the two types of ossification.

Intramembranous Ossification

At approximately the fifth week of development, embryonic mesenchyme condenses around the developing brain to form a membrane of connective tissue with randomly oriented, delicate collagen fibers. Intramembranous ossification of the membrane begins at approximately the eighth week of development and is completed by approximately 2 years of age. Many skull bones, part of the mandible (lower jaw), and the diaphyses of the clavicles (collarbones) develop by intramembranous ossification (figure 6.10).

Centers of ossification are the locations in the membrane where ossification begins. The centers of ossification expand to form a bone by gradually ossifying the membrane. Thus, the centers of ossification have the oldest bone and the expanding edges the youngest bone. The larger membrane-covered spaces between the developing skull bones that have not yet been ossified are called **fontanels,** or soft spots (figure 6.11; see chapter 8). The bones eventually grow together, and all the fontanels have usually closed by 2 years of age.

1. Intramembranous ossification begins when some of the mesenchymal cells in the membrane become osteochondral progenitor cells, which specialize to become osteoblasts. The osteoblasts produce bone matrix that surrounds the collagen fibers of the connective tissue membrane, and the osteoblasts become osteocytes. As a result of this process, many tiny trabeculae of woven bone develop (see figure 6.10, step 1).
2. Additional osteoblasts gather on the surfaces of the trabeculae and produce more bone, thereby causing the trabeculae to become larger and longer (see figure 6.10, step 2). Cancellous bone forms as the trabeculae join together, resulting in an interconnected network of trabeculae separated by spaces.
3. Cells within the spaces of the cancellous bone specialize to form red bone marrow, and cells surrounding the developing bone specialize to form the periosteum. Osteoblasts from the periosteum lay down bone matrix to form an outer surface of compact bone (see figure 6.10, step 3).

Thus, the end products of intramembranous bone formation are bones with outer compact bone surfaces and cancellous centers (see figure 6.9). Remodeling converts woven bone to lamellar bone and contributes to the final shape of the bone.

Endochondral Ossification

The formation of cartilage begins at approximately the end of the fourth week of development. Endochondral ossification of some of this cartilage starts at approximately the eighth week of development. Endochondral ossification of some cartilage might not begin until as late as age 18–20 years. Bones of the base of the skull, part of the mandible, the epiphyses of the clavicles, and most

TABLE 6.2 Comparison of Intramembranous and Endochondral Ossification

Intramembranous Ossification	Endochondral Ossification
Embryonic mesenchyme forms a collegen membrane containing osteochondral progenitor cells.	Embryonic mesenchymal cells become chondroblasts, which produce a cartilage template surrounded by the perichondrium.
No stage is comparable.	Chondrocytes hypertrophy, the cartilage, matrix becomes calcified, and the chondrocytes die.
Embryonic mesenchyme forms the periosteum, which contains osteoblasts.	The perichondrium becomes the periosteum when osteochondral progenitor cells within the periosteum become osteoblasts.
Osteochondral progenitor cells become osteoblasts at centers of ossification; internally, the osteoblasts form cancellous bone; externally, the periosteal osteoblasts form compact bone.	Blood vessels and osteoblasts from the periosteum invade the calcified cartilage template; internally, these osteoblasts form cancellous bone at primary ossification centers (and later at secondary ossification centers); externally, the periosteal osteoblasts form compact bone.
Intramembranous bone is remodeled and becomes indistinguishable from endochondral bone.	Endochondral bone is remodeled and becomes indistinguishable from intramembranous bone.

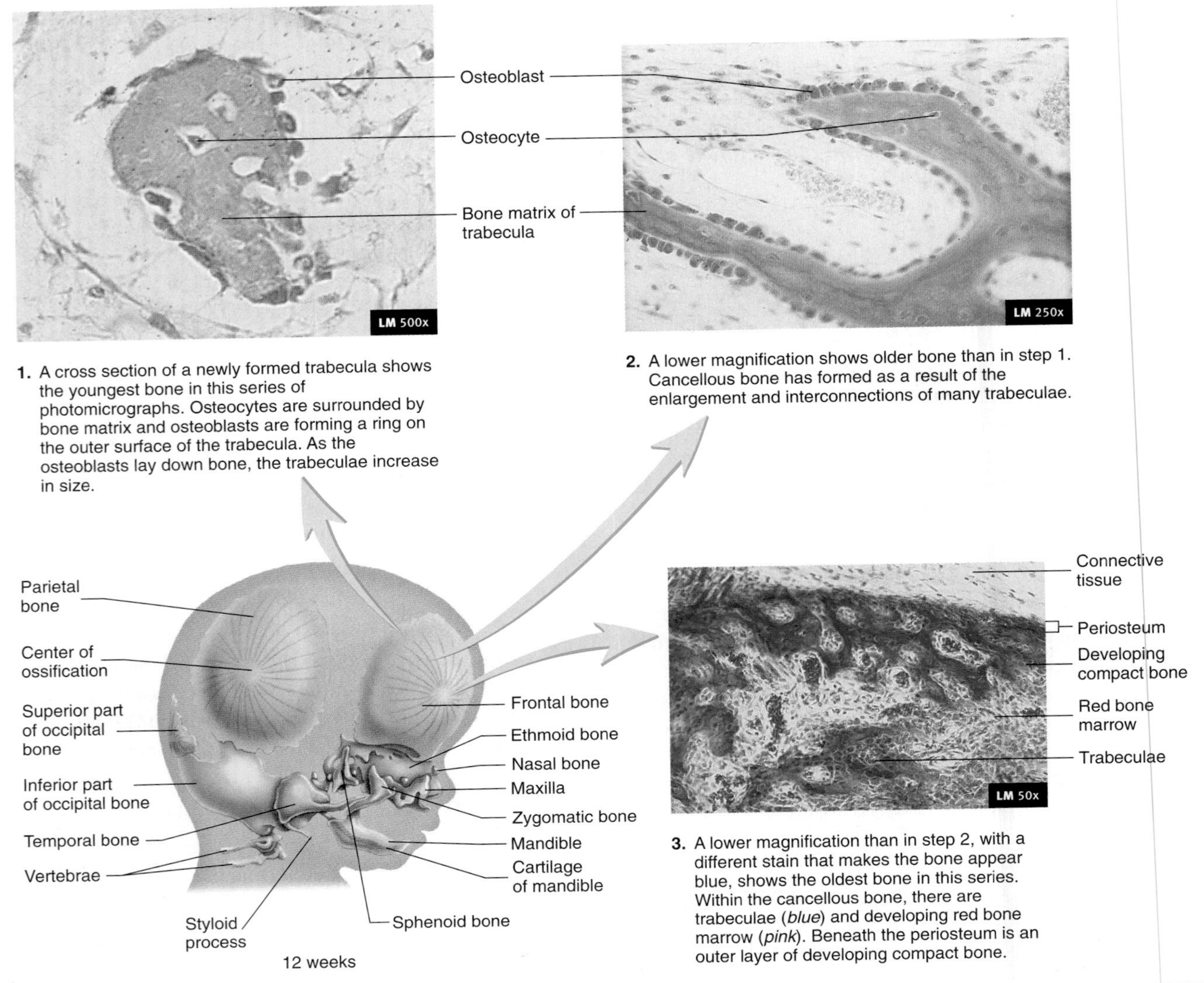

PROCESS FIGURE 6.10 Intramembranous Ossification

The inset shows a 12-week-old fetus. Bones formed by intramembranous ossification are *yellow* and bones formed by endochondral ossification are *blue*. Intramembranous ossification starts at a center of ossification and expands outward. Therefore, the youngest bone is at the edge of the expanding bone, and the oldest bone is at the center of ossification.

of the remaining skeletal system develop through the process of endochondral ossification (see figures 6.10 and 6.11).

1. Endochondral ossification begins as mesenchymal cells aggregate in regions of future bone formation. The mesenchymal cells become osteochondral progenitor cells that become chondroblasts. The chondroblasts produce a hyaline **cartilage model** having the approximate shape of the bone that will later be formed (figure 6.12, step 1). As the chondroblasts become surrounded by cartilage matrix, they become chondrocytes. The cartilage model is surrounded by perichondrium, except where a joint will form connecting one bone to another bone. Not shown in figure 6.12, the perichondrium is continuous with tissue that will become the joint capsule (see chapter 8).
2. When blood vessels invade the perichondrium surrounding the cartilage model (figure 6.12, step 2), osteochondral progenitor cells within the perichondrium become osteoblasts. The perichondrium becomes the periosteum when the osteoblasts begin to produce bone. The osteoblasts produce compact bone on the surface of the cartilage model, forming a **bone collar.** Two other events are occurring at the same time that the bone collar is forming. First, the cartilage

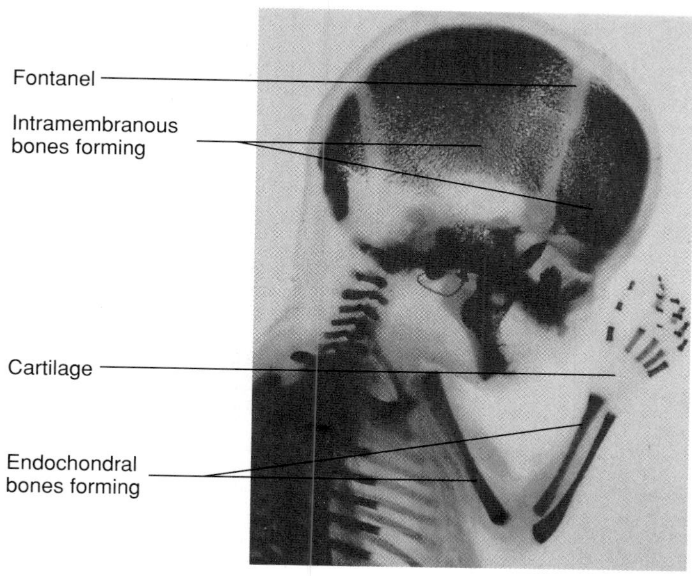

FIGURE 6.11 Bone Formation

Eighteen-week-old fetus showing intramembranous and endochondral ossification. Intramembranous ossification occurs at centers of ossification in the flat bones of the skull. Endochondral ossification has formed bones in the diaphyses of long bones. The epiphyses are still cartilage at this stage of development.

model increases in size as a result of interstitial and appositional cartilage growth. Second, the chondrocytes in the center of the cartilage model absorb some of the cartilage matrix and **hypertrophy** (hī-per′trō-fē), or enlarge. The chondrocytes also release matrix vesicles, which initiate the formation of hydroxyapatite crystals in the cartilage matrix. At this point, the cartilage is referred to as **calcified cartilage.** The chondrocytes in this calcified area eventually die, leaving enlarged lacunae with thin walls of calcified matrix.

3. Blood vessels grow into the enlarged lacunae of the calcified cartilage (figure 6.12, step 3). The connective tissue surrounding the blood vessels brings in osteoblasts and osteoclasts from the periosteum. The osteoblasts produce bone on the surface of the calcified cartilage, forming bone trabeculae, which changes the calcified cartilage of the diaphysis into cancellous bone. This area of bone formation is called the **primary ossification center.**
4. As bone development proceeds, the cartilage model continues to grow, more perichondrium becomes periosteum, the bone collar thickens and extends farther along the diaphysis, and additional cartilage within the diaphysis is calcified and transformed into cancellous bone (figure 6.12, step 4). Remodeling converts woven bone to lameller bone and contributes to the final shape of the bone. Osteoclasts remove bone from the center of the diaphysis to form the medullary cavity, and cells within the medullary cavity specialize to form red bone marrow.
5. In long bones, the diaphysis is the primary ossification center, and additional sites of ossification, called **secondary ossification centers,** appear in the epiphyses (figure 6.12, step 5). The events occurring at the secondary ossification centers are the same as those occurring at the primary ossification centers, except that the spaces in the epiphyses do not enlarge to form a medullary cavity as in the diaphysis. Primary ossification centers appear during early fetal development, whereas secondary ossification centers appear in the proximal epiphysis of the femur, humerus, and tibia about 1 month before birth. A baby is considered full-term if one of these three ossification centers can be seen on radiographs at the time of birth. At about 18–20 years of age, the last secondary ossification center appears in the medial epiphysis of the clavicle.
6. Replacement of cartilage by bone continues in the cartilage model until all the cartilage, except that in the epiphyseal plate and on articular surfaces, has been replaced by bone (figure 6.12, step 6). The epiphyseal plate, which exists throughout an individual's growth, and the articular cartilage, which is a permanent structure, are derived from the original embryonic cartilage model.
7. In mature bone, cancellous and compact bone are fully developed and the epiphyseal plate has become the epiphyseal line. The only cartilage present is the articular cartilage at the ends of the bone (figure 6.12, step 7). All the original perichondrium that surrounded the cartilage model has become periosteum.

18 ***Describe the formation of cancellous and compact bone during intramembranous ossification. What are centers of ossification? What are fontanels?***

19 ***For the process of endochondral ossification, describe the formation of these structures: cartilage model, bone collar, calcified cartilage, primary ossification center, medullary cavity, secondary ossification center, epiphyseal plate, epiphyseal line, and articular cartilage.***

20 ***When do primary and secondary ossification centers appear during endochondral ossification?***

PREDICT 4

During endochondral ossification, calcification of cartilage results in the death of chondrocytes. However, ossification of the bone matrix does not result in the death of osteocytes. Explain.

BONE GROWTH

Unlike cartilage, bones cannot grow by interstitial growth. Bones increase in size only by appositional growth, the formation of new bone on the surface of older bone or cartilage. For example, trabeculae grow in size by the deposition of new bone matrix by osteoblasts onto the surface of the trabeculae (see figure 6.10).

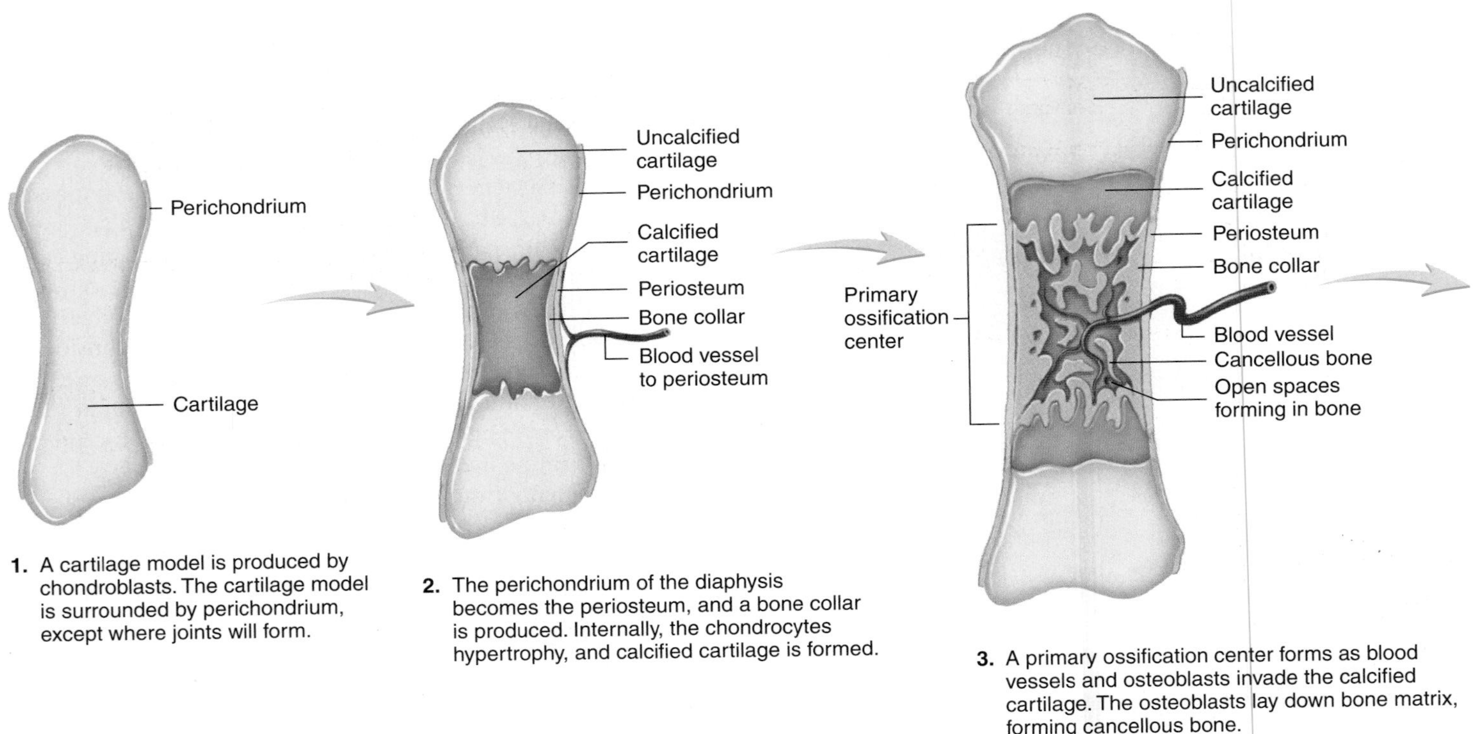

PROCESS FIGURE 6.12 Endochondral Ossification
Endochondral ossification begins with the formation of a cartilage model. See successive steps as indicated by the *blue arrows*.

PREDICT 5

Explain why bones cannot undergo interstitial growth, as does cartilage.

Growth in Bone Length

Long bones and bony projections increase in length because of growth at the epiphyseal plate. In a long bone, the epiphyseal plate separates the epiphysis from the diaphysis (figure 6.13*a*). Long projections of bones, such as the processes of vertebrae (see chapter 7), also have epiphyseal plates.

Growth at the epiphyseal plate involves the formation of new cartilage by interstitial cartilage growth followed by appositional bone growth on the surface of the cartilage. The epiphyseal plate is organized into four zones (figure 6.13*b*). The **zone of resting cartilage** is nearest the epiphysis and contains randomly arranged chondrocytes that do not divide rapidly. The chondrocytes in the **zone of proliferation** produce new cartilage through interstitial cartilage growth. The chondrocytes divide and form columns resembling stacks of plates or coins. In the **zone of hypertrophy,** the chondrocytes produced in the zone of proliferation mature and enlarge. Thus, a maturation gradient exists in each column: The cells nearer the epiphysis are younger and are actively proliferating, whereas the cells progressively nearer the diaphysis are older and are undergoing hypertrophy. The **zone of calcification** is very thin and contains hypertrophied chondrocytes and calcified cartilage matrix. The hypertrophied chondrocytes die, and blood vessels from the diaphysis grow into the area. The connective tissue surrounding the blood vessels contains osteoblasts from the endosteum. The osteoblasts line up on the surface of the calcified cartilage and, through appositional bone growth, deposit new bone matrix, which is later remodeled. The process of cartilage calcification and ossification in the epiphyseal plate occurs by the same basic process as the calcification and ossification of the cartilage model during endochondral bone formation.

As new cartilage cells form in the zone of proliferation, and as these cells enlarge in the zone of hypertrophy, the overall length of the diaphysis increases (figure 6.14). The thickness of the epiphyseal plate does not increase, however, because the rate of cartilage growth on the epiphyseal side of the plate is equal to the rate at which cartilage is replaced by bone on the diaphyseal side of the plate.

As the bones achieve normal adult size, growth in bone length ceases because the epiphyseal plate is ossified and becomes the epiphyseal line. This event, called closure of the epiphyseal plate, occurs between approximately 12 and 25 years of age, depending on the bone and the individual.

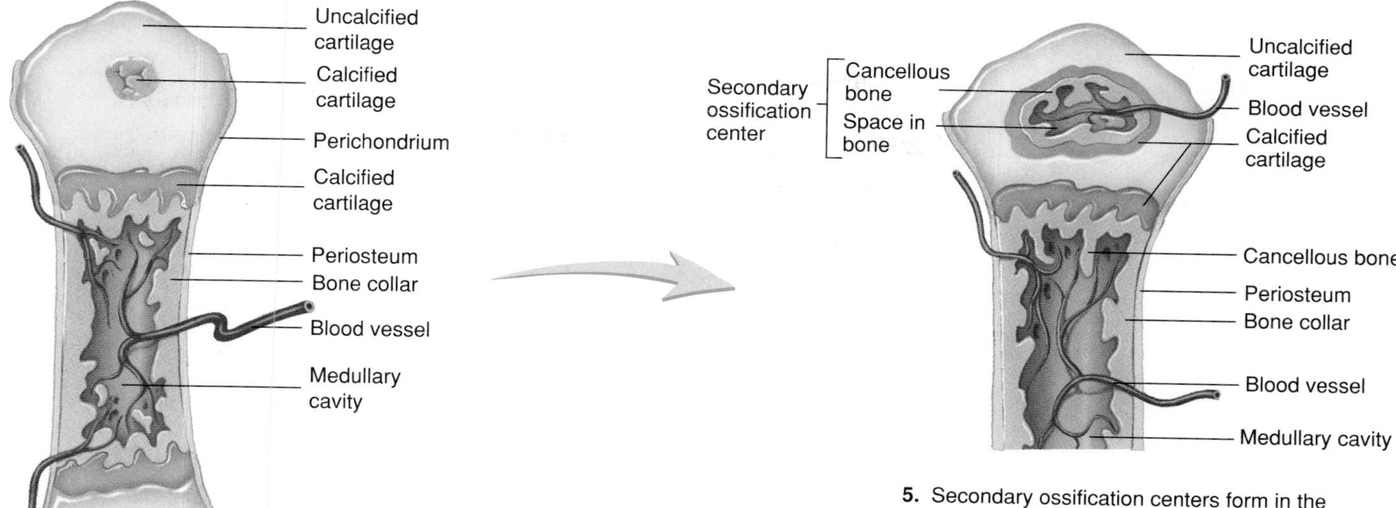

4. The process of bone collar formation, cartilage calcification, and cancellous bone production continues. Calcified cartilage begins to form in the epiphyses. A medullary cavity begins to form in the center of the diaphysis.

5. Secondary ossification centers form in the epiphyses of long bones.

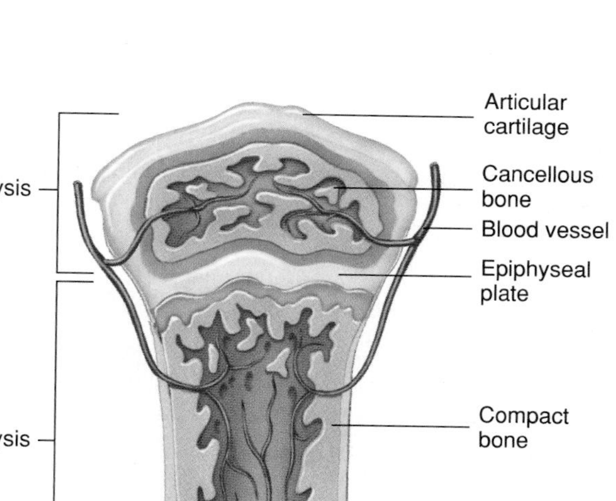

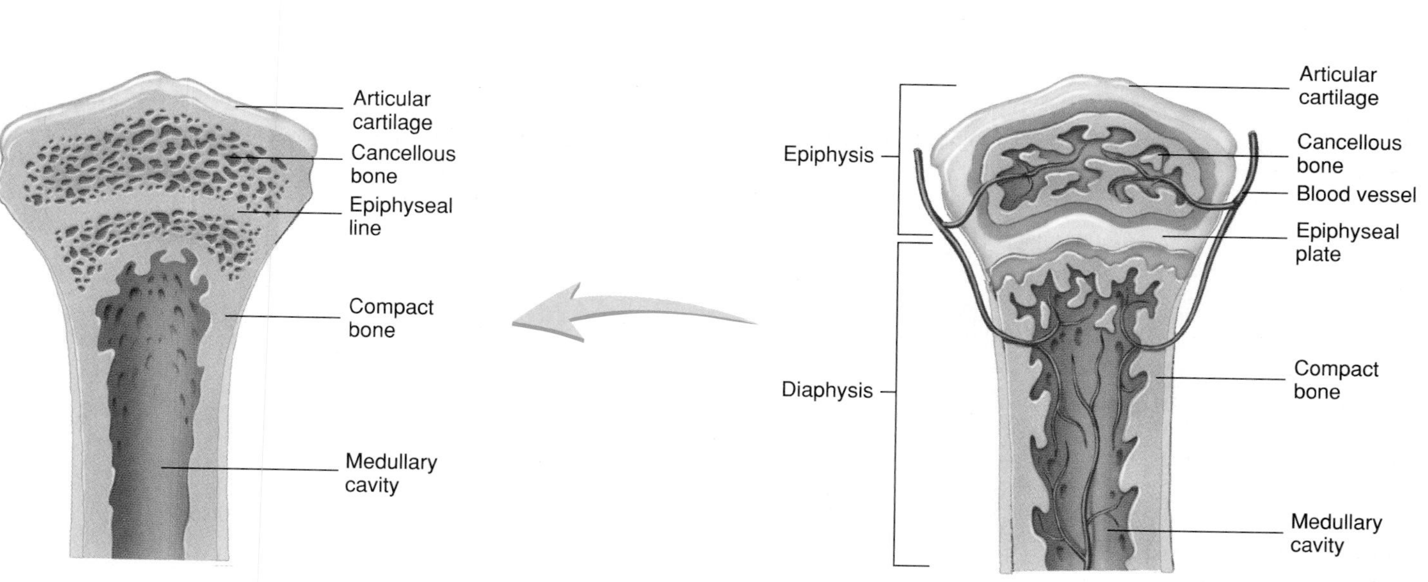

6. The original cartilage model is almost completely ossified. Unossified cartilage becomes the epiphyseal plate and the articular cartilage.

7. In a mature bone, the epiphyseal plate has become the epiphyseal line and all the cartilage in the epiphysis, except the articular cartilage, has become bone.

PROCESS FIGURE 6.12 ***(continued)***

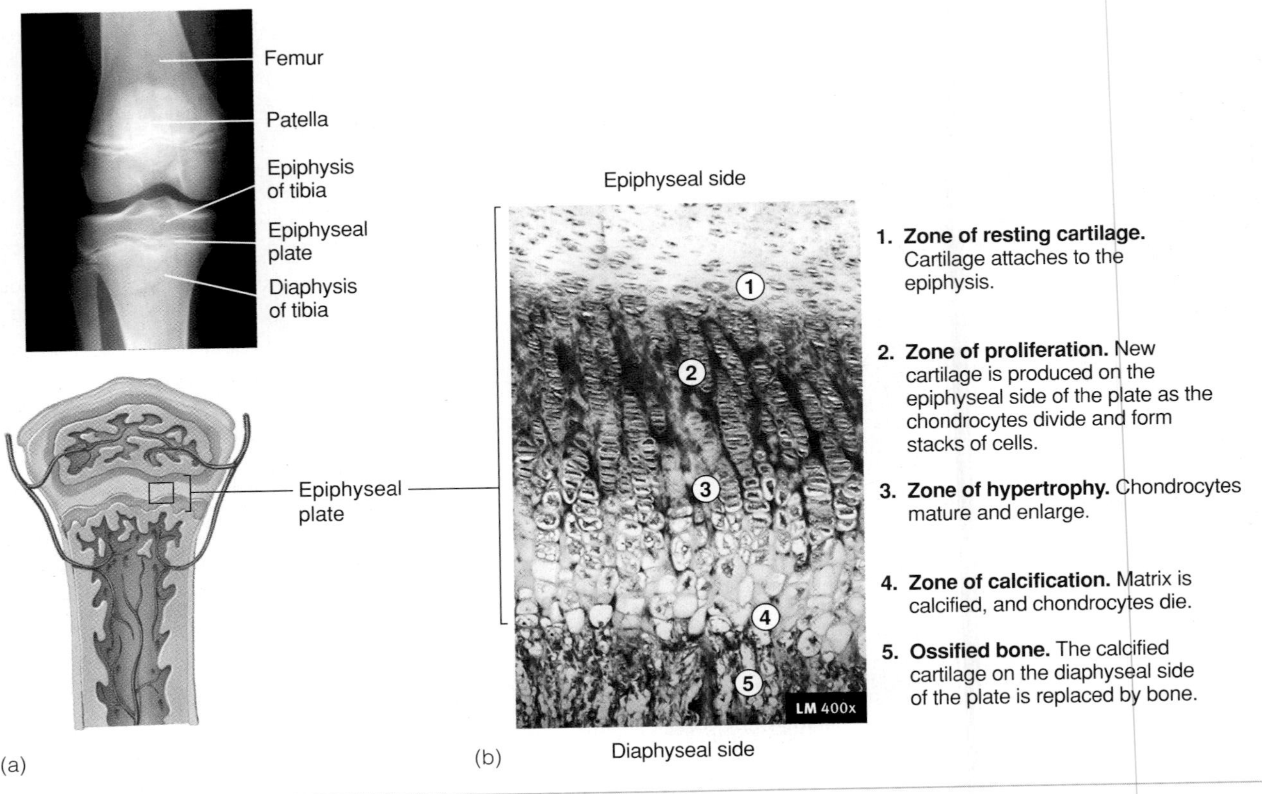

PROCESS FIGURE 6.13 Epiphyseal Plate

(*a*) Radiograph of the knee, showing the epiphyseal plate of the tibia (shinbone). Because cartilage does not appear readily on x-ray film, the epiphyseal plate appears as a black area between the white diaphysis and the epiphyses. (*b*) Zones of the epiphyseal plate.

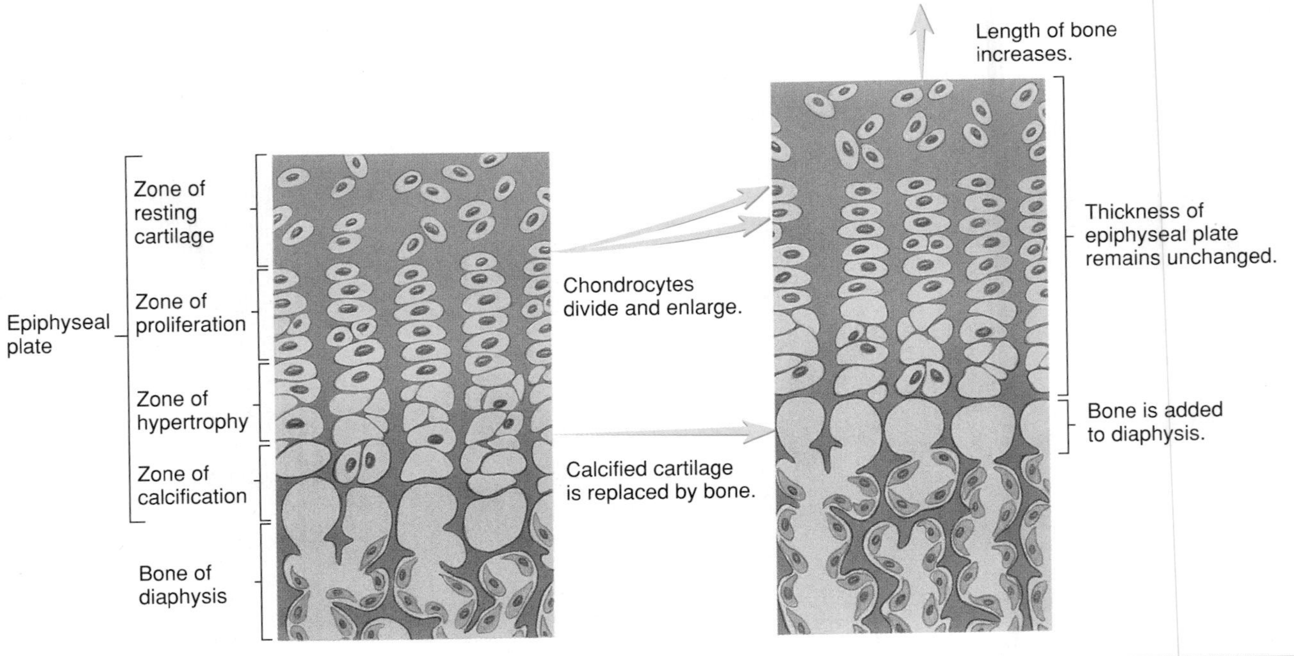

FIGURE 6.14 Bone Growth in Length at the Epiphyseal Plate

New cartilage is formed on the epiphyseal side of the plate at the same rate that new bone is formed on the diaphyseal side of the plate. Consequently, the epiphyseal plate remains the same thickness, but the length of the diaphysis increases.

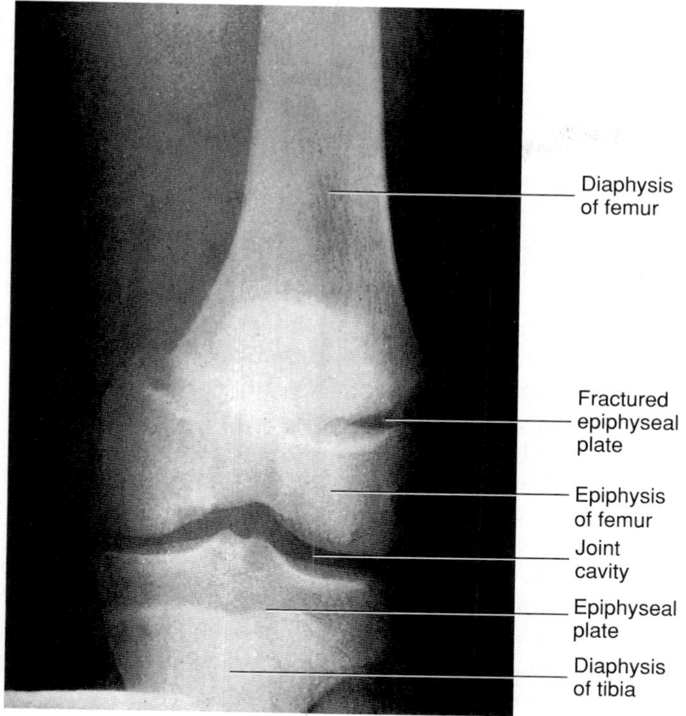

FIGURE 6.15 Fracture of the Epiphyseal Plate
Radiograph of an adolescent's knee. The femur (thighbone) is separated from the tibia (leg bone) by a joint cavity. The epiphyseal plate of the femur is fractured, thereby separating the diaphysis from the epiphysis.

PREDICT 6

A 15-year-old football player is tackled during a game, and the epiphyseal plate of the left femur is damaged (figure 6.15). What are the results of such an injury, and why is recovery difficult?

Growth at Articular Cartilage

Epiphyses increase in size because of growth at the articular cartilage. In addition, growth at the articular cartilage increases the size of bones that do not have an epiphysis, such as short bones. The process of growth in articular cartilage is similar to that occurring in the epiphyseal plate, except that the chondrocyte columns are not as obvious. The chondrocytes near the surface of the articular cartilage are similar to those in the zone of resting cartilage of the epiphyseal plate. In the deepest part of the articular cartilage, nearer bone tissue, the cartilage is calcified and ossified to form new bone.

When the epiphyses reach their full size, the growth of cartilage and its replacement by bone cease. The articular cartilage, however, persists throughout life and does not become ossified as does the epiphyseal plate.

PREDICT 7

Growth at the epiphyseal plate stops when the epiphyseal cartilage becomes ossified. The articular cartilage, however, does not become ossified when growth of the epiphysis ceases. Explain why it is advantageous for the articular cartilage not to be ossified.

Growth in Bone Width

Long bones increase in width (diameter) and other bones increase in size or thickness because of appositional bone growth beneath the periosteum. When bone growth in width is rapid, as in young bones or during puberty, osteoblasts from the periosteum lay down bone to form a series of ridges with grooves between them (figure 6.16, step 1). The periosteum covers the bone ridges and extends down into the bottom of the grooves, and one or more blood vessels of the periosteum lie within each groove. As the osteoblasts continue to produce bone, the ridges increase in size, extend toward each other, and meet to change the groove into a tunnel (figure 6.16, step 2). The name of the periosteum in the tunnel changes to *endosteum* because the membrane now lines an internal bone surface. Osteoblasts from the endosteum lay down bone to form a concentric lamella (figure 6.16, step 3). The production of additional lamellae fills in the tunnel, encloses the blood vessel, and produces an osteon (figure 6.16, step 4).

When bone growth in width is slow, the surface of the bone becomes smooth as osteoblasts from the periosteum lay down even layers of bone to form circumferential lamellae. The circumferential lamellae are broken down during remodeling to form osteons (see "Bone Remodeling," p. 191).

21 ***Name and describe the events occurring in the four zones of the epiphyseal plate. Explain how the epiphyseal plate remains the same thickness while the bone increases in length.***

22 ***Describe the process of growth at the articular cartilage. What happens to the epiphyseal plate and the articular cartilage when bone growth ceases?***

23 ***Describe how new osteons are produced as a bone increases in width.***

Factors Affecting Bone Growth

The bones of an individual's skeleton usually reach a certain length, thickness, and shape through the processes described in the previous sections. The potential shape and size of a bone and an individual's final adult height are determined genetically, but factors such as nutrition and hormones can greatly modify the expression of those genetic factors.

Nutrition

Because bone growth requires chondroblast and osteoblast proliferation, any metabolic disorder that affects the rate of cell proliferation or the production of collagen and other matrix components affects bone growth, as does the availability of calcium or other minerals needed in the mineralization process.

The long bones of a child sometimes exhibit lines of arrested growth, which are transverse regions of greater bone density crossing an otherwise normal bone. These lines are caused by greater calcification below the epiphyseal plate of a bone, where it has grown at a slower rate during an illness or severe nutritional deprivation. They demonstrate that illness or malnutrition during the time of bone growth can cause a person to be shorter than he or she would have been otherwise.

Certain vitamins are important in very specific ways to bone growth. **Vitamin D** is necessary for the normal absorption of

1. Osteoblasts beneath the periosteum lay down bone (*dark brown*) to form ridges separated by grooves. Blood vessels of the periosteum lie in the grooves.

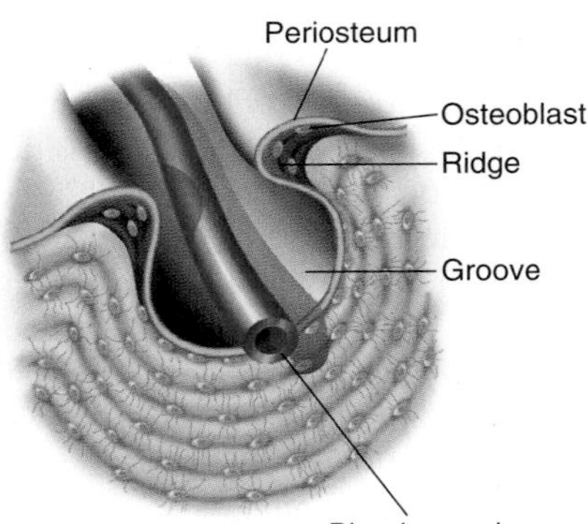

2. The groove is transformed into a tunnel when the bone built on adjacent ridges meets. The periosteum of the groove becomes the endosteum of the tunnel.

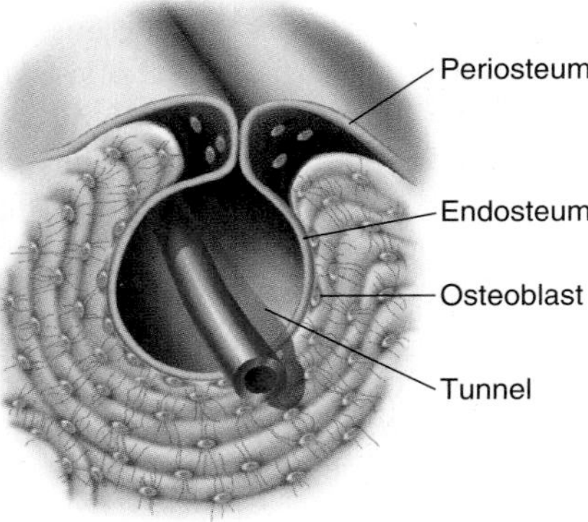

3. Appositional growth by osteoblasts from the endosteum results in the formation of a new concentric lamella.

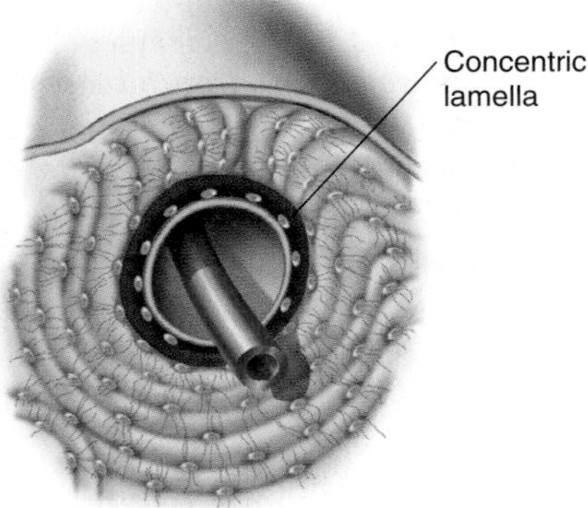

4. The production of additional concentric lamellae fills in the tunnel and completes the formation of the osteon.

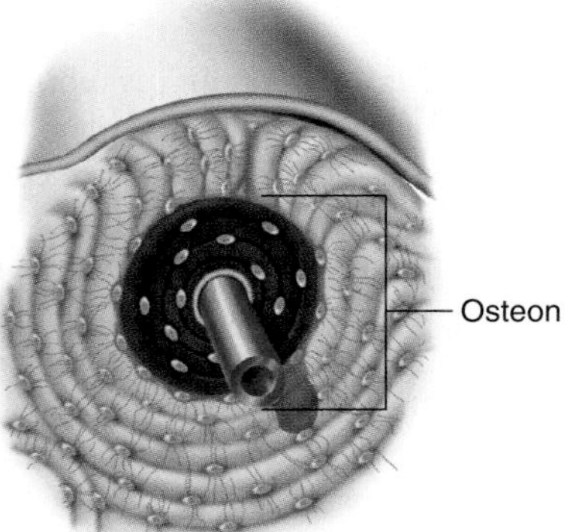

PROCESS FIGURE 6.16 Bone Growth in Width
Bones can increase in width by the formation of new osteons beneath the periosteum.

calcium from the intestines (see chapters 5 and 24). The body can either synthesize or ingest vitamin D. Its rate of synthesis increases when the skin is exposed to sunlight.

Insufficient vitamin D in children causes **rickets,** a disease resulting from reduced mineralization of the bone matrix. Children with rickets can have bowed bones and inflamed joints. During the winter in northern climates, if children are not exposed to sufficient sunlight, they can take vitamin D as a dietary supplement to prevent rickets. The body's inability to absorb fats in which vitamin D is soluble can also result in vitamin D deficiency. This condition can occur in adults who suffer from digestive disorders and can be one cause of "adult rickets," or **osteomalacia** (os′tē-ō-mă-lā′shē-ă), which is a softening of the bones as a result of calcium depletion.

Vitamin C is necessary for collagen synthesis by osteoblasts. Normally, as old collagen breaks down, new collagen is synthesized to replace it. Vitamin C deficiency results in bones and cartilage that are deficient in collagen because collagen synthesis is impaired. In children, vitamin C deficiency can cause growth retardation. In children and adults, vitamin C deficiency can result in **scurvy,** which is marked by ulceration and hemorrhage in almost any area of the body because of a lack of normal collagen synthesis in connective tissues. Wound healing, which requires collagen synthesis, is hindered in patients with vitamin C deficiency. In extreme cases, the teeth can fall out because the ligaments that hold them in place break down.

Hormones

Hormones are very important in bone growth. **Growth hormone** from the anterior pituitary increases general tissue growth (see chapters 17 and 18), including overall bone growth, by stimulating interstitial cartilage growth and appositional bone growth. **Thyroid hormone** is also required for normal growth of all tissues, including cartilage; therefore, a decrease in this hormone can result in decreased size of the individual. **Sex hormones** also influence bone growth. Estrogen (a class of female sex hormones) and testosterone (a male sex hormone) initially stimulate bone growth, which accounts for the burst of growth at puberty, when production of these hormones increases. Both hormones also stimulate ossification of epiphyseal plates, however, and thus the cessation of growth. Females usually stop growing earlier than males because estrogens cause a quicker closure of the epiphyseal plate than does testosterone. Because their entire growth period is somewhat shorter, females usually do not reach the same height as males. Decreased levels of testosterone or estrogen can prolong the growth phase of the epiphyseal plates, even though the bones grow more slowly. Growth is very complex, however, and is influenced by many factors in addition to sex hormones, such as other hormones, genetics, and nutrition.

24 ***Explain how illness or malnutrition can affect bone growth. How do vitamins D and C affect bone growth?***

25 ***What are the effects of growth hormone and thyroid hormone on bone growth?***

26 ***What effects do estrogen and testosterone have on bone growth? How do these effects account for the average height difference observed in men and women?***

PREDICT 8

A 12-year-old female has an adrenal tumor that produces large amounts of estrogen. If untreated, what effect will this condition have on her growth for the next 6 months? On her height when she is 18?

BONE REMODELING

Just as we renovate or remodel our homes when they become outdated, when bone becomes old it is replaced with new bone in a process called **bone remodeling.** In this process, osteoclasts remove old bone and osteoblasts deposit new bone. Bone remodeling converts woven bone into lamellar bone, and it is involved in bone growth, changes in bone shape, the adjustment of the bone to stress, bone repair, and calcium ion (Ca^{2+}) regulation in the body. For example, as a long bone increases in length and diameter, the size of the medullary cavity also increases (figure 6.17). Otherwise, the bone would consist of nearly solid bone matrix and would be very heavy. A cylinder with the same height, weight, and composition as a solid rod but with a greater diameter can support much more weight than the rod without bending. Bone therefore has a mechanical advantage as a cylinder rather than as a rod. The relative thickness of compact bone is maintained by the removal of bone on the inside by osteoclasts and the addition of bone to the outside by osteoblasts.

A **basic multicellular unit (BMU)** is a temporary assembly of osteoclasts and osteoblasts that travels through or across the surface of bone, removing old bone matrix and replacing it with new bone matrix. The average life span of a BMU is approximately 6 months, and BMU activity renews the entire skeleton every 10 years. In compact bone, the osteoclasts of a BMU break down bone matrix, forming a tunnel. Interstitial lamellae (see figure 6.6*a*) are remnants of osteons that were not completely removed when a BMU formed a tunnel. Blood vessels grow into the tunnel and osteoblasts of the BMU move in and lay down a layer of bone on the tunnel wall, forming a concentric lamella. Additional concentric lamellae are produced, filling in the tunnel from the outside to the inside, until an osteon is formed in which the center of the tunnel becomes a central canal containing blood vessels. In cancellous bone, the BMU removes bone matrix from the surface of a trabecula, forming a cavity, which the BMU then fills in with new bone matrix.

Mechanical Stress and Bone Strength

Remodeling, the formation of additional bone, alteration in trabecular alignment to reinforce the scaffolding, or other changes can modify the strength of the bone in response to the amount of stress applied to it. Mechanical stress applied to bone increases osteoblast activity in bone tissue, and removal of mechanical stress decreases osteoblast activity. Under conditions of reduced stress, such as when a person is bedridden or paralyzed, osteoclast activity continues at a nearly normal rate, but osteoblast activity is reduced, resulting in a decrease in bone density. In addition, pressure in bone causes an electrical change that increases the activity of osteoblasts. Applying weight (pressure) to a broken bone therefore speeds the healing process. Weak pulses of electric current applied to a broken bone sometimes are used clinically to speed the healing process.

27 ***What cells are involved in bone remodeling? Describe how the medullary cavity of a long bone can increase in size as the width of the bone increases.***

28 ***Explain two ways that remodeling is responsible for the formation of new osteons in compact bone.***

29 ***How does bone adjust to stress? Describe the role of osteoblasts and osteoclasts in this process. What happens to bone that is not subjected to stress?***

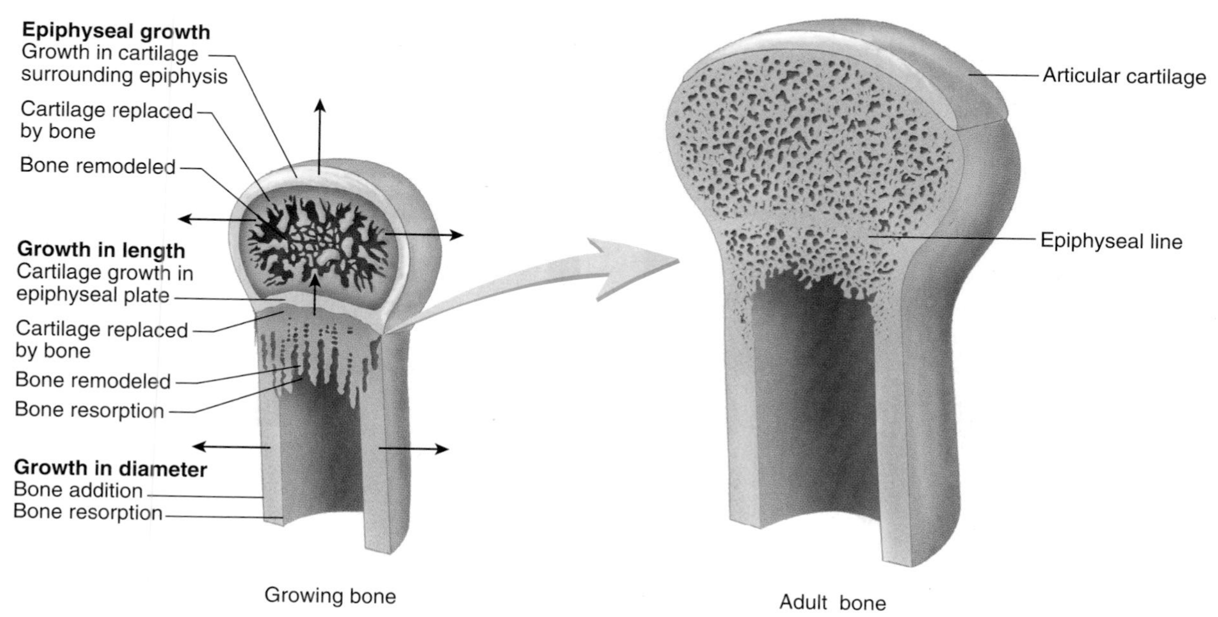

FIGURE 6.17 Remodeling of a Long Bone

The epiphysis enlarges and the diaphysis increases in length as new cartilage is formed and replaced by bone, which is remodeled. The diameter of the bone increases as a result of bone growth on the outside of the bone, and the size of the medullary cavity increases because of bone resorption.

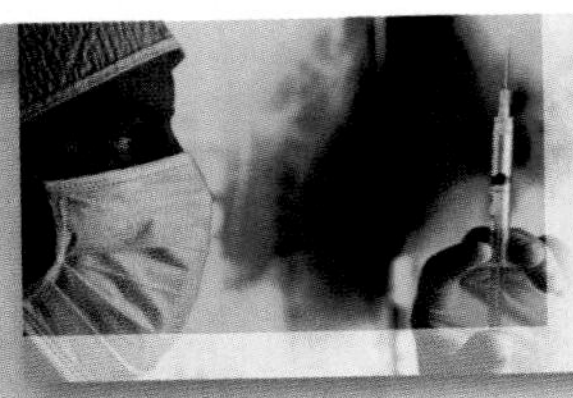

CLINICAL FOCUS

Bone Disorders

GROWTH AND DEVELOPMENT DISORDERS

Giantism is a condition of abnormally increased height that usually results from excessive cartilage and bone formation at the epiphyseal plates of long bones (figure B). The most common type of giantism, **pituitary giantism,** results from excess secretion of pituitary growth hormone. The large stature of some individuals, however, can result from genetic factors rather than from abnormal levels of growth hormone. **Acromegaly** (ak-rō-meg′ă-lē) is also caused by excess pituitary growth hormone secretion; however, acromegaly involves the growth of connective tissue, including bones, after the epiphyseal plates have ossified. The effect mainly involves increased diameter of all bones and is most strikingly apparent in the face and hands. Many pituitary giants also develop acromegaly later in life.

Dwarfism, the condition in which a person is abnormally short, is the opposite of giantism (see figure B). **Pituitary dwarfism** results when abnormally low levels of pituitary growth hormone affect the whole body, thus producing a small person who is normally proportioned. **Achondroplasia** (ā-kon-drō-plā′zē-ă), or **achondroplastic** (ā-kon-drō-plas′tik) **dwarfism,** is the most common type of dwarfism; it produces a person with a nearly normal-sized trunk and head but shorter than normal limbs. Achondroplasia is an autosomal-dominant trait. Approximately 80% of cases result from a spontaneous mutation of the fibroblast growth factor receptor gene on chromosome 4 during the formation of sperm cells or oocytes. Thus, the parents of most achondroplastic dwarfs are of normal height and proportions. The normal effect of the gene is to slow bone growth by inhibiting chondrocyte division at the epiphyseal plate. Mutation of the gene results in a "gain of function," in which the normal inhibitory effect is increased, resulting in severely reduced bone growth in length.

Giant and dwarf

FIGURE B Bone Disorders

BACTERIAL INFECTIONS

Osteomyelitis (os′tē-ō-mī-ĕ-lī′tis) is bone inflammation that often results from bacterial infection. It can lead to complete destruction of the bone. *Staphylococcus aureus,* often introduced into the body through wounds, is a common cause of osteomyelitis. Bone tuberculosis, a specific type of osteomyelitis, results from spread of the tubercular bacterium (*Mycobacterium tuberculosis*) from the initial site of infection, such as the lungs to the bones through the circulatory system.

TUMORS

Many types of tumors cause a wide range of resultant bone defects with varying prognoses. Tumors can be benign or malignant. Malignant bone tumors can metastasize to other parts of the body, or they can spread to bone from metastasizing tumors elsewhere in the body.

DECALCIFICATION

Osteomalacia (os′t-ē-ō-mă-lā′shē-ă), or the softening of bones, results from calcium depletion from bones. If the body has an unusual need for calcium—such as during pregnancy, when growth of the fetus requires large amounts of calcium—it can be removed from the mother's bones, which become soft and weakened. **Osteoporosis,** which is a major disorder of decalcification, is discussed in the "Systems Pathology" section on p. 196.

BONE REPAIR

Bone is a living tissue that can undergo repair following damage to it. This process has four major steps.

1. *Hematoma formation* (figure 6.18, step 1). When bone is fractured, the blood vessels in the bone and surrounding periosteum are damaged, and a hematoma forms. A **hematoma** (hē-mă-tō′mă, hem-ă-tō′mă) is a localized mass of blood released from blood vessels but confined within an organ or a space. Usually, the blood in a hematoma forms a clot, which consists of fibrous proteins that stop the bleeding. Disruption of blood vessels in the central canals results

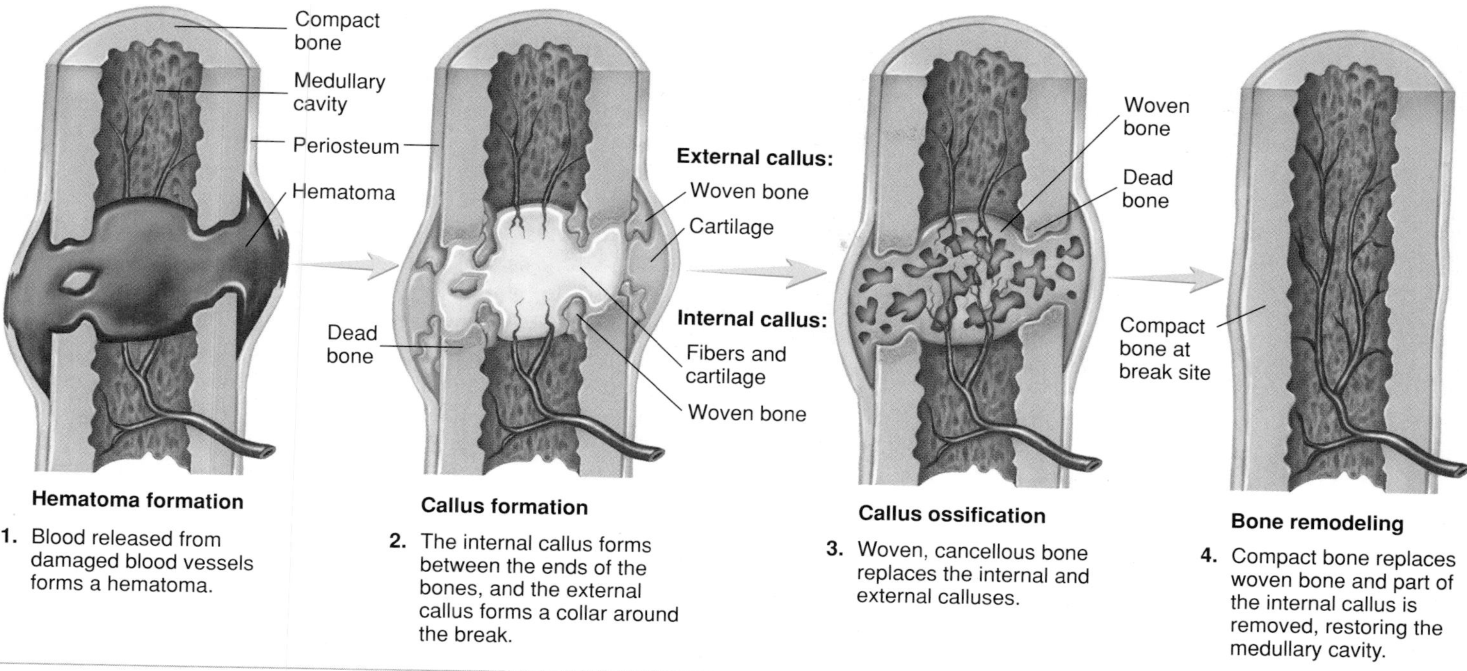

PROCESS FIGURE 6.18 Bone Repair

in inadequate blood delivery to osteocytes, and bone tissue adjacent to the fracture site dies. Inflammation and swelling of tissues around the bone often occur following the injury.

2. *Callus formation* (figure 6.18, step 2). A **callus** (kal′ŭs) is a mass of tissue that forms at a fracture site, connecting the broken ends of the bone. An **internal callus** forms *between* the ends of the broken bone, as well as in the marrow cavity if the fracture occurs in the diaphysis of a long bone. Several days after the fracture, blood vessels grow into the clot. As the clot dissolves (see chapter 19), macrophages clean up cell debris, osteoclasts break down dead bone tissue, and fibroblasts produce collagen and other extracellular materials to form granulation tissue (see chapter 4). As the fibroblasts continue to produce collagen fibers, a denser fibrous network, which helps hold the bone together, is produced. Chondroblasts derived from osteochondral progenitor cells of the periosteum and endosteum begin to produce cartilage in the fibrous network. As these events are occurring, osteochondral progenitor cells in the endosteum become osteoblasts and produce new bone that contributes to the internal callus.

   The **external callus** forms a collar *around* the opposing ends of the bone fragments. Osteochondral progenitor cells from the periosteum become osteoblasts, which produce bone, and chondroblasts, which produce cartilage. Cartilage production is more rapid than bone production, and the cartilage from each side of the break eventually grows together. The external callus is a bone–cartilage collar that stabilizes the ends of the broken bone. In modern medical practice, stabilization of the bone is assisted with a cast or surgical implantation of metal supports.

3. *Callus ossification* (figure 6.18, step 3). Like the cartilage models formed during fetal development, the cartilage in the external callus is replaced by woven, cancellous bone through endochondral ossification. The result is a stronger external callus. Even as the internal callus is forming and replacing the hematoma, osteoblasts from the periosteum and endosteum enter the internal callus and begin to produce bone. Eventually, the fibers and cartilage of the internal callus are replaced by woven, cancellous bone, which further stabilizes the broken bone.
4. *Bone remodeling* (figure 6.18, step 4). Filling the gap between bone fragments with an internal callus of woven bone is not the end of the repair process because woven bone is not as structurally strong as the original lamellar bone. Repair is not complete until the woven bone of the internal callus and the dead bone adjacent to the fracture site are replaced by compact bone. In this compact bone, osteons from both sides of the break extend across the fracture line to "peg" the bone fragments together. This remodeling process takes time and may not be complete even after a year. As the internal callus is remodeled and becomes stronger, the external callus is reduced in size by osteoclast activity. Eventually, repair may be so complete that no evidence of the break remains; however, the repaired zone usually remains slightly thicker than the adjacent bone. If the fracture occurred in the diaphysis of a long bone, remodeling also restores the medullary cavity.

30 ***How does breaking a bone result in hematoma formation?***

31 ***Distinguish between the location and the composition of the internal and external callus.***

32 ***Why is remodeling of the ossified callus necessary?***

Uniting Broken Bones

Before the formation of compact bone between the broken ends of a bone can take place, the appropriate substrate must be present. Normally, this is the woven, cancellous bone of the internal callus. If formation of the internal callus is prevented by infections, bone movements, or the nature of the injury, nonunion of the bone occurs. This condition can be treated by surgically implanting an appropriate substrate, such as living bone from another site in the body or dead bone from cadavers. Other substrates have also been used. For example, a specific marine coral calcium phosphate is converted into a predominantly hydroxyapatite biomatrix that is very much like cancellous bone.

CALCIUM HOMEOSTASIS

Bones play an important role in regulating blood Ca^{2+} levels, which must be maintained within narrow limits for functions such as muscle contraction and membrane potentials to occur normally (see chapters 9 and 11). Bone is the major storage site for calcium in the body, and movement of Ca^{2+} into and out of bone helps determine blood Ca^{2+} levels. Calcium ions move into bone as osteoblasts build new bone and out of bone as osteoclasts break down bone (figure 6.19). When osteoblast and osteoclast activity is balanced, the movement of Ca^{2+} into and out of a bone is equal.

When blood Ca^{2+} levels are too low, osteoclast activity increases. More Ca^{2+} are released by osteoclasts from bone into the blood than are removed by osteoblasts from the blood to make new bone. Consequently, a net movement of Ca^{2+} occurs from bone into blood, and blood Ca^{2+} levels increase. Conversely, if blood Ca^{2+} levels are too high, osteoclast activity decreases. Fewer Ca^{2+} are released by osteoclasts from bone into the blood than are taken from the blood by osteoblasts to produce new bone. As a result, a net movement of Ca^{2+} occurs from the blood to bone, and blood Ca^{2+} levels decrease.

Parathyroid hormone (PTH) from the parathyroid glands (see figure 18.11) is the major regulator of blood Ca^{2+} levels. If the blood Ca^{2+} level decreases, the secretion of PTH increases, resulting in increased numbers of osteoclasts, which causes increased bone breakdown and increased blood Ca^{2+} levels (see figure 6.19). In addition, osteoblasts respond to PTH by releasing enzymes, resulting in the breakdown of the layer of unmineralized organic bone matrix covering bone, thereby making the mineralized bone matrix available to osteoclasts.

Osteoclast numbers are regulated by the interactions of osteoblasts, red bone marrow stromal (stem) cells, and osteoclast precursor cells. Osteoblasts and stromal cells have receptors for PTH. When PTH binds to these receptors, these cells respond by producing **receptor for activation of nuclear factor kappaB ligand (RANKL).** RANKL is expressed on the surface of the osteoblasts and stromal cells and can combine with **receptor for activation of nuclear factor kappaB (RANK)** found on the cell surfaces of osteoclast precursor cells. In a cell-to-cell interaction, RANKL on osteoblasts or stromal cells binds to RANK on osteoclast precursor cells, stimulating them to become osteoclasts.

Osteoclast production is inhibited by **osteoprotegerin** (os′tē-ō-prō-teg′er-in) **(OPG),** which is secreted by osteoblasts and stromal cells. OPG inhibits osteoclast production by binding

1. Osteoclasts break down bone and release Ca^{2+} into the blood, and osteoblasts remove Ca^{2+} from the blood to make bone. (*Blue arrows represent the movement of Ca^{2+}.*) Parathyroid hormone (PTH) regulates blood Ca^{2+} levels by indirectly stimulating osteoclast activity, resulting in increased Ca^{2+} release into the blood. Calcitonin plays a minor role in Ca^{2+} maintenance by inhibiting osteoclast activity.
2. In the kidneys, PTH increases Ca^{2+} reabsorption from the urine.
3. In the kidneys, PTH also promotes the formation of active vitamin D (*green arrows*), which increases Ca^{2+} absorption from the small intestine.

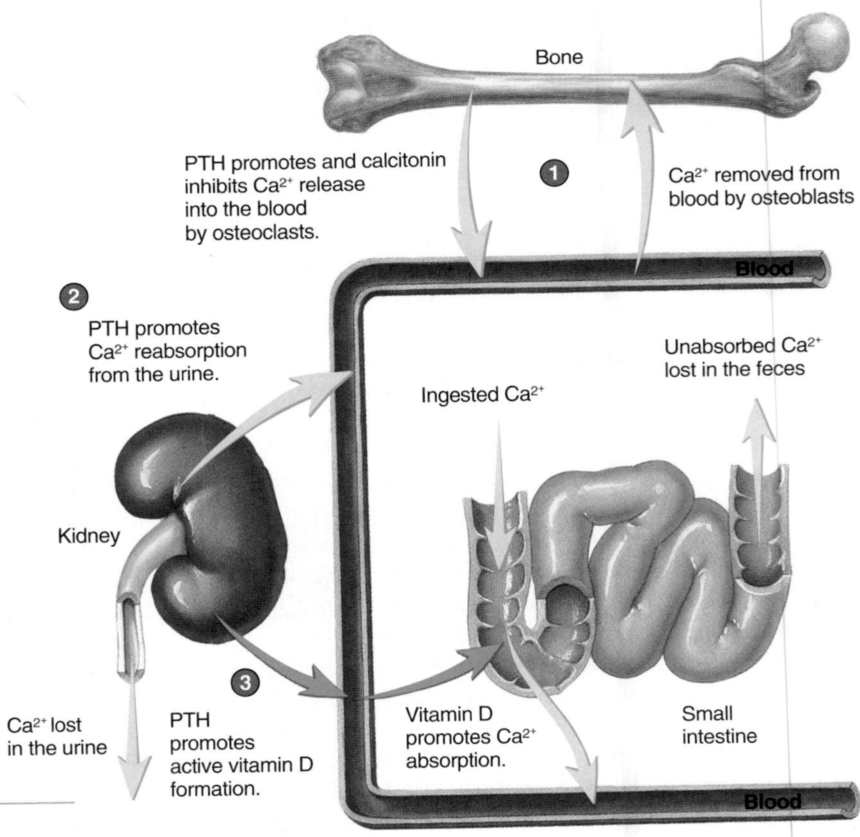

PROCESS FIGURE 6.19 Calcium Homeostasis

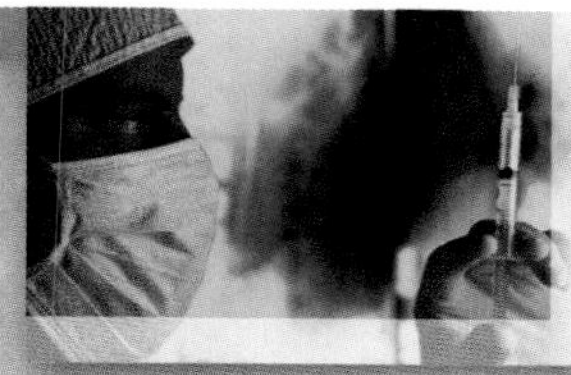

CLINICAL FOCUS

Classification of Bone Fractures

Bone fractures are classified in several ways. The most commonly used classification involves the severity of the injury to the soft tissues surrounding the bone. An **open fracture** (formerly called compound) occurs when an open wound extends to the site of the fracture or when a fragment of bone protrudes through the skin. If the skin is not perforated, the fracture is called a **closed fracture** (formerly called simple). If the soft tissues around a closed fracture are damaged, the fracture is called a **complicated fracture.**

Two other terms to designate fractures are **incomplete,** in which the fracture does not extend completely across the bone, and **complete,** in which the bone is broken into at least two fragments (figure C*a*). An incomplete fracture that occurs on the convex side of the curve of the bone is a **greenstick fracture. Hairline fractures** are incomplete fractures in which the two sections of bone do not separate; they are common in skull fractures.

Comminuted (kom′i-noo-ted) **fractures** are complete fractures in which the bone breaks into more than two pieces—usually two major fragments and a smaller fragment (figure C*b*). **Impacted fractures** are those in which one fragment is driven into the cancellous portion of the other fragment (figure C*c*).

Fractures are also classified according to the direction of the fracture within a bone. **Linear fractures** run parallel to the long axis of the bone, and **transverse fractures** are at right angles to the long axis (see figure C*b*). **Spiral fractures** have a helical course around the bone, and **oblique fractures** run obliquely in relation to the long axis (figure C*d*). **Dentate fractures** have rough, toothed, broken ends, and **stellate fractures** have breakage lines radiating from a central point.

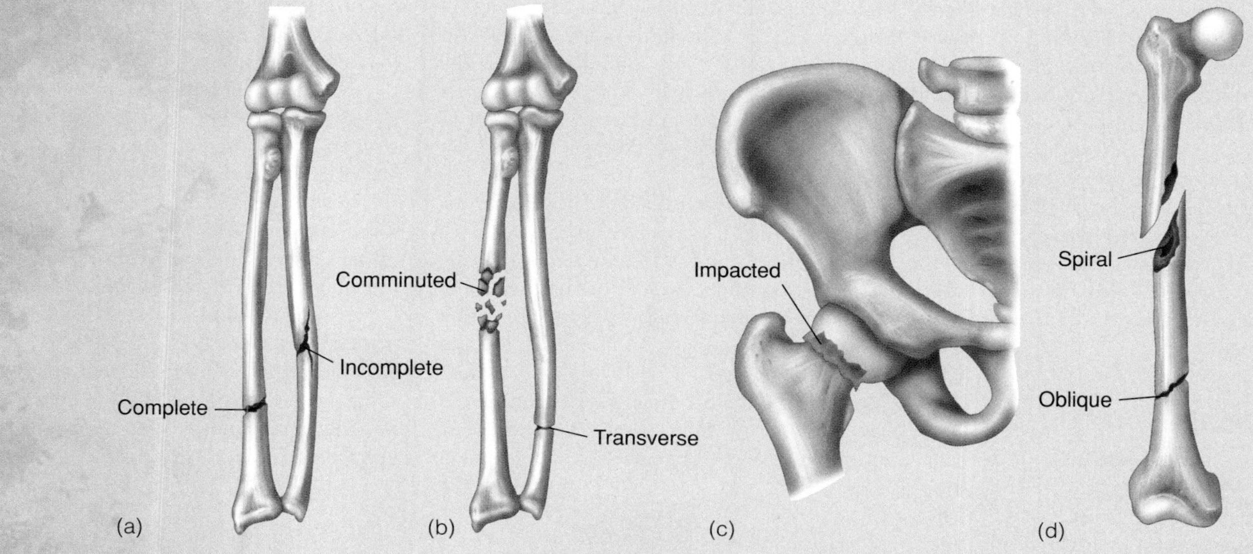

FIGURE C **Bone Fractures**

(*a*) Complete and incomplete. (*b*) Transverse and comminuted. (*c*) Impacted. (*d*) Spiral and oblique.

to RANKL and preventing RANKL from binding to its receptor on osteoclast precursor cells. Increased PTH causes decreased secretion of OPG from osteoblasts and stromal cells. Thus, increased PTH promotes an increase in osteoclast numbers by increasing RANKL and decreasing OPG. The increased RANKL stimulates osteoclast precursor cells and the decreased OPG results in less inhibition of osteoclast precursor cells. Conversely, decreased PTH results in fewer osteoclasts by decreasing RANKL and increasing OPG.

PTH also regulates blood Ca^{2+} levels by increasing Ca^{2+} uptake in the small intestine (see figure 6.19). Increased PTH promotes the formation of vitamin D in the kidneys, and vitamin D increases the absorption of Ca^{2+} from the small intestine. PTH also increases the reabsorption of Ca^{2+} from urine in the kidneys, which reduces Ca^{2+} lost in the urine.

Tumors that secrete large amounts of PTH can cause so much bone breakdown that bones become weakened and fracture easily. On the other hand, an increase in blood Ca^{2+} levels results in less PTH secretion, decreased osteoclast activity, reduced Ca^{2+} release from bone, and decreased blood Ca^{2+} levels.

Calcitonin (kal-si-tō′nin), secreted from the thyroid gland (see figure 18.8), decreases osteoclast activity (see figure 6.19) by binding to receptors on the osteoclasts. An increase in blood Ca^{2+} levels stimulates the thyroid gland to secrete calcitonin, which

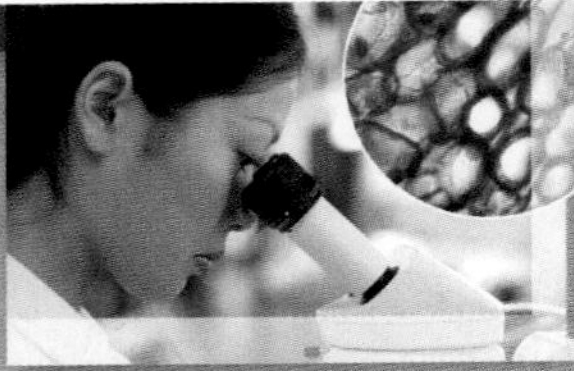

SYSTEMS PATHOLOGY

Osteoporosis

Mrs. B is a 70-year-old grandmother. Since she was a teenager, she has been a heavy smoker. She is typically sedentary, seldom goes outside, has not had the best dietary habits, and is underweight. One of her favorite yearly events is the family picnic on the Fourth of July. During one picnic, misfortune struck when Mrs. B tripped on a lawn sprinkler and fell. She was unable to stand because of severe hip pain, so she was rushed to the hospital, where a radiograph revealed that her femur was broken (figure D*a*) and that she had osteoporosis (figure D*b*).

It was decided that hip replacement surgery was indicated. Before the surgery could be performed, however, a fat embolism from the fracture site lodged in her lungs, making it difficult for her to breathe. The surgery was postponed and the fracture immobilized until she recovered from the fat embolism. Three weeks after the accident, Mrs. B had a successful hip transplant and began physical therapy. She appeared to be on the road to recovery, but 6 weeks after the surgery she developed persistent pain and edema in her hip. A bone biopsy confirmed a postoperative infection, which was successfully treated with antibiotics.

BACKGROUND INFORMATION

Osteoporosis (os′tē-ō-pō-rō′sis), or porous bone, results from a reduction in the overall quantity of bone tissue. It occurs when the rate of bone resorption exceeds the rate of bone formation. The loss of bone mass makes bones so porous and weakened that they become deformed and prone to fracture. The occurrence of osteoporosis increases with age. In both men and women, bone mass starts to decrease at about age 35 and continually decreases thereafter. Women can eventually lose approximately half, and men a quarter, of their cancellous bone. Osteoporosis is two and a half times more common in women than in men.

Osteoporosis has a strong genetic component. It is estimated that approximately 60% of a person's peak bone mass is genetically determined and that 40% is attributed to environmental factors, such as diet and physical activity. A woman whose mother has osteoporosis is more likely to develop osteoporosis than is a woman whose mother does not have the disorder. The genetic component of osteoporosis is complex and probably involves variations in a number of genes, such as those encoding vitamin D, calcitonin and estrogen receptors, collagen type I alpha 1 chain (COL1A1), and others.

In postmenopausal women, the decreased production of the female sex hormone estrogen can cause osteoporosis. Estrogen is secreted by the ovaries, and it normally contributes to the maintenance of normal bone mass by inhibiting the stimulatory effects of PTH on osteoclast activity. Following menopause, estrogen production decreases, resulting in degeneration of cancellous bone, especially in the vertebrae of the spine and the bones of the forearm. Collapse of the vertebrae can cause a decrease in height or, in more severe cases, can produce kyphosis, or a "dowager's hump," in the upper back.

Conditions that result in decreased estrogen levels, other than menopause, can also cause osteoporosis. Examples include removal of the ovaries before menopause, extreme exercise to the point of amenorrhea (lack of menstrual flow), anorexia nervosa (self-starvation), and cigarette smoking.

In males, reduction in testosterone levels can cause loss of bone tissue. Decreasing testosterone levels are usually less of a problem for men than decreasing estrogen levels are for women for two reasons. First, because males have denser bones than females, a loss of some bone tissue has less of an effect. Second, testosterone levels generally do not decrease significantly until after age 65, and even then the rate of decrease is often slow.

Overproduction of PTH, which results in overstimulation of osteoclast activity, can also cause osteoporosis.

Inadequate dietary intake or absorption of calcium can contribute to osteoporosis. Absorption of calcium from the small intestine decreases with age, and individuals with osteoporosis often have insufficient intake of calcium or vitamin D. Drugs that interfere with calcium uptake or use can also increase the risk for osteoporosis.

Finally, osteoporosis can result from inadequate exercise or disuse caused by fractures or paralysis. Significant amounts of bone are lost after 8 weeks of immobilization.

Treatments for osteoporosis are designed to reduce bone loss, increase bone formation, or both. Increased dietary calcium and vitamin D can increase calcium uptake and promote bone formation. Daily doses of 1000–1500 mg of calcium and 800 IU (20 μg) of vitamin D are recommended. Exercise, such as walking or using light weights, also appears to be effective not only in reducing bone loss but in increasing bone mass as well.

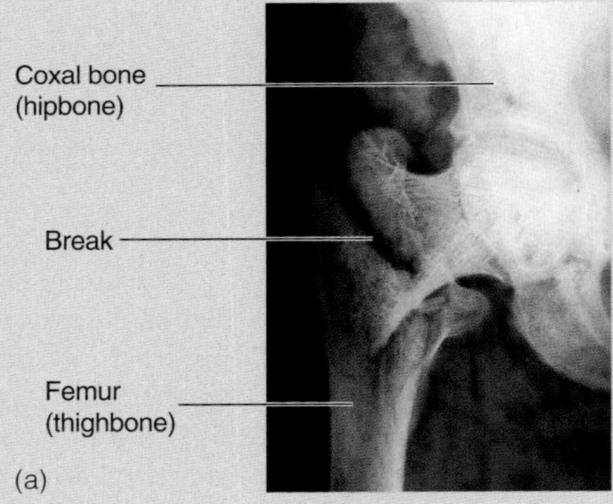

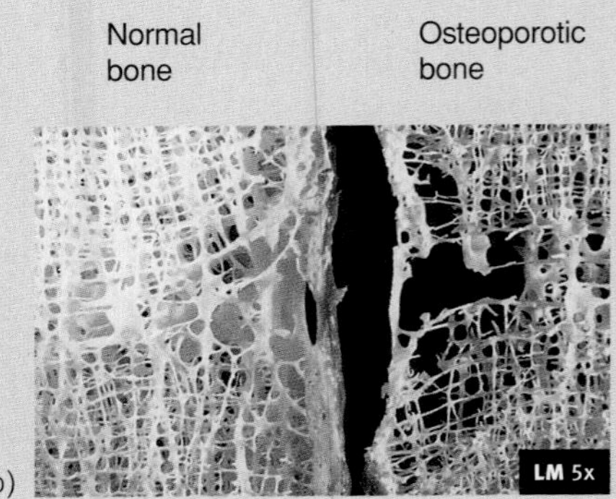

FIGURE D Osteoporosis
(*a*) Radiograph of a "broken hip," which is actually a break of the femur (thighbone) in the hip region. (*b*) Photomicrograph of normal bone and osteoporotic bone.

In postmenopausal women, **hormone replacement therapy (HRT)** with estrogen decreases osteoclast numbers by inhibiting the production of RANKL (see p. 194). This reduces bone loss but does not result in an increase in bone mass because osteoclast activity still exceeds osteoblast activity. However, the use of HRT to prevent bone loss is now discouraged because of the results of a study sponsored by the Women's Health Initiative. The study examined HRT in over 16,000 women, and found that HRT increased the risk for breast cancer, uterine cancer, heart attacks, strokes, and blood clots but decreased the risk for hip fractures and colorectal cancer. **Selective estrogen receptor modulators (SERMs)** are a class of drugs that bind to estrogen receptors. They may be able to protect against bone loss without increasing the risk of breast cancer. For example, raloxifene (ral-ox'ĭ-fēn) stimulates estrogen receptors in bone but inhibits them in the breast and uterus.

Osteoprotegerin, which prevents RANKL from binding to its receptors, is under consideration as a treatment for osteoporosis. Calcitonin (Miacalcin), which inhibits osteoclast activity, is now available as a nasal spray. Calcitonin can be used to treat osteoporosis in men and women and has been shown to produce a slight increase in bone mass. **Statins** (stat'ins) are drugs that inhibit cholesterol synthesis; they also stimulate osteoblast activity, and there is some evidence that statins can reduce the risk for fractures. Alendronate (Fosamax) belongs to a class of drugs called **bisphosphonates** (bis-fos'fō-nāts). Bisphosphonates concentrate in bone; when osteoclasts break down bone, the bisphosphonates are taken up by the osteoclasts. The bisphosphonates interfere with certain enzymes, leading to inactivation and lysis of the osteoclasts. Alendronate increases bone mass and reduces fracture rates even more effectively than calcitonin. Slow-releasing sodium fluoride (Slow Fluoride) in combination with calcium citrate (Citracal) also appears to increase bone mass.

Early diagnosis of osteoporosis may lead to the use of more preventive treatments. Instruments that measure the absorption of photons (particles of light) by bone are currently used, of which dual-energy x-ray absorptiometry (DEXA) is considered the best.

PREDICT 9

What advice should Mrs. B give to her granddaughter so that the granddaughter will be less likely to develop osteoporosis when she is Mrs. B's age?

SYSTEM INTERACTIONS — The Effects of Osteoporosis on Other Systems

SYSTEM	INTERACTIONS
Integumentary	Decreased exposure to sunlight because of an indoor lifestyle reduces vitamin D production an decreases Ca^{2+} absorption. Surgical wounds through the skin can allow the entry of bacteria, resulting in postoperative infections.
Muscular	A sedentary lifestyle and decreased body weight reduces stress on bone and contributes to osteoporosis. Muscle atrophy and weakness make it difficult to maintain balance, which increases the likelihood of falling and injury. Following hip replacement surgery, physical therapy places stress on the bones and improves muscular strength.
Nervous	Pain sensations following the injury and during rehabilitation help prevent further injury.
Endocrine	Although not a factor in this case of osteoporosis, elevated PTH (usually from a benign parathyroid tumor) or elevated thyroid hormone (Graves disease) can result in excessive osteoclast activity. Calcitonin is being used to treat osteoporosis.
Cardiovascular	Blood clotting following the injury starts the process of tissue repair. Blood cells are carried to the injury site to fight infections and remove cell debris. Blood vessels grow into the recovering tissue, providing nutrients and removing waste products.
Lymphatic and Immune	Immune cells resist infections and release chemicals that promote tissue repair. New immune cells are produced in bone marrow.
Respiratory	Excessive smoking lowers estrogen levels, which increases bone loss. A fat embolism from a fractured bone can impair respiration.
Digestive	Inadequate calcium and vitamin D in the inadequate calcium absorption by the digestive system can contribute to osteoporosis.
Urinary	Calcium ions released from the bones are excreted through the urinary system.
Reproductive	Decreased estrogen levels following menopause contribute to osteoporosis.

inhibits osteoclast activity. PTH and calcitonin are described more fully in chapters 18 and 27.

33 ***Name the hormone that is the major regulator of Ca^{2+} levels in the body. What stimulates the secretion of this hormone?***

34 ***Describe how PTH controls the number of osteoclasts. What are the effects of PTH on the formation of vitamin D, Ca^{2+} uptake in the small intestine, and reabsorption of Ca^{2+} from urine?***

35 ***What stimulates calcitonin secretion? How does calcitonin affect osteoclast activity?***

CASE STUDY

Henry is a 65-year-old man who was admitted to the emergency room after a fall. A radiograph confirmed that he had fractured the proximal part of his arm bone (surgical neck of the humerus). The radiograph also revealed that his bone matrix was not as dense as it should be for a man his age. A test for blood Ca^{2+} levels was normal. On questioning, Henry confessed that he is a junk food addict with poor eating habits. He eats few vegetables and never consumes dairy products. In addition, Henry never exercises and seldom goes outdoors except at night.

PREDICT 10

Use your knowledge of bone physiology and figure 6.19 to answer the following questions.

a. Why is Henry more likely than most men his age to break a bone?
b. How has Henry's eating habits contributed to his low bone density?
c. Would Henry's PTH levels be lower than normal, normal, or higher than normal?
d. What effect has Henry's nocturnal lifestyle had on his bone density?
e. How has lack of exercise affected his bone density?

EFFECTS OF AGING ON THE SKELETAL SYSTEM

The most significant age-related changes in the skeletal system affect the quality and quantity of bone matrix. Recall that a mineral (hydroxyapatite) in the bone matrix gives bone compression (weight-bearing) strength, but collagen fibers make the bone flexible. The bone matrix in an older bone is more brittle than in a younger bone because decreased collagen production results in a matrix that has relatively more mineral and fewer collagen fibers. With aging, the amount of matrix also decreases because the rate of matrix formation by osteoblasts becomes slower than the rate of matrix breakdown by osteoclasts.

Bone mass is at its highest around age 30, and men generally have denser bones than women because of the effects of testosterone and greater body weight. Race also affects bone mass. African-Americans and Hispanics have higher bone masses than Caucasians and Asians. After age 35, both men and women have an age-related loss of bone of 0.3%–0.5% a year. This loss can increase by 10 times in women after menopause, and women can have a bone loss of 3%–5% a year for approximately 5–7 years (see "Osteoporosis," p. 196).

Cancellous bone is lost at first as the trabeculae become thinner and weaker. The ability of the trabeculae to provide support also decreases as they become disconnected from each other. Eventually, some of the trabeculae completely disappear. Trabecular bone loss is greatest in the trabeculae that are under the least stress.

A slow loss of compact bone begins about age 40 and increases after age 45. The rate of compact bone loss, however, is approximately half the rate of trabecular bone loss. Bones become thinner, but their outer dimensions change little, because most loss of compact bone occurs under the endosteum on the inner surfaces of bones. In addition, the remaining compact bone becomes weaker as a result of incomplete bone remodeling. In a young bone, when osteons are removed, the resulting spaces are filled with new osteons. With aging, the new osteons fail to completely fill in the spaces produced when the older osteons are removed.

Significant loss of bone increases the likelihood of having bone fractures. For example, loss of trabeculae greatly increases the risk of compression fractures of the vertebrae (backbones) because the weight-bearing body of the vertebrae consists mostly of cancellous bone. In addition, loss of bone can cause deformity, loss of height, pain, and stiffness. For example, compression fractures of the vertebrae can cause an exaggerated curvature of the spine, resulting in a bent-forward, stooped posture. Loss of bone from the jaws can also lead to tooth loss.

36 ***What effect does aging have on the quality and quantity of bone matrix?***

37 ***Describe how cancellous and compact bone change with age. How do these changes affect a person's health?***

SUMMARY

Functions of the Skeletal System (p. 174)

1. The skeletal system consists of bones, cartilage, tendons, and ligaments.
2. The skeletal system supports the body, protects the organs it surrounds, allows body movements, stores minerals and fats, and is the site of blood cell production.

Cartilage (p. 174)

1. Chondroblasts produce cartilage and become chondrocytes. Chondrocytes are located in lacunae surrounded by matrix.
2. The matrix of cartilage contains collagen fibers (for strength) and proteoglycans (trap water).
3. The perichondrium surrounds cartilage.
   - The outer layer contains fibroblasts.
   - The inner layer contains chondroblasts.
4. Cartilage grows by appositional and interstitial growth.

Bone Histology (p. 175)

Bone Matrix

1. Collagen provides flexible strength.
2. Hydroxyapatite provides compressional strength.

Bone Cells

1. Osteoblasts produce bone matrix and become osteocytes.
   - Osteoblasts connect to one another through cell processes and surround themselves with bone matrix to become osteocytes.
   - Osteocytes are located in lacunae and are connected to one another through canaliculi.
2. Osteoclasts break down bone (with assistance from osteoblasts).
3. Osteoblasts originate from osteochondral progenitor cells, whereas osteoclasts originate from stem cells in red bone marrow.
4. Ossification, the formation of bone, occurs through appositional growth.

Woven and Lamellar Bone

1. Woven bone has collagen fibers oriented in many different directions. It is remodeled to form lamellar bone.
2. Lamellar bone is arranged in thin layers, called lamellae, which have collagen fibers oriented parallel to one another.

Cancellous and Compact Bone

1. Cancellous bone has many spaces.
   - Lamellae combine to form trabeculae, beams of bone that interconnect to form a latticelike structure with spaces filled with bone marrow and blood vessels.
   - The trabeculae are oriented along lines of stress and provide structural strength.
2. Compact bone is dense with few spaces.
   - Compact bone consists of organized lamellae: Circumferentia lamellae form the outer surface of compact bones; concentric lamellae surround central canals, forming osteons; interstitial lamellae are remnants of lamellae left after bone remodeling.
   - Canals within compact bone provide a means for the exchange of gases, nutrients, and waste products. From the periosteum or endosteum, perforating canals carry blood vessels to central canals, and canaliculi connect central canals to osteocytes.

Bone Anatomy (p. 180)

Bone Shapes

Individual bones can be classified as long, short, flat, or irregular.

Structure of a Long Bone

1. The diaphysis is the shaft of a long bone, and the epiphyses are the ends.
2. The epiphyseal plate is the site of bone growth in length.
3. The medullary cavity is a space within the diaphysis.
4. Red marrow is the site of blood cell production, and yellow marrow consists of fat.
5. The periosteum covers the outer surface of bone.
   - The outer layer contains blood vessels and nerves.
   - The inner layer contains osteoblasts, osteoclasts, and osteochondral progenitor cells.
   - Perforating fibers hold the periosteum, ligaments, and tendons in place.
6. The endosteum lines cavities inside bone and contains osteoblasts, osteoclasts, and osteochondral progenitor cells.

Structure of Flat, Short, and Irregular Bones

Flat, short, and irregular bones have an outer covering of compact bone surrounding cancellous bone.

Bone Development (p. 183)

Intramembranous Ossification

1. Some skull bones, part of the mandible, and the diaphyses of the clavicles develop from membranes.
2. Within the membrane at centers of ossification, osteoblasts produce bone along the membrane fibers to form cancellous bone.
3. Beneath the periosteum, osteoblasts lay down compact bone to form the outer surface of the bone.
4. Fontanels are areas of membrane that are not ossified at birth.

Endochondral Ossification

1. Most bones develop from a cartilage model.
2. The cartilage matrix is calcified, and chondrocytes die. Osteoblasts form bone on the calcified cartilage matrix, producing cancellous bone.
3. Osteoblasts build an outer surface of compact bone beneath the periosteum.
4. Primary ossification centers form in the diaphysis during fetal development. Secondary ossification centers form in the epiphyses.
5. Articular cartilage on the ends of bones and the epiphyseal plate is cartilage that does not ossify.

Bone Growth (p. 185)

1. Bones increase in size only by appositional growth, the adding of new bone on the surface of older bone or cartilage.
2. Trabeculae grow by appositional growth.

Growth in Bone Length

1. Epiphyseal plate growth involves the interstitial growth of cartilage followed by appositional bone growth on the cartilage.
2. Epiphyseal plate growth results in an increase in the length of the diaphysis and bony processes. Bone growth in length ceases when the epiphyseal plate becomes ossified and forms the epiphyseal line.

Growth at Articular Cartilage

1. Articular cartilage growth involves the interstitial growth of cartilage followed by appositional bone growth on the cartilage.
2. Articular cartilage growth results in larger epiphyses and an increase in the size of bones that do not have epiphyseal plates.

Growth in Bone Width

1. Appositional bone growth beneath the periosteum increases the diameter of long bones and the size of other bones.
2. Osteoblasts from the periosteum form ridges with grooves between them. The ridges grow together, converting the grooves into tunnels filled with concentric lamellae to form osteons.
3. Osteoblasts from the periosteum lay down circumferential lamellae, which can be remodeled.

Factors Affecting Bone Growth

1. Genetic factors determine bone shape and size. The expression of genetic factors can be modified.
2. Factors that alter the mineralization process or the production of organic matrix, such as deficiencies in vitamins D and C, can affect bone growth.
3. Growth hormone, thyroid hormone, estrogen, and testosterone stimulate bone growth.
4. Estrogen and testosterone cause increased bone growth and closure of the epiphyseal plate.

Bone Remodeling (p. 191)

1. Remodeling converts woven bone to lamellar bone and allows bone to change shape, adjust to stress, repair itself, and regulate body calcium levels.
2. Bone adjusts to stress by adding new bone and by realigning bone through remodeling.

Bone Repair (p. 192)

1. Fracture repair begins with the formation of a hematoma.
2. The hematoma is replaced by an internal callus consisting of fibers and cartilage.
3. The external callus is a bone–cartilage collar that stabilizes the ends of the broken bone.
4. The internal and external calluses are ossified to become woven bone.
5. Woven bone is remodeled.

Calcium Homeostasis (p. 194)

PTH increases blood Ca^{2+} levels by increasing bone breakdown, Ca^{2+} absorption from the small intestine, and reabsorption of Ca^{2+} from the urine. Calcitonin decreases blood Ca^{2+} by decreasing bone breakdown.

Effects of Aging on the Skeletal System (p. 198)

1. With aging, bone matrix is lost and the matrix becomes more brittle.
2. Cancellous bone loss results from a thinning and a loss of trabeculae. Compact bone loss mainly occurs from the inner surface of bones and involves less osteon formation.
3. Loss of bone increases the risk for fractures and causes deformity, loss of height, pain, stiffness, and loss of teeth.

REVIEW AND COMPREHENSION

1. Which of these is *not* a function of bone?
   a. internal support and protection
   b. attachment for the muscles
   c. calcium and phosphate storage
   d. blood cell production
   e. vitamin D storage
2. The extracellular matrix for hyaline cartilage
   a. is produced by chondroblasts.
   b. contains collagen.
   c. contains proteoglycans.
   d. is usually covered by the perichondrium.
   e. all of the above.
3. Chondrocytes are mature cartilage cells found within the ______________, and they are derived from ______________.
   a. perichondrium, fibroblasts
   b. perichondrium, chondroblasts
   c. lacunae, fibroblasts
   d. lacunae, chondroblasts
4. Which of these statements concerning cartilage is correct?
   a. Cartilage often occurs in thin plates or sheets.
   b. Chondrocytes receive nutrients and oxygen from blood vessels in the matrix.
   c. Articular cartilage has a thick perichondrium layer.
   d. The perichondrium has both chondrocytes and osteocytes.
   e. Appositional growth of cartilage occurs when chondrocytes within the tissue add more matrix from the inside.
5. Which of these substances makes up the major portion of bone?
   a. collagen
   b. hydroxyapatite
   c. proteoglycan aggregates
   d. osteocytes
   e. osteoblasts
6. The flexible strength of bone occurs because of
   a. osteoclasts.
   b. ligaments.
   c. hydroxyapatite.
   d. collagen fibers.
   e. periosteum.
7. The prime function of osteoclasts is to
   a. prevent osteoblasts from forming.
   b. become osteocytes.
   c. break down bone.
   d. secrete calcium salts and collagen fibers.
   e. form the periosteum.
8. Osteochondral progenitor cells
   a. can become osteoblasts or chondroblasts.
   b. are derived from mesenchymal stem cells.
   c. are located in the perichondrium, periosteum, and endosteum.
   d. do not produce osteoclasts.
   e. all of the above.
9. Lamellar bone
   a. is mature bone.
   b. is remodeled to form woven bone.
   c. is the first type of bone formed during early fetal development.
   d. has collagen fibers randomly oriented in many directions.
   e. all of the above.

10. Central canals
    a. connect perforating canals to canaliculi.
    b. connect cancellous bone to compact bone.
    c. are where blood cells are produced.
    d. are found only in cancellous bone.
    e. are lined with periosteum.
11. The type of lamellae found in osteons is ____________ lamellae.
    a. circumferential b. concentric c. interstitial
12. Cancellous bone consists of interconnecting rods or plates of bone called
    a. osteons.
    b. canaliculi.
    c. circumferential lamellae.
    d. a haversian system.
    e. trabeculae.
13. A fracture in the shaft of a bone is a break in the
    a. epiphysis.
    b. perichondrium.
    c. diaphysis.
    d. articular cartilage.
14. Yellow marrow is
    a. found mostly in children's bones.
    b. found in the epiphyseal plate.
    c. important for blood cell production.
    d. mostly adipose tissue.
15. The periosteum
    a. is an epithelial tissue membrane.
    b. covers the outer and internal surfaces of bone.
    c. contains only osteoblasts.
    d. becomes continuous with collagen fibers of tendons or ligaments.
    e. has a single fibrous layer.
16. Given these events:
    1. Osteochondral progenitor cells become osteoblasts.
    2. Connective tissue membrane is formed.
    3. Osteoblasts produce woven bone.

    Which sequence best describes intramembranous bone formation?
    a. 1,2,3 b. 1,3,2 c. 2,1,3 d. 2,3,1 e. 3,2,1
17. Given these processes:
    1. Chondrocytes die.
    2. Cartilage matrix calcifies.
    3. Chondrocytes hypertrophy.
    4. Osteoblasts deposit bone.
    5. Blood vessels grow into lacunae.

    Which sequence best represents the order in which they occur during endochondral bone formation?
    a. 3,2,1,4,5
    b. 3,2,1,5,4
    c. 5,2,3,4,1
    d. 3,2,5,1,4
    e. 3,5,2,4,1
18. Intramembranous bone formation
    a. occurs at the epiphyseal plate.
    b. is responsible for growth in diameter of a bone.
    c. gives rise to the flat bones of the skull.
    d. occurs within a hyaline cartilage model.
    e. produces articular cartilage in the long bones.
19. The ossification regions formed during early fetal development
    a. are secondary ossification centers.
    b. become articular cartilage.
    c. become medullary cavities.
    d. become the epiphyses.
    e. are primary ossification centers.
20. Growth in the length of a long bone occurs
    a. at the primary ossification center.
    b. beneath the periosteum.
    c. at the center of the diaphysis.
    d. at the epiphyseal plate.
    e. at the epiphyseal line.
21. During growth in length of a long bone, cartilage is formed and then ossified. The location of the ossification is the zone of
    a. calcification.
    b. hypertrophy.
    c. proliferation.
    d. resting cartilage.
22. Given these processes:
    1. An osteon is produced.
    2. Osteoblasts from the periosteum form a series of ridges.
    3. The periosteum becomes the endosteum.
    4. Osteoblasts lay down bone to produce a concentric lamella.
    5. Grooves are changed into tunnels.

    Which sequence best represents the order in which these processes occur during growth in width of a long bone?
    a. 1,4,2,3,5
    b. 2,5,3,4,1
    c. 3,4,2,1,5
    d. 4,2,1,5,3
    e. 5,4,2,1,3
23. Chronic vitamin D deficiency results in which of these consequences?
    a. Bones become brittle.
    b. The percentage of bone composed of hydroxyapatite increases.
    c. Bones become soft and pliable.
    d. Scurvy occurs.
    e. both a and b
24. Osteomalacia occurs as a result of a deficiency of
    a. growth hormone.
    b. sex hormones.
    c. thyroid hormone.
    d. vitamin C.
    e. vitamin D.
25. Estrogen
    a. stimulates a burst of growth at puberty.
    b. causes a later closure of the epiphyseal plate than does testosterone.
    c. causes a longer growth period in females than testosterone causes in males.
    d. tends to prolong the growth phase of the epiphyseal plates.
    e. all of the above.
26. Bone remodeling can occur
    a. when woven bone is converted into lamellar bone.
    b. as bones are subjected to varying patterns of stress.
    c. as a long bone increases in diameter.
    d. when new osteons are formed in compact bone.
    e. all of the above.
27. Given these processes:
    1. cartilage ossification
    2. external callus formation
    3. hematoma formation
    4. internal callus formation
    5. remodeling of woven bone into compact bone

    Which sequence best represents the order in which the processes occur during repair of a fracture?
    a. 1,2,3,4,5
    b. 2,4,3,1,5
    c. 3,4,2,1,5
    d. 4,1,5,2,3
    e. 5,3,4,2,1
28. Which of these processes during bone repair requires the longest period of time?
    a. cartilage ossification
    b. external callus formation
    c. hematoma formation
    d. internal callus formation
    e. remodeling of woven bone into compact bone
29. If the secretion of parathyroid hormone (PTH) increases, osteoclast activity ____________, and blood Ca^{2+} levels ____________.
    a. decreases, decrease
    b. decreases, increase
    c. increases, decrease
    d. increases, increase
30. Osteoclast activity is inhibited by
    a. calcitonin.
    b. growth hormone.
    c. parathyroid hormone.
    d. sex hormones.
    e. thyroid hormone.

Answers in Appendix E

CRITICAL THINKING

1. When a person develops Paget disease, for unknown reasons the collagen fibers in the bone matrix run randomly in all directions. In addition, the amount of trabecular bone decreases. What symptoms would you expect to observe?
2. When closure of the epiphyseal plate occurs, the cartilage of the plate is replaced by bone. Does this occur from the epiphyseal side of the plate or the diaphyseal side?
3. Assume that two patients have identical breaks in the femur (thighbone). If one is bedridden and the other has a walking cast, which patient's fracture heals faster? Explain.
4. Explain why running helps prevent osteoporosis in the elderly. Does the benefit include all bones or mainly those of the legs and spine?
5. Astronauts can experience a dramatic decrease in bone density while in a weightless environment. Explain how this happens, and suggest a way to slow the loss of bone tissue.
6. Would a patient suffering from kidney failure be more likely to develop osteomalacia or osteoporosis? Explain.
7. In some cultures, eunuchs were responsible for guarding harems, which are the collective wives of one male. Eunuchs are males who were castrated as boys. Castration removes the testes, the major site of testosterone production in males. Because testosterone is responsible for the sex drive in males, the reason for castration is obvious. As a side effect of this procedure, the eunuchs grew to above-normal heights. Can you explain why?
8. When a long bone is broken, blood vessels at the fracture line are severed. The formation of blood clots stops the bleeding. Within a few days, bone tissue on both sides of the fracture site dies. The bone only dies back a certain distance from the fracture line, however. Explain.
9. A patient has hyperparathyroidism because of a tumor in the parathyroid gland that produces excessive amounts of PTH. What effect does this hormone have on bone? Would administration of large doses of vitamin D help the situation? Explain.

Answers in Appendix F

Visit this book's website at www.mhhe.com/seeley8 for chapter quizzes, interactive learning exercises, and other study tools.

Skeletal System

Gross Anatomy

If the body had no skeleton, it would look somewhat like a poorly stuffed rag doll. Without a skeletal system, we would have no framework to help maintain shape and we would not be able to move normally. Most muscles act on bones to produce movement, often pulling on the bones with considerable force. Without the skeleton, muscles would not make the body move. Human bones are very strong and can resist tremendous bending and compression forces without breaking. Nonetheless, each year nearly 2 million Americans break a bone.

The skeletal system includes the bones, cartilage, ligaments, and tendons. To study skeletal gross anatomy, however, dried, prepared bones are used. This allows the major features of individual bones to be seen clearly without being obstructed by associated soft tissues, such as muscles, tendons, ligaments, cartilage, nerves, and blood vessels. As a consequence, however, it is easy to ignore the important relationships among bones and soft tissues and the fact that living bones have soft tissue, such as the periosteum (see chapter 6).

This chapter includes a discussion of *general considerations* (p. 204). It then discusses the two categories of the named bones: the *axial skeleton* (p. 206) and the *appendicular skeleton* (p. 233).

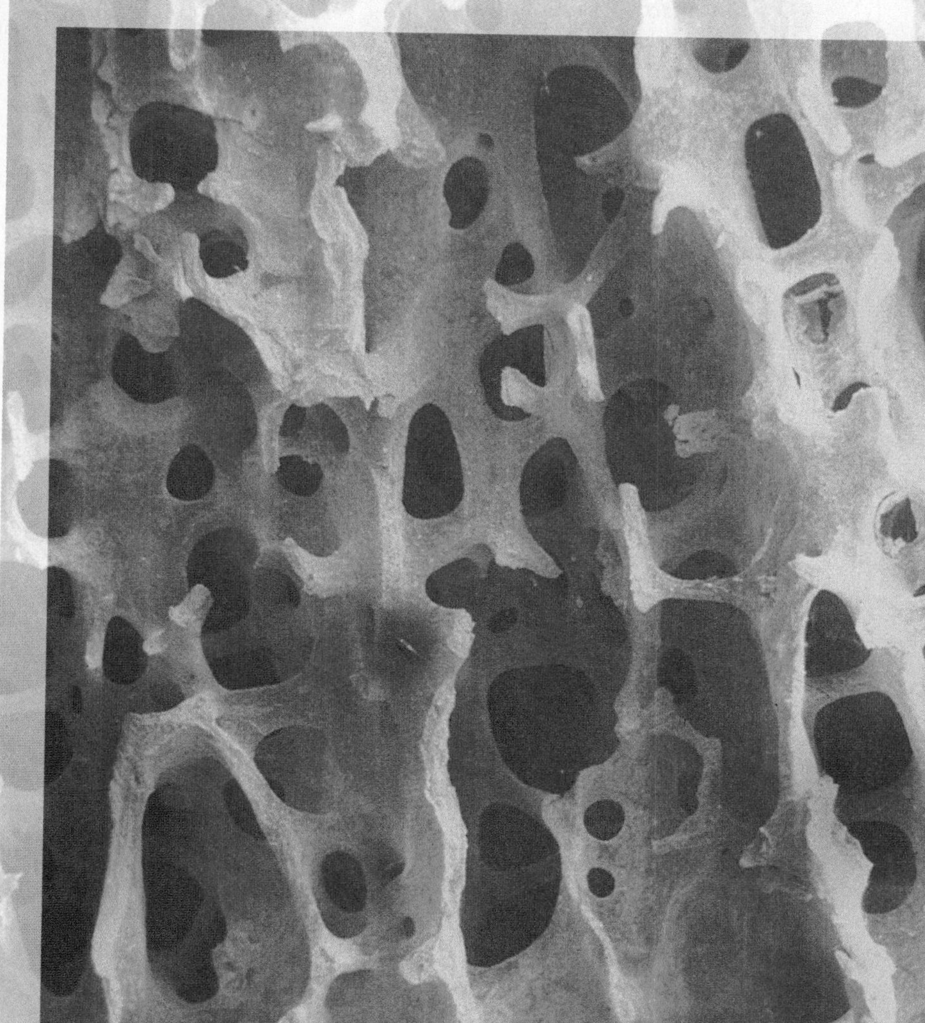

Colorized SEM of bone trabeculae.

Skeletal System

GENERAL CONSIDERATIONS

The average adult skeleton has 206 bones (table 7.1 and figure 7.1). Although this is the traditional number, the actual number of bones varies from person to person and decreases with age as some bones become fused.

Many of the anatomical terms used to describe the features of bones are listed in table 7.2. Most of these features are based on the relationship between the bones and associated soft tissues. If a bone possesses a **tubercle** (too′ber-kl; lump) or **process** (projection), such structures usually exist because a ligament or tendon was attached to that tubercle or process during life. If a bone has

TABLE 7.1 Number of Named Bones Listed by Category

Bones		Number
Axial Skeleton		
Skull (Cranium)		
Neurocranium (braincase)		
Paired	Parietal	2
	Temporal	2
Unpaired	Frontal	1
	Sphenoid	1
	Occipital	1
	Ethmoid	1
Viscerocranium (face)		
Paired	Maxilla	2
	Zygomatic	2
	Palatine	2
	Lacrimal	2
	Nasal	2
	Inferior nasal concha	2
Unpaired	Mandible	1
	Vomer	1
	Total skull	22
Bones Associated with the Skull		
Auditory ossicles		
Malleus		2
Incus		2
Stapes		2
Hyoid		1
	Total associated	7
Vertebral Column		
Cervical vertebrae		7
Thoracic vertebrae		12
Lumbar vertebrae		5
Sacrum		1
Coccyx		1
	Total vertebral column	26
Thoracic Cage (Rib Cage)		
Ribs		24
Sternum		1
	Total thoracic cage	25
	Total axial skeleton	80

Bones		Number
Appendicular Skeleton		
Pectoral Girdle		
Scapula		2
Clavicle		2
Upper Limb		
Humerus		2
Ulna		2
Radius		2
Carpals		16
Metacarpals		10
Phalanges		28
	Total upper limb and girdle	64
Pelvic Girdle		
Coxal bone		2
Lower Limb		
Femur		2
Tibia		2
Fibula		2
Patella		2
Tarsals		14
Metatarsals		10
Phalanges		28
	Total lower limb and girdle	62
	Total appendicular skeleton	126
	Total axial skeleton	80
	Total appendicular skeleton	126
	Total bones	206

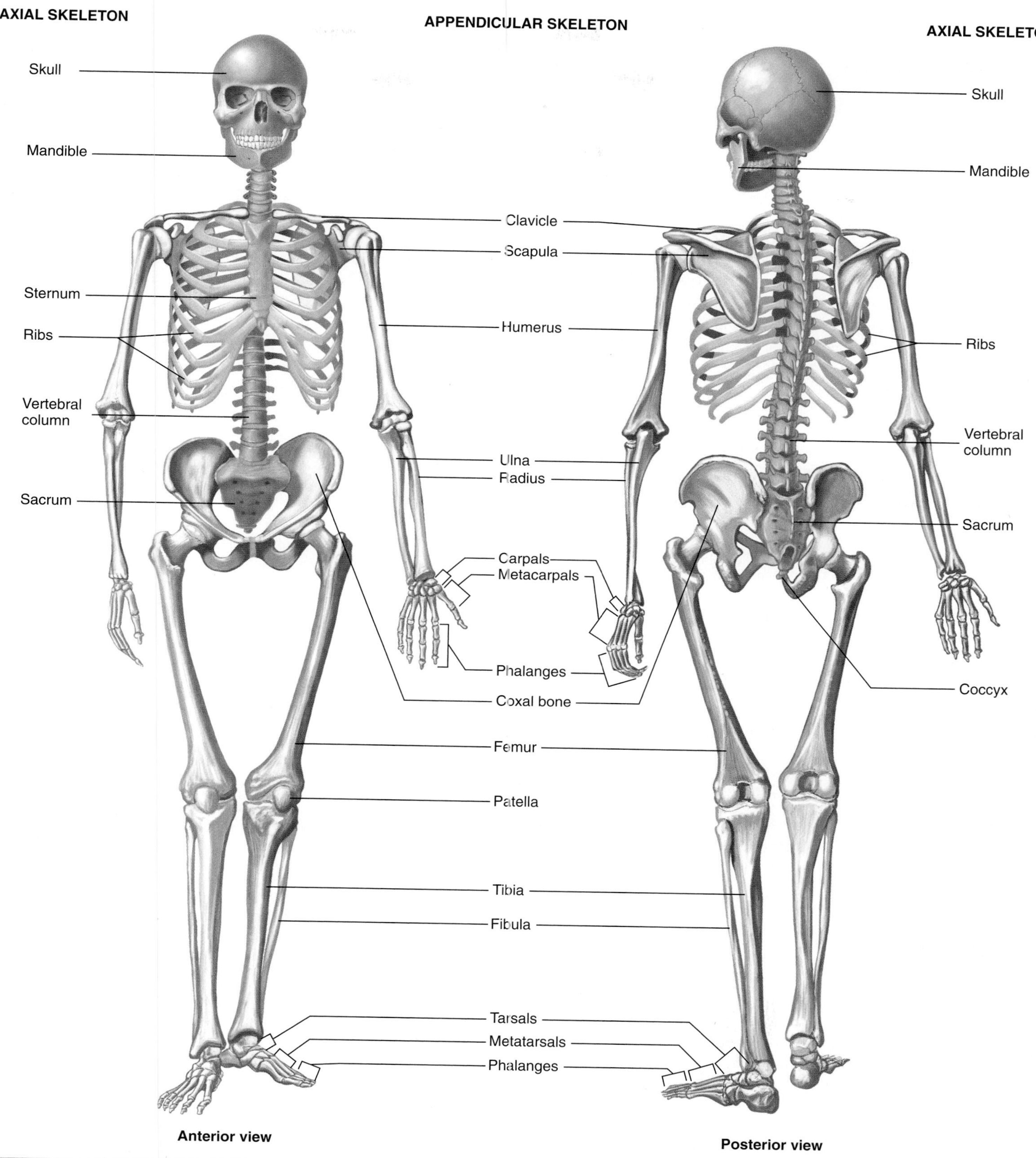

FIGURE 7.1 Complete Skeleton

The skeleton is not shown in the anatomical position.

TABLE 7.2 General Anatomical Terms for Various Features of Bones

Term	Description
Body	Main part
Head	Enlarged (often rounded) end
Neck	Constriction between head and body
Margin or border	Edge
Angle	Bend
Ramus	Branch off the body (beyond the angle)
Condyle	Smooth, rounded articular surface
Facet	Small, flattened articular surface
Ridges	
Line or linea	Low ridge
Crest or crista	Prominent ridge
Spine	Very high ridge
Projections	
Process	Prominent projection
Tubercle	Small, rounded bump
Tuberosity or tuber	Knob; larger than a tubercle
Trochanter	Tuberosities on the proximal femur
Epicondyle	Upon a condyle
Lingula	Flat, tongue-shaped process
Hamulus	Hook-shaped process
Cornu	Horn-shaped process
Openings	
Foramen	Hole
Canal or meatus	Tunnel
Fissure	Cleft
Sinus or labyrinth	Cavity
Depressions	
Fossa	General term for a depression
Notch	Depression in the margin of a bone
Fovea	Little pit
Groove or sulcus	Deeper, narrow depression

a smooth, articular surface, that surface was part of a joint and was covered with articular cartilage. If the bone has a **foramen** (fō-rā′men; pl. *foramina;* fō-ram′i-nă; a hole) in it, that foramen was occupied by something such as a nerve or blood vessel. Some skull bones contain mucous membrane–lined air spaces called **sinuses.** Regions of some skull bones are composed of paper-thin, translucent compact bone only and have little or no cancellous center (see chapter 6).

1 ***How many bones are in an average adult skeleton?***

2 ***How are lumps, projections, and openings in bones related to soft tissues?***

AXIAL SKELETON

The **axial skeleton** is divided into the skull, auditory ossicles, hyoid bone, vertebral column, and thoracic cage, or rib cage. The axial skeleton forms the upright axis of the body. It protects the brain, the spinal cord, and the vital organs housed within the thorax.

3 ***List the parts of the axial skeleton and its functions.***

Skull

The **skull,** or **cranium** (krā′nē-ŭm), protects the brain; supports the organs of vision, hearing, smell, and taste; and provides a foundation for the structures that take air, food, and water into the body. When the skull is disassembled, the mandible is easily separated from the rest of the skull, which remains intact. Special effort is needed to separate the other bones. For this reason, it is convenient to think of the skull, except for the mandible, as a single unit. The top of the skull is called the **calvaria** (kal-vā′rē-ă), or skullcap. It is usually cut off to reveal the skull's interior. The exterior and interior of the skull have ridges, lines, processes, and plates. These structures are important for the attachment of muscles or for articulations between the bones of the skull. Selected features of the intact skull are listed in table 7.3.

Superior View of the Skull

The skull appears quite simple when viewed from above. Only four bones are seen from this view: the frontal bone, two parietal bones, and a small part of the occipital bone. The paired **parietal bones** are joined at the midline by the **sagittal suture,** and the parietal bones are connected to the **frontal bone** by the **coronal suture** (figure 7.2).

Posterior View of the Skull

The parietal and occipital bones are the major structures seen from the posterior view (figure 7.3). The parietal bones are joined to the **occipital bone** by the **lambdoid** (lam′doyd; the shape resembles the Greek letter lambda) **suture.** Occasionally, extra small bones called **sutural** (soo′choor-ăl), or wormian, **bones** form along the lambdoid suture.

PREDICT 1

Explain the basis for the names *sagittal* and *coronal sutures.*

Inca Bone

Sutural bones are usually small, bilateral, and apparently genetically determined. A large midline bone, called an Inca bone, may form at the junction of the lambdoid and sagittal sutures. The bone was common in the skulls of Incas and is still present in their Andean descendants.

An **external occipital protuberance** is present on the posterior surface of the occipital bone (see figure 7.3). It can be felt through the scalp at the base of the head and varies considerably in size from person to person. The external occipital protuberance is the site of attachment of the **ligamentum nuchae** (noo′kē; nape of neck),

TABLE 7.3 Processes and Other Features of the Skull

Feature	Bone on Which Feature is Found	Description
External Features		
Alveolar process	Mandible, maxilla	Ridges on the mandible and maxilla containing the teeth
Angle	Mandible	The posterior, inferior corner of the mandible
Coronoid process	Mandible	Attachment point for the temporalis muscle
Genu	Mandible	Chin (resembles a bent knee)
Horizontal plate	Palatine	The posterior third of the hard palate
Mandibular condyle	Mandible	Region where the mandible articulates with the skull
Mandibular fossa	Temporal	Depression where the mandible articulates with the skull
Mastoid process	Temporal	Enlargement posterior to the ear; attachment site for several muscles that move the head
Nuchal lines	Occipital	Attachment points for several posterior neck muscles
Occipital condyle	Occipital	Point of articulation between the skull and the vertebral column
Palatine process	Maxilla	Anterior two-thirds of the hard palate
Pterygoid hamulus	Sphenoid	Hooked process on the inferior end of the medial pterygoid plate, around which the tendon of one palatine muscle passes; an important dental landmark
Pterygoid plates (medial and lateral)	Sphenoid	Bony plates on the inferior aspect of the sphenoid bone; the lateral pterygoid plate is the site of attachment for two muscles of mastication (chewing)
Ramus	Mandible	Portion of the mandible superior to the angle
Styloid process	Temporal	Attachment site for three muscles (to the tongue, pharynx, and hyoid bone) and some ligaments
Temporal lines	Parietal	Where the temporalis muscle, which closes the jaw, attaches
Internal Features		
Crista galli	Ethmoid	Process in the anterior part of the cranial vault to which one of the connective tissue coverings of the brain (dura mater) connects
Petrous portion	Temporal	Thick, interior part of temporal bone containing the middle and inner ears and the auditory ossicles
Sella turcica	Sphenoid	Bony structure resembling a saddle in which the pituitary gland is located

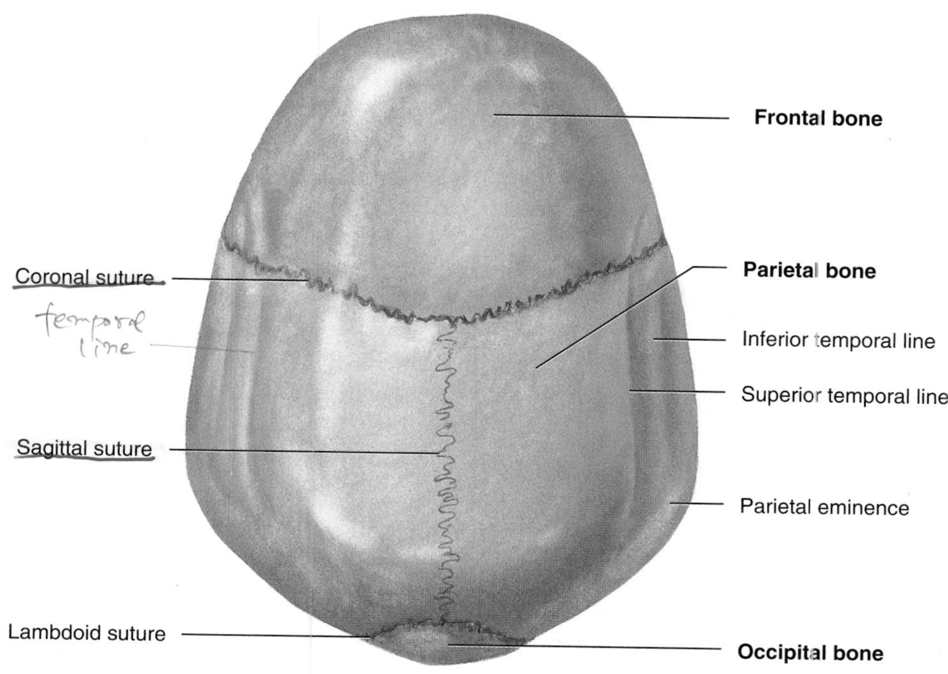

FIGURE 7.2 Skull as Seen from the Superior View
The names of the bones are bolded.

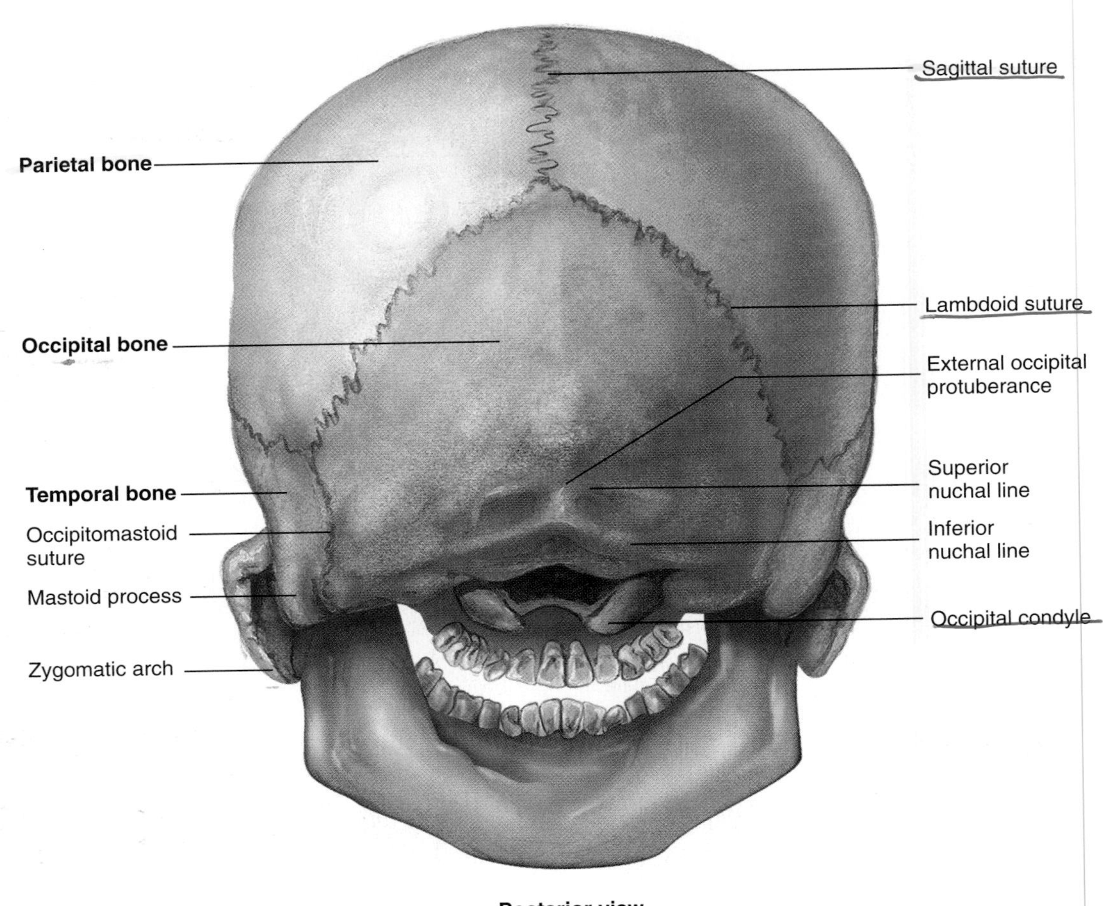

FIGURE 7.3 Skull as Seen from the Posterior View
The names of the bones are bolded.

an elastic ligament that extends down the neck and helps keep the head erect by pulling on the occipital region of the skull. **Nuchal lines** are a set of small ridges that extend laterally from the protuberance and are the points of attachment for several neck muscles.

> **Nuchal Lines**
>
> The ligamentum nuchae and neck muscles in humans are not as strong as comparable structures in other animals; therefore, the human bony prominence and lines of the posterior skull are not as well developed as they are in those animals. The location of the human foramen magnum allows the skull to balance above the vertebral column and allows for an upright posture. Thus human skulls require less ligamental and muscular effort to balance the head on the spinal column than do the skulls of other animals, including other primates, such as chimpanzees, whose skulls are not balanced over the vertebral column. The presence of small nuchal lines in hominids (i.e., animals with an upright stance, such as humans) reflects this decreased musculature and is one way paleontologists establish probable upright posture in hominids.

Lateral View of the Skull

The parietal bone and the squamous part of the temporal bone form a large part of the side of the head (figure 7.4). The term *temporal* means related to time, and the temporal bone is so named because the hair of the temples is often the first to turn white, indicating the passage of time. The **squamous suture** joins these bones. A prominent feature of the temporal bone is a large hole, the **external acoustic,** or **auditory, meatus** (mē-ā′tŭs; passageway or tunnel), which transmits sound waves toward the eardrum. The external ear, or auricle, surrounds the meatus. Just posterior and inferior to the external auditory meatus is a large inferior projection, the **mastoid** (mas′toyd; resembling a breast) **process.** The process can be seen and felt as a prominent lump just posterior to the ear. The process is not solid bone but is filled with cavities called **mastoid air cells,** which are connected to the middle ear. Important neck muscles involved in rotation of the head attach to the mastoid process. The superior and inferior **temporal lines,** which are attachment points of the temporalis muscle, one of the major muscles of mastication, arch across the lateral surface of the parietal bone.

> **Temporal Lines**
>
> The temporal lines are important to anthropologists because a heavy temporal line suggests a strong temporalis muscle supporting a heavy jaw. In a male gorilla, the temporalis muscles are so large that the temporal lines meet in the midline of the skull to form a heavy sagittal crest. The temporal lines are much smaller in humans.

Lateral view

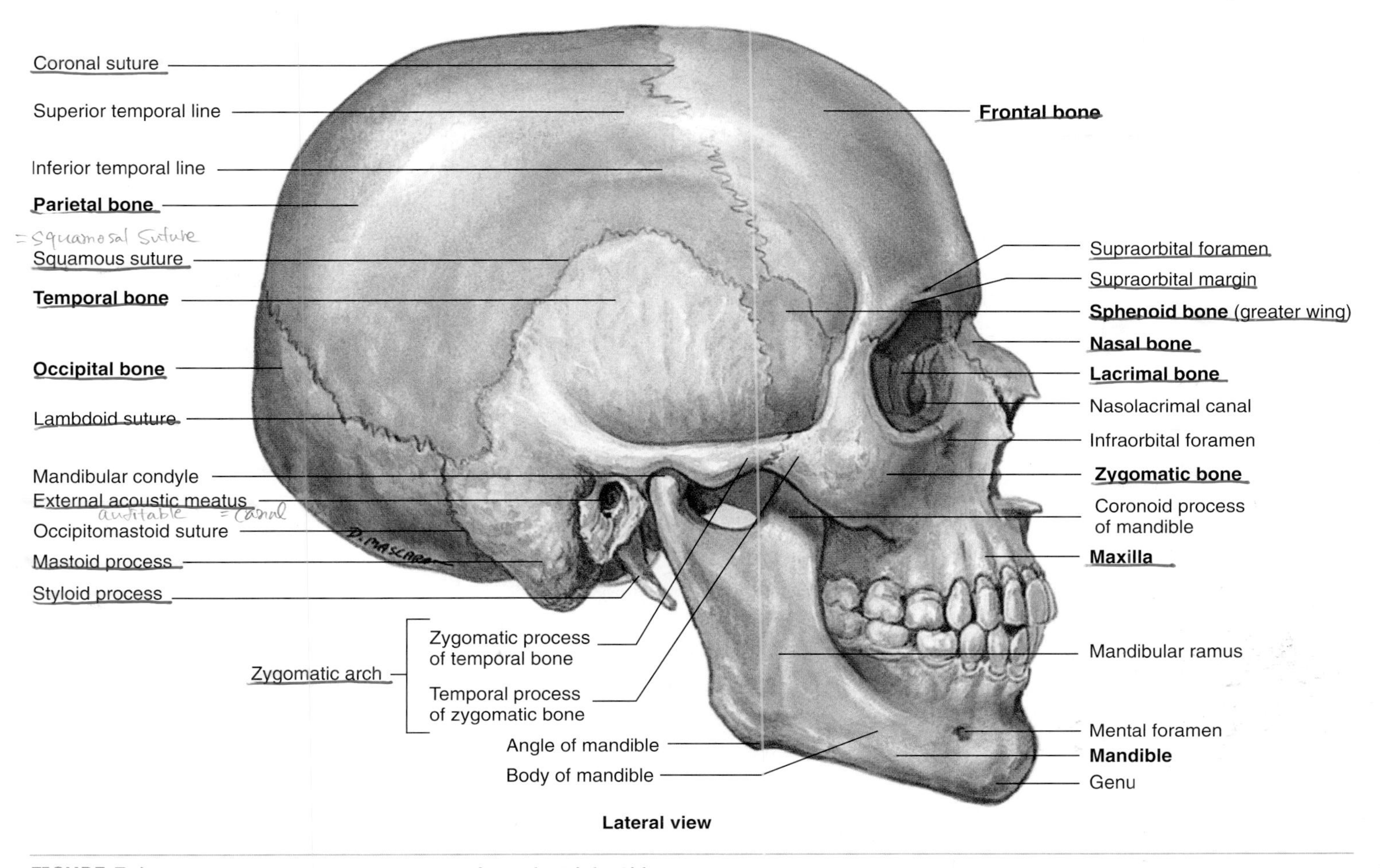

FIGURE 7.4 Lateral View of the Skull as Seen from the Right Side
The names of the bones are bolded.

The lateral surface of the **greater wing** of the **sphenoid** (sfē′noyd; wedge-shaped) **bone** is immediately anterior to the temporal bone (see figure 7.4). Although appearing to be two bones, one on each side of the skull, the sphenoid bone is actually a single bone that extends completely across the skull. Anterior to the sphenoid bone is the **zygomatic** (zī′gō-mat′ik; a bar or yoke) **bone,** or cheekbone, which can be easily seen and felt on the face (figure 7.5).

The **zygomatic arch,** which consists of joined processes from the temporal and zygomatic bones, forms a bridge across the side of the skull (see figure 7.4). The zygomatic arch is easily felt on the side of the face, and the muscles on each side of the arch can be felt as the jaws are opened and closed (see figure 7.5).

The **maxilla** (mak-sil′ă; upper jaw) is anterior and inferior to the zygomatic bone to which it is joined. The **mandible** (lower jaw) is inferior to the maxilla and articulates posteriorly with the temporal bone (see figure 7.4). The mandible consists of two main portions: the **body,** which extends anteroposteriorly, and the **ramus** (branch), which extends superiorly from the body toward the temporal bone. The superior end of the ramus has a mandibular **condyle,** which articulates with the mandibular fossa of the temporal bone, and the **coronoid** (kōr′ŏ-noyd; shaped like a crow's beak) **process,** to which the powerful temporalis muscle, one of the chewing muscles, attaches. The alveolar process of the maxilla contains the superior set of teeth, and the alveolar process of the mandible contains the inferior teeth.

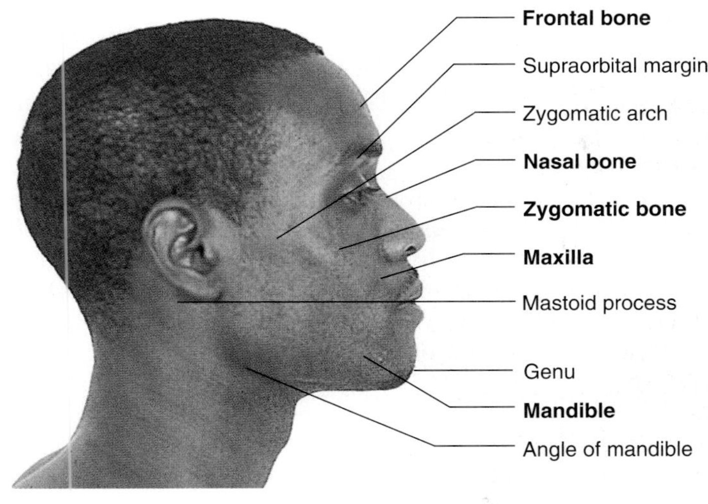

FIGURE 7.5 Lateral View of Bony Landmarks on the Face

Frontal bone
Parietal bone
Supraorbital foramen
Optic canal
Orbital plate of frontal bone
Sphenoid bone (greater wing)
Superior orbital fissure
Lacrimal bone
Infraorbital foramen
Middle nasal concha
Inferior nasal concha
Anterior nasal spine
Oblique line of mandible
Mandible
Mandibular symphysis
Genu
Coronal suture
Glabella
Supraorbital margin
Temporal bone
Nasal bone
Infraorbital margin
Zygomatic bone
Nasal septum
Perpendicular plate of ethmoidal bone
Vomer
Nasal cavity
Maxilla
Alveolar processes
Body of mandible
Mental foramen
Anterior view

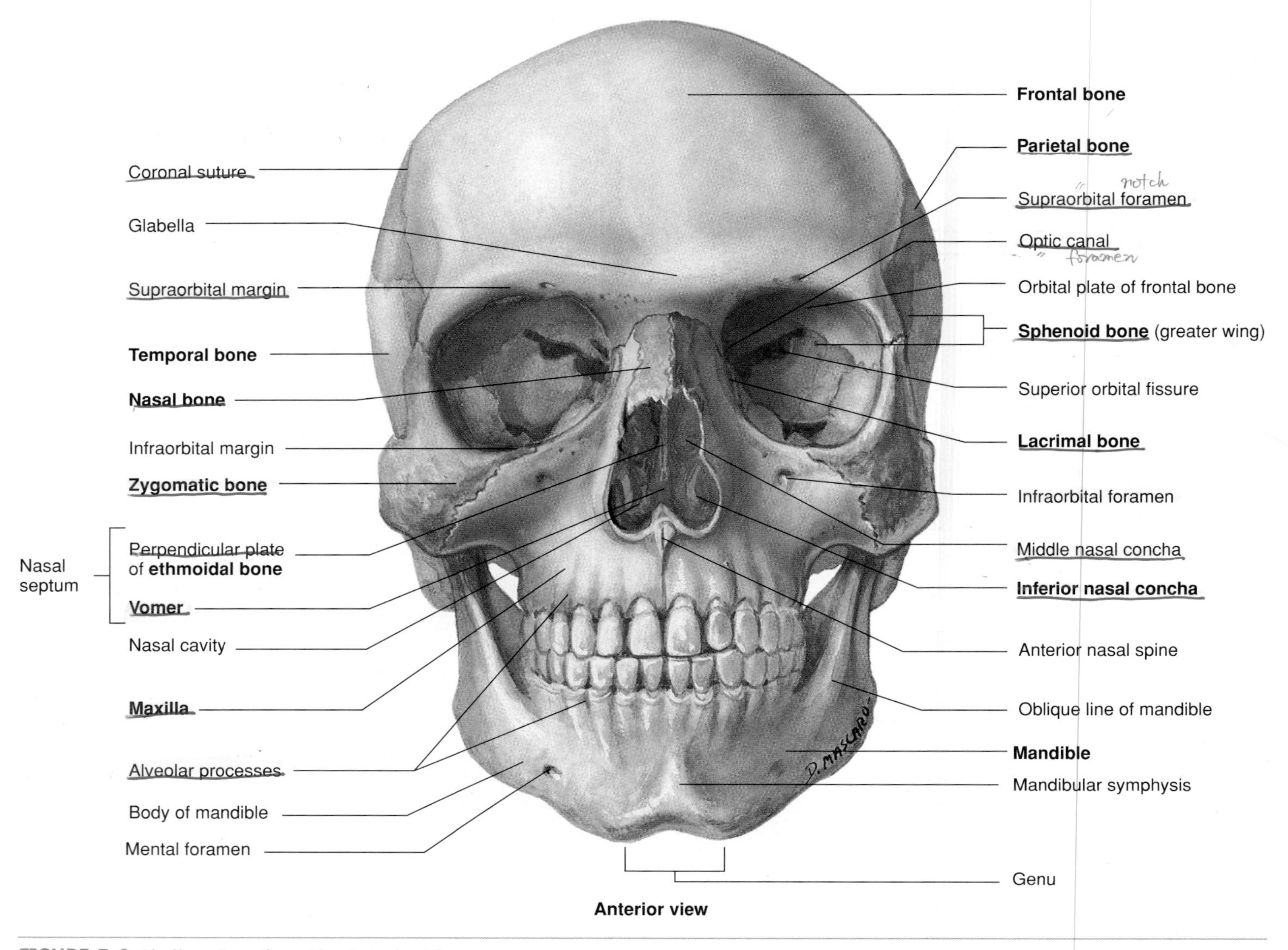

FIGURE 7.6 Skull as Seen from the Anterior View
The names of the bones are bolded.

Anterior View of the Skull

The major structures seen from the anterior view are the frontal bone (forehead), the zygomatic bones (cheekbones), the maxillae, and the mandible (figure 7.6). The teeth, which are very prominent in this view, are discussed in chapter 24. Many bones of the face can be easily felt through the skin of the face (figure 7.7).

From this view, the most prominent openings into the skull are the orbits and the nasal cavity. The **orbits** are cone-shaped fossae with their apices directed posteriorly (see figure 7.6; figure 7.8). They are called orbits because of the rotation of the eyes within the fossae. The bones of the orbits provide both protection for the eyes and attachment points for the muscles that move the eyes. The major portion of each eyeball is within the orbit, and the portion of the eye visible from the outside is relatively small. Each orbit contains blood vessels, nerves, and fat, as well as the eyeball and the muscles that move it. The bones forming the orbit are listed in table 7.4.

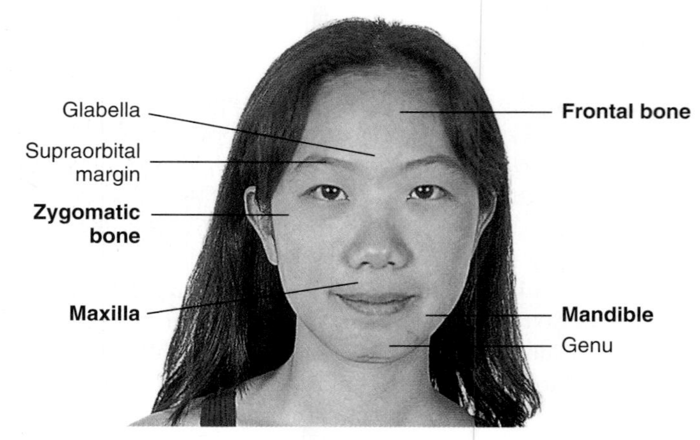

FIGURE 7.7 Anterior View of Bony Landmarks on the Face
The names of the bones are bolded.

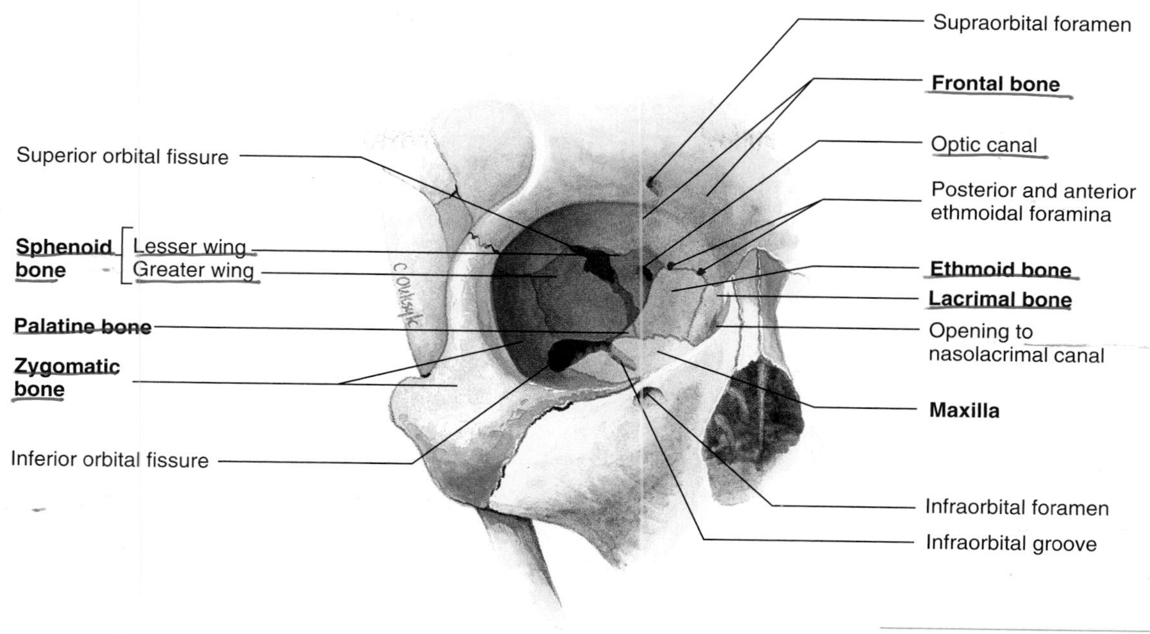

FIGURE 7.8 Bones of the Right Orbit
The names of the bones are bolded.

Orbit Weak Point

The superolateral corner of the orbit, where the zygomatic and frontal bones join, is a weak point in the skull that is easily fractured by a blow to that region of the head. The bone tends to collapse into the orbit, resulting in an injury that is difficult to repair.

The orbit has several openings through which structures communicate between the orbit and other cavities. The nasolacrimal duct passes from the orbit into the nasal cavity through the **nasolacrimal canal,** carrying tears from the eyes to the nasal cavity. The optic nerve for the sense of vision passes from the eye through the **optic canal** at the posterior apex of the orbit and enters the cranial cavity. Superior and inferior fissures in the posterior region of the orbit provide openings through which nerves and blood vessels communicate with structures in the orbit or pass to the face.

TABLE 7.4 Bones Forming the Orbit (See Figures 7.6 and 7.8)

Bone	Part of Orbit
Frontal	Roof
Sphenoid	Roof and posterolateral wall
Zygomatic	Lateral wall
Maxilla	Floor
Lacrimal	Medial wall
Ethmoid	Medial wall
Palatine	Medial wall

The **nasal cavity** (table 7.5 and figure 7.9; see figure 7.6) has a pear-shaped opening anteriorly and is divided into right and left halves by a **nasal septum** (sep′tŭm; wall). The bony part of the nasal septum consists primarily of the vomer and the perpendicular plate of the ethmoid bone. Hyaline cartilage forms the anterior part of the nasal septum.

Deviated Nasal Septum

The nasal septum usually is located in the midsagittal plane until a person is 7 years old. Thereafter, it tends to deviate, or bulge slightly to one side. The septum can also deviate abnormally at birth or, more commonly, as a result of injury. Deviations can be severe enough to block one side of the nasal passage and interfere with normal breathing. The repair of severe deviations requires surgery.

TABLE 7.5 Bones Forming the Nasal Cavity (See Figures 7.6 and 7.9)

Bone	Part of Nasal Cavity
Frontal	Roof
Nasal	Roof
Sphenoid	Roof
Ethmoid	Roof, septum, lateral wall
Inferior nasal concha	Lateral wall
Lacrimal	Lateral wall
Maxilla	Floor
Palatine	Floor and lateral wall
Vomer	Septum

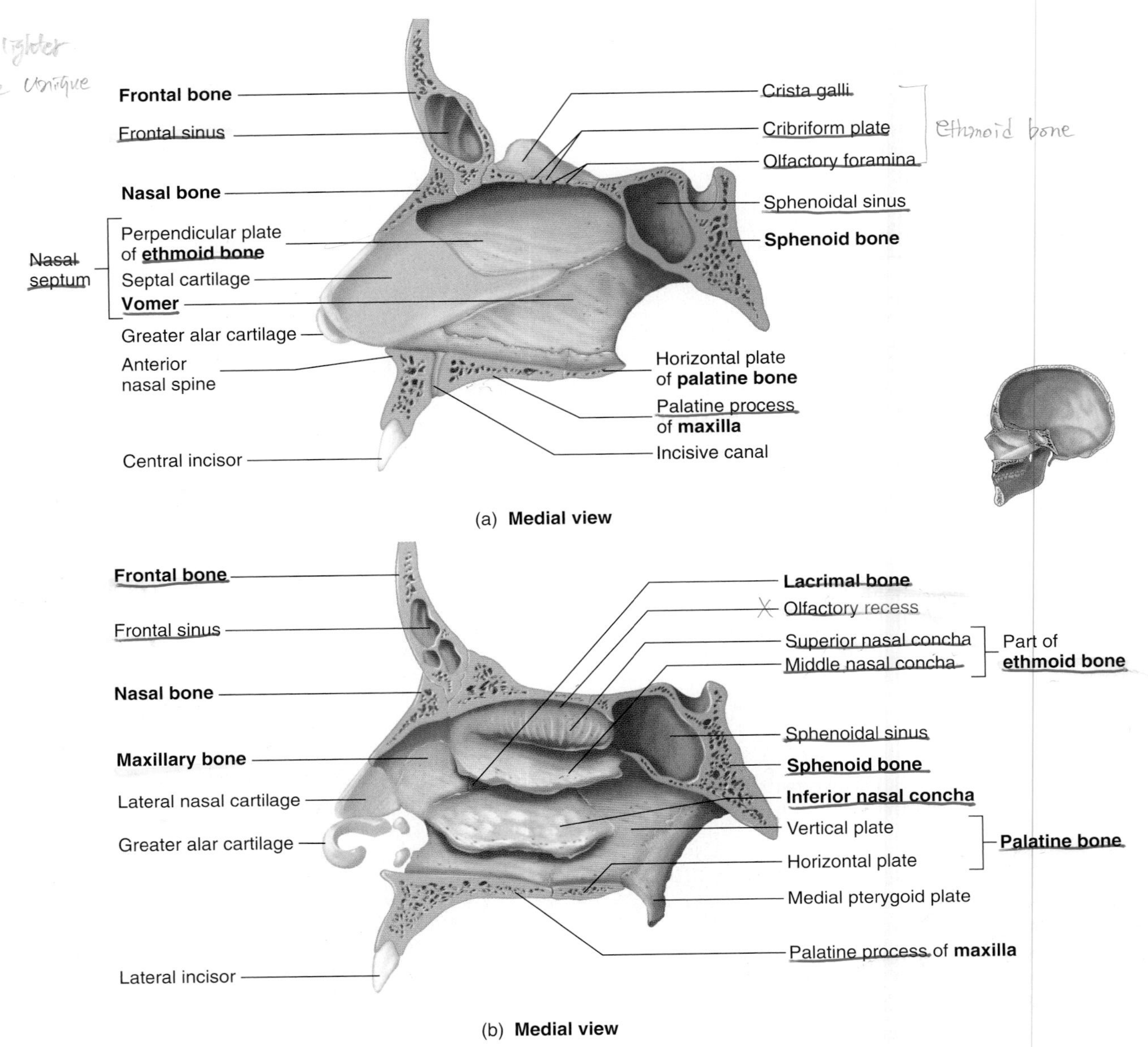

FIGURE 7.9 Bones of the Nasal Cavity

The names of the bones are bolded. (*a*) Nasal septum as seen from the left nasal cavity. (*b*) Right lateral nasal wall as seen from inside the nasal cavity (nasal septum removed).

The external part of the nose, formed mostly of hyaline cartilage, is almost entirely absent in the dried skeleton and is represented mainly by the nasal bones and the frontal processes of the maxillary bones, which form the bridge of the nose.

PREDICT 2

A direct blow to the nose may result in a "broken nose." List at least three bones that may be broken.

The lateral wall of the nasal cavity has three bony shelves, the **nasal conchae** (kon'kē; resembling a conch shell), which are directed inferiorly (see figure 7.9). The inferior nasal concha is a separate bone, and the middle and superior nasal conchae are projections from the ethmoid bone. The conchae increase the surface area in the nasal cavity, thereby facilitating moistening, the removal of particles, and warming of the air inhaled through the nose.

Several of the bones associated with the nasal cavity have large cavities within them called the **paranasal sinuses,** which open into the nasal cavity (figure 7.10). The sinuses decrease the weight of the skull and act as resonating chambers during voice production. Compare a normal voice with the voice of a person who has a cold and whose sinuses are "stopped up." The sinuses are named for the bones in which they are located and include the frontal, maxillary, and sphenoidal sinuses. The cavities within

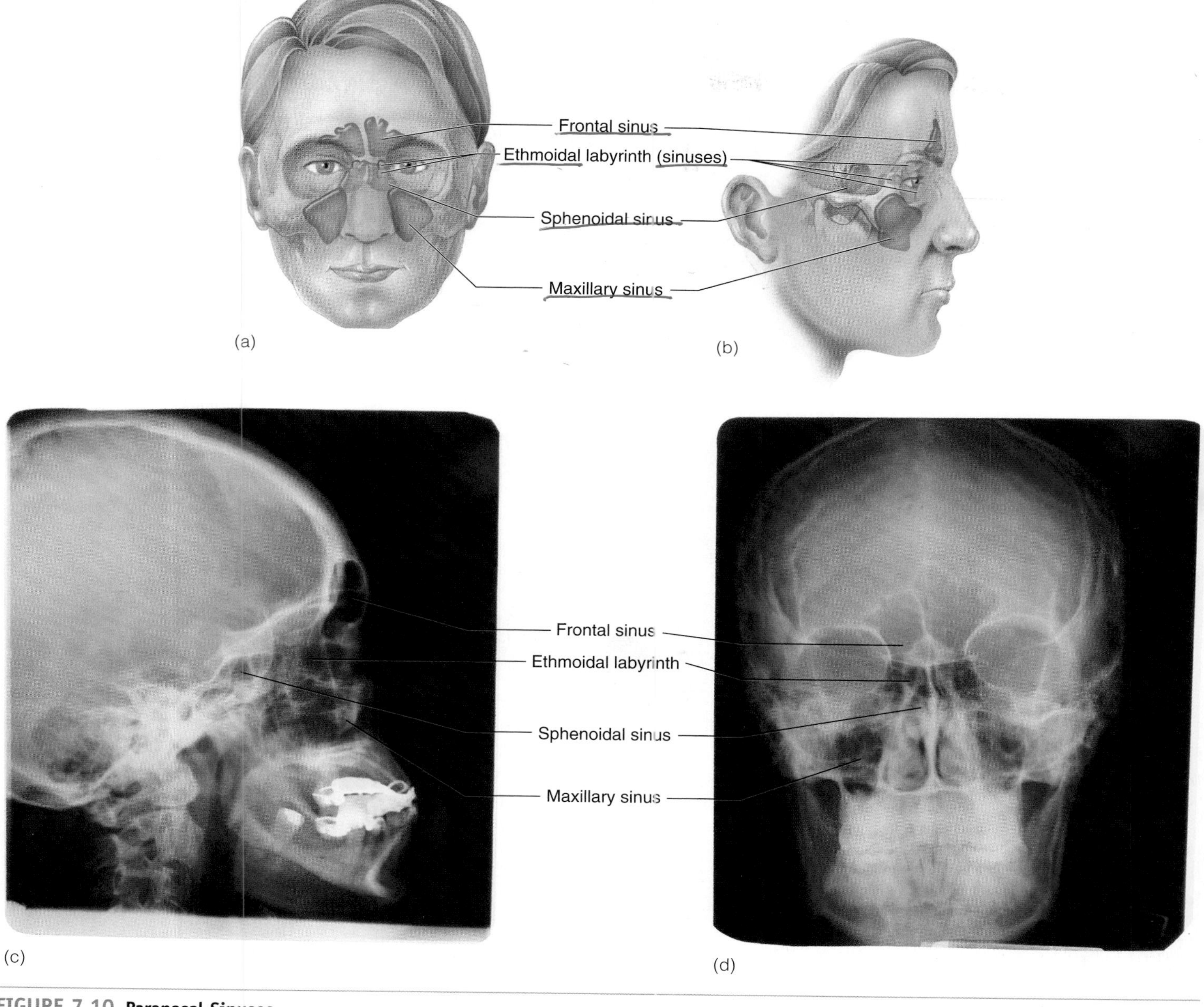

FIGURE 7.10 Paranasal Sinuses
(*a*) Lateral view. (*b*) Anterior view. (*c*) X-ray of the sinuses, lateral view. (*d*) X-ray of the sinuses, anterior view.

each ethmoid bone are collectively called the ethmoidal labyrinth because it consists of a maze of interconnected ethmoidal air cells. The ethmoidal labyrinth is often referred to as the ethmoidal sinuses.

Interior of the Cranial Cavity

The **cranial cavity** is the cavity in the skull occupied by the brain. The cranial cavity can be exposed by cutting away the calvaria, the upper, domelike portion of the skull. With the calvaria removed, the floor of the cranial cavity can be seen (figure 7.11). That floor can be divided roughly into anterior, middle, and posterior cranial fossae, which are formed as the developing neurocranium conforms to the shape of the brain.

A prominent ridge, the **crista galli** (kris′tă găl′ē; rooster's comb), is located in the center of the anterior fossa. The crista galli is a point of attachment for one of the **meninges** (mĕ-nin′jēz), a thick connective tissue membrane that supports and protects the brain (see chapter 13). On each side of the crista galli is an olfactory fossa. An olfactory bulb rests in each fossa and receives the olfactory nerves for the sense of smell. The **cribriform** (krib′ri-fōrm; sievelike) **plate** of the ethmoid bone forms the floor of each olfactory fossa. The olfactory

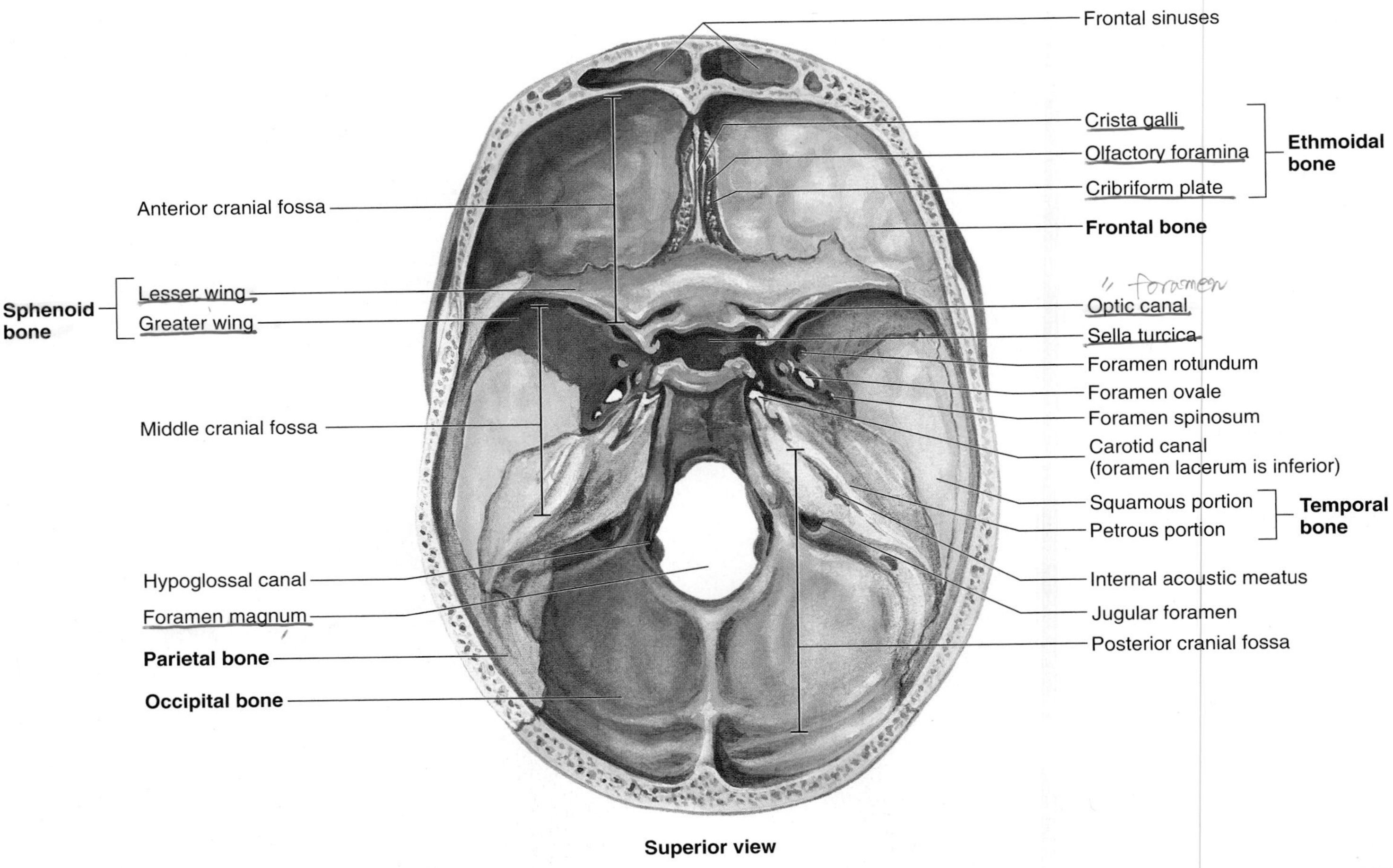

FIGURE 7.11 Floor of the Cranial Cavity

The names of the bones are bolded. The roof of the skull has been removed, and the floor is seen from a superior view.

nerves extend from the cranial cavity into the roof of the nasal cavity through sievelike perforations in the cribriform plate called **olfactory foramina** (see figure 7.9*a* and chapter 15).

> **Fracture of the Cribriform Plate**
>
> The cribriform plate may be fractured in an automobile accident involving a car without air bags, if the driver's nose strikes the steering wheel. Cerebrospinal (ser'ĕ-brō-spī-năl, sĕ-rē'brō-spī-năl) fluid from the cranial cavity may leak through the fracture into the nose. This leakage is a dangerous sign and requires immediate medical attention because risk of infection is very high.

The body of the sphenoid bone forms a central prominence located within the floor of the cranial cavity. This prominence is modified into a structure resembling a saddle, the **sella turcica** (sel'ă tŭr'si-kă; Turkish saddle), which is occupied by the pituitary gland. An **optic canal** is located on each side just anterior to the sella turcica. The lesser wings of the sphenoid bone form a ridge to each side of the optic canals. This ridge separates the **anterior cranial fossa** from the **middle cranial fossa.**

The petrous (rocky) part of the temporal bone extends posterolaterally from each side of the sella turcica. This thick, bony ridge is hollow and contains the middle and inner ears. The petrous ridge separates the middle cranial fossa from the **posterior cranial fossa.**

The **superior orbital fissure, foramen rotundum, foramen ovale, foramen spinosum,** and internal opening of the **carotid canal** are all important openings in the floor of the middle cranial fossa (table 7.6). The **foramen lacerum** (lă-ser'um), in the floor of the carotid canal, is an artifact of the dried skull. In life, it is filled with cartilage.

The prominent **foramen magnum,** through which the spinal cord and brain are connected, is located in the posterior fossa. A **hypoglossal canal** is located on the anterolateral sides of the foramen magnum. **Jugular** (jŭg'ū-lar, throat) **foramina** are located on each side of the foramen magnum near the base of the petrous ridge. An **internal acoustic meatus** is located about midway up the face of each ridge (see table 7.6).

Inferior View of the Skull

Seen from below with the mandible removed, the base of the skull is complex, with a number of foramina and specialized surfaces (figure 7.12). The foramen magnum passes through the occipital bone just slightly posterior to the center of the skull base. **Occipital condyles,** smooth points of articulation between the skull and the

TABLE 7.6 Skull Foramina, Fissures, and Canals (See Figures 7.11 and 7.12)

Opening	Bone Containing the Opening	Structures Passing Through Openings
Carotid canal	Temporal	Carotid artery and carotid sympathetic nerve plexus
Ethmoidal foramina, anterior and posterior	Between frontal and ethmoid	Anterior and posterior ethmoidal nerves
External acoustic meatus	Temporal	Sound waves enroute to the eardrum
Foramen lacerum	Between temporal, occipital, and sphenoid	The foramen is filled with cartilage during life; the carotid canal and pterygoid canal cross its superior part but do not actually pass through it
Foramen magnum	Occipital	Spinal cord, accessory nerves, and vertebral arteries
Foramen ovale	Sphenoid	Mandibular division of trigeminal nerve
Foramen rotundum	Sphenoid	Maxillary division of trigeminal nerve
Foramen spinosum	Sphenoid	Middle meningeal artery
Hypoglossal canal	Occipital	Hypoglossal nerve
Incisive foramen (canal)	Between maxillae	Incisive nerve
Inferior orbital fissure	Between sphenoid and maxilla	Infraorbital nerve and blood vessels and zygomatic nerve
Infraorbital foramen	Maxilla	Infraorbital nerve
Internal acoustic meatus	Temporal	Facial nerve and vestibulocochlear nerve
Jugular foramen	Between temporal and occipital	Internal jugular vein, glossopharyngeal nerve, vagus nerve, and accessory nerve
Mandibular foramen	Mandible	Inferior alveolar nerve to the mandibular teeth
Mental foramen	Mandible	Mental nerve
Nasolacrimal canal	Between lacrimal and maxilla	Nasolacrimal (tear) duct
Olfactory foramina	Ethmoid	Olfactory nerves
Optic canal	Sphenoid	Optic nerve and ophthalmic artery
Palatine foramina, anterior and posterior	Palatine	Palatine nerves
Pterygoid canal	Sphenoid	Sympathetic and parasympathetic nerves to the face
Sphenopalatine foramen	Between palatine and sphenoid	Nasopalatine nerve and sphenopalatine blood vessels
Stylomastoid foramen	Temporal	Facial nerve
Superior orbital fissures	Sphenoid	Oculomotor nerve, trochlear nerve, ophthalmic division of trigeminal nerve, abducens nerve, and ophthalmic veins
Supraorbital foramen or notch	Frontal	Supraorbital nerve and vessels
Zygomaticofacial foramen	Zygomatic	Zygomaticofacial nerve
Zygomaticotemporal foramen	Zygomatic	Zygomaticotemporal nerve

vertebral column, are located on the lateral and anterior margins of the foramen magnum.

The major entry and exit points for blood vessels that supply the brain can be seen from this view. Blood reaches the brain through the internal carotid arteries, which pass through the **carotid** (ka-rot′id; put to sleep) **canals,** and the vertebral arteries, which pass through the foramen magnum. An internal carotid artery enters the inferior opening of each carotid canal (see figure 7.11) and passes through the carotid canal, which runs anteriomedially through the temporal bone. A thin plate of bone separates the carotid canal from the middle ear, making it possible for a person to hear his or her own heartbeat—for example, when frightened or after running. Most blood leaves the brain through the internal jugular veins, which exit through the jugular foramina located lateral to the occipital condyles.

Two long, pointed **styloid** (stī′loyd; stylus- or pen-shaped) **processes** project from the floor of the temporal bone (see figures 7.4 and 7.12). Three muscles involved in movement of the tongue, hyoid bone, and pharynx attach to each process. The **mandibular fossa,** where the mandible articulates with the rest of the skull, is anterior to the mastoid process at the base of the zygomatic arch.

The posterior opening of the nasal cavity is bounded on each side by the vertical bony plates of the sphenoid bone: the **medial pterygoid** (ter′i-goyd; wing-shaped) **plate** and the **lateral pterygoid plate.** The medial and lateral pterygoid muscles, which help move the mandible, attach to the lateral plate (see chapter 10). The **vomer** forms most of the posterior portion of the nasal septum and can be seen between the medial pterygoid plates in the center of the nasal cavity.

The **hard palate,** or **bony palate,** forms the floor of the nasal cavity. Sutures join four bones to form the hard palate; the palatine

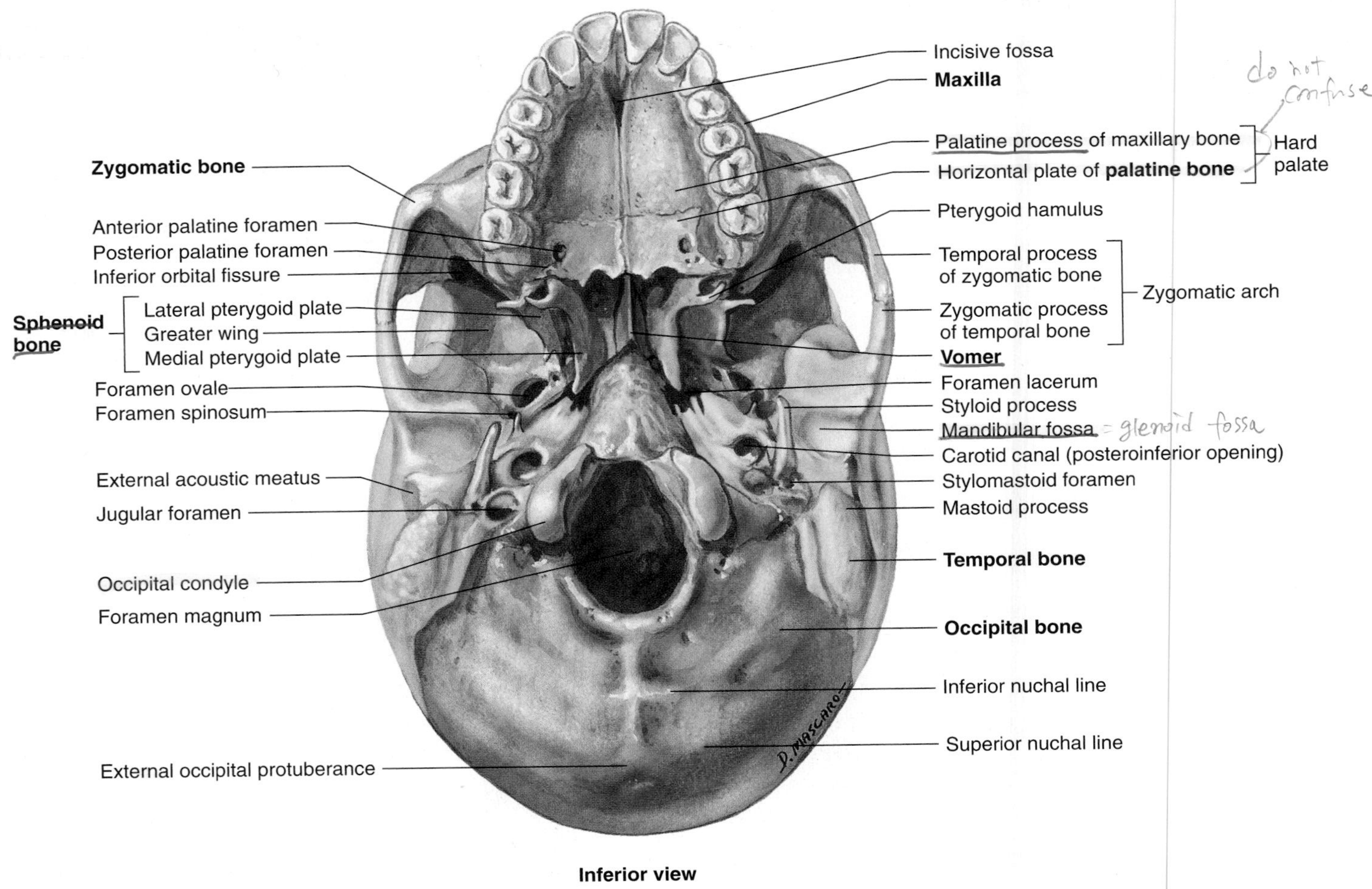

FIGURE 7.12 Inferior View of the Skull
The names of the bones are bolded.

processes of the two maxillary bones form the anterior two-thirds of the palate, and the horizontal plates of the two palatine bones form the posterior one-third of the palate. The tissues of the soft palate extend posteriorly from the hard palate. The hard and soft palates separate the nasal cavity from the mouth, enabling humans to chew and breathe at the same time.

> ### Cleft Lip or Palate
> During development, the facial bones sometimes fail to fuse with one another. A **cleft lip** results if the maxillae do not form normally, and a **cleft palate** occurs when the palatine processes of the maxillae do not fuse with one another. A cleft palate produces an opening between the nasal and oral cavities, making it difficult to eat or drink or to speak distinctly. An artificial palate may be inserted into a newborn's mouth until the palate can be repaired. Approximately 1 in 700 babies in the United States is born with a facial cleft. A cleft lip alone, or cleft lip and palate, occurs approximately once in every 1000 births and is more common in males than in females. A cleft palate alone occurs approximately once in every 2000 births and is more common in females than in males. A cleft lip and cleft palate may also occur in the same person.

4 ***List the seven bones that form the orbit of the eye.***

5 ***Describe the bones and cartilage found in the nasal septum.***

6 ***What is a sinus? What are the functions of sinuses? Give the location of the paranasal sinuses.***

7 ***Name the bones that form the hard palate. What is the function of the hard palate?***

8 ***Through what foramen does the brainstem connect to the spinal cord? Name the foramina that contain nerves for the senses of vision (optic nerve), smell (olfactory nerves), and hearing (vestibulocochlear nerve).***

9 ***Name the foramina through which the major blood vessels enter and exit the skull.***

10 ***List the places where these muscles attach to the skull: neck muscles, throat muscles, muscles of mastication, muscles of facial expression, and muscles that move the eyeballs.***

Bones of the Skull

The skull, or cranium, is composed of 22 separate bones (see table 7.1; table 7.7). In addition, the skull contains six **auditory ossicles,** which function in hearing (see chapter 15). Each

TABLE 7.7 Skull Bones

(a) Parietal Bone (Right)—Lateral View

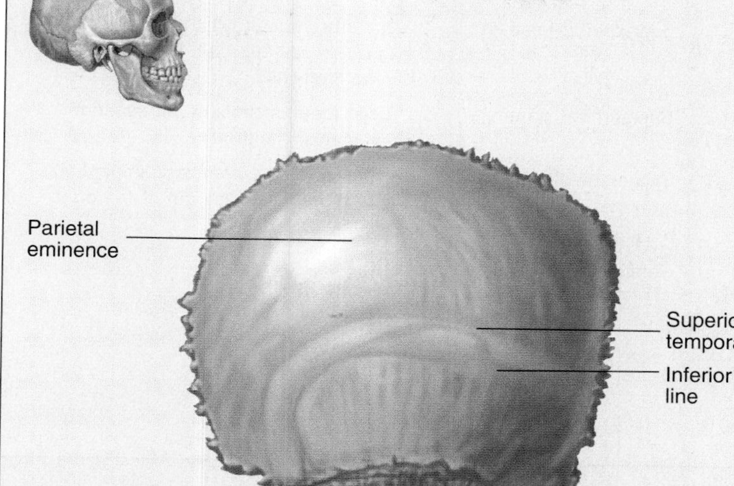

Landmark	Description
Parietal eminence	The widest part of the head is from one parietal eminence to the other.
Superior and inferior temporal lines	Attachment point for temporalis muscle

Special Feature

Forms lateral wall of skull

(b) Temporal Bone (Right)—Lateral and Medial Views

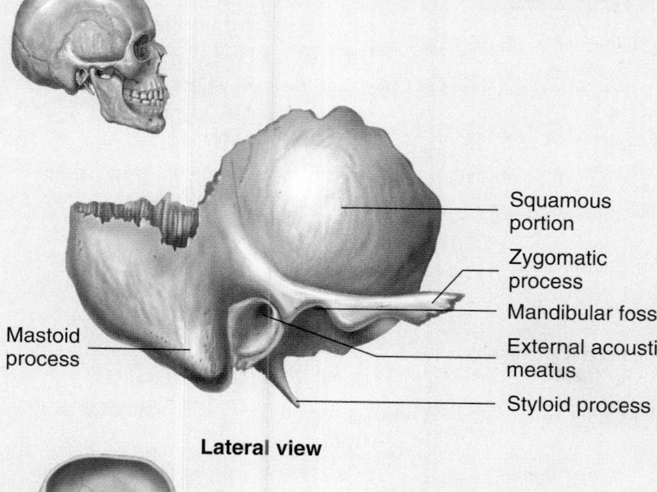

Lateral view

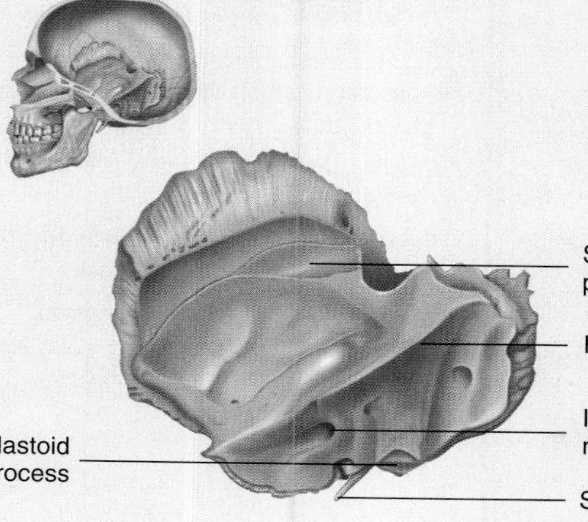

Medial view

Landmark	Description
Carotid canal (shown in figures 7.11 and 7.12)	Canal through which the internal carotid artery enters the cranial cavity
External acoustic meatus	External canal of the ear; carries sound to the ear
Internal acoustic meatus (shown in figure 7.11)	Opening through which the facial (cranial nerve VII) and vestibulocochlear (cranial nerve VIII) nerves enter the petrous portion of the temporal bone
Forms one side of jugular foramen (shown in figures 7.11 and 7.12)	Foramen through which the internal jugular vein exits the cranial cavity
Mandibular fossa	Articulation point between the mandible and skull
Mastoid process	Attachment point for muscles moving the head and for a hyoid muscle
Middle cranial fossa (shown in figure 7.11)	Depression in the floor of the cranial cavity formed by the temporal lobes of the brain
Petrous portion (shown in figure 7.11)	Thick, "rocky" portion of the temporal bone
Squamous portion	Flat, lateral portion of the temporal bone
Styloid process	Attachment for muscles of the tongue, throat, and hyoid bone
Stylomastoid foramen (shown in figure 7.12)	Foramen through which the facial nerve (cranial nerve VII) exits the skull
Zygomatic process	Helps form the bony bridge from the cheek to just anterior to the ear; attachment for a muscle moving the mandible

Special Features

Contains the middle and inner ear and the mastoid air cells

Place where the mandible articulates with the rest of the skull

TABLE 7.7 Skull Bones—Continued

(c) Frontal Bone—Anterior View

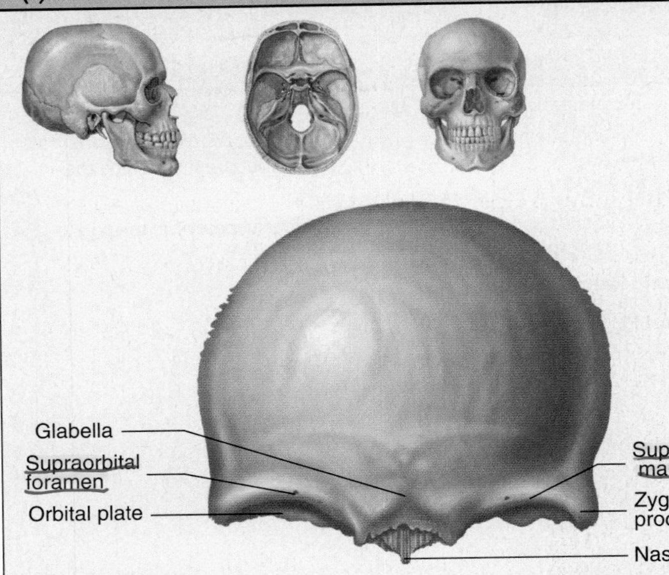

Landmark	Description
Glabella	Area between the supraorbital margins
Nasal spine	Superior part of the nasal bridge
Orbital plate	Roof of the orbit
Supraorbital foramen	Opening through which nerves and vessels exit the skull to the skin of the forehead
Supraorbital margin	Ridge forming the anterior superior border of the orbit
Zygomatic process	Connects to the zygomatic bone; helps form the lateral margin of the orbit

Special Features

Forms the forehead and roof of the orbit

Contains the frontal sinus

(d) Sphenoid Bone—Superior and Posterior Views

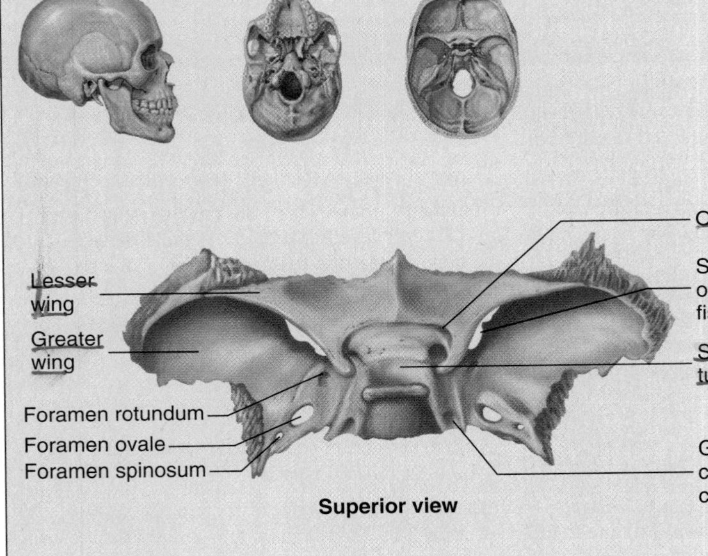

Superior view

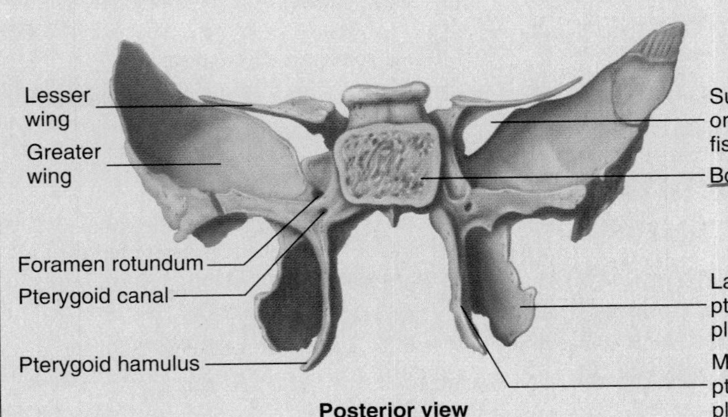

Posterior view

Landmark	Description
Body	Thickest part of the bone which articulates with the occipital bone
Foramen ovale	Opening through which a branch of the trigeminal nerve (cranial nerve V) exits the cranial cavity
Foramen rotundum	Opening through which a branch of the trigeminal nerve (cranial nerve V) exits the cranial cavity
Foramen spinosum	Opening through which a major artery to the meninges (membranes around the brain) enters the cranial cavity
Greater wing	Forms the floor of the middle cranial fossa; several foramina pass through this wing
Lateral pterygoid plate	Attachment point for muscles of mastication (chewing)
Lesser wing	Superior border of the superior orbital fissure
Medial pterygoid plate	Posterolateral walls of the nasal cavity
Optic canal	Opening through which the optic nerve (cranial nerve II) passes from the orbit to the cranial cavity
Pterygoid canal	Opening through which nerves and vessels exit the cranial cavity
Pterygoid hamulus	Process around which the tendon from a muscle to the soft palate passes
Sella turcica	Fossa containing the pituitary gland
Superior orbital fissure	Opening through which nerves and vessels enter the orbit from the cranial cavity

Special Feature

Contains the sphenoidal sinus

TABLE 7.7 Skull Bones—Continued

(e) Occipital Bone—Inferior View

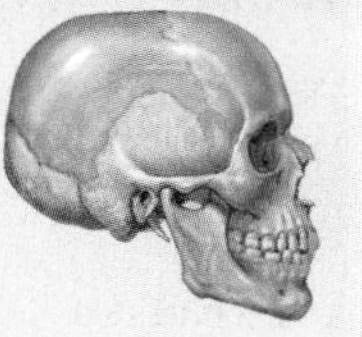
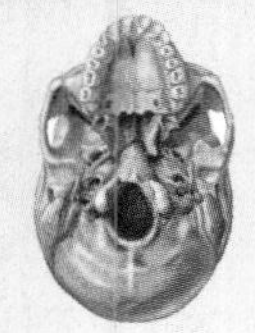
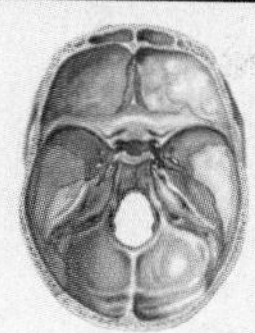

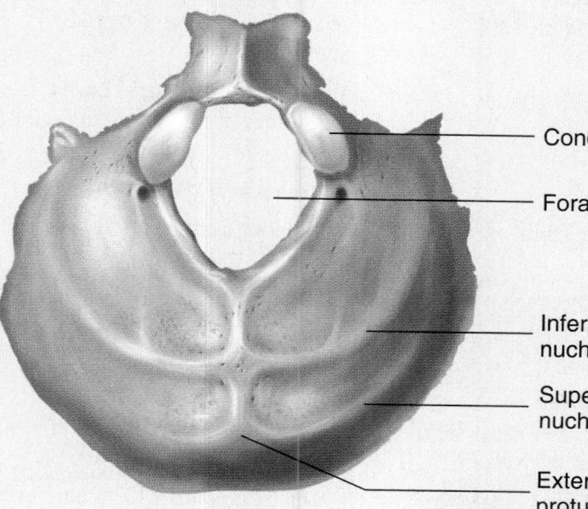

Landmark	Description
Condyle	Articulation point between the skull and first vertebra
External occipital protuberance	Attachment point for a strong ligament (nuchal ligament) in the back of the neck
Foramen magnum	Opening around the point where the brain and spinal cord connect
Hypoglossal canal (shown in figure 7.11)	Opening through which the hypoglossal nerve (cranial nerve XII) passes
Inferior nuchal line	Attachment point for neck muscles
Posterior cranial fossa (shown in figure 7.11)	Depression in the posterior of the cranial cavity formed by the cerebellum
Superior nuchal line	Attachment point for neck muscles

Special Feature

Forms the base of the skull

(f) Zygomatic Bone (Right)—Lateral View

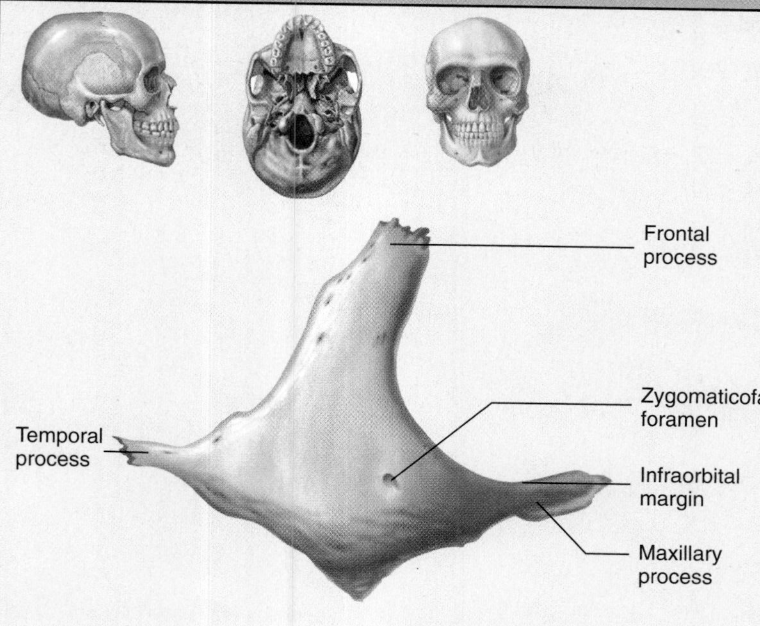

Landmark	Description
Frontal process	Connection to the frontal bone; helps form the lateral margin of the orbit
Infraorbital margin	Ridge forming the inferior border of the orbit
Temporal process	Helps form the bony bridge from the cheek to just anterior to the ear
Zygomaticofacial foramen	Opening through which a nerve and vessels exit the orbit to the face

Special Features

Forms the prominence of the cheek

Forms the anterolateral wall of the orbit

TABLE 7.7 Skull Bones—Continued

(g) Ethmoid Bone—Superior, Lateral, and Anterior Views

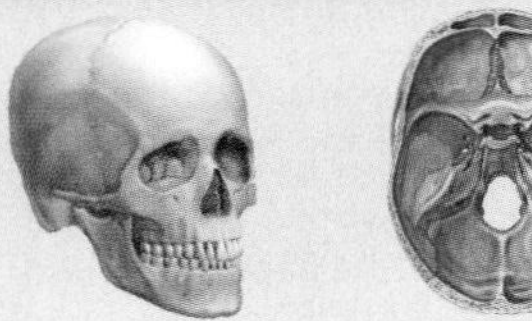

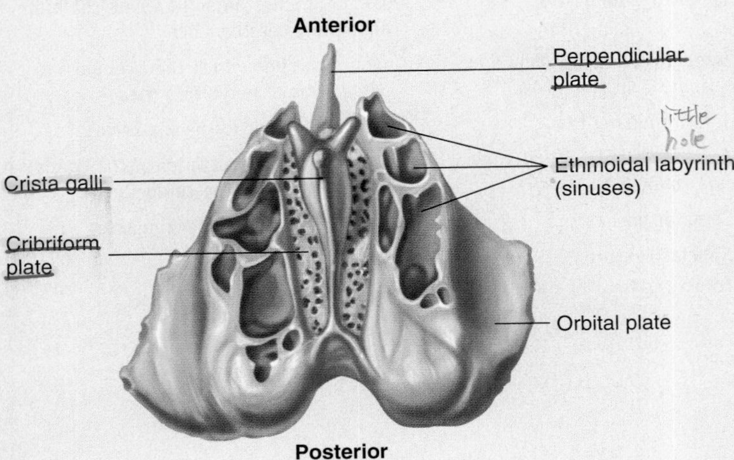

Superior view

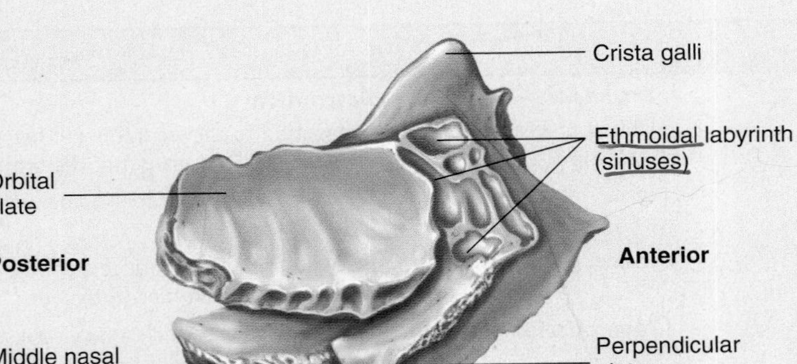

Lateral view

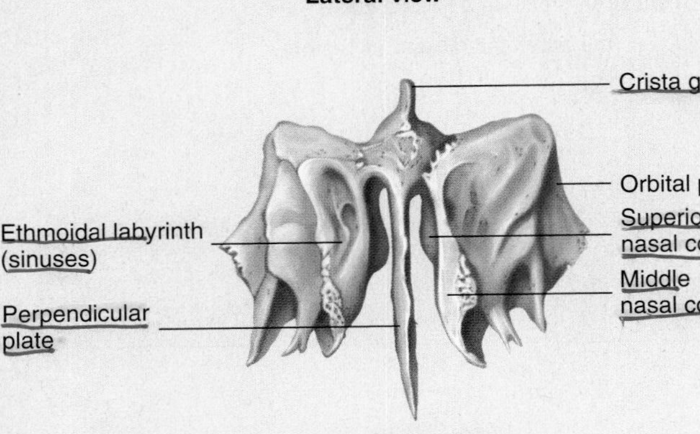

Anterior view

Landmark	Description
Cribriform plate	Contains numerous olfactory foramina through which branches of the olfactory nerve (cranial nerve I) enter the cranial cavity from the nasal cavity
Crista galli	Attachment for meninges (membranes around brain)
Ethmoidal foramina (shown in figure 7.8)	Openings through which nerves and vessels pass from the orbit to the nasal cavity
Middle nasal concha	Ridge extending into the nasal cavity; increases surface area, helps warm and moisten air in the cavity
Orbital plate	Forms the medial wall of the orbit
Perpendicular plate	Forms the superior portion of the nasal septum
Superior nasal concha	Ridge extending into the nasal cavity; increases surface area, helps warm and moisten air in the cavity

Special Features

Forms part of the nasal septum and part of the lateral walls and roof of the nasal cavity

Contains the ethmoidal labyrinth, or ethmoidal sinuses. The labyrinth is divided into anterior, middle, and posterior ethmoidal cells.

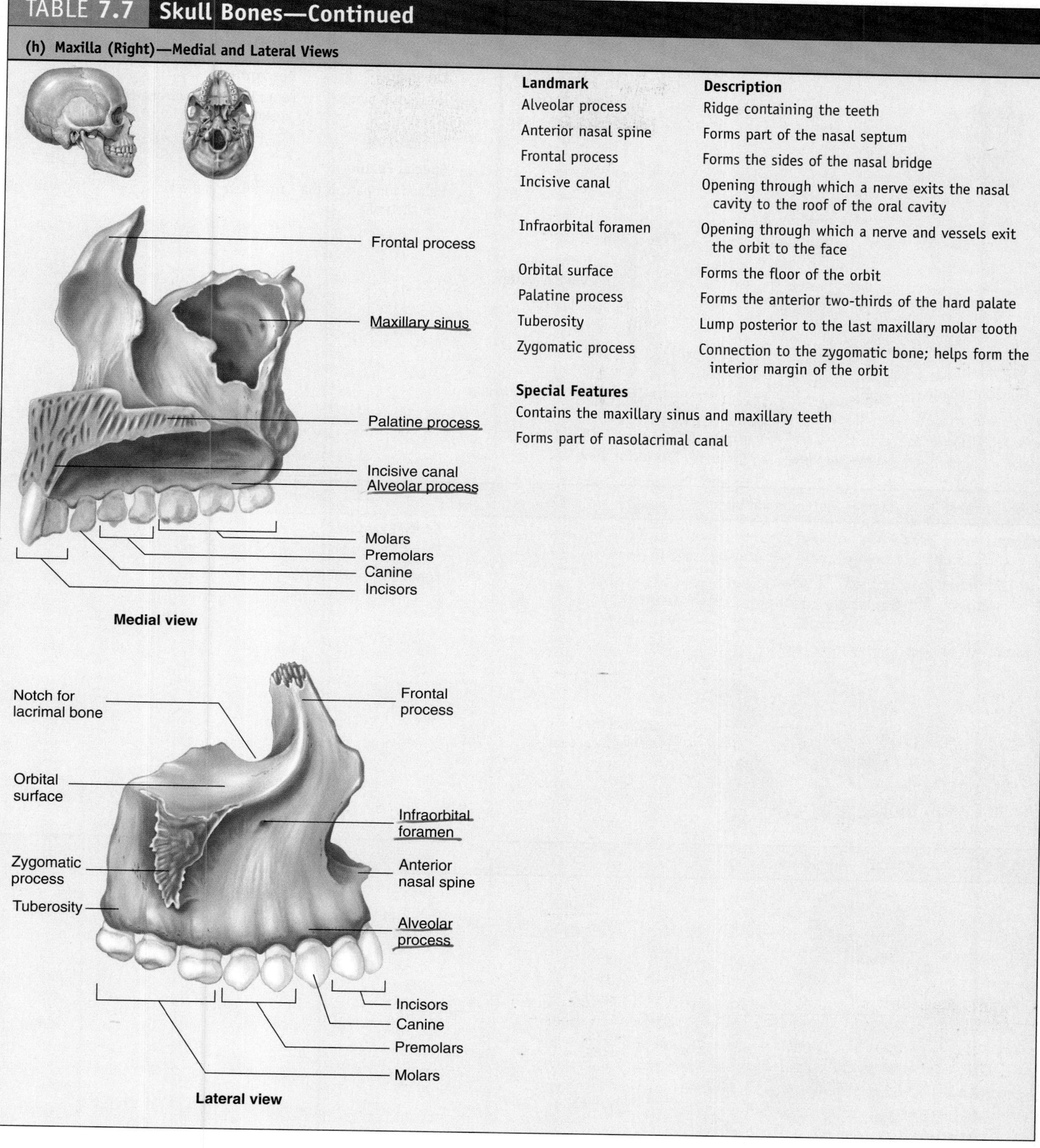

TABLE 7.7 Skull Bones—Continued

(h) Maxilla (Right)—Medial and Lateral Views

Landmark	Description
Alveolar process	Ridge containing the teeth
Anterior nasal spine	Forms part of the nasal septum
Frontal process	Forms the sides of the nasal bridge
Incisive canal	Opening through which a nerve exits the nasal cavity to the roof of the oral cavity
Infraorbital foramen	Opening through which a nerve and vessels exit the orbit to the face
Orbital surface	Forms the floor of the orbit
Palatine process	Forms the anterior two-thirds of the hard palate
Tuberosity	Lump posterior to the last maxillary molar tooth
Zygomatic process	Connection to the zygomatic bone; helps form the interior margin of the orbit

Special Features

Contains the maxillary sinus and maxillary teeth

Forms part of nasolacrimal canal

TABLE 7.7 Skull Bones—Continued

(i) Palatine Bone (Right)—Medial and Anterior Views

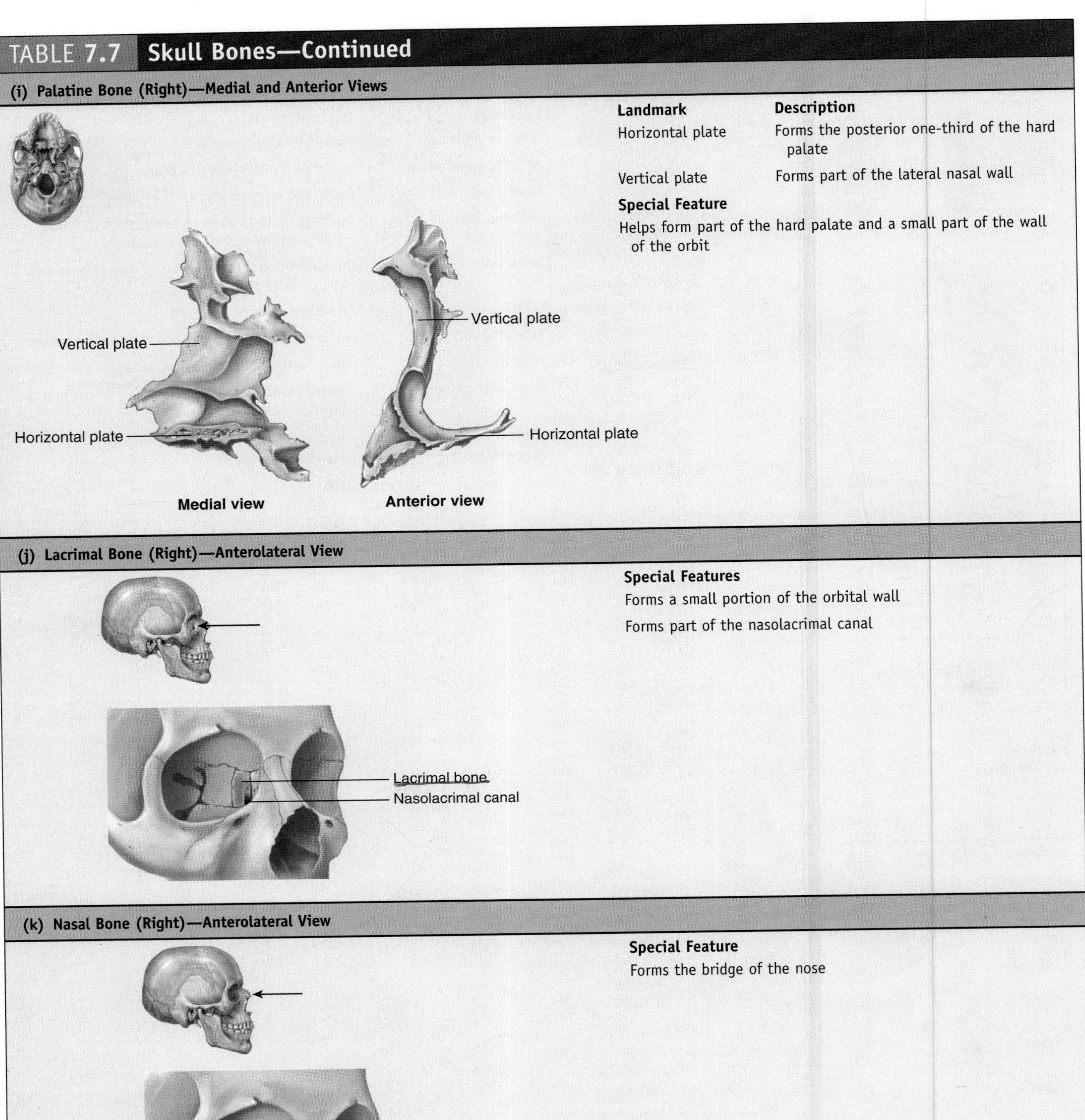

Landmark	Description
Horizontal plate	Forms the posterior one-third of the hard palate
Vertical plate	Forms part of the lateral nasal wall

Special Feature

Helps form part of the hard palate and a small part of the wall of the orbit

(j) Lacrimal Bone (Right)—Anterolateral View

Special Features

Forms a small portion of the orbital wall

Forms part of the nasolacrimal canal

(k) Nasal Bone (Right)—Anterolateral View

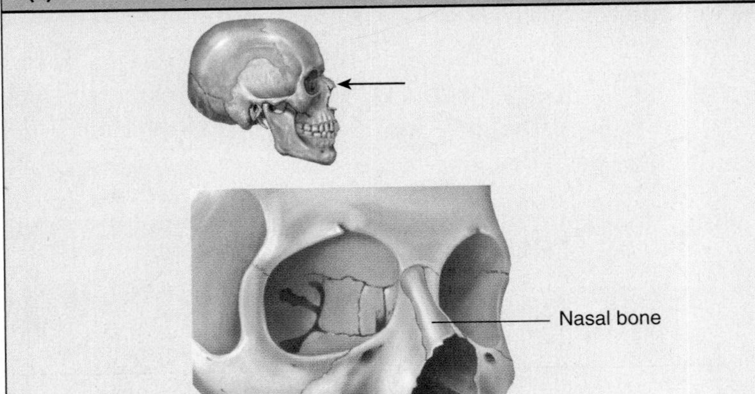

Special Feature

Forms the bridge of the nose

TABLE 7.7 Skull Bones—Continued

(l) Mandible (Right Half)—Medial and Lateral Views

Coronoid process
Mandibular notch
Molars
Premolars
Canine
Incisors
Mandibular condyle (head)
Condylar process
Ramus
Mandibular foramen
Lingula
Alveolar process
Mylohyoid line
Angle
Body

Medial view

Mandibular notch
Mandibular condyle
Condylar process
Oblique line
Ramus
Body
Angle
Coronoid process
Molars
Premolars
Canine
Incisors
Alveolar process
Mental foramen

Lateral view

Landmark	Description
Alveolar process	Ridge containing the teeth
Angle	Corner between the body and ramus
Body	Major, horizontal portion of the bone
Condylar process	Extension containing the mandibular condyle
Coronoid process	Attachment for a muscle of mastication
Mandibular condyle	Helps form the temporomandibular joint (the point of articulation between the mandible and the rest of the skull)
Mandibular foramen	Opening through which nerves and vessels to the mandibular teeth enter the bone
Mandibular notch	Depression between the condylar process and the coronoid process
Mental foramen	Opening through which a nerve and vessels exit the mandible to the skin of the chin
Mylohyoid line	Attachment point of the mylohyoid muscle
Oblique line	Ridge from the anterior edge of the ramus onto the body of the mandible
Ramus	Major, nearly vertical portion of the bone

Special Features

The only bone in the skull that is freely movable relative to the rest of the skull bones

Holds the lower teeth

(m) Vomer—Anterior and Lateral Views

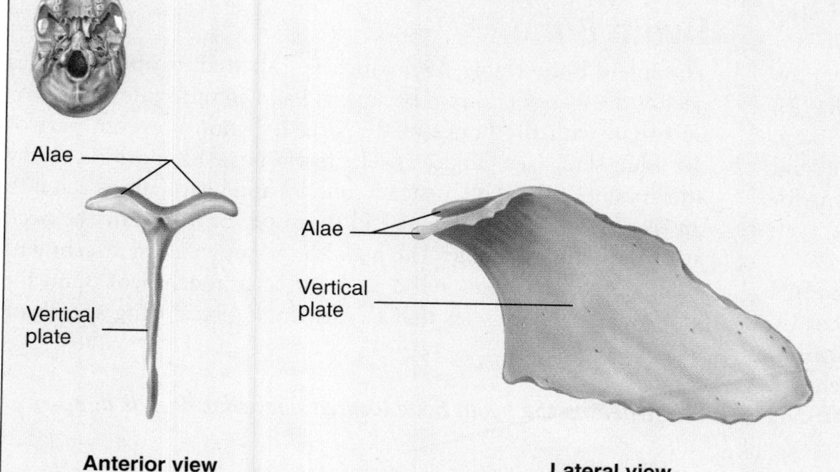

Anterior view

Lateral view

Landmark	Description
Alae	Attachment points between the vomer and sphenoid
Vertical plate	Forms part of the nasal septum

Special Feature

Forms most of the posterior and inferior portions of the nasal septum

TABLE 7.8 Hyoid Bone—Anterior and Lateral Views

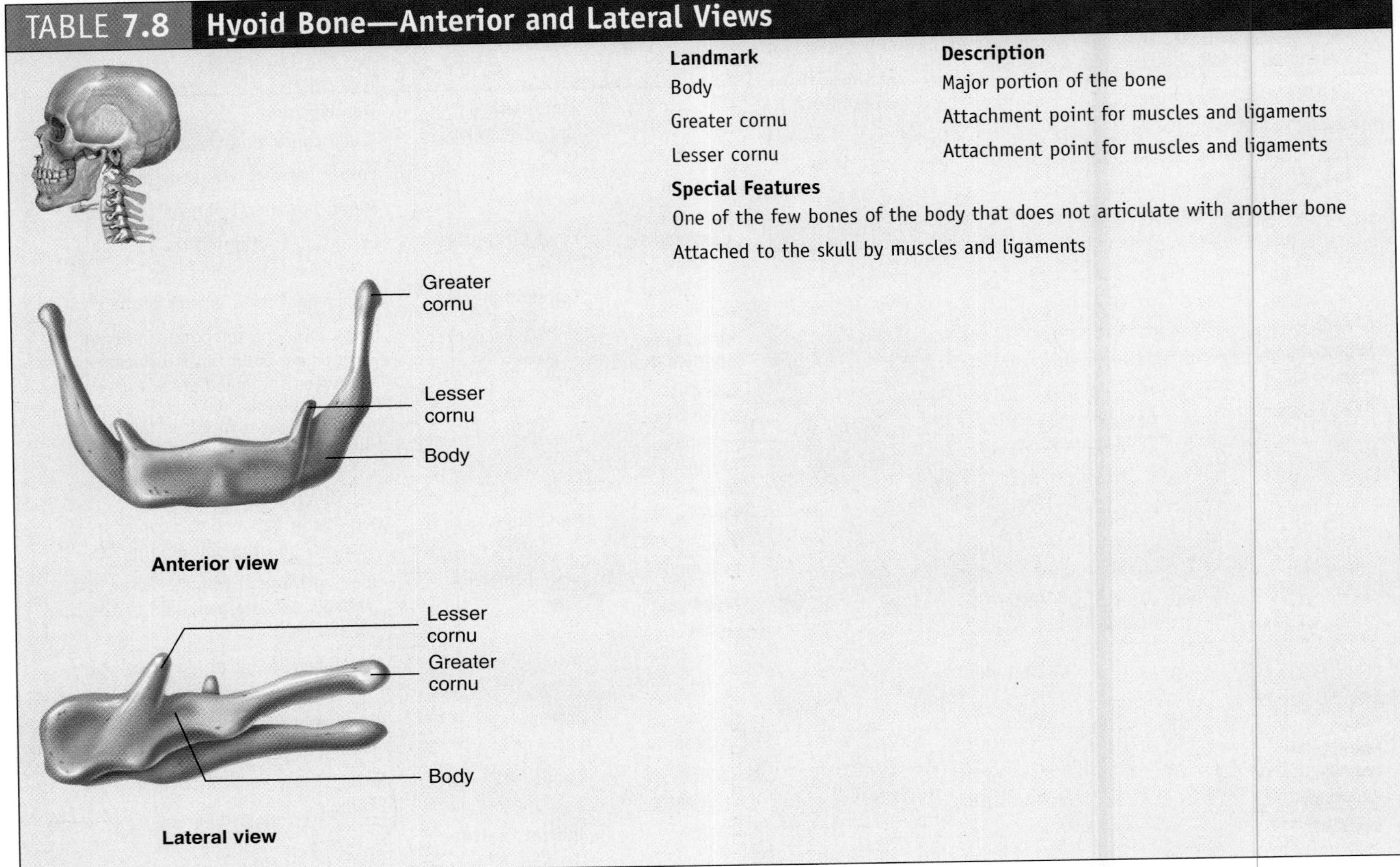

Landmark	Description
Body	Major portion of the bone
Greater cornu	Attachment point for muscles and ligaments
Lesser cornu	Attachment point for muscles and ligaments

Special Features

One of the few bones of the body that does not articulate with another bone

Attached to the skull by muscles and ligaments

temporal bone holds one set of auditory ossicles, which consists of the malleus, incus, and stapes. These bones cannot be observed unless the temporal bones are cut open.

The 22 bones of the skull are divided into two portions: the neurocranium and the viscerocranium. The **neurocranium,** or **braincase,** consists of 8 bones that immediately surround and protect the brain. They include the paired parietal and temporal bones and the unpaired frontal, occipital, sphenoid, and ethmoid bones.

The 14 bones of the **viscerocranium,** or **facial bones,** form the structure of the face in the anterior skull. They are the maxilla (2), zygomatic (2), palatine (2), lacrimal (2), nasal (2), inferior nasal concha (2), mandible (1), and vomer (1) bones. The frontal and ethmoid bones, which are part of the neurocranium, also contribute to the face. The mandible is often listed as a facial bone, even though it is not part of the intact skull.

The facial bones protect the major sensory organs located in the face: the eyes, nose, and tongue. The bones of the face also provide attachment points for the muscles involved in **mastication** (mas-ti-kā′shŭn; chewing), facial expression, and eye movement. The jaws (mandible and maxillae) possess **alveolar** (al-vē′ō-lăr) **processes** with sockets for the attachment of the teeth. The bones of the face and their associated soft tissues determine the unique facial features of each individual.

11 ***Name the bones of the neurocranium and viscerocranium. What functions are accomplished by each group?***

Hyoid Bone

The **hyoid bone** (table 7.8), which is unpaired, is often listed as part of the viscerocranium because it has a common developmental origin with the bones of the face. It is not, however, part of the adult skull (see table 7.1). The hyoid bone has no direct bony attachment to the skull. Instead, muscles and ligaments attach it to the skull, so the hyoid "floats" in the superior aspect of the neck just below the mandible. The hyoid bone provides an attachment point for some tongue muscles, and it is an attachment point for important neck muscles that elevate the larynx during speech or swallowing.

12 ***Where is the hyoid bone located and what does it do?***

Vertebral Column

The **vertebral column** usually consists of 26 bones, which can be divided into five regions (figure 7.13). Seven **cervical vertebrae** (ver′tĕ-brē), 12 **thoracic vertebrae,** 5 **lumbar vertebrae,** 1 **sacral bone,** and 1 **coccygeal** (kok-sij′ē-ăl) bone make up the vertebral column. The cervical vertebrae are designated "C," thoracic "T," and lumbar "L." A number after the letter indicates the number of the vertebra, from superior to inferior, within each vertebral region. The developing embryo has about 33–34 vertebrae, but the 5 sacral vertebrae fuse to form 1 bone, and the 4 or 5 coccygeal bones usually fuse to form 1 bone.

The five regions of the adult vertebral column have four major curvatures (see figure 7.13). Two of the curves appear during embryonic development and reflect the C-shaped curve of the embryo and fetus within the uterus. When the infant raises its head in the first few months after birth, a secondary curve, which is convex anteriorly, develops in the neck. Later, when the infant learns to sit and then walk, the lumbar portion of the column also becomes convex anteriorly. Thus, in the adult vertebral column, the cervical region is convex anteriorly, the thoracic region is concave anteriorly, the lumbar region is convex anteriorly, and the sacral and coccygeal regions together are concave anteriorly.

Abnormal Spinal Curvatures

Lordosis (lōr-dō′sis; hollow back) is an exaggeration of the convex curve of the lumbar region. **Kyphosis** (kī-fō′sis; hump back) is an exaggeration of the concave curve of the thoracic region. **Scoliosis** (skō′lē-ō′sis) is an abnormal lateral and rotational curvature of the vertebral column, which is often accompanied by secondary abnormal curvatures, such as kyphosis (figure A).

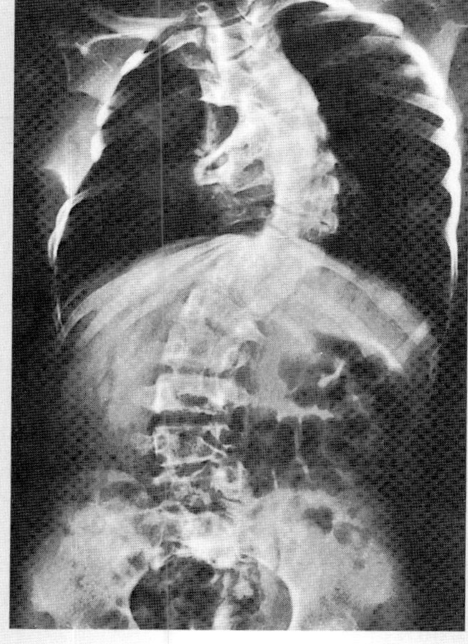

FIGURE A Scoliosis

Cervical region (convex anteriorly)
Thoracic region (concave posteriorly)
Lumbar region (convex anteriorly)
Sacral and coccygeal regions (concave posteriorly)
First cervical vertebra (atlas)
Second cervical vertebra (axis)
Seventh cervical vertebra
First thoracic vertebra
Intervertebral foramina
Body
Intervertebral disk
Twelfth thoracic vertebra
First lumbar vertebra
Transverse process
Spinous process
Fifth lumbar vertebra
Sacral promontory
Sacrum
Coccyx
Lateral view

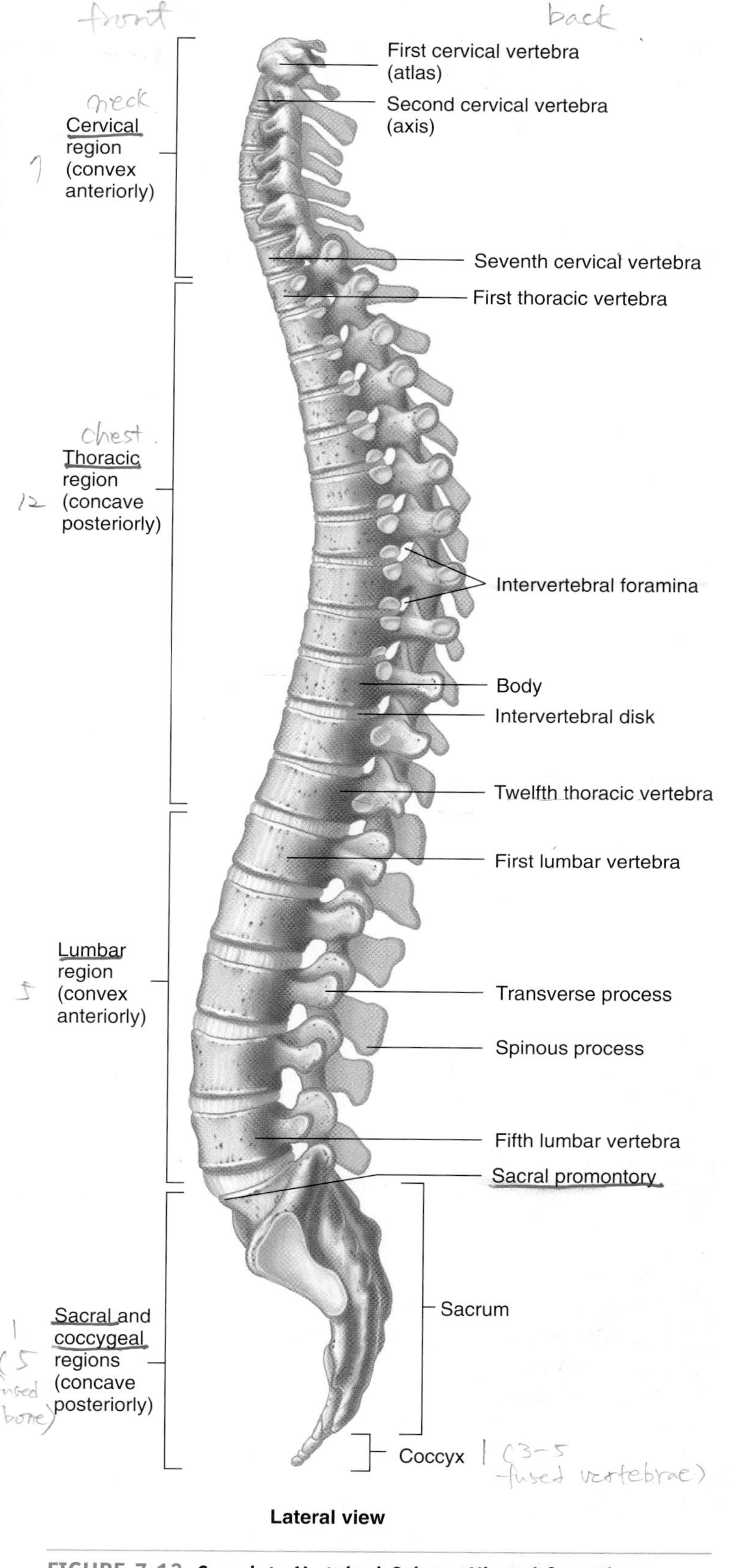

FIGURE 7.13 Complete Vertebral Column Viewed from the Left Side

TABLE 7.9 General Structure of a Vertebra

Feature	Description
Body	Disk-shaped; usually the largest part with flat surfaces directed superiorly and inferiorly; forms the anterior wall of the vertebral foramen; intervertebral disks are located between the bodies
Vertebral foramen	Hole in each vertebra through which the spinal cord passes; adjacent vertebral foramina form the vertebral canal
Vertebral arch	Forms the lateral and posterior walls of the vertebral foramen; possesses several processes and articular surfaces
Predicle	Foot of the arch with one on each side; forms the lateral walls of the vertebral foramen
Lamina	Posterior part of the arch; forms the posterior wall of the vertebral foramen
Transverse process	Process projecting laterally from the junction of the lamina and pedicle; a site of muscle attachment
Spinous process	Process projecting posteriorly at the point where the two laminae join; a site of muscle attachment; strengthens the vertebral column and allows for movement
Articular processes	Superior and inferior projections containing articular facets where vertebrae articulate with each other; strengthen the vertebral column and allow for movement
Intervertebral notches	Form intervertebral foramina between two adjacent vertebrae through which spinal nerves exit the vertebral canal

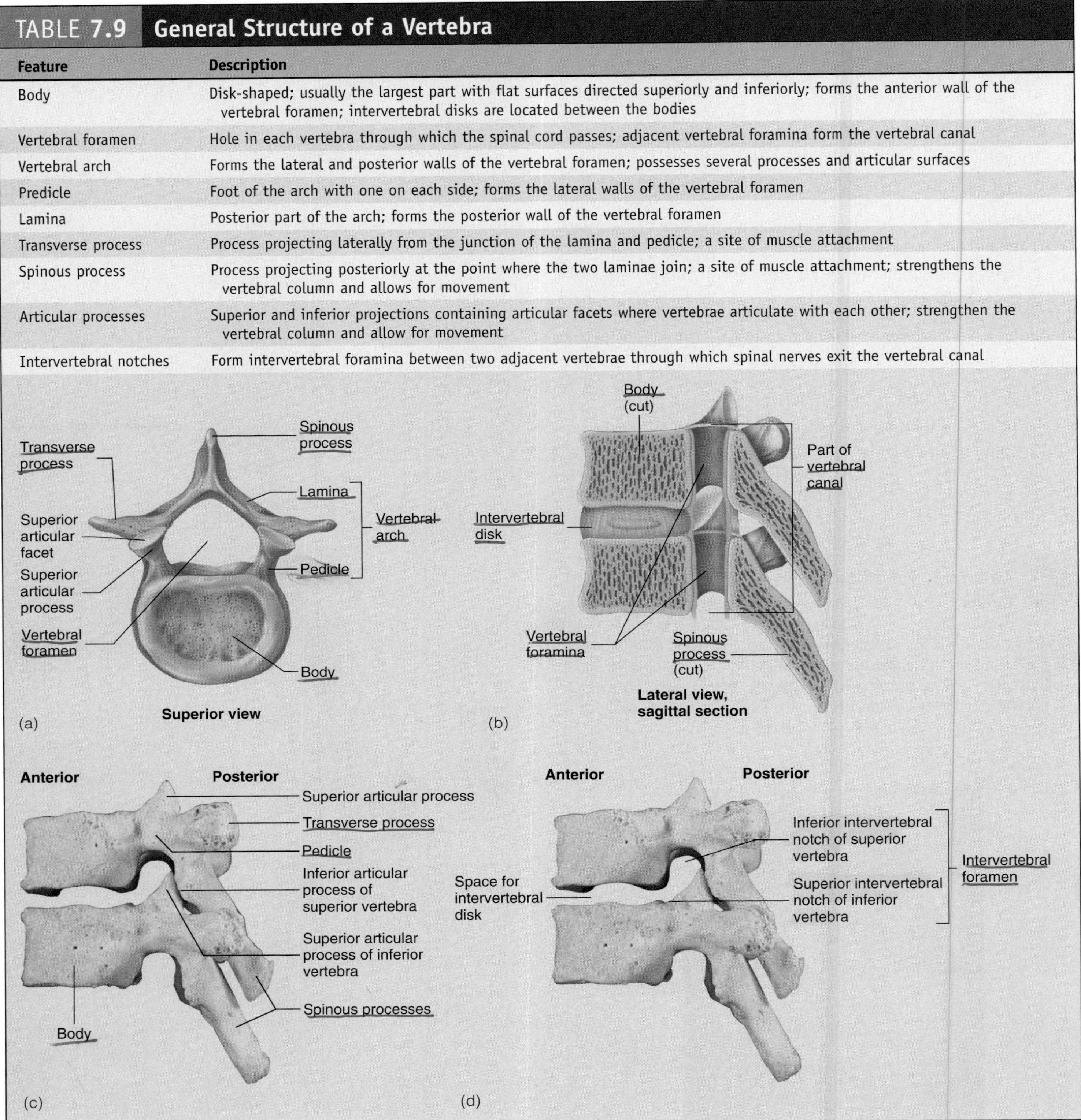

General Plan of the Vertebrae

The vertebral column performs five major functions: (1) It supports the weight of the head and trunk, (2) it protects the spinal cord, (3) it allows spinal nerves to exit the spinal cord, (4) it provides a site for muscle attachment, and (5) it permits movement of the head and trunk. The general structure of a vertebra is outlined in table 7.9. Each vertebra consists of a body, an arch, and various processes. The weight-bearing portion of the vertebra is a bony disk called the **body.**

The **vertebral arch** projects posteriorly from the body. The arch is divided into left and right halves, and each half has two parts: the **pedicle** (ped′i-kl; foot), which is attached to the body, and the **lamina** (lam′i-na; thin plate), which joins the lamina from the opposite half of the arch. The vertebral arch and the posterior part of the body surround a large opening called the **vertebral foramen.** The vertebral foramina of adjacent vertebrae combine to form the **vertebral canal,** which contains the spinal cord or cauda equina (see figure 12.1). The vertebral arches and bodies protect the spinal cord.

A **transverse process** extends laterally from each side of the arch between the lamina and pedicle, and a single **spinous process** is present at the junction between the two laminae. The spinous processes can be seen and felt as a series of lumps down the midline of the back (figure 7.14). Much vertebral movement is accomplished by the contraction of the skeletal muscles attached to the transverse and spinous processes (see chapter 10).

Spinal nerves exit the spinal cord through the **intervertebral foramina** (see table 7.9 and figure 7.13). Each intervertebral foramen is formed by **intervertebral notches** in the pedicles of adjacent vertebrae.

Movement and additional support of the vertebral column are made possible by the vertebral processes. Each vertebra has two **superior** and two **inferior articular processes,** with the superior processes of one vertebra articulating with the inferior processes of the next superior vertebra (see table 7.9*c* & *d*). Overlap of these processes increases the rigidity of the vertebral column. The region of overlap and articulation between the superior and inferior articular processes creates a smooth **articular facet** (fas′et; little face) on each articular process.

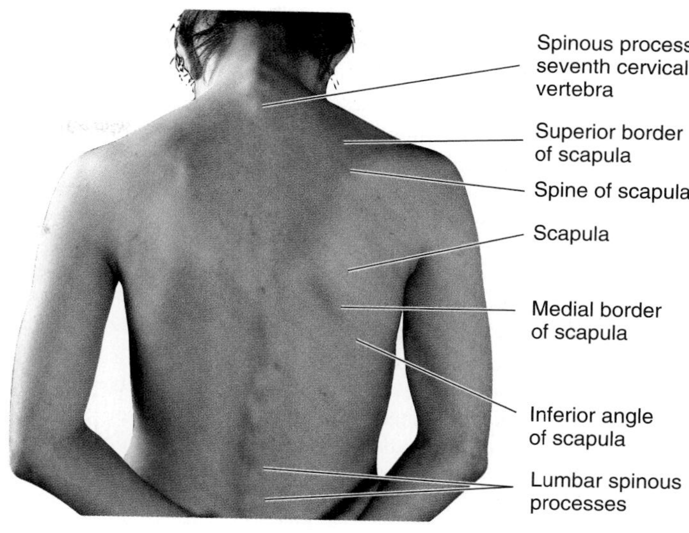

FIGURE 7.14 Surface View of the Back Showing the Scapula and Vertebral Spinous Processes

Laminectomy and Spina Bifida

Sometimes vertebral laminae partly or completely fail to fuse (or even fail to form) during fetal development, resulting in a condition called **spina bifida** (spī′nă bif′i-dă; split spine). This defect is most common in the lumbar region. If the defect is severe and involves the spinal cord (figure B), it may interfere with normal nerve function below the point of the defect.

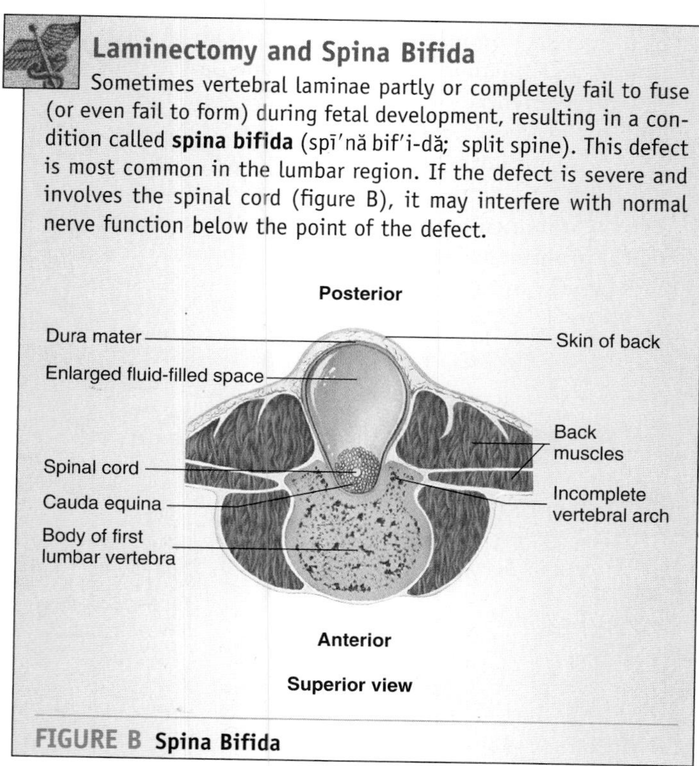

FIGURE B Spina Bifida

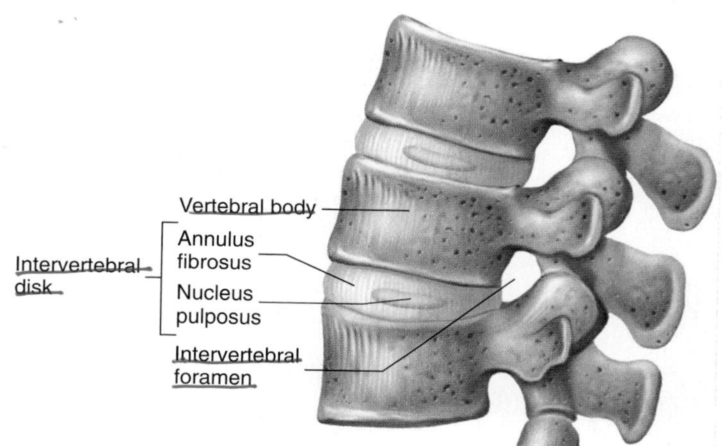

(a) **Lateral view**

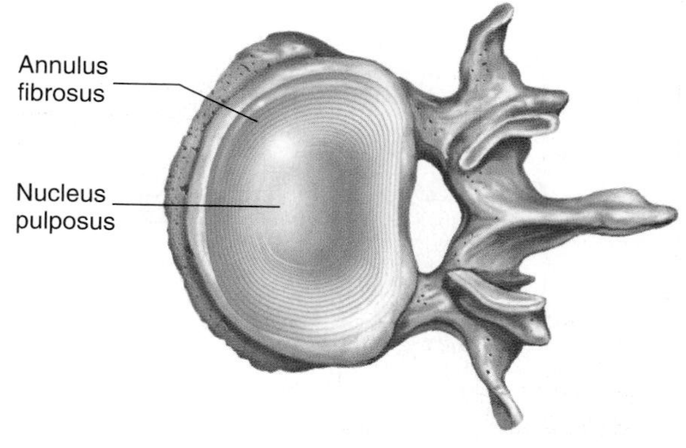

(b) **Superior view**

FIGURE 7.15 Intervertebral Disk

Intervertebral Disks

During life, **intervertebral disks** of fibrocartilage, which are located between the bodies of adjacent vertebrae (see table 7.9, figure 7.13; figure 7.15), provide additional support and prevent the vertebral

Herniated, or Ruptured, Intervertebral Disk

A **herniated,** or **ruptured, disk** results from the breakage or ballooning of the annulus fibrosus with a partial or complete release of the nucleus pulposus (figure C). The herniated part of the disk may push against and compress the spinal cord, cauda equina, or spinal nerves, compromising their normal function and producing pain. Herniation of the inferior lumbar intervertebral disks is most common, but herniation of the inferior cervical disks is almost as common.

Herniated, or ruptured, disks can be repaired in one of several ways. One procedure uses prolonged bed rest and is based on the tendency for the herniated part of the disk to recede and the annulus fibrosus to repair itself. In many cases, however, surgery is required. A **laminectomy** is the removal of a vertebral lamina, or vertebral arch. A **hemilaminectomy** is the removal of a portion of a vertebral lamina. These procedures are performed to reduce the compression of the spinal nerve or spinal cord. **Fenestration** involves removal of the nucleus pulposus, leaving the annulus fibrosus intact. In extreme cases, the entire damaged disk is removed. To enhance the stability of the vertebral column, a piece of hipbone is sometimes inserted into the space previously occupied by the disk, and the adjacent vertebrae become fused by bone across the gap.

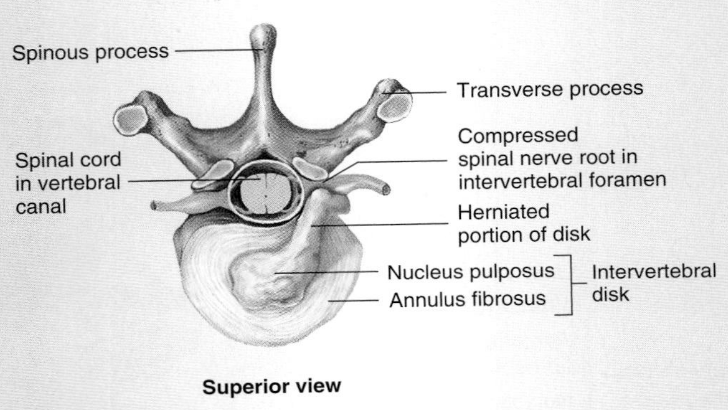

FIGURE C Herniated Disk
Part of the annulus fibrosus has been removed to reveal the nucleus pulposus in the center of the disk and in the intervertebral foramen.

bodies from rubbing against each other. The intervertebral disks consist of an external **annulus fibrosus** (an′ū-lŭs fī-brō′sŭs; fibrous ring) and an internal, gelatinous **nucleus pulposus** (pŭl-pō′sŭs; pulp). The disk becomes more compressed with increasing age so that the distance between vertebrae and therefore the overall height of the individual decreases. The annulus fibrosus also becomes weaker with age and more susceptible to herniation.

Regional Differences in Vertebrae

The vertebrae of each region of the vertebral column have specific characteristics that tend to blend at the boundaries between regions (table 7.10). The **cervical vertebrae** (see figure 7.13; figure 7.16*a–g*) have very small bodies; most have **bifid** (bī′fid; split) **spinous processes,** and a **transverse foramen** in each transverse process through which the vertebral arteries extend toward the head. Only cervical vertebrae have transverse foramina.

The first cervical vertebra is called the **atlas** (see figure 7.16*a*) because it holds up the head, just as Atlas in classical mythology held up the world. The atlas vertebra has no body and no spinous process, but it has large superior articular facets, where it articulates with the occipital condyles on the base of the skull. This joint allows the head to move in a "yes" motion or to tilt from side to side. The second cervical vertebra is called the **axis** (figure 7.16*c* and *d*) because a considerable amount of rotation occurs at this vertebra to produce a "no" motion of the head. The axis has a highly modified process on the superior side of its small body called the **dens,** or **odontoid** (ō-don′toyd; tooth-shaped), **process** (both *dens* and *odontoid* mean tooth-shaped). The dens fits into the enlarged vertebral foramen of the atlas, and the latter rotates around this process. The spinous process of the seventh cervical vertebra, which is not bifid, is quite pronounced and often can be seen and felt as a lump between the shoulders (see figure 7.14). The most prominent spinous process in this area is called the **vertebral prominens.** This is usually the spinous process of the seventh cervical vertebra, but it may be that of the sixth cervical

TABLE 7.10 Comparison of Vertebral Regions

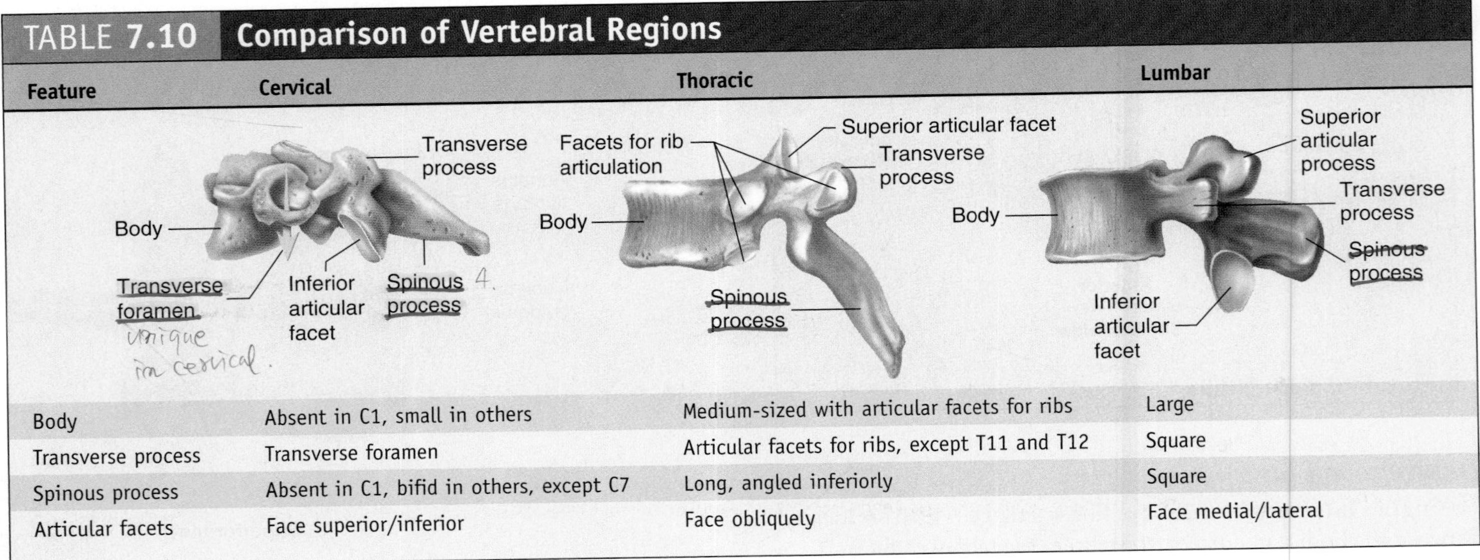

Feature	Cervical	Thoracic	Lumbar
Body	Absent in C1, small in others	Medium-sized with articular facets for ribs	Large
Transverse process	Transverse foramen	Articular facets for ribs, except T11 and T12	Square
Spinous process	Absent in C1, bifid in others, except C7	Long, angled inferiorly	Square
Articular facets	Face superior/inferior	Face obliquely	Face medial/lateral

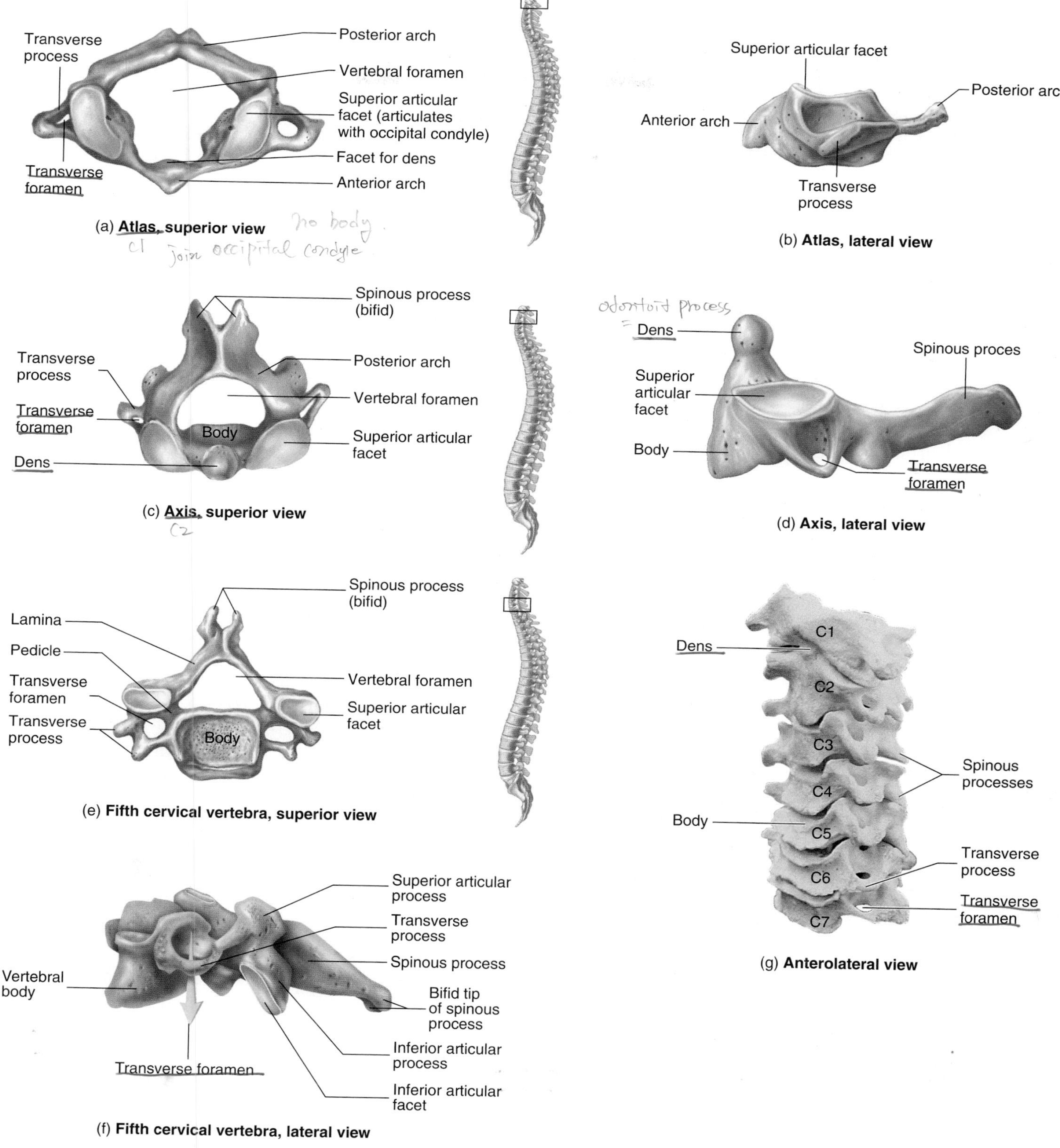

FIGURE 7.16 Cervical Vertebrae

(*a*) Atlas (first cervical vertebra), superior view. (*b*) Atlas, lateral view. (*c*) Axis (second cervical vertebra), superior view. (*d*) Axis, lateral view. (*e*) Fifth cervical vertebra, superior view. (*f*) Fifth cervical vertebra, lateral view. (*g*) Cervical vertebrae together from an anterolateral view.

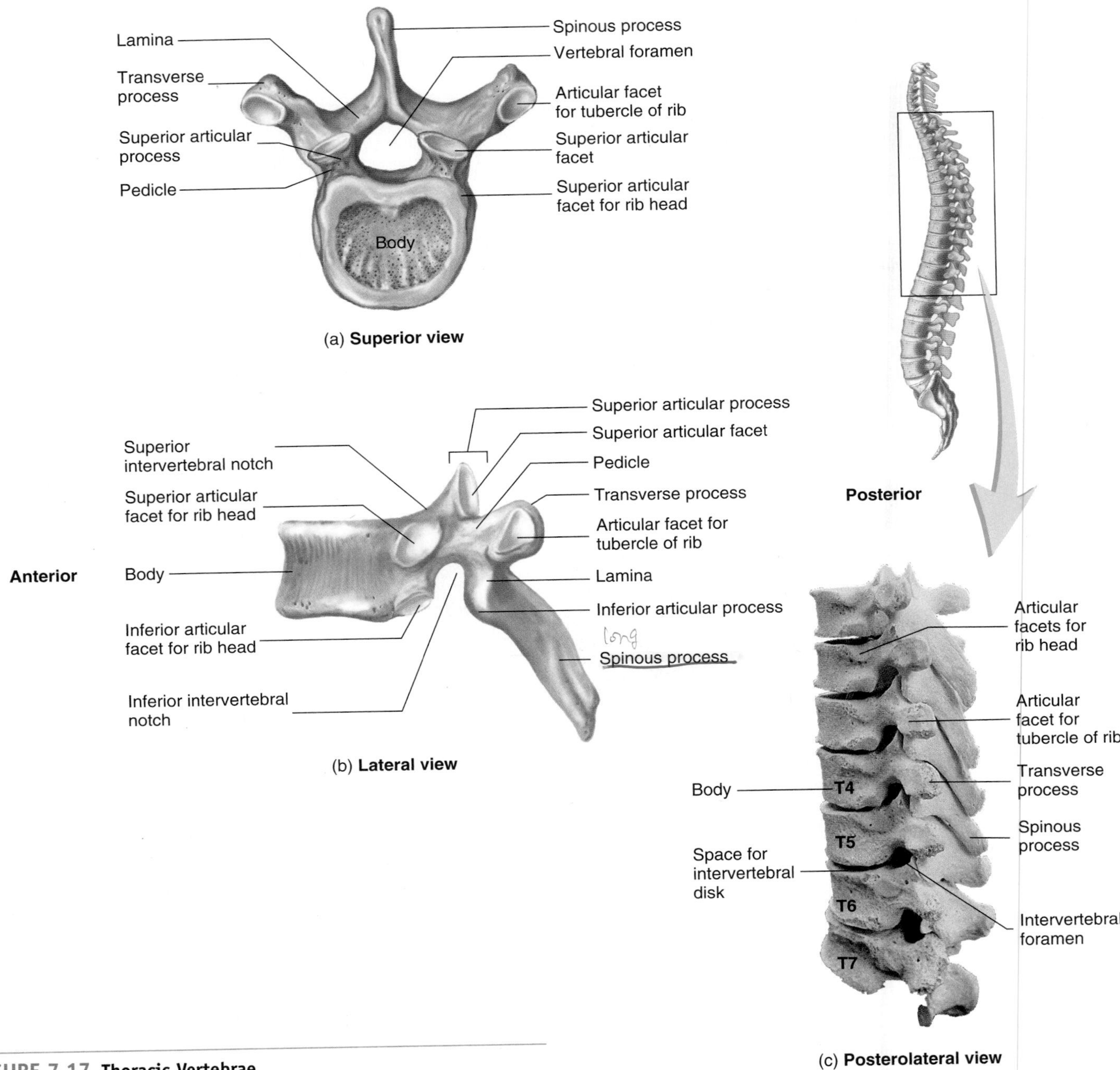

FIGURE 7.17 Thoracic Vertebrae

vertebra or even the first thoracic. The superior articular facets face superiorly and the inferior articular facets face inferiorly.

The **thoracic vertebrae** (see figure 7.13; figure 7.17) possess long, thin spinous processes, which are directed inferiorly, and they have relatively long transverse processes. The first 10 thoracic vertebrae have articular facets on their transverse processes, where they articulate with the tubercles of the ribs. Additional articular facets are on the superior and inferior margins of the body where the heads of the ribs articulate. The head of most ribs articulates with the inferior articular facet of one vertebra and the superior articular facet for the rib head on the next vertebra down.

The **lumbar vertebrae** (see figure 7.13; figure 7.18) have large, thick bodies and heavy, rectangular transverse and spinous processes. The superior articular facets face medially, and the inferior articular facets face laterally. When the superior articular surface of one lumbar vertebra joins the inferior articulating surface of another lumbar vertebra, the resulting arrangement adds strength to the inferior portion of the vertebral column and limits rotation of the lumbar vertebrae.

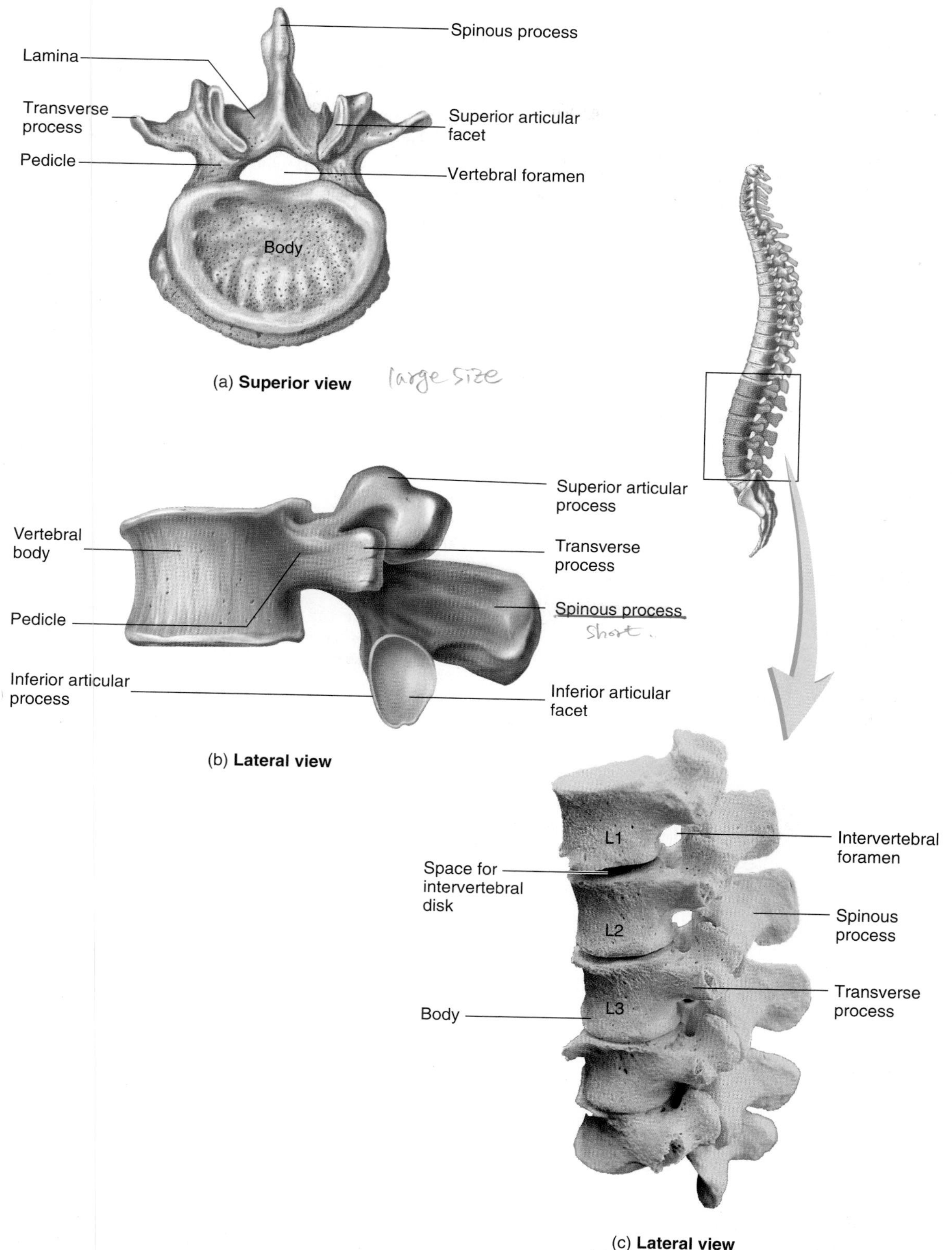

FIGURE 7.18 Lumbar Vertebrae

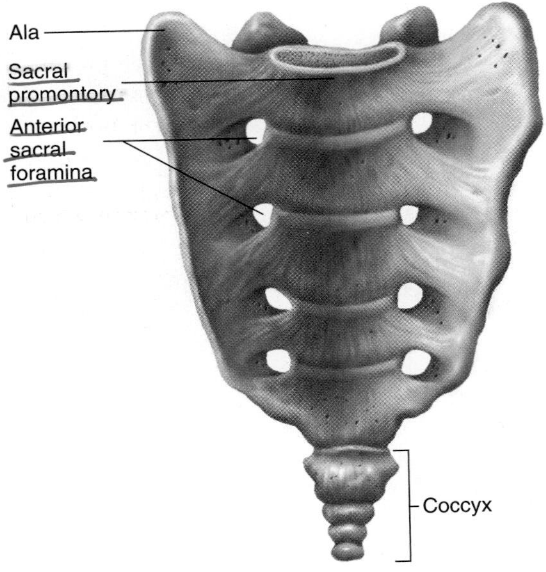

(a) **Anterior view**

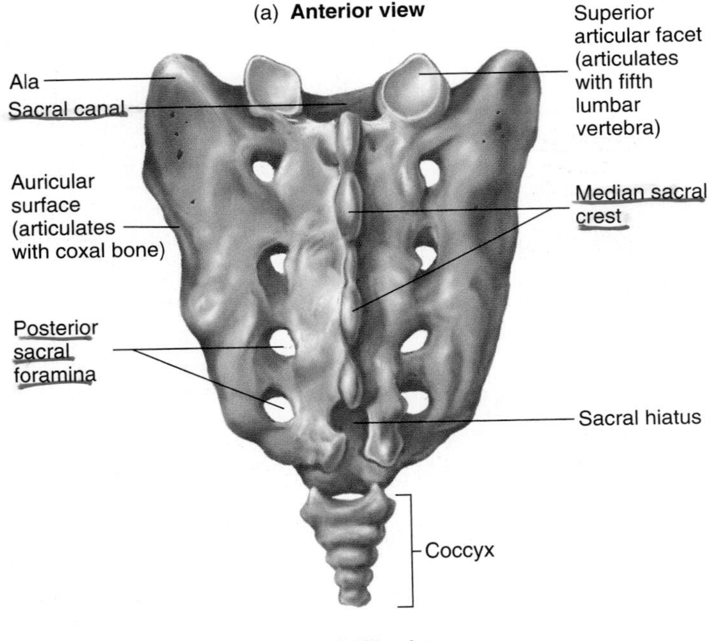

(b) **Posterior view**

FIGURE 7.19 Sacrum

PREDICT 3

Why are the lumbar vertebrae more massive than the cervical vertebrae?

The **sacral** (sā′krăl) **vertebrae** (see figure 7.13; figure 7.19) are highly modified, compared with the others. These five vertebrae are fused into a single bone called the **sacrum** (sā′krŭm).

The transverse processes of the sacral vertebrae fuse to form the lateral parts of the sacrum. The superior surfaces of the lateral parts are wing-shaped areas called the **alae** (ā′lē; wings). Much of the lateral surfaces of the sacrum are ear-shaped **auricular surfaces,** which join the sacrum to the pelvic bones. The spinous processes of the first four sacral vertebrae partially fuse to form projections, called the **median sacral crest,** along the dorsal surface of the sacrum. The spinous process of the fifth vertebra does not form, thereby leaving a **sacral hiatus** (hī-ā′tŭs) at the inferior end of the sacrum, which is often the site of anesthetic injections. The intervertebral foramina are divided into anterior and posterior foramina, called the **sacral foramina,** which are lateral to the midline. Anterior and posterior branches of the spinal nerves pass through these foramina. The anterior edge of the body of the first sacral vertebra bulges to form the **sacral promontory** (see figure 7.13), a landmark that separates the abdominal cavity from the pelvic cavity. The sacral promontory can be felt during a vaginal examination, and it is used as a reference point during measurement of the pelvic inlet.

The **coccyx** (kok′siks; shaped like a cuckoo's bill; see figures 7.13 and 7.19), or tailbone, is the most inferior portion of the vertebral column and usually consists of three to five more or less fused vertebrae that form a triangle, with the apex directed inferiorly. The coccygeal vertebrae are greatly reduced in size relative to the other vertebrae and have neither vertebral foramina nor well-developed processes.

13 ***Describe the four major curvatures of the vertebral column; explain what causes them and when they develop. Define scoliosis, kyphosis, and lordosis.***

14 ***Describe the structures forming the vertebral foramen. Where do spinal nerves exit the vertebral column?***

15 ***Describe how superior and inferior articular processes help support and allow movement of the vertebral column.***

16 ***Name and give the number of each type of vertebra. Describe the characteristics that distinguish the different types of vertebrae.***

Whiplash

Whiplash is a traumatic hyperextension of the cervical vertebrae. The head is a heavy object at the end of a flexible column, and it may become hyperextended when the head "snaps back" as a result of a sudden acceleration of the body. This commonly occurs in "rear-end" automobile accidents, or athletic injuries, in which the body is quickly forced forward while the head remains stationary. Common injuries resulting from whiplash are fracture of the spinous processes of the cervical vertebrae and herniated disks, with an anterior tear of the annulus fibrosus. These injuries can cause posterior pressure on the spinal cord or spinal nerves and strained or torn muscles, tendons, and ligaments.

Variation in Lumbar Vertebrae

The fifth lumbar vertebra or first coccygeal vertebra may become fused into the sacrum. Conversely, the first sacral vertebra may fail to fuse with the rest of the sacrum, resulting in six lumbar vertebrae.

Vertebral Column Injuries

Because the cervical vertebrae are rather delicate and have small bodies, dislocations and fractures are more common in this area than in other regions of the column. Because the lumbar vertebrae have massive bodies and carry a large amount of weight, fractures are less common, but ruptured intervertebral disks are more common in this area than in other regions of the column. The coccyx is easily broken in a fall in which a person sits down hard on a solid surface.

Thoracic Cage

The **thoracic cage,** or **rib cage,** protects the vital organs within the thorax and forms a semi-rigid chamber that can increase and decrease in volume during respiration. It consists of the thoracic vertebrae, the ribs with their associated costal (rib) cartilages, and the sternum (figure 7.20*a*).

Ribs and Costal Cartilages

The 12 pairs of **ribs** are classified as either true or false ribs. The superior 7 pairs are called **true ribs,** or **vertebrosternal** (ver′tĕ′brō-ster′năl) **ribs;** they articulate with the thoracic vertebrae and attach directly through their **costal cartilages** to the sternum. The inferior 5 pairs, or **false ribs,** articulate with the thoracic vertebrae but do not attach directly to the sternum. The false ribs consist of two groups. The eighth, ninth, and tenth ribs, the **vertebrochondral** (ver′tĕ-brō-kon′drăl) **ribs,** are joined by a common cartilage to the costal cartilage of the seventh rib, which, in turn, is attached to the sternum. Two of the false ribs, the eleventh and twelfth ribs, are also called **floating,** or **vertebral, ribs** because they do not attach to the sternum. The costal cartilages are flexible and permit the thoracic cage to expand during respiration.

Most ribs have two points of articulation with the thoracic vertebrae (figure 7.20*b* and *c*). First, the **head** articulates with the bodies of two adjacent vertebrae and the intervertebral disk between them. The head of each rib articulates with the inferior articular facet of the superior vertebra and the superior articular facet of the inferior vertebra. Second, the **tubercle** articulates with the transverse process of the inferior vertebra. The **neck** is between the head and tubercle, and the **body,** or shaft, is the main part of the rib. The **angle** of the rib is located just lateral to the tubercle and is the point of greatest curvature.

Rib Defects

A **separated rib** is a dislocation between a rib and its costal cartilage. As a result of the dislocation, the rib can move, override adjacent ribs, and cause pain. Separation of the tenth rib is the most common.

The angle is the weakest part of the rib and may be fractured in a crushing accident, such as an automobile accident.

The transverse processes of the seventh cervical vertebra may form separate bones called **cervical ribs.** These ribs may be just tiny pieces of bone or may be long enough to reach the sternum. The first lumbar vertebra may develop lumbar ribs.

Sternum

The **sternum,** or breastbone, has been described as being sword-shaped and has three parts (see figure 7.20*a*). The **manubrium** (mă-noo′brē-ŭm; handle) is the sword handle, the **body,** or gladiolus (sword), is the blade, and the **xiphoid** (zi′foyd; sword) **process** is the sword tip. The superior margin of the manubrium has a **jugular** (neck), or **suprasternal, notch** in the midline, which can be easily felt at the anterior base of the neck (figure 7.21). The first rib and the clavicle articulate with the manubrium. The point at which the manubrium joins the body of the sternum can be felt as a prominence on the anterior thorax called the **sternal angle** (see figure 7.20*a*). The cartilage of the second rib attaches to the sternum at the sternal angle, the third through seventh ribs attach to the body of the sternum, and no ribs attach to the xiphoid process.

Sternal Angle and Thoracic Landmarks

The sternal angle is important clinically because the second rib is found lateral to it and can be used as a starting point for counting the other ribs. Counting ribs is important because they are landmarks used to locate structures in the thorax, such as areas of the heart. The sternum often is used as a site for taking red bone marrow samples because it is readily accessible. Because the xiphoid process of the sternum is attached only at its superior end, it may be broken during cardiopulmonary resuscitation (CPR) and then may lacerate the liver.

17 ***What is the function of the thoracic (rib) cage? Distinguish among true, false, and floating ribs, and give the number of each type.***

18 ***Describe the articulation of the ribs with thoracic vertebrae.***

19 ***Describe the parts of the sternum. Name the structures that attach to, or articulate with, the sternum.***

APPENDICULAR SKELETON

The appendicular skeleton (see figure 7.1) consists of the bones of the **upper** and **lower limbs** and the **girdles** by which they are attached to the body. The term *girdle* means a belt or a zone and refers to the two zones, pectoral and pelvic, where the limbs are attached to the body.

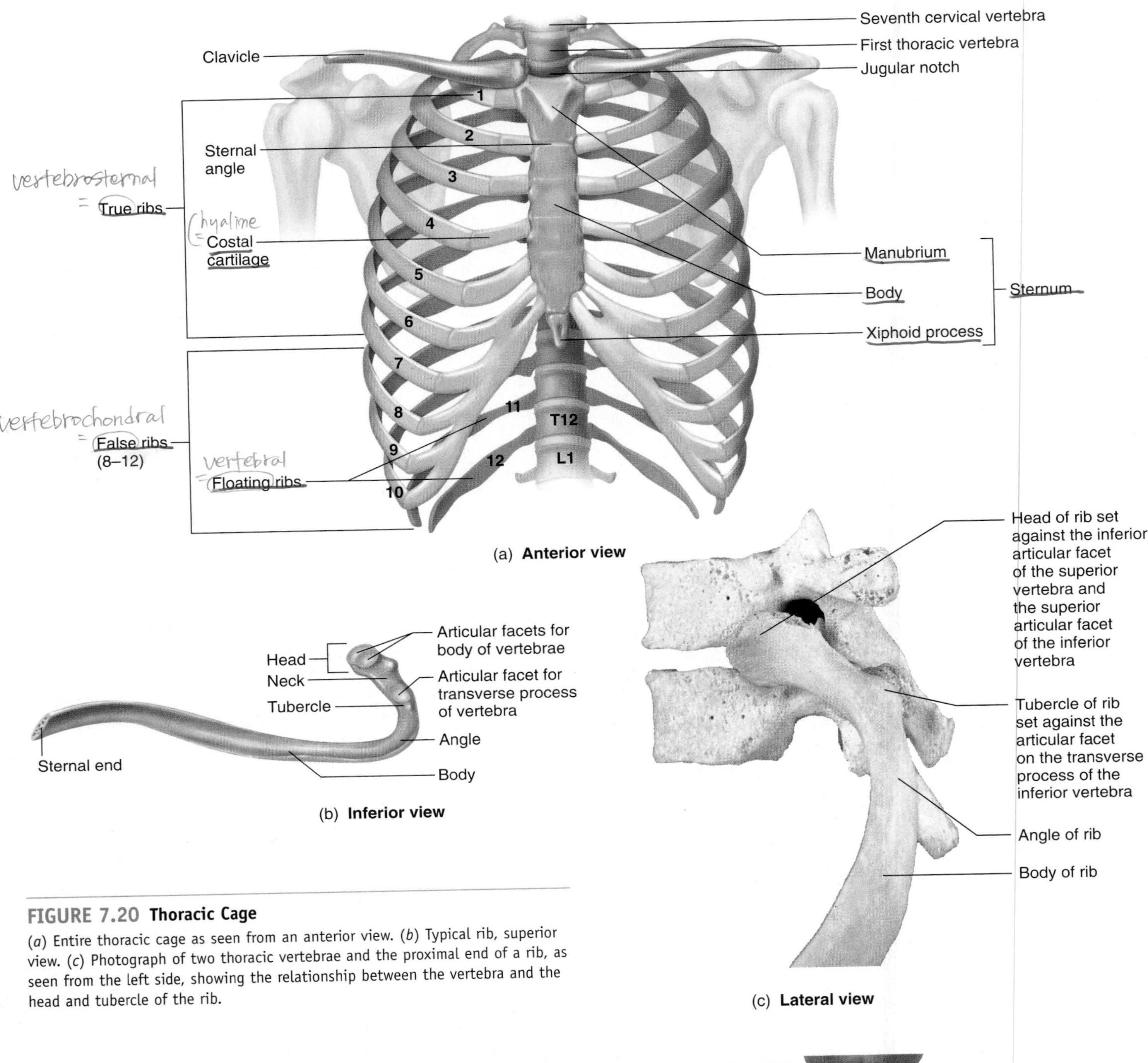

FIGURE 7.20 Thoracic Cage

(*a*) Entire thoracic cage as seen from an anterior view. (*b*) Typical rib, superior view. (*c*) Photograph of two thoracic vertebrae and the proximal end of a rib, as seen from the left side, showing the relationship between the vertebra and the head and tubercle of the rib.

FIGURE 7.21 Surface Anatomy Showing Bones of the Upper Thorax

Pectoral Girdle and Upper Limb

The human upper limb (figure 7.22) is extremely mobile. It is capable of a wide range of movements, including lifting, grasping, pulling, and touching. Many structural characteristics of the upper limb reflect these functions. The upper limb and its girdle are attached rather loosely by muscles to the rest of the body, an arrangement that allows considerable freedom of movement of this extremity. This freedom of movement allows placement of the hand in a wide range of positions to accomplish its functions.

Pectoral Girdle

The **pectoral** (pek'tō-răl), or **shoulder, girdle** consists of two pairs of bones that attach the upper limb to the body: Each pair is composed of a **scapula** (skap'ū-lă), or shoulder blade (figure 7.23), and a **clavicle** (klav'i-kl), or collarbone (see figures 7.20, 7.22, and 7.23). The scapula is a flat, triangular bone that can easily be seen and felt in a living person (see figure 7.14). The base of the triangle, the superior border, faces superiorly; the apex, the inferior angle, is directed inferiorly.

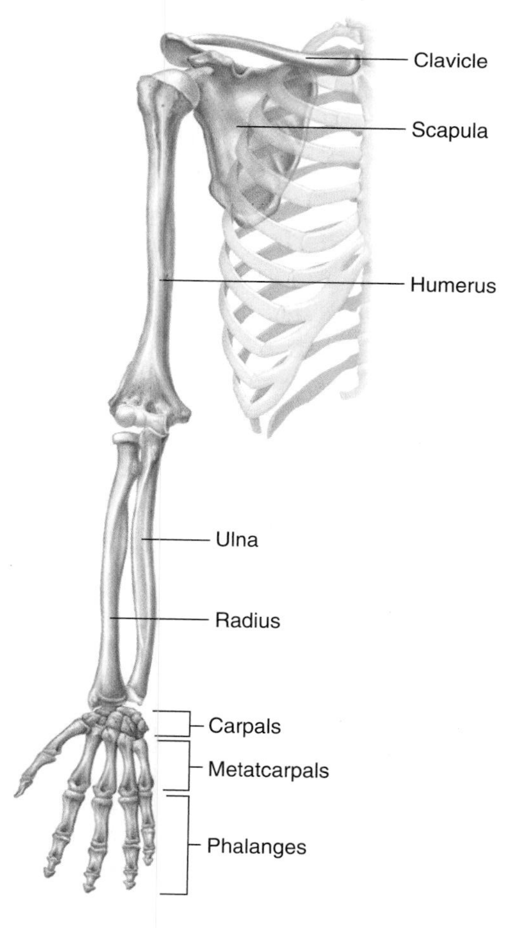

FIGURE 7.22 Bones of the Pectoral Girdle and Right Upper Limb

The large **acromion** (ă-krō'mē-on; shoulder tip) **process** of the scapula, which can be felt at the tip of the shoulder, has three functions: (1) to form a protective cover for the shoulder joint, (2) to form the attachment site for the clavicle, and (3) to provide attachment points for some of the shoulder muscles. The **scapular spine** extends from the acromion process across the posterior surface of the scapula and divides that surface into a small **supraspinous fossa** superior to the spine and a larger **infraspinous fossa** inferior to the spine. The deep, anterior surface of the scapula constitutes the **subscapular fossa.** The smaller **coracoid** (meaning shaped like a crow's beak) **process** provides attachments for some shoulder and arm muscles. A **glenoid** (glē'noyd, glen'oyd) **cavity,** located in the superior lateral portion of the bone, articulates with the head of the humerus.

The clavicle (see figures 7.20, 7.22, and 7.23*c*) is a long bone with a slight sigmoid (S-shaped) curve and is easily seen and felt in the living human (see figure 7.21). The lateral end of the clavicle articulates with the acromion process, and its medial end articulates with the manubrium of the sternum. These articulations form the only bony connections between the pectoral girdle and the axial skeleton. Because the clavicle holds the upper limb away from the body, it facilitates the limb's mobility.

20 ***Name the bones that make up the pectoral girdle. Describe their functions.***

21 ***What are the functions of the acromion process and the coracoid process of the scapula?***

PREDICT 4

A broken clavicle changes the position of the upper limb in what way?

Arm

The arm, the part of the upper limb from the shoulder to the elbow, contains only one bone, the **humerus** (figure 7.24). The humeral **head** articulates with the glenoid cavity of the scapula. The **anatomical neck,** immediately distal to the head, is almost nonexistent; thus, a surgical neck has been designated. The **surgical neck** is so-named because it is a common fracture site that often requires surgical repair. If it becomes necessary to remove the humeral head because of disease or injury, it is removed down to the surgical neck. The **greater tubercle** is located on the lateral surface, and the **lesser tubercle** is located on the anterior surface of the proximal end of the humerus, where they are sites of muscle attachment. The groove between the two tubercles contains one tendon of the biceps brachii muscle and is called the **intertubercular,** or **bicipital** (bī-sip'i-tăl), **groove.** The **deltoid tuberosity** is located on the lateral surface of the humerus a little more than a third of the way along its length and is the attachment site for the deltoid muscle.

The articular surfaces of the distal end of the humerus exhibit unusual features where the humerus articulates with the two forearm bones. The lateral portion of the articular surface is very rounded, articulates with the radius, and is called the **capitulum** (kă-pit'ū-lŭm; head-shaped). The medial portion somewhat resembles a spool or pulley, articulates with the ulna, and is called

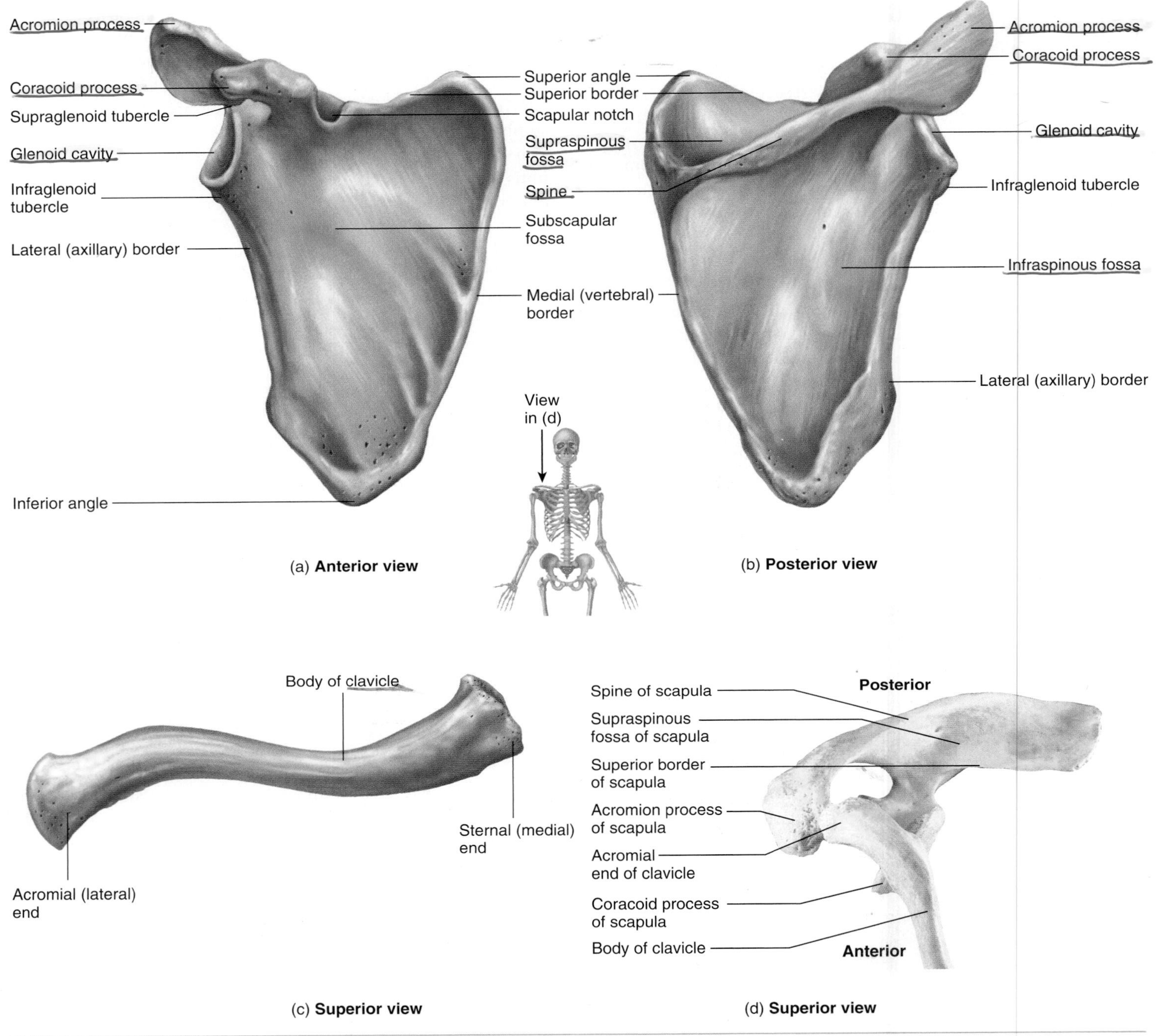

FIGURE 7.23 Right Scapula and Clavicle

(*a*) Right scapula, anterior view. (*b*) Right scapula, posterior view. (*c*) Right clavicle, superior view. (*d*) Photograph of the right scapula and clavicle from a superior view, showing the relationship between the distal end of the clavicle and the acromion process of the scapula.

the **trochlea** (trok′lē-ă; spool). Proximal to the capitulum and the trochlea are the **medial** and **lateral epicondyles,** which are points of muscle attachment for the muscles of the forearm.

Forearm

The forearm has two bones. The **ulna** is on the medial side of the forearm, the side with the little finger. The **radius** is on the lateral, or thumb side, of the forearm (figure 7.25).

The proximal end of the ulna has a C-shaped articular surface, called the **trochlear,** or **semilunar, notch** that fits over the trochlea of the humerus. The trochlear notch is bounded by two processes. The larger, posterior process is the **olecranon** (ō-lek′ră-non; the point of the elbow) **process.** It can easily be felt and is commonly referred to as "the elbow" (see figure 7.27). Posterior arm muscles attach to the olecranon process. The smaller, anterior process is the **coronoid** (kōr′ŏ-noyd; crow's beak) **process.**

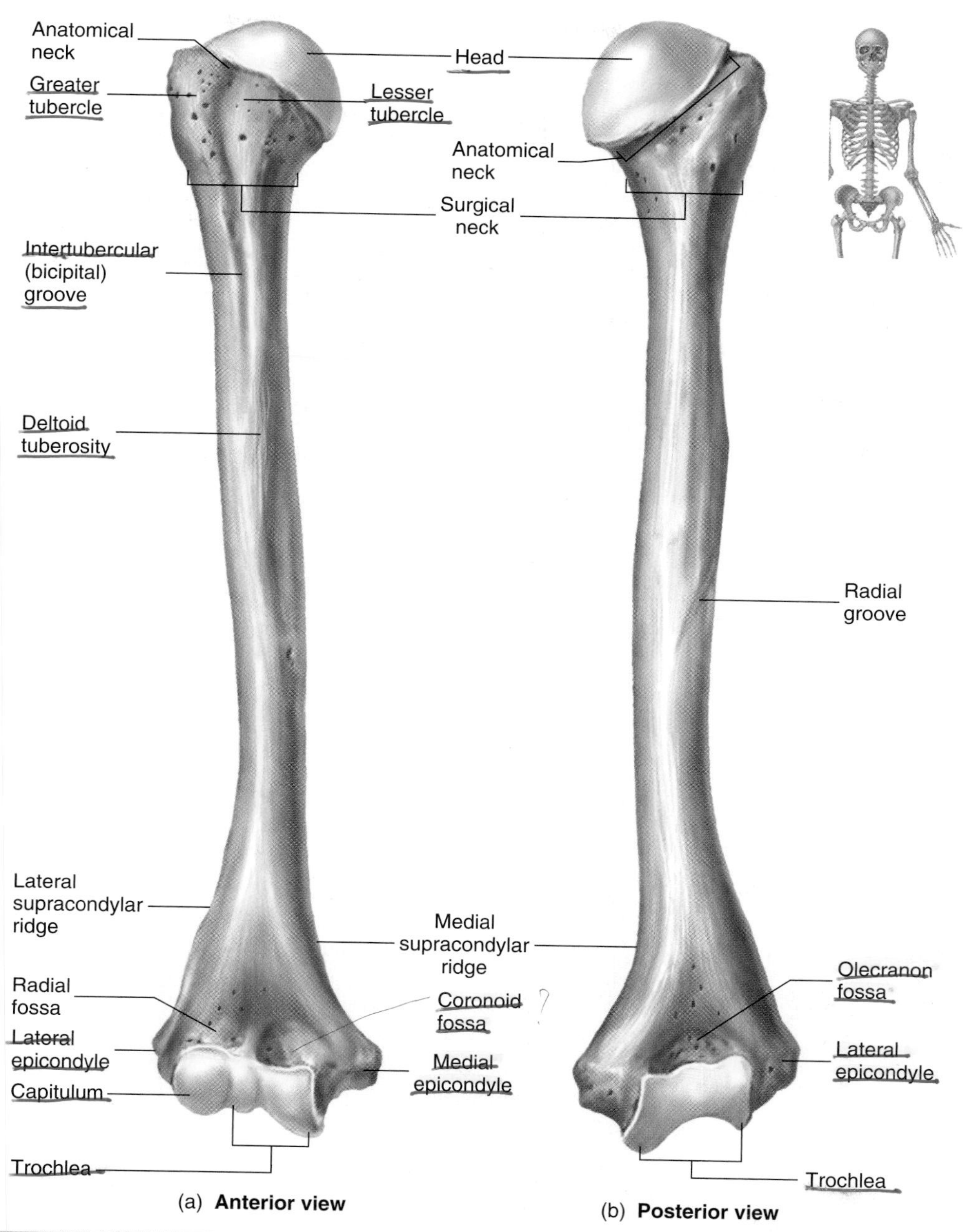

FIGURE 7.24 Right Humerus

PREDICT 5

Explain the function of the olecranon and coronoid fossae on the distal humerus (see figure 7.25).

The distal end of the ulna has a small **head,** which articulates with both the radius and the wrist bones (see figures 7.25 and 7.27). The head can be seen on the posterior, medial (ulnar) side of the distal forearm. The posteromedial side of the head has a small **styloid** (stī′loyd; shaped like a stylus or writing instrument) **process,** to which ligaments of the wrist are attached.

The proximal end of the radius is the **head.** It is concave and articulates with the capitulum of the humerus. The lateral surfaces of the head constitute a smooth cylinder, where the radius rotates against the **radial notch** of the ulna. As the forearm rotates (supination and pronation; see chapter 8), the proximal end of the ulna stays in place, and the radius rotates. The **radial tuberosity** is the point at which a major anterior arm muscle, the biceps brachii, attaches.

The distal end of the radius, which articulates with the ulna and the carpals, is somewhat broadened, and a **styloid process** to which wrist ligaments are attached is located on the lateral side of the distal radius.

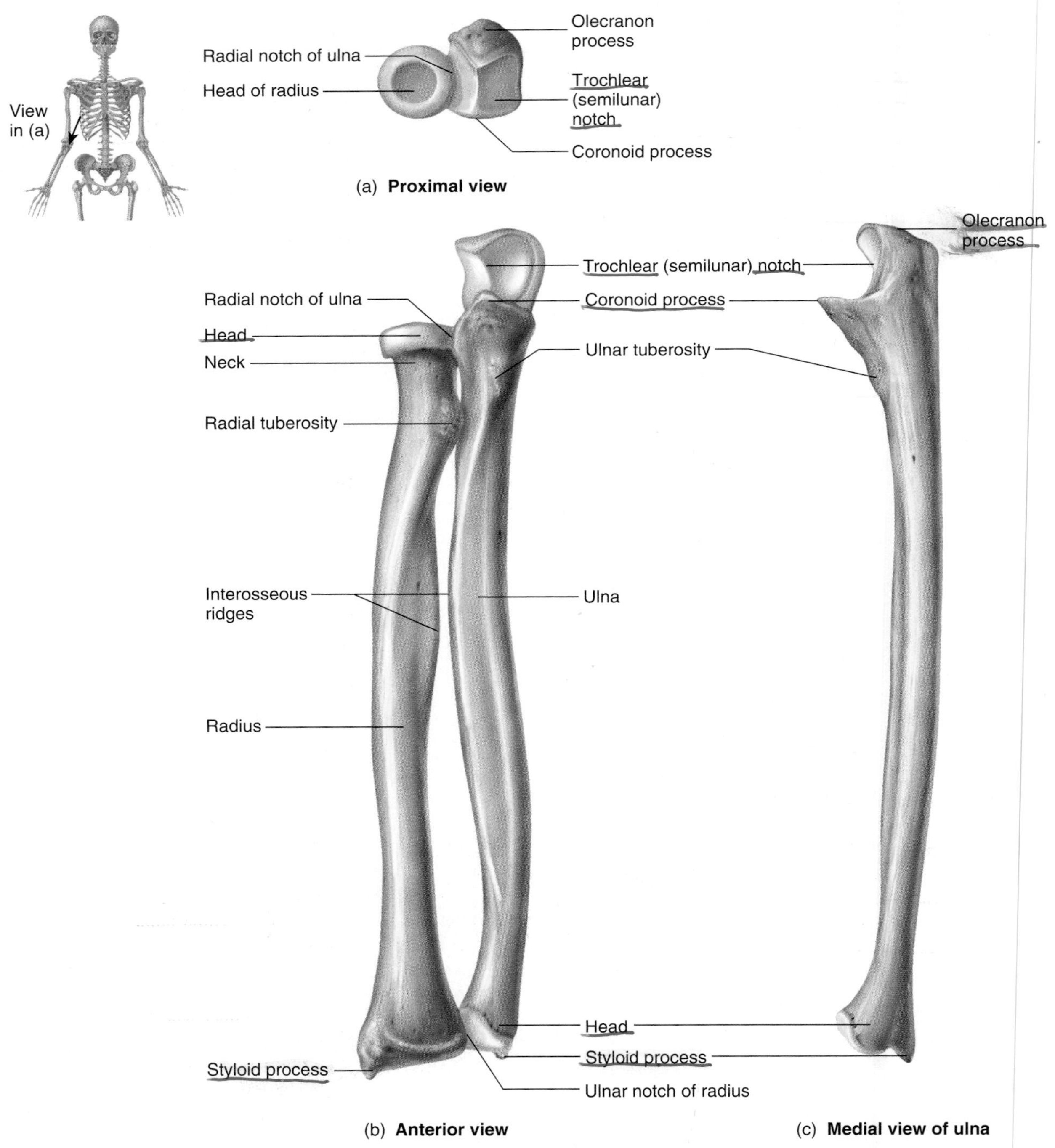

FIGURE 7.25 **Right Ulna and Radius**

> **Radius Fractures**
>
> The radius is the most commonly fractured bone in people over 50 years old. It is often fractured as the result of a fall on an outstretched hand. The most common site of fracture is 2.5 cm proximal to the wrist, and the fracture is often comminuted, or impacted. Such a fracture, which results in posterior displacement of the hand (extension), is called a Colles fracture.

Wrist

The wrist is a relatively short region between the forearm and hand; it is composed of eight **carpal** (kar′păl) **bones** arranged into two rows of four each (figure 7.26). The proximal row of carpals, lateral to medial, includes the **scaphoid** (skaf′oyd), which is boat-shaped; the **lunate** (loo′nāt), which is moon-shaped; the three-cornered **triquetrum** (trī-kwē′trŭm, trī-kwet′rŭm); and

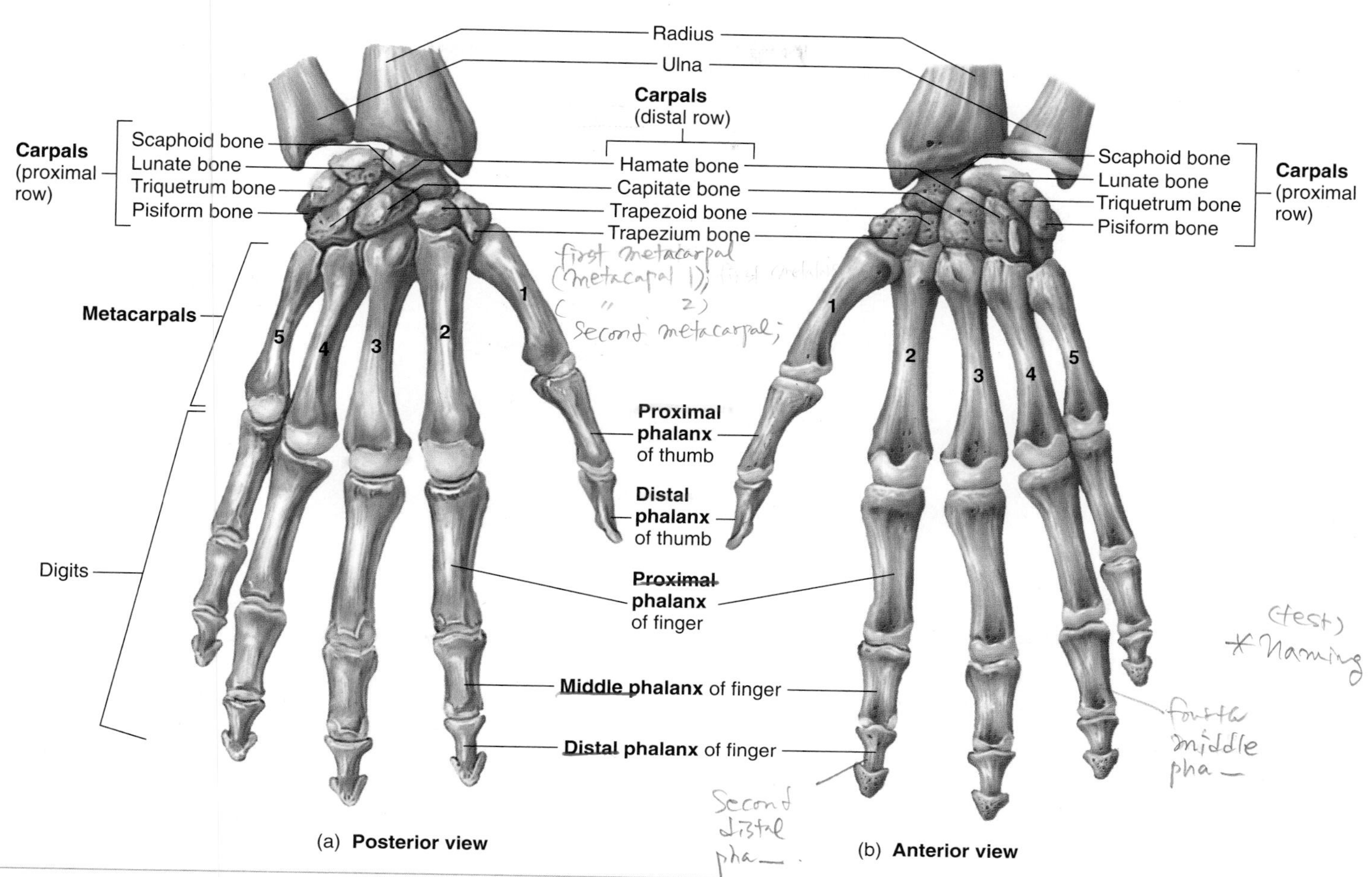

FIGURE 7.26 Bones of the Right Wrist and Hand

the pea-shaped **pisiform** (pis′i-fōrm), which is located on the palmar surface of the triquetrum. The distal row of carpals, from medial to lateral, includes the **hamate** (ha′māt), which has a hooked process on its palmar side, called the hook of the hamate; the head-shaped **capitate** (kap′i-tāt); the **trapezoid** (trap′ē-zoyd), which is shaped like a trapezoid (a four-sided geometric form with two parallel sides); and the **trapezium** (tra-pē′zē-ŭm), which is shaped like a trapezium (a four-sided geometric form with no two sides parallel). A number of mnemonics have been developed to help students remember the carpal bones. This one allows students to remember them in order from lateral to medial for the proximal row (top) and from medial to lateral (by the thumb) for the distal row: So Long Top Part, Here Comes The Thumb, that is Scaphoid, Lunate, Triquetrum, Pisiform, Hamate, Capitate, Trapezoid, and Trapezium.

The eight carpals, taken together, are convex posteriorly and concave anteriorly. The anterior concavity of the carpals is accentuated by the tubercle of the trapezium at the base of the thumb and the hook of the hamate at the base of the little finger. A ligament stretches across the wrist from the tubercle of the trapezium to the hook of the hamate to form a tunnel on the anterior surface of the wrist called the **carpal tunnel.** Tendons, nerves, and blood vessels pass through this tunnel to enter the hand.

Carpal Tunnel Syndrome

The bones and ligaments that form the walls of the carpal tunnel do not stretch. Edema (fluid buildup) or connective tissue deposition may occur within the carpal tunnel as the result of trauma or some other problem. The edema or connective tissue may apply pressure against the nerve and vessels passing through the tunnel, causing carpal tunnel syndrome, which consists of tingling, burning, and numbness in the hand. Carpal tunnel syndrome occurs more frequently in people whose work involves extending the wrist and flexing the fingers, as in typing. The number of cases has increased in recent decades among people who perform repetitive tasks, such as computer keyboarding.

Hand

Five **metacarpals** are attached to the carpal bones and constitute the bony framework of the hand (see figure 7.26). They are numbered one to five, starting with the most lateral metacarpal, at the base of the thumb. The metacarpals form a curve so that, in the resting position, the palm of the hand is concave. The distal ends of the metacarpals help form the knuckles of the hand (figure 7.27). The spaces between the metacarpals are occupied by soft tissue.

The five **digits** of each hand include one thumb and four fingers. Each digit consists of small long bones called **phalanges**

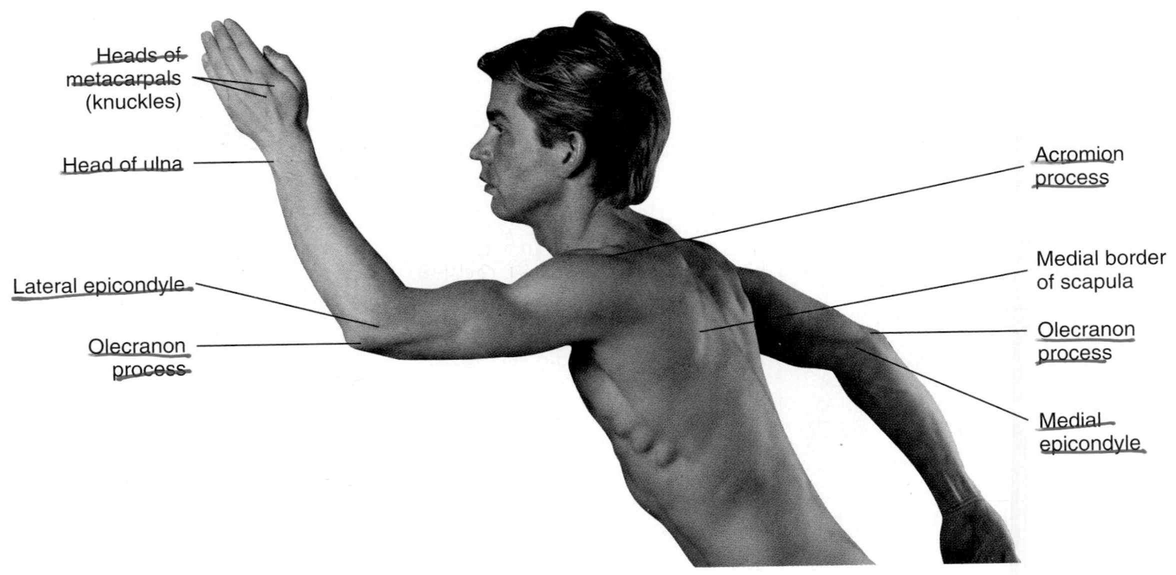

FIGURE 7.27 **Surface Anatomy Showing Bones of the Pectoral Girdle and Upper Limb**

(fă-lan′jēz; sing. *phalanx,* a line or wedge of soldiers holding their spears, tips outward, in front of them). The thumb has two phalanges, called proximal and distal. Each finger has three phalanges designated proximal, middle, and distal. One or two **sesamoid** (ses′ă-moyd; resembling a sesame seed) **bones** (not shown in figure 7.26) often form near the junction between the proximal phalanx and the metacarpal of the thumb. Sesamoid bones are small bones located within some tendons increasing their mechanical advantage where they cross joints.

PREDICT 6

Explain why the dried, articulated skeleton appears to have much longer "fingers" than are seen in the hand with the soft tissue intact.

22 ***Name the important sites of muscle attachment on the humerus.***

23 ***Give the points of articulation between the scapula, humerus, radius, ulna, and wrist bones.***

24 ***What is the function of the radial tuberosity? Of the styloid processes? Name the part of the ulna commonly referred to as "the elbow."***

25 ***List the eight carpal bones. What is the carpal tunnel?***

26 ***What bones form the hand? The knuckles? How many phalanges are in each finger and in the thumb?***

Pelvic Girdle and Lower Limb

The lower limbs support the body and are essential for normal standing, walking, and running. The general pattern of the lower limb (figure 7.28) is very similar to that of the upper limb, except that the pelvic girdle is attached much more firmly to the body than is the pectoral girdle, and the bones in general are thicker, heavier, and longer than those of the upper limb. These structures

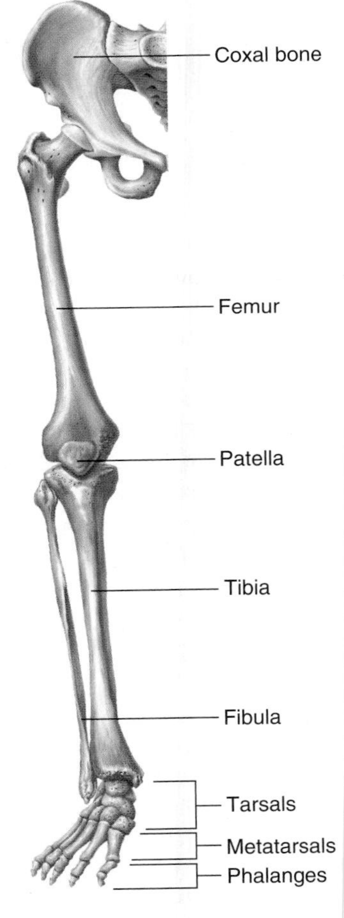

FIGURE 7.28 **Bones of the Right Half of the Pelvic Girdle and Right Lower Limb**

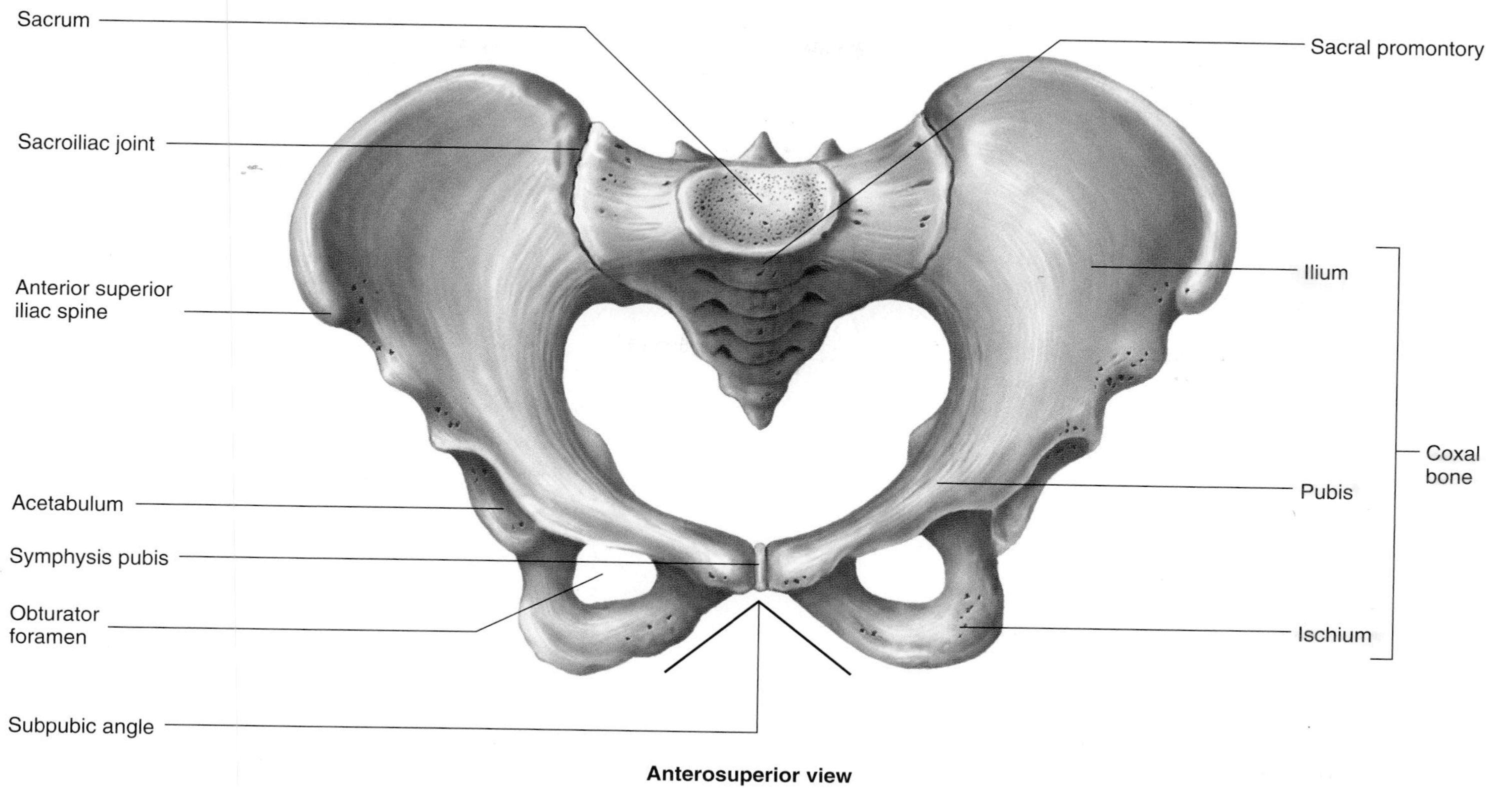

FIGURE 7.29 Pelvis

reflect the function of the lower limb in the support and movement of the body.

Pelvic Girdle

The right and left **coxal** (kok′sul) **bones,** os coxae, or hipbones, join each other anteriorly and the **sacrum** posteriorly to form a ring of bone called the **pelvic girdle.** The **pelvis** (pel′vis; basin; figure 7.29) includes the pelvic girdle and the coccyx. Each coxal bone consists of a large, concave bony plate superiorly, a slightly narrower region in the center, and an expanded bony ring inferiorly, which surrounds a large **obturator** (ob′too-rā-tŏr; to occlude or close up, indicating that the foramen is occluded by soft tissue) **foramen.** A fossa called the **acetabulum** (as-ĕ-tab′ū-lŭm; a shallow vinegar cup—a common household item in ancient times) is located on the lateral surface of each coxal bone and is the point of articulation of the lower limb with the girdle. The articular surface of the acetabulum is crescent-shaped and occupies only the superior and lateral aspects of the fossa. The pelvic girdle is the place of attachment for the lower limbs, supports the weight of the body, and protects internal organs. Because the pelvic girdle is a complete bony ring, it provides more stable support but less mobility than the incomplete ring of the pectoral girdle. In addition, the pelvis in a woman protects the developing fetus and forms a passageway through which the fetus passes during delivery.

The Sacroiliac Joint

Each sacroiliac joint is formed by the junction of the auricular surface of a coxal bone and one articular surface of the sacrum. The sacroiliac joints receive most of the weight of the upper body and are strongly supported by ligaments. Excessive strain on the joints, however, can cause slight joint movement and can stretch connective tissue and associated nerve endings in the area and cause pain. Thus is derived the expression "My aching sacroiliac!" This problem sometimes develops in pregnant women because of the forward weight distribution of the fetus.

Each coxal bone is formed by the fusion of three bones during development: the **ilium** (il′ē-ŭm; groin), the **ischium** (is′kē-ŭm; hip), and the **pubis** (pū′bis; genital hair). All three bones join near the center of the acetabulum (figure 7.30*a*). The superior portion of the ilium is called the **iliac crest** (figure 7.30*b* and *c*). The crest ends anteriorly as the **anterior superior iliac spine** and posteriorly as the **posterior superior iliac spine.** The crest and anterior spine can be felt and even seen in thin individuals (figure 7.31). The anterior superior iliac spine is an important anatomical landmark used, for example, to find the correct location for giving gluteal injections into the hip. A dimple overlies the posterior superior iliac spine just superior to the buttocks. The **greater sciatic** (sī-at′ik) **notch** is on the posterior side of the ilium, just inferior to the posterior inferior iliac spine. The sciatic

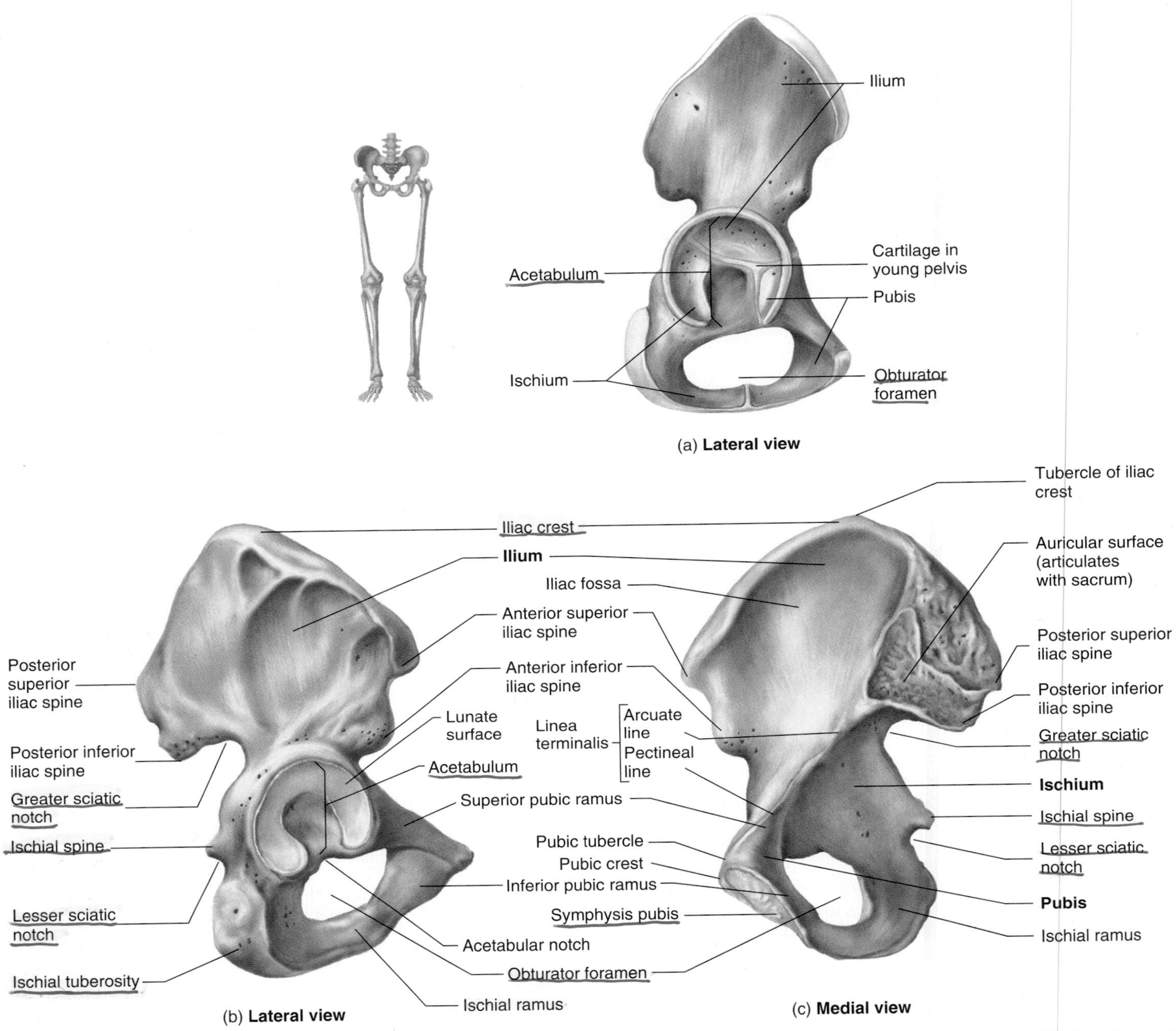

FIGURE 7.30 Coxal Bone

The three bones forming the coxal bone are bolded. (*a*) Right coxal bone of a growing child. Each coxal bone is formed by fusion of the ilium, ischium, and pubis. The three bones can be seen joining near the center of the acetabulum, separated by lines of cartilage. (*b*) Right coxal bone, lateral view. (*c*) Right coxal bone, medial view.

nerve passes through the greater sciatic notch. The **auricular surface** of the ilium joins the sacrum to form the **sacroiliac joint** (see figure 7.29). The medial side of the ilium consists of a large depression called the **iliac fossa.**

The ischium possesses a heavy **ischial** (is′kē-ăl) **tuberosity,** where posterior thigh muscles attach and on which a person sits (see figure 7.30*b*). The pubis possesses a **pubic crest,** medially and a **pubic tubercle** laterally where abdominal muscles attach (see figure 7.30*c*). The pubic crest can be felt anteriorly. Just inferior to the pubic crest is the point of junction, the **symphysis** (sim′fi-sis; a coming together) **pubis,** or **pubic symphysis,** between the two coxal bones (see figure 7.29).

The pelvis can be thought of as having two parts divided by an imaginary plane passing from the sacral promontory along the **linea terminalis** of the ilium to the pubic crest (see figure 7.32). The bony boundary of this plane is the **pelvic brim.** The **false,** or **greater, pelvis** is superior to the pelvic brim and is partially surrounded by bone on the posterior and lateral sides. During life,

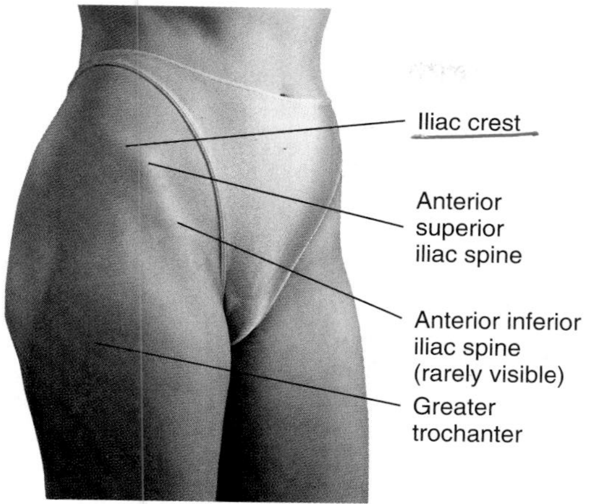

FIGURE 7.31 Surface Anatomy Showing an Anterior View of the Hipbones

the abdominal muscles form the anterior wall of the false pelvis. The **true pelvis** is inferior to the pelvic brim and is completely surrounded by bone. The superior opening of the true pelvis, at the level of the pelvic brim, is the **pelvic inlet.** The inferior opening of the true pelvis, bordered by the inferior margin of the pubis, the ischial spines and tuberosities, and the coccyx, is the **pelvic outlet** (see figure 7.32*c*).

Comparison of the Male and Female Pelvis

The male pelvis usually is more massive than the female pelvis as a result of the greater weight and size of the male, and the female pelvis is broader and has a larger, more rounded pelvic inlet and outlet (figure 7.32*a* and *b*), consistent with the need to allow the fetus to pass through these openings in the female pelvis during delivery. Table 7.11 lists additional differences between the male and female pelvis.

27 ***Define the pelvic girdle. What bones fuse to form each coxal bone? Where and with what bones does each coxal bone articulate?***

28 ***Name the important sites of muscle attachment on the pelvis.***

29 ***Distinguish between the true pelvis and the false pelvis.***

30 ***Describe the differences between a male and a female pelvis.***

Gluteal Injections

The large gluteal muscles (hip muscles; see chapter 10) are a common site for intramuscular injections. Gluteal injections are made in the superolateral region of the hip (figure D) because a large nerve (the sciatic nerve; see chapter 12) lies deep to the other gluteal regions. The landmarks for such an injection are the anterior superior iliac spine and the tubercle of the iliac crest, which lies about one-third of the way along the iliac crest from anterior to posterior.

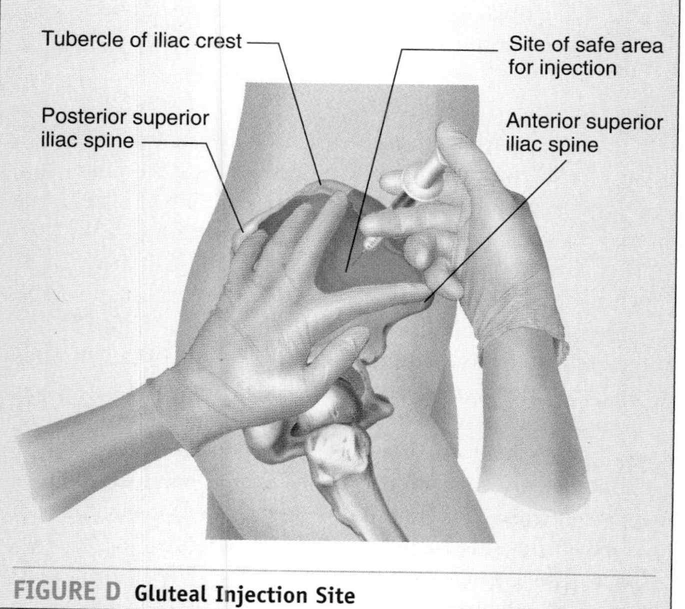

FIGURE D Gluteal Injection Site

Pelvic Outlet and Birth

A wide, circular pelvic inlet and a pelvic outlet with widely spaced ischial spines can facilitate delivery. A smaller pelvic outlet can cause problems during delivery; thus, the size of the pelvic inlet and outlet is routinely measured during prenatal pelvic examinations of pregnant women. If the pelvic outlet is too small for normal delivery, delivery can be accomplished by **cesarean section,** which is the surgical removal of the fetus through the abdominal wall.

Thigh

The thigh, like the arm, contains a single bone, which is called the **femur.** The femur has a prominent, rounded **head,** where it articulates with the acetabulum, and a well-defined **neck;** both are located at an oblique angle to the shaft of the femur (figure 7.33). The proximal shaft exhibits two projections: a **greater trochanter** (trō-kan′ter; runner) lateral to the neck and a smaller, or **lesser, trochanter** inferior and posterior to the neck. Both trochanters are attachment sites for muscles that fasten the hip to the thigh. The greater trochanter and its attached muscles form a bulge that can be seen as the widest part of the hips (see figure 7.31). The distal end of the femur has **medial** and **lateral condyles,** smooth, rounded surfaces that articulate with the tibia. Located proximally to the condyles are the **medial** and **lateral epicondyles,** important sites of ligament attachment. An **adductor tubercle,** to which muscles attach, is located just proximal to the medial epicondyle.

The **patella,** or kneecap, is a large sesamoid bone located within the tendon of the quadriceps femoris muscle group, which is the major muscle group of the anterior thigh (figure 7.34). The patella articulates with the patellar groove of the femur to create a smooth articular surface over the anterior distal end of the femur. The patella holds the tendon away from the distal end of the femur and therefore changes the angle of the tendon between the quadriceps femoris muscle and the tibia, where the tendon attaches. This

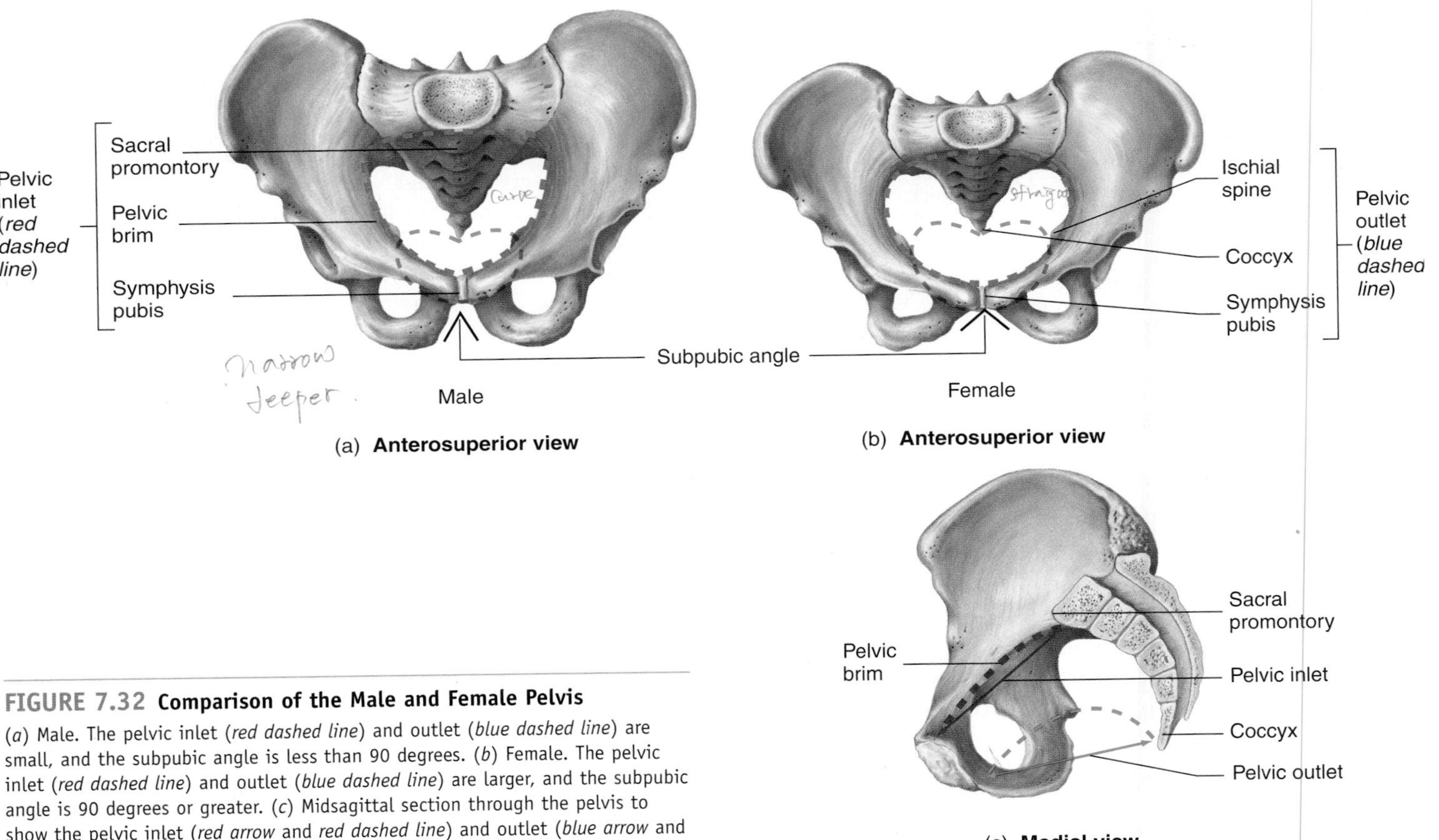

FIGURE 7.32 Comparison of the Male and Female Pelvis

(*a*) Male. The pelvic inlet (*red dashed line*) and outlet (*blue dashed line*) are small, and the subpubic angle is less than 90 degrees. (*b*) Female. The pelvic inlet (*red dashed line*) and outlet (*blue dashed line*) are larger, and the subpubic angle is 90 degrees or greater. (*c*) Midsagittal section through the pelvis to show the pelvic inlet (*red arrow* and *red dashed line*) and outlet (*blue arrow* and *blue dashed line*).

TABLE 7.11 Differences Between the Male and Female Pelvis (See Figure 7.32)

Area	Description
General	In females, somewhat lighter in weight and wider laterally but shorter superiorly to inferiorly and less funnel-shaped; less obvious muscle attachment points in females than in males
Sacrum	Broader in females with the inferior part directed more posteriorly; the sacral promontory does not project as far anteriorly in females
Pelvic inlet	Heart-shaped in males; oval in females
Pelvic outlet	Broader and more shallow in females
Subpubic angle	Less than 90 degrees in males; 90 degrees or more in females
Ilium	More shallow and flared laterally in females
Ischial spines	Farther apart in females
Ischial tuberosities	Turned laterally in females and medially in males

change in angle increases the force that can be applied from the muscle to the tibia. As a result of this increase in applied force, less muscle contraction force is required to move the tibia.

Leg

The leg is the part of the lower limb between the knee and the ankle. Like the forearm, it consists of two bones: the **tibia** (tib′ē-ă), or shinbone, and the **fibula** (fib′ū-lă; resembling a clasp or buckle; figure 7.35). The tibia is by far the larger of the two and supports most of the weight of the leg. A **tibial tuberosity,** which is the attachment point for the quadriceps femoris muscle group, can easily be seen and felt just inferior to the patella (figure 7.36). The **anterior crest** forms the shin. The proximal end of the tibia has flat **medial** and **lateral condyles,** which articulate with the condyles of the femur. Located between the condyles is the **intercondylar eminence,** which is a ridge between the two articular surfaces of the proximal tibia. The distal end of the tibia is enlarged to form the **medial malleolus** (ma-lē′ō-lŭs; mallet-shaped), which helps form the medial side of the ankle joint.

The fibula does not articulate with the femur but has a small proximal head where it articulates with the tibia. The distal end of

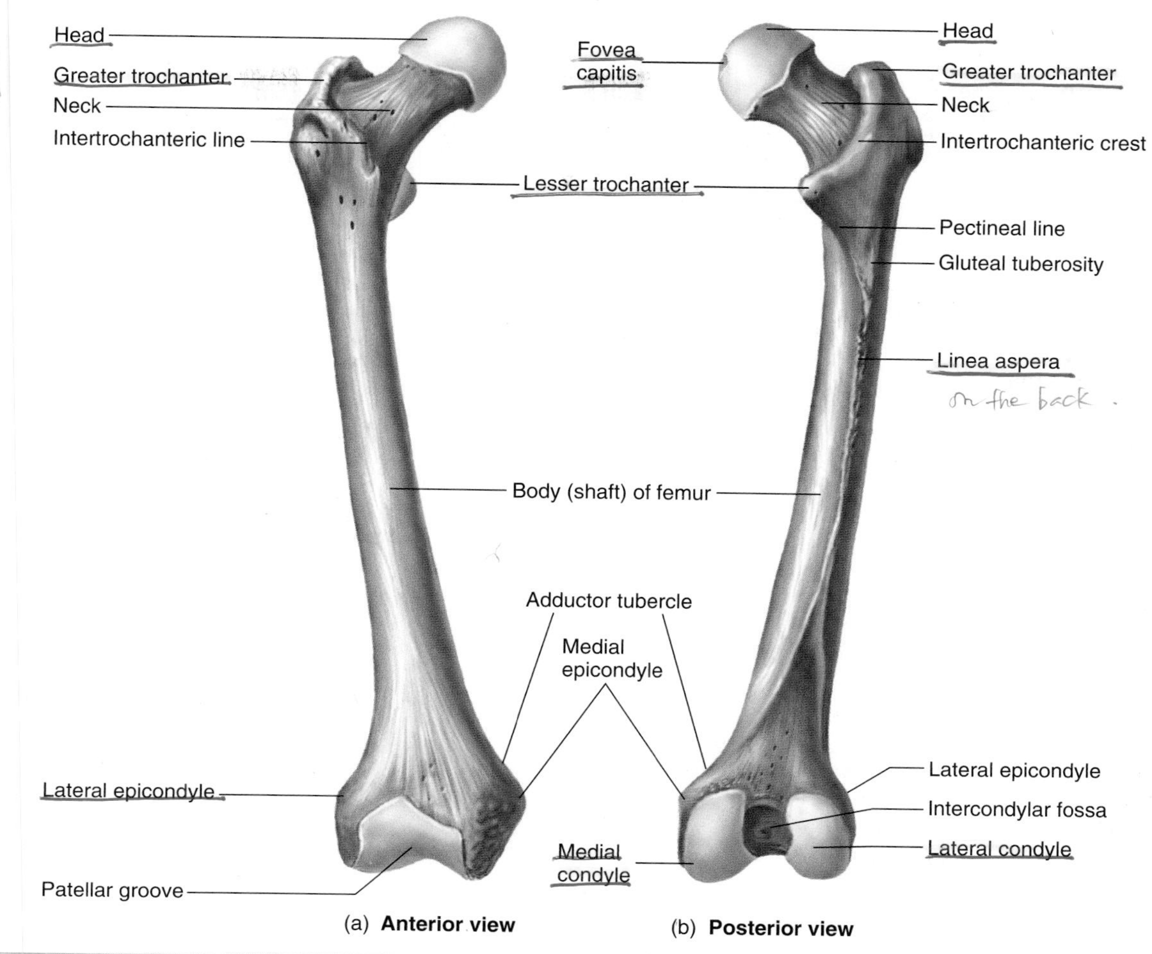

FIGURE 7.33 **Right Femur**

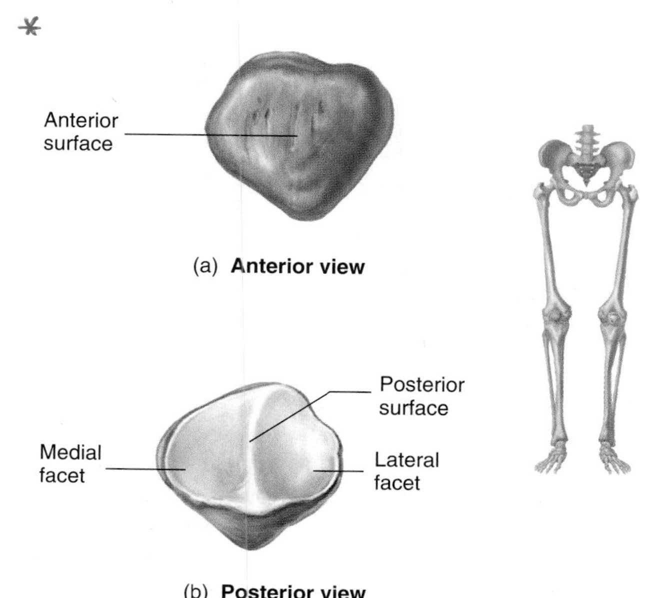

FIGURE 7.34 **Right Patella**

Patellar Defects

If the patella is severely fractured, the tendon from the quadriceps femoris muscle group may be torn, resulting in a severe decrease in muscle function. In extreme cases, it may be necessary to remove the patella to repair the tendon. Patella removal results in a decrease in the amount of power the quadriceps femoris muscle can generate at the tibia.

The patella normally tracks in the patellar groove on the anterodistal end of the femur. Abnormal tracking of the patella can become a problem in some teenagers, especially females. As a young woman's hips widen during puberty, the angles at the joints between the hips and the tibia may change considerably. As the knee becomes located more medially relative to the hip, the patella may be forced to track more laterally than normal. This lateral tracking may result in pain in the knees of some young athletes.

CASE STUDY

Fracture of the Femoral Neck

An 85-year-old woman who lived alone at home was found lying on her kitchen floor by her daughter, who had come to check on her mother. The woman could not rise, even with help, and when she tried she experienced extreme pain in her right hip. Her daughter immediately dialed 911 and her mother was taken by paramedics to the hospital.

The elderly woman's hip was x-rayed in the emergency room, and it was determined that she had a fracture of the right femoral neck. A femoral neck fracture is commonly, but incorrectly, called a broken hip. Two days later, she received a partial hip replacement in which the head and neck of the femur were replaced, but not the acetabulum.

In the case of falls involving femoral neck fracture, it is not always clear whether the fall caused the femoral neck to fracture or whether a fracture of the femoral neck caused the fall. Femoral neck fractures are among the most common injuries resulting in morbidity (disease) and mortality (death) in older adults. Despite treatment with anticoagulants and antibiotics, about 5% of patients with femoral neck fractures develop deep vein thrombosis (blood clot) and about 5% develop wound infections, either of which can be life-threatening. Hospital mortality is 1%–7% among patients with femoral neck fractures, and nearly 20% of femoral neck fracture victims die within 3 months of the fracture.

Four percent of women over age 85 experience femoral neck fractures each year. Only about 25% of victims ever fully recover from the injury. Forty percent of all deaths caused by injury in men over age 65 are from falls and 58% of injury deaths result from falls among women of the same age group. Many of those falls result in, or from, femoral neck fractures.

PREDICT 7

Fracture of the femoral neck increases dramatically with age, and 81% of victims are women. The average age of such injury is 82. Why is the femoral neck so commonly injured (*hint:* see figure 7.1), and why are elderly women the group most commonly affected?

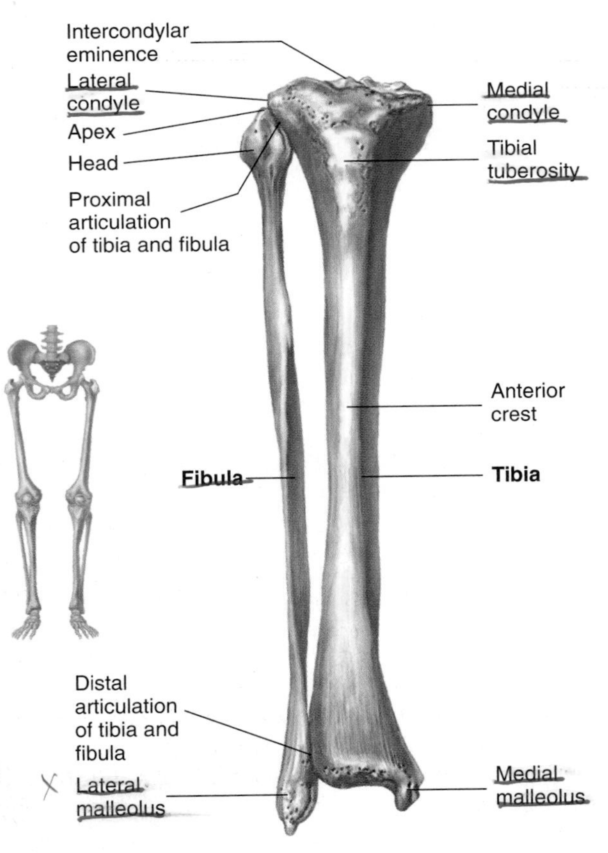

FIGURE 7.35 Right Tibia and Fibula

the fibula is also slightly enlarged as the **lateral malleolus** to create the lateral wall of the ankle joint. The lateral and medial malleoli can be felt and seen as prominent lumps on both sides of the ankle (see figure 7.36). The thinnest, weakest portion of the fibula is just proximal to the lateral malleolus.

PREDICT 8

Explain why modern ski boots are designed with high tops that extend partway up the leg.

Foot

The proximal portion of the foot consists of seven **tarsal** (tar′săl; the sole of the foot) **bones** (figure 7.37). The **talus** (tā′lŭs), or ankle bone, articulates with the tibia and the fibula to form the ankle joint. It also articulates with the calcaneous and navicular. The **calcaneus** (kal-kā′nē-us; heel), the heel bone, is the largest and strongest bone in the foot. It is located inferior to the talus and supports that bone. The calcaneus protrudes posteriorly, is the important attachment point for the large calf muscles, and can be easily felt as the heel of the foot. The **navicular** (nă-vik′ū-lar), which is boat-shaped, lies between the talus posteriorly and the cuneiforms anteriorly. The talus, calcaneus, and navicular are proximal in the foot but do not form a row. The proximal foot is relatively much larger than the wrist. The distal four bones do form a row. The medial, wedge-shaped three bones of the distal row are called **cuneiforms** (kū′-nē-i-fōrmz). They are the **medial cuneiform, intermediate cuneiform,** and **lateral cuneiform.** The **cuboid** (kū′boyd), which is cube-shaped, is the most lateral of

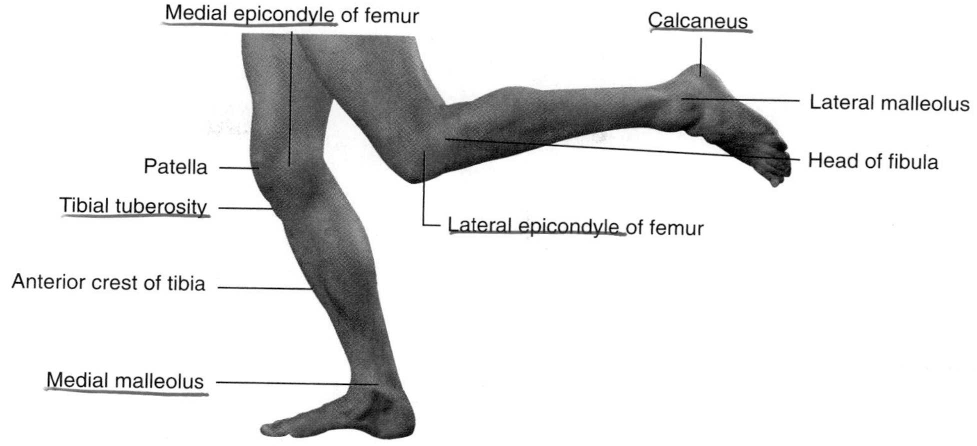

FIGURE 7.36 Surface Anatomy Showing Bones of the Lower Limb

Fractures of the Malleoli

Turning the plantar surface of the foot outward so that it faces laterally is called eversion. Forceful eversion of the foot, such as when a person slips and twists the ankle or jumps and lands incorrectly on the foot, may cause the distal ends of the tibia and/or fibula to fracture (figure E*a*). Such fractures are not uncommon in soccer, football, and basketball players. When the foot is forcefully everted, the medial malleolus moves inferiorly toward the ground or floor and the talus slides laterally, forcing the medial and lateral malleoli to separate. The ligament holding the medial malleolus to the tarsal bones is stronger than the bones it connects, and often it does not tear as the bones separate because of the eversion. Instead, the medial malleolus breaks. Also, as the talus slides laterally, the force can shear off the lateral malleolus or, more commonly, can cause the fibula to break superior to the lateral malleolus. This type of injury to the tibia and fibula is often called a Pott's fracture.

Turning the plantar surface of the foot inward so that it faces medially is called inversion. Forceful inversion of the foot can fracture the fibula just proximal to the lateral malleolus (figure E*b*). More often, because the ligament holding the medial malleolus to the tarsal bones is weaker than the bones it connects, the inversion of the foot causes a sprain in which ligaments are damaged.

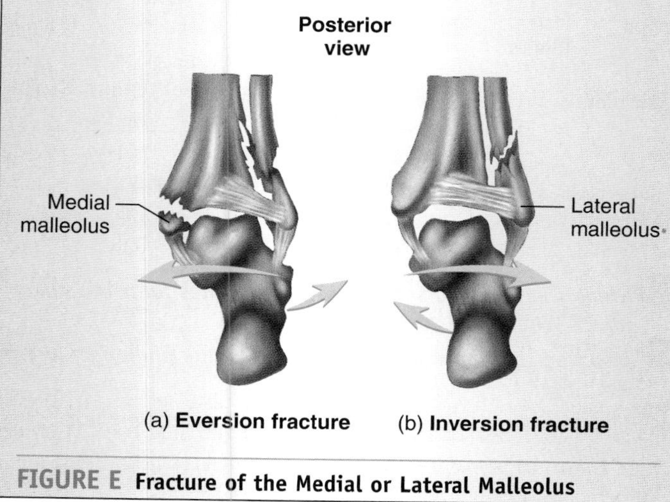

FIGURE E Fracture of the Medial or Lateral Malleolus

the distal row. A mnemonic for the distal row is **MILC**—that is, **M**edial, **I**ntermediate, and **L**ateral cuneiforms and the **C**uboid; A mnemonic for the proximal three bones is **No T**hanks **C**ow, that is **N**avicular, **T**alus, and **C**alcaneus.

The **metatarsals** and **phalanges** of the foot are arranged in a manner very similar to that of the metacarpals and phalanges of the hand, with the great toe analogous to the thumb (see figure 7.37). Small sesamoid bones often form in the tendons of muscles attached to the great toe. The ball of the foot is the junction between the metatarsals and phalanges. The foot as a unit is convex dorsally and concave ventrally to form the arches of the foot (described more fully in chapter 8).

PREDICT 9

A decubitus ulcer is a chronic ulcer that appears in pressure areas of skin overlying a bony prominence in bedridden or otherwise immobilized patients. Where are likely sites for decubitus ulcers to occur?

The foot has three major **arches,** which distribute the weight of the body between the heel and the ball of the foot during standing and walking (see figure 7.37*b*). As the foot is placed on the ground, weight is transferred from the tibia and the fibula to the talus. From there, the weight is distributed first to the heel (calcaneus) and then through the arch system along the lateral side of the foot to the ball of the foot (head of the metatarsals). This effect can be observed when a person with wet, bare feet walks across a dry surface; the print of the heel, the lateral border of the foot, and the ball of the foot can be seen, but the middle of the plantar surface and the medial border leave no impression. The medial side leaves no mark because the arches on this side of the foot are higher than those on the lateral side. The shape of the arches is maintained by the configuration of the bones, the ligaments connecting them, and the muscles acting on the foot.

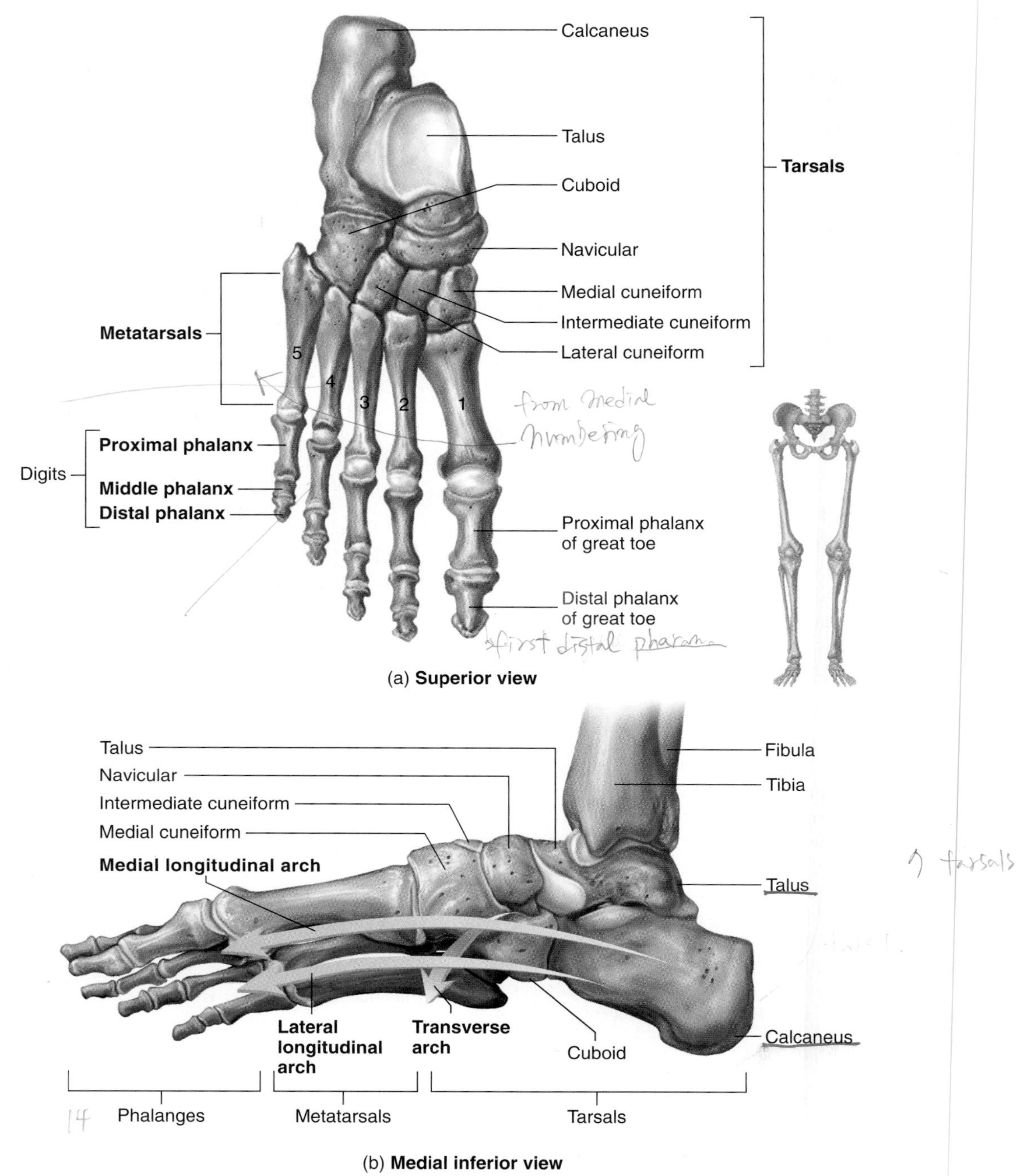

FIGURE 7.37 Bones of the Right Foot

The medial longitudinal arch is formed by the calcaneus, talus, navicular, cuneiforms, and three medial metatarsals. The lateral longitudinal arch is formed by the calcaneus, cuboid, and two lateral metatarsals. The transverse arch is formed by the cuboid and cuneiforms.

31 ***What is the function of the greater trochanter and the lesser trochanter?***

32 ***Describe the function of the patella.***

33 ***Name the bones of the leg.***

34 ***Give the points of articulation among the pelvis, femur, leg, and ankle.***

35 ***What is the function of the tibial tuberosity?***

36 ***Name the seven tarsal bones. Which bones form the ankle joint? What bone forms the heel?***

37 ***Describe the bones of the foot. How many phalanges are in each toe?***

38 ***List the three arches of the foot and describe their function.***

SUMMARY

The gross anatomy of the skeletal system considers the features of bone, cartilage, tendons, and ligaments that can be seen without the use of a microscope. Dried, prepared bones display the major features of bone but obscure the relationship between bone and soft tissue.

General Considerations (p. 204)

Bones have processes, smooth surfaces, and holes that are associated with ligaments, muscles, joints, nerves, and blood vessels.

Axial Skeleton (p. 206)

The axial skeleton consists of the skull, hyoid bone, vertebral column, and thoracic cage.

Skull

1. The skull, or cranium, can be thought of as a single unit.
2. The parietal bones are joined at the midline by the sagittal suture; they are joined to the frontal bone by the coronal suture, to the occipital bone by the lambdoid suture, and to the temporal bone by the squamous suture.
3. Nuchal lines are the points of attachment for neck muscles.
4. Several skull features are seen from a lateral view.
   - The external auditory meatus transmits sound waves toward the eardrum.
   - Important neck muscles attach to the mastoid process.
   - The temporal lines are attachment points of the temporalis muscle.
   - The zygomatic arch, from the temporal and zygomatic bones, forms a bridge across the side of the skull.
5. Several skull features are seen from a frontal view.
   - The orbits contain the eyes.
   - The nasal cavity is divided by the nasal septum, and the hard palate separates the nasal cavity from the oral cavity.
   - Sinuses within bone are air-filled cavities. The paranasal sinuses, which connect to the nasal cavity, are the frontal, sphenoidal, and maxillary sinuses and the ethmoidal labyrinth.
   - The mandible articulates with the temporal bone.
6. Several skull features are inside the cranial cavity.
   - The crista galli is a point of attachment for one of the meninges.
   - The olfactory nerves extend into the roof of the nasal cavity through the cribriform plate.
   - The sella turcica is occupied by the pituitary gland.
   - The spinal cord and brain are connected through the foramen magnum.
7. Several features are on the inferior surface of the skull.
   - Occipital condyles are points of articulation between the skull and the vertebral column.
   - Blood reaches the brain through the internal carotid arteries, which pass through the carotid canals, and the vertebral arteries, which pass through the foramen magnum.
   - Most blood leaves the brain through the internal jugular veins, which exit through the jugular foramina.
   - Styloid processes provide attachment points for three muscles involved in movement of the tongue, hyoid bone, and pharynx.
   - The hard palate forms the floor of the nasal cavity.
8. The skull is composed of 22 bones.
   - The auditory ossicles, which function in hearing, are located inside the temporal bones.
   - The braincase protects the brain.
   - The facial bones protect the sensory organs of the head and are muscle attachment sites (mastication, facial expression, and eye muscles).
   - The mandible and maxillae possess alveolar processes with sockets for the attachment of the teeth.

Hyoid Bone

The hyoid bone, which "floats" in the neck, is the attachment site for throat and tongue muscles.

Vertebral Column

1. The vertebral column provides flexible support and protects the spinal cord.
2. The vertebral column has four major curvatures: cervical, thoracic, lumbar, and sacral/coccygeal. Abnormal curvatures are lordosis (lumbar), kyphosis (thoracic), and scoliosis (lateral).
3. A typical vertebra consists of a body, a vertebral arch, and various processes.
   - Part of the body and the vertebral arch (pedicle and lamina) form the vertebral foramen, which contains and protects the spinal cord.
   - Spinal nerves exit through the intervertebral foramina.
   - The transverse and spinous processes are points of muscle and ligament attachment.
   - Vertebrae articulate with one another through the superior and inferior articular processes.
4. Adjacent bodies are separated by intervertebral disks. The disk has a fibrous outer covering (annulus fibrosus) surrounding a gelatinous interior (nucleus pulposus).
5. Several types of vertebrae can be distinguished.
   - All seven cervical vertebrae have transverse foramina, and most have bifid spinous processes.
   - The 12 thoracic vertebrae are characterized by long, downward-pointing spinous processes and demifacets.
   - The five lumbar vertebrae have thick, heavy bodies and processes.
   - The sacrum consists of five fused vertebrae and attaches to the coxal bones to form the pelvis.
   - The coccyx consists of four fused vertebrae attached to the sacrum.

Thoracic Cage

1. The thoracic cage (consisting of the ribs, their associated costal cartilages, and the sternum) protects the thoracic organs and changes volume during respiration.
2. Twelve pairs of ribs attach to the thoracic vertebrae. They are divided into seven pairs of true ribs and five pairs of false ribs. Two pairs of false ribs are floating ribs.
3. The sternum is composed of the manubrium, the body, and the xiphoid process.

Appendicular Skeleton (p. 233)

The appendicular skeleton consists of the upper and lower limbs and the girdles that attach the limbs to the body.

Pectoral Girdle and Upper Limb

1. The upper limb is attached loosely and functions in grasping and manipulation.
2. The pectoral girdle consists of the scapulae and clavicles.
   - The scapula articulates with the humerus and the clavicle. It is an attachment site for shoulder, back, and arm muscles.
   - The clavicle holds the shoulder away from the body, permitting free movement of the arm.
3. The arm bone is the humerus.
   - The humerus articulates with the scapula (head), the radius (capitulum), and the ulna (trochlea).
   - Sites of muscle attachment are the greater and lesser tubercles, the deltoid tuberosity, and the epicondyles.
4. The forearm contains the ulna and radius.
   - The ulna and radius articulate with each other and with the humerus and wrist bones.
   - The wrist ligaments attach to the styloid processes of the radius and ulna.
5. Eight carpal, or wrist, bones are arranged in two rows.
6. The hand consists of five metacarpal bones.
7. The phalanges are digital bones. Each finger has three phalanges, and the thumb has two phalanges.

Pelvic Girdle and Lower Limb

1. The lower limb is attached solidly to the coxal bone and functions in support and movement.
2. The pelvic girdle consists of the right and left coxal bones and sacrum. Each coxal bone is formed by the fusion of the ilium, the ischium, and the pubis.
   - The coxal bones articulate with each other (symphysis pubis) and with the sacrum (sacroiliac joint) and the femur (acetabulum).
   - Important sites of muscle attachment are the iliac crest, the iliac spines, and the ischial tuberosity.
   - The female pelvis has a larger pelvic inlet and outlet than the male pelvis.
3. The thighbone is the femur.
   - The femur articulates with the coxal bone (head), the tibia (medial and lateral condyles), and the patella (patellar groove).
   - Sites of muscle attachment are the greater and lesser trochanters, as well as the adductor tubercle.
   - Sites of ligament attachment are the lateral and medial epicondyles.
4. The leg consists of the tibia and the fibula.
   - The tibia articulates with the femur, the fibula, and the talus. The fibula articulates with the tibia and the talus.
   - Tendons from the thigh muscles attach to the tibial tuberosity.
5. Seven tarsal bones form the proximal portion of the foot.
6. The foot consists of five metatarsal bones.
7. The toes have three phalanges each, except for the big toe, which has two.
8. The bony arches transfer weight from the heels to the toes and allow the foot to conform to many different positions.

REVIEW AND COMPREHENSION

1. Which of these is part of the appendicular skeleton?
   a. cranium
   b. ribs
   c. clavicle
   d. sternum
   e. vertebra
2. A knoblike lump on a bone is called a
   a. spine.
   b. facet.
   c. tuberosity.
   d. sulcus.
   e. ramus.
3. The superior and middle nasal conchae are formed by projections of the
   a. sphenoid bone.
   b. vomer bone.
   c. palatine process of maxillae.
   d. palatine bone.
   e. ethmoid bone.
4. The perpendicular plate of the ethmoid and the ____________ form the nasal septum.
   a. palatine process of the maxilla
   b. horizontal plate of the palatine
   c. vomer
   d. nasal bone
   e. lacrimal bone
5. Which of these bones does *not* contain a paranasal sinus?
   a. ethmoid
   b. sphenoid
   c. frontal
   d. temporal
   e. maxilla
6. The mandible articulates with the skull at the
   a. styloid process.
   b. occipital condyle.
   c. mandibular fossa.
   d. zygomatic arch.
   e. medial pterygoid.
7. The nerves for the sense of smell pass through the
   a. cribriform plate.
   b. nasolacrimal canal.
   c. internal acoustic meatus.
   d. optic canal.
   e. orbital fissure.
8. The major blood supply to the brain enters through the
   a. foramen magnum.
   b. carotid canals.
   c. jugular foramina.
   d. both a and b.
   e. all of the above.
9. The site of the sella turcica is the
   a. sphenoid bone.
   b. maxillae.
   c. frontal bone.
   d. ethmoid bone.
   e. temporal bone.
10. Which of these bones is *not* in contact with the sphenoid bone?
    a. maxilla
    b. inferior nasal concha
    c. ethmoid
    d. parietal
    e. vomer
11. A herniated disk occurs when
    a. the annulus fibrosus ruptures.
    b. the intervertebral disk slips out of place.
    c. the spinal cord ruptures.
    d. too much fluid builds up in the nucleus pulposus.
    e. all of the above.
12. The weight-bearing portion of a vertebra is the
    a. vertebral arch.
    b. articular process.
    c. body.
    d. transverse process.
    e. spinous process.
13. Transverse foramina are found only in
    a. cervical vertebrae.
    b. thoracic vertebrae.
    c. lumbar vertebrae.
    d. the sacrum.
    e. the coccyx.
14. Articular facets on the bodies and transverse processes are found only on
    a. cervical vertebrae.
    b. thoracic vertebrae.
    c. lumbar vertebrae.
    d. the sacrum.
    e. the coccyx.
15. Which of these statements concerning ribs is true?
    a. The true ribs attach directly to the sternum with costal cartilage.
    b. There are five pairs of floating ribs.
    c. The head of the rib attaches to the transverse process of the vertebra.
    d. Vertebrochondral ribs are classified as true ribs.
    e. Floating ribs do not attach to vertebrae.

16. The point where the scapula and clavicle connect is the
    a. coracoid process. c. glenoid cavity. e. capitulum.
    b. styloid process. d. acromion process.
17. The distal medial process of the humerus to which the ulna joins is the
    a. epicondyle. c. malleolus. e. trochlea.
    b. deltoid tuberosity. d. capitulum.
18. Which of these is *not* a point of muscle attachment on the pectoral girdle or upper limb?
    a. epicondyles c. radial tuberosity e. greater tubercle
    b. mastoid process d. spine of scapula
19. The bone/bones of the foot on which the tibia rests is the
    a. talus. c. metatarsals. e. phalanges.
    b. calcaneus. d. navicular.
20. A place where nerves or blood vessels pass from the trunk to the lower limb is the
    a. pubic crest. d. iliac crest.
    b. greater sciatic notch. e. pubis symphysis.
    c. ischial tuberosity.
21. The projection on the coxal bone of the pelvic girdle that is used as a landmark for finding an injection site is the
    a. ischial tuberosity. d. posterior inferior iliac spine.
    b. iliac crest. e. ischial spine.
    c. anterior superior iliac spine.
22. When comparing the pectoral girdle with the pelvic girdle, which of these statements is true?
    a. The pectoral girdle has greater mass than the pelvic girdle.
    b. The pelvic girdle is more firmly attached to the body than the pectoral girdle.
    c. The pectoral girdle has the limbs more securely attached than the pelvic girdle.
    d. The pelvic girdle allows greater mobility than the pectoral girdle.
23. When comparing a male pelvis with a female pelvis, which of these statements is true?
    a. The pelvic inlet in males is larger and more circular.
    b. The subpubic angle in females is less than 90 degrees.
    c. The ischial spines in males are closer together.
    d. The sacrum in males is broader and less curved.
24. A site of muscle attachment on the proximal end of the femur is the
    a. greater trochanter. d. intercondylar eminence.
    b. epicondyle. e. condyle.
    c. greater tubercle.
25. The process that forms the large lateral bump in the ankle is the lateral
    a. malleolus. c. epicondyle. e. tubercle.
    b. condyle. d. tuberosity.

Answers in Appendix E

CRITICAL THINKING

1. A patient has an infection in the nasal cavity. Name seven adjacent structures to which the infection could spread.
2. A patient is unconscious. Radiographic films reveal that the superior articular process of the atlas has been fractured. Which of the following could have produced this condition: falling on the top of the head or being hit in the jaw with an uppercut? Explain.
3. If the vertebral column is forcefully rotated, what part of the vertebra is most likely to be damaged? In what area of the vertebral column is such damage most likely?
4. An asymmetric weakness of the back muscles can produce which of the following: scoliosis, kyphosis, or lordosis? Which can result from pregnancy? Explain.
5. What might be the consequences of a broken forearm involving both the ulna and the radius when the ulna and radius fuse to each other during repair of the fracture?
6. Suppose you need to compare the length of one lower limb with the length of the other in an individual. Using bony landmarks, suggest an easy way to accomplish the measurements.
7. A paraplegic individual develops decubitus ulcers (pressure sores) on the buttocks from sitting in a wheelchair for extended periods. Name the bony protuberance responsible.
8. Why are more women than men knock-kneed?
9. On the basis of bone structure of the lower limb, explain why it is easier to turn the foot medially (sole of the foot facing toward the midline of the body) than laterally. Why is it easier to cock the wrist medially than laterally?
10. Justin Time leaped from his hotel room to avoid burning to death in a fire. If he landed on his heels, what bone was likely fractured? Unfortunately for Justin, a 240 lb firefighter ran by and stepped heavily on the proximal part of Justin's foot (not the toes). What bones could have been broken?

Answers in Appendix F

Visit this book's website at www.mhhe.com/seeley8 for chapter quizzes, interactive learning exercises, and other study tools.

8 Articulations and Movement

Muscles pull on bones to make them move, but movement would not be possible without joints between the bones. Humans would resemble statues, were it not for the joints between bones that allow bones to move once the muscles have provided the pull. Machine parts most likely to wear out are those that rub together and they require the most maintenance. Movable joints are places in the body where the bones rub together, yet we tend to pay little attention to them. Fortunately, our joints are self-maintaining, but damage to or disease of a joint can make movement very difficult. We realize then how important the movable joints are for normal function.

An **articulation,** or **joint,** is a place where two or more bones come together. We usually think of joints as being movable, but that is not always the case. Many joints allow only limited movement, and others allow no apparent movement. The structure of a given joint is directly correlated with its degree of movement. Fibrous joints have much less movement than joints containing fluid and having smooth articulating surfaces.

Joints develop between adjacent bones or areas of ossification, and movement is important in determining the type of joint that develops. If movement is restricted—even in a highly movable joint—at any time during an individual's life, the joint may be transformed into a nonmovable joint.

This chapter presents a scheme for *naming joints* (p. 253) and an explanation of *classes of joints* (p. 253) *types of movement* (p. 259), and *range of motion* (p. 263). It then presents a *description of selected joints* (p. 263) and summarizes the *effects of aging on the joints* (p. 272).

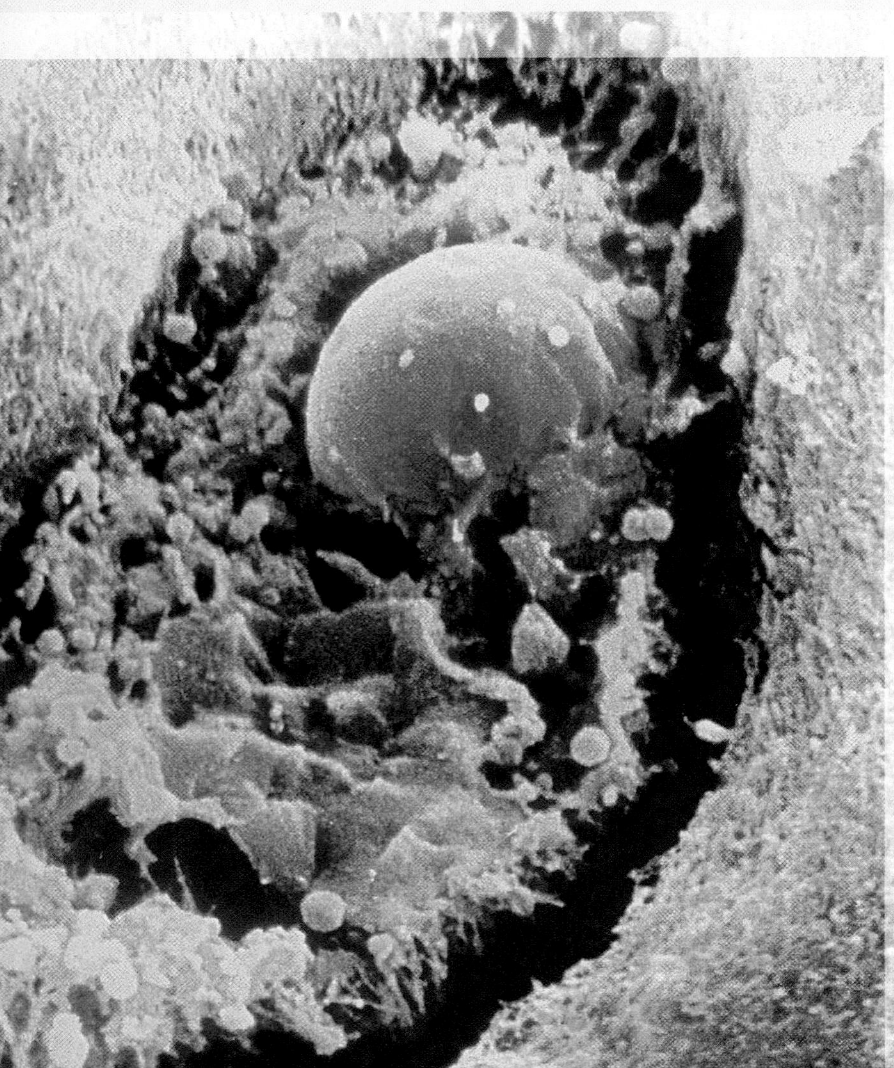

Colorized scanning electron micrograph of a chondrocyte within a lacuna surrounded by cartilage matrix. Cartilage provides the durable, resilient cushion within joints.

Skeletal System

NAMING JOINTS

Joints are commonly named according to the bones or portions of bones that are united at the joint, such as the temporomandibular joint between the temporal bone and the mandible.

PREDICT 1

What is the name of the joint between the metacarpals and the phalanges?

Some joints are simply given the Greek or Latin equivalent of the common name, such as **cubital** (kū'bi-tăl; cubit, elbow or forearm) **joint** for the elbow joint.

1 ***What criteria are used to name joints?***

CLASSES OF JOINTS

The three major kinds of joints are classified structurally as fibrous, cartilaginous, and synovial. In this classification scheme, joints are categorized according to the major connective tissue type that binds the bones together, and whether or not a fluid-filled joint capsule is present. Joints may also be classified according to their function. This classification is based on the degree of motion at each joint and includes the terms *synarthrosis* (nonmovable joint), *amphiarthrosis* (slightly movable joint), and *diarthrosis* (freely movable joint). This functional classification is somewhat limited and is not used in this book. The structural classification scheme with its various subclasses allows for a more precise classification and is the scheme used in this book.

Fibrous Joints

Fibrous joints consist of two bones that are united by fibrous connective tissue, have no joint cavity, and exhibit little or no movement. Joints in this group are classified further as sutures, syndesmoses, or gomphoses (table 8.1), based on their structure.

Sutures

Sutures (soo'choorz) are seams between the bones of the skull (figure 8.1). Some sutures may become completely immovable in older adults. Few sutures are smooth, and the opposing bones often interdigitate (have interlocking, fingerlike processes). This interdigitation adds considerable stability to sutures. The tissue between the bones is dense regular collagenous connective tissue, and the periosteum on the inner and outer surfaces of the adjacent bones continues over the joint. The two layers of periosteum plus the dense fibrous connective tissue in between form a **sutural ligament.**

In a newborn, a membranous area called a **fontanel** (fon'tă-nel'; little fountain, so-named because the membrane can be seen to move with the pulse; soft spot) is present within some of the sutures. A fontanel makes the skull flexible during the birth process and allows for growth of the head after birth (figure 8.2).

The margins of bones within sutures are sites of continuous intramembranous bone growth, and many sutures eventually become ossified. For example, ossification of the suture between the two frontal bones occurs shortly after birth so that they usually form a single frontal bone in the adult skull. In most normal adults, the coronal, sagittal, and lambdoid sutures are not fused. In some very old adults, however, even these sutures become ossified. A **synostosis** (sin-os-tō'sis) results when two bones grow together across a joint to form a single bone.

TABLE 8.1 Fibrous and Cartilaginous Joints

Class and Example of Joint	Bones or Structures Joined	Movement
Fibrous Joints		
Sutures		
Coronal	Frontal and parietal	None
Lambdoid	Occipital and parietal	None
Sagittal	The two parietal bones	None
Squamous	Parietal and temporal	Slight
Syndesmoses		
Radioulnar	Ulna and radius	Slight
Stylohyoid	Styloid process and hyoid bone	Slight
Stylomandibular	Styloid process and mandible	Slight
Tibiofibular	Tibia and fibula	Slight
Gomphoses		
Dentoalveolar	Tooth and alveolar process	Slight
Cartilaginous Joints		
Synchondroses		
Epiphyseal plate	Diaphysis and epiphysis of a long bone	None
Sternocostal	Anterior cartilaginous part of first rib; between rib and sternum	Slight
Sphenooccipital	Sphenoid and occipital	None
Symphyses		
Intervertebral	Bodies of adjacent vertebrae	Slight
Manubriosternal	Manubrium and body of sternum	None
Symphysis pubis	The two coxal bones	None except during childbirth
Xiphisternal	Xiphoid process and body of sternum	None

PREDICT 2

Predict the result of a sutural synostosis that occurs prematurely in a child's skull before the brain has reached its full size.

Syndesmoses

A **syndesmosis** (sin'dez-mō'sis; to fasten or bind) is a fibrous joint in which the bones are farther apart than in a suture and are joined by ligaments. Some movement may occur at

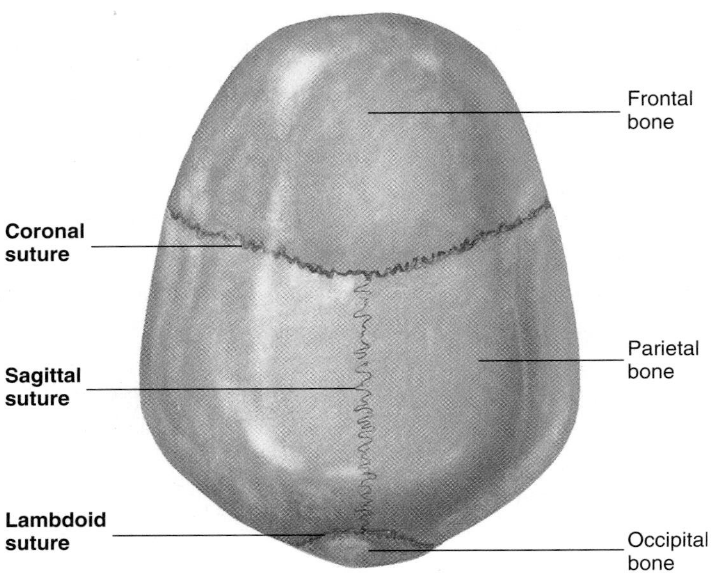

FIGURE 8.1 **Sutures, Superior View**

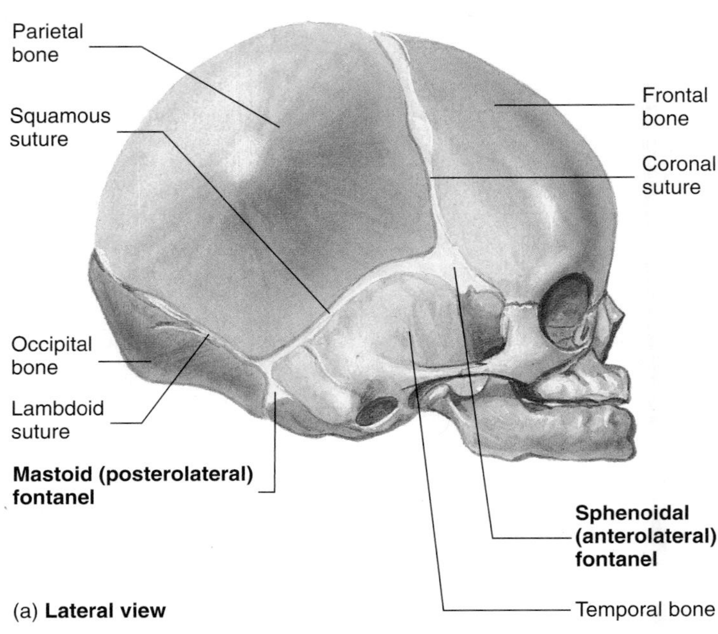

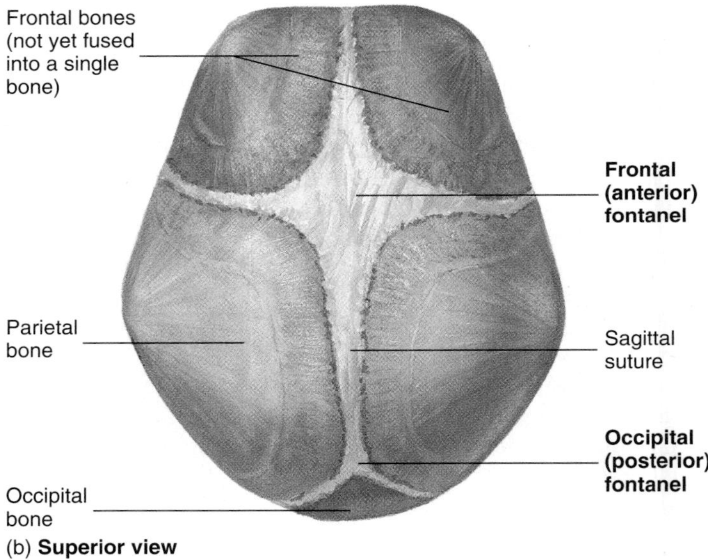

FIGURE 8.2 **Fetal Skull Showing Fontanels and Sutures**

syndesmoses because of flexibility of the ligaments, such as in the radioulnar syndesmosis, which binds the radius and ulna together (figure 8.3).

Gomphoses

Gomphoses (gom-fō′sēz) are specialized joints consisting of pegs that fit into sockets and that are held in place by fine bundles of regular collagenous connective tissue. The joints between the teeth and the sockets (alveoli) of the mandible and maxillae are gomphoses (figure 8.4). The connective tissue bundles between the teeth and their sockets are called **periodontal** (per′ē-ō-don′tăl) **ligaments;** they allow a slight amount of "give" to the teeth during mastication.

> **Gingivitis**
>
> The gingiva, or gums, are the soft tissues covering the alveolar process. Neglect of the teeth can result in **gingivitis,** an inflammation of the gingiva, often resulting from bacterial infection. Left untreated, gingivitis can spread to the tooth socket, resulting in periodontal disease, the leading cause of tooth loss in the United States. Periodontal disease involves an accumulation of plaque and bacteria and the resulting inflammation, which gradually destroys the periodontal ligaments and the bone. As a result, teeth may become so loose that they come out of their sockets. Proper brushing, flossing, and professional cleaning to remove plaque can usually prevent gingivitis and periodontal disease.

2 ***Define fibrous joint, describe three types, and give an example of each. What is a synostosis? Where are periodontal ligaments found?***

Cartilaginous Joints

Cartilaginous joints unite two bones by means of either hyaline cartilage or fibrocartilage (see table 8.1). Joints containing hyaline cartilage are called synchondroses; joints containing fibrocartilage are called symphyses.

Synchondroses

A **synchondrosis** (sin′kon-drō′sis; union through cartilage) consists of two bones joined by hyaline cartilage where little or no movement occurs (figure 8.5*a*). The epiphyseal plates of growing bones are synchondroses (figure 8.5*b*). Most synchondroses are temporary, with bone eventually replacing them to form synostoses. On the other hand,

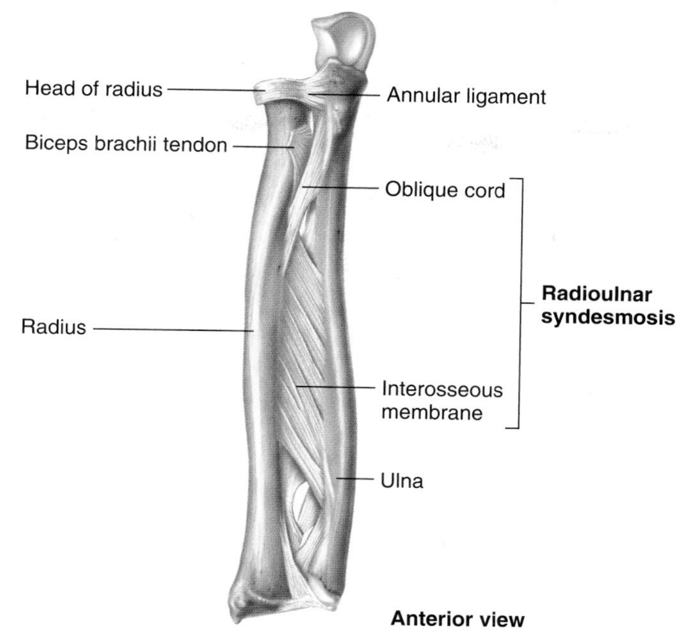

FIGURE 8.3 Right Radioulnar Syndesmosis

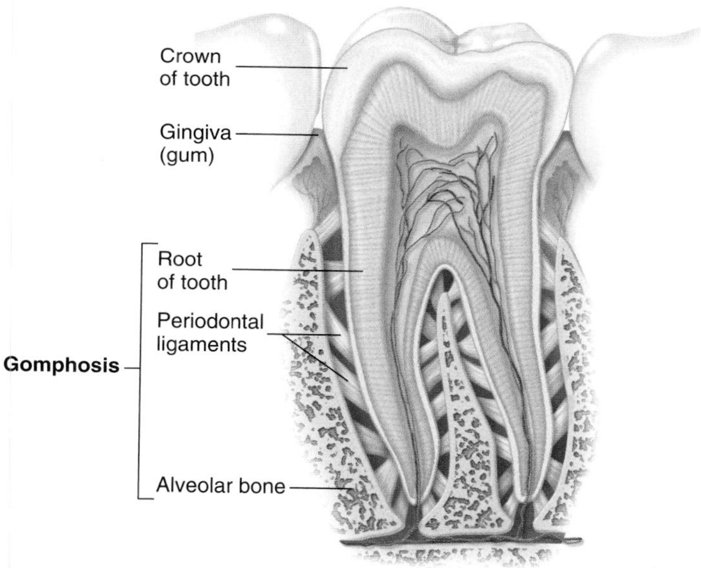

FIGURE 8.4 Gomphosis Between a Tooth and Alveolar Bone of the Mandible

some synchondroses persist throughout life. An example is the sternocostal synchondrosis between the first rib and the sternum by way of the first costal cartilage (figure 8.5*c*). All the costal cartilages begin as synchondroses, but, because of the movement that occurs between them and the sternum, all but the first usually develop synovial joints at those junctions. As a result, even though the **costochondral joints** (between the ribs and the costal cartilages) are retained, most costal cartilages no longer qualify as synchondroses because one end of the cartilage attaches to bone (the sternum) by a synovial joint.

Symphyses

A **symphysis** (sim′fi-sis; a growing together) consists of fibrocartilage uniting two bones. Symphyses include the junction between the manubrium and the body of the sternum (see figure 8.5*c*), the symphysis pubis (figure 8.6), and the intervertebral disks (see figures 7.13 and 7.14). Some of these joints are slightly movable because of the somewhat flexible nature of fibrocartilage.

Joint Changes During Pregnancy

During pregnancy certain hormones, such as estrogen, progesterone, and relaxin, act on the connective tissue of joints, such as the symphysis pubis, causing them to become more stretchable and allowing the joints to loosen. This change allows the pelvic opening to enlarge at the time of delivery. After delivery, the connective tissue of the symphysis pubis returns to its original condition. The enlarged pelvic opening, however, may not return completely to its original size, and the woman may have slightly wider hips after the birth of the child.

The same hormones may act on the connective tissue of other joints in the body, such as the arches of the feet, causing them to relax, which may result in fallen arches (see "Arch Problems," p. 272). They may also act on some of a baby's joints, such as the hip, causing the joints to become more mobile than normal. Increased mobility of the hip can result in congenital (appearing at birth) subluxation, or congenital dislocation, of the hip. Congenital hip dislocation occurs approximately once in every 670 births.

3 ***Define cartilaginous joints, describe two types, and give an example of each. Why are costochondral joints unique?***

Synovial Joints

Synovial (si-nō′vē-ăl; joint fluid; *syn,* coming together, *ovia,* resembling egg albumin) **joints** contain synovial fluid and allow considerable movement between articulating bones (figure 8.7). These joints are anatomically more complex than fibrous and cartilaginous joints. Most joints that unite the bones of the appendicular skeleton are synovial joints, reflecting the far greater mobility of the appendicular skeleton, compared with that of the axial skeleton.

The articular surfaces of bones within synovial joints are covered with a thin layer of hyaline cartilage called **articular cartilage,** which provides a smooth surface where the bones meet. In some synovial joints, a flat plate or pad of fibrocartilage, called an **articular disk,** is located between the articular cartilages of bones. The circumference of the disk is attached to the fibrous capsule. Examples of joints with articular disks are the temporomandibular, sternoclavicular, and acromioclavicular joints. A **meniscus** is an incomplete, crescent-shaped fibrocartilage pad found in joints such as the knee and wrist. A meniscus is much like an articular disk with a hole in the center. The circumference of the meniscus is attached to the fibrous joint capsule.

The articular surfaces of the bones that meet at a synovial joint are enclosed within a synovial **joint cavity,** which is surrounded by a **joint capsule.** This capsule helps hold the bones together while allowing for movement. The joint capsule consists of two layers: an outer **fibrous capsule** and an inner **synovial membrane** (see

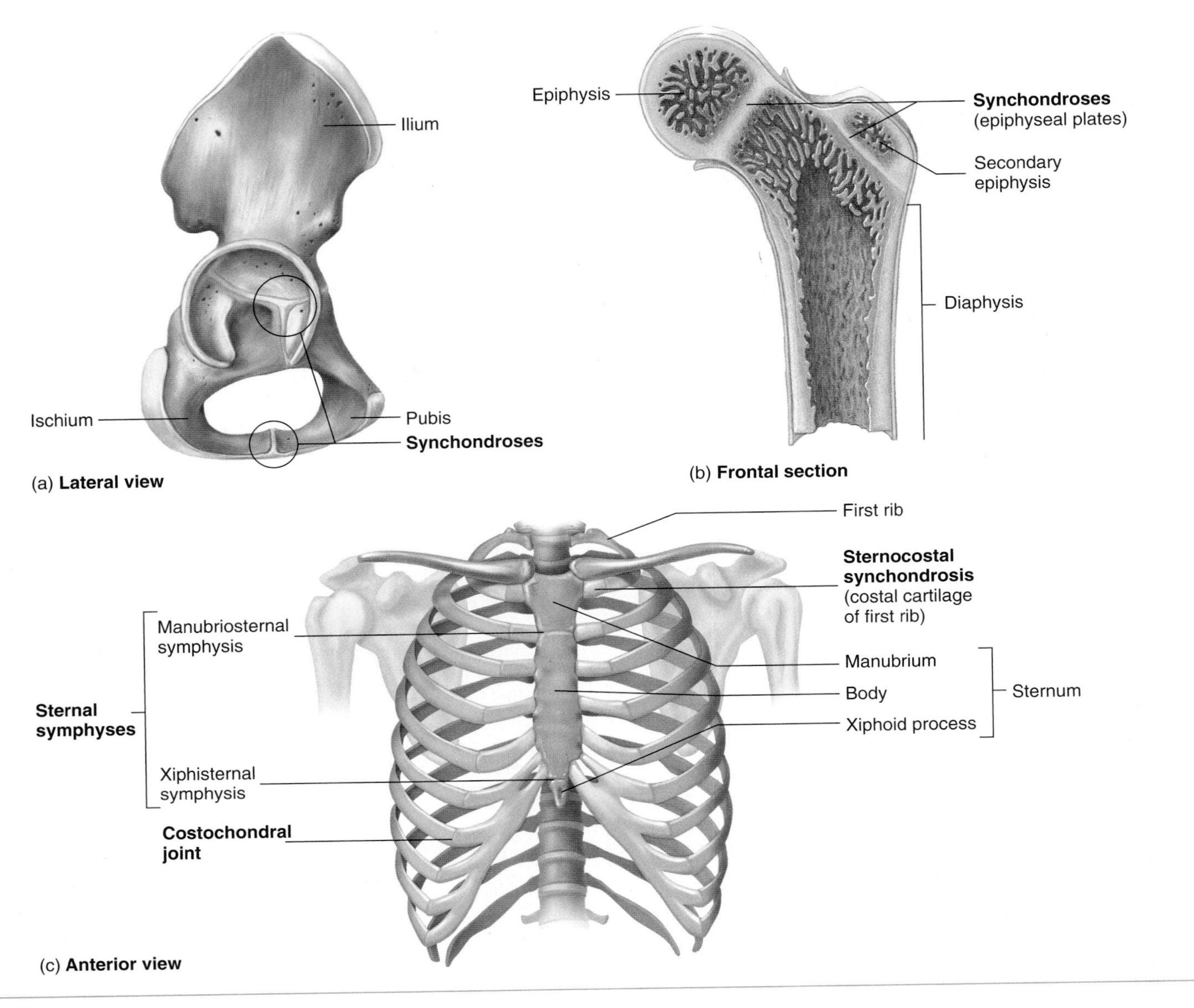

FIGURE 8.5 Synchondroses

(*a*) Synchondroses (epiphyseal plates) between the developing bones of the coxal bone. (*b*) Epiphyseal plates (frontal section of proximal femur). (*c*) Sternocostal synchondroses.

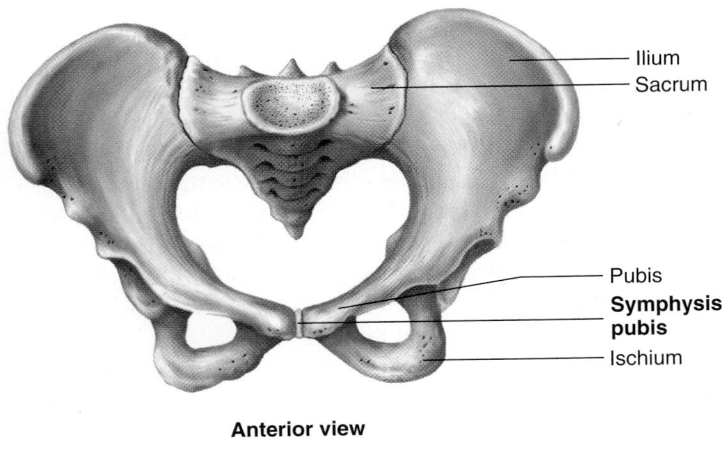

FIGURE 8.6 Symphysis Pubis

figure 8.7). The fibrous capsule consists of dense irregular connective tissue and is continuous with the fibrous layer of the periosteum that covers the bones united at the joint. Portions of the fibrous capsule may thicken and the collagen fibers become regularly arranged to form ligaments. In addition, ligaments and tendons may be present outside the fibrous capsule, thereby contributing to the strength and stability of the joint while limiting movement in some directions.

The synovial membrane lines the joint cavity, except over the articular cartilage and articular disks. It is a thin, delicate membrane consisting of a collection of modified connective tissue cells either intermixed with part of the fibrous capsule or separated from it by a layer of areolar tissue or adipose tissue. The membrane produces synovial fluid, which consists of a serum (blood fluid) filtrate and secretions from the synovial cells. Synovial fluid is a complex mixture of polysaccharides, proteins, fat, and cells. The major polysaccharide is hyaluronic acid, which provides much of the slippery consistency

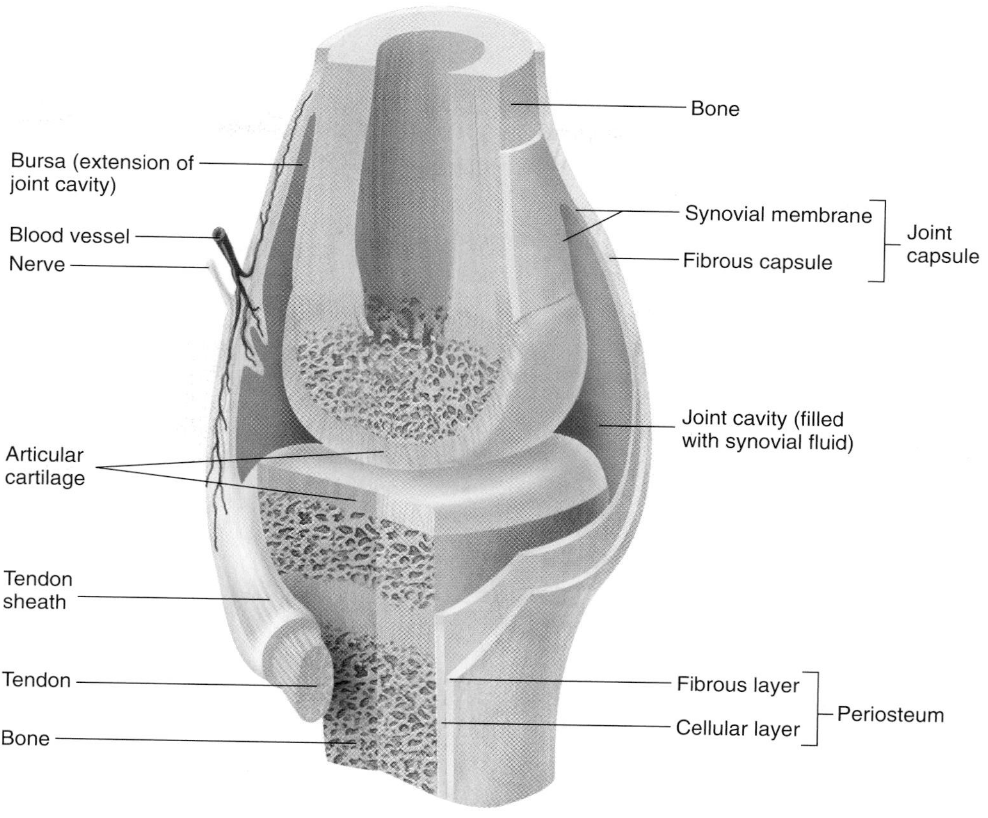

FIGURE 8.7 Structure of a Synovial Joint

and lubricating qualities of synovial fluid. Synovial fluid forms an important thin lubricating film that covers the surfaces of a joint.

PREDICT 3

What would happen if a synovial membrane covered the articular cartilage?

In certain synovial joints, such as the shoulder and knee, the synovial membrane extends as a pocket, or sac, called a **bursa** (ber′să; pocket) for a distance away from the rest of the joint cavity (see figure 8.7). Bursae contain synovial fluid, providing a cushion between structures that otherwise would rub against each other, such as tendons rubbing on bones or other tendons. Some bursae, such as the subcutaneous olecranon bursae, are not associated with joints but provide a cushion between the skin and underlying bony prominences, where friction could damage the tissues. Other bursae extend along tendons for some distance, forming **tendon sheaths. Bursitis** (ber-sī′tis) is an inflammation of a bursa and may cause considerable pain around the joint and restrict movement.

At the peripheral margin of the articular cartilage, blood vessels form a vascular circle that supplies the cartilage with nourishment, but no blood vessels penetrate the cartilage or enter the joint cavity. Additional nourishment to the articular cartilage comes from the underlying cancellous bone and from the synovial fluid covering the articular cartilage. Sensory nerves enter the fibrous capsule and, to a lesser extent, the synovial membrane. They not only supply information to the brain about pain in the joint but also furnish constant information to the brain about the position of the joint and its degree of movement (see chapter 14). Nerves do not enter the cartilage or joint cavity.

4. ***Describe the structure of a synovial joint. How do the different parts of the joint permit joint movement? What are articular disks and where are they found?***
5. ***Define bursa and tendon sheath. What is their function?***

Types of Synovial Joints

Synovial joints are classified according to the shape of the adjoining articular surfaces. The six types of synovial joints are plane, saddle, hinge, pivot, ball-and-socket, and ellipsoid. These joints are listed in table 8.2. Movements at synovial joints are described as **uniaxial** (occurring around one axis), **biaxial** (occurring around two axes situated at right angles to each other), or **multiaxial** (occurring around several axes).

Plane, or **gliding, joints** consist of two opposed flat surfaces of about equal size in which a slight amount of gliding motion can occur between the bones (figure 8.8). These joints are considered uniaxial because some rotation is also possible but is limited by ligaments and adjacent bone. Examples are the articular processes between vertebrae.

TABLE 8.2 Types of Joints

Class	Example	Structures Joined	Movement
Plane (uniaxial)	Acromioclavicular	Acromion process of scapula and clavicle	Slight
	Carpometacarpal	Carpals and metacarpals 2–5	Slight
	Costovertebral	Ribs and vertebrae	Slight
	Intercarpal	Between carpals	Slight
	Intertarsal	Between tarsals	Slight
	Intervertebral	Between articular processes of adjacent vertebrae	Slight
	Sacroiliac	Between sacrum and coxal bone (complex joint with several planes and synchondroses)	Slight
	Tarsometatarsal	Tarsals and metatarsals	Slight
Saddle (biaxial)	Carpometacarpal pollicis	Carpal and metacarpal of thumb	Two axes
	Intercarpal	Between carpals	Slight
	Sternoclavicular	Manubrium of sternum and clavicle	Slight
Hinge (uniaxial)	Cubital (elbow)	Humerus, ulna, and radius	One axis
	Genu (knee)	Femur and tibia	One axis
	Interphalangeal	Between phalanges	One axis
	Talocrural (ankle)	Talus, tibia, and fibula	Multiple axes, one predominates
Pivot (uniaxial)	Atlantoaxial	Atlas and axis	Rotation
	Proximal radioulnar	Radius and ulna	Rotation
	Distal radioulnar	Radius and ulna	Rotation
Ball-and-socket (multiaxial)	Coxal (hip)	Coxal bone and femur	Multiple axes
	Glenohumeral (shoulder)	Scapula and humerus	Multiple axes
Ellipsoid (biaxial or multiaxial)	Atlantooccipital	Atlas and occipital bone	Two axes
	Metacarpophalangeal (knuckles)	Metacarpals and phalanges	Two axes
	Metatarsophalangeal	Metatarsals and phalanges	Two axes
	Radiocarpal (wrist)	Radius and carpals	Multiple axes
	Temporomandibular	Mandible and temporal bone	Multiple axes, one predominates

Saddle joints consist of two saddle-shaped articulating surfaces oriented at right angles to each other so that complementary surfaces articulate with each other (figure 8.9). Saddle joints are biaxial joints. The carpometacarpal joint of the thumb is an example.

Hinge joints are uniaxial joints (figure 8.10). They consist of a convex cylinder in one bone applied to a corresponding concavity in the other bone. Examples include the elbow and knee joints.

Pivot joints are uniaxial joints that restrict movement to rotation around a single axis (figure 8.11). A pivot joint consists of a relatively cylindrical bony process that rotates within a ring composed partly of bone and partly of ligament. The articulation between the head of the radius and the proximal end of the ulna

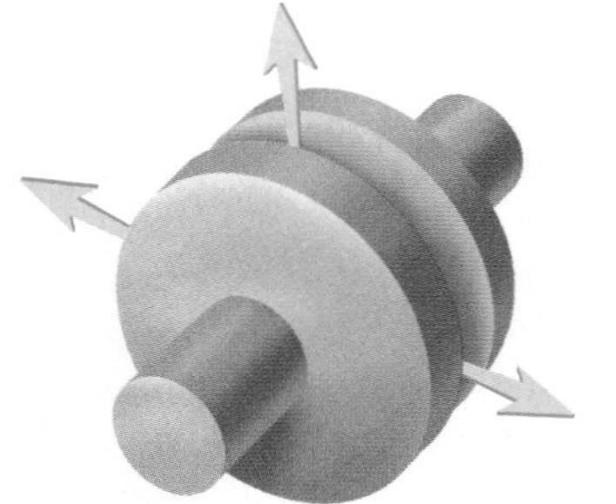

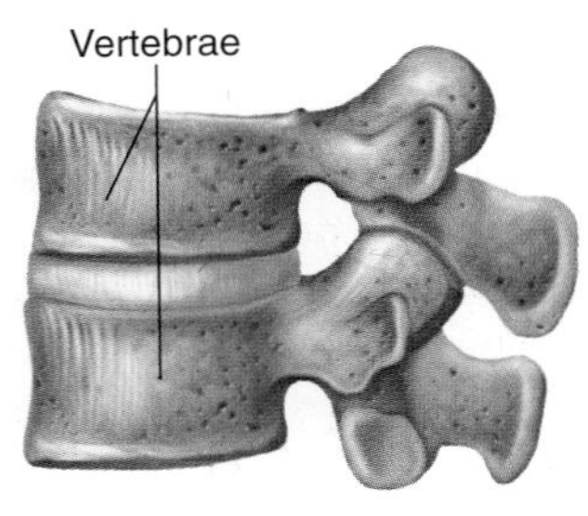

FIGURE 8.8 **Plane Joint**

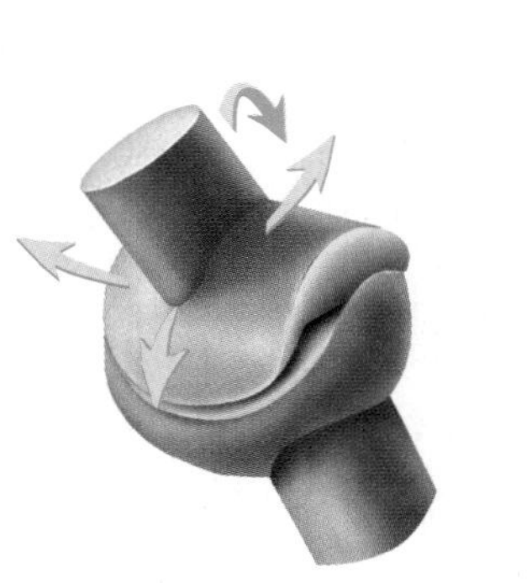

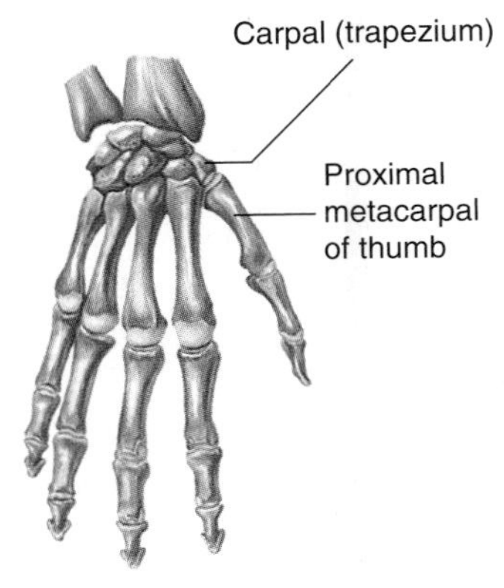

FIGURE 8.9 **Saddle Joint**

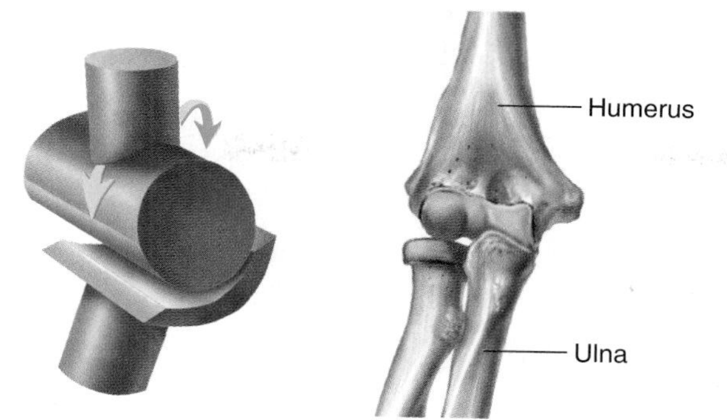

FIGURE 8.10 **Hinge Joint**

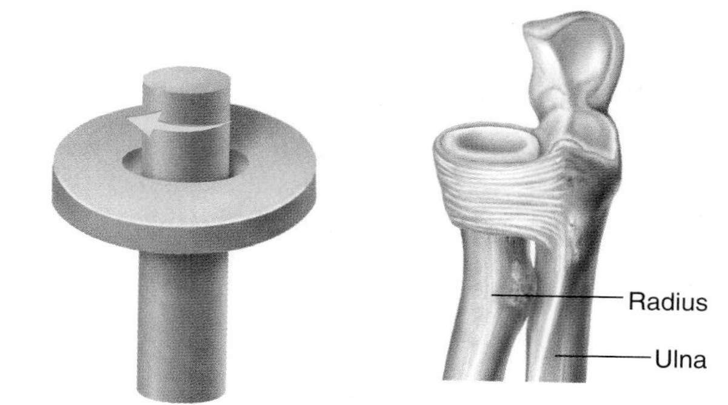

FIGURE 8.11 **Pivot Joint**

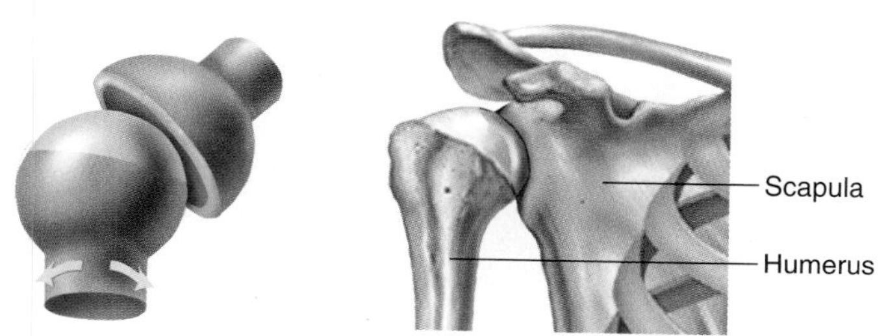

FIGURE 8.12 **Ball-and-Socket Joint**

is an example. The articulation between the dens, a process on the axis (see chapter 7), and the atlas is another example.

Ball-and-socket joints consist of a ball (head) at the end of one bone and a socket in an adjacent bone into which a portion of the ball fits (figure 8.12). This type of joint is multiaxial, allowing a wide range of movement in almost any direction. Examples are the shoulder and hip joints.

Ellipsoid joints (or condyloid joints) are modified ball-and-socket joints (figure 8.13). The articular surfaces are ellipsoid in shape rather than spherical as in regular ball-and-socket joints. Ellipsoid joints are biaxial, because the shape of the joint limits its range of movement almost to a hinge motion in two axes and restricts rotation. The atlantooccipital joint is an example.

6 ***On what basis are synovial joints classified? Describe the types of synovial joints, and give examples of each. What movements does each type of joint allow?***

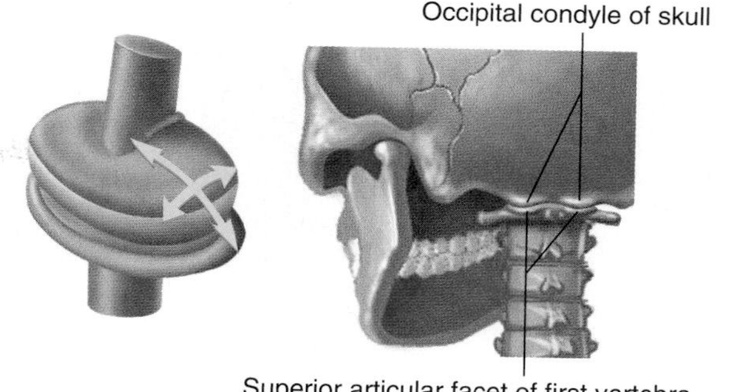

FIGURE 8.13 **Ellipsoid Joint**

TYPES OF MOVEMENT

A joint's structure relates to the movements that occur at that joint. Some joints are limited to only one type of movement; others can move in several directions. With few exceptions, movement is best described in relation to the anatomical position: (1) movement away from the anatomical position and (2) movement returning a structure toward the anatomical position. Most movements are accompanied by movements in the opposite direction and therefore are listed in pairs.

Gliding Movements

Gliding movements are the simplest of all the types of movement. These movements occur in plane joints between two flat or nearly flat surfaces where the surfaces slide or glide over each other. These joints often give only slight movement, such as between carpal bones.

Angular Movements

Angular movements are those in which one part of a linear structure, such as the body as a whole or a limb, is bent relative to another part of the structure, thereby changing the angle between the two parts. Angular movements also involve the movement of a solid rod, such as a limb, that is attached at one end to the body so that the angle at which it meets the body changed. The most common angular movements are flexion and extension and abduction and adduction.

Flexion and Extension

Flexion and extension can be defined in a number of ways, but in each case exceptions to the definition exist. The literal definition is to bend and straighten, respectively. This bending and straightening can easily be seen in the elbow (figure 8.14) or knee (see figure 8.16). The following definition, however, has more utility and fewer exceptions, with reference to the coronal plane. **Flexion** is movement of a body part anterior to the coronal plane, or in the anterior direction. **Extension** is movement of a body part posterior to the coronal plane, or in the posterior direction (figure 8.15).

The exceptions to defining flexion and extension according to the coronal plane are the knee and foot. At the knee, flexion moves the leg in a posterior direction and extension moves it in an anterior direction (figure 8.16). Movement of the foot toward the plantar surface, such as when standing on the toes, is commonly

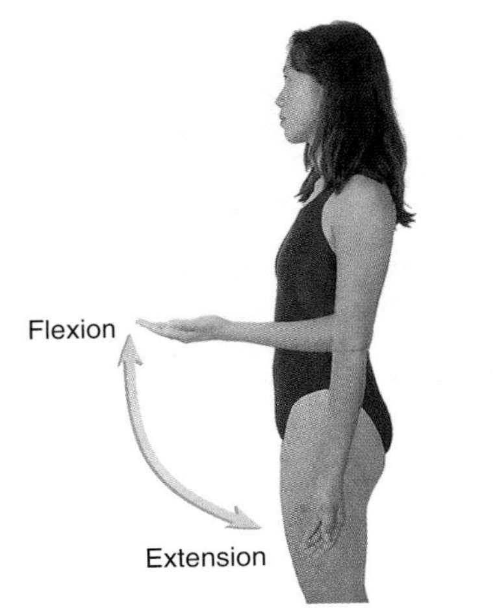

FIGURE 8.14 **Flexion and Extension of the Elbow**

called **plantar flexion;** movement of the foot toward the shin, such as when walking on the heels, is called **dorsiflexion** (figure 8.17).

> **Hyperextension**
>
> **Hyperextension** is usually defined as an abnormal, forced extension of a joint beyond its normal range of motion. For example, if a person falls and attempts to break the fall by putting out a hand, the force of the fall directed into the hand and wrist may cause hyperextension of the wrist, which may result in sprained joints or broken bones. Some health professionals, however, define hyperextension as the normal movement of structures, except the leg, into the space posterior to the anatomical position.

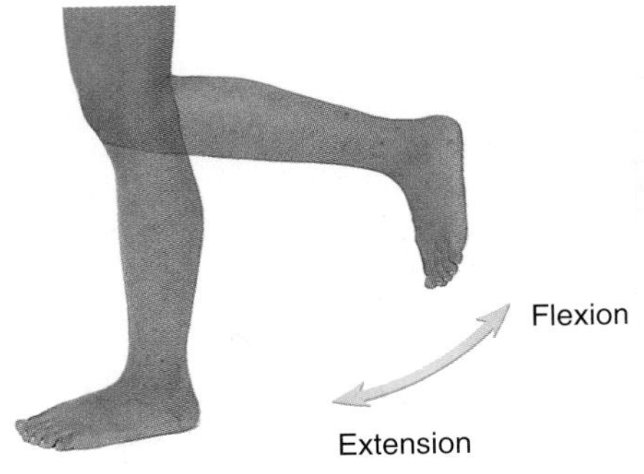

FIGURE 8.16 **Flexion and Extension of the Knee**

Abduction and Adduction

Abduction (to take away) is movement away from the median plane; **adduction** (to bring together) is movement toward the midline. Moving the upper limbs away from the body, such as in the outward portion of doing jumping jacks, is abduction, and bringing the upper limbs back toward the body is adduction (figure 8.18*a*). Abduction of the fingers involves spreading the fingers apart, away from the midline of the hand, and adduction is bringing them back together (figure 8.18*b*). Abduction of the thumb moves it anteriorly, away from the palm. Abduction of the wrist, which is sometimes called radial deviation, is movement of the hand away from the midline of the body, and adduction of the wrist, which is sometimes called ulnar deviation, results in movement of the hand toward the midline of the body. Abduction of the head is tilting the

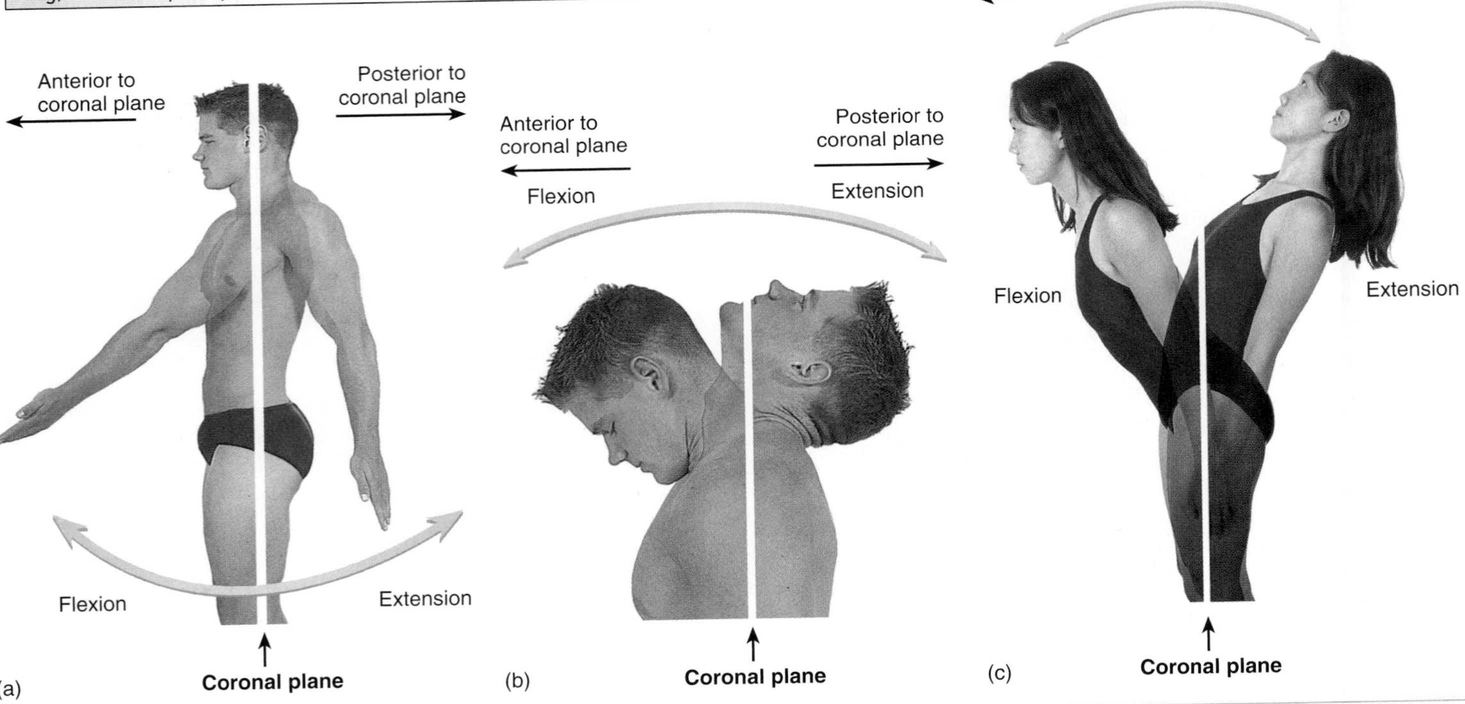

FIGURE 8.15 **Flexion and Extension Defined According to the Coronal Plane**

Flexion and extension of (*a*) the shoulder, (*b*) the neck, and (*c*) the trunk.

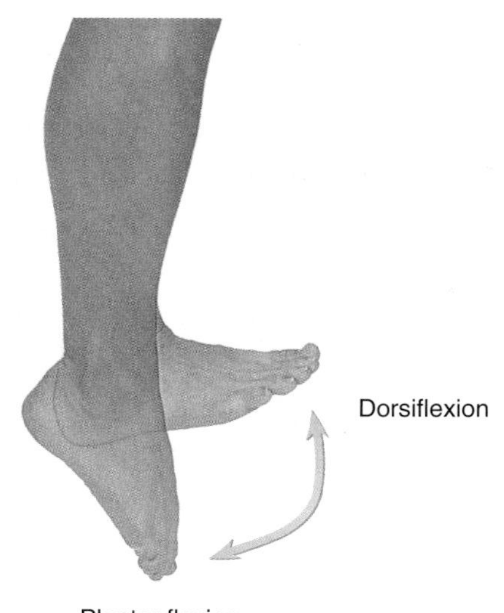

FIGURE 8.17 Dorsiflexion and Plantar Flexion of the Foot

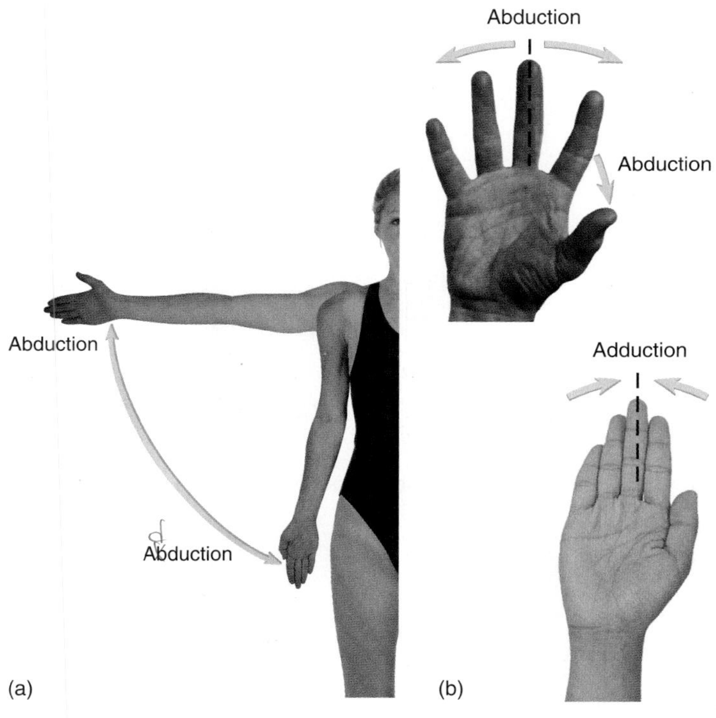

FIGURE 8.18 Abduction and Adduction
Abduction and adduction of (*a*) the upper limb and (*b*) the fingers.

head to one side or the other and is commonly called lateral flexion of the neck. Bending at the waist to one side is usually called **lateral flexion** of the vertebral column, rather than abduction.

7 ***Define flexion and extension. How are they different for the upper and lower limbs? What is hyperextension?***

8 ***Contrast abduction and adduction. Describe these movements for the head, upper limbs, wrist, fingers, lower limbs, and toes. For what part of the body is the term* lateral flexion *used?***

Circular Movements

Circular movements involve the rotation of a structure around an axis or movement of the structure in an arc.

Rotation

Rotation is the turning of a structure around its long axis, such as rotation of the head, the humerus, or the entire body (figure 8.19). Medial rotation of the humerus with the forearm flexed brings the hand toward the body. Rotation of the humerus so that the hand moves away from the body is lateral rotation.

Pronation and Supination

Pronation (prō-nā′shŭn) and **supination** (soo′pi-nā′shūn) refer to the unique rotation of the forearm (figure 8.20). The word *prone* means lying facedown; the word *supine* means lying faceup. Pronation is rotation of the forearm so that the palm faces posteriorly in relation to the anatomical position. The palm of the hand faces inferiorly if the elbow is flexed to 90 degrees. Supination is rotation of the forearm so that the palm faces anteriorly in relation to the anatomical position. The palm of the hand faces superiorly if the elbow is flexed to 90 degrees. In pronation, the radius and ulna cross; in supination, they are in a parallel position. The head of the radius rotates against the radial notch of the ulna during supination and pronation.

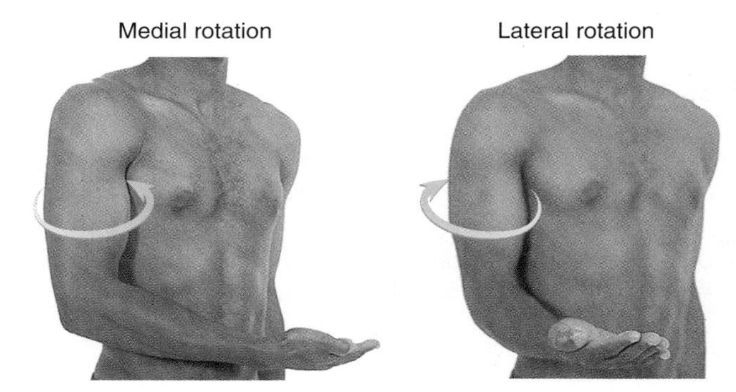

FIGURE 8.19 Medial and Lateral Rotation of the Arm

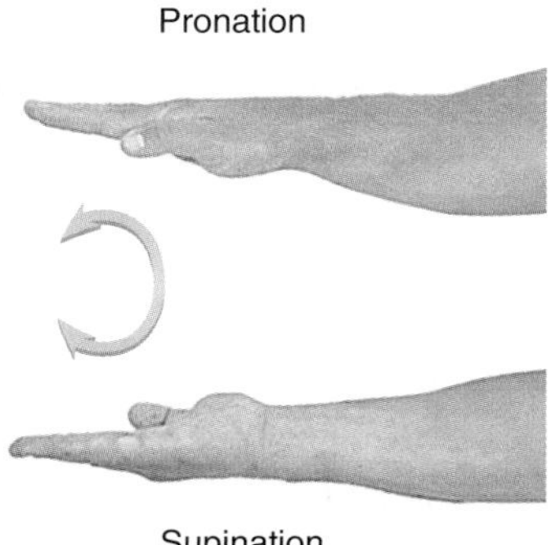

FIGURE 8.20 Pronation and Supination of the Hand

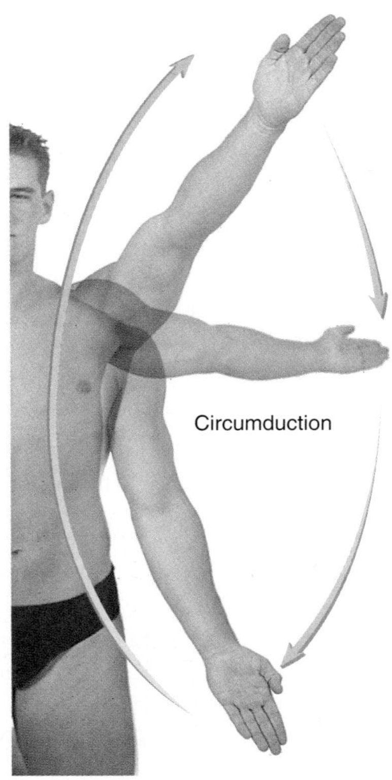

FIGURE 8.21 Circumduction

Circumduction

Circumduction is a combination of flexion, extension, abduction, and adduction (figure 8.21). It occurs at freely movable joints, such as the shoulder. In circumduction, the arm moves so that it describes a cone with the shoulder joint at the apex.

9 ***Distinguish among rotation, circumduction, pronation, and supination. Give an example of each.***

Special Movements

Special movements are those movements unique to only one or two joints; they do not fit neatly into one of the other categories.

Elevation and Depression

Elevation moves a structure superiorly; **depression** moves it inferiorly (figure 8.22). The scapulae and mandible are primary examples. Shrugging the shoulders is an example of scapular elevation. Depression of the mandible opens the mouth, and elevation closes it.

Protraction and Retraction

Protraction consists of moving a structure in a gliding motion in an anterior direction (figure 8.23). **Retraction** moves the structure back to the anatomical position or even more posteriorly. As with elevation and depression, the mandible and scapulae are primary examples. Pulling the scapulae back toward the vertebral column is retraction.

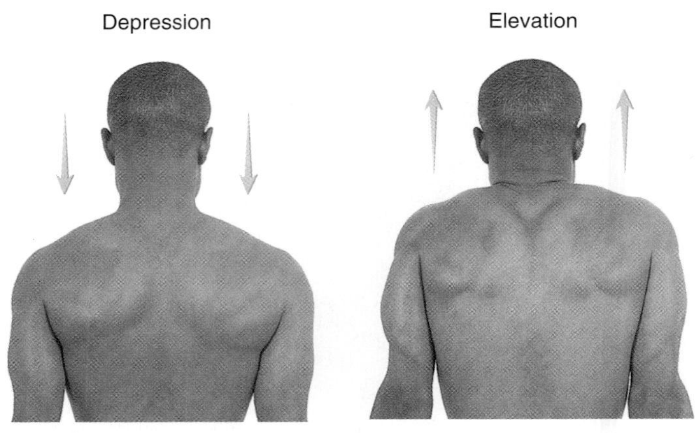

FIGURE 8.22 Elevation and Depression of the Scapula

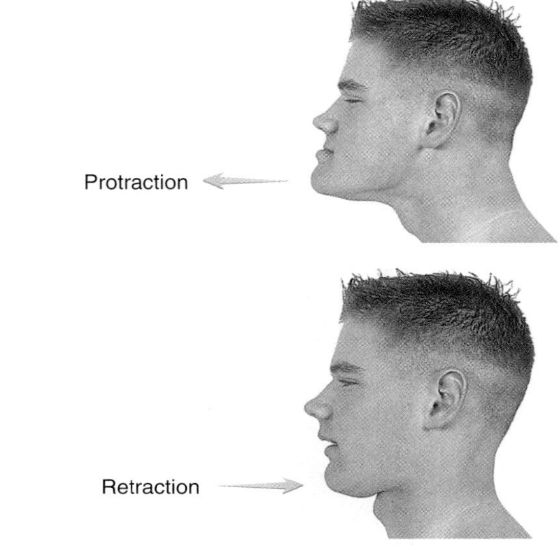

FIGURE 8.23 Protraction and Retraction of the Mandible

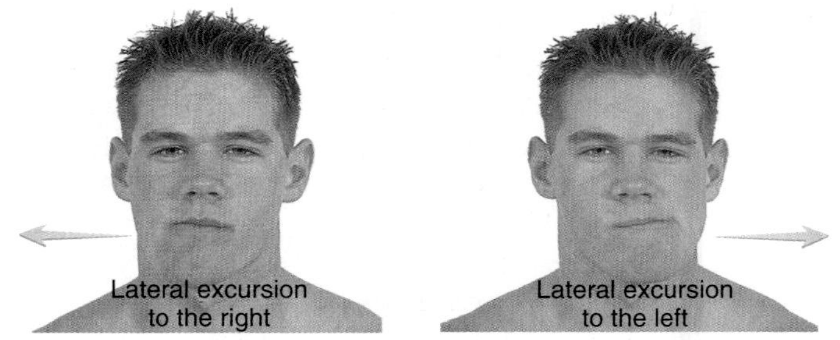

FIGURE 8.24 Excursion of the Mandible

Excursion

Lateral excursion is moving the mandible to either the right or left of the midline (figure 8.24), such as in grinding the teeth or chewing. **Medial excursion** returns the mandible to the neutral position.

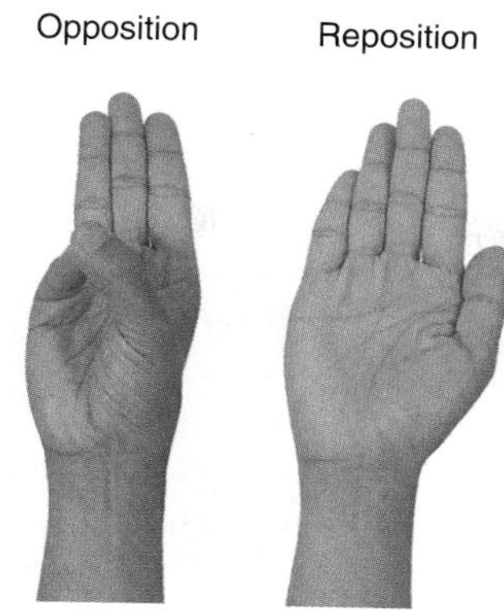

FIGURE 8.25 **Opposition and Reposition of the Thumb and Little Finger**

Opposition and Reposition

Opposition is a unique movement of the thumb and little finger (figure 8.25). It occurs when these two digits are brought toward each other across the palm of the hand. The thumb can also oppose the other digits, but those digits flex to touch the tip of the opposing thumb. **Reposition** is the movement returning the thumb and little finger to the neutral, anatomical position.

Inversion and Eversion

Inversion consists of turning the ankle so that the plantar surface of the foot faces medially, toward the opposite foot. **Eversion** is turning the ankle so that the plantar surface faces laterally (figure 8.26). Inversion of the foot is sometimes called supination, and eversion is called pronation.

10 ***Define the following jaw movements: protraction, retraction, lateral excursion, medial excursion, elevation, and depression.***

11 ***Define opposition and reposition.***

12 ***What terms are used for flexion and extension of the foot? For turning the side of the foot medially or laterally?***

Combination Movements

Most movements that occur in the course of normal activities are combinations of the movements named previously and are described by naming the individual movements involved in the combined movement. For example, when a person steps forward at a 45-degree angle to the side, the movement at the hip is a combination of flexion and abduction.

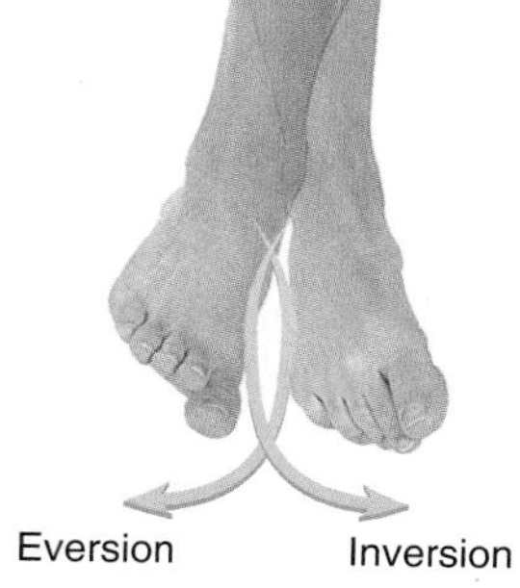

FIGURE 8.26 **Inversion and Eversion of the Foot**

PREDICT 4

What combination of movements is required at the shoulder and elbow joints for a person to move the right upper limb from the anatomical position to touch the right side of the head with the fingertips?

RANGE OF MOTION

Range of motion is an expression of the amount of mobility that can be demonstrated in a given joint. The **active range of motion** is the amount of movement that can be accomplished by contraction of the muscles that normally act across a joint. The **passive range of motion** is the amount of movement that can be accomplished at a joint when the structures that meet at the joint are moved by an outside force, such as when a therapist holds onto the forearm of a patient and moves it toward the patient's arm, flexing the elbow joint. The active and passive ranges of motion for normal joints are usually about equal.

The range of motion for a given joint is influenced by a number of factors, including the shape of the articular surfaces of the bones forming the joint, the amount and shape of cartilage covering those articular surfaces, the strength and location of ligaments and tendons surrounding the joint, the strength and location of the muscles associated with the joint, the amount of fluid in and around the joint, the amount of pain in and around the joint, and the amount of use or disuse the joint has received over time. Abnormalities in the range of motion can occur when any of those components changes. For example, damage to a ligament associated with a given joint may increase the range of motion of that joint. A torn piece of cartilage within a joint can limit its range of motion. If the nerve supply to a muscle is damaged so that the muscle is weakened, the active range of motion for the joint acted on by that muscle may decrease, but the passive range of motion for the joint should remain unchanged. Fluid buildup and/or pain in or around a joint (such as when the soft tissues around the joint develop edema following an injury) can severely limit both the active and passive ranges of motion for that joint. With disuse, both the active and passive ranges of motion for a given joint decrease.

13 ***Define range of motion. Contrast active range of motion with passive range of motion. What factors influence range of motion?***

DESCRIPTION OF SELECTED JOINTS

It is impossible in a limited space to describe all the joints of the body; therefore, only selected joints are described in this chapter, and they have been chosen because of their representative structure, important function, or clinical significance.

Temporomandibular Joint

The mandible articulates with the temporal bone to form the **temporomandibular joint (TMJ).** The mandibular condyle fits into the mandibular fossa of the temporal bone. A fibrocartilage

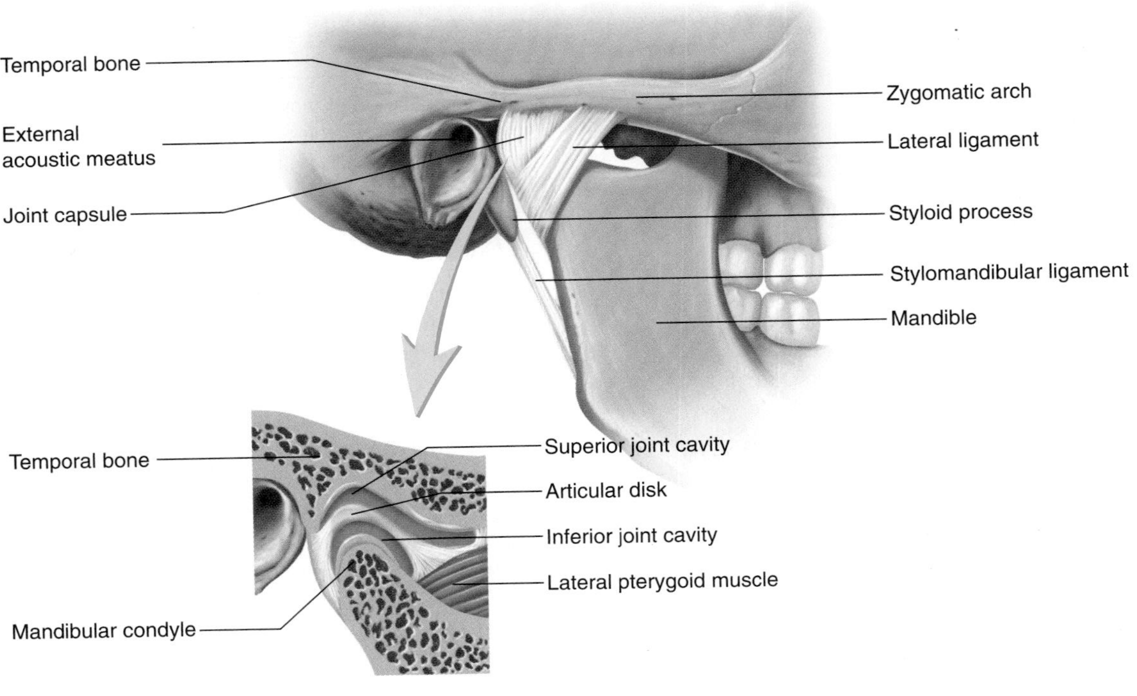

FIGURE 8.27 Right Temporomandibular Joint, Lateral View

articular disk is located between the mandible and the temporal bone, dividing the joint into superior and inferior joint cavities (figure 8.27). The joint is surrounded by a fibrous capsule, to which the articular disk is attached at its margin, and is strengthened by lateral and accessory ligaments.

The temporomandibular joint is a combination plane and ellipsoid joint, with the ellipsoid portion predominating. Depression of the mandible to open the mouth involves an anterior gliding motion of the mandibular condyle and articular disk relative to the temporal bone, which is about the same motion that occurs in protraction of the mandible; it is followed by a hinge motion between the articular disk and the mandibular head. The mandibular condyle is also capable of slight mediolateral movement, allowing excursion of the mandible.

Shoulder Joint

The **shoulder,** or **glenohumeral, joint** is a ball-and-socket joint (figure 8.28) in which stability is reduced and mobility is increased compared with the other ball-and-socket joint, the hip. Flexion, extension, abduction, adduction, rotation, and circumduction can all occur at the shoulder joint. The rounded head of the humerus articulates with the shallow glenoid cavity of the scapula. The rim of the glenoid cavity is built up slightly by a fibrocartilage ring, the **glenoid labrum,** to which the joint capsule is attached. A **subscapular bursa** (not shown in figure 8.28) opens into the joint cavity. A **subacromial bursa** is located near the joint cavity but is separated from the cavity by the joint capsule (see figure 8.28).

TMJ Disorders

TMJ disorders are conditions that cause most chronic orofacial pain. The conditions include joint noise; pain in the muscles, joint, or face; headache; and reduction in the range of joint movement. TMJ pain is often felt as referred pain in the ear. Patients may go to a physician complaining of an earache and are then referred to a dentist. As many as 65%–75% of people between ages 20 and 40 experience some of these symptoms. Symptoms appear to affect men and women about equally, but only about 10% of the symptoms are severe enough to cause people to seek medical attention. Women experience severe pain eight times more often than do men.

TMJ disorders are classified as those involving the joint, with or without pain; those involving only muscle pain; and those involving both the joint disorder and muscle pain. TMJ disorders are also classified as acute or chronic. Acute cases are usually self-limiting and have an identifiable cause. Chronic cases are not self-limiting, may be permanent, and often have no apparent cause. Chronic TMJ disorders are not easily treated, and chronic TMJ pain has much in common with other types of chronic pain. Whereas some people learn to live with the pain, others experience psychological problems, such as a sense of helplessness and hopelessness, high tension, and loss of sleep and appetite. Drug dependency may occur if strong drugs are used to control the pain, and relationships, lifestyle, vocation, and social interactions may be disrupted. Many of these problems may make the pain worse through positive feedback. Treatment includes teaching the patient to reduce jaw movements that aggravate the problem and to reduce stress and anxiety. Physical therapy may help relax the muscles and restore function. Analgesic and anti-inflammatory drugs may be used, and oral splints may be helpful, especially at night.

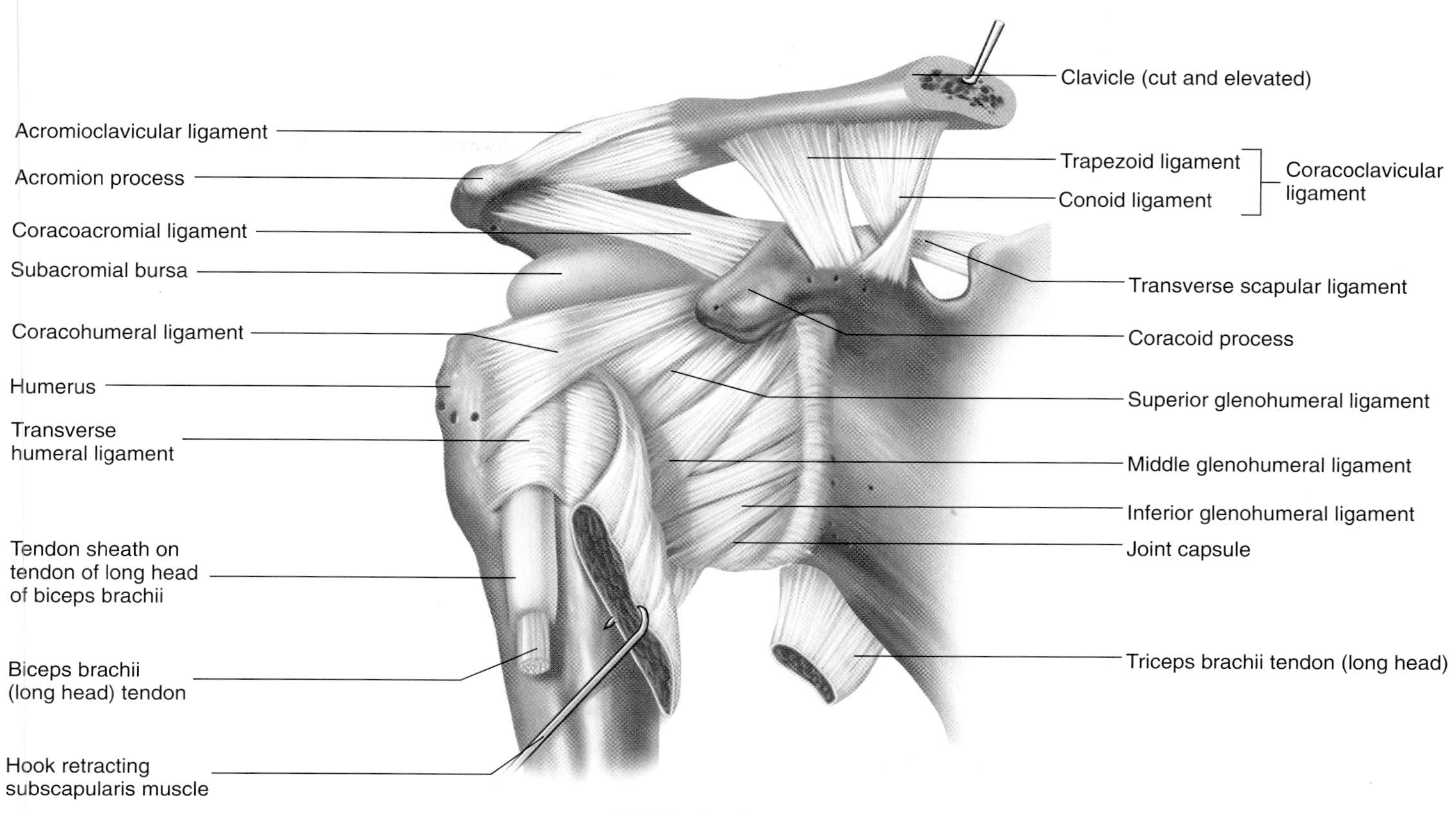

(a) **Anterior view**

Acromion process (articular surface)
Subacromial bursa
Joint cavity
Articular cartilage over head of humerus
Articular cartilage over glenoid cavity
Tendon sheath on tendon of long head of biceps brachii
Scapula (cut surface)
Glenoid labrum
Joint capsule
Biceps brachii (long head) tendon
Humerus
Biceps brachii (long head) muscle

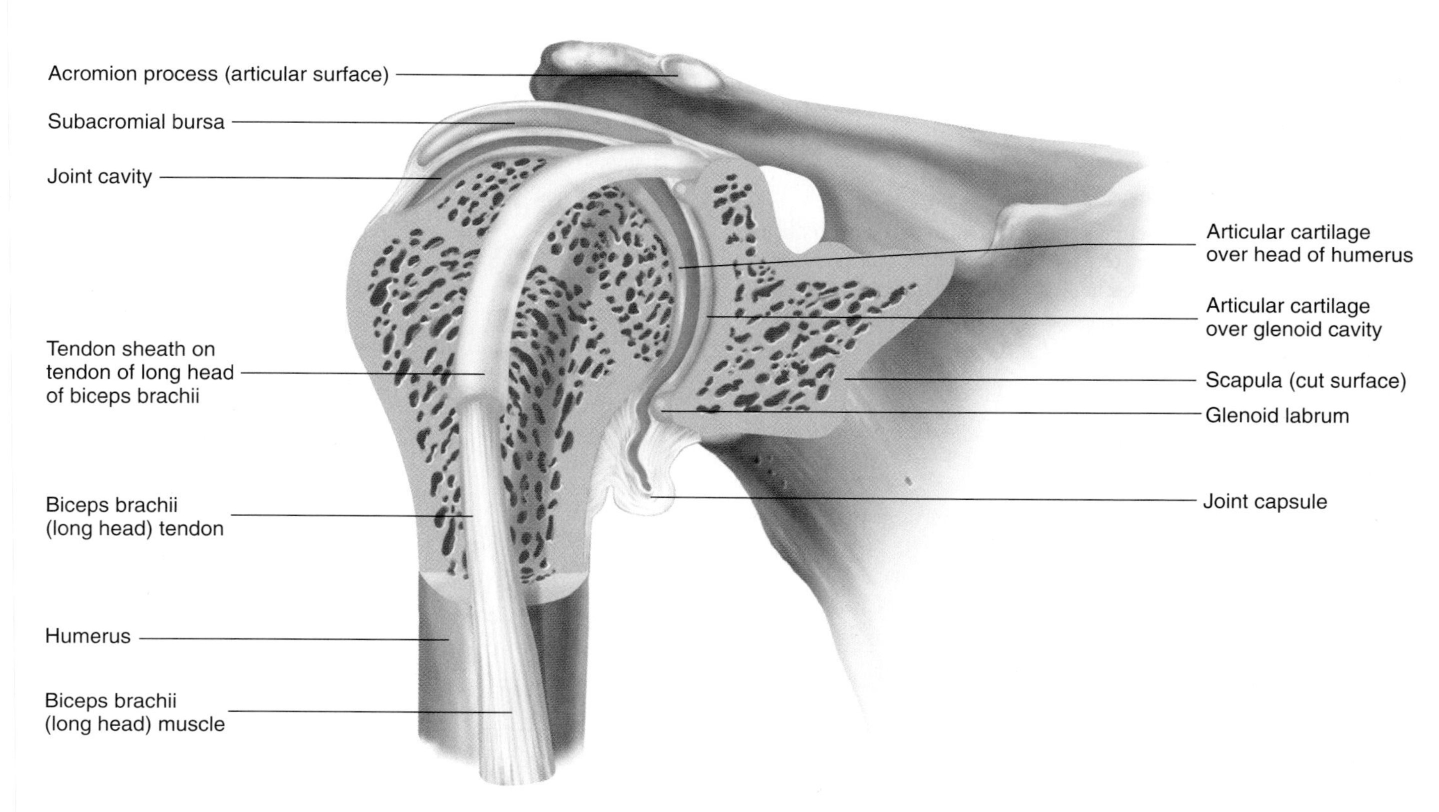

(b) **Frontal section**

FIGURE 8.28 Right Shoulder Joint

TABLE 8.3 Ligaments of the Shoulder Joint (See Figure 8.28)

Ligament	Description
Glenohumeral (superior, middle, and inferior)	Three slightly thickened longitudinal sets of fibers on the anterior side of the capsule; extend from the humerus to the margin of the glenoid cavity
Transverse humeral	Lateral, transverse fibrous thickening of the joint capsule; crosses between the greater and lesser tubercles and holds down the tendon from the long head of the biceps brachii muscle
Coracohumeral	Crosses from the root of the coracoid process to the humeral neck
Coracoacromial	Crosses above the joint between the coracoid process and the acromion process; an accessory, protective ligament

The stability of the joint is maintained primarily by three sets of ligaments and four muscles. The ligaments of the shoulder are listed in table 8.3. The four muscles, referred to collectively as the **rotator cuff,** hold the humeral head tightly within the glenoid cavity. These muscles are discussed in more detail in chapter 10. The head of the humerus is also supported against the glenoid cavity by the tendon from the biceps brachii muscle in the anterior part of the arm. This tendon is unusual in that it passes through the articular capsule of the shoulder joint before crossing the head of the humerus and attaching to the scapula at the supraglenoid tubercle (see figure 7.24*a*).

Shoulder Disorders

The most common traumatic shoulder disorders are **dislocation** and muscle or tendon **tears.** The shoulder is the most commonly dislocated joint in the body. The major ligaments cross the superior part of the shoulder joint, and no major ligaments or muscles are associated with the inferior side. As a result, dislocation of the humerus is most likely to occur inferiorly into the axilla. Because the axilla contains very important nerves and arteries, severe and permanent damage may occur when the humeral head dislocates inferiorly. The axillary nerve is the most commonly damaged (see chapter 12).

Chronic shoulder disorders include tendonitis (inflammation of tendons), bursitis (inflammation of bursae), and arthritis (inflammation of joints). Bursitis of the subacromial bursa can become very painful when the large shoulder muscle, called the deltoid muscle, compresses the bursa during shoulder movement.

PREDICT 5

Separation of the shoulder consists of stretching or tearing the ligaments of the acromioclavicular joint (acromioclavicular, or AC, separation). Using figure 8.28*a* and your knowledge of the articulated skeleton for assistance, explain the nature of a shoulder separation and predict the problems that may follow a separation.

Elbow Joint

The **elbow joint** (figure 8.29) is a compound hinge joint consisting of the **humeroulnar joint,** between the humerus and ulna, and the **humeroradial joint,** between the humerus and radius. The **proximal radioulnar joint,** between the proximal radius and

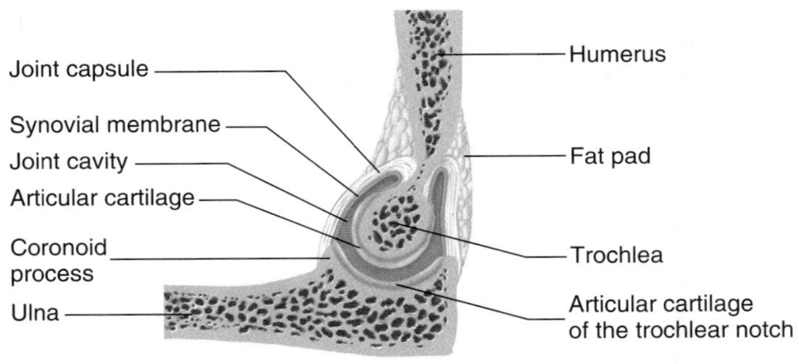

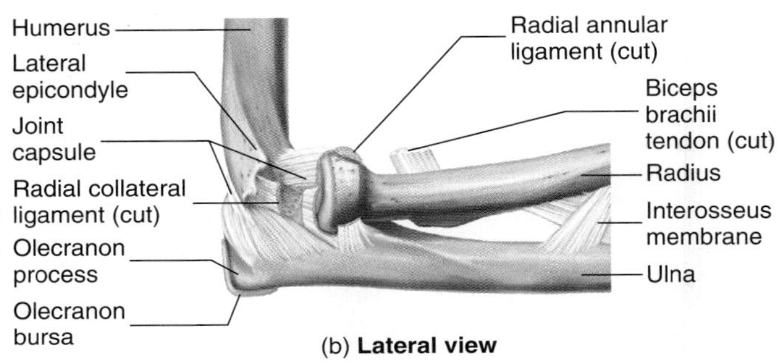

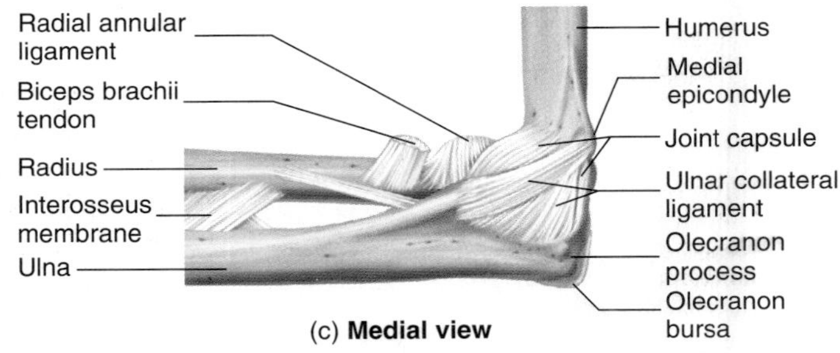

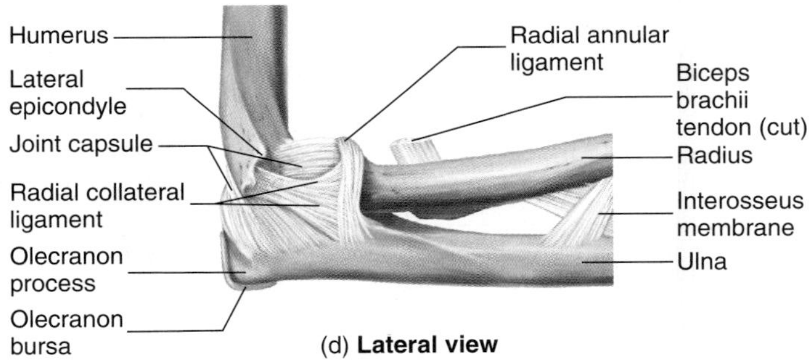

FIGURE 8.29 Right Elbow Joint

(*a*) Sagittal section showing the relation between the ulna and humerus. (*b*) Lateral view with ligaments cut to show the relation between the radial head, ulna, and humerus. (*c*) Medial view. (*d*) Lateral view.

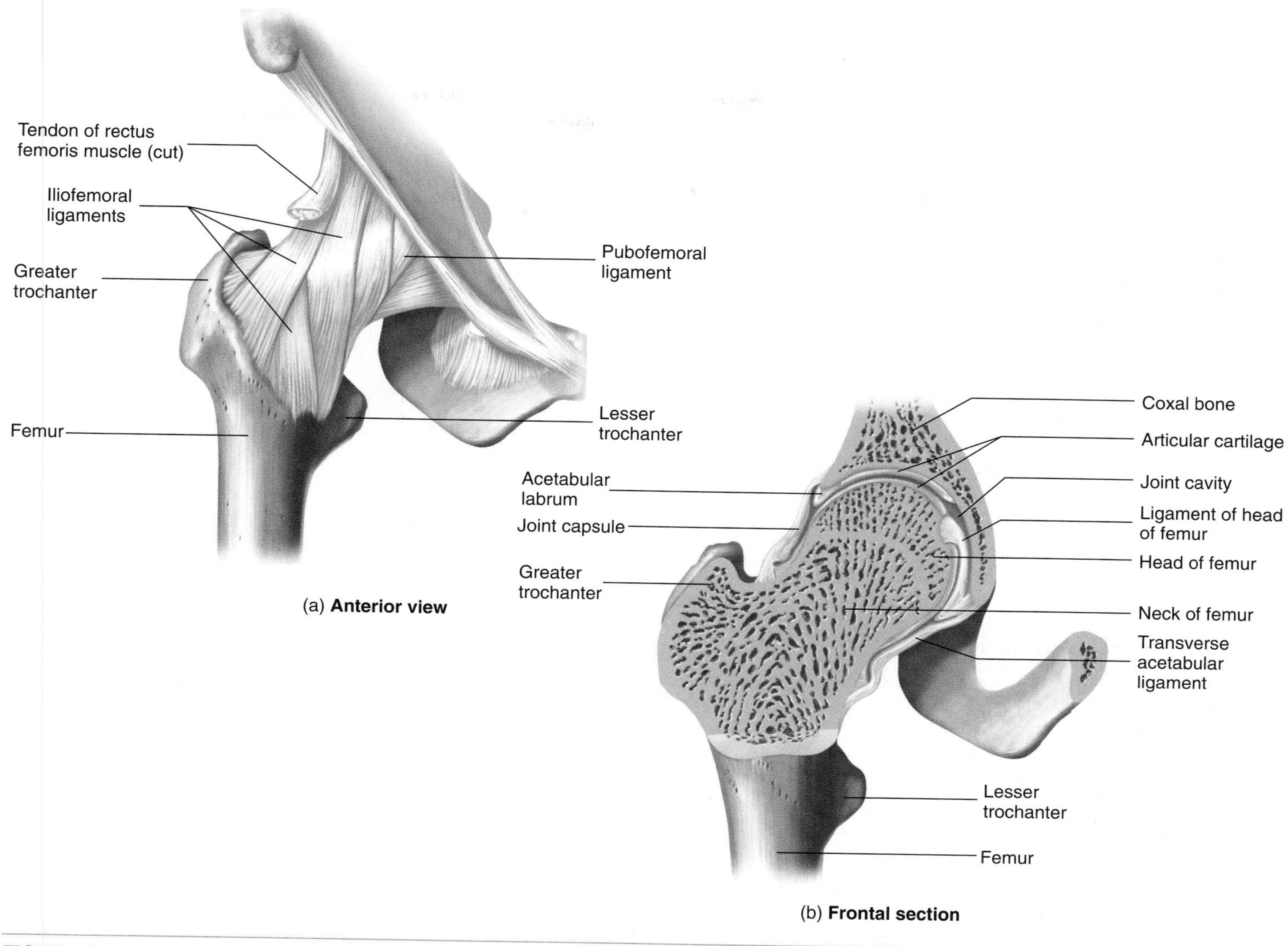

FIGURE 8.30 Right Coxal, or Hip, Joint

ulna, is also closely related. The shape of the trochlear notch and its association with the trochlea of the humerus (see figure 8.29*a*) limit movement at the elbow joint to flexion and extension. The rounded radial head, however, rotates in the radial notch of the ulna and against the capitulum of the humerus (see figure 8.29*b*), allowing pronation and supination of the hand.

Elbow Problems

Olecranon bursitis is an inflammation of the olecranon bursa. This inflammation can be caused by excessive rubbing of the elbow against a hard surface and is sometimes referred to as **student's elbow.** The radial head can become subluxated (partial joint separation) from the annular ligament of the radius. This condition is called **nursemaid's elbow.** If a child is lifted by one hand, the action may subluxate (partially dislocate) the radial head.

The elbow joint is surrounded by a joint capsule. The humeroulnar joint is reinforced by the **ulnar collateral ligament** (see figure 8.29*c*). The humeroradial and proximal radioulnar joints are reinforced by the **radial collateral ligament** and **radial annular ligament** (see figure 8.29*d*). A subcutaneous **olecranon bursa** covers the proximal and posterior surfaces of the olecranon process.

Hip Joint

The femoral head articulates with the relatively deep, concave acetabulum of the coxal bone to form the **coxal,** or **hip, joint** (figure 8.30). The head of the femur is more nearly a complete ball than the articulating surface of any other bone of the body. The acetabulum is deepened and strengthened by a lip of fibrocartilage called the **acetabular labrum,** which is incomplete inferiorly, and by a **transverse acetabular ligament,** which crosses the acetabular notch on the inferior edge of the acetabulum. The hip is capable of a wide range of movement, including flexion, extension, abduction, adduction, rotation, and circumduction.

Hip Dislocation

Dislocation of the hip may occur when the hip is flexed and the femur is driven posteriorly, such as when a person sitting in an automobile is involved in an accident. The head of the femur usually dislocates posterior to the acetabulum, tearing the acetabular labrum, the fibrous capsule, and the ligaments. Fracture of the femur and the coxal bone often accompanies hip dislocation.

An extremely strong joint capsule, reinforced by several ligaments, extends from the rim of the acetabulum to the neck of the femur (table 8.4). The **iliofemoral ligament** is especially strong. When standing, most people tend to thrust the hips anteriorly. This position is relaxing because the iliofemoral ligament supports much of the body's weight. The **ligament of the head of the femur** (round ligament of the femur) is located inside the hip joint between the femoral head and the acetabulum. This ligament does not contribute much toward strengthening the hip joint; however, it does carry a small nutrient artery to the head of the femur in about 80% of the population. The acetabular labrum, ligaments of the hip, and the surrounding muscles make the hip joint much more stable but less mobile than the shoulder joint.

TABLE 8.4 Ligaments of the Hip Joint (See Figure 8.30)

Ligament	Description
Transverse acetabular	Bridges gap in the inferior margin of the fibrocartilage acetabular labrum
Iliofemoral	Strong, thick band between the anterior inferior iliac spine and the intertrochanteric line of the femur
Pubofemoral	Extends from the pubic portion of the acetabular rim to the inferior portion of the femoral neck
Ischiofemoral	Bridges the ischial acetabular rim and the superior portion of the femoral neck; less well defined
Ligament of the head of the femur	Weak, flat band from the margin of the acetabular notch and the transverse ligament to a fovea in the center of the femoral head

Knee Joint

The **knee joint** traditionally is classified as a modified hinge joint located between the femur and the tibia (figure 8.31). Actually, it is a complex ellipsoid joint that allows flexion, extension, and a small amount of rotation of the leg. The distal end of the femur

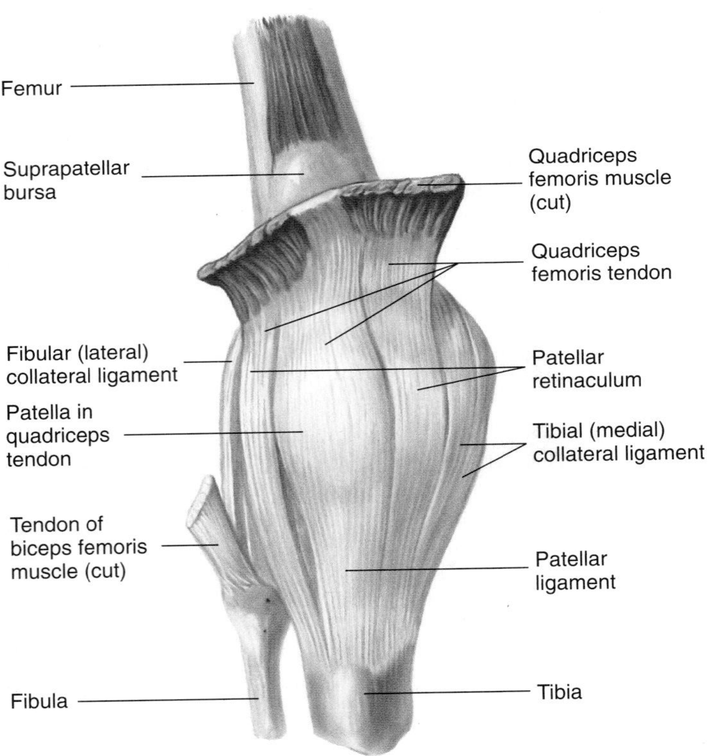

(a) **Anterior view**

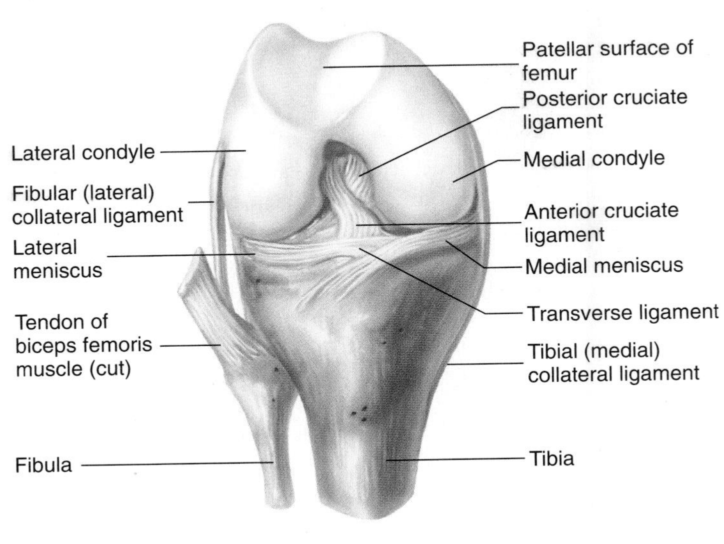

(b) **Anterior view**

FIGURE 8.31 **Right Knee Joint**
(*a*) Anterior superficial view. (*b*) Anterior deep view (knee flexed).

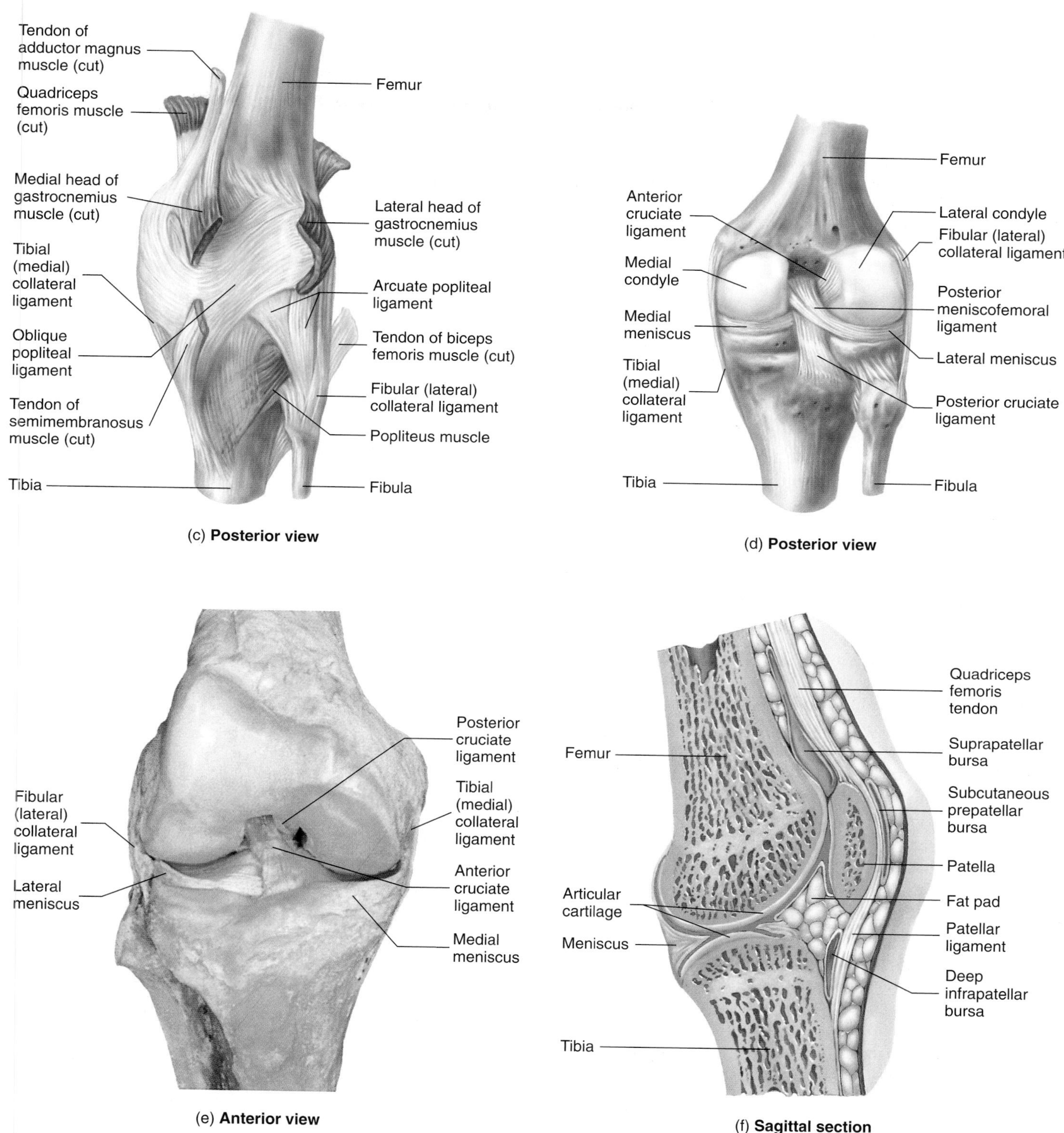

FIGURE 8.31 ***(continued)***
(*c*) Posterior superficial view. (*d*) Posterior deep view. (*e*) Photograph of anterior deep view. (*f*) Sagittal section.

TABLE 8.5 Ligaments of the Knee Joint (See Figure 8.31)

Ligament	Description
Patellar	Thick, heavy, fibrous band between the patella and the tibial tuberosity; actually part of the quadriceps femoris tendon
Patellar retinaculum	Thin band from the margins of the patella to the sides of the tibial condyles
Oblique popliteal	Thickening of the posterior capsule; extension of the semimembranous tendon
Arcuate popliteal	Extends from the posterior fibular head to the posterior fibrous capsule
Tibial collateral	Thickening of the lateral capsule from the medial epicondyle of the femur to the medial surface of the tibia; also called the medial collateral ligament
Fibular collateral	Round ligament extending from the lateral femoral epicondyle to the head of the fibula; also called the lateral collateral ligament
Anterior cruciate	Extends obliquely, superiorly, and posteriorly from the anterior intercondylar eminence of the tibia to the medial side of the lateral femoral condyle
Posterior cruciate	Extends superiorly and anteriorly from the posterior intercondylar eminence to the lateral side of the medial condyle
Coronary (medial and lateral)	Attaches the menisci to the tibial condyles (not illustrated)
Transverse	Connects the anterior portions of the medial and lateral menisci
Meniscofemoral (anterior and posterior)	Joins the posterior part of the lateral menisci to the medial condyle of the femur, passing anterior and posterior to the posterior cruciate ligament

has two large ellipsoid surfaces and a deep fossa between them. The femur articulates with the proximal end of the tibia, which is flattened and smooth laterally, with a crest called the intercondylar eminence in the center (see figure 7.36). The margins of the tibia are built up by thick fibrocartilage articular disks, called **menisci** (mĕ-nis′sī; crescent-shaped; see figure 8.31*b* and *d*), that deepen the articular surface. The fibula does not articulate with the femur but articulates only with the lateral side of the tibia.

Two **cruciate** (kroo′shē-āt; crossed) **ligaments** extend between the intercondylar eminence of the tibia and the fossa of the femur (see figure 8.31*b*, *d*, and *e*). The anterior cruciate ligament prevents anterior displacement of the tibia relative to the femur, and the posterior cruciate ligament prevents posterior displacement of the tibia. The joint is also strengthened by **collateral** and **popliteal ligaments** and by the tendons of the thigh muscles, which extend around the knee (table 8.5).

A number of bursae surround the knee (see figure 8.31*f*). The largest is the **suprapatellar bursa,** which is a superior extension of the joint capsule that allows for movement of the anterior thigh muscles over the distal end of the femur. Other knee bursae include the subcutaneous prepatellar bursa and the deep infrapatellar bursa, as well as the popliteal bursa, the gastrocnemius bursa, and the subcutaneous infrapatellar bursa (not shown in figure 8.31).

Ankle Joint and Arches of the Foot

The distal tibia and fibula form a highly modified hinge joint with the talus called the **ankle,** or **talocrural** (tā′lō-kroo′răl), **joint** (figure 8.32). The medial and lateral malleoli of the tibia and fibula, which form the medial and lateral margins of the ankle, are rather extensive, whereas the anterior and posterior margins are almost nonexistent. As a result, a hinge joint is created. A fibrous

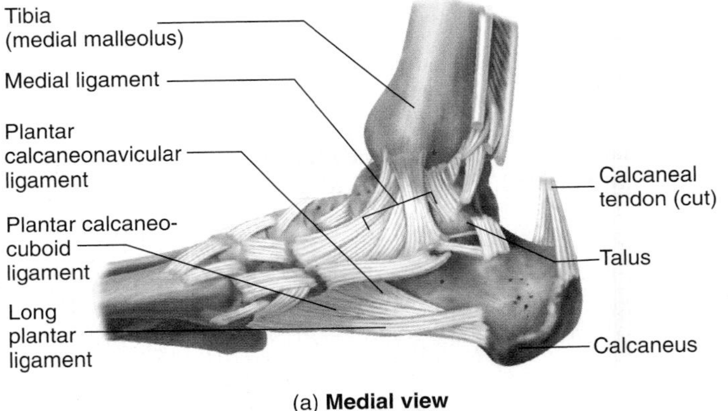

(a) **Medial view**

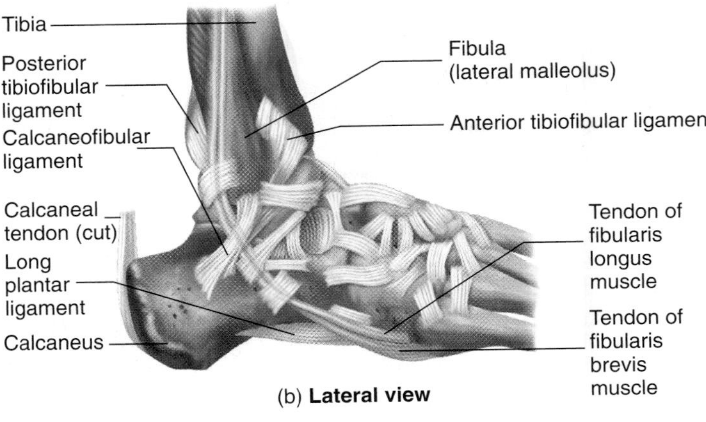

(b) **Lateral view**

FIGURE 8.32 Ligaments of the Right Ankle Joint

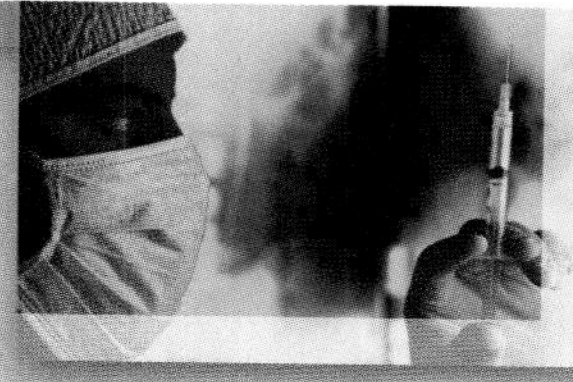

CLINICAL FOCUS

Knee Injuries and Disorders

Injuries to the medial side of the knee are much more common than injuries to the lateral side. The **fibular** (lateral) **collateral ligament** strengthens the joint laterally and is stronger than the tibial (medial) collateral ligament. Damage to the collateral ligaments occurs as a result of blows to the opposite side of the knee. Severe blows to the medial side of the knee, which would damage the fibular collateral ligament, are far less common than blows to the lateral side of the knee, which damage the **tibial collateral ligament.**

In addition, the **medial meniscus** is fairly tightly attached to the tibial collateral ligament and is damaged 20 times more often in a knee injury than the lateral meniscus, which is thinner and not attached to the fibular collateral ligament. A **torn meniscus** may result in a "clicking" sound during extension of the leg; if the damage is more severe, the torn piece of cartilage may move between the articulating surfaces of the tibia and femur, causing the knee to "lock" in a partially flexed position.

If the knee is driven anteriorly or if the knee is hyperextended, the **anterior cruciate ligament** may be torn, which causes the knee joint to be very unstable. If the knee is flexed and the tibia is driven posteriorly, the **posterior cruciate ligament** may be torn. Surgical replacement of a cruciate ligament with a transplanted or an artificial ligament repairs the damage.

A common type of football injury results from a block or tackle to the lateral side of the knee, which can cause the knee to bend inward, tearing the tibial (medial) collateral ligament and opening the medial side of the joint. The **medial meniscus,** which is strongly attached to this ligament, often is torn as well. In severe medial knee injuries, the anterior cruciate ligament is also damaged (figure A). Tearing of the tibial collateral ligament, medial meniscus, and anterior cruciate ligament is often referred to as "the unhappy triad of injuries."

Bursitis in the **subcutaneous prepatellar bursa** (see figure 8.31*f*), commonly called "housemaid's knee," may result from prolonged work performed while on the hands and knees. Another bursitis, "clergyman's knee," results from excessive kneeling and affects the **subcutaneous infrapatellar bursa.** This type of bursitis is common among carpet layers and roofers.

Other common knee problems are **chondromalacia** (kon′drō-mă-lā′shē-ă), or softening of the cartilage, which results from abnormal movement of the patella within the patellar groove, and **fat pad syndrome,** which consists of an accumulation of fluid in the fat pad posterior to the patella. An acutely swollen knee appearing immediately after an injury is usually a sign of blood accumulation within the joint cavity and is called a **hemarthrosis** (hē′mar-thrō′sis, hem′ar-thrō′sis). A slower accumulation of fluid, "water on the knee," may be caused by bursitis.

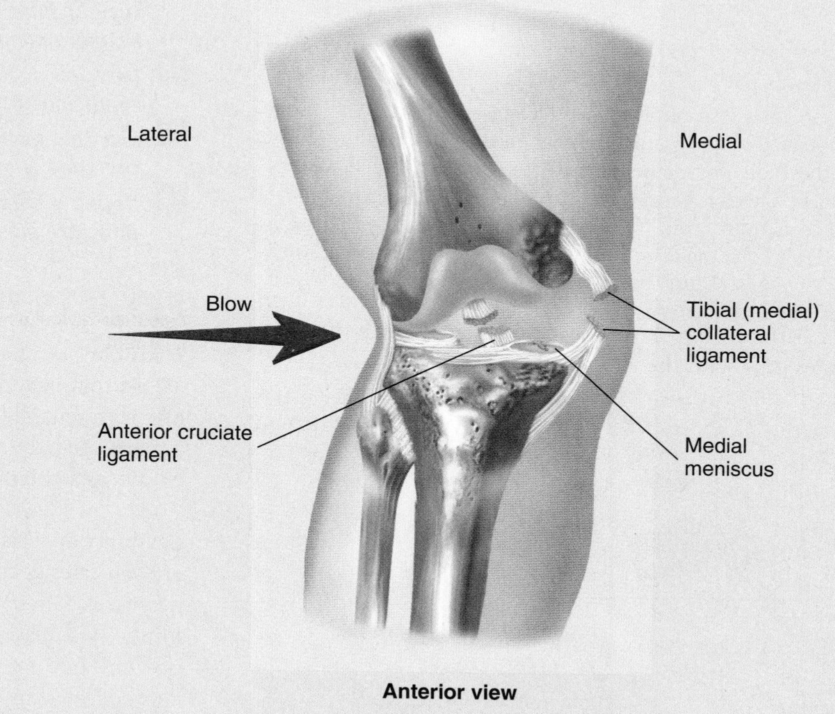

FIGURE A Injury to the Right Knee

capsule surrounds the joint, with the medial and lateral parts thickened to form ligaments. Other ligaments also help stabilize the joint (table 8.6). Dorsiflexion, plantar flexion, and limited inversion and eversion can occur at this joint.

The ligaments of the arches (see figure 7.39) serve two major functions: to hold the bones in their proper relationship as segments of the arch and to provide ties across the arch somewhat like a bowstring. As weight is transferred through the arch system, some of the ligaments are stretched, giving the foot more flexibility and allowing it to adjust to uneven surfaces. When weight is removed from the foot, the ligaments recoil and restore the arches to their unstressed shape.

TABLE 8.6 Ligaments of the Ankle and Arch (See Figure 8.32)

Ligament	Description
Medial	Thickening of the medial fibrous capsule that attaches the medial malleolus to the calcaneus, navicular, and talus; also called the deltoid ligament
Calcaneofibular	Extends from the lateral malleolus to the lateral surface of the calcaneus; separate from the capsule
Anterior talofibular	Extends from the lateral malleolus to the neck of the talus; fused with the joint capsule
Long plantar	Extends from the calcaneus to the cuboid and bases of metatarsals 2–5
Plantar calcaneocuboid	Extends from the calcaneus to the cuboid
Plantar calcaneonavicular (short plantar)	Extends from the calcaneus to the navicular

Ankle Injury

The ankle is the most frequently injured major joint in the body. The most common ankle injuries result from forceful inversion of the foot. A **sprained ankle** results when the ligaments of the ankle are torn partially or completely. The calcaneofibular ligament tears most often (figure B), followed in frequency by the anterior talofibular ligament. A fibular fracture can occur with severe inversion because the talus can slide against the lateral malleolus and break it (see chapter 7).

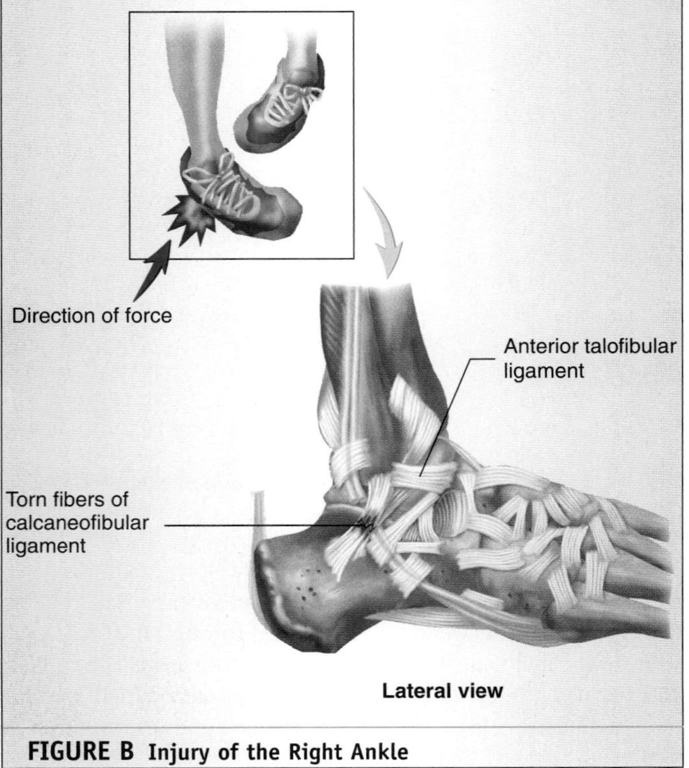

FIGURE B Injury of the Right Ankle

Arch Problems

The arches of the foot normally form early in fetal life. Failure to form results in congenital **flat feet,** or fallen arches, a condition in which the arches, primarily the medial longitudinal arch, are depressed or collapsed (see figure 7.39). This condition is not always painful. Flat feet may also occur when the muscles and ligaments supporting the arch fatigue and allow the arch, usually the medial longitudinal arch, to collapse. During prolonged standing, the plantar calcaneonavicular ligament may stretch, flattening the medial longitudinal arch. The transverse arch may also become flattened. The strained ligaments can become painful.

The plantar fascia is the deep connective tissue superficial to the ligaments in the central plantar surface of the foot and the thinner fascia on the medial and lateral sides of the plantar surface (see figure 8.32). **Plantar fasciitis,** which is an inflammation of the plantar fascia, can be a problem for distance runners as a result of continuous stretching.

14 ***For each of the following joints, name the bones of the joint, the specific part of the bones that form the joint, the type of joint, and the possible movement(s) at the joint: temporomandibular, shoulder, elbow, hip, knee, and ankle.***

15 ***Describe dislocations of the shoulder and hip. What conditions are most likely to cause each type?***

16 ***List the most common knee injuries, and tell which part of the knee is most often damaged in each type.***

17 ***Define a sprain, and describe which portions of the ankle joint are most commonly damaged when it is sprained.***

EFFECTS OF AGING ON THE JOINTS

A number of changes occur within many joints as a person ages. Those that occur in synovial joints have the greatest impact and often present major problems for elderly people. In general, as a person ages, the tissues of the body become less flexible and less elastic as protein cross-linking, especially in fibrous connective tissue, increases. The most important proteins related to tissue flexibility are elastin and collagen. Tissue repair slows as cell proliferation rates decline and the rate of new blood vessel development decreases. These general changes can significantly affect synovial joints. With use, the cartilage covering articular surfaces can wear down. When a person is young, production of new, resilient matrix compensates for the wear. As a person ages, the rate of replacement declines and the matrix becomes more rigid, thus adding to its rate of wear. The production rate of lubricating synovial fluid also declines with age, further contributing to the wear of the articular cartilage. In addition, the ligaments and tendons surrounding a joint shorten and become less flexible with age, resulting in a decrease in the range of motion of the joint. With age, muscles, which strengthen the joints, tend to weaken. Older people often experience a general decrease in activity, which causes the joints to become less flexible and their range of motion to decrease.

18 ***List the age-related factors that contribute to cartilage wear in synovial joints. List the age-related factors that cause a loss of flexibility and loss of range of motion in synovial joints.***

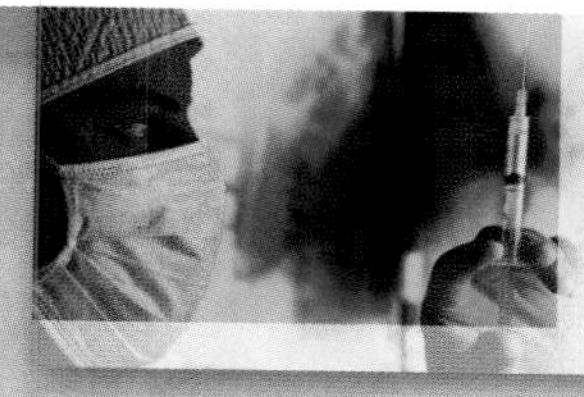

CLINICAL FOCUS

Joint Disorders

ARTHRITIS

Arthritis, an inflammation of any joint, is the most common and best known of the joint disorders, affecting 10% of the world's population and 14% of the U.S. population. There are over 37 million cases of arthritis in the United States alone. More than 100 types of arthritis exist. Classification is often based on the cause and progress of the arthritis. Its causes include infectious agents, metabolic disorders, trauma, and immune disorders. Mild exercise retards joint degeneration and enhances mobility. Swimming and walking are recommended for people with arthritis, but running, tennis, and aerobics are not recommended. Therapy depends on the type of arthritis but usually includes the use of anti-inflammatory drugs. Current research is focusing on the possible development of antibodies against the cells that initiate the inflammatory response in the joints or against cell surface markers on those cells.

Osteoarthritis (OA), or **degenerative arthritis,** is the most common type of arthritis, affecting 10% of people in the United States (85% of those over age 70). OA may begin as a molecular abnormality in articular cartilage, with heredity and normal wear-and-tear of the joint important contributing factors. Slowed metabolic rates with increased age also seem to contribute to OA. Inflammation is usually secondary in this disorder. It tends to occur in the weight-bearing joints, such as the knees, and is more common in overweight individuals. OA is becoming more common in younger people as a result of increasing rates of childhood obesity.

The first line of treatment for osteoarthritis is to change the lifestyle to reduce stress on affected joints. Synovial joints require movement to remain healthy. Long periods of inactivity may cause joints to stiffen. **Moderate exercise** helps reduce pain and increase flexibility. Exercising also helps people reduce excess weight, which can place stress on joints of the lower limbs. Older people should avoid high-impact sports, such as jogging, tennis, and racquetball, which place stress on the joints. Cycling and walking are recommended, but swimming is the best for people with osteoarthritis, as it exercises the muscles and joints without stressing the joints. Wearing shock-absorbing shoes can help. Splints or braces worn over an affected joint may be necessary to align the joint properly and distribute weight around it.

Applying heat, such as with hot soaks, warm paraffin, heating pads, low-power infrared light, or diathermy (mild electric currents that produce heat), directly over the joint may be helpful. Moving to a warmer climate, however, does not seem to make much difference.

The American Geriatrics Society has released guidelines for managing chronic pain in elderly patients with osteoarthritis. They recommend acetaminophen (Tylenol) or other **nonsteroidal anti-inflammatory drugs (NSAIDs),** such as aspirin and ibuprofen (Advil), for mild to moderate pain. Capsaicin, a component of hot red peppers, may help relieve pain when applied as a skin cream (Zostrix). Capsaicin seems to reduce levels of a chemical, substance P, that contributes both to joint inflammation and to the conduction of pain sensations to the brain. If pain becomes a major problem and over-the-counter pain relievers appear ineffective, physicians may inject corticosteroids directly into the affected joint.

Synvisc and Hyalgan are two drugs derived from **hyaluronic acid,** a natural substance that lubricates joints. They may be administered by injection into the joint when standard medication and exercise programs fail to relieve pain. **Glucosamine** and **chondroitin sulfate** are also natural substances associated with joints. If taken orally or by injection, they may help affected joints. However, glucosamine may also raise blood sugar levels, so people with diabetes should not use it without consulting their physician. Injections of genetically treated cells from synovial fluid, which are able to block the immune factors thought to cause the breakdown of joint cartilage, are currently under investigation. An immune system protein called **transforming growth factor beta (TGF-β),** introduced by gene therapy, is showing some promise in repairing cartilage damaged by osteoarthritis.

If other treatments fail, **surgical procedures** may be used to relieve pain and increase function in osteoarthritis patients. Using arthroscopy, a surgeon can examine the joint and clean out bone and cartilage fragments that stimulate pain and inflammation. In **osteotomy,** the bones of the joint are reshaped to better align the joint. In a procedure called **chondroplasty,** a small amount of healthy cartilage is removed and grown in the laboratory. The newly grown cartilage is then implanted into the joint, where it may stimulate the regeneration of damaged tissue. Research is currently underway to grow new cartilage from stem cells.

Joint replacement is discussed at the end of this "Clinical Focus." If the affected joint cannot be replaced, surgeons may perform a procedure called **arthrodesis,** in which the bones meeting at the joint are fused together. This procedure is intended to eliminate the pain, but the joint is eliminated and movement at that point becomes impossible.

Rheumatoid arthritis (RA) is the second most common type of arthritis. It affects about 3% of all women and about 1% of all men in the United States. It is a general connective tissue disorder that affects the skin, vessels, lungs, and other organs, but it is most pronounced in the joints. It is severely disabling and most commonly destroys small joints, such as those in the hands and feet (figure C). The initial cause is unknown but may involve a transient **infection** or an **autoimmune disease** (an immune reaction to one's own tissues; see chapter 22) that develops against collagen. A genetic predisposition may also exist. Whatever the cause, the ultimate course appears to be immunologic. People with classic RA have a protein, **rheumatoid factor,** in their blood. In RA, the synovial fluid and associated connective tissue cells proliferate, forming a **pannus** (clothlike layer), which causes the joint capsule to become thickened and which destroys the articular cartilage. In advanced stages, opposing joint surfaces can become fused. **Juvenile rheumatoid arthritis** is similar to the adult type in many ways, but no rheumatoid factor is found in the serum.

Hemophilic arthritis may result from bleeding into the joint cavity caused by hemophilia, a hereditary disease characterized by a deficient clotting mechanism in the blood. Some evidence exists that the iron in the blood is toxic to the chondrocytes, resulting in degeneration of the articular cartilage.

Continued

Continued

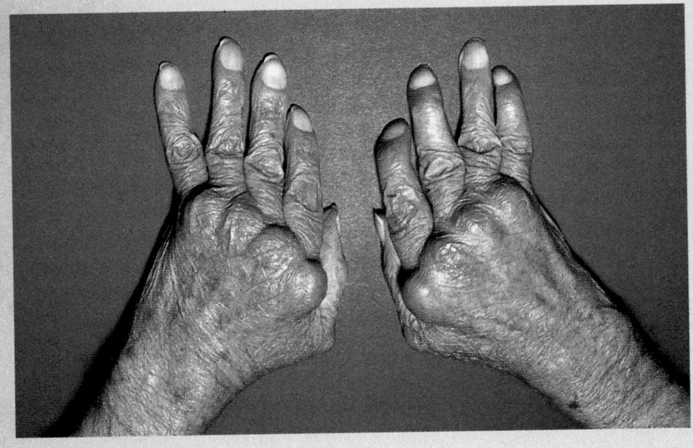

(a)

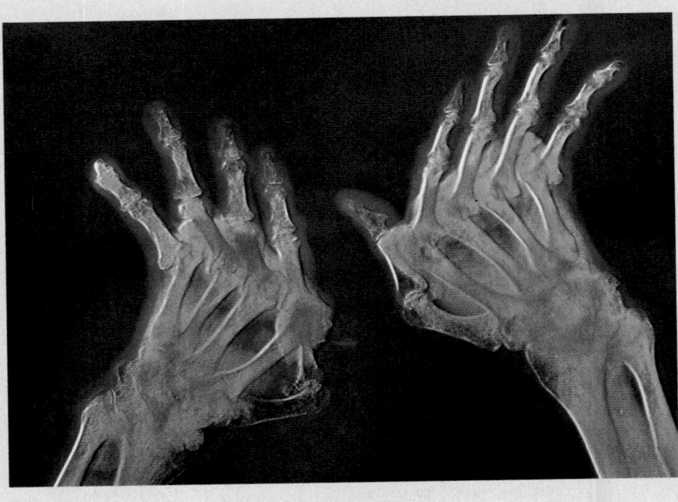

(b)

FIGURE C Rheumatoid Arthritis

(*a*) Photograph of hands with rheumatoid arthritis. (*b*) Radiographs of the same hands shown in (*a*).

JOINT INFECTIONS

Lyme disease is the result of a bacterial infection (*Borrelia burgdorferi*) transmitted to humans by a **tick** vector (usually *Ixodes sp.*) that affects the brain, nerves, eyes, heart, and joints. The chronic arthritis and central nervous system dysfunction that are symptoms of the disease are severely disabling but rarely fatal.

The disease is named for an epidemic of childhood arthritis occurring in Lyme, Connecticut, in 1975. Lyme disease has probably existed in Europe for many years and in North America since before the first European colonization, but it was not recognized as a distinct disease before 1975. Humans and domestic animals are only incidental hosts to the ticks, which normally infect wild mammals and birds. Deer are of particular concern as carriers of the ticks. The northeastern United States was greatly deforested during the eighteenth and nineteenth centuries, and deer and other wildlife populations declined dramatically. The more recent abandonment and reforestation of farms in New England has lead to an increase in the deer and tick populations, with a resurgence of the associated joint and nervous system disease.

Nearly 300,000 cases of Lyme disease have been reported in the United States since 1982 and over 20,000 new cases are reported each year. Although the disease is most common in the northeastern United States, it is relatively rare in Canada. In the United States, scattered cases have been reported in the north central states, along the West Coast, and scattered throughout the eastern, southeastern, and central states.

Early manifestations of the disease include flulike symptoms, with a localized skin rash. If untreated, the bacterium can spread to the nervous system, heart, and joints within a few weeks to months. A human vaccine against Lyme disease is currently being used for high-risk individuals.

Suppurative (pus-forming) **arthritis** may result from a number of infectious agents. These joint infections may be transferred from another infected site in the body or may be systemic (i.e., throughout the body). Usually, only one joint, normally, one of the larger joints, is affected, and the course of suppurative arthritis, if treated early, is transitory. With prolonged infection, however, the articular surfaces may degenerate. **Tuberculous arthritis** can occur as a secondary infection from pulmonary tuberculosis and is more damaging than typical suppurative arthritis. It usually affects the spine or large joints and causes ulceration of the articular cartilages and even erosion of the underlying bone. Transient arthritis of multiple joints is a common symptom of rheumatic fever, but permanent damage seldom occurs in joints with this disorder.

GOUT

Gout is a group of metabolic disorders involving joints. These disorders are largely idiopathic (of unknown cause), although some cases of gout seem to be familial (occur in families and therefore are probably genetic). Gout is more common in males than in females. The ultimate problem in gout patients is an increase in **uric acid** in the blood because of too much synthesis or decreased removal through the kidneys. The limited solubility of uric acid salts in the body results in the precipitation of **monosodium urate crystals** in various tissues, including the kidneys and joint capsules.

The earliest symptom of gout is transient arthritis resulting from urate crystal accumulation in a joint, causing irritation of the synovial membrane. This irritation can ultimately lead to an inflammatory response in the joints, and both the crystal deposition and the inflammation can become chronic. This condition is called **gouty arthritis.** Normally, only one or two joints are affected. The most commonly affected joints (85% of the cases) are the base of the **great toe** and other foot and leg joints to a lesser extent. Any joint may ultimately be involved, and damage to the kidneys from crystal formation occurs in almost all advanced cases. **Kidney failure** may occur in untreated cases. With modern medications, these complications seldom occur. Weight control and reduced alcohol consumption can help prevent gout.

Pseudogout is a disorder that causes pain and swelling similar to that seen in gout, but it is characterized by calcium hypophosphate crystal deposits in joints.

HALLUX VALGUS AND BUNION

In people who wear pointed shoes, the great toe can be deformed and displaced laterally, a condition called **hallux valgus** (hal′ŭks val′gŭs). Bunions are often associated with hallux valgus. A **bunion** is a bursitis that develops over the first metatarsophalangeal joint because of pressure and rubbing by shoes.

JOINT REPLACEMENT

As a result of recent advancements in biomedical technology, many joints of the body can now be replaced by artificial joints. Joint replacement, called **arthroplasty,** was first developed in the late 1950s. One of the major reasons for its use is to eliminate unbearable pain in patients near ages 55 to 60 with joint disorders. Osteoarthritis is the leading disease requiring joint replacement and accounts for two-thirds of the patients. Rheumatoid arthritis accounts for more than half of the remaining cases.

The major objectives in the design of joint prostheses (artificial replacements) include the development of stable articulations, low friction, solid fixation to the bone, and normal range of motion. Synthetic replacement materials are being designed by biomedical engineers to accomplish these objectives. Prosthetic joints usually are composed of metal, such as stainless steel, titanium alloys, or cobalt–chrome alloys, in combination with modern plastics, such as high-density polyethylene, silastic, or elastomer.

The bone of the articular area is removed on one side (a procedure called **hemireplacement**) or both sides (**total replacement**) of the joint, and the artificial articular areas are glued to the bone with a synthetic adhesive, such as methylmethacrylate. The smooth metal surface rubbing against the smooth plastic surface provides a low-friction contact, with a range of movement that depends on the design.

The success of joint replacement depends on the joint being replaced, the age and condition of the patient, the state of the technology, and the definition of success. Success is usually defined as minimizing pain while maintaining movement. Most reports are based on examinations of patients 2–10 years after joint replacement. The technology is improving constantly, so current reports do not adequately reflect the effect of the most recent improvements. Still, reports indicate a success rate of 80%–90% in hip replacements and 60% or more in ankle and elbow replacements. The major reason for the failure of prosthetic joints is loosening of the artificial joint from the bone to which it is attached. New prostheses with porous surfaces help overcome this problem.

SUMMARY

An articulation, or joint, is a place where two bones come together.

Naming Joints (p. 253)

Joints are named according to the bones or parts of bones involved.

Classes of Joints (p. 253)

Joints are classified according to function or type of connective tissue that binds them together and whether fluid is present between the bones.

Fibrous Joints

1. Fibrous joints are those in which bones are connected by fibrous tissue with no joint cavity. They are capable of little or no movement.
2. Sutures involve interdigitating bones held together by dense fibrous connective tissue. They occur between most skull bones.
3. Syndesmoses are joints consisting of fibrous ligaments.
4. Gomphoses are joints in which pegs fit into sockets and are held in place by periodontal ligaments (teeth in the jaws).
5. Some sutures and other joints can become ossified (synostosis).

Cartilaginous Joints

1. Synchondroses are immovable joints in which bones are joined by hyaline cartilage. Epiphyseal plates are examples.
2. Symphyses are slightly movable joints made of fibrocartilage.

Synovial Joints

1. Synovial joints are capable of considerable movement. They consist of the following.
   - Articular cartilage on the ends of bones provides a smooth surface for articulation. Articular disks can provide additional support.
   - A joint cavity is surrounded by a joint capsule of fibrous connective tissue, which holds the bones together while permitting flexibility, and a synovial membrane produces synovial fluid that lubricates the joint.
2. Bursae are extensions of synovial joints that protect skin, tendons, or bone from structures that could rub against them.
3. Synovial joints are classified according to the shape of the adjoining articular surfaces: plane (two flat surfaces), saddle (two saddle-shaped surfaces), hinge (concave and convex surfaces), pivot (cylindrical projection inside a ring), ball-and-socket (rounded surface into a socket), and ellipsoid (ellipsoid concave and convex surfaces).

Types of Movement (p. 259)

1. Gliding movements occur when two flat surfaces glide over one another.
2. Angular movements include flexion and extension, plantar and dorsiflexion, and abduction and adduction.
3. Circular movements include rotation, pronation and supination, and circumduction.
4. Special movements include elevation and depression, protraction and retraction, excursion, opposition and reposition, and inversion and eversion.
5. Combination movements involve two or more of the previously mentioned movements.

Range of Motion (p. 263)

Range of motion is the amount of movement, active or passive, that can occur at a joint.

Description of Selected Joints (p. 263)

1. The temporomandibular joint is a complex hinge and gliding joint between the temporal and mandibular bones. It is capable of elevation and depression, protraction and retraction, and lateral and medial excursion movements.
2. The shoulder joint is a ball-and-socket joint between the head of the humerus and the glenoid cavity of the scapula that permits a wide range of movement. It is strengthened by ligaments and the muscles of the rotator cuff. The tendon of the biceps brachii passes through the joint capsule. The shoulder joint is capable of flexion and extension, abduction and adduction, rotation, and circumduction.
3. The elbow joint is a compound hinge joint between the humerus, ulna, and radius. Movement at this joint is limited to flexion and extension.
4. The hip joint is a ball-and-socket joint between the head of the femur and the acetabulum of the coxal bone. It is greatly strengthened by ligaments and is capable of a wide range of movement, including flexion, extension, abduction, adduction, rotation, and circumduction.
5. The knee joint is a complex ellipsoid joint between the femur and the tibia that is supported by many ligaments. The joint allows flexion and extension and slight rotation of the leg.
6. The ankle joint is a special hinge joint of the tibia, fibula, and talus that allows dorsiflexion and plantar flexion and inversion and eversion.
7. Ligaments of the foot arches hold the bones in an arch and transfer weight in the foot.

Effects of Aging on the Joints (p. 272)

With age, the connective tissue of the joints becomes less flexible and less elastic. The resulting joint rigidity increases the rate of wear in the articulating surfaces. The change in connective tissue also reduces the range of motion.

REVIEW AND COMPREHENSION

1. Given these types of joints:
   1. gomphosis
   2. suture
   3. symphysis
   4. synchondrosis
   5. syndesmosis

   Choose the types that are held together by fibrous connective tissue.
   a. 1,2,3
   b. 1,2,5
   c. 2,3,5
   d. 3,4,5
   e. 1,2,3,4,5
2. Which of these joints is *not* matched with the correct joint type?
   a. parietal bone to occipital bone—suture
   b. between the coxal bones—symphysis
   c. humerus and scapula—synovial
   d. shafts of the radius and ulna—synchondrosis
   e. teeth in alveolar process—gomphosis
3. Which joint is the most movable?
   a. sutures
   b. syndesmoses
   c. symphyses
   d. synovial
   e. gomphoses
4. In which of these joints are periodontal ligaments found?
   a. sutures
   b. syndesmoses
   c. symphyses
   d. synovial
   e. gomphoses
5. The intervertebral disks are an example of
   a. sutures.
   b. syndesmoses.
   c. symphyses.
   d. synovial joints.
   e. gomphoses.
6. Joints containing hyaline cartilage are called __________, and joints containing fibrocartilage are called __________.
   a. sutures, synchondroses
   b. syndesmoses, symphyses
   c. symphyses, syndesmoses
   d. synchondroses, symphyses
   e. gomphoses, synchondroses
7. The inability to produce the fluid that keeps most joints moist would likely be caused by a disorder of the
   a. cruciate ligaments.
   b. synovial membrane.
   c. articular cartilage.
   d. bursae.
   e. tendon sheath.
8. Which of these is *not* associated with synovial joints?
   a. perichondrium on the surface of articular cartilage
   b. fibrous capsule
   c. synovial membrane
   d. synovial fluid
   e. bursae
9. Assume that a sharp object penetrated a synovial joint. From this list of structures:
   1. tendon or muscle
   2. ligament
   3. articular cartilage
   4. fibrous capsule (of joint capsule)
   5. skin
   6. synovial membrane (of joint capsule)

   Choose the order in which they would most likely be penetrated.
   a. 5,1,2,6,4,3
   b. 5,2,1,4,3,6
   c. 5,1,2,6,3,4
   d. 5,1,2,4,3,6
   e. 5,1,2,4,6,3
10. Which of these do hinge joints and saddle joints have in common?
    a. Both are synovial joints.
    b. Both have concave surfaces that articulate with a convex surface.
    c. Both are uniaxial joints.
    d. both a and b
    e. all of the above
11. Which of these joints is correctly matched with the type of joint?
    a. atlas to occipital condyle—pivot
    b. tarsals to metatarsals—saddle
    c. femur to coxal bone—ellipsoid
    d. tibia to talus—hinge
    e. scapula to humerus—plane
12. Once a doorknob is grasped, what movement of the forearm is necessary to unlatch the door—that is, turn the knob in a clockwise direction? (Assume using the right hand.)
    a. pronation
    b. rotation
    c. supination
    d. flexion
    e. extension
13. After the door is unlatched, what movement of the elbow is necessary to open it? (Assume the door opens in, and you are on the inside.)
    a. pronation
    b. rotation
    c. supination
    d. flexion
    e. extension
14. After the door is unlatched, what movement of the shoulder is necessary to open it? (Assume the door opens in, and you are on the inside.)
    a. pronation
    b. rotation
    c. supination
    d. flexion
    e. extension

15. When grasping a doorknob, the thumb and little finger undergo
    a. opposition. c. lateral excursion. e. dorsiflexion.
    b. reposition. d. medial excursion.
16. Spreading the fingers apart is
    a. rotation. c. abduction. e. flexion.
    b. depression. d. lateral excursion.
17. A runner notices that the lateral side of her right shoe is wearing much more than the lateral side of her left shoe. This could mean that her right foot undergoes more ____________ than her left foot.
    a. eversion c. plantar flexion e. lateral excursion
    b. inversion d. dorsiflexion
18. For a ballet dancer to stand on her toes, her feet must
    a. evert. c. plantar flex. e. abduct.
    b. invert. d. dorsiflex.
19. A meniscus is found in the
    a. shoulder joint. c. hip joint. e. ankle joint.
    b. elbow joint. d. knee joint.
20. A lip (labrum) of fibrocartilage deepens the joint cavity of the
    a. temporomandibular joint. d. knee joint.
    b. shoulder joint. e. ankle joint.
    c. elbow joint.
21. Which of these joints has a tendon inside the joint cavity?
    a. temporomandibular joint d. knee joint
    b. shoulder joint e. ankle joint
    c. elbow joint
22. Which of these structures helps stabilize the shoulder joint?
    a. rotator cuff muscles
    b. cruciate ligaments
    c. medial and lateral collateral ligaments
    d. articular disk
    e. all of the above
23. Bursitis of the subacromial bursa could result from
    a. flexing the wrist. d. running a long distance.
    b. kneeling. e. extending the elbow.
    c. overusing the shoulder joint.
24. Which of these does *not* occur with the aging of joints?
    a. decrease in production of new cartilage matrix
    b. decline in synovial fluid production
    c. stretching of ligaments and tendons and increase in range of motion
    d. weakening of muscles
    e. increase in protein cross-linking in tissues

Answers in Appendix E

CRITICAL THINKING

1. What would be the result if the sternal synchondroses and the sternocostal synchondrosis of the first rib were to become synostoses?
2. Using an articulated skeleton, examine the following list of joints. Describe the type of joint and the movement(s) possible.
   a. the joint between the zygomatic bone and the maxilla
   b. the ligamentous connection between the coccyx and the sacrum
   c. the elbow joint
3. For each of the following muscles, describe the motion(s) produced when the muscle contracts. It may be helpful to use an articulated skeleton.
   a. The biceps brachii muscle attaches to the coracoid process of the scapula (one head) and the radial tuberosity of the radius. Name two movements that the muscle accomplishes in the forearm.
   b. The rectus femoris muscle attaches to the anterior inferior iliac spine and the tibial tuberosity. How does contraction move the thigh? The leg?
   c. The supraspinatus muscle is located in and attached to the supraspinatus fossa of the scapula. Its tendon runs over the head of the humerus to the greater tubercle. When it contracts, what movement occurs at the glenohumeral (shoulder) joint?
   d. The gastrocnemius muscle attaches to the medial and lateral condyles of the femur and to the calcaneus. What movement of the leg results when this muscle contracts? Of the foot?
4. Crash McBang hurt his knee in an auto accident by ramming it into the dashboard. The doctor tested the knee for ligament damage by having Crash sit on the edge of a table with his knee flexed at a 90-degree angle. The doctor attempted to pull the tibia in an anterior direction (the anterior drawer test) and then tried to push the tibia in a posterior direction (the posterior drawer test). No unusual movement of the tibia occurred in the anterior drawer test but did occur during the posterior drawer test. Explain the purpose of each test, and tell Crash which ligament he has damaged.

Answers in Appendix F

Visit this book's website at www.mhhe.com/seeley8 for chapter quizzes, interactive learning exercises, and other study tools.

9 Muscular System

Histology and Physiology

Movements of the limbs, the heart, and other parts of the body are made possible by muscle cells that function like tiny motors. Muscle cells use energy extracted from nutrient molecules much as motors use energy provided by electric current. The nervous system regulates and coordinates muscle cells, so that smooth, coordinated movements are produced much as a computer regulates and coordinates several motors in robotic machines that perform assembly line functions.

This chapter presents the *functions of the muscular system* (p. 279), *general functional characteristics of muscle* (p. 279), and *skeletal muscle structure* (p. 279). The *sliding filament model* (p. 285) of muscle contraction is explained. The *physiology of skeletal muscle fibers* (p. 285), the *physiology of skeletal muscle* (p. 295), the *types of muscle contractions* (p. 299), *fatigue* (p. 301), *energy sources* (p. 303), *slow and fast fibers* (p. 305), and *heat production* (p. 307) are presented. The structure and function of *smooth muscle* (p. 307) and *cardiac muscle* (p. 311) are introduced, but cardiac muscle is discussed in greater detail in chapter 20. Finally, the *effects of aging on skeletal muscle* (p. 312) are presented. Because skeletal muscle is more abundant and more is known about it, skeletal muscle is examined in the greatest detail.

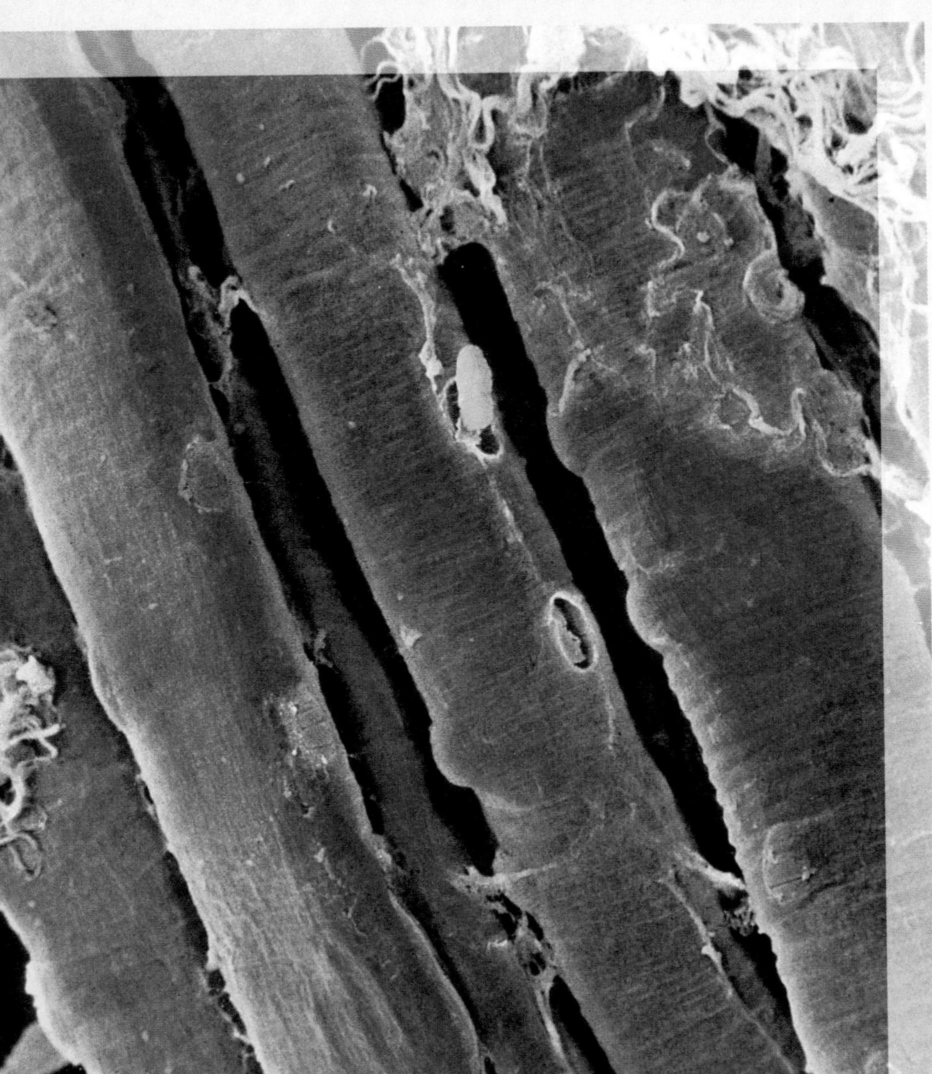

Color-enhanced scanning electron micrograph of skeletal muscle fibers.

Muscular System

FUNCTIONS OF THE MUSCULAR SYSTEM

Movements within the body are accomplished by cilia or flagella on the surface of some cells, by the force of gravity, or by the contraction of muscles. Most of the body's movements, from the beating of the heart to the running of a marathon, result from muscle contractions. As described in chapter 4, there are three types of muscle tissue: skeletal, smooth, and cardiac. The following are the major functions of muscles:

1. *Body movement.* Most skeletal muscles are attached to bones, are typically under conscious control, and are responsible for most body movements, including walking, running, chewing and manipulating objects with the hands.
2. *Maintenance of posture.* Skeletal muscles constantly maintain tone, which keeps us sitting or standing erect.
3. *Respiration.* Skeletal muscles of the thorax are responsible for the movements necessary for respiration.
4. *Production of body heat.* When skeletal muscles contract, heat is given off as a by-product. This released heat is critical to the maintenance of body temperature.
5. *Communication.* Skeletal muscles are involved in all aspects of communication, such as speaking, writing, typing, gesturing, and facial expression.
6. *Constriction of organs and vessels.* The contraction of smooth muscle within the walls of internal organs and vessels causes constriction of those structures. This constriction can help propel and mix food and water in the digestive tract, propel secretions from organs, and regulate blood flow through vessels.
7. *Heart beat.* The contraction of cardiac muscle causes the heart to beat, propelling blood to all parts of the body.

1 ***List the functions of skeletal, smooth, and cardiac muscles and explain how each is accomplished.***

GENERAL FUNCTIONAL CHARACTERISTICS OF MUSCLE

Muscle tissue is highly specialized to contract, or shorten, forcefully. The process of metabolism extracts energy from nutrient molecules. Part of that energy is used for muscle contraction, and the remainder is used for other cell processes or is released as heat.

Properties of Muscle

Muscle has four major functional properties: contractility, excitability, extensibility, and elasticity.

1. **Contractility** is the ability of muscle to shorten forcefully. When muscle contracts, it causes movement of the structures to which it is attached, or it may increase pressure inside hollow organs or vessels. Although muscle shortens forcefully during contraction, it lengthens passively; that is, gravity, contraction of an opposing muscle, or the pressure of fluid in a hollow organ or vessel produces a force that acts on the shortened muscle, causing it to lengthen.
2. **Excitability** is the capacity of muscle to respond to a stimulus. Normally, skeletal muscle contracts as a result of stimulation by nerves. Smooth muscle and cardiac muscle can contract without outside stimuli, but they also respond to stimulation by nerves and hormones.
3. **Extensibility** means that muscle can be stretched beyond its normal resting length and is still able to contract.
4. **Elasticity** is the ability of muscle to recoil to its original resting length after it has been stretched.

Types of Muscle Tissue

Table 9.1 provides a comparison of the major characteristics of skeletal, smooth, and cardiac muscle. Skeletal muscle with its associated connective tissue constitutes about 40% of the body's weight and is responsible for locomotion, facial expressions, posture, respiratory movements, and many other body movements. The nervous system voluntarily, or consciously, controls the functions of the skeletal muscles.

Smooth muscle is the most widely distributed type of muscle in the body, and it has the greatest variety of functions. It is in the walls of hollow organs and tubes, the interior of the eye, the walls of blood vessels, and other areas. Smooth muscle performs a variety of functions, including propelling urine through the urinary tract, mixing food in the stomach and intestine, dilating and constricting the pupils, and regulating the flow of blood through blood vessels.

Cardiac muscle is found only in the heart, and its contractions provide the major force for moving blood through the circulatory system. Unlike skeletal muscle, cardiac muscle and many smooth muscles are autorhythmic; that is, they contract spontaneously at somewhat regular intervals, and nervous or hormonal stimulation is not always required for them to contract. Furthermore, unlike skeletal muscle, smooth muscle and cardiac muscle are not consciously controlled by the nervous system. Rather, they are controlled involuntarily, or unconsciously, by the endocrine and autonomic nervous systems (see chapters 16 and 18).

2 ***Define contractility, excitability, extensibility, and elasticity of muscle tissue.***

3 ***Describe the structure, function, location, and control of the three major muscle tissue types.***

SKELETAL MUSCLE STRUCTURE

Skeletal muscles are composed of skeletal **muscle fibers** associated with smaller amounts of connective tissue, blood vessels, and nerves. Skeletal muscle cells are often called skeletal muscle fibers. Each skeletal muscle fiber is a single, long, cylindrical cell containing several nuclei located around the periphery of the fiber near the plasma membrane (figure 9.1). Muscle fibers develop from less mature multinucleated cells called **myoblasts**

TABLE 9.1 Comparison of Muscle Types

Features	Skeletal Muscle	Smooth Muscle	Cardiac Muscle
Location	Attached to bones	Walls of hollow organs, blood vessels, eyes, glands, and skin	Heart
Cell shape	Very long and cylindrical (1 mm–4 cm); extends the length of muscle fasciculi, which in some cases is the length of the muscle	Spindle-shaped (15–200 μm in length, 5–8 μm in diameter)	Cylindrical and branched (100–500 μm in length, 12–20 μm in diameter)
Nucleus	Multiple, peripherally located	Single, centrally located	Single, centrally located
Special cell–cell attachments	None	Gap junctions join some visceral smooth muscle cells together	Intercalated disks join cells to one another
Striations	Yes	No	Yes
Control	Voluntary and involuntary (reflexes)	Involuntary	Involuntary
Capable of spontaneous contraction	No	Yes (some smooth muscle)	Yes
Function	Body movement	Food movement through the digestive tract, emptying of the urinary bladder, regulation of blood vessel diameter, change in pupil size, contraction of many gland ducts, movement of hair, and many other functions	Pumps blood; contractions provide the major force for propelling blood through blood vessels

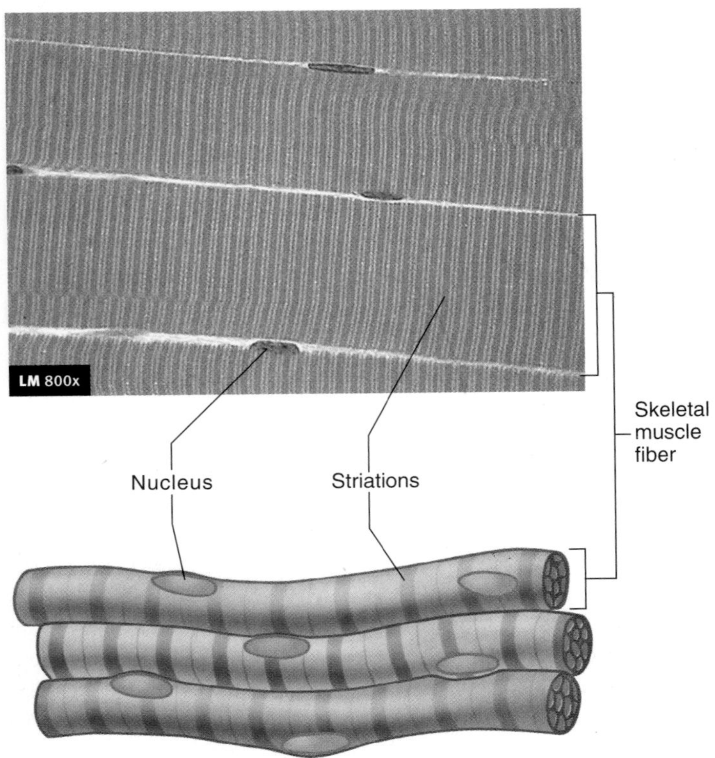

FIGURE 9.1 Skeletal Muscle Fibers
Skeletal muscle fibers in longitudinal section.

(mī′ō-blasts). Their multiple nuclei result from the fusion of myoblast precursor cells, not from the division of nuclei within myoblasts. Myoblasts are converted to muscle fibers as contractile proteins accumulate within their cytoplasm. Shortly after the myoblasts form, nerves grow into the area and innervate the developing muscle fibers.

The number of skeletal muscle fibers remains relatively constant after birth. Enlargement, or hypertrophy, of muscles after birth therefore results not from a substantial increase in the number of muscle fibers but from an increase in their size. Similarly, hypertrophy of muscles in response to exercise is due mainly to an increase in muscle fiber size, rather than an increase in number.

As seen in longitudinal section, alternating light and dark bands give the muscle fiber a **striated** (strī′āt-ed; banded), or striped, appearance (see figure 9.1). A single fiber can extend from one end of a muscle to the other. In most muscles, the fibers range from approximately 1 mm to about 4 cm in length and from 10 μm to 100 μm in diameter. Large muscles contain a large percentage of large-diameter fibers, whereas small, delicate muscles contain a large percentage of small-diameter fibers. However, any given muscle contains a mixture of small- and large-diameter fibers.

4 ***Define skeletal muscle fiber. Do the number of muscle fibers increase significantly after birth?***

Connective Tissue Coverings of Muscle

Connective tissue associated with skeletal muscle (figure 9.2) is critical to its proper function. Connective tissue provides muscle

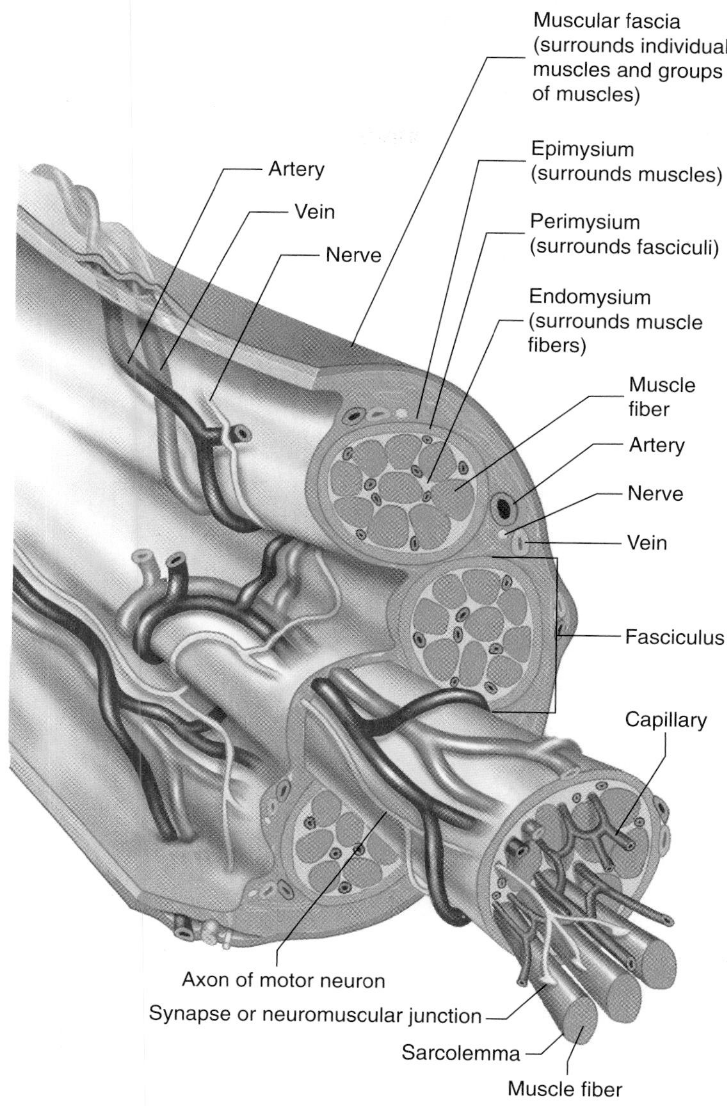

FIGURE 9.2 Skeletal Muscle Structure: Connective Tissue, Innervation, and Blood Supply

This figure shows the relationship between muscle fibers, fasciculi, and associated connective tissue layers: the epimysium, perimysium, and endomysium. Arteries, veins, and nerves course together through the connective tissue of muscles. They branch frequently as they approach individual muscle fibers. At the level of the perimysium, axons of neurons branch, and each branch extends to a muscle fiber.

fibers with a solid structure, to which they attach. The connective tissue fibers that surround a muscle and muscle fascicles extend beyond the belly of the muscle to become tendons, which connect muscles to bones or to the dermis of the skin.

The plasma membrane of a muscle fiber (cell) is called a **sarcolemma** (sar′kō-lem′ă). Two delicate connective tissue layers are located just outside the sarcolemma. The deeper and thinner of the two is the **external lamina.** It consists mostly of reticular (collagen) fibers and is so thin that it cannot be distinguished from the sarcolemma when viewed under a light microscope. The second layer also consists mostly of reticular fibers but is a much thicker layer, called the **endomysium** (en′dō-miz′ē-ŭm, en′dō-mis′ē-ŭm; G. *mys,* muscle). A bundle of muscle fibers with their endomysium is surrounded by another, heavier connective tissue layer called the **perimysium** (per′i-mis′ē-ŭm, per′i-miz′ē-ŭm). Each bundle ensheathed by perimysium is called a muscle **fasciculus** (fă-sik′u-lus). A muscle consists of many fasciculi grouped together and surrounded by a third layer of connective tissue, called the **epimysium** (ep-ĭ-mis′ē-ŭm). The epimysium is composed of dense collagenous connective tissue and covers the entire surface of the muscle. Fascia (fash′ē-ă) is a general term for connective tissue sheets within the body. **Muscular fascia** (formerly *deep fascia*), located superficial to the epimysium, separates and compartmentalizes individual muscles or groups of muscles. It consists of dense irregular collagenous connective tissue.

5 ***Name the connective tissue layers that surround muscle fibers, muscle fasciculi, and whole muscles. Define sarcolemma and muscular fascia.***

Nerves and Blood Vessels

The nerves and blood vessels that extend to skeletal muscles are abundant. **Motor neurons** are specialized nerve cells that stimulate muscles to contract. Their cell bodies are located in the brain or spinal cord, and their axons extend to skeletal muscle fibers through nerves. An artery and either one or two veins extend together with a nerve through the connective tissue layers of skeletal muscles (see figure 9.2). Numerous branches of the arteries supply the extensive capillary beds surrounding the muscle fibers, and blood is carried away from the capillary beds by branches of the veins. At the level of the perimysium, the axons of motor neurons branch repeatedly, each branch projecting toward the center of one muscle fiber. The contact between the axons and the muscle fibers, called synapses, or neuromuscular junctions, are described later in the chapter (see "Neuromuscular Junction," p. 288). Each motor neuron innervates more than one muscle fiber, and every muscle fiber receives a branch of an axon.

6 ***What are motor neurons? How do the axons of motor neurons and blood vessels extend to muscle fibers?***

Muscle Fibers

The many nuclei of each muscle fiber lie just inside the sarcolemma, whereas most of the interior of the fiber is filled with myofibrils (figure 9.3). Other organelles, such as the numerous mitochondria and glycogen granules, are packed between the myofibrils. The cytoplasm without the myofibrils is called **sarcoplasm** (sar′kō-plazm). Each **myofibril** (mī-ō-fī′bril) is a threadlike structure approximately 1–3 μm in diameter that extends from one end of the muscle fiber to the other. Two kinds of protein filaments, called **myofilaments** (mī-ō-fil′ă-ments), are major components of myofibrils. **Actin** (ak′tin) **myofilaments,** or thin myofilaments, are approximately 8 nanometers (nm) in diameter and 1000 nm in length, whereas **myosin** (mī′ō-sin) **myofilaments,** or thick myofilaments, are approximately 12 nm in diameter and

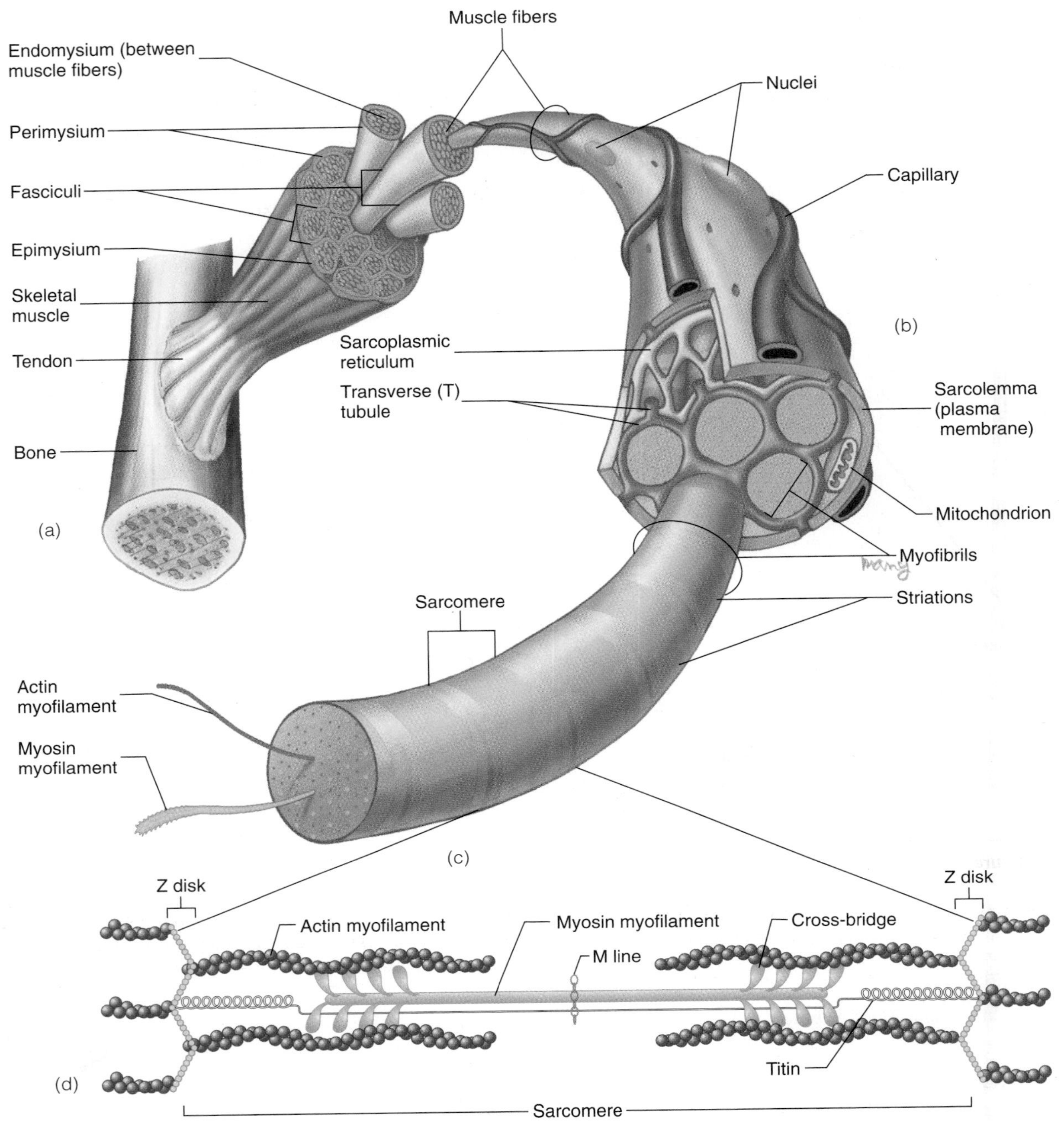

FIGURE 9.3 Parts of a Muscle
(*a*) Part of a muscle attached by a tendon to a bone. A muscle is composed of muscle fasciculi, each surrounded by perimysium. The fasciculi are composed of bundles of individual muscle fibers (muscle cells), each surrounded by endomysium. (*b*) Enlargement of one muscle fiber. The muscle fiber contains several myofibrils. (*c*) A myofibril extended out the end of the muscle fiber. The banding patterns of the sarcomeres are shown in the myofibril. (*d*) A single sarcomere of a myofibril is composed of actin myofilaments and myosin myofilaments. The Z disk anchors the actin myofilaments, and the myosin myofilaments are held in place by titin molecules and the M line.

1800 nm in length. The actin and myosin myofilaments form highly ordered units called **sarcomeres** (sar′kō-mērz), which are joined end to end to form the myofibrils (figure 9.4*a*).

Actin and Myosin Myofilaments

Each actin myofilament is composed of two strands of **fibrous actin (F actin),** a series of **tropomyosin** (trō-pō-mī′ō-sin) **molecules,** and a series of **troponin** (trō′pō-nin) **molecules** (figure 9.4*b* and *c*). The two strands of F actin are coiled to form a double helix, which extends the length of the actin myofilament. Each F actin strand is a polymer of approximately 200 small, globular units called **globular actin (G actin)** monomers. Each G actin monomer has an active site, to which myosin molecules can bind during muscle contraction. Tropomyosin is an elongated

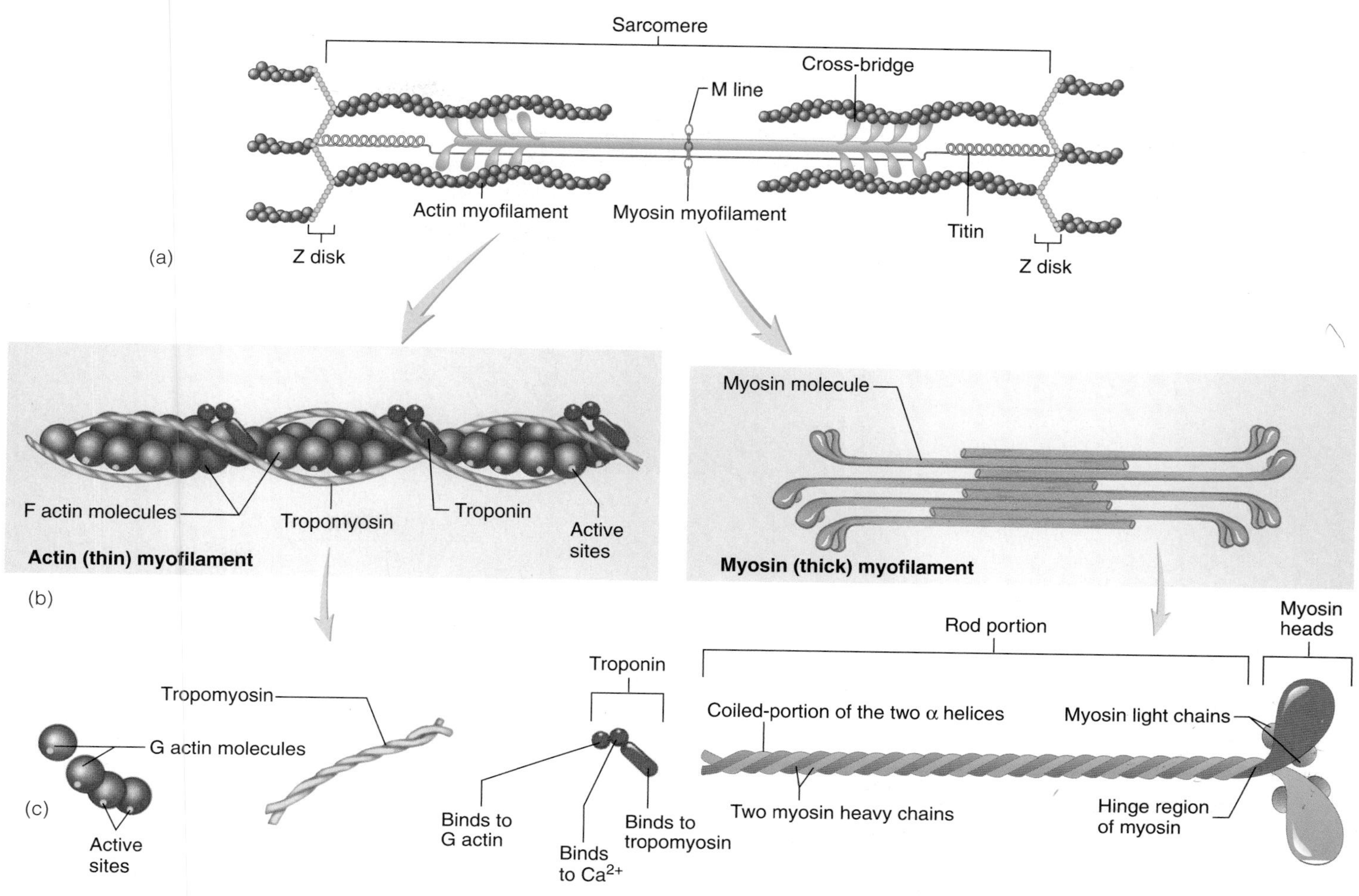

FIGURE 9.4 Structure of Actin and Myosin

(*a*) The sarcomere consists of actin (thin) myofilaments, attached to the Z disks, and myosin (thick) myofilaments, suspended between the actin myofilaments. (*b*) Actin myofilaments are made up of F actin (*chains of purple spheres*), tropomyosin, (*blue strands*), and troponin (*red spheres and rod*). Myosin myofilaments are made up of many golf-club-shaped myosin molecules, with all the heads pointing in one direction at one end and the opposite direction at the other end. (*c*) G actin molecules (*purple spheres*), with their active sites (*yellow*), tropomyosin, and troponin, make up actin myofilaments. Myosin molecules (*green*) are golf-club-shaped structures composed of two molecules of heavy myosin wound together to form the rod portion and double globular heads. Four small light myosin molecules are located on the heads of each of the myosin molecules.

protein that winds along the groove of the F actin double helix. Each tropomyosin molecule is sufficiently long to cover seven G actin active sites. Troponin is composed of three subunits: one that binds to actin, a second that binds to tropomyosin, and a third that has a binding site for Ca^{2+}. The troponin molecules are spaced between the ends of the tropomyosin molecules in the groove between the F actin strands. The complex of tropomyosin and troponin regulates the interaction between active sites on G actin and myosin.

Myosin myofilaments are composed of many elongated **myosin molecules** shaped like golf clubs (see figure 9.4*b* and *c*). Each myosin molecule consists of two **heavy myosin molecules** wound together to form a **rod portion** lying parallel to the myosin myofilament and two **heads** that extend laterally (see figure 9.4*b* and figure 9.3*d*). Four light myosin chains are attached to the heads of each myosin molecule. Each myosin myofilament consists of about 300 myosin molecules arranged so that about 150 of them have their heads projecting toward each end. The centers of the myosin myofilaments consist of only the rod portions of the myosin molecules. The myosin heads have three important properties: (1) The heads can bind to active sites on the actin molecules to form **cross-bridges;** (2) the heads are attached to the rod portion by a hinge region that can bend and straighten during contraction; and (3) the heads have ATPase activity, the enzymatic activity that breaks down adenosine triphosphate (ATP), releasing energy. Part of the energy is used to bend the hinge region of the myosin molecule during contraction.

Sarcomeres

Each sarcomere extends from one Z disk to an adjacent Z disk (see figure 9.4; figure 9.5). A **Z disk** is a filamentous network of protein forming a disklike structure for the attachment of actin

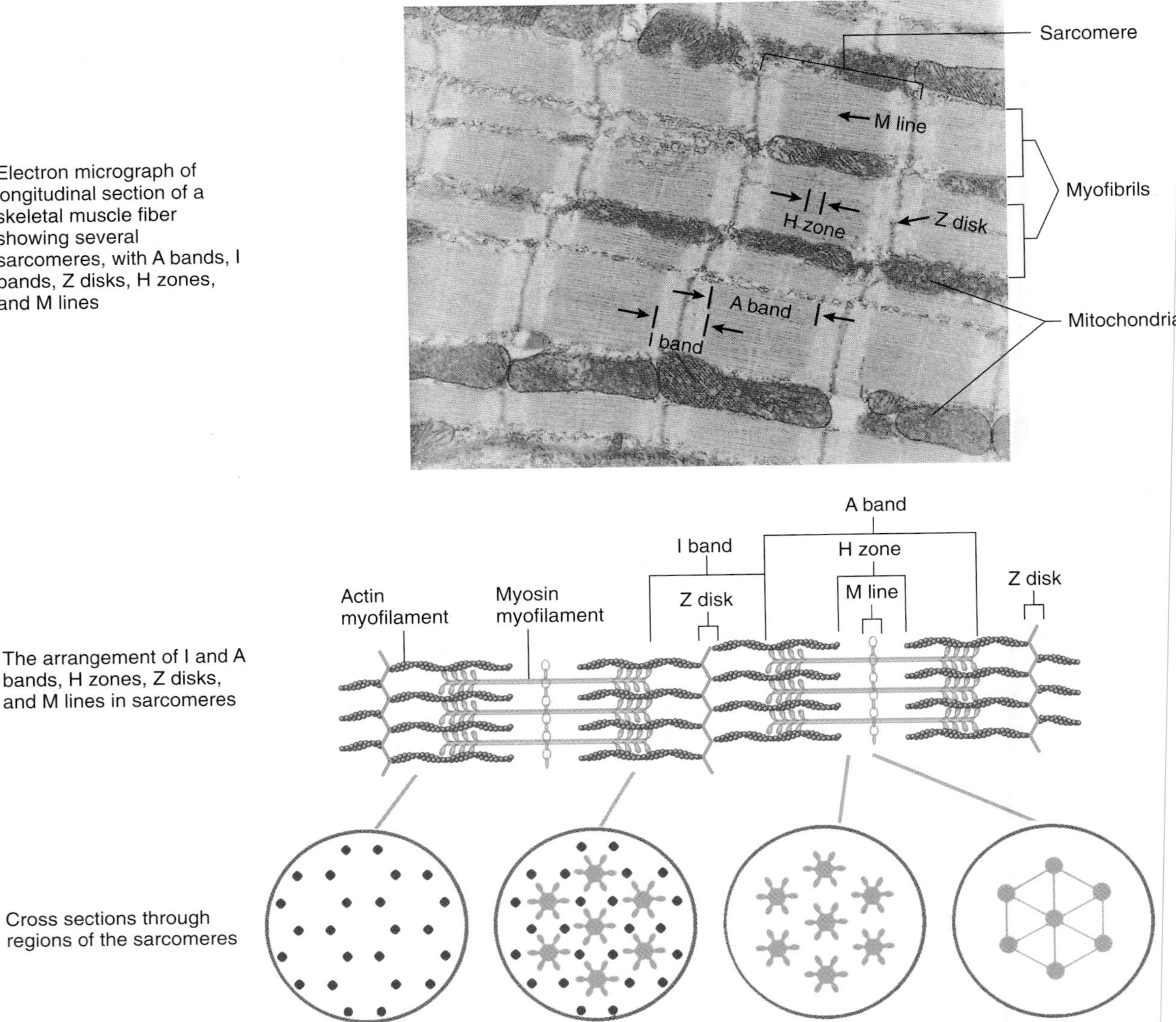

FIGURE 9.5 Components of Sarcomeres

(*a*) A sarcomere is the part of a myofibril between two adjacent Z disks. Actin myofilaments are attached to the Z disks. (*b*) The I band is between the ends of myosin myofilaments on each side of a Z disk. The A band is formed by the myosin myofilaments within a sarcomere. The H zone is between the ends of the actin myofilaments within a sarcomere. Myosin myofilaments are attached to the M line. (*c*) Cross sections through regions of the sarcomeres show the arrangement of proteins in three dimensions.

myofilaments (see figure 9.5). The arrangement of the actin myofilaments and myosin myofilaments gives the myofibril a banded, or striated, appearance when viewed longitudinally. Each **isotropic** (ī-sō-trop′ik), or **I, band** includes a Z disk and extends from each side of the Z disk to the ends of the myosin myofilaments. When seen in longitudinal and cross sections, the I band on each side of the Z disk consists only of actin myofilaments. Each **anisotropic** (an-ī-sō-trop′ik), or **A, band** extends the length of the myosin myofilaments within a sarcomere. The actin and myosin myofilaments overlap for some distance at both ends of the A band. In a cross section of the A band where actin and myosin myofilaments overlap, each myosin myofilament is visibly surrounded by six actin myofilaments. In the center of each A band is a smaller band called the **H zone,** where the actin and myosin myofilaments do not overlap and only myosin myofilaments are present. A dark line, called the **M line,** is in the middle of the H zone and consists of delicate filaments that attach to the center of the myosin myofilaments. The M line helps hold the myosin myofilaments in place, similar to the way the Z disk holds actin myofilaments in place (see figure 9.5*b* and *c*). The numerous myofibrils are oriented within each muscle fiber so that A bands and I bands of parallel myofibrils are aligned and thus produce the striated pattern seen through a microscope.

What Makes Muscles Extensible and Elastic?

In addition to actin and myosin, there are other, less visible proteins within sarcomeres. These proteins help hold actin and myosin in place, and one of them accounts for muscle's ability to stretch (extensibility) and recoil (elasticity). **Titin** (tī'tin; see figure 9.3) is one of the largest known proteins, consisting of a single chain of nearly 27,000 amino acids. It attaches to Z disks and extends along myosin myofilaments to the M line. The myosin myofilaments are attached to the titin molecules, which help hold them in position. Part of the titin molecule in the I band functions like a spring, allowing the sarcomere to stretch and recoil. Another large protein, called **nebulin** (neb'ū-lin), appears to hold the actin myofilaments in place. These proteins extend from each side of the Z disk and along the actin myofilaments. Each nebulin molecule is as long as an actin myofilament.

7 ***Define sarcoplasm, myofibril, and myofilament.***

8 ***How do G actin, tropomyosin, and troponin combine to form an actin myofilament? Name the molecule or ion to which each of the three subunits of troponin binds.***

9 ***Describe the structure of myosin molecules and how they combine to form a myosin myofilament.***

10 ***List three important properties of the myosin head. What is a cross-bridge?***

11 ***How are Z disks, actin myofilaments, myosin myofilaments, and M lines arranged to form a sarcomere? Describe how this arrangement produces the I band, A band, and H zone.***

SLIDING FILAMENT MODEL

The **sliding filament model** of muscle contraction includes all the events that result in actin myofilaments sliding over myosin myofilaments to shorten the sarcomeres of muscle fibers. Actin and myosin myofilaments do not change length during contraction of skeletal muscle. Instead, the actin and myosin myofilaments slide past one another, causing the sarcomeres to shorten (figure 9.6). The shortening of sarcomeres is responsible for the contraction of skeletal muscles. When sarcomeres shorten, the myofibrils also shorten because the myofibrils consist of sarcomeres joined end to end. The myofibrils extend the length of the muscle fibers and, when they shorten, the muscle fibers shorten. Muscle bundles are made up of muscle fibers and muscles are made up of muscle bundles. Therefore, when sarcomeres shorten, myofibrils, muscle fibers, muscle bundles, and muscles shorten to produce muscle contractions.

During muscle relaxation, the sarcomeres lengthen. For this to happen, an external force must be applied to a muscle to cause sarcomeres to lengthen, such as forces produced by other muscles or by gravity. For example, contraction of a muscle causes a joint, such as the elbow or knee, to flex. Extension of the joint and lengthening of the muscle result from the contraction of other, antagonistic muscles (see chapter 10).

12 ***Why do the I bands and H zones shorten during muscle contraction but the length of the A band is unchanged?***

13 ***How does shortening of sarcomeres explain muscle contraction?***

PREDICT 1

Explain the events that influence the width of each band of a sarcomere when a muscle goes through the sequence of being stretched, contracted, and relaxed.

PHYSIOLOGY OF SKELETAL MUSCLE FIBERS

Axons of nerve cells extend from the brain and spinal cord to skeletal muscle fibers. The nervous system controls the contraction of skeletal muscles through these axons. Electrical signals, called **action potentials,** travel from the brain or spinal cord along the axons to muscle fibers and cause them to contract.

Membrane Potentials

Plasma membranes are **polarized**—that is, there is a voltage difference, or electric charge difference, across each plasma membrane. This charge difference across the plasma membrane of an unstimulated cell is called the **resting membrane potential** (figure 9.7). Action potentials cannot be produced without the resting membrane potential. The negative charge at the internal surface of the plasma membrane compared with the outer surface results mainly from the concentration differences of ions and charged molecules across the plasma membrane and to its permeability characteristics.

There is a difference in the ion concentrations across the plasma membrane. The Na^+-K^+ pump moves Na^+ to the outside of the cell and K^+ to the inside of the cell. This results in a higher concentration of Na^+ in the extracellular fluid and a lower concentration of Na^+ in the intracellular fluid, and it results in a higher concentration of K^+ in the intracellular fluid and a lower concentration of K^+ in the extracellular fluid.

The plasma membrane is relatively permeable to K^+ and much less permeable to Na^+ and to negatively charged molecules found inside of the cell. Consequently, positively charged K^+ tend to diffuse out of the cell through K^+ leak channels, leaving the negatively charged molecules behind, thus polarizing the membrane. The membrane is at equilibrium when the tendency for K^+ to diffuse out of the cell is resisted by the negative charge of the molecules inside the cell. The details of the resting membrane potential are more fully described in chapter 11.

The resting membrane potential can be measured in units called **millivolts** (mV; mV = 1/1000 volt). The potential differences across the plasma membranes of nerve cells and muscle fibers are between −70 and −90 mV. The potential difference is reported as a negative number because the inner surface of the plasma membrane is negative, compared with the outside.

Ion Channels

The increased permeability of the plasma membrane to K^+ is due to K^+ leak channels. These channels are specific for K^+ and they are open continuously. Once the resting membrane potential is established, action potentials can be produced. An action potential is a reversal of the resting membrane potential such

1. Actin and myosin myofilaments in a relaxed muscle (*right*) and a contracted muscle (*#4 below*) are the same length. Myofilaments do not change length during muscle contraction.

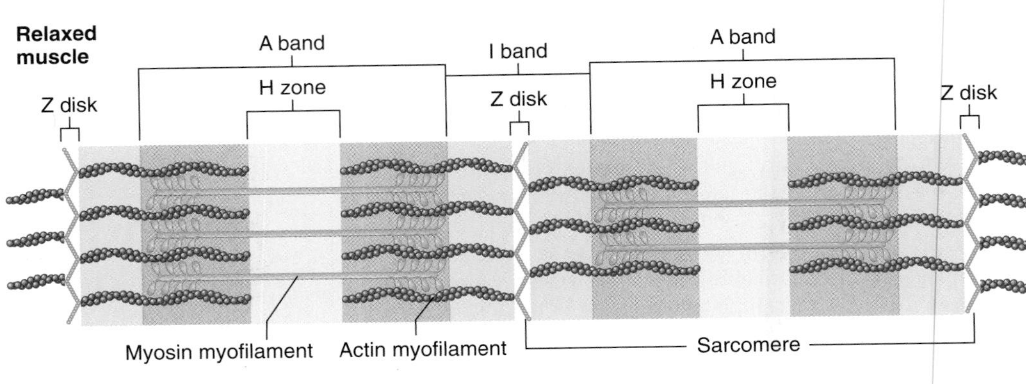

2. During contraction, actin myofilaments at each end of the sarcomere slide past the myosin myofilaments toward each other. As a result, the Z disks are brought closer together, and the sarcomere shortens.

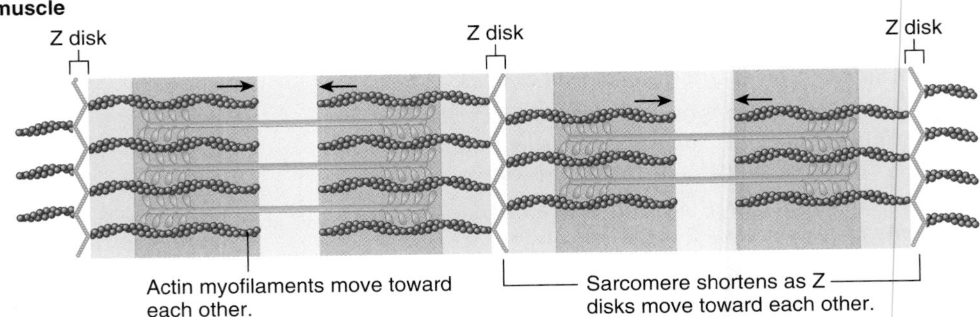

3. As the actin myofilaments slide over the myosin myofilaments, the H zones (*yellow*) and the I bands (*blue*) narrow. The A bands, which are equal to the length of the myosin myofilaments, do not narrow, because the length of the myosin myofilaments does not change.

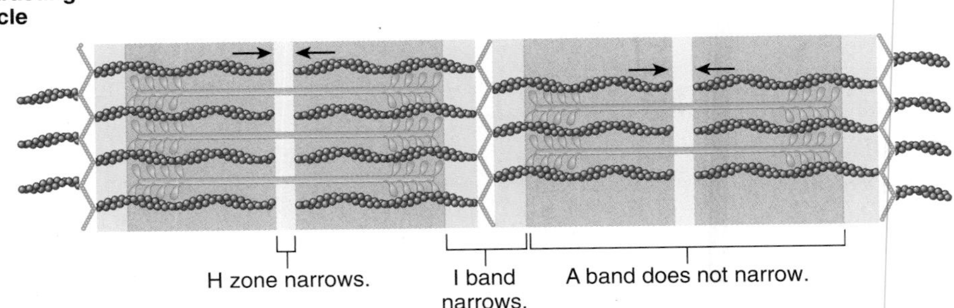

4. In a fully contracted muscle, the ends of the actin myofilaments overlap at the center of the sarcomere and the H zone disappears.

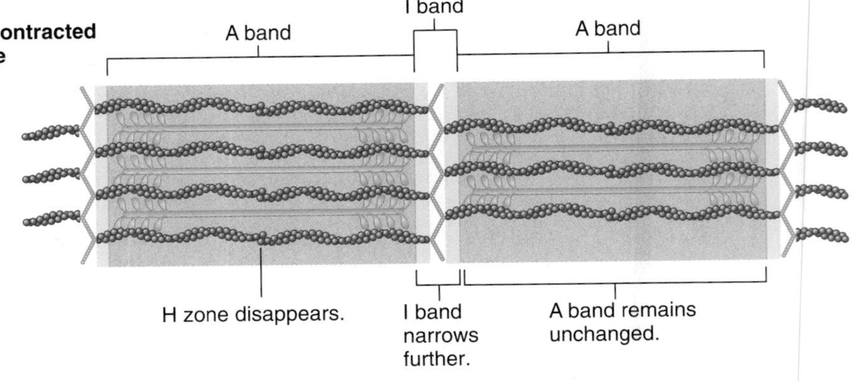

PROCESS FIGURE 9.6 **Sarcomere Shortening**

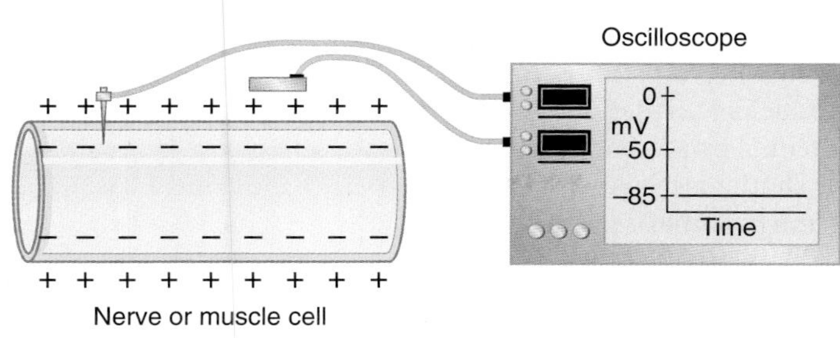

FIGURE 9.7 Measuring the Resting Membrane Potential in Skeletal Muscle

The recording electrode is inside the membrane, the reference electrode is outside, and a potential difference of about −85 mV is recorded by the recording device (oscilloscope), with the inside of the membrane negative with respect to the outside of the membrane.

that the inside of the plasma membrane becomes positively charged, compared with the outside. The permeability characteristics of the plasma membrane change as a result of the opening of certain ion channels, when a cell is stimulated. The diffusion of ions through these channels changes the charge across the plasma membrane and produces an action potential. Two types of gated ion channels play important roles in producing action potentials:

1. *Ligand-gated ion channels.* A **ligand** (lī′gand) is a molecule that binds to a receptor. A **receptor** is a protein or glycoprotein that has a receptor site to which a ligand can bind. **Ligand-gated ion channels** are channels that open in response to a ligand binding to a receptor that is part of the ion channel (see figure 3.10). For example, the axons of nerve cells supplying skeletal muscle fibers release ligands, called **neurotransmitters** (noor′ō-trans-mit′erz), which bind to ligand-gated Na^+ channels in the membranes of the muscle fibers. As a result, the Na^+ channels open, allowing Na^+ to enter the cell.
2. *Voltage-gated ion channels.* These channels are gated membrane channels that open and close in response to small voltage (charge) changes across the plasma membrane, or changes in the membrane potential. When a nerve or muscle cell is stimulated, the charge difference changes and that change causes voltage-gated ion channels to open or close. The major voltage-gated channels that play important roles in an action potential are voltage-gated Na^+ and K^+ channels.

Both ligand-gated and voltage-gated ion channels are specific for the type of ion that passes through them. The channels that open determine what ions move across the plasma membrane. For example, opening voltage-gated Na^+ channels allow Na^+ to cross the plasma membrane, whereas opening voltage-gated K^+ channels allow K^+ to cross.

The concentration gradient for an ion determines whether that ion enters or leaves the cell after the ion channel, specific for that ion, opens (see chapter 3). For example, there is a higher concentration of Na^+ outside the cell than inside it. Consequently, when voltage-gated Na^+ channels open, Na^+ move through them into the cell. In a similar fashion, when gated K^+ channels open, K^+ move out of the cell.

14 ***Define resting membrane potential.***

15 ***Describe the differences in Na^+ and K^+ concentrations across the plasma membrane.***

16 ***Describe the permeability characteristics of the plasma membrane.***

17 ***List the two major categories of gated ion channels in the plasma membrane.***

18 ***List the two major types of voltage-gated ion channels that play important roles in the resting membrane potential.***

PREDICT 2

When ligand-gated K^+ channels open, how does this affect the resting membrane potential?

Action Potentials

An action potential takes approximately 1 millisecond to a few milliseconds to occur, and it has two phases, called depolarization and repolarization. Stimulation of a cell can cause depolarization of its plasma membrane (figure 9.8*a*). Depolarization occurs when the inside of the plasma membrane becomes less negative, which is indicated by movement of the curve upward toward zero. If the depolarization changes the membrane potential to a value called **threshold** (figure 9.8*b*), an action potential is triggered. The **depolarization phase** of the action potential is a brief period during which further depolarization occurs and the inside of the cell becomes positively charged. The charge difference across the plasma membrane is said to be reversed when the membrane potential becomes a positive value. The **repolarization phase** is the return of the membrane potential to its resting value.

The opening and closing of voltage-gated ion channels change the permeability of the plasma membrane to ions, resulting in depolarization and repolarization. Before a nerve or muscle cell is stimulated, these voltage-gated ion channels are closed (figure 9.9, step 1). When the cell is stimulated, gated Na^+ channels open, and Na^+ diffuse into the cell. The positively charged Na^+ make the inside of the cell membrane less negative. If this depolarization reaches threshold, many voltage-gated Na^+ channels open rapidly, and Na^+ diffuse into the cell until the inside of the membrane becomes positive for a brief time (figure 9.9, step 2). As the inside of the cell becomes positive, this voltage change causes additional permeability changes in the plasma membrane, which stop depolarization and start repolarization. Repolarization results from the closing of voltage-gated Na^+ channels and the opening of voltage-gated K^+ channels (figure 9.9, step 3). Thus, the movement of Na^+ into the cell stops, and the movement of K^+ out of the cell increases. These changes cause the inside of the plasma membrane to become more negative and the outside to become more positive. The action potential ends and the resting membrane potential is reestablished when the voltage-gated K^+ channels close.

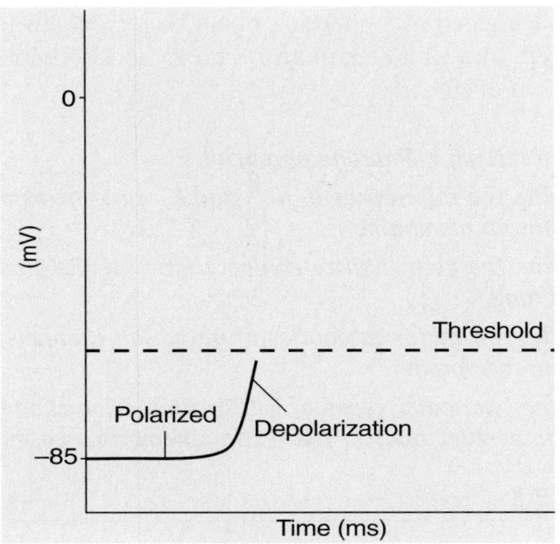

(a) Depolarization is a change of the charge difference across the plasma membrane, making the charge inside of the cell less negative and the outside of the plasma membrane less positive. Once threshold is reached, an action potential is produced.

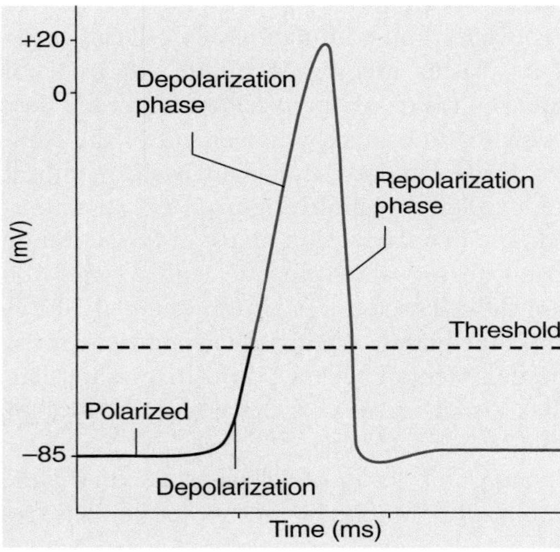

(b) During the depolarization phase of the action potential, the membrane potential changes from approximately –85 mV to approximately +20 mV. During the repolarization phase, the inside of the plasma membrane changes from approximately +20 mV back to –85 mV.

FIGURE 9.8 Depolarization and the Action Potential in Skeletal Muscle

Action potentials occur according to the **all-or-none principle.** If a stimulus is strong enough to produce a depolarization that reaches threshold, or even if it exceeds threshold by a substantial amount, all of the permeability changes responsible for an action potential proceed without stopping. Consequently, all of the action potentials in a given cell are alike (the "all" part). If a stimulus is so weak that the depolarization does not reach threshold, few of the permeability changes occur. The membrane potential returns to its resting level after a brief period without producing an action potential (the "none" part). An action potential can be compared to the flash system of a camera. Once the shutter is triggered (reaches threshold), the camera flashes (an action potential is produced), and each flash is the same brightness (the "all" part) as the previous flashes. If the shutter is depressed but not triggered (does not reach threshold), no flash results (the "none" part).

An action potential occurs in a very small area of the plasma membrane and does not affect the entire plasma membrane at one time. Action potentials can travel, or **propagate,** across the plasma membrane because an action potential produced at one location in the plasma membrane can stimulate the production of an action potential in an adjacent location (figure 9.10). Note that an action potential does not actually move along the plasma membrane. Rather, an action potential at one location stimulates the production of another action potential in an adjacent location, which in turn stimulates the production of another and so on. It is much like a long row of toppling dominos in which each domino knocks down the next. Each domino falls, but no single domino actually travels the length of the row.

The **action potential frequency** is the number of action potentials produced per unit of time. As the strength of the stimulus applied to a nerve or muscle cell increases (once threshold is reached), the action potential frequency increases as the strength of the stimulus increases. All the action potentials are identical. The action potential frequency can affect the strength of a muscle contraction (see "Stimulus Frequency and Muscle Contraction," p. 298).

In summary, the resting membrane potential results from a charge difference across the plasma membrane. An action potential, which is a reversal of that charge difference, stimulates cells to respond by contracting. The nervous system controls muscle contractions by sending action potentials along axons to muscle cells and stimulating action potentials in them. An increased frequency of action potentials sent to the muscle cells can result in an increased strength of muscle contraction.

19 ***What types of gated ion channels are responsible for producing action potentials?***

20 ***What value must depolarization reach in a cell to trigger an action potential?***

21 ***Describe the changes that occur during the depolarization and repolarization phases of an action potential.***

22 ***What is the all-or-none principle of action potentials and what is its significance?***

23 ***Describe the propagation of an action potential.***

24 ***How does the frequency of action potentials affect muscle contractions?***

Neuromuscular Junction

Action potentials carried by motor neurons cause action potentials to be produced in muscle fibers because of events that occur in the neuromuscular junction. Axons of motor neurons carry action

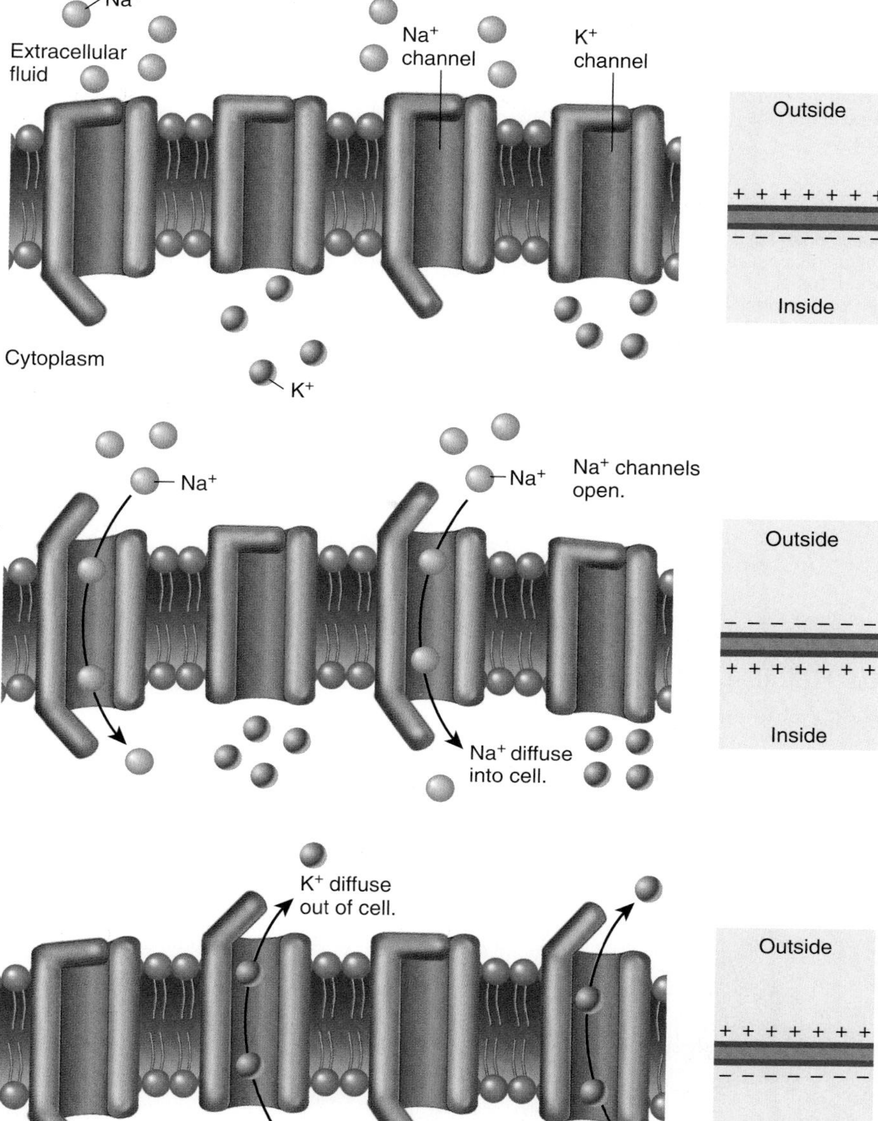

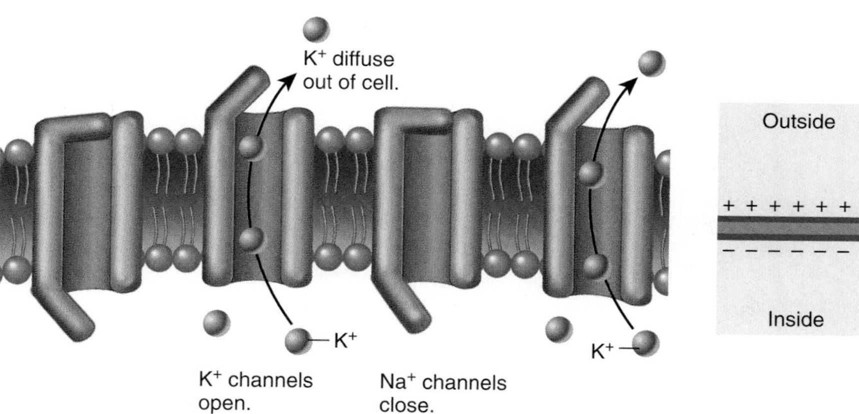

PROCESS FIGURE 9.9 Gated Ion Channels and the Action Potential

Step 1 illustrates the status of gated Na^+ and K^+ channels in a resting cell. Steps 2 and 3 show how the channels open and close to produce an action potential. Next to each step, the charge difference across the plasma membrane is illustrated.

potentials at a high velocity from the brain and spinal cord to skeletal muscle fibers. The axons branch repeatedly, and each branch projects toward one muscle fiber to innervate it. Thus, each muscle fiber receives a branch of an axon, and each axon innervates more than one muscle fiber (see figure 9.2).

Near the muscle fiber it innervates, each axon branch forms a cluster of enlarged axon terminals that rests in an invagination of the sarcolemma to form a **synapse,** or **neuromuscular junction,** which consists of the axon terminals and the area of the muscle fiber sarcolemma they innervate. Each axon terminal is the **presynaptic** (prē′si-nap′tik) **terminal.** The space between the presynaptic terminal and the muscle fiber is the **synaptic** (si-nap′tik) **cleft,** and the muscle plasma membrane in the area of the junction is the **postsynaptic** (pōst-si-nap′tik) **membrane,** or **motor end-plate** (figure 9.11).

Each presynaptic terminal contains numerous mitochondria and many small, spherical sacs approximately 45 μm in diameter, called **synaptic vesicles.** The vesicles contain **acetylcholine**

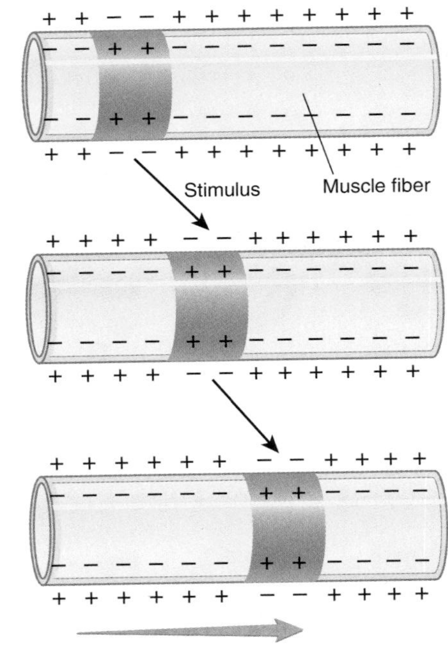

PROCESS FIGURE 9.10 Action Potential Propagation in a Muscle Fiber

(**ACh;** as-e-til-kō′lēn), an organic molecule composed of acetic acid and choline, which functions as a neurotransmitter. A **neurotransmitter** (noor′ō-trans-mit′er) is a substance released from a presynaptic membrane that diffuses across the synaptic cleft and stimulates (or inhibits) the production of an action potential in the postsynaptic membrane by binding to ligand-gated ion channels.

When an action potential reaches the presynaptic terminal, it causes voltage-gated calcium ion (Ca^{2+}) channels in the plasma membrane of the axon to open; as a result, Ca^{2+} diffuse into the cell (figure 9.12). Once inside the cell, the Ca^{2+} cause the contents of a few synaptic vesicles to be secreted by exocytosis from the presynaptic terminal into the synaptic cleft. The acetylcholine molecules released from the synaptic vesicles then diffuse across the cleft and bind to receptor molecules located within the postsynaptic membrane. This causes ligand-gated Na^+ channels to open, increasing the permeability of the membrane to Na^+. Sodium ions then diffuse into the cell, causing depolarization. In skeletal muscle, each action potential in the motor neuron causes a depolarization that exceeds threshold, which causes changes in voltage-gated ion channels that result in the production of an action potential in the muscle fiber.

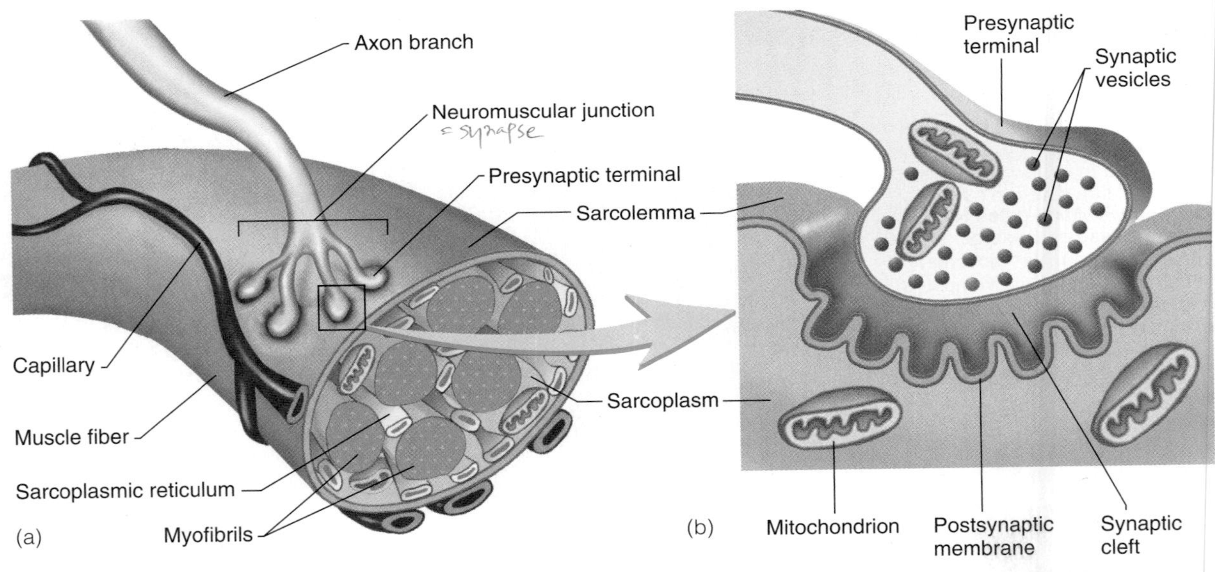

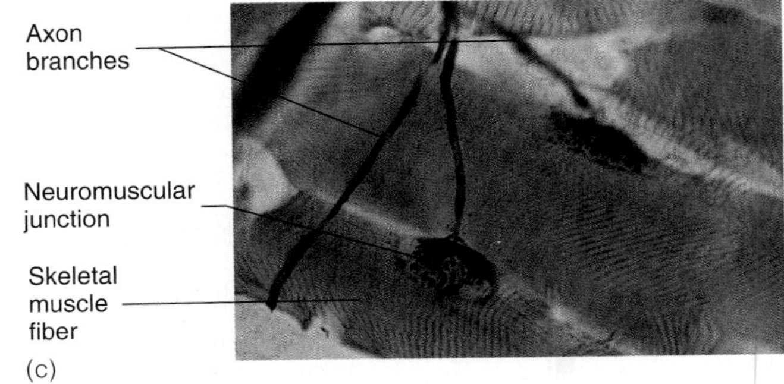

FIGURE 9.11 Neuromuscular Junction

(*a*) Several branches of an axon form the neuromuscular junction with a single muscle fiber. (*b*) The presynaptic terminal containing synaptic vesicles is separated from the postsynaptic membrane by the synaptic cleft. (*c*) Photomicrograph of neuromuscular junctions.

1. An action potential (orange arrow) arrives at the presynaptic terminal and causes voltage-gated Ca^{2+} channels in the presynaptic membrane to open.

2. Calcium ions enter the presynaptic terminal and initiate the release of the neurotransmitter acetylcholine (ACh) from synaptic vesicles.

3. ACh is released into the synaptic cleft by exocytosis.

4. ACh diffuses across the synaptic cleft and binds to ligand-gated Na^+ channels on the postsynaptic membrane.

5. Ligand-gated Na^+ channels open and Na^+ enters the postsynaptic cell, causing the postsynaptic membrane to depolarize. If depolarization passes threshold, an action potential is generated along the postsynaptic membrane.

6. ACh unbinds from the ligand-gated Na^+ channels, which then close.

7. The enzyme acetylcholinesterase, which is attached to the postsynaptic membrane, removes acetylcholine from the synaptic cleft by breaking it down into acetic acid and choline.

8. Choline is symported with Na^+ into the presynaptic terminal, where it can be recycled to make ACh. Acetic acid diffuses away from the synaptic cleft.

9. ACh is reformed within the presynaptic terminal using acetic acid generated from metabolism and from choline recycled from the synaptic cleft. ACh is then taken up by synaptic vesicles.

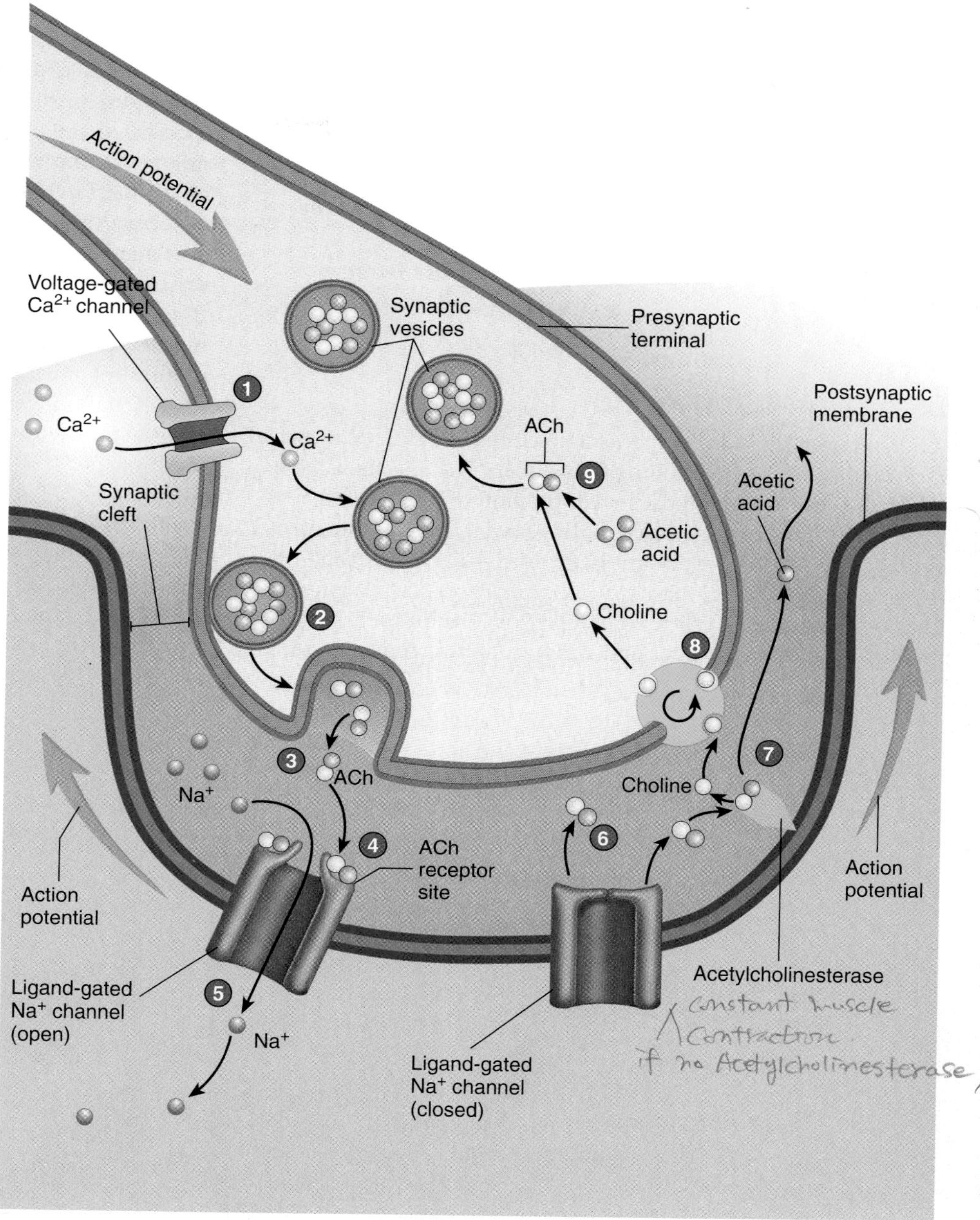

PROCESS FIGURE 9.12 Function of the Neuromuscular Junction

PREDICT 3

Predict the consequence if presynaptic action potentials in an axon could not release sufficient acetylcholine to cause depolarization to threshold in a skeletal muscle fiber.

Acetylcholine released into the synaptic cleft is rapidly broken down to acetic acid and choline by the enzyme **acetylcholinesterase** (as′e-til-kō-lin-es′ter-ās; figure 9.12). Acetylcholinesterase keeps acetylcholine from accumulating within the synaptic cleft, where it would act as a constant stimulus at the postsynaptic terminal, producing many action potentials and continuous contraction in the muscle fiber. The release of acetylcholine and its rapid degradation in the synaptic cleft ensures that one presynaptic action potential yields only one postsynaptic action potential. Choline molecules are actively reabsorbed by the presynaptic terminal and then combined with the acetic acid produced within the cell to form acetylcholine. Recycling choline molecules requires less energy and is more rapid than completely synthesizing new acetylcholine molecules each time they are released from the presynaptic terminal. Acetic acid is an intermediate in the process of glucose metabolism (see chapter 25). A variety of cells can take it up and use it after it diffuses from the area of the neuromuscular junction.

25 ***Describe the neuromuscular junction. How does an action potential in the neuron produce an action potential in the muscle cell?***

26 ***What is the importance of acetylcholinesterase in the synaptic cleft?***

Excitation–Contraction Coupling

Action potentials produced in the sarcolemma of a skeletal muscle fiber can lead to contraction of the fiber. The mechanism by which an action potential causes contraction of a muscle fiber is called **excitation–contraction coupling,** and it involves the sarcolemma, T tubules, sarcoplasmic reticulum, Ca^{2+}, and troponin. The sarcolemma has, along its surface, many regularly arranged, tubelike invaginations called **transverse,** or **T, tubules.** T tubules project into the muscle fiber and wrap around sarcomeres in the region where actin and myosin myofilaments overlap (see figure 9.3; figure 9.13). The lumen of each T tubule is filled with extracellular fluid and is continuous with the exterior of the muscle fiber. Suspended in the sarcoplasm between the T tubules is a highly specialized smooth endoplasmic reticulum called the **sarcoplasmic reticulum** (sar-kō-plaz′mik re-tik′ū-lŭm). Near the T tubules, the sarcoplasmic reticulum is enlarged to form **terminal cisternae** (sis-ter′nē). A T tubule and the two adjacent terminal cisternae together are called a **triad** (trī′ad; see figure 9.13). The sarcoplasmic reticulum actively transports Ca^{2+} into its lumen; thus, the concentration of Ca^{2+} is approximately 2000 times higher within the sarcoplasmic reticulum than in the sarcoplasm of a resting muscle.

Excitation–contraction coupling begins at the neuromuscular junction with the production of an action potential in the sarcolemma. The action potential is propagated along the entire sarcolemma of the muscle fiber and into the T tubules. The T tubules carry action potentials into the interior of the muscle fiber, where they cause Ca^{2+} channels in the terminal cisternae of the sarcoplasmic reticulum to open. When the Ca^{2+} channels open, Ca^{2+} rapidly diffuse into the sarcoplasm surrounding the myofibrils (figure 9.14).

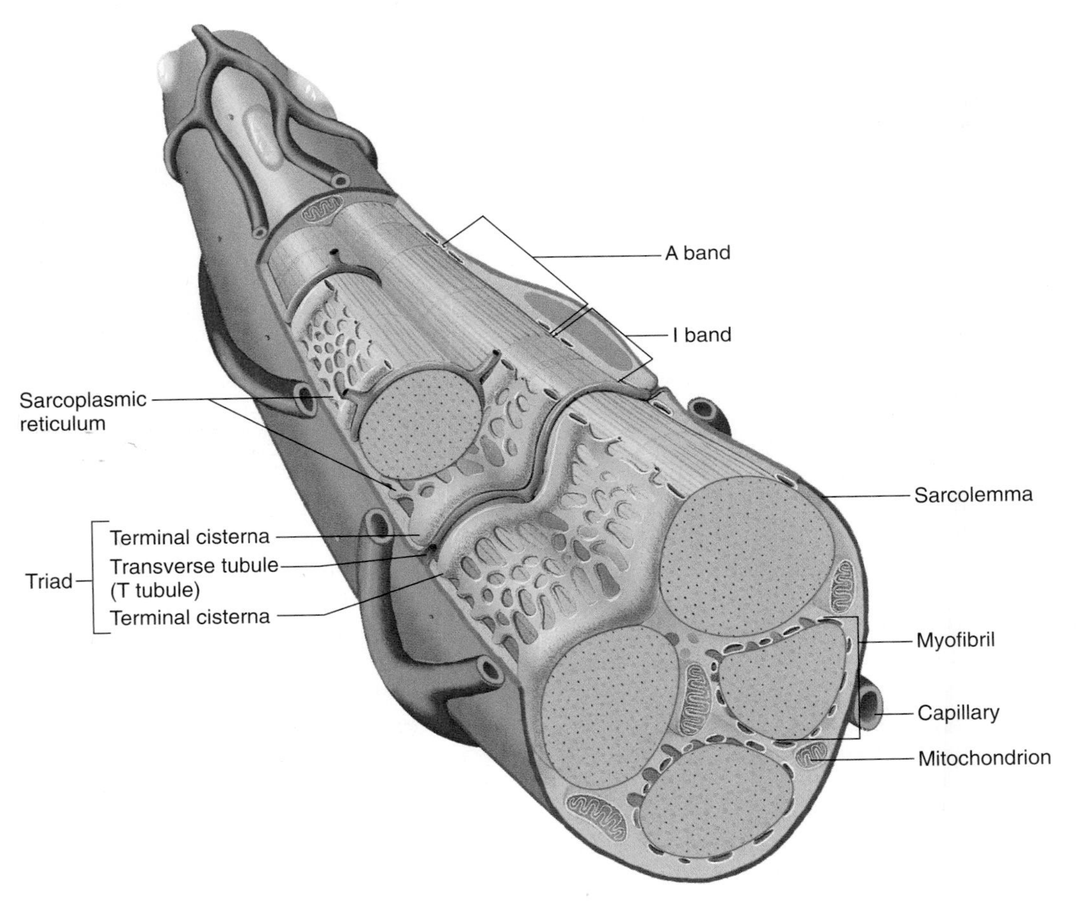

FIGURE 9.13 T Tubules and Sarcoplasmic Reticulum

A T tubule and the sarcoplasmic reticulum on each side of the T tubule form a triad.

1. An action potential produced at the neuromuscular junction is propagated along the sarcolemma of the skeletal muscle. The depolarization also spreads along the membrane of the T tubules.

2. The depolarization of the T tubule causes gated Ca^{2+} channels in the sarcoplasmic reticulum to open, resulting in an increase in the permeability of the sarcoplasmic reticulum to Ca^{2+}, especially in the terminal cisternae. Calcium ions then diffuse from the sarcoplasmic reticulum into the sarcoplasm.

3. Calcium ions released from the sarcoplasmic reticulum bind to troponin molecules. The troponin molecules bound to G actin molecules are released, causing tropomyosin to move, exposing the active sites on G actin.

4. Once active sites on G actin molecules are exposed, the heads of the myosin myofilaments bind to them to form cross-bridges.

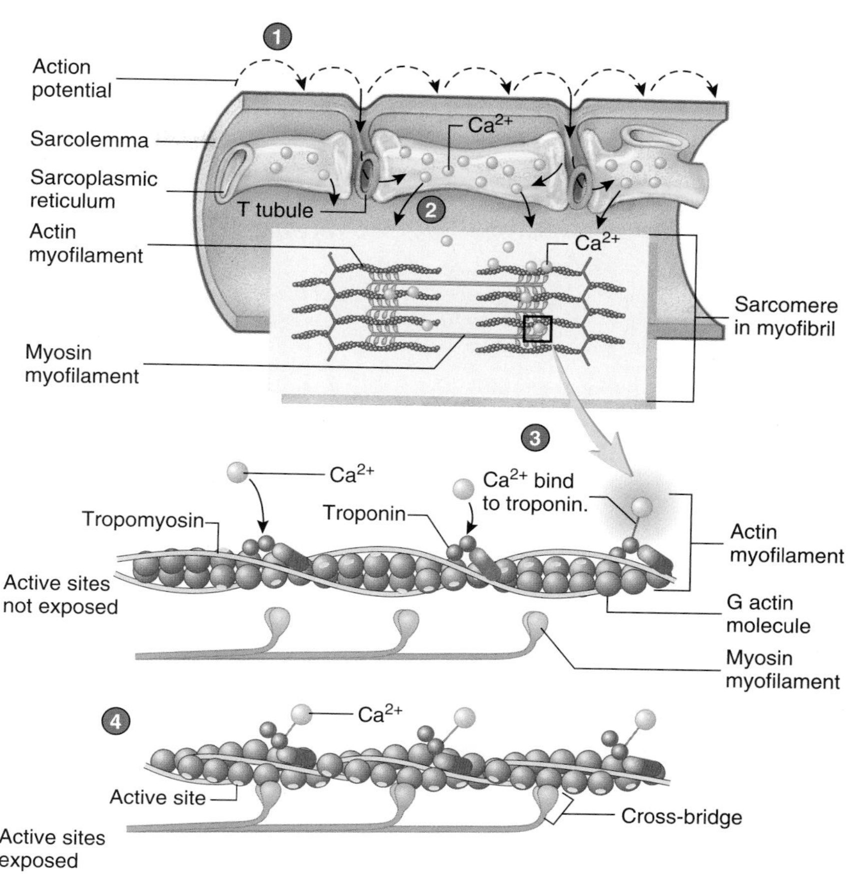

PROCESS FIGURE 9.14 Action Potentials and Muscle Contraction

Calcium ions bind to the Ca^{2+} binding sites on the troponin molecules of the actin myofilaments. The combination of Ca^{2+} with troponin causes the troponin–tropomyosin complex to move deeper into the groove between the two F actin strands, which exposes active sites on the actin myofilaments. The heads of the myosin molecules then bind to the exposed active sites to form cross-bridges (see figure 9.14). Movement of the cross-bridges results in contraction.

Cross-Bridge Movement

A cycle of events resulting in contraction proceeds very rapidly when the heads of the myosin molecules bind to actin to form cross-bridges (figure 9.15). The heads of myosin molecules move at their hinged region, resulting in cross-bridge movement, which forces the actin myofilament, to which the heads of the myosin molecules are attached, to slide over the surface of the myosin myofilament. After cross-bridge movement, each myosin head releases from the actin and returns to its original position. It can then form another cross-bridge at a different site on the actin myofilament, followed by movement, release of the cross-bridge, and return to its original position. During a single contraction, each myosin molecule undergoes the cycle of cross-bridge formation, movement, release, and return to its original position many times. This process is called **cross-bridge cycling.**

The energy from one ATP molecule is required for each cycle of cross-bridge formation, movement, and release. Before a myosin head binds to the active site on an actin myofilament, the head of the myosin molecule is in its resting position, and ADP and phosphate are bound to the head of the myosin molecule (see figure 9.15). Once Ca^{2+} bind to troponin and the tropomyosin moves, the active sites on actin myofilaments are then exposed. The head of a myosin molecule can then bind to an exposed active site, and the phosphate is released from the head of the myosin molecule. Energy stored in the head of the myosin molecule causes the head of the myosin molecule to move. Movement of the head causes the actin myofilament to slide past the myosin myofilament and ADP is released from the myosin head. ATP must then bind to the head of the myosin before the cross-bridge can release. As the ATP molecule binds to the head of the myosin molecule, the ATP is broken down to ADP and phosphate, and the myosin head releases from the active site on actin. The ADP and phosphate remain attached to the head of the myosin molecule, and the head of the myosin returns to its resting position. Energy

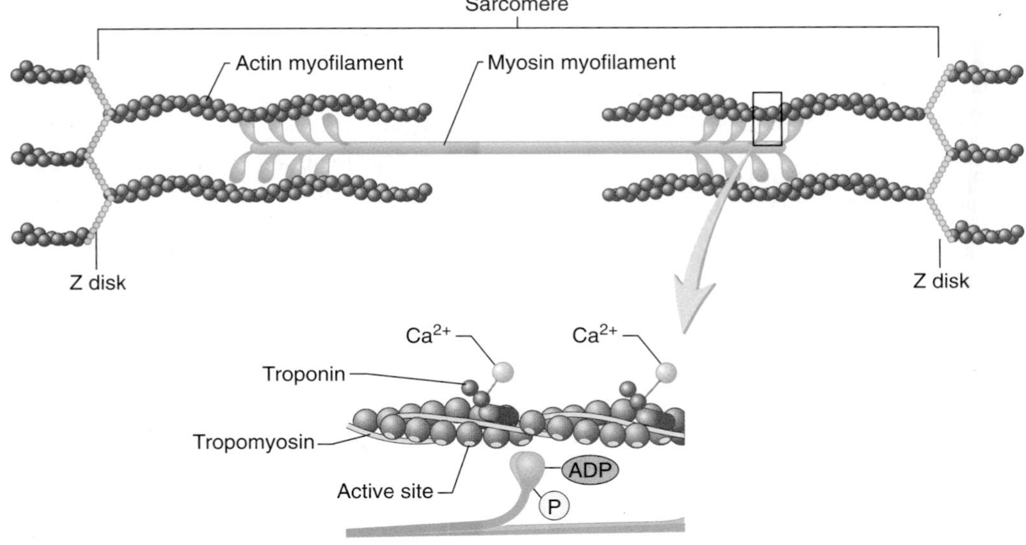

1. **Exposure of active sites.** Before cross-bridges cycle, Ca^{2+} bind to the troponins and the tropomyosins move, causing the exposure of active sites on actin myofilaments.

Cross-bridge
ADP
P

2. **Cross-bridge formation.** The myosin heads bind to the exposed active sites on the actin myofilaments to form cross-bridges, and phosphates are released from the myosin heads.

ADP

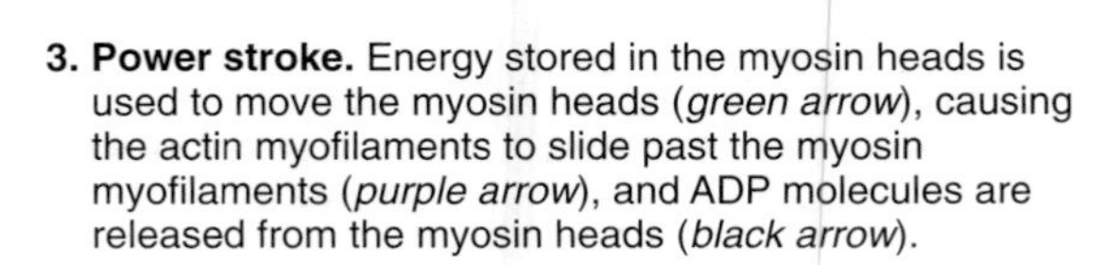

3. **Power stroke.** Energy stored in the myosin heads is used to move the myosin heads (*green arrow*), causing the actin myofilaments to slide past the myosin myofilaments (*purple arrow*), and ADP molecules are released from the myosin heads (*black arrow*).

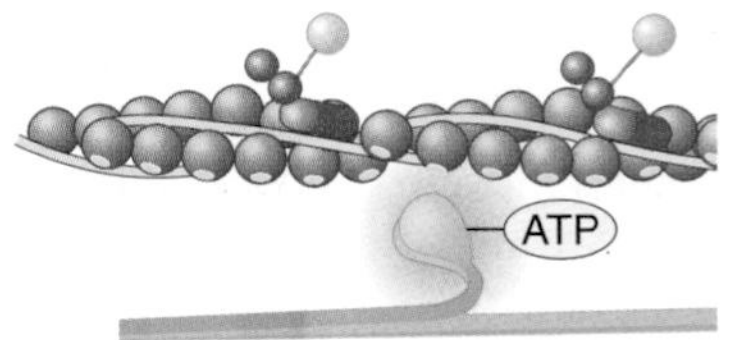

4. **Cross-bridge release.** An ATP molecule binds to each of the myosin heads, causing them to detach from the actin.

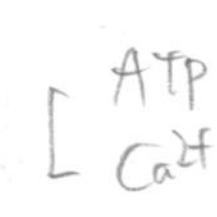

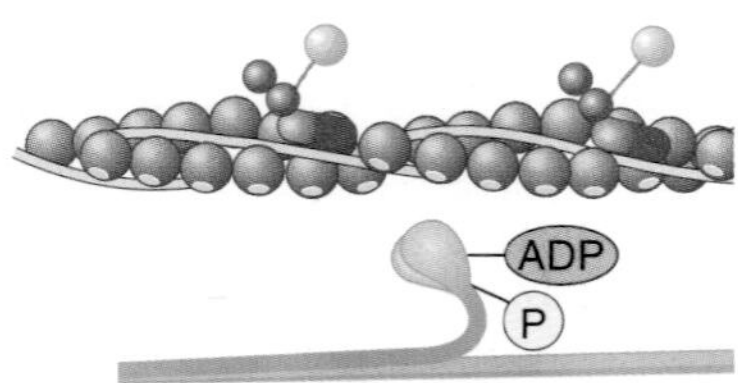
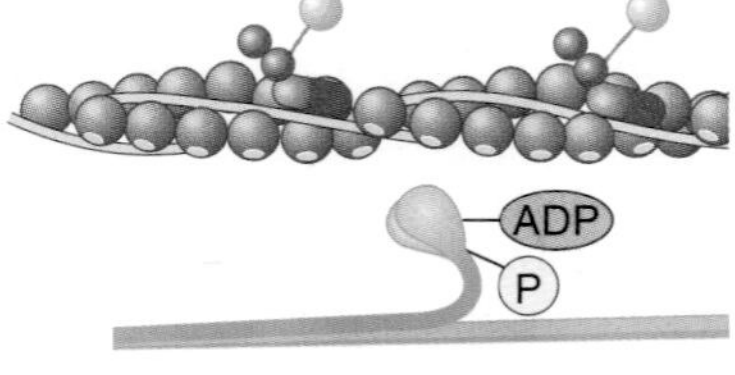

5. **Hydrolysis of ATP.** The myosin ATPase portion of the myosin heads split ATP in to ADP and phosphate (P), which remain attached to the myosin heads.

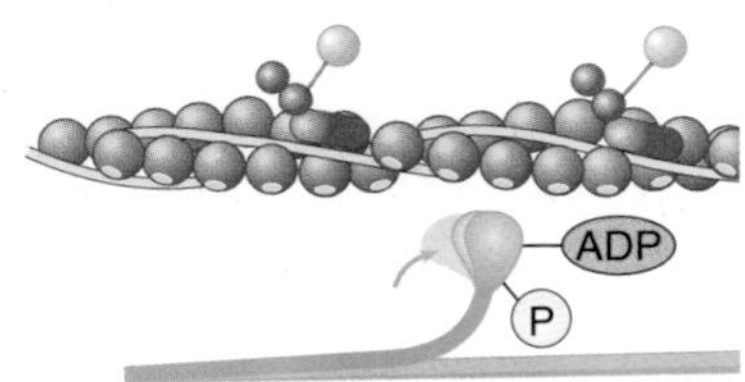

6. **Recovery stroke.** The heads of the myosin molecules return to their resting position (*green arrow*), and energy is stored in the heads of the myosin molecules. If Ca^{2+} are still attached to troponins, cross-bridge formation and movement are repeated (return to step 2). This cycle occurs many times during a muscle contraction. Not all cross-bridges form and release simultaneously.

PROCESS FIGURE 9.15 Breakdown of ATP and Cross-Bridge Movement During Muscle Contraction

The Effect of Blocking Acetylcholine Receptors and Acetylcholinesterase

Anything that affects the production, release, and degradation of acetylcholine or its ability to bind to its receptor molecule also affects the transmission of action potentials across the neuromuscular junction. For example, some insecticides contain organophosphates that bind to and inhibit the function of acetylcholinesterase. As a result, acetylcholine is not degraded and accumulates in the synaptic cleft, where it acts as a constant stimulus to the muscle fiber. Insects exposed to such insecticides die partly because their muscles contract and cannot relax—a condition called **spastic paralysis** (spas′tik pă-ral′i-sis), which is followed by fatigue of the muscles.

In humans, a similar response to these insecticides occurs. The skeletal muscles responsible for respiration cannot undergo their normal cycle of contraction and relaxation. Instead, they remain in a state of spastic paralysis until they become fatigued. Victims die of respiratory failure. Other organic poisons, such as curare, bind to the acetylcholine receptors, preventing acetylcholine from binding to them. Curare does not allow activation of the receptors, and therefore the muscle is incapable of contracting in response to nervous stimulation—a condition called **flaccid** (flak′sid, flas′id) **paralysis.** Curare is not a poison to which people are commonly exposed, but it has been used to investigate the role of acetylcholine in the neuromuscular synapse and is sometimes used in small doses to relax muscles during certain kinds of surgery.

Myasthenia gravis (mī-as-thē′nē-ă grăv′is) results from the production of antibodies that bind to acetylcholine receptors, eventually causing the destruction of the receptor and thus reducing the number of receptors. As a consequence, muscles exhibit a degree of flaccid paralysis or are extremely weak. A class of drugs that includes neostigmine partially blocks the action of acetylcholinesterase and sometimes is used to treat myasthenia gravis. The drugs cause acetylcholine levels to increase in the synaptic cleft and combine more effectively with the remaining acetylcholine receptors.

released from the breakdown of ATP is stored in the head of the myosin molecule.

Movement of the myosin molecule while the cross-bridge is attached is called the **power stroke,** whereas return of the myosin head to its original position after cross-bridge release is called the **recovery stroke.** Many cycles of power and recovery strokes occur during each muscle contraction. While muscle is relaxed, energy stored in the heads of the myosin molecules is held in reserve until the next contraction. When Ca^{2+} are released from the sarcoplasmic reticulum in response to an action potential, the cycle of cross-bridge formation, movement, and release, which results in contraction, begins (see figures 9.14 and 9.15).

Muscle Relaxation

Relaxation occurs as a result of the active transport of Ca^{2+} back into the sarcoplasmic reticulum. As the Ca^{2+} concentration decreases in the sarcoplasm, the ions diffuse away from the troponin molecules. The troponin–tropomyosin complex then reestablishes its position, which blocks the active sites on the actin molecules. As a consequence, cross-bridges cannot re-form once they have been released, and relaxation occurs.

Energy is needed to make muscles contract, but it is also needed to make muscles relax. The active transport of Ca^{2+} into the sarcoplasmic reticulum requires ATP. The active transport processes that maintain the normal concentrations of Na^+ and K^+ across the sarcolemma also require ATP. The amount of ATP required for cross-bridge formation during contraction is much greater than the other energy requirements in a skeletal muscle. The diffusion of Ca^{2+} from the sarcoplasmic reticulum is rapid, compared with the reuptake of Ca^{2+} into the sarcoplasmic reticulum by active transport. Consequently, a muscle takes at least twice as long to relax as it does to contract.

27 ***How does an action potential produced in the postsynaptic membrane of the neuromuscular junction eventually result in contraction of the muscle fiber?***

28 ***Where in the contraction and relaxation processes is ATP required? Define power stroke and recovery stroke.***

29 ***What conditions are required for relaxation of the muscle fiber?***

PREDICT 4

Predict the consequences of having the following conditions develop in a muscle in response to a stimulus: (a) Na^+ cannot enter the skeletal muscle through voltage-gated Na^+ channels, (b) very little ATP is present in the muscle fiber before a stimulus is applied, and (c) adequate ATP is present within the muscle fiber, but action potentials occur at a frequency so great that calcium is not transported back into the sarcoplasmic reticulum between individual action potentials.

PHYSIOLOGY OF SKELETAL MUSCLE

Muscle Twitch

A **muscle twitch** is the contraction of a muscle in response to a stimulus that causes an action potential in one or more muscle fibers. Even though the normal function of muscles is more complex, an understanding of the muscle twitch makes the function of muscles in living organisms easier to comprehend.

A hypothetical contraction of a single muscle fiber in response to a single action potential is illustrated in figure 9.16. The time between application of the stimulus to the motor neuron and the beginning of contraction is the **lag,** or **latent, phase;** the time during which contraction occurs is the **contraction phase;** and the time during which relaxation occurs is the **relaxation phase** (table 9.2). An action potential is an electrochemical event, but contraction is a mechanical event. An action potential is measured in millivolts and is completed in less than 2 milliseconds. Muscle contraction is measured as a force, also called tension. It is reported as the number of grams lifted, or the distance the muscle shortens, and requires up to 1 second to occur.

30 ***Define the phases of a muscle twitch and describe the events responsible for each phase.***

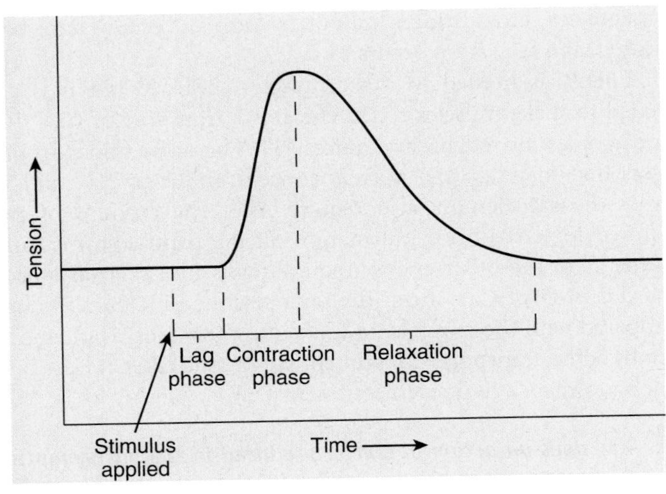

FIGURE 9.16 Phases of a Muscle Twitch in a Single Muscle Fiber

There is a short lag phase after stimulus application, followed by a contraction phase and a relaxation phase.

Stimulus Strength and Muscle Contraction

An isolated skeletal muscle fiber produces contractions of equal force in response to each action potential. This relationship, called the **all-or-none law of skeletal muscle contraction,** can be explained on the basis of action potential production in the skeletal muscle fiber. When brief electric stimuli of increasing strength are applied to the muscle fiber sarcolemma, the following events occur: (1) A subthreshold stimulus does not produce an action potential, and no muscle contraction occurs; (2) a threshold stimulus produces an action potential and results in contraction of the muscle cell; or (3) a stronger-than-threshold stimulus produces an action potential of the same magnitude as the threshold stimulus and therefore produces an identical contraction. Thus, for a given condition, once an action potential is generated, the skeletal muscle fiber contracts to produce a constant force. If internal conditions change, the force of contraction can change as well. For example, increasing the amount of calcium available to the muscle cell results in a stronger force of contraction; conversely, muscle fatigue can result in a weaker force of contraction.

Within a given skeletal muscle, such as the biceps brachii (see chapter 10), skeletal muscle fibers form **motor units,** each of

TABLE 9.2 Events That Occur During Each Phase of a Muscle Twitch*

Events
Lag Phase
1. An action potential is propagated to the presynaptic terminal of the motor neuron.
2. The action potential causes voltage-gated Ca^{2+} channels in the presynaptic terminal to open.
3. Calcium ions diffuse into the presynaptic terminal, causing acetylcholine contained within several synaptic vesicles to be released by exocytosis into the synaptic cleft.
4. Acetylcholine released from the presynaptic terminal diffuses across the synaptic cleft and binds to acetylcholine receptor sites on ligand-gated Na^+ channels in the postsynaptic membrane of the sarcolemma.
5. The binding of acetylcholine to its receptor site causes the ligand-gated Na^+ channels to open, and the postsynaptic membrane becomes more permeable to Na^+.
6. Sodium ions diffuse into the muscle fiber, causing a local depolarization that exceeds threshold and produces an action potential.
7. Acetycholine is rapidly degraded in the synaptic cleft to acetic acid and choline by acetylcholinesterase, thus limiting the length of time acetylcholine is bound to its receptor site. The result is that one presynaptic action potential produces one postsynaptic action potential in each muscle fiber.
8. The action potential produced in a muscle fiber is propagated from the postsynaptic membrane near the middle of the fiber toward both ends and into the T tubules.
9. The depolarization that occurs in the T tubule in response to the action potential causes gated Ca^{2+} channels of the membrane of the sarcoplasmic reticulum to open, and the membrane of the sarcoplasmic reticulum becomes very permeable to Ca^{2+}.
10. Calcium ions diffuse from the sarcoplasmic reticulum into the sarcoplasm.
11. Calcium ions bind to troponin; the troponin–tropomyosin complex changes its position and exposes active sites on the actin myofilaments.
Contraction Phase
12. Cross-bridges between actin molecules and myosin molecules form, move, release, and re-form many times, causing the sarcomeres to shorten. Energy stored in the heads of the myosin molecules allows cross-bridge formation and movement. After cross-bridge movement has occurred, ATP must bind to the myosin heads. The ATP is broken down to ADP, and some of the energy is used to release the cross-bridges and cause the heads of the myosin molecules to move back to their resting position, where they are ready to form other cross-bridges. Some of the energy from the ATP is stored in the myosin heads and is used for the next cross-bridge formation and movement (see figure 9.15). Energy is also released as heat.
Relaxation Phase
13. Calcium ions are actively transported back into the sarcoplasmic reticulum (requires ATP).
14. The troponin–tropomyosin complexes move to inhibit cross-bridge formation.
15. The muscle fibers lengthen passively.

*Assuming that the process begins with a single action potential in the motor neuron.

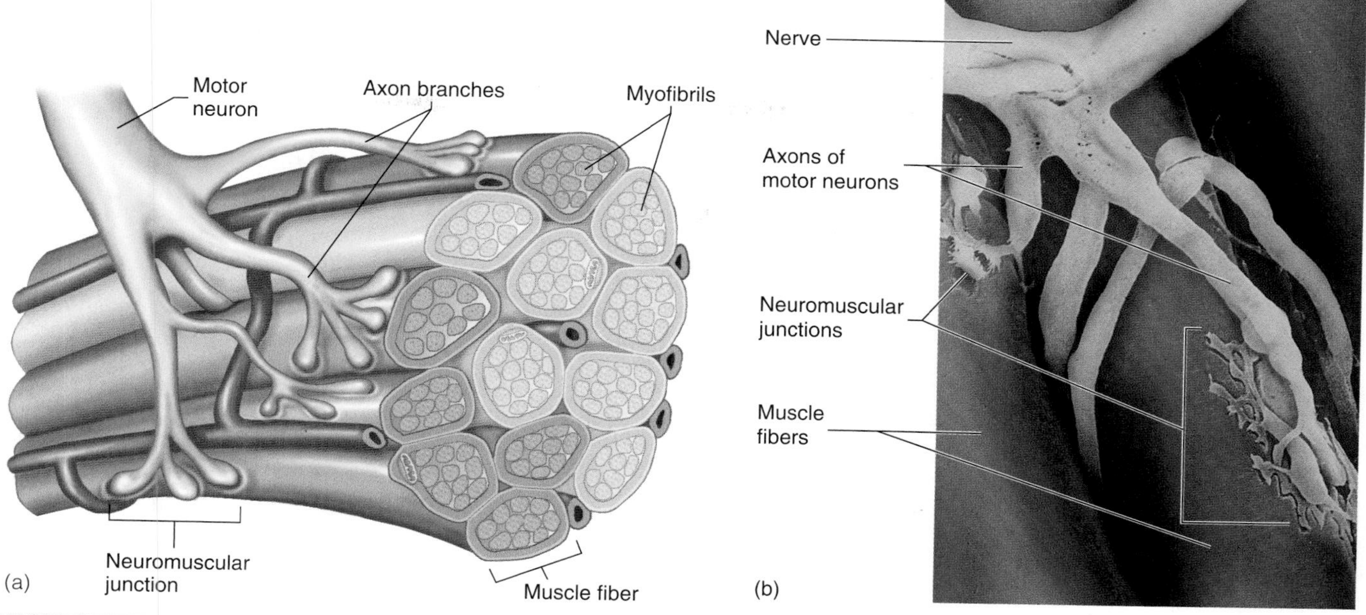

FIGURE 9.17 Motor Unit

(*a*) A motor unit consists of a single motor neuron and all the muscle fibers its branches innervate. (*b*) Photomicrograph of motor units.

which consists of a single motor neuron and all of the muscle fibers it innervates (figure 9.17). Motor units respond as a single unit. All the muscle fibers of a motor unit contract to produce a constant force in response to a threshold stimulus because an action potential in a motor neuron initiates an action potential in each of the muscle fibers it innervates.

Whole muscles exhibit characteristics that are more complex than those of individual muscle fibers or motor units. Instead of responding in an all-or-none fashion, whole muscles respond to stimuli in a **graded fashion,** which means that the strength of the contractions can range from weak to strong.

A muscle is composed of many motor units, and the axons of the motor units combine to form a nerve. A whole muscle contracts with either a small force or a large force, depending on the number of motor units stimulated to contract. This relationship is called **multiple motor unit summation** because the force of contraction increases as more and more motor units are stimulated. Multiple motor unit summation resulting in graded responses can be demonstrated by applying brief electric stimuli of increasing strength to the nerve supplying a muscle (figure 9.18). A **subthreshold stimulus** is not strong enough to cause an action potential in any of the axons in a nerve and causes no contraction. As the stimulus strength increases, it eventually becomes a **threshold stimulus,** which is strong enough to produce an action potential in a single motor unit axon, causing all the muscle fibers of that motor unit to contract. Progressively stronger stimuli, called **submaximal stimuli,** produce action potentials in axons of additional motor units. A **maximal stimulus** produces action potentials in the axons of all motor units of that muscle. Consequently, even greater stimulus strengths, or **supramaximal stimuli,** have no additional effect. As the stimulus strength increases between threshold and maximum values, motor units are **recruited**—the number of motor units responding to the stimuli increases and the force of contraction produced by the muscle increases in a graded fashion.

Motor units in different muscles do not always contain the same number of muscle fibers. Muscles performing delicate and precise movements have motor units with a small number of muscle fibers, whereas muscles performing more powerful but less precise contractions have motor units with many muscle fibers. For example, in very delicate muscles, such as those that move the eye, the number of muscle fibers per motor unit can be less than 10, whereas, in the heavy muscles of the thigh, the number can be several hundred.

31 ***Why does a single muscle fiber either not contract or contract with the same force in response to stimuli of different strengths?***

32 ***Why does a motor unit either not contract or contract with the same force in response to stimuli of different strengths?***

33 ***How does increasing the strength of a stimulus cause a whole muscle to respond in a graded fashion?***

34 ***Define multiple motor unit summation.***

PREDICT 5

In patients with poliomyelitis (pō′lē-ō-mī′ĕ-lī′tis), motor neurons are destroyed, causing loss of muscle function and even flaccid paralysis. Sometimes recovery occurs because of the formation of axon branches from the remaining motor neurons. These branches innervate the paralyzed muscle fibers to produce motor units with many more muscle fibers than usual, resulting in the recovery of muscle function. What effect would this reinnervation of muscle fibers have on the degree of muscle control in a person who has recovered from poliomyelitis?

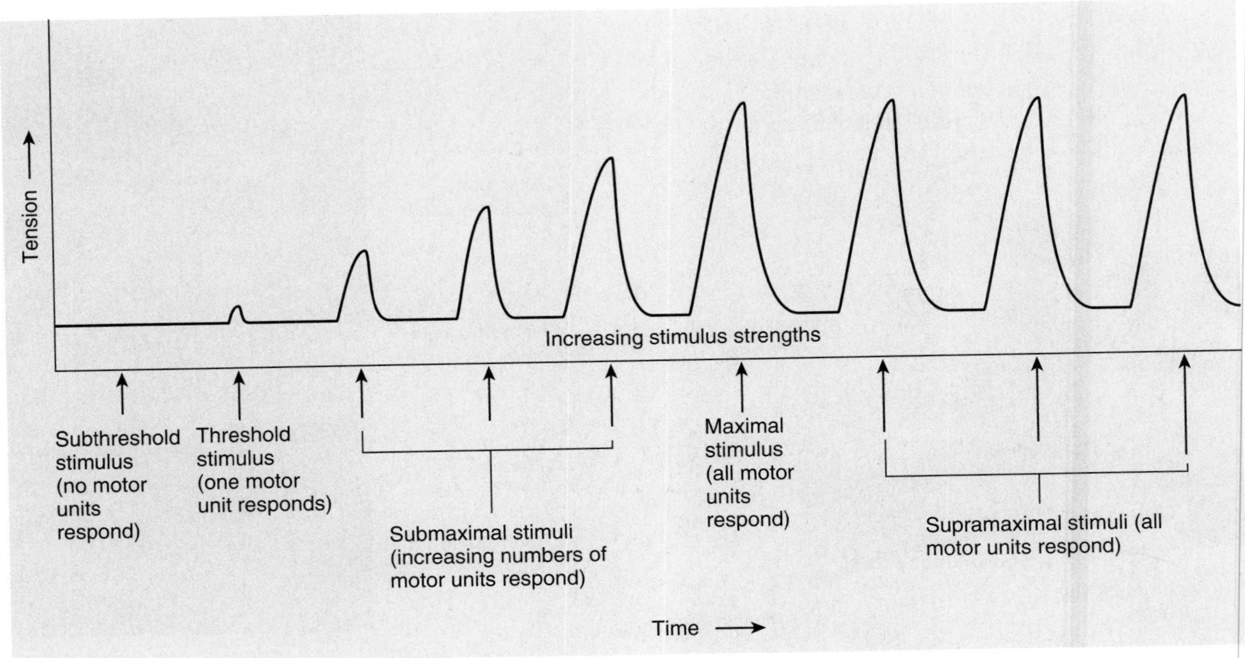

FIGURE 9.18 Multiple Motor Unit Summation in a Muscle
Multiple motor unit summation occurs as stimuli of increasing strength are applied to a nerve that innervates a muscle. The amount of tension (height of peaks) is influenced by the number of motor units responding.

Stimulus Frequency and Muscle Contraction

An action potential in a single muscle fiber causes it to contract. Although the action potential triggers contraction of the muscle fiber, the action potential is completed long before the contraction phase is completed. In addition, the contractile mechanism in a muscle fiber exhibits no refractory period. That is, relaxation of a muscle fiber is not required before a second action potential can stimulate a second contraction. As the frequency of action potentials in a skeletal muscle fiber increases, the frequency of contraction also increases. In **incomplete tetanus** (tet'ă-nŭs), muscle fibers partially relax between the contractions, but, in **complete tetanus,** action potentials are produced so rapidly in muscle fibers that no muscle relaxation occurs between them. The tension produced by a muscle increases as the frequency of contractions increases. This increased tension is called **multiple-wave summation** (figure 9.19).

Tetanus of a muscle caused by stimuli of increasing frequency can be explained by the effect of the action potentials on Ca^{2+} release from the sarcoplasmic reticulum. The first action potential causes Ca^{2+} release from the sarcoplasmic reticulum, the Ca^{2+} diffuse to the myofibrils, and contraction occurs. Relaxation begins as the Ca^{2+} are pumped back into the sarcoplasmic reticulum. If the next action potential occurs before relaxation is complete, two things happen. First, because not enough time has passed for all the Ca^{2+} to reenter the sarcoplasmic reticulum, Ca^{2+} levels around the myofibrils remain elevated. Second, the next action potential causes the release of additional Ca^{2+} from the sarcoplasmic reticulum. Thus, the elevated Ca^{2+} levels in the sarcoplasm produce continued contraction of the muscle fiber. Action potentials at a high frequency can increase Ca^{2+} concentrations in the sarcoplasm to an extent that the muscle fiber is contracted completely and does not relax at all.

At least two factors play roles in the increased tension observed during multiple-wave summation. First, as the action potential frequency increases, the concentration of Ca^{2+} around the myofibrils becomes greater than during a single muscle twitch, thereby causing a much greater degree of contraction. The additional Ca^{2+} cause the exposure of additional active sites on the actin myofilaments. Second, the sarcoplasm and the connective tissue components of

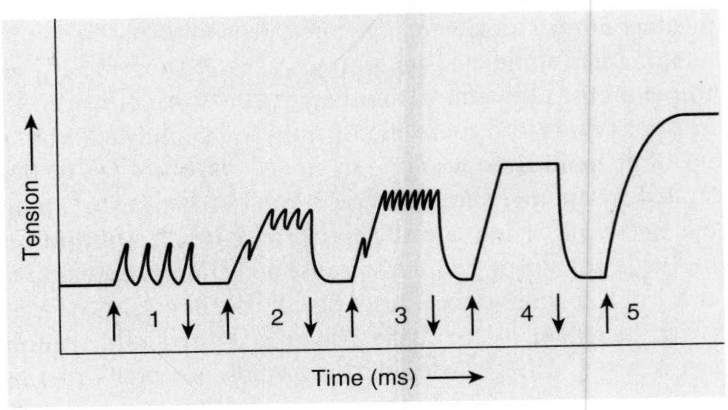

FIGURE 9.19 Multiple-Wave Summation
Stimuli 1–5 are stimuli of increasing frequency. For each stimulus, the up arrow indicates the start of stimulation, and the down arrow indicates the end of stimulation. Stimulus frequency 1 produces successive muscle twitches with complete relaxation between stimuli. Stimuli frequencies 2–4 do not allow complete relaxation between stimuli. Stimulus frequency 5 causes tetanus—no relaxation between stimuli.

muscle have some elasticity. During each separate muscle twitch, some of the tension produced by the contracting muscle fibers is used to stretch those elastic elements, and the remaining tension is applied to the load to be lifted. In a single muscle twitch, relaxation begins before the elastic components are totally stretched. The maximum tension produced during a single muscle twitch is therefore not applied to the load to be lifted. In a muscle stimulated at a high frequency, the elastic elements are stretched during the very early part of the prolonged contraction. After the elastic components are stretched, all of the tension produced by the muscle is applied to the load to be lifted, and the observed tension produced by the muscle is increased.

Another example of a graded response is **treppe** (trep'eh; staircase), which occurs in muscle that has rested for a prolonged period (figure 9.20). If the muscle is stimulated with a maximal stimulus at a low frequency, which allows complete relaxation between the stimuli, the contraction triggered by the second stimulus produces a slightly greater tension than the first. The contraction triggered by the third stimulus produces a contraction with a greater tension than the second. After only a few stimuli, the levels of tension produced by all the contractions are equal.

A possible explanation for treppe is an increase in Ca^{2+} levels around the myofibrils. The Ca^{2+} released in response to the first stimulus is not taken up completely by the sarcoplasmic reticulum before the second stimulus causes the release of additional Ca^{2+}, even though the muscle completely relaxes between the muscle twitches. As a consequence, during the first few contractions of the muscle, the Ca^{2+} concentration in the sarcoplasm increases slightly, making contraction more efficient because of the increased number of ions available to bind to troponin. Treppe achieved during warm-up exercises can contribute to improved muscle efficiency during athletic events. Factors such as increased blood flow to the muscle and increased muscle temperature probably are involved as well. Increased muscle temperature causes the enzymes responsible for muscle contraction to function at a more rapid rate.

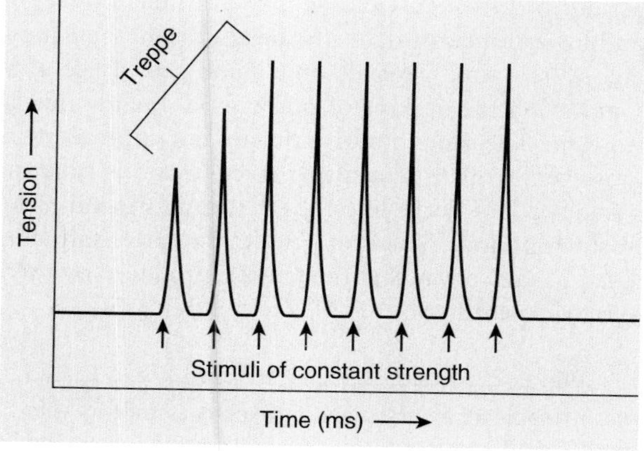

FIGURE 9.20 Treppe

When a rested muscle is stimulated repeatedly with maximal stimuli at a frequency that allows complete relaxation between stimuli, the second contraction produces a slightly greater tension than the first, and the third contraction produces greater tension than the second. After a few contractions, the levels of tension produced by all contractions are equal.

CASE STUDY

Organophosphate Poisoning

John has a number of prize apple trees in his back yard. To prevent them from becoming infested with insects, he sprayed them with an organophosphate insecticide. He was in a rush to spray the trees before leaving town on vacation, and he failed to pay attention to the safety precautions on the packaging. He sprayed the trees without using any skin or respiratory protection. Soon he experienced severe stomach cramps, double vision, difficulty breathing, and spastic contractions of his skeletal muscles. His wife took John to the emergency room where he was diagnosed with organophosphate poisoning. While in the emergency room, his physician administered a drug, and soon many of John's symptoms subsided.

Organophosphate insecticides exert their effects by binding to the enzyme acetylcholinesterase within synaptic clefts, rendering it ineffective. Thus, the organophosphate poison and acetylcholine "compete" for the acetylcholinesterase as the organophosphate poison increases in concentration the enzyme is less effective in degrading acetylcholine. Organophosphate poisons affect synapses in which acetylcholine is the neurotransmitter, including skeletal muscle synapses and some smooth muscle synapses, such as in the walls of the stomach, intestines, and air passageways.

PREDICT 6

Organophosphate insecticides exert their effects by binding to the enzyme acetylcholinesterase within synaptic clefts, rendering it ineffective. Use figure 9.12 to help answer the following question.

a. Explain the spastic contractions that occurred in John's skeletal muscles.

b. Propose as many mechanisms as you can by which a drug could counteract the effects of organophosphate poisoning.

35 ***How does the lack of a refractory period in skeletal muscle fiber contraction explain multiple-wave summation?***

36 ***Define incomplete tetanus and complete tetanus.***

37 ***Give two reasons why rapid, repeated stimulation of a muscle fiber increases its force of contraction.***

38 ***Describe treppe and explain how it occurs.***

TYPES OF MUSCLE CONTRACTIONS

Muscle contractions are classified based on the type of contraction that predominates (table 9.3). In **isometric** (ī-sō-met'rik) **contractions,** the length of the muscle does not change, but the amount of tension increases during the contraction process. Isometric contractions are responsible for the constant length of the postural muscles of the body, such as muscles that hold the spine erect while a person is sitting or standing. In **isotonic** (ī-sō-ton'ik) **contractions,** the amount of tension produced by the

TABLE 9.3 Types of Muscle Contractions

Contraction Types	Characteristics
Multiple motor unit summation	Each motor unit responds in an all-or-none fashion. A whole muscle is capable of producing an increasing amount of tension as the number of motor units stimulated increases.
Multiple-wave summation	Summation results when many action potentials are produced in a muscle fiber. Contraction occurs in response to the first action potential, but there is not enough time for relaxation to occur between action potentials. Because each action potential cause the release of Ca^{2+} from the sarcoplasmic reticulum, the ion levels remain elevated in the sarcoplasm to produce a tetanic contraction. The tension produced as a result of multiple-wave summation is greater than the tension produced by a single muscle twitch. The increased tension results from the greater concentration of Ca^{2+} in the sarcoplasm and the stretch of the elastic components of the muscle early in contraction.
Tetanus of muscles	Tetanus of muscles results from multiple-wave summation. Incomplete tetanus occurs when the action potential frequency is low enough to allow partial relaxation of the muscle fibers. Complete tetanus occurs when the action potential frequency is high enough that no relaxation of the muscle fibers occurs.
Treppe	Tension produced increases for the first few contractions in response to a maximal stimulus at a low frequency in a muscle that has been at rest for some time. Increased tension may result from the accumulation of small amounts of Ca^{2+} in the sarcoplasm for the first few contractions or from an increasing rate of enzyme activity.
Isometric contractions	A muscle produces increasing tension as it remains at a constant length. This type of contraction is characteristic of postural muscles that maintain a constant tension without changing their length.
Isotonic contractions	A muscle produces a constant tension during contraction. A muscle shortens during contraction. This type of contraction is characteristic of finger and hand movements.
Concentric contractions	A muscle produces increasing tension as it shortens.
Eccentric contractions	A muscle produces increasing tension as it lengthens.

muscle is constant during contraction, but the length of the muscle changes. Movements of the upper limbs or fingers are predominantly isotonic contractions. Examples include waving and using a computer keyboard. Most muscle contractions are not strictly isometric or isotonic contractions. For example, both the length and tension of muscles change when a person walks or opens a heavy door. Although some mechanical differences exist, both types of contractions result from the same contractile process within muscle cells.

Concentric (kon-sen′trik) **contractions** are isotonic contractions in which tension in the muscle is great enough to overcome the opposing resistance, and the muscle shortens. Concentric contractions include contractions that result in an increasing tension as the muscle shortens. Many of the movements performed by muscle contractions are concentric contractions. **Eccentric** (ek-sen′trik) **contractions** are isotonic contractions in which tension is maintained in a muscle, but the opposing resistance is great enough to cause the muscle to increase in length (see table 9.3). Eccentric contractions are performed when a person slowly lowers a heavy weight. During eccentric contractions, muscles produce substantial force. During exercise, eccentric contractions often produce greater tension than concentric contractions. Eccentric contractions are of clinical interest because repetitive eccentric contractions, such as seen in the lower limbs of people who run downhill for long distances, tend to injure muscle fibers and the connective tissue of muscles.

Muscle tone is the constant tension produced by muscles for long periods of time. Muscle tone is responsible for keeping the back and lower limbs straight, the head upright, and the abdomen flat. Muscle tone depends on a small percentage of all the motor units contracting out of phase with one another at any point in time. The same motor units are not contracting all the time, however. A small percentage of motor units is stimulated with a frequency of nerve impulses that causes incomplete tetanus for short periods. The motor units that are contracting are stimulated in such a way that the tension produced by the whole muscle remains constant.

PREDICT 7

Mary Myosin overheard an argument between two students who could not decide if a weight lifter who lifts a weight above the head and then holds it there before lowering it is using isometric, isotonic, concentric, or eccentric muscle contractions. Mary is an expert on muscle contractions, so she settles the debate. What is her explanation?

Movements of the body are usually smooth and occur at widely differing rates—some very slowly and others quite rapidly. Most movements are produced by muscle contractions, but very few of the movements resemble the rapid contractions of individual muscle twitches. Smooth, slow contractions result from an increasing number of motor units contracting out of phase as the muscles shorten, as well as from a decreasing number of motor units contracting out of phase as muscles lengthen. Each motor unit exhibits either incomplete or complete tetanus, but, because the contractions are out of phase and because the number of motor units activated varies at each point in time, a smooth contraction results. Consequently, muscles are capable of contracting either slowly or rapidly, depending on the number of motor units stimulated and the rate at which that number increases or decreases.

Length Versus Tension

Active tension is the force applied to an object to be lifted when a muscle contracts. The initial length of a muscle has a strong influence on the amount of active tension it produces. As the length of a muscle increases, its active tension also increases, to a point. If the muscle is stretched farther than that optimum length, the active tension it produces begins to decline. The muscle length plotted against the tension produced by the muscle in response to maximal stimuli is the **active tension curve** (figure 9.21).

If a muscle is stretched so that the actin and myosin myofilaments within the sarcomeres do not overlap or overlap to a very small extent, the muscle produces very little active tension when it is stimulated. Also, if the muscle is not stretched at all, the myosin myofilaments touch each of the Z disks in each sarcomere, and very little contraction of the sarcomeres can occur. If the muscle is stretched to its optimum length, optimal overlap of the actin and myosin myofilaments takes place. When the muscle is stimulated, cross-bridge formation results in maximal contraction.

Passive tension is the tension applied to the load when a muscle is stretched but not stimulated. It is similar to the tension produced if the muscle is replaced with an elastic band. Passive tension exists because the muscle and its connective tissue have some elasticity. The sum of active and passive tension is called **total tension.**

39 ***Define isometric, isotonic, concentric, and eccentric contractions.***

40 ***What is muscle tone, and how is it maintained?***

41 ***How are smooth, slow contractions produced in muscles?***

42 ***Draw an active tension curve. How does the overlap of actin and myosin explain the shape of the curve?***

Weight Lifters and Muscle Length

Weight lifters and others who lift heavy objects usually assume positions so that their muscles are stretched close to their optimum length before lifting. For example, the position a weight lifter assumes before power lifting stretches the upper limb and lower limb muscles to a near-optimum length for muscle contraction, and the stance a lineman assumes in a football game stretches most muscle groups in the lower limbs so they are near their optimum length for suddenly moving the body forward.

FATIGUE

Fatigue (fă-tēg′) is the decreased capacity to do work and the reduced efficiency of performance that normally follows a period of activity. The rate at which individuals develop fatigue is highly variable, but it is a phenomenon that everyone has experienced. Fatigue can develop at three possible sites: the nervous system, the muscles, and the neuromuscular junction.

Psychologic fatigue, the most common type, involves the central nervous system. The muscles are capable of functioning, but the individual "perceives" that additional muscular work is not possible. A burst of activity in a tired athlete in response to encouragement from spectators shows how psychologic fatigue can be overcome. The onset and duration of psychologic fatigue vary greatly and depend on the emotional state of the individual.

The second most common type of fatigue occurs in the muscle fiber. **Muscular fatigue** results from ATP depletion. Without adequate ATP levels in muscle fibers, cross-bridges and ion transport cannot function normally. As a consequence, the tension that a muscle is capable of producing declines. Fatigue in the lower limbs of marathon runners or in the upper and lower limbs of swimmers are examples.

The least common type of fatigue, called **synaptic fatigue,** occurs in the neuromuscular junction. If the action potential frequency in motor neurons is great enough, the release of acetylcholine from the presynaptic terminals is greater than the rate of acetylcholine synthesis. As a result, the synaptic vesicles become depleted, and insufficient acetylcholine is released to stimulate the muscle fibers. Under normal physiologic conditions, fatigue of neuromuscular junctions is rare; however, it may occur under conditions of extreme exertion.

Muscle Soreness Resulting from Exercise

Pain frequently develops after 1 or 2 days in muscles that are vigorously exercised, and the pain can last for several days. The pain is more common in untrained people who exercise vigorously. In addition, highly repetitive eccentric contractions of muscles produce pain more readily than concentric contractions. The pain is associated with damage to skeletal muscle fibers and with connective tissue surrounding the skeletal muscle fibers. In people with muscle soreness induced by exercise, enzymes that are normally found inside muscle fibers can be detected in the extracellular fluid. Enzymes are able to leave the muscle fibers because injury causes an increase in the permeability or rupture of plasma membranes. In addition, fragments of collagen molecules can be found in the extracellular fluid of muscles. These observations indicate that injury occurs to both muscle fibers and the connective tissue of muscles. The pain produced appears to be the result of inflammation resulting from damage to muscle fibers and the connective tissue. Exercise schedules that alternate exercise with periods of rest, such as lifting weights every other day, provide time for the repair of muscle tissue.

As a result of extreme muscular fatigue, muscles occasionally become incapable of either contracting or relaxing—a condition called **physiologic contracture** (kon-trak′choor), which is caused by a lack of ATP within the muscle fibers. ATP can decline

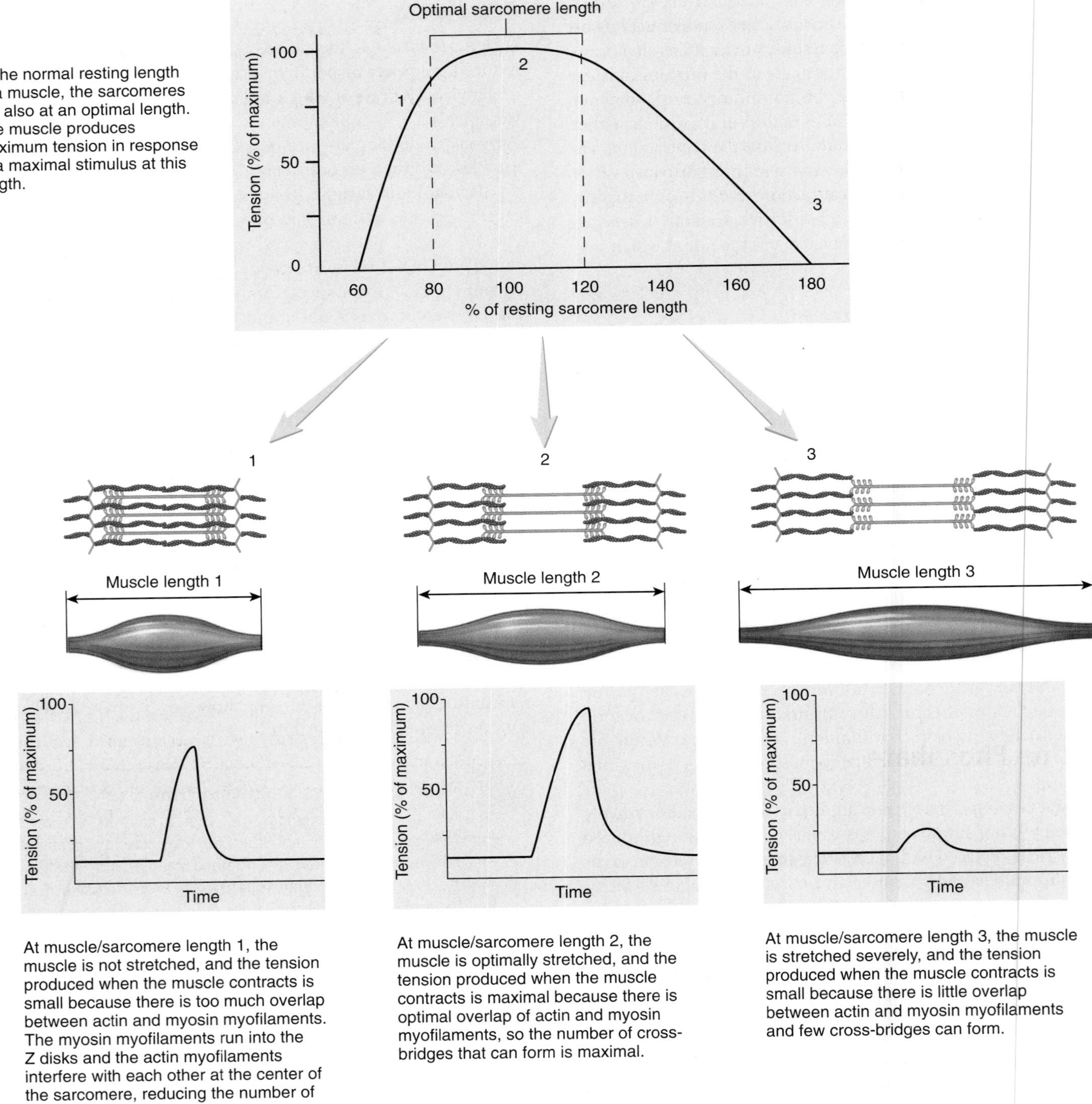

FIGURE 9.21 Muscle Length and Tension

The length of a muscle before it is stimulated influences the force of contraction of the muscle. As the muscle changes length, the sarcomeres also change length.

to very low levels when a muscle is stimulated strongly, such as under conditions of extreme exercise. When ATP levels are very low, active transport of Ca^{2+} into the sarcoplasmic reticulum slows, Ca^{2+} accumulate within the sarcoplasm, and ATP is unavailable to bind to the myosin molecules that have formed cross-bridges with the actin myofilaments. As a consequence, the previously formed cross-bridges cannot release, resulting in physiologic contracture.

Rigor mortis (rig′er mōr′tĭs), the development of rigid muscles several hours after death, is similar to physiologic contracture.

ATP production stops shortly after death, and ATP levels within muscle fibers decline. Because of low ATP levels, active transport of Ca^{2+} into the sarcoplasmic reticulum stops, and Ca^{2+} leak from the sarcoplasmic reticulum into the sarcoplasm. Calcium ions can also leak from the sarcoplasmic reticulum as a result of the breakdown of the sarcoplasmic reticulum membrane after cell death. As Ca^{2+} levels increase in the sarcoplasm, cross-bridges form. Too little ATP is available to bind to the myosin molecules, however, so the cross-bridges are unable to release and re-form in a cyclic fashion to produce contractions. As a consequence, the muscles remain stiff until tissue degeneration occurs.

43 ***Define fatigue, and list three locations in which fatigue can develop.***

44 ***Define and explain the cause of physiologic contracture and rigor mortis.***

ENERGY SOURCES

ATP is the immediate source of energy for muscle contractions. As long as adequate amounts of ATP are present, muscles can contract repeatedly for a long time. ATP must be synthesized continuously to sustain muscle contractions, and ATP synthesis must be equal to ATP breakdown because the small amount of ATP stored in muscle fibers is sufficient to support vigorous muscle contractions for only a few seconds. The energy required to produce ATP comes from three sources: (1) creatine phosphate, (2) anaerobic respiration, and (3) aerobic respiration (table 9.4). Only the main points of anaerobic respiration and aerobic respiration are considered here (a more detailed discussion can be found in chapter 25).

Creatine Phosphate

During resting conditions, energy from aerobic respiration is used to synthesize **creatine** (krē′ă-tēn, krē′ă-tin) **phosphate.** Creatine phosphate accumulates in muscle cells and functions to store energy, which can be used to synthesize ATP. As ATP levels begin to fall, ADP reacts with creatine phosphate to produce ATP and creatine.

$$\text{ADP} + \text{Creatine phosphate} \xrightarrow{\text{Creatine kinase}} \text{Creatine} + \text{ATP}$$

The reaction is catalyzed by the enzyme **creatine kinase,** occurs very rapidly, and is able to maintain ATP levels as long as creatine phosphate is available in the cell. During intense muscular contraction, however, creatine phosphate levels are quickly exhausted. ATP and creatine phosphate present in the cell provide enough energy to sustain maximum contractions for about 8–10 seconds.

Anaerobic Respiration

Anaerobic (an-ār-ō′bik) **respiration** does not require oxygen and results in the breakdown of glucose to yield ATP and lactic acid. For each molecule of glucose metabolized, a net production of two ATP molecules and two molecules of lactic acid occurs. The first part of anaerobic respiration and aerobic respiration share an enzymatic pathway called **glycolysis** (glī-kol′-ī-sis). In glycolysis, a glucose molecule is broken down into two molecules of pyruvic acid. Two molecules of ATP are used in this process, but four molecules of ATP are produced, resulting in a net gain of two ATP molecules for each glucose molecule metabolized. In anaerobic respiration, the pyruvic acid is then converted to lactic acid. Unlike pyruvic acid, much of the lactic acid diffuses out of the muscle fibers into the bloodstream.

Anaerobic respiration is less efficient than aerobic respiration, but it is much faster, especially when oxygen availability limits aerobic respiration. By using many glucose molecules, anaerobic respiration can rapidly produce ATP for a short time. During short periods of intense exercise, such as sprinting, anaerobic respiration combined with the breakdown of creatine phosphate provides enough ATP to support intense muscle contraction for up to 3 minutes. ATP formation from creatine phosphate and anaerobic respiration is limited by the depletion of creatine phosphate and glucose and by the brief time anaerobic respiration can produce ATP.

Aerobic Respiration

Aerobic (ār-ō′bik) **respiration** requires oxygen and breaks down glucose to produce ATP, carbon dioxide, and water. Compared with anaerobic respiration, aerobic respiration is much more efficient. The metabolism of a glucose molecule by anaerobic respiration produces a net gain of 2 ATP molecules for each glucose molecule. In contrast, aerobic respiration can produce up to 38 ATP molecules for each glucose molecule. In addition, aerobic respiration uses a greater variety of molecules as energy sources, such as fatty acids and amino acids. Some glucose is used as an energy source in skeletal muscles, but fatty acids provide a more important source of energy during sustained exercise and during resting conditions.

In aerobic respiration, pyruvic acid is metabolized by chemical reactions within mitochondria. Two closely coupled sequences of reactions in mitochondria, called the citric acid cycle and the electron-transport chain, produce many ATP molecules. Carbon dioxide molecules are produced and, in the last step, oxygen atoms combine with hydrogen atoms to form water. Thus, carbon dioxide, water, and ATP are major end products of aerobic respiration. The following equation represents the aerobic respiration of one molecule of glucose:

$$\text{Glucose} + 6\ O_2 + 38\ \text{ADP} + 38\ \text{P} \rightarrow$$
$$6\ CO_2 + 6\ H_2O + \text{About } 38\ \text{ATP}$$

Although aerobic respiration produces many more ATP molecules for each glucose molecule metabolized than does anaerobic respiration, the rate at which the ATP molecules are produced is slower. Resting muscles or muscles undergoing long-term exercise, such as long-distance running or other endurance exercises, depend primarily on aerobic respiration for ATP synthesis.

Oxygen Deficit and Recovery Oxygen Consumption

There is a lag time between when exercise begins and when an individual begins to breathe more heavily because of the exercise. After exercise stops, there is a lag time before breathing returns to preexercise levels. These changes in breathing patterns reflect changes in oxygen consumption by muscle cells as they increase their demand for ATP supplied through aerobic respiration. The

TABLE 9.4 Sources of ATP in Muscles

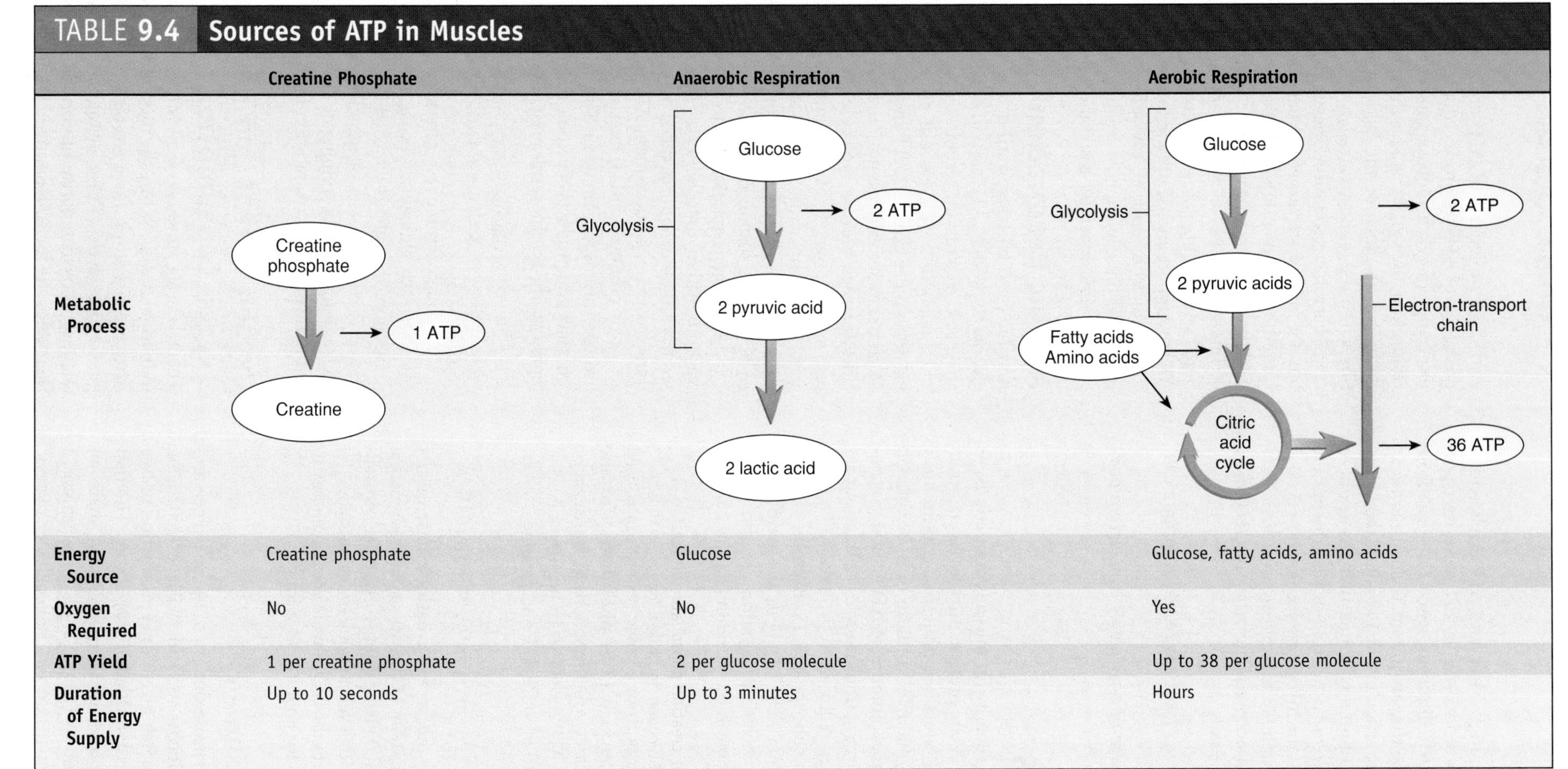

	Creatine Phosphate	Anaerobic Respiration	Aerobic Respiration
Metabolic Process			
Energy Source	Creatine phosphate	Glucose	Glucose, fatty acids, amino acids
Oxygen Required	No	No	Yes
ATP Yield	1 per creatine phosphate	2 per glucose molecule	Up to 38 per glucose molecule
Duration of Energy Supply	Up to 10 seconds	Up to 3 minutes	Hours

lack of increased oxygen consumption relative to increased activity at the onset of exercise creates an **oxygen deficit,** or **debt.** This deficit must be repaid during and after exercise once oxygen consumption catches up with the increased activity level. At the onset of exercise, the ATP required by muscles for the increased level of activity is primarily supplied by the creatine phosphate system and anaerobic respiration: two systems that can supply ATP relatively quickly and without a requirement for oxygen (see table 9.4). The ability of aerobic respiration to supply ATP at the onset of exercise lags behind that of the creatine phosphate system and anaerobic respiration. Consequently, there is a lag between the onset of exercise and the need for increased oxygen consumption.

The elevated oxygen consumption relative to activity level after exercise has ended is called **recovery oxygen consumption.** A portion of the recovery oxygen consumption is used to "repay" the oxygen deficit that was incurred at the onset of exercise, but most of the recovery oxygen consumption is used to support metabolic processes that restore homeostasis after the disturbances that occurred during exercise. Such disturbances include exercise-related increases in body temperature, changes in intra- and extracellular ion concentrations, and changes in metabolite and hormone levels. The duration of the recovery oxygen consumption generally lasts only minutes to hours, and it is dependent on the individual's physical conditioning and on the length and intensity of the exercise session. In extreme cases, such as following a marathon, recovery oxygen consumption can last as long as 15 hours.

Anaerobic Exercise and Oxygen Deficit

During brief but intense exercise, such as sprinting or lifting a heavy weight, much of the ATP used by exercising muscles comes from the conversion of creatine phosphate to creatine and from anaerobic respiration. Glycogen is broken down to glucose in the skeletal muscle fibers and in the liver. Glucose is released from the liver into the circulatory system and can be taken up by skeletal muscle fibers. Anaerobic respiration within the skeletal muscle fibers converts the glucose molecules to ATP and lactic acid. Increased breathing and elevated aerobic respiration after exercise partially result from the oxygen deficit that occurs during such exercise. The increased aerobic respiration pays back a portion of the oxygen deficit by converting creatine to creatine phosphate and converting the excess lactic acid to glucose, which is once again stored as glycogen in muscles and in the liver. The magnitude of the oxygen deficit depends on the intensity of the exercise, the length of time it was sustained, and the physical condition of the individual. Those who are in poor physical condition do not have as great a capacity as well-trained athletes to perform aerobic respiration.

45 ***List the energy sources used to synthesize ATP for muscle contraction.***

46 ***Contrast the efficiency of aerobic and anaerobic respiration. When is each type used by cells?***

47 ***What is the function of creatine phosphate and when is it used?***

48 ***When does lactic acid increase in a muscle cell?***

49 ***Define oxygen deficit. Explain the factors that contribute to oxygen deficit.***

PREDICT 8

Eric is a highly trained cross-country runner and his brother John is a computer programmer who almost never exercises. While the two brothers were working on a remodeling project in the basement of their house, the doorbell upstairs rang: A package they were both very excited about was being delivered. They raced each other up the stairs to the front door to see which one would get the package first. When they reached the front door, both were breathing heavily. However, John continued to breathe heavily for several minutes while Eric was opening the package. Explain why John continued to breathe heavily longer than Eric, even though they both had run the same distance.

SLOW AND FAST FIBERS

Not all skeletal muscles have identical functional capabilities. They differ in several respects, including having muscle fibers that contain slightly different forms of myosin. The myosin of slow-twitch muscle fibers causes these fibers to contract more slowly, and these cells are more resistant to fatigue, whereas the myosin of fast-twitch muscle fibers cause these fibers to contract quickly and these cells fatigue quickly (table 9.5). The proportion of muscle fiber types differs within individual muscles.

Slow-Twitch Oxidative Muscle Fibers

Slow-twitch oxidative (SO), or **type I, muscle fibers** contract more slowly, are smaller in diameter, have a better developed blood supply, have more mitochondria, and are more fatigue-resistant than fast-twitch muscle fibers. Slow-twitch muscle fibers respond relatively slowly to nervous stimulation. The enzymes on the myosin heads responsible for the breakdown of ATP are called **myosin ATPase.** Slow-twitch fibers break down ATP slowly because their myosin heads have a slow form of myosin ATPase. The relatively slow breakdown of ATP means that cross-bridge movement occurs slowly, which is what causes the muscle to contract slowly. Aerobic respiration is the primary source for ATP synthesis in slow-twitch muscles, and their capacity to perform aerobic respiration is enhanced by a plentiful blood supply and the presence of numerous mitochondria. They are called oxidative muscle fibers because of their enhanced capacity to carry out aerobic respiration. Slow-twitch fibers also contain large amounts of **myoglobin** (mī-ō-glō′bin), a dark pigment similar to hemoglobin in red blood cells, which binds oxygen and acts as a muscle reservoir for oxygen when the blood does not supply an adequate amount. Myoglobin thus enhances the capacity of the muscle fibers to perform aerobic respiration.

Fast-Twitch Muscle Fibers

Fast-twitch, or **type II, muscle fibers** respond rapidly to nervous stimulation and their myosin heads have a fast form of myosin ATPase, which allows them to break down ATP more rapidly than slow-twitch muscle fibers. This allows their cross-bridges to release and form more rapidly than those in slow-twitch muscle fibers. Muscles containing a high percentage of these fibers have a less well-developed blood supply than muscles containing a high percentage of slow-twitch muscle fibers. In addition, fast-twitch

TABLE 9.5 Characteristics of Skeletal Muscle Fiber Types

	Slow-Twitch Oxidative (SO) Type I	Fast-Twitch Oxidative Glycolytic (FOG) Type IIa	Fast-Twitch Glycolytic (FG) Type IIb
Fiber Diameter	Smallest	Intermediate	Largest
Myoglobin Content	High	High	Low
Mitochondria	Many	Many	Few
Capillaries	Many	Many	Few
Metabolism	High aerobic capacity Low anaerobic capacity	Intermediate aerobic capacity High anaerobic capacity	Low aerobic capacity Highest anaerobic capacity
Fatigue Resistance	High	Intermediate	Low
Myosin ATPase Activity	Slow	Fast	Fast
Glycogen Concentration	Low	High	High
Location Where Fibers Are Most Abundant	Generally in postural muscles and more in lower limbs than upper limbs	Generally predominate in lower limbs	Generally predominate in upper limbs
Functions	Maintenance of posture and performance in endurance activities	Endurance activities in endurance-trained muscles	Rapid, intense movements of short duration

muscle fibers have very little myoglobin and fewer and smaller mitochondria. Fast-twitch muscle fibers have large deposits of glycogen and are well adapted to perform anaerobic respiration. The anaerobic respiration of fast-twitch muscle fibers, however, is not adapted for supplying a large amount of ATP for a prolonged period. The muscle fibers tend to contract rapidly for a shorter time and fatigue relatively quickly. Fast-twitch muscle fibers come in two forms, which differ mainly in their metabolic capacities. The two types are type IIa, or fast-twitch oxidative glycolytic (FOG) fibers, and type IIb, or fast-twitch glycolytic (FG) fibers (see table 9.5). Type IIa fibers rely on both anaerobic and aerobic ATP production, whereas type IIb fibers rely almost exclusively on anaerobic glycolysis for ATP production.

Distribution of Fast-Twitch and Slow-Twitch Muscle Fibers

The muscles of many animals are composed primarily of either fast-twitch or slow-twitch muscle fibers. The white meat of a chicken or pheasant breast, which is composed mainly of fast-twitch fibers, appears whitish because of its relatively poor blood supply and lack of myoglobin. The muscles are adapted to contract rapidly for a short time but fatigue quickly. The red, or dark, meat of a chicken leg or of a duck breast is composed of slow-twitch fibers and appears darker because of the relatively well-developed blood supply and a large amount of myoglobin. These muscles are adapted to contract slowly for a longer time and to fatigue slowly. The distribution of slow-twitch and fast-twitch muscle fibers is consistent with the behavior of these animals. For example, pheasants can fly relatively fast for short distances, and ducks fly more slowly for long distances.

Humans exhibit no clear separation of slow-twitch and fast-twitch muscle fibers in individual muscles. Most muscles have both types of fibers, although the number of each varies for each muscle. The large postural muscles contain more slow-twitch fibers, whereas the muscles of the upper limbs contain more fast-twitch fibers.

The distribution of slow-twitch and fast-twitch muscle fibers in a given muscle is fairly constant for each individual and apparently is established developmentally. People who are good sprinters have a greater percentage of fast-twitch muscle fibers, whereas good long-distance runners have a higher percentage of slow-twitch muscle fibers in their lower limb muscles. Athletes who are able to perform a variety of anaerobic and aerobic exercises tend to have a more balanced mixture of fast-twitch and slow-twitch muscle fibers.

50 ***Contrast the structural and functional differences between slow-twitch and fast-twitch muscle fibers.***

51 ***Explain the functions for which each type is best adapted and how they are distributed.***

Effects of Exercise

Neither fast-twitch nor slow-twitch muscle fibers can be easily converted to muscle fibers of the other type without specialized training. Training can increase the size and the capacity of both types of muscle fibers to perform more efficiently. Intense exercise resulting in anaerobic respiration, such as weight lifting, increases muscular strength and mass and results in an enlargement of fast-twitch muscle fibers more than slow-twitch muscle fibers. Aerobic exercise increases the vascularity of muscle and causes an enlargement of slow-twitch muscle fibers. Aerobic exercise training can convert some fast-twitch muscle fibers that fatigue readily (type IIb) to fast-twitch muscle fibers that resist fatigue (type IIa). In addition to changes in myosin, there is an increase in the number of mitochondria in the muscle cells and an increase in their blood supply. Weight training followed by periods of rest can convert some muscle fibers from type IIa to type IIb. However, a type I muscle fiber cannot be converted to a type II fiber, and vice versa. Through specific training, a person with more fast-twitch muscle fibers can run long distances, and a person with more slow-twitch muscle fibers can increase the speed at which he or she runs.

PREDICT 9

Susan recently began racing her bicycle. Her training consists entirely of long rides on her bicycle at a steady pace. When she entered her first race, she was excited to find that she was able to keep pace with the rest of the riders. However, at the final sprint to the finish line, the other riders left her in their dust, and she finished in last place. Why was she unable to keep pace during the finishing sprint? As her coach, what advice would you give Susan about her training to prepare her better for her next race?

A muscle increases in size, or **hypertrophies** (hī-per′trō-fēz), and increases in strength and endurance in response to exercise. Conversely, a muscle that is not used decreases in size, or **atrophies** (at′rō-fēz). The muscular atrophy that occurs in limbs placed in casts for several weeks is an example. Because muscle cell numbers do not change appreciably during most of a person's life, atrophy and hypertrophy of muscles result from changes in the size of individual muscle fibers. As fibers increase in size, the number of myofibrils and sarcomeres increases within each muscle fiber. The number of nuclei in each muscle cell increases in response to exercise, but the nuclei of muscle cells cannot divide. New nuclei are added to muscle fibers because small satellite cells near skeletal muscle cells increase in number in response to exercise and then fuse with the skeletal muscle cells. Other elements, such as blood vessels, connective tissue, and mitochondria, also increase. Atrophy of muscles due to lack of exercise results from a decrease in all of these elements without a decrease in muscle cell number. Severe atrophy, such as occurs in elderly people who cannot readily move their limbs, however, does involve an irreversible decrease in the number of muscle cells and can lead to paralysis.

The increased strength of trained muscle is greater than would be expected if that strength were based only on the change in muscle size. Part of the increase in strength results from the ability of the nervous system to recruit a large number of motor units simultaneously in a trained person to perform movements with better neuromuscular coordination. In addition, trained muscles usually are restricted less by excess fat. Metabolic enzymes increase in hypertrophied muscle fibers, resulting in a greater capacity for nutrient uptake and ATP production. Improved endurance in trained muscles is in part a result of improved metabolism, increased circulation to the exercising muscles, increased numbers of capillaries, more efficient respiration, and a greater capacity for the heart to pump blood.

Anabolic Steroids and Growth Hormone

Some people take synthetic hormones called **anabolic steroids** (an-ă-bol′ik stēr′oydz, ster′oydz) to increase the size and strength of their muscles. Anabolic steroids are related to testosterone, a reproductive hormone secreted by the testes, except that they have been altered so that the reproductive effects of these compounds are minimized but their effect on skeletal muscles is maintained. Testosterone and anabolic steroids cause skeletal muscle tissue to hypertrophy. People who take large doses of an anabolic steroid exhibit an increase in body weight and total skeletal muscle mass, and many athletes believe that anabolic steroids improve performance that depends on strength. Unfortunately, evidence indicates that harmful side effects are associated with taking anabolic steroids, including periods of irritability, testicular atrophy and sterility, cardiovascular diseases (such as heart attack or stroke), and abnormal liver function. Most athletic organizations prohibit the use of anabolic steroids, and some even analyze urine samples either randomly or periodically for evidence of their use. Penalties exist for athletes who have evidence of anabolic steroid metabolites in their urine.

Some individuals use growth hormone inappropriately to increase muscle size. Growth hormone increases protein synthesis in muscle tissue, although it does not produce the same kinds of side effects as those produced by anabolic steroids. The large doses of growth hormone used by athletes, however, can cause harmful side effects if taken over a long period (see chapter 18).

52 ***What factors contribute to an increase in muscle size, strength, and endurance?***

53 ***How does anaerobic versus aerobic exercise affect muscles?***

HEAT PRODUCTION

The rate of metabolism in skeletal muscle differs before, during, and after exercise. As chemical reactions occur within cells, some energy is released in the form of heat. Normal body temperature results in large part from this heat. Because the rate of chemical reactions increases in muscle fibers during contraction, the rate of heat production also increases, causing an increase in body temperature. After exercise, elevated metabolism resulting from the oxygen deficit helps keep the body temperature elevated. If the body temperature increases as a result of increased contraction of skeletal muscle, vasodilation of blood vessels in the skin and sweating speed heat loss and keep body temperature within its normal range (see chapter 25).

When body temperature declines below a certain level, the nervous system responds by inducing **shivering,** which involves rapid skeletal muscle contractions that produce shaking rather than coordinated movements. The muscle movement increases heat production up to 18 times that of resting levels, and the heat produced during shivering can exceed that produced during moderate exercise. The elevated heat production during shivering helps raise body temperature to its normal range.

54 ***How do muscles contribute to the heat responsible for body temperature before, during, and after exercise? What is accomplished by shivering?***

SMOOTH MUSCLE

Smooth muscle is distributed widely throughout the body and is more variable in function than other muscle types. Smooth muscle cells (figure 9.22) are smaller than skeletal muscle cells, ranging from 15 to 500 μm in length and from 5 to 10 μm in diameter. They are spindle-shaped, with a single nucleus located in the middle of the cell. Compared with skeletal muscle, fewer actin and myosin myofilaments are present, and there are more actin than myosin myofilaments. The actin and myosin myofilaments overlap, but they are organized as loose bundles. Consequently, smooth muscle does not have a striated appearance. Actin myofilaments are attached to **dense bodies,** which are scattered through the cell cytoplasm, and to **dense areas,** which are in the plasma membrane. Dense bodies and dense areas are considered to be equivalent to the Z disks in skeletal muscle. Noncontractile **intermediate filaments** also attach to the

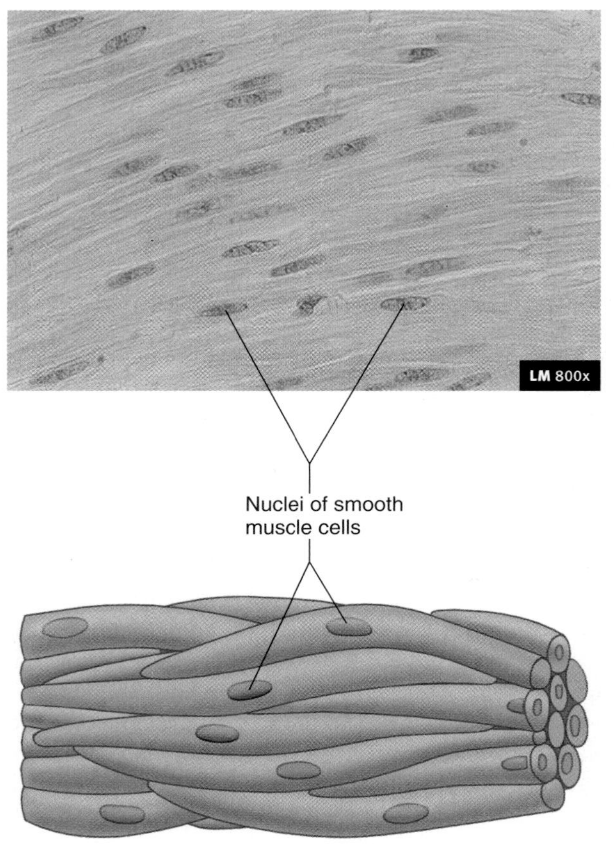

FIGURE 9.22 **Smooth Muscle Histology**

Smooth muscle tissue is made up of sheets or bundles of spindle-shaped cells, with a single nucleus located in the middle of each cell.

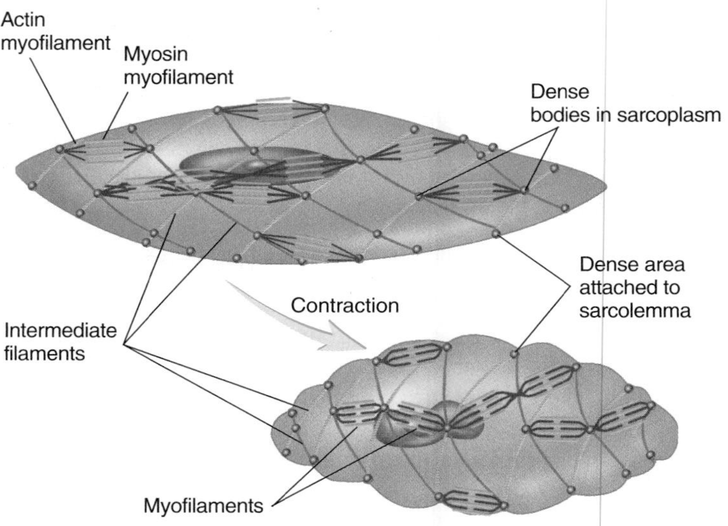

FIGURE 9.23 **Contractile Proteins in a Smooth Muscle Cell**

Bundles of contractile myofilaments containing actin and myosin are anchored at one end to dense areas in the plasma membrane and at the other end, through dense bodies, to intermediate filament. The contractile myofilaments are oriented with the long axis of the cell; when actin and myosin slide over one another during contraction, the cell shortens.

dense bodies. The intermediate filaments and dense bodies form an intracellular cytoskeleton, which has a longitudinal or spiral organization. The smooth muscle cells shorten when the actin and myosin slide over one another during contraction (figure 9.23).

Sarcoplasmic reticulum is present in smooth muscle cells, but no T tubule system exists. Some shallow invaginated areas called **caveolae** (kav-ē-ō′lē) are along the surface of the plasma membrane. The function of caveolae is not well known, but it may be similar to that of both the T tubules and the sarcoplasmic reticulum of skeletal muscle.

Some of the Ca^{2+} required to initiate contractions in smooth muscle enters the cell from the extracellular fluid and from the sarcoplasmic reticulum. The greater distance that Ca^{2+} must diffuse, the rate at which action potentials are propagated between smooth muscle cells, and the slower rate of cross-bridge formation between actin and myosin myofilaments are responsible for the slower contraction of smooth muscle, compared with skeletal muscle.

Calcium ions regulate contraction in smooth muscle cells (figure 9.24). The role of Ca^{2+} in smooth muscle differs from that in skeletal muscle cells because there are no troponin molecules associated with actin fibers of smooth muscle cells. Calcium ions that enter the cytoplasm bind to a protein called **calmodulin** (kal-mod′ū-lin). Calmodulin molecules with Ca^{2+} bound to them activate an enzyme called **myosin kinase** (kī′nās), which transfers a phosphate group from ATP to light myosin molecules on the heads of myosin molecules. Cross-bridge formation occurs when myosin myofilaments have phosphate groups bound to them. The enzymes responsible for cross-bridge cycling are slower than the enzymes in skeletal muscle, resulting in slower cross-bridge formation. Once activated, cross-bridge formation has energy requirements very similar to those for cross-bridge formation in skeletal muscle fibers.

Relaxation of smooth muscle results because of the activity of another enzyme, called **myosin phosphatase** (fos′fă-tās). This enzyme removes the phosphate group from the myosin molecules (see figure 9.24). If the phosphate is removed from myosin while the cross-bridges are attached to actin, the cross-bridges release very slowly. This explains how smooth muscle is able to sustain tension for long periods and without extensive energy expenditure. This is often called the **latch state** of smooth muscle contraction. If myosin phosphatase removes the phosphate from myosin molecules while the cross-bridges are not attached, relaxation occurs much more rapidly.

Elevated Ca^{2+} levels in the sarcoplasm of smooth muscle cells result in the activation of myosin molecules and cross-bridge formation. Also, the action of myosin phosphatase results in a high percentage of myosin molecules having their phosphates removed while bound to actin. This process favors sustained contractions, or the latch state, and a low rate of energy consumption because of the slow release of cross-bridges. As long as Ca^{2+} are present, cross-bridges re-form quickly after they are released. Consequently, many cross-bridges are intact at any given time in contracted smooth muscle.

Calcium ion levels in the sarcoplasm of smooth muscle are reduced as Ca^{2+} are actively transported across the plasma membrane, including the plasma membrane of caveolae, and into the sarcoplasmic reticulum. Relaxation occurs in response to lower intracellular levels of Ca^{2+}.

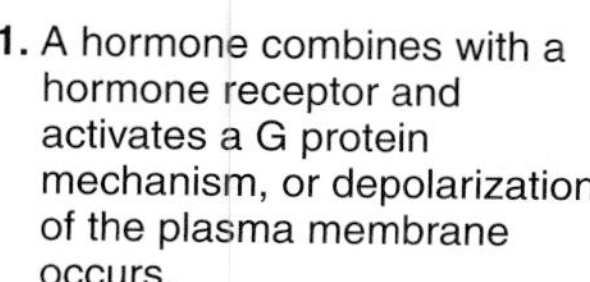

1. A hormone combines with a hormone receptor and activates a G protein mechanism, or depolarization of the plasma membrane occurs.

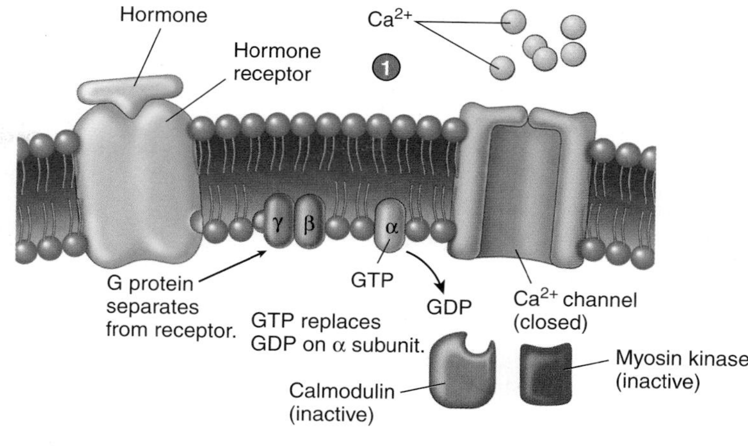

2. An α subunit opens the Ca^{2+} channel in the plasma membrane, or depolarization opens Ca^{2+} channels. Calcium ions diffuse through the Ca^{2+} channels and combine with calmodulin.

3. Calmodulin with a Ca^{2+} bound to it binds with myosin kinase and activates it.

4. Activated myosin kinase attaches phosphate from ATP to myosin heads to activate the contractile process.

5. A cycle of cross-bridge formation, movement, detachment, and cross-bridge formation occurs.

6. Relaxation occurs when myosin phosphatase removes phosphate from myosin.

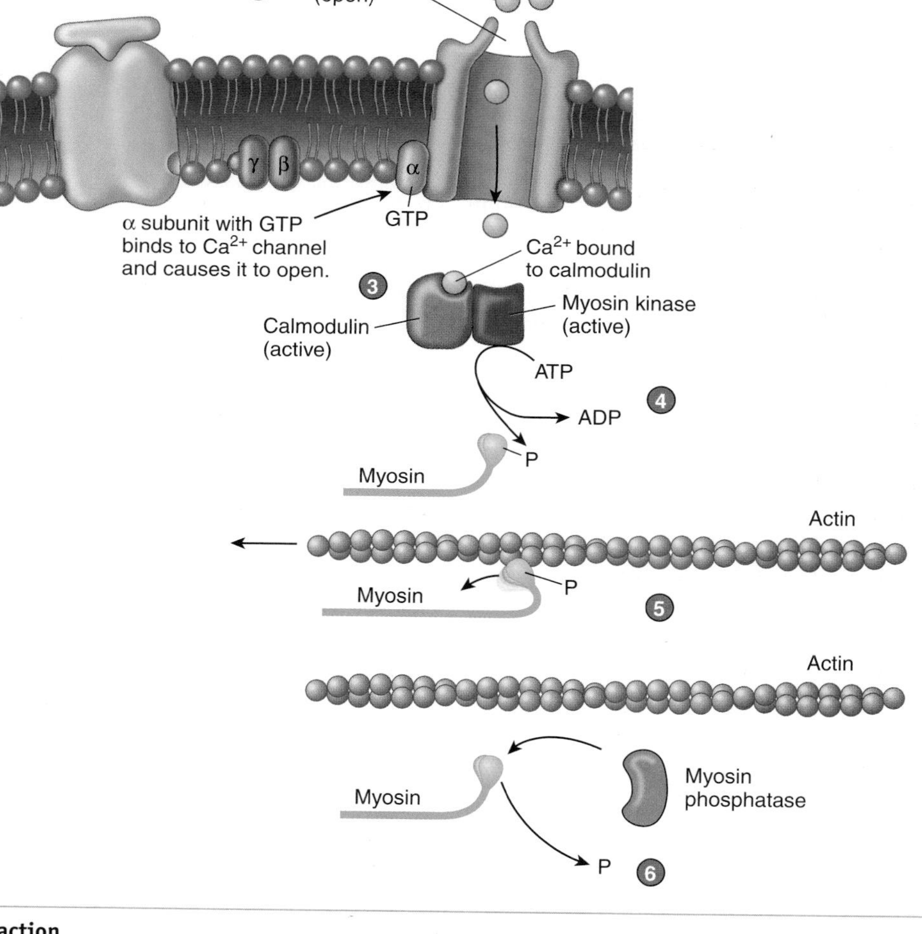

PROCESS FIGURE 9.24 Smooth Muscle Contraction

Types of Smooth Muscle

Smooth muscle can be either visceral or multiunit. **Visceral** (vis′er-ăl), or **unitary, smooth muscle** is more common than multiunit smooth muscle. It occurs in sheets and includes smooth muscle of the digestive, reproductive, and urinary tracts. Visceral smooth muscle has numerous gap junctions (see chapter 4), which allow action potentials to pass directly from one cell to another. As a consequence, sheets of smooth muscle cells function as a unit, and a wave of contraction traverses the entire smooth muscle sheet. Visceral smooth muscle is often autorhythmic, but some contracts only when stimulated. For example, visceral smooth muscles of the digestive tract contract spontaneously and at relatively regular intervals, whereas the visceral smooth muscle of the urinary bladder contracts when stimulated by the nervous system.

Multiunit smooth muscle occurs as sheets, such as in the walls of blood vessels; as small bundles, such as in the arrector pili muscles and the iris of the eye; and as single cells, such as in the capsule of the spleen. Multiunit smooth muscle has fewer gap junctions than visceral smooth muscle, and cells or groups of cells act as independent units. It normally contracts only when stimulated by nerves or hormones.

Visceral smooth muscle tissue has a different arrangement between neurons and smooth muscle fibers than in skeletal muscle tissue. The synapses are more diffuse than in skeletal muscle. Axons of nerve cells terminate in a series of dilations along the branching axons within the connective tissue among the smooth muscle cells. These dilations have vesicles containing neurotransmitter molecules. Once released, the neurotransmitter molecules diffuse among the smooth muscle cells and bind to receptors on their surfaces. Multiunit smooth muscle has synapses more like those found in skeletal muscle tissue.

55 ***Describe a typical smooth muscle cell. How does its structure and its contraction process differ from those of a skeletal muscle cell?***

56 ***Compare visceral smooth muscle and multiunit smooth muscle. Explain why visceral smooth muscle contracts as a single unit.***

Electrical Properties of Smooth Muscle

The resting membrane potential of smooth muscle cells is usually not as negative as that of skeletal muscle fibers. It normally ranges from −55 to −60 mV, compared with approximately −85 mV in skeletal muscle fibers. Furthermore, the resting membrane potential of some smooth muscle types fluctuates, with slow depolarization and repolarization phases occurring in many visceral smooth muscle cells. These slow waves of depolarization and repolarization are propagated from cell to cell for short distances (figure 9.25*a*). More "classic" action potentials can be triggered by the slow waves of depolarization and usually are propagated for longer distances (figure 9.25*b*). In addition, some smooth muscle types have action potentials with a plateau, or prolonged, depolarization (figure 9.25*c*). The slow waves in the resting membrane potential may result from a spontaneous and progressive increase in the permeability of the plasma membrane to Na^+ and Ca^{2+}, or they may be controlled by neurons. Sodium ions and Ca^{2+} diffuse into the cell through their respective channels and produce the depolarization.

Smooth muscle does not respond in an all-or-none fashion to action potentials. A series of action potentials in smooth muscle can result in a single, slow contraction followed by slow relaxation instead of individual contractions in response to each action potential, as occurs in skeletal muscle. A slow wave of depolarization that has one to several more classic-appearing action potentials superimposed on it is common in many types of smooth muscle. After the wave of depolarization, the smooth muscle undergoes contraction. Action potentials with plateaus are common in smooth muscle that exhibits periods of sustained contraction.

Spontaneously generated action potentials that lead to contractions are characteristic of visceral smooth muscle in the uterus, the ureter, and the digestive tract. Certain smooth muscle cells in these organs function as **pacemaker cells,** which tend to develop action potentials more rapidly than other cells.

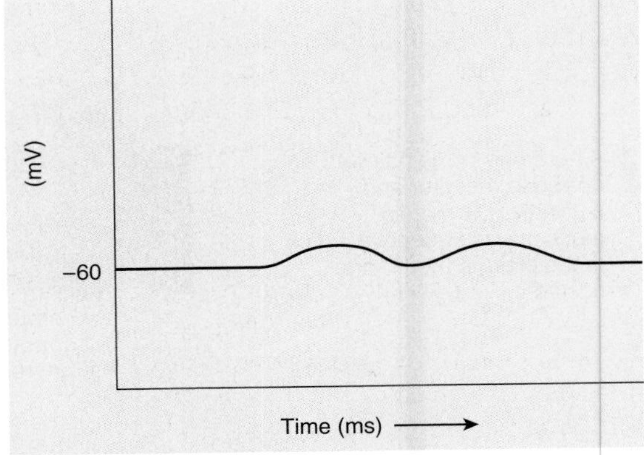

(a) **Slow waves of depolarization**

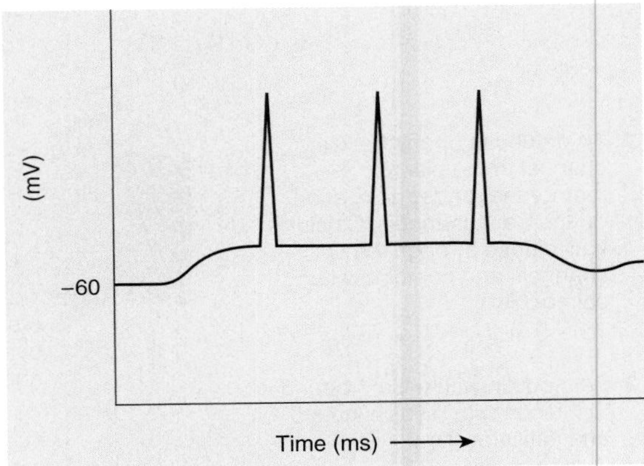

(b) **Action potentials in smooth muscle superimposed on a slow wave of depolarization**

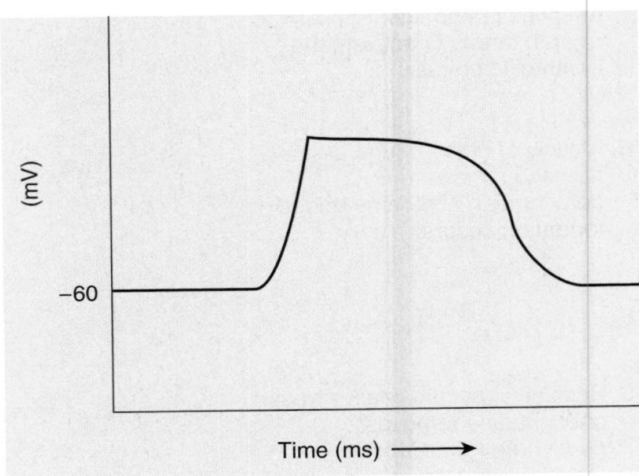

(c) **Action potential with prolonged depolarization (plateau)**

FIGURE 9.25 Membrane Potential from a Smooth Muscle Preparation

The nervous system can regulate smooth muscle contractions by increasing or decreasing action potentials carried by nerve cell axons to smooth muscle. Responses of smooth muscle cells result in depolarization and increased contraction or hyperpolarization and decreased contraction. The nervous system can also regulate the pacemaker cells.

Hormones and ligands produced locally in tissues can bind to receptors on some smooth muscle plasma membranes. The combination of a hormone or other ligands with a receptor causes ligand-gated Ca^{2+} channels in the plasma membrane to open (see figure 9.24). Calcium ions then enter the cell and cause smooth muscle contractions to occur without a major change in the membrane potential. For example, some smooth muscles contract when exposed to the hormone epinephrine because epinephrine combines with epinephrine receptors. Epinephrine combined with its receptors activates G proteins in the plasma membrane (see figure 9.24). The α subunit of the G complex can produce intracellular mediator molecules, which open the ligand-gated Ca^{2+} channels in the plasma membrane or sarcoplasmic reticulum.

PREDICT 10

Explain how a ligand could bind to a membrane-bound receptor in a smooth muscle cell and cause a sustained contraction of the smooth muscle cell for a prolonged period without a large increase in ATP breakdown.

Functional Properties of Smooth Muscle

Smooth muscle has four functional properties not seen in skeletal muscle: (1) Some visceral smooth muscle exhibits autorhythmic contractions; (2) smooth muscle tends to contract in response to being stretched but a slow increase in length produces less of a response than a more rapid increase in length; (3) smooth muscle exhibits a relatively constant tension, called **smooth muscle tone,** over a long period and maintains that same tension in response to a gradual increase in the smooth muscle length; (4) the amplitude of contraction produced by smooth muscle also remains constant, although the muscle length varies. Smooth muscle is therefore well adapted for lining the walls of hollow organs, such as the stomach and the urinary bladder. As the volume of the stomach or urinary bladder increases, only a small increase develops in the tension applied to its contents. Also, as the volume of the large and small intestines increases, the contractions that move food through them do not change dramatically in amplitude.

Regulation of Smooth Muscle

Nerves that innervate smooth muscle are part of the autonomic division of the nervous system, whereas skeletal muscle is innervated by the somatic motor nervous system (see chapter 11). The regulation of smooth muscle is therefore involuntary, and the regulation of skeletal muscle is voluntary.

The most important neurotransmitters released from nerves that innervate smooth muscle cells are acetylcholine and norepinephrine. Acetylcholine stimulates some smooth muscle types to contract and inhibits others.

Hormones are also important in regulating smooth muscle. Epinephrine, a hormone from the adrenal medulla, stimulates some smooth muscles, such as those in the blood vessels of the intestine, and inhibits other smooth muscles, such as those in the intestinal wall. Oxytocin stimulates contractions of uterine smooth muscle, especially during delivery of a baby. These and other hormones are discussed more thoroughly in chapters 17 and 18. Other chemical substances produced locally by surrounding tissues—such as histamine, prostaglandins, and by-products of metabolism—also influence smooth muscle function. For example, blood flow through capillaries is dramatically influenced by these substances (see chapter 21).

The type of receptors present on the plasma membrane to which the neurotransmitters or hormones bind determines the response of the smooth muscle. Some smooth muscle types have receptors to which acetylcholine binds, and the receptor's response is to stimulate contractions; other smooth muscle types have receptors to which acetylcholine binds, and the receptor's response is to inhibit contractions. A similar relationship exists for smooth muscle receptors for norepinephrine and certain hormones.

The receptor molecules that result in stimulation of smooth muscle contractions often open either Na^+ or Ca^{2+} channels. When these channels open, Na^+ and Ca^{2+} pass through their respective channels into the cell and cause depolarization of the plasma membrane. It is also possible for the receptor to open Ca^{2+} channels in the plasma membrane and sarcoplasmic reticulum. As a result, Ca^{2+} can diffuse into the cytoplasm of the smooth muscle cells without depolarization of the membrane potential to its threshold level and therefore not produce action potentials.

The receptor molecules that result in the inhibition of smooth muscle contractions often close Na^+ and Ca^{2+} channels or open K^+ channels. The result is hyperpolarization of the smooth muscle cells and inhibition. It is also possible for the receptors to increase the activity of the Ca^{2+} pump that transports Ca^{2+} out of the cell or into the sarcoplasmic reticulum. As a result, relaxation may occur without a change in the resting membrane potential.

The response of specific smooth muscle types to either neurotransmitters or hormones is presented in chapters dealing with the smooth muscle types.

57 ***How are spontaneous contractions produced in smooth muscle?***

58 ***List four functional properties of smooth muscle that are not seen in skeletal muscle. Can smooth muscle develop an oxygen deficit?***

59 ***How do the nervous system and hormones regulate smooth muscle contraction? How are ion channels affected by receptors that stimulate smooth muscle contractions? How are ion channels affected by receptors that inhibit smooth muscle contractions?***

CARDIAC MUSCLE

Cardiac muscle is found only in the heart; it is discussed in detail in chapter 20. Cardiac muscle tissue is striated, like skeletal muscle, but each cell usually contains one nucleus located near the center. Adjacent cells join together to form branching fibers by specialized cell-to-cell attachments called **intercalated** (in-ter′kă-lā-ted) **disks,** which have gap junctions that allow action potentials to pass from cell to cell. Some cardiac muscle cells are autorhythmic, and one part of the heart normally acts as the pacemaker. The action potentials of cardiac muscle are similar to those of nerve and skeletal muscle but have a much longer duration and refractory period. The depolarization of cardiac muscle results from the

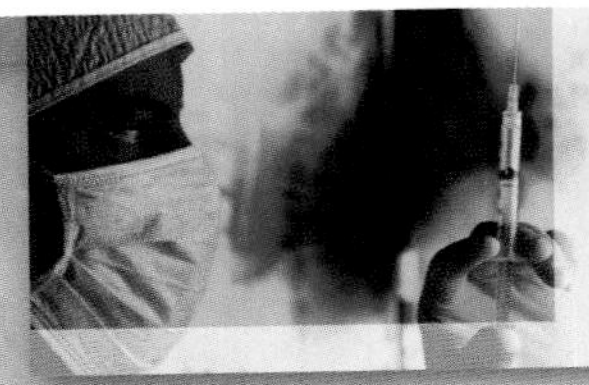

CLINICAL FOCUS

Disorders of Muscle Tissue

Muscle disorders are caused by disruption of normal innervation, degeneration and replacement of muscle cells, injury, lack of use, and disease.

ATROPHY

Muscular atrophy is a decrease in the size of muscles. Individual muscle fibers decrease in size, and a progressive loss of myofibrils occurs.

Disuse atrophy is muscular atrophy that results from a lack of muscle use. Bedridden people, people with limbs in casts, and those who are inactive for other reasons experience disuse atrophy in the unused muscles. Disuse atrophy is temporary if a muscle is exercised after it is taken out of a cast. Extreme disuse of a muscle, however, results in muscular atrophy in which skeletal muscle fibers are permanently lost and replaced by connective tissue. Immobility that occurs in bedridden elderly people can lead to permanent and severe muscular atrophy.

Denervation (dē-ner-vā′shŭn) **atrophy** results when nerves that supply skeletal muscles are severed. When motor neurons innervating skeletal muscle fibers are severed, the result is flaccid paralysis. If the muscle is reinnervated, muscle function is restored, and atrophy is stopped. If skeletal muscle is permanently denervated, however, it atrophies and exhibits permanent flaccid paralysis. Eventually, muscle fibers are replaced by connective tissue and the condition cannot be reversed.

Transcutaneous stimulators are used to supply electric stimuli to muscles that have had their nerves temporarily damaged or to muscles that are put in casts for a prolonged period. The electric stimuli keep the muscles functioning and prevent them from permanently atrophying while the nerves resupply the muscles or until the cast is removed.

FIBROSIS

Fibrosis (fī-brō′sis), or scarring, is the replacement of damaged cardiac muscle or skeletal muscle by connective tissue. Fibrosis is associated with severe trauma to skeletal muscle and with heart attack (myocardial infarction) in cardiac muscle.

FIBROSITIS

Fibrositis (fī-brō-sī′tis) is an inflammation of fibrous connective tissue, resulting in stiffness, pain, or soreness. It is not progressive, nor does it lead to tissue destruction. Fibrositis can be caused by repeated muscular strain or prolonged muscular tension.

CRAMPS

Cramps are painful, spastic contractions of muscles that usually result from an irritation within a muscle that causes a reflex contraction (see chapter 12). Local inflammation resulting from a buildup of lactic acid and fibrositis causes reflex contraction of muscle fibers surrounding the irritated region.

Fibromyalgia (fī-brō-mī-al′ja), or chronic muscle pain syndrome, has muscle pain as its main symptom. Fibromyalgia has no known cure, but it is not progressive, crippling, or life-threatening. The pain occurs in muscles or where muscles join their tendons, but not in joints. The pain is chronic, widespread, and distinguished from other causes of chronic pain by the identification of tender points in muscles, by the length of time the pain persists, and by failure to identify any other cause of the condition.

influx of both Na^+ and Ca^{2+} across the plasma membrane. The regulation of contraction in cardiac muscle by Ca^{2+} is similar to that of skeletal muscle.

60 ***Compare the structural and functional characteristics of cardiac muscle with those of skeletal muscle.***

EFFECTS OF AGING ON SKELETAL MUSCLE

Several changes occur in aging skeletal muscle that reduce muscle mass, increase the time that muscle takes to contract in response to nervous stimuli, reduce stamina, and increase recovery time. There is a loss of muscle fibers as aging occurs, and the loss begins as early as 25 years of age. By 80 years of age, 50% of the muscle mass is gone, and this is due mainly to the loss of muscle fibers. Weight-lifting exercises help slow the loss of muscle mass, but they do not prevent the loss of muscle fibers. In addition, fast-twitch muscle fibers decrease in number more rapidly than slow-twitch fibers. Most of the loss of strength and speed is due to the loss of fast-twitch muscle fibers. Also, the surface area of the synapse decreases. Consequently, action potentials in neurons stimulate action potential production in muscle cells more slowly, and action potentials may not be produced in muscle cells consistently. The number of motor neurons also decreases. Some of the muscle fibers that lose their innervation when a neuron dies are reinnervated by a branch of another motor neuron. This makes motor units in skeletal muscle fewer in number, with a greater number of muscle fibers for each neuron. This may result in less precise control of muscles. Aging is associated with a decrease in the density of capillaries in skeletal muscles, and after exercise a longer period of time is required to recover.

Many of the age-related changes in skeletal muscle can be dramatically slowed if people remain physically active. As people age, they often assume a sedentary lifestyle. Age-related changes develop more rapidly in these people. It has been demonstrated that elderly people who are sedentary can become stronger and more mobile in response to exercise.

61 ***Describe the changes in muscle mass and response time that occur in aging skeletal muscle.***

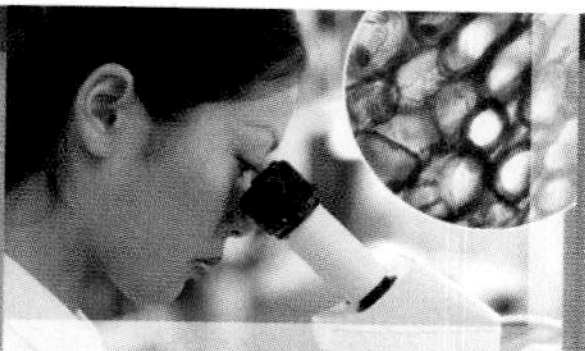

SYSTEMS PATHOLOGY

Duchenne Muscular Dystrophy

A couple became concerned about their 3-year-old son when they noticed that he was much weaker than other boys his age. The differences appeared to become more pronounced as time passed. Eventually, the boy had difficulty sitting, standing, and walking. He seemed clumsy and fell often. He had difficulty climbing stairs, and he often moved from a sitting position on the floor to a standing position by using his hands and arms to climb up his legs. His muscles appeared to be poorly developed. The couple took their son to a physician and, after several tests, they were informed that he had Duchenne muscular dystrophy.

BACKGROUND INFORMATION

Duchenne muscular dystrophy (DMD) is usually identified in children around 3 years of age when their parents notice slow motor development with progressive weakness and muscle wasting (atrophy). Typically, muscular weakness begins in the hip muscles, which causes a waddling gait. Temporary enlargement of the calf muscles is apparent in 80% of cases. The enlargement is paradoxical because the muscle fibers are actually getting smaller, but there is an increase in fibrous connective tissue and fat between the muscle fibers (figures A and B). Rising from the floor by climbing up the legs is characteristic and is caused by weakness of the lumbar and hip muscles figure C. Muscles of the shoulder are affected by 6–8 years of age. Wasting of the muscles with replacement by connective tissue results in shortened, inflexible muscles called contractures. The contractures limit movements and can cause severe deformities of the skeleton. People with DMD are usually unable to walk by 10–12 years of age, and few live beyond age 20. No effective treatment exists to prevent the progressive deterioration of muscles in DMD. Therapy primarily involves exercises to help strengthen muscles and prevent contractures. Braces and corrective surgery sometimes help correct abnormal posture caused by the advanced disease. There are other types of muscular dystrophies that are less severe than DMD. Most of these are also inherited conditions, and all are characterized by degeneration of muscle fibers, leading to atrophy and replacement by connective tissues. Research is directed at identifying the genes responsible for all types of muscular dystrophy, exploring the mechanism that leads to the disease condition, and finding an effective treatment once the mechanism is known.

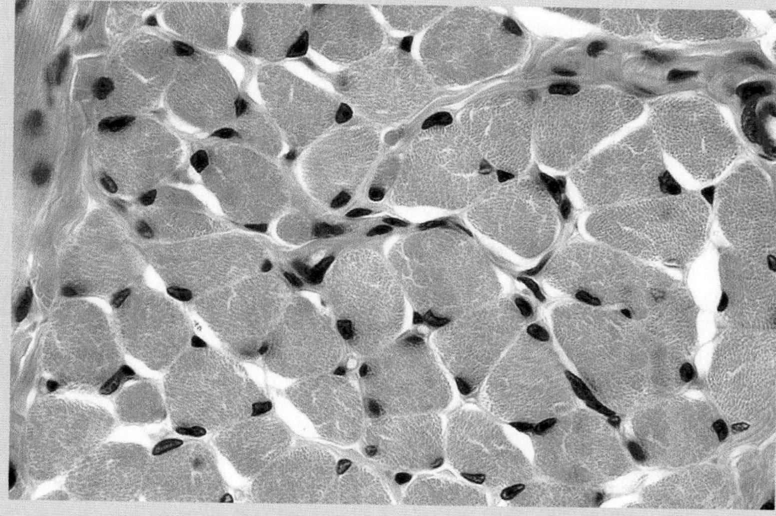

FIGURE A

Normal appearance of skeletal muscle in cross section. The muscle fibers are of normal size, and a small amount of connective tissue is located between them.

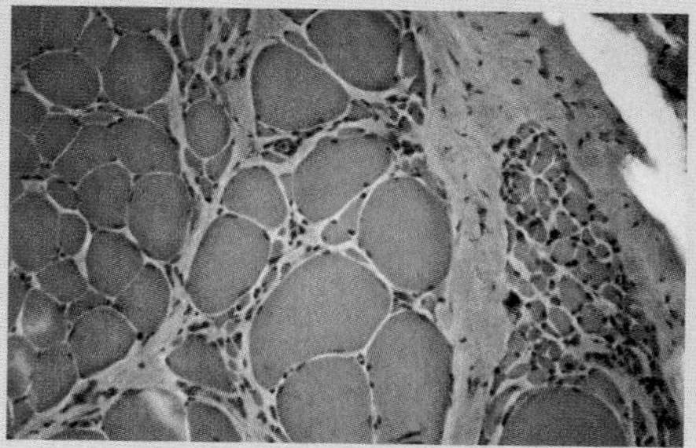

FIGURE B

Skeletal muscle biopsy of a patient with DMD demonstrates muscle fibers of varying sizes, most of them smaller than normal (atrophic). An abundance of fibrous connective tissue and fat fills the spaces previously occupied by muscle tissue.

FIGURE C

DMD is characterized by progressive muscle weakness. Rising from the floor to a standing position becomes difficult and use of the arms to push against the thighs while rising is common.

Continued

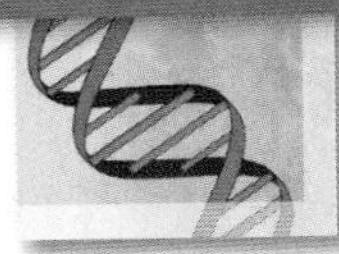

CLINICAL GENETICS

Mutations in the *dystrophin* (dis-trō′-fin; *DMD*) gene on the X chromosome are responsible for Duchenne muscular dystrophy. The dystrophin gene is responsible for producing a protein called **dystrophin.** Dystrophin plays a role in attaching actin myofilaments of myofibrils to, and regulating the activity of, other proteins in the sarcolemma (figure D). Dystrophin is thought to protect muscle fibers against mechanical stress in a normal individual. In DMD, part of the dystrophin gene is missing, and the protein it produces is nonfunctional, resulting in progressive muscular weakness and muscle contractures. DMD is an X-linked recessive disorder. Thus, although the gene is carried by females, DMD affects males almost exclusively.

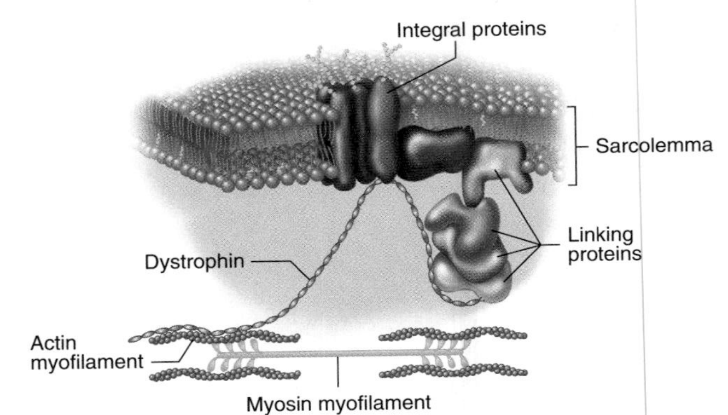

FIGURE D

The protein dystrophin links actin myofilaments within myofibrils to integral proteins in the sarcolemma of the muscle fiber and to linking proteins.

PREDICT 11

A boy with advanced Duchenne muscular dystrophy developed pulmonary edema, which is an accumulation of fluid in the lungs, and pneumonia caused by a bacterial infection. His physician diagnosed the condition in the following way: The pulmonary edema was the result of heart failure and the increased fluid in the lungs acted as a site where bacteria invaded and grew. The fact that the boy could not breathe deeply or cough effectively made the condition worse. How would the muscle tissues in a boy with advanced DMD with heart failure and ineffective respiratory movements differ from the muscle tissues in a boy with less advanced DMD?

SYSTEM INTERACTIONS — **Effect of Duchenne Muscular Dystrophy on Other Systems**

SYSTEM	INTERACTION
Skeletal	The shortened, inflexible muscles (contractures) cause severe deformities of the skeletal system. Kyphoscoliosis, severe curvature of the spinal column laterally and anteriorly, can be so severe that normal respiratory movements are impaired. Surgery is sometimes required to prevent contractures from making it impossible for the individual to sit in a wheelchair.
Nervous	Some degree of mental retardation occurs in a large percentage of people with DMD, although the specific cause is unknown. Research suggests that dystrophin is important in the formation of synapses between nerve cells.
Cardiovascular	Cardiac muscle is affected by DMD; consequently, heart failure occurs in a large number of people with advanced DMD. Cardiac involvement becomes serious in as many as 95% of cases and is one of the leading causes of death for individuals with DMD.
Lymphatic and immune	No obvious direct effects occur to the lymphatic system, but damaged muscle fibers are phagocytized by macrophages.
Respiratory	Deformity of the thorax and increasing weakness of the respiratory muscles result in inadequate respiratory movements, which causes an increase in respiratory infections, such as pneumonia. Insufficient movement of air into and out of the lungs due to weak respiratory muscles is a major contributing factor in many deaths.
Digestive	Smooth muscle tissue is affected by DMD and the reduced ability of smooth muscle to contract can result in disorders of the digestive system. Digestive system disorders include an enlarged colon diameter and twisting of the small intestine, resulting in intestinal obstruction, cramping, and reduced absorption of nutrients.
Urinary	Reduced smooth muscle function and wheelchair dependency increase the frequency of urinary tract infections.

SUMMARY

Functions of the Muscular System (p. 279)

Muscle is responsible for movement of the arms, legs, heart, and other parts of the body; maintenance of posture; respiration; production of body heat; communication; contraction of organs and vessels; and heart beat.

General Functional Characteristics of Muscle (p. 279)

Properties of Muscle

1. Muscle exhibits contractility (shortens forcefully), excitability (responds to stimuli), extensibility (can be stretched and still contract), and elasticity (recoils to resting length).
2. Muscle tissue shortens forcefully but lengthens passively.

Types of Muscle Tissue

1. The three types of muscle are skeletal, smooth, and cardiac.
2. Skeletal muscle is responsible for most body movements, smooth muscle is found in the wall of hollow organs and tubes and moves substances through them, and cardiac muscle is found in the heart and pumps blood.

Skeletal Muscle Structure (p. 279)

Skeletal muscle fibers are multinucleated and appear striated.

Connective Tissue Coverings of Muscle

1. The connective tissue of muscle is bound firmly to the connective tissue of tendons and bone.
2. Endomysium surrounds each muscle fiber.
3. Muscle fibers are covered by the external lamina and the endomysium.
4. Muscle fasciculi, bundles of muscle fibers, are covered by the perimysium.
5. Muscle consisting of fasciculi is covered by the epimysium, or muscular fascia.

Nerves and Blood Vessels

1. Motor neurons extend together with arteries and veins through the connective tissue of skeletal muscles.
2. At the level of the perimysium, axons of motor neurons branch and each branch projects to a muscle fiber to form a synapse.

Muscle Fibers

1. A muscle fiber is a single cell consisting of a plasma membrane (sarcolemma), cytoplasm (sarcoplasm), several nuclei, and myofibrils.
2. Myofibrils are composed of two major protein fibers: actin and myosin.
   - Actin myofilaments consist of a double helix of F actin (composed of G actin monomers), tropomyosin, and troponin.
   - Myosin molecules, consisting of two globular heads and a rodlike portion, constitute myosin myofilaments.
   - A cross-bridge is formed when the myosin binds to the actin.
3. Actin and myosin are organized to form sarcomeres.
   - Sarcomeres are bound by Z disks that hold actin myofilaments.
   - Six actin myofilaments (thin filaments) surround a myosin myofilament (thick filament).
   - Myofibrils appear striated because of A bands and I bands.

Sliding Filament Model (p. 285)

1. Actin and myosin myofilaments do not change in length during contraction.
2. Actin and myosin myofilaments slide past one another in a way that causes sarcomeres to shorten.
3. The I band and H zones become narrower during contraction, and the A band remains constant in length.

Physiology of Skeletal Muscle Fibers (p. 285)

Membrane Potentials

Plasma membranes are polarized, which means there is a charge difference, called the resting membrane potential, across the plasma membrane. The membrane becomes polarized because the tendency for K^+ to diffuse out of the cell is resisted by the negative charges of ions and molecules inside the cell.

Ion Channels

1. An action potential is a reversal of the resting membrane potential so that the inside of the plasma membrane becomes positive.
2. Ion channels are responsible for producing action potentials.
3. Two types of membrane channels produce action potentials, ligand-gated and voltage-gated channels.

Action Potentials

1. The charge difference across the plasma membrane of cells is the resting membrane potential.
2. Depolarization results from an increase in the permeability of the plasma membrane to Na^+.
3. An all-or-none action potential is produced if depolarization reaches threshold.
4. The depolarization phase of the action potential results from many Na^+ channels opening in an all-or-none fashion.
5. The repolarization phase of the action potential occurs when the Na^+ channels close and K^+ channels open briefly.
6. Propagation of action potentials along the plasma membrane of neurons and skeletal muscle fibers occurs in an all-or-none fashion.

Neuromuscular Junction

1. The presynaptic terminal of the axon is separated from the postsynaptic membrane of the muscle fiber by the synaptic cleft.
2. Acetylcholine released from the presynaptic terminal binds to receptors of the postsynaptic membrane, thereby changing membrane permeability and producing an action potential.
3. After an action potential occurs, acetylcholinesterase splits acetylcholine into acetic acid and choline. Choline is reabsorbed into the presynaptic terminal to re-form acetylcholine.

Excitation–Contraction Coupling

1. Invaginations of the sarcolemma form T tubules, which wrap around the sarcomeres.
2. A triad is a T tubule and two terminal cisternae (an enlarged area of sarcoplasmic reticulum).
3. Action potentials move into the T tubule system, causing Ca^{2+} channels to open to release Ca^{2+} from the sarcoplasmic reticulum.
4. Calcium ions diffuse from the sarcoplasmic reticulum to the myofilaments and bind to troponin, causing tropomyosin to move and expose active sites on actin to myosin.

5. Contraction occurs when myosin heads bind to active sites on actin, myosin changes shape, and actin is pulled past the myosin.
6. Relaxation occurs when calcium is taken up by the sarcoplasmic reticulum, ATP binds to myosin, and tropomyosin moves back so that active sites on actin are no longer exposed to myosin.

Cross-Bridge Movement

1. ATP is required for the cycle of cross-bridge formation, movement, and release.
2. ATP is also required to transport Ca^{2+} into the sarcoplasmic reticulum and to maintain normal concentration gradients across the plasma membrane.

Muscle Relaxation

1. Calcium ions are transported into the sarcoplasmic reticulum.
2. Calcium ions diffuse away from troponin, preventing further cross-bridge formation.

Physiology of Skeletal Muscle (p. 295)

Muscle Twitch

1. A muscle twitch is the contraction of a single muscle fiber or a whole muscle in response to a stimulus.
2. A muscle twitch has lag, contraction, and relaxation phases.

Stimulus Strength and Muscle Contraction

1. For a given condition, a muscle fiber or motor unit contracts with a consistent force in response to each action potential, which is called the all-or-none law of skeletal muscle contraction.
2. For a whole muscle, a stimulus of increasing magnitude results in graded contractions of increased force as more motor units are recruited (multiple motor unit summation).

Stimulus Frequency and Muscle Contraction

1. A stimulus of increasing frequency increases the force of contraction (multiple-wave summation).
2. Incomplete tetanus is partial relaxation between contractions, and complete tetanus is no relaxation between contractions.
3. The force of contraction of a whole muscle increases with increased frequency of stimulation because of an increasing concentration of Ca^{2+} around the myofibrils and because of complete stretching of muscle elastic elements.
4. Treppe is an increase in the force of contraction during the first few contractions of a rested muscle.

Types of Muscle Contractions (p. 299)

1. Isometric contractions cause a change in muscle tension but no change in muscle length.
2. Isotonic contractions cause a change in muscle length but no change in muscle tension.
3. Concentric contractions cause muscles to shorten and tension to increase.
4. Eccentric contractions cause muscles to increase in length and the tension to decrease gradually.
5. Muscle tone is maintenance of a steady tension for long periods.
6. Asynchronous contractions of motor units produce smooth, steady muscle contractions.

Length Versus Tension

Muscle contracts with less than maximum force if its initial length is shorter or longer than optimum.

Fatigue (p. 301)

Fatigue, the decreased ability to do work, can be caused by the central nervous system, depletion of ATP in muscles, or depletion of acetylcholine in the neuromuscular synapse. Physiologic contracture (the inability of muscles to contract or relax) and rigor mortis (stiff muscles after death) result from inadequate amounts of ATP.

Energy Sources (p. 303)

Energy for muscle contraction comes from ATP.

Creatine Phosphate

ATP can be synthesized when ADP reacts with creatine phosphate to form creatine and ATP. ATP from this source provides energy for a short time during intense exercise.

Anaerobic Respiration

ATP is synthesized by anaerobic respiration and is used to provide energy for a short time during intense exercise. Anaerobic respiration produces ATP less efficiently but more rapidly than aerobic respiration. Lactic acid levels increase because of anaerobic respiration.

Aerobic Respiration

ATP is synthesized by aerobic respiration. Although ATP is produced more efficiently, it is produced more slowly. Aerobic respiration produces energy for muscle contractions under resting conditions or during exercises such as long-distance running.

Oxygen Deficit and Recovery Oxygen Consumption

After anaerobic respiration, aerobic respiration is higher than normal, as the imbalances of homeostasis that occurred during exercise become rectified.

Slow and Fast Fibers (p. 305)

Slow-Twitch Oxidative Muscle Fibers

Slow-twitch muscle fibers split ATP slowly and have a well-developed blood supply, many mitochondria, and myoglobin.

Fast-Twitch Muscle Fibers

Fast-twitch muscle fibers split ATP rapidly.

1. Type IIa fibers have a well-developed blood supply, more mitochondria, and more myoglobin.
2. Type IIb fibers have large amounts of glycogen, a poor blood supply, fewer mitochondria, and little myoglobin.

Distribution of Fast-Twitch and Slow-Twitch Muscle Fibers

People who are good sprinters have a greater percentage of fast-twitch muscle fibers, and people who are good long-distance runners have a higher percentage of slow-twitch muscle fibers in their leg muscles.

Effects of Exercise

1. Muscles increase (hypertrophy) or decrease (atrophy) in size because of a change in the size of muscle fibers.
2. Anaerobic exercise develops Type IIb fibers. Aerobic exercise develops Type I fibers and changes Type IIb fibers into Type IIa fast-twitch fibers.

Heat Production (p. 307)

1. Heat is produced as a by-product of chemical reactions in muscles.
2. Shivering produces heat to maintain body temperature.

Smooth Muscle (p. 307)

1. Smooth muscle cells are spindle-shaped with a single nucleus. They have actin myofilaments and myosin myofilaments but are not striated.
2. The sarcoplasmic reticulum is poorly developed, and caveolae may function as a T tubule system.
3. Calcium ions enter the cell to initiate contraction; calmodulin binds to Ca^{2+} and activates an enzyme that transfers a phosphate group from ATP to myosin. When phosphate groups are attached to myosin, cross-bridges form.
4. Relaxation results when myosin phosphatase removes a phosphate group from the myosin molecule.
   - If phosphate is removed while the cross-bridges are attached, relaxation occurs very slowly, and this is referred to as the catch phase.
   - If phosphate is removed while the cross-bridges are not attached, relaxation occurs rapidly.

Types of Smooth Muscle

1. Visceral smooth muscle fibers contract slowly, have gap junctions (and thus function as a single unit), and can be autorhythmic.
2. Multiunit smooth muscle fibers contract rapidly in response to stimulation by neurons and function independently.

Electrical Properties of Smooth Muscle

1. Spontaneous contractions result from Na^+ and Ca^{2+} leakage into cells; Na^+ and Ca^{2+} movement into the cell is involved in depolarization.
2. The autonomic nervous system, hormones, and chemicals produced locally can inhibit or stimulate action potentials (and thus contractions). Hormones can also stimulate or inhibit contractions without affecting membrane potentials.

Functional Properties of Smooth Muscle

1. Smooth muscle can contract autorhythmically in response to stretch or when stimulated by the autonomic nervous system or hormones.
2. Smooth muscle maintains a steady tension for long periods.
3. The force of smooth muscle contraction remains nearly constant, despite changes in muscle length.
4. Smooth muscle does not develop an oxygen deficit.

Regulation of Smooth Muscle

1. Smooth muscle is innervated by the autonomic nervous system and is involuntary.
2. Hormones are important in regulating smooth muscle. Some hormones can increase the Ca^{2+} permeability of some smooth muscle membranes and therefore cause contraction without a change in the resting membrane potential.

Cardiac Muscle (p. 311)

Cardiac muscle fibers are striated, have a single nucleus, are connected by intercalated disks (thus function as a single unit), and are capable of autorhythmicity.

Effects of Aging on Skeletal Muscle (p. 312)

1. Aging skeletal muscle is associated with reduced muscle mass, increased response time, and increased time that muscle takes to contract in response to nervous stimuli.
2. Muscle fibers decrease in number, motor units decrease in number, and recovery time increases.

REVIEW AND COMPREHENSION

1. Which of these is true of skeletal muscle?
   a. spindle-shaped cells
   b. under involuntary control
   c. many peripherally located nuclei per muscle cell
   d. forms the walls of hollow internal organs
   e. may be autorhythmic
2. Which of these is *not* a major functional characteristic of muscle?
   a. contractility
   b. elasticity
   c. excitability
   d. extensibility
   e. secretability
3. The connective tissue sheath that surrounds a muscle fasciculus is the
   a. perimysium.
   b. endomysium.
   c. epimysium.
   d. hypomysium.
   e. external lamina.
4. Given these structures:
   1. whole muscle
   2. muscle fiber (cell)
   3. myofilament
   4. myofibril
   5. muscle fasciculus

   Choose the arrangement that lists the structures in the correct order from the largest to the smallest structure.
   a. 1,2,5,3,4
   b. 1,2,5,4,3
   c. 1,5,2,3,4
   d. 1,5,2,4,3
   e. 1,5,4,2,3
5. Each myofibril
   a. is made up of many muscle fibers.
   b. contains sarcoplasmic reticulum.
   c. is made up of many sarcomeres.
   d. contains T tubules.
   e. is the same thing as a muscle fiber.
6. Myosin myofilaments are
   a. attached to the Z disk.
   b. found primarily in the I band.
   c. thinner than actin myofilaments.
   d. absent from the H zone.
   e. attached to filaments that form the M line.
7. Which of these statements about the molecular structure of myofilaments is true?
   a. Tropomyosin has a binding site for Ca^{2+}.
   b. The head of the myosin molecule binds to an active site on G actin.
   c. ATPase is found on troponin.
   d. Troponin binds to the rodlike portion of myosin.
   e. Actin molecules have a hingelike portion, which bends and straightens during contraction.
8. The part of the sarcolemma that invaginates into the interior of the skeletal muscle cells is the
   a. T tubule system.
   b. sarcoplasmic reticulum.
   c. myofibrils.
   d. terminal cisternae.
   e. mitochondria.
9. During the depolarization phase of an action potential, the permeability of the plasma membrane to
   a. Ca^{2+} increases.
   b. Na^+ increases.
   c. K^+ increases.
   d. Ca^{2+} decreases.
   e. Na^+ decreases.
10. During depolarization, the inside of the membrane
    a. becomes more negative than the outside of the membrane.
    b. becomes more positive than the outside of the membrane.
    c. is unchanged.

11. During repolarization of the plasma membrane,
   a. Na^+ move to the inside of the cell.
   b. Na^+ move to the outside of the cell.
   c. K^+ move to the inside of the cell.
   d. K^+ move to the outside of the cell.
12. Given these events:
   1. Acetylcholine is broken down into acetic acid and choline.
   2. Acetylcholine diffuses across the synaptic cleft.
   3. Action potential reaches the terminal branch of the motor neuron.
   4. Acetylcholine combines with a ligand-gated ion channel.
   5. Action potential is produced on the muscle fiber's plasma membrane.

   Choose the arrangement that lists the events in the order they occur at a neuromuscular junction.
   a. 2,3,4,1,5 c. 3,4,2,1,5 e. 5,1,2,4,3
   b. 3,2,4,5,1 d. 4,5,2,1,3
13. Acetylcholinesterase is an important molecule in the neuromuscular junction because it
   a. stimulates receptors on the presynaptic terminal.
   b. synthesizes acetylcholine from acetic acid and choline.
   c. stimulates receptors within the postsynaptic membrane.
   d. breaks down acetylcholine.
   e. causes the release of Ca^{2+} from the sarcoplasmic reticulum.
14. Given these events:
   1. Sarcoplasmic reticulum releases Ca^{2+}.
   2. Sarcoplasmic reticulum takes up Ca^{2+}.
   3. Calcium ions diffuse into myofibrils.
   4. Action potential moves down the T tubule.
   5. Sarcomere shortens.
   6. Muscle relaxes.

   Choose the arrangement that lists the events in the order they occur following a single stimulation of a skeletal muscle cell.
   a. 1,3,4,5,2,6 c. 4,1,3,5,2,6 e. 5,1,4,3,2,6
   b. 2,3,5,4,6,1 d. 4,2,3,5,1,6
15. Given these events:
   1. Calcium ions combine with tropomyosin.
   2. Calcium ions combine with troponin.
   3. Tropomyosin pulls away from actin.
   4. Troponin pulls away from actin.
   5. Tropomyosin pulls away from myosin.
   6. Troponin pulls away from myosin.
   7. Myosin binds to actin.

   Choose the arrangement that lists the events in the order they occur during muscle contraction.
   a. 1,4,7 c. 1,3,7 e. 2,3,7
   b. 2,5,6 d. 2,4,7
16. Which of these regions shortens during skeletal muscle contraction?
   a. A band c. H zone e. both b and c
   b. I band d. both a and b
17. With stimuli of increasing strength, which of these is capable of a graded response?
   a. nerve axon c. motor unit
   b. muscle fiber d. whole muscle
18. Considering the force of contraction of a skeletal muscle cell, multiple-wave summation occurs because of
   a. increased strength of action potentials on the plasma membrane.
   b. a decreased number of cross-bridges formed.
   c. an increase in Ca^{2+} concentration around the myofibrils.
   d. an increased number of motor units recruited.
   e. increased permeability of the sarcolemma to Ca^{2+}.
19. Which of these events occurs during the lag (latent) phase of muscle contraction?
   a. cross-bridge movement
   b. active transport of Ca^{2+} into the sarcoplasmic reticulum
   c. Ca^{2+} binding to troponin
   d. sarcomere shortening
   e. ATP broken down to ADP
20. A weight lifter attempts to lift a weight from the floor, but the weight is so heavy that he is unable to move it. The type of muscle contraction the weight lifter is using is mostly
   a. isometric. c. isokinetic. e. eccentric.
   b. isotonic. d. concentric.
21. An active tension curve illustrates
   a. how isometric contractions occur.
   b. that the greatest force of contraction occurs if a muscle is not stretched at all.
   c. that passive tension can create active tension.
   d. that optimal overlap of actin and myosin produces the greatest force of contraction.
   e. that the greatest force of contraction occurs with little or no overlap of actin and myosin.
22. Which of these types of fatigue is the most common?
   a. muscular fatigue c. synaptic fatigue
   b. psychologic fatigue d. army fatigue
23. Given these conditions:
   1. low ATP levels
   2. little or no transport of Ca^{2+} into the sarcoplasmic reticulum
   3. release of cross-bridges
   4. Na^+ accumulation in the sarcoplasm
   5. formation of cross-bridges

   Choose the conditions that occur in both physiologic contracture and rigor mortis.
   a. 1,2,3 c. 1,2,3,4 e. 1,2,3,4,5
   b. 1,2,5 d. 1,2,4,5
24. Jerry Jogger's 3-mile run every morning takes about 30 minutes. Which of these sources provides most of the energy for his run?
   a. aerobic respiration c. creatine phosphate
   b. anaerobic respiration d. stored ATP
25. Which of these conditions would one expect to find within the leg muscle cells of a world-class marathon runner?
   a. myoglobin-poor
   b. contract very quickly
   c. primarily anaerobic
   d. numerous mitochondria
   e. large deposits of glycogen
26. Which of these increases the least as a result of muscle hypertrophy?
   a. number of sarcomeres
   b. number of myofibrils
   c. number of fibers
   d. blood vessels and mitochondria
   e. connective tissue
27. Relaxation in smooth muscle occurs when
   a. myosin kinase attaches phosphate to the myosin head.
   b. Ca^{2+} bind to calmodulin.
   c. myosin phosphatase removes phosphate from myosin.
   d. Ca^{2+} channels open.
   e. Ca^{2+} are released from the sarcoplasmic reticulum.

28. Compared with skeletal muscle, visceral smooth muscle
    a. has the same ability to be stretched.
    b. when stretched loses the ability to contract forcefully.
    c. maintains about the same tension, even when stretched.
    d. cannot maintain long, steady contractions.
    e. can accumulate a substantial oxygen deficit.
29. Which of these often has spontaneous contractions?
    a. multiunit smooth muscle
    b. visceral smooth muscle
    c. skeletal muscle
    d. both a and b
    e. both b and c
30. Which of these statements concerning aging and skeletal muscle is correct?
    a. There is a loss of muscle fibers with aging.
    b. Slow-twitch fibers decrease in number faster than fast-twitch fibers.
    c. Loss of strength and speed is due mainly to loss of neuromuscular junctions.
    d. There is an increase in density of capillaries in skeletal muscle.
    e. The number of motor neurons remains constant.

Answers in Appendix E

CRITICAL THINKING

1. Bob Canner improperly canned some homegrown vegetables. As a result, he contracted botulism poisoning after eating the vegetables. His symptoms included difficulty in swallowing and breathing. Eventually, he died of respiratory failure (his respiratory muscles relaxed and would not contract). Assuming that botulism toxin affects the neuromuscular synapse, propose the ways that the toxin produced the observed symptoms.
2. A patient is thought to be suffering from either muscular dystrophy or myasthenia gravis. How would you distinguish between the two conditions?
3. Under certain circumstances, the actin and myosin myofilaments can be extracted from muscle cells and placed in a beaker. They subsequently bind together to form long filaments of actin and myosin. The addition of what cell organelle or molecule to the beaker would make the actin and myosin myofilaments unbind?
4. Explain the effect of a lower than normal temperature on each of the processes that occur in the lag phase of muscle contraction.
5. Design an experiment to test the following hypothesis: Muscle A has the same number of motor units as muscle B. Assume you could stimulate the nerves that innervate skeletal muscles with an electronic stimulator and monitor the tension produced by the muscles.
6. Explain what is happening at the level of individual sarcomeres when an individual is using his or her biceps brachii muscle to hold a weight in a constant position. Contrast this with what is happening at the level of individual sarcomeres when an individual lowers the weight, as well as when the individual raises the weight.
7. Predict the shape of an active tension curve for visceral smooth muscle. How does it differ from the active tension curve for skeletal muscle?
8. A researcher is investigating the composition of muscle tissue in the gastrocnemius muscles (in the calf of the leg) of athletes. A needle biopsy is taken from the muscle, and the concentration (or enzyme activity) of several substances is determined. Describe the major differences this researcher sees when comparing the muscles from athletes who perform in the following events: 100 m dash, weight lifting, and 10,000 m run.
9. Harvey Leche milked cows by hand each morning before school. One morning, he slept later than usual and had to hurry to get to school on time. As he was milking the cows as fast as he could, his hands became very tired, and for a short time he could neither release his grip nor squeeze harder. Explain what happened.
10. Explain how the force produced by smooth muscle contractions in the stomach change between 5 minutes after eating a meal and 1 hour after eating a meal.
11. Shorty McFleet noticed that his rate of respiration was elevated after running a 100 m race but was not as elevated after running slowly for a much longer distance. How would you explain this?
12. High blood K^+ concentrations cause depolarization of the resting membrane potential. Predict the effect of high blood K^+ levels on smooth muscle function. Explain.
13. Predict and explain the response if the ATP concentration in a muscle that was exhibiting rigor mortis could be instantly increased.
14. A hormone stimulates the smooth muscle of a blood vessel to contract. Although the hormone causes a small change in membrane potential, the smooth muscle contracts substantially. Explain.
15. Experiments were performed in an anatomy and physiology laboratory. The rate and depth of respiration for a resting student were determined. In experiment A, students ran in place for 30 seconds and then immediately sat down and relaxed, and respiration rate and depth were again determined. Experiment B was just like experiment A, except that the students held their breath while running in place. What differences in respiration would you expect for the two different experiments? Explain the basis for your predictions.
16. After learning about muscle fiber types in his anatomy and physiology class, Alex started to notice differences in the color of the meat in the turkey he ate for lunch. Some of the meat was very white and some of it was much darker. From the color of the meat, Alex guessed which muscles the bird used for maintenance of posture and/or slow movements such as walking, and which muscles were used for quicker movements, such as running or flying. What type of muscle fiber predominates in white meat and in dark meat? Explain how the color of the meat relates to the function of the muscle.

Answers in Appendix F

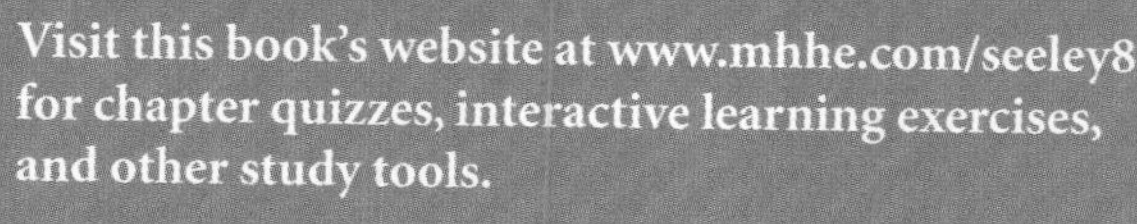

10 Muscular System

Gross Anatomy

Mannequins are rigid, expressionless, immobile recreations of the human form. They cannot walk or talk. One of the major characteristics of a living human being is our ability to move about. Without muscles, humans would be little more than mannequins. We would not be able to hold this book. We would not be able to blink, so our eyes would dry out. None of these inconveniences would bother us for long because we would not be able to breathe, either.

We use our skeletal muscles all the time—even when we are not "moving." Postural muscles are constantly contracting to keep us sitting or standing upright. Respiratory muscles are constantly functioning to keep us breathing, even when we sleep. Communication of any kind requires skeletal muscles, whether we are writing, typing, or speaking. Even silent communication with hand signals or facial expression requires skeletal muscle function.

This chapter explains the *general principles* (p. 321) of the muscular system and describes in detail the *head muscles* (p. 327), *trunk muscles* (p. 340), *upper limb muscles* (p. 346), and *lower limb muscles* (p. 359).

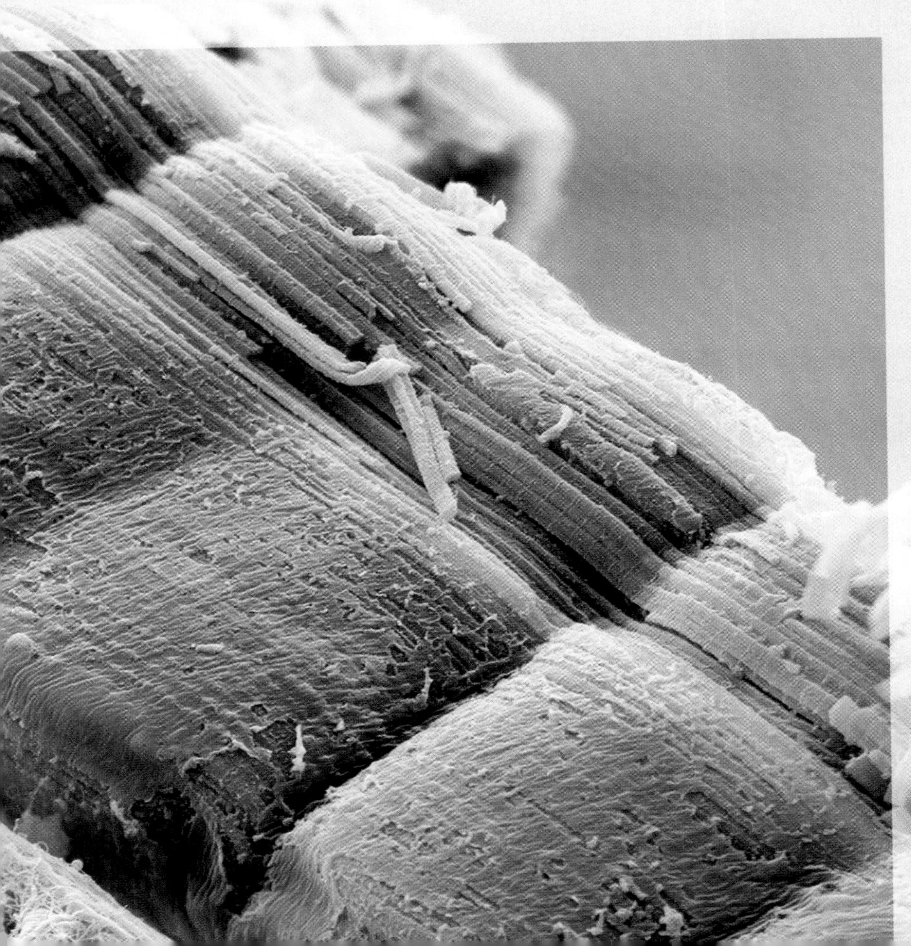

Colorized SEM of skeletal muscle.

Muscular System

GENERAL PRINCIPLES

This chapter is devoted to the description of the major named skeletal muscles. The structure and function of cardiac and smooth muscle are considered in other chapters. Most skeletal muscles extend from one bone to another and cross at least one joint. Muscle contraction causes most body movements by pulling one of the bones toward the other across a movable joint. The **action** of a muscle is the movement accomplished when it contracts. Some muscles are not attached to bone at both ends. For example, some facial muscles attach to the skin, which moves as the muscles contract.

The two points of attachment of each muscle are its origin and insertion. The **origin,** also called the **fixed end** or the head, is usually the most stationary, usually proximal end of the muscle. The **insertion,** also called the **mobile end,** is usually the distal end of the muscle attached to the bone undergoing the greatest movement. The part of the muscle between the origin and the insertion is the **belly** (figure 10.1). Some muscles, such as the biceps brachii with two heads and the triceps brachii with three heads, have multiple origins, or heads. At the attachment point, each muscle is connected to bone by **tendons.** Tendons are long, cablelike structures; broad, sheetlike structures called **aponeuroses** (ap′ō-noo-rō′sēz); or short, almost nonexistent structures.

Muscles are typically grouped so that the action of one muscle or group of muscles is opposed by that of another muscle or group of muscles. For example, the biceps brachii flexes the elbow and the triceps brachii extends the elbow. A muscle that accomplishes a certain movement, such as flexion, is called the **agonist** (ag′ō-nist; a contest). A muscle acting in opposition to an agonist is called an **antagonist** (an-tag′ō-nist). The biceps brachii is the agonist in elbow flexion, whereas the triceps brachii is the antagonist, which extends the elbow.

Muscles also tend to function in groups to accomplish specific movements. For example, the deltoid, biceps brachii, and pectoralis major all help flex the shoulder. Furthermore, many muscles are members of more than one group, depending on the type of movement being considered. For example, the anterior part of the deltoid muscle functions with the flexors of the shoulder, whereas the posterior part functions with the extensors of the shoulder. Members of a group of muscles working together to produce a movement are called **synergists** (sin′er-jistz). For example, the biceps brachii and brachialis are synergists in elbow flexion. Among a group of synergists, if one muscle plays the major role in accomplishing the desired movement, it is the **prime mover.** The brachialis is the prime mover in flexing the elbow. **Fixators** are muscles that hold one bone in place relative to the body while a usually more distal bone is moved. The origin of a prime mover is often stabilized by fixators so that its action occurs at its point of insertion. The muscles of the scapula act as fixators to hold the scapula in place while other muscles contract to move the humerus.

1 ***Define origin and insertion; agonist and antagonist; and synergist, prime mover, and fixator.***

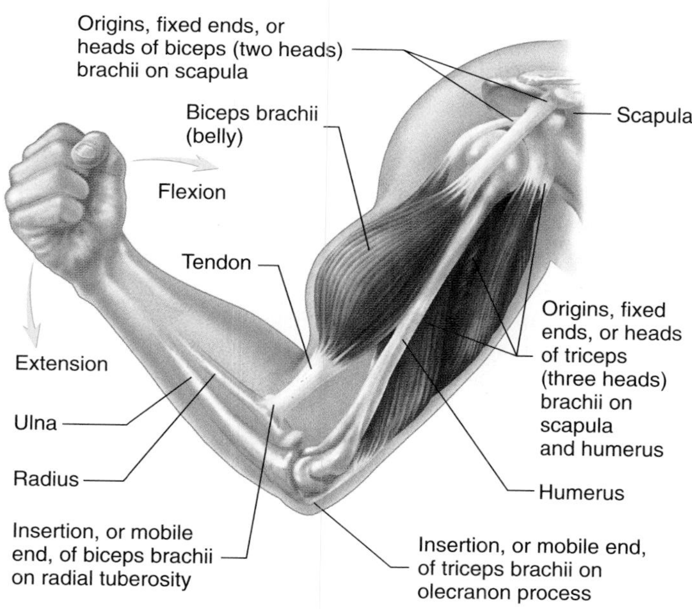

FIGURE 10.1 **Muscle Attachment**

Muscles are attached to bones by tendons. The biceps brachii has two heads, which originate on the scapula. The triceps brachii has three heads, which originate on the scapula and humerus. The biceps brachii inserts onto the radial tuberosity and onto nearby connective tissue. The triceps brachii inserts onto the olecranon process of the ulna.

Muscle Shapes

Muscles come in a wide variety of shapes. The shape and size of any given muscle greatly influence the degree to which it can contract and the amount of force it can generate. Muscle shapes are grouped into three classes, according to the orientation of the fasciculi: pennate, straight, and orbicular.

Some muscles have their fasciculi (bundles of muscle fibers that can be distinguished by the unaided eye) arranged like the barbs of a feather on two sides of a common tendon and therefore are called **pennate** (pen′āt; *pennatus* is Latin, meaning feather), or **bipennate,** muscles. A muscle with all fasciculi on one side of the tendon is **semipennate,** or **unipennate,** and a muscle with fasciculi arranged at many places around the central tendon is **multipennate** (figure 10.2*a*). The long tendons of pennate muscles can extend for some distance between a muscle belly and its insertion. The pennate arrangement allows a large number of fasciculi to attach to a single tendon with the force of contraction concentrated at the tendon. The muscles that extend the knee are multipennate muscles.

In other muscles, called **straight** muscles, fasciculi are organized parallel to the long axis of the muscle (figure 10.2*b*). As a consequence, the muscles shorten to a greater degree than do pennate muscles because the fasciculi are in a direct line with the tendon; however, they contract with less force because fewer total fascicles are attached to the tendon. The hyoid muscles are an example of straight muscles.

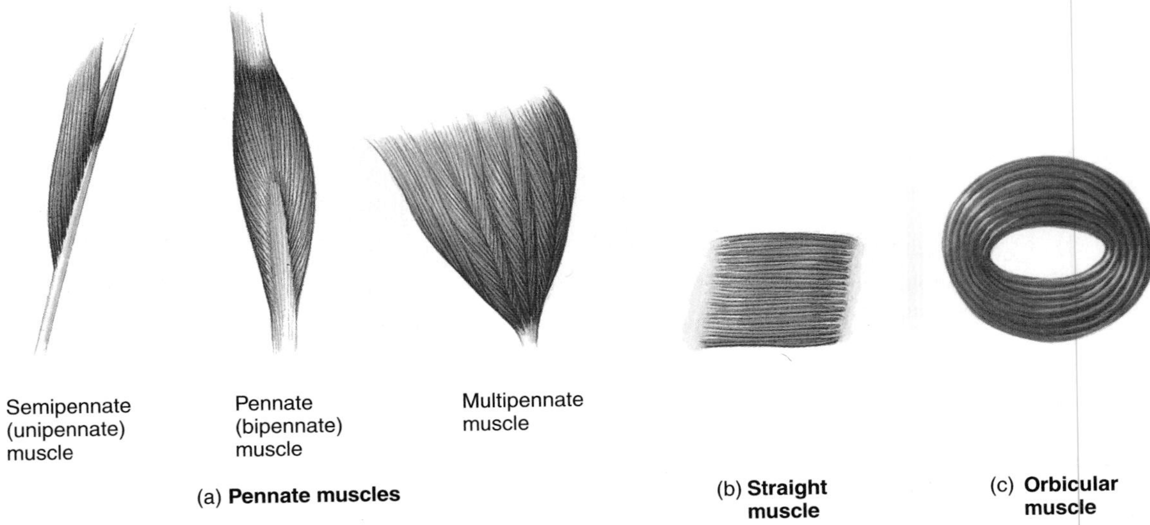

FIGURE 10.2 Examples of Muscle Types
Three classes of muscle shapes based on fasciculi arrangement.

Orbicular, or **circular,** muscles, such as the orbicularis oris and orbicularis oculi (figure 10.2*c*) have their fasciculi arranged in a circle around an opening and act as sphincters to close the opening.

Muscles can also have specific shapes, such as quadrate, triangular, trapezium, rhomboidal, or fusiform (figure 10.3*a*). Muscles also can have multiple components, such as two bellies or two heads. A digastric muscle has two bellies separated by a tendon, whereas a bicipital muscle has two origins (heads) and a single insertion (figure 10.3*b*).

2 ***Describe the different shapes of muscles. How are the shapes related to the muscle's force of contraction and the range of movement the contraction produces?***

Nomenclature

Muscles are named according to their location, size, shape, orientation of fasciculi, origin and insertion, number of heads, and function. Recognizing the descriptive nature of muscle names makes learning those names much easier.

1. *Location.* A pectoralis (chest) muscle is located in the chest, a gluteus (buttock) muscle is located in the buttock, and a brachial (arm) muscle is located in the arm.
2. *Size.* The gluteus maximus (large) is the largest muscle of the buttock, and the gluteus minimus (small) is the smallest. A longus (long) muscle is longer than a brevis (short) muscle.
3. *Shape.* The deltoid (triangular) muscle is triangular, a quadratus (quadrate) muscle is rectangular, and a teres (round) muscle is round.
4. *Orientation of fasciculi.* A rectus (straight) muscle has muscle fasciculi running straight with the axis of the structure to which the muscle is associated, whereas the fasciculi of an oblique muscle lie oblique to the longitudinal axis of the structure.
5. *Origin and insertion.* The sternocleidomastoid originates on the sternum and clavicle and inserts onto the mastoid process of the temporal bone. The brachioradialis originates in the arm (brachium) and inserts onto the radius.
6. *Number of heads.* A biceps muscle has two heads, and a triceps muscle has three heads.
7. *Function.* An abductor moves a structure away from the midline, and an adductor moves a structure toward the midline. The masseter (a chewer) is a chewing muscle.

3 ***List the criteria used to name muscles, and give an example of each.***

Movements Accomplished by Muscles

When muscles contract, the **pull** (P), or force, of muscle contraction is applied to levers, such as bones, resulting in movement of the levers (figure 10.4). A **lever** is a rigid shaft capable of turning about a hinge, or pivot point, called a **fulcrum** (F) and transferring a force applied at one point along the lever to a **weight** (W), or resistance, placed at another point along the lever. The joints

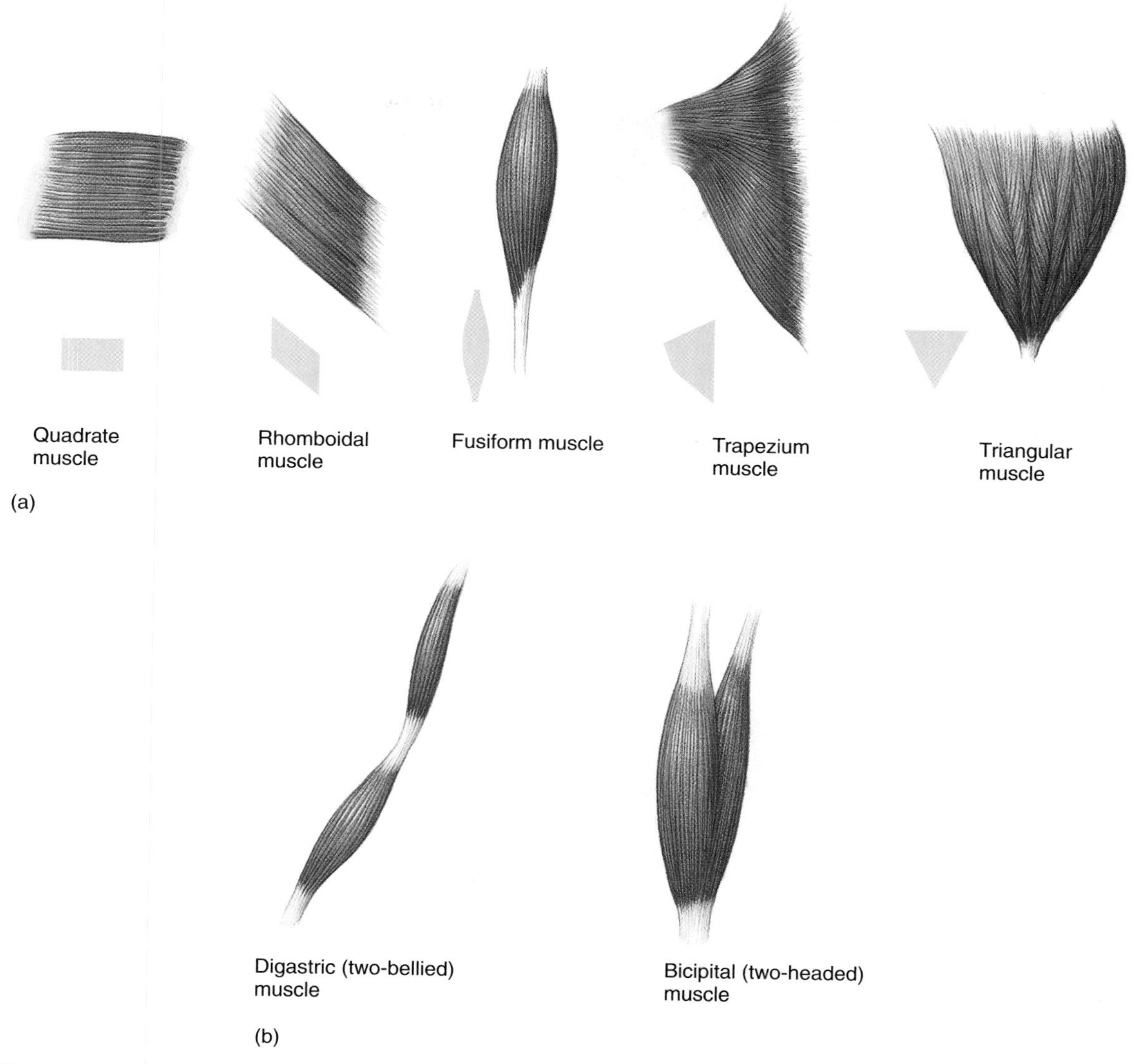

FIGURE 10.3 Examples of Muscle Shapes

(*a*) Muscles with various shapes: quadrate (four-sided with right angles), rhomboidal (four-sided with no right angles), fusiform (spindle-shaped), trapezium (four-sided with no right angles and no parallel sides), and triangular (three sided). (*b*) Muscles with various components: digastric (two bellied) and bicipital (two headed).

function as fulcrums, the bones function as levers, and the muscles provide the pull to move the levers. Three classes of levers exist based on the relative positions of the levers, weights, fulcrums, and forces.

Class I Lever

In a **class I lever system,** the fulcrum is located between the pull and the weight (see figure 10.4*a*). A child's seesaw is an example of this type of lever. The children on the seesaw alternate between being the weight and the pull across a fulcrum in the center of the board. The head is an example of this type of lever in the body. The atlanto-occipital joint is the fulcrum, the posterior neck muscles provide the pull depressing the back of the head, and the face, which is elevated, is the weight. With the weight balanced over the fulcrum, only a small amount of pull is required to lift a weight. For example, only a very small shift in weight is needed for one child to lift the other on a seesaw. This system is quite limited, however, as to how much weight can be lifted and how high it can be lifted. For example, consider what happens when the child on one end of the seesaw is much larger than the child on the other end.

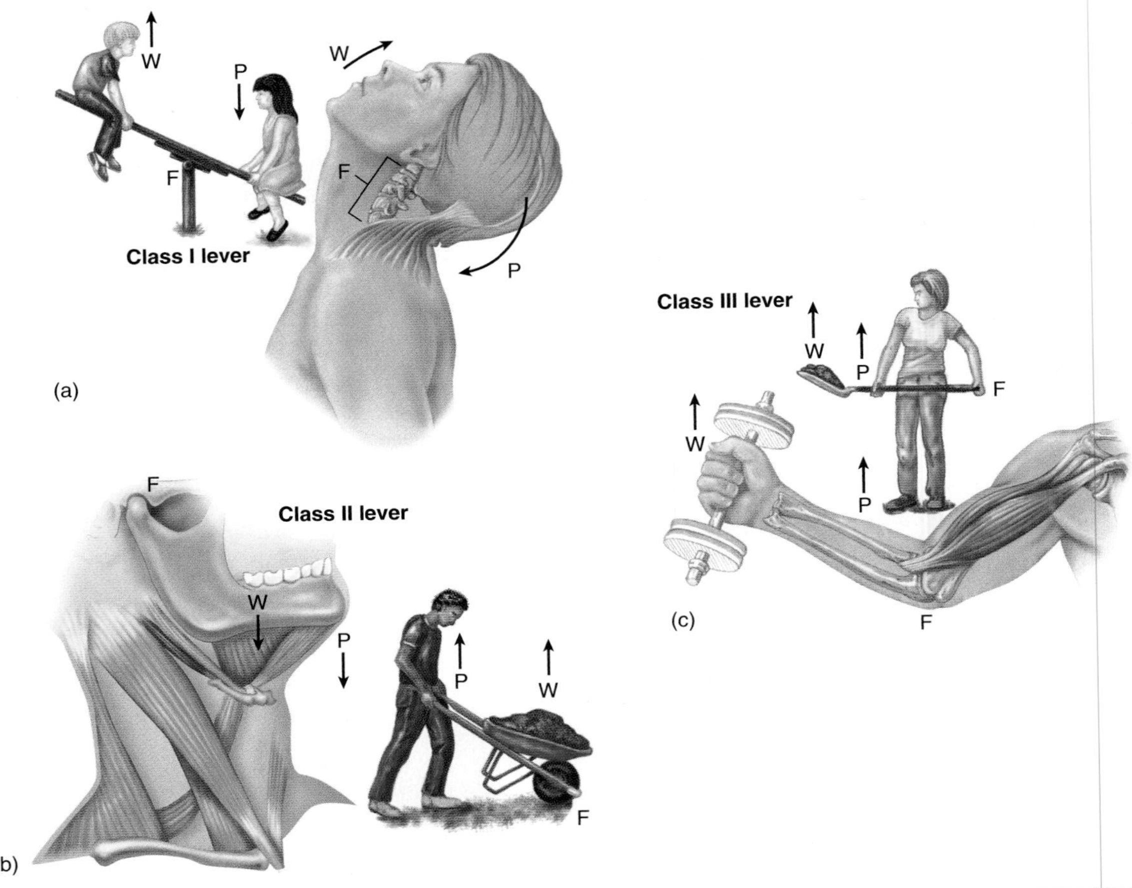

FIGURE 10.4 Lever Classes
(*a*) **Class I:** The fulcrum (*F*) is located between the weight (*W*) and the pull (*P*), or force. The pull is directed downward, and the weight, on the opposite side of the fulcrum, is lifted. In the example in the body, the fulcrum extends through several cervical vertebrae. (*b*) **Class II:** The weight (*W*) is located between the fulcrum (*F*) and the pull (*P*), or force. The upward pull lifts the weight. The movement of the mandible is easier to compare to a wheelbarrow if the head is considered upside down. (*c*) **Class III:** The pull (*P*), or force, is located between the fulcrum (*F*) and the weight (*W*). The upward pull lifts the weight.

Class II Lever

In a **class II lever system,** the weight is located between the fulcrum and the pull (see figure 10.4*b*). An example is a wheelbarrow; the wheel is the fulcrum and the person lifting on the handles provides the pull. The weight, or load, carried in the wheelbarrow is placed between the wheel and the operator. In the body, an example of a class II lever is depression of the mandible (as in opening the mouth; in this case, the whole system, the wheelbarrow and the person lifting it, is upside down).

Class III Lever

In a **class III lever system,** the most common type in the body, the pull is located between the fulcrum and the weight (see figure 10.4*c*). An example is a person using a shovel. The hand placed on the part of the handle closest to the blade provides the pull to lift the weight, such as a shovelful of dirt, and the hand placed near the end of the handle acts as the fulcrum. In the body, the action of the biceps brachii muscle (force) pulling on the radius (lever) to flex the elbow (fulcrum) and elevate the hand (weight) is an example of a class III lever. This type of lever system does not allow as great a weight to be lifted, but the weight can be lifted a greater distance.

4 ***Using the terms* fulcrum, lever, *and* force, *explain how contraction of a muscle results in movement. Define the three classes of levers, and give an example of each in the body.***

Muscle Anatomy

An overview of the superficial skeletal muscles is presented in figure 10.5. Muscles of the head, neck, trunk, and limbs are described in the following sections.

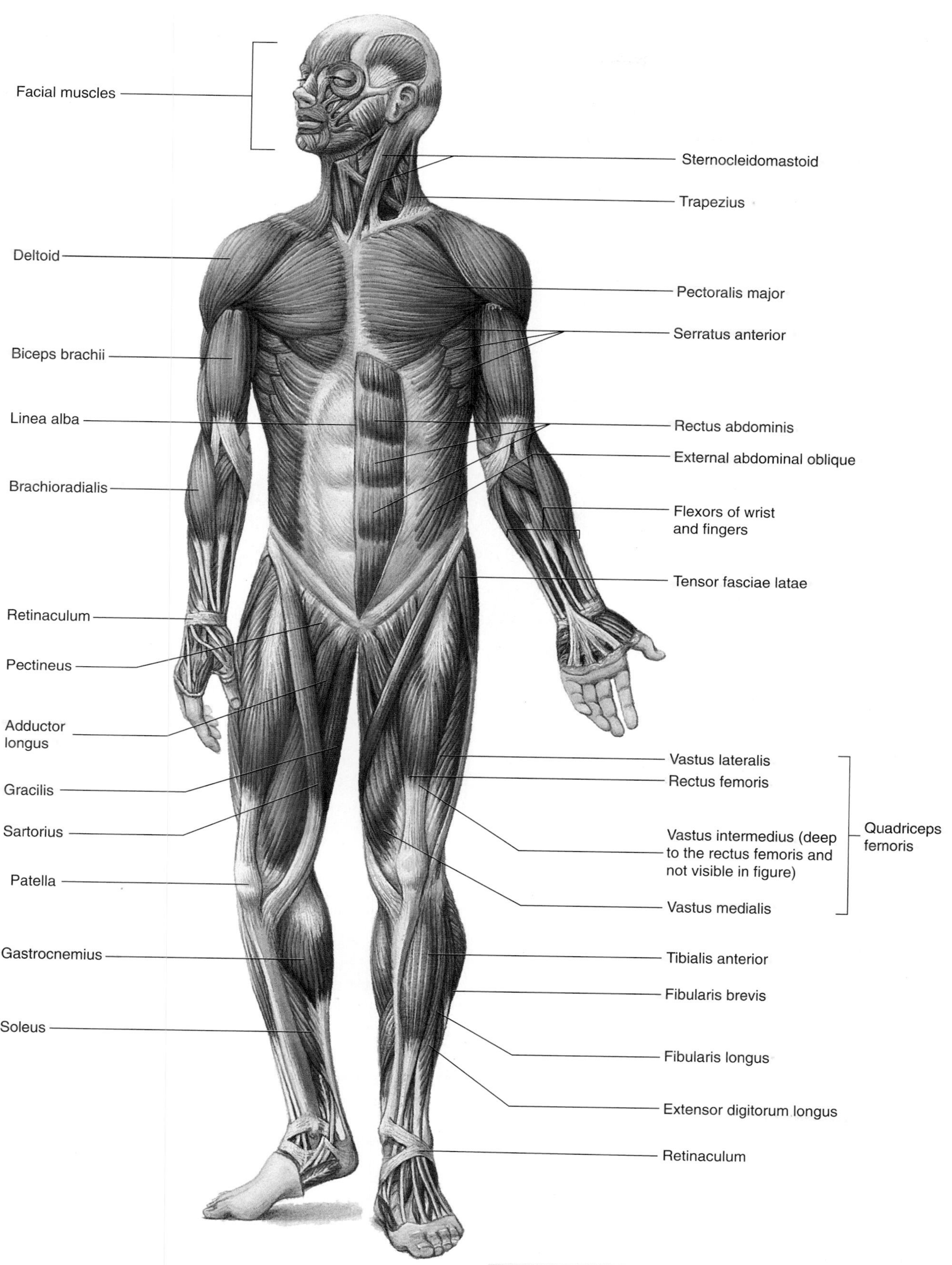

(a) **Anterior view**

FIGURE 10.5 General Overview of the Superficial Body Musculature

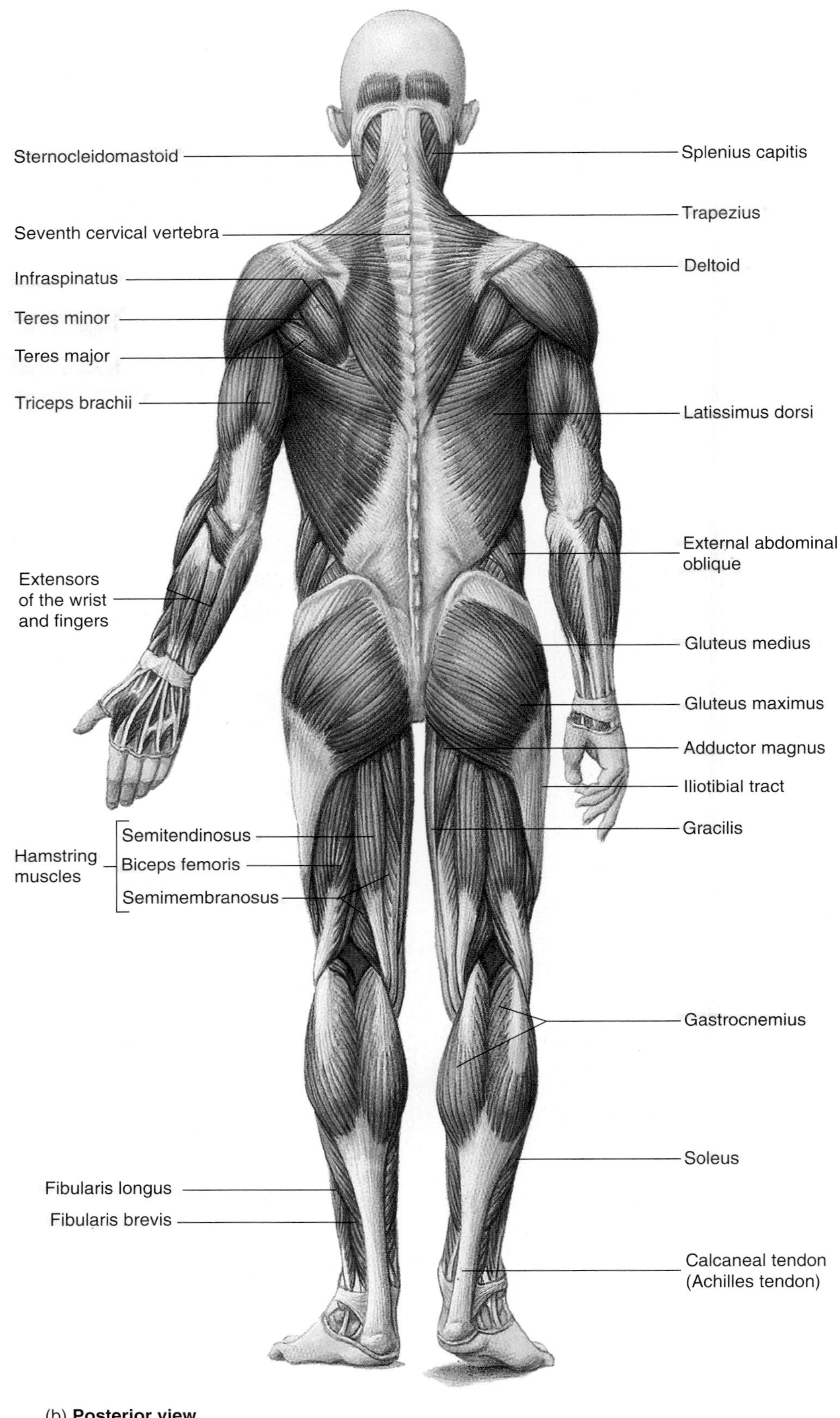

(b) **Posterior view**

FIGURE 10.5 ***(continued)***

HEAD MUSCLES

Head and Neck Muscles

The muscles of the neck are listed in table 10.1 and are illustrated in figures 10.6–10.8. The anterior superficial neck muscles are illustrated in figure 10.6. Most of the flexors of the head and neck lie deep within the neck along the anterior margins of the vertebral bodies (not illustrated). Extension of the neck is accomplished by the posterior neck muscles that attach to the occipital bone and mastoid process of the temporal bone (see figures 10.7 and 10.8) and function as the force of a class I lever system. They also rotate and laterally flex the neck.

The muscular ridge seen superficially in the posterior part of the neck and lateral to the midline is composed of the trapezius muscle overlying the splenius capitis (see figures 10.6*b* and 10.7*b*). The fasciculi of the trapezius muscles are shorter at the base of the neck and leave a diamond-shaped area over the inferior cervical and superior thoracic vertebral spines (see figure 10.7*b*).

Rotation and lateral flexion of the neck are accomplished by muscles of both the lateral and posterior groups (see table 10.1). The **sternocleidomastoid** (ster′nō-klī′dō-mas′toyd) muscle is the prime mover of the lateral group. It is easily seen on the anterior and lateral sides of the neck, especially if the head is extended slightly and rotated to one side (see figure 10.6*b*). If the sternocleidomastoid muscle on only one side of the neck contracts, the neck is rotated toward the opposite side. If both contract together, they flex the neck. Lateral flexion of the neck (moving the head back to the midline after it has been tilted to one side) is accomplished by the lateral flexors of the opposite side.

TABLE 10.1* **Muscles Moving the Head (See Figures 10.6–10.8)**

Muscle	Origin	Insertion	Nerve	Action
Anterior				
Longus capitis (lon′gŭs ka′pi-tis) (not illustrated)	C3–C6	Occipital bone	C1–C3	Flexes neck
Rectus capitis anterior (rek′tŭs ka′pi-tis) (not illustrated)	Atlas	Occipital bone	C1–C2	Flexes neck
Posterior				
Longissimus capitis (lon-gis′ĭ-mŭs kă′pi-tis)	Upper thoracic and lower cervical vertebrae	Mastoid process	Dorsal rami of cervical nerves	Extends, rotates, and laterally flexes neck
Oblique capitis superior (ka′pi-tis)	Atlas	Occipital bone (inferior nuchal line)	Dorsal ramus of C1	Extends and laterally flexes neck
Rectus capitis posterior (rek′tŭs ka′pi-tis)	Axis, atlas	Occipital bone	Dorsal ramus of C1	Extends and rotates neck
Semispinalis capitis	C4–T6	Occipital bone	Dorsal rami of cervical nerves	Extends and rotates neck
Splenius capitis	C4–T6	Superior nuchal line and mastoid process	Dorsal rami of cervical nerves	Extends, rotates, and laterally flexes neck
Trapezius	Occipital protuberance, nuchal ligament, spinous processes of C7–T12	Clavicle, acromion process, and scapular spine	Accessory (cranial nerve XI)	Extends and laterally flexes neck
Lateral				
Rectus capitis lateralis (not illustrated)	Atlas	Occipital bone	C1	Laterally flexes neck
Sternocleidomastoid	Manubrium and medial clavicle	Mastoid process and superior nuchal line	Accessory (cranial nerve XI)	One contracting alone: rotates and extends neck Both contracting together: flex neck and elevate face

*The tables in this chapter are to be used as references. As you study the muscular system, first locate the muscle on the figure, and then find its description in the corresponding table.

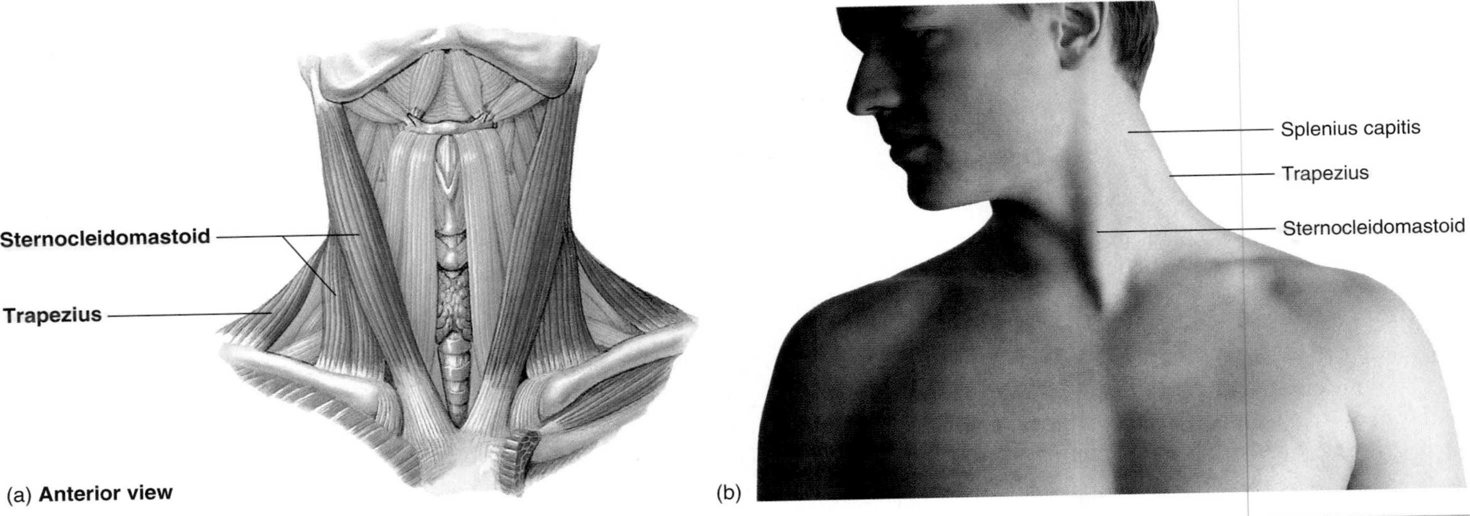

FIGURE 10.6 Anterior Neck Muscles
(*a*) Anterior superficial neck muscles. (*b*) Surface anatomy of anterior neck muscles.

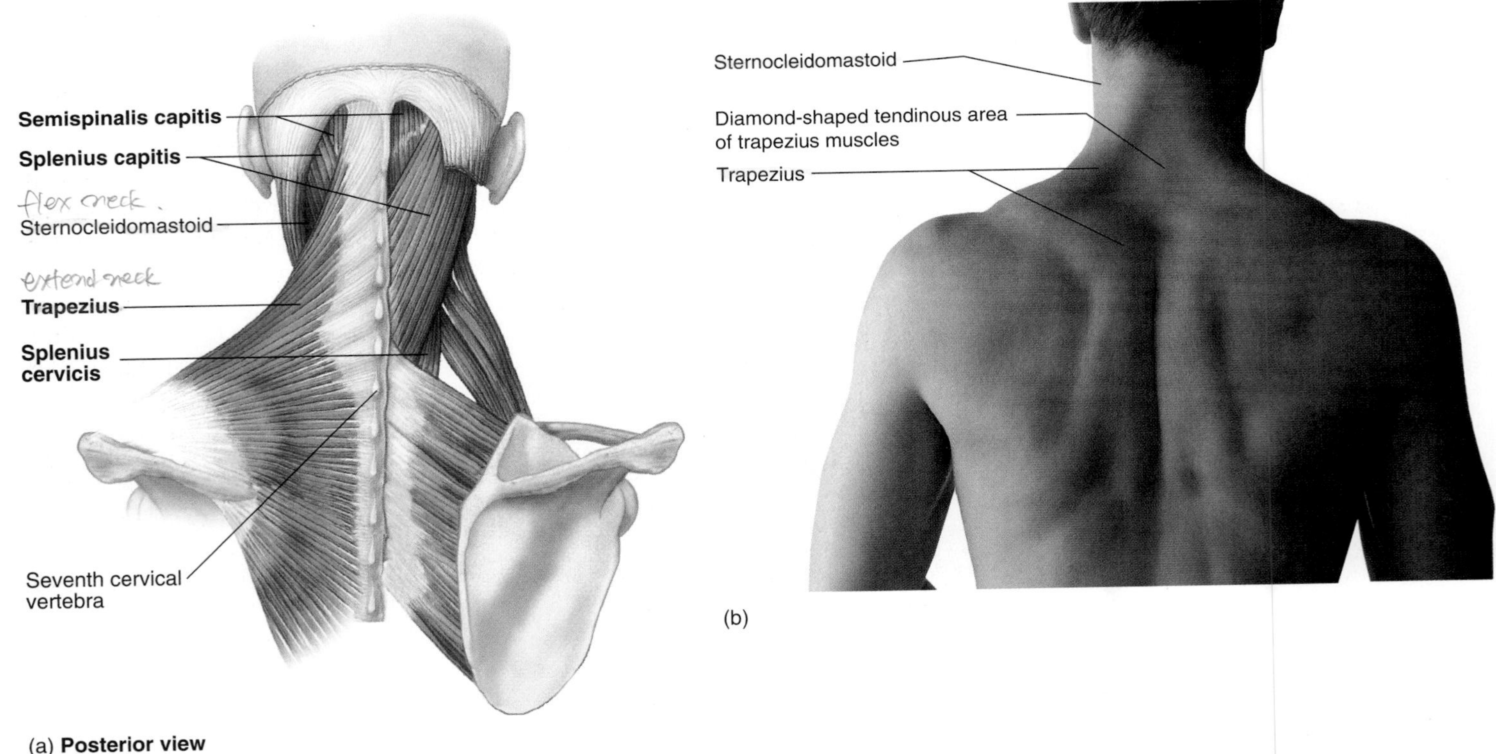

FIGURE 10.7 Posterior Neck Muscles
(*a*) Posterior superficial neck muscles. (*b*) Surface anatomy of anterior neck muscles.

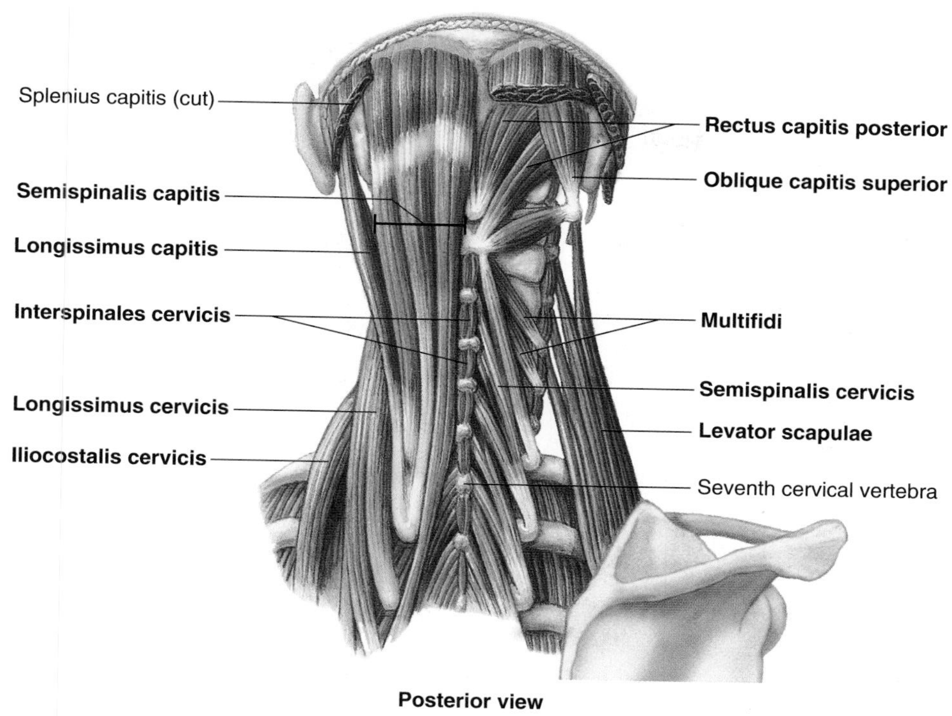

FIGURE 10.8 **Posterior Deep Neck Muscles**

Torticollis

Torticollis (tōr-ti-kol′is; twisted neck or wry neck) may result from damage to one of the sternocleidomastoid muscles or the accessory nerve (cranial nerve XI) supplying it. Torticollis may also result from a tumor forming in a baby's sternocleidomastoid before birth. Damage to an infant's neck muscles during a difficult birth sometimes causes torticollis and can usually be corrected by exercising the muscle. **Spasmodic torticollis** can occur in adults. It is characterized by intermittent contraction of neck muscles, especially the sternocleidomastoid and trapezius, resulting in rotation, flexion, and extention of the neck and elevation of the shoulder.

PREDICT 1

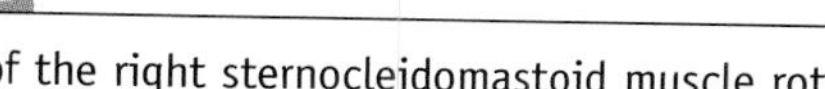

Shortening of the right sternocleidomastoid muscle rotates the head in which direction?

5. ***Name the major movements of the head caused by contraction of the anterior, posterior, and lateral neck muscles.***
6. ***Name the movements of the head and neck caused by contraction of the sternocleidomastoid muscle. What is torticollis (wry neck)?***

Facial Expression

The skeletal muscles of the face (table 10.2 and figure 10.9) are cutaneous muscles attached to the skin. Many animals have cutaneous muscles over the trunk that allow the skin to twitch to remove irritants, such as insects. In humans, facial expressions are important components of nonverbal communication, and the cutaneous muscles are confined primarily to the face and neck.

Several muscles act on the skin around the eyes and eyebrows (figure 10.10; see figure 10.9). The **occipitofrontalis** (ok-sip′i-tō-frŭn-tă′lis) raises the eyebrows and furrows the skin of the forehead. The **orbicularis oculi** (ōr-bik′ū-lā′ris ok′ū-lī) closes the eyelids and causes "crow's-feet" wrinkles in the skin at the lateral corners of the eyes. The **levator palpebrae** (le-vā′ter, lē-vā′tōr pal-pē′brē; the palpebral fissure is the opening between the eyelids) **superioris** raises the upper lids (see figure 10.10*a*). A droopy eyelid on one side, called **ptosis** (tō′sis), usually indicates that the nerve to the levator palpebrae superioris, or the part of the brain controlling that nerve, has been damaged. The **corrugator supercilii** (kōr′ŭ-gā′ter, kōr′ŭ-gā′tōr soo′per-sil′ē-ī) draws the eyebrows inferiorly and medially, producing vertical corrugations (furrows) in the skin between the eyes (see figures 10.9 and 10.10*c*).

Several muscles function in moving the lips and the skin surrounding the mouth (see figures 10.9 and 10.10). The **orbicularis oris** (ōr-bik′ū-lā′ris ōr′is) and **buccinator** (buk′si-nā-tōr), the kissing muscles, pucker the mouth. Smiling is accomplished by the **zygomaticus** (zī′gō-mat′i-kŭs) **major** and **minor,** the **levator anguli** (ang′gū-lī) **oris,** and the **risorius** (rī-sōr′ē-ŭs). Sneering is accomplished by the **levator labii** (lā′bē-ī) **superioris** and frowning or pouting by the **depressor anguli oris,** the **depressor labii inferioris,** and the **mentalis** (men-tā′lis). If the mentalis muscles are well developed on each side of the chin, a chin dimple, where

TABLE 10.2 Muscles of Facial Expression (See Figure 10.9)

Muscle	Origin	Insertion	Nerve	Action
Auricularis (aw-rik′ū-lăr′is)				
Anterior	Aponeurosis over head	Cartilage of auricle	Facial	Draws auricle superiorly and anteriorly
Posterior	Mastoid process	Posterior root of auricle	Facial	Draws auricle posteriorly
Superior	Aponeurosis over head	Cartilage of auricle	Facial	Draws auricle superiorly and posteriorly
Buccinator (buk′sĭ-nā′tōr)	Mandible and maxilla	Orbicularis oris at angle of mouth	Facial	Retracts angle of mouth; flattens cheek
Corrugator supercilii (kōr′ŭ′gā′ter, soo′per-sil′ē-ī)	Nasal bridge and orbicularis oculi	Skin of eyebrow	Facial	Depresses medial portion of eyebrow; draws eyebrows together, as in frowning
Depressor anguli oris (dē-pres′ŏr ang′gū-lī ōr′ŭs)	Lower border of mandible	Skin of lip near angle of mouth	Facial	Depresses angle of mouth
Depressor labii inferioris (dē-pres′ŏr lā′bē-ī in-fēr′ē-ōr-is)	Lower border of mandible	Skin of lower lip and orbicularis oris	Facial	Depresses lower lip
Levator anguli oris (lē-vā′tor, le-vā′ter ang′gū-lī ōr′ŭs)	Maxilla	Skin at angle of mouth and orbicularis oris	Facial	Elevates angle of mouth
Levator labii superioris (lē-vā′tor, le-vā′ter lā′bē-ī sū-pēr′ē-ōr-is)	Maxilla	Skin and orbicularis oris of upper lip	Facial	Elevates upper lip
Levator labii superioris alaeque nasi (lē-vā′tor, le-vā′ter lā′bē-ī sū-pēr′ē-ōr-is ă-lak′ă nā′zī)	Maxilla	Ala at nose and upper lip	Facial	Elevates ala of nose and upper lip
Levator palpebrae superioris (lē-vā′tor, le-vā′ter pal-pē′brē sū-pēr′ē-ōr-is)	Lesser wing of sphenoid	Skin of eyelid	Oculomotor	Elevates upper eyelid
Mentalis (men-tā′lis)	Mandible	Skin of chin	Facial	Elevates and wrinkles skin over chin; protrudes lower lip
Nasalis (nā′ză-lis)	Maxilla	Bridge and ala of nose	Facial	Dilates nostril
Occipitofrontalis (ok-sip′i-tō-frŭn′tā′lis)	Occipital bone	Skin of eyebrow and nose	Facial	Moves scalp; elevates eyebrows
Orbicularis oculi (ōr-bik′ū-lā′ris ok′ū-lī)	Maxilla and frontal bones	Circles orbit and inserts near origin	Facial	Closes eye
Orbicularis oris (ōr-bik′ū-lā′ris ōr′is)	Nasal septum, maxilla, and mandible	Fascia and other muscles of lips	Facial	Closes lip
Platysma (plă-tiz′mă)	Fascia of deltoid and pectoralis major	Skin over inferior border of mandible	Facial	Depresses lower lip; wrinkles skin of neck and upper chest
Procerus (prō-sē′rŭs)	Bridge of nose	Frontalis	Facial	Creates horizontal wrinkles between eyes, as in frowning
Risorius (ri-sōr′ē-ŭs)	Platysma and masseter fascia	Orbicularis oris and skin at corner of mouth	Facial	Abducts angle of mouth
Zygomaticus major (zī′gō-mat′i-kŭs)	Zygomatic bone	Angle of mouth	Facial	Elevates and abducts upper lip
Zygomaticus minor (zī′gō-mat′i-kŭs)	Zygomatic bone	Orbicularis oris of upper lip	Facial	Elevates and abducts upper lip

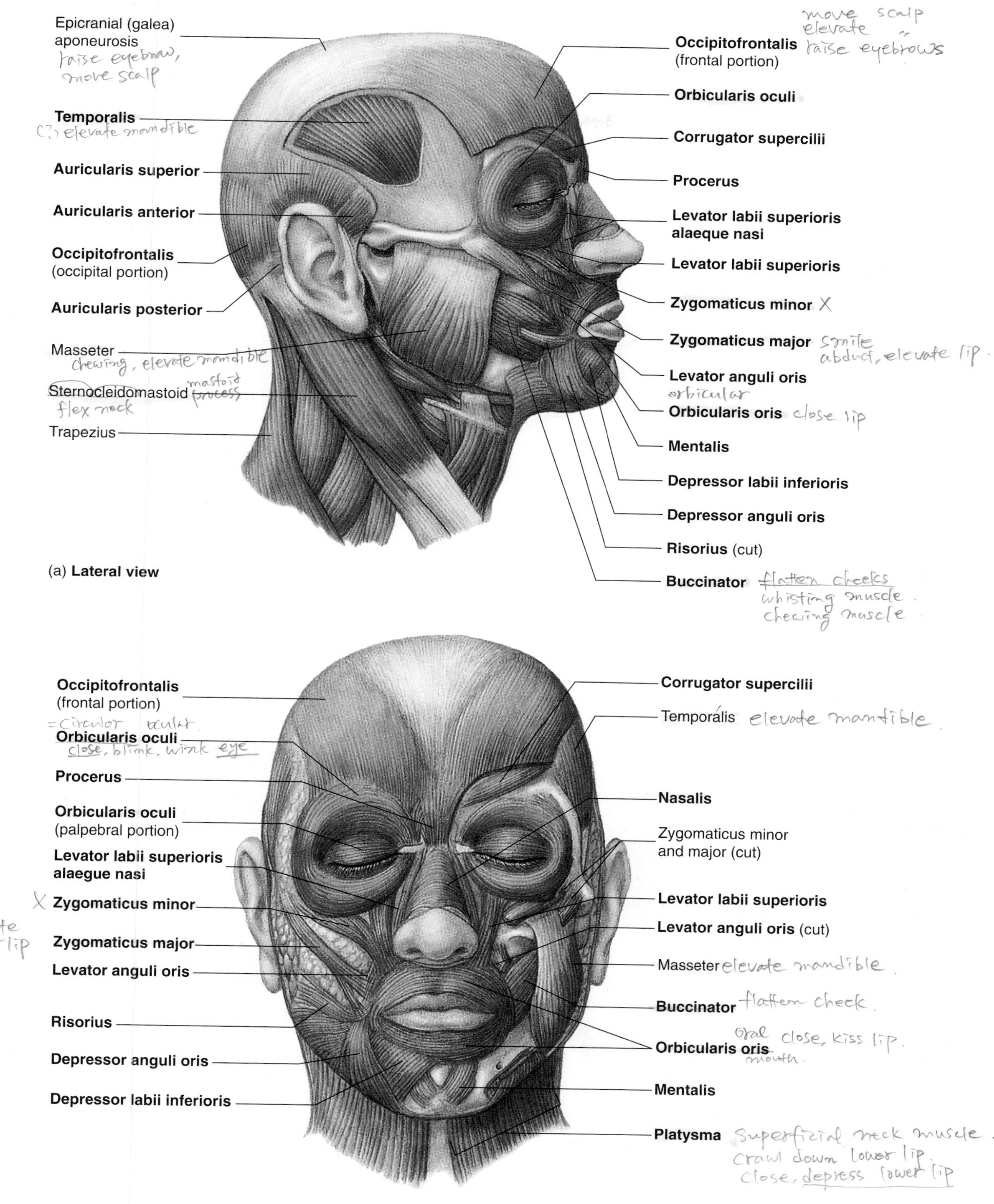

FIGURE 10.9 Muscles of Facial Expression

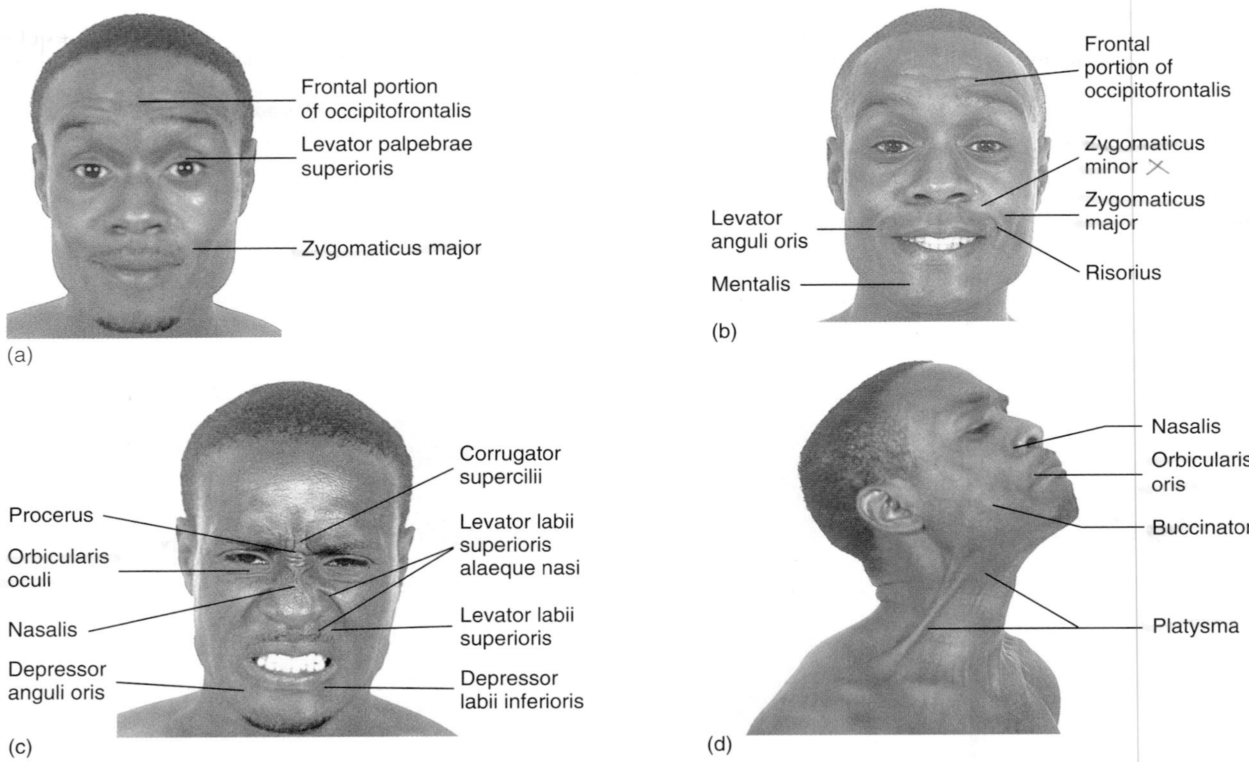

FIGURE 10.10 Surface Anatomy, Muscles of Facial Expression

the skin is tightly attached to the underlying bone or other connective tissue, may appear between the two muscles.

7 ***What is unusual about the insertion (and sometimes the origin) of facial muscles?***

8 ***Which muscles are responsible for moving the ears, the eyebrows, the eyelids, and the nose? For puckering the lips, smiling, sneering, and frowning? What causes a dimple on the chin? What usually causes ptosis on one side?***

PREDICT 2

Harry Wolf, a notorious flirt, on seeing Sally Gorgeous, raises his eyebrows, winks, whistles, and smiles. Name the facial muscles he uses to carry out this communication. Sally, thoroughly displeased with this exhibition, frowns and flares her nostrils in disgust. What muscles does she use?

Mastication

Chewing, or **mastication** (mas-ti-kā′shŭn), involves forcefully closing the mouth (elevating the mandible: temporalis, masseter, and medial pterygoid) and grinding food between the teeth (medial and lateral excursion of the mandible; involving all muscles of mastication). The **muscles of mastication** and the **hyoid muscles** move the mandible (tables 10.3 and 10.4; figures 10.11 and 10.12). The elevators of the mandible are some of the strongest muscles of the body and bring the mandibular teeth forcefully against the maxillary teeth to crush food. Slight mandibular depression involves relaxation of the mandibular elevators and the pull of gravity. Opening the mouth wide requires the action of the depressors of the mandible (lateral pterygoid, digastric, geniohyoid, mylohyoid); even though the muscles of the tongue and the buccinator (see table 10.2; table 10.5) are not involved in chewing, they help move the food in the mouth and hold it in place between the teeth.

9 ***Name the muscles responsible for opening and closing the jaw and for lateral and medial excursion of the jaw.***

Tongue Movements

The tongue is very important in mastication and speech: (1) It moves food around in the mouth; (2) with the buccinator, it holds food in place while the teeth grind it; (3) it pushes food up to the palate and back toward the pharynx to initiate swallowing; and (4) it changes shape to modify sound during speech. The tongue consists of a mass of **intrinsic muscles** (entirely within the tongue), which are involved in changing the shape of the tongue, and **extrinsic muscles** (outside of the tongue but attached to it), which help change the shape and move the tongue

TABLE 10.3 Muscles of Mastication (See Figures 10.9 and 10.11)

Muscle	Origin	Insertion	Nerve	Action
Temporalis (tem-pŏ-rā′lis)	Temporal fossa	Anterior portion of mandibular ramus and coronoid process	Mandibular division of trigeminal	Elevates and retracts mandible; involved in excursion
Masseter (ma′se-ter)	Zygomatic arch	Lateral side of mandibular ramus	Mandibular division of trigeminal	Elevates and protracts mandible; involved in excursion
Pterygoids (ter′i-goydz)				
Lateral	Lateral side of lateral pterygoid plate and greater wing of sphenoid	Condylar process of mandible and articular disk	Mandibular division of trigeminal	Protracts and depresses mandible; involved in excursion
Medial	Medial side of lateral pterygoid plate and tuberosity of maxilla	Medial surface of mandible	Mandibular division of trigeminal	Protracts and elevates mandible; involved in excursion

TABLE 10.4 Hyoid Muscles (See Figure 10.12)

Muscle	Origin	Insertion	Nerve	Action
Suprahyoid Muscles				
Digastric (dī-gas′trik)	Mastoid process (posterior belly)	Mandible near midline (anterior belly)	Posterior belly—facial; anterior belly—mandibular division of trigeminal	Depresses and retracts mandible; elevates hyoid
Geniohyoid (jĕ-nī-ō-hī′oyd)	Genu of mandible	Body of hyoid	Fibers of C1 and C2 with hypoglossal	Protracts hyoid; depresses mandible
Mylohyoid (mī′lō-hī′oyd)	Body of mandible	Hyoid	Mandibular division of trigeminal	Elevates floor of mouth and tongue; depresses mandible when hyoid is fixed
Stylohyoid (stī-lō-hī′oyd)	Styloid process	Hyoid	Facial	Elevates hyoid
Infrahyoid Muscles				
Omohyoid (ō-mō-hī′oyd)	Superior border of scapula	Hyoid	Upper cervical through ansa cervicalis	Depresses hyoid; fixes hyoid in mandibular depression
Sternohyoid (ster′nō-hī′oyd)	Manubrium and first costal cartilage	Hyoid	Upper cervical through ansa cervicalis	Depresses hyoid; fixes hyoid in mandibular depression
Sternothyroid (ster′nō-thī′royd)	Manubrium and first or second costal cartilage	Thyroid cartilage	Upper cervical through ansa cervicalis	Depresses larynx; fixes hyoid in mandibular depression
Thyrohyoid (thī-rō-hī′oyd)	Thyroid cartilage	Hyoid	Upper cervical, passing with hypoglossal	Depresses hyoid and elevates thyroid cartilage of larynx; fixes hyoid in mandibular depression

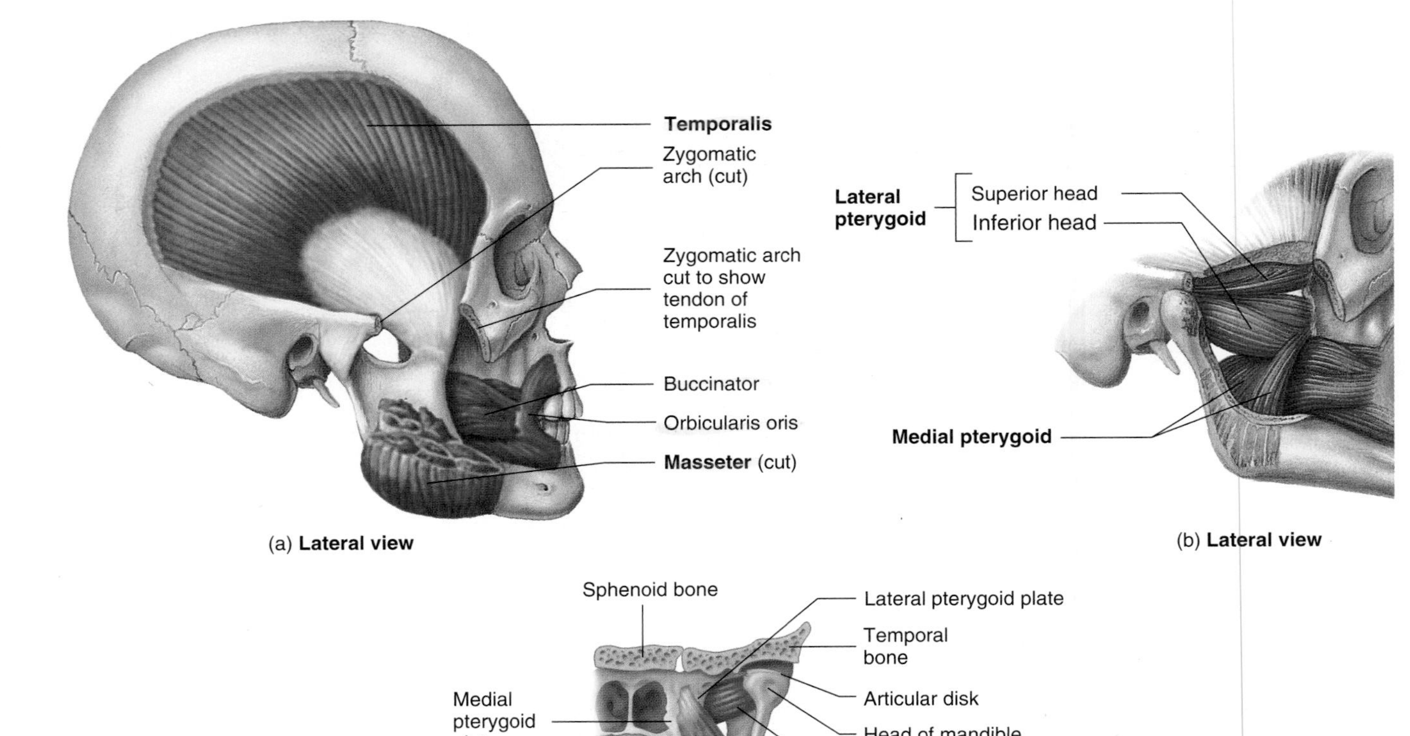

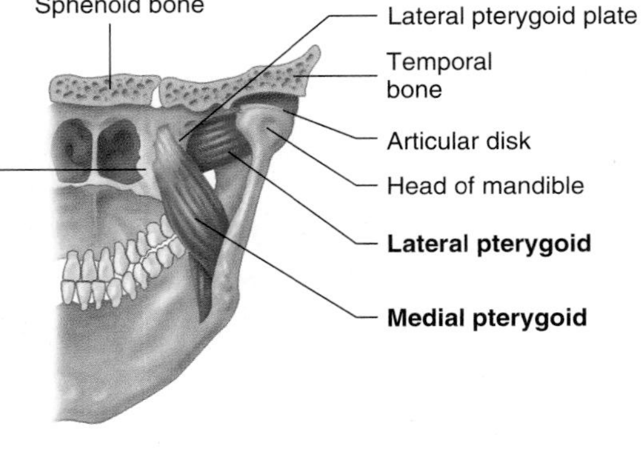

FIGURE 10.11 Muscles of Mastication

(*a*) Masseter and zygomatic arch are cut away to expose the temporalis. (*b*) Masseter and temporalis muscles are removed, and the zygomatic arch and part of the mandible are cut away to reveal the deeper muscles. (*c*) Frontal section of the head, showing the pterygoid muscles.

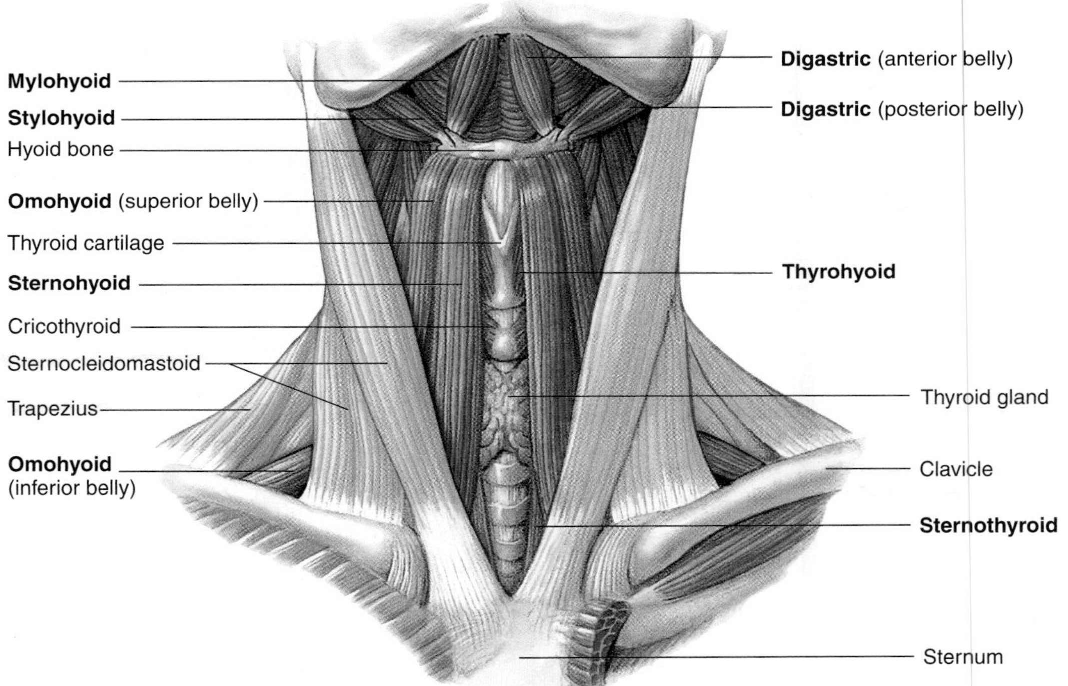

FIGURE 10.12 Hyoid Muscles

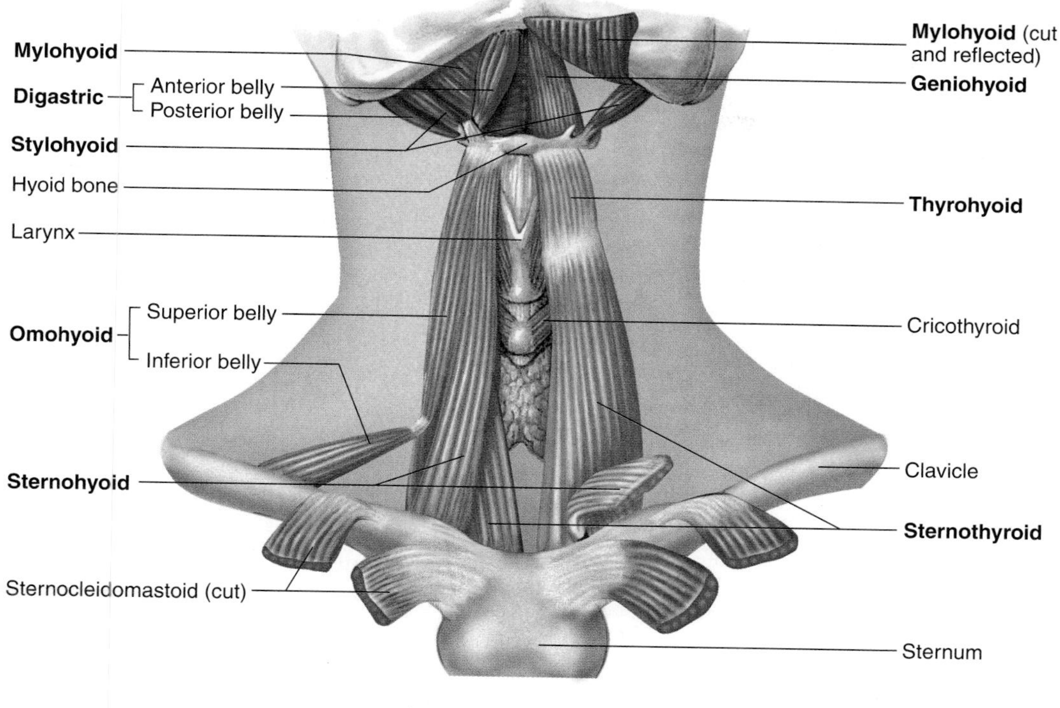

(b) **Anterior deep view**

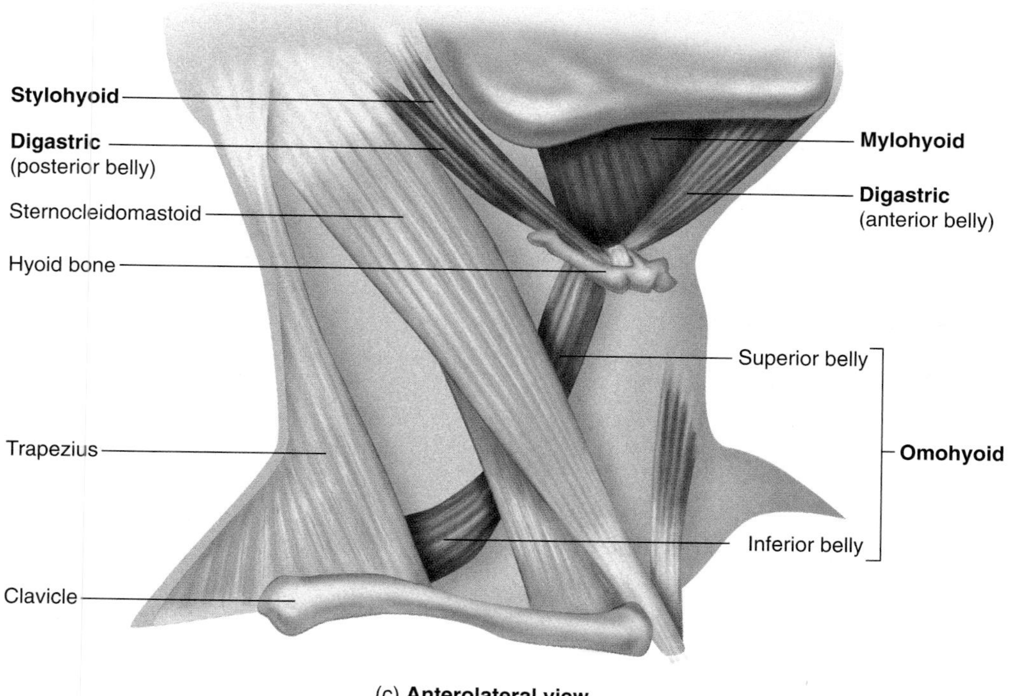

(c) **Anterolateral view**

FIGURE 10.12 ***(continued)***

TABLE 10.5 Tongue Muscles

Muscle	Origin	Insertion	Nerve	Action
Intrinsic Muscles				
Longitudinal, transverse, and vertical (not illustrated)	Within tongue	Within tongue	Hypoglossal	Change tongue shape
Extrinsic Muscles				
Genioglossus (jĕ′nī-ō-glos′ŭs)	Genu of mandible	Tongue	Hypoglossal	Depresses and protrudes tongue
Hyoglossus (hī′ō-glos′ŭs)	Hyoid	Side of tongue	Hypoglossal	Retracts and depresses side of tongue
Styloglossus (stī′lō-glos′ŭs)	Styloid process of temporal bone	Tongue (lateral and inferior)	Hypoglossal	Retracts tongue
Palatoglossus (pal-ă-tō-glos′ŭs)	Soft palate	Tongue	Pharyngeal plexus	Elevates posterior tongue

(see table 10.5; figure 10.13). The intrinsic muscles are named for their fiber orientation in the tongue. The extrinsic muscles are named for their origin and insertion.

Tongue Rolling

Everyone can change the shape of the tongue, but not everyone can roll the tongue into the shape of a tube. This ability apparently is partially controlled genetically, but other factors are involved. In some cases, one of a pair of identical twins can roll the tongue but the other twin cannot. It is not known exactly what tongue muscles are involved in **tongue rolling,** and no anatomical differences are reported between tongue rollers and nonrollers.

10 ***Contrast the movements produced by the extrinsic and intrinsic tongue muscles.***

Swallowing and the Larynx

The hyoid muscles (see table 10.4 and figures 10.12 and 10.13) are divided into a **suprahyoid group** superior to the hyoid bone and an **infrahyoid group** inferior to it. When the hyoid bone is fixed by the infrahyoid muscles so that the bone is stabilized from below, the suprahyoid muscles can help depress the mandible. If the suprahyoid muscles fix the hyoid and thus stabilize it from above, the thyrohyoid muscle (an infrahyoid muscle) can elevate the larynx. To observe this effect, place your hand on your larynx (Adam's apple) and swallow.

The soft palate, pharynx, and larynx contain several muscles involved in swallowing and speech (table 10.6 and figure 10.14). The muscles of the soft palate close the posterior opening to the nasal cavity during swallowing.

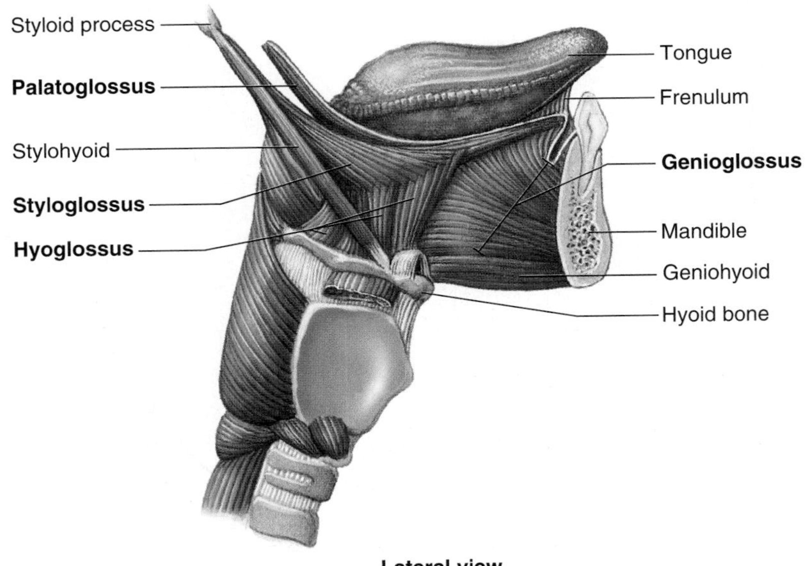

FIGURE 10.13 Tongue Muscles
As seen from the right side.

TABLE 10.6 Muscles of Swallowing and the Larynx (See Figure 10.14)

Muscle	Origin	Insertion	Nerve	Action
Larynx				
Arytenoids (ar-i-tē′noydz)				
Oblique (not illustrated)	Arytenoid cartilage	Opposite arytenoid cartilage	Recurrent laryngeal	Narrows opening to larynx
Transverse (not illustrated)	Arytenoid cartilage	Opposite arytenoid cartilage	Recurrent laryngeal	Narrows opening to larynx
Cricoarytenoids (krī′kō-ar-i-tē′noydz)				
Lateral (not illustrated)	Lateral side of cricoid cartilage	Arytenoid cartilage	Recurrent laryngeal	Narrows opening to larynx
Posterior (not illustrated)	Posterior side of cricoid cartilage	Arytenoid cartilage	Recurrent laryngeal	Widens opening of larynx
Cricothyroid (krī-kō-thī′royd)	Anterior cricoid cartilage	Thyroid cartilage	Superior laryngeal	Lengthens (tenses) vocal cords
Thyroarytenoid (thī′rō-ar′i-tē′noyd) (not illustrated)	Thyroid cartilage	Arytenoid cartilage	Recurrent laryngeal	Shortens (relaxes) vocal cords
Vocalis (vō-kal′ĭs) (not illustrated)	Thyroid cartilage	Arytenoid cartilage	Recurrent laryngeal	Shortens (relaxes) vocal cords
Soft Palate				
Levator veli palatini (lē-vā′tor, le-vā′ter vel′ī pal′ă-tē′nī)	Temporal bone and pharyngotympanic	Soft palate	Pharyngeal plexus	Elevates soft palate
Palatoglossus (pal-ă-tō-glos′ŭs)	Soft palate	Tongue	Pharyngeal plexus	Narrows fauces; elevates posterior tongue
Palatopharyngeus (pal′ă-tō-far-in-jē′ŭs)	Soft palate	Pharynx	Pharyngeal plexus	Narrows fauces; depresses palate; elevates pharynx
Tensor veli palatini (ten′sōr vel′ī pal′ă-tē′nī)	Sphenoid and auditory tube	Soft palate division of auditory tube	Mandibular, division of trigeminal	Tenses soft palate; opens auditory tube
Uvulae (ū′vū-lē)	Posterior nasal spine	Uvula	Pharyngeal plexus	Elevates uvula
Pharynx				
Pharyngeal constrictors (fă-rin′jē-ăl)				
Inferior	Thyroid and cricoid cartilages	Pharyngeal raphe	Pharyngeal plexus and external laryngeal nerve	Narrows inferior portion of pharynx in swallowing
Middle	Stylohyoid ligament and hyoid	Pharyngeal raphe	Pharyngeal plexus	Narrows pharynx in swallowing
Superior	Medial pterygoid plate, mandible, floor of mouth, and side of tongue	Pharyngeal raphe	Pharyngeal plexus	Narrows superior portion of pharynx in swallowing
Salpingopharyngeus (sal-ping′gō-far-in-jē′ŭs)	Auditory tube	Pharynx	Pharyngeal plexus	Elevates pharynx; opens auditory tube in swallowing
Stylopharyngeus (stī′lō-far-in-jē′ŭs)	Styloid process	Pharynx	Glossopharyngeus	Elevates pharynx

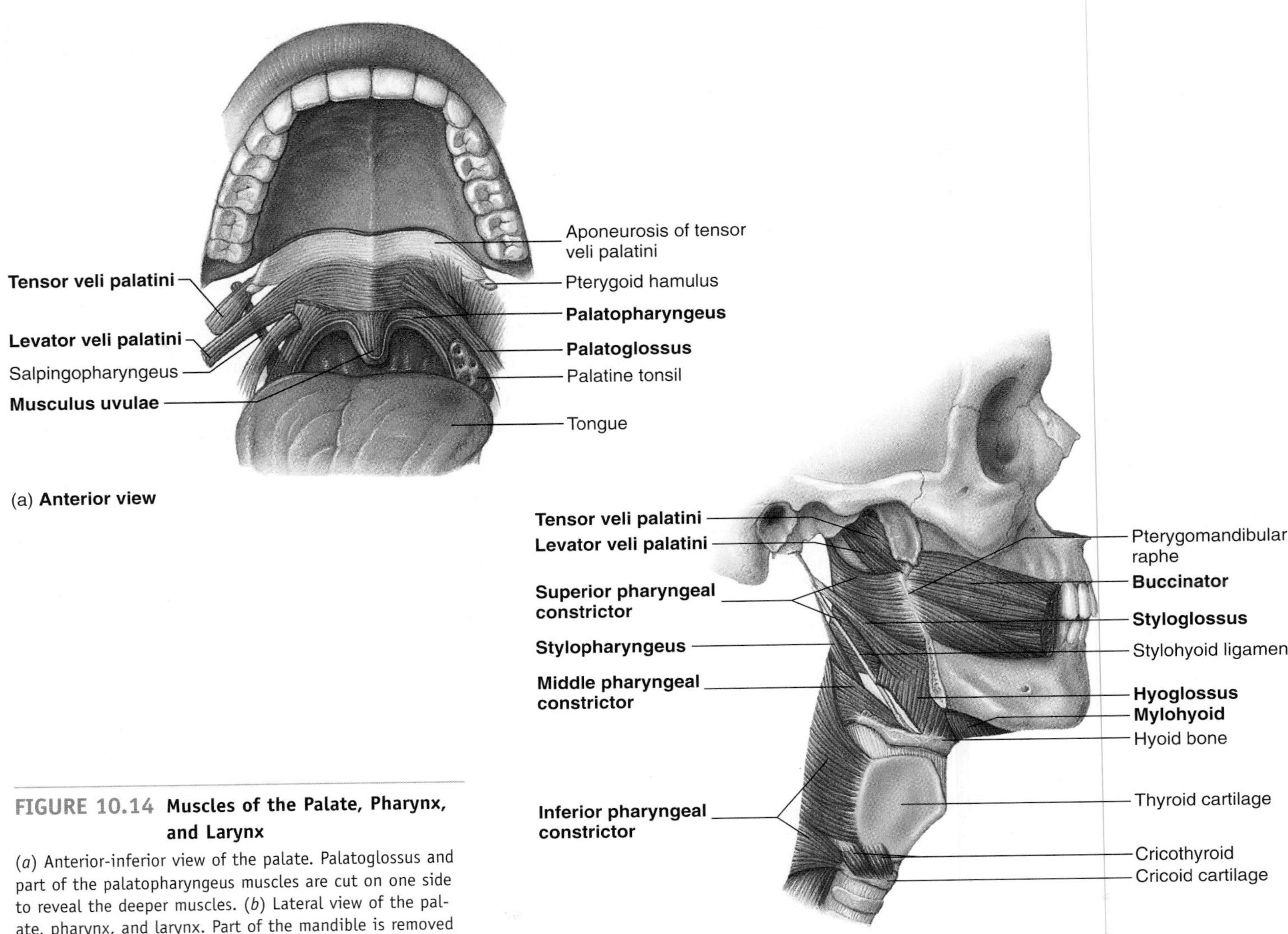

FIGURE 10.14 Muscles of the Palate, Pharynx, and Larynx

(*a*) Anterior-inferior view of the palate. Palatoglossus and part of the palatopharyngeus muscles are cut on one side to reveal the deeper muscles. (*b*) Lateral view of the palate, pharynx, and larynx. Part of the mandible is removed to reveal the deeper structures.

Swallowing (see chapter 24) is accomplished by elevation of the pharynx, which in turn is accomplished by elevation of the larynx, to which the pharynx is attached, and constriction of the **palatopharyngeus** (pal′ă-tō-far-in-jē′ŭs) and **salpingopharyngeus** (sal-pin′gō-far-in-jē′ŭs; *salpingo* means trumpet and refers to the trumpet-shaped opening of the pharyngotympanic, or auditory, tube). The pharyngeal constrictor muscles then constrict from superior to inferior, forcing food into the esophagus.

The salpingopharyngeus also opens the pharyngotympanic tube, which connects the middle ear with the pharynx. Opening the pharyngotympanic tube equalizes the pressure between the middle ear and the atmosphere; this is why it is sometimes helpful to chew gum or swallow when ascending or descending a mountain in a car or when changing altitudes in an airplane.

The muscles of the larynx are listed in table 10.6 and are illustrated in figure 10.14*b*. Most of the laryngeal muscles help narrow or close the laryngeal opening so food does not enter the larynx when a person swallows. The remaining muscles shorten (relax) the vocal cords to lower the pitch of the voice or lengthen (tense) the vocal cords to raise the pitch of the voice.

Snoring and Laryngospasm

Snoring is a rough, raspy noise that can occur when a sleeping person inhales through the mouth and nose. The noise usually is made by vibration of the soft palate but also may occur as a result of vocal cord vibration.

Laryngospasm is a tetanic contraction of the muscles that narrows the opening of the larynx (arytenoids, lateral cricoarytenoids). In severe cases, the opening is closed completely, air can no longer pass through the larynx into the lungs, and the victim may die of asphyxiation. Laryngospasm can develop as a result of severe allergic reactions, tetanus infections, or hypocalcemia.

Movements of the Eyeball

The eyeball rotates within the orbit to allow vision in a wide range of directions. The movements of each eye are accomplished by six

TABLE 10.7 Muscles Moving the Eye (See Figure 10.15)

Muscle	Origin	Insertion	Nerve	Action
Oblique				
Inferior	Orbital plate of maxilla	Sclera of eye	Oculomotor	Elevates and laterally deviates gaze
Superior	Common tendinous ring	Sclera of eye	Trochlear	Depresses and laterally deviates gaze
Rectus				
Inferior	Common tendinous ring	Sclera of eye	Oculomotor	Depresses and medially deviates gaze
Lateral	Common tendinous ring	Sclera of eye	Abducent	Laterally deviates gaze
Medial	Common tendinous ring	Sclera of eye	Oculomotor	Medially deviates gaze
Superior	Common tendinous ring	Sclera of eye	Oculomotor	Elevates and medially deviates gaze

muscles named for the orientation of their fasciculi relative to the spherical eye (table 10.7; figure 10.15).

Each rectus muscle (so-named because the fibers are nearly straight with the axis of the eye) attaches to the eyeball anterior to the center of the sphere. The superior rectus rotates the anterior portion of the eyeball superiorly so that the pupil, and thus the gaze, is directed superiorly (looking up). The inferior rectus depresses the gaze, the lateral rectus laterally deviates (abducts) the gaze (looking to the side), and the medial rectus medially deviates (adducts) the gaze (looking toward the nose). The superior rectus and inferior rectus are not completely straight in their orientation to the eye; thus, they also medially deviate the gaze as they contract.

The oblique muscles (so-named because their fibers are oriented obliquely to the axis of the eye) insert onto the posterolateral margin of the eyeball so that both muscles laterally deviate the gaze as they contract (see chapter 15, figure 15.10). The superior oblique

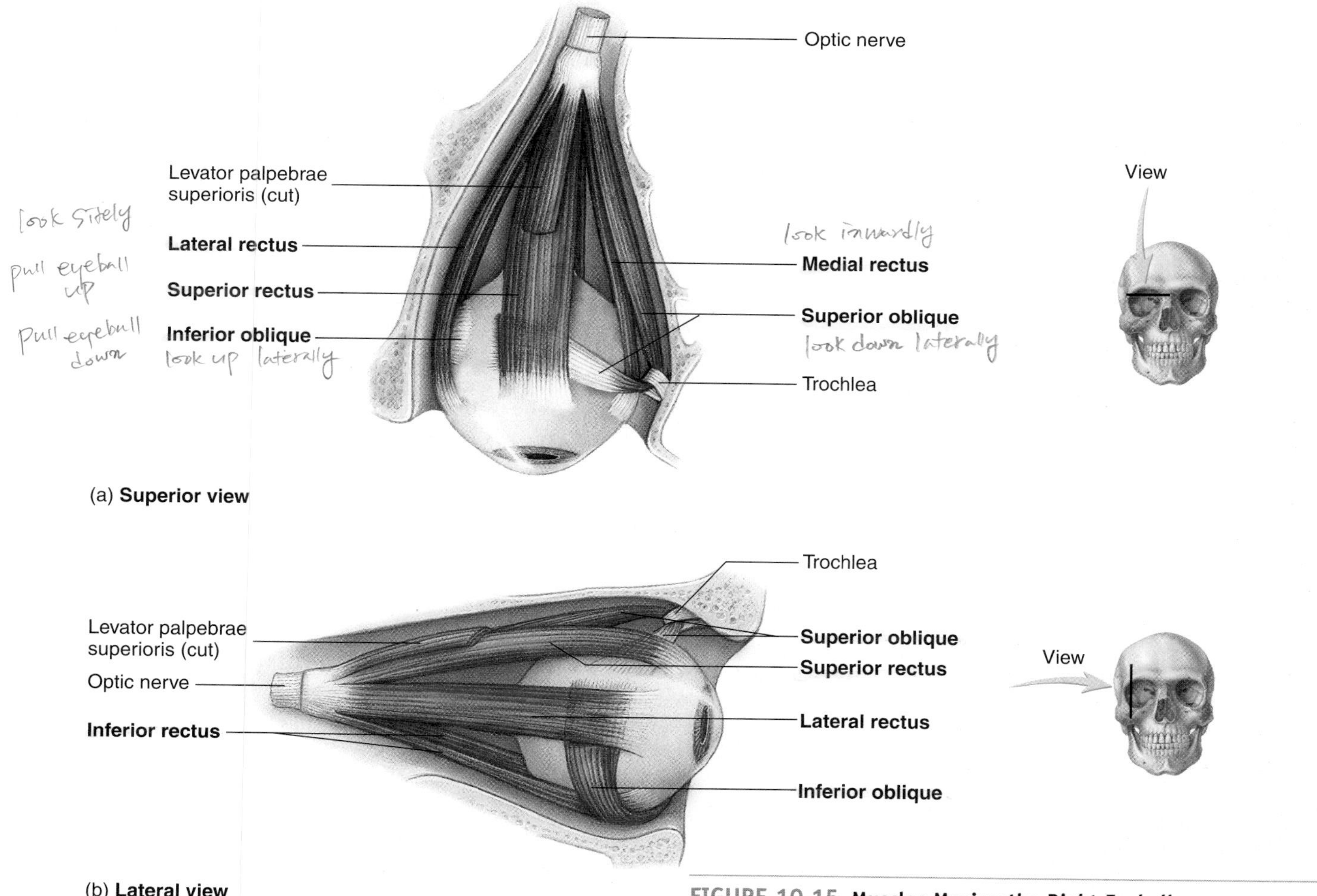

FIGURE 10.15 Muscles Moving the Right Eyeball

elevates the posterior part of the eye, thus directing the pupil inferiorly and depressing the gaze. The inferior oblique elevates the gaze.

11 ***Explain the interaction of the suprahyoid and infrahyoid muscles to depress the mandible and to elevate the larynx.***

12 ***Which muscles open and close the openings to the auditory tube and larynx?***

13 ***Describe the muscles of the eye and the movements they cause.***

PREDICT 3

Strabismus (stra-biz′mŭs) is a condition in which one or both eyes deviate in a medial or lateral direction. In some cases, the condition is caused by a weakness in either the medial or the lateral rectus muscle. If the lateral rectus of the right eye is weak, in which direction does the eye deviate?

TRUNK MUSCLES

Muscles Moving the Vertebral Column

The muscles that extend, laterally flex, and rotate the vertebral column are divided into superficial and deep groups (table 10.8). In general, the muscles of the deep group extend from vertebra to vertebra, whereas the muscles of the superficial group extend from the vertebrae to the ribs. In humans, these back muscles are very strong to maintain erect posture. Comparable muscles in cattle are relatively delicate, although quite large. The **erector spinae** (spī′nē) group of muscles on each side of the back consists of three subgroups: the **iliocostalis** (il′ē-ō-kos-tā′lis), the **longissimus** (lon-gis′i-mŭs), and the **spinalis** (sp-ī-nā′lis). The longissimus group accounts for most of the muscle mass in the lower back (figure 10.16). The deepest

TABLE 10.8 Muscles Acting on the Vertebral Column (See Figures 10.7, 10.8, 10.16, and 10.17)

Muscle	Origin	Insertion	Nerve	Action
Superficial				
Erector spinae (ē-rek′tŏr, ē-rek′tōr spī′nē; divides into three columns)				
Iliocostalis (il′ē-ō-kos-tā′lis)	Sacrum, ilium, and lumbar spines	Ribs and vertebrae	Dorsal rami of spinal nerves	Extends vertebral column
Cervicis (ser-vī′sis)	Superior six ribs	Transverse processes of middle cervical vertebrae	Dorsal rami of thoracic nerves	Extends, laterally flexes, and rotates vertebral column
Thoracis (thō-ra′sis)	Inferior six ribs	Superior six ribs	Dorsal rami of thoracic nerves	Extends, laterally flexes, and rotates vertebral column
Lumborum (lum-bōr′ŭm)	Sacrum, ilium, and lumbar vertebrae	Inferior six ribs	Dorsal rami of thoracic and lumbar nerves	Extends, laterally flexes, and rotates vertebral column
Longissimus (lon-gis′i-mŭs)				
Capitis (ka′pĭ-tis)	Upper thoracic and lower cervical vertebrae	Mastoid process	Dorsal rami of cervical nerves	Extends head
Cervicis (ser-vī′sis)	Upper thoracic vertebrae	Transverse processes of upper cervical vertebrae	Dorsal rami of cervical nerves	Extends neck
Thoracis (thō-ra′sis)	Ribs and lower thoracic vertebrae	Transverse processes of upper lumbar vertebrae and ribs	Dorsal rami of thoracic and lumbar nerves	Extends vertebral column
Spinalis (spī-nā′lis)				
Cervicis (ser-vī′sis) (not illustrated)	C6–C7	Spinous processes of C2–C3	Dorsal rami of cervical nerves	Extends neck
Thoracis (thō-ra′sis)	T11–L2	Spinous processes of middle and upper thoracic vertebrae	Dorsal rami of thoracic nerves	Extends vertebral column
Semispinalis (sem′ē-spī-nā′lis)				
Cervicis (ser-vī′sis)	Transverse processes of T2–T5	Spinous processes of C2–C5	Dorsal rami of cervical nerves	Extends neck
Thoracis (thō-ra′sis)	Transverse processes of T5–T11	Spinous processes of C5–T4	Dorsal rami of thoracic nerves	Extends vertebral column
Splenius cervicis (splē′nē-ŭs ser-vī′sis)	Spinous processes of C3–C5	Transverse processes of C1–C3	Dorsal rami of cervical nerves	Rotates and extends neck
Longus colli (lon′gŭs kō′lī) (not illustrated)	Bodies of C3–T3	Bodies of C1–C6	Ventral rami of cervical nerves	Flexes neck

TABLE 10.8 Muscles Acting on the Vertebral Column—Continued

Muscle	Origin	Insertion	Nerve	Action
Deep				
Interspinales (in-ter-spī-nā′lēz)	Spinous processes of all vertebrae	Next superior spinous process	Dorsal rami of spinal nerves	Extends back and neck
Intertransversarii (in-ter-trans′ver-săr′ē-ī)	Transverse processes of all vertebrae	Next superior transverse process	Dorsal rami of spinal nerves	Laterally flexes vertebral column
Multifidus (mŭl-tif′i-dŭs)	Transverse processes of vertebrae; posterior surface of sacrum and ilium	Spinous processes of superior vertebrae	Dorsal rami of spinal nerves	Extends and rotates vertebral column
Psoas minor (sō′as mī′ner)	T12–L1	Pectineal line near pubic crest	L1	Flexes vertebral column
Rotatores (rō-tā′tōrz)	Transverse processes of all vertebrae	Base of spinous process of superior vertebrae	Dorsal rami of spinal nerves	Extends and rotates vertebral column

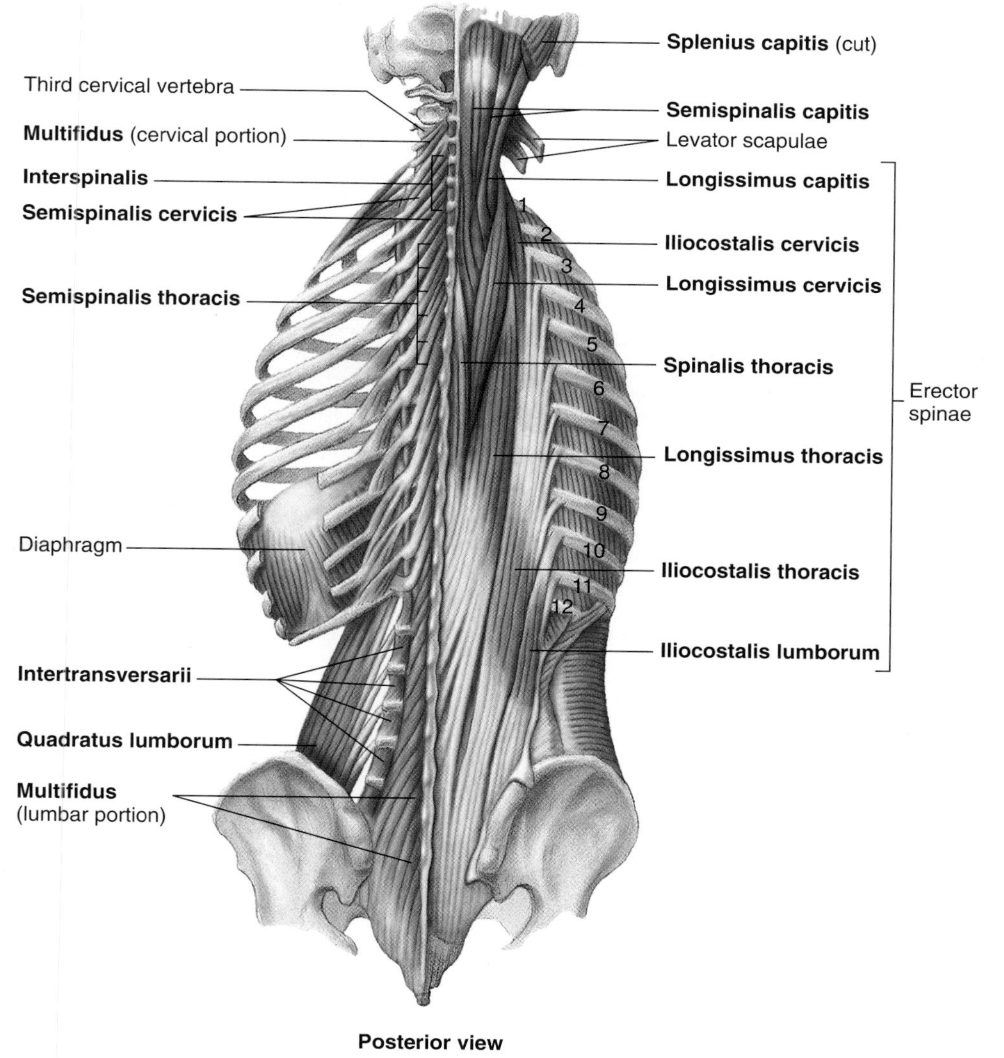

FIGURE 10.16 Deep Back Muscles

On the right, the erector spinae group of muscles is shown. On the left, these muscles are removed to reveal the deeper back muscles.

muscles of the back attach between the spinous and transverse processes of individual vertebra (figure 10.17).

14 ***List the actions of the group of back muscles that attaches to the vertebrae or ribs (or both). What is the name of the superficial group?***

> **Back Pain**
>
> **Low back pain** can result from poor posture, from being overweight, or from having a poor fitness level. A few changes may help: sitting and standing up straight; using a low-back support when sitting; losing weight; exercising, especially the back and abdominal muscles; and sleeping on your side on a firm mattress. Sleeping on your side all night, however, may be difficult because most people change position over 40 times during the night.

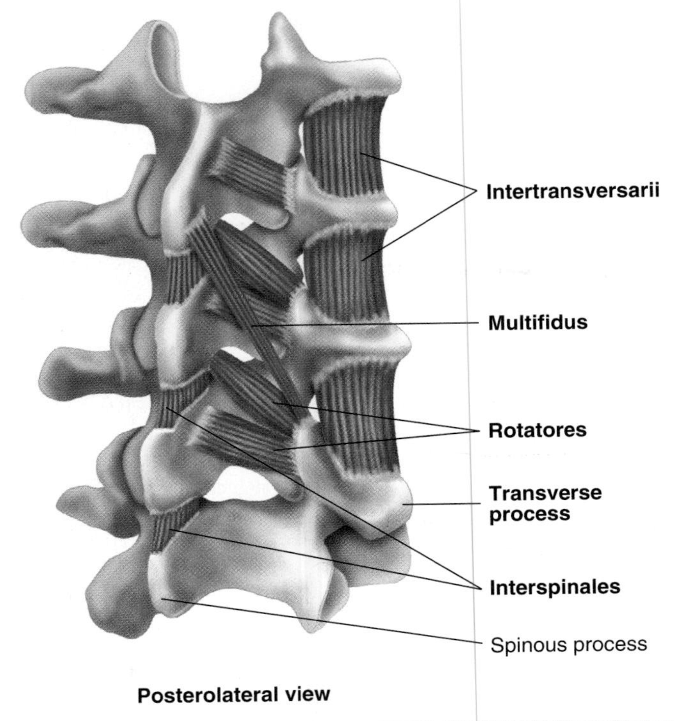

FIGURE 10.17 **Deep Muscles Associated with the Vertebrae**

Thoracic Muscles

The muscles of the thorax are involved mainly in the process of breathing (see chapter 23). Four major groups of muscles are associated with the rib cage (table 10.9 and figure 10.18). The **scalene** (skā′lēn) muscles elevate the first two ribs during inspiration. The **external intercostals** (in-ter-kos′tălz) also

TABLE 10.9 Muscles of the Thorax (See Figure 10.18)

Muscle	Origin	Insertion	Nerve	Action
Diaphragm	Interior of ribs, sternum, and lumbar vertebrae	Central tendon of diaphragm	Phrenic	Inspiration; depresses floor of thorax
Intercostalis (in′ter-kos-ta′lis)				
External	Inferior margin of each rib	Superior border of next rib below	Intercostal	Inspiration; elevates ribs
Internal	Superior margin of each rib	Inferior border of next rib above	Intercostal	Expiration; depresses ribs
Scalenus (skā-lē′nŭs)				
Anterior	Transverse processes of C3–C6	First rib	Cervical plexus	Elevates first rib
Medial	Transverse processes of C2–C6	First rib	Cervical plexus	Elevates first rib
Posterior	Transverse processes of C4–C6	Second rib	Cervical and brachial plexuses	Elevates second rib
Serratus posterior (sĕr-ā′tŭs)				
Inferior (not illustrated)	Spinous processes of T11–L2	Inferior four ribs	Ninth to eleventh intercostals and subcostal	Depresses inferior ribs and extends vertebral column
Superior (not illustrated)	Spinous processes of C6–T2	Second to fifth ribs	First to fourth intercostals	Elevates superior ribs
Transversus thoracis (trans-ver′sus thō-ra′sis) (not illustrated)	Sternum and xiphoid process	Second to sixth costal cartilages	Intercostal	Decreases diameter of thorax

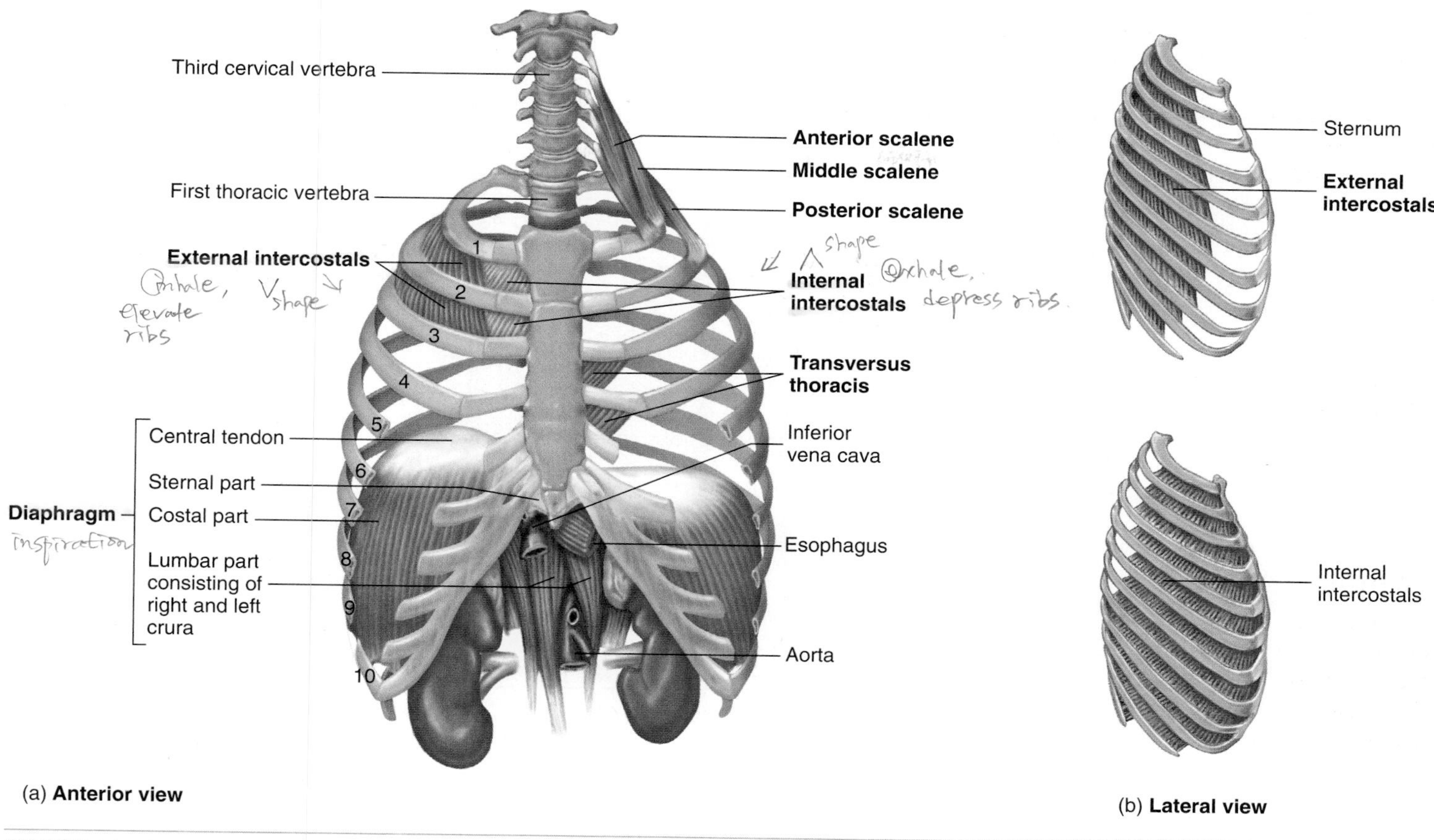

FIGURE 10.18 **Muscles of the Thorax**

elevate the ribs during inspiration. The **internal intercostals** and **transversus thoracis** (thō-ra′sis) muscles depress the ribs during forced expiration.

The **diaphragm** (dī′ă-fram; see figure 10.18*a*) causes the major movement produced during quiet breathing. It is a dome-shaped structure and, when it contracts, the dome flattens slightly, causing the volume of the thoracic cavity to increase, resulting in inspiration. If this dome of skeletal muscle or the phrenic nerve supplying it is severely damaged, the amount of air moving into and out of the lungs may be so small that the individual is likely to die without the aid of an artificial respirator.

15 ***Name the muscle that is mainly responsible for respiratory movements. How do other muscles aid this movement?***

Abdominal Wall

The muscles of the anterior abdominal wall (table 10.10 and figures 10.19 and 10.20) flex and rotate the vertebral column. Contraction of the abdominal muscles when the vertebral column is fixed decreases the volume of the abdominal cavity and the thoracic cavity and can aid in such functions as forced expiration, vomiting, defecation, urination, and childbirth. The crossing pattern of the abdominal muscles creates a strong anterior wall, which holds in and protects the abdominal viscera.

In a relatively muscular person with little fat, a vertical line is visible, extending from the area of the xiphoid process of the sternum through the navel to the pubis. This tendinous area of the abdominal wall is devoid of muscle; the **linea alba** (lin′ē-ă al′bă), or white line, is so-named because it consists of white connective tissue rather than muscle (see figure 10.19). On each side of the linea alba is the **rectus abdominis** (see figures 10.19 and 10.20), surrounded by a **rectus sheath. Tendinous intersections** (tendinous inscriptions) transect the rectus abdominis at three, or sometimes more, locations, causing the abdominal wall of a well-muscled person to appear segmented. Lateral to the rectus abdominis is the **linea semilunaris** (sem-ē-loo-nar′is; a crescent- or half-moon-shaped line); lateral to it are three layers of muscle

TABLE 10.10 Muscles of the Abdominal Wall (See Figures 10.5*a*, 10.19, and 10.20)

Muscle	Origin	Insertion	Nerve	Action
Anterior				
Rectus abdominis (rek′tŭs ab-dom′i-nis)	Pubic crest and symphysis pubis	Xiphoid process and inferior ribs	Branches of lower thoracic	Flexes vertebral column; compresses abdomen
External abdominal oblique	Fifth to twelfth ribs	Iliac crest, inguinal ligament, and rectus sheath	Branches of lower thoracic	Flexes and rotates vertebral column; compresses abdomen; depresses thorax
Internal abdominal oblique	Iliac crest, inguinal ligament, and lumbar fascia	Tenth to twelfth ribs and rectus sheath	Lower thoracic	Flexes and rotates vertebral column; compresses abdomen; depresses thorax
Transversus abdominis (trans-ver′sŭs ab-dom′i-nis)	Seventh to twelfth costal cartilages, lumbar fascia, iliac crest, and inguinal ligament	Xiphoid process, linea alba, and pubic tubercle	Lower thoracic	Compresses abdomen
Posterior				
Quadratus lumborum (kwah-drā′tŭs lŭm-bōr′ŭm)	Iliac crest and lower lumbar vertebrae	Twelfth rib and upper lumbar vertebrae	Upper lumbar	Laterally flexes vertebral column and depresses twelfth rib

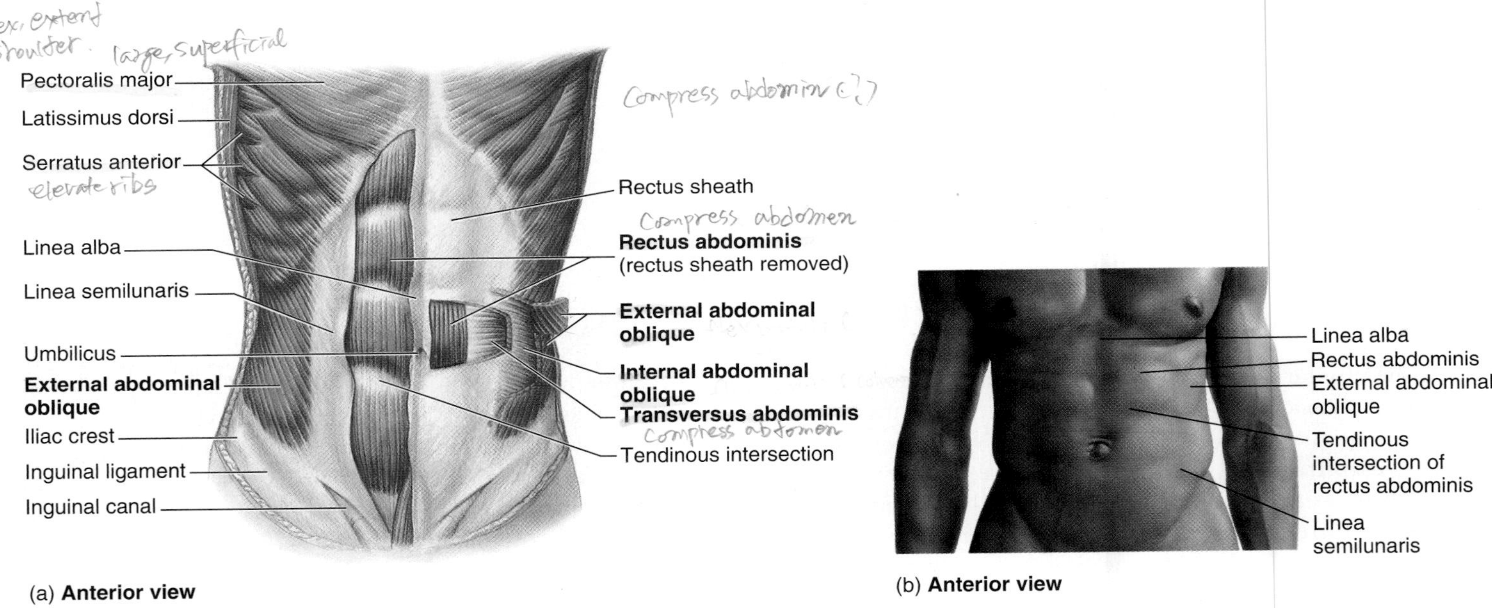

FIGURE 10.19 Anterior Abdominal Wall Muscles
(*a*) Anterior abdominal muscles. Windows in the side reveal the various muscle layers. (*b*) Surface anatomy of anterior abdominal muscle.

(see figures 10.19 and 10.20). From superficial to deep, these muscles are the **external abdominal oblique, internal abdominal oblique,** and **transversus abdominis.**

16 ***Explain the anatomical basis for the segments ("cuts") seen on a well-muscled individual's abdomen. What are the functions of the abdominal muscles? List the muscles of the anterior abdominal wall.***

Pelvic Floor and Perineum

The pelvis is a ring of bone (see chapter 7) with an inferior opening that is closed by a muscular wall, through which the anus and the urogenital openings penetrate (table 10.11). Most of the pelvic floor is formed by the **coccygeus** (kok-si′jē-ŭs) muscle and the **levator ani** (a′nī) muscle, referred to jointly as the **pelvic diaphragm.** The area inferior to the pelvic floor is the **perineum**

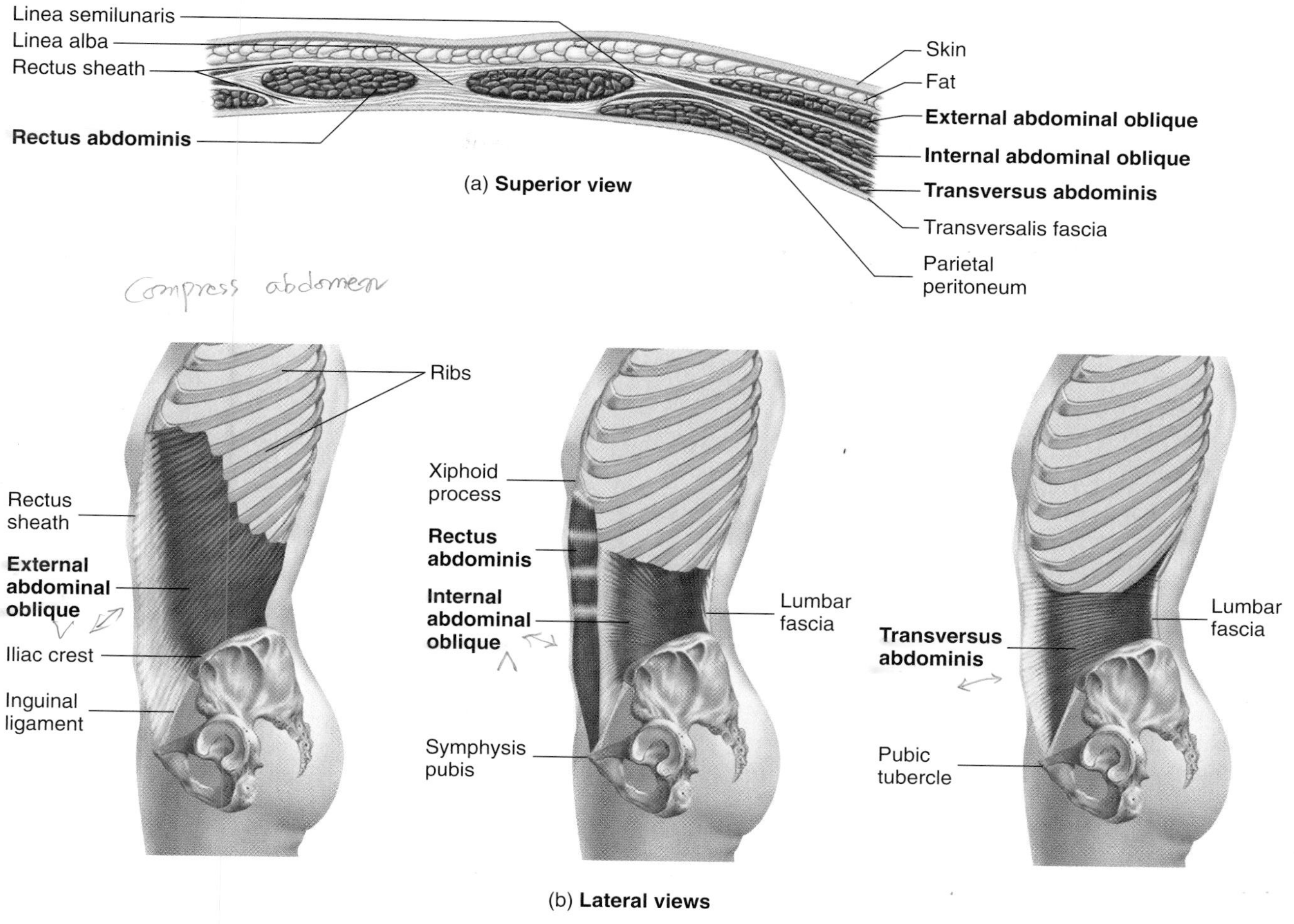

FIGURE 10.20 Anterior Abdominal Wall Muscles
(*a*) Cross section superior to the umbilicus. (*b*) Abdominal muscles shown individually.

TABLE 10.11 Muscles of the Pelvic Floor and Perineum

Muscle	Origin	Insertion	Nerve	Action
Bulbospongiosus (bul′bō-spŭn′jē-ō′sŭs)	Male—central tendon of perineum and median raphe of penis	Dorsal surface of penis and bulb of penis	Pudendal	Constricts urethra; erects penis
	Female—central tendon of perineum	Base of clitoris	Pudendal	Erects clitoris
Coccygeus (kok-si′jē-ŭs) (not illustrated)	Ischial spine	Coccyx	S3 and S4	Elevates and supports pelvic floor
Ischiocavernosus (ish′ē-ō-kav′er-nō′sŭs)	Ischial ramus	Corpus cavernosum	Perineal	Compresses base of penis or clitoris
Levator ani (lē-vā′tor, le-vā′ter ā′nī)	Posterior pubis and ischial spine	Sacrum and coccyx	Fourth sacral	Elevates anus; supports pelvic viscera
External anal sphincter (ā′năl sfingk′ter)	Coccyx	Central tendon of perineum	Fourth sacral and pudenda	Keeps orifice of anal canal closed
External urethral sphincter (ū-rē′thrăl sfingk′ter) (not illustrated)	Pubic ramus	Median raphe	Pudendal	Constricts urethra
Transverse perinei (pĕr′i-nē′ī)				
Deep	Ischial ramus	Median raphe	Pudendal	Supports pelvic floor
Superficial	Ischial ramus	Central perineal	Pudendal	Fixes central tendon

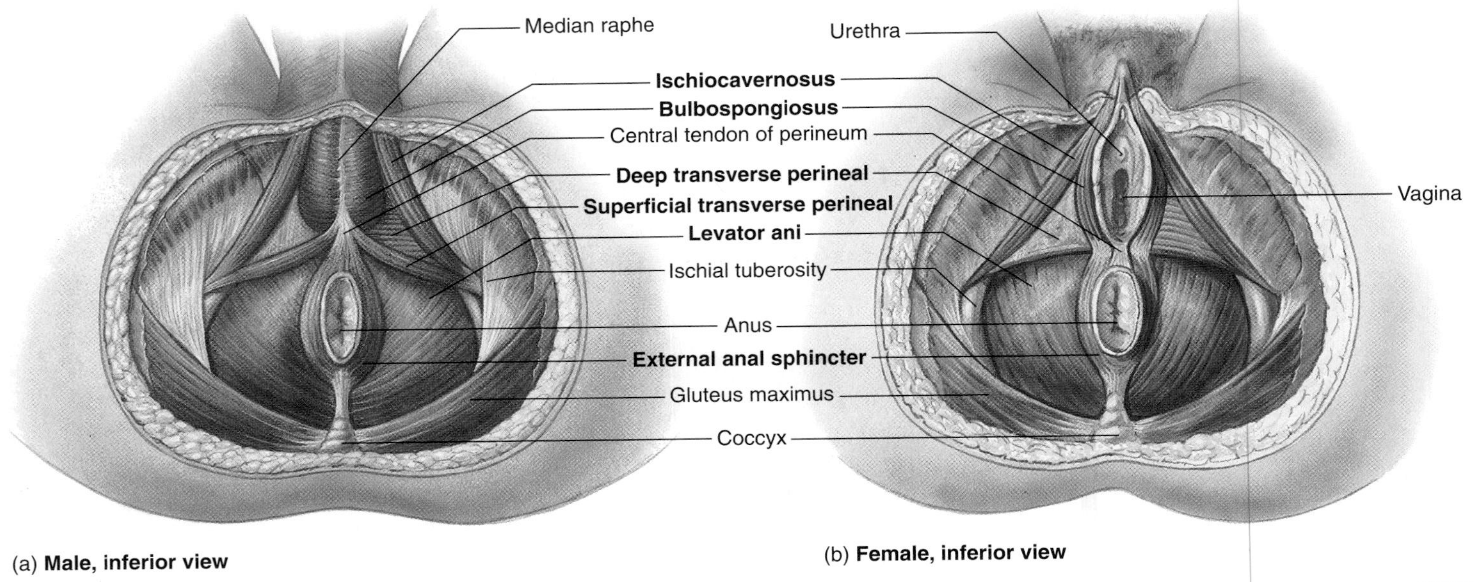

FIGURE 10.21 Muscles of the Pelvic Floor and Perineum

(per′i-nē′ŭm), which is somewhat diamond-shaped (figure 10.21). The anterior half of the diamond is the urogenital triangle, and the posterior half is the anal triangle (see chapter 28). The urogenital triangle contains the **urogenital diaphragm,** which forms a "subfloor" to the pelvis in that area and consists of the **deep transverse perineal** (pĕr′ĭ-nē′ăl) muscle and the **external urethral sphincter** muscle. During pregnancy, the muscles of the pelvic diaphragm and urogenital diaphragm may be stretched by the extra weight of the fetus, and specific exercises are designed to strengthen them.

17 ***What openings penetrate the pelvic floor muscles? Name the area inferior to the pelvic floor.***

UPPER LIMB MUSCLES

The muscles of the upper limb include those that move the scapula and those that move the arm, forearm, and hand.

Scapular Movements

The major connection of the upper limb to the body is accomplished by muscles (table 10.12 and figure 10.22). The muscles attaching the scapula to the thorax include the **trapezius, levator scapulae** (skap′ū-lē), **rhomboideus** (rom-bō-id′ē-ŭs) **major** and **minor, serratus** (sĕr-ā′tŭs) **anterior,** and **pectoralis** (pek′tō-ra′lis) **minor.**

TABLE 10.12 Muscles Acting on the Scapula (See Figure 10.22)

Muscle	Origin	Insertion	Nerve	Action
Levator scapulae (lē-vā′tor, le-vā′ter skap′ū-lē)	Transverse processes of C1–C4	Superior angle of scapula	Dorsal scapular	Elevates, retracts, and rotates scapula; laterally flexes neck
Pectoralis minor (pek′tō-ra′lis)	Third to fifth ribs	Coracoid process of scapula	Medial pectoral	Depresses scapula or elevates ribs
Rhomboideus (rom-bō-id′ē-ŭs)				
Major	Spinous processes of T1–T4	Medial border of scapula	Dorsal scapular	Retracts, rotates, and fixes scapula
Minor	Spinous processes of C6–C7	Medial border of scapula	Dorsal scapular	Retracts, slightly elevates, rotates, and fixes scapula
Serratus anterior (ser-ā′tŭs)	First to eighth or ninth ribs	Medial border of scapula	Long thoracic	Rotates and protracts scapula; elevates ribs
Subclavius (sŭb-klā′vē-ŭs)	First rib	Clavicle	Subclavian	Fixes clavicle or elevates first rib
Trapezius (tra-pē′zē-ŭs)	External occipital protuberance, ligamentum nuchae, and spinous processes of C7–T12	Clavicle, acromion process, and scapular spine	Accessory and cervical plexus	Elevates, depresses, retracts, rotates, and fixes scapula; extends neck

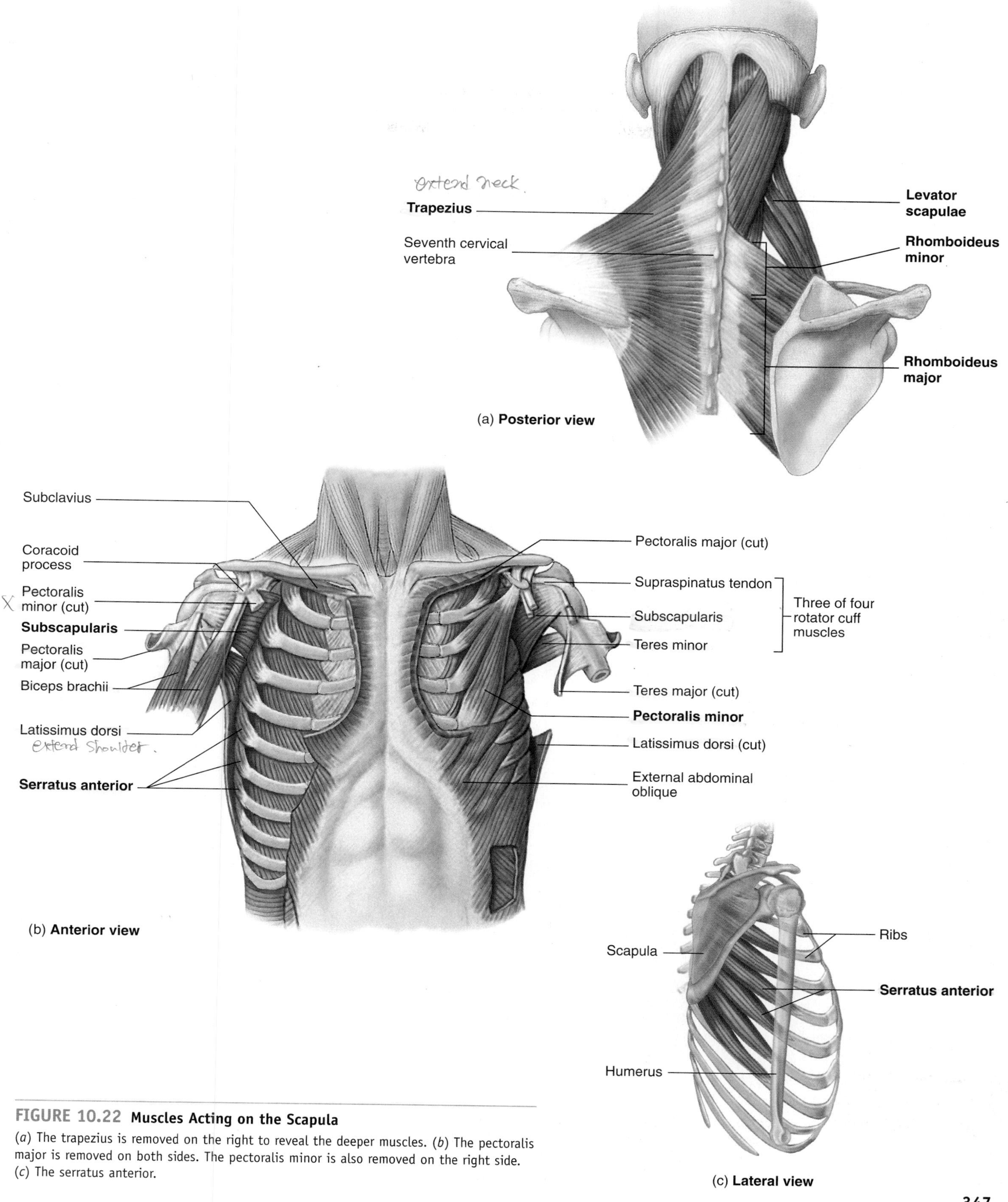

FIGURE 10.22 Muscles Acting on the Scapula

(*a*) The trapezius is removed on the right to reveal the deeper muscles. (*b*) The pectoralis major is removed on both sides. The pectoralis minor is also removed on the right side. (*c*) The serratus anterior.

TABLE 10.13 Muscles Acting on the Arm (See Figures 10.22–10.24)

Muscle	Origin	Insertion	Nerve	Action
Coracobrachialis (kōr′ă-kō-brā-kē-ā′lis)	Coracoid process of scapula	Midshaft of humerus	Musculocutaneous	Adducts arm and flexes shoulder
Deltoid (del′toyd)	Clavicle, acromion process, and scapular spine	Deltoid tuberosity	Axillary	Flexes and extends shoulder; abducts and medially and laterally rotates arm
Latissimus dorsi (lă-tis′i-mŭs dōr′sī)	Spinous processes of T7–L5; sacrum and iliac crest; inferior angle of scapula in some people	Medial crest of intertubercular groove	Thoracodorsal	Adducts and medially rotates arm; extends shoulder
Pectoralis major (pek′tō-rā′lis)	Clavicle, sternum, superior six costal cartilages, and abdominal aponeurosis	Lateral crest of intertubercular groove	Medial and lateral pectoral	Flexes shoulder; adducts and medially rotates arm; extends shoulder from flexed position
Teres major (ter′ēz, tēr-ēz)	Lateral border of scapula	Medial crest of intertubercular groove	Lower subscapular C5 and C6	Extends shoulder; adducts and medially rotates arm
Rotator Cuff				
Infraspinatus (in-fră-spī-nā′tŭs)	Infraspinous fossa of scapula	Greater tubercle of humerus	Suprascapular C5 and C6	Laterally rotates arm; holds head of humerus in place
Subscapularis (sŭb-skap-ū-lā′ris)	Subscapular fossa	Lesser tubercle of humerus	Upper and lower subscapular C5 and C6	Medially rotates arm; holds head of humerus in place
Supraspinatus (soo-pră-spī-nā′tŭs)	Supraspinous fossa	Greater tubercle of humerus	Suprascapular C5 and C6	Abducts arm; holds head of humerus in place
Teres minor (ter′ēz, tēr-ēz)	Lateral border of scapula	Greater tubercle of humerus	Axillary C5 and C6	Laterally rotates and adducts arm; holds head of humerus in place

These muscles move the scapula, permitting a wide range of movements of the upper limb, or act as fixators to hold the scapula firmly in position when the arm muscles contract. The superficial muscles that act on the scapula can easily be seen on a living person (see figure 10.7*b*): The trapezius forms the upper line from each shoulder to the neck, and the origin of the serratus anterior from the first eight or nine ribs can be seen along the lateral thorax. The serratus anterior inserts onto the medial border of the scapula (see figure 10.22*c*).

Arm Movements

The arm is attached to the thorax by several muscles, including the **pectoralis major** and the **latissimus dorsi** (lă-tis′i-mŭs dōr′sī) muscles (table 10.13 and figures 10.23 and 10.24; see figure 10.22*b*). Notice that the pectoralis major muscle is listed in table 10.13 as both a flexor and an extensor. The muscle flexes the extended shoulder and extends the flexed shoulder. Try these

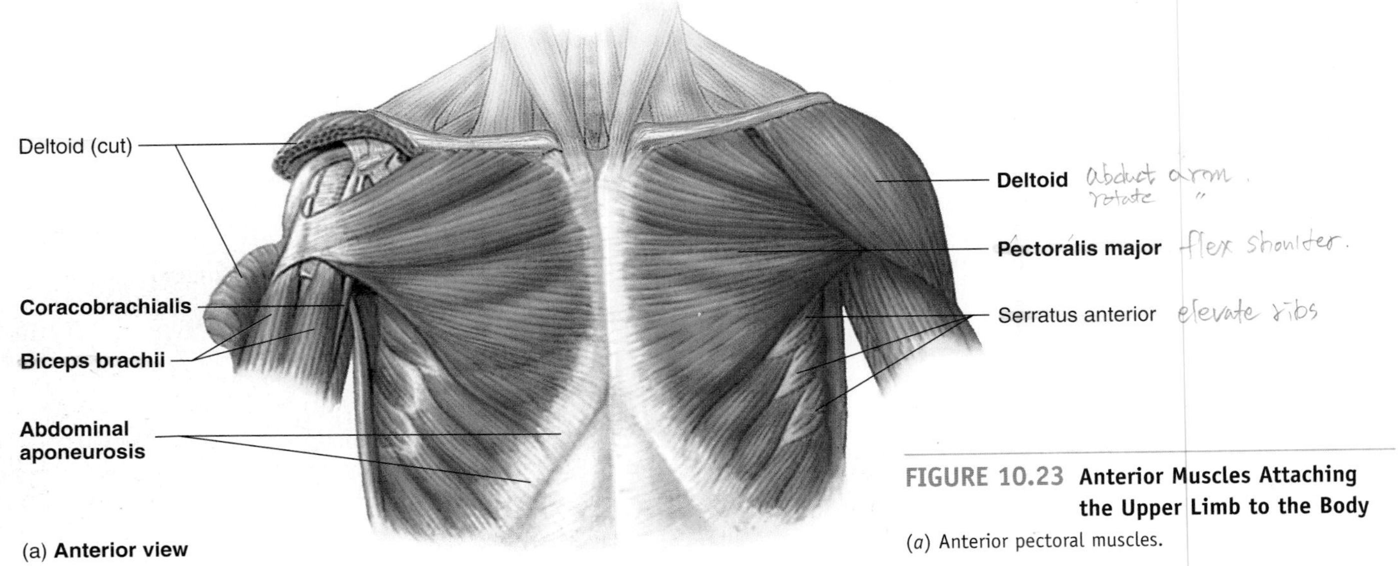

FIGURE 10.23 Anterior Muscles Attaching the Upper Limb to the Body

(*a*) Anterior pectoral muscles.

Sternocleidomastoid
Deltoid
Pectoralis major
Biceps brachii
Serratus anterior
(b) **Anterior view**

Trapezius
Deltoid
Sternocleidomastoid
Biceps brachii
Pectoralis major
Serratus anterior
Teres major and latissimus dorsi
(c) **Anterior view**

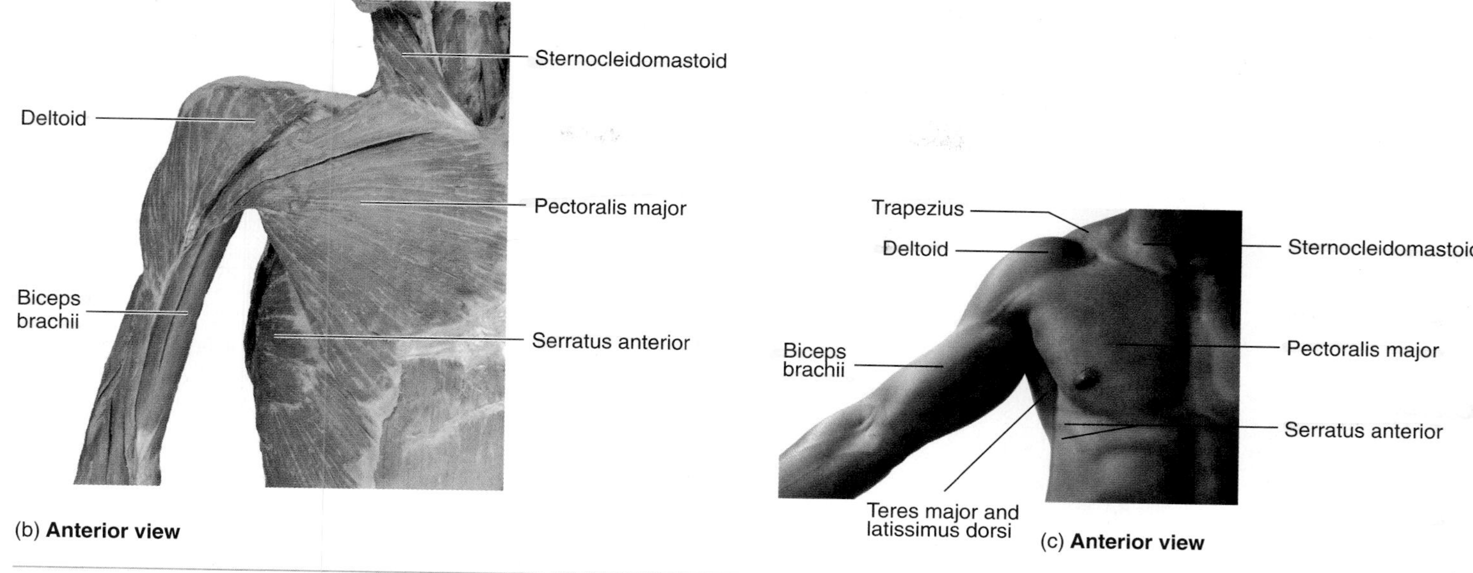

FIGURE 10.23 ***(continued)***
(*b*) The right pectoral region of a cadaver. (*c*) Surface anatomy of the right anterior pectoral region.

Levator scapulae
Rhomboideus minor
Rhomboideus major
spine of scapula
Rotator cuff
stabilize shoulder joint
Supraspinatus
Infraspinatus
Subscapularis (deep to scapula and not visible in figure)
adduct arm
Teres minor
Teres major
extend shoulder
Latissimus dorsi
extend shoulder, adduct arm
Twelfth thoracic vertebra
External abdominal oblique
(a) **Posterior view**

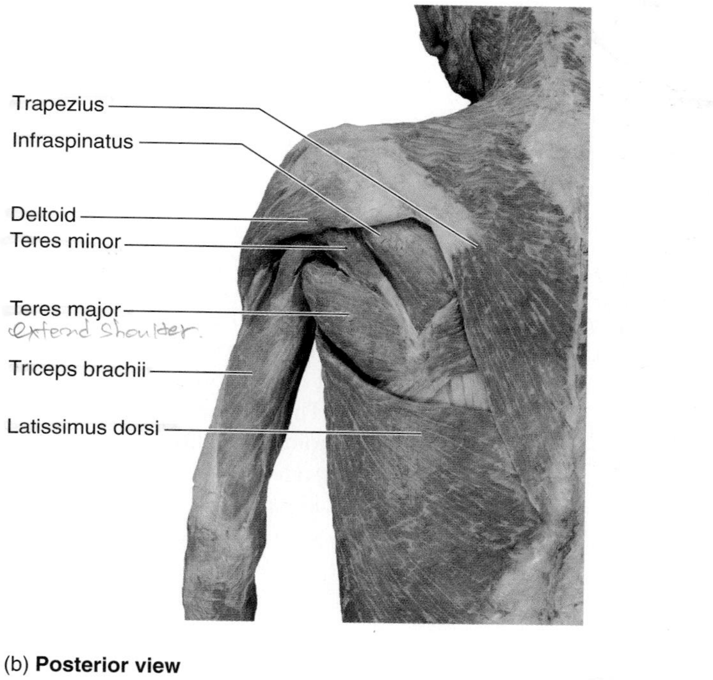

Trapezius
Infraspinatus
Deltoid
Teres minor
Teres major
extend shoulder
Triceps brachii
Latissimus dorsi
(b) **Posterior view**

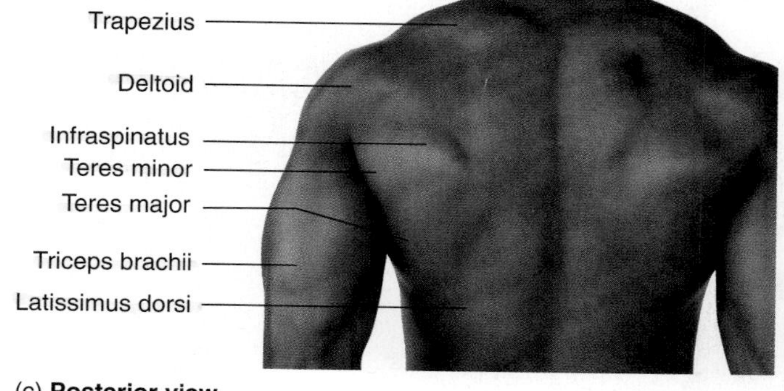

Trapezius
Deltoid
Infraspinatus
Teres minor
Teres major
Triceps brachii
Latissimus dorsi
(c) **Posterior view**

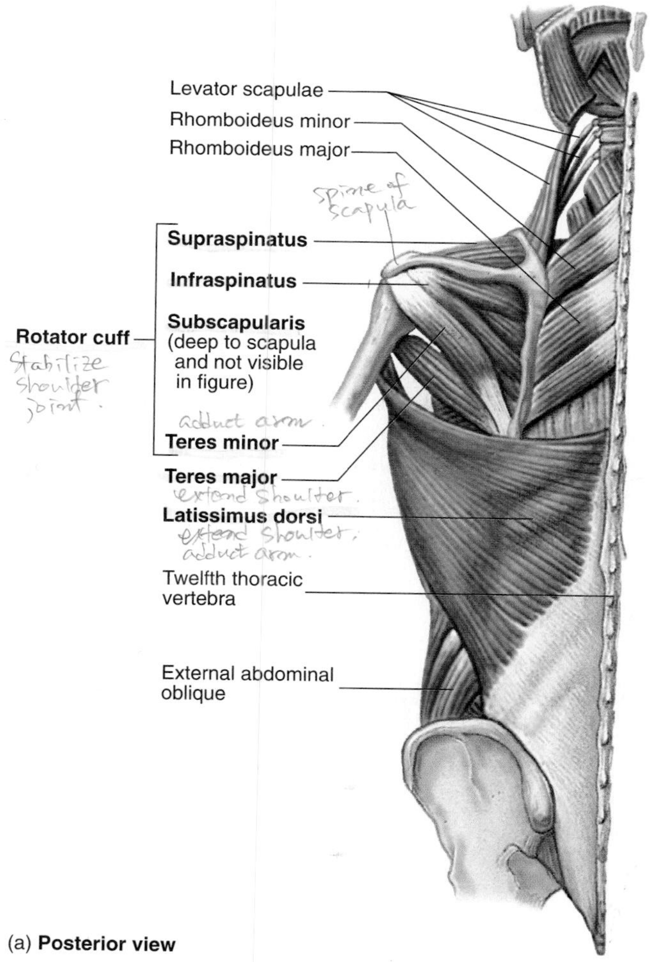

FIGURE 10.24 Posterior Muscles Attaching the Upper Limb to the Body
(*a*) Posterior view of muscles. (*b*) Posterior view of the left pectoral region of a cadaver. (*c*) Surface anatomy of the left posterior pectoral region.

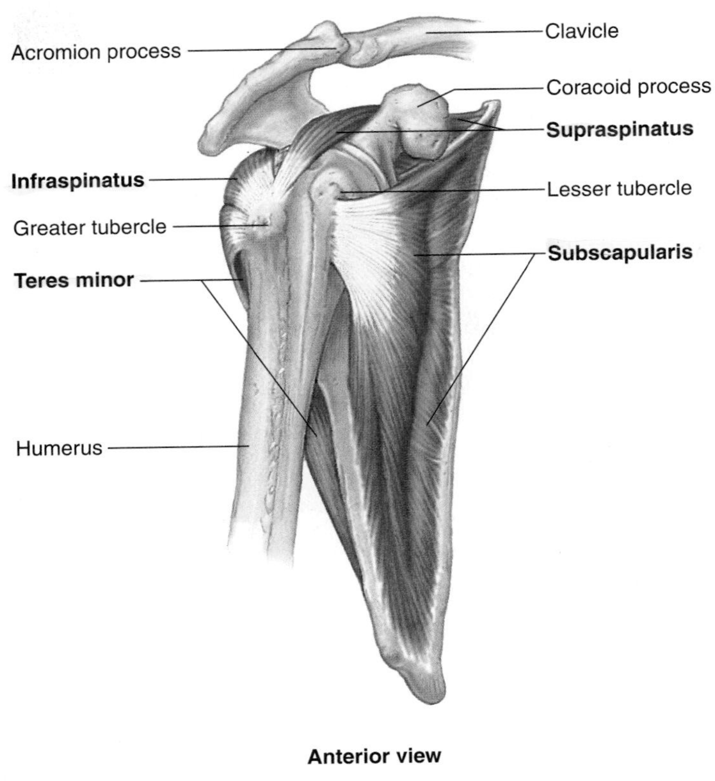

FIGURE 10.25 Right Rotator Cuff Muscles

movements and notice the position and action of the muscle. The **deltoid** (deltoideus) muscle (see figure 10.23) also is listed in table 10.13 as a flexor and an extensor. The deltoid muscle is like three muscles in one: The anterior fibers flex the shoulder, the lateral fibers abduct the arm, and the posterior fibers extend the shoulder. The deltoid muscle is part of the group of muscles that binds the humerus to the scapula. The primary muscles holding the head of the humerus in the glenoid cavity, however, are called the **rotator cuff** muscles (listed separately in table 10.13) because they form a cuff or cap over the proximal humerus (figure 10.25).

A rotator cuff injury involves damage to one or more of these muscles or their tendons, usually the supraspinatus muscle. The muscles moving the arm are involved in flexion, extension, abduction, adduction, rotation, and circumduction (table 10.14).

Raising the upper limb so as to place the hand above the head can be accomplished by abduction of the arm and then rotation of the scapula. Abduction of the arm from the anatomical position through 90 degrees (to the point at which the hand is level to the shoulder) involves movement of the humerus and is accomplished by the deltoid muscle assisted by the rotator cuff muscles, which hold the head of the humerus tightly in place. In the initial phase of abduction (first 15 degrees), the deltoid is assisted by the supraspinatus. Place your hand on your deltoid and feel it contract as you abduct 90 degrees. Moving the arm from 90 degrees to 180 degrees, so that the hand is held high above the head, primarily involves rotation of the scapula, which is accomplished by the trapezius and serratus anterior muscles. Feel the inferior angle of your scapula as you abduct the arm to 90 degrees and then rotate to 180 degrees. Do you notice a big difference? Movement of the arm from 90 degrees to 180 degrees, however, cannot occur unless the head of the humerus is held tightly in the glenoid cavity by the rotator cuff muscles, especially the supraspinatus. Damage to the supraspinatus muscle can prevent abduction past 90 degrees.

PREDICT 4

A tennis player complains of pain in the shoulder when abducting the arm in attempting to serve or when attempting an overhead volley (during extreme abduction). In extreme abduction, the supraspinatus muscle rises superiorly and may be damaged by compression against what bony structure?

Several muscles acting on the arm can be seen very clearly in the living individual (see figures 10.23*c* and 10.24*c*). The pectoralis major forms the upper chest, and the deltoids are prominent over the shoulders. The deltoid is a common site for administering injections.

TABLE 10.14 Summary of Muscle Actions on the Shoulder and Arm

Flexion	Extension	Abduction	Adduction	Medial Rotation	Lateral rotation
Deltoid	Deltoid	Deltoid	Pectoralis major	Pectoralis major	Deltoid
Pectoralis major	Teres major	Supraspinatus	Latissimus dorsi	Teres major	Infraspinatus
Coracobrachialis	Lattissimus dorsi		Teres major	Lattissimus dorsi	Teres minor
Biceps brachii	Pectoralis major		Teres minor	Deltoid	
	Triceps brachii		Triceps brachii	Subscapularis	
			Coracobrachialis		

TABLE 10.15 Muscles Acting on the Forearm (See Figures 10.26 and 10.27)

Muscle	Origin	Insertion	Nerve	Action
Arm				
Biceps brachii (bī′seps brā′kē-ī)	Long head—supraglenoid tubercle Short head—coracoid process	Radial tuberosity and aponeurosis of biceps brachii	Musculocutaneous	Flexes shoulder and elbow; supinates forearm and hand
Brachialis (brā′kē-al′is)	Anterior surface of humerus	Ulnar tuberosity and coronoid process of ulna	Musculocutaneous and radial	Flexes elbow
Triceps brachii (trī′seps brā′kē-ī)	Long head—infraglenoid tubercle on the lateral border of scapula Lateral head—lateral and posterior surface of humerus Medial head—posterior humerus	Olecranon process of ulna	Radial	Extends elbow; extends shoulder and adducts arm
Forearm				
Anconeus (ang-kō′nē-ŭs)	Lateral epicondyle of humerus	Olecranon process and posterior ulna	Radial	Extends elbow
Brachioradialis (brā′kē-ō-rā′dē-al′is)	Lateral supracondylar ridge of humerus	Styloid process of radius	Radial	Flexes elbow
Pronator quadratus (prō-nā-ter, prō-nā-tōr kwah-drā′tŭs)	Distal ulna	Distal radius	Anterior interosseous	Pronates forearm (and hand)
Pronator teres (prō-nā-tōr ter′ēz, tĕr-ēz)	Medial epicondyle of humerus and coronoid process of ulna	Radius	Median	Pronates forearm (and hand)
Supinator (soo′pi-nā-ter, soo′pi-nā-tōr)	Lateral epicondyle of humerus and ulna	Radius	Radial	Supinates forearm (and hand)

Shoulder Pain

Baseball pitchers, because they throw very hard, may tear their rotator cuffs. Such tears result in pain in the anterosuperior part of the shoulder. Older people may also develop such pain because of **degenerative tendonitis** of the rotator cuff. The supraspinatus tendon is the most commonly affected part of the rotator cuff in either trauma or degeneration, probably because it has a relatively poor blood supply.

Pain in the shoulder can also result from **subacromial bursitis,** which is inflammation of the subacromial bursa. **Biceps tendinitis,** inflammation of the biceps brachii long head tendon, can also cause shoulder pain. This inflammation is also commonly caused by throwing a baseball or football.

18 ***Name seven muscles that attach the humerus to the scapula. What two muscles attach the humerus directly to the trunk?***

19 ***List the muscles forming the rotator cuff, and describe their function.***

20 ***What muscles cause flexion and extension of the shoulder? Abduction and adduction of the arm? What muscle is involved in abduction of the arm to 90 degrees? Above 90 degrees? What muscles cause rotation of the arm?***

21 ***List the muscles that cause flexion and extension of the elbow. Where are these muscles located?***

Forearm Movements

Extension and Flexion of the Elbow

Extension of the elbow is accomplished by the **triceps brachii** (brā′kē-ī) and **anconeus** (ang-kō′nē-ŭs); flexion of the elbow is accomplished by the **brachialis** (brā′-kē-al′is), **biceps brachii,** and **brachioradialis** (brā′kē-ō-rā′dē-al′is; table 10.15; figures 10.26 and 10.27). The triceps brachii constitute the main mass visible on the posterior aspect of the arm (see figures 10.24*c* and 10.26*c*). The biceps brachii is readily visible on the anterior aspect of the arm (see figures 10.23*c* and 10.26*c*). The brachialis lies deep to the biceps brachii and can be seen only as a mass on the medial and lateral sides of the arm. The brachioradialis forms a bulge on the anterolateral side of the forearm just distal to the elbow (see figure 10.26*a*; figure 10.28*b* and *d*). If the elbow is forcefully flexed in the midprone position (midway between pronation and supination), the brachioradialis stands out clearly on the forearm (see figure 10.28*d*).

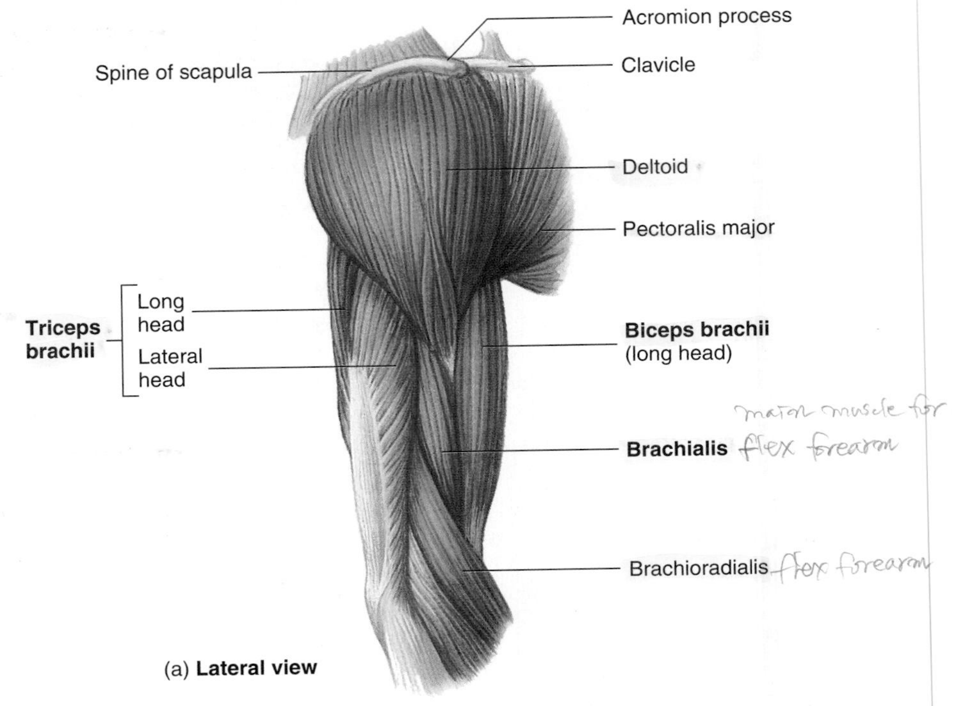

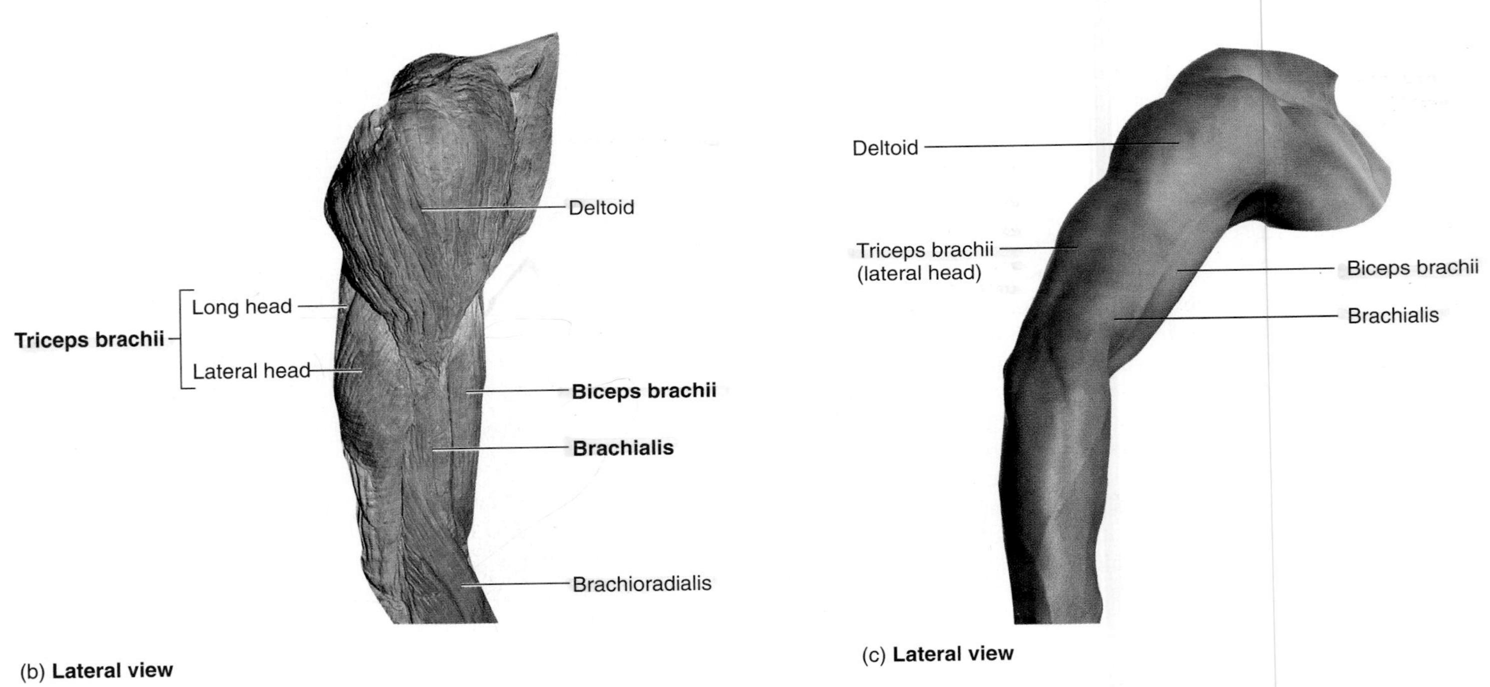

FIGURE 10.26 Lateral Right Arm Muscles

(*a*) The right shoulder and arm. (*b*) The right shoulder and arm muscles of a cadaver. (*c*) Surface anatomy of the right shoulder and arm.

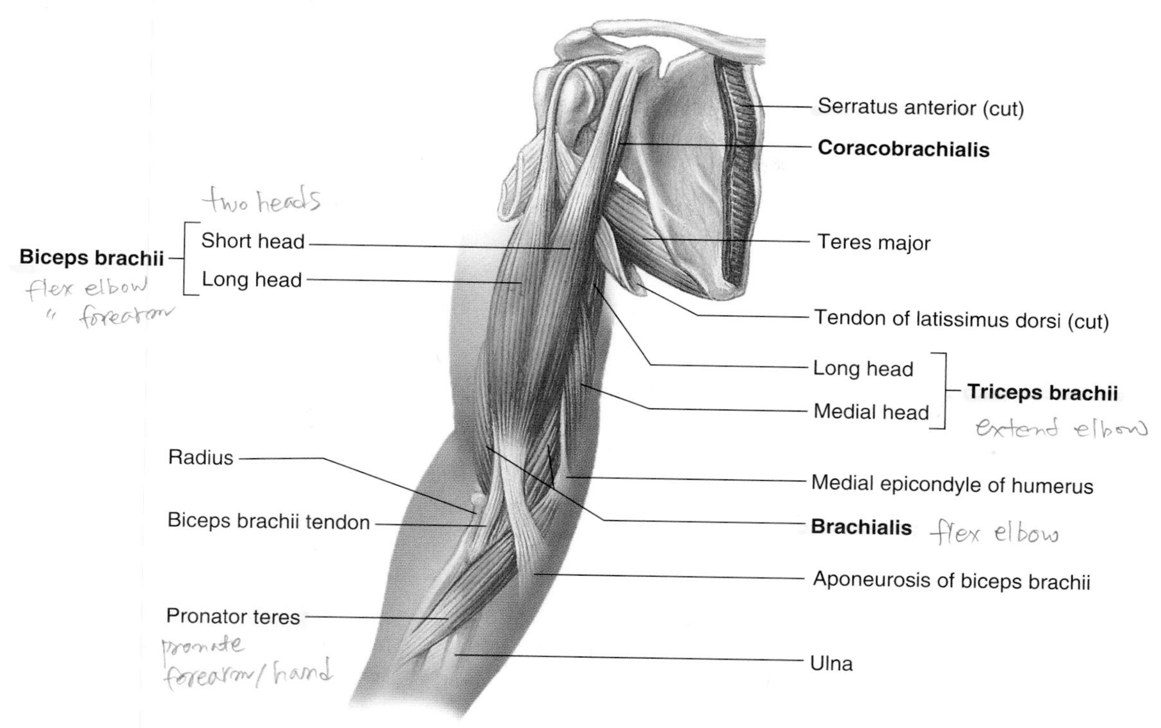

FIGURE 10.27 Anterior Right Arm Muscles

Medial epicondyle of humerus

Pronator teres

Flexor carpi radialis

Palmaris longus

Flexor carpi ulnaris

Radius

Ulna

Palmar aponeurosis

(a) **Anterior view**

Brachioradialis

Flexor digitorum superficialis

(b) **Anterior view**

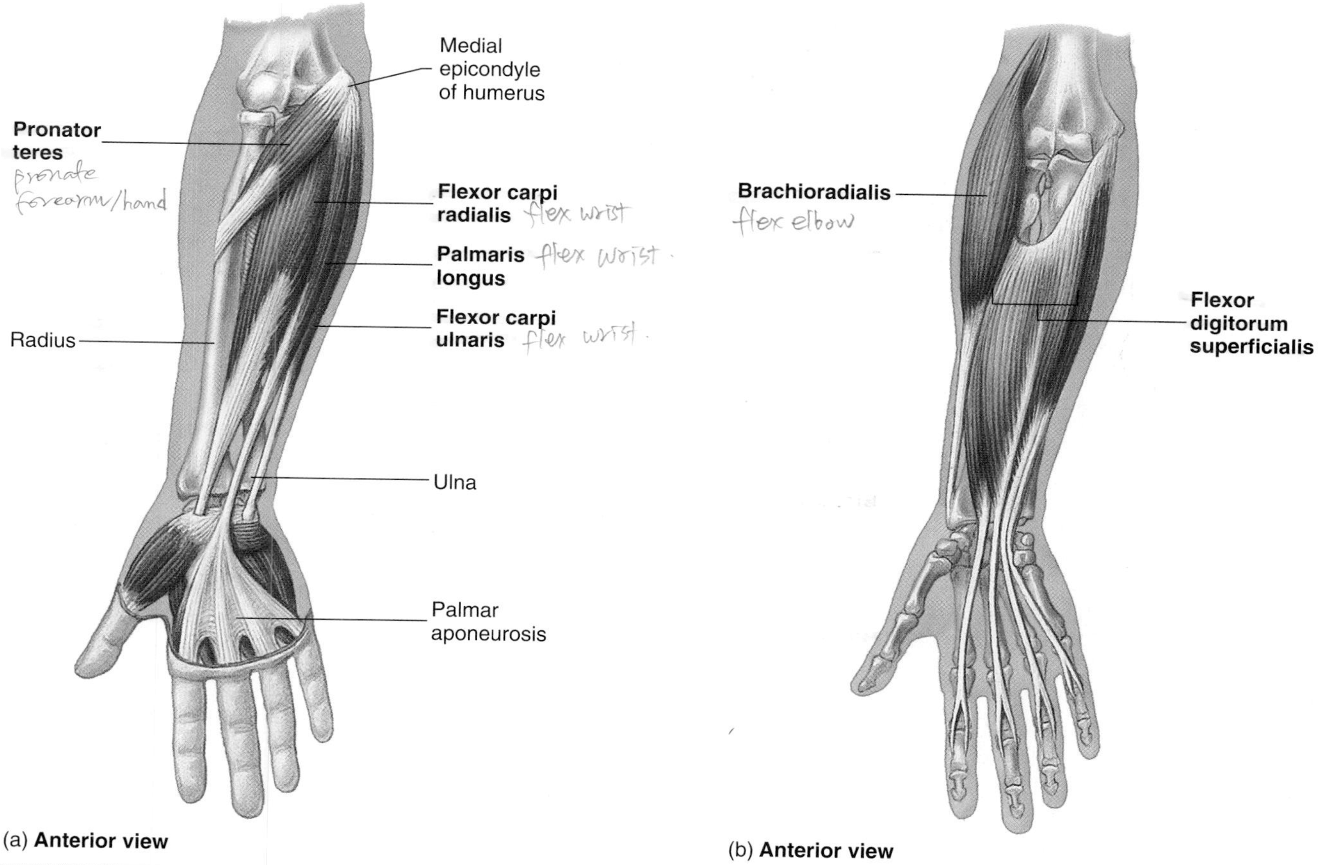

FIGURE 10.28 Anterior Right Forearm Muscles

(*a*) Right forearm (superficial). Brachioradialis muscle is removed. (*b*) Right forearm (deeper than *a*). Pronator teres, flexor carpi radialis and ulnaris, and palmaris longus muscles are removed.

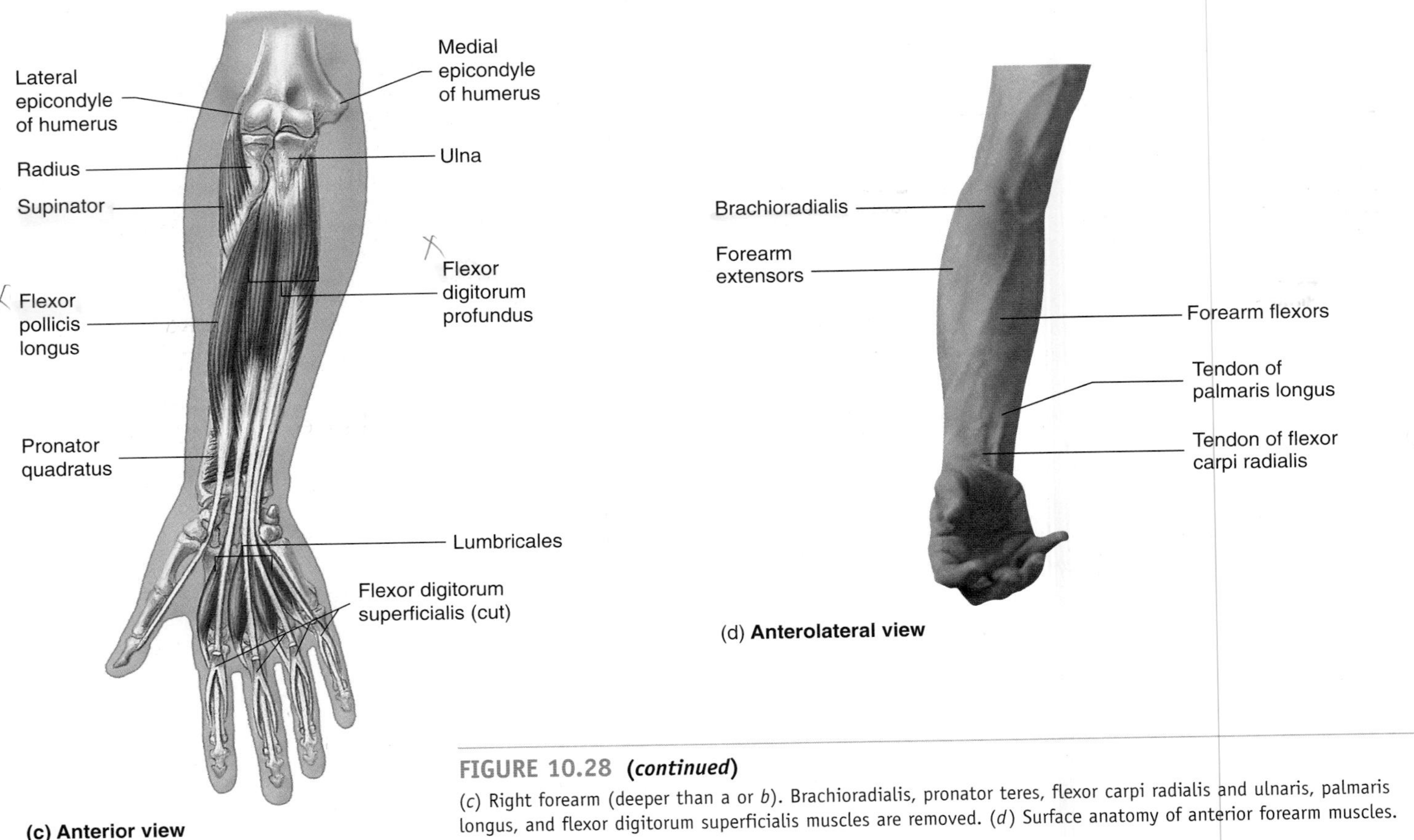

FIGURE 10.28 ***(continued)***

(*c*) Right forearm (deeper than a or *b*). Brachioradialis, pronator teres, flexor carpi radialis and ulnaris, palmaris longus, and flexor digitorum superficialis muscles are removed. (*d*) Surface anatomy of anterior forearm muscles.

Supination and Pronation

Supination of the forearm and hand is accomplished by the **supinator** and the **biceps brachii** (see figure 10.27; figure 10.28*c* and figure 10.29*b*). Pronation is a function of the **pronator quadratus** (kwah-drā′tŭs) and the **pronator teres** (ter′ēz, tēr-ēz; figure 10.28*a*; see figures 10.27 and 10.28*c*).

22 ***Supination and pronation of the forearm and hand are produced by what muscles? Where are these muscles located?***

PREDICT 5

Explain the difference between doing chin-ups with the forearm supinated versus pronated. The action of which muscle predominates in each type of chin-up? Which type is easier? Why?

Wrist, Hand, and Finger Movements

The forearm muscles are divided into anterior and posterior groups (table 10.16; see figures 10.28 and 10.29). Most of the anterior forearm muscles are responsible for flexion of the wrist and fingers. Most of the posterior forearm muscles cause extension of the wrist and fingers.

Extrinsic Hand Muscles

The **extrinsic hand muscles** are in the forearm but have tendons that extend into the hand. A strong band of fibrous connective tissue, the **retinaculum** (ret-i-nak′ū-lŭm; bracelet), covers the flexor and extensor tendons and holds them in place around the wrist so that they do not "bowstring" during muscle contraction (see figures 10.29*a* and *c*).

Two major anterior muscles, the **flexor carpi radialis** (kar′pī-rā-dē-ā′lis) and the **flexor carpi ulnaris** (ŭl-nā′ris), flex the wrist; and three posterior muscles, the **extensor carpi radialis longus,** the **extensor carpi radialis brevis,** and the **extensor carpi ulnaris,** extend the wrist. The wrist flexors and extensors are visible on the anterior and posterior surfaces of the forearm (see figures 10.28*d* and 10.29*d*). The tendon of the flexor carpi radialis is an important landmark because the radial pulse can be felt just lateral to the tendon (see figure 10.28*d*).

Tennis Elbow

Forceful, repetitive use of the forearm extensor muscles can damage them where they attach to the lateral epicondyle. This condition is often called **tennis elbow** because it can result from playing tennis. It is also called **lateral epicondylitis** because it can result from other sports and activities, such as shoveling snow.

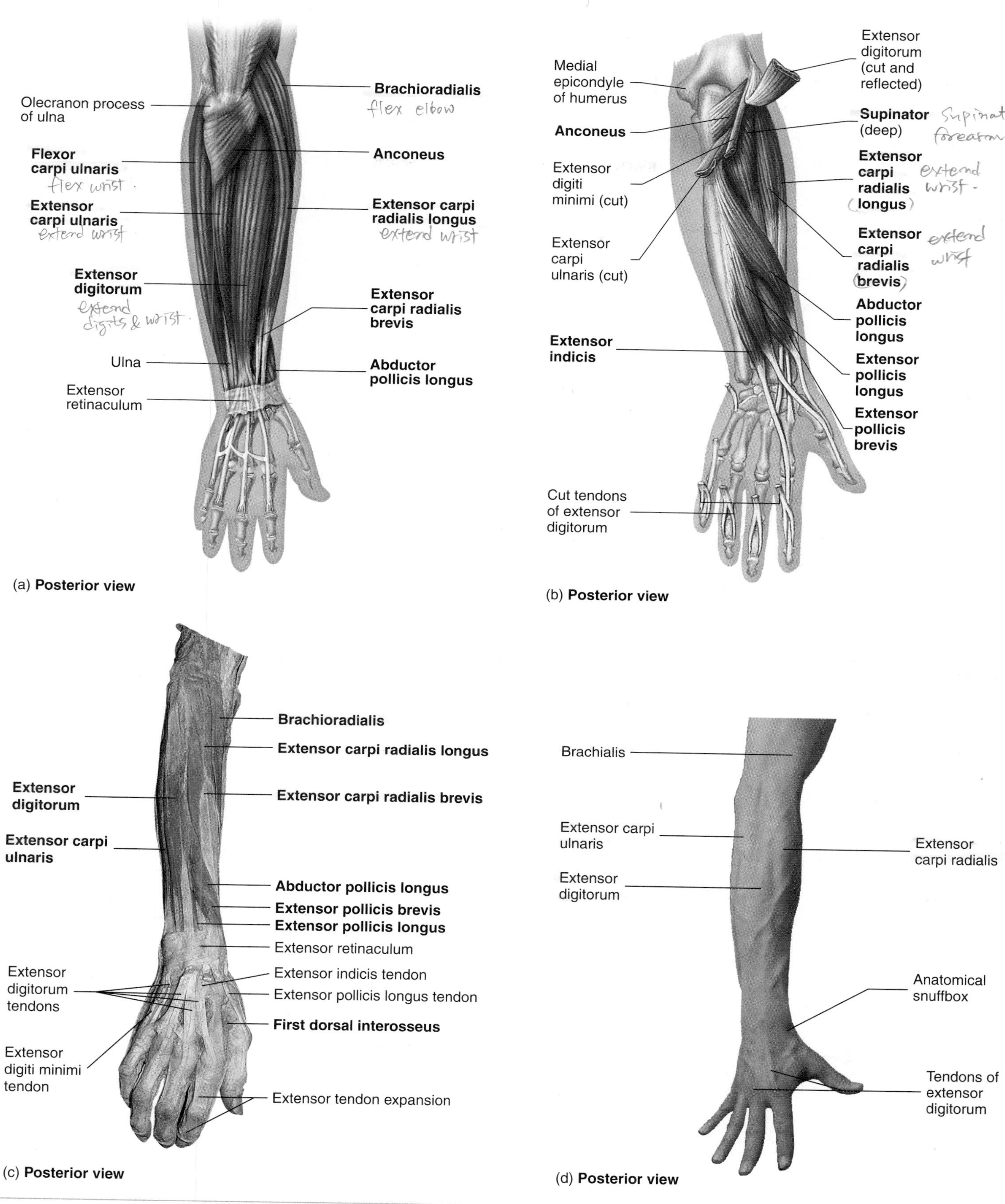

FIGURE 10.29 Posterior Right Forearm Muscles

(*a*) Right forearm (superficial). (*b*) Deep muscles of the right posterior forearm. Extensor digitorum, extensor digiti minimi, and extensor carpi ulnaris muscles are cut to reveal deeper muscles. (*c*) Photograph showing dissection of the posterior right forearm and hand. (*d*) Surface anatomy of posterior forearm.

TABLE 10.16 Muscles of the Forearm Acting on the Wrist, Hand, and Fingers (See Figures 10.28 and 10.29)

Muscle	Origin	Insertion	Nerve	Action
Anterior Forearm				
Flexor carpi radialis (kar′pī rā-dē-ā′lis)	Medial epicondyle of humerus	Second and third metacarpals	Median	Flexes and abducts wrist
Flexor carpi ulnaris (kar′pī ŭl-nā′ris)	Medial epicondyle of humerus and ulna	Pisiform, hamate, and fifth metacarpal	Ulnar	Flexes and adducts wrist
Flexor digitorum profundus (dij′i-tōr′ŭm prō-fŭn′dŭs)	Ulna	Distal phalanges of digits 2–5	Ulnar and median	Flexes fingers at metacarpal-phalangeal joints and interphalangeal joints and wrist
Flexor digitorum superficialis (dij′i-tōr′ŭm soo′per-fish-ē-ā′lis)	Medial epicondyle of humerus, coronoid process, and radius	Middle phalanges of digits 2–5	Median	Flexes fingers at interphalangeal joints and wrist
Flexor pollicis longus (pol′i-sis lon′gŭs)	Radius	Distal phalanx of thumb	Median	Flexes thumb
Palmaris longus (pawl-mār′is lon′gŭs)	Medial epicondyle of humerus	Palmar fascia	Median	Tenses palmar fascia; flexes wrist
Posterior Forearm				
Abductor pollicis longus (pol′i-sis lon′gŭs)	Posterior ulna and radius and interosseous membrane	Base of first metacarpal	Radial	Abducts and extends thumb; abducts wrist
Extensor carpi radialis brevis (kar′pī rā-dē-ā′lis brev′is)	Lateral epicondyle of humerus	Base of third metacarpal	Radial	Extends and abducts wrist
Extensor carpi radialis longus (kar′pī rā-dē-ā′lis lon′gus)	Lateral supracondylar ridge of humerus	Base of second metacarpal	Radial	Extends and abducts wrist
Extensor carpi ulnaris (kar′pī ŭl-nā′ris)	Lateral epicondyle of humerus and ulna	Base of fifth metacarpal	Radial	Extends and adducts wrist
Extensor digiti minimi (dij′i-tī mi′nĭ-mī)	Lateral epicondyle of humerus	Phalanges of fifth digit	Radial	Extends little finger and wrist
Extensor digitorum (dij′i-tōr′ŭm)	Lateral epicondyle of humerus	Extensor tendon expansion over phalanges of digits 2–5	Radial	Extends fingers and wrist
Extensor indicis (in′di-sis)	Ulna	Extensor tendon expansion over second digit	Radial	Extends forefinger and wrist
Extensor pollicis brevis (pol′i-sis brev′is)	Radius	Proximal phalanx of thumb	Radial	Extends and abducts thumb; abducts wrist
Extensor pollicis longus (pol′i-sis lon′gŭs)	Ulna	Distal phalanx of thumb	Radial	Extends thumb

TABLE 10.17 Intrinsic Hand Muscles (See Figure 10.30)

Muscle	Origin	Insertion	Nerve	Action
Midpalmar Muscles				
Interossei (in′ter-os′e-ī)				
Dorsal	Sides of metacarpal bones	Proximal phalanges of second, third, and fourth digits	Ulnar	Abducts second, third, and fourth digits
Palmar	Second, fourth, and fifth metacarpals	Second, fourth, and fifth digits	Ulnar	Adducts second, fourth, and fifth digits
Lumbricals (lum′bră-kălz)	Tendons of flexor digitorum profundis	Second through fifth digits	Two on radial side—median; two on ulnar side—ulnar	Flexes proximal and extends middle and distal phalanges
Thenar Muscles				
Abductor pollicis brevis (ab-dŭk-ter, ab-dŭk-tōr pol′i-sis brev′is)	Flexor retinaculum, trapezium, and scaphoid	Proximal phalanx of thumb	Median	Abducts thumb
Adductor pollicis (ab-dŭk-ter, ab-dŭk-tōr pol′i-sis)	Third metacarpal, second metacarpal, trapezoid, and capitate	Proximal phalanx of thumb	Ulnar	Adducts thumb
Flexor pollicis brevis (pol′i-sis brev′is)	Flexor retinaculum and first metacarpal	Proximal phalanx of thumb	Median and ulnar	Flexes thumb
Opponens pollicis (ŏ-pō′nens pol′i-sis)	Trapezium and flexor retinaculum	First metacarpal	Median	Opposes thumb
Hypothenar Muscles				
Abductor digiti minimi (ab-dŭk-ter, ab-dŭk-tōr dij′i-tī min′imī)	Pisiform	Base of fifth digit	Ulnar	Abducts and flexes little finger
Flexor digiti minimi brevis (dij′i-tī min′ĭ-mī brev′is)	Hamate	Base of proximal phalanx of fifth digit	Ulnar	Flexes little finger
Opponens digiti minimi (ŏ-pō′nens dij′i-tī min′i-mī)	Hamate and flexor retinaculum	Fifth metacarpal	Ulnar	Opposes little finger

Flexion of the four medial digits is a function of the **flexor digitorum** (dij′i-tor′ŭm) **superficialis** and **flexor digitorum profundus** (prō-fŭn′dŭs; deep). Extension is accomplished by the **extensor digitorum.** The tendons of this muscle are very visible on the dorsum of the hand (see figure 10.29*d*). The little finger has an additional extensor, the **extensor digiti minimi** (dij′i-tī min′i-mī). The index finger also has an additional extensor, the **extensor indicis** (in′di-sis).

Movement of the thumb is caused in part by the **abductor pollicis** (pol′i-sis) **longus,** the **extensor pollicis longus,** and the **extensor pollicis brevis.** These tendons form the sides of a depression on the posterolateral side of the wrist called the "anatomical snuffbox" (see figure 10.29). When snuff was in use, a small pinch could be placed into the anatomical snuffbox and inhaled through the nose.

Intrinsic Hand Muscles

The **intrinsic hand muscles** are entirely within the hand (table 10.17 and figure 10.30). Abduction of the fingers is accomplished by the **dorsal interossei** (in′ter-os′e-ī) and the **abductor digiti minimi,** whereas adduction is a function of the **palmar interossei.**

The **flexor pollicis brevis,** the **abductor pollicis brevis,** and the **opponens pollicis** form a fleshy prominence at the base of the thumb called the **thenar** (thē′nar) **eminence** (see figure 10.30). The **abductor digiti minimi, flexor digiti minimi brevis,** and **opponens digiti minimi** constitute the **hypothenar eminence** on the ulnar side of the hand. The thenar and hypothenar muscles are involved in the control of the thumb and little finger.

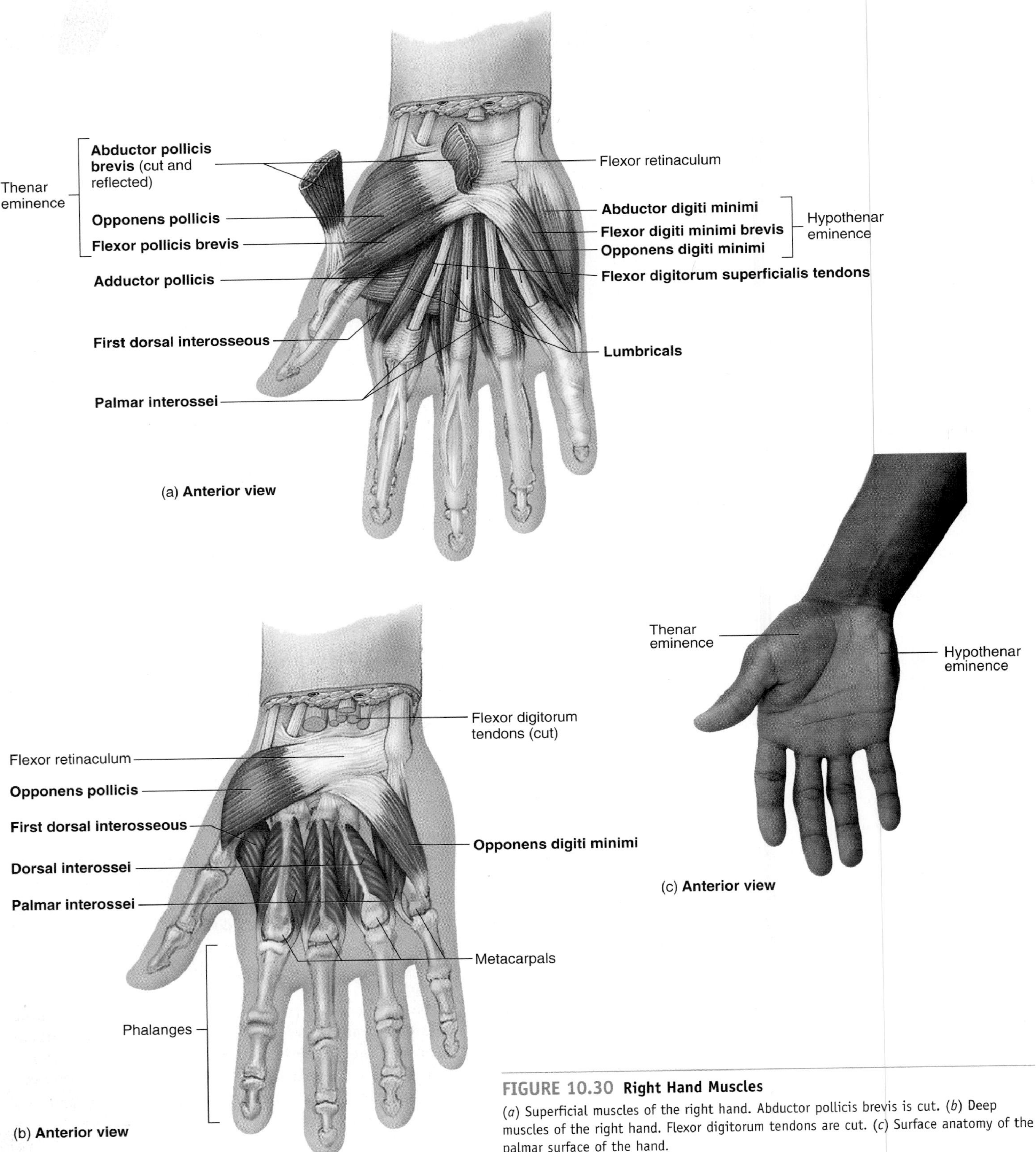

FIGURE 10.30 Right Hand Muscles

(*a*) Superficial muscles of the right hand. Abductor pollicis brevis is cut. (*b*) Deep muscles of the right hand. Flexor digitorum tendons are cut. (*c*) Surface anatomy of the palmar surface of the hand.

TABLE 10.18 Muscles Acting on the Thigh (See Figures 10.31–10.35)

Muscle	Origin	Insertion	Nerve	Action
Anterior				
Iliopsoas (il′ē-ō-sō′as)				
Iliacus (il-ī′ă-kus)	Iliac fossa	Lesser trochanter of femur and capsule of hip joint	Lumbar plexus	Flexes hip
Psoas major (sō′as)	T12–L5	Lesser trochanter of femur	Lumbar plexus	Flexes hip
Posterior and Lateral				
Gluteus maximus (gloo-tē′ŭs mak′si-mŭs)	Posterior surface of ilium, sacrum, and coccyx	Gluteal tuberosity of femur and iliotibial tract	Inferior gluteal	Extends hip; abducts and laterally rotates thigh
Gluteus medius (gloo-tē′ŭs mē′dē-ŭs)	Posterior surface of ilium	Greater trochanter of femur	Superior gluteal	Abducts and medially rotates thigh; tilts pelvis toward supported side
Gluteus minimus (gloo-tē′ŭs min-i-mŭs)	Posterior surface of ilium	Greater trochanter of femur	Superior gluteal	Abducts and medially rotates thigh; tilts pelvis toward supported side
Tensor fasciae latae (ten′sōr fash′ē-ē lā′tē)	Anterior superior iliac spine	Through iliotibial tract to lateral condyle of tibia	Superior gluteal	Tenses lateral fascia and stabilizes femur on tibia when standing; flexes hip; abducts and medially rotates thigh; tilts pelvis
Deep Thigh Rotators				
Gemellus (jĕ-mel′ŭs)				
Inferior	Ischial tuberosity	Obturator internus tendon	L5 and S1	Laterally rotates and abducts thigh
Superior	Ischial spine	Obturator internus tendon	L5 and S1	Laterally rotates and abducts thigh
Obturator (ob′too-rā-tōr)				
Externus (eks-ter′nŭs)	Inferior margin of obturator foramen	Greater trochanter of femur	Obturator	Laterally rotates thigh
Internus (in-ter′nŭs)	Interior margin of obturator foramen	Greater trochanter of femur	L5 and S1	Laterally rotates thigh
Piriformis (pir′i-fōr′mis)	Sacrum and ilium	Greater trochanter of femur	S1 and S2	Laterally rotates and abducts thigh
Quadratus femoris (kwah′-drā′tŭs fem′ō-ris)	Ischial tuberosity	Intertrochanteric ridge of femur	L5 and S1	Laterally rotates thigh

23 ***Describe the muscle groups that cause flexion and extension of the wrist.***

24 ***Contrast the location and actions of the extrinsic and intrinsic hand muscles. What is the retinaculum? What is the location and action of the thenar and hypothenar muscles?***

25 ***Describe the muscles that move the thumb. The tendons of what muscles form the anatomical snuffbox?***

LOWER LIMB MUSCLES

Thigh Movements

Several hip muscles originate on the coxal bone and insert onto the femur (table 10.18 and figures 10.31–10.35). These muscles are divided into three groups: anterior, posterolateral, and deep.

The anterior muscles (see figure 10.31), the **iliacus** (il-ī′ă-kŭs) and the **psoas** (sō′as) **major,** flex the hip. Because these muscles share an insertion and produce the same movement, they often are referred to as the **iliopsoas** (il′ē-ō-sō′as). When the thigh is fixed, the iliopsoas flexes the trunk on the thigh. For example, the iliopsoas does most of the work when a person does sit-ups.

The posterolateral hip muscles consist of the gluteal muscles and the **tensor fasciae latae** (fash′ē-ē lā′tē). The **gluteus** (gloo-tē′ŭs) **maximus** (see figure 10.32) contributes most of the mass that can be seen as the buttocks (see figure 10.32*c*); the **gluteus medius,** a common site for injections, creates a smaller mass just superior and lateral to the gluteus maximus. The gluteus maximus functions at its maximum force in extension of the thigh when the hip is flexed at a 45-degree angle so that the muscle is optimally stretched, which accounts for both the sprinter's stance and the bicycle racing posture.

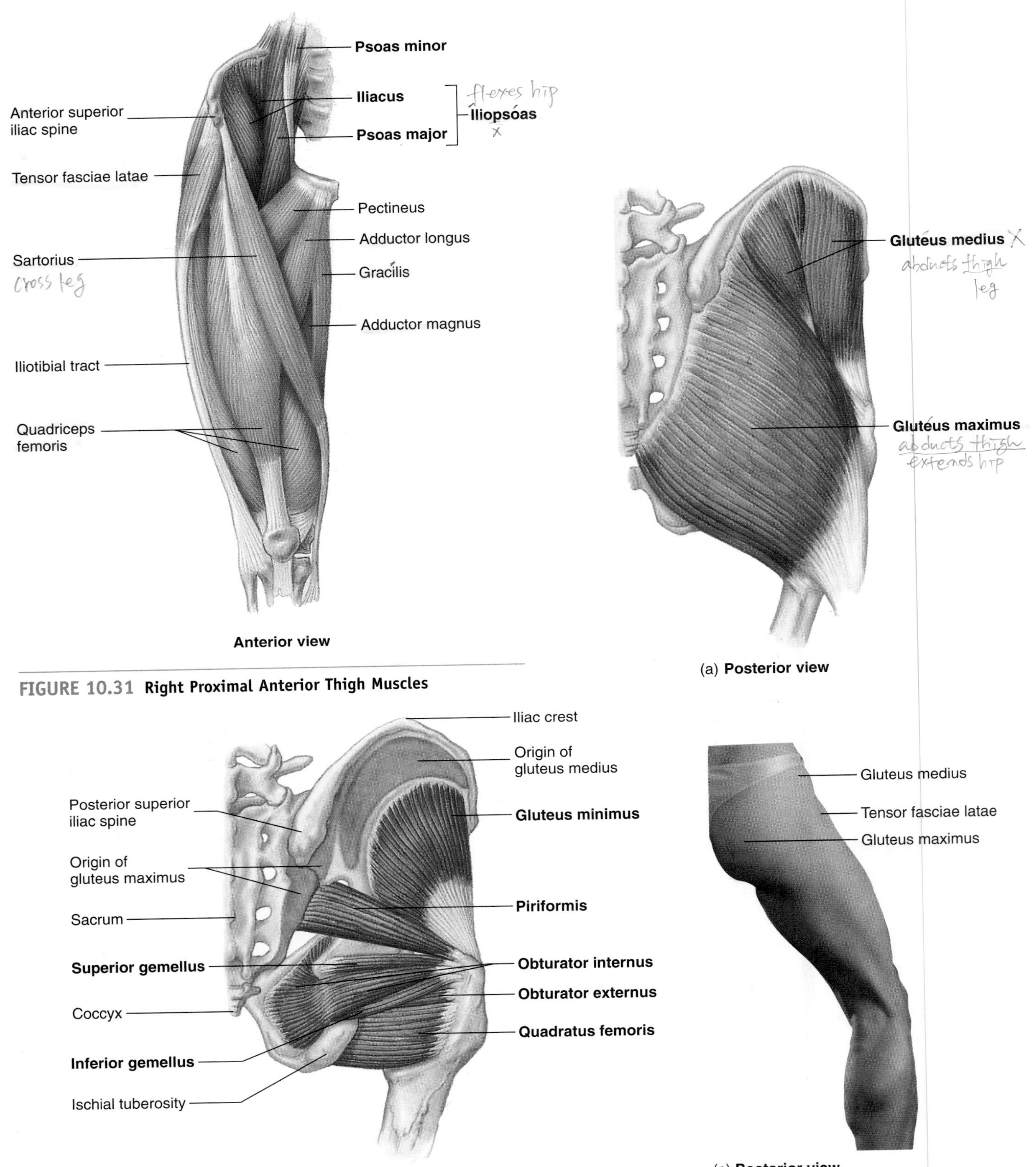

FIGURE 10.31 Right Proximal Anterior Thigh Muscles

FIGURE 10.32 Right Posterior Hip Muscles
(*a*) Right hip, superficial. (*b*) Right hip, deep. Gluteus maximus and medius are removed to reveal deeper muscles. (*c*) Surface anatomy of the right posterior hip muscles.

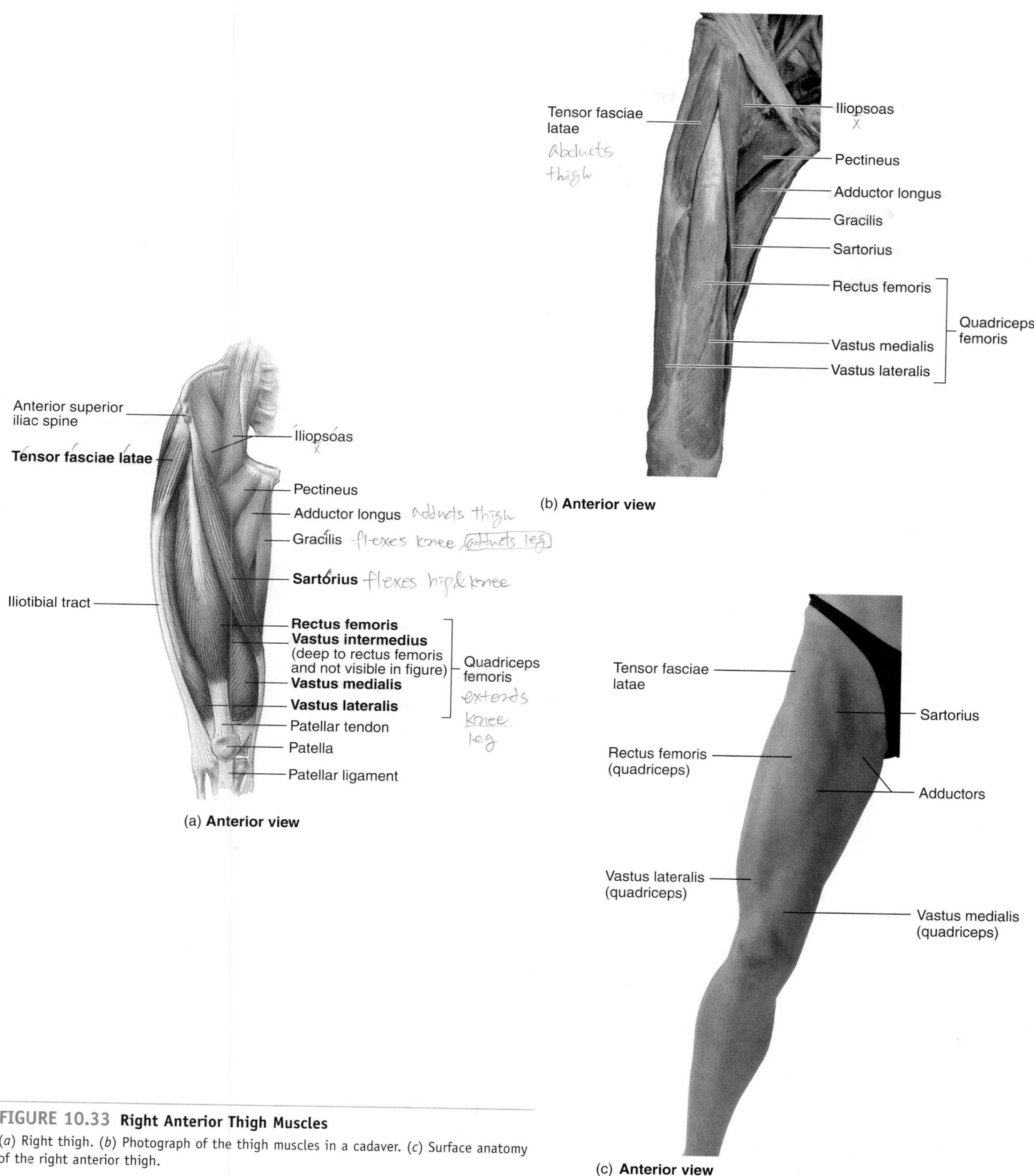

FIGURE 10.33 Right Anterior Thigh Muscles

(*a*) Right thigh. (*b*) Photograph of the thigh muscles in a cadaver. (*c*) Surface anatomy of the right anterior thigh.

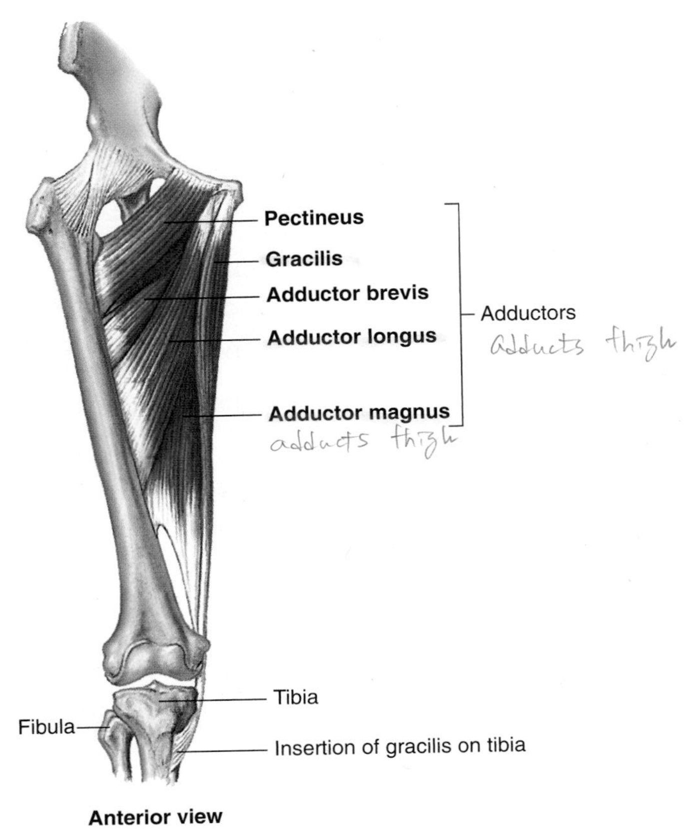

FIGURE 10.34 Right Medial Thigh Muscles

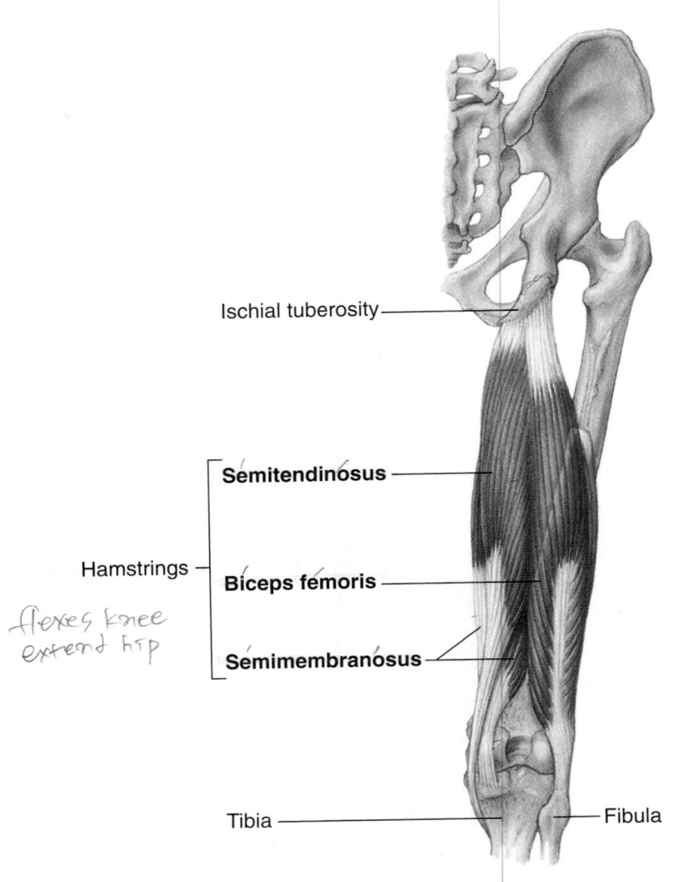

FIGURE 10.35 Right Posterior Thigh Muscles
Hip muscles are removed.

The deep hip muscles, as well as the gluteus maximus, function as lateral thigh rotators. The gluteus medius, gluteus minimus, and tensor fasciae latae are medial hip rotators (see table 10.18 and figure 10.32*b*). The gluteus medius and minimus muscles help tilt the pelvis and maintain the trunk in an upright posture during walking, as the foot of the opposite limb is raised from the ground. Without the action of these muscles, the pelvis tends to sag downward on the unsupported side.

In addition to the hip muscles, some of the muscles located in the thigh originate on the coxal bone and can cause movement of the thigh (tables 10.19 and 10.20). Three groups of thigh muscles have been identified based on their location in the thigh: the anterior, which flex the hip and/or extend the knee (see figure 10.33*a*); the medial, which adduct the thigh (see figure 10.34); and the posterior, which extend the hip and flex the knee (see figure 10.35).

26 ***Name the anterior hip muscle that flexes the hip. What muscles act as synergists to this muscle?***

27 ***Describe the movements produced by the gluteus muscles.***

Leg Movements

The anterior thigh muscles are the **quadriceps femoris** (fem′ŏ-ris) and the **sartorius** (sar-tōr′ē-ŭs; see table 10.20 and figure 10.33*a*). The quadriceps femoris is actually four muscles: the **rectus femoris,** the **vastus lateralis,** the **vastus medialis,** and the **vastus intermedius.** The quadriceps group extends the knee. The rectus femoris also flexes the hip because it crosses both the hip and knee joints.

The quadriceps femoris makes up the large mass on the anterior thigh (see figure 10.33*c*). The vastus lateralis sometimes is used as an injection site, especially in infants who may not have well-developed deltoid or gluteal muscles. The muscles of the quadriceps femoris have a common insertion, the patellar tendon, on and around the patella. The patellar ligament is an extension of the patellar tendon onto the tibial tuberosity. The patellar ligament is the point that is tapped with a rubber hammer when testing the knee-jerk reflex in a physical examination.

The sartorius is the longest muscle of the body, crossing from the lateral side of the hip to the medial side of the knee. As the muscle contracts, it flexes the hip and knee and laterally rotates the thigh. This movement is the action required for crossing the legs.

Sartorius—the Tailor's Muscle

The term *sartorius* means tailor. The sartorius muscle is so-named because its action is to cross the legs, a common position traditionally preferred by tailors because they can hold their sewing in their lap as they sit and sew by hand.

TABLE 10.19 Summary of Muscle Actions on the Hip and Thigh

Flexion	Extension	Abduction	Adduction	Medial Rotation	Lateral Rotation
Iliopsoas	Gluteus maximus	Gluteus maximus	Adductor magnus	Tensor fasciae latae	Gluteus maximus
Tensor fasciae latae	Semitendinosus	Gluteus medius	Adductor longus	Gluteus medius	Obturator internus
Rectus femoris	Semimembranosus	Gluteus minimus	Adductor brevis	Gluteus minimus	Obturator externus
Sartorius	Biceps femoris	Tensor fasciae latae	Pectineus		Superior gemellus
Adductor longus	Adductor magnus	Obturator internus	Gracilis		Inferior gemellus
Adductor brevis		Gemellus superior and inferior			Quadratus femoris
Pectineus		Piriformis			Piriformis
					Adductor magnus
					Adductor longus
					Adductor brevis

TABLE 10.20 Muscles of the Thigh (See Figures 10.33 and 10.35)

Muscle	Origin	Insertion	Nerve	Action
Anterior Compartment				
Quadriceps femoris (kwah′dri-seps fem′ŏ-ris)	Rectus femoris—anterior inferior iliac spine	Patella and onto tibial tuberosity through patellar ligament	Femoral	Extends knee; rectus femoris also flexes hip
	Vastus lateralis—greater trochanter and linea aspera of femur			
	Vastus intermedius—body of femur			
	Vastus medialis—linea aspera of femur			
Sartorius (sar-tōr′ē-ŭs)	Anterior superior iliac spine	Medial side of tibial tuberosity	Femoral	Flexes hip and knee; rotates thigh laterally and leg medially
Medial Compartment				
Adductor brevis (a-dŭk′ter, a-dŭk′tōr brev′is)	Pubis	Pectineal line and linea aspera of femur	Obturator	Adducts, laterally rotates thigh; flexes hip
Adductor longus (a-dŭk′ter, a-dŭk′tōr lon′gŭs)	Pubis	Linea aspera of femur	Obturator	Adducts, laterally rotates thigh; flexes hip
Adductor magnus (a-dŭk′ter, a-dŭk′tōr mag′nŭs)	Adductor part: pubis and ischium	Adductor part: linea aspera of femur	Adductor part: obturator	Adductor part: adducts thigh and flexes hip
	Hamstring part: ischial tuberosity	Hamstring part: adductor tubercle of femur	Hamstring part: tibial	Hamstring part: extends hip and adducts thigh
Gracilis (gras′i-lis)	Pubis near symphysis	Tibia	Obturator	Adducts thigh; flexes knee
Pectineus (pek′ti-nē′ŭs)	Pubic crest	Pectineal line of femur	Femoral and obturator	Adducts thigh; flexes hip
Posterior Compartment				
Biceps femoris (bī′seps fem′ŏ-ris)	Long head—ischial tuberosity	Head of fibula	Long head—tibial	Flexes knee; laterally rotates leg; extends hip
	Short head—femur		Short head—common fibular	
Semimembranosus (sem′ē-mem-bră-nō′sŭs)	Ischial tuberosity	Medial condyle of tibia and collateral ligament	Tibial	Flexes knee; medially rotates leg; tenses capsule of knee joint; extends hip
Semitendinosus (sem′ē-ten-di-nō′sŭs)	Ischial tuberosity	Tibia	Tibial	Flexes knee; medially rotates leg; extends hip

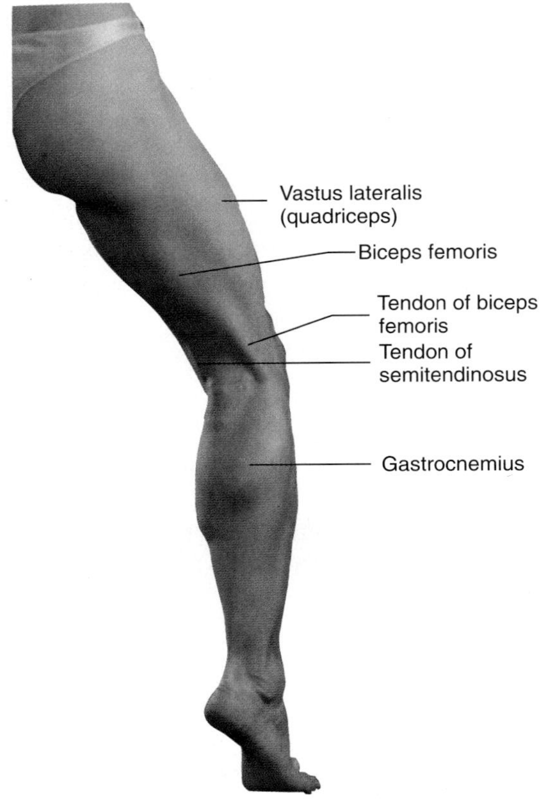

FIGURE 10.36 **Surface Anatomy of the Posterior Lower Limb**

The medial thigh muscles (see figure 10.34) are involved primarily in adduction of the thigh. Some of these muscles also laterally rotate the thigh and/or flex or extend the hip. The gracilis also flexes the knee.

The posterior thigh muscles (see figure 10.35), collectively called the hamstring muscles, consist of the **biceps femoris, semimembranosus** (sem′ē-mem-bră-nō′sŭs), and **semitendinosus** (sem′ē-ten-di-nō′sŭs; see table 10.20). Their tendons are easily seen or felt on the medial and lateral posterior aspect of a slightly bent knee (figure 10.36).

Hamstrings

The **hamstrings** are so-named because in pigs these tendons can be used to suspend hams during curing. Some animals, such as wolves, often bring down their prey by biting through the hamstrings; therefore, "to hamstring" someone is to render the person helpless. A "pulled hamstring" results from tearing one or more of these muscles or their tendons, usually near the origin of the muscle.

28 ***Name the muscle compartments of the thigh and the movements produced by the muscles of each compartment. List the muscles of each compartment and the individual action of each muscle.***

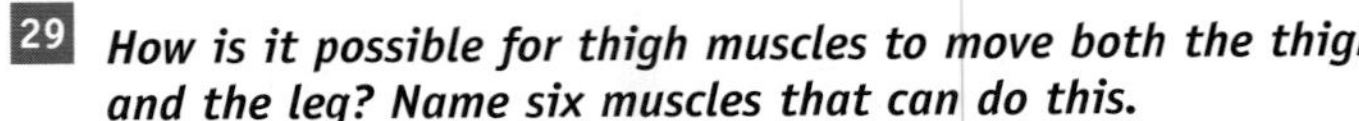

29 ***How is it possible for thigh muscles to move both the thigh and the leg? Name six muscles that can do this.***

Ankle, Foot, and Toe Movements

The muscles of the leg that move the ankle and the foot are listed in table 10.21 and are illustrated in figures 10.37–10.39. These **extrinsic foot muscles** are divided into three groups, each located within a separate compartment of the leg: anterior, posterior, and lateral (see figure 10.37). The anterior leg muscles (see figure 10.38*a*) are extensor muscles involved in dorsiflexion and eversion or inversion of the foot and extension of the toes.

Shinsplints

Shinsplints is a catchall term involving any one of the following four conditions associated with pain in the anterior portion of the leg:

1. Excessive stress on the tibialis anterior, resulting in pain along the origin of the muscle
2. Tibial periostitis, an inflammation of the tibial periosteum
3. Anterior compartment syndrome. During hard exercise, the anterior compartment muscles may swell with blood. The overlying fascia is very tough and does not expand; thus, the nerves and vessels are compressed, causing pain.
4. Stress fracture of the tibia 2–5 cm distal to the knee

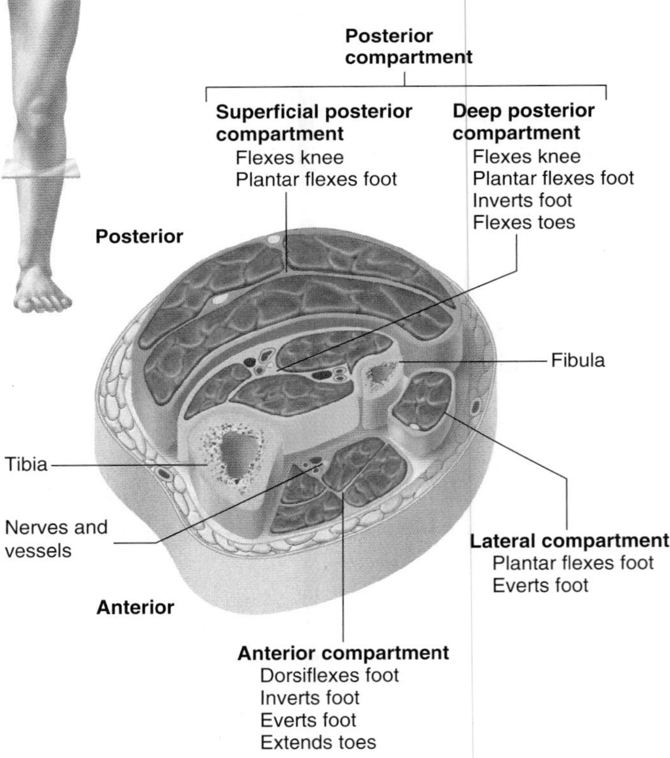

FIGURE 10.37 **Cross Section Through the Left Leg**
Drawing of the muscular compartments.

TABLE 10.21 Muscles of the Leg Acting on the Leg, Ankle, and Foot (See Figures 10.37–10.39)

Muscle	Origin	Insertion	Nerve	Action
Anterior Compartment				
Extensor digitorum longus (dij′i-tōr-ŭm lon′gŭs)	Lateral condyle of tibia and fibula	Four tendons to phalanges of four lateral toes	Deep fibular*	Extends four lateral toes; dorsiflexes and everts foot
Extensor hallucis longus (hal′i-sis lon′gŭs)	Middle fibula and interosseous membrane	Distal phalanx of great toe	Deep fibular*	Extends great toe; dorsiflexes and inverts foot
Tibialis anterior (tib-ē-a′lis)	Proximal, lateral tibia and interosseous membrane	Medial cuneiform and first metatarsal	Deep fibular*	Dorsiflexes and inverts foot
Fibularis tertius (peroneus tertius) (per′ō-nē′ŭs ter′shē-ŭs)	Fibula and interosseous membrane	Fifth metatarsal	Deep fibular*	Dorsiflexes and everts foot
Lateral Compartment				
Fibularis brevis (peroneus brevis) (fib-ū-lā′ris brev′is)	Inferior two-thirds of lateral fibula	Fifth metatarsal	Superficial fibular*	Everts and plantar flexes foot
Fibularis longus (peroneus longus) (fib-ū-lā′ris lon′gŭs)	Superior two-thirds of lateral fibula	First metatarsal and medial cuneiform	Superficial fibular*	Everts and plantar flexes foot
Posterior Compartment				
Superficial				
Gastrocnemius (gas-trok-nē′mē-ŭs)	Medial and lateral condyles of femur	Through calcaneal (Achilles) tendon to calcaneus	Tibial	Plantar flexes foot; flexes knee
Plantaris (plan-tār′is)	Femur	Through calcaneal tendon to calcaneus	Tibial	Plantar flexes foot; flexes knee
Soleus (sō-lē′ŭs)	Fibula and tibia	Through calcaneal tendon to calcaneus	Tibial	Plantar flexes foot
Deep				
Flexor digitorum longus (dij′i-tōr′ŭm lon′gŭs)	Tibia	Four tendons to distal phalanges of four lateral toes	Tibial	Flexes four lateral toes; plantar flexes and inverts foot
Flexor hallucis longus (hal′i-sis lon′gŭs)	Fibula	Distal phalanx of great toe	Tibial	Flexes great toe; plantar flexes and inverts foot
Popliteus (pop-li-tē′ŭs)	Lateral femoral condyle	Posterior tibia	Tibial	Flexes knee; medially rotates leg
Tibialis posterior (tib-ē-a′lis)	Tibia, interosseous membrane, and fibula	Navicular, cuneiforms, cuboid, and second through fourth metatarsals	Tibial	Plantar flexes and inverts foot

*Also referred to as the peroneal nerve.

The lateral muscles (see figure 10.38*b*) are primarily everters of the foot, but they also aid plantar flexion. The **fibularis brevis** inserts onto the fifth metatarsal and everts and plantar flexes the foot. The **fibularis longus** crosses under the lateral four metatarsals to insert onto the first metatarsal and medial cuneiform. The fibular muscle was formerly referred to as *peroneal,* which means relating to the fibula; the term has now been simplified by using the term *fibular*. The tendons of the fibularis muscles can be seen on the lateral side of the ankle (see figure 10.38*d*).

The superficial muscles of the posterior compartment, the **gastrocnemius** (gas-trok-nē′mē-ŭs) and **soleus,** form the bulge of the calf (posterior leg; see figures 10.39*a*, *b*, and *d*). They join with the small **plantaris** muscle to form the common **calcaneal** (kal-kā′nē-al), or Achilles, **tendon.** These muscles are involved in plantar flexion of the foot. The deep muscles of the posterior compartment plantar flex and invert the foot and flex the toes.

Achilles Tendon

The **Achilles tendon** derives its name from a hero of Greek mythology. When Achilles was a baby, his mother dipped him into magic water, which made him invulnerable to harm everywhere the water touched his skin. His mother, however, held him by the heel and failed to submerge this part of his body under the water. Consequently, his heel was vulnerable and proved to be his undoing; he was shot in the heel with an arrow at the battle of Troy and died. Thus, saying that someone has an "Achilles heel" means that the person has a weak spot that can be attacked.

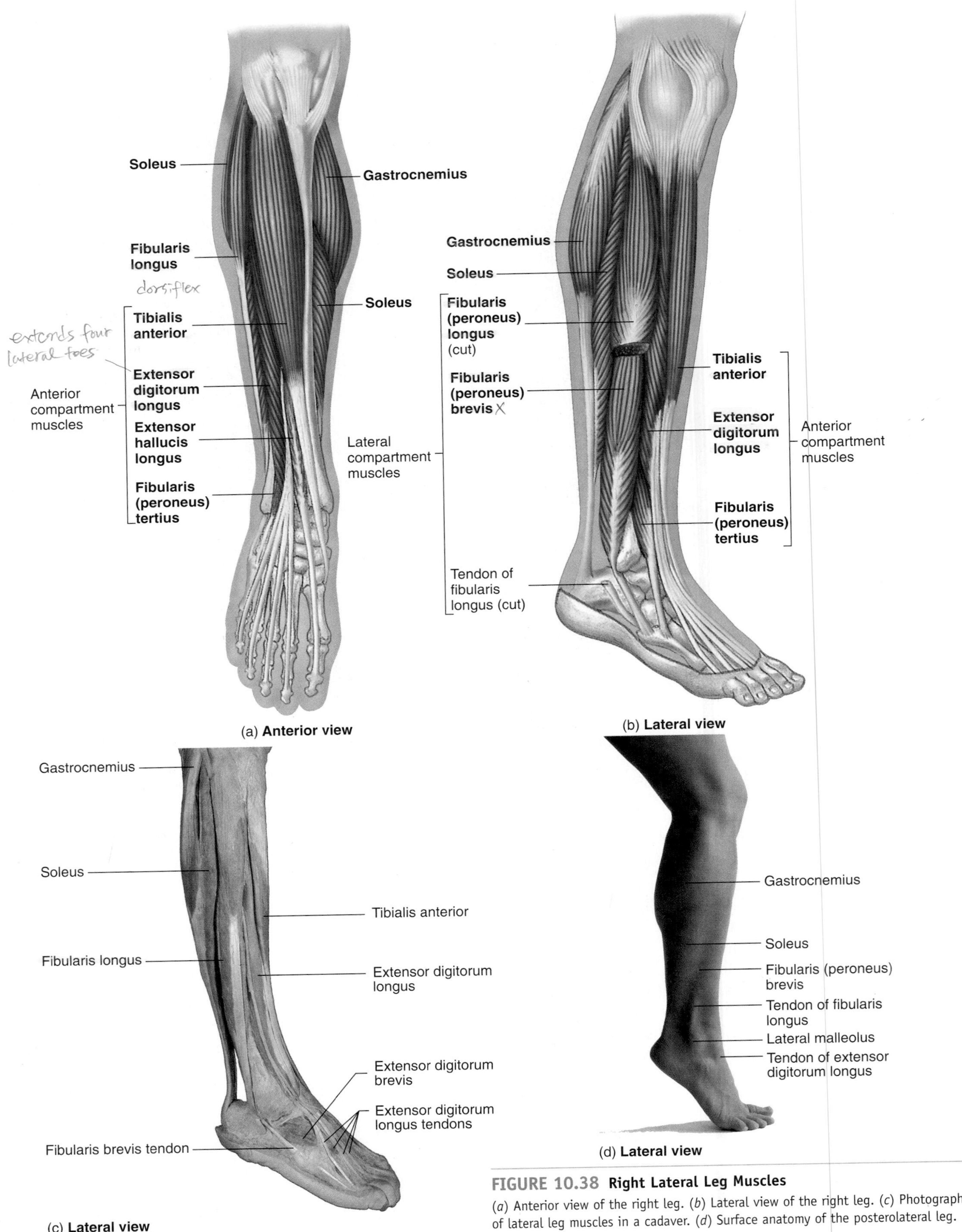

FIGURE 10.38 Right Lateral Leg Muscles

(*a*) Anterior view of the right leg. (*b*) Lateral view of the right leg. (*c*) Photograph of lateral leg muscles in a cadaver. (*d*) Surface anatomy of the posterolateral leg.

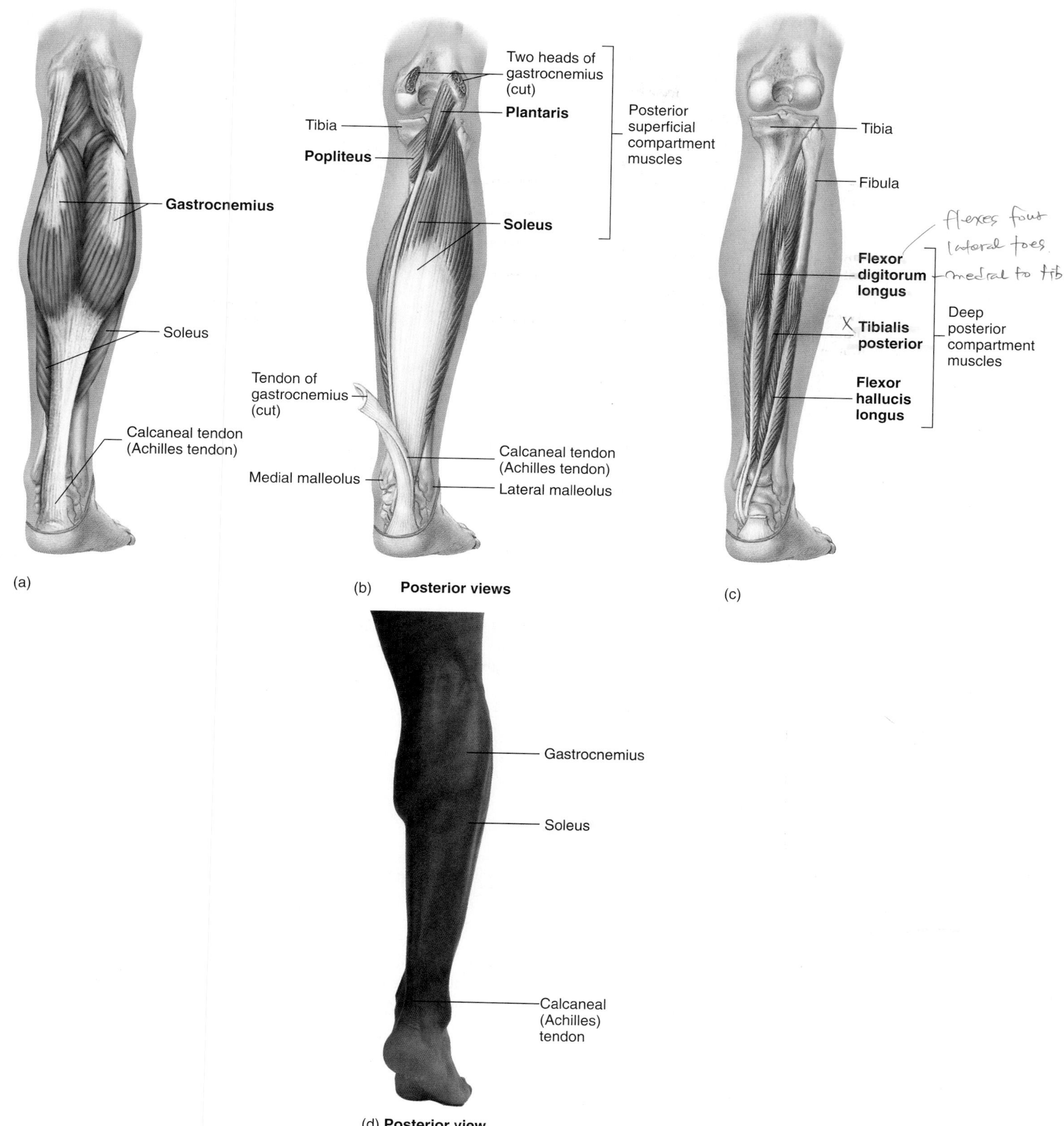

FIGURE 10.39 Right Posterior Leg Muscles

(*a*) Superficial muscles. (*b*) Posterior view of the right calf, superficial. Gastrocnemius is removed. (*c*) Posterior view of the right calf, deep. Gastrocnemius, plantaris, and soleus muscles are removed. (*d*) Surface anatomy of the posterior right leg.

TABLE 10.22 Intrinsic Muscles of the Foot (See Figure 10.40)

Muscle	Origin	Insertion	Nerve	Action
Abductor digiti minimi (ab-dŭk′ter, ab-dŭk′tōr dij′i-tī min′ĭ-mī)	Calcaneus	Proximal phalanx of fifth toe	Lateral plantar	Abducts and flexes little toe
Abductor hallucis (ab-dŭk′ter, ab-dŭk′tōr hal′i-sis)	Calcaneus	Base of proximal phalanx of great toe	Medial plantar	Abducts great toe
Adductor hallucis (a-dŭk′ter, a-dŭk′tōr hal′i-sis) (not illustrated)	Lateral four metatarsals	Proximal phalanx of great toe	Lateral plantar	Adducts great toe
Extensor digitorum brevis (dij′i-tōr′ŭm brev′is) (not illustrated)	Calcaneus	Four tendons fused with tendons of extensor digitorum longus	Deep fibular*	Extends toes
Flexor digiti minimi brevis (dij′i-tī min′ĭ-mī brev′is)	Fifth metatarsal	Proximal phalanx of fifth digit	Lateral plantar	Flexes little toe (proximal phalanx)
Flexor digitorum brevis (dij′i-tōr′ŭm brev′is)	Calcaneus and plantar fascia	Four tendons to middle phalanges of four lateral toes	Medial plantar	Flexes lateral four toes
Flexor hallucis brevis (hal′i-sis brev′is)	Cuboid; medial and lateral cuneiforms	Two tendons to proximal phalanx of great toe	Medial and lateral plantar	Flexes great toe
Dorsal interossei (in′ter-os′e-ī) (not illustrated)	Metatarsal bones	Proximal phalanges of second, third, and fourth digits	Lateral plantar	Abduct second, third, and fourth toes
Plantar interossei (plan′tăr in′ter-os′e-ī)	Third, fourth, and fifth metatarsals	Proximal phalanges of third, fourth, and fifth digits	Lateral plantar	Adduct third, fourth, and fifth toes
Lumbricales (lum′bri-kā-lēz)	Tendons of flexor digitorum longus	Extensor expansion of second through fifth digits	Lateral and medial plantar	Flex proximal and extend middle and distal phalanges
Quadratus plantae (kwah′drā′tŭs plan′tē)	Calcaneus	Tendons of flexor digitorum longus	Lateral plantar	Assists flexor digitorum longus in flexing lateral four toes

*Also referred to as the peroneal nerve.

Intrinsic foot muscles, located within the foot itself (table 10.22 and figure 10.40), flex, extend, abduct, and adduct the toes. They are arranged in a manner similar to that of the intrinsic muscles of the hand.

Plantar Fasciitis

The muscles in the plantar region of the foot are covered with thick fascia and the plantar aponeurosis. Running on a hard surface wearing poorly fitting or worn-out shoes can result in inflammation of the plantar aponeurosis. The pain resulting from **plantar fasciitis** occurs in the fascia over the heel and along the medial-inferior side of the foot.

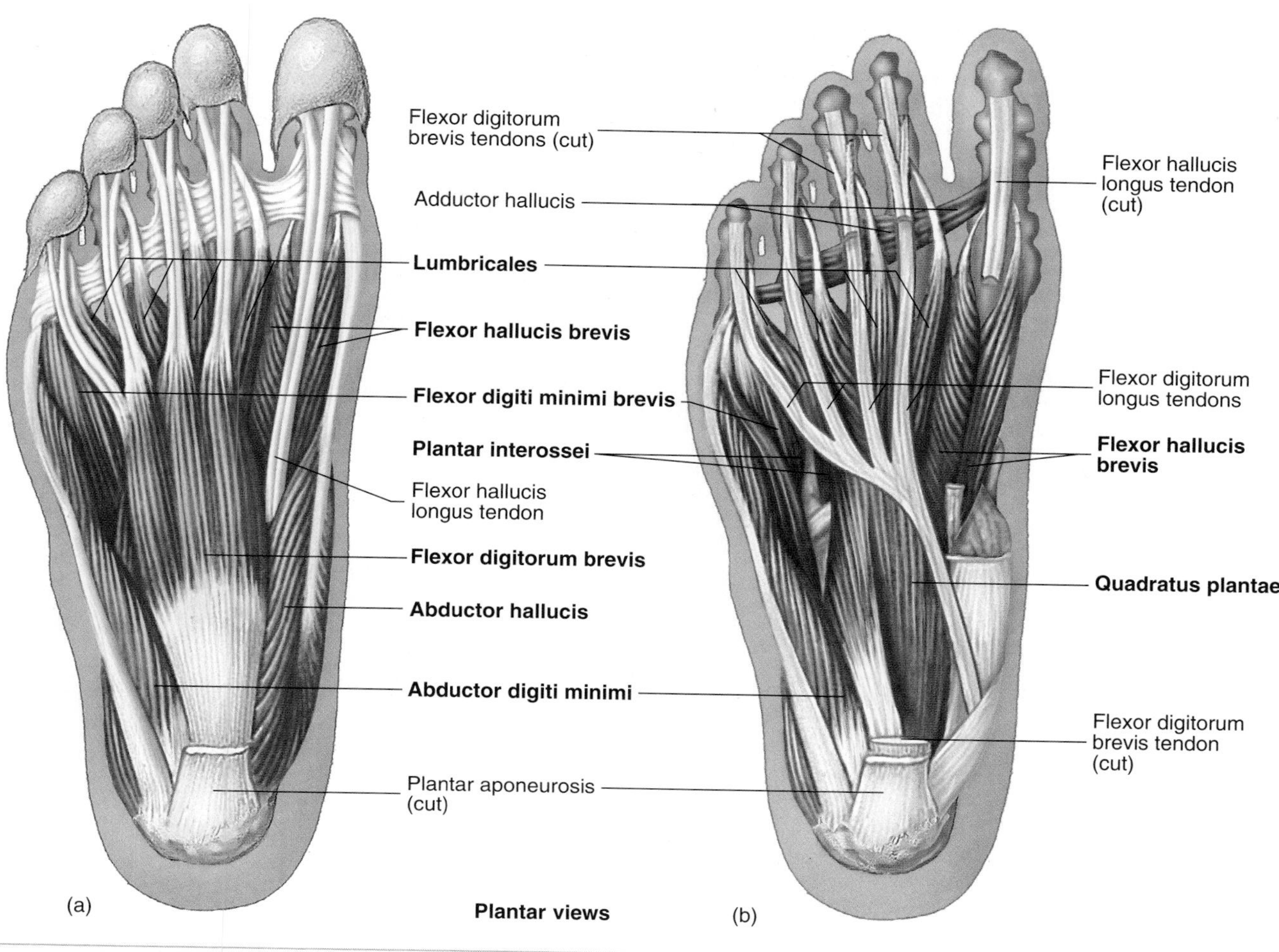

FIGURE 10.40 Right Foot Muscles

(*a*) Superficial muscles of the right foot. Plantar aponeurosis is cut. (*b*) Deep muscles of the right foot. Flexor digitorum brevis and flexor hallucis longus are cut.

30 ***What movements are produced by the three muscle compartments of the leg? Name the muscles of each compartment, and describe the movements for which each muscle is responsible.***

31 ***What movement do the fibularis (peroneus) muscles have in common? The tibialis muscles?***

32 ***Name the leg muscles that flex the knee. Which of them can also plantar flex the foot?***

33 ***List the general actions performed by the intrinsic foot muscles.***

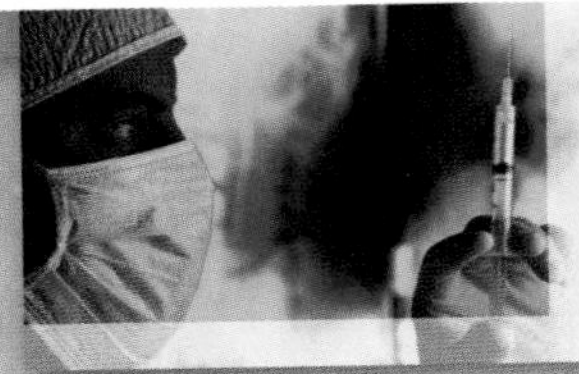

CLINICAL FOCUS

Bodybuilding

Bodybuilding is a popular sport worldwide. Participants in this sport combine diet and specific weight training to develop maximum muscle mass and minimum body fat, with their major goal being a well-balanced, complete physique. An uninformed, untrained muscle builder can build some muscles and ignore others; the result is a disproportioned body. Skill, training, and concentration are required to build a well-proportioned, muscular body and to know which exercises build a large number of muscles and which are specialized to build certain parts of the body. Is the old adage "no pain, no gain" correct? Not really. Overexercising can cause small tears in muscles and soreness. Torn muscles are weaker, and it may take up to 3 weeks to repair the damage, even though the soreness may last only 5–10 days.

Bodybuilders concentrate on increasing skeletal muscle mass. Endurance tests conducted years ago demonstrated that the cardiovascular and respiratory abilities of bodybuilders were similar to the abilities of normal, healthy persons untrained in a sport. More recent studies, however, indicate that the cardiorespiratory fitness of bodybuilders is similar to that of other well-trained athletes. The difference between the results of the new studies and those of the older ones is attributed to modern bodybuilding techniques that include aerobic exercise and running, as well as "pumping iron."

FIGURE A **Bodybuilders**

Bodybuilding has its own language. Bodybuilders refer to the "lats," "traps," and "delts" rather than the latissimus dorsi, trapezius, and deltoids. The exercises also have special names, such as "lat pulldowns," "preacher curls," and "triceps extensions."

Photographs of bodybuilders are very useful in the study of anatomy because they enable the easy identification of the surface anatomy of muscles that cannot usually be seen in untrained people (figure A).

SUMMARY

Body movements result from the contraction of skeletal muscles.

General Principles (p. 321)

1. The less movable end of a muscle attachment is the origin; the more movable end is the insertion.
2. An agonist causes a certain movement and an antagonist acts in opposition to the agonist.
3. Synergists are muscles that function together to produce movement.
4. Prime movers are mainly responsible for a movement. Fixators stabilize the action of prime movers.

Muscle Shapes

Muscle shape is determined primarily by the arrangement of muscle fasciculi.

Nomenclature

Muscles are named according to their location, size, shape, orientation of fasciculi, origin and insertion, number of heads, and function.

Movements Accomplished by Muscles

Contracting muscles generate a force that acts on bones (levers) across joints (fulcrums) to create movement. Three classes of levers have been identified.

Muscle Anatomy

Muscles of the head, neck, trunk and limbs are presented.

Head Muscles (p. 327)

Head and Neck Muscles

Origins of these muscles are mainly on the cervical vertebrae (except for the sternocleidomastoid); insertions are on the occipital bone or mastoid process. They cause flexion, extension, rotation, abduction, and adduction of the head.

Facial Expression

Origins of facial muscles are on skull bones or fascia; insertions are into the skin, causing movement of the facial skin, lips, and eyelids.

Mastication

Three pairs of muscles close the jaw; gravity opens the jaw. Forced opening is caused by the lateral pterygoids and the hyoid muscles.

Tongue Movements

Intrinsic tongue muscles change the shape of the tongue; extrinsic tongue muscles move the tongue.

Swallowing and the Larynx

1. Hyoid muscles can depress the jaw and assist in swallowing.
2. Muscles open and close the openings to the nasal cavity, pharyngotympanic tubes, and larynx.

Movements of the Eyeball

Six muscles with their origins on the orbital bones insert on the eyeball and cause it to move within the orbit.

Trunk Muscles (p. 340)

Muscles Moving the Vertebral Column

1. These muscles extend, abduct, rotate, or flex the vertebral column.
2. A deep group of muscles connects adjacent vertebrae.
3. A more superficial group of muscles runs from the pelvis to the skull, extending from the vertebrae to the ribs.

Thoracic Muscles

1. Most respiratory movement is caused by the diaphragm.
2. Muscles attached to the ribs aid in respiration.

Abdominal Wall

Abdominal wall muscles hold and protect abdominal organs and cause flexion, rotation, and lateral flexion of the vertebral column.

Pelvic Floor and Perineum

These muscles support the abdominal organs inferiorly.

Upper Limb Muscles (p. 346)

Scapular Movements

Six muscles attach the scapula to the trunk and enable the scapula to function as an anchor point for the muscles and bones of the arm.

Arm Movements

Seven muscles attach the humerus to the scapula. Two additional muscles attach the humerus to the trunk. These muscles cause flexion and extension of the shoulder and abduction, adduction, rotation, and circumduction of the arm.

Forearm Movements

1. Flexion and extension of the elbow are accomplished by three muscles located in the arm and two in the forearm.
2. Supination and pronation are accomplished primarily by forearm muscles.

Wrist, Hand, and Finger Movements

1. Forearm muscles that originate on the medial epicondyle are responsible for flexion of the wrist and fingers. Muscles extending the wrist and fingers originate on the lateral epicondyle.
2. Extrinsic hand muscles are in the forearm. Intrinsic hand muscles are in the hand.

Lower Limb Muscles (p. 359)

Thigh Movements

1. Anterior pelvic muscles cause flexion of the hip.
2. Muscles of the buttocks are responsible for extension of the hip and abduction and rotation of the thigh.
3. The thigh can be divided into three compartments.
   - The anterior compartment muscles flex the hip and extend the knee.
   - The medial compartment muscles adduct the thigh.
   - The posterior compartment muscles extend the hip and flex the knee.

Leg Movements

Some muscles of the thigh also act on the leg. The anterior thigh muscles extend the leg, and the posterior thigh muscles flex the leg.

Ankle, Foot, and Toe Movements

1. The leg is divided into three compartments.
   - Muscles in the anterior compartment cause dorsiflexion, inversion, or eversion of the foot and extension of the toes.
   - Muscles of the lateral compartment plantar flex and evert the foot.
   - Muscles of the posterior compartment flex the leg, plantar flex and invert the foot, and flex the toes.
2. Intrinsic foot muscles flex or extend and abduct or adduct the toes.

REVIEW AND COMPREHENSION

1. Muscles that oppose one another are
   a. synergists.
   b. levers.
   c. hateful.
   d. antagonists.
   e. fixators.
2. The most movable attachment of a muscle is its
   a. origin.
   b. insertion.
   c. fascia.
   d. fulcrum.
   e. belly.
3. Which of these muscles is correctly matched with its type of fascicle orientation?
   a. pectoralis major—pennate
   b. transversus abdominis—circular
   c. temporalis—convergent
   d. biceps femoris—straight
   e. orbicularis oris—straight

4. The muscle whose name means it is larger and round is the
   a. gluteus maximus.
   b. vastus lateralis.
   c. teres major.
   d. latissimus dorsi.
   e. adductor magnus.
5. In a class III lever system, the
   a. fulcrum is located between the pull and the weight.
   b. weight is located between the fulcrum and the pull.
   c. pull is located between the fulcrum and the weight.
6. A prominent lateral muscle of the neck that can cause flexion of the neck or rotate the head is the
   a. digastric.
   b. mylohyoid.
   c. sternocleidomastoid.
   d. buccinator.
   e. platysma.
7. Harry Wolf has just picked up his date for the evening. She is wearing a stunning new outfit. Harry shows his appreciation by moving his eyebrows up and down, winking, smiling, and finally kissing her. Given the muscles listed:
   1. zygomaticus
   2. levator labii superioris
   3. occipitofrontalis
   4. orbicularis oris
   5. orbicularis oculi

   In which order did Harry use these muscles?
   a. 2,3,4,1
   b. 2,5,3,1
   c. 2,5,4,3
   d. 3,5,1,4
   e. 3,5,2,4
8. An aerial circus performer who supports herself only by her teeth while spinning around should have strong
   a. temporalis muscles.
   b. masseter muscles.
   c. buccinator muscles.
   d. both a and b.
   e. all of the above.
9. The tongue curls and folds *primarily* because of the action of the
   a. extrinsic tongue muscles.
   b. intrinsic tongue muscles.
10. The infrahyoid muscles
    a. elevate the mandible.
    b. move the mandible from side to side.
    c. fix (prevent movement of) the hyoid.
    d. both a and b.
    e. all of the above.
11. The soft palate muscles
    a. prevent food from entering the nasal cavity.
    b. close the pharyngotympanic tube.
    c. force food into the esophagus.
    d. prevent food from entering the larynx.
    e. elevate the mandible.
12. Which of these movements is *not* caused by contraction of the erector spinae muscles?
    a. flexion of the vertebral column
    b. lateral flexion of the vertebral column
    c. extension of the vertebral column
    d. rotation of the vertebral column
13. Which of these muscles is (are) responsible for flexion of the vertebral column (below the neck)?
    a. deep back muscles
    b. superficial back muscles (erector spinae)
    c. rectus abdominis
    d. both a and b
    e. all of the above
14. Which of these muscles is *not* involved with the inspiration of air?
    a. diaphragm
    b. external intercostals
    c. scalene
    d. transversus thoracis
15. Given these muscles:
    1. external abdominal oblique
    2. internal abdominal oblique
    3. transversus abdominis

    Choose the arrangement that lists the muscles from most superficial to deepest.
    a. 1,2,3
    b. 1,3,2
    c. 2,1,3
    d. 2,3,1
    e. 3,1,2
16. Tendinous intersections
    a. attach the rectus abdominis muscles to the xiphoid process.
    b. divide the rectus abdominis muscles into segments.
    c. separate the abdominal wall from the thigh.
    d. are the site of exit of blood vessels from the abdomen into the thigh.
    e. are the central point of attachment for all the abdominal muscles.
17. Which of these muscles can both elevate and depress the scapula?
    a. rhomboideus major and minor
    b. levator scapulae
    c. serratus anterior
    d. trapezius
    e. pectoralis minor
18. Which of these muscles does *not* adduct the arm (humerus)?
    a. latissimus dorsi
    b. deltoid
    c. teres major
    d. pectoralis major
    e. coracobrachialis
19. Which of these muscles is most directly involved in abduction of the arm (humerus)?
    a. supraspinatus
    b. infraspinatus
    c. teres minor
    d. teres major
    e. subscapularis
20. Which of these muscles would you expect to be especially well developed in a boxer known for his powerful jab (punching straight ahead)?
    a. biceps brachii
    b. brachialis
    c. trapezius
    d. triceps brachii
    e. supinator
21. Which of these muscles is an antagonist of the triceps brachii?
    a. biceps brachii
    b. anconeus
    c. latissimus dorsi
    d. brachioradialis
    e. supinator
22. The posterior group of forearm muscles is responsible for
    a. flexion of the wrist.
    b. flexion of the fingers.
    c. extension of the fingers.
    d. both a and b.
    e. all of the above.
23. Which of these muscles is an intrinsic hand muscle that moves the thumb?
    a. flexor pollicis brevis
    b. flexor digiti minimi brevis
    c. flexor pollicis longus
    d. extensor pollicis longus
    e. all of the above
24. Which of these muscles can extend the hip?
    a. gluteus maximus
    b. gluteus medius
    c. gluteus minimus
    d. tensor fasciae latae
    e. sartorius
25. Given these muscles:
    1. iliopsoas
    2. rectus femoris
    3. sartorius

    Which of the muscles flex the hip?
    a. 1
    b. 1,2
    c. 1,3
    d. 2,3
    e. 1,2,3
26. Which of these muscles is found in the medial compartment of the thigh?
    a. rectus femoris
    b. sartorius
    c. gracilis
    d. vastus medialis
    e. semitendinosus
27. Which of these is *not* a muscle that can flex the knee?
    a. biceps femoris
    b. vastus medialis
    c. gastrocnemius
    d. gracilis
    e. sartorius

28. The ____________ muscles evert the foot, whereas the ____________ muscles invert the foot.
    a. fibularis (longus and brevis), gastrocnemius and soleus
    b. fibularis (longus and brevis), tibialis anterior and extensor hallucis longus
    c. tibialis anterior and extensor hallucis longus, fibularis longus and brevis
    d. tibialis anterior and extensor hallucis longus, flexor digitorum longus and flexor hallucis longus
    e. flexor digitorum longus and flexor hallucis longus, gastrocnemius and soleus
29. Which of these muscles causes plantar flexion of the foot?
    a. tibialis anterior
    b. extensor digitorum longus
    c. fibularis (peroneus) tertius
    d. soleus
    e. sartorius

Answers in Appendix E

CRITICAL THINKING

1. For each of the following muscles, (1) describe the movement that the muscle produces and (2) name the muscles that act as synergists and antagonists for them: longus capitis, erector spinae, coracobrachialis.
2. Propose an exercise that would benefit each of the following muscles specifically: biceps brachii, triceps brachii, deltoid, rectus abdominis, quadriceps femoris, and gastrocnemius.
3. Consider only the effect of the brachioradialis muscle for this question. If a weight is held in the hand and the forearm is flexed, what type of lever system is in action? If the weight is placed on the forearm? Which system can lift more weight, and how far?
4. A patient was involved in an automobile accident in which the car was "rear-ended," resulting in whiplash injury of the head (hyperextension). What neck muscles might be injured in this type of accident? What is the easiest way to prevent such injury in an automobile accident?
5. During surgery, a branch of a patient's facial nerve was accidentally cut on one side of the face. As a result, after the operation, the lower eyelid and the corner of the patient's mouth drooped on that side of the face. What muscles were apparently affected?
6. When a person becomes unconscious, the tongue muscles relax and the tongue tends to retract or fall back and obstruct the airway. Which tongue muscle is responsible? How can this be prevented or reversed?
7. The mechanical support of the head of the humerus in the glenoid fossa is weakest in the inferior direction. What muscles help prevent dislocation of the shoulder when a heavy weight, such as a suitcase, is carried?
8. How would paralysis of the quadriceps femoris of the left leg affect a person's ability to walk?
9. Speedy Sprinter started a 200 m dash and fell to the ground in pain. Examination of her right leg revealed the following symptoms: inability to plantar flex the foot against resistance, normal ability to evert the foot, dorsiflexion of the foot more than normal, and abnormal bulging of the calf muscles. Explain the nature of her injury.
10. What muscles are required to turn this page?

Answers in Appendix F

Visit this book's website at www.mhhe.com/seeley8 for chapter quizzes, interactive learning exercises, and other study tools.

11 Functional Organization of Nervous Tissue

A hungry man prepares to drink a cup of hot soup. He smells the aroma and anticipates the taste of the soup. Feeling the warmth of the cup in his hands, he quickly raises the cup to his lips and takes a sip. The soup is so hot that it burns his lips and tongue. He jerks the cup away from his lips and gasps in pain. None of these sensations, thoughts, emotions, and movements would be possible without the nervous system, which is responsible for sensations, mental activities, and control of muscles and many glands. The nervous system is made up of the brain, spinal cord, nerves and sensory receptors.

This chapter explains the *functions of the nervous system* (p. 375), the *divisions of the nervous system* (p. 375), the *cells of the nervous system* (p. 377), the *organization of nervous tissue* (p. 382), *electric signals* (p. 382), *the synapse* (p. 394), and *neuronal pathways and circuits* (p. 404).

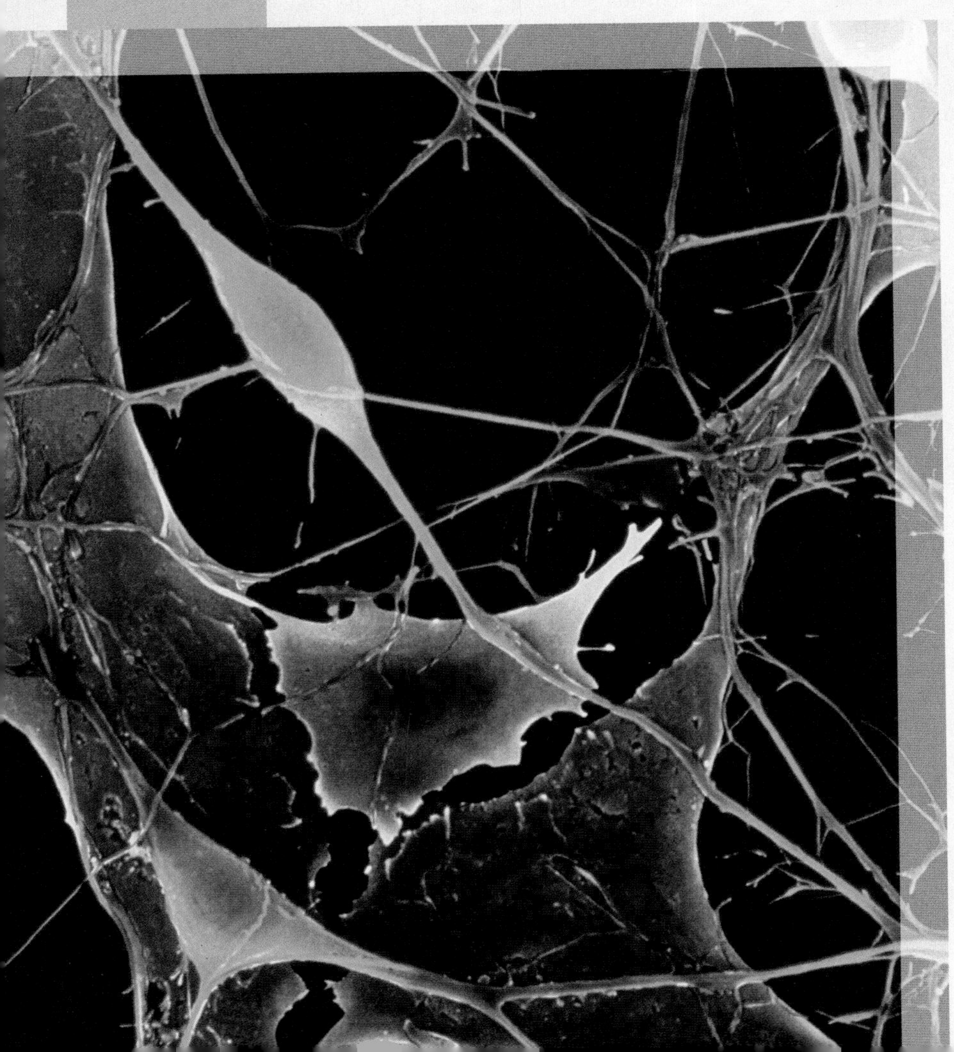

Light photomicrograph of pyramid-shaped neurons (*green*) growing on a fibrous network (*yellow*) in the central nervous system.

Nervous System

FUNCTIONS OF THE NERVOUS SYSTEM

The nervous system is involved in most body functions. Some of the major functions of the nervous system are

1. *Sensory input.* Sensory receptors monitor numerous external and internal stimuli. Some stimuli result in sensations we are aware of, such as sight, vision, hearing, taste, smell, touch, pain, body position, and temperature. Other stimuli, such as blood pH, blood gases, or blood pressure, are processed at an unconscious level.
2. *Integration.* The brain and spinal cord are the major organs for processing sensory input and initiating responses. The input may produce an immediate response, may be stored as memory, or may be ignored.
3. *Control of muscles and glands.* Skeletal muscles normally contract only when stimulated by the nervous system, and the nervous system controls the major movements of the body through the control of skeletal muscle. Some smooth muscle, such as in the walls of blood vessels, contracts only when stimulated by the nervous system or hormones (see chapter 18). Cardiac muscle and some smooth muscle, such as in the wall of the stomach, contract autorhythmically—that is, no external stimulation is necessary for contraction to occur. Although the nervous system does not initiate contraction in these muscles, it can cause the contractions to occur more rapidly or slowly. Finally, the nervous system controls the secretions from many glands, such as sweat glands, salivary glands, and the glands of the digestive system.
4. *Homeostasis.* The regulatory and coordinating activities of the nervous system are necessary for maintaining homeostasis. The trillions of cells in the human body do not function independently of each other but must work together to maintain homeostasis. For example, heart cells must contract at a rate that ensures adequate delivery of blood, skeletal muscles of respiration must contract at a rate that ensures oxygenation of blood, and kidney cells must regulate blood volume and remove waste products. The nervous system can stimulate or inhibit the activities of these and other structures to help maintain homeostasis.
5. *Mental activity.* The brain is the center of mental activities, including consciousness, thinking, memory, and emotions.

1 ***List and give examples of the general functions of the nervous system.***

DIVISIONS OF THE NERVOUS SYSTEM

Humans have only one nervous system, even though some of its subdivisions are referred to as separate systems (figure 11.1). Thus, the central nervous system and the peripheral nervous system are subdivisions of the nervous system, instead of separate organ systems, as their names suggest. The **central nervous system (CNS)** consists of the brain and the spinal cord. The brain is located within the skull, and the spinal cord is located within the vertebral canal, formed by the vertebrae (see chapter 7). The brain and spinal cord are continuous with each other at the foramen magnum.

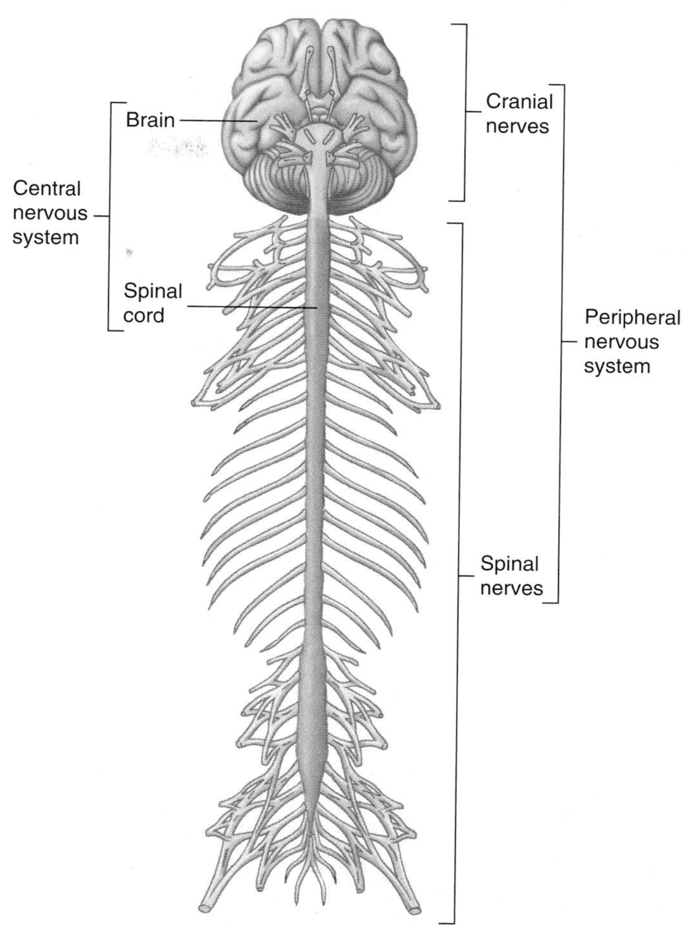

FIGURE 11.1 Nervous System
The central nervous system (CNS) consists of the brain and spinal cord. The peripheral nervous system (PNS) consists of cranial nerves, which arise from the brain, and spinal nerves, which arise from the spinal cord. The nerves, which are shown cut in the illustration, actually extend throughout the body.

The **peripheral nervous system (PNS)** is external to the CNS. It consists of sensory receptors, nerves, ganglia, and plexuses. **Sensory receptors** are the endings of nerve cells or separate, specialized cells that detect temperature, pain, touch, pressure, light, sound, odors, and other stimuli. Sensory receptors are located in the skin, muscles, joints, internal organs, and specialized sensory organs, such as the eyes and ears. A **nerve** is a bundle of axons and their sheaths; it connects the CNS to sensory receptors, muscles, and glands. Twelve pairs of **cranial nerves** originate from the brain, and 31 pairs of **spinal nerves** originate from the spinal cord (see figure 11.1). A **ganglion** (gang′glē-on; pl. *ganglia,* gang′glē-ă; knot) is a collection of neuron cell bodies located outside the CNS. A **plexus** (plek′sus; braid) is an extensive network of axons and, in some cases, neuron cell bodies, located outside the CNS.

The PNS can be divided into two subcategories. The **sensory,** or **afferent, division** transmits electric signals, called **action potentials,** from the sensory receptors to the CNS. The cell bodies of sensory neurons are located in dorsal root ganglia near the spinal cord (figure 11.2*a*) or in ganglia near the origin of certain cranial nerves. The **motor,** or **efferent, division** transmits action potentials from the CNS to effector organs, such as muscles and glands.

The motor division is divided into the **somatic nervous system** and the **autonomic nervous system (ANS).** The somatic nervous system transmits action potentials from the CNS to skeletal muscles (figure 11.2*b*). Skeletal muscles are voluntarily controlled through the somatic nervous system. The cell bodies of somatic motor neurons are located within the CNS, and their axons extend through nerves to form synapses with skeletal muscle cells. A **synapse** is the junction of a nerve cell with another cell. The neuromuscular junction, the synapse between a neuron and a skeletal muscle cell, is discussed in detail in chapter 9. Nerve cells can also form synapses with other nerve cells, smooth muscle cells, cardiac muscle cells, and gland cells.

The ANS transmits action potentials from the CNS to smooth muscle, cardiac muscle, and certain glands. Subconscious, or involuntary, control of smooth muscle, cardiac muscle, and glands occurs through the ANS. The ANS has two sets of neurons in a series between the CNS and the effector organs (figure 11.2*c*). Cell bodies of the first neurons are within the CNS and send their axons to autonomic ganglia, where neuron cell bodies of the second neurons are located. Synapses exist between the first and second neurons within the autonomic ganglia, and the axons of the second neurons extend from the autonomic ganglia to the effector organs.

The ANS is subdivided into the sympathetic and the parasympathetic divisions and the enteric nervous system. In general, the **sympathetic division** is most active during physical activity, whereas the **parasympathetic division** regulates resting or vegetative functions, such as digesting food or emptying the urinary bladder. The **enteric nervous system** consists of plexuses within the wall of the digestive tract (see figure 24.4). Although the enteric nervous system is capable of controlling the digestive tract independently of the CNS, it is considered part of the ANS because of the parasympathetic and sympathetic neurons that contribute to the plexuses. See chapters 16 and 24 for details on the enteric nervous system.

The sensory division of the PNS detects stimuli and transmits information in the form of action potentials to the CNS (figure 11.3). The CNS is the major site for processing information, initiating responses, and integrating mental processes. It is much like a computer, with the ability to receive input, process and store information, and generate responses. The motor division of the PNS conducts action potentials from the CNS to muscles and glands.

2 ***Define CNS and PNS.***

3 ***What is a sensory receptor, nerve, ganglion, and plexus?***

4 ***Based on the direction they transmit action potentials, what are the two subcategories of the PNS?***

5 ***Based on the structures they supply, what are the two subcategories of the motor division?***

6 ***Where are the cell bodies of sensory, somatic motor, and autonomic neurons located? What is a synapse?***

7 ***What are the subcategories of the ANS?***

8 ***Compare the general functions of the CNS and the PNS.***

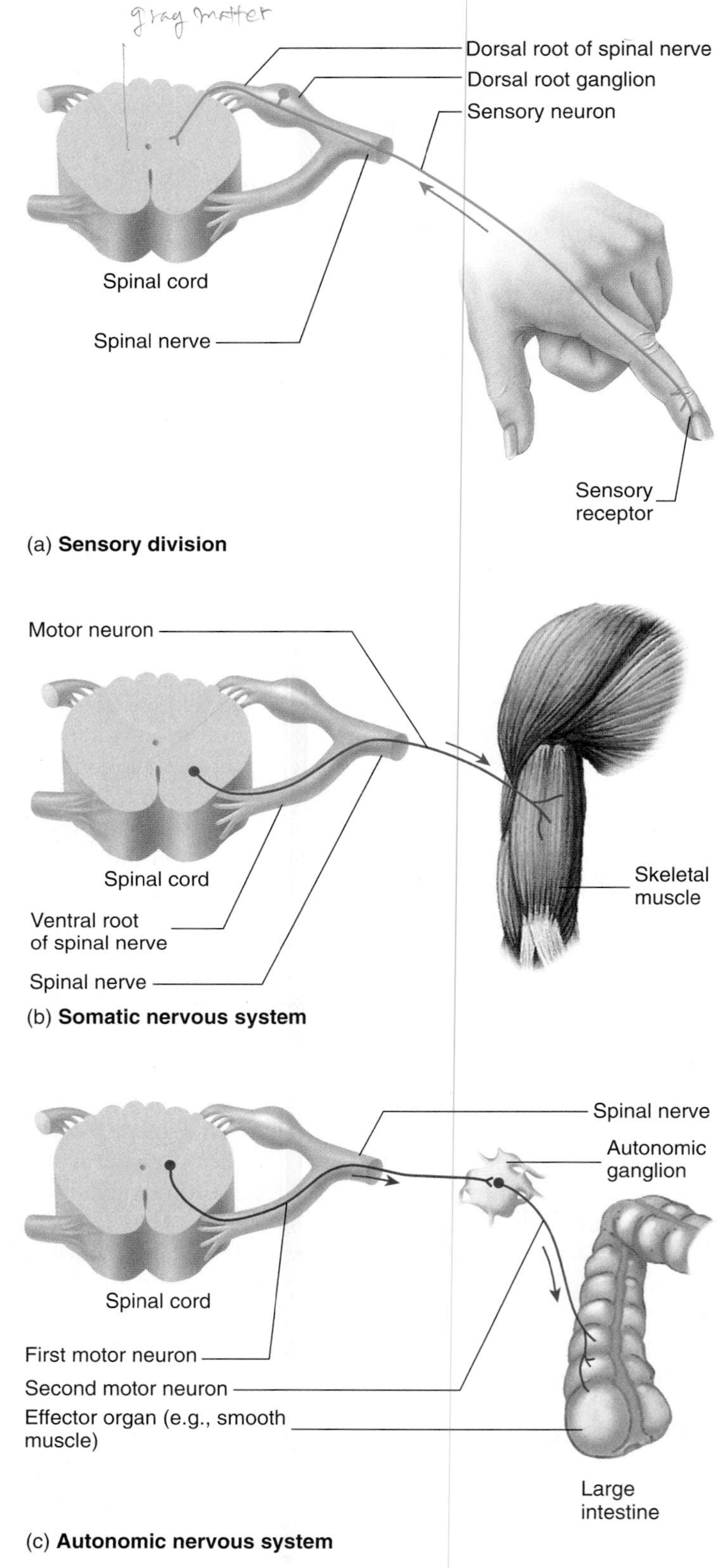

FIGURE 11.2 Divisions of the Peripheral Nervous System

(*a*) Sensory division. A neuron with its cell body in a dorsal root ganglion. (*b*) Somatic nervous system. The neuron extends from the CNS to skeletal muscle. (*c*) Autonomic nervous system. Two neurons are in series between the CNS and the effector cells (smooth muscle or glands). The first neuron has its cell body in the CNS, and the second neuron has its cell body in an autonomic ganglion.

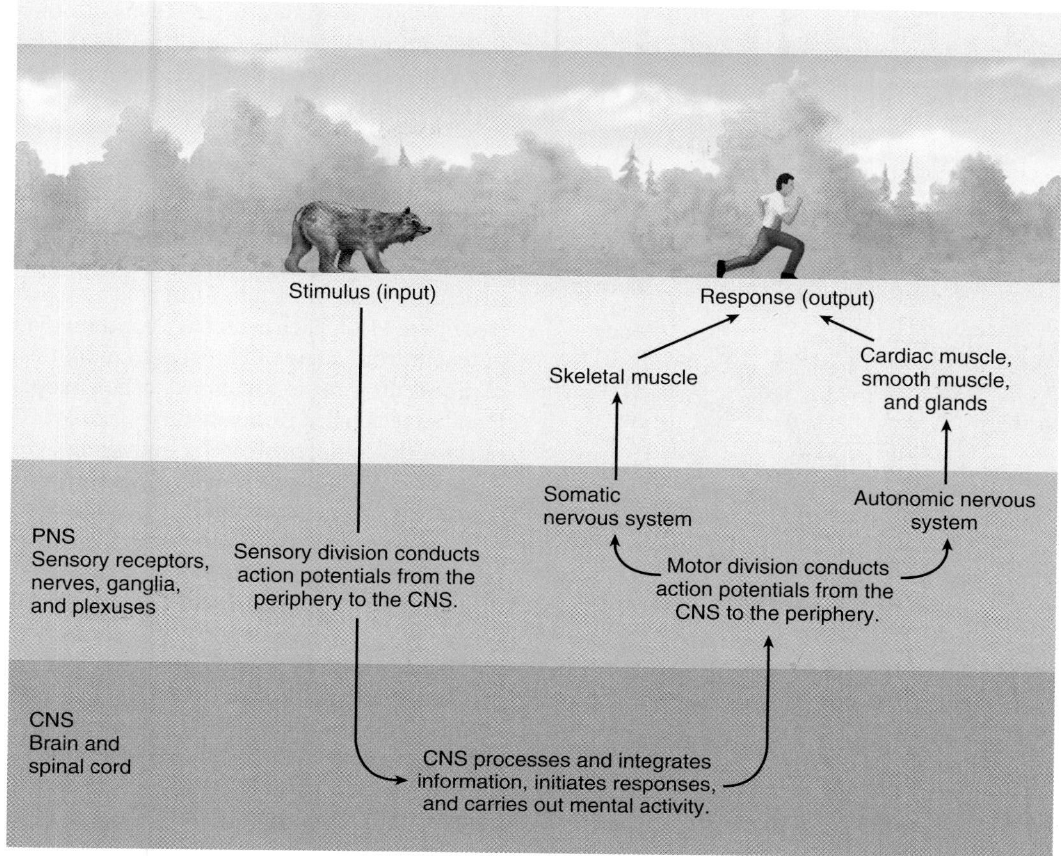

FIGURE 11.3 Organization of the Nervous System

The sensory division of the peripheral nervous system (PNS) detects stimuli and conducts action potentials to the central nervous system (CNS). The CNS interprets incoming action potentials and initiates action potentials that are conducted through the motor division to produce a response. The motor division is divided into the somatic nervous system and the autonomic nervous system.

CELLS OF THE NERVOUS SYSTEM

The nervous system is made up of neurons and nonneural cells. Neurons receive stimuli and conduct action potentials. Nonneural cells are called **glial** (glī′ăl, glē′ăl) **cells,** or **neuroglia** (noo-rog′lē-ă; nerve glue), and they support and protect neurons and perform other functions. Glial cells account for over half of the brain's weight, and there can be 10 to 50 times more glial cells than neurons in various parts of the brain.

Neurons

Neurons, or **nerve cells,** receive stimuli and transmit action potentials to other neurons or to effector organs. They are organized to form complex networks that perform the functions of the nervous system. Each neuron consists of a cell body and two types of processes. The cell body is called the **neuron cell body,** or **soma** (sō′mă; body), and the processes are called **dendrites** (den′drītz) and **axons** (ak′sonz; figure 11.4). *Dendrite* means tree and refers to the branching organization of dendrites. *Axon* means axis and refers to the straight alignment and uniform diameter of most axons. Axons are also referred to as **nerve fibers.**

Neuron Cell Body

Each neuron cell body contains a single relatively large and centrally located nucleus with a prominent nucleolus. Extensive rough endoplasmic reticulum and Golgi apparatuses surround the nucleus, and a moderate number of mitochondria and other organelles are present. Randomly arranged lipid droplets and melanin pigments accumulate in the cytoplasm of some neuron cell bodies. The lipid droplets and melanin pigments increase as humans age, but their functional significance is unknown. Large numbers of intermediate filaments (neurofilaments) and microtubules form bundles that course through the cytoplasm in all directions. The neurofilaments separate abundant rough endoplasmic reticulum (ER), called **Nissl** (nis′l) **substance,** which is located in the cell body and dendrites but not the axon. Nissl substance is the primary site of protein synthesis in neurons.

PREDICT 1

Predict the effect on the part of a severed axon that is no longer connected to its neuron cell body. Explain your prediction.

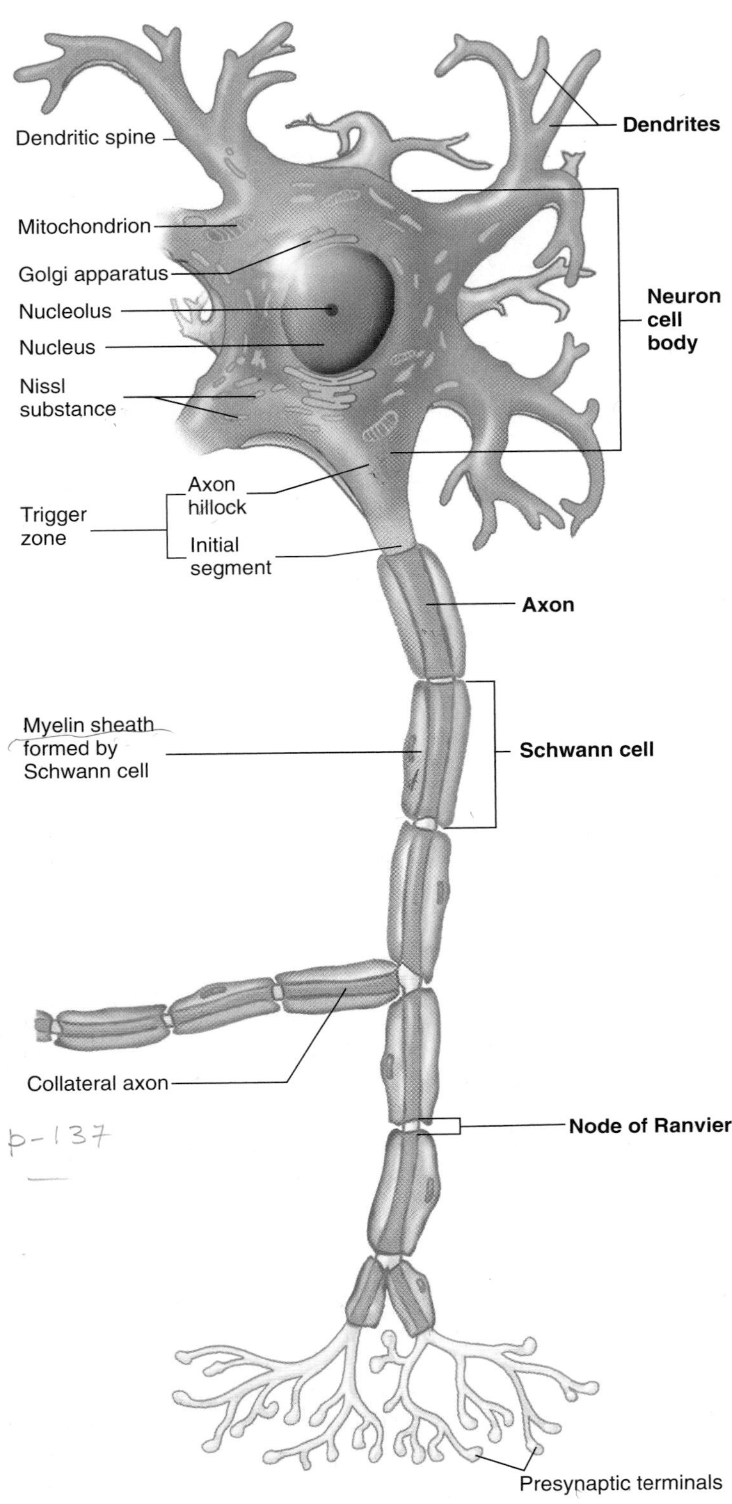

FIGURE 11.4 Neuron

The structural features of a neuron include a cell body and two types of cell processes: dendrites and an axon.

Dendrites

Dendrites are short, often highly branched cytoplasmic extensions that are tapered from their bases at the neuron cell body to their tips (see figure 11.4). Many dendrite surfaces have small extensions called **dendritic spines,** where axons of other neurons form synapses with the dendrites. Dendrites are the input part of the neuron. When stimulated, they generate small electric currents, which are conducted to the neuron cell body.

Axons

In most neurons, a single axon arises from a cone-shaped area of the neuron cell body called the **axon hillock.** The beginning of the axon is called the **initial segment.** An axon can remain as a single structure or can branch to form collateral axons, or side branches (see figure 11.4). Each axon has a constant diameter, but it can vary in length from a few millimeters to more than 1 meter. The cytoplasm of an axon is sometimes called **axoplasm,** and its plasma membrane is called the **axolemma** (*lemma* is Greek, meaning husk or sheath). Axons terminate by branching to form small extensions with enlarged ends called **presynaptic terminals.** Numerous small vesicles containing neurotransmitters are present in the presynaptic terminals. The presynaptic terminals release the **neurotransmitters** which are chemicals that cross the synapse to stimulate or inhibit the postsynaptic cell. Functionally, action potentials are generated at the **trigger zone,** which consists of the axon hillock and the part of the axon nearest the cell body. Action potentials are conducted along the axon to the presynaptic terminal, where they stimulate the release of neurotransmitters.

Axon transport mechanisms can move cytoskeletal proteins (see chapter 3), organelles (such as mitochondria), and vesicles containing neurohormones to be secreted (see chapter 17) down the axon to the presynaptic terminals. In addition, damaged organelles, recycled plasma membrane, and substances taken in by endocytosis can be transported up the axon to the neuron cell body. The movement of materials within the axon is necessary for its normal function, but it also provides a way for infectious agents and harmful substances to be transported from the periphery to the CNS. For example, rabies and herpes viruses can enter damaged axons in the skin and be transported within the axons to the CNS.

9 ***Compare the functions of glial cells and neurons.***

10 ***Describe and give the function of a neuron cell body, a dendrite, and an axon.***

11 ***Define trigger zone, presynaptic terminal, and neurotransmitter.***

Types of Neurons

Neurons are classified according to their function or structure. The functional classification is based on the direction in which action potentials are conducted. **Sensory,** or **afferent, neurons** conduct action potentials toward the CNS; **motor,** or **efferent, neurons** conduct action potentials away from the CNS toward muscles or glands. **Interneurons,** or **association neurons,** conduct action potentials from one neuron to another within the CNS.

The structural classification scheme is based on the number of processes that extend from the neuron cell body. The three major categories of neurons are multipolar, bipolar, and unipolar.

Multipolar neurons have many dendrites and a single axon. The dendrites vary in number and in their degree of branching

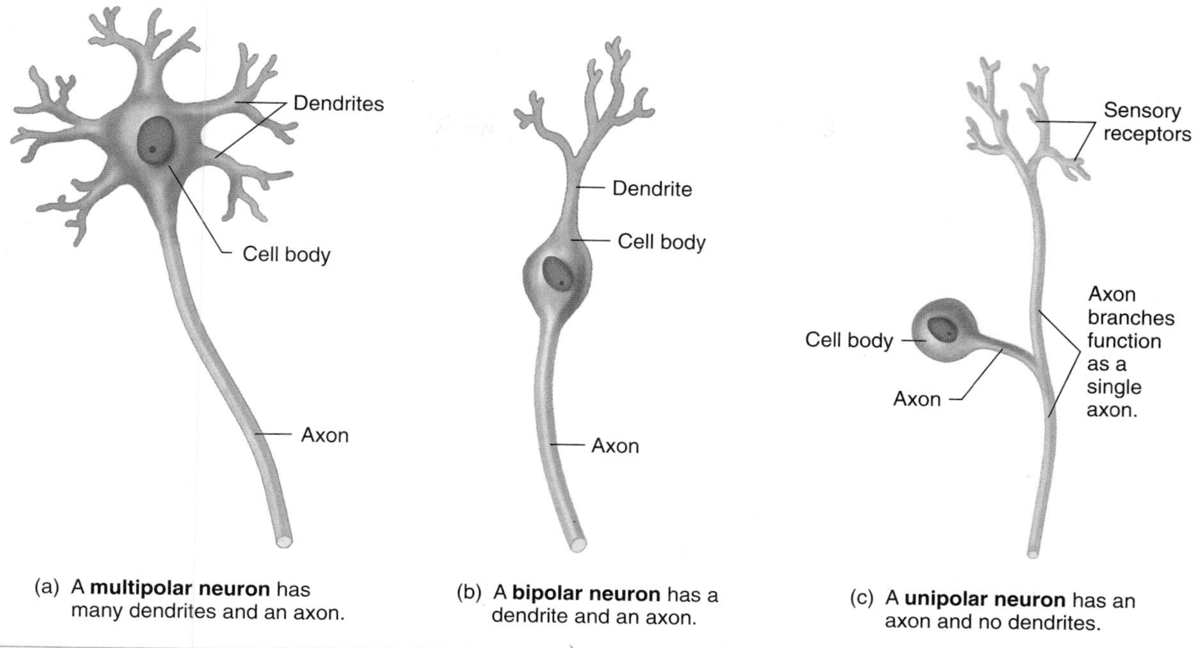

(a) A **multipolar neuron** has many dendrites and an axon.

(b) A **bipolar neuron** has a dendrite and an axon.

(c) A **unipolar neuron** has an axon and no dendrites.

FIGURE 11.5 Types of Neurons

Neurons are classified structurally by the number of cell processes extending from their cell bodies. Dendrites and sensory receptors are specialized to receive stimuli and axons are specialized to conduct action potentials.

(figure 11.5*a*). Most of the neurons within the CNS and motor neurons are multipolar.

Bipolar neurons have two processes: one dendrite and one axon (figure 11.5*b*). The dendrite often is specialized to receive the stimulus, and the axon conducts action potentials to the CNS. Bipolar neurons are located in some sensory organs, such as in the retina of the eye and in the nasal cavity.

Unipolar neurons have a single process extending from the cell body (figure 11.5*c*). This process divides into two branches a short distance from the cell body. One branch extends to the CNS, and the other branch extends to the periphery and has dendritelike sensory receptors. The two branches function as a single axon. The sensory receptors respond to stimuli, resulting in the production of action potentials that are transmitted to the CNS. Most sensory neurons are unipolar neurons. According to a functional definition of a dendrite, the branch of a unipolar neuron that extends from the periphery to the neuron cell body can be classified as a dendrite because it conducts action potentials toward the neuron cell body. This branch is usually referred to as an axon, however, for two reasons: It cannot be distinguished from an axon on the basis of its structure, and it conducts action potentials in the same fashion as an axon.

12 ***Describe the three types of neurons based on their function.***

13 ***Describe the three types of neurons based on their structure, and give an example of where each type is found.***

Glial Cells of the CNS

Glial cells are the major supporting cells in the CNS; they participate in the formation of a permeability barrier between the blood and the neurons, phagocytize foreign substances, produce cerebrospinal fluid, and form myelin sheaths around axons. There are four types of CNS glial cells, each with unique structural and functional characteristics.

Astrocytes

Astrocytes (as′trō-sītz, *aster* is Greek, meaning star) are glial cells that are star-shaped because of cytoplasmic processes that extend from the cell body. These extensions widen and spread out to form foot processes, which cover the surfaces of blood vessels (figure 11.6), neurons, and the pia mater. (The pia mater is a membrane covering the outside of the brain and spinal cord.) Astrocytes have an extensive cytoskeleton of microfilaments (see chapter 3), which enables them to form a supporting framework for blood vessels and neurons.

Astrocytes play a role in regulating the extracellular composition of brain fluid. They release chemicals that promote the formation of tight junctions (see chapter 4) between the endothelial cells of capillaries. The endothelial cells with their tight junctions form the **blood–brain barrier,** which determines what substances can pass from the blood into the nervous tissue of the brain and spinal cord. The blood–brain barrier protects neurons from toxic substances in the blood, allows the exchange of nutrients and waste products between neurons and the blood, and prevents fluctuations in the composition of the blood from affecting the functions of the brain.

Astrocytes contribute to beneficial and detrimental responses to tissue damage in the CNS. Almost all injuries to CNS tissue induce **reactive astrocytosis,** in which astrocytes participate in walling off the injury site and limiting the spread of inflammation to the surrounding healthy tissue. Reactive scar-forming astrocytes also limit the regeneration of the axons of injured neurons.

Other astrocyte functions include the release of chemicals that promote the development of synapses and assistance in the

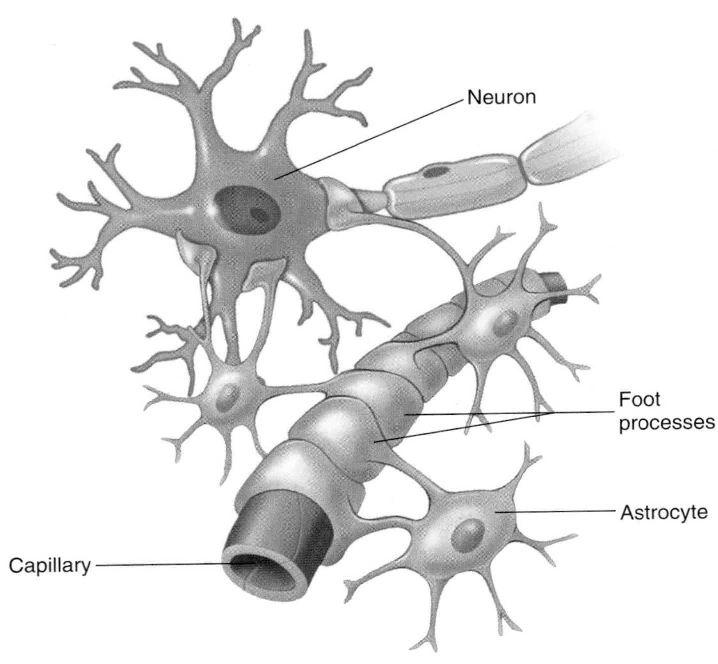

FIGURE 11.6 Astrocytes
Astrocyte processes form foot processes, which cover the surfaces of neurons, blood vessels, and the pia mater. The astrocytes provide structural support and play a role in regulating what substances from the blood reach the neurons.

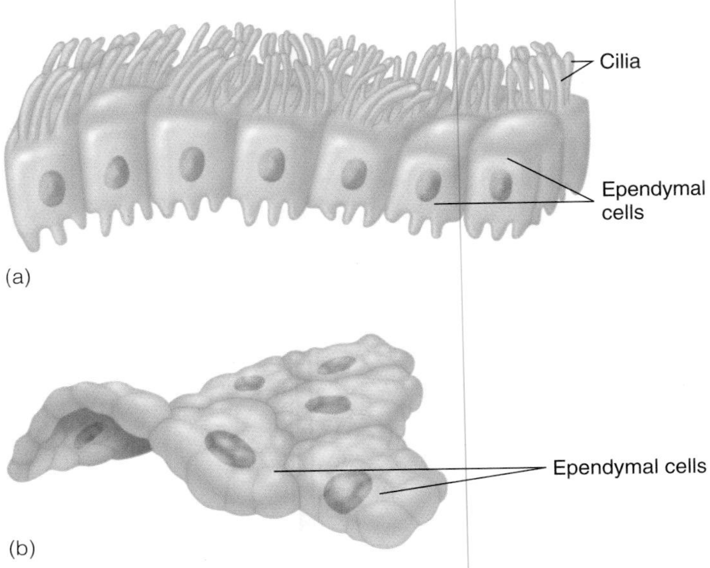

FIGURE 11.7 Ependymal Cells
(*a*) Ciliated ependymal cells lining the ventricles of the brain and the central canal of the spinal cord help move cerebrospinal fluid. (*b*) Ependymal cells on the surface of the choroid plexus secrete cerebrospinal fluid.

regulation of synaptic activity through the synthesis, absorption, and recycling of neurotransmitters.

Ependymal Cells

Ependymal (ep-en′di-măl) cells line the ventricles (cavities) of the brain and the central canal of the spinal cord (figure 11.7*a*). Specialized ependymal cells and blood vessels form the **choroid plexuses** (ko′royd plek′sŭs-ez; figure 11.7*b*), which are located within certain regions of the ventricles. The choroid plexuses secrete the cerebrospinal fluid that circulates through the ventricles of the brain (see chapter 13). The free surface of the ependymal cells frequently has patches of cilia that help move cerebrospinal fluid through the cavities of the brain. Ependymal cells also have long processes at their basal surfaces that extend deep into the brain and the spinal cord and seem, in some cases, to have astrocyte-like functions.

Microglia

Microglia (mī-krog′lē-ă) are glial cells in the CNS that become mobile and phagocytic in response to inflammation. They phagocytize necrotic tissue, microorganisms, and other foreign substances that invade the CNS (figure 11.8).

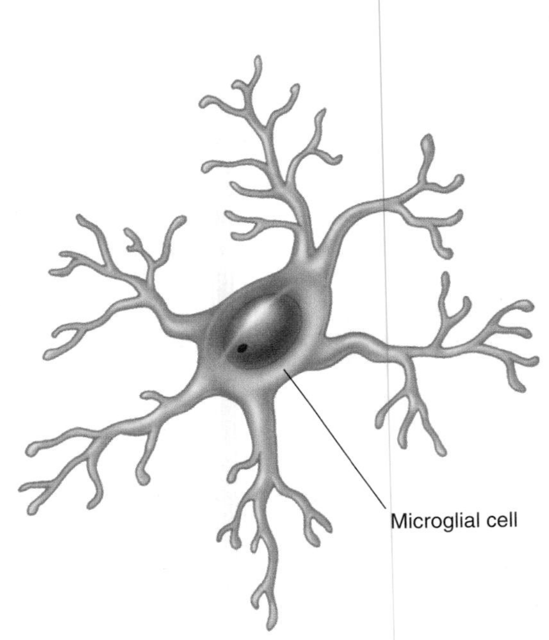

FIGURE 11.8 Microglia
Microglia are phagocytic cells within the CNS.

Microglia and Brain Damage

Numerous microglia migrate to areas damaged by infection, trauma, or stroke and perform phagocytosis. A pathologist can identify these damaged areas in the CNS during an autopsy because large numbers of microglia are found in them.

Oligodendrocytes

Oligodendrocytes (ol′i-gō-den′drō-sītz) have cytoplasmic extensions that can surround axons. If the cytoplasmic extensions wrap many times around the axons, they form **myelin** (mī′ĕ-lin) **sheaths.** A single oligodendrocyte can form myelin sheaths around portions of several axons (figure 11.9).

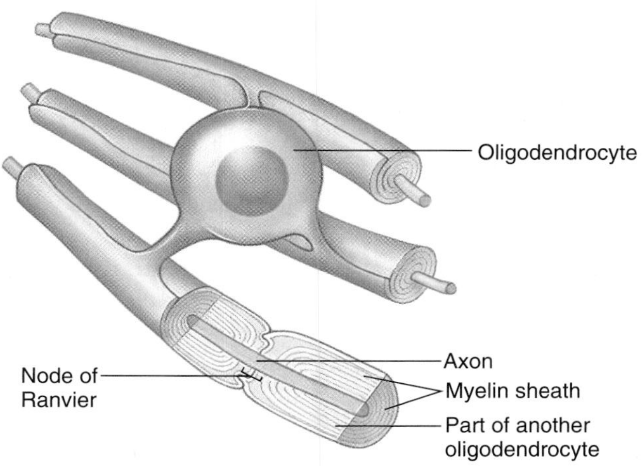

FIGURE 11.9 Oligodendrocytes

Extensions from oligodendrocytes form part of the myelin sheaths of several axons within the CNS.

Glial Cells of the PNS

Schwann cells, or **neurolemmocytes** (noor-ō-lem′mō-sītz), are glial cells in the PNS that wrap around axons. If a Schwann cell wraps many times around an axon, it forms a myelin sheath. Unlike oligodendrocytes, however, each Schwann cell forms a myelin sheath around a portion of only one axon (figure 11.10).

Satellite cells surround neuron cell bodies in sensory ganglia (see figure 11.10). They provide support and nutrition to the neuron cell bodies, and they protect neurons from heavy metal poisons, such as lead and mercury, by absorbing them and reducing their access to the neuron cell bodies.

14 ***Which type of glial cell supports neurons and blood vessels and promotes the formation of the blood–brain barrier? What is the blood–brain barrier, and what is its function?***

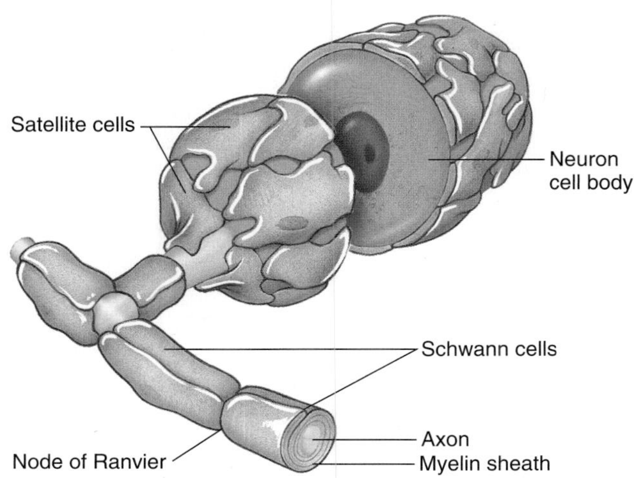

FIGURE 11.10 Glial Cells of the PNS

Neuron cell bodies within ganglia are surrounded by satellite cells. Schwann cells form the myelin sheath of an axon within the PNS.

15 ***Name the different kinds of glial cells responsible for the following functions: production of cerebrospinal fluid, phagocytosis, production of myelin sheaths in the CNS, production of myelin sheaths in the PNS, support of neuron cell bodies in the PNS.***

Myelinated and Unmyelinated Axons

Cytoplasmic extensions of the Schwann cells in the PNS and of the oligodendrocytes in the CNS surround axons to form either myelinated or unmyelinated axons. Myelin protects and electrically insulates axons from one another. In addition, action potentials travel along myelinated axons more rapidly than along unmyelinated axons (see "Propagation of Action Potentials," p. 391).

In **myelinated axons,** the extensions from Schwann cells or oligodendrocytes repeatedly wrap around a segment of an axon to form a series of tightly wrapped membranes rich in phospholipids with little cytoplasm sandwiched between the membrane layers (figure 11.11*a*). The tightly wrapped membranes constitute the myelin sheath and give myelinated axons a white appearance because of the high lipid concentration. The myelin sheath is not continuous but is interrupted every 0.3–1.5 mm. At these locations, there are slight constrictions where the myelin sheaths of adjacent cells dip toward the axon but do not cover it, leaving a bare area 2–3 μm in length. These interruptions in the myelin sheath are the **nodes of Ranvier** (ron′vē-ā). Although the axon at a node of Ranvier is not covered with myelin, Schwann cells or oligodendrocytes extend across the node and connect to each other.

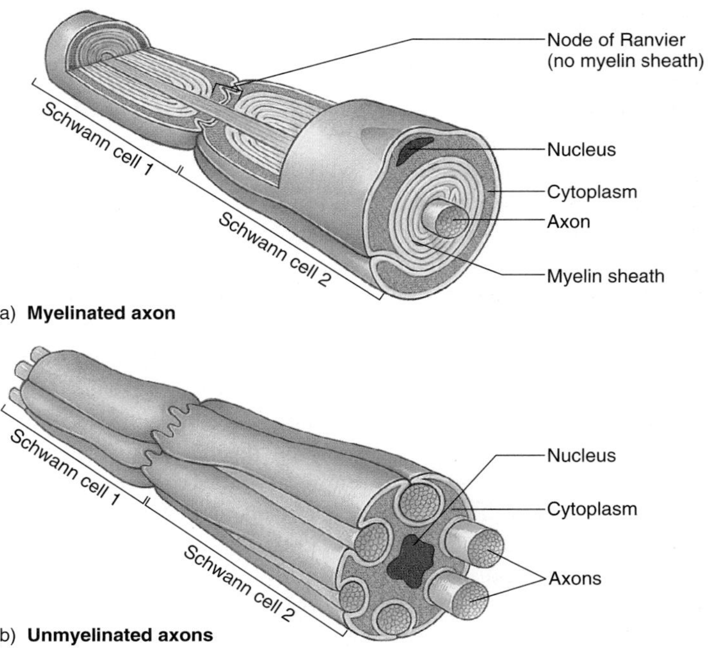

FIGURE 11.11 Comparison of Myelinated and Unmyelinated Axons

(*a*) Myelinated axon with two Schwann cells forming part of the myelin sheath around a single axon. Each Schwann cell surrounds part of one axon. (*b*) Unmyelinated axons with two Schwann cells surrounding several axons in parallel formation. Each Schwann cell surrounds part of several axons.

Unmyelinated axons rest in invaginations of the Schwann cells or oligodendrocytes (figure 11.11*b*). The cell's plasma membrane surrounds each axon but does not wrap around it many times. Thus, each axon is surrounded by a series of cells, and each cell can simultaneously surround more than one unmyelinated axon.

16 ***Define myelin sheath and node of Ranvier.***

17 ***How are myelinated and unmyelinated axons different from each other?***

ORGANIZATION OF NERVOUS TISSUE

Nervous tissue is organized so that axons form bundles, and neuron cell bodies and their relatively short dendrites are grouped together. Bundles of parallel axons with their associated myelin sheaths are whitish in color, which accounts for their name, **white matter.** Collections of neuron cell bodies and unmyelinated axons are grayer in color and are called **gray matter.**

The axons that make up the white matter of the CNS form **nerve tracts,** which propagate action potentials from one area in the CNS to another. The gray matter of the CNS performs integrative functions or acts as relay areas in which axons synapse with the cell bodies of other neurons. The central area of the spinal cord is gray matter, and the outer surface of much of the brain consists of gray matter called **cortex.** Within the brain are other collections of gray matter called **nuclei.**

In the PNS, bundles of axons form nerves, which conduct action potentials to and from the CNS. Most nerves contain myelinated axons, but some consist of unmyelinated axons. Collections of neuron cell bodies in the PNS are called ganglia.

18 ***What is white and gray matter?***

19 ***Define and state the location of nerve tracts, nerves, brain cortex, nuclei, and ganglia.***

ELECTRIC SIGNALS

Like computers, humans depend on electric signals to communicate and process information. The electric signals produced by cells are called **action potentials.** They are an important means by which cells transfer information from one part of the body to another. For example, stimuli, such as light, sound, and pressure, act on specialized sensory cells in the eye, ear, and skin to produce action potentials, which are conducted from these cells to the spinal cord and brain. Action potentials originating within the brain and spinal cord are conducted to muscles and certain glands to regulate their activities.

The ability to perceive our environment, perform complex mental activities, and act depends on action potentials. For example, interpreting the action potentials received from sensory cells results in the sensations of sight, hearing, and touch. Complex mental activities, such as conscious thought, memory, and emotions, result from action potentials. The contraction of muscles and the secretion of certain glands occur in response to action potentials generated in them.

A basic knowledge of the electrical properties of cells is necessary for understanding many of the normal functions and pathologies of the body. These properties result from the ionic concentration differences across the plasma membrane and from the permeability characteristics of the plasma membrane.

Concentration Differences Across the Plasma Membrane

Table 11.1 lists the concentration differences for positively charged ions (cations) and negatively charged ions (anions) between the intracellular and extracellular fluids. The concentration of sodium ions (Na^+) and chloride ions (Cl^-) is much greater outside the cell than inside. The concentration of potassium ions (K^+) and negatively charged molecules, such as proteins and other molecules containing phosphate, is much greater inside the cell than outside. Note that a steep concentration gradient (see chapter 3) exists for Na^+ from outside the cell to the inside. Also, a steep concentration gradient exists for K^+ from the inside to the outside of the cell.

Differences in intracellular and extracellular concentrations of ions result primarily from (1) the Na^+–K^+ pump and (2) the permeability characteristics of the plasma membrane.

The Na^+–K^+ Pump

The differences in K^+ and Na^+ concentrations across the plasma membrane are maintained primarily by the action of the **Na^+–K^+pump** (see figure 3.21). Through active transport, the Na^+–K^+ pump moves K^+ and Na^+ through the plasma membrane against their concentration gradients. Potassium ions are transported into the cell, increasing the concentration of K^+ inside the cell, and Na^+ are transported out of the cell, increasing the concentration of Na^+ outside the cell. Approximately three Na^+ are transported out of the cell and two K^+ are transported into the cell for each ATP molecule used.

TABLE 11.1 Representative Concentrations of the Principal Cations and Anions in Extracellular and Intracellular Fluids of Vertebrates

Ions	Intracellular Fluid (mEq/L)*	Extracellular Fluid (mEq/L)
Cations (Positive)		
Potassium (K^+)	148	5
Sodium (Na^+)	10	142
Calcium (Ca^{2+})	<1	5
Others	41	3
TOTAL	200	155
Anions (Negative)		
Proteins	56	16
Chloride (Cl^-)	4	103
Others	140	36
TOTAL	200	155

*See Appendix C for an explanation of milliequivalents (mEq).

Permeability Characteristics of the Plasma Membrane

As noted in chapter 3, the plasma membrane is selectively permeable, thus allowing some, but not all, substances to pass through it. Negatively charged proteins are synthesized inside the cell, and because of their large size and their solubility characteristics they cannot readily diffuse across the plasma membrane (figure 11.12). Negatively charged Cl^- are repelled by the negatively charged proteins and other negatively charged ions inside the cell. Chloride ions diffuse through the plasma membrane and accumulate outside it, resulting in a higher concentration of Cl^- outside the cell than inside.

Ions pass through the plasma membrane through ion channels. The two major types of ion channels are leak channels and gated ion channels.

Leak Channels

Leak channels, or **nongated ion channels,** are always open and are responsible for the permeability of the plasma membrane to ions when the plasma membrane is unstimulated, or at rest (see figure 11.12). Each ion channel is specific for one type of ion, although the specificity is not absolute. The number of each type of leak channels in the plasma membrane determines the permeability characteristics of the resting plasma membrane to different types of ions. The plasma membrane is more permeable to K^+ and Cl^- and much less permeable to Na^+ because there are many more K^+ and Cl^- leak channels than Na^+ leak channels in the plasma membrane.

Gated Ion Channels

Gated ion channels open and close in response to stimuli. By opening and closing, these channels can change the permeability characteristics of the plasma membrane. The major types of gated ion channels are

1. *Ligand-gated ion channels.* A **ligand** is a molecule that binds to a receptor. A **receptor** is a protein or glycoprotein that has a **receptor site** to which a ligand can bind. Most receptors are located in the plasma membrane. **Ligand-gated ion channels** are receptors that have an extracellular receptor site and a membrane-spanning part that forms an ion channel. When a ligand binds to the receptor site, the ion channel opens or closes. For example, the neurotransmitter acetylcholine released from the presynaptic terminal of a neuron is a chemical signal that can bind to a ligand-gated Na^+ channel in the membrane of a muscle cell. As a result, the Na^+ channel opens, allowing Na^+ to enter the cell (see figure 3.10). Ligand-gated ion channels exist for Na^+, K^+, Ca^{2+}, and Cl^-, and these channels are common in tissues such as nervous and muscle tissues, as well as glands.
2. *Voltage-gated ion channels.* These channels open and close in response to small voltage changes across the plasma membrane. In an unstimulated cell, the inside of the plasma membrane is negatively charged relative to the outside. This charge difference can be measured in units called **millivolts** (mV; 1 mV = 1/1000 V). When a cell is stimulated, the permeability of the plasma membrane changes because gated ion channels open or close. The movement of ions into or out of the cell changes the charge difference across the plasma membrane, which causes voltage-gated ion channels to open or close. Voltage-gated channels specific for Na^+ and K^+ are most numerous in electrically excitable tissues, but voltage-gated Ca^{2+} channels are also important, especially in smooth muscle and cardiac muscle cells (see chapters 9 and 20).
3. *Other gated ion channels.* Gated ion channels that respond to stimuli other than ligands or voltage changes are present in specialized electrically excitable tissues. Examples include touch receptors, which respond to mechanical stimulation of the skin, and temperature receptors, which respond to temperature changes in the skin.

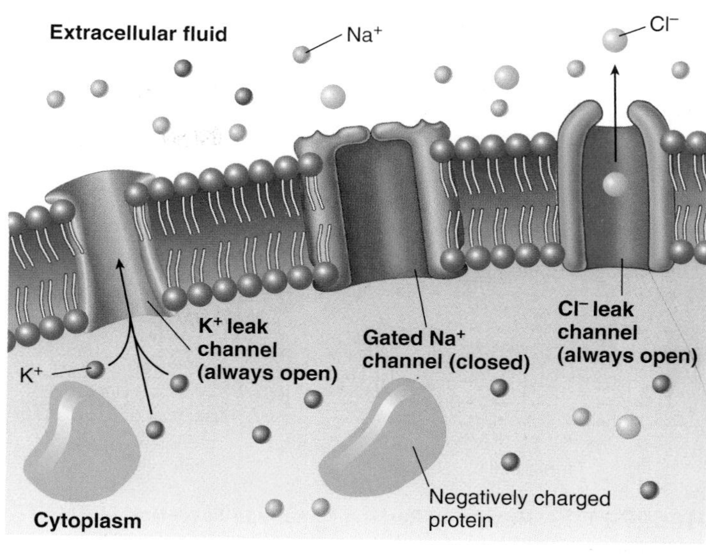

FIGURE 11.12 Membrane Permeability and Ion Channels

The permeability of the membrane to K^+ and Cl^- is greater than its permeability to Na^+ because some K^+ and Cl^- leak channels are open, whereas most gated Na^+ channels are closed. The membrane is not permeable to the negatively charged proteins inside the cell because they are too large to pass through membrane channels.

20 ***Describe the concentration differences for Na^+ and K^+ that exist across the plasma membrane.***

21 ***In what direction, into or out of cells, does the Na^+–K^+ pump move Na^+ and K^+?***

22 ***Define leak channels and gated ion channels. How are they responsible for the permeability characteristics of a resting versus a stimulated plasma membrane?***

23 ***Define ligand, receptor, and receptor site.***

24 ***What kinds of stimuli cause gated ion channels to open or close?***

Establishing the Resting Membrane Potential

Intracellular fluid is electrically neutral because the number of positively charged cations is equal to the number of negatively charged anions (see table 11.1). Similarly, extracellular fluid is electrically neutral. There is, however, a difference in charge across the plasma membrane. The immediate inside of the membrane is negative, compared with the immediate outside of

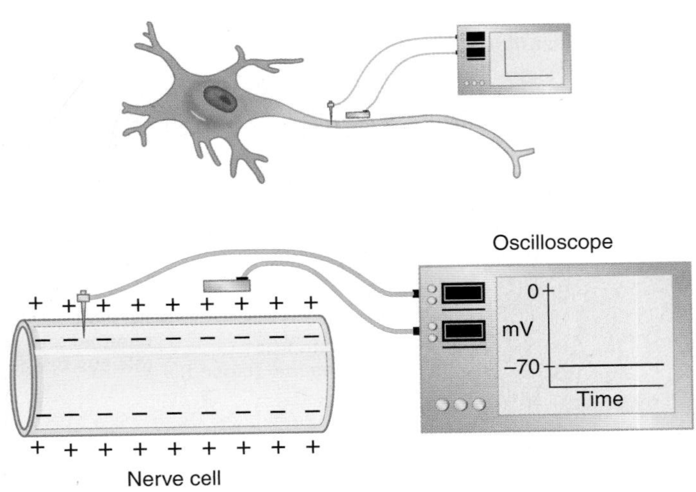

FIGURE 11.13 Measuring Resting Membrane Potential
The recording electrode is inside the cell; the reference electrode is outside. A potential difference of about −70 mV is recorded, with the inside of the plasma membrane negative with respect to the outside of the membrane.

the membrane (figure 11.13). The plasma membrane, therefore, is said to be **polarized** because there are opposite charges, or poles, across the membrane.

The electric charge difference across the plasma membrane is called a **potential difference.** In an unstimulated, or resting, cell, the potential difference is called the **resting membrane potential.** It can be measured using a voltmeter or an oscilloscope connected to microelectrodes positioned just inside and outside the plasma membrane (see figure 11.13). The resting membrane potential of nerve cells is approximately −70 mV and of skeletal muscle cells is approximately −85 mV (see chapter 9). The potential difference is reported as a negative number because the inside of the plasma membrane is negative, compared with the outside. The greater the charge difference across the plasma membrane, the greater the potential difference. A resting membrane potential of −85 mV has a greater charge difference than a resting membrane potential of −70 mV.

The resting membrane potential results from the permeability characteristics of the resting plasma membrane and the difference in concentration of ions between the intracellular and the extracellular fluids. The plasma membrane is somewhat permeable to K^+ because of K^+ leak channels. Positively charged K^+ can therefore diffuse down their concentration gradient from inside to just outside the cell (figure 11.14). Negatively charged proteins and other molecules cannot diffuse through the plasma membrane with the K^+. As K^+ diffuse out of the cell, the loss of positive charges makes the inside of the plasma membrane more negative. Because opposite charges attract, the K^+ are attracted back toward the cell. The accumulation of K^+ just outside of the plasma membrane makes the outside of the plasma membrane positive relative to the inside. The resting membrane potential is an equilibrium that is established when the tendency for K^+ to diffuse out of the cell, because of the K^+ concentration gradient, is equal to the tendency for K^+ to move into the cell because of the attraction of the positively charged K^+ to negatively charged proteins and other molecules.

PREDICT 2

Given that tissue A has significantly more K^+ leak channels than tissue B, which tissue has the larger resting membrane potential?

Other ions, such as Na^+, Cl^-, and Ca^{2+}, have a minor influence on the resting membrane potential, but the major influence

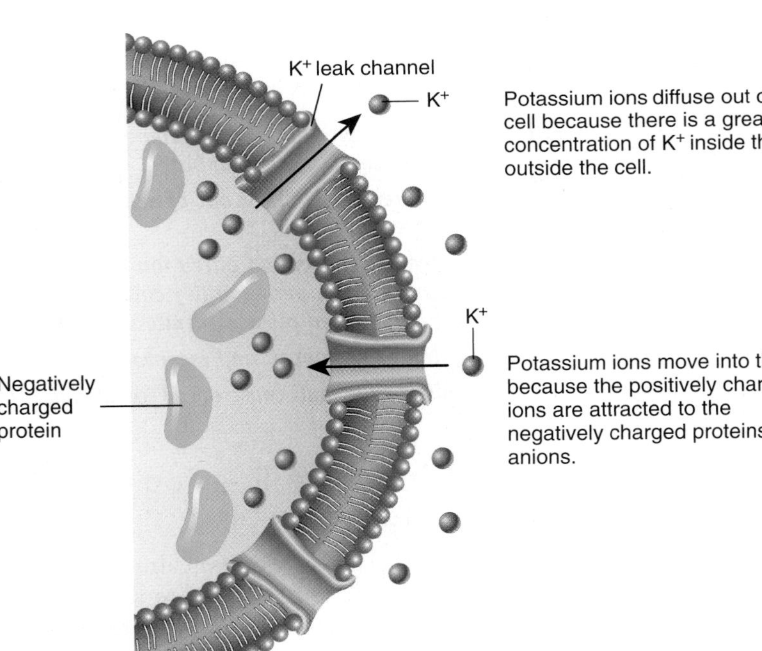

FIGURE 11.14 Potassium Ions and the Resting Membrane Potential

is from K^+. Because the resting plasma membrane is 50–100 times less permeable to Na^+ than to K^+, very few Na^+ can diffuse from the outside to the inside of the resting cell. The resting plasma membrane is not very permeable to Ca^{2+}, either. The plasma membrane is relatively permeable to Cl^-, but these negatively charged ions are repelled by the negative charge inside the cell.

The resting membrane potential is proportional to the tendency for K^+ to diffuse out of the cell, not to the actual rate of flow for K^+. At equilibrium, very few of these ions pass through the plasma membrane because their movement out of the cell is opposed by the negative charge inside the cell. Still, some Na^+ and K^+ diffuse continuously across the plasma membrane, although at a low rate. The large concentration gradients for Na^+ and K^+ would eventually disappear without the continuous activity of the Na^+–K^+ pump.

As already noted, the function of the Na^+–K^+ pump is to maintain the normal concentration gradients for Na^+ and K^+ across the plasma membrane. The pump is also responsible for a small portion of the resting membrane potential, usually less than 15 mV, because it transports approximately three Na^+ out of the cell and two K^+ into the cell for each ATP molecule used (see figure 3.21). The outside of the plasma membrane becomes more positively charged than the inside because more positively charged ions are pumped out of the cell than are pumped into it.

The characteristics responsible for a resting membrane potential are summarized in table 11.2.

25 ***Define resting membrane potential. Is the outside of the plasma membrane positively or negatively charged relative to the inside?***

26 ***Explain the role of K^+ and the Na^+–K^+ pump in establishing the resting membrane potential.***

Changing the Resting Membrane Potential

The resting membrane potential can decrease or increase (figure 11.15). **Depolarization** (dē-pō′lăr-i-zā′shŭn) is a decrease in the membrane potential in which the charge difference, or polarity, across the plasma membrane decreases. **Hyperpolarization** (hī′per-pō′lăr-i-zā′shŭn) is an increase in the membrane potential caused by an increase in the charge difference across the plasma membrane. The charge difference can be changed by decreasing or increasing the movement of ions across the plasma membrane. Ion movement can be changed by altering ion concentration gradients or ion permeability.

TABLE 11.2 Characteristics Responsible for the Resting Membrane Potential

1. The number of charged molecules and ions inside and outside the cell is nearly equal.
2. The concentration of K^+ is higher inside than outside the cell, and the concentration of Na^+ is higher outside than inside the cell.
3. The plasma membrane is 50–100 times more permeable to K^+ than to other positively charged ions, such as Na^+.
4. The plasma membrane is impermeable to large, intracellular, negatively charged molecules, such as proteins.
5. Potassium ions tend to diffuse across the plasma membrane from the inside to the outside of the cell.
6. Because negatively charged molecules cannot follow the positively charged K^+, a small negative charge develops just inside the plasma membrane.
7. The negative charge inside the cell attracts positively charged K^+. When the negative charge inside the cell is great enough to prevent additional K^+ from diffusing out of the cell through the plasma membrane, an equilibrium is established.
8. The charge difference across the plasma membrane at equilibrium is reflected as a difference in potential, which is measured in millivolts (mV).
9. The resting membrane potential is proportional to the potential for K^+ to diffuse out of the cell but not to the actual rate of flow for K^+.
10. At equilibrium, there is very little movement of K^+ or other ions across the plasma membrane.

Potassium Ions

Changes in the K^+ concentration gradient and K^+ permeability can change the resting membrane potential. An increase in the extracellular concentration of K^+ decreases the K^+ concentration gradient because the concentration of K^+ is lower outside than inside a cell. As a consequence, the tendency for K^+ to diffuse out

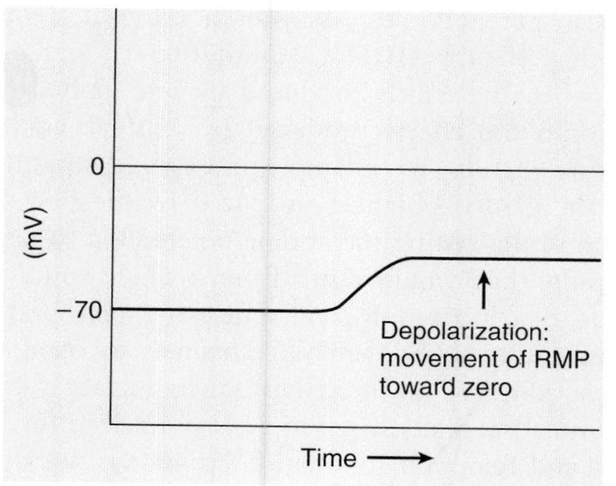

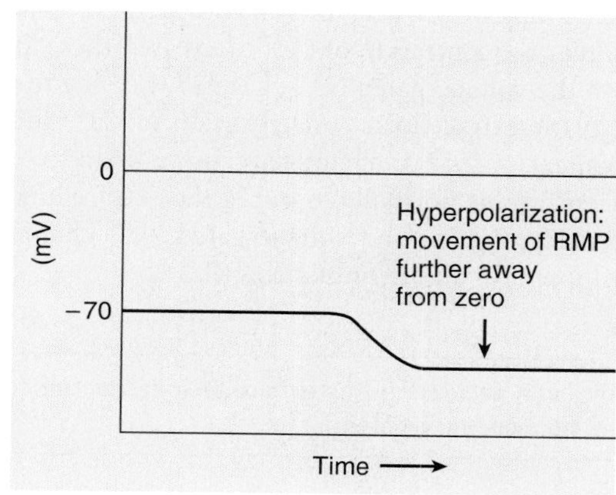

FIGURE 11.15 Depolarization and Hyperpolarization of the Resting Membrane Potential
(*a*) In depolarization, membrane potential decreases, becoming less negative. (*b*) In hyperpolarization, membrane potential increases, becoming more negative.

of the cell decreases, and a smaller negative charge inside the cell is required to oppose the diffusion of K^+ out of the cell. At this new equilibrium, the smaller charge difference across the plasma membrane is a depolarization. On the other hand, a decrease in the extracellular concentration of K^+ increase the K^+ concentration gradient. As a result, the tendency for K^+ to diffuse out of the cell increases, and a larger negative charge inside the cell is required to resist that diffusion. At this new equilibrium, the larger charge difference across the plasma membrane is a hyperpolarization.

Although K^+ leak channels allow some K^+ to diffuse across the plasma membrane, the resting membrane is not freely permeable to K^+. There are gated K^+ channels in the plasma membrane, however; if they open, membrane permeability to K^+ increases, and more K^+ diffuse out of the cell. The increased tendency for K^+ to diffuse out of the cell is opposed by the greater negative charge that develops inside the plasma membrane, resulting in hyperpolarization.

PREDICT 3

Does the resting membrane potential increase or decrease when the intracellular concentration of K^+ is increased by the injection of a potassium succinate solution into a cell? Explain.

Sodium Ions

In an unstimulated cell, the membrane is not very permeable to Na^+ because there are few Na^+ leak channels. Changes in the concentration of Na^+ on either side of the plasma membrane do not influence the resting membrane potential very much because of this low permeability. There are gated Na^+ channels in the plasma membrane, however; if they open, membrane permeability to Na^+ increases (see figure 3.10). Sodium ions then diffuse into the cell because the concentration gradient for Na^+ is from the outside to the inside of the cell. As Na^+ diffuse into the cell, the inside of the membrane becomes more positive, resulting in depolarization.

Calcium Ions

Calcium ions alter membrane potentials in two ways: (1) by affecting voltage-gated Na^+ channels and (2) by entering cells through gated Ca^{2+} channels. Voltage-gated Na^+ channels are sensitive to changes in the extracellular concentration of Ca^{2+}. Positively charged Ca^{2+} in the extracellular fluid are attracted to the negatively charged groups of proteins in voltage-gated Na^+ channels. If the extracellular concentration of Ca^{2+} decreases, these ions diffuse away from the voltage-gated Na^+ channels, causing the channels to open. If the extracellular concentration of Ca^{2+} increases, they bind to voltage-gated Na^+ channels, causing them to close. At the Ca^{2+} concentrations normally found in the extracellular fluid, only a small percentage of the voltage-gated Na^+ channels are open at any moment in an unstimulated cell.

PREDICT 4

Predict the effect of a decrease in the extracellular concentration of Ca^{2+} on the resting membrane potential.

Changes in the permeability of Ca^{2+} can change the resting membrane potential. If the plasma membrane becomes permeable to Ca^{2+} by the opening of gated Ca^{2+} channels, Ca^{2+} diffuse into the cell and the inside of the membrane becomes more positive, resulting in depolarization.

Chloride Ions

Changes in the permeability of Cl^- can change the resting membrane potential. If the permeability of the plasma membrane to Cl^- increases by the opening of gated Cl^- channels, then Cl^- diffuse into the cell and the inside of the membrane becomes more negative, resulting in hyperpolarization.

27 ***Define depolarization and hyperpolarization. How do alterations in the K^+ concentration gradient, changes in membrane permeability to K^+, Na^+, Ca^{2+}, or Cl^- and changes in extracellular Ca^{2+} concentration affect depolarization and hyperpolarization?***

Graded Potentials

A stimulus applied at one location on the plasma membrane of a cell normally causes a change in the resting membrane potential called a **graded potential.** Graded potentials are socalled because the potential change can vary from small to large. Graded potentials are also called **local potentials** because they are confined to a small region of the plasma membrane. Graded potentials can result from (1) chemical signals binding to their receptors, (2) changes in the voltage across the plasma membrane, (3) mechanical stimulation, (4) temperature changes, or (5) spontaneous changes in membrane permeability.

A graded potential can be either a depolarization or a hyperpolarization (see figure 11.5). A change in membrane permeability to Na^+, K^+, or other ions can produce a graded potential. For example, if a stimulus causes gated Na^+ channels to open, the diffusion of a few Na^+ into cells results in depolarization. If a stimulus causes gated K^+ channels to open, the diffusion of a few K^+ out of the cell results in hyperpolarization.

The magnitude of graded potentials can vary from small to large, depending on the stimulus strength or on summation. For example, a weak stimulus can cause a few gated Na^+ channels to open. A few Na^+ diffuse into the cell and cause a small depolarization. A stronger stimulus can cause a greater number of gated Na^+ channels to open. A greater number of Na^+ diffusing into the cell causes a larger depolarization (figure 11.16*a*).

Local potentials can **summate** (sŭm-āt'), or add onto each other (figure 11.16*b*). **Summation** of graded potentials occurs when the effects produced by one graded potential are added onto the effects produced by another graded potential (see figure 11.16*b*). For example, if a second stimulus is applied before the graded potential produced by the first stimulus has returned to the resting membrane potential, a larger depolarization results than would result from a single stimulus. The first stimulus causes gated Na^+ channels to open, and the second stimulus causes additional Na^+ channels to open. As a result, more Na^+ diffuse into the cell, producing a larger graded potential. Summation is discussed in greater detail in this chapter (see "Spatial and Temporal Summation," p. 404).

Graded potentials spread, or are conducted, over the plasma membrane in a decremental fashion. That is, they rapidly decrease in magnitude as they spread over the surface of the plasma membrane,

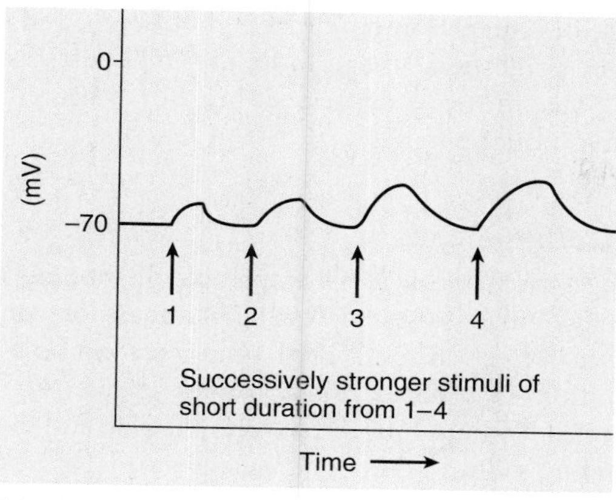

(a)

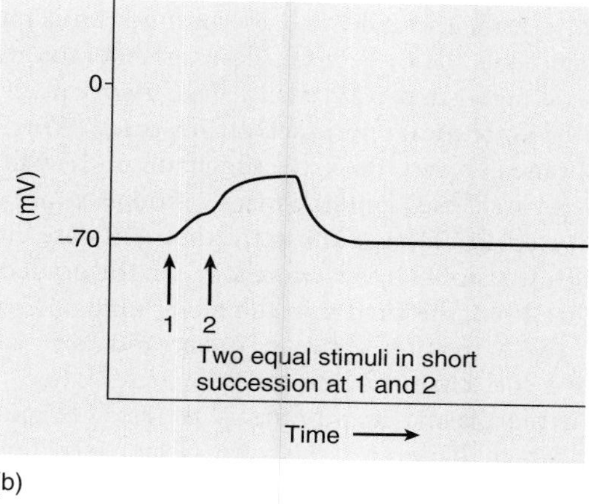

(b)

FIGURE 11.16 Graded Potentials

(*a*) Graded potentials are proportional to the stimulus strength. A weak stimulus applied briefly causes a small depolarization, which quickly returns to the resting membrane potential (*1*). Progressively stronger stimuli result in larger depolarizations (*2* to *4*). (*b*) A stimulus applied to a cell causes a small depolarization. When a second stimulus is applied before the depolarization disappears, the depolarization caused by the second stimulus is added to the depolarization caused by the first to result in a larger depolarization.

much as a teacher talks to a large class. At the front of the class, the teacher's voice can be easily heard, but, the farther away a student sits, the more difficult it is to hear. Normally, a graded potential cannot be detected more than a few millimeters from the site of stimulation. As a consequence, a graded potential cannot transfer information over long distances from one part of the body to another.

Graded potentials are important because of their effect on the generation of action potentials. The characteristics of graded potentials are summarized in table 11.3.

28 ***Define a graded potential. What does it mean to say a graded potential can summate and spreads in a decremental fashion?***

PREDICT 5

Given two cells that are identical in all ways except that the extracellular concentration of Na^+ is less for cell A than for cell B, how would the magnitude of the graded potential in cell A differ from that in cell B if stimuli of identical strength were applied to each?

TABLE 11.3 Characteristics of Graded Potentials
1. A stimulus causes ion channels to open, resulting in increased permeability of the membrane to Na^+, K^+, or Cl^-.
2. Increased permeability of the membrane to Na^+ results in depolarization. Increased permeability of the membrane to K^+ or Cl^- results in hyperpolarization.
3. The size of the graded potential is proportional to the strength of the stimulus. Graded potentials can also summate. Thus, a graded potential produced in response to several stimuli is larger than one produced in response to a single stimulus.
4. Graded potentials are conducted in a decremental fashion, meaning that their magnitude decreases as they spread over the plasma membrane. Graded potentials cannot be measured a few millimeters from the point of stimulation.
5. A depolarizing graded potential can cause an action potential.

Action Potentials

When a graded potential causes depolarization of the plasma membrane to a level called **threshold,** a series of permeability changes occur that results in an action potential (figure 11.17). The action potential has a **depolarization phase,** in which the membrane potential moves away from the resting membrane potential and becomes more positive, and a **repolarization phase,** in which the membrane potential returns toward the resting membrane state and becomes more negative. After the repolarization phase, the plasma membrane may be slightly hyperpolarized for a short period called the **afterpotential.** An action potential is a large change in the membrane potential that propagates, without changing its magnitude, over long distances along the plasma membrane. Thus, action potentials can transfer information from one part of the body to another.

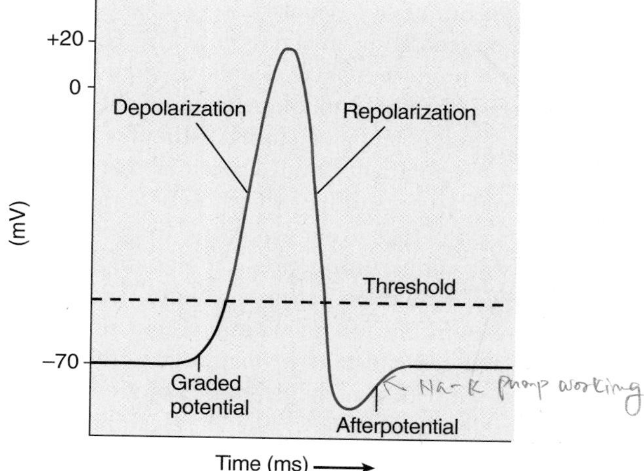

FIGURE 11.17 Action Potential

The action potential consists of a depolarization phase and a repolarization phase, often followed by a short period of hyperpolarization, called the afterpotential.

TABLE 11.4 Characteristics of Action Potentials

1. Action potentials are produced when a graded potential reaches threshold.
2. Action potentials are all-or-none.
3. Depolarization is a result of increased membrane permeability to Na^+ and movement of Na^+ into the cell. Activation gates of the voltage-gated Na^+ channels open.
4. Repolarization is a result of decreased membrane permeability to Na^+ and increased membrane permeability to K^+, which stops Na^+ movement into the cell and increases K^+ movement out of the cell. The inactivation gates of the voltage-gated Na^+ channels close, and the voltage-gated K^+ channels open.
5. No action potential is produced by a stimulus, no matter how strong, during the absolute refractory period. During the relative refractory period, a stronger-than-threshold stimulus can produce an action potential.
6. Action potentials are propagated, and for a given axon or muscle fiber the magnitude of the action potential is constant.
7. Stimulus strength determines the frequency of action potentials.

It generally takes 1–2 milliseconds (ms; 1 ms = 0.001 s) for an action potential to occur. The characteristics of action potentials are summarized in table 11.4.

All-or-None Principle

The generation of action potentials is dependent on graded potentials. Depolarizing graded potentials that reach threshold produce an action potential. Hyperpolarizing graded potentials can never reach threshold and do not produce action potentials. Thus, depolarizing graded potentials can potentially activate a cell by causing the production of an action potential, whereas hyperpolarizing graded potentials can prevent the production of an action potential.

The magnitude of a depolarizing graded potential affects the likelihood of generating an action potential. For example, a weak stimulus can produce a small depolarizing graded potential that does not reach threshold and therefore does not cause the production of an action potential. A stronger stimulus, however, can produce a larger depolarizing graded potential that reaches threshold, resulting in the production of an action potential.

Action potentials occur according to the **all-or-none principle.** If a stimulus produces a depolarizing graded potential that is large enough to reach threshold, all the permeability changes responsible for an action potential proceed without stopping and are constant in magnitude (the "all" part). If a stimulus is so weak that the depolarizing graded potential does not reach threshold, few of the permeability changes occur. The membrane potential returns to its resting level after a brief period without producing an action potential (the "none" part). An action potential can be compared to the flash system of a camera. When the shutter is triggered (reaches threshold), the camera flashes (an action potential is produced), and each flash is the same brightness (magnitude; the "all" part) as previous flashes. If the shutter is depressed, but not triggered, no flash results (the "none" part).

Depolarization Phase

The change in charge across the plasma membrane caused by a graded potential causes increasing numbers of voltage-gated Na^+ channels to open for a brief time. As soon as a threshold depolarization is reached, many voltage-gated Na^+ channels begin to open. Sodium ions diffuse into the cell, and the resulting depolarization causes additional voltage-gated Na^+ channels to open. As a consequence, more Na^+ diffuse into the cell, causing a greater depolarization of the membrane, which in turn causes still more voltage-gated Na^+ channels to open. This is an example of a positive-feedback cycle, and it continues until most of the voltage-gated Na^+ channels in the plasma membrane are open.

Each voltage-gated Na^+ channel has two voltage-sensitive gates, called **activation** and **inactivation gates.** When the plasma membrane is at rest, the activation gates of the voltage-gated Na^+ channel are closed, and the inactivation gates are open (figure 11.18, step 1). Because the activation gates are closed, Na^+ cannot diffuse through the channels. When the graded potential reaches threshold, the change in the membrane potential causes many of the activation gates to open, and Na^+ can diffuse through the Na^+ channels into the cell.

When the plasma membrane is at rest, voltage-gated K^+ channels, which have one gate, are closed (see figure 11.18, step 1). When the graded potential reaches threshold, the voltage-gated K^+ channels begin to open at the same time as the voltage-gated Na^+ channels, but they open more slowly (figure 11.18, step 2). Only a small number of voltage-gated K^+ channels are open, compared with the number of voltage-gated Na^+ channels because the voltage-gated K^+ channels open slowly. Depolarization occurs because more Na^+ diffuse into the cell than K^+ diffuse out of it.

PREDICT 6

Predict the effect of a reduced extracellular concentration of Na^+ on the magnitude of the action potential in an electrically excitable cell.

Repolarization Phase

As the membrane potential approaches its maximum depolarization, the change in the potential difference across the plasma membrane causes the inactivation gates in the voltage-gated Na^+ channels to begin closing, and the permeability of the plasma membrane to Na^+ decreases. During the repolarization phase, the voltage-gated K^+ channels, which started to open along with the voltage-gated Na^+ channels, continue to open (figure 11.18, step 3). Consequently, the permeability of the plasma membrane to Na^+ decreases, and the permeability to K^+ increases. The decreased diffusion of Na^+ into the cell and the increased diffusion of K^+ out of the cell causes repolarization.

1. Resting membrane potential. Voltage-gated Na^+ channels (*pink*) are closed (the activation gates are closed and the inactivation gates are open). Voltage-gated K^+ channels (*purple*) are closed.

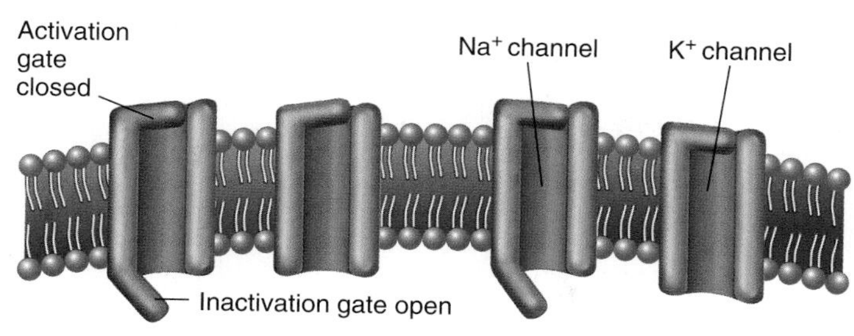

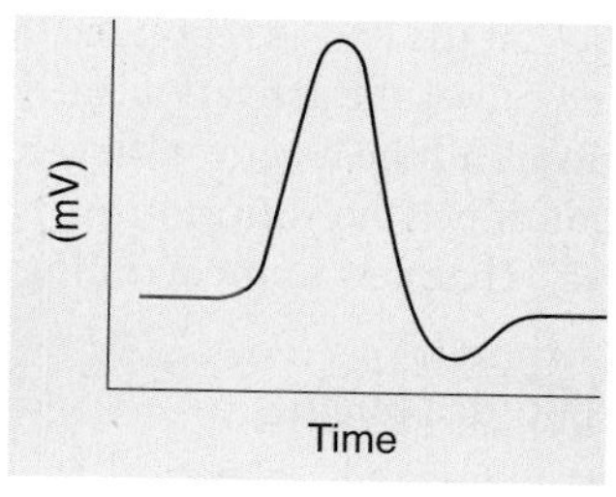

2. Depolarization. Voltage-gated Na^+ channels open because the activation gates open. Voltage-gated K^+ channels start to open. Depolarization results because the inward diffusion of Na^+ is much greater than the outward diffusion of K^+.

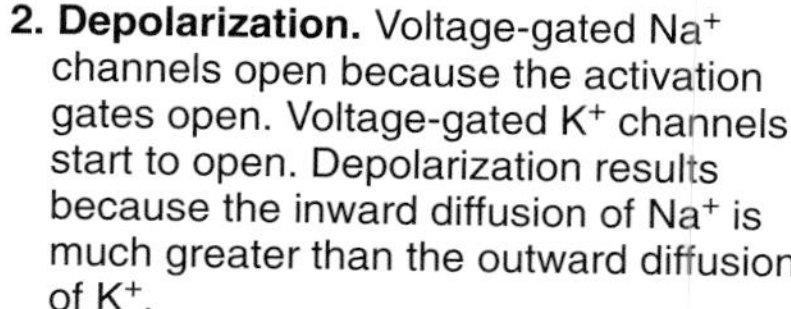

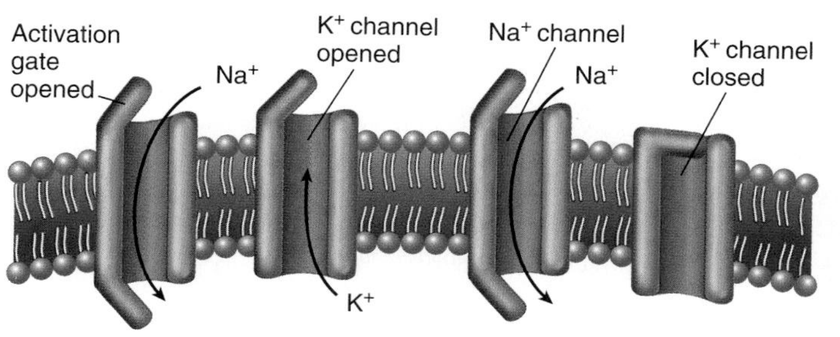

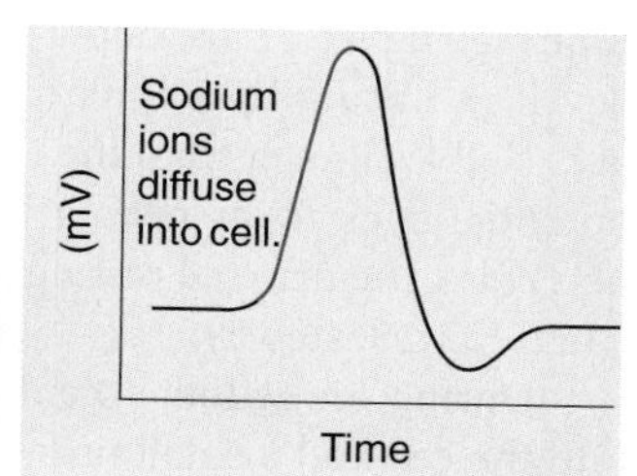

3. Repolarization. Voltage-gated Na^+ channels are closed because the inactivation gates close. Voltage-gated K^+ channels are now open. Sodium ions diffusion into the cell stops and K^+ diffuse out of the cell, causing repolarization.

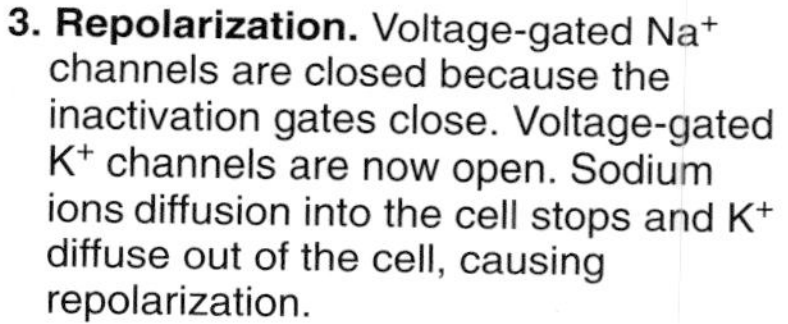

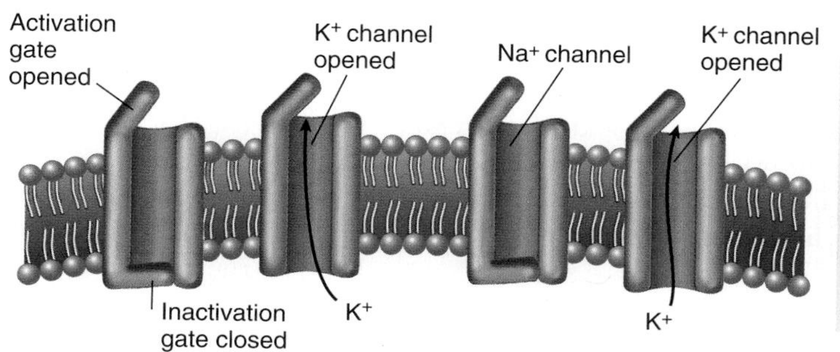

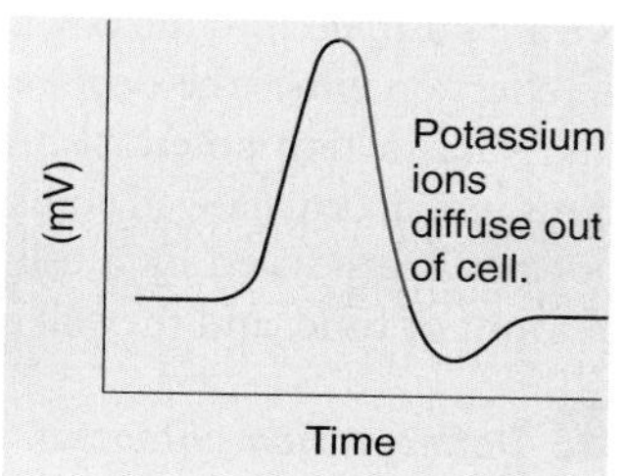

4. End of repolarization and afterpotential. Voltage-gated Na^+ channels are closed. Closure of the activation gates and opening of the inactivation gates reestablish the resting condition for Na^+ channels (see step 1). Diffusion of K^+ through voltage-gated channels produces the afterpotential.

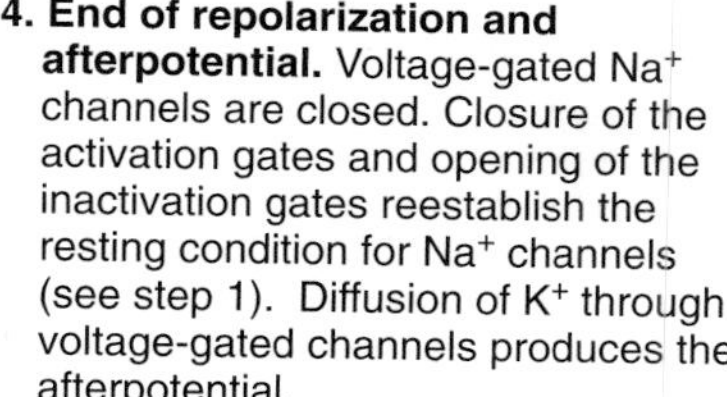

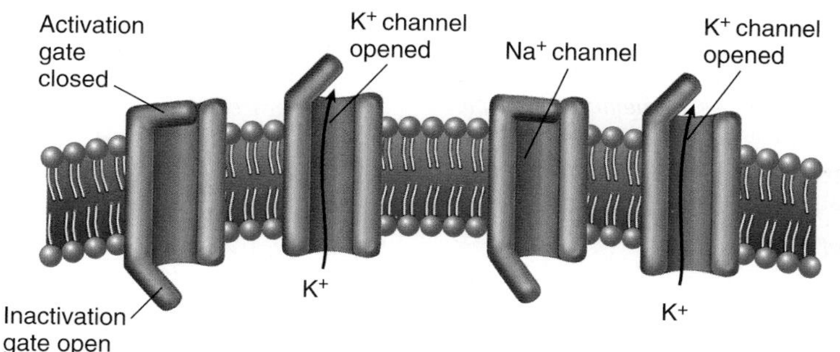

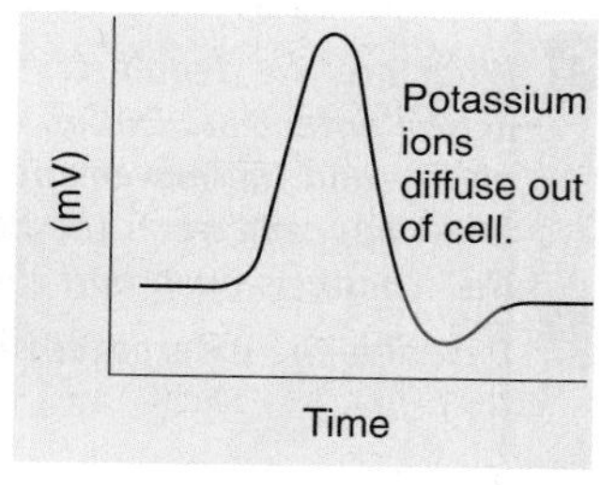

5. Resting membrane potential. The resting membrane potential is reestablished after the voltage-gated K^+ channels close.

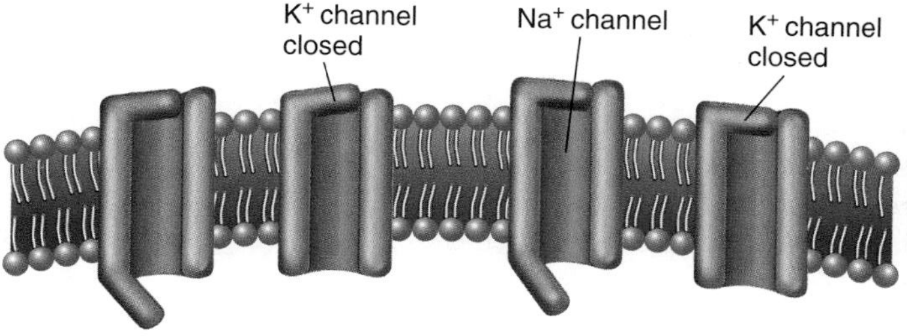

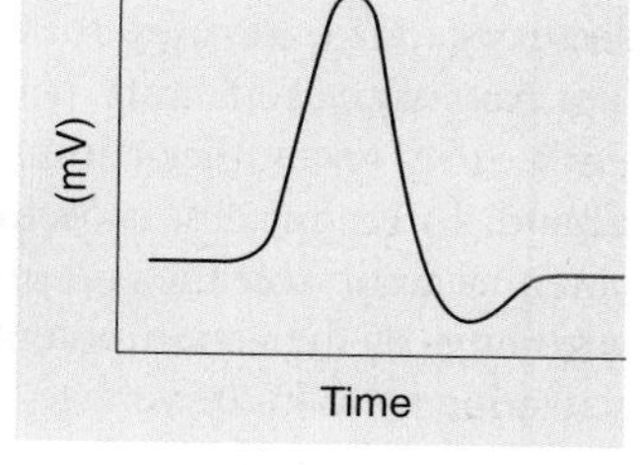

PROCESS FIGURE 11.18 Voltage-Gated Ion Channels and the Action Potential

Step 1 illustrates the status of voltage-gated Na^+ and K^+ channels in a resting cell. Steps 2–5 show how the channels open and close to produce an action potential. Next to each step, a graph shows in *red* the membrane potential resulting from the condition of the ion channels.

At the end of repolarization, the decrease in membrane potential causes the activation gates in the voltage-gated Na^+ channels to close and the inactivation gates to open. Although this change does not affect the diffusion of Na^+, it does return the voltage-gated Na^+ channels to their resting state (figure 11.18, step 4).

Afterpotential

In many cells, a period of hyperpolarization, or afterpotential, exists following each action potential. The afterpotential exists because the voltage-gated K^+ channels remain open for a short time (see figure 11.18, step 4). The increased K^+ permeability that develops during the repolarization phase of the action potential lasts slightly longer than the time required to bring the membrane potential back to its resting level. As the voltage-gated K^+ channels close, the original resting membrane potential is reestablished (figure 11.18, step 5).

During an action potential, a small number of Na^+ diffuse into the cell and a small number of K^+ diffuse out of the cell. The Na^+–K^+ pump restores normal resting ion concentrations by transporting these ions in the opposite direction of their movement during the action potential. That is, Na^+ are pumped out of the cell and K^+ are pumped into the cell. The Na^+–K^+ pump is too slow to have an effect on either the depolarization or the repolarization phase of individual action potentials. As long as the Na^+ and K^+ concentrations remain unchanged across the plasma membrane, all the action potentials produced by a cell are identical. They all take the same amount of time, and they all exhibit the same magnitude.

29 ***Define action potential. How do depolarizing and hyperpolarizing graded potentials affect the likelihood of generating an action potential?***

30 ***Explain the "all" and the "none" parts of the all-or-none principle of action potentials.***

31 ***What are the depolarization and repolarization phases of an action potential? Explain how changes in membrane permeability and the movement of Na^+ and K^+ cause each phase. What happens when the activation gates in the voltage-gated Na^+ channels open and the inactivation gates close?***

32 ***Describe the afterpotential and its cause.***

Refractory Period

Once an action potential is produced at a given point on the plasma membrane, the sensitivity of that area to further stimulation decreases for a time called the **refractory** (rē-frak′tōr-ē) **period.** The first part of the refractory period, during which complete insensitivity exists to another stimulus, is called the **absolute refractory period.** In many cells, it occurs from the beginning of the action potential until near the end of repolarization (figure 11.19). At the beginning of the action potential, depolarization occurs when the activation gates in the voltage-gated Na^+ channel open. At this time, the inactivation gates in the voltage-gated Na^+ channels are already open (see figure 11.18, step 2). Depolarization ends as the inactivation gates close (see figure 11.18, step 3). As long as the inactivation gates are closed, further depolarization cannot occur. When the inactivation gates open and the activation gates close near the end of repolarization (see figure 11.18, step 4), once again it is possible to stimulate the production of another action potential.

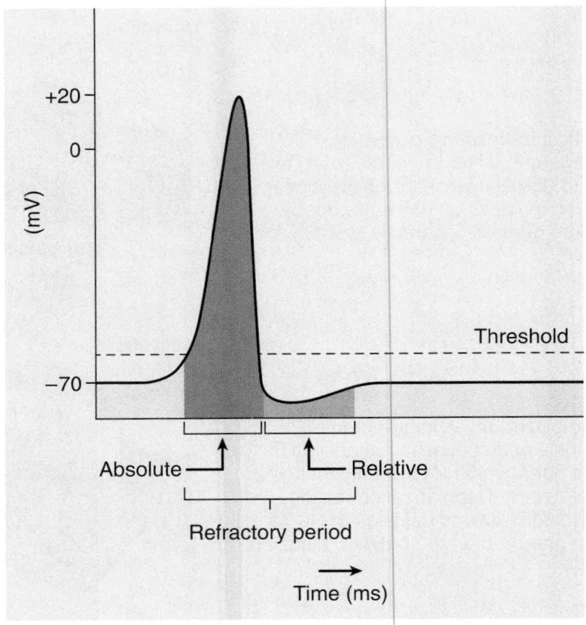

FIGURE 11.19 Refractory Period

The absolute and relative refractory periods of an action potential. In some cells, the absolute refractory period ends during the repolarization phase of the action potential.

The existence of the absolute refractory period guarantees that, once an action potential is begun, both the depolarization and the repolarization phases will be completed, or nearly completed, before another action potential can begin and that a strong stimulus cannot lead to prolonged depolarization of the plasma membrane. The absolute refractory period has important consequences for the rate at which action potentials can be generated and for the propagation of action potentials.

The second part of the refractory period, called the **relative refractory period,** follows the absolute refractory period. A stronger-than-threshold stimulus can initiate another action potential during the relative refractory period. Thus, after the absolute refractory period, but before the relative refractory period is completed, a sufficiently strong stimulus can produce another action potential. During the relative refractory period, the membrane is more permeable to K^+ because many voltage-gated K^+ channels are open (see figure 11.18, step 4). The relative refractory period ends when the voltage-gated K^+ channels close (see figure 11.18, step 5).

33 ***What are the absolute and relative refractory periods? Relate them to the depolarization and repolarization phases of the action potential.***

PREDICT 7

Does a prolonged threshold stimulus or a prolonged stronger-than-threshold stimulus of the same duration produce the most action potentials? Explain.

Action Potential Frequency

The **action potential frequency** is the number of action potentials produced per unit of time in response to a stimulus. Action potential frequency is directly proportional to stimulus strength and to the size of the graded potential. A **subthreshold stimulus** is any stimulus not strong enough to produce a graded potential that reaches threshold. Therefore, no action potential is produced (figure 11.20). A **threshold stimulus** produces a graded potential that is just strong enough to reach threshold and cause the production of a single action potential. A **maximal stimulus** is just strong enough to produce a maximum frequency of action potentials. A **submaximal stimulus** includes all stimuli between threshold and the maximal stimulus strength. For submaximal stimuli, the action potential frequency increases in proportion to the strength of the stimulus because the size of the graded potential increases with stimulus strength. A **supramaximal stimulus** is any stimulus stronger than a maximal stimulus. These stimuli cannot produce a greater frequency of action potentials than a maximal stimulus.

The duration of the absolute refractory period determines the maximum frequency of action potentials generated in an excitable cell. During the absolute refractory period, a second stimulus, no matter how strong, cannot stimulate an additional action potential. As soon as the absolute refractory period ends, however, it is possible for a second stimulus to cause the production of an action potential.

PREDICT 8

If the duration of the absolute refractory period of a nerve cell is 1 millisecond (ms), how many action potentials are generated by a maximal stimulus in 1 second?

The frequency of action potentials provides information about the strength of a stimulus. For example, a weak pain stimulus generates a low frequency of action potentials, whereas a stronger pain stimulus generates a higher frequency of action potentials. The ability to interpret a stimulus as mildly painful versus very painful depends, in part, on the frequency of action potentials generated by individual pain receptors. Communication regarding the strength of stimuli cannot depend on the magnitudes of action potentials because, according to the all-or-none principle, the magnitudes are always the same. The magnitudes of action potentials produced by weak and strong pain stimuli are the same.

The ability to stimulate muscle or gland cells also depends on action potential frequency. A low frequency of action potentials produces a weaker muscle contraction or less secretion than does a higher frequency. For example, a low frequency of action potentials in a muscle results in incomplete tetanus and a high frequency in complete tetanus (see chapter 9).

In addition to the frequency of action potentials, how long the action potentials are produced provides important information.

For example, a pain stimulus of 1 second is interpreted differently than the same stimulus applied for 30 seconds.

34 ***Define action potential frequency. What two factors determine action potential frequency?***

35 ***Define subthreshold, threshold, maximal, submaximal, and supramaximal stimuli. What determines the maximum frequency of action potential generation?***

Propagation of Action Potentials

An action potential occurs in a very small area of the plasma membrane and does not affect the entire membrane at one time. Action

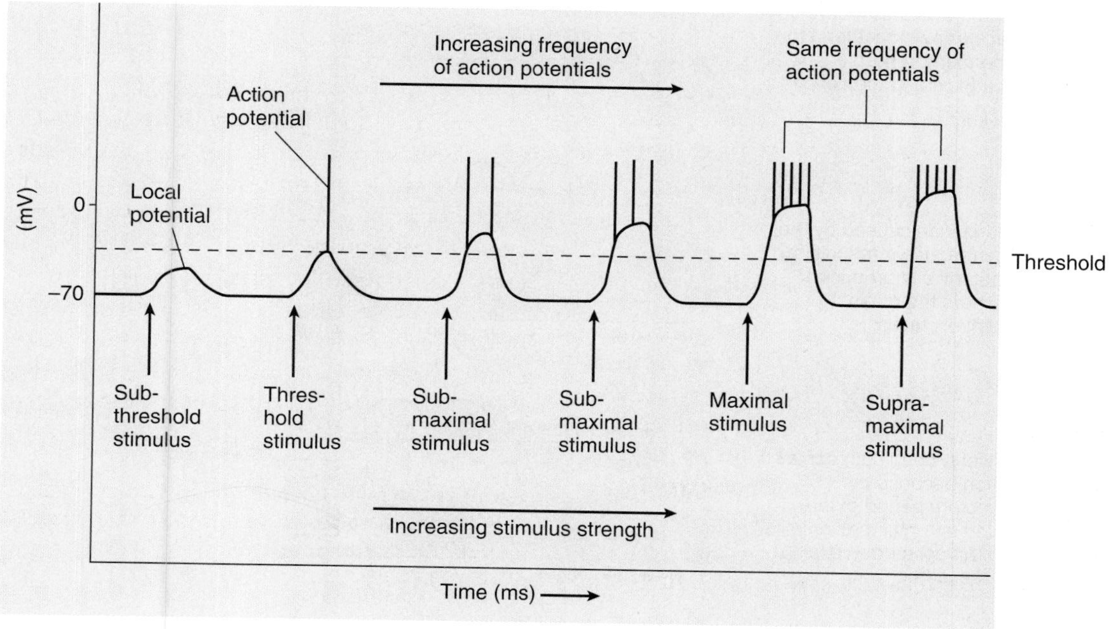

FIGURE 11.20 Stimulus Strength and Action Potential Frequency

From left to right, each stimulus in the figure is stronger than the previous one. As stimulus strength increases, the frequency of action potentials increases until a maximal rate is produced. Thereafter, increasing stimulus strength does not increase action potential frequency.

potentials can, however, **propagate,** or spread, across the plasma membrane because an action potential produced at one location in the plasma membrane can stimulate the production of an action potential at an adjacent area of the plasma membrane. Note that an action potential does not actually move along an axon. Rather, an action potential at one location stimulates the production of another action potential at an adjacent location, which in turn stimulates the production of another, and so on, like a long row of toppling dominos in which each domino knocks down the next one. Each domino falls, but no one domino actually travels the length of the row.

In a neuron, action potentials are normally produced at the trigger zone and propagate in one direction along the axon (figure 11.21, step 1). The location at which the next action potential is generated is different for unmyelinated and myelinated axons (see figure 11.11). In an unmyelinated axon, the next action potential is generated immediately adjacent to the previous action potential. When an action potential is produced, the inside of the membrane becomes more positive than the outside (figure 11.21, step 2). On the outside of the membrane, positively charged ions from the adjacent area are attracted to the negative charges at the site of the action potential. On the inside of the plasma membrane, positively charged ions at the site of the action potential are attracted to the adjacent negatively charged part of the membrane. The movement of positively charged ions is called an **ionic current,** or **local current.** As a result of the ionic current, the part of the membrane immediately adjacent to the action potential depolarizes. That is, the outside of the membrane immediately adjacent to the action potential becomes more negative because of the loss of positive charges and the inside becomes more positive because of the gain of positive charges. When the depolarization reaches threshold, an action potential is produced (figure 11.21, step 3).

If an action potential is initiated at one end of an axon, it is propagated in one direction down the axon. The absolute refractory period ensures one-way propagation of an action

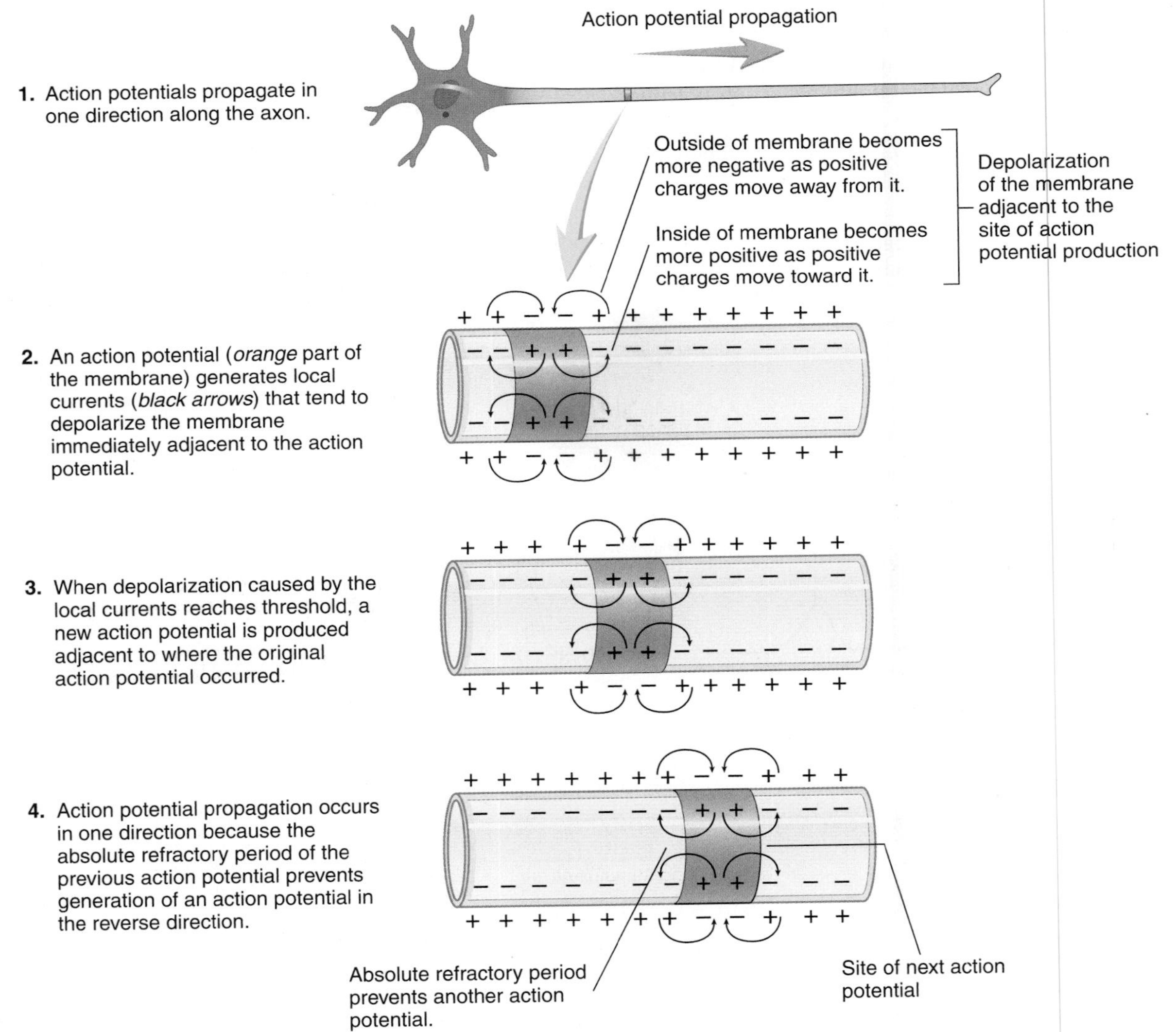

PROCESS FIGURE 11.21 Action Potential Propagation in an Unmyelinated Axon

CLINICAL FOCUS

Examples of Abnormal Membrane Potentials

Several conditions are examples of the physiology of membrane potentials and the consequence of abnormal ones. **Hypokalemia** (hī-pō-ka-lē′mē-ă) is a lower than normal concentration of K^+ in the blood or extracellular fluid. Reduced extracellular K^+ concentrations cause hyperpolarization of the resting membrane potential (see figure 11.15*b*). Thus, a greater than normal stimulus is required to depolarize the membrane to its threshold level and to initiate action potentials in neurons, skeletal muscle, and cardiac muscle. Symptoms of hypokalemia include muscular weakness, an abnormal electrocardiogram, and sluggish reflexes. The symptoms result from the reduced sensitivity of the excitable tissues to stimulation. The causes of hypokalemia include potassium depletion during starvation, alkalosis, and certain kidney diseases.

Hypocalcemia (hī-pō-kal-sē′mē-ă) is a lower than normal concentration of Ca^{2+} in blood or extracellular fluid. Symptoms of hypocalcemia include nervousness and uncontrolled contraction of skeletal muscles, called **tetany** (tet′ă-nē). The symptoms are due to an increased membrane permeability to Na^+ that results because low blood levels of Ca^{2+} cause voltage-gated Na^+ channels in the membrane to open. Sodium ions diffuse into the cell, causing depolarization of the plasma membrane to threshold and initiating action potentials. The tendency for action potentials to occur spontaneously in nervous tissue and muscles accounts for the symptoms. A lack of dietary calcium, a lack of vitamin D, and a reduced secretion rate of a parathyroid gland hormone are conditions that cause hypocalcemia.

potential because it prevents the ionic current from stimulating the production of an action potential in the reverse direction (figure 11.21, step 4).

In a myelinated axon, an action potential is conducted from one node of Ranvier to another in a process called **saltatory conduction** (*saltare* is Latin, meaning to leap). An action potential at one node of Ranvier generates ionic currents that flow toward the next node of Ranvier (figure 11.22, step 1). The lipids within the membranes of the myelin sheath act as a layer of insulation, forcing the ionic currents to flow from one node of Ranvier to the next. In addition, voltage-gated Na^+ channels are highly concentrated at the nodes of Ranvier. Therefore, the ionic current quickly flows

1. An action potential (*orange*) at a node of Ranvier generates local currents (*black arrows*). The local currents flow to the next node of Ranvier because the myelin sheath of the Schwann cell insulates the axon of the internode.

2. When the depolarization caused by the local currents reaches threshold at the next node of Ranvier, a new action potential is produced (*orange*).

3. Action potential propagation is rapid in myelinated axons because the action potentials are produced at successive nodes of Ranvier (*1–5*) instead of at every part of the membrane along the axon.

PROCESS FIGURE 11.22 Saltatory Conduction: Action Potential Propagation in a Myelinated Axon
The gaps between the Schwann cells are exaggerated for clarity.

to a node and stimulates the voltage-gated Na^+ channels to open, resulting in the production of an action potential (figure 11.22, step 2).

The speed of action potential conduction along an axon depends on the myelination of the axon. Action potentials are conducted more rapidly in myelinated than unmyelinated axons because they are formed quickly at each successive node of Ranvier (figure 11.22, step 3), instead of being propagated more slowly through every part of the axon's membrane, as in unmyelinated axons (see figure 11.21). Action potential conduction in a myelinated fiber is like a grasshopper jumping; in an unmyelinated axon, it is like a grasshopper walking. The grasshopper (action potential) moves more rapidly by jumping. The generation of action potentials at nodes of Ranvier occurs so rapidly that as many as 30 successive nodes of Ranvier are simultaneously in some phase of an action potential.

The speed of action potential conduction is also affected by the thickness of the myelin sheath, which is determined by how many times oligodendrocytes or Schwann cells wrap around the axon. Heavily myelinated axons have a thicker myelin sheath and conduct action potentials more rapidly than lightly myelinated axons.

In addition to myelination, the diameter of axons affects the speed of action potential conduction. Large-diameter axons conduct action potentials more rapidly than small-diameter axons because large-diameter axons have a greater surface area. Consequently, at a given site on an axon, more voltage-gated Na^+ channels open during depolarization, resulting in a greater ionic current flow, which more rapidly stimulates adjacent membrane areas.

Nerve fibers (axons) are classified according to their size and myelination. It is not surprising that the structure of nerve fibers reflects their functions. Type A fibers are large-diameter, myelinated axons that conduct action potentials at 15–120 m/s. Motor neurons supplying skeletal muscles and most sensory neurons have type A fibers. Rapid response to the external environment is possible because of the rapid input of sensory information to the CNS and rapid output of action potentials to skeletal muscles.

Type B fibers are medium-diameter, lightly myelinated axons that conduct action potentials at 3–15 m/s, and type C fibers are small-diameter, unmyelinated axons that conduct action potentials at 2 m/s or less. Type B and C fibers are primarily part of the ANS, which stimulates internal organs, such as the stomach, intestines, and heart. The responses necessary to maintain internal homeostasis, such as digestion, need not be as rapid as responses to the external environment.

36 ***What is an ionic current? How do ionic currents cause the propagation of action potentials in unmyelinated axons?***

37 ***What prevents an action potential from reversing its direction of propagation?***

38 ***Describe saltatory conduction of an action potential.***

39 ***Compare the speed of action potential conduction in (a) heavily myelinated, lightly myelinated, and unmyelinated axons and (b) large-diameter and small-diameter axons.***

40 ***Compare the functions of type A nerve fibers with that of type B and C nerve fibers.***

Importance of Myelin Sheaths

Myelin sheaths begin to form late in fetal development. The process continues rapidly until the end of the first year after birth and continues more slowly thereafter. The development of myelin sheaths is associated with the infant's continuing development of more rapid and better coordinated responses.

The importance of myelinated fibers is dramatically illustrated in diseases in which the myelin sheath is gradually destroyed. Action potential transmission is slowed, resulting in impaired control of skeletal and smooth muscles. In severe cases, complete blockage of action potential transmission can occur. Multiple sclerosis and some cases of diabetes mellitus result in myelin sheath destruction.

THE SYNAPSE

Just as the fire from one lit torch can light another torch, action potentials in one cell can stimulate action potentials in another cell, thereby allowing communication between the cells. For example, if your finger touches a hot pan, the heat is a stimulus that produces action potentials in sensory nerve fibers. The action potentials are propagated along the sensory fibers from the finger toward the CNS. For the CNS to get this information, the action potentials of the sensory neurons must produce action potentials in CNS neurons. After the CNS has received the information, it produces a response. One response is the contraction of the appropriate skeletal muscles, causing the finger to move away from the hot pan. CNS action potentials cause motor neurons to produce action potentials that are then transmitted by the motor neurons toward skeletal muscles. The action potentials of the motor neuron produce skeletal muscle action potentials, which are the stimuli that cause muscle fibers to contract (see chapter 9).

The **synapse** (sin′aps), which is the junction between two cells, is where two cells communicate with each other. The cell that transmits a signal toward a synapse is called the **presynaptic cell,** and the cell that receives the signal is called the **postsynaptic cell.**

The average presynaptic neuron synapses with about 1000 other neurons, but the average postsynaptic neuron has up to 10,000 synapses. Some postsynaptic neurons in the part of the brain called the cerebellum have up to 100,000 synapses. There are two types of synapses: electrical and chemical.

Electrical Synapses

Electrical synapses are gap junctions (see chapter 4) that allow an ionic current to flow between adjacent cells (figure 11.23). At these gap junctions, the membranes of adjacent cells are separated by a 2 nm gap spanned by tubular proteins called **connexons.** The movement of ions through the connexons can generate an ionic current. Thus, an action potential in one cell produces an ionic current that generates an action potential in the adjacent cell almost as if the two cells had the same membrane. As a result, action potentials are conducted rapidly between cells, allowing the cells' activity to be synchronized. Electrical synapses are not common in the nervous system of vertebrates, but there are some in humans.

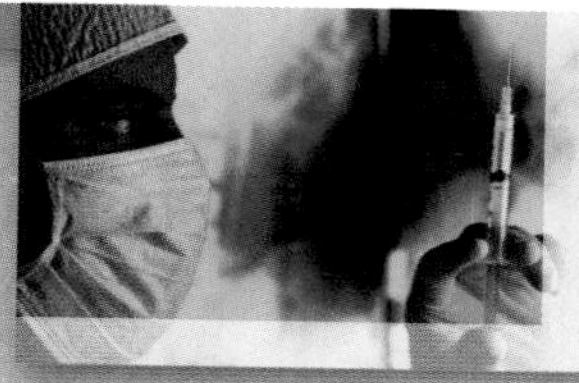

CLINICAL FOCUS

Nervous Tissue Response to Injury

When a nerve is cut, either healing or permanent interruption of the neural pathways occurs. The final outcome depends on the severity of the injury and on its treatment.

Several degenerative changes result when a nerve is cut (figure A). Within about 3–5 days, the axons in the part of the nerve distal to the cut break into irregular segments and degenerate. This occurs because the neuron cell body produces the substances essential to maintain the axon, and these substances have no way of reaching parts of the axon distal to the point of damage. Eventually, the distal part of the axon completely degenerates. As the axons degenerate, the myelin part of the Schwann cells around them also degenerates, and macrophages invade the area to phagocytize the myelin. The Schwann cells then enlarge, undergo mitosis, and finally form a column of cells along the regions once occupied by the axons. The columns of Schwann cells are essential for the growth of new axons. If the ends of the regenerating axons encounter a Schwann cell column, their rate of growth increases, and reinnervation of peripheral structures is likely. If the ends of the axons do not encounter the columns, they fail to reinnervate the peripheral structures.

The end of each regenerating axon forms several axonal sprouts. It normally takes about 2 weeks for the axonal sprouts to enter the Schwann cell columns. Only one of the sprouts from each severed neuron forms an axon, however. The other branches degenerate. After the axons grow through the Schwann cell columns, new myelin sheaths are formed, and the neurons reinnervate the structures they previously supplied.

Treatment strategies that increase the probability of reinnervation include bringing the ends of the severed nerve close together surgically. In some cases in which sections of nerves are destroyed as a result of trauma, nerve transplants are performed to replace damaged segments. The transplanted nerve eventually degenerates, but it does provide Schwann cell columns through which axons can grow.

The regeneration of damaged nerve tracts within the CNS is very limited and is poor in comparison with the regeneration of nerves in the PNS. In part, the difference may result from the oligodendrocytes, which exist only in the CNS. Each oligodendrocyte has several processes, each of which forms part of a myelin sheath. The cell bodies of the oligodendrocytes are a short distance from the axons they ensheathe, and fewer oligodendrocytes than Schwann cells are present. Consequently, when the myelin degenerates following damage, no column of cells remains in the CNS to act as a guide for the growing axons.

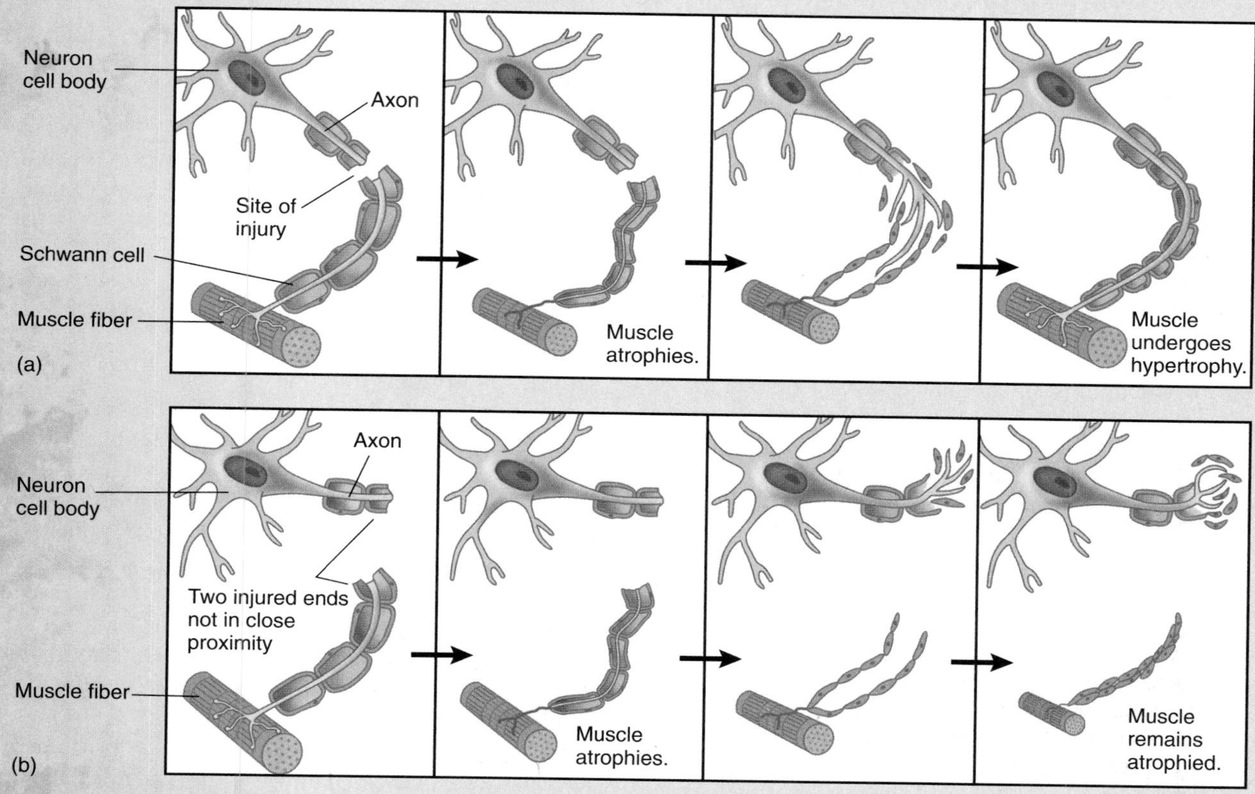

FIGURE A Changes That Occur in an Injured Nerve Fiber

(*a*) When the two ends of an injured nerve fiber are aligned in close proximity, healing and regeneration of the axon are likely to occur. Without stimulation from the nerve, the muscle is paralyzed and atrophies (shrinks in size). After reinnervation, the muscle can become functional and hypertrophy (increase in size). (*b*) When the two ends of an injured nerve fiber are not aligned in close proximity, regeneration is unlikely to occur. Without innervation from the nerve, muscle function is completely lost, and the muscle remains atrophied.

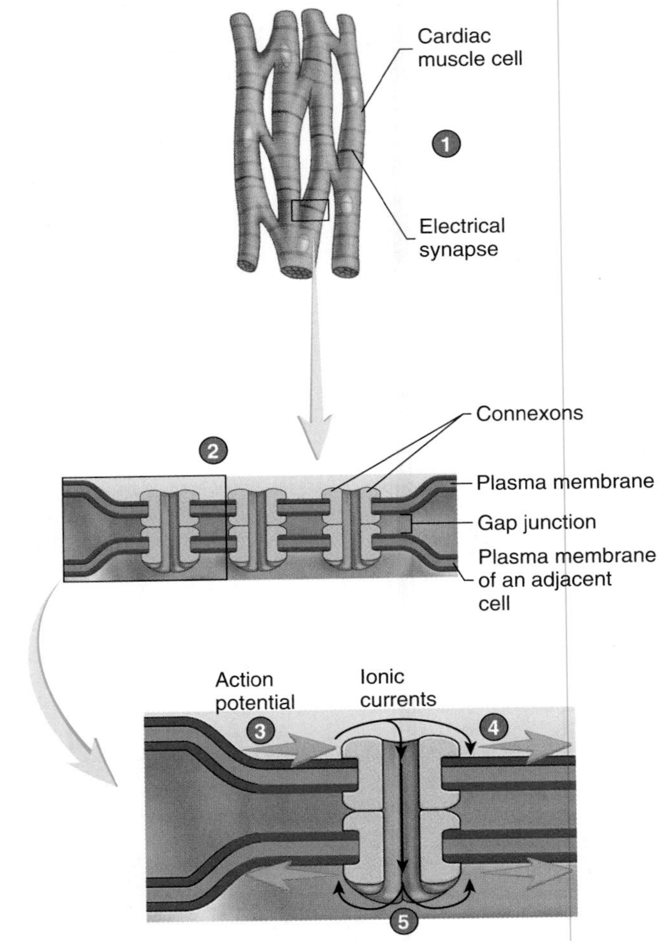

1. Electrical synapses connect cardiac muscle cells.
2. An electrical synapse has gap junctions in which the membranes of two cells are separated by a gap but are connected by proteins called connexons.
3. An action potential (*orange arrow*) in the plasma membrane generates ionic currents (*black arrows*) that flow to adjacent parts of the plasma membrane and through the gap junction.
4. An ionic current stimulates the production of another action potential. Thus, the action potential propagates along the plasma membrane.
5. An ionic current flows through a gap junction and stimulates the production of an action potential in the adjacent cardiac muscle cell. Thus, the action potential propagates to the adjacent cell.
Cardiac muscle cell
Electrical synapse
Connexons
Plasma membrane
Gap junction
Plasma membrane of an adjacent cell
Action potential
Ionic currents

PROCESS FIGURE 11.23 Electrical Synapse

Electrical synapses are important, however, in cardiac muscle and in many types of smooth muscle. Coordinated contractions of these muscle cells occur when action potentials in one cell propagate to adjacent cells because of electrical synapses (see chapters 9 and 20).

41 ***What is an electrical synapse? Describe its operation. In what kinds of tissue are electrical synapses found?***

Chemical Synapses

The essential components of a **chemical synapse** are the presynaptic terminal, the synaptic cleft, and the postsynaptic membrane (figure 11.24). The **presynaptic terminal** is formed from the end of an axon, and the space separating the axon ending and the cell with which it synapses is the **synaptic cleft.** The membrane of the postsynaptic cell opposed to the presynaptic terminal is the **postsynaptic membrane.** Postsynaptic cells are typically other neurons, muscle cells, or gland cells.

Neurotransmitter Release

In chemical synapses, action potentials do not pass directly from the presynaptic terminal to the postsynaptic membrane. Instead, the action potentials in the presynaptic terminal cause the release of neurotransmitters from the terminal.

Presynaptic terminals are specialized to produce and release neurotransmitters. The major cytoplasmic organelles within presynaptic terminals are mitochondria and numerous membrane-bound **synaptic vesicles,** which contain neurotransmitters, such as acetylcholine (see figure 11.24). Each action potential arriving at the presynaptic terminal initiates a series of specific events, which result in the release of neurotransmitters. In response to an action potential, voltage-gated Ca^{2+} channels open, and Ca^{2+} diffuse into the presynaptic terminal. These ions cause synaptic vesicles to fuse with the presynaptic membrane and release their neurotransmitters by exocytosis into the synaptic cleft.

Once neurotransmitters are released from the presynaptic terminal, they diffuse rapidly across the synaptic cleft, which is about 20 nm wide, and bind in a reversible fashion to specific receptors in the postsynaptic membrane (see figure 11.24). Depending on the receptor type, this binding produces a depolarizing or hyperpolarizing graded potential in the postsynaptic membrane. For example, the binding of acetylcholine to ligand-gated Na^+ channels causes them to open, allowing the diffusion of Na^+ into the postsynaptic cell. If the resulting depolarizing graded potential reaches threshold, an action

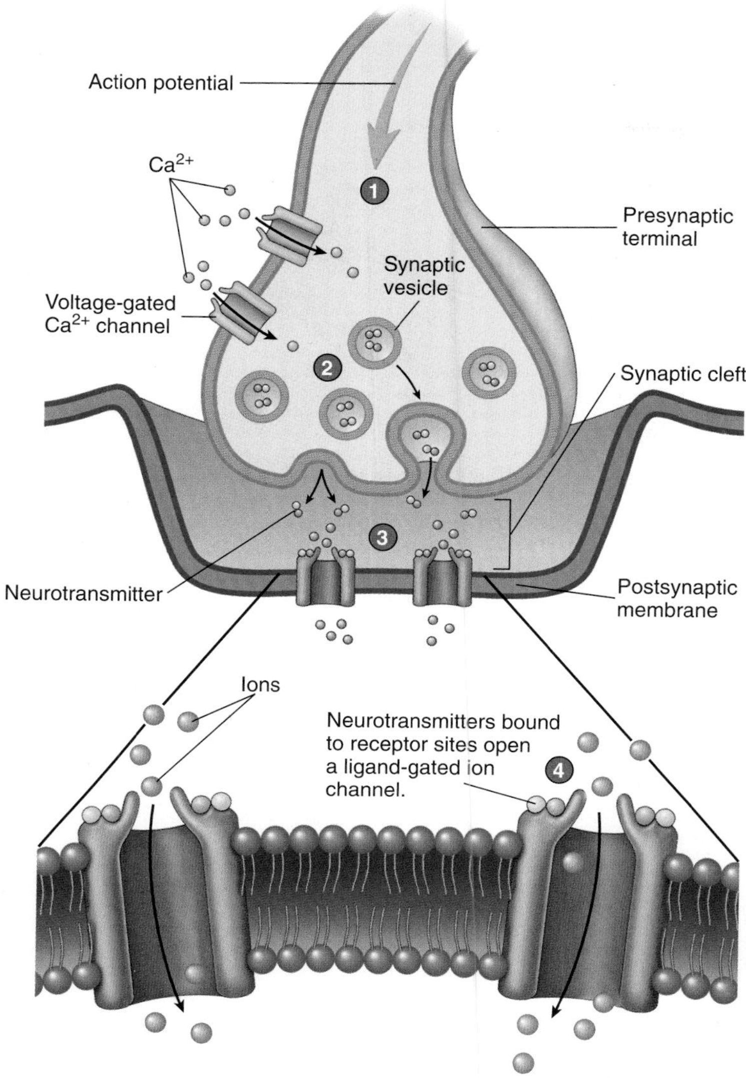

1. Action potentials arriving at the presynaptic terminal cause voltage-gated Ca^{2+} channels to open.
2. Calcium ions diffuse into the cell and cause synaptic vesicles to release neurotransmitters.
3. Neurotransmitters diffuse from the presynaptic terminal across the synaptic cleft.
4. Neurotransmitters combine with their receptor sites and cause ligand-gated ion channels to open. Ions diffuse into the cell (*shown in illustration*) or out of the cell (*not shown*) and cause a change in membrane potential.

PROCESS FIGURE 11.24 Chemical Synapse

A chemical synapse consists of the end of a neuron (presynaptic terminal), a small space (synaptic cleft), and the postsynaptic membrane of another neuron or an effector cell, such as a muscle or gland cell.

potential is produced. On the other hand, the opening of K^+ or Cl^- channels results in a hyperpolarizing graded potential.

42 ***What are the three parts of a chemical synapse?***

43 ***Describe the release of a neurotransmitter in a chemical synapse.***

PREDICT 9

Is an action potential transmitted fastest between cells connected by electrical or chemical synapses? Explain.

Neurotransmitter Removal

The interaction between a neurotransmitter and a receptor represents an equilibrium.

$$\text{Neurotransmitter} + \text{Receptor} \rightleftarrows \text{Neurotransmitter–receptor complex}$$

When the neurotransmitter concentration in the synaptic cleft is high, many of the receptor molecules have neurotransmitter molecules bound to them; when the neurotransmitter concentration declines, the neurotransmitter molecules diffuse away from the receptor molecules.

Neurotransmitters have short-term effects on postsynaptic membranes because the neurotransmitter is rapidly destroyed or removed from the synaptic cleft. For example, in the neuromuscular junction (see chapter 9), the neurotransmitter acetylcholine is broken down by the enzyme **acetylcholinesterase** (as′e-til-kō-lin-es′ter-ās) to acetic acid and choline (figure 11.25*a*). Choline is then transported back into the presynaptic terminal and reacts with acetyl-CoA to form acetylcholine. Acetyl-CoA is synthesized by mitochondria as foods are metabolized to produce ATP (see chapter 25). Acetic acid can be absorbed from the synaptic cleft into the presynaptic terminal, or it can diffuse out of the synaptic cleft and be taken up by a variety of cells. Acetic acid can be used to synthesize acetyl-CoA.

When the neurotransmitter norepinephrine is released into the synaptic cleft, most of it is transported back into the presynaptic terminal, where most of it is repackaged into synaptic vesicles for reuse (figure 11.25*b*). The enzyme **monoamine oxidase (MAO;** mon-ō-am′īn ok′si-dās) inactivates some of the norepinephrine.

Diffusion of neurotransmitter molecules away from the synapse and into the extracellular fluid also limits the length of time the neurotransmitter molecules remain bound to their receptors. Norepinephrine in the circulation is taken up primarily by liver and kidney cells, where the enzymes monoamine oxidase and **catechol-*O*-methyltransferase** (kat′ĕ-kol-ō-meth-il-trans′fer-ās) convert it into inactive metabolites.

44 ***Name three ways to stop the effect of a neurotransmitter on the postsynaptic membrane. Give an example of each way.***

Receptor Molecules in Synapses

Receptor molecules in synapses are membrane-bound, ligand-activated receptors with highly specific receptor sites. Consequently, only neurotransmitter molecules or very closely related substances normally bind to their receptors. For example, acetylcholine binds to acetylcholine receptors, but not to norepinephrine receptors, whereas norepinephrine binds to norepinephrine receptors, but not to acetylcholine receptors. Any given cell does not have all possible receptors. Therefore, a neurotransmitter affects only the cells with receptors for that neurotransmitter.

A neurotransmitter can stimulate some cells but inhibit others. More than one type of receptor molecule exists for some neurotransmitters. Different cells respond differently to a neurotransmitter when

1. Acetylcholine molecules bind to their receptors.
2. Acetylcholine molecules unbind from their receptors.
3. Acetylcholinesterase splits acetylcholine into choline and acetic acid, which prevents acetylcholine from again binding to its receptors. Choline is taken up by the presynaptic terminal.
4. Choline is used to make new acetylcholine molecules that are packaged into synaptic vesicles.

(a)

1. Norepinephrine binds to its receptor.
2. Norepinephrine unbinds from its receptor.
3. Norepinephrine is taken up by the presynaptic terminal, which prevents norepinephrine from again binding to its receptor.
4. Norepinephrine is repackaged into synaptic vesicles or is broken down by monoamine oxidase (MAO).

(b)

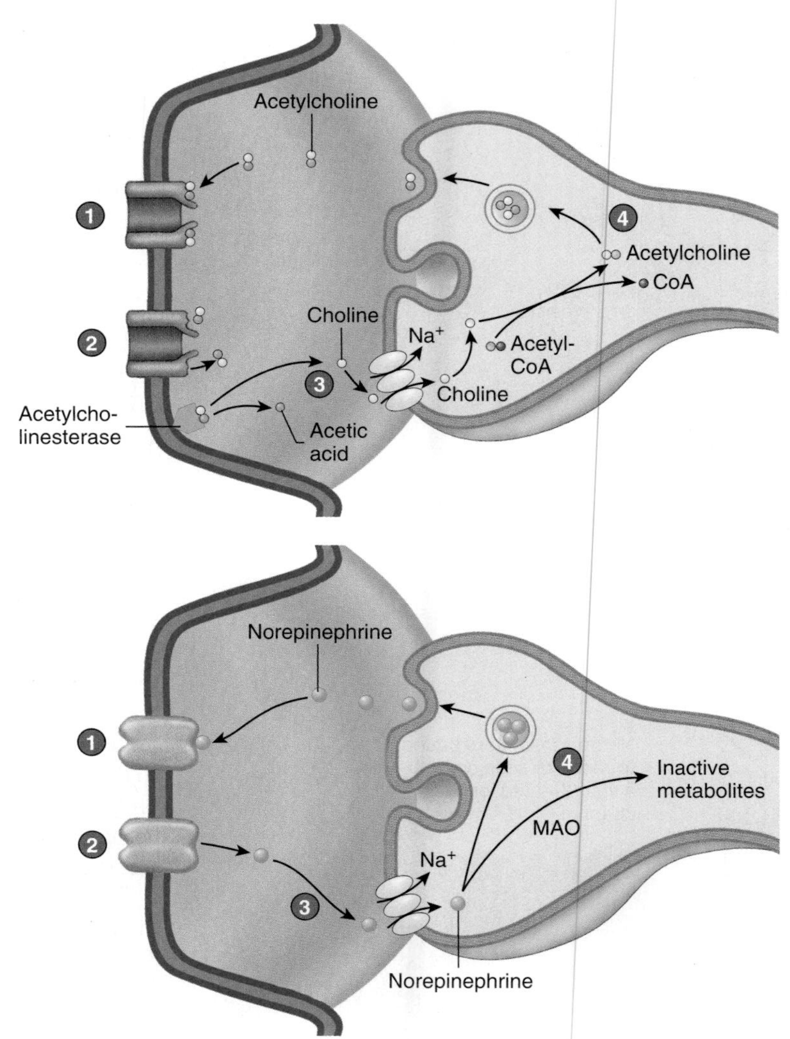

PROCESS FIGURE 11.25 Removal of Neurotransmitter from the Synaptic Cleft

(*a*) In some synapses, neurotransmitters are broken down by enzymes and recycled into the presynaptic terminal. (*b*) In some synapses, neurotransmitters are taken up whole into the presynaptic terminal.

these cells have different receptors. For example, norepinephrine can bind to one type of norepinephrine receptor to cause depolarization in one synapse and to another type of norepinephrine receptor to cause hyperpolarization in another synapse. Thus, norepinephrine is either stimulatory or inhibitory, depending on the type of norepinephrine receptor to which it binds and on the effect that receptor has on the permeability of the postsynaptic membrane.

Although neurotransmitter receptors are in greater concentrations on postsynaptic membranes, some receptors exist on presynaptic membranes. For example, norepinephrine released from the presynaptic membrane binds to receptors on both the presynaptic and postsynaptic membranes. Its binding to the receptors of the presynaptic membrane decreases the release of additional synaptic vesicles. Norepinephrine can therefore modify its own release by binding to presynaptic receptors. A high frequency of presynaptic action potentials results in the release of fewer synaptic vesicles in response to later action potentials.

Neurotransmitters and Neuromodulators

Several substances have been identified as neurotransmitters, and others are suspected neurotransmitters. It was once thought that each neuron contains only one type of neurotransmitter; however, it is now known that some neurons can secrete more than one type. If a neuron does produce more than one neurotransmitter, it secretes all of them from each of its presynaptic terminals. The physiologic significance of presynaptic terminals that secrete more than one type of neurotransmitter has not been clearly established.

Neuromodulators are substances released from neurons that can presynaptically or postsynaptically influence the likelihood that an action potential in the presynaptic terminal will result in the production of an action potential in the postsynaptic cell. For example, a neuromodulator that decreases the release of an excitatory neurotransmitter from a presynaptic terminal decreases the likelihood of action potential production in the postsynaptic cell. A list of neurotransmitters and neuromodulators is presented in table 11.5.

TABLE 11.5 Clinical Examples of Synaptic Function

<table>
<tr><th>Neurotransmitter/Neuromodulator</th><th>Clinical Examples</th></tr>
<tr><td>

Acetylcholine

Structure:

Site of release: CNS synapses, ANS synapses, and neuromuscular junctions

Effect: excitatory in the CNS and neuromuscular junctions; inhibitory or excitatory in ANS synapses

</td><td>

Myasthenia Gravis

Myasthenia gravis is a disease in which the ability of skeletal muscle to respond to nervous system stimulation decreases, resulting in muscle weakness and even paralysis. Antibodies are proteins produced by the immune system that can attach to foreign substances, such as bacteria (see chapter 22). In myasthenia gravis, antibodies inappropriately attach to acetylcholine receptors. The antibodies link the receptors together, which cause them to be removed from the plasma membrane faster than normal, decreasing the number of receptors. The antibodies also stimulate immune responses that lead to destruction of the postsynaptic membrane, which decreases the number of Na^+ channels in the synapse. Thus, the ability of ACh to stimulate action potential production decreases because there are fewer ligand-gated ACh receptors, and the ability to generate an action potential is reduced because there are fewer voltage-gated Na^+ channels.

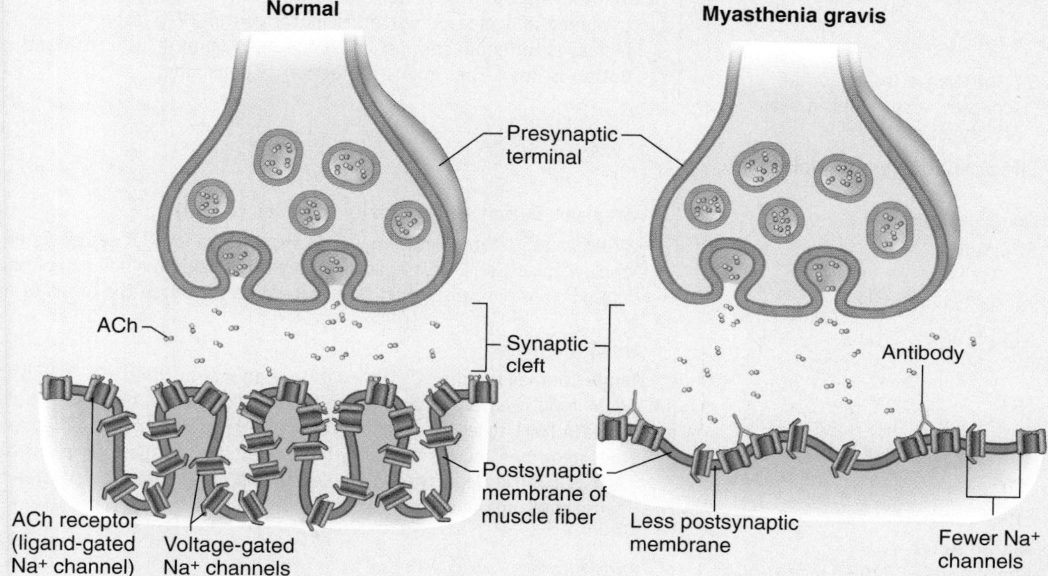

</td></tr>
<tr><td>

Biogenic Amines

Serotonin

Structure:

Site of release: CNS synapses

Effect: generally inhibitory

</td><td>

Antidepressant Therapy

Selective serotonin reuptake inhibitors (SSRIs), such as Prozac and Zoloft, are drugs commonly used to treat depression. They temporarily block serotonin transporters (symporters), which decreases serotonin transport back into presynaptic terminals, resulting in increased serotonin levels in synaptic clefts. In some people, the increased stimulation of the postsynaptic neuron by serotonin relieves depression.

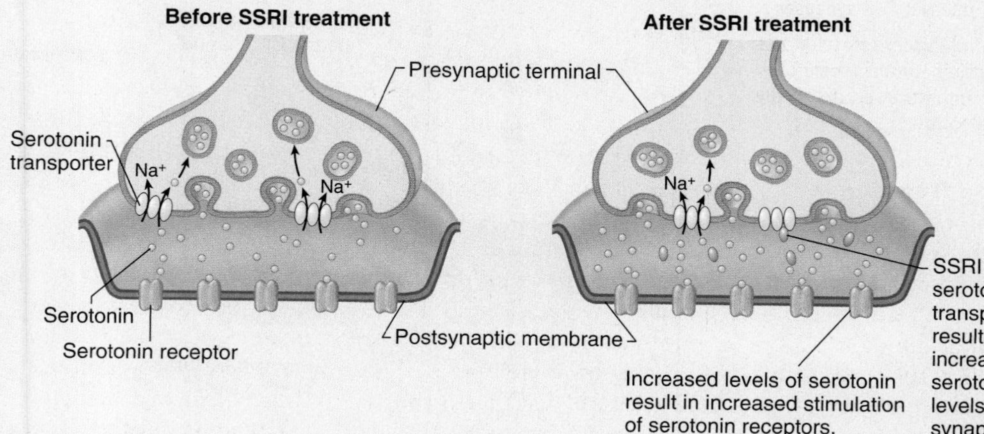

</td></tr>
</table>

TABLE 11.5 Clinical Examples of Synaptic Function—Continued

Neurotransmitter/Neuromodulator	Clinical Examples
Serotonin (continued)	**Anxiety Disorders** SSRIs are also used to treat panic disorders, such as obsessive-compulsive disorder (OCD), a fact that makes researchers believe that OCD might be linked to abnormalities in serotonin function. **Hallucinogens** The hallucinogenic drug D-lysergic acid diethylamide (LSD) blocks serotonin transporters in specific areas of the brain and produces hallucinogenic effects. Other drugs, such as ecstasy, are also hallucinogens that block serotonin transporters.
Dopamine Structure: Site of release: selected CNS synapses; also found in some ANS synapses Effect: excitatory or inhibitory	**Drug Addiction** Cocaine blocks dopamine transporters (symporters), which increases dopamine levels in synaptic clefts, resulting in overstimulation of postsynaptic neurons. Although moderate levels of dopamine can cause euphoria, high levels of dopamine can produce psychotic effects. **Parkinson Disease** Parkinson disease results from the destruction of dopamine-producing neurons, and it is characterized by tremors and decreased voluntary motor control. Parkinson disease is treated with the drug L-Dopa, which increases the production of dopamine in the presynaptic terminals of remaining neurons. Another treatment option is drugs that mimic the action of dopamine.
Norepinephrine Structure: Site of release: selected CNS synapses and some ANS synapses Effect: excitatory	**Attention-Deficit Hyperactivity Disorder (ADHD)** ADHD is often treated with drugs that increase the level of excitatory neurotransmitters, such as norepinephrine, in the synaptic clefts. This is achieved by using selective norepinephrine reuptake inhibitors (SNRIs) that block norepinephrine transporters (symporters) and increase the levels of norepinephrine in the synaptic clefts. **Amphetamines** Amphetamines are drugs with excitatory effects on the CNS. They increase the levels of norepinephrine and dopamine in synaptic clefts by either blocking the reuptake of these neurotransmitters or promoting their release from synaptic vesicles. The CNS effects of amphetamines include decreased appetite, increased alertness, and increased ability to concentrate and perform physical tasks. Amphetamines are used to treat ADHD, clinical depression, narcolepsy, and chronic fatigue syndrome. An overdose of amphetamines can cause insomnia, tremors, anxiety, and panic.
Amino Acids *Gamma-Amino Butyric Acid (GABA)* Structure: Site of release: CNS synapses Effect: inhibitory effect on postsynaptic nerurons; some presynaptic inhibition occurs in the spinal cord	**Barbiturates** Certain GABA receptors are ligand-gated channels that permit the inflow of Cl^- when stimulated. GABA produces an inhibitory (or a hyperpolarizing) effect by binding to these receptors and promoting Cl^- inflow. Barbiturates enhance the binding of GABA to their receptors, resulting in the prolonged inhibition of postsynaptic neurons. These drugs are used as sedatives and anesthetics and as a treatment for epilepsy, which is characterized by excessive neuronal discharge.

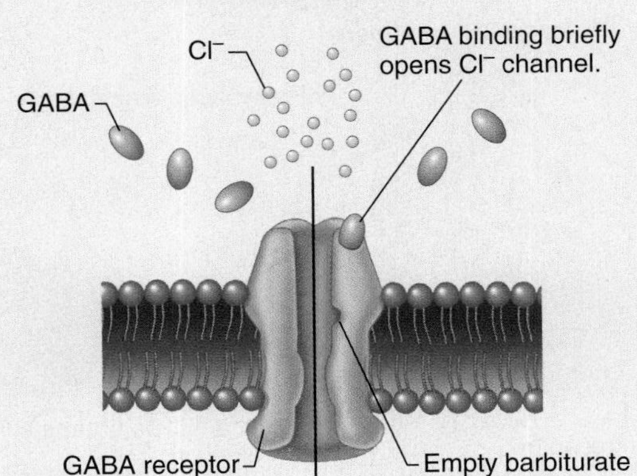

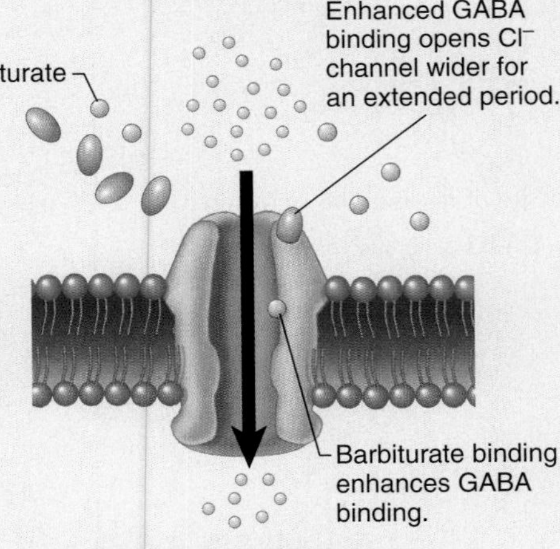

TABLE 11.5 Clinical Examples of Synaptic Function—Continued

Neurotransmitter/Neuromodulator	Clinical Examples
GABA (*continued*)	**Benzodiazepines** Benzodiazepines that are used in anti-anxiety drugs also have binding sites on certain GABA receptors. Their action is similar to that of barbiturates in that they enhance the binding of GABA to its receptor, producing an inhibitory effect. **Alcohol Dependence** Alcohol acts in ways similar to barbiturates to enhance the effect of GABA. As a result, the ligand-gated Cl^- channel becomes more permeable to Cl^-, producing an inhibitory effect. Chronic consumption of alcohol renders the GABA receptor less sensitive to both alcohol and GABA, resulting in increased alcohol dependence and alcohol withdrawal symptoms, such as anxiety, tremors, and insomnia. Alcohol withdrawal symptoms are often treated with benzodiazepines.
Glycine Structure: H_2N-CH_2-COOH Site of release: CNS synapses Effect: inhibitory	**Strychnine Poisoning** Glycine receptors are similar to GABA receptors in that they act as ligand-gated channels permitting the inflow of Cl^-. The poison strychnine blocks glycine receptors, which increases the excitability of certain neurons by preventing their inhibition. Strychnine poisoning results in powerful muscle contractions and convulsions. Tetanus of respiratory muscles can cause death.
Glutamate Structure: $H_2N-CH(COOH)-CH_2-CH_2-COOH$ Site of release: CNS synapses; in areas of the brain that are involved in learning and memory Effect: excitatory	**Stroke and Excitotoxicity** Glutamate is the major excitatory neurotransmitter of the CNS. Glutamate receptors are ligand-gated Ca^{2+} channels. When stimulated, Ca^{2+} channels open, causing depolarization of postsynaptic membranes. Some glutamate is removed from the synapse by transporters in presynaptic terminals, whereas the bulk of it is removed by transporters (symporters) in neighboring astrocytes. When a person suffers a stroke, brain tissue is deprived of oxygen and ATP levels decrease. This causes the secondary active tranport of glutamate by the glutamate transporters to fail temporarily. As a result, glutamate accumulates in the synaptic clefts and causes excessive stimulation of postsynaptic neurons. Excessive movement of Ca^{2+} into neurons activates a variety of destructive processes, which can cause cell death. 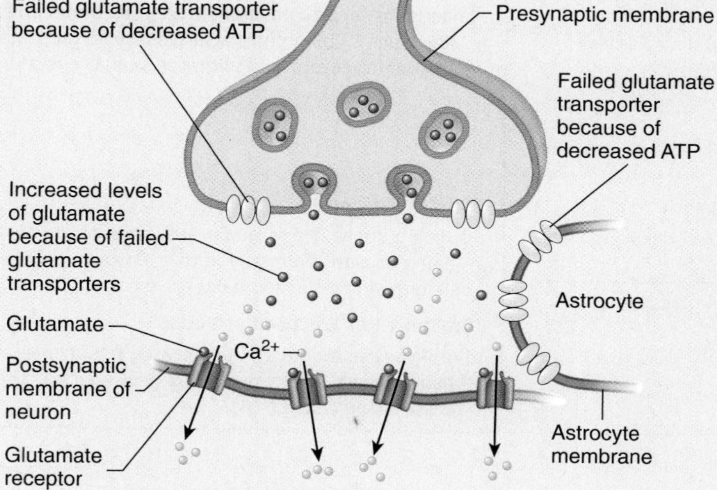 **Cognition** Glutamate is implicated in learning and memory. Drugs that target specific glutamate receptors are often used in the treatment of Alzheimer disease.

TABLE 11.5 Clinical Examples of Synaptic Function—Continued

Neurotransmitter/Neuromodulator	Clinical Examples
Purines *Adenosine* Structure: 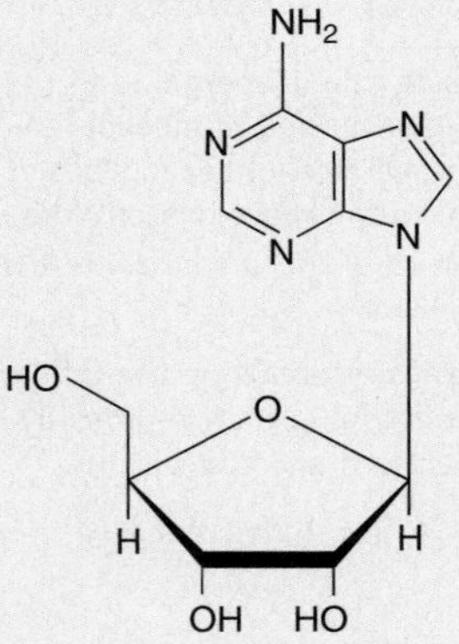 Site of release: CNS synapses; in the areas of the brain that are involved in learning and memory Effect: inhibitory	**Neuroprotective Agent** Adenosine acts both as a neurotransmitter and a neuromodulator. Adenosine receptors are linked to G-proteins (see figure 3.11). As a neurotransmitter, adenosine stimulates the opening of Cl^- and K^+ channels on postsynaptic membranes, thereby producing a hyperpolarizing effect. Acting as a neuromodulator, adenosine stimulates the closing of Ca^{2+} channels on presynaptic neurons, causing the inhibition of neurotransmitter release. Adenosine production greatly increase during a stroke. It prevents the release of glutamate from presynaptic vesicles, which reduces the level of glutamate in synaptic clefts. It also hyperpolarizes the postsynaptic membranes of glutamate synapses, thereby countering the excitatory effects of glutamate. As a result, the damaging effects of glutamate during a stroke are diminished. The possibility of using adenosine as an antistroke agent is being investigated. **Caffeine** Adenosine produces drowsiness, which is countered by caffeine, which blocks adenosine receptors and promotes alertness. Caffeine also promotes cognition by blocking adenosine's inhibitory effect on glutamate function.
Neuropeptides *Substance P* Structure: polypeptide (10 amino acids) Site of release: descending pain pathways Effect: excitatory	**Pain Therapy** Substance P acts as a neurotransmitter and neuromodulator. The receptor for substance P is called a neurokinin receptor, which is linked to a G-protein complex (see figure 3.11). Drugs such as morphine reduce pain by blocking the release of substance P.
Endorphins Structure: Polypeptide (30 amino acids) Site of release: descending pain pathways Effect: inhibitory	**Opiates** Endorphins bind to endorphin receptors on presynaptic neurons and reduce pain by blocking the release of substance P. Endorphins also produce feelings of euphoria. Opiates such as morphine and heroin also bind to endorphin receptors, resulting in similar effects.
Gases *Nitric Oxide (NO)* Structure: N=O Site of release: CNS, nerves supplying the adrenal gland, penis Effect: excitatory	**Stroke Damage** During a stroke, rising glutamate levels act on postsynaptic neurons and cause the release of NO, which in high concentrations can be toxic to cells. Nitric oxide also diffuses out of postsynaptic neurons, enters neighboring cells, and damages them. **Treatment of Erectile Dysfunction** During sexual arousal, NO is released from nerves, causing the vasodilation of the blood vessels supplying the penis. Viagra, which is used to treat erectile dysfunction, acts by prolonging the effect of NO on these blood vessels (see chapter 28).

45 ***Why does a given type of neurotransmitter affect only certain types of cells? How can a neurotransmitter stimulate one type of cell but inhibit another type?***

46 ***What is a neuromodulator?***

Excitatory and Inhibitory Postsynaptic Potentials

The combination of neurotransmitters with their specific receptors causes either depolarization or hyperpolarization of the postsynaptic membrane. When depolarization occurs, the response is stimulatory, and the graded potential is called an **excitatory postsynaptic potential** (**EPSP;** figure 11.26*a*). EPSPs are important because the depolarization might reach threshold, thereby producing an action potential and a response from the cell. Neurons releasing neurotransmitter substances that cause EPSPs are **excitatory neurons.** In general, an EPSP occurs because of an increase in the membrane's permeability to Na^+. For example, glutamate in the brain and acetylcholine in skeletal muscle can

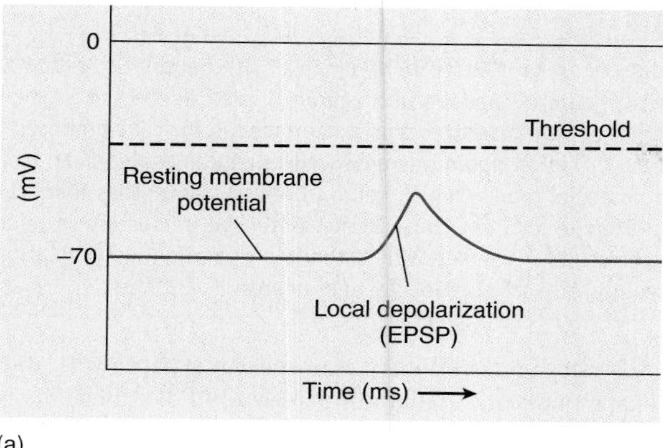

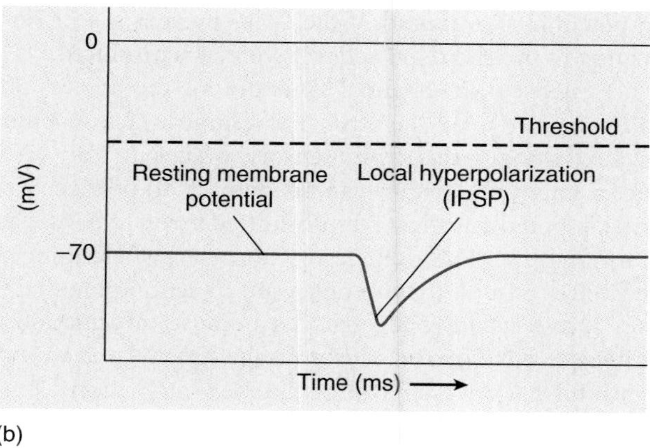

FIGURE 11.26 Postsynaptic Potentials
(*a*) Excitatory postsynaptic potential (EPSP) is closer to threshold. (*b*) Inhibitory postsynaptic potential (IPSP) is farther from threshold.

bind to their receptors, causing Na^+ channels to open. Because the concentration gradient is large for Na^+ and because the negative charge inside the cell attracts the positively charged Na^+, they diffuse into the cell and cause depolarization. If EPSPs cause a depolarizing graded potential that reaches threshold, an action potential is produced.

How Local Anesthetics Work

Awareness of pain can occur only if action potentials generated by sensory neurons stimulate the production of action potentials in CNS neurons. Local anesthetics, such as procaine (Novacain), act at their site of application to prevent pain sensations. They do so by blocking voltage-gated Na^+ channels, which prevents the propagation of action potentials along sensory neurons. Consequently, neurotransmitters are not released from the presynaptic terminals of the sensory neurons and EPSPs are not produced in CNS neurons.

When the combination of a neurotransmitter with its receptor results in hyperpolarization of the postsynaptic membrane, the response is inhibitory, and the local hyperpolarization is called an **inhibitory postsynaptic potential** (**IPSP;** figure 11.26*b*). IPSPs are important because they decrease the likelihood of producing action potentials by moving the membrane potential farther from threshold. Neurons releasing neurotransmitter substances that cause IPSPs are called **inhibitory neurons.** The IPSP is the result of an increase in the permeability of the plasma membrane to Cl^- or K^+. For example, in the spinal cord, glycine binds to its receptors, directly causing Cl^- channels to open. Because Cl^- are more concentrated outside the cell than inside, when the membrane's permeability to Cl^- increases, they diffuse into the cell, causing the inside of the cell to become more negative, resulting in hyperpolarization. Acetylcholine can bind to its receptors in the heart, causing G protein–mediated opening of K^+ channels (see chapter 3). The concentration of K^+ is greater inside the cell than outside, and increased permeability of the membrane to K^+ results in diffusion of K^+ out of the cell. Consequently, the outside of the cell becomes more positive than the inside, resulting in hyperpolarization.

Presynaptic Inhibition and Facilitation

Many of the synapses of the CNS are **axoaxonic synapses,** meaning that the axon of one neuron synapses with the presynaptic terminal (axon) of another (figure 11.27). Through axoaxonic synapses, one neuron can release a neuromodulator that influences the release of a neurotransmitter from the presynaptic terminal of another neuron.

In **presynaptic inhibition,** there is a reduction in the amount of neurotransmitter released from the presynaptic terminal. For example, sensory neurons for pain can release neurotransmitters from their presynaptic terminals and stimulate the postsynaptic membranes of neurons in the brain or spinal cord. Awareness of pain occurs only if action potentials are produced in the postsynaptic membranes of the CNS neurons. Enkephalins and endorphins released from inhibitory neurons of axoaxonic

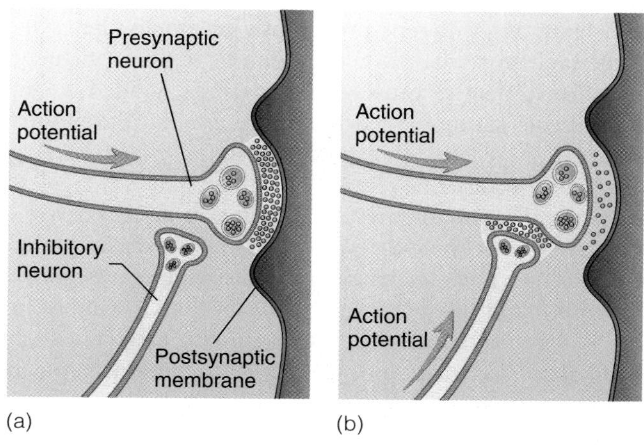

FIGURE 11.27 Presynaptic Inhibition at an Axoaxonic Synapse
(*a*) The inhibitory neuron of the axoaxonic synapse is inactive and has no effect on the release of neurotransmitter from the presynaptic terminal.
(*b*) The release of a neuromodulator from the inhibitory neuron of the axoaxonic synapse reduces the amount of neurotransmitter released from the presynaptic terminal.

synapses can reduce or eliminate pain sensations by inhibiting the release of neurotransmitter from the presynaptic terminals of sensory neurons (see figure 11.27). Enkephalins and endorphins can block voltage-gated Ca^{2+} channels. Consequently, when action potentials reach the presynaptic terminal, the influx of Ca^{2+} that normally stimulates neurotransmitter release is blocked.

In **presynaptic facilitation,** there is an increase in the amount of neurotransmitter released from the presynaptic terminal. For example, serotonin, released in certain axoaxonic synapses, functions as a neuromodulator that increases the release of neurotransmitter from the presynaptic terminal by causing voltage-gated Ca^{2+} channels to open.

47 ***Define and explain the production of EPSPs and IPSPs. Why are they important?***

48 ***What is presynaptic inhibition and facilitation?***

Spatial and Temporal Summation

Depolarizations produced in postsynaptic membranes are graded potentials. Within the CNS and in many PNS synapses, a single presynaptic action potential does not cause a graded potential in the postsynaptic membrane sufficient to reach threshold and produce an action potential. Instead, many presynaptic action potentials cause many graded potentials in the postsynaptic neuron. The graded potentials combine in summation at the trigger zone of the postsynaptic neuron, which is the normal site of action potential generation for most neurons. If summation results in a graded potential that exceeds threshold at the trigger zone, an action potential is produced. Action potentials are readily produced at the trigger zone because the concentration of voltage-gated Na^+ channels is approximately seven times greater there than at the rest of the cell body.

Two types of summation, called spatial summation and temporal summation, are possible. The simplest type of **spatial summation** occurs when two action potentials arrive simultaneously at two different presynaptic terminals that synapse with the same postsynaptic neuron. In the postsynaptic neuron, each action potential causes a depolarizing graded potential that undergoes summation at the trigger zone. If the summated depolarization reaches threshold, an action potential is produced (figure 11.28*a*).

Temporal summation results when two or more action potentials arrive in very close succession at a single presynaptic terminal. The first action potential causes a depolarizing graded potential in the postsynaptic membrane that remains for a few milliseconds before it disappears, although its magnitude decreases through time. Temporal summation results when another action potential initiates another graded depolarization before the depolarization caused by the previous action potential returns to its resting value (see figure 11.16*b*). Subsequent action potentials cause depolarizations that summate with previous depolarizations. If the summated depolarizing graded potentials reach threshold at the trigger zone, an action potential is produced in the postsynaptic neuron (figure 11.28*b*).

PREDICT 10

Excitatory neurons A and B both synapse with neuron C. Neuron A releases a neurotransmitter, and neuron B releases the same type and amount of neurotransmitter plus a neuromodulator that produces EPSPs in neuron C. Action potentials produced in neuron A alone can result in action potential production in neuron C. Action potentials produced in neuron B alone also can cause action potential production in neuron C. Which results in more action potentials in neuron C, stimulation by only neuron A or stimulation by only neuron B? Explain.

Excitatory and inhibitory neurons can synapse with the same postsynaptic neuron. Spatial summation of EPSPs and IPSPs occurs in the postsynaptic neuron, and whether a postsynaptic action potential is initiated or not depends on which type of graded potential has the greatest influence on the postsynaptic membrane potential (figure 11.28*c*). If the EPSPs (local depolarizations) cancel the IPSPs (local hyperpolarizations) and summate to threshold, an action potential is produced. If the IPSPs prevent the EPSPs from summating to threshold, no action potential is produced.

The synapse is an essential structure for the process of integration carried out by the CNS. For example, action potentials propagated along axons from sensory organs to the CNS can produce a sensation, or they can be ignored. To produce a sensation, action potentials must be transmitted across synapses as they travel through the CNS to the cerebral cortex, where information is interpreted. Stimuli that do not result in action potential transmission across synapses are ignored because information never reaches the cerebral cortex. The brain can ignore large amounts of sensory information as a result of complex integration.

49 ***Define spatial and temporal summation. In what part of the neuron does summation take place?***

50 ***How do EPSPs and IPSPs affect the likelihood that summation will result in an action potential?***

NEURONAL PATHWAYS AND CIRCUITS

The organization of neurons within the CNS varies from relatively simple to extremely complex patterns. The axon of a neuron can branch repeatedly to form synapses with many other neurons, and hundreds or even thousands of axons can synapse with the cell body and dendrites of a single neuron. Although their complexity varies, three basic patterns can be recognized: convergent pathways, divergent pathways, and oscillating circuits.

In **convergent pathways,** many neurons converge and synapse with a smaller number of neurons (figure 11.29*a*). Convergence makes it possible for different parts of the nervous system to activate or inhibit the activity of neurons. For example, one part of the nervous system can stimulate the neurons responsible for making a muscle contract, whereas another part can inhibit those neurons. Through summation, action potentials resulting in muscle contraction can be produced if the converging neurons stimulate the production of more EPSPs than IPSPs. Conversely, muscle contraction is inhibited if the converging neurons stimulate the production of more IPSPs than EPSPs.

(a) **Spatial summation.** Action potentials 1 and 2 cause the production of graded potentials at two different dendrites. These graded potentials summate at the trigger zone to produce a graded potential that exceeds threshold, resulting in an action potential.

(b) **Temporal summation.** Two action potentials arrive in close succession at the presynaptic membrane. The first action potential causes the production of a graded potential that does not reach threshold at the trigger zone. The second action potential results in the production of a second graded potential that summates with the first to reach threshold, resulting in the production of an action potential.

(c) **Combined spatial and temporal summation with both excitatory postsynaptic potentials (EPSPs) and inhibitory postsynaptic potentials (IPSPs).** An action potential is produced at the trigger zone when the graded potentials produced as a result of the EPSPs and IPSPs summate to reach threshold.

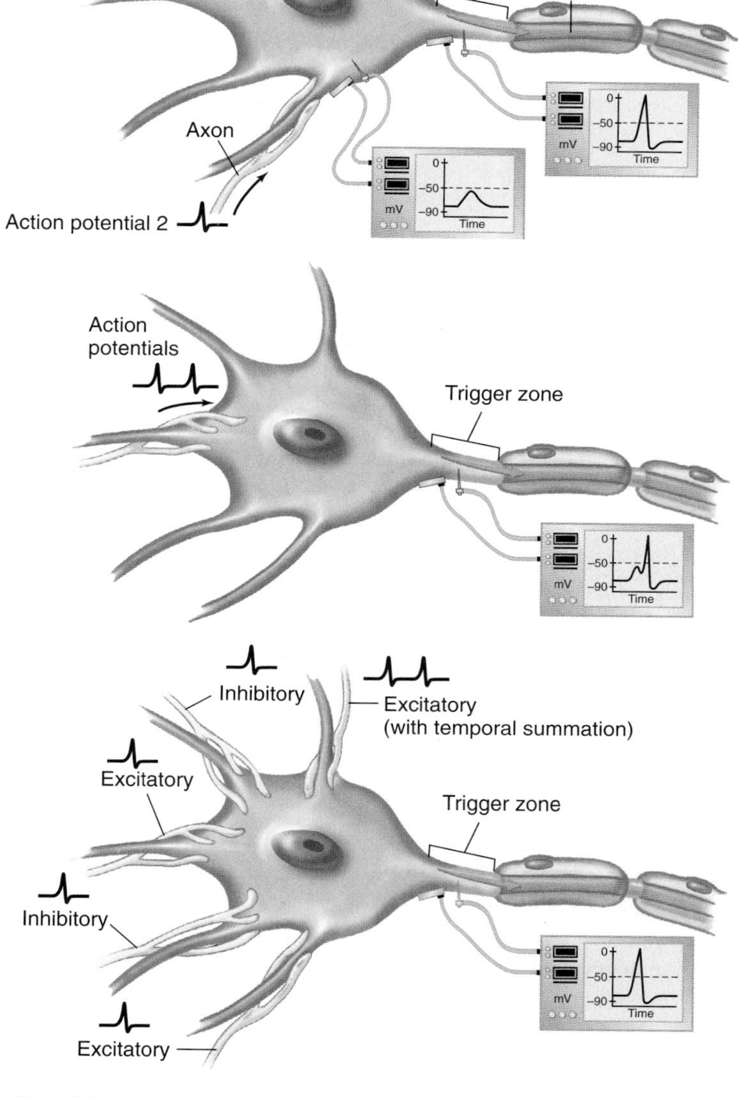

FIGURE 11.28 Summation

In **divergent pathways,** a smaller number of presynaptic neurons synapse with a larger number of postsynaptic neurons to allow information transmitted in one neuronal pathway to diverge into two or more pathways (figure 11.29*b*). Diverging pathways allow one part of the nervous system to affect more than one other part of the nervous system. For example, sensory input to the central nervous system can go to both the spinal cord and the brain.

Oscillating circuits have neurons arranged in a circular fashion, which allows action potentials entering the circuit to cause a neuron farther along in the circuit to produce an action potential more than once (figure 11.29*c*). This response is called **afterdischarge,** and its effect is to prolong the response to a stimulus. Oscillating circuits are similar to positive-feedback systems. Once an oscillating circuit is stimulated, it continues to discharge until the synapses involved become fatigued or until they are inhibited by other neurons. Oscillating circuits play a role in neuronal circuits that are periodically active. Respiration may be controlled by an oscillating circuit that controls inspiration and another that controls expiration.

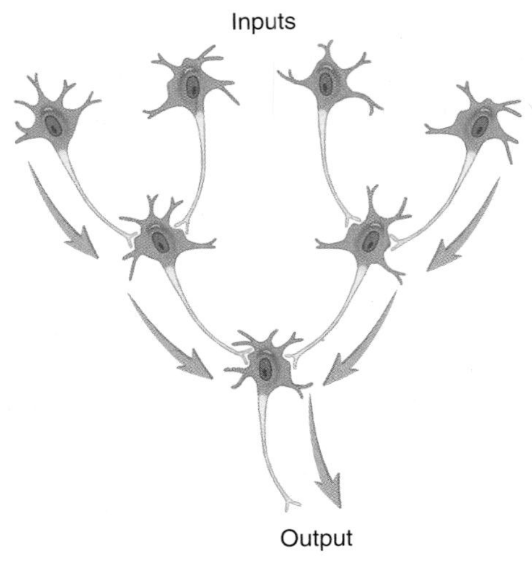

(a) **Convergent pathway**

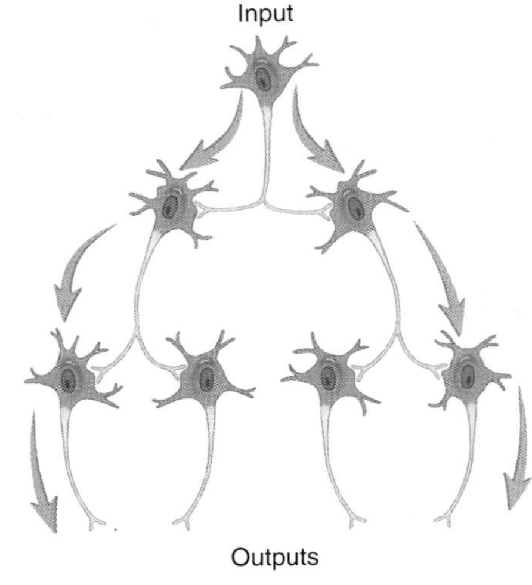

(b) **Divergent pathway**

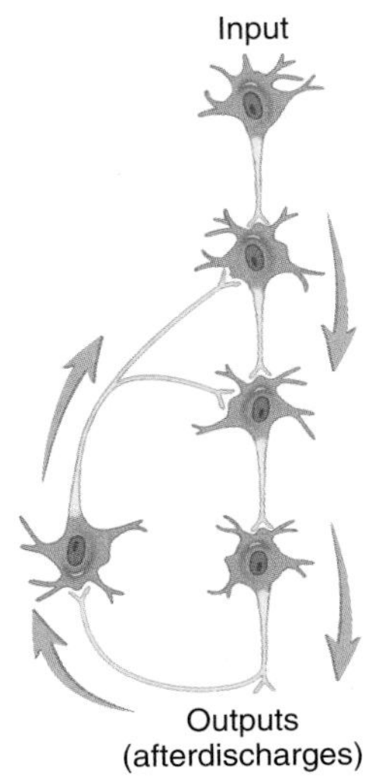

(c) **Oscillating circuit**

FIGURE 11.29 Neuronal Pathways and Circuits

The direction of action potential propagation is represented by the *gold arrows*. (*a*) General model of a convergent pathway; many neurons converge and synapse with a smaller number of neurons. (*b*) General model of a divergent pathway; a few neurons synapse with a larger number of neurons. (*c*) Simple model of an oscillating circuit; input action potentials result in the production of a larger number of output action potentials because neurons within the circuit are repeatedly stimulated to produce action potentials.

Neurons that spontaneously produce action potentials are common in the CNS and may activate oscillating circuits, which remain active awhile. The cycle of wakefulness and sleep may involve circuits of this type. Spontaneously active neurons can also influence the activity of other circuit types. The complex functions carried out by the CNS are affected by the numerous circuits operating together and influencing the activity of one another.

51 ***Diagram a convergent pathway, a divergent pathway, and an oscillating circuit, and describe what is accomplished in each.***

SUMMARY

Functions of the Nervous System (p. 375)

The nervous system detects external and internal stimuli (sensory input), processes and responds to sensory input (integration), controls body movements through skeletal muscles, maintains homeostasis by regulating other systems, and is the center for mental activities.

Divisions of the Nervous System (p. 375)

1. The nervous system has two anatomical divisions.
   - The central nervous system (CNS) consists of the brain and spinal cord and is encased in bone.
   - The peripheral nervous system (PNS), the nervous tissue outside of the CNS, consists of sensory receptors, nerves, ganglia, and plexuses.
2. The PNS has two divisions.
   - The sensory division transmits action potentials to the CNS and usually consists of single neurons that have their cell bodies in ganglia.
   - The motor division carries action potentials away from the CNS in cranial or spinal nerves.
3. The motor division has two subdivisions.
   - The somatic nervous system innervates skeletal muscle and is mostly under voluntary control. It consists of single neurons that have their cell bodies located within the CNS.
   - The autonomic nervous system (ANS) innervates cardiac muscle, smooth muscle, and glands. It has two sets of neurons between the CNS and effector organs. The first set has its cell bodies within the CNS, and the second set has its cell bodies within autonomic ganglia.
   - The ANS is subdivided into the sympathetic division, which is most active during physical activity; the parasympathetic division, which regulates resting functions; and the enteric nervous system, which controls the digestive system.
4. The anatomical divisions perform different functions.
   - The PNS detects stimuli and transmits information to and receives information from the CNS.
   - The CNS processes, integrates, stores, and responds to information from the PNS.

Cells of the Nervous System (p. 377)

Neurons

1. Neurons receive stimuli and transmit action potentials.
2. Neurons have three components.
   - The cell body is the primary site of protein synthesis.
   - Dendrites are short, branched cytoplasmic extensions of the cell body that usually conduct electric signals toward the cell body.
   - An axon is a cytoplasmic extension of the cell body that transmits action potentials to other cells.

Types of Neurons

1. Multipolar neurons have several dendrites and a single axon. Interneurons and motor neurons are multipolar.

2. Bipolar neurons have a single axon and dendrite and are found as components of sensory organs.
3. Unipolar neurons have a single axon. Most sensory neurons are unipolar.

Glial Cells of the CNS

1. Glial cells are nonneural cells that support and aid the neurons of the CNS and PNS.
2. Astrocytes provide structural support for neurons and blood vessels. Astrocytes influence the functioning of the blood–brain barrier and process substances that pass through it. Astrocytes isolate damaged tissue and limit the spread of inflammation. Astrocytes also help maintain synaptic function.
3. Ependymal cells line the ventricles and the central canal of the spinal cord. Some are specialized to produce cerebrospinal fluid.
4. Microglia phagocytize microorganisms, foreign substances, and necrotic tissue.
5. An oligodendrocyte forms myelin sheaths around the axons of several CNS neurons.

Glial Cells of the PNS

1. A Schwann cell forms a myelin sheath around part of the axon of a PNS neuron.
2. Satellite cells support and nourish neuron cell bodies within ganglia.

Myelinated and Unmyelinated Axons

1. Myelinated axons are wrapped by several layers of plasma membrane from Schwann cells (PNS) or oligodendrocytes (CNS). Spaces between the wrappings are the nodes of Ranvier. Myelinated axons conduct action potentials rapidly.
2. Unmyelinated axons rest in invaginations of Schwann cells (PNS) or oligodendrocytes (CNS). They conduct action potentials slowly.

Organization of Nervous Tissue (p. 382)

1. Nervous tissue can be grouped into white and gray matter.
   - White matter consists of myelinated axons; it propagates action potentials.
   - Gray matter consists of collections of neuron cell bodies or unmyelinated axons. Axons synapse with neuron cell bodies, which are functionally the site of integration in the nervous system.
2. White matter forms nerve tracts in the CNS and nerves in the PNS. Gray matter forms cortex and nuclei in the CNS and ganglia in the PNS.

Electric Signals (p. 382)

Electrical properties of cells result from the ionic concentration differences across the plasma membrane and from the permeability characteristics of the plasma membrane.

Concentration Differences Across the Plasma Membrane

1. The Na^+–K^+ pump moves ions by active transport. Potassium ions are moved into the cell, and Na^+ are moved out of it.
2. The concentration of K^+ and negatively charged proteins and other molecules is higher inside, and the concentrations of Na^+ and Cl^- are higher outside the cell.
3. Negatively charged proteins and other negatively charged ions are synthesized inside the cell and cannot diffuse out of it, and they repel negatively charged Cl^-.
4. The permeability of the plasma membrane to ions is determined by leak channels and gated ion channels.
   - Potassium ion leak channels are more numerous than Na^+ leak channels, thus the plasma membrane is more permeable to K^+ than to Na^+ when at rest.
   - Gated ion channels in the plasma membrane include ligand-gated ion channels, voltage-gated ion channels, and other gated ion channels.

Establishing the Resting Membrane Potential

1. The resting membrane potential is a charge difference across the plasma membrane when the cell is in an unstimulated condition. The inside of the cell is negatively charged, compared with the outside of the cell.
2. The resting membrane potential is due mainly to the tendency of positively charged K^+ to diffuse out of the cell, which is opposed by the negative charge that develops inside the plasma membrane.

Changing the Resting Membrane Potential

1. Depolarization is a decrease in the resting membrane potential and can result from a decrease in the K^+ concentration gradient, a decrease in membrane permeability to K^+, an increase in membrane permeability to Na^+, an increase in membrane permeability to Ca^{2+}, or a decrease in extracellular Ca^{2+} concentration.
2. Hyperpolarization is an increase in the resting membrane potential that can result from an increase in the K^+ concentration gradient, an increase in membrane permeability to K^+, an increase in membrane permeability to Cl^-, a decrease in membrane permeability to Na^+, or an increase in extracellular Ca^{2+} concentration.

Graded Potentials

1. A graded potential is a small change in the resting membrane potential that is confined to a small area of the plasma membrane.
2. An increase in membrane permeability to Na^+ can cause graded depolarization, and an increase in membrane permeability to K^+ or Cl^- can result in graded hyperpolarization.
3. The term *graded potential* is used because a stronger stimulus produces a greater potential change than a weaker stimulus.
4. Graded potentials can summate, or add together.
5. A graded potential decreases in magnitude as the distance from the stimulation increases.

Action Potentials

1. An action potential is a larger change in the resting membrane potential that spreads over the entire surface of the cell.
2. Threshold is the membrane potential at which a graded potential depolarizes the plasma membrane sufficiently to produce an action potential.
3. Action potentials occur in an all-or-none fashion. If action potentials occur, they are of the same magnitude, no matter how strong the stimulus.
4. Depolarization occurs as the inside of the membrane becomes more positive because Na^+ diffuse into the cell through voltage-gated ion channels.
5. Repolarization is a return of the membrane potential toward the resting membrane potential because voltage-gated Na^+ channels close and Na^+ diffusion into the cell slows to resting levels and because voltage-gated K^+ channels continue to open and K^+ diffuse out of the cell.
6. The afterpotential is a brief period of hyperpolarization following repolarization.

Refractory Period

1. The absolute refractory period is the time during an action potential when a second stimulus, no matter how strong, cannot initiate another action potential.
2. The relative refractory period follows the absolute refractory period and is the time during which a stronger-than-threshold stimulus can evoke another action potential.

Action Potential Frequency

1. The strength of stimuli affects the frequency of action potentials.
   - A subthreshold stimulus produces only a graded potential.
   - A threshold stimulus causes a graded potential that reaches threshold and results in a single action potential.
   - A submaximal stimulus is greater than a threshold stimulus and weaker than a maximal stimulus. The action potential frequency increases as the strength of the submaximal stimulus increases.
   - A maximal or a supramaximal stimulus produces a maximum frequency of action potentials.
2. A low frequency of action potentials represents a weaker stimulus than a high frequency.

Propagation of Action Potentials

1. An action potential generates ionic currents, which stimulate voltage-gated Na^+ channels in adjacent regions of the plasma membrane to open, producing a new action potential.
2. In an unmyelinated axon, action potentials are generated immediately adjacent to previous action potentials.
3. In a myelinated axon, action potentials are generated at successive nodes of Ranvier.
4. Reversal of the direction of action potential propagation is prevented by the absolute refractory period.
5. Action potentials propagate most rapidly in myelinated, large-diameter axons.

The Synapse (p. 394)

Electrical Synapses

1. Electrical synapses are gap junctions in which tubular proteins called connexons allow ionic currents to move between cells.
2. At an electrical synapse, an action potential in one cell generates an ionic current that causes an action potential in an adjacent cell.

Chemical Synapses

1. Anatomically, a chemical synapse has three components.
   - The enlarged ends of the axon are the presynaptic terminals containing synaptic vesicles.
   - The postsynaptic membranes contain receptors for the neurotransmitter.
   - The synaptic cleft, a space, separates the presynaptic and postsynaptic membranes.
2. An action potential arriving at the presynaptic terminal causes the release of a neurotransmitter, which diffuses across the synaptic cleft and binds to the receptors of the postsynaptic membrane.
3. The effect of the neurotransmitter on the postsynaptic membrane is stopped in several ways.
   - The neurotransmitter is broken down by an enzyme.
   - The neurotransmitter is taken up by the presynaptic terminal.
   - The neurotransmitter diffuses out of the synaptic cleft.
4. Neurotransmitters are specific for their receptors. A neurotransmitter can be stimulatory in one synapse and inhibitory in another, depending on the type of receptor present.
5. Neuromodulators influence the likelihood that an action potential in a presynaptic terminal will result in an action potential in a postsynaptic cell.
6. An excitatory postsynaptic potential (EPSP) is a depolarizing graded potential of the postsynaptic membrane. It can be caused by an increase in membrane permeability to Na^+.
7. An inhibitory postsynaptic potential (IPSP) is a hyperpolarizing graded potential of the postsynaptic membrane. It can be caused by an increase in membrane permeability to K^+ or Cl^-.
8. Presynaptic inhibition decreases neurotransmitter release. Presynaptic facilitation increases neurotransmitter release.

Spatial and Temporal Summation

1. Presynaptic action potentials through neurotransmitters produce graded potentials in postsynaptic neurons. The graded potential can summate to produce an action potential at the trigger zone.
2. Spatial summation occurs when two or more presynaptic terminals simultaneously stimulate a postsynaptic neuron.
3. Temporal summation occurs when two or more action potentials arrive in succession at a single presynaptic terminal.
4. Inhibitory and excitatory presynaptic neurons can converge on a postsynaptic neuron. The activity of the postsynaptic neuron is determined by the integration of the EPSPs and IPSPs produced in the postsynaptic neuron.

Neuronal Pathways and Circuits (p. 404)

1. Convergent pathways have many neurons synapsing with a few neurons.
2. Divergent pathways have a few neurons synapsing with many neurons.
3. Oscillating circuits have collateral branches of postsynaptic neurons synapsing with presynaptic neurons.

REVIEW AND COMPREHENSION

1. The peripheral nervous system includes the
   a. somatic nervous system.
   b. brain.
   c. spinal cord.
   d. nuclei.
   e. all of the above.
2. The part of the nervous system that controls smooth muscle, cardiac muscle, and glands is the
   a. somatic nervous system.
   b. autonomic nervous system.
   c. skeletal division.
   d. sensory division.
3. Neurons have cytoplasmic extensions that connect one neuron to another neuron. Given these structures:
   1. axon
   2. dendrite
   3. dendritic spine
   4. presynaptic terminal

   Choose the arrangement that lists the structures in the order they are found between two neurons.
   a. 1,4,2,3
   b. 1,4,3,2
   c. 4,1,2,3
   d. 4,1,3,2
   e. 4,3,2,1

4. A neuron with many short dendrites and a single long axon is a ______ neuron.
   a. multipolar b. unipolar c. bipolar
5. Motor neurons and interneurons are ______ neurons.
   a. unipolar c. multipolar
   b. bipolar d. afferent
6. Cells found in the choroid plexuses that secrete cerebrospinal fluid are
   a. astrocytes. c. ependymal cells. e. Schwann cells.
   b. microglia. d. oligodendrocytes.
7. Glial cells that are phagocytic within the central nervous system are
   a. oligodendrocytes. d. astrocytes.
   b. microglia. e. Schwann cells.
   c. ependymal cells.
8. Unmyelinated axons within nerves may have which of these associated with them?
   a. Schwann cells c. oligodendrocytes
   b. nodes of Ranvier d. all of the above
9. Action potentials are conducted more rapidly
   a. in small-diameter axons than in large-diameter axons.
   b. in unmyelinated axons than in myelinated axons.
   c. along axons that have nodes of Ranvier.
   d. all of the above.
10. Clusters of neuron cell bodies within the peripheral nervous system are
   a. ganglia. c. nuclei.
   b. fascicles. d. laminae.
11. Gray matter contains primarily
   a. myelinated fibers. c. Schwann cells.
   b. neuron cell bodies. d. oligodendrocytes.
12. Concerning concentration difference across the plasma membrane, there are
   a. more K^+ and Na^+ outside the cell than inside.
   b. more K^+ and Na^+ inside the cell than outside.
   c. more K^+ outside the cell than inside and more Na^+ inside the cell than outside.
   d. more K^+ inside the cell than outside and more Na^+ outside the cell than inside.
13. Compared with the inside of the resting plasma membrane, the outside surface of the membrane is
   a. positively charged.
   b. electrically neutral.
   c. negatively charged.
   d. continuously reversing so that it is positive 1 second and negative the next.
   e. negatively charged whenever the Na^+–K^+ pump is operating.
14. Leak channels
   a. open in response to small voltage changes.
   b. open when a chemical signal binds to its receptor.
   c. are responsible for the ion permeability of the resting plasma membrane.
   d. allow substances to move into the cell but not out.
   e. all of the above.
15. The resting membrane potential results when the tendency for ______ to diffuse out of the cell is balanced by their attraction to opposite charges inside the cell.
   a. Na^+ c. Cl^-
   b. K^+ d. negatively charged proteins
16. If the permeability of the plasma membrane to K^+ increases, resting membrane potential ______. This is called ______.
   a. increases, hyperpolarization c. decreases, hyperpolarization
   b. increases, depolarization d. decreases, depolarization
17. Decreasing the extracellular concentration of K^+ affects the resting membrane potential by causing
   a. hyperpolarization. b. depolarization. c. no change.
18. Which of these terms are correctly matched with their definition or description?
   a. depolarization: membrane potential becomes more negative
   b. hyperpolarization: membrane potential becomes more negative
   c. hypopolarization: membrane potential becomes more negative
19. Which of these statements about ion movement through the plasma membrane is true?
   a. Movement of Na^+ out of the cell requires energy (ATP).
   b. When Ca^{2+} binds to proteins in ion channels, the diffusion of Na^+ into the cell is inhibited.
   c. Specific ion channels regulate the diffusion of Na^+ through the plasma membrane.
   d. All of the above are true.
20. The *major* function of the Na^+–K^+ pump is to
   a. pump Na^+ into and K^+ out of the cell.
   b. generate the resting membrane potential.
   c. maintain the concentration gradients of Na^+ and K^+ across the plasma membrane.
   d. oppose any tendency of the cell to undergo hyperpolarization.
21. Graded potentials
   a. spread over the plasma membrane in decremental fashion.
   b. are not propagated for long distances.
   c. are confined to a small region of the plasma membrane.
   d. can summate.
   e. all of the above.
22. During the depolarization phase of an action potential, the permeability of the membrane
   a. to K^+ is greatly increased. c. to Ca^{2+} is greatly increased.
   b. to Na^+ is greatly increased. d. is unchanged.
23. During repolarization of the plasma membrane,
   a. Na^+ diffuse into the cell. c. K^+ diffuse into the cell.
   b. Na^+ diffuse out of the cell. d. K^+ diffuse out of the cell.
24. The absolute refractory period
   a. limits how many action potentials can be produced during a given period of time.
   b. prevents an action potential from starting another action potential at the same point on the plasma membrane.
   c. is the period of time when a strong stimulus can initiate a second action potential.
   d. both a and b.
   e. all of the above.
25. A subthreshold stimulus
   a. produces an afterpotential.
   b. produces a graded potential.
   c. causes an all-or-none response.
   d. produces more action potentials than a submaximal stimulus.
26. Neurotransmitter substances are stored in vesicles located in specialized portions of the
   a. neuron cell body. c. dendrite.
   b. axon. d. postsynaptic membrane.

27. In a chemical synapse,
    a. action potentials in the presynaptic terminal cause voltage-gated Ca^{2+} channels to open.
    b. neurotransmitters can cause ligand-gated Na^+ channels to open.
    c. neurotransmitters can be broken down by enzymes.
    d. neurotransmitters can be taken up by the presynaptic terminal.
    e. all of the above.
28. An inhibitory presynaptic neuron can affect a postsynaptic neuron by
    a. producing an IPSP in the postsynaptic neuron.
    b. hyperpolarizing the plasma membrane of the postsynaptic neuron.
    c. causing K^+ to diffuse out of the postsynaptic neuron.
    d. causing Cl^- to diffuse into the postsynaptic neuron.
    e. all of the above.
29. Summation
    a. is caused by combining two or more graded potentials.
    b. occurs at the trigger zone of the postsynaptic neuron.
    c. results in an action potential if it reaches the threshold potential.
    d. can occur when two action potentials arrive in close succession at a single presynaptic terminal.
    e. all of the above.
30. In convergent pathways,
    a. the response of the postsynaptic neuron depends on the summation of EPSPs and IPSPs.
    b. a smaller number of presynaptic neurons synapse with a larger number of postsynaptic neurons.
    c. information transmitted in one neuronal pathway can go into two or more pathways.
    d. all of the above.

Answers in Appendix E

CRITICAL THINKING

1. Predict the consequence of a reduced intracellular K^+ concentration on the resting membrane potential.
2. A child eats a whole bottle of salt (NaCl) tablets. What effect does this have on action potentials?
3. Lithium ions reduce the permeability of plasma membranes to Na^+. Predict the effect lithium ions in the extracellular fluid would have on the response of a neuron to stimuli.
4. Some smooth muscle has the ability to contract spontaneously—that is, it contracts without any external stimulation. Propose an explanation for the ability of smooth muscle to contract spontaneously based on what you know about membrane potentials. Assume that an action potential in a smooth muscle cell causes it to contract.
5. Assume that you have two nerve fibers of the same diameter, but one is myelinated and the other is unmyelinated. The conduction of an action potential is most energy-efficient along which type of fiber? (*Hint:* ATP.)
6. Explain the consequences when an inhibitory neuromodulator is released from a presynaptic terminal and a stimulatory neurotransmitter is released from another presynaptic terminal, both of which synapse with the same neuron.
7. The speed of action potential propagation and synaptic transmission decreases with aging. List possible explanations.
8. Students in a veterinary school are given the following hypothetical problem. A dog ingests organophosphate poison, and the students are responsible for saving the animal's life. Organophosphate poisons bind to and inhibit acetylcholinesterase. Several substances they could inject include the following: acetylcholine, curare (which blocks acetylcholine receptors), and potassium chloride. If you were a student in the class, what would you do to save the animal?
9. Strychnine blocks receptor sites for inhibitory neurotransmitter substances in the CNS. Explain how strychnine could produce tetanus in skeletal muscles.
10. The venom of many cobras contains a potent neurotoxin that binds to ligand-gated Na^+ channels, causing them to open. Unlike ACh, which binds to and then rapidly unbinds from ligand-gated Na^+ channels, the neurotoxin tends to remain bound to ligand-gated Na^+ channels. What effect would this neurotoxin have on the ability of the nervous system to stimulate skeletal muscle contraction? What effect would it have on the ability of skeletal muscle fibers to respond to stimulation?
11. Epilepsy is a chronic disorder characterized by seizures resulting from excessive production of action potentials by neurons in the brain. Channelopathies are genetic disorders caused by mutations in ion channel genes, which result in ion channels that do not function normally. What kind of Na^+ or Ca^{2+} channelopathies might contribute to epilepsy?

Answers in Appendix F

Visit this book's website at www.mhhe.com/seeley8 for chapter quizzes, interactive learning exercises, and other study tools.

Spinal Cord and Spinal Nerves

The **central nervous system (CNS)** consists of the brain and spinal cord, with the division between these two parts of the CNS placed at the level of the foramen magnum. The **peripheral nervous system (PNS)** consists of nerves and ganglia outside the CNS (see chapter 11). **Nerves** consist of axons, Schwann cells, and connective tissue sheaths. **Ganglia** are accumulations of cell bodies in the PNS. The PNS includes 12 pairs of cranial nerves and 31 pairs of spinal nerves and their ganglia.

The CNS receives sensory information, integrates and evaluates that information, stores some information, and initiates reactions. The PNS collects information from numerous sources both inside and outside the body and relays it through axons of sensory neurons to the CNS. Axons of motor neurons in the PNS relay information from the CNS to various parts of the body, primarily to muscles and glands, thereby regulating activity in those structures.

The spinal cord and spinal nerves are described in this chapter. The brain and cranial nerves are considered in chapter 13. The specific topics of chapter 12 are the *spinal cord* (p. 412), *reflexes* (p. 415), *interactions with spinal cord reflexes* (p. 421), the *structure of peripheral nerves* (p. 421), and *spinal nerves* (p. 422).

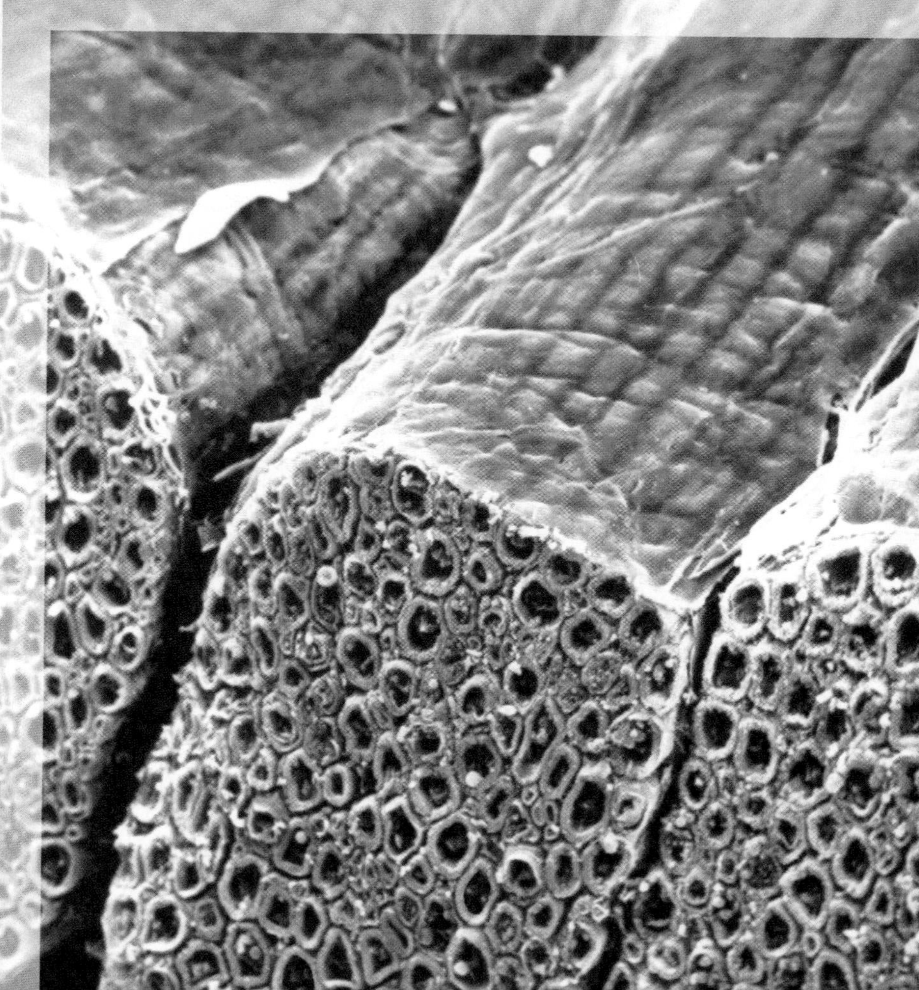

Colorized SEM of nerve fascicles containing bundles of axons.

Nervous System

SPINAL CORD

The **spinal cord** is extremely important to the overall function of the nervous system. It is the major communication link between the brain and the PNS inferior to the head; it participates in the integration of incoming information and produces responses through reflex mechanisms.

General Structure

The spinal cord (figure 12.1) extends from the foramen magnum to the level of the second lumbar vertebra. It is considerably shorter than the vertebral column because it does not grow as rapidly as the vertebral column during development. The spinal cord is composed of cervical, thoracic, lumbar, and sacral segments, named according to the portion of the vertebral column from which their nerves enter and exit. The spinal cord gives rise to 31 pairs of spinal nerves, which exit the vertebral column through intervertebral and sacral foramina (see figure 7.13). The nerves from the lower segments descend some distance in the vertebral canal before they exit because the spinal cord is shorter than the vertebral column.

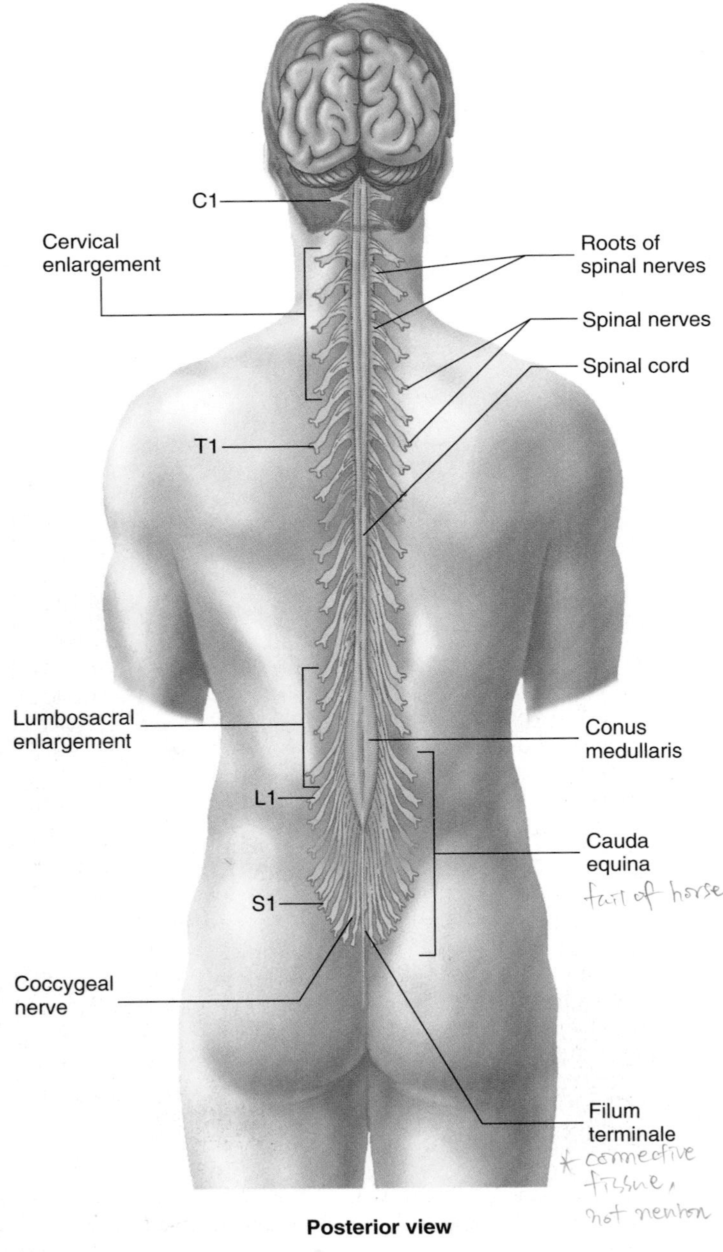

FIGURE 12.1 Spinal Cord and Spinal Nerve Roots

The spinal cord is not uniform in diameter throughout its length. It is larger in diameter at its superior end, and it gradually decreases in diameter toward its inferior end. Two enlargements occur where nerves supplying the upper and lower limbs enter and leave the spinal cord (see figure 12.1). The **cervical enlargement** in the inferior cervical region corresponds to the location where axons that supply the upper limbs enter and leave the spinal cord. The **lumbosacral enlargement** in the inferior thoracic, lumbar, and superior sacral regions is the site where the axons supplying the lower limbs enter or leave the spinal cord.

PREDICT 1

Why is the spinal cord enlarged in the cervical and lumbosacral regions?

Immediately inferior to the lumbosacral enlargement, the spinal cord tapers to form a conelike region called the **conus medullaris.** Its tip is the inferior end of the spinal cord; it extends to the level of the second lumbar vertebra. The nerves supplying the lower limbs and other inferior structures of the body arise from the lumbar and sacral nerves. They exit the lumbosacral enlargement and conus medullaris, course inferiorly through the vertebral canal, and exit through the intervertebral foramina from the second lumbar to the fifth sacral vertebrae. The numerous roots (origins) of spinal nerves extending inferiorly from the lumbosacral enlargement and conus medullaris resemble a horse's tail and are therefore called the **cauda** (kaw′dă; tail) **equina** (ē-kwī′nă; horse; see figure 12.1).

1 ***Describe the cervical and lumbar enlargements of the spinal cord, the conus medullaris, and the cauda equina. How many pairs of spinal nerves exit the spinal cord?***

Meninges of the Spinal Cord

The spinal cord and brain are surrounded by connective tissue membranes called **meninges** (mĕ-nin′jēz; figure 12.2). The most superficial and thickest membrane is the **dura mater** (doo′ră mā′ter; tough mother). The dura mater forms a sac, often called the **thecal** (the′kal) **sac,** that surrounds the spinal cord. The thecal sac attaches to the rim of the foramen magnum and ends at the level of the second sacral vertebra. The spinal dura mater is continuous with the dura mater surrounding the brain and the connective tissue surrounding the spinal nerves. The dura mater around the spinal cord is separated from the periosteum of the vertebral canal by the **epidural space.** This is a true space between the walls of the vertebral canal and the dura mater of the spinal cord that contains spinal nerve roots, blood vessels, areolar connective tissue, and fat. In contrast, the epidural space around the brain is only a potential space. **Epidural anesthesia** of the spinal nerves is induced by injecting anesthetics into this space. Epidural anesthesia is often given to women during childbirth.

Dura mater
Subdural space
Denticulate ligament
Arachnoid mater
Subarachnoid space
Pia mater
Epineurium of spinal nerve
Dorsal root ganglion
Spinal nerve
Ventral root

(a) **Anterolateral view**

Posterior
Spinous process of vertebra
Transverse process of vertebra
Spinal cord
Denticulate ligament
Dorsal root
Dorsal root ganglion
Spinal nerve
Ventral root
Body of vertebra
Pia mater
Subarachnoid space
Periosteum
Arachnoid mater
Epidural space filled with fat
Subdural space
Dura mater
Anterior

(b) **Superior view**

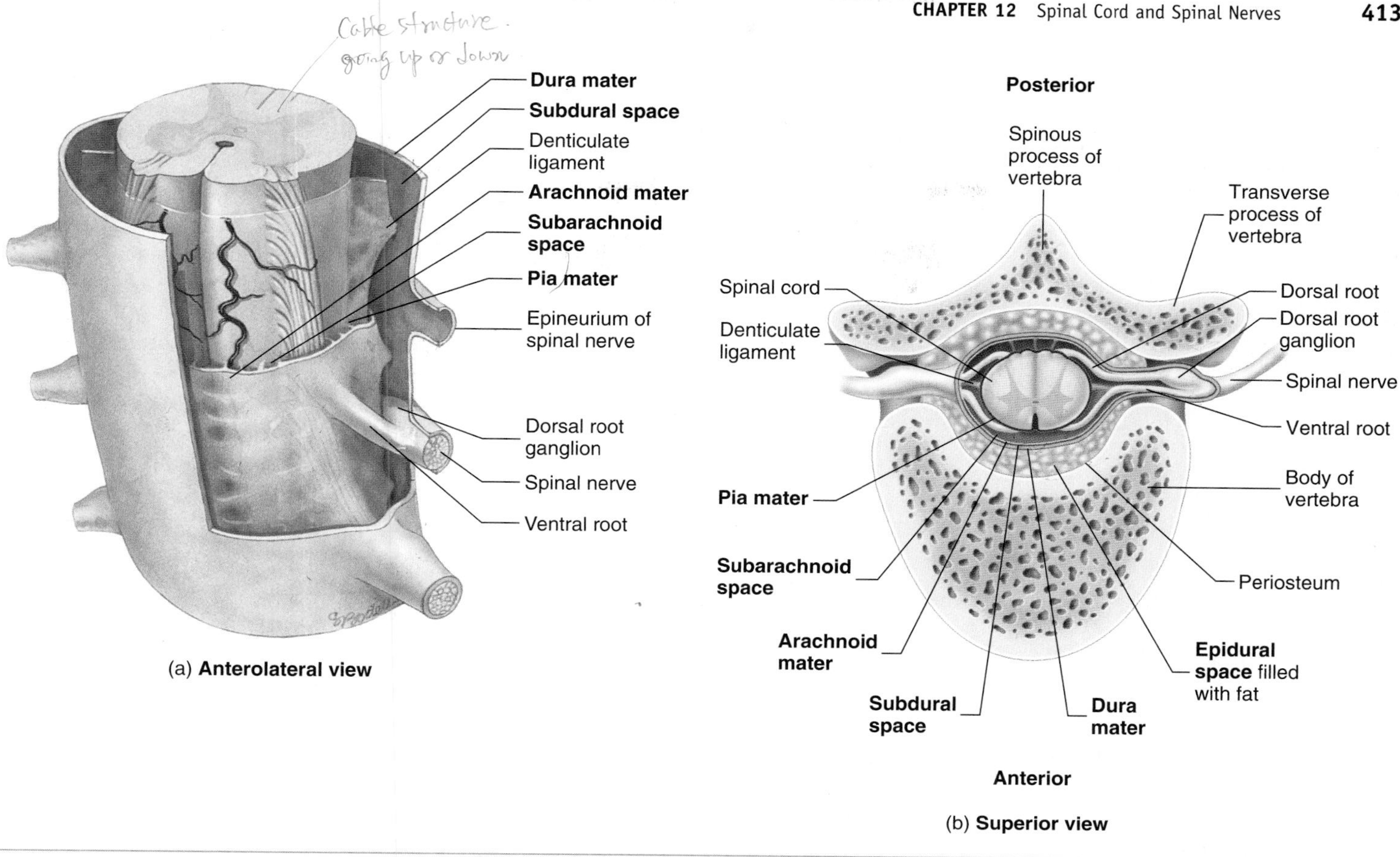

FIGURE 12.2 **Meningeal Membranes Surrounding the Spinal Cord**

PREDICT 2

Epidural anesthesia is usually induced by injection at what level of the vertebral column?

The next deeper meningeal membrane is a very thin, wispy **arachnoid** (ă-rak′noyd; spiderlike—i.e., cobwebs) **mater.** The space between this membrane and the dura mater is the **subdural space;** it contains only a very small amount of serous fluid.

The third, deepest, meningeal layer, the **pia** (pī′ă; affectionate) **mater** is bound very tightly to the surface of the spinal cord. The spinal cord is held in place within the thecal sac by the denticulate ligaments and the filum terminale. The paired **denticulate** (den-tik′ū-lāt) **ligaments** are connective tissue septa extending from the lateral sides of the spinal cord to the dura mater (see figure 12.2*b*). The term *denticulate* refers to having small teeth, and the denticulate ligaments attach to the dura mater by 21 triangular, or toothlike, processes between the exits of the cervical and thoracic spinal nerves. The denticulate ligaments limit the lateral movement of the spinal cord. The **filum terminale** (fī′lŭm ter′mi-nal′ē) is a connective tissue strand that anchors the conus medullaris and the thecal sac to the first coccygeal vertebra, limiting their superior movement.

Between the arachnoid mater and the pia mater is the **subarachnoid space,** which contains weblike strands of the arachnoid mater, blood vessels, and **cerebrospinal** (ser′ĕ-brō-spī-năl, sĕ-rē′brō-spī-nal) **fluid (CSF),** which is described in chapter 13.

Introduction of Needles into the Subarachnoid Space

Several clinical procedures involve the insertion of a needle into the subarachnoid space inferior to the level of the second lumbar vertebra. The needle is introduced into either the L3/L4 or the L4/L5 intervertebral space. The needle does not contact the spinal cord because it extends only approximately to the second lumbar vertebra of the vertebral column, but the subarachnoid space extends to level S2 of the vertebral column. Nor does the needle damage the nerve roots of the cauda equina located in the subarachnoid space because the needle quite easily pushes them aside. In **spinal anesthesia,** or spinal block, drugs that block action potential transmission are introduced into the subarachnoid space to prevent pain sensations in the lower half of the body. A **spinal tap** is the removal of CSF from the subarachnoid space. A spinal tap may be performed to examine the CSF for infectious agents (meningitis), for the presence of blood (hemorrhage), or for the measurement of CSF pressure. A radiopaque substance may also be injected into this area, and a **myelogram** (radiograph of the spinal cord) may be taken to visualize spinal cord defects or damage.

2 ***Name the meninges surrounding the spinal cord. What is found within the epidural, subdural, and subarachnoid spaces?***

3 ***How is the spinal cord held within the vertebral canal?***

Cross Section of the Spinal Cord

A cross section of the spinal cord reveals that the cord consists of a peripheral white portion and a central gray portion (figure 12.3*a* and *b*). The white matter consists of myelinated axons forming nerve tracts, and the gray matter consists of neuron cell bodies, dendrites, and axons. An **anterior median fissure** and a **posterior median sulcus** are deep clefts partially separating the two halves of the cord. The white matter in each half of the spinal cord is organized into three **columns,** or **funiculi** (fū-nik′ū-lī), called the **ventral** (anterior), **dorsal** (posterior), and **lateral columns.** Each column of the spinal cord is subdivided into **tracts,** or **fasciculi** (fă-sik′ū-lī), also referred to as **pathways.** Individual axons ascending to the brain or descending from the brain are usually grouped together within the tracts (figure 12.3*c*). Axons within a given tract carry basically the same type of information, although they may overlap to some extent. For example, one ascending tract carries action potentials related to pain and temperature sensations, whereas another carries action potentials related to light touch.

The central gray matter is organized into horns. Each half of the central gray matter of the spinal cord consists of a relatively thin **posterior** (dorsal) **horn** and a larger **anterior** (ventral) **horn.** Small **lateral horns** exist in levels of the cord associated with the autonomic nervous system (see chapter 16). The two halves of the spinal cord are connected by **gray** and **white commissures** (see figure 12.3). The white and gray commissures contain axons

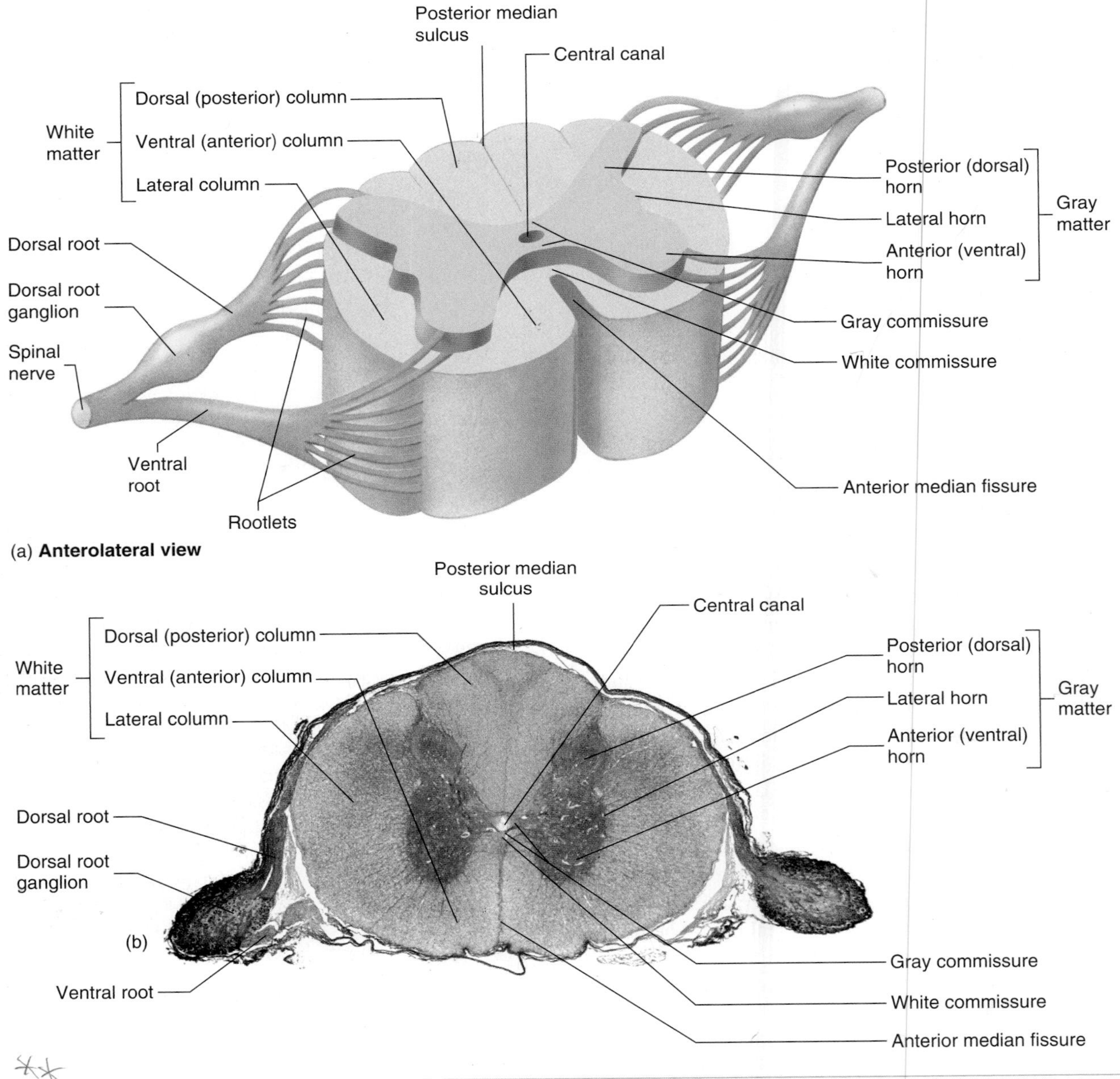

FIGURE 12.3 Cross Section of the Spinal Cord

(*a*) A segment of the spinal cord showing one dorsal and one ventral root on each side and the rootlets that form them. (*b*) Photograph of a cross section through the midlumbar region. The *lighter tan areas* are white matter, where tracts are located. The *darker area* is gray matter, where neuron cell bodies are located.

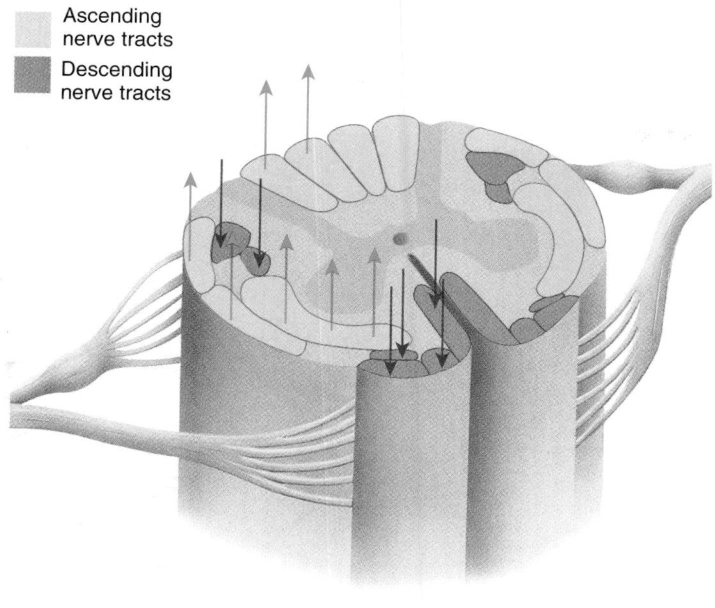

(c) **Anterolateral view**

FIGURE 12.3 **(*continued*)**

(*c*) Ascending and descending tracts in the spinal cord. Ascending nerve tracts are *blue*; descending nerve tracts are *pink*. The arrows indicate the direction of each pathway.

that cross from one side of the spinal cord to the other. The **central canal** is in the center of the gray commissure.

Spinal nerves arise from numerous rootlets along the dorsal and ventral surfaces of the spinal cord (see figure 12.3*a*). Six to eight of these rootlets combine to form each **ventral root** on the ventral (anterior) side of the spinal cord, and another six to eight form each **dorsal root** on the dorsal (posterior) side of the cord at each segment. The ventral and dorsal roots extend laterally from the spinal cord, passing through the subarachnoid space, piercing the arachnoid mater and dura mater, and joining one another to form a spinal nerve. Each dorsal root contains a ganglion, called the **dorsal root,** or **spinal, ganglion** (gang′glē-on; a swelling or knot).

The dorsal root ganglia are collections of cell bodies of the sensory neurons forming the dorsal roots of the spinal nerves (figure 12.4). The axons of these unipolar neurons extend from various parts of the body and pass through spinal nerves to the dorsal root ganglia. The axons do not synapse in the dorsal root ganglion but pass through the dorsal root into the posterior horn of the spinal cord gray matter. The axons either synapse with interneurons in the posterior horn or pass into the white matter and ascend or descend in the spinal cord.

Motor neurons innervate muscles and glands. The cell bodies of the motor neurons are located in the anterior and lateral horns of the spinal cord gray matter (see figure 12.4). The cell bodies of multipolar somatic motor neurons are in the anterior horn, also called the motor horn, and autonomic neuron cell bodies are in the lateral horn. Axons of the motor neurons form the ventral roots and pass into the spinal nerves. Thus, dorsal roots contain sensory axons, ventral roots contain motor axons, and spinal nerves have both sensory and motor axons.

4 ***Describe the arrangement of white and gray matter in the spinal cord.***

5 ***What are tracts and commissures?***

6 ***Where are the cell bodies of sensory, somatic motor, and autonomic neurons located in the gray matter?***

7 ***Where do dorsal and ventral roots exit the spinal cord? What kinds of axons are in the dorsal and ventral roots and in the spinal nerves?***

PREDICT 3

Explain why the dorsal root ganglia are larger in diameter than the dorsal roots, and describe the direction of action potential propogation in the spinal nerves, dorsal roots, and ventral roots.

REFLEXES

The basic structural unit of the nervous system is the neuron. The **reflex arc** is the basic functional unit of the nervous system because it is the smallest, simplest portion capable of receiving a stimulus and producing a response. Because the reflex arc is the basic functional unit of the nervous system, it is possible to learn

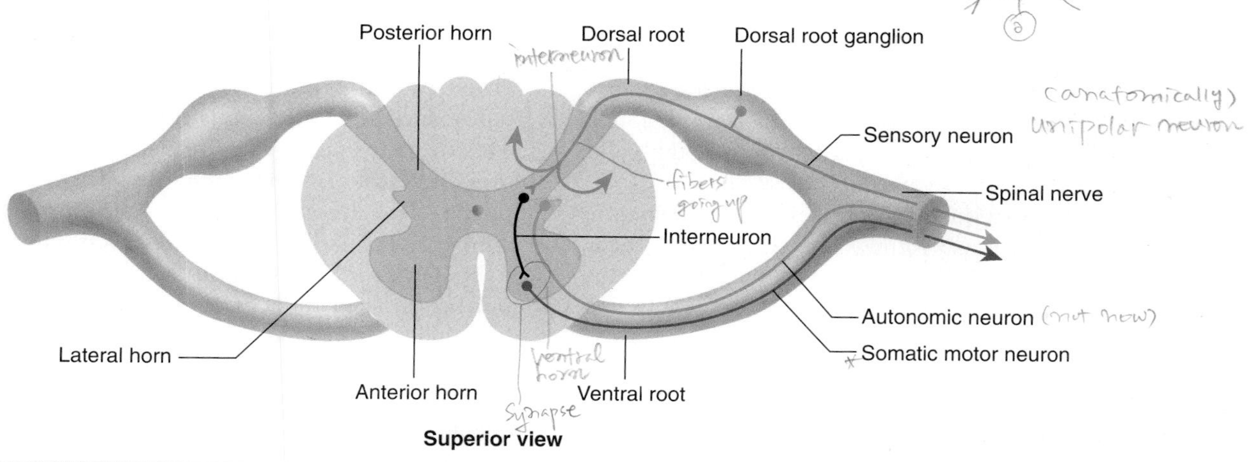

FIGURE 12.4 **Relationship of Sensory and Motor Neurons to the Spinal Cord**

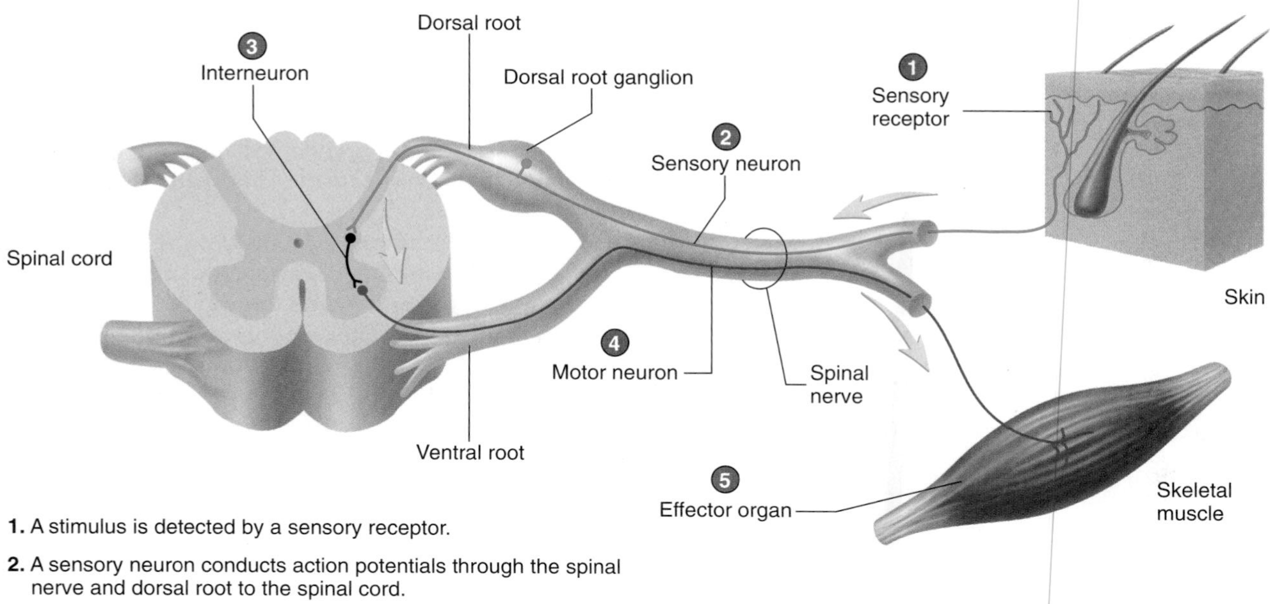
Dorsal root
3
Interneuron
Dorsal root ganglion
1
Sensory receptor
2
Sensory neuron
Spinal cord
Skin
4
Motor neuron
Spinal nerve
Ventral root
5
Effector organ
Skeletal muscle
1. A stimulus is detected by a sensory receptor.
2. A sensory neuron conducts action potentials through the spinal nerve and dorsal root to the spinal cord.
3. In the spinal cord, the sensory neuron synapses with an interneuron.
4. The interneuron synapses with a motor neuron.
5. A motor neuron axon conducts action potentials through the ventral root and spinal nerve to an effector organ.

PROCESS FIGURE 12.5 Reflex Arc

The parts of a reflex arc are labeled in the order in which action potentials pass through them.

much about nervous system functions by examining how reflex arcs receive stimuli and produce responses. The reflex arc has five basic components: (1) a sensory receptor, (2) a sensory neuron, (3) an interneuron, (4) a motor neuron, and (5) an effector organ (figure 12.5).

A **reflex** is an automatic response to a stimulus produced by a reflex arc. It occurs without conscious thought. Action potentials initiated in sensory receptors are transmitted along the axons of sensory neurons to the CNS, where the axons usually synapse with interneurons. Interneurons synapse with motor neurons, which send axons out of the spinal cord and through the PNS to muscles or glands, where the action potentials of the motor neurons cause these effector organs to respond.

Reflexes are homeostatic. Some reflexes, called somatic reflexes, remove the body from painful stimuli that would cause tissue damage or keep the body from suddenly falling or moving because of external forces. A number of other reflexes, called autonomic reflexes, are responsible for maintaining relatively constant blood pressure, blood carbon dioxide levels, and water intake.

Individual reflexes vary in their complexity. **Monosynaptic reflexes** involve simple neuronal pathways in which sensory neurons synapse directly with motor neurons. **Polysynaptic reflexes** involve more complex pathways and integrative centers with one or more interneurons between the sensory and motor neurons.

Many reflexes are integrated within the spinal cord, and others are integrated within the brain. Some reflexes involve excitatory neurons and result in a response, such as when a muscle contracts (see chapter 11). Other reflexes involve inhibitory neurons and result in the inhibition of a response, such as when a muscle is inhibited and relaxes. In addition, higher brain centers influence reflexes by either suppressing or exaggerating them. Major spinal cord reflexes include the stretch reflex, the Golgi tendon reflex, and the withdrawal reflex.

8 ***Name the parts of a reflex arc and explain how they produce a reflex.***

Stretch Reflex

The simplest reflex is the **stretch reflex** (figure 12.6), a reflex in which muscles contract in response to a stretching force applied to them. The sensory receptor of this reflex is the **muscle spindle,** which consists of 3–10 small, specialized skeletal muscle cells. The cells are contractile only at their ends and are innervated by specific motor neurons called **gamma motor neurons** (the term *gamma* refers to motor neurons with small-diameter axons) originating from the spinal cord and controlling the sensitivity of the muscle spindle cells. Sensory neurons innervate the noncontractile centers of the muscle spindle cells. Axons of these sensory neurons extend to the spinal cord and synapse directly with motor neurons in the spinal cord called **alpha motor neurons,** which in turn innervate the muscle in which the muscle spindle is embedded. Neurons can be classified by the diameter of their axons, with alpha motor neurons having the largest diameter. The stretch reflex is a monosynaptic reflex because there is no interneuron between the sensory neuron and the alpha motor neuron.

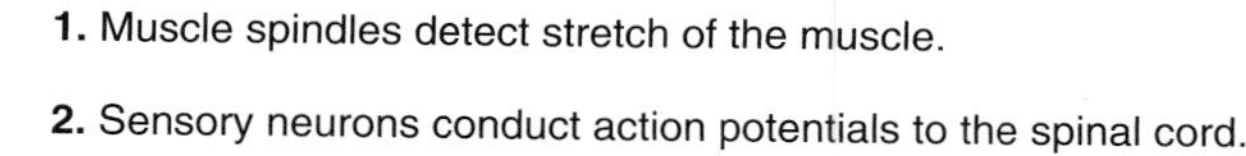

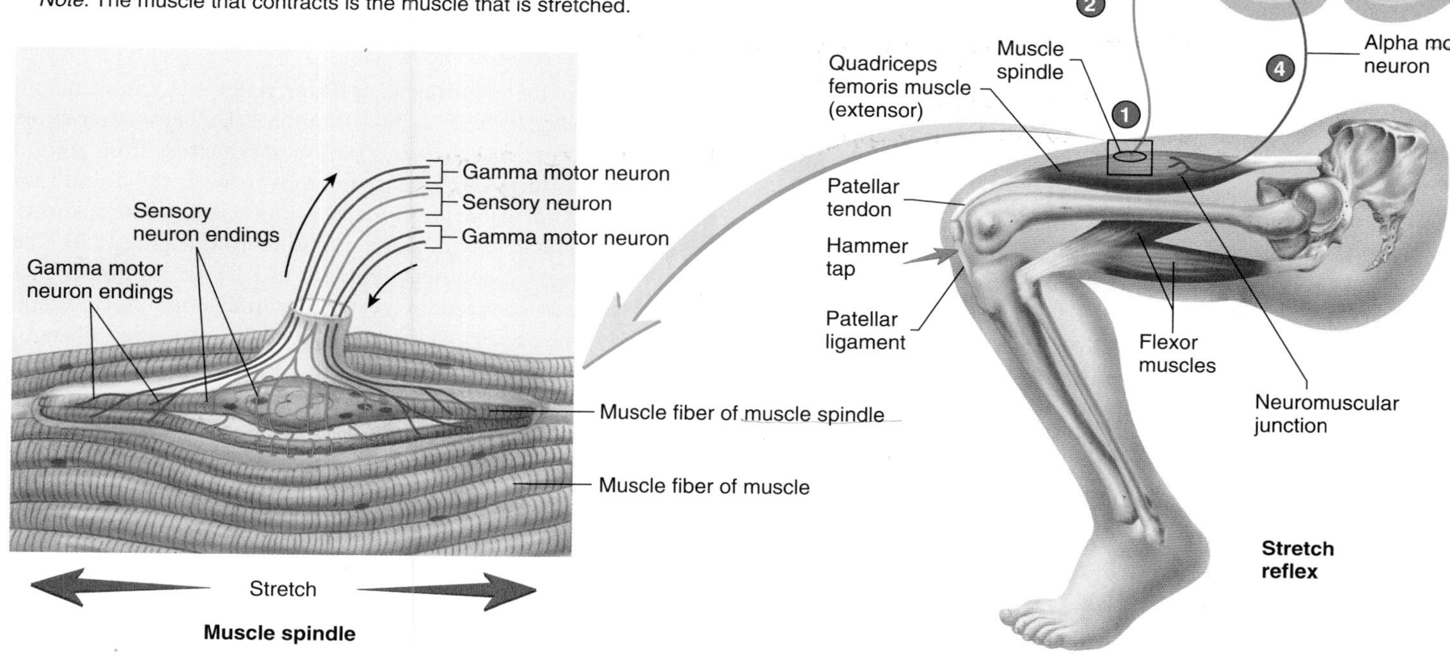

PROCESS FIGURE 12.6 Stretch Reflex

Stretching a muscle also stretches muscle spindles located among the muscle fibers. The stretch stimulates the sensory neurons that innervate the center of each of the muscle spindles. The increased frequency of action potentials carried to the spinal cord by sensory neurons stimulates the alpha motor neurons in the spinal cord. The alpha motor neurons transmit action potentials to skeletal muscle, causing a rapid contraction of the stretched muscle, which opposes the stretch of the muscle. The postural muscles demonstrate the adaptive nature of this reflex. If a person is standing upright and then begins to tip slightly to one side, the postural muscles associated with the vertebral column on the other side are stretched. As a result, stretch reflexes are initiated in those muscles, which cause them to contract and reestablish normal posture.

Collateral axons from the sensory neurons of the muscle spindles also synapse with neurons whose axons contribute to ascending nerve tracts, which enable the brain to perceive that a muscle has been stretched (see figure 12.3*c*). Descending neurons within the spinal cord synapse with the neurons of the stretch reflex modifying their activity. This activity is important in maintaining posture and in coordinating muscular activity.

Gamma motor neurons are responsible for regulating the sensitivity of the muscle spindles. As a skeletal muscle contracts, the tension on the centers of muscle spindles within the muscle decreases because the muscle spindles passively shorten as the muscle shortens. The decrease in tension in the centers of the muscle spindles cause them to be less sensitive to stretch. Sensitivity is maintained because while alpha motor neurons are stimulating the muscle to contract gamma motor neurons stimulate the muscle spindles to contract. The contraction of the muscle fibers at the ends of the muscle spindles pulls on the center part of the muscle spindles and maintains the proper tension. The activity of the muscle spindles help control posture, muscle tension, and muscle length.

Knee-Jerk Reflex

The **knee-jerk reflex,** or **patellar reflex,** is a classic example of the stretch reflex. Clinicians use this reflex to determine whether the higher CNS centers that normally influence this reflex are functional. When the patellar ligament is tapped, the tendons and muscles of the quadriceps femoris muscle group are stretched. The muscle spindle fibers within these muscles are also stretched, and the stretch reflex is activated. Consequently, contraction of the muscles extends the leg, thus producing the characteristic knee-jerk response. When the stretch reflex is greatly exaggerated, it indicates that the neurons within the brain that innervate the gamma motor neurons and enhance the stretch reflex are overly active. On the other hand, if the neurons that innervate the gamma motor neurons are depressed, the stretch reflex can be suppressed or absent. Absence of the stretch reflex may also indicate that the reflex pathway is not intact.

Golgi Tendon Reflex

The **Golgi tendon reflex** prevents contracting muscles from applying excessive tension to tendons. **Golgi tendon organs** are encapsulated nerve endings that have at their ends numerous terminal branches with small swellings associated with bundles of collagen fibers in tendons. The Golgi tendon organs are located within tendons near the muscle–tendon junction (figure 12.7). As a muscle contracts, the attached tendons are stretched, resulting in increased tension in the tendon. The increased tension stimulates action potentials in the sensory neurons from the Golgi tendon organs. Golgi tendon organs have a high threshold and are sensitive only to intense stretch.

The sensory neurons of the Golgi tendon organs pass through the dorsal root to the spinal cord and enter the posterior gray matter, where they branch and synapse with inhibitory interneurons. The interneurons synapse with alpha motor neurons that innervate the muscle to which the Golgi tendon organ is attached. When a great amount of tension is applied to the tendon, the sensory neurons of the Golgi tendon organs are stimulated. The sensory neurons stimulate the interneurons to release inhibitory neurotransmitters, which inhibit the alpha motor neurons of the associated muscle and cause it to relax. This reflex protects muscles and tendons from damage caused by excessive tension. The sudden relaxation of the muscle reduces the tension applied to the muscle and tendons. A weight lifter who suddenly drops a heavy weight after straining to lift it does so, in part, because of the effect of the Golgi tendon reflex.

Tremendous amounts of tension can be applied to muscles and tendons in the legs. Frequently, an athlete's Golgi tendon reflex is inadequate to protect muscles and tendons from excessive tension. The large muscles and sudden movements of football players and sprinters can make them vulnerable to relatively frequent hamstring pulls and calcaneal (Achilles) tendon injuries.

Withdrawal Reflex

The function of the **withdrawal,** or **flexor, reflex** is to remove a limb or another body part from a painful stimulus. The sensory receptors are pain receptors (see chapter 15). Action potentials from painful stimuli are conducted by sensory neurons through the dorsal root to the spinal cord, where they synapse with excitatory interneurons, which in turn synapse with alpha motor neurons (figure 12.8). The alpha motor neurons stimulate muscles, usually flexor muscles, that remove the limb from the source of the painful stimulus. Collateral branches of the sensory neurons synapse with ascending fibers to the brain, providing conscious awareness of the painful stimuli.

Reciprocal Innervation

Reciprocal innervation is associated with the withdrawal reflex and reinforces its efficiency (figure 12.9). Collateral axons of sensory

1. Golgi tendon organs detect tension applied to a tendon.
2. Sensory neurons conduct action potentials to the spinal cord.
3. Sensory neurons synapse with inhibitory interneurons that synapse with alpha motor neurons.
4. Inhibition of the alpha motor neurons causes muscle relaxation, relieving the tension applied to the tendon. *Note:* The muscle that relaxes is attached to the tendon to which tension is applied.

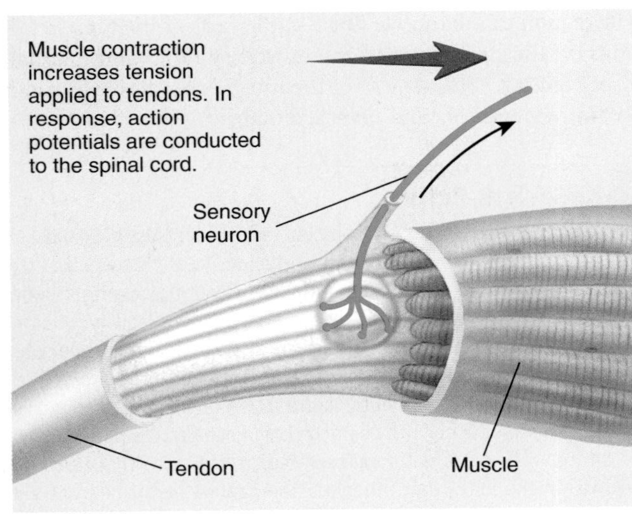

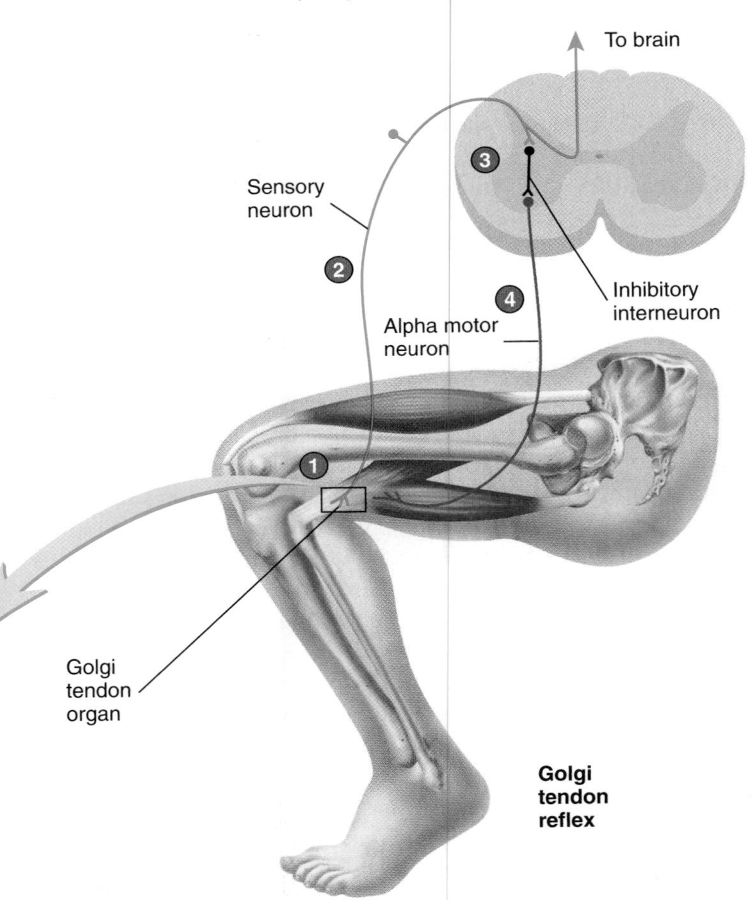

PROCESS FIGURE 12.7 **Golgi Tendon Reflex**

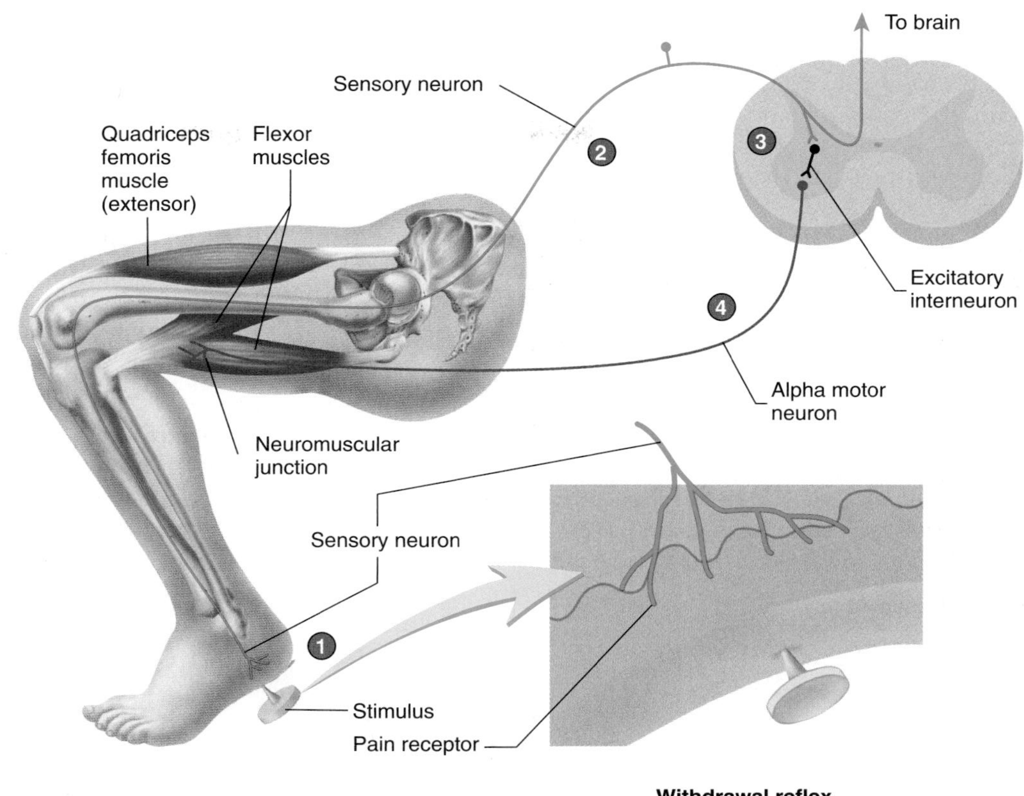

Stimulation of pain receptors results in:
1. Pain receptors detect a painful stimulus.
2. Sensory neurons conduct action potentials to the spinal cord.
3. Sensory neurons synapse with excitatory interneurons that synapse with alpha motor neurons.
4. Excitation of the alpha motor neurons results in contraction of the flexor muscles and withdrawal of the limb from the painful stimulus.
To brain
Sensory neuron
Quadriceps femoris muscle (extensor)
Flexor muscles
Excitatory interneuron
Alpha motor neuron
Neuromuscular junction
Sensory neuron
Stimulus
Pain receptor
Withdrawal reflex

PROCESS FIGURE 12.8 **Withdrawal Reflex**

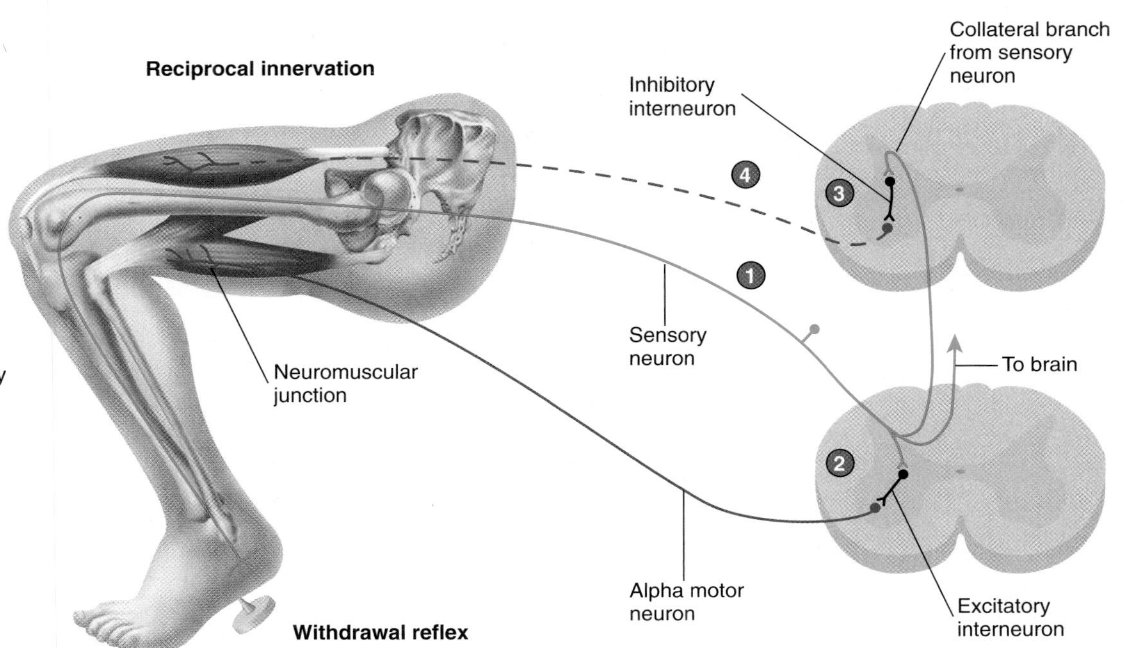

Reciprocal innervation
1. During the withdrawal reflex, sensory neurons conduct action potentials from pain receptors to the spinal cord.
2. Sensory neurons synapse with excitatory interneurons that are part of the withdrawal reflex.
3. Collateral branches of the sensory neurons also synapse with inhibitory interneurons that are part of reciprocal innervation.
4. The inhibitory interneurons synapse with alpha motor neurons supplying the extensor muscles, causing them to relax and not oppose the flexor muscles of the withdrawal reflex, which are contracting.
Reciprocal innervation
Collateral branch from sensory neuron
Inhibitory interneuron
Sensory neuron
Neuromuscular junction
To brain
Alpha motor neuron
Excitatory interneuron
Withdrawal reflex

PROCESS FIGURE 12.9 **Withdrawal Reflex with Reciprocal Innervation**

neurons that carry action potentials from pain receptors synapse with inhibitory interneurons in the dorsal horn of the spinal cord. The inhibitory interneurons synapse with and inhibit alpha motor neurons of extensor (antagonist) muscles. When the withdrawal reflex is initiated, flexor muscles contract, and reciprocal innervation causes relaxation of the extensor muscles. This reduces the resistance to movement that the extensor muscles would otherwise generate.

Reciprocal innervation is also involved in the stretch reflex. When the stretch reflex causes a muscle to contract, reciprocal innervation causes opposing muscles to relax. In the patellar reflex, for example, the quadriceps femoris muscle contracts and the hamstring muscles relax.

Crossed Extensor Reflex

The **crossed extensor reflex** is another reflex associated with the withdrawal reflex (figure 12.10). Interneurons that stimulate alpha motor neurons, resulting in withdrawal of a limb, have collateral axons that extend through the white commissure to the opposite side of the spinal cord and synapse with alpha motor neurons that innervate extensor muscles in the opposite side of the body. When a withdrawal reflex is initiated in one lower limb, the crossed extensor reflex causes extension of the opposite lower limb.

The crossed extensor reflex is adaptive in that it helps prevent falls by shifting the weight of the body from the affected to the unaffected limb. For example, when a person steps on a sharp object, the affected limb is withdrawn from the stimulus (withdrawal reflex) while the other limb is extended (crossed extensor reflex). Therefore, when a person steps on a sharp object with the right foot, the body weight is shifted from the right to the left lower limb. Initiating a withdrawal reflex in both legs at the same time would cause one to fall.

9 ***Contrast and give the functions of a stretch reflex and a Golgi tendon reflex. Describe the sensory receptors for each.***

10 ***Describe the operation of gamma motor neurons. What do they accomplish?***

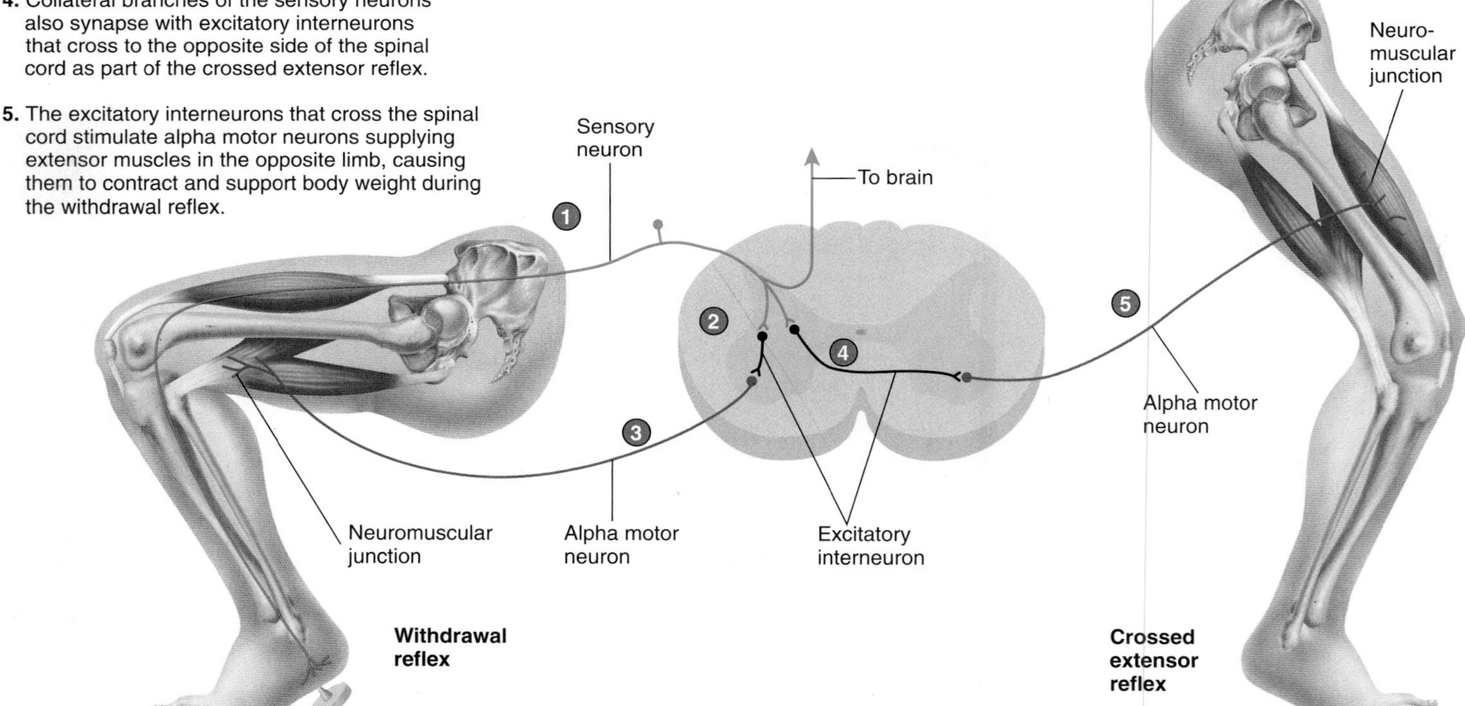
Crossed extensor reflex
1. During the withdrawal reflex, sensory neurons from pain receptors conduct action potentials to the spinal cord.
2. Sensory neurons synapse with excitatory interneurons that are part of the withdrawal reflex.
3. The excitatory interneurons that are part of the withdrawal reflex stimulate alpha motor neurons that innervate flexor muscles, causing withdrawal of the limb from the painful stimulus.
4. Collateral branches of the sensory neurons also synapse with excitatory interneurons that cross to the opposite side of the spinal cord as part of the crossed extensor reflex.
5. The excitatory interneurons that cross the spinal cord stimulate alpha motor neurons supplying extensor muscles in the opposite limb, causing them to contract and support body weight during the withdrawal reflex.
Sensory neuron
To brain
Neuro-muscular junction
Alpha motor neuron
Neuromuscular junction
Alpha motor neuron
Excitatory interneuron
Withdrawal reflex
Crossed extensor reflex

PROCESS FIGURE 12.10 Withdrawal Reflex with Crossed Extensor Reflex

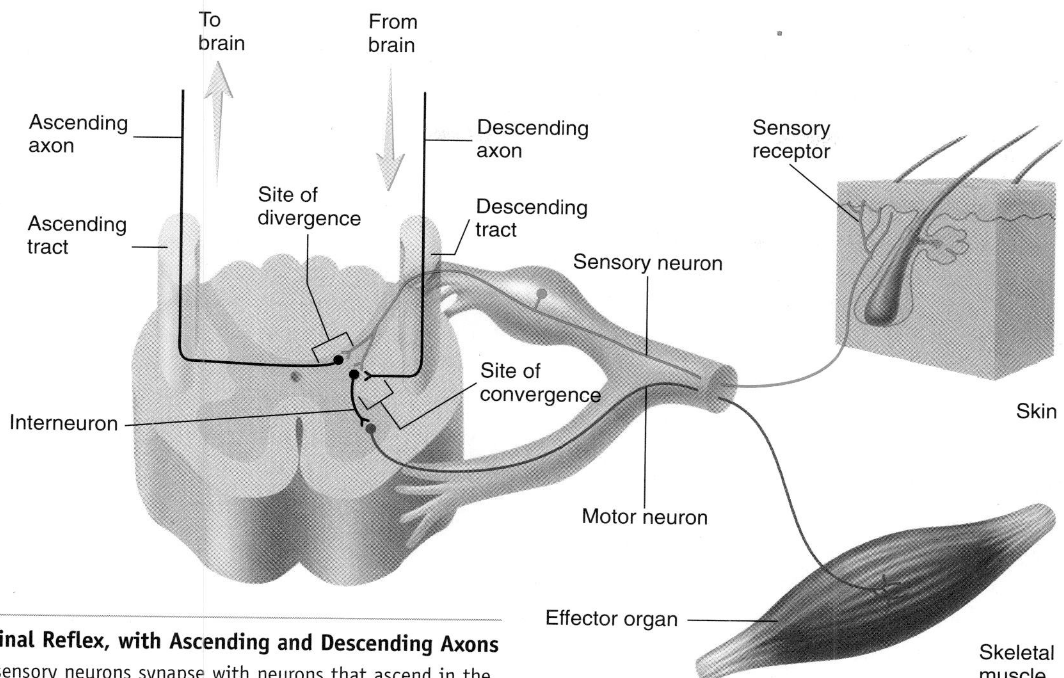

FIGURE 12.11 **Spinal Reflex, with Ascending and Descending Axons**
Diverging branches of sensory neurons synapse with neurons that ascend in the spinal cord to the brain. Axons from the brain descend in the spinal cord and synapse with interneurons, influencing their action on motor neurons.

11 ***What is a withdrawal reflex? How do reciprocal innervation and the crossed extensor reflex assist the withdrawal reflex?***

INTERACTIONS WITH SPINAL CORD REFLEXES

Reflexes do not operate as isolated entities within the nervous system because of divergent and convergent pathways (see chapter 11). Diverging branches of the sensory neurons or interneurons in a reflex arc send action potentials along ascending nerve tracts to the brain (figure 12.11). A pain stimulus, for example, not only initiates a withdrawal reflex, which removes the affected part of the body from the painful stimulus, but also causes perception of the pain sensation as a result of action potentials sent to the brain.

Axons within descending tracts from the brain carry action potentials to motor neurons in the anterior horn of the spinal cord, converging with neurons of reflex arcs. The neurotransmitters released from the axons of these tracts either stimulate or inhibit the motor neurons in the anterior horn. Neurotransmitters change the sensitivity of the reflex by stimulating (EPSP) or inhibiting (IPSP) the motor neurons.

12 ***How do ascending and descending pathways relate to reflexes and other neuron functions?***

STRUCTURE OF PERIPHERAL NERVES

Peripheral nerves consist of axons, Schwann cells, and connective tissue (figure 12.12). Each axon, or nerve fiber, and its Schwann cell sheath are surrounded by a delicate connective

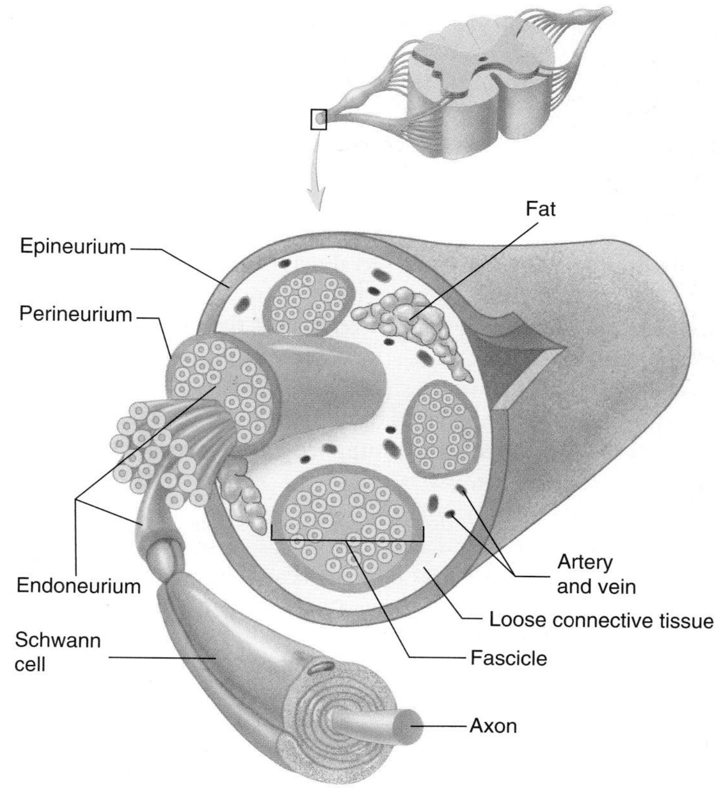

FIGURE 12.12 **Structure of a Peripheral Nerve**
Nerve structure illustrating axons surrounded by various layers of connective tissue: epineurium surrounds the whole nerve, perineurium surrounds nerve fascicles, and endoneurium surrounds Schwann cells and axons. Loose connective tissue also surrounds the nerve fascicles.

tissue layer, the **endoneurium** (en-dō-noo′rē-ŭm). A heavier connective tissue layer, the **perineurium** (per-i-noo′rē-ŭm), surrounds groups of axons to form nerve **fascicles** (fas′i-klz). A third layer of dense connective tissue, the **epineurium** (ep-i-noo′rē-ŭm), binds the nerve fascicles together to form a nerve. The connective tissue of the epineurium is continuous with the dura mater surrounding the CNS. An analogy of this relationship is a coat (the dura) and its sleeve (epineurium). The connective tissue layers of nerves make them tougher than the nerve tracts in the CNS.

13 ***Describe the structure of peripheral nerves.***

SPINAL NERVES

All of the 31 pairs of spinal nerves, except the first pair and those in the sacrum, exit the vertebral column through intervertebral foramina located between adjacent vertebrae (see figure 7.13). The first pair of spinal nerves exits between the skull and the first cervical vertebra. The nerves of the sacrum exit from the single bone of the sacrum through the sacral foramina (see chapter 7). Eight spinal nerve pairs exit the vertebral column in the cervical region, 12 in the thoracic region, 5 in the lumbar region, 5 in the sacral region, and 1 in the coccygeal region (figure 12.13). For convenience, each of the spinal nerves is designated by a letter and number. The letter indicates the region of the vertebral column from which the nerve emerges: C, cervical; T, thoracic; L, lumbar; and S, sacral. The single coccygeal nerve is often not designated, but when it is the symbol often used is Co. The number indicates the location in each region where the nerve emerges from the vertebral column, with the smallest number always representing the most superior origin. For example, the most superior nerve exiting from the thoracic region of the vertebral column is designated T1. The cervical nerves are designated C1–C8, the thoracic nerves T1–T12, the lumbar nerves L1–L5, and the sacral nerves S1–S5.

The nerves arising from each region of the spinal cord and vertebral column supply specific regions of the body. Each of the spinal nerves except C1 has a specific cutaneous sensory distribution. Figure 12.14 illustrates the **dermatomal** (der-mă-tō′măl) **map** for the sensory cutaneous distribution of the spinal nerves. A **dermatome** is the area of skin supplied with sensory innervation by a pair of spinal nerves.

PREDICT 4

The dermatomal map is important in clinical considerations of nerve damage. Loss of sensation in a dermatomal pattern can provide valuable information about the location of nerve damage. Predict the possible site of nerve damage for a patient who suffered whiplash in an automobile accident and subsequently developed anesthesia (no sensations) in the left arm, forearm, and hand (see figure 12.14 for help).

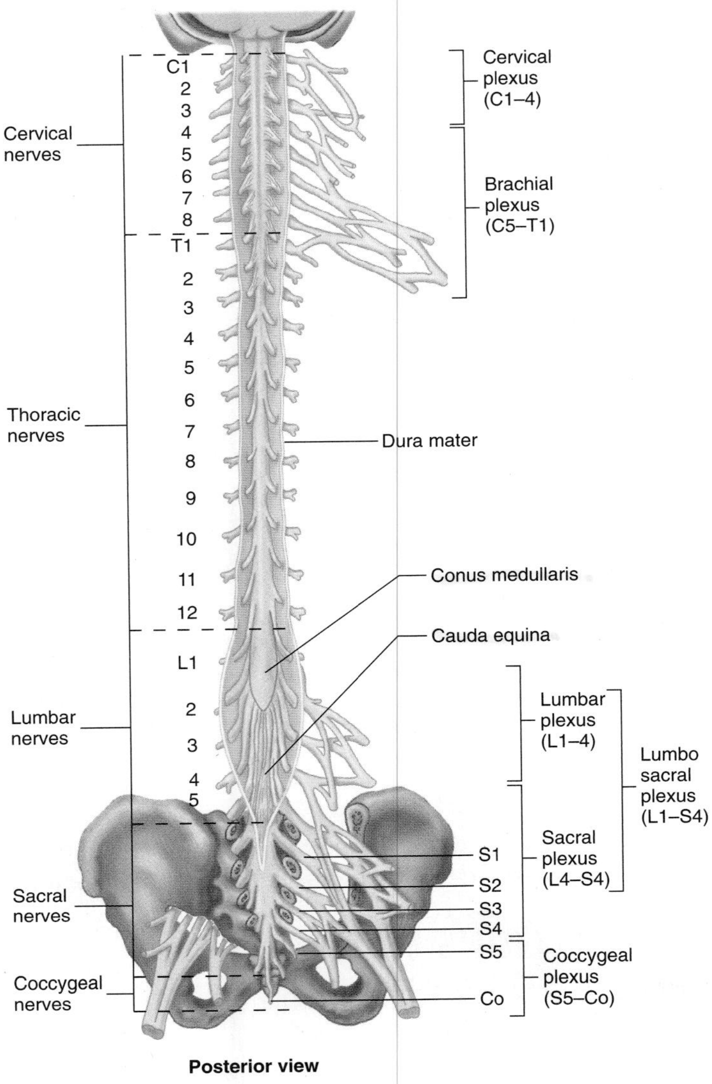

FIGURE 12.13 Spinal Nerves

The regional designations and the numbers of the spinal nerves are shown on the left. The plexuses formed by the spinal nerves are shown on the right.

Figure 12.15 depicts an idealized section through the trunk. Each spinal nerve has a dorsal and a ventral **ramus** (rā′mŭs; branch). Additional rami (rā′mī), called communicating rami, from the thoracic and upper lumbar spinal cord regions carry axons associated with the sympathetic division of the autonomic nervous system (see chapter 16). The **dorsal rami** (rā′mī) innervate most of the deep muscles of the dorsal trunk responsible for movement of the vertebral column. They also innervate the connective tissue and skin near the midline of the back.

The **ventral rami** are distributed in two ways. In the thoracic region, the ventral rami form **intercostal** (between ribs) **nerves** (see figure 12.15), which extend along the inferior margin of each rib and innervate the intercostal muscles and the skin over the thorax. The ventral rami of the remaining spinal nerves form five major **plexuses** (plek′sŭs-ēz). The term *plexus* means braid and describes the organization produced by the intermingling of the nerves. The ventral rami of different spinal nerves, called the **roots** of the plexus, join with each other to form a plexus. These roots should not be confused with the dorsal and ventral roots from

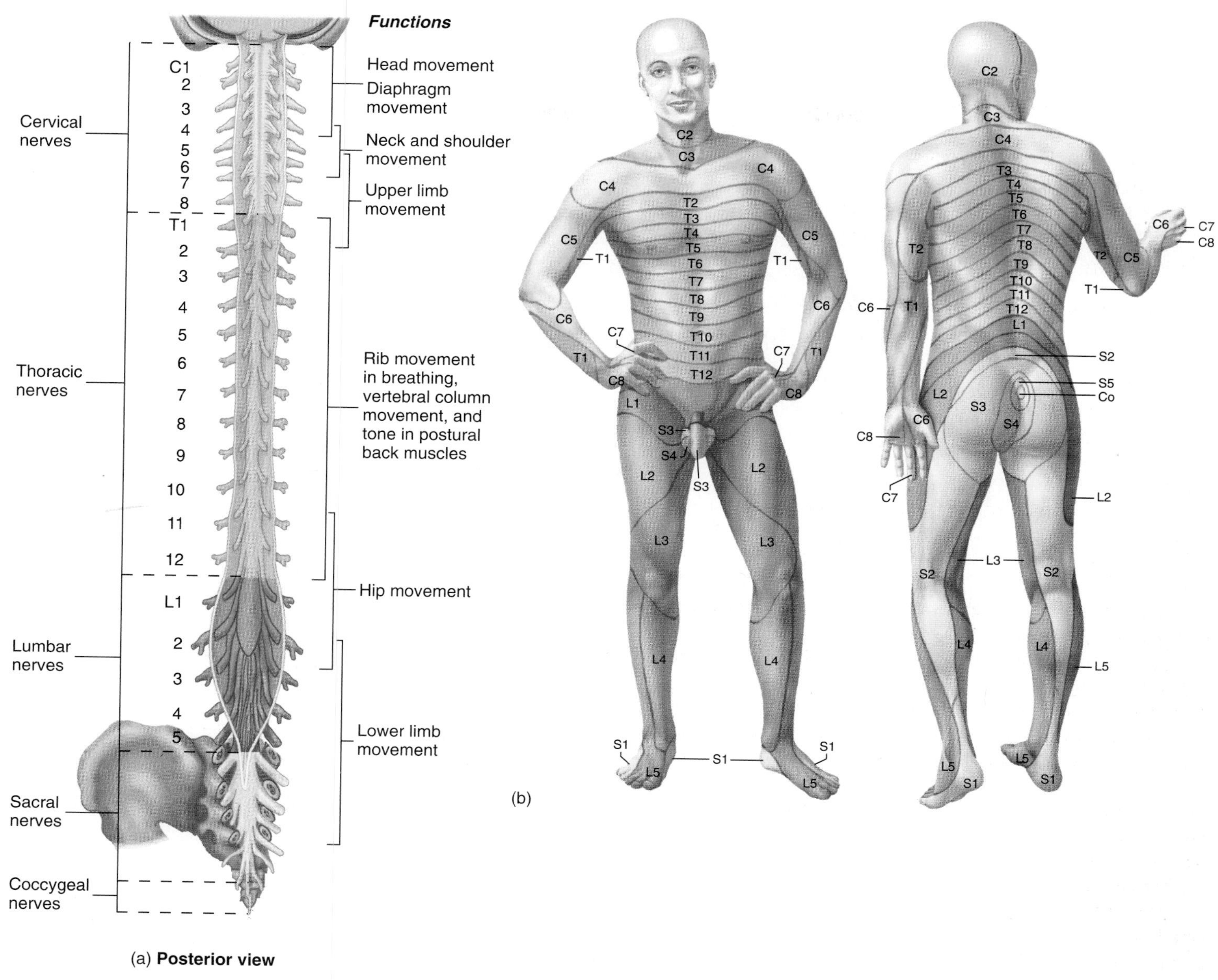

FIGURE 12.14 Spinal Cord and Dermatomal Map

(*a*) Nerves and functions of the spinal cord. Regions are color-coded. (*b*) Letters and numbers indicate the spinal nerves innervating a given region of skin.

the spinal cord, which are more medial. Nerves that arise from plexuses usually have axons from more than one spinal nerve and thus more than one level of the spinal cord. The ventral rami of spinal nerves C1–C4 form the cervical plexus, C5–T1 form the brachial plexus, L1–L4 form the lumbar plexus, L4–S4 form the sacral plexus, and S5 and the coccygeal nerve (Co) form the coccygeal plexus.

Several smaller somatic plexuses, such as the pudendal plexus in the pelvis, are derived from more distal branches of the spinal nerves. Some of the somatic plexuses are mentioned where appropriate in this chapter. Autonomic plexuses (described in chapter 16) also exist in the thorax and abdomen.

The five major plexuses derived from the ventral rami of spinal nerves are described more fully in the following sections.

14 ***Describe the connective tissue layers within and surrounding spinal nerves.***

15 ***Differentiate among rootlet, dorsal root, ventral root, and spinal nerve. Indicate whether each contains motor fibers, sensory fibers, or both.***

16 ***List all of the spinal nerves by name and number. Where do they exit the vertebral column?***

17 ***What is a dermatome? Why are dermatomes clinically important?***

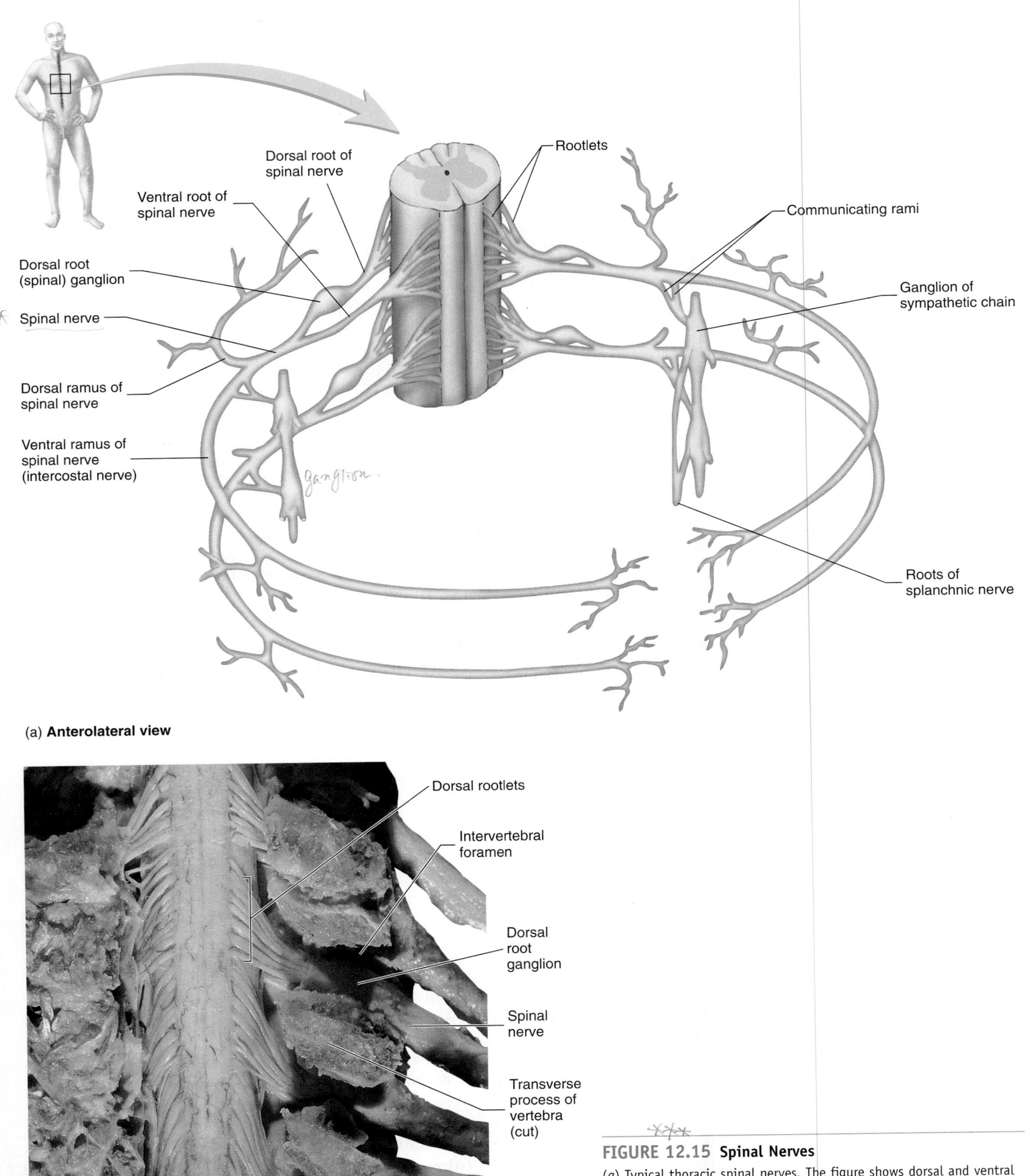

FIGURE 12.15 Spinal Nerves

(*a*) Typical thoracic spinal nerves. The figure shows dorsal and ventral roots, as well as dorsal, ventral, and communicating rami. Communicating rami connect to the sympathetic chain (see chapter 16). (*b*) Photograph of four dorsal roots in place along the vertebral column.

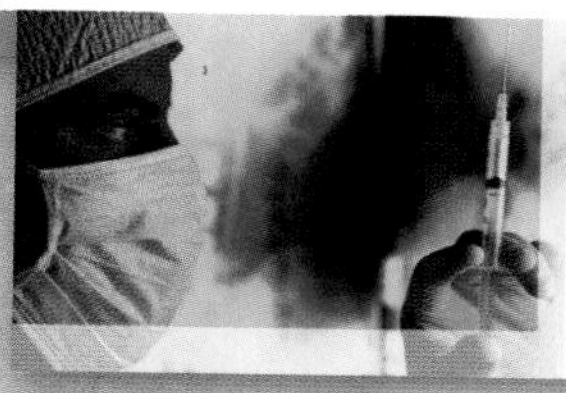

CLINICAL FOCUS

Spinal Cord Injury

Damage to the spinal cord can disrupt ascending tracts from the spinal cord to the brain, resulting in the loss of sensation, and/or descending tracts from the brain to motor neurons in the spinal cord, resulting in the loss of motor functions. About 10,000 new cases of **spinal cord injury** occur each year in the United States. Automobile and motorcycle accidents are leading causes, followed by gunshot wounds, falls, and swimming accidents. Spinal cord injury is classified according to the vertebral level at which the injury occurred, whether the entire cord is damaged at that level or only a portion of the cord, and the mechanism of injury. Most spinal cord injuries occur in the cervical region or at the thoracolumbar junction and are incomplete. The primary mechanisms include concussion (an injury caused by a blow), contusion (an injury resulting in hemorrhage), and laceration (a tear or cut) and involve excessive flexion, extension, rotation, or compression of the vertebral column. The majority of spinal cord injuries are acute contusions of the cord due to bone or disk displacement into the cord and involve a combination of excessive directional movements, such as simultaneous flexion and compression.

At the time of spinal cord injury, two types of tissue damage occur: (1) primary, mechanical damage and (2) secondary, tissue damage. Secondary spinal cord damage, which begins within minutes of the primary damage, is caused by ischemia, edema, ion imbalances, the release of "excitotoxins" (such as glutamate), and inflammatory cell invasion. Secondary damage extends into a much larger region of the cord than the primary damage. It is the primary focus of current research in spinal cord injury. The only treatment for primary damage is prevention, such as wearing seat belts when riding in automobiles and not diving into shallow water. Once an accident occurs, however, little can be done about the primary damage. On the other hand, much of the secondary damage can be prevented or reversed.

Until the 1950s, many spinal cord injuries were ultimately fatal. Now, with quick treatment directed at the mechanisms of secondary tissue damage, much of the total damage to the spinal cord can be prevented. Treatment of the damaged spinal cord with large doses of anti-inflammatory steroids, such as methylprednisolone, within 8 hours of the injury can dramatically reduce the secondary damage to the cord. The objectives of this treatment are to reduce inflammation and edema. Current treatment includes anatomical realignment and stabilization of the vertebral column, decompression of the spinal cord, and administration of anti-inflammatory steroids. Rehabilitation is based on retraining the patient to use whatever residual connections exist across the site of damage.

It had long been thought that the spinal cord was incapable of regeneration following severe damage. However, it is now known that, following injury, most neurons of the adult spinal cord survive and begin to regenerate, growing about 1 mm into the site of damage, but then they regress to an inactive, atrophic state. In addition, fetuses and newborns exhibit considerable regenerative ability and functional improvement. A major block to adult spinal cord regeneration is the formation of a scar, consisting mainly of myelin and astrocytes, at the site of injury. Myelin in the scar is apparently the primary inhibitor of regeneration. Implantation of peripheral nerves, Schwann cells, or fetal CNS tissue can bridge the scar and stimulate some regeneration. Certain growth factors can also stimulate some regeneration. Current research continues to look for the right combination of chemicals and other factors to stimulate regeneration of the spinal cord following injury.

18 ***Contrast dorsal and ventral rami of spinal nerves. What muscles do the dorsal rami innervate?***

19 ***Describe the distribution of the ventral rami of the thoracic region.***

20 ***What is a plexus? What happens to the axons of spinal nerves as they pass through a plexus?***

21 ***Name the main spinal plexuses and the spinal nerves associated with each one.***

Cervical Plexus

The **cervical plexus** is a relatively small plexus originating from spinal nerves C1–C4 (figure 12.16). Branches derived from this plexus innervate superficial neck structures, including several of the muscles attached to the hyoid bone. The cervical plexus innervates the skin of the neck and posterior portion of the head (see figure 12.14). An unusual part of the cervical plexus, the **ansa** (an′sah; bucket handle) **cervicalis,** is a loop between C1 and C3. Nerves to the infrahyoid muscles branch from the ansa cervicalis (see chapter 10).

One of the most important derivatives of the cervical plexus is the **phrenic** (fren′ik) **nerve,** which originates from spinal nerves C3–C5, derived from both the cervical and brachial plexus. The phrenic nerves descend along each side of the neck to enter the thorax. They descend along the sides of the mediastinum to reach the diaphragm, which they innervate. Contraction of the diaphragm is largely responsible for the ability to breathe.

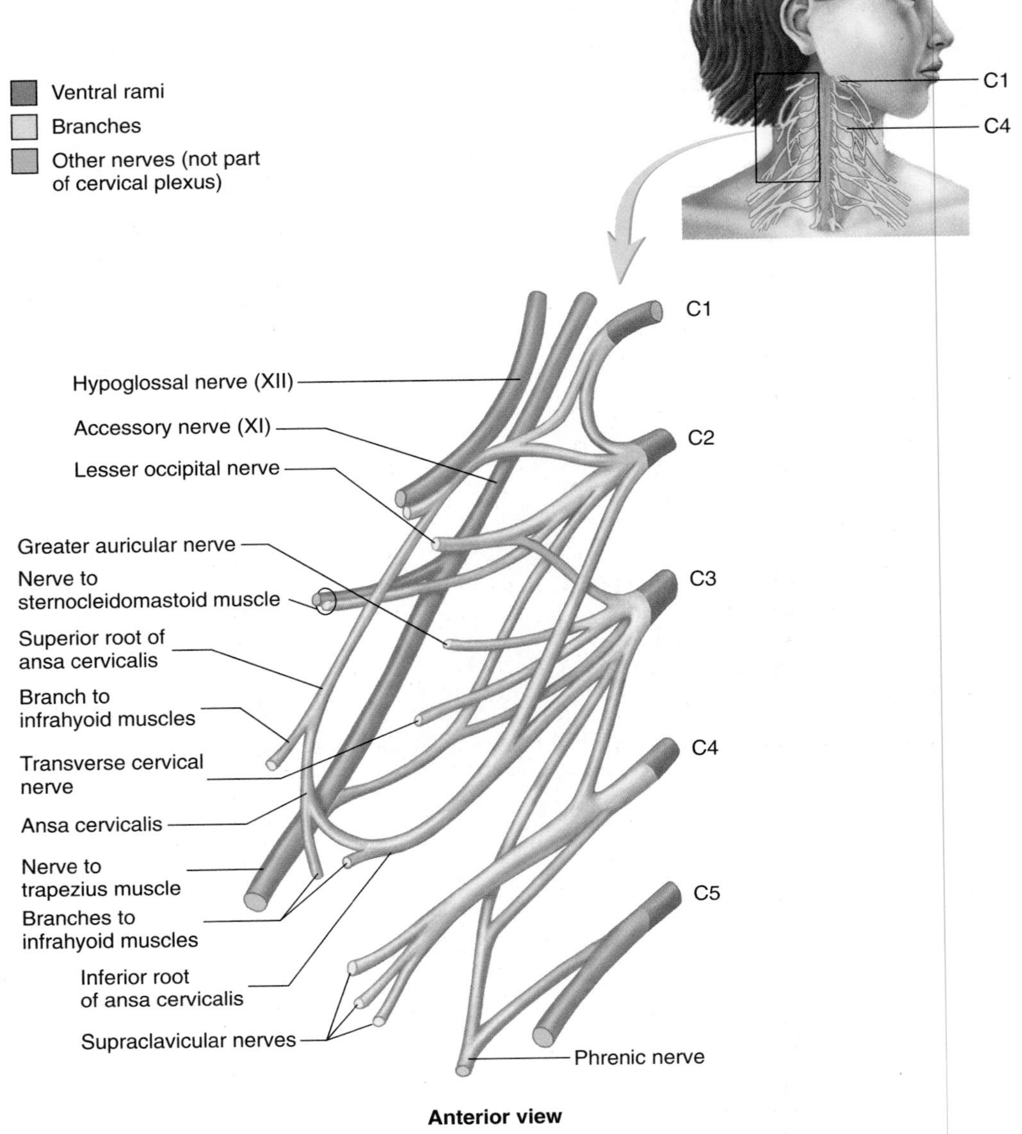

FIGURE 12.16 Cervical Plexus

The roots of the plexus are formed by the ventral rami of the spinal nerves C1–C4. Branches of the cervical plexus innervate the muscles (infrahyoid) and skin of the neck. The phrenic nerve (C3–C5) innervates the diaphragm.

Phrenic Nerve Damage

Damage to the phrenic nerve severely limits a person's ability to breathe. Care must be taken not to damage the phrenic nerve during open thoracic surgery or open heart surgery. Cancer of the bronchus is the most common type of cancer in men, accounting for about 30% of all male cancers, and it most often occurs in men who smoke cigarettes. Tumors at the base of the lung can compress the phrenic nerve.

PREDICT 5

Explain how damage to or compression of the right phrenic nerve affects the diaphragm. Describe the effect on breathing of a completely severed spinal cord at the level of C2 versus at the level of C6.

22 ***Name the structures innervated by the cervical plexus. Describe the innervation of the phrenic nerve.***

Brachial Plexus

The **brachial plexus** originates from spinal nerves C5–T1 (figure 12.17). The five ventral rami that constitute the brachial plexus join to form three **trunks,** which separate into six **divisions** and then join again to create three **cords** (posterior, lateral, and medial) from which five **branches,** or nerves of the upper limb, emerge.

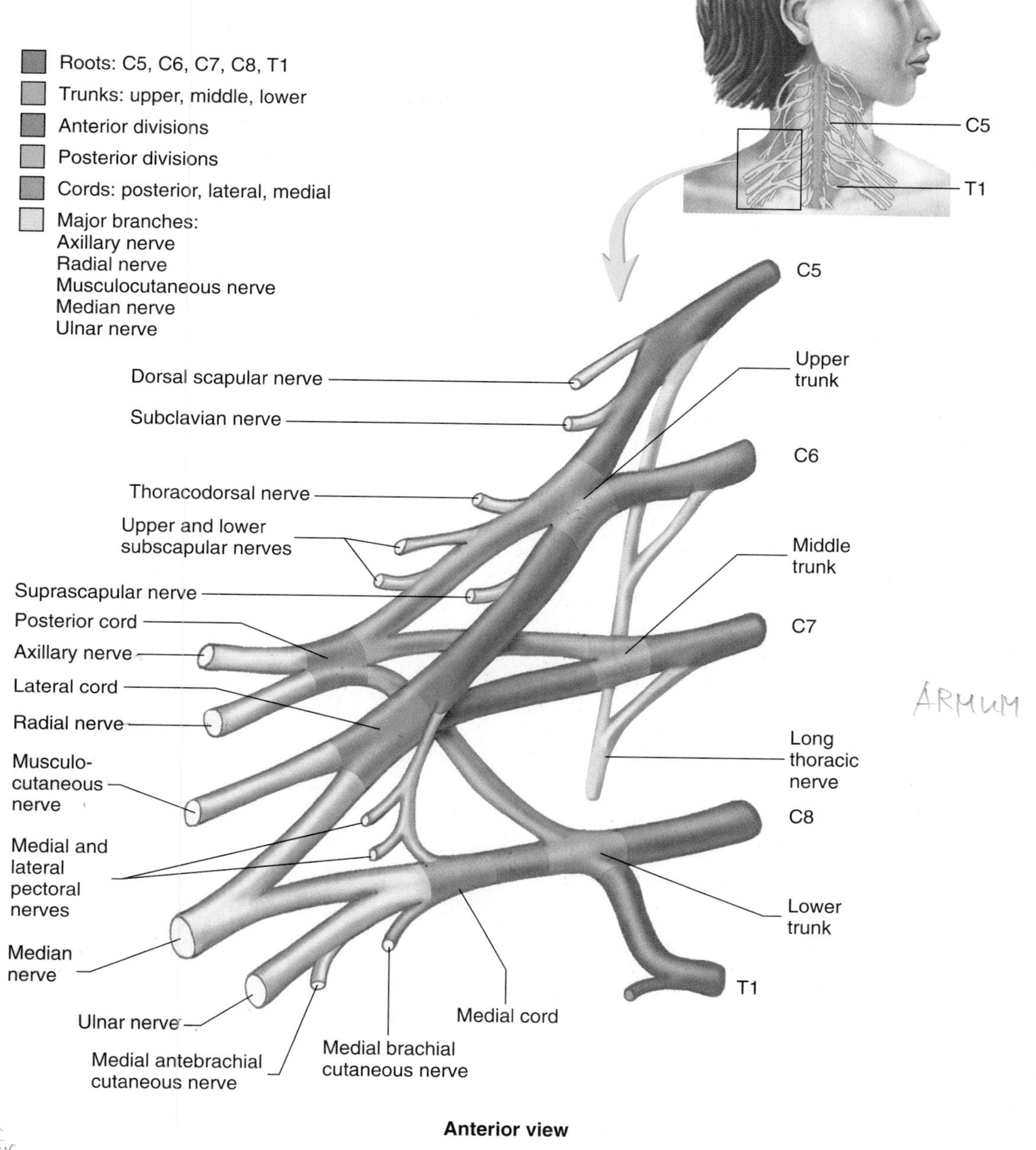

FIGURE 12.17 Brachial Plexus

The roots of the plexus are formed by the ventral rami of the spinal nerves C5–T1 and join to form an upper, middle, and lower trunk. Each trunk divides into anterior and posterior divisions. The divisions join together to form the posterior, lateral, and medial cords from which the major brachial plexus nerves arise. The major brachial plexus nerves include the axillary, radial, musculocutaneous, median, and ulnar nerves. These nerves innervate the muscles and skin of the upper limb.

The five major nerves emerging from the brachial plexus to supply the upper limb are the axillary, radial, musculocutaneous, ulnar, and median nerves. The axillary nerve innervates part of the shoulder; the radial nerve innervates the posterior arm, forearm, and hand; the musculocutaneous nerve innervates the anterior arm; and the ulnar and median nerves innervate the anterior forearm and hand. Smaller nerves from the brachial plexus innervate the shoulder and pectoral muscles.

Brachial Anesthesia

The entire upper limb can be anesthetized by injecting an anesthetic near the brachial plexus. This is called **brachial anesthesia.** The anesthetic can be injected between the neck and the shoulder posterior to the clavicle.

Axillary Nerve

Origin
Posterior cord of brachial plexus, C5–C6

Movements/Muscles Innervated
Laterally rotates arm
- *Teres minor*

Abducts arm
- *Deltoid*

Cutaneous (Sensory) Innervation
Inferior lateral shoulder

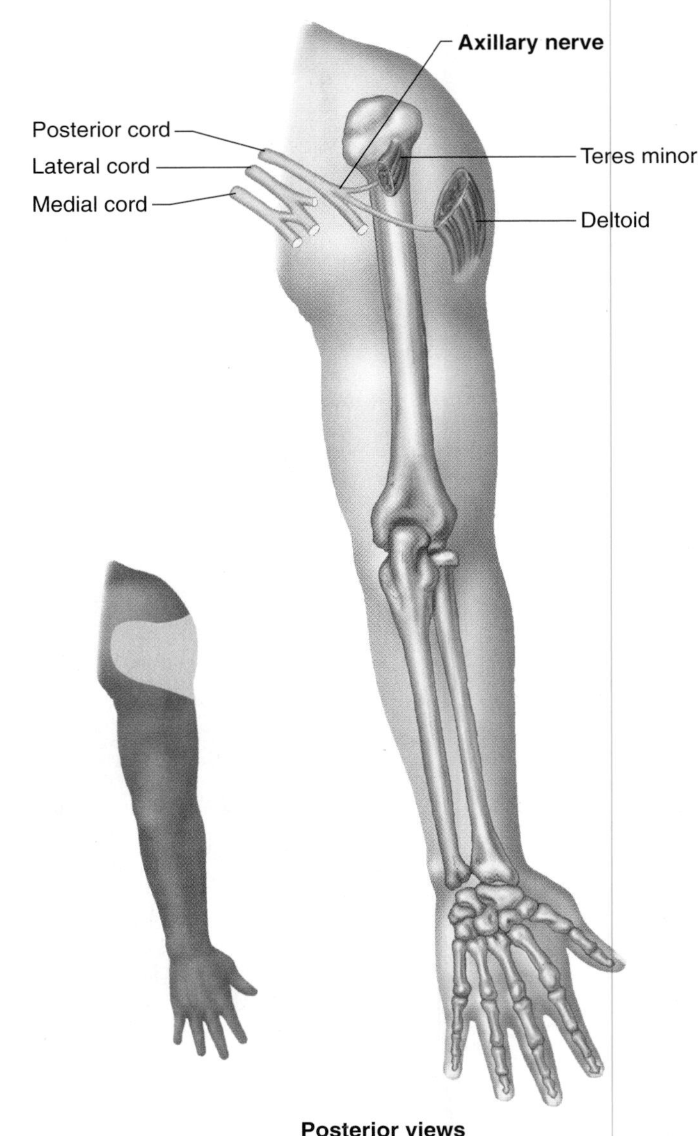

FIGURE 12.18 **Axillary Nerve**
The route of the axillary nerve and the muscles it innervates are illustrated and listed. The inset depicts the cutaneous (sensory) distribution of the nerve (*blue area*).

Axillary Nerve

The **axillary** (ak′sil-ār-ē) **nerve** innervates the deltoid and teres minor muscles (figure 12.18). It also provides sensory innervation to the shoulder joint and to the skin over part of the shoulder.

Radial Nerve

Origin
Posterior cord of brachial plexus, C5–T1

Movements/Muscles Innervated

Extends elbow
- *Triceps brachii*
- *Anconeus*

Flexes elbow
- *Brachialis (part; sensory only)*
- *Brachioradialis*

Extends and abducts wrist
- *Extensor carpi radialis longus*
- *Extensor carpi radialis brevis*

Supinates forearm and hand
- *Supinator*

Extends fingers
- *Extensor digitorum*
- *Extensor digiti minimi*
- *Extensor indicis*

Extends and adducts wrist
- *Extensor carpi ulnaris*

Abducts thumb
- *Abductor pollicis longus*

Extends thumb
- *Extensor pollicis longus*
- *Extensor pollicis brevis*

Cutaneous (Sensory) Innervation
Posterior surface of arm and forearm, lateral two-thirds of dorsum of hand

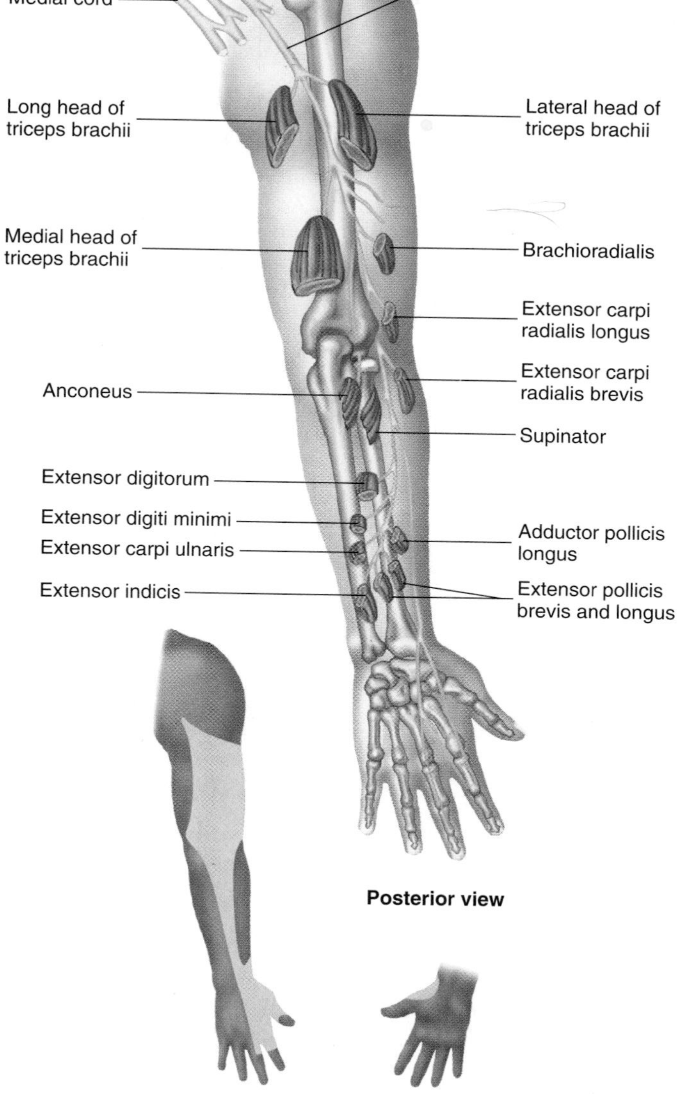

FIGURE 12.19 Radial Nerve
The route of the radial nerve and the muscles it innervates are illustrated and listed. The insets depict the cutaneous (sensory) distribution of the nerve (*blue area*).

Radial Nerve

The **radial nerve** emerges from the posterior cord of the brachial plexus and descends within the deep aspect of the posterior arm (figure 12.19). About midway down the shaft of the humerus, it lies against the bone in the radial groove. The radial nerve innervates all of the extensor muscles of the upper limb, the supinator muscle, and the brachioradialis. Its cutaneous sensory distribution is to the posterior portion of the upper limb, including the posterior surface of the hand.

Radial Nerve Damage

Because the radial nerve lies near the humerus in the axilla, it can be damaged if it is compressed against the humerus. Improper use of crutches (i.e., when the crutch is pushed tightly into the axilla) can result in **crutch paralysis.** In this disorder, the radial nerve is compressed between the top of the crutch and the humerus. As a result, the radial nerve is damaged, and the muscles it innervates lose their function. The major symptom of crutch paralysis is **wrist drop,** in which the extensor muscles of the wrist and fingers, which are innervated by the radial nerve, fail to function; as a result, the elbow, wrist, and fingers are constantly flexed.

Musculocutaneous Nerve

Origin

Lateral cord of brachial plexus, C5–C7

Movements/Muscles Innervated

Flexes shoulder
- *Biceps brachii*
- *Coracobrachialis*

Flexes elbow and supinates forearm and-hand
- *Biceps brachii*

Flexes elbow
- *Brachialis (also small amount of innervation from radial nerve)*

Cutaneous (Sensory) Innervation

Lateral surface of forearm

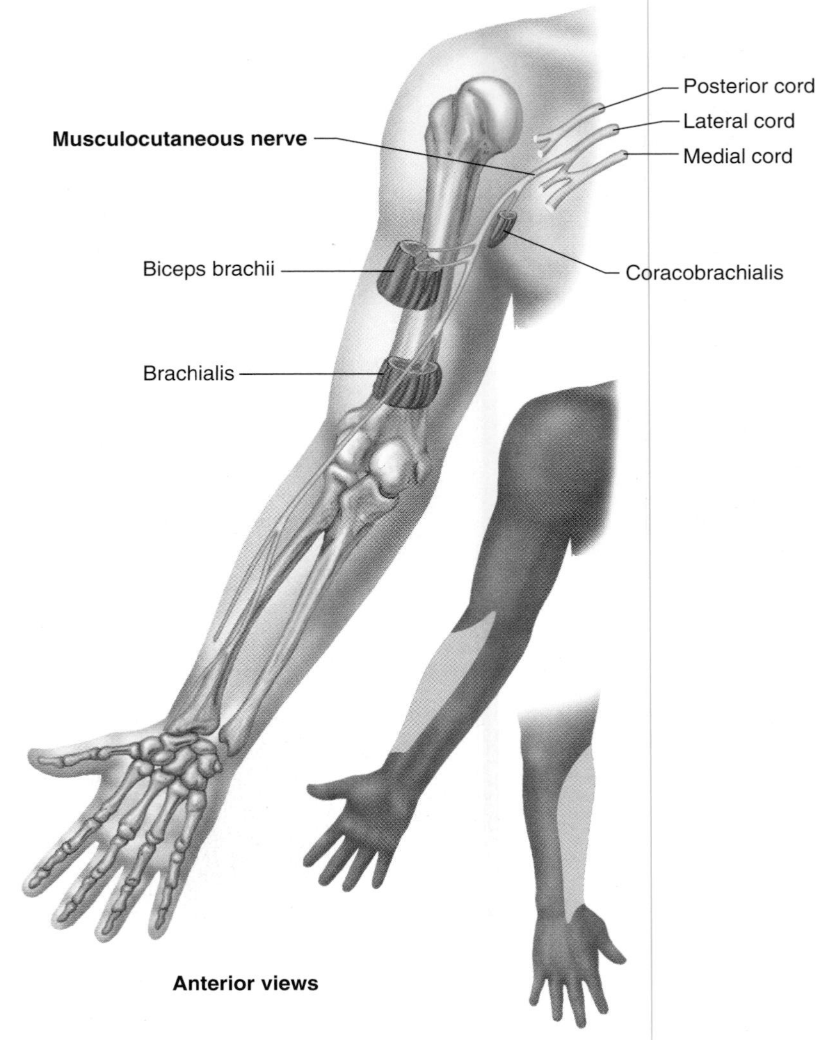

FIGURE 12.20 Musculocutaneous Nerve

The route of the musculocutaneous nerve and the muscles it innervates are illustrated and listed. The insets depict the cutaneous (sensory) distribution of the nerve (*blue area*).

PREDICT 6

Wrist drop can also result from a compound fracture of the humerus. Explain how and where damage to the nerve can occur.

Musculocutaneous Nerve

The **musculocutaneous** (mŭs′kū-lō-kū-tā′nē-ŭs) **nerve** provides motor innervation to the anterior muscles of the arm, as well as cutaneous sensory innervation to part of the forearm (figure 12.20).

Ulnar Nerve

Origin
Medial cord of brachial plexus, C8–T1

Movements/Muscles Innnervated

Flexes and adducts wrist
- *Flexor carpi ulnaris*

Flexes fingers
- *Part of the flexor digitorum profundus controlling the distal phalanges of little and ring fingers*

Adducts thumb
- *Adductor pollicis*

Controls hypothenar muscles
- *Flexor digiti minimi brevis*
- *Abductor digiti minimi*
- *Opponens digiti minimi*

Flexes metacarpophalangeal joints and extends interphalangeal joints
- *Two medial (ulnar) lumbricales*

Abducts and adducts fingers
- *Interossei*

Cutaneous (Sensory) Innervation
Medial third of hand, little finger, and medial half of ring finger

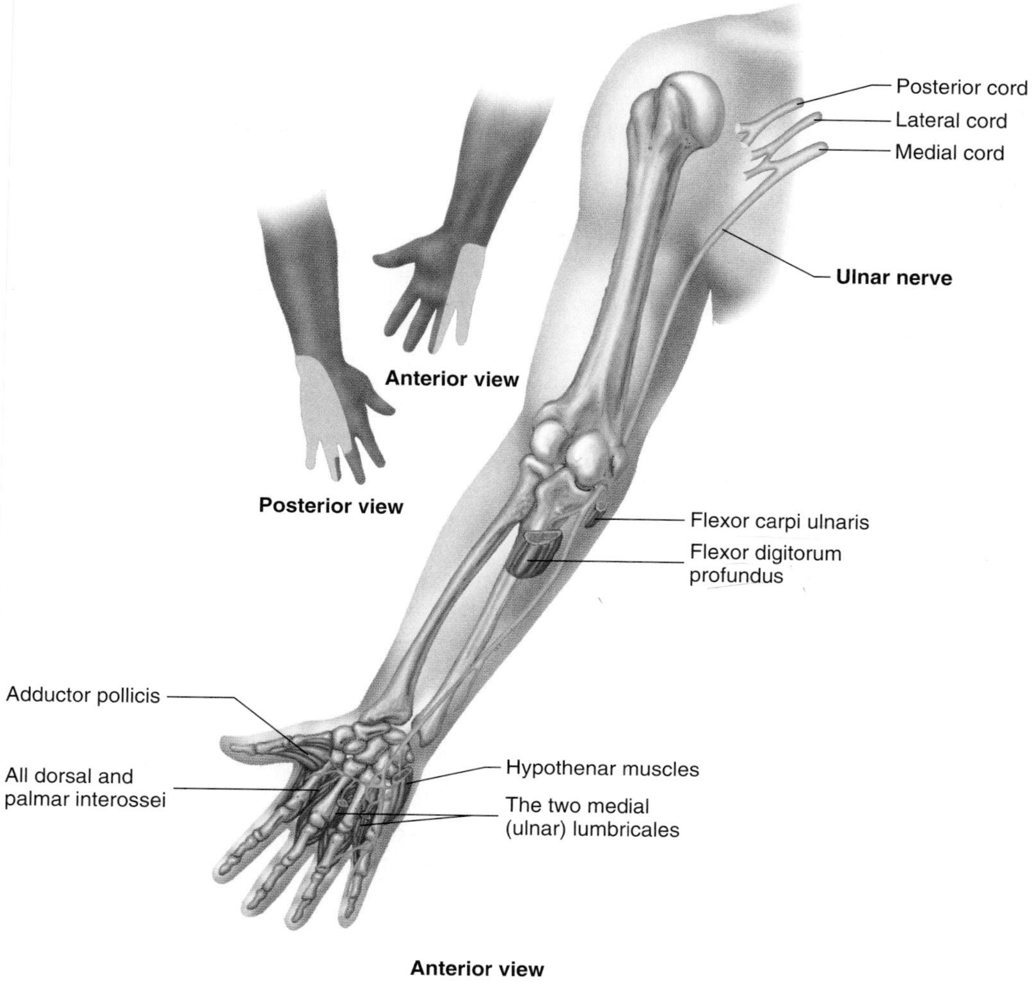

FIGURE 12.21 Ulnar Nerve
The route of the ulnar nerve and the muscles it innervates are illustrated and listed. The insets depict the cutaneous (sensory) distribution of the nerve (*blue area*).

Ulnar Nerve

The **ulnar nerve** innervates two forearm muscles plus most of the intrinsic hand muscles, except some associated with the thumb. Its sensory distribution is to the ulnar side of the hand (figure 12.21).

Ulnar Nerve Damage

The ulnar nerve is the most easily damaged of all the peripheral nerves, but such damage is almost always temporary. Slight damage to the ulnar nerve may occur where it passes posterior to the medial epicondyle of the humerus. The nerve can be felt just below the skin at this point, and, if this region of the elbow is banged against a hard object, temporary ulnar nerve damage may occur. This damage results in painful tingling sensations radiating down the ulnar side of the forearm and hand. Because of this sensation, this area of the elbow is often called the **funny bone** or **crazy bone.**

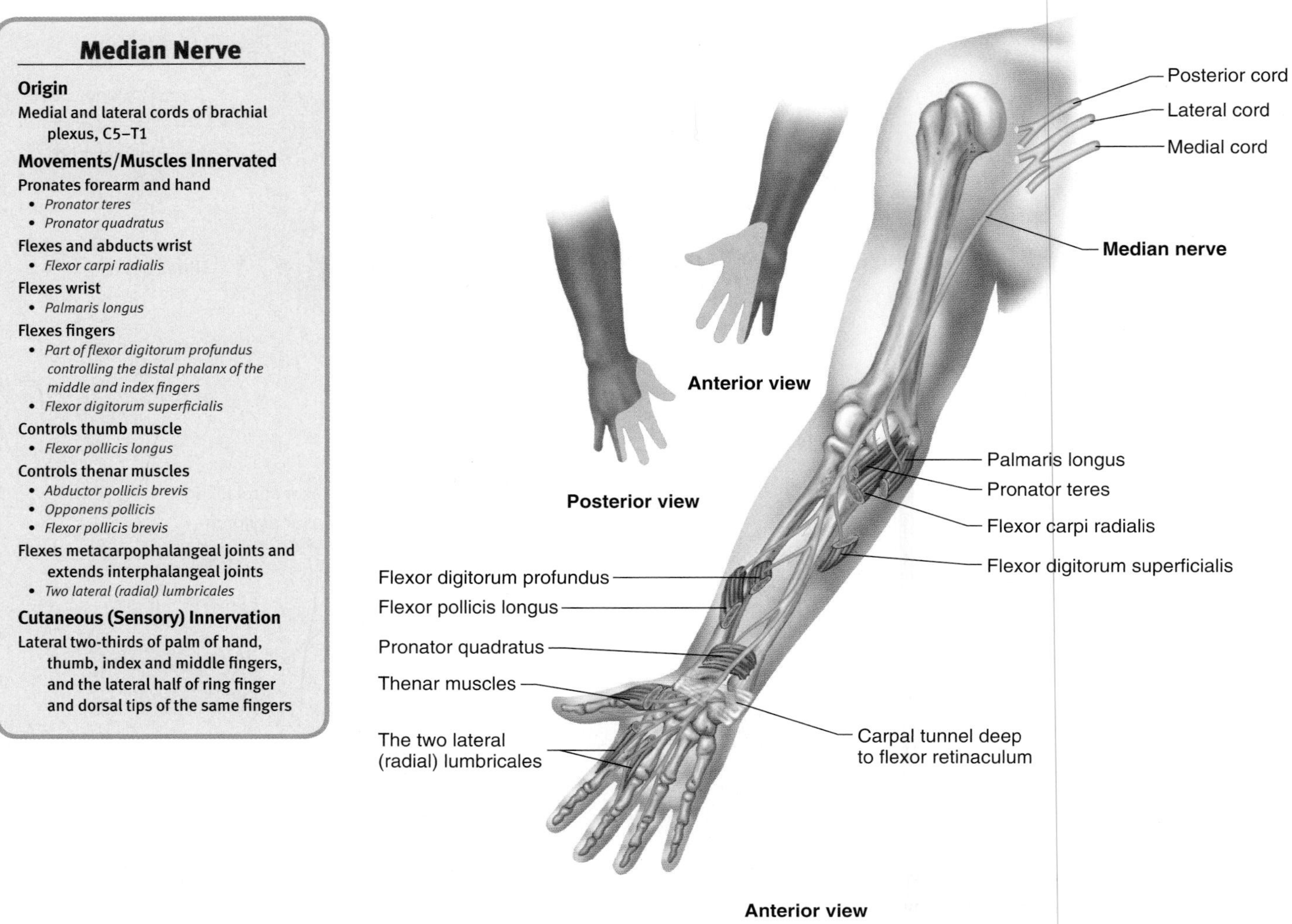

FIGURE 12.22 Median Nerve
The route of the median nerve and the muscles it innervates are illustrated and listed. The insets depict the cutaneous (sensory) distribution of the nerve (*blue area*).

Median Nerve

The **median nerve** innervates all but one of the flexor muscles of the forearm and most of the hand muscles at the base of the thumb, called the thenar area of the hand. Its cutaneous sensory distribution is to the radial portion of the palm of the hand (figure 12.22).

Median Nerve Damage

Damage to the median nerve occurs most commonly where it enters the wrist through the **carpal tunnel.** This tunnel is created by the concave organization of the carpal bones and the flexor retinaculum on the anterior surface of the wrist. None of the connective tissue components of the carpal tunnel expands readily. The tendons passing through the carpal tunnel may become inflamed and enlarged as a result of repetitive movements. This inflammation can produce pressure within the carpal tunnel, thereby compressing the median have and resulting in numbness, tingling, and pain in the fingers. The thenar muscles, innervated by the median nerve, have reduced function, resulting in weakness in thumb flexion and opposition. This condition is referred to as **carpal tunnel syndrome.** Carpal tunnel syndrome is common among people who perform repetitive movements of the wrists and fingers, such as keyboard operators. Surgery is often required to relieve the pressure.

People attempting suicide by cutting the wrists commonly cut the median nerve proximal to the carpal tunnel.

CASE STUDY
Cervical Rib Syndrome

Sarah, who is 26, noticed that over a period of time she experienced pain, tingling, and numbness in the ring finger and little finger of her right hand. She also had pain in her elbow, which radiated down the anterolateral portion of her forearm and hand. She made an appointment with her physician. After careful examination of Sarah's upper limb, her physician ordered an x-ray of her neck. The x-ray disclosed a cervical rib on the right, which was attached to the C7 vertebra. Cervical ribs are not uncommon, occurring in about 1% of the population. Most people exhibit no symptoms, but symptoms may develop in some people. If the extra rib compresses the inferior roots of the brachial plexus (see figure 12.17), the condition is called **cervical rib syndrome**—one of several **thoracic outlet syndromes.** An alternative thoracic outlet syndrome may involve the brachial artery rather than the brachial plexus.

PREDICT 7

Use figures 12.14, 12.17, 12.19, 12.20, and 12.21 to answer these questions.

a. Name the brachial plexus nerves supplying the skin of the hand.
b. Damage to which of these nerves could produce the symptoms seen in Sarah's hand?
c. What made Sarah's physician suspect cervical rib syndrome rather than damage to an individual nerve?
d. Cervical rib syndrome can also affect muscles, producing muscle weakness and paralysis. Muscles supplied by what nerves could be affected by a cervical rib?

Other Nerves of the Brachial Plexus

Several nerves, other than the five just described, arise from the brachial plexus (see figure 12.17). They supply most of the muscles acting on the scapula and arm and include the pectoral, long thoracic, thoracodorsal, subscapular, and suprascapular nerves. In addition, brachial plexus nerves supply the cutaneous innervation of the medial arm and forearm.

23 ***Name the five major nerves that emerge from the brachial plexus. List the muscles they innervate and the areas of the skin they supply. In addition to these five nerves, name the muscles and skin areas supplied by the remaining brachial plexus nerves.***

Lumbar and Sacral Plexuses

The **lumbar plexus** originates from the ventral rami of spinal nerves L1–L4 and the **sacral plexus** from L4–S4. Because of their close, overlapping relationship and their similar distribution, however, the two plexuses often are considered together as a single **lumbosacral plexus** (L1–S4; figure 12.23). Four major

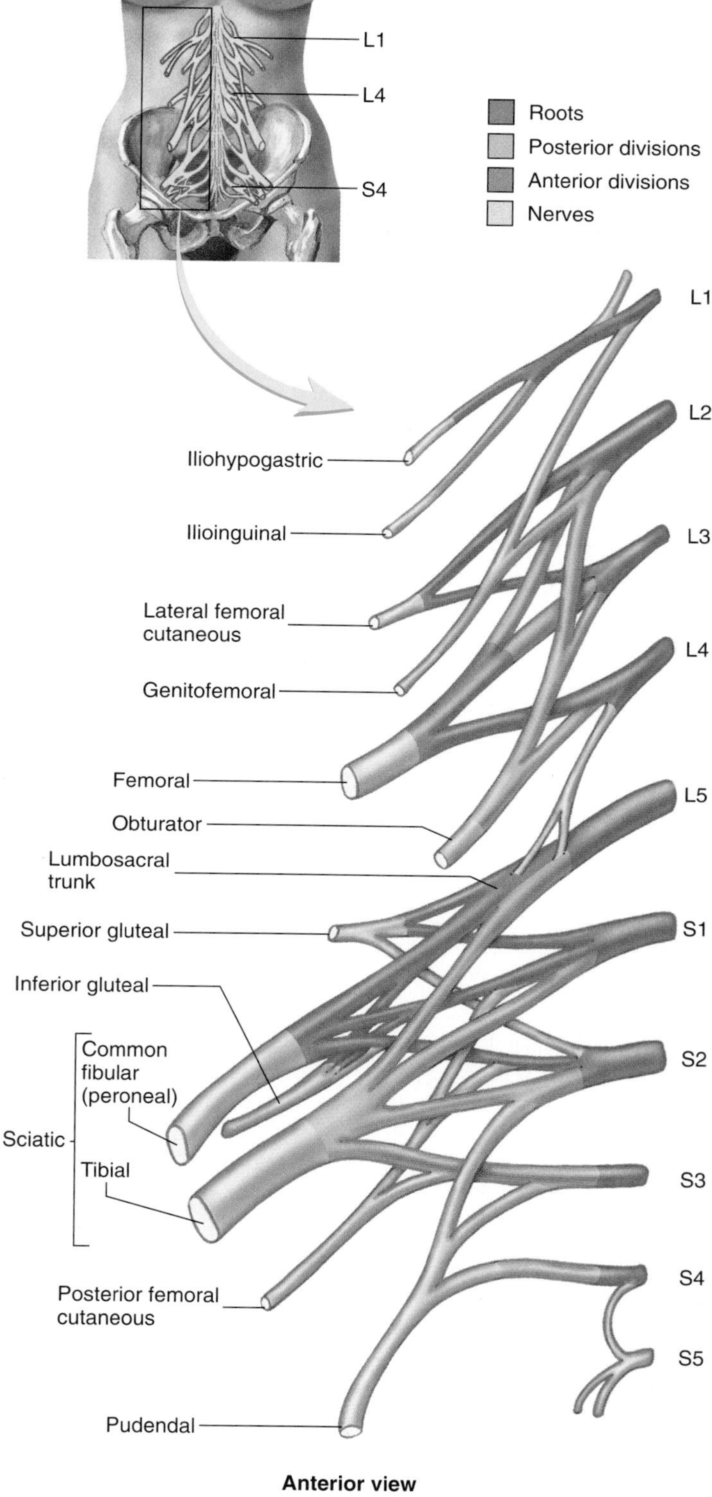

FIGURE 12.23 Lumbosacral Plexus

The roots of the plexus are formed by the ventral rami of the spinal nerves L1–S4 and form anterior and posterior divisions, which give rise to the lumbrosacral nerves. The lumbosacral trunk joins the lumbar and sacral plexuses.

Obturator Nerve

Origin
Lumbosacral plexus, L2–L4

Movements/Muscles Innervated

Rotates thigh laterally
- *Obturator externus*

Adducts thigh
- *Adductor magnus (adductor part)*
- *Adductor longus*
- *Adductor brevis*

Adducts thigh and flexes knee
- *Gracilis*

Cutaneous (Sensory) Innervation
Superior medial side of thigh

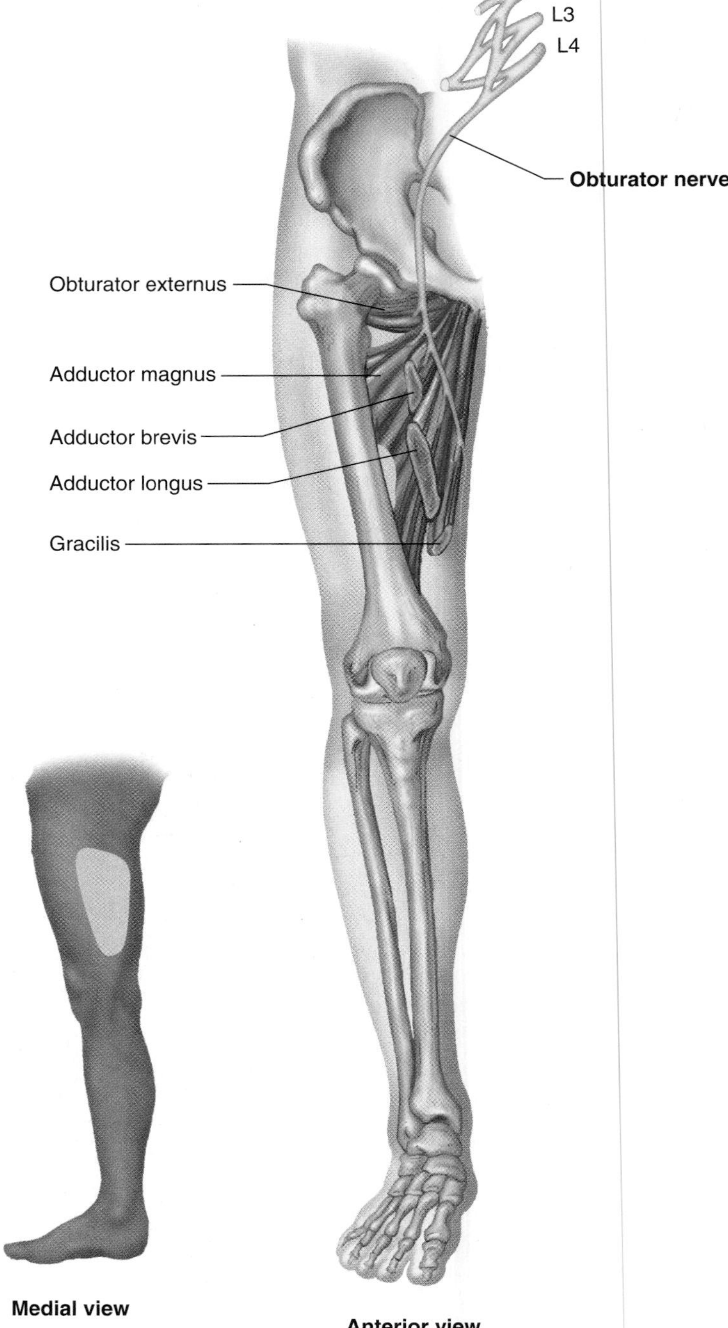

FIGURE 12.24 Obturator Nerve
The route of the obturator nerve and the muscles it innervates are illustrated and listed. The inset depicts the cutaneous (sensory) distribution of the nerve (*blue area*).

nerves exit the lumbosacral plexus and enter the lower limb: the obturator, femoral, tibial, and common fibular (peroneal). The obturator nerve innervates the medial thigh; the femoral nerve innervates the anterior thigh; the tibial nerve innervates the posterior thigh, leg, and foot; and the common fibular nerve innervates the posterior thigh, anterior and lateral leg, and foot. Other lumbosacral nerves supply the lower back, hip, and lower abdomen.

Obturator Nerve

The **obturator** (ob′too-rā-tōr) **nerve** supplies the muscles that adduct the thigh. Its cutaneous sensory distribution is to the medial side of the thigh (figure 12.24).

Femoral Nerve

Origin

Lumbosacral plexus, L2–L4

Movements/Muscles Innervated

Flexes hip
- *Psoas major*
- *Iliacus*
- *Pectineus*

Flexes hip and flexes knee
- *Sartorius*

Extends knee
- *Vastus lateralis*
- *Vastus intermedius*
- *Vastus medialis*

Extends knee and flexes hip
- *Rectus femoris*

Cutaneous (Sensory) Innervation

Anterior and lateral branches supply the anterior and lateral thigh; saphenous branch supplies the medial leg and foot

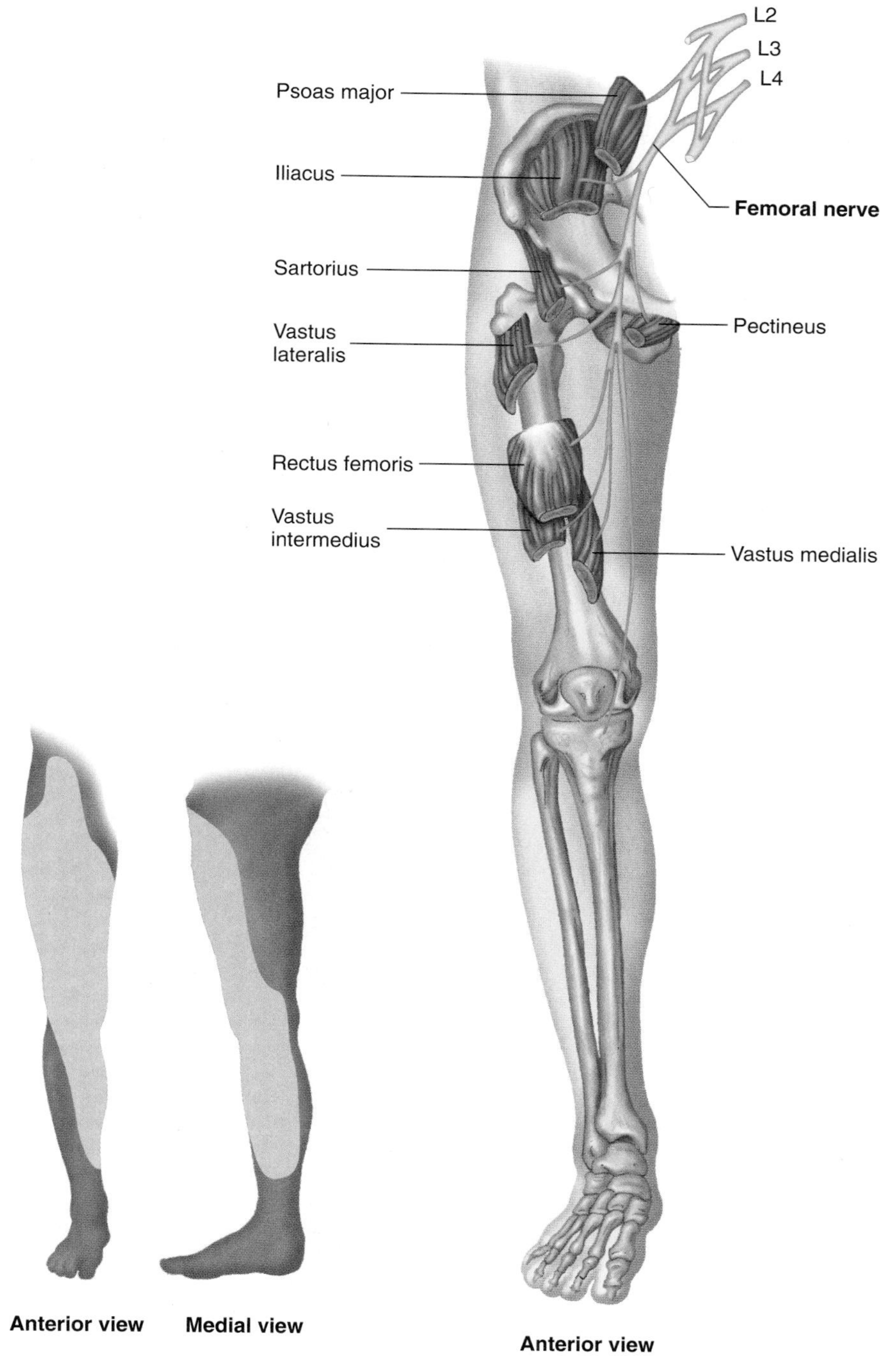

FIGURE 12.25 Femoral Nerve

The route of the femoral nerve and the muscles it innervates are illustrated and listed. The insets depict the cutaneous (sensory) distribution of the nerve (*blue area*).

Femoral Nerve

The **femoral nerve** innervates the iliopsoas and sartorius muscles and the quadriceps femoris group. Its cutaneous sensory distribution is the anterior and lateral thigh and the medial leg and foot (figure 12.25).

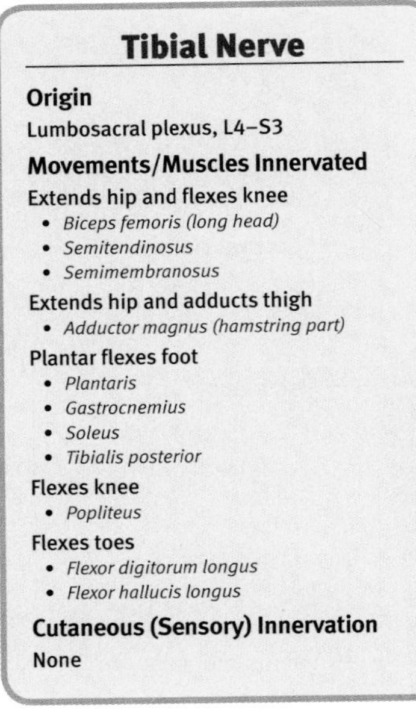

Tibial Nerve

Origin
Lumbosacral plexus, L4–S3

Movements/Muscles Innervated

Extends hip and flexes knee
- *Biceps femoris (long head)*
- *Semitendinosus*
- *Semimembranosus*

Extends hip and adducts thigh
- *Adductor magnus (hamstring part)*

Plantar flexes foot
- *Plantaris*
- *Gastrocnemius*
- *Soleus*
- *Tibialis posterior*

Flexes knee
- *Popliteus*

Flexes toes
- *Flexor digitorum longus*
- *Flexor hallucis longus*

Cutaneous (Sensory) Innervation
None

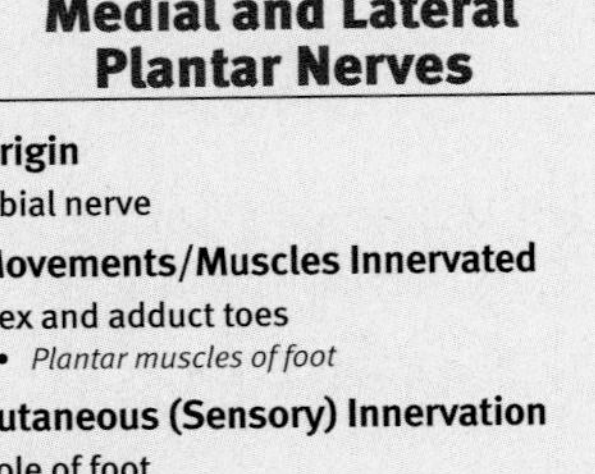

Medial and Lateral Plantar Nerves

Origin
Tibial nerve

Movements/Muscles Innervated

Flex and adduct toes
- *Plantar muscles of foot*

Cutaneous (Sensory) Innervation
Sole of foot

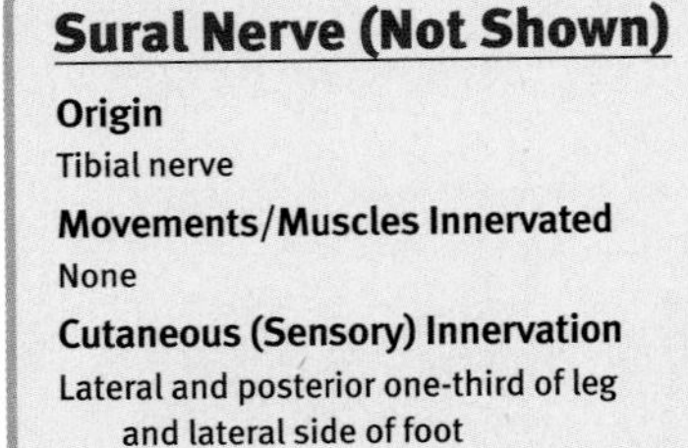

Sural Nerve (Not Shown)

Origin
Tibial nerve

Movements/Muscles Innervated
None

Cutaneous (Sensory) Innervation
Lateral and posterior one-third of leg and lateral side of foot

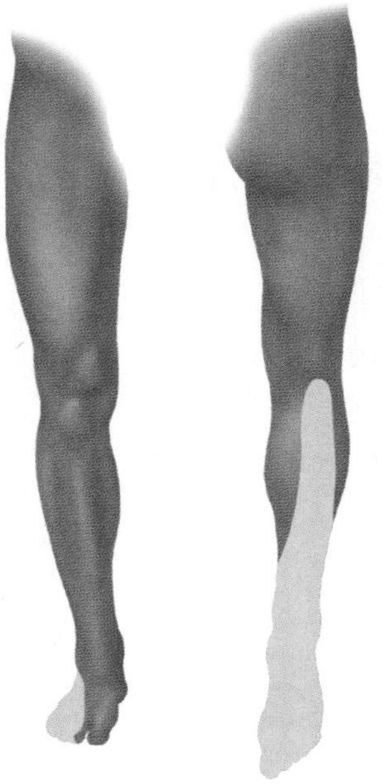

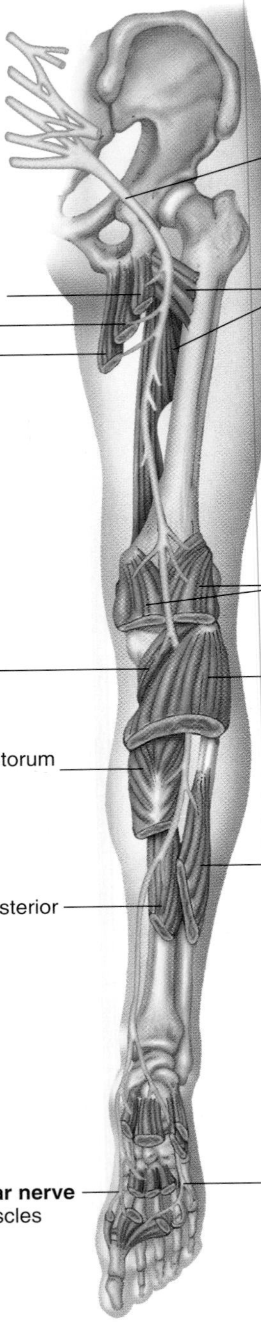

FIGURE 12.26 Tibial Nerve
The route of the tibial nerve and the muscles it innervates are illustrated and listed. The insets depict the cutaneous (sensory) distribution of the nerve (*blue area*).

Tibial and Common Fibular Nerves

The **tibial** and **common fibular (peroneal)** (per-ō-nē′ăl) **nerves** originate from spinal segments L4–S3 and are bound together within a connective tissue sheath for the length of the thigh (figures 12.26 and 12.27; see figure 12.23). These two nerves, combined within the same sheath, are referred to jointly as the **sciatic** (sī-at′ik; hip), or **ischiadic** (is-kē-ad′ik) **nerve** (see figure 12.23). The sciatic nerve, by far the largest peripheral nerve in the body, passes through the greater sciatic notch in the pelvis and descends in the posterior thigh to the popliteal fossa, where the two portions of the sciatic nerve separate.

Common Fibular (Peroneal) Nerve

Origin
Lumbosacral plexus, L4–S2

Movements/Muscles Innervated
Extends hip and flexes knee
- *Biceps femoris (short head)*

Cutaneous (Sensory) Innervation
Lateral surface of knee

Deep Fibular (Peroneal) Nerve

Origin
Common fibular (peroneal) nerve

Movements/Muscles Innervated
Dorsiflexes foot
- *Tibialis anterior*
- *Fibularis tertius*

Extends toes
- *Extensor digitorum longus*
- *Extensor hallucis longus*
- *Extensor digitorum brevis*

Cutaneous (Sensory) Innervation
Great and second toe

Superficial Fibular (Peroneal) Nerve

Origin
Common fibular (peroneal) nerve

Movements/Muscles Innervated
Plantar flexes and everts foot
- *Fibularis longus*
- *Fibularis brevis*

Cutaneous (Sensory) Innervation
Dorsal anterior third of leg and dorsum of foot

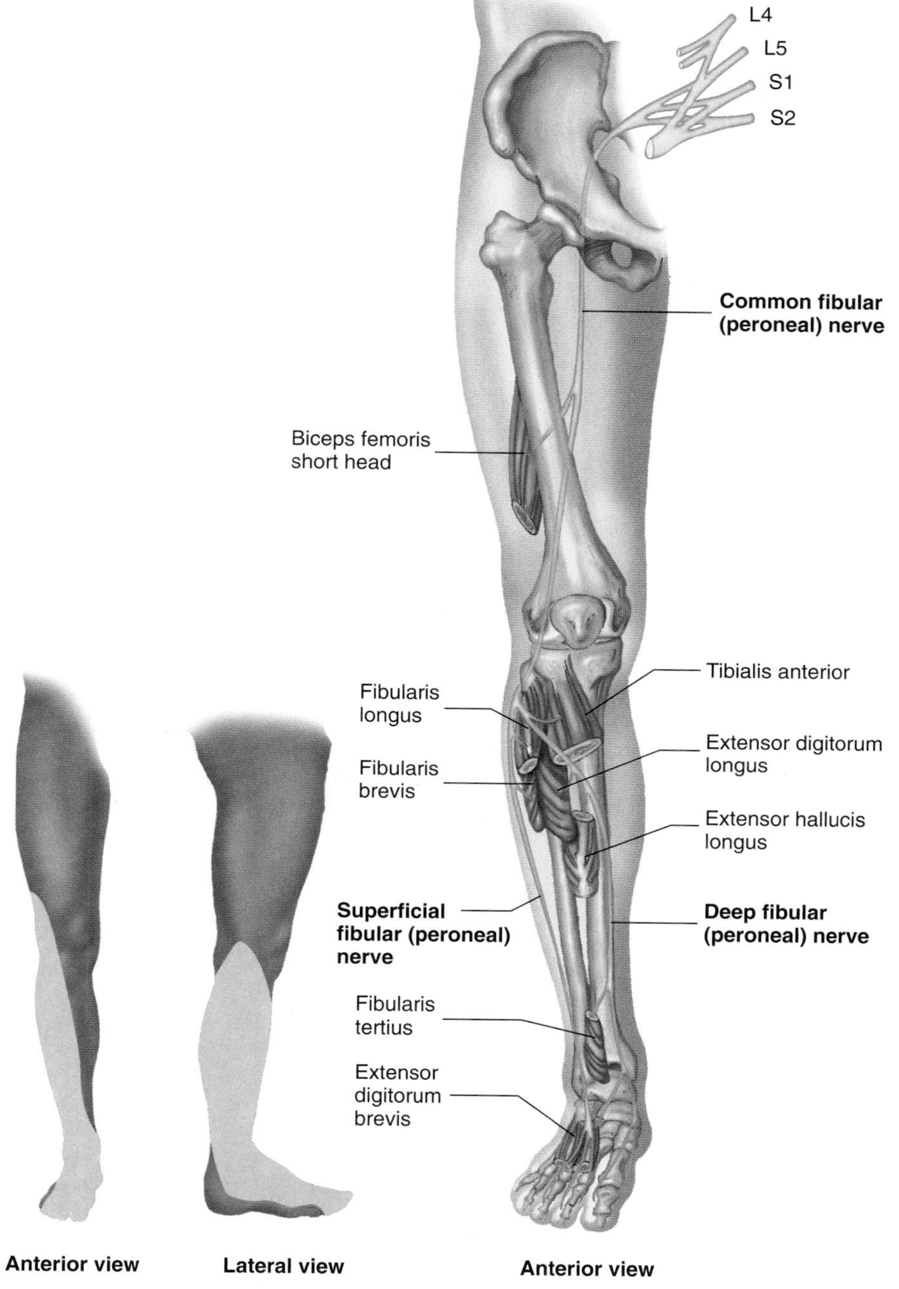

FIGURE 12.27 Fibular Nerve
The route of the common fibular (peroneal) nerve and the muscles it innervates are illustrated and listed. The insets depict the cutaneous (sensory) distribution of the nerve (*blue area*).

The tibial nerve innervates most of the posterior thigh and leg muscle (see figure 12.26). It branches in the foot to form the **medial** and **lateral plantar** (plan′tăr) **nerves,** which innervate the plantar muscles of the foot and the skin over the sole of the foot. Another branch, the **sural** (soo′răl) **nerve,** supplies part of the cutaneous innervation over the calf of the leg and the plantar surface of the foot.

The common fibular nerve divides into the **deep** and **superficial fibular (peroneal) nerves.** These branches innervate the anterior and lateral muscles of the leg and foot. The cutaneous distribution of the common fibular nerve and its branches is the lateral and anterior leg and the dorsum of the foot (see figure 12.27).

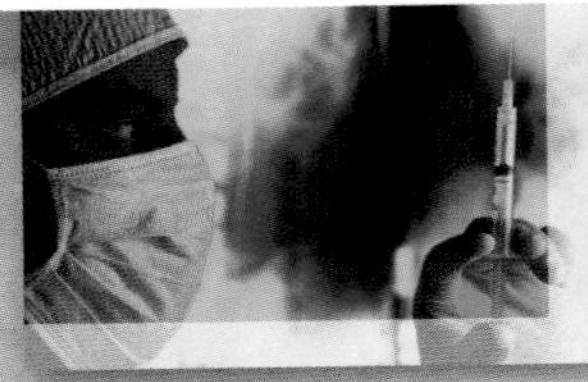

CLINICAL FOCUS

Nerve Replacement

Patients paralyzed by strokes or spinal cord lesions are able to regain certain limited functions through the use of microcomputers that stimulate certain programmed activities, such as grasping and walking. The microcomputer initiates electric impulses, which are conveyed through fine wire leads either to peripheral nerves or directly to the muscles responsible for the desired movement. The subtle movement of muscles not affected by the paralysis initiate the program. Sensors connected to the microcomputer are attached to the skin overlying functional muscles and are able to detect electrical activity associated with movement of the underlying muscles. For example, a person with both legs paralyzed may have such a sensor attached to the abdomen. The abdominal muscles normally involved in stabilizing and moving the pelvis during walking are stimulated by descending tracts when walking is initiated by CNS centers. The resultant abdominal muscle activity is detected by the sensor, which activates the program that stimulates the appropriate sequence of muscles in the lower limbs, and the paralyzed person walks. Similarly, a quadriplegic using subtle movements of the shoulder, neck, or face, where specific sensors can be placed, can initiate certain upper limb and grasping actions.

Sciatic Nerve Damage

If a person sits on a hard surface for a considerable time, the sciatic nerve may be compressed against the ischial portion of the coxal bone. When the person stands up, a tingling sensation, described as "pins and needles," can be felt throughout the lower limb, and the limb is said to have "gone to sleep."

The sciatic nerve may be seriously injured in a number of ways. A ruptured intervertebral disk or pressure from the uterus during pregnancy may compress the roots of the sciatic nerve. Other causes of sciatic nerve damage include hip injury and an improperly administered injection in the hip region (see p. 243).

Other Lumbosacral Plexus Nerves

In addition to the nerves just described, the lumbosacral plexus gives rise to **gluteal nerves,** which supply the hip muscles that act on the femur, and the **pudendal** (pū-den′dăl) **nerves,** which supply the muscles of the abdominal floor (see figure 12.23). The **iliohypogastric** (il′ē-ō-hī-pō-gas′trik), **ilioinguinal** (il′ē-ō-ing′gwi-năl), **genitofemoral** (jen′i-tō-fem′ŏ-răl), **cutaneous femoral,** and pudendal nerves innervate the skin of the suprapubic area, the external genitalia, the superior medial thigh, and the posterior thigh. The pudendal nerve plays a vital role in sexual stimulation and response.

Pudendal Nerve Anesthesia

Branches of the pudendal nerve are anesthetized before a doctor performs an episiotomy for childbirth. An **episiotomy** (e-piz-ē-ot′ō-mē, epis-ē-ot′ō-mē) is a cut in the perineum that makes the opening of the birth canal larger.

Coccygeal Plexus

The **coccygeal** (kok-sij′ē-ăl) **plexus** is a very small plexus formed from the ventral rami of spinal nerve S5 and the coccygeal nerve (Co). This small plexus supplies motor innervation to the muscles of the pelvic floor and sensory cutaneous innervation to the skin over the coccyx. The dorsal rami of the coccygeal nerves innervate some skin over the coccyx.

24 ***Name the four major nerves that arise from the lumbosacral plexus, and describe the muscles and skin area they supply. What is the name applied to the tibial and common fibular nerves bound together?***

25 ***Describe the structures innervated by the remaining lumbosacral nerves.***

26 ***What structures are innervated by the coccygeal plexus?***

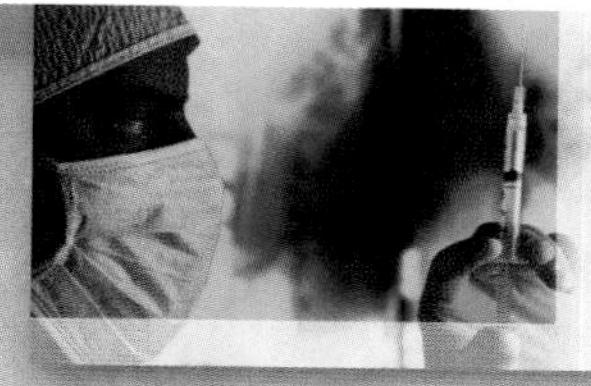

CLINICAL FOCUS

Peripheral Nervous System Disorders—Spinal Nerves

GENERAL TYPES OF PNS DISORDERS

Anesthesia is the loss of sensation (the Greek word *esthesis* means sensation). It may be a pathologic condition if it happens spontaneously, or it may be induced to facilitate surgery or some other medical treatment.

Hyperesthesia (hī′per-es-thē′zē-ă) is an abnormal acuteness to sensation, especially an increased sensitivity to pain, pressure, or light.

Paresthesia (par-es-the′zē-ă) is an abnormal spontaneous sensation, such as tingling, prickling, or burning.

Neuralgia (noo-ral′jē-ă) consists of severe spasms of throbbing or stabbing pain resulting from inflammation or damage along the pathway of a nerve.

Sciatica, or **ischiadica** (is′kē-ad′i-kă), is a neuralgia of the sciatic nerve, with pain radiating down the back of the thigh and leg. The most common cause is a herniated lumbar disk, resulting in pressure on the spinal nerves contributing to the lumbar plexus. Sciatica may also be produced by sciatic neuritis arising from a number of causes, including mechanical stretching of the nerve during exertion, vitamin deficiency, or metabolic disorders (such as gout or diabetes).

Neuritis (noo-rī′tis) is a general term referring to inflammation of a nerve that has a wide variety of causes, including mechanical injury or pressure, viral or bacterial infection, poisoning by drugs or other chemicals, and vitamin deficiencies. Neuritis in sensory nerves is characterized by neuralgia or may result in anesthesia and loss of reflexes in the affected area. Neuritis in motor nerves results in loss of motor function.

INFECTIONS

Herpes is a family of diseases characterized by skin lesions, which are caused by a group of closely related viruses (the herpes viruses). The term is derived from the Greek word *herpo,* meaning to creep, and indicates a spreading skin eruption. The viruses apparently reside in the ganglia of sensory nerves and cause lesions along the course of the nerves. **Herpes simplex II,** or genital herpes, is usually responsible for a sexually transmitted disease causing lesions on the external genitalia.

The varicella-zoster virus causes the diseases chicken pox in children and **shingles** in older adults, a disease also called **herpes zoster.** Normally, this virus first enters the body in childhood to cause chicken pox. The virus then lies dormant in the spinal ganglia for many years and can become active during a time of reduced resistance to cause shingles, a unilateral patch of skin blisters and discoloration along the path of one or more spinal nerves, most commonly around the waist. The symptoms can persist for 3–6 months.

Poliomyelitis (pō′lē-ō-mī′ĕ-lī′tis; "polio" or infantile paralysis; the Greek word *polio* means gray matter) is a disease caused by an *Enterovirus.* It is actually a CNS infection, but its major effect is on the peripheral nerves and the muscles they supply. The virus infects the motor neurons in the anterior horn of the spinal cord. The infection causes degeneration of the motor neurons, which results in paralysis and atrophy of the muscles innervated by those nerves.

Anesthetic leprosy is a bacterial infection of the peripheral nerves caused by *Mycobacterium leprae.* The infection results in anesthesia, paralysis, ulceration, and gangrene.

GENETIC AND AUTOIMMUNE DISORDERS

Myotonic dystrophy is an autosomal dominant hereditary disease characterized by muscle weakness, dysfunction, and atrophy and by visual impairment as a result of nerve degeneration.

Myasthenia (mī-as-thē′nē-ă) **gravis** is an autoimmune disorder resulting in a reduction in the number of functional acetylcholine receptors in the postsynaptic terminals and morphological changes in the neuromuscular synapse. T cells of the immune system break down acetylcholine receptor proteins into two fragments, which trigger antibody production by the immune system. Myasthenia gravis results in fatigue and progressive muscular weakness because of the neuromuscular dysfunction.

SUMMARY

Spinal Cord (p. 412)

General Structure

1. The spinal cord gives rise to 31 pairs of spinal nerves. The spinal cord has cervical and lumbosacral enlargements where nerves of the limbs enter and leave.
2. The spinal cord is shorter than the vertebral column. Nerves from the end of the spinal cord form the cauda equina.

Meninges of the Spinal Cord

Three meningeal layers surround the spinal cord: the dura mater, arachnoid mater, and pia mater.

Cross Section of the Spinal Cord

1. The spinal cord consists of peripheral white matter and central gray matter.
2. White matter is organized into funiculi, which are subdivided into fasciculi, or nerve tracts, which carry action potentials to and from the brain.
3. Gray matter is divided into horns.
   - The dorsal horns contain sensory axons that synapse with interneurons. The ventral horns contain the neuron cell bodies of somatic motor neurons, and the lateral horns contain the neuron cell bodies of autonomic neurons.
   - The gray and white commissures connect each half of the spinal cord.

4. The dorsal root conveys sensory input into the spinal cord, and the ventral root conveys motor output away from the spinal cord.

Reflexes (p. 415)

1. A reflex arc is the functional unit of the nervous system.
   - Sensory receptors respond to stimuli and produce action potentials in sensory neurons.
   - Sensory neurons propagate action potentials to the CNS.
   - Interneurons in the CNS synapse with sensory neurons and with motor neurons.
   - Motor neurons carry action potentials from the CNS to effector organs.
   - Effector organs, such as muscles or glands, respond to the action potentials.
2. Reflexes do not require conscious thought, and they produce a consistent and predictable result.
3. Reflexes are homeostatic.
4. Reflexes are integrated within the brain and spinal cord. Higher brain centers can suppress or exaggerate reflexes.

Stretch Reflex

Muscle spindles detect the stretch of skeletal muscles and cause the muscle to shorten reflexively.

Golgi Tendon Reflex

Golgi tendon organs respond to increased tension within tendons and cause skeletal muscles to relax.

Withdrawal Reflex

1. Activation of pain receptors causes contraction of muscles and the removal of some part of the body from a painful stimulus.
2. Reciprocal innervation causes relaxation of muscles that would oppose the withdrawal movement.
3. In the crossed extensor reflex, during flexion of one limb caused by the withdrawal reflex, the opposite limb is stimulated to extend.

Interactions with Spinal Cord Reflexes (p. 421)

Convergent and divergent pathways interact with reflexes.

Structure of Peripheral Nerves (p. 421)

In the PNS, individual axons are surrounded by the endoneurium. Groups of axons, called fascicles, are bound together by the perineurium. The fascicles form the nerve and are held together by the epineurium.

Spinal Nerves (p. 422)

1. Eight cervical, 12 thoracic, 5 lumbar, 5 sacral pairs, and 1 coccygeal pair make up the spinal nerves.
2. Spinal nerves have specific cutaneous distributions called dermatomes.
3. Spinal nerves branch to form rami.
   - The dorsal rami supply the muscles and skin near the midline of the back.
   - The ventral rami in the thoracic region form intercostal nerves, which supply the thorax and upper abdomen. The remaining ventral rami join to form plexuses. Communicating rami supply sympathetic nerves.

Cervical Plexus

Spinal nerves C1–C4 form the cervical plexus, which supplies some muscles and the skin of the neck and shoulder. The phrenic nerves innervate the diaphragm.

Brachial Plexus

1. Spinal nerves C5–T1 form the brachial plexus, which supplies the upper limb.
2. The axillary nerve innervates the deltoid and teres minor muscles and the skin of the shoulder.
3. The radial nerve supplies the extensor muscles of the arm and forearm and the skin of the posterior surface of the arm, forearm, and hand.
4. The musculocutaneous nerve supplies the anterior arm muscles and the skin of the lateral surface of the forearm.
5. The ulnar nerve innervates most of the intrinsic hand muscles and the skin on the ulnar side of the hand.
6. The median nerve innervates the pronator and most of the flexor muscles of the forearm, most of the thenar muscles, and the skin of the radial side of the palm of the hand.
7. Other nerves supply most of the muscles that act on the arm, the scapula, and the skin of the medial arm and forearm.

Lumbar and Sacral Plexuses

1. Spinal nerves L1–S4 form the lumbosacral plexus.
2. The obturator nerve supplies the muscles that adduct the thigh and the skin of the medial thigh.
3. The femoral nerve supplies the muscles that flex the thigh and extend the leg and the skin of the anterior and lateral thigh and the medial leg and foot.
4. The tibial nerve innervates the muscles that extend the thigh and flex the leg and the foot. It also supplies the plantar muscles and the skin of the posterior leg and the sole of the foot.
5. The common fibular nerve supplies the short head of the biceps femoris, the muscles that dorsiflex and plantar flex the foot, and the skin of the lateral and anterior leg and the dorsum of the foot.
6. In the thigh, the tibial nerve and the common fibular nerve are combined as the sciatic nerve.
7. Other lumbosacral nerves supply the lower abdominal muscles, the hip muscles, and the skin of the suprapubic area, external genitalia, and upper medial thigh.

Coccygeal Plexus

Spinal nerve S5 and the coccygeal nerve form the coccygeal plexus, which supplies the muscles of the pelvic floor and the skin over the coccyx.

REVIEW AND COMPREHENSION

1. The spinal cord extends from the
   a. medulla oblongata to the coccyx.
   b. level of the third cervical vertebra to the coccyx.
   c. level of the axis to the lowest lumbar vertebra.
   d. medulla oblongata to the level of the second lumbar vertebra.
   e. axis to the sacral hiatus.
2. The structure that anchors the inferior end of the spinal cord to the coccyx is the
   a. conus medullaris.
   b. cauda equina.
   c. filum terminale.
   d. lumbar enlargement.
   e. posterior median sulcus.
3. Axons of sensory neurons synapse with the cell bodies of interneurons in the ____________ of spinal cord gray matter.
   a. anterior horn
   b. lateral horn
   c. posterior horn
   d. gray commissure
   e. lateral funiculi
4. Cell bodies for spinal sensory neurons are located in the
   a. anterior horn of spinal cord gray matter.
   b. lateral horn of spinal cord gray matter.
   c. posterior horn of spinal cord gray matter.
   d. dorsal root ganglia.
   e. posterior columns.
5. Given these components of a reflex arc:
   1. effector organ
   2. interneuron
   3. motor neuron
   4. sensory neuron
   5. sensory receptor

   Choose the correct order an action potential follows after a sensory receptor is stimulated.
   a. 5,4,3,2,1
   b. 5,4,2,3,1
   c. 5,3,4,1,2
   d. 5,2,4,3,1
   e. 5,3,2,1,4
6. A reflex response accompanied by the conscious sensation of pain is possible because of
   a. convergent pathways.
   b. divergent pathways.
   c. a reflex arc that contains only one neuron.
   d. sensory perception in the spinal cord.
7. Several of the events that occurred between the time that a physician struck a patient's patellar tendon with a rubber hammer and the time the quadriceps femoris contracted (knee-jerk reflex) are listed below:
   1. increased frequency of action potentials in sensory neurons
   2. stretch of the muscle spindles
   3. increased frequency of action potentials in the alpha motor neurons
   4. stretch of the quadriceps femoris
   5. contraction of the quadriceps femoris

   Which of these lists most closely describes the sequence of events as they normally occur?
   a. 4,1,2,3,5
   b. 4,1,3,2,5
   c. 1,4,3,2,5
   d. 4,2,1,3,5
   e. 4,2,3,1,5
8. ____________ are responsible for regulating the sensitivity of the muscle spindle.
   a. Alpha motor neurons
   b. Sensory neurons
   c. Gamma motor neurons
   d. Golgi tendon organs
   e. Inhibitory interneurons
9. Which of these events occurs when a person steps on a tack with the right foot?
   a. The right foot is pulled away from the tack because of the Golgi tendon reflex.
   b. The left leg is extended to support the body because of the stretch reflex.
   c. The flexor muscles of the right thigh contract, and the extensor muscles of the right thigh relax because of reciprocal innervation.
   d. Extensor muscles contract in both thighs because of the crossed extensor reflex.
10. Which of these is a correct count of the spinal nerves?
    a. 9 cervical, 12 thoracic, 5 lumbar, 5 sacral, 1 coccygeal
    b. 8 cervical, 12 thoracic, 5 lumbar, 5 sacral, 1 coccygeal
    c. 7 cervical, 12 thoracic, 5 lumbar, 5 sacral, 1 coccygeal
    d. 8 cervical, 11 thoracic, 4 lumbar, 6 sacral, 1 coccygeal
    e. 7 cervical, 11 thoracic, 5 lumbar, 6 sacral, 1 coccygeal
11. Given these structures:
    1. dorsal ramus
    2. dorsal root
    3. plexus
    4. ventral ramus
    5. ventral root

    Choose the arrangement that lists the structures in the order that an action potential passes through them, given that the action potential originates in the spinal cord and propagates to a peripheral nerve.
    a. 2,1,3
    b. 2,3,1
    c. 3,4,5
    d. 5,3,4
    e. 5,4,3
12. Damage to the dorsal ramus of a spinal nerve results in
    a. loss of sensation.
    b. loss of motor function.
    c. both a and b.
13. A collection of spinal nerves that join together after leaving the spinal cord is called a
    a. ganglion.
    b. nucleus.
    c. projection nerve.
    d. plexus.
14. A dermatome
    a. is the area of skin supplied by a pair of spinal nerves.
    b. exists for each spinal nerve except C1.
    c. can be used to locate the site of spinal cord or nerve root damage.
    d. all of the above.
15. Which of these nerves arises from the cervical plexus?
    a. median
    b. musculocutaneous
    c. phrenic
    d. obturator
    e. ulnar
16. The skin on the posterior surface of the hand is supplied by the
    a. median nerve.
    b. musculocutaneous nerve.
    c. ulnar nerve.
    d. axillary nerve.
    e. radial nerve.
17. Most thenar muscles and most of the flexor muscles in the forearm are supplied by the
    a. musculocutaneous nerve.
    b. radial nerve.
    c. median nerve.
    d. ulnar nerve.
    e. axillary nerve.
18. Most intrinsic hand muscles are supplied by the
    a. musculocutaneous nerve.
    b. radial nerve.
    c. median nerve.
    d. ulnar nerve.
    e. axillary nerve.

19. The sciatic nerve is actually two nerves combined within the same sheath. The two nerves are the
    a. femoral and obturator.
    b. femoral and gluteal.
    c. common fibular (peroneal) and tibial.
    d. common fibular (peroneal) and obturator.
    e. tibial and gluteal.
20. The muscles of the anterior compartment of the thigh are supplied by the
    a. obturator nerve.
    b. gluteal nerve.
    c. sciatic nerve.
    d. femoral nerve.
    e. ilioinguinal nerve.

Answers in Appendix E

CRITICAL THINKING

1. Describe how stimulation of a neuron that has its cell body in the cerebrum could inhibit a reflex that is integrated within the spinal cord.
2. A cancer patient has his left lung removed. To reduce the space remaining where the lung is removed, the diaphragm on the left side is paralyzed to allow the abdominal viscera to push the diaphragm upward. What nerve is cut? Where is a good place to cut it, and when would the surgery be done?
3. Based on sensory response to pain in the skin of the hand, how could you distinguish between damage to the ulnar, median, and radial nerves?
4. During a difficult delivery, the baby's arm delivered first. The attending physician grasped the arm and forcefully pulled it. Later a nurse observed that the baby could not abduct or adduct the medial four fingers and flexion of the wrist was impaired. What nerve was damaged?
5. Two patients are admitted to the hospital. According to their charts, both have herniated disks that are placing pressure on the roots of the sciatic nerve. One patient has pain in the buttocks and the posterior aspect of the thigh. The other patient experiences pain in the posterior and lateral aspects of the leg and the lateral part of the ankle and foot. Explain how the same condition, a herniated disk, could produce such different symptoms.
6. In an automobile accident a woman suffers a crushing hip injury. For each of the conditions given here, state what nerve is damaged.
    a. Unable to adduct the thigh
    b. Unable to extend the leg
    c. Unable to flex the leg
    d. Loss of sensation from the skin of the anterior thigh
    e. Loss of sensation from the skin of the medial thigh
7. A skier breaks his ankle. As part of his treatment, the ankle and leg are placed in a plaster cast. Unfortunately, the cast is too tight around the proximal portion of the leg and presses in against the neck of the fibula. Where would you predict the patient would experience tingling or numbness in the leg? Explain.

Answers in Appendix F

Visit this book's website at www.mhhe.com/seeley8 for chapter quizzes, interactive learning exercises, and other study tools.

Brain and Cranial Nerves 13

The brain is that part of the CNS contained within the cranial cavity (figure 13.1). It is the control center for many of the body's functions. The brain is much like a complex central computer but with additional functions that no computer can as yet match. Indeed, one goal in computer technology is to make computers that can function more like the human brain. The brain consists of the brainstem, the cerebellum, the diencephalon, and the cerebrum (table 13.1). The brainstem includes the medulla oblongata, pons, midbrain, and reticular formation. The structure of the brain is described in this chapter. Its functions are primarily discussed in chapter 14.

Twelve pairs of cranial nerves, which are part of the PNS, arise directly from the brain. Two pairs arise from the cerebrum, nine pairs arise from the brainstem, and one pair arises from the spinal cord.

This chapter describes the *development of the CNS* (p. 445), *brainstem* (p. 445), *cerebellum* (p. 449), *diencephalon* (p. 449), *cerebrum* (p. 453), *meninges, ventricles, and cerebrospinal fluid* (p. 456), *blood supply to the brain* (p. 461), and *cranial nerves* (p. 462).

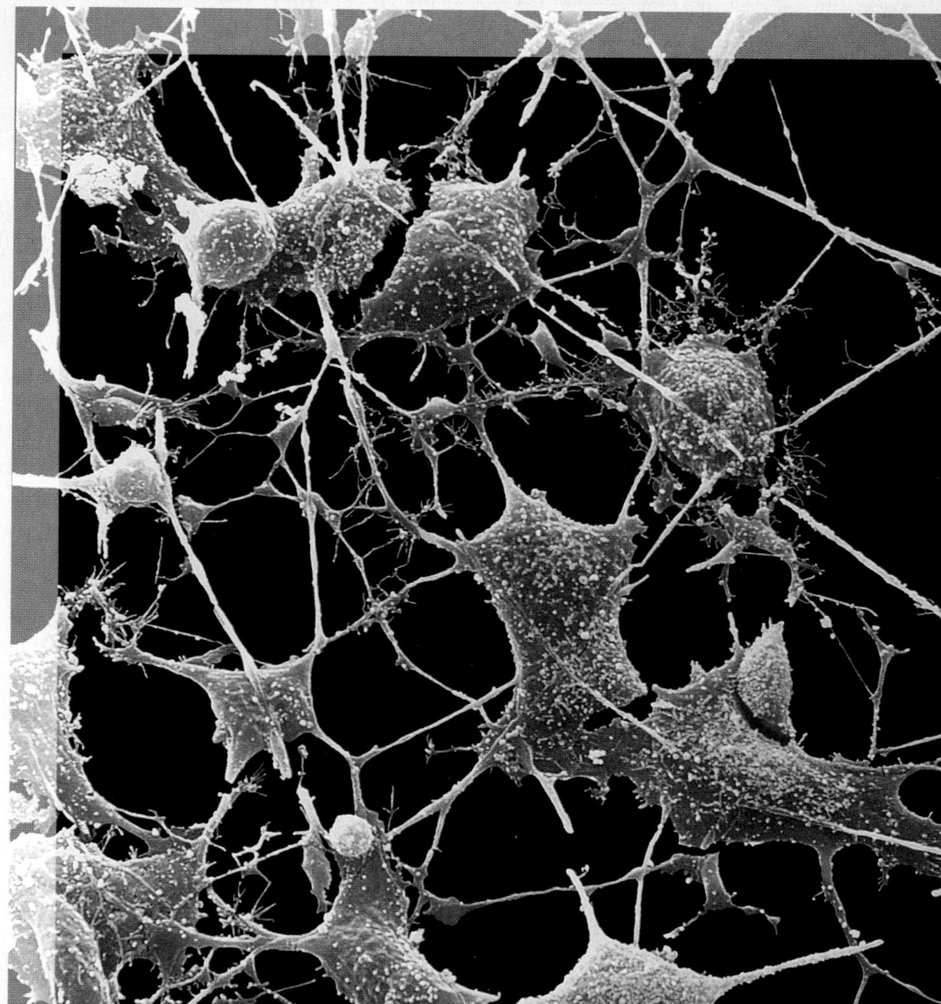

Colorized SEM of a neuron network.

Nervous System

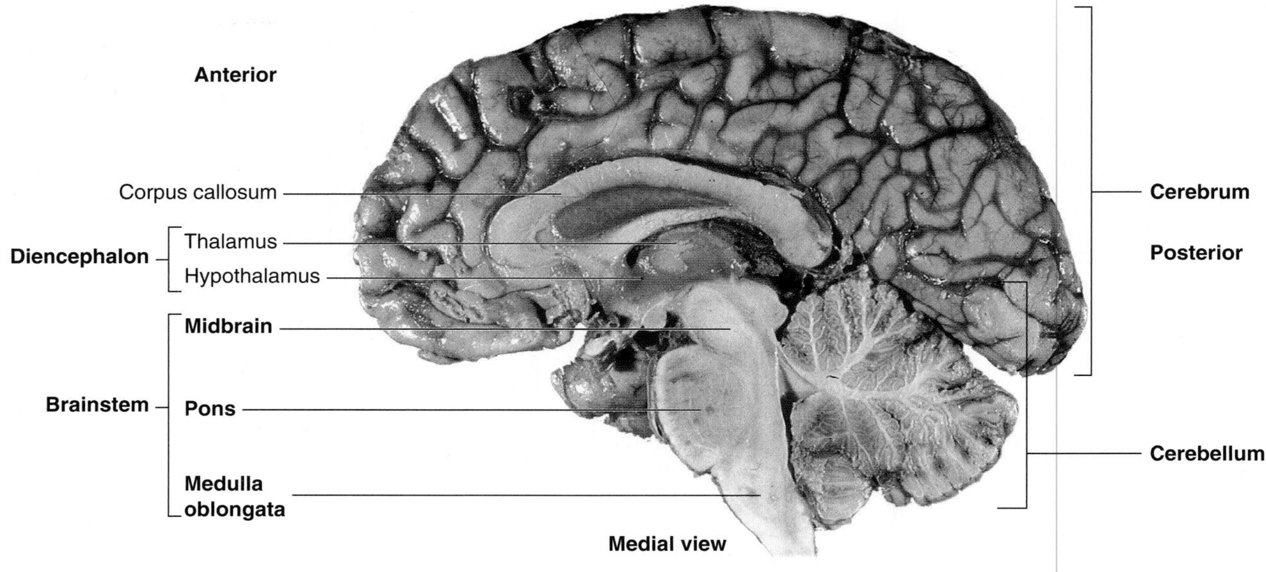

FIGURE 13.1 **Regions of the Right Half of the Brain (a Median Section)**

TABLE 13.1 Divisions and Functions of the Brain

Division	Function
Brainstem	Connects the spinal cord to the cerebrum; consists of the medulla oblongata, pons, and midbrain, with the reticular formation scattered throughout the three regions; has many important functions as listed under each subdivision; is the location of cranial nerve nuclei
Medulla oblongata	Pathway for ascending and descending nerve tracts; center for several important reflexes (e.g., heart rate, breathing, swallowing, vomiting)
Pons	Contains ascending and descending nerve tracts; relay between cerebrum and cerebellum; reflex centers
Midbrain	Contains ascending and descending nerve tracts; visual reflex center; part of auditory pathway
Reticular formation	Scattered throughout brainstem; controls cyclic activities, such as the sleep-wake cycle
Cerebellum	Control of muscle movement and tone; balance; regulates extent of intentional movement; involved in learning motor skills
Diencephalon	
Thalamus	Major sensory relay center; influences mood and movement
Subthalamus	Contains nerve tracts and nuclei
Epithalamus	Contains nuclei responding to olfactory stimulation and contains pineal body
Hypothalamus	Major control center for maintaining homeostasis and regulating endocrine function
Cerebrum	Conscious perception, thought, and conscious motor activity; can override most other systems
Basal nuclei	Control of muscle activity and posture; largely inhibit unintentional movement when at rest
Limbic system	Autonomic response to smell, emotion, mood, memory, and other such functions

DEVELOPMENT OF THE CNS

The CNS develops from a flat plate of ectodermal tissue, the **neural plate,** on the dorsal surface of the embryo, influence in part by the underlying rod-shaped **notochord** (figure 13.2). The lateral sides of the neural plate become elevated as waves, forming **neural folds.** The crest of each fold is called a **neural crest,** and the center of the neural plate becomes the **neural groove.** The neural folds move toward each other in the midline, and the crests fuse to create a **neural tube** (see figure 13.2*b*). The cephalic portion of the neural tube becomes the brain, and the caudal portion becomes the spinal cord. **Neural crest cells** are cells that separate from the neural crests and give rise to sensory and autonomic neurons of the peripheral nervous system. They also give rise to all pigmented cells of the body, the adrenal medulla, facial bones, and the dentin of the teeth.

A series of pouches develops in the anterior part of the neural tube (figure 13.3). The pouch walls become the various portions of the adult brain (table 13.2), and the pouch cavities become fluid-filled **ventricles** (ven′tri-klz). The ventricles are continuous with each other and with the **central canal** of the spinal cord. The neural tube develops flexures that cause the brain to be oriented almost 90 degrees to the spinal cord.

Three brain regions can be identified in the early embryo (see table 13.2 and figure 13.3*a*): a forebrain, or **prosencephalon** (pros-en-sef′ă-lon); a midbrain, or **mesencephalon** (mez-en-sef′ă-lon); and a hindbrain, or **rhombencephalon** (rom-ben-sef′ă-lon). During development, the forebrain divides into the **telencephalon** (tel-en-sef′ă-lon), which becomes the cerebrum, and the **diencephalon** (dī-en-sef′ă-lon). The midbrain remains as a single structure, but the hindbrain divides into the **metencephalon** (met′en-sef′ă-lon), which becomes the pons and cerebellum, and the **myelencephalon** (mī′el-en-sef′ă-lon), which becomes the medulla oblongata (see figure 13.3*b* and *c*).

1 ***Explain how the neural tube forms. Name the five divisions of the neural tube and the part of the brain that each division becomes.***

2 ***What do the cavities of the neural tube become in the adult brain?***

BRAINSTEM

The medulla oblongata, pons, and midbrain constitute the **brainstem** (figure 13.4). The brainstem connects the spinal cord to the remainder of the brain and is responsible for many essential functions. Damage to small brainstem areas often causes death because many reflexes essential for survival are integrated in the brainstem, whereas relatively large areas of the cerebrum or cerebellum may be damaged without life-threatening consequences.

Medulla Oblongata

The **medulla oblongata** (ob-long-gah′tă), often called the medulla, is about 3 cm long, is the most inferior part of the brainstem, and is continuous inferiorly with the spinal cord. It contains sensory and motor tracts; cranial nerve nuclei; other, related nuclei; and part of the reticular formation (see chapter 14). Superficially, the spinal cord blends into the medulla, but internally several

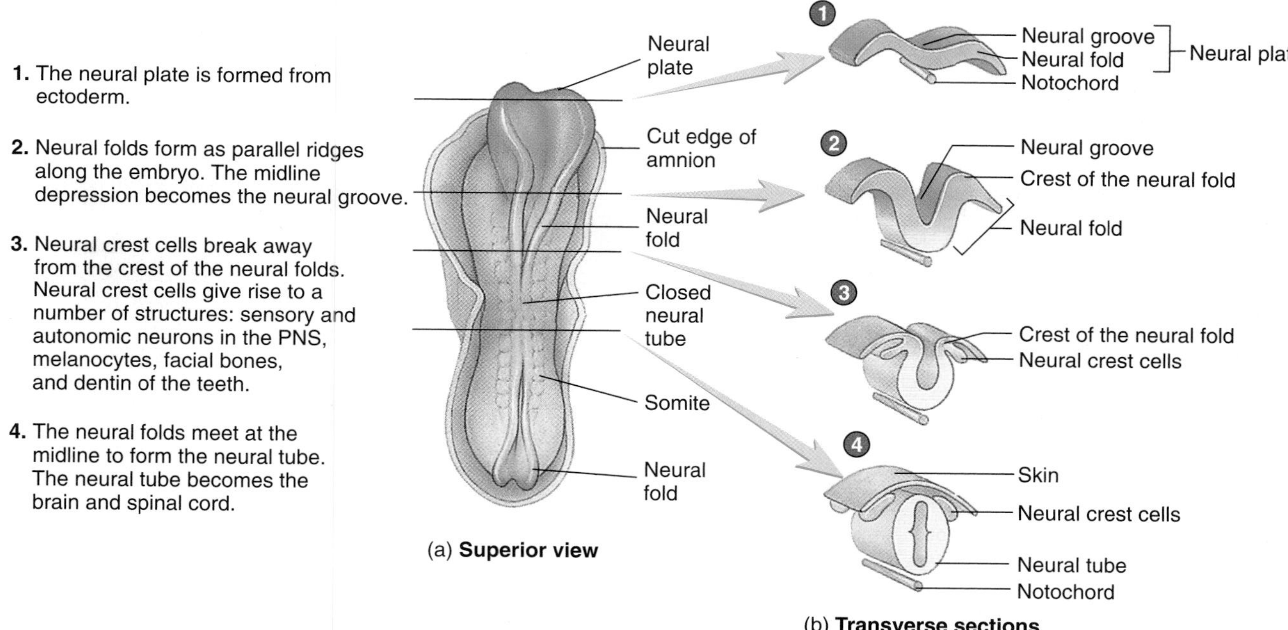

PROCESS FIGURE 13.2 Formation of the Neural Tube

(*a*) A 21-day-old human embryo. (*b*) Cross sections through the embryo. The level of each section is indicated by a line in part (*a*).

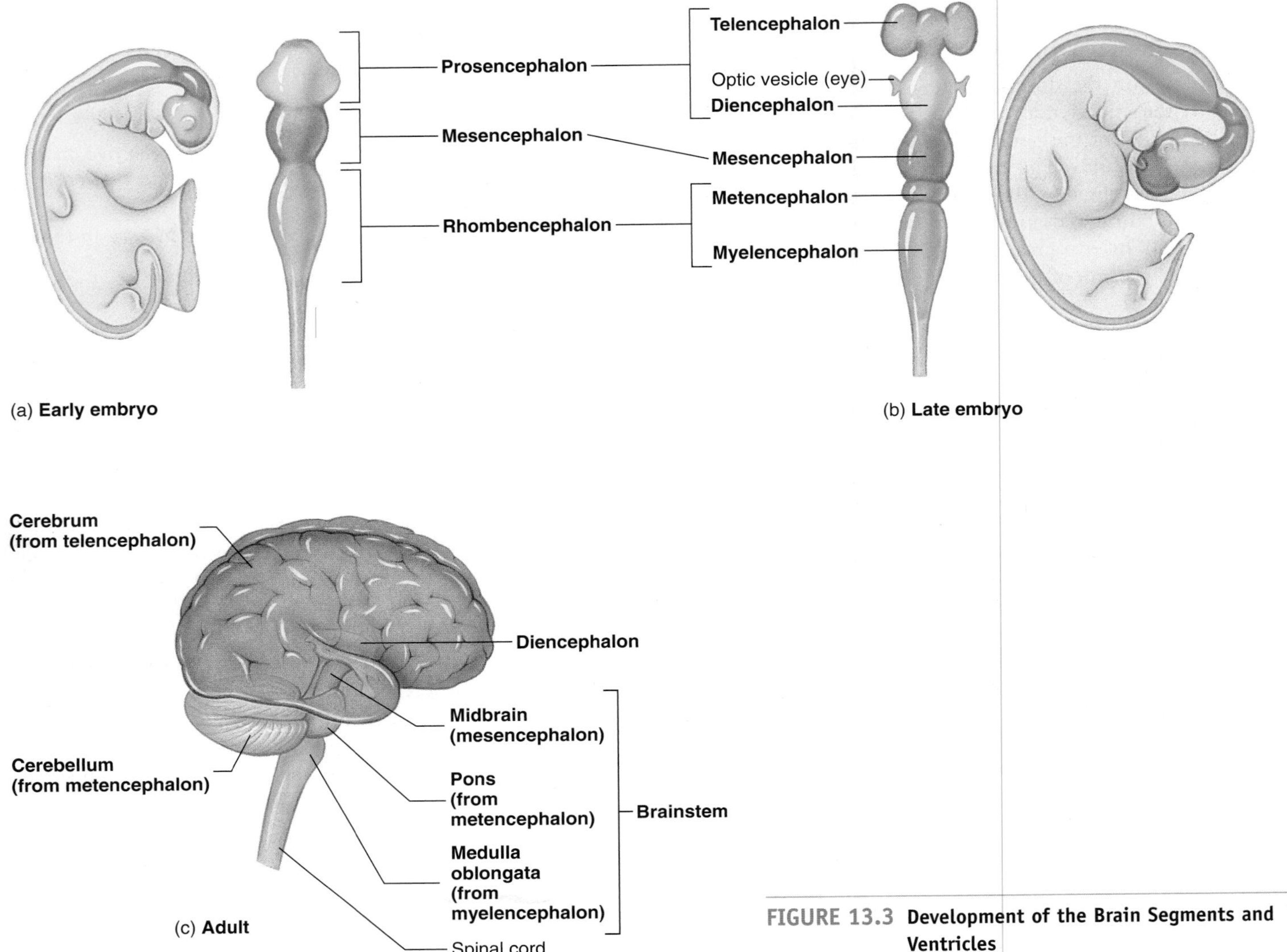

FIGURE 13.3 Development of the Brain Segments and Ventricles

TABLE 13.2 Development of the Central Nervous System (See Figure 13.3)

Early Embryo	Late Embryo	Adult	Cavity	Function
Prosencephalon (forebrain)	**Telencephalon**	**Cerebrum**	Lateral ventricles	Higher brain functions
	Diencephalon	**Diencephalon** (thalamus, subthalamus, epithalamus, hypothalamus)	Third ventricle	Relay center, autonomic nerve control, endocrine control
Mesencephalon (midbrain)	**Mesencephalon**	**Mesencephalon** (midbrain)	Cerebral aqueduct	Nerve pathways, reflex centers
Rhombencephalon (hindbrain)	**Metencephalon**	**Pons** and **cerebellum**	Fourth ventricle	Nerve pathways, reflex centers, muscle coordination, balance
	Myelencephalon	**Medulla oblongata**	Central canal	Nerve pathways, reflex centers

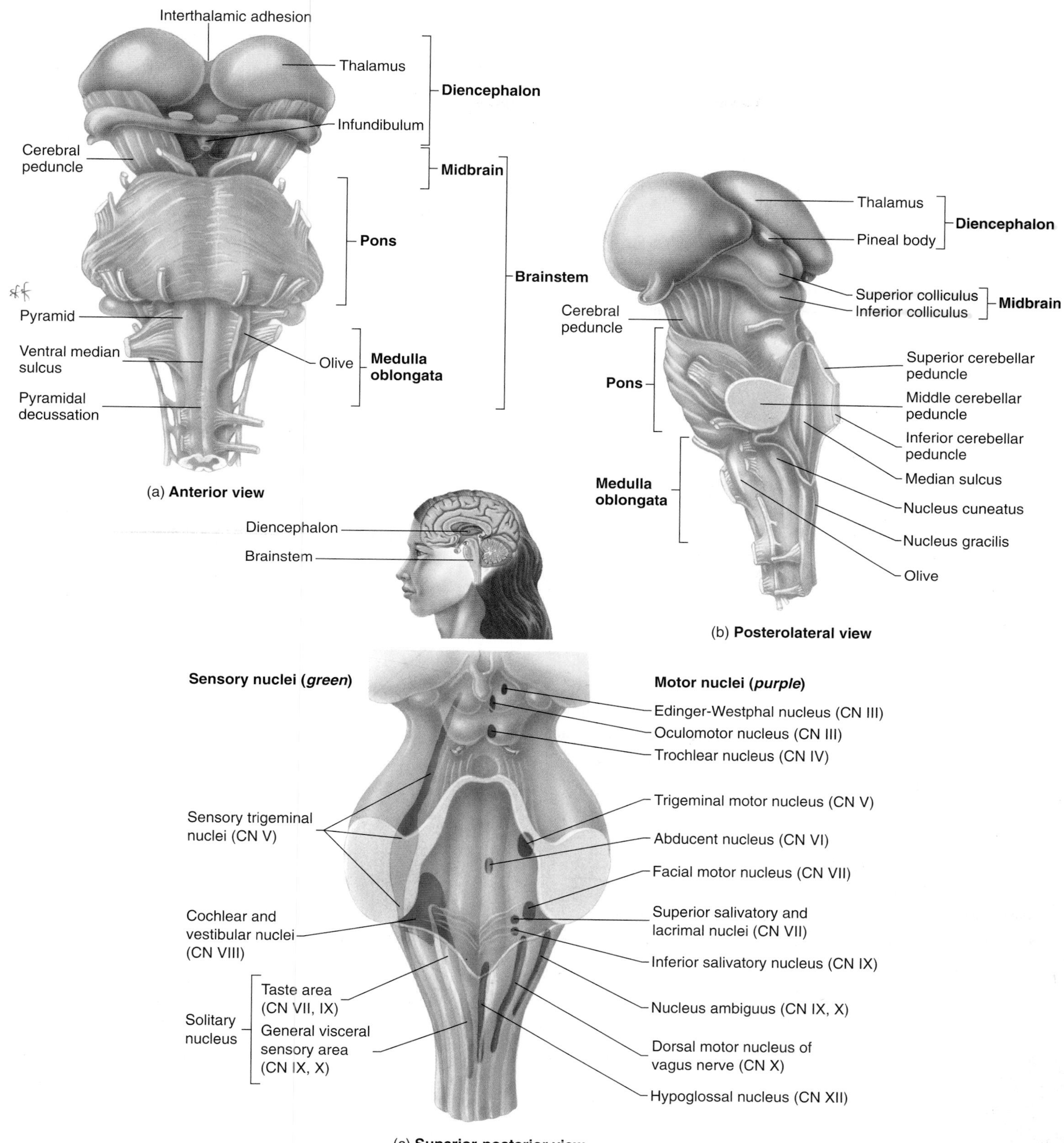

FIGURE 13.4 Brainstem and Diencephalon

(*a*) Anterior view. (*b*) Posterolateral view. (*c*) Brainstem nuclei. The sensory nuclei are shown on the left (*green*). The motor nuclei are shown on the right (*purple*). Even though the nuclei are shown on only one side, each half of the brainstem has both sensory and motor nuclei. The inset shows the location of the diencephalon (*red*) and brainstem (*blue*). (*CN* = cranial nerve.)

differences exist. Discrete **nuclei** (see figure 13.4), clusters of gray matter composed mostly of neuron cell bodies and having specific functions, are found in the medulla oblongata, whereas the gray matter of the spinal cord extends as a continuous mass in the center of the cord. In addition, the tracts within the medulla do not have the same organization as those of the spinal cord. Several medullary nuclei function as centers for reflexes, such as those involved in the regulation of heart rate, blood vessel diameter, respiration, swallowing, vomiting, hiccuping, coughing, and sneezing.

Two prominent enlargements on the anterior surface of the medulla oblongata are called **pyramids** because they are broader near the pons and taper toward the spinal cord (see figure 13.4*a*). The pyramids are descending tracts involved in the conscious control of skeletal muscles. Near their inferior ends, most of the fibers of the descending tracts cross to the opposite side, or **decussate** (dē′kŭ-sāt, dē-kŭs′āt; the Latin word *decussatus* means to form an X, as in the Roman numeral X). This decussation accounts, in part, for the fact that each half of the brain controls the opposite half of the body. Its role as a conduction pathway is discussed in the description of ascending and descending tracts (see chapter 14).

Two rounded, oval structures, called **olives,** protrude from the anterior surface of the medulla oblongata just lateral to the superior ends of the pyramids (see figure 13.4*a* and *b*). The olives are nuclei involved in functions such as balance, coordination, and modulation of sound from the inner ear (see chapter 15). Nuclei of cranial nerves V (trigeminal), IX (glossopharyngeal), X (vagus), XI (accessory), and XII (hypoglossal) also are located within the medulla (see figure 13.4*c*).

Pons

The part of the brainstem just superior to the medulla oblongata is the **pons** (see figure 13.4*a*), which contains ascending and descending tracts and several nuclei. The pontine nuclei, located in the anterior portion of the pons, relay information from the cerebrum to the cerebellum.

Nuclei for cranial nerves V (trigeminal), VI (abducent), VII (facial), VIII (vestibulocochlear), and IX (glossopharyngeal) are contained within the posterior pons. Other important pontine areas include the pontine sleep center and respiratory center, which work with the respiratory centers in the medulla to help control respiratory movements (see chapter 23).

Midbrain

The **midbrain,** or **mesencephalon,** is the smallest region of the brainstem (see figure 13.4*b*). It is located just superior to the pons and contains the nuclei of cranial nerves III (oculomotor), IV (trochlear), and V (trigeminal).

The **tectum** (tek′tŭm; roof; figure 13.5) of the midbrain consists of four nuclei that form mounds on the dorsal surface, collectively called **corpora** (kōr′pōr-ă; bodies) **quadrigemina** (kwah′dri-jem′i-nă; four twins). Each mound is called a **colliculus** (ko-lik′ū-lŭs; hill); the two superior mounds are called **superior colliculi,** and the two inferior mounds are called **inferior colliculi** (see figure 13.4*b*). The superior colliculi are involved in visual reflexes. These reflexes control the movement of the head, eyes, and body toward visual, auditory, or tactile stimuli such as loud noises, flashing lights, or startling pain. The superior colliculi receive input from the eyes, the inferior colliculi, the skin, and the cerebrum. The inferior colliculi are involved in hearing and are an integral part of the auditory pathways in the CNS. Neurons conducting action potentials from the structures of the inner ear (see chapter 15) to the brain synapse in the inferior colliculi. Collateral fibers from the inferior colliculi to the superior colliculi provide auditory input that stimulates visual reflexes.

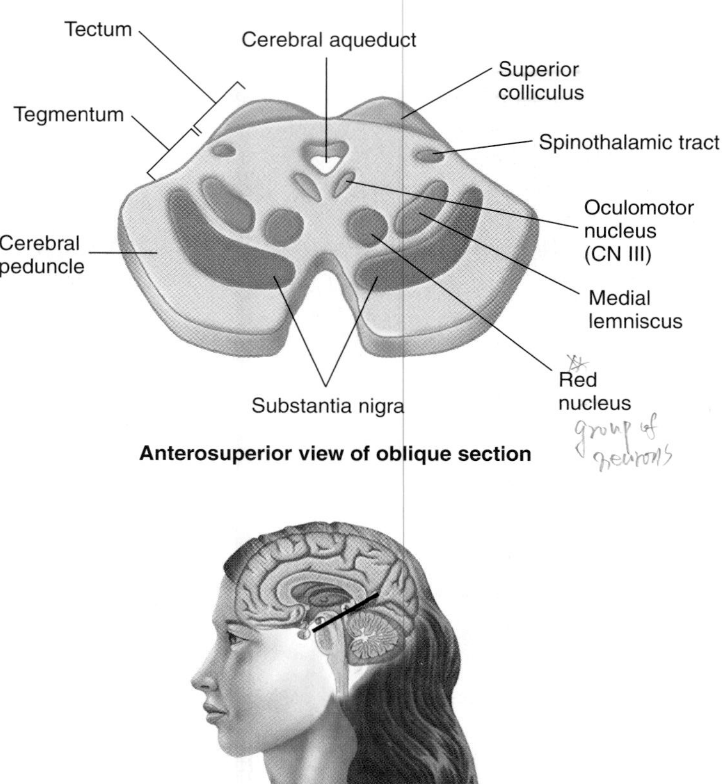

FIGURE 13.5 Oblique Section Through the Midbrain
Inset shows the level of section.

Reflex Movements of the Eyes and Head

The superior colliculi regulate the reflex movements of the eyes and head in response to various stimuli. When a bright object suddenly appears in a person's field of vision, a reflex turns the eyes to focus on it. When a person hears a sudden, loud noise, a reflex turns the head and eyes toward it. When a part of the body, such as the shoulder, is touched, a reflex turns the person's head and eyes toward that part of the body. In each situation, the pathway involves the superior colliculus.

The **tegmentum** (teg-men′tŭm; floor) of the midbrain largely consists of ascending tracts, such as the spinothalamic tract and the medial lemniscus, which carry sensory information from the spinal cord to the brain. The tegmentum also contains the paired **red nuclei** (see figure 13.5), which are so-named because in fresh brain specimens they are pinkish in color as a result of an abundant blood supply. The red nuclei aid in the unconscious regulation and coordination of motor activities. **Cerebral peduncles** (pe-dŭng′klz, pē′dŭng-klz; the foot of a column) constitute the portion of the midbrain ventral to the tegmentum. They consist primarily of descending tracts, which carry motor information from the cerebrum to the brainstem and spinal cord. The **substantia nigra** (nī′gră; black substance) is a nuclear mass between the tegmentum and cerebral peduncles, containing cytoplasmic melanin granules that give it a dark gray or black color (see figure 13.5). The substantia nigra is interconnected with other basal nuclei of the cerebrum, described later in this chapter, and it is involved in maintaining muscle tone and in coordinating movements.

Reticular Formation

The **reticular formation** (see table 13.1) is a diffuse system consisting of several loosely packed nuclei scattered throughout the length of the brainstem. The reticular formation receives axons from a large number of sources and especially from nerves that innervate the face. The functions of the reticular formation involve "cycles" of activity, such as the sleep–wake cycle, and are discussed in chapter 14.

Describe the major components of the medulla oblongata, pons, midbrain, and reticular formation. What are the general functions of each region?

CEREBELLUM

The term **cerebellum** (ser-e-bel′ŭm; figure 13.6) means little brain. The cerebellum is attached to the brainstem posterior to the pons. It communicates with other regions of the CNS through three large tracts: the **superior, middle,** and **inferior cerebellar peduncles** (see figure 13.4*b*), which connect the cerebellum to the midbrain, pons, and medulla oblongata, respectively. The cerebellum has a gray cortex and nuclei, with white medulla in between. The cerebellar cortex has ridges called **folia.** The white matter of the medulla resembles a branching tree and is called the **arbor vitae** (ar′bōr vī′te; tree of life). The **nuclei** of the cerebellum are located in the deep inferior center of the white matter.

The cerebellar cortex contains several cell types, including stellate cells, basket cells, granule cells, Golgi cells, and Purkinje cells. It also contains mossy fibers, which are afferent axons that branch extensively within the cerebellum. **Purkinje** (pŭr-kin′jē; Johannes Purkinje, a Bohemian anatomist and physiologist, 1787–1869) **cells** are the largest and probably most interesting cells in the CNS. Purkinje cells receive 200,000 synapses, are inhibitory neurons, and are the only cerebellar cortex neurons that send axons to the cerebellar nuclei. The cerebellar cortex contains more than 10^{12} neurons, more neurons than the entire cerebral cortex.

The cerebellum consists of three parts: a small inferior part, the **flocculonodular** (flok′ū-lō-nod′ū-lăr; floccular, a tuft of wool) **lobe;** a narrow central **vermis** (worm-shaped); and two large **lateral hemispheres** (see figure 13.6*b* and *c*).

The flocculonodular lobe, the simplest part of the cerebellum, helps control balance and eye movements. The vermis and medial portion of the lateral hemispheres are involved in the control of posture, locomotion, and fine motor coordination, thereby producing smooth, flowing movements. The major portions of the lateral hemispheres of the cerebellum function in concert with the frontal lobes of the cerebral cortex in planning, practicing, and learning complex movements.

Each lateral hemisphere is divided by a **primary fissure** into an **anterior lobe** and a **posterior lobe.** The lobes are subdivided into **lobules,** which contain the folia.

4 ***What are the major regions of the cerebellum? Describe the major functions of each.***

DIENCEPHALON

The **diencephalon** (dī-en-sef′ă-lon) is the part of the brain located between the brainstem and the cerebrum (see figure 13.1; figure 13.7). Its main components are the thalamus, subthalamus, epithalamus, and hypothalamus.

Thalamus

The **thalamus** (thal′ă-mŭs; see figure 13.7*a* and *b*) is by far the largest part of the diencephalon, constituting about four-fifths of its weight. It consists of a cluster of nuclei shaped somewhat like a yo-yo, with two large, lateral portions connected in the center by a small stalk called the **interthalamic adhesion,** or **intermediate mass.** The space surrounding the interthalamic adhesion and separating the two large portions of the thalamus is the third ventricle of the brain.

Except for the olfactory neurons, all sensory neurons that project to the cerebrum first synapse in the thalamus. Thalamic neurons then send projections to the appropriate areas of the cerebral cortex where sensory input is localized (see chapter 14). For this reason, the thalamus is considered the **sensory relay center** of the brain. Axons carrying auditory information synapse in the **medial geniculate** (je-nik′ū-lāt; Latin, *genu,* meaning bent like a knee) **nucleus** of the thalamus. Axons carrying visual information synapse in the **lateral geniculate nucleus.** Most other sensory impulses synapse in the **ventral posterior nucleus.** Axons originating in the ventral posterior nucleus project to the **dorsal tier** of nuclei, which register pain (see figure 13.7*b*). Other axons project to the cerebral cortex where sensory input is localized (see chapter 14). The **ventral anterior** and **ventral lateral nuclei** are involved with motor functions, communicating between the basal nuclei, cerebellum, and the motor cortex (these areas are described later in this chapter).

The thalamus also influences mood and actions associated with strong emotions, such as fear and rage. The **anterior** and **medial nuclei** are connected to the limbic system and to the prefrontal cortex (described later in this chapter and in chapter 14). These nuclei are involved in mood modification. The **lateral dorsal nuclei** are connected to other thalamic nuclei and to the cerebral cortex and are involved in regulating emotions. The **lateral posterior nuclei** and the **pulvinar** (pŭl-vī′năr; pillow) also have connections to other thalamic nuclei and are involved in sensory integration.

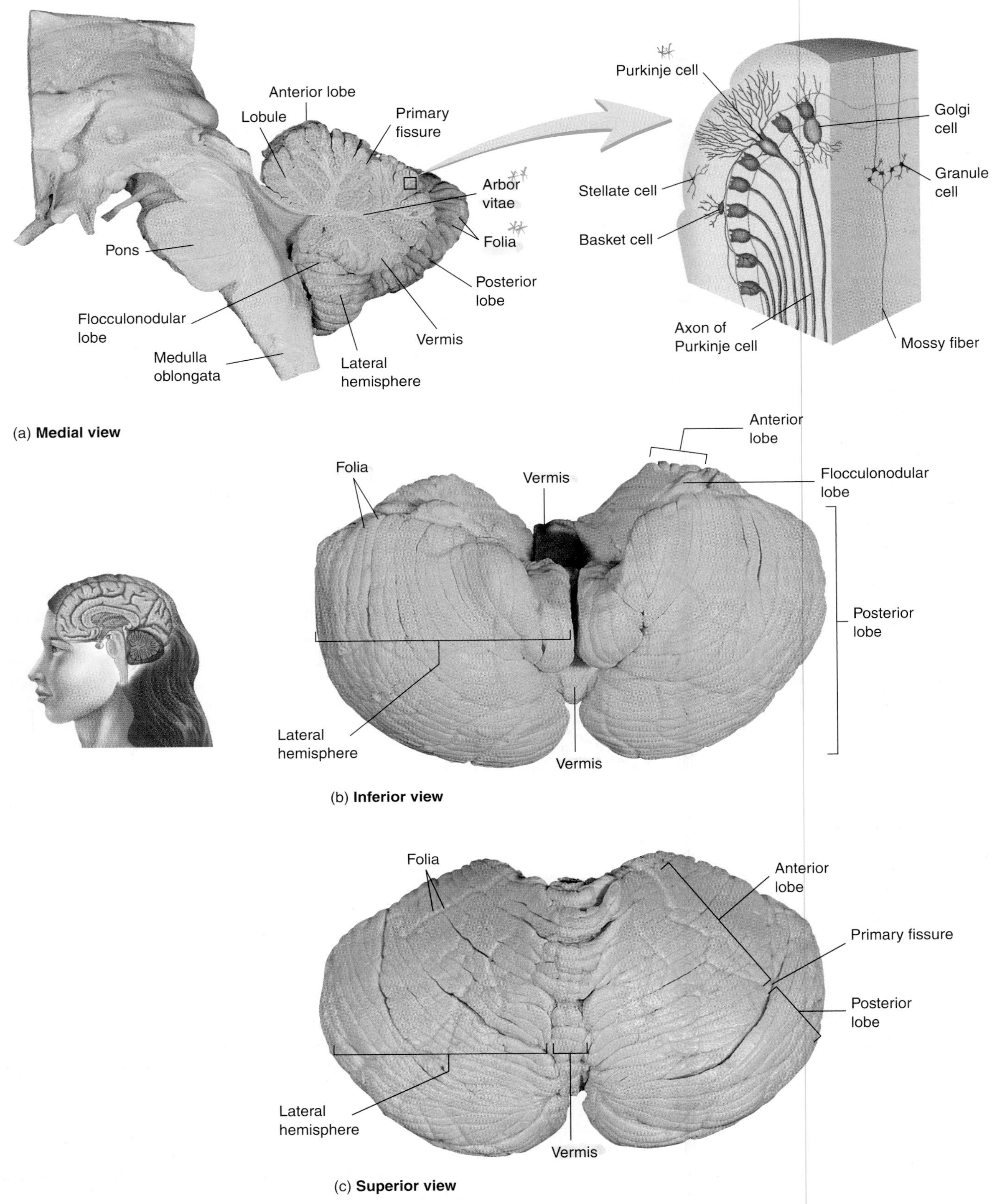

FIGURE 13.6 Cerebellum

(*a*) Right half of the cerebellum as seen in a median section. Inset shows the histology of the cerebellum. (*b*) Inferior view of the cerebellum. (*c*) Superior view of the cerebellum.

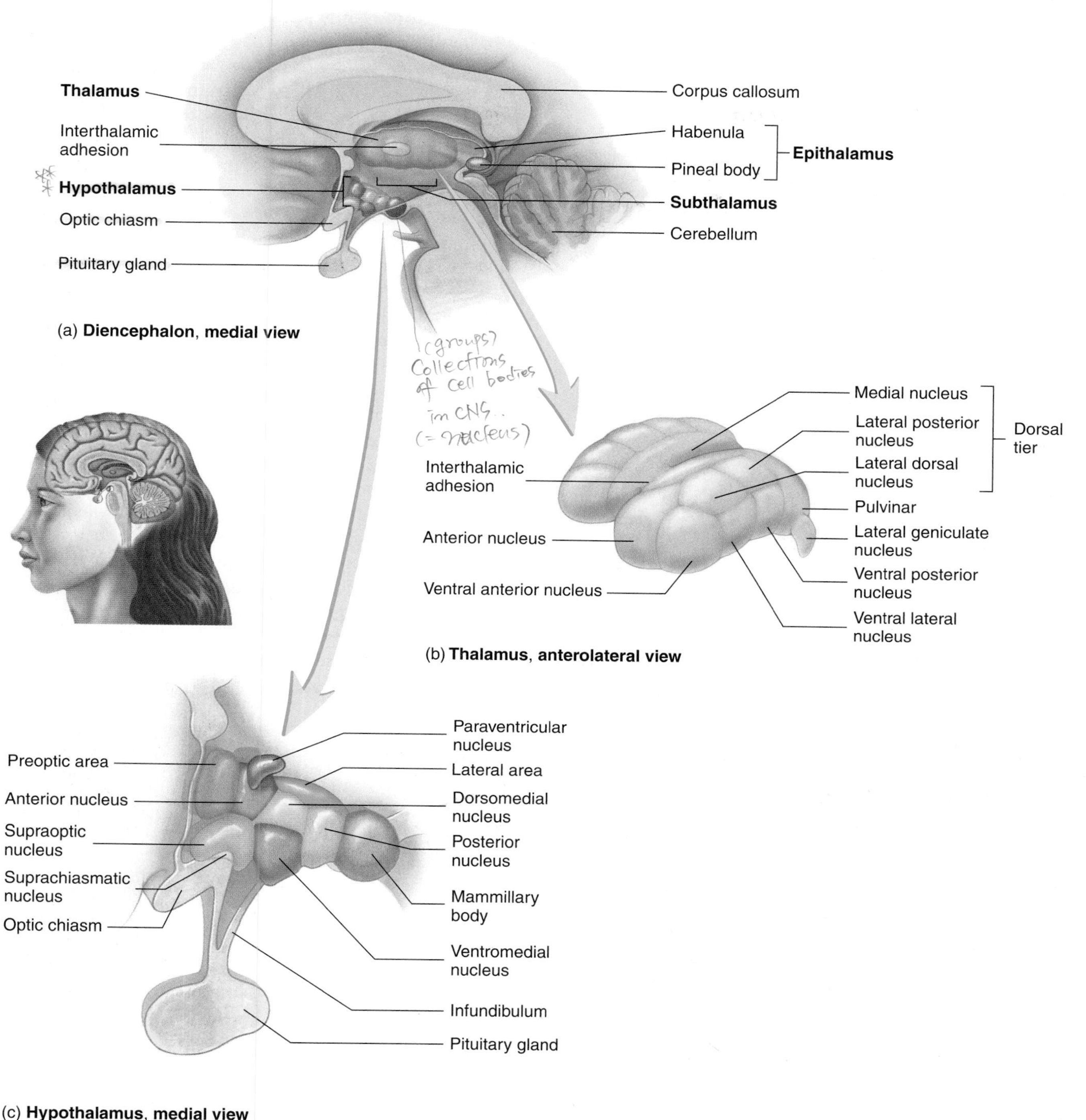

FIGURE 13.7 Diencephalon

(*a*) General overview of the right half of the diencephalon as seen in a median section. (*b*) Thalamus showing the nuclei. (*c*) Hypothalamus showing the nuclei and right half of the pituitary.

Subthalamus

The **subthalamus** is a small area immediately inferior to the thalamus (see figure 13.7*a*) that contains several ascending and descending tracts and the **subthalamic nuclei.** A small portion of the red nucleus and substantia nigra of the midbrain extend into this area.

The subthalamic nuclei are associated with the basal nuclei and are involved in controlling motor functions.

Epithalamus

The **epithalamus** is a small area superior and posterior to the thalamus (see figure 13.7*a*). It consists of habenular nuclei and the pineal body. The **habenular** (hă-ben′ū-lăr) **nuclei** are influenced by the sense of smell and are involved in emotional and visceral responses to odors. The **pineal** (pin′ē-ăl) **gland,** or **pineal body,** is shaped somewhat like a pinecone, from which the name *pineal* is derived. It appears to play a role in controlling the onset of puberty, but data are inconclusive, so active research continues in this field. The pineal body also may influence the sleep–wake cycle and other biorhythms.

Brain Sand in the Pineal

In about 75% of adults, the pineal body contains granules of calcium and magnesium salts called "brain sand." These granules can be seen on radiographs and are useful as landmarks in determining whether or not the pineal body has been displaced by a pathologic enlargement of a part of the brain, such as a tumor or a hematoma.

Hypothalamus

very important

The **hypothalmus,** the most inferior portion of the diencephalon (see figure 13.7*a* and *c*), contains several small nuclei and tracts. The most conspicuous nuclei, called the **mammillary bodies,** appear as bulges on the ventral surface of the diencephalon. They are involved in olfactory reflexes and emotional responses to odors. They may also be involved in memory. A funnel-shaped stalk, the **infundibulum** (in-fŭn-dib′ū-lŭm), extends from the floor of the hypothalamus and connects it to the **posterior pituitary gland,** or **neurohypophysis** (noor′ō-hī-pof′i-sis). The hypothalamus plays an important role in controlling the endocrine system because it regulates the pituitary gland's secretion of hormones, which influence functions as diverse as metabolism, reproduction, responses to stressful stimuli, and urine production (table 13.3; see chapter 18).

Sensory neurons that terminate in the hypothalamus provide input from (1) internal organs; (2) taste receptors of the tongue; (3) the limbic system, which is involved in responses to smell; (4) specific cutaneous areas, such as the nipples and external genitalia; (5) the eyes; and (6) the prefrontal cortex of the cerebrum carrying information relative to "mood" through the thalamus. Efferent fibers from the hypothalamus extend into the brainstem and the spinal cord, where they synapse with neurons of the autonomic nervous system (see chapter 16). These connections regulate functions such as heart rate and digestive functions (see table 13.3). Other fibers extend through the infundibulum to the posterior portion of the pituitary gland (see chapter 18); some extend to trigeminal and facial nerve nuclei to help control the muscles involved in swallowing; and some extend to motor neurons of the spinal cord to stimulate shivering.

Hypothalamic nuclei directly control temperature by stimulating sweating or shivering. Other hypothalamic nuclei are involved in the control of thirst, hunger, and sex drive. The hypothalamus is also very important in a number of functions related to mood, motivation, and emotion. This is one reason

TABLE 13.3 Hypothalamic Functions

Function	Hypothalamic Nuclei	Description
Autonomic	Preoptic area and anterior nucleus (parasympathetic) Lateral area and posterior nucleus (sympathetic)	Helps control heart rate, urine release from the bladder, movement of food through the digestive tract, and blood vessel diameter
Endocrine	Paraventricular nucleus Supraoptic nucleus	Helps regulate pituitary gland secretions and influences metabolism, ion concentration, sexual development, and sexual functions; site of antidiuretic hormone and oxytocin production (see chapter 18)
Muscle control	Lateral area	Controls the muscles involved in swallowing and stimulates shivering
Temperature regulation	Preoptic area Anterior nucleus Posterior nucleus	Promotes heat loss when the hypothalamic temperature increases by increasing sweat production (anterior nucleus) and promotes heat production when the hypothalamic temperature decreases by promoting shivering (posterior nucleus); aspirin affects the preoptic area to reduce fever
Regulation of food and water intake	Ventromedial nucleus Lateral area	Hunger center promotes eating and satiety center inhibits eating; thirst center promotes water intake
Emotions	Lateral area Medial area	Large range of emotional influences over body functions; directly involved in stress-related and psychosomatic illnesses and with feelings of fear and rage
Regulation of the sleep–wake cycle	Lateral area Suprachiasmatic nucleus	Coordinates responses to the sleep-wake cycle with the other areas of the brain (e.g., the reticular activating system); suprachiasmatic nucleus receives direct input from the eyes concerning light/dark cycles and is implicated in jet lag
Sexual development and behavior	Preoptic area Dorsomedial nucleus Ventromedial nucleus	Stimulates sexual development, sexual arousal, and sexual behavior; the preoptic area is larger in males than in females

that strong emotional experiences may affect a person's desire or ability to eat, drink, or experience sexual pleasure, and vice versa. Sensations such as sexual pleasure, relaxation after a meal, rage, and fear are related to hypothalamic functions. The hypothalamus interacts with the reticular activating system in the brainstem to coordinate the sleep–wake cycle (see table 13.3)

5 ***Name the four main components of the diencephalon.***

6 ***What are the functions of the thalamus and hypothalamus? Explain why the hypothalamus is an important link between the nervous system and the endocrine system.***

7 ***List the general functions of the subthalamus. Name the parts of the epithalamus and give their functions.***

CEREBRUM

The cerebrum (figure 13.8) is the part of the brain that most people think of when the term *brain* is mentioned. The cerebrum accounts for the largest portion of total brain weight, which is about 1200 g in females and 1400 g in males. Brain size is related to body size; larger brains are associated with larger bodies, not with greater intelligence.

The cerebrum is divided into left and right hemispheres by a **longitudinal fissure** (see figure 13.8*a*). The most conspicuous features on the surface of each hemisphere are numerous folds called **gyri** (jī′rī; sing. *gyrus*), which greatly increase the surface area of the cortex. The grooves between the gyri are called **sulci**

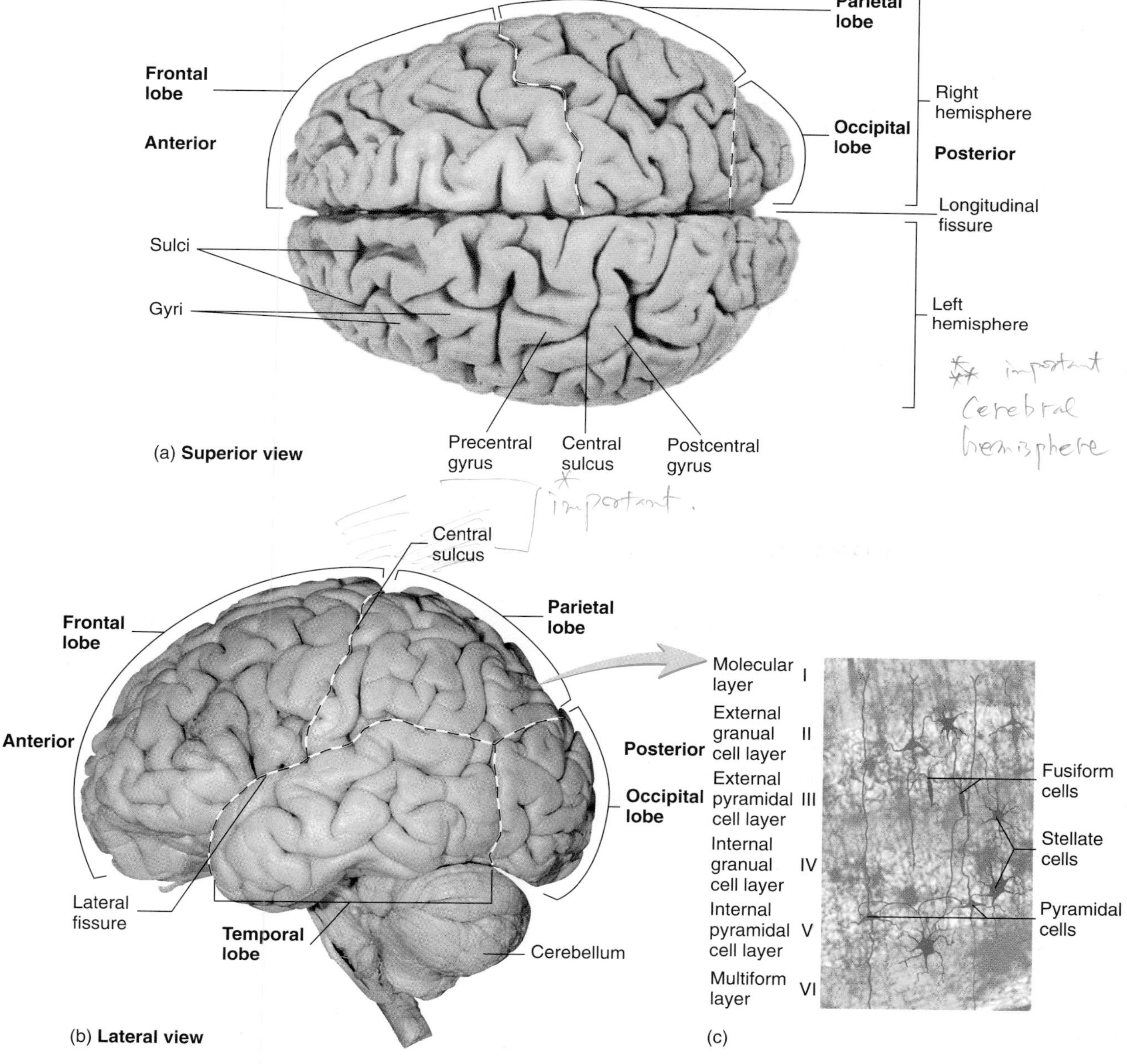

FIGURE 13.8 Brain

(*a*) Superior view. (*b*) Lateral view of the left cerebral hemisphere. (*c*) Histology of the cerebral cortex.

(sŭl′sī; sing. *sulcus*). The **central sulcus,** which extends across the lateral surface of the cerebrum from superior to inferior, is located about midway along the length of the brain. The central sulcus is located between the **precentral gyrus** anteriorly, which is the **primary motor cortex,** and a **postcentral gyrus** posteriorly, which is the **primary somatic sensory cortex** (see chapter 14). The general pattern of the gyri is similar in all normal human brains, but some variation exists between individuals and even between the two hemispheres of the same cerebrum.

Each cerebral hemisphere is divided into lobes, which are named for the skull bones overlying each one (see figure 13.8*b*). The **frontal lobe** is important in voluntary motor function, motivation, aggression, the sense of smell, and mood. The **parietal lobe** is the major center for the reception and evaluation of most sensory information, except for smell, hearing, and vision. The frontal and parietal lobes are separated by the central sulcus. The **occipital lobe** functions in the reception and integration of visual input and is not distinctly separate from the other lobes. The **temporal lobe** receives and evaluates input for smell and hearing and plays an important role in memory. Its anterior and inferior portions are referred to as the "psychic cortex," and they are associated with such brain functions as abstract thought and judgment. The temporal lobe is separated from the rest of the cerebrum by a **lateral fissure,** and deep within the fissure is the **insula** (in′soo-lă; island), often referred to as a fifth lobe.

The gray matter on the outer surface of the cerebrum is the **cortex,** and clusters of gray matter deep inside the brain are nuclei. The cerebral cortex contains a number of neuron types, named largely for their shape, such as fusiform cells, stellate cells, and pyramidal cells (see figure 13.8*c*). These cells are distributed in layers within the cerebral cortex. The thickness of the cortex is not uniform throughout the cerebrum but ranges from two or three layers in the most "primitive" parts of the cortex to six layers in the more "advanced" regions.

The white matter of the brain between the cortex and nuclei is the **cerebral medulla.** This term should not be confused with the medulla oblongata; *medulla* is a general term meaning the center of a structure. The cerebral medulla consists of tracts that connect areas of the cerebral cortex to each other or to other parts of the CNS (figure 13.9). The fibers in these tracts fall into three main categories: (1) **Association fibers** connect areas of the cerebral cortex within the same hemisphere. (2) **Commissural fibers** connect one cerebral hemisphere to the other. The largest bundle of commissural fibers connecting the two cerebral hemispheres is the **corpus callosum** (kōr′pŭs ka-lō′sŭm; see figures 13.1 and 13.9). (3) **Projection fibers** are between the cerebrum and other parts of the brain and spinal cord (see figure 13.9). The projection fibers form the **internal capsule.**

8 ***Define gyri and sulci. What structures do the longitudinal fissure, central sulcus, and lateral fissure separate?***

9 ***Define cerebral cortex and cerebral medulla.***

10 ***Name the five lobes of the cerebrum, and describe their locations and functions.***

11 ***List three categories of tracts in the cerebral medulla.***

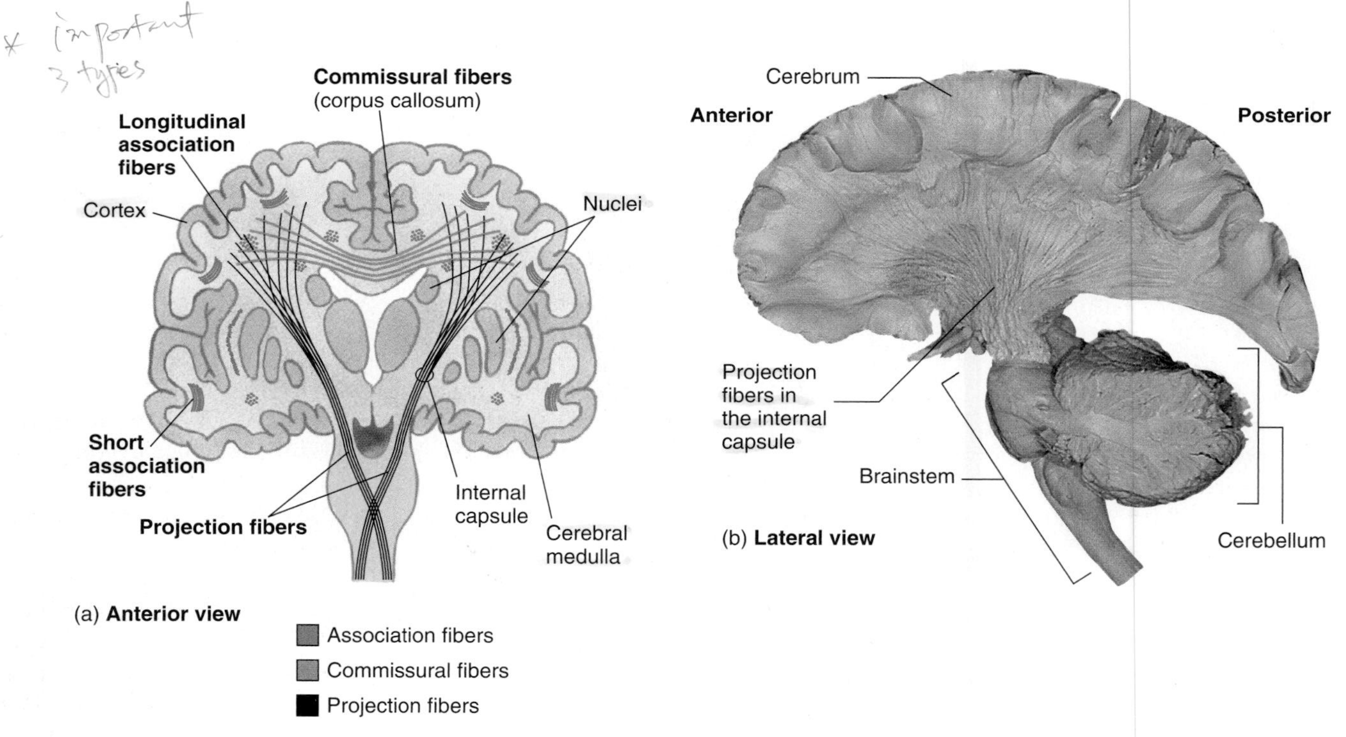

FIGURE 13.9 Cerebral Medullary Tracts

(*a*) Coronal section of the brain showing commissural, association, and projection fibers. (*b*) Photograph of the left cerebral hemisphere from a lateral view with the cortex and association fibers removed to reveal the projection fibers of the internal capsule deep within the brain.

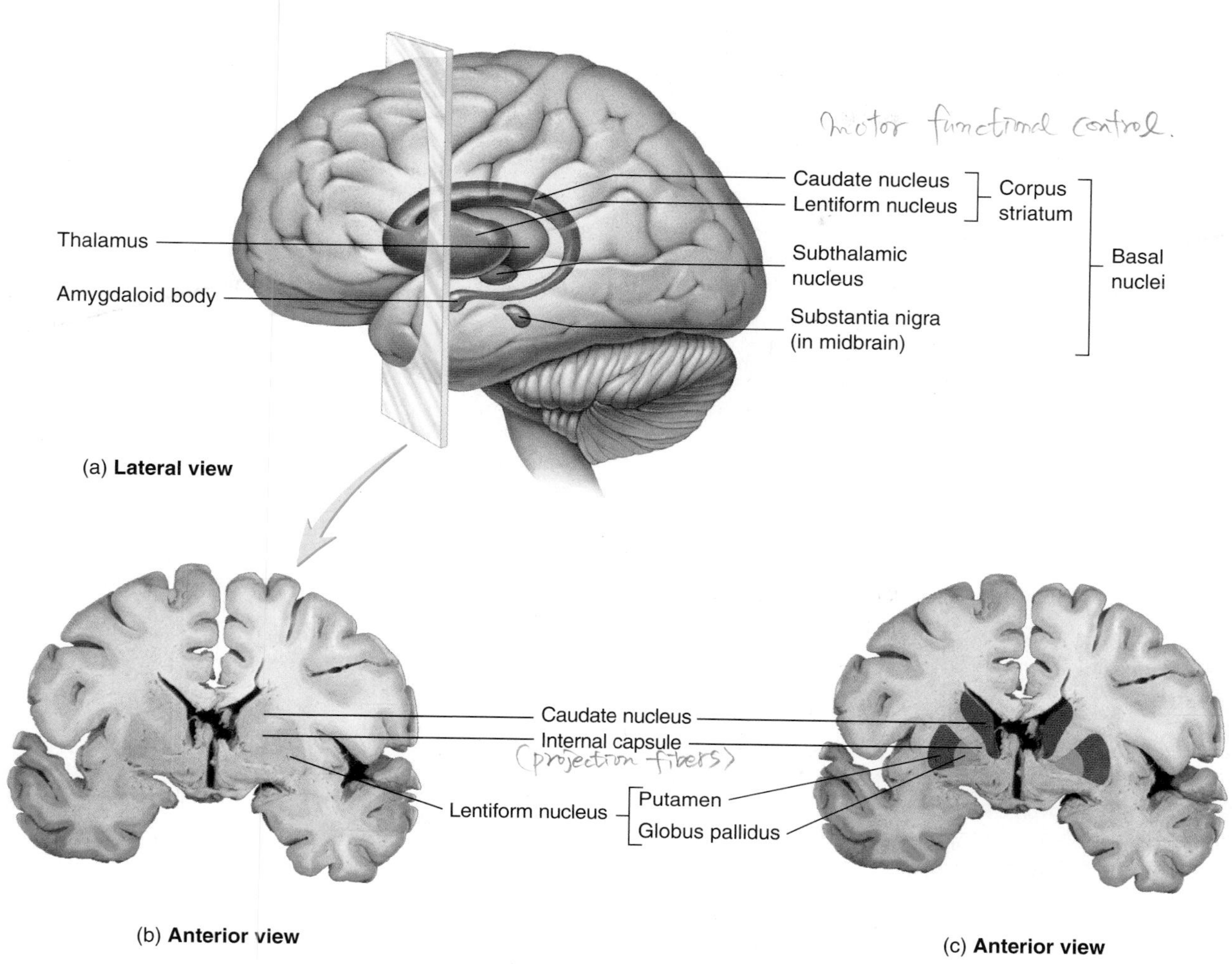

FIGURE 13.10 Basal Nuclei of the Left Hemisphere
(*a*) "Transparent 3-D" drawing of the basal nuclei inside the left hemisphere. (*b*) Photograph of a frontal section of the brain showing the basal nuclei and other structures. (*c*) The same photograph as in (*b*) with nuclei highlighted.

Basal Nuclei

The **basal nuclei** are a group of functionally related nuclei located bilaterally in the inferior cerebrum, diencephalon, and midbrain (figure 13.10). These nuclei are involved in the control of motor functions (see chapter 14). The nuclei in the cerebrum are collectively called the **corpus striatum** (kōr′pŭs strī-ā′tŭm; striped body) and include the **caudate** (kaw′dāt; having a tail) **nucleus** and **lentiform** (len′ti-fōrm; lens-shaped) **nucleus.** The lentiform nucleus, in turn, is divided into a lateral **putamen** (pū-tā′men; that which falls off in pruning) and the medial **globus pallidus** (glō′bŭs pal′li-dŭs; pale globe). The basal nuclei are the largest nuclei of the brain and occupy a large part of the cerebrum. The **subthalamic nucleus** and the **substantia nigra** function in conjunction with the caudate and lentiform nuclei in the control of movement. The subthalamic nucleus is located in the diencephalon, and the substantia nigra is located in the midbrain.

12 ***List the basal nuclei and state their general function.***

Limbic System

Parts of the cerebrum and diencephalon are grouped together under the title **limbic** (lim′bik) **system** (figure 13.11). The limbic system plays a central role in basic survival functions, such as memory, reproduction, and nutrition. It is also involved in emotional interpretation of sensory input and emotions in general. *Limbus* means border, and the term *limbic* refers to deep portions of the cerebrum that form a ring around the diencephalon. Structurally, the limbic system consists of (1) certain cerebral cortical areas, including the **cingulate** (sin′gū-lāt; to surround) **gyrus,** located along the inner surface of the longitudinal fissure just above the corpus callosum, and the **parahippocampal gyrus,** located on the medial side of the temporal lobe; (2) various nuclei, such as anterior nuclei of the thalamus, the habenula in the epithalamus, and the **dentate nucleus** of the **hippocampus;** (3) parts of the basal nuclei, such as the **amygdala,** or **amygdaloid nuclear complex;** (4) the hypothalamus, especially the mammillary bodies; (5) the **olfactory cortex;** and (6) tracts connecting the various cortical areas and nuclei, such as the

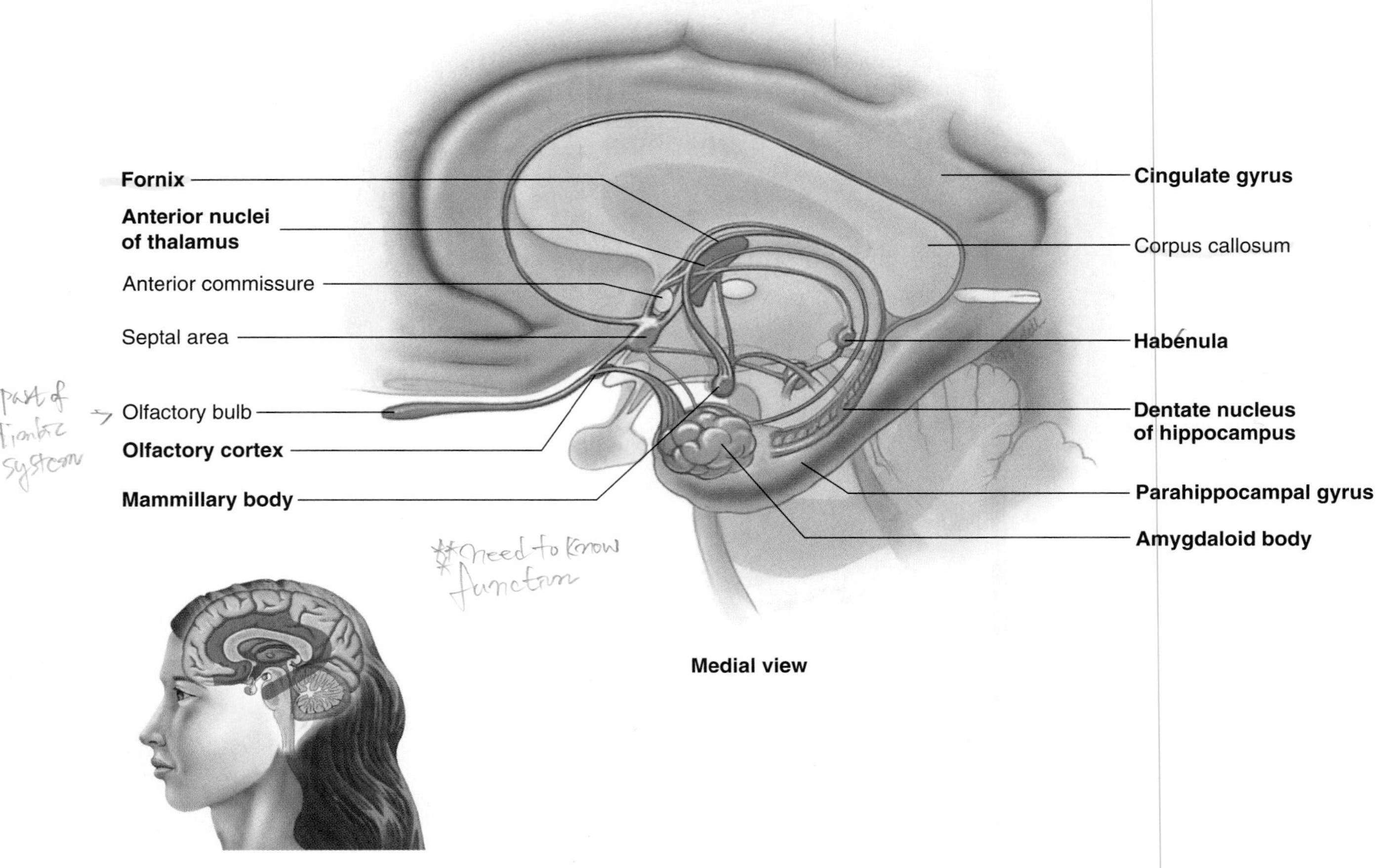

FIGURE 13.11 Limbic System and Associated Structures of the Right Hemisphere as Seen in a Median Section

fornix, which connects the hippocampus to the thalamus and mammillary bodies.

13 ***List the parts of the limbic system. What are the functions of the limbic system?***

MENINGES, VENTRICLES, AND CEREBROSPINAL FLUID

Meninges

Three connective tissue membranes, the **meninges** (mĕ-nin′jēz), surround and protect the brain and spinal cord (figure 13.12). The most superficial and thickest membrane is the **dura mater** (doo′ră mā′ter; tough mother). The dura mater is composed of dense irregular connective tissue. Within the vertebral canal, the dura mater is distinctly separate from the vertebrae, forming an **epidural space.** Within the cranial cavity, the dura mater is tightly adherent to the cranial bones, so the epidural space there is only a potential space. The dura mater within the cranial cavity consists of two layers. The outer layer is called the **periosteal dura,** and it is the inner periosteum of the cranial bones. The inner layer is the **meningeal dura.** The meningeal dura is continuous with the dura of the spinal cord. The meningeal dura is separated from the periosteal dura in several regions to form structures called dural folds and dural venous sinuses.

Dural folds are tough connective tissue partitions that extend into the major brain fissures. The dura mater and dural folds help hold the brain in place within the skull and keep it from moving around too freely. The largest of the dural folds is the **falx cerebri** (falks se-rē′brē; sickle-shaped), which lies in the longitudinal fissure and is anchored anteriorly to the crista galli of the ethmoid bone. The **tentorium cerebelli** (ten-tō′rē-ŭm, ser′ĕ-bel′ē; tent) is oriented horizontally between the cerebrum and the cerebellum. The **falx cerebelli** lies between the two cerebellar hemispheres.

Dural venous sinuses are spaces that form where the two layers of the dura mater are separated from each other. The largest of the sinuses is the **superior sagittal sinus,** which forms between the falx cerebri and the periosteal dura and runs along the median plane (see figure 13.12*b*). The dural venous sinuses are lined with endothelium; they transport venous blood and cerebrospinal

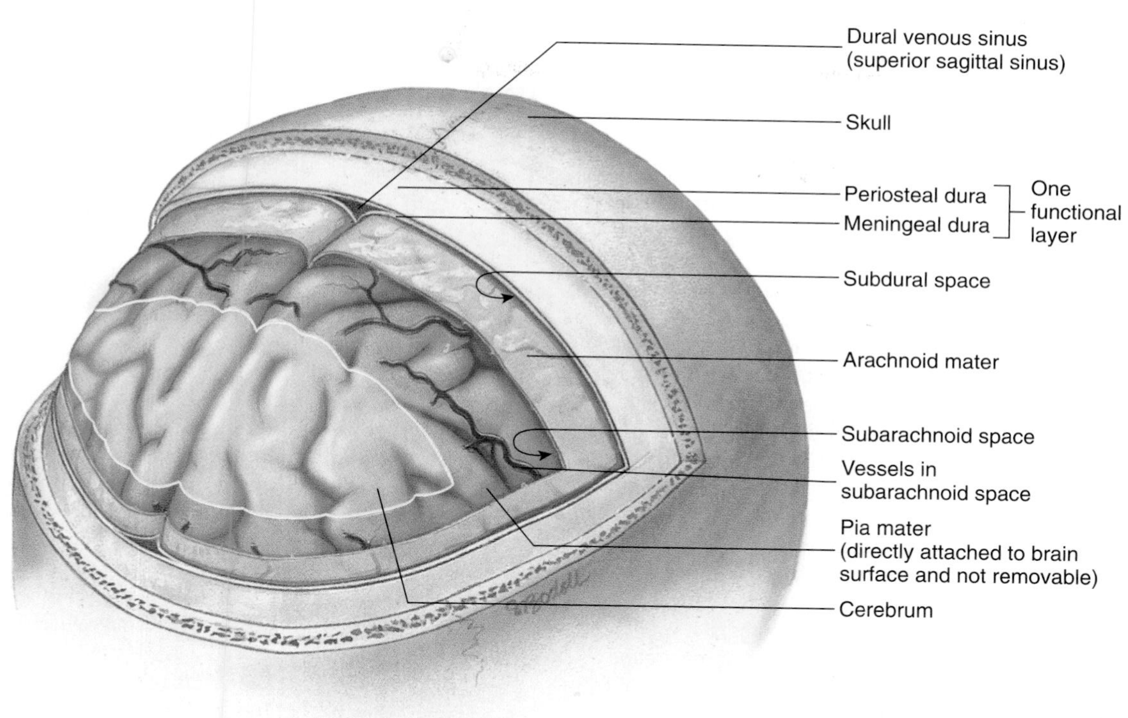

(a) **Anterosuperior view**

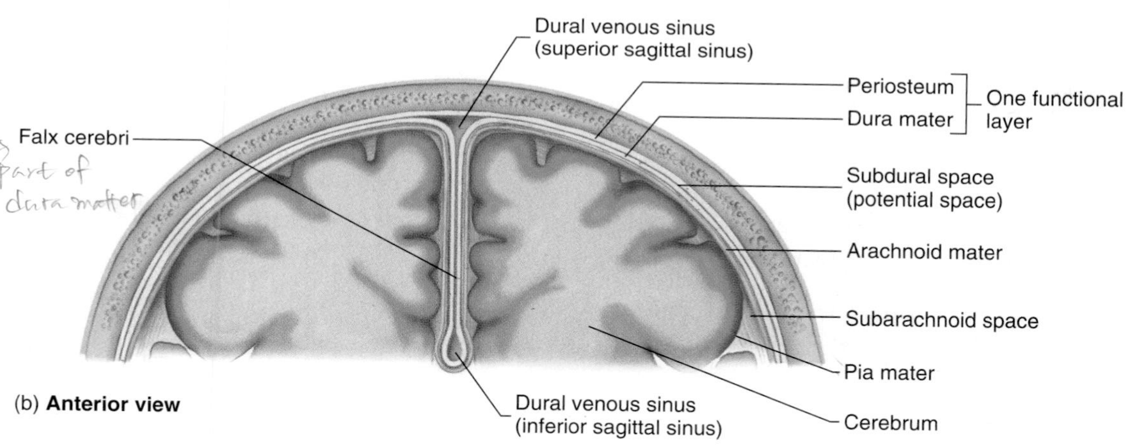

(b) **Anterior view**

FIGURE 13.12 Meninges

(*a*) Meningeal membranes surrounding the brain. (*b*) Frontal section of the head to show the meninges.

fluid (CSF; see "Cerebrospinal Fluid," p. 459) away from the brain. All veins draining blood from the brain empty into dural venous sinuses. The dural sinuses subsequently drain into the internal jugular veins, which are the major veins that exit the cranial cavity to carry blood back to the heart (see chapter 21).

The next meningeal membrane is a very thin, wispy **arachnoid** (ă-rak′noyd; spiderlike; i.e., cobwebs) **mater.** The space between this membrane and the dura mater is the **subdural space;** it contains only a very small amount of serous fluid. The third meningeal layer, the **pia** (pī′ă, pē′ă; affectionate) **mater** is bound very tightly to the surface of the brain. Between the arachnoid mater and the pia mater is the **subarachnoid space,** which contains weblike strands of arachnoid mater and the blood vessels supplying the brain and is filled with CSF.

Traumatic Brain Injuries and Hematomas

Head injuries are classified as **open** (when cranial cavity contents are exposed to the outside) or **closed** (when the cranial cavity remains intact) (see "CSF and Skull Fractures"). Closed injuries are more common and involve the head striking a hard surface or an object striking the head. Such injuries may result in brain trauma. The main trauma may be **coup** (kū), occurring at the site of impact, or **contrecoup** (kon′tra-kū), occurring on the opposite side of the brain from the impact and resulting from movement of the brain within the skull. Traumatic brain injury is three times more common in males than in females. The most common type of traumatic brain injury (75%–90%) is a **concussion,** which is defined as an immediate, but transient, impairment of neural function, such as a loss of consciousness or blurred vision.

Traumatic brain injury may be diffuse or focal. **Diffuse brain injury** usually results from shaking, such as when a person shakes a child or when a person is thrown about in an automobile accident. As the name suggests, such injury is not localized to one focal point but involves damage to many small vessels and nerves, especially around the brainstem. **Focal** traumatic brain injury may involve cortical **contusions** (bruisings), involving direct impact injury to the brain, or hemorrhages in or around the brain. Contusions are usually superficial and involve only the gyri.

Hemorrhagic brain injury results from bleeding outside the dura (extradural or epidural), between the dura and the brain (subdural), or within the brain (intracerebral). A hemorrhage, which is bleeding, results in a hematoma, an accumulation of blood. **Extradural hemorrhages,** or **epidural hematomas,** occur in about 1%–2% of major head injuries. They usually occur in the middle cranial fossa and involve a tear of the middle meningeal artery (85%) or in the middle meningeal vein or dural sinus (15%). **Subdural hematomas** are much more common, occurring in 10%–20% of major head injuries. They most commonly involve tears of the cortical veins or dural venous sinuses, occur in the superior portion of the cranial cavity, and appear within hours of the head injury. **Chronic subdural hematomas,** which involve slow bleeding over an extended period of time (weeks to months), are common in elderly people and in people who abuse alcohol. **Intracerebral hematomas** occur in about 2%–3% of major head injuries and are often associated with contusions. They involve the damage of small vessels within the brain itself and are most common in the frontal and temporal lobes. They occur within 3–10 days of the head trauma.

CASE STUDY

Traumatic Brain Injury

The body of an 80-year-old female was being examined by the hospital pathologist to determine the cause of death. Her medical records indicated no significant history of cardiovascular disease, stroke, Alzheimer disease, or cancer. She had been taken to the emergency room after having been found lying in her bathtub and not breathing. Bleeding over the occipital region of her scalp led the pathologist to hypothesize that she had slipped and fallen while getting into the bath and had hit the back of her head on the edge of the tub. Because the pathologist suspected a traumatic brain injury as the cause of death, he focused most of his attention on the contents of the cranial cavity. The pathologist noted superficial bruising and bleeding of the scalp in the occipital region of the woman's head and then proceeded to open the cranial cavity. Inside the cranial cavity, the pathologist noted a large subdural hematoma above the right frontal lobe of the woman's brain. In addition, he noted that the brain had shifted as a result of the extensive bleeding such that the medulla oblongata had been pushed inferiorly (herniated) through the foramen magnum into the vertebral canal. After completing a thorough inspection of the remainder of the woman's body, the pathologist determined the cause of death to be a subdural hematoma caused by a traumatic brain injury.

PREDICT 1

a. Explain why the subdural hematoma was found in the frontal region of the brain when the blow to the head occurred over the occipital region.

b. What were the consequences of the herniation of the medulla oblongata into the vertebral canal, and how did this contribute to the woman's death?

Ventricles

The CNS is formed as a hollow tube that may be quite reduced in some areas of the adult CNS and expanded in other areas (see "Development of the CNS" p. 445). The interior of the tube is lined with a single layer of epithelial cells called **ependymal** (ep-en′di-măl; see chapter 11) **cells.** Each cerebral hemisphere contains a relatively large cavity, the **lateral ventricle** (figure 13.13). The lateral ventricles are separated from each other by thin **septa pellucida** (sep′tă pe-loo′sid-ă; sing. *septum pellucidum;* translucent walls), which lie in the midline just inferior to the corpus callosum and usually are fused with each other. A smaller midline cavity, the **third ventricle,** is located in the center of the diencephalon between the two halves of the thalamus. The two lateral ventricles communicate with the third ventricle through two **interventricular foramina** (foramina of Monro). The lateral ventricles can be thought of as the first and second ventricles in the numbering scheme, but they are not designated as such. The **fourth ventricle** is in the inferior part of the pontine region and the superior region of the medulla oblongata at the base of the cerebellum. The third ventricle communicates with the fourth ventricle through a narrow canal, the **cerebral aqueduct** (aqueduct of Sylvius), which passes through the midbrain. The fourth ventricle is continuous with the central canal of the spinal cord, which extends nearly the full length of the cord. The fourth ventricle is also continuous with the subarachnoid space through two lateral apertures and one median aperture.

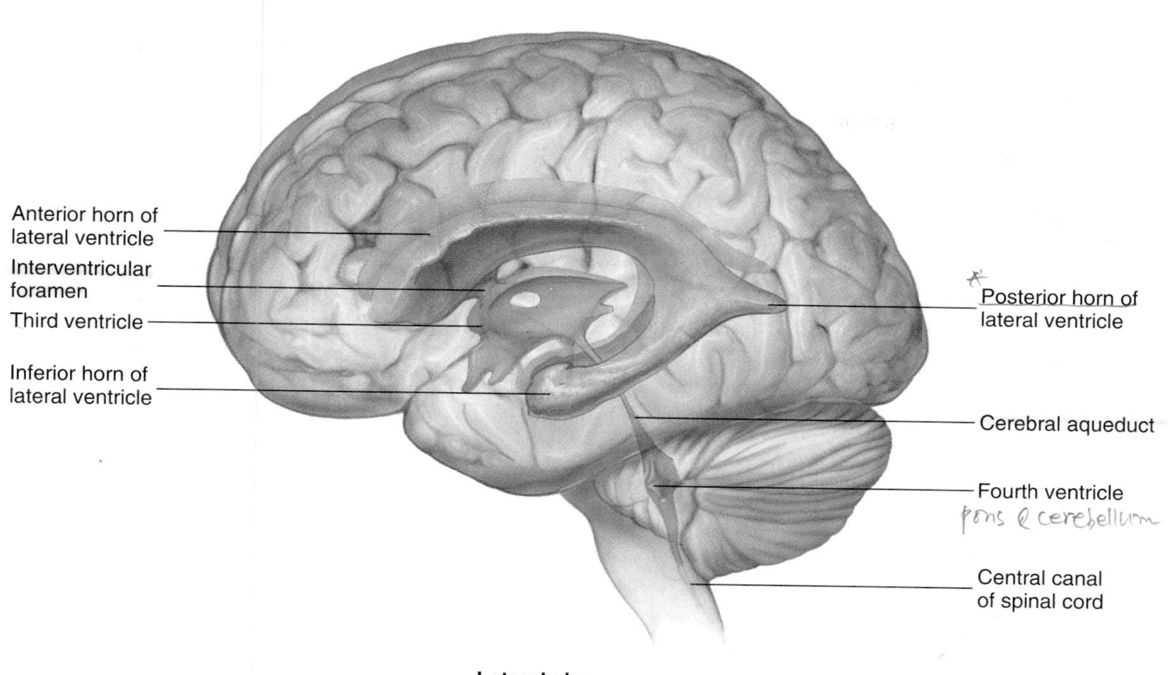

FIGURE 13.13 Ventricles of the Brain Viewed from the Left

Cerebrospinal Fluid

Cerebrospinal (ser′ĕ-brō-spī-năl; sĕ-rē′brō-spī-năl) **fluid (CSF)** is a fluid similar to blood serum with most of the proteins removed. It bathes the brain and the spinal cord and provides a protective cushion around the CNS. CSF allows the brain to float within the cranial cavity so it does not rest directly on the surface of the skull or dura mater. In addition, it protects the brain against the shock of rapid movements of the head. It also provides some nutrients to CNS tissues. About 80%–90% of the CSF is produced by specialized ependymal cells within the lateral ventricles, with the remainder produced by similar cells in the third and fourth ventricles. These specialized ependymal cells, their support tissue, and the associated blood vessels are collectively called the **choroid** (kō′royd; lacy) **plexuses** (plek′sŭs-ez; figure 13.14). The choroid plexuses are formed by invaginations of the vascular pia mater into the ventricles, thus producing a vascular connective tissue core covered by ependymal cells.

CSF and Skull Fractures

Open head trauma involves a fracture or hole in the skull, which exposes the contents of the cranial cavity (brain, blood, and/or CSF) to the exterior. Head injuries involving scalp lacerations or damage to the eye, ear, or nose should be carefully evaluated for the possibility of open head trauma involving the cranial cavity. In skull fractures in which the meninges are torn, CSF may leak from the nose if the fracture is in the frontal area or from the ear if the fracture is in the temporal area. Leakage of CSF indicates serious mechanical damage to the head and presents a risk for meningitis because bacteria may pass from the nose or ear through the tear and into the meninges. A skull fracture involving the base of the skull may result in cranial nerves damage where the nerves exit the skull.

CSF is formed through a variety of mechanisms. The majority of the fluid enters the ventricles by following a Na^+ concentration gradient. Ependymal cells of the choroid plexus actively transport Na^+ into the ventricles and water passively follows. Large molecules are transported by pinocytosis. The precise mechanisms for the transport of glucose and other substances into CSF remain unknown.

Endothelial cells of the blood vessels in the choroid plexuses, which are joined by tight junctions (see chapter 4), form the **blood–brain barrier,** or, more correctly, the **blood–cerebrospinal fluid barrier.** Consequently, substances do not pass between the cells but must pass through the cells.

CSF fills the ventricles, the subarachnoid space of the brain and spinal cord, and the central canal of the spinal cord. Approximately 23 mL of fluid fills the ventricles, and 117 mL fills the subarachnoid space. The route taken by the CSF from its origin in the choroid plexuses to its return to the circulation is depicted in figure 13.14. The flow rate of CSF from its origin to the point at which it enters the bloodstream is about 0.4 mL/min. CSF passes from the lateral ventricles through the interventricular foramina into the third ventricle and then through the cerebral aqueduct into the fourth ventricle. It can exit the interior of the brain only from the fourth ventricle. One **median aperture** (foramen of Magendie), which opens through the roof of the fourth ventricle, and two **lateral apertures** (foramina of Luschka), which open through the walls, allow the CSF to pass from the fourth ventricle to the subarachnoid space. Some CSF continues to flow inferiorly into the central canal of the spinal cord. However, parts of the central canal are closed off in adults, so the amount circulating there is very small. Masses of arachnoid tissue, **arachnoid granulations,** penetrate into the dural venous sinuses, especially the superior sagittal sinus. CSF passes into the blood of the dural venous sinuses through these granulations. The sinuses

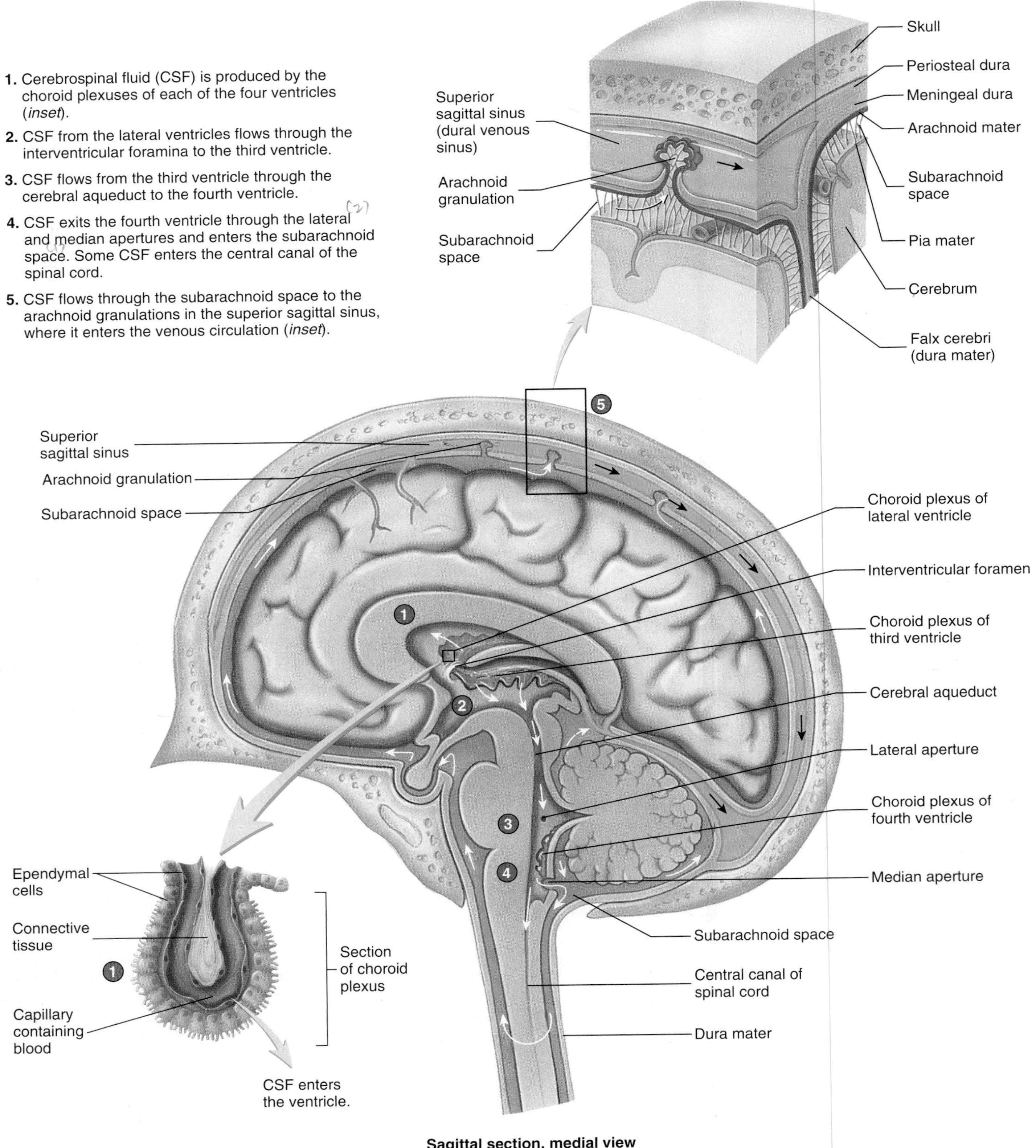
1. Cerebrospinal fluid (CSF) is produced by the choroid plexuses of each of the four ventricles (*inset*).
2. CSF from the lateral ventricles flows through the interventricular foramina to the third ventricle.
3. CSF flows from the third ventricle through the cerebral aqueduct to the fourth ventricle.
4. CSF exits the fourth ventricle through the lateral and median apertures and enters the subarachnoid space. Some CSF enters the central canal of the spinal cord.
5. CSF flows through the subarachnoid space to the arachnoid granulations in the superior sagittal sinus, where it enters the venous circulation (*inset*).
Superior sagittal sinus (dural venous sinus)
Arachnoid granulation
Subarachnoid space
Skull
Periosteal dura
Meningeal dura
Arachnoid mater
Subarachnoid space
Pia mater
Cerebrum
Falx cerebri (dura mater)
5
Superior sagittal sinus
Arachnoid granulation
Subarachnoid space
Choroid plexus of lateral ventricle
Interventricular foramen
Choroid plexus of third ventricle
Cerebral aqueduct
Lateral aperture
Choroid plexus of fourth ventricle
Median aperture
Subarachnoid space
Central canal of spinal cord
Dura mater
1
2
3
4
Ependymal cells
Connective tissue
1
Capillary containing blood
Section of choroid plexus
CSF enters the ventricle.
Sagittal section, medial view

PROCESS FIGURE 13.14 Flow of CSF

CSF flows through the ventricles and subarachnoid space is shown by *white arrows*. Those going through the foramina in the wall and roof of the fourth ventricle depict the CSF entering the subarachnoid space. CSF passes back into the blood through the arachnoid granulations (*white* and *black arrows*), which penetrate the dural sinus. The *black arrows* show the direction of blood flow in the sinuses.

are blood-filled; thus, it is within these dural sinuses that the CSF reenters the bloodstream. From the dural venous sinuses, the blood drains into the internal jugular veins to enter the veins of the general circulation.

Hydrocephalus

If the apertures of the fourth ventricle or the cerebral aqueduct are blocked, CSF can accumulate within the ventricles. This condition is called **internal,** or noncommunicating, **hydrocephalus,** and it results in increased CSF pressure. The production of CSF continues, even when the passages that normally allow it to exit the brain are blocked. Consequently, fluid builds inside the brain, causing pressure, which compresses the nervous tissue and dilates the ventricles. Compression of the nervous tissue usually results in irreversible brain damage. If the skull bones are not completely ossified when the hydrocephalus occurs, the pressure may also severely enlarge the head. The cerebral aqueduct may be blocked at the time of birth or may become blocked later in life because of a tumor growing in the brainstem.

Internal hydrocephalus can be successfully treated by placing a drainage tube (shunt) between the brain ventricles and abdominal cavity to eliminate the high internal pressures. There is some risk of infection being introduced into the brain through these shunts, however, and the shunts must be replaced as the person grows. A subarachnoid hemorrhage may block the return of CSF to the circulation. If CSF accumulates in the subarachnoid space, the condition is called **external,** or communicating, **hydrocephalus.** In this condition, pressure is applied to the brain externally, compressing neural tissues and causing brain damage.

14 ***Describe the three meninges that surround the CNS. What are the falx cerebri, tentorium cerebelli, and falx cerebelli?***

15 ***Describe and list the contents of the dural sinuses subdural space, and subarachnoid space.***

16 ***Name the four ventricles of the brain, and describe their locations and the connections between them. What are the septa pellucida?***

17 ***Describe the production and circulation of CSF. Where does the CSF return to the blood?***

BLOOD SUPPLY TO THE BRAIN

The brain requires a tremendous amount of blood to maintain its normal functions. The brain has a very high metabolic rate and brain cells are not capable of storing high-energy molecules for any length of time. In addition, brain cells depend almost entirely on glucose as their energy source (see chapter 25). Thus, the brain requires a constant blood supply to meet the demands of brain cells for both glucose and oxygen. Even though the brain accounts for only about 2% of the total weight of the body, it receives approximately 15%–20% of blood pumped by the heart. Interruption of the brain's blood supply for only seconds can cause unconsciousness, and interruption of the blood supply for minutes can cause irreversible brain damage.

The brain's blood supply is illustrated in chapter 21 (see figures 21.10 and 21.11). Blood reaches the brain through the **internal carotid arteries,** which ascend to the head along the anterior-lateral part of the neck, and the **vertebral arteries,** which ascend along the posterior part of the neck, through the transverse foramina of the cervical vertebrae. The internal carotid arteries enter the cranial cavity through the carotid canals, and the vertebral arteries enter through the foramen magnum. The vertebral arteries join together to form the **basilar artery,** which lies on the ventral surface of the pons. The basilar artery and internal carotid arteries contribute to the **cerebral arterial circle** (circle of Willis). Branches from this circle and from the basilar artery supply blood to the brain.

The cerebral cortex on each side of the brain is supplied by three branches that arise from the cerebral arterial circle: the **anterior, middle,** and **posterior cerebral arteries.** The middle cerebral artery supplies most of the lateral surface of each cerebral hemisphere. The anterior cerebral artery supplies the medial portion of the parietal and frontal lobes. The posterior cerebral artery supplies the occipital lobe and the medial surface of the temporal lobe.

The arteries to the brain and their larger branches are located in the subarachnoid space. Small cortical arterial branches leave the subarachnoid space and enter the pia mater, where they branch extensively. Precapillary branches leave the pia mater and enter the substance of the brain. Most of these branches are short and remain in the cortex. Fewer, longer branches extend into the medulla.

The arteries within the substance of the brain quickly divide into capillaries. The epithelial cells of these capillaries are surrounded by the foot processes of astrocytes (see figure 11.6). The astrocytes promote the formation of tight junctions between the epithelial cells. The epithelial cells with their tight junctions, form the **blood–brain barrier,** which regulates the movement of materials from the blood into the brain. Materials that would enter many tissues by passing between the epithelial cells of capillaries can not pass through the blood–brain barrier because of the tight junctions. Most materials that enter the brain pass through the epithelial cells. Lipid-soluble substances, such as nicotine, ethanol, and heroin, can diffuse through the plasma membranes of the epithelial cells and enter the brain. Water-soluble molecules, such as amino acids and glucose, move across the plasma membranes of the epithelial cells by mediated transport (see chapter 3).

Drugs and the Blood–Brain Barrier

The permeability characteristics of the blood–brain barrier must be considered when developing drugs designed to affect the CNS. For example, Parkinson disease is caused by a lack of the neurotransmitter dopamine, which normally is produced by certain neurons of the brain. A lack of dopamine results in decreased muscle control and shaking movements. Administering dopamine is not helpful because dopamine cannot cross the blood–brain barrier. Levodopa (L-dopa), a precursor to dopamine, is administered instead because it can cross the blood–brain barrier. CNS neurons then convert levodopa to dopamine, which helps reduce the symptoms of Parkinson disease.

18 ***Describe the blood supply to the brain. List the arteries supplying each part of the cerebral cortex.***

19 ***Describe the blood–brain barrier.***

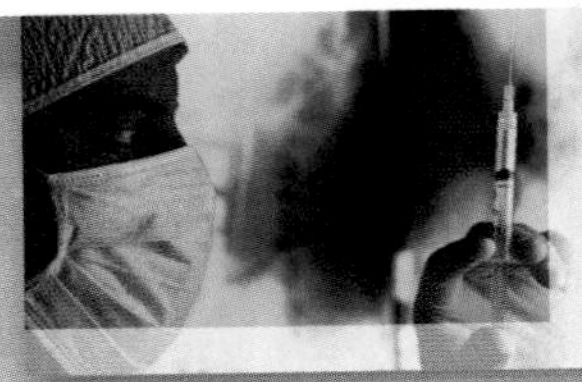

CLINICAL FOCUS

General CNS Disorders

INFECTIONS

Encephalitis (en-sef-ă-lī′tis) is an inflammation of the brain; it is most often caused by a virus and less often by bacteria or other agents. A large variety of symptoms may result, including fever, paralysis, coma, or even death.

Myelitis (mī-ĕ-lī′tis) is an inflammation of the spinal cord caused by trauma, multiple sclerosis, or a number of infectious agents, including viruses, bacteria, or other agents. A large variety of symptoms may result depending on the extent and level of injury or infection.

Meningitis (men-in-jī′tis) is an inflammation of the meninges. It may be virally induced but is more often bacterial. Symptoms usually include stiffness in the neck, headache, and fever. Pus may accumulate in the subarachnoid space, block CSF flow, and result in hydrocephalus. In severe cases, meningitis may also cause paralysis, coma, or death.

OTHER CNS DISORDERS

An **aneurysm** (an′ū-rizm) is a dilation, or ballooning, of an artery. The arteries around the brain are common sites for aneurysms, and hypertension can cause one of these "balloons" to burst or leak, causing a hemorrhage around the brain. With hemorrhaging, blood may enter the epidural space (epidural hematoma), subdural space (subdural hematoma), subarachnoid space, or the brain tissue. Blood in the subdural or subarachnoid space can apply pressure to the brain, causing damage to brain tissue. Blood is toxic to brain tissue, so blood entering the brain can directly damage brain tissue.

A **concussion** is a blow to the head producing momentary loss of consciousness without immediate detectable damage to the brain. Often, no more problems occur after the person regains consciousness; however, in some cases, **postconcussion syndrome** may occur a short time after the injury. The syndrome includes increased muscle tension or migraine headaches, reduced alchol tolerance, difficulty in learning new things, reduction in creativity; as well as motivation, fatigue, and personality changes. The symptoms may be gone in a month or may persist for as much as a year. In some cases, postconcussion syndrome is the result of a slowly occurring subdural hematoma that may be missed by an early examination. The blood may accumulate from small leaks in the dural sinuses.

Cerebral compression may occur as a result of hematomas, hydrocephalus, tumors, or edema of the brain, which can occur as a result of a severe blow to the head. The intracranial pressure increases, which may directly damage brain tissue. The cerebellum may compress the fourth ventricle, blocking the foramina and causing internal hydrocephalus, which further increases intracranial pressure. The greatest problem comes from compression of the brainstem. Compression of the midbrain can kink the oculomotor nerves, resulting in dilation of the pupils with no light response. Compression of the medulla oblongata may disrupt cardiovascular and respiratory centers, which can cause death. Compression of any part of the CNS that results in ischemia for as little as 3–5 minutes can result in local neuronal cell death. This is a major problem in spinal cord injuries.

CRANIAL NERVES

The 12 pairs of cranial nerves by convention are indicated by Roman numerals (I–XII) from anterior to posterior (figure 13.15). The first two pairs of cranial nerves connect directly to the cerebrum (I) or diencephalon (II). Nine pairs of the cranial nerves connect to the brainstem. One pair of cranial nerves (XI) is connected to the spinal cord and has no direct connection to brain structures. A given cranial nerve may have one or more of three functions: (1) sensory, (2) somatic motor, and (3) parasympathetic (table 13.4). **Sensory** functions include the special senses, such as vision, and the more general senses, such as touch and pain. **Somatic** (sō-mat′ik) **motor** functions are the control of skeletal muscles through motor neurons. **Proprioception** (prō-prē-ō-sep′shun) informs the brain about the position of various body parts, including joints and muscles. The cranial nerves innervating skeletal muscles also contain proprioceptive sensory fibers, which convey action potentials to the CNS from those muscles. Because proprioception is the only sensory function of several otherwise somatic motor cranial nerves, however, that function is usually ignored, and the nerves are designated by convention as motor only. **Parasympathetic** function involves the regulation of glands, smooth muscles, and cardiac muscle. These functions are part of the autonomic nervous system and are discussed in chapter 16. Several of the cranial nerves have associated ganglia, and these ganglia are of two types: parasympathetic and sensory. Table 13.5 lists specific information about each cranial nerve.

The **olfactory (I)** and **optic (II) nerves** are exclusively sensory and are involved in the special senses of smell and vision, respectively. The functions of these nerves are discussed in chapter 15.

The **oculomotor nerve (III)** innervates four of the six muscles that move the eyeball (the superior, inferior, and medial rectus

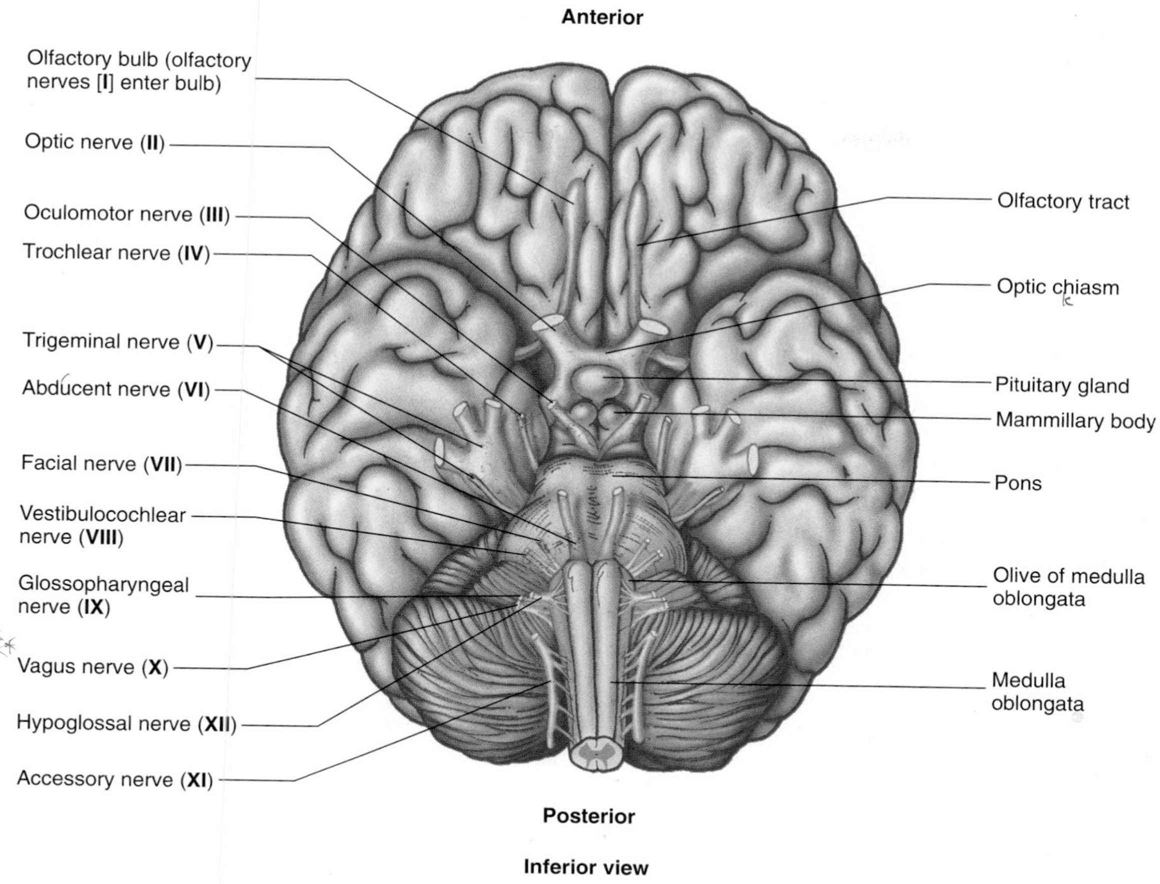

FIGURE 13.15 Inferior Surface of the Brain Showing the Origin of the Cranial Nerves

TABLE 13.4 Functional Organization of the Cranial Nerves

Nerve Function	Cranial Nerve	
Sensory	I	Olfactory
	II	Optic
	VIII	Vestibulocochlear
Somatic motor	IV	Trochlear
	VI	Abducent
	XI	Accessory
	XII	Hypoglossal
Somatic motor and sensory	V	Trigeminal
Somatic motor and parasympathetic	III	Oculomotor
Somatic motor, sensory, and parasympathetic	VII	Facial
	IX	Glossopharyngeal
	X	Vagus

muscles and the inferior oblique muscle; see chapter 10) and the levator palpebrae superioris muscle, which raises the superior eyelid. In addition, parasympathetic nerve fibers in the oculomotor nerve innervate smooth muscles in the eye and regulate the size of the pupil and the shape of the lens of the eye.

The **trochlear** (trōk′lē-ar) **nerve (IV)** is a somatic motor nerve that innervates one of the six eye muscles responsible for moving the eyeball (superior oblique).

The **trigeminal** (trī-jem′i-năl) **nerve (V)** has somatic motor, proprioceptive, and cutaneous sensory functions. It supplies motor innervation to the muscles of mastication, one middle ear muscle, one palatine muscle, and two throat muscles. In addition to proprioception associated with its somatic motor functions, the trigeminal nerve also carries proprioception from the temporomandibular joint, tongue, and cheek, which allow you to chew food without biting your tongue or cheek. Damage to the trigeminal nerve may impede chewing.

The trigeminal nerve has the greatest general sensory function of all the cranial nerves and is the only cranial nerve involved in **sensory cutaneous innervation** of the head. All other cutaneous innervation comes from spinal nerves (see figure 12.14). *Trigeminal* means three twins, and the sensory

TABLE 13.5	Cranial Nerves and Their Functions		
Cranial Nerve	**Foramen or Fissure***	**Function**	**Consequences of Lesions to Nerve**
I: Olfactory	Cribriform plate	Sensory Special sense of smell	Inability to smell
II: Optic	Optic foramen	Sensory Special sense of vision	Blindness on the affected side
III: Oculomotor	Superior orbital fissure	Motor† and parasympathetic Motor to eye muscles (superior, medial, and inferior rectus; inferior oblique) and upper eyelid (levator palpebrae superioris) Proprioceptive from those muscles Parasympathetic to the sphincter of the pupil (causing constriction) and the ciliary muscle of the lens (causing accommodation)	Pupil dilation; eye deviates inferiorly and laterally from muscle paralysis, resulting in double vision; eyelid droops (ptosis); blurred vision from loss of accommodation

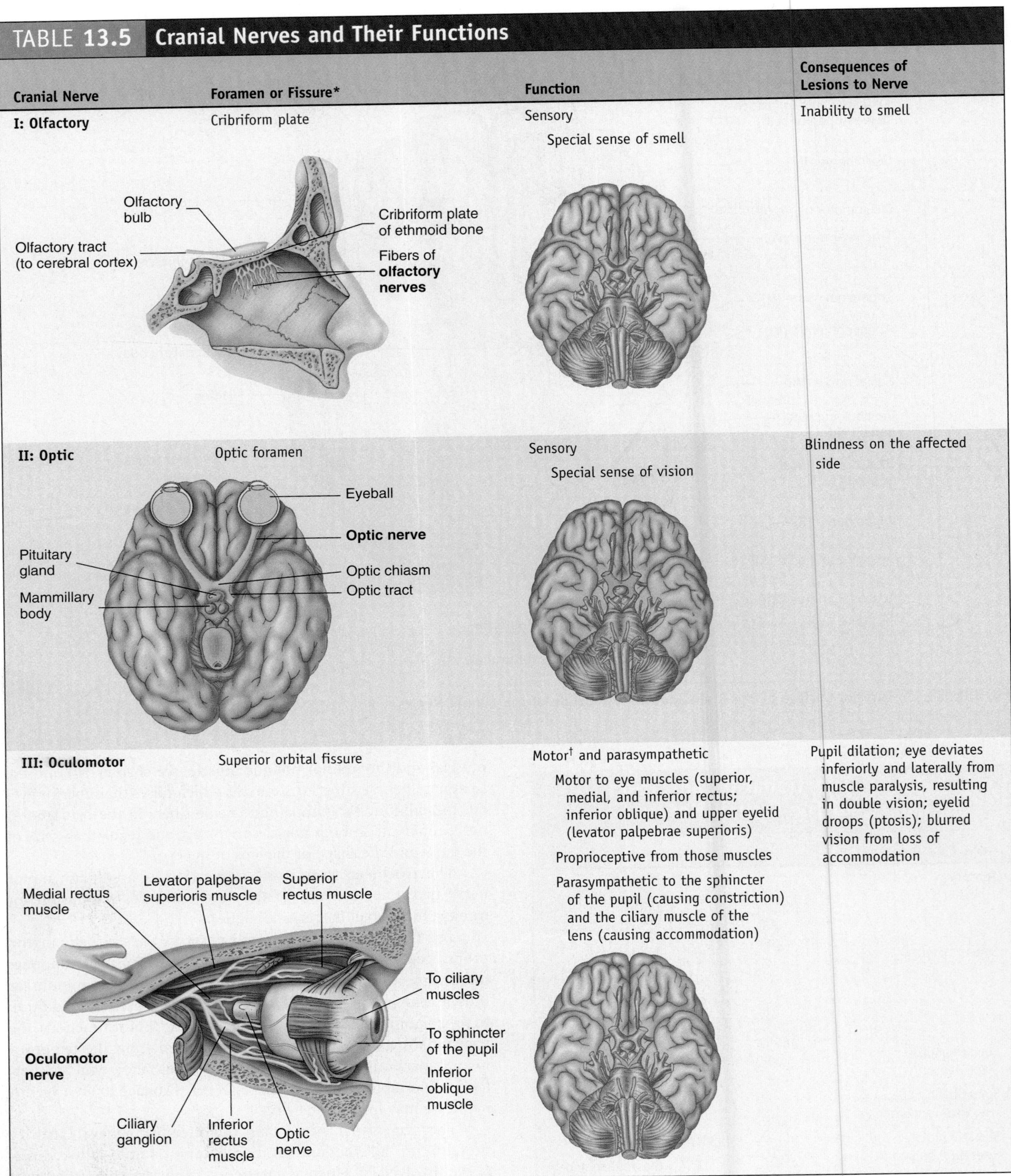

*Route of entry or exit from the skull.
†Proprioception is a sensory function, not a motor function; however, motor nerves to muscles also contain some proprioceptive afferent fibers from those muscles. Because proprioception is the only sensory information carried by some cranial nerves, these nerves still are considered "motor."

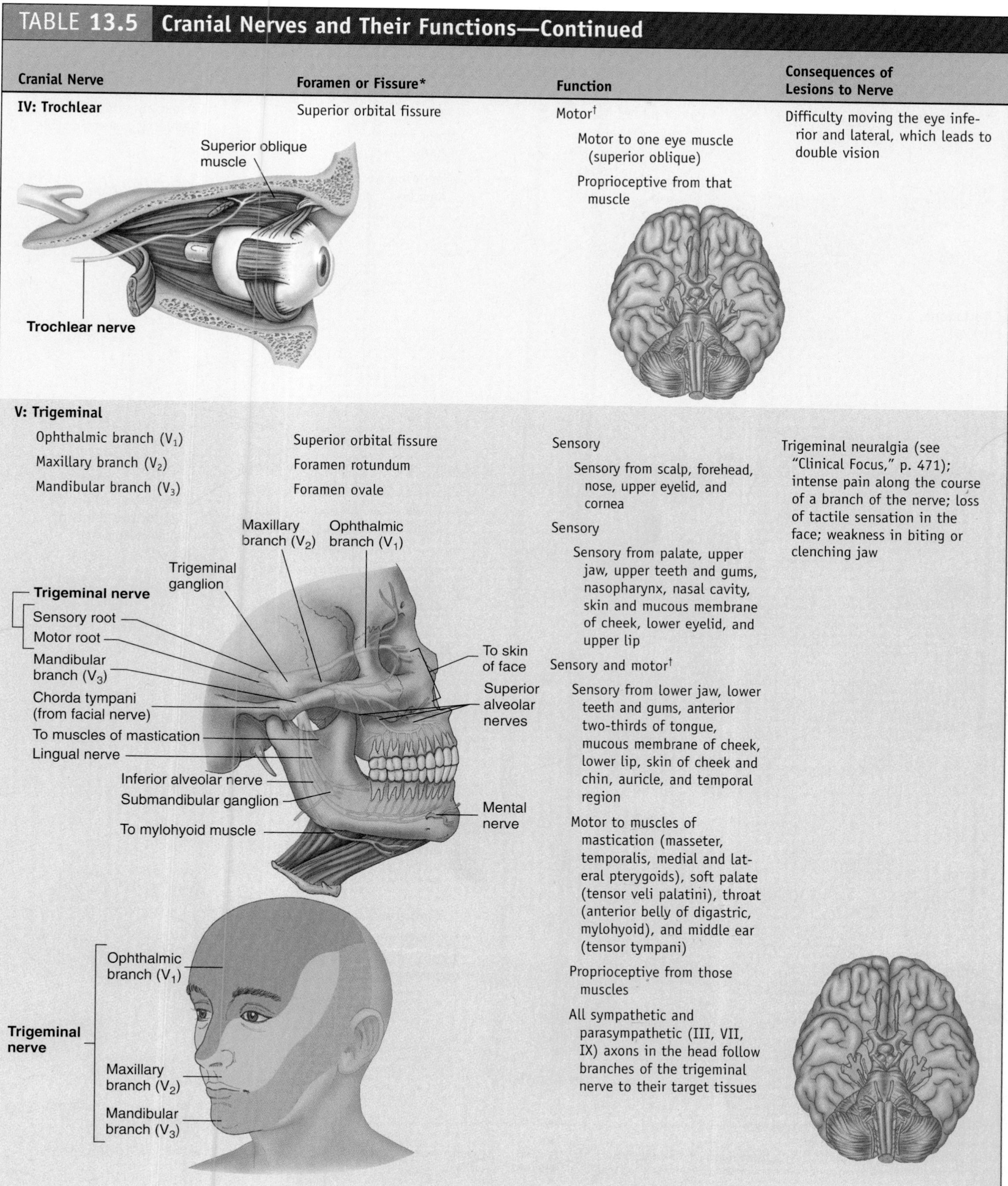

TABLE 13.5 Cranial Nerves and Their Functions—Continued

Cranial Nerve	Foramen or Fissure*	Function	Consequences of Lesions to Nerve
IV: Trochlear	Superior orbital fissure	Motor[†] Motor to one eye muscle (superior oblique) Proprioceptive from that muscle	Difficulty moving the eye inferior and lateral, which leads to double vision
V: Trigeminal			Trigeminal neuralgia (see "Clinical Focus," p. 471); intense pain along the course of a branch of the nerve; loss of tactile sensation in the face; weakness in biting or clenching jaw
Ophthalmic branch (V_1)	Superior orbital fissure	Sensory Sensory from scalp, forehead, nose, upper eyelid, and cornea	
Maxillary branch (V_2)	Foramen rotundum	Sensory Sensory from palate, upper jaw, upper teeth and gums, nasopharynx, nasal cavity, skin and mucous membrane of cheek, lower eyelid, and upper lip	
Mandibular branch (V_3)	Foramen ovale	Sensory and motor[†] Sensory from lower jaw, lower teeth and gums, anterior two-thirds of tongue, mucous membrane of cheek, lower lip, skin of cheek and chin, auricle, and temporal region Motor to muscles of mastication (masseter, temporalis, medial and lateral pterygoids), soft palate (tensor veli palatini), throat (anterior belly of digastric, mylohyoid), and middle ear (tensor tympani) Proprioceptive from those muscles All sympathetic and parasympathetic (III, VII, IX) axons in the head follow branches of the trigeminal nerve to their target tissues	

TABLE 13.5 Cranial Nerves and Their Functions—Continued

Cranial Nerve	Foramen or Fissure*	Function	Consequences of Lesions to Nerve
VI: Abducent	Superior orbital fissure	Motor† Motor to one eye muscle (lateral rectus) Proprioceptive from that muscle	Eye deviates medially (adducts) causing double vision
VII: Facial	Internal auditory meatus Stylomastoid foramen	Sensory, motor,† and parasympathetic Sense of taste from anterior two-thirds of tongue, sensory from some of external ear and palate Motor to muscles of facial expression, throat (posterior belly of digastric, stylohyoid), and middle ear (stapedius) Proprioceptive from those muscles Parasympathetic to submandibular and sublingual salivary glands, lacrimal gland, and glands of the nasal cavity and palate	Facial palsy (see "Clinical Focus," p. 471); loss of taste sensation on the anterior two-thirds of tongue; decreased salivation

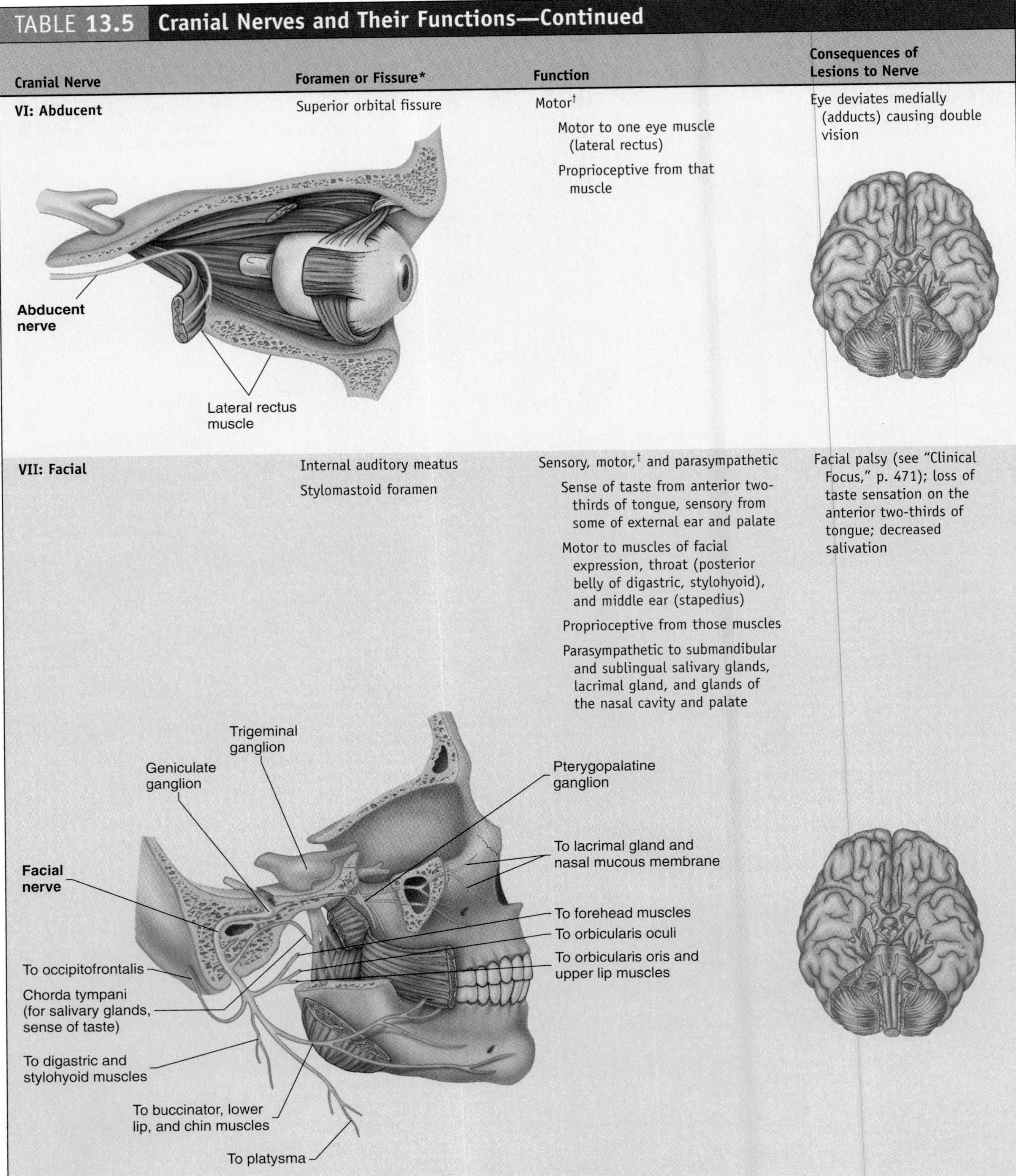

TABLE 13.5 Cranial Nerves and Their Functions—Continued

Cranial Nerve	Foramen or Fissure*	Function	Consequences of Lesions to Nerve
VIII: Vestibulocochlear	Internal auditory meatus	Sensory Special senses of hearing and balances	Loss of balance and equilibrium; nausea; vertigo; vomiting; and dizziness and/or inability to hear
IX: Glossopharyngeal	Jugular foramen	Sensory, motor,† and parasympathetic Sense of taste from posterior third of tongue, sensory from pharynx, palatine tonsils, posterior third of tongue, middle ear, carotid sinus and carotid body Motor to pharyngeal muscle (stylopharyngeus) Proprioceptive from that muscle Parasympathetic to parotid salivary gland and the glands of the posterior third of tongue	Difficulty swallowing; loss of taste sensation in the posterior one-third of tongue; decreased salivation

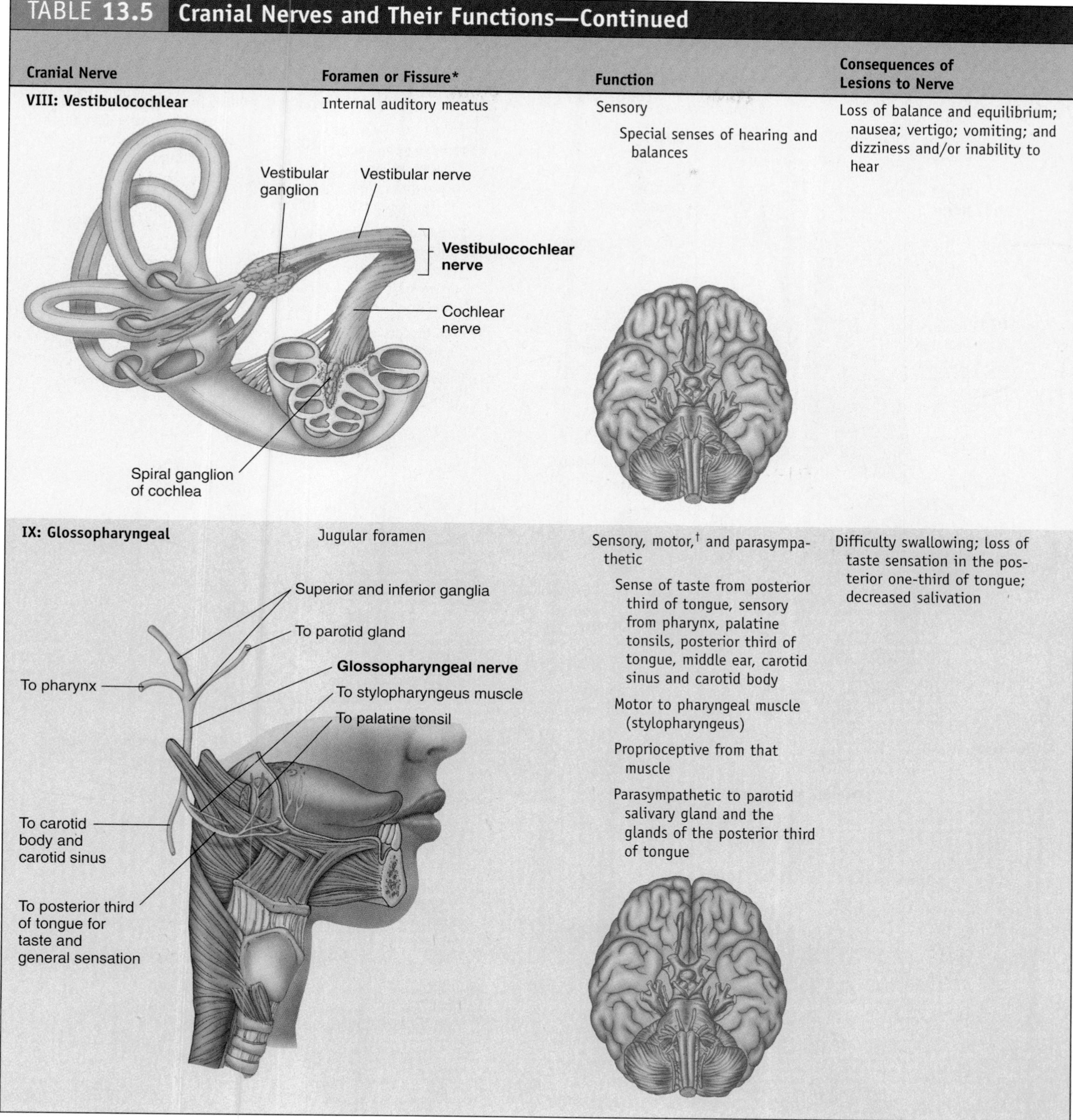

TABLE 13.5 Cranial Nerves and Their Functions—Continued

Cranial Nerve	Foramen or Fissure*	Function	Consequences of Lesions to Nerve
X: Vagus	Jugular foramen	Sensory, motor,† and parasympathetic Sensory from inferior pharynx, larynx, thoracic and abdominal organs, sense of taste from posterior tongue Motor to soft palate, pharynx, intrinsic laryngeal muscles (voice production), and an extrinsic tongue muscle (palatoglossus) Proprioceptive from those muscles Parasympathetic to thoracic and abdominal viscera	Difficulty swallowing and/or hoarseness; uvula deviates away from side of the disfunction
XI: Accessory	Foramen magnum Jugular foramen	Motor† Motor to sternocleidomastoid and trapezius	Difficulty elevating the scapula or rotating the neck

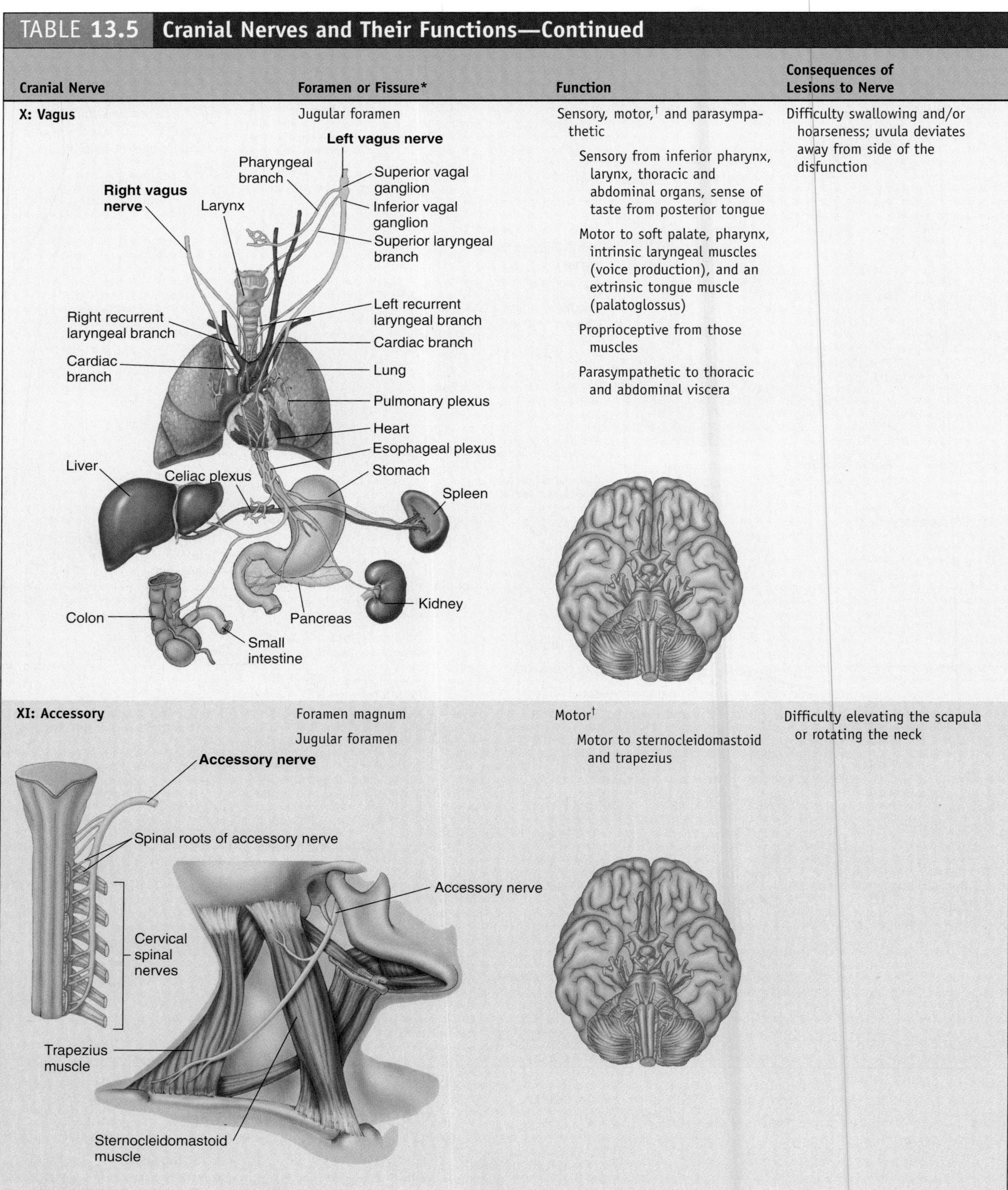

TABLE 13.5 Cranial Nerves and Their Functions—Continued

Cranial Nerve	Foramen or Fissure*	Function	Consequences of Lesions to Nerve
XII: Hypoglossal	Hypoglossal canal	Motor† Motor to intrinsic and extrinsic tongue muscles (styloglossus, hypoglossus, genioglossus) and throat muscles (thyrohyoid and geniohyoid) Proprioceptive from those muscles	When sticking out the tongue, it deviates toward the side of the damaged nerve

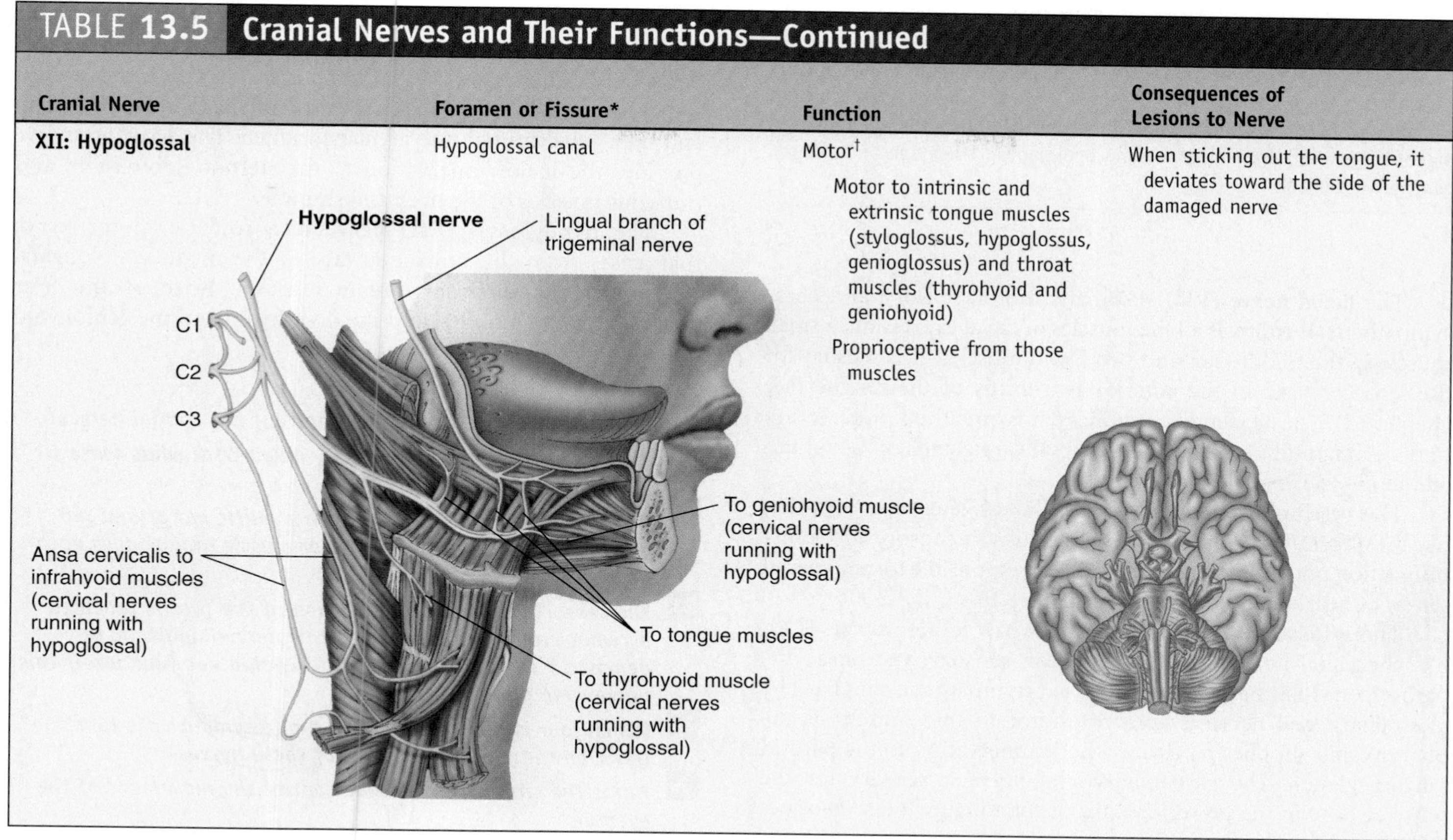

distribution of the trigeminal nerve in the face is divided into three regions, each supplied by a branch of the nerve. The three branches—ophthalmic, maxillary, and mandibular—arise directly from the trigeminal ganglion, which serves the same function as the dorsal root ganglia of the spinal nerves. Only the mandibular branch contains motor axons, which bypass the trigeminal ganglion, much as the ventral root of a spinal nerve bypasses a dorsal root ganglion.

In addition to these cutaneous functions, the maxillary and mandibular branches are important in dentistry. The maxillary nerve supplies sensory innervation to the maxillary teeth, palate, and gingiva (jin′jī-vă; gum). The mandibular branch supplies sensory innervation to the mandibular teeth, tongue, and gingiva. The various nerves innervating the teeth are referred to as **alveolar nerves** (al-vē′ō-lăr; the sockets in which the teeth are located). The **superior alveolar nerves** to the maxillary teeth are derived from the maxillary branch of the trigeminal nerve, and the **inferior alveolar nerves** to the mandibular teeth are derived from the mandibular branch of the trigeminal nerve.

The **abducent** (ab-doo′sent) **nerve (VI),** like the trochlear nerve, is a somatic motor nerve that innervates one of the six eye muscles responsible for moving the eyeball (lateral rectus).

Dental Anesthesia

Dentists inject anesthetic to block sensory transmission by the alveolar nerves. The superior alveolar nerves are not usually anesthetized directly because they are difficult to approach with a needle. For this reason, the maxillary teeth are usually anesthetized locally by inserting the needle beneath the oral mucosa surrounding the teeth. The inferior alveolar nerve probably is anesthetized more often than any other nerve in the body. To anesthetize this nerve, the dentist inserts the needle somewhat posterior to the patient's last molar and extends the needle near where the mandibular branch of the trigeminal nerve enters the mandibular foramen.

Several nondental nerves are usually anesthetized during an inferior alveolar block. The mental nerve, which supplies cutaneous innervation to the anterior lip and chin, is a distal branch of the inferior alveolar nerve. When the inferior alveolar nerve is blocked, the mental nerve is blocked also, resulting in a numb lip and chin. Nerves lying near the point where the inferior alveolar nerve enters the mandible often are also anesthetized during inferior alveolar anesthesia. For example, the lingual nerve can be anesthetized to produce a numb tongue. The facial nerve lies some distance from the inferior alveolar nerve, but in rare cases anesthetic can diffuse far enough posteriorly to anesthetize that nerve. The result is a temporary facial palsy (paralysis or paresis), with the injected side of the face drooping because of flaccid muscles, which disappears when the anesthesia wears off. If the facial nerve is cut by an improperly inserted needle, permanent facial palsy may occur.

PREDICT 2

A drooping upper eyelid on one side of the face is a sign of possible oculomotor nerve damage. Describe how this could be evaluated by examining other oculomotor nerve functions. Describe the eye movements that would distinguish among oculomotor, trochlear, and abducent nerve damage.

The **facial nerve (VII)** is somatic motor, sensory, and parasympathetic. It controls all the muscles of facial expression, a small muscle in the middle ear, and two hyoid muscles. It is sensory for the sense of taste in the anterior two-thirds of the tongue (see chapter 15). The facial nerve supplies parasympathetic innervation to the submandibular and sublingual salivary glands of the mouth and to the lacrimal glands of the eye.

The **vestibulocochlear** (ves-tib′ū-lō-kok′lē-ăr) **nerve (VIII),** like the olfactory and optic nerves, is exclusively sensory and transmits action potentials from the inner ear responsible for the special senses of hearing and balance (see chapter 15).

The **glossopharyngeal** (glos′ō-fă-rin′jē-ăl) **nerve (IX),** like the facial nerve, is somatic motor, sensory, and parasympathetic and has both sensory and parasympathetic ganglia. The glossopharyngeal nerve is somatic motor to one muscle of the pharynx and supplies parasympathetic innervation to the parotid salivary glands. The glossopharyngeal nerve is sensory for the sense of taste in the posterior third of the tongue. It also supplies tactile sensory innervation from the posterior tongue, middle ear, and pharynx and transmits sensory stimulation from receptors in the carotid arteries, which monitor blood pressure and blood carbon dioxide, oxygen, and pH levels (see chapter 21).

The **vagus** (vā′gŭs) **nerve (X),** like the facial and glossopharyngeal nerves, is somatic motor, sensory, and parasympathetic and has both sensory and parasympathetic ganglia. Most muscles of the soft palate, pharynx, and larynx are innervated by the vagus nerve. Damage to the laryngeal branches of the vagus nerve can interfere with normal speech. The vagus nerve is sensory for taste from the root of the tongue (see chapter 15). It is sensory for the inferior pharynx and the larynx and transmits sensory input from receptors in the aortic arch, which monitor blood pressure and carbon dioxide, oxygen, and pH levels in the blood (see chapter 21). In addition, the vagus nerve conveys sensory information from the thoracic and abdominal organs. The parasympathetic part of the vagus nerve is very important in regulating the functions of the thoracic and abdominal organs. It carries parasympathetic fibers to the heart and lungs in the thorax and to the digestive organs and kidneys in the abdomen.

The **accessory (XI) nerve** is a somatic motor nerve. The accessory nerve has been thought for many years to have both a cranial and a spinal component. The cranial component was thought to join the vagus nerve (hence the name *accessory*) and participate in its function. An understanding of the accessory nerve's structure was based on dissections conducted nearly 150 years ago and, apparently, not repeated until very recently. Extensive dissections of 15 human brainstems have failed to find any connection between the brainstem and the accessory nerve. These data suggest that the accessory nerve is not actually a cranial nerve but originates entirely from the superior part of the cervical spinal cord, enters the cranial cavity through the foramen magnum, and then exits through the jugular foramen. The accessory nerve provides the major innervation to the sternocleidomastoid and trapezius muscles of the neck and shoulder.

The **hypoglossal (XII) nerve** is a somatic motor nerve that arises from the ventral surface of the medulla oblongata. It supplies the intrinsic tongue muscles, three of the four extrinsic tongue muscles, and the thyrohyoid and the geniohyoid muscles.

20 ***What are the three major functions of the cranial nerves?***

21 ***Which cranial nerves are sensory only? With what sense is each of these nerves associated?***

22 ***Name the cranial nerves that are somatic motor and proprioceptive only. What muscles or muscle groups does each nerve supply?***

23 ***The sensory cutaneous innervation of the face is provided by what cranial nerve? How is this nerve important in dentistry? Name the muscles that would not function if this nerve were damaged.***

24 ***Which four cranial nerves have a parasympathetic function? Describe the function of each of these nerves.***

25 ***Name the cranial nerves that control the movement of the eyeball.***

26 ***Which cranial nerves are involved in the sense of taste? What part of the tongue does each supply?***

27 ***Speech production involves which cranial nerves? Describe the branches of these nerves.***

PREDICT 3

Injury to the accessory nerve may result in sternocleidomastoid muscle dysfunction, a condition called "wry neck." If the head of a person with wry neck is turned to the left, does this position indicate injury to the left or right spinal component of the accessory nerve?

PREDICT 4

Unilateral damage to the hypoglossal nerve results in loss of tongue movement on one side, which is most obvious when the tongue is protruded. If the tongue is deviated to the right, is the left or right hypoglossal nerve damaged?

Reflexes integrated within the spinal cord were discussed in chapter 12. Many of the body's functions, especially those involved in maintaining homeostasis, involve reflexes that are integrated within the brain. Some of these reflexes, such as those involved in the control of heart rate (see chapter 20), blood pressure (see chapter 21), and respiration (see chapter 23), are integrated in the brainstem and many involve cranial nerve X (vagus nerve).

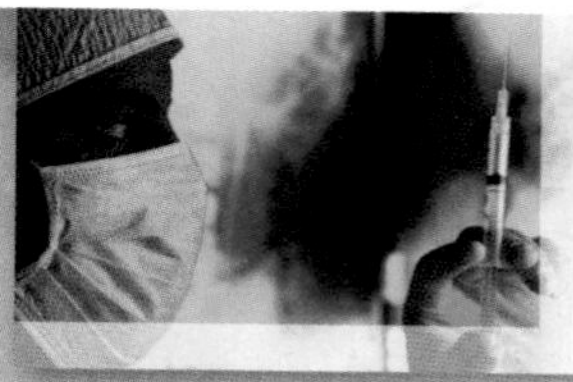

CLINICAL FOCUS

Peripheral Nervous System Disorders—Cranial Nerves

General issues of PNS disorders are described in chapter 12. This chapter addresses only those specific to the cranial nerves.

Trigeminal neuralgia, also called tic douloureux, involves one or more of the trigeminal nerve branches and consists of sharp bursts of pain in the face. This disorder often has a trigger point in or around the mouth, which, when touched, elicits the pain response in some other part of the face. The cause of trigeminal neuralgia is unknown.

Facial palsy (called Bell palsy) is a unilateral paralysis of the facial muscles. The affected side of the face droops because of the absence of muscle tone. Facial palsy involves the facial nerve and may result from facial nerve neuritis. The facial nerve passes from deep to superficial through the parotid gland. Although the cause of facial palsy is often unknown, it can result from a stroke or tumor in the cerebral cortex or brainstem (see chapter 14). Temporary facial palsy can result from inflammation of the parotid gland or from anesthesia accidently introduced into the gland during dental anesthesia. Temporary facial palsy can even result from extreme cold in the face, where the superficial branches of the facial nerve are located.

INFECTIONS

Herpes simplex I is usually characterized by one or more lesions (sores) on the lips or nose. The virus apparently remains dormant in the trigeminal ganglion. Eruptions are usually recurrent and often occur in times of stress or of reduced resistance, such as during a case of the common cold. For this reason, they are called cold sores or fever blisters.

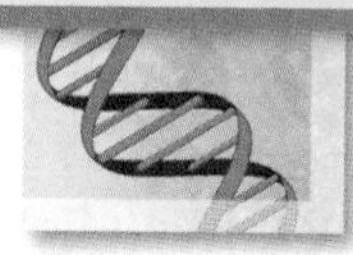

CLINICAL GENETICS

Neurofibromatosis

Neurofibromatosis type 1 (noor′ō-fī-brō-mă-tō′sis; von Recklinghausen disease) is an autosomal-dominant trait localized to chromosome 17. Neurofibromatosis type 1 is characterized by hyperpigmented skin present at birth and by multiple benign tumors (neurofibromas), which grow on nerves just under the skin and along nerves throughout the body. The skin of affected individuals is covered with characteristic large, flat, dark brown spots called café au lait spots, which are a type of birthmark. The neurofibromas increase in size and abundance with age and can cause severe disfigurement. Neurofibromatosis type 1 results from a mutation in the *neurofibromin 1* (NF1) gene, which encodes the protein neurofibromin. Neurofibromin is produced by neurons and glial cells, such as Schwann cells and oligodendrocytes. Neurofibromin is a tumor suppressor protein, which means it plays a role in keeping cells from growing uncontrollably. A mutated NF1 gene produces a defective neurofibromin protein, which then allows the glial cells to grow unchecked and form neurofibromas. Neurofibromatosis type 1 has an incidence of 1 : 2500–3300 individuals. Approximately 50% of the mutated genes are inherited from an affected parent, but the remaining 50% are caused by new mutations. Two mutated copies of the gene are necessary to cause the disorder, but most individuals born with one NF1 mutation eventually develop a second mutation and begin to develop neurofibromas and café au lait spots.

A second form of neurofibromatosis, **neurofibromatosis type 2** (NF2), has the same inheritance characteristics of NF1, but is much rarer (incidence of 1 : 50,000–120,000 individuals). NF2 is localized to chromosome 22 and encodes the protein merlin (schwannomin), which is also a tumor suppressor protein. Individuals with neurofibromatosis type 2 do not develop café au lait spots. Instead, the disease is characterized by the development of bilateral tumors surrounding the vestibular division of CN VIII (vestibular schwannomas). Tumors on CN VIII cause ringing in the ears (tinnitus), hearing loss, and vertigo from the pressure of the tumor on the nerve as it travels through the internal acoustic meatus. Treatment of neurofibromatosis consists of surgery to remove tumors that are causing the patient severe pain, loss of function, or disfigurement or to remove tumors that are thought to have become malignant.

Many of the brainstem reflexes are associated with cranial nerve function. The circuitry of most of these reflexes is too complex for our discussions. In general, these reflexes involve sensory input from the cranial nerves or spinal cord, and the motor output of the cranial nerves.

Turning of the eyes toward a flash of light, sudden noise, or a touch on the skin are examples of brainstem reflexes. Moving the eyes to track a moving object is another complex brainstem reflex. Some of the sensory neurons from cranial nerve VIII form a reflex arc with neurons of cranial nerves V and VII, which send axons to muscles of the middle ear and dampen the effects of very loud, sustained noises on delicate inner ear structures (see chapter 15). Reflexes that occur during the process of chewing allow the jaws to react to foods of various hardness and protect the teeth from breakage from very hard food items. Both the sensory and motor components of the reflex arc are carried by cranial nerve V. Reflexes involving input through cranial nerve V and output through cranial nerve XII move the tongue about to position food between the teeth for chewing and then move the tongue out of the way so it is not bitten.

SUMMARY

Development of the CNS (p. 445)

The brain and spinal cord develop from the neural tube. The ventricles and central canal develop from the lumen of the neural tube.

Brainstem (p. 445)

Medulla Oblongata

1. The medulla oblongata is continuous with the spinal cord and contains ascending and descending tracts.
2. The pyramids are tracts controlling voluntary muscle movement.
3. The olives are nuclei that function in equilibrium, coordination, and modulation of sound from the inner ear.
4. Medullary nuclei regulate the heart, blood vessels, respiration, swallowing, vomiting, coughing, sneezing, and hiccuping. The nuclei of cranial nerves V and IX–XII are in the medulla.

Pons

1. The pons is superior to the medulla.
2. Ascending and descending tracts pass through the pons.
3. Pontine nuclei regulate sleep and respiration. The nuclei of cranial nerves V–IX are in the pons.

Midbrain

1. The midbrain is superior to the pons.
2. The midbrain contains the nuclei for cranial nerves III, IV, and V.
3. The tectum consists of four colliculi. The two inferior colliculi are involved in hearing and the two superior colliculi in visual reflexes.
4. The tegmentum contains ascending tracts and the red nuclei, which are involved in motor activity.
5. The cerebral peduncles are the major descending motor pathway.
6. The substantia nigra connects to other basal nuclei and is involved with muscle tone and movement.

Reticular Formation

The reticular formation consists of nuclei scattered throughout the brainstem. The reticular system functions in the sleep–wake cycle and other cycles of activity.

Cerebellum (p. 449)

1. The cerebellar cortex contains more neurons than does the cerebral cortex. The Purkinje cells are the largest cells in the CNS.
2. The cerebellum has three parts that control balance, gross motor coordination, and fine motor coordination.
3. The cerebellum corrects discrepancies between intended movements and actual movements.
4. The cerebellum can "learn" highly specific complex motor activities.

Diencephalon (p. 449)

The diencephalon is located between the brainstem and the cerebrum.

Thalamus

1. The thalamus consists of two lobes connected by the intermediate mass. The thalamus functions as an integration center.
2. Most sensory input synapses in the thalamus. Pain is registered in the thalamus.
3. The thalamus also has some motor functions.

Subthalamus

The subthalamus is inferior to the thalamus and is involved in motor function.

Epithalamus

The epithalamus is superior and posterior to the thalamus and contains the habenular nuclei, which influence emotions through the sense of smell. The pineal body may play a role in the onset of puberty.

Hypothalamus

1. The hypothalamus, the most inferior portion of the diencephalon, contains several nuclei and tracts.
2. The mammillary bodies are reflex centers for olfaction.
3. The hypothalamus regulates many endocrine functions (e.g., metabolism, reproduction, response to stress, and urine production). The pituitary gland attaches to the hypothalamus.
4. The hypothalamus regulates body temperature, hunger, thirst, satiety, swallowing, and emotions.

Cerebrum (p. 453)

1. The cortex of the cerebrum is folded into ridges called gyri and grooves called sulci, or fissures.
2. The longitudinal fissure divides the cerebrum into left and right hemispheres. Each hemisphere has five lobes.
   - The frontal lobes are involved in smell, voluntary motor function, motivation, aggression, and mood.
   - The parietal lobes contain the major sensory areas receiving general sensory input, taste, and balance.
   - The occipital lobes contain the visual centers.
   - The temporal lobes receive olfactory and auditory input and are involved in memory, abstract thought, and judgment.
3. Tracts connect areas of the cortex within the same hemisphere (association fibers), between hemispheres (commissural fibers), and with other parts of the brain and the spinal cord (projection fibers).

Basal Nuclei

1. Basal nuclei include the subthalamic nuclei, substantia nigra, and corpus striatum.
2. The basal nuclei are important in controlling motor functions.

Limbic System

1. The limbic system includes parts of the cerebral cortex, basal nuclei, the thalamus, the hypothalamus, and the olfactory cortex.
2. The limbic system controls visceral functions through the autonomic nervous system and the endocrine system and is involved in emotions and memory.

Meninges, Ventricles, and Cerebrospinal Fluid (p. 456)

Meninges

1. The brain and spinal cord are covered by the dura, arachnoid, and pia mater.
2. The dura mater attaches to the skull and has two layers that can separate to form dural sinuses.
3. Beneath the arachnoid mater the subarachnoid space contains CSF that helps cushion the brain.
4. The pia mater attaches directly to the brain.

Ventricles

1. The lateral ventricles in the cerebrum are connected to the third ventricle in the diencephalon by the interventricular foramen.
2. The third ventricle is connected to the fourth ventricle in the pons by the cerebral aqueduct. The central canal of the spinal cord is connected to the fourth ventricle.

Cerebrospinal Fluid

1. CSF is produced from the blood in the choroid plexus of each ventricle. CSF moves from the lateral to the third and then to the fourth ventricle.
2. From the fourth ventricle, CSF enters the subarachnoid space through three apertures.
3. CSF leaves the subarachnoid space through arachnoid granulations and returns to the blood in the dural venous sinuses.

Blood Supply to the Brain (p. 461)

1. The brain receives blood from the internal carotid and vertebral arteries. The latter form the basilar artery. The basilar and internal carotid arteries contribute to the cerebral arterial circle. Branches from the circle and basilar artery supply the brain.
2. The blood–brain barrier is formed from the endothelial cells of the capillaries in the brain, the astrocytes in the brain tissue, and the basement membrane in between.

Cranial Nerves (p. 462)

1. Cranial nerves perform sensory, somatic motor, proprioceptive, and parasympathetic functions.
2. The olfactory (I) and optic (II) nerves are involved in the sense of smell and vision.
3. The oculomotor nerve (III) innervates four of six extrinsic eye muscles and the upper eyelid. The oculomotor nerve also provides parasympathetic supply to the iris and lens of the eye.
4. The trochlear nerve (IV) controls an extrinsic eye muscle.
5. The trigeminal nerve (V) supplies the muscles of mastication, as well as a middle ear muscle, a palatine muscle, and two throat muscles. The trigeminal nerve has the greatest cutaneous sensory distribution of any cranial nerve. The trigeminal nerve has three branches. Two of the three trigeminal nerve branches innervate the teeth.
6. The abducent nerve (VI) controls an extrinsic eye muscle.
7. The facial nerve (VII) supplies the muscles of facial expression, an inner ear muscle, and two throat muscles. It is involved in the sense of taste. It is parasympathetic to two sets of salivary glands and to the lacrimal glands.
8. The vestibulocochlear nerve (VIII) is involved in the sense of hearing and balance.
9. The glossopharyngeal nerve (IX) is involved in taste and supplies tactile sensory innervation from the posterior tongue, middle ear, and pharynx. It is also sensory for receptors that monitor blood pressure and gas levels in the blood. The glossopharyngeal nerve is parasympathetic to the parotid salivary glands.
10. The vagus nerve (X) innervates the muscles of the pharynx, palate, and larynx. It is also involved in the sense of taste. The vagus nerve is sensory for the pharynx and larynx and for receptors that monitor blood pressure and gas levels in the blood. The vagus nerve is sensory for thoracic and abdominal organs. The vagus nerve provides parasympathetic innervation to the thoracic and abdominal organs.
11. The accessory nerve (XI) has only a spinal component. It supplies the sternocleidomastoid and trapezius muscles.
12. The hypoglossal nerve (XII) supplies the intrinsic tongue muscles, three of four extrinsic tongue muscles, and two throat muscles.
13. Many reflexes involved in homeostasis involve the cranial nerves and occur in the brainstem.

REVIEW AND COMPREHENSION

1. Which of these parts of the embryonic brain is correctly matched with the structure it becomes in the adult brain?
   a. mesencephalon—midbrain
   b. metencephalon—medulla oblongata
   c. myelencephalon—cerebrum
   d. telencephalon—pons and cerebellum
   e. atom.
2. If a section is made that separates the brainstem from the rest of the brain, the cut is between the
   a. medulla oblongata and pons.
   b. pons and midbrain.
   c. midbrain and diencephalon.
   d. thalamus and cerebrum.
   e. medulla oblongata and spinal cord.
3. Important centers for heart rate, blood pressure, respiration, swallowing, coughing, and vomiting are located in the
   a. cerebrum. b. medulla oblongata. c. midbrain. d. pons. e. cerebellum.
4. In which of these parts of the brain does decussation of descending tracts involved in the conscious control of skeletal muscles occur?
   a. cerebrum
   b. diencephalon
   c. midbrain
   d. pons
   e. medulla oblongata
5. Important respiratory centers are located in the
   a. cerebrum.
   b. cerebellum.
   c. pons and medulla oblongata.
   d. midbrain.
   e. limbic system.
6. The cerebral peduncles are a major descending motor pathway found in the
   a. cerebrum. b. cerebellum. c. pons. d. midbrain. e. medulla oblongata.
7. The superior colliculi are involved in __________, whereas the inferior colliculi are involved in __________.
   a. hearing, visual reflexes
   b. visual reflexes, hearing
   c. balance, motor pathways
   d. motor pathways, balance
   e. respiration, sleep
8. The cerebellum communicates with other regions of the CNS through the
   a. flocculonodular lobe.
   b. cerebellar peduncles.
   c. vermis.
   d. lateral hemispheres.
   e. folia.
9. The major relay station for sensory input that projects to the cerebral cortex is the
   a. hypothalamus. b. thalamus. c. pons. d. cerebellum. e. midbrain.

10. Which part of the brain is involved with olfactory reflexes and emotional responses to odors?
    a. inferior colliculi
    b. superior colliculi
    c. mammillary bodies
    d. pineal body
    e. pituitary gland
11. The part of the diencephalon directly connected to the pituitary gland is the
    a. hypothalamus.
    b. epithalamus.
    c. subthalamus.
    d. thalamus.
12. Which of the following is a function of the hypothalamus?
    a. regulates autonomic nervous system functions
    b. regulates the release of hormones from the posterior pituitary
    c. regulates body temperature
    d. regulates food intake (hunger) and water intake (thirst)
    e. all of the above
13. The grooves on the surface of the cerebrum are called the
    a. nuclei.
    b. commissures.
    c. tracts.
    d. sulci.
    e. gyri.
14. Which of these areas is located in the postcentral gyrus of the cerebral cortex?
    a. olfactory cortex
    b. visual cortex
    c. primary motor cortex
    d. primary somatic sensory cortex
    e. primary auditory cortex
15. Which of these cerebral lobes is important in voluntary motor function, motivation, aggression, sense of smell, and mood?
    a. frontal
    b. insula
    c. occipital
    d. parietal
    e. temporal
16. Fibers that connect areas of the cerebral cortex within the same hemisphere are
    a. projection fibers.
    b. commissural fibers.
    c. association fibers.
    d. all of the above.
17. The basal nuclei are located in the
    a. inferior cerebrum.
    b. diencephalon.
    c. midbrain.
    d. all of the above.
18. The most superficial of the meninges is a thick, tough membrane called the
    a. pia mater.
    b. dura mater.
    c. arachnoid mater.
    d. epidural mater.
19. The ventricles of the brain are interconnected. Which of these ventricles are *not* correctly matched with the structures that connect them?
    a. lateral ventricle to the third ventricle—interventricular foramina
    b. left lateral ventricle to right lateral ventricle—central canal
    c. third ventricle to fourth ventricle—cerebral aqueduct
    d. fourth ventricle to subarachnoid space—median and lateral apertures
20. Cerebrospinal fluid is produced by the ______________, circulates through the ventricles, and enters the subarachnoid space. The cerebrospinal fluid leaves the subarachnoid space through the ______________.
    a. choroid plexuses, arachnoid granulations
    b. arachnoid granulations, choroid plexuses
    c. dural venous sinuses, dura mater
    d. dura mater, dural venous sinuses
21. Given these spaces:
    1. third ventricle
    2. epidural space
    3. subarachnoid space
    4. subdural space
    5. superior sagittal sinus

    Which of these spaces contains only cerebrospinal fluid (CSF)?
    a. 1, 3
    b. 1,2,3
    c. 1,3,5
    d. 1,2,3,5
    e. 2,3,4,5
22. Some portions of the blood plasma move across the blood–brain barrier by
    a. diffusion.
    b. endocytosis.
    c. exocytosis.
    d. symport.
    e. filtration.
23. The cranial nerve involved in chewing food is the
    a. trochlear (IV).
    b. trigeminal (V).
    c. abducent (VI).
    d. facial (VII).
    e. vestibulocochlear (VIII).
24. The cranial nerve responsible for focusing the eye (innervates the ciliary muscle of the eye) is the
    a. optic (II).
    b. oculomotor (III).
    c. trochlear (IV).
    d. abducent (VI).
    e. facial (VII).
25. The cranial nerve involved in moving the tongue is the
    a. trigeminal (V).
    b. facial (VII).
    c. glossopharyngeal (IX).
    d. accessory (XI).
    e. hypoglossal (XII).
26. The cranial nerve involved in feeling a toothache is the
    a. trochlear (IV).
    b. trigeminal (V).
    c. abducent (VI).
    d. facial (VII).
    e. vestibulocochlear (VIII).
27. From this list of cranial nerves:
    1. olfactory (I)
    2. optic (II)
    3. oculomotor (III)
    4. abducent (VI)
    5. vestibulocochlear (VIII)

    Select the nerves that are sensory only.
    a. 1,2,3
    b. 2,3,4
    c. 1,2,5
    d. 2,3,5
    e. 3,4,5
28. From this list of cranial nerves:
    1. optic (II)
    2. oculomotor (III)
    3. trochlear (IV)
    4. trigeminal (V)
    5. abducent (VI)

    Select the nerves that are involved in moving the eyes.
    a. 1,2,3
    b. 1,2,4
    c. 2,3,4
    d. 2,4,5
    e. 2,3,5
29. From this list of cranial nerves:
    1. trigeminal (V)
    2. facial (VII)
    3. glossopharyngeal (IX)
    4. vagus (X)
    5. hypoglossal (XII)

    Select the nerves that are involved in the sense of taste.
    a. 1,2,3
    b. 1,4,5
    c. 2,3,4
    d. 2,3,5
    e. 3,4,5
30. From this list of cranial nerves:
    1. trigeminal (V)
    2. facial (VII)
    3. glossopharyngeal (IX)
    4. vagus (X)
    5. hypoglossal (XII)

    Select the nerves that innervate the salivary glands.
    a. 1,2
    b. 2,3
    c. 3,4
    d. 4,5
    e. 3,5
31. From this list of cranial nerves:
    1. oculomotor (III)
    2. trigeminal (V)
    3. facial (VII)
    4. vestibulocochlear (VIII)
    5. glossopharyngeal (IX)
    6. vagus (X)

    Select the nerves that are part of the parasympathetic division of the ANS.
    a. 1,2,4,5
    b. 1,3,5,6
    c. 1,4,5,6
    d. 2,3,4,5
    e. 2,3,5,6

Answers in Appendix E

CRITICAL THINKING

1. A patient loses all sense of feeling in the left side of the back, below the upper limb, and extending in a band around to the chest, as well as below the upper limb. All sensation on the right is normal. The line between normal and absent sensation is the anterior and posterior midline. Explain this condition.
2. What happens to the developing brain if the CSF is not properly drained, resulting in early hydrocephalus?
3. A patient exhibits enlargement of the lateral and third ventricles, but no enlargement of the fourth ventricle. What do you conclude?
4. During a spinal tap of a patient, blood is discovered in the CSF. What does this finding suggest?
5. Describe a clinical test to evaluate each of the 12 cranial nerves.
6. A patient presented to her physician with a loss of muscle tone and coordination. She was unable to perform coordinated tasks, such as touching the tip of her finger to her nose. The physician concluded that a part of her brain had been damaged. Which part of the brain was damaged?
7. A baseball player was accidentally hit with a baseball, which struck him on the bridge of his nose. He suffered fractures of several facial bones as a result. Soon after sustaining the injury, he noticed he had lost his sense of smell. Explain how this probably happened.

Answers in Appendix F

Visit this book's website at www.mhhe.com/seeley8 for chapter quizzes, interactive learning exercises, and other study tools.

14 Integration of Nervous System Functions

The nervous system is involved in almost all bodily functions. Although humans have larger, more complex brains than other animals, most human nervous system functions are similar to those of other animals. The sensory input we receive and most of the ways we respond to that input are similar to that in other animals, yet the human brain is capable of unique and complex functions, such as recording history, reasoning, and planning, to a degree unparalleled in the animal kingdom. Many of these functions can be studied only in humans. That is why much of human brain function remains elusive and is one of the most challenging frontiers of anatomy and physiology.

This chapter presents the concept of *sensation* (p. 477) and then discusses the *control of skeletal muscles* (p. 490), the *brainstem functions* (p. 498), *other brain functions* (p. 500), and the *effects of aging on the nervous system* (p. 506).

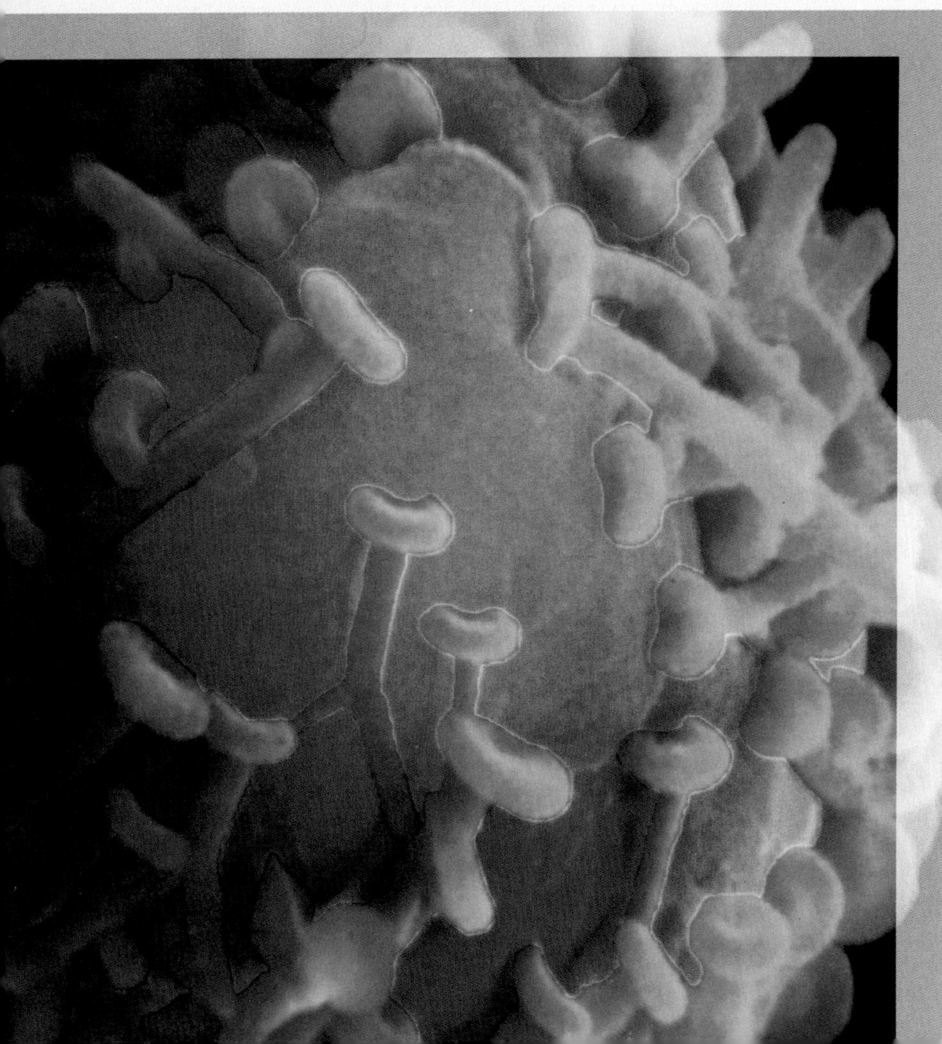

Colorized SEM of presynaptic terminals associated with a postsynaptic neuron.

Nervous System

SENSATION

Sensation, or **perception,** is the conscious awareness of the effects of stimuli on sensory receptors. The brain constantly receives action potentials from a wide variety of sensory receptors both inside and outside the body. Sensory receptors respond to stimuli by generating action potentials that are propagated along nerves to the spinal cord and brain. Sensations result when action potentials reach the cerebral cortex. Some other parts of the brain are involved in sensation. For example, the thalamus and amygdala are involved in the sensation of pain.

The **senses** are the means by which the brain perceives information about the environment and the body. Historically, five senses were recognized: smell, taste, sight, hearing, and touch. Today, the senses are divided into two basic groups: general and special senses. The **general senses** are those with receptors distributed over a large part of the body. They are divided into two groups: the somatic and the visceral senses (table 14.1 and figure 14.1). The **somatic senses,** which provide sensory information about the body and the environment, include touch, pressure, itch, vibration, temperature, proprioception, and pain. The **visceral senses,** which provide information about various internal organs, consist primarily of pain and pressure.

The **special senses** are more specialized in structure, have specialized nerve endings, and are localized to specific organs (see table 14.1). The special senses are smell, taste, sight, hearing, and balance. Chapter 15 considers the special senses in detail.

Awareness

As you read this paragraph, you are probably not aware of the weight of the book in your hands, if you are holding it, or the weight of your forearms on the desk or on your lap, if you are reading at a desk. It is also unlikely that you are aware of the small noises around you or the clothes touching your body until your attention is drawn to them. You certainly are not aware of changes in your blood pressure and your body fluid pH and blood glucose levels. This is because not all of the sensory information detected by sensory receptors results in sensory perception. Some action potentials reach areas of the brain where they are not consciously perceived.

Although we can be consciously aware of body position and movements, much of this sensory information is propagated to the cerebellum, where is it processed at an unconscious level. Sensory information from receptors that monitor blood pressure, blood oxygen, and pH levels are processed unconsciously in the medulla oblongata. For example, blood pressure must be regulated constantly to maintain homeostasis. If we had to regulate blood pressure consciously, we could not be able to think of much else. Homeostasis, therefore, is controlled largely without our conscious involvement.

Even the cerebral cortex screens much of what it receives, ignoring many of the action potentials that reach it. In addition, humans exhibit selective awareness. We are more aware of sensations on which we focus our attention than on other sensations. If we were aware of all the sensations that arrived at the cerebral cortex, we would probably not be able to function.

Sensation requires the following steps:

1. Stimuli originating either inside or outside the body are detected by sensory receptors and action potentials are produced, which are propagated to the CNS by nerves (see "Sensory Receptors").
2. Within the CNS, tracts conduct action potentials to the cerebral cortex and to other areas of the CNS (see "Sensory Tracts," p. 482).
3. Many action potentials reaching the cerebral cortex are ignored. Others are translated so the person becomes aware of the stimulus (see "Sensory Areas of the Cerebral Cortex," p. 486).

Sensory Receptors

Types of Sensory Receptors

The various senses depend on sensory receptors specialized to respond to specific types of stimuli (see table 14.1). **Mechanoreceptors** respond to mechanical stimuli, such as compression, bending, or stretching of cells. The senses of touch, tickle, itch, vibration, pressure, proprioception, hearing, and balance all depend on a variety of mechanoreceptors. **Chemoreceptors** respond to chemicals that become attached to receptors on their membranes. Smell and taste depend on chemoreceptors. **Thermoreceptors** respond to changes in temperature at the site of the receptor and are necessary for the sense of temperature. **Photoreceptors** respond to light striking the receptor cells and are necessary for vision. **Nociceptors** (nō-si-sep′ters; Latin, *noceo* means hurt), or **pain receptors,** respond to painful mechanical, chemical, and thermal stimuli. Most sensory receptors typically respond to one type of stimulus, but some nociceptors respond to more than one.

At least eight major types of sensory receptors, which differ in their structure and the types of stimuli to which they are most sensitive, are involved in general sensation (table 14.2 and figure 14.2). Many of these sensory receptors are associated with the skin; others are associated with deeper structures, such as tendons, ligaments, and muscles; and some can be found in both the skin and deeper structures. In general, sensory receptors are classified into three groups based on their location: **Exteroreceptors** (cutaneous receptors) are associated with the skin, **visceroreceptors** are associated with the viscera or organs, and **proprioceptors** are associated with joints, tendons, and other connective tissue. Exteroreceptors provide information about the external environment, visceroreceptors provide information about the internal environment, and proprioceptors provide information about body position, movement, and the extent of stretch or the force of muscular contractions.

Structurally, the simplest and most common sensory receptors are the **free nerve endings** (see figure 14.2), which are relatively unspecialized neuronal branches similar to dendrites. Free nerve endings are distributed throughout almost all parts of the body. Most visceroreceptors consist of free nerve endings, which are responsible for a number of sensations, including pain, temperature, itch, and movement. The free nerve endings responsible for temperature detection respond to three types of sensations. One type, the **cold receptors,** increases its rate of action potential

TABLE 14.1 **Classification of the Senses**

Types of Sense	Nerve Endings	Receptor Type	Initiation of Response
Somatic			
Touch		Mechanoreceptors	Compression of receptors
Stroking	Meissner corpuscle		
	Hair follicle receptor		
Texture	Merkel's disk		
Vibration	Pacinian corpuscle		
Skin stretch	Ruffini end organ		
Itch	Free nerve endings		
Pressure	Merkel disk	Mechanoreceptors	Compression of receptors
Proprioception	Pacinian corpuscle	Mechanoreceptors	Compression of receptors
Temperature	Free nerve endings	Thermoreceptors	Temperature around nerve endings
	Cold receptors		
	Warm receptors		
Pain	Free nerve endings	Nociceptors	Irritation of nerve endings (e.g., mechanical, chemical, or thermal)
Visceral			
Pain	Free nerve endings	Nociceptors	Irritation of nerve endings
Pressure	Pacinian corpuscle	Mechanoreceptors	Compression of receptors
Special			
Smell	Specialized	Chemoreceptors	Binding of molecules to membrane receptors
Taste	Specialized	Chemoreceptors	Binding of molecules to membrane receptors
Sight	Specialized	Photoreceptors	Chemical change in receptors initiated by light
Hearing	Specialized	Mechanoreceptors	Bending of microvilli on receptor cells
Balance	Specialized	Mechanoreceptors	Bending of microvilli on receptor cells

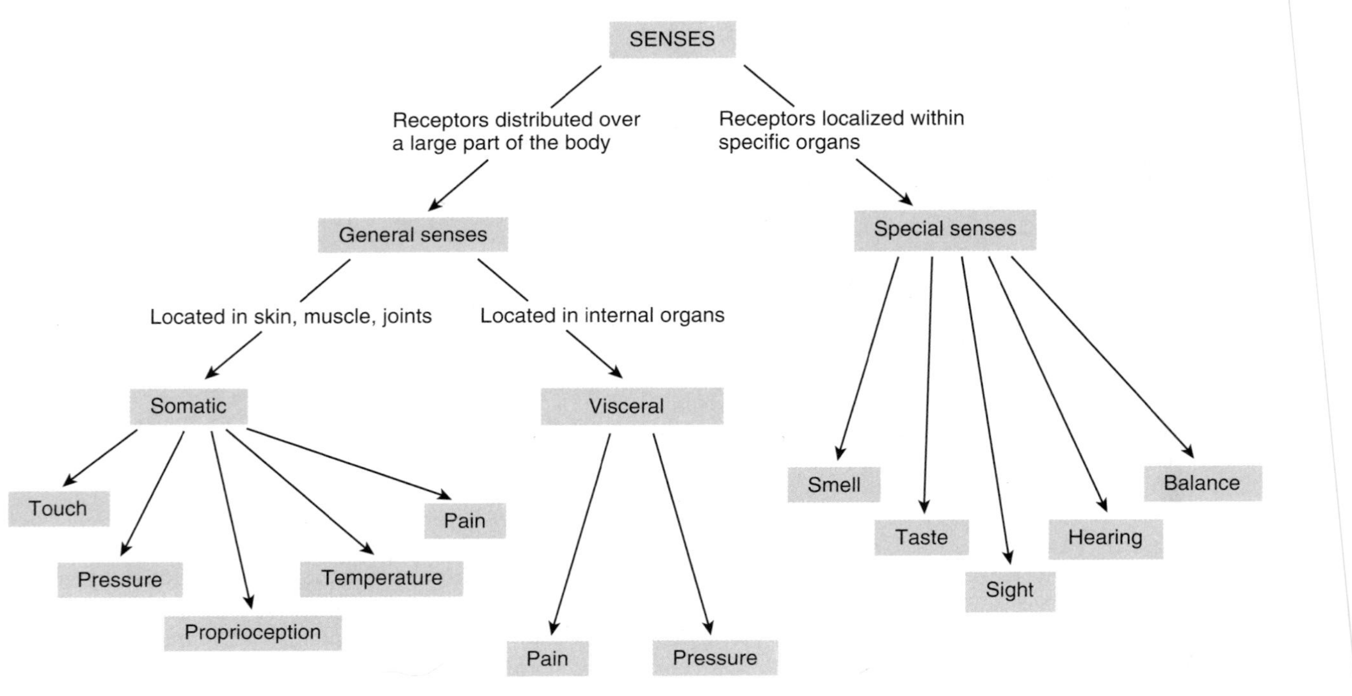

FIGURE 14.1 Classification of Senses

TABLE 14.2 Sensory Receptors

Type of Receptor	Structure	Function
Free nerve ending	Branching, no capsule	Pain, itch, tickle, temperature, joint movement, and proprioception
Merkel disk	Flattened expansions at the end of axons; each expansion associated with a Markel cell	Light touch and superficial pressure
Hair follicle receptor	Wrapped around hair follicles or extending along the hair axis, each axon supplies several hairs, and each hair receives branches from several neurons, resulting in considerable overlap.	Light touch; responds to very slight bending of the hair
Pacinian corpuscle	Onion-shaped capsule of several cell layers with a single central nerve process	Deep cutaneous pressure, vibration, and propriception
Meissner corpuscle	Several branches of a single axon associated with wedge-shaped epitheloid cells and surrounded by a connective tissue capsule	Two-point discrimination
Ruffini end organ	Branching axon with numerous small, terminal knobs surrounded by a connective tissue capsule	Continuous touch or pressure; responds to depression or stretch of the skin
Muscle spindle	Three to 10 striated muscle fibers enclosed by a loose connective tissue capsule, striated only at the ends, with sensory nerve endings in the center	Proprioception associated with detection of muscle stretch; important for control of muscle tone
Golgi tendon organ	Surrounds a bundle of tendon fascicles and is enclosed by a delicate connective tissue capsule; nerve terminations are branched with small swellings applied to individual tendon fascicles	Proprioception associated with the stretch of a tendon; important in the control of muscle contraction

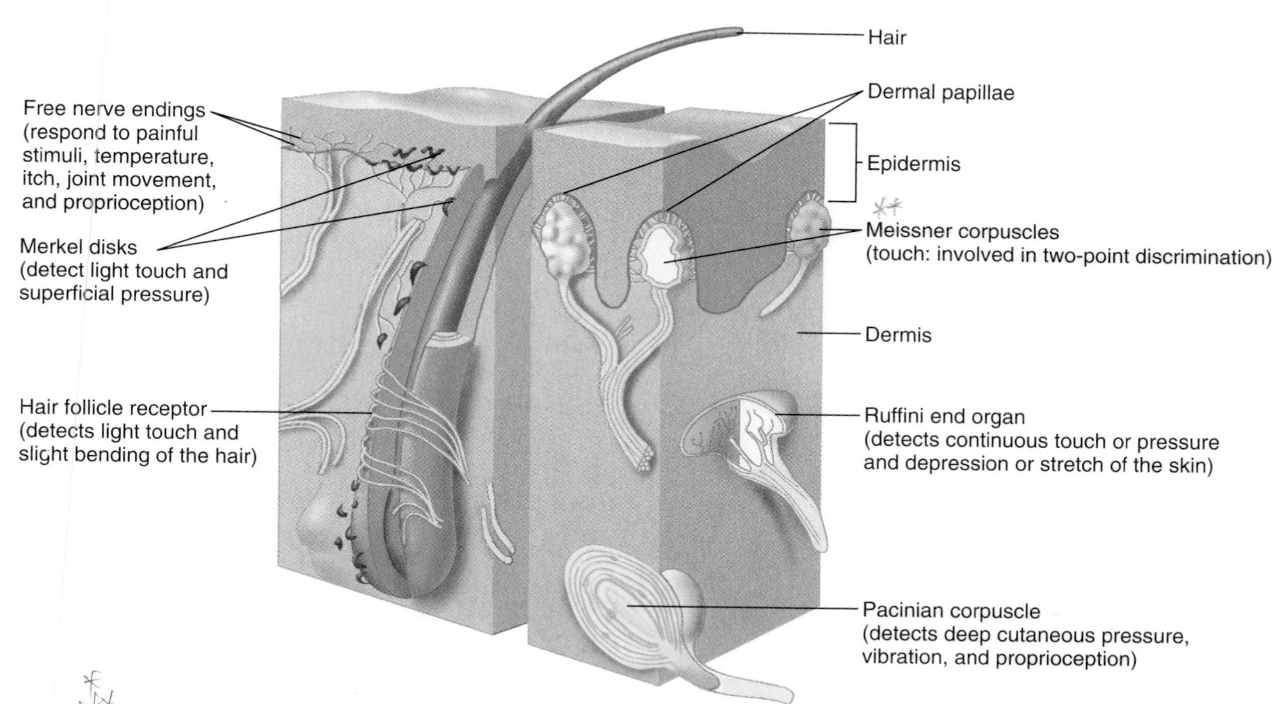

FIGURE 14.2 Sensory Nerve Endings in the Skin

production as the skin is cooled. The second type, **warm receptors,** increases its rate of action potential production as skin temperature increases. Both cold and warm receptors respond most strongly to changes in temperature. Cold receptors are 10 to 15 times more numerous in any given area of skin than warm receptors. The third type is a pain receptor, which is stimulated by extreme cold or heat. At very cold temperatures (0°–12°C), only pain receptors are stimulated. The pain sensation ends as the temperature increases above 15°C. Between 12° and 35°C, cold fibers are stimulated. Nerve fibers from warm receptors are stimulated between 25° and 47°C. "Comfortable" temperatures, between 25° and 35°C, therefore stimulate both warm and cold receptors. Temperatures above 47°C stimulate cold and pain receptors but do not stimulate warm receptors.

PREDICT 1

How might a very cold object placed in the hand be misperceived as being hot?

Merkel (mer'kĕl), or **tactile, disks** are more complex than free nerve endings (see figure 14.2) and consist of axonal branches that end as flattened expansions, each associated with a specialized epithelial cell. They are distributed throughout the basal layers of the epidermis just superficial to the basement membrane and are associated with dome-shaped mounds of thickened epidermis in hairy skin. Merkel disks are involved with the sensations of light touch and superficial pressure. These receptors can detect a skin displacement of less than 1 mm (1/25 inch).

Hair follicle receptors, or **hair end organs,** respond to very slight bending of the hair and are involved in light touch (see figure 14.2). These receptors are extremely sensitive and require very little stimulation to elicit a response. The sensation, however, is not very well localized. The dendritic tree at the distal end of a sensory axon has several hair follicle receptors. The field of hairs innervated by these receptors overlaps with the fields of hair follicle receptors of adjacent axons. The considerable overlap that exists in the endings of sensory neurons helps explain why light touch is not highly localized, yet, because of converging signals within the CNS, it is very sensitive.

Pacinian (pa-sin'ē-an, pa-chin'ē-an), or **lamellated, corpuscles** are complex receptors that resemble an onion (see figure 14.2). A single dendrite extends to the center of each lamellated corpuscle. The corpuscles are located within the deep dermis or hypodermis, where they are responsible for deep cutaneous pressure and vibration. Pacinian corpuscles associated with the joints help relay **proprioceptive** (prō-prē-ō-sep'tiv; perception of position) information about joint positions.

Meissner (mīs'ner), or **tactile, corpuscles** are distributed throughout the dermal papillae (see figure 14.2 and chapter 5) and are involved in two-point discrimination. **Two-point discrimination** (fine touch) is the ability to detect simultaneous stimulation at two points on the skin. The distance between two points that a person can detect as separate points of stimulation differs for various regions of the body. This sensation is important in evaluating the texture of objects. Meissner corpuscles are numerous and close together in the tongue and fingertips but are less numerous and more widely separated in other areas, such as the back (figure 14.3).

Ruffini (rū-fē'nē) **end organs** are located in the dermis of the skin (see figure 14.2), primarily in the fingers. They respond to pressure on the skin directly superficial to the receptor and to stretch of adjacent skin. These receptors are important in responding to continuous touch or pressure.

Muscle spindles (figure 14.4) consist of 3–10 specialized skeletal muscle fibers. They are located in skeletal muscles and provide information about the length of the muscle (see "Stretch

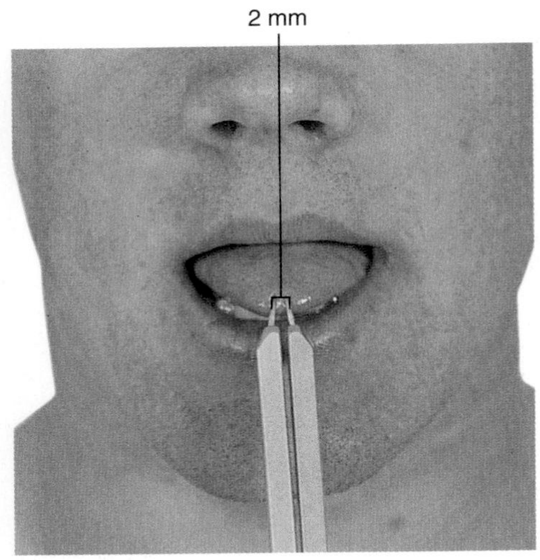

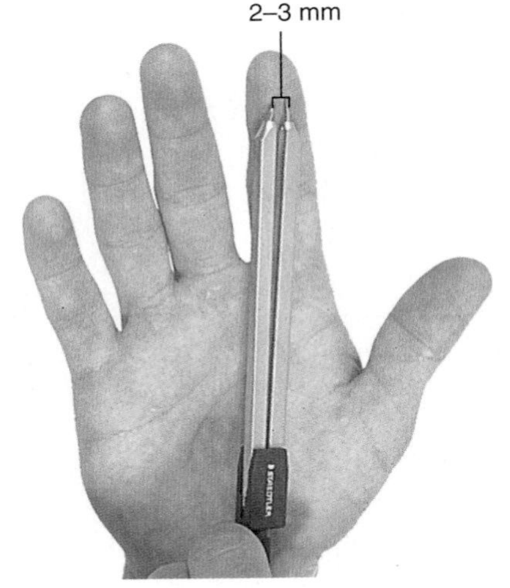

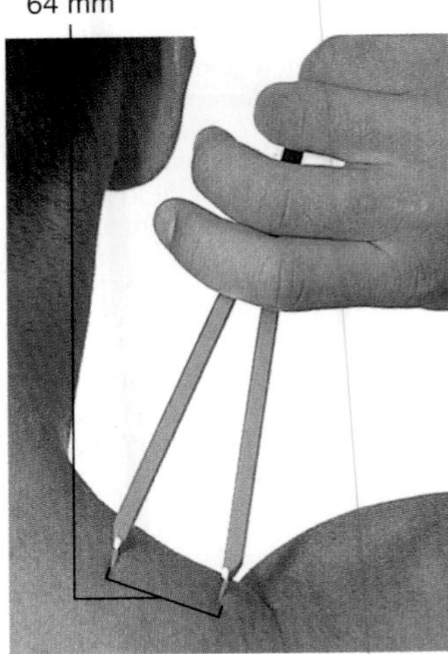

FIGURE 14.3 Two-Point Discrimination

Two-point discrimination can be demonstrated by touching a person's skin with the two points of a compass. When the two points are closer together than the receptor field, the individual perceives only one point. When the two points of the compass are opened wider, the person becomes aware of two points.

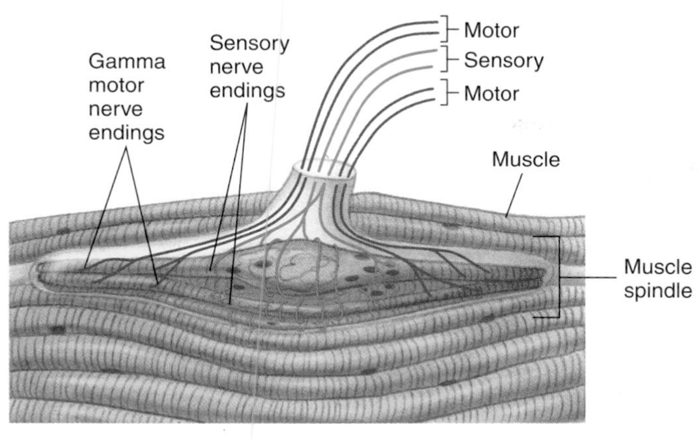

FIGURE 14.4 **Muscle Spindle**

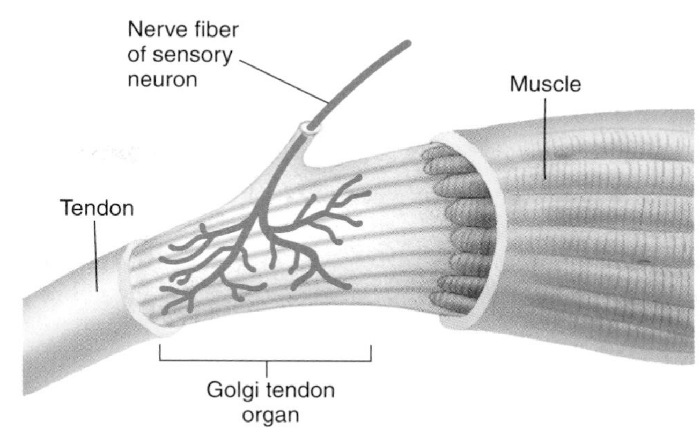

FIGURE 14.5 **Golgi Tendon Organ**

Reflex," p. 417). Muscle spindles are important to the control and tone of postural muscles. Brain centers act through descending tracts to either increase or decrease action potentials in gamma motor neurons. Stimulation of the gamma motor system, caused by stretch of the muscle, activates the stretch reflex, which in turn increases the tone of the muscles involved.

Golgi tendon organs are proprioceptive receptors associated with the fibers of a tendon near the junction between the muscle and tendon (figure 14.5). They are activated by an increase in tendon tension, caused either by contraction of the muscle or by passive stretch of the tendon.

1 ***In general, into what three groups can sensory receptors be classified?***

2 ***List the eight major types of sensory receptors, indicate where they are located, and state the functions they perform.***

Responses of Sensory Receptors

Interaction of a stimulus with a sensory receptor produces a graded potential called a **receptor potential.** Some sensory receptor cells, called **primary receptors,** have axons that conduct action potentials in response to the receptor potential (figure 14.6*a*). When

1. A mechanoreceptor (Pacinian corpuscle) is subjected to a pressure stimulus.

2. The sensory receptor responds by generating a graded potential, also known as the receptor potential.

3. When the graded potential reaches threshold, action potentials are generated in the axon. The action potentials are propagated towards the CNS.

(a)

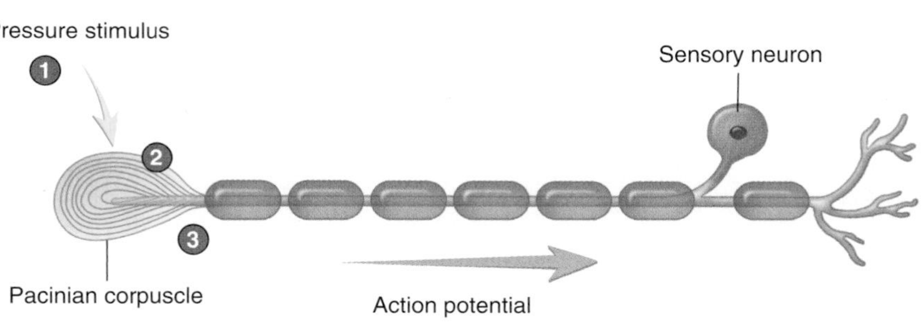

1. A chemoreceptor (taste cell) is subjected to a chemical stimulus in the form of a salty substance.

2. The sensory receptor responds by generating a graded potential, also known as the receptor potential.

3. The sensory receptor releases a neurotransmitter into the synaptic cleft, which in turn stimulates the sensory neuron.

4. When the stimulation reaches threshold, the neuron generates action potentials, which are propagated toward the CNS.

(b)

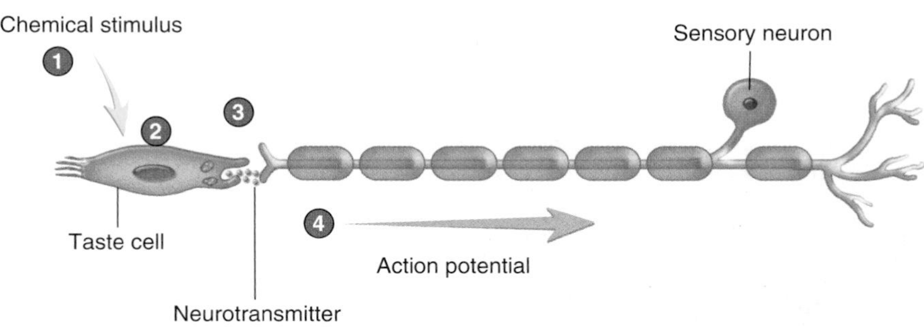

PROCESS FIGURE 14.6 **Comparison of Primary and Secondary Receptors**

(*a*) A primary receptor has an axon that conducts action potentials in response to the receptor potential. (*b*) A secondary receptor has no axon, but the receptor potential results in the release of a neurotransmitter, which in turn stimulates a postsynaptic neuron.

the ends of these neurons are stimulated, a receptor potential is produced. If it reaches threshold, an action potential is produced and is propagated toward the CNS. Most sensory neurons, including all those in table 14.2, belong to this category. Other receptor cells, called **secondary receptors,** have no axons or have short, axonlike projections and the receptor potentials produced in those cells do not result in action potentials (figure 14.6*b*). Instead, the receptor potentials cause the release of neurotransmitter molecules from the receptor cell that bind to receptors on the membrane of a neuron. This causes a receptor potential in the neuron, which produces an action potential if threshold is reached. The receptor cells of the special senses of taste, hearing, and balance belong to this category.

Some sensations have the quality of **accommodation,** or **adaptation,** a decreased sensitivity to a continued stimulus. After exposure to a stimulus for a time, the response of the receptors or the sensory pathways to a certain stimulus strength lessens from that which occurs when the stimulus was first applied. The local graded depolarization that produces a receptor potential accommodates, or returns to, its resting level even though the stimulus is still applied. For example, when a person first gets dressed, tactile receptors and pathways relay information to the brain that create an awareness that the clothes are touching the skin. After a time, the action potentials from the skin decrease, and the clothes are ignored.

Another way that sensations change through time occurs in proprioception. Proprioception provides information about the precise position and the rate of movement of various body parts, the weight of an object being held in the hand, and the range of movement of a joint. This information is involved in activities such as walking, climbing stairs, shooting a basketball, driving a car, eating, and writing. Receptors for this system are located around joints and in muscles. Two types of proprioceptors are involved in providing positional information: tonic receptors and phasic receptors. **Tonic receptors,** or **slowly adapting receptors,** generate action potentials as long as a stimulus is applied and accommodate very slowly. Information from slowly adapting receptors allows a person to know, for example, where the little finger is at all times without having to look for it. **Phasic receptors,** or **rapidly adapting receptors,** by contrast, accommodate rapidly and are most sensitive to changes in stimuli. For example, information from rapidly adapting receptors allows a person to know where the little finger is as it moves; thus, we can control its movement through space and predict where it will be in the next moment.

We are usually not conscious of tonic or phasic input because most of this peripheral input is ignored by higher brain centers most of the time. Through selective awareness, however, we can call up the information when we wish. For example, where is the thumb of your right hand at this moment? Were you aware of its position a few seconds ago?

3 ***What are primary and secondary receptors? What effect does a receptor potential have on them?***

4 ***Define adaptation. Describe tonic and phasic receptors.***

Sensory Tracts

The spinal cord and brainstem contain a number of sensory pathways that transmit action potentials from the periphery to various parts of the brain (table 14.3). Each pathway is involved with specific modalities (the type of information transmitted). The

TABLE 14.3 Ascending Spinal Pathways

Pathway	Modality (Information Transmitted)	Origin	Primary Cell Body	Secondary Cell Body
Anterolateral				
Spinothalamic	Pain, temperature, light touch, pressure, tickle, and itch	Cutaneous receptors	Dorsal root ganglion	Posterior horn of spinal cord
Spinoreticular	Pain	Cutaneous receptors	Dorsal root ganglion	Posterior horn of spinal cord
Spinomesencephalic (including spinotectal)	Pain and touch	Cutaneous receptors	Dorsal root ganglion	Posterior horn of spinal cord
Dorsal-column/ medial-lemniscal system	Proprioception, two-point discrimination, pressure, and vibration	Joints, tendons, muscles	Dorsal root ganglion	Medulla oblongata
Spinocerebellar				
Posterior	Proprioception	Joints and tendons	Dorsal root ganglion	Posterior horn of spinal cord
Anterior	Proprioception	Joints and tendons	Dorsal root ganglion	Posterior horn of spinal cord
Spinoolivary	Proprioception relating to balance	Joints and tendons	Dorsal root ganglion	Posterior horn of spinal cord

neurons that make up each pathway are associated with specific types of sensory receptors. For example, thermoreceptors located in the skin generate action potentials that are propagated along the sensory pathway for pain and temperature, whereas Golgi tendon organs located in tendons generate action potentials that are propagated along the sensory pathway involved with proprioception.

The names of many ascending pathways, or tracts, in the CNS indicate their origin and termination (figure 14.7). Others indicate their location in the spinal cord. Each pathway usually is given a composite name in which the first half of the word indicates its origin and the second half indicates its termination. Ascending pathways therefore usually begin with the prefix *spino-*, indicating that they originate in the spinal cord. For example, a spinocerebellar (spī′nō-ser-e-bel′ar) tract is one that originates in the spinal cord and terminates in the cerebellum. An example of location nomenclature is the dorsal-column/medial-lemniscal system, whose name is a combination of the pathway names in the spinal cord and brainstem. The specific function of each ascending tract, however, is not suggested by its name.

The two major ascending systems of tracts involved in the conscious perception of external stimuli are the anterolateral system and the dorsal-column/medial-lemniscal system (see table 14.3). Those pathways carrying sensory input that we are not consciously aware of include the spinocerebellar, spinoolivary, spinomesencephalic, and spinoreticular tracts, the last two of which are part of the anterolateral system.

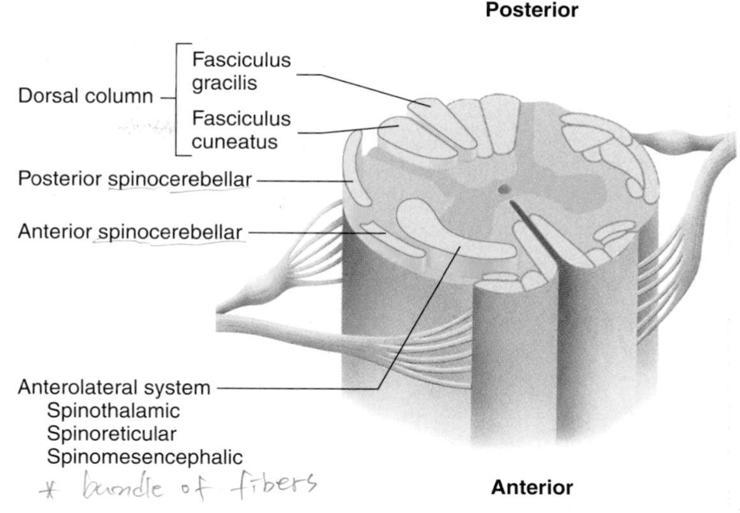

FIGURE 14.7 Cross Section of the Spinal Cord at the Cervical Level Depicting the Ascending Pathways

Ascending pathways are labeled on the left side of the figure only (*blue*), although they exist on both sides.

Anterolateral System

The **anterolateral system** is one of the two major systems that convey cutaneous sensory information to the brain. Historically, the anterolateral system was divided into anterior and lateral spinothalamic tracts. It is now known that dividing the anterolateral system into anterior and lateral spinothalamic tracts was artificial and not clinically relevant. The anterolateral system includes the spinothalamic, spinoreticular, and spinomesencephalic tracts. There is, however, considerable overlap among these three tracts

Crossover	Tertiary Cell Body	Termination	Side of Body Where Fibers Terminate
Level at which primary neuron enters spinal cord for pain and temperature or 8 to 10 segments from where primary neuron enters spinal cord for light touch	Thalamus	Cerebral cortex	Contralateral
Reticular formation	Reticular formation	Reticular formation and thalamus	Contralateral
At point of origin	Superior colliculus	Mesencephalon and superior colliculus	Contralateral
Medulla oblongata	Thalamus	Cerebral cortex and cerebellum	Contralateral
Uncrossed	Cerebellum	Cerebellum	Ipsilateral
Some uncrossed; some cross at point of entry; recross at cerebellum	Cerebellum	Cerebellum	Ipsilateral
Crossed at point of entry; recross to reach cerebellum	Accessory olivary nucleus	Accessory olivary nucleus, then cerebellum	Ipsilateral

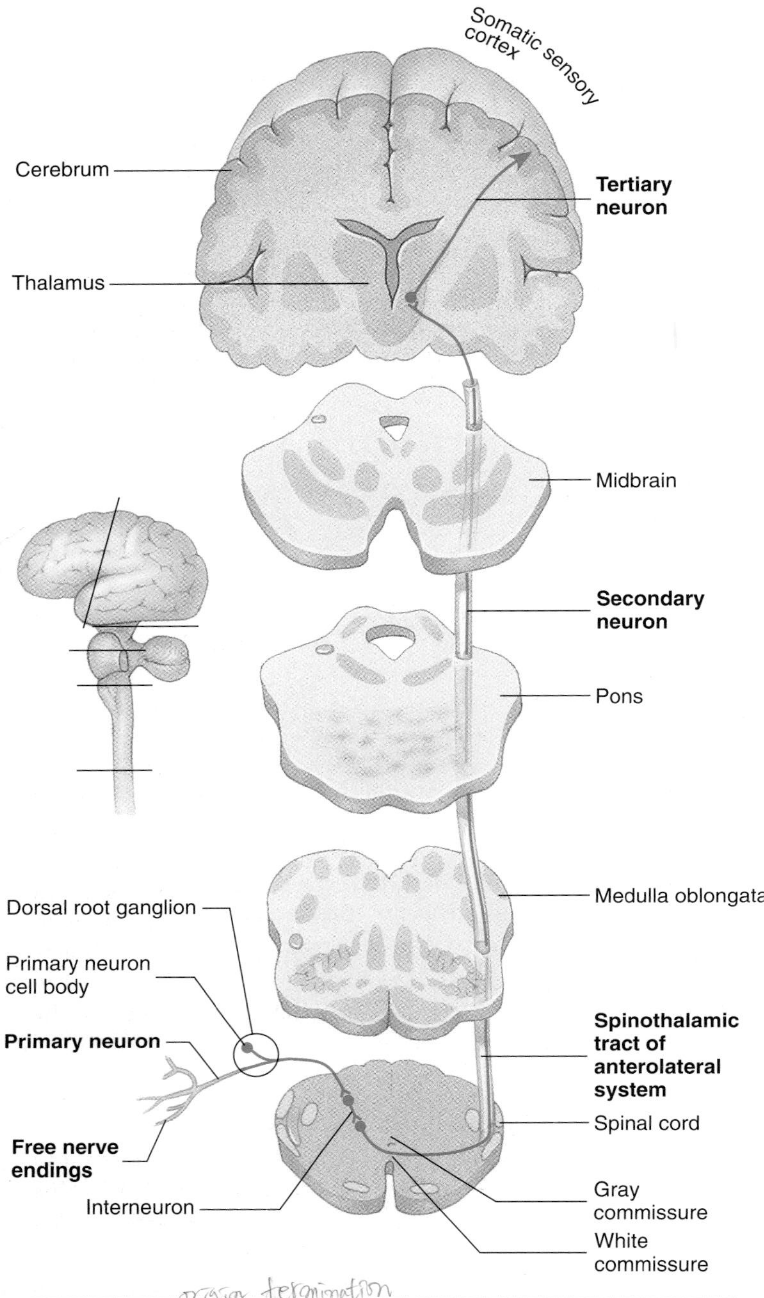

FIGURE 14.8 Spinothalamic Tract of the Anterolateral System

Some axons of the spinothalamic tract in the anterolateral system transmit action potentials for pain and temperature. Lines on the inset indicate levels of section.

within the anterolateral system. The **spinothalamic tract** carries pain and temperature information, as well as light touch and pressure, tickle and itch sensations (figure 14.8).

Three neurons in sequence—the primary, secondary, and tertiary—are involved in the spinothalamic tract from the peripheral receptor to the cerebral cortex. The **primary neuron** cell bodies of the spinothalamic tract are in the dorsal root ganglia. The primary neurons relay sensory input from the periphery to the posterior horn of the spinal cord, where they synapse with interneurons. The interneurons, which are not specifically named in the three-neuron sequence, synapse with secondary neurons. Axons from the **secondary neurons** cross to the **contralateral,** or opposite side, of the spinal cord through the anterior portion of the gray and white commissures and enter the spinothalamic tract, where they ascend to the thalamus. The secondary neurons synapse with cell bodies of tertiary neurons in the thalamus. **Tertiary neurons** from the thalamus project to the somatic sensory cortex.

Some neurons in the spinoreticular tracts do not cross over but ascend on the **ipsilateral,** or same side, of the spinal cord on which they enter.

PREDICT 2

Describe the clinical results of a lesion on one side of the spinal cord that interrupts the spinothalamic part of the anterolateral system.

Syringomyelia

Syringomyelia (sĭ-ring′gō-mī-ē′lē-ă) is a degenerative cavitation of the central canal of the spinal cord, often caused by a cord tumor. Symptoms include neuralgia (pain along the course of the nerve), paresthesia (abnormal sensations, burning, tingling, increased sensitivity to pain), specific loss of pain and temperature sensation, and paresis (partial paralysis). This defect is unusual in that it occurs in a distinct band that includes both sides of the body because commissural tracts are destroyed.

The defect causing syringomyelia occurs where the fibers giving rise to the anterolateral system cross over in the center of the spinal cord. As a result of damage to crossing fibers from both sides, there is loss of pain and temperature on both sides of the body in a segment at the level of the cord damage. The loss of function is only at that level. There is no loss of function below the level of the damage.

Dorsal-Column/Medial-Lemniscal System

The **dorsal-column/medial-lemniscal** (lem-nis′kăl) **system** carries the sensations of two-point discrimination, proprioception, pressure, and vibration (figure 14.9). This system is named for the dorsal column of the spinal cord and the medial lemniscus, which is the continuation of the dorsal column in the brainstem. The term *lemniscus* means ribbon and refers to the thin, ribbonlike appearance of the pathway as it passes through the brainstem.

Primary neurons of the dorsal-column/medial-lemniscal system are located in the dorsal root ganglia. They are the largest cell bodies in the dorsal root ganglia, especially those for two-point discrimination. Many axons of the primary neurons of the dorsal-column/medial-lemniscal system enter the spinal cord and ascend the entire length of the spinal cord, without crossing to its opposite side, and synapse with secondary neurons located in the medulla oblongata. Others synapse in the thoracic portion of the spinal cord.

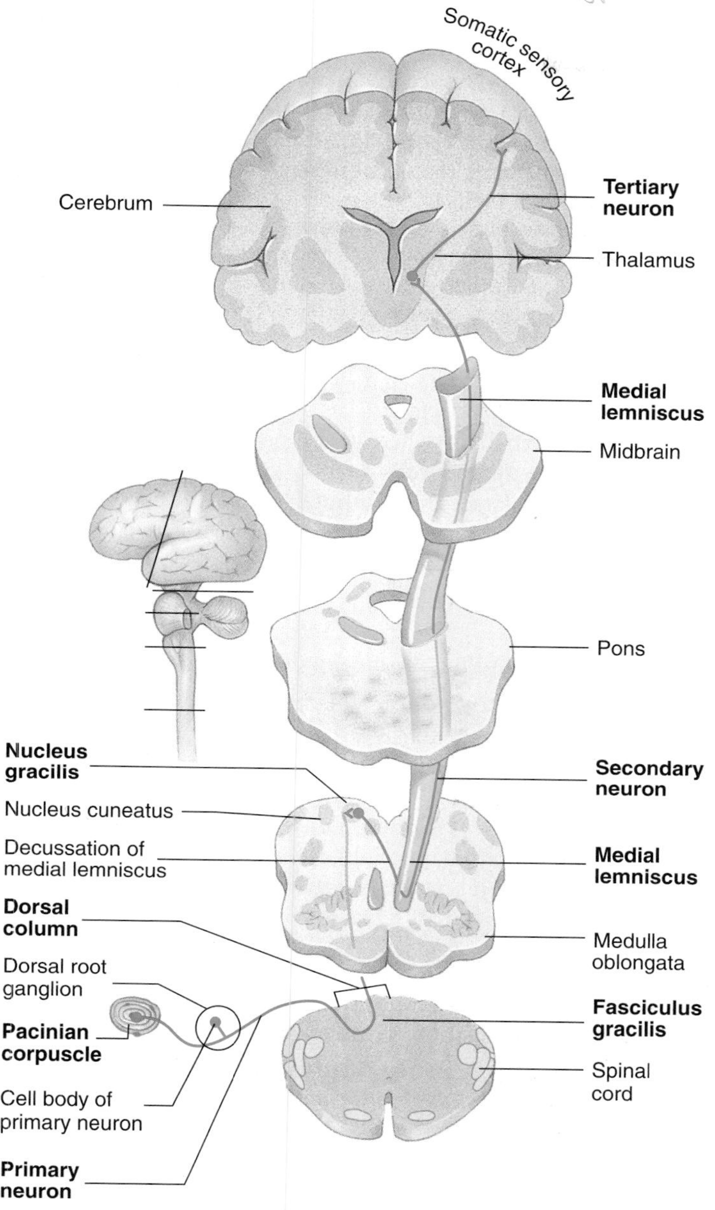

FIGURE 14.9 Dorsal-Column/Medial-Lemniscal System

The fasciculus gracilis and fasciculus cuneatus (not illustrated) convey proprioception, pressure, vibration, and two-point discrimination. Lines on the inset indicate levels of section.

In the spinal cord, the dorsal-column/medial-lemniscal system is divided into two tracts (see figure 14.7) based on the source of the stimulus. The **fasciculus gracilis** (gras′i-lis; thin) conveys sensations from nerve endings below the midthoracic level, and the **fasciculus cuneatus** (kū′nē-ā′tŭs; wedge-shaped) conveys impulses from nerve endings above the midthorax. The fasciculus gracilis terminates by synapsing with secondary neurons in the **nucleus gracilis** or with neurons of the posterior spinocerebellar tracts. The fasciculus cuneatus primarily terminates by synapsing with secondary neurons in the **nucleus cuneatus.** Both the nucleus gracilis and the nucleus cuneatus are in the medulla oblongata. Secondary neurons exit the nucleus gracilis and the nucleus cuneatus, cross to the opposite side of the medulla through the decussations of the medial lemniscus, and ascend through the medial lemniscus to terminate in the thalamus. Tertiary neurons from the thalamus project to the primary somatic sensory cortex (see page 486).

PREDICT 3

Two people, Bill and Mary, were involved in an accident and each experienced a loss of proprioception, fine touch, and vibration on the left side of the body below the waist. It was determined that Bill had damage to his spinal cord as a result of the accident and that Mary had damage to her brainstem. Explain which side of the spinal cord was damaged in Bill and which side of the brainstem was damaged in Mary.

Trigeminothalamic Tract

As the fibers of the spinothalamic tracts pass through the brainstem, they are joined by fibers of the **trigeminothalamic tract.** The trigeminothalamic tract is made up primarily of afferent fibers from the trigeminal nerve, joined by a few tactile afferent fibers from the ear and tongue carried by cranial nerves VII, IX, and sometimes X. This tract carries the same sensory information as the spinothalamic tracts and dorsal-column/medial-lemniscal system but from the face, nasal cavity, and oral cavity, including the teeth. The trigeminothalamic tract is similar to the spinothalamic tracts and dorsal-column/medial-lemniscal system in that primary neurons from one side of the face synapse with secondary neurons, which cross to the opposite side of the brainstem. The secondary neurons synapse with tertiary neurons in the thalamus. Tertiary neurons from the thalamus project to the somatic sensory cortex.

Spinocerebellar System and Other Tracts

The spinocerebellar tracts (see figure 14.7) carry proprioceptive information to the cerebellum so that information concerning actual movements can be monitored and compared with cerebral information representing intended movements.

Two spinocerebellar tracts extend through the spinal cord: (1) the **posterior spinocerebellar tract** (figure 14.10), which originates in the thoracic and upper lumbar regions and contains uncrossed nerve fibers that enter the cerebellum through the inferior cerebellar peduncles, and (2) the **anterior spinocerebellar tract,** which carries information from the lower trunk and lower limbs and contains both crossed and uncrossed nerve fibers that enter the cerebellum through the superior cerebellar peduncle. The crossed fibers recross in the cerebellum. Both spinocerebellar tracts transmit proprioceptive information to the cerebellum from the same side of the body as the cerebellar hemisphere to which they project. Why the anterior spinocerebellar tract crosses twice to accomplish this feat is unknown. Much of the proprioceptive information carried from the lower limbs by the fasciculus gracilis of the dorsal-column/medial-lemniscal

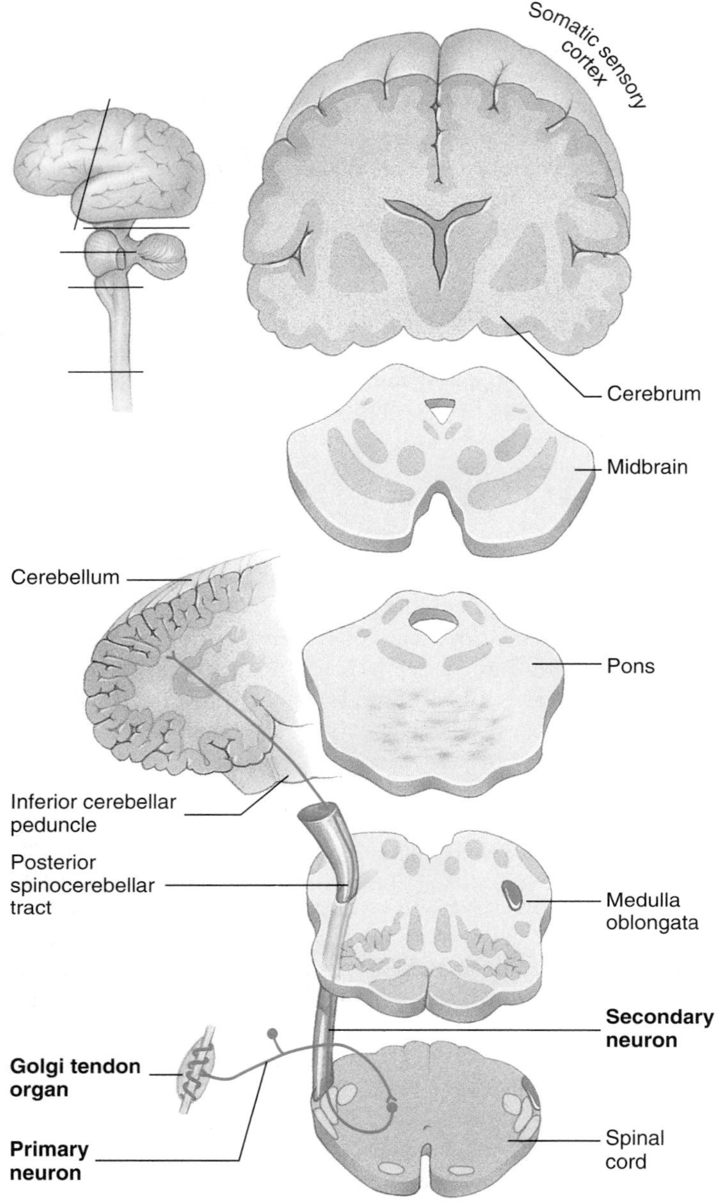

FIGURE 14.10 Posterior Spinocerebellar Tract

This tract transmits proprioceptive information from the thorax, upper limbs, and upper lumbar region to the cerebellum. Lines on the inset indicate levels of section.

system is transferred by synapses in the inferior thorax to the spinocerebellar system and enters the cerebellum as unconscious proprioceptive information. The spinocerebellar tracts convey very little information from the upper limbs to the cerebellum. Proprioception from the upper limbs is projected to the thalamus. This input enters the cerebellum through the inferior peduncle from the cuneate nucleus of the dorsal-column/medial-lemniscal system. The dorsal-column/medial-lemniscal system, therefore, is involved not only in conscious awareness of proprioception but also unconscious neuromuscular functions.

PREDICT 4

Most of the neurons from the fasciculus gracilis synapse in the inferior thorax and enter the spinocerebellar system, whereas most of the neurons from the fasciculus cuneatus synapse in the nucleus cuneatus and then continue to the thalamus and cerebrum. It can therefore be deduced that most of the proprioception from the lower limbs is unconscious and most of the proprioception from the upper limbs is conscious. Explain why this difference in the two sets of limbs is of value.

The **spinoolivary tracts** project to the accessory olivary nucleus and to the cerebellum, where action potentials carried by these tracts contribute to the coordination of movement associated primarily with balance. The **spinotectal** (spī-nō-tek′tăl) **tracts** end in the superior colliculi of the midbrain and transmit action potentials involved in reflexes that turn the head and eyes toward a point of cutaneous stimulation.

Descending Pathways Modifying Sensation

The corticospinal (see p. 493) and other descending tracts send collateral branches to the thalamus, reticular formation, trigeminal nuclei, and spinal cord. Neuromodulators, such as endorphins and enkephalin, released from axons originating in these CNS regions decrease the frequency of action potentials in sensory tracts (see the discussion of presynaptic inhibition in chapter 11). Through this route, the cerebral cortex or other brain regions may reduce the conscious perception of sensations.

5 ***What are the functions of the anterolateral and dorsal-column/medial-lemniscal systems? Describe where the neurons of these systems cross over and synapse.***

6 ***What kind of information is carried in the spinocerebellar tracts? Where do the anterior and posterior spinocerebellar tracts originate? Do these tracts terminate on the same or opposite side of the body from where they originate?***

7 ***What are the functions of the spinoolivary and spinomesencephalic tracts?***

8 ***How do descending pathways modulate sensation?***

Sensory Areas of the Cerebral Cortex

Figure 14.11 depicts a lateral view of the left cerebral cortex with some of its functional areas labeled. Sensory pathways project to specific regions of the cerebral cortex, called **primary sensory areas,** where these sensations are perceived.

Most of the postcentral gyrus is called the **primary somatic sensory cortex,** or **general sensory area.** The terms *area* and *cortex* are often used interchangeably for the same functional region of the cerebral cortex. Fibers carrying general sensory input, such as pain, pressure, and temperature, synapse in the thalamus, and thalamic neurons relay the information to the primary somatic sensory cortex.

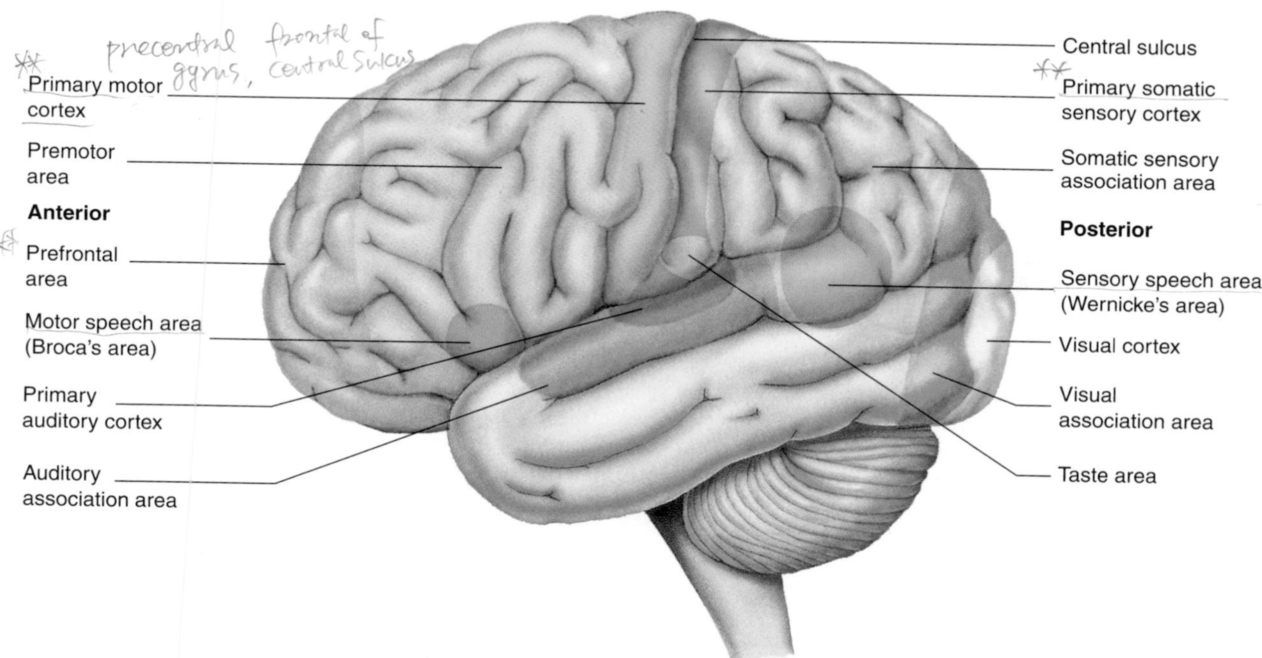

FIGURE 14.11 Functional Regions of the Lateral Side of the Left Cerebral Cortex

The somatic sensory cortex is organized topographically relative to the general plan of the body (figure 14.12). Sensory impulses conducting input from the feet project to the most superior portion of the somatic sensory cortex, and sensory impulses from the face project to the most inferior portion. The pattern of the somatic sensory cortex in each hemisphere is arranged in the form of an upside-down half homunculus (hō-mŭngk′ū-lŭs; little human) representing the opposite side of the body, with the feet located superiorly and the head located inferiorly. The size of various regions of the somatic sensory cortex is related to the number of sensory receptors in that area of the body. The density of sensory receptors is much greater in the face than in the legs; therefore, a greater area of the somatic sensory cortex contains sensory neurons associated with the face, and the homunculus has a disproportionately large face.

There are other primary sensory areas of the cerebral cortex (see figure 14.11). The **taste area,** where taste sensations are consciously perceived in the cortex, is located at the inferior end of the postcentral gyrus. The **olfactory cortex** (not shown in figure 14.11) is on the inferior surface of the frontal lobe and is where both conscious and unconscious responses to odor are initiated (see chapter 15). The **primary auditory cortex,** where auditory stimuli are processed by the brain, is located in the superior part of the temporal lobe. The **visual cortex,** where portions of visual images are processed, is located in the occipital lobe. In the visual cortex, color, shape, and movement are processed separately rather than as a complete "color motion picture." These sensory areas are discussed more fully in chapter 15.

The primary sensory areas of the cerebral cortex must be intact for conscious perception, localization, and identification of a stimulus. Cutaneous sensations, although integrated within the cerebrum, are perceived as though they were on the surface of the body. This is called **projection** and indicates that the brain refers a cutaneous sensation to the superficial site at which the stimulus interacts with the sensory receptors.

PREDICT 5

Using the visual association areas as an example, explain the general functions of the association areas around the other primary cortical areas (see figure 14.11).

Sensory Processing

Cortical areas immediately adjacent to the primary sensory centers, called **association areas,** are involved in the process of recognition. The **somatic sensory association area** is posterior to the primary somatic sensory cortex, and the **visual association area** is anterior to the visual cortex (see figure 14.11). Sensory action potentials originating in the retina of the eye reach the visual cortex, where the image is "perceived." Action potentials then pass from the visual cortex to the visual association area, where the present visual information is compared with past visual experience ("Have I seen this before?"). On the basis of this comparison, the visual association area "decides" whether or not the visual input is recognized and passes judgment concerning the significance of the input. For

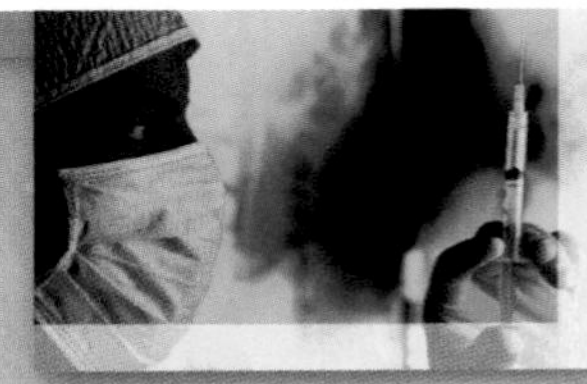

CLINICAL FOCUS

Pain

Pain is a sensation characterized by a group of unpleasant and complex perceptual and emotional experiences that trigger autonomic, psychologic, and somatic motor responses. Pain sensation has two components: (1) rapidly conducted action potentials carried by large-diameter, myelinated axons, resulting in sharp, well-localized, pricking, or cutting pain, followed by (2) more slowly propagated action potentials, carried by smaller, less heavily myelinated axons, resulting in diffuse burning or aching pain. Research indicates that pain receptors have very uniform sensitivity, which does not change dramatically from one instant to another. Variations in pain sensation result from the mechanisms by which pain receptors are stimulated, differences in the integration of action potentials from the pain receptors, and complex interactions in the cerebral cortex, cingulate gyrus, and thalamus, where the emotional component of pain is registered. Neurons in the cerebral cortex respond to pain stimuli selectively based on prior experience and context. Stress, for example, can reduce pain perception.

Although the dorsal-column/medial-lemniscal system contains no pain fibers, tactile and mechanoreceptors are often activated by the same stimuli that affect pain receptors. Action potentials from tactile receptors provide information that allows the pain sensation to be localized. Superficial pain is highly localized, in part, because of the simultaneous stimulation of pain receptors and mechanoreceptors in the skin. Deep or visceral pain is not highly localized (diffuse) because of fewer mechanoreceptors in the deeper structures, and it is normally perceived as a diffuse pain.

Dorsal-column/medial-lemniscal system neurons are involved in what is called the **gate-control theory** of pain control. Primary neurons of the dorsal-column/medial-lemniscal system send out collateral branches that synapse with interneurons in the posterior horn of the spinal cord. These interneurons have an inhibitory effect on secondary neurons of the spinothalamic tract. Thus, pain action potentials traveling through the spinothalamic tract can be suppressed by action potentials that originate in neurons of the dorsal-column/medial-lemniscal system. The arrangement may act as a "gate" for pain action potentials transmitted in the spinothalamic tract. Increased activity in the dorsal-column/medial-lemniscal system tends to close the gate, thereby reducing pain action potentials transmitted in the spinothalamic tract. Descending pathways from the cerebral cortex or other brain regions can also regulate this "gate."

The gate-control theory may explain the physiologic basis for the following methods that have been used to reduce the intensity of chronic pain: electric stimulation of the dorsal-column/medial-lemniscal neurons, transcutaneous electric stimulation (applying a weak electric stimulus to the skin), acupuncture, massage, and exercise. The frequency of action potentials that are transmitted in the dorsal-column/medial-lemniscal system is increased when the skin is rubbed vigorously and when the limbs are moved and may explain why vigorously rubbing a large area around a source of pricking pain tends to reduce the intensity of the painful sensation. Exercise normally decreases the sensation of pain, and exercise programs are important components in the management of chronic pain not associated with illness. Action potentials initiated by acupuncture procedures may act through a gating mechanism in which inhibition of action potentials in neurons that transmit pain action potentials upward in the spinal cord is influenced by activity in sensory cells that send collateral branches to the posterior horn.

Analgesics act in much the same way as gate control. Some analgesics block the transmission of nociception in the spinal cord from primary neurons to neurons of the ascending pathways. Other analgesics function at the level of the cerebral cortex to modulate pain.

REFERRED PAIN

Referred pain is a painful sensation in a region of the body that is not the source of the pain stimulus. Most commonly, referred pain is sensed in the skin or other superficial structures when internal organs are damaged or inflamed. This sensation usually occurs because both the area of skin to which the pain is referred and the visceral area that is damaged are innervated by neurons that project to the same area of the cerebral cortex. The brain cannot distinguish between the two sources of painful stimuli, and the painful sensation is referred to the most superficial structures innervated by the converging neurons. This referral may occur because the number of receptors is much greater in superficial structures than in deep structures and the brain is more "accustomed" to dealing with superficial stimuli.

Referred pain is clinically useful in diagnosing the actual cause of a painful stimulus. Heart attack victims often feel cutaneous pain radiating from the left shoulder down the arm. Other examples of referred pain are shown in figure A.

PHANTOM PAIN

Phantom pain occurs in people who have had appendages amputated or a structure such as a tooth removed. Many of these people perceive pain, which can be intense, or other sensations, in the amputated structure as if it were still in place. If a neuron pathway that transmits action potentials is stimulated at any point along that pathway, action potentials are initiated and propagated toward the CNS. Integration results in the perception of pain that is projected to the site of the sensory receptors, even if those sensory receptors are no longer present. A similar phenomenon can be easily demonstrated by bumping the ulnar nerve as it crosses the elbow (the funny bone). A sensation of pain is often felt in the fourth and fifth digits, even though the neurons were stimulated at the elbow.

A factor that may be important in phantom pain is the lack of touch, pressure, and proprioceptive impulses from the amputated limb. Those action potentials suppress the transmission of pain action potentials in the pain pathways, as explained by the gate-control theory of pain. When a limb is amputated, the inhibitory effect of sensory information is removed. As a consequence, the intensity of phantom pain may be increased. Another factor in phantom

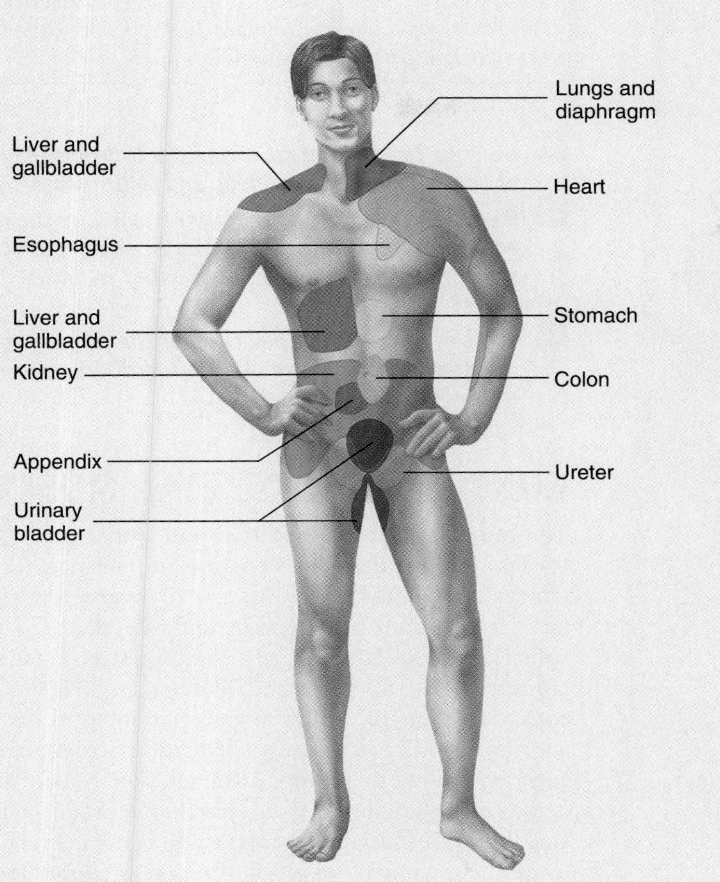

FIGURE A Areas of Referred Pain on the Body Surface
Pain from the indicated internal organs is referred to the surface areas shown.

pain may be that the cerebral cortex retains an image of the amputated body part.

CHRONIC PAIN

Chronic pain is long-lasting. Some chronic pain has a known cause, such as tissue damage, as in the case of arthritis. Other chronic pain cannot be associated with tissue damage and has no known cause.

Pain is important in warning us of potentially injurious conditions because pain receptors are stimulated when tissues are injured. Pain itself, however, can become a problem. Chronic pain, such as migraine headaches, localized facial pain, or back pain, can be very debilitating, and pain loses its value of providing information about the condition of the body. People suffering from chronic pain often feel helpless and hopeless, and they may become dependent on drugs. The pain can interfere with vocational pursuits, and many victims are unemployed or even housebound and socially isolated. They are easily frustrated or angered, and they suffer symptoms of major depression. These qualities are associated with what is called **chronic pain syndrome.** Over 2 million people in the United States at any given time suffer chronic pain sufficient to impair activity.

Chronic pain may originate with acute pain associated with an injury or may develop for no apparent reason. How sensory signals are processed in the thalamus and cerebrum may determine if the input is evaluated as only a discomfort, a minor pain, or a severe pain and how much distress is associated with the sensation. The brain actively regulates the amount of pain information that gets through to the level of perception, thereby suppressing much of the input. If this dampening system becomes less functional, pain perception may increase. Other nervous system factors, such as a loss of some sensory modalities from an area, or habituation of pain transmission, which may remain even after the stimulus is removed, may actually intensify otherwise normal pain sensations. Treatment often requires a multidisciplinary approach, including such interventions as surgery or psychotherapy. Some sufferers respond well to drug therapy, but some drugs, such as opiates, have a diminishing effect and may become addictive.

SENSITIZATION IN CHRONIC PAIN

Tissue damage within an area of injury, such as the skin, can cause an increase in the sensitivity of nerve endings in the area of damage, a condition called **peripheral sensitization.** One class of pain receptors is not activated by traditional noxious stimuli but is recruited only when tissues become inflamed. These receptors, once activated, add to the total barrage of sensory signals to the brain and intensify the sensation of pain.

The CNS may also respond to tissue damage by decreasing its threshold and increasing its sensitivity to pain. This condition is called **central sensitization.** Central sensitization apparently results from a specific subset of receptors that is only recruited during repetitive neuron firing, such as when intense pain sensations are experienced. These receptors maintain a hyperexcitable state in the CNS cells. This chronic, hyperexcitable state can result in persistent, chronic pain states.

This information concerning peripheral and central sensitization and the knowledge that sensitization involves neuronal and chemical receptors not normally involved in sensation may lead to the discovery of new drugs for treating chronic pain. Rather than searching for new analgesics, which may decrease a broad range of sensations, an opportunity is now available to develop a new class of drugs, the "antihyperanalgesics," which may block sensitization without diminishing other sensations, including normal pain.

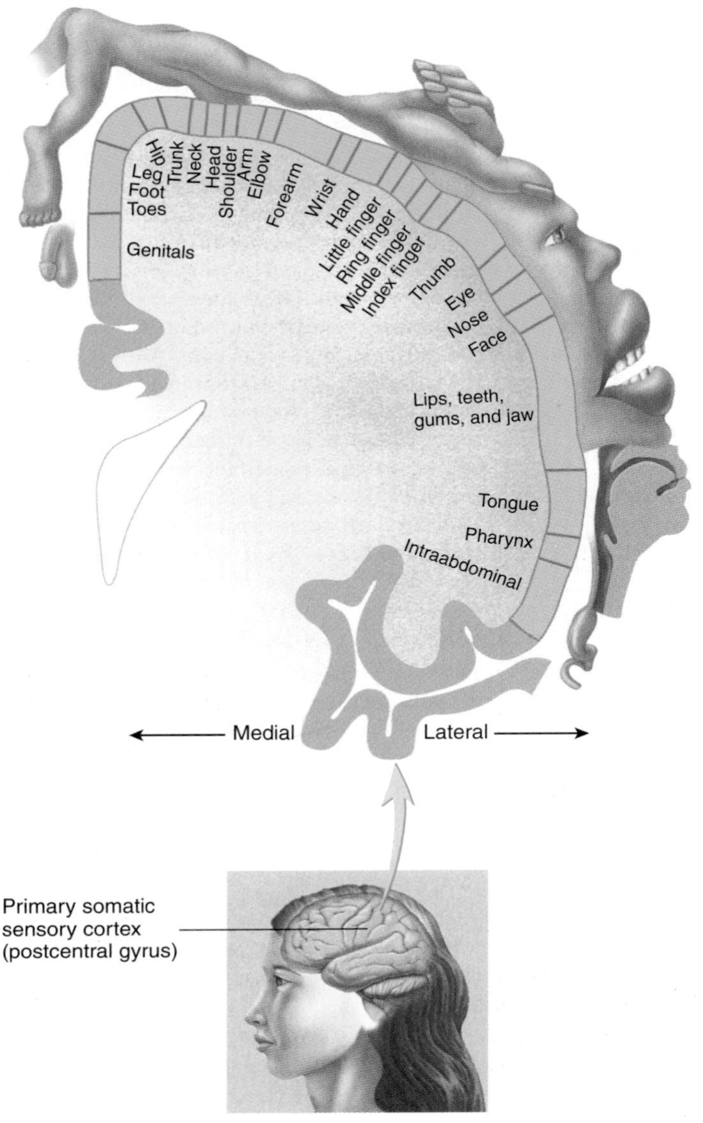

FIGURE 14.12 Topography of the Somatic Sensory Cortex

Cerebral cortex seen in coronal section on the left side of the brain. The figure of the body (homunculus) depicts the nerve distributions; the size of each body region shown indicates relative innervation. The cortex occurs on both sides of the brain but appears on only one side in this illustration. The inset shows the somatic sensory region of the left hemisphere (*green*).

example, we generally pay less attention to people in a crowd we have never seen before than to someone we know.

The visual association area, like other association areas of the cortex, has reciprocal connections with other parts of the cortex that influence decisions. For example, the visual association area has input from the frontal lobe, where emotional value is placed on the visual input. Because of these numerous connections, visual information is judged several times as it passes beyond the visual association area. This may be one of the reasons that, when two people view the same painting, their individual perceptions of it can be vastly different.

PREDICT 6

A man has constipation, which causes distention and painful cramping in his colon. What kind of pain does he experience (local or diffuse) and where is it perceived? Explain.

9 ***Describe in the cerebral cortex the locations and functions of the primary sensory areas and their association areas.***

10 ***Describe the topographic organization of the general body plan in the primary somatic sensory cortex. Why are some areas of the body represented as larger than other areas?***

11 ***What are the related functions of the primary motor area, the premotor area, and the prefrontal area of the cerebral cortex?***

CONTROL OF SKELETAL MUSCLES

The motor system of the brain and spinal cord is responsible for maintaining the body's posture and balance; for moving the trunk, head, limbs, and eyes; and for communicating through facial expressions and speech. Reflexes mediated through the spinal cord (see chapter 12) and brainstem (see chapter 13) are responsible for some body movements. They occur without conscious thought. **Voluntary movements,** on the other hand, are movements consciously activated to achieve a specific goal, such as walking or typing. Although consciously activated, the details of most voluntary movements occur automatically. After walking begins, it is not necessary to think about the moment-to-moment control of every muscle because neural circuits automatically control the limbs. After learning how to do complex tasks, such as typing, people can perform them relatively automatically.

Voluntary movements depend on upper and lower motor neurons. **Upper motor neurons** directly or through interneurons connect to lower motor neurons. The cell bodies of upper motor neurons are in the cerebral cortex. **Lower motor neurons** have axons that leave the central nervous system and extend through peripheral nerves to supply skeletal muscles. The cell bodies of lower motor neurons are located in the anterior horns of the spinal cord gray matter and in cranial nerve nuclei of the brainstem.

Voluntary movements depend on the following:

1. The initiation of most voluntary movement begins in the premotor areas of the cerebral cortex and involves stimulation of upper motor neurons (see "Motor Areas of the Cerebral Cortex," p. 491).
2. The axons of the upper motor neurons form the descending tracts. They stimulate lower motor neurons, which stimulate skeletal muscles to contract (see "Motor Tracts," p. 491).
3. The cerebral cortex interacts with the basal nuclei and cerebellum in the planning, coordination, and execution of movements (see "Modifying and Refining Motor Activities," p. 496).

12 ***Distinguish between upper and lower motor neurons.***

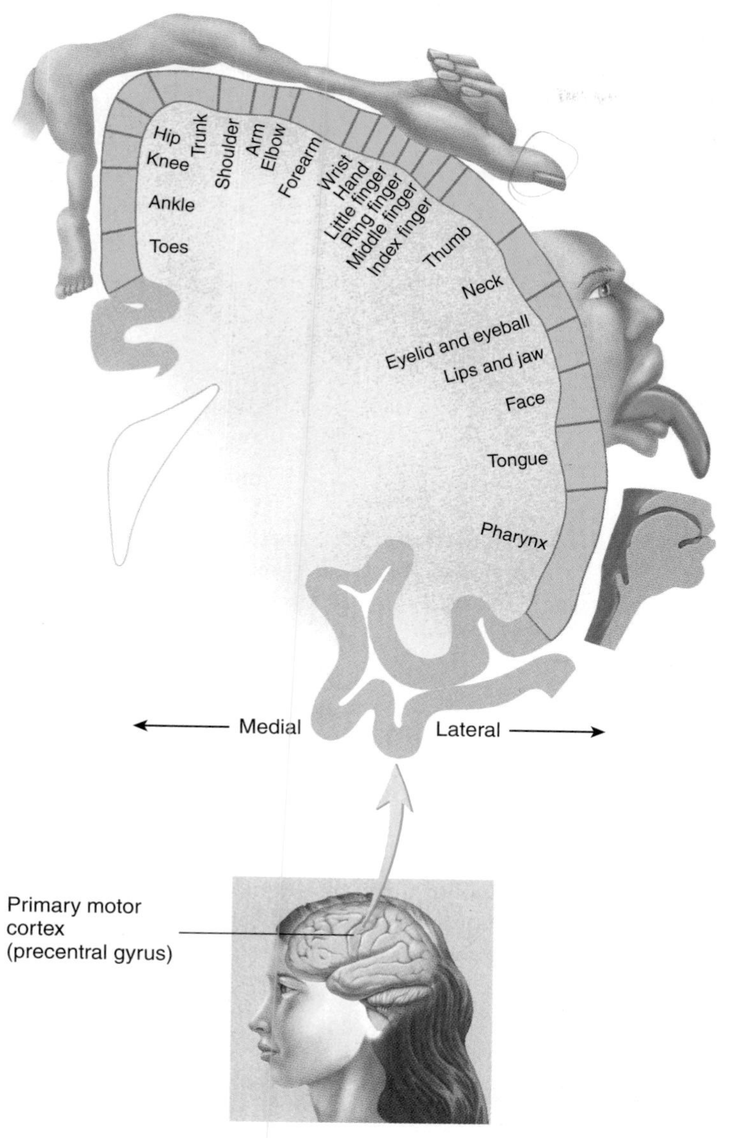

FIGURE 14.13 Topography of the Primary Motor Cortex

Cerebral cortex seen in coronal section on the left side of the brain. The figure of the body (homunculus) depicts the nerve distributions; the size of each body region shown indicates relative innervation. The cortex occurs on both sides of the brain but appears on only one side in this illustration. The inset shows the motor region of the left hemisphere (*pink*).

Motor Areas of the Cerebral Cortex

The **precentral gyrus** is also called the **primary motor cortex** or **primary motor area** (see figure 14.11). Action potentials initiated in this region control many voluntary movements, especially the fine motor movements of the hands. Upper motor neurons are not confined to the precentral gyrus—only about 30% of them are located there. Another 30% are in the premotor area, and the rest are in the somatic sensory cortex.

The cortical functions of the precentral gyrus are arranged topographically according to the general plan of the body—similar to the topographic arrangement of the postcentral gyrus (figure 14.13). The neuron cell bodies controlling motor functions of the feet are in the most superior and medial portions of the precentral gyrus, whereas those for the face are in the inferior region. Muscle groups with many motor units are represented by relatively large areas of the precentral gyrus. For example, muscles performing precise movements, such as those controlling the hands and face, have many motor units, each of which has a small number of muscle fibers. Multiple-motor unit summation (see chapter 9) can precisely control the force of contraction of these muscles because only a few muscle fibers at a time are recruited. Muscle groups with few motor units are represented by relatively small areas of the precentral gyrus, even if the muscles innervated are quite large. Muscles such as those controlling movements of the thigh and leg have proportionately fewer motor units than hand muscles but have many more and much larger muscle fibers per motor unit. They are less precisely controlled because the activation of a motor unit stimulates the contraction of many large muscle fibers.

The **premotor area,** located anterior to the primary motor cortex (see figure 14.11), is the staging area in which motor functions are organized before they are initiated in the motor cortex. For example, if a person decides to take a step, the neurons of the premotor area are stimulated first. The determination is made in the premotor area as to which muscles must contract, in what order, and to what degree. Action potentials are then passed to the upper motor neurons in the motor cortex, which actually initiate the planned movements.

Apraxia

The premotor area must be intact for a person to carry out complex, skilled, or learned movements, especially ones related to manual dexterity—for example, a surgeon's use of a scalpel or a student's use of a pencil. Impairment in the performance of learned movements, called **apraxia** (ă-prak′sē-ă), can result from a lesion in the premotor area. Apraxia is characterized by hesitancy and reduced dexterity in performing these movements.

The motivation and foresight to plan and initiate movements occur in the next most anterior portion of the brain, the **prefrontal area,** an association area that is well developed only in primates and especially in humans. It is involved in the motivation and regulation of emotional behavior and mood. The large size of this area of the brain in humans may account for their relatively well-developed forethought and motivation and for the emotional complexity of humans.

13 ***Where are the primary motor, premotor, and prefrontal areas of the cerebral cortex located?***

14 ***Why are some areas of the body represented as larger than other areas on the topographic map of the body on the primary motor areas?***

Motor Tracts

Motor tracts are descending pathways containing axons that carry action potentials from regions of the cerebrum or cellebellum to the brainstem or spinal cord. The names of descending tracts are based on their origin and termination (figure 14.14 and table 14.4). Much like the names of ascending tracts, the prefix indicates its origin and the suffix indicates its destination. For example, the corticospinal tract is a motor tract that originates in the cerebral cortex and terminates in the spinal cord.

Amyotrophic Lateral Sclerosis

Amyotrophic (ă-mī-ō-trō′fik) **lateral sclerosis (ALS),** also called Lou Gehrig's disease, usually affects people between the ages of 40 and 70. About 10% of the cases of ALS are inherited. It begins with weakness and clumsiness and progresses within 2–5 years to loss of muscle control. The disease selectively destroys both upper and lower motor neurons. The inherited form of ALS apparently results from a mutation in DNA coding for the enzyme **superoxide dismutase (SOD)** and is located on chromosome 21. SOD is involved in eliminating a free radical, known as superoxide, from the body. Free radicals are molecules that readily accept electrons, which makes them highly reactive. They can strip electrons from proteins, lipids, or nucleic acids, thereby destroying their functions and resulting in cell dysfunction or death. Free-radical damage has been implicated in ALS, arteriosclerosis, arthritis, cancer, and aging. Superoxide is one of the most important and toxic free radicals. It forms as the result of oxygen reacting with other free radicals. Although oxygen is critical for aerobic metabolism, it is also dangerous to tissues. SOD catalyzes the conversion of superoxide to hydrogen peroxide, which is then converted by catalase to oxygen and water. Apparently, if SOD is defective, superoxide is not degraded and can destroy cells. Motor neurons appear to be particularly sensitive to superoxide attack.

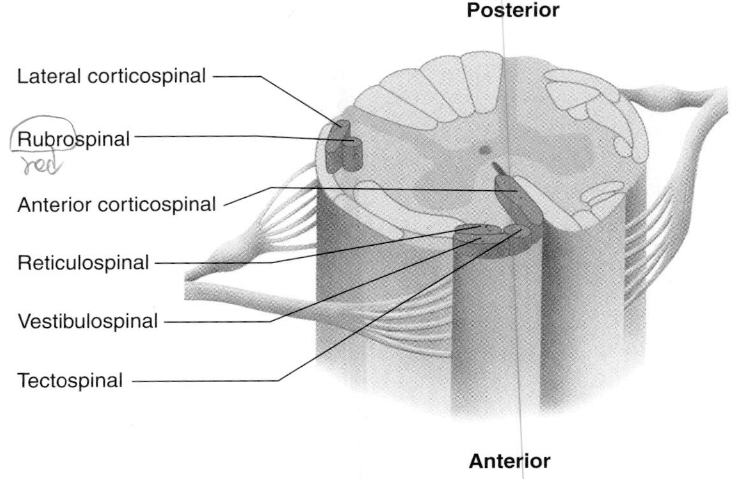

FIGURE 14.14 Cross Section of the Spinal Cord at the Cervical Level Depicting the Descending Pathways

Descending pathways are labeled on the left side of the figure only (*pink*), though they exist on both sides.

The descending motor fibers are divided into two groups: direct pathways and indirect pathways (figure 14.15). The **direct pathways,** also called the **pyramidal** (pi-ram′i-dal) **system,** are involved in the maintenance of muscle tone and in controlling the speed and precision of skilled movements, primarily fine movements involved in dexterity. Most of the **indirect pathways,** sometimes called the **extrapyramidal system,** are involved in less precise control of motor functions, especially those associated with overall body coordination and cerebellar function, such as posture. Many of the indirect pathways are phylogenetically older

TABLE 14.4 Descending Spinal Pathways

Pathway	Functions Controlled	Examples of Movements Controlled	Origin
Direct	Conscious skilled movements		
Corticospinal tract	Movements below the head, especially of the hands		
Lateral	Movements of the neck, trunk, and limbs	Push-ups, moving with a hula hoop	Cerebral cortex
Anterior	Movements of the neck and upper limb extremities	Typing	Cerebral cortex
Corticobulbar tract	Movements of the head and face	Facial expression, chewing	Cerebral cortex
Indirect	Unconscious movements		
Rubrospinal	Movement coordination	Positioning of digits and palm of the hand when reaching out to grasp	Red nucleus
Vestibulospinal	Maintenance of upright posture, balance	Extension of upper limbs when falling	Vestibular nucleus
Reticulospinal	Posture adjustment, walking	Maintenance of posture when standing on one foot	Reticular formation
Tectospinal	Movements of head and neck in response to visual and auditory reflexes	Movement of head and neck away from a sudden flash of light	Superior colliculus

and control more "primitive" movements of the trunk and proximal portions of the limbs. The direct pathways, which exist only in mammals, may be thought of as overlying the indirect pathways and are more involved in finely controlled movements of the face and distal portions of the limbs. Some indirect pathways, such as those from the basal nuclei and cerebellum, help in fine control of the direct pathways.

Direct Pathways

Direct pathways are so-named because upper motor neurons in the cerebral cortex, whose axons form these pathways, synapse directly with lower motor neurons in the brainstem or spinal cord. They are also called the pyramidal system because the fibers of these pathways form the medullary **pyramids.** The direct pathways include groups of nerve fibers arrayed into two tracts: the **corticospinal tract,** which is involved in direct cortical control of movements below the head, and the **corticobulbar tract,** which is involved in direct cortical control of movements in the head and neck.

The corticospinal tract consists of axons of upper motor neurons located in the primary motor and premotor areas of the frontal lobes and the somatic sensory parts of the parietal lobes. They descend through the **internal capsules** and the cerebral peduncles of the midbrain to the pyramids of the medulla oblongata. At the inferior end of the medulla 75%–85% of the corticospinal fibers cross to the opposite side of the CNS through the **pyramidal decussation,** which is visible on the anterior surface of the inferior medulla. The crossed fibers descend in the **lateral corticospinal tracts** of the spinal cord (figure 14.16). The remaining 15%–25% descend uncrossed in the **anterior corticospinal tracts** and decussate near the level where they synapse with lower motor neurons. The anterior corticospinal tracts supply the neck and upper limbs, and the lateral corticospinal tracts supply all levels of the body (see table 14.4).

Most of the corticospinal fibers synapse with interneurons in the lateral portions of the spinal cord central gray matter. The interneurons, in turn, synapse with the lower motor neurons of the anterior horn that innervate primarily distal limb muscles.

Damage to the corticospinal tracts results in reduced muscle tone, clumsiness, and weakness but not in complete paralysis, even if the damage is bilateral. Experiments with monkeys have demonstrated that bilateral sectioning of the medullary pyramids results in (1) loss of contact-related activities, such as tactile placing of the foot and grasping, (2) defective fine movements, and (3) hypotonia (reduced tone). These and other experimental data support the conclusion that the corticospinal system is superimposed over the older, indirect pathways and that it has many parallel functions. The main function of the direct pathways is to add speed and agility to conscious movements, especially of the hands, and to provide a high degree of fine motor control, such as in movements of individual fingers. Spinal cord lesions that affect both the direct and indirect pathways result in complete paralysis.

The corticobulbar tracts are analogous to the corticospinal tracts. The former innervate the head, and the latter innervate the rest of the body. Cells that contribute to the corticobulbar tracts are in regions of the cortex similar to those of the corticospinal tracts, except that they are more laterally and inferiorly located. Corticobulbar tracts follow the same basic route as the corticospinal system down to the level of the brainstem. At that point, most corticobulbar fibers terminate in the reticular formation near the **cranial nerve nuclei.** Interneurons from the reticular formation then enter the cranial nerve nuclei, where they synapse with lower motor neurons. These nuclei give rise to the nerves that control

Crossover	Termination	Side of Body Where Fibers Terminate
Inferior end of medulla oblongata	Anterior horn of spinal cord	Contralateral
Level of lower motor neuron	Anterior horn of spinal cord	Contralateral
Varies for the different cranial nerves	Cranial nerve nuclei in brainstem (lower motor neuron)	Contralateral
Midbrain	Anterior horn of spinal cord	Contralateral
Uncrossed	Anterior horn of spinal cord	Ipsilateral
Some uncrossed; some cross at termination	Anterior horn of spinal cord	Ipsilateral or contralateral
Midbrain	Cranial nerve nucleus in medulla oblongata and anterior horn of upper levels of spinal cord (lower motor neurons that turn head and neck)	Contralateral

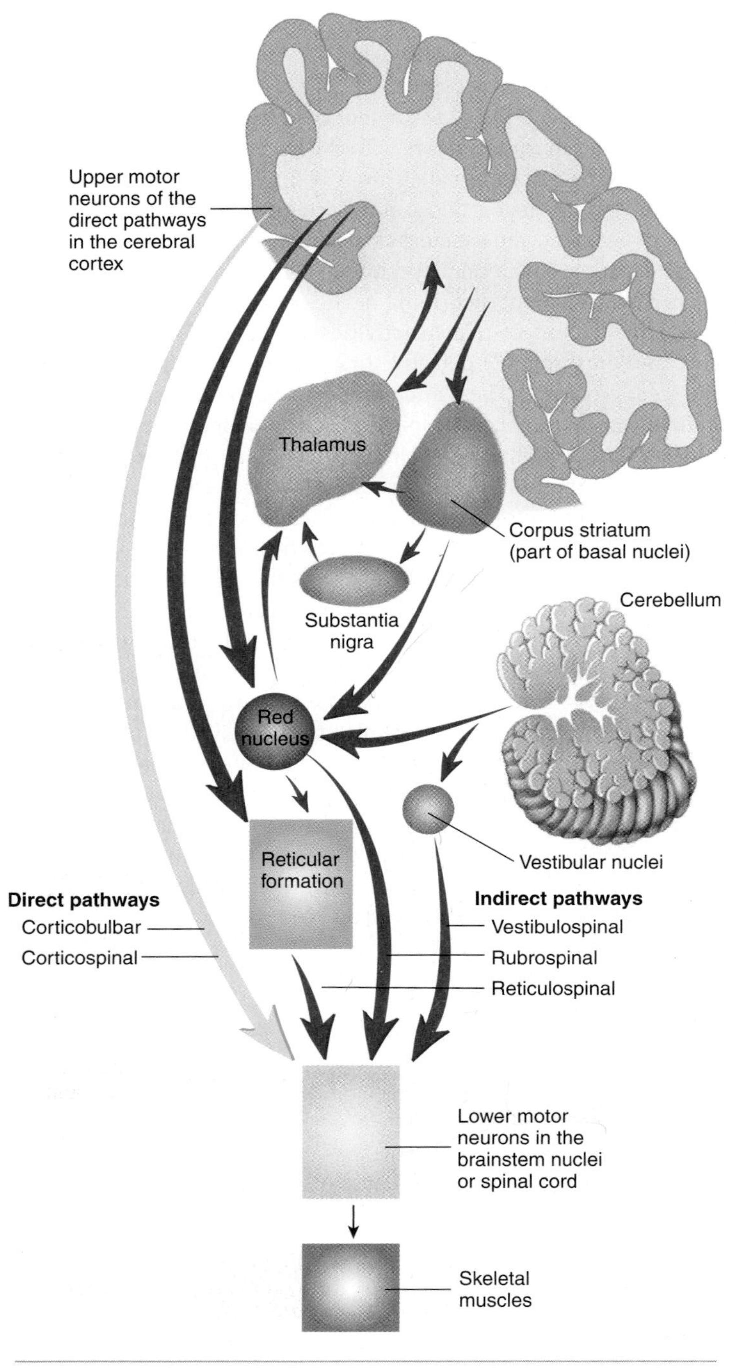

FIGURE 14.15 Descending Pathways

The direct pathways (corticobulbar and corticospinal) are indicated by the *blue arrow.* The indirect pathways and their interconnections are indicated by the *red arrows.*

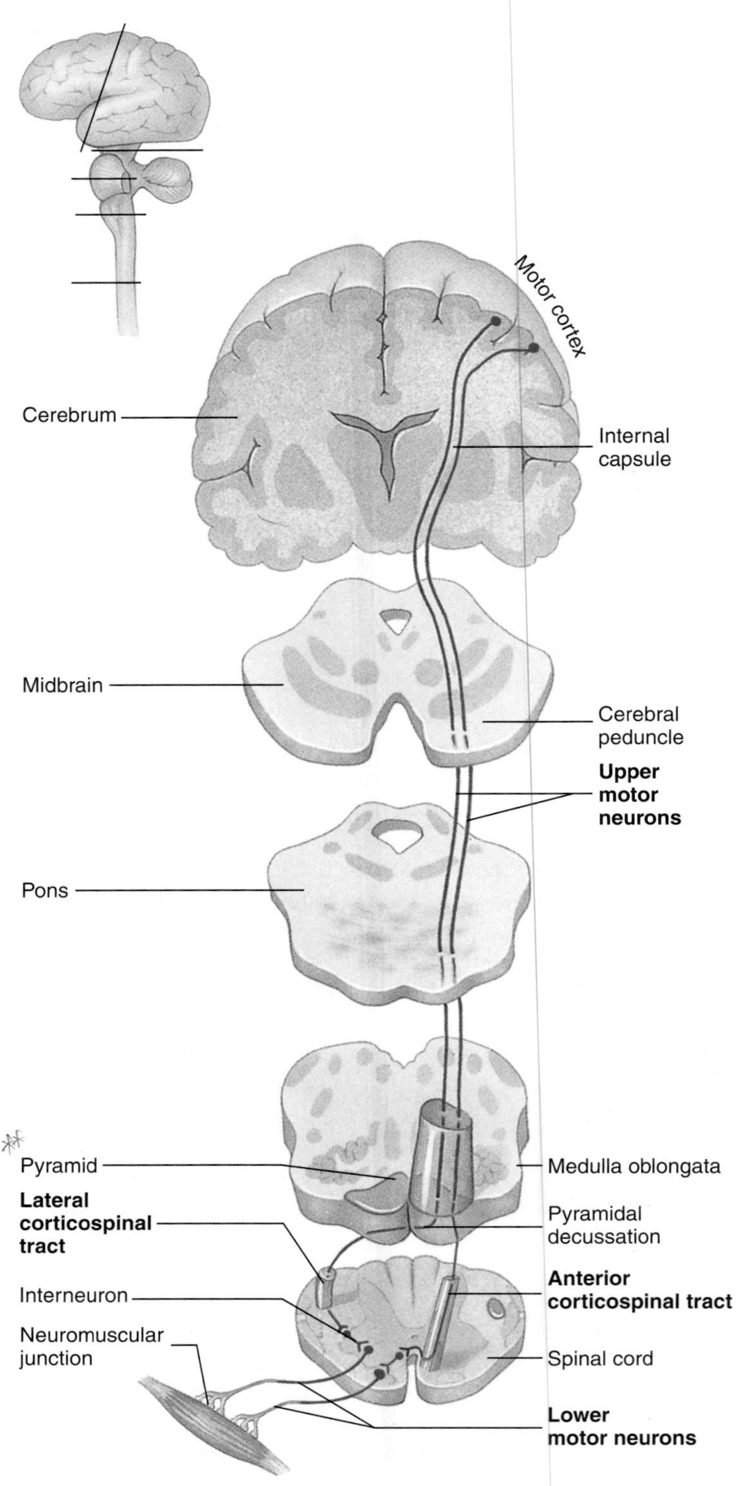

FIGURE 14.16 Direct Pathways

Lateral and anterior corticospinal tract, which is responsible for movement below the head. Lines on the inset indicate levels of section.

eye and tongue movements, mastication, facial expression, and palatine, pharyngeal, and laryngeal movements.

Brown-Séquard Syndrome

A lesion of the spinal cord that destroys half the cord at a specific level (hemisection) results in a very specific syndrome, called the **Brown-Séquard syndrome.** The syndrome includes contralateral loss of pain and temperature because fibers entering the anterolateral system cross over where they enter the spinal cord, as well as ipsilateral loss of proprioception, two-point discrimination, and most upper motor control because the dorsal-column/medial-lemniscal system ascends ipsilaterally and the lateral corticospinal tract descends ipsilaterally.

15 ***What two tracts form the direct pathways? What area of the body is supplied by each tract? Describe the location of the neurons in each tract, where they cross over, and where they synapse.***

Indirect Pathways

The indirect pathways (figure 14.17) originate in neurons of the cerebrum and cerebellum whose axons synapse in some intermediate nucleus rather than directly with lower motor neurons. Axons from the neurons in those nuclei form the indirect pathways. They do not pass through the pyramids or through the corticobulbar tracts and, therefore, are sometimes called extrapyramidal. The major tracts are the rubrospinal, vestibulospinal, and reticulospinal tracts. Many interconnections and feedback loops are present in this system.

Neurons of the **rubrospinal tract** begin in the red nucleus, which is located at the boundary between the diencephalon and midbrain. The tract decussates in the midbrain and descends in the lateral column of the spinal cord. The red nucleus receives input from both the motor cortex and the cerebellum. Lesions in the red nucleus result in intention, or action, tremors similar to those seen in cerebellar lesions (see the Clinical Focus "Dyskinesias," p. 496). The function of the red nucleus, therefore, is related closely to cerebellar function. The rubrospinal tract is the one indirect tract that is very closely related to the direct, corticospinal tract. It terminates in the lateral portion of the spinal cord central gray matter with the corticospinal tract, and it transmits action potentials involved in the comparator function of the cerebellum (see p. 497). It plays a major role in regulating fine motor control of muscles in the distal part of the upper limbs. Damage to the rubrospinal tract impairs forearm and hand movements but does not greatly affect general body movements.

The **vestibulospinal tracts** (see figure 14.14) originate in the vestibular nuclei, descend in the anterior column, and synapse with interneurons and lower motor neurons in the ventromedial portion of the spinal cord central gray matter. Their fibers preferentially influence neurons innervating extensor muscles in the trunk and proximal portion of the lower limbs and are involved primarily in the maintenance of upright posture. The vestibular nuclei receive major input from the vestibular nerve (see chapter 15) and the cerebellum.

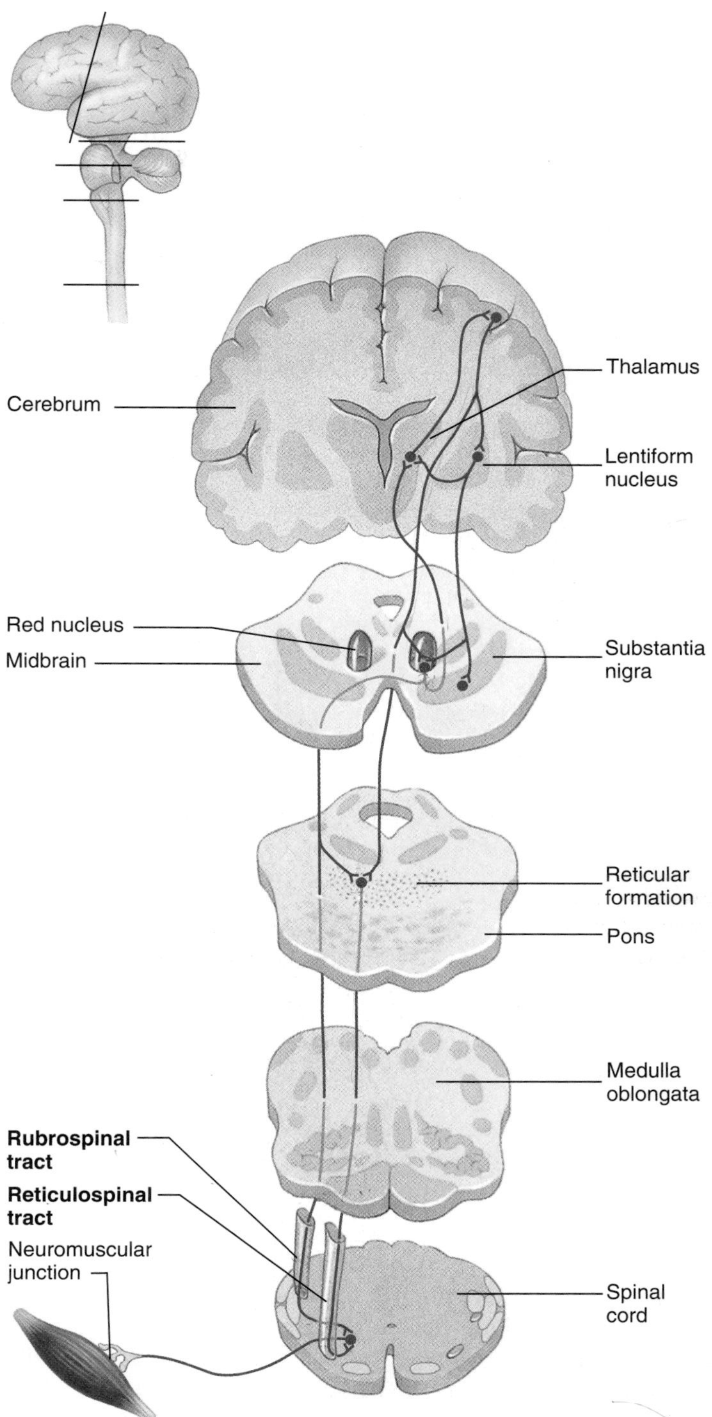

FIGURE 14.17 Indirect Pathways

Examples of indirect pathways are rubrospinal and reticulospinal tracts. Lines on the inset indicate levels of section.

Neuron cell bodies of the **reticulospinal tract** (see figure 14.14) are in the reticular formation of the pons and medulla oblongata. Their axons descend in the anterior portion of the lateral column and synapse with interneurons and lower motor neurons in the ventromedial portion of the spinal cord central gray matter. This

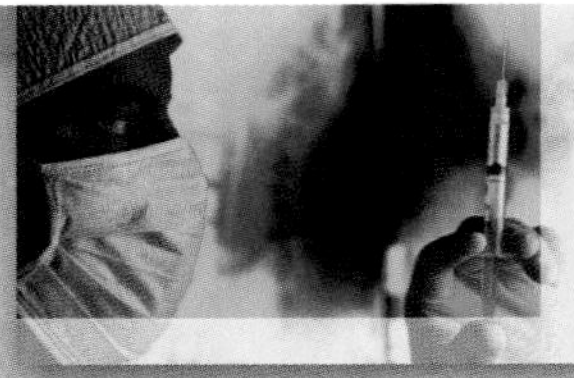

CLINICAL FOCUS

Dyskinesias

Dyskinesias (dis-ki-nē′zē-ăs) are a group of disorders often involving the basal nuclei in which unwanted, superfluous movements occur. Defects in the basal nuclei may result in brisk, jerky, purposeless movements that resemble fragments of voluntary movements. **Sydenham chorea** (kōr-ē′ă; also called St. Vitus dance) is a disease usually associated with a toxic or an infectious disorder that apparently causes temporary dysfunction of the corpus striatum. It usually affects children. **Huntington chorea** is a dominant hereditary disorder that begins in middle life, causing mental deterioration and progressive degeneration of the corpus striatum in affected individuals.

Cerebral palsy (pawl′zē) is a general term referring to defects in motor functions or coordination resulting from several types of brain damage, which may be caused by abnormal brain development or birth-related injury. Some symptoms of cerebral palsy, such as increased muscle tension, are related to basal nuclei dysfunction. **Athetosis** (ath-ĕ-tō′sis), often one of the features of cerebral palsy, is characterized by slow, sinuous, aimless movements. When the face, neck, and tongue muscles are involved, grimacing, protrusion, and writhing of the tongue and difficulty in speaking and swallowing are characteristics.

Damage to the subthalamic nucleus can result in **hemiballismus** (hem-ē-bal-iz′mŭs), an uncontrolled, purposeless, and forceful throwing or flailing of the arm. Forceful twitching of the face and neck may also result from subthalamic nuclear damage.

Parkinson disease—characterized by muscular rigidity; loss of facial expression; tremor; a slow, shuffling gait; and general lack of movement—is caused by a dysfunction in the substantia nigra. The disease usually occurs after age 55 and is not contagious or inherited. A resting tremor, called "pill-rolling," is characteristic of Parkinson disease; it consists of circular movement of the opposed thumb and index fingertips. The increased muscular rigidity in Parkinson disease results from defective inhibition of some of the basal nuclei by the substantia nigra. In this disease, dopamine, an inhibitory neurotransmitter produced by the substantia nigra is deficient. The melanin-containing cells of the substantia nigra degenerate, resulting in a loss of pigment.

Parkinson disease can be treated with levodopa (lē-vō-dō′pă, L-dopa), a precursor to dopamine, or more effectively with Sinemet, a combination of L-dopa and carbidopa (kar-bi-dō′pă). Carbidopa is a decarboxylase inhibitor, which prevents the breakdown of L-dopa before it can reach the brain. Because of the long-term side effects, including dyskinesias, associated with levodopa, other dopamine agonists, such as ropinirole and pramipexole, are being examined. A protein called **glial cell line–derived neurotrophic factor** (GDNF) has been discovered that selectively promotes the survival of dopamine-secreting neurons. Chronic stimulation of the globus pallidus (part of the lentiform nucleus) with an electrical pulse generator has shown some success. Treatment of the disorder by transplanting fetal tissues, or stem cells from adult tissues, capable of producing dopamine is also under investigation.

Cerebellar lesions result in a spectrum of characteristic functional disorders. Movements tend to be **ataxic** (jerky) and **dysmetric** (overshooting—for example, pointing past or deviating from a mark that one tries to touch with the finger). Alternating movements, such as supination and pronation of the hand, are performed in a clumsy manner. **Nystagmus** (nis-tag′mŭs), which is a constant motion of the eyes, may also occur. A cerebellar tremor is an intention tremor (i.e., the more carefully one tries to control a given movement, the greater the tremor becomes). For example, when a person with a cerebellar tremor attempts to drink a glass of water, the closer the glass comes to the mouth the shakier the movement becomes. This type of tremor is in direct contrast to basal nuclei tremors, in which the resting tremor largely or completely disappears during purposeful movement.

tract maintains posture through the action of trunk and proximal upper and lower limb muscles during certain movements. For example, when a person who is standing lifts one foot off the ground, the weight of the body is shifted to the other limb. The reticulospinal tract apparently enhances the functions of the alpha motor neurons in the crossed extensor reflex during this type of movement so that balance is maintained.

Another major portion of the indirect pathways involves the basal nuclei (see figure 14.15). The basal nuclei have a number of connections with each other, as well as with the thalamus and cerebrum. The basal nuclei also interact with other indirect pathways, such as the rubrospinal tract, by which they modulate motor functions.

16 ***Name the structures and the tracts that form the indirect pathways. What functions do they control? Contrast them with the functions of the direct pathways.***

Modifying and Refining Motor Activities

Basal Nuclei

The **basal nuclei** (see figure 13.10) are important in planning, organizing, and coordinating motor movements and posture. Complex neural circuits link the basal nuclei with each other, with the thalamus, and with the cerebral cortex. These connections form several feedback loops, some of which are stimulatory and others inhibitory. The stimulatory circuits facilitate muscle activity, especially at the beginning of a voluntary movement, such as rising from a sitting position or beginning to walk. The inhibitory circuits facilitate the actions of the stimulatory circuits by inhibiting muscle activity in antagonist muscles. Inhibitory circuits also decrease muscle tone when the body, limbs, and head are at rest (eliminating "unwanted" movement). Disorders of the basal nuclei result in difficulty in rising from a sitting position and difficulty in initiating walking. People with basal nuclei disorders exhibit increased muscle tone and exaggerated, uncontrolled movements when they are at rest. A specific feature of some basal nuclei disorders is a "resting tremor," a slight shaking of the hands when a person is not performing a task. Parkinson disease and cerebral palsy are basal nuclei disorders.

Cerebellum

The **cerebellum** (see figure 13.6) consists of three functional parts: the vestibulocerebellum, or flocculonodular lobe; the spinocerebellum; and the cerebrocerebellum. The vestibulocerebellum receives direct input from the vestibular structures, especially the semicircular canals (see chapter 15), and sends axons to the vestibular nuclei of the brainstem. It helps maintain muscle tone in postural muscles. It also helps control balance, especially during movements, and it helps coordinate eye movement.

The **vermis** and medial portion of the **lateral hemisphere,** referred to jointly as the **spinocerebellum,** helps accomplish fine motor coordination of simple movements by means of its **comparator** function. Action potentials from the motor cortex descend into the spinal cord to initiate voluntary movements. At the same time, action potentials are carried from the motor cortex to the cerebellum to give the cerebellar neurons information representing the intended movement (figure 14.18). Simultaneously, action potentials from proprioceptive neurons ascend through the spinocerebellar tracts to the cerebellum. Proprioceptive neurons innervate the joints and tendons of the structure being moved, such as the elbow or knee, and provide information about the position of the body or body parts. These action potentials give the cerebellar neurons information from the periphery about the actual movements. The cerebellum compares the action potentials from the motor cortex with those from the moving structures. That is, it compares the intended movement with the actual movement. If a difference is detected, the cerebellum sends action potentials through the thalamus to the motor cortex and to the spinal cord to correct the discrepancy. The result is smooth and coordinated movements.

The comparator function coordinates simple movements, such as touching your nose. Rapid, complex movements, however, require much greater coordination and training. The **cerebrocerebellum** consists of the lateral two-thirds of the lateral hemispheres. It communicates with the motor, premotor, and prefrontal portions of the cerebral cortex to help in planning and practicing rapid, complex motor actions. The connections from the cerebrum to the cerebellum constitute a large portion of the axons in the cerebral peduncles. Because of the cerebrocerebellum, with training, a person can learn highly skilled and rapid movements that are accomplished more rapidly than can be accounted for by the comparator function of the cerebellum. In these cases, the cerebellum participates with the cerebrum in learning highly specialized movements, such as playing the piano or swinging a baseball bat. The cerebrocerebellum is also involved in cognitive functions, such as rhythm, conceptualization of time intervals, some word associations, and solutions to pegboard puzzles—tasks once thought to occur only in the cerebrum.

17 ***What are the functions of the basal nuclei?***

18 ***Explain the comparator activities of the spinocerebellum.***

19 ***Describe the role of the cerebrocerebellum in rapid and skilled motor movements, such as playing the piano.***

1. The motor cortex sends action potentials to lower motor neurons in the spinal cord.
2. Action potentials from the motor cortex inform the cerebellum of the intended movement.
3. Lower motor neurons in the spinal cord send action potentials to skeletal muscles, causing them to contract.
4. Proprioceptive signals from the skeletal muscles and joints to the cerebellum convey information concerning the status of the muscles and the structure being moved during contraction.
5. The cerebellum compares the information from the motor cortex to the proprioceptive information from the skeletal muscles and joints.
6. Action potentials from the cerebellum to the spinal cord modify the stimulation from the motor cortex to the lower motor neurons.
7. Action potentials from the cerebellum are sent to the motor cortex, which modify its motor activity.

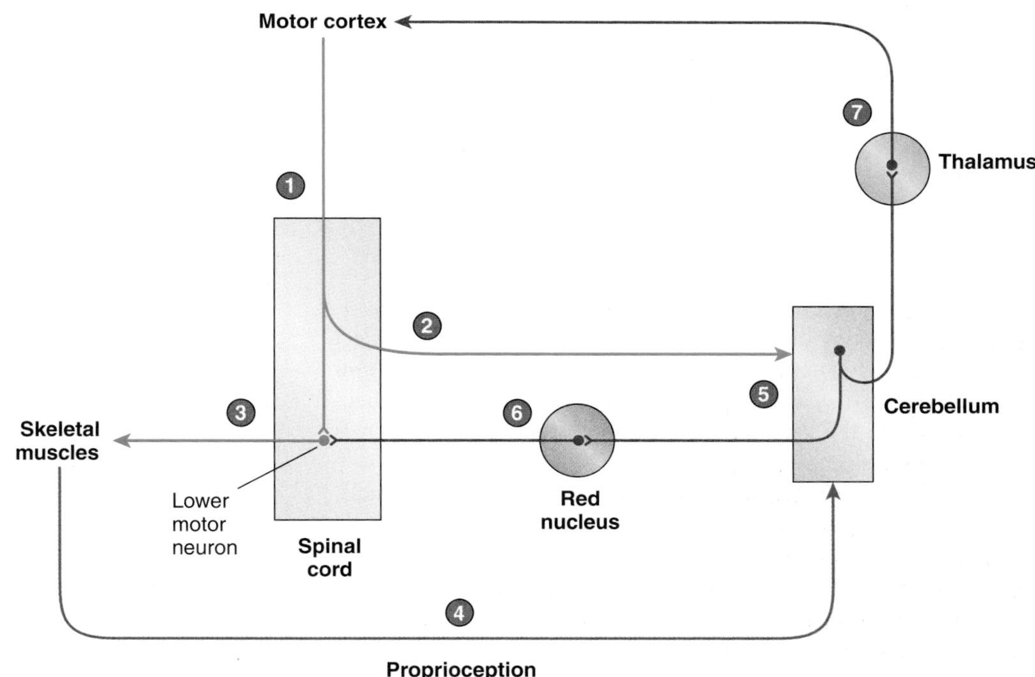

PROCESS FIGURE 14.18 Cerebellar Comparator Function

Cerebellar Dysfunction

Cerebellar dysfunction results in (1) decreased muscle tone, (2) balance impairment, (3) a tendency to overshoot when reaching for or touching an object, and (4) an intention tremor, which is a shaking in the hands that occurs only while attempting to perform a task. Although the cerebellum and basal nuclei both control motor functions, they have opposite effects, and they exhibit opposite symptoms when injured. For example, cerebellar dysfunction results in decreased muscle tone and an intention tremor, whereas basal nuclear dysfunction often results in increased muscle tone and a resting tremor.

BRAINSTEM FUNCTIONS

The major ascending and descending pathways project through the brainstem. In addition, the brainstem contains nuclei of cranial nerves III–X and XII (see table 13.4 and figure 13.2), and nuclei of the reticular formation. Only collateral branches of cranial nerve II (optic nerve) project to brainstem nuclei. Cranial nerves I (olfactory nerve) and XI (spinal accessory nerve) do not have projections to or nuclei in the brainstem.

Sensory Input Projecting Through the Brainstem

The brainstem receives sensory input from collateral branches of ascending spinal cord pathways and from the axons of cranial nerves II (vision), V (tactile sensation from the face, nasal cavity, and oral cavity), VII (taste), VIII (hearing and balance), IX (taste and tactile sensation in the throat), and X (taste, tactile sensation in the larynx, and visceral sensation in the thorax and abdomen). Cranial nerves V, VII, VIII, IX, and X have sensory nuclei in the brainstem. Many of these nuclei are involved in the special senses (see chapter 15). The brainstem nuclei associated with cranial nerve II are involved in visual reflexes.

As noted earlier, fibers of the spinothalamic tracts passing through the brainstem are joined by fibers of the trigeminothalamic tract (see p. 485). This tract carries tactile sensations, such as pain and temperature, two-point discrimination, and light touch from the face, nasal cavity and oral cavity, including the teeth.

RAS Functions of the Brainstem

Collateral branches of trigeminothalamic tract neurons project to the reticular formation, where they stimulate wakefulness and consciousness. This part of the reticular formation and its connections constitute the **reticular activation system (RAS),** which is involved in the sleep–wake cycle.

Collateral branches of cranial nerves II (optic), V (trigeminal), and VIII (vestibulocochlear); ascending tactile sensory pathways; and descending neurons from the cerebrum also project to the RAS. Visual and acoustic stimuli, as well as mental activites, stimulate the RAS to help maintain alertness and attention. Ringing alarm clocks, sudden flashes of bright lights, and cold water being splashed on the face can all arouse consciousness (figure 14.19). The removal of auditory, visual, and other stimuli may lead to drowsiness or sleep. For example, consider what happens to students during a monotonous lecture in a dark lecture hall. Damage to RAS cells of the reticular formation can result in coma.

PREDICT 7

Luke tried to go to sleep, but a dripping bathroom faucet kept him awake. Explain.

Drugs and the Reticular Activating System

Certain drugs can either stimulate or depress the RAS. General anesthetics suppress this system, and many tranquilizers depress it. On the other hand, ammonia (smelling salts) and other irritants stimulate trigeminal nerve endings in the nose. As a result, action potentials are sent to the reticular formation and the cerebral cortex to arouse an unconscious patient.

Motor Output and Reflexes Projecting Through the Brainstem

Descending pathways in the brainstem pass to the spinal cord, pass into the cerebellum, or synapse with cranial nerve motor nuclei and other nuclei in the brainstem. Some of the descending pathways originate in the cerebral cortex and pass directly through the brainstem (direct spinal pathways). Many direct descending pathways synapse in cranial nerve motor nuclei to initiate movements within the head, such as eye movements. Others synapse with brainstem nuclei, which in turn send descending fibers into the spinal cord (indirect pathways). Descending fibers from the reticular formation constitute one of the body's most important motor pathways. Fibers from the reticular formation are critical in controlling many functions, such as respiratory movements and cardiac rhythms. Cranial nerves III, IV, V, VI, VII, IX, X, and XII all have motor nuclei in the brainstem. Functionally, the motor output projecting through the brainstem can be classified into two categories, somatic motor and parasympathetic.

Somatic Motor Output and Reflexes

Cranial nerves III (oculomotor), IV (trochlear), and VI (abducent) control the eye muscles. The eyes can be controlled by the cerebrum to move and look toward an object. Action potentials reaching the superior colliculi from the cerebrum are involved in the visual tracking of moving objects. Visual tracking with both eyes to the right involves the lateral rectus muscle and abducent nerve of the right eye and the medial rectus muscle and oculomotor nerve of the left eye. Coordination of these two nerves and muscles requires nuclei of the reticular formation. The eye muscles along with the muscles of the neck (mainly the trapezius and sternocleidomastoid, innervated by cranial nerve IX) can also be involved in reflexes that are initiated in the superior colliculi in response to visual and audioty stimuli. Collateral branches from the optic tract (II) synapse in the superior colliculi of the midbrain (see figure 13.7). Axons from the superior colliculi project to cranial nerve nuclei III, IV, and VI and to the upper cervical part of the spinal cord (motor neurons of XI), where they stimulate the motor neurons involved in turning the eyes and

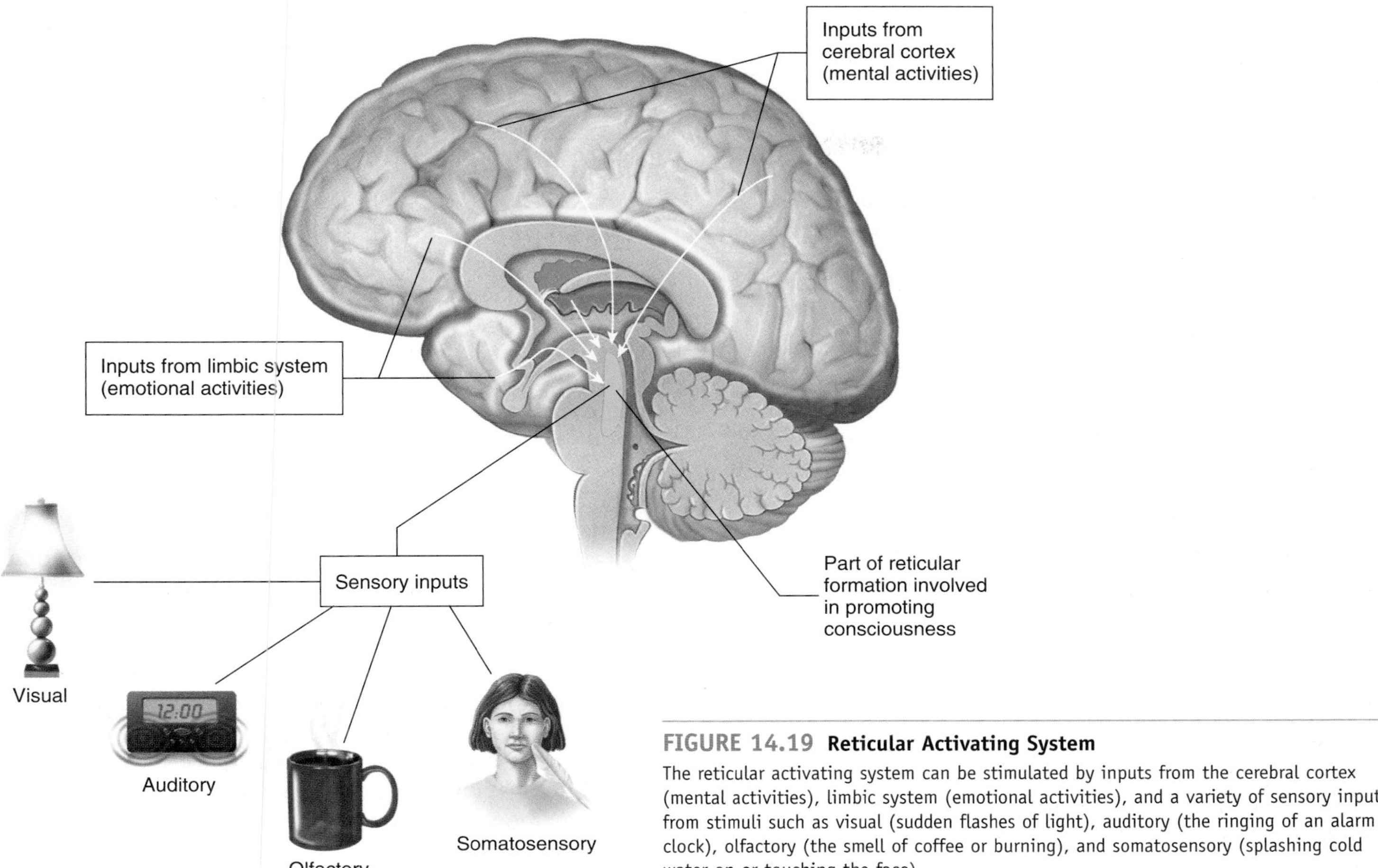

FIGURE 14.19 Reticular Activating System

The reticular activating system can be stimulated by inputs from the cerebral cortex (mental activities), limbic system (emotional activities), and a variety of sensory inputs from stimuli such as visual (sudden flashes of light), auditory (the ringing of an alarm clock), olfactory (the smell of coffee or burning), and somatosensory (splashing cold water on or touching the face).

head toward a visual stimulus. Similarly, the superior colliculi also receive input from auditory pathways, which can initiate a reflex that turns the eyes and head toward a sudden noise. Axons in the spinomesencephalic pathway of the spinal cord (see table 14.3) also project to the superior colliculi. Input from this pathway initiates a reflex that turns the eyes and head toward a tactile stimulus on the body.

Motor fibers from cranial nerve V (trigeminal) innervate the muscles of mastication and control chewing; however, once the chewing cycle is initiated, the reticular formation regulates the cycle. Even though the initiation of chewing can be under conscious control, the presence of food in the mouth initiates a reflex between the sensory nuclei and the motor nucleus of cranial nerve V, thereby starting the chewing cycle. Other reflexes in the trigeminal nerve system detect how hard or soft an item is in the mouth and adjusts the bite accordingly. Reflexes between the trigeminal sensory nuclei and the motor nucleus of cranial nerve XII (hypoglossal) control the tongue to help place the food between the teeth for chewing, while keeping the tongue out of harm's way.

Cranial nerve VII (facial), which innervates the muscles of facial expression, is controlled by the cerebrum and is very important in communication.

Somatic motor fibers from cranial nerves IX (glossopharyngeal) and X (vagus) supply muscles of the pharynx and larynx associated with swallowing and speech. Swallowing, once initiated under conscious control, continues as a reflex. The pharyngeal muscles of swallowing are largely innervated by the vagus nerve and, to a smaller extent, by the glossopharyngeal nerve.

Unlike swallowing, which is largely a reflex, speech is highly controlled by the cerebrum. The vagus nerve innervates the muscles of the larynx responsible for voice production and controls the pharyngeal and most palatine muscles responsible for the movement of the soft palate during speech. The complex movements of the tongue during speech are controlled by the hypoglossal nerve (XII), which innervates nearly all the muscles of the tongue.

Parasympathetic Output and Reflexes

The constriction of the pupil involves parasympathetic stimulation through the oculomotor (III) nerve. The visual reflexes resulting in pupil constriction are coordinated through nuclei in the reticular formation. These reflexes are also coordinated by a nuclear region in the diencephalon called the pretectal area (in front of the tectum, the roof of the midbrain).

The Brain's Canary

Function of the pretectal area is critical for normal pupillary constriction in response to light. This area of the brain can be thought of as the "brain's canary." For many years, miners carried caged canaries with them into deep mine shafts to detect poisonous gas. An unconscious or dead canary warned the miners of methane gas before enough had accumulated to kill them. Like a canary, the pretectal area is more sensitive to brain damage, or symptoms of the damage are more easily observed in functions coordinated by the pretectal area, than many other parts of the brain. Fixed, dilated pupils in a patient are a sign that the pretectal area is not functioning normally and that the patient may have experienced damage to the brain.

Tactile sensory input from the nasal cavity via the trigeminal (V) nerve can initiate a sneeze reflex. Tactile sensory input from the oral cavity via the trigeminal (V) nerve informs the cerebrum of the presence of food or another object in the mouth. The presence of an object in the mouth—even a nonfood item, such as a marble—stimulates a reflex between the trigeminal sensory nuclei and the motor nuclei of VII (facial) and IX (glossopharyngeal), which innervate the salivary glands to stimulate salivation.

Sensory input from the glossopharyngeal nerve (IX) conveys tactile information from the back of the tongue, the soft palate, and the throat (pharynx) to the brainstem. Mechanical stimulation of these areas can initiate a gag reflex, whereas other stimulation of the throat can initiate a cough reflex. Sensory input from the vagus nerve (X) conveys tactile information from the larynx (voice box) and thoracic and abdominal viscera. Tactile input from the larynx can also initiate a cough reflex. In addition, the vagus nerve (X) is involved in many complex reflexes associated with vital functions such as heart rate, respiration, and digestion. Many of these involve the reticular formation and are discussed in later chapters.

Vital Functions Controlled in the Brainstem

Because many vital functions, such as heart rate, blood pressure, and respiration are regulated by nuclei in the brainstem, many medical emergency evaluations are of brainstem function. When a person is involved in a serious accident or is extremely ill, these vital functions may be affected. Such signs of brainstem damage can be assessed by medical personnel.

20 ***List the motor nuclei in the brainstem.***

21 ***Describe the motor reflexes that occur in the brainstem.***

22 ***Describe the parasympathetic reflexes that occur in the brainstem.***

23 ***What are some of the vital functions that are regulated in the brainstem?***

PREDICT 8

Some types of epilepsy are treated by placing an electronic device inside the neck to stimulate the vagus nerve. Minor injury to the vagus nerve during implantation of the device can lead to hoarseness. Why is this so? Can you predict two other consequences of minor injury to the vagus nerve?

OTHER BRAIN FUNCTIONS

The human brain is capable of many functions besides awareness of sensory input and the control of skeletal muscles. Speech, mathematical and artistic abilities, sleep, memory, emotions, and judgment are functions of the brain.

Speech

In most people, the speech area is in the left cerebral cortex. Two major cortical areas are involved in speech: **Wernicke's area** (sensory speech area), a portion of the parietal lobe, and **Broca's area** (motor speech area) in the inferior part of the frontal lobe (see figure 14.11). Wernicke's area is necessary for understanding and formulating coherent speech. Broca's area initiates the complex series of movements necessary for speech. The two areas are connected by a bundle of neurons known as the **arcuate fasciculus** (figure 14.20*a*).

For someone to speak a word that he or she sees, such as when reading out loud, the following sequence of events must take place. Action potentials from the eyes reach the primary visual cortex, where the word is seen. The word is then recognized in the visual association area and understood in parts of Wernicke's area. Then action potentials representing the word are conducted through association fibers that connect Wernicke's and Broca's areas. In Broca's area, the word is formulated as it will be spoken. Action potentials are then propagated to the premotor area, where the movements are programmed, and finally to the primary motor cortex, where the proper movements are triggered (figure 14.20*b*).

To repeat a word that has been heard is similar. The information passes from the ears to the primary auditory cortex and then passes to the auditory association area, where the word is recognized, and continues to Wernicke's area, where the word is understood and formulated as it will be spoken. From Wernicke's area, it follows the same route as followed for speaking words that are seen.

Aphasia

Aphasia (ă-fā′zē-ă), absent or defective speech or language comprehension, results from a lesion in the language areas of the cortex. The several types of aphasia depend on the site of the lesion. **Receptive aphasia** (Wernicke's aphasia), which includes defective auditory and visual comprehension of language, defective naming of objects, and repetition of spoken sentences, is caused by a lesion in Wernicke's area. Both **jargon aphasia,** in which a person may speak fluently but unintelligibly, and **conduction aphasia,** in which a person has poor repetition but relatively good comprehension, can result from a lesion in the tracts between Wernicke's and Broca's areas. **Anomic** (ă-nō′mik) **aphasia,** caused by the isolation of Wernicke's area from the parietal or temporal association areas, is characterized by fluent but circular speech resulting from poor word-finding ability. **Expressive aphasia** (Broca's aphasia), caused by a lesion in Broca's area, is characterized by hesitant and distorted speech.

24 ***List the sequence of events that must occur for a person to repeat a word that he or she hears.***

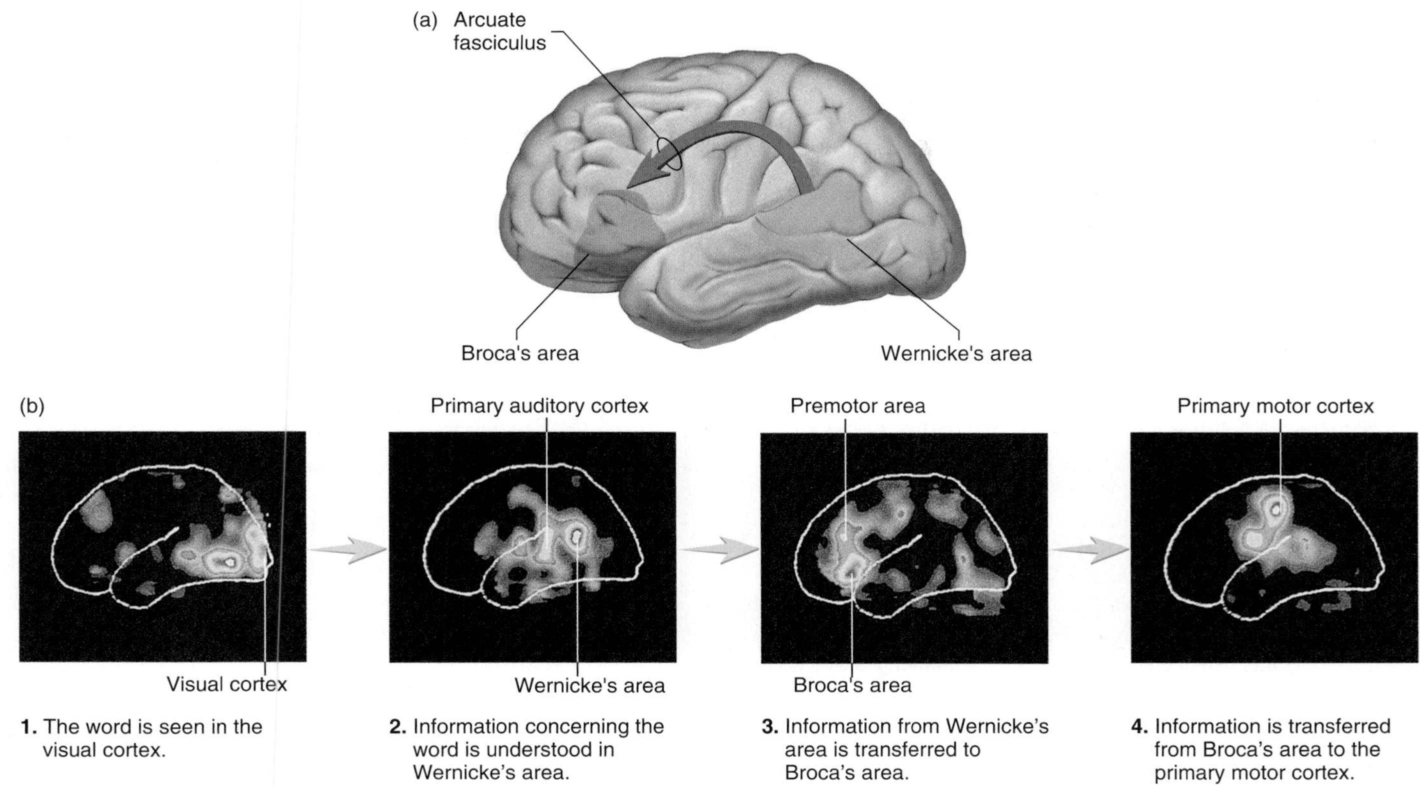

FIGURE 14.20 Demonstration of Cortical Activities During Speech

(*a*) The arcuate fasciculus connects the two key areas involved in speech. (*b*) The figures show the pathway for reading and naming something that is seen, such as reading aloud. PET scans show the areas of the brain that are most active during various phases of speech. *Red* indicates the most active areas; *blue* indicates the least active areas.

PREDICT 9

Propose the sequence needed for a blindfolded person to name an object placed in the right hand.

Right and Left Cerebral Cortex

The cortex of the right cerebral hemisphere controls muscular activity in and receives sensory input from the left half of the body. The left cerebral hemisphere controls muscles in and receives sensory input from the right half of the body. Sensory information received by the cortex of one hemisphere is shared with the other through connections between the two hemispheres called **commissures** (kom′i-shūrz; a joining together). The largest of these commissures is the **corpus callosum** (kōr′pūs kă-lō′sŭm; callous body), which is a broad band of tracts at the base of the longitudinal fissure (see figure 13.1).

Language and perhaps other functions, such as artistic activities, are not shared equally between the left and right cerebral hemispheres. The left hemisphere is more involved in such skills as mathematics and speech. The right hemisphere is involved in activities such as spatial perception, the recognition of faces, and musical ability.

25 ***Name the largest pathway that connects the right and left cerebral hemispheres.***

26 ***What are the functions localized in the left cerebral hemisphere? In the right cerebral hemisphere?***

Hemisphere Dominance and Amorphosynthesis

Dominance of one cerebral hemisphere over the other, for most functions, is probably not very important in most people because the two hemispheres are in constant communication through the corpus callosum, literally allowing the right hand to know what the left hand is doing. Surgical cutting of the corpus callosum has been successful in treating a limited number of epilepsy cases. Under certain conditions, however, interesting functional defects develop in people who have had their corpus callosum severed. For example, if a patient with a severed corpus callosum is asked to reach behind a screen to touch one of several items with one hand without being able to see it and then is asked to point out the same object with the other hand, the person cannot do it. Tactile information from the left hand enters the right somatic sensory cortex, but that information is not transferred to the left hemisphere, which controls the right hand. As a result, the left hemisphere cannot direct the right hand to the correct object.

A person suffering a stroke in the right parietal lobe may lose the ability to recognize faces while retaining essentially all other brain functions. A more severe lesion can cause a person to lose the ability to identify simple objects. This defect is called **amorphosynthesis** (ă-mōr′fō-sin′thĕ-sis). Some people with a similar lesion in the right cerebral hemisphere may tend to ignore the left half of the world, including the left half of their own bodies. These people may completely ignore a person who is to their left but react normally when the person moves to their right. They may also fail to dress the left half of their bodies or to eat the food on the left half of their plates.

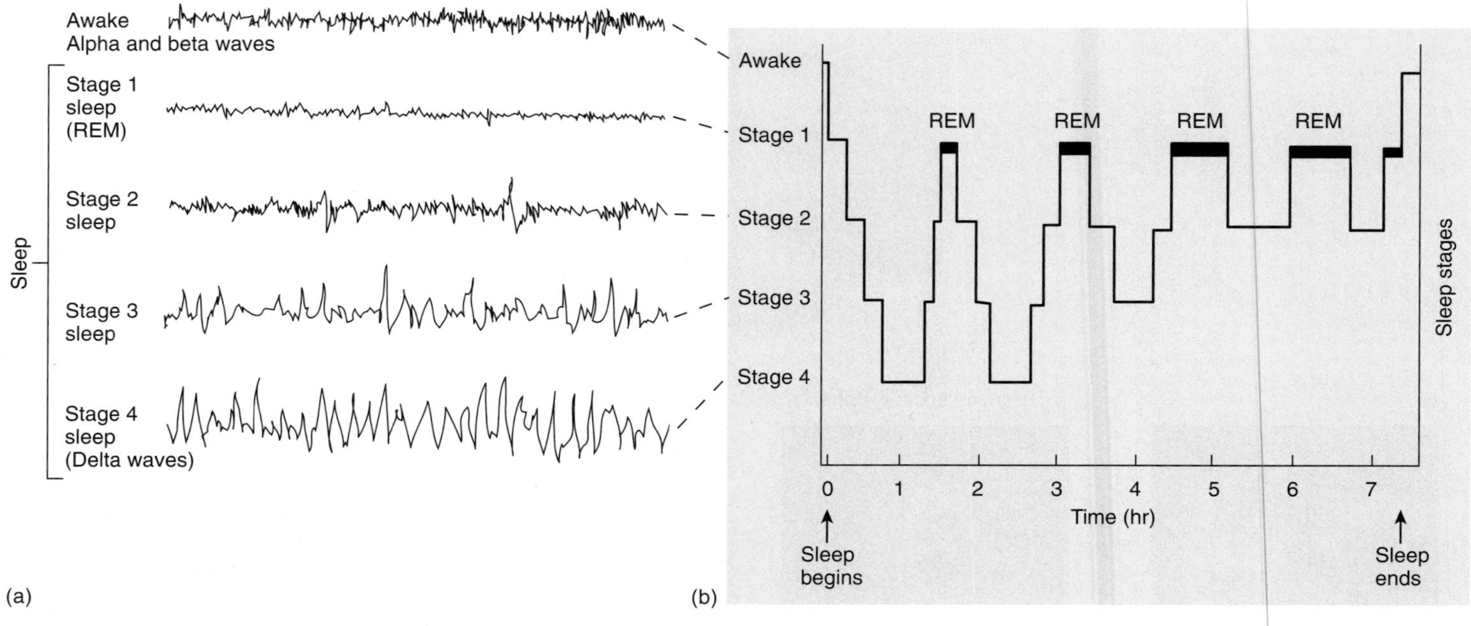

FIGURE 14.21 Electroencephalograms (EEGs) Showing Brain Waves

(*a*) EEG tracings when a person is awake and during four stages of sleep. (*b*) A typical night's sleep pattern in a young adult. The time spent in REM sleep is labeled and shown by *dark bars.*

Brain Waves and Sleep

Electrodes placed on a person's scalp and attached to a recording device can record the electrical activity of the brain, producing an **electroencephalogram** (ē-lek′trō-en-sef′ă-lō-gram; **EEG;** figure 14.21). These electrodes are not sensitive enough to detect individual action potentials, but they can detect the simultaneous action potentials in large numbers of neurons. As a result, the EEG displays wavelike patterns known as **brain waves.** Brain waves are produced continuously, but their intensity and frequency differ from time to time based on the state of brain activity.

Most of the time, EEG patterns from a given individual are irregular, with no particular pattern, because, although the normal brain is active, most of its electrical activity is not synchronous. At other times, however, specific patterns can be detected. These regular patterns are classified as alpha, beta, theta, or delta waves (see figure 14.21). **Alpha waves** are observed in a normal person who is awake but in a quiet, resting state with the eyes closed. **Beta waves** have a higher frequency than alpha waves and occur during intense mental activity. **Theta waves** usually occur in children, but they can also occur in adults who are experiencing frustration or who have certain brain disorders. **Delta waves** occur in deep sleep, in infancy, and in patients with severe brain disorders.

Brain wave patterns vary during the four stages of sleep (see figure 14.21). A sleeping person arouses several times during a period of sleep. Dreaming occurs during periods when eye movement can be observed in a sleeping person, called **rapid eye movement (REM) sleep.**

Distinct types of EEG patterns can be detected in patients with specific brain disorders, such as epileptic seizures. Neurologists use these patterns to diagnose the disorders and determine the appropriate treatment.

27 ***What is an EEG? What four conditions produce alpha, beta, theta, and delta waves, respectively?***

Memory

Memory is divided into three major types: sensory, short-term (or primary), and long-term (figure 14.22). **Sensory memory** is the very short-term retention of sensory input received by the brain while something is scanned, evaluated, and acted on. This type of memory lasts less than a second and apparently involves transient changes in membrane potentials.

If a given piece of data held in sensory memory is considered valuable enough, it is moved into **short-term memory,** where it is retained for a few seconds to a few minutes. This memory is limited primarily by the number of bits of information (usually about seven or eight) that can be stored at any one time, although the amount varies from person to person. More bits can be stored when the numbers are grouped into specific segments separated by spaces. When new information is presented, or when the person is

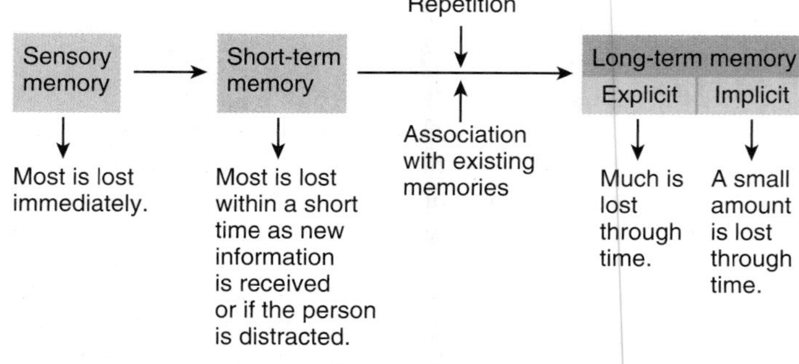

FIGURE 14.22 Memory Processing

CASE STUDY

Multiple Sclerosis

Betty, a 32-year-old woman, woke up one morning with weakness in her lower limbs. By that afternoon, she had become completely exhausted with a familiar ache in her left eye and tingling sensations in her extremities. At the end of the day, she could barely stand up, and her left eye was blurry and increasingly painful. Betty's family physician made an appointment for her with a neurologist. The neurologist suspected **multiple sclerosis,** an autoimmune disease resulting in the demyelination of CNS neurons, which become sclerotic, or hard. He ordered an MRI scan and a **visual evoked potential (VEP) test,** which is a recording of action potentials in the optic nerve "evoked" in response to visual stimuli. During the VEP test, electrodes were placed on Betty's scalp, and she was asked to respond to visual stimuli with the normal eye covered by an eye-patch. The normal eye was also subjected to the same treatment.

The MRI scan showed lesions in the optic nerve and white matter of the brain, and the VEP test showed optic nerve dysfunction. These results confirmed the diagnosis of multiple sclerosis accompanied by **optic neuritis,** or inflammation of the optic nerve. Betty's neurologist explained that, in multiple sclerosis, sensory neurons that conduct sensory information to the brain and motor neurons that conduct motor stimulation from the brain to the muscles are damaged. Betty asked if she would be cured, to which the neurologist replied that she would experience periods of remission, which may be interrupted by symptomatic periods. He also warned that, with each successive episode, neurons may become progressively more damaged.

PREDICT 10

a. What effect would demyelination have on action potential propagation along axons? How might this explain blurred vision?

b. How would you explain the weakness and tingling that Betty experienced?

distracted, information previously stored in short-term memory is eliminated; therefore, if a person is given a second telephone number or if the person's attention is drawn to something else, the first number usually is forgotten.

Long-term memory lasts for hours or longer. Some long-term memory may last a lifetime. Two types of long-term memory exist: declarative, or explicit, and procedural, or implicit. **Declarative** or **explicit, memory** involves the retention of facts, such as names, dates, and places. Explicit memory is accessed by part of the temporal lobe called the **hippocampus** (hip-ō-kam′pŭs; shaped like a seahorse) and the **amygdala,** or **amygdaloid** (ă-mig′dă-loyd; almond-shaped) **nuclear complex.** The hippocampus is involved in retrieving the actual memory, such as recalling a person's name, and the amygdala is involved in the emotional overtones of that memory, such as feelings of like or dislike, and the recollection of good or bad memories associated with that person. A lesion in the temporal lobe affecting the hippocampus can prevent the brain from moving information from short-term to long-term memory.

Fear

Some aspects of fearful responses appear to be "hardwired" in the brain and do not require learning. For example, infant rodents are terrified when exposed to a cat, even though they have never seen a cat. Loud sounds seem to be particularly effective in eliciting fear responses. A direct collateral branch runs from the auditory pathway to the amygdala, which does not involve the cerebral cortex. Fear can be evoked by a loud sound acting directly on the amygdala. Overcoming fear, however, requires the involvement of the cerebral cortex; therefore, the stimulation of fear appears to involve one process, its suppression another. Flaws in either process could result in fear-related disorders, such as anxiety, depression, panic, phobias, and posttraumatic stress disorder.

Emotion and mood apparently serve as gates in the brain and determine what is or is not stored in long-term explicit memory. The amygdala is also a key to the development of fear, which also involves the prefrontal cortex and the hypothalamus.

Parts of explicit memory appear to be stored separately in various parts of the cerebrum, especially in the parietal lobe, much like storing items in separate "pigeonholes." Memories of people appear to be stored separately from memories of places. People's faces may be stored in yet other pigeonholes. Family members appear to be stored together. Items that are recognized by sight, such as an animal, are stored separately from items that are recognized by feel, such as tools. Damage, such as a stroke, to one part of the brain can remove certain memories without affecting others.

Retrieval of a complete memory requires accessing parts of the memory from different pigeonholes. A complex memory requires accessing and reassembling segments of memory each time the memory is recalled. The memory of an experience, for example, may be stored in at least four different pigeonholes. Where you were is stored in one place, who you were with in another, what happened in another, and how you felt in yet another place. The complexity of this process may be responsible for the changes in what is recalled over time. On occasion, parts of unrelated memories may be pulled out and put together incorrectly to create a "false memory." Much of what is stored as explicit memory is gradually lost through time.

Procedural, or **implicit, memory,** also called **reflexive memory,** involves the development of skills like riding a bicycle or playing a piano. Implicit memory is stored primarily in the cerebellum and the premotor area of the cerebrum. Conditioned, or Pavlovian, reflexes are also implicit and can be eliminated in experimental animals by producing cerebellar lesions in the animals. The most famous example of a conditioned reflex is that of Ivan Pavlov's experiments with dogs. Each time he fed the dogs, a bell was rung; soon the dogs

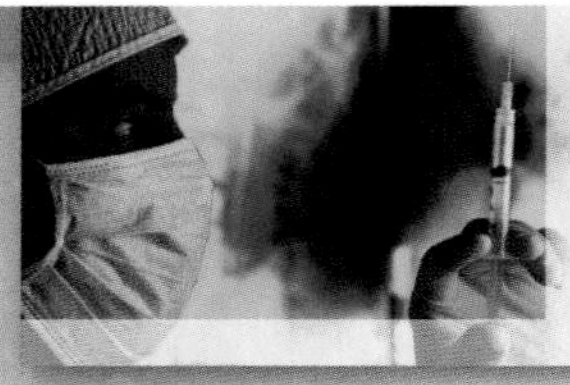

CLINICAL FOCUS

General CNS Disorders

INFECTIONS

Reye syndrome may develop in children following a viral infection, especially influenza or chicken pox. The use of aspirin to treat viral infections has been linked to development of the syndrome in the United States. A predisposing disorder in fat metabolism may also be present in some cases. In children affected by the syndrome, the brain cells swell, and the liver and kidneys accumulate fat. Symptoms include vomiting, lethargy, and loss of consciousness and may progress to coma and death or to permanent brain damage.

Rabies is a viral disease transmitted by the bite of an infected mammal. The rabies virus infects the brain, salivary glands (through which it is transmitted), muscles, and connective tissue. When the patient attempts to swallow, the effort can produce pharyngeal muscle spasms; sometimes even the thought of swallowing water or the sight of water can induce pharyngeal spasms. Thus, the term *hydrophobia,* fear of water, is applied to the disease. The virus also infects the brain and results in abnormal excitability, aggression, and in later stages paralysis and death.

Tabes dorsalis (tā'bēz dōr-sā'lis) is a progressive disorder occurring as a result of untreated syphilis. *Tabes* means a wasting away, and *dorsalis* refers to a degeneration of the dorsal roots and dorsal columns of the spinal cord. Symptoms include ataxia, resulting from lack of proprioceptive input; anesthesia, resulting from dorsal root damage; and eventually paralysis as the infection spreads.

OTHER DISORDERS

Tumors of the brain develop from neuroglial cells. Symptoms vary widely, depending on the location of the tumor, but include headaches, neuralgia (pain along the distribution of a peripheral nerve), paralysis, seizures, coma, and death. **Meningiomas** (mĕ-nin'jē-ō'măz), tumors of the meninges, account for 25% of all primary intracranial tumors.

Alzheimer disease is a severe type of mental deterioration, or dementia, usually affecting older people but occasionally affecting people younger than 60. Alzheimer disease is estimated to affect 10% of all people older than 65 and nearly half of those older than 85. Alzheimer disease involves a general decrease in brain size, resulting from a loss of neurons in the cerebral cortex. The gyri become narrower, and sulci widen. The frontal lobes and specific regions of the temporal lobes are affected most severely. Alzheimer symptoms include general intellectual deficiency, memory loss, short attention span, moodiness, disorientation, and irritability.

Alzheimer disease is characterized by the appearance of **amyloid** (am'i-loyd) **plaques** and **neurofibrillary tangles.** Amyloid plaques are localized axonal enlargements of degenerating nerve fibers, containing large amounts of β-amyloid protein, and neurofibrillary tangles are filaments inside the cell bodies of dead or dying neurons.

The gene for B-amyloid protein has been mapped to chromosome 21; however, it is thought that only the rare, inherited, early-onset (beginning before age 60) form of Alzheimer maps to chromosomes 21. The more common, late-onset form (beginning after age 65), which makes up more than three-fourths of all cases, maps to chromosome 19.

Another protein, **apolipoprotein E** (ap'ō-lip-ō-prō'tēn; **apoE**), which binds to β-amyloid protein and is known to transport cholesterol in the blood, has also been associated with Alzheimer disease. *apoE-III* is the normal protein and *apoE-IV* is an abnormal, mutant form. *apoE-IV* has been found in amyloid plaques and neurofibrillary tangles and has been mapped to the same region of chromosome 19 as the late-onset form of Alzheimer. People with two copies of the *apoE-IV* gene are eight times more likely to develop the disease than people with two copies of the *apoE-III* gene. *apoE-IV* apparently binds to β-amyloid more rapidly and more tightly than does *apoE-III. ApoE* also may be involved with regulating phosphorylation of another protein, called τ (tau), which in turn is involved in microtuble formation inside neurons. If τ is overphosphorylated, microtubules are not properly constructed, and the τ proteins intertwine to form neurofibrillary tangles.

Chronic mercury poisoning can cause brain disorders, such as intention tremor, exaggerated reflexes, and emotional instability. Mercury causes oxidative damage and apoptosis in several organs, including the kidneys and CNS.

Lead poisoning is a serious problem, particularly among urban children. Lead is taken into the body from contaminated air, food, and water. Flaking lead paint in older houses and soil contamination can be major sources of lead poisoning in children. Lead usually accumulates slowly in the body until toxic levels are reached.

Brain damage caused by lead poisoning in children includes edema, demyelination, and cortical neuron necrosis with astrocyte proliferation. This damage appears to be permanent and can result in reduced intelligence, learning disabilities, poor psychomotor development, and blindness. In severe cases, psychoses, seizures, coma,

salivated when the bell rang, even if no food was presented. Only a small amount of implicit memory is lost through time.

Several physiologic explanations have been proposed for short-term memory, most of which involve short-term changes in membrane potentials. The changes in membrane potentials are transitory but are longer than those involved in sensory memory, and they can be eliminated by new information reaching the cells.

Certain pieces of information are transferred from short-term to long-term memory. Long-term memory involves changes in neurons, called **long-term potentiation,** which facilitates the future transmission of action potentials (figure 14.23). The amount of the neurotransmitter glutamate produced and released by the presynaptic neuron increases. The number of glutamate receptors in the postsynaptic neuron also increases, and the reaction of the postsynaptic neuron to glutamate is potentiated. Long-term memory storage in a single neuron also involves Ca^{2+} influx into the postsynaptic cell. Calcium ions associate with and activate **calmodulin** (kal-mod'ū-lin) inside the cell. Calmodulin, through a cAMP mechanism, stimulates the synthesis of specific proteins, which are involved in changing the shape of the cell. The change in shape is stabilized by the creation of a new cytoskeleton, and the memory becomes more or less permanent.

or death may occur. Adults exhibit more mild PNS symptoms, including demyelination with decreased neuromuscular function. Other symptoms include abdominal pain and renal disease.

Epilepsy is a group of brain disorders that have seizure episodes in common. The seizure, a sudden massive neuronal discharge, can be either partial or complete, depending on the amount of brain involved and whether or not consciousness is impaired. A seizure involves a change in sensation, consciousness, or behavior due to brief electrical discharge in the brain.

Normally, a balance exists between excitation and inhibition in the brain. When this balance is disrupted by increased excitation or decreased inhibition, a seizure may result. The neuronal discharges may stimulate muscles innervated by the neurons involved, resulting in involuntary muscle contractions, or convulsions.

Epilepsy occurs in 1% of all people age 20 and 3% of those over age 75. A number of factors can increase the incidence of epilepsy: 2.4% of children whose fathers have epilepsy develop epilepsy and 8.7% of children whose mothers have epilepsy develop the disorder. Ten percent of children with cerebral palsy develop epilepsy, and 10% of those with mental retardation develop epilepsy. Half of children with both cerebral palsy and mental retardation develop epilepsy. Twenty-two percent of stroke patients develop epilepsy. Seventy percent of those with epilepsy will eventually experience remission.

Depression may cause more "grief and misery" than any other single disease. Although the illness has been known for over 2000 years, its medical status is still uncertain. Is depression a disease state caused by a chemical excess or deficiency, or is it a psychologic condition that a person can decide to snap out of? The answer is probably that both types of depression exist. Depression is a complex, multifaceted group of disorders. Some types of "endogenous" depression can be treated with antidepressants, of which there are five groups: tricyclic antidepressants, nontricyclic compounds, MAO inhibitors, serotonin agonists, and lithium. Many people with depression also have epilepsy. Recent research in which "pacemaker-like" stimulation of the vagus nerve to treat epilepsy has shown some promise in treating depression that does not respond to drugs.

Headaches have a variety of causes, which can be grouped into two basic classes: extracranial and intracranial. Extracranial headaches can be caused by inflammation of the sinuses, dental irritations, temporomandibular joint disorders, ophthalmologic disorders, and tension in the muscles moving the head and neck. Intracranial headaches may result from inflammation of the brain or meninges, vascular problems, mechanical damage, or tumors.

Tension headaches are extracranial muscle tension, stress headaches, consisting of a dull, steady pain in the forehead, temples, and neck or throughout the head. Tension headaches are associated with stress, fatigue, and posture.

Migraine headaches (*migraine* means half a skull) occur in only one side of the head and appear to involve the abnormal dilation and constriction of blood vessels. They often start with distorted vision, shooting spots, and blind spots. Migraines consist of severe throbbing, pulsating pain. About 80% of migraine sufferers have a family history of the disorder, and women are affected four times more often than men. Those suffering migraines are usually women younger than 35. The severity and frequency usually decrease with age.

A **concussion** is a blow to the head, producing momentary loss of consciousness without immediate detectable damage to the brain. Often, no more problems occur after the person regains consciousness; however, in some cases, **postconcussion syndrome** occurs shortly after the injury. The syndrome includes increased muscle tension or migraine headaches, reduced alcohol tolerance, difficulty in learning new things, reduction in creativity and motivation, fatigue, and personality changes. The symptoms may be gone in a month or may persist for as much as a year. In some cases, postconcussion syndrome is the result of a slowly occurring subdural hematoma, which is missed by an early examination. The blood may accumulate from small leaks in the dural sinuses.

Alexia (ă-lek′sē-ă), loss of the ability to read, may result from a lesion in the visual association cortex. **Dyslexia** (dĭs-lek′sē-ă) is a defect in which the reading level is below that expected on the basis of an individual's overall intelligence. Most people with dyslexia have normal or above-normal intelligence quotients. The term means reading deficiency and is also called partial alexia. It is three times more common in males than females. As many as 10% of males in the United States suffer from the disorder. The symptoms vary considerably from person to person and include the transposition of letters in a word, confusion between the letters *b* and *d*, and a lack of orientation in three-dimensional space. The brains of some dyslexics have abnormal cellular arrangements, including cortical disorganization and the appearance of bits of gray matter in medullary areas. Dyslexia apparently results from abnormal brain development.

Children with **attention-deficit disorder (ADD)** are easily distractible, have short attention spans, and may shift from one uncompleted task to another. Children with **attention-deficit/hyperactivity disorder (ADHD)** exhibit the characteristics of ADD, but they are also fidgety, have difficulty remaining seated and waiting their turn, engage in excessive talking, and commonly interrupt others. About 3% of all children exhibit ADHD, more boys than girls. Symptoms usually occur before age 7. The neurologic basis of both ADD and ADHD is as yet unknown.

A whole series of neurons and their pattern of activity, called a **memory engram,** or memory trace, probably is involved in the long-term retention of information, a thought, or an idea. Repetition of the information and association of the new information with existing memories assist in the transfer of information from short-term to long-term memory.

28 ***Name the three types of memory, and describe the processes that result in the transfer of information from short-term to long-term memory.***

29 ***Distinguish between implicit and explicit memory.***

Sleep and Memory

Is "pulling an all nighter" a good idea when preparing for an exam, or is it better to quit early and get some sleep? Does sleep actually help in memory processing? The results of studies seem to disagree. A number of studies indicate that REM sleep plays a critical role in learning. Other research, however, indicates that REM sleep is critical to the consolidation of procedural memory but not of declarative memory. There is no clear evidence that REM sleep helps consolidate declarative memory, the type of memory used in an examination. Some data suggest that deep, slow-wave sleep (SWS), which occurs early in the night, may be important in consolidating declarative memory.

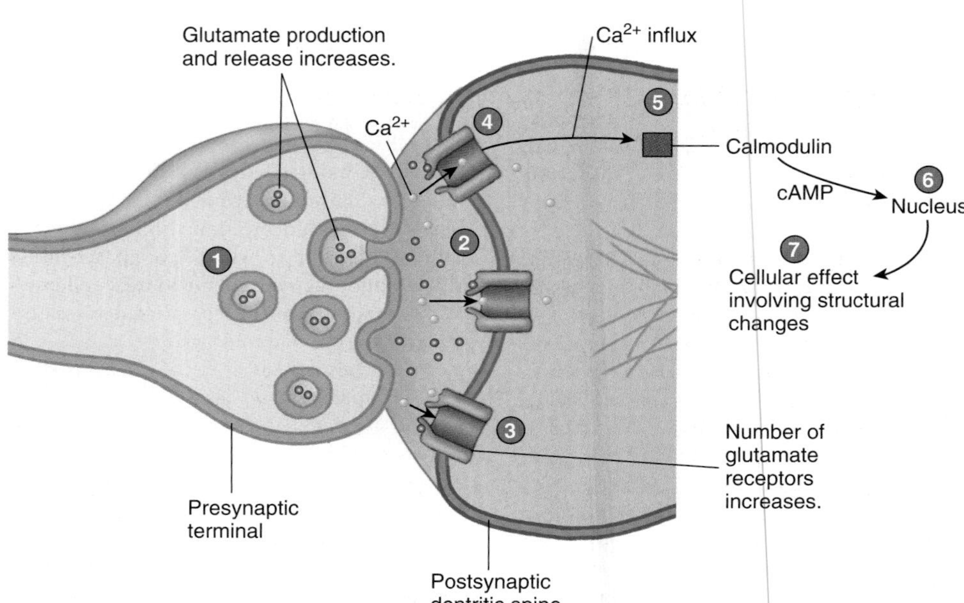
1. The amount of the neurotransmitter glutamate produced by the presynaptic neuron increases.
2. The amount of glutamate released by the presynaptic neuron also increases.
3. The number of glutamate receptors on the postsynaptic neuron membrane increases.
4. Calcium ion channels in the postsynaptic membrane open, allowing Ca2+ to enter the cell.
5. The Ca2+ that enter the cell associate with the intracellular molecule calmodulin.
6. Activated calmodulin activates a cAMP intracellular mediator, which stimulates synthesis of specific proteins.
7. The cellular effect may involve structural changes in the cell.
Glutamate production and release increases.
Ca2+ influx
Ca2+
Calmodulin
cAMP
Nucleus
Cellular effect involving structural changes
Number of glutamate receptors increases.
Presynaptic terminal
Postsynaptic dentritic spine

PROCESS FIGURE 14.23 Cellular Mechanisms of Long-Term Potentiation

Limbic System

The limbic system (see figure 13.9) influences emotions, the visceral responses to emotions, motivation, mood, and the sensations of pain and pleasure. This system is associated with basic survival instincts: reproduction and the acquisition of food and water. One of the major sources of sensory input into the limbic system is the olfactory nerves. The smell or thought of food stimulates the sense of hunger in the hypothalamus, which motivates us to seek food. Many animals can also smell water, even over great distances. In animals such as dogs and cats, olfactory detection of **pheromones** (fer′ō-mōnz) is important in reproduction. Pheromones are molecules released into the air by one animal that attract another animal of the same species, usually of the opposite sex. Pheromones released by human females can influence the menstrual cycles of other women.

Apparently, the cingulate gyrus is a "satisfaction center" for the brain and is associated with the feeling of satisfaction after a meal or after sexual intercourse. The relationship of the hippocampus with the limbic system and with memory is probably important to survival. For example, it is very important for an animal to remember where to obtain food. Once a person has eaten, the satiety center in the hypothalamus is stimulated, the hunger center is inhibited, and the person feels satiated. The hypothalamus interacts with the cingulate gyrus and other parts of the limbic system, causing a sense of satisfaction associated with the satiation.

Lesions in the limbic system can result in a voracious appetite; increased sexual activity, which is often inappropriate; and docility, including the loss of normal fear and anger responses. Because the hippocampus is part of the temporal lobe, damage to that portion can also result in a loss of memory formation.

30 ***What are the functions of the limbic system? Which of the special senses has a major input into the limbic system?***

31 ***Define pheromones.***

Addiction, Obsession, and Compulsion

The "satisfaction center" of the brain, the cingulate gyrus, can be directly affected by chemicals—alcohol and other drugs. These substances cause a physical dependence. The brain becomes dependent on these substances for a quick sense of satisfaction. Obsessions and compulsions, such as compulsive gambling, kleptomania, fire-setting, and even compulsive overeating, exhibit similar neurologic patterns to addiction. The same brain areas, such as the cingulate gyrus, show activity during obsessive and compulsive behavior as in addictive behavior.

EFFECTS OF AGING ON THE NERVOUS SYSTEM

As a person ages, sensory function gradually declines because the number of sensory neurons declines, the function of remaining neurons decreases, and CNS processing decreases. In the skin, free nerve endings and hair follicle receptors remain largely unchanged with age. Meissner corpuscles and Pacinian corpuscles, however, decrease in number. The capsules of those that remain become thicker and structurally distorted and therefore exhibit reduced function. As a result of these changes in Meissner corpuscles and Pacinian corpuscles, elderly people are less conscious of something touching or pressing on the skin, have a decreased sense of two-point discrimination, and have a more difficult time identifying objects by touch. These functional changes leave elderly people more prone to skin injuries and with a greater sense of isolation.

A loss of Pacinian corpuscles also results in a decreased sense of position of the limbs and in the joints, which can affect balance and coordination. The functions of Golgi tendon organs and muscle spindles also decline with increasing age. As a result, information on the position, tension, and length of tendons and

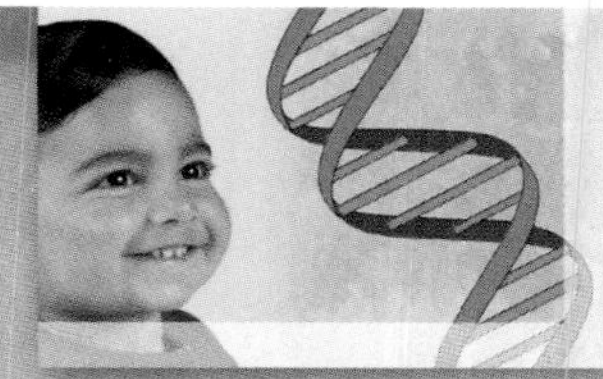

CLINICAL GENETICS

Tay-Sachs Disease

Neurons contain specialized membrane lipids known as **gangliosides.** Lysosomes contain a variety of hydrolytic enzymes, including those that digest gangliosides. **Hexoseaminidase A** is a lysosomal enzyme that breaks down a specific type of ganglioside called **GM2 ganglioside.** The gene that encodes hexoseaminidase A is called *hex A* and is located on chromosome 15.

Tay-Sachs disease is an autosomal-recessive lysosomal enzyme disorder (see chapter 3) caused by mutations to the *hex A* gene. With *hex A* not functioning properly, GM2 ganglioside accumulates in cells and damages them, especially in neurons of the CNS. Hexoseaminidase A also breaks down gangliosides in the light-sensitive cells in the eye. Therefore, in the absence of this enzyme, gangliosides accumulate in these cells, leading to blindness.

The classic form of Tay-Sachs disease is a hereditary disorder of infants, mostly of Eastern European Jewish descent. Symptoms include paralysis, blindness, and death, usually before age 5. The disease is named for Warren Tay (1843–1927), a British ophthalmologist, and Bernard Sachs (1858–1944), an American neurologist, who first described aspects of the disease. Damage to the CNS begins in the fetus, but clinical manifestations do not appear until several months after birth. The infant appears to develop normally but then begins to regress, losing motor functions, exhibiting seizures, becoming blind and mentally retarded, and eventually dying by the age of 4 or 5.

So far, more than 100 mutations of the *hex A* gene have been reported. The most common mutation found among those of Eastern European Jewish descent is a four base insertion in one of the 14 exons (exon 11) of the *hex A* gene. Heterozygous carriers of Tay-Sachs disease produce lowered levels of the enzyme. This does not, however, lead to the clinical manifestation of Tay-Sachs disease because the level of enzyme produced is sufficient to remove excess GM2 gangliosides in neurons.

Another form of Tay-Sachs disease is called late-onset Tay-Sachs (LOTS), which includes juvenile-onset and adult-onset Tay-Sachs disease. LOTS is also caused by mutations in the *hex A* gene and, consequently, deficiencies in the production of hexoseaminidase A. Affected individuals often have a combination of one severely affected allele that fails to code for hexoseaminidase A and another allele that results in reduced production of hexoseaminidase A. LOTS is characterized by problems in gait, cognition, and speech and patients often experience psychiatric problems. Juvenile-onset Tay-Sachs manifests between 5 and 15 years of age, leading to progressive neural degeneration. In adult-onset Tay-Sachs disease, symptoms appear during the teens or much later in life and tend to be milder. Similar to the infantile form of Tay-Sachs, LOTS occurs more often in the Jewish population of Eastern European descent.

Genetic counseling of high-risk couples, accompanied by prenatal screening for level of hexoseaminidase A or mutations in the *hex A* gene, has greatly reduced the incidence of Tay-Sachs disease.

muscles decreases, resulting in additional reduction in the senses of movement, posture, and position, as well as reduced control and coordination of movement.

Other sensory neurons with reduced function include those that monitor blood pressure, thirst, objects in the throat, the amount of urine in the urinary bladder, and the amount of feces in the rectum. As a result, elderly people are more prone to high blood pressure, dehydration, swallowing and choking problems, urinary incontinence, and constipation or bowel incontinence.

There is also a general decline in the number of motor neurons. As many as 50% of the lower motor neurons in the lumbar region of the spinal cord may be lost by age 60. Muscle fibers innervated by the lost motor neurons are also lost, resulting in a general decline in muscle mass. The remaining motor units can compensate for some of the lost function. This, however, often results in a feeling that one must work harder to perform activities that were previously not so difficult. Loss of motor units also leads to more rapid fatigue, as the remaining units must perform compensatory work.

Reflexes slow as people age because both the generation and the conduction of action potentials and synaptic functions slow. The number of neurotransmitters and receptors declines. Age-related changes in the CNS also slow reflexes. The more complicated the reflex, the more it is affected by age. As reflexes slow, older people are less able to react automatically, quickly, and accurately to changes in internal and external conditions.

The size and weight of the brain decrease as a person ages. At least some of these changes result from the loss of neurons within the cerebrum. The remaining neurons can apparently compensate for much of this loss. In addition to loss of neurons, structural changes occur in the remaining neurons. Neuron plasma membranes become more rigid, the endoplasmic reticulum becomes more irregular in structure, neurofibrillar tangles develop in the cells, and amyloid plaques form in synapses. All these changes decrease the ability of neurons to function. Age-related changes in brain function include decreased voluntary movement, conscious sensations, reflexes, memory, and sleep. Short-term memory is decreased in most older people. This change varies greatly among individuals but, in general, such changes are slow until about age 60 and then become more rapid, especially after age 70. However, the total amount of memory loss is normally not great for most people. The most difficult information for older people to assimilate is that which is unfamiliar and presented verbally and rapidly. Some of these problems may occur as older people are required to deal with new information in the face of existing, contradictory memories. Long-term memory appears to be unaffected or even improved in older people.

As with short-term memory, thinking, which includes problem solving, planning, and intelligence, in general declines slowly

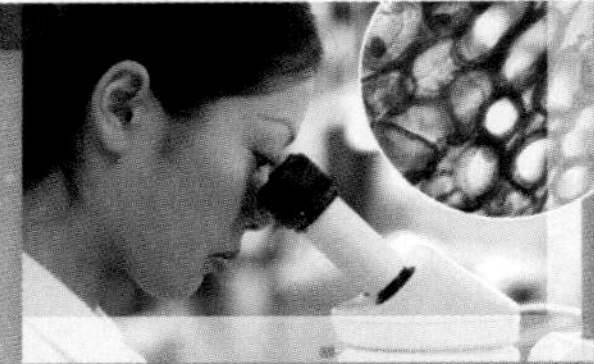

SYSTEMS PATHOLOGY

Stroke

Mr. S, who is approaching middle age, is somewhat overweight and has high blood pressure. He was seated on the edge of his couch, surrounded by empty pizza boxes, bowls of chips and salsa, empty beer cans, and full ashtrays. As Mr. S cheered on his favorite team in a hotly contested big game, he noticed that he felt drowsy and that the television screen seemed blurry. He began to feel dizzy. As he tried to stand up, he suddenly vomited and collapsed to the floor, unconscious.

Mr. S was rushed to the local hospital, where the following signs and symptoms were observed. He exhibited weakness in his limbs, especially on the right, and ataxia (inability to walk). He had loss of pain and temperature sensation in his right lower limb and loss of all sensation in the left side of his face. The dizziness persisted, and he appeared disoriented and lacked attentiveness. He also exhibited dysphagia (the inability to swallow) and hoarseness. He had nystagmus (rhythmic oscillation of the eyes). His pupils were slightly dilated, his respiration was short and shallow, and his pulse rate and blood pressure were elevated.

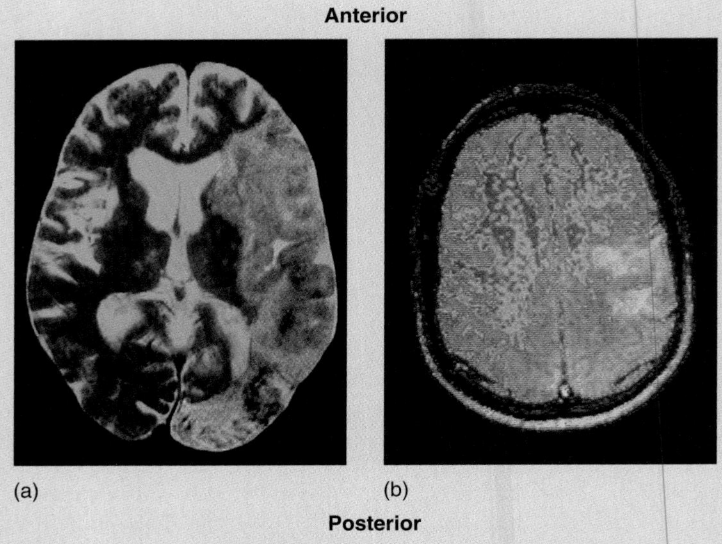

FIGURE B Images of Stroke
(*a*) MRI (magnetic resonance imaging) of a massive stroke in the left side (the viewer's right) of the brain. (*b*) Colorized MMR (nuclear magnetic resonance) showing disruption of blood flow to the left side (the viewer's right) of the brain (*yellow*). This disruption could cause a stroke.

BACKGROUND INFORMATION

Mr. S suffered a stroke, also referred to as a cerebrovascular accident (CVA). The term *stroke* describes a heterogeneous group of conditions involving the death of brain tissue resulting from a disruption of its vascular supply. Two types of stroke exist: **hemorrhagic stroke,** which results from the bleeding of arteries supplying brain tissue, and **ischemic stroke,** which results from blockage of the arteries supplying brain tissue (figure B). The blockage in ischemic stroke can result from a thrombus (a clot that develops in place within an artery) or an embolism (a plug, composed of a detached thrombus or other foreign body, such as a fat globule or gas bubble, that becomes lodged in an artery, blocking it). Mr. S was at high risk for developing a stroke. He was approaching middle age, was overweight, did not exercise enough, smoked, was under stress, and had a poor diet.

The combination of motor loss, which was seen as weakness in his limbs, and sensory loss, seen as loss of pain and temperature sensation in his right lower limb and loss of all sensation in the left side of his face, along with the ataxia, dizziness, nystagmus, and hoarseness, suggests that the stroke affected the brainstem and cerebellum. Blockage of the vertebral artery, a major artery supplying the brain, or its branches can result in what is called a lateral medullary infarction (an area of dead tissue resulting from a loss of blood supply to an area). Damage to the descending motor pathways in that area, above the medullary decussation, results in muscle weakness. Damage to ascending pathways can result in loss of pain and temperature sensation

to age 60 but more rapidly thereafter. These changes, however, are slight and quite variable. Many older people show no change, and about 10% show an increase in thinking ability. Many of these changes are impacted by a person's background, education, health, motivation, and experience.

Among older people, more time is required to fall asleep, there are more periods of waking during the night, and the wakeful periods are of greater duration. Factors that can affect sleep include pain, indigestion, rhythmic leg movements, sleep apnea, decreased urinary bladder capacity, and circulatory problems. There is, on the average, an increase in stage 1 sleep, which is the least restful, and less time spent in stage 4 and REM sleep, which are the most restful.

32 ***How does aging affect sensory function? How does loss of motor neurons affect muscle mass?***

33 ***Does aging always produce memory loss?***

(or other sensory modalities, depending on the affected tract). Damage to cranial nerve nuclei results in the loss of pain and temperature sensation in the face, dizziness, blurred vision, nystagmus, vomiting, and hoarseness. These signs and symptoms are not observed unless the lesion is in the brainstem, where these nuclei are located. Some damage to the cerebellum, also supplied by branches of the vertebral artery, can account for the ataxia.

Drowsiness, disorientation, inattentiveness, and loss of consciousness are examples of generalized neurologic response to damage. Seizures may also result from severe neurologic damage. Depression from neurologic damage or from discouragement is also common. Slight dilation of the pupils; short, shallow respiration; and increased pulse rate and blood pressure are all signs of Mr. S's anxiety, not about the outcome of the game but about his current condition and his immediate future. With a loss of consciousness, Mr. S would not remember the last few minutes of what he saw in the game he was watching. People in these circumstances are often worried about how they are going to deal with work tomorrow. They often have no idea that the motor and sensory losses may be permanent, or that they have a long period of therapy ahead.

PREDICT 11

Given that Mr. S exhibited weakness in his right limbs and loss of pain and temperature sensation in his right lower limb and the left side of his face, state which side of the brainstem was most severely affected by the stroke. Explain your answer.

SYSTEM INTERACTIONS Effect of Stroke on Other Systems

SYSTEM	INTERACTIONS
Integumentary	Decubitus ulcers (bedsores) from immobility; loss of motor function following a stroke leads to immobility.
Skeletal	Loss of bone mass, if muscles are dysfunctional for a prolonged time; in the absence of muscular activity, the bones to which those muscle are attached begin to be resorbed by osteoclasts.
Muscular	Major area of effect; absence of stimulation due to damaged pathways or neurons leads to decreased motor function and may result in muscle atrophy.
Endocrine	Strokes in other parts of the brain could involve the hypothalamus, pineal body, or pituitary gland functions.
Cardiovascular	The following are risks: Phlebothrombosis (blood clot in a vein) can occur from inactivity. Edema around the brain can apply pressure to the cardioregulatory and vasomotor centers of the brain. This pressure can stimulate these centers, which result in elevated blood pressure, and congestive heart failure can result. If the cardioregulatory center in the brain is damaged, death may occur rapidly. Bleeding is due to the use of anticoagulants. Hypotension results from use of antihypertensives.
Respiratory	Pneumonia from aspiration of the vomitus or hypoventilation results from decreased function in the respiratory center. If the respiratory center is severely damaged, death may occur rapidly.
Digestive	Vomiting; dysphagia (difficulty swallowing): hypovolemia (decreased blood volume resulting from decreased fluid intake), which occurs because of dysphagia; possibly a loss of bowel control.
Urinary	Control of the micturition reflex may be inhibited. Urinary tract infection results from catheter implanation or from urinary bladder distension.
Reproductive	Loss of libido; innervation of the reproductive organs is often affected.

SUMMARY

Sensation (p. 477)

1. The senses include general senses and special senses.
2. The somatic senses include touch, pressure, temperature, proprioception, and pain.
3. The visceral senses are primarily pain and pressure.
4. The special senses are smell, taste, sight, hearing, and balance.
5. Sensation, or perception, is the conscious awareness of stimuli received by sensory receptors.
6. Sensation requires a stimulus, a receptor, the conduction of an action potential to the CNS, translation of the action potential, and processing of the action potential in the CNS so that the person is aware of the sensation.

Sensory Receptors

1. Receptors include mechanoreceptors, chemoreceptors, thermoreceptors, photoreceptors, and nociceptors.
2. Free nerve endings detect light touch, pain, itch, tickle, and temperature.
3. Merkel disks respond to light touch and superficial pressure.
4. Hair follicle receptors wrap around the hair follicle and are involved in the sensation of light touch when the hair is bent.
5. Pacinian corpuscles, located in the dermis and hypodermis, detect pressure. In joints, they serve a proprioceptive function.
6. Meissner corpuscles, located in the dermis, are responsible for two-point discriminative touch.
7. Ruffini end organs are involved in continuous touch or pressure.
8. Muscle spindles, located in skeletal muscle, are proprioceptors.
9. Golgi tendon organs, embedded in tendons, respond to changes in tension.
10. A stimulus produces a receptor potential in a sensory receptor. Primary receptors have axons that transmit action potentials toward the CNS. Secondary receptors have no axons but release neurotransmitters.
11. Adaptation is decreased sensitivity to a continued stimulus. Tonic receptors accommodate slowly; phasic receptors accommodate rapidly.

Sensory Tracts

1. Ascending pathways carry conscious and unconscious sensations. The two major ascending systems are the anterolateral and dorsal-column/medial-lemniscal systems.
2. In the anterolateral system,
   - The spinothalamic tract carries pain, temperature, light touch, pressure, tickle, and itch sensations.
   - The spinothalamic tracts are formed by primary neurons that enter the spinal cord and synapse with secondary neurons. The secondary neurons conducting pain and temperature sensations cross the spinal cord and ascend to the thalamus, where they synapse with tertiary neurons that project to the somatic sensory cortex. Axons conducting other sensations, such as light touch or itch, may ascend several spinal cord levels before crossing.
   - The ascending neurons of the spinoreticular tract ascend both contralaterally and ipsilaterally, project to the reticular formation, and influence the level of consciousness.
   - The ascending neurons of the spinomesencephalic tract carry action potentials from cutaneous pain receptors. The spinomesencephalic tract also contributes to eye reflexes.
3. The dorsal-column/medial-lemniscal system carries the sensations of two-point discrimination, proprioception, pressure, and vibration. Primary neurons enter the spinal cord and ascend to the medulla, where they synapse with secondary neurons. Secondary neurons cross over and project to the thalamus. Tertiary neurons extend from there to the somatic sensory cortex.
4. The trigeminothalamic tract carries sensory information from the face, nose, and mouth.
5. In the spinocerebellar system and other tracts,
   - The spinocerebellar tracts carry unconscious proprioception to the cerebellum from the same side of the body.
   - Neurons of the dorsal-column/medial-lemniscal system synapse with the neurons that carry proprioception information to the cerebellum.
   - The spinoolivary tract contributes to coordination of movement.
6. Descending pathways can reduce conscious perception of sensations.

Sensory Areas of the Cerebral Cortex

1. Sensory pathways project to primar ory areas in the cerebral cortex.
2. Sensory areas are organized topographically in the somatic sensory cortex.

Sensory Processing

Association areas of the cerebral cortex process sensory input from the primary sensory areas.

Control of Skeletal Muscles (p. 490)

1. Upper motor neurons are located in the cerebral cortex, cerebellum, and brainstem. Lower motor neurons are found in the cranial nuclei or the anterior horn of the spinal cord gray matter.
2. Upper motor neurons in the cerebral cortex and other brain areas project to lower motor neurons in the brainstem and spinal cord.

Motor Areas of the Cerebral Cortex

1. The primary motor cortex is the precentral gyrus. The premotor and prefrontal areas are staging areas for motor function.
2. The motor cortex is organized topographically.

Motor Tracts

1. The direct pathways maintain muscle tone and control fine, skilled movements in the face and distal limbs. The indirect pathways control conscious and unconscious muscle movements in the trunk and proximal limbs.
2. The corticospinal tracts control muscle movements below the head.
   - About 75%–85% of the upper motor neurons of the corticospinal tracts cross over in the medulla to form the lateral corticospinal tracts in the spinal cord.
   - The remaining upper motor neurons pass through the medulla to form the anterior corticospinal tracts, which cross over in the spinal cord.
   - The upper motor neurons of both tracts synapse with interneurons that then synapse with lower motor neurons in the spinal cord.
3. The corticobulbar tracts innervate the head muscles. Upper motor neurons synapse with interneurons in the reticular formation that, in turn, synapse with lower motor neurons in the cranial nerve nuclei.
4. The indirect pathways include the rubrospinal, vestibulospinal, and reticulospinal tracts and fibers from the basal nuclei.
5. The indirect pathways are involved in conscious and unconscious trunk and proximal limb muscle movements, posture, and balance.

Modifying and Refining Motor Activities

1. Basal nuclei are important in planning, organizing, and coordinating motor movements and posture.
2. The cerebellum has three parts.
   - The vestibulocerebellum controls balance and eye movement.
   - The spinocerebellum corrects discrepancies between intended movements and actual movements.
   - The cerebrocerebellum can "learn" highly specific complex motor activities.

Brainstem Functions (p. 498)

The brainstem contains nuclei for cranial nerves III–X and XII and nuclei of the reticular formation.

Sensory Input Projecting Through the Brainstem

The brainstem receives sensory from ascending spinal cord pathways and from the axons of cranial nerves.

RAS Functions of the Brainstem

Collateral branches of cranial nerves II, V, and VIII project to the reticular activating system (RAS) of the brainstem where they stimulate wakefulness and consciousness.

Motor Output and Reflexes Projecting Through the Brainstem

1. Descending spinal pathways either pass directly through the brainstem or synapse with brainstem nuclei.
2. Several somatic motor and parasympathetic reflexes are controlled by the brainstem.

Vital Functions Controlled by the Brainstem

Many vital functions are controlled by the brainstem.

Other Brain Functions (p. 500)

Speech

1. Speech is located only in the left cortex in most people.
2. Wernicke's area comprehends and formulates speech.
3. Broca's area receives input from Wernicke's area and sends impulses to the premotor and motor areas, which cause the muscle movements required for speech.

Right and Left Cerebral Cortex

1. Each cerebral hemisphere controls and receives input from the opposite side of the body.
2. The right and left hemispheres are connected by commissures. The largest commissure is the corpus callosum, which allows the sharing of information between hemispheres.
3. In most people, the left hemisphere is dominant, controlling speech and analytic skills. The right hemisphere controls spatial and musical abilities.

Brain Waves and Sleep

1. Electroencephalograms (EEGs) record the electrical activity of the brain as alpha, beta, theta, and delta waves.
2. Some brain disorders can be detected with EEGs.
3. Sleep patterns are characterized by specific EEGs.

Memory

At least three kinds of memory exist: sensory, short-term, and long-term.

Limbic System

The limbic system controls visceral functions through the autonomic nervous system and the endocrine system and is involved in emotions and memory.

Effects of Aging on the Nervous System (p. 506)

1. There is a general decline in sensory and motor functions as a person ages.
2. Short-term memory is decreased in most older people.
3. Thinking ability does not decrease in most older people.

REVIEW AND COMPREHENSION

1. Nociceptors respond to
   a. changes in temperature at the site of the receptor.
   b. compression, bending, or stretching of cells.
   c. painful mechanical, chemical, or thermal stimuli.
   d. light striking a receptor cell.
2. Which of these types of sensory receptors respond to pain, itch, tickle, and temperature?
   a. Merkel disks
   b. Meissner corpuscles
   c. Ruffini end organs
   d. free nerve endings
   e. Pacinian corpuscles
3. Which of these types of sensory receptors are involved with proprioception?
   a. free nerve endings
   b. Golgi tendon organs
   c. muscle spindles
   d. Pacinian corpuscles
   e. all of the above
4. The sensory receptor in the dermis and hypodermis responsible for sensing continuous touch or pressure are
   a. Merkel disks.
   b. Meissner corpuscles.
   c. Ruffini end organs.
   d. free nerve endings.
   e. Pacinian corpuscles.
5. Decreased sensitivity to a continued stimulus is called
   a. adaptation.
   b. projection.
   c. translation.
   d. conduction.
   e. phantom pain.
6. Secondary neurons in the spinothalamic tracts synapse with tertiary neurons in the
   a. medulla oblongata.
   b. gray matter of the spinal cord.
   c. cerebellum.
   d. thalamus.
   e. midbrain.
7. If the spinothalamic tract on the right side of the spinal cord is severed,
   a. pain sensations below the damaged area on the right side are eliminated.
   b. pain sensations below the damaged area on the left side are eliminated.
   c. temperature sensations are unaffected.
   d. neither pain sensations nor temperature sensations are affected.
8. Fibers of the dorsal-column/medial-lemniscal system
   a. carry the sensations of two-point discrimination, proprioception, pressure, and vibration.
   b. cross to the opposite side in the medulla oblongata.
   c. are divided into the fasciculus gracilis and fasciculus cuneatus in the spinal cord.
   d. include secondary neurons that exit the medulla and synapse in the thalamus.
   e. all of the above.

9. Tertiary neurons in both the anterolateral system and the dorsal-column/medial-lemniscal system
   a. project to the somatic sensory cortex.
   b. cross to the opposite side in the medulla oblongata.
   c. are found in the spinal cord.
   d. connect to quaternary neurons in the thalamus.
   e. are part of a descending pathway.
10. Unlike the anterolateral and dorsal-column/medial-lemniscal systems, the spinocerebellar tracts
    a. are descending tracts.
    b. transmit information from the same side of the body as the side of the CNS to which they project.
    c. have four neurons in each pathway.
    d. carry only pain sensations.
    e. have primary neurons that synapse in the thalamus.
11. General sensory inputs (pain, pressure, temperature) to the cerebrum end in the
    a. precentral gyrus.
    b. postcentral gyrus.
    c. central sulcus.
    d. corpus callosum.
    e. arachnoid mater.
12. Neurons from which area of the body occupy the greatest area of the somatic sensory cortex?
    a. foot
    b. leg
    c. torso
    d. arm
    e. face
13. A cutaneous nerve to the hand is severed at the elbow. The distal end of the nerve at the elbow is then stimulated. The person reports
    a. no sensation because the receptors are gone.
    b. a sensation only in the region of the elbow.
    c. a sensation "projected" to the hand.
    d. a vague sensation on the side of the body containing the cut nerve.
14. Which of these areas of the cerebral cortex is involved in the motivation and foresight to plan and initiate movements?
    a. primary motor cortex
    b. somatic sensory cortex
    c. prefrontal area
    d. premotor area
    e. basal nuclei
15. Which of these pathways is *not* an ascending (sensory) pathway?
    a. spinothalamic tract
    b. corticospinal tract
    c. dorsal-column/medial-lemniscal system
    d. trigeminothalamic tract
    e. spinocerebellar tract
16. The ______________ tracts innervate the head muscles.
    a. corticospinal
    b. rubrospinal
    c. vestibulospinal
    d. corticobulbar
    e. dorsal-column/medial-lemniscal
17. Most fibers of the direct (pyramidal) system
    a. decussate in the medulla oblongata.
    b. synapse in the pons.
    c. descend in the rubrospinal tract.
    d. begin in the cerebellum.
18. A person with a spinal cord injury is suffering from paresis (partial paralysis) in the right lower limb. Which of these pathways is probably involved?
    a. left lateral corticospinal tract
    b. right lateral corticospinal tract
    c. left dorsal-column/medial-lemniscal system
    d. right dorsal-column/medial-lemniscal system
19. Which of these pathways is *not* an indirect (extrapyramidal) pathway?
    a. reticulospinal tract
    b. corticobulbar tract
    c. rubrospinal tract
    d. vestibulospinal tract
20. The indirect (extrapyramidal) system is concerned with
    a. posture.
    b. trunk movements.
    c. proximal limb movements.
    d. all of the above.
21. The major effect of the basal nuclei is
    a. to act as a comparator for motor coordination.
    b. to decrease muscle tone and inhibit unwanted muscular activity.
    c. to affect emotions and emotional responses to odors.
    d. to modulate pain sensations.
22. Which of the parts of the cerebellum is correctly matched with its function?
    a. vestibulocerebellum—planning and learning rapid, complex movements
    b. spinocerebellum—comparator function
    c. cerebrocerebellum—balance
    d. none of the above
23. Given the following events:
    1. Action potentials from the cerebellum go to the motor cortex and spinal cord.
    2. Action potentials from the motor cortex go to lower motor neurons and the cerebellum.
    3. Action potentials from proprioceptors go to the cerebellum.

    Arrange the events in the order they occur in the cerebellar comparator function.
    a. 1,2,3
    b. 1,3,2
    c. 2,1,3
    d. 2,3,1
    e. 3,2,1
24. The brainstem
    a. consists of ascending and descending pathways.
    b. contains cranial nerve nuclei III–X and XII.
    c. has nuclei and connections that form the reticular activating system.
    d. has many important reflexes, some of which are necessary for survival.
    e. has all of the above.
25. Given these areas of the cerebral cortex:
    1. Broca's area
    2. premotor area
    3. primary motor cortex
    4. Wernicke's area

    If a person hears and understands a word and then says the word out loud, in what order are the areas used?
    a. 1,4,2,3
    b. 1,4,3,2
    c. 3,1,4,2
    d. 4,1,2,3
    e. 4,1,3,2
26. The main connection between the right and left hemispheres of the cerebrum is the
    a. intermediate mass.
    b. corpus callosum.
    c. vermis.
    d. unmyelinated nuclei.
    e. thalamus.
27. Which of these activities is mostly associated with the left cerebral hemisphere in most people?
    a. sensory input from the left side of the body
    b. mathematics and speech
    c. spatial perception
    d. recognition of faces
    e. musical ability

28. The limbic system is involved in the control of
    a. sleep and wakefulness.
    b. posture.
    c. higher intellectual processes.
    d. emotion, mood, and sensations of pain or pleasure.
    e. hearing.
29. Long-term memory involves
    a. a change in the cytoskeleton of neurons.
    b. the movement of calcium into the neuron.
    c. an increase in glutamate release by presynaptic neurons.
    d. specific protein synthesis.
    e. all of the above.
30. Concerning long-term memory,
    a. declarative (explicit) memory involves the development of skills, such as riding a bicycle.
    b. procedural (implicit) memory involves the retention of facts, such as names, dates, or places.
    c. much of declarative (explicit) memory is lost through time.
    d. declarative (explicit) memory is stored primarily in the cerebellum and premotor area of the cerebrum.
    e. all of the above.

Answers in Appendix E

CRITICAL THINKING

1. Describe all the sensations involved when a woman picks up an apple and bites into it. Explain which of those sensations are special and which are general. What types of receptors are involved? Which aspects of the taste of the apple are actually taste and which are olfaction?
2. Some student nurses are at a party. Because they love anatomy and physiology so much, they are discussing adaptation of the special senses. They make the following observations:
    a. When entering a room, an odor like brewing coffee is easily noticed. A few minutes later, the odor might be barely, if at all, detectable, no matter how hard one tries to smell it.
    b. When entering a room, the sound of a ticking clock can be detected. Later the sound is not noticed until a conscious effort is made to hear it. Then it is easily heard.

    Explain the basis for each of these observations.
3. A patient suffered a loss of two-point discrimination and proprioception on the right side of the body. Voluntary movement of muscles was not affected, and pain and temperature sensations were normal. Is it possible to conclude that the right side of the spinal cord was damaged?
4. A patient is suffering from the loss of two-point discrimination and proprioceptive sensations on the right side of the body resulting from a lesion in the pons. What tract is affected, and which side of the pons is involved?
5. A patient suffers a lesion in the central core of the spinal cord. It is suspected that the fibers that decussate and that are associated with the lateral spinothalamic tracts are affected in the area of the lesion. What observations would be consistent with that diagnosis?
6. A person in a car accident exhibits the following symptoms: extreme paresis on the right side, including the arm and leg; reduction of pain sensation on the left side; and normal tactile sensation on both sides. Which tracts are damaged? Where did the patient suffer tract damage?
7. If the right side of the spinal cord is completely transected, what symptoms do you expect to observe with regard to motor function, two-point discrimination, light touch, and pain perception?
8. A patient with a cerebral lesion exhibits a loss of fine motor control of the left hand, arm, forearm, and shoulder. All other motor and sensory functions appear to be intact. Describe the location of the lesion as precisely as possible.
9. A patient suffers brain damage in an automobile accident. It is suspected that the cerebellum is the part of the brain that is affected. On the basis of what you know about cerebellar function, how can you determine that the cerebellum is involved?
10. Woody Knothead was accidentally struck in the head with a baseball bat. He fell to the ground, unconscious. Later, when he regained consciousness, he was not able to remember any of the events that happened 10 minutes before the accident. Explain. What complications might be looked for at a later time?
11. Pamela Frail is a 40-year-old woman experiencing limb weakness on her left side. A brain MRI shows lesions in her cerebral cortex. Which of the descending pathways are affected by these lesions? Would you expect the brain lesions to appear ipsilateral or contralateral to her limb weakness? Explain why.

Answers in Appendix F

Visit this book's website at www.mhhe.com/seeley8 for chapter quizzes, interactive learning exercises, and other study tools.

15 The Special Senses

Historically, it was thought that we had just five senses: smell, taste, sight, hearing, and touch. Today we recognize many more. Some specialists suggest that there are at least 20, or perhaps as many as 40, different senses. Most of these senses are part of what was originally classified as "touch." These "general senses" were discussed in chapter 14. The sense of balance is now recognized as a "special sense," making a total of 5 special senses: smell, taste, sight, hearing, and balance. **Special senses** are defined as senses with highly localized receptors that provide specific information about the environment. This chapter describes *olfaction* (p. 515), *taste* (p. 518), the *visual system* (p. 521), and *hearing and balance* (p. 542). The chapter concludes with a look at the *effects of aging on the special senses* (p. 556).

Photograph of an isolated cochlea from the inner ear.

Nervous System

OLFACTION

Olfaction (ol-fak′shŭn), the sense of smell, occurs in response to odors that stimulate sensory receptors located in the extreme superior region of the nasal cavity, called the **olfactory region** (figure 15.1*a*). Most of the nasal cavity is involved in respiration, with only a small superior part devoted to olfaction. The major anatomical features of the nasal cavity are described in chapter 23 in relation to respiration. The specialized nasal epithelium of the olfactory region is called the **olfactory epithelium.**

PREDICT 1

Explain why it sometimes helps to inhale slowly and deeply through the nose when trying to identify an odor.

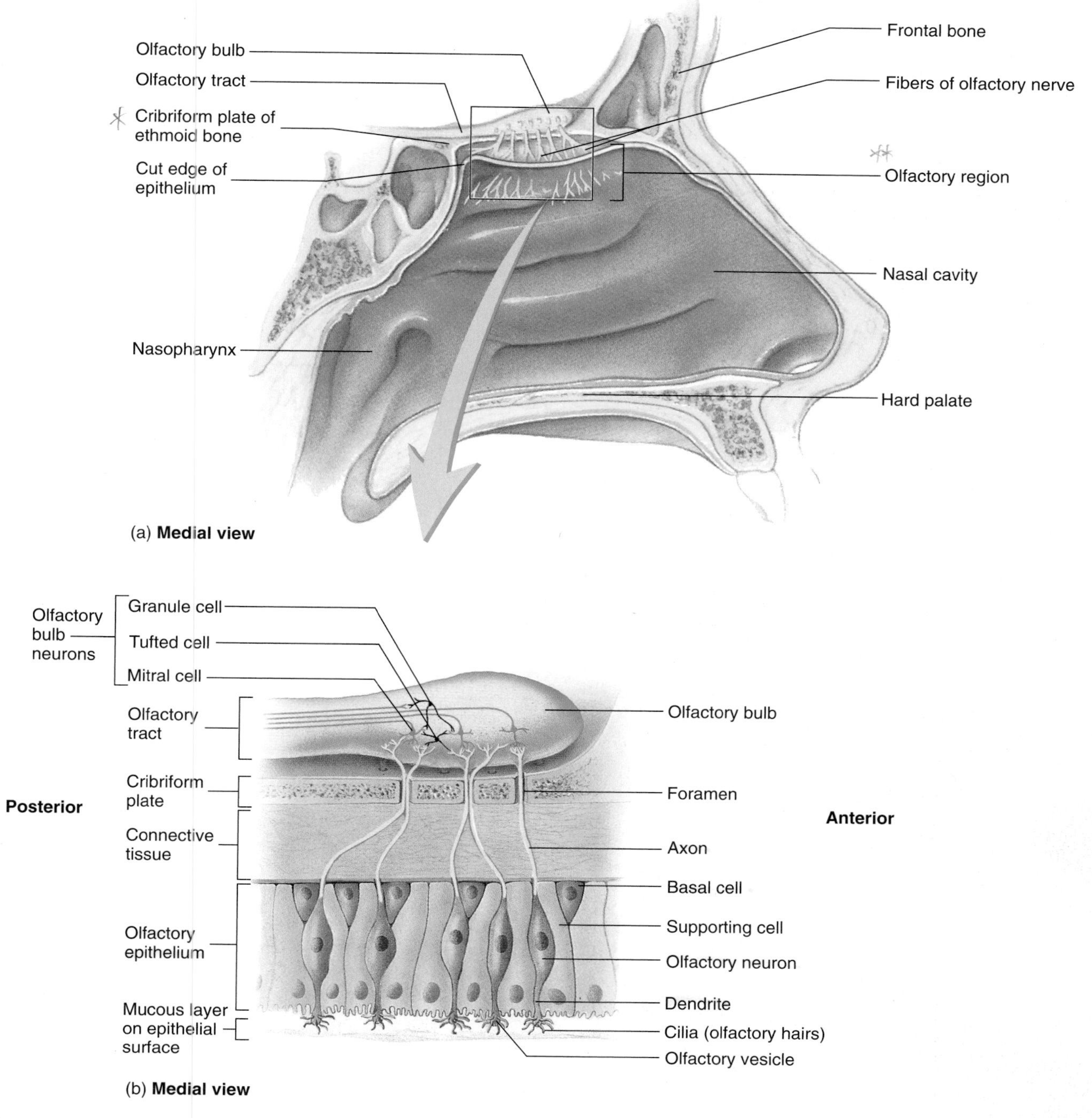

FIGURE 15.1 Olfactory Region, Epithelium, and Bulb

(*a*) The lateral wall of the nasal cavity (cut in sagittal section), showing the olfactory region and olfactory bulb. (*b*) The olfactory cells within the olfactory epithelium, the olfactory nerve processes passing through the cribriform plate, and the fine structure of the olfactory bulb.

Olfactory Epithelium and Bulb

Ten million **olfactory neurons** are present within the olfactory epithelium (figure 15.1*b*). Axons of these bipolar neurons project through numerous, small foramina of the bony cribriform plate (see chapter 7) to the **olfactory bulbs,** where they synapse with secondary neurons. **Olfactory tracts** project from the bulbs to the cerebral cortex.

The dendrites of olfactory neurons extend to the epithelial surface of the nasal cavity, and their ends are modified into bulbous enlargements called **olfactory vesicles** (see figure 15.1*b*). These vesicles possess cilia called **olfactory hairs,** which lie in a thin mucous film on the epithelial surface.

Airborne molecules enter the nasal cavity and are dissolved in the fluid covering the olfactory epithelium. Some of these molecules, referred to as **odorants** (ō′dŏr-ants; molecules with an odor), bind to transmembrane odorant receptor molecules (chemoreceptors) of the olfactory hair membranes (figure 15.2). A G protein, associated with each odorant receptor, is activated by the binding of the odorant. The α subunit of the activated G protein binds to and activates adenylate cyclase, which in turn catalyzes the formation of cyclic AMP (cAMP) from ATP. cAMP in these cells causes Na^+ and Ca^{2+} channels to open. The influx of ions into the olfactory hairs results in depolarization and the production of action potentials in the olfactory neurons.

Odorant receptors are composed of seven transmembrane polypeptide subunits produced by a large family of genes. Combinations of subunits from approximately 1000 different odorant receptors can be produced. These receptors can react to odorants of different sizes, shapes, and functional groups. These capabilities, together with multiple intracellular pathways involving G proteins, adenylate cyclase, and ion channels allow for a wide variety of detectable smells, which is about 4000 for the average person. With such a wide range of smells, however, only seven primary classes of odors have been proposed: (1) camphoraceous (e.g., moth balls), (2) musky, (3) floral, (4) pepperminty, (5) ethereal (e.g., fresh pears), (6) pungent, and (7) putrid. It is very unlikely, however, that this list is an accurate representation of all primary odors, and some studies point to the possibility of as many as 50 primary odors.

The threshold for the detection of odors is very low, so very few odorant molecules are required to trigger the response. Apparently, there is rather low specificity in the olfactory epithelium. A given

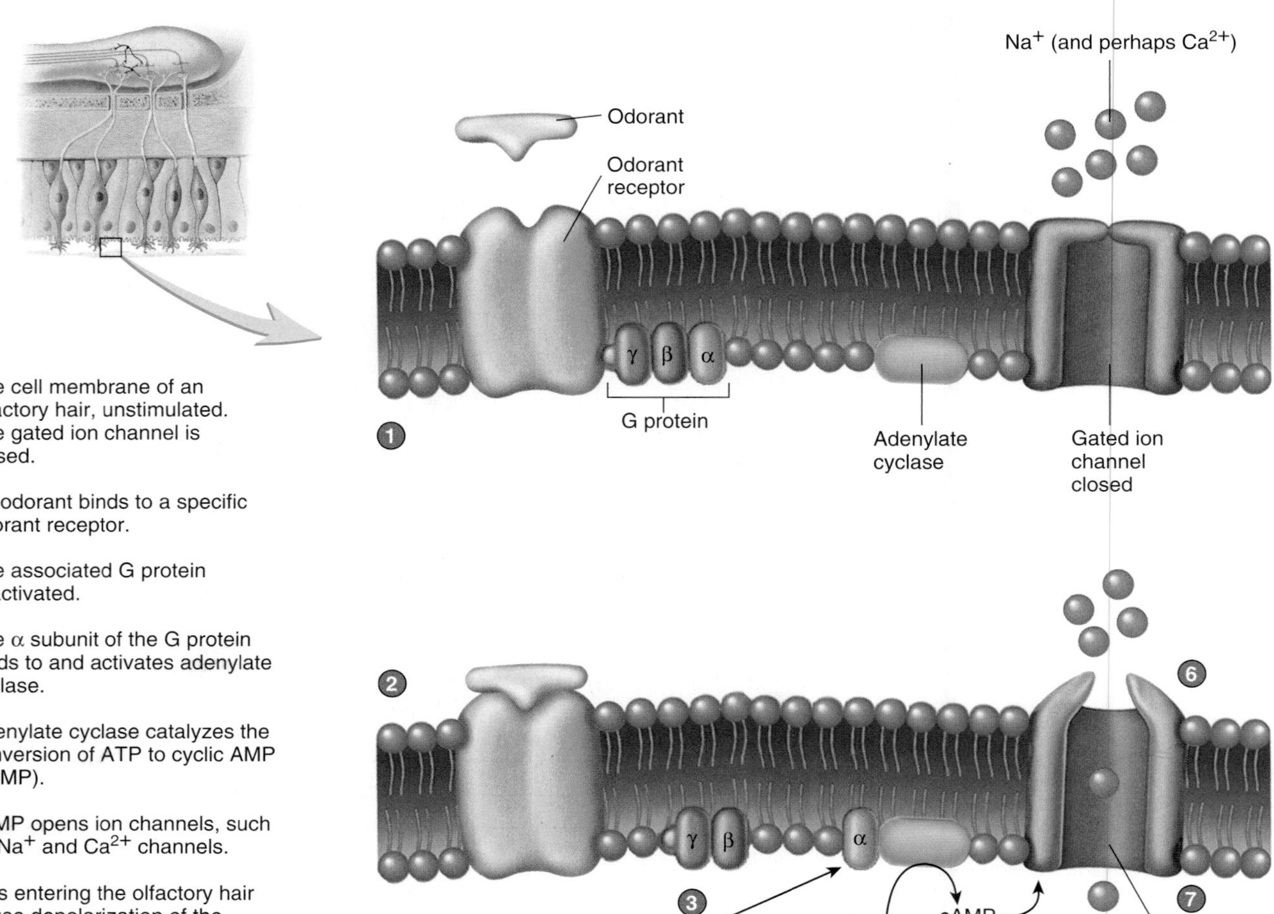

PROCESS FIGURE 15.2 Action of Odorant Binding to Membrane of Olfactory Hair

receptor may react to more than one type of odorant. Odorant receptors may become saturated with odorants and no longer respond to odorant molecules. This accommodation makes a person less sensitive to an odorant after being exposed to it for a short time. For example, when a person first enters a room with a distinctive odor, that odor is very noticeable, but it becomes almost unnoticeable after the person remains in the room for a short time.

The "Odor" of Natural Gas

Methylmercaptan, which has a nauseating odor similar to that of rotten cabbage, is added to natural gas at a concentration of about one part per million. A person can detect the odor of about 1/25 billionth of a milligram of the substance and therefore is aware of the presence of the more dangerous but odorless natural gas.

Odor Survey Results

The National Geographic Society conducted a smell survey in 1986, which was the largest sampling of its kind ever conducted. One and a half million people participated. Of six odors studied, 98%–99% of those responding could smell isoamyl acetate (banana), eugenol (cloves), mercaptans, and rose, but 29% could not smell galaxolide (musk), and 35% could not smell androsterone (contained in sweat). Of those responding to the survey, 1.2% could not smell at all, a disorder called **anosmia** (an-oz′mē-ă).

The primary olfactory neurons have the most exposed nerve endings of any neurons, and they are constantly being replaced. The entire olfactory epithelium, including the neurosensory cells, is lost about every 2 months as the olfactory epithelium degenerates and is lost from the surface. Lost olfactory cells are replaced by a proliferation of **basal cells** in the olfactory epithelium. This replacement of olfactory neurons is unique among neurons, most of which are permanent cells that have a very limited ability to replicate (see chapter 4).

Neuronal Pathways for Olfaction

Axons from the olfactory neurons form the olfactory nerves (cranial nerve I), which enter the olfactory bulbs (see figure 15.1*b*), where they synapse with **mitral** (mī′trăl; triangular; shaped like a bishop's miter or hat) **cells** or **tufted cells.** The mitral and tufted cells relay olfactory information to the brain through the olfactory tracts and synapse with granule cells in the olfactory bulb. Olfactory bulb neurons also receive input from nerve cell processes entering the olfactory bulb from the brain. As a result of input from both mitral cells and the brain, olfactory bulb neurons can modify olfactory information before it leaves the olfactory bulb. This modification enhances the accommodation occurring in the odorant receptors.

Olfaction is the only major sensation that is relayed directly to the cerebral cortex without first passing through the thalamus. Each olfactory tract terminates in an area of the brain called the **olfactory cortex** (figure 15.3). The olfactory cortex is in the frontal and temporal lobes, within the lateral fissure of the cerebrum, and it can be divided structurally and functionally into three areas: lateral, medial, and intermediate. The **lateral olfactory area** is involved in the conscious perception of smell. The **medial olfactory area** is responsible for visceral and emotional reactions to odors and has connections to the limbic system, through which it connects to the hypothalamus. Axons extend from the **intermediate olfactory area** along the olfactory tract to the bulb and synapse with the olfactory bulb neurons, thus constituting a major mechanism by which sensory information is modulated within the

1. Axons of the olfactory neurons in the olfactory epithelium project through foramina in the cribriform plate to the olfactory bulb.
2. Axon of neurons in the olfactory bulb project through the olfactory tract to the olfactory cortex.
3. The lateral olfactory area is involved in the conscious perception of smell.
4. The medial olfactory area is involved in the visceral and emotional reaction to odors.
5. The intermediate olfactory area receives input from the medial and lateral olfactory areas.
6. Axons from the intermediate olfactory area project along the olfactory tract to the olfactory bulb. Action potentials carried by those axons modulate the activity of the neurons in the olfactory bulb.

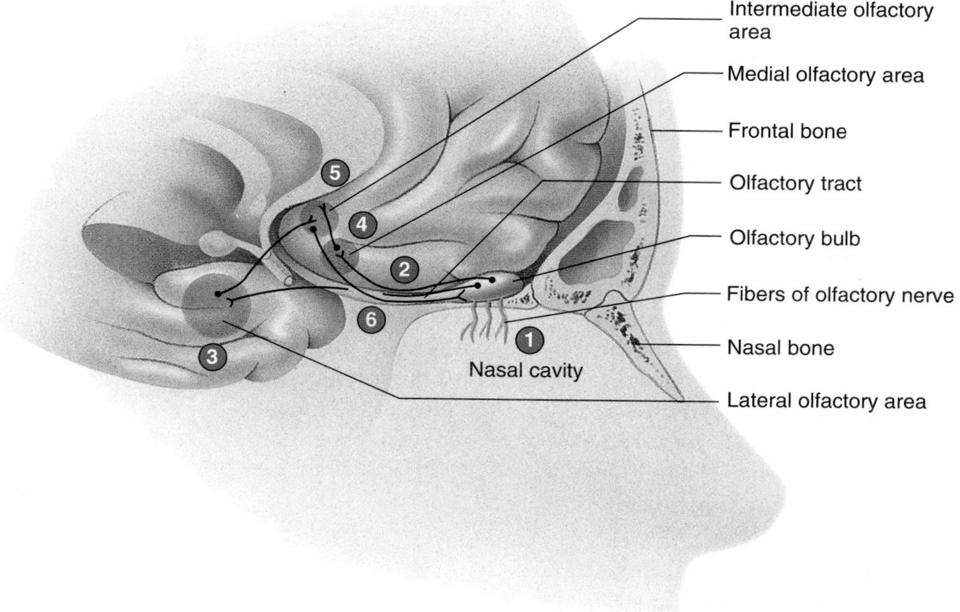

PROCESS FIGURE 15.3 Olfactory Neuronal Pathways and Cortex

olfactory bulb. The feedback from the intermediate olfactory area is mostly inhibitory, further enhancing the rapid accommodation of the olfactory system. Accommodation at the receptor level, the olfactory bulb level, and the CNS level make the olfactory system insensitive to an odorant after a brief exposure.

1. ***Describe the initiation of an action potential in an olfactory neuron. Name all the structures and cells that the action potential encounters on the way to the olfactory cortex.***
2. ***What is a primary odor? Name seven examples. How do the primary odors relate to the ability to smell many different odors?***
3. ***What type of neurons are olfactory neurons? What is unique about olfactory neurons with respect to replacement?***
4. ***How is the sense of smell modified in the olfactory bulb?***
5. ***Name the three areas of the olfactory cortex, and give their functions.***
6. ***Explain how the CNS connections elicit various visceral and conscious responses to smell.***

PREDICT 2

The olfactory system quickly adapts to continued stimulation, and a particular odor becomes unnoticed before very long, even though the odor molecules are still present in the air. Describe as many sites as you can in the olfactory pathways where such adaptation can occur.

TASTE

The sensory structures that detect **gustatory,** or **taste,** stimuli are the **taste buds.** Most taste buds are associated with specialized portions of the tongue called **papillae** (pă-pil′ē). Taste buds, however, are also located on other areas of the tongue, the palate, and even the lips and throat, especially in children. The four major types of papillae are named according to their shape (figure 15.4): **vallate** (val′āt; surrounded by a wall), **fungiform** (fŭn′ji-fōrm; mushroom-shaped), **foliate** (fō′lē-āt; leaf-shaped), and **filiform** (fil′i-fōrm; filament-shaped). Taste buds (see figure 15.4*c–e*) are associated with vallate, fungiform, and foliate papillae. Filiform papillae are the most numerous papillae on the surface of the tongue but have no taste buds. Rather, filiform papillae provide a rough surface on the tongue, allowing it to manipulate food more easily.

Vallate papillae are the largest but least numerous of the papillae. Eight to 12 of these papillae form a V-shaped row along the border between the anterior and posterior parts of the tongue (figure 15.4*a*). Fungiform papillae are scattered irregularly over the entire superior surface of the tongue, appearing as small, red dots interspersed among the far more numerous filiform papillae. Foliate papillae are distributed in folds on the sides of the tongue and contain the most sensitive of the taste buds. They are most numerous in young children and decrease with age. They are located mostly posteriorly in adults.

Histology of Taste Buds

Taste buds are oval structures embedded in the epithelium of the tongue and mouth (figure 15.4*g*). Each of the 10,000 taste buds on a person's tongue consists of three major types of specialized epithelial cells. The sensory cells of each taste bud consist of about 50 **taste,** or **gustatory, cells.** The remaining two cell types, which are nonsensory cells, are **basal cells** and **supporting cells.** Like olfactory cells, the cells of the taste buds are replaced continuously, each having a normal life span of about 10 days. Each taste cell has several microvilli, called **gustatory hairs,** extending from its apex into a tiny opening in the epithelium called the **taste,** or **gustatory, pore.**

Function of Taste

Substances called **tastants** (tās′tants), dissolved in saliva, enter the taste pores and, by various mechanisms, cause the taste cells to depolarize. These cells do not have classic axons but have short connections with secondary sensory neurons (see chapter 14). Those connections have some characteristics of chemical synapses. Neurotransmitters (apparently including ATP) are released from the taste cells and stimulate action potentials in the axons of sensory neurons associated with them.

The taste of **salt** results when Na^+ diffuse through Na^+ channels (figure 15.5*a*) of the gustatory hairs or other cell surfaces of taste cells, resulting in depolarization of the cells. Hydrogen ions (H^+) of **acids,** which result in **sour** taste, can cause depolarization of taste cells by one of three mechanisms (figure 15.5*b*): (1) They can enter the cell directly through H^+ channels, (2) they can bind to ligand-gated K^+ channels and block the exit of K^+ from the cell, or (3) they can open ligand-gated channels for other positive ions and allow them to diffuse into the cell. **Sweet** and **bitter** tastants bind to receptors (figure 15.5*c* and *d*) on the gustatory hairs of taste cells and cause depolarization through a G protein mechanism. A taste called **umami** (ū-ma′mē; a Japanese term loosely translated as savory) results when amino acids, such as glutamate, bind to receptors (figure 15.5*e*) on gustatory hairs of taste cells and cause depolarization through a G protein mechanism.

The texture of food in the oral cavity also affects the perception of taste. Hot or cold food may interfere with the ability of the taste buds to function in tasting food. If a cold fluid is held in the mouth, the body warms the fluid and the taste becomes enhanced. On the other hand, adaptation is very rapid for taste. This adaptation apparently occurs both at the level of the taste bud and within the CNS. Adaptation may begin within 1 or 2 seconds after a taste sensation is perceived, and complete adaptation may occur within 5 minutes.

Even though only five primary tastes have been identified, humans can perceive a fairly large number of different tastes, presumably by combining the five basic taste sensations. As with olfaction, the specificity of the receptor molecules is not perfect. For example, artificial sweeteners have different chemical structures than the sugars they are designed to replace and some are many times more powerful than natural sugars in stimulating taste sensations.

Many of the sensations thought of as being taste are strongly influenced by olfactory sensations. This phenomenon can be demonstrated by pinching one's nose to close the nasal passages while trying to taste something. With olfaction blocked, it is difficult to distinguish between the taste of a piece of apple and the taste of potato. Much of the "taste" is lost by this action. This is one reason that a person with a cold has a reduced sensation of taste. Although all taste buds are able to detect all five of the basic tastes, each taste cell is usually most sensitive to one.

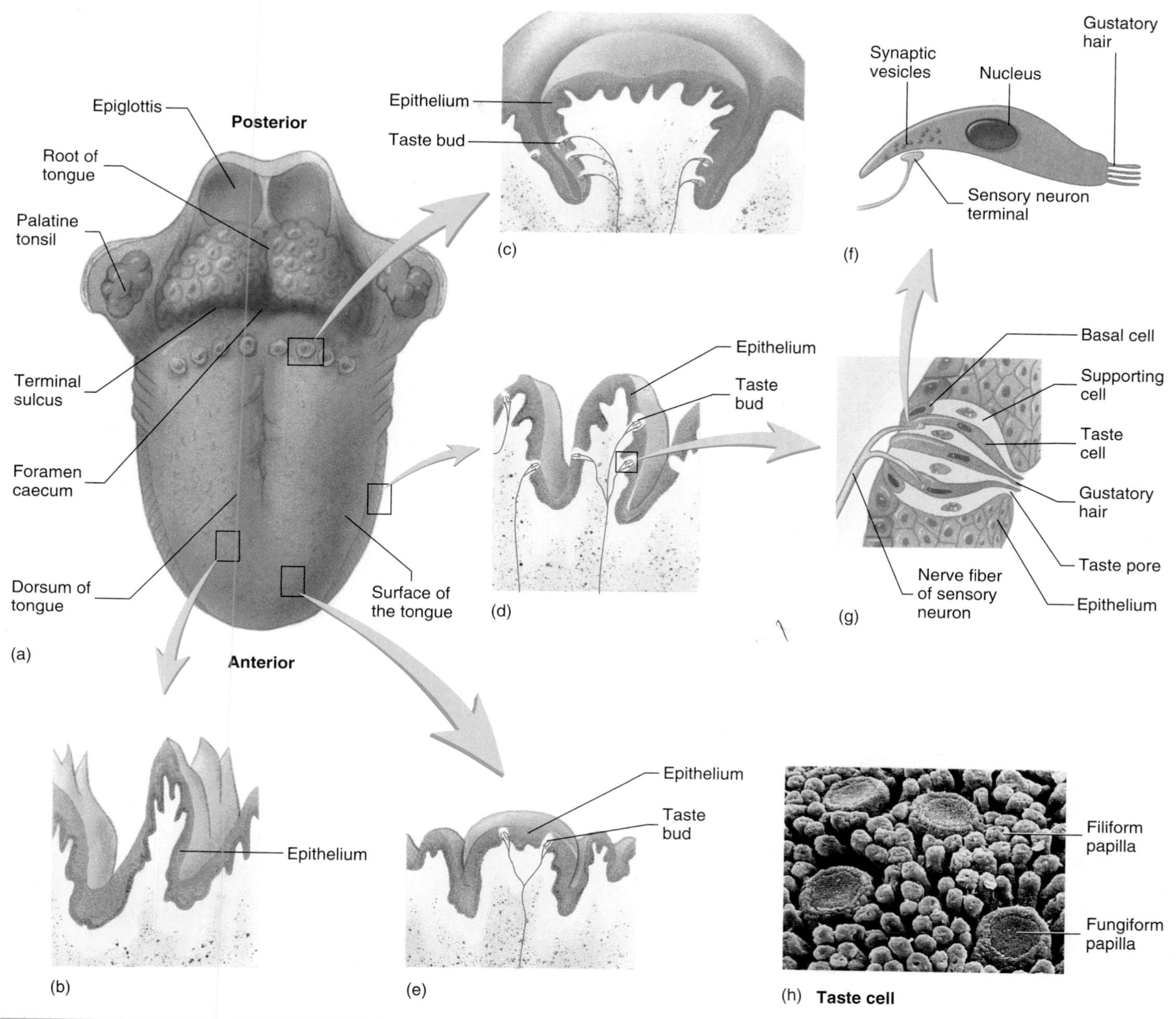

FIGURE 15.4 Papillae and Taste Buds

(*a*) Surface of the tongue. (*b*) Filiform papilla. (*c*) Vallate papilla. (*d*) Foliate papilla. (*e*) Fungiform papilla. (*f*) Taste cell. (*g*) A taste bud. (*h*) Scanning electron micrograph of taste buds (fungiform and filiform papillae) on the surface of the tongue.

Thresholds vary for the five primary tastes. Sensitivity for bitter substances is the highest; sensitivities for sweet and salty tastes are the lowest. Sugars, some other carbohydrates, and some proteins produce sweet tastes; many proteins and amino acids produce umami tastes; acids produce sour tastes; metal ions tend to produce salty tastes; and alkaloids (bases) produce bitter tastes. Many alkaloids are poisonous; thus, the high sensitivity for bitter tastes may be protective. On the other hand, humans tend to crave sweet, salty, and umami tastes, perhaps in response to the body's need for sugars, carbohydrates, proteins, and minerals.

Neuronal Pathways for Taste

Taste from the anterior two-thirds of the tongue, except from the circumvallate papillae, is carried by means of a branch of the facial nerve (VII) called the **chorda tympani** (kōr′dă tim′pă-nē; so-named because it crosses over the surface of the tympanic membrane of the middle ear). Taste from the posterior one-third of the tongue, the circumvallate papillae, and the superior pharynx is carried by means of the glossopharyngeal nerve (IX). In addition to these two major nerves, the vagus nerve (X) carries a few fibers for taste sensation from the epiglottis.

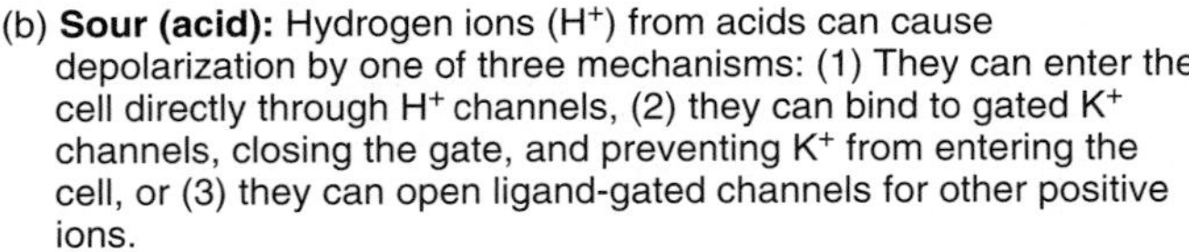

(a) **Salt:** Sodium ions diffuse through Na^+ channels, resulting in depolarization.

(b) **Sour (acid):** Hydrogen ions (H^+) from acids can cause depolarization by one of three mechanisms: (1) They can enter the cell directly through H^+ channels, (2) they can bind to gated K^+ channels, closing the gate, and preventing K^+ from entering the cell, or (3) they can open ligand-gated channels for other positive ions.

(c) **Sweet:** Sugars, such as glucose, or artificial sweeteners bind to receptors and cause the cell to depolarize by means of a G protein mechanism. The α subunit of the G protein activates adenylate cyclase, which produces cAMP. cAMP activates a kinase that phosphorylates K^+ channels. The K^+ channels close, resulting in depolarization.

(d) **Bitter:** Bitter tastants, such as quinine, bind to receptors and cause depolarization of the cell through a G protein mechanism. The α subunit of the G protein activates phospholipase C, which converts phosphoinositol (PIP_2) to inositol triphosphate (IP_3). IP_3 causes Ca^{2+} release from intracellular stores and depolarization of the cell.

(e) **Glutamate (umami):** Amino acids, such as glutamate, bind to receptors and cause depolarization through a G protein mechanism. The α subunit of the G protein activates adenylate cyclase, which catalyzes the conversion of ATP to cAMP. cAMP opens Ca^{2+} channals. The influx of Ca^{2+} causes depolarization of the cell.

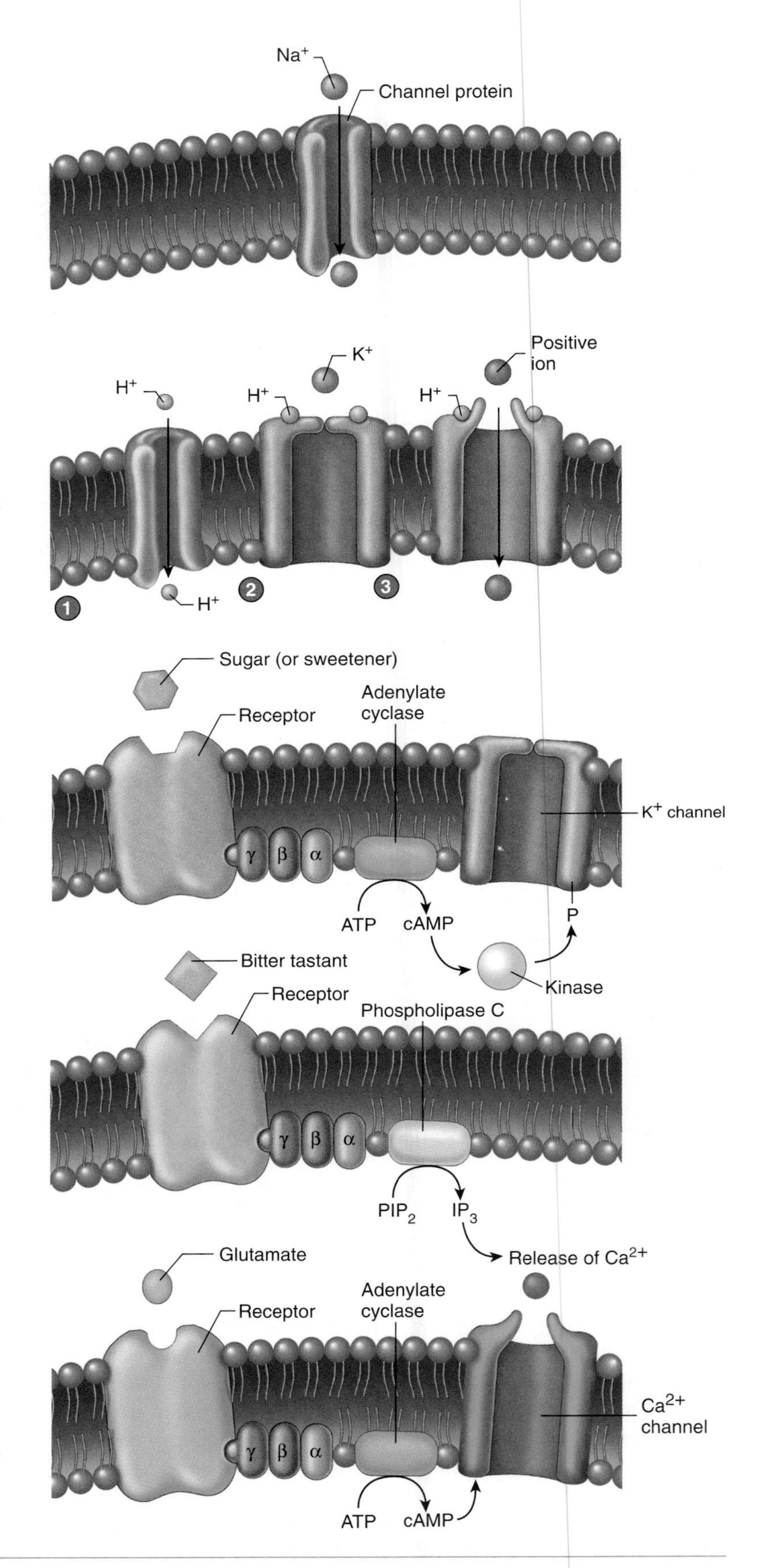

FIGURE 15.5 Actions of the Major Tastants

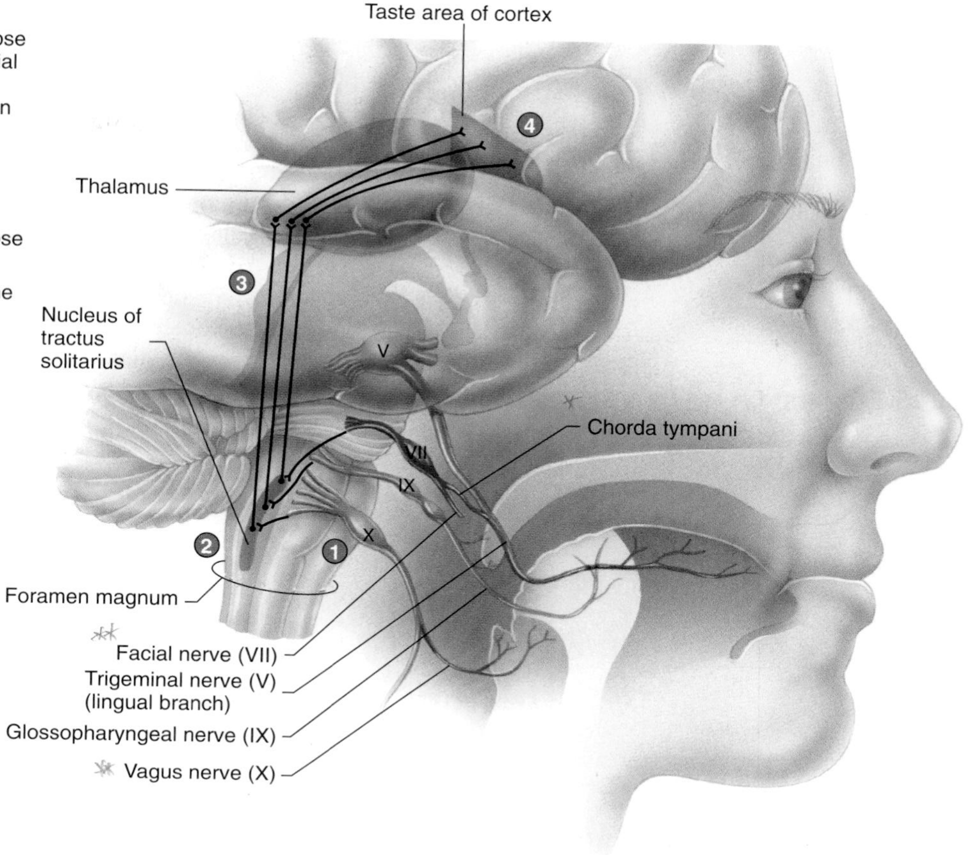
1. Axons of sensory neurons, which synapse with taste receptors, pass through cranial nerves VII, IX, and X and through the ganglion of each nerve (enlarged portion of each nerve).
2. The axons enter the brainstem and synapse in the nucleus of the tractus solitarius.
3. Axons from the nucleus solitarius synapse in the thalamus.
4. Axons from the thalamus terminate in the taste area of the cortex.
Taste area of cortex
Thalamus
Nucleus of tractus solitarius
V
Chorda tympani
VII
IX
X
Foramen magnum
Facial nerve (VII)
Trigeminal nerve (V) (lingual branch)
Glossopharyngeal nerve (IX)
Vagus nerve (X)

PROCESS FIGURE 15.6 Pathways for the Sense of Taste

The facial nerve (anterior two-thirds of the tongue), glossopharyngeal nerve (posterior one-third of the tongue), and vagus nerve (root of the tongue) all carry taste sensations. The trigeminal nerve carries tactile sensations from the anterior two-thirds of the tongue. The chorda tympani from the facial nerve (carrying taste input) joins the trigeminal nerve.

These nerves extend from the taste buds to the tractus solitarius of the medulla oblongata (figure 15.6). Fibers from this nucleus decussate and extend to the thalamus. Neurons from the thalamus project to the taste area of the cortex, which is at the extreme inferior end of the postcentral gyrus.

7 ***Name and describe the four kinds of papillae found on the tongue. Which ones have taste buds associated with them?***

8 ***Starting with the gustatory hair, name the structures and cells that an action potential would encounter on the way to the taste area of the cerebral cortex.***

9 ***What is the life span of a normal gustatory cell?***

10 ***What are the five primary tastes? Describe how each type of tastant causes depolarization of a taste cell.***

11 ***How is the sense of taste related to the sense of smell?***

VISUAL SYSTEM

The visual system includes the eyes, the accessory structures, and the optic nerves (II), tracts, and pathways. The **eye** includes the eyeball (the globe of the eye) and the lens. The eyes respond to light and initiate afferent action potentials, which are transmitted from the eyes to the brain by the optic nerves and tracts. The accessory structures, such as eyebrows, eyelids, eyelashes, and tear glands, help protect the eyes from direct sunlight and damaging particles. Much of the information about the world around us is detected by the visual system. Our education is largely based on visual input and depends on our ability to read words and numbers. Visual input includes information about light and dark, color and hue. Visual stimuli can come from far greater distances than can stimuli for any other sense. The eye can detect light originating from stars billions of miles away.

Accessory Structures

Accessory structures protect, lubricate, move, and in other ways aid in the function of the eye. These structures include the eyebrows, eyelids, conjunctiva, lacrimal apparatus, and extrinsic eye muscles.

Eyebrows

The **eyebrows** (figure 15.7) protect the eyes by preventing perspiration, which can irritate the eyes, from running down the forehead and into them, and they help shade the eyes from direct sunlight.

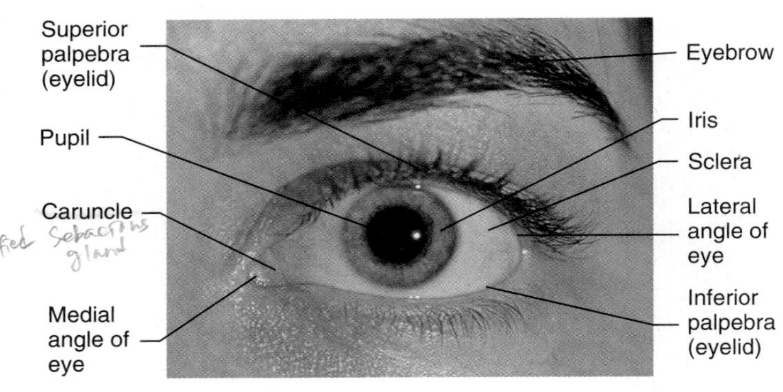

FIGURE 15.7 The Left Eye and Its Accessory Structures

Eyelids

The **eyelids,** also called **palpebrae** (pal-pē′brē), with their associated lashes, protect the eyes from foreign objects. The space between the two eyelids is called the **palpebral fissure,** and the angles where the eyelids join at the medial and lateral margins of the eye are called **canthi** (kan′thī; corners of the eye; see figure 15.7). The medial canthus contains a small, reddish-pink mound called the **caruncle** (kar′ŭng-kl; mound of tissue). The caruncle contains some modified sebaceous and sweat glands.

The eyelids consist of five layers of tissue (figure 15.8). From the outer to the inner surface, they are (1) a thin layer of integument on the external surface; (2) a thin layer of areolar connective tissue; (3) a layer of skeletal muscle consisting of the orbicularis

Sagittal section

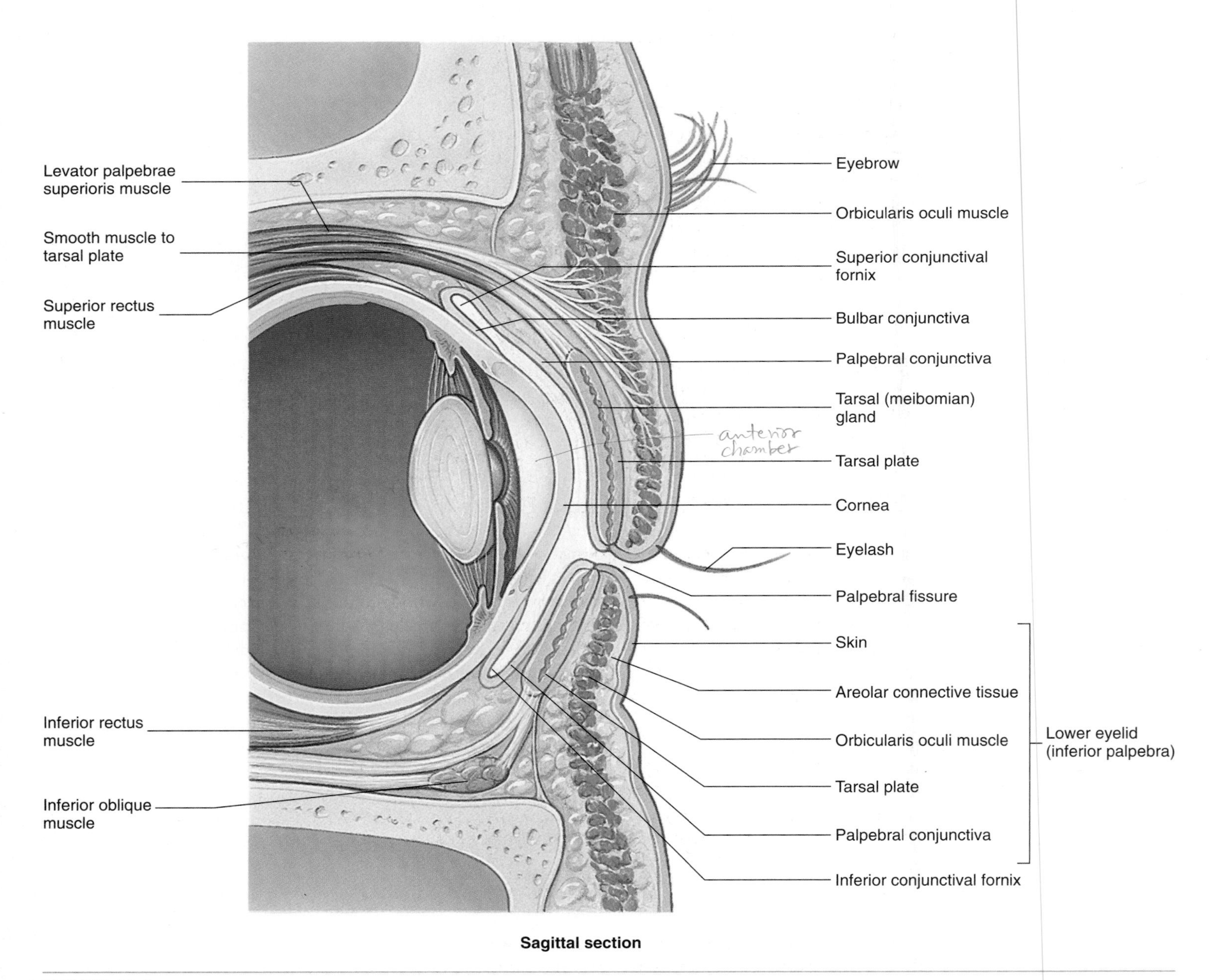

FIGURE 15.8 Sagittal Section Through the Eye, Showing Its Accessory Structures

oculi and levator palpebrae superioris muscles; (4) a crescent-shaped layer of dense connective tissue called the **tarsal** (tar′săl) **plate,** which helps maintain the shape of the eyelid; and (5) the palpebral conjunctiva (described in the next section), which lines the inner surface of the eyelid and the anterior surface of the eyeball.

If an object suddenly approaches the eye, the eyelids protect the eye by rapidly closing and then opening (blink reflex). Blinking, which normally occurs about 25 times per minute, also helps keep the eye lubricated by spreading tears over the surface. Movements of the eyelids are a function of skeletal muscles. The orbicularis oculi muscle closes the lids, and the levator palpebrae superioris elevates the upper lid (see chapter 10). The eyelids also help regulate the amount of light entering the eye.

Eyelashes (see figures 15.7 and 15.8) are attached as a double or triple row of hairs to the free edges of the eyelids. **Ciliary glands** are modified sweat glands that open into the follicles of the eyelashes to keep them lubricated. When one of these glands becomes inflamed, it is called a **sty. Meibomian** (mī-bō′mē-an), or tarsal, **glands** are sebaceous glands near the inner margins of the eyelids; they produce **sebum** (sē′bŭm; an oily semifluid substance), which lubricates the lids and restrains tears from flowing over the margin of the eyelids. An infection or a blockage of a meibomian gland is called a **chalazion** (ka-lā′zē-on), or **meibomian cyst.**

Conjunctiva

The **conjunctiva** (kon-jŭnk-tī′vă; see figure 15.8) is a thin, transparent mucous membrane. The **palpebral conjunctiva** covers the inner surface of the eyelids, and the **bulbar conjunctiva** covers the anterior white surface of the eye. The points at which the palpebral and bulbar conjunctivae meet are the superior and inferior **conjunctival fornices.**

Conjunctivitis

Conjunctivitis is an inflammation of the conjunctiva caused by an infection or another irritation. An example of conjunctivitis caused by a bacterium is **acute contagious conjunctivitis,** also called **pinkeye.**

Lacrimal Apparatus

The **lacrimal** (lak′ri-măl) **apparatus** (figure 15.9) consists of a lacrimal gland situated in the superolateral corner of the orbit and a nasolacrimal duct beginning in the inferomedial corner of the orbit. The **lacrimal gland** is innervated by parasympathetic fibers from the facial nerve (VII). The gland produces tears, which leave the gland through several **lacrimal ducts** and pass over the anterior surface of the eyeball. The gland produces tears constantly at the rate of about 1 mL/day to moisten the surface of the eye, lubricate the eyelids, and wash away foreign objects. Tears are mostly water, with some salts, mucus, and lysozyme, an enzyme that kills certain bacteria. Most of the fluid produced by the lacrimal glands evaporates from the surface of the eye, but excess tears are collected in the medial corner of the eye by the **lacrimal canaliculi.** The opening of each lacrimal canaliculus is called a **punctum** (pŭngk′tŭm). The upper and lower eyelids each have a punctum near the medial canthus. Each punctum is located on a small lump called a **lacrimal papilla.** The lacrimal canaliculi open into a **lacrimal sac,** which in turn continues into the **nasolacrimal duct** (see figure 15.9). The nasolacrimal duct opens into the inferior meatus of the nasal cavity beneath the inferior nasal concha (see chapter 23).

1. Tears are produced in the lacrimal gland and exit the gland through several lacrimal ducts.
2. The tears pass over the surface of the eye.
3. Tears enter the lacrimal canaliculi.
4. Tears are carried through the nasolacrimal duct.
5. Tears enter the nasal cavity from the nasolacrimal duct.

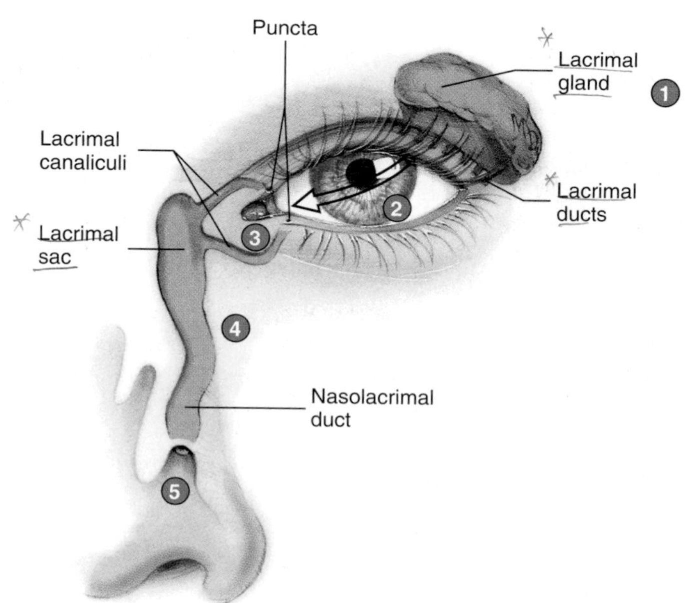

PROCESS FIGURE 15.9 Lacrimal Apparatus

Facial Nerve Damage

Facial nerve damage results in an inability to close the eyelid on the affected side. With the ability to blink being lost, tears cannot be washed across the eye, and the conjunctiva and cornea become dry. A dry cornea may become ulcerated; if not treated, eyesight may be lost.

PREDICT 3

Explain why it is often possible to "taste" medications, such as eyedrops, that have been placed into the eyes. Why does a person's nose "run" when he or she cries?

Extrinsic Eye Muscles

Six **extrinsic muscles** of the eye (figures 15.10 and 15.11; see chapter 10) cause the eyeball to move. Four of these muscles run more or less straight anteroposteriorly. They are the superior, inferior, medial, and lateral **rectus muscles.** Two muscles, the superior and inferior **oblique muscles,** are placed at an angle to the globe of the eye.

The movements of the eye can be described graphically by a figure resembling the letter *H.* The clinical test for normal eye movement is therefore called the **H test.** A person's inability to move the eye toward one part of the *H* may indicate dysfunction of an extrinsic eye muscle or the cranial nerve to the muscle (the actions of the eye muscles are listed in table 10.7).

The superior oblique muscle is innervated by the trochlear nerve (IV). The nerve is so-named because the superior oblique muscle goes around a little pulley, or trochlea, in the superomedial corner of the orbit. The lateral rectus muscle is innervated by the abducent nerve (VI), so-named because the lateral rectus muscle abducts the eye. The other four extrinsic eye muscles are innervated by the oculomotor nerve (III).

12 ***Describe and state the functions of the eyebrows, eyelids, conjunctiva, lacrimal apparatus, and extrinsic eye muscles.***

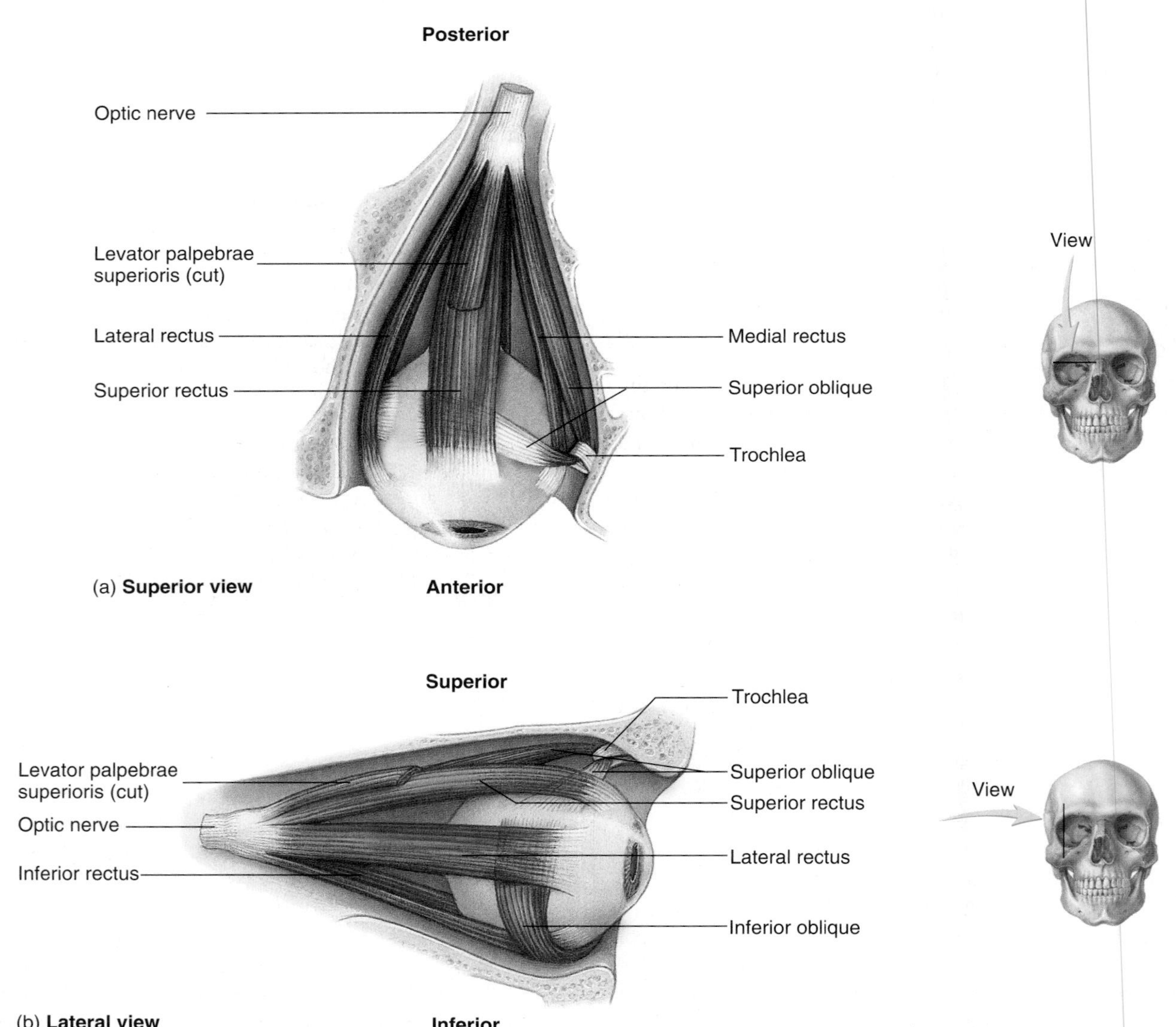

FIGURE 15.10 Extrinsic Muscles of the Eye

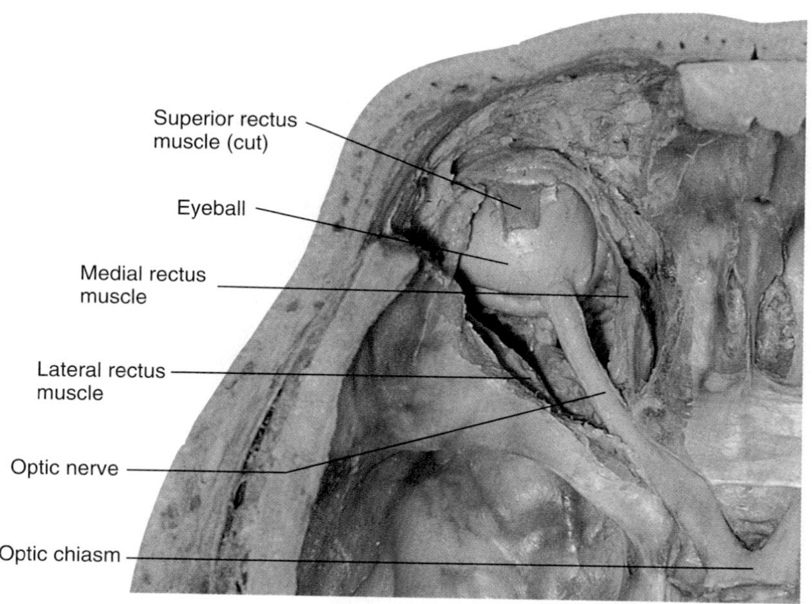

FIGURE 15.11 Photograph of the Eye and Its Associated Structures

Anatomy of the Eye

The eyeball is composed of three layers (figure 15.12). The outer, or **fibrous, layer** consists of the sclera and cornea; the middle, or **vascular, layer** consists of the choroid, ciliary body, and iris; and the inner, or **nervous, layer** consists of the retina.

Fibrous Layer

The **sclera** (sklēr′ă) is the firm, opaque, white outer layer of the posterior five-sixths of the eyeball. It consists of dense collagenous connective tissue with elastic fibers. The sclera helps maintain the shape of the eyeball, protects its internal structures, and provides an attachment point for the muscles that move it. Usually, a small portion of the sclera can be seen as the "white of the eye" when the eye and its surrounding structures are intact (see figure 15.7). The bulbar conjunctiva is loosely attached to the sclera.

The sclera is continuous anteriorly with the cornea. The **cornea** (kōr′nē-ă) is an avascular, transparent structure that permits light to enter the eye and bends, or refracts, that light as part of the focusing system of the eye. The cornea consists of a connective tissue matrix containing collagen, elastic fibers, and proteoglycans, with a layer of stratified squamous epithelium covering the outer surface and a layer of simple squamous epithelium on the inner surface. The outer epithelium is continuous with the bulbar conjunctiva over the sclera. Large collagen fibers are white, whereas smaller collagen fibers and proteoglycans are transparent. The cornea is transparent, rather than white, like the sclera, in part because fewer large collagen fibers and more proteoglycans are present in the cornea than in the sclera. The transparency of the cornea also results from its low water content. In the presence of water, proteoglycans trap water and expand, which scatters light. In the absence of water, the proteoglycans decrease in size and do not interfere with the passage of light through the matrix.

PREDICT 4

Predict the effect of corneal inflammation on vision.

Cornea

The central part of the cornea receives oxygen from the outside air. Soft plastic contact lenses worn for long periods must therefore be permeable so that air can reach the cornea.

The most common eye injuries are cuts or tears of the cornea caused by foreign objects, stones or sticks, hitting the cornea. Extensive injury to the cornea may cause connective tissue deposition, thereby making the cornea opaque.

The cornea was one of the first organs to be transplanted. Several characteristics make it relatively easy to transplant: It is easily accessible and relatively easily removed; it is avascular and therefore does not require extensive circulation, as do other tissues; and it is less immunologically active and therefore less likely to be rejected than are other tissues.

Vascular Layer

The middle tunic of the eyeball is called the vascular layer because it contains most of the blood vessels of the eyeball (see figure 15.12). The arteries of the vascular layer are derived from a number of arteries called **short ciliary arteries,** which pierce the sclera in a circle around the optic nerve. These arteries are branches of the **ophthalmic** (of-thal′mik) **artery,** which is a branch of the internal carotid artery. The vascular layer contains a large number of melanin-containing pigment cells and appears black in color. The portion of the vascular layer associated with the sclera of the eye is the **choroid** (ko′royd). The term *choroid* means membrane and suggests that this layer is relatively thin (0.1–0.2 mm thick). Anteriorly, the vascular layer consists of the ciliary body and iris.

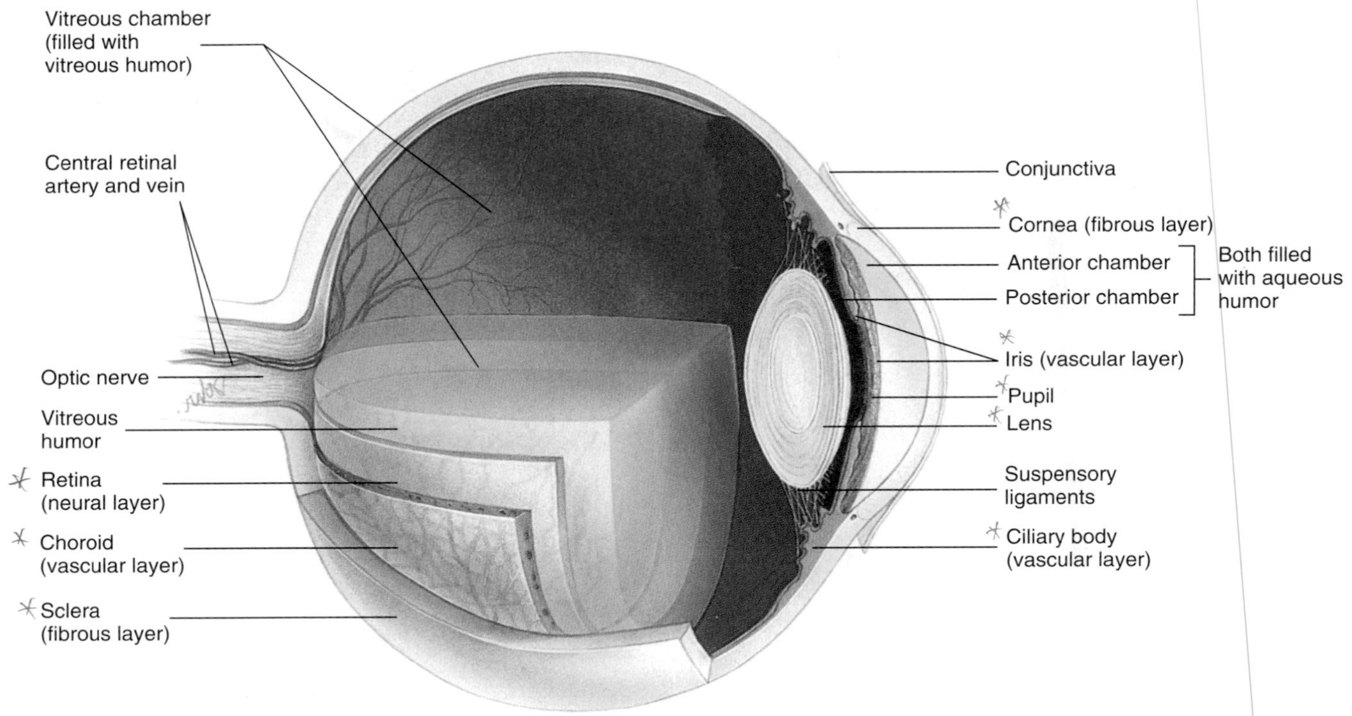

FIGURE 15.12 Sagittal Section of the Eye, Demonstrating the Layers of the Eyeball

The **ciliary** (sil′ē-ar-ē) **body** is continuous with the choroid, and the **iris** is attached at its lateral margins to the ciliary body (figure 15.13*a* and *b*). The ciliary body consists of an outer **ciliary ring** and an inner group of **ciliary processes,** which are attached to the lens by **suspensory ligaments.** The ciliary body contains smooth muscles called the **ciliary muscles,** which are arranged with the outer muscle fibers oriented radially and the central fibers oriented circularly. The ciliary muscles function as a sphincter, and contraction of these muscles can change the shape of the lens. (This function is described in more detail on p. 516.) The ciliary processes are a complex of capillaries and cuboidal epithelium that produces aqueous humor.

The iris is the "colored part" of the eye, and its color differs from person to person. A large amount of melanin in the iris causes it to appear brown or even black. Less melanin results in light brown, green, or gray irises. Even less melanin causes the eyes to appear blue. If there is no pigment in the iris, as in albinism, the iris is pink because blood vessels in the eye reflect light back to the iris.

The genetics of eye color is quite complex. A gene on chromosome 15 and another on chromosome 19 affect eye color, but other genes, not yet identified, are postulated to explain complex eye colors and patterns.

The iris is a contractile structure, consisting mainly of smooth muscle, surrounding an opening called the **pupil.** Light enters the eye through the pupil, and the iris regulates the amount of light by controlling the size of the pupil. The iris contains two groups of smooth muscles: a circular group called the **sphincter pupillae** (pū-pil′ē), and a radial group called the **dilator pupillae** (figure 15.13*c* and *d*). The sphincter pupillae are innervated by parasympathetic fibers from the oculomotor nerve (III). When they contract, the iris decreases, or constricts, the size of the pupil. The dilator pupillae are innervated by sympathetic fibers. When they contract, the pupil is dilated. The ciliary muscles, sphincter pupillae, and dilator pupillae are sometimes referred to as the intrinsic eye muscles.

Retina

The **retina** is the **inner layer** of the eyeball (see figure 15.12). It consists of the outer **pigmented layer,** which is pigmented simple cuboidal epithelium, and the inner **neural layer,** which responds to light. The neural layer contains 120 million photoreceptor cells called **rods** and 6 or 7 million **cones,** as well as numerous relay neurons. The retina covers the inner surface of the eyeball posterior to the ciliary body. A more detailed description of the histology and function of the retina is presented on page 531.

Eye Pigment

The pupil appears black because of the pigment in the choroid and the pigmented portion of the retina. The eyeball is a closed chamber, which allows light to enter only through the pupil. Light is absorbed by the pigmented inner lining of the eyeball; thus, looking into it is like looking into a dark room. If a bright light is directed into the pupil, however, the reflected light is red because of the blood vessels on the surface of the retina. This is why the pupils of a person looking directly at a flash camera often appear red in a photograph. People with albinism lack the pigment melanin, and the pupil always appears red because no melanin is present to absorb light and prevent it from being reflected from the back of the eyeball. The diffusely lighted blood vessels in the interior of the eyeball contribute to the red color of the pupil.

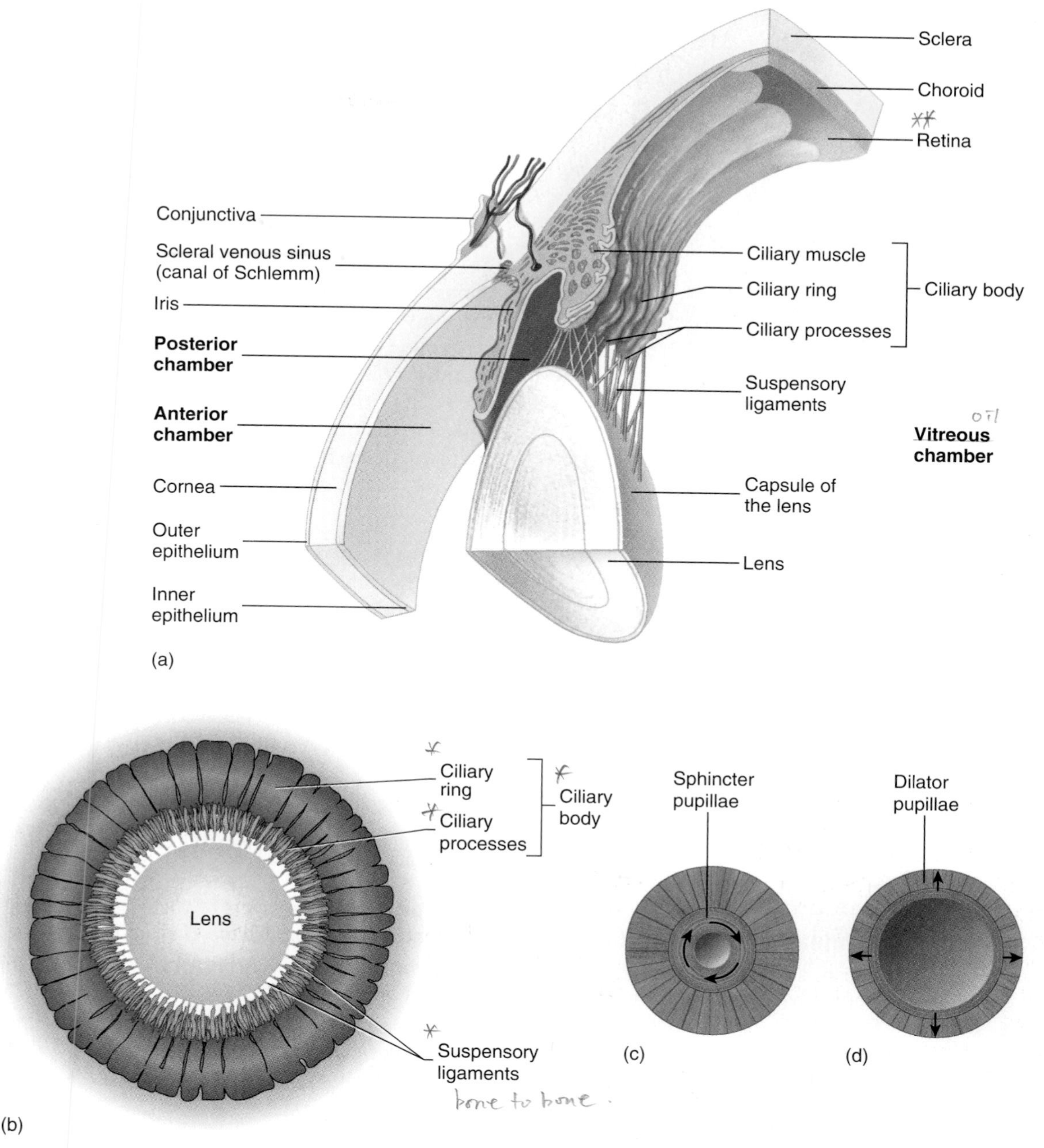

FIGURE 15.13 Lens, Cornea, Iris, and Ciliary Body

(*a*) The orientation is the same as in figure 15.12. (*b*) The lens and ciliary body. (*c*) The sphincter pupillae muscles of the iris constrict the pupil. (*d*) The dilator pupillae muscles of the iris dilate the pupil.

When the posterior region of the retina is examined with an **ophthalmoscope** (of-thal′mō-skōp; figure 15.14), several important features can be observed. Near the center of the posterior retina is a small yellow spot approximately 4 mm in diameter, the **macula** (mak′ū-lă). In the center of the macula is a small pit, the **fovea** (fō′vē-ă) **centralis.** The fovea and macula make up the region of the retina where light is focused. The fovea is the portion of the retina with the greatest visual acuity (the ability to see fine images) because the photoreceptor cells are more tightly packed in that portion of the retina than anywhere else. Just medial to the macula is a white spot, the **optic disc,** through which the central retinal artery enters and the central retinal vein exits the eyeball. Branches from these vessels spread over the surface of the retina. This is also the spot where nerve processes from the neural layer meet, pass through the outer two layers, and exit the eye as the optic nerve. The optic disc contains no photoreceptor cells and does not respond to light; therefore, it is called the **blind spot** of the eye.

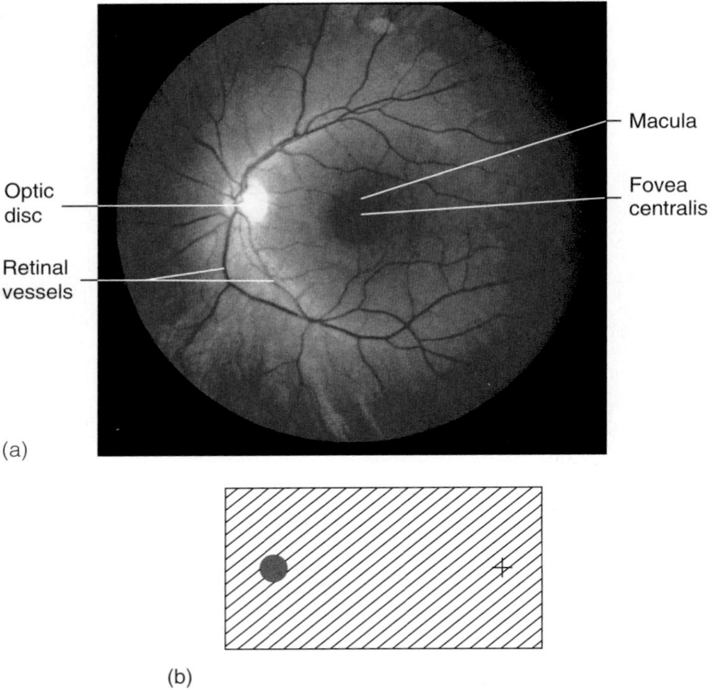

FIGURE 15.14 Ophthalmoscopic View of the Left Retina

(*a*) The posterior wall of the retina as seen when looking through the pupil. Notice the vessels entering the eye through the optic disc (the optic nerve) and the macula with the fovea (the part of the retina with the greatest visual acuity). (*b*) Demonstration of the blind spot. Close your right eye. Hold the figure in front of your left eye and stare at the +. Move the figure toward your eye. At a certain point, when the image of the spot is over the optic disc, the red spot seems to disappear.

Ophthalmoscopic Examination of the Retina

Ophthalmoscopic examination of the posterior retina can reveal some general disorders of the body. **Hypertension,** or high blood pressure, results in "nicking" (compression) of the retinal veins where the abnormally pressurized arteries cross them. **Increased cerebrospinal fluid (CSF) pressure** associated with hydrocephalus may cause swelling of the optic disc. This swelling is referred to as **papilledema** (pă-pil-e-dē′mă). Furthermore, **cataracts** (opacity of the lens) are usually discovered or confirmed by ophthalmoscopic examination.

Chambers of the Eye

The interior of the eye is divided into three chambers: **anterior chamber, posterior chamber,** and **vitreous** (vit′rē-ŭs), or **postremal, chamber** (see figure 15.12). The anterior chamber lies between the cornea and the iris, and a smaller posterior chamber lies between the iris and the lens (see figure 15.13). These two chambers are filled with **aqueous humor,** which helps maintain intraocular pressure. The pressure within the eyeball keeps it inflated and is largely responsible for maintaining the shape of the eyeball. The aqueous humor also refracts light and provides nutrition for the structures of the anterior chamber, such as the cornea, which has no blood vessels. Aqueous humor is produced by the ciliary processes as a blood filtrate and is returned to the circulation through a venous ring at the base of the cornea called the scleral venous sinus (canal of Schlemm; shlem; see figure 15.13). The production and removal of aqueous humor result in the "circulation" of aqueous humor and maintenance of a constant intraocular pressure. If circulation of the aqueous humor is inhibited, a defect called **glaucoma** (glaw-kō′mă), which is an abnormal increase in intraocular pressure, can result.

The vitreous chamber of the eye is much larger than the anterior and posterior chambers. It is surrounded almost completely by the retina and is filled with a transparent, jellylike substance, **vitreous** (vit′rē-ŭs) **humor.** Vitreous humor is not produced as rapidly as is the aqueous humor, and its turnover is extremely slow. The vitreous humor helps maintain intraocular pressure and therefore the shape of the eyeball, and it holds the lens and retina in place. It also functions in the refraction of light in the eye.

Lens

The **lens** is an unusual biologic structure. It is transparent and biconvex, with the greatest convexity on its posterior side. The lens consists of a layer of cuboidal epithelial cells on its anterior surface and a posterior region of very long, columnar epithelial cells called **lens fibers.** Cells from the anterior epithelium proliferate and give rise to the lens fibers at the equator of the lens. The lens fibers lose their nuclei and other cellular organelles and accumulate a set of proteins called **crystallines.** This crystalline lens is covered by a highly elastic, transparent **capsule.**

The lens is suspended between the two eye compartments by the suspensory ligaments of the lens, which are connected from the ciliary body to the lens capsule.

13 ***Name the three layers of the eyeball, describe the parts each forms, and explain their functions.***

14 ***How does the pupil constrict? How does it dilate? What is the blind spot?***

15 ***Name the three chambers of the eye and the substances that fill each chamber.***

16 ***What is the function of the scleral venous sinus and the ciliary processes?***

17 ***Describe the lens of the eye, and explain how the lens is held in place.***

Functions of the Complete Eye

The iris of the eye allows light into the eye, and the lens, cornea, and humors focus the light onto the retina. The light striking the retina is converted into action potentials, which are relayed to the brain.

The electromagnetic spectrum is the entire range of wavelengths, or frequencies, of electromagnetic radiation, from very short gamma waves at one end of the spectrum to the longest radio waves at the other end (figure 15.15). **Visible light** is the portion of the electromagnetic spectrum that can be detected by the human eye. Light has characteristics of both particles (photons)

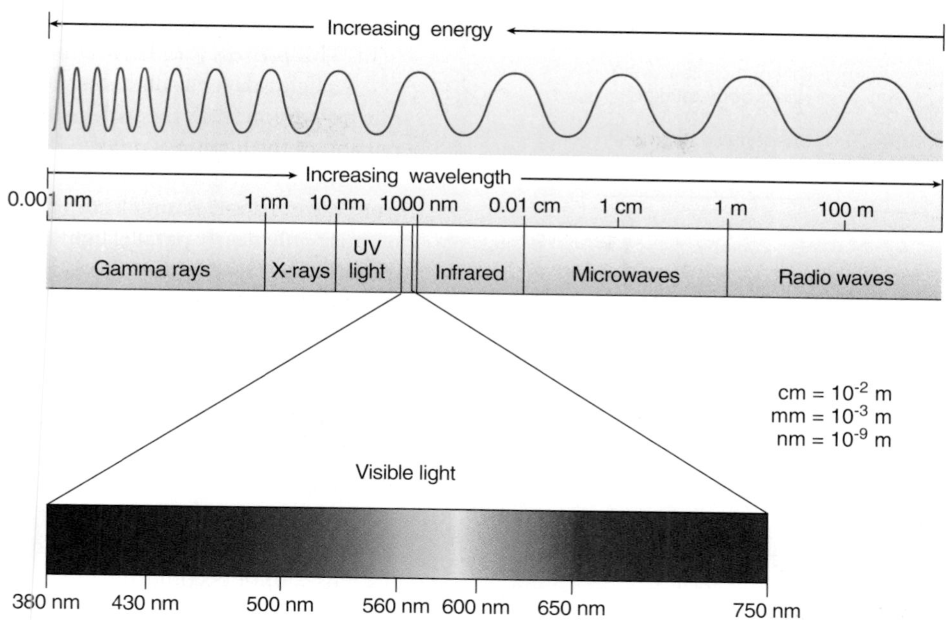

FIGURE 15.15 Electromagnetic Spectrum
The spectrum of visible light is pulled out and expanded. The wavelengths of the various colors are also depicted.

and waves, with a wavelength between 400 and 700 nm. This range sometimes is called the range of visible light or, more correctly, the **visible spectrum.** Within the visible spectrum, each color has a different wavelength.

An important characteristic of light is that it can be refracted (bent). As light passes from air to a denser substance, such as glass or water, its speed is reduced. If the surface of that substance is at an angle other than 90 degrees to the direction the light rays are traveling, the rays are bent as a result of variation in the speed of light as it encounters the new medium. This bending of light is called **refraction.**

If the surface of a lens is concave, with the lens thinnest in the center, the light rays diverge as a result of refraction. If the surface is convex, with the lens thickest in the center, the light rays tend to converge. As light rays converge, they finally reach a point at which they cross. This point is called the **focal point (FP),** and causing light to converge is called **focusing.** No image is formed exactly at the focal point, but an inverted, focused image can form on a surface located some distance past the focal point. How far past the focal point the focused image forms depends on a number of factors. A biconvex lens causes light to focus closer to the lens than does a lens with a single convex surface. Furthermore, the more nearly spherical the lens, the closer to the lens the light is focused; the more flattened the biconcave lens, the more distant is the point where the light is focused.

If light rays strike an object that is not transparent, they bounce off the surface. This phenomenon is called **reflection.** If the surface is very smooth, such as the surface of a mirror, the light rays bounce off in a specific direction. If the surface is rough, the light rays are reflected in several directions and produce a more diffuse reflection.

The focusing system of the eye projects a clear image on the retina. Light rays converge as they pass from the air through the convex cornea. Additional convergence occurs as light encounters the aqueous humor, lens, and vitreous humor. The greatest contrast in media density is between the air and the cornea; therefore, the greatest amount of convergence occurs at that point. The shape of the cornea and its distance from the retina are fixed, however, so that no adjustment in the location of the focal point can be made by the cornea. Fine adjustment in focal point location is accomplished by changing the shape of the lens. In general, focusing can be accomplished in two ways. One is to keep the shape of the lens constant and move it nearer or farther from the point at which the image will be focused, such as occurs in a camera, microscope, or telescope. The second way is to keep the distance constant and to change the shape of the lens, which is the technique used in the eye.

As light rays enter the eye and are focused, the image formed just past the focal point is inverted (figure 15.16). Action potentials that represent the inverted image are passed to the visual cortex of the cerebrum, where the brain interprets them as being right side up.

Visual Image Inversion

Because the visual image is inverted when it reaches the retina, the image of the world focused on the retina is upside down. The brain processes information from the retina so that the world is perceived the way "it really is." If, as an experiment, a person wears glasses that invert the image entering the eye, he or she will see the world upside down for a few days, after which time the brain adjusts to the new input to set the world right side up again. If the glasses are then removed, another adjustment period is required before the world is made right by the brain.

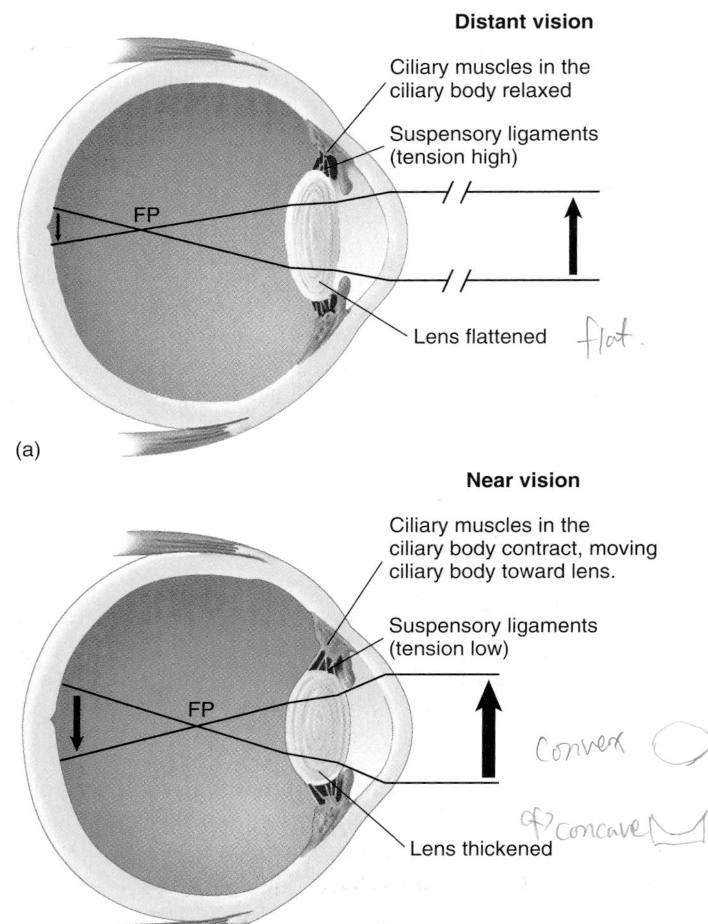

FIGURE 15.16 Focus and Accommodation by the Eye
The focal point (FP) is where light rays cross. (*a*) Distant image. The lens is flattened, and the image is focused on the retina. (*b*) Accommodation for near vision. The lens is more rounded, and the image is focused on the retina.

When the ciliary muscles are relaxed, the suspensory ligaments of the ciliary body maintain elastic pressure on the lens, thereby keeping it relatively flat and allowing for distant vision (see figure 15.16*a*). The condition in which the lens is flattened so that nearly parallel rays from a distant object are focused on the retina is referred to as **emmetropia** (em-ĕ-trō′pē-ă; measure) and is the normal resting condition of the lens. The point at which the lens does not have to thicken for focusing to occur is called the **far point of vision** and normally is 20 feet or more from the eye.

When an object is brought closer than 20 feet to the eye, three events occur to bring the image into focus on the retina: accommodation by the lens, constriction of the pupil, and convergence of the eyes.

1. *Accommodation.* When the eye focuses on a nearby object, the ciliary muscles contract as a result of parasympathetic stimulation from the oculomotor nerve (III). This sphincterlike contraction pulls the choroid toward the lens to reduce the tension on the suspensory ligaments. This allows the lens to assume a more spherical form because of its own elastic nature (see figure 15.16b). The more spherical lens then has a more convex surface, causing greater refraction of light. This process is called accommodation.

   As light strikes a solid object, the rays are reflected in every direction from the surface of the object. Only a small portion of the light rays reflected from a solid object, however, pass through the pupil and enter the eye. An object far away from the eye appears small, compared with a nearby object, because only nearly parallel light rays enter the eye from a distant object (see figure 15.16*a*). Converging rays leaving an object closer to the eye can also enter the eye (see figure 15.16*b*), and the object appears larger.

   When rays from a distant object reach the lens, they do not have to be refracted to any great extent to be focused on the retina, and the lens can remain fairly flat. When an object is closer to the eye, the more obliquely directed rays must be refracted to a greater extent to be focused on the retina.

   As an object is brought closer and closer to the eye, accommodation becomes more and more difficult because the lens cannot become any more convex. At some point, the eye no longer can focus the object, and it is seen as a blur. The point at which this blurring occurs is called the **near point of vision,** which is usually about 2–3 inches from the eye for children, 4–6 inches for a young adult, 20 inches for a 45-year-old adult, and 60 inches for an 80-year-old adult. This increase in the near point of vision, called **presbyopia,** occurs because the lens becomes more rigid with increasing age, which is primarily why some older people say they could read with no problem if they only had longer arms.

Vision Charts

When a person's visual acuity is tested, a chart is placed 20 feet from the eye, and the person is asked to read a line of letters that is standardized for normal vision at 20 feet. If the person can read the line, the vision is considered to be 20/20, which means that the person can see at 20 feet what people with normal vision can see at 20 feet. If, on the other hand, the person can see words only at 20 feet that people with normal vision can see at 40 feet, the vision is considered 20/40.

2. *Pupil constriction.* **Depth of focus** is the greatest distance through which an object can be moved and still remain in focus on the retina. The main factor affecting depth of focus is the size of the pupil. If the pupillary diameter is small, the depth of focus is greater than if the pupillary diameter is large. With a smaller pupillary opening, an object may therefore be moved slightly nearer or farther from the eye without disturbing its focus. This is particularly important when viewing an object at close range because the interest in detail is much greater, and therefore the acceptable margin for error is smaller. When the pupil is constricted, the light entering the eye tends to pass more nearly through the center of the lens and is more accurately focused than light passing through the edges of the lens. Pupillary diameter also regulates the amount of light entering the eye. The dimmer the light, the greater the pupil diameter must be. As the pupil constricts during close vision, therefore, more light is required on the object being observed.

3. *Convergence.* Because the light rays entering the eyes from a distant object are nearly parallel, both pupils can pick up the light rays when the eyes are directed more or less straight ahead. As an object moves closer, however, the eyes must be rotated medially so that the object is kept focused on corresponding areas of each retina. Otherwise, the object appears blurry. This medial rotation of the eyes is accomplished by a reflex that stimulates the medial rectus muscle of each eye. This movement of the eyes is called convergence. Convergence can easily be observed. Have someone stand facing you. Have the person reach out one hand and extend an index finger as far in front of his or her face as possible. While the person keeps the gaze fixed on the finger, have the person slowly bring the finger in toward his or her nose until finally touching it. Notice the movement of the person's pupils. What happens?

PREDICT 5

Explain how several hours of reading can cause eyestrain, or eye fatigue. Describe what structures are involved.

18 ***What causes light to refract? What is a focal point? What is emmetropia?***

19 ***Describe the changes that occur in the lens, pupil, and extrinsic eye muscles as an object moves from 25 feet away to 6 inches away. Define near point and far point of vision.***

Structure and Function of the Retina

Leonardo da Vinci, in speaking of the eye, said, "Who would believe that so small a space could contain the images of all the universe?" The retina of each eyeball, which gives us the potential to see the whole world, is about the size and thickness of a postage stamp.

The retina consists of a neural layer and a pigmented layer. The neural layer contains three layers of neurons: photoreceptor, bipolar, and ganglionic. The cell bodies of these neurons form nuclear layers separated by plexiform layers, where the neurons of adjacent layers synapse with each other (figure 15.17). The outer plexiform (plexus-like) layer is between the photoreceptor and bipolar cell layers. The inner plexiform layer is between the bipolar and ganglionic cell layers.

The pigmented layer, or pigmented epithelium, consists of a single layer of cells. This layer of cells is filled with melanin pigment and, together with the pigment in the choroid, provides a black matrix, which enhances visual acuity by isolating individual photoreceptors and reducing light scattering. This pigmentation is not strictly necessary for vision, however. People with albinism (lack of pigment) can see, although their visual acuity is reduced because of some light scattering.

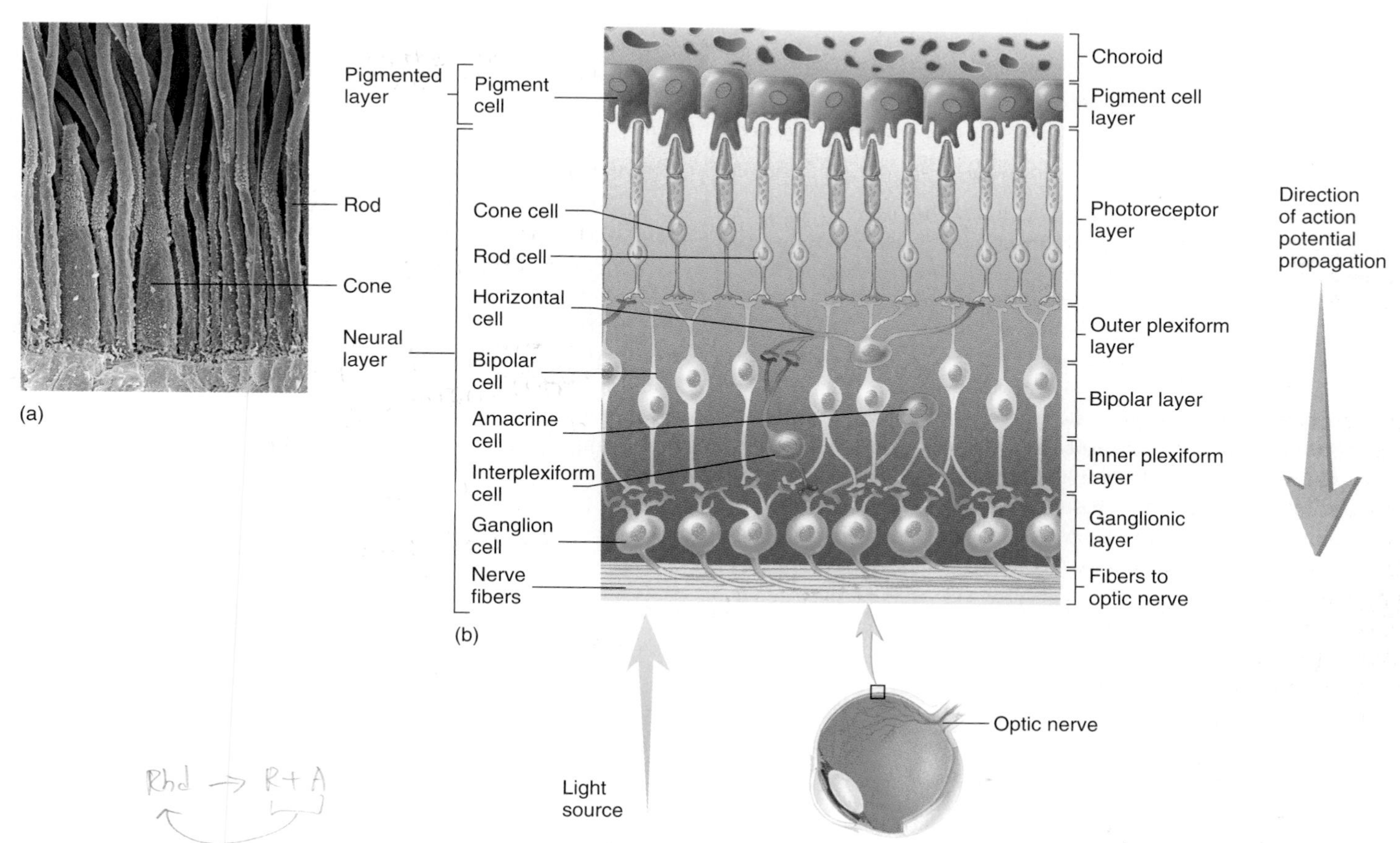

FIGURE 15.17 Retina

(*a*) Colorized electron micrograph of rods and cones. (*b*) Section through the retina with its major layers labeled.

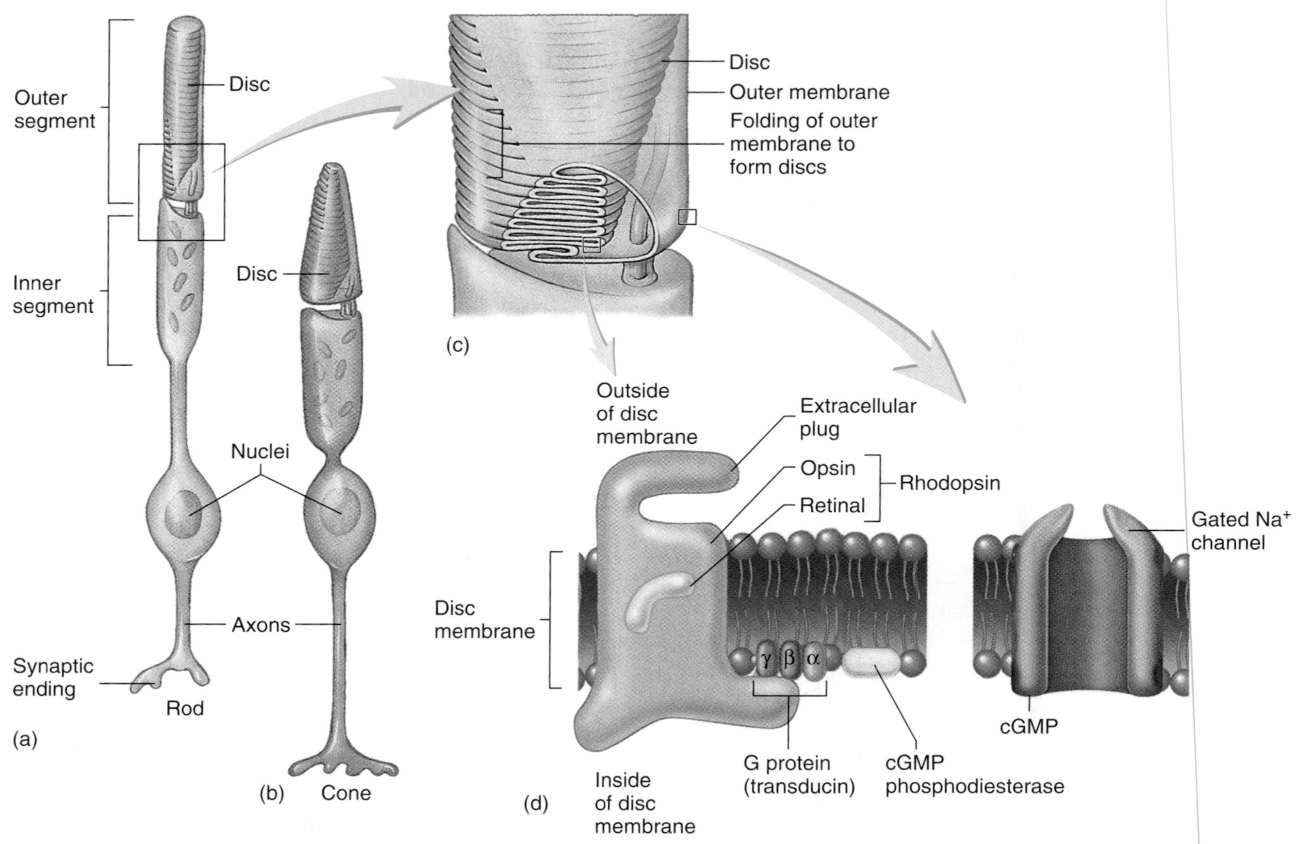

FIGURE 15.18 Sensory Receptor Cells of the Retina

(*a*) Rod cell. (*b*) Cone cell. (*c*) An enlargement of the discs in the outer segment. (*d*) An enlargement of one of the discs, showing the relation of rhodopsin and a cGMP gated Na^+ channel in the membrane.

The portion of the neural layer nearest the pigmented layer consists of rods and cones. The rods and cones are photoreceptor cells, which are sensitive to stimulation from visible light. The light-sensitive portion of each photoreceptor cell is adjacent to the pigmented layer.

Rods

Rods are bipolar photoreceptor cells involved in noncolor vision and are responsible for vision under conditions of reduced light (table 15.1). The modified, dendritic, light-sensitive part of rod cells is cylindrical, with no taper from base to apex (figure 15.18*a*). This rod-shaped photoreceptive part of the rod cell contains about 700 double-layered membranous discs. The discs contain **rhodopsin** (rō-dop′sin), a purple pigment, which consists of the protein **opsin** covalently bound to a yellow photosensitive pigment called **retinal** (derived from vitamin A). Opsin is a protein much like a channel protein, consisting of seven transmembrane helical regions. An extracellular plug closes the external opening of the opsin "channel." G proteins and cyclic GMP (cGMP) phosphodiesterase enzymes are also associated with the disc membranes. Sodium ion channels are located in the outer membrane of the rod cell outer segment (figure 15.18*d*).

Figure 15.19 depicts the changes that rhodopsin undergoes in response to light. In the resting (dark) state, 11-*cis*-retinal is tightly bound to the internal surface of opsin. The extracellular plug helps keep retinal from being lost from the cell out of the opsin "channel." Cyclic GMP is attached to the Na^+ channel, keeping it

TABLE 15.1 Rods and Cones

Photoreceptive End	Photoreceptive Molecule	Function	Location
Rod			
Cylindrical	Rhodopsin	Noncolor vision; vision under conditions of low light	Over most of retina; none in fovea
Cone			
Conical	Iodopsin	Color vision; visual acuity	Numerous in fovea and macula; sparse over rest of retina

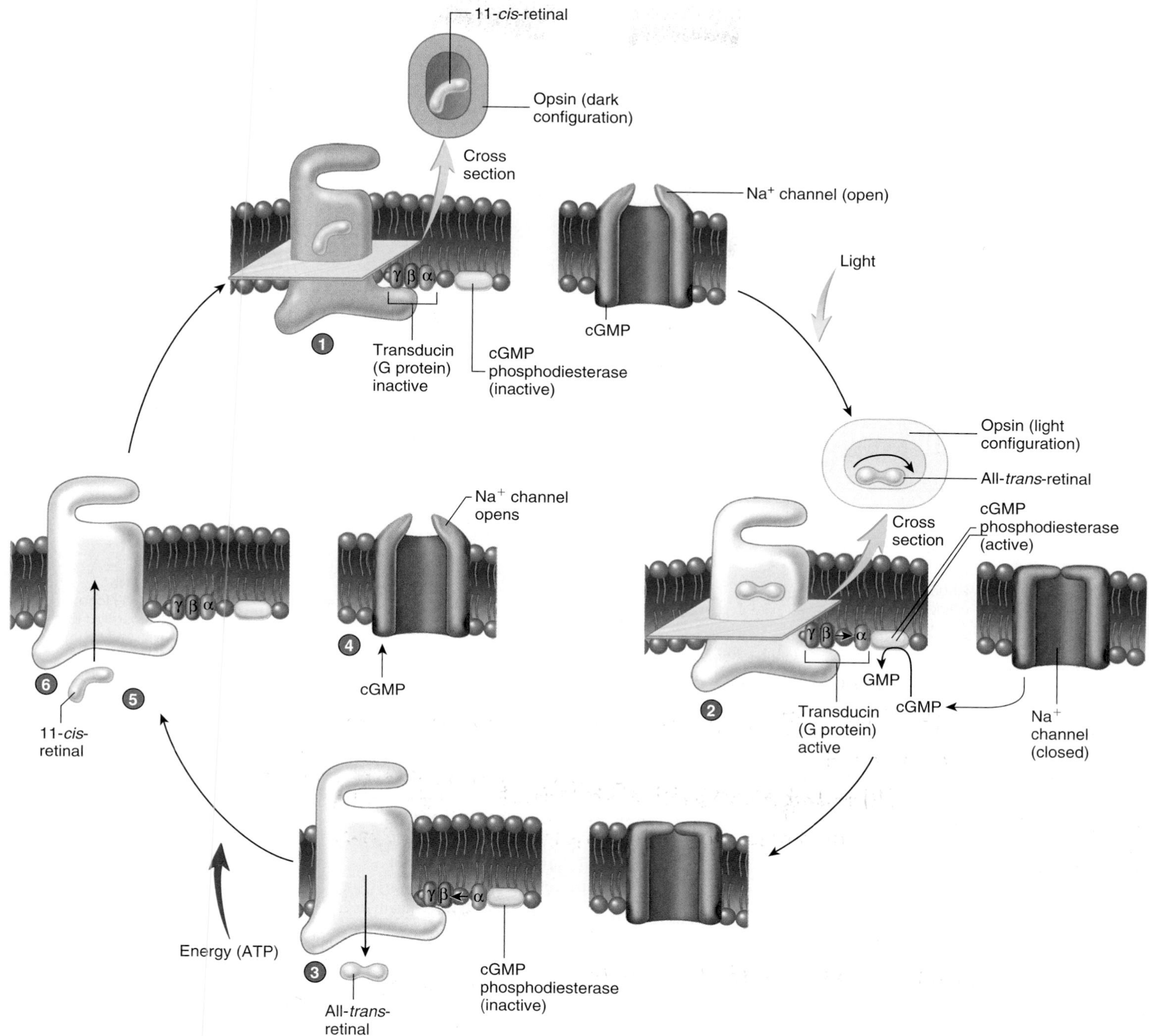

1. Retinal (in an inactive configuration called 11-*cis*-) is attached inside opsin to make rhodopsin. Cyclic GMP attached to a Na^+ channel keeps it open.
2. Light causes retinal to change shape from 11-*cis*-retinal to all-*trans*-retinal. This change causes opsin also to change shape (from its dark to light configuration). This activated rhodopsin activates a G protein (called transducin), which activates a cGMP phosphodiesterase. The cGMP phosphodiesterase catalyzes the conversion of cGMP to GMP, which decreases cGMP concentration and causes the cGMP to diffuse away from the Na^+ channels, closing these channels.
3. All-*trans*-retinal detaches from opsin. Transducin is reassociated and cGMP phosphodiesterase is inactivated.
4. Cyclic GMP concentration increases and cGMP attaches to Na^+ channels, causing them to open.
5. All-*trans*-retinal is converted to 11-*cis*-retinal, a process that requires energy.
6. 11-*cis*-retinal attaches to opsin, which returns to its original (*dark*) configuration.

PROCESS FIGURE 15.19 Rhodopsin Cycle X not for exam

open. As light is absorbed by rod cells, retinal changes shape from 11-*cis*-retinal to all-*trans*-retinal. This change causes opsin to also change shape (dark to light state). The changes in opsin activate a G protein, called **transducin** (trans-doo′sin), which activates a cyclic GMP phosphodiesterase. The cGMP phosphodiesterase catalyzes the conversion of cGMP to GMP. This reaction causes cGMP to diffuse away from Na^+ channels, resulting in closure of the Na^+ channels and hyperpolarization of the cell (figure 15.20).

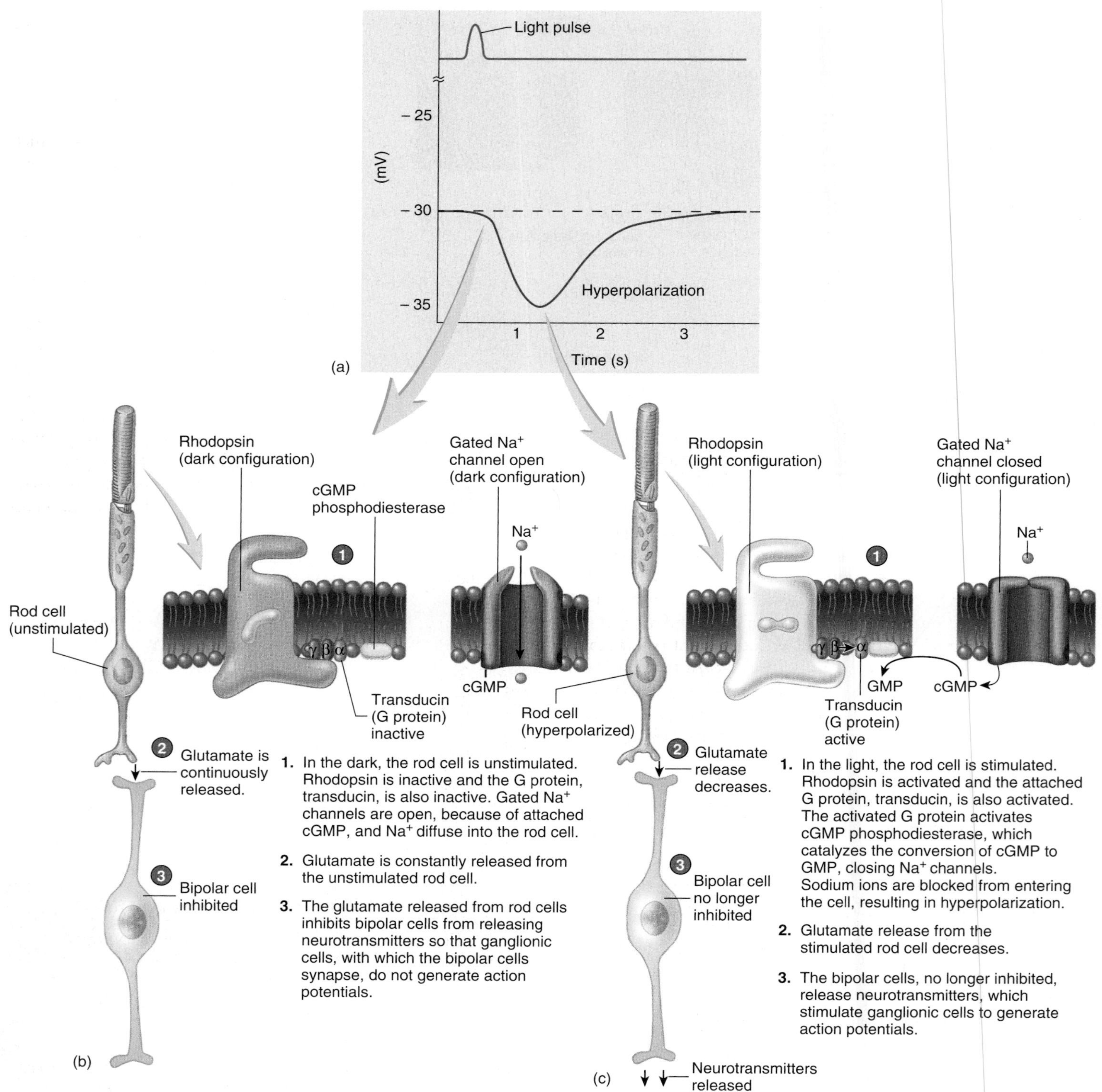

PROCESS FIGURE 15.20 Rod Cell Hyperpolarization

(*a*) Changes in the rod cell membrane potential following the opsin and retinal cell shape changes is a hyperpolarization. (*b*) Unstimulated rod cell (*dark*). (*c*) Stimulated rod cell (*light*).

This hyperpolarization in the photoreceptor cells is somewhat remarkable because most neurons respond to stimuli by depolarizing. When photoreceptor cells are not exposed to light and are in a resting, nonactivated state, gated Na^+ channels in their membranes are open, and Na^+ flow into the cell. This influx of Na^+, referred to as the dark current, causes the photoreceptor cells to release the neurotransmitter glutamate from their presynaptic terminals (see figure 15.20). Glutamate binds to receptors on the postsynaptic membranes of the bipolar cells of the retina, causing them to hyperpolarize. Thus, glutamate causes an inhibitory postsynaptic potential (IPSP) in the bipolar cells. The influx of Na^+ is offset by the efflux of K^+ through nongated K^+ channels. Equilibrium of Na^+ and K^+ in the cell is maintained by a Na^+–K^+ pump.

When photoreceptor cells are exposed to light, the Na^+ channels close, fewer Na^+ enter the cell, and the amount of glutamate released from the presynaptic terminals decreases. As a result, the hyperpolarization in the bipolar cells decreases, and they depolarize sufficiently to release neurotransmitters, which stimulate ganglionic cells to generate action potentials. The number of Na^+ channels that close and the degree to which they close is proportional to the amount of light exposure.

At the final stage of this light-initiated reaction, retinal is completely released from the opsin. This free retinal may then be converted back to vitamin A, from which it was originally derived. The total vitamin A/retinal pool is in equilibrium so that under normal conditions the amount of free retinal is relatively constant. To create more rhodopsin, the altered retinal must be converted back to its original shape, a reaction that requires energy. Once the retinal resumes its original shape, its recombination with opsin is spontaneous, and the newly formed rhodopsin can again respond to light.

Light and **dark adaptation** is the adjustment of the eyes to changes in light. Adaptation to light or dark conditions, which occurs when a person goes out of a darkened building into the sunlight or vice versa, is accomplished by changes in the amount of available rhodopsin. In bright light, excess rhodopsin is broken down so that not as much is available to initiate action potentials, and the eyes become adapted to bright light. Conversely, in a dark room more rhodopsin is produced, making the retina more light-sensitive.

PREDICT 6

If the breakdown of rhodopsin occurs rapidly and production is slow, do eyes adapt more rapidly to light or dark conditions?

Light and dark adaptation also involves pupil reflexes. The pupil enlarges in dim light to allow more light into the eye and contracts in bright light to allow less light into the eye. In addition, rod function decreases and cone function increases in light conditions, and vice versa during dark conditions. This occurs because rod cells are more sensitive to light than cone cells and because rhodopsin is depleted more rapidly in rods than in cones.

Cones

Color vision and visual acuity are functions of cone cells. Color is a function of the wavelength of light, and each color results from a certain wavelength within the visible spectrum. Even though rods are very sensitive to light, they cannot detect color, and sensory input that ultimately reaches the brain from these cells is interpreted by the brain as shades of gray. Cones require relatively bright light to function. As a result, as the light decreases, so does the color of objects that can be seen until, under conditions of very low illumination, the objects appear gray. This occurs because, as the light decreases, fewer cone cells respond to the dim light.

Cones are bipolar photoreceptor cells with a conical light-sensitive part that tapers slightly from base to apex (see figure 15.18*b*). The outer segments of the cone cells, like those of the rods, consist of double-layered discs. The discs are slightly more numerous and more closely stacked in the cones than in the rods. Cone cells contain a visual pigment, **iodopsin** (ī-ō-dop′sin), which consists of retinal combined with a photopigment opsin protein. Three major types of color-sensitive opsin exist—blue, red, and green; each closely resembles the opsin proteins of rod cells but with somewhat different amino acid sequences. These color photopigments function in much the same manner as rhodopsin; however, whereas rhodopsin responds to the entire spectrum of visible light, each iodopsin is sensitive to a much narrower spectrum.

Most people have one red pigment gene and one or more green pigment genes located in a tandem array on each X chromosome. An enhancer gene on the X chromosome apparently determines that only one color opsin gene is expressed in each cone cell. Only the first or second gene in the tandem array is expressed in each cone cell so that some cone cells express only the red pigment gene and others express only one of the green pigment genes.

As can be seen in figure 15.21, although considerable overlap occurs in the wavelength of light to which these pigments

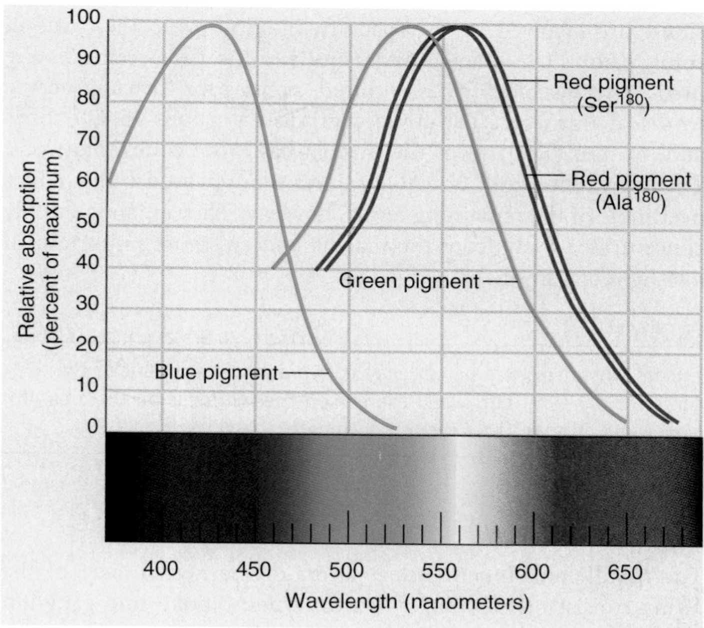

FIGURE 15.21 Wavelengths to Which Each of the Three Visual Pigments are Sensitive: Blue, Green, Red

There are actually two forms of the red pigment. One, found in 60% of the population, has a serine at position 180; the other, found in 40% of the population, has an alanine at position 180. Each red pigment has a slightly different wavelength sensitivity.

are sensitive, each pigment absorbs light of a certain range of wavelengths. As light of a given wavelength, representing a certain color, strikes the retina, all cone cells containing photopigments capable of responding to that wavelength generate action potentials. Because of the overlap among the three types of cones, especially between the green and red pigments, different proportions of cone cells respond to each wavelength, thus allowing color perception over a wide range. Color is interpreted in the visual cortex as combinations of sensory input originating from cone cells. For example, when orange light strikes the retina, 99% of the red-sensitive cones respond, 42% of the green-sensitive cones respond, and no blue cones respond. When yellow light strikes the retina, the response is shifted so that a greater number of green-sensitive cones respond. The variety of combinations created allows humans to distinguish several million gradations of light and shades of color.

Seeing Red

Not everyone sees the same red. Two forms of the red photopigment are common in humans. Approximately 60% of people have the amino acid serine in position 180 of the red opsin protein, whereas 40% have alanine in that position. That subtle difference in the protein results in slightly different absorption characteristics (see figure 15.21). Even though we were each taught to recognize red when we see a certain color, we apparently do not see that color in quite the same way. This difference may contribute to people having different favorite colors.

Distribution of Rods and Cones in the Retina

Cones are involved in visual acuity, in addition to their role in color vision. The macula and especially the fovea centralis are used when visual acuity is required, such as for focusing on the words of this page. The fovea centralis has about 35,000 cones and no rods. The rest of the macula has more cones than rods. The 120 million rods are 10–20 times more plentiful than cones over most of the remaining retina, however. They are more highly concentrated away from the macula and are more important in low-light conditions.

PREDICT 7

Explain why at night a person may notice a movement "out of the corner of the eye," but, when the person tries to focus on the area of movement, it appears as though nothing is there.

Inner Layers of the Retina

The middle and inner nuclear layers of the neural layer of the retina consist of two major neuron types: bipolar and ganglion cells. The rod and cone photoreceptor cells synapse with **bipolar cells,** which in turn synapse with **ganglion cells.** Axons from the ganglion cells pass over the inner surface of the retina (see figure 15.17), except in the area of the fovea centralis; converge at the **optic disc;** and exit the eye as the **optic nerve** (II). The fovea centralis is devoid of ganglion cell processes, resulting in a small depression in this area—thus the name *fovea,* meaning small pit. As a result of the absence of ganglion cell processes in addition to the concentration of cone cells mentioned previously, visual acuity is further enhanced in the fovea centralis because light rays do not have to pass through as many tissue layers before reaching the photoreceptor cells.

Rod and cone cells differ in the way they interact with bipolar and ganglion cells. One bipolar cell receives input from numerous rods, and one ganglion cell receives input from several bipolar cells so that spatial summation of the signal occurs and the signal is enhanced, thereby allowing awareness of stimulus from very dim light sources but decreasing visual acuity in these cells. Cones, on the other hand, exhibit little or no convergence on bipolar cells so that one cone cell may synapse with only one bipolar cell. This system reduces light sensitivity but enhances visual acuity.

The **receptive field** of each ganglion cell, the area from which a ganglion cell receives input, is roughly circular, with a smaller portion in the center, called the receptive field center, and a larger surrounding area. The receptive fields in the fovea centralis are very small, compared with receptive fields of more peripheral parts of the retina. The small size of these receptive fields is consistent with the visual acuity of the fovea centralis.

Two types of receptive fields exist: on-center ganglion cells and off-center ganglion cells. On-center cells generate more action potentials when light is directed onto the receptive field center. Off-center neurons generate more action potentials when light is turned off in the receptive field center or when light shines on the surrounding area. These receptive fields respond primarily to contrasts in light (edges) rather than the absolute intensity of light.

Within the inner layers of the retina, interneurons modify the signals from the photoreceptor cells before the signal ever leaves the retina (see figure 15.17). **Horizontal cells** form the outer plexiform layer and synapse with photoreceptor cells and bipolar cells. **Amacrine** (am′ă-krin) **cells** form the inner plexiform layer and synapse with bipolar and ganglion cells. **Interplexiform cells** form the bipolar layer and synapse with amacrine, bipolar, and horizontal cells to form a feedback loop. The interneurons are either excitatory or inhibitory on the cells with which they synapse. These interneurons enhance borders and contours, thereby increasing the intensity at boundaries, such as the edge of a dark object against a light background.

20 ***What is the function of the pigmented retina and of the choroid?***

21 ***Describe the changes that occur in a rod cell after light strikes rhodopsin. How does rhodopsin re-form? Why is the response of a rod cell to a stimulus unusual?***

22 ***How does dark and light adaptation occur?***

23 ***What are the three types of cone cells? How do they produce the colors we see?***

24 ***Describe the arrangement of rods and cones in the fovea, the macula, and the periphery of the eye.***

25 ***Starting with a rod or cone cell, name the cells or structures that an action potential encounters while traveling to the visual cortex.***

Motion Pictures

Action potentials pass from the retina through the optic nerve at the rate of 20–25/s. We "see" a given image for a fraction of a second longer than it actually appears. Motion pictures take advantage of these two facts. When still photographs are flashed on a screen at the rate of 24 frames per second, they appear to flow into each other, and a motion picture results.

Neuronal Pathways for Vision

The optic nerve (II; figure 15.22) leaves the eye and exits the orbit through the optic foramen to enter the cranial cavity. Just inside the vault and just anterior to the pituitary, the optic nerves are connected to each other at the **optic chiasm** (kī′azm). Ganglion cell axons from the nasal retina (the medial portion of the retina) cross through the optic chiasm and project to the

1. Each visual field is divided into a temporal and nasal half.
2. After passing through the lens, light from each half of a visual field projects to the opposite side of the retina.
3. An optic nerve consists of axons extending from the retina to the optic chiasm.
4. In the optic chiasm, axons from the nasal part of the retina cross and project to the opposite side of the brain. Axons from the temporal part of the retina do not cross.
5. An optic tract consists of axons that have passed through the optic chiasm (with or without crossing) to the thalamus.
6. The axons synapse in the lateral geniculate nuclei of the thalamus. Collateral branches of the axons in the optic tracts synapse in the superior colliculi.
7. An optic radiation consists of axons from thalamic neurons that project to the visual cortex.
8. The right part of each visual field (*dark green* and *light blue*) projects to the left side of the brain, and the left part of each visual field projects to the right side of the brain (*light green* and *dark blue*).

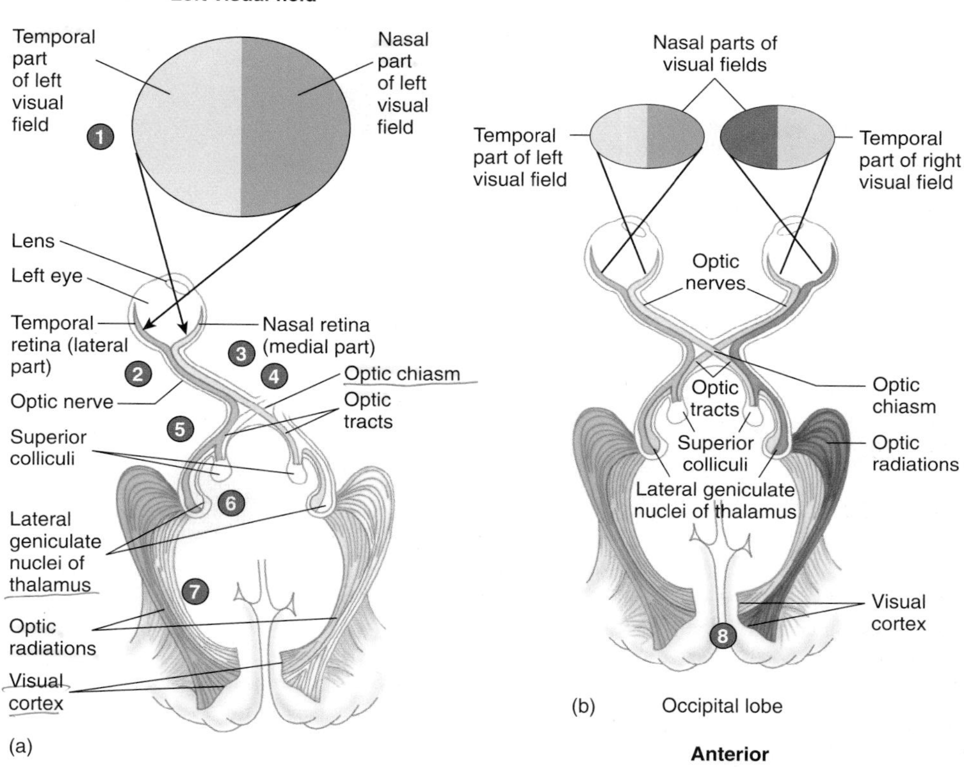

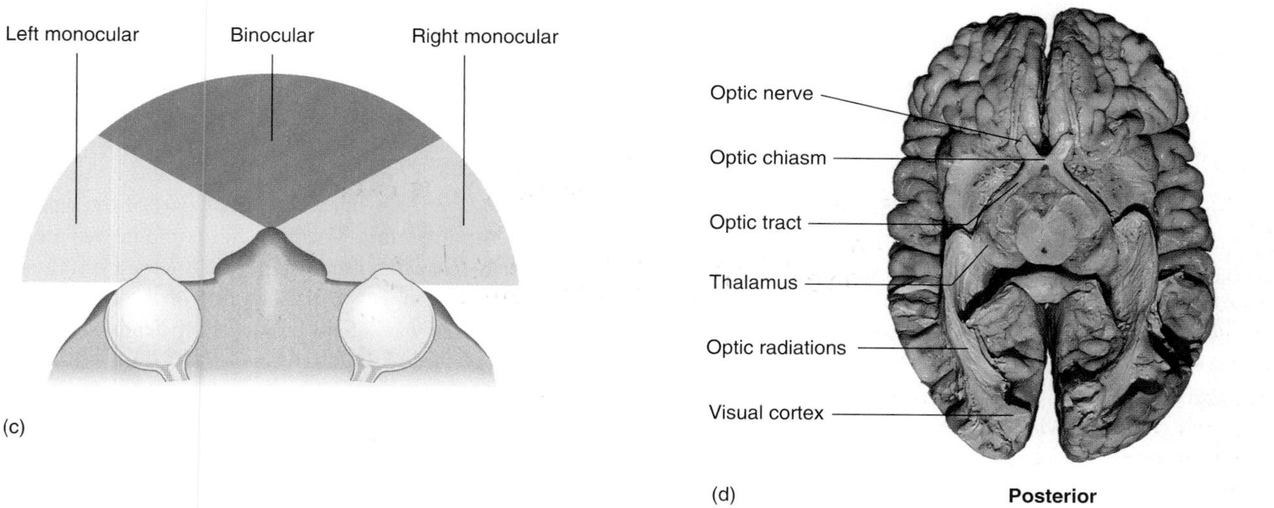

PROCESS FIGURE 15.22 Visual Pathways

(*a*) Pathways for the left eye (superior view). (*b*) Pathways for both eyes (superior view). (*c*) Overlap of the fields of vision (superior view). (*d*) Photograph of the visual nerves, tracts, and pathways (inferior view).

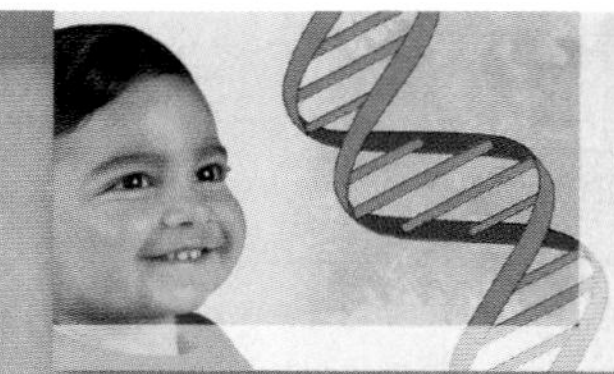

CLINICAL GENETICS

Retinitis Pigmentosa

Retinitis pigmentosa (RP) is an inherited disorder characterized by the progressive degeneration of the retina, accompanied by the accumulation of pigment islands in the retina. Diagnosis of the disease can usually be made on the basis of family history and the presence of dark spots in the retina. RP results in the degeneration of rod and/or cone cells. The degeneration and death of photoreceptor cells result in progressive vision loss. In most forms of RP, called **rod-cone dystrophies,** the rod cells degenerate first. Progression of the disease includes night blindness and loss of peripheral vision as more rod cells are lost, because rod cells are most important for those two functions. Night blindness is often the first symptom of RP. Some patients retain a small amount of central vision throughout life (figure A). Other forms of RP, called **cone-rod dystrophies,** first affect cone cells, resulting in loss of visual acuity and color vision, because cone cells are most important for those two functions.

Symptoms of RP most often appear first in young adults, and the disease progresses throughout the person's life. RP is a fairly uncommon disease, affecting about 1 in 4000 people worldwide. There is no cure for this condition, although the use of sunglasses and/or vitamin therapy (such as vitamin A) may delay the progression of the disease. A Swiss company has tested an electronic retinal implant. In addition, a French biotech company is testing the delivery of a genetically engineered protein, ciliary neurotrophic factor (CNTF), into the vitreous body of RP patients, using the company's patented encapsulated cell technology (ECT). The latest clinical trials of this technique indicate that some patients experience improvement in visual acuity. Stem cell research is also currently in progress in an attempt to correct RP.

FIGURE A Visual Image as Seen by a Person with RP

RP may result from defects in any one of at least 36 different genes. It may be inherited as an autosomal-recessive, autosomal-dominant, or X-linked condition. The heterozygous autosomal-dominant forms of RP generally have a milder course than the autosomal-recessive or X-linked forms. A single mutation in the opsin gene, affecting one amino acid in opsin, apparently is sufficient to cause photoreceptor cell death. A gene for opsin has been mapped to chromosome 3 and many different mutations (multiple alleles) of the opsin gene have been found in cases of autosomal-dominant RP. The opsin protein is composed of 338 amino acids. A mutation resulting in a change in either amino acid 23 or 28, within the opsin external plug, can cause RP. Because of the mutation, the external plug changes shape and does not adequately close the extracellular opening of opsin, thereby allowing retinal to leak out of the cells. This leaking retinal pigment infiltrates the sensory retina, causing it to degenerate.

At least one autosomal-recessive rhodopsin mutation has also been found. Mutations in the cGMP phosphodiesterase gene also have been linked to some cases of autosomal-recessive RP. Other forms of RP have been mapped to chromosomes 1, 6, 7, and 17. Autosomal-dominant cone-rod dystrophy has been mapped to chromosome 19.

opposite side of the brain. Ganglion cell axons from the temporal retina (the lateral portion of the retina) pass through the optic nerves and project to the brain on the same side of the body without crossing.

Beyond the optic chiasm, the route of the ganglionic axons is called the **optic tract** (see figure 15.22). Most of the optic tract axons terminate in the **lateral geniculate nucleus** of the thalamus. Some axons do not terminate in the thalamus but separate from the optic tract to terminate in the **superior colliculi,** the center for visual reflexes (see chapter 13). Neurons of the lateral geniculate nucleus form the fibers of the **optic radiations,** which project to the **visual cortex** in the **occipital lobe.** Neurons of the visual cortex integrate the messages coming from the retina into a single message, translate that message into a mental image, and then transfer the image to other parts of the brain, where it is evaluated and either ignored or acted on.

The projections of ganglion cells from the retina can be related to the **visual fields** (see figure 15.22). The visual field of one eye can be evaluated by closing the other eye. Everything that can be seen with the one open eye is the visual field of that eye. The visual field of each eye can be divided into temporal (lateral) and nasal (medial) parts. In each eye, the temporal part of the visual field projects onto the nasal retina, whereas the nasal part of the visual field projects to the temporal retina. The projections and nerve pathways are arranged in such a way that images entering the eye from the right part of each visual field project to the left side of the brain. Conversely, the left part of each visual field projects to the right side of the brain.

Tunnel Vision

Because the optic chiasm lies just anterior to the pituitary, a pituitary tumor can put pressure on the optic chiasm and may result in visual defects. Because the nerve fibers crossing in the optic chiasm are carrying information from the temporal halves of the visual fields, a person with optic chiasm damage cannot see objects in the temporal halves of the visual fields, a condition called **tunnel vision.** Tunnel vision is often an early sign of a pituitary tumor.

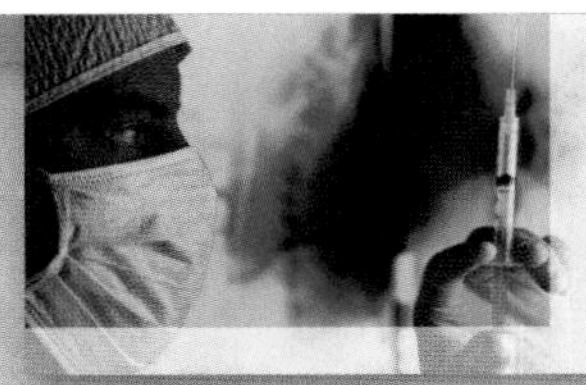

CLINICAL FOCUS

Eye Disorders

MYOPIA

Myopia (mī-ō′pē-ă), or nearsightedness, is the ability to see close objects clearly but distant objects appear blurry. Myopia is a defect of the eye in which the focusing system, the cornea and lens, is optically too powerful, or the eyeball is too long (axial myopia). As a result, the focal point is too near the lens, and the image is focused in front of the retina (figure B*a*).

Myopia is corrected by a concave lens that counters the refractive power of the eye. Concave lenses cause the light rays coming to the eye to diverge and are therefore called "minus" lenses (figure B*b*).

Another technique for correcting myopia is **radial keratotomy** (ker′ă-tot′ō-mē), which consists of making a series of four to eight radiating cuts in the cornea. The cuts are intended to slightly weaken the dome of the cornea so that it becomes more flattened and eliminates the myopia. One problem with the technique is that it is difficult to predict exactly how much flattening will occur. In one study of 400 patients 5 years after the surgery, 55% had normal vision, 28% were still somewhat myopic, and 17% had become hyperopic. Another problem is that some patients are bothered by glare following radial keratotomy because the slits apparently do not heal evenly.

An alternative procedure is **lasix,** or **laser corneal sculpturing,** in which a thin portion of the cornea is etched away to make the cornea less convex. The advantage of this procedure is that the results can be more accurately predicted than those from radial keratotomy.

HYPEROPIA

Hyperopia (hī-per-ō′pē-ă), or farsightedness, is the ability to see distant objects clearly but close objects appear blurry. Hyperopia is a disorder in which the cornea and lens system is optically too weak or the eyeball is too short. The image is focused behind the retina (figure B*c*).

Hyperopia can be corrected by convex lenses that cause light rays to converge as they approach the eye (figure B*d*). Such lenses are called "plus" lenses.

PRESBYOPIA

Presbyopia (prez-bē-ō′pē-ă) is the normal, presently unavoidable degeneration of the accommodation power of the eye that occurs as a consequence of aging. It occurs because the lens becomes sclerotic and less flexible. The eye is presbyopic when the near point of vision has increased beyond 9 inches. The average age for onset of presbyopia is the midforties. Avid readers and people engaged in fine, close work may develop the symptoms earlier.

Presbyopia can be corrected by the use of "reading glasses," which are worn only for

Continued

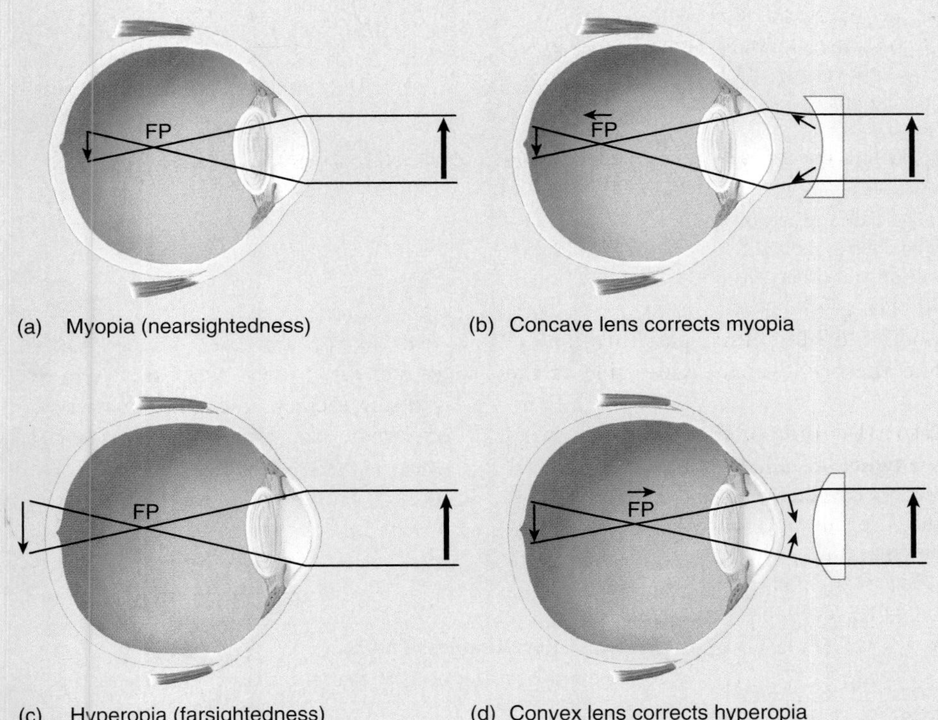

FIGURE B Visual Disorders and Their Correction by Various Lenses

Continued

close work and are removed when the person wants to see at a distance. It is sometimes annoying to keep removing and replacing glasses because reading glasses hamper vision of only a few feet away. This problem may be corrected by the use of half glasses, or by **bifocals,** which have a different lens in the top and the bottom, or by **progressive lenses,** in which the lens is graded.

ASTIGMATISM

Astigmatism (ă-stig′mă-tizm) is a type of refractive error in which the quality of focus is affected. If the cornea or lens is not uniformly curved, the light rays do not focus at a single point but fall as a blurred circle. Regular astigmatism can be corrected by glasses that are formed with the opposite curvature gradation. Irregular astigmatism is a situation in which the abnormal form of the cornea fits no specific pattern and is very difficult to correct with glasses.

STRABISMUS

Strabismus (stra-biz′mŭs) is a lack of parallelism of light paths through the eyes. Strabismus can involve one or both eyes, and the eyes may turn in (convergent) or out (divergent). In **concomitant strabismus,** the most common congenital type, the angle between visual axes remains constant, regardless of the direction of the gaze. In **noncomitant strabismus,** the angle varies, depending on the direction of the gaze, and deviates as the gaze changes.

In some cases, the image that appears on the retina of one eye may be considerably different from that appearing on the other eye. This problem is called **diplopia** (di-plō′pē-ă; double vision) and is often the result of weak or abnormal eye muscles.

RETINAL DETACHMENT

Retinal detachment is a relatively common problem that can result in complete blindness. The integrity of the retina depends on the vitreous humor, which keeps the retina pushed against the other layers of the eye. If a hole or tear occurs in the retina, fluid may accumulate between the neural and pigmented layers, thereby separating them. This separation, or detachment, may continue until the neural layer has become totally detached from the pigmented layer and has folded into a funnel-like form around the optic nerve. When the neural layer becomes separated from its nutrient supply in the choroid, it degenerates, and blindness follows. Causes of retinal detachment include a severe blow to the eye or head; a shrinking of the vitreous humor, which may occur with aging; and diabetes. The space between the sensory and pigmented retina, called the subretinal space, is also important in keeping the retina from detaching, as well as in maintaining the health of the retina. The space contains a gummy substance, which glues the neural layer to the pigmented layer.

COLOR BLINDNESS

Color blindness results from the dysfunction of one or more of the three photopigments involved in color vision. If one pigment is dysfunctional and the other two are functional, the condition is called **dichromatism.** An example of dichromatism is red-green color blindness (figure C).

The genes for the red and green photopigments are arranged in tandem on the X chromosome, which explains why color blindness is over eight times more common in males than in females (see chapter 3).

Six exons exist for each gene. The red and green genes are 96%–98% identical; as a result, the exons may be shuffled to form hybrid genes in some people. Some of the hybrid genes produce proteins with nearly normal function, but others do not. Exon 5 is the most critical for determining normal red-green function. If the fifth exon from a green gene replaces a red pigment gene that has the fifth exon, the protein made from the gene responds to wavelengths more toward the green pigment range. The person has a red perception deficiency and is not able to distinguish between red and green. If the fifth exon from a red gene replaces a green pigment gene that has the fifth exon, the protein made from the gene responds to wavelengths more toward the red pigment range. The person has a green perception deficiency and is not able to distinguish between red and green.

Apparently, only about 3 of the over 360 amino acids in the color opsin proteins (those at positions 180 in exon 3 and those at 277 and 285 in exon 5) are key to determining their wavelength absorption characteristics. If those amino acids are altered by hydroxylation, the absorption shifts toward the red end of the spectrum. If they are not hydroxylated, the absorption shifts toward the green end.

NIGHT BLINDNESS

Everyone sees less clearly in the dark than in the light. A person with **night blindness,** or nyctalopia, however, may not see well enough in a dimly lit environment to function adequately. **Progressive night blindness** results from general retinal degeneration. This form of night blindness is the type associated with retinitis pigmentosa (RP, see p. 538). Another hereditary form of progressive night blindness

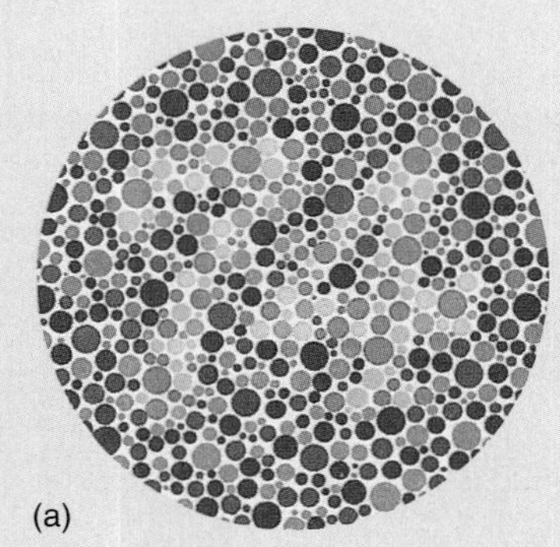

(a)

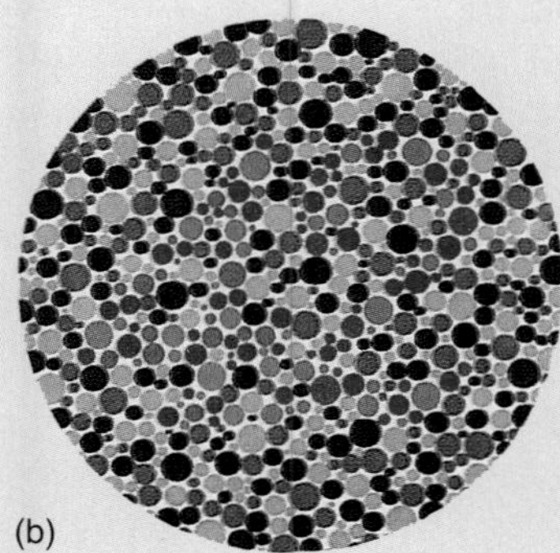

(b)

FIGURE C Color Blindness Charts

(*a*) A person with normal color vision can see the number 74, whereas a person with red-green color blindness sees the number 21. (*b*) A person with normal color vision can see the number 42. A person with red color blindness sees the number 2, and a person with green color blindness sees the number 4.

involves a mutation affecting amino acid 90 in the opsin protein, located in one of the seven tran-membrane helical regions of the protein. This amino acid change changes the shape of opsin and affects the attachment of retinal to opsin. **Stationary night blindness** results from nonprogressive abnormal rod function. Temporary night blindness can result from a vitamin A deficiency.

Patients with night blindness can be helped with electronic optical devices, including monocular pocket scopes and binocular goggles that electronically amplify light.

GLAUCOMA

Glaucoma is a disease of the eye involving increased intraocular pressure caused by a buildup of aqueous humor. It usually results from blockage of the aqueous veins or the scleral venous sinus, restricting drainage of the aqueous humor, or from overproduction of aqueous humor. If untreated, glaucoma can lead to retinal, optic disc, and optic nerve damage. The damage results from the increased intraocular pressure, which is sufficient to close off the blood vessels, causing the starvation and death of the retinal cells.

Glaucoma is one of the leading causes of blindness in the United States, affecting 2% of people over age 35, and accounting for 15% of all blindness. Fifty thousand people in the United States are blind as the result of glaucoma, and it occurs three times more often in black people than in white people. The symptoms include a slow closing in of the field of vision. No pain or redness occurs, nor do light flashes occur.

Glaucoma has a strong hereditary tendency but may develop after surgery or with the use of certain eyedrops containing cortisone. Everyone older than 40 should be checked every 2–3 years for glaucoma; those older than 40 who have relatives with glaucoma should have an annual checkup. During a checkup, the field of vision and the optic nerve are examined. Ocular pressures can also be measured. Glaucoma is usually treated with eyedrops, which do not cure the problem but keep it from advancing. In some cases, laser or conventional surgery may be used.

CATARACT

Cataract (figure D*a*) is a clouding of the lens resulting from a buildup of proteins. The lens relies on the aqueous humor for its nutrition. Any loss of this nutrient source leads to degeneration of the lens and, ultimately, opacity of the lens (i.e., a cataract). A cataract may occur with advancing age, infection, or trauma.

A certain amount of lens clouding occurs in 65% of patients older than 50 and 95% of patients older than 65. The decision of whether to remove the cataract depends on the extent to which light passage is blocked. Over 400,000 cataracts are removed in the United States each year. Surgery to remove a cataract is actually the removal of the lens. The posterior portion of the lens capsule is left intact. Although the cornea can still accomplish light convergence, with the lens gone, the rays cannot be focused as well, and an artificial lens must be supplied to help accomplish focusing. In most cases, an artificial lens is implanted into the remaining portion of the lens capsule at the time that the natural lens is removed. The implanted lens helps restore normal vision, but glasses may be required for near vision.

MACULAR DEGENERATION

Macular degeneration (figure D*b*) is very common in older people. It does not cause total blindness but results in the loss of acute vision. This degeneration has a variety of causes, including hereditary disorders, infections, trauma, tumor, and most often poorly understood degeneration associated with aging. Because no satisfactory medical treatment has been developed, optical aids, such as magnifying glasses, are used to improve visual function.

DIABETES

Loss of visual function is one of the most common consequences of diabetes because a major complication of the disease is dysfunction of the peripheral circulation. Defective circulation to the eye may result in retinal degeneration or detachment. Diabetic retinopathy (figure D*c*) is one of the leading causes of blindness in the United States.

INFECTIONS

Trachoma (trā-kō′mă) is the leading cause of blindness worldwide. It is caused by an intracellular microbial infection (*Chlamydia trachomatis*) of the conjunctiva (conjunctivitis). As inflammation resulting from the disease progresses and spreads, the eyelids may turn inward, causing the eyelashes to abrade the cornea, resulting in scar tissue formation in the cornea. The bacteria are spread from one eye to the other by towels, fingers, and other objects. Five hundred million cases of trachoma exist in the world, and 7 million people are blind or otherwise visually impaired as a result of it.

Neonatal gonorrheal ophthalmia (of-thal′-mē-ă) is a bacterial infection (*Neisseria gonorrhoeae*) of the eye that causes blindness. If the mother has gonorrhea, which is a sexually transmitted disease of the reproductive tract, the bacteria can infect the newborn during delivery. The disease can be prevented by treating the infant's eyes with silver nitrate, tetracycline, or erythromycin drops.

(a)

(c)

(b)

FIGURE D Defects in Vision

Visual images as seen with various defects in vision. (*a*) Cataract. (*b*) Macular degeneration. (*c*) Diabetic retinopathy.

PREDICT 8

The lines at *A* and *B* in figure E depict two lesions in the visual pathways. The effect of a lesion at *A* in the optic radiations on the visual fields is depicted (with the right and left fields separated) in the ovals at the bottom of the figure. The black areas indicate what parts of the visual fields are defective. Describe the effect that the lesion at *B* has on the visual fields (see figure 15.22 for help).

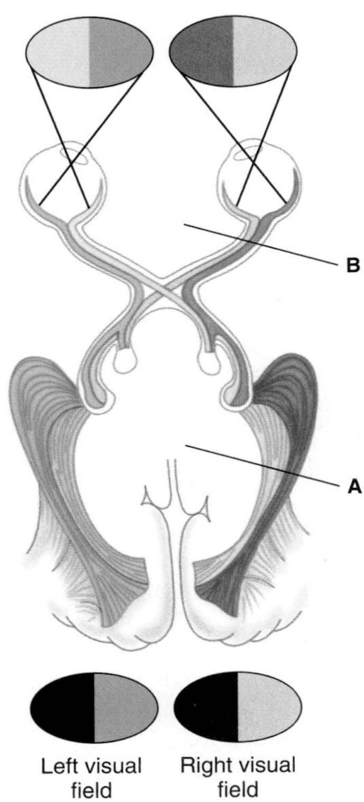

FIGURE E Lesions of Visual Pathways

The visual fields of the eyes partially overlap (see figure 15.22). The region of overlap is the area of **binocular vision,** seen with two eyes at the same time, and it is responsible for **depth perception,** the ability to distinguish between near and far objects and to judge their distance. Because humans see the same object with both eyes, the image of the object reaches the retina of one eye at a slightly different angle from that of the other. With experience, the brain can interpret these differences in angle so that distance can be judged quite accurately.

26 ***What is a visual field? How do the visual fields project to the brain?***

27 ***Explain how binocular vision allows for depth perception.***

HEARING AND BALANCE

The organs of hearing and balance are divided into three parts: the external, middle, and inner ears (figure 15.23). The external and middle ears are involved in hearing only, whereas the inner ear functions in both hearing and balance.

The **external ear** includes the **auricle** (aw′ri-kl; ear) and the **external acoustic meatus** (mē-ā′tŭs; passageway from the outside to the eardrum). The external ear terminates medially at the **eardrum,** or **tympanic** (tim-pan′ik) **membrane.** The **middle ear** is an air-filled space within the petrous portion of the temporal bone, which contains the **auditory ossicles.** The **inner ear** contains the sensory organs for hearing and balance. It consists of interconnecting, fluid-filled tunnels and chambers within the petrous portion of the temporal bone.

Auditory Structures and Their Functions

External Ear

The **auricle,** or **pinna** (pin′ă), is the fleshy part of the external ear on the outside of the head; it consists primarily of elastic cartilage covered with skin (figure 15.24). Its shape helps collect sound waves and direct them toward the external acoustic meatus. The external acoustic meatus is lined with **hairs** and **ceruminous** (sĕ-roo′mi-nŭs) **glands,** which produce **cerumen,** a modified sebum commonly called earwax. The hairs and cerumen help prevent foreign objects from reaching the delicate eardrum. Overproduction of cerumen, however, may block the meatus.

The tympanic membrane, or eardrum, is a thin, semitransparent, nearly oval, three-layered membrane that separates the external ear from the middle ear. It consists of a low, simple cuboidal epithelium on the inner surface and a thin stratified squamous epithelium on the outer surface, with a layer of connective tissue between. Sound waves reaching the tympanic membrane through the external acoustic meatus cause it to vibrate.

Tympanic Membrane Rupture

Rupture of the tympanic membrane may result in hearing impairment. Foreign objects thrust into the ear, pressure, and infections of the middle ear can rupture the tympanic membrane. Sufficient differential pressure between the middle ear and the outside air can also rupture the tympanic membrane. This can occur in flyers, divers, or individuals who are hit on the side of the head by an open hand.

Middle Ear

Medial to the tympanic membrane is the air-filled cavity of the middle ear (see figure 15.23). Two covered openings, the round and oval windows, on the medial side of the middle ear separate it from the inner ear. Two openings provide air passages from the middle ear. One passage opens into the **mastoid air cells** in the mastoid process of the temporal bone. The other passageway, the **auditory,** or **pharyngotympanic, tube** (also called the eustachian ū-stā′shŭn tube), opens into the pharynx and equalizes air pressure between the outside air and the middle ear cavity. Unequal pressure between the middle ear and the outside environment can distort the eardrum, dampen its vibrations, and make hearing difficult. Distortion of the eardrum, which occurs under these conditions, also stimulates pain fibers associated with it. Because of this distortion, when a person changes altitude, sounds seem muffled and the eardrum may become painful. These symptoms can be relieved by opening the auditory tube to allow air to pass through the auditory tube to equalize air pressure. Swallowing, yawning, chewing, and holding the nose and mouth shut while gently trying to force air out of the lungs are methods used to open the auditory tube.

External ear
Auricle
Temporal bone
External acoustic meatus
Tympanic membrane
Chorda tympani
Semicircular canals
Oval window
Facial nerve
Vestibulocochlear nerve
Inner ear
Cochlear nerve
Vestibule
Cochlea
Round window
Auditory tube
Malleus Incus Stapes
Auditory ossicles in the middle ear

Frontal section

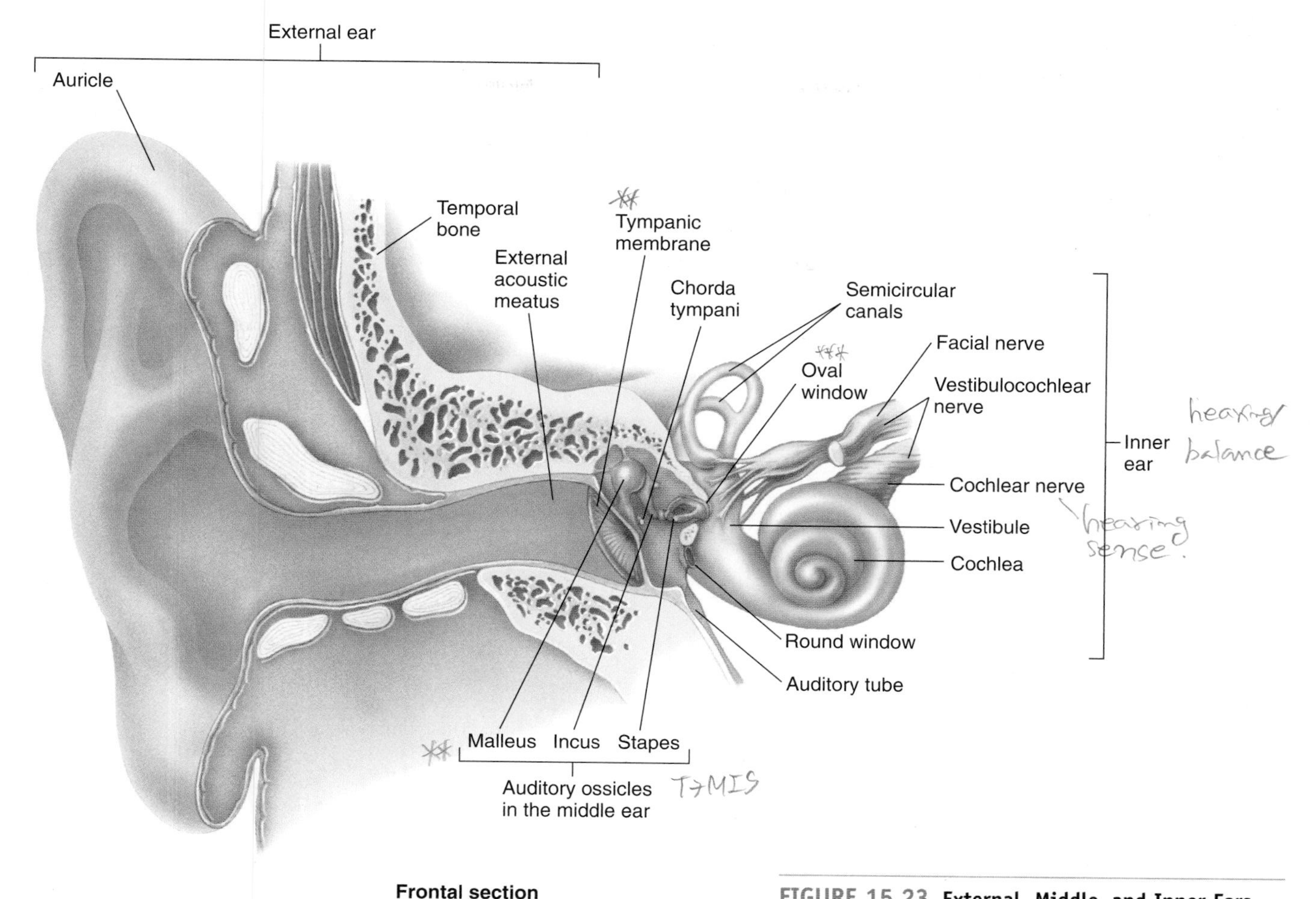

FIGURE 15.23 External, Middle, and Inner Ears

The middle ear contains three auditory ossicles (see figure 15.23—the **malleus** (mal'ē-ŭs; hammer), **incus** (ing'kŭs; anvil), and **stapes** (stā'pēz; stirrup)—which transmit vibrations from the tympanic membrane to the **oval window.** The handle of the malleus is attached to the inner surface of the tympanic membrane, and vibration of the membrane causes the malleus to vibrate as well. The head of the malleus is attached by a very small synovial joint to the incus, which in turn is attached by a small synovial joint to the stapes. The foot plate of the stapes fits into the oval window and is held in place by a flexible **annular ligament.**

Two small skeletal muscles originate from bone around the middle ear and insert onto auditory ossicles (figure 15.25). The **tensor tympani** (ten'sōr tim'păn-ē) muscle is attached to the malleus and is innervated by the trigeminal nerve (V). The **stapedius** (stā-pē'dē-ŭs) muscle is attached to the stapes and is innervated by the facial nerve (VII).

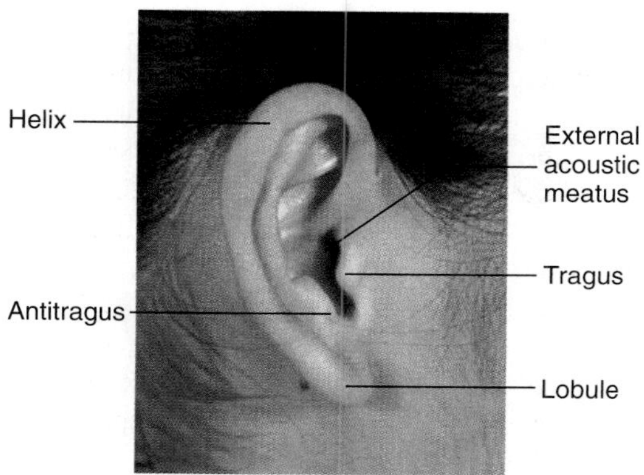

FIGURE 15.24 Structures of the Auricle (the Right Ear)

Chorda Tympani

A structure you might be somewhat surprised to find in the middle ear is the **chorda tympani,** a branch of the facial nerve carrying taste impulses from the anterior two-thirds of the tongue. It crosses over the inner surface of the tympanic membrane (see figure 15.23). The chorda tympani has nothing to do with hearing but is just passing through. This nerve can be damaged, however, during ear surgery or by a middle ear infection, resulting in loss of taste sensation from the anterior two-thirds of the tongue on the side innervated by that nerve.

Inner Ear

The tunnels and chambers inside the temporal bone are called the **bony labyrinth** (lab'i-rinth; maze; figure 15.26). Because the

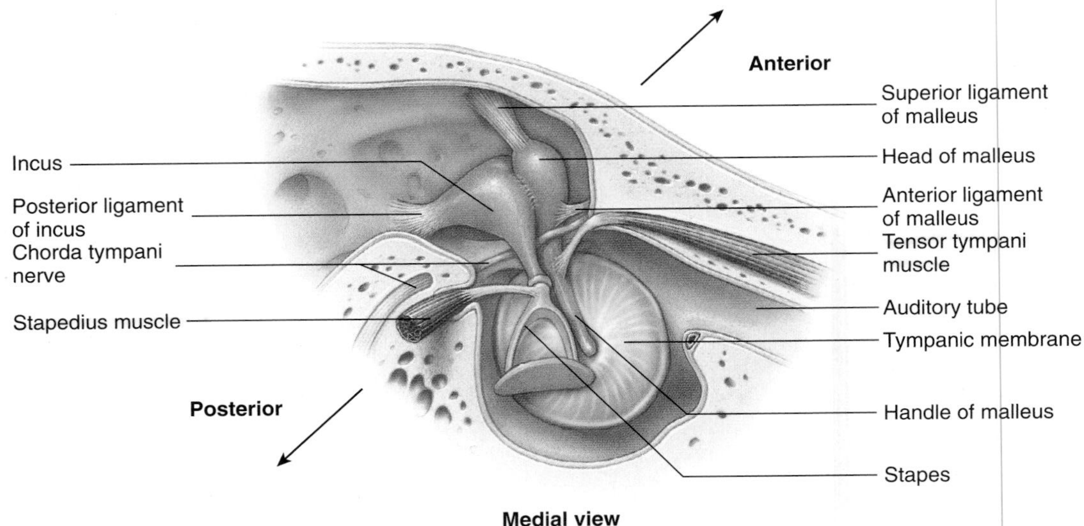

FIGURE 15.25 Muscles of the Middle Ear
Medial view of the middle ear (as though viewed from the inner ear), showing the three ear ossicles with their ligaments and the two muscles of the middle ear: the tensor tympani and the stapedius.

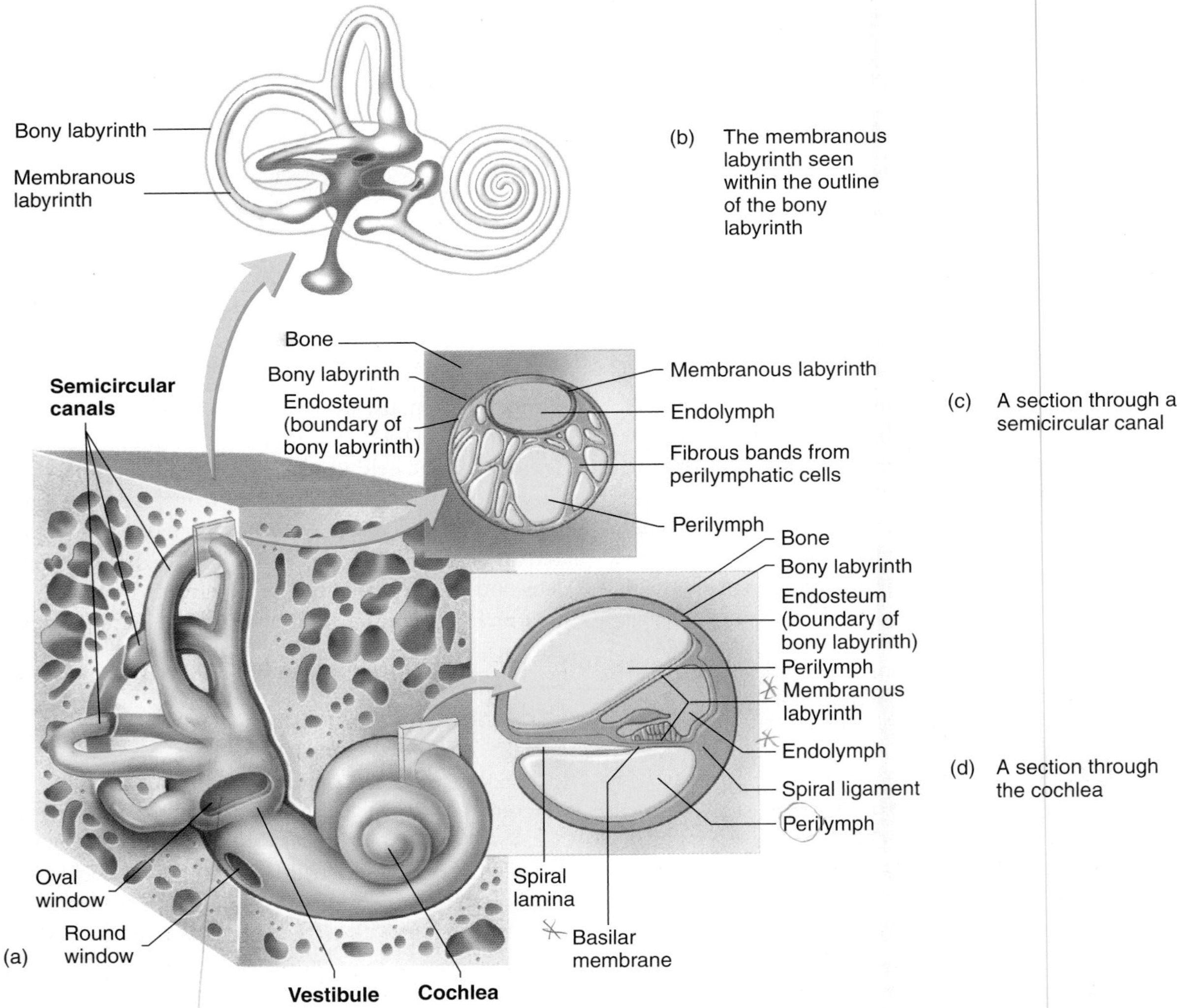

Bony labyrinth
Membranous labyrinth
(b) The membranous labyrinth seen within the outline of the bony labyrinth
Bone
Bony labyrinth
Endosteum (boundary of bony labyrinth)
Membranous labyrinth
Endolymph
Fibrous bands from perilymphatic cells
Perilymph
(c) A section through a semicircular canal
Semicircular canals
Bone
Bony labyrinth
Endosteum (boundary of bony labyrinth)
Perilymph
Membranous labyrinth
Endolymph
Spiral ligament
Perilymph
(d) A section through the cochlea
Oval window
Round window
(a)
Vestibule
Cochlea
Spiral lamina
Basilar membrane

FIGURE 15.26 Inner Ear: Bony and Membranous Labyrinths
The cross sections are taken through a semicircular canal and the cochlea to show the relationship between the bony and membranous labyrinths.

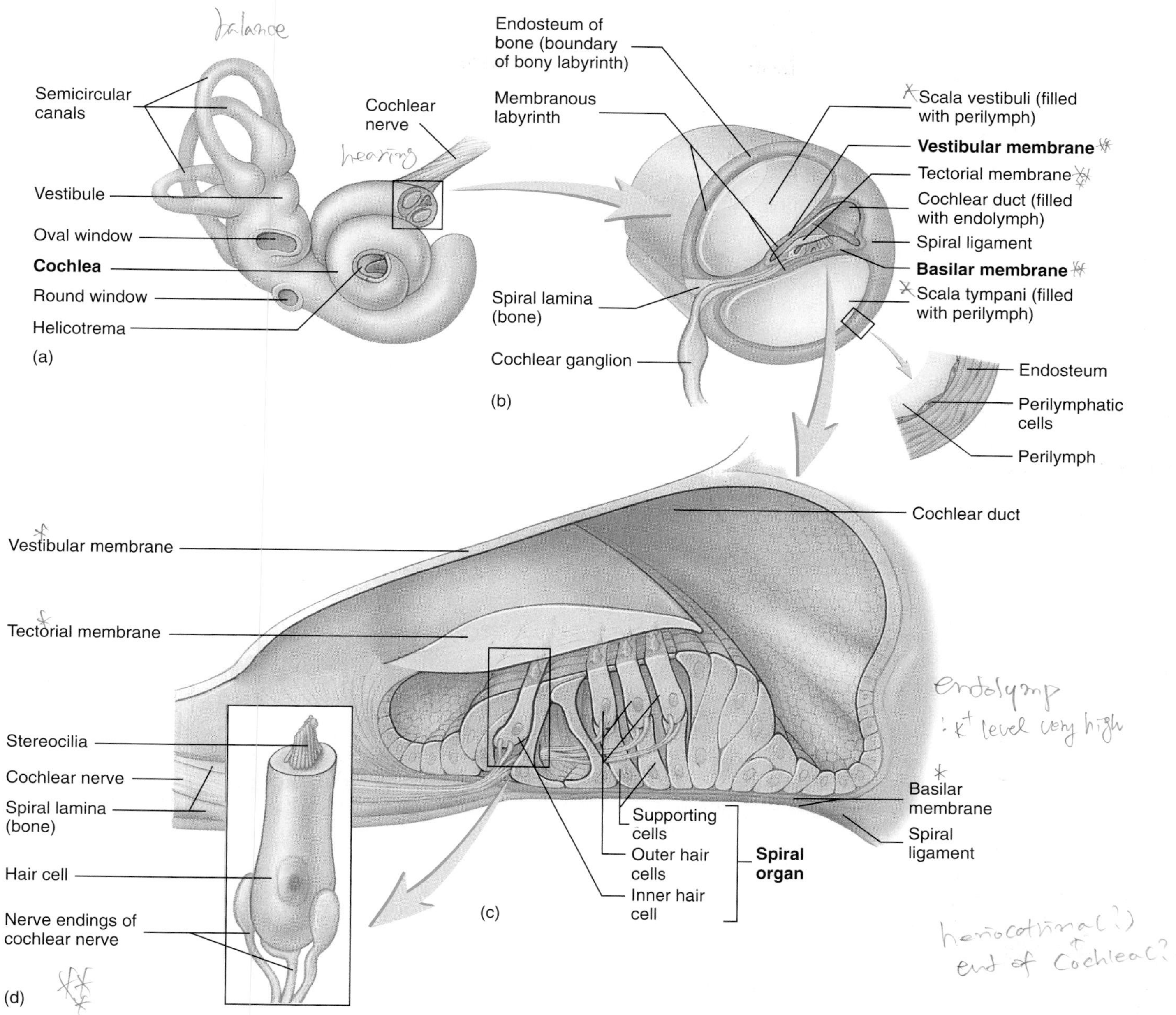

FIGURE 15.27 Structure of the Cochlea

(*a*) *The inner ear.* The outer surface (*gray*) is the endosteum lining the inner surface of the bony labyrinth. (*b*) *A cross section of the cochlea.* The outer layer is the endosteum lining the inner surface of the bony labyrinth. The membranous labyrinth is very small in the cochlea and consists of the vestibular and basilar membranes. The space between the membranous and bony labyrinths consists of two parallel tunnels: the scala vestibuli and scala tympani. (*c*) An enlarged section of the cochlear duct (membranous labyrinth). (*d*) A greatly enlarged individual sensory hair cell.

bony labyrinth consists of tunnels within the bone, it cannot easily be removed and examined separately. The bony labyrinth is lined with endosteum, and, when the inner ear is shown separately (figure 15.27*a*), the endosteum is what is depicted. Inside the bony labyrinth is a similarly shaped but smaller set of membranous tunnels and chambers called the **membranous labyrinth.** The inner surface of the endosteum and the outer surface of the membranous labyrinth are covered with a very thin layer of cells called the **perilymphatic cells** (figure 15.27*b*). The membranous labyrinth is filled with a clear fluid called **endolymph,** and the space between the membranous and bony labyrinths is filled with a fluid called **perilymph.** Perilymph is very similar to cerebrospinal fluid, but endolymph has a high concentration of potassium and a low concentration of sodium, which is opposite from perilymph and cerebrospinal fluid.

The bony labyrinth is divided into three regions: cochlea, vestibule, and semicircular canals. The **vestibule** (ves′ti-bool) and **semicircular canals** are involved primarily in balance, and the

cochlea (kok′lē-ă) is involved in hearing. The membranous labyrinth of the cochlea is divided into three parts: the scala vestibuli, the scala tympani, and the cochlear duct.

The oval window communicates with the vestibule of the inner ear, which in turn communicates with a cochlear chamber, the **scala vestibuli** (skā′lă ves-tib′ū-lē; see figure 15.27*a*). The scala vestibuli extends from the oval window to the **helicotrema** (hel′i-kō-trē′mă; a hole at the end of a helix or spiral) at the apex of the cochlea; a second cochlear chamber, the **scala tympani** (tim′pă-nē), extends from the helicotrema, back from the apex, parallel to the scala vestibuli, to the membrane of the **round window.**

The scala vestibuli and the scala tympani are the perilymph-filled spaces between the walls of the bony and membranous labyrinths. The wall of the membranous labyrinth that bounds the scala vestibuli is called the **vestibular membrane** (Reissner membrane); the wall of the membranous labyrinth bordering the scala tympani is the **basilar membrane** (see figure 15.27*b* and *c*). The space between the vestibular membrane and the basilar membrane is the interior of the membranous labyrinth and is called the **cochlear duct,** or **scala media,** which is filled with endolymph.

The vestibular membrane consists of a double layer of squamous epithelium and is the simplest region of the membranous labyrinth. The vestibular membrane is so thin that it has little or no mechanical effect on the transmission of sound waves through the inner ear; therefore, the perilymph and endolymph on the two sides of the vestibular membrane can be thought of mechanically as one fluid. The role of the vestibular membrane is to separate the two chemically different fluids. The basilar membrane is somewhat more complex and is of much greater physiologic interest in relation to the mechanics of hearing. It consists of an acellular portion with collagen fibers, ground substance, and sparsely dispersed elastic fibers and a cellular part with a thin layer of vascular connective tissue that is overlaid with simple squamous epithelium.

The basilar membrane is attached at one side to the bony **spiral lamina,** which projects from the sides of the **modiolus** (mō′dī′ō-lus), the bony core of the cochlea, like the threads of a screw. At the other side, the basilar membrane is attached to the lateral wall of the bony labyrinth by the **spiral ligament,** a local thickening of the endosteum. The distance between the spiral lamina and the spiral ligament (i.e., the width of the basilar membrane) increases from 0.04 mm near the oval window to 0.5 mm near the helicotrema. The collagen fibers of the basilar membrane are oriented across the membrane between the spiral lamina and the spiral ligament, somewhat like the strings of a piano. The collagen fibers near the oval window are both shorter and thicker than those near the helicotrema. The diameter of the collagen fibers in the membrane decreases as the basilar membrane widens. As a result, the basilar membrane near the oval window is short and stiff, and it responds to high-frequency vibrations, whereas the part near the helicotrema is wide and limber and responds to low-frequency vibrations.

The cells inside the cochlear duct are highly modified to form a structure called the **spiral organ,** or **organ of Corti** (see figure 15.27*b* and *c*). The spiral organ contains supporting epithelial cells and specialized sensory cells called **hair cells,** which have hairlike projections at their apical ends. In children, these projections consist of one cilium (kinocilium) and about 80 very long microvilli, often referred to as **stereocilia,** but in adults the cilium is absent from most hair cells (figures 15.27*d* and 15.28). The hair cells are arranged in four long rows extending the length of the cochlear duct. Each row contains 3500–4000 hair cells. The inner row consists of hair cells, called **inner hair cells,** which are the hair cells primarily responsible for hearing. The outer three rows contain **outer hair cells.** The outer hair cells are involved in regulating the tension of the basilar membrane and are separated from the inner hair cells by a gap in the basilar membrane (see figure 15.27). The stereocilia

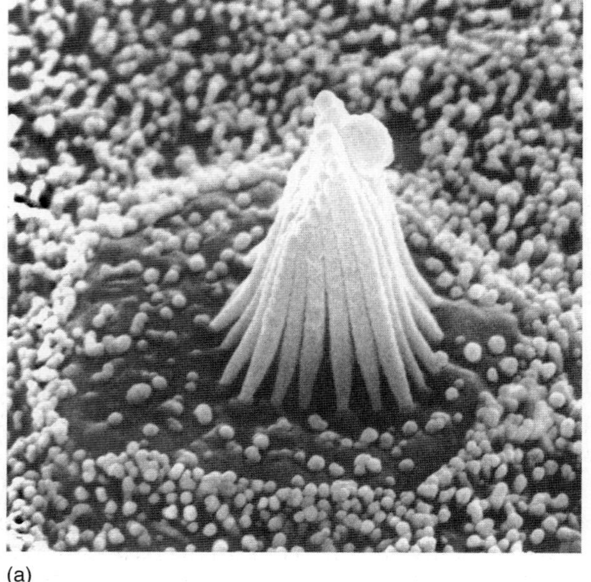
(a)

(b)

FIGURE 15.28 Scanning Electron Micrograph of Cochlear Hair Cell Stereocilia
(*a*) The hair bundle of one inner hair cell. (*b*) The stereocilia of three outer hair cells.

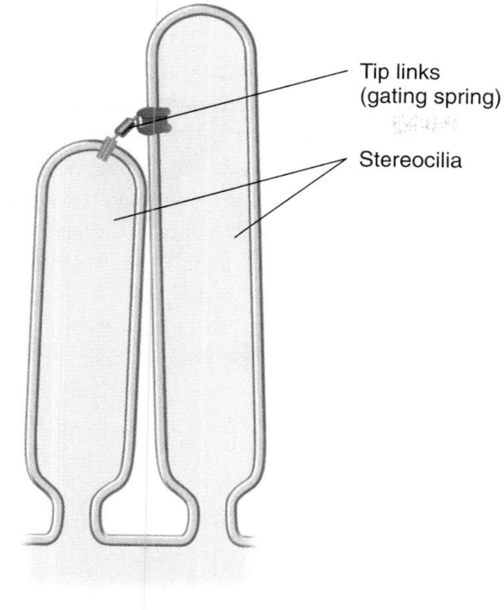

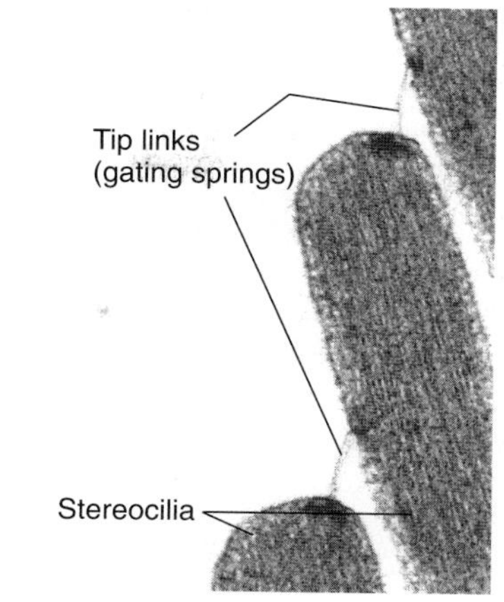

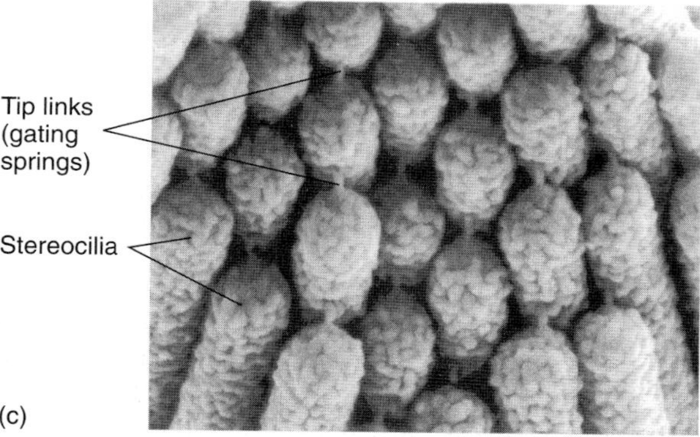

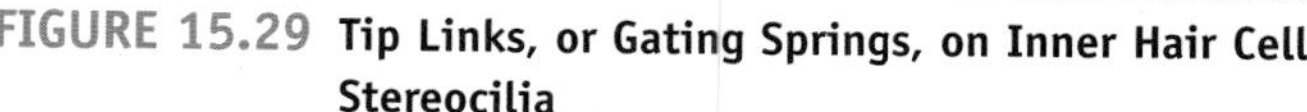

FIGURE 15.29 Tip Links, or Gating Springs, on Inner Hair Cell Stereocilia

(*a*) Diagram of two stereocilia connected by one tip link. (*b*) Transmission electron micrograph of three stereocilia and two tip links. (*c*) Scanning electron micrograph of the tops of stereocilia showing tip links.

of one inner hair cell form a conical group called a **hair bundle** (see figure 15.28*a*). The length of each stereocilium within a hair bundle increases gradually from one side of the hair cell to the other. The stereocilia of an outer hair cell are arranged in a curved line (see figure 15.28*b*). The tips of the longest stereocilia of the outer hair cells are embedded within an acellular gelatinous shelf called the **tectorial** (tek-tōr′ē-ăl) **membrane,** which is attached to the spiral lamina.

A **tip link** connects the tip of each stereocilium in a hair bundle to the side of the next longer stereocilium (figure 15.29). Each tip link is a **gating spring,** a pair of microtubule strands that attaches to the gate of a gated K^+ channel. The gated K^+ channels of hair cells open mechanically. As the stereocilia bend, the gating spring pulls the K^+ gate open (figure 15.30). The response time for such a mechanism is very brief and functions much faster than for a gating mechanism involving the synthesis of intracellular chemical signals such as cAMP.

Hair cells have no axons, but the basilar regions of each hair cell are covered by synaptic terminals of sensory neurons. The cell bodies of afferent neurons are located within the cochlear modiolus and are grouped into a **cochlear,** or **spiral, ganglion** (see figure 15.27*b*). Afferent fibers of these neurons join to form the **cochlear nerve.** This nerve then joins the vestibular nerve to become the **vestibulocochlear nerve** (VIII), which traverses the internal acoustic meatus and enters the cranial cavity.

28 ***Name the three regions of the ear, and list each region's parts.***

29 ***Describe the relationship among the tympanic membrane, the ear ossicles, and the oval window of the ear.***

30 ***What are the functions of the external acoustic meatus and of the auditory tube?***

31 ***Explain how the cochlear duct is divided into three compartments. What is found in each compartment?***

32 ***Explain the differences between inner and outer hair cells.***

33 ***What is the function of tip links, or gating springs?***

Auditory Function

Vibration of matter, such as air, water, or a solid material, creates sound. No sound occurs in a vacuum. When a person speaks, the vocal cords vibrate, causing the air passing out of the lungs to vibrate. The vibrations consist of bands of compressed air followed by bands of less compressed air (figure 15.31*a*). These vibrations are propagated through the air as sound waves, somewhat as ripples are propagated over the surface of water. **Volume,** or loudness, is a function of wave amplitude, or height, measured in decibels (figure 15.31*b*). The greater the amplitude, the louder the sound. **Pitch** is a function of the wave frequency (i.e., the number of waves or cycles per second) measured in hertz (Hz; figure 15.31*c*). The higher the frequency, the higher the pitch. The normal range of human hearing is 20–20,000 Hz and 0 or more decibels (db). Sounds louder than 125 db are painful to the ear.

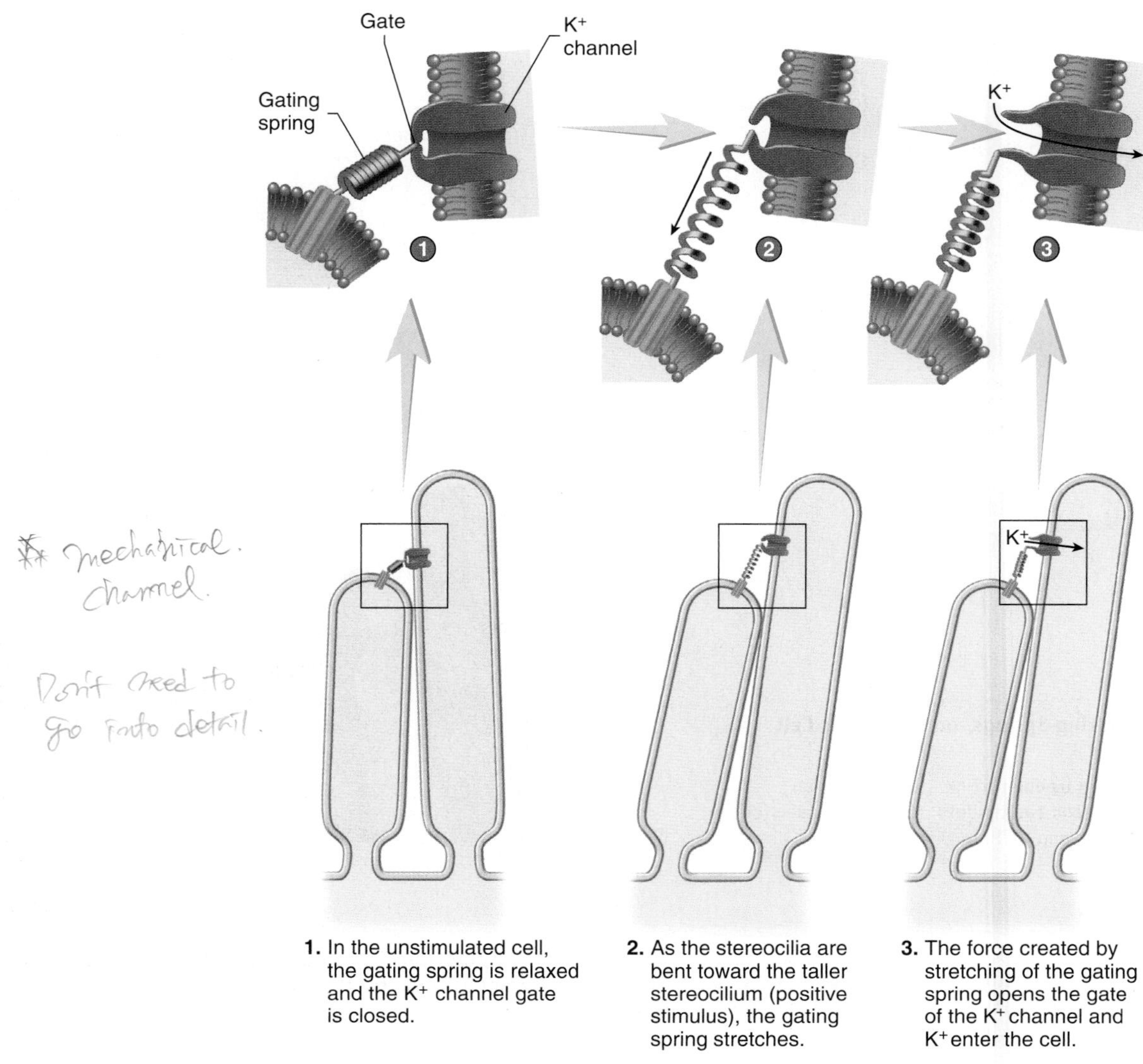

Gate
K+ channel
Gating spring
K+
1. In the unstimulated cell, the gating spring is relaxed and the K+ channel gate is closed.
2. As the stereocilia are bent toward the taller stereocilium (positive stimulus), the gating spring stretches.
3. The force created by stretching of the gating spring opens the gate of the K+ channel and K+ enter the cell.
* mechanical channel.
Don't need to go into detail.

PROCESS FIGURE 15.30 Action of the Gating Spring to Open a K^+ Channel When Two Stereocilia Bend

> **Human Speech and Hearing Impairment**
>
> The range of normal human speech is 250–8000 Hz. This is the range that is tested for the possibility of hearing impairment because it is the most important for communication.

Timbre (tam′br, tim′br) is the resonance quality or overtones of a sound. A smooth sigmoid curve is the image of a "pure" sound wave, but such a wave almost never exists in nature. The sounds made by musical instruments and the human voice are not smooth sigmoid curves but, rather, are rough, jagged curves formed by numerous, superimposed curves of various amplitudes and frequencies. The roughness of the curve accounts for the timbre. Timbre allows one to distinguish between, for example, an oboe and a French horn playing a note at the same pitch and volume. The steps involved in the mechanical part of hearing are illustrated in figure 15.32.

External Ear

The auricle collects sound waves, which are then conducted through the external acoustic meatus toward the tympanic membrane. Sound waves travel relatively slowly in air, 332 m/s, and a significant time interval may elapse between the time a sound wave reaches one ear and the time that it reaches the other. The brain can interpret this interval to determine the direction from which a sound is coming.

Middle Ear

Sound waves strike the tympanic membrane and cause it to vibrate. This vibration causes vibration of the three ossicles of the middle ear, and by this mechanical linkage vibration is transferred to the oval window. More force is required to cause vibration in a liquid, such as the perilymph of the inner ear, than is required in air; thus, the vibrations reaching the perilymph must be amplified

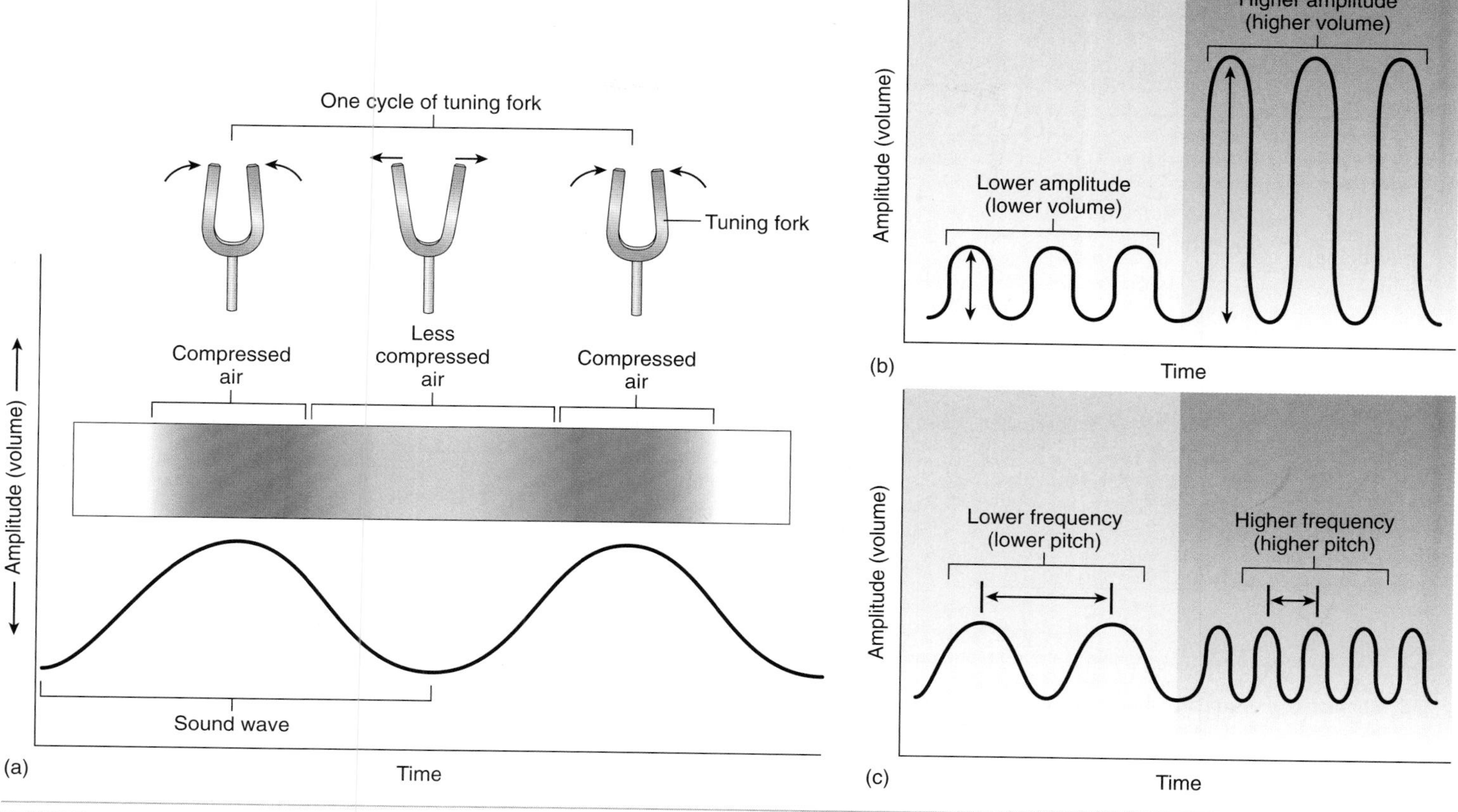

FIGURE 15.31 Sound Waves

(*a*) When something, such as a tuning fork or vocal cords, vibrates, the movements of the object alternate between compressing the air and decompressing the air, or making the air less compressed, thus producing sound. Each sound wave consists of a region of compressed air between two regions of less compressed air (*blue bars*). The sigmoid waves correspond to the regions of more compressed air (peaks) and less compressed air (troughs). The *green shadowed area* represents the width of one cycle (distance between peaks). (*b*) Low- and high-volume sound waves. Compare the relative lengths of the *arrows* indicating the wave height (amplitude). (*c*) Lower- and higher-pitch sound. Compare the relative number of peaks (frequency) within a given time interval (*between arrows*).

as they cross the middle ear. The footplate of the stapes and its annular ligament, which occupy the oval window, are much smaller than the tympanic membrane. Because of this size difference, the mechanical force of vibration is amplified about 20-fold as it passes from the tympanic membrane through the ossicles and to the oval window. This amplification occurs because the area of the tympanic membrane is roughly 20 times that of the oval window.

The tensor tympani and stapedius muscles, attached to auditory ossicles, reflexively dampen excessively loud sounds (see figure 15.25). This **sound attenuation reflex** protects the delicate ear structures from damage by loud noises. The sound attenuation reflex responds most effectively to low-frequency sounds and can reduce by a factor of 100 the energy reaching the oval window. The facial nerve and stapedius are primarily involved in the sound attenuation reflex. The trigeminal nerve and tensor tympani are only stimulated by extremely loud noise. The reflex is too slow to prevent damage from a sudden noise, such as a gunshot, and it cannot function effectively for longer than about 10 minutes, in response to prolonged noise.

PREDICT 9

What effect does facial nerve damage have on hearing?

Inner Ear

As the stapes vibrates, it produces waves in the perilymph of the scala vestibuli (see figure 15.32). Vibrations of the perilymph are transmitted through the thin vestibular membrane and cause simultaneous vibrations of the endolymph. The mechanical effect is as though the perilymph and endolymph were a single fluid. Vibration of the endolymph causes distortion of the basilar membrane. Waves in the perilymph of the scala vestibuli are transmitted also through the helicotrema and into the scala tympani. Because the helicotrema is very small, however, this transmitted vibration is probably of little consequence. Distortions of the basilar membrane, together with weaker waves coming through the helicotrema, cause waves in the scala tympani perilymph and ultimately result in vibration of the membrane of the round window. Vibration of the round window membrane is important

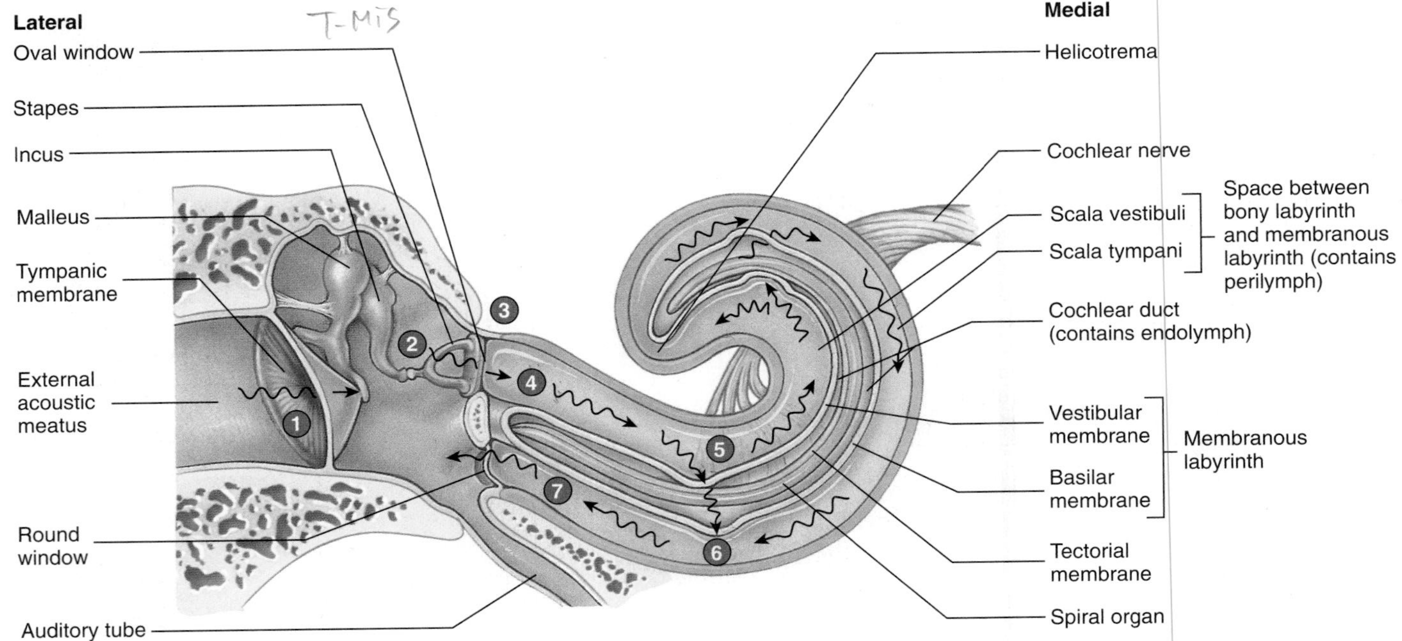

1. Sound waves strike the tympanic membrane and cause it to vibrate.
2. Vibration of the tympanic membrane causes the three bones of the middle ear—the malleus, incus, and stapes—to vibrate.
3. The foot plate of the stapes vibrates in the oval window.
4. Vibration of the foot plate causes the perilymph in the scala vestibuli to vibrate.
5. Vibration of the perilymph causes the vestibular membrane to vibrate, which causes vibrations in the endolymph.
6. Vibration of the endolymph causes displacement of the basilar membrane. Short waves (high pitch) cause displacement of the basilar membrane near the oval window, and longer waves (low pitch) cause displacement of the basilar membrane some distance from the oval window. Movement of the basilar membrane is detected in the hair cells of the spiral organ, which are attached to the basilar membrane. Vibrations of the perilymph in the scala vestibuli and of the basement membrane are transferred to the perilymph of the scala tympani.
7. Vibrations in the perilymph of the scala tympani are transferred to the round window, where they are dampened.

PROCESS FIGURE 15.32 Effect of Sound Waves on Cochlear Structures

to hearing because it acts as a mechanical release for waves from within the cochlea. If this window were solid, it would reflect the waves, which would interfere with and dampen later waves. The round window also allows the relief of pressure in the perilymph because fluid is not compressible, thereby preventing compression damage to the spiral organ.

The distortion of the basilar membrane is most important to hearing. As this membrane distorts, the hair cells resting on the basilar membrane move relative to the tectorial membrane, which remains stationary. The inner hair cell microvilli are bent as they move against the tectorial membrane.

The apical portion of each hair cell is surrounded by endolymph and the basal portion of the cell is surrounded by perilymph. Endolymph has a high K^+ concentration similar to the intracellular K^+ concentration of hair cells. Perilymph has a low concentration of K^+ similar to that of other extracellular fluid. The intracellular charge of hair cells compared with perilymph is -60 mV. The charge of the endolymph is $+80$ mV compared with the perilymph. This charge difference is called the **endocochlear potential.** Consequently, the intracellular charge of hair cells compared with endolymph is -140 mV, which is a large charge difference. Consequently, when K^+ channels open, K^+ flow into the hair cells because they are attracted to the negative charge inside of the hair cell, even though the intracellular concentration of K^+ is about the same as in the endolymph. The movement of the K^+ causes depolarization of the hair cells. This is a rare instance in which an increase in K^+ permeability of the plasma membrane of a cell results in depolarization.

In the unstimulated hair cell, approximately 15% of the gated K^+ channels are open and the resting membrane potential of the cell is approximately -60 mV. If the hair bundle is bent toward the shortest stereocilium (negative stimulus) the gating springs attached to the K^+ channel gates slacken, allowing the open K^+ channels to close and the cell hyperpolarizes. If the hair bundle is bent toward the longest stereocilium (positive stimulus), gating springs pull additional K^+ channel gates open, and K^+ rush into the cell. The influx of K^+ into the hair cell causes a slight depolarization of the cell. This depolarization causes voltage-gated Ca^{2+} channels to open. Calcium ions rush into the cell, causing a further depolarization. The cell depolarizes by a total of about 10 mV. Depolarization of hair cells results in the increased release of neurotransmitters, which increases the action potential frequency in the afferent neurons. Hyperpolarization decreases neurotransmitter release and decreases action potential frequency in afferent neurons. Depolarization also opens voltage-gated K^+ channels in the basal portion of the hair cell. Potassium ions tend to leave the cell, causing the cell to repolarize.

The neurotransmitter released by the inner hair cells is apparently glutamate, but other neurotransmitters may also be involved. The release of neurotransmitters from the inner hair cells induces action potentials in the cochlear neurons that synapse on the hair cells. The cell bodies of those neurons are located in the cochlear ganglion.

The part of the basilar membrane that distorts as a result of endolymph vibration depends on the pitch of the sound that created the vibration and, as a result, on the vibration frequency within the endolymph. The width of the basilar membrane and the length and diameter of the collagen fibers stretching across the membrane at each level along the cochlear duct determine the location of the optimal amount of basilar membrane vibration produced by a given pitch (figure 15.33). Higher-pitched tones cause optimal vibration near the base, and lower-pitched tones cause optimal vibration near the apex of the basilar membrane. As the basilar membrane vibrates, hair cells along a large part of the basilar membrane are stimulated. In areas of minimum vibration, the amount of stimulation may not reach threshold. In other areas, a low frequency of afferent action potentials may be transmitted, whereas, in the optimally vibrating regions of the basilar membrane, a high frequency of action potentials is initiated.

There are approximately twice as many nerve cells in the cochlear ganglion as there are hair cells. Over 90% of the afferent axons synapse with inner hair cells, 10 to 30 axons per hair cell. Only a few, small-diameter afferent axons synapse with the three rows of outer hair cells. The outer cells, however, receive input from efferent axons. Action potentials from those efferent axons stimulate the contraction of actin filaments within the hair cells, causing them to shorten. This adjustment in the height of the outer hair cells, attached to both the basilar membrane and the tectorial membrane, fine tunes the tension of the basilar membrane and the distance between the basilar membrane and tectorial membrane. Additional sensitivity can be adjusted within the inner hair cells. The mechanically gated K^+ channels are attached to actin filaments inside the cell. Those filaments can move the K^+ channels along the cell membrane,

Loud Noises and Hearing Loss

Prolonged or frequent exposure to excessively loud noises can cause degeneration of the spiral organ at the base of the cochlea, resulting in high-frequency deafness. The actual amount of damage can vary greatly from person to person. High-frequency loss can cause a person to miss hearing consonants in a noisy setting. Loud music, amplified to 120 db, can impair hearing. The defects may not be detectable on routine diagnosis, but they include decreased sensitivity to sound in specific narrow frequency ranges and a decreased ability to discriminate between two pitches. Loud music, however, is not as harmful as is the sound of a nearby gunshot, which is a sudden sound occurring at 140 db. The sound is too sudden for the attenuation reflex to protect the inner ear structures, and the intensity is great enough to cause auditory damage. In fact, gunshot noise is the most common recreational cause of serious hearing loss.

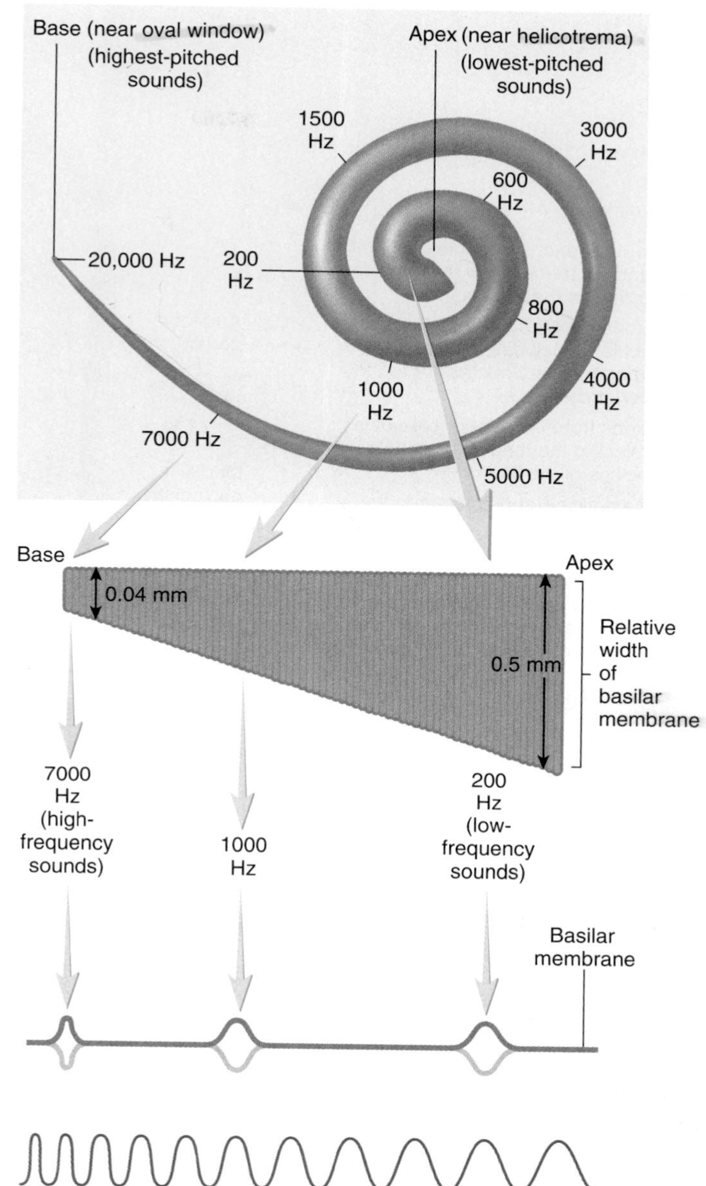

FIGURE 15.33 Effect of Sound Waves on Points Along the Basilar Membrane

Points of maximum vibration along the basilar membrane resulting from stimulation by sounds of various frequencies (in hertz).

tightening or loosening the gating spring. By this means, hair cells are tuned to very specific frequencies. Likewise, the response of an inner hair cell to stimulation is graded and can increase up to a point of saturation (when all the K^+ channels are open maximally). Inner hair cells are also very sensitive. A 100 nm (1 degree) deflection of the stereocilia results in a response that is 90% of maximum.

Afferent action potentials conducted by cochlear nerve fibers from all along the spiral organ terminate in the **superior olivary nucleus** in the medulla oblongata (figure 15.34; see chapter 13). These action potentials are compared with one another, and the strongest action potential, corresponding to the area of maximum

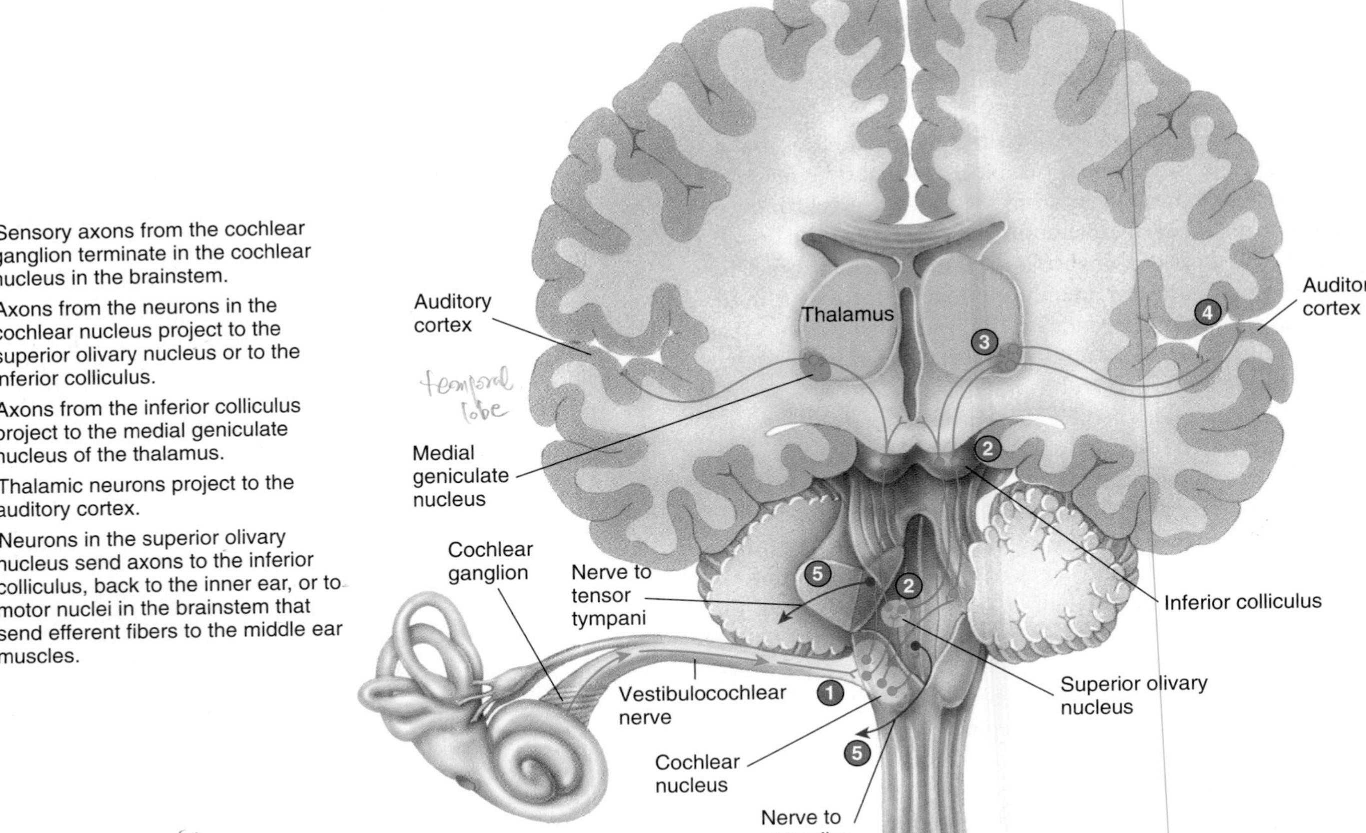

PROCESS FIGURE 15.34 Central Nervous System Pathways for Hearing

basilar membrane vibration, is taken as standard. Efferent action potentials then are sent from the superior olivary nucleus back to the spiral organ to all regions where the maximum vibration did not occur. These action potentials inhibit the hair cells from initiating additional action potentials in the sensory neurons. Thus, only action potentials from regions of maximum vibration are received by the cortex, where they become consciously perceived.

By this process, tones are localized along the cochlea. As a result of this localization, neurons along a given portion of the cochlea send action potentials only to the cerebral cortex in response to specific pitches. Action potentials near the base of the basilar membrane stimulate neurons in a certain part of the auditory cortex, which interpret the stimulus as a high-pitched sound, whereas action potentials from the apex stimulate a different part of the cortex, which interprets the stimulus as a low-pitched sound.

PREDICT 10

Suggest some possible sites and mechanisms within the auditory structures to explain why certain people have "perfect pitch" and other people are "tone deaf."

Sound volume, or loudness, is a function of sound wave amplitude. As high-amplitude sound waves reach the ear, the perilymph, endolymph, and basilar membrane vibrate more intensely, and the hair cells are stimulated more intensely. As a result of the increased stimulation, more hair cells send action potentials at a higher frequency to the cerebral cortex, where this information is perceived as a greater sound volume.

34 ***Starting with the auricle, trace a sound wave into the inner ear to the point at which action potentials are generated in the cochlear nerve.***

35 ***Explain how pulling on a gating spring results in the generation of an action potential.***

PREDICT 11

Explain why it is much easier to perceive subtle musical tones when music is played somewhat softly, as opposed to very loudly.

Neuronal Pathways for Hearing

The special senses of hearing and balance are both transmitted by the vestibulocochlear (VIII) nerve. The term *vestibular* refers to the vestibule of the inner ear, which is involved in balance. The term *cochlear* refers to the cochlea and is that portion of the inner ear involved in hearing. The vestibulocochlear nerve functions as two separate nerves carrying information from two separate but closely related structures.

The auditory pathways within the CNS are very complex, with both crossed and uncrossed tracts (see figure 15.34). Unilateral CNS damage therefore usually has little effect on hearing. The neurons from the cochlear ganglion synapse with CNS neurons in the dorsal or ventral **cochlear nucleus** in the superior medulla near the inferior cerebellar peduncle. These neurons in turn either synapse in or pass through the superior olivary nucleus. Neurons terminating in this nucleus may synapse with efferent neurons returning to the cochlea to modulate pitch perception. Nerve fibers from the superior olivary nucleus also project to the trigeminal (V) nucleus, which controls the tensor tympani, and the facial (VII) nucleus, which controls the stapedius muscle. This reflex pathway dampens loud sounds by initiating contractions of these muscles, the sound attenuation reflex. Neurons synapsing in the superior olivary nucleus may also join other ascending neurons to the cerebral cortex.

Ascending neurons from the superior olivary nucleus travel in the **lateral lemniscus.** All ascending fibers synapse in the **inferior colliculi,** and neurons from there project to the **medial geniculate nucleus** of the **thalamus,** where they synapse with neurons that project to the cortex. These neurons terminate in the **auditory cortex** in the dorsal portion of the temporal lobe within the lateral fissure and, to a lesser extent, on the superolateral surface of the temporal lobe (see chapter 13). Neurons from the inferior colliculus also project to the **superior colliculus,** where reflexes that turn the head and eyes in response to loud sounds are initiated.

36 ***Describe the neuronal pathways for hearing from the cochlear nerve to the cerebral cortex.***

Balance

The organs of balance are divided structurally and functionally into two parts. The first, the **static labyrinth,** consists of the **utricle** (oo′tri-kl) and the **saccule** (sak′ūl) of the vestibule and is primarily involved in evaluating the position of the head relative to gravity, although the system also responds to linear acceleration or deceleration, such as when a person is in a car that is increasing or decreasing speed. The second, the **kinetic labyrinth,** is associated with the semicircular canals and is involved in evaluating movements of the head.

Most of the utricular and saccular walls consist of simple cuboidal epithelium. The utricle and saccule, however, each contain a specialized patch of epithelium about 2–3 mm in diameter called the **macula** (mak′ū-lă; figure 15.35*a* and *b*). The macula of the utricle is oriented parallel to the base of the skull, and the macula of the saccule is perpendicular to the base of the skull.

The maculae resemble the spiral organ and consist of columnar supporting cells and hair cells. The "hairs" of these cells, which consist of numerous microvilli, called **stereocilia,** and one cilium, called a **kinocilium** (kī-nō-sil′ē-ŭm), are embedded in a **gelatinous mass** weighted by the presence of **otoliths** (ō′tō-liths) composed of protein and calcium carbonate (see figure 15.35*b*). The gelatinous mass moves in response to gravity, bending the hair cells and initiating action potentials in the associated neurons. The stereocilia function much as do the stereocilia of cochlear hair cells, with tip links connected to gated K^+ channels. Deflection of the hairs toward the kinocilium results in depolarization of the hair cell, whereas deflection of the hairs away from the kinocilium results in hyperpolarization of the hair cell. If the head is tipped, otoliths move in response to gravity and stimulate certain hair cells (figure 15.36).

The hair cells have no axons but synapse directly with neurons of the vestibulocochlear nerve (VIII). Several neurotransmitters are released from the hair cells. One of those neurotransmitters is glutamate. Others have not yet been identified.

The hair cells are constantly being stimulated at a low level by the presence of the otolith-weighted covering of the macula;

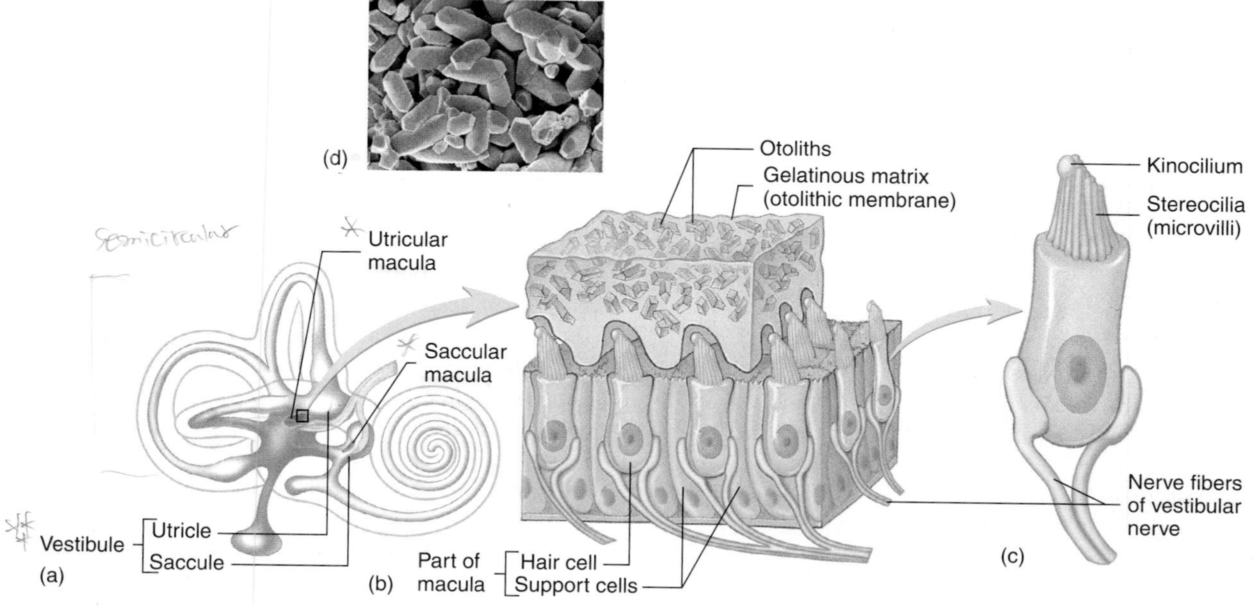

FIGURE 15.35 Structure of the Macula

(*a*) Vestibule showing the location of the utricular and saccular maculae. (*b*) Enlargement of the utricular macula, showing hair cells and otoliths in the macula. (*c*) An enlarged hair cell, showing the kinocilium and stereocilia. (*d*) Colorized scanning electron micrograph of otoliths.

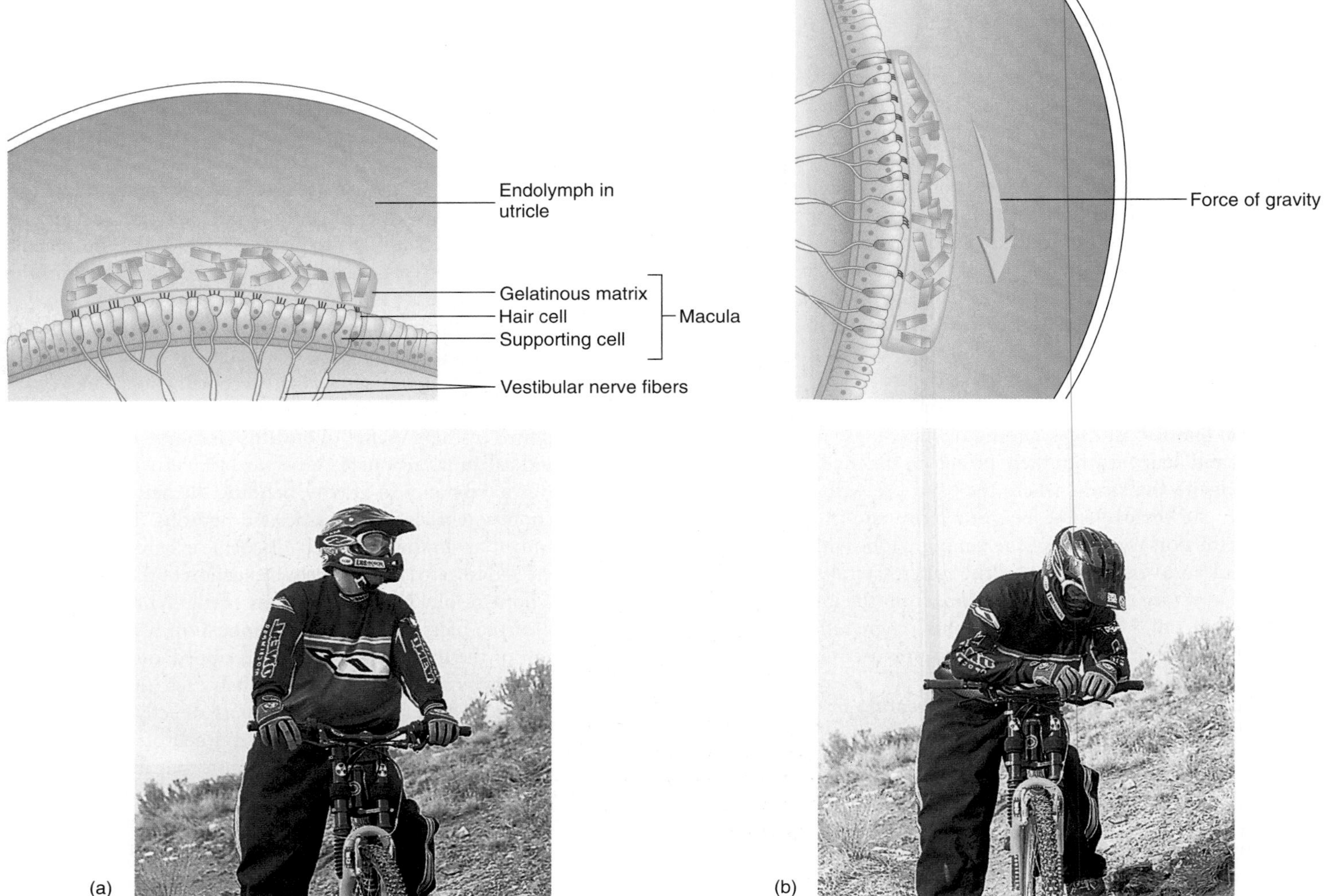

FIGURE 15.36 Function of the Vestibule in Maintaining Balance

(*a*) As the position of the head changes, such as when a person bends over, the maculae respond to changes in position of the head relative to gravity by moving in the direction of gravity. (*b*) In an upright position, the maculae do not move.

however, as this covering moves in response to gravity, the pattern of intensity of hair cell stimulation changes. This pattern of stimulation and the subsequent pattern of action potentials from the numerous hair cells of the maculae can be translated by the brain into specific information about head position or acceleration. Much of this information is not perceived consciously but is dealt with subconsciously. The body responds by making subtle tone adjustments in the muscles of the back and neck, which are intended to restore the head to its proper neutral, balanced position.

The kinetic labyrinth (figure 15.37) consists of three **semicircular canals** placed at nearly right angles to one another, one lying nearly in the transverse plane, one in the coronal plane, and one in the sagittal plane (see chapter 1). The arrangement of the semicircular canals enables a person to detect movement in all directions. The base of each semicircular canal is expanded into an **ampulla** (see figure 15.37*a*). Within each ampulla, the epithelium is specialized to form a **crista ampullaris** (kris′tă am-pū-lar′ŭs). This specialized sensory epithelium is structurally and functionally very similar to the sensory epithelium of the maculae. Each crista consists of a ridge or crest of epithelium with a curved gelatinous mass, the **cupula** (koo′poo-lă), suspended over the crest. The hairlike processes of the crista hair cells, which are stereocilia similar to those in the maculae, and cochlear hair cells are embedded in the cupula (see figure 15.37*b*). The cupula contains no otoliths and therefore does not respond to gravitational pull. Instead, the cupula is a float that is displaced by fluid movements within the semicircular canals. Endolymph movement within each semicircular canal moves the cupula, bends the hairs, and initiates action potentials (figure 15.38).

As the head begins to move, the endolymph does not move at the same rate as the semicircular canals (see figure 15.38). This difference causes displacement of the cupula in a direction opposite to that of the movement of the head, resulting in relative movement between the cupula and the endolymph. As movement continues,

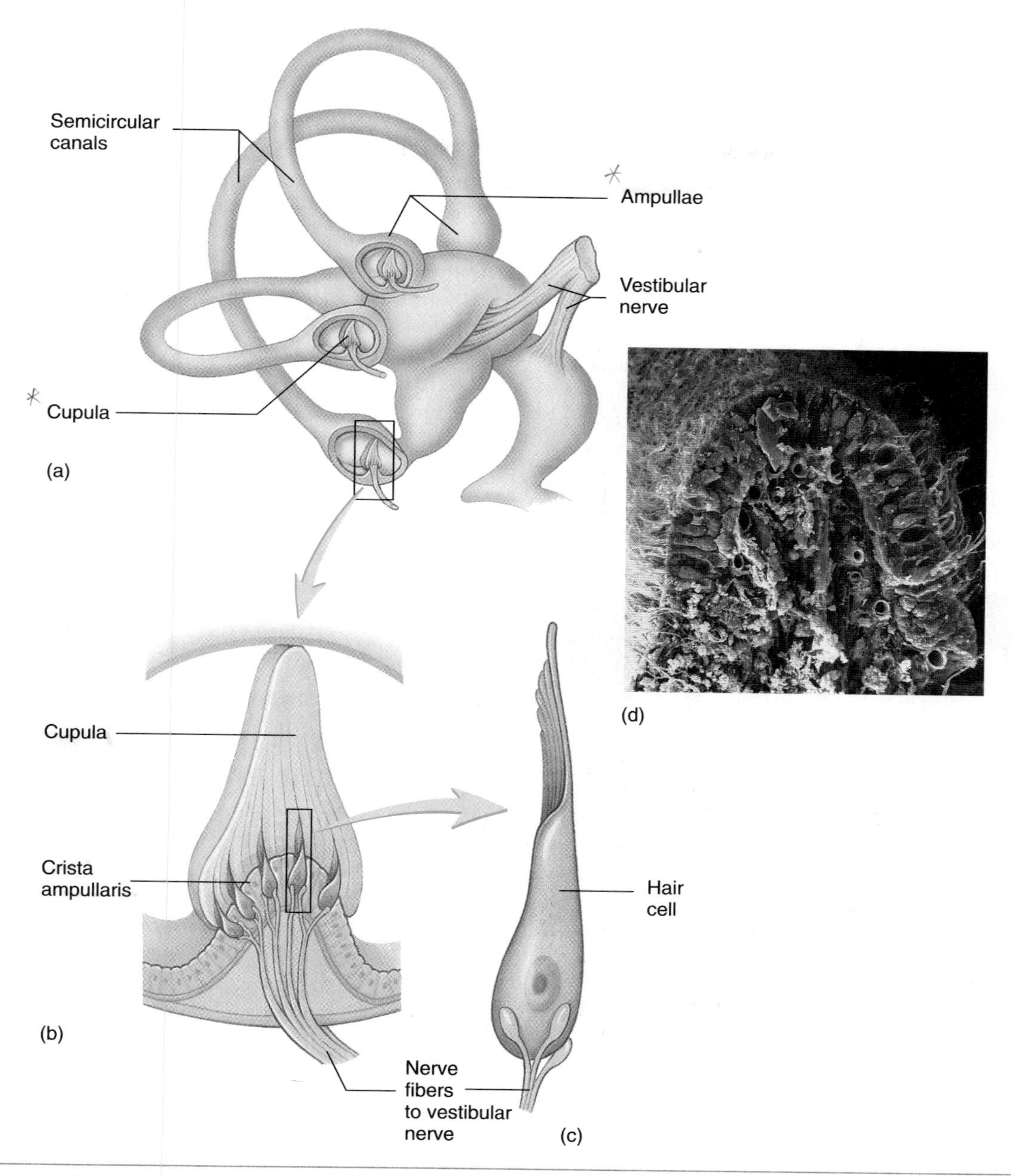

FIGURE 15.37 Semicircular Canals
(*a*) Semicircular canals showing the location of the crista ampullaris in the ampullae of the semicircular canals. (*b*) Enlargement of the crista ampullaris, showing the cupula and hair cells. (*c*) Enlargement of a hair cell. (*d*) SEM of a crista ampullaris with hair cells.

the fluid of the semicircular canals begins to move and catches up with the cupula, and stimulation is stopped. As movement of the head ceases, the endolymph continues to move because of its momentum, causing displacement of the cupula in the same direction as the head had been moving. Because displacement of the cupula is most intense when the rate of head movement changes, this system detects changes in the rate of movement rather than movement alone. As with the static labyrinth, the information the brain obtains from the kinetic labyrinth is largely subconscious.

37 ***What are the functions of the saccule and utricle? Describe the macula and its functions.***

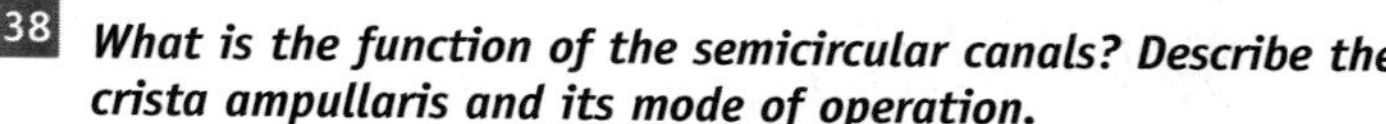

38 ***What is the function of the semicircular canals? Describe the crista ampullaris and its mode of operation.***

Space Sickness

Space sickness is a balance disorder occurring in zero gravity and resulting from unfamiliar sensory input to the brain. The brain must adjust to these unusual signals, or severe symptoms, such as headaches and dizziness, may result. Space sickness is unlike motion sickness in that motion sickness results from an excessive stimulation of the brain, whereas space sickness results from too little stimulation as a result of weightlessness.

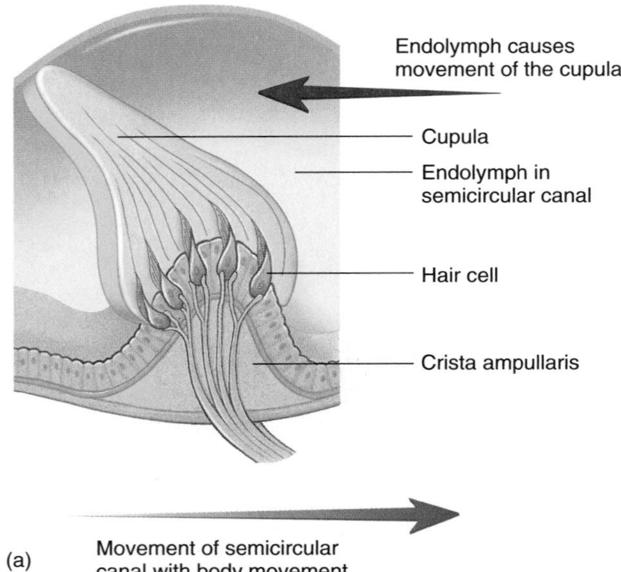

(b)

FIGURE 15.38 Function of the Semicircular Canals

The crista ampullaris responds to fluid movements within the semicircular canals. (*a*) When a person is at rest, the crista ampullaris does not move. (*b*) As a person begins to move, the semicircular canals begin to move with the body (*blue arrow*), but the endolymph tends to remain stationary relative to the movement (momentum force: *red arrow* pointing in the opposite direction of body and semicircular canal movement), and the crista ampullaris is displaced by the endolymph in a direction opposite to the direction of movement.

Neuronal Pathways for Balance

Neurons synapsing on the hair cells of the maculae and cristae ampullares converge into the **vestibular ganglion,** where their cell bodies are located (figure 15.39). Sensory fibers from these neurons join sensory fibers from the cochlear ganglion to form the vestibulocochlear nerve (VIII) and terminate in the **vestibular nucleus** within the medulla oblongata. Axons run from this nucleus to numerous areas of the CNS, such as the spinal cord, the cerebellum, the cerebral cortex, and the nuclei controlling extrinsic eye muscles.

Balance is a complex process not simply confined to one type of input. In addition to vestibular sensory input, the vestibular nucleus receives input from proprioceptive neurons throughout the body, as well as from the visual system. People are asked to close their eyes while balance is evaluated in a sobriety test because alcohol affects the proprioceptive and vestibular components of balance (cerebellar function) to a greater extent than it does the visual portion.

Reflex pathways exist between the kinetic part of the vestibular system and the nuclei controlling the extrinsic eye muscles (oculomotor, trochlear, and abducent). A reflex pathway allows the maintenance of visual fixation on an object while the head is in motion. This function can be demonstrated by spinning a person around about 10 times in 20 seconds, stopping him or her, and observing eye movements. The reaction is most pronounced if the individual's head is tilted forward about 30 degrees while spinning, thus bringing the lateral semicircular canals into the horizontal plane. A slight oscillatory movement of the eyes occurs. The eyes track in the direction of motion and return with a rapid recovery movement before repeating the tracking motion. This oscillation of the eyes is called **nystagmus** (nis-tag′mŭs). If asked to walk in a straight line, the individual deviates in the direction of rotation; if asked to point to an object, his or her finger deviates in the direction of rotation.

39 ***Describe the neuronal pathways for balance.***

EFFECTS OF AGING ON THE SPECIAL SENSES

Elderly people experience only a slight loss in the ability to detect odors. However, the ability to identify specific odors correctly is decreased, especially in men over age 70.

In general, the sense of taste decreases as people age. The number of sensory receptors decreases and the brain's ability to interpret taste sensations declines.

Responses to taste change in some elderly people who are fighting cancer. One side effect of radiation treatment and chemotherapy is the gastrointestinal discomfort resulting from the treatments. The patients experience a loss of appetite because of conditioned taste aversions resulting from treatment.

The lenses of the eyes lose flexibility as a person ages because the connective tissue of the lenses becomes more rigid. Consequently, there is first a reduction in and then an eventual loss of the lenses' ability to change shape. Recall that this condition, called presbyopia, is the most common age-related change in the eyes.

The most common visual problem in older people requiring medical treatment, such as surgery, is the development of cataracts. Macular degeneration is the second most common defect, glaucoma is third, and diabetic retinopathy is fourth.

The number of cones decreases, especially in the fovea centralis. These changes cause a gradual decline in visual acuity and color perception.

As people age, the number of hair cells in the cochlea decreases, leading to age-related hearing loss, called presbyacusis (prez′bē-ă-koo′sis). This decline does not occur equally in both ears, however. As a result, because direction is determined by comparing sounds

CASE STUDY

Motion Sickness

Earl booked his first trip on a charter fishing boat. Crossing the bar was refreshing and invigorating, and Earl was surprised that all those warnings about his becoming sea sick did not seem to apply to him. At last, the boat arrived at the fishing site, the engine was cut, and the sea anchor was set. As the boat drifted, it began to roll and pitch. Earl noticed, for the first time, that the bait smelled oily and unpleasant. The smell mixed poorly with the diesel fumes. Earl felt a little light-headed and a bit drowsy. He then noticed that he was definitely nauseated and felt pale. "I'm sea sick," he realized. Trying to fish seemed to worsen his condition. Eventually, his nausea intensified and he leaned over the boat rail and vomited into the ocean. "Improves the fishing," the ship owner shouted to him cheerily. Earl noticed that he felt distinctly better. His condition continued to improve. He found that looking at the horizon rather than at the water helped his condition. He enjoyed the rest of the trip and even caught a couple of fish.

Sea sickness is a form of **motion sickness,** which consists of nausea, weakness, and other dysfunctions caused by stimulation of the semicircular canals during motion, such as in a boat, an automobile, or an airplane, on a swing, or on an amusement park ride. It can progress to vomiting and incapacitation. The brain compares sensory input from the semicircular canals, eyes, and proprioceptors in the lower limbs. Perceived differences in the input may result in motion sickness, which can be decreased by closing the eyes or looking at a distant object, such as the horizon. Antiemetics, such as anticholinergic or antihistamine medications, can be taken to counter the nausea and vomiting. Scopolamine is an anticholinergic drug that blocks acetylcholine-mediated transmission in the parasympathetic nervous system. It depresses parasympathetic activity within areas of the CNS, such as the hypothalamus, in response to vestibular stimulation. It also depresses other CNS functions, which may cause side effects such as restlessness, agitation, psychosis, mania, and Parkinson-like tremors. Cyclizine (Marezine), dimenhydrinate (Dramamine), and diphenhydramine (Benadryl) are antihistamines that affect the neural pathways from the vestibule. Scopolamine can be administered transdermally in the form of a patch placed on the skin behind the ear (Transdermal-Scop). A patch lasts about 3 days.

PREDICT 12

Explain why closing your eyes can help decrease motion sickness. Why might looking at the horizon help?

1. Sensory axons from the vestibular ganglion pass through the vestibular nerve to the vestibular nucleus, which also receives input from several other sources, such as proprioception from the legs.
2. Vestibular neurons send axons to the cerebellum, which influences postural muscles.
3. Vestibular neurons also send axons to motor nuclei (oculomotor, trochlear, and abducens), which control extrinsic eye muscles.
4. Vestibular neurons also send axons to the posterior ventral nucleus of the thalamus.
5. Thalamic neurons project to the vestibular area of the cortex.

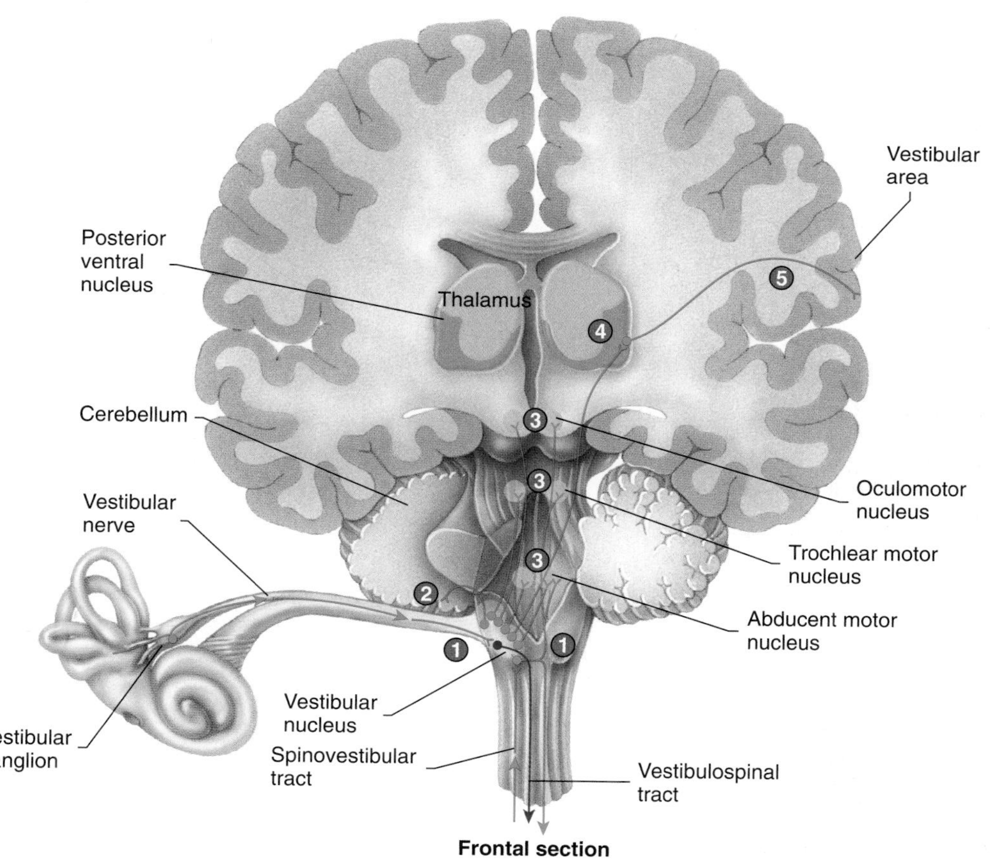

PROCESS FIGURE 15.39 **Central Nervous System Pathways for Balance**

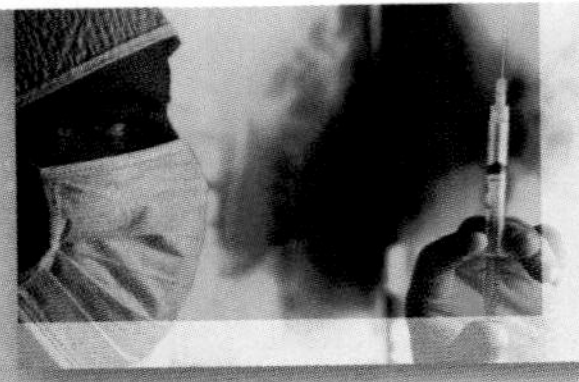

CLINICAL FOCUS

Hearing Impairment and Functional Replacement of the Ear

The term **hearing-impaired** refers to any type or degree of hearing loss; the hearing loss can be conductive, sensorineural, or a combination of both. **Conduction deafness** involves a mechanical deficiency in the transmission of sound waves from the external ear to the spiral organ. The spiral organ and neuronal pathways for hearing function normally. Conductive hearing loss often can be treated—for example, by removing ear wax blocking the external acoustic meatus or by replacing or repairing the auditory ossicles. If the degree of conductive hearing loss does not justify surgical treatment, or if treatment does not resolve the hearing loss, a hearing aid may be beneficial because the amplified (louder) sound waves it produces are transmitted through the conductive blockage and may provide normal stimulation to the spiral organ.

Sensorineural hearing loss involves the spiral organ or neuronal pathways. Sound waves are transmitted normally to the spiral organ, but the nervous system's ability to respond to the sound waves is impaired. Hearing aids are commonly used by people with sensorineural hearing loss because the amplified sound waves they produce result in greater than normal stimulation of the spiral organ, helping overcome the perception of reduced sound volume. Sound clarity also improves with sound amplification but may never be perceived as normal.

The term **deaf** refers to sensorineural hearing loss so profound that the sense of hearing is nonfunctional, with or without amplification, for ordinary purposes of life. Stimulation of the spiral organ or hearing nerve pathways can help deaf people hear.

Research is currently being conducted on ways to replace the hearing pathways with electric circuits. One approach involves the direct stimulation of nerves by electric impulses. There has been considerable success in the area of cochlear nerve stimulation. Certain types of sensorineural deafness in which the hair cells of the spiral organ are impaired can now be partially corrected. Prostheses are available that consist of a microphone for picking up the initial sound waves, a microelectronic processor for converting the sound into electric signals, a transmission system for relaying the signals to the inner ear, and a long, slender electrode that is threaded into the cochlea. This electrode delivers electric signals directly to the endings of the cochlear nerve (figure F). High-frequency sounds are picked up by the microphone and transmitted through specific circuits to terminate near the oval window, whereas low-frequency sounds are transmitted farther up the cochlea to cochlear nerve endings near the helicotrema.

Research is currently underway to develop implants directly into the cochlear nucleus of the brainstem for patients with vestibulocochlear nerve damage. These implants have electrodes of various lengths to stimulate parts of the cochlear nucleus, at various depths from the surface, which respond to sounds of different frequencies.

1. A receiver, a transmitter, and an antenna are implanted under the skin near the auricle.
2. A small lead from the transmitter is fed through the external acoustic meatus, tympanic membrane, and middle ear into the cochlea.
3. In the cochlea, the cochlear nerve can be directly stimulated by electric impulses from the receiver.

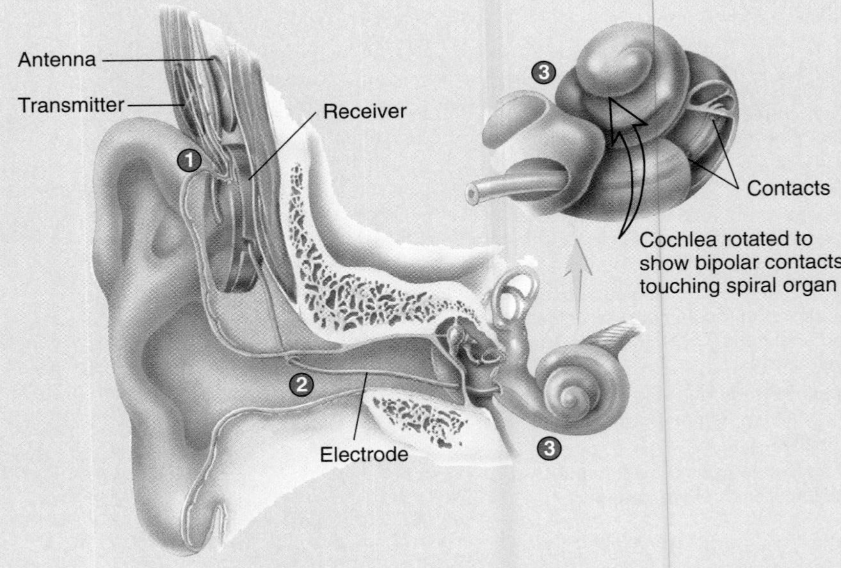

FIGURE F **Cochlear Implant**

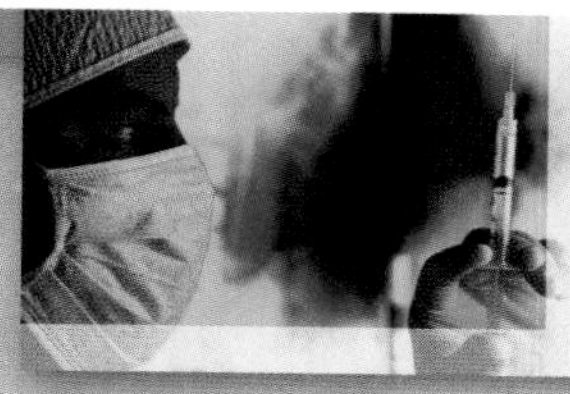

CLINICAL FOCUS

Ear Disorders

OTOSCLEROSIS

Otosclerosis (ō′tō-sklē-rō′sis) is an ear disorder in which spongy bone grows over the oval window and immobilizes the stapes, leading to progressive loss of hearing. This disorder can be surgically corrected by breaking away the bony growth and the immobilized stapes. During surgery, the stapes is replaced by a small rod connected by a fat pad or a synthetic membrane to the oval window at one end and to the incus at the other end.

TINNITUS

Tinnitus (ti-nī′tŭs) consists of phantom noises, such as ringing, clicking, whistling, buzzing, or booming, in the ears. These noises may occur as a result of disorders in the middle or inner ear or along the central neuronal pathways. Tinnitus is a common problem, affecting 17% of the world's population. It is treated primarily by training people to ignore the sounds.

MENIERE DISEASE

Meniere disease is the most common disease involving dizziness from the inner ear. Its cause is unknown but it appears to involve a fluid abnormality in one (usually) or both ears. Symptoms include vertigo, hearing loss, tinnitus, and a feeling of "fullness" in the affected ear. Treatment includes a low-salt diet and diuretics (water pills). Symptoms may also be treated with medications for motion sickness.

OTITIS MEDIA

Infections of the middle ear, called **otitis media** (ō-tī′tis mē′dē-ă), are quite common in young children. These infections usually result from the spread of infection from the mucous membrane of the pharynx through the auditory tube to the mucous lining of the middle ear. The symptoms of otitis media, consisting of low-grade fever, lethargy, irritability, and pulling at the ear, are often not easily recognized by the parent as signs of middle ear infection. The infection can also cause a temporary decrease or loss of hearing because fluid buildup has dampened the tympanic membrane or ossicles. In extreme cases, the infection can damage or rupture the tympanic membrane.

Chronic middle ear infections increase the chances of inner ear infectios. Inner ear infections can decrease the inner ear's detection of sound and maintenance of equilibrium.

EARACHE

Earache can result from otitis media, otitis externa (inflammation of the external acoustic meatus), dental abscesses, or temporomandibular joint pain.

coming into each ear, elderly people may experience a decreased ability to localize the origin of certain sounds. In some people, this leads to a general sense of disorientation. In addition, CNS defects in the auditory pathways can result in difficulty understanding sounds with echoes or background noise. Such deficit makes it difficult for elderly people to understand rapid or broken speech.

With age, the number of hair cells in the saccule, utricle, and ampullae decreases. The number of otoliths also declines. As a result, elderly people experience a decreased sensitivity to gravity, acceleration, and rotation. Because of these decreases, elderly people experience dizziness (instability) and vertigo (a feeling of spinning). They often feel that they cannot maintain posture and are prone to fall.

40 ***Explain the changes in taste, vision, hearing, and balance that occur with aging.***

SUMMARY

Olfaction (p. 515)

Olfaction is the sense of smell.

Olfactory Epithelium and Bulb

1. Olfactory neurons in the olfactory epithelium are bipolar neurons. Their distal ends are enlarged as olfactory vesicles, which have long cilia.
2. The cilia have receptors that respond to dissolved substances. There are approximately 1000 different odorant receptors.
3. The receptors activate a G protein complex, which opens ion channels.
4. At least 7 (perhaps 50) primary odors exist. The olfactory neurons have a very low threshold and accommodate rapidly.

Neuronal Pathways for Olfaction

1. Axons from the olfactory neurons extend as olfactory nerves to the olfactory bulb, where they synapse with mitral and tufted cells. Axons from these cells form the olfactory tracts and synapse with granule cells. The olfactory bulb neurons can modulate output to the olfactory tracts.
2. The olfactory tracts terminate in the olfactory and temporal cortex. The lateral olfactory area is involved in the conscious perception of smell, the intermediate area with modulating smell, and the medial area with visceral and emotional responses to smell.

Taste (p. 518)

Taste buds usually are associated with circumvallate, fungiform, and foliate papillae. Filiform papillae do not have taste buds.

Histology of Taste Buds

1. Taste buds consist of basilar, supporting, and gustatory cells.
2. The gustatory cells have gustatory hairs that extend into taste pores.

Function of Taste

1. Receptors on the hairs detect dissolved substances.
2. Five basic types of taste exist: sour, salty, bitter, sweet, and umami.

Neuronal Pathways for Taste

1. The facial nerve carries taste sensations from the anterior two-thirds of the tongue, the glossopharyngeal nerve from the posterior one-third of the tongue, and the vagus nerve from the epiglottis.
2. The neural pathways for taste extend from the medulla oblongata to the thalamus and to the cerebral cortex.

Visual System (p. 521)

Accessory Structures

1. The eyebrows prevent perspiration from entering the eyes and help shade the eyes.
2. The eyelids consist of five tissue layers. They protect the eyes from foreign objects and help lubricate the eyes by spreading tears over their surface.
3. The conjunctiva covers the inner eyelid and the anterior part of the eye.
4. Lacrimal glands produce tears, which flow across the surface of the eye. Excess tears enter the lacrimal canaliculi and reach the nasal cavity through the nasolacrimal canal. Tears lubricate and protect the eye.
5. The extrinsic eye muscles move the eyeball.

Anatomy of the Eye

1. The fibrous layer is the outer layer of the eyeball. It consists of the sclera and cornea.
   - The sclera is the posterior four-fifths of the eyeball. It is white connective tissue that maintains the shape of the eyeball and provides a site for muscle attachment.
   - The cornea is the anterior one-fifth of the eye. It is transparent and refracts light that enters the eye.
2. The vascular layer is the middle layer of the eyeball.
   - The iris is smooth muscle regulated by the autonomic nervous system. It controls the amount of light entering the pupil.
   - The ciliary muscles control the shape of the lens. They are smooth muscles regulated by the autonomic nervous system. The ciliary process produces aqueous humor.
3. The retina is the inner layer of the eyeball and contains neurons sensitive to light.
   - The macula (fovea centralis) is the area of greatest visual acuity.
   - The optic disc is the location through which nerves exit and blood vessels enter the eye. It has no photosensory cells and is therefore the blind spot of the eye.
4. The eyeball has three chambers: anterior, posterior, and vitreus.
   - The anterior and posterior chambers are filled with aqueous humor, which circulates and leaves by way of the scleral venous sinus.
   - The vitreus chamber is filled with vitreous humor.
5. The lens is held in place by the suspensory ligaments, which are attached to the ciliary muscles.

Functions of the Complete Eye

1. Light is the portion of the electromagnetic spectrum that humans can see.
2. When light travels from one medium to another, it can bend, or refract. Light striking a concave surface refracts outward (divergence). Light striking a convex surface refracts inward (convergence).
3. Converging light rays meet at the focal point and are said to be focused.
4. The cornea, aqueous humor, lens, and vitreous humor all refract light. The cornea is responsible for most of the convergence, whereas the lens can adjust the focal point by changing shape.
   - Relaxation of the ciliary muscles causes the lens to flatten, producing the emmetropic eye.
   - Contraction of the ciliary muscles causes the lens to become more spherical. This change in lens shape enables the eye to focus on objects that are less than 20 feet away, a process called accommodation.
5. The far point of vision is the distance at which the eye no longer has to change shape to focus on an object. The near point of vision is the closest an object can come to the eye and still be focused.
6. The pupil becomes smaller during accommodation, increasing the depth of focus.

Structure and Function of the Retina

1. The pigmented layer of the retina provides a black backdrop for increasing visual acuity.
2. Rods are responsible for vision in low illumination (night vision).
   - A pigment, rhodopsin, is split by light into retinal and opsin, producing hyperpolarization in the rod.
   - Light adaptation is caused by a reduction of rhodopsin; dark adaptation is caused by rhodopsin production.
3. Cones are responsible for color vision and visual acuity.
   - Cones are of three types, each with a different type of iodopsin photopigment. The pigments are most sensitive to blue, red, and green lights.
   - Perception of many colors results from mixing the ratio of the different types of cones that are active at a given moment.
4. Most visual images are focused on the fovea centralis, which has a very high concentration of cones. Moving away from the fovea, fewer cones (the macula) are present; mostly rods are in the periphery of the retina.
5. The rods and the cones synapse with bipolar cells that in turn synapse with ganglion cells, which form the optic nerves.
6. Bipolar and ganglion cells in the retina can modify information sent to the brain.
7. Ganglion cells have receptive fields with on-centers or off-centers. This arrangement enhances contrast.

Neuronal Pathways for Vision

1. Ganglia cell axons extend to the lateral geniculate ganglion of the thalamus, where they synapse. From there, neurons form the optic radiations that project to the visual cortex.
2. Neurons from the nasal visual field (temporal retina) of one eye and the temporal visual field (nasal retina) of the opposite eye project to the same cerebral hemisphere. Axons from the nasal retina cross in the optic chiasm, and axons from the temporal retina remain uncrossed.
3. Depth perception is the ability to judge relative distances of an object from the eyes and is a property of binocular vision. Binocular vision results because a slightly different image is seen by each eye.

Hearing and Balance (p. 542)

The osseous labyrinth is a canal system within the temporal bone that contains perilymph and the membranous labyrinth. Endolymph is inside the membranous labyrinth.

Auditory Structures and Their Functions

1. The external ear consists of the auricle and external acoustic meatus.
2. The middle ear connects the external and inner ears.
   - The tympanic membrane is stretched across the external acoustic meatus.
   - The malleus, incus, and stapes connect the tympanic membrane to the oval window of the inner ear.
   - The auditory tube connects the middle ear to the pharynx and functions to equalize pressure.
   - The middle ear is connected to the mastoid air cells.
3. The inner ear has three parts: the semicircular canals; the vestibule, which contains the utricle and the saccule; and the cochlea.
4. The cochlea is a spiral-shaped canal within the temporal bone.
   - The cochlea is divided into three compartments by the vestibular and basilar membranes. The scala vestibuli and scala tympani contain perilymph. The cochlear duct contains endolymph and the spiral organ (organ of Corti).
   - The spiral organ consists of inner and outer hair cells that attach to the tectorial membrane.

Auditory Function

1. Sound waves are funneled by the auricle down the external acoustic meatus, causing the tympanic membrane to vibrate.
2. The tympanic membrane vibrations are passed along the auditory ossicles to the oval window of the inner ear.
3. Movement of the stapes in the oval window causes the perilymph, vestibular membrane, and endolymph to vibrate, producing movement of the basilar membrane.
4. Movement of the basilar membrane causes bending of the stereocilia of inner hair cells in the spiral organ.
5. Bending of the stereocilia pulls on gating springs, which open K^+ channels.
6. Potassium ions entering the hair cell depolarize the cell, which open Ca^{2+} channels. Calcium ions entering the cell cause further depolarization.
7. Depolarization causes the release of glutamate, generating action potentials in the vestibulocochlear nerve.
8. Some vestibulocochlear nerve axons synapse in the superior olivary nucleus. Efferent neurons from this nucleus project back to the cochlea, where they regulate the perception of pitch.
9. The round window protects the inner ear from pressure buildup and dissipates sound waves.

Neuronal Pathways for Hearing

1. Axons from the vestibulocochlear nerve synapse in the medulla. Neurons from the medulla project axons to the inferior colliculi, where they synapse. Neurons from this point project to the thalamus and synapse. Thalamic neurons extend to the auditory cortex.
2. Efferent neurons project to cranial nerve nuclei responsible for controlling muscles that dampen sound in the middle ear.

Balance

1. Static balance evaluates the position of the head relative to gravity and detects linear acceleration and deceleration.
   - The utricle and saccule in the inner ear contain maculae. The maculae consist of hair cells with the hairs embedded in a gelatinous mass that contains otoliths.
   - The gelatinous mass moves in response to gravity.
2. Kinetic balance evaluates movements of the head.
   - Three semicircular canals at right angles to one another are present in the inner ear. The ampulla of each semicircular canal contains the crista ampullaris, which has hair cells with hairs embedded in a gelatinous mass, the cupula.
   - When the head moves, endolymph within the semicircular canal moves the cupula.

Neuronal Pathways for Balance

1. Axons from the maculae and the cristae ampullares extend to the vestibular nucleus of the medulla. Fibers from the medulla run to the spinal cord, cerebellum, cortex, and nuclei that control the extrinsic eye muscles.
2. Balance also depends on proprioception and visual input.

Effects of Aging on the Special Senses (p. 556)

Elderly people experience a decline in function of all special functions: olfaction, taste, vision, hearing, and balance. These declines can result in loss of appetite, visual impairment, disorientation, and risk of falling.

REVIEW AND COMPREHENSION

1. Olfactory neurons
   a. have projections called cilia.
   b. have axons that combine to form the olfactory nerves.
   c. connect to the olfactory bulb.
   d. have receptors that react with odorants dissolved in fluid.
   e. all of the above.
2. Which of these statements is *not* true with respect to olfaction?
   a. Olfactory sensation is relayed directly to the cerebral cortex without passing through the thalamus.
   b. Olfactory neurons are replaced about every 2 months.
   c. The lateral olfactory area of the cortex is involved in the conscious perception of smell.
   d. The medial olfactory area of the cortex is responsible for visceral and emotional reactions to odors.
   e. The olfactory cortex is in the occipital lobe of the cerebrum.
3. Gustatory (taste) cells
   a. are found only on the tongue.
   b. extend through tiny openings called taste buds.
   c. have no axons but release neurotransmitters when stimulated.
   d. have axons that extend directly to the taste area of the cerebral cortex.
4. Which of these is *not* one of the basic tastes?
   a. spicy c. bitter e. sour
   b. salt d. umami
5. Which of these types of papillae have no taste buds associated with them?
   a. circumvallate c. foliate
   b. filiform d. fungiform

6. Tears
   a. are released onto the surface of the eye near the medial corner of the eye.
   b. in excess are removed by the scleral venous sinus.
   c. in excess can cause a sty.
   d. can pass through the nasolacrimal duct into the oral cavity.
   e. contain water, salts, mucus, and lysozyme.
7. The fibrous layer of the eye includes the
   a. conjunctiva. c. choroid. e. retina.
   b. sclera. d. iris.
8. The ciliary body
   a. contains smooth muscles that attach to the lens by suspensory ligaments.
   b. produces the vitreous humor.
   c. is part of the iris of the eye.
   d. is part of the sclera.
   e. all of the above.
9. The lens normally focuses light onto the
   a. optic disc. c. macula. e. ciliary body.
   b. iris. d. cornea.
10. Given these structures:
   1. lens
   2. aqueous humor
   3. vitreous humor
   4. cornea

   Choose the arrangement that lists the structures in the order that light entering the eye encounters them.
   a. 1,2,3,4 c. 4,1,2,3 e. 4,3,2,1
   b. 1,4,2,3 d. 4,2,1,3
11. Aqueous humor
   a. is the pigment responsible for the black color of the choroid.
   b. exits the eye through the scleral venous sinus.
   c. is produced by the iris.
   d. can cause cataracts if overproduced.
   e. is composed of proteins called crystallines.
12. Contraction of the smooth muscle in the ciliary body causes the
   a. lens to flatten. c. pupil to constrict.
   b. lens to become more spherical. d. pupil to dilate.
13. Given these events:
   1. medial rectus contracts
   2. lateral rectus contracts
   3. pupils dilate
   4. pupils constrict
   5. lens of the eye flattens
   6. lens of the eye becomes more spherical

   Assume you are looking at an object 30 feet away. If you suddenly look at an object that is 1 foot away, which events occur?
   a. 1,3,6 c. 1,4,6 e. 2,4,5
   b. 1,4,5 d. 2,3,6
14. Given these events:
   1. bipolar cells depolarize
   2. decrease in glutamate released from presynaptic terminals of photoreceptor cells
   3. light strikes photoreceptor cells
   4. photoreceptor cells depolarized
   5. photoreceptor cells hyperpolarized

   Choose the arrangement that lists the correct order of events, starting with the photoreceptor cells in the resting, nonactivated state.
   a. 1,2,3,4,5 c. 3,4,2,5,1 e. 5,3,4,1,2
   b. 2,4,3,5,1 d. 4,3,5,2,1
15. Given these neurons in the retina:
   1. bipolar cells
   2. ganglionic cells
   3. photoreceptor cells

   Choose the arrangement that lists the correct order of the cells encountered by light as it enters the eye and travels toward the pigmented retina.
   a. 1,2,3 c. 2,1,3 e. 3,1,2
   b. 1,3,2 d. 2,3,1
16. Which of these photoreceptor cells is *not* correctly matched with its function?
   a. rods—vision in low light
   b. rods—visual acuity
   c. cones—color vision
17. Concerning dark adaptation,
   a. the amount of rhodopsin increases.
   b. the pupils constrict.
   c. it occurs more rapidly than light adaptation.
   d. all of the above.
18. In the retina, there are cones that are most sensitive to a particular color. Given this list of colors:
   1. red 3. green
   2. yellow 4. blue

   Indicate which colors correspond to specific types of cones.
   a. 2,3 c. 1,2,3 e. 1,2,3,4
   b. 3,4 d. 1,3,4
19. Given these areas of the retina:
   1. macula 3. optic disc
   2. fovea centralis 4. periphery of the retina

   Choose the arrangement that lists the areas according to the density of cones, starting with the area that has the highest density of cones.
   a. 1,2,3,4 c. 2,1,4,3 e. 3,4,1,2
   b. 1,3,2,4 d. 2,4,1,3
20. Concerning axons in the optic nerve from the right eye,
   a. they all go to the right occipital lobe.
   b. they all go to the left occipital lobe.
   c. they all go to the thalamus.
   d. most go to the thalamus but some go to the superior colliculus.
   e. some go to the right occipital lobe, but some go to the left occipital lobe.
21. A lesion that has destroyed the left optic tract of a boy eliminates vision in his
   a. left nasal visual field. d. both a and b.
   b. left temporal visual field. e. both a and c.
   c. right temporal visual field
22. A person with an abnormally powerful focusing system is ______________ and uses a ______________ to correct his or her vision.
   a. nearsighted, concave lens c. farsighted, concave lens
   b. nearsighted, convex lens d. farsighted, convex lens
23. Which of these structures is found within or is a part of the external ear?
   a. oval window d. external acoustic meatus
   b. auditory tube e. cochlear duct
   c. ossicles
24. Given these ear bones:
   1. incus 2. malleus 3. stapes

   Choose the arrangement that lists the ear bones in order from the tympanic membrane to the inner ear.
   a. 1,2,3 c. 2,1,3 e. 3,2,1
   b. 1,3,2 d. 2,3,1

25. Given these structures:
    1. perilymph
    2. endolymph
    3. vestibular membrane
    4. basilar membrane

    Choose the arrangement that lists the structures in the order sound waves coming from the outside encounter them in producing sound.
    a. 1,3,2,4
    b. 1,4,2,3
    c. 2,3,1,4
    d. 2,4,1,3
    e. 3,4,2,1
26. The spiral organ is found within the
    a. cochlear duct.
    b. scala vestibuli.
    c. scala tympani.
    d. vestibule.
    e. semicircular canals.
27. An increase in the loudness of sound occurs as a result of an increase in the ____________ of the sound wave.
    a. frequency
    b. amplitude
    c. resonance
    d. both a and b
28. Interpretation of different sounds is possible because of the ability of the ____________ to vibrate at different frequencies and stimulate the ____________.
    a. vestibular membrane, vestibular nerve
    b. vestibular membrane, spiral organ
    c. basilar membrane, vestibular nerve
    d. basilar membrane, spiral organ
29. Which structure is a specialized receptor found within the utricle?
    a. macula
    b. crista ampullaris
    c. spiral organ
    d. cupula
30. Damage to the semicircular canals affects the ability to detect
    a. sound.
    b. the position of the head relative to the ground.
    c. the movement of the head in all directions.
    d. all of the above.

Answers in Appendix E

CRITICAL THINKING

1. An elderly man with normal vision develops cataracts. He is surgically treated by removing the lenses of his eyes. What kind of glasses would you recommend he wear to compensate for the removal of his lenses?
2. Some animals have a reflective area in the choroid called the tapetum lucidum. Light entering the eye is reflected back instead of being absorbed by the choroid. What would be the advantage of this arrangement? The disadvantage?
3. Perhaps you have heard someone say that eating carrots is good for the eyes. What is the basis for this claim?
4. On a camping trip, Jean Tights rips her pants. That evening, she is going to repair the rip. As the sun goes down, the light becomes dimmer and dimmer. When she tries to thread the needle, it is obvious that she is not looking directly at the needle but is looking a few inches to the side. Why does she do this?
5. A man stares at a black clock on a white wall for several minutes. Then he shifts his view and looks at only the blank white wall. Although he is no longer looking at the clock, he sees a light clock against a dark background. Explain what is happening.
6. Describe the results of a lesion of the optic chiasm.
7. Persistent exposure to loud noise can cause loss of hearing, especially for high-frequency sounds. What part of the ear is probably damaged? Be as specific as possible.
8. Professional divers are subject to increased pressure as they descend to the bottom of the ocean. Sometimes this pressure can lead to damage to the ear and loss of hearing. Describe the normal mechanisms that adjust for changes in pressure, suggest some conditions that might interfere with pressure adjustment, and explain how the increased pressure might cause loss of hearing.
9. If a vibrating tuning fork is placed against the mastoid process of the temporal bone, the vibrations are perceived as sound, even if the external acoustic meatus is plugged. Explain how this happens.

Answers in Appendix F

Visit this book's website at www.mhhe.com/seeley8 for chapter quizzes, interactive learning exercises, and other study tools.

16 Autonomic Nervous System

During a picnic on a sunny spring day, it is easy to concentrate on the delicious food and the pleasant surroundings. The maintenance of homeostasis requires no conscious thought. The autonomic nervous system (ANS) helps keep body temperature at a constant level by controlling the activity of sweat glands and the amount of blood flowing through the skin. The ANS helps regulate the complex activities necessary for the digestion of food. The movement of absorbed nutrients to tissues is possible because the ANS controls heart rate, which helps maintain the blood pressure necessary to deliver blood to tissues. Without the ANS, all of the activities necessary to maintain homeostasis would be overwhelming.

A functional knowledge of the ANS enables you to predict general responses to a variety of stimuli, explain responses to changes in environmental conditions, comprehend symptoms that result from abnormal autonomic functions, and understand how drugs affect the ANS. This chapter examines the autonomic nervous system by *contrasting the somatic and autonomic nervous systems* (p. 565); describing the *anatomy of the autonomic nervous system* (p. 565), the *physiology of the autonomic nervous system* (p. 572), and the *regulation of the autonomic nervous system* (p. 576); and examining *functional generalizations about the autonomic nervous system* (p. 578).

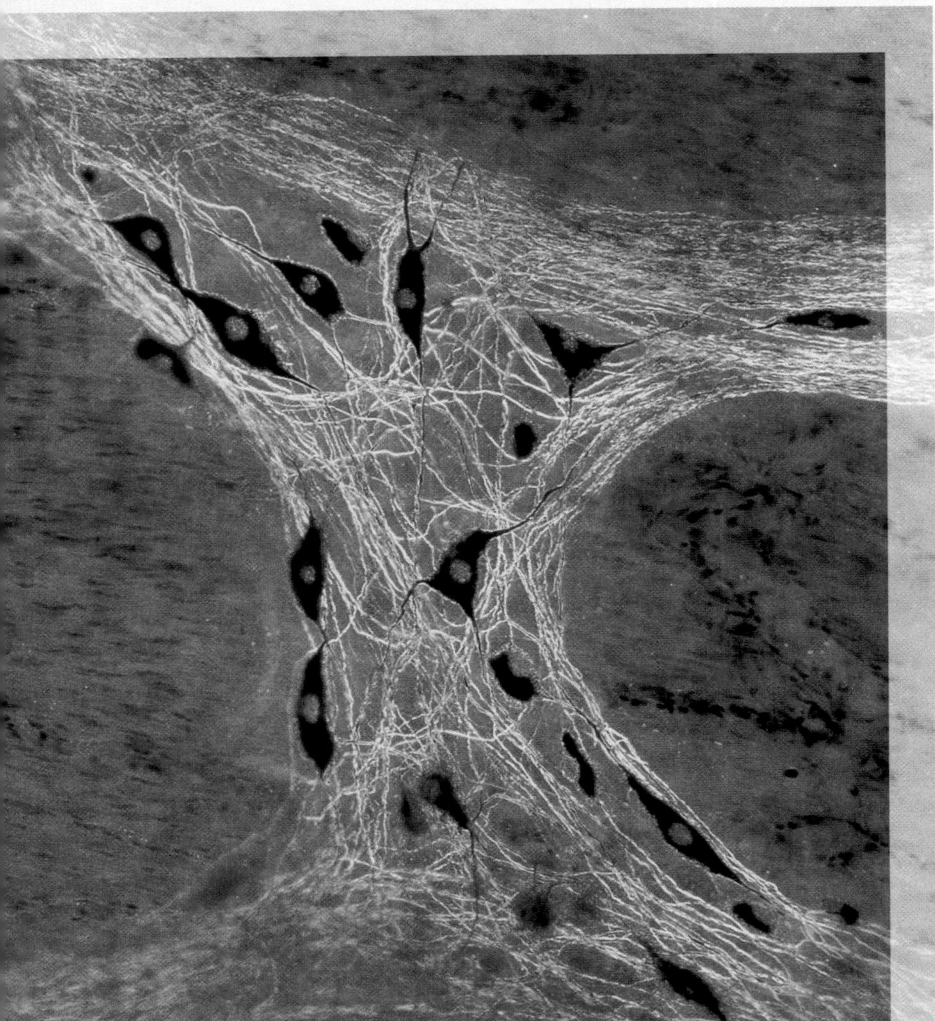

Light photomicrograph from a section of the small intestine, showing the nerve cells of the enteric plexus. These nerve cells regulate the contraction of smooth muscle and the secretion of glands within the intestinal wall.

Nervous System

CONTRASTING THE SOMATIC AND AUTONOMIC NERVOUS SYSTEMS

The peripheral nervous system (PNS) is composed of sensory and motor neurons. Sensory neurons carry action potentials from the periphery to the central nervous system (CNS), and motor neurons carry action potentials from the CNS to the periphery. Motor neurons are either somatic motor neurons, which innervate skeletal muscle, or autonomic motor neurons, which innervate smooth muscle, cardiac muscle, and glands.

Although axons of autonomic, somatic, and sensory neurons are in the same nerves, the proportion varies from nerve to nerve. For example, nerves innervating smooth muscle, cardiac muscle, and glands, such as the vagus nerves, consist primarily of axons of autonomic motor neurons and sensory neurons. Nerves innervating skeletal muscles, such as the sciatic nerves, consist primarily of axons of somatic motor neurons and sensory neurons. Some cranial nerves, such as the olfactory, optic, and vestibulocochlear nerves, are composed entirely of axons of sensory neurons.

The cell bodies of somatic motor neurons are in the CNS, and their axons extend from the CNS to skeletal muscle (figure 16.1*a*). The ANS, on the other hand, has two neurons in a series extending between the CNS and the organs innervated (figure 16.1*b*). The first neurons of the series are called **preganglionic neurons.** Their cell bodies are located in the CNS within either the brainstem or the lateral part of the spinal cord gray matter, and their axons extend to autonomic ganglia located outside the CNS. The **autonomic ganglia** contain the cell bodies of the second neurons of the series, which are called **postganglionic neurons.** The preganglionic neurons synapse with the postganglionic neurons in the autonomic ganglia. The axons of the postganglionic neurons extend from autonomic ganglia to effector organs, where they synapse with their target tissues.

Many movements controlled by the somatic nervous system are conscious, whereas ANS functions are unconsciously controlled. The effect of somatic motor neurons on skeletal muscle is always excitatory, but the effect of the ANS on target tissues can be excitatory or inhibitory. For example, after a meal, the ANS can stimulate stomach activities but, during exercise, the ANS can inhibit those activities. Table 16.1 summarizes the differences between the somatic nervous system and the ANS.

Sensory neurons are not classified as somatic or autonomic. These neurons propagate action potentials from sensory receptors to the CNS and can provide information for reflexes mediated through the somatic nervous system or the ANS. For example, stimulation of pain receptors can initiate somatic reflexes, such as the withdrawal reflex, and autonomic reflexes, such as an increase in heart rate. Although some sensory neurons primarily affect somatic functions and others primarily influence autonomic functions, functional overlap makes attempts to classify sensory neurons as either somatic or autonomic meaningless.

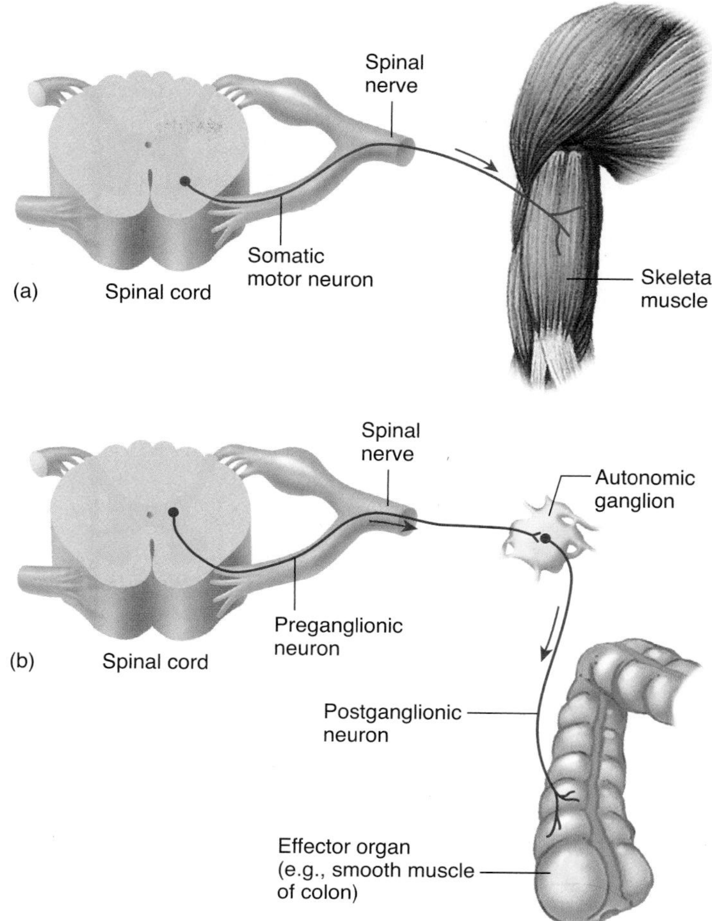

FIGURE 16.1 Organization of Somatic and Autonomic Nervous System Neurons

(*a*) The cell body of the somatic neuron is in the CNS, and its axon extends to the skeletal muscle. (*b*) The cell body of the preganglionic neuron is in the CNS, and its axon extends to the autonomic ganglion and synapses with the postganglionic neuron. The postganglionic neuron extends to and synapses with its effector organ.

1 ***Contrast the somatic nervous system with the ANS for each of the following:***

a. the number of neurons between the CNS and effector organ
b. the location of neuron cell bodies
c. the structures each innervates
d. inhibitory or excitatory effects
e. conscious or unconscious control

2 ***Why are sensory neurons not classified as somatic or autonomic?***

3 ***Define preganglionic neuron, postganglionic neuron, and autonomic ganglia.***

ANATOMY OF THE AUTONOMIC NERVOUS SYSTEM

The ANS is subdivided into the **sympathetic** and the **parasympathetic divisions** and the **enteric** (en-ter′ik; bowels) **nervous system (ENS).** The sympathetic and parasympathetic divisions differ structurally in (1) the location of their preganglionic neuron cell bodies within the CNS and (2) the location of their autonomic ganglia.

The enteric nervous system is a complex network of neuron cell bodies and axons within the wall of the digestive tract. An important part of this network is sympathetic and parasympathetic

TABLE 16.1 Comparison of the Somatic and Autonomic Nervous Systems

Features	Somatic Nervous System	Autonomic Nervous System
Target tissues	Skeletal muscle	Smooth muscle, cardiac muscle, and glands
Regulation	Controls all conscious and unconscious movements of skeletal muscle	Unconscious regulation, although influenced by conscious mental functions
Response to stimulation	Skeletal muscle contracts.	Target tissues are stimulated or inhibited.
Neuron arrangement	One neuron extends from the central nervous system (CNS) to skeletal muscle.	There are two neurons in series; the preganglionic neuron extends from the CNS to an autonomic ganglion, and the postganglionic neuron extends from the autonomic ganglion to the target tissue.
Neuron cell body location	Neuron cell bodies are in motor nuclei of the cranial nerves and in the ventral horn of the spinal cord.	Preganglionic neuron cell bodies are in autonomic nuclei of the cranial nerves and in the lateral part of the spinal cord; postganglionic neuron cell bodies are in autonomic ganglia.
Number of synapses	One synapse between the somatic motor neuron and the skeletal muscle	Two synapses; first is in the autonomic ganglia; second is at the target tissue
Axon sheaths	Myelinated	Preganglionic axons are myelinated; postganglionic axons are unmyelinated.
Neurotransmitter substance	Acetylcholine	Acetylcholine is released by preganglionic neurons; either acetylcholine or norepinephrine is released by postganglionic neurons.
Receptor molecules	Receptor molecules for acetylcholine are nicotinic.	In autonomic ganglia, receptor molecules for acetylcholine are nicotinic; in target tissues, receptor molecules for acetylcholine are muscarinic, whereas receptor molecules for norepinephrine are either α- or β-adrenergic.

neurons. For this reason, the enteric nervous system is considered to be part of the ANS.

Sympathetic Division

Cell bodies of sympathetic preganglionic neurons are in the lateral horns of the spinal cord gray matter between the first thoracic (T1) and the second lumbar (L2) segments (figure 16.2). Because of the location of the preganglionic cell bodies, the sympathetic division is sometimes called the **thoracolumbar division.** The axons of the preganglionic neurons pass through the ventral roots of spinal nerves T1–L2, course through the spinal nerves for a short distance, leave these nerves, and project to sympathetic ganglia. There are two types of sympathetic ganglia: sympathetic chain ganglia and collateral ganglia. **Sympathetic chain ganglia** are ganglia connected to each other to form a chain. A set of sympathetic chain ganglia is located along the left and right sides of the vertebral column. They are also called **paravertebral** (alongside the vertebral column) **ganglia** because of their location. Although the sympathetic division originates in the thoracic and lumbar vertebral regions, the sympathetic chain ganglia extend into the cervical and sacral regions. As a result of ganglia fusion during fetal development, there are typically 3 pairs of cervical ganglia, 11 pairs of thoracic ganglia, 4 pairs of lumbar ganglia, and 4 pairs of sacral ganglia. The **collateral** (meaning accessory) **ganglia** are unpaired ganglia located in the abdominopelvic cavity. They are also called **prevertebral** (in front of the vertebral column) **ganglia** because they are anterior to the vertebral column.

The axons of preganglionic neurons are small in diameter and myelinated. The short connection between a spinal nerve and a sympathetic chain ganglion through which the preganglionic axons pass is called a **white ramus communicans** (rā′mĭs kŏ-mū′ni-kans; pl. *rami communicantes,* rā′mī kŏ-mū-ni-kan′tēz) because of the whitish color of the myelinated axons (figure 16.3).

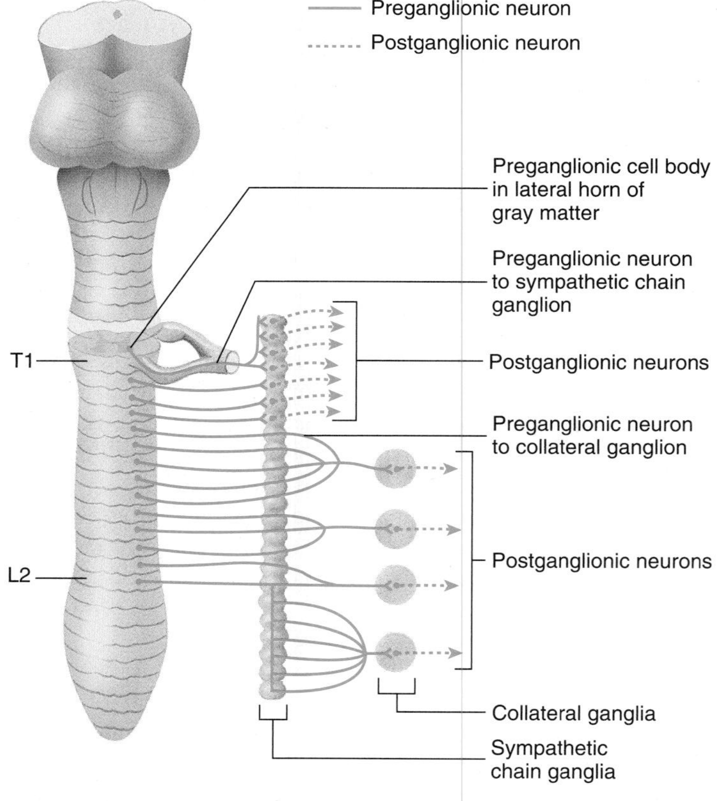

FIGURE 16.2 Sympathetic Division
The location of sympathetic preganglionic (*solid blue*) and postganglionic (*dotted blue*) neurons. The preganglionic cell bodies are in the lateral gray matter of the thoracic and lumbar parts of the spinal cord. The cell bodies of the postganglionic neurons are within the sympathetic chain ganglia or within collateral ganglia.

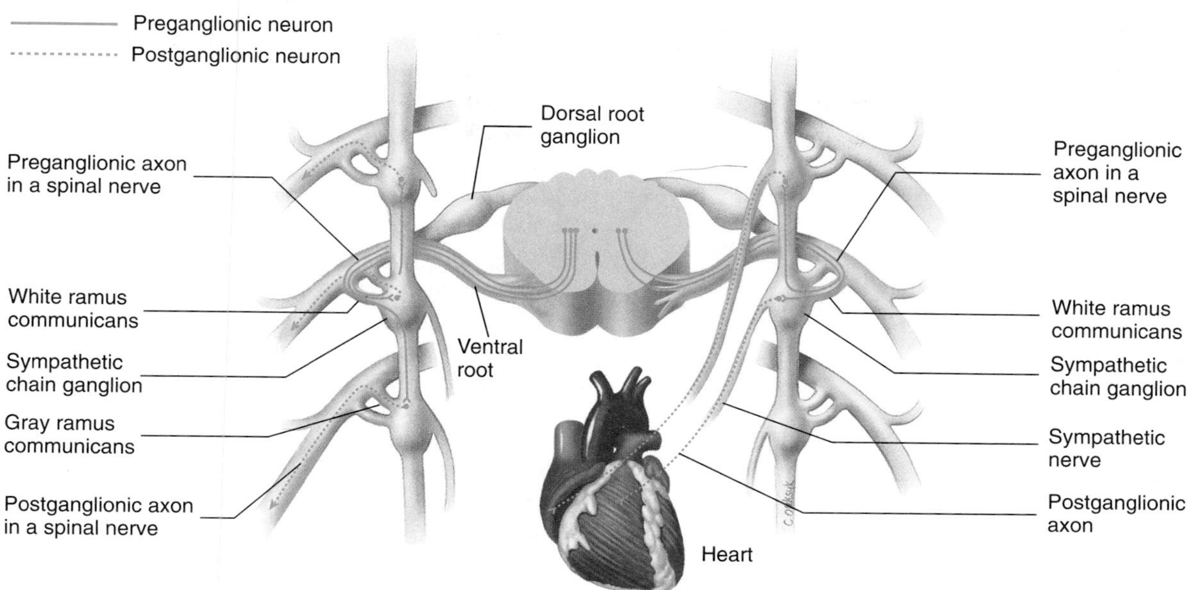

(a) Preganglionic axons from a spinal nerve pass through a white ramus communicans into a sympathetic chain ganglion. Some axons synapse with a postganglionic neuron at the level of entry; others ascend or descend to other levels before synapsing. Each postganglionic axon exits the sympathetic chain through a gray ramus communicans and enters a spinal nerve.

(b) Part (*b*) is like part (*a*), except that each postganglionic neuron exits a sympathetic chain ganglion through a sympathetic nerve.

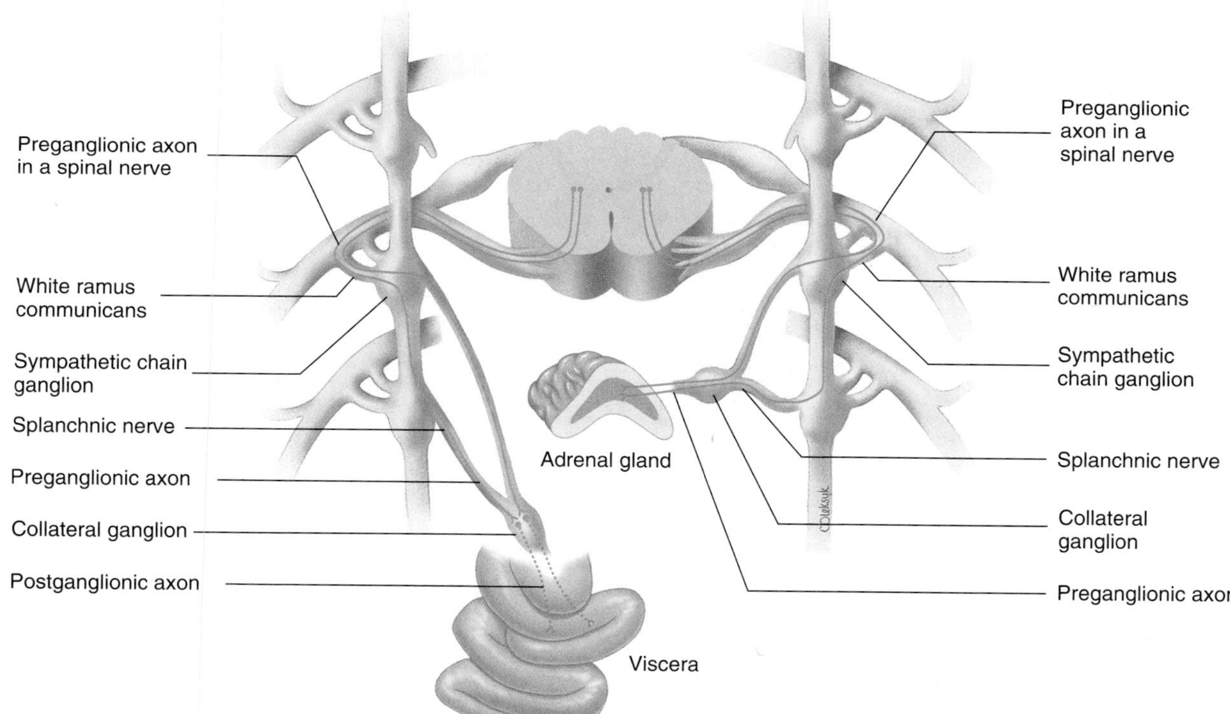

(c) Preganglionic neurons do not synapse in the sympathetic chain ganglia, but exit in splanchnic nerves and extend to a collateral ganglion, where they synapse with postganglionic neurons.

(d) Part (*d*) is like part (*c*), except that the preganglionic axons extend to the adrenal medulla, where they synapse with specialized adrenal medullary cells.

FIGURE 16.3 Routes Taken by Sympathetic Axons

Preganglionic axons are illustrated as solid lines and postganglionic axons as dashed lines.

Sympathetic axons exit the sympathetic chain ganglia by the following four routes:

1. *Spinal nerves* (see figure 16.3*a*). Preganglionic axons synapse with postganglionic neurons in sympathetic chain ganglia. They can synapse at the same level that the preganglionic axons enter the sympathetic chain, or they can pass superiorly or inferiorly through one or more ganglia and synapse with postganglionic neurons in a sympathetic chain ganglion at a different level. Axons of the postganglionic neurons pass through a **gray ramus communicans** and reenter a spinal nerve. Postganglionic axons are not myelinated, thereby giving the gray ramus communicans its grayish color. All spinal nerves receive postganglionic axons from a gray ramus communicans. The postganglionic axons then project through the spinal nerve to the skin and skeletal muscles.
2. *Sympathetic nerves* (see figure 16.3*b*). Preganglionic axons enter the sympathetic chain and synapse in a sympathetic chain ganglion at the same or a different level with postganglionic neurons. The postganglionic axons leaving the sympathetic chain ganglion form **sympathetic nerves,** which supply organs in the thoracic cavity.
3. *Splanchnic* (splangk′nik) *nerves* (see figure 16.3*c*). Some preganglionic axons enter sympathetic chain ganglia and, without synapsing, exit at the same or a different level to form **splanchnic nerves.** Those preganglionic axons extend to collateral ganglia, where they synapse with postganglionic neurons. Axons of the postganglionic neurons leave the collateral ganglia through small nerves that extend to target organs in the abdominopelvic cavity.
4. *Innervation to the adrenal gland* (see figure 16.3*d*). The splanchnic nerve innervation to the adrenal glands is different from other ANS nerves because it consists of only preganglionic neurons. Axons of the preganglionic neurons do not synapse in sympathetic chain ganglia or in collateral ganglia. Instead, the axons pass through those ganglia and synapse with cells in the adrenal medulla. The **adrenal medulla** (me-dool′ă) is the inner portion of the adrenal gland and consists of specialized cells derived during embryonic development from neural crest cells (see figure 13.13), which are the same population of cells that gives rise to the postganglionic cells of the ANS. Adrenal medullary cells are round, have no axons or dendrites, and are divided into two groups. About 80% of the cells secrete **epinephrine** (ep′i-nef′rin), also called **adrenaline** (ă-dren′ă-lin), and about 20% secrete **norepinephrine** (nōr′ep-i-nef′rin), also called **noradrenaline** (nōr-ă-dren′ă-lin). Stimulation of these cells by preganglionic axons causes the release of epinephrine and norepinephrine. These substances circulate in the blood and affect all tissues having receptors to which they can bind. The general response to epinephrine and norepinephrine released from the adrenal medulla is to prepare the individual for physical activity. Secretions of the adrenal medulla are considered hormones because they are released into the general circulation and travel some distance to the tissues in which they have their effect (see chapters 17 and 18).

4 ***Where are the cell bodies of sympathetic preganglionic neurons located?***

5 ***What types of axon (preganglionic or postganglionic, myelinated or unmyelinated) are found in white and gray rami communicantes?***

6 ***Where do preganglionic neurons synapse with postganglionic neurons that are found in spinal and sympathetic nerves?***

7 ***Where do preganglionic axons that form splanchnic nerves (except those to the adrenal gland) synapse with postganglionic neurons?***

8 ***What is unusual about the splanchnic nerve innervation to the adrenal gland? What do the specialized cells of the adrenal medulla secrete, and what is the effect of these substances?***

Parasympathetic Division

The cell bodies of parasympathetic preganglionic neurons are either within cranial nerve nuclei in the brainstem or within the lateral parts of the gray matter in the sacral region of the spinal cord from S2–S4 (figure 16.4). For that reason, the parasympathetic division is sometimes called the **craniosacral** (krā′nē-ō-sā′krăl) **division.**

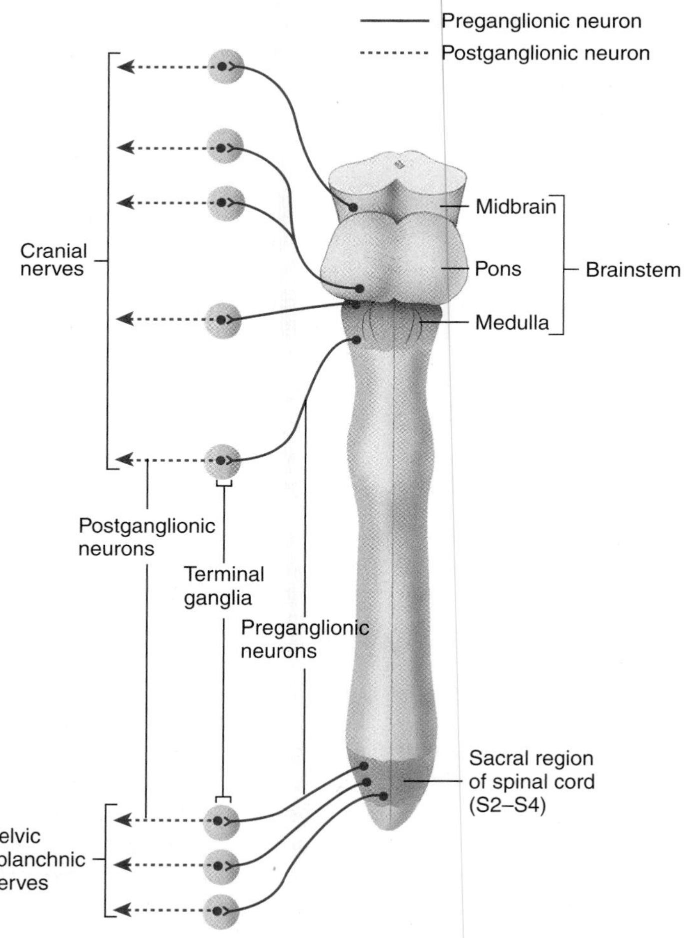

FIGURE 16.4 Parasympathetic Division

The location of parasympathetic preganglionic (*solid red*) and postganglionic (*dotted red*) neurons. The preganglionic neuron cell bodies are in the brainstem and the lateral gray matter of the sacral part of the spinal cord, and the postganglionic neuron cell bodies are within terminal ganglia.

TABLE 16.2 Comparison of the Sympathetic and Parasympathetic Divisions

Features	Sympathetic Division	Parasympathetic Division
Location of preganglionic cell body	Lateral horns of spinal cord gray matter (T1–L2)	Brainstem and lateral parts of spinal gray matter (S2–S4)
Outflow from the CNS	Spinal nerves Sympathetic nerves Splanchnic nerves	Cranial nerves Pelvic splanchnic nerves
Ganglia	Sympathetic chain ganglia along spinal cord for spinal and sympathetic nerves; collateral ganglia for splanchnic nerves	Terminal ganglia near or on effector organ
Number of postganglionic neurons for each preganglionic neuron	Many (much divergence)	Few (less divergence)
Relative length of neurons	Short preganglionic Long postganglionic	Long preganglionic Short postganglionic

Axons of the parasympathetic preganglionic neurons from the brain are in **cranial nerves** III, VII, IX, and X and from the spinal cord in **pelvic splanchnic nerves.** The preganglionic axons course through these nerves to **terminal ganglia,** where they synapse with postganglionic neurons. The axons of the postganglionic neurons extend relatively short distances from the terminal ganglia to the target organs. The terminal ganglia are either near or embedded within the walls of the organs innervated by the parasympathetic neurons. Many of the parasympathetic ganglia are small but some, such as those in the wall of the digestive tract, are large.

Table 16.2 summarizes the structural differences between the sympathetic and parasympathetic divisions.

9 ***Where are the cell bodies of parasympathetic preganglionic neurons located? In what structure do parasympathetic preganglionic neurons synapse with postganglionic neurons?***

10 ***What nerves are formed by the axons of parasympathetic preganglionic neurons?***

Enteric Nervous System

The enteric nervous system consists of nerve plexuses within the wall of the digestive tract (see chapter 24). The plexuses have contributions from three sources: (1) sensory neurons that connect the digestive tract to the CNS, (2) ANS motor neurons that connect the CNS to the digestive tract, and (3) enteric neurons, which are confined to the enteric plexuses. The CNS is capable of monitoring the digestive tract and controlling its smooth muscle and glands through autonomic reflexes (see "Regulation of the Autonomic Nervous System," p. 576). For example, stretch of the digestive tract is detected by sensory neurons and action potentials are transmitted to the CNS. In response, the CNS sends action potentials to glands in the digestive tract, causing them to secrete.

There are several major types of enteric neurons: (1) Enteric sensory neurons can detect changes in the chemical composition of the contents of the digestive tract or detect stretch of the digestive tract wall, (2) enteric motor neurons can stimulate or inhibit smooth muscle contraction and gland secretion, and (3) enteric interneurons connect enteric sensory and motor neurons to each other. Enteric neurons are capable of monitoring and controlling the digestive tract independently of the CNS through local reflexes (see "Regulation of the Autonomic Nervous System," p. 576). For example, stretch of the digestive tract is detected by enteric sensory neurons, which stimulate enteric interneurons. The enteric interneurons stimulate enteric motor neurons, which stimulate glands to secrete. Although the enteric nervous system is capable of controlling the activities of the digestive tract completely independently of the CNS, normally the two systems work together.

11 ***What is the enteric nervous system and where is it located?***

12 ***How does the CNS monitor and control the digestive tract?***

13 ***Name three major types of enteric neurons. How do the enteric neurons monitor and control the digestive tract?***

PREDICT 1

Are the ANS ganglia found in the enteric plexus chain ganglia, collateral ganglia, or terminal ganglia? What type (preganglionic or postganglionic) of sympathetic and parasympathetic axons contribute to the enteric plexus?

Distribution of Autonomic Nerve Fibers

Sympathetic Division

Sympathetic axons pass from the sympathetic chain ganglia to their target tissues through spinal, sympathetic, and splanchnic nerves. The sympathetic and splanchnic nerves can join **autonomic nerve plexuses,** which are complex, interconnected neural networks formed by neurons of the sympathetic and parasympathetic divisions. In addition, the axons of sensory neurons contribute to these plexuses.

The autonomic nerve plexuses typically are named according to organs they supply or blood vessels along which they are found. For example, the cardiac plexus supplies the heart and the thoracic aortic plexus is found along the thoracic aorta. Plexuses following the route of blood vessels are a major means by which autonomic axons are distributed throughout the body.

The major means by which sympathetic axons reach organs include the following:

1. *Spinal nerves.* From all levels of the sympathetic chain, some postganglionic axons project through gray rami communicates

to spinal nerves. The axons extend to the same structures innervated by the spinal nerves and supply sweat glands in the skin, smooth muscle in skeletal and skin blood vessels, and the smooth muscle of the arrector pili. See figure 12.14 for the distribution of spinal nerves to the skin.

2. *Head and neck nerve plexuses.* Most of the sympathetic nerve supply to the head and neck is derived from the superior cervical sympathetic chain ganglion (figure 16.5). Postganglionic axons of sympathetic nerves form plexuses that extend superiorly to the head and inferiorly to the neck. The plexuses give off branches to supply sweat glands in the skin, smooth muscle in skeletal and skin blood vessels, and the smooth muscle of the arrector pili. Axons from the plexuses also join branches of the trigeminal nerves (cranial nerve V) to supply the skin of the face, the salivary glands, the iris, and the ciliary muscles of the eye.
3. *Thoracic nerve plexuses.* The sympathetic supply for organs of the thorax is mainly derived from the cervical and upper five thoracic sympathetic chain ganglia. Postganglionic axons in sympathetic nerves contribute to the **cardiac plexus,** supplying the heart, the **pulmonary plexus,** supplying the lungs, and other thoracic plexuses (see figure 16.5).
4. *Abdominopelvic nerve plexuses.* Sympathetic chain ganglia from T5 and below mainly supply the abdominopelvic organs. The preganglionic axons of splanchnic nerves synapse with postganglionic neurons in the collateral ganglia of abdominopelvic nerve plexuses. Postganglionic axons from the collateral ganglia innervate smooth muscle and glands in the abdominopelvic organs. There are several abdominopelvic nerve plexuses (see figure 16.5). The **celiac** (sē′lē-ak) **plexus** has two large celiac ganglia and other, smaller ganglia. It supplies the diaphragm, stomach, spleen, liver, gallbladder, adrenal glands, kidneys, testes, and ovaries. The **superior mesenteric** (mez-en-ter′ik) **plexus** includes the superior mesenteric ganglion and supplies the pancreas, small intestine, ascending colon, and transverse colon. The **inferior mesenteric plexus** includes the inferior mesenteric ganglion and supplies the transverse colon to the rectum. The **superior** and **inferior hypogastric plexuses** supply the descending colon to the rectum, urinary bladder, and reproductive organs in the pelvis.

Parasympathetic Division

Parasympathetic outflow is through cranial and pelvic splanchnic nerves. Branches of these nerves either supply organs or join nerve plexuses to be distributed to organs. The major means by which parasympathetic axons reach organs include the following:

1. *Cranial nerves supplying the head and neck.* Three pairs of cranial nerves have parasympathetic preganglionic axons that extend to terminal ganglia in the head. Postganglionic neurons from the terminal ganglia supply nearby structures. The following are the parasympathetic cranial nerves, their terminal ganglia, and the structures innervated (see figure 16.5 and table 13.5):
   a. The **oculomotor (III) nerve,** through the **ciliary** (sil′ē-ar-ē) **ganglion,** supplies the ciliary muscles and the iris of the eye.
   b. The **facial (VII) nerve,** through the **pterygopalatine** (ter′i-gō-pal′ă-tīn) **ganglion,** supplies the lacrimal gland and mucosal glands of the nasal cavity and palate. The facial nerve, through the **submandibular ganglion,** also supplies the submandibular and sublingual salivary glands.
   c. The **glossopharyngeal (IX) nerve,** through the **otic** (ō′tik) **ganglion,** supplies the parotid salivary gland.
2. *The vagus nerve and thoracic nerve plexuses.* Although cranial nerve X, the **vagus nerve,** has somatic motor and sensory functions in the head and neck, its parasympathetic distribution is to the thorax and abdomen. Preganglionic axons extend through the vagus nerves to the thorax, where they pass through branches of the vagus nerves to contribute to the cardiac plexus, which supplies the heart, and the pulmonary plexus, which supplies the lungs. The vagus nerves continue down the esophagus and give off branches to form the **esophageal plexus.**
3. *Abdominal nerve plexuses.* After the esophageal plexus passes through the diaphragm, some of the vagal preganglionic axons supply terminal ganglia in the wall of the stomach, whereas others contribute to the celiac and superior mesenteric plexuses. Through these plexuses, the preganglionic axons supply terminal ganglia in the walls of the gallbladder, biliary ducts, pancreas, small intestine, ascending colon, and transverse colon.
4. *Pelvic splanchnic nerves and pelvic nerve plexuses.* Parasympathetic preganglionic axons whose cell bodies are in the S2–S4 region of the spinal cord pass to the ventral rami of spinal nerves and enter the pelvic splanchnic nerves. The pelvic splanchnic nerves supply terminal ganglia in the transverse colon to the rectum, and they contribute to the hypogastric plexus. The hypogastric plexus and its derivatives supply the lower colon, rectum, urinary bladder, and organs of the reproductive system in the pelvis.

Sensory Neurons in Autonomic Nerve Plexuses

Although not strictly part of the ANS, the axons of sensory neurons run alongside ANS axons within ANS nerves and plexuses. Some of these sensory neurons are part of reflex arcs regulating organ activities. Sensory neurons also transmit pain and pressure sensations from organs to the CNS. The cell bodies of these sensory neurons are found in the dorsal root ganglia and in the sensory ganglia of certain cranial nerves, which are swellings on the nerves close to their attachment to the brain.

14 ***Define autonomic nerve plexuses. How are they typically named?***

15 ***Describe the four major ways by which sympathetic axons pass from sympathetic chain ganglia to reach organs. Name four thoracic and four abdominopelvic autonomic nerve plexuses.***

16 ***List the four major means by which parasympathetic axons reach organs. List the cranial nerves and ganglia that supply the head and neck. What cranial nerve supplies the thoracic and abdominal nerve plexuses? To what plexus do pelvic splanchnic nerves contribute?***

PREDICT 2

Starting in the small intestine and ending with the ganglia where their cell bodies are located, trace the route for sensory axons passing alongside sympathetic axons. Name all of the plexuses, nerves, ganglia, and so on, that the sensory axon passes through. Also trace the route for sensory neurons passing alongside parasympathetic axons.

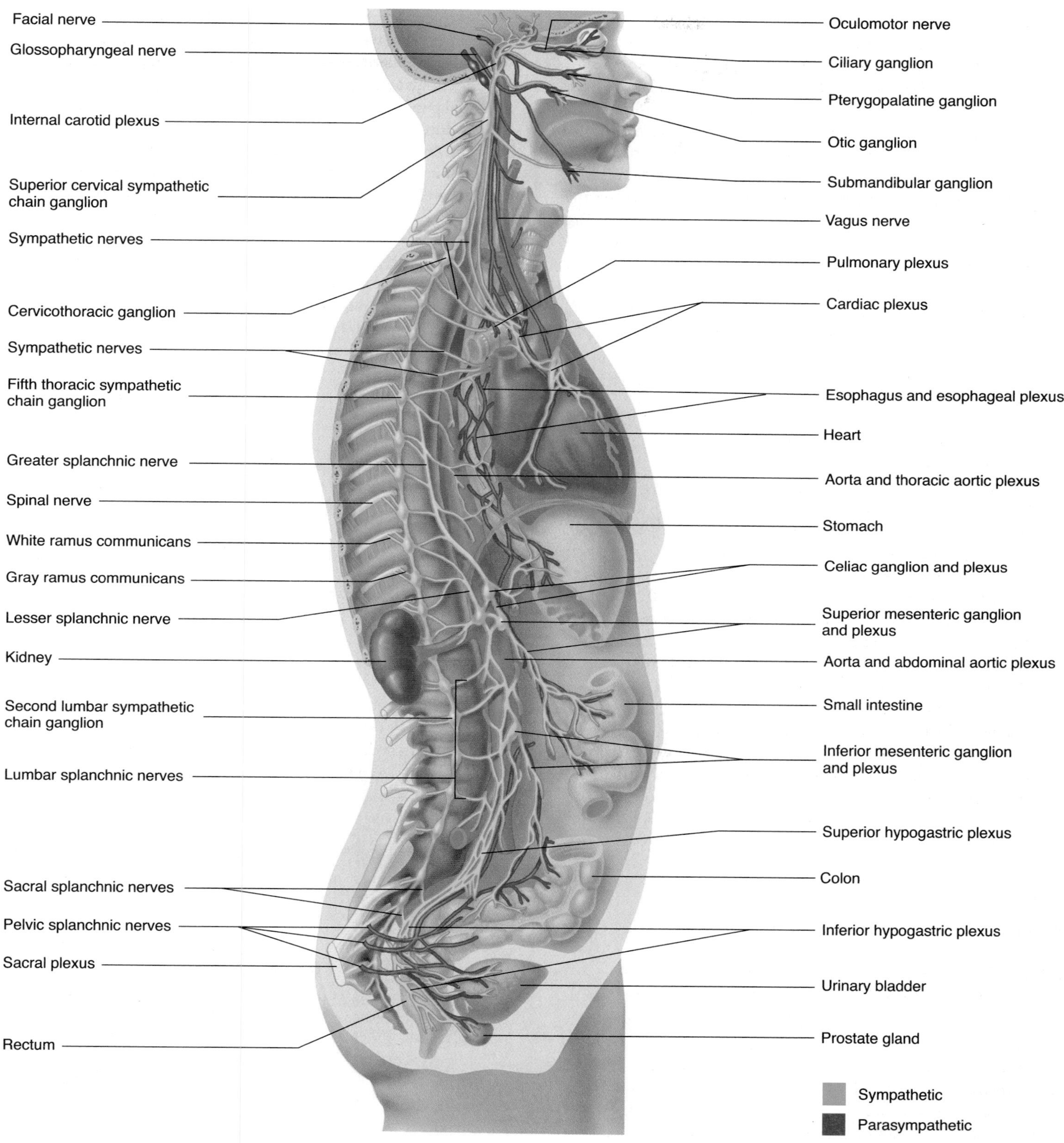

FIGURE 16.5 Distribution of Autonomic Nerve Fibers

PHYSIOLOGY OF THE AUTONOMIC NERVOUS SYSTEM

Neurotransmitters

Sympathetic and parasympathetic nerve endings secrete one of two neurotransmitters. If the neuron secretes acetylcholine, it is a **cholinergic** (kol-in-er'jik) **neuron,** and, if it secretes norepinephrine (or epinephrine), it is an **adrenergic** (ad-rĕ-ner'jik) **neuron.** Adrenergic neurons are so-named because at one time it was believed that they secreted adrenalin, which is another name for epinephrine. All preganglionic neurons of the sympathetic and parasympathetic divisions and all postganglionic neurons of the parasympathetic division are cholinergic. Almost all postganglionic neurons of the sympathetic division are adrenergic, but a few postganglionic neurons that innervate thermoregulatory sweat glands are cholinergic (figure 16.6).

In recent years, substances in addition to the regular neurotransmitters have been extracted from ANS neurons. These substances include nitric oxide; fatty acids, such as eicosanoids; peptides, such as gastrin, somatostatin, cholecystokinin, vasoactive intestinal peptide, enkephalins, and substance P; and monoamines, such as dopamine, serotonin, and histamine. The specific role that many of these compounds play in the regulation of the ANS is unclear, but they appear to function as either neurotransmitters or neuromodulator substances (see chapter 11).

Receptors

Receptors for acetylcholine and norepinephrine are located in the plasma membrane of certain cells (table 16.3). The combination of neurotransmitter and receptor functions as a signal to cells, causing them to respond. Depending on the type of cell, the response is excitatory or inhibitory.

Sympathetic division

Most target tissues innervated by the sympathetic division have adrenergic receptors. When norepinephrine (NE) binds to adrenergic receptors, some target tissues are stimulated, and others are inhibited. For example, smooth muscle cells in blood vessels are stimulated to constrict, and stomach glands are inhibited.

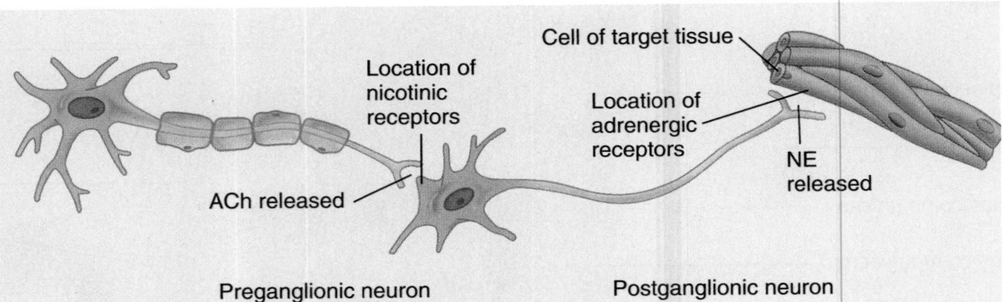

Sympathetic division

Some sympathetic target tissues, such as sweat glands, have muscarinic receptors, which respond to acetylcholine (ACh). Stimulation of sweat glands results in increased sweat production.

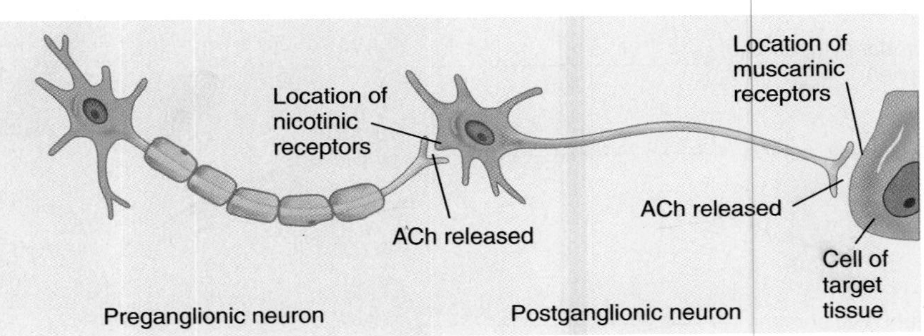

Parasympathetic division

All parasympathetic target tissues have muscarinic receptors. The general response to ACh is excitatory, but some target tissues, such as the heart, are inhibited.

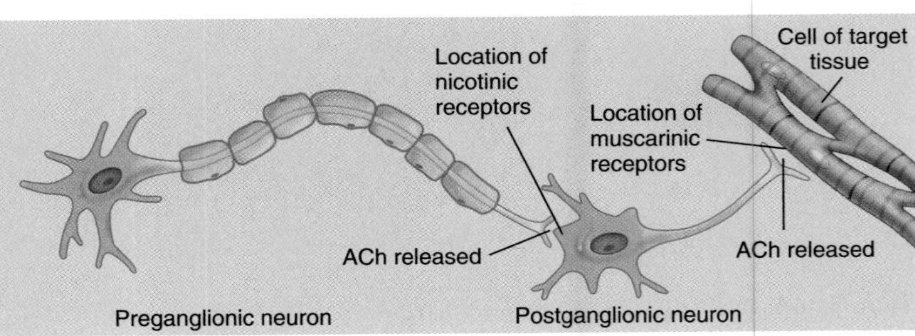

FIGURE 16.6 Location of ANS Receptors

Nicotinic receptors are on the cell bodies of both sympathetic and parasympathetic postganglionic cells in the autonomic ganglia. *Abbreviations: NE,* norepinephrine; *ACh,* acetylcholine.

TABLE 16.3 Effects of the Sympathetic and Parasympathetic Divisions on Various Tissues

Organ	Sympathetic Effects and Receptor Type*	Parasympathetic Effects and Receptor Type*
Adipose tissue	Fat breakdown and release of fatty acids (α_2, β_1)	None
Arrector pili muscle	Contraction (α_1)	None
Blood (platelets)	Increased coagulation (α_2)	None
Blood vessels		
Arterioles (carry blood to tissues)		
Digestive organs	Constriction (α_1)	None
Heart	Constriction (α_1), dilation (β_2)†	None
Kidneys	Constriction (α_1, α_2), dilation (β_1, β_2)	None
Lungs	Constriction (α_1), dilation (β_2)	None
Skeletal muscle	Constriction (α_1), dilation (β_2)	None
Skin	Constriction (α_1, α_2)	None
Veins (carry blood away from tissues)	Constriction (α_1, α_2), dilation (β_2)	
Eye		
Ciliary muscle	Relaxation for far vision (β_2)	Contraction for near vision (m)
Pupil	Dilated (α_1)‡	Constriction (m)‡
Gallbladder	Relaxation (β_2)	Contraction (m)
Glands		
Adrenal	Release of epinephrine and norepinephrine (n)	None
Gastric	Decreased gastric secretion (α_2)	Increased gastric secretion (m)
Lacrimal	Slight tear production (α)	Increased tear secretion (m)
Pancreas	Decreased insulin secretion (α_2)	Increased insulin secretion (m)
	Decreased exocrine secretion (α)	Increased exocrine secretion (m)
Salivary	Constriction of blood vessels and slight production of a thick, viscous saliva (α_1)	Dilation of blood vessels and thin, copious saliva (m)
Sweat		
Apocrine	Thick, organic secretion (m)	None
Merocrine	Watery sweat from most of the skin (m), sweat from the palms and soles (α_1)	None
Heart	Increased rate and force of contraction (β_1, β_2)	Decreased rate of contraction (m)
Liver	Glucose released into blood (α_1, β_2)	None
Lungs	Dilated air passageways (β_2)	Constricted air passageways (m)
Metabolism	Increased up to 100% (α, β)	None
Sex organs	Ejaculation (α_1), erection§	Erection (m)
Skeletal muscles	Breakdown of glycogen to glucose (β_2)	None
Stomach and intestines		
Wall	Decreased tone (α_1, α_2, β_2)	Increased motility (m)
Sphincter	Increased tone (α_1)	Decreased tone (m)
Urinary bladder		
Wall (detrusor)	None	Contraction (m)
Neck of bladder	Contraction (α_1)	Relaxation (m)
Internal urinary sphincter	Contraction (α_1)	Relaxation (m)

*Receptor subtypes are indicated. The receptors are α_1- and α_2-adrenergic, β_1- and β_2-adrenergic, nicotinic cholinergic (n), and muscarinic cholinergic (m).

†Normally, blood-flow increases through coronary arteries because of increased demand by cardiac tissue for oxygen (local control of blood flow is discussed in chapter 21). In experiments that isolate the coronary arteries, sympathetic nerve stimulation, acting through α-adrenergic receptors, causes vasoconstriction. The β-adrenergic receptors are relatively insensitive to sympathetic nerve stimulation but can be activated by epinephrine released from the adrenal gland and by drugs. As a result, coronary arteries vasodilate.

‡Contraction of the radial muscles of the iris causes the pupil to dilate. Contraction of the circular muscles causes the pupil to constrict (see chapter 15).

§Decreased stimulation of alpha receptors by the sympathetic division can cause vasodilation of penile blood vessels, resulting in an erection.

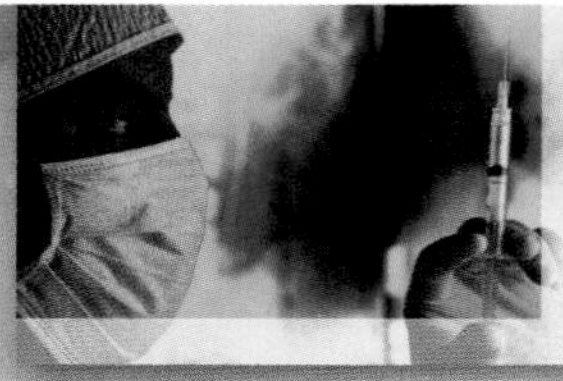

CLINICAL FOCUS

Influence of Drugs on the Autonomic Nervous System

Some drugs that affect the ANS have important therapeutic value in treating certain diseases because they can increase or decrease activities normally controlled by the ANS. Chemicals that affect the ANS are also found in medically hazardous substances, such as tobacco and insecticides.

Direct-acting and indirect-acting drugs influence the ANS. Direct-acting drugs bind to ANS receptors to produce their effects. For example, **agonists,** or **stimulating agents,** bind to specific receptors and activate them, and **antagonists,** or **blocking agents,** bind to specific receptors and prevent them from being activated. The main topic of this Clinical Focus is direct-acting drugs. It should be noted, however, that some indirect-acting drugs produce a stimulatory effect by causing the release of neurotransmitters or by preventing the metabolic breakdown of neurotransmitters. Other indirect-acting drugs produce an inhibitory effect by preventing the biosynthesis or release of neurotransmitters.

DRUGS THAT BIND TO NICOTINIC RECEPTORS

Drugs that bind to nicotinic receptors and activate them are **nicotinic agents.** Although these agents have little therapeutic value and are mainly of interest to researchers, nicotine is medically important because of its presence in tobacco. Nicotinic agents bind to the nicotinic receptors on all postganglionic neurons within autonomic ganglia and produce stimulation. Responses to nicotine are variable and depend on the amount taken into the body. Because nicotine stimulates the postganglionic neurons of both the sympathetic and parasympathetic divisions, much of the variability of its effects results from the opposing actions of these divisions. For example, in response to the nicotine contained in a cigarette, the heart rate may either increase or decrease. Heart rate rhythm tends to become less regular as a result of the simultaneous actions on the sympathetic division, which increase the heart rate, and the parasympathetic division, which decrease the heart rate. Blood pressure tends to increase because of the constriction of blood vessels, which are almost exclusively innervated by sympathetic neurons. In addition to its influence on the ANS, nicotine also affects the CNS; therefore, not all of its effects can be explained on the basis of action on the ANS. Nicotine is extremely toxic, and small amounts can be lethal.

Drugs that bind to and block nicotinic receptors are called **ganglionic blocking agents** because they block the effect of acetylcholine on both parasympathetic and sympathetic postganglionic neurons. The effect of these substances on the sympathetic division, however, overshadows the effect on the parasympathetic division. For example, trimethaphan camsylate (trī-meth'ă-fan kam'sil-āt), used to treat high blood pressure, blocks sympathetic stimulation of blood vessels, causing the blood vessels to dilate, which decreases blood pressure. Ganglionic blocking agents have limited uses because they affect both sympathetic and parasympathetic ganglia. Whenever possible, more selective drugs, which affect receptors of target tissues, are now used.

DRUGS THAT BIND TO MUSCARINIC RECEPTORS

Drugs that bind to and activate muscarinic receptors are **muscarinic,** or **parasympathomimetic** (par-ă-sim'pă-thō-mi-met'ik), **agents.** These drugs activate the muscarinic receptors of target tissues of the parasympathetic division and the muscarinic receptors of sweat glands, which are innervated by the sympathetic division. Muscarine causes in creased sweating, increased secretion of glands in the digestive system, decreased heart rate, constriction of the pupils, and contraction of respiratory, digestive, and urinary system smooth muscles. Bethanechol (be-than'ē-kol) chloride is a parasympathomimetic agent used to stimulate the urinary bladder following surgery because the general anesthetics used for surgery can temporarily inhibit a person's ability to urinate.

Drugs that bind to and block the action of muscarinic receptors are **muscarinic,** or **parasympathetic, blocking agents.** For example, the activation of muscarinic receptors causes the constriction of air passageways. Ipratropium (i-pră-trō'pē-ŭm) is used to treat chronic obstructive pulmonary disease because it blocks muscarinic receptors, which promotes the relaxation of air passageways. Atropine (at'rō-pēn) is used to block parasympathetic reflexes associated with the surgical manipulation of organs.

DRUGS THAT BIND TO ALPHA AND BETA RECEPTORS

Drugs that activate adrenergic receptors are **adrenergic,** or **sympathomimetic** (sim'pă-thō-mi-met'ik), **agents.** Drugs such as phenylephrine (fen-il-ef'rin) stimulate alpha receptors, which are numerous in the smooth muscle cells of certain blood vessels, especially in the digestive tract and the skin. These drugs increase blood pressure by causing vasoconstriction. On the other hand, albuterol (al-bū'ter-ol) is a drug that selectively activates beta receptors in bronchiolar smooth muscle. β-adrenergic-stimulating agents are sometimes used to dilate bronchioles in respiratory disorders such as asthma.

Drugs that bind to and block the action of alpha receptors are **α-adrenergic-blocking agents.** For example, prazosin (prā'zō-sin) hydrochloride is used to treat hypertension. By binding to alpha receptors in the smooth muscle of blood vessel walls, prazosin hydrochloride blocks the normal effects of norepinephrine released from sympathetic postganglionic neurons. Thus, the blood vessels relax, and blood pressure decreases.

Propranolol (prō-pran'ō-lōl) is an example of a **β-adrenergic-blocking agent.** These drugs are sometimes used to treat high blood pressure, some types of cardiac arrhythmias, and patients recovering from heart attacks. Blockage of the beta receptors within the heart prevents sudden increases in heart rate and thus decreases the probability of arrhythmic contractions.

FUTURE RESEARCH

Present knowledge of the ANS is more complicated than the broad outline presented here. In fact, each of the major receptor types has subtype receptors. For example, α-adrenergic receptors are subdivided into the following subgroups: α_{1A}-, α_{1B}-, α_{2A}-, and α_{2B}-adrenergic receptors. The exact number of subtypes in humans is not yet known; however, their existence suggests the possibility of designing drugs that affect only one subtype. For example, a drug that affects the blood vessels of the heart but not other blood vessels might be developed. Such drugs could produce specific effects yet would not produce undesirable side effects because they would act only on specific target tissues.

Cholinergic Receptors

Cholinergic receptors are receptors to which acetylcholine binds. They have two major, structurally different forms. **Nicotinic** (nik-ō-tin′ik) **receptors** bind to nicotine, an alkaloid substance found in tobacco, and **muscarinic** (mŭs-kă-rin′ik) **receptors** bind to muscarine, an alkaloid extracted from some poisonous mushrooms. Although nicotine and muscarine are not naturally in the human body, they demonstrate the differences between the two classes of cholinergic receptors. Nicotine binds to nicotinic receptors but not to muscarinic receptors, whereas muscarine binds to muscarinic receptors but not to nicotinic receptors. On the other hand, nicotinic and muscarinic receptors are very similar because acetylcholine binds to and activates both types of receptors.

The membranes of all postganglionic neurons in autonomic ganglia and the membranes of skeletal muscle cells have nicotinic receptors. The membranes of effector cells that respond to acetylcholine released from postganglionic neurons have muscarinic receptors (see figure 16.6).

PREDICT 3

Would structures innervated by the sympathetic division or the parasympathetic division be affected after the consumption of nicotine? After the consumption of muscarine? Explain.

Acetylcholine binding to nicotinic receptors has an excitatory effect because it results in the direct opening of Na^+ channels and the production of action potentials. When acetylcholine binds to muscarinic receptors, the cell's response is mediated through G proteins (see chapters 3 and 17). The response is either excitatory or inhibitory, depending on the target tissue in which the receptors are found. For example, acetylcholine binds to muscarinic receptors in cardiac muscle, thereby reducing heart rate, and acetylcholine binds to muscarinic receptors in smooth muscle cells of the stomach, thus increasing its rate of contraction.

17 ***Define cholinergic and adrenergic neurons. Which neurons of the ANS are cholinergic and adrenergic?***

18 ***Name the two major subtypes of cholinergic receptors. Where are they located? When acetylcholine binds to each subtype, does it result in an excitatory or inhibitory cell response?***

Adrenergic Receptors

Adrenergic receptors are receptors to which norepinephrine or epinephrine bind. They are located in the plasma membranes of target tissues innervated by the sympathetic division (see figure 16.6). The response of cells to norepinephrine or epinephrine binding to adrenergic receptors is mediated through G proteins (see chapters 3 and 17). Depending on the target tissue, the activation of G proteins can result in excitatory or inhibitory responses.

Adrenergic receptors are subdivided into two major categories: **alpha (α) receptors** and **beta (β) receptors.** Epinephrine has a greater effect than norepinephrine on most α and β receptors. The main subtypes for alpha receptors are α_1- and α_2-adrenergic receptors and for beta receptors are β_1- and β_2-adrenergic receptors.

Adrenergic receptors can be stimulated in two ways: by the nervous system and by epinephrine and norepinephrine released from the adrenal gland. Sympathetic postganglionic neurons release norepinephrine, which stimulates adrenergic receptors within synapses (see figure 16.6). For example, blood vessels are continually stimulated to contract through the release of norepinephrine. Increasing or decreasing this stimulation regulates blood flow. Increased stimulation causes further constriction and reduces blood flow, whereas decreased stimulation results in dilation and increases blood flow. The control of blood vessel diameter plays an important role in the regulation of blood flow and blood pressure (see chapter 20).

Epinephrine and norepinephrine released from the adrenal glands and carried to effector organs by the blood can bind to adrenergic receptors located in the plasma membrane away from synapses. For example, during exercise epinephrine and norepinephrine bind to β_2 receptors and cause blood vessel dilation in skeletal muscles.

CASE STUDY

Eye Drops

Sally is a 50-year-old woman with diabetes. Every year, she has her eyes examined by her ophthalmologist because damage to the retina can be associated with diabetes. In addition to checking her vision using an eye chart, her doctor examines the insides of her eyes using an ophthalmoscope (see figure 15.14). To see inside Sally's eyes better, the doctor applies pupil-dilating eye drops that cause the pupils to enlarge. The diameter of a typical pupil in normal room light is approximately 3–4 mm, whereas a dilated pupil is 7–8 mm. Because Sally has complained in the past about light sensitivity as a result of the pupil-dilating eye drops, after the eye exam the doctor applies pupil-constricting eye drops to cause the pupil diameter to decrease.

PREDICT 4

a. Review the anatomy of the iris of the eye in chapter 15 (see p. 526). Do the radial muscles or circular muscles of the iris cause the pupil to dilate?
b. Which division of the ANS controls the radial muscles? The circular muscles?
c. Four types of drugs act on the receptors of target organs of the ANS (see Clinical Focus "Influence of Drugs on the Autonomic Nervous System," p. 574). Which of these drugs could explain dilation of Sally's pupils? (*Hint:* See table 16.3.)
d. A side effect of the pupil-dilating eye drops is blurred vision—that is, an inability to see close-up objects clearly. Based on this observation, which type of drug is in the pupil-dilating eye drops?
e. Which type of drug could be in the pupil-constricting eye drops that reversed the dilation of Sally's pupils?
f. A side effect of the pupil-constricting eye drops is bloodshot eyes. Based on this observation, which type of drug is in the pupil-constricting eye drops?

Dopamine and the Treatment of Shock

Norepinephrine is produced from a precursor molecule called dopamine. Certain sympathetic neurons release dopamine, which binds to dopamine receptors. Dopamine is structurally similar to norepinephrine and it binds to beta receptors. Dopamine hydrochloride has been used successfully to treat circulatory shock because it can bind to dopamine receptors in kidney blood vessels. The resulting vasodilation increases blood flow to the kidneys and prevents kidney damage. At the same time, dopamine can bind to beta receptors in the heart, causing stronger contractions.

19 ***Where are adrenergic receptors located? Name the two major types of adrenergic receptors.***

20 ***In what two ways are adrenergic receptors stimulated?***

REGULATION OF THE AUTONOMIC NERVOUS SYSTEM

Much of the regulation of structures by the ANS occurs through autonomic reflexes, but input from the cerebrum, the hypothalamus, and other areas of the brain allows conscious thoughts and

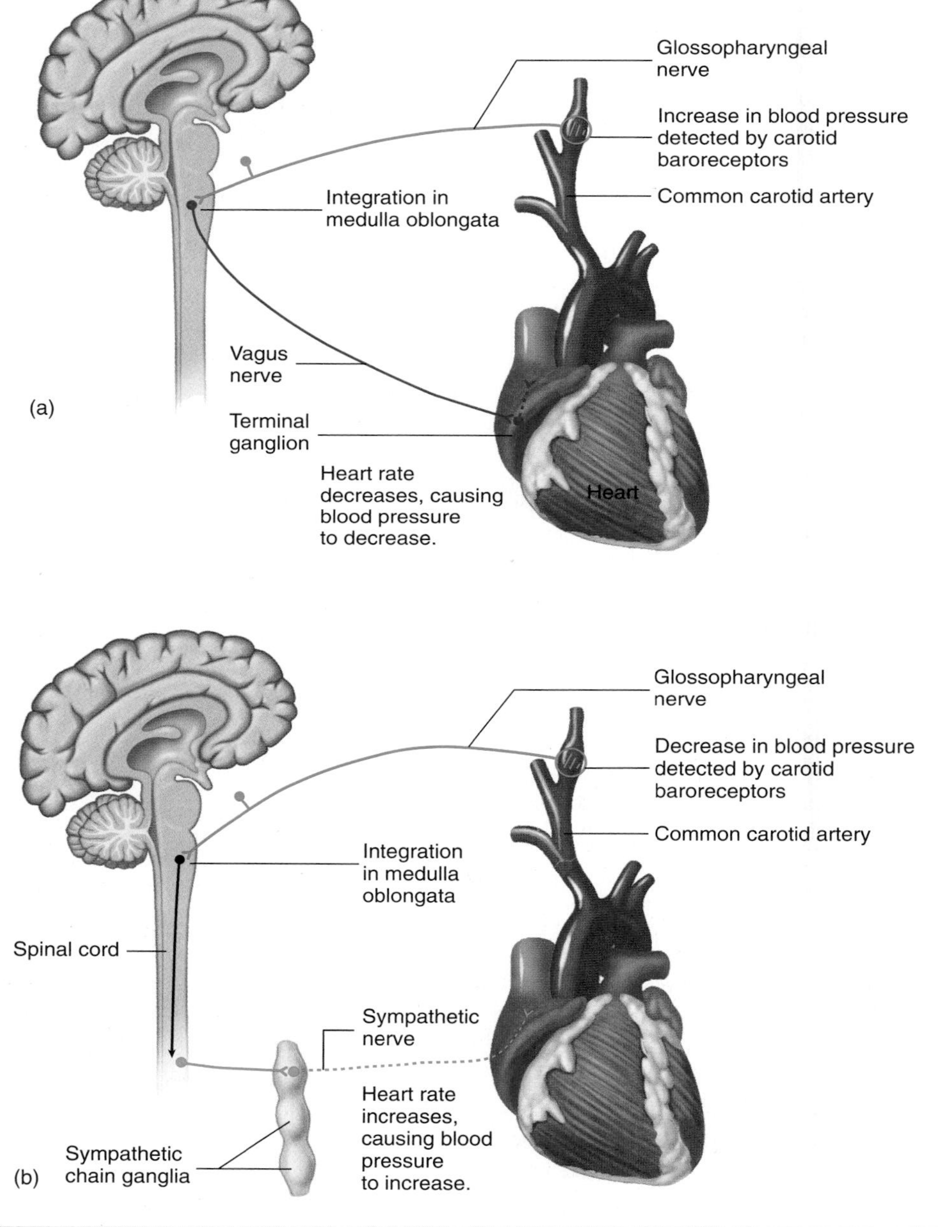

FIGURE 16.7 Autonomic Reflexes

Sensory input from the carotid baroreceptors is sent along the glossopharyngeal nerves to the medulla oblongata. The input is integrated in the medulla, and motor output is sent to the heart. (*a*) *Parasympathetic reflex.* Increased blood pressure results in increased stimulation of the heart by the vagus nerves, which increases inhibition of the heart and lowers heart rate. (*b*) *Sympathetic reflex.* Decreased blood pressure results in increased stimulation of the heart by sympathetic nerves, which in turn increases stimulation of the heart and increases heart rate and the force of contraction.

actions, emotions, and other CNS activities to influence autonomic functions. Without the regulatory activity of the ANS, an individual has limited ability to maintain homeostasis.

Autonomic reflexes, like other reflexes, involve sensory receptors, sensory neurons, interneurons, motor neurons, and effector cells (figure 16.7; see chapter 12). For example, **baroreceptors** (stretch receptors) in the walls of large arteries near the heart detect changes in blood pressure, and sensory neurons transmit information from the baroreceptors through the glossopharyngeal and vagus nerves to the medulla oblongata. Interneurons in the medulla oblongata integrate the information, and action potentials are produced in autonomic neurons that extend to the heart. If baroreceptors detect a change in blood pressure, autonomic reflexes change heart rate, which returns blood pressure to normal. A sudden increase in blood pressure initiates a parasympathetic reflex, which inhibits cardiac muscle cells and reduces heart rate, thus bringing blood pressure down toward its normal value. Conversely, a sudden decrease in blood pressure initiates a sympathetic reflex, which stimulates the heart to increase its rate and force of contraction, thus increasing blood pressure.

PREDICT 5

Sympathetic neurons stimulate sweat glands in the skin. Predict how they function to control body temperature during exercise and during exposure to cold temperatures.

Other autonomic reflexes participate in the regulation of blood pressure (see chapter 21). For example, numerous sympathetic neurons transmit a low but relatively constant frequency of action potentials that stimulate blood vessels throughout the body, keeping them partially constricted. If the vessels constrict further, blood pressure increases; if they dilate, blood pressure decreases. Thus, altering the frequency of action potentials delivered to blood vessels along sympathetic neurons can either raise or lower blood pressure.

PREDICT 6

How would sympathetic reflexes that control blood vessels respond to a sudden decrease and a sudden increase in blood pressure?

The brainstem and the spinal cord contain important autonomic reflex centers responsible for maintaining homeostasis (figure 16.8). The hypothalamus, however, is in overall control of the ANS. Almost any type of autonomic response can be evoked by stimulating a part of the hypothalamus, which in turn stimulates ANS centers in the brainstem or spinal cord. Although there is overlap, stimulation of the posterior hypothalamus produces sympathetic responses, whereas stimulation of the anterior hypothalamus produces parasympathetic responses. In addition, the hypothalamus monitors and controls body temperature.

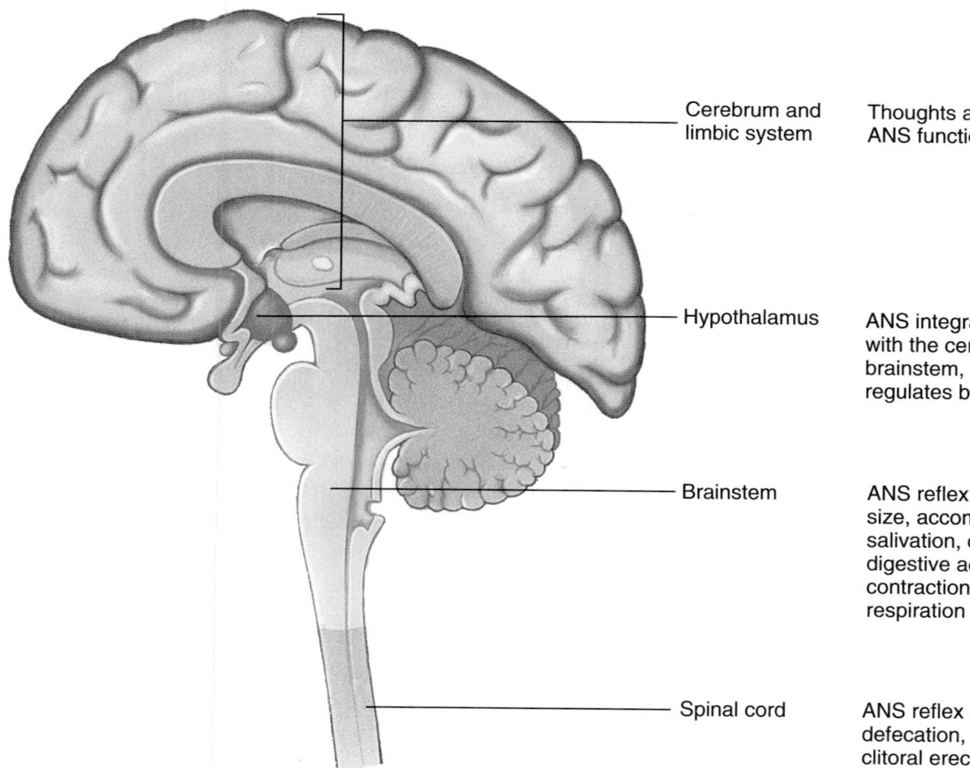

FIGURE 16.8 Influence of Higher Parts of the Brain on Autonomic Functions

The hypothalamus and cerebrum influence the ANS. Neural pathways extend from the cerebrum to the hypothalamus and from the hypothalamus to neurons of the ANS.

The hypothalamus has connections with the cerebrum and is an important part of the limbic system, which plays an important role in emotions. The hypothalamus integrates thoughts and emotions to produce ANS responses. Pleasant thoughts of a delicious banquet initiate increased secretion by salivary glands and by glands within the stomach and increased smooth muscle contractions within the digestive system. These responses are controlled by parasympathetic neurons. Emotions such as anger increase blood pressure by increasing heart rate and constricting blood vessels through sympathetic stimulation.

The enteric nervous system is involved with autonomic and local reflexes that regulate the activity of the digestive tract. Autonomic reflexes help control the digestive tract because sensory neurons of the enteric plexuses supply the CNS with information about intestinal contents and ANS neurons to the enteric plexuses affect the responses of smooth muscle and glands within the digestive tract wall. For example, sensory neurons detecting stretch of the digestive tract wall send action potentials to the CNS. In response, the CNS sends action potentials out the ANS, causing smooth muscle in the digestive tract wall to contract.

The neurons of the enteric nervous system also operate independently of the CNS to produce local reflexes. A **local reflex** does not involve the CNS, but it does produce an involuntary, unconscious, stereotypic response to a stimulus. For example, sensory neurons not connected to the CNS detect stretch of the digestive tract wall. These sensory neurons send action potentials through the enteric plexuses to motor neurons, causing smooth muscle contraction or relaxation. See chapter 24 for more information on local reflexes.

Effects of Spinal Cord Injury on ANS Functions

Spinal cord injury can damage nerve tracts, resulting in the loss of sensation and motor control below the level of the injury. Spinal cord injury also interrupts the control of autonomic neurons by ANS centers in the brain. For the parasympathetic division, effector organs innervated through the sacral region of the spinal cord are affected, but most effector organs still have normal parasympathetic function because they are innervated by the vagus nerve. For the sympathetic division, brain control of sympathetic neurons is lost below the site of the injury. The higher the level of injury, the greater the number of body parts affected.

Immediately after spinal cord injury, spinal cord reflexes below the level of the injury are lost, including ANS reflexes. With time, the reflex centers in the spinal cord become functional again. This recovery is particularly important for reflexes involving urination and defecation. Autonomic reflexes mediated through the vagus nerves or the enteric nervous system are not affected by spinal cord injury.

21 ***Name the components of an autonomic reflex. Describe the autonomic reflex that maintains blood pressure by altering heart rate or the diameter of blood vessels.***

22 ***What part of the CNS stimulates ANS reflexes and integrates thoughts and emotions to produce ANS responses?***

23 ***Define local reflex. Explain how the enteric nervous system operates to produce local reflexes.***

FUNCTIONAL GENERALIZATIONS ABOUT THE AUTONOMIC NERVOUS SYSTEM

Generalizations can be made about the function of the ANS on effector organs, but most of the generalizations have exceptions.

Stimulatory Versus Inhibitory Effects

Both divisions of the ANS produce stimulatory and inhibitory effects. For example, the parasympathetic division stimulates contraction of the urinary bladder and inhibits the heart, causing a decrease in heart rate. The sympathetic division causes vasoconstriction by stimulating smooth muscle contraction in blood vessel walls and produces dilation of lung air passageways by inhibiting smooth muscle contraction in the walls of the passageways. Thus, it is *not* true that one division of the ANS is always stimulatory and the other is always inhibitory.

Dual Innervation

Most organs that receive autonomic neurons are innervated by both the parasympathetic and the sympathetic division (figure 16.9). The gastrointestinal tract, heart, urinary bladder, and reproductive tract are examples (see table 16.3). Dual innervation of organs by both divisions of the ANS is not universal, however. Sweat glands and blood vessels, for example, are innervated by sympathetic neurons almost exclusively. In addition, most structures receiving dual innervation are not regulated equally by both divisions. For example, parasympathetic innervation of the gastrointestinal tract is more extensive and exhibits a greater influence than does sympathetic innervation.

Opposite Effects

When a *single* structure is innervated by both autonomic divisions, the two divisions usually produce opposite effects on the structure. As a consequence, the ANS is capable of both increasing and decreasing the activity of the structure. In the gastrointestinal tract, for example, parasympathetic stimulation increases secretion from glands, whereas sympathetic stimulation decreases secretion. In a few instances, however, the effect of the two divisions is not clearly opposite. For example, both divisions of the ANS increase salivary secretion: The parasympathetic division initiates the production of a large volume of thin, watery saliva, and the sympathetic division causes the secretion of a small volume of viscous saliva.

Cooperative Effects

One autonomic division can coordinate the activities of *different* structures. For example, the parasympathetic division stimulates the pancreas to release digestive enzymes into the small intestine and stimulates contractions to mix the digestive enzymes with food within the small intestine. These responses enhance the digestion and absorption of the food.

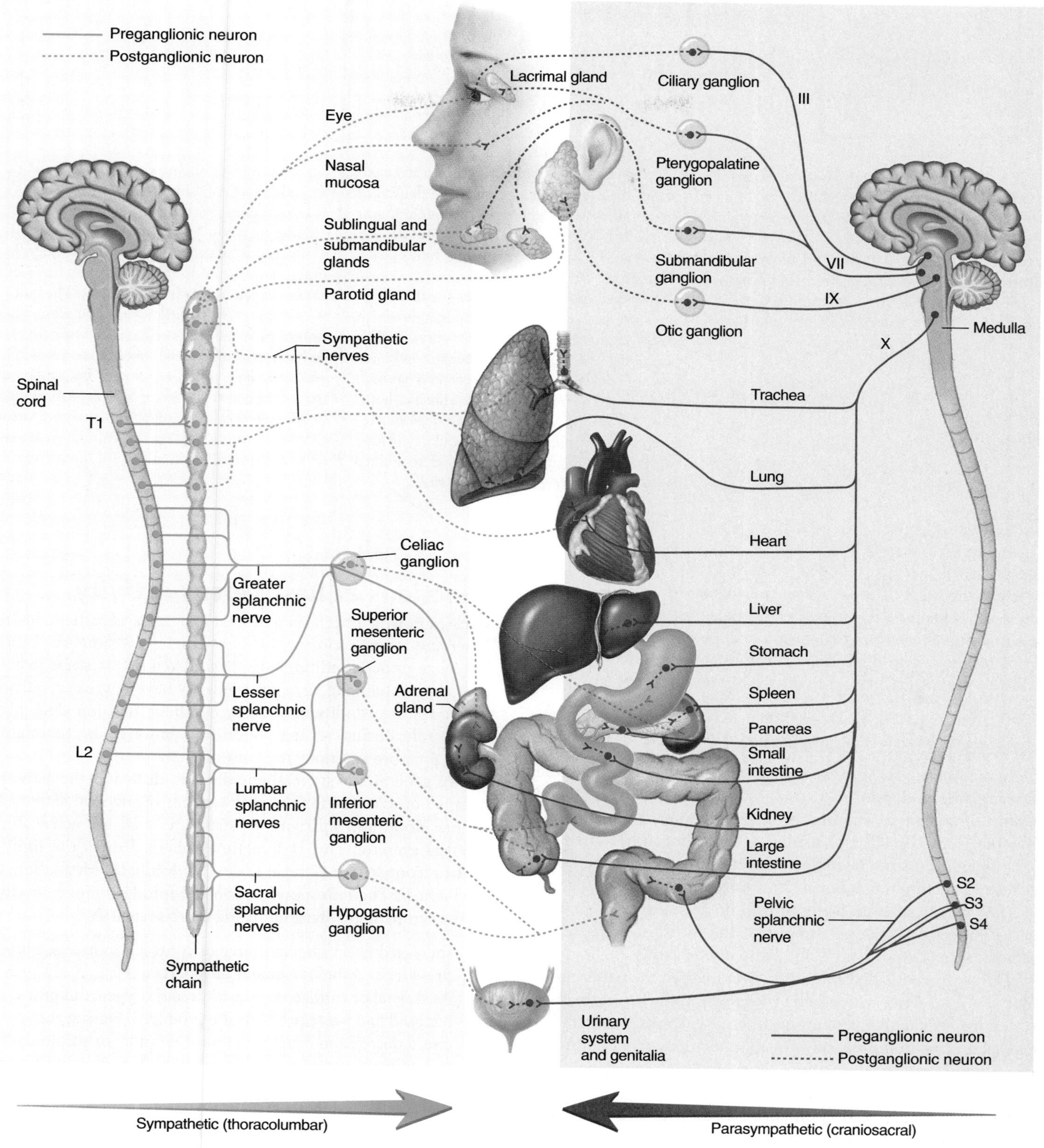

FIGURE 16.9 Innervation of Organs by the ANS

Preganglionic fibers are indicated by *solid lines,* and postganglionic fibers are indicated by *dashed lines.*

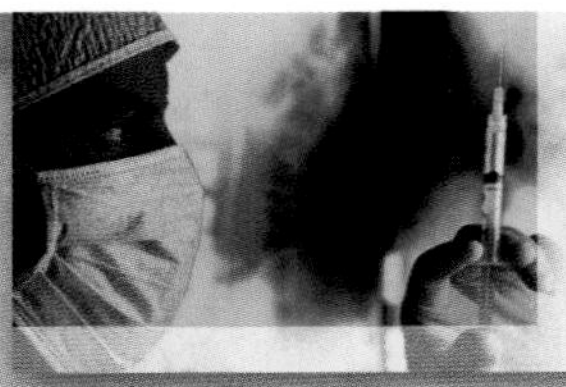

CLINICAL FOCUS

Biofeedback, Meditation, and the Fight-or-Flight Response

Biofeedback takes advantage of electronic instruments or other techniques to monitor and change subconscious activities, many of which are regulated by the ANS. Skin temperature, heart rate, and brain waves are monitored electronically. By watching the monitor and using biofeedback techniques, a person can learn how to reduce heart rate and blood pressure consciously and regulate blood flow in the limbs. For example, people claim that they can prevent the onset of migraine headaches or reduce their intensity by learning to dilate blood vessels in the skin of their forearms and hands. Increased blood vessel dilation increases skin temperature, which is correlated with a decrease in the severity of the migraine. Some people use biofeedback methods to relax by learning to reduce their heart rate or change the pattern of their brain waves. The severity of some stomach ulcers, high blood pressure, anxiety, and depression may be reduced by using biofeedback techniques.

Meditation is another technique that influences autonomic functions. Although numerous claims about the value of meditation include improving one's spiritual well-being, consciousness, and holistic view of the universe, it has been established that meditation does influence autonomic functions. Meditation techniques are useful in some people in reducing heart rate, blood pressure, severity of ulcers, and other symptoms that are frequently associated with stress.

The fight-or-flight response occurs when an individual is subjected to stress, such as a threatening, frightening, embarrassing, or exciting situation. Whether a person confronts or avoids a stressful situation, the nervous system and the endocrine system are involved either consciously or unconsciously. The autonomic part of the fight-or-flight response results in a general increase in sympathetic activity, including heart rate, blood pressure, sweating, and other responses, that prepare the individual for physical activity. The fight-or-flight response is adaptive because it also enables the individual to resist or move away from a threatening situation.

Both divisions of the ANS can act together to coordinate the activity of *different* structures. In the male, the parasympathetic division initiates erection of the penis, and the sympathetic division stimulates the release of secretions from male reproductive glands and helps initiate ejaculation in the male reproductive tract.

General Versus Localized Effects

The sympathetic division has a more general effect than the parasympathetic division because activation of the sympathetic division often causes secretion of both epinephrine and norepinephrine from the adrenal medulla. These hormones circulate in the blood and stimulate effector organs throughout the body. Because circulating epinephrine and norepinephrine can persist for a few minutes before being broken down, they can also produce an effect for a longer time than the direct stimulation of effector organs by postganglionic sympathetic axons.

The sympathetic division diverges more than the parasympathetic division. Each sympathetic preganglionic neuron synapses with many postganglionic neurons, whereas each parasympathetic preganglionic neuron synapses with about two postganglionic neurons. Consequently, stimulation of sympathetic preganglionic neurons can result in greater stimulation of an effector organ.

Sympathetic stimulation often activates many different kinds of effector organs at the same time as a result of CNS stimulation or epinephrine and norepinephrine release from the adrenal medulla. It is possible, however, for the CNS to selectively activate effector organs. For example, vasoconstriction of cutaneous blood vessels in a cold hand is not always associated with an increased heart rate or other responses controlled by the sympathetic division.

Functions at Rest Versus Activity

In cases in which both sympathetic and parasympathetic neurons innervate a single organ, the sympathetic division has a major influence under conditions of physical activity or stress, whereas the parasympathetic division tends to have a greater influence under resting conditions. The sympathetic division does play a major role during resting conditions, however, by maintaining blood pressure and body temperature.

In general, the sympathetic division decreases the activity of organs not essential for the maintenance of physical activity and shunts blood and nutrients to structures that are active during physical exercise. This is sometimes referred to as the **fight-or-flight response** (see Clinical Focus "Biofeedback, Meditation, and the Fight-or-Flight Response," p. 580). Typical responses produced by the sympathetic division during exercise include

1. Increased heart rate and force of contraction increase blood pressure and the movement of blood.
2. As skeletal or cardiac muscle contracts, oxygen and nutrients are used and waste products are produced. During exercise, a decrease in oxygen and nutrients and an accumulation of waste products are stimuli that cause vasodilation of muscle blood vessels (see "Local Control of Blood Flow by the Tissues," chapter 21). Vasodilation is beneficial because it increases blood flow, bringing needed oxygen and nutrients and removing waste products. Too much vasodilation, however, can cause a decrease in blood pressure that decreases blood flow. Increased stimulation of skeletal muscle blood vessels by sympathetic nerves during exercise causes vasoconstriction, which prevents a drop in blood pressure (see chapter 21).

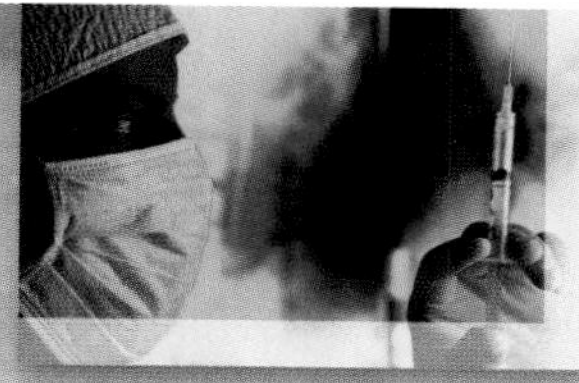

CLINICAL FOCUS

Disorders of the Autonomic Nervous System

Normal function of all the components of the ANS is not required to maintain life, as long as environmental conditions are constant and optimal. Abnormal autonomic functions, however, markedly affect an individual's ability to respond to changing conditions. **Sympathectomy,** the removal of sympathetic ganglia, demonstrates this. The normal regulation of body temperature is lost following sympathectomy. In a hot environment, the ability to lose heat by increasing blood flow to the skin and by sweating is decreased. When one is exposed to the cold, the ability to reduce blood flow to the skin and conserve heat is decreased. Sympathectomy also results in low blood pressure caused by dilation of peripheral blood vessels and in the inability to increase blood pressure during periods of physical activity.

Raynaud disease involves the spasmodic contraction of blood vessels in the periphery of the body, especially in the digits, and results in pale, cold hands that are prone to ulcerations and gangrene because of poor circulation. This condition can be caused by exaggerated sensitivity of blood vessels to sympathetic innervation. Preganglionic denervation (cutting the preganglionic neurons) is occasionally performed to alleviate the condition.

Hyperhidrosis (hī′per-hī-drō′sis), or excessive sweating, is caused by exaggerated sympathetic innervation of the sweat glands.

Achalasia (ak-ă-lā′zē-ă) is characterized by difficulty in swallowing and in controlling contraction of the esophagus where it enters the stomach, therefore interrupting normal peristaltic contractions of the esophagus. The swallowing reflex is controlled partly by somatic reflexes and partly by parasympathetic reflexes. The cause of achalasia can be abnormal parasympathetic regulation of the swallowing reflex. The condition is aggravated by emotions.

Dysautonomia (dis′aw-tō-nō′mē-ă), an inherited condition involving an autosomal-recessive gene, causes reduced tear gland secretion, poor vasomotor control, trouble swallowing, and other symptoms. It is the result of poorly controlled autonomic reflexes.

Hirschsprung disease, or **megacolon,** is caused by a functional obstruction in the lower colon and rectum. Ineffective parasympathetic stimulation and a predominance of sympathetic stimulation of the colon inhibit peristaltic contractions, causing feces to accumulate above the inhibited area. The resulting dilation of the colon can be so great that surgery is required to alleviate the condition.

3. Increased heart rate and force of contraction potentially increase blood flow through tissues. Vasoconstriction of blood vessels in tissues not involved in exercise, such as abdominopelvic organs, reduces blood flow through them, thus making more blood available for the exercising tissues.
4. Dilation of air passageways increases air flow into and out of the lungs.
5. The availability of energy sources increases. Skeletal muscle cells and liver cells (hepatocytes) are stimulated to break down glycogen to glucose. Skeletal muscle cells use the glucose and liver cells release it into the blood for use by other tissues. Fat cells (adipocytes) break down triglycerides and release fatty acids into the blood, which are used as an energy source by skeletal and cardiac muscle.
6. As exercising muscles generate heat, body temperature increases. Vasodilation of blood vessels in the skin brings warm blood close to the surface, where heat is lost to the environment. Sweat gland activity increases, resulting in increased sweat production, and evaporation of the sweat removes additional heat.
7. The activities of organs not essential for exercise decrease. For example, the process of digesting food slows as digestive glands decrease their secretions and the contractions of smooth muscle that mix and move food through the gastrointestinal tract decrease.

Increased activity of the parasympathetic division is generally consistent with resting conditions. The parasympathetic division regulates digestion by increasing the secretions of glands, promoting the mixing of food with digestive enzymes and bile, and moving materials through the digestive tract. Defecation and urination are also controlled by the parasympathetic division. Increased parasympathetic stimulation lowers heart rate, which lowers blood pressure, and constricts air passageways, which decreases air movement through them.

24 ***What kinds of effects, excitatory or inhibitory, are produced by the sympathetic and parasympathetic divisions?***

25 ***Give two exceptions to the generalization that organs are innervated by both divisions of the ANS.***

26 ***When a single organ is innervated by both ANS divisions, do they usually produce opposite effects?***

27 ***Explain how the ANS coordinates the activities of different organs.***

28 ***Which ANS division produces the most general effects? How does this happen?***

29 ***Use the fight-or-flight response to describe the responses produced by the sympathetic division.***

PREDICT 7

Bethanechol (be-than′ĕ-kol) chloride is a drug that binds to muscarinic receptors. Explain why this drug can be used to promote emptying of the urinary bladder. Which of the following side effects would you predict: abdominal cramps, asthmatic attack, decreased tear production, decreased salivation, dilation of the pupils, or sweating?

SUMMARY

Contrasting the Somatic and Autonomic Nervous Systems (p. 565)

1. The cell bodies of somatic neurons are located in the CNS, and their axons extend to skeletal muscles, where they have an excitatory effect that usually is controlled consciously.
2. The cell bodies of the preganglionic neurons of the ANS are located in the CNS and extend to ganglia, where they synapse with postganglionic neurons. The postganglionic axons extend to smooth muscle, cardiac muscle, or glands and have an excitatory or inhibitory effect, which usually is controlled unconsciously.

Anatomy of the Autonomic Nervous System (p. 565)

Sympathetic Division

1. Preganglionic cell bodies are in the lateral horns of the spinal cord gray matter from T1–L2.
2. Preganglionic axons pass through the ventral roots to the white rami communicantes to the sympathetic chain ganglia. From there, four courses are possible:
   - Preganglionic axons synapse (at the same or a different level) with postganglionic neurons, which exit the ganglia through the gray rami communicantes and enter spinal nerves.
   - Preganglionic axons synapse (at the same or a different level) with postganglionic neurons, which exit the ganglia through sympathetic nerves.
   - Preganglionic axons pass through the chain ganglia without synapsing to form splanchnic nerves. Preganglionic axons then synapse with postganglionic neurons in collateral ganglia.
   - Preganglionic axons synapse with the cells of the adrenal medulla.

Parasympathetic Division

1. Preganglionic cell bodies are in nuclei in the brainstem or the lateral parts of the spinal cord gray matter from S2–S4.
   - Preganglionic axons from the brain pass to ganglia through cranial nerves.
   - Preganglionic axons from the sacral region pass through the pelvic splanchnic nerves to the ganglia.
2. Preganglionic axons pass to terminal ganglia within the wall of or near the organ that is innervated.

Enteric Nervous System

1. The enteric nerve plexus is within the wall of the digestive tract.
2. The enteric plexus consists of sensory neurons, ANS motor neurons, and enteric neurons.

Distribution of Autonomic Nerve Fibers

1. Sympathetic axons reach organs through spinal nerves, head and neck nerve plexuses, thoracic nerve plexuses, and abdominopelvic nerve plexuses.
2. Parasympathetic axons reach organs through cranial nerves, thoracic nerve plexuses, abdominopelvic nerve plexuses, and pelvic splanchnic nerves.
3. Sensory neurons run alongside sympathetic and parasympathetic neurons within nerves and nerve plexuses.

Physiology of the Autonomic Nervous System (p. 572)

Neurotransmitters

1. Acetylcholine is released by cholinergic neurons (all preganglionic neurons, all parasympathetic postganglionic neurons, and some sympathetic postganglionic neurons).
2. Norepinephrine is released by adrenergic neurons (most sympathetic postganglionic neurons).

Receptors

1. Acetylcholine binds to nicotinic receptors (found in all postganglionic neurons) and muscarinic receptors (found in all parasympathetic and some sympathetic effector organs).
2. Norepinephrine and epinephrine bind to alpha and beta receptors (found in most sympathetic effector organs).
3. Activation of nicotinic receptors is excitatory, whereas activation of muscarinic, alpha, or beta receptors is either excitatory or inhibitory.

Regulation of the Autonomic Nervous System (p. 576)

1. Autonomic reflexes control most of the activity of visceral organs, glands, and blood vessels.
2. Autonomic reflex activity can be influenced by the hypothalamus and higher brain centers.
3. The sympathetic and parasympathetic divisions can influence the activities of the enteric nervous system through autonomic reflexes. The enteric nervous system can function independently of the CNS through local reflexes.

Functional Generalizations About the Autonomic Nervous System (p. 578)

1. Both divisions of the ANS produce stimulatory and inhibitory effects.
2. Most organs are innervated by both divisions. Usually, each division produces an opposite effect on a given organ.
3. Either division alone or both working together can coordinate the activities of different structures.
4. The sympathetic division produces more generalized effects than the parasympathetic division.
5. Sympathetic activity generally prepares the body for physical activity, whereas parasympathetic activity is more important for vegetative functions.

REVIEW AND COMPREHENSION

1. Given these phrases:
   1. neuron cell bodies in the nuclei of cranial nerves
   2. neuron cell bodies in the lateral gray matter of the spinal cord (S2–S4)
   3. two synapses between the CNS and effector organs
   4. regulates smooth muscle

   Which of the phrases are true for the autonomic nervous system?
   a. 1,3
   b. 2,4
   c. 1,2,3
   d. 2,3,4
   e. 1,2,3,4
2. Given these structures:
   1. gray ramus communicans
   2. white ramus communicans
   3. sympathetic chain ganglion

   Choose the arrangement that lists the structures in the order an action potential passes through them from a spinal nerve to an effector organ.
   a. 1,2,3
   b. 1,3,2
   c. 2,1,3
   d. 2,3,1
   e. 3,2,1
3. Given these structures:
   1. collateral ganglion
   2. sympathetic chain ganglion
   3. white ramus communicans
   4. splanchnic nerve

   Choose the arrangement that lists the structures in the order an action potential travels through them on the way from a spinal nerve to an effector organ.
   a. 1,3,2,4
   b. 1,4,2,3
   c. 3,1,4,2
   d. 3,2,4,1
   e. 4,3,1,2
4. The white ramus communicans contains
   a. preganglionic sympathetic fibers.
   b. postganglionic sympathetic fibers.
   c. preganglionic parasympathetic fibers.
   d. postganglionic parasympathetic fibers.
5. The cell bodies of the postganglionic neurons of the sympathetic division are located in the
   a. sympathetic chain ganglia.
   b. collateral ganglia.
   c. terminal ganglia.
   d. dorsal root ganglia.
   e. both a and b.
6. Splanchnic nerves
   a. are part of the parasympathetic division.
   b. have preganglionic neurons that synapse in the collateral ganglia.
   c. exit from the cervical region of the spinal cord.
   d. travel from the spinal cord to the sympathetic chain ganglia.
   e. all of the above.
7. Which of the following statements regarding the adrenal gland is true?
   a. The parasympathetic division stimulates the adrenal gland to release acetylcholine.
   b. The parasympathetic division stimulates the adrenal gland to release epinephrine.
   c. The sympathetic division stimulates the adrenal gland to release acetylcholine.
   d. The sympathetic division stimulates the adrenal gland to release epinephrine.
8. The parasympathetic division
   a. is also called the craniosacral division.
   b. has preganglionic axons in cranial nerves.
   c. has preganglionic axons in pelvic splanchnic nerves.
   d. has ganglia near or in the wall of effector organs.
   e. all of the above.
9. Which of these is *not* a part of the enteric nervous system?
   a. ANS motor neurons
   b. neurons located only in the digestive tract
   c. sensory neurons
   d. somatic neurons
10. Sympathetic axons reach organs through all of the following *except*
    a. abdominopelvic nerve plexuses.
    b. head and neck nerve plexuses.
    c. thoracic nerve plexuses.
    d. pelvic splanchnic nerves.
    e. spinal nerves.
11. Which of these cranial nerves does *not* contain parasympathetic fibers?
    a. oculomotor (III)
    b. facial (VII)
    c. glossopharyngeal (IX)
    d. trigeminal (V)
    e. vagus (X)
12. Which of the following statements concerning the preganglionic neurons of the ANS is true?
    a. All parasympathetic preganglionic neurons secrete acetylcholine.
    b. Only parasympathetic preganglionic neurons secrete acetylcholine.
    c. All sympathetic preganglionic neurons secrete norepinephrine.
    d. Only sympathetic preganglionic neurons secrete norepinephrine.
13. A cholinergic neuron
    a. secretes acetylcholine.
    b. has receptors for acetylcholine.
    c. secretes norepinephrine.
    d. has receptors for norepinephrine.
    e. secretes both acetylcholine and norepinephrine.
14. When acetylcholine binds to nicotinic receptors,
    a. the cell's response is mediated by G proteins.
    b. the response can be excitatory or inhibitory.
    c. Na^+ channels open.
    d. the binding occurs at the effector organ.
    e. all of the above.
15. Nicotinic receptors are located in
    a. postganglionic neurons of the parasympathetic division.
    b. postganglionic neurons of the sympathetic division.
    c. membranes of skeletal muscle cells.
    d. both a and b.
    e. all of the above.
16. The activation of α and β receptors
    a. can produce an excitatory or inhibitory response.
    b. can be caused by the sympathetic division.
    c. can be caused by epinephrine released from the adrenal gland.
    d. can be caused by norepinephrine
    e. all of the above.
17. The sympathetic division
    a. is always stimulatory.
    b. is always inhibitory.
    c. is usually under conscious control.
    d. generally opposes the actions of the parasympathetic division.
    e. both a and c.
18. A sudden increase in blood pressure
    a. initiates a sympathetic reflex that decreases heart rate.
    b. initiates a local reflex that decreases heart rate.
    c. initiates a parasympathetic reflex that decreases heart rate.
    d. both a and b.
    e. both b and c.

19. Which of these structures is innervated almost exclusively by the sympathetic division?
    a. gastrointestinal tract
    b. heart
    c. urinary bladder
    d. reproductive tract
    e. blood vessels
20. Which of these is expected if the sympathetic division is activated?
    a. Secretion of watery saliva increases.
    b. Tear production increases.
    c. Air passageways dilate.
    d. Glucose release from the liver decreases.
    e. All of the above are true.

Answers in Appendix E

CRITICAL THINKING

1. When a person is startled or sees a "pleasurable" object, the pupils of the eyes may dilate. What division of the ANS is involved in this reaction? Describe the nerve pathway involved.
2. Reduced secretion from salivary and lacrimal glands could indicate damage to what nerve?
3. In a patient with Raynaud disease, blood vessels in the skin of the hand may become chronically constricted, thereby reducing blood flow and producing gangrene. These vessels are supplied by nerves that originate at levels T2 and T3 of the spinal cord and eventually exit through the first thoracic and inferior cervical sympathetic ganglia. Surgical treatment for Raynaud disease severs this nerve supply. At which of the following locations would you recommend that the cut be made: white rami of T2–T3, gray rami of T2–T3, spinal nerves T2–T3, or spinal nerves C1–T1? Explain.
4. Patients with diabetes mellitus can develop autonomic neuropathy, which is damage to parts of the autonomic nerves. Given the following parts of the ANS—vagus nerve, oculomotor nerve, splanchnic nerve, pelvic splanchnic nerve, outflow of gray ramus—match the part with the symptom it would produce if the part were damaged:
    a. impotence
    b. subnormal sweat production
    c. stomach muscles relaxed and delayed emptying of the stomach
    d. diminished pupil reaction (constriction) to light
    e. bladder paralysis with urinary retention
5. Explain why methacholine, a drug that acts like acetylcholine, is effective for treating tachycardia (heart rate faster than normal). Which of the following side effects would you predict: increased salivation, dilation of the pupils, sweating, or difficulty in taking a deep breath?
6. A patient has been exposed to the organophosphate pesticide malathion, which inactivates acetylcholinesterase. Which of the following symptoms would you predict: blurring of vision, excess tear formation, frequent or involuntary urination, pallor (pale skin), muscle twitching, or cramps? Would atropine be an effective drug to treat the symptoms (see p. 574 for the action of atropine)? Explain.
7. Epinephrine is routinely mixed with local anesthetic solutions. Why?
8. A drug blocks the effect of the sympathetic division on the heart. Careful investigation reveals that, after administration of the drug, normal action potentials are produced in the sympathetic preganglionic and postganglionic neurons. Also, injection of norepinephrine produces a normal response in the heart. Explain, in as many ways as you can, the mode of action of the unknown drug.
9. A drug is known to decrease heart rate. After cutting the white rami of T1–T4, the drug still causes heart rate to decline. After cutting the vagus nerves, the drug no longer affects heart rate. Which division of the ANS does the drug affect? Does the drug have its effect at the synapse between preganglionic and postganglionic neurons, at the synapse between postganglionic neurons and effector organs, or in the CNS? Is the effect of the drug excitatory or inhibitory?
10. Make a list of the responses controlled by the ANS in (a) a person who is extremely angry and (b) a person who has just finished eating and is relaxing.

Answers in Appendix F

Visit this book's website at www.mhhe.com/seeley8 for chapter quizzes, interactive learning exercises, and other study tools.

Functional Organization of the Endocrine System

17

The nervous and endocrine systems are the two major regulatory systems of the body. Together, they regulate and coordinate the activities of essentially all other body structures. The nervous system functions something like telephone messages sent along many telephone wires to their specific destinations. It transmits information in the form of action potentials along the axons of nerve cells. Chemical signals in the form of neurotransmitters are released at synapses between neurons and the cells they control. The endocrine system is more like satellite radio or television signals broadcast widely so that every radio or television set, with its receiver adjusted properly, can receive the signals. It sends information to the cells it controls in the form of chemical signals, called **hormones,** that are released from endocrine glands. Hormones are carried to all parts of the body by the circulatory system. Cells that are able to recognize the hormones respond to them, whereas other cells do not.

This chapter introduces the general characteristics of the endocrine system. It compares some of the functions of the nervous and endocrine systems, emphasizes the role of the endocrine system in the maintenance of homeostasis, and illustrates the means by which the endocrine system regulates the functions of cells. This chapter explains the *general characteristics of the endocrine system* (p. 586), the *chemical structure of hormones* (p. 587), the *control of secretion rate* (p. 587), *transport and distribution in the body* (p. 593), *metabolism and excretion* (p. 594), *interaction of hormones with their target tissues* (p. 595), and *classes of receptors* (p. 597). The structure and function of the endocrine glands, the chemicals they secrete, and the means by which activities are regulated are described in chapter 18.

Colorized transmission electron micrograph of a growth hormone-secreting cell from the anterior pituitary gland. The secretory vesicles (*brown*) contain growth hormone.

Endocrine System

GENERAL CHARACTERISTICS OF THE ENDOCRINE SYSTEM

The term **endocrine** (en′dō-krin) is derived from the Greek words *endo,* meaning within, and *krinō,* to secrete. The term implies that cells of endocrine glands secrete chemical signals that influence tissues that are separated from the endocrine glands by some distance. The **endocrine system** is composed of glands that typically secrete chemical signals into the circulatory system (figure 17.1). In contrast, exocrine glands have ducts that carry their secretions to surfaces (see chapter 4). The chemical signals secreted by endocrine glands are called hormones (hōr′mōnz), a term derived from the Greek word *hormon,* meaning to set into motion. Traditionally, a hormone is defined as a chemical signal, or **ligand,** that (1) is produced in minute amounts by a collection of cells; (2) is secreted into the interstitial spaces; (3) enters the circulatory system, where it is transported some distance; and (4) acts on specific tissues, called **target tissues,** at another site in the body to influence the activity of those tissues in a specific fashion. All hormones exhibit most components of this definition, but some components do not apply to every hormone.

Both the endocrine system and the nervous system regulate the activities of structures in the body, but they do so in different ways. For example, the hormones secreted by most endocrine glands can be described as **amplitude-modulated signals,** which consist mainly of increases or decreases in the concentration of hormones in the body fluids (figure 17.2), over periods ranging usually from minutes to hours. The responses to hormones either increase or decrease as a function of the hormone concentration. For example, antidiuretic hormone released from the posterior pituitary gland acts on the kidney to decrease the volume of urine produced. It produces a response within minutes, and the urine volume produced is a function of the amount of hormone released. On the other hand, the all-or-none action potentials carried along axons can be described as **frequency-modulated signals** (see figure 17.2), which vary in frequency but not in amplitude. A low frequency of action potentials is a weak stimulus, whereas a high frequency of action potentials is a strong stimulus (see chapter 11). Each action potential usually lasts from one to a few milliseconds (msec). For example, action potentials carried by axons to a muscle cause a response within msec. The muscle response is a product of the action potential frequency and the length of time the action potential frequency is maintained. The responses of the endocrine system are usually slower and of longer duration, and its effects are usually more generally distributed than those of the nervous system.

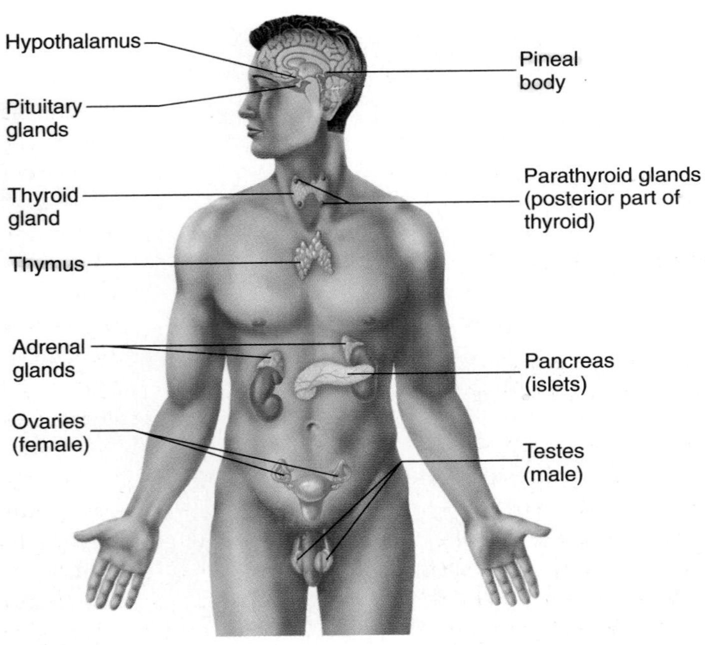

FIGURE 17.1 Major Endocrine Glands

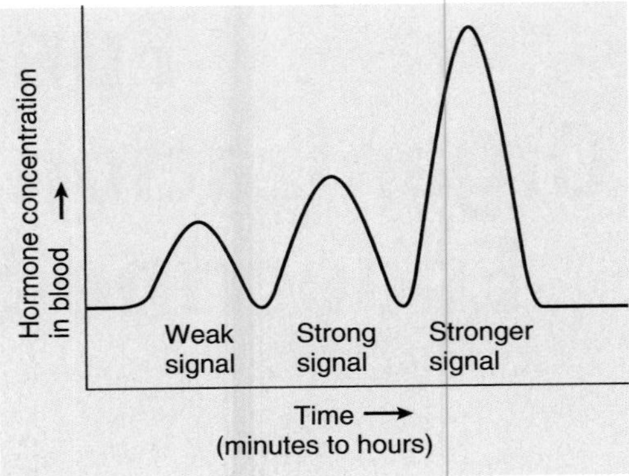

Amplitude-modulated system. The concentration of the hormone determines the strength of the signal and the magnitude of the response. For most hormones, a small concentration of a hormone is a weak signal and produces a small response, whereas a larger concentration is a stronger signal and results in a greater response.

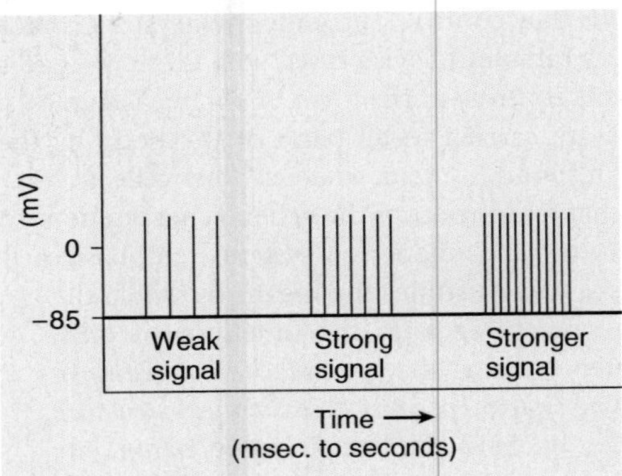

Frequency-modulated system. The strength of the signal depends on the frequency, not the size, of the action potentials. All action potentials are the same size in a given tissue. A low frequency of action potentials is a weak stimulus, and a higher frequency is a stronger stimulus.

FIGURE 17.2 Regulatory Systems

Although the stated differences between the endocrine and nervous systems are generally true, exceptions exist. Some endocrine responses are more rapid than some neural responses, and some endocrine responses have a shorter duration than some neural responses. In addition, a few hormones act as both amplitude- and frequency-modulated signals, in which the concentrations of the hormones and the frequencies at which the increases in hormone concentrations occur are important.

At one time, the endocrine system was believed to be relatively independent and different from the nervous system; however, an intimate relationship between these systems is now recognized, and the two systems cannot be completely separated either anatomically or functionally. Some neurons secrete chemical signals called **neurohormones** (noor-ō-hōr′mōnz) into the circulatory system, which function as hormones and commonly are referred to as hormones. Also, some neurons directly innervate endocrine glands and influence their secretory activity. Neurons release chemical signals at synapses in the form of neurotransmitters and neuromodulators, and the membrane potentials of some endocrine glands undergo depolarization or hyperpolarization, which results in either an increase or a decrease in the rate of hormone secretion. Conversely, some hormones secreted by endocrine glands affect the nervous system and markedly influence its activity.

Intercellular chemical signals allow one cell to communicate with other cells. These signals coordinate and regulate the activities of most cells. Several categories of intercellular chemical signals exist. Neurotransmitters and neuromodulators play essential roles in the function of the nervous system (see chapter 11). Hormones are secreted by endocrine glands.

Autocrine (aw′tō-krin) **chemical signals** are released by cells and have a local effect on the same cell type from which the chemical signals are released. Examples include prostaglandin-like chemicals released from smooth muscle cells and platelets in response to inflammation. These chemicals cause the relaxation of blood vessel smooth muscle cells and the aggregation of platelets. As a result, the blood vessels dilate and blood clots.

Paracrine (par′ă-krin) **chemical signals** are released by cells and affect other cell types locally without being transported in the blood. For example, a peptide called somatostatin is released by cells in the pancreas and functions locally to inhibit the secretion of insulin from other cells of the pancreas (see chapter 18).

Pheromones (fer′ō-mōnz) are special chemical signals because they are secreted into the environment and modify the behavior and the physiology of other individuals of the same species. For example, pheromones released in the urine of cats and dogs at certain times are olfactory signals that indicate fertility. Evidence supports the existence of pheromones produced by women that influence the timing of menstrual cycles in other women (table 17.1).

Many intercellular chemical signals consistently fit one specific definition, but others do not. For example, norepinephrine functions as both a neurotransmitter and a neurohormone; prostaglandins function as neurotransmitters, neuromodulators, paracrine, and autocrine chemical signals. The schemes used to classify chemicals on the basis of their functions are useful, but they do not indicate that a specific molecule always performs as the same type of chemical signal in every place it is found. For that reason, the study of endocrinology often includes the study of autocrine and paracrine chemical signals in addition to hormones.

1 ***Define endocrine gland, endocrine system, and hormone. Explain why a simple definition for hormone is difficult to create.***

2 ***Contrast the endocrine system and the nervous system for the following: amplitude versus frequency modulation; speed and duration of target cell response.***

3 ***Explain why, despite their differences, the nervous and endocrine systems cannot be completely separated.***

4 ***Name and describe five categories of intercellular chemical signals, other than hormones.***

CHEMICAL STRUCTURE OF HORMONES

Hormones, including neurohormones, can be either proteins or short sequences of amino acids called polypeptides, derivatives of amino acids, or lipids. Some protein hormones, called glycoprotein hormones, are composed of one or more polypeptide chains combined with carbohydrate molecules. Lipid hormones are either steroids or derivatives of fatty acids. Table 17.2 presents the major categories and subcategories of hormones based on their chemical structures, and figure 17.3 provides examples.

5 ***List three categories of hormones and their subcategories, based on chemical structure, and give an example of each.***

CONTROL OF SECRETION RATE

Most hormones are not secreted at a constant rate. Instead, the secretion of most hormones increases and decreases dramatically over time. The specific mechanisms regulating the secretion rates of hormones are presented in chapter 18, but the general patterns of regulation are introduced in this chapter. Hormones regulate the rates of many activities in the body. Generally, the secretion rate of each hormone is controlled by a negative-feedback mechanism (see chapter 1) so that the body activity it regulates is maintained within a normal range and homeostasis is maintained. In a few instances, positive-feedback systems play a role in the regulation of hormone secretion. In these instances, the positive-feedback system is limited.

Hormones have three major patterns of regulation. One pattern involves changes in the extracellular concentration of a substance other than a hormone and the effect of those changes on the endocrine gland. An increasing blood glucose level causes an increase in insulin secretion from the pancreas. Insulin increases glucose movement into cells, resulting in a decrease in blood glucose levels, which in turn causes a decrease in insulin secretion. Thus, insulin levels increase and decrease in response to changes in

TABLE 17.1 Functional Classification of Intercellular Chemical Signals

Intercellular Chemical Signal	Description	Example
Autocrine	Secreted by cells in a local area and influences the activity of the same cell type from which it was secreted	Eicosanoids (prostaglandins, thromboxanes, prostacyclins, leukotrienes) 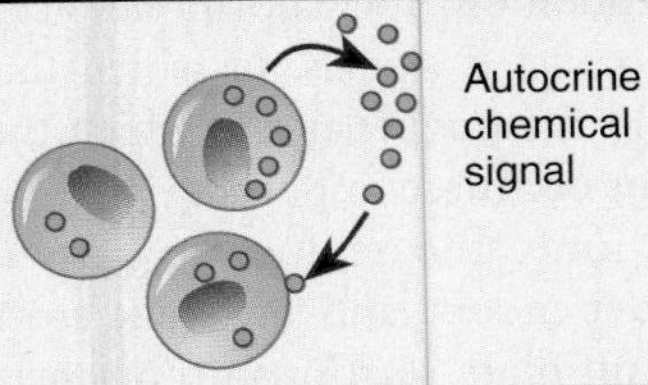
Paracrine	Produced by a wide variety of tissues and secreted into tissue spaces; usually has a localized effect on other tissues	Somatostatin, histamine, eicosanoids 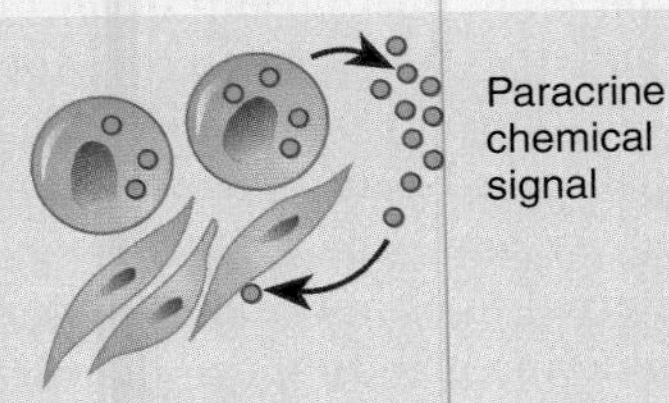
Hormone	Secreted into the blood by specialized cells; travels some distance to target tissues; influences specific activities	Thyroid hormones, growth hormone, insulin, epinephrine, estrogen, progesterone, testosterone 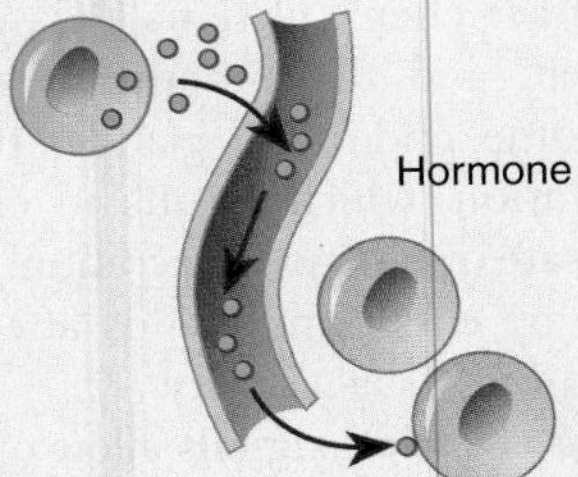
Neurohormone (often referred to as hormone)	Produced by neurons and functions as hormones	Oxytocin, antidiuretic hormone, hypothalamic-releasing and inhibiting hormones 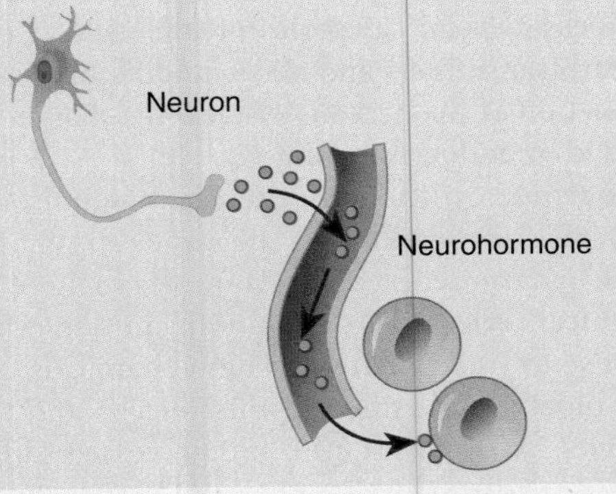
Neurotransmitter or neuromodulator	Produced by neurons and secreted into extracellular spaces by presynaptic nerve terminals; travels short distances; influences postsynaptic cells	Acetylcholine, epinephrine 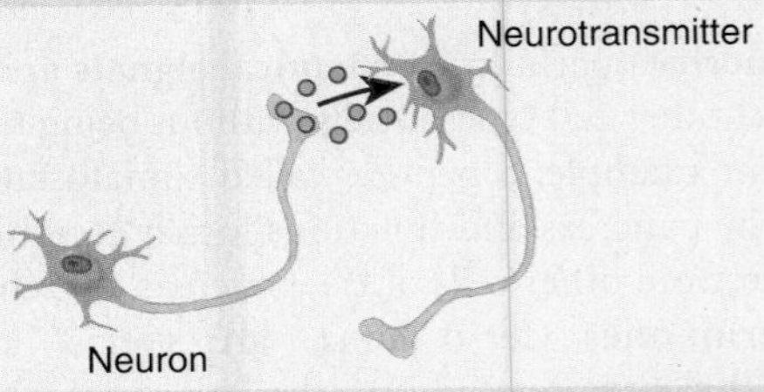
Pheromone	Secreted into the environment; modifies the physiology and behavior of other individuals of the same species	Released by humans and many other animals; released in the urine of animals such as dogs and cats; pheromones produced by women influence the timing of the menstrual cycle of other women 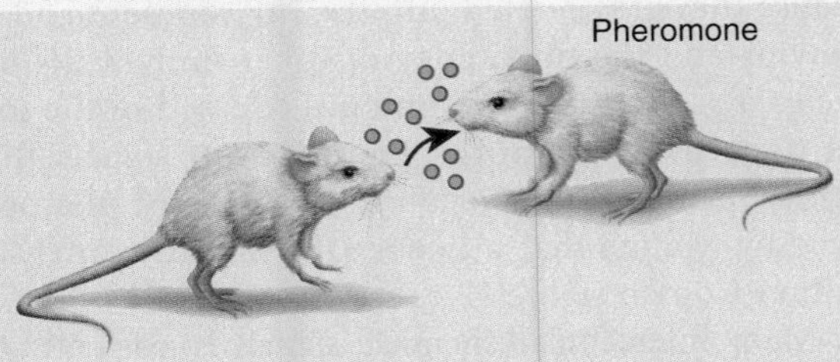

TABLE 17.2 Structural Categories of Hormones

Structural Category	Examples	Structural Category	Examples
Proteins	Growth hormone Prolactin Insulin	Amino acid derivatives	Epinephrine Norepinephrine Thyroid hormones (both T_4 and T_3) Melatonin
Glycoproteins (protein and carbohydrate)	Follicle-stimulating hormone Luteinizing hormone Thyroid-stimulating hormone	Lipids Steroids (cholesterol is a precursor for all steroids)	Estrogens Progestins (progesterone) Testosterone Mineralocorticoids (aldosterone) Glucocorticoids (cortisol)
Polypeptides*	Parathyroid hormone Thyrotropin-releasing hormone Oxytocin Antidiuretic hormone Calcitonin Glucagon Adrenocorticotropic hormone Endorphins Thymosin Melanocyte-stimulating hormone Hypothalamic hormones Lipotropins Somatostatin	Fatty acid derivatives	Prostaglandins Thromboxanes Prostacyclins Leukotrienes

Abbreviations: T_4 = tetraiodothyronine or thyroxine; T_3 = triiodothyronine.

*Polypeptides consisting of short sequences of amino acids are sometimes referred to as peptides.

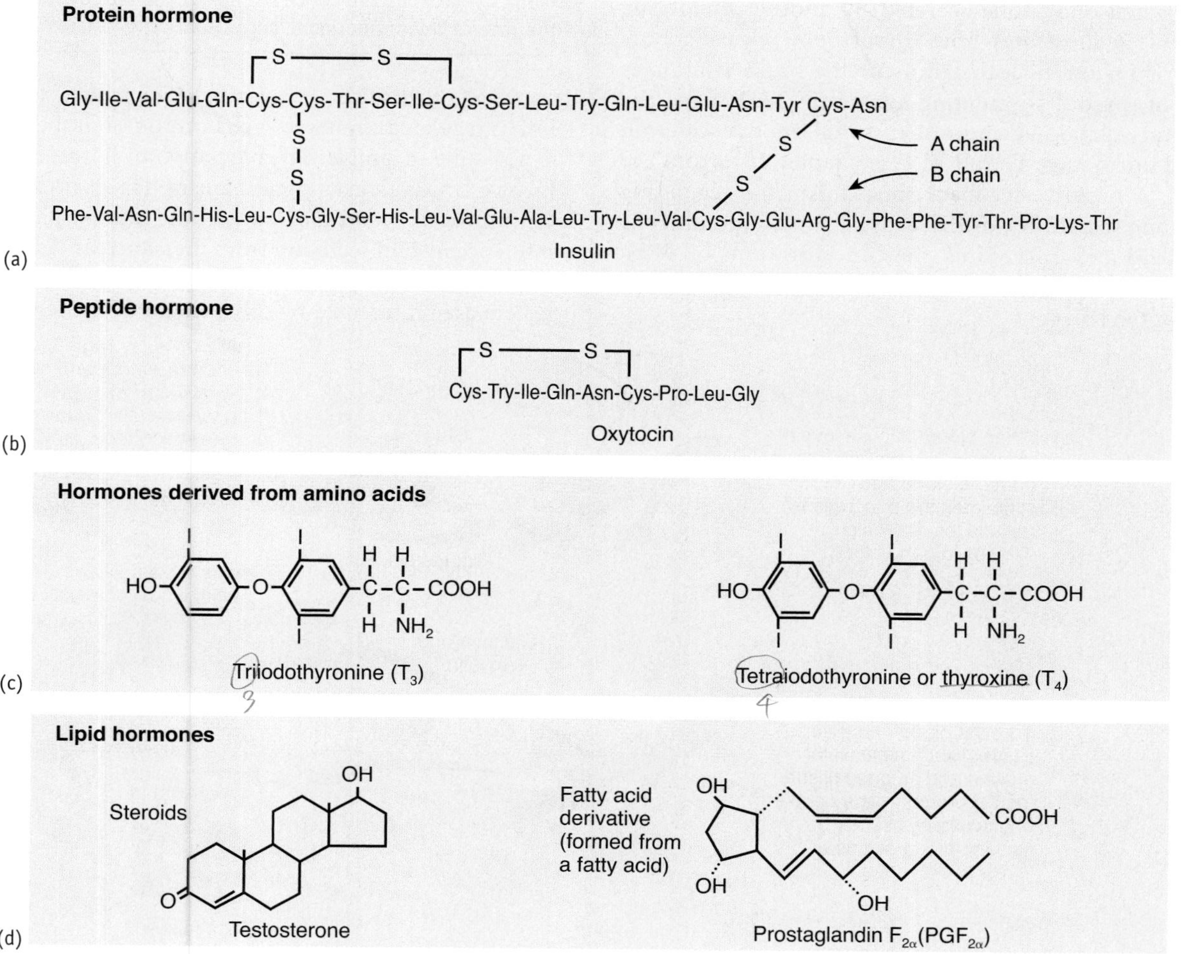

FIGURE 17.3 Examples of Hormone Chemical Structure

(*a*) Insulin is an example of a protein hormone. (*b*) Oxytocin is an example of a peptide hormone. (*c*) The thyroid hormones triiodothyronine (T_3) and tetraiodothyronine (T_4) are examples of amino acid derivatives. (*d*) Testosterone, a steroid, and prostaglandin $F_{2\alpha}$ are examples of lipid hormones.

blood glucose levels. Figure 17.4 describes the influence of blood glucose on insulin secretion from the pancreas.

A second pattern of hormone regulation involves neural control of the endocrine gland. Neurons synapse with the cells that produce the hormone and, when action potentials result, the neurons release a neurotransmitter. In some cases, the neurotransmitter is stimulatory and causes the cells to increase hormone secretion. In other cases, the neurotransmitter is inhibitory and decreases hormone secretion. Thus, sensory input and emotions acting through the nervous system can influence hormone secretion. Figure 17.5 illustrates the neural control of epinephrine and norepinephrine secretion from the adrenal gland. In response to stimuli, such as stress or exercise, the sympathetic division of the autonomic nervous system (see chapter 16) stimulates the adrenal gland to secrete epinephrine and norepinephrine, which help the body respond. Responses include an increased heart rate and increased blood flow through exercising muscles. When the stimuli are no longer present, secretion of epinephrine and norepinephrine decreases.

A third pattern of hormone regulation involves the control of the secretory activity of one endocrine gland by a hormone or a neurohormone secreted by another endocrine gland. Figure 17.6 illustrates how thyrotropin-releasing hormone (TRH) from the hypothalamus of the brain stimulates the secretion of thyroid-stimulating hormone (TSH) from the anterior pituitary gland, which in turn stimulates the secretion of the thyroid hormones, T_3 and T_4 (see chapter 18), from the thyroid gland. A negative-feedback mechanism for regulating thyroid hormone secretion exists because T_3 and T_4 inhibit the secretion of TRH and TSH. Thus, the concentrations of TRH, TSH, and T_3 and T_4 increase and decrease within a normal range (see chapter 18).

Neural Control of Insulin Secretion

Blood glucose levels regulate insulin secretion, but insulin secretion is also regulated by the nervous system. When action potentials in parasympathetic neurons that innervate the pancreas increase, the neurotransmitter acetylcholine is released. Acetylcholine binds to membrane-bound receptors and opens Na^+ channels, resulting in depolarization of pancreatic cells and insulin secretion. When action potentials in sympathetic neurons that innervate the pancreas increase, the neurotransmitter norepinephrine is released. Norepinephrine binds to membrane-bound receptors and activates intracellular chemical signals, causing hyperpolarization of pancreatic cells, and insulin secretion decreases. Thus, nervous stimulation of the pancreas can either increase or decrease insulin secretion.

PREDICT 1

For a person having normal thyroid function, the rate at which TSH and thyroid hormones are secreted remains within a normal range of concentrations. In some people, however, the immune system begins to produce large amounts of an abnormal substance that functions as TSH. Predict what that substance will do to the rate of TSH secretion and the rate of thyroid hormone secretion.

One of these three major patterns by which hormone secretion is regulated applies to each hormone, but the complete picture is not quite so simple. The regulation of hormone secretion often involves more than one mechanism. For example, both the concentration of blood glucose and the autonomic nervous system influence insulin secretion from the pancreas.

Examples of positive-feedback regulation in the endocrine system are represented by components of the female reproductive

1. When blood glucose levels increase, glucose diffuses from the blood and stimulates the pancreas to release insulin, which diffuses into the blood.

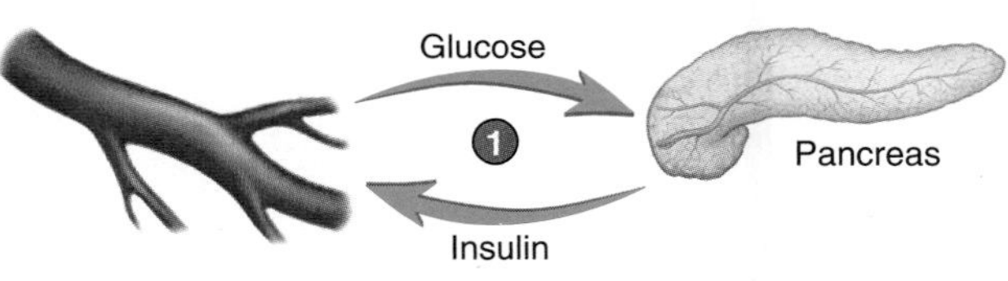

2. Sugar and insulin diffuse from the blood and enter tissues, such as skeletal muscle and adipose tissue. Insulin promotes the uptake of glucose by tissues, causing blood glucose levels to decrease.

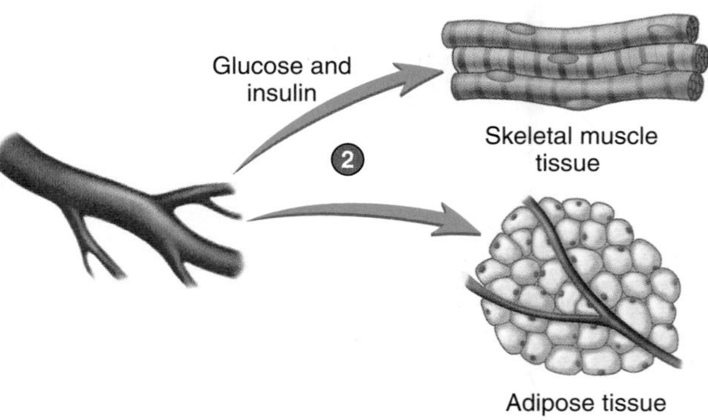

PROCESS FIGURE 17.4 Nonhormonal Regulation of Hormone Secretion
Glucose, which is not a hormone, regulates the secretion of insulin from the pancreas.

1. Stimuli, such as stress or exercise, activate the sympathetic division of the autonomic nervous system.

2. Sympathetic neurons stimulate the release of epinephrine and smaller amounts of norepinephrine from the adrenal medulla. Epinephrine and norepinephrine prepare the body to respond to stressful conditions.

Once the stressful stimuli are removed, less epinephrine is released as a result of decreased stimulation from the autonomic nervous system.

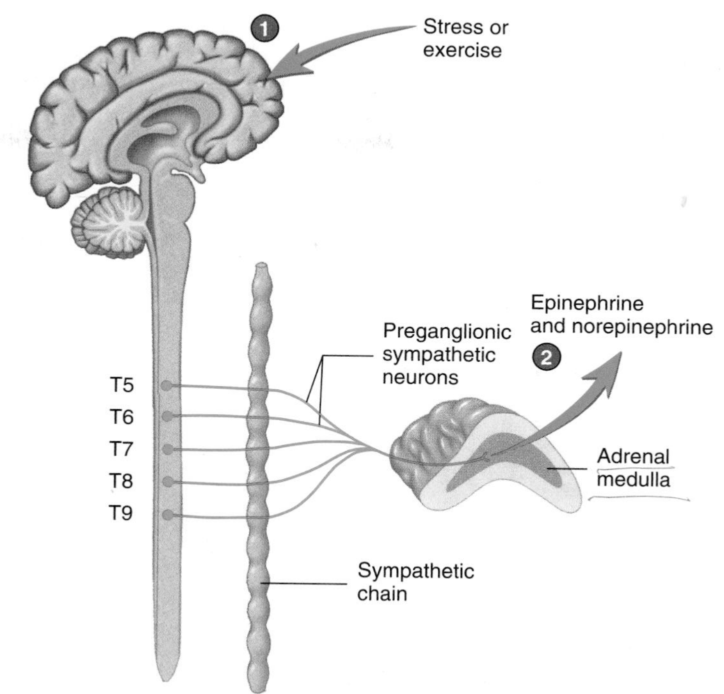

PROCESS FIGURE 17.5 Nervous System Regulation of Hormone Secretion
The sympathetic division of the autonomic nervous system stimulates the adrenal gland to secrete epinephrine and norepinephrine.

1. Thyrotropin-releasing hormone (TRH) is released from neurons in the hypothalamus and travels in the blood to the anterior pituitary gland.
2. TRH stimulates the release of thyroid-stimulating hormone (TSH) from the anterior pituitary gland. TSH travels in the blood to the thyroid gland.
3. TSH stimulates the secretion of thyroid hormones (T_3 and T_4) from the thyroid gland into the blood.
4. T_3 and T_4 act on tissues to produce responses.
5. T_3 and T_4 also have a negative-feedback effect on the hypothalamus and the anterior pituitary to inhibit both TRH secretion and TSH secretion. The negative feedback helps keep blood thyroid hormone levels within a narrow range.

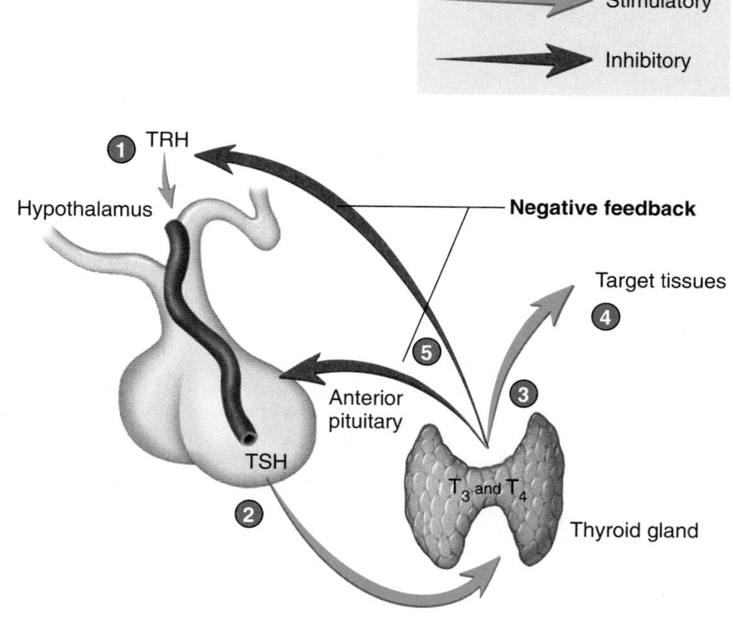

PROCESS FIGURE 17.6 Hormonal Regulation of Hormone Secretion
Hormones can stimulate or inhibit the secretion of other hormones.

system. Prior to ovulation, estrogen from the ovary stimulates luteinizing hormone (LH) secretion from the anterior pituitary gland. LH, in turn, stimulates estrogen secretion from the ovary. Consequently, blood levels of estrogen and LH increase prior to ovulation (figure 17.7*a*). The release of oxytocin during delivery of an infant is another example (see chapters 28 and 29). In cases of positive feedback, negative feedback limits the degree to which positive feedback proceeds (figure 17.7*b*). For example, after ovulation the ovary secretes progesterone, which inhibits LH secretion.

Some hormones are in the circulatory system at relatively constant levels, some change suddenly in response to certain stimuli, and others change in relatively constant cycles (figure 17.8). For example, thyroid hormones in the blood vary within a small range of concentrations that remain relatively constant over long periods of time. Epinephrine is released in large amounts in response to stress or physical exercise; thus its concentration can change suddenly. Reproductive hormones increase and decrease in a cyclic fashion in women during their reproductive years.

(a) **Positive feedback**

1. During the menstrual cycle, before ovulation, small amounts of estrogen are secreted from the ovary.
2. Estrogen stimulates the release of gonadotropin-releasing hormone (GnRH) from the hypothalamus and luteinizing hormone (LH) from the anterior pituitary.
3. GnRH also stimulates the release of LH from the anterior pituitary.
4. LH causes the release of additional estrogen from the ovary. The GnRH and LH levels in the blood increase because of this positive-feedback effect.

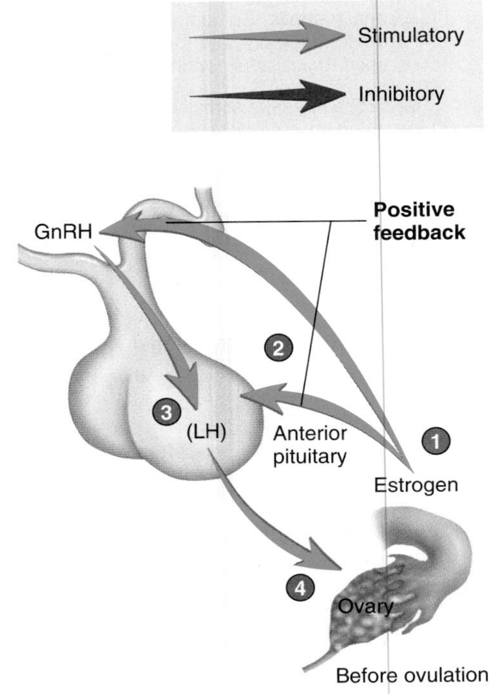

(b) **Negative feedback**

1. During the menstrual cycle, after ovulation, the ovary begins to secrete progesterone in response to LH.
2. Progesterone inhibits the release of GnRH from the hypothalamus and LH from the anterior pituitary.
3. Decreased GnRH release from the hypothalamus reduces LH secretion from the anterior pituitary. GnRH, LH, and estrogen levels in the blood decrease because of this negative-feedback effect.

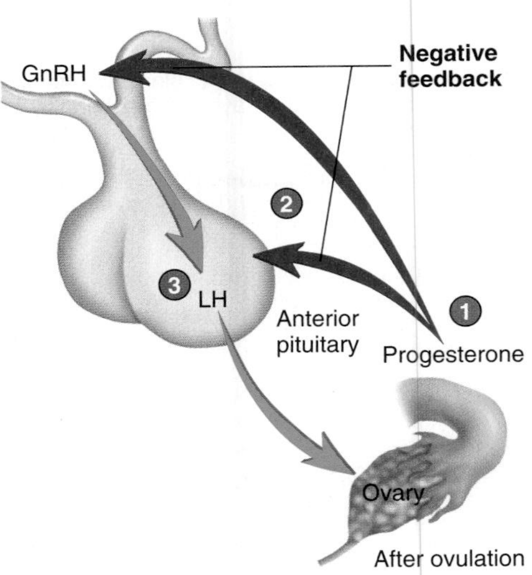

PROCESS FIGURE 17.7 Positive and Negative Feedback

A positive feedback mechanism increases GnRH, LH, and estrogen secretion prior to ovulation. A negative-feedback mechanism decreases GnRH, LH, and estrogen secretion after ovulation.

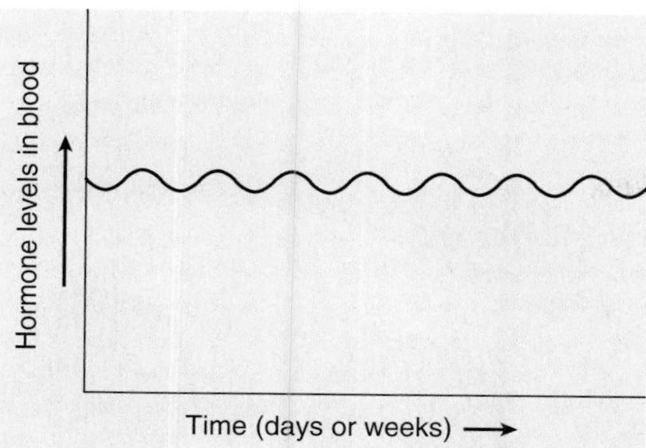

(a) **Chronic hormone regulation.** A relatively constant concentration of hormone is maintained in the circulating blood over a relatively long period.

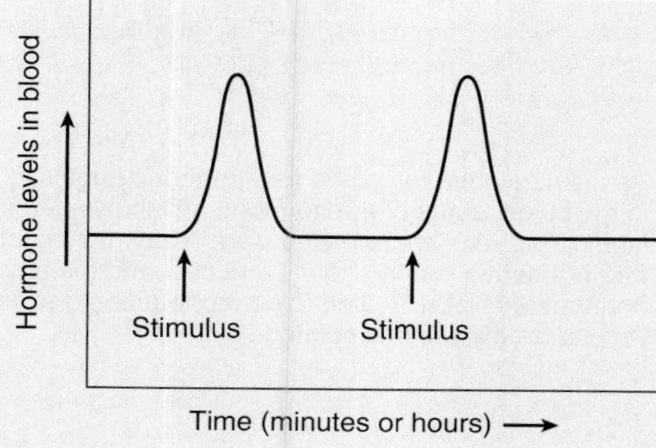

(b) **Acute hormone regulation.** A hormone rapidly increases in the blood for a short time in response to a stimulus.

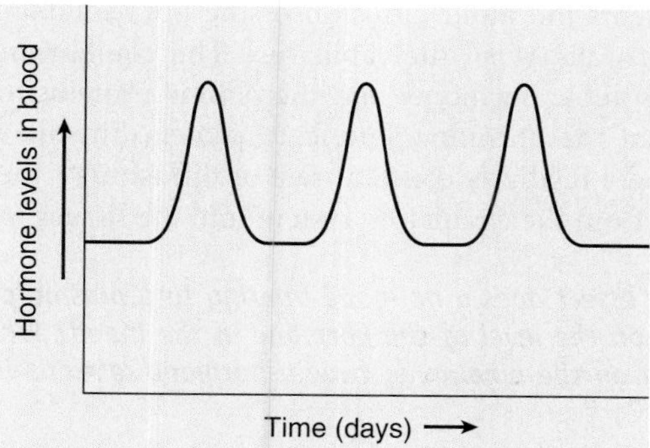

(c) **Cyclic hormone regulation.** A hormone is regulated so that it increases and decreases in the blood at a relatively constant time and to roughly the same amount.

FIGURE 17.8 Changes in Hormone Secretion Through Time
Three basic patterns of hormone secretion exist.

CASE STUDY

Negative Feedback and Hyperthyroidism

Josie owns a business, has several employees, and works hard to manage her business and make time for her family. Over several months, she slowly recognized that she felt warm when others did not; she sweat excessively and her skin was often flushed. She often felt as if her heart were pounding, she was much more nervous than usual, and it was difficult for her to concentrate. She began to feel weak and lose weight, even though her appetite was greater than normal. Her family recognized some of these changes and that her eyes seemed larger than usual. They encouraged her to see her physician. Based on the symptoms, her physician suspected that Josie was suffering from hyperthyroidism (see chapter 18). A blood sample was taken and the results indicated that her blood levels of T_3 and T_4 were elevated and her blood levels of TSH were very low. In addition, a specific immune globulin, called a long-acting thyroid stimulator (LATS), was present in significant concentrations in her blood. The physician concluded that Josie was suffering from Grave disease (see chapter 18).

Josie was treated with radioactive iodine (^{131}I) atoms, which were actively transported into Josie's thyroid cells, where they destroyed a substantial portion of the thyroid gland. Subsequently, Josie had to take T_3 and T_4 in the form of a pill to keep her blood levels of T_3 and T_4 within their normal range of values.

PREDICT 2

a. Prior to treatment, explain why the blood levels of TSH were low and blood levels of T_3 and T_4 were high.
b. Explain why Josie's condition was not the result of a TSH-secreting tumor of the anterior pituitary.
c. What role did LATS play in causing her blood levels of T_3 and T_4 to be high?
d. After a large portion of the thyroid gland was destroyed, predict how the blood levels of T_3 and T_4, as well as TSH, changed in Josie.
e. Explain why Josie had to take T_3 and T_4 in the form of a pill to keep her blood levels of T_3 and T_4 within their normal range of values.

6 ***Describe and give examples of the three major patterns by which hormone secretion is regulated. Give an example of a hormone that is controlled by more than one mechanism.***

7 ***Is hormone secretion generally regulated by negative-feedback or positive-feedback mechanisms?***

8 ***Describe chronic, acute, and cyclic patterns of hormone secretion.***

TRANSPORT AND DISTRIBUTION IN THE BODY

Hormones in blood plasma are transported either as unbound hormones or hormones bound to plasma proteins called **binding proteins.** Hormones bind to binding proteins in a reversible fashion. An equilibrium is established between the unbound hormones and hormones bound to binding proteins.

H	+	BP	⟷	HBP
Hormone		Binding protein		Hormone bound to binding protein

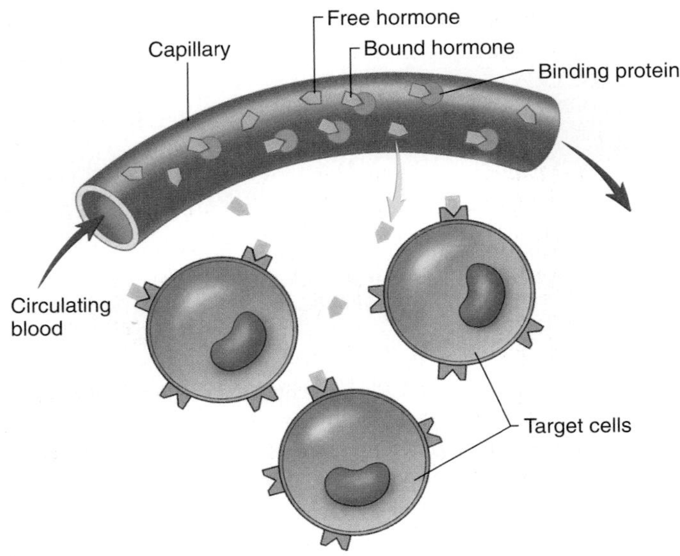

(a) Some hormones in the blood bind to binding proteins so that some hormone molecules are free and some are bound to the binding proteins. The binding proteins function as a reservoir.

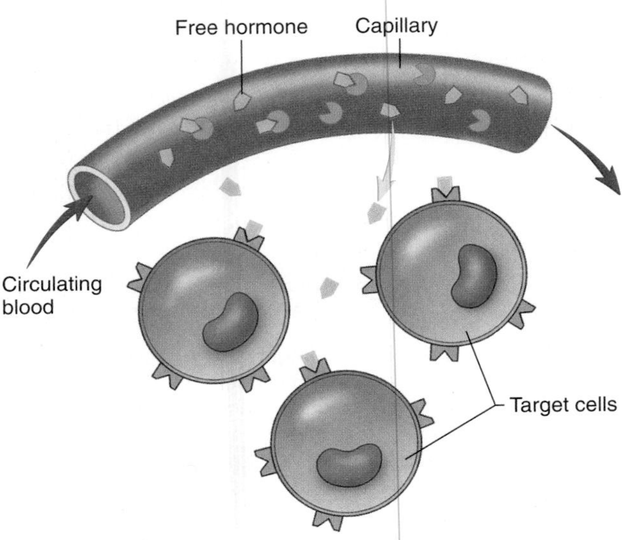

(b) As the concentration of the free hormone begins to decrease in the blood, some of the hormone molecules bound to the binding proteins are released. As a result, the concentration of free hormones is only slightly reduced, and the amount of free hormone available to bind to hormone receptors in target tissues is only slightly reduced.

FIGURE 17.9 Effect of Plasma Binding Proteins

Hormones bind only to certain types of binding proteins. For example, a specific type of binding protein binds to thyroid hormones, and different types of binding proteins bind to hormones such as testosterone and progesterone. The equilibrium between the unbound hormones and bound hormones is important because only an unbound hormone is able to diffuse through capillary walls and bind to target tissues. The hormone molecules bound to binding proteins act as a reservoir for the hormone. If the amount of unbound hormone begins to decrease, some of the bound hormones are released from the binding proteins so that the concentration of unbound hormones decreases less than it would without the equilibrium between the unbound and bound hormones (figure 17.9). Consequently hormones that bind to binding proteins can remain at a relatively constant level in the blood for longer periods of time than unbound hormones.

Hormones circulate in the blood and are, therefore, distributed quickly throughout the body. They diffuse through the capillary wall and enter the interstitial spaces, although the rate at which this movement occurs varies from one hormone to the next. Lipid-soluble hormones readily diffuse through the walls of capillaries. In contrast, water-soluble hormones, such as proteins, must pass through pores called fenestrae (see chapter 21) in the capillary wall. The capillaries of organs that are regulated by protein hormones have large fenestrae. Large pores exist in the capillaries of the endocrine glands that secrete these hormones as well. Because the large water-soluble hormones do not readily diffuse through the wall of all capillaries, they tend to diffuse from the blood into tissue spaces more slowly.

Lipid-soluble hormones typically bind to binding proteins. The combination of the lipid-soluble hormones with plasma proteins prevents the rapid diffusion of the lipid-soluble hormones across the walls of all the capillaries. The equilibrium between the lipid-soluble hormones and the plasma proteins maintains a reservoir of the hormone, bound to proteins, in the circulatory system and a relatively constant rate of diffusion of the unbound hormone from the circulatory system into the tissues.

9 ***What effect does a hormone binding to a plasma protein have on the level of the hormone in the blood? What is the effect on the amount of time a hormone remains in the blood?***

10 ***Why does the wall of capillaries or organs regulated by protein hormones have large pores?***

METABOLISM AND EXCRETION

The destruction and elimination of hormones limit the length of time they are active, and body activities can increase and decrease quickly when hormones are secreted and remain active for only short periods. The length of time it takes for half a dose of a substance to be eliminated from the circulatory system is called its **half-life.** The half-life of a hormone is a standard measurement used by endocrinologists because it allows them to predict the rate at which hormones are eliminated from the body. The length of time required for the total removal of a hormone from the body is not as useful because that measurement is influenced dramatically by the starting concentration. Water-soluble hormones, such as proteins, glycoproteins, epinephrine, and norepinephrine, have relatively short half-lives because they are degraded rapidly by enzymes within the circulatory system of organs, such as the kidneys, liver, or lungs. Hormones with short half-lives normally have concentrations that increase and decrease rapidly within the blood. They generally regulate activities that have a rapid onset and a short duration.

Hormones that are lipid-soluble, such as the steroids and T_3 and T_4 secreted by the thyroid gland, commonly circulate in the blood in combination with binding proteins. The rate at which hormones are degraded, or are eliminated from the circulation, is greatly reduced when the hormones bind to binding proteins. The combination increases their half-life by protecting them and reducing the rate at which they diffuse through the wall of blood vessels. Without the binding proteins, the lipid-soluble hormones would diffuse out of capillaries and be eliminated by the kidneys and enzymes of the liver and lungs. Hormones with a long half-life have blood levels that are maintained at a relatively constant level through time. Table 17.3 outlines the ways the half-life of a hormone is shortened or lengthened.

Hormones are removed from the blood in four major ways: excretion, metabolism, active transport, and conjugation. The kidneys excrete hormones into the urine, or the liver excretes them into the bile. Enzymes in the blood or in tissues, such as the liver, kidneys, lungs, or other target cells, metabolize or chemically modify hormones. The end products can be excreted in the urine or bile, or they can be taken up by cells and used in metabolic processes. For example, epinephrine is modified enzymatically and then excreted by the kidneys. Protein hormones are broken down to their amino acid building blocks. The amino acids can then be taken up by cells and used to synthesize new proteins. Some hormones can be actively transported into cells and recycled. For example, both epinephrine and norepinephrine can be actively transported into cells and secreted again.

The liver conjugates some hormones. **Conjugation** (kon-jŭ-gā′shŭn) is accomplished when cells in the liver attach water-soluble molecules to the hormone. These molecules are usually sulfate or glucuronic acid. Once they are conjugated, hormones are excreted by the kidneys and liver at a greater rate.

TABLE 17.3 Factors That Influence the Half-Life of Hormones
A. Means by which the half-life of hormones is shortened:
1. **Excretion** Hormones are excreted by the kidney into the urine or excreted by the liver into the bile.
2. **Metabolism** Hormones are enzymatically degraded in the blood, liver, kidney, lungs, or target tissues. End products of metabolism are either excreted in urine or bile or are used in other metabolic processes by cells in the body.
3. **Active Transport** Some hormones are actively transported into cells and are used again as either hormones or neurotransmitter substances.
4. **Conjugation** Substances such as sulfate or glucuronic acid groups are attached to hormones primarily in the liver, normally making them less active as hormones and increasing the rate at which they are excreted in the urine or bile.
B. Means by which the half-life of hormones is lengthened:
1. Some hormones are protected from rapid excretion or metabolism by binding reversibly with binding proteins in the plasma.
2. Some hormones are protected by their structure. The carbohydrate components of the glycoprotein hormones protect them from proteolytic enzymes in the circulatory system.

11 ***Define the half-life of a hormone. What happens to this half-life when a hormone binds to a plasma protein? What kinds of hormones bind to plasma proteins?***

12 ***What kinds of activities do hormones with a short half-life regulate? With a long half-life?***

13 ***What are the ways by which the half-life of a hormone is shortened or lengthened?***

PREDICT 3

How is the half-life of a hormone affected by a decrease in the concentration of the specific binding protein to which that hormone binds?

INTERACTION OF HORMONES WITH THEIR TARGET TISSUES

Hormones bind to proteins or glycoproteins called **receptors.** The portion of each protein or glycoprotein molecule where a hormone binds is called a **receptor site,** or **binding site.** The shape and chemical characteristics of each receptor site allow only a specific type of chemical signal to bind to it (figure 17.10). The tendency for each type of chemical signal to bind to a specific type of receptor, and not to others, is called **specificity.** Insulin therefore binds to insulin receptors but not to receptors for growth hormone. However, some hormones can bind to a number of different receptors that are closely related. For example, epinephrine

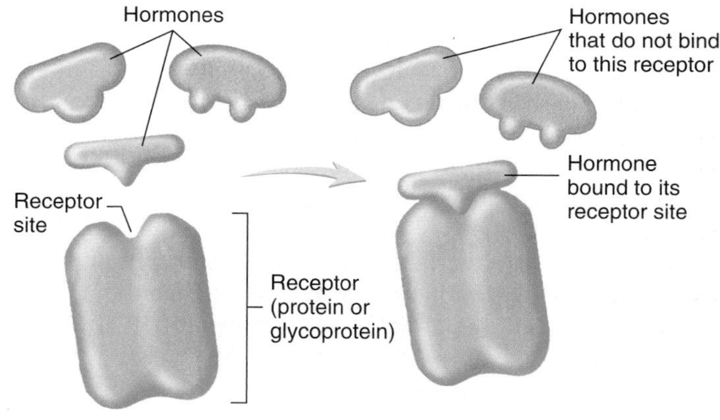

FIGURE 17.10 Receptors and Specificity of Receptor Sites
Hormones bind to receptor molecules. The shape and chemical characteristics of each receptor site allow certain hormones that have a compatible shape and compatible chemical characteristics to bind to it, not others. This relationship is called specificity.

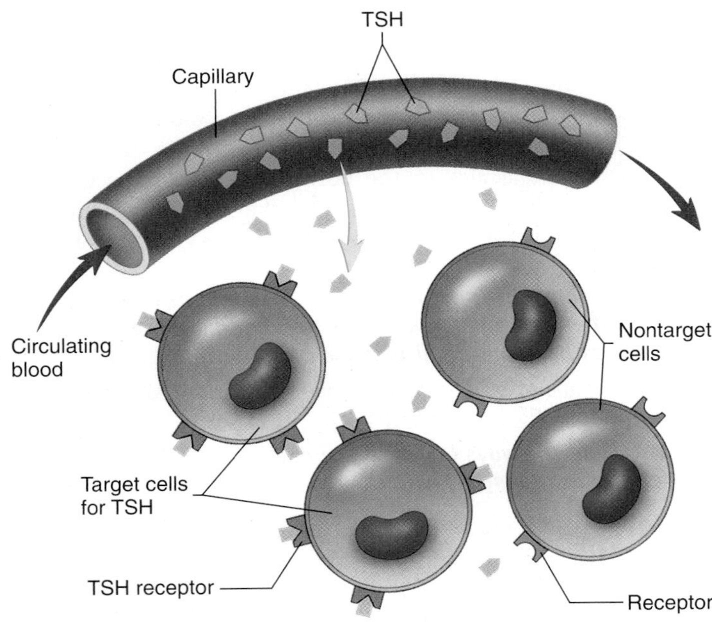

FIGURE 17.11 Response of Target Cells to Hormones

TSH is secreted into the blood and distributed throughout the body, where TSH diffuses from the blood into the interstitial fluid (*blue arrow*). Only target cells have receptors for TSH; therefore, although TSH is distributed throughout the body, only target cells for that hormone can respond to it.

can bind to more than one type of epinephrine receptor. Hormone receptors have a high affinity for the hormones that bind to them, so only a small concentration of a given hormone results in a significant number of receptors with hormones bound to them.

Hormones are secreted and distributed throughout the body by the circulatory system, but the presence or absence of specific receptor molecules in cells determines which cells will or will not respond to each hormone (figure 17.11). For example, there are receptors for TSH in cells of the thyroid gland, but there are no such receptors in most other cells of the body. Consequently, cells of the thyroid gland produce a response when exposed to TSH, but cells without receptor molecules do not respond to it.

Drugs with structures similar to specific hormones may compete with those hormones for their receptors (see chapter 3). A drug that binds to a hormone receptor and activates it is called an **agonist** for that hormone. A drug that binds to a hormone receptor and inhibits its action is called an **antagonist** for that hormone. For example, drugs exist that compete with epinephrine for its receptor. Some of these drugs, called epinephrine agonists, activate epinephrine receptors and others, called epinephrine antagonists, inhibit them.

The response to a given concentration of a hormone is constant in some cases but variable in others. In some cells, the response rapidly decreases through time. Fatigue of the target cells after prolonged stimulation explains some decreases in responsiveness. Also, the number of receptors can rapidly decrease after exposure to certain hormones—a phenomenon called **down-regulation** (figure 17.12*a*). Two mechanisms are commonly

Down-regulation

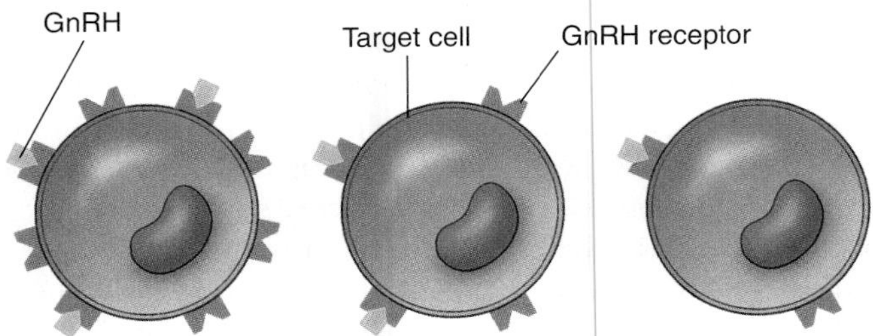

(a) Down-regulation occurs when the number of receptors for a hormone decreases within target cells. For example, gonadotropin-releasing hormone (GnRH) released from the hypothalamus binds to GnRH receptors in the anterior pituitary. GnRH bound to its receptors causes down-regulation of the GnRH receptors so that eventually the target cells become less sensitive to the GnRH.

Up-regulation

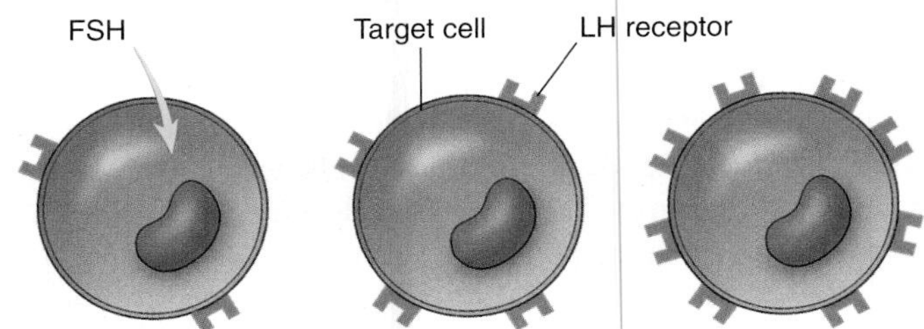

(b) Up-regulation occurs when some stimulus causes the number of receptors for a hormone to increase within a target cell. For example, FSH acts on cells of the ovary to up-regulate the number of receptors for LH. Thus the ovary becomes more sensitive to the effect of LH.

FIGURE 17.12 Down-Regulation and Up-Regulation

responsible for down-regulation. First, the rate at which receptors are synthesized decreases in some cells after the cells are exposed to a hormone. Because most receptor molecules are degraded after a time, a decrease in the synthesis rate reduces the total number of receptor molecules in a cell. Second, the combination of hormones and receptors can increase the rate at which receptor molecules are degraded. In some cases, when a hormone binds to a receptor, both the hormone and the receptor are taken into the cell by endocytosis. Once the hormone and receptor are inside the cell, the cell can break them down.

Gonadotropin-releasing hormone (GnRH), which is released from neurons of the hypothalamus, causes the secretion of LH and follicle-stimulating hormone (FSH) from the anterior pituitary cells. In addition, exposure of the anterior pituitary cells to GnRH causes the number of receptor molecules for GnRH in the pituitary gland cells to decrease dramatically several hours after exposure to the hormone. The down-regulation of GnRH receptors causes the pituitary gland to become less sensitive to additional GnRH. The normal response of the pituitary gland cells to GnRH, therefore, depends on periodic rather than constant exposure of the gland to the hormone.

In general, tissues that exhibit down-regulation of receptors are adapted to respond to short-term increases in hormone concentrations, and tissues that respond to hormones maintained at constant levels normally do not exhibit down-regulation.

In addition to down-regulation, periodic increases in the sensitivity of some cells to certain hormones also occur. This is called **up-regulation,** and it results from an increase in the rate of receptor molecule synthesis (figure 17.12*b*). An example of up-regulation is the increased number of receptors for LH in cells of the ovary during each menstrual cycle. FSH molecules secreted by the pituitary gland increase the rate of LH receptor synthesis in cells of the ovary. Thus, exposure of a tissue to one hormone can increase its sensitivity to a second by causing up-regulation of hormone receptors.

14 ***What characteristics of a hormone receptor make it specific for one type of hormone?***

15 ***What is down-regulation? What two mechanisms are commonly responsible for down-regulation? Give an example of down-regulation in the body.***

16 ***What is up-regulation? Give an example of up-regulation in the body.***

PREDICT 4

Estrogen is a hormone secreted by the ovary. It is secreted in greater amounts after menstruation and a few days before ovulation. Among its many effects is causing up-regulation of receptors in the uterus for another hormone secreted by the ovaries called progesterone. Progesterone is secreted after ovulation. A major effect of progesterone is to cause the uterus to become ready for an embryo to attach to its wall following ovulation. Pregnancy cannot occur unless the embryo attaches to the wall of the uterus. Predict the consequence if the ovary secretes too little estrogen.

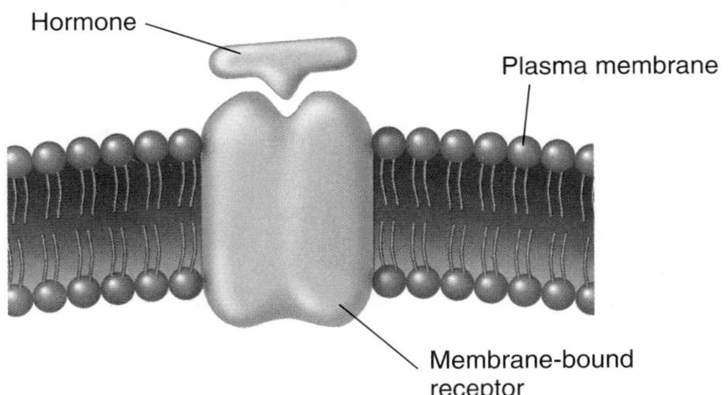

(a) Hormones that are large and water-soluble interact with membrane-bound receptors that extend across the plasma membrane and are exposed on the outer surface of the plasma membrane. The portion of the receptor inside the cell produces the response.

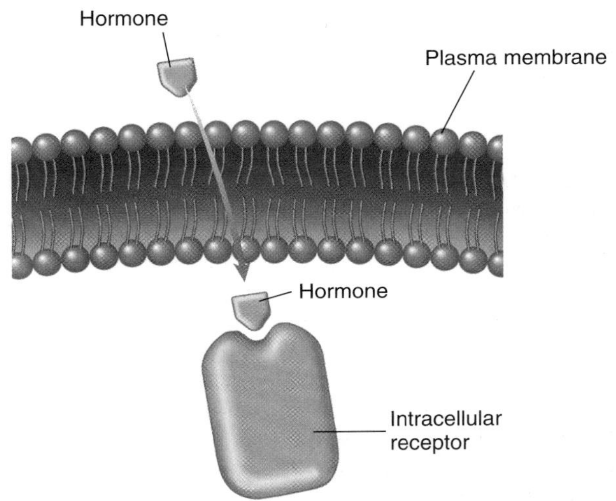

(b) Small lipid-soluble hormones diffuse through the plasma membrane and combine with intracellular receptors of intracellular receptors. The combination of hormones and intracellular receptors produces a response.

FIGURE 17.13 Membrane-Bound and Intracellular Receptors

CLASSES OF RECEPTORS

Hormones can be placed into two major categories, and they bind to two major categories of receptors.

1. *Hormones that bind to membrane-bound receptors.* These hormones include large molecules and water-soluble molecules that cannot pass through the plasma membrane. They interact with **membrane-bound receptors,** which are receptors that extend across the plasma membrane and have their receptor sites exposed to the outer surface of the plasma membrane (figure 17.13*a*). When a hormone binds to a receptor on the outside of the plasma membrane, the receptor initiates a response inside the cell. Hormones that bind to membrane-bound receptors are proteins, glycoproteins, polypeptides, and some smaller molecules, such as epinephrine and norepinephrine.
2. *Hormones that bind to intracellular receptors.* These hormones are lipid-soluble and relatively small. They diffuse through the plasma membrane and bind to **intracellular receptors,** which are receptors in the cytoplasm or in the nucleus of the cell (figure 17.13*b*). Subsequently, the receptors, with the hormones bound to them, interact with DNA in the nucleus of the cell or interact with existing enzymes to produce a response. Thyroid hormones and steroid hormones, such as testosterone, estrogen, progesterone, aldosterone, and cortisol, are examples.

17 ***Define membrane-bound receptor and intracellular receptor. Describe the types of hormones that bind to each type of receptor.***

TABLE 17.4 Overview of Responses to Hormones Binding to Their Receptors

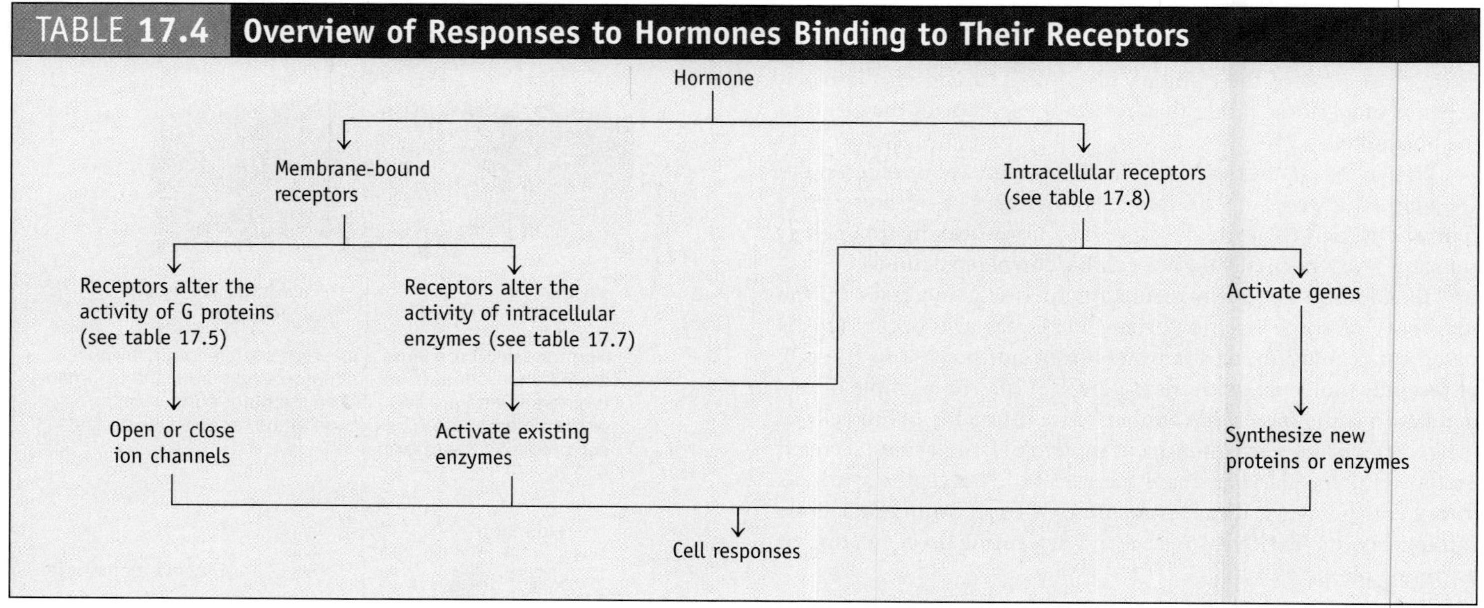

Membrane-Bound Receptors

Hormones bind in a reversible fashion to membrane-bound receptors. Hormone receptors have peptide chains that cross the membrane once in the case of some receptors and several times for other receptors (see chapter 3). After a hormone binds to its receptor, the intracellular part of the receptor initiates events that lead to a response. The mechanisms by which all membrane-bound receptors produce an intracellular response is not known, but many hormone receptors involve two major categories of mechanisms. Many hormones bind to membrane-bound receptors that (1) alter the activity of G proteins at the inner surface of the plasma membrane or (2) alter the activity of intracellular enzymes (table 17.4). These changes produce specific responses in cells.

Receptors That Activate G Proteins

Many membrane-bound receptors produce responses through the action of G proteins (see chapter 3; table 17.5 and figure 17.14). G proteins consist of three subunits; from the largest to smallest, they are called alpha (α), beta (β), and gamma (γ). The G proteins are so-named because one of the subunits binds to guanine nucleotides. In the inactive state, a guanine diphosphate (GDP) molecule is bound to the α subunit of each G protein.

G proteins can bind with receptors at the inner surface of the plasma membrane. After a hormone binds to the receptor on the outside of a cell, the receptor changes shape. As a result, the receptor combines with a G protein on the inner surface of the plasma membrane, and GDP is released from the α subunit. Guanine

TABLE 17.5 Examples of Hormones That Bind to Membrane-Bound Receptors and Activate G Proteins

Hormone	Source	Target Tissue
Luteinizing hormone	Anterior pituitary	Ovary or testis
Follicle-stimulating hormone	Anterior pituitary	Ovary or testis
Prolactin	Anterior pituitary	Mammary gland and ovary or testis
Thyroid-stimulating hormone	Anterior pituitary	Thyroid gland
Adrenocorticotropic hormone	Anterior pituitary	Adrenal cortex
Oxytocin	Posterior pituitary	Uterus
Vasopressin	Posterior pituitary	Kidney
Calcitonin	Thyroid gland (parafollicular cells)	Bone
Parathyroid hormone	Parathyroid gland	Bone
Glucagon	Pancreas	Liver
Epinephrine	Medulla of adrenal gland	Cardiac muscle

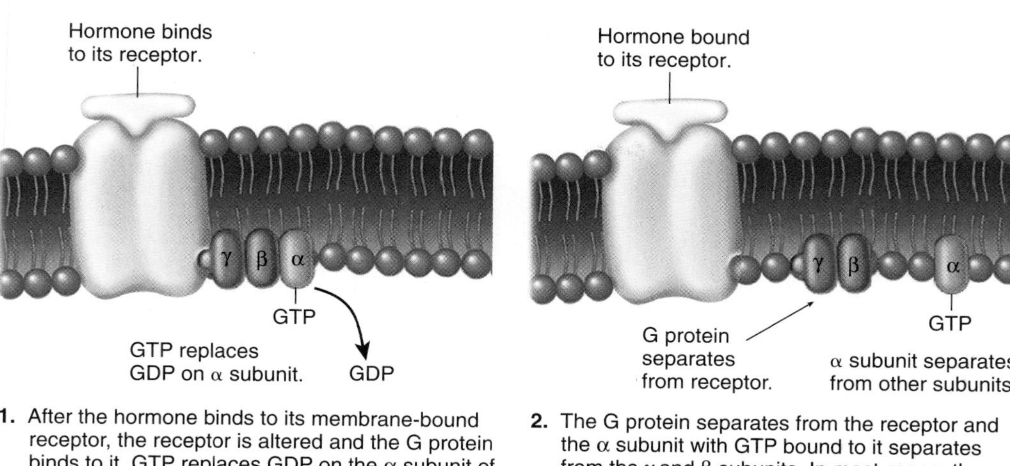

1. After the hormone binds to its membrane-bound receptor, the receptor is altered and the G protein binds to it. GTP replaces GDP on the α subunit of the G protein.

2. The G protein separates from the receptor and the α subunit with GTP bound to it separates from the γ and β subunits. In most cases, the α subunit is the active component of the G protein.

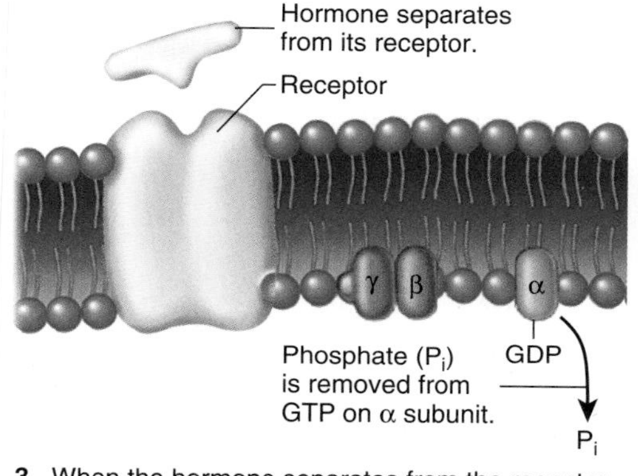

3. When the hormone separates from the receptor, additional G proteins are no longer activated. Inactivation of the α subunit occurs when phosphate (P_i) is removed from the GTP, leaving GDP bound to the α subunit.

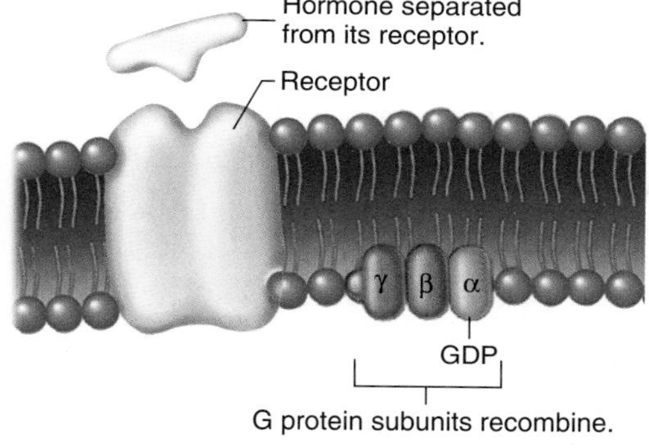

4. The subunits of the G protein recombine.

PROCESS FIGURE 17.14 Membrane-Bound Receptors That Activate G Proteins

triphosphate (GTP), which is more abundant than GDP, binds to the α subunit, thereby activating it. The G proteins separate from the receptor and the activated α subunit separates from the β and γ subunits (see figure 17.14, steps 1 and 2). The activated α subunit can alter the activity of molecules within the plasma membrane or inside the cell, thus producing cellular responses. After a short time, the activated α subunit is turned off because a phosphate group is removed from GTP and converts it to GDP. The α subunit then recombines with the β and γ subunits (see figure 17.14, steps 3 and 4).

Some activated α subunits of G proteins can combine with ion channels, causing them to open or close (figure 17.15). For example, activated α subunits can open Ca^{2+} channels in smooth muscle cells, resulting in the movement of Ca^{2+} into those cells. The Ca^{2+} function as **intracellular mediators.** Intracellular chemical mediators are ions or molecules that either enter the cell or are synthesized in the cell and regulate enzyme activities inside of the cell. The Ca^{2+} combine with calmodulin (kal-mod′ū-lin) molecules, and the calcium-calmodulin complexes activate enzymes that cause contraction of the smooth muscle cells (see figure 17.15, steps 1 and 2 and figure 9.24). After a short time, the activated α subunit is inactivated because GTP is converted to GDP. The α subunit then recombines with the β and γ subunits (see 17.15, steps 3 and 4).

Activated α subunits of G proteins can also alter the activity of enzymes inside the cell. For example, activated α subunits can influence the rate of cyclic adenosine monophosphate (cAMP) formation (figure 17.16). **Adenylate cyclase** (a-den′i-lāt sī′klās), an enzyme activated by G proteins, converts ATP to cAMP. The cAMP molecules act as intracellular mediator molecules. Cyclic AMP binds to protein kinases and activate them. **Protein kinases** are enzymes that regulate the activity of other enzymes by attaching phosphates to them, a process called **phosphorylation.** Depending on the enzyme, phosphorylation increases or decreases the activity of the enzyme. The amount of time cAMP is present to produce a response in a cell is limited. An enzyme in the cytoplasm, called **phosphodiesterase** (fos′fō-dī-es′ter-ās), breaks down cAMP to AMP. The response of the cell is terminated after cAMP levels are reduced below a certain level.

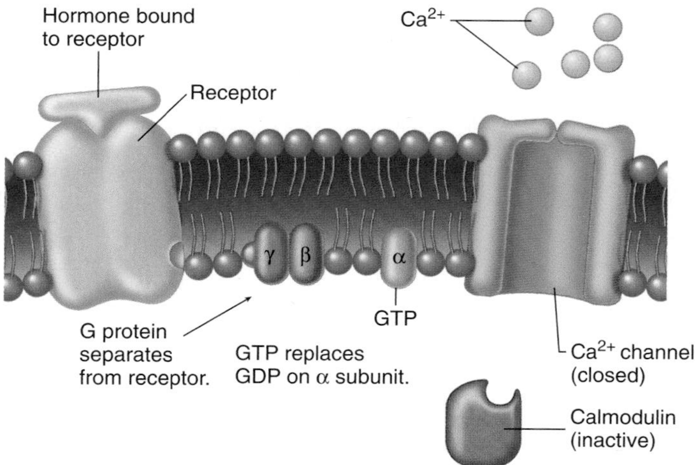

1. After a hormone binds to its receptor, the G protein is activated (see figure 17.14). The activated α subunit with GTP bound to it separates from the γ and β subunits.

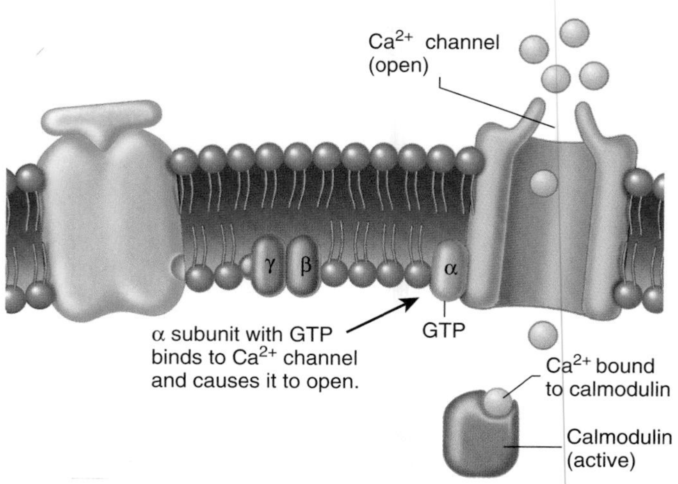

2. The α subunit, with GTP bound to it, combines with the Ca^{2+} channel, and the combination causes the Ca^{2+} channel to open. The Ca^{2+} diffuse into the cell and combine with calmodulin. The combination of Ca^{2+} with calmodulin produces the cell's response to the hormone.

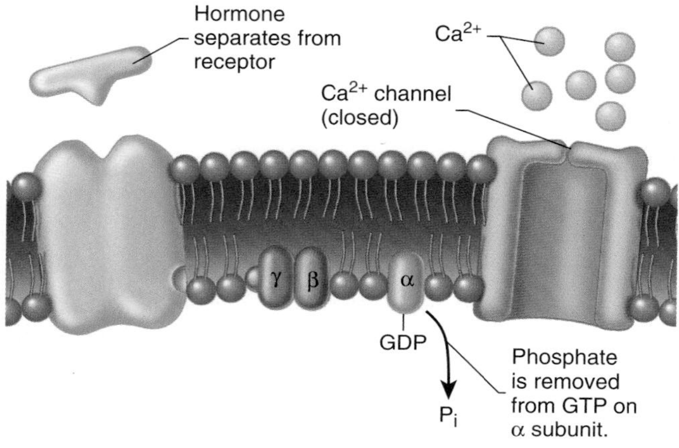

3. Phosphate is removed from the GTP bound to the α subunit, leaving GDP bound to the α subunit. The α subunit can no longer stimulate a cellular response, it separates from the Ca^{2+} channel, and the channel closes.

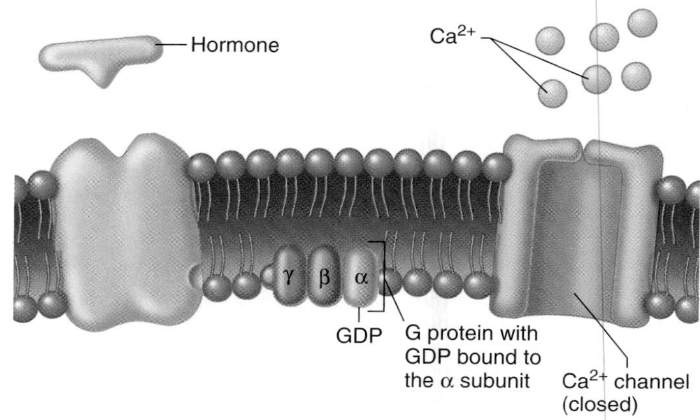

4. The α subunit recombines with the γ and β subunits.

PROCESS FIGURE 17.15 Membrane-Bound Receptors, G Proteins, and Ca^{2+} Channels

Cyclic AMP acts as an intracellular mediator in many cell types. The response in each cell type is different because the enzymes activated by cAMP in each cell type are different. For example, glucagon combines with receptors on the surface of liver cells, activating G proteins and causing an increase in cAMP synthesis, which increases the release of glucose from liver cells (see figure 17.16). In contrast, LH combines with receptors on the surface of cells of the ovary, activating G proteins, and increasing cAMP synthesis. The major response to the increased cAMP is ovulation.

The combination of hormones with their receptors does not always result in increased cAMP synthesis. There are other common intracellular mediators (table 17.6). In some cell types, the combination of hormones with their receptors causes the G proteins to inhibit the synthesis of cAMP, producing a response.

G proteins can also alter the concentration of intracellular mediators other than Ca^{2+} or cAMP (see table 17.6). For example, **diacylglycerol** (dī′as-il-glis′er-ol; **DAG**) and **inositol** (in-ō′si-tōl, in-ō′si-tol) **triphosphate (IP_3)** are intracellular mediator molecules that are influenced by G proteins (figure 17.17). Epinephrine binds to certain membrane-bound receptors in some types of smooth muscle. The combination activates a G protein mechanism, which in turn increases the activity of phospholipase C. Phospholipase C converts phosphoinositol diphosphate (PIP_2) to DAG and IP_3. DAG activates enzymes that synthesize prostaglandins. Prostaglandins increase smooth muscle contraction. IP_3 releases Ca^{2+} from the endoplasmic reticulum or opens Ca^{2+} channels in the plasma membrane. The ions enter the cytoplasm and increase the contraction of the smooth muscle cells.

1. After glucagon binds to a glucagon receptor, the G protein is activated (see figure 17.14).
2. The activated α subunit, with GTP bound to it, binds to and activates an adenylate cyclase enzyme so that it converts ATP to cAMP.
3. The cAMP activates protein kinase enzymes, which phosphorylate specific enzymes that break down glycogen to glucose molecules and the glucose is released from liver cells.
4. Phosphodiesterase enzymes inactivate cAMP by converting cAMP to AMP.

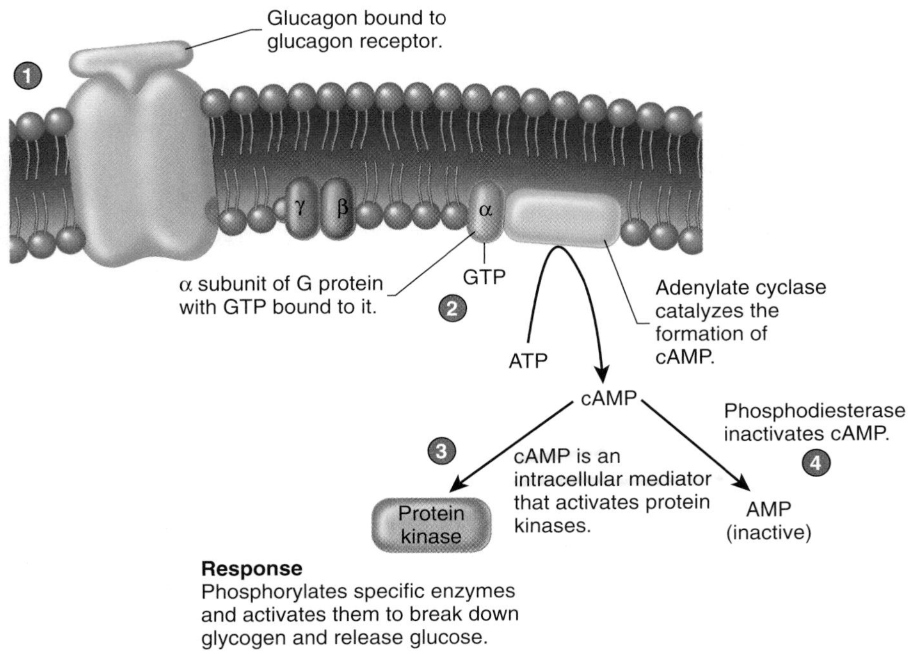

PROCESS FIGURE 17.16 **Membrane-Bound Receptors That Activate G Proteins and Increase the Synthesis of cAMP**

Receptors That Directly Alter the Activity of Intracellular Enzymes

Some hormones bind to membrane-bound receptors and directly change the activity of an intracellular enzyme. The altered enzyme activity either increases or decreases the synthesis of intracellular mediator molecules, or it results in the phosphorylation of intracellular proteins. The intracellular mediators or phosphorylated proteins activate processes that produce the response of cells to the chemical signals.

Intracellular enzymes that are controlled by membrane-bound receptors can be part of the membrane-bound receptor, or they may be separate molecules. The intracellular mediator molecules act as chemical signals that move from the enzymes that produce them into the cytoplasm of the cell, where they activate processes that produce the response of the cell.

Cyclic guanine (gwahn′ēn) **monophosphate (cGMP)** is an intracellular mediator molecule that is synthesized in response to a hormone binding with a membrane-bound receptor (figure 17.18). The hormone binds to its receptor, and the combination activates an enzyme called **guanylate cyclase** (gwahn′i-lāt sī′klās) located at the inner surface of the plasma membrane. The guanylate cyclase enzyme converts guanine triphosphate (GTP) to cGMP and two inorganic phosphate groups. The cGMP molecules then combine with specific enzymes in the cytoplasm of the cell and activate them. The activated enzymes, in turn, produce the cell's response to the hormone. For example, atrial natriuretic hormone is a hormone that combines with its

TABLE 17.6 Common Intracellular Mediators

Intracellular Mediator	Example of Cell Type	Example of Response
Cyclic guanine monophosphate (cGMP)	Kidney cells	Increases Na^+ and water excretion by the kidney
Cyclic adenosine monophosphate (cAMP)	Liver cells	Increases the breakdown of glycogen and the release of glucose into the circulatory system
Calcium ions (Ca^{2+})	Smooth muscle cells	Contraction of smooth muscle cells
Inositol triphosphate (IP_3)	Smooth muscle cells	Contraction of certain smooth muscle cells in response to epinephrine
Diacylglycerol (DAG)	Smooth muscle cells	Contraction of certain smooth muscle cells in response to epinephrine
Nitric oxide (NO)	Smooth muscle cells	Relaxation of smooth muscle cells of blood vessels, resulting in vasodilation

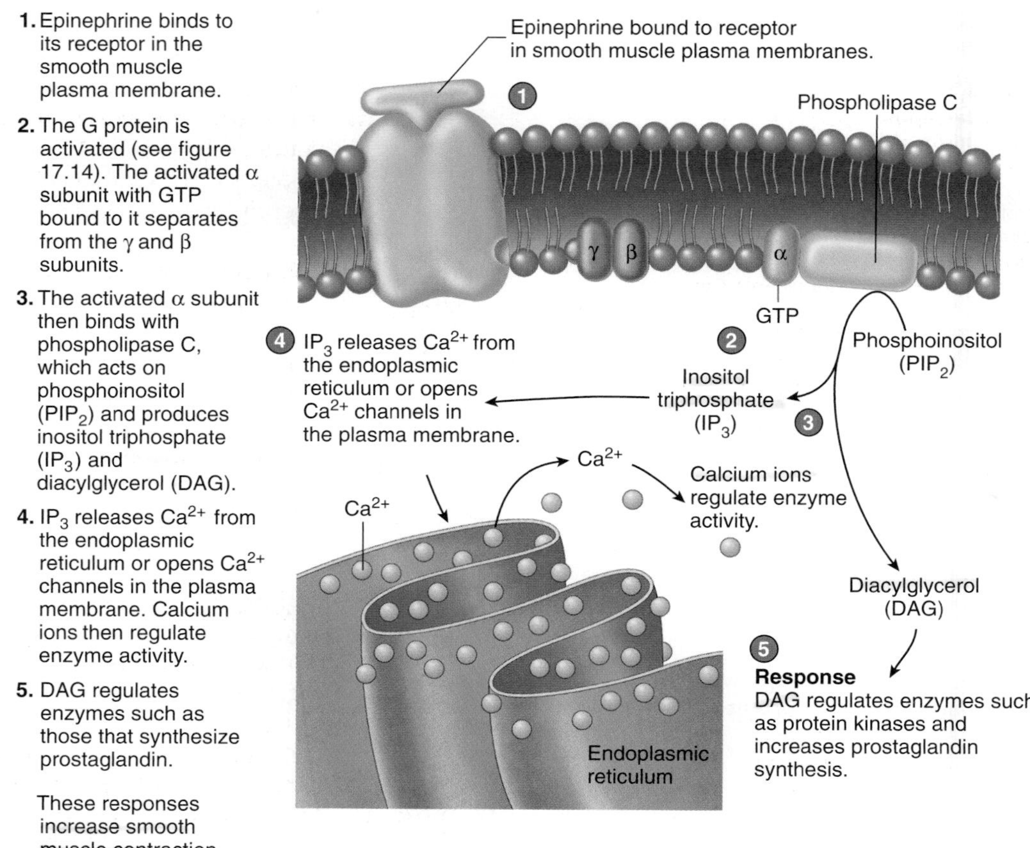

PROCESS FIGURE 17.17 Membrane-Bound Receptors That Activate G Proteins and Increase the Synthesis of IP_3 and DAG
Epinephrine receptors in some smooth muscle cells are associated with G proteins.

1. Atrial natriuretic hormone binds with its receptor.
2. At the inner surface of the plasma membrane, guanylate cyclase is activated to produce cGMP from GTP.
3. Cyclic GMP is an intracellular mediator that mediates the response of the cell.
4. Phosphodiesterase is an enzyme that converts cGMP to inactive GMP.

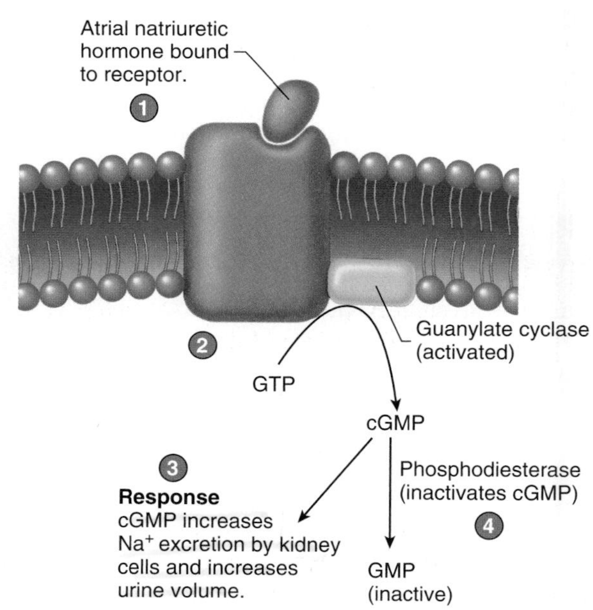

PROCESS FIGURE 17.18 Membrane-Bound Receptor That Directly Activates an Intracellular Mediator

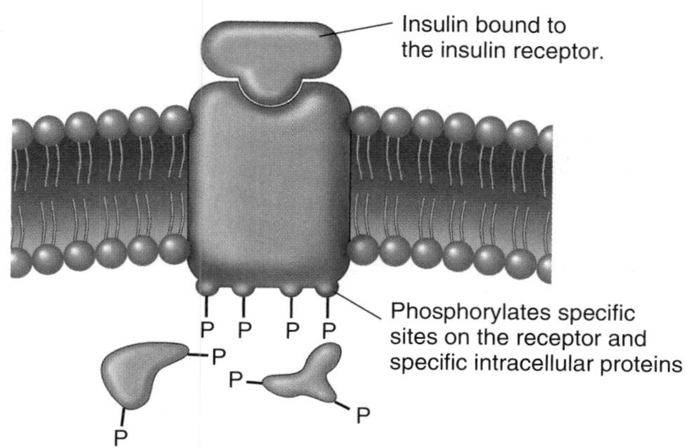

FIGURE 17.19 Membrane-Bound Receptors That Phosphorylate Intracellular Proteins

Insulin receptors are membrane-bound receptors. When insulin binds to the insulin receptor, the receptor acts as an enzyme that phosphorylates specific sites on the receptor and on intracellular proteins. The phosphorylated proteins produce the normal response to insulin.

receptor in the plasma membrane of kidney cells. The result is an increase in the rate of cGMP synthesis at the inner surface of the plasma membranes (see figure 17.18). Cyclic GMP influences the action of enzymes in the kidney cells, which increase the rate of Na^+ and water excretion by the kidney (see chapter 26). The amount of time the cGMP is present to produce a response in the cell is limited. Phosphodiesterase breaks down cGMP to GMP. Consequently, the length of time a hormone increases cGMP synthesis and has an effect on a cell is brief after the hormone is no longer present.

Some hormones bind to membrane-bound receptors, and the portion of the receptor on the inner surface of the plasma membrane acts as a phosphorylase enzyme that phosphorylates several specific proteins. Some of the phosphorylated proteins are part of the membrane-bound receptor, and others are in the cytoplasm of the cell (figure 17.19). The phosphorylated proteins influence the activity of other enzymes in the cytoplasm of the cell. For example, insulin binds to its membrane-bound receptor, resulting in the phosphorylation of parts of the receptor on the inner surface of the plasma membrane and the phosphorylation of certain other intracellular proteins. The phosphorylated proteins produce the cells' responses to insulin. Some receptors for hormones that phosphorylate intracellular proteins are listed in table 17.7.

Hormones that stimulate the synthesis of an intracellular mediator molecule often produce rapid responses. This is possible because the mediator influences already existing enzymes and causes a **cascade effect,** which results when a few mediator molecules activate several enzymes, and each of the activated enzymes in turn activates several other enzymes that produce the final response. Thus, an amplification system exists in which a few molecules, such as cAMP, cGMP, or phosphorylated proteins, can control the activity of many enzymes within a cell (figure 17.20)

18 ***Explain how the combination of a hormone and its receptor can alter the G proteins on the inner surface of the plasma membrane. Which activated subunit of the G protein alters the activity of molecules inside the plasma membrane or inside the cell?***

19 ***Describe how G proteins can alter the permeability of the plasma membrane and how they can alter the synthesis of an intracellular mediator molecule, such as cAMP. Give examples.***

20 ***Other than cAMP and Ca^{2+}, list two additional intracellular mediators affected by G proteins.***

21 ***Describe how a hormone can combine with a membrane-bound receptor, change enzyme activity inside the cell, and increase phosphorylation of intracellular proteins. Give examples.***

22 ***What limits the activity of intracellular mediator molecules, such as cAMP, and phosphorylated proteins?***

23 ***Explain the cascade effect for the intracellular mediator model of hormone action. Does the intracellular mediator mechanism produce a slow or rapid response?***

TABLE 17.7 Hormones That Bind to Receptors Linked to Intracellular Enzymes That Phosphorylate Intracellular Proteins

Hormone	Source	Target Tissue and Effect
Insulin	Pancreatic islets	Most cells; increases glucose and amino acid uptake
Growth hormone	Anterior pituitary gland	Most cells; increases protein synthesis and resists protein breakdown
Prolactin	Anterior pituitary gland	Mammary glands and ovary; initiates milk production following pregnancy
Growth factors	Various tissues	Stimulate growth in certain cell types
Some intercellular immune signal molecules	Cells of the immune system	Immune-competent cells; help mediate responses of the immune system
Atrial nutriuretic hormone	Cells of kidney tubules	Increases Na^+ excretion by kidney tubule cells and increases urine volume

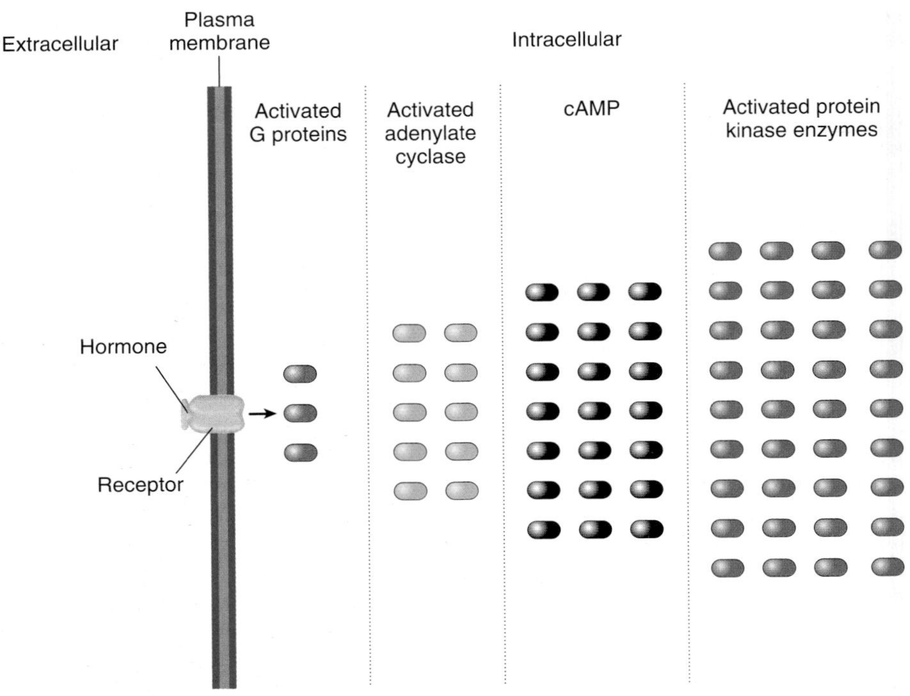

FIGURE 17.20 Cascade Effect

The combination of a hormone with a membrane-bound receptor activates several G proteins. The G proteins, in turn, activate many inactive adenylate cyclase enzymes, which cause the synthesis of a large number of cAMP molecules. The large number of cAMP molecules, in turn, activate many inactive protein kinase enzymes, which produce a rapid and amplified response.

PREDICT 5

When smooth muscle cells in the airways of the lungs contract, as in asthma, breathing becomes very difficult, whereas breathing is easy if the smooth muscle cells are relaxed. During asthma attacks, the smooth muscle cells in the airways of the lungs contract. Some of the drugs used to treat asthma increase cAMP in smooth muscle cells. Explain as many ways as possible how these drugs might work.

Intracellular Receptors

Intracellular receptors are either in the cytoplasm or in the nucleus of cells. Lipid-soluble hormones cross the plasma membrane into the cytoplasm or into the nucleus and bind to intracellular receptors by the process of diffusion (figure 17.21). After a hormone binds with an intracellular receptor, the receptor can alter the activity of enzymes in the cell, or it can bind to DNA to produce a response (see table 17.4). Some intracellular receptors that influence the expression of DNA are located in the cytoplasm. Once a hormone binds to its receptor, the receptor and hormone diffuse into the nucleus and bind to DNA. Other intracellular receptors are located in the nucleus. A hormone diffuses into the nucleus and binds to its receptor, and the receptor then binds to DNA.

Receptors that interact with DNA have specific "fingerlike" projections that interact with specific parts of a DNA molecule. The combination of the chemical signal and its receptor with DNA increases the synthesis of specific **messenger ribonucleic acid (mRNA)** molecules. The mRNA molecules then move to the cytoplasm and increase the synthesis of specific proteins at the ribosomes. The newly synthesized proteins produce the cell response to the hormone. For example, testosterone from the testes and estrogen from the ovaries stimulate the synthesis of proteins that are responsible for the secondary sex characteristics of males and females. The effect of the steroid aldosterone on its target cells in the kidney is to stimulate the synthesis of proteins that increase the rate of Na^+ and K^+ transport. The result is an increase in the reabsorption of Na^+ from, and increased secretion of K^+ into, the filtrate in the kidney, and a reduction in the amount of Na^+ and an increase in K^+ lost in the urine. Other hormones that produce responses through intracellular receptor mechanisms include T_3 and T_4 from the thyroid follicles and vitamin D, which has some hormonelike characteristics (table 17.8).

Cells that synthesize new protein molecules in response to hormonal stimuli normally have a latent period of several hours between the time the hormones bind to their receptors and the time responses are observed. During this latent period, mRNA and new proteins are synthesized. Receptor–hormone complexes normally are degraded within the cell, limiting the length of time hormones influence the cells' activities, and the cells slowly return to their previous functional states.

1. Aldosterone is a lipid-soluble hormone and can easily diffuse through the plasma membrane.
2. Once inside the cell, aldosterone binds with an aldosterone receptor in the cytoplasm.
3. The aldosterone–receptor complex moves into the nucleus and binds to DNA.
4. The binding of the aldosterone–receptor complex to DNA stimulates the synthesis of messenger RNA (mRNA), which codes for specific proteins.
5. The mRNA leaves the nucleus, passes into the cytoplasm of the cell, and binds to ribosomes, where it directs the synthesis of the specific proteins.
6. The proteins synthesized on the ribosomes produce the cell's response to aldosterone. For example, aldosterone increases the synthesis of proteins that transport Na^+ and K^+ across epithelial plasma membranes in the kidney (see chapter 26).

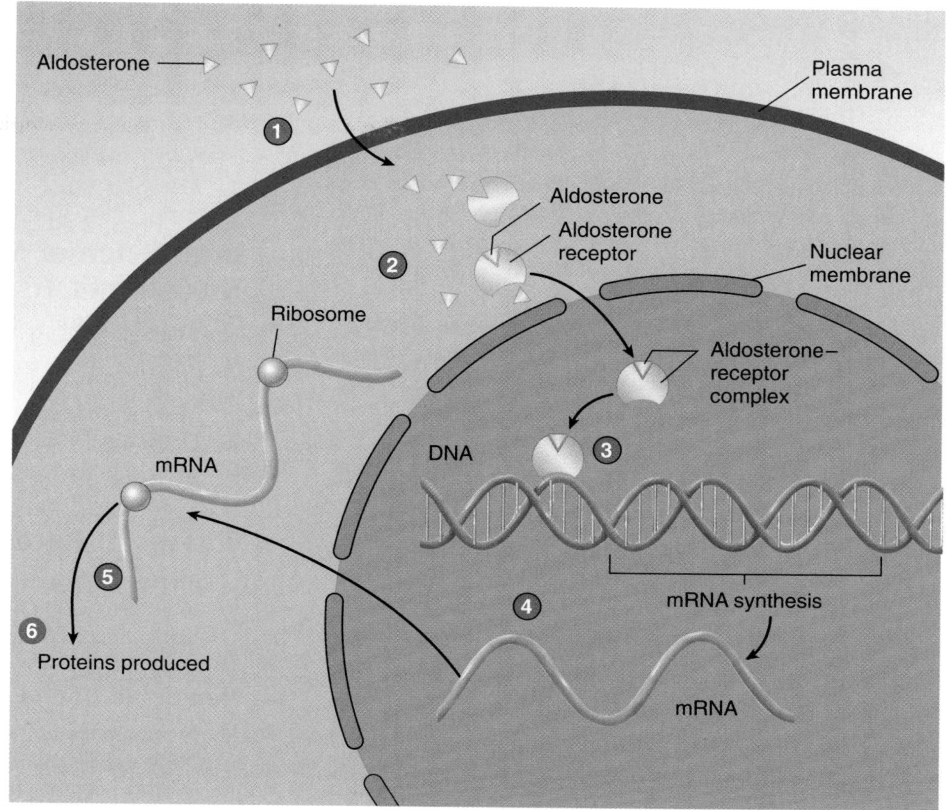

PROCESS FIGURE 17.21 Intracellular Receptor Model

24 ***Describe how a hormone that crosses the plasma membrane interacts with its receptor and how it alters the rate of protein synthesis. Why is there normally a latent period between the time hormones bind to their receptors and the time responses are observed?***

25 ***What finally limits the processes activated by the intracellular receptor mechanism?***

PREDICT 6

Of membrane-bound receptors and intracellular receptors, which is better adapted for mediating a response that lasts a considerable length of time and which is better for mediating a response with a rapid onset and a short duration? Explain why.

TABLE 17.8 Major Hormones That Combine with Intracellular Receptors

Category of Hormone	Hormone	Source	Target Tissue and Effect
Sex steroids	Testosterone	Testis	Development of the reproductive structures and development of male secondary sex characteristics
	Progesterone	Ovary	Increased size of cells lining the uterus
	Estrogen	Ovary	Increased cell division in the lining of the uterus
Mineralocorticoid steroids	Aldosterone	Adrenal cortex	Increased reabsorption of Na^+ and increased secretion of K^+ in the kidney
Glucocorticoid steroid hormones	Cortisol	Adrenal cortex	Increased breakdown of proteins and fats and increased blood levels of glucose
Thyroid hormones	Triiodothyronine (T_3)	Thyroid gland	Regulation of development and metabolism
Vitamin D*	1,25-dihydroxycholecalciferol	Combination of the skin, liver, and kidney	Increased reabsorption of Ca^{2+} in the kidney and absorption of Ca^{2+} in the gastrointestinal tract

*Although vitamin D is not a hormone, it has hormonelike characteristics.

SUMMARY

General Characteristics of the Endocrine System (p. 586)

1. Endocrine glands produce hormones that are released into the interstitial fluid and diffuse into the blood. Hormones act on target tissues, which produce specific responses.
2. Endocrine glands produce other chemical signals, including neurohormones, neurotransmitters, neuromodulators, parahormones, and pheromones.
3. Generalizations about the differences between the endocrine and nervous systems include the following: (a) the endocrine system is amplitude-modulated, whereas the nervous system is frequency-modulated, and (b) the response of target tissues to hormones is usually slower and of longer duration than that to neurons.

Chemical Structure of Hormones (p. 587)

Hormones are proteins, glycoproteins, peptides, derivatives of amino acids, or lipids (steroids or derivatives of fatty acids).

Control of Secretion Rate (p. 587)

1. Most hormones are not secreted at a constant rate.
2. Negative-feedback mechanisms, which maintain homeostasis, control the secretion of most hormones.
3. Hormone secretion from an endocrine tissue is regulated by one or more of three mechanisms: changes in the extracellular concentration of a nonhormone substance, stimulation by the nervous system, or stimulation by a hormone from another endocrine tissue.

Transport and Distribution in the Body (p. 593)

Hormones are dissolved in plasma or bind to plasma proteins. The blood quickly distributes hormones throughout the body.

Metabolism and Excretion (p. 594)

1. Nonpolar, readily diffusible hormones bind to plasma proteins and have an increased half-life.
2. Water-soluble hormones, such as proteins, epinephrine, and norepinephrine, do not bind to plasma proteins or readily diffuse out of the blood. Instead, they are broken down by enzymes or are taken up by tissues. They have a short half-life.
3. Hormones with a short half-life regulate activities that have a rapid onset and a short duration.
4. Hormones with a long half-life regulate activities that remain at a constant rate through time.
5. Hormones are eliminated from the blood by excretion from the kidney and liver, enzymatic degradation, conjugation, or active transport.

Interaction of Hormones with Their Target Tissues (p. 595)

1. Target tissues have receptor molecules that are specific for a particular hormone.
2. Hormones bound with receptors affect the rate at which already existing processes occur.
3. Down-regulation is a decrease in the number of receptors in a target tissue, and up-regulation is an increase in the number of receptors.

Classes of Receptors (p. 597)

1. Membrane-bound receptors bind to water-soluble or large-molecular-weight hormones.
2. Intracellular receptors bind to lipid-soluble hormones.

Membrane-Bound Receptors

1. Membrane-bound receptors are proteins or glycoproteins that have polypeptide chains that are folded to cross the cell several times.
2. When a hormone binds to a membrane-bound receptor,
   - G proteins are activated. The α subunit of the G protein can bind to ion channels and cause them to open or change the rate of synthesis of intracellular mediator molecules, such as cAMP, cGMP, IP3, and DAG.
   - Intracellular enzymes can be directly activated, which in turn synthesizes intracellular mediators, such as cGMP, or adds a phosphate group to intracellular enzymes, which alters their activity.
3. Intracellular mediator mechanisms are rapid-acting because they act on already existing enzymes and produce a cascade effect.

Intracellular Receptors

1. Intracellular receptors are proteins in the cell cytoplasm or nucleus.
2. Hormones bind with the intracellular receptor, and the receptor–hormone complex activates genes. Consequently, DNA is activated to produce mRNA. The mRNA initiates the production of certain proteins (enzymes) that produce the response of the target cell to the hormone.
3. Intracellular receptor mechanisms are slow-acting because time is required to produce the mRNA and the protein.
4. Intracellular receptor–activated processes are limited by the breakdown of the receptor–hormone complex.

REVIEW AND COMPREHENSION

1. When comparing the endocrine system and the nervous system, generally speaking, the endocrine system
   a. is faster-acting than the nervous system.
   b. produces effects that are of shorter duration.
   c. uses amplitude-modulated signals.
   d. produces more localized effects.
   e. relies less on chemical signals.
2. A chemical signal released from a cell that has a local effect on the same cell type from which the chemical signal is released is a(n)
   a. paracrine chemical signal.
   b. pheromone.
   c. autocrine chemical signal.
   d. hormone.
   e. intracellular mediator.

3. Given this list of molecule types:
   1. nucleic acid derivatives
   2. fatty acid derivatives
   3. polypeptides
   4. proteins
   5. phospholipids

   Which could be hormone molecules?
   a. 1,2,3 b. 2,3,4 c. 1,2,3,4 d. 2,3,4,5 e. 1,2,3,4,5
4. Which of these regulates the secretion of a hormone from an endocrine tissue?
   a. other hormones
   b. negative-feedback mechanisms
   c. nonhormone substance in the blood
   d. the nervous system
   e. all of the above
5. Hormones are released into the blood
   a. at relatively constant levels.
   b. in large amounts in response to a stimulus.
   c. in a cyclic fashion.
   d. all of the above.
6. Lipid-soluble hormones readily diffuse through capillary walls, whereas water-soluble hormones, such as proteins, must
   a. pass through capillary cells.
   b. pass through pores in the capillary endothelium.
   c. be moved out of the capillary by active transport.
   d. remain in the blood.
   e. be broken down to amino acids before leaving the blood.
7. Concerning the half-life of hormones,
   a. lipid-soluble hormones generally have a longer half-life.
   b. hormones with shorter half-lives regulate activities with a slow onset and long duration.
   c. hormones with a shorter half-life are maintained at more constant levels in the blood.
   d. lipid-soluble hormones are degraded rapidly by enzymes in the circulatory system.
   e. water-soluble hormones usually combine with plasma proteins.
8. Given these observations:
   1. A hormone affects only a specific tissue (not all tissues).
   2. A tissue can respond to more than one hormone.
   3. Some tissues respond rapidly to a hormone, whereas others take many hours to respond.

   Which of these observations can be explained by the characteristics of hormone receptors?
   a. 1 b. 1,2 c. 2,3 d. 1,3 e. 1,2,3
9. Which of these is *not* a means by which hormones are eliminated from the circulatory system?
   a. excreted into urine or bile
   b. bound to plasma proteins
   c. metabolism (enzymatically degraded in the blood)
   d. actively transported into cells
   e. conjugated with sulfate or glucuronic acid
10. Down-regulation
    a. produces a decrease in the number of receptors in the target cells.
    b. produces an increase in the sensitivity of the target cells to a hormone.
    c. is found in target cells that respond to hormones that are maintained at constant levels.
    d. occurs partly because of an increase in receptor synthesis by the target cell.
    e. all of the above.
11. A hormone
    a. can function as an enzyme.
    b. is also a G protein.
    c. can bind to a receptor.
    d. is an intracellular mediator.
    e. all of the above.
12. Activated G proteins can
    a. cause ion channels to open or close.
    b. activate adenylate cyclase.
    c. inhibit the synthesis of cAMP.
    d. alter the activity of IP3.
    e. all of the above.
13. Given these events:
    1. GTP is converted to GDP.
    2. The α subunit separates from the β and γ units.
    3. GDP is released from the α subunit.

    List the order in which the events occur after a hormone binds to a membrane-bound receptor.
    a. 1,2,3 b. 1,3,2 c. 2,3,1 d. 3,2,1 e. 3,1,2
14. Which of these can limit the response of a cell to a hormone?
    a. phosphodiesterase
    b. converting GTP to GDP
    c. decreasing the number of receptors
    d. blocking binding sites
    e. all of the above
15. Given these events:
    1. The α subunit of a G protein interacts with Ca^{2+} channels.
    2. Calcium ions diffuse into the cell.
    3. The α subunit of a G protein is activated.

    Choose the arrangement that lists the events in the order they occur after a hormone combines with a receptor on a smooth muscle cell.
    a. 1,2,3 b. 1,3,2 c. 2,1,3 d. 3,1,2 e. 3,2,1
16. Given these events:
    1. cAMP is synthesized.
    2. The α subunit of G protein is activated.
    3. Phosphodiesterase breaks down cAMP.

    Choose the arrangement that lists the events in the order they occur after a hormone binds to a receptor.
    a. 1,2,3 b. 1,3,2 c. 2,1,3 d. 2,3,1 e. 3,2,1
17. Which of these events can occur after a G protein activates phospholipase C?
    a. DAG and IP3 are synthesized from PIP2.
    b. IP_3 causes Ca^{2+} channels to open.
    c. DAG activates enzymes that synthesize prostaglandins.
    d. All of the above are true.

18. When a hormone binds to an intracellular receptor,
    a. DNA produces mRNA.
    b. G proteins are activated.
    c. the receptor–hormone complex causes ion channels to open or close.
    d. the cell's response is faster than when a hormone binds to a membrane-bound receptor.
    e. the hormone is usually a large, water-soluble molecule.

19. Given these events:
    1. activation of cAMP
    2. activation of genes
    3. enzyme activity altered

    Which of these events can occur when a hormone binds to an intracellular hormone receptor?
    a. 1
    b. 1,2
    c. 2,3
    d. 1,2,3

Answers in Appendix E

CRITICAL THINKING

1. Consider a hormone that is secreted in large amounts at a given interval, modified chemically by the liver, and excreted by the kidney at a rapid rate, thus making the half-life of the hormone in the circulatory system very short. The hormone therefore rapidly increases in the blood and then decreases rapidly. Predict the consequences of liver and kidney disease on the blood levels of that hormone.
2. Consider a hormone that increases the concentration of a substance in the circulatory system. If a tumor begins to produce that substance in large amounts in an uncontrolled fashion, predict the effect on the secretion rate for the hormone.
3. How could you determine whether or not a hormone-mediated response resulted from the intracellular mediator mechanism or the intracellular receptor mechanism?
4. If the effect of a hormone on a target tissue is through a membrane-bound receptor that has a G protein associated with it, predict the consequences if a genetic disease causes the α subunit of the G protein to have a structure that prevents it from binding to GTP.
5. Prostaglandins are a group of hormones produced by many cells of the body. Unlike other hormones, they do not circulate but usually have their effect at or very near their site of production. Prostaglandins apparently affect many body functions, including blood pressure, inflammation, induction of labor, vomiting, fever, and inhibition of the clotting process. Prostaglandins also influence the formation of cAMP. Explain how an inhibitor of prostaglandin synthesis could be used as a therapeutic agent. Inhibitors of prostaglandin synthesis can produce side effects. Why?
6. For a hormone that binds to a membrane-bound receptor and has cAMP as the intracellular mediator, predict and explain the consequences if a drug is taken that strongly inhibits phosphodiesterase.
7. When an individual is confronted with a potentially harmful or dangerous situation, epinephrine (adrenaline) is released from the adrenal gland. Epinephrine prepares the body for action by increasing the heart rate and blood glucose levels. Explain the advantages or disadvantages associated with a short half-life for epinephrine and those associated with a long half-life.
8. Thyroid hormones are important in regulating the basal metabolic rate of the body. What are the advantages or disadvantages of the following?
    a. a long half-life for thyroid hormones
    b. a short half-life
9. An increase in thyroid hormones causes an increase in metabolic rate. If liver disease results in reduced production of the plasma proteins to which thyroid hormones normally bind, what is the effect on metabolic rate? Explain.
10. Predict the effect on LH and FSH secretion if a small tumor in the hypothalamus of the brain secretes large concentrations of GnRH continuously. Given that LH and FSH regulate the function of the male and female reproductive systems, predict whether the condition increases or decreases the activity of these systems.
11. Predict some consequences of trying to use a skin patch to administer insulin to a person who has diabetes mellitus.

Answers in Appendix F

Visit this book's website at www.mhhe.com/seeley8 for chapter quizzes, interactive learning exercises, and other study tools.

Endocrine Glands 18

Homeostasis depends on the precise regulation of the organs and organ systems of the body. Together, the nervous and endocrine systems regulate and coordinate the activity of nearly all other body structures. When either the nervous or the endocrine system fails to function properly, conditions can rapidly deviate from homeostasis.

Disorders of the endocrine system can result in diseases such as insulin-dependent diabetes and Addison disease. Early in the 1900s, people who developed those diseases died. No effective treatments were available for those and other diseases of the endocrine system, such as diabetes insipidus, Cushing syndrome, and many reproductive abnormalities. Advances have been made in understanding the endocrine system, so the outlook for people with these and other endocrine diseases has improved.

The endocrine system is small, compared with its importance to healthy body functions. It consists of several small glands distributed throughout the body that could escape notice if not for the importance of the small amounts of hormones they secrete.

This chapter first explains the *functions of the endocrine system* (p. 610) and then profiles the *pituitary gland and hypothalamus* (p. 610), *thyroid gland* (p. 619), *parathyroid glands* (p. 624), *adrenal glands* (p. 627), and *pancreas* (p. 632). It then moves to discussions about *hormonal regulation of nutrients* (p. 638), *hormones of the reproductive system* (p. 640), *hormones of the pineal body* (p. 641), *hormones of the thymus* (p. 642), and *hormones of the gastrointestinal tract* (p. 642), and *hormonelike substances* (p. 642). The chapter concludes with a look at the *effects of aging on the endocrine system* (p. 643).

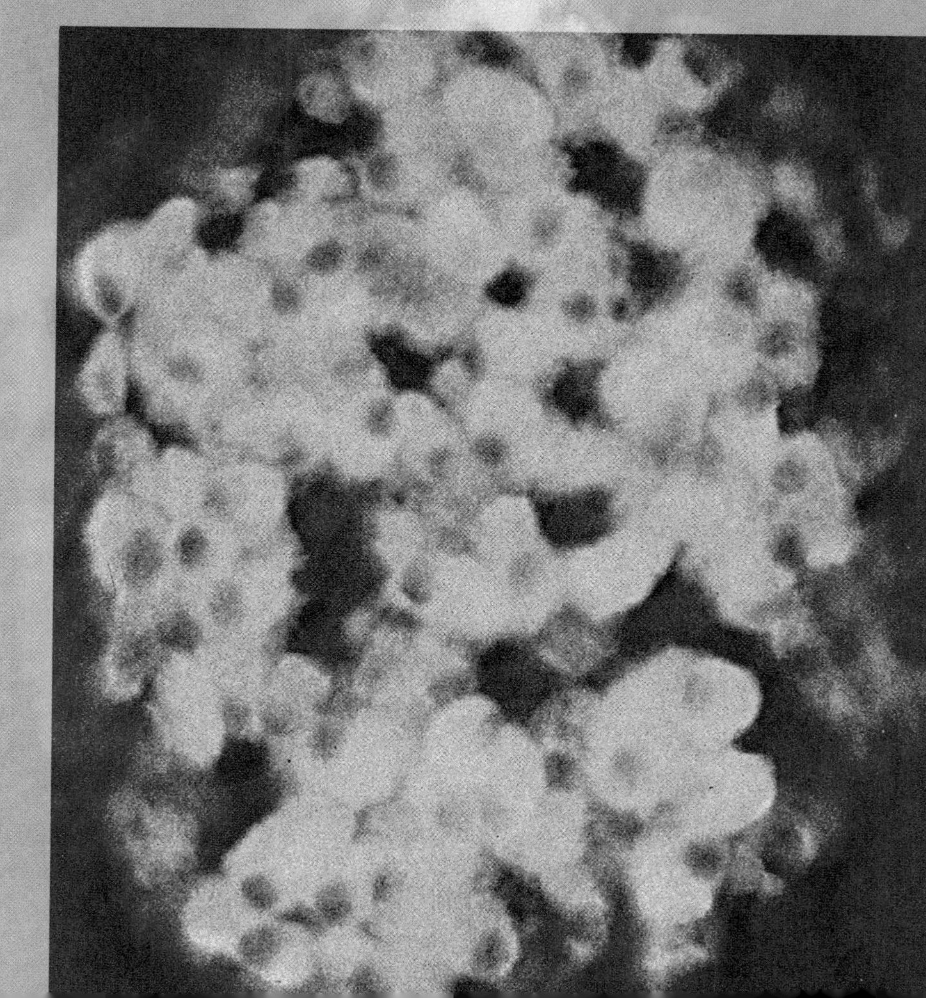

Light micrograph of a pancreatic islet showing insulin-secreting beta cells (*green*) and the glucagon-secreting cells (red).

Endocrine System

FUNCTIONS OF THE ENDOCRINE SYSTEM

* ductless.

Several pieces of information are needed to understand how the endocrine system regulates body functions:

1. the anatomy of each gland and its location
2. the hormone secreted by each gland
3. the target tissues and the response of target tissues to each hormone
4. the means by which the secretion of each hormone is regulated
5. the consequences and causes, if known, of hypersecretion and hyposecretion of the hormone

The main regulatory functions of the endocrine system are the following:

1. *Metabolism and tissue maturation.* The endocrine system regulates the rate of metabolism and influences the maturation of tissues, such as those of the nervous system.
2. *Ion regulation.* The endocrine system helps regulate blood pH, as well as Na^+, K^+, and Ca^{2+} concentrations in the blood.
3. *Water balance.* The endocrine system regulates water balance by controlling the solute concentration of the blood.
4. *Immune system regulation.* The endocrine system helps control the production of immune cells.
5. *Heart rate and blood pressure regulation.* The endocrine system helps regulate the heart rate and blood pressure and helps prepare the body for physical activity.
6. *Control of blood glucose and other nutrients.* The endocrine system regulates blood glucose levels and other nutrient levels in the blood.
7. *Control of reproductive functions.* The endocrine system controls the development and functions of the reproductive systems in males and females.
8. *Uterine contractions and milk release.* The endocrine system regulates uterine contractions during delivery and stimulates milk release from the breasts in lactating females.

1 ***What pieces of information are needed to understand how the endocrine system regulates body functions?***

2 ***List eight regulatory functions of the endocrine system.***

PITUITARY GLAND AND HYPOTHALAMUS

The **pituitary** (pi-too′i-tār-rē) **gland,** or **hypophysis** (hī-pof′i-sis; an undergrowth), secretes nine major hormones that regulate numerous body functions and the secretory activity of several other endocrine glands.

The **hypothalamus** (hī′pō-thal′ă-mŭs) of the brain and the pituitary gland are major sites where the nervous and endocrine systems interact (figure 18.1). The hypothalamus regulates the secretory activity of the pituitary gland. Indeed, the posterior pituitary is an extension of the hypothalamus. Hormones, sensory information that enters the central nervous system, and emotions, in turn, influence the activity of the hypothalamus.

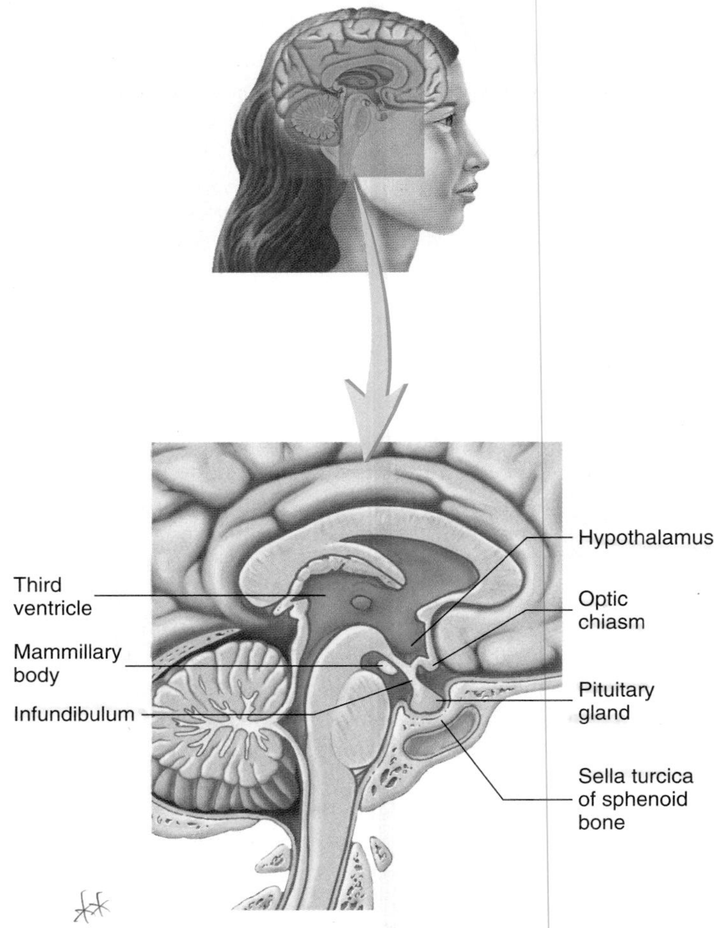

FIGURE 18.1 Hypothalamus and Pituitary Gland

A midsagittal section of the head through the pituitary gland showing the location of the hypothalamus of the brain and the pituitary gland. The pituitary gland is in a depression called the sella turcica in the floor of the skull. It is connected to the hypothalamus by the infundibulum.

Structure of the Pituitary Gland

The pituitary gland is roughly 1 cm in diameter, weighs 0.5–1.0 g, and rests in the sella turcica of the sphenoid bone (see figure 18.1). It is located inferior to the hypothalamus and is connected to it by a stalk of tissue called the **infundibulum** (in-fŭn-dib′ū-lŭm).

The pituitary gland is divided functionally into two parts: the **posterior pituitary,** or **neurohypophysis** (noor′ō-hī-pof′i-sis), and the **anterior pituitary,** or **adenohypophysis** (ad′ĕ-nō-hī-pof′i-sis).

Posterior Pituitary, or Neurohypophysis

The posterior pituitary is called the neurohypophysis because it is continuous with the brain (*neuro-* refers to the nervous system). It is formed during embryonic development from an outgrowth of the inferior part of the brain in the area of the hypothalamus (see chapter 29). The outgrowth of the brain forms the infundibulum, and the distal end of the infundibulum enlarges to form the posterior pituitary (figure 18.2). Secretions of the posterior pituitary are **neurohormones** (noor-ōhōr′mōnz) because the posterior pituitary is an extension of the nervous system.

Anterior Pituitary, or Adenohypophysis

The anterior pituitary, or adenohypophysis (*adeno-* means gland), arises as an outpocketing of the roof of the embryonic oral cavity called the pituitary diverticulum, or Rathke's pouch, which grows toward the posterior pituitary. As it nears the posterior pituitary, the pituitary diverticulum loses its connection with the oral cavity and becomes the anterior pituitary. The anterior pituitary is subdivided into three areas with indistinct boundaries: the pars tuberalis, which extends superiorly and partially wraps around the infundibulum; the pars distalis, which is the enlarged distal portion of the anterior pituitary; and the pars intermedia, which is adjacent to the posterior pituitary (see figure 18.2). The hormones secreted from the anterior pituitary, in contrast to those from the posterior pituitary, are not neurohormones because the anterior pituitary is derived from epithelial tissue of the embryonic oral cavity, not from neural tissue.

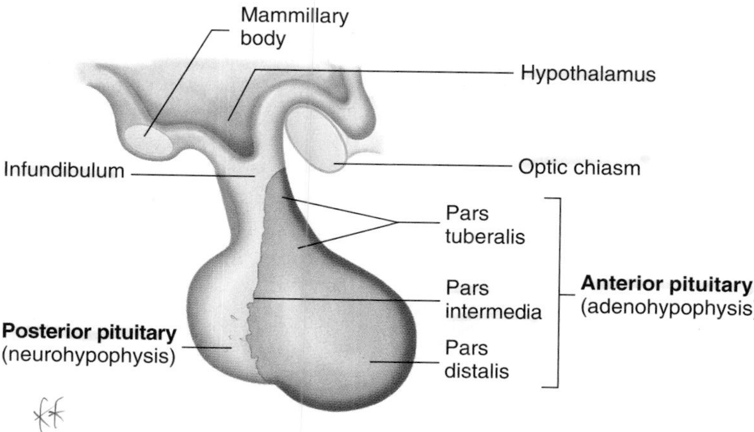

FIGURE 18.2 Subdivisions of the Pituitary Gland

The pituitary gland is divided into the anterior pituitary, or adenohypophysis, and the posterior pituitary, or neurohypophysis. The anterior pituitary is subdivided further into the pars distalis, pars intermedia, and pars tuberalis. The posterior pituitary consists of the enlarged distal end of the infundibulum, which connects the posterior pituitary to the hypothalamus.

Relationship of the Pituitary to the Brain

Portal vessels are blood vessels that begin in a primary capillary network, extend some distance, and end in a secondary capillary network. The **hypothalamohypophysial** (hī′pō-thal′ă-mō-hī′pō-fiz′ē-ăl) **portal system** is one of two major portal systems. The other is the hepatic portal system (see chapter 21). The hypothalamohypophysial portal system extends from a part of the hypothalamus to the anterior pituitary (figure 18.3). The primary capillary network in the hypothalamus is supplied with blood from

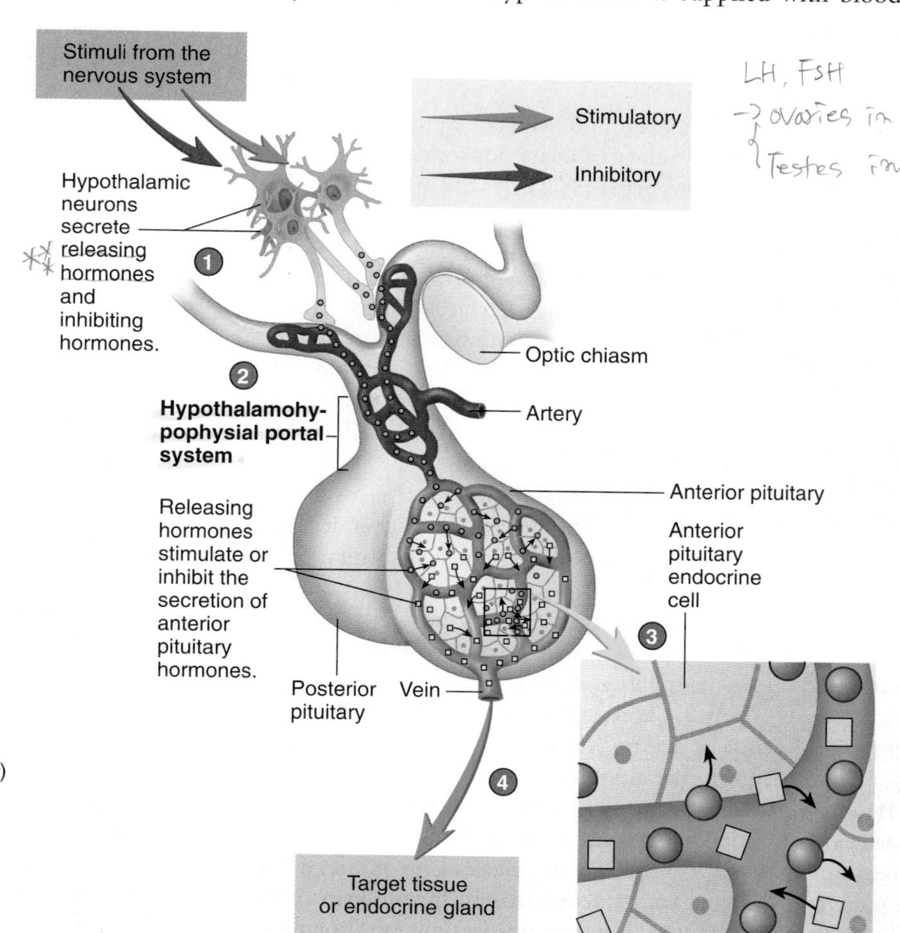

1. Stimuli within the nervous system increase or decrease the secretion of releasing hormones and inhibiting hormones (*blue balls*) from neurons of the hypothalamus.
2. Releasing hormones and inhibiting hormones pass through the hypothalamohypophysial portal system to the anterior pituitary.
3. Releasing hormones and inhibiting hormones leave capillaries, bind to membrane-bound receptors, and stimulate or inhibit the release of hormones (*yellow squares*) from anterior pituitary cells.
4. Anterior pituitary hormones (*yellow squares*) are carried in the blood to their target tissues (*green arrow*), which, in some cases, are other endocrine glands.

PROCESS FIGURE 18.3 Relationships Among the Hypothalamus, Anterior Pituitary, and Target Tissues

TABLE 18.1 Hormones of the Hypothalamus

Hormones	Structure	Target Tissue	Response
Growth hormone–releasing hormone (GHRH)	Peptide	Anterior pituitary cells that secrete growth hormone	Increased growth hormone secretion
Growth hormone–inhibiting hormone (GHIH), or somatostatin	Small peptide	Anterior pituitary cells that secrete growth hormone	Decreased growth hormone secretion
Thyrotropine-releasing hormone (TRH)	Small peptide	Anterior-pituitary cells that secrete thyroid-stimulating hormone	Increased thyroid-stimulating hormone secretion
Corticotropin-releasing hormone (CRH)	Peptide	Anterior pituitary cells that secrete adrenocorticotropic hormone	Increased adrenocorticotropic hormone secretion
Gonadotropin-releasing hormone (GnRH)	Small peptide	Anterior pituitary cells that secrete luteinizing hormone and follicle-stimulating hormone	Increased secretion of luteinizing hormone and follicle-stimulating hormone
Prolactin-inhibiting hormone (PIH)	Unknown (possibly dopamine)	Anterior pituitary cells that secrete prolactin	Decreased prolactin secretion
Prolactin-releasing hormone (PRH)	Unknown	Anterior pituitary cells that secrete prolactin	Increased prolactin secretion

arteries that deliver blood to the hypothalamus. From the primary capillary network, the hypothalamohypophysial portal vessels carry blood to a secondary capillary network in the anterior pituitary. Veins from the secondary capillary network eventually merge with the general circulation.

Neurohormones, produced and secreted by neurons of the hypothalamus, enter the primary capillary network and are carried to the secondary capillary network. There the neurohormones leave the blood and act on cells of the anterior pituitary. They act either as **releasing hormones,** increasing the secretion of anterior pituitary hormones, or as **inhibiting hormones,** decreasing the secretion of anterior pituitary hormones. Each releasing hormone stimulates and each inhibiting hormone inhibits the production and secretion of a specific hormone by the anterior pituitary. In response to the releasing hormones, anterior pituitary cells secrete hormones that enter the secondary capillary network and are carried by the general circulation to their target tissues. Thus, the hypothalamohypophysial portal system provides a means by which the hypothalamus, using neurohormones as chemical signals, regulates the secretory activity of the anterior pituitary (see figure 18.3).

Several major releasing and inhibiting hormones are released from hypothalamic neurons. **Growth hormone–releasing hormone (GHRH)** is a small peptide that stimulates the secretion of growth hormone from the anterior pituitary gland, and **growth hormone–inhibiting hormone (GHIH),** also called **somatostatin,** is a small peptide that inhibits growth hormone secretion. **Thyrotropin-releasing hormone (TRH)** is a small peptide that stimulates the secretion of thyroid-stimulating hormone from the anterior pituitary gland. **Corticotropin-releasing hormone (CRH)** is a peptide that stimulates adrenocorticotropic hormone from the anterior pituitary gland. **Gonadotropin-releasing hormone (GnRH)** is a small peptide that stimulates luteinizing hormone and follicle-stimulating hormone from the anterior pituitary gland. **Prolactin-releasing hormone (PRH)** and **prolactin-inhibiting hormone (PIH)** regulate the secretion of prolactin from the anterior pituitary gland (table 18.1). Secretions of the anterior pituitary gland are described in the section "Posterior Pituitary Hormones" (p. 613).

There is no portal system to carry hypothalamic neurohormones to the posterior pituitary. Neurohormones released from the posterior pituitary are produced by neurosecretory cells with their cell bodies located in the hypothalamus. The axons of these cells extend from the hypothalamus through the infundibulum into the posterior pituitary and form a tract called the **hypothalamohypophysial tract** (figure 18.4). Neurohormones produced in the hypothalamus pass down these axons in tiny vesicles and are stored in secretory vesicles in the enlarged ends of the axons. Action potentials originating in the neuron cell bodies in the hypothalamus are propagated along the axons to the axon terminals in the posterior pituitary. The action potentials cause the release of neurohormones from the axon terminals, and they enter the circulatory system. Secretions of the posterior pituitary gland are described in the section "Posterior Pituitary Hormones" (p. 613).

3 ***Where is the pituitary gland located? Contrast the embryonic origin of the anterior pituitary and the posterior pituitary.***

4 ***Name the parts of the pituitary gland and the function of each part.***

5 ***Define portal system. Describe the hypothalamohypo-physial portal system. How does the hypothalamus regulate the secretion of the anterior pituitary hormones?***

1. Stimuli within the nervous system stimulate hypothalamic neurons to either increase or decrease their action potential frequency.
2. Action potentials are carried by axons of the hypothalamic neurons through the hypothalamohypophysial tract to the posterior pituitary. The axons of neurons store hormones in the posterior pituitary.
3. In the posterior pituitary gland, action potentials cause the release of neurohormones (*red balls*) from axon terminals into the circulatory system.
4. The neurohormones pass through the circulatory system and influence the activity of their target tissues (*green arrow*).

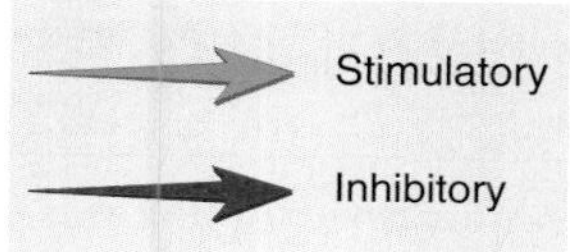

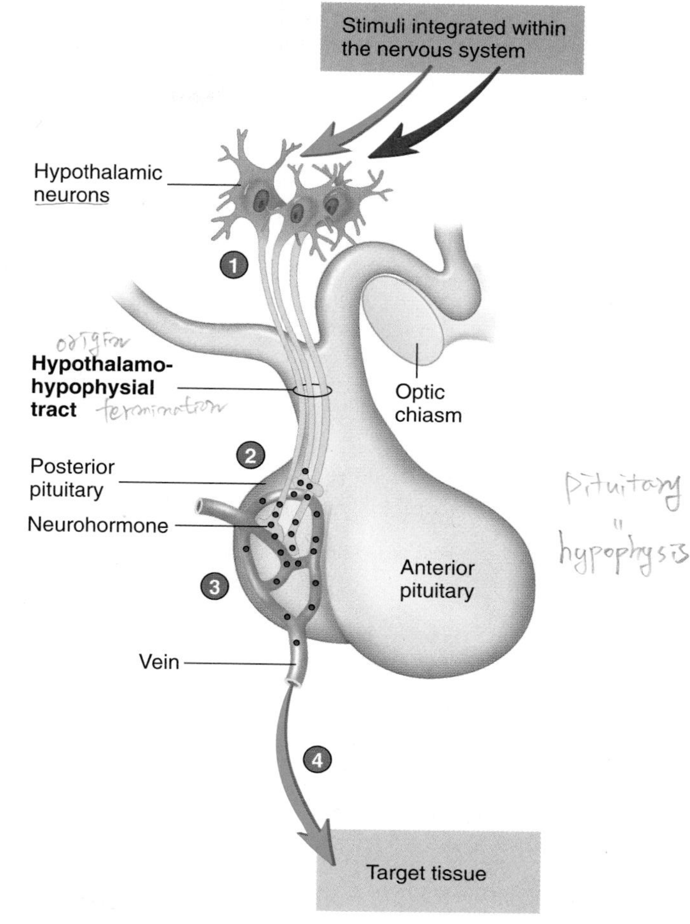

PROCESS FIGURE 18.4 Relationship Among the Hypothalamus, Posterior Pituitary, and Target Tissues

6 ***List the releasing and inhibiting hormones that are released from hypothalamic neurons.***

7 ***Describe the hypothalamohypophysial tract, including the production of neurohormones in the hypothalamus and their release from the posterior pituitary.***

PREDICT 1

Surgical removal of the posterior pituitary in experimental animals results in marked symptoms, but these symptoms associated with hormone shortage are temporary. Explain these results.

Hormones of the Pituitary Gland

This section describes the hormones secreted from the pituitary gland (table 18.2), their effects on the body, and the mechanisms that regulate their secretion rate. In addition, some major consequences of abnormal hormone secretion are stressed.

Posterior Pituitary Hormones

The posterior pituitary stores and secretes two polypeptide neurohormones called antidiuretic hormone and oxytocin. A separate population of cells secretes each neurohormone.

Antidiuretic Hormone

Antidiuretic (an′tē-d-ī-ū-ret′ik) **hormone (ADH)** is so-named because it prevents (*anti-*) the output of large amounts of urine (*diuresis*). ADH is sometimes called **vasopressin** (vā-sō-pres′in, vas-ō-pres′in) because it constricts blood vessels and raises blood pressure when large amounts are released. ADH molecules are synthesized predominantly by neuron cell bodies in the supraoptic nuclei of the hypothalamus and are transported within the axons of the hypothalamohypophysial tract from the supraoptic nuclei to the posterior pituitary, where they are stored in axon terminals. ADH molecules are released in response to action potentials from these axon terminals into the blood and are carried to the primary target tissue for ADH, the kidney tubules. Kidney tubules are the

TABLE 18.2 Hormones of the Pituitary Gland

Hormones	Structure	Target Tissue	Response
Posterior Pituitary (Neurohypophysis)			
Antidiuretic hormone (ADH)	Small peptide	Kidney	Increased water reabsorption (less water is lost in the form of urine)
Oxytocin	Small peptide	Uterus; mammary glands	Increased uterine contractions; increased milk expulsion from mammary glands; unclear function in males
Anterior Pituitary (Adenohypophysis)			
Growth hormone (GH), or somatotropin	Protein	Most tissues	Increased growth in tissues; increased amino acid uptake and protein synthesis; increased breakdown of lipids and release of fatty acids from cells; increased glycogen synthesis and increased blood glucose levels; increased somatomedin production
Thyroid-stimulating hormone (TSH)	Glycoprotein	Thyroid gland	Increased thyroid hormone secretion
Adrenocorticotropic hormone (ACTH)	Peptide	Adrenal cortex	Increased glucocorticoid hormone secretion
Lipotropins	Peptides	Fat tissues	Increased fat breakdown
β endorphins	Peptides	Brain, but not all target tissues are known	Analgesia in the brain; inhibition of gonadotropin-releasing hormone secretion
Melanocyte-stimulating hormone (MSH)	Peptide	Melanocytes in the skin	Increased melanin production in melanocytes to make the skin darker in color
Luteinizing hormone (LH)	Glycoprotein	Ovaries in females; testes in males	Ovulation and progesterone production in ovaries; testosterone synthesis and support for sperm cell production in testes
Follicle-stimulating hormone (FSH)	Glycoprotein	Follicles in ovaries in females; seminiferous tubes in males	Follicle maturation and estrogen secretion in ovaries; sperm cell production in testes
Prolactin	Protein	Ovaries and mammary glands in females	Milk production in lactating women; increased response of follicle to LH and FSH; unclear function in males

sites of urine production in the kidneys. ADH molecules promote the retention of water in kidney tubules and reduce urine volume (see chapter 26).

The secretion rate for ADH changes in response to alterations in blood osmolality and blood volume (figure 18.5). The osmolality of a solution increases as the concentration of solutes in the solution increases. Specialized neurons, called **osmoreceptors** (os′mō-rē-sep′terz, os′mō-rē-sep′tōrz), synapse with the ADH neurosecretory cells in the hypothalamus. When blood osmolality increases, the frequency of action potentials in the osmoreceptors increases, resulting in a greater frequency of action potentials in the axons of ADH neurosecretory cells. As a consequence, ADH secretion increases. In addition, an increase in blood osmolality can directly stimulate the ADH neurosecretory cells. Because ADH stimulates the kidney tubules to retain water, it reduces blood osmolality and resists any further increase in the osmolality of body fluids.

As the osmolality of the blood decreases, the action potential frequency in the osmoreceptors and the neurosecretory cells decreases. Thus, less ADH is secreted from the posterior pituitary gland, and the decreased ADH secretion causes an increased volume of water to be eliminated in the form of urine.

Urine volume increases within minutes to a few hours in response to the consumption of a large volume of water. In contrast, urine volume decreases and urine concentration increases within hours if little water is consumed. ADH plays a major role in these changes in urine formation. The effect is to maintain the osmolality and the volume of the extracellular fluid within a normal range of values.

Sensory receptors that detect changes in blood pressure send action potentials through sensory nerve fibers of the vagus nerve that eventually synapse with the ADH neurosecretory cells. A decrease in blood pressure, which normally accompanies a decrease in blood volume, causes an increased action potential frequency in the neurosecretory cells and increased ADH secretion, which stimulates the kidneys to retain water. Because the water in urine is derived from blood as it passes through the kidneys, ADH slows any further reduction in blood volume.

An increase in blood pressure decreases the action potential frequency in neurosecretory cells. This leads to the secretion of less ADH from the posterior pituitary. As a result, the volume of urine produced by the kidneys increases (see figure 18.5). Even small changes in blood osmolality influence ADH secretion. Larger changes in blood pressure are required to influence ADH secretion. The effect of ADH on the kidney and its role in the regulation of extracellular osmolality and volume are described in greater detail in chapters 26 and 27.

1. Osmoreceptors in the hypothalamus detect changes in blood osmolality, and sensory neurons that send action potentials through the vagus nerves to the hypothalamus detect changes in blood pressure.
2. An increase in osmolality and a decrease in blood pressure increase action potentials in ADH-secreting neurons.
3. Action potentials are carried by axons of ADH-secreting neurons through the hypothalamohypophysial tract to the posterior pituitary.
4. In the posterior pituitary, action potentials cause the release of ADH from the axon terminals into the circulatory system.
5. Increasing ADH acts on the kidney tubules to increase water reabsorption, resulting in a reduced urine volume, increased urine osmolality, and decreased blood osmolality. This helps maintain blood osmolality and volume.

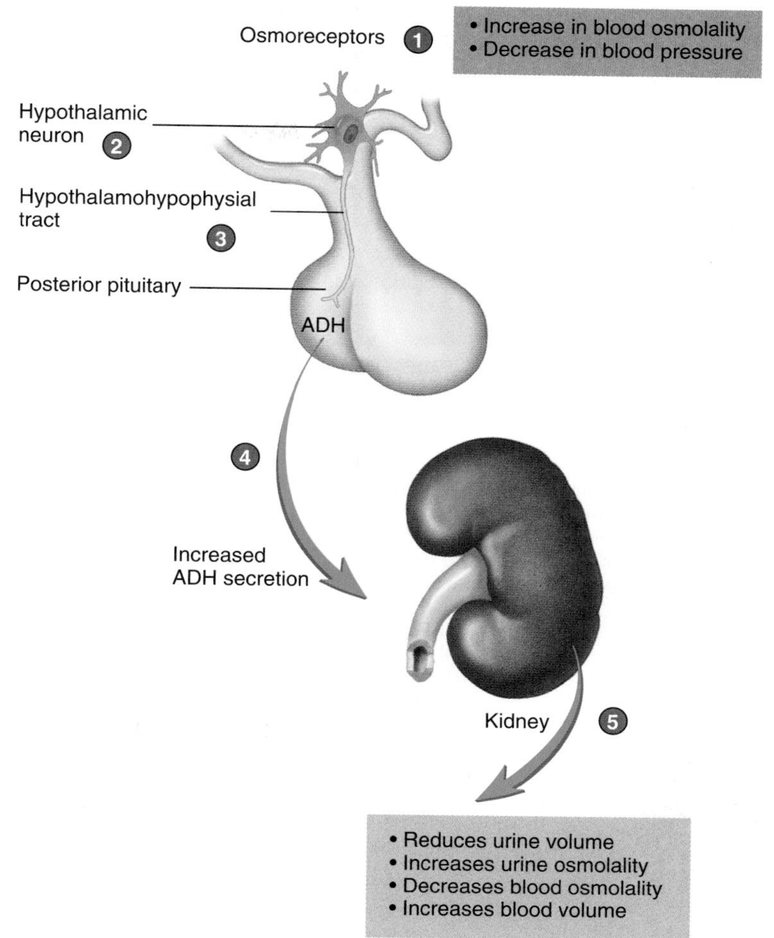

PROCESS FIGURE 18.5 Control of Antidiuretic Hormone (ADH) Secretion
Small changes in blood osmolality are important in regulating ADH secretion. Larger changes in blood pressure are required to influence ADH secretion.

8 ***Where is ADH produced, from where is it secreted, and what is its target tissue? When ADH levels increase, how are urine volume, blood osmolality, and blood volume affected?***

9 ***The secretion rate for ADH changes in response to alterations in what two factors? Name the types of sensory cells that respond to alterations in those factors.***

Oxytocin

Oxytocin (ok-sē-tō′sin) is synthesized by neuron cell bodies in the paraventricular nuclei of the hypothalamus and then is transported through axons to the posterior pituitary, where it is stored in the axon terminals.

Oxytocin stimulates smooth muscle cells of the uterus. This neurohormone plays an important role in the expulsion of the fetus from the uterus during delivery by stimulating uterine smooth muscle contraction. It also causes contraction of uterine smooth muscle in nonpregnant women, primarily during menses and sexual intercourse. The uterine contractions help expel the uterine epithelium and small amounts of blood during menses and can participate in the movement of sperm cells through the uterus after sexual intercourse. Oxytocin is also responsible for milk ejection in lactating females by promoting contraction of smooth musclelike cells surrounding the alveoli of the mammary glands (see chapter 29). Little is known about the effect of oxytocin in males.

Stretch of the uterus, mechanical stimulation of the cervix, and stimulation of the nipples of the breast when a baby nurses activate nervous reflexes that stimulate oxytocin release. Action potentials are carried by sensory neurons from the uterus and from the nipples to the spinal cord. Action potentials are then carried up the spinal cord to the hypothalamus, where they stimulate an increase in the action potential frequency in the axons of oxytocin-secreting neurons. The action potentials pass along the axons in the hypothalamohypophysial tract to the posterior pituitary, where they cause the axon terminals to secrete oxytocin (figure 18.6). The role of oxytocin in the reproductive system is described in greater detail in chapter 29.

10 ***Where is oxytocin produced and secreted, and what effects does it have on its target tissues? What factors stimulate the secretion of oxytocin?***

1. Action potentials are carried by sensory neurons from the uterus and breast to the spinal cord and up spinal nerve tracts to the hypothalamus.
2. Stretch of the uterus and the uterine cervix or nipples of the breasts increases action potentials in axons of oxytocin-secreting neurons.
3. Action potentials are carried by axons of oxytocin-secreting neurons in the hypothalamohypophysial tract to the posterior pituitary, where they increase oxytocin secretion.
4. Oxytocin enters the circulation. It increases contractions of the uterus and milk ejection from the lactating breast.

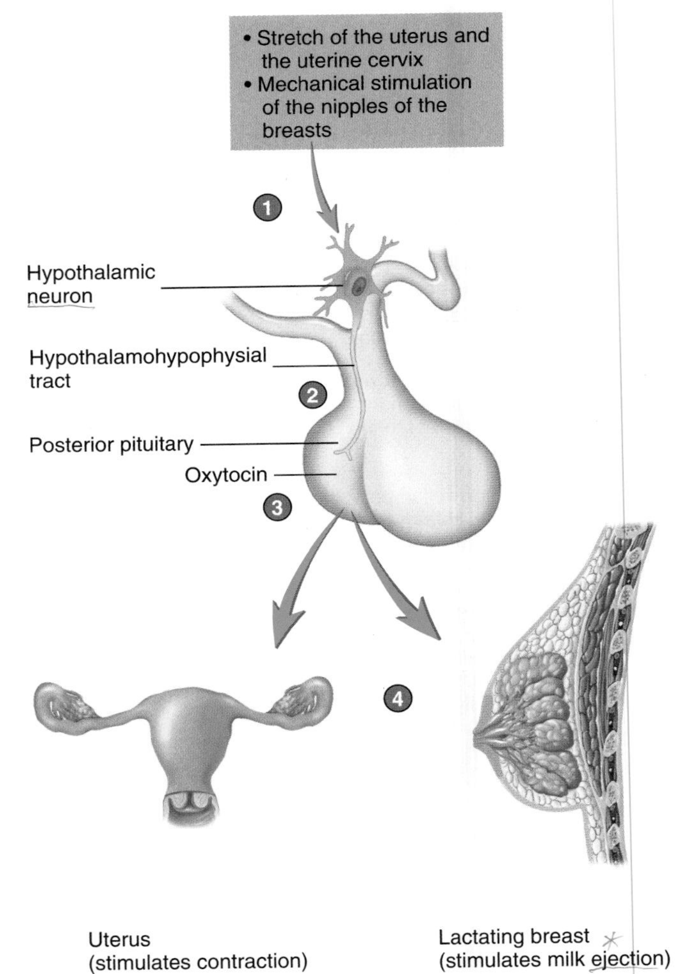

PROCESS FIGURE 18.6 Control of Oxytocin Secretion

Anterior Pituitary Hormones

Releasing and inhibiting hormones that pass from the hypothalamus through the hypothalamohypophysial portal system to the anterior pituitary influence anterior pituitary secretions. For some anterior pituitary hormones, the hypothalamus produces both releasing hormones and inhibiting hormones. For others, regulation is primarily by releasing hormones (see table 18.1).

The hormones secreted from the anterior pituitary are proteins, glycoproteins, or polypeptides. They are transported in the circulatory system, have a half-life measured in minutes, and bind to membrane-bound receptor molecules on their target cells. For the most part, each hormone is secreted by a separate cell type. Adrenocorticotropic hormone and lipotropin are exceptions because these hormones are derived from the same precursor protein.

Anterior pituitary hormones are called **tropic** (trop′ik, trō′ pik) **hormones.** They are released from the anterior pituitary gland and regulate target tissues. This regulation includes controlling the secretion of hormones, as well as growth of, target tissues. The anterior pituitary hormones include growth hormone, adrenocorticotropic hormone and related substances, luteinizing hormone, follicle-stimulating hormone, prolactin, and thyroid-stimulating hormone.

11 ***Structurally, what kind of hormones are released from the posterior pituitary and the anterior pituitary? Do these hormones bind to plasma proteins, how long is their half-life, and how do they activate their target tissues?***

Growth Hormone

Growth hormone (GH), sometimes called **somatotropin** (sō′ mă-tō-trō′ pin), stimulates growth in most tissues, plays a major role in regulating growth, and therefore plays an important role in determining how tall a person becomes. It also regulates metabolism. GH increases the movement of amino acids into cells, favors their incorporation into proteins, and slows protein breakdown. GH increases lipolysis, or the breakdown of lipids, and the release of fatty acids from fat cells into the blood. Fatty acids then can be used as energy sources to drive chemical reactions, including anabolic reactions, by other cells. GH increases glucose synthesis by the liver, which releases glucose into the blood. The increased use of fats as an energy source accompanies a decrease in glucose usage. GH plays an important role in regulating blood nutrient levels after a meal and during periods of fasting.

GH binds directly to membrane-bound receptors on target cells (see chapter 17), such as fat cells, to produce responses. These

1. Stress and decreased blood glucose levels increase the release of growth hormone–releasing hormone (GHRH) and decrease the release of growth hormone-inhibiting hormone (GHIH).
2. GHRH and GHIH travel through the hypothalamohypophysial portal system to the anterior pituitary.
3. Increased GHRH and reduced GHIH act on the anterior pituitary and result in increased GH secretion.
4. GH acts on target tissues.
5. Increasing GH has a negative-feedback effect on the hypothalamus, resulting in decreased GHRH and decreased GHIH release.

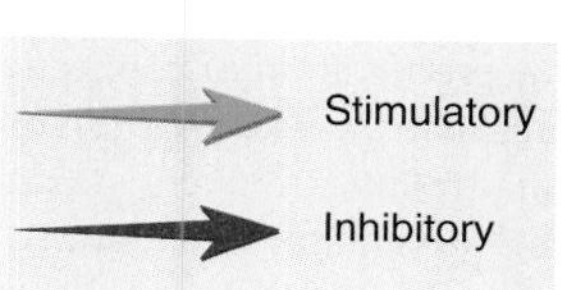

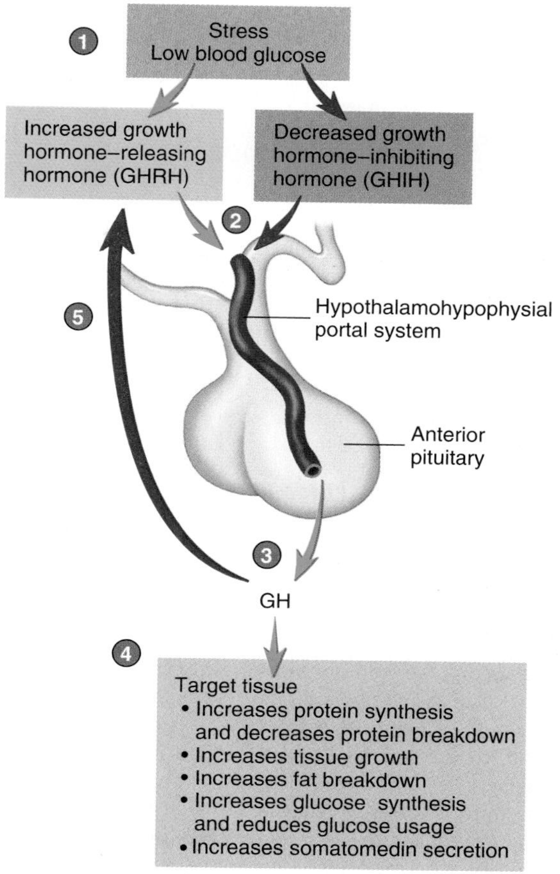

PROCESS FIGURE 18.7 Control of Growth Hormone (GH) Secretion

Secretion of GH is controlled by two neurohormones released from the hypothalamus: growth hormone–releasing hormone (GHRH), which stimulates GH secretion, and growth hormone–inhibiting hormone (GHIH), which inhibits GH secretion. Stress increases GHRH secretion and inhibits GHIH secretion. High levels of GH have a negative-feedback effect on the production of GHRH by the hypothalamus.

responses are called the direct effects of GH and include the increased breakdown of lipids and decreased use of glucose as an energy source.

GH also has indirect effects on some tissues. It increases the production of a number of polypeptides, primarily by the liver but also by skeletal muscle and other tissues. These polypeptides, called **somatomedins** (sō′mă-tō-mē′dinz), circulate in the blood and bind to receptors on target tissues. Somatomedins stimulate growth in cartilage and bone and increases the synthesis of protein in skeletal muscles. The best known somatomedins are two polypeptide hormones produced by the liver called **insulin-like growth factor I** and **II** because of the similarity of their structure to insulin and because the receptor molecules function through a mechanism similar to that of the receptors for insulin. Growth hormone and growth factors, like somatomedins, bind to membrane-bound receptors that phosphorylate intracellular proteins (see chapter 17).

Two neurohormones released from the hypothalamus regulate the secretion of GH (figure 18.7). One factor, growth hormone–releasing hormone (GHRH), stimulates the secretion of GH, and the other, growth hormone–inhibiting hormone (GHIH), or **somatostatin** (sō′mă-tō-stat′in), inhibits the secretion of GH. Stimuli that influence GH secretion act on the hypothalamus to increase or decrease the secretion of the releasing and inhibiting hormones. Low blood glucose levels and stress stimulate secretion of GH, and high blood glucose levels inhibit secretion of GH. Rising blood levels of certain amino acids also increases GH secretion.

In most people, a rhythm of GH secretion occurs. Daily peak levels of GH are correlated with deep sleep. A chronically elevated blood GH level during periods of rapid growth does not occur, although children tend to have somewhat higher blood levels of GH than adults. In addition to GH, factors such as genetics, nutrition, and sex hormones influence growth.

Several pathologic conditions are associated with abnormal GH secretion. In general, the causes for hypersecretion or hyposecretion of GH are the result of tumors in the hypothalamus or pituitary, the synthesis of structurally abnormal GH, the liver's inability to produce somatomedins, and the lack of functional receptors in target tissues. The consequences of hypersecretion and hyposecretion of growth hormone are described in Clinical Focus "Growth Hormone and Growth Disorders" (see chapter 6).

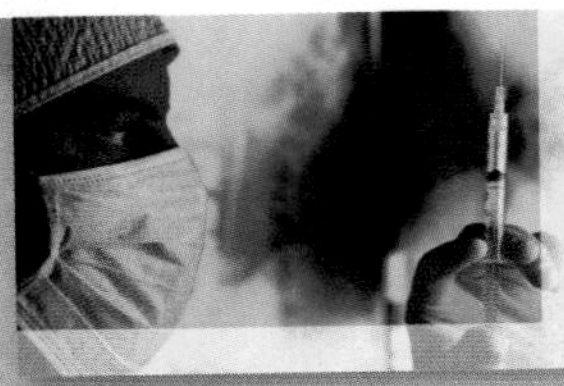

CLINICAL FOCUS

Growth Hormone and Growth Disorders

Chronic hyposecretion of GH in infants and children leads to a type of **dwarfism** (dwōrf'izm) sometimes called pituitary dwarfism. It results in a short stature due to delayed bone growth. The bones usually have a normal shape, and people with this condition exhibit normal intelligence, in contrast to dwarfism caused by hyposecretion of thyroid hormones. Other symptoms resulting from the lack of GH include mild obesity and retarded development of adult reproductive functions. Two types of dwarfism result from a lack of GH secretion: (1) In approximately two-thirds of the cases, GH and other anterior pituitary hormones are secreted in reduced amounts. The decrease in other anterior pituitary hormones can result in additional disorders caused by reduced secretion of thyroid hormones, adrenal cortex hormones, and reproductive hormones. (2) In the remaining approximately one-third of cases, a reduced amount of GH is observed, and the secretion of other anterior pituitary hormones is closer to normal. No disorders caused by a lack of thyroid hormones or adrenal cortex hormones are observed, and normal reproduction is possible for these individuals. No obvious pathology is associated with hyposecretion of GH in adults, although some evidence suggests that lack of GH can lead to reduced bone mineral content in adults.

The gene responsible for determining the structure of GH has been transferred successfully from human cells to bacterial cells, which produce GH that is identical to human GH. The GH produced in this fashion is available to treat patients who suffer from a lack of GH secretion. The availability of GH is especially important for children who suffer from reduced GH secretion.

Chronic hypersecretion of GH leads to **giantism** (jī'an-tizm) or **acromegaly** (ak-rō-meg'ă-lē), depending on whether the hypersecretion occurs before or after complete ossification of the epiphysial plates in the skeletal system. Chronic hypersecretion of GH before the epiphysial plates have ossified causes exaggerated and prolonged growth in long bones, resulting in giantism. Some individuals thus affected have grown to be 8 feet tall or more.

In adults, chronically elevated GH levels result in acromegaly. No increase in height occurs because of the ossified epiphysial plates. The condition does result in an increased diameter of fingers, toes, hands, and feet; the deposition of heavy bony ridges above the eyes; and a prominent jaw. The influence of GH on soft tissues results in a bulbous or broad nose, an enlarged tongue, thickened skin, and sparse subcutaneous adipose tissue. Nerves frequently are compressed as a result of the proliferation of connective tissue. Because GH spares glucose usage, chronic hyperglycemia results, frequently leading to diabetes mellitus and the development of severe atherosclerosis. Treatment for chronic hypersecretion of GH often involves the surgical removal or irradiation of a GH-producing tumor.

12 ***What effects do stress, amino acid levels in the blood, and glucose levels in the blood have on GH secretion?***

13 ***Describe the effects of GH on its target tissues. What stimulates somatomedin production, where is it produced, and what are its effects?***

PREDICT 2

Mr. Hoops has a son who wants to be a basketball player almost as much as Mr. Hoops wants him to be one. Mr. Hoops knows a little bit about growth hormone and asks his son's doctor if she would prescribe some for his son, so he can grow tall. What do you think the doctor tells Mr. Hoops?

Thyroid-Stimulating Hormone

Thyroid-stimulating hormone (TSH), also called **thyrotropin** (thī-rot'rō-pin, thī-rō-trō'pin), stimulates the synthesis and secretion of thyroid hormones from the thyroid gland. TSH is a glycoprotein consisting of α and β subunits, which bind to membrane-bound receptors of the thyroid gland. The receptors respond through a G protein mechanism that increases the intracellular chemical signal cAMP. In higher concentrations, TSH also increases the activity of phospholipase. Phospholipase activates mechanisms that open Ca^{2+} channels and increases the Ca^{2+} concentration in cells of the thyroid gland (see chapter 17).

TSH secretion is controlled by TRH from the hypothalamus and thyroid hormones from the thyroid gland. TRH binds to membrane-bound receptors in cells of the anterior pituitary gland and activates G proteins, which results in increased TSH secretion. In contrast, thyroid hormones inhibit both TRH and TSH secretion. TSH is secreted in a pulsatile fashion and its blood levels are highest at night, but the blood levels of thyroid hormones are maintained within a narrow range of values (see "Thyroid Hormones," p. 619).

Adrenocorticotropic Hormone and Related Substances

Adrenocorticotropic (ă-drē'nō-kōr'ti-kō-trō'pik) **hormone (ACTH)** is one of several peptide hormones derived from a precursor protein called **proopiomelanocortin** (prō-ō'pē-ō-mel'ă-nō-kōr'tin) and secreted from the anterior pituitary. This precursor protein gives rise to ACTH, lipotropins, β endorphin, and melanocyte-stimulating hormone.

ACTH binds to membrane-bound receptors and activates a G protein mechanism that increases cAMP, which produces a response. ACTH increases the secretion of hormones, primarily cortisol, from the adrenal cortex. ACTH and melanocyte-stimulating hormone also bind to melanocytes in the skin and increase skin pigmentation (see chapter 5). In pathologic conditions such as chronic adrenocortical insufficiency (Addison disease), the adrenal cortex degenerates, usually due to an autoimmune condition that destroys the adrenal cortex (see chapter 22). Blood levels of ACTH and related hormones (page 618), are chronically elevated, and the skin becomes markedly darker. Regulation of ACTH secretion and the effect of hypersecretion and hyposecretion of ACTH are described in the section "Adrenal Glands" on p. 627.

The **lipotropins** (li-pō-trō′pinz) secreted from the anterior pituitary bind to membrane-bound receptor molecules on adipose tissue cells. They cause fat breakdown and the release of fatty acids into the circulatory system.

The **β endorphins** (en′dōr-finz) have the same effects as opiate drugs, such as morphine, and they can play a role in analgesia in response to stress and exercise. Other functions have been proposed for the β endorphins, including regulation of body temperature, food intake, and water balance. Both ACTH and β-endorphin secretions increase in response to stress and exercise.

Melanocyte-stimulating hormone (MSH) binds to membrane-bound receptors on skin melanocytes and stimulates increased melanin deposition in the skin. The regulation of MSH secretion and its function in humans is not well understood, although it is an important regulator of skin pigmentation in some other vertebrates.

Luteinizing Hormone, Follicle-Stimulating Hormone, and Prolactin

Gonadotropins (gō′nad-ō-trō′pinz) are glycoprotein hormones capable of promoting the growth and function of the **gonads,** the ovaries and testes. The two major gonadotropins secreted from the anterior pituitary are **luteinizing** (loo′tē-ĭ-nīz-ing) **hormone (LH)** and **follicle-stimulating hormone (FSH).** LH, FSH, and another anterior pituitary hormone called **prolactin** (prō-lak′tin) play important roles in regulating reproduction.

LH and FSH secreted into the blood bind to membrane-bound receptors, increase the intracellular synthesis of cAMP through G protein mechanisms, and stimulate the production of **gametes** (gam′ēts)—sperm cells in the testes and oocytes in ovaries. LH and FSH also control the production of reproductive hormones—estrogens and progesterone in the ovaries and testosterone in the testes.

LH and FSH are released from anterior pituitary cells under the influence of the hypothalamic-releasing hormone **gonadotropin-releasing hormone (GnRH).** GnRH is also called **luteinizing hormone–releasing hormone (LHRH).**

Prolactin plays an important role in milk production in the mammary glands of lactating females. It binds to a membrane-bound receptor that is linked to a kinase that phosphorylates intracellular proteins. The phosphorylated proteins produce the response in the cell. Prolactin also can enhance progesterone secretion of the ovary after ovulation. No role for this hormone has been clearly established in males. Several hypothalamic neurohormones can be involved in the complex regulation of prolactin secretion. One neurohormone is prolactin-releasing hormone (PRH), and another is prolactin-inhibiting hormone (PIH). The regulation of gonadotropin and prolactin secretion and their specific effects are explained more fully in chapter 28.

14 ***For each of the following hormones secreted by the anterior pituitary—GH, TSH, ACTH, LH, FSH, and prolactin—name its target tissue and the hormone's effect on its target tissue.***

15 ***How are ACTH, MSH, lipotropins, and β endorphins related? What are the functions of these hormones?***

16 ***Define gonadotropins, and name two gonadotropins produced by the anterior pituitary.***

THYROID GLAND

The **thyroid gland** is composed of two lobes connected by a narrow band of thyroid tissue called the **isthmus.** The lobes are lateral to the upper portion of the trachea just inferior to the larynx, and the isthmus extends across the anterior aspect of the trachea (figure 18.8*a*). The thyroid gland is one of the largest endocrine glands, with a weight of approximately 20 g. It is highly vascular and appears more red than its surrounding tissues.

The thyroid gland contains numerous **follicles,** which are small spheres whose walls are composed of a single layer of cuboidal epithelial cells (figure 18.8*b* and *c*). The center, or lumen, of each thyroid follicle is filled with proteins called **thyroglobulin** (thī-rō-glob′ū-lin) synthesized and secreted by cells of the thyroid follicles. Large amounts of the thyroid hormones are stored in the thyroid follicles as part of the thyroglobulin molecules.

Between the follicles, a delicate network of loose connective tissue contains numerous capillaries. Scattered **parafollicular** (par-ă-fo-lik′ū-lăr) **cells** are found between the follicles and among the cells that make up the walls of the follicle. **Calcitonin** (kal-si-tō′nin) is secreted from the parafollicular cells and plays a role in reducing the concentration of calcium in the body fluids when calcium levels become elevated.

Thyroid Hormones

The thyroid hormones include both **triiodothyronine** (trī-ī′ō-dō-thī′rō-nēn; T_3) and **tetraiodothyronine** (tet′ră-ī′ō-dō-thī′rō-nēn; T_4). T_4 is also called **thyroxine** (thī-rok′sēn, thī-rok′sin). T_3 and T_4 constitute major secretory products of the thyroid gland, consisting of 10% T_3 and 90% T_4 (table 18.3). Although calcitonin is secreted by the parafollicular cells of the thyroid gland, T_3 and T_4 are considered to be the thyroid hormones because they are more clinically important and because they are secreted from the thyroid follicles.

T_3 and T_4 Synthesis

Thyroid-stimulating hormone (TSH) from the anterior pituitary stimulates thyroid hormone synthesis and secretion. TSH causes an increase in synthesis of T_3 and T_4, which are then stored inside the thyroid follicles as part of thyroglobulin. TSH also causes T_3 and T_4 to be released from thyroglobulin and enter the circulatory system. Because iodine is a component of T_3 and T_4, an adequate amount of iodine in the diet is required for thyroid hormone

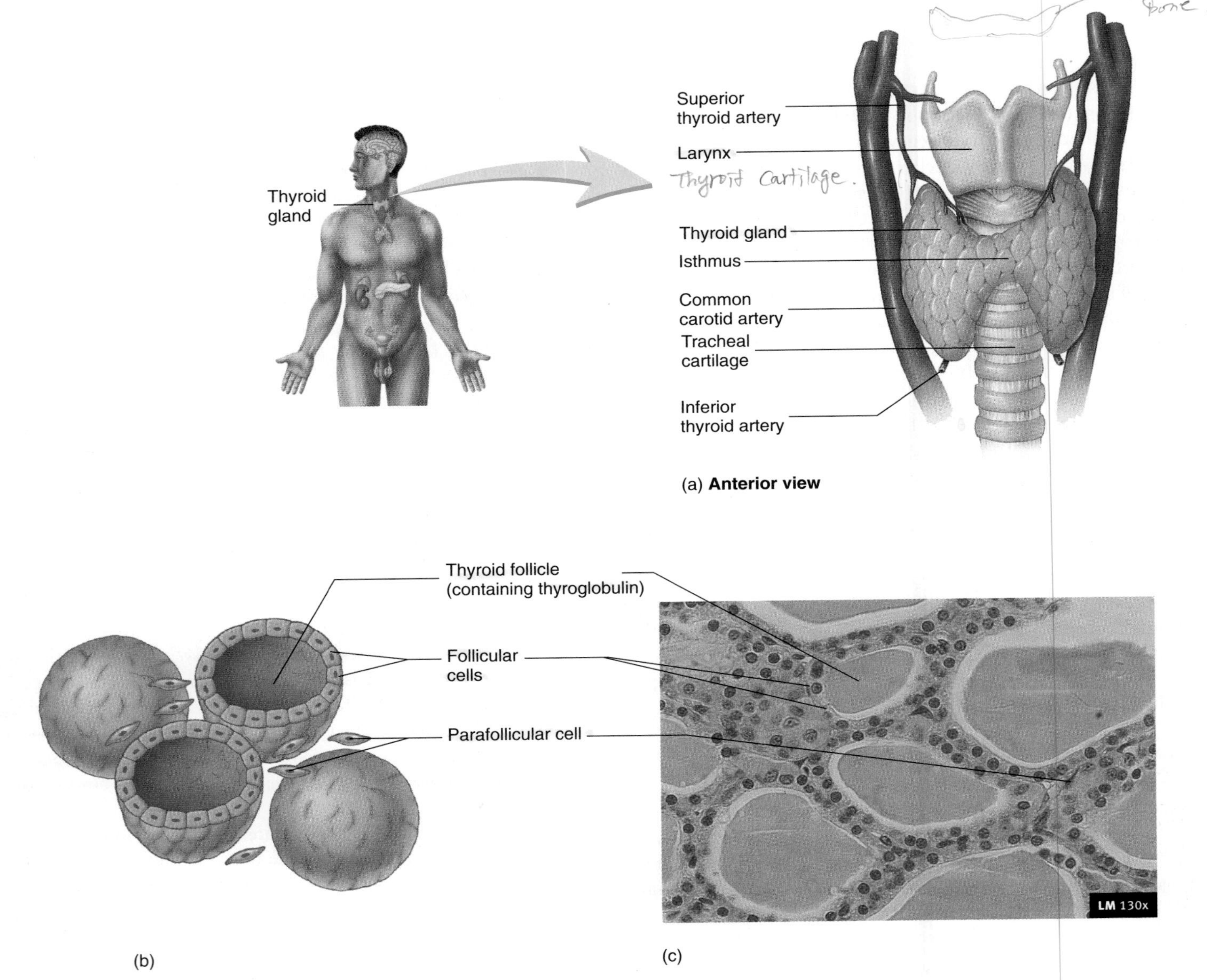

FIGURE 18.8 Anatomy and Histology of the Thyroid Gland

(*a*) Anterior view of the thyroid gland. (*b*) Histology of the thyroid gland. The gland is made up of many spheric thyroid follicles containing thyroglobulin. Parafollicular cells are in the tissue between the thyroid follicles. (*c*) Low-power photomicrograph of thyroid follicles.

TABLE 18.3 Hormones of the Thyroid and Parathyroid Glands

Hormones	Structure	Target Tissue	Response
Thyroid Gland			
Thyroid Follicles			
Thyroid hormones (triiodothyronine and tetraiodothyronine)	Amino acid derivative	Most cells of the body	Increased metabolic rate; essential for normal process of growth and maturation
Parafollicular Cells			
Calcitonin	Polypeptide	Bone	Decreased rate of breakdown of bone by osteoclasts; prevention of a large increase in blood Ca^{2+} levels
Parathyroid			
Parathyroid hormone	Polypeptide	Bone; kidney; small intestine	Increased rate of breakdown of bone by osteoclasts; increased reabsorption of Ca^{2+} in kidneys; increased absorption of Ca^{2+} from the small intestine; increased vitamin D synthesis; increased blood Ca^{2+} levels

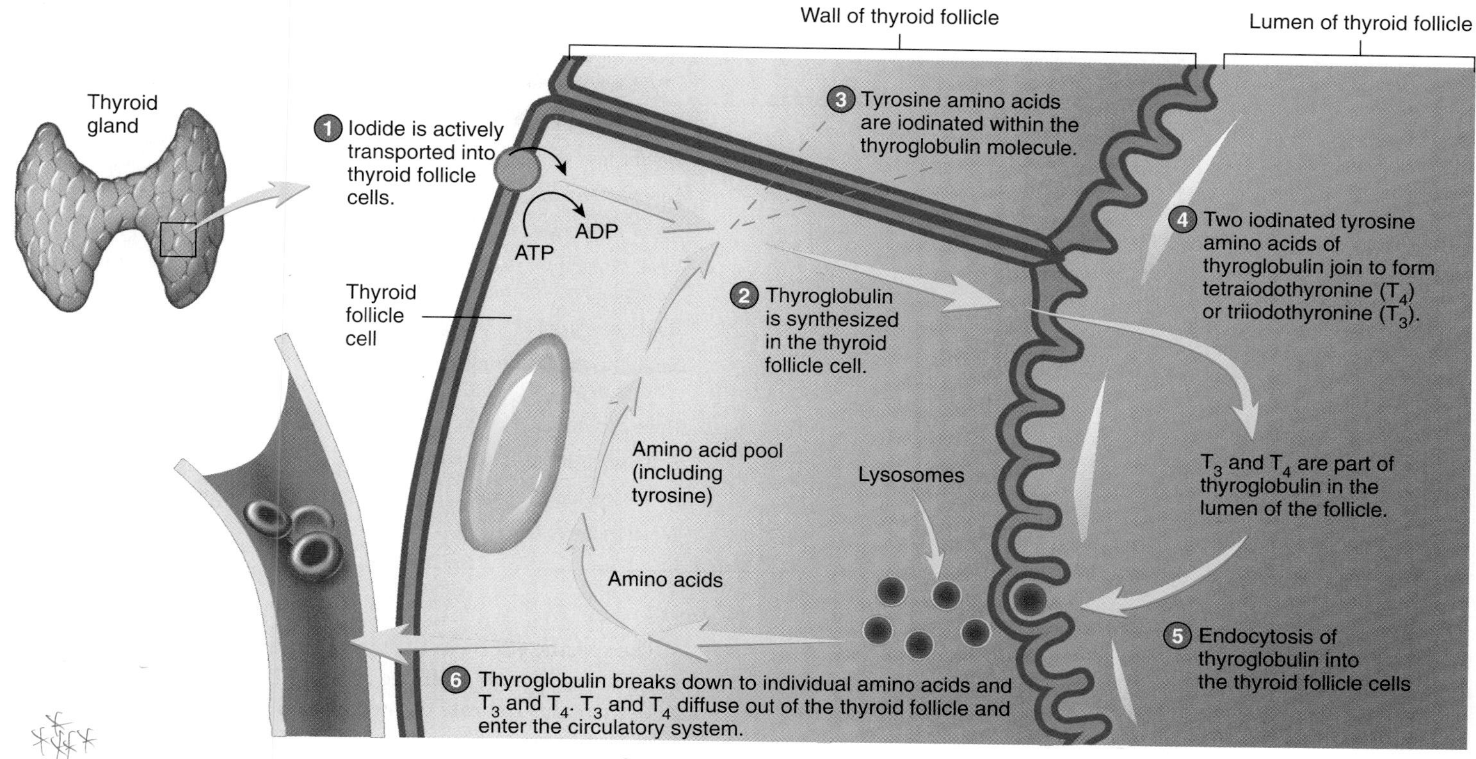

PROCESS FIGURE 18.9 Biosynthesis of Thyroid Hormones

The numbered steps describe the synthesis and secretion of T_3 and T_4 from the thyroid gland.

synthesis. The following events in the thyroid follicles result in T_3 and T_4 synthesis and secretion (figure 18.9):

1. Iodide ions (I^-) are taken up by thyroid follicle cells by active transport. The active transport of the I^- is against a concentration gradient of approximately 30-fold in healthy individuals.
2. Thyroglobulins, which contain numerous tyrosine amino acid molecules, are synthesized within the cells of the follicles.
3. Nearly simultaneously, the I^- are oxidized to form iodine (I) and either one or two iodine atoms are bound to each of the tyrosine molecules of thyroglobulin. This occurs close to the time the thyroglobulin molecules are secreted by the process of exocytosis into the lumen of the follicle. As a result, the secreted thyroglobulin contains many iodinated tyrosine amino acids with either one iodine atom (monoiodotyrosine) or two iodine atoms (diiodotyrosine).
4. In the lumen of the follicle, two diiodotyrosine molecules of thyroglobulin combine to form tetraiodothyronine (T_4), or one monoiodotyrosine and one diiodotyrosine molecule combine to form triiodothyronine (T_3). Large amounts of T_3 and T_4 are stored within the thyroid follicles as part of thyroglobulin. A reserve sufficient to supply thyroid hormones for approximately 2–4 months is stored in this form.
5. Thyroglobulin is taken into the thyroid follicle cells by endocytosis, where lysosomes fuse with the endocytotic vesicles.
6. Proteolytic enzymes break down thyroglobulin to release T_3 and T_4. T_3 and T_4 are lipid-soluble, so they diffuse through the plasma membranes of the follicular cells into the interstitial spaces and finally into the capillaries of the thyroid gland. The remaining amino acids of thyroglobulin are used again to synthesize more thyroglobulin.

Transport in the Blood

T_3 and T_4 are transported in combination with plasma proteins in the circulatory system. Approximately 70%–75% of the circulating T_3 and T_4 are bound to **thyroxine-binding globulin (TBG),** which is synthesized by the liver and 20%–30% are bound to other plasma proteins, including albumen. T_3 and T_4, bound to these plasma proteins, form a large reservoir of circulating thyroid hormones, and the half-life of these hormones is increased greatly because of this binding. After thyroid gland removal in experimental animals, it takes approximately 1 week for T_3 and T_4 levels in the blood to decrease by 50%. As free T_3 and T_4 levels decrease in the interstitial spaces, additional T_3 and T_4 dissociate from the plasma proteins to maintain the levels in the tissue spaces. When sudden secretion of T_3 and T_4 occurs, the excess binds to the plasma proteins. As a consequence, the concentration of thyroid hormones in the tissue spaces fluctuates very little.

Approximately 33%–40% of the T_4 is converted to T_3 in the body tissues. This conversion can be important in the action of thyroid hormones on their target tissues because T_3 is the major hormone that interacts with target cells. In addition, T_3 is several times more potent than T_4.

Much of the circulating T_4 is eliminated from the body by being converted to tetraiodothyroacetic acid and then excreted in

TABLE 18.4 Effects of Hyposecretion and Hypersecretion of Thyroid Hormones

Hypothyroidism	Hyperthyroidism
Decreased metabolic rate, low body temperature, cold intolerance	Increased metabolic rate, high body temperature, heat intolerance
Weight gain, reduced appetite	Weight loss, increased appetite
Reduced activity of sweat and sebaceous glands, dry and cold skin	Copious sweating, warm and flushed skin
Reduced heat rate, reduced blood pressure, dilated and enlarged heart	Rapid heart rate, elevated blood pressure, abnormal electrocardiogram
Weak, flabby skeletal muscles; sluggish movements	Weak skeletal muscles that exhibit tremors, quick movements with exaggerated reflexes
Constipation	Bouts of diarrhea
Myxedema (swelling of the face and body) as a result of mucoprotein deposits	Exophthalmos (protruding of the eyes) as a result of mucoprotein and other deposits behind the eye
Apathy, somnolence	Hyperactivity, insomnia, restlessness, irritability, short attention span
Coarse hair, rough and dry skin	Soft, smooth hair and skin
Decreased iodide uptake	Increased iodide uptake
Possible goiter (enlargement of the thyroid gland)	Almost always a goiter

the urine or bile. In addition, a large amount is converted to an inactive form of T_3, rapidly metabolized, and excreted.

Mechanism of Action of T_3 and T_4

T_3 and T_4 interact with their target tissues in a fashion similar to that of the steroid hormones. They readily diffuse through plasma membranes into the cytoplasm of cells. Within cells, they bind to receptor molecules in the nuclei. Thyroid hormones combined with their receptor molecules interact with DNA in the nucleus to influence regulatory genes and initiate new protein synthesis. The newly synthesized proteins within the target cells mediate the cells' response to thyroid hormones. It takes up to a week after the administration of thyroid hormones for a maximal response to develop, and new protein synthesis occupies much of that time.

Effects of T_3 and T_4

T_3 and T_4 affect nearly every tissue in the body, but not all tissues respond identically. Metabolism is primarily affected in some tissues, and growth and maturation are influenced in others.

The normal rate of metabolism for an individual depends on an adequate supply of thyroid hormone, which increases the rate at which glucose, fat, and protein are metabolized. The increased rate of metabolism produces heat. Blood levels of cholesterol decline. Thyroid hormones increase the activity of Na^+–K^+ pump, which helps increase the body temperature. T_3 and T_4 can alter the number and activity of mitochondria, resulting in greater ATP synthesis and heat production. The metabolic rate can increase 60%–100% when blood T_3 and T_4 are elevated. Low levels of T_3 and T_4 lead to the opposite effect. Normal body temperature depends on an adequate amount of T_3 and T_4.

Normal growth and maturation of organs also depend on T_3 and T_4. For example, bone, hair, teeth, connective tissue, and nervous tissue require thyroid hormone for normal growth and development. Both normal growth and normal maturation of the brain require thyroid hormones. Also, T_3 and T_4 play a permissive role for GH, and GH does not have its normal effect on target tissues if T_3 and T_4 are not present.

The specific effects of the hyposecretion and hypersecretion of thyroid hormones are outlined in table 18.4. Hypersecretion of T_3 and T_4 increases the rate of metabolism. High body temperature, weight loss, increased appetite, rapid heart rate, and an enlarged thyroid gland are major symptoms.

Hyposecretion of T_3 and T_4 decreases the rate of metabolism. Low body temperature, weight gain, reduced appetite, reduced heart rate, reduced blood pressure, weak skeletal muscles, and apathy are major symptoms. If hyposecretion of T_3 and T_4 occurs during development, there is a decreased metabolic rate, abnormal nervous system development, abnormal growth, and abnormal maturation of tissues. The consequence is a mentally retarded person of short stature and distinctive form called a **cretin** (krē′tin).

Regulation of Thyroid Hormone Secretion

Thyrotropin-releasing hormone (TRH) from the hypothalamus and TSH from the anterior pituitary function together to increase T_3 and T_4 secretion from the thyroid gland. Exposure to cold and stress cause increased TRH secretion and prolonged fasting decreases TRH secretion. TRH stimulates the secretion of TSH from the anterior pituitary. Thus, when TRH release increases, TSH secretion from the anterior pituitary gland increases. When TRH release decreases, TSH secretion decreases. Small fluctuations in blood levels of TSH occur on a daily basis, with a small nocturnal increase. TSH stimulates T_3 and T_4 secretion from the thyroid gland. TSH also increases the synthesis of T_3 and T_4, as well as causing hypertrophy (increased cell size) and hyperplasia (increased cell number) of the thyroid gland. Decreased blood levels of TSH lead to decreased T_3 and T_4 secretion and thyroid gland atrophy. Figure 18.10 illustrates the regulation of T_3 and T_4 secretion. T_3 and T_4 have a negative-feedback effect on the hypothalamus and anterior pituitary gland. As T_3 and T_4 levels increase in the circulatory system, they inhibit TRH and TSH secretion. Also, if the thyroid gland is removed or if T_3 and T_4 secretion declines, TSH levels in the blood increase dramatically.

Abnormal thyroid conditions are outlined in table 18.5. Hypothyroidism, or reduced secretion of thyroid hormones, can

1. TRH is released from neurons within the hypothalamus. It passes through the hypothalamohypophysial portal system to the anterior pituitary.
2. TRH causes cells of the anterior pituitary to secrete TSH, which passes through the general circulation to the thyroid gland.
3. TSH causes increased release of T_3 and T_4 into the general circulation.
4. T_3 and T_4 act on target tissues to produce a response.
5. T_3 and T_4 also have an inhibitory effect on the secretion of TRH from the hypothalamus and TSH from the anterior pituitary.

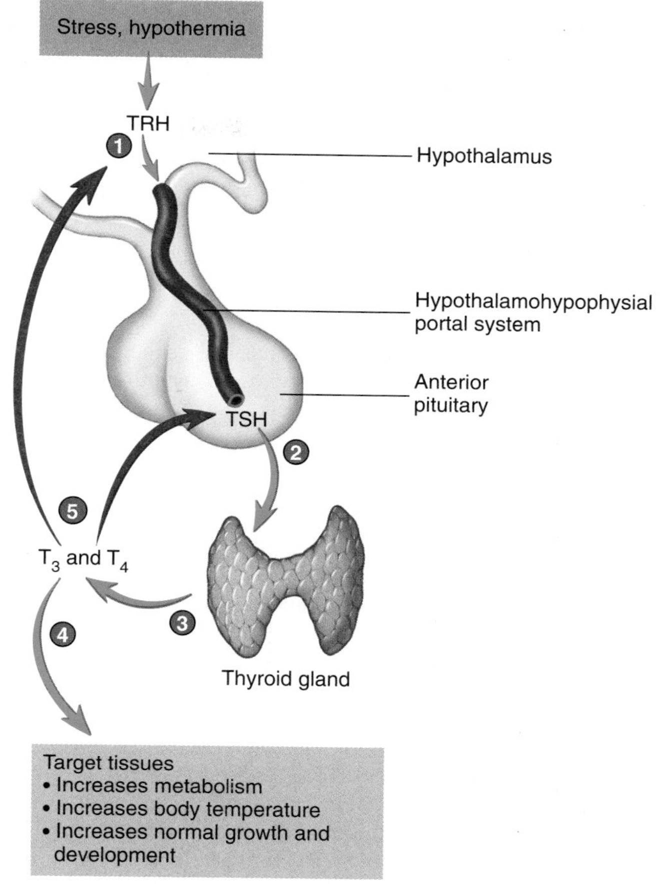

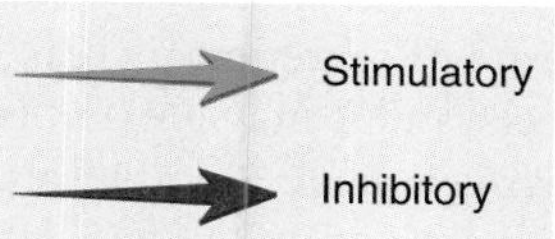

PROCESS FIGURE 18.10 Regulation of Thyroid Hormone (T_3 and T_4) Secretion

TABLE 18.5 Abnormal Thyroid Conditions

Cause	Description
Hypothyroidism	
Iodine deficiency	Causes inadequate T_3 and T_4 synthesis, which results in elevated thyroid-stimulating hormone (TSH) secretion; thyroid gland enlarges (goiter) as a result of TSH stimulation; T_3 and T_4 frequently remain in the low to normal range
Goiterogenic substances	Found in certain drugs and in small amounts in certain plants, such as cabbage; inhibit T_3 and T_4 synthesis
Cretinism	Caused by maternal iodine deficiency or congenital errors in thyroid hormone synthesis; results in mental retardation and a short, grotesque appearance
Lack of thyroid gland	Partial or complete surgical removal or drug-induced destruction of the thyroid gland as a treatment for Graves disease (hyperthyroidism)
Pituitary insufficiency	Results from lack of TSH secretion; often associated with inadequate secretion of other anterior pituitary hormones
Hashimoto disease	Autoimmune disease in which thyroid hormone secretion can be normal or depressed
Hyperthyroidism	
Graves disease	Characterized by goiter and exophthalmos; apparently an autoimmune disease; most patients have a TSH-like immunoglobulin in their plasma
Tumors—benign adenoma or cancer	Result in either normal secretion or hypersecretion of thyroid hormones (rarely hyposecretion)
Thyroiditis—a viral infection	Produces painful swelling of the thyroid gland with normal or slightly increased T_3 and T_4 production
Elevated TSH levels	Result from a pituitary tumor
Thyroid storm	Sudden release of large amounts of T_3 and T_4; caused by surgery, stress, infections, and unknown reasons

result from iodine deficiency, the taking of certain drugs, and exposure to other chemicals that inhibit T_3 and T_4 synthesis. It can also be due to inadequate secretion of TSH, an autoimmune disease that depresses thyroid hormone function, or surgical removal of the thyroid gland. Hypersecretion of T_3 and T_4 can result from the synthesis of an immune globulin that can bind to TSH receptors and acts like TSH, and from TSH-secreting tumors of the pituitary gland.

Goiter and Exophthalmos

An abnormal enlargement of the thyroid gland is called a **goiter.** Goiters can result from conditions that cause hypothyroidism as well as conditions that cause hyperthyroidism. An **iodine-deficiency goiter** results when dietary iodine intake is very low and there is too little iodine to synthesize T_3 and T_4 (see table 18.5). As a result, blood levels of T_3 and T_4 decrease and the person may exhibit symptoms of hypothyroidism. The reduced negative feedback of T_3 and T_4 on the anterior pituitary and hypothalamus result in elevated TSH secretion. TSH causes hypertrophy and hyperplasia of the thyroid gland and increased thyroglobulin synthesis, even though there is not enough iodine to synthesize T_3 and T_4. Consequently, the thyroid gland enlarges. Historically, iodine-deficiency goiters were common in people from areas where the soil was depleted of iodine. Consequently, plants grown in these areas, called goiter belts, had little iodine in them and caused iodine-deficient diets. Iodized salt has nearly eliminated iodine-deficiency goiters. However, it remains a problem in some developing countries. **Toxic goiter** secretes excess T_3 and T_4, and it can result from elevated TSH secretion or elevated TSH-like immunoglobulin (see Grave disease in table 18.5). Toxic goiter results in elevated T_3 and T_4 secretion and symptoms of hyperthyroidism. **Exophthalmos** often accompanies hyperthyroidism and is caused by the deposition of excess connective tissue proteins behind the eyes. The excess tissue makes the eyes move anteriorly, and consequently they appear to be larger than normal.

Grave disease is the most common cause of hyperthyroidism. Elevated T_3 and T_4 resulting from this condition suppresses TSH and TRH, but the T_3 and T_4 levels remain elevated. Exophthalmos is common. Treatment often involves the use of antithyroid drugs that reduce T_3 and T_4 secretion, partial destruction of the thyroid gland using radioactive iodine, or partial or complete surgical removal of the thyroid gland. These treatments are followed by the oral administration of the appropriate amount of T_3 and T_4. Unfortunately removal of the thyroid gland normally does not reverse exophthalmos.

17 ***Where is the thyroid gland located? Describe the follicles and the parafollicular cells within the thyroid. What hormones do they produce?***

18 ***Starting with the uptake of iodide by the follicles, describe the production and secretion of thyroid hormones.***

19 ***How are the thyroid hormones transported in the blood? What effect does this transportation have on their half-life?***

20 ***What are the target tissues of thyroid hormone? By what mechanism do thyroid hormones alter the activities of their target tissues? What effects are produced?***

21 ***Starting in the hypothalamus, explain how chronic exposure to cold, food deprivation, or stress can affect thyroid hormone production.***

22 ***Diagram two negative-feedback mechanisms involving hormones that regulate the production of thyroid hormones.***

PREDICT 3

Predict the effect of surgical removal of the thyroid gland on blood levels of TRH, TSH, T_3, and T_4. Predict the effect of oral administration of T_3 and T_4 on TRH and TSH.

Calcitonin

The parafollicular cells, or C cells, of the thyroid gland, which secrete calcitonin, are dispersed between the thyroid follicles throughout the thyroid gland. The major stimulus for increased calcitonin secretion is an increase in calcium levels in the blood.

The primary target tissue for calcitonin is bone (see chapter 6). Calcitonin binds to membrane-bound receptors, decreases osteoclast activity, and lengthens the life span of osteoblasts. The result is a decrease in blood calcium and phosphate levels caused by increased bone deposition.

The importance of calcitonin in the regulation of blood Ca^{2+} levels is unclear. Its rate of secretion increases in response to elevated blood Ca^{2+} levels, and it may prevent large increases in blood Ca^{2+} levels following a meal. Blood levels of calcitonin decrease with age to a greater extent in females than males. Also, osteoporosis increases with age and occurs to a greater degree in females than males. Complete thyroidectomy does not result in high blood Ca^{2+} levels, however. It is possible that the regulation of blood Ca^{2+} levels by other hormones, such as parathyroid hormone, and vitamin D compensates for the loss of calcitonin in individuals who have undergone a thyroidectomy. No pathologic condition is associated directly with a lack of calcitonin secretion. Some evidence suggests that calcitonin may play a role in regulating food intake by decreasing appetite.

23 ***What effect does calcitonin have on osteoclasts, osteoblasts, and blood calcium levels? What stimulus can cause an increase in calcitonin secretion?***

PARATHYROID GLANDS

The **parathyroid** (par-ă-thī′royd) **glands** are usually embedded in the posterior part of each lobe of the thyroid gland and are made up of two cell types, the chief cells and oxyphils. The chief cells secrete parathyroid hormone, but the function of the oxyphils is unknown. Usually, four parathyroid glands are present, with their cells organized in densely packed masses or cords rather than in follicles (figure 18.11). In some cases, one or more of the parathyroid glands does not become embedded in the thyroid gland and remains in the nearby connective tissue.

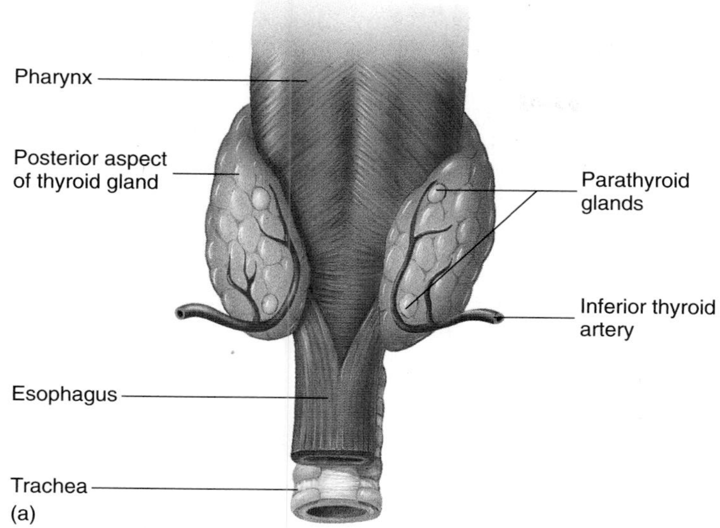

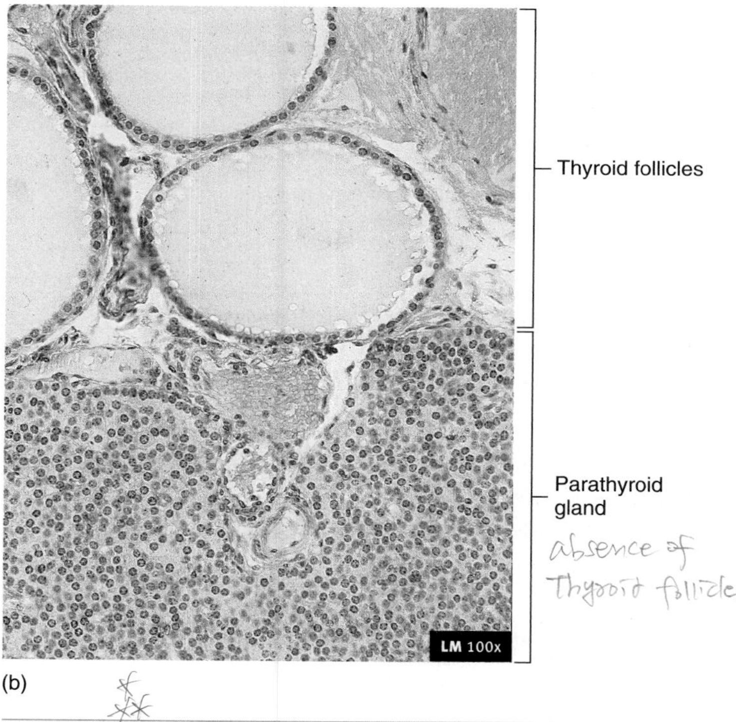

FIGURE 18.11 Anatomy and Histology of the Parathyroid Glands
(*a*) The parathyroid glands are embedded in the posterior part of the thyroid gland. (*b*) The parathyroid glands are composed of densely packed cords of cells.

Parathyroid hormone (PTH), also called parathormone, is a polypeptide hormone that is important in the regulation of calcium levels in body fluids (see table 18.3). Bone, the kidneys, and the intestine are its major target tissues. PTH binds to membrane-bound receptors and activates a G protein mechanism that increases intracellular cAMP levels in target tissues. Without functional parathyroid glands, the ability to regulate blood calcium levels adequately is lost.

PTH stimulates osteoclast activity in bone and can cause the number of osteoclasts to increase. The increased osteoclast activity results in bone resorption and the release of calcium and phosphate, causing an increase in blood calcium levels. PTH receptors are not present on osteoclasts but are present on osteoblasts and on red bone marrow stromal (stem) cells. PTH binds to receptors on osteoblasts, which then promote an increase in osteoclast activity (see chapter 6).

PTH induces calcium reabsorption within the kidneys so that less calcium leaves the body in urine. It also increases the enzymatic formation of active vitamin D in the kidneys. Calcium is actively absorbed by the epithelial cells of the small intestine, and the synthesis of transport proteins in the intestinal cells requires active vitamin D. PTH increases the rate of active vitamin D synthesis, which in turn increases the rate of calcium and phosphate absorption in the intestine, thereby elevating blood levels of calcium.

Although PTH increases the release of phosphate ions (PO_4^{3-}) from bone and increases PO_4^{3-} absorption in the gut, it increases PO_4^{3-} excretion in the kidney. The overall effect of PTH is to decrease blood phosphate levels. A simultaneous increase in both Ca^{2+} and PO_4^{3-} results in the precipitation of calcium phosphate in soft tissues of the body, where they cause irritation and inflammation.

The regulation of PTH secretion and the role of PTH and calcitonin in regulating blood Ca^{2+} levels are outlined in figure 18.12. The primary stimulus for the secretion of PTH is a decrease in blood Ca^{2+} levels, whereas elevated blood Ca^{2+} levels inhibit PTH secretion. This regulation keeps blood Ca^{2-} levels fluctuating within a normal range of values. Both hypersecretion and hyposecretion of PTH cause serious symptoms (table 18.6). The regulation of blood Ca^{2+} levels is discussed more thoroughly in chapter 27.

PREDICT 4

Predict the effect of an inadequate dietary intake of calcium on PTH secretion and on target tissues for PTH.

Inactive parathyroid glands result in hypocalcemia. Reduced extracellular calcium levels cause voltage-gated Na^+ channels in plasma membranes to open, which increases the permeability of plasma membranes to Na^+. As a consequence, Na^+ diffuse into cells and cause depolarization (see chapter 11). Symptoms of hypocalcemia are nervousness, muscle spasms, cardiac arrhythmias, and convulsions. In extreme cases, tetany of skeletal muscles results and tetany of the respiratory muscles can cause death.

24. ***Where are the parathyroid glands located, and what hormone do they produce?***
25. ***What effect does PTH have on osteoclasts, osteoblasts, the kidneys, the small intestine, and blood calcium and blood phosphate levels? What stimulus can cause an increase in PTH secretion?***

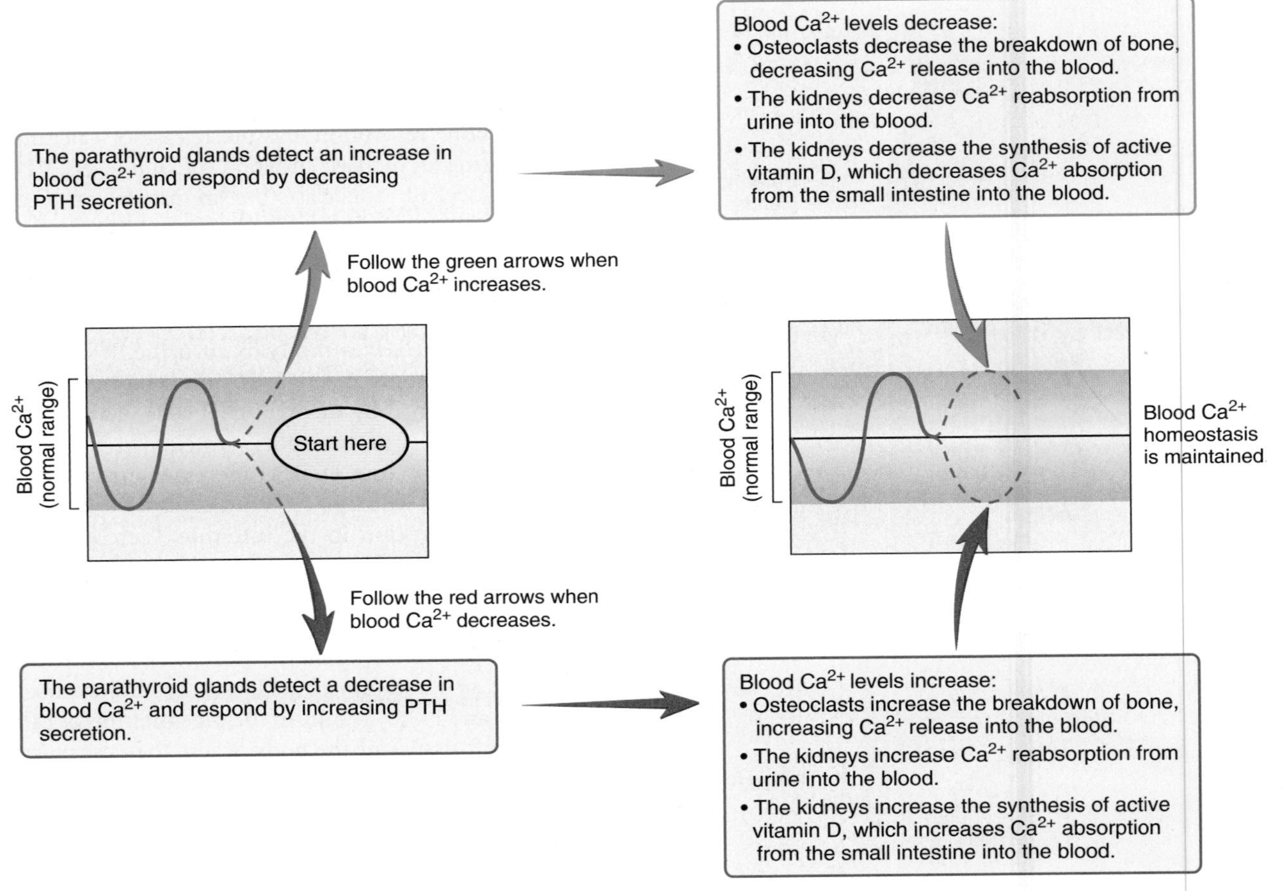

HOMEOSTASIS FIGURE 18.12 **Summary of Regulation of Blood Ca^{2+} Concentration Regulation by Parathyroid Hormone (PTH)**

TABLE 18.6 Causes and Symptoms of Hypersecretion and Hyposecretion of Parathyroid Hormone

Hypoparathyroidism	Hyperparathyroidism
Causes	
Accidental removal during thyroidectomy	Primary hyperparathyroidism: a result of abnormal parathyroid function—adenomas of the parathyroid gland (90%), hyperplasia of parathyroid idiopathic (unknown cause) cells (9%), and carcinomas (1%) Secondary hyperparathyroidism: caused by conditions that reduce blood Ca^{2+} levels, such as inadequate Ca^{2+} in the diet, inadequate levels of vitamin D, pregnancy, or lactation
Symptoms	
Hypocalcemia	Hypercalcemia or normal blood Ca^{2+} levels; calcium carbonate salts may be deposited throughout the body, especially in the renal tubules (kidney stones), lungs, blood vessels, and gastric mucosa
Normal bone structure	Bones weaken and are eaten away as a result of resorption; some cases are first diagnosed when a radiograph is taken of a broken bone
Increased neuromuscular excitability; tetany, laryngospasm, and death from asphyxiation can result	Neuromuscular system less excitable; muscular weakness may be present
Flaccid heart muscle; cardiac arrhythmia may develop	Increased force of contraction of cardiac muscle; at very high blood Ca^{2+} levels, cardiac arrest during contraction is possible
Diarrhea	Constipation

PREDICT 5

A patient with a malignant tumor had his thyroid gland removed. What effect would this removal have on blood levels of Ca^{2+}? If the parathyroid glands are inadvertently removed along with the thyroid gland during surgery, death can result because the muscles of respiration undergo sustained contractions. Explain.

ADRENAL GLANDS

The **adrenal** (ă-drē′năl) **glands,** also called the **suprarenal** (soo′pră-rē′năl) **glands,** are near the superior poles of the kidneys. Like the kidneys, they are retroperitoneal, and they are surrounded by abundant adipose tissue. The adrenal glands are enclosed by a connective tissue capsule and have a well-developed blood supply (figure 18.13*a*).

The adrenal glands are composed of an inner **medulla** and an outer **cortex,** which are derived from two separate embryonic tissues. The adrenal medulla arises from neural crest cells, which also give rise to postganglionic neurons of the sympathetic division of the autonomic nervous system (see chapters 16 and 29). Unlike most glands of the body, which develop from invaginations of epithelial tissue, the adrenal cortex is derived from mesoderm.

Trabeculae of the connective tissue capsule penetrate into the adrenal gland in several locations, and numerous small blood

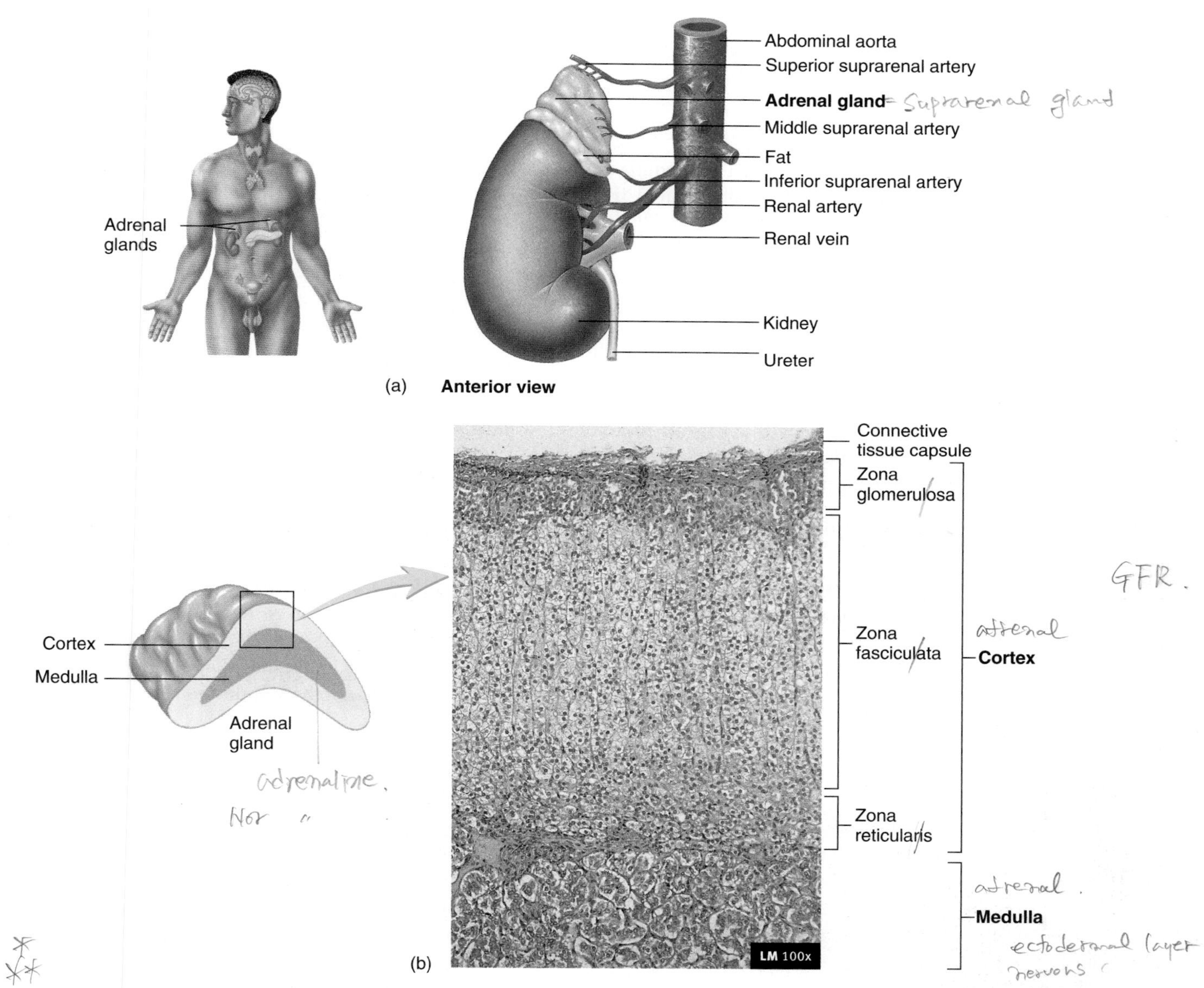

FIGURE 18.13 Anatomy and Histology of Adrenal Glands

(*a*) An adrenal gland is at the superior pole of each kidney. (*b*) The adrenal glands have an outer cortex and an inner medulla. The cortex is surrounded by a connective tissue capsule and consists of three layers: the zona glomerulosa, the zona fasciculata, and the zona reticularis.

TABLE 18.7 Hormones of the Adrenal Gland

Hormones	Structure	Target Tissue	Response
Adrenal Medulla			
Epinephrine primarily; norepinephrine	Amino acid derivatives	Heart, blood vessels, liver, fat cells	Increased cardiac output; increased blood flow to skeletal muscles and increased blood flow to the heart (see chapter 20); vasoconstriction of blood vessels, especially visceral and skin blood vessels; increased release of glucose and fatty acids into blood; in general, preparation for physical activity
Adrenal Cortex			
Glucocorticoids (cortisol)	Steroids	Most tissues	Increased protein and fat breakdown; increased glucose production; inhibition of immune response and decreased inflammation
Mineralocorticoids (aldosterone)	Steroids	Kidney	Increased Na^+ reabsorption and K^+ and H^+ excretion; enhances water reabsorption
Androgens	Steroids	Many tissues	Minor importance in males; in females, development of some secondary sexual characteristics, such as axillary and pubic hair

vessels course with them to supply the gland. The medulla consists of closely packed polyhedral cells centrally located in the gland (figure 18.13*b*). The cortex is composed of smaller cells and forms three indistinct layers: the **zona glomerulosa** (glō-mār′ū-lōs-ă), the **zona fasciculata** (fa-sik′ū-lă-tă), and the **zona reticularis** (re-tik′ū-lăr′is). These three layers are functionally and structurally specialized. The zona glomerulosa is immediately beneath the capsule and is composed of small clusters of cells. Beneath the zona glomerulosa is the thickest part of the adrenal cortex, the zona fasciculata. In this layer, the cells form long columns, or fascicles, of cells that extend from the surface toward the medulla of the gland. The deepest layer of the adrenal cortex is the zona reticularis, which is a thin layer of irregularly arranged cords of cells.

Hormones of the Adrenal Medulla

The adrenal medulla secretes two major hormones: **epinephrine** (**adrenaline;** ă-dren′ă-lin), 80%, and **norepinephrine** (**noradrenaline;** nor-ă-dren′ă-lin), 20% (table 18.7). Epinephrine and norepinephrine are closely related to each other. In fact, norepinephrine is a precursor to the formation of epinephrine. Because the adrenal medulla consists of cells derived from the same cells that give rise to postganglionic sympathetic neurons, its secretory products are neurohormones.

Epinephrine and norepinephrine combine with adrenergic receptors, which are membrane-bound receptors in target cells. They are classified as either α-adrenergic or β-adrenergic receptors, and each of these categories has subcategories. All of the adrenergic receptors function through G protein mechanisms. The α-adrenergic receptors cause Ca^{2+} channels to open, cause the release of Ca^{2+} from endoplasmic reticulum by activating phospholipase enzymes, open K^+ channels, decrease cAMP synthesis, or stimulate the synthesis of eicosanoid molecules, such as prostaglandins. The β-adrenergic receptors all increase cAMP synthesis.

In this book, the effects of epinephrine and norepinephrine released from the adrenal medulla are described when the systems these hormones affect are discussed (see chapters 16, 20, 21, 24, and 26).

Epinephrine increases blood glucose levels. It combines with membrane-bound receptors in liver cells. Cyclic AMP, in turn, activates enzymes that catalyze the breakdown of glycogen to glucose and the release of glucose from the liver cells into the blood. Epinephrine also increases the breakdown of glycogen in muscle cells, but muscle cells do not release glucose into the blood. The glucose in muscle cells is metabolized in the muscle cells. Epinephrine also increases fat breakdown in adipose tissue, and fatty acids are released into the blood. The fatty acids can be taken up and metabolized by tissues such as skeletal and cardiac muscle. Epinephrine and norepinephrine increase the heart's rate and force of contraction and cause blood vessels to constrict in the skin, kidneys, gastrointestinal tract, and other viscera. Also, epinephrine causes dilation of blood vessels in skeletal muscles and cardiac muscle.

Secretion of adrenal medullary hormones prepares the individual for physical activity and is a major component of the fight-or-flight response (see chapter 16). The response results in reduced activity in organs not essential for physical activity and in increased blood flow and metabolic activity in organs that participate in physical activity. In addition, it mobilizes nutrients that can be used to sustain physical exercise.

The effects of epinephrine and norepinephrine are short-lived because they are rapidly metabolized, excreted, or taken up by tissues. Their half-life in the circulatory system is measured in minutes.

The release of adrenal medullary hormones primarily occurs in response to stimulation by sympathetic neurons because the adrenal medulla is a specialized part of the autonomic nervous system. Several conditions, including emotional excitement, injury, stress, exercise, and low blood glucose levels, lead to the release of adrenal medullary neurohormones (figure 18.14).

1. Stress, physical activity, and low blood glucose levels act as stimuli to the hypothalamus, resulting in increased sympathetic nervous system activity.
2. An increased frequency of action potentials conducted through the sympathetic division of the autonomic nervous system stimulates the adrenal medulla to secrete epinephrine and some norepinephrine into the circulatory system.
3. Epinephrine and norepinephrine act on their target tissues to produce responses.

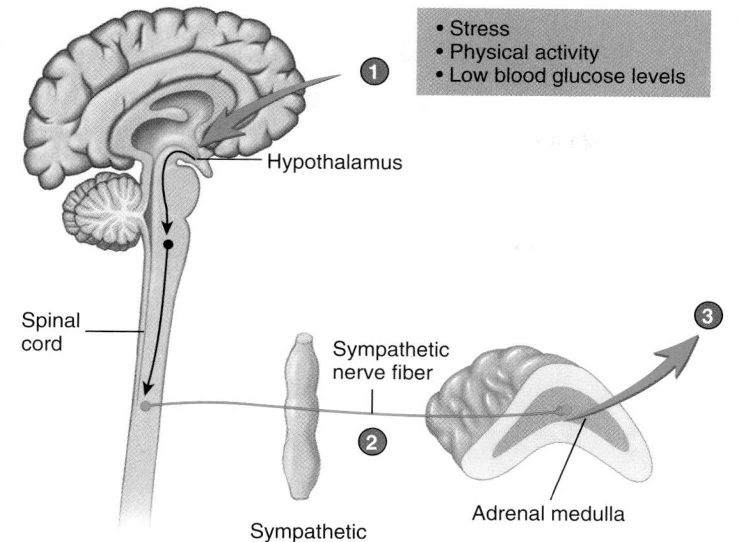

- Increases the release of glucose from the liver into the blood
- Increases the release of fatty acids from adipose tissue into the blood
- Increases heart rate
- Decreases blood flow through blood vessels of internal organs
- Increases blood flow through blood vessels of skeletal muscle and the heart
- Increases blood pressure
- Decreases the function of visceral organs
- Increases the metabolic rate of skeletal muscles

PROCESS FIGURE 18.14 Regulation of Adrenal Medullary Secretions
Stress, physical exercise, and low blood glucose levels cause increased activity of the sympathetic nervous system, which increases epinephrine and norepinephrine secretion from the adrenal medulla.

Pheochromocytoma and Neuroblastoma

The two major disorders of the adrenal medulla are both tumors: pheochromocytoma (fē′ō-krō′mō-sī-tō′mă), a benign tumor, and neuroblastoma (noor′ō-blas-tō′mă), a malignant tumor. Symptoms result from the release of large amounts of epinephrine and norepinephrine and include hypertension (high blood pressure), sweating, nervousness, pallor, and tachycardia (rapid heart rate). The high blood pressure results from the effect of these hormones on the heart and blood vessels and is correlated with an increased chance of heart disease and stroke.

26 ***Where are the adrenal glands located? Describe the embryonic origin of the adrenal medulla and adrenal cortex.***

27 ***Name two hormones secreted by the adrenal medulla, and list the effects of these hormones.***

28 ***List several conditions that can stimulate the production of adrenal medullary hormones. What role does the nervous system play in the release of adrenal medullary hormones? How does this role relate to the embryonic origin of the adrenal medulla?***

Hormones of the Adrenal Cortex

The adrenal cortex secretes three hormone types: **mineralocorticoids** (min′er-al-ō-kōr′ti-koydz), **glucocorticoids** (gloo-kō-kōr′ti-koydz), and **androgens** (an′drō-jenz; see table 18.7). All are similar in structure in that they are steroids, highly specialized lipids that are derived from cholesterol. Because they are lipid-soluble, they are not stored in the adrenal gland cells but diffuse from the cells as they are synthesized. Adrenal cortical hormones are transported in the blood in combination with specific plasma proteins; they are metabolized in the liver and excreted in the bile and urine. The hormones of the adrenal cortex bind to intracellular receptors and stimulate the synthesis of specific proteins that are responsible for producing the cell's responses.

Mineralocorticoids

The major secretory products of the zona glomerulosa are the mineralocorticoids. **Aldosterone** (al-dos′ter-ōn) is produced in the greatest amounts, although other closely related mineralocorticoids are also secreted. Aldosterone increases the rate of sodium reabsorption by the kidneys, thereby increasing blood levels of sodium. Sodium reabsorption can result in increased water reabsorption by the kidneys and an increase in blood volume providing ADH is also secreted. Aldosterone increases K^+ excretion into the urine by the kidneys, thereby decreasing blood levels of K^+. It also increases the rate of H^+ excretion into the urine. When aldosterone is secreted in high concentrations, it can result in reduced blood levels of K^+ and alkalosis (elevated pH of body fluids). The details of the effects of aldosterone and the mechanisms controlling aldosterone secretion are discussed along with kidney functions in chapters 26 and 27 and with the cardiovascular system in chapter 21.

TABLE 18.8 Target Tissues and Their Responses to Glucocorticoid Hormones

Target Tissues	Responses
Peripheral tissues, such as skeletal muscle, liver, and adipose tissue	Inhibits glucose use; stimulates the formation of glucose from amino acids and, to some degree, from fats (gluconeogenesis) in the liver, which results in elevated blood glucose levels; stimulates glycogen synthesis in cells; mobilizes fats by increasing lipolysis, which results in the release of fatty acids into the blood and an increased rate of fatty acid metabolism; increases protein breakdown and decreases protein synthesis
Immune tissues	Anti-inflammatory—depresses antibody production, white blood cell production, and the release of inflammatory components in response to injury; the immune system is suppressed
Target cells for epinephrine	Receptor molecules for epinephrine and norepinephrine decrease without adequate amounts of glucocorticoid hormone

PREDICT 6

Alterations in blood levels of sodium and potassium have profound effects on the electrical properties of cells. Because high blood levels of aldosterone cause retention of sodium and excretion of potassium, predict and explain the effects of high aldosterone levels on nerve and muscle function. Conversely, because low blood levels of aldosterone cause low blood levels of sodium and elevated blood levels of potassium, predict the effects of low aldosterone levels on nerve and muscle function.

Glucocorticoids

The zona fasciculata of the adrenal cortex primarily secretes **glucocorticoid hormones,** the major one of which is **cortisol** (kōr′ti-sol). The target tissues and responses to the glucocorticoids are numerous (table 18.8). The responses are classified as metabolic, developmental, or anti-inflammatory. Glucocorticoids increase fat catabolism, decrease glucose and amino acid uptake in skeletal muscle, increase **gluconeogenesis** (gloo′kō-nē-ō-jen′ĕ-sis; the synthesis of glucose from precursor molecules, such as amino acids in the liver), and increase protein degradation. Thus, a major effect of glucocorticoids is an increased use of fats and proteins to provide energy for cells. There is an increase in blood glucose levels and glycogen deposits in cells. The glucose and glycogen are a reservoir of molecules that can be metabolized rapidly. Glucocorticoids are also required for the maturation of tissues, such as fetal lungs, and for the development of receptor molecules in target tissues for epinephrine and norepinephrine. Glucocorticoids decrease the intensity of the inflammatory and immune responses by decreasing both the number of white blood cells and the secretion of inflammatory chemicals from tissues. This anti-inflammatory effect is important under conditions of stress, when the rate of glucocorticoid secretion is relatively high. Synthetic glucocorticoids are often used to suppress the immune response in people suffering from autoimmune conditions and in people who are transplant recipients (see chapter 22).

ACTH is necessary to maintain the secretory activity of the adrenal cortex, which rapidly atrophies without this hormone. Corticotropin-releasing hormone (CRH) is released from the hypothalamus and stimulates the anterior pituitary to secrete ACTH. The zona fasciculata is very sensitive to ACTH, and it responds by increasing cortisol secretion. At high concentrations, ACTH acts on the zona glomerulosa to enhance aldosterone secretion. The regulation of ACTH and cortisol secretion is outlined in figure 18.15. Both ACTH and cortisol inhibit CRH secretion from the hypothalamus and thus constitute a negative-feedback influence on CRH secretion. In addition, high concentrations of cortisol in the blood inhibit ACTH secretion from the anterior pituitary, and low concentrations stimulate it. This negative-feedback loop is important in maintaining blood cortisol levels within a narrow range of concentrations. In response to stress or hypoglycemia, blood levels of cortisol increase rapidly because these stimuli trigger a large increase in CRH release from the hypothalamus. Table 18.9 outlines several abnormalities associated with the hyposecretion and hypersecretion of adrenal hormones.

PREDICT 7

Cortisone, a drug similar to cortisol, is sometimes given to people who have severe allergies or extensive inflammation or who suffer from autoimmune diseases. Taking this substance chronically can damage the adrenal cortex. Explain how this damage can occur.

Adrenal Androgens

Some adrenal steroids, including **androstenedione** (an-drō-stēn′dī-ōn), are weak androgens. Androgens are steroid hormones that cause the development of male secondary sexual characteristics (see chapter 28). Androgens are secreted by the zona reticularis and converted by peripheral tissues to the more potent androgen testosterone. Adrenal androgens stimulate pubic and axillary hair growth and sexual drive in females. Their effects in males are negligible, in comparison with testosterone secreted by the testes. Chapter 28 presents additional information about androgens.

29 ***Describe the three layers of the adrenal cortex, and name the hormones produced by each layer.***

30 ***Name the target tissue of aldosterone, and list the effects of an increase in aldosterone secretion on the concentration of ions in the blood.***

31 ***Describe the effects produced by an increase in cortisol secretion. Starting in the hypothalamus, describe how stress or low blood sugar levels can stimulate cortisol release.***

32 ***What effects do adrenal androgens have on males and females?***

1. Cortiocotropin-releasing hormone (CRH) is released from hypothalamic neurons in response to stress or low blood glucose and passes, by way to the hypothalamohypophysial portal system, to the anterior pituitary.
2. In the anterior pituitary, CRH binds to and stimulates cells that secrete adrenocorticotropic hormone (ACTH).
3. ACTH binds to membrane-bound receptors on cells of the adrenal cortex and stimulates the secretion of glucocorticoids, primarily cortisol.
4. Cortisol acts on target tissues, resulting in increased fat and protein breakdown, increased glucose levels, and anti-inflammatory effects.
5. Cortisol has a negative-feedback effect because it inhibits CRH release from the hypothalamus and ACTH secretion from the anterior pituitary.

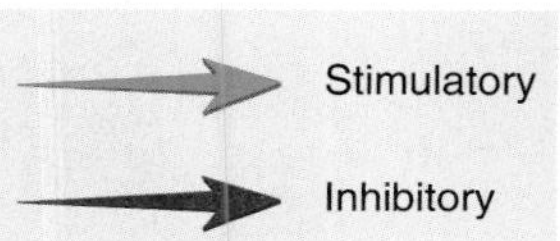

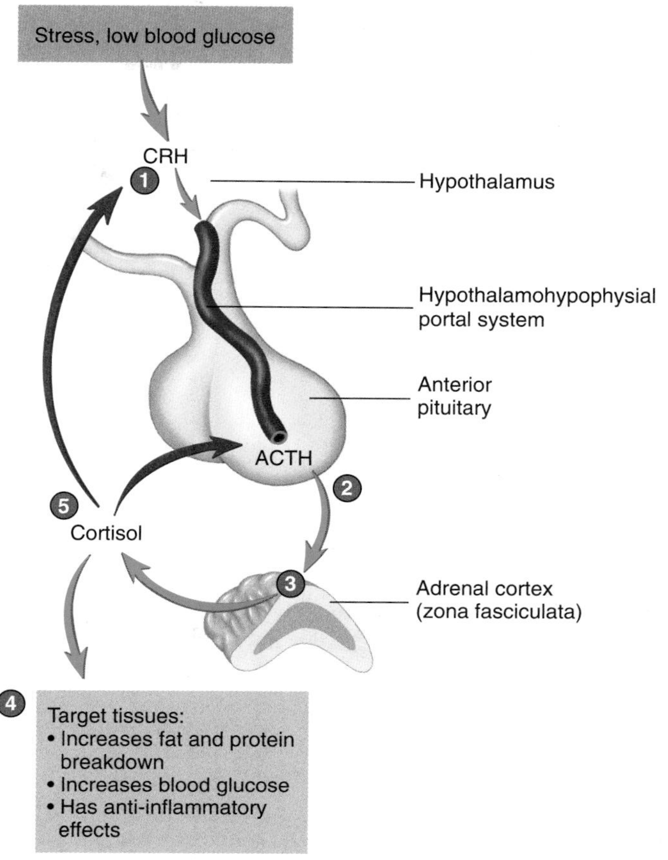

PROCESS FIGURE 18.15 **Regulation of Cortisol Secretion**

TABLE 18.9 Symptoms of Hyposecretion and Hypersecretion of Adrenal Cortex Hormones

Hyposecretion	Hypersecretion
Mineralocorticoids (Aldosterone)	
Hyponatremia (low blood levels of sodium)	Slight hypernatremia (high blood levels of sodium)
Hyperkalemia (high blood levels of potassium)	Hypokalemia (low blood levels of potassium)
Acidosis	Alkalosis
Low blood pressure	High blood pressure
Tremors and tetany of skeletal muscles	Weakness of skeletal muscles
Polyuria	Acidic urine
Glucocorticoids (Cortisol)	
Hypoglycemia (low blood glucose levels)	Hyperglycemia (high blood glucose levels; adrenal diabetes)—leads to diabetes mellitus
Depressed immune system	Depressed immune system
Protein and fats from diet unused, resulting in weight loss	Destruction of tissue proteins, causing muscle atrophy and weakness osteoporosis, weak capillaries (easy bruising), thin skin, and impaired wound healing; mobilization and redistribution of fats, causing depletion of fat from limbs and deposition in face (moon face), neck (buffalo hump), and abdomen
Loss of appetite, nausea, and vomiting	Emotional effects, including euphoria and depression
Increased skin pigmentation (occurs if ACTH levels are elevated)	
Androgens	
In women, reduction of pubic and axillary hair	In women, hirsutism (excessive facial and body hair), acne, increased sex drive, regression of breast tissue, and loss of regular menses

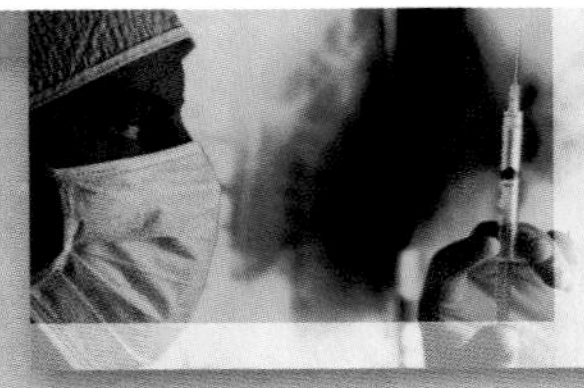

CLINICAL FOCUS

Hormone Pathologies of the Adrenal Cortex

Several pathologies are associated with abnormal secretion of adrenal cortex hormones. **Chronic adrenocortical insufficiency,** often called **Addison disease,** results from abnormally low levels of aldosterone and cortisol. The cause of many cases of chronic adrenocortical insufficiency is unknown, but it frequently results from an autoimmune disease in which the body's defense mechanisms inappropriately destroy the adrenal cortex. Bacteria such as tuberculosis bacteria, acquired immunodeficiency syndrome (AIDS), fungal infections, adrenal hemorrhage, and cancer can also damage the adrenal cortex, thus causing some cases of chronic adrenocortical insufficiency. Prolonged treatment with glucocorticoids, which suppresses pituitary gland function, can cause chronic adrenocortical insufficiency. Tumors that damage the hypothalamus can result in reduced secretion of ACTH from the anterior pituitary gland, thus resulting in chronically reduced secretion of hormones from the adrenal cortex. Symptoms of chronic adrenocortical insufficiency include weakness, fatigue, weight loss, anorexia, and in many cases increased pigmentation of the skin. Reduced blood pressure results from the loss of Na^+ and water through the kidneys. Reduced blood pressure is the most critical manifestation and requires immediate treatment. Low blood levels of Na^+, high blood levels of K^+, and reduced blood pH are consistent with the condition.

Aldosteronism (al-dos′ter-on-izm) is caused by excess production of aldosterone. Primary aldosteronism results from an adrenal cortex tumor, and secondary aldosteronism occurs when an extraneous factor, such as an overproduction of renin, a substance produced by the kidneys, increases aldosterone secretion. Major symptoms of aldosteronism include reduced blood levels of K^+, increased blood pH, and elevated blood pressure. Elevated blood pressure is a result of the retention of water and Na^+ by the kidneys.

Cushing syndrome (figure A) is a disorder characterized by hypersecretion of cortisol and androgens and possibly by excess aldosterone production. Most cases are caused by excess ACTH production by nonpituitary tumors, which usually result from a type of lung cancer. Some cases of increased ACTH secretion do result from pituitary tumors. Sometimes adrenal tumors or unidentified causes can be responsible for hypersecretion of the adrenal cortex without increases in ACTH secretion. Elevated secretion of glucocorticoids results in muscle wasting, the accumulation of adipose tissue in the face and trunk of the body, and increased blood glucose levels.

Hypersecretion of androgens from the adrenal cortex causes a condition called **adrenogenital** (ă-drē′nō-jen′i-tăl) **syndrome,** in which secondary sexual characteristics develop early in male children, and female children are masculinized. If the condition develops before birth in females, the external genitalia can be masculinized to the extent that the infant's reproductive structures are neither clearly female nor male. Hypersecretion of adrenal androgens in male children before puberty results in rapid and early development of the reproductive system. If not treated, early sexual development and short stature result. The short stature results from the effect of androgens on skeletal growth (see chapter 6). In women, partial development of male secondary sexual characteristics, such as facial hair and a masculine voice, occurs.

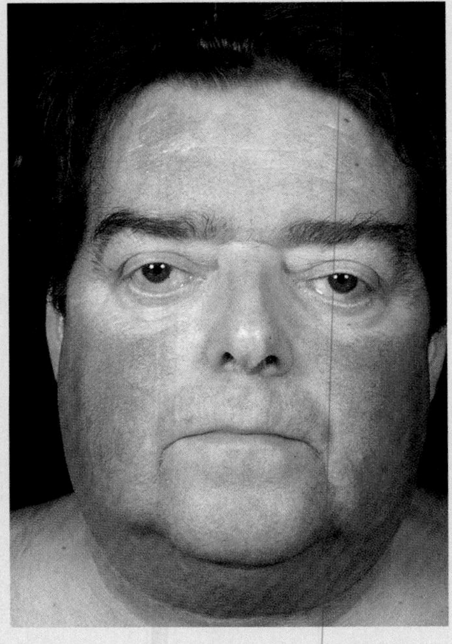

FIGURE A Male Patient with Cushing Syndrome

PANCREAS

The **pancreas** (pan′krē-us) lies behind the peritoneum between the greater curvature of the stomach and the duodenum. It is an elongated structure approximately 15 cm long weighing approximately 85–100 g. The head of the pancreas lies near the duodenum, and its body and tail extend toward the spleen.

The pancreas is both an exocrine gland and an endocrine gland. The exocrine portion consists of **acini** (as′i-nī), which produce pancreatic juice, and a duct system, which carries the pancreatic juice to the small intestine (see chapter 24). The endocrine part, consisting of **pancreatic islets** (islets of Langerhans),

CASE STUDY

Cushing Syndrome

Fred noticed that he had gained a substantial amount of weight over the past few months and that he was feeling weak. His physician observed that the fat distribution was mainly in his trunk, face, and neck. There was also evidence of decreased muscle mass, and Fred had several bruises on his upper and lower limbs. Results of a routine blood sample showed elevated blood glucose levels and low blood K^+ levels. There was no observable evidence that Fred had cancer. His physician suspected that Fred was suffering from Cushing syndrome.

A second blood sample was taken and blood cortisol and ACTH were measured. Fred's blood cortisol levels were very high and his blood ACTH levels were very low. Based on these data, Freds' physician explained to him that he was probably suffering from an adrenal gland tumor, which was secreting large amounts of cortisol, and that the tumor was not responding to the normal negative-feedback mechanisms that normally control cortisol secretion. Subsequently, imaging techniques indicated that a tumor was present in Fred's left adrenal gland. His left adrenal gland was surgically removed and his symptoms decreased dramatically over the next few weeks.

PREDICT 8

a. Why did the physician suspect Cushing syndrome?

b. Explain why Fred's physician concluded that a hormone-secreting tumor was likely to be located in one of Fred's adrenal glands.

c. After surgical removal of his left adrenal gland, how did Fred's blood cortisol and ACTH levels change?

d. How would the data from the second blood sample have been different if Fred's condition had been due to a hormone-secreting tumor in his anterior pituitary gland?

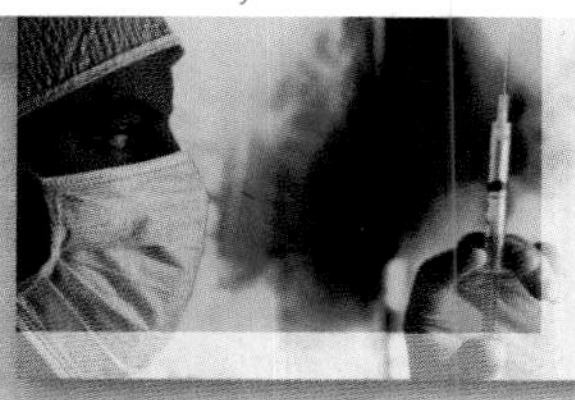

CLINICAL FOCUS

Stress

The adrenal cortex and the adrenal medulla play major roles in response to stress.

In general, stress activates nervous and endocrine responses that prepare the body for physical activity, even when physical activity is not the most appropriate response to the stressful conditions, such as during an examination or other mentally stressful situations. The endocrine response to stress involves increased CRH release from the hypothalamus and increased sympathetic stimulation of the adrenal medulla. CRH stimulates ACTH secretion from the anterior pituitary, which in turn stimulates cortisol from the adrenal cortex. Increased sympathetic stimulation of the adrenal medulla increases epinephrine and norepinephrine secretion.

Together, epinephrine and cortisol increase blood glucose levels and the release of fatty acids from adipose tissue and the liver. Sympathetic innervation of the pancreas decreases insulin secretion. Consequently, most tissues do not readily take up and use glucose. Thus, glucose is available primarily to the nervous system, and fatty acids are used by skeletal muscle, cardiac muscle, and other tissues.

Epinephrine and sympathetic stimulation also increase cardiac output, increase blood pressure, and act on the central nervous system to increase alertness and aggressiveness. Cortisol also decreases the initial inflammatory response.

Responses to stress illustrate the close relationship between the nervous and endocrine systems and provide an example of their integrated functions. Our ability to respond to stressful conditions depends on the nervous and endocrine responses to stress.

Although responses to stress are adaptive under many circumstances, they can become harmful. For example, if stress is chronic, the elevated secretion of cortisol and epinephrine produces harmful effects.

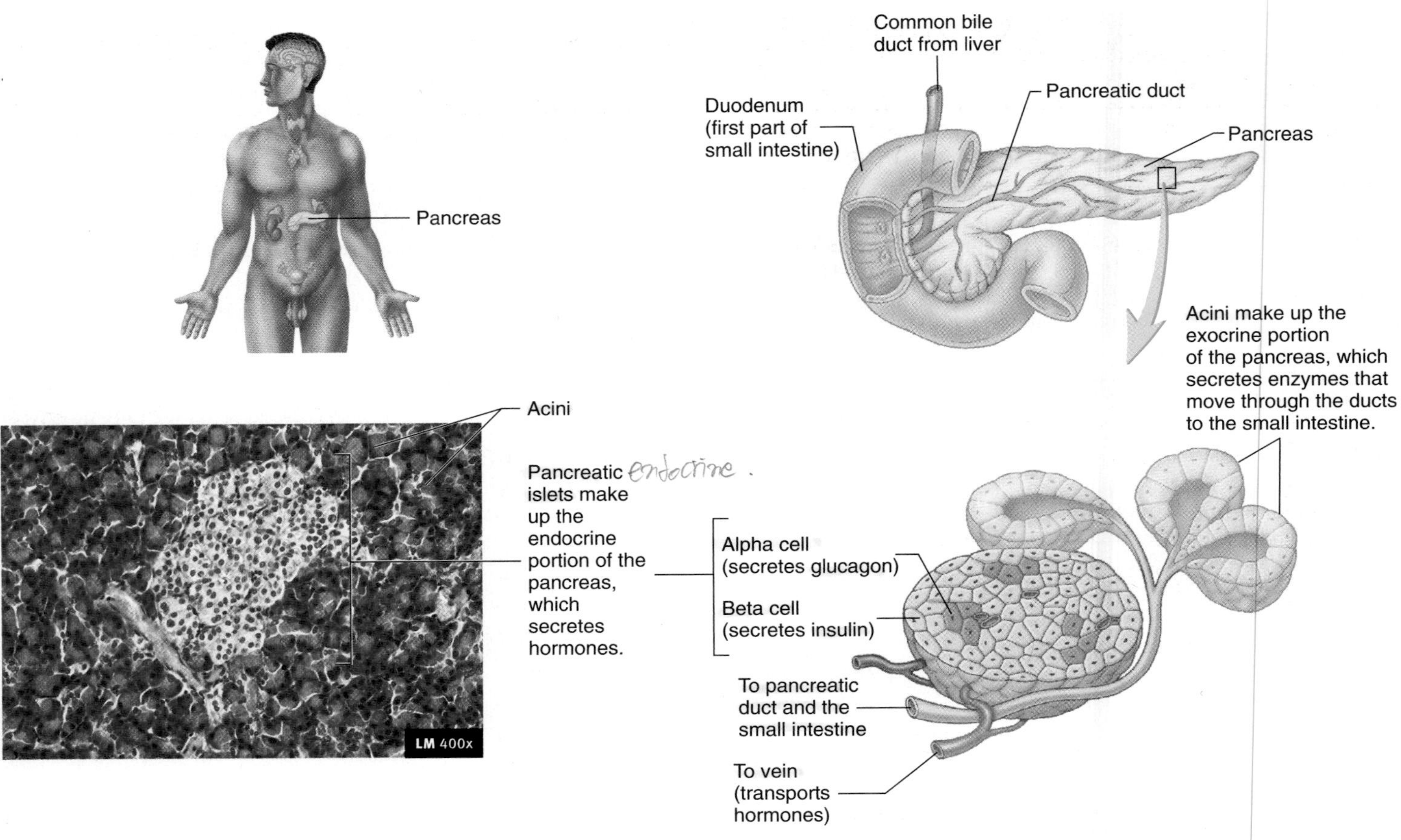

FIGURE 18.16 Histology of the Pancreatic Islets

A pancreatic islet consists of clusters of specialized cells among the acini of the exocrine portion of the pancreas. The stain used for this slide does not distinguish between alpha and beta cells.

which (figure 18.16) secrete hormones that enter the circulatory system.

Between 500,000 and 1 million pancreatic islets are dispersed among the ducts and acini of the pancreas. Each islet is composed of **alpha (α) cells** (20%), which secrete **glucagon,** a small polypeptide hormone; **beta (β) cells** (75%), which secrete **insulin,** a small protein hormone consisting of two polypeptide chains bound together; and other cell types (5%). The remaining cells are either immature cells of questionable function or **delta (δ) cells,** which secrete somatostatin, a small polypeptide hormone. Nerves from both divisions of the autonomic nervous system innervate the pancreatic islets, and a well-developed capillary network surrounds each islet.

Effect of Insulin and Glucagon on Their Target Tissues

The pancreatic hormones play an important role in regulating the concentration of critical nutrients in the circulatory system, especially glucose and amino acids (table 18.10). The major target tissues of insulin are the liver, adipose tissue, the skeletal muscles, and the satiety center within the hypothalamus of the brain. The **satiety** (sa′-tī-ĕ-tē) **center** is a collection of neurons in the hypothalamus that controls appetite, but insulin does not directly affect most areas of the nervous system. The specific effects of insulin on these target tissues are listed in table 18.11.

Insulin molecules bind to membrane-bound receptors on target cells. Once insulin molecules bind their receptors, the receptors cause specific proteins in the membrane to become phosphorylated. Part of the cells' response to insulin is to increase the number of transport proteins in the membrane of cells for glucose and amino acids. Finally, insulin and its receptors are taken by endocytosis into the cell. The insulin molecules are released from the insulin receptors and broken down within the cell, and the insulin receptor can once again become associated with the plasma membrane.

In general, the target tissue response to insulin is an increase in its ability to take up and use glucose and amino acids. Glucose molecules that are not needed immediately as an energy source to maintain cell metabolism are stored as glycogen in skeletal muscle,

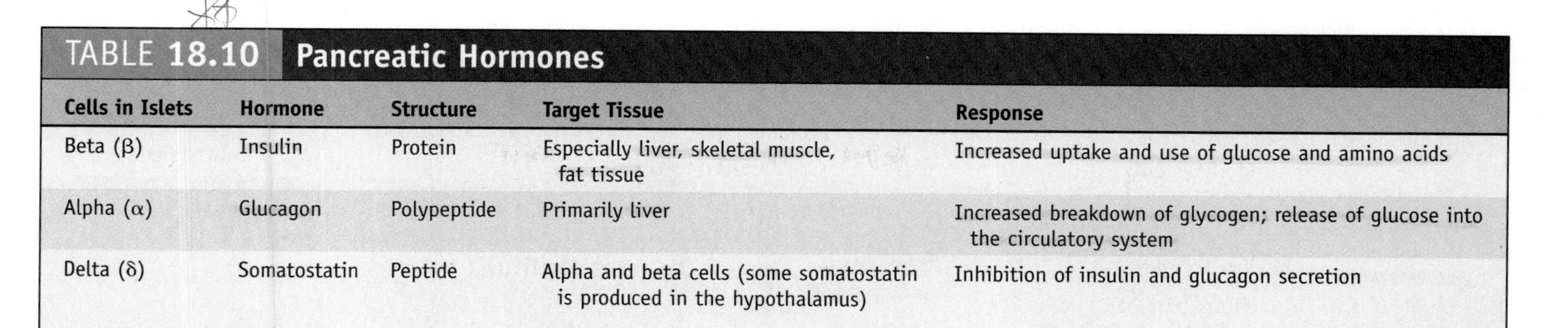

TABLE 18.10 Pancreatic Hormones

Cells in Islets	Hormone	Structure	Target Tissue	Response
Beta (β)	Insulin	Protein	Especially liver, skeletal muscle, fat tissue	Increased uptake and use of glucose and amino acids
Alpha (α)	Glucagon	Polypeptide	Primarily liver	Increased breakdown of glycogen; release of glucose into the circulatory system
Delta (δ)	Somatostatin	Peptide	Alpha and beta cells (some somatostatin is produced in the hypothalamus)	Inhibition of insulin and glucagon secretion

TABLE 18.11 Effect of Insulin and Glucagon on Target Tissues

Target Tissue	Response to Insulin	Response to Glucagon
Skeletal muscle, cardiac muscle, cartilage, bone, fibroblasts, leukocytes, and mammary glands	Increased glucose uptake and glycogen synthesis; increased uptake of certain amino acids	Little effect
Liver	Increased glycogen synthesis; increased use of glucose for energy (glycolysis)	Rapid increase in the breakdown of glycogen to glucose (glycogenolysis) and release of glucose into the blood; increased formation of glucose (gluconeogenesis) from amino acids and, to some degree, from fats; increased metabolism of fatty acids, resulting in increased ketones in the blood
Adipose cells	Increased glucose uptake, glycogen synthesis, fat synthesis, and fatty acid uptake; increased glycolysis	High concentrations cause breakdown of fats (lipolysis); probably unimportant under most conditions
Nervous system	Little effect except to increase glucose uptake in the satiety center	No effect

the liver, and other tissues and are converted to fat in adipose tissue. Amino acids can be broken down and used as an energy source or to synthesize glucose, or they can be converted to protein. Without insulin, the ability of these tissues to take up glucose and amino acids and use them is minimal.

The normal regulation of blood glucose levels requires insulin. Blood glucose levels can increase dramatically when too little insulin is secreted or when insulin receptors do not respond to it (see Clinical Focus "Diabetes Mellitus," p. 636). In the absence of insulin, the movement of glucose and amino acids into cells declines dramatically, even though blood levels of these molecules may increase to very high levels. The satiety center requires insulin to take up glucose. In the absence of insulin, the satiety center cannot detect the presence of glucose in the extracellular fluid, even when high levels are present. The result is an intense sensation of hunger in spite of high blood glucose levels, a condition called polyphagia. High blood glucose levels cause increased urine volume (polyuria) and loss of water in the urine because glucose enters kidney tubules. The high glucose concentration attracts water because of osmosis and increases the amount of water in urine (see chapter 26). Increased blood glucose levels also cause an increased sensation of thirst (polydipsia; see chapter 27).

Blood glucose levels can fall very low when too much insulin is secreted. When too much insulin is present, target tissues rapidly take up glucose from the blood, causing blood levels of glucose to decline to very low levels. Although the nervous system, except for cells of the satiety center, is not a target tissue for insulin, the nervous system depends primarily on blood glucose for a nutrient source. Consequently, low blood glucose levels cause the central nervous system to malfunction.

Glucagon primarily influences the liver, although it has some effect on skeletal muscle and adipose tissue (see table 18.11). Glucagon binds to membrane-bound receptors, activates G proteins, and increases cAMP synthesis. In general, glucagon causes the breakdown of glycogen and increases glucose synthesis in the liver. It also increases the breakdown of fats. The amount of glucose released from the liver into the blood increases dramatically after glucagon secretion increases. Because glucagon is secreted into the hepatic portal circulation, which carries blood from the intestine and pancreas to the liver, glucagon is delivered in a relatively high concentration to the liver, where it has its major effect. The liver also rapidly metabolizes it. Thus, glucagon has less of an effect on skeletal muscles and adipose tissue than on the liver.

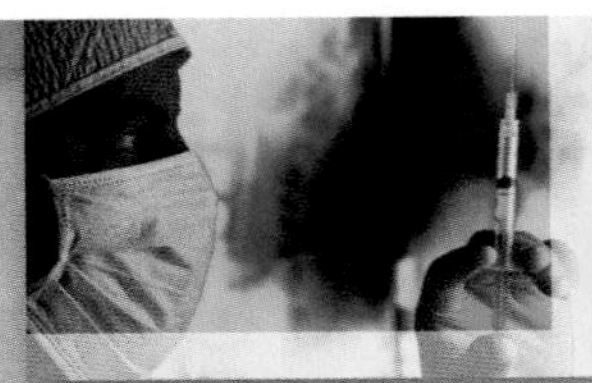

CLINICAL FOCUS

Diabetes Mellitus

Diabetes mellitus results primarily from the inadequate secretion of insulin or the inability of tissues to respond to insulin. **Type 1 diabetes mellitus,** also called **insulin-dependent diabetes mellitus (IDDM),** affects approximately 3% of people with diabetes mellitus and results from diminished insulin secretion. It develops as a result of autoimmune destruction of the pancreatic islets, and symptoms appear after approximately 90% of the islets have been destroyed. Type 1 diabetes mellitus most commonly develops in young people. Heredity may play a role in the condition, although the initiation of pancreatic islet destruction may involve a viral infection of the pancreas (see "Systems Pathology," p. 644).

Type 2 diabetes mellitus, also called **noninsulin-dependent diabetes mellitus (NIDDM),** results from the inability of tissues to respond to insulin. Type 2 diabetes mellitus usually develops in people older than 40–45 years of age, although the age of onset varies considerably.

Type 2 diabetes mellitus is more common than type 1 diabetes mellitus. Approximately 80% of people who have diabetes mellitus have type 2 diabetes mellitus. People with type 2 diabetes mellitus have a reduced number of functional receptors for insulin, or one or more of the enzymes activated by the insulin receptor are defective. Thus, the uptake of glucose by cells is very slow, which results in elevated blood glucose levels after a meal. Obesity is common, although not universal, in patients with type 2 diabetes mellitus. Elevated blood glucose levels cause fat cells to convert glucose to fat, even though the rate at which adipose cells take up glucose is impaired. Increased blood glucose and increased urine production lead to hyperosmolality of blood and dehydration of cells. The poor use of nutrients and dehydration of cells lead to lethargy, fatigue, and periods of irritability. The elevated blood glucose levels lead to recurrent infections and prolonged wound healing.

Patients with type 2 diabetes mellitus do not suffer sudden, large increases in blood glucose and severe tissue wasting because a slow rate of glucose uptake does occur, even though the insulin receptors are defective. In some people with type 2 diabetes mellitus, insulin production eventually decreases because pancreatic islet cells atrophy and type 2 diabetes mellitus develops. Approximately 25%–30% of patients with type 2 diabetes mellitus take insulin, 50% take oral medication to increase insulin secretion and increase the efficiency of glucose utilization, and the remainder control blood glucose levels with exercise and diet.

Glucose tolerance tests are used to diagnose diabetes mellitus. In general, the test involves feeding the patient a large amount of glucose after a period of fasting. Blood samples are collected for a few hours, and a sustained increase in blood glucose levels strongly indicates that the person is suffering from diabetes mellitus.

Too much insulin relative to the amount of glucose ingested leads to insulin shock. The high levels of insulin cause target tissues to take up glucose at a very high rate. As a result, blood glucose levels rapidly fall to a low level. Because the nervous system depends on glucose as its major source of energy, neurons malfunction because of a lack of metabolic energy. The result is a series of nervous system responses that include disorientation, confusion, and convulsions. Too much insulin, too little food intake after an injection of insulin, or increased metabolism of glucose due to excess exercise by a diabetic patient can cause insulin shock.

It appears that damage to blood vessels and reduced nerve function can be reduced in diabetic patients suffering from either type 1 or type 2 diabetes mellitus by keeping blood glucose well within normal levels at all times. Doing so, however, requires increased attention to diet, frequent blood glucose testing, and an increased chance of suffering from low blood glucose levels, which leads to symptoms of insulin shock. A strict diet and routine exercise are often effective components of a treatment strategy for diabetes mellitus, and in many cases diet and exercise are adequate to control type 2 diabetes mellitus.

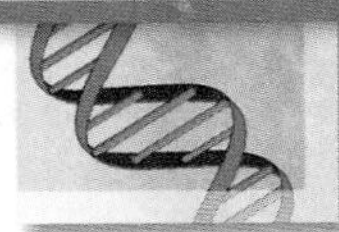

CLINICAL GENETICS

Type 2 Diabetes Mellitus

Most people who suffer from diabetes mellitus have type 2 diabetes mellitus. There is a genetic basis for type 2 diabetes mellitus, and it appears to involve several genes that can make people more susceptible to developing type 2 diabetes. For example, individuals who have close relatives who have type 2 diabetes mellitus have an increased risk for the development of the condition. In addition, type 2 diabetes mellitus is more prevalent among certain populations. For example, it is more prevalent in Native American populations than it is in Caucasians, African-Americans, and Hispanics.

The pathways activated by the insulin receptor are complex, and several genes that code for proteins in these pathways can be involved in type 2 diabetes mellitus. Genes on 10 different chromosomes have been associated with the development of type 2 diabetes mellitus. Antibodies that bind to insulin receptors and make the receptors nonfunctional, or decreases in the number of functional insulin receptors, can cause type 2 diabetes mellitus. In most cases, however, the insulin receptor is normal but mutations in genes that code for enzymes activated by the combination of insulin and its receptors result in a reduced response to insulin.

Type 2 diabetes mellitus involves a gradual failure of cells to respond to insulin and to take up glucose. Therefore, people who inherit genes that make them susceptible to type 2 diabetes mellitus are likely to develop the condition later in life. Also, symptoms are more likely to develop in people with an unhealthy lifestyle, which includes a diet high in calories and simple sugars and a sedentary tendency. A high percentage of people who have type 2 diabetes mellitus are obese, and the severity of the type 2 diabetes may decrease in response to weight loss. An unhealthy lifestyle is also associated with a trend to develop type 2 diabetes mellitus in people at younger ages than in the past.

The "thrifty genotype" hypothesis suggests that type 2 diabetes mellitus may be more common today because the genes that make people susceptible to type 2 diabetes mellitus might once have been beneficial. The tendency to store fat in times of plenty, and altered glucose metabolism, may have been advantageous during periods of famine, but today, when food is abundant, these genes increase the likelihood of type 2 diabetes mellitus.

Regulation of Pancreatic Hormone Secretion

Blood levels of nutrients, neural stimulation, and hormones control the secretion of insulin. Hyperglycemia, or elevated blood levels of glucose, directly affects the β cells and stimulates insulin secretion. Hypoglycemia, or low blood levels of glucose, directly inhibits insulin secretion. Thus, blood glucose levels play a major role in the regulation of insulin secretion. Certain amino acids also stimulate insulin secretion by acting directly on the β cells. After a meal, when glucose and amino acid levels increase in the circulatory system, insulin secretion increases. During periods of fasting, when blood glucose levels are low, the rate of insulin secretion declines (figure 18.17).

The autonomic nervous system also controls insulin secretion. Parasympathetic stimulation is associated with food intake, and its stimulation acts with the elevated blood glucose levels to increase insulin secretion. Sympathetic innervation inhibits insulin secretion and helps prevent a rapid fall in blood glucose levels. Because most tissues, except nervous tissue, require insulin to take up glucose, sympathetic stimulation maintains blood glucose levels in a normal range during periods of physical activity or excitement. This response is important for maintaining normal functioning of nervous system.

Gastrointestinal hormones involved with the regulation of digestion, such as gastrin, secretin, and cholecystokinin (see chapter 24), increase insulin secretion. Somatostatin inhibits insulin and glucagon secretion, but the factors that regulate somatostatin secretion are not clear. It can be released in response to food intake, in which case somatostatin may prevent the oversecretion of insulin.

PREDICT 9

Explain why the increase in insulin secretion in response to parasympathetic stimulation and gastrointestinal hormones is consistent with the maintenance of blood glucose levels in the circulatory system.

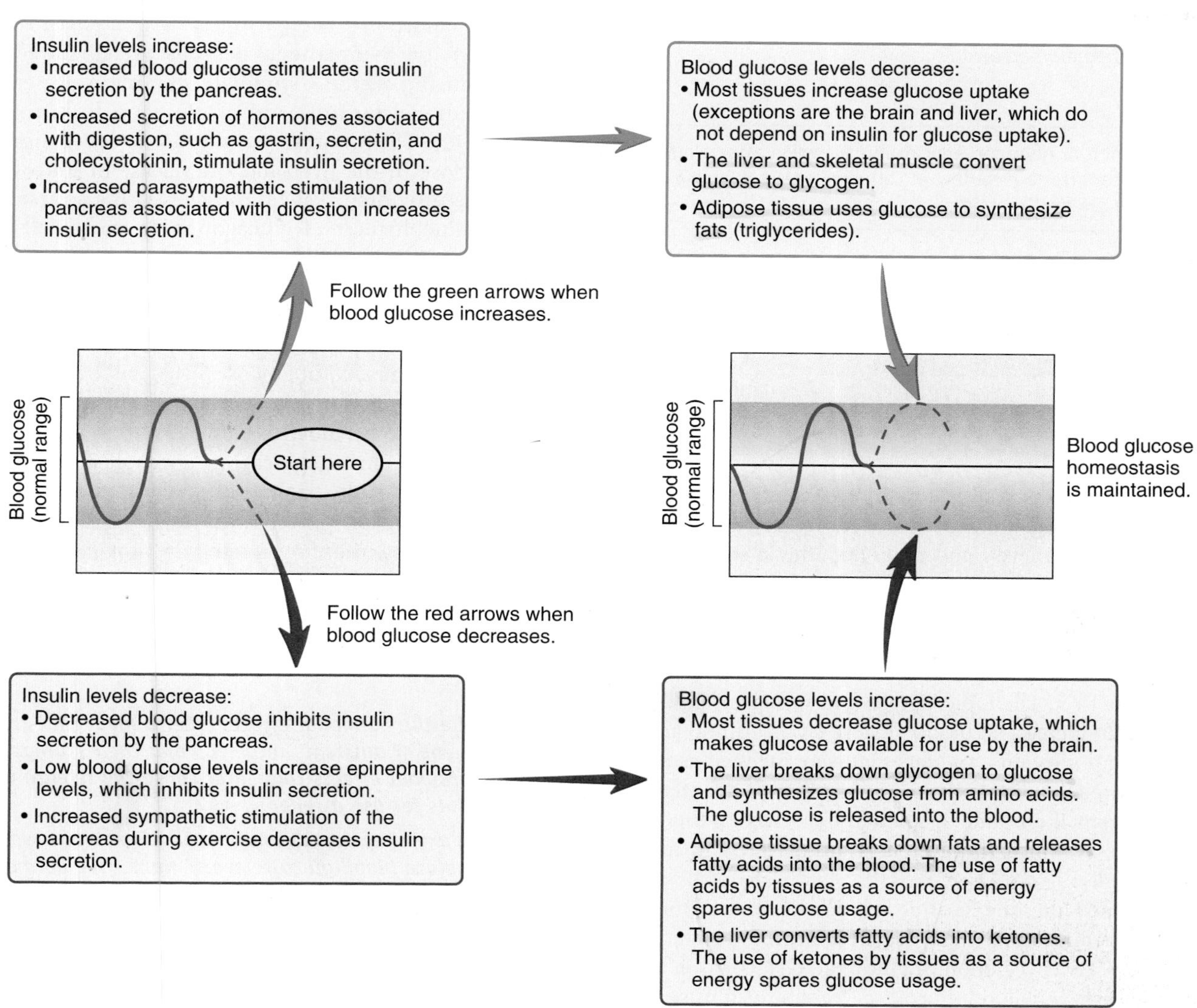

HOMEOSTASIS FIGURE 18.17 Summary of Insulin Secretion Regulation

Low blood glucose levels stimulate glucagon secretion, and high blood glucose levels inhibit it. Certain amino acids and sympathetic stimulation also increase glucagon secretion. After a high-protein meal, amino acids increase both insulin and glucagon secretion. Insulin causes target tissues to accept the amino acids for protein synthesis, and glucagon increases the process of glucose synthesis from amino acids in the liver (gluconeogenesis). Both protein synthesis and the use of amino acids to maintain blood glucose levels result from the low, but simultaneous, secretion of insulin and glucagon induced by meals high in protein content.

33 ***Where is the pancreas located? Describe the exocrine and endocrine parts of this gland and the secretions produced by each portion.***

34 ***Name the target tissues for insulin and glucagon, and list the effects they have on their target tissues.***

35 ***How does insulin affect the nervous system in general and the satiety center in the hypothalamus in particular?***

36 ***What effect do blood glucose levels, blood amino acid levels, the autonomic nervous system, and somatostatin have on insulin and glucagon secretion?***

PREDICT 10

Compare the regulation of glucagon and insulin secretion after a meal high in carbohydrates, after a meal low in carbohydrates but high in proteins, and during physical exercise.

HORMONAL REGULATION OF NUTRIENTS

Two situations—after a meal and during exercise—can illustrate how several hormones function together to regulate blood nutrient levels.

After a meal and under resting conditions, secretion of glucagon, cortisol, GH, and epinephrine is reduced (figure 18.18). Both increasing blood glucose levels and parasympathetic stimulation elevate insulin secretion to increase the uptake of glucose, amino acids, and fats by target tissues. Substances not immediately used for cell metabolism are stored. Glucose is converted to glycogen in skeletal muscle and the liver, and it is used for fat synthesis in adipose tissue and the liver. The rapid uptake and storage of glucose prevent too large an increase in blood glucose levels. Amino acids are incorporated into proteins, and fats that were ingested as part of the meal are stored in adipose tissue and the liver. If the meal is high in protein, a small amount of glucagon is secreted, thereby increasing the rate at which the liver uses amino acids to form glucose.

Within 1–2 hours after the meal, absorption of digested materials from the gastrointestinal tract declines, and blood glucose levels decline. As a result, secretion of glucagon, GH, cortisol, and epinephrine increases. As blood glucose decreases, insulin secretion decreases, and the rate of glucose entry into the target tissues for insulin also decreases. Glycogen, stored in cells, is converted back to glucose and is used as an energy source. Glucose is released into the blood by the liver. The decreased uptake of glucose by most tissues, combined with its release from the liver, helps maintain blood glucose at levels necessary for normal brain function. Cells that use less glucose start using more fats and proteins. Adipose tissue releases fatty acids, and the liver releases triglycerides (in lipoproteins) and ketones into the blood. Tissues take up these substances from the blood and use them for energy. Fat molecules are a major source of energy for most tissues when blood glucose levels are low.

The interactions of insulin, GH, glucagon, epinephrine, and cortisol are excellent examples of negative-feedback mechanisms. When blood glucose levels are high, these hormones cause the rapid uptake and storage of glucose, amino acids, and fats. When blood glucose levels are low, they cause the release of glucose and a switch to fat and protein metabolism as a source of energy for most tissues.

During exercise, skeletal muscles require energy to support the contraction process (see chapter 9). Although metabolism of intracellular nutrients can sustain muscle contraction for a short time, additional energy sources are required during prolonged activity. Sympathetic nervous system activity, which increases during exercise, stimulates the release of epinephrine from the adrenal medulla and of glucagon from the pancreas (figure 18.19). These hormones induce the conversion of glycogen to glucose in the liver and the release of glucose into the blood, thus providing skeletal muscles with a source of energy. Because epinephrine and glucagon have short half-lives, they can rapidly adjust blood glucose levels for varying conditions of activity.

During sustained activity, glucose released from the liver and other tissues is not adequate to support muscle activity, and a danger exists that blood glucose levels will become too low to support brain function. A decrease in insulin prevents the uptake of glucose by most tissues, thus conserving glucose for the brain. Epinephrine, glucagon, cortisol, and GH cause an increase of fatty acids, triglycerides, and ketones in the blood. Because GH increases protein synthesis and it slows the breakdown of proteins, muscle proteins are not used as an energy source. Consequently, glucose metabolism decreases, and fat metabolism in skeletal muscles increases. At the end of a long race, for example, muscles rely to a large extent on fat metabolism for energy.

37 ***Describe the hormonal effects after a meal that result in the movement of nutrients into cells and their storage. Describe the hormonal effects that later cause the release of stored materials for use as energy.***

38 ***During exercise, how does sympathetic nervous system activity regulate blood glucose levels? Name five hormones that interact to ensure that both the brain and muscles have adequate energy sources.***

PREDICT 11

Explain why long-distance runners may not have much of a "kick" left when they try to sprint to the finish line.

Soon after a meal, blood levels of nutrients, such as glucose, amino acids, and fatty acids, increase.

Stomach

Nutrients enter the blood.

Parasympathetic stimulation and increasing blood glucose levels cause increased insulin secretion from the pancreas.

Pancreas

Increased insulin secretion

Nutrients move into cells.

Within 1–2 hours after a meal, nutrient levels decline.

Decreasing blood glucose levels and increasing sympathetic stimulation cause insulin secretion to decrease.

Decreasing blood glucose levels and increasing sympathetic stimulation cause increased glucagon secretion.

Decreasing blood nutrient levels cause increased GH secretion.

Decreasing blood levels cause increased cortisol secretion.

Most cells

Glucose uptake decreases and switches to fat and protein metabolism.

Liver

Releases glucose, ketones, and triglycerides into circulation

Adipose tissue

Releases fatty acids into circulation

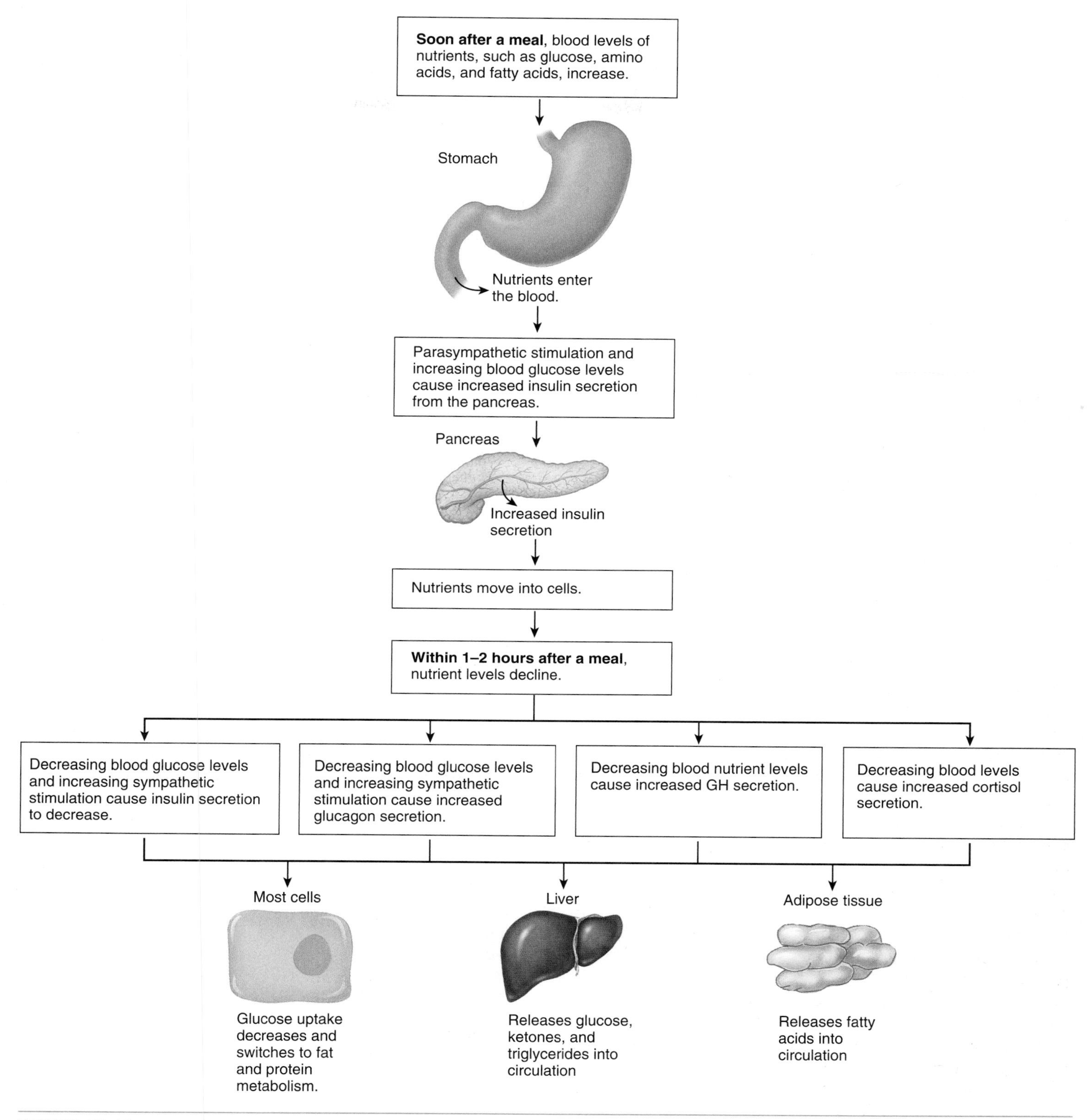

FIGURE 18.18 Regulation of Blood Nutrient Levels

Blood nutrient levels are maintained immediately after a meal and several hours after a meal.

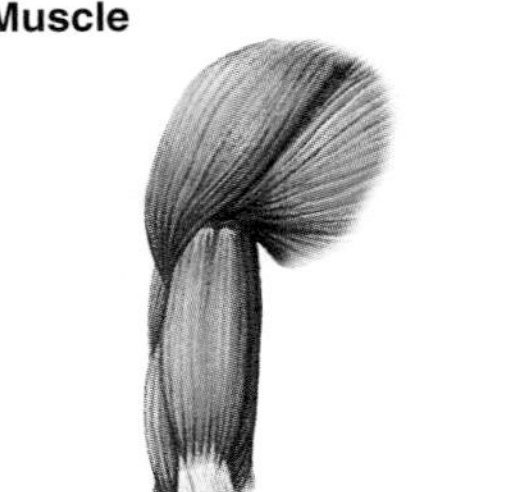

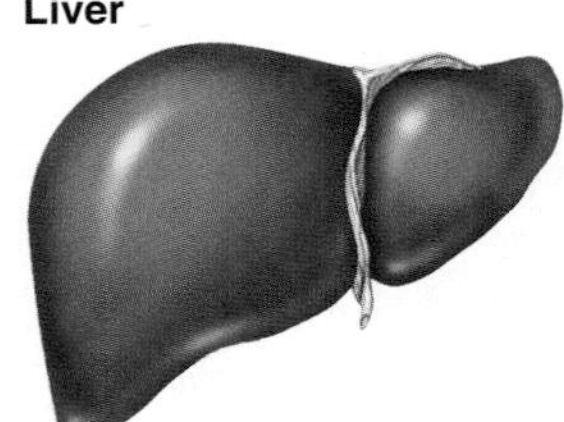

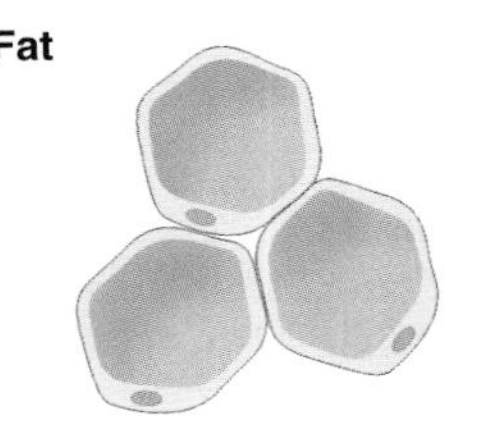

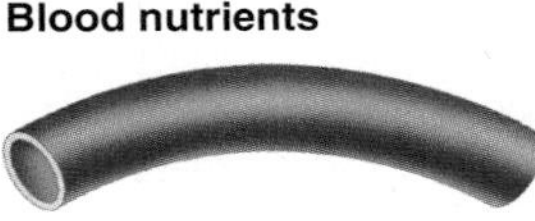

FIGURE 18.19 **Regulation of Blood Nutrient Levels During Exercise**

HORMONES OF THE REPRODUCTIVE SYSTEM

Reproductive hormones are secreted primarily from the ovaries, testes, placenta, and pituitary gland (table 18.12). These hormones are discussed in chapter 28. The main endocrine glands of the male reproductive system are the testes. The functions of the testes depend on the secretion of FSH and LH from the anterior pituitary gland. The main hormone secreted by the testes is testosterone, an androgen. **Testosterone** regulates the production of sperm cells by the testes and the development and maintenance of male reproductive organs and secondary sex characteristics. The testes secrete another hormone, called **inhibin,** which inhibits the secretion of FSH from the anterior pituitary.

The main endocrine glands of the female reproductive system are the ovaries. Like the testes, the functions of the ovaries depend on the secretion of FSH and LH from the anterior pituitary gland. The main hormones secreted by the ovaries are **estrogen** and **progesterone.** These hormones, along with FSH and LH, control the female reproductive cycle, prepare the mammary glands for lactation, and maintain pregnancy. Estrogen and progesterone are also responsible for the development of the female reproductive organs and female secondary sex characteristics. The ovaries also secrete inhibin, which inhibits FSH secretion.

During the first one-third of pregnancy, the placenta secretes an LH-like substance that is necessary to maintain pregnancy (see chapter 28). Throughout most of pregnancy, the ovaries and placenta secrete increasing amounts of estrogen and progesterone, which are also necessary to maintain pregnancy. In addition, the ovaries secrete **relaxin,** which increases the flexibility of connective tissue of the symphysis pubis and helps

TABLE 18.12 Hormones of the Reproductive Organs

Hormones	Structure	Target Tissue	Response
Testis			
Testosterone	Steroid	Most cells	Aids in spermatogenesis; development of genitalia; maintenance of functional reproductive organs; secondary sex characteristics; sexual behavior
Inhibin	Polypeptide	Anterior pituitary gland	Inhibits FSH secretion
Ovary			
Estrogens	Steroids	Most cells	Uterine and mammary gland development and function; maturation of genitalia; secondary sex characteristics; sexual behavior and menstrual cycle
Progesterone	Steroid	Most cells	Uterine and mammary gland development and function; maturation of genitalia; secondary sex characteristics; menstrual cycle
Inhibin	Polypeptide	Anterior pituitary gland	Inhibits FSH secretion
Relaxin	Polypeptide	Connective tissue cells	Increases flexibility of connective tissue in the pelvic area, especially the symphysis pubis

TABLE 18.13 Other Hormones and Hormonelike Substances

Chemical Signal	Structure	Target Tissue	Response
Pineal Gland			
Melatonin	Amino acid derivative	At least the hypothalamus	Inhibition of gonadotropin-releasing hormone-secretion, thereby inhibiting reproduction; significance is not clear in humans; may help regulate sleep–wake cycles
Arginine vasotocin	Peptide	Possibly the hypothalamus	Possible inhibition of gonadotropin-releasing hormone secretion
Thymus			
Thymosin	Peptide	Immune tissues	Development and function of the immune system
Several Tissues (Autocrine and Paracrine Regulatory Substances)			
Eicosanoids			
Prostaglandins	Modified fatty acid	Most tissues	Mediation of the inflammatory response; increases uterine contractions; involved in ovulation, possible inhibition of progesterone synthesis; blood coagulation; and other functions
Prostacyclins	Modified fatty acid	Most tissues	Mediation of the inflammatory response and other functions
Thromboxanes	Modified fatty acid	Most tissues	Mediation of the inflammatory response and other functions
Leukotrienes	Modified fatty acid	Most tissues	Mediation of the inflammatory response and other functions
Enkephalins and endorphins	Peptides	Nervous system	Reduction of pain sensation and other functions
Epidermal growth factor	Protein	Many tissues	Stimulation of division in many cell types; embryonic development
Fibroblast growth factor	Protein	Many tissues	Stimulation of cell division in many cell types; embryonic development
Interleukin-2	Protein	Certain immune competent cells	Stimulation of cell division of T lymphocytes

dilate the cervix of the uterus. This facilitates delivery by making the birth canal larger.

39 ***List the hormones secreted by the testes, and give their functions. What hormones regulate the testes?***

40 ***List the hormones secreted by the ovaries, and give their functions. During pregnancy, what other organ, in addition to the ovaries, secretes hormones? On what hormones does ovarian function depend?***

HORMONES OF THE PINEAL BODY

The **pineal** (pin′ē-ăl) **gland** (pineal body) in the epithalamus of the brain secretes hormones that act on the hypothalamus or the gonads to inhibit reproductive functions. Two substances have been proposed as secretory products: **melatonin** (mel-ă-tōn′in) and **arginine vasotocin** (ar′ji-nēn vā-sō-tō′sin, vas-ō-tos′in; table 18.13). Melatonin can decrease GnRH secretion from the hypothalamus and may inhibit reproductive functions through this

1. Light entering the eye stimulates neurons in the retina of the eye to produce action potentials.
2. Action potentials are transmitted to the hypothalamus in the brain.
3. Action potentials from the hypothalamus are transmitted through the sympathetic division to the pineal body.
4. A decrease in light (dark) results in increased sympathetic stimulation of the pineal gland and increased melatonin secretion. An increase in light results in decreased sympathetic stimulation of the pineal gland and decreased melatonin secretion.
5. Melatonin inhibits GnRH secretion from the hypothalamus and may help regulate sleep cycles.

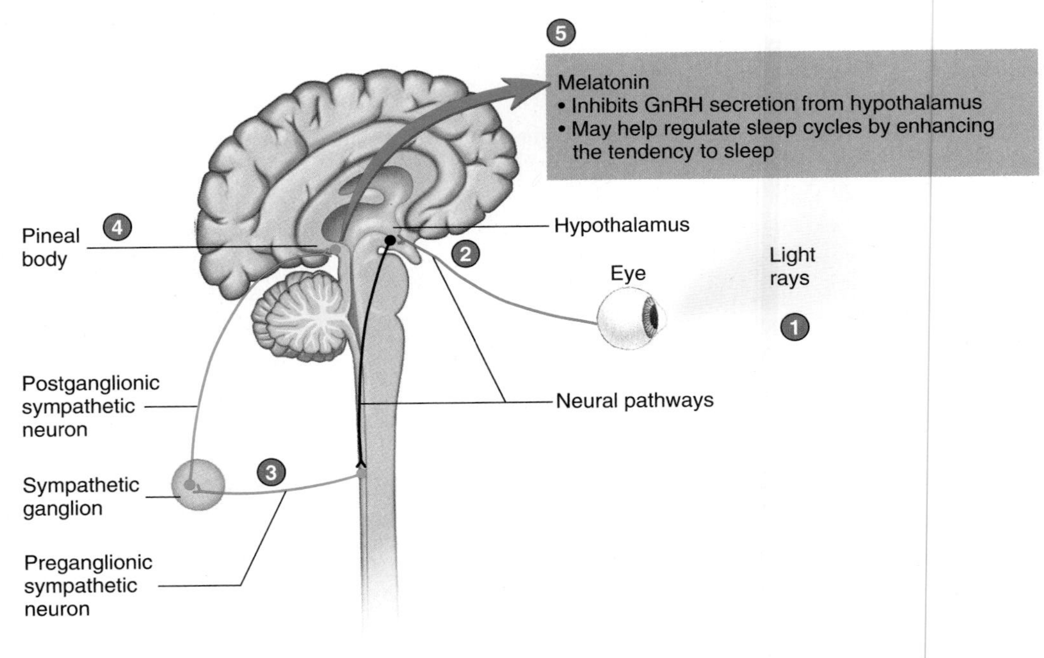

PROCESS FIGURE 18.20 Regulation of Melatonin Secretion from the Pineal Body
Light entering the eye inhibits and dark stimulates the release of melatonin from the pineal body.

mechanism. It may also help regulate sleep cycles by increasing the tendency to sleep.

The **photoperiod** is the amount of daylight and darkness that occurs each day and changes with the seasons of the year. In some animals, the photoperiod regulates pineal secretions (figure 18.20). For example, increased daylight initiates action potentials in the retina of the eye, which are propagated to the brain and cause a decrease in the action potentials sent first to the spinal cord and then through sympathetic neurons to the pineal body. Decreased pineal secretion results. In the dark, action potentials delivered by sympathetic neurons to the pineal body increase, thereby stimulating the secretion of pineal hormones. Humans secrete larger amounts of melatonin at night than in the daylight. In animals that breed in the spring, the increased length of a day decreases pineal secretions. Because pineal secretions inhibit reproductive functions in these species, the increased length of a day results in hypertrophy of the reproductive structures.

Melatonin's role in the regulation of reproductive functions in humans is not clear, but it is recommended by some to enhance sleep. Because melatonin causes atrophy of reproductive structures in some species, there may be undesirable side effects on the reproductive system.

The function of the pineal body in humans is not clear, but tumors that destroy the pineal body correlate with early sexual development, and tumors that result in pineal hormone secretion correlate with retarded development of the reproductive system. It is not clear, however, if the pineal body controls the onset of puberty.

Arginine vasotocin works with melatonin to regulate the function of the reproductive system in some animals. Evidence for the role of melatonin is more extensive, however.

41 ***Where is the pineal body located? Name the hormones it produces and their possible effects.***

HORMONES OF THE THYMUS

The **thymus** (thī′mŭs) is in the neck and superior to the heart in the thorax; it secretes a hormone called **thymosin** (thī′mō-sin; see table 18.13). Both the thymus and thymosin play an important role in the development and maturation of the immune system, as discussed in chapter 22.

HORMONES OF THE GASTROINTESTINAL TRACT

Several hormones are released from the gastrointestinal tract. They regulate digestive functions by influencing the activity of the stomach, intestines, liver, and pancreas. They are discussed in chapter 24.

HORMONELIKE SUBSTANCES

Autocrine chemical signals are released from cells that influence the same cell type from which they are released. **Paracrine chemical signals** are released from one cell type, diffuse short distances, and influence the activity of another cell type, which is its target tissue. Autocrine and paracrine chemical signals differ from hormones in that they are not secreted from discrete endocrine glands, they have local effects rather than systemic effects, or they have functions that are not understood adequately to explain their role in the body. Examples of autocrine chemical signals include chemical mediators of inflammation derived from the fatty acid **arachidonic** (ă-rak-i-don′ik) **acid,** such as eicosanoids and modified phospholipids. The **eicosanoids** include **prostaglandins** (pros′-stă-glandinz), **thromboxanes** (throm′bok-zānz), **prostacyclins** (pros-tă-sī′klinz), and **leukotrienes** (loo′kō-trī′ēnz). Modified phospholipids include platelet activating factor (see chapter 19). Paracrine chemical signals include substances that play a role in modulating the sensation of pain, such as **endorphins** (en′dōr-finz) and **enkephalins** (en-kef′ă-linz), and several peptide growth factors, such as **epidermal growth factor, fibroblast growth factor,** and **interleukin-2** (in-ter-loo′kin; see table 18.13).

Prostaglandins, thromboxanes, prostacyclins, and leukotrienes are released from injured cells and are responsible for initiating some of the symptoms of inflammation (see chapter 22), in addition to being released from certain healthy cells. For example, prostaglandins are involved in the regulation of uterine contractions during menstruation and childbirth, the process of ovulation, the inhibition of progesterone synthesis by the corpus luteum, the regulation of coagulation, kidney function, and the modification of the effect of other hormones on their target tissues. Pain receptors are stimulated directly by prostaglandins and other inflammatory compounds, or prostaglandins cause vasodilation of blood vessels, which is associated with headaches. Anti-inflammatory drugs, such as aspirin, inhibit prostaglandin synthesis and, as a result, reduce inflammation and pain. These examples are paracrine regulatory substances because they are synthesized and secreted by the cells near their target cells. Once prostaglandins enter the circulatory system, they are metabolized rapidly.

Three classes of peptide molecules, which are endogenously produced analgesics, bind to the same receptors as morphine. They include enkephalins, endorphins, and **dynorphins** (dī′nōr-finz). They are produced in several sites in the body, such as parts of the brain, pituitary, spinal cord, and gut. They act as neurotransmitters in some neurons of both the central and peripheral nervous systems and as hormones or paracrine regulatory substances. In general, they moderate the sensation of pain (see chapter 14). Decreased sensitivity to painful stimuli during exercise and stress may result from the increased secretion of these substances.

Several proteins can be classified as growth factors. They generally function as paracrine chemical signals because they are secreted near their target tissues. Epidermal growth factor stimulates cell divisions in a number of tissues and plays an important role in embryonic development. Interleukin-2 stimulates the proliferation of T lymphocytes and plays a very important role in immune responses (see chapter 22). The number of hormonelike substances in the body is large, and only a few of them have been mentioned here. Chemical communication among cells in the body is complex, well developed, and necessary for the maintenance of homeostasis. Investigations into chemical regulation increase our knowledge of body functions—knowledge that can be used in the development of techniques for the treatment of pathologic conditions.

42 ***Define autocrine chemical signals. List eicosanoids and modified phospholipids that function as autocrine chemical signals, and explain their function.***

43 ***Define paracrine chemical signals. List examples of substances that play a role in modulating pain or are peptide growth factors. How can prostaglandins function as both autocrine and paracrine chemical signals?***

EFFECTS OF AGING ON THE ENDOCRINE SYSTEM

Age-related changes in the endocrine system are not the same for all of the endocrine glands. There is a gradual decrease in the secretory activity of some endocrine glands, but not in all of them. In addition, some decreases in secretory activity of endocrine glands appear to be secondary to a decrease in physical activity as people age.

There is a decrease in the secretion of GH as people age. The decrease is greater in people who do not exercise, and it may not occur in people who exercise regularly. Decreasing GH secretion may explain the gradual decrease in lean body mass. For example, bone mass and muscle mass decrease as GH levels decline. At the same time, adipose tissue increases.

Melatonin secretion decreases in aging people. The decrease may influence age-related changes in sleep patterns and the secretory patterns of other hormones, such as GH and testosterone.

The secretion of thyroid hormones decreases slightly with increasing age, and there is a decrease in the T_3/T_4 ratio. This may be less of a decrease in the secretory activity of the thyroid gland than it is compensating for the decrease in the lean body mass in aging people. Age-related damage to the thyroid gland by the immune system can occur. This change occurs in women more than in men. The result is that approximately 10% of elderly women have thyroid glands that do not produce enough T_3 and T_4.

Parathyroid hormone secretion does not appear to decrease with age. Blood levels of Ca^{2+} may decrease slightly because of reduced dietary calcium intake and vitamin D levels. The greatest risk is a loss of bone matrix as parathyroid hormone increases to maintain blood levels of Ca^{2+} within their normal range.

The kidneys of the elderly secrete less renin. Consequently, there is a reduced ability to respond to decreases in blood pressure by activating the renin-angiotensin-aldosterone mechanism (see chapter 26).

Reproductive hormone secretion gradually declines in elderly men, and women experience menopause. These age-related changes are described in chapter 28.

There are no age-related decreases in the ability to regulate blood glucose levels. However, there is an age-related tendency to develop type 2 diabetes mellitus in those who have a familial tendency to do so, and it is correlated with age-related increases in body weight.

Thymosin from the thymus decreases with age. Fewer immature lymphocytes are able to mature and become functional, and the immune system becomes less effective in protecting the body. There is an increased susceptibility to infection and to cancer.

44 ***Describe age-related changes in the secretion and the consequences of these changes in the following: GH, melatonin, thyroid hormones, renin, and reproductive hormones. Name one hormone that does not appear to decrease with age.***

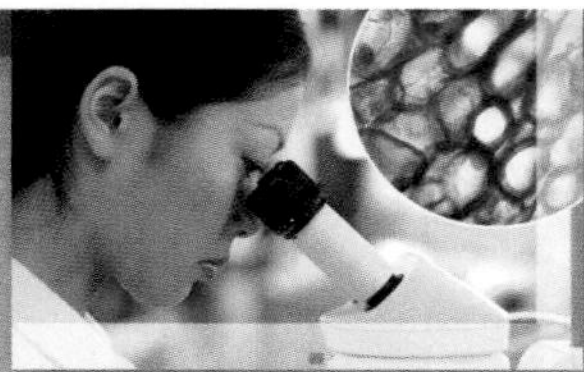

SYSTEMS PATHOLOGY

Type 1 Diabetes Mellitus

Billy, a 10-year-old boy, was diagnosed as having type 1 diabetes mellitus. Billy's mother took him to a physician after noticing that he was constantly hungry and was losing weight rapidly in spite of his unusually large food intake. More careful observation made it clear that Billy was constantly thirsty and that he urinated frequently. In addition, he felt weak and lethargic, and his breath occasionally had a distinctive sweet, or acetone, odor. Diagnostic tests confirmed that he had type 1 diabetes mellitus.

BACKGROUND INFORMATION

Type 1 diabetes mellitus is caused by diminished insulin secretion. In patients with type 1 diabetes mellitus, nutrients are absorbed from the intestine after a meal, but skeletal muscle, adipose tissue, and other target tissues do not readily take glucose into their cells, and liver cells cannot convert glucose to glycogen. Consequently, blood levels of glucose increase dramatically. Glucagon and glucocorticoid secretion increase because the glucose in the blood cannot enter the cells that produce these hormones, so their rate of secretion is similar to when blood glucose levels are low. Epinephrine secretion also increases. In response to these hormones, glycogen, fats, and proteins are broken down and metabolized to produce the ATP required by cells.

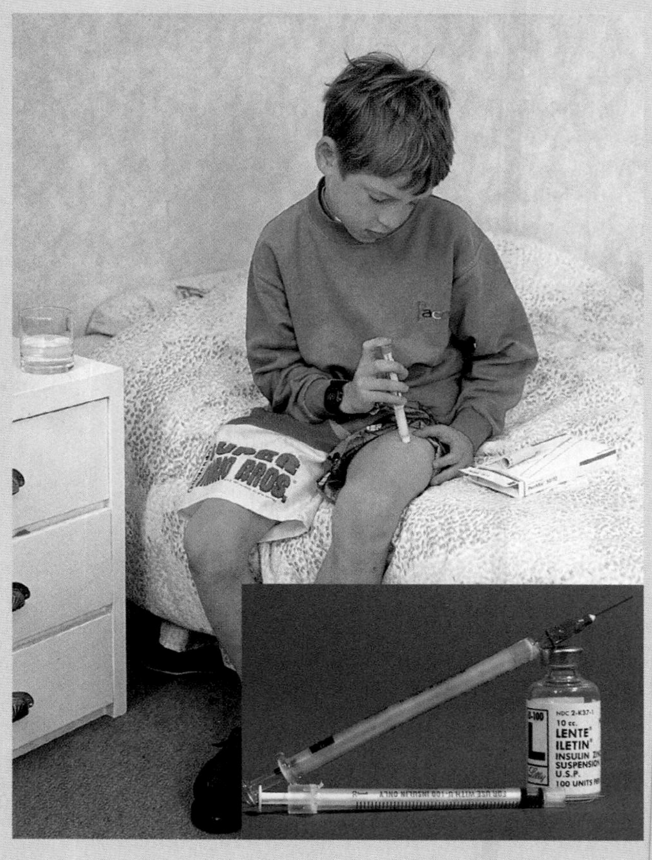

FIGURE B **A 10-Year-Old Boy Giving Himself Insulin**

When blood glucose levels are very high, glucose molecules enter the kidney tubules and increase the solute concentration of the urine. Water is attracted to the glucose because of osmosis. Therefore, the urine volume increases. The loss of water and the high blood glucose levels increase the osmotic concentration of blood, which increases the sensation of thirst. The increased osmolality of blood and the ionic imbalances caused by the loss of Ca^{2+} and K^{+} in the large amount of urine produced cause neurons to malfunction and result in diabetic coma in severe cases. When insulin levels in the blood are low and the cells of the nervous system that control appetite appear to be unable to take up glucose even when blood glucose levels are high, the result is an increased appetite. **Polyuria** (pol-ē-ū'rē-ă; increased urine volume), **polydipsia** (pol-ē-dip'sē-ă; increased thirst), and **polyphagia** (pol-ē-fā'jē-ă; increased appetite) are major symptoms of type 1 diabetes mellitus. Acidosis is caused by rapid fat catabolism, which results in increased levels of **acetoacetic** (as'e-tō-a-sē'tik) **acid,** which is converted to **acetone** (as'e-tōn) and **β-hydroxybutyric** (bā'tă hī-drok'sē-bū-tir'ik) **acid.** These three substances collectively are referred to as **ketone** (kē'tōn) **bodies.** The presence of excreted ketone bodies in the urine and in expired air ("acetone breath") suggests that the person has diabetes mellitus.

Billy's physician explained that, prior to the late 1920s, people with his condition always died in a relatively short time. They suffered from massive weight loss and appeared to starve to death in spite of eating a large amount of food. The physician explained that, because of the discovery of insulin, many people with his type of diabetes mellitus are able to live nearly normal lives. Taking insulin injections (figure B), monitoring blood glucose levels, and following a strict diet to keep blood glucose levels within a normal range of values are the major treatments for type 1 diabetes mellitus.

PREDICT 12

After Billy was diagnosed with diabetes mellitus, he followed a strict diet and took insulin for a few months. He began to feel much better than before. In fact, he felt so well that he began to sneak candy and soft drinks when his parents were not around. Predict the consequences of his actions on his health.

SYSTEM INTERACTIONS	Effect of Type 1 Diabetes Mellitus on Other Systems
SYSTEM	**INTERACTION**
Muscular	Untreated diabetes mellitus, especially type 1 diabetes mellitus, results in severe muscle atrophy because glycogen, stored fat, and proteins of muscles are broken down and used as energy sources. Ionic imbalances can also lead to muscular weakness.
Nervous	Untreated type 1 diabetes mellitus can have dramatic effects on the nervous system. When the blood glucose reaches very high levels, the osmolality of the extracellular fluid is increased. Thus, water diffuses from the neurons of the brain. In addition, acidosis develops because of the rapid metabolism of fats. As a result, the nervous system cannot function normally, and diabetic coma can result. A long-term effect is the degeneration of the myelin sheaths of neurons, resulting in abnormal nerve functions.
Cardiovascular	Atherosclerosis develops more rapidly in diabetics than in the healthy population. Changes in the capillary structure and high blood glucose levels increase the probability of reduced circulation and gangrene.
Lymphatic and immune	The tendency to develop infections increases, and the rate of healing is slower. In some cases, an allergic reaction to the injected insulin occurs.
Respiratory	Acidosis causes hyperventilation, which increases blood pH back toward normal levels by decreasing blood CO_2 levels.
Urinary	High blood glucose levels cause polyuria, the urine contains glucose and has a high osmolality, and people with diabetes are more likely to develop urinary tract infections.
Reproductive	Pregnant women with diabetes mellitus may have babies with a larger than normal birth weight because the blood glucose levels may be high in the mother and fetus, and the fetus's pancreas produces insulin. Glucose is therefore taken up by the cells of the fetus, where it is converted to fat.

SUMMARY

Functions of the Endocrine System (p. 610)

Main regulatory functions include water balance, uterine contractions and milk release, metabolism and tissue maturation, ion regulation, heart rate and blood pressure regulation, control of blood glucose and other nutrients, immune system regulation, and control of reproductive functions.

Pituitary Gland and Hypothalamus (p. 610)

1. The pituitary gland secretes at least nine hormones that regulate numerous body functions and other endocrine glands.
2. The hypothalamus regulates pituitary gland activity through neurohormones and action potentials.

Structure of the Pituitary Gland

1. The posterior pituitary develops from the floor of the brain and consists of the infundibulum and pars nervosa.
2. The anterior pituitary develops from the roof of the mouth and consists of the pars distalis, pars intermedia, and pars tuberalis.

Relationship of the Pituitary to the Brain

1. The hypothalamohypophysial portal system connects the hypothalamus and the anterior pituitary.
   - Neurohormones are produced in hypothalamic neurons.
   - Through the portal system, the neurohormones inhibit or stimulate hormone production in the anterior pituitary.
2. The hypothalamohypophysial tract connects the hypothalamus and the posterior pituitary.
   - Neurohormones are produced in hypothalamic neurons.
   - The neurohormones move down the axons of the tract and are secreted from the posterior pituitary.

Hormones of the Pituitary Gland

1. ADH promotes water retention by the kidneys.
2. Oxytocin promotes uterine contractions during delivery and causes milk ejection in lactating women.
3. GH is sometimes called somatotropin.
   - GH stimulates growth in most tissues and is a regulator of metabolism.
   - GH stimulates the uptake of amino acids and their conversion into proteins and stimulates the breakdown of fats and the synthesis of glucose.
   - GH stimulates the production of somatomedins; together, they promote bone and cartilage growth.
   - GH secretion increases in response to an increase in blood amino acids, low blood glucose, or stress.
   - GH is regulated by GHRH and GHIH or by somatostatin.
4. TSH, or thyrotropin, causes the release of thyroid hormones.
5. ACTH is derived from proopiomelanocortin; it stimulates cortisol secretion from the adrenal cortex and increases skin pigmentation.

6. Several hormones in addition to ACTH are derived from proopiomelanocortin.
   - Lipotropins cause fat breakdown.
   - β endorphins play a role in analgesia.
   - MSH increases skin pigmentation.
7. LH and FSH are major gonadotropins.
   - Both hormones regulate the production of gametes and reproductive hormones (testosterone in males, estrogen and progesterone in females).
   - GnRH from the hypothalamus stimulates LH and FSH secretion.
8. Prolactin stimulates milk production in lactating females. Prolactin-releasing hormone and prolactin-inhibiting hormone from the hypothalamus affect prolactin secretion.

Thyroid Gland (p. 619)

1. The thyroid gland is just inferior to the larynx.
2. The thyroid gland is composed of small, hollow balls of cells called follicles, which contain thyroglobulin.
3. Parafollicular cells are scattered throughout the thyroid gland.

Thyroid Hormones

1. T_3 and T_4 synthesis occurs in thyroid follicles.
   - Iodide ions are taken into the follicles by active transport, are oxidized, and are bound to tyrosine molecules in thyroglobulin.
   - Thyroglobulin is secreted into the follicle lumen. Tyrosine molecules with iodine combine to form T_3 and T_4 thyroid hormones.
   - Thyroglobulin is taken into the follicular cells and is broken down; T_3 and T_4 diffuse from the follicles to the blood.
2. T_3 and T_4 are transported in the blood.
   - T_3 and T_4 bind to thyroxine-binding globulin and other plasma proteins.
   - The plasma proteins prolong the half-life of T_3 and T_4 and regulate the levels of T_3 and T_4 in the blood.
   - Approximately one-third of the T_4 is converted into functional T_3.
3. T_3 and T_4 bind with intracellular receptor molecules and initiate new protein synthesis.
4. T_3 and T_4 affect nearly every tissue in the body.
   - T_3 and T_4 increase the rate of glucose, fat, and protein metabolism in many tissues, thus increasing body temperature.
   - Normal growth of many tissues is dependent on T_3 and T_4.
5. TRH and TSH regulate T_3 and T_4 secretion.
   - Increased TSH from the anterior pituitary increases T_3 and T_4 secretion.
   - TRH from the hypothalamus increases TSH secretion. TRH increases as a result of chronic exposure to cold, food deprivation, and stress.
   - T_3 and T_4 inhibit TSH and TRH secretion.

Calcitonin

1. The parafollicular cells secrete calcitonin.
2. An increase in blood calcium levels stimulates calcitonin secretion.
3. Calcitonin decreases blood cal cium and phosphate levels by inhibiting osteoclasts.

Parathyroid Glands (p. 624)

1. The parathyroid glands are embedded in the thyroid gland.
2. PTH increases blood calcium levels.
   - PTH stimulates osteoclasts.
   - PTH promotes calcium reabsorption by the kidneys and the formation of active vitamin D by the kidneys.
   - Active vitamin D increases calcium absorption by the intestine.
3. A decrease in blood calcium levels stimulates PTH secretion.

Adrenal Glands (p. 627)

1. The adrenal glands are near the superior poles of the kidneys.
2. The adrenal medulla arises from neural crest cells and functions as part of the sympathetic nervous system. The adrenal cortex is derived from mesoderm.
3. The adrenal medulla is composed of closely packed cells.
4. The adrenal cortex is divided into three layers: the zona glomerulosa, the zona fasciculata, and the zona reticularis.

Hormones of the Adrenal Medulla

1. Epinephrine accounts for 80% and norepinephrine for 20% of the adrenal medulla hormones.
   - Epinephrine increases blood glucose levels, the use of glycogen and glucose by skeletal muscle, and heart rate and force of contraction, and it causes vasoconstriction in the skin and viscera and vasodilation in skeletal and cardiac muscle.
   - Norepinephrine stimulates cardiac muscle and causes the constriction of most peripheral blood vessels.
2. The adrenal medulla hormones prepare the body for physical activity.
3. Release of adrenal medulla hormones is mediated by the sympathetic nervous system in response to emotions, injury, stress, exercise, and low blood glucose levels.

Hormones of the Adrenal Cortex

1. The zona glomerulosa secretes the mineralocorticoids, especially aldosterone. Aldosterone acts on the kidneys to increase sodium and to decrease potassium and hydrogen levels in the blood.
2. The zona fasciculata secretes glucocorticoids, especially cortisol.
   - Cortisol increases fat and protein breakdown, increases glucose synthesis from amino acids, decreases the inflammatory response, and is necessary for the development of some tissues.
   - ACTH from the anterior pituitary stimulates cortisol secretion. CRH from the hypothalamus stimulates ACTH release. Low blood glucose levels and stress stimulate CRH secretion.
3. The zona reticularis secretes androgens. In females, androgens stimulate axillary and pubic hair growth and sexual drive.

Pancreas (p. 632)

1. The pancreas is located along the small intestine and the stomach. It is both an exocrine and an endocrine gland.
2. The exocrine portion of the pancreas consists of a complex duct system, which ends in small sacs, called acini, that produce pancreatic digestive juices.
3. The endocrine portion consists of the pancreatic islets. Each islet is composed of alpha cells, which secrete glucagon; beta cells, which secrete insulin; and delta cells, which secrete somatostatin.

Effect of Insulin and Glucagon on Their Target Tissues

1. Insulin's target tissues are the liver, adipose tissue, muscle, and the satiety center in the hypothalamus. The nervous system is not a target tissue, but it does rely on blood glucose levels maintained by insulin.
2. Insulin increases the uptake of glucose and amino acids by cells. Glucose is used for energy or is stored as glycogen. Amino acids are used for energy or are converted to glucose or proteins.
3. Glucagon's target tissue is mainly the liver.
4. Glucagon causes the breakdown of glycogen and fats for use as an energy source.

Regulation of Pancreatic Hormone Secretion

1. Insulin secretion increases because of elevated blood glucose levels, an increase in some amino acids, parasympathetic stimulation, and gastrointestinal hormones. Sympathetic stimulation decreases insulin secretion.
2. Glucagon secretion is stimulated by low blood glucose levels, certain amino acids, and sympathetic stimulation.
3. Somatostatin inhibits insulin and glucagon secretion.

Hormonal Regulation of Nutrients (p. 638)

1. After a meal, the following events take place:
   - High glucose levels inhibit glucagon, cortisol, GH, and epinephrine, which reduces the release of glucose from tissues.
   - Insulin secretion increases as a result of the high blood glucose levels, thereby increasing the uptake of glucose, amino acids, and fats, which are used for energy or are stored.
   - Sometime after the meal, blood glucose levels drop. Glucagon, GH, cortisol, and epinephrine levels increase, insulin levels decrease, and glucose is released from tissues.
   - Adipose tissue releases fatty acids, triglycerides, and ketones, which most tissues use for energy.
2. During exercise, the following events occur:
   - Sympathetic activity increases epinephrine and glucagon secretion, causing a release of glucose into the blood.
   - Low blood sugar levels, caused by the uptake of glucose by skeletal muscles, stimulate epinephrine, glucagon, GH, and cortisol secretion, causing an increase in fatty acids, triglycerides, and ketones in the blood, all of which are used for energy.

Hormones of the Reproductive System (p. 640)

The ovaries, testes, placenta, and pituitary gland secrete reproductive hormones.

Hormones of the Pineal Body (p. 641)

The pineal body produces melatonin and arginine vasotocin, which can inhibit reproductive maturation and may regulate sleep–wake cycles.

Hormones of the Thymus (p. 642)

The thymus produces thymosin, which is involved in the development of the immune system.

Hormones of the Gastrointestinal Tract (p. 642)

The gastrointestinal tract produces several hormones that regulate digestive functions.

Hormonelike Substances (p. 642)

1. Autocrine and paracrine chemical signals are produced by many cells of the body and usually have a local effect. They affect many body functions.
2. Eicosanoids, such as prostaglandins, prostacyclins, thromboxanes, and leukotrienes, are derived from fatty acids and mediate inflammation and other functions. Endorphins, enkephalins, and dynorphins are analgesic substances. Growth factors influence cell division and growth in many tissues, and interleukin-2 influences cell division in T cells of the immune system.

Effects of Aging on the Endocrine System (p. 643)

There is a gradual decrease in the secretion rate of most, but not all, hormones. Some decreases are secondary to gradual decreases in physical activity.

REVIEW AND COMPREHENSION

1. The pituitary gland
   a. develops from the floor of the brain.
   b. develops from the roof of the mouth.
   c. is stimulated by neurohormones produced in the midbrain.
   d. secretes only three major hormones.
   e. both a and b.
2. The hypothalamohypophysial portal system
   a. contains one capillary bed.
   b. carries hormones from the anterior pituitary to the body.
   c. carries hormones from the posterior pituitary to the body.
   d. carries hormones from the hypothalamus to the anterior pituitary.
   e. carries hormones from the hypothalamus to the posterior pituitary.
3. Which of these hormones is *not* a hormone that is secreted into the hypothalamohypophysial portal system?
   a. GHRH c. PIH e. ACTH
   b. TRH d. GnRH
4. Hormones secreted from the posterior pituitary
   a. are produced in the anterior pituitary.
   b. are transported to the posterior pituitary within axons.
   c. include GH and TSH.
   d. are steroids.
   e. all of the above.
5. Which of these stimulates the secretion of ADH?
   a. elevated blood osmolality
   b. decreased blood osmolality
   c. the release of hormones from the hypothalamus
   d. ACTH
   e. increased blood pressure
6. Oxytocin is responsible for
   a. preventing milk release from the mammary glands.
   b. preventing goiter.
   c. causing contraction of the uterus.
   d. maintaining normal calcium levels.
   e. increasing metabolic rate.
7. Growth hormone
   a. increases the usage of glucose.
   b. increases the breakdown of lipids.
   c. decreases the synthesis of proteins.
   d. decreases the synthesis of glycogen.
   e. all of the above.
8. Which of these hormones stimulates somatomedin secretion?
   a. FSH c. LH e. TSH
   b. GH d. prolactin

9. Hypersecretion of growth hormone
   a. results in giantism if it occurs in children.
   b. causes acromegaly in adults.
   c. increases the probability that one will develop diabetes.
   d. can lead to severe atherosclerosis.
   e. all of the above.
10. LH and FSH
    a. are produced in the hypothalamus.
    b. production is increased by TSH.
    c. promote the production of gametes and reproductive hormones.
    d. inhibit the production of prolactin.
    e. all of the above.
11. T_3 and T_4
    a. require iodine for their production.
    b. are made from the amino acid tyrosine.
    c. are transported in the blood bound to thyroxine-binding globulin.
    d. all of the above.
12. Which of these symptoms is associated with hyposecretion of thyroid hormones?
    a. hypertension
    b. nervousness
    c. diarrhea
    d. weight loss with a normal or an increased food intake
    e. decreased metabolic rate
13. Which of these conditions most likely occurs if a healthy person receives an injection of T_3 and T_4?
    a. The secretion rate of TSH declines.
    b. The person develops symptoms of hypothyroidism.
    c. The person develops hypercalcemia.
    d. The person secretes more TRH.
14. Which of these occurs as a response to a thyroidectomy (removal of the thyroid gland)?
    a. increased calcitonin secretion
    b. increased T_3 and T_4 secretion
    c. decreased TRH secretion
    d. increased TSH secretion
15. Choose the statement that most accurately predicts the long-term effect of a substance that prevents the active transport of iodide by the thyroid gland.
    a. Large amounts of T_3 and T_4 accumulate within the thyroid follicles, but little is released.
    b. The person exhibits hypothyroidism.
    c. The anterior pituitary secretes smaller amounts of TSH.
    d. The circulating levels of T_3 and T_4 increase.
16. Calcitonin
    a. is secreted by the parathyroid glands.
    b. levels increase when blood calcium levels decrease.
    c. causes blood calcium levels to decrease.
    d. insufficiency results in weak bones and tetany.
17. Parathyroid hormone secretion increases in response to
    a. a decrease in blood calcium levels.
    b. increased production of parathyroid-stimulating hormone from the anterior pituitary.
    c. increased secretion of parathyroid-releasing hormone from the hypothalamus.
    d. increased secretion of calcitonin.
    e. a decrease in secretion of ACTH.
18. If parathyroid hormone levels increase, which of these conditions is expected?
    a. Osteoclast activity is increased.
    b. Calcium absorption from the small intestine is inhibited.
    c. Calcium reabsorption from the urine is inhibited.
    d. Less active vitamin D is formed in the kidneys.
    e. All of the above are true.
19. The adrenal medulla
    a. produces steroids.
    b. has cortisol as its major secretory product.
    c. decreases its secretions during exercise.
    d. is formed from a modified portion of the sympathetic division of the ANS.
    e. all of the above.
20. Pheochromocytoma is a condition in which a benign tumor results in hypersecretion of hormones from the adrenal medulla. The symptoms that one would expect include
    a. hypotension.
    b. bradycardia.
    c. pallor (decreased blood flow to the skin).
    d. lethargy.
    e. hypoglycemia.
21. Which of these is *not* a hormone secreted by the adrenal cortex?
    a. aldosterone
    b. androgens
    c. cortisol
    d. epinephrine
22. If aldosterone secretions increase,
    a. blood potassium levels increase.
    b. blood hydrogen levels increase.
    c. acidosis results.
    d. blood sodium levels decrease.
    e. blood volume increases.
23. Glucocorticoids (cortisol)
    a. increase the breakdown of fats.
    b. increase the breakdown of proteins.
    c. increase blood glucose levels.
    d. decrease inflammation.
    e. all of the above.
24. The release of cortisol from the adrenal cortex is regulated by other hormones. Which of these hormones is correctly matched with its origin and function?
    a. CRH—secreted by the hypothalamus; stimulates the adrenal cortex to secrete cortisol
    b. CRH—secreted by the anterior pituitary; stimulates the adrenal cortex to secrete cortisol
    c. ACTH—secreted by the hypothalamus; stimulates the adrenal cortex to secrete cortisol
    d. ACTH—secreted by the anterior pituitary; stimulates the adrenal cortex to produce cortisol
25. Which of these is/are expected in Cushing syndrome?
    a. loss of hair in women
    b. deposition of fat in the face, neck, and abdomen
    c. low blood glucose
    d. low blood pressure
    e. all of the above
26. Within the pancreas, the pancreatic islets produce
    a. insulin.
    b. glucagon.
    c. digestive enzymes.
    d. both a and b.
    e. all of the above.

27. Insulin increases
    a. the uptake of glucose by its target tissues.
    b. the breakdown of protein.
    c. the breakdown of fats.
    d. glycogen breakdown in the liver.
    e. all of the above.
28. Which of these tissues is least affected by insulin?
    a. adipose tissue c. skeletal muscle e. liver
    b. heart d. brain
29. Glucagon
    a. primarily affects the liver.
    b. causes glycogen to be stored.
    c. causes blood glucose levels to decrease.
    d. decreases fat metabolism.
    e. all of the above.
30. When blood glucose levels increase, the secretion of which of these hormones increases?
    a. glucagon c. GH e. epinephrine
    b. insulin d. cortisol
31. If a person who has diabetes mellitus has forgotten to take an insulin injection, the symptoms that may soon appear include
    a. acidosis. d. lethargy and fatigue.
    b. hyperglycemia. e. all of the above.
    c. increased urine production.
32. Which of these is *not* a hormone produced by the ovaries?
    a. estrogen c. prolactin e. relaxin
    b. progesterone d. inhibin
33. Melatonin
    a. is produced by the posterior pituitary.
    b. production increases as day length increases.
    c. inhibits the development of the reproductive system.
    d. increases GnRH secretion from the hypothalamus.
    e. decreases the tendency to sleep.
34. Which of these substances, produced by many tissues of the body, can promote inflammation, pain, and vasodilation of blood vessels?
    a. endorphin d. epidermal growth factor
    b. enkephalin e. prostaglandin
    c. thymosin
35. Which of these changes does *not* decrease with aging of the endocrine system?
    a. GH secretion d. parathyroid hormone secretion
    b. melatonin secretion e. renin secretion by the kidneys
    c. thyroid hormone secretion

Answers in Appendix E

CRITICAL THINKING

1. The hypothalamohypophysial portal system connects the hypothalamus with the anterior pituitary. Why is such a special circulatory system advantageous?
2. The secretion of ADH can be affected by exposure to hot or cold environmental temperatures. Predict the effect of a hot environment on ADH secretion, and explain why it is advantageous. Propose a mechanism by which temperature produces a change in ADH secretion.
3. A patient exhibits polydipsia (thirst), polyuria (excess urine production), and urine with a low specific gravity (contains few ions and no glucose). If you wanted to reverse the symptoms, would you administer insulin, glucagon, ADH, or aldosterone? Explain.
4. A patient complains of headaches and visual disturbances. A casual glance reveals that the patient's finger bones are enlarged in diameter, a heavy deposition of bone exists over the eyes, and the patient has a prominent jaw. The doctor tells you that the headaches and visual disturbances result from increased pressure within the skull and that the patient is suffering from a pituitary tumor that is affecting hormone secretion. Name the hormone that is causing the problem, and explain why an increase in pressure exists within the skull.
5. Most laboratories have the ability to determine blood levels of TSH, T_3, and T_4. Given that ability, design a method of determining whether hyperthyroidism in a patient results from a pituitary abnormality or from the production of a nonpituitary thyroid stimulatory substance.
6. An anatomy and physiology instructor asks two students to predict a patient's response to chronic vitamin D deficiency. One student claims that the person would suffer from hypocalcemia and the symptoms associated with that condition. The other student claims that calcium levels would remain within their normal range, although at the low end of the range, and that bone resorption would occur to the point that advanced osteomalacia might be seen. With whom do you agree, and why?
7. Given the ability to measure blood glucose levels, design an experiment that distinguishes among a person with diabetes, a healthy person, and a person who has a pancreatic tumor that secretes large amounts of insulin.
8. A patient arrives in an unconscious condition. A medical emergency bracelet reveals that he has diabetes. The patient can be in diabetic coma or insulin shock. How can you tell which, and what treatment do you recommend for each condition?
9. Diabetes mellitus can result from a lack of insulin, which results in hyperglycemia. Adrenal diabetes and pituitary diabetes also produce hyperglycemia. What hormones produce the last two conditions?
10. Predict some of the consequences of exposure to intense and prolonged stress.

Answers in Appendix F

Visit this book's website at www.mhhe.com/seeley8 for chapter quizzes, interactive learning exercises, and other study tools.

19 Cardiovascular System

Blood

Historically, many cultures around the world, both ancient and modern, shared beliefs in the magical qualities of blood. Blood is considered the "essence of life" because the uncontrolled loss of it can result in death. Blood was also thought to define character and emotions. People of a noble bloodline were described as "blue bloods," whereas criminals were considered to have "bad" blood. It was said that anger caused the blood to "boil," and fear resulted in blood "curdling." The scientific study of blood reveals characteristics as fascinating as any of these fantasies. Blood performs many functions essential to life and often can reveal much about our health.

Blood is a type of connective tissue, consisting of cells and cell fragments surrounded by a liquid matrix. The cells and cell fragments are the formed elements, and the liquid is the plasma. The formed elements make up about 45%, and plasma makes up about 55% of the total blood volume (figure 19.1). The total blood volume in the average adult is about 4–5 L in females and 5–6 L in males. Blood makes up about 8% of the total weight of the body.

Cells require constant nutrition and waste removal because they are metabolically active. The cardiovascular system, which consists of the heart, blood vessels, and blood, connects the various tissues of the body. The heart pumps blood through blood vessels, and the blood delivers nutrients and picks up waste products.

This chapter explains the *functions of blood* (p. 651), *plasma* (p. 652), and the *formed elements* (p. 653) of blood. *Hemostasis* (p. 662), *blood grouping* (p. 667), and *diagnostic blood tests* (p. 671) are also described.

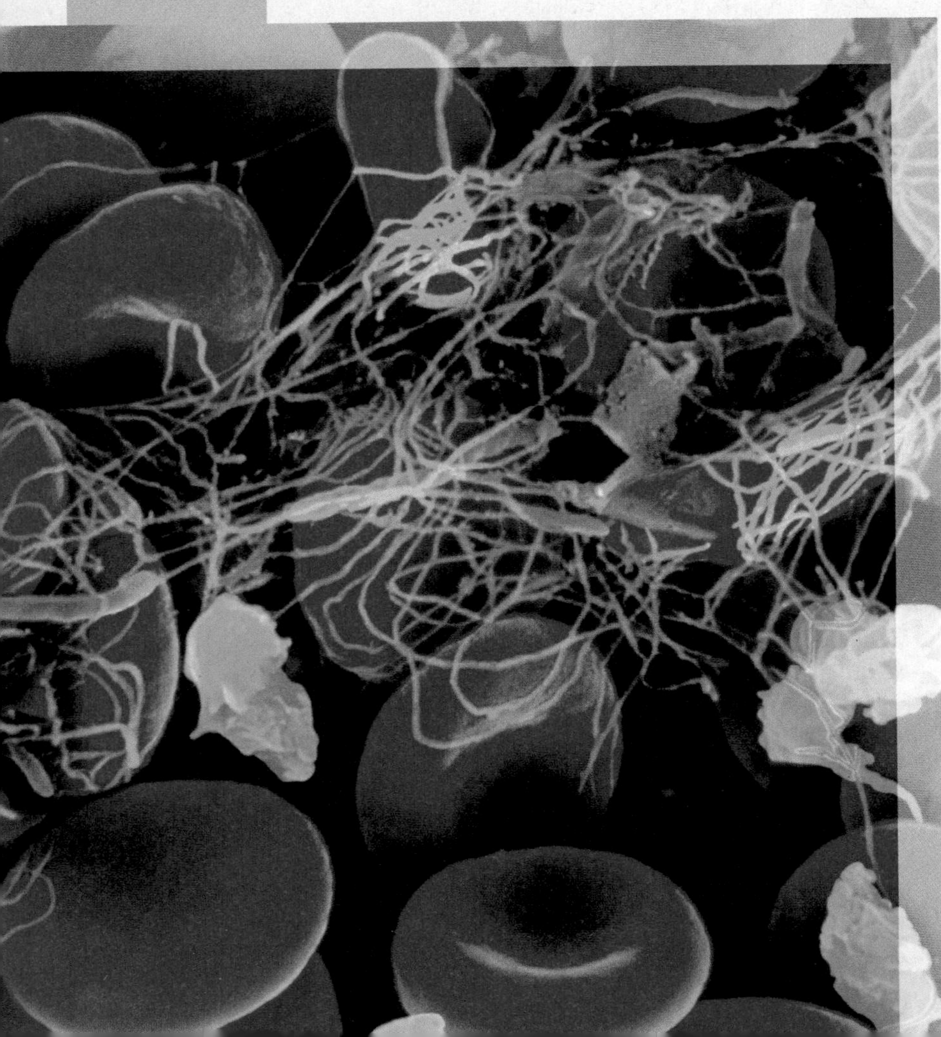

Colorized scanning electron micrograph of a blood clot. The red discs are red blood cells, the blue particles are platelets, and the yellow strands are fibrin.

Cardiovascular System

FUNCTIONS OF BLOOD

The heart pumps blood through blood vessels, which extend throughout the body. Blood helps maintain homeostasis in several ways:

1. *Transport of gases, nutrients, and waste products.* Oxygen enters blood in the lungs and is carried to cells. Carbon dioxide, produced by cells, is carried in the blood to the lungs, from which it is expelled. The blood transports ingested nutrients, ions, and water from the digestive tract to cells, and the blood transports the waste products of cells to the kidneys for elimination.
2. *Transport of processed molecules.* Many substances are produced in one part of the body and transported in the blood to another part, where they are modified. For example, the precursor to vitamin D is produced in the skin (see chapter 5) and transported by the blood to the liver and then to the kidneys for processing into active vitamin D. Active vitamin D is transported in the blood to the small intestines, where it promotes the uptake of calcium. Another example is lactic acid produced by skeletal muscles during anaerobic respiration (see chapter 9). The blood carries lactic acid to the liver, where it is converted into glucose.
3. *Transport of regulatory molecules.* The blood carries many of the hormones and enzymes that regulate body processes from one part of the body to another.
4. *Regulation of pH and osmosis.* Buffers (see chapter 2), which help keep the blood's pH within its normal limits of 7.35–7.45, are in the blood. The osmotic composition of blood is also critical for maintaining normal fluid and ion balance.
5. *Maintenance of body temperature.* Blood is involved with body temperature regulation because warm blood is

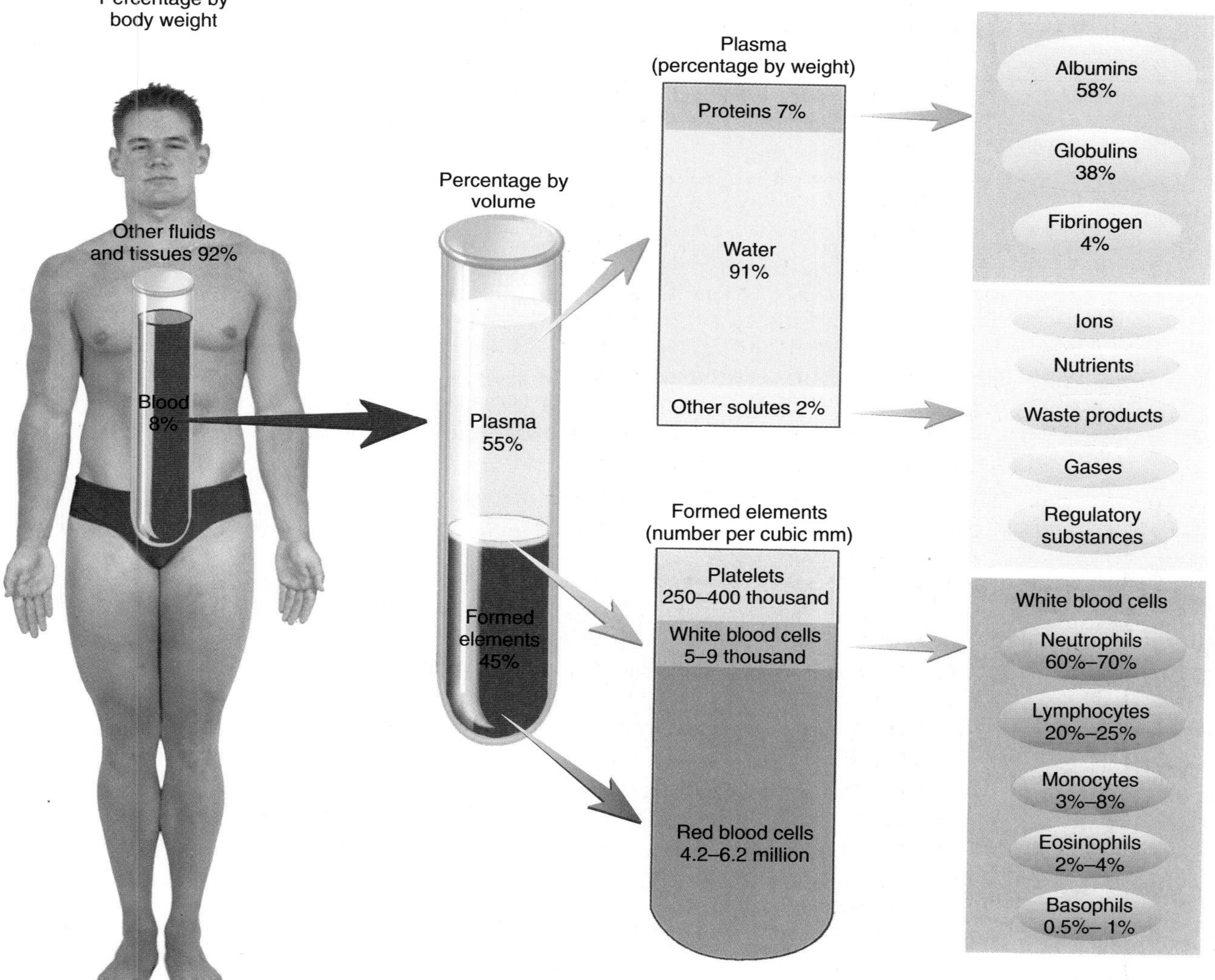

FIGURE 19.1 Composition of Blood

Approximate values for the components of blood in a normal adult.

transported from the interior to the surface of the body, where heat is released from the blood.

6. *Protection against foreign substances.* The cells and chemicals of the blood make up an important part of the immune system, protecting against foreign substances, such as microorganisms and toxins.
7. *Clot formation.* Blood clotting protects against excessive blood loss when blood vessels are damaged. When tissues are damaged, the blood clot that forms is also the first step in tissue repair and the restoration of normal function (see chapter 4).

1 ***List the ways that blood helps maintain homeostasis in the body.***

PLASMA

Plasma (plaz′mă) is the liquid part of blood. It is a pale yellow fluid that consists of about 91% water and 9% other substances, such as proteins, ions, nutrients, gases, and waste products (table 19.1). Plasma is a **colloid** (kol′oyd), which is a liquid containing suspended substances that do not settle out of solution. Most of the suspended substances are plasma proteins. Based on molecular size and charge, the plasma proteins can be classified as albumin, globulins, and fibrinogen. Almost all of the plasma proteins are produced by the liver or blood cells, a notable exception being protein hormones. **Albumin** (al-bū′min) makes up 58% of the plasma proteins and is important in the regulation of water movement between tissues and blood. Because albumin does not easily pass

TABLE 19.1 Composition of Plasma

Plasma Components	Function
Water	Acts as a solvent and suspending medium for blood components
Plasma Proteins	
Albumin	Is partly responsible for blood viscosity and osmotic pressure; acts as a buffer; transports fatty acids, free bilirubin, and thyroid hormones
Globulins	
α	Protect tissues from damage by inflammation (alpha-1 antitrypsin); transport thyroid hormones (thyroid-binding globulin), cortisol (transcortin), and testosterone and estrogen (sex hormone–binding globulin); transport lipids (e.g., cholesterol in high-density lipoproteins); convert ferrous iron (Fe^{2+}) to ferric iron (Fe^{3+}), which promotes iron transport by transferrin (ceruloplasmin); transport hemoglobin released from damaged red blood cells (haptoglobin)
β	Transport iron (transferrin); transport lipids (beta-lipoproteins), especially cholesterol in low-density lipoproteins; are involved with immunity (complement); prevent blood loss (coagulation proteins)
γ	Are involved in immunity (most antibodies are γ globulins, but some are β or α globulins)
Fibrinogen	Functions in blood clotting
Ions	
Sodium, potassium, calcium, magnesium, chloride, iron, phosphate, hydrogen, hydroxide, bicarbonate	Is involved in osmosis, membrane potentials, and acid–base balance
Nutrients	
Glucose, amino acids, triacylglycerol, cholesterol	Are a source of energy and basic "building blocks" of more complex molecules
Vitamins	Promote enzyme activity
Waste Products	
Urea, uric acid, creatinine, ammonia salts	Are breakdown products of protein metabolism; are excreted by the kidneys
Bilirubin	Is a breakdown product of red blood cells; is excreted as part of the bile from the liver into the intestine
Lactic acid	Is an end product of anaerobic respiration; is converted to glucose by the liver
Gases	
Oxygen	Is necessary for aerobic respiration; terminal electron acceptor in electron-transport chain
Carbon dioxide	Is a waste product of aerobic respiration; as bicarbonate, helps buffer blood
Nitrogen	Is inert
Regulatory Substances	Enzymes catalyze chemical reactions; hormones stimulate or inhibit many body functions

from the blood into tissues, it plays an important role in maintaining blood colloid osmotic pressure (see chapters 21 and 26). Other molecules, such as fatty acids, bilirubin, and thyroid hormones, are transported in the blood attached to albumin. **Globulins** (glob'ū-linz) account for 38% of the plasma proteins. The globulins are subdivided into α, β, and γ globulins. Many substances in the blood are transported by globulins and, as part of immunity, globulins provide protection against microorganisms (see chapter 22 and table 19.1). **Fibrinogen** (fī-brin'ō-jen) constitutes 4% of the plasma proteins and is responsible for the formation of blood clots (see "Coagulation," p. 664). **Serum** (ser'um; whey) is plasma without the clotting factors.

The water, proteins, and other substances in the blood, such as ions, nutrients, waste products, gases, and regulatory substances, are maintained within narrow limits. Normally, water intake through the digestive tract closely matches water loss through the kidneys, lungs, digestive tract, and skin. Therefore, plasma volume remains relatively constant. Suspended or dissolved substances in the blood come from the liver, kidneys, intestines, endocrine glands, and immune tissues, such as the lymph nodes and spleen. Oxygen enters blood in the lungs and leaves the blood as it flows through tissues. Carbon dioxide enters blood from the tissues and leaves the blood as it flows through the lungs.

Define plasma. What are the functions of albumin, globulins, and fibrinogen in plasma? What other substances are found in plasma?

FORMED ELEMENTS

About 95% of the volume of the **formed elements** consists of red blood cells, or erythrocytes (ĕ-rith'rō-sītz). The remaining 5% consists of white blood cells, or leukocytes (loo'kō-sītz), and cell fragments called platelets, or thrombocytes (throm'bō-sītz). The formed elements of the blood are outlined and illustrated in table 19.2. In healthy adults, white blood cells are the only formed elements possessing nuclei, whereas red blood cells and platelets lack nuclei.

White blood cells are named according to their appearance in stained preparations. **Granulocytes** (gran'yū-lō-sītz) are white blood cells with large cytoplasmic granules and lobed nuclei (see table 19.2). Their granules stain with dyes that make the cells more visible when viewed through a light microscope. The three types of granulocytes are named according to the staining characteristics of their granules: **neutrophils** (nu'trō-filz) stain with acidic and basic dyes, **eosinophils** (ē-ō-sin'ō-filz) stain red with acidic dyes, and **basophils** (bā'sō-filz) stain dark purple with basic

TABLE 19.2 Formed Elements of the Blood

Cell Type	Illustration	Description	Function
Red Blood Cell		Biconcave disk; no nucleus; contains hemoglobin, which colors the cell red; 7.5 μm in diameter	Transports oxygen and carbon dioxide
White Blood Cells		Spherical cells with a nucleus	Five types of white blood cells, each with specific functions
Granulocytes			
Neutrophil		Nucleus with two to four lobes connected by thin filaments; cytoplasmic granules stain a light pink or reddish purple; 10–12 μm in diameter	Phagocytizes microorganisms and other substances
Basophil		Nucleus with two indistinct lobes; cytoplasmic granules stain blue-purple; 10–12 μm in diameter	Releases histamine, which promotes inflammation, and heparin, which prevents clot formation
Eosinophil		Nucleus often bilobed; cytoplasmic granules stain orange red or bright red; 11–14 μm in diameter	Releases chemicals that reduce inflammation; attacks certain worm parasites
Agranulocytes			
Lymphocyte		Round nucleus; cytoplasm forms a thin ring around the nucleus; 6–14 μm in diameter	Produces antibodies and other chemicals responsible for destroying microorganisms; contributes to allergic reactions, graft rejection, tumor control, and regulation of the immune system
Monocyte		Nucleus round, kidney-shaped, or horseshoe-shaped; contains more cytoplasm than does lymphocyte; 12–20 μm in diameter	Phagocytic cell in the blood; leaves the blood and becomes a macrophage, which phagocytizes bacteria, dead cells, cell fragments, and other debris within tissues
Platelet		Cell fragment surrounded by a plasma membrane and containing granules; 2–4 μm in diameter	Forms platelet plugs; releases chemicals necessary for blood clotting

dyes. **Agranulocytes** (ă-gran′yū-lō-sītz) are white blood cells that appear to have no granules when viewed in the light microscope. Agranulocytes actually have granules, but they are so small that they cannot be seen easily with the light microscope. The two types of agranulocytes are **lymphocytes** (lim′fō-sītz) and **monocytes** (mon′ō-sītz). They have nuclei that are not lobed.

Production of Formed Elements

The process of blood cell production, called **hematopoiesis** (hē′mă-tō-poy-ē′sis, hem′ă-to-poy-ē′sis) or **hemopoiesis** (hē′mō-poy-ē′sis), occurs in the embryo and fetus in tissues such as the yolk sac, liver, thymus, spleen, lymph nodes, and red bone marrow. After birth, hematopoiesis is confined primarily to red bone marrow, with some lymphoid tissue helping in the production of lymphocytes (see chapter 22). In young children, nearly all the marrow is red bone marrow. In adults, however, red marrow is confined to the ribs, sternum, vertebrae, pelvis, proximal femur, and proximal humerus. Yellow marrow replaces red marrow in other locations in the body (see chapter 6).

All the formed elements of the blood are derived from a single population of **stem cells** called **hemocytoblasts,** located in the red bone marrow. Hemopoietic stem cells are precursor cells capable of dividing to produce daughter cells that can differentiate into various types of blood cells (figure 19.2): **proerythroblasts** (prō-ĕ-rith′rō-blastz), from which red blood cells develop; **myeloblasts** (mī′ĕ-lō-blastz), from which basophils, eosinophils, and neutrophils develop; **lymphoblasts** (lim′fō-blastz), from which lymphocytes develop; **monoblasts** (mon′ō-blastz), from which monocytes develop; and **megakaryoblasts** (meg-ă-kar′ē-ō-blastz), from which platelets develop. The development of the cell lines is regulated by growth factors. That is, the type of formed element derived from the stem cells and how many formed elements are produced are determined by different growth factors.

3 ***Name the three general types of formed elements in the blood.***

4 ***Define hematopoiesis. What is a stem cell? What types of formed elements develop from proerythroblasts, myeloblasts, lymphoblasts, monoblasts, and megakaryoblasts?***

Stem Cells and Cancer Therapy

Many cancer therapies affect dividing cells, such as those found in tumors. An undesirable side effect of such therapies, however, can be the destruction of nontumor cells that are dividing, such as the stem cells and their derivatives in red bone marrow. After treatment for cancer, growth factors are used to stimulate the rapid regeneration of the red bone marrow. Although not a cure for cancer, the use of growth factors can speed recovery from the cancer therapy.

Some types of leukemia and genetic immune deficiency diseases can be treated with a bone marrow or stem cell transplant. To avoid the problems of tissue rejection, families with a history of these disorders can freeze the umbilical cord blood of their newborn children. The cord blood, which contains many stem cells, can be used instead of bone marrow.

Red Blood Cells

Red blood cells, or **erythrocytes,** are about 700 times more numerous than white blood cells and 17 times more numerous than platelets in the blood (figure 19.3*a*). Males have about 5.4 million red blood cells per microliter (μL; 1 mm^3 or 10^{-6} L) of blood (range: 4.6–6.2 million), whereas females have about 4.8 million/μL (range: 4.2–5.4 million). Red blood cells cannot move of their own accord; they are passively moved by forces that cause the blood to circulate.

Structure

Normal red blood cells are biconcave discs about 7.5 μm in diameter, with edges that are thicker than the center of the cell (figure 19.3*b*). The biconcave shape increases the surface area of the red blood cell, compared with a flat disc of the same size. The greater surface area makes the movement of gases into and out of the red blood cell more rapid. In addition, the red blood cell can bend or fold around its thin center, thereby decreasing its size and enabling it to pass more easily through small blood vessels.

Red blood cells are derived from specialized cells that lose their nuclei and nearly all their cellular organelles during maturation. The main component of the red blood cell is the pigmented protein **hemoglobin** (hē-mō-glō′bin), which occupies about one-third of the total cell volume and accounts for its red color. Other red blood cell contents include lipids, adenosine triphosphate (ATP), and the enzyme carbonic anhydrase.

5 ***How does the shape of red blood cells contribute to their ability to exchange gases and move through blood vessels?***

Function

The primary functions of red blood cells are to transport oxygen from the lungs to the various tissues of the body and to transport carbon dioxide from the tissues to the lungs. Approximately 98.5% of the oxygen in the blood is transported in combination with the hemoglobin in the red blood cells, and the remaining 1.5% is dissolved in the water part of the plasma. If red blood cells rupture, the hemoglobin leaks out into the plasma and becomes nonfunctional because the shape of the molecule changes as a result of denaturation (see chapter 2). Red blood cell rupture followed by hemoglobin release is called **hemolysis** (hē-mol′i-sis). Hemolysis occurs in hemolytic anemia (see Clinical Focus "Disorders of the Blood," p. 673), transfusion reactions, hemolytic disease of the newborn, and malaria.

Carbon dioxide is transported in the blood in three major ways: Approximately 7% is transported as carbon dioxide dissolved in the plasma, approximately 23% is transported in combination with hemoglobin, and 70% is transported in the form of bicarbonate ions. The bicarbonate ions (HCO_3^-) are produced when carbon dioxide (CO_2) and water (H_2O) combine to form carbonic acid (H_2CO_3), which dissociates to form hydrogen (H^+) and bicarbonate ions. The combination of carbon dioxide and

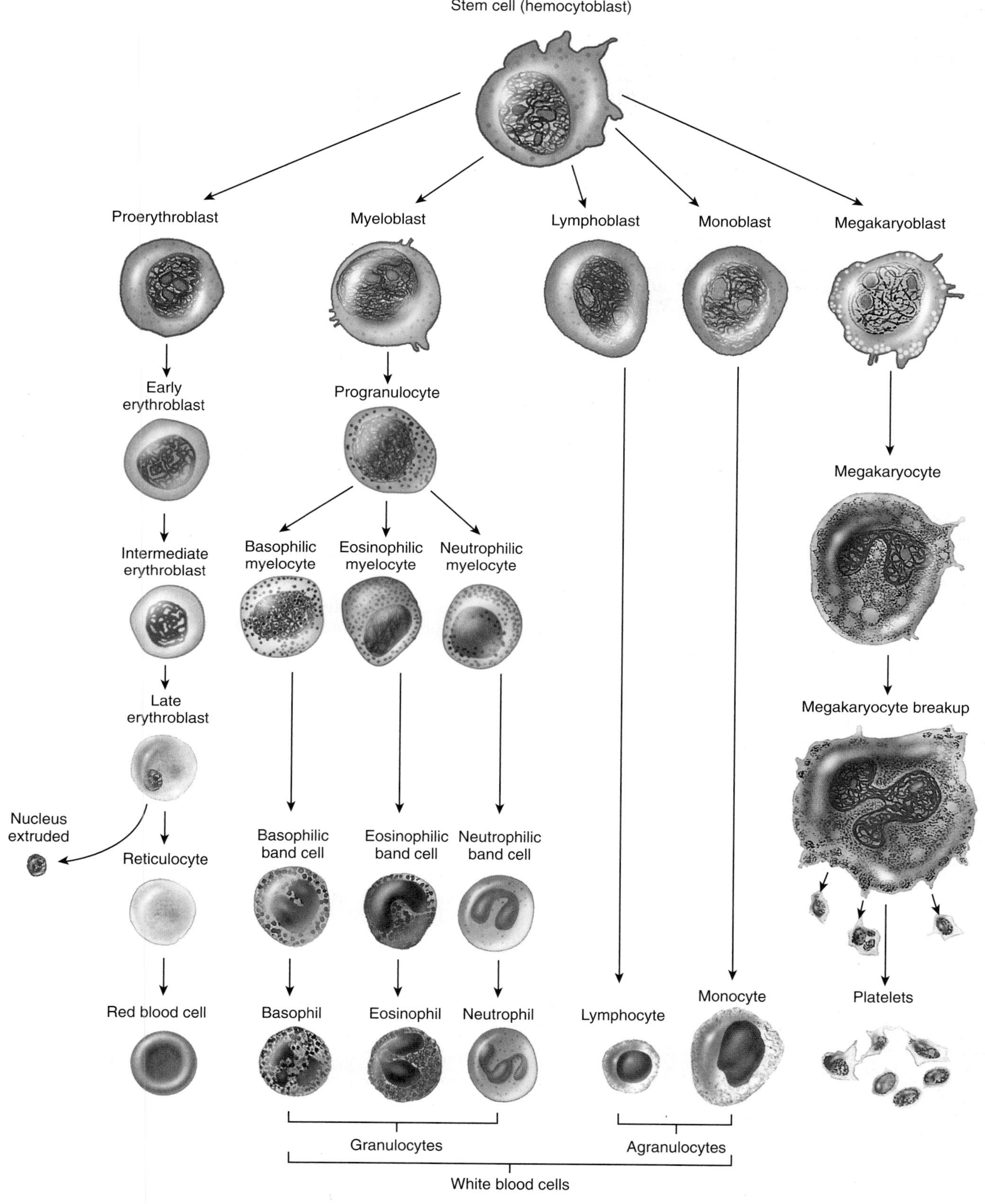

FIGURE 19.2 Hematopoiesis

Stem cells give rise to the cell lines that produce the formed elements.

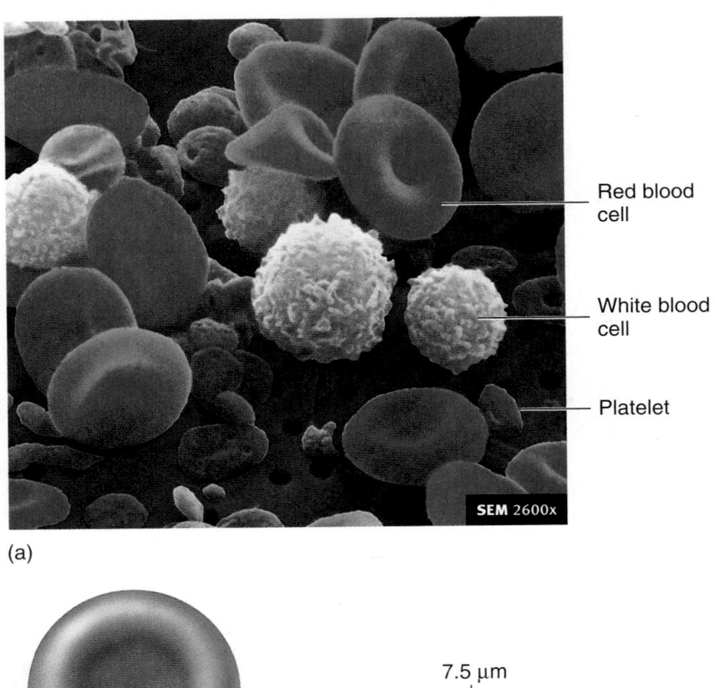

FIGURE 19.3 Red Blood Cells and White Blood Cells

(*a*) Scanning electron micrograph of formed elements: red blood cells (*red doughnut shapes*) and white blood cells (*yellow*). (*b*) Shape and dimensions of a red blood cell.

water is catalyzed by an enzyme, **carbonic anhydrase,** which is located primarily within red blood cells.

$$\underset{\text{Carbon dioxide}}{CO_2} + \underset{\text{Water}}{H_2O} \xrightleftharpoons{\text{Carbonic anhydrase}} \underset{\text{Carbonic acid}}{H_2CO_3} \rightleftharpoons \underset{\text{Hydrogen ion}}{H^+} + \underset{\text{Bicarbonate ion}}{HCO_3^-}$$

6 ***Give the percentage for each of the ways that oxygen and carbon dioxide are transported in the blood. What is the function of carbonic anhydrase?***

Hemoglobin

Hemoglobin consists of four polypeptide chains and four heme groups. Each polypeptide chain, called a **globin** (glō′bin), is bound to one **heme** (hēm). Each heme is a red-pigment molecule containing one **iron atom** (figure 19.4). There are three kinds of hemoglobin: embryonic, fetal, and adult. The first type of hemoglobin produced during development is embryonic hemoglobin. By the third month of development, embryonic hemoglobin has been replaced with fetal hemoglobin. At birth, 60%–90% of the hemoglobin is adult hemoglobin. At 2 to 4 years of age, fetal hemoglobin is less than 2% of the hemoglobin and, in adulthood, only traces of fetal hemoglobin can be found.

The different kinds of hemoglobin have different affinities for, or abilities to bind with, oxygen. Embryonic and fetal hemoglobins have a higher affinity for oxygen than does adult hemoglobin. In the embryo and fetus, hemoglobin picks up oxygen from the mother's blood at the placenta. After birth, hemoglobin picks up oxygen from the air in the baby's lungs. Even though placental blood contains less oxygen than air, adequate amounts of oxygen are picked up because of the higher affinity of embryonic and fetal hemoglobins for oxygen.

Although embryonic, fetal, and adult hemoglobins each have four globins, the types of globins are different. There are nine types

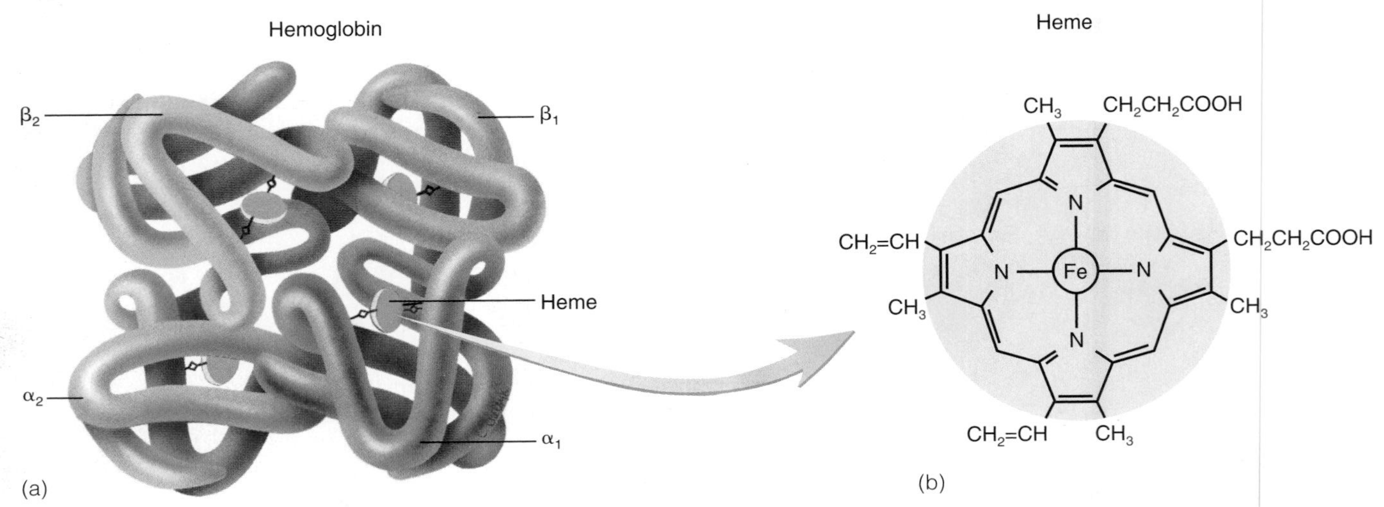

FIGURE 19.4 Hemoglobin

(*a*) Four polypeptide chains, each with a heme, form a hemoglobin molecule. (*b*) Each heme contains one iron atom.

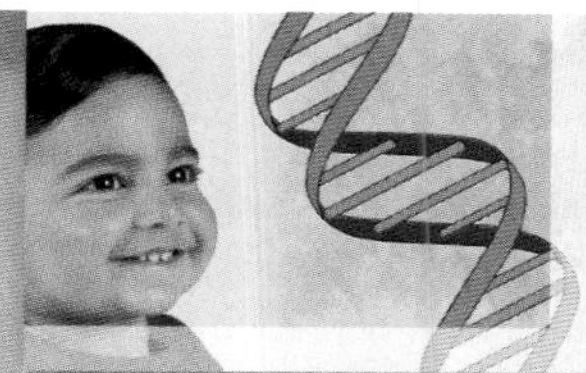

CLINICAL GENETICS

Sickle-Cell Disease

Sickle-cell disease is a disorder in which red blood cells become sickle-shaped. It results from a mutation in the gene on chromosome 11 that codes for the beta globin of hemoglobin. Within this gene is a DNA triplet, CTC, which codes for the mRNA codon GAG (see "Genes and Gene Expression," chapter 3). At the ribosome, GAG specifies that glutamic acid is the sixth amino acid inserted into the beta globin as it is synthesized. The mutation substitutes adenine for thymine in the DNA triplet. Thus, CTC becomes CAC, which codes for the mRNA codon GUG. At the ribosome, GUG specifies that valine is inserted into the beta globulin instead of glutamic acid. The change of this one amino acid in the beta globulin has a dramatic effect on hemoglobin. When blood oxygen levels decrease, as when oxygen diffuses away from the hemoglobin in tissue capillaries, the abnormal hemoglobin molecules join together, causing a change in red blood cell shape (figure A). When blood oxygen levels increase, as in the lungs, the abnormal hemoglobin molecules separate and red blood cells can resume their normal shape. Thus, red blood cells can undergo repeated cycles of sickling as they circulate between the lungs and tissues.

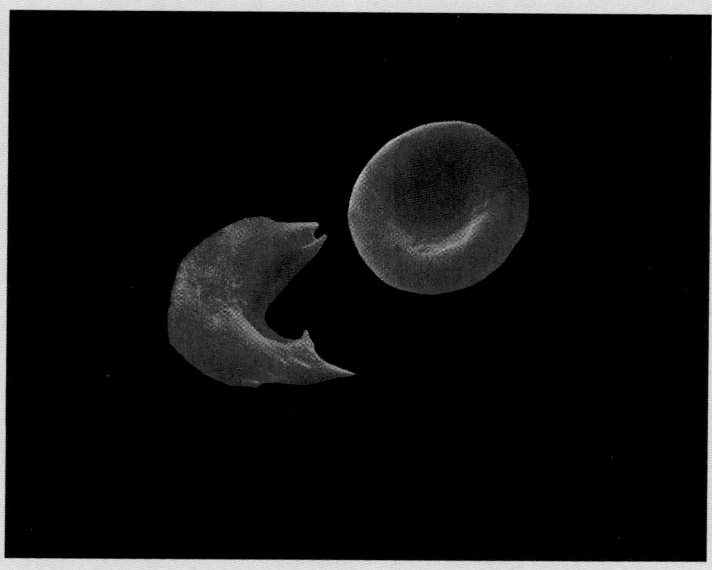

FIGURE A Sickle-Cell Disease

Red blood cells in a person with sickle-cell disease appear normal in oxygenated blood. In deoxygenated blood, hemoglobin changes shape and causes the cells to become sickle-shaped and rigid.

Sickle-shaped red blood cells are less able to squeeze through small capillaries. Consequently, they become lodged in capillaries, blocking blood flow through them. This causes a further decrease in oxygen levels, which promotes more sickling. As oxygen levels decrease further, more capillary blockage is promoted, and so on. After repeated cycles of sickling, red blood cells lose their ability to resume their normal shape. This increases the number of sickled cells.

The major consequence of sickle-cell disease is tissue damage resulting from reduced blood flow through tissues. The most common symptom is pain, often severe pain, associated with the tissues deprived of blood. Spleen and liver enlargement, kidney and lung damage, and stroke can occur. Priapism (prī′ă-prizm), a prolonged, painful erection due to venous blockage, can develop in men.

Sickle-shaped red blood cells are also likely to rupture, and they have a life span of 10–20 days, compared with 120 days for normal red blood cells. Rupture of red blood cells can result in hemolytic anemia (see "Anemia," p. 673).

Sickle-cell disease is an autosomal-recessive disorder. Only individuals who have two mutated beta globin alleles express the disease. Individuals who are heterozygous have a normal beta globin allele and produce sufficient amounts of normal beta globin that their red blood cells do not usually become sickle-shaped. Heterozygotes are carriers (see chapter 3) and are said to have **sickle-cell trait.** They usually do not have disease symptoms, but mild anemia may occur. Under stressful situations, such as low oxygen levels at high altitude, severe infections, or exhaustion, complications caused by sickling may develop.

Sickle-cell disease is an example of **balanced polymorphism,** in which the heterozygote has a better ability to survive under certain circumstances than either the homozygous dominant or recessive. Carriers (heterozygotes) with sickle-cell trait are healthier than those with sickle-cell disease. They also have increased resistance to malaria, compared with individuals who are homozygous for the normal beta globin alleles. Malaria is a disease caused by a parasitic protozoan that reproduces inside red blood cells. The parasite is usually transmitted from one person to another through the bite of a mosquito. The red blood cells of people with sickle-cell trait tend to rupture before the parasite successfully reproduces. Those people are, therefore, less likely to contract malaria and have a much milder disease if they do.

The highest percentage of people with sickle-cell trait occurs in populations exposed to malaria or whose ancestors were exposed to malaria. In certain parts of Africa where malaria is rampant, the percentage of sickle-cell carriers can be as high as 50%. In the United States, 8% of African-Americans are sickle-cell carriers and 0.8% have sickle-cell disease. Sickle-cell disease, however, is not a disorder of only Africans and African-Americans. It is hypothesized that the sickle-cell mutation independently arose four times in Africa and once in India. Through population movements and interracial relationships, the mutant gene can also be found in other groups, but at lower frequencies.

Treatment for sickle-cell disease attempts to reduce the blockage of blood vessels, alleviate pain, and prevent infections. Hydroxyurea (hī-drok′sē-ū-rē′ă) stimulates the production of gamma (fetal) globins. When the gamma globins combine with defective beta globins, the formation of sickle-cell cells is slowed. Bone marrow transplants can cure sickle-cell disease, but such transplants can be dangerous and even fatal. Gene therapy is under investigation.

of globins, each with a slightly different amino acid composition. For example, there are two types of alpha globins, each of which differs from the other by one amino acid. Because they are so similar, they are usually referred to simply as alpha globins. There are also a beta globin, two kinds of gamma globins, a delta globin, and three kinds of embryonic globins. Most adult hemoglobin has two alpha globins (one of each type) and two beta globins (see figure 19.4). Fetal hemoglobin has two alpha globins (one of each type) and two gamma globins (one of each type).

There are nine globin genes, each of which codes for one of the globins. Five of these genes are on chromosome 11 and four are on chromosome 16. The globin genes are active during different stages of development. The embryonic globin genes are active first, but they become inactive as fetal globin genes become active. In a similar fashion, the fetal globin genes become inactive as the adult globin genes become active.

PREDICT 1

What would happen to a fetus if maternal blood had an equal or a greater affinity for oxygen than does fetal blood?

Iron is necessary for the normal function of hemoglobin because each oxygen molecule that is transported is associated with an iron atom. The adult human body normally contains about 4 g of iron, two-thirds of which is associated with hemoglobin. Small amounts of iron are regularly lost from the body in waste products, such as urine and feces. Females lose additional iron as a result of menstrual bleeding and, therefore, require more dietary iron than do males. Dietary iron is absorbed into the circulation from the upper part of the intestinal tract. Stomach acid and vitamin C in food increase the absorption of iron by converting ferric iron (Fe^{3+}) to ferrous iron (Fe^{2+}), which is more readily absorbed.

Effect of Carbon Monoxide on Oxygen Transport

Various types of poisons affect the hemoglobin molecule. Carbon monoxide (CO), which is produced by the incomplete combustion of gasoline, binds very strongly to the iron of hemoglobin to form the relatively stable compound **carboxyhemoglobin** (kar-bok'sē-hē-mō-glō'bin). As a result of the stable binding of carbon monoxide, hemoglobin cannot transport oxygen, and death may occur. Carbon monoxide is found in cigarette smoke, and the blood of smokers can contain 5%–15% carboxyhemoglobin.

When hemoglobin is exposed to oxygen, one oxygen molecule can become associated with each heme group. This oxygenated form of hemoglobin is called **oxyhemoglobin** (ok'sē-hē-mō-glō'bin). The oxyhemoglobin in one red blood cell transports about 1 billion molecules of oxygen. Each heme molecule binds to one oxygen molecule; there are four heme molecules per hemoglobin and 280 million hemoglobin molecules per red blood cell. Hemoglobin containing no oxygen is called **deoxyhemoglobin.** Oxyhemoglobin is bright red, whereas deoxyhemoglobin has a darker red color.

Hemoglobin transports carbon dioxide, which does not combine with the iron atoms but is attached to amino groups of the globin molecule. This hemoglobin form is **carbaminohemoglobin** (kar-bam'i-nō-hē-mō-glō'bin). The transport of oxygen and carbon dioxide by the blood is discussed more fully in chapter 23.

Hemoglobin also transports nitric oxide, which is produced by the endothelial cells lining blood vessels. In the lungs, at the same time that heme picks up oxygen, in each β-globin a sulfur-containing amino acid, cysteine, binds with a nitric oxide molecule to form *S*-nitrosothiol (nī-trōs'ō-thī-ol; SNO). When oxygen is released in tissues, so is the nitric oxide, which functions as a chemical signal that induces the smooth muscle of blood vessels to relax. By affecting the amount of nitric oxide in tissues, hemoglobin may play a role in regulating blood pressure because the relaxation of blood vessels results in a decrease in blood pressure (see chapter 21).

Hemoglobin-Based Oxygen Carriers

Hemoglobin-based oxygen carriers (HBOCs) are being studied as an alternative to blood used in transfusions for people who have inadequate delivery of oxygen to tissues as a result of blood loss, anemia, and so on. The preparation of HBOCs involves removing and purifying the hemoglobin in red blood cells. These hemoglobin molecules are chemically cross-linked and placed in an isotonic salt solution. HBOCs have been produced using human hemoglobin (PolyHeme and HemoLink) and cow hemoglobin (HemoPure). The use of HBOCs for blood transfusions has several benefits, compared with using blood. They have a longer shelf life than blood and can be used when blood is not available. The free oxygen-carrying hemoglobin molecules are smaller than red blood cells, allowing them to flow past partially blocked arteries. There are no transfusion reactions because HBOCs do not have red blood cell membranes, which are necessary for transfusion reactions (see "Blood Grouping," p. 667). HBOCs prepared from cow blood can be used by those who oppose transfusions of human blood on religious grounds.

7 ***Describe the two basic parts of a hemoglobin molecule. Which part is associated with iron? What gases are transported by each part?***

Life History of Red Blood Cells

Under normal conditions, about 2.5 million red blood cells are destroyed every second. This loss seems staggering, but it represents only 0.00001% of the total 25 trillion red blood cells contained in the normal adult circulation. Homeostasis is maintained by replacing the 2.5 million cells lost every second with an equal number of new red blood cells. Thus, approximately 1% of the total number of red blood cells is replaced each day.

The process by which new red blood cells are produced is called **erythropoiesis** (ĕ-rith'rō-poy-ē'sis; see figure 19.2), and the time required for the production of a single red blood cell is about 4 days. Stem cells, from which all blood cells originate, give rise to **proerythroblasts.** After several mitotic divisions, proerythroblasts become **early (basophilic) erythroblasts** (ĕ-rith'rō-blastz), which stain with a basic dye. The dye stains the cytoplasm a purplish color because it binds to the large numbers of ribosomes, which are sites of synthesis for the protein hemoglobin. Early erythroblasts give rise to **intermediate (polychromatic) erythroblasts,** which stain different colors with basic and acidic dyes. As hemoglobin is synthesized and accumulates in the cytoplasm, it is stained a reddish color by an acidic dye. Intermediate erythroblasts continue to produce hemoglobin, and then most of their ribosomes and other organelles degenerate. The resulting **late erythroblasts** have a reddish color because about one-third of the cytoplasm is hemoglobin.

The late erythroblasts lose their nuclei by a process of extrusion to become immature red blood cells, which are called **reticulocytes** (re-tik′ū-lō-sītz), because a reticulum, or network, can be observed in the cytoplasm when a special staining technique is used. The reticulum is artificially produced by the reaction of the dye with the few remaining ribosomes in the reticulocyte. Reticulocytes are released from the bone marrow into the circulating blood, which normally consists of mature red blood cells and 1%–3% reticulocytes. Within 1 to 2 days, reticulocytes become mature red blood cells when the ribosomes degenerate.

PREDICT 2

What does an elevated reticulocyte count indicate? Would the reticulocyte count change during the week after a person had donated a unit (about 500 mL) of blood?

Cell division requires the B vitamins folate and B_{12}, which are necessary for the synthesis of DNA (see chapter 3). Hemoglobin production requires iron. Consequently, adequate amounts of folate, vitamin B_{12}, and iron are necessary for normal red blood cell production.

Red blood cell production is stimulated by low blood oxygen levels, typical causes of which are decreased numbers of red blood cells, decreased or defective hemoglobin, diseases of the lungs, high altitude, inability of the cardiovascular system to deliver blood to tissues, and increased tissue demands for oxygen—for example, during endurance exercises.

Low blood oxygen levels stimulate red blood cell production by increasing the formation of the glycoprotein **erythropoietin** (ĕ-rith-rō-poy′ĕ-tin), which is a hormone produced mostly by the kidneys (figure 19.5). Erythropoietin stimulates red bone marrow to produce more red blood cells by increasing the number of proerythroblasts formed and by decreasing the time required for red blood cells to mature. Thus, when oxygen levels in the blood decrease, erythropoietin production increases, which increases red blood cell production. The increased number of red blood cells increases the blood's ability to transport oxygen. This mechanism returns blood oxygen levels to normal and maintains homeostasis by increasing the delivery of oxygen to tissues. Conversely, if blood oxygen levels increase, less erythropoietin is released, and red blood cell production decreases.

PREDICT 3

Cigarette smoke produces carbon monoxide. If a nonsmoker smoked a pack of cigarettes a day for a few weeks, what would happen to the number of red blood cells in the person's blood? Explain.

Red blood cells normally stay in the circulation for about 120 days in males and 110 days in females. These cells have no nuclei

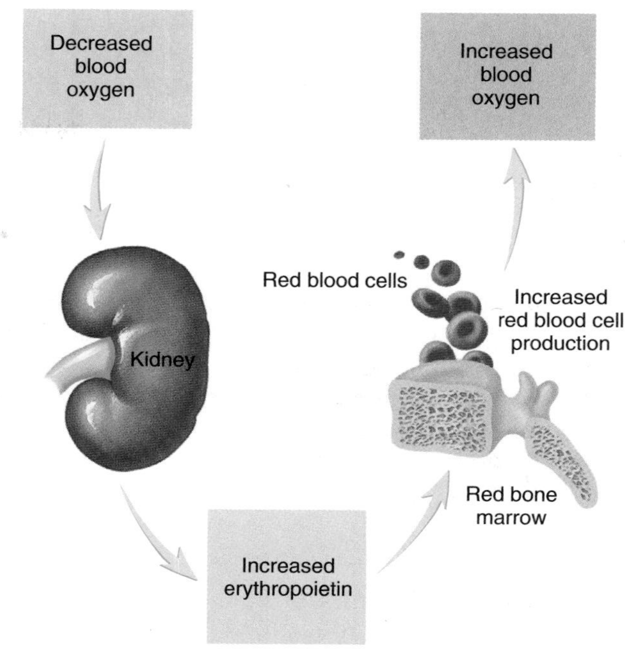

FIGURE 19.5 Red Blood Cell Production

In response to decreased blood oxygen, the kidneys release erythropoietin into the general circulation. The increased erythropoietin stimulates red blood cell production in the red bone marrow. This process increases blood oxygen levels.

and therefore cannot produce new proteins or divide. As their existing proteins, enzymes, plasma membrane components, and other structures degenerate, the red blood cells are less able to transport oxygen and their plasma membranes become more fragile. Eventually, the red blood cells rupture as they squeeze through a tight spot in the circulation.

Macrophages located in the spleen, liver, and other lymphatic tissue (figure 19.6) take up the hemoglobin released from ruptured red blood cells. Within a macrophage, lysosomal enzymes digest the hemoglobin to yield amino acids, iron, and bilirubin. The globin part of hemoglobin is broken down into its component amino acids, most of which are reused in the production of other proteins. Iron atoms released from heme can be carried by the blood to red bone marrow, where they are incorporated into new hemoglobin molecules. The non-iron part of the heme groups are converted to **biliverdin** (bil-i-ver′din) and then to **bilirubin** (bil-i-roo′bin), which is released into the plasma. Bilirubin binds to albumin and is transported to liver cells. This bilirubin is called **free bilirubin** because it is not yet conjugated. Free bilirubin is taken up by the liver cells and is conjugated, or joined, to glucuronic acid to form **conjugated bilirubin,** which is more water-soluble than free bilirubin. The conjugated bilirubin becomes part of the **bile,** which is the fluid secreted from the liver into the small intestine. In the intestine, bacteria convert bilirubin into the pigments that give feces its characteristic brownish color. Some of these pigments are absorbed from the intestine, modified in the kidneys, and excreted in the urine, thus contributing to the characteristic yellowish color of urine. **Jaundice** (jawn′dis) is a yellowish staining of the skin and sclerae caused by a buildup of bile

1. Hemoglobin is broken down into heme and globin chains.
2. The globin chains of hemoglobin are broken down to individual amino acids (*pink arrow*) and are metabolized or used to build new proteins.
3. Iron is released from the heme of hemoglobin. The heme is converted into biliverdin, which is converted into bilirubin.
4. Iron is transported in combination with transferrin in the blood to various tissues for storage or transported to the red bone marrow and used in the production of new hemoglobin (*green arrows*).
5. Free bilirubin (*blue arrow*) is transported in the blood to the liver.
6. Conjugated bilirubin is excreted as part of the bile into the small intestine.
7. Bilirubin derivatives contribute to the color of feces or are reabsorbed from the intestine into the blood and excreted from the kidneys in the urine.

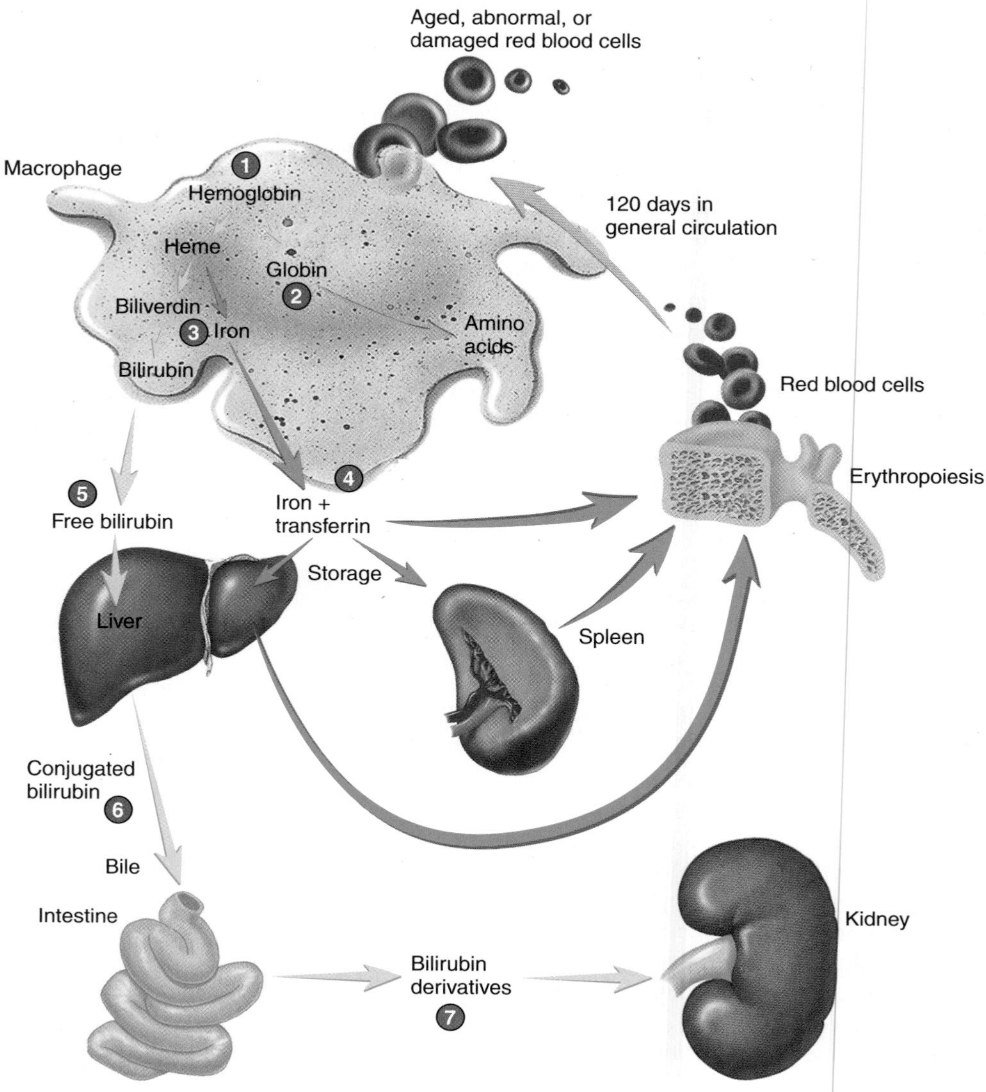

PROCESS FIGURE 19.6 Hemoglobin Breakdown
Hemoglobin is broken down in macrophages, and the breakdown products are used or excreted.

pigments in the circulation and interstitial spaces. Any process that causes increased destruction of red blood cells can cause jaundice, such as damage by toxins, genetic defects in red blood cell plasma membranes, infections, and immune reactions. Other causes of jaundice include dysfunction or destruction of liver tissue and blockage of the duct system that drains bile from the liver (see chapter 24).

8 ***Define erythropoiesis. Describe the formation of red blood cells, starting with the stem cells in the red bone marrow.***

9 ***What is erythropoietin, where is it produced, what causes it to be produced, and what effect does it have on red blood cell production?***

10 ***Where are red blood cells removed from the blood? List the three breakdown products of hemoglobin and explain what happens to them.***

White Blood Cells

White blood cells, or **leukocytes,** form a thin, white layer of cells between plasma and red blood cells when the components of blood are separated from each other (see figure 19.1). White blood cells lack hemoglobin but have a nucleus. In stained preparations, white blood cells attract stain, whereas red blood cells remain relatively unstained (figure 19.7; see table 19.2).

White blood cells protect the body against invading microorganisms and remove dead cells and debris from the body. Most white blood cells are motile, exhibiting **ameboid movement,** which is the ability to move as an ameba does, by putting out irregular cytoplasmic projections. White blood cells leave the circulation and enter tissues by **diapedesis** (dī′ă-pĕ-dē′sis), a process in which they become thin and elongated and slip

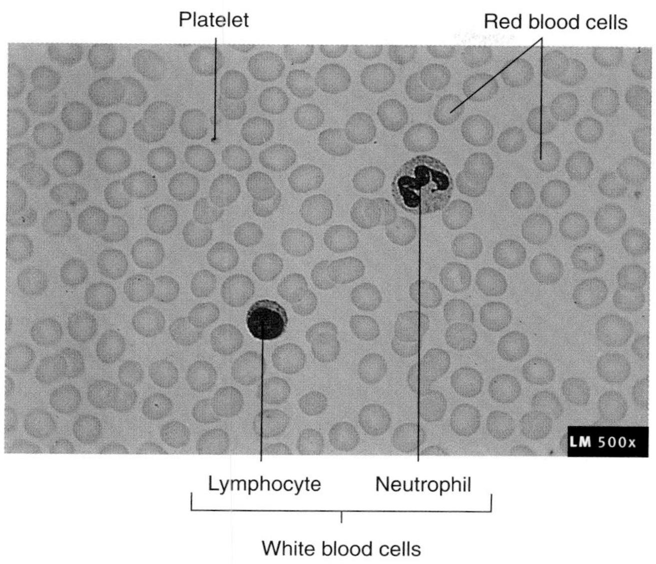

FIGURE 19.7 Standard Blood Smear

A thin film of blood is spread on a microscope slide and stained. The white blood cells have pink-colored cytoplasm and purple-colored nuclei. The red blood cells do not have nuclei. The center of a red blood cell appears whitish because light more readily shines through the thin center of the disc than through the thicker edges. The platelets are purple cell fragments.

between or, in some cases, through the cells of blood vessel walls. The white blood cells can then be attracted to foreign materials or dead cells within the tissue by **chemotaxis** (kē-mō-tak′sis; see chapter 22). At the site of an infection, white blood cells accumulate and phagocytize bacteria, dirt, and dead cells; then they die. The accumulation of dead white blood cells and bacteria, along with fluid and cell debris, is called **pus.**

The five types of white blood cells are neutrophils, eosinophils, basophils, lymphocytes, and monocytes.

Neutrophils

Neutrophils comprise 60%–70% of white blood cells (see table 19.2). They have small cytoplasmic granules that stain with both acidic and basic dyes. Their nuclei are commonly lobed, with the number of lobes varying from two to five. Neutrophils are often called **polymorphonuclear** (pol′ē-mōr-fō-noo′klē-ăr) **neutrophils,** or PMNs, to indicate that their nuclei can occur in more than one (*poly*) form (*morph*). Neutrophils usually remain in the circulation for about 10–12 hours and then move into other tissues, where they become motile and seek out and phagocytize bacteria, antigen–antibody complexes (antigens and antibodies bound together), and other foreign matter. Neutrophils also secrete a class of enzymes called **lysozymes** (lī′sō-zīmz), which are capable of destroying certain bacteria. Neutrophils usually survive for 1–2 days after leaving the circulation.

Eosinophils

Eosinophils comprise 2%–4% of white blood cells (see table 19.2). They contain cytoplasmic granules that stain bright red with eosin, an acidic stain. Eosinophils are motile cells that leave the circulation to enter the tissues during an inflammatory reaction. They are most common in tissues undergoing an allergic response, and their numbers are elevated in the blood of people with allergies. Eosinophils apparently reduce the inflammatory response by producing enzymes that destroy inflammatory chemicals, such as histamine. Eosinophils also release toxic chemicals that attack certain worm parasites, such as tapeworms, flukes, pinworms, and hookworms.

Basophils

Basophils comprise 0.5%–1% of white blood cells (see table 19.2). They contain large cytoplasmic granules that stain blue or purple with basic dyes. Basophils, like eosinophils and neutrophils, leave the circulation and migrate through the tissues. They increase in number in both allergic and inflammatory reactions. Basophils contain large amounts of **histamine** (see chapter 22), which they release within tissues to increase inflammation. They also release **heparin,** which inhibits blood clotting.

Lymphocytes

Lymphocytes comprise 20%–25% of white blood cells (see table 19.2). They are the smallest white blood cells, most of which are slightly larger in diameter than red blood cells. The lymphocytic cytoplasm consists of only a thin, sometimes imperceptible ring around the nucleus. Although lymphocytes originate in red bone marrow, they migrate through the blood to lymphatic tissues, where they can proliferate and produce more lymphocytes. The majority of the body's total lymphocyte population is in the lymphatic tissues: the lymph nodes, spleen, tonsils, lymphatic nodules, and thymus.

Although they cannot be identified by standard microscopic examination, a number of different kinds of lymphocytes play important roles in immunity (see chapter 22). For example, **B cells** can be stimulated by bacteria or toxins to divide and form cells that produce proteins called **antibodies.** Antibodies can attach to bacteria and activate mechanisms that result in the destruction of the bacteria. **T cells** protect against viruses and other intracellular microorganisms by attacking and destroying the cells in which they are found. In addition, T cells are involved in the destruction of tumor cells and tissue graft rejections.

Monocytes

Monocytes comprise 3%–8% of white blood cells (see table 19.2). They are typically the largest of the white blood cells. Monocytes normally remain in the circulation for about 3 days, leave the circulation, become transformed into macrophages, and migrate through various tissues. They phagocytize bacteria, dead cells, cell fragments, and other debris within the tissues. An increase in the number of monocytes is often associated with chronic infections. Macrophages also stimulate responses from other cells in two ways: (1) by the release of chemical signals and (2) by phagocytizing and processing foreign substances, which are presented to lymphocytes. The responses of these other cells help protect against microorganisms and other foreign substances (see chapter 22).

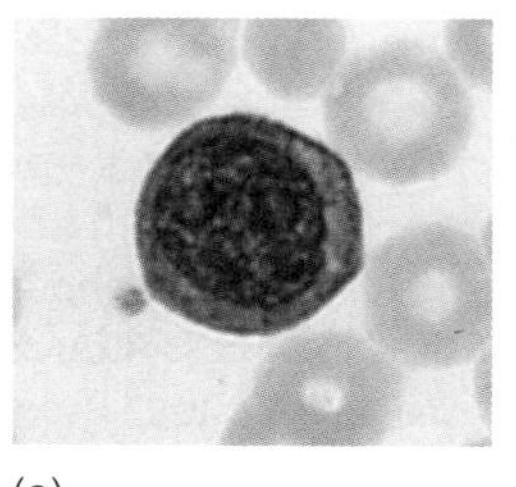
(a)

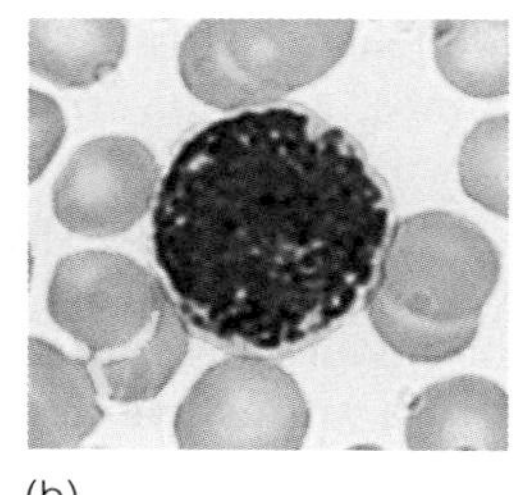
(b)

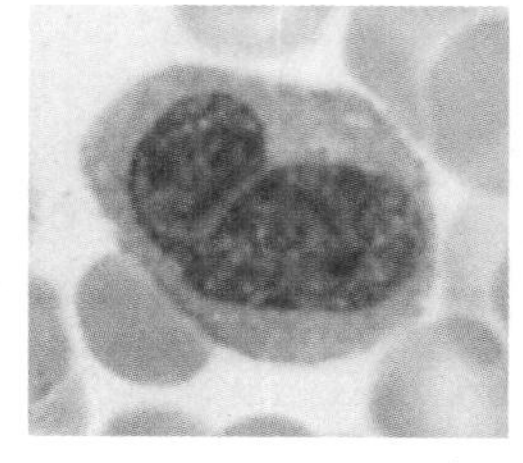
(c)

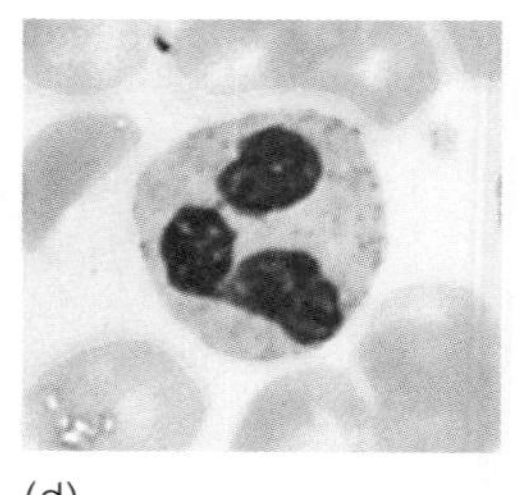
(d)

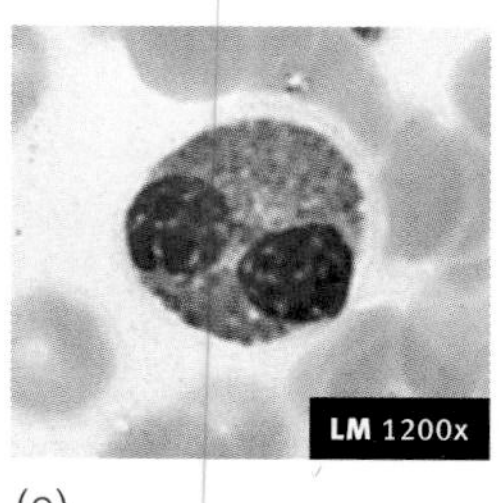

(e)

FIGURE 19.8 Identification of White Blood Cells
See Predict question 4.

11 ***What are the two major functions of white blood cells? Define ameboid movement, diapedesis, and chemotaxis.***

12 ***Describe the morphology of the five types of white blood cells.***

13 ***Name the two white blood cells that function primarily as phagocytic cells. Define lysozymes.***

14 ***Which white blood cell reduces the inflammatory response? Which white blood cell releases histamine and promotes inflammation?***

15 ***B and T cells are examples of what type of white blood cell? How do these cells protect against bacteria and viruses?***

PREDICT 4

Based on their morphology, identify each of the white blood cells shown in figure 19.8.

Platelets

Platelets, or **thrombocytes** (see figure 19.7 and table 19.2), are minute fragments of cells consisting of a small amount of cytoplasm surrounded by a plasma membrane. Platelets are roughly disk-shaped and average about 3 μm in diameter. The surface of platelets has glycoproteins and proteins that allow platelets to attach to other molecules, such as collagen in connective tissue. Some of these surface molecules, as well as molecules released from granules in the platelet cytoplasm, play important roles in controlling blood loss. The platelet cytoplasm also contains actin and myosin, which can cause contraction of the platelet (see "Clot Retraction and Dissolution," p. 666).

The life expectancy of platelets is about 5–9 days. They are produced within the red marrow and are derived from **megakaryocytes** (meg-ă-kar′ē-ō-sītz), which are extremely large cells with diameters up to 100 μm. Small fragments of these cells break off and enter the circulation as platelets.

Platelets play an important role in preventing blood loss by (1) forming platelet plugs, which seal holes in small vessels, and (2) promoting the formation and contraction of clots, which help seal off larger wounds in the vessels.

16 ***What is a platelet? How are platelets formed?***

17 ***What are the two major roles of platelets in preventing blood loss?***

HEMOSTASIS

Hemostasis (hē′mō-stā-sis, hē-mos′tă-sis), the stoppage of bleeding, is very important to the maintenance of homeostasis. If not stopped, excessive bleeding from a cut or torn blood vessel can result in a positive-feedback cycle, consisting of ever-decreasing blood volume and blood pressure, which disrupts homeostasis and results in death. Fortunately, when a blood vessel is damaged, a number of events occur that help prevent excessive blood loss. Vascular spasm, platelet plug formation, and coagulation can cause hemostasis.

Vascular Spasm

Vascular spasm is the immediate but temporary constriction of a blood vessel resulting from contraction of smooth muscle within the wall of the vessel. This constriction can close small vessels completely and stop the flow of blood through them. Damage to blood vessels can activate nervous system reflexes that cause vascular spasms. Chemicals also produce vascular spasms. For example, during the formation of a platelet plug, platelets release **thromboxanes** (throm′bok-zānz), which are derived from certain prostaglandins, and endothelial cells release the peptide **endothelin** (en-dō′thē-lin).

18 ***What is a vascular spasm? Name two factors that produce it. What is the source of thromboxanes and endothelin?***

Platelet Plug Formation

A **platelet plug** is an accumulation of platelets that can seal up small breaks in blood vessels. Platelet plug formation is very important in maintaining the integrity of the circulatory system because small tears occur in the smaller vessels and capillaries many times each day, and platelet plug formation quickly closes them. People who lack the normal number of platelets tend to develop numerous small hemorrhages in their skin and internal organs.

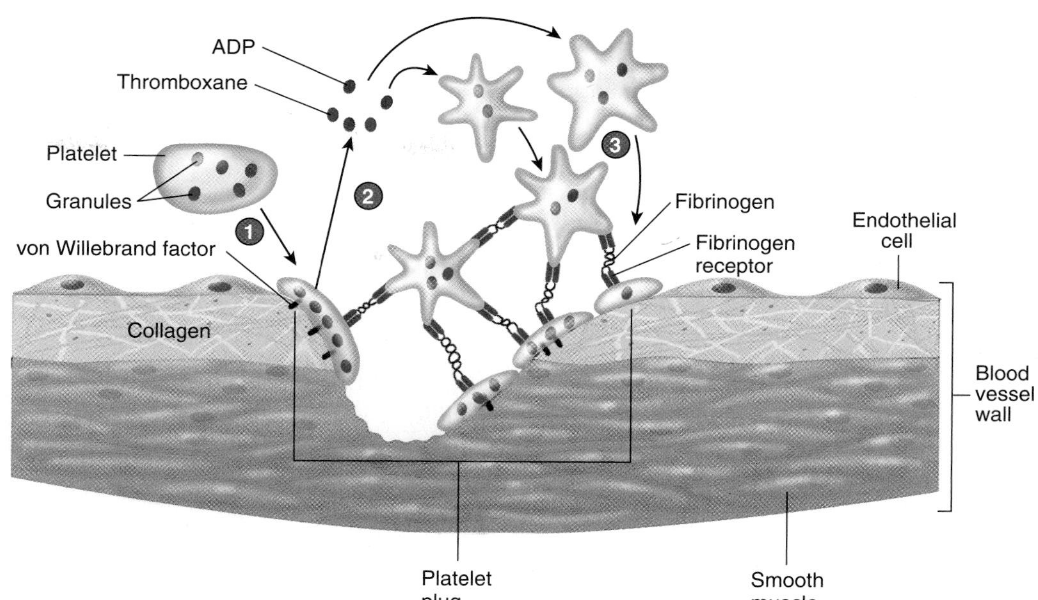

1. Platelet adhesion occurs when von Willebrand factor connects collagen and platelets.
2. The platelet release reaction results in the release of ADP, thromboxanes, and other chemicals that activate other platelets.
3. Platelet aggregation occurs when fibrinogen receptors on activated platelets bind to fibrinogen, connecting the platelets to one another. A platelet plug is formed by the accumulating mass of platelets.

PROCESS FIGURE 19.9 Platelet Plug Formation

The formation of a platelet plug can be described as a series of steps, but in actuality many of the steps take place simultaneously (figure 19.9):

1. **Platelet adhesion** occurs when platelets bind to collagen exposed by blood vessel damage. Most platelet adhesion is mediated through **von Willebrand factor (vWF),** which is a protein produced and secreted by blood vessel endothelial cells. Von Willebrand factor forms a bridge between collagen and platelets by binding to platelet surface receptors and collagen. In addition, other platelet surface receptors can bind directly to collagen.
2. After platelets adhere to collagen, they become activated; in the **platelet release reaction,** adenosine diphosphate (ADP), thromboxanes, and other chemicals are extruded from the platelets by exocytosis. The ADP and thromboxane bind to their respective receptors on the surfaces of other platelets, resulting in their activation. These activated platelets release additional chemicals, thereby producing a cascade of chemical release by the platelets. Thus, more and more platelets become activated.
3. As platelets become activated, they change shape and express fibrinogen receptors that can bind to fibrinogen, a plasma protein. In **platelet aggregation,** fibrinogen forms a bridge between the fibrinogen receptors of different platelets, resulting in the formation of a platelet plug.
4. Activated platelets express phospholipids (platelet factor III) and coagulation factor V, which are important in clot formation (see "Coagulation," p. 664).

Clinical Importance of Activating Platelets

Eicosanoids (ī-kō'să-noydz; *eicosa-,* twenty + *eidos,* form) are a group of related compounds derived from 20-carbon essential fatty acids (see chapter 25). Examples of eicosanoids are prostaglandins, thromboxanes, and leukotrienes. In humans, arachidonic acid is the most common precursor molecule for the eicosanoids. The enzyme cyclooxygenase (COX) converts arachidonic acid into a prostaglandin that can be converted into thromboxane. Aspirin inhibits COX, which inhibits prostaglandin synthesis and therefore thromboxane synthesis. As a result, aspirin reduces platelet activation.

If an expectant mother ingests aspirin near the end of pregnancy, thromboxane synthesis is inhibited and several effects are possible. The mother can experience excessive bleeding after delivery because of decreased platelet function, and the baby can exhibit numerous localized hemorrhages called **petechiae** (pe-tē'kē-ē) over the surface of its body as a result of decreased platelet function. If the quantity of ingested aspirin is large, the infant, the mother, or both may die as a result of hemorrhage.

Platelet plug and clot formation can cause the blockage of blood vessels, producing heart attacks and strokes. Suspected heart attack victims are routinely given aspirin en route to the emergency room as part of their treatment. The United States Preventive Services Task Force (USPSTF) and the American Heart Association (AHA) recommend low-dose aspirin therapy (75–160 mg/day) for the prevention of cardiovascular disease for all men and women at high risk for cardiovascular disease. Determining high risk involves analyzing many risk factors and should be done in consultation with a physician. Decreased risk for cardiovascular disease from aspirin therapy must be weighed against the increased risk for hemorrhagic stroke and gastrointestinal bleeding. Risk factors for cardiovascular disease include age (men over 40 and postmenopausal women), high blood cholesterol levels, high blood pressure, a history of smoking, diabetes, a family history of cardiovascular disease, and a previous clotting event, such as a heart attack, transient ischemic attack, or occlusive stroke (see Systems Pathology, "Stroke," chapter 14).

Plavix (clopidogrel bisulfate) reduces the activation of platelets by blocking the ADP receptors on the surface of platelets. It is used to prevent clotting and, with other anticlotting drugs, to treat heart attacks.

19 ***What is the function of a platelet plug? Describe the process of platelet plug formation. How are platelets an important part of clot formation?***

Coagulation

Vascular spasms and platelet plugs alone are not sufficient to close large tears or cuts. When a blood vessel is severely damaged, **coagulation** (kō-ag-ū-lā′shŭn), or blood clotting, results in the formation of a clot. A **blood clot** is a network of threadlike protein fibers, called **fibrin,** that traps blood cells, platelets, and fluid (figure 19.10).

The formation of a blood clot depends on a number of proteins, called **coagulation factors,** found within plasma (table 19.3). Normally, the coagulation factors are in an inactive state and do not cause clotting. After injury, the clotting factors are activated to produce a clot. This activation is a complex process involving

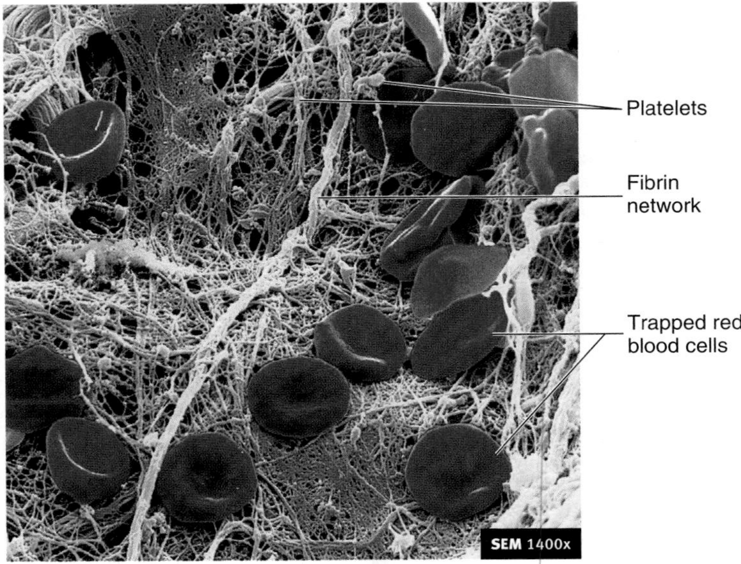

FIGURE 19.10 Blood Clot
A blood clot consists of fibrin fibers that trap red blood cells, platelets, and fluid.

TABLE 19.3 Coagulation Factors

Factor Number	Name (Synonym)	Description and Function
I	Fibrinogen	Plasma protein synthesized in liver; converted to fibrin in stage 3
II	Prothrombin	Plasma protein synthesized in liver (requires vitamin K); converted to thrombin in stage 2
III	Thromboplastin (tissue factor)	Mixture of lipoproteins released from damaged tissue; required in extrinsic stage 1
IV	Calcium ion	Required throughout entire clotting sequence
V	Proaccelerin (labile factor)	Plasma protein synthesized in liver; activated form functions in stages 1 and 2 of both intrinsic and extrinsic clotting pathways
VI		Once thought to be involved but no longer accepted as playing a role in coagulation; apparently the same as activated factor V
VII	Serum prothrombin conversion accelerator (stable factor, proconvertin)	Plasma protein synthesized in liver (requires vitamin K); functions in extrinsic stage 1
VIII	Antihemophilic factor (antihemophilic globulin)	Plasma protein synthesized in megakaryocytes and endothelial cells; required for intrinsic stage 1
IX	Plasma thromboplastin component (Christmas factor)	Plasma protein synthesized in liver (requires vitamin K); required for intrinsic stage 1
X	Stuart factor (Stuart-Prower factor)	Plasma protein synthesized in liver (requires vitamin K); required in stages 1 and 2 of both intrinsic and extrinsic clotting pathways
XI	Plasma thromboplastin antecedent	Plasma protein synthesized in liver; required for intrinsic stage 1
XII	Hageman factor	Plasma protein required for intrinsic stage 1
XIII	Fibrin-stabilizing factor	Protein found in plasma and platelets; required for stage 3
Platelet Factors		
I	Platelet accelerator	Same as plasma factor V
II	Thrombin accelerator	Accelerates thrombin (intrinsic clotting pathway) and fibrin production
III		Phospholipids necessary for the intrinsic and extrinsic clotting pathways
IV		Binds heparin, which prevents clot formation

many chemical reactions, some of which require calcium ions (Ca^{2+}) and molecules on the surface of activated platelets, such as phospholipids and coagulation factor V.

PREDICT 5

Why is it advantageous for clot formation to involve molecules on the surface of activated platelets?

The activation of clotting proteins begins with the extrinsic and intrinsic pathways (figure 19.11). These pathways converge to form the common pathway, which results in the formation of a fibrin clot.

Extrinsic Pathway

The extrinsic pathway is so-named because it begins with chemicals that are outside of, or extrinsic to, the blood (see figure 19.11). Damaged tissues release a mixture of lipoproteins and phospholipids called **thromboplastin** (throm-bō-plas′tin), also known as **tissue factor (TF),** or factor III. Thromboplastin, in the presence of Ca^{2+}, forms a complex with factor VII, which activates factor X, which is the beginning of the common pathway.

Intrinsic Pathway

The intrinsic pathway is so-named because it begins with chemicals that are inside, or intrinsic to, the blood (see figure 19.11). Damage to blood vessels can expose collagen in the connective tissue beneath the epithelium lining the blood vessel. When plasma factor XII comes into contact with collagen, factor XII is activated and it stimulates factor XI, which in turn activates factor IX. Activated factor IX joins with factor VIII, platelet phospholipids, and Ca^{2+} to activate factor X, which is the beginning of the common pathway.

Although once considered distinct pathways, it is now known that the extrinsic pathway can activate the clotting proteins in the

1. The extrinsic pathway of clotting starts with thromboplastin, which is released outside of the plasma in damaged tissue.
2. The intrinsic pathway of clotting starts when inactive factor XII, which is in the plasma, is activated by coming into contact with a damaged blood vessel.
3. Activation of the extrinsic or intrinsic clotting pathway results in the production of activated factor X.
4. Activated factor X, factor V, phospholipids, and Ca^{2+} form prothrombinase.
5. Prothrombin is converted to thrombin by prothrombinase.
6. Fibrinogen is converted to fibrin (the clot) by thrombin.
7. Thrombin activates factor XIII, which stabilizes the fabrin clot.

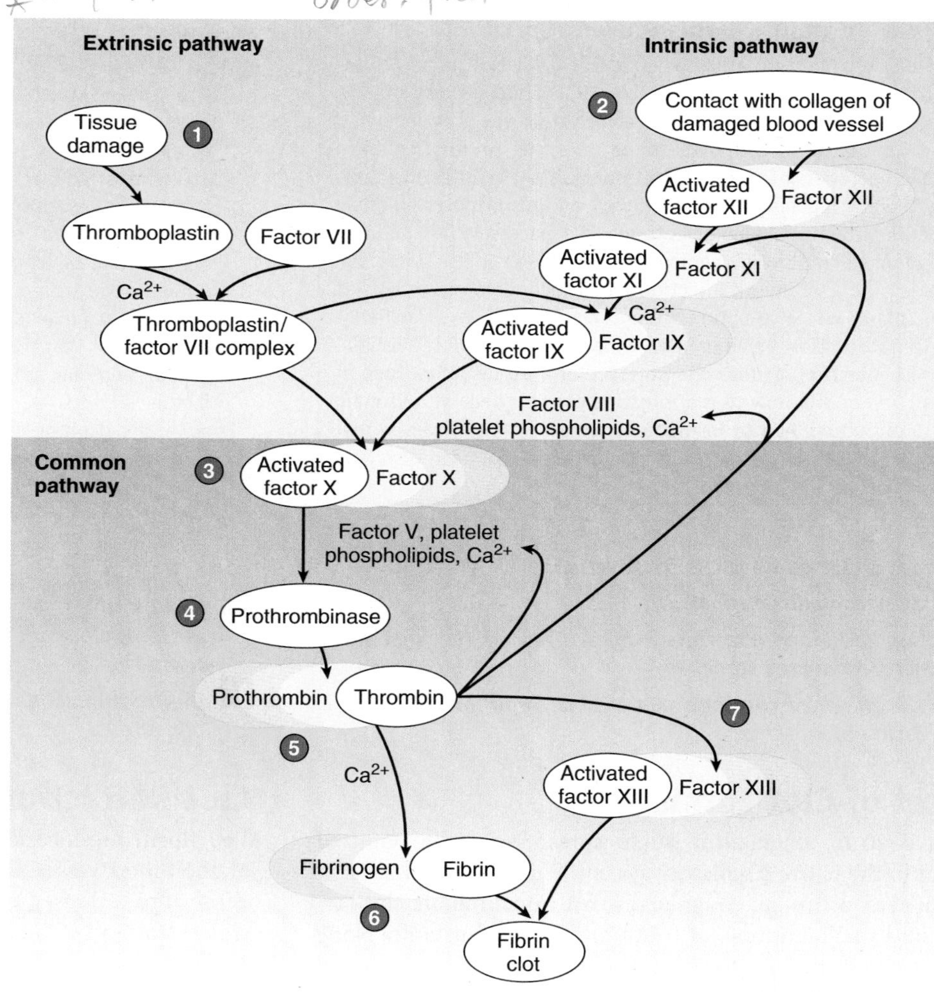

PROCESS FIGURE 19.11 Clot Formation

A sequence of chemical reactions occurs as activated coagulation factors (*white ovals*) activate inactive coagulation factors (*yellow ovals*). Clot formation begins through the extrinsic or intrinsic pathway. The common pathway starts with factor X and results in a fibrin clot.

intrinsic pathway. The thromboplastin/factor VII complex from the extrinsic pathway can stimulate the formation of activated factors IX in the intrinsic pathway.

Common Pathway

On the surface of platelets, activated factor X, factor V, platelet phospholipids, and Ca^{2+} complex to form **prothrombinase.** Prothrombinase converts the soluble plasma protein **prothrombin** into the enzyme **thrombin.** Thrombin converts the soluble plasma protein fibrinogen into the insoluble protein **fibrin.** Fibrin forms the fibrous network of the clot. Thrombin stimulates factor XIII activation, which is necessary to stabilize the clot.

Thrombin can activate many of the clotting proteins, such as factor XI and prothrombinase. Thus, thrombin is part of a positive-feedback system in which thrombin production stimulates the production of additional thrombin. Thrombin also has a positive-feedback effect on coagulation by stimulating platelet activation.

How Vitamin K Helps Prevent Bleeding

Many of the factors involved in clot formation require vitamin K for their production (see table 19.3). Humans rely on two sources for vitamin K. About half comes from the diet, and half comes from bacteria within the large intestine. Antibiotics taken to fight bacterial infections sometimes kill these intestinal bacteria, thereby reducing vitamin K levels and resulting in bleeding problems. Vitamin K supplements may be necessary for patients on prolonged antibiotic therapy. Newborns lack these intestinal bacteria, and a vitamin K injection is routinely given to infants at birth. Infants can also obtain vitamin K from food, such as milk.

The absorption of vitamin K, which is a fat-soluble vitamin, from the intestine requires the presence of bile. Disorders such as obstruction of bile flow to the intestine can interfere with vitamin K absorption and lead to insufficient clotting. Liver diseases that result in the decreased synthesis of clotting factors can also lead to insufficient clot formation.

20 ***What is a clot and what is its function?***

21 ***What are coagulation factors?***

22 ***Clotting is divided into three stages. Describe the final event that occurs in each stage.***

23 ***What is the difference between extrinsic and intrinsic activation of clotting?***

Control of Clot Formation

Without control, coagulation would spread from the point of initiation to the entire circulatory system. Furthermore, vessels in a healthy person contain rough areas that can stimulate clot formation, and small amounts of prothrombin are constantly being converted into thrombin. To prevent unwanted clotting, the blood contains several **anticoagulants** (an′tē-kō-ag′ū-lantz), which prevent coagulation factors from initiating clot formation. Only when coagulation factor concentrations exceed a given threshold in a local area does coagulation occur. At the site of injury, so many coagulation factors are activated that the anticoagulants are unable to prevent clot formation. Away from the injury site, however, the activated coagulation factors are diluted in the blood, anticoagulants neutralize them, and clotting is prevented.

Examples of anticoagulants in the blood are antithrombin, heparin, and prostacyclin. **Antithrombin,** a plasma protein produced by the liver, slowly inactivates thrombin. **Heparin,** produced by basophils and endothelial cells, increases the effectiveness of antithrombin because heparin and antithrombin together rapidly inactivate thrombin. **Prostacyclin** (pros-tă-sī′klin) is a prostaglandin derivative produced by endothelial cells. It counteracts the effects of thrombin by causing vasodilation and inhibiting the release of coagulation factors from platelets.

Anticoagulants are also important when blood is outside the body. They prevent the clotting of blood used in transfusions and laboratory blood tests. Examples include heparin, **ethylenediaminetetraacetic** (eth′il-ēn-dī′ă-mēn-tet-ră-ă-sē′tik) **acid (EDTA),** and sodium citrate. EDTA and sodium citrate prevent clot formation by binding to Ca^{2+}, thus making the ions inaccessible for clotting reactions.

The Danger of Unwanted Clots

When platelets encounter damaged or diseased areas on the walls of blood vessels or the heart, an attached clot called a **thrombus** (throm′bŭs) may form. A thrombus that breaks loose and begins to float through the circulation is called an **embolus** (em′bō-lŭs). Both thrombi and emboli can result in death if they block vessels that supply blood to essential organs, such as the heart, brain, or lungs. Abnormal coagulation can be prevented or hindered by the injection of anticoagulants, such as heparin, which acts rapidly. Coumadin (koo′mă-din), or warfarin (war′fă-rin), acts more slowly than heparin. Coumadin prevents clot formation by suppressing the production of vitamin K–dependent coagulation factors (II, VII, IX, and X) by the liver. Coumadin was first used as a rat poison by causing rats to bleed to death. In small doses, warfarin is a proven, effective anticoagulant in humans. Caution is necessary with anticoagulant treatment, however, because the patient can hemorrhage internally or bleed excessively when cut.

24 ***What is the function of anticoagulants in blood? Name three anticoagulants in blood, and explain how they prevent clot formation.***

25 ***Define thrombus and embolus, and explain why they are dangerous.***

Clot Retraction and Dissolution

The fibrin meshwork constituting a clot adheres to the walls of the blood vessel. Once a clot has formed, it begins to condense into a denser, compact structure through the process of **clot retraction.** Platelets contain the contractile proteins actin and myosin, which operate in a similar fashion to that of actin and myosin in smooth muscle (see chapter 9). Platelets form extensions, which attach to fibrinogen through fibrinogen receptors (see figure 19.9). Contraction of the extensions pulls on the fibrinogen and is responsible for clot retraction. As the clot condenses, a fluid called **serum** (sēr′ŭm) is squeezed out of

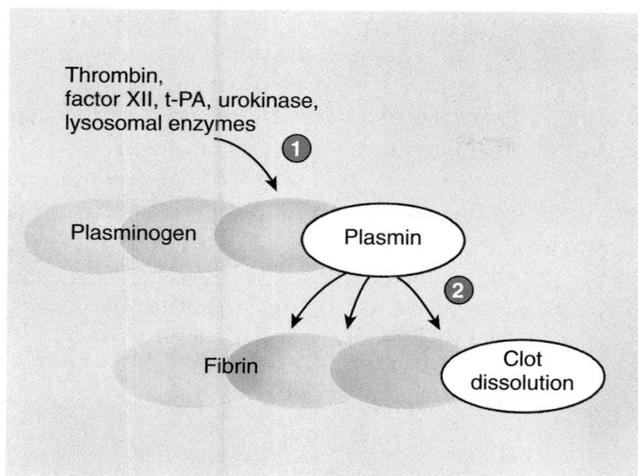

1. Inactive plasminogen is converted to the active enzyme plasmin.

2. Plasmin breaks the fibrin molecules, and therefore the clot, into smaller pieces, which are washed away in the blood or phagocytized.

PROCESS FIGURE 19.12 Fibrinolysis

it. Serum is plasma from which fibrinogen and some of the clotting factors have been removed.

Consolidation of the clot pulls the edges of the damaged blood vessel together, which can help stop blood flow, reduce infection, and enhance healing. The damaged vessel is repaired by the movement of fibroblasts into the damaged area and the formation of new connective tissue. In addition, epithelial cells around the wound proliferate and fill in the torn area.

The clot usually is dissolved within a few days after clot formation by a process called **fibrinolysis** (fī-bri-nol′i-sis), which involves the activity of **plasmin** (plaz′min), an enzyme that hydrolyzes fibrin. Plasmin is formed from inactive plasminogen, which is a normal blood protein produced by the liver. Plasmin becomes part of the clot as it is formed. Plasmin is activated by thrombin, factor XII, tissue plasminogen activator (t-PA), urokinase, and lysosomal enzymes released from damaged tissues (figure 19.12). In disorders that are caused by the blockage of a vessel by a clot, such as a heart attack, dissolving the clot can restore blood flow and reduce damage to tissues. For example, streptokinase (a bacterial enzyme), t-PA, or urokinase can be injected into the blood or introduced at the clot site by means of a catheter. These substances convert plasminogen to plasmin, which breaks down the clot.

26 ***Describe clot retraction. What is serum?***

27 ***What is fibrinolysis? How does it occur?***

BLOOD GROUPING

If large quantities of blood are lost during surgery or in an accident, the patient can go into shock and die unless a transfusion or an infusion is performed. A **transfusion** is the transfer of blood or blood components from one individual to another. When large quantities of blood are lost, red blood cells must be replaced to restore the blood's oxygen-carrying capacity. An **infusion** is the introduction of a fluid other than blood, such as a saline or glucose solution, into the blood. In many cases, the return of blood volume to normal levels is all that is necessary to prevent shock. Eventually, the body produces red blood cells to replace those that were lost.

Early attempts to transfuse blood from one person to another were often unsuccessful because they resulted in transfusion reactions, which included clotting within blood vessels, kidney damage, and death. It is now known that transfusion reactions are caused by interactions between antigens and antibodies (see chapter 22). In brief, the surfaces of red blood cells have molecules called **antigens** (an′ti-jenz) and, in the plasma, molecules called **antibodies** are present. Antibodies are very specific, meaning that each antibody can combine only with a certain antigen. When the antibodies in the plasma bind to the antigens on the surfaces of the red blood cells, they form molecular bridges that connect the red blood cells. As a result, **agglutination** (ă-gloo-ti-nā′shŭn), or clumping, of the cells occurs. The combination of the antibodies with the antigens can also initiate reactions that cause hemolysis. Because the antigen–antibody combinations can cause agglutination, the antigens are often called **agglutinogens** (ă-gloo-tin′ō-jenz), and the antibodies are called **agglutinins** (ă-gloo′ti-ninz).

The antigens on the surface of red blood cells have been categorized into **blood groups,** and more than 35 blood groups, most of which are rare, have been identified. For transfusions, the ABO and Rh blood groups are among the most important. Other well-known groups include the Lewis, Duffy, MNSs, Kidd, Kell, and Lutheran groups.

28 ***What are blood groups, and how do they cause transfusion reactions? Define agglutination.***

ABO Blood Group

In the **ABO blood group,** type A blood has type A antigens, type B blood has type B antigens, type AB blood has both A and B antigens, and type O blood has neither A nor B antigens on the surface of red blood cells (figure 19.13). The genes for the ABO blood group are on chromosome 9. The ABO blood group is an example of codominance in that the A and B antigens can be expressed at the same time (see chapter 3).

Plasma from type A blood contains anti-B antibodies, which act against type B antigens, whereas plasma from type B blood contains anti-A antibodies, which act against type A antigens. Type AB blood has neither type of antibody, and type O blood has both anti-A and anti-B antibodies.

The ABO blood types are not found in equal numbers. In Caucasians in the United States, the distribution is type O, 47%; type A, 41%; type B, 9%; and type AB, 3%. Among African-Americans, the distribution is type O, 46%; type A, 27%; type B, 20%; and type AB, 7%.

Antibodies normally do not develop against an antigen unless the body is exposed to that antigen. One possible explanation

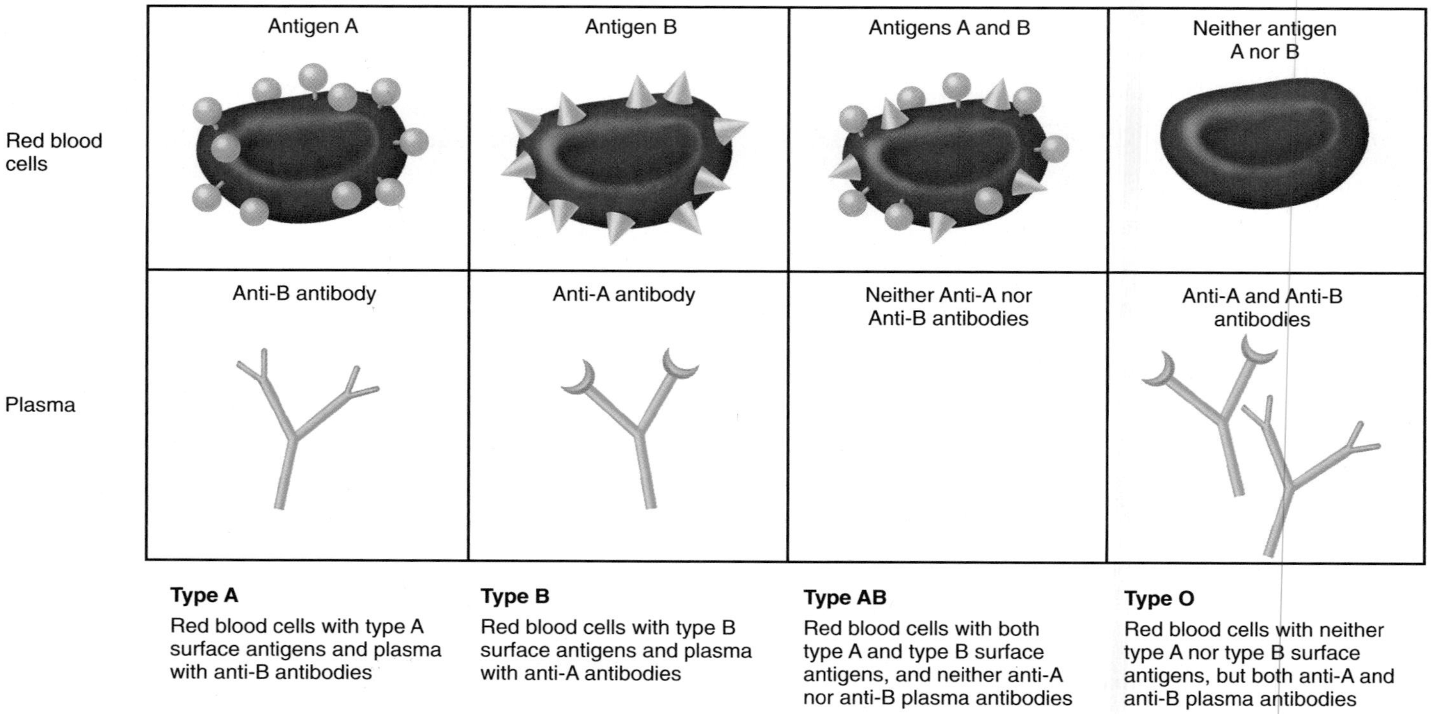

	Type A	Type B	Type AB	Type O
Red blood cells	Antigen A	Antigen B	Antigens A and B	Neither antigen A nor B
Plasma	Anti-B antibody	Anti-A antibody	Neither Anti-A nor Anti-B antibodies	Anti-A and Anti-B antibodies

Type A
Red blood cells with type A surface antigens and plasma with anti-B antibodies

Type B
Red blood cells with type B surface antigens and plasma with anti-A antibodies

Type AB
Red blood cells with both type A and type B surface antigens, and neither anti-A nor anti-B plasma antibodies

Type O
Red blood cells with neither type A nor type B surface antigens, but both anti-A and anti-B plasma antibodies

FIGURE 19.13 ABO Blood Groups
For simplicity, only parts of the anti-A and anti-B antibodies are illustrated. Each antibody has 5 identical Y-shaped arms (see IgM in chapter 22).

for the production of anti-A and/or anti-B antibodies is that type A or B antigens on bacteria or food in the digestive tract stimulate the formation of antibodies against antigens that are different from one's own antigens. In support of this explanation is the observation that anti-A and anti-B antibodies are not found in the blood until about 2 months after birth. For example, an infant with type A blood produces anti-B antibodies against the B antigens on bacteria or food. An infant with A antigens does not produce antibodies against the A antigen on bacteria or food because mechanisms exist in the body to prevent the production of antibodies that react with the body's own antigens (see chapter 22).

A blood **donor** gives blood, and a **recipient** receives blood. Usually, a recipient can receive blood from a donor if they have the same blood type. For example, a person with type A blood can receive blood from a person with type A blood. No ABO transfusion reaction would occur because the recipient has no anti-A antibodies against the type A antigen. On the other hand, if type A blood were donated to a person with type B blood, a transfusion reaction would occur because the person with type B blood has anti-A antibodies against the type A antigen, and agglutination would result (figure 19.14).

Historically, people with type O blood have been called universal donors because they usually can give blood to the other ABO blood types without causing an ABO transfusion reaction. Their red blood cells have no ABO surface antigens and, therefore, do not react with a recipient's anti-A or anti-B antibodies. For example, if type O blood is given to a person with type A blood, the type O red blood cells do not react with the anti-B antibodies in the recipient's blood. In a similar fashion, if type O blood is given to a person with type B blood, no reaction occurs to the recipient's anti-A antibodies.

The term *universal donor* is misleading, however. The transfusion of type O blood, in some cases, produces a transfusion reaction for two reasons. First, other blood groups can cause a transfusion reaction. Second, antibodies in the donor's blood can react with antigens in the blood of the recipient. For example, type O blood has anti-A and anti-B antibodies. If type O blood is transfused into a person with type A blood, the anti-A antibodies (in the type O blood) react against the A antigens (in the type A blood). Usually, such reactions are not serious because the antibodies in the donor's blood are diluted in the larger volume of the recipient's blood, and few reactions take place. Blood banks separate donated blood into several products, such as packed red blood cells; plasma; platelets; and cryoprecipitate, which contains von Willebrand factor, clotting factors, and fibrinogen. This process allows the donated blood to be used by multiple recipients, each of whom may need only one of the blood components. Type O packed red blood cells are unlikely to cause an ABO transfusion reaction when given to a person with a different blood type because it has very little plasma with anti-A and anti-B antibodies.

29 ***What kinds of antigens and antibodies are found in each of the four ABO blood types?***

30 ***Why is a person with type O blood considered to be a universal donor?***

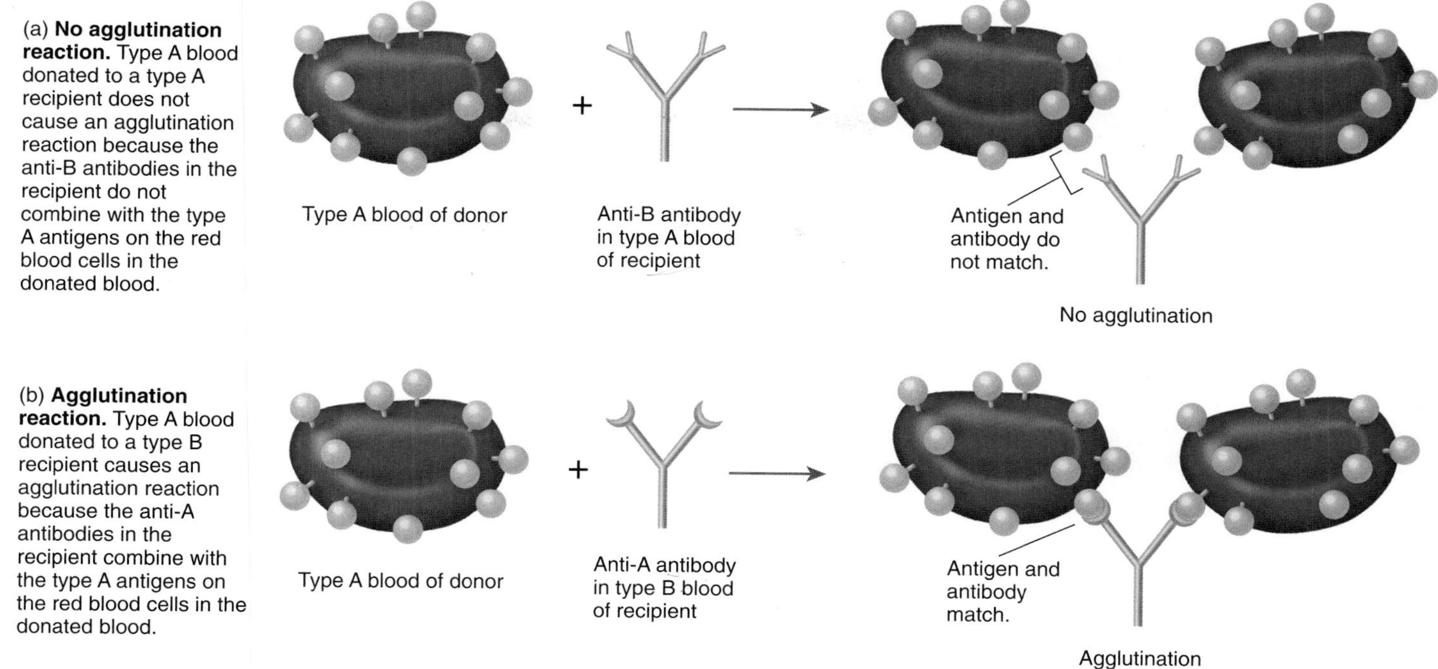

FIGURE 19.14 Agglutination Reaction

For simplicity, only parts of the anti-A and anti-B antibodies are illustrated. Each antibody has 5 identical Y-shaped arms (see IgM in chapter 22).

PREDICT 6

Historically, people with type AB blood were called universal recipients. What is the rationale for this term? Explain why the term is misleading.

Rh Blood Group

The **Rh blood group** is so-named because it was first studied in rhesus monkeys. People are Rh-positive if they have a certain Rh antigen (the D antigen) on the surface of their red blood cells, and people are Rh-negative if they do not have this Rh antigen. The gene for the D antigen is on chromosome 1. About 85% of Caucasians in the United States and 88% of African-Americans are Rh-positive. The ABO blood type and the Rh blood type usually are designated together. For example, a person designated as A positive is type A in the ABO blood group and Rh-positive. The rarest combination in the United States is AB negative, which occurs in less than 1% of all Americans.

Antibodies against the Rh antigen do not develop unless an Rh-negative person is exposed to Rh-positive blood. This can occur through a transfusion or if blood crosses the placenta to a mother from her fetus. When an Rh-negative person is exposed to Rh-positive blood, the person can become sensitized to the Rh antigens and produce anti-Rh antibodies.

Rh incompatibility can pose a major problem in some pregnancies when the mother is Rh-negative and the fetus is Rh-positive. If fetal blood leaks through the placenta and mixes with the mother's blood, the mother becomes sensitized to the Rh antigen. The mother produces anti-Rh antibodies that cross the placenta and cause agglutination and hemolysis of fetal red blood cells. This disorder is called **hemolytic** (hē-mō-lit'ik) **disease of the newborn (HDN),** or **erythroblastosis fetalis** (ĕ-rith'rō-blas-tō'sis fē-ta'lis), and it may be fatal to the fetus (figure 19.15). In the mother's first pregnancy, there is often no problem. The leakage of fetal blood is usually the result of a tear in the placenta that takes place either late in the pregnancy or during delivery. Thus, there is not enough time for the mother to produce enough anti-Rh antibodies to harm the fetus. If sensitization occurs, however, it can cause problems in a subsequent pregnancy in two ways. First, once a woman is sensitized and produces anti-Rh antibodies, she may continue to produce the antibodies through her life. Thus, in a subsequent pregnancy, anti-Rh antibodies may already be present. Second, and especially dangerous in a subsequent pregnancy with an Rh-positive fetus, if any leakage of fetal blood into the mother's blood occurs, she rapidly produces large amounts of anti-Rh antibodies, and HDN develops. Therefore, the levels of anti-Rh antibodies in the mother should be tested. If they are too high, the fetus should be tested to determine the severity of the HDN. In severe cases, a transfusion to replace lost red blood cells can be performed through the umbilical cord, or the baby can be delivered if mature enough.

After birth, treatment of HDN consists of slowly removing the newborn's blood and replacing it. The newborn can also be exposed to fluorescent light, because the light helps break down the large amounts of bilirubin formed as a result of red blood cell destruction. High levels of bilirubin are toxic to the nervous system and can damage brain tissue.

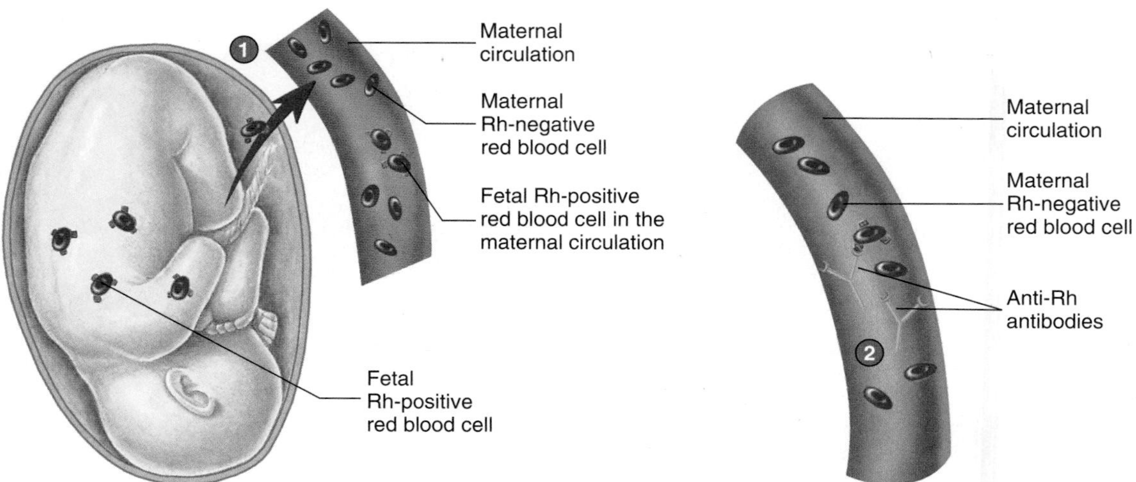

1. Before or during delivery, Rh-positive red blood cells from the fetus enter the blood of an Rh-negative woman through a tear in the placenta.

2. The mother is sensitized to the Rh antigen and produces anti-Rh antibodies. Because this usually happens after delivery, there is no effect on the fetus in the first pregnancy.

3. During a subsequent pregnancy with an Rh-positive fetus, Rh-positive red blood cells cross the placenta, enter the maternal circulation, and stimulate the mother to produce antibodies against the Rh antigen. Antibody production is rapid because the mother has been sensitized to the Rh antigen.

4. The anti-Rh antibodies from the mother cross the placenta, causing agglutination and hemolysis of fetal red blood cells, and hemolytic disease of the newborn (HDN) develops.

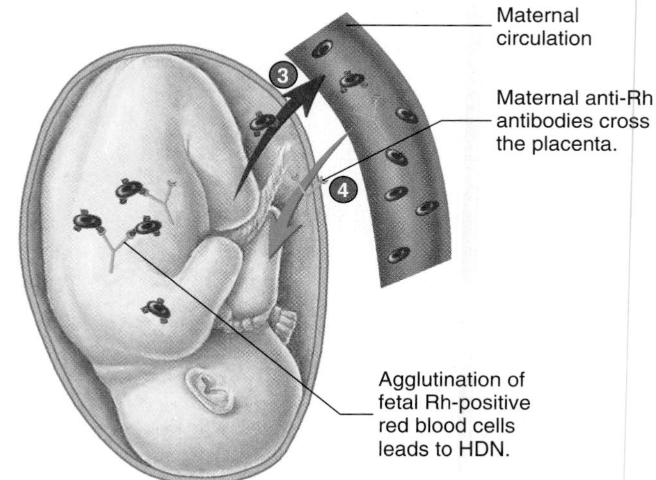

FIGURE 19.15 Hemolytic Disease of the Newborn (HDN)

Prevention of HDN is often possible if the Rh-negative mother is given an injection of a specific type of antibody preparation, called Rh_0(D) immune globulin (RhoGAM). The injection can be given during the pregnancy, before delivery, or immediately after each delivery, miscarriage, or abortion. The injection contains antibodies against Rh antigens. The injected antibodies bind to the Rh antigens of any fetal red blood cells that may have entered the mother's blood. This treatment inactivates the fetal Rh antigens and prevents sensitization of the mother. However, if sensitization of the mother has already occurred, the treatment is ineffective.

31 ***Define Rh-positive.***

32 ***What Rh blood types must the mother and fetus have before HDN can occur?***

33 ***How is HDN harmful to the fetus?***

34 ***Why doesn't HDN usually develop in the first pregnancy?***

35 ***How can HDN be prevented? How is HDN treated?***

CASE STUDY

Treatment of Hemolytic Disease of the Newborn

Billy was born with HDN. He was treated with exchange transfusion, phototherapy, and erythropoietin. An exchange transfusion replaced Billy's blood with donor blood. In this procedure, as the donor's blood was transfused into Billy, his blood was withdrawn. Phototherapy, in which blood passing through the skin is exposed to blue or white lights, results in the break down of bilirubin to less toxic compounds that are removed by the newborn's liver. During fetal development, the increased rate of red blood cell destruction caused by the mother's anti-Rh antibodies results in lower than normal numbers of red blood cells, a condition called **anemia** (ă-nē′mē-ă). It also results in increased levels of bilirubin. Although high levels of bilirubin can damage the brain by killing nerve cells, this is not usually a problem in the fetus because the bilirubin is removed by the placenta. Following birth, bilirubin levels can increase because of the continued lysis of red blood cells and the inability of the newborn's liver to handle the large bilirubin load.

PREDICT 7

Answer the following questions about Billy's treatment for HDN.

a. What was the purpose of giving Billy a transfusion?
b. What was the benefit of an exchange transfusion?
c. Explain the reason for giving Billy erythropoietin.
d. Just before birth, would Billy's erythropoietin levels have been higher or lower than those of a fetus without HDN?
e. After birth, but before treatment, would Billy's erythropoietin levels have increased or decreased?
f. When treating HDN with an exchange transfusion, should the donor's blood be Rh-positive or Rh-negative? Explain.
g. Does giving an Rh-positive newborn a transfusion of Rh-negative blood change the newborn's blood type? Explain.

DIAGNOSTIC BLOOD TESTS

Type and Crossmatch

To prevent transfusion reactions, blood is typed and a **crossmatch** is made. **Blood typing** determines the ABO and Rh blood groups of the blood sample. Typically, the cells are separated from the serum. The cells are tested with known antibodies to determine the type of antigen on the cell surface. For example, if a patient's blood cells agglutinate when mixed with anti-A antibodies but do not agglutinate when mixed with anti-B antibodies, it is concluded that the cells have type A antigen. In a similar fashion, the serum is mixed with known cell types (antigens) to determine the type of antibodies in the serum. Normally, donor blood must match the ABO and Rh type of the recipient.

The International Society of Blood Transfusion recognizes 29 blood groups, including the ABO and Rh groups. Because any of these blood groups can cause a transfusion reaction, a crossmatch is performed. In a crossmatch, the donor's blood cells are mixed with the recipient's serum, and the donor's serum is mixed with the recipient's cells. The donor's blood is considered safe for transfusion only if no agglutination occurs in either match.

Complete Blood Count

A **complete blood count (CBC),** an analysis of blood, provides much information. It consists of a red blood count, hemoglobin and hematocrit measurements, a white blood count, and a differential white blood count.

Red Blood Count

Blood cell counts usually are done automatically with an electronic instrument, but they can also be done manually with a microscope. The normal range for a **red blood count (RBC)** is the number (expressed in millions) of red blood cells per microliter of blood. It is 4.6–6.2 million/μL of blood for a male and 4.2–5.4 million/μL of blood for a female. **Erythrocytosis** (ĕ-rith′rō-sī-tō′sis) is an overabundance of red blood cells (see Clinical Focus "Disorders of the Blood," p. 673).

Hemoglobin Measurement

A **hemoglobin measurement** determines the amount of hemoglobin in a given volume of blood, usually expressed as grams of hemoglobin per 100 mL of blood. The normal hemoglobin count for a male is 14–18 g/100 mL of blood, and for a female it is 12–16 g/100 mL of blood. Abnormally low hemoglobin is an indication of anemia (see Clinical Focus "Disorders of the Blood," p. 673).

Hematocrit Measurement

The percentage of total blood volume composed of red blood cells is the **hematocrit** (hē′mă-tō-krit, hem′ă-tō-krit). One way to determine hematocrit is to place blood in a tube and spin it in a centrifuge. The formed elements are heavier than the plasma and are forced to one end of the tube (figure 19.16). White blood cells and platelets form a thin, whitish layer, called the **buffy coat,** between the plasma and the red blood cells. The red blood cells account for 40%–54% of the total blood volume in males and 38%–47% in females.

The number and size of red blood cells affect the hematocrit measurement. **Normocytes** (nōr′mō-sītz) are normal-sized red blood cells with a diameter of 7.5 mm. **Microcytes** (mī′krō-sītz) are smaller than normal, with a diameter of 6 μm or less, and **macrocytes** (mak′krō-sītz) are larger than normal, with a diameter 9 μm or greater. Blood disorders can result in an abnormal hematocrit measurement because they cause red blood cells numbers to be abnormally high or low or cause red blood cells to be abnormally small or large (see Clinical Focus "Disorders of the Blood," p. 673). A decreased hematocrit indicates that the volume of red blood cells is less than normal. It can result from a decreased number of normocytes or a normal number of microcytes. For example, inadequate iron in the diet can impair hemoglobin production. Consequently, during their formation red blood cells do not fill with hemoglobin, and they remain smaller than normal.

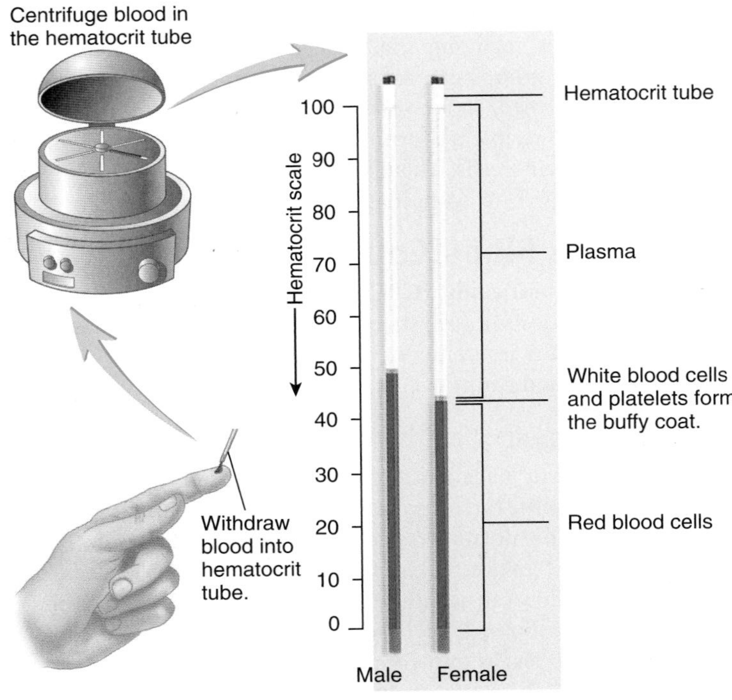

FIGURE 19.16 Hematocrit

Blood is withdrawn into a capillary tube and placed in a centrifuge. The blood is separated into plasma, red blood cells, and a small amount of white blood cells and platelets, which rest on the red blood cells. The hematocrit measurement is the percentage of the total blood volume that is red blood cells. It does not measure the white blood cells and platelets. Normal hematocrits for a male and a female are shown.

White Blood Count

A **white blood count (WBC)** measures the total number of white blood cells in the blood. Normally, 5000–10,000 white blood cells are present in each microliter of blood. **Leukopenia** (loo-kō-pē′nē-ă) is a lower than normal WBC resulting from depression or destruction of the red marrow. Viral infections, radiation, drugs, tumors, or vitamin deficiencies (B_{12} or folate) can cause leukopenia. **Leukocytosis** (loo′kō-sī-tō′sis) is an abnormally high WBC. **Leukemia** (loo-kē′mē-ă; a cancer of the red marrow) often results in leukocytosis, but the white blood cells have an abnormal structure and function. Bacterial infections also can cause leukocytosis by stimulating neutrophils to increase in number.

Differential White Blood Count

A **differential white blood count** determines the percentage of each of the five kinds of white blood cells in the WBC. Normally, neutrophils account for 60%–70%; lymphocytes, 20%–30%; monocytes, 2%–8%; eosinophils, 1%–4%; and basophils, 0.5%–1%. A differential WBC can provide insight into a patient's condition. For example, in patients with bacterial infections the neutrophil count is often greatly increased, whereas in patients with allergic reactions the eosinophil and basophil counts are elevated.

Clotting

Measurements that test the blood's ability to clot are platelet count and prothrombin time measurement.

Platelet Count

A normal **platelet count** is 250,000–400,000 platelets per microliter of blood. **Thrombocytopenia** (throm′bō-sī-tō-pē′nē-ă) is a condition in which the platelet count is greatly reduced, resulting in chronic bleeding through small vessels and capillaries. It can be caused by decreased platelet production as a result of hereditary disorders, lack of vitamin B_{12}, drug therapy, or radiation therapy.

Prothrombin Time Measurement

Prothrombin time measurement is a measure of how long it takes for the blood to start clotting, which normally is 9–12 seconds. Prothrombin time is determined by adding thromboplastin to whole plasma. Thromboplastin is a chemical released from injured tissues that starts the process of clotting (see figure 19.11). Prothrombin time is officially reported as the International Normalized Ratio (INR), which standardizes the time it takes to clot based on the slightly different thromboplastins used by different labs. Because many clotting factors must be activated to form fibrin, a deficiency of any one of them can cause an abnormal prothrombin time. Vitamin K deficiency, certain liver diseases, and drug therapy can cause an increased prothrombin time.

Blood Chemistry

The composition of materials dissolved or suspended in the plasma can be used to assess the functioning of many of the body's systems (Appendix E). For example, high blood glucose levels can indicate that the pancreas is not producing enough insulin; high blood urea nitrogen (BUN) can be a sign of reduced kidney function; increased bilirubin can indicate liver dysfunction or hemolysis; and high cholesterol levels can indicate an increased risk of developing cardiovascular disease. A number of blood chemistry tests are routinely done when a blood sample is taken, and additional tests are available.

36 ***For each of the following tests, define the test and give an example of a disorder that would cause an abnormal test result:***

a. red blood count
b. hemoglobin measurement
c. hematocrit measurement
d. white blood count
e. differential white blood count
f. platelet count
g. prothrombin time measurement
h. blood chemistry tests

PREDICT 8

When a patient complains of acute pain in the abdomen, the physician suspects appendicitis, which is often caused by a bacterial infection of the appendix. What blood test should be done to support the diagnosis?

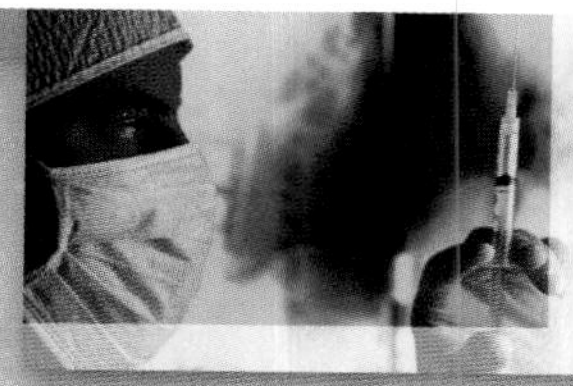

CLINICAL FOCUS

Disorders of the Blood

ERYTHROCYTOSIS

Erythrocytosis (ĕ-rith′rō-sī-tō′sis) is an overabundance of red blood cells, resulting in increased blood viscosity, reduced flow rates, and, if severe, plugging of the capillaries. **Relative erythrocytosis** results from decreased plasma volume, such as that caused by dehydration, diuretics, and burns. **Primary erythrocytosis,** often called **polycythemia vera** (pol′ē-sī-thē′mē-ă ve′ra), is a stem cell defect of unknown cause that results in the overproduction of red blood cells, granulocytes, and platelets. Erythropoietin levels are low and the spleen can be enlarged. **Secondary erythrocytosis (polycythemia)** results from a decreased oxygen supply, such as that which occurs at high altitudes, in chronic obstructive pulmonary disease, and in congestive heart failure. The resulting decrease in oxygen delivery to the kidneys stimulates erythropoietin secretion and causes an increase in red blood cell production. In primary and secondary erythrocytosis, the increased number of red blood cells increases blood viscosity and blood volume. There can be clogging of capillaries and the development of hypertension.

ANEMIA

Anemia (ă-nē′mē-ă) is a deficiency of hemoglobin in the blood. It can result from a decrease in the number of red blood cells, a decrease in the amount of hemoglobin in each red blood cell, or both. The decreased hemoglobin reduces the blood's ability to transport oxygen. Anemic patients suffer from a lack of energy and feel excessively tired and listless. They can appear pale and quickly become short of breath with only slight exertion.

One general cause of anemia is nutritional deficiency. **Iron-deficiency anemia** results from a deficient intake or absorption of iron or from excessive iron loss. Consequently, not enough hemoglobin is produced, and the red blood cells are smaller than normal (microcytic). **Folate deficiency** can also cause anemia. An inadequate amount of folate in the diet is the usual cause of folate deficiency, with the disorder developing most often in the poor, in pregnant women, and in chronic alcoholics. Because folate helps in the synthesis of DNA, folate deficiency results in fewer cell divisions. There is decreased red blood cell production, but the cells grow larger than normal (macrocytic). Another type of nutritional anemia is **pernicious** (per-nish′ŭs) **anemia,** which is caused by inadequate amounts of vitamin B_{12}. A 2- to 3-year supply of vitamin B_{12} can be stored in the liver. Because vitamin B_{12} is important for folate synthesis, inadequate amounts of vitamin B_{12} causes a secondary folate deficiency. Thus, a deficiency of vitamin B_{12} also causes the decreased production of red blood cells that are larger than normal. Although inadequate levels of vitamin B_{12} in the diet can cause pernicious anemia, the usual cause is insufficient absorption of the vitamin. Normally, the stomach produces intrinsic factor, a protein that binds to vitamin B_{12}. The combined molecules pass into the small intestine, where intrinsic factor facilitates the absorption of the vitamin. Without adequate levels of intrinsic factor, insufficient vitamin B_{12} is absorbed, and pernicious anemia develops. Present evidence suggests that the inability to produce intrinsic factor is due to an autoimmune disease in which the body's immune system damages the cells in the stomach that produce intrinsic factor.

Another general cause of anemia is the loss or destruction of red blood cells. **Hemorrhagic** (hem-ŏ-raj′ik) **anemia** results from a loss of blood, such as can result from trauma, ulcers, or excessive menstrual bleeding. Chronic blood loss, in which small amounts of blood are lost over time, can result in iron-deficiency anemia. **Hemolytic** (hē-mō-lit′ik) **anemia** is a disorder in which red blood cells rupture or are destroyed at an excessive rate. It can be caused by inherited defects. For example, hereditary spherocytosis (sfēr′ō-sī-tō′sis) results from a mutation of a gene on chromosome 14 that codes for a cytoskeleton protein called spectrin (spek′-trin). Spectrin normally stabilizes the plasma membrane, but when defective, red blood cells assume a spherical shape and rupture easily. Many kinds of hemolytic anemia result from unusual damage to the red blood cells by drugs, snake venom, artificial heart valves, autoimmune disease, or hemolytic disease of the newborn.

Aplastic anemia is caused by an inability of the red bone marrow to produce red blood cells and often white blood cells and platelets. It is usually acquired as a result of damage to stem cells in the red marrow by chemicals (e.g., benzene), drugs (e.g., certain antibiotics and sedatives), or radiation. Aplastic anemia is usually associated with an inability to produce granulocytes and platelets.

Some anemias result from inadequate or defective hemoglobin production. **Thalassemia** (thal-ă-sē′mē-ă) is an autosomal-recessive hereditary disease found predominantly in people of Mediterranean, Asian, and African ancestry. It is caused by insufficient production of the alpha or beta globin in hemoglobin (see figure 19.4). The gene for alpha globin is on chromosome 16 and the gene for beta globin is on chromosome 11. Depending on the mutation of these genes, the effects of thalassemia can vary from minor changes to hemoglobin to severe or fatal anemia.

VON WILLEBRAND DISEASE

Von Willebrand disease is an autosomal-dominant hereditary disorder caused by a mutation of the von Willebrand factor gene on chromosome 12. It is the most common inherited bleeding disorder, occurring as frequently as in 1 in 1000 individuals. Von Willebrand factor (vWF) helps platelets stick to collagen (platelet adhesion) and is the plasma carrier for factor VIII (see "Coagulation," p. 664 and table 19.3). One treatment for von Willebrand disease involves injections of vWF or concentrates of factor VIII to which vWF is attached. Another therapeutic approach is to administer a drug that increases vWF levels in the blood.

HEMOPHILIA

Hemophilia (hē-mō-fil′ē-ă) is a genetic disorder in which clotting is abnormal or absent. It is most often found in people from northern Europe and their descendants. Because hemophilia is most often an X-linked trait (see chapter 3), it occurs almost exclusively

Continued

Continued

in males. **Hemophilia A** (classic hemophilia) results from a deficiency of plasma coagulation factor VIII, and **hemophilia B** is caused by a deficiency of plasma factor IX. Hemophilia A occurs in approximately 1 in 10,000 male births, and hemophilia B occurs in approximately 1 in 100,000 male births. Treatment of hemophilia involves injection of the missing clotting factor.

DISSEMINATED INTRAVASCULAR COAGULATION

Disseminated intravascular coagulation (DIC) is a complex disorder involving clotting throughout the vascular system followed by bleeding. Normally, excessive clotting is prevented by anticoagulants. DIC can develop when these control mechanisms are overwhelmed. Many conditions can cause DIC by overstimulating blood clotting. Examples include massive tissue damage, such as burns, and the alteration of the lining of blood vessels caused by infections or snake bites. If DIC occurs slowly, the predominant effect is thrombosis and the blockage of blood vessels. If DIC occurs rapidly, massive clot formation occurs, quickly using up available blood clotting factors and platelets. The result is continual bleeding around wounds, intravenous lines, and catheters, as well as internal bleeding. The best therapy for DIC is to treat the condition that is causing it.

THROMBOCYTOPENIA

Thrombocytopenia (throm′bō-sī-tō-pē′nē-ă) is a condition in which the number of platelets is greatly reduced, resulting in chronic bleeding through small vessels and capillaries. Thrombocytopenia has several causes, including increased platelet destruction, caused by autoimmune disease (see chapter 22) or infections, and decreased platelet production, resulting from pernicious anemia, drug therapy, radiation therapy, or leukemias. Hereditary thrombocytopenia can be X-linked or autosomal-dominant (chromosome 10).

LEUKEMIA

The **leukemias** are cancers of the red bone marrow in which abnormal production of one or more of the white blood cell types occur. Because these cells are usually immature or abnormal and lack their normal immunologic functions, patients are very susceptible to infections. The excess production of white blood cells in the red marrow can also interfere with red blood cell and platelet formation and thus lead to anemia and bleeding.

INFECTIOUS DISEASES OF THE BLOOD

Microorganisms do not normally survive in the blood. Blood can transport microorganisms, however, and they can multiply in the blood. Microorganisms can enter the body and be transported by the blood to the tissues they infect. For example, the poliomyelitis virus enters through the gastrointestinal tract and is carried to nervous tissue. After microorganisms are established at a site of infection, some can enter the blood. They can then be transported to other locations in the body, multiply within the blood, or be eliminated by the body's immune system.

Septicemia (sep-ti-sē′mē-ă), or blood poisoning, is the spread of microorganisms and their toxins by the blood. Often, septicemia results from the introduction of microorganisms by a medical procedure, such as the insertion of an intravenous tube into a blood vessel. The release of toxins by microorganisms can cause septic shock, which is a decrease in blood pressure that can result in death.

In a few diseases, microorganisms actually multiply within blood cells. **Malaria** (mă-lār′ē-ă) is caused by a protozoan (*Plasmodium*) that is introduced into the blood by the bite of the *Anopheles* mosquito. Part of the development of the protozoan occurs inside red blood cells. The symptoms of chills and fever in malaria are produced by toxins released when the protozoan causes the red blood cells to rupture. **Infectious mononucleosis** (mon′ō-noo-klē-ō′sis) is caused by a virus (Epstein-Barr virus) that infects lymphocytes (B cells). The virus alters the lymphocytes, and the immune system attacks and destroys the lymphocytes. The immune system response is believed to produce the symptoms of fever, sore throat, and swollen lymph nodes. **Acquired immunodeficiency syndrome (AIDS)** is caused by the human immunodeficiency virus (HIV), which infects lymphocytes and suppresses the immune system (see chapter 22).

The presence of microorganisms in the blood is a concern with blood transfusions, because it is possible to infect the blood recipient. Blood is routinely tested, especially for AIDS and hepatitis, in an effort to eliminate this risk. One cause of **hepatitis** (hep-ă-tī′tis) is an infection of the liver caused by viruses (see chapter 24). After recovering, hepatitis victims can become virus carriers. Although they show no signs of the disease, they release the virus into their blood or bile. To prevent the infection of others, anyone who has had hepatitis is asked not to donate blood products.

SUMMARY

Functions of Blood (p. 651)

1. Blood transports gases, nutrients, waste products, and hormones.
2. Blood is involved in the regulation of homeostasis and the maintenance of pH, body temperature, fluid balance, and electrolyte levels.
3. Blood protects against disease and blood loss.

Plasma (p. 652)

1. Plasma is mostly water (91%) and contains proteins, such as albumin (maintains osmotic pressure), globulins (function in transport and immunity), fibrinogen (involved in clot formation), and hormones and enzymes (involved in regulation).
2. Plasma contains ions, nutrients, waste products, and gases.

Formed Elements (p. 653)

The formed elements include red blood cells (erythrocytes), white blood cells (leukocytes), and platelets (cell fragments).

Production of Formed Elements

1. In the embryo and fetus, the formed elements are produced in a number of locations.
2. After birth, red bone marrow becomes the source of the formed elements.
3. All formed elements are derived from stem cells.

Red Blood Cells

1. Red blood cells are biconcave discs containing hemoglobin and carbonic anhydrase.
   - A hemoglobin molecule consists of four heme and four globin molecules. The heme molecules transport oxygen, and the globin molecules transport carbon dioxide and nitric oxide. Iron is required for oxygen transport.
   - Carbonic anhydrase is involved with the transport of carbon dioxide.
2. Erythropoiesis is the production of red blood cells.
   - Stem cells in red bone marrow eventually give rise to late erythroblasts, which lose their nuclei and are released into the blood as reticulocytes. Loss of the endoplasmic reticulum by a reticulocyte produces a red blood cell.
   - In response to low blood oxygen, the kidneys produce erythropoietin, which stimulates erythropoiesis.
3. Hemoglobin from ruptured red blood cells is phagocytized by macrophages. The hemoglobin is broken down, and heme becomes bilirubin, which is secreted in bile.

White Blood Cells

1. White blood cells protect the body against microorganisms and remove dead cells and debris.
2. Five types of white blood cells exist.
   - Neutrophils are small phagocytic cells.
   - Eosinophils reduce inflammation.
   - Basophils release histamine and are involved with increasing the inflammatory response.
   - Lymphocytes are important in immunity, including the production of antibodies.
   - Monocytes leave the blood, enter tissues, and become large phagocytic cells called macrophages.

Platelets

Platelets, or thrombocytes, are cell fragments pinched off from megakaryocytes in the red bone marrow.

Hemostasis (p. 662)

Hemostasis, the stoppage of bleeding, is very important to the maintenance of homeostasis.

Vascular Spasm

Vasoconstriction of damaged blood vessels reduces blood loss.

Platelet Plug Formation

1. Platelets repair minor damage to blood vessels by forming platelet plugs.
   - In platelet adhesion, platelets bind to collagen in damaged tissues.
   - In the platelet release reaction, platelets release chemicals that activate additional platelets.
   - In platelet aggregation, platelets bind to one another to form a platelet plug.
2. Platelets also release chemicals involved with coagulation.

Coagulation

1. Coagulation is the formation of a blood clot.
2. The first stage of coagulation occurs through the extrinsic or intrinsic pathway. Both pathways end with the production of activated factor X.
   - The extrinsic pathway begins with the release of thromboplastin from damaged tissues.
   - The intrinsic pathway begins with the activation of factor XII.
3. Activated factor X, factor V, phospholipids, and Ca^{2+} form prothrombinase.
4. Prothrombin is converted to thrombin by prothrombinase.
5. Fibrinogen is converted to fibrin by thrombin. The insoluble fibrin forms the clot.

Control of Clot Formation

1. Heparin and antithrombin inhibit thrombin activity. Fibrinogen is, therefore, not converted to fibrin, and clot formation is inhibited.
2. Prostacyclin counteracts the effects of thrombin.

Clot Retraction and Dissolution

1. Clot retraction results from the contraction of platelets, which pull the edges of damaged tissue closer together.
2. Serum, which is plasma minus fibrinogen and some clotting factors, is squeezed out of the clot.
3. Factor XII, thrombin, tissue plasminogen activator, and urokinase activate plasmin, which dissolves fibrin (the clot).

Blood Grouping (p. 667)

1. Blood groups are determined by antigens on the surface of red blood cells.
2. Antibodies can bind to red blood cell antigens, resulting in agglutination or hemolysis of red blood cells.

ABO Blood Group

1. Type A blood has A antigens, type B blood has B antigens, type AB blood has A and B antigens, and type O blood has neither A nor B antigens.
2. Type A blood has anti-B antibodies, type B blood has anti-A antibodies, type AB blood has neither anti-A nor anti-B antibodies, and type O blood has both anti-A and anti-B antibodies.
3. Mismatching the ABO blood group results in transfusion reactions.

Rh Blood Group

1. Rh-positive blood has certain Rh antigens (the D antigen), whereas Rh-negative blood does not.
2. Antibodies against the Rh antigen are produced by an Rh-negative person when the person is exposed to Rh-positive blood.
3. The Rh blood group is responsible for hemolytic disease of the newborn.

Diagnostic Blood Tests (p. 671)

Type and Crossmatch

Blood typing determines the ABO and Rh blood groups of a blood sample. A crossmatch tests for agglutination reactions between donor and recipient blood.

Complete Blood Count

A complete blood count consists of the following: red blood count, hemoglobin measurement (grams of hemoglobin per 100 mL of blood), hematocrit measurement (percent volume of red blood cells), and white blood count.

Differential White Blood Count

A differential white blood count determines the percentage of each type of white blood cell.

Clotting

Platelet count and prothrombin time measurement measure the blood's ability to clot.

Blood Chemistry

The composition of materials dissolved or suspended in plasma (e.g., glucose, urea nitrogen, bilirubin, and cholesterol) can be used to assess the functioning and status of the body's systems.

REVIEW AND COMPREHENSION

1. Which of these is a function of blood?
   a. clot formation
   b. protection against foreign substances
   c. maintenance of body temperature
   d. regulation of pH and osmosis
   e. all of the above
2. Which of these is *not* a component of plasma?
   a. nitrogen c. platelets e. urea
   b. sodium ions d. water
3. Which of these plasma proteins plays an important role in maintaining the osmotic concentration of the blood?
   a. albumin c. platelets e. globulins
   b. fibrinogen d. hemoglobin
4. Red blood cells
   a. are the least numerous formed element in the blood.
   b. are phagocytic cells.
   c. are produced in the yellow marrow.
   d. do not have a nucleus.
   e. all of the above.
5. Given these ways of transporting carbon dioxide in the blood:
   1. bicarbonate ions
   2. combined with blood proteins
   3. dissolved in plasma

   Choose the arrangement that lists them in the correct order from largest to smallest percentage of carbon dioxide transported.
   a. 1, 2, 3 c. 2, 3, 1 e. 3, 1, 2
   b. 1, 3, 2 d. 2, 1, 3
6. Which of these components of a red blood cell is correctly matched with its function?
   a. heme group of hemoglobin—oxygen transport
   b. globin portion of hemoglobin—carbon dioxide transport
   c. carbonic anhydrase—carbon dioxide transport
   d. cysteine on β-globin—nitric oxide transport
   e. all of the above
7. Each hemoglobin molecule can become associated with ____________ oxygen molecules.
   a. one d. four
   b. two e. an unlimited number of
   c. three
8. Which of these substances is *not* required for normal red blood cell production?
   a. folate c. iron
   b. vitamin K d. vitamin B_{12}
9. Erythropoietin
   a. is produced mainly by the heart.
   b. inhibits the production of red blood cells.
   c. production increases when blood oxygen decreases.
   d. production is inhibited by testosterone.
   e. all of the above.
10. Which of these changes occurs in the blood in response to the initiation of a vigorous exercise program?
    a. increased erythropoietin production
    b. increased concentration of reticulocytes
    c. decreased bilirubin formation
    d. both a and b
    e. all of the above
11. Which of the components of hemoglobin is correctly matched with its fate following the destruction of a red blood cell?
    a. heme—reused to form a new hemoglobin molecule
    b. globin—broken down into amino acids
    c. iron—mostly secreted in bile
    d. all of the above
12. If you lived near sea level and were training for a track meet in Denver (5280 ft elevation), you would want to spend a few weeks before the meet training at
    a. sea level.
    b. an altitude similar to Denver's.
    c. a facility with a hyperbaric chamber.
    d. any location—it does not matter.
13. The blood cells that inhibit inflammation are
    a. eosinophils. c. neutrophils. e. lymphocytes.
    b. basophils. d. monocytes.
14. The most numerous type of white blood cell, whose primary function is phagocytosis, is
    a. eosinophils. c. neutrophils. e. lymphocytes.
    b. basophils. d. monocytes.
15. Monocytes
    a. are the smallest white blood cells.
    b. increase in number during chronic infections.
    c. give rise to neutrophils.
    d. produce antibodies.
16. The white blood cells that release large amounts of histamine and heparin are
    a. eosinophils. c. neutrophils. e. lymphocytes.
    b. basophils. d. monocytes.

17. The smallest white blood cells, which include B cells and T cells, are
    a. eosinophils.
    b. basophils.
    c. neutrophils.
    d. monocytes.
    e. lymphocytes.
18. Platelets
    a. are derived from megakaryocytes.
    b. are cell fragments.
    c. have surface molecules that attach to collagen.
    d. play an important role in clot formation.
    e. all of the above.
19. Given these processes in platelet plug formation:
    1. platelet adhesion
    2. platelet aggregation
    3. platelet release reaction

    Choose the arrangement that lists the processes in the correct order after a blood vessel is damaged.
    a. 1, 2, 3
    b. 1, 3, 2
    c. 3, 1, 2
    d. 3, 2, 1
    e. 2, 3, 1
20. A constituent of blood plasma that forms the network of fibers in a clot is
    a. fibrinogen.
    b. tissue factor.
    c. platelets.
    d. thrombin.
    e. prothrombinase.
21. Given these chemicals:
    1. activated factor XII
    2. fibrinogen
    3. prothrombinase
    4. thrombin

    Choose the arrangement that lists the chemicals in the order they are used during clot formation.
    a. 1, 3, 4, 2
    b. 2, 3, 4, 1
    c. 3, 2, 1, 4
    d. 3, 1, 2, 4
    e. 3, 4, 2, 1
22. The extrinsic pathway
    a. begins with the release of thromboplastin (tissue factor).
    b. leads to the production of activated factor X.
    c. requires Ca^{2+}.
    d. all of the above.
23. The chemical involved in the breakdown of a clot (fibrinolysis) is
    a. antithrombin.
    b. fibrinogen.
    c. heparin.
    d. plasmin.
    e. sodium citrate.
24. A person with type A blood
    a. has anti-A antibodies.
    b. has type B antigens.
    c. will have a transfusion reaction if given type B blood.
    d. all of the above.
25. In the United States, the most common blood type is
    a. A positive.
    b. B positive.
    c. O positive.
    d. O negative.
    e. AB negative.
26. Rh-negative mothers who receive a RhoGAM injection are given that injection to
    a. initiate the synthesis of anti-Rh antibodies in the mother.
    b. initiate anti-Rh antibody production in the baby.
    c. prevent the mother from producing anti-Rh antibodies.
    d. prevent the baby from producing anti-Rh antibodies.

Answers in Appendix E

CRITICAL THINKING

1. In hereditary hemolytic anemia, massive destruction of red blood cells occurs. Would you expect the reticulocyte count to be above or below normal? Explain why one of the symptoms of the disease is jaundice. In 1910, it was discovered that hereditary hemolytic anemia can be successfully treated by removing the spleen. Explain why this treatment is effective.
2. Red Packer, a physical education major, wanted to improve his performance in an upcoming marathon race. About 6 weeks before the race, 500 mL of blood was removed from his body, and the formed elements were separated from the plasma. The formed elements were frozen, and the plasma was reinfused into his body. Just before the competition, the formed elements were thawed and injected into his body. Explain why this procedure, called blood doping or blood boosting, would help Red's performance. Can you suggest any possible bad effects?
3. Chemicals such as benzene and chloramphenicol can destroy red bone marrow and cause aplastic anemia. What symptoms develop as a result of the lack of (a) red blood cells, (b) platelets, and (c) white blood cells?
4. Some people habitually use barbiturates to depress feelings of anxiety. Barbiturates cause hypoventilation, which is a slower than normal rate of breathing, because they suppress the respiratory centers in the brain. What happens to the red blood count of a habitual user of barbiturates? Explain.
5. What blood problems would you expect to observe in a patient after total gastrectomy (removal of the stomach)? Explain.
6. According to an old saying, "good food makes good blood." Name three substances in the diet that are essential for "good blood." What blood disorders develop if these substances are absent from the diet?
7. Why do anemic patients often have gray feces?
8. Reddie Popper has a cell membrane defect in her red blood cells that makes them more susceptible to rupturing. Her red blood cells are destroyed faster than they can be replaced. Are her RBC, hemoglobin, hematocrit, and bilirubin levels below normal, normal, or above normal? Explain.

Answers in Appendix F

Visit this book's website at www.mhhe.com/seeley8 for chapter quizzes, interactive learning exercises, and other study tools.

20 Cardiovascular System
The Heart

Approximately 370 years ago, it was established that the heart's pumping action is essential to maintain the continuous circulation of blood throughout the body. The current understanding of the detailed function of this amazing pump, its regulation, and modern treatments for heart disease is, in comparison, very recent.

The heart is actually two pumps in one. The right side of the heart receives blood from the body and pumps blood through the **pulmonary** (pŭl′mō-nār-ē) **circulation,** which carries blood to the lungs and returns it to the left side of the heart. In the lungs, carbon dioxide diffuses from the blood into the lungs, and oxygen diffuses from the lungs into the blood. The left side of the heart pumps blood through the **systemic circulation,** which delivers oxygen and nutrients to all the remaining tissues of the body. From those tissues, carbon dioxide and other waste products are carried back to the right side of the heart (figure 20.1).

The heart of a healthy 70 kg person pumps approximately 7200 L (approximately 1900 gallons) of blood each day at a rate of 5 L/min. For most people, the heart continues to pump for more than 75 years. During periods of vigorous exercise, the amount of blood pumped per minute increases severalfold, but the life of the individual is in danger if the heart loses its ability to pump blood for even a few minutes. **Cardiology** (kar-dē-ol′ō-jē) is the medical specialty concerned with the diagnosis and treatment of heart disease.

This chapter describes the *functions of the heart* (p. 679), *size, shape, and location of the heart* (p. 679), the *anatomy of the heart* (p. 681), the *route of blood flow through the heart* (p. 687), and its *histology* (p. 689) and *electrical properties* (p. 692). The *cardiac cycle* (p. 695), the *mean arterial blood pressure* (p. 703), the *regulation of the heart* (p. 705), and the *heart and homeostasis* (p. 707) are described. The chapter ends with the *effects of aging on the heart* (p. 711).

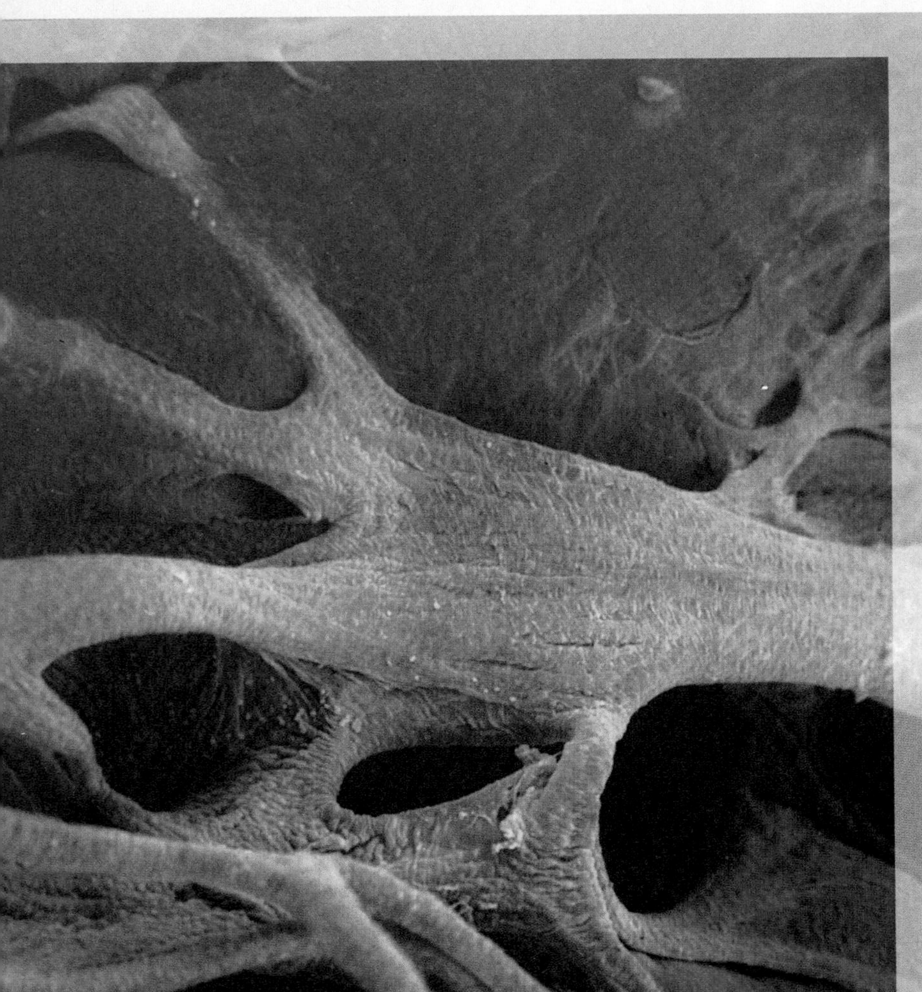

Colorized scanning electron micrograph of Purkinje fibers of the heart.

Cardiovascular System

* Composition of whole blood
* Cardiac cycle (5 periods)
* renin-angiotensin-aldosterone mechanism
* essay test questions

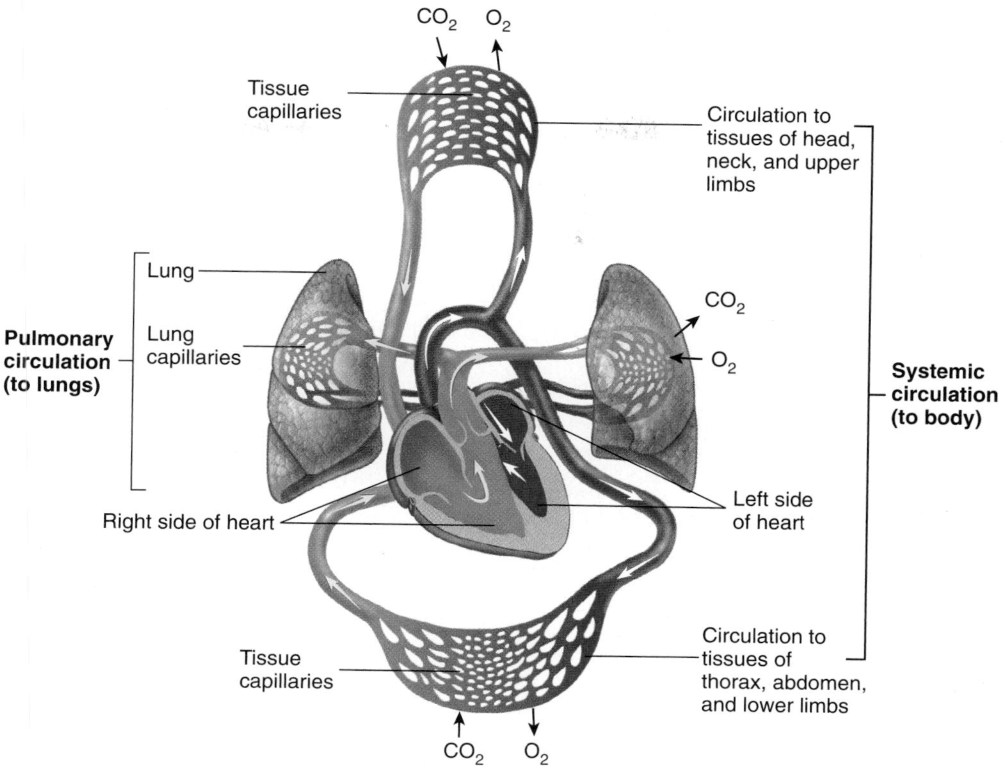

FIGURE 20.1 Systemic and Pulmonary Circulation
The circulatory system consists of the pulmonary and systemic circulations. The right side of the heart pumps blood through vessels to the lungs and back to the left side of the heart through the pulmonary circulation. The left side of the heart pumps blood through vessels to the tissues of the body and back to the right side of the heart through the systemic circulation.

FUNCTIONS OF THE HEART

The functions of the heart include

1. *Generating blood pressure.* Contractions of the heart generate blood pressure, which is responsible for blood movement through the blood vessels.
2. *Routing blood.* The heart separates the pulmonary and systemic circulations and ensures better oxygenation of blood flowing to tissues.
3. *Ensuring one-way blood flow.* The valves of the heart ensure a one-way flow of blood through the heart and blood vessels.
4. *Regulating blood supply.* Changes in the rate and force of contraction match blood delivery to the changing metabolic needs of the tissues, such as during rest, exercise, and changes in body position.

1 ***List the four functions of the heart.***

SIZE, SHAPE, AND LOCATION OF THE HEART

The adult heart is shaped like a blunt cone and is approximately the size of a closed fist. It is larger in physically active adults than in other healthy adults, and it generally decreases in size after approximately age 65, especially in those who are not physically active. The blunt, rounded point of the cone is the **apex;** the larger, flat part at the opposite end of the cone is the **base.**

The heart is located in the thoracic cavity between the lungs. The heart, trachea, esophagus, and associated structures form a midline partition, the **mediastinum** (me′dē-as-tī′nŭm; see figure 1.15).

It is important for clinical reasons to know the location of the heart in the thoracic cavity. Positioning a stethoscope to hear the heart sounds and positioning electrodes to record an **electrocardiogram** (ē-lek-trō-kar′dē-ō-gram; **ECG** or **EKG**) from chest leads depend on this knowledge. Effective **cardiopulmonary resuscitation** (kar′dē-ō-pŭl′mo-nār-ē rē-sŭs′i-tā-shŭn; **CPR**) also depends on a reasonable knowledge of the position and shape of the heart. The heart lies obliquely in the mediastinum, with its base directed posteriorly and slightly superiorly and the apex directed anteriorly and slightly inferiorly. The apex is also directed to the left so that approximately two-thirds of the heart's mass lies to the left of the midline of the sternum (figure 20.2). The base of the heart is located deep to the sternum and extends to the second intercostal space. The apex is located deep to the fifth intercostal space, approximately 7–9 centimeters (cm) to the left of the sternum and medial to the midclavicular line, which is a perpendicular line that extends down from the middle of the clavicle.

2 ***Describe the approximate size and shape of the heart. Where is it located?***

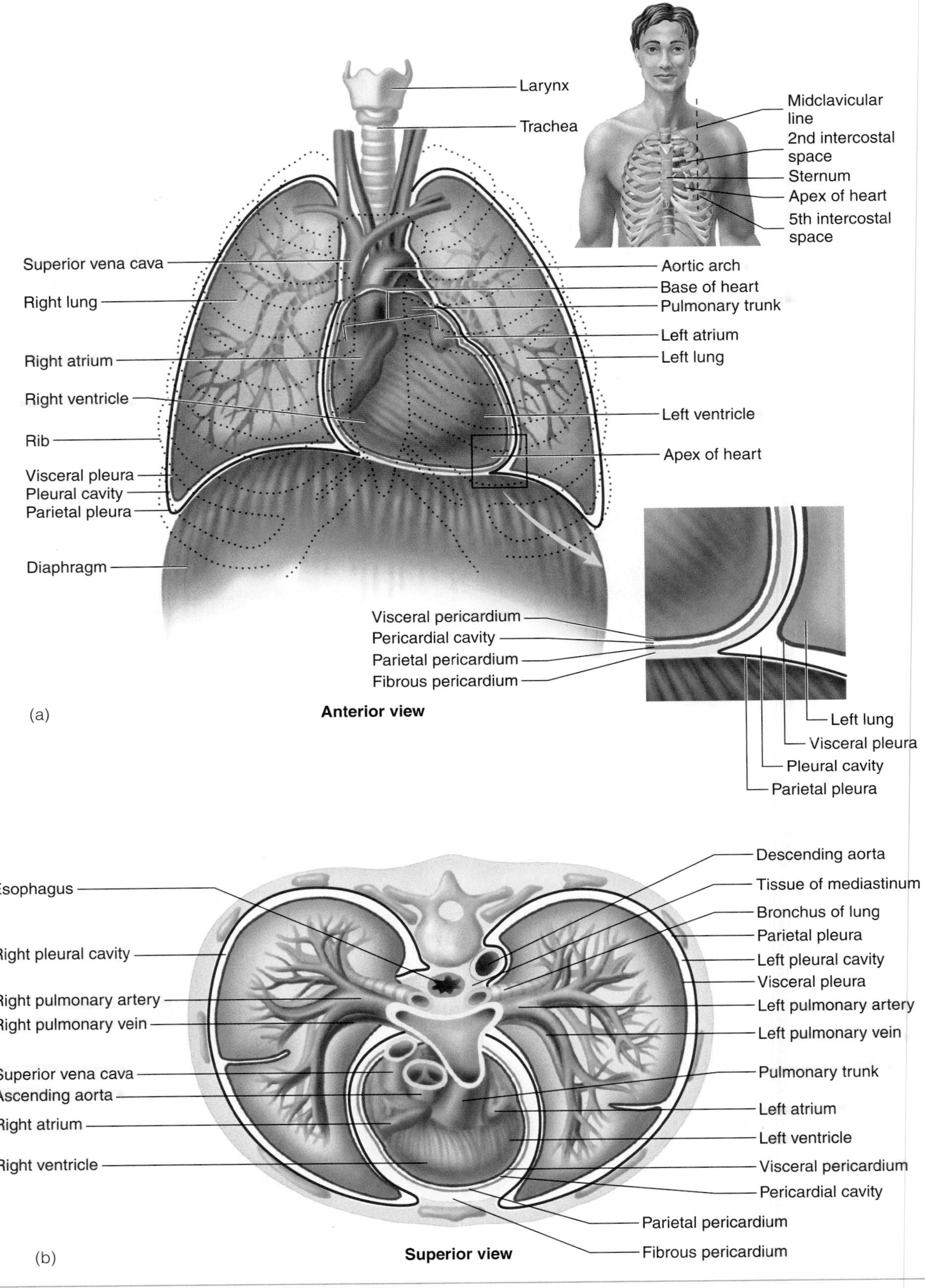

FIGURE 20.2 Location of the Heart in the Thorax

(*a*) The heart is located in the thoracic cavity between the lungs, deep and slightly to the left of the sternum. The base of the heart, located deep to the sternum, extends to the second intercostal space, and the apex of the heart is located deep to the fifth intercostal space, approximately 7–9 cm to the left of the sternum, or where the midclavicular line intersects with the fifth intercostal space (see inset). (*b*) Cross section of the thorax, showing the position of the heart in the mediastinum and its relationship to other structures.

Cardiopulmonary Resuscitation (CPR)

When the heart suddenly stops beating, CPR can save lives. CPR involves firm and rhythmic compression of the chest combined with artificial ventilation of the lungs. Applying pressure to the sternum compresses the chest wall, which also compresses the heart and causes it to pump blood. CPR can provide an adequate blood supply to the heart wall and brain until emergency medical assistance arrives.

ANATOMY OF THE HEART

Pericardium

The **pericardium** (per-i-kar′dē-ŭm), or **pericardial sac,** is a double-layered, closed sac that surrounds the heart (figure 20.3). It consists of a tough, fibrous connective tissue outer layer called the **fibrous pericardium** and a thin, transparent inner layer of simple squamous epithelium called the **serous pericardium.** The fibrous pericardium prevents overdistention of the heart and anchors it within the mediastinum. Superiorly, the fibrous pericardium is continuous with the connective tissue coverings of the great vessels, and inferiorly it is attached to the surface of the diaphragm.

The part of the serous pericardium lining the fibrous pericardium is the **parietal pericardium,** and the part covering the heart surface is the **visceral pericardium,** or epicardium (see figure 20.3). The parietal and visceral portions of the serous pericardium are continuous with each other where the great vessels enter or leave the heart. The **pericardial cavity,** between the visceral and parietal pericardia, is filled with a thin layer of serous **pericardial fluid,** which helps reduce friction as the heart moves within the pericardial sac.

Even though the pericardium contains fibrous connective tissue, it can accommodate changes in heart size by gradually enlarging. The pericardial cavity can also increase in volume to hold a significant volume of pericardial fluid.

Pericarditis and Cardiac Tamponade

Pericarditis (per′i-kar-dī′tis) is an inflammation of the serous pericardium. The cause is frequently unknown, but it can result from infection, diseases of connective tissue, or damage due to radiation treatment for cancer. It can be extremely painful, with sensations of pain referred to the back and chest, which can be confused with the pain of a myocardial infarction (heart attack). Pericarditis can result in fluid accumulation within the pericardial sac.

Cardiac tamponade (tam-pŏ-nād′) is a potentially fatal condition in which a large volume of fluid or blood accumulates in the pericardial sac. The fluid compresses the heart from the outside. Although the heart is a powerful muscle, it relaxes passively. When it is compressed by fluid within the pericardial sac, it cannot dilate when the cardiac muscle relaxes. Consequently, it cannot fill with blood during relaxation, which makes it impossible for it to pump blood. Cardiac tamponade can cause a person to die quickly unless the fluid is removed. Causes of cardiac tamponade include rupture of the heart wall following a myocardial infarction, rupture of blood vessels in the pericardium after a malignant tumor invades the area,damage to the pericardium resulting from radiation therapy, and trauma (e.g., a traffic accident).

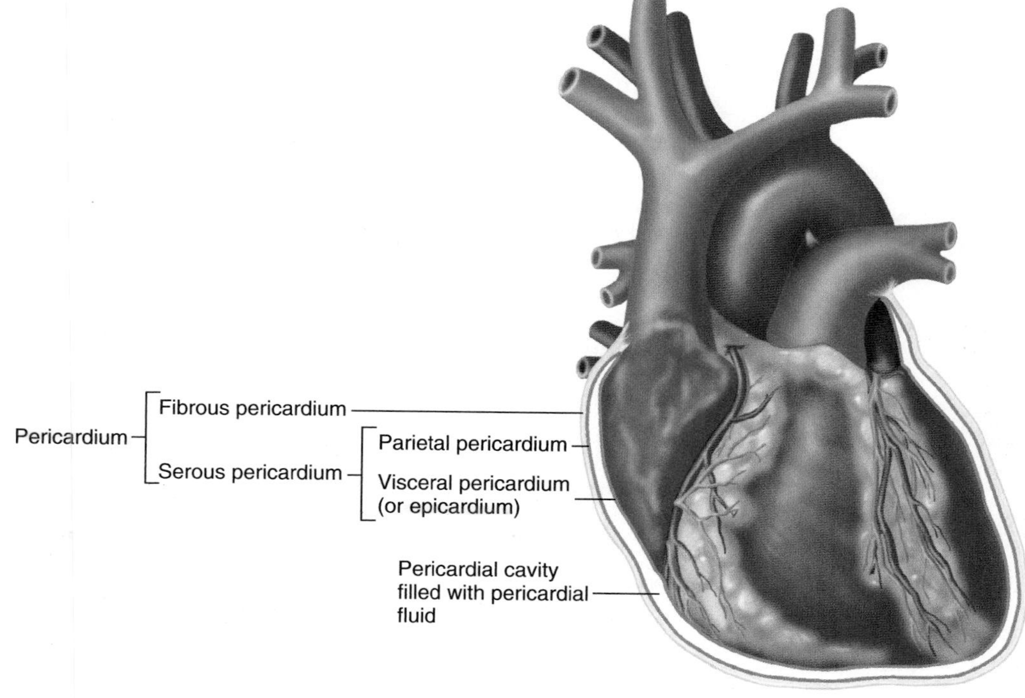

FIGURE 20.3 Heart in the Pericardium

The heart is located in the pericardium, which consists of an outer fibrous pericardium and an inner serous pericardium. The serous pericardium has two parts: The parietal pericardium lines the fibrous pericardium, and the visceral pericardium (epicardium) covers the surface of the heart. The pericardial cavity, between the parietal and visceral pericardia, is filled with a small amount of pericardial fluid.

3 ***What is the pericardium?***

4 ***Describe the parts of the pericardium and their functions.***

5 ***Define pericarditis.***

6 ***Explain the effects of cardiac tamponade on the heart.***

Heart Wall

The heart wall is composed of three layers of tissue: the epicardium, myocardium, and endocardium (figure 20.4). The **epicardium** (ep-i-kar′dē-ŭm), or visceral pericardium, is a thin serous membrane that constitutes the smooth outer surface of the heart. The serous pericardium is called the epicardium when considered a part of the heart and the visceral pericardium when considered a part of the pericardium. The thick middle layer of the heart, the **myocardium** (mī-ō-kar′dē-ŭm), is composed of cardiac muscle cells and is responsible for the heart's ability to contract. The smooth inner surface of the heart chambers is the **endocardium** (en-dō-kar′dē-ŭm), which consists of simple squamous epithelium over a layer of connective tissue. The smooth inner surface allows blood to move easily through the heart. The heart valves are modified folds of the endocardium that consist of a double layer of endocardium with connective tissue in between.

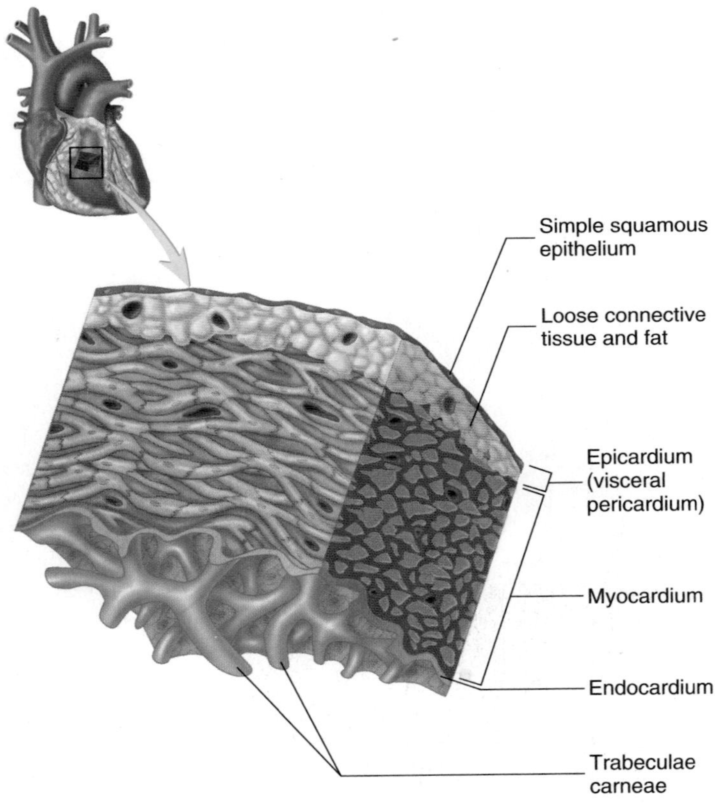

FIGURE 20.4 Heart Wall

Part of the wall of the heart has been removed, enlarged, and rotated so that its inner surface is visible. The enlarged section illustrates the epicardium (visceral pericardium), myocardium, and endocardium.

The interior surfaces of the atria are mainly flat, but the interior of both auricles and a part of the right atrial wall contain muscular ridges called **pectinate muscles** (pek′ti-nāt; hair comb–shaped). The pectinate muscles of the right atrium are separated from the larger, smooth portions of the atrial wall by a ridge called the **crista terminalis** (kris′tă ter′mi-nal′is; terminal crest). The interior walls of the ventricles contain larger, muscular ridges and columns called **trabeculae** (tră-bek′ū-lē; beams) **carneae** (kar′nē-ē; flesh).

7 ***Describe the three layers of the heart, and state their functions.***

8 ***Name the muscular ridges found on the interior of the auricles, and name the ridges and columns found on the interior walls of the ventricles.***

External Anatomy and Coronary Circulation

The heart consists of four chambers: two **atria** (ā′trē-ă; entrance chamber) and two **ventricles** (ven′tri-klz; belly). The thin-walled atria form the superior and posterior parts of the heart, and the thick-walled ventricles form the anterior and inferior portions (figure 20.5). Flaplike **auricles** (aw′ri-klz; ears) are extensions of the atria that can be seen anteriorly between each atrium and ventricle. The entire atrium used to be called the auricle, and some medical personnel still refer to it as such.

Several large veins carry blood to the heart. The **superior vena cava** (vē′nă kā′vă) and the **inferior vena cava** carry blood from the body to the right atrium, and four **pulmonary veins** carry blood from the lungs to the left atrium. In addition, the smaller coronary sinus carries blood from the walls of the heart to the right atrium.

Two arteries, the **aorta** and the **pulmonary trunk,** exit the heart. The aorta carries blood from the left ventricle to the body, and the pulmonary trunk carries blood from the right ventricle to the lungs.

A large **coronary** (kōr′o-nār-ē; circling like a crown) **sulcus** (sool′kŭs; ditch) runs obliquely around the heart, separating the atria from the ventricles. Two more sulci extend inferiorly from the coronary sulcus, indicating the division between the right and left ventricles. The **anterior interventricular sulcus,** or **groove,** is on the anterior surface of the heart, and the **posterior interventricular sulcus,** or **groove,** is on the posterior surface of the heart. In a healthy, intact heart, the sulci are covered by fat, and only after this fat is removed can the sulci be seen.

The major arteries supplying blood to the tissue of the heart lie within the coronary sulcus and interventricular sulci on the surface of the heart. The **right** and **left coronary arteries** exit the aorta just above the point where the aorta leaves the heart and lie within the coronary sulcus (figure 20.6*a*). The right coronary artery is usually smaller in diameter than the left one, and it does not carry as much blood as the left coronary artery.

A major branch of the left coronary artery, called the **anterior interventricular artery,** or the **left anterior descending artery,** extends inferiorly in the anterior interventricular sulcus and supplies blood to most of the anterior part of the heart. The **left marginal artery** branches from the left coronary artery to supply blood to the lateral wall of the left ventricle. The **circumflex** (ser'kŭm-fleks) **artery** branches from the left coronary artery and extends around to the posterior side of the heart in the coronary sulcus. Its branches supply blood to much of the posterior wall of the heart.

The right coronary artery lies within the coronary sulcus and extends from the aorta around to the posterior part of the heart. A larger branch of the right coronary artery, called the **right marginal artery,** and other branches supply blood to the lateral wall of the right ventricle. A branch of the right coronary artery, called the **posterior interventricular artery,** lies in the posterior interventricular sulcus and supplies blood to the posterior and inferior part of the heart.

PREDICT 1

Predict the effect on the heart if blood flow through a coronary artery, such as the anterior interventricular artery, is restricted or completely blocked.

Most of the myocardium receives blood from more than one arterial branch. Furthermore, there are many anastamoses, or direct connections, between the arterial branches. The anastamoses are either between branches of a given artery or between branches of different arteries. In the event that one artery is blocked, the areas primarily supplied by that artery may still receive some blood through other arterial branches and anastamoses. Aerobic exercise tends to increase the density of blood vessels supplying blood to the myocardium and the number and extent of the anastamoses increase. Consequently, aerobic exercise increases the chance that a person will survive the blockage of a small coronary artery. The blockage of larger coronary blood vessels still has the potential to permanently damage large areas of the heart wall.

The major vein draining the tissue on the left side of the heart is the **great cardiac vein,** and a **small cardiac vein** drains the right margin of the heart (figure 20.6*b*). These veins converge toward the posterior part of the coronary sulcus and empty into a large venous cavity called the **coronary sinus,** which in turn empties into the right atrium. A number of smaller veins empty into the cardiac veins, into the coronary sinus, or directly into the right atrium.

Blood flow through the coronary blood vessels is not continuous. When the cardiac muscle contracts, blood vessels in the wall of the heart are compressed and blood does not readily flow through them. When the cardiac muscle is relaxing, the blood vessels are not compressed and blood flow through the coronary blood vessels resumes.

In a resting person, blood flowing through the coronary arteries of the heart gives up approximately 70% of its oxygen. In comparison, blood flowing through arteries to skeletal muscle gives up only about 25% of its oxygen. The percentage of oxygen the blood releases to skeletal muscle can increase to 70% or more during exercise. Because the percentage of oxygen the blood releases to cardiac muscle is near its maximum at rest, it cannot increase substantially during exercise. Cardiac muscle is therefore very dependent on an increased rate of blood flow through the coronary arteries above its resting level to provide an adequate oxygen supply during exercise.

9. ***Name the chambers of the heart and describe their structure.***
10. ***Name the major blood vessels that enter and leave the heart. Which chambers of the heart do they enter or exit?***
11. ***Describe the flow of blood through the coronary arteries and their branches.***
12. ***Is blood flow through the coronary vessels continuous?***
13. ***Describe the flow of blood through the cardiac veins.***

Heart Chambers and Valves

Right and Left Atria

The **right atrium** has three major openings: The openings from the superior vena cava and the inferior vena cava receive blood from the body, and the opening of the coronary sinus receives blood from the heart itself (figure 20.7). The **left atrium** has four relatively uniform openings that receive blood from the four pulmonary veins from the lungs. The two atria are separated from each other by the **interatrial septum.** A slight, oval depression, the **fossa ovalis** (fos'ă ō-va'lis), on the right side of the septum marks the former location of the **foramen ovale** (ō-va'lē), an opening between the right and left atria in the embryo and the fetus. This opening allows blood to flow from the right to the left atrium in the fetus to bypass the pulmonary circulation (see chapter 29).

Right and Left Ventricles

The atria open into the ventricles through **atrioventricular canals** (see figure 20.7). Each ventricle has one large, superiorly placed outflow route near the midline of the heart. The **right ventricle** opens into the pulmonary trunk, and the **left ventricle** opens into the aorta. The two ventricles are separated from each other by the **interventricular septum,** which has a thick, muscular part toward the apex and a thin, membranous part toward the atria.

Atrioventricular Valves

An **atrioventricular valve** is in each atrioventricular canal and is composed of cusps, or flaps. These valves allow blood to flow from the atria into the ventricles but prevent blood from flowing back into the atria. The atrioventricular valve between the right atrium and the right ventricle has three cusps and is therefore called the **tricuspid** (trī-kŭs'pid) **valve.** The atrioventricular valve between the left atrium and left ventricle has two cusps and is therefore called the **bicuspid** (bī-kŭs'pid), or **mitral** (mī'trăl; resembling a bishop's miter, a two-pointed hat), **valve** (see figure 20.7; figure 20.8*a*).

Each ventricle contains cone-shaped, muscular pillars called **papillary** (pap'i-lār-ē; nipple, or pimple-shaped) **muscles.** These muscles are attached by thin, strong connective tissue strings called **chordae tendineae** (kōr'dē ten'di-nē-ē; heart strings) to the cusps of the atrioventricular valves (see figure 20.7; figure 20.8*b*). The papillary muscles contract when the ventricles contract and

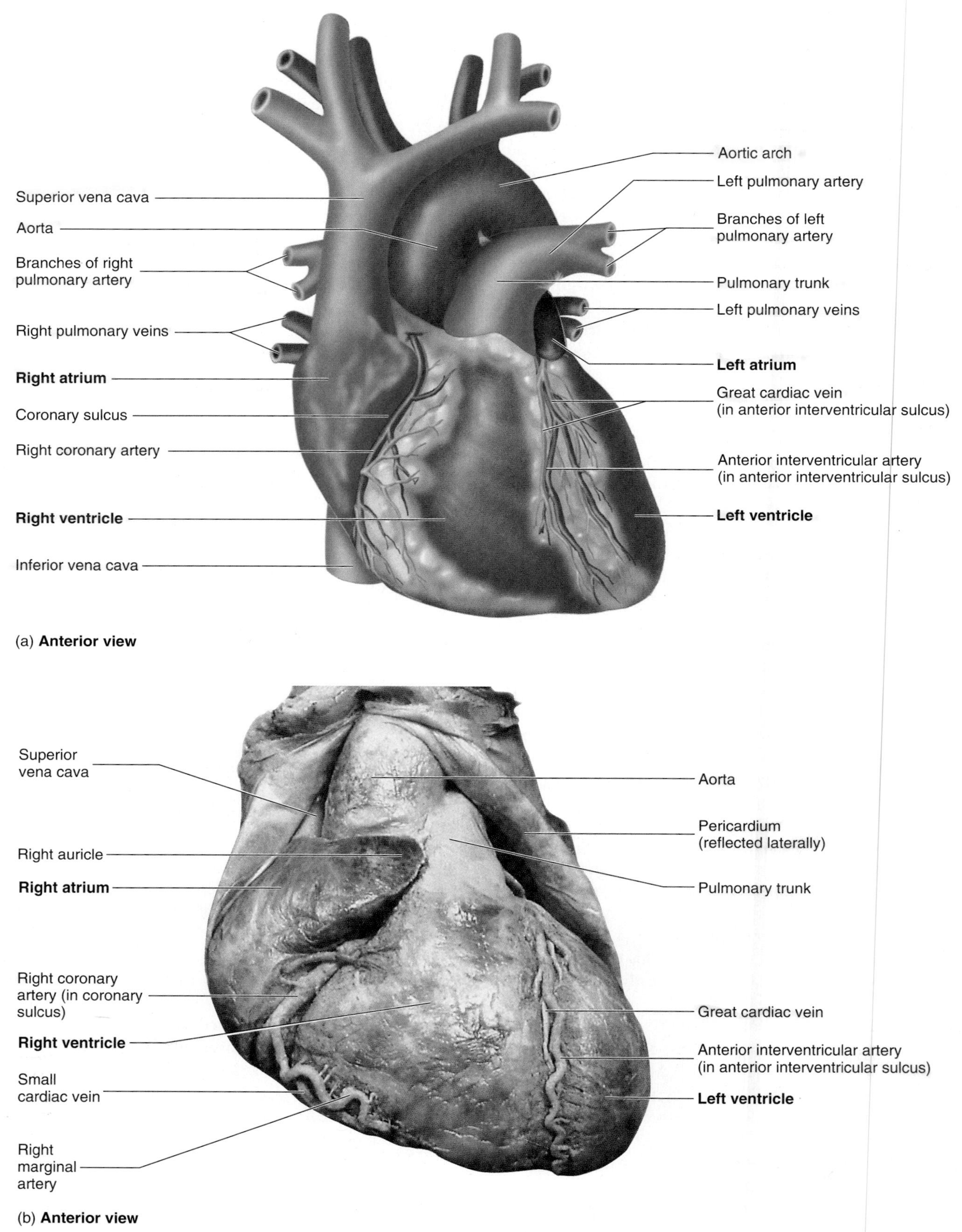

FIGURE 20.5 Surface View of the Heart

(*a*) The two atria (right and left) are located superiorly, and the two ventricles (right and left) are located inferiorly. The superior and inferior venae cava enter the right atrium. The pulmonary veins enter the left atrium. The pulmonary trunk exits the right ventricle, and the aorta exits the left ventricle. (*b*) Photograph surface.

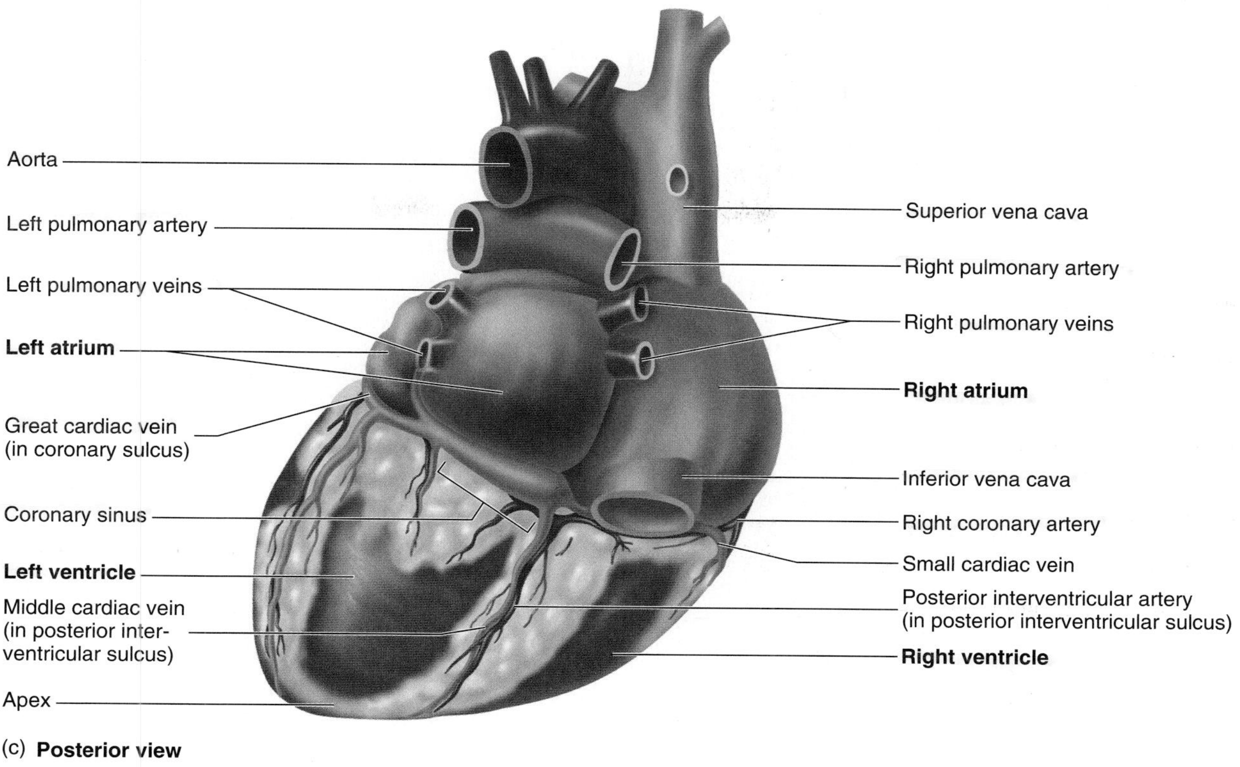

(c) **Posterior view**

FIGURE 20.5 ***(continued)***

(*c*) The two atria (right and left) are located superiorly, and the two ventricles (right and left) are located inferiorly. The superior and inferior venae cava enter the right atrium, and the four pulmonary veins enter the left atrium.

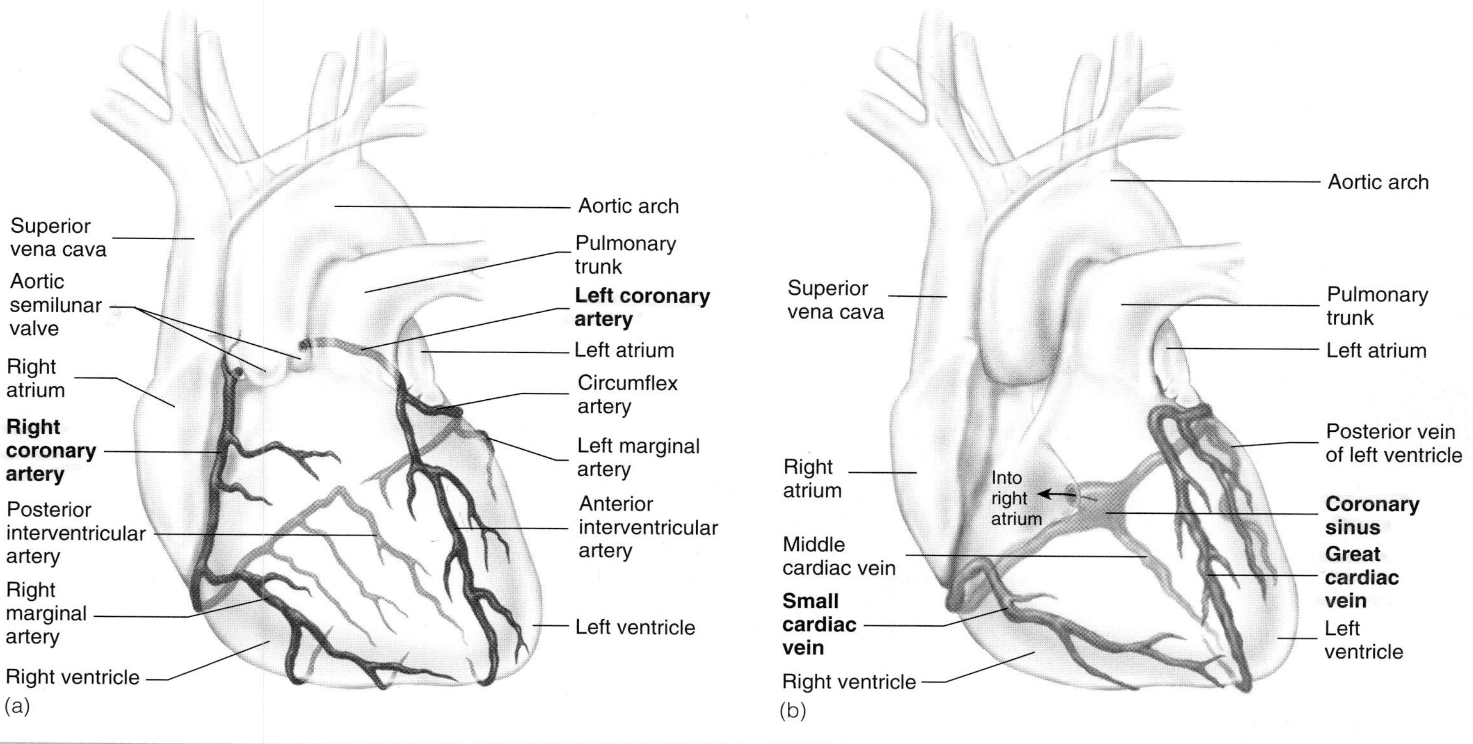

FIGURE 20.6 Coronary Circulation

(*a*) Arteries supplying blood to the heart. The arteries of the anterior surface are seen directly and are darker in color; the arteries of the posterior surface are seen through the heart and are lighter in color. (*b*) Veins draining blood from the heart. The veins of the anterior surface are seen directly and are darker in color; the veins of the posterior surface are seen through the heart and are lighter in color.

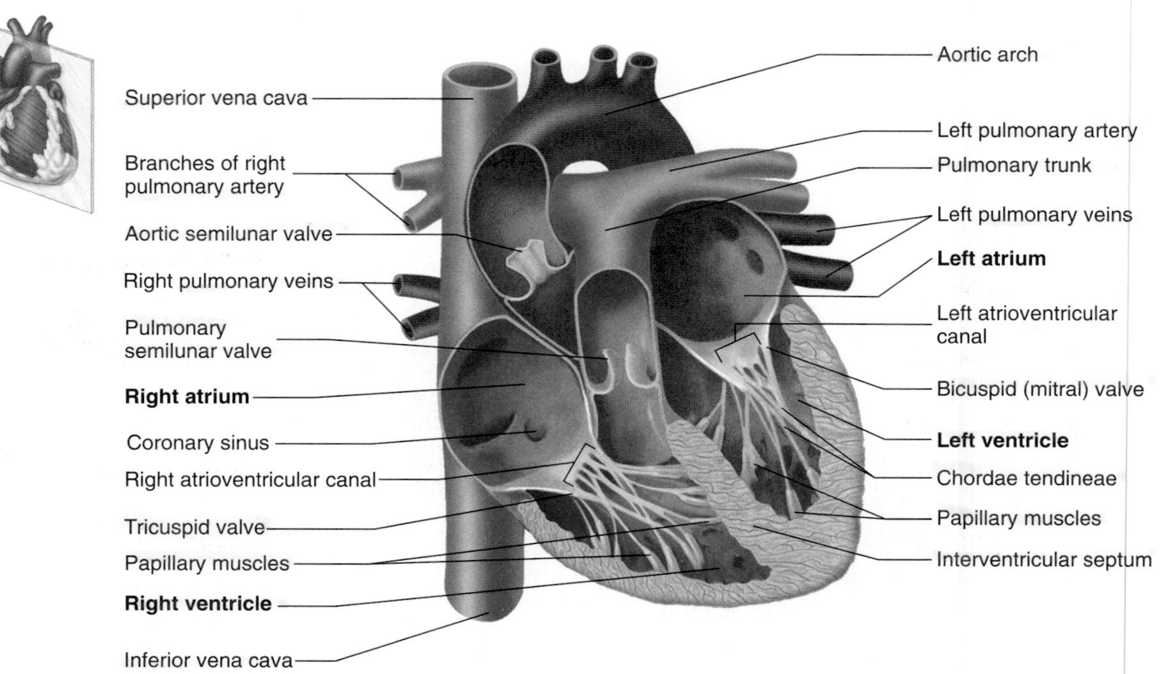

FIGURE 20.7 Internal Anatomy of the Heart
The heart is cut in a frontal plane to show the internal anatomy.

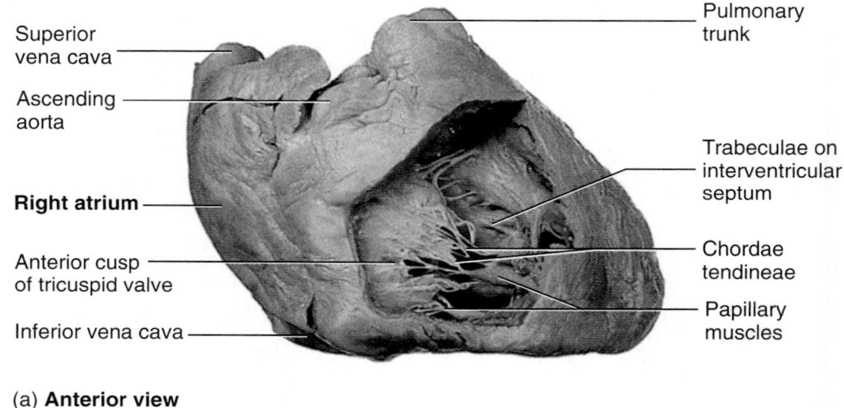

(a) **Anterior view**

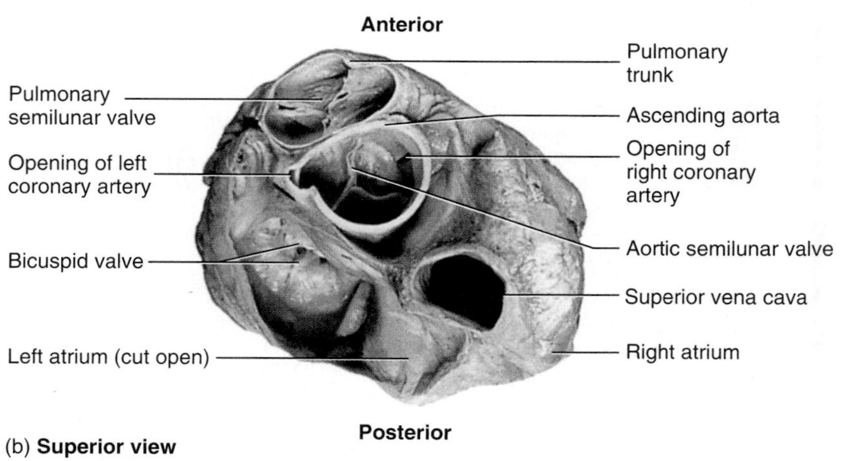

(b) **Superior view**

FIGURE 20.8 Heart Valves
(*a*) Tricuspid valve, chordae tendineae, and papillary muscles. (*b*) Heart valves. Note the three cusps of each semilunar valve meeting to prevent the backflow of blood.

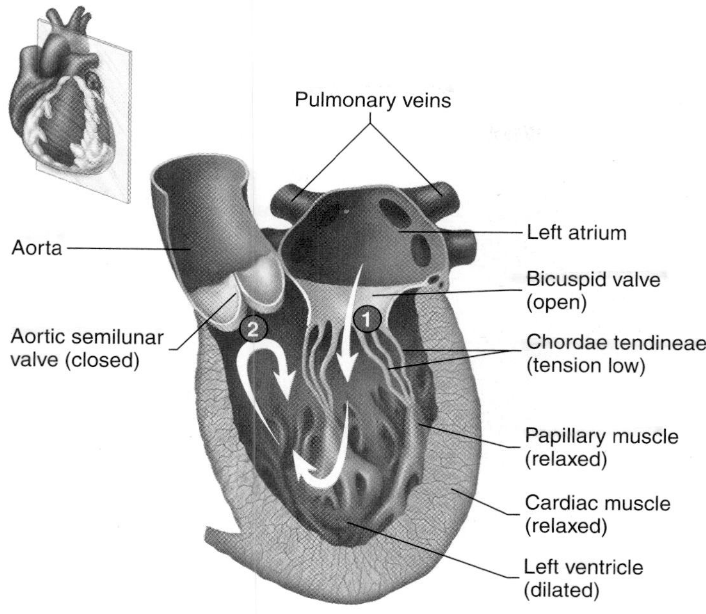

(a) Valve positions when blood is flowing into the left ventricle.

1. The bicuspid valve is open. The cusps of the valve are pushed by the blood into the ventricle.
2. The aortic semilunar valve is closed. The cusps of the valve overlap as they are pushed by the blood in the aorta toward the ventricle.

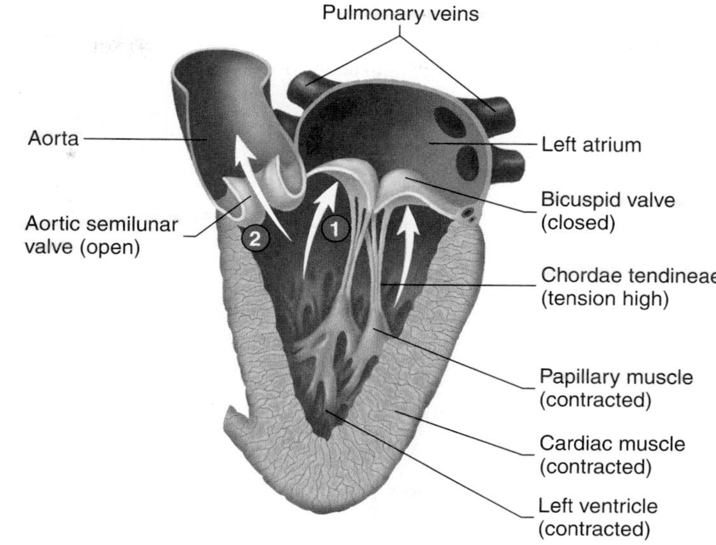

(b) Valve positions when blood is flowing out of the left ventricle.

1. The bicuspid valve is closed. The cusps of the valves overlap as they are pushed by the blood toward the left atrium.
2. The aortic semilunar valve is open. The cusps of the valve are pushed by the blood toward the aorta.

PROCESS FIGURE 20.9 Function of the Heart Valves

(*a*) Valve positions when blood is flowing into the left ventricle. (*b*) Valve positions when blood is flowing out of the left ventricle. Illustrations of the bicuspid and aortic semilunar valves show the function of those valves, as well as the tricuspid and pulmonary semilunar valves.

prevent the valves from opening into the atria by pulling on the chordae tendineae attached to the valve cusps. Blood flowing from the atrium into the ventricle pushes the valve open into the ventricle, but, when the ventricle contracts, blood pushes the valve back toward the atrium. The atrioventricular canal is closed as the valve cusps meet (figure 20.9).

Semilunar Valves

Within the aorta and pulmonary trunk are **aortic** and **pulmonary semilunar** (sem-ē-loo′năr; half-moon–shaped) **valves,** respectively. Each valve consists of three pocketlike semilunar cusps, the free inner borders of which meet in the center of the artery to block blood flow (see figures 20.7 and 20.8*b*). Blood flowing out of the ventricles pushes against each valve, forcing it open; however, when blood flows back from the aorta or pulmonary trunk toward the ventricles, it enters the pockets of the cusps, causing them to meet in the center of the aorta or pulmonary trunk, thus closing them and keeping blood from flowing back into the ventricles (see figure 20.9).

14 ***Describe the openings into the right and left atria. What structure separates the atria from each other?***

15 ***Describe the openings of the right and left ventricles. What structure separates the ventricles from each other?***

16 ***Name the valves that separate the right atrium from the right ventricle and the left atrium from the left ventricle. What are the functions of the papillary muscles and the chordae tendineae?***

17 ***Name the valves found in the aorta and pulmonary trunk.***

ROUTE OF BLOOD FLOW THROUGH THE HEART

Blood flow through the heart is depicted in figure 20.10. Even though it is more convenient to discuss blood flow through the heart one side at a time, it is important to understand that both atria contract at about the same time and both ventricles contract at about the same time. This concept is particularly important when considering electrical activity, pressure changes, and heart sounds.

Blood enters the right atrium from the systemic circulation, which returns blood from all the tissues of the body. Blood flows from an area of higher pressure in the systemic circulation to the right atrium, which has a lower pressure. Most of the blood in the right atrium then passes into the right ventricle as the ventricle relaxes following the previous contraction. The

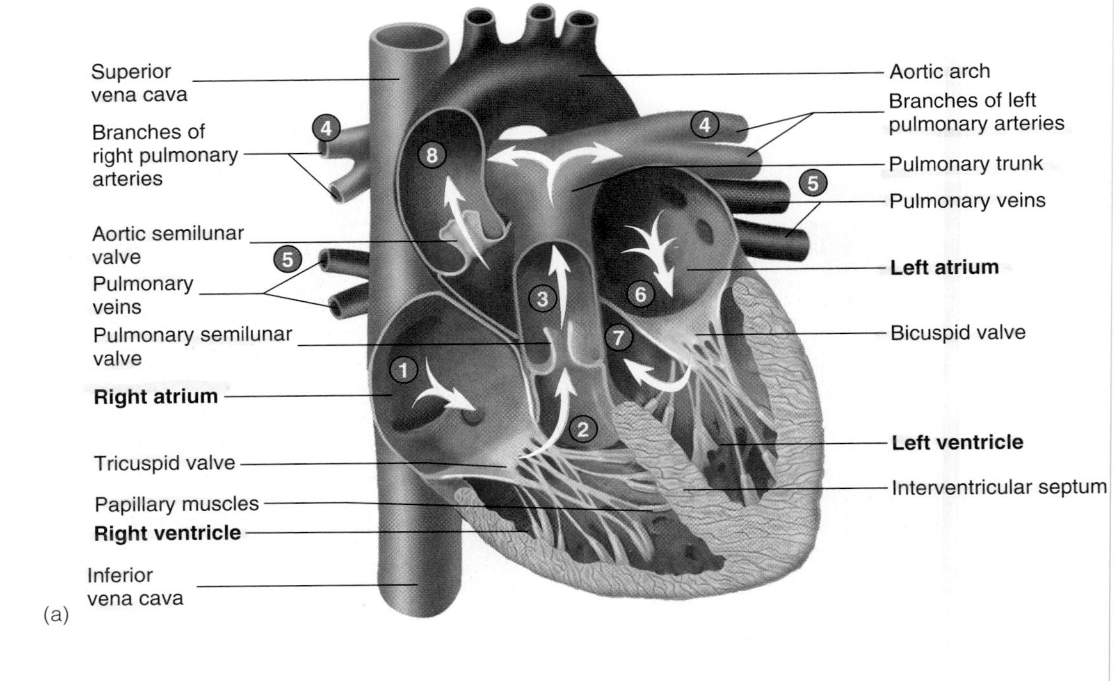

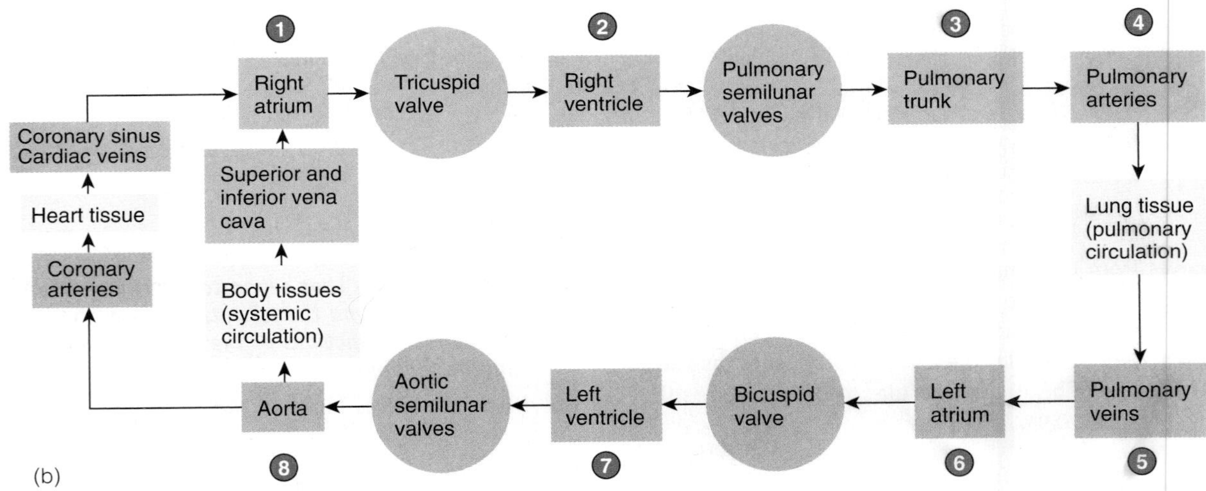

FIGURE 20.10 Blood Flow Through the Heart

(*a*) Frontal section of the heart revealing the four chambers and the direction of blood flow through the heart. The numbers identify the direction of blood flow. (*b*) Diagram listing in order the structures through which blood flows in the systemic and pulmonary circulations. The heart valves are indicated by *circles:* deoxygenated blood (*blue*), oxygenated blood (*red*).

right atrium then contracts, and most of the blood remaining in the atrium is pushed into the ventricle to complete right ventricular filling.

Contraction of the right ventricle pushes blood against the tricuspid valve, forcing it closed, and against the pulmonary semilunar valve, forcing it open, thus allowing blood to enter the pulmonary trunk.

The pulmonary trunk branches to form the **pulmonary arteries** (see figure 20.5), which carry blood to the lungs, where carbon dioxide is released and oxygen is picked up (see chapters 21 and 23). Blood returning from the lungs enters the left atrium through the four pulmonary veins. The blood passing from the left atrium to the left ventricle opens the bicuspid valve, and contraction of the left atrium completes left ventricular filling.

Contraction of the left ventricle pushes blood against the bicuspid valve, closing it, and against the aortic semilunar valve, opening it and allowing blood to enter the aorta. Blood flowing through the aorta is distributed to all parts of the body, except to the parts of the lungs supplied by the pulmonary blood vessels (see chapter 23).

18 ***Starting at the venae cavae and ending at the aorta, describe the flow of blood through the heart.***

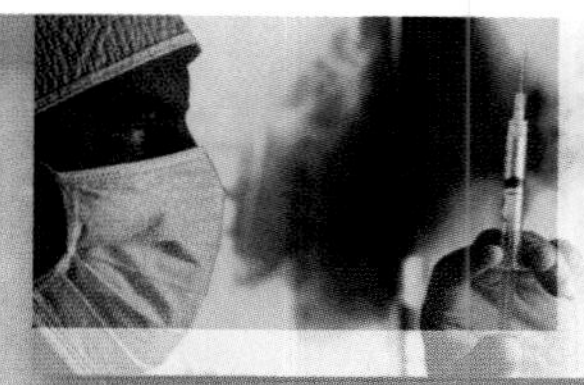

CLINICAL FOCUS

Angina, Infarctions, and the Treatment of Blocked Coronary Arteries

Angina pectoris (an′ji-nă, an-jī′nă pek′tō-ris) is pain that results from a reduction in blood supply to cardiac muscle. The pain is temporary and, if blood flow is restored, little permanent change or damage results. Angina pectoris is characterized by chest discomfort deep to the sternum, often described as heaviness, pressure, or moderately severe pain. It is often mistaken for indigestion. The pain can also be referred to the neck, lower jaw, left arm, and left shoulder (see chapter 14, p. 489).

Most often, angina pectoris results from narrowed and hardened coronary arterial walls. The reduced blood flow results in a reduced supply of oxygen to cardiac muscle cells. As a consequence, the limited anaerobic metabolism of cardiac muscle results in a reduced pH in affected areas of the heart. Pain receptors are stimulated by the decreased pH. The pain is predictably associated with exercise because the increased pumping activity of the heart requires more oxygen, and the narrowed blood vessels cannot supply it. Rest and drugs, such as nitroglycerin, frequently relieve angina pectoris. Nitroglycerin dilates the blood vessels, including the coronary arteries. Consequently, the drug increases the oxygen supply to cardiac muscle and reduces the heart's workload. Because peripheral arteries are dilated, the heart has to pump blood against a smaller pressure, and the need for oxygen decreases. The heart also pumps less blood because blood tends to remain in the dilated blood vessels and less blood is returned to the heart.

Myocardial infarction (mī-ō-kar′dē-ăl in-fark′shŭn) results from a prolonged lack of blood flow to a part of the cardiac muscle, resulting in a lack of oxygen and ultimately cellular death. Myocardial infarctions vary with the amount of cardiac muscle and the part of the heart that is affected. If blood supply to cardiac muscle is reestablished within 20 minutes, no permanent damage occurs. If the lack of oxygen lasts longer, cell death results. Within 30–60 seconds after blockage of a coronary blood vessel, however, functional changes are obvious. The electrical properties of the cardiac muscle are altered, and the heart's ability to function properly is lost. The most common cause of myocardial infarction is thrombus formation that blocks a coronary artery. Coronary arteries narrowed by **atherosclerotic** (ath′er-ō-skler-ot′ik) **lesions** increase the risk for myocardial infarction. Atherosclerotic lesions partially block blood vessels, resulting in turbulent blood flow, and the surfaces of the lesions are rough. These changes increase the probability of thrombus formation.

Angioplasty (an′jē-ō-plas-tē) is a process whereby a small balloon is inserted, usually into the femoral artery (see chapter 21), and is threaded through the aorta and into a coronary artery. After the balloon has entered the partially occluded coronary artery, it is inflated, thereby flattening the atherosclerotic deposits against the vessel walls and opening the occluded blood vessel. This technique improves the function of cardiac muscle in patients suffering from an inadequate blood flow to the cardiac muscle through the coronary arteries. Some controversy exists about its effectiveness, however. At least in some patients, dilation of the coronary arteries can be reversed within a few weeks or months and blood clots can form in coronary arteries following angioplasty. To help prevent future blockage, a metal-mesh tube sometimes called a **stent,** is inserted into the vessel. Although the stent is better able to hold the vessel open, it, too, can eventually become blocked. Small, rotating blades and lasers are also used to remove lesions from coronary vessels.

A **coronary bypass** is a surgical procedure that relieves the effects of obstructions in the coronary arteries. The technique involves taking healthy segments of blood vessels from other parts of the patient's body and using them to bypass obstructions in the coronary arteries. The technique is common for those who suffer from severe occlusion in specific parts of coronary arteries.

Special enzymes are used to break down blood clots that form in the coronary arteries and cause heart attacks. The major enzymes used are **streptokinase** (strep-tō-kī′nās), **tissue plasminogen** (plaz-min′ō-jen) **activator (t-PA),** and **urokinase** (ūr-ō-kī′nās). These enzymes activate plasminogen, which is an inactive form of an enzyme in the body that breaks down the fibrin of clots. The strategy is to administer these drugs to people suffering from myocardial infarctions as soon as possible following the onset of symptoms. Removal of the occlusions produced by clots reestablishes blood flow to the cardiac muscle and reduces the amount of cardiac muscle permanently damaged by the occlusions.

HISTOLOGY

 ***Describe and list the functions of the skeleton of the heart.***

Heart Skeleton

The **heart skeleton** consists of a plate of fibrous connective tissue between the atria and ventricles. This connective tissue plate forms **fibrous rings** around the atrioventricular and semilunar valves and provides a solid support for them (figure 20.11). The fibrous connective tissue plate also serves as electrical insulation between the atria and the ventricles and provides a rigid site for attachment of the cardiac muscles.

Cardiac Muscle

Cardiac muscle cells are elongated, branching cells that contain one, or occasionally two, centrally located nuclei. Cardiac muscle cells contain actin and myosin myofilaments organized to form sarcomeres, which join end-to-end to form myofibrils (see chapter 9). The actin and myosin myofilaments are responsible for muscle contraction, and their organization gives cardiac muscle a striated (banded) appearance. The striations are less

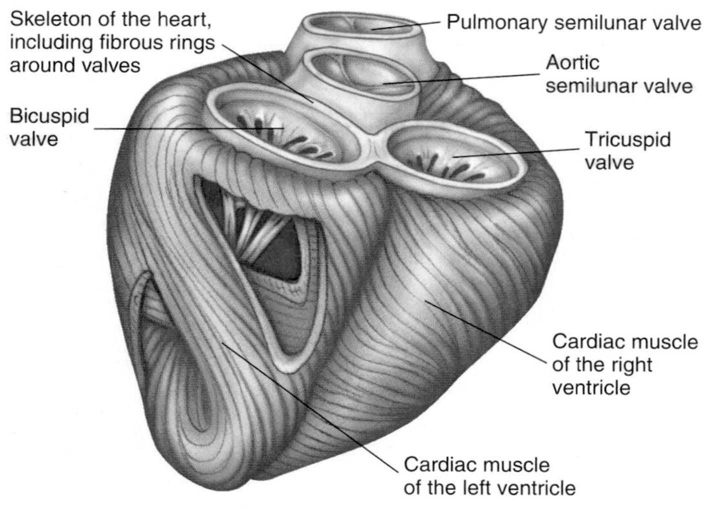

FIGURE 20.11 Skeleton of the Heart

The skeleton of the heart consists of fibrous connective tissue rings, which surround the heart valves and separate the atria from the ventricles. Cardiac muscle attaches to the fibrous connective tissue. The muscle fibers are arranged so that, when the ventricles contract, a wringing motion is produced and the distance between the apex and the base of the heart shortens.

regularly arranged and less numerous than in skeletal muscle (figure 20.12).

Cardiac muscle has a **smooth sarcoplasmic reticulum,** but it is not as regularly arranged as it is in skeletal muscle fibers, and there are no dilated cisternae, as in skeletal muscle. The sarcoplasmic reticulum comes into close association at various points with membranes of **transverse (T) tubules.** The T tubules of cardiac muscle are found near the Z disks of the sarcomeres instead of where the actin and myosin overlap as in skeletal muscle. The T tubules in cardiac muscle are larger in diameter than in skeletal muscle, and there are extensions of T tubules that are parallel with the sarcoplasmic reticulum. The loose association between the sarcoplasmic reticulum and the T tubules is partly responsible for the slow onset of contraction and the prolonged contraction phase in cardiac muscle. Depolarizations of the cardiac muscle plasma membrane are not carried from the surface of the cell to the sarcoplasmic reticulum as efficiently as they are in skeletal muscles, and calcium must diffuse a greater distance from the sarcoplasmic reticulum to the actin myofilaments. In addition, some Ca^{2+} enter the cardiac muscle cells from the extracellular fluid and from the T tubules.

Adenosine triphosphate (ATP) provides the energy for cardiac muscle contraction, and, as in other tissues, ATP production depends on oxygen availability. Cardiac muscle, however, cannot develop a large oxygen debt, a characteristic that is consistent with the function of the heart. Development of a large oxygen debt would result in muscular fatigue and cessation of cardiac muscle contraction. Cardiac muscle cells are rich in mitochondria, which perform oxidative metabolism at a rate rapid enough to sustain normal myocardial energy requirements. The extensive capillary network provides an adequate oxygen supply to the cardiac muscle cells.

PREDICT 2

Under resting conditions, most of the ATP produced in cardiac muscle is derived from the metabolism of fatty acids. During heavy exercise, however, cardiac muscle cells use lactic acid as an energy source. Explain why this arrangement is an advantage.

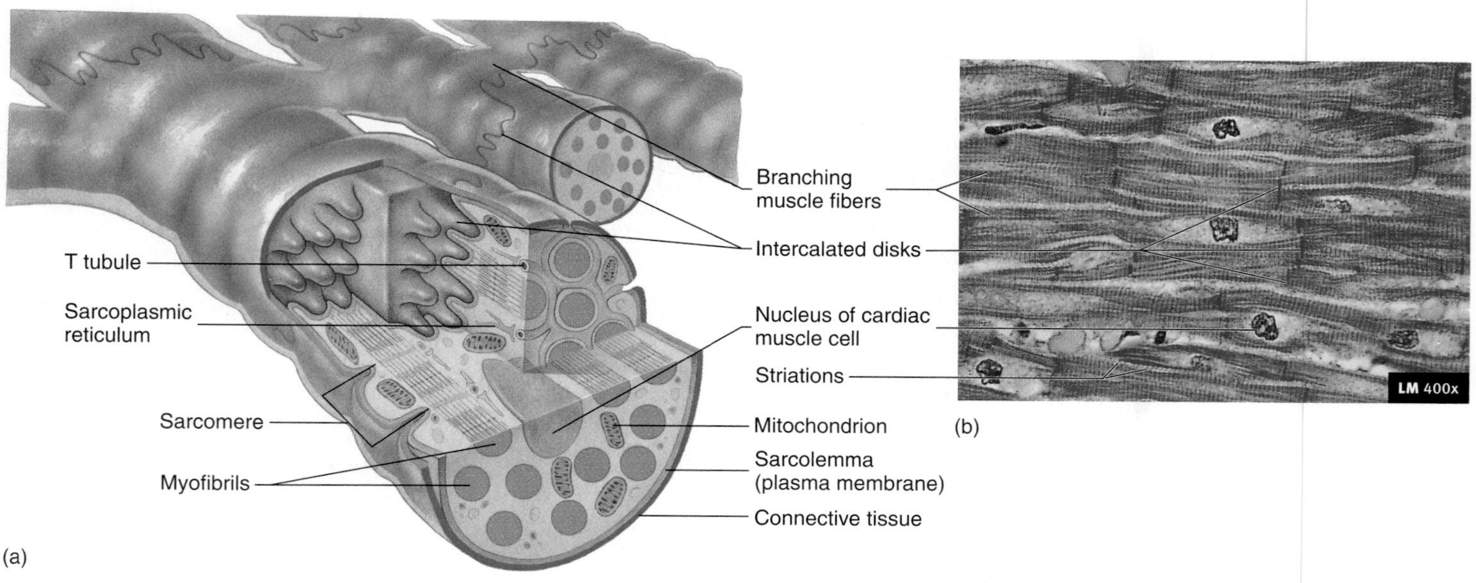

FIGURE 20.12 Histology of the Heart

(*a*) Cardiac muscle cells are branching cells with centrally located nuclei. As in skeletal muscle, sarcomeres join end-to-end to form myofibrils, and mitochondria provide ATP for contraction. The cells are joined to one another by intercalated disks. Gap junctions in the intercalated disks allow action potentials to pass from one cardiac muscle cell to the next. Sarcoplasmic reticulum and T tubules are visible but are not as numerous as they are in skeletal muscle. (*b*) A light micrograph of cardiac muscle tissue. The cardiac muscle fibers appear to be striated because of the arrangement of the individual myofilaments.

Cardiac muscle cells are organized in spiral bundles or sheets. The cells are bound end-to-end and laterally to adjacent cells by specialized cell–cell contacts called **intercalated** (in-ter′kă-lā-ted) **disks** (see figure 20.12). The membranes of the intercalated disks have folds, and the adjacent cells fit together, thus greatly increasing contact between them. Specialized plasma membrane structures called **desmosomes** (dez′mō-sōmz) hold the cells together, and **gap junctions** function as areas of low electric resistance between the cells, allowing action potentials to pass from one cell to adjacent cells (see figure 4.2). Electrically, the cardiac muscle cells behave as a single unit, and the highly coordinated contractions of the heart depend on this functional characteristic.

20 ***Describe the similarities and differences between cardiac muscle and skeletal muscle.***

21 ***Why does cardiac muscle have a slow onset of contraction and a prolonged contraction?***

22 ***What substances do cardiac muscle cells use as an energy source? Do cardiac muscle cells develop an oxygen debt?***

23 ***What anatomical features are responsible for the ability of cardiac muscle cells to contract as a unit?***

Conducting System

The conducting system of the heart, which relays action potentials through the heart, consists of modified cardiac muscle cells that form two nodes (knots or lumps) and a conducting bundle (figure 20.13). The two nodes are contained within the walls of the right atrium and are named according to their position in the atrium. The **sinoatrial (SA) node** is medial to the opening of the superior vena cava, and the **atrioventricular (AV) node** is medial to the right atrioventricular valve. The AV node gives rise to a conducting bundle of the heart, the **atrioventricular (AV) bundle** (bundle of His). This bundle passes through a small opening in the fibrous skeleton to reach the interventricular septum, where it divides to form the **right** and **left bundle branches,** which extend beneath the endocardium on each side of the interventricular septum to the apices of the right and left ventricles, respectively.

The inferior terminal branches of the bundle branches are called **Purkinje** (per-kin′jē) **fibers,** which are large-diameter cardiac muscle fibers. They have fewer myofibrils than most cardiac muscle cells and do not contract as forcefully. Intercalated disks are well developed between the Purkinje fibers and contain numerous gap junctions. As a result of these structural modifications, action potentials travel along the Purkinje fibers much more rapidly than through other cardiac muscle tissue.

Cardiac muscle cells have the capacity to generate spontaneous action potentials, but cells of the SA node do so at a greater frequency. As a result, the SA node is called the **pacemaker** of the heart. The SA node is made up of specialized, small-diameter cardiac muscle cells that merge with the other cardiac muscle cells of the right

1. Action potentials originate in the sinoatrial (SA) node (the pacemaker) and travel across the wall of the atrium (*arrows*) from the SA node to the atrioventricular (AV) node.

2. Action potentials pass through the AV node and along the atrioventricular (AV) bundle, which extends from the AV node, through the fibrous skeleton, into the interventricular septum.

3. The AV bundle divides into right and left bundle branches, and action potentials descend to the apex of each ventricle along the bundle branches.

4. Action potentials are carried by the Purkinje fibers from the bundle branches to the ventricular walls.

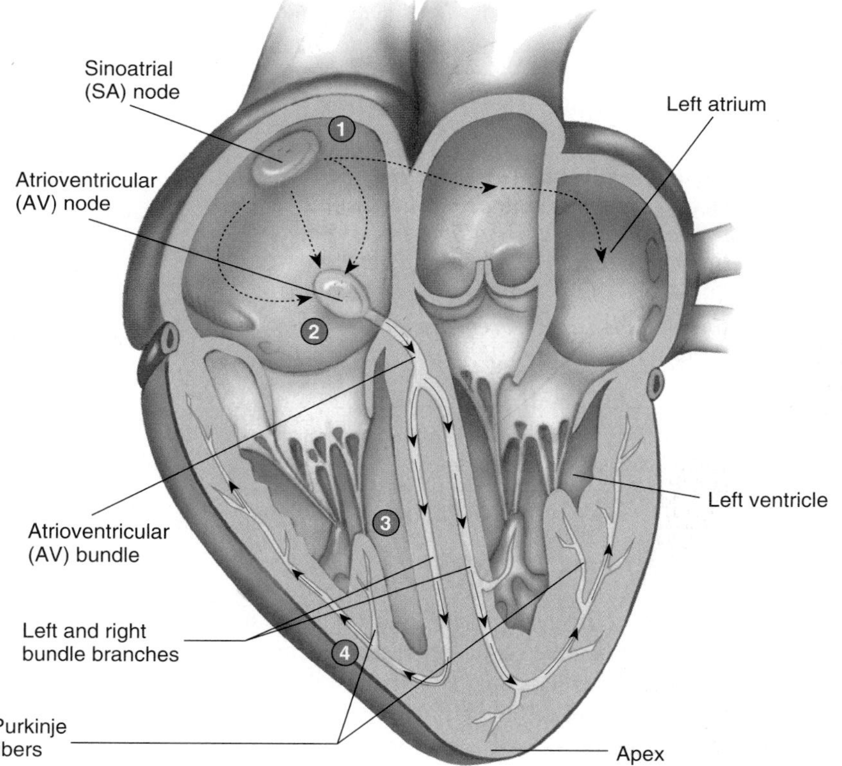

PROCESS FIGURE 20.13 Conducting System of the Heart

atrium. Thus, the heart contracts spontaneously and rhythmically. Once action potentials are produced, they spread from the SA node to adjacent cardiac muscle fibers of the atrium. Preferential pathways conduct action potentials from the SA node to the AV node at a greater velocity than they are transmitted in the remainder of the atrial muscle fibers, although such pathways cannot be distinguished structurally from the remainder of the atrium.

When the heart beats under resting conditions, approximately 0.04 second is required for action potentials to travel from the SA node to the AV node. Within the AV node, action potentials are propagated slowly, compared with the remainder of the conducting system. The slow rate of action potential conduction in the AV node is due, in part, to the smaller-diameter muscle fibers and fewer gap junctions in their intercalated disks. Like the other specialized conducting fibers in the heart, they have fewer myofibrils than most cardiac muscle cells. As a consequence, a delay occurs of 0.11 second from the time action potentials reach the AV node until they pass to the AV bundle. The total delay of 0.15 second allows completion of the atrial contraction before ventricular contraction begins.

After action potentials pass from the AV node to the highly specialized conducting bundles, the velocity of conduction increases dramatically. The action potentials pass through the left and right bundle branches and through the individual Purkinje fibers that penetrate into the myocardium of the ventricles (see figure 20.13).

Because of the arrangement of the conducting system, the first part of the myocardium that is stimulated is the inner wall of the ventricles near the apex. Thus, ventricular contraction begins at the apex and progresses throughout the ventricles. The spiral arrangement of muscle layers in the wall of the heart results in a wringing action, which proceeds from the apex toward the base of the heart as contraction proceeds. During the process, the distance between the apex and the base of the heart decreases.

24 ***List the parts of the conducting system of the heart. Explain how the conducting system coordinates contraction of the atria and ventricles. Explain why Purkinje fibers conduct action potentials more rapidly than other cardiac muscle cells.***

25 ***Explain why the SA node is the pacemaker of the heart.***

PREDICT 3

Explain why it is more efficient for contraction of the ventricles to begin at the apex of the heart than at the base.

ELECTRICAL PROPERTIES

Cardiac muscle cells—like other electrically excitable cells, such as neurons and skeletal muscle fibers—have a **resting membrane potential (RMP).** The RMP depends on a low permeability of the plasma membrane to Na^+ and Ca^{2+} and a higher permeability to K^+. When neurons, skeletal muscle cells, and cardiac muscle cells are depolarized to their threshold level, action potentials result (see chapter 11).

Action Potentials

Like action potentials in skeletal muscle, those in cardiac muscle exhibit depolarization followed by repolarization of the RMP. Alterations in membrane channels are responsible for the changes in the permeability of the plasma membrane that produce the action potentials. Action potentials in cardiac muscle last longer than those in skeletal muscle, and the membrane channels differ somewhat from those in skeletal muscle. In contrast to action potentials in skeletal muscle, which take less than 2 milliseconds (ms) to complete, action potentials in cardiac muscle take approximately 200–500 ms to complete (figure 20.14).

In cardiac muscle, the action potential consists of a rapid **depolarization phase,** followed by a rapid but partial **early repolarization** phase. Then a prolonged period of slow repolarization occurs, called the **plateau phase.** At the end of the plateau phase, a more rapid **final repolarization phase** takes place, during which the membrane potential returns to its resting level (see figure 20.14).

Depolarization is the result of changes in membrane permeability to Na^+, K^+, and Ca^{2+}. Membrane channels, called **voltage-gated Na^+ channels,** open, bringing about the depolarization phase of the action potential. As the voltage-gated Na^+ channels open, Na^+ diffuse into the cell, causing rapid depolarization until the cell is depolarized to approximately +20 millivolts (mV).

The voltage change occurring during depolarization affects other ion channels in the plasma membrane. Several types of **voltage-gated K^+ channels** exist, each of which opens and closes at different membrane potentials, causing changes in membrane permeability to K^+. For example, at rest, the movement of K^+ through open voltage-gated K^+ channels is primarily responsible for establishing the resting membrane potential in cardiac muscle cells. Depolarization causes these voltage-gated K^+ channels to close, thereby decreasing membrane permeability to K^+. Depolarization also causes **voltage-gated Ca^{2+}** channels to begin to open. These changes contribute to depolarization. Compared with sodium channels, the calcium channels open and close slowly.

Repolarization is also the result of changes in membrane permeability to Na^+, K^+, and Ca^{2+}. Early repolarization occurs when the voltage-gated Na^+ channels and some voltage-gated Ca^{2+} channels close, and a small number of voltage-gated K^+ channels open. Sodium ion movement into the cell slows, and some K^+ move out of the cell. The plateau phase occurs as voltage-gated Ca^{2+} channels remain open, and the movement of Ca^{2+} and some Na^+ through the voltage-gated Ca^{2+} channels into the cell counteracts the potential change produced by the movement of K^+ out of the cell. The plateau phase ends and final repolarization begins as the voltage-gated Ca^{2+} channels close and many more voltage-gated K^+ channels open. Thus, Ca^{2+} and Na^+ stop diffusing into the cell, and the tendency for K^+ to diffuse out of the cell increases. These permeability changes cause the membrane potential to return to its resting level.

Action potentials in cardiac muscle are conducted from cell to cell, whereas action potentials in skeletal muscle fibers are conducted along the length of a single muscle fiber, but not from fiber to fiber. Also, the rate of action potential propagation is slower

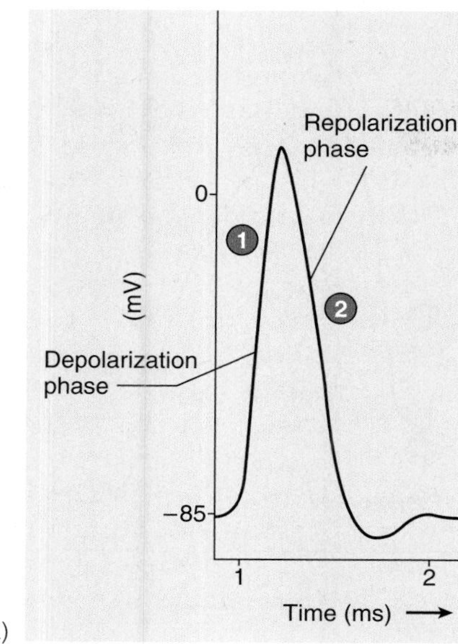

(a)

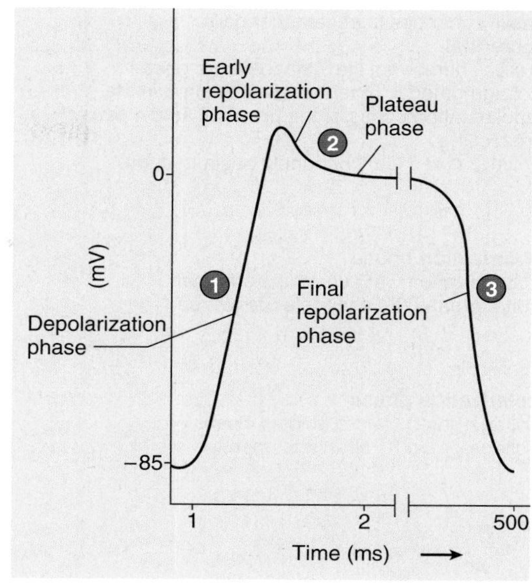

(b)

Permeability changes during an action potential in skeletal muscle:

1. **Depolarization phase**
   - Voltage-gated Na^+ channels open.
   - Voltage-gated K^+ channels begin to open.
2. **Repolarization phase**
   - Voltage-gated Na^+ channels close.
   - Voltage-gated K^+ channels continue to open.
   - Voltage-gated K^+ channels close at the end of repolarization and return the membrane potential to its resting value.

Permeability changes during an action potential in cardiac muscle:

1. **Depolarization phase**
   - Voltage-gated Na^+ channels open.
   - Voltage-gated K^+ channels close.
   - Voltage-gated Ca^{2+} channels begin to open.
2. **Early repolarization and plateau phases**
   - Voltage-gated Na^+ channels close.
   - Some voltage-gated K^+ channels open, causing early repolarization.
   - Voltage-gated Ca^{2+} channels are open, producing the plateau by slowing further repolarization.
3. **Final repolarization phase**
   - Voltage-gated Ca^{2+} channels close.
   - Many voltage-gated K^+ channels open.

PROCESS FIGURE 20.14 **Comparison of Action Potentials in Skeletal and Cardiac Muscle**

(*a*) An action potential in skeletal muscle consists of depolarization and repolarization phases. (*b*) An action potential in cardiac muscle consists of depolarization, early repolarization, plateau, and final repolarization phases. Cardiac muscle does not repolarize as rapidly as skeletal muscle (*indicated by the break in the curve*) because of the plateau phase.

in cardiac muscle than in skeletal muscle because cardiac muscle cells are smaller in diameter and much shorter than skeletal muscle fibers. Although the gap junctions of intercalated disks allow the transfer of action potentials between cardiac muscle cells, they do slow the rate of action potential conduction between the cardiac muscle cells.

The movement of Ca^{2+} through the plasma membrane, including the membranes of the T tubules, into cardiac muscle cells stimulates the release of Ca^{2+} from the sarcoplasmic reticulum, a process called **calcium-induced calcium release (CICR).** When an action potential occurs in a cardiac muscle cell, Ca^{2+} enter the cell and bind to receptors in the membranes of sarcoplasmic reticulum, resulting in the opening of Ca^{2+} channels. Calcium ions then move out of the sarcoplasmic reticulum and activate the interaction between actin and myosin to produce contraction of the cardiac muscle cells.

Autorhythmicity of Cardiac Muscle

The heart is said to be **autorhythmic** (aw′tō-rith′mik) because it stimulates itself (*auto*) to contract at regular intervals (*rhythmic*). If the heart is removed from the body and maintained under physiologic conditions with the proper nutrients and temperature, it will continue to beat autorhythmically for a long time.

In the SA node, pacemaker cells generate action potentials spontaneously and at regular intervals. These action potentials spread through the conducting system of the heart to other cardiac muscle cells, causing voltage-gated Na^+ channels to open. As a result, action potentials are produced and the cardiac muscle cells contract.

The generation of action potentials in the SA node results when a spontaneously developing local potential, called the **prepotential,** reaches threshold (figure 20.15). Changes in ion movement into and out of the pacemaker cells cause the prepotential. Sodium

Permeability changes in pacemaker cells

1. **Prepotential**
   - A small number of Na^+ channels are open.
   - Voltage-gated K^+ channels that opened in the repolarization phase of the previous action potential are closing.
   - Voltage-gated Ca^{2+} channels begin to open.
2. **Depolarization phase**
   - Voltage-gated Ca^{2+} channels are open.
   - Voltage-gated K^+ channels are closed.
3. **Repolarization phase**
   - Voltage-gated Ca^{2+} channels close.
   - Voltage-gated K^+ channels open.

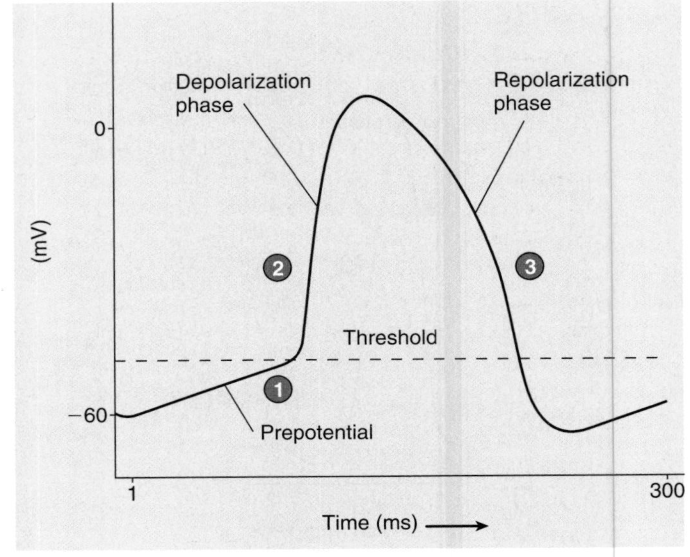

PROCESS FIGURE 20.15 SA Node Action Potential

The production of action potentials by the SA node is responsible for the autorhythmicity of the heart.

ions cause depolarization by moving into the cells through specialized nongated Na^+ channels. A decreasing permeability to K^+ also causes depolarization as fewer K^+ move out of the cells. The decreasing K^+ permeability occurs due to the voltage changes at the end of the previous action potential. As a result of the depolarization, voltage-gated Ca^{2+} channels open, and the movement of Ca^{2+} into the pacemaker cells causes further depolarization. When the prepotential reaches threshold, many voltage-gated Ca^{2+} channels open. Unlike other cardiac muscle cells, the movement of Ca^{2+} into the pacemaker cells is primarily responsible for the depolarization phase of the action potential. Repolarization occurs, as in other cardiac muscle cells, when the voltage-gated Ca^{2+} channels close and the voltage-gated K^+ channels open. After the RMP is reestablished, production of another prepotential starts the generation of the next action potential.

Drugs That Block Calcium Channels

Various chemical agents, such as manganese ions (Mn^{2+}) and verapamil (ver-ap'ă-mil), block voltage-gated Ca^{2+} channels. Voltage-gated Ca^{2+} channel-blocking agents prevent the movement of Ca^{2+} through voltage-gated Ca^{2+} channels into the cell; for that reason, they are called **calcium channel blockers.** Some calcium channel blockers are widely used clinically in the treatment of various cardiac disorders, including tachycardia and certain arrhythmias. Calcium channel blockers slow the development of the prepotential and thus reduce the heart rate. If action potentials arise prematurely within the SA node or other areas of the heart, calcium channel blockers reduce that tendency. Calcium channel blockers also reduce the amount of work performed by the heart because less calcium enters cardiac muscle cells to activate the contractile mechanism. On the other hand, epinephrine and norepinephrine increase the heart rate and its force of contraction by opening voltage-gated Ca^{2+} channels.

Although most cardiac muscle cells respond to action potentials produced by the SA node, some cardiac muscle cells in the conducting system can also generate spontaneous action potentials. Normally, the SA node controls the rhythm of the heart because its pacemaker cells generate action potentials at a faster rate than other potential pacemaker cells to produce a heart rate of 70–80 beats per minute (bpm). An **ectopic focus** (eek-top'ik fō'kŭs; pl. *foci,* fō'sī) is any part of the heart other than the SA node that generates a heartbeat. For example, if the SA node does not function properly, the part of the heart to produce action potentials at the next highest frequency is the AV node, which produces a heart rate of 40–60 bpm. Another cause of an ectopic focus is blockage of the conducting pathways between the SA node and other parts of the heart. For example, if action potentials do not pass through the AV node, an ectopic focus can develop in an AV bundle, resulting in a heart rate of 30 bpm.

Ectopic foci can also appear when the rate of action potential generation in cardiac muscle cells outside of the SA node becomes enhanced. For example, when cells are injured their plasma membranes become more permeable, resulting in depolarization. Inflammation or lack of adequate blood flow to cardiac muscle tissue can injure cardiac muscle cells. These injured cells can be the source of ectopic action potentials. Also, alterations in blood levels of K^+ and Ca^{2+} can change the cardiac muscle membrane potential, and certain drugs, such as those that mimic the effect of epinephrine on the heart, can alter cardiac muscle membrane permeability. Changes in cardiac muscle cells' membrane potentials or permeability can produce ectopic foci.

PREDICT 4

Predict the consequences for the heart's pumping effectiveness if numerous ectopic foci in the ventricles produce action potentials at the same time.

Refractory Periods of Cardiac Muscle

Cardiac muscle, like skeletal muscle, has **refractory** (rē-frak′tōr-ē) **periods** associated with its action potentials. During the **absolute refractory period,** the cardiac muscle cell is completely insensitive to further stimulation. During the **relative refractory period,** the cell is sensitive to stimulation, but a greater stimulation than normal is required to cause an action potential. Because the plateau phase of the action potential in cardiac muscle delays repolarization to the RMP, the refractory period is prolonged. The long refractory period ensures that contraction and most of relaxation are complete before another action potential can be initiated. This prevents tetanic contractions from occurring in cardiac muscle and is responsible for rhythmic contractions.

PREDICT 5

Predict the consequences if cardiac muscle could undergo tetanic contraction.

Electrocardiogram

The conduction of action potentials through the myocardium during the cardiac cycle produces electric currents that can be measured at the surface of the body. Electrodes placed on the surface of the body and attached to an appropriate recording device can detect small voltage changes resulting from action potentials in the cardiac muscle. The electrodes detect a summation of all the action potentials that are transmitted by the cardiac muscle cells through the heart at a given time. Electrodes do not detect individual action potentials. The summated record of the cardiac action potentials is an electrocardiogram (ECG or EKG).

The ECG is not a direct measurement of mechanical events in the heart, and neither the force of contraction nor blood pressure can be determined from it. Each deflection in the ECG record, however, indicates an electrical event within the heart that is correlated with a subsequent mechanical event. Consequently, it is an extremely valuable diagnostic tool in identifying a number of abnormal cardiac rhythms (arrythmias; table 20.1) and other abnormalities, particularly because it is painless, easy to record, and noninvasive (it does not require surgical procedures). Abnormal heart rates or rhythms, abnormal conduction pathways, hypertrophy or atrophy of portions of the heart, and the approximate location of damaged cardiac muscle can be determined from analysis of an ECG (see table 20.1).

The normal ECG consists of a P wave, a QRS complex, and a T wave (figure 20.16). The **P wave,** which is the result of action potentials that cause depolarization of the atrial myocardium, signals the onset of atrial contraction. The **QRS complex** is composed of three individual waves: the Q, R, and S waves. The QRS complex results from ventricular depolarization and signals the onset of ventricular contraction. The **T wave** represents repolarization of the ventricles and precedes ventricular relaxation. A wave representing repolarization of the atria cannot be seen because it occurs during the QRS complex.

The time between the beginning of the P wave and the beginning of the QRS complex is the PQ interval, commonly called the PR interval because the Q wave is often very small. During the PR interval, which lasts approximately 0.16 second, the atria contract and begin to relax. The ventricles begin to depolarize at the end of the PR interval. The QT interval extends from the beginning of the QRS complex to the end of the T wave, lasts approximately 0.36 second, and represents the approximate length of time required for the ventricles to contract and begin to relax.

Alterations in the Electrocardiogram

Elongation of the PR interval can result from (1) a delay in action potential conduction through the atrial muscle because of damage, such as that caused by **ischemia** (is-kē′mē-ă), which is the obstruction of the blood supply to the walls of the heart; (2) a delay of action potential conduction through atrial muscle because of a dilated atrium; or (3) a delay of action potential conduction through the AV node and bundle because of ischemia, compression, or necrosis of the AV node or bundle. These conditions result in slow conduction of action potentials through the bundle branches. An unusually long QT interval reflects the abnormal conduction of action potentials through the ventricles, which can result from myocardial infarctions or from an abnormally enlarged left or right ventricle.

Examples of alteration in the form of the electrocardiogram due to cardiac abnormalities are illustrated in figure 20.17. Examples include complete heart block, premature ventricular contraction, bundle branch block, atrial fibrillation, and ventricular fibrillation.

26 ***For cardiac muscle action potentials, describe ion movement during the depolarization, early repolarization, plateau, and final repolarization phases. What ions are associated with fast channels and slow channels?***

27 ***Why is cardiac muscle referred to as autorhythmic? What are ectopic foci?***

28 ***How does the depolarization of pacemaker cells differ from the depolarization of other cardiac muscle cells? What is the prepotential?***

29 ***Why does cardiac muscle have a prolonged refractory period? What is the advantage of a prolonged refractory period?***

30 ***What does an ECG measure? Name the waves produced by an ECG, and state what events occur during each wave.***

CARDIAC CYCLE

The heart is actually two separate pumps that work together, one in the right half and the other in the left half of the heart. Each pump consists of a primer pump—the atrium—and a power pump—the ventricle. Both atrial primer pumps complete the filling of the ventricles with blood, and both ventricular power pumps produce the major force that causes blood to flow through the pulmonary and systemic arteries. The term **cardiac cycle** refers to the repetitive pumping process that begins with the onset of cardiac muscle

TABLE 20.1 Major Cardiac Arrhythmias

Conditions	Symptoms	Possible Causes
Abnormal Heart Rhythms		
Tachycardia	Heart rate in excess of 100 bpm	Elevated body temperature; excessive sympathetic stimulation; toxic conditions
Paroxysmal atrial tachycardia	Sudden increase in heart rate to 95–150 bpm for a few seconds or even for several hours; P wave precedes every QRS complex; P wave inverted and superimposed on T wave	Excessive sympathetic stimulation; abnormally elevated permeability of slow channels
Ventricular tachycardia	Frequently causes fibrillation	Often associated with damage to AV node or ventricular muscle
Abnormal Rhythms Resulting from Ectopic Action Potentials		
Atrial flutter	300 P waves/min; 125 QRS complexes/min, resulting in two or three P waves (atrial contraction) for every QRS complex (ventricular contraction)	Ectopic action potentials in the atria
Atrial fibrillation	No P waves; normal QRS complexes; irregular timing; ventricles constantly stimulated by atria; reduced pumping effectiveness and filling time	Ectopic action potentials in the atria
Ventricular fibrillation	No QRS complexes; no rhythmic contraction of the myocardium; many patches of asynchronously contracting ventricular muscle	Ectopic action potentials in the ventricles
Bradycardia	Heart rate less than 60 bpm	Elevated stroke volume in athletes; excessive vagal stimulation; carotid sinus syndrome
Sinus Arrhythmia	Heart rate varies 5% during respiratory cycle and up to 30% during deep respiration.	Cause not always known; occasionally caused by ischemia or inflammation or associated with cardiac failure
SA Node Block	Cessation of P wave; new low heart rate due to AV node acting as pacemaker; normal QRS complex and T wave	Ischemia; tissue damage due to infarction; causes unknown
AV Node Block		
First-degree	PR interval greater than 0.2 second	Inflammation of AV bundle
Second-degree	PR interval 0.25–0.45 second; some P waves trigger QRS complexes and others do not; 2:1, 3:1, and 3:2 P wave/QRS complex ratios may occur	Excessive vagal stimulation
Third-degree (complete heart block)	P wave dissociated from QRS complex; atrial rhythm approximately 100 bpm; ventricular rhythm less than 40 bpm	Ischemia of AV nodal fibers or compression of AV bundle
Premature Atrial Contractions	Occasional shortened intervals between one contraction and the succeeding contraction; frequently occurs in healthy people P wave superimposed on QRS complex	Excessive smoking; lack of sleep; too much caffeine; alcoholism
Premature Ventricular Contractions (PVCs)	Prolonged QRS complex; exaggerated voltage because only one ventricle may depolarize; inverted T wave; increased probability of fibrillation	Ectopic foci in ventricles; lack of sleep; too much caffeine, irritability; occasionally occurs with coronary thrombosis

Abbreviations: SA = sinoatrial; AV = atrioventricular.

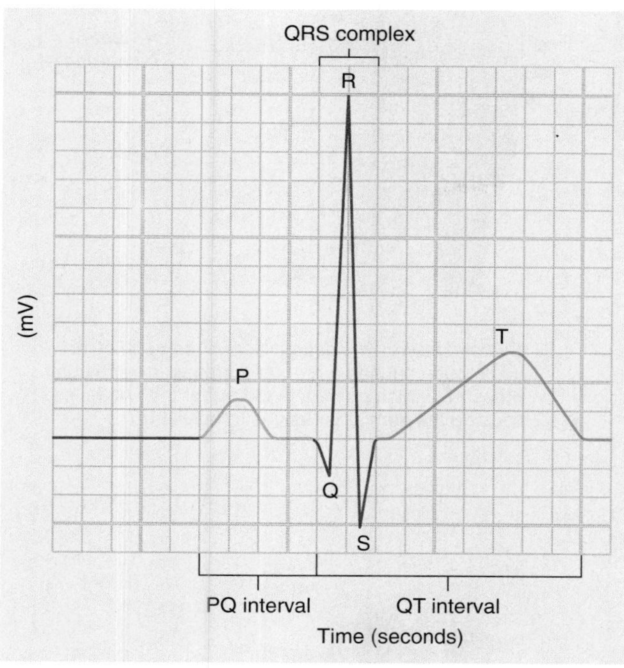

FIGURE 20.16 Electrocardiogram

The major waves and intervals of an electrocardiogram. Each thin horizontal line on the ECG recording represents 1 mV, and each thin vertical line represents 0.04 second.

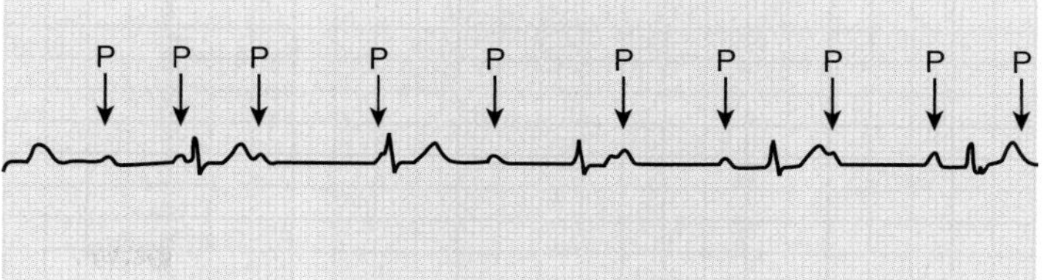

Complete heart block (P waves and QRS complexes are not coordinated)

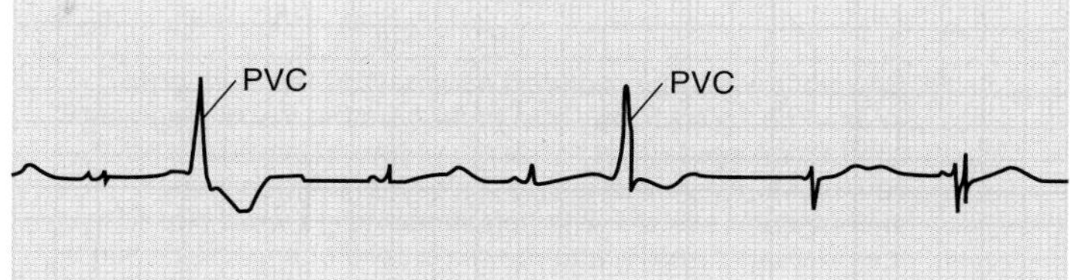

Premature ventricular contraction (PVC) (no P waves precede PVCs)

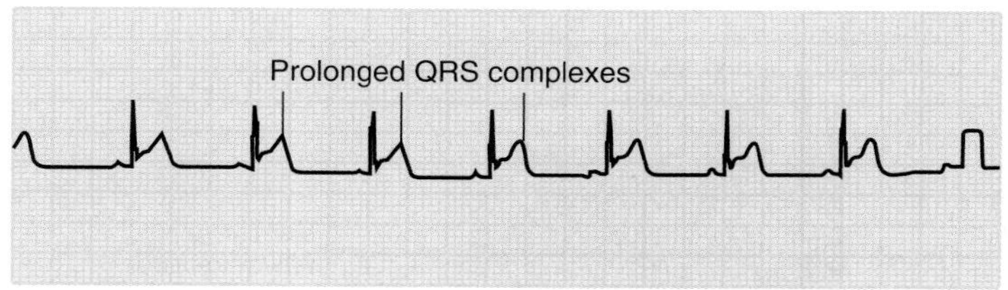

Bundle branch block

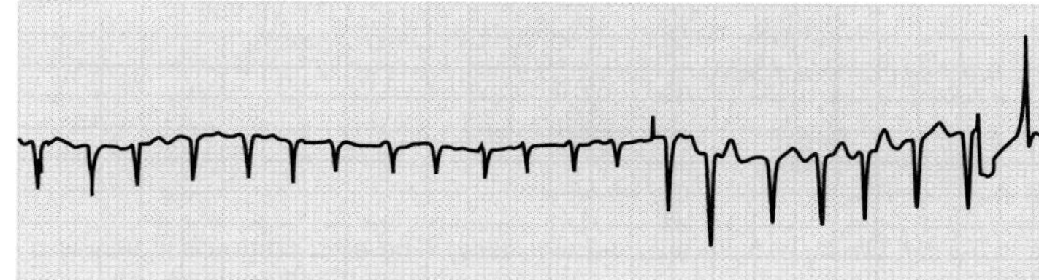

Atrial fibrillation (no clear P waves and rapid QRS complexes)

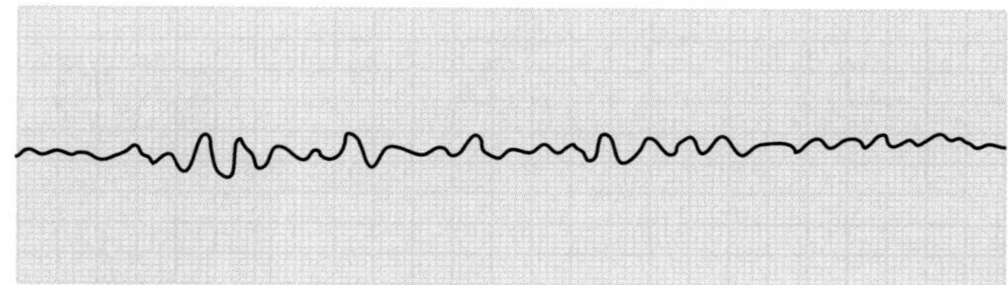

Ventricular fibrillation (no P, QRS, or T waves)

FIGURE 20.17 Alterations in an Electrocardiogram

contraction and ends with the beginning of the next contraction (figures 20.18 and 20.19; table 20.2). Pressure changes produced within the heart chambers as a result of cardiac muscle contraction are responsible for blood movement because blood moves from areas of higher pressure to areas of lower pressure.

The duration of the cardiac cycle varies considerably among humans and during an individual's lifetime. It can be as short as 0.25–0.3 second in a newborn or as long as 1 or more seconds in a well-trained athlete. The normal cardiac cycle of 0.7–0.8 second depends on the capability of cardiac muscle to contract and on the functional integrity of the conducting system.

The term **systole** (sis′tō-lē) means to contract, and **diastole** (dī-as′tō-lē) means to dilate. **Atrial systole** is contraction of the atrial myocardium, and **atrial diastole** is relaxation of the atrial myocardium. Similarly, **ventricular systole** is contraction of the ventricular myocardium, and **ventricular diastole** is relaxation of the ventricular myocardium. When the terms *systole* and *diastole* are used without reference to specific chambers, however, they mean ventricular systole or diastole.

Just before systole begins, the atria and ventricles are relaxed, the ventricles are filled with blood, the semilunar valves are closed, and the AV valves are open. As systole begins, contraction of the ventricles increases ventricular pressures, causing blood to flow toward the atria and close the AV valves. As contraction proceeds, ventricular pressures continue to rise, but no blood flows from the ventricles because all the valves are closed. This brief interval is called the **period of isovolumic** (ī′sō-vol-ū′mik) **contraction** because the volume of blood in the ventricles does not change, even though the ventricles are contracting (see figure 20.18, step 1). As the ventricles continue to contract, ventricular pressures become greater than the pressures in the pulmonary trunk and aorta. As a result, during the **period of ejection,** the semilunar valves are pushed open and blood flows from the ventricles into those arteries (see figure 20.18, step 2).

As diastole begins, the ventricles relax and ventricular pressures decrease below the pressures in the pulmonary trunk and

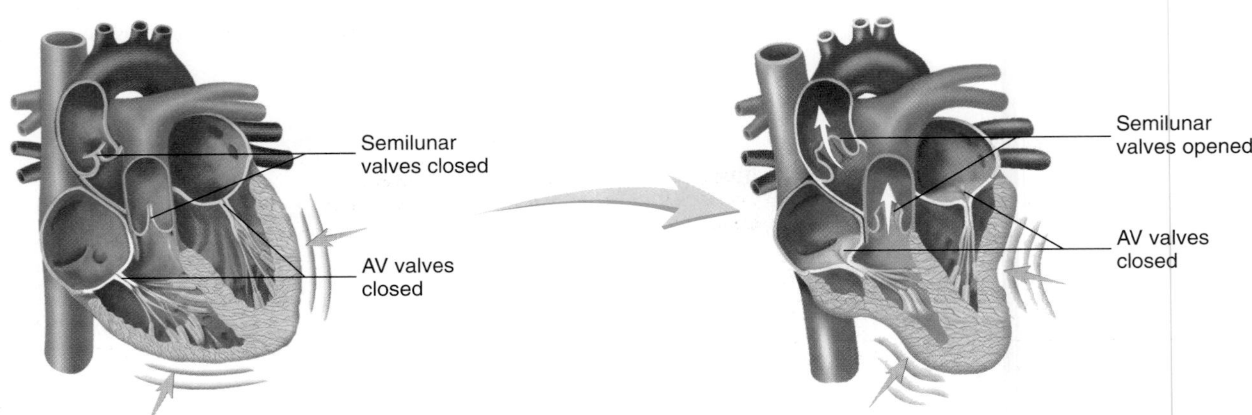

1. *Systole: period of isovolumic contraction.* The atria are relaxed and blood flows into them from the veins. Ventricular contraction causes ventricular pressure to increase and causes the AV valves to close, which is the beginning of ventricular systole. The semilunar valves were closed in the previous diastole and remain closed during this period.

2. *Systole: period of ejection.* Continued ventricular contraction causes a greater increase in ventricular pressure, which pushes blood out of the ventricles, causing the semilunar valves to open.

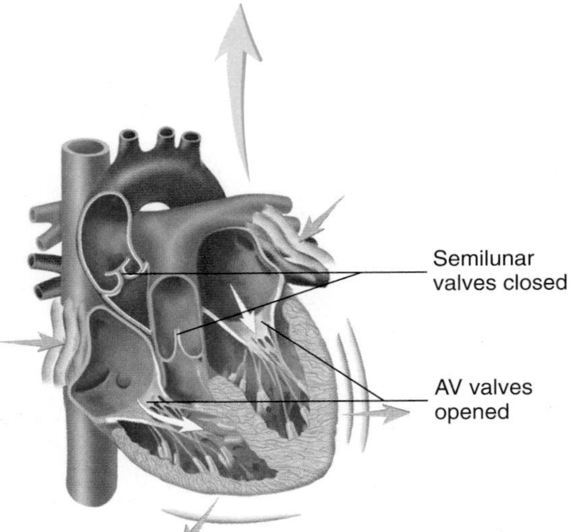

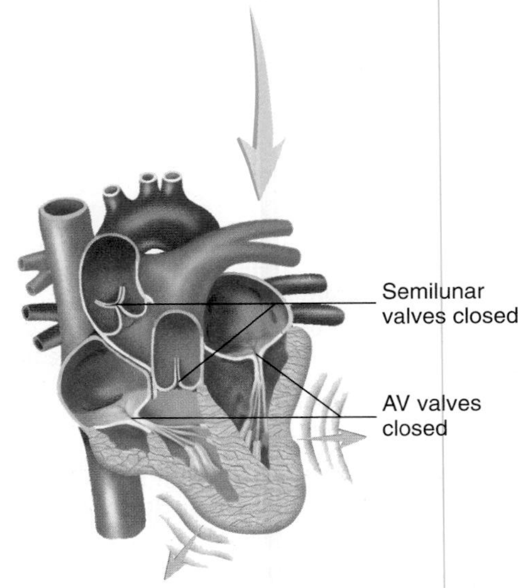

5. *Diastole: active ventricular filling.* The atria contract, increasing atrial pressure and completing ventricular filling while the ventricles are relaxed.

3. *Diastole: period of isovolumic relaxation.* As the ventricles begin to relax at the beginning of ventricular diastole, blood flowing back from the aorta and pulmonary trunk toward the relaxing ventricles causes the semilunar valves to close. Note that the AV valves are closed also.

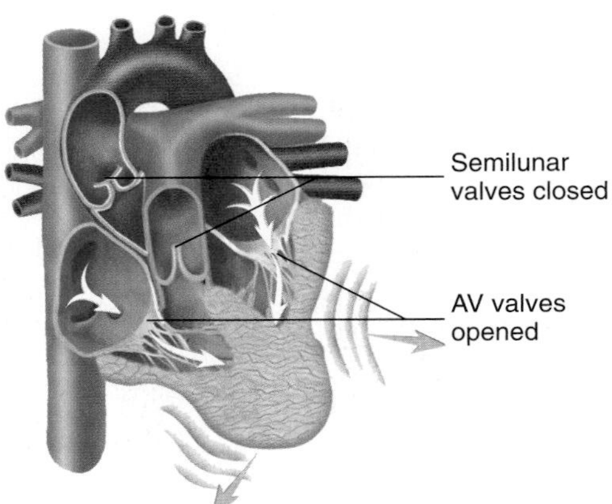

4. *Diastole: passive ventricular filling.* As ventricular relaxation continues, the AV valves open and blood flows from the atria into the relaxing ventricles, accounting for most of the ventricular filling.

PROCESS FIGURE 20.18 Cardiac Cycle

The cardiac cycle is a repeating series of contraction and relaxation that moves blood through the heart (*AV* = atrioventricular).

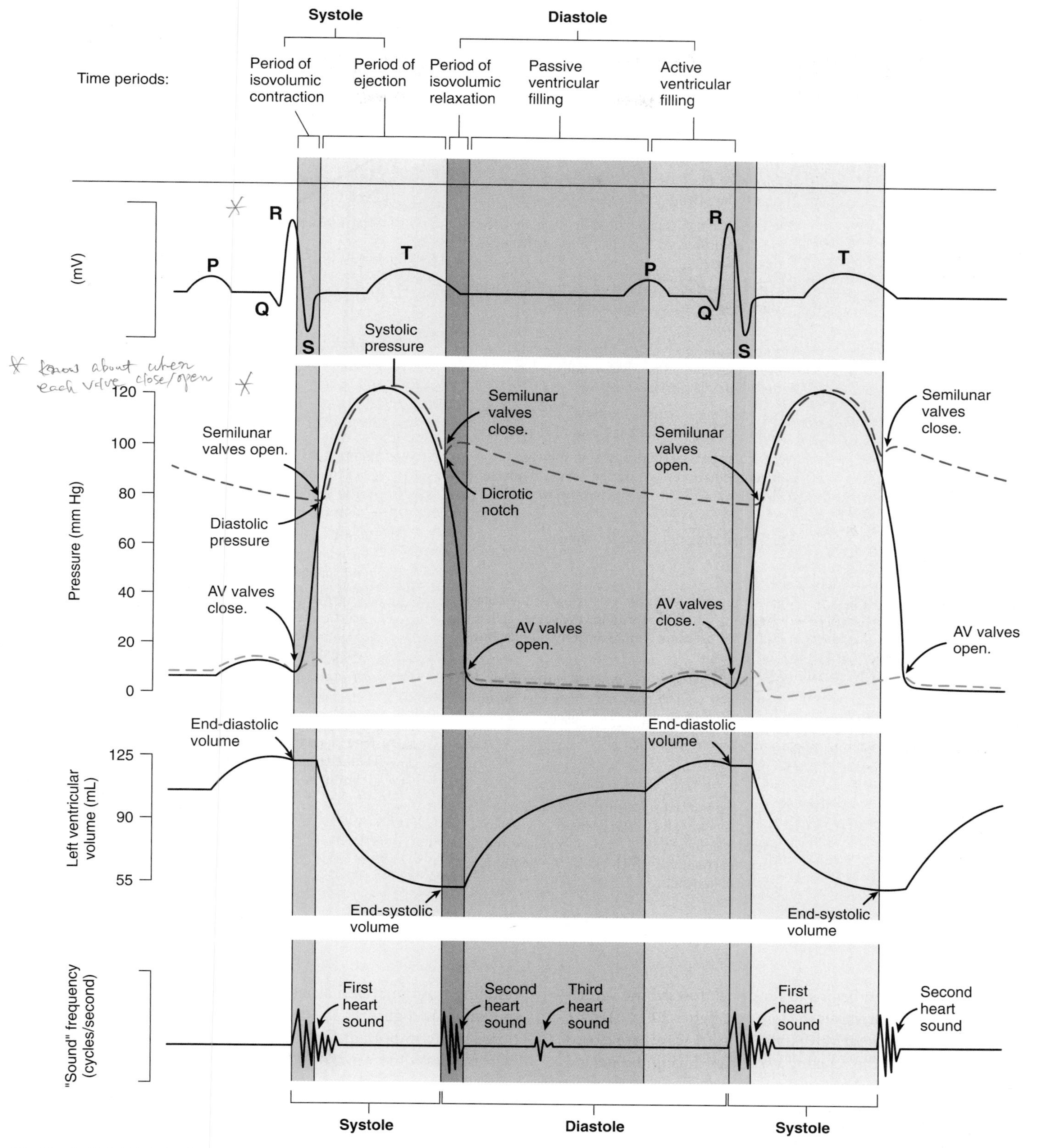

FIGURE 20.19 Events Occurring During the Cardiac Cycle

The cardiac cycle is divided into five periods (*top*). *From top to bottom,* the electrocardiogram; pressure changes for the left atrium (*blue line*), left ventricle (*black line*), and aorta (*red line*); left ventricular volume curve; and heart sounds.

TABLE 20.2 Summary of the Events of the Cardiac Cycle

	Ventricular Systole	
	Period of Isovolumic Contraction	**Period of Ejection**
Time Period	The ventricles begin to contract, but ventricular volume does not change.	The ventricles continue to contract and blood is pumped out of the ventricles.
Condition of Valves	The semilunar valves are closed; the AV valves are closed (see figure 20.18,*1*).	The semilunar valves are opened; the AV valves are closed (see figure 20.18,*2*).
ECG	The QRS complex is completed and the ventricles are depolarized. As a result, the ventricles begin to contract. Atrial repolarization is masked by the QRS complex. The atria are relaxed (atrial diastole).	The **T wave** results from ventricular repolarization.
Atrial Pressure Graph	Atrial pressure decreases in the relaxed atria. When atrial pressure is less than venous pressure, blood flows into the atria. Atrial pressure increases briefly as the contracting ventricles push blood back toward the atria.	Atrial pressure increases gradually as blood flows from the veins into the relaxed atria.
Ventricular Pressure Graph	Ventricular contraction causes an increase inventricular pressure, which causes blood to flow toward the atria, closing the AV valves. Ventricular pressure increases rapidly.	Ventricular pressure becomes greater than pressure in the aorta as the ventricles continue to contract. The semilunar valves are pushed open as blood flows out of the ventricles. Ventricular pressure peaks as the ventricles contract maximally; then pressure decreases as blood flow out of the ventricles decreases.
Aortic Pressure Graph	Just before the semilunar valves open, pressure in the aorta decreases to its lowest value, called the **diastolic pressure** (approximately 80 mm Hg).	As ventricular contraction forces blood into the aorta, pressure in the aorta increases to its highest value, called the **systolic pressure** (approximately 120 mm Hg).
Volume Graph	During the **period of isovolumic contraction,** ventricular volume does not change because the semilunar and AV valves are closed.	After the semilunar valves open, blood volume decreases as blood flows out of the ventricles during the **period of ejection.** The amount of blood left in a ventricle at the end of the period of ejection is called the **end-systolic volume.**
Sound Graph	Blood flowing from the ventricles toward the atria closes the AV valves. Vibrations of the valves and the turbulent flow of blood produce the **first heart sound,** which marks the beginning of ventricular systole.	

Abbreviation: AV = atrioventricular.

aorta. Consequently, blood begins to flow back toward the ventricles, causing the semilunar valves to close (see figure 20.18, step 3). With closure of the semilunar valves, all the heart valves are closed and no blood flows into the relaxing ventricles during the **period of isovolumic relaxation.**

Throughout ventricular systole and the period of isovolumic relaxation, the atria relax and blood flows into them from the veins. As the ventricles continue to relax, ventricular pressures become lower than atrial pressures, the AV valves open, and blood flows from the atria into the relaxed ventricles (see figure 20.18, step 4). At rest, most ventricular filling is a passive process resulting from the greater pressure of blood in the veins and atria than in the completely relaxed ventricles. Completion of ventricular filling is an active process resulting from increased atrial pressure produced by contraction of the atria (see figure 20.18, step 5). During exercise, atrial contraction is more important for ventricular filling because, as heart rate increases, less time is available for passive ventricular filling.

Events Occurring During Ventricular Systole

An ECG indicates the electrical events that cause contraction and relaxation of the atria and ventricles. (See figure 20.19, which displays the main events of the cardiac cycle in graphic form and should be examined from top to bottom for each period of the cardiac cycle.) The pressure graph shows the pressure changes within the left

Ventricular Diastole		
Period of Isovolumic Relaxation The ventricles relax, but ventricular volume does not change.	**Passive Ventricular Filling** Blood flows into the ventricles because blood pressure is higher in the veins and atria than in the relaxed ventricles.	**Active Ventricular Filling** Contraction of the atria pumps blood into the ventricles.
The semilunar valves are closed; the AV valves are closed (see figure 20.18,*3*).	The semilunar valves are closed; the AV valves are opened (see figure 20.18,*4*).	The semilunar valves are closed; the AV valves are opened (see figure 20.18,*5*).
The T wave is completed and the ventricles are repolarized. The ventricles relax.	The **P wave** is produced when the SA node generates action potentials and a wave of depolarization begins to propagate across the atria.	The P wave is completed and the atria are stimulated to contract. Action potentials are delayed in the AV node for 0.11 second, allowing time for the atria to contract. The **QRS complex** begins as action potentials are propagated from the AV node to the ventricles.
Atrial pressure continues to increase gradually as blood flows from the veins into the relaxed atria.	After the AV valves open, atrial pressure decreases as blood flows out of the atria into the relaxed ventricles.	Atrial contraction (systole) causes an increase in atrial pressure, and blood is forced to flow from the atria into the ventricles.
Elastic recoil of the aorta pushes blood back toward the heart, causing the semilunar valves to close. After closure of the semilunar valves, the pressure in the relaxing ventricles rapidly decreases.	No significant change occurs in ventricular pressure during this time period.	Atrial contraction (systole) and the movement of blood into the ventricles cause a slight increase in ventricular pressure.
After the semilunar valves close, elastic recoil of the aorta causes a slight increase in aortic pressure, producing the **dicrotic notch,** or **incisura.**	Aortic pressure gradually decreases as blood runs out of the aorta into other systemic blood vessels.	Aortic pressure gradually decreases as blood runs out of the aorta into other systemic blood vessels.
During the **period of isovolumic relaxation,** ventricular volume does not change because the semilunar and AV valves are closed.	After the AV valves open, blood flows from the atria and veins into the ventricles because of pressure differences. Most ventricular filling occurs during the first one-third of diastole. Little ventricular filling occurs during the middle one-third of diastole.	Atrial contraction (systole) completes ventricular filling during the last one-third of diastole.
Blood flowing from the ventricles toward the aorta and pulmonary trunk closes the semilunar valves. Vibrations of the valves and the turbulent flow of blood produce the **second heart sound,** which marks the beginning of ventricular diastole.	Sometimes the turbulent flow of blood into the ventricles produces a **third heart sound.**	The amount of blood in a ventricle at the end of ventricular diastole is called the **end-diastolic volume.**

atrium, left ventricle, and aorta resulting from atrial and ventricular contraction and relaxation. Although pressure changes in the right side of the heart are not shown, they are similar to those in the left side, only lower. The volume graph presents the changes in left ventricular volume as blood flows into and out of the left ventricle as a result of the pressure changes. The sound graph records the closing of valves caused by blood flow. See also figure 20.18 for illustrations of the valves and blood flow and table 20.2 for a summary of the events occurring during each period.

Period of Isovolumic Contraction

Completion of the QRS complex initiates contraction of the ventricles. Ventricular pressure rapidly increases, resulting in closure of the AV valves. During the previous ventricular diastole, the ventricles were filled with 120–130 mL of blood, which is called the **end-diastolic volume.** Ventricular volume does not change during the period of isovolumic contraction because all the heart valves are closed.

PREDICT 6

Is the cardiac muscle contracting isotonically or isometrically during the period of isovolumic contraction?

Period of Ejection

As soon as ventricular pressures exceed the pressures in the aorta and pulmonary trunk, the semilunar valves open. The aortic semilunar valve opens at approximately 80 mm Hg ventricular pressure, whereas the pulmonary semilunar valve opens at approximately 8 mm Hg. Although the pressures are different, both valves open at nearly the same time.

As blood flows from the ventricles during the period of ejection, the left ventricular pressure continues to climb to approximately 120 mm Hg, and the right ventricular pressure increases to approximately 25 mm Hg. The larger left ventricular pressure causes blood to flow throughout the body (systemic circulation), whereas the lower right ventricle pressure causes blood to flow through the lungs (pulmonary circuit). Even though the pressure generated by the left ventricle is much higher than that of the right ventricle, the amount of blood pumped by each is almost the same.

PREDICT 7

Which ventricle has the thickest wall? Why is it important for each ventricle to pump approximately the same volume of blood?

During the first part of ejection, blood flows rapidly out of the ventricles. Toward the end of ejection, very little blood flow occurs, which causes the ventricular pressure to decrease despite continued ventricular contraction. By the end of ejection, the volume has decreased to 50–60 mL, which is called the **end-systolic volume.**

Events Occurring During Ventricular Diastole

Period of Isovolumic Relaxation

Completion of the T wave results in ventricular repolarization and relaxation. The already decreasing ventricular pressure falls very rapidly as the ventricles suddenly relax. When the ventricular pressures fall below the pressures in the aorta and pulmonary trunk, the recoil of the elastic arterial walls, which were stretched during the period of ejection, forces the blood to flow back toward the ventricles, thereby closing the semilunar valves. Ventricular volume does not change during the period of isovolumic relaxation because all the heart valves are closed at this time.

Passive Ventricular Filling

During ventricular systole and the period of isovolumic relaxation, the relaxed atria fill with blood. As ventricular pressure drops below atrial pressure, the atrioventricular valves open and allow blood to flow from the atria into the ventricles. Blood flows from the area of higher pressure in the veins and atria toward the area of lower pressure in the relaxed ventricles. Most ventricular filling occurs during the first one-third of ventricular diastole. At the end of passive ventricular filling, the ventricles are approximately 70% filled.

PREDICT 8

Fibrillation is abnormal, rapid contractions of different parts of the heart that prevent the heart muscle from contracting as a single unit. Explain why atrial fibrillation does not immediately cause death but ventricular fibrillation does.

Active Ventricular Filling

Depolarization of the SA node generates action potentials that spread over the atria, producing the P wave and stimulating both atria to contract (atrial systole). The atria contract during the last one-third of ventricular diastole and complete ventricular filling.

Under most conditions, the atria function primarily as reservoirs, and the ventricles can pump sufficient blood to maintain homeostasis even if the atria do not contract at all. During exercise, however, the heart pumps 300%–400% more blood than during resting condition. It is under these conditions that the pumping action of the atria becomes important in maintaining the pumping efficiency of the heart.

Heart Sounds

Distinct sounds are heard when a stethoscope is used to listen to the heart (see figure 20.19; figure 20.20). The **first heart sound** is a low-pitched sound, often described as a "lubb" sound. It is caused by vibration of the atrioventricular valves and surrounding fluid as the valves close at the beginning of ventricular systole. The **second heart sound** is a higher-pitched sound often described as a "dupp" sound. It results from closure of the aortic and pulmonary semilunar valves, at the beginning of ventricular diastole. Systole is, therefore, approximately the time between the first and second heart sounds. Diastole, which lasts somewhat longer, is approximately the time between the second heart sound and the next first heart sound.

Occasionally, a faint **third heart sound** can be heard in some normal people, particularly in those who are thin and young. It is

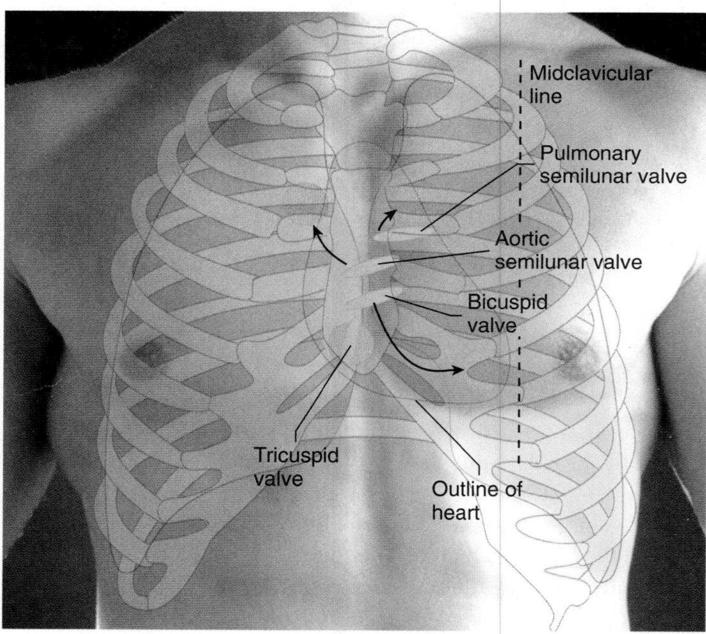

FIGURE 20.20 Location of the Heart Valves in the Thorax
Surface markings of the heart in the male. The positions of the four heart valves are indicated by *blue ellipses,* and the sites where the sounds of the valves are best heard with the stethoscope are indicated by *pink circles.*

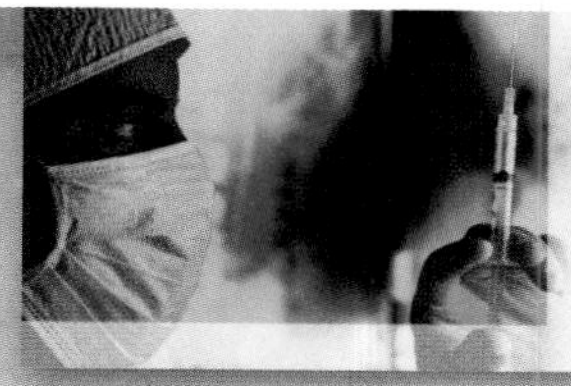

CLINICAL FOCUS

Abnormal Heart Sounds

Heart sounds provide important information to clinicians about the normal function of the heart and help them diagnose cardiac abnormalities. Abnormal heart sounds are called **murmurs** (mer′merz), and certain murmurs are important indicators of specific cardiac abnormalities. For example, an **incompetent valve** (also called a valvular insufficiency) leaks significantly. After an incompetent valve closes, blood flows through it but in a reverse direction. The flow through a valve in reverse direction is called regurgitation. Regurgitation results in turbulence, which causes a gurgling or swishing sound immediately after the valve closes. An incompetent tricuspid valve or bicuspid valve makes a swish sound immediately after the first heart sound, and the first heart sound may be muffled. An incompetent aortic or pulmonary semilunar valve results in a swish sound immediately after the second heart sound.

Stenosed (sten′ōzd) valves have an abnormally narrow opening and produce abnormal heart sounds. Blood flows through stenosed valves in a very turbulent fashion and produces a rushing sound before the valve closes. A stenosed atrioventricular valve, therefore, results in a rushing sound immediately before the first heart sound, and a stenosed semilunar valve results in a rushing sound immediately before the second heart sound.

Inflammation of the heart valves, resulting from conditions such as rheumatic fever, can cause valves to become either incompetent or stenosed. In addition, myocardial infarctions that make papillary muscles nonfunctional can cause bicuspid or tricuspid valves to be incompetent. Heart murmurs also result from congenital abnormalities in the hearts of infants. For example, septal defects in the heart and patent ductus arteriosus result in distinct heart murmurs (see chapter 29).

Either incompetent or stenosed valves increase the amount of work the cardiac muscle must perform. Consequently, these conditions can lead to heart failure.

caused by blood flowing in a turbulent fashion into the ventricles, and it can be detected near the end of the first one-third of diastole.

Aortic Pressure Curve

The elastic walls of the aorta are stretched as blood is ejected into the aorta from the left ventricle. Aortic pressure remains slightly below ventricular pressure during this period of ejection. As ventricular pressure drops below that in the aorta, blood flows back toward the ventricle because of the elastic recoil of the aorta. Consequently, the aortic semilunar valve closes, and pressure within the aorta increases slightly, producing a **dicrotic** (dī-krot′ik) **notch** in the aortic pressure curve (see figure 20.19). The term *dicrotic* means double-beating; when increased pressure caused by recoil is large, a double pulse can be felt. The dicrotic notch is also called an **incisura** (in′sī-soo′ră; a cutting into). Aortic pressure then gradually falls throughout the rest of ventricular diastole as blood flows through the peripheral vessels. When aortic pressure has fallen to approximately 80 mm Hg, the ventricles again contract, forcing blood once more into the aorta.

Blood pressure measurements performed for clinical purposes reflect the pressure changes that occur in the aorta rather than in the left ventricle (see chapter 21). The blood pressure in the aorta fluctuates between systolic pressure, which is about 120 mm Hg, and diastolic pressure, which is about 80 mm Hg for the average young adult at rest.

31 ***Define systole and diastole.***

32 ***List the five periods of the cardiac cycle (see figure 20.19 and table 20.2), and state whether the AV and semilunar valves are open or closed during each period.***

33 ***Define isovolumic. When does most ventricular filling occur?***

34 ***Define end-diastolic volume and end-systolic volume.***

35 ***What produces the first heart sound, the second heart sound, and the third heart sound?***

36 ***Explain the production in the aorta of systolic pressure, diastolic pressure, and the dicrotic notch, or incisura.***

MEAN ARTERIAL BLOOD PRESSURE

Blood pressure is necessary for blood movement and, therefore, is critical to the maintenance of homeostasis. Blood flows from areas of higher to areas of lower pressure. For example, during one cardiac cycle, blood flows from the higher pressure in the aorta, resulting from contraction of the left ventricle, toward the lower pressure in the relaxed right atrium.

Mean arterial pressure (MAP) is slightly less than the average of the systolic and diastolic pressure in the aorta. It is proportional to **cardiac output (CO)** times **peripheral resistance (PR).** Cardiac output, or **minute volume,** is the amount of blood pumped by the heart per minute, and peripheral resistance is the total resistance against which blood must be pumped.

$$\text{MAP} = \text{CO} \times \text{PR}$$

Changes in cardiac output and peripheral resistance (figure 20.21) can alter mean arterial pressure. Cardiac output is discussed in this chapter, and peripheral resistance is explained in chapter 21.

Cardiac output is equal to heart rate times stroke volume. **Heart rate (HR)** is the number of times the heart beats (contracts)

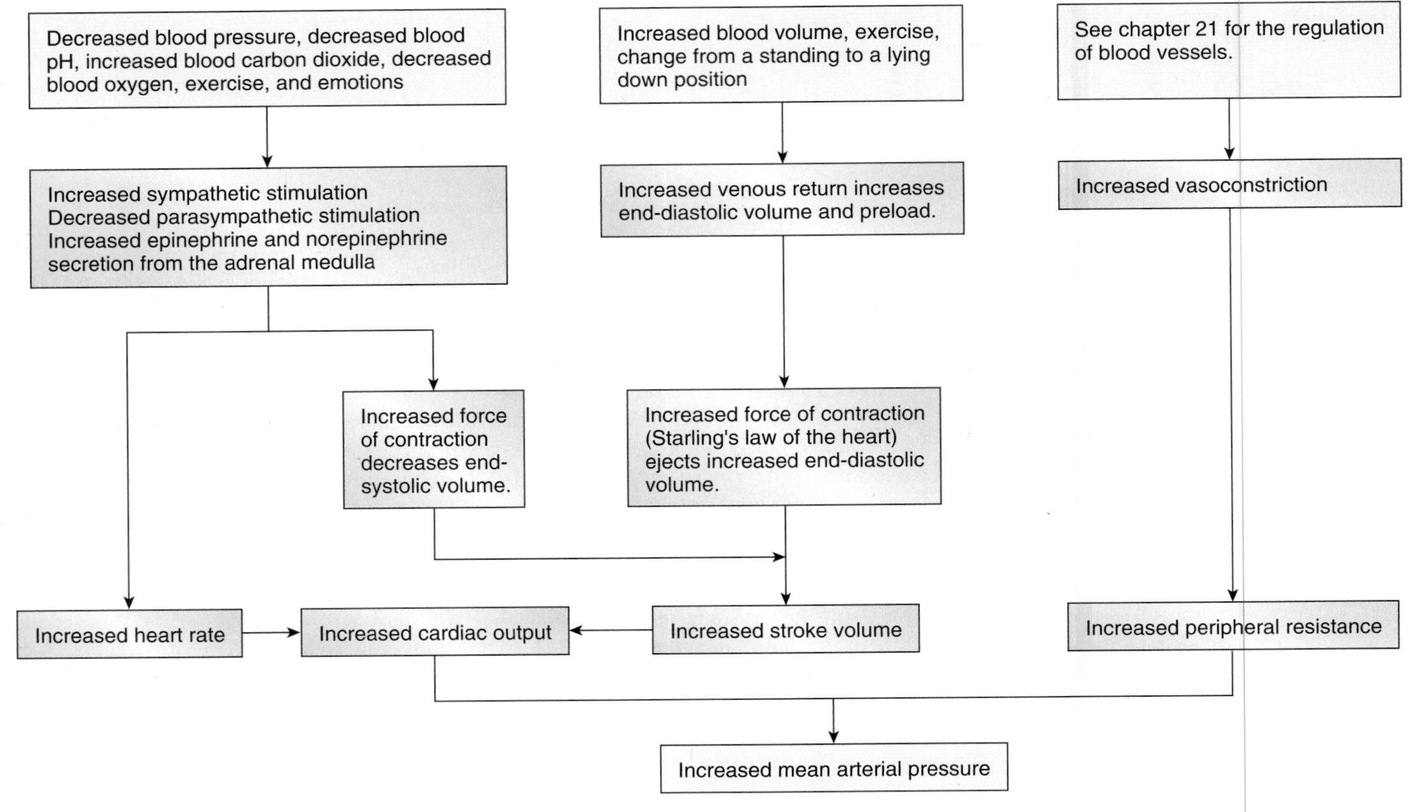

FIGURE 20.21 Factors Affecting Mean Arterial Pressure
Mean arterial pressure is regulated by controlling cardiac output and peripheral resistance.

per minute. **Stroke volume (SV),** which is the volume of blood pumped during each heartbeat (cardiac cycle), is equal to end-diastolic volume minus end-systolic volume. During diastole, blood flows from the atria into the ventricles, and end-diastolic volume normally increases to approximately 125 mL. After the ventricles partially empty during systole, end-systolic volume decreases to approximately 55 mL. The stroke volume is therefore equal to 70 mL (125 − 55).

To better understand stroke volume, imagine that you are rinsing out a sponge under a running water faucet. As you relax your hand, the sponge fills with water; as your fingers contract, water is squeezed out of the sponge; and, after you have squeezed it, some water is left in the sponge. In this analogy, the amount of water you squeeze out of the sponge (stroke volume) is the difference between the amount of water in the sponge when your hand is relaxed (end-diastolic volume) and the amount that is left in the sponge after you squeeze it (end-systolic volume).

Stroke volume can be increased by increasing end-diastolic volume or by decreasing end-systolic volume (see figure 20.21). During exercise, end-diastolic volume increases because of an increase in **venous return,** which is the amount of blood returning to the heart from the peripheral circulation. End-systolic volume decreases because the heart contracts more forcefully. For example, stroke volume could increase from a resting value of 70 mL to an exercising value of 115 mL by increasing end-diastolic volume to 145 mL and decreasing end-systolic volume to 30 mL.

Under resting conditions, the heart rate is approximately 72 bpm, and the stroke volume is approximately 70 mL/beat, although these values can vary considerably from person to person. The cardiac output is therefore

$$\begin{aligned} CO &= HR \times SV \\ &= 72 \text{ bpm} \times 70 \text{ mL/beat} \\ &= 5040 \text{ mL/min (approximately 5 L/min)} \end{aligned}$$

During exercise, heart rate can increase to 190 bpm, and the stroke volume can increase to 115 mL. Consequently, cardiac output is

$$\begin{aligned} CO &= 190 \text{ bpm} \times 115 \text{ mL/beat} \\ &= 21{,}850 \text{ mL/min (approximately 22 L/min)} \end{aligned}$$

The difference between cardiac output when a person is at rest and maximum cardiac output is called **cardiac reserve.** The greater a person's cardiac reserve, the greater his or her capacity for doing exercise. Lack of exercise and cardiovascular diseases can reduce cardiac reserve and affect a person's quality of life. Exercise training can greatly increase cardiac reserve by increasing cardiac output. In well-trained athletes, stroke volume during exercise can increase to over 200 mL/beat, resulting in cardiac outputs of 40 L/min or more.

PREDICT 9

Predict the effect of an incompetent bicuspid valve on the stroke volume, the volume of the left atrium at the end of ventricular systole, and the volume of the left ventricle at the end of ventricular diastole.

37 ***Define mean arterial pressure, cardiac output, and peripheral resistance. Explain the role of mean arterial pressure in causing blood flow.***

38 ***Define stroke volume, and state two ways to increase stroke volume.***

39 ***What is cardiac reserve? How can exercise training influence cardiac reserve?***

REGULATION OF THE HEART

To maintain homeostasis, the amount of blood pumped by the heart must vary dramatically. For example, during exercise cardiac output can increase several times over resting values. Intrinsic and extrinsic regulatory mechanisms control cardiac output. **Intrinsic regulation** results from the normal functional characteristics of the heart and does not depend on either neural or hormonal regulation. It functions when the heart is in place in the body or is removed and maintained outside the body under proper conditions. On the other hand, **extrinsic regulation** involves neural and hormonal control. Neural regulation of the heart results from sympathetic and parasympathetic reflexes, and the major hormonal regulation comes from epinephrine and norepinephrine secreted from the adrenal medulla.

Intrinsic Regulation

As venous return increases, end-diastolic volume increases (see figure 20.21). A greater end-diastolic volume increases the stretch of the ventricular walls. The extent to which the ventricular walls are stretched is sometimes called the **preload.** An increased preload causes an increase in cardiac output, and a decreased preload causes a decrease in cardiac output.

Cardiac muscle exhibits a length-versus-tension relationship similar to that of skeletal muscle. Skeletal muscle, however, is normally stretched to nearly its optimal length before contraction, whereas cardiac muscle fibers are not stretched to the point at which they contract with a maximal force (see chapter 9). An increased preload, therefore, causes the cardiac muscle fibers to contract with a greater force and produce a greater stroke volume. This relationship between preload and stroke volume is commonly referred to as **Starling's law of the heart,** which describes the relationship between changes in the pumping effectiveness of the heart and changes in preload (see figure 20.21). Venous return can decrease to a value as low as 2 L/min or increase to as much as 24 L/min, which has a major effect on the preload.

Afterload is the pressure the contracting ventricles must produce to overcome the pressure in the aorta and move blood into the aorta. Although the pumping effectiveness of the heart is greatly influenced by relatively small changes in the preload, it is very insensitive to large changes in afterload. Aortic blood pressure must increase to more than 170 mm Hg before it hampers the ventricles' ability to pump blood.

During physical exercise, blood vessels in exercising skeletal muscles dilate and allow an increased flow of blood through the vessels. The increased blood flow increases oxygen and nutrient delivery to the exercising muscles. In addition, skeletal muscle contractions repeatedly compress veins and cause an increased rate of blood flow from the skeletal muscles toward the heart. As blood rapidly flows through skeletal muscles and back to the heart, venous return to the heart increases, resulting in an increased preload. The increased preload causes an increased force of cardiac muscle contraction, which increases stroke volume. The increase in stroke volume results in increased cardiac output, and the volume of blood flowing to the exercising muscles increases. When a person rests, venous return to the heart decreases because arteries in the skeletal muscles constrict and because muscular contractions no longer repeatedly compress the veins. As a result, blood flow through skeletal muscles decreases, and preload and cardiac output decrease.

Extrinsic Regulation

The heart is innervated by both **parasympathetic** and **sympathetic** nerve fibers (figure 20.22). They influence the pumping action of the heart by affecting both heart rate and stroke volume. The influence of parasympathetic stimulation on the heart is much less than that of sympathetic stimulation. Sympathetic stimulation can increase cardiac output by 50%–100% over resting values, whereas parasympathetic stimulation can cause only a 10%–20% decrease.

Extrinsic regulation of the heart keeps blood pressure, blood oxygen levels, blood carbon dioxide levels, and blood pH within their normal ranges of values. For example, if blood pressure suddenly decreases, extrinsic mechanisms detect the decrease and initiate responses that increase cardiac output to bring blood pressure back to its normal range.

Parasympathetic Control

Parasympathetic nerve fibers are carried to the heart through the **vagus nerves.** Preganglionic fibers of the vagus nerve extend from the brainstem to terminal ganglia within the wall of the heart, and postganglionic fibers extend from the ganglia to the SA node, AV node, coronary vessels, and atrial myocardium.

Parasympathetic stimulation has an inhibitory influence on the heart, primarily by decreasing the heart rate. When a person is at rest, continuous parasympathetic stimulation inhibits the heart to some degree. An increase in heart rate during exercise results, in part, from decreased parasympathetic stimulation. Strong parasympathetic stimulation can decrease the heart rate below resting levels by at least 20–30 bpm, but it has little effect on stroke volume. In fact, if venous return remains constant while the

1. Sensory (*green*) neurons carry action potentials from baroreceptors and carotid body chemoreceptors to the cardioregulatory center. Chemoreceptors in the medulla oblongata also influence the cardioregulatory center.
2. The cardioregulatory center controls the frequency of action potentials in the parasympathetic (*red*) neurons extending to the heart through the vagus nerves. The parasympathetic neurons decrease the heart rate.
3. The cardioregulatory center controls the frequency of action potential in the sympathetic (*blue*) neurons. The sympathetic neurons extend through the cardiac nerves and increase the heart rate and the stroke volume.
4. The cardioregulatory center influences the frequency of action potentials in the sympathetic (*blue*) neurons extending to the adrenal medulla. The sympathetic neurons increase the secretion of epinephrine and some norepinephrine into the general circulation. Epinephrine and norepinephrine increase the heart rate and stroke volume.

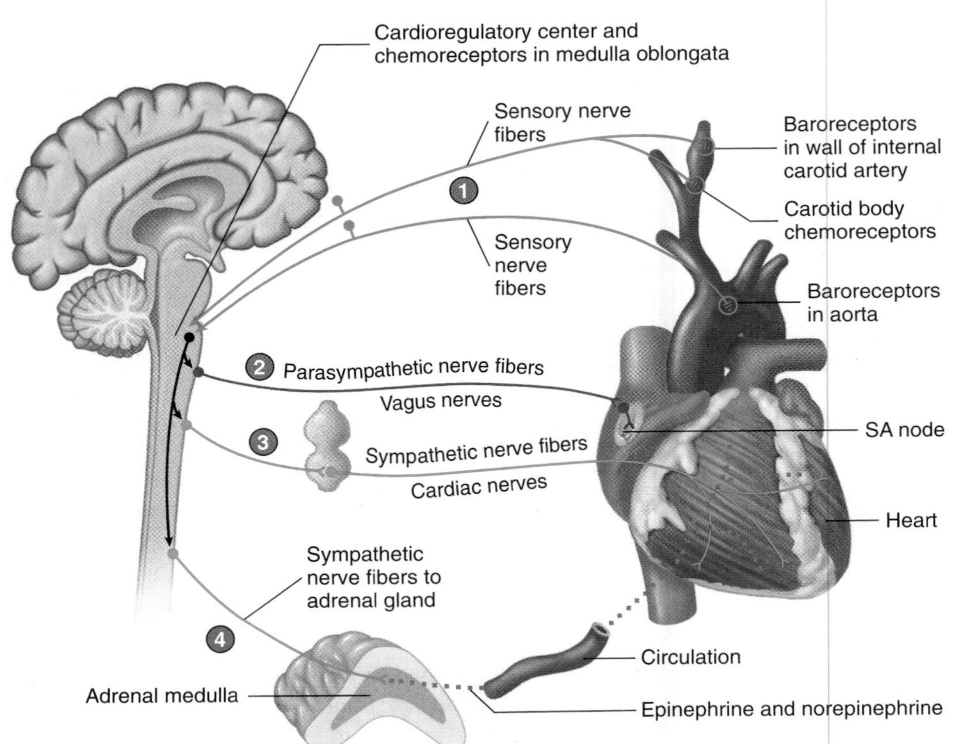

PROCESS FIGURE 20.22 Baroreceptor and Chemoreceptor Reflexes

Sensory (*green*) nerves carry action potentials from sensory receptors to the medulla oblongata. Sympathetic (*blue*) and parasympathetic (*red*) nerves exit the spinal cord or medulla oblongata and extend to the heart to regulate its function. Epinephrine and norepinephrine from the adrenal gland also help regulate the heart's action (*SA* = sinoatrial).

heart is inhibited by parasympathetic stimulation, stroke volume actually can increase. The longer time between heartbeats allows the heart to fill to a greater capacity, resulting in an increased preload, which increases stroke volume because of Starling's law of the heart.

Acetylcholine, the neurotransmitter produced by postganglionic parasympathetic neurons, binds to ligand-gated channels that cause cardiac plasma membranes to become more permeable to K^+. As a consequence, the membrane hyperpolarizes. Heart rate decreases because the hyperpolarized membrane takes longer to depolarize and cause an action potential.

Sympathetic Control

Sympathetic nerve fibers originate in the thoracic region of the spinal cord as preganglionic neurons. These neurons synapse with postganglionic neurons of the inferior **cervical** and upper **thoracic sympathetic chain ganglia,** which project to the heart as **cardiac nerves** (see figure 20.22 and chapter 16). The postganglionic sympathetic nerve fibers innervate the SA and AV nodes, the coronary vessels, and the atrial and ventricular myocardium.

Sympathetic stimulation increases both the heart rate and the force of muscular contraction. In response to strong sympathetic stimulation, the heart rate can increase to 250 or, occasionally, 300 bpm. Stronger contractions also can increase stroke volume. The increased force of contraction resulting from sympathetic stimulation causes a lower end-systolic volume in the heart; therefore, the heart empties to a greater extent (see figure 20.21).

PREDICT 10

What effect does sympathetic stimulation have on stroke volume if the venous return remains constant? Dilation of the coronary blood vessels occurs in response to an increased heart rate and stroke volume. Explain the functional advantage of that effect.

Limitations exist, however, to the relationship between increased heart rate and cardiac output. If the heart rate becomes too fast, ventricular diastole is not long enough to allow complete ventricular filling, end-diastolic volume decreases, and stroke volume actually decreases. In addition, if heart rate increases beyond a critical level, the strength of contraction decreases, probably as a result of the accumulation of metabolites in cardiac muscle cells. The limit of the heart's ability to increase the volume of blood pumped is 170–250 bpm in response to intense sympathetic stimulation.

Sympathetic stimulation of the ventricular myocardium plays a significant role in regulation of its contraction force when a person is at rest. Sympathetic stimulation maintains the strength of ventricular contraction at a level approximately 20% greater than it would be with no sympathetic stimulation.

Norepinephrine, the postganglionic sympathetic neurotransmitter, increases the rate and degree of cardiac muscle depolarization so that both the frequency and the amplitude of the action potentials are increased. The effect of norepinephrine on the heart involves the association between norepinephrine and cell surface β-adrenergic receptors. This combination causes a G protein–mediated synthesis and accumulation of cAMP in the cytoplasm of cardiac muscle cells. Cyclic AMP increases the permeability of the plasma membrane to Ca^{2+}, primarily by opening calcium channels in the plasma membrane.

Increased sympathetic stimulation causes coronary arteries to constrict, to some degree. However, increased metabolism of cardiac muscle, in response to sympathetic stimulation, results in the accumulation of metabolic by-products in cardiac muscle that cause dilation of coronary blood vessels. The dilation effect of these metabolites predominates (see chapter 21).

Hormonal Control

Epinephrine and norepinephrine released from the adrenal medulla can markedly influence the pumping effectiveness of the heart. Epinephrine has essentially the same effect on cardiac muscle as norepinephrine and, therefore, increases the rate and force of heart contractions (see figure 20.21).

The secretion of epinephrine and norepinephrine from the adrenal medulla is controlled by sympathetic stimulation of the adrenal medulla and occurs in response to increased physical activity, emotional excitement, or stressful conditions. Many stimuli that increase sympathetic stimulation of the heart also increase release of epinephrine and norepinephrine from the adrenal gland (see chapter 18). Epinephrine and norepinephrine are transported in the blood through the vessels of the heart to the cardiac muscle cells, where they bind to β-adrenergic receptors and stimulate cAMP synthesis. Epinephrine takes a longer time to act on the heart than sympathetic stimulation does, but the effect lasts longer.

40 ***Define venous return, and explain how it affects preload. How does preload affect cardiac output? State Starling's law of the heart.***

41 ***Define afterload, and describe its effect on the pumping effectiveness of the heart.***

42 ***What part of the brain regulates the heart? Describe the autonomic nerve supply to the heart.***

43 ***What effect do parasympathetic stimulation and sympathetic stimulation have on heart rate, force of contraction, and stroke volume?***

44 ***What neurotransmitters are released by the parasympathetic and sympathetic postganglionic neurons of the heart? What effects do they have on membrane permeability and excitablity?***

45 ***Name the two main hormones that affect the heart. Where are they produced, what causes their release, and what effects do they have on the heart?***

HEART AND HOMEOSTASIS

The pumping efficiency of the heart plays an important role in the maintenance of homeostasis. Blood pressure in the systemic vessels must be maintained at a level that is high enough to achieve nutrient and waste product exchange across the walls of the capillaries that meets metabolic demands. The heart's activity must be regulated because the metabolic activities of the tissues change under such conditions as exercise and rest.

Effect of Blood Pressure

Baroreceptor (bar′ō-rē-sep′ter, bar′ō-rē-sep′tōr) **reflexes** detect changes in blood pressure and result in changes in heart rate and in the force of contraction. The sensory receptors of the baroreceptor reflexes are stretch receptors. They are in the walls of certain large arteries, such as the internal carotid arteries and the aorta, and they measure blood pressure (see figure 20.22). The anatomy of these sensory structures and their afferent pathways are described in chapter 21.

Afferent neurons project primarily through the glossopharyngeal (cranial nerve IX) and vagus (cranial nerve X) nerves from the baroreceptors to an area in the medulla oblongata called the **cardioregulatory center,** where sensory action potentials are integrated (see figure 20.22). The part of the cardioregulatory center that increases heart rate is called the **cardioacceleratory center,** and the part that decreases heart rate is called the **cardioinhibitory center.** Efferent action potentials then are sent from the cardioregulatory center to the heart through both the sympathetic and the parasympathetic divisions of the autonomic nervous system.

Increased blood pressure within the internal carotid arteries and aorta causes their walls to stretch, thereby stimulating an increase in action potential frequency in the baroreceptors (figure 20.23). At normal blood pressures (80–120 mm Hg), afferent action potentials are sent from the baroreceptors to the medulla oblongata at a relatively constant frequency. When blood pressure increases, the arterial walls are stretched farther, and the afferent action potential frequency increases. When blood pressure decreases, the arterial walls are stretched to a lesser extent, and the afferent action potential frequency decreases. In response to increased blood pressure, the baroreceptor reflexes

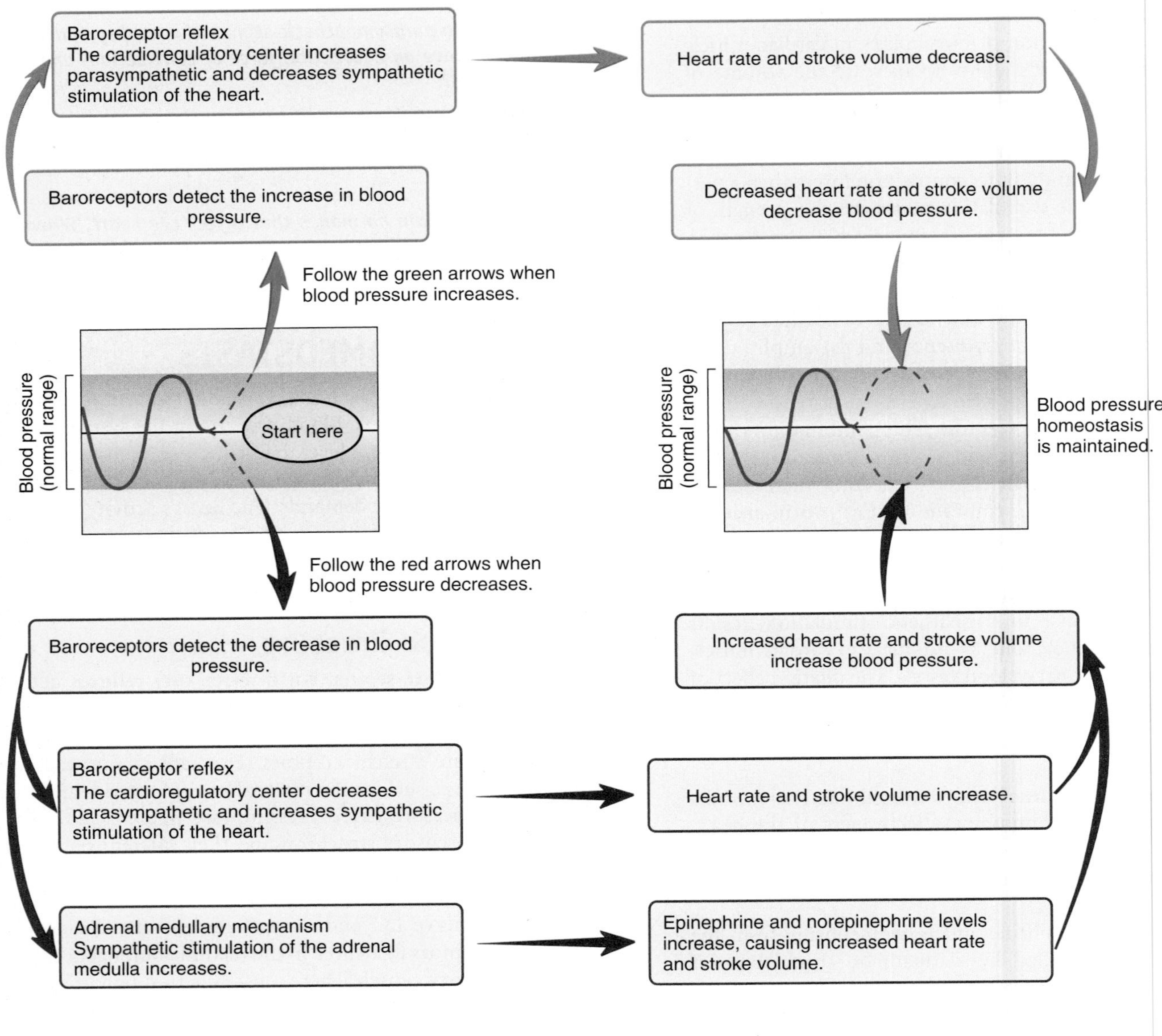

HOMEOSTASIS FIGURE 20.23 Summary of the Baroreceptor Reflex
The baroreceptor reflex maintains homeostasis in response to changes in blood pressure.

decrease sympathetic stimulation and increase parasympathetic stimulation of the heart, causing the heart rate to decrease. Decreased blood pressure causes decreased parasympathetic and increased sympathetic stimulation of the heart, resulting in an increased heart rate and force of contraction. Withdrawal of parasympathetic stimulation is primarily responsible for increases in heart rate up to approximately 100 bpm. Larger increases in heart rate, especially during exercise, result from sympathetic stimulation. The baroreceptor reflexes are homeostatic because they keep the blood pressure within a narrow range of values, which is adequate to maintain blood flow to the tissues.

Effect of pH, Carbon Dioxide, and Oxygen

Chemoreceptor (kē′mō-rē-sep′tor) **reflexes** help regulate the heart's activity. Chemoreceptors sensitive to changes in pH

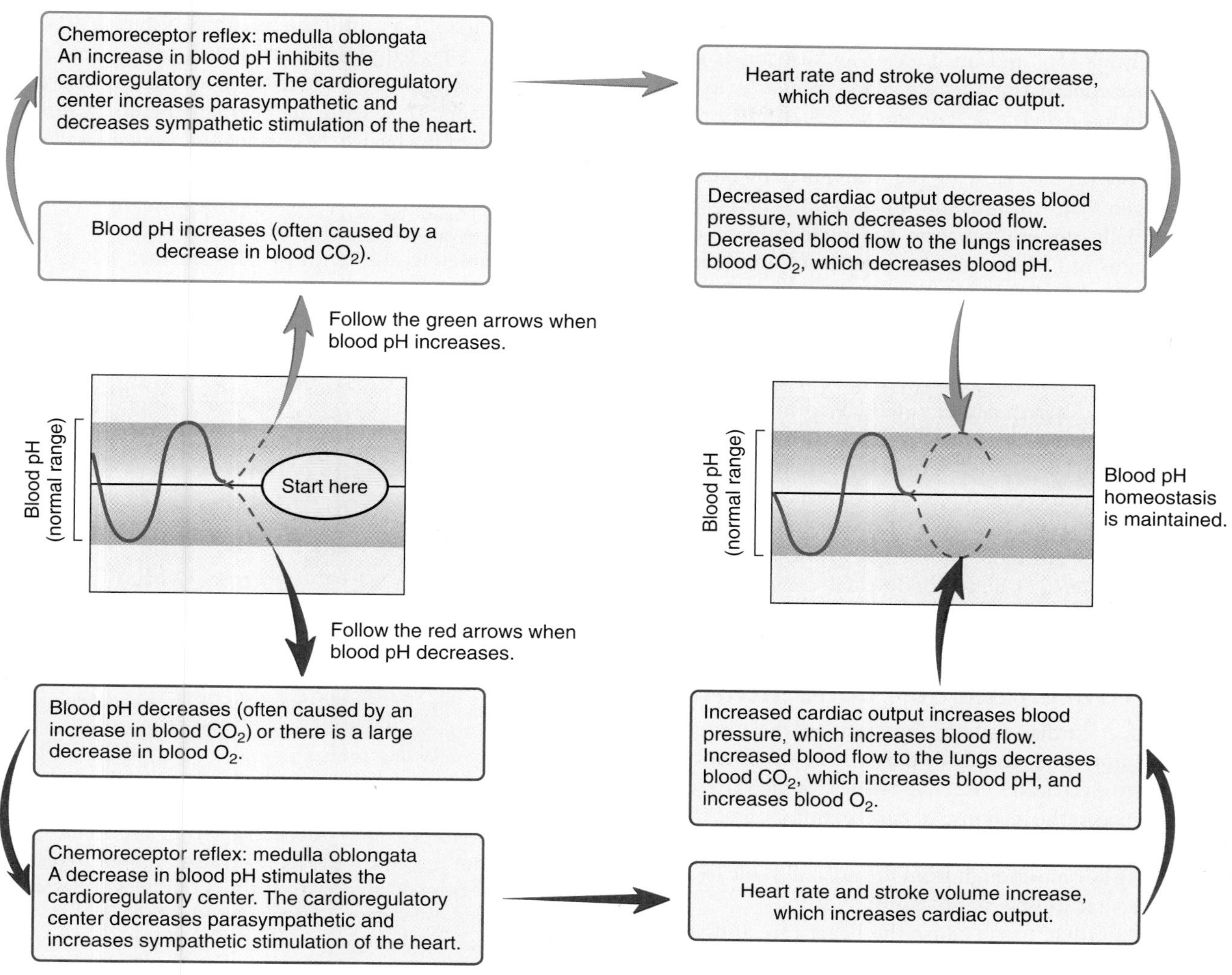

HOMEOSTASIS FIGURE 20.24 Summary of Chemoreceptor Reflex

The chemoreceptor reflex maintains homeostasis in response to changes in blood CO_2 and H^+ concentrations.

and carbon dioxide levels exist within the medulla oblongata. A drop in pH and a rise in carbon dioxide decrease parasympathetic and increase sympathetic stimulation of the heart, resulting in an increased heart rate and force of contraction (figure 20.24).

The increased cardiac output causes greater blood flow through the lungs, where carbon dioxide is eliminated from the body. This helps bring the blood carbon dioxide level down to its normal range of values and helps increase blood pH.

Chemoreceptors primarily sensitive to blood oxygen levels are found in the carotid and aortic bodies. These small structures are located near large arteries close to the brain and heart, and they monitor blood flowing to the brain and to the rest of the body. A dramatic decrease in blood oxygen levels, such as during asphyxiation, activates the carotid and aortic body chemoreceptor reflexes. In carefully controlled experiments, it is possible to isolate the effects of the carotid and aortic body chemoreceptor reflexes from other reflexes, such as the medullary chemoreceptor reflexes.

These experiments indicate that a decrease in blood oxygen results in a decrease in heart rate and an increase in vasoconstriction. The increased vasoconstriction causes blood pressure to rise, which promotes blood delivery despite the decrease in heart rate. The carotid and aortic body chemoreceptor reflexes may protect the heart for a short time by slowing the heart, thereby reducing its need for oxygen. The carotid and aortic body chemoreceptor reflexes normally do not function independently of other regulatory mechanisms. When all the regulatory mechanisms function together, the effect of large, prolonged decreases in blood oxygen levels is to increase the heart rate. Low blood oxygen levels also result in increased stimulation of respiratory movements (see chapter 23). Increased inflation of the lungs stimulates stretch receptors in the lungs. Afferent action potentials from these stretch receptors influence the cardioregulatory center, which causes an increase in heart rate. The reduced oxygen levels that exist at high altitudes can cause an increase in heart rate even when blood carbon dioxide levels remain low. The carotid and aortic body chemoreceptor reflexes are more important in the regulation of respiration (see chapter 23) and blood vessel constriction (see chapter 21) than in the regulation of heart rate.

Effect of Extracellular Ion Concentration

The ions that affect cardiac muscle function are the same ions (potassium, calcium, and sodium) that influence membrane potentials in other electrically excitable tissues. Some differences exist, however, between the response of cardiac muscle and that of nerve or muscle tissue to these ions. For example, the extracellular levels of Na^+ rarely deviate enough from the normal value to affect the function of cardiac muscle significantly.

Excess K^+ in cardiac tissue cause the heart rate and stroke volume to decrease. A twofold increase in extracellular K^+ results in **heart block,** which is loss of the functional conduction of action potentials through the conducting system of the heart. The excess K^+ in the extracellular fluid cause partial depolarization of the resting membrane potential, resulting in a decreased amplitude of action potentials and, because of the decreased amplitude, a decreased rate at which action potentials are conducted along cardiac muscle fibers. As the conduction rates decrease, ectopic action potentials can occur. In many cases, partially depolarized cardiac muscle cells spontaneously produce action potentials because the membrane potential reaches threshold. Increased blood levels of K^+ can produce enough ectopic action potentials to cause fibrillation. The reduced action potential amplitude also results in less calcium entering the sarcoplasm of the cell; thus, the strength of cardiac muscle contraction decreases.

Although the extracellular concentration of K^+ normally is small, a decrease in extracellular K^+ results in a decrease in the heart rate because the resting membrane potential is hyperpolarized; as a consequence, it takes longer for the membrane to depolarize to threshold. The force of contraction is not affected, however.

An increase in the extracellular concentration of Ca^{2+} produces an increase in the force of cardiac contraction because of a greater influx of Ca^{2+} into the sarcoplasm during action potential generation. Elevated plasma Ca^{2+} levels have an indirect effect on heart rate because they reduce the frequency of action potentials in nerve fibers, thus reducing sympathetic and parasympathetic stimulation of the heart (see chapter 11). Generally, elevated blood Ca^{2+} levels reduce the heart rate.

A low blood Ca^{2+} level increases the heart rate, although the effect is imperceptible until blood Ca^{2+} levels are reduced to approximately one-tenth of their normal value. The reduced extracellular Ca^{2+} levels cause Na^+ channels to open, which allows Na^+ to diffuse more readily into the cell, resulting in

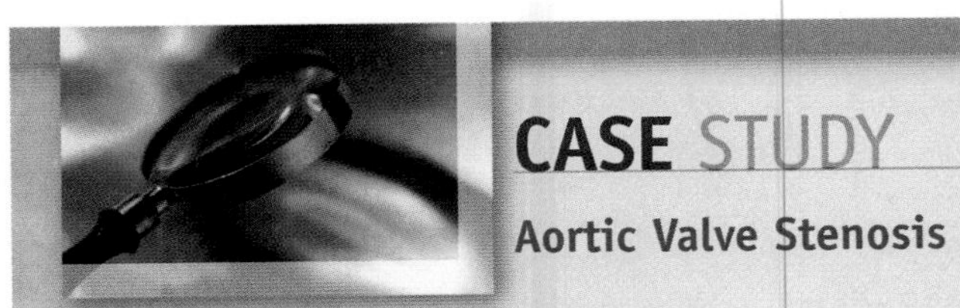

CASE STUDY
Aortic Valve Stenosis

Norma is a 62-year-old woman who had rheumatic fever when she was 12 years old. She has had a heart murmur since then. Norma went to her doctor complaining of fatigue; dizziness, especially on rising from a sitting or lying position; and pain in her chest when she exercises. Her doctor listened to Norma's heart sounds and determined she has a systolic murmur (see Clinical Focus "Abnormal Heart Sounds," p. 703). Norma's blood pressure (90/65 mm Hg) and heart rate (55 beats/min) were lower than normal. Norma's doctor referred her to a cardiologist, who did additional tests. An electrocardiogram indicated she has left ventricular hypertrophy and imaging techniques confirmed the left ventricular hypertrophy and a stenosed aortic semilunar valve. The cardiologist explained to Norma that the rheumatic fever she had as a child damaged her aortic semilunar valve and that the valve's condition had gradually become worse. The cardiologist recommended surgical replacement of Norma's aortic semilunar valve. Otherwise, she is likely to develop heart failure.

PREDICT 11

a. What effect does Norma's stenosed valve have on stroke volume?
b. Norma has left ventricular hypertrophy, which means the left ventricle is enlarged and has thicker walls than normal. Explain how that condition developed.
c. Explain Norma's low blood pressure.
d. Explain why Norma becomes dizzy on rising from a sitting or lying position (*Hint:* venous return).
e. Predict how Norma's heart rate changes on standing (see figure 20.23).
f. Norma experiences chest pain when she exercises, a condition called angina pectoris (see p. 689). Why doesn't she feel this pain at rest?

depolarization and action potential generation. Reduced Ca^{2+} levels, however, usually cause death as a result of tetany of skeletal muscles before they decrease enough to markedly influence the heart's function.

Effect of Body Temperature

Under resting conditions, the temperature of cardiac muscle normally does not change dramatically in humans, although alterations in temperature influence the heart rate. Small increases in cardiac muscle temperature cause the heart rate to increase, and decreases in temperature cause the heart rate to decrease. For example, during exercise or fever, increased heart rate and force of contraction accompany temperature increases, but the heart rate decreases under conditions of hypothermia. During heart surgery, body temperature sometimes is reduced dramatically to slow the heart rate and other metabolic functions.

46 ***How does the nervous system detect and respond to (a) a decrease in blood pressure, (b) an increase in carbon dioxide levels, (c) a decrease in blood pH, and (d) a decrease in blood oxygen levels?***

47 ***Describe the baroreceptor reflex and the response of the heart to an increase in venous return.***

48 ***What effect does an increase or a decrease in extracellular potassium, calcium, and sodium ions have on heart rate and the force of contraction of the heart?***

49 ***What effect does temperature have on heart rate?***

EFFECTS OF AGING ON THE HEART

Aging results in gradual changes in the function of the heart, which are minor under resting conditions but become more significant in response to exercise and when age-related diseases develop. The mechanisms that regulate the heart compensate effectively for most of the changes under resting conditions.

Hypertrophy of the left ventricle is a common age-related change. This appears to result from a gradual increase in the pressure in the aorta against which the left ventricle must pump blood and a gradual increase in the stiffness of cardiac muscle tissue. The increased pressure in the aorta results from a gradual decrease in arterial elasticity, resulting in an increased stiffness of the aorta and other large arteries. Myocardial cells accumulate lipid, and collagen fibers increase in cardiac tissue. These changes make the cardiac muscle tissue stiffer and less compliant. The increased volume of the left ventricle can sometimes result in an increase in left atrial pressure and increased pulmonary capillary pressure. This can cause pulmonary edema and a tendency for people to feel out of breath when they exercise strenuously.

There is a gradual decrease in the maximum heart rate. This can be roughly predicted by the following formula: Maximum heart rate = 220 − age of the individual. There is an increase in the rate at which ATP is broken down by cardiac muscle and a decrease in the rate of Ca^{2+} transport. There is a decrease in the maximum rate at which cardiac muscle can carry out aerobic metabolism. In addition, there is a decrease in the degree to which epinephrine and norepinephrine can increase the heart rate. These changes are consistent with longer contraction and relaxation times for cardiac muscle and a decrease in the maximum heart rate. Both the resting and maximum cardiac output slowly decrease as people age and, by 85 years of age, the cardiac output may have decreased by 30%–60%.

Age-related changes in the connective tissue of the heart valves occur. The connective tissue becomes less flexible and Ca^{2+} deposits increase. The result is an increased tendency for heart valves to function abnormally. There is especially an increased tendency for the aortic semilunar valve to become stenosed, but other heart valves, such as the bicuspid valve, may become either stenosed or incompetent.

Atrophy and replacement of cells of the left bundle branch and a decrease in the number of SA node cells alter the electrical conduction system of the heart and lead to a higher rate of cardiac arrhythmias in elderly people.

The enlarged and thickened cardiac muscle, especially in the left ventricle, consumes more oxygen to pump the same amount of blood pumped by a younger heart. This change is not significant unless the coronary circulation is decreased by coronary artery disease. However, the development of coronary artery disease is age-related. Congestive heart disease is also age-related. Approximately 10% of elderly people over 80 have congestive heart failure, and a major contributing factor is coronary artery disease. Because of the age-related changes in the heart, many elderly people are limited in their ability to respond to emergencies, infections, blood loss, and stress.

Exercise has many beneficial effects on the heart. Regular aerobic exercise improves the functional capacity of the heart at all ages, providing no conditions develop that cause the increased workload of the heart to be harmful.

50 ***Explain how age-related changes affect the function of the left ventricle.***

51 ***Describe the age-related changes in the heart rate.***

52 ***Describe how increasing age affects the function of the conduction system and the heart valves.***

53 ***Describe the effect of two age-related heart diseases on the functions of the aging heart.***

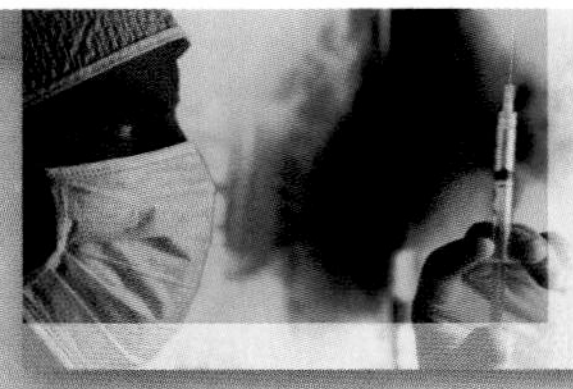

CLINICAL FOCUS

Conditions and Diseases Affecting the Heart

INFLAMMATION OF HEART TISSUES

Endocarditis (en′dō-kar-dī′tis) is inflammation of the endocardium. It affects the valves more severely than other areas of the heart and can lead to deposition of scar tissue, causing valves to become stenosed or incompetent.

Myocarditis (mī′ō-kar-dī′tis) is inflammation of the myocardium and can lead to heart failure.

Pericarditis is inflammation of the pericardium. Pericarditis can result from bacterial or viral infections and can be extremely painful.

Rheumatic (roo-mat′ik) **heart disease** can result from a streptococcal infection in young people. Toxin produced by the bacteria can cause an immune reaction called rheumatic fever about 2–4 weeks after the infection. The immune reaction can cause inflammation of the endocardium, called **rheumatic endocarditis.** The inflamed valves, especially the bicuspid valve, can become stenosed or incompetent. The effective treatment of streptococcal infections with antibiotics has reduced the frequency of rheumatic heart disease.

REDUCED BLOOD FLOW TO CARDIAC MUSCLE

Coronary heart disease reduces the amount of blood that the coronary arteries are able to deliver to the myocardium. The reduction in blood flow damages the myocardium. The degree of damage depends on the size of the arteries involved, whether occlusion (blockage) is partial or complete, and whether occlusion is gradual or sudden. As the walls of the arteries thicken and harden with age, the volume of blood they can supply to the heart muscle declines, and the heart's ability to pump blood decreases. Inadequate blood flow to the heart muscle can result in angina pectoris, which is a poorly localized sensation of pain in the region of the chest, left arm, and left shoulder.

Degenerative changes in the artery wall can cause the inside surface of the artery to become roughened. The chance of platelet aggregation increases at the rough surface, which increases the chance of **coronary thrombosis** (throm-bō′sis; formation of a blood clot in a coronary vessel). Inadequate blood flow can cause an **infarct** (in′farkt), an area of damaged cardiac tissue. A heart attack is often referred to as a coronary thrombosis, or a **myocardial infarct.** The outcome of coronary thrombosis depends on the extent of the damage to heart muscle caused by inadequate blood flow and whether other blood vessels can supply enough blood to maintain the heart's function. Death can occur swiftly if the infarct is large; if the infarct is small, the heart can continue to function. In most cases, scar tissue replaces damaged cardiac muscle in the area of the infarct.

People who survive myocardial infarctions often lead fairly normal lives if they take precautions. Most cases call for moderate exercise, adequate rest, a disciplined diet, and reduced stress.

CONGENITAL CONDITIONS AFFECTING THE HEART

Congenital heart disease is the result of abnormal development of the heart. The following conditions are common congenital defects.

Septal defect is a hole in a septum between the left and right sides of the heart. The hole may be in the interatrial or interventricular septum. These defects allow blood to flow from one side of the heart to the other and, as a consequence, greatly reduce the pumping effectiveness of the heart (see chapter 29).

Patent ductus arteriosus (dŭk′tŭs ar-tēr′ē-ō-sŭs) results when a blood vessel called the **ductus arteriosus,** which is present in the fetus, fails to close after birth. The ductus arteriosus extends between the pulmonary trunk and the aorta. It allows blood to pass from the pulmonary trunk to the aorta, thus bypassing the lungs. This is normal before birth because the lungs are not functioning (see chapter 29). If the ductus arteriosus fails to close after birth, blood flows in the opposite direction, from the aorta to the pulmonary trunk. As a consequence, blood flows through the lungs under higher pressure, causing damage to the lungs. In addition, the amount of work required of the left ventricle to maintain adequate systemic blood pressure increases.

Stenosis (ste-nō′sis) **of a heart valve** is a narrowed opening through one of the heart valves. In aortic or pulmonary valve stenosis, the heart's workload is increased because the ventricles must contract with a much greater force to pump blood from the ventricles. Stenosis of the bicuspid valve prevents the flow of blood into the left ventricle, causing blood to back up in the left atrium and in the lungs, resulting in congestion of the lungs. Stenosis of the tricuspid valve causes blood to back up in the right atrium and systemic veins, causing swelling in the periphery.

An **incompetent heart valve** is one that leaks. Blood, therefore, flows through the valve when it is closed. The heart's workload is increased because incompetent valves reduce the heart's pumping efficiency. For example, an incompetent aortic semilunar valve allows blood to flow from the aorta into the left ventricle during diastole. Thus, the left ventricle fills with blood to a greater degree than normal. The increased filling of the left ventricle results in a greater stroke volume because of Starling's law of the heart. The pressure produced by the contracting ventricle and the pressure in the aorta are greater than normal during ventricular systole. The pressure in the aorta, however, decreases very rapidly as blood leaks into the left ventricle during diastole.

An incompetent bicuspid valve allows blood to flow back into the left atrium from the left ventricle during ventricular systole. This increases the pressure in the left atrium and pulmonary veins, which results in pulmonary edema. Also, the stroke volume of the left ventricle is reduced, which causes a decrease in systemic blood pressure. Similarly, an incompetent tricuspid valve allows blood to flow back into the right atrium and systemic veins, causing edema in the periphery.

Cyanosis (sī-ă-nō′sis) is a symptom of inadequate heart function in babies suffering from congenital heart disease. The term *blue baby* is sometimes used to refer to infants with cyanosis. Low blood oxygen levels in the peripheral blood vessels cause the skin to look blue.

HEART FAILURE

Heart failure is the result of progressive weakening of the heart muscle and the failure of the heart to pump blood effectively. Hypertension (high blood pressure) increases the afterload on the heart, can produce significant enlargement of the heart, and can finally result in heart failure. Advanced age, malnutrition, chronic infections, toxins, severe anemias, and hyperthyroidism can cause degeneration of the heart muscle, resulting in heart failure. Hereditary factors can also be responsible for increased susceptibility to heart failure.

HEART MEDICATIONS

Digitalis (dij-i-tal′is, dij-i-ta′lis) slows and strengthens contractions of the heart muscle by increasing the amount of Ca^{2+} that enters cardiac muscle cells and by prolonging the action potentials' refractory period. This drug is frequently given to people who suffer from heart failure, although it also can be used to treat atrial tachycardia.

Nitroglycerin (nī-trō-glis′er-in) causes dilation of all of the veins and arteries, including coronary arteries, without an increase in heart rate or stroke volume. When all blood vessels dilate, a greater volume of blood pools in the dilated blood vessels, causing a decrease in the venous return to the heart. The flow of blood through coronary arteries also increases. The reduced preload causes cardiac output to decrease, resulting in a decreased amount of work performed by the heart. Nitroglycerin is frequently given to people who suffer from coronary artery disease, which restricts coronary blood flow. The decreased work performed by the heart reduces the amount of oxygen required by the cardiac muscle. Consequently, the heart does not suffer from a lack of oxygen, and angina pectoris does not develop.

Beta-adrenergic-blocking agents reduce the rate and strength of cardiac muscle contractions, thus reducing the heart's demand for oxygen. These blocking agents bind to receptors for norepinephrine and epinephrine and prevent these substances from having their normal effects. They are often used to treat people who suffer from rapid heart rates, certain types of arrhythmias, and hypertension.

Calcium channel blockers reduce the rate at which Ca^{2+} diffuse into cardiac muscle cells and smooth muscle cells. Because the action potentials that produce cardiac muscle contractions depend in part on the flow of Ca^{2+} into cardiac muscle cells, calcium channel blockers can be used to control the force of heart contractions and to reduce arrhythmia, tachycardia, and hypertension. Because the entry of Ca^{2+} into smooth muscle cells causes contraction, calcium channel blockers cause dilation of coronary blood vessels and can be used to treat angina pectoris.

Antihypertensive (an′tē-hī-per-ten′siv) **agents** comprise several drugs used specifically to treat hypertension. These drugs reduce blood pressure and, therefore, reduce the work required by the heart to pump blood. In addition, the reduction of blood pressure reduces the risk for heart attacks and strokes. Drugs used to treat hypertension include those that reduce the activity of the sympathetic nervous system, dilate arteries and veins, increase urine production (diuretics), and block the conversion of angiotensinogen to angiotensin I.

Anticoagulants (an′tē-kō-ag′ū-lantz) prevent clot formation in persons with damage to heart valves or blood vessels or in persons who have had a myocardial infarction. Aspirin functions as a weak anticoagulant.

INSTRUMENTS AND SELECTED PROCEDURES

An **artificial pacemaker** is an instrument placed beneath the skin, equipped with an electrode that extends to the heart. The instrument provides an electric stimulus to the heart at a set frequency. Artificial pacemakers are used in patients in whom the natural pacemaker of the heart does not produce a heart rate high enough to sustain normal physical activity. Artificial pacemakers can increase the heart rate as increases in physical activity occur. Pacemakers can also detect cardiac arrest, extreme arrythmias, or fibrillation. In response, strong stimulation of the heart by the pacemaker may restore heart function.

A **heart lung machine** serves as a temporary substitute for a patient's heart and lungs. It oxygenates the blood, removes carbon dioxide, and pumps blood throughout the body. It has made possible many surgeries on the heart and lungs.

Heart valve replacement or repair is a surgical procedure performed on those who have diseased valves that are so deformed and scarred from conditions such as endocarditis that the valves are severely incompetent or stenosed. Substitute valves made of synthetic materials, such as plastic or Dacron, are effective; valves transplanted from pigs are also used.

A **heart transplant** is a surgical procedure made possible when the immune characteristics of a donor and the recipient are closely matched (see chapter 22). The heart of a recently deceased donor is transplanted to the recipient, and the diseased heart of the recipient is removed. People who have received heart transplants must continue to take drugs that suppress their immune responses for the rest of their lives. If they do not, their immune system will reject the transplanted heart.

An **artificial heart** is a mechanical pump that replaces the heart. It is still experimental and cannot be viewed as a permanent substitute for the heart. It has been used to keep a patient alive until a donor heart can be found.

Cardiac assistance involves temporarily implanting a mechanical device that assists the heart in pumping blood. In some cases, the decreased workload on the heart provided by the device appears to promote recovery of failing hearts, and the device has been successfully removed. In **cardiomyoplasty,** a piece of a back muscle (latissimus dorsi) is wrapped around the heart and stimulated to contract in synchrony with the heart.

PREVENTION OF HEART DISEASE

Heart disease is a major cause of death. Several precautions can be taken to help prevent heart disease. Proper nutrition is important in reducing the risk for heart disease (see chapter 25). A recommended diet is low in fats, especially saturated fats and cholesterol, and low in refined sugar. Diets should be high in fiber, whole grains, fruits, and vegetables. Total food intake should be limited to avoid obesity, and sodium chloride intake should be reduced.

Smoking and the excessive use of alcohol should be avoided. Smoking increases the risk for heart disease at least 10-fold, and excessive alcohol use also substantially increases the risk for heart disease.

Chronic stress, frequent emotional upsets, and a lack of physical exercise can increase the risk for cardiovascular disease. Remedies include relaxation techniques and aerobic exercise programs involving gradual increases in duration and difficulty in activities, such as walking, swimming, jogging, and aerobic dancing.

Hypertension (hī′per-ten′shŭn) is abnormally high systemic blood pressure. It affects about one-fifth of the U.S. population. Regular blood pressure measurements are important because hypertension does not produce obvious symptoms. If hypertension cannot be controlled by diet and exercise, it is important to treat the condition with prescribed drugs. The cause of hypertension in most cases is unknown.

Some data suggest that taking an aspirin daily reduces the chance of a heart attack. Aspirin inhibits the synthesis of prostaglandins in platelets, thereby helping prevent clot formation (see chapter 19).

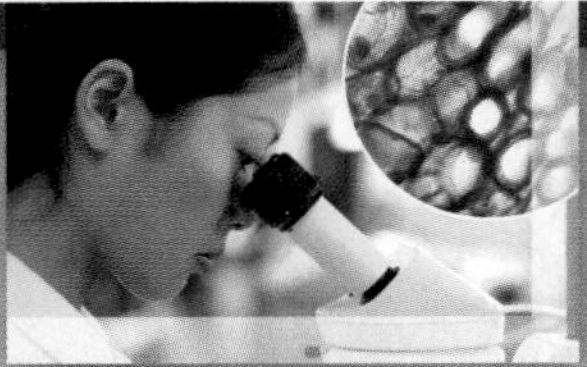

SYSTEMS PATHOLOGY

Myocardial Infarction

Mr. P was an overweight, out-of-shape executive who regularly smoked and consumed food with a high fat content. He viewed his job as frustrating because he was frequently confronted with stressful deadlines. He had not had a physical examination for several years, so he was not aware that his blood pressure was high. One evening, Mr. P was walking to his car after work when he began to feel chest pain that radiated down his left arm. Shortly after the onset of pain, he felt out of breath, developed marked pallor, became dizzy, and had to lie down on the sidewalk. The pain in his chest and arm was poorly localized but intense, and he became anxious and then disoriented. Mr. P lost consciousness, although he did not stop breathing. After a short delay, one of his coworkers noticed him and called for help. When paramedics arrived, they determined that Mr. P's blood pressure was low and he exhibited arrhythmia and tachycardia. The paramedics transmitted the electrocardiogram they took to a physician by way of their electronic communications system, and they discussed Mr. P's symptoms with the physician who was at the hospital. The paramedics were directed to administer oxygen and medication to control arrhythmias and transport him to the hospital. At the hospital, tissue plasminogen activator (t-PA) was administered, which improved blood flow to the damaged area of the heart by activating plasminogen, which dissolves blood clots. Enzymes, such as creatine phosphokinase, increased in Mr. P's blood over the next few days, which confirmed that damage to cardiac muscle had resulted from an infarction.

In the hospital, Mr. P began to experience shortness of breath because of pulmonary edema, and, after a few days in the hospital, he developed pneumonia. He was treated for pneumonia and gradually improved over the next few weeks. An angiogram (an′jē-ō-gram) performed several days after Mr. P's infarction indicated that he had suffered damage to a significant part of the lateral wall of his left ventricle and that neither angioplasty nor bypass surgery was necessary, although Mr. P has some serious restrictions to blood flow in his coronary arteries.

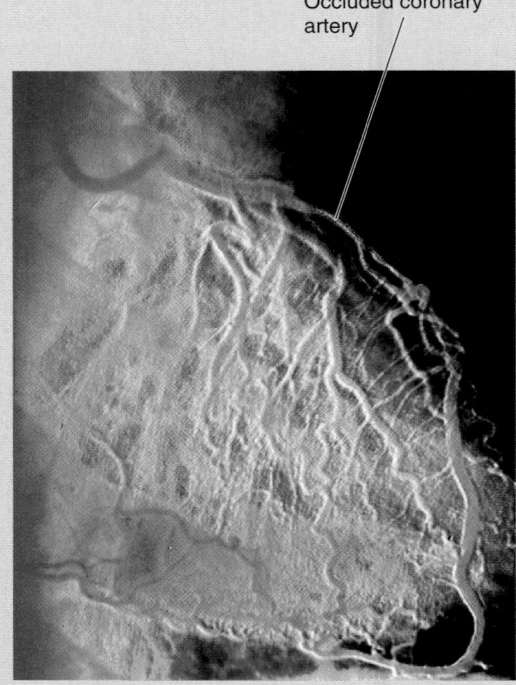

FIGURE A Angiogram

An angiogram is usually obtained by placing a catheter into a blood vessel and injecting a dye that can be detected with x-rays. Note the occluded (blocked) coronary blood vessel in this angiogram, which has been computer-enhanced to show colors.

BACKGROUND INFORMATION

Mr. P experienced a myocardial infarction. A thrombosis in one of the branches of the left coronary artery reduces the blood supply to the lateral wall of the left ventricle, resulting in ischemia of the left ventricle wall. That t-PA is effective in treating a heart attack is consistent with the conclusion that the infarction was caused by a thrombosis. An ischemic area of the heart wall is not able to contract normally and, therefore, the pumping effectiveness of the heart is dramatically reduced. The reduced pumping capacity of the heart is responsible for the low blood pressure, which causes the blood flow to the brain to decrease, resulting in confusion, disorientation, and unconsciousness.

Low blood pressure, increasing blood carbon dioxide levels, pain, and anxiousness increase sympathetic stimulation of the heart and adrenal glands. Increased sympathetic stimulation of the adrenal medulla results in the release of epinephrine. Increased parasympathetic stimulation of the heart results from pain sensations. In such cases, the heart is periodically arrhythmic due to the combined effects of parasympathetic stimulation, epinephrine and norepinephrine from the adrenal gland, and sympathetic stimulation. In addition, ectopic beats are produced by the ischemic areas of the left ventricle.

Pulmonary edema results from the increased pressure in the pulmonary veins because of the inability of the left ventricle to pump blood. The edema allows bacteria to infect the lungs and cause pneumonia.

Mr. P's heart began to beat rhythmically in response to medication because the infarction did not damage the conducting system of the heart, which is an indication that no permanent arrhythmias developed. Permanent arrhythmias are indications of damage done to cardiac muscle specialized to conduct action potentials in the heart.

Analysis of the electrocardiogram, blood pressure measurements, and the angiogram (figure A) indicate that the infarction was located on the left side of Mr. P's heart. Mr. P exhibited several characteristics that are correlated with an increased probability of myocardial infarction: lack of physical exercise, overweight, smoking, and stress.

Mr. P's physician made it very clear to him that he was lucky to have survived a myocardial infarction, and the physician recommended a weight-loss program and a low-sodium and low-fat diet. The physician also suggested that Mr. P stop smoking. She explained that Mr. P would have to take medication for high blood pressure if his blood pressure did not decrease in response to the recommended changes. After a period of recovery, Mr. P's physician recommended an aerobic exercise program to him. She advised Mr. P to seek ways to reduce the stress associated with his job. She also recommended that Mr. P regularly take a small amount of aspirin to reduce the probability of thrombosis. Because aspirin inhibits prostaglandin synthesis, it reduces the tendency for blood to clot. Mr. P followed the doctor's recommendations, and, after several months, he began to feel better than he had in years, and his blood pressure was normal.

PREDICT 12

Severe ischemia in the wall of a ventricle can result in the death of cardiac muscle cells. Inflammation around the necrotic tissue results, and macrophages invade the necrotic tissue and phagocytize dead cells. At the same time, blood vessels and connective tissue grow into the necrotic area and begin to deposit connective tissue to replace the necrotic tissue. Assume that Mr. P had a myocardial infarction and was recovering. After about a week, however, his blood pressure suddenly decreased to very low levels and he died within a very short time. At autopsy, a large amount of blood was found in the pericardial sac, and the wall of the left ventricle was ruptured. Explain.

SYSTEM INTERACTIONS **Effect of Myocardial Infarction on Other Systems**

SYSTEM	INTERACTIONS
Integumentary	Pallor of the skin results from intense constriction of peripheral blood vessels, including those in the skin.
Muscular	Reduced skeletal muscle activity required for activities such as walking results from lack of blood flow to the brain and because blood is shunted from blood vessels that supply skeletal muscles to those that supply the heart and brain.
Nervous	Decreased blood flow to the brain, decreased blood pressure, and pain due to ischemia of heart muscle result in increased sympathetic and decreased parasympathetic stimulation of the heart. Loss of consciousness occurs when the blood flow to the brain decreases enough to result in too little oxygen to maintain normal brain function, especially in the reticular activating system.
Endocrine	When blood pressure decreases to low values, antidiuretic hormone (ADH) is released from the posterior pituitary gland and renin, released from the kidney, activates the renin-angiotensinogen-aldosterone mechanism. ADH, secreted in large amounts, and angiotensin II cause vasoconstriction of peripheral blood vessels. ADH and aldosterone act on the kidneys to retain water and electrolytes. An increased blood volume increases venous return, which results in an increased stroke volume of the heart and an increase in blood pressure unless damage to the heart is very severe.
Lymphatic and Immune	White blood cells, including macrophages, move to the area of cardiac muscle damaged and phagocytize any dead cardiac muscle cells.
Respiratory	Decreased blood pressure results in a decreased blood flow to the lungs. The decrease in gas exchange results in increased blood CO_2 levels, acidosis, and decreased blood O_2 levels. Initially, respiration becomes deep and labored because of the elevated CO_2 levels, decreased blood pH, and depressed O_2 levels. If the blood O_2 levels decrease too much, the person loses consciousness. Pulmonary edema can result when the pumping effectiveness of the left ventricle is substantially reduced.
Digestive	Decreased blood flow to the digestive system to very low levels often results in increased nausea and vomiting.
Urinary	Blood flow to the kidney decreases dramatically in response to sympathetic stimulation. If the kidney becomes ischemic, damage to the kidney tubules can occur, resulting in acute renal failure. Acute renal failure reduces urine production. Increased blood urea nitrogen, increased blood levels of K^+, and edema are indications that the kidneys cannot eliminate waste products and excess water. If damage is not too great, the period of reduced urine production may last up to 3 weeks and then the rate of urine production slowly returns to normal as the kidney tubules heal.

SUMMARY

Functions of the Heart (p. 679)

The heart produces the force that causes blood circulation.

Size, Shape, and Location of the Heart (p. 679)

The heart is approximately the size of a closed fist and is shaped like a blunt cone. The heart lies obliquely in the mediastinum, with its base directed posteriorly and slightly superiorly and the apex directed anteriorly, inferiorly, and to the left. The base is deep to the second intercostal space and the apex extends to the fifth intercostal space. It is in the mediastinum.

Anatomy of the Heart (p. 681)

The heart consists of two atria and two ventricles.

Pericardium

1. The pericardium is a sac that surrounds the heart and consists of the fibrous pericardium and the serous pericardium.
2. The fibrous pericardium helps hold the heart in place.
3. The serous pericardium reduces friction as the heart beats. It consists of the following parts:
   - The parietal pericardium lines the fibrous pericardium.
   - The visceral pericardium lines the exterior surface of the heart.
   - The pericardial cavity lies between the parietal and visceral pericardium and is filled with pericardial fluid.

Heart Wall

1. The heart wall has three layers:
   - The outer epicardium (visceral pericardium) provides protection against the friction of rubbing organs.
   - The middle myocardium is responsible for contraction.
   - The inner endocardium reduces the friction resulting from the blood's passing through the heart.
2. The inner surfaces of the atria are mainly smooth. The auricles have raised areas called musculi pectinati.
3. The ventricles have ridges called trabeculae carneae.

External Anatomy and Coronary Circulation

1. Each atrium has a flap called the auricle.
2. The coronary sulcus separates the atria from the ventricles. The interventricular grooves separate the right and left ventricles.
3. The inferior and superior venae cavae and the coronary sinus enter the right atrium. The four pulmonary veins enter the left atrium.
4. The pulmonary trunk exits the right ventricle, and the aorta exits the left ventricle.
5. Coronary arteries branch off the aorta to supply the heart. Blood returns from the heart tissues to the right atrium through the coronary sinus and cardiac veins.

Heart Chambers and Valves

1. The interatrial septum separates the atria from each other, and the interventricular septum separates the ventricles.
2. The tricuspid valve separates the right atrium and ventricle. The bicuspid valve separates the left atrium and ventricle. The chordae tendineae attach the papillary muscles to the atrioventricular valves.
3. The semilunar valves separate the aorta and pulmonary trunk from the ventricles.

Route of Blood Flow Through the Heart (p. 687)

1. Blood from the body flows through the right atrium into the right ventricle and then to the lungs.
2. Blood returns from the lungs to the left atrium, enters the left ventricle, and is pumped back to the body.

Histology (p. 689)

Heart Skeleton

The fibrous heart skeleton supports the openings of the heart, electrically insulates the atria from the ventricles, and provides a point of attachment for heart muscle.

Cardiac Muscle

1. Cardiac muscle cells are branched and have a centrally located nucleus. Actin and myosin are organized to form sarcomeres. The sarcoplasmic reticulum and T tubules are not as organized as in skeletal muscle.
2. Cardiac muscle cells are joined by intercalated disks, which allow action potentials to move from one cell to the next. Thus, cardiac muscle cells function as a unit.
3. Cardiac muscle cells have a slow onset of contraction and a prolonged contraction time caused by the length of time required for calcium to move to and from the myofibrils.
4. Cardiac muscle is well supplied with blood vessels that support aerobic respiration.
5. Cardiac muscle aerobically uses glucose, fatty acids, and lactic acid to produce ATP for energy. Cardiac muscle does not develop a significant oxygen debt.

Conducting System

1. The SA node and the AV node are in the right atrium.
2. The AV node is connected to the bundle branches in the interventricular septum by the AV bundle.
3. The bundle branches give rise to Purkinje fibers, which supply the ventricles.
4. The SA node is made up of small-diameter cardiac muscle cells that initiate action potentials, which spread across the atria and cause them to contract.
5. Action potentials are slowed in the AV node, allowing the atria to contract and blood to move into the ventricles. Then the action potentials travel through the AV bundles and bundle branches to the Purkinje fibers, causing the ventricles to contract, starting at the apex. The AV node is also made up of small-diameter cardiac muscle fibers.

Electrical Properties (p. 692)

Action Potentials

1. After depolarization and partial repolarization, a plateau is reached, during which the membrane potential only slowly repolarizes.
2. The movement of Na^+ through the voltage-gated Na^+ channels causes depolarization.

3. During depolarization, voltage-gated K^+ channels close and voltage-gated Ca^{2+} channels begin to open.
4. Early repolarization results from closure of the voltage-gated Na^+ channels and the opening of some voltage-gated K^+ channels.
5. The plateau exists because voltage-gated Ca^{2+} channels remain open.
6. The rapid phase of repolarization results from closure of the voltage-gated Ca^+ channels and the opening of many voltage-gated K^+ channels.
7. The entry of Ca^{2+} into cardiac muscle cells causes Ca^{2+} to be released from the sarcoplasmic reticulum to trigger contractions.

Autorhythmicity of Cardiac Muscle

1. Cardiac pacemaker muscle cells are autorhythmic because of the spontaneous development of a prepotential.
2. The prepotential results from the movement of Na^+ and Ca^{2+} into the pacemaker cells.
3. Ectopic foci are areas of the heart that regulate heart rate under abnormal conditions.

Refractory Periods of Cardiac Muscle

Cardiac muscle has a prolonged depolarization and thus a prolonged refractory period, which allows time for the cardiac muscle to relax before the next action potential causes a contraction.

Electrocardiogram

1. The ECG records only the electrical activities of the heart.
   - Depolarization of the atria produces the P wave.
   - Depolarization of the ventricles produces the QRS complex. Repolarization of the atria occurs during the QRS complex.
   - Repolarization of the ventricles produces the T wave.
2. Based on the magnitude of the ECG waves and the time between waves, ECGs can be used to diagnose heart abnormalities.

Cardiac Cycle (p. 695)

1. The cardiac cycle is repetitive contraction and relaxation of the heart chambers.
2. Blood moves through the circulatory system from areas of higher pressure to areas of lower pressure. Contraction of the heart produces the pressure.
3. The cardiac cycle is divided into five periods.
   - Although the heart is contracting, during the period of isovolumic contraction ventricular volume does not change because all the heart valves are closed.
   - During the period of ejection, the semilunar valves open and blood is ejected from the heart.
   - Although the heart is relaxing, during the period of isovolumic relaxation, ventricular volume does not change because all the heart valves are closed.
   - Passive ventricular filling results when blood flows from the higher pressure in the veins and atria to the lower pressure in the relaxed ventricles.
   - Active ventricular filling results when the atria contract and pump blood into the ventricles.

Events Occurring During Ventricular Systole

1. Contraction of the ventricles closes the AV valves, opens the semilunar valves, and ejects blood from the heart.
2. The volume of blood in a ventricle just before it contracts is the end-diastolic volume. The volume of blood after contraction is the end-systolic volume.

Events Occurring During Ventricular Diastole

1. Relaxation of the ventricles results in the closing of the semilunar valves, the opening of the AV valves, and the movement of blood into the ventricles.
2. Most ventricular filling occurs when blood flows from the higher pressure in the veins and atria to the lower pressure in the relaxed ventricles.
3. Contraction of the atria completes ventricular filling.

Heart Sounds

1. Closure of the atrioventricular valves produces the first heart sound.
2. Closure of the semilunar valves produces the second heart sound.
3. Turbulent flow of blood into the ventricles that can be heard in some people produces the third heart sound.

Aortic Pressure Curve

1. Contraction of the ventricles forces blood into the aorta, producing the peak systolic pressure.
2. Blood pressure in the aorta falls to the diastolic level as blood flows out of the aorta.
3. Elastic recoil of the aorta maintains pressure in the aorta and produces the dicrotic notch.

Mean Arterial Blood Pressure (p. 703)

1. Mean arterial pressure is the average blood pressure in the aorta. Adequate blood pressure is necessary to ensure delivery of blood to the tissues.
2. Mean arterial pressure is proportional to cardiac output (amount of blood pumped by the heart per minute) times peripheral resistance (total resistance to blood flow through blood vessels).
3. Cardiac output is equal to stroke volume times heart rate.
4. Stroke volume, the amount of blood pumped by the heart per beat, is equal to end-diastolic volume minus end-systolic volume.
   - Venous return is the amount of blood returning to the heart. Increased venous return increases stroke volume by increasing end-diastolic volume.
   - Increased force of contraction increases stroke volume by decreasing end-systolic volume.
5. Cardiac reserve is the difference between resting and exercising cardiac output.

Regulation of the Heart (p. 705)

Intrinsic Regulation

1. Venous return is the amount of blood that returns to the heart during each cardiac cycle.
2. Starling's law of the heart describes the relationship between preload and the stroke volume of the heart. An increased preload causes the cardiac muscle fibers to contract with a greater force and produce a greater stroke volume.

Extrinsic Regulation

1. The cardioregulatory center in the medulla oblongata regulates the parasympathetic and sympathetic nervous control of the heart.
2. Parasympathetic stimulation is supplied by the vagus nerve.
   - Parasympathetic stimulation decreases heart rate.
   - Postganglionic neurons secrete acetylcholine, which increases membrane permeability to K^+, producing hyperpolarization of the membrane.

3. Sympathetic stimulation is supplied by the cardiac nerves.
   - Sympathetic stimulation increases heart rate and the force of contraction (stroke volume).
   - Postganglionic neurons secrete norepinephrine, which increases membrane permeability to Na^+ and Ca^{2+} and produces depolarization of the membrane.
4. Epinephrine and norepinephrine are released into the blood from the adrenal medulla as a result of sympathetic stimulation.
   - The effects of epinephrine and norepinephrine on the heart are long-lasting, compared with those of neural stimulation.
   - Epinephrine and norepinephrine increase the rate and force of heart contraction.

Heart and Homeostasis (p. 707)

Effect of Blood Pressure

1. Baroreceptors monitor blood pressure.
2. In response to a decrease in blood pressure, the baroreceptor reflexes increase sympathetic stimulation and decrease parasympathetic stimulation of the heart, resulting in an increase in heart rate and force of contraction.

Effect of pH, Carbon Dioxide, and Oxygen

1. Chemoreceptors monitor blood carbon dioxide, pH, and oxygen levels.
2. In response to increased carbon dioxide and decreased pH, medullary chemoreceptor reflexes increase sympathetic stimulation and decrease parasympathetic stimulation of the heart.
3. Carotid body chemoreceptor receptors stimulated by low oxygen levels result in a decreased heart rate and vasoconstriction.
4. All regulatory mechanisms functioning together in response to low blood pH, high blood carbon dioxide, and low blood oxygen levels usually produce an increase in heart rate and vasoconstriction. Decreased oxygen levels stimulate an increase in heart rate indirectly by stimulating respiration, and the stretch of the lungs activates a reflex that increases sympathetic stimulation of the heart.

Effect of Extracellular Ion Concentration

1. An increase or decrease in extracellular K^+ decreases heart rate.
2. Increased extracellular Ca^{2+} increase the force of contraction of the heart and decrease the heart rate. Decreased Ca^{2+} levels produce the opposite effect.

Effect of Body Temperature

Heart rate increases when body temperature increases, and it decreases when body temperature decreases.

Effects of Aging on the Heart (p. 711)

1. Aging results in gradual changes in the function of the heart, which are minor under resting conditions but are more significant during exercise.
2. Hypertrophy of the left ventricle is a common age-related condition.
3. The maximum heart rate decreases and by age 85 the cardiac output may be decreased by 30%–60%.
4. There is an increased tendency for valves to function abnormally and for arrhythmias to occur.
5. An increased oxygen consumption, required to pump the same amount of blood, makes age-related coronary artery disease more severe.
6. Exercise improves the functional capacity of the heart at all ages.

REVIEW AND COMPREHENSION

1. The fibrous pericardium
   a. is in contact with the heart.
   b. is a serous membrane.
   c. is also known as the epicardium.
   d. forms the outer layer of the pericardial sac.
   e. all of the above.
2. Which of these structures returns blood to the right atrium?
   a. coronary sinus
   b. inferior vena cava
   c. superior vena cava
   d. both b and c
   e. all of the above
3. The valve located between the right atrium and the right ventricle is the
   a. aortic semilunar valve.
   b. pulmonary semilunar valve.
   c. tricuspid valve.
   d. bicuspid (mitral) valve.
4. The papillary muscles
   a. are attached to chordae tendineae.
   b. are found in the atria.
   c. contract to close the foramen ovale.
   d. are attached to the semilunar valves.
   e. surround the openings of the coronary arteries.
5. Given these blood vessels:
   1. aorta
   2. inferior vena cava
   3. pulmonary trunk
   4. pulmonary vein

   Choose the arrangement that lists the vessels in the order a red blood cell would encounter them in going from the systemic veins to the systemic arteries.
   a. 1,3,4,2
   b. 2,3,4,1
   c. 2,4,3,1
   d. 3,2,1,4
   e. 3,4,2,1
6. Which of these does *not* correctly describe the skeleton of the heart?
   a. electrically insulates the atria from the ventricles
   b. provides a rigid source of attachment for the cardiac muscle
   c. reinforces or supports the valve openings
   d. is composed mainly of cartilage
7. The bulk of the heart wall is
   a. epicardium.
   b. pericardium.
   c. myocardium.
   d. endocardium.
   e. exocardium.

8. Muscular ridges on the interior surface of the auricles are called
   a. trabeculae carneae.
   b. crista terminalis.
   c. musculi pectinati.
   d. endocardium.
   e. papillary muscles.
9. Cardiac muscle has
   a. sarcomeres.
   b. a sarcoplasmic reticulum.
   c. transverse tubules.
   d. many mitochondria.
   e. all of the above.
10. Action potentials pass from one cardiac muscle cell to another
    a. through gap junctions.
    b. by a special cardiac nervous system.
    c. because of the large voltage of the action potentials.
    d. because of the plateau phase of the action potentials.
    e. by neurotransmitters.
11. During the transmission of action potentials through the conducting system of the heart, there is a temporary delay at the
    a. bundle branches.
    b. Purkinje fibers.
    c. AV node.
    d. SA node.
    e. AV bundle.
12. Given these structures of the conduction system of the heart:
    1. atrioventricular bundle
    2. AV node
    3. bundle branches
    4. Purkinje fibers
    5. SA node

    Choose the arrangement that lists the structures in the order an action potential passes through them.
    a. 2,5,1,3,4
    b. 2,5,3,1,4
    c. 2,5,4,1,3
    d. 5,2,1,3,4
    e. 5,2,4,3,1
13. Purkinje fibers
    a. are specialized cardiac muscle cells.
    b. conduct impulses much more slowly than ordinary cardiac muscle.
    c. conduct action potentials through the atria.
    d. connect between the SA node and the AV node.
    e. ensure that ventricular contraction starts at the base of the heart.
14. T waves on an ECG represent
    a. depolarization of the ventricles.
    b. repolarization of the ventricles.
    c. depolarization of the atria.
    d. repolarization of the atria.
15. Which of these conditions observed in an electrocardiogram suggests that the AV node is not conducting action potentials?
    a. a complete lack of the P wave
    b. a complete lack of the QRS complex
    c. more QRS complexes than P waves
    d. a prolonged PR interval
    e. P waves and QRS complexes that are not synchronized
16. The greatest amount of ventricular filling occurs during
    a. the first one-third of diastole.
    b. the middle one-third of diastole.
    c. the last one-third of diastole.
    d. ventricular systole.
17. While the semilunar valves are open during a normal cardiac cycle, the pressure in the left ventricle is
    a. greater than the pressure in the aorta.
    b. less than the pressure in the aorta.
    c. the same as the pressure in the left atrium.
    d. less than the pressure in the left atrium.
18. The pressure within the left ventricle fluctuates between
    a. 120 and 80 mm Hg.
    b. 120 and 0 mm Hg.
    c. 80 and 0 mm Hg.
    d. 20 and 0 mm Hg.
19. Blood flows neither into nor out of the ventricles during
    a. the period of isovolumic contraction.
    b. the period of isovolumic relaxation.
    c. diastole.
    d. systole.
    e. both a and b.
20. Stroke volume is the
    a. amount of blood pumped by the heart per minute.
    b. difference between end-diastolic and end-systolic volume.
    c. difference between the amount of blood pumped at rest and that pumped at maximum output.
    d. amount of blood pumped from the atria into the ventricles.
21. Cardiac output is defined as
    a. blood pressure times peripheral resistance.
    b. peripheral resistance times heart rate.
    c. heart rate times stroke volume.
    d. stroke volume times blood pressure.
    e. blood pressure minus peripheral resistance.
22. Pressure in the aorta is at its lowest
    a. at the time of the first heart sound.
    b. at the time of the second heart sound.
    c. just before the AV valves open.
    d. just before the semilunar valves open.
23. Just after the dicrotic notch on the aortic pressure curve,
    a. the pressure in the aorta is greater than the pressure in the left ventricle.
    b. the pressure in the left ventricle is greater than the pressure in the aorta.
    c. the pressure in the left atrium is greater than the pressure in the left ventricle.
    d. the pressure in the left atrium is greater than the pressure in the aorta.
    e. blood pressure in the aorta is 0 mm Hg.
24. The "lubb" sound (first heart sound) of the heart is caused by the
    a. closing of the AV valves.
    b. closing of the semilunar valves.
    c. blood rushing out of the ventricles.
    d. filling of the ventricles.
    e. ventricular contraction.
25. Increased venous return results in
    a. increased stroke volume.
    b. increased cardiac output.
    c. decreased heart rate.
    d. both a and b.
26. Parasympathetic nerve fibers are found in the ______________ nerves and release ______________ at the heart.
    a. cardiac, acetylcholine
    b. cardiac, norepinephrine
    c. vagus, acetylcholine
    d. vagus, norepinephrine
27. Increased parasympathetic stimulation of the heart
    a. increases the force of ventricular contraction.
    b. increases the rate of depolarization in the SA node.
    c. decreases the heart rate.
    d. increases cardiac output.
28. Because of the baroreceptor reflex, when normal arterial blood pressure decreases, the
    a. heart rate decreases.
    b. stroke volume decreases.
    c. frequency of afferent action potentials from baroreceptors decreases.
    d. cardioregulatory center stimulates parasympathetic neurons.

29. A decrease in blood pH and an increase in blood carbon dioxide levels result in
    a. increased heart rate.
    b. increased stroke volume.
    c. increased sympathetic stimulation of the heart.
    d. increased cardiac output.
    e. all of the above.

30. An increase in extracellular potassium levels could cause
    a. an increase in stroke volume.
    b. an increase in the force of contraction.
    c. a decrease in heart rate.
    d. both a and b.

Answers in Appendix E

CRITICAL THINKING

1. Explain why the walls of the ventricles are thicker than the walls of the atria.
2. In most tissues, peak blood flow occurs during systole and decreases during diastole. In heart tissue, however, the opposite is true, and peak blood flow occurs during diastole. Explain why this difference occurs.
3. A patient has tachycardia. Would you recommend a drug that prolongs or shortens the plateau of cardiac muscle cell action potentials?
4. Endurance-trained athletes often have a decreased heart rate, compared with that of a nonathlete when both are resting. Explain why an endurance-trained athlete's heart rate decreases rather than increases.
5. A doctor lets you listen to a patient's heart with a stethoscope at the same time that you feel the patient's pulse. Once in a while, you hear two heartbeats very close together, but you feel only one pulse beat. Later, the doctor tells you that the patient has an ectopic focus in the right atrium. Explain why you hear two heartbeats very close together. The doctor also tells you that the patient exhibits a pulse deficit (i.e., the number of pulse beats felt is fewer than the number of heartbeats heard). Explain why a pulse deficit occurs.
6. Heart rate and cardiac output were measured in a group of nonathletic students. After 2 months of aerobic exercise training, their measurements were repeated. It was found that heart rate had decreased, but cardiac output remained the same for many activities. Explain these findings.
7. Explain why it is sufficient to replace the ventricles, but not the atria, in artificial heart transplantation.
8. During an experiment in a physiology laboratory, a student named Cee Saw was placed on a table that could be tilted. The instructor asked the students to predict what would happen to Cee Saw's heart rate if the table were tilted so that her head was lower than her feet. Some students predicted an increase in heart rate, and others claimed it would decrease. Can you explain why both predictions might be true?
9. After Cee Saw, the student in question 8 is tilted so that her head is lower than her feet for a few minutes. Predict how her blood pressure will change in response to raising her head above her feet.
10. A friend tells you that her son had an ECG and it revealed that he has a slight heart murmur. Should you be convinced that he has a heart murmur? Explain.
11. Predict the effect of an incompetent aortic semilunar valve on stroke volume, on the volume of blood in the left ventricle at the end of diastole, and on heart sounds.
12. An experiment on a dog was performed in which the mean arterial blood pressure was monitored before and after the common carotid arteries were partially clamped (at time A). The results are graphed below:

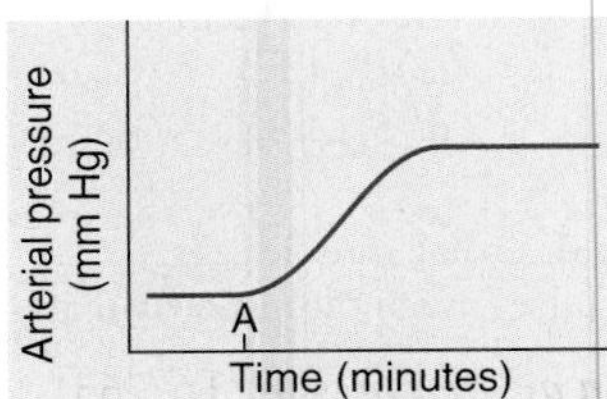

    Explain the change in mean arterial blood pressure (*Hint:* Baroreceptors are located in the internal carotid arteries, which are superior to the site of clamping of the common carotid arteries).
13. During hemorrhagic shock (caused by loss of blood), blood pressure may fall dramatically, although the heart rate is elevated. Explain why blood pressure falls despite the increase in heart rate.

Answers in Appendix F

Visit this book's website at www.mhhe.com/seeley8 for chapter quizzes, interactive learning exercises, and other study tools.

Cardiovascular System

Peripheral Circulation and Regulation

21

Complex urban water systems seem rather simple when compared with the intricacy and coordinated functions of blood vessels. The heart is the pump that provides the major force causing blood to circulate, and the blood vessels are the pipes that carry blood to the tissues of the body and back to the heart. In addition, the blood vessels participate in the regulation of blood pressure and help direct blood flow to the tissues that are most active.

The peripheral circulatory system comprises two sets of blood vessels: systemic and pulmonary vessels. **Systemic vessels** transport blood through all parts of the body from the left ventricle and back to the right atrium. **Pulmonary vessels** transport blood from the right ventricle through the lungs and back to the left atrium (see figure 20.1). Both the blood vessels and the heart are regulated to ensure that the blood pressure is high enough to cause blood flow in sufficient quantities to meet the metabolic needs of the tissues. The cardiovascular system ensures the survival of the tissues by supplying nutrients to and removing waste products from them.

This chapter explains the *functions of the peripheral circulation* (p. 722) *general features of blood vessel structure* (p. 722), *pulmonary circulation* (p. 728), *systemic circulation: arteries* (p. 728), *systemic circulation: veins* (p. 739), *dynamics of blood circulation* (p. 751), *physiology of systemic circulation* (p. 755), *control of blood flow in tissues* (p. 761), and *regulation of mean arterial pressure* (p. 765).

Color enhanced scanning electron micrograph of an artery. The inner surface of the artery appears to be scalloped, or folded, because the circular smooth muscle cells in the wall of the artery partially contract when tissues are prepared for microscopic examination.

Cardiovascular System

FUNCTIONS OF THE PERIPHERAL CIRCULATION

The heart provides the major force that causes blood to circulate, and the peripheral circulation has five functions:

1. *Carries blood.* Blood vessels carry blood from the heart to all the tissues of the body and back to the heart.
2. *Exchanges nutrients, waste products, and gases with tissues.* Nutrients and oxygen diffuse from blood vessels to cells in all areas of the body. Waste products and carbon dioxide diffuse from the cells, where they are produced, to blood vessels.
3. *Transports substances.* Hormones, components of the immune system, molecules required for coagulation, enzymes, nutrients, gases, waste products, and other substances are transported in the blood to all areas of the body.
4. *Helps regulate blood pressure.* The peripheral circulatory system and the heart work together to regulate blood pressure within a normal range of values.
5. *Directs blood flow to tissues.* The peripheral circulatory system directs blood to tissues when increased blood flow is required to maintain homeostasis.

GENERAL FEATURES OF BLOOD VESSEL STRUCTURE

The ventricles pump blood from the heart into large, elastic arteries that branch repeatedly to form many progressively smaller arteries. As they become smaller, the arteries undergo a gradual transition from having walls that contain a large amount of elastic tissue and a smaller amount of smooth muscle to having walls with a smaller amount of elastic tissue and a relatively large amount of smooth muscle. Although the arteries form a continuum from the largest to the smallest branches, they normally are classified as (1) elastic arteries, (2) muscular arteries, or (3) arterioles.

Blood flows from arterioles into capillaries. Most of the exchange that occurs between the blood and interstitial spaces occurs across the walls of capillaries. Their walls are the thinnest of all the blood vessels, blood flows through them slowly, and a greater number of them exist than any other blood vessel type.

From the capillaries, blood flows into the venous system. When compared with arteries, the walls of the veins are thinner and contain less elastic tissue and fewer smooth muscle cells. The veins increase in diameter and decrease in number, and their walls increase in thickness as they project toward the heart. They are classified as (1) venules, (2) small veins, or (3) medium or large veins.

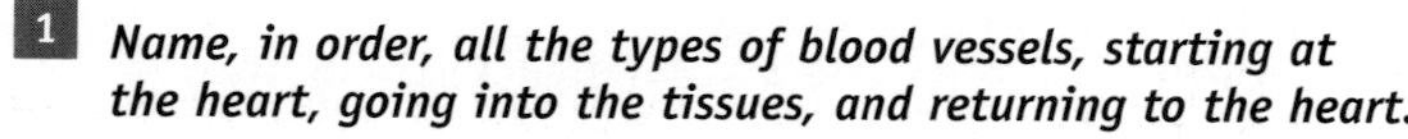

1 ***Name, in order, all the types of blood vessels, starting at the heart, going into the tissues, and returning to the heart.***

Capillaries

All blood vessels have an internal lining of simple squamous epithelial cells called the **endothelium** (en-dō-thē′lē-ŭm), which is continuous with the endocardium of the heart.

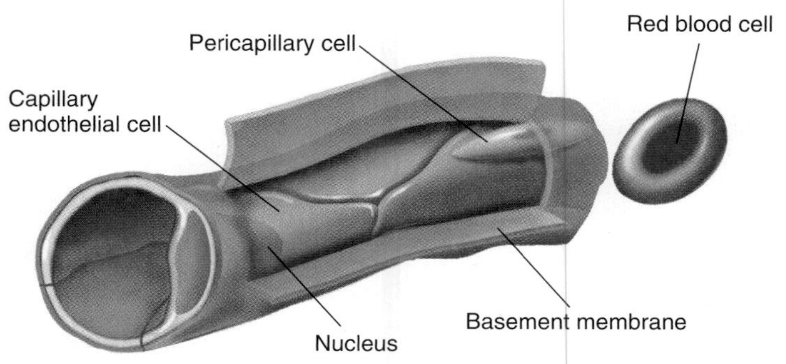

FIGURE 21.1 Capillary
Section of a capillary, showing that it is composed of flattened endothelial cells.

The capillary wall consists primarily of endothelial cells (figure 21.1), which rest on a basement membrane. Outside the basement membrane is a delicate layer of loose connective tissue that merges with the connective tissue surrounding the capillary.

Along the length of the capillary are some scattered cells that are closely associated with the endothelial cells. These scattered cells lie between the basement membrane and the endothelial cells and are called **pericapillary cells.** They are apparently fibroblasts, macrophages, or undifferentiated smooth muscle cells.

Most capillaries range from 7–9 μm in diameter, and they branch without a change in their diameter. Capillaries are variable in length, but in general they are approximately 1 mm long. Red blood cells flow through most capillaries in a single file and frequently are folded as they pass through the smaller-diameter capillaries.

Types of Capillaries

Capillaries are classified as continuous, fenestrated, or sinusoidal, depending on their diameter and permeability characteristics. **Continuous capillaries** are approximately 7–9 μm in diameter, and their walls exhibit no gaps between the endothelial cells (figure 21.2*a*). Continuous capillaries are less permeable to large molecules than are other capillary types and exist in muscle, nervous tissue, and many other locations.

In **fenestrated** (fen′es-trā′ted) **capillaries,** endothelial cells have numerous fenestrae (figure 21.2*b*). The **fenestrae** (fe-nes′trē; windows) are areas approximately 70–100 nm in diameter in which the cytoplasm is absent and the plasma membrane consists of a porous diaphragm that is thinner than the normal plasma membrane. In some capillaries, the diaphragm is not present. Fenestrated capillaries are in tissues where capillaries are highly permeable, such as in the intestinal villi, ciliary processes of the eyes, choroid plexuses of the central nervous system, and glomeruli of the kidneys.

Sinusoidal (sī-nŭ-soy′dăl) **capillaries** are larger in diameter than either continuous or fenestrated capillaries, and their basement membrane is less prominent (figure 21.2*c*) or absent. Their fenestrae are larger than those in fenestrated capillaries and gaps may exist between endothelial cells. The sinusoidal capillaries occur in such places as endocrine glands, where large molecules cross their walls.

Sinusoids are large-diameter sinusoidal capillaries. Their basement membrane is sparse and often missing, and their structure suggests that large molecules and sometimes cells can move

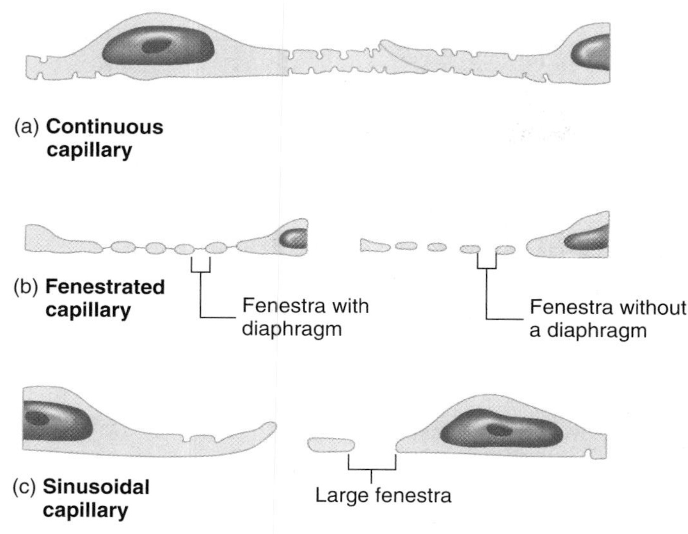

FIGURE 21.2 Structure of Capillary Walls

(*a*) Continuous capillaries have no gaps between endothelial cells and no fenestrae. They are common in muscle, nervous, and connective tissue. (*b*) Fenestrated capillaries have fenestrae 7–100 nm in diameter, covered by thin porous diaphragms, which are not present in some capillaries. They are found in intestinal villi, ciliary processes of the eyes, choroid plexuses of the central nervous system, and glomeruli of the kidneys. (*c*) Sinusoidal capillaries have larger fenestrae without diaphragms and can have gaps between endothelial cells. They are found in bone marrow, the liver, the spleen, and the lymphoid organs.

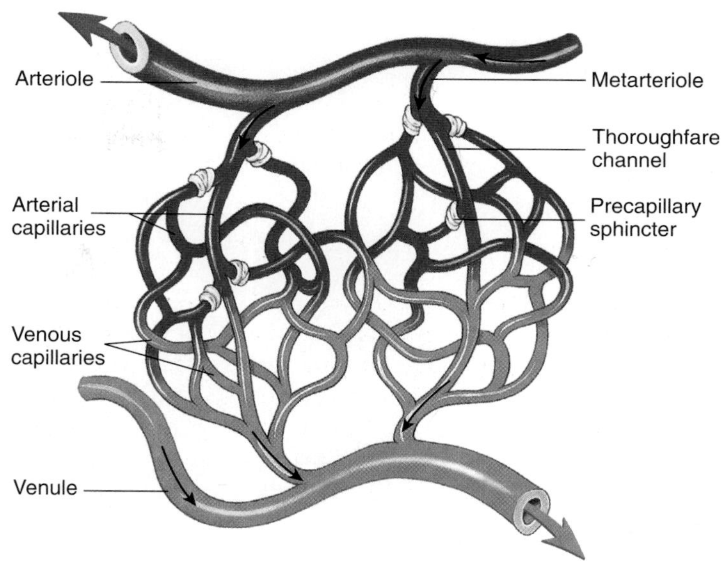

FIGURE 21.3 Capillary Network

The metarteriole giving rise to the network feeds directly from an arteriole into the thoroughfare channel, which feeds into the venule. The network forms numerous branches, which transport blood from the thoroughfare channel and can return to the channel. Smooth muscle cells, called precapillary sphincters, regulate blood flow through the capillaries. Blood flow decreases when the precapillary sphincters constrict and increases when they dilate.

readily across their walls between the endothelial cells (see figure 21.2*c*). Sinusoids are common in the liver and the bone marrow. Macrophages are closely associated with the endothelial cells of the liver sinusoids. **Venous sinuses** are similar in structure to the sinusoidal capillaries but are even larger in diameter. They exist primarily in the spleen, and they have large gaps between the endothelial cells that make up their walls.

Substances cross capillary walls by diffusing through the endothelial cells, through fenestrae, or between the endothelial cells. Lipid-soluble substances, such as oxygen and carbon dioxide, and small, water-soluble molecules readily diffuse through the plasma membrane. Larger water-soluble substances must pass through the fenestrae or gaps between the endothelial cells. In addition, transport by pinocytosis occurs, but little is known about its role in the capillaries. The walls of the capillaries are effective permeability barriers because red blood cells and large, water-soluble molecules, such as proteins, cannot readily pass through them.

Capillary Network

Arterioles supply blood to each capillary network (figure 21.3). Blood then flows through the capillary network and into the venules. The ends of capillaries closest to the arterioles are **arterial capillaries,** and the ends closest to venules are **venous capillaries.**

Blood flows from arterioles through **metarterioles** (met′ar-tēr′ē-ōlz), which have isolated smooth muscle cells along their walls. Blood flows from a metarteriole into a **thoroughfare channel,** which extends in a relatively direct fashion from a metarteriole to a venule. Blood flow through thoroughfare channels is relatively continuous. Several capillaries branch from the thoroughfare channels, and in these branches blood flow is intermittent. Smooth muscle cells called **precapillary sphincters,** which are located at the origin of the branches (see figure 21.3), regulate flow in these capillaries.

Capillary networks are more numerous and more extensive in highly metabolic tissues, such as in the lung, liver, kidney, skeletal muscle, and cardiac muscle. Capillary networks in the skin have many more thoroughfare channels than capillary networks in cardiac or skeletal muscle. Capillaries in the skin function in thermoregulation, and heat loss results from the flow of a large volume of blood through them. In muscle, however, nutrient and waste product exchange is the major function of the capillaries.

2 ***Describe the structure of a capillary.***

3 ***Compare the structure of the three types of capillaries. Explain the ways that materials pass through capillary walls.***

4 ***Describe a capillary network. Where is the smooth muscle that regulates blood flow into and through the capillary network located? What is the function of a thoroughfare channel?***

5 ***Contrast the function of capillaries in the skin with the function of capillaries in muscle tissue.***

Structure of Arteries and Veins

General Features

Except for the capillaries and the venules, the blood vessel walls consist of three relatively distinct layers, which are most apparent in the muscular arteries and least apparent in the veins. From the lumen to the outer wall of the blood vessels, the layers, or **tunics** (too′niks), are (1) the tunica intima, (2) the tunica media, (3) and the tunica adventitia, or tunica externa (figure 21.4 and figure 21.5).

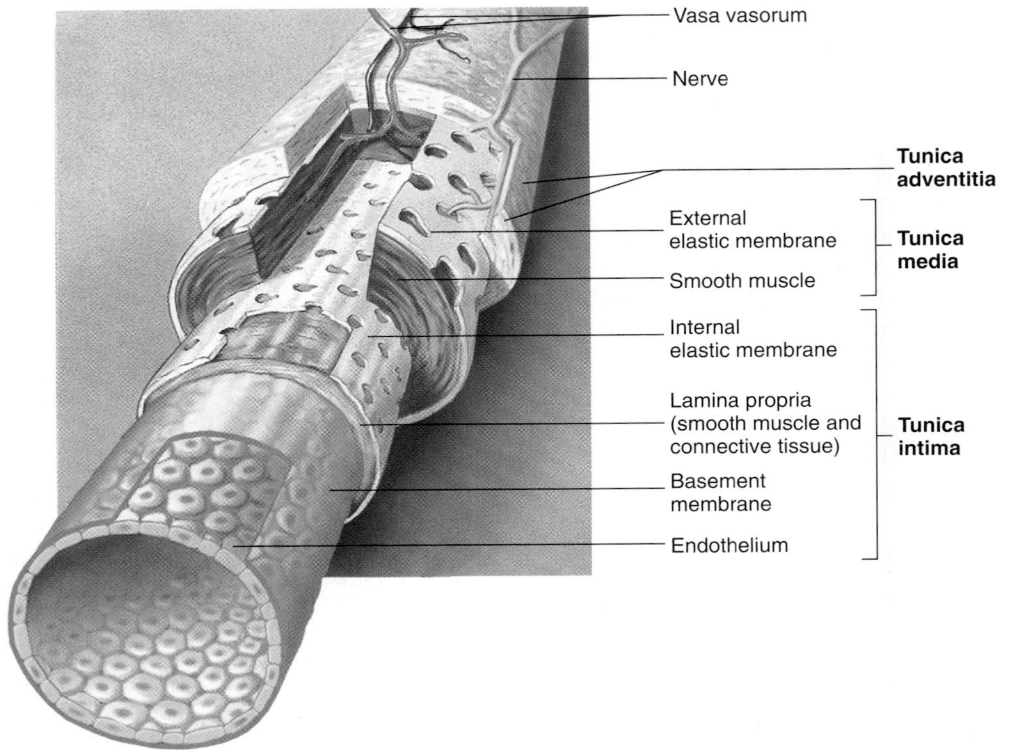

FIGURE 21.4 Histology of a Blood Vessel

The layers, or tunics, of the blood vessel wall include the intima, media, and adventitia. A vasa vasorum is a blood vessel that supplies blood to the wall of the blood vessel.

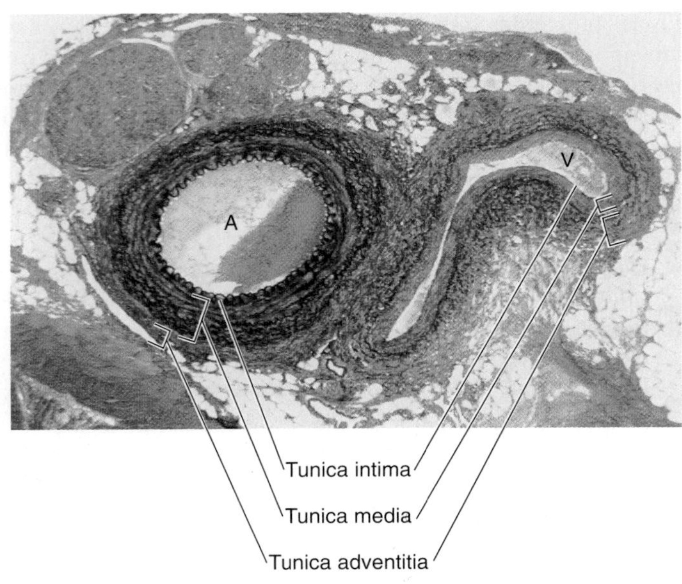

FIGURE 21.5 Photomicrograph of an Artery and a Vein

The typical structure of a medium-sized artery (A) and vein (V). Note that the artery has a thicker wall than the vein. The tunica intima, tunica media, and tunica adventitia make up the walls of the blood vessels. The predominant layer in the wall of the artery is the tunica media, with its circular layers of smooth muscle. The predominant layer in the wall of the vein is the tunica adventitia, and the tunica media is thinner.

The **tunica intima** consists of endothelium, a delicate connective tissue basement membrane, a thin layer of connective tissue called the lamina propria, and a fenestrated layer of elastic fibers called the **internal elastic membrane.** The internal elastic membrane separates the tunica intima from the next layer, the tunica media.

The **tunica media,** or middle layer, consists of smooth muscle cells arranged circularly around the blood vessel. The amount of blood flowing through a blood vessel can be regulated by contraction or relaxation of the smooth muscle in the tunica media. A decrease in blood flow results from **vasoconstriction** (vā′sō-kon-strik′shŭn, vas′ō-kon-strik′shŭn), a decrease in blood vessel diameter caused by smooth muscle contraction, whereas an increase in blood flow is produced by **vasodilation** (vā′sō-dī-lā′shŭn, vas-ō-dī-lā′shŭn), an increase in blood vessel diameter because of smooth muscle relaxation. The tunica media also contains variable amounts of elastic and collagen fibers, depending on the size of the vessel. An **external elastic membrane,** which separates the tunica media from the tunica adventitia, can be identified at the outer border of the tunica media in some arteries. A few longitudinally oriented smooth muscle cells occur in some arteries near the tunica intima.

The **tunica adventitia** (too′ni-kă ad-ven-tish′ă) is composed of connective tissue, which varies from dense connective tissue near the tunica media to loose connective tissue that merges with the connective tissue surrounding the blood vessels.

The relative thickness and composition of each layer varies with the diameter of the blood vessel and its type. The transition from one artery type or from one vein type to another is gradual, as are the structural changes.

Large Elastic Arteries

Elastic arteries have the largest diameters (figure 21.6*a*) and often are called **conducting arteries.** Pressure is relatively high in these vessels, and it fluctuates between systolic and diastolic values. A greater amount of elastic tissue and a smaller amount of smooth muscle occur in their walls, compared with the elastic tissue and smooth muscle of other arteries. The elastic fibers are responsible for the elastic characteristics of the blood vessel wall, but collagenous connective tissue determines the degree to which the arterial wall can be stretched.

The tunica intima is relatively thick. The elastic fibers of the internal and external elastic membranes merge and are not recognizable as distinct layers. The tunica media consists of a meshwork of elastic fibers with interspersed circular smooth muscle cells and some collagen fibers. The tunica adventitia is relatively thin.

Muscular Arteries

The larger **muscular arteries,** often called **medium arteries,** can be observed in a gross dissection. They include most of the smaller *unnamed* arteries. Their walls are relatively thick, compared with their diameter, mainly because the tunica media contains 25–40 layers of smooth muscle (figure 21.6*b*). The tunica intima of the muscular arteries has a well-developed internal elastic membrane. The tunica adventitia is composed of a relatively thick layer of collagenous connective tissue that blends with the surrounding connective tissue. Muscular arteries frequently are called **distributing arteries** because the smooth muscle cells allow these vessels to partially regulate blood supply to different regions of the body by either constricting or dilating.

Smaller muscular arteries range from 40–300 μm in diameter, and those that are 40 μm in diameter have approximately three or four layers of smooth muscle in their tunica media, whereas arteries that are 300 μm across have essentially the same structure as the larger muscular arteries. The small muscular arteries are adapted for vasodilation and vasoconstriction.

Arterioles

The **arterioles** (ar-tēr′ē-ōlz) transport blood from small arteries to capillaries and are the smallest arteries in which the three tunics can be identified. They range from approximately 40 μm to as small as 9 μm in diameter. The tunica intima has no observable internal elastic membrane, and the tunica media consists of one or two layers of circular smooth muscle cells. The arterioles, like the small arteries, are capable of vasodilation and vasoconstriction.

Venules and Small Veins

Venules (ven′-oolz, vē′noolz), with a diameter of up to 40–50 μm, are tubes composed of endothelium resting on a delicate basement membrane. Their structure, except for their diameter, is very similar to that of capillaries. A few isolated smooth muscle cells exist outside the endothelial cells, especially in the larger venules. As the vessels increase to 0.2–0.3 mm in diameter, the smooth muscle cells form a continuous layer; the vessels then are called **small veins.** The small veins also have a tunica adventitia composed of collagenous connective tissue.

The venules collect blood from the capillaries and transport it to small veins, which in turn transport it to medium-sized veins. Nutrient exchange occurs across the venule walls, but, as the walls of the small veins increase in thickness, the degree of nutrient exchange decreases.

Medium and Large Veins

Most of the veins observed in gross anatomical dissections, except for the large veins, are **medium veins.** They collect blood from small veins and deliver it to large veins. The **large veins** transport blood from the medium veins to the heart. Their tunica intima is thin and consists of endothelial cells, a relatively thin layer of collagenous connective tissue, and a few scattered elastic fibers. The tunica media is also thin and is composed of a thin layer of circularly arranged smooth muscle cells containing some collagen fibers and a few sparsely distributed elastic fibers. The tunica adventitia, which is composed of collagenous connective tissue, is the predominant layer (figure 21.6*c*).

Valves

Veins having diameters greater than 2 mm contain **valves** that allow blood to flow toward the heart but not in the opposite direction (figure 21.6*d*). The valves consist of folds in the tunica intima that form two flaps that are shaped and function like the semilunar valves of the heart. The two folds overlap in the middle of the vein so that, when blood attempts to flow in a reverse direction, the valves occlude the vessel. Many valves are present in the medium veins, and the number is greater in veins of the lower extremities than in veins of the upper extremities.

Varicose Veins, Phlebitis, and Gangrene

Stretching of the vein walls in the lower limbs causes valves to become incompetent, which results in **varicose veins.** The veins become so dilated that the flaps of the venous valves no longer overlap and prevent the backflow of blood. As a consequence, the venous pressure is greater than normal in the veins of the lower limbs, resulting in edema. Blood flow in the veins can become sufficiently stagnant that the blood clots. The condition can result in **phlebitis** (fle-bī′tis), which is inflammation of the veins. If the inflammation is severe and blood flow becomes stagnant in a large area, it can lead to **gangrene** (gang′grēn), which is tissue death caused by a reduction in or loss of blood supply. Some people have a genetic propensity for the development of varicose veins. For women with that genetic propensity, some conditions increase the pressure in veins, causing them to stretch, and varicose veins can develop. One such condition is pregnancy, in which the venous pressure in the veins that drain the lower limbs increases because of compression of the veins by the expanded uterus.

6 ***Name the three layers of a blood vessel. What kinds of tissue are in each layer?***

7 ***Compare the amount of elastic fibers and smooth muscle found in each type of artery and vein.***

8 ***What is the function of valves in blood vessels? In which blood vessels are they found?***

(a) **Elastic arteries.** The tunica media is mostly elastic connective tissue. Elastic arteries recoil when stretched, which prevents blood pressure from falling rapidly.

(b) **Muscular arteries.** The tunica media is a thick layer of smooth muscle. Muscular arteries regulate blood flow to different regions of the body.

(c) **Medium and large veins.** All three tunics are present. The tunica media is thin but can regulate vessel diameter because blood pressure in the venous system is low. The predominant layer is the tunica adventitia.

(d) Folds in the endothelium form the valves of veins, which allow blood to flow toward the heart but not in the opposite direction.

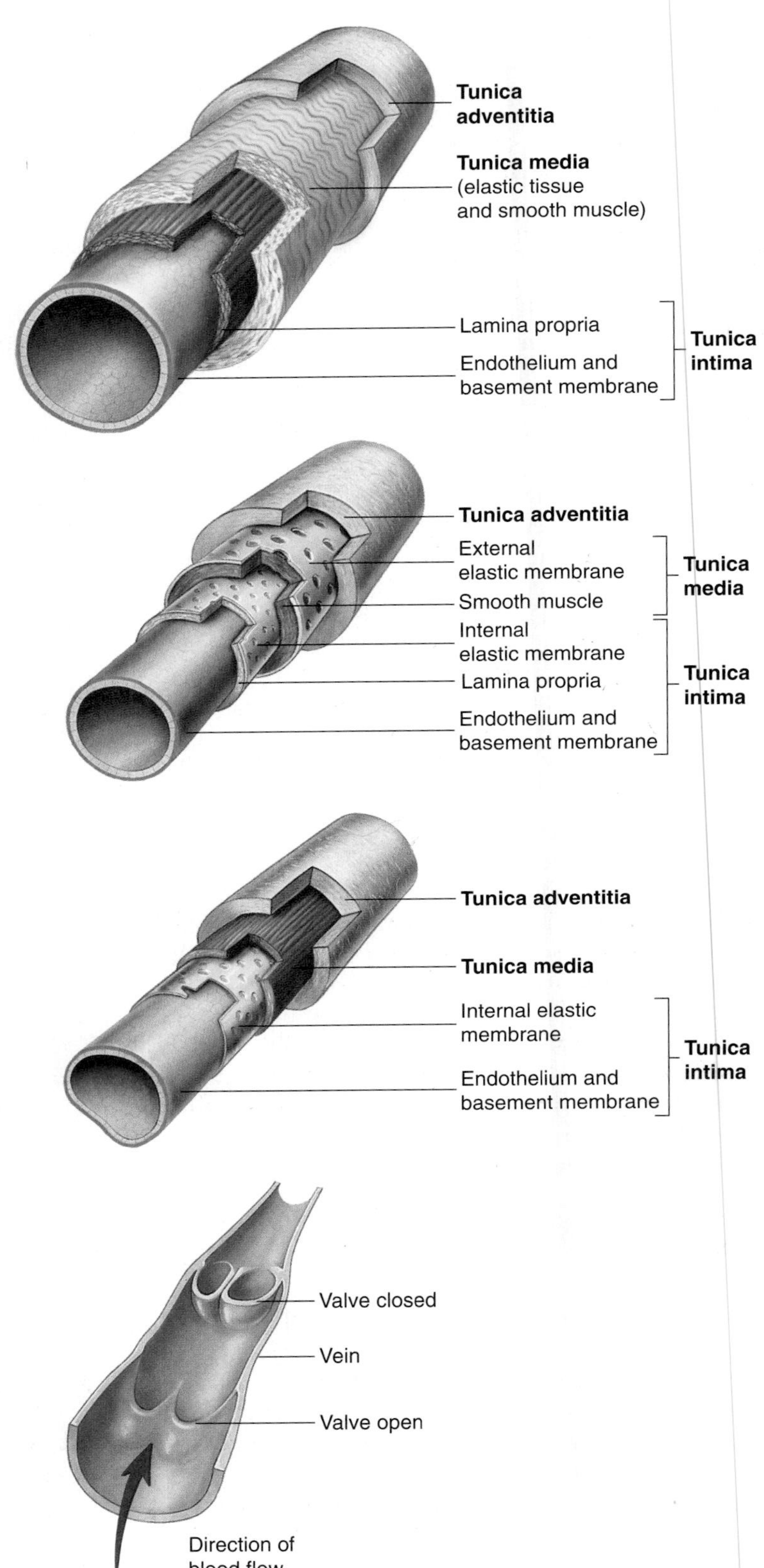

FIGURE 21.6 Structural Comparison of Blood Vessel Types

(*a*) Elastic arteries are large-diameter arteries with thick walls that contain a large amount of elastic connective tissue in the tunica media. (*b*) Muscular arteries have a distinctive layer of smooth muscle cells in the tunica media, and they are capable of constriction and dilation. (*c*) Medium veins have thinner walls. The tunica media is thinner than the tunica media in arteries and contains fewer smooth muscle cells. The dominant layer in the veins is the tunica adventitia. (*d*) The valves in veins are folds in the endothelium that allow blood to flow toward the heart but not in the opposite direction.

Vasa Vasorum

For arteries and veins greater than 1 mm in diameter, nutrients cannot diffuse from the lumen of the vessel to all of the layers of the wall. Nutrients are, therefore, supplied to their walls by way of small blood vessels called **vasa vasorum** (vā′să vā′sor-ŭm), which penetrate from the exterior of the vessel to form a capillary network in the tunica adventitia and the tunica media (see figure 21.4).

Arteriovenous Anastomoses

Arteriovenous anastomoses (ă-nas′tō-mō′sez) allow blood to flow from arterioles to small veins without passing through capillaries. A **glomus** (glō′mŭs, pl. *glomera,* glom′er-ă) is an arteriovenous anastomosis that consists of arterioles with abundant smooth muscle in their walls. The vessels are branched and coiled and are surrounded by connective tissue sheaths.

Glomera are present in large numbers in the sole of the foot, the palm of the hand, the terminal phalanges, and the nail beds. The glomera function in body temperature regulation. As the body temperature decreases, glomera constrict and less blood flows through them. As the body temperature increases, glomera dilate and more blood flows through them, increasing the rate of heat loss from the body. **Pathologic arteriovenous anastomoses** can result from injury or tumors. They cause the direct flow of blood from arteries to veins and can, if they are sufficiently large, lead to heart failure because of the tremendous increase in venous return to the heart.

Portal Veins

Portal (pōr′tăl, door) **veins** are veins that begin in a primary capillary network, extend some distance, and end in a secondary capillary network. There is no pumping mechanism like the heart between the two capillary networks. There are two systems of portal veins in humans. The hepatic portal veins carry blood from the capillaries in the gastrointestinal tract and spleen to dilated capillaries, called sinusoids, in the liver (see figure 21.26). The hypothalamohypophysial portal veins carry blood from the hypothalamus of the brain to the anterior pituitary gland (see figure 18.3).

9 ***Define vasa vasorum and arteriovenous anastamosis, and give their function. Define portal veins and name two examples.***

Nerves

Unmyelinated sympathetic nerve fibers (see figure 21.4) richly innervate the walls of most blood vessels. Some blood vessels, such as those in the penis and clitoris, are innervated by parasympathetic fibers. The nerve fibers branch to form plexuses in the tunica adventitia, and nerve terminals containing neurotransmitter vesicles project among the smooth muscle cells of the tunica media. Synapses consist of several enlargements of each of the nerve fibers among the smooth muscle cells. Small arteries and arterioles are innervated to a greater extent than other blood vessel types. The response of the blood vessels to sympathetic stimulation is vasoconstriction. Parasympathetic stimulation of blood vessels in the penis and clitoris results in vasodilation.

The smooth muscle cells of blood vessels act to some extent as a functional unit. Frequent gap junctions exist between adjacent smooth muscle cells; as a consequence, stimulation of a few smooth muscle cells in the vessel wall results in constriction of a relatively large segment of the blood vessel.

A few myelinated sensory neurons innervate some blood vessels and function as baroreceptors. They monitor stretch in the blood vessel wall and detect changes in blood pressure.

10 ***Describe the innervation of the walls of blood vessels. Which types of vessels have the greatest innervation?***

Aging of the Arteries

The walls of all arteries undergo changes as they age, although some arteries change more rapidly than others and some individuals are more susceptible to change than others. The most significant change occurs in the large elastic arteries, such as the aorta, the large arteries that carry blood to the brain, and the coronary arteries. The age-related changes described here refer to these blood vessel types. Changes in muscular arteries do occur, but they are less dramatic and they often do not result in the disruption of normal blood vessel function.

Arteriosclerosis (ar-tēr′ē-ō-skler-o′sis; hardening of the arteries) consists of degenerative changes in arteries that make them less elastic. These changes occur in many individuals, and they become more severe with advancing age. Arteriosclerosis greatly increases resistance to blood flow. Therefore, advanced arteriosclerosis reduces the normal circulation of blood and greatly increases the work performed by the heart.

Arteriosclerosis involves general hypertrophy of the tunica intima, including the internal elastic membrane, and hypertrophy of the tunica media. For example, when arteriosclerosis is associated with hypertension, there is an increase in smooth muscle and elastic tissue of the arterial walls. The elastic tissue can form concentric layers in the tunica intima, and it becomes less elastic, and some of the smooth muscle cells of the tunica media can ultimately be replaced by collagen fibers. Arteriosclerosis in some older people can involve arteries, primarily of the lower limbs, in which calcium deposits form in the tunica media of the arteries with little or no encroachment on their lumens.

Atherosclerosis (ath′er-ō-skler-ō′sis) is the deposition of material in the walls of arteries to form distinct plaques. It is a common type of arteriosclerosis, and, like other types of arteriosclerosis, it is age-related. It also develops more rapidly in people who exhibit risk factors for atherosclerosis. Atherosclerosis affects primarily medium and larger arteries, including the coronary arteries. The plaques form as macrophages containing cholesterol accumulate in the tunica intima, and smooth muscle cells of the tunica media proliferate (figure 21.7). After the plaques enlarge, they consist of smooth muscle cells; leukocytes; lipids, including cholesterol; and, in the largest plaques, fibrous connective tissue and calcium deposits. The plaques narrow the lumens of blood vessels and make their walls less elastic. Atherosclerotic plaques can become so large that they severely restrict, or block, blood flow through arteries, and the plaques are sites where thromboses and emboli form.

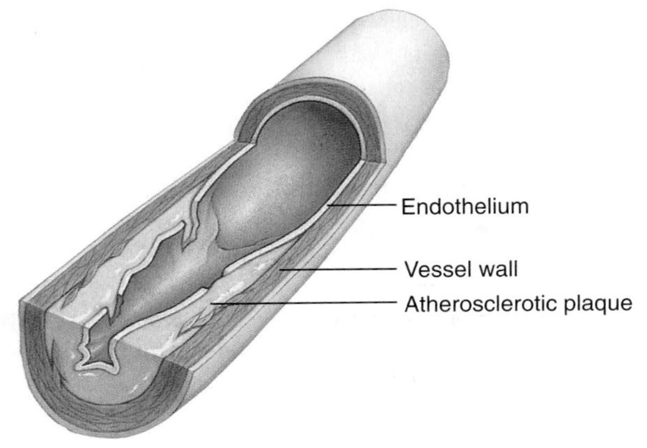

FIGURE 21.7 Atherosclerotic Plaque in an Artery
Atherosclerotic plaques develop within the tissue of the artery wall.

Some investigators propose that arteriosclerosis may not be a pathologic process. Instead, it may be an aging or wearing out process. Evidence also suggests that arteriosclerosis may result from inflammation, which, in some cases, may be the result of an autoimmune disease. Atherosclerosis has been studied extensively, and there are many risk factors associated with the development of atherosclerotic plaques. These factors include being at an advanced age, being a male, being a postmenopausal woman, having a family history of atherosclerosis, smoking cigarettes, having hypertension, having diabetes mellitus, having increased blood LDL and cholesterol levels, being overweight, leading a sedentary lifestyle, and having high blood triglyceride levels. Avoiding the environmental factors that influence atherosclerosis slows the development of atherosclerotic plaques. In some cases, the severity of the plaques can be reduced. For example, treatments such as regulating blood glucose levels in people with diabetes mellitus, and taking drugs that lower blood cholesterol levels, especially in people with high blood cholesterol levels that cannot be controlled by dietary changes, can provide some protection.

11 ***Describe the changes that occur in arteries due to aging. In which vessels do the most significant changes occur?***

12 ***Name the factors associated with premature arteriosclerosis.***

PULMONARY CIRCULATION

The heart pumps blood from the right ventricle into the **pulmonary** (pŭl′mō-nār-ē; relating to the lungs) **trunk** (figure 21.8). This short vessel, 5 cm long, branches into the right and left **pulmonary arteries,** one transporting blood to each lung. Within the lungs, gas exchange occurs between air in the lungs and the blood. Two **pulmonary veins** exit each lung and enter the left atrium (see figure 20.10).

13 ***For the vessels of the pulmonary circulation, give their starting point, ending point, and function.***

SYSTEMIC CIRCULATION: ARTERIES

Oxygenated blood entering the heart from the pulmonary veins passes through the left atrium into the left ventricle and from the left ventricle into the aorta. Blood flows from the aorta to all parts of the body (see figure 21.8).

Aorta

All **arteries** of the systemic circulation are derived either directly or indirectly from the **aorta** (ā-ōr′tă), which usually is divided into three general parts: the ascending aorta, the aortic arch, and the descending aorta. The descending aorta is divided further into a thoracic aorta and an abdominal aorta (see figure 21.14).

At its origin from the left ventricle, the aorta is approximately 2.8 cm in diameter. Because it passes superiorly from the heart, this part is called the **ascending aorta.** It is approximately 5 cm long and has only two arteries branching from it: the right and left **coronary arteries,** which supply blood to the cardiac muscle (see figure 20.6*a*).

The aorta then arches posteriorly and to the left as the **aortic arch.** Three major branches, which carry blood to the head and upper limbs, originate from the aortic arch: the brachiocephalic artery, the left common carotid artery, and the left subclavian artery.

Trauma and the Aorta

Trauma that ruptures the aorta is almost immediately fatal. Trauma can also lead to an **aneurysm** (an′ū-rizm), however, which is a bulge caused by a weakened spot in the aortic wall. If the weakened aortic wall leaks blood slowly into the thorax, the aneurysm must be corrected surgically. Also, once the aneurysm is formed it is likely to enlarge and becomes more likely to rupture. The majority of traumatic aortic arch ruptures occur during automobile accidents and result from the great force with which the body is thrown into the steering wheel, dashboard, or other objects. Waist-type safety belts alone do not prevent this type of injury as effectively as shoulder-type safety belts and air bags.

The next part of the aorta is the longest part, the **descending aorta.** It extends through the thorax in the left side of the mediastinum and through the abdomen to the superior margin of the pelvis. The **thoracic aorta** is the portion of the descending aorta located in the thorax. It has several branches that supply various structures between the aortic arch and the diaphragm. The **abdominal aorta** is the part of the descending aorta between the diaphragm and the point at which the aorta ends by dividing into the two **common iliac** (il′ē-ak; relating to the flank area) **arteries.** The abdominal aorta has several branches that supply the abdominal wall and organs. Its terminal branches, the common iliac arteries, supply blood to the pelvis and lower limbs.

14 ***Name the parts of the aorta.***

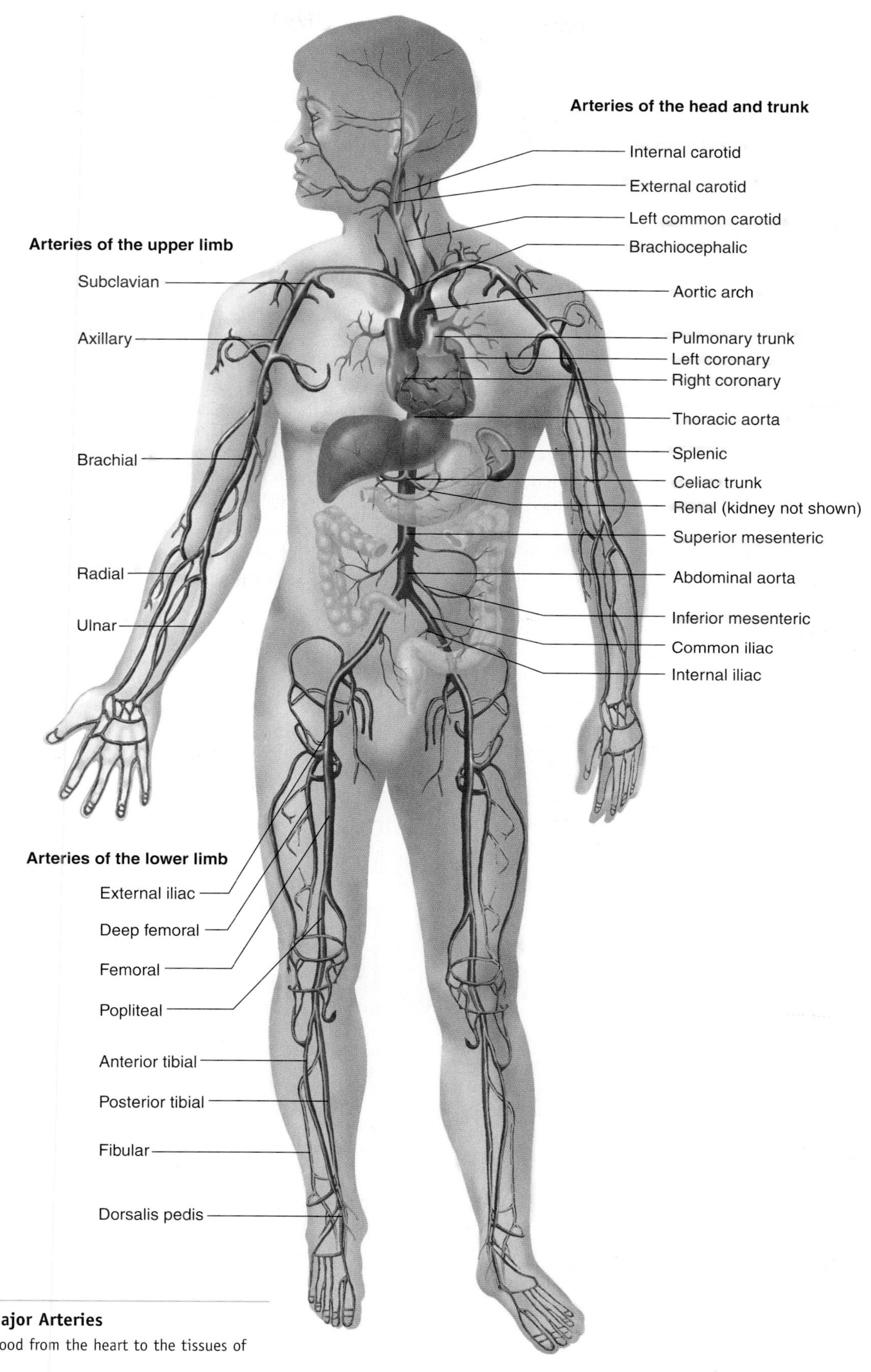

FIGURE 21.8 Major Arteries

The arteries carry blood from the heart to the tissues of the body.

Coronary Arteries

The **coronary** (kōr′o-năr-ē; encircling the heart like a crown) **arteries,** which are the only branches of the ascending aorta, are described in chapter 20.

15 ***Name the arteries that branch from the aorta to supply the heart.***

Arteries to the Head and the Neck

The first vessel to branch from the aortic arch is the **brachiocephalic** (brā′kē-ō-se-fal′ik; arm and head) **artery** (figure 21.9). It is a very short artery, and it branches at the level of the clavicle to form the **right common carotid** (ka-rot′id) **artery,** which transports blood to the right side of the head and neck, and the **right subclavian** (sŭb-klā′vē-an; below the clavicle) **artery,** which transports blood to the right upper limb (see figures 21.8, 21.9, 21.11, 21.12, and 21.13).

The second and third branches of the aortic arch are the **left common carotid artery,** which transports blood to the left side of the head and neck, and the **left subclavian artery,** which transports blood to the left upper limb.

The common carotid arteries extend superiorly, without branching, along either side of the neck from their base to the inferior angle of the mandible, where each common carotid artery branches into **internal** and **external carotid arteries** (see figures 21.9 and 21.11). At the point of bifurcation on each side of the neck, the common carotid artery and the base of the internal carotid artery are dilated slightly to form the **carotid sinus,** which is important in monitoring blood pressure (baroreceptor reflex). The external carotid arteries have several branches that supply the structures of the neck and face (table 21.1; see figures 21.9 and 21.11). The internal carotid arteries, together with the vertebral arteries, which are branches of the subclavian arteries, supply the brain (see table 2.1 and figures 21.9, 21.10, and 21.11).

PREDICT 1

The term *carotid* means to put to sleep, implying that, if the carotid arteries are occluded for even a short time, the patient can lose consciousness (go to sleep). The blood supply to the brain is extremely important to its function. Elimination of this supply for even a relatively short time can result in permanent brain damage because the brain is dependent on oxidative metabolism and quickly malfunctions in the absence of oxygen. What is the physiologic significance of arteriosclerosis, which slowly reduces blood flow through the carotid arteries?

Branches of the subclavian arteries, the **left** and **right vertebral arteries,** pass through the transverse foramina of the cervical

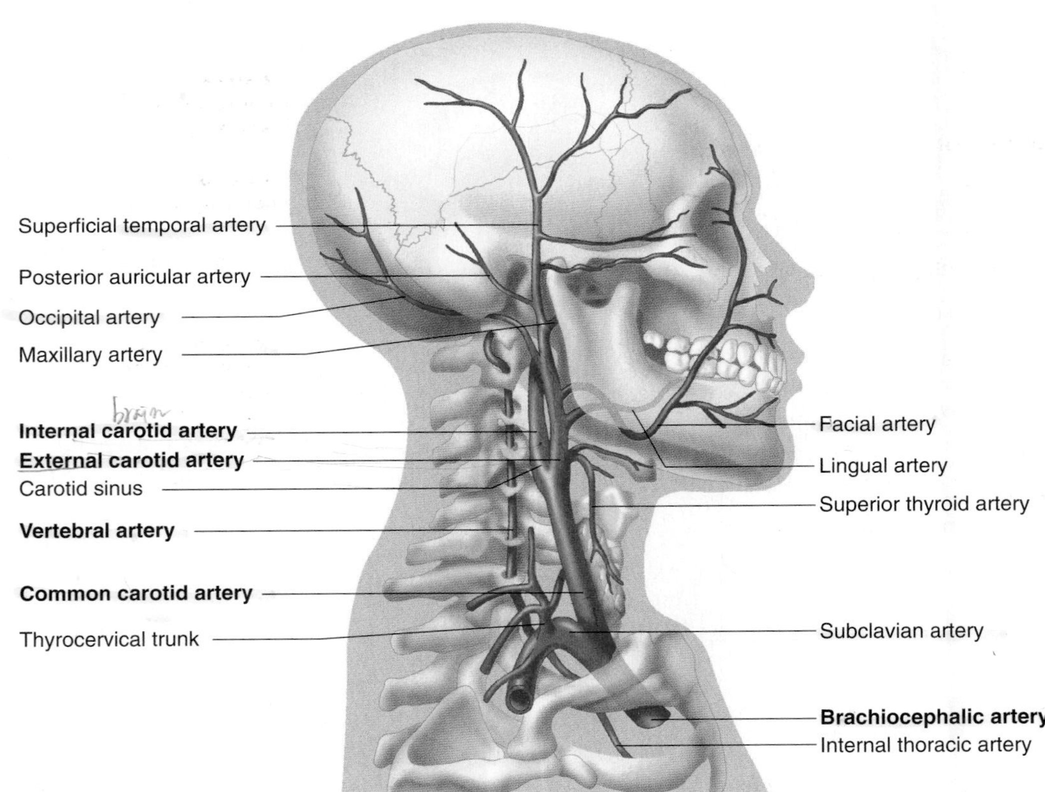

FIGURE 21.9 Arteries of the Head and Neck

The brachiocephalic artery, the right common carotid artery, and the right vertebral artery are the major arteries that supply the head and neck. The right common carotid artery branches from the brachiocephalic artery, and the vertebral artery branches from the subclavian artery.

TABLE 21.1 Arteries of the Head and Neck (Figures 21.9 and 21.10; see Figure 21.11)

Arteries	Tissues Supplied
Common Carotid Arteries	Head and neck by branches listed below
External Carotid	
Superior thyroid	Neck, larynx, and thyroid gland
Lingual	Tongue, mouth, and submandibular and sublingual glands
Facial	Mouth, pharynx, and face
Occipital	Posterior head and neck and meninges around posterior brain
Posterior auricular	Middle and inner ear, head, and neck
Ascending pharyngeal	Deep neck muscles, middle ear, pharynx, soft palate, and meninges around posterior brain
Superficial temporal	Temple, face, and anterior ear
Maxillary	Middle and inner ears, meninges, lower jaw and teeth, upper jaw and teeth, temple external eye structures, face, palate, and nose
Internal Carotid	
Posterior communicating	Joins the posterior cerebral artery
Anterior cerebral	Anterior portions of the cerebrum and forms the anterior communicating arteries
Middle cerebral	Most of the lateral surface of the cerebrum
Vertebral Arteries (Branches of the Subclavian Arteries)	
Anterior spinal	Anterior spinal cord
Posterior inferior cerebellar	Cerebellum and fourth ventricle
Basilar Artery (Formed by Junction of Vertebral Arteries)	
Anterior interior cerebellar	Cerebellum
Superior cerebellar	Cerebellum and midbrain
Posterior cerebral	Posterior portions of the cerebrum

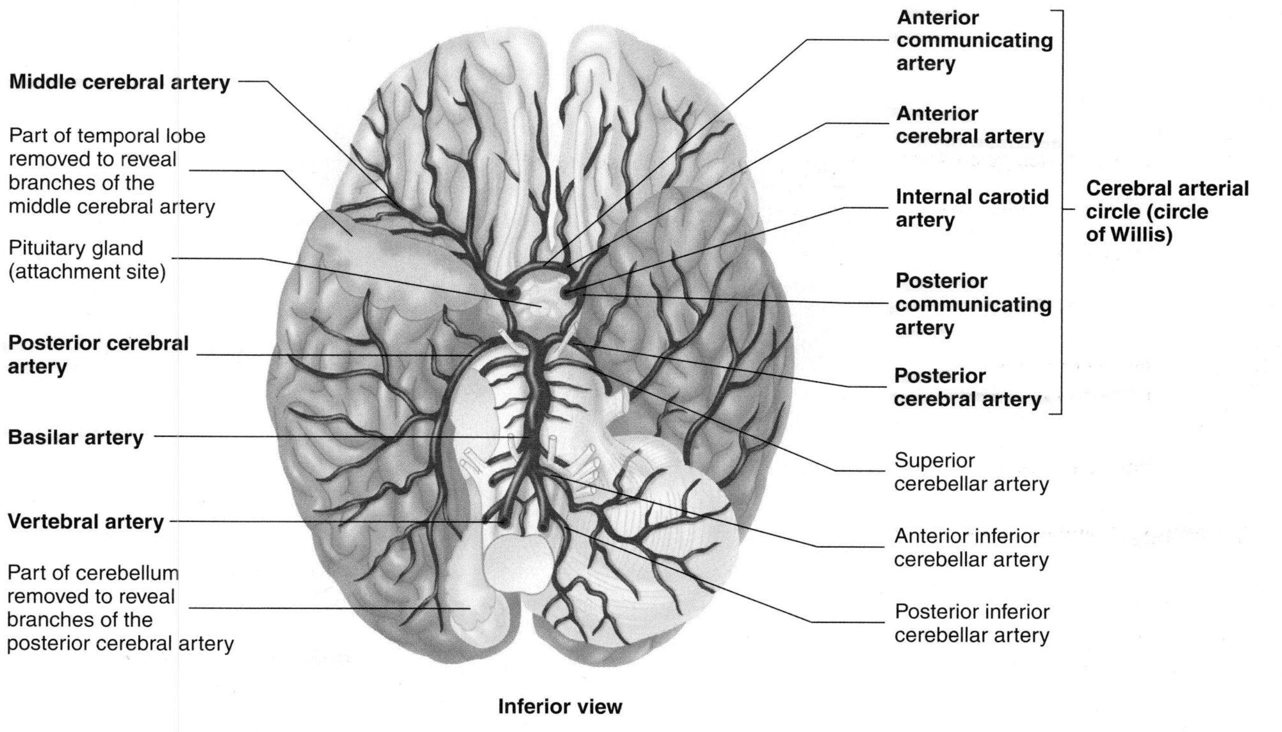

FIGURE 21.10 Cerebral Arterial Circle (Circle of Willis)

The blood supply to the brain is carried by the internal carotid and vertebral arteries. The vertebral arteries join to form the basilar artery. Branches of the internal carotid arteries and basilar artery supply blood to the brain and complete a circle of arteries around the pituitary gland and the base of the brain called the cerebral arterial circle (circle of Willis).

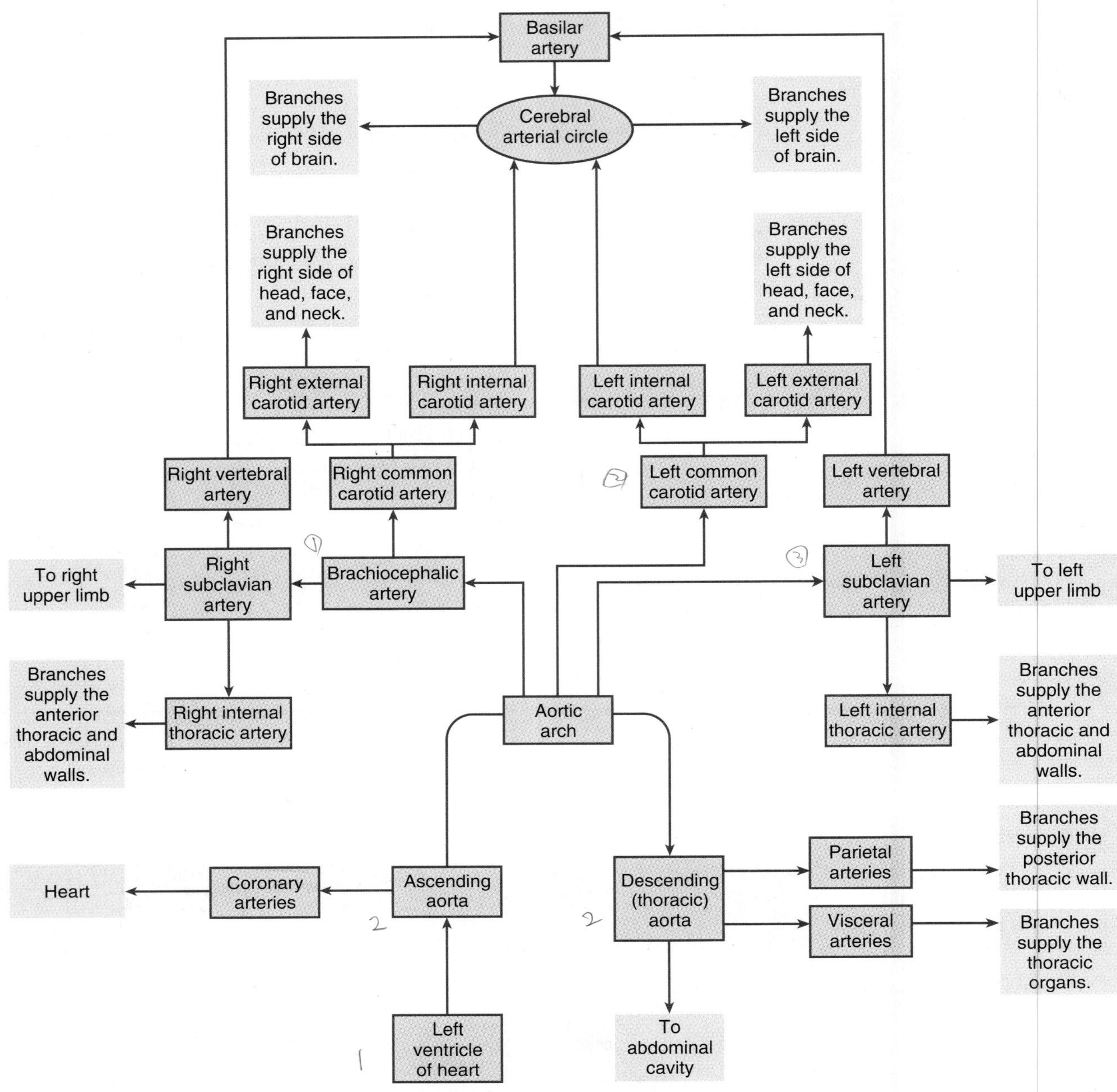

FIGURE 21.11 Major Arteries of the Head and Thorax

vertebrae and enter the cranial cavity through the foramen magnum. They give off arteries to the cerebellum and then unite to form a single, midline **basilar** (bas′i-lăr) **artery** (see figures 21.10 and 21.11; see table 21.1). The basilar artery gives off branches to the pons and the cerebellum and then branches to form the **posterior cerebral arteries,** which supply the posterior part of the cerebrum (see figure 21.10).

The internal carotid arteries enter the cranial vault through the carotid canals and terminate by forming the **middle cerebral arteries,** which supply large parts of the lateral cerebral cortex (see figure 21.10). Posterior branches of these arteries, called the **posterior communicating arteries,** unite with the posterior cerebral arteries. Anterior branches, called the **anterior cerebral arteries,** supply blood to the frontal lobes of the brain. The anterior cerebral arteries are in turn connected by an **anterior communicating artery,** which completes a circle around the pituitary gland and the base of the brain called the **cerebral arterial circle** (circle of Willis; see figures 21.10 and 21.11).

Stroke

A **stroke** is a sudden neurologic disorder often caused by a decreased blood supply to a part of the brain. It can occur as a result of a **thrombosis** (throm-bō′sis; a stationary clot), an **embolism** (em′bō-lizm; a floating clot that becomes lodged in smaller vessels), or a **hemorrhage** (hem′ŏ-rij; a rupture or leaking of blood from vessels). Any one of these conditions can result in a loss of blood supply or in trauma to a part of the brain. As a result, the tissue normally supplied by the arteries becomes **necrotic** (nĕ-krot′ik; dead). The affected area is called an **infarct** (in′farkt; to stuff into, an area of cell death). The neurologic results of a stroke are described in chapter 14.

16 ***Name the arteries that branch from the aorta to supply the head and neck.***

17 ***List the arteries that are part of, and branch from, the cerebral arterial circle.***

Arteries of the Upper Limb

The three major arteries of the upper limb, called the **subclavian, axillary,** and **brachial arteries,** are a continuum rather than a branching system. The axillary artery is the continuation of the subclavian artery, and the brachial artery is the continuation of the axillary artery. The subclavian artery is located deep to the clavicle, the axillary artery is within the axilla, and the brachial artery lies within the arm itself (table 21.2 and figures 21.12 and 21.13).

The brachial artery divides at the elbow into **ulnar** and **radial arteries,** which form two arches within the palm of the hand, referred to as the superficial and deep palmar arches. The **superficial palmar arch** is formed by the ulnar artery and is completed by anastomosing with the radial artery. The **deep palmar arch** is formed by the radial artery and is completed by anastomosing with the ulnar artery. This arch is not only deep to the superficial arch but is proximal as well.

Digital (dij′i-tăl; relating to the digits—the fingers and the thumb) **arteries** branch from each of the two palmar arches and unite to form single arteries on the medial and lateral sides of each digit.

18 ***Name the arteries that branch from the aorta to supply the upper limbs.***

19 ***List, in order, the arteries that travel through the upper limb to the digits.***

Thoracic Aorta and Its Branches

The branches of the thoracic aorta are divided into two groups: the **visceral branches** supplying the thoracic organs and the **parietal branches** supplying the thoracic wall (figure 21.14*a* and *b* and table 21.3). The visceral branches supply a portion of the lungs, including the bronchi and bronchioles (see chapter 23), the esophagus, and the pericardium. Even though a large quantity of blood flows to the lungs through the pulmonary arteries, the bronchi and bronchioles require a separate oxygenated blood supply through small bronchial branches from the thoracic aorta.

The walls of the thorax are supplied with blood by the **intercostal** (in-ter-kos′tăl; between the ribs) **arteries,** which consist of two sets: the anterior intercostals and the posterior intercostals. The **anterior intercostals** are derived from the **internal thoracic arteries,** which are branches of the subclavian arteries and lie on

TABLE 21.2 Arteries of the Upper Limbs (See Figure 21.12)

Arteries	Tissues Supplied
Subclavian Arteries (Right Subclavian Originates from the Brachiocephalic Artery, and Left Subclavian Originates Directly from the Aorta)	
Vertebral	Spinal cord and cerebellum form the basilar artery (see table 21.1)
Internal thoracic	Diaphragm, mediastinum, pericardium, anterior thoracic wall, and anterior abdominal wall
Thyrocervical trunk	Inferior neck and shoulder
Axillary Arteries (Continuation of Subclavian)	
Thoracoacromial	Pectoral region and shoulder
Lateral thoracic	Pectoral muscles, mammary gland, and axilla
Subscapular	Scapular muscles
Brachial Arteries (Continuation of Axillary Arteries)	
Deep brachial	Arm and humerus
Radial	Forearm
Deep palmar arch	Hand and fingers
Digital arteries	Fingers
Ulnar	Forearm
Superficial palmar arch	Hand and fingers
Digital arteries	Fingers

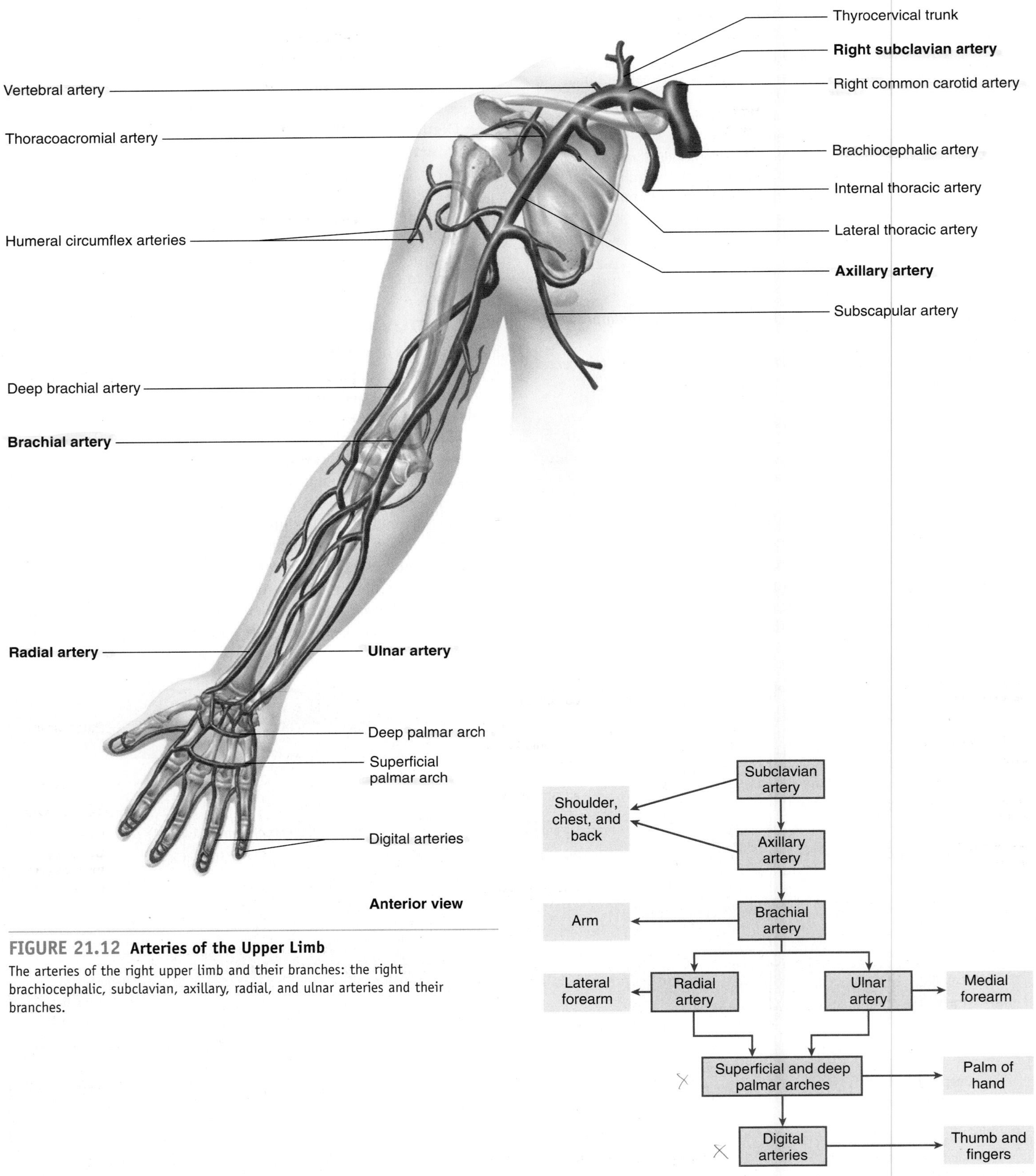

FIGURE 21.12 Arteries of the Upper Limb
The arteries of the right upper limb and their branches: the right brachiocephalic, subclavian, axillary, radial, and ulnar arteries and their branches.

FIGURE 21.13 Major Arteries of the Shoulder and Upper Limb

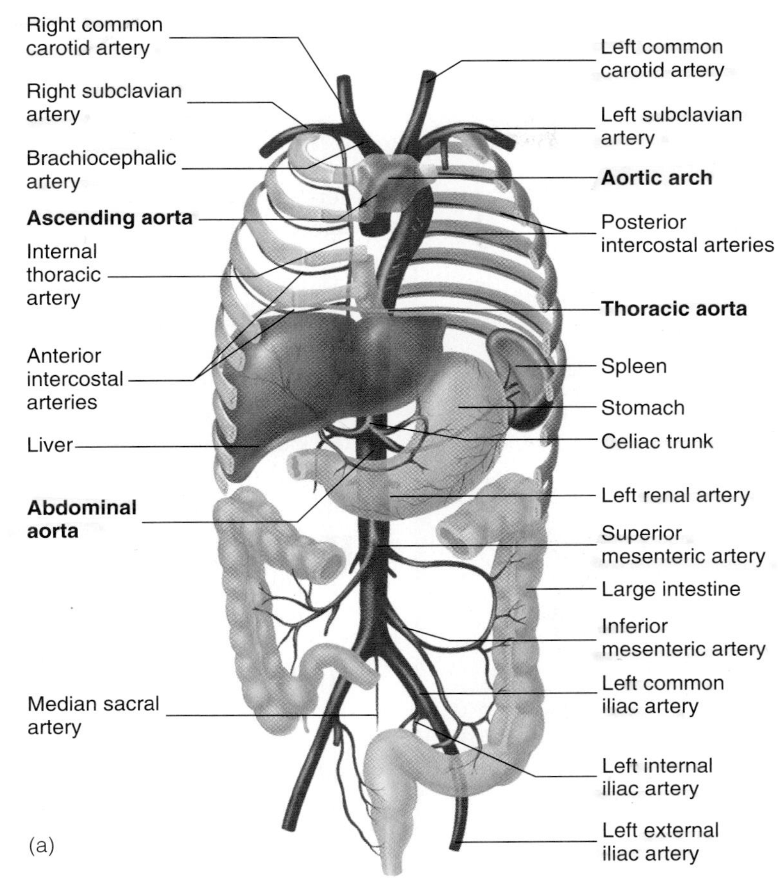

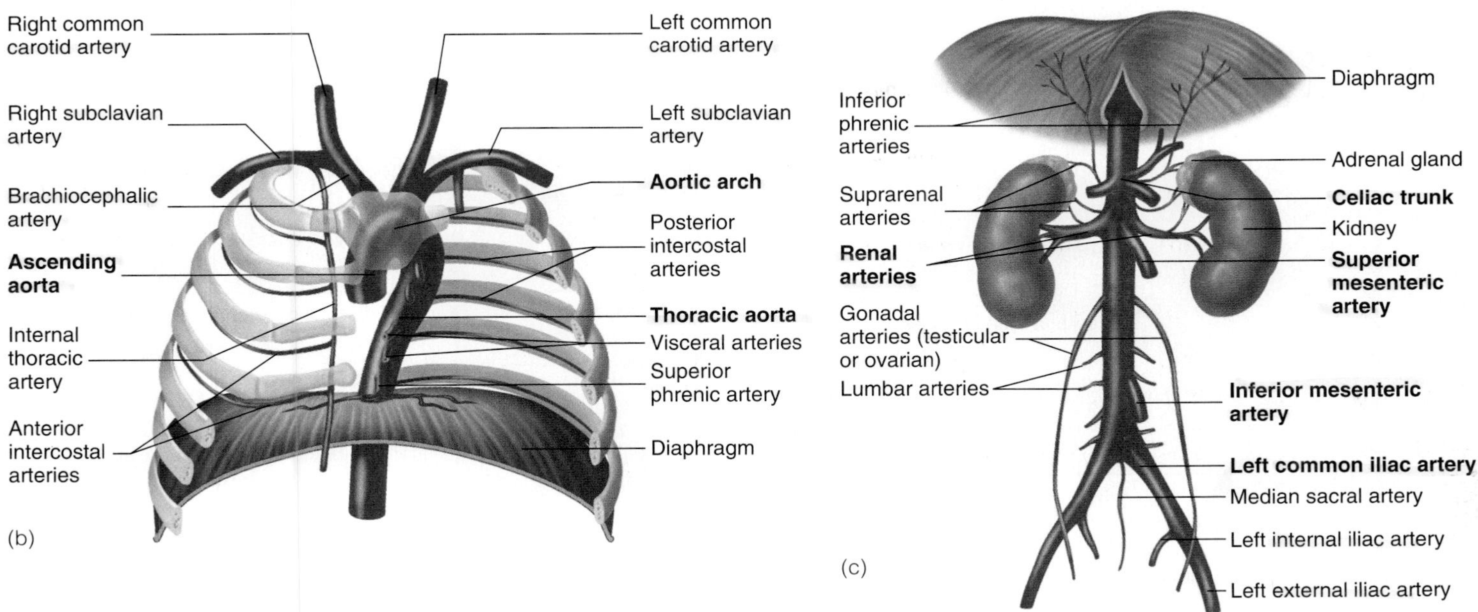

FIGURE 21.14 Branches of the Aorta

(*a*) The aorta is considered in three portions: the ascending aorta, the aortic arch, and the descending aorta. The descending aorta consists of the thoracic and abdominal aorta. (*b*) The thoracic aorta. (*c*) The abdominal aorta.

TABLE 21.3 Thoracic and Abdominal Aorta (Figures 21.14 and 21.15)	
Arteries	**Tissues Supplied**
Thoracic Aorta	
Visceral Branches	
Bronchial	Lung tissue
Esophageal	Esophagus
Parietal Branches	
Intercostal	Thoracic wall
Superior phrenic	Superior surface of diaphragm
Abdominal Aorta	
Visceral Branches	
Unpaired	
Celiac trunk	
Left gastric	Stomach and esophagus
Common hepatic	
Gastroduodenal	Stomach and duodenum
Right gastric	Stomach
Hepatic	Liver
Splenic	Spleen and pancreas
Left gastroepiploic	Stomach
Superior mesenteric	Pancreas, small intestine, and colon
Inferior mesenteric	Descending colon and rectum
Paired	
Suprarenal	Adrenal gland
Renal	Kidney
Gonadal	
Testicular (male)	Testis and ureter
Ovarian (female)	Ovary, ureter, and uterine tube
Parietal Branches	
Inferior phrenic	Adrenal gland and inferior surface of diaphragm
Lumbar	Lumbar vertebrae and back muscles
Median sacral	Inferior vertebrae
Common iliac	
External iliac	Lower limb (see table 21.5)
Internal iliac	Lower back, hip, pelvis, urinary bladder, vagina, uterus, rectum, and external genitalia (see table 21.4)

the inner surface of the anterior thoracic wall (see table 21.3 and figure 21.14*a* and *b*). The **posterior intercostals** are derived as bilateral branches directly from the descending aorta. The anterior and posterior intercostal arteries lie along the inferior margin of each rib and anastomose with each other approximately midway between the ends of the ribs. **Superior phrenic** (fren′ik; to the diaphragm) **arteries** supply blood to the diaphragm.

20 ***Name the two types of branches arising from the thoracic aorta. What structures are supplied by each group?***

Abdominal Aorta and Its Branches

The branches of the abdominal aorta, like those of the thoracic aorta, are divided into visceral and parietal parts (see table 21.3 and figures 21.14*a* and *c* and 21.15). The visceral arteries are in turn divided into paired and unpaired branches. Three major unpaired branches exist: the **celiac** (sē′lē-ak; belly) **trunk,** the **superior mesenteric artery** (mez-en-ter′ik; relating to the mesenteries), and the **inferior mesenteric artery** (see figure 21.14*a* and *c*). Each has several major branches supplying the abdominal organs.

The paired visceral branches of the abdominal aorta supply the kidneys, adrenal glands, and gonads (testes and ovaries). The parietal arteries of the abdominal aorta supply the diaphragm and abdominal wall (see figure 21.15).

21 ***What areas of the body are supplied by the paired arteries that branch from the abdominal aorta? The unpaired arteries? Name the three major unpaired branches.***

Arteries of the Pelvis

The abdominal aorta divides at the level of the fifth lumbar vertebra into two **common iliac arteries.** They divide to form the **external iliac arteries,** which enter the lower limbs, and the **internal iliac arteries,** which supply the pelvic area. Visceral branches supply the pelvic organs, such as the urinary bladder, rectum, uterus, and vagina. Parietal branches supply blood to the walls and floor of the pelvis; the lumbar, gluteal, and proximal thigh muscles; and the external genitalia (table 21.4; see figure 21.15; figure 21.16).

22 ***Name the arteries that branch from the aorta to supply the pelvic area. List the organs of the pelvis that are supplied by branches of these arteries.***

TABLE 21.4 Arteries of the Pelvis (See Figures 21.15 and 21.16)	
Arteries	**Tissues Supplied**
Internal Iliac	Pelvis through the branches listed below
Visceral Branches	
Middle rectal	Rectum
Vaginal	Vagina and uterus
Uterine	Uterus, vagina, uterine tube, and ovary
Parietal Branches	
Lateral sacral	Sacrum
Superior gluteal	Muscles of the gluteal region
Obturator	Public region, deep groin muscles, and hip joint
Internal pudendal	Rectum, external genitalia, and floor of pelvis
Inferior gluteal	Inferior gluteal region, coccyx, and proximal thigh

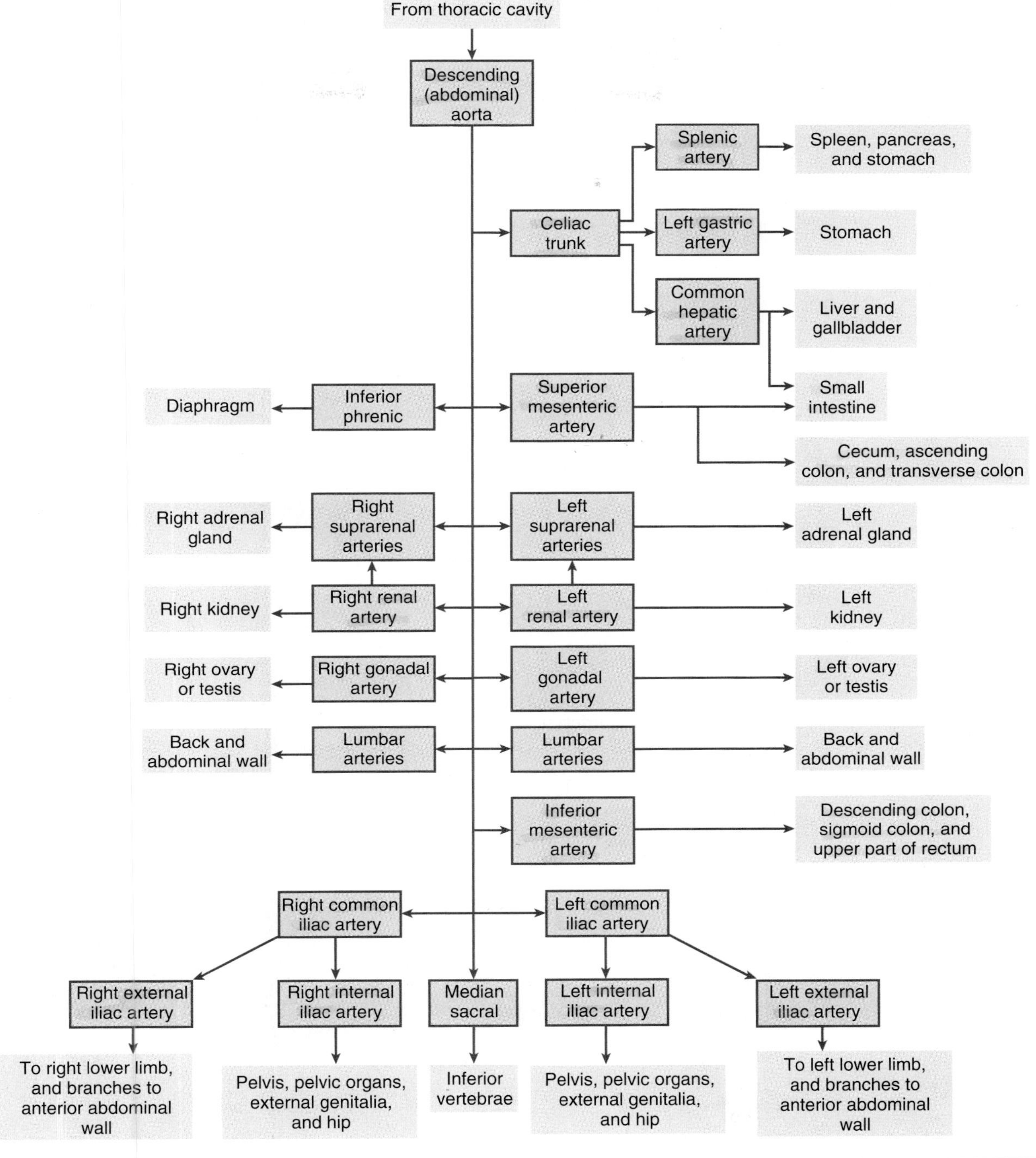

FIGURE 21.15 Major Arteries of the Abdomen and Pelvis

Visceral branches include those that are unpaired (celiac trunk, superior mesenteric, inferior mesenteric) and those that are paired (renal, suprarenal, testicular, ovarian). Parietal branches include inferior phrenic, lumbar, and median sacral.

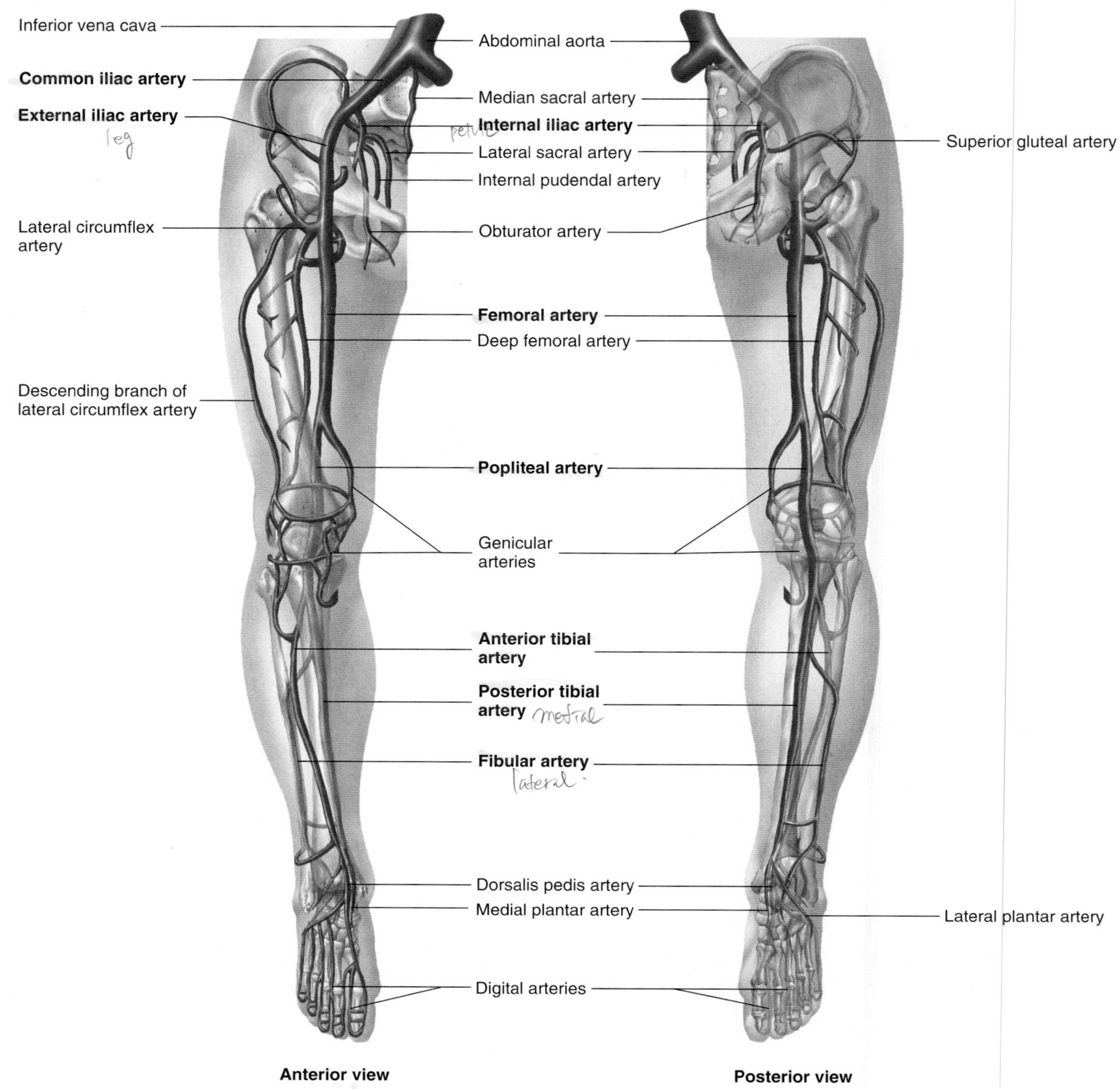

FIGURE 21.16 Arteries of the Pelvis and Lower Limb

The internal and external iliac arteries and their branches. The internal iliac artery supplies the pelvis and hip, and the external iliac artery supplies the lower limb through the femoral artery.

Arteries of the Lower Limb

The arteries of the lower limb form a continuum similar to that of the arteries of the upper limb. The **external iliac artery** becomes the **femoral** (fem′ŏ-răl; relating to the thigh) **artery** in the thigh, which becomes the **popliteal** (pop-lit′ē-ăl, pop-li-tē′ăl; ham, the hamstring area posterior to the knee) **artery** in the popliteal space. The popliteal artery gives off the **anterior tibial artery** just inferior to the knee and then continues as the **posterior tibial artery.** The anterior tibial artery becomes the **dorsalis pedis artery** at the foot. The posterior tibial artery gives off the **fibular,** or **peroneal, artery** and then gives rise to **medial** and **lateral plantar** (plan′tăr; the sole of the foot) **arteries,** which in turn give off **digital branches** to the toes. The arteries of the lower limb are illustrated in figures 21.16 and 21.17 and are listed in table 21.5.

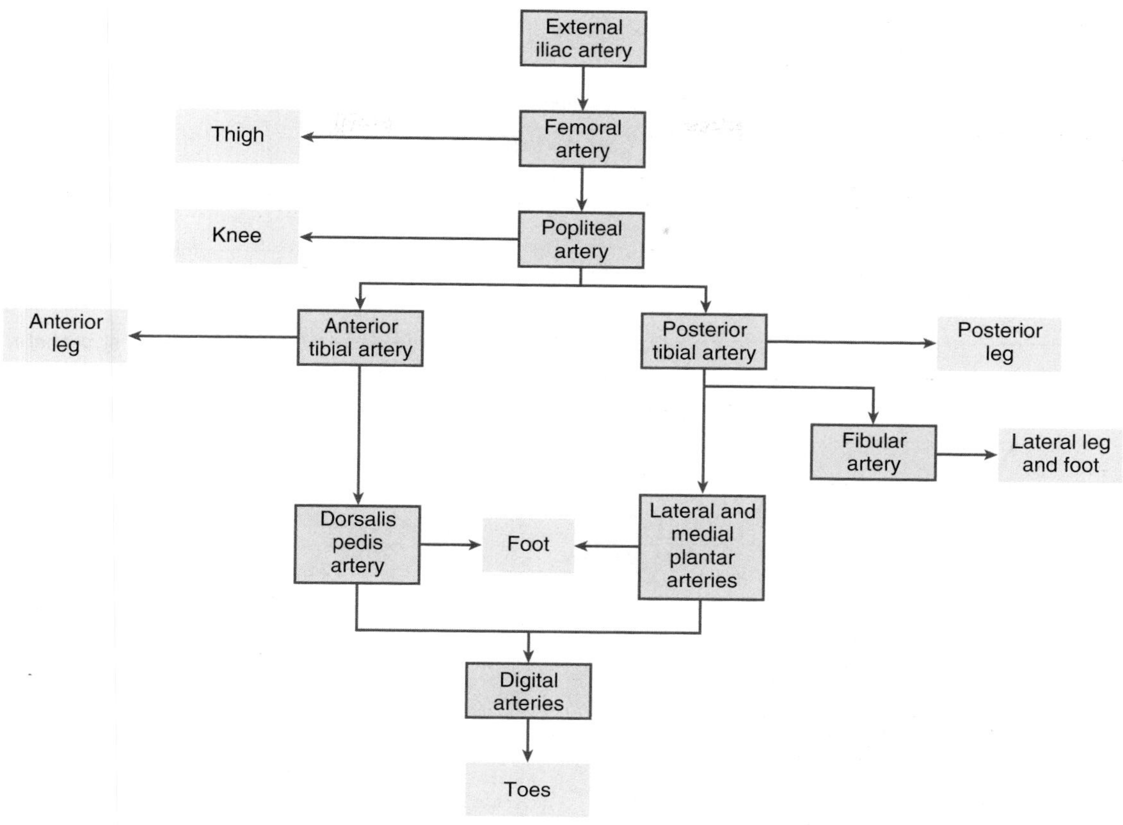

FIGURE 21.17 **Major Arteries of the Lower Limb**

TABLE 21.5 Arteries of the Lower Limb (See Figures 21.16 and 21.17)

Arteries	Tissues Supplied
Femoral	Thigh, external genitalia, and anterior abodominal wall
Deep femoral	Thigh, knee, and femur
Popliteal (Continuation of the Femoral Artery)	
Posterior tibial	Knee and leg
Fibular (peroneal)	Calf and peroneal muscles and ankle
Medial plantar	Plantar region of foot
Digital arteries	Digits of foot
Lateral plantar	Plantar region of foot
Digital arteries	Digits of foot
Anterior tibial	Knee and leg
Dorsalis pedis	Dorsum of foot
Digital arteries	Digits of foot

23 ***List, in order, the arteries that travel from the aorta to the digits of the lower limbs.***

SYSTEMIC CIRCULATION: VEINS

Three major veins return blood from the body to the right atrium: the **coronary sinus,** returning blood from the walls of the heart (see figures 20.6*b* and 20.7); the **superior vena cava** (vē′nă kā′vă, kā′vă; venous cave), returning blood from the head, neck, thorax, and upper limbs; and the **inferior vena cava,** returning blood from the abdomen, pelvis, and lower limbs (figure 21.18).

In a very general way, the smaller veins follow the same course as the arteries and many are given the same names. The veins, however, are more numerous and more variable. The larger veins often follow a very different course and have names different from the arteries.

24 ***List the veins that carry blood from each of the major body areas.***

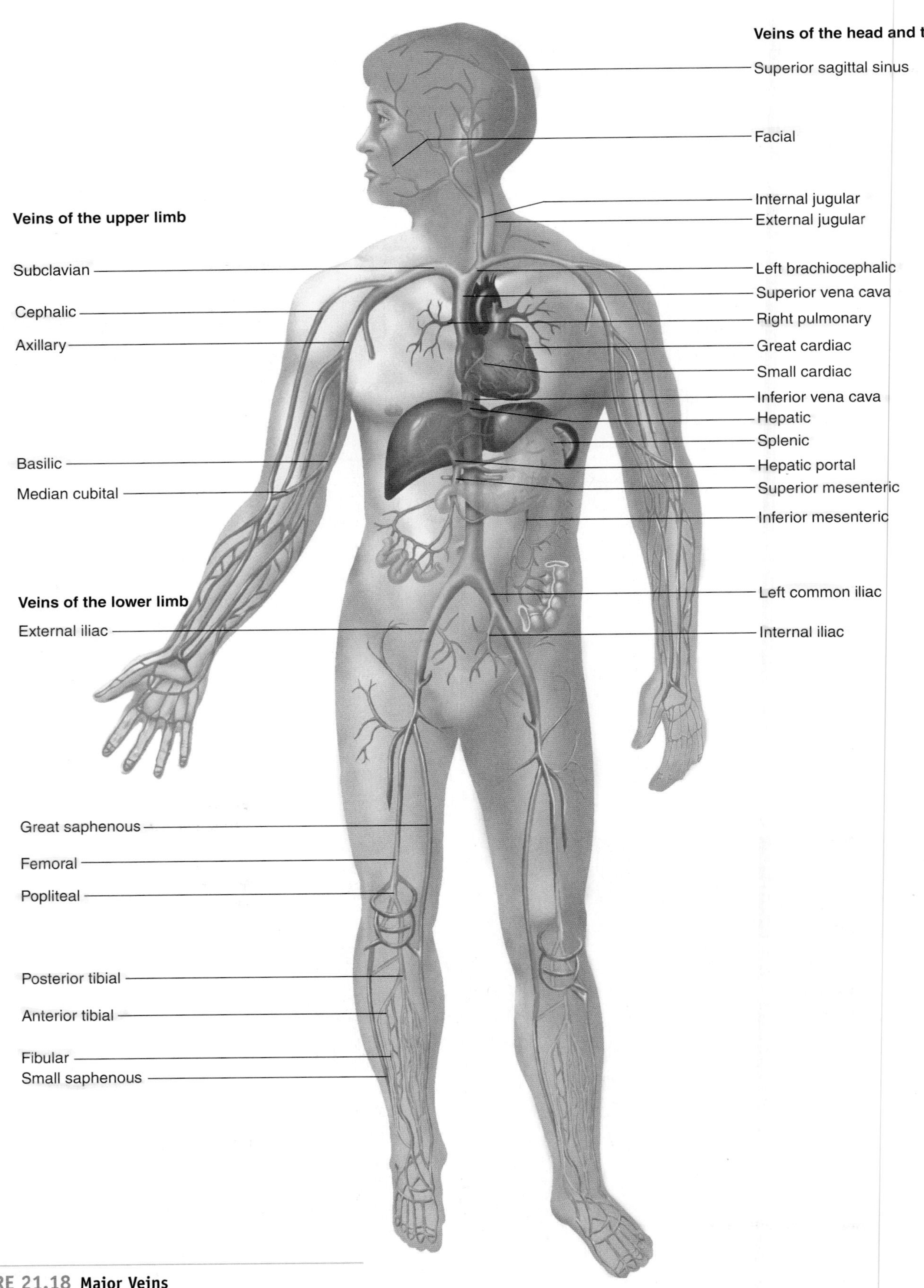

FIGURE 21.18 Major Veins
The veins carry blood to the heart from the tissues of the body.

The three major types of veins are the superficial veins, deep veins, and sinuses. The superficial veins of the limbs are, in general, larger than the deep veins, whereas in the head and trunk the opposite is the case. Venous sinuses occur primarily in the cranial cavity and the heart.

Veins Draining the Heart

The **cardiac veins,** which transport blood from the walls of the heart and return it through the coronary sinus to the right atrium, are described in chapter 20.

Veins of the Head and Neck

The two pairs of major veins that drain blood from the head and neck are the **external** and **internal jugular** (jŭg′ū-lar; neck) **veins.** The external jugular veins are the more superficial of the two sets, and they drain blood primarily from the posterior head and neck. The external jugular vein drains into the subclavian vein. The internal jugular veins are much larger and deeper than the external jugular veins. They drain blood from the cranial cavity and the anterior head, face, and neck.

The internal jugular vein is formed primarily as the continuation of the **venous sinuses** of the cranial cavity. The venous sinuses are actually spaces within the dura mater surrounding the brain (see chapter 13). They are depicted in figure 21.19 and are listed in table 21.6.

Facial Pimples and Meningitis

Because venous communication exists between the facial veins and venous sinuses through the ophthalmic veins, infections can be introduced into the cranial cavity through this route. A superficial infection of the face on either side of the nose can enter the facial vein. The infection can then pass through the ophthalmic veins to the venous sinuses and result in meningitis. For this reason, people are warned not to aggravate pimples or boils on the face on either side of the nose.

TABLE 21.6 Venous Sinuses of the Cranial Cavity (See Figure 21.19)

Veins	Tissues Drained
Internal Jugular Vein	
Sigmoid sinus	
Superior and inferior petrosal sinuses	Anterior portion of cranial cavity
Cavernous sinus	
Ophthalmic veins	Orbit
Transverse sinus	
Occipital sinus	Central floor of posterior fossa of skull
Superior sagittal sinus	Superior portion of cranial cavity and brain
Straight sinus	
Inferior sagittal sinus	Deep portion of longitudinal fissure

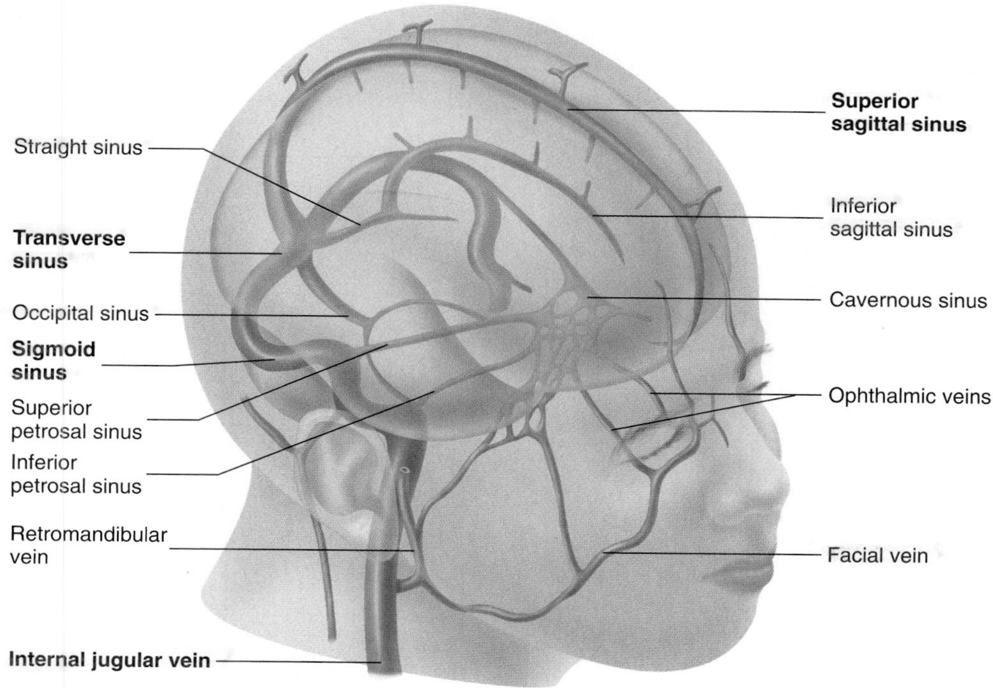

FIGURE 21.19 **Venous Sinuses Associated with the Brain**

Once the internal jugular veins exit the cranial cavity, they receive several venous tributaries that drain the external head and face (table 21.7 and figures 21.20 and 21.21). The **internal jugular veins** join the **subclavian veins** on each side of the body to form the **brachiocephalic veins.**

25 ***List the two pairs of major veins that drain blood from the head and neck. Describe the venous sinuses. To what large vein do the venous sinuses connect?***

TABLE 21.7 Veins Draining the Head and Neck (See Figures 21.20 and 21.21)

Veins	Tissues Drained
Brachiocephallc	
Internal jugular	Brain
Lingual	Tongue and mouth
Superior thyroid	Thyroid and deep posterior facial structures (also empties into external jugular)
Facial	Superficial and anterior facial structures
External jugular	Superficial surface of posterior head and neck

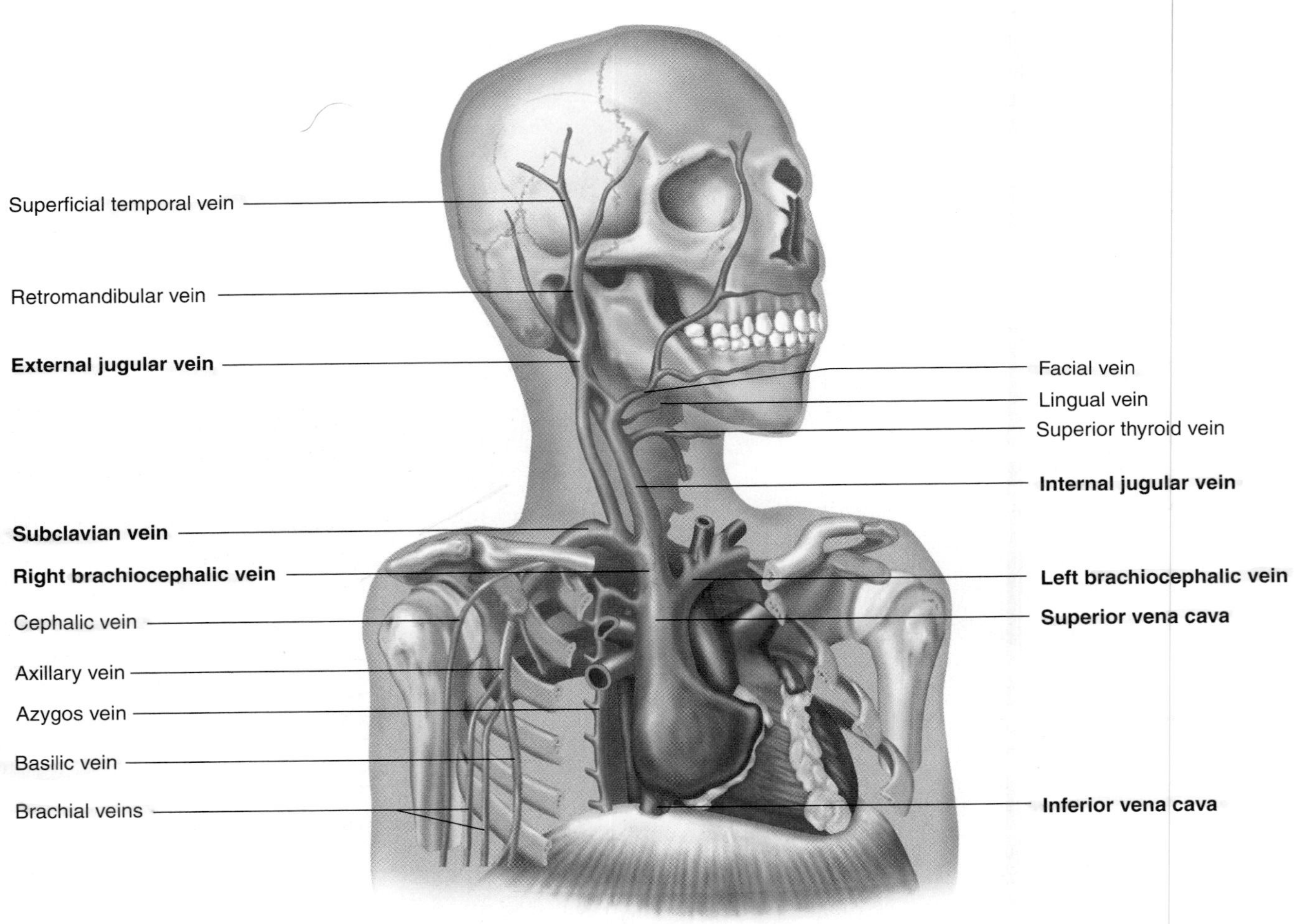

FIGURE 21.20 Veins of the Head and Neck
The right brachiocephalic vein and its tributaries. The major veins draining the head and neck are internal and external jugular veins.

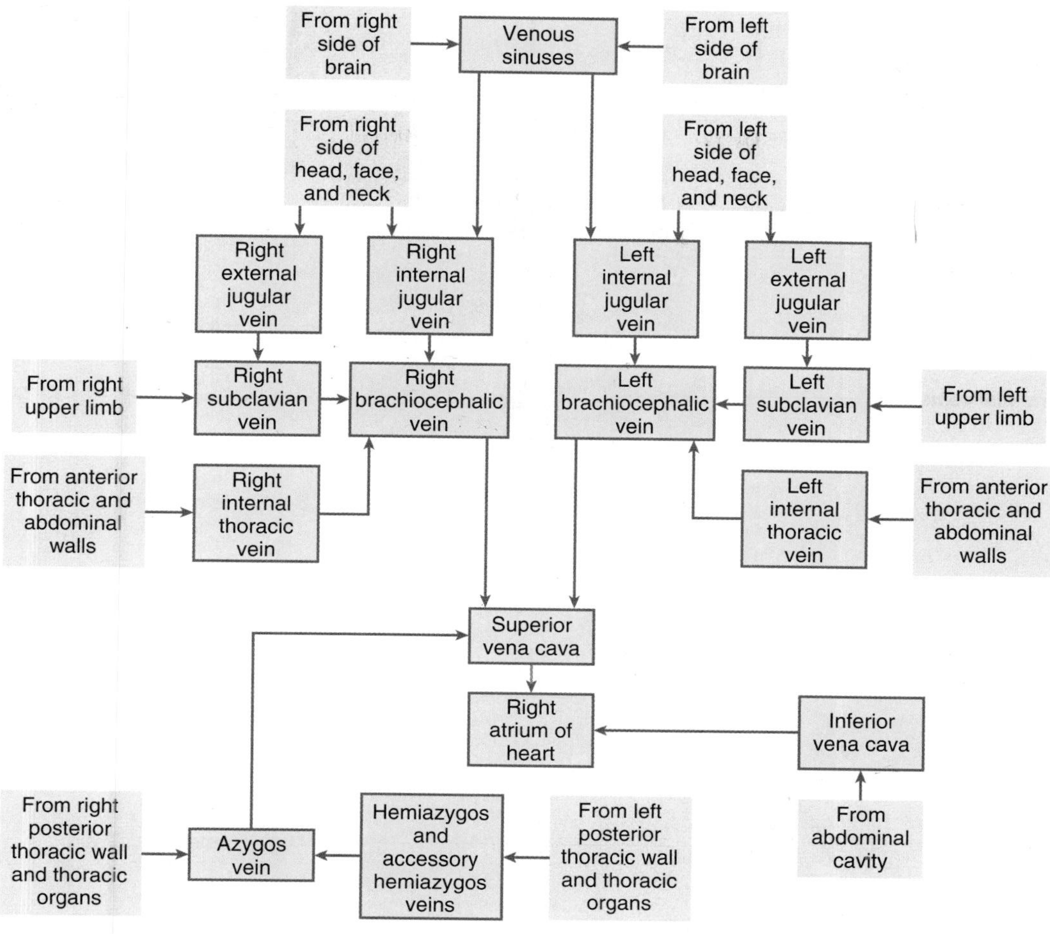

FIGURE 21.21 Major Veins of the Head and Thorax

Veins of the Upper Limb

The **cephalic** (se-fal′ik; toward the head), **basilic** (ba-sil′ik), and **brachial veins** are responsible for draining most of the blood from the upper limbs (table 21.8 and figures 21.22 and 21.23). Many of the tributaries of the cephalic and basilic veins in the forearm and hand can be seen through the skin. Because of the considerable variation in the tributary veins of the forearm and hand, they often are left unnamed. The basilic vein of the arm becomes the **axillary vein** as it courses through the axillary region. The axillary vein then becomes the **subclavian vein** at the margin of the first rib. The cephalic vein enters the axillary vein.

The **median cubital** (kū′bi-tăl; pertaining to the elbow) **vein** is a variable vein that usually connects the cephalic vein or its tributaries with the basilic vein. In many people, this vein is quite prominent on the anterior surface of the upper limb at the level of the elbow (cubital fossa) and is, therefore, often used as a site for drawing blood from a patient.

The deep veins draining the upper limb follow the same course as the arteries. The **radial** and **ulnar veins,** therefore, are named for the arteries they attend. They usually are paired, with one small vein lying on each side of the artery, and they have

TABLE 21.8 Veins of the Upper Limb (See Figures 21.22 and 21.23)

Veins	Tissues Drained
Subclavian (Continuation of the Axillary Vein)	
Axillary (Continuation of the Basilic Vein)	
Cephalic	Lateral arm, forearm, and hand (superficial veins of the forearm and hand are variable)
Brachial (paired, deep veins)	Deep structures of the arm
Radial vein	Deep forearm
Ulnar vein	Deep forearm
Basilic	Medial arm, forearm, and hand (superficial veins of the forearm and hand are variable)
Median cubital	Connects basilic and cephalic veins
Deep and superficial palmar venous arches	Drain into superficial and deep veins of the forearm
Digital veins	Fingers

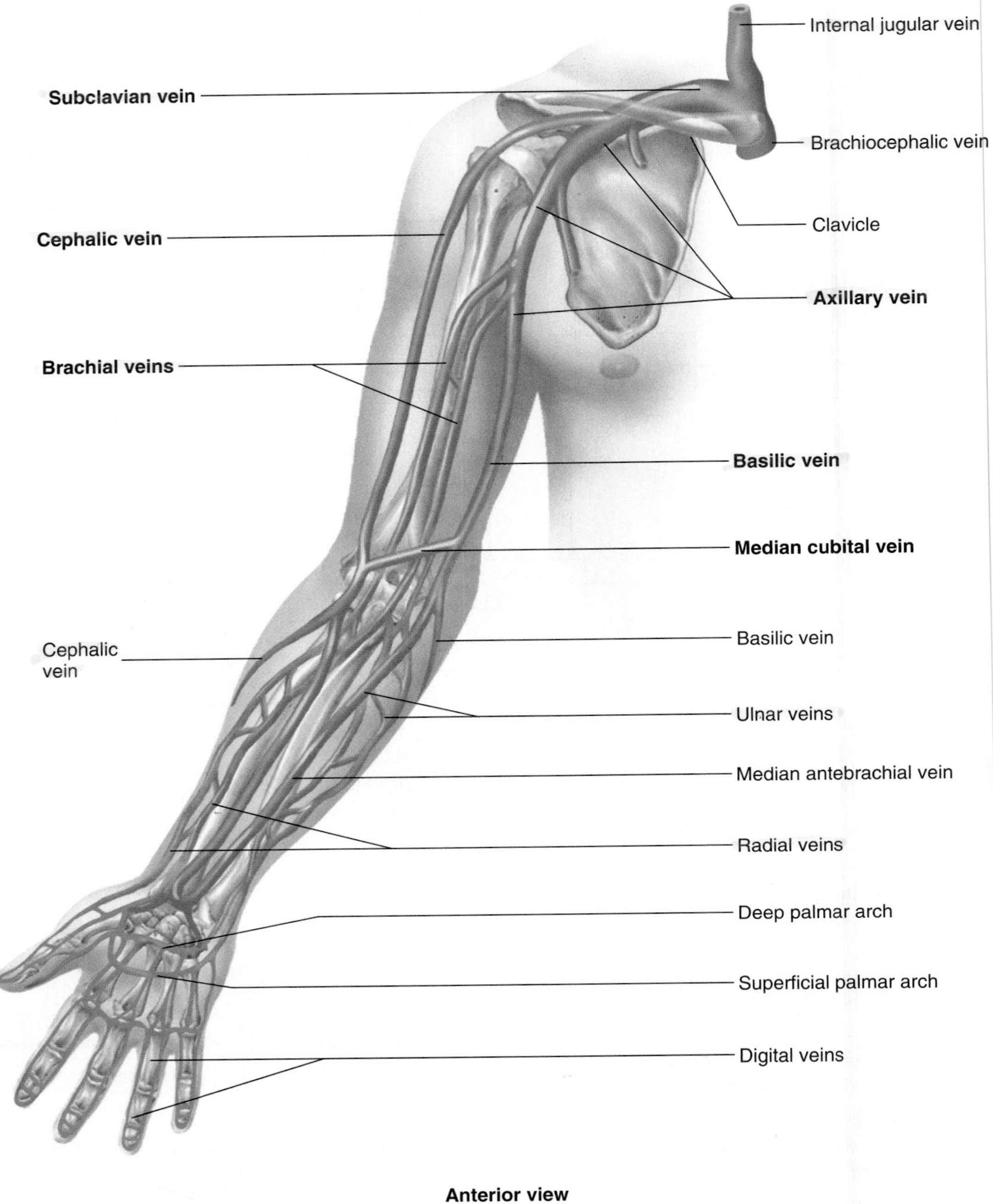

FIGURE 21.22 Veins of the Upper Limb
The subclavian vein and its tributaries. The major veins draining the superficial structures of the limb are the cephalic and basilic veins. The brachial veins drain the deep structures.

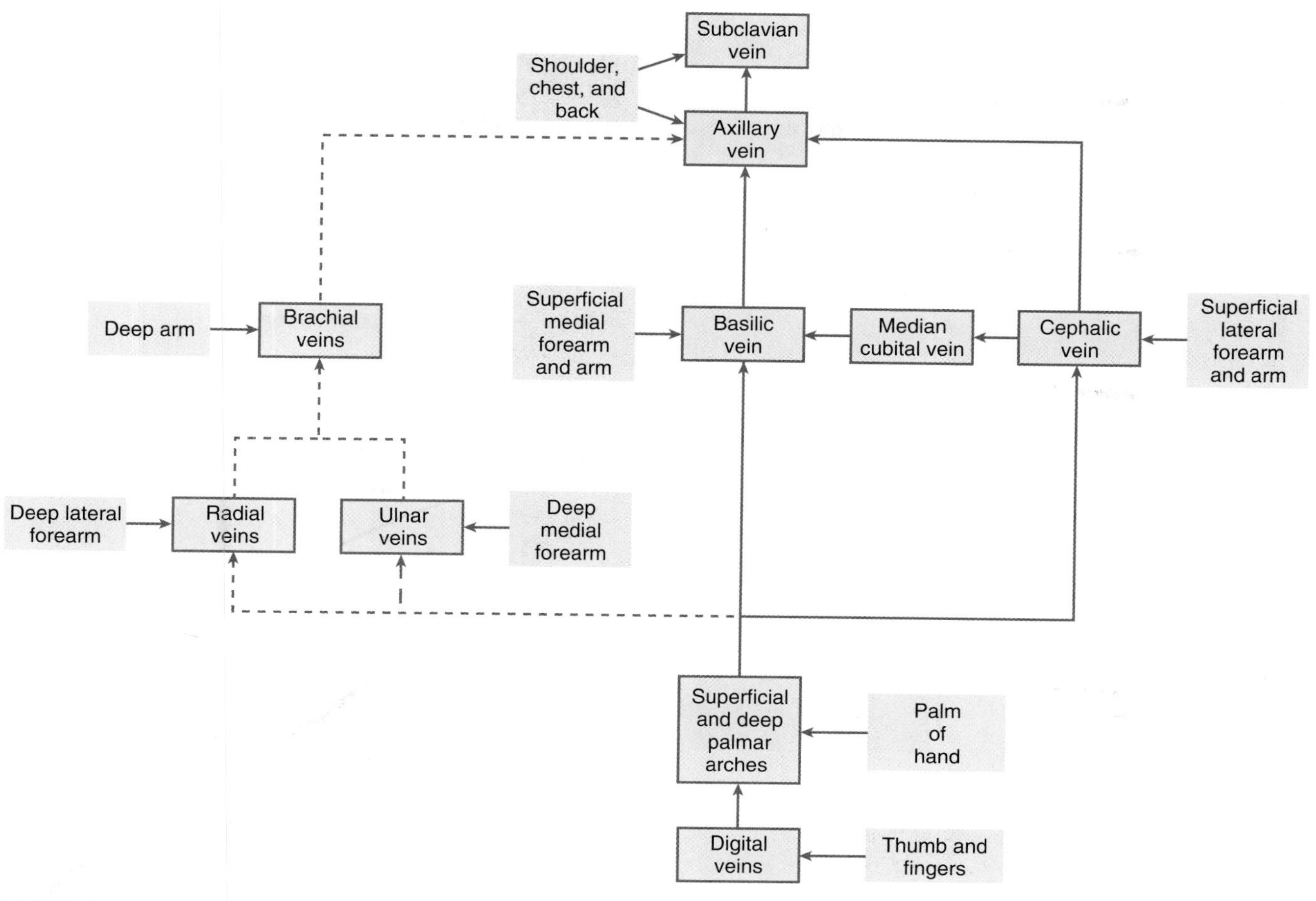

FIGURE 21.23 Major Veins of the Shoulder and Upper Limb
The deep veins, which carry far less blood than the superficial veins, are indicated by *dashed lines*.

numerous connections with one another and with the superficial veins. The radial and ulnar veins empty into the **brachial veins,** which accompany the brachial artery and empty into the axillary vein (see figures 21.22 and 21.23).

26 ***List the major deep and superficial veins of the upper limb.***

Veins of the Thorax

Three major veins return blood from the thorax to the superior vena cava: the right and left brachiocephalic veins and the **azygos** (az′ī-gos; unpaired) **vein.** The thoracic drainage to the brachiocephalic veins is through the anterior thoracic wall by way of the **internal thoracic veins.** They receive blood from the **anterior intercostal veins.** Blood from the posterior thoracic wall is collected by **posterior intercostal veins** that drain into the azygos vein on the right and the **hemiazygos** (hem′ē-az′ī-gos) or **accessory hemiazygos vein** on the left. The hemiazygos and accessory hemiazygos veins empty into the azygos vein, which drains into the superior vena cava. The thoracic veins are listed in table 21.9 and illustrated in figure 21.24 (see also figure 21.21).

27 ***List the three major veins that return blood from the thorax to the superior vena cava.***

TABLE 21.9 Veins of the Thorax (See Figure 21.24)

Veins	Tissues Drained
Superior Vena Cava	
Brachiocephalic	
Azygos vein	Right side, posterior thoracic wall and posterior abdominal wall; esophagus, bronchi, pericardium, and mediastinum
Hemiazygos	Left side, inferior posterior thoracic wall and posterior abdominal wall; esophagus and mediastinum
Accessory hemiazygos	Left side, superior posterior thoracic wall

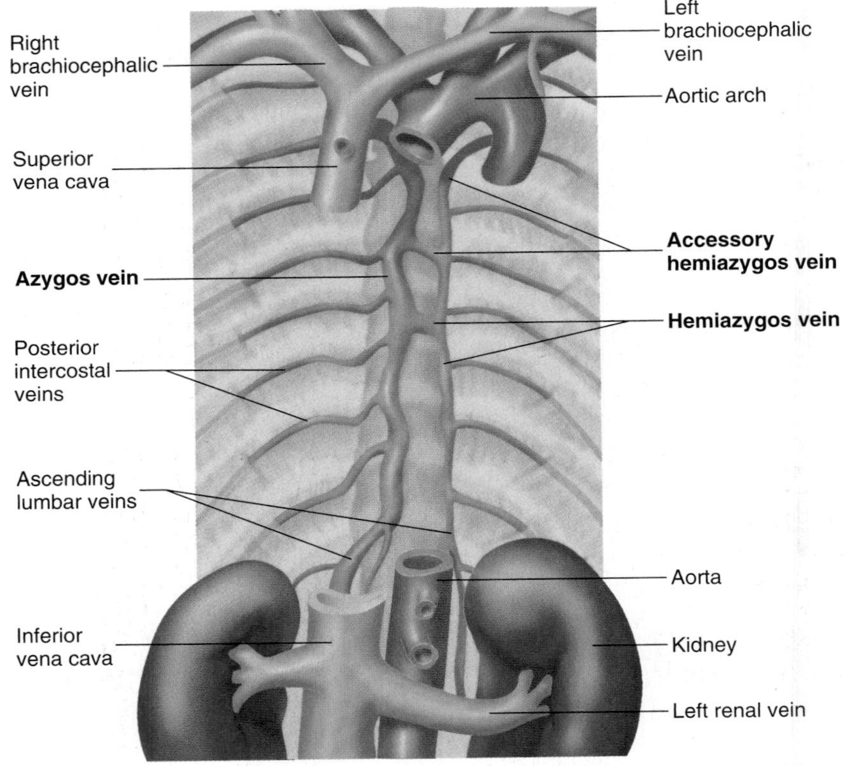

FIGURE 21.24 Veins of the Thorax
Anterior view of the azygos and hemiazygos veins and their tributaries.

Veins of the Abdomen and Pelvis

Blood from the posterior abdominal wall drains into the **ascending lumbar veins.** These veins are continuous superiorly with the hemiazygos on the left and the azygos on the right. Blood from the rest of the abdomen, pelvis, and lower limbs returns to the heart through the inferior vena cava. The gonads (testes and ovaries), kidneys, and adrenal glands are the only abdominal organs outside the pelvis that drain directly into the inferior vena cava. The **internal iliac veins** drain the pelvis and join the **external iliac veins** from the lower limbs to form the **common iliac veins,** which unite to form the inferior vena cava. The major abdominal and pelvic veins are listed in table 21.10 and illustrated in figure 21.25; see also figure 21.27.

TABLE 21.10 Veins Draining the Abdomen and Pelvis (See Figures 21.25 and 21.27)

Veins	Tissues Drained
Inferior Vena Cava	
Hepatic veins	Liver (see hepatic portal system)
Common iliac	
External iliac	Lower limb (see table 21.12)
Internal iliac	Pelvis and its viscera
Ascending lumbar	Posterior abdominal wall (empties into common iliac, azygos, and hemiazygos veins)
Renal	Kidney
Suprarenal	Adrenal gland
Gonadal	
Testicular (male)	Testis
Ovarian (female)	Ovary
Phrenic	Diaphragm

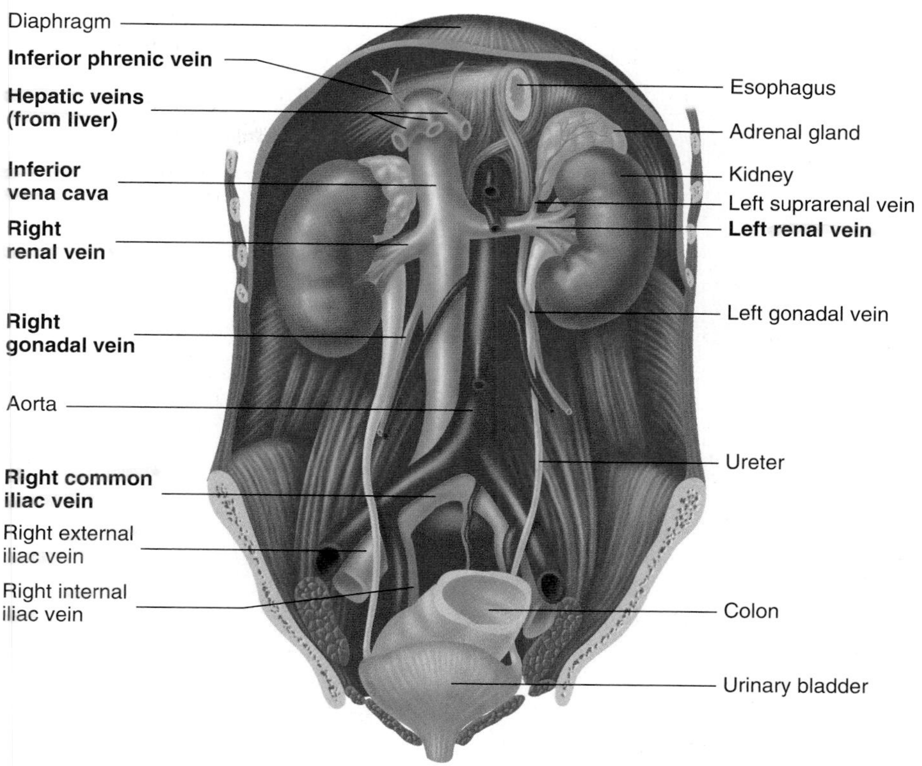

FIGURE 21.25 Inferior Vena Cava and Its Tributaries
The hepatic veins transport blood to the inferior vena cava from the hepatic portal system, which ends as a series of blood sinusoids in the liver (see figure 21.26).

Hepatic Portal System

The **hepatic** (he-pat′ik; relating to the liver) **portal system** (table 21.11 and figures 21.26 and 21.27) carries blood through veins from capillaries within most of the abdominal viscera, such as the stomach, intestines, and spleen, to a series of dilated capillaries, called **sinusoids,** in the liver. Nutrients and other substances absorbed from the stomach or intestine are delivered to the liver by the hepatic portal system (see chapter 24).

The **hepatic portal vein,** the largest vein of the system, is formed by the union of the **superior mesenteric vein,** which drains the small intestine, and the **splenic vein,** which drains the spleen. The splenic vein receives the **inferior mesenteric vein,** which drains part of the large intestine, and the **pancreatic veins,** which drain the pancreas. The hepatic portal vein also receives gastric veins before entering the liver.

TABLE 21.11 Hepatic Portal System (See Figures 21.26 and 21.27)

Veins	Tissues Drained
Hepatic Portal	
Superior mesenteric	Small intestine and most of the colon
Splenic	Spleen
Inferior mesenteric	Descending colon and rectum
Pancreatic	Pancreas
Gastroomental	Stomach
Gastric	Stomach
Cystic	Gallbladder

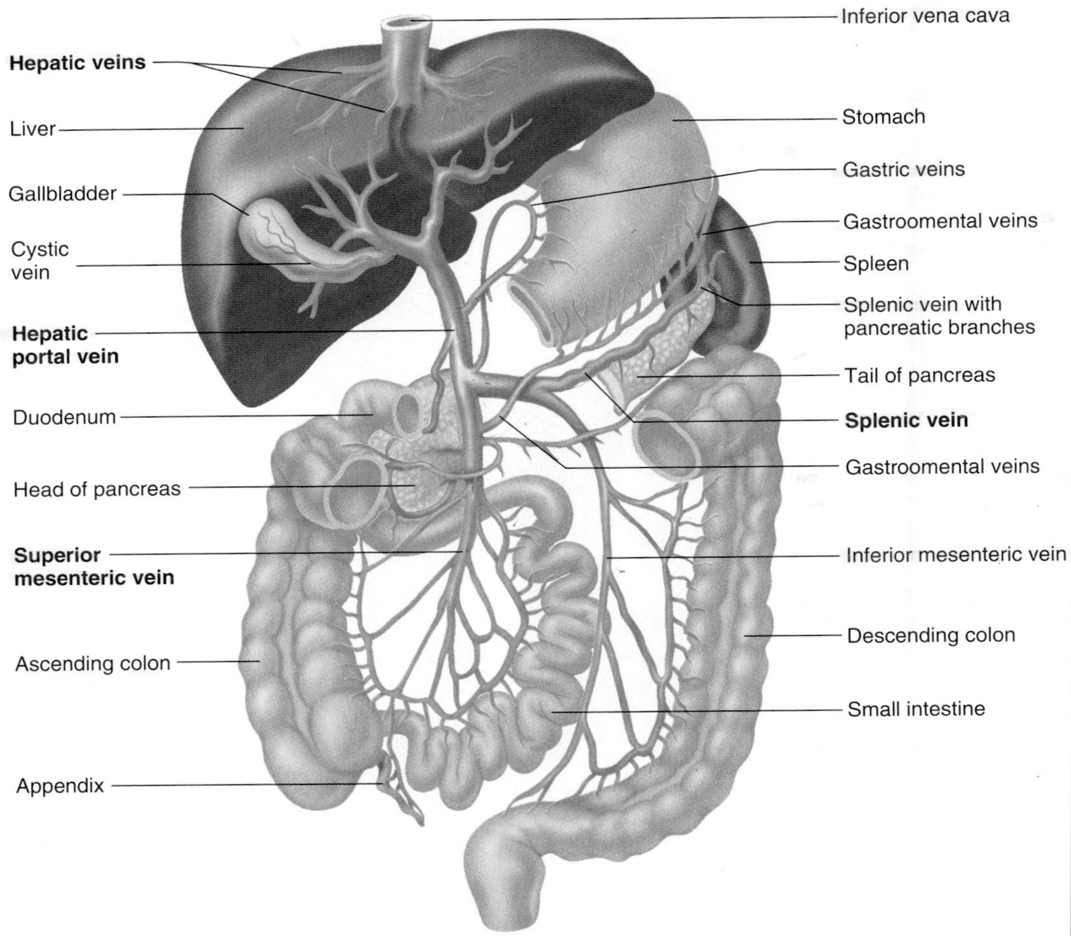

FIGURE 21.26 Veins of the Hepatic Portal System

The hepatic portal system begins as capillary beds in the stomach, pancreas, spleen, small intestine, and large intestine. The veins of the hepatic portal system converge on the hepatic portal vein, which carries blood to a series of capillaries (sinusoids) in the liver. Hepatic veins carry blood from capillaries in the liver to the inferior vena cava (see figure 21.25).

Blood from the liver sinusoids is collected into **central veins,** which empty into **hepatic veins.** Blood from the cystic veins, which drain the gallbladder, also enters the hepatic veins. The hepatic veins join the inferior vena cava. Blood entering the liver through the hepatic portal vein is rich with nutrients collected from the intestines, but it also can contain a number of toxic substances harmful to the tissues of the body. Within the liver, the nutrients are either taken up and stored or are modified chemically and used by other cells of the body (see chapter 24). The cells of the liver also help remove toxic substances by altering their structure or making them water-soluble, a process called **biotransformation.** The water-soluble substances can then be transported in the blood to the kidneys, from which they are excreted in the urine (see chapter 26).

28 ***Explain the three ways that blood from the abdomen returns to the heart.***

29 ***List the vessels that carry blood from the abdominal organs to the hepatic portal vein.***

Veins of the Lower Limb

The veins of the lower limb, like those of the upper limb, consist of superficial and deep groups. The distal deep veins of each limb are paired and follow the same path as the arteries, whereas the proximal deep veins are unpaired. The **anterior** and **posterior tibial veins** are paired and accompany the anterior and posterior tibial arteries. They unite just inferior to the knee to form the single **popliteal vein,** which ascends through the thigh and becomes the **femoral vein.** The femoral vein becomes the external iliac vein. **Fibular,** or **peroneal** (per-ō-nē′ăl), **veins** also are paired in each leg and accompany the fibular arteries. They empty into the posterior tibial veins just before those veins contribute to the popliteal vein.

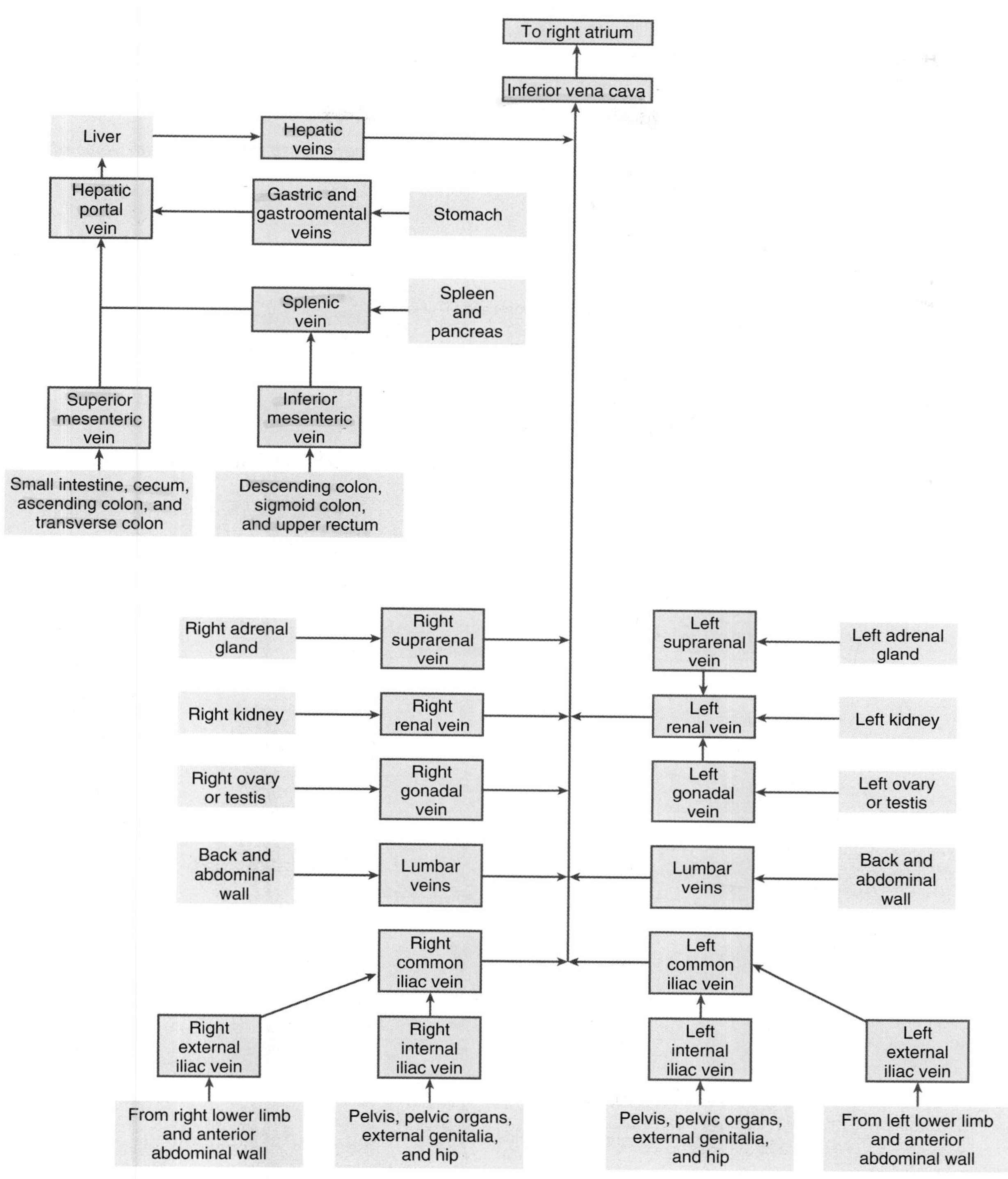

FIGURE 21.27 Major Veins of the Abdomen and Pelvis

The superficial veins consist of the great and small saphenous veins. The **great saphenous** (să-fē′nŭs; visible) **vein,** the longest vein of the body, originates over the dorsal and medial side of the foot and ascends along the medial side of the leg and thigh to empty into the femoral vein. The **small saphenous vein** begins over the lateral side of the foot and ascends along the posterior leg to the popliteal space, where it empties into the popliteal vein. The veins of the lower limb are illustrated in figures 21.28 and 21.29 and are listed in table 21.12. The saphenous veins can be removed and used as a source of blood vessels for coronary bypass surgery (see chapter 20).

30 ***List the major deep and superficial veins of the lower limbs.***

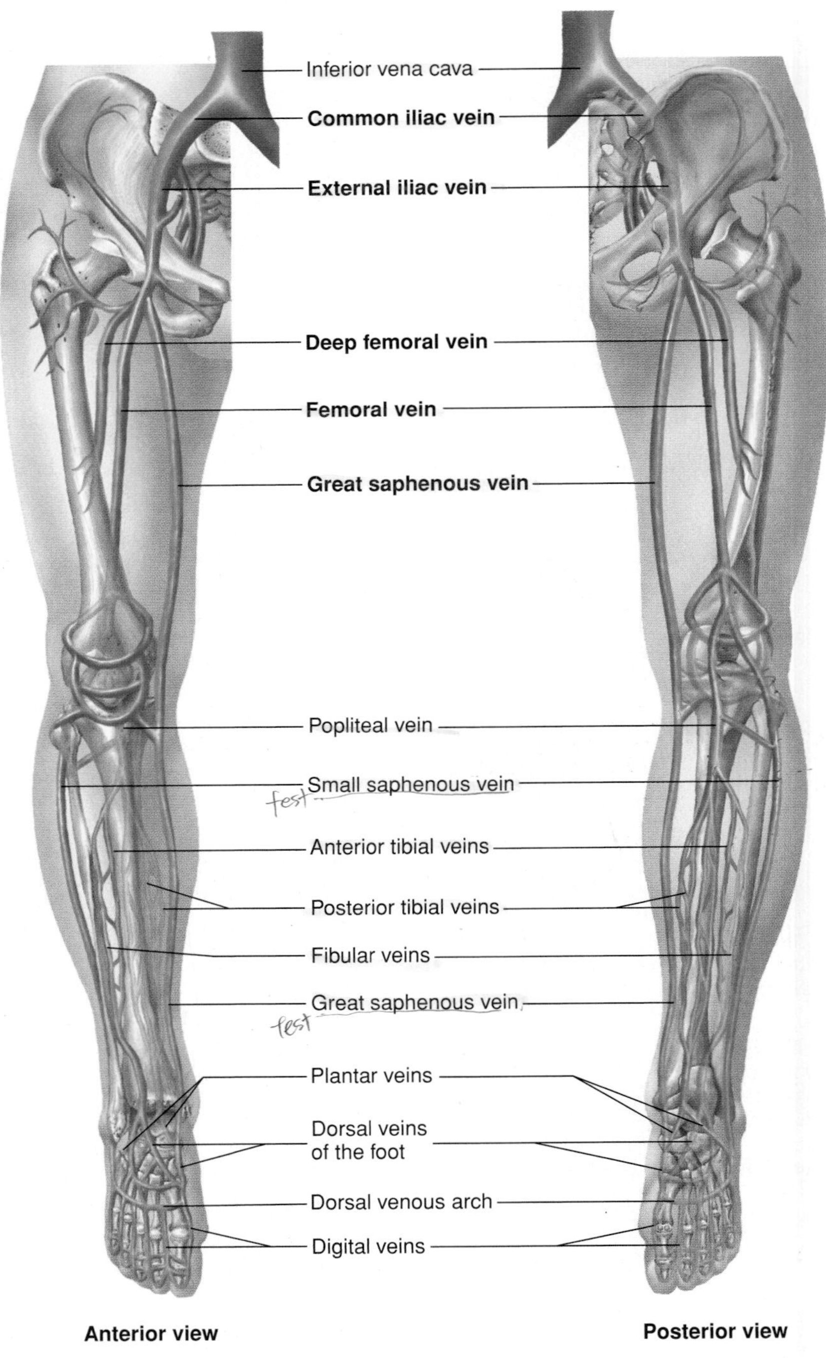

FIGURE 21.28 Veins of the Pelvis and Lower Limb

The right common iliac vein and its tributaries.

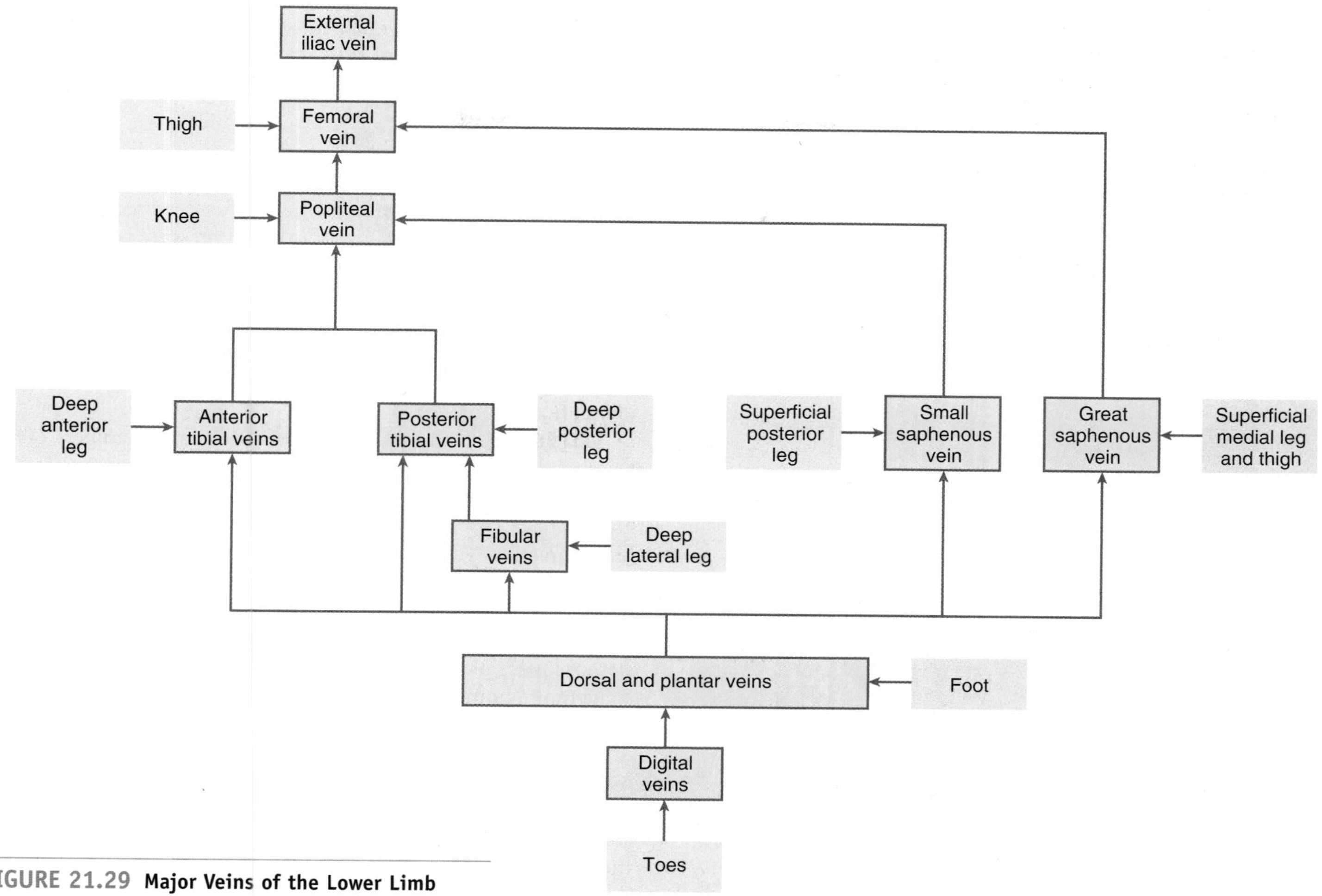

FIGURE 21.29 **Major Veins of the Lower Limb**

TABLE 21.12 Veins of the Lower Limb (See Figures 21.28 and 21.29)

Veins	Tissues Drained
External Iliac Vein (Continuation of the Femoral Vein)	
Femoral (Continuation of the Popliteal Vein)	Thigh
Popliteal	
Anterior tibial	Deep anterior leg
Dorsal vein of foot	Dorsum of foot
Posterior tibial	Deep posterior leg
Plantar veins	Plantar region of foot
Fibular (peroneal)	Deep lateral leg and foot
Small saphenous	Superficial posterior leg and lateral side of foot
Great saphenous	Superficial anterior and medial leg, thigh, and dorsum of foot
Dorsal vein foot	Dorsum of foot
Dorsal venous arch	Foot
Digital veins	Toes

DYNAMICS OF BLOOD CIRCULATION

The dynamics of blood circulation through blood vessels are the same as those affecting the flow of water or other liquids through pipes. The interrelationships among pressure, flow, resistance, and the control mechanisms that regulate blood pressure and blood flow through vessels play critical roles in the functions of the circulatory system. Many of these interrelationships are clinically significant.

Laminar and Turbulent Flow in Vessels

Fluid, including blood, tends to flow through long, smooth-walled tubes in a streamlined fashion called **laminar flow** (figure 21.30*a*). Fluid behaves as if it were composed of a large number of concentric layers. The layer nearest the wall of the tube experiences the greatest resistance to flow because it moves against the stationary wall. The innermost layers slip over the surface of the outermost layers and experience less resistance to movement. Thus, flow in a vessel consists of movement of concentric layers, with the outermost layer moving slowest and the layer at the center moving fastest.

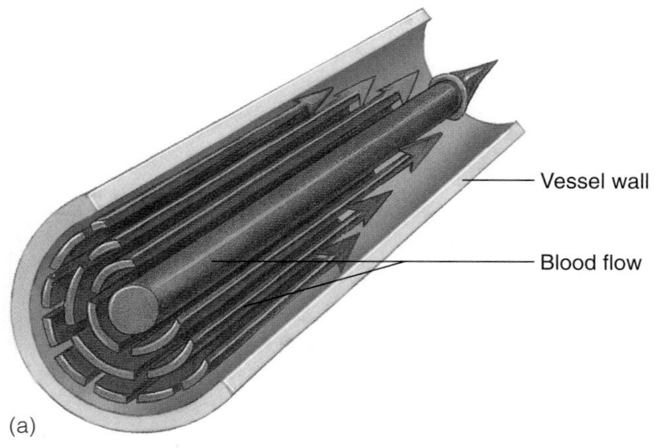

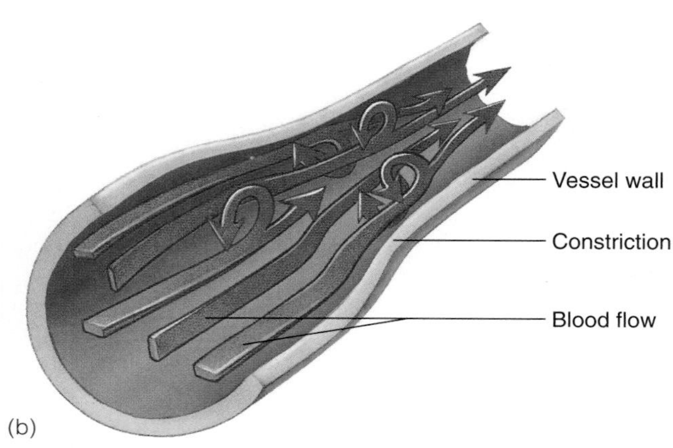

FIGURE 21.30 Laminar and Turbulent Flow

(*a*) In laminar flow, fluid flows in long, smooth-walled tubes as if it were composed of a large number of concentric layers. (*b*) Turbulent flow is caused by numerous small currents flowing crosswise or obliquely to the long axis of the vessel, resulting in flowing whorls and eddy currents.

Laminar flow is interrupted and becomes **turbulent flow** when the rate of flow exceeds a critical velocity or when the fluid passes a constriction, a sharp turn, or a rough surface (figure 21.30*b*). Vibrations of the liquid and blood vessel walls during turbulent flow cause the sounds produced when blood pressure is measured using a blood pressure cuff. Turbulent flow is also common as blood flows past the valves in the heart and is partially responsible for the heart sounds (see chapter 20).

Turbulent flow of blood through vessels occurs primarily in the heart and to a lesser extent where arteries branch. Sounds caused by turbulent blood flow in arteries are not normal and usually indicate that the artery is constricted abnormally. In addition, turbulent flow in abnormally constricted arteries increases the probability that thromboses will develop.

31 ***Describe laminar flow and turbulent flow through a tube. What conditions cause turbulent flow of blood?***

Blood Pressure

Blood pressure is a measure of the force blood exerts against blood vessel walls. An instrument for measuring blood pressure is the **mercury (Hg) manometer,** which measures pressure in millimeters of mercury (mm Hg). If the blood pressure is 100 mm Hg, the pressure is great enough to lift a column of mercury 100 mm.

Blood pressure can be measured directly by inserting a **cannula** (or tube) into a blood vessel and connecting a manometer or an electronic pressure transducer to it. Electronic transducers are very sensitive to changes in pressure and can precisely detect rapid fluctuation in pressure.

Placing catheters into blood vessels or into chambers of the heart to monitor pressure changes is possible but not appropriate for routine clinical determinations of systemic blood pressure. The **auscultatory** (aws-kŭl′tă-tō′rē) **method** can be used to measure blood pressure without surgical procedures or discomfort, so it is used under most clinical conditions. A blood pressure cuff connected to a **sphygmomanometer** (sfig′mō-mă-nom′ĕ-ter) is placed around the patient's arm just above the elbow, and a stethoscope is placed over the brachial artery (figure 21.31). Some sphygmomanometers have mercury manometers, and others have digital manometers, but they all measure pressure in terms of millimeters of mercury. The blood pressure cuff is inflated until the brachial artery is completely collapsed. Because no blood flows through the constricted area, no sounds can be heard. The pressure in the cuff is gradually lowered. As soon as it declines below the systolic pressure, blood flows through the constricted area during systole. The blood flow is turbulent and produces vibrations in the blood and surrounding tissues that can be heard through the stethoscope. These sounds are called **Korotkoff** (kō-rot′kof) **sounds,** and the pressure at which a Korotkoff sound is first heard represents the **systolic pressure.**

As the pressure in the blood pressure cuff is lowered still more, the Korotkoff sounds change tone and loudness. When the pressure has dropped until continuous laminar blood flow is reestablished, the sound disappears completely. The pressure at which continuous laminar flow is reestablished is the **diastolic pressure.** This method for determining systolic and diastolic pressures is not entirely accurate, but its results are within 10% of methods that are more direct. In healthy people, blood pressure values are maintained within a normal range of values.

32 ***Define blood pressure. Describe how blood pressure can be measured.***

Blood Flow and Poiseuille's Law

The **rate** at which blood or any other liquid flows through a tube can be expressed as the volume that passes a specific point per unit of time. Blood flow usually is reported in either milliliters (mL) per minute or liters (L) per minute. For example, when a person is resting, the **cardiac output** of the heart is approximately 5 L/min; thus, blood flow through the aorta is approximately 5 L/min.

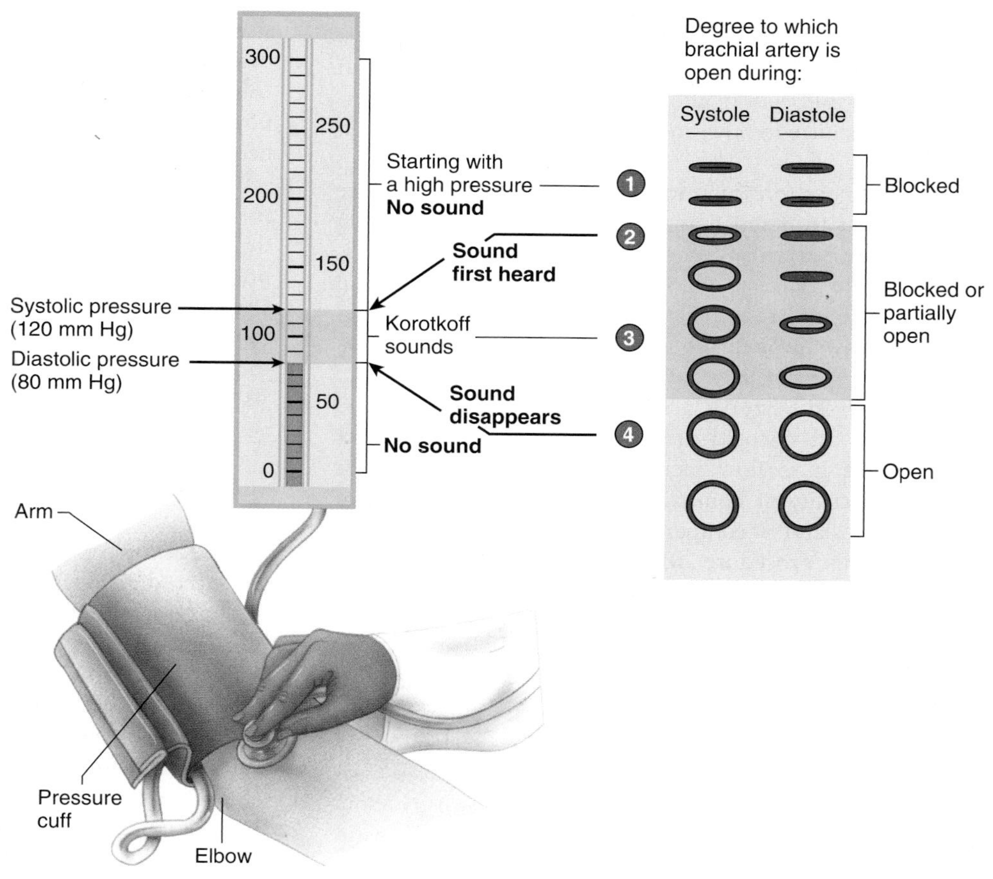

1. No sound is heard because there is no blood flow when the cuff pressure is high enough to keep the brachial artery closed.
2. Systolic pressure is the pressure at which a Korotkoff sound is first heard. When cuff pressure decreases and is no longer able to keep the brachial artery closed during systole, blood is pushed through the partially opened brachial artery to produce turbulent blood flow and a sound. The brachial artery remains closed during diastole.
3. As cuff pressure continues to decrease, the brachial artery opens even more during systole. At first, the artery is closed during diastole, but, as cuff pressure continues to decrease, the brachial artery partially opens during diastole. Turbulent blood flow during systole produces Korotkoff sounds, although the pitch of the sounds changes as the artery becomes more open.
4. Diastolic pressure is the pressure at which the sound disappears. Eventually, cuff pressure decreases below the pressure in the brachial artery and it remains open during systole and diastole. Nonturbulent flow is reestablished and no sounds are heard.
300
250
200
150
100
50
0
Systolic pressure (120 mm Hg)
Diastolic pressure (80 mm Hg)
Starting with a high pressure
No sound
Sound first heard
Korotkoff sounds
Sound disappears
No sound
Degree to which brachial artery is open during:
Systole
Diastole
Blocked
Blocked or partially open
Open
Arm
Pressure cuff
Elbow

PROCESS FIGURE 21.31 **Blood Pressure Measurement**

The rate of blood flow in a vessel can be described by the following:

$$\text{Flow} = \frac{P_1 - P_2}{R}$$

where P_1 and P_2 are the pressures in the vessel at points one and two, respectively, and R is the resistance to flow. Blood always flows from an area of higher pressure to an area of lower pressure, and, the greater the pressure difference, the greater the rate of flow. For example, the average blood pressure in the aorta (P_1) is greater than the blood pressure in the relaxed right atrium (P_2). Therefore, blood flows from the aorta to tissues and from tissues to the right atrium. If the heart should stop contracting, the pressure in the aorta would become equal to that in the right atrium and blood would no longer flow.

The flow of blood, resulting from a pressure difference between the two ends of a blood vessel, is opposed by a resistance to flow. As the resistance increases, blood flow decreases; as the resistance decreases, blood flow increases. Factors that affect resistance can be represented as follows:

$$\text{Resistance} = \frac{8vl}{\pi r^4}$$

where v is the viscosity of blood, l is the length of the vessel, and r is the radius of the vessel. Both 8 and π are constants and, for practical purposes, the length of the blood vessel is constant. Thus, the radius of the blood vessel and the viscosity of the blood determine resistance. The viscosity of blood changes slowly.

When the equation for resistance is combined with the equation for flow, the following relationship, called **Poiseuille's** (pwah-zuh′yes) **law** results:

$$\text{Flow} = \frac{P_1 - P_2}{R} = \frac{\pi(P_1 - P_2)r^4}{8vl}$$

According to Poiseuille's law, a small change in the radius of a vessel dramatically changes the resistance to flow, and therefore the amount of blood that flows through the vessel, because the radius is raised to the fourth power. Vasoconstriction decreases the radius of a vessel, increases resistance to flow, and decreases blood flow through the vessel. For example, decreasing the radius of a vessel by half increases the resistance to flow 16-fold and decreases flow 16-fold. Vasodilation increases the radius of a vessel, decreases resistance to flow, and increases blood flow through the vessel.

Changes in blood pressure and blood vessel radius produce the major changes in blood flow through blood vessels. During exercise, heart rate and stroke volume increase, causing blood pressure in the aorta to increase. In addition, blood vessels in skeletal muscles vasodilate and resistance to flow decreases. As a consequence, blood flow through blood vessels in exercising skeletal muscles increases dramatically.

Viscosity (vis-kos′i-tē) is a measure of the resistance of a liquid to flow. As the viscosity of a liquid increases, the pressure required to force it to flow increases. A common means for reporting the viscosity of liquids is to consider the viscosity of distilled water as 1 and to compare the viscosity of other liquids with it. Using this procedure, whole blood has a viscosity of 3.0–4.5, which means that about three times as much pressure is required to force whole blood to flow through a given tube at the same rate as water.

The viscosity of blood is influenced largely by **hematocrit** (hē′mă-tō-krit, hem′ă-tō-krit), which is the percentage of the total blood volume composed of red blood cells (see chapter 19). As the hematocrit increases, the viscosity of blood increases logarithmically. Blood with a hematocrit of 45% has a viscosity about three times that of water, whereas blood with a very high hematocrit of 65% has a viscosity about seven to eight times that of water. The plasma proteins have only a minor effect on the viscosity of blood. Dehydration or uncontrolled production of erythrocytes can increase hematocrit and the viscosity of blood substantially. Viscosity above its normal range of values increases the workload on the heart; if this workload is great enough, heart failure can result.

33 ***Define blood flow and resistance.***

34 ***Describe the relationship among blood flow, blood pressure, and resistance.***

35 ***According to Poiseuille's law, what effect do viscosity, blood vessel diameter, and blood vessel length have on resistance? On blood flow?***

36 ***Define viscosity, and state the effect of hematocrit on viscosity.***

PREDICT 2

Predict the effect of each of the following conditions on blood flow: (a) vasoconstriction of blood vessels in the skin in response to cold exposure, (b) vasodilation of the blood vessels in the skin in response to an elevated body temperature, and (c) erythrocytosis, which results in a greatly increased hematocrit.

Critical Closing Pressure and Laplace's Law

Each blood vessel exhibits a **critical closing pressure,** the pressure below which the vessel collapses and blood flow through the vessel stops. Under conditions of shock, blood pressure can decrease below the critical closing pressure in vessels (see Clinical Focus "Shock," p. 775). As a consequence, the blood vessels collapse, and flow ceases. Tissues supplied by those vessels can become necrotic because of the lack of blood supply.

Laplace's (la-plas′ez) **law** states that the force that stretches the vascular wall is proportional to the diameter of the vessel times the blood pressure. Laplace's law helps explain the critical closing pressure. As the pressure in a vessel decreases, the force that stretches the vessel wall also decreases. Some minimum force is required to keep the vessel open. If the pressure decreases so that the force is below that minimum requirement, the vessel will close. As the pressure in a vessel increases, the force that stretches the vessel wall also increases.

Laplace's law is expressed by the following formula:

$$F = D \times P$$

where F is force, D is vessel diameter, and P is pressure.

According to Laplace's law, as the diameter of a vessel increases, the force applied to the vessel wall increases, even if the pressure remains constant. If a part of an arterial wall becomes weakened so that a bulge forms in it, the force applied to the weakened part is greater than at other points along the blood vessel because its diameter is greater. The greater force causes the weakened vessel wall to bulge even more, further increasing the force applied to it. This series of events can proceed until the vessel finally ruptures. As the bulges in weakened blood vessel walls, called **aneurysms,** enlarge, the danger of their rupturing increases. Ruptured aneurysms in the blood vessels of the brain or in the aorta often result in death.

37 ***State Laplace's law. How does it explain critical closing pressure and aneurysms?***

Vascular Compliance

Compliance (kom-plī′ans) is the tendency for blood vessel volume to increase as the blood pressure increases. The more easily the vessel wall stretches, the greater is its compliance. The less easily the vessel wall stretches, the smaller is its compliance.

Compliance is expressed by the following formula:

$$\text{Compliance} = \frac{\text{Increase in volume (mL)}}{\text{Increase in pressure (mm Hg)}}$$

Vessels with a large compliance exhibit a large increase in volume when the pressure increases a small amount. Vessels with a small compliance do not show a large increase in volume when the pressure increases.

Venous compliance is approximately 24 times greater than the compliance of arteries. As venous pressure increases, the volume of the veins increases greatly. Consequently, veins act as

TABLE 21.13 Distribution of Blood Volume in Blood Vessels

Vessels		Total Blood Volume (%)
Systemic		
Veins		64
Large veins	(39%)	
Small veins	(25%)	
Arteries		15
Large arteries	(8%)	
Small arteries	(5%)	
Arterioles	(2%)	
Capillaries		5
	TOTAL IN SYSTEMIC VESSELS	84
Pulmonary vessels		9
Heart		7
	TOTAL BLOOD VOLUME	100

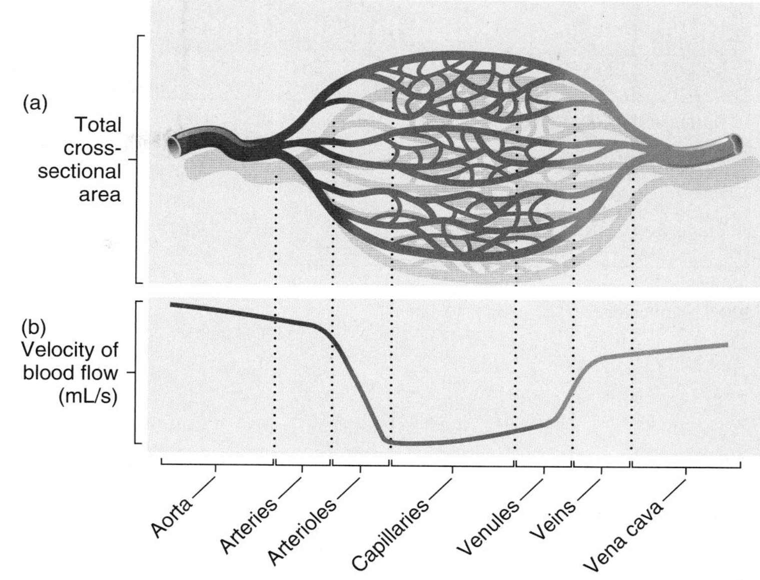

FIGURE 21.32 **Blood Vessel Types and Velocity of Blood Flow**
(*a*) The total cross-sectional area for each of the major blood vessel types is represented. The total cross-sectional areas of all the capillaries is much greater (2500 cm^2) than that of the aorta (5 cm^2), although the cross-sectional area of each capillary is much smaller than that of the aorta. (*b*) Blood velocity decreases dramatically in arterioles, capillaries, and venules and is greater in the aorta and the large veins. As the total cross-sectional areas increases, the velocity of blood flow decreases.

storage areas, or reservoirs, for blood because their large compliance allows them to hold much more blood than other areas of the vascular system (table 21.13).

38 ***Define vascular compliance. Do veins or arteries have greater compliance?***

PHYSIOLOGY OF SYSTEMIC CIRCULATION

The anatomy of the circulatory system, the dynamics of blood flow, and the regulatory mechanisms that control the heart and blood vessels determine the physiologic characteristics of the circulatory system. The entire circulatory system maintains adequate blood flow to all tissues.

Approximately 84% of the total blood volume is contained in the systemic blood vessels. Most of the blood volume is in the veins (64%), which are the vessels with the greatest compliance. Smaller volumes of blood are in the arteries (15%) and capillaries (5%; see table 21.13).

39 ***List the percent distribution of blood in the large arteries, small arteries, arterioles, capillaries, small veins, and large veins.***

Cross-Sectional Area of Blood Vessels

If the cross-sectional area of each blood vessel type is determined and multiplied by the number of each type of blood vessel, the result is the total cross-sectional area for each blood vessel type. For example, only one aorta exists, and it has a cross-sectional area of 5 square centimeters (cm^2). On the other hand, millions of capillaries exist, and each has a very small cross-sectional area. The total cross-sectional area of all capillaries, however, is 2500 cm^2, which is much greater than the cross-sectional area of the aorta (figure 21.32).

The velocity of blood flow is greatest in the aorta, but the total cross-sectional area is small. In contrast, the total cross-sectional area for the capillaries is large, but the velocity of blood flow is low. As the veins become larger in diameter, their total cross-sectional area decreases, and the velocity of blood flow increases. The relationship between blood vessel diameter and velocity of blood flow is much like a stream that flows rapidly through a narrow gorge but flows slowly through a broad plane (see figure 21.32).

40 ***Describe the total cross-sectional areas of the aorta, arteries, arterioles, capillaries, venules, veins, and vena cava.***

41 ***Describe how the velocity of blood flow changes as blood flows through the aorta to the superior vena cava and inferior vena cava.***

TABLE 21.14 Blood Pressure Classification (Adults)

	Systolic Blood Pressure (mm Hg)	Diastolic Blood Pressure (mm Hg)
Normal Blood Pressure	<120	<80
Prehypertension	120–139	80–89
Stage 1 Hypertension	140–159	90–99
Stage 2 Hypertension	≥160	≥100

Source: The seventh Report of the Joint National Committee on Prevention, Detection, and Evaluation, and Treatment of High Blood Pressure, U.S. Dept. of Health and Human Services, NIH Publication No. 03-5233, May 2003.

Pressure and Resistance

The left ventricle of the heart forcefully ejects blood from the heart into the aorta. Because the heart's pumping action is pulsatile, the aortic pressure fluctuates between a systolic pressure of 120 mm Hg and a diastolic pressure of 80 mm Hg (table 21.14 and figure 21.33). As blood flows from arteries through the capillaries and the veins, the pressure falls progressively to a minimum of approximately 0 mm Hg or even slightly lower by the time it returns to the right atrium.

The decrease in arterial pressure in each part of the systemic circulation is directly proportional to the resistance to blood flow. Resistance is small in the aorta, so the average pressure at the end of the aorta is nearly the same as at the beginning of the aorta, about 100 mm Hg. The resistance in medium arteries, which are as small as 3 mm in diameter, is also small, so that their average pressure is only decreased to 95 mm Hg. In the smaller arteries, however, the resistance to blood flow is greater; by the time blood reaches the arterioles, the average pressure is approximately 85 mm Hg. Within the arterioles, the resistance to flow is higher than in any other part of the systemic circulation, and, at their ends, the average pressure is only approximately 30 mm Hg. The resistance is also fairly high in the capillaries. The blood pressure at the arterial end of the capillaries is approximately 30 mm Hg, and it decreases to approximately 10 mm Hg at the venous end. Resistance to blood flow in the veins is low because of their relatively large diameter; by the time the blood reaches the right atrium in the venous system, the average pressure has decreased from 10 mm Hg to approximately 0 mm Hg.

The muscular arteries and arterioles are capable of constricting or dilating in response to autonomic and hormonal stimulation. If constriction occurs, the resistance to blood flow increases, less blood flows through the constricted blood vessels, and blood is shunted to other, nonconstricted areas of the body. Muscular arteries help control the amount of blood flowing to each region of the body, and arterioles regulate blood flow through specific tissues. Constriction of an arteriole decreases blood flow through the local area it supplies, and vasodilation increases the blood flow.

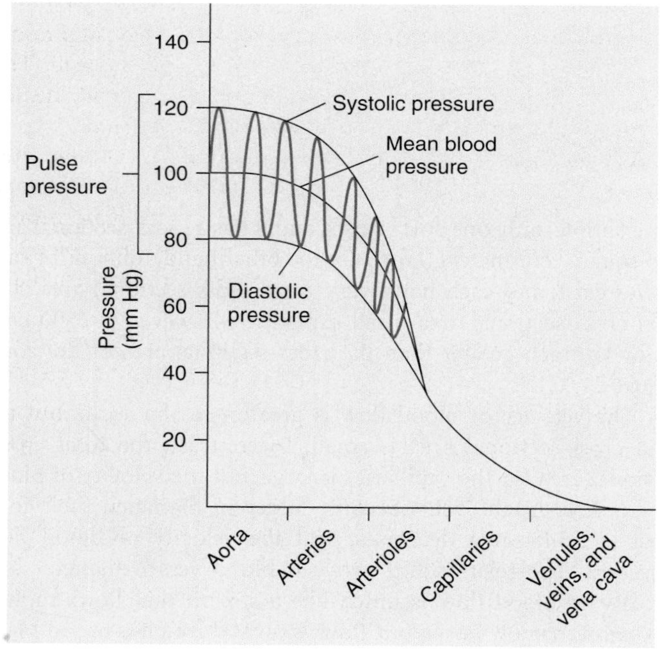

FIGURE 21.33 Blood Pressure in the Major Blood Vessel Types

Blood pressure fluctuations between systole and diastole are damped in small arteries and arterioles. There are no large fluctuations in blood pressure in capillaries and veins.

42 ***Describe the changes in resistance and blood pressure as blood flows through the aorta to the superior vena cava and/or inferior vena cava.***

43 ***Explain how constriction and dilation of muscular arteries shunts blood from one area of the body to another and constriction and dilation of arterioles changes blood flow through local areas.***

Pulse Pressure

The difference between systolic and diastolic pressures is called **pulse pressure** (see figure 21.33). In a healthy, young adult at rest, systolic pressure is approximately 120 mm Hg, and diastolic pressure is approximately 80 mm Hg; thus, the pulse pressure is approximately 40 mm Hg. Two major factors influence pulse pressure: stroke volume of the heart and vascular compliance. When stroke volume decreases, pulse pressure also decreases; when stroke volume increases, pulse pressure increases. For example, during exercise, such as running, the stroke volume increases; as a consequence, the pulse pressure also increases. After running, the pulse pressure gradually returns to its resting value as the stroke volume of the heart decreases. The compliance of blood vessels decreases as arteries age. Arteries in older people become less elastic, or arteriosclerotic, and the resulting decrease in compliance causes the pressure in the aorta to rise more rapidly and to a greater degree during systole and to fall more rapidly to its diastolic value. Thus, for a given stroke volume, systolic pressure and pulse pressure are higher as vascular compliance decreases.

PREDICT 3

Explain the consequences of arteriosclerosis that is getting progressively more severe on a large aortic aneurysm.

The pulse pressure caused by the ejection of blood from the left ventricle into the aorta produces a pressure wave, or pulse, that travels rapidly along the arteries. Its rate of transmission is approximately 15 times greater in the aorta (7–10 m/s) and 100 times greater (15–35 m/s) in the distal arteries than the velocity of blood flow. The pulse is monitored frequently, especially in the radial artery, where it is called the **radial pulse,** to determine heart rate and rhythm. Also, weak pulses usually indicate a decreased stroke volume or increased constriction of the arteries as a result of intense sympathetic stimulation of the arteries.

As the pulse passes through the smallest arteries and arterioles, it is gradually damped so that there is a smaller fluctuation between the systolic and diastolic pressure. This difference is almost absent at the end of the arterioles (see figure 21.33). At the beginning of the capillary, there is a steady pressure of close to 30 mm Hg. As a consequence, there is adequate pressure to cause blood to flow through capillaries if the precapillary sphincters dilate.

44 ***What is pulse pressure? How do stroke volume and vascular compliance affect pulse pressure?***

PREDICT 4

Explain each of the following: weak pulses in response to ectopic and premature beats of the heart, strong bounding pulses in a person who received too much saline solution intravenously, and weak pulses in a person who is suffering from hemorrhagic shock.

Capillary Exchange and Regulation of Interstitial Fluid Volume

Approximately 10 billion capillaries exist in the body. The heart and blood vessels all function to maintain blood flow through those capillaries and to support **capillary exchange,** which is the movement of substances into and out of capillaries. Capillary exchange is the process by which cells receive everything they need to survive and to eliminate metabolic waste products. If blood flow through capillaries is not maintained, cells cannot survive.

By far, the most important means by which capillary exchange occurs is **diffusion.** Nutrients, such as glucose and amino acids; O_2; and hormones diffuse from a higher concentration in capillaries to a lower concentration in the interstitial fluid. Waste products, including CO_2, diffuse from a higher concentration in the interstitial fluid to a lower concentration in the capillaries. Lipid-soluble molecules cross capillary walls by diffusing through the plasma membranes of the endothelial cells of the capillaries. Examples include O_2, CO_2, steroid hormones, and fatty acids. Water-soluble substances, such as glucose and amino acids, diffuse through intercellular spaces or through fenestrations of capillaries. In a few areas of the body, such as the spleen and liver, the spaces between the endothelial cells are large enough to allow proteins to pass through them. In other areas, the connections between endothelial cells are extensive and few molecules pass between the endothelial cells, such as in the capillaries of the brain that form the blood–brain barrier. In these capillaries, mediated transport processes move water-soluble substances across the capillary walls (see chapter 13 for a description of the blood–brain barrier).

Endothelial cells of capillaries appear to take up small pinocytotic vesicles and transport them across the capillary wall. The pinocytotic vesicles, however, do not appear to be a major means by which molecules move across the wall of the capillary.

A small amount of fluid moves out of capillaries at their arterial ends, and most of that fluid reenters capillaries at their venous ends (figure 21.34). The remaining fluid enters lymphatic vessels, which eventually return it to the venous circulation (see chapter 22). Alterations in the forces affecting fluid movement across capillary walls are responsible for edema.

Net filtration pressure (NFP) is the force responsible for moving fluid across capillary walls. It is the difference between net hydrostatic pressure and net osmotic pressure:

$$\text{NFP} = \text{Net hydrostatic pressure} - \text{Net osmotic pressure}$$

Net hydrostatic pressure is the difference in pressure between the blood and interstitial fluid. **Blood pressure (BP)** at the arterial end of a capillary is about 30 mm Hg. It results mainly from the force of contraction of the heart, but it can be modified by the effect of gravity on fluids within the body (see "Blood Pressure and the Effect of Gravity," on p. 760).

Interstitial fluid pressure (IFP) is the pressure of interstitial fluid within the tissue spaces. It is −3 mm Hg. IFP is a negative number because of the suction effect produced by the lymphatic vessels as they pump excess fluid from the tissue spaces. The lymphatic system

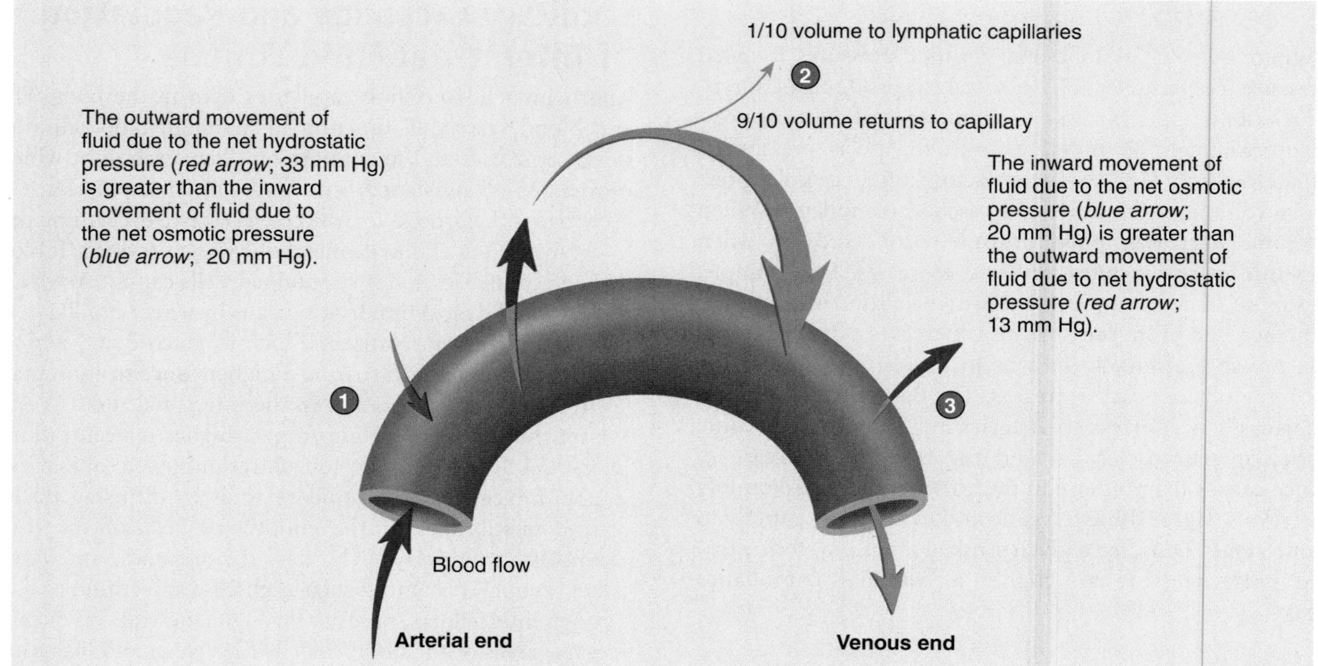

1. At the arterial end of the capillary, the **net hydrostatic pressure** is greater than the **net osmotic pressure.** When the net osmotic pressure is subtracted from the net hydrostatic pressure, it results in a positive **net filtration pressure** that causes fluid to move out of the capillary.

   33 mm Hg (Net hydrostatic pressure)
   −20 mm Hg (Net osmotic pressure)
   13 mm Hg (Net filtration pressure)

2. Approximately nine-tenths of the fluid that leaves the capillary at its arterial end reenters the capillary at its venous end. About one-tenth of the fluid passes into the lymphatic capillaries.

3. At the venous end of the capillary, the **net hydrostatic pressure** is less than the **net osmotic pressure.** When the net osmotic pressure is subtracted from the net hydrostatic pressure, it results in a negative **net filtration pressure** that causes fluid to move into the capillary.

   13 mm Hg (Net hydrostatic pressure)
   −20 mm Hg (Net osmotic pressure)
   − 7 mm Hg (Net filtration pressure)

PROCESS FIGURE 21.34 Fluid Exchange Across the Walls of Capillaries

The total pressure differences between the inside and the outside of the capillary at its arterial and venous ends.

is described in chapter 22. Here, it is only necessary to understand that excess interstitial fluid enters lymphatic capillaries and is eventually returned to the blood.

At the arterial end of capillaries, the net hydrostatic pressure that moves fluid across capillary walls into the tissue spaces is the difference between BP and IFP.

$$\begin{aligned}\text{Net hydrostatic pressure} &= \text{BP} - \text{IFP}\\ &= 30 - (-3)\\ &= 33 \text{ mm Hg}\end{aligned}$$

Net osmotic pressure is the difference in osmotic pressure between the blood and the interstitial fluid. The osmotic pressure caused by the plasma proteins is called the **blood colloid osmotic pressure (BCOP),** and the osmotic pressure caused by proteins in the interstitial fluid is called the **interstitial colloid osmotic pressure (ICOP).** Large proteins do not pass freely through the capillary walls, and the difference in protein concentrations between the blood and interstitial fluid is responsible for osmosis. Ions and small molecules do not make a significant contribution to osmosis across the capillary wall because they pass freely through it and their concentrations are approximately the same in the blood as in the interstitial fluid.

The BCOP (28 mm Hg) is several times larger than the ICOP (8 mm Hg) because of the presence of albumin and other proteins in the plasma (see chapter 19). Therefore, the net osmotic pressure is equal to BCOP − ICOP.

$$\begin{aligned}\text{Net osmotic pressure} &= \text{BCOP} - \text{ICOP}\\ &= 28 - 8\\ &= 20 \text{ mm Hg}\end{aligned}$$

CLINICAL FOCUS

Pulse

The pulse is important clinically because one can determine heart rate, rhythmicity, and other characteristics by feeling it. A pulse can be felt at 10 major locations on each side of the body where large arteries are close to the surface.

On the head and neck, a pulse can be felt in three arteries: the common carotid artery in the neck, the superficial temporal artery immediately anterior to the ear, and the facial artery at the point where it crosses the inferior border of the mandible approximately midway between the angle and the genu (figure A).

On the upper limb, a pulse can also be felt in three arteries: the axillary artery in the axilla, the brachial artery on the medial side of the arm slightly proximal to the elbow, and the radial artery on the lateral side of the anterior forearm just proximal to the wrist. The radial artery is traditionally the most common site for taking a pulse, because it is the most easily accessible artery in the body.

In the lower part of the body, a pulse can be felt in four locations: the femoral artery in the groin, the popliteal artery just proximal to the knee, and the dorsalis pedis artery and posterior tibial artery at the ankle.

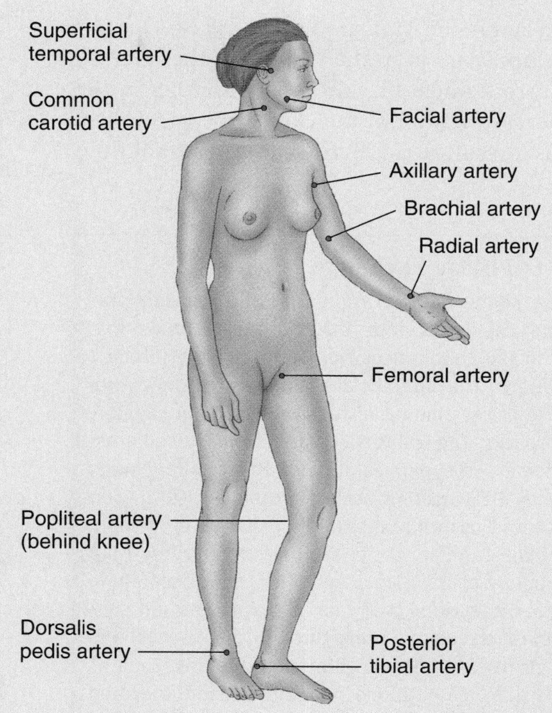

FIGURE A Location of Major Points at Which the Pulse Can Be Monitored
Each pulse point is named after the artery on which it occurs.

The greater the osmotic pressure of a fluid, the greater the tendency for water to move into that fluid (see chapter 3). The net osmotic pressure results in the osmosis of water into the capillary because there is a greater tendency for water to move into the blood than into the interstitial fluid.

The net filtration pressure at the arterial end of the capillary is equal to the net hydrostatic pressure, which moves fluid out of the capillary, minus the net osmotic pressure, which moves fluid into the capillary:

$$\begin{aligned} \text{NFP} &= \text{Net hydrostatic pressure} - \text{Net osmotic pressure} \\ &= 33 - 20 \\ &= 13 \text{ mm Hg} \end{aligned}$$

Between the arterial ends of capillaries and their venous ends, the blood pressure decreases from about 30 mm Hg to 10 mm Hg, which reduces the net hydrostatic pressure moving fluid out of the venous end of the capillary:

$$\begin{aligned} \text{Net hydrostatic pressure} &= \text{BP} - \text{IFP} \\ &= 10 - (-3) \\ &= 13 \text{ mm Hg} \end{aligned}$$

The concentration of proteins within capillaries and the concentration of proteins within interstitial fluid do not change significantly because only a small amount of fluid passes from the capillaries into the tissue spaces. Therefore, the net osmotic pressure moving fluid into capillaries by osmosis is still approximately 20 mm Hg. At the venous end of capillaries, the NFP now causes fluid to reenter the capillary:

$$\begin{aligned} \text{NFP} &= \text{Net hydrostatic pressure} - \text{Net osmotic pressure} \\ &= 13 - 20 \\ &= -7 \text{ mm Hg} \end{aligned}$$

Exchange of fluid across the capillary wall and movement of fluid into lymphatic capillaries keep the volume of the interstitial fluid within a narrow range of values. Disruptions in the movement of fluid across the wall of the capillary can result in edema, or swelling, as a result of an increase in interstitial fluid volume.

Edema and Capillary Exchange

Increases in the permeability of capillaries allow plasma proteins to move from capillaries into the interstitial fluid. This causes an increase in the colloid osmotic pressure of the interstitial fluid. An increase in the interstitial colloid osmotic pressure causes a net increase in the amount of fluid moving from capillaries into interstitial spaces. The result is edema. Chemical mediators of inflammation increase the permeability of the capillary walls and can cause edema. Decreases in plasma protein concentration reduce the blood colloid osmotic pressure, which results in more fluid moving out of the capillary at its arterial end and less fluid moving into the capillary at its venous end. The result once again is edema. Severe liver infections that reduce plasma protein synthesis, loss of protein molecules in urine through the kidneys, and protein starvation all result in edema. Blockage of veins, such as in venous thrombosis, increases blood pressure in capillaries and can result in edema. Either blockage or removal of lymphatic vessels causes fluid to accumulate in the interstitial spaces and results in edema. Removal of lymphatic vessels occurs when lymph nodes that are suspected to be cancerous are removed.

45 ***What is the most important means by which capillary exchange occurs?***

46 ***Describe the factors that influence the movement of fluid from capillaries into the tissues.***

47 ***What happens to the fluid in the tissues? What is edema?***

PREDICT 5

Edema often results from a disruption in the normal inwardly and outwardly directed pressures across the capillary wall. On the basis of what you know about fluid movement across the wall of the capillary and the regulation of capillary blood pressure, explain why large fluctuations in arterial blood pressure occur without causing significant edema and why small increases in venous pressure can lead to edema.

Functional Characteristics of Veins

Cardiac output depends on the preload, which is determined by the volume of blood that enters the heart from the veins (see chapter 20). The factors that affect flow in the veins are, therefore, of great importance to the overall function of the cardiovascular system. If the volume of blood is increased because of a rapid transfusion, the amount of blood flow to the heart through the veins increases. This increases the preload, which causes the cardiac output to increase because of Starling's law of the heart. On the other hand, a rapid loss of a large volume of blood decreases venous return to the heart, which decreases the preload and cardiac output.

Venous tone is a continual state of partial contraction of the veins as a result of sympathetic stimulation. Increased sympathetic stimulation increases venous tone by causing constriction of the veins, which forces the large venous volume to flow toward the heart. Consequently, venous return and preload increase, causing an increase in cardiac output. Conversely, decreased sympathetic stimulation decreases venous tone, allowing veins to relax and dilate. As the veins fill with blood, venous return to the heart, preload, and cardiac output decrease.

The periodic muscular compression of veins forces blood to flow more rapidly through them toward the heart. The valves in the veins prevent flow away from the heart so that, when veins are compressed, blood is forced to flow toward the heart. The combination of arterial dilation and compression of the veins by muscular movements during exercise causes blood to return to the heart more rapidly than under conditions of rest.

48 ***How do blood volume and venous tone affect cardiac output?***

Blood Pressure and the Effect of Gravity

Blood pressure is approximately 0 mm Hg in the right atrium, and it averages approximately 100 mm Hg in the aorta. The pressure in vessels above and below the heart, however, is affected by gravity. While a person is standing, the pressure in the venules of the feet can be as much as 90 mm Hg, instead of its usual 10 mm Hg pressure. Arterial pressure is influenced by gravity to the same degree; thus, the arterial ends of the capillaries can have a pressure of 110 mm Hg rather than 30 mm Hg. The normal pressure difference between the arterial and the venous ends of capillaries still remains the same so that flow continues through the capillaries. The major effect of the high pressure in the feet and legs when a person stands for a prolonged time without moving is edema. Without muscular movement, the pressure at the venous end of the capillaries increases. Up to 15%–20% of the total blood volume can pass through the walls of the capillaries into the interstitial spaces of the lower limbs during 15 minutes of standing still.

49 ***What effect does standing have on blood pressure in the feet and the head? Explain why this effect occurs.***

PREDICT 6

Explain why people who are suffering from edema in the legs are told to keep them elevated.

CASE STUDY

A Venous Thrombosis

Harry is a 55-year-old college professor who teaches a night class in a small town about 50 miles from his home. One night, as he walked to his car after class, Harry noticed that his right leg was uncomfortable. By the time he reached home, about 90 minutes later, the calf of his right leg had become very swollen. When he extended his knee and plantar flexed his foot, the pain in his right leg increased. Harry thought this might be a serious condition, so he drove to the emergency room.

In the emergency room, a Doppler test, which monitors the flow of blood through blood vessels, was performed on Harry's right leg. The test confirmed that a thrombus had formed in one of the deep veins of his right leg. His pain and edema were consistent with the presence of a venous thrombosis.

Harry was admitted to the hospital and his physician prescribed intravenous (IV) heparin. About 4 A.M., Harry experienced an increase in his respiratory rate and his breathing became labored. He experienced pain in his chest and back, and there was a decrease in his arterial oxygen levels. In response to these changes, Harry's physician increased the amount of heparin. The chest pain and changes in Harry's respiratory movements improved over the next 24 hours. The next day, a CT scan was performed and pulmonary emboli were identified, but infarctions of the lung were not apparent. The edema in Harry's leg also slowly improved. Harry was told that he would have to remain in the hospital for several days. Heparin was continued for several days and then oral coumadin was prescribed. Frequent blood samples were taken to determine Harry's prothrombin time. After about a week, Harry was released from the hospital, but his physician prescribed oral coumadin for at least several months. Harry was required to have his 'pro-time' checked periodically.

PREDICT 7

a. Explain why edema and pain developed in response to a thrombus in one of the deep veins of Harry's right leg.

b. If a thrombus in the posterior tibial vein gave rise to an embolus, name in order the parts of the circulatory system the embolus would pass through before lodging in a blood vessel in the lungs. Explain why the lungs are the most likely places the embolus will lodge.

c. Predict the effect of pulmonary emboli on the ability of the right ventricle to pump blood.

d. Predict the effect of pulmonary emboli on blood oxygen levels, on the ability of the left ventricle to pump blood, and on systemic blood pressure. What responses would be activated by this change in blood pressure? (*Hint:* See figure 21.38.)

e. Explain why Harry's physician prescribed heparin and coumadin, and explain why coumadin was prescribed long after the venous thrombosis and lung emboli were dissolved.

CONTROL OF BLOOD FLOW IN TISSUES

Blood flow provided to the tissues by the cardiovascular system is highly controlled and matched closely to the metabolic needs of tissues. Mechanisms that control blood flow through tissues are classified as (1) local control and (2) nervous and hormonal control.

Local Control of Blood Flow by the Tissues

Blood flow is much greater in some organs than in others. For example, blood flow through the brain, kidneys, and liver is relatively high. The muscle mass of the body is large so that flow through resting skeletal muscles, although not high, is greater than that through other tissue types because skeletal muscle constitutes 35%–40% of the total body mass. Flow through exercising skeletal muscles can increase up to 20-fold, however, and the blood flow through the viscera, including the kidneys and liver, either remains the same or decreases.

In most tissues, blood flow is proportional to the metabolic needs of the tissue; therefore, as the activity of skeletal muscle increases, blood flow increases to supply the increased need for oxygen and other nutrients. Blood flow also increases in response to a buildup of metabolic end products.

In some tissues, however, blood flow serves purposes other than the delivery of nutrients and the removal of waste products. In the skin, blood flow also dissipates heat from the body. In the kidneys, it eliminates metabolic waste products, regulates water balance, and controls the pH of body fluids. Among other functions, blood flow delivers nutrients that enter the blood from the small intestine to the liver for processing.

Functional Characteristics of the Capillary Bed

The innervation of the metarterioles and the precapillary sphincters in capillary beds is sparse (table 21.15). Local factors regulate these structures primarily. As the rate of metabolism increases in a tissue, blood flow through its capillaries increases. The precapillary sphincters relax, allowing blood to flow into the local capillary bed. Blood flow can increase sevenfold to eightfold as a result of vasodilation of the metarterioles and the precapillary sphincters in response to an increased rate of metabolism.

Vasodilator substances are produced as the rate of metabolism increases. The vasodilator substances then diffuse from the tissues supplied by the capillary to the area of the precapillary sphincter, the metarterioles, and the arterioles to cause vasodilation (figure 21.35*a*). Several chemicals, including carbon dioxide, lactic acid, adenosine, adenosine monophosphate, adenosine

TABLE 21.15 Homeostasis: Local Control of Blood Flow

Stimulus	Response
Regulation by Metabolic Need of Tissues	
Increased vasodilator substances (e.g., CO_2, lactic acid, adenosine, adenosine monophosphate, adenosine diphosphate, endothelium-derived relaxation factor, K^+, decreased pH) or decreased nutrients (e.g., O_2, glucose, amino acids, fatty acids, and other nutrients) as a result of increased metabolism	Relaxation of precapillary sphincters and subsequent increase in blood flow through capillaries
Decreased vasodilator substances and a reduced need for O_2 and other nutrients	Contraction of precapillary sphincters and subsequent decrease in blood flow through capillaries
Regulation by Nervous Mechanisms	
Increased physical activity or increased sympathetic activity	Construction of blood vessels in skin and viscera
Increased body temperature detected by neurons of the hypothalamus	Dilation of blood vessels in skin (see chapter 5)
Decreased body temperature detected by neurons of the hypothalamus	Constriction of blood vessels in skin (see chapter 5)
Decrease in skin temperature below a critical value	Dilation of blood vessels in skin (protects skin from extreme cold)
Anger or embarrassment	Dilation of blood vessels in skin of face and upper thorax
Regulation by Hormonal Mechanisms (Reinforces Increased Activity of the Sympathetic Nervous System)	
Increased physical activity and increased sympathetic activity causing release of epinephrine and small amounts of norepinephrine from the adrenal medulla	Constriction of blood vessels in skin and viscera; dilation of blood vessels in skeletal and cardiac muscle
Autoregulation	
Increased blood pressure	Contraction of precapillary sphincters to maintain constant capillary blood flow
Decreased blood pressure	Relaxation of precapillary sphincters to maintain constant capillary blood flow
Long-Term Local Blood Flow	
Increased metabolic activity of tissues over a long period	Increased diameter and number of capillaries
Decreased metabolic activity of tissues over a long period	Decreased diameter and number of capillaries

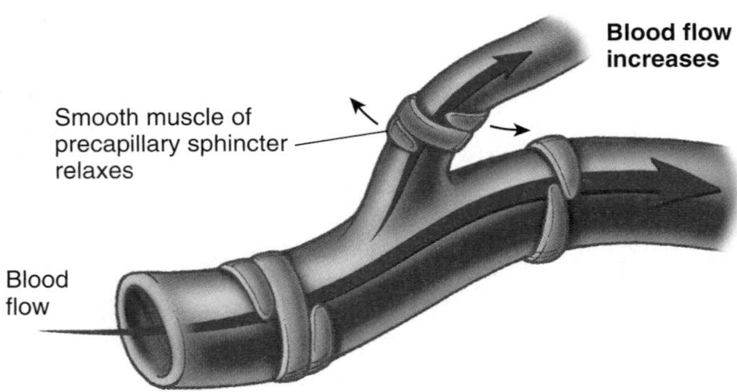

(a) Precapillary sphincters relax as the tissue concentration of nutrients such as O_2, glucose, amino acids, and fatty acids decreases.

Precapillary sphincters relax as the concentration of tissue metabolic by-products, such as CO_2, lactic acid, adenosine, adenosine monophosphate, adenosine diphosphate, nitric oxide, and K^+ increase, and as the pH decreases.

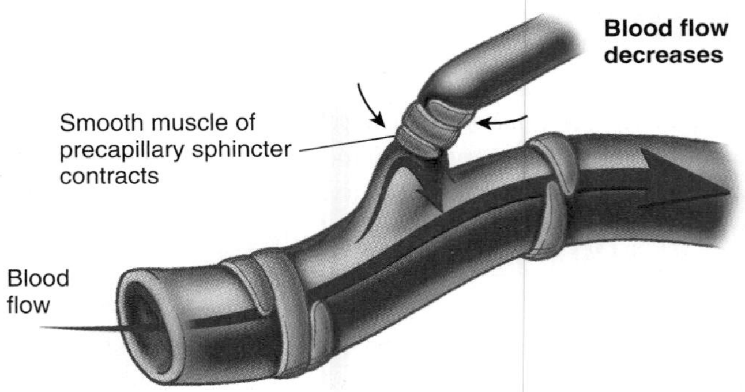

(b) Precapillary sphincters contract as the tissue concentration of nutrients such as O_2, glucose, amino acids, and fatty acids increases.

Precapillary sphincters contract as the tissue concentration of metabolic by-products, such as CO_2, lactic acid, adenosine, adenosine monophosphate, adenosine diphosphate, nitric oxide, and K^+ decrease, and as the pH increases.

FIGURE 21.35 Control of Local Blood Flow Through Capillary Beds

(*a*) Dilation of precapillary sphincters. (*b*) Constriction of precapillary sphincters.

CLINICAL FOCUS

Hypertension

Hypertension, or high blood pressure, affects approximately 20% of the human population at sometime in their lives. Generally, a person is considered hypertensive if the systolic blood pressure is greater than 140 mm Hg and the diastolic pressure is greater than 90 mm Hg. Current methods of evaluation, however, take into consideration diastolic and systolic blood pressures in determining whether a person is suffering from hypertension (see table 21.14). In addition, normal blood pressure is age-dependent, so classification of an individual as hypertensive depends on the person's age.

Chronic hypertension has an adverse effect on the function of both the heart and blood vessels. Hypertension requires the heart to work harder than normal. This extra work leads to hypertrophy of the cardiac muscle, especially in the left ventricle, and can lead to heart failure. Hypertension also increases the rate at which arteriosclerosis develops. Arteriosclerosis, in turn, increases the probability that blood clots, or thromboemboli (throm′bō-em′bō-lī), may form and that blood vessels will rupture. Common medical problems associated with hypertension are cerebral hemorrhage, coronary infarction, hemorrhage of renal blood vessels, and poor vision caused by burst blood vessels in the retina.

Some conditions leading to hypertension include a decrease in functional kidney mass, excess aldosterone or angiotensin production, and increased resistance to blood flow in the renal arteries. All of these conditions cause an increase in total blood volume, which causes cardiac output to increase. Increased cardiac output forces blood to flow through tissue capillaries, causing the precapillary sphincters to constrict. Thus, increased blood volume increases cardiac output and peripheral resistance, both of which result in greater blood pressure.

Although these conditions result in hypertension, roughly 90% of the diagnosed cases of hypertension are called **idiopathic,** or **essential, hypertension,** which means the cause of the condition is unknown. Drugs that dilate blood vessels (called vasodilators), drugs that increase the rate of urine production (called diuretics), and drugs that decrease cardiac output normally are used to treat essential hypertension. The vasodilator drugs increase the rate of blood flow through the kidneys and thus increase urine production, and the diuretics also increase urine production. Increased urine production reduces blood volume, which reduces blood pressure. Substances that decrease cardiac output, such as β-adrenergic-blocking agents, decrease the heart rate and force of contraction. In addition to these treatments, low-salt diets normally are recommended to reduce the amount of sodium chloride and water absorbed from the intestine into the bloodstream.

diphosphate, endothelium-derived relaxation factor (EDRF), potassium ions, and hydrogen ions, cause vasodilation, and they increase in concentration in the extracellular fluid as the rate of metabolism in tissues increases.

Lack of nutrients can also be important in regulating local blood flow. For example, oxygen and other nutrients are required to maintain vascular smooth muscle contraction. An increased rate of metabolism decreases the amount of oxygen and other nutrients in the tissues. Smooth muscle cells of the precapillary sphincter relax in response to a lack of oxygen and other nutrients, resulting in vasodilation (see figure 21.35*a*).

Blood flow through capillaries is not continuous but cyclic. The cyclic fluctuation is the result of periodic contraction and relaxation of the precapillary sphincters called **vasomotion** (vā-sō-mō′shŭn, vas-ō-mō′shŭn). Blood flows through the capillaries until the by-products of metabolism are reduced in concentration and until nutrient supplies to precapillary smooth muscles are replenished. Then the precapillary sphincters constrict and remain constricted until the by-products of metabolism increase and nutrients decrease (figure 21.35*b*).

Autoregulation of Blood Flow

Arterial pressure can change over a wide range, whereas blood flow through tissues remains relatively constant. The maintenance of blood flow by tissues is called **autoregulation** (aw′tō-reg-ū-lā′shŭn). Between arterial pressures of approximately 75 mm Hg and 175 mm Hg, blood flow through tissues remains within 10%–15% of its normal value. The mechanisms responsible for autoregulation are the same as those for vasomotion. The need for nutrients and the buildup of metabolic by-products cause precapillary sphincters to dilate, and blood flow through tissues increases if a minimum blood pressure exists. On the other hand, once the supply of nutrients and oxygen to tissues is adequate, the precapillary sphincters constrict, and blood flow through the tissues decreases, even if blood pressure is very high.

PREDICT 8

When blood flow to a tissue has been blocked for a short time, the blood flow through that tissue increases to as much as five times its normal value after the removal of the blockade. The response is called reactive hyperemia. Explain this phenomenon based on the local control of blood flow.

Long-Term Local Blood Flow

The long-term regulation of blood flow through tissues is matched closely to the metabolic requirements of the tissue. If the metabolic activity of a tissue increases and remains elevated for an extended period, the diameter and the number of capillaries in the tissue increase, and local blood flow increases. The increased density of capillaries in the well-trained skeletal muscles of athletes, compared with that in poorly trained skeletal muscles, is an example.

The availability of oxygen to a tissue can be a major factor in determining the adjustment of the vascularity of a tissue to its long-term metabolic needs. If oxygen is scarce, capillaries increase in diameter and in number and, if the oxygen levels remain elevated in a tissue, the vascularity decreases.

Occlusion of Blood Vessels and Collateral Circulation

Blockage, or occlusion, of a blood vessel leads to an increase in the diameter of smaller blood vessels that bypass the occluded vessel. In many cases, the development of these collateral vessels is marked. For example, if a vessel such as the femoral artery becomes occluded, the small vessels that bypass the occluded vessel become greatly enlarged. An adequate blood supply to the lower limb is often reestablished over a period of weeks. If the occlusion is sudden and so complete that tissues supplied by a blood vessel suffer from ischemia (lack of blood flow), cell death (necrosis) can occur. In this instance, collateral circulation does not have a chance to develop before necrosis occurs.

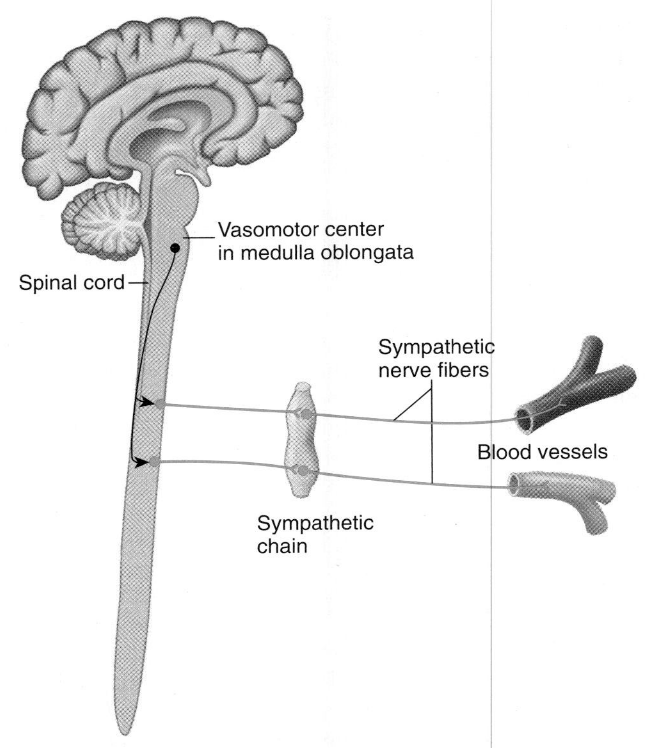

FIGURE 21.36 Nervous Regulation of Blood Vessels

Most blood vessels are innervated by sympathetic nerve fibers. The vasomotor center within the medulla oblongata plays a major role in regulating the frequency of action potentials in nerve fibers that innervate blood vessels.

Nervous and Hormonal Regulation of Local Circulation

Nervous control of arterial blood pressure is important in the minute-to-minute regulation of local circulation. The blood pressure must be adequate to cause blood flow through capillaries while at rest, during exercise, or in response to circulatory shock, during which blood pressure decreases to a very low value. For example, during exercise increased arterial blood pressure is needed to sustain increased blood flow through the capillaries of skeletal muscles, which have dilated precapillary sphincters. The increased blood flow is needed to supply oxygen and nutrients to the exercising skeletal muscles.

Nervous regulation also provides a means by which blood can be shunted from one large area of the peripheral circulatory system to another. For example, in response to blood loss, blood flow to the viscera and the skin is reduced dramatically. This helps maintain the arterial blood pressure within a range sufficient to allow adequate blood flow through the capillaries of the brain and cardiac muscle.

Nervous regulation, by the autonomic nervous system, can function rapidly (within 1–30 seconds). The most important part of the autonomic nervous system for this regulation is the sympathetic division (figure 21.36). Sympathetic vasomotor fibers innervate all the blood vessels of the body except the capillaries, precapillary sphincters, and most metarterioles. The innervation of the small arteries and arterioles allows the sympathetic nervous system to increase or decrease resistance to blood flow.

PREDICT 9

A strong athlete just finished a 1-mile run and sat down to have a drink with her friends. Her blood pressure was not dramatically elevated during the run, but her cardiac output was greatly increased. After the run, her cardiac output decreased dramatically, but her blood pressure only decreased to its resting level. Predict how sympathetic stimulation of her large veins, arteries in her digestive system, and arteries in her skeletal muscles change while she is relaxing. Explain why this is consistent with the decrease in her cardiac output.

Sympathetic vasoconstrictor fibers extend to most parts of the circulatory system, but they are less prominent in skeletal muscle, cardiac muscle, and the brain and more prominent in the kidneys, gastrointestinal tract, spleen, and skin.

An area of the lower pons and upper medulla oblongata, called the **vasomotor** (vā-sō-mō′ter, vas-ō-mō′ter) **center** (see figure 21.36), is tonically active. A low frequency of action potentials is transmitted continually through the sympathetic vasoconstrictor fibers. As a consequence, the peripheral blood vessels are partially constricted, a condition called **vasomotor tone.**

Part of the vasomotor center inhibits vasomotor tone. Thus, the vasomotor center consists of an excitatory part, which is tonically active, and an inhibitory part, which can induce vasodilation.

Vasoconstriction results from an increase and vasodilation from a decrease in vasomotor tone.

Areas throughout the pons, midbrain, and diencephalon can either stimulate or inhibit the vasomotor center. For example, the hypothalamus can exert either strong excitatory or inhibitory effects on the vasomotor center. Increased body temperature detected by temperature receptors in the hypothalamus causes vasodilation of blood vessels in the skin (see chapter 5). The cerebral cortex also can either excite or inhibit the vasomotor center. For example, action potentials that originate in the cerebral cortex during periods of emotional excitement activate hypothalamic centers, which in turn increase vasomotor tone (see table 21.15).

The neurotransmitter for the vasoconstrictor fibers is norepinephrine, which binds to α-adrenergic receptors on vascular smooth muscle cells to cause vasoconstriction. Sympathetic action potentials also cause the release of epinephrine and norepinephrine into the blood from the adrenal medulla. These hormones are transported in the blood to all parts of the body. In most vessels, they cause vasoconstriction, but in some vessels, especially those in skeletal muscle, epinephrine binds to β-adrenergic receptors, which are present in larger numbers, and can cause the skeletal muscle blood vessels to dilate.

50 ***Explain how vasodilator substances and nutrients are involved with local control of blood flow. What is vasomotion? What is autoregulation of local blood flow?***

51 ***How is long-term regulation of blood flow through tissues accomplished?***

52 ***Describe nervous and hormonal control of blood flow. Under what conditions is nervous control of blood flow important? Define vasomotor tone.***

REGULATION OF MEAN ARTERIAL PRESSURE

Blood flow to all areas of the body depends on the maintenance of an adequate pressure in the arteries. As long as arterial blood pressure is adequate, local control of blood flow through tissues is appropriately matched to their metabolic needs. Blood flow through tissues cannot be adequate if arterial blood pressure is too low, and damage, including heart and blood vessel damage, can result if arterial blood pressure is too high.

Mean arterial pressure (MAP) is slightly less than the average of systolic and diastolic pressures because diastole lasts longer than systole. MAP is approximately 70 mm Hg at birth, is slightly less than 100 mm Hg from adolescence to middle age, and reaches 110 mm Hg in healthy older persons, but it can be as high as 130 mm Hg (see table 21.14).

Cardiac output (CO) is the volume of blood pumped by the heart each minute. It is equal to the **heart rate (HR)** times the **stroke volume (SV)**. **Peripheral resistance (PR)** is the resistance to blood flow in all the blood vessels. MAP in the body is proportional to the cardiac output times the peripheral resistance: Blood flow through the entire circulatory system is determined by the cardiac output (*CO*), which is equal to the heart rate (*HR*) times the stroke volume (*SV*) and peripheral resistance (*PR*), which is the resistance to blood flow in all the blood vessels:

$$\text{MAP} = CO \times PR \quad \text{or} \quad \text{MAP} = HR \times SV \times PR$$

This equation expresses the effect of heart rate, stroke volume, and peripheral resistance on blood pressure. An increase in any one of them results in an increase in blood pressure. Conversely, a decrease in any one of them produces a decrease in blood pressure. The mechanisms that control blood pressure do so by changing peripheral resistance, heart rate, or stroke volume. Because stroke volume depends on the amount of blood entering the heart, regulatory mechanisms that control blood volume also affect blood pressure. For example, an increase in blood volume increases venous return, which increases preload, and the increased preload increases stroke volume.

When blood pressure suddenly drops because of hemorrhage or some other cause, the control systems respond by increasing blood pressure to a value consistent with life and by increasing blood volume to its normal value. Two major types of control systems operate to achieve these responses: (1) those that respond in the short term and (2) those that respond in the long term.

The regulatory mechanisms that control pressure on a short-term basis respond quickly but begin to lose their capacity to regulate blood pressure a few hours to a few days after blood pressure is maintained at higher or lower values. This occurs because sensory receptors adapt to the altered pressures. Long-term regulation of blood pressure is controlled primarily by mechanisms that influence kidney function, and those mechanisms do not adapt rapidly to altered blood pressures.

53 ***Define mean arterial pressure, cardiac output, and peripheral resistance.***

54 ***Describe the factors that determine mean arterial pressure.***

Short-Term Regulation of Blood Pressure

The short-term, rapidly acting mechanisms controlling blood pressure include the baroreceptor reflexes, the adrenal medullary mechanism, chemoreceptor reflexes, and the central nervous system ischemic response. Some of these reflex mechanisms operate on a minute-to-minute basis and help regulate blood pressure within a narrow range of values. Other mechanisms respond primarily to emergency situations (see figures 21.39 and 21.41).

Baroreceptor Reflexes

Baroreceptor reflexes are very important in regulating blood pressure on a minute-to-minute basis. They detect even small changes in blood pressure and respond quickly. However, they are not as important as other mechanisms in regulating blood pressure over long periods of time.

Baroreceptors, or **pressoreceptors,** are sensory receptors sensitive to stretch. They are scattered along the walls of most of the large arteries of the neck and thorax and are most numerous in the area of the carotid sinus at the base of the internal carotid artery and in the walls of the aortic arch. Action potentials are transmitted from the carotid sinus baroreceptors through the glossopharyngeal (IX) nerves to the cardioregulatory and vasomotor centers in the medulla oblongata and from the aortic arch

1. Baroreceptors in the carotid sinus and aortic arch monitor blood pressure.
2. Action potentials are conducted by the glossopharyngeal and vagus nerves to the cardioregulatory and vasomotor centers in the medulla oblongata.
3. Increased parasympathetic stimulation of the heart decreases the heart rate.
4. Increased sympathetic stimulation of the heart increases the heart rate and stroke volume.
5. Increased sympathetic stimulation of blood vessels increases vasoconstriction.

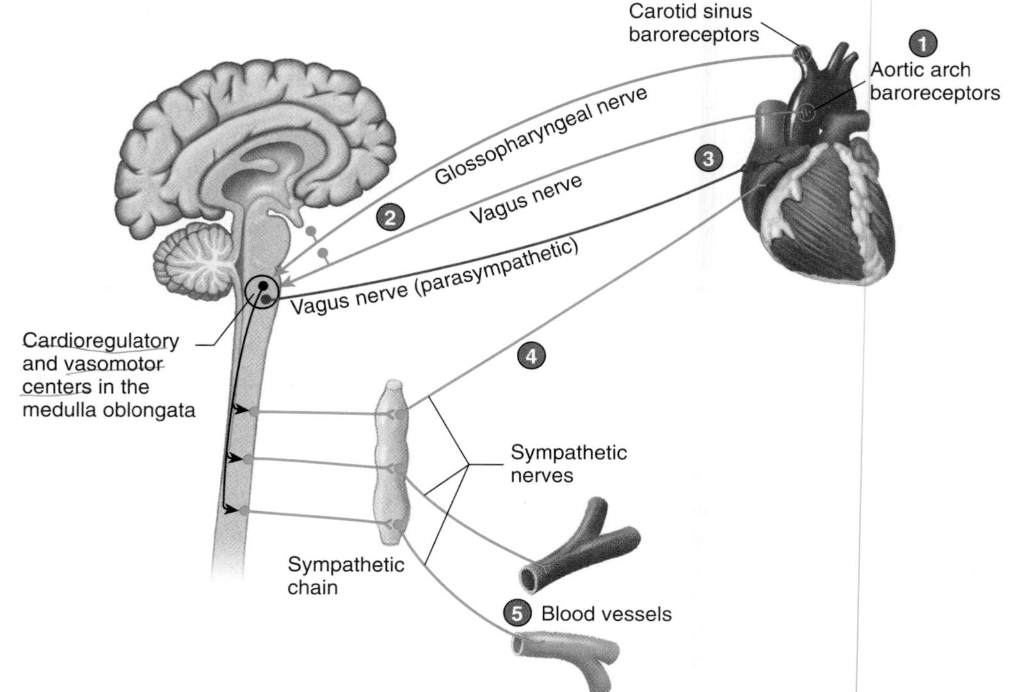

PROCESS FIGURE 21.37 Baroreceptor Reflex Control of Blood Pressure

An increase in blood pressure increases parasympathetic stimulation of the heart and decreases sympathetic stimulation of the heart and blood vessels, resulting in a decrease in blood pressure. A decrease in blood pressure decreases parasympathetic stimulation of the heart and increases sympathetic stimulation of the heart and blood vessels, resulting in an increase in blood pressure.

through the vagus (X) nerves to the medulla oblongata (figure 21.37). Stimulation of baroreceptors in the carotid sinus activates the **carotid sinus reflex,** and stimulation of baroreceptors in the aortic arch activates the **aortic arch reflex.** Both of these reflexes are baroreceptor reflexes, and they help keep blood pressure within a narrow range of values.

In the carotid sinus and the aortic arch, normal blood pressure partially stretches the arterial wall so that the baroreceptors produce a constant but low frequency of action potentials. Increased pressure in the blood vessels stretches the vessel walls and causes the baroreceptors to increase the frequency of action potentials. Conversely, a decrease in blood pressure reduces the stretch of the arterial wall and causes the baroreceptors to decrease the frequency of action potentials.

A sudden increase in blood pressure increases the frequency of action potentials produced in the baroreceptors. The increase in action potentials influences the vasomotor and cardioregulatory centers of the medulla oblongata. The vasomotor center responds by decreasing sympathetic stimulation of blood vessels, and the cardioregulatory center responds by increasing parasympathetic stimulation of the heart. As a result, peripheral blood vessels dilate, the heart rate decreases, and the blood pressure decreases (figure 21.38).

A sudden decrease in blood pressure results in a decreased frequency of action potentials produced by the baroreceptors. The decreased action potentials influence the vasomotor center and cardioregulatory centers of the medulla oblongata. The vasomotor center responds by increasing sympathetic stimulation of blood vessels, and the cardioregulatory center responds by increasing sympathetic stimulation of the heart. As a result, peripheral blood vessels constrict, the heart rate and stroke volume increase, and the blood pressure increases. This increase is accompanied by a decrease in parasympathetic stimulation of the heart (see figures 21.37 and 21.38).

The carotid sinus and aortic arch baroreceptor reflexes are important in regulating blood pressure moment to moment. When a person rises rapidly from a sitting or lying position to a standing position, a dramatic drop in blood pressure in the neck and thoracic regions occurs because of the pull of gravity on the blood. This reduction can be so great that blood flow to the brain becomes sufficiently sluggish to cause dizziness or loss of consciousness. The falling blood pressure activates the baroreceptor reflexes, which reestablish normal blood pressure within a few seconds. A healthy person may experience only a temporary sensation of dizziness.

PREDICT 10

Explain how the baroreceptor reflex responds when a person does a headstand.

The baroreceptor reflexes are short-term and rapid-acting. They do not change the average blood pressure in the long run. The baroreceptors adapt within 1–3 days to any new sustained blood pressure to which they are exposed. If the blood pressure is elevated for more than a few days, the baroreceptors adapt to the elevated pressure and the baroreceptor reflex does not reduce the blood pressure to its original value. This adaptation is common in people who have hypertension.

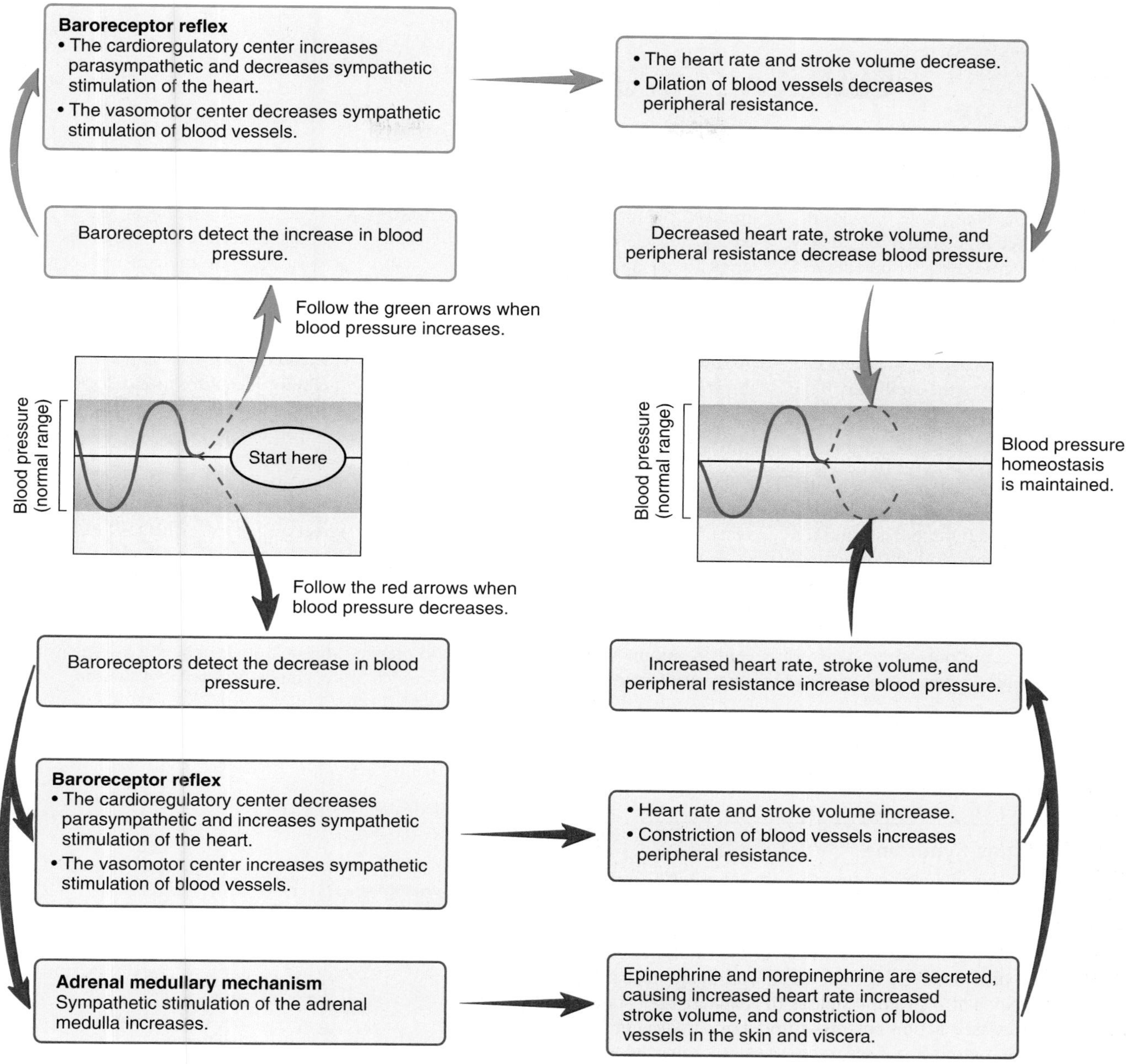

HOMEOSTASIS FIGURE 21.38 **Summary of the Baroreceptor Effects on Blood Pressure**

For more information on the baroreceptor reflex, see figure 21.37 (p. 766); on the adrenal medullary mechanism, see figure 21.39 (p. 768).

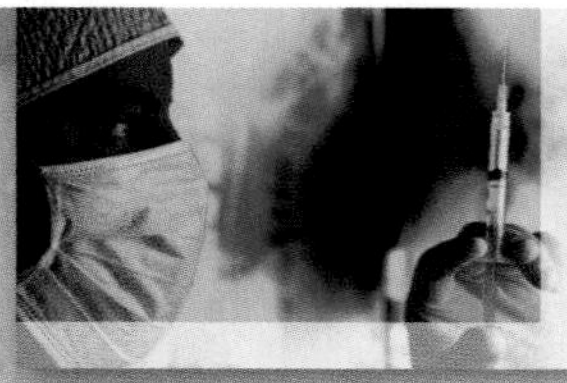

CLINICAL FOCUS

Blood Flow Through Tissues During Exercise

During exercise, blood flow through tissues is changed dramatically. Blood flow through exercising skeletal muscles can be 15–20 times greater than through resting muscles. Local, nervous, and hormonal regulatory mechanisms are responsible for the increased blood flow. When skeletal muscle is resting, only 20%–25% of the capillaries in the skeletal muscle are open, whereas during exercise 100% of the capillaries are open.

Low oxygen tensions resulting from greatly increased muscular activity and the release of vasodilator substances, such as lactic acid, carbon dioxide, and potassium ions, cause the dilation of precapillary sphincters. Increased sympathetic stimulation and epinephrine released from the adrenal medulla cause vasoconstriction in the blood vessels of the skin and viscera, but only some vasoconstriction in the blood vessels of skeletal muscles. The total resistance to blood flow in skeletal muscle decreases because the capillaries are all open, even though some vasoconstriction in skeletal muscle blood vessels occurs. The increased resistance to blood flow in the skin and viscera causes blood to be shunted from the viscera and the skin to the skeletal muscles.

The movement of skeletal muscles compresses veins in a cyclic fashion and greatly increases the venous return to the heart. In addition, veins undergo some constriction, which reduces the total volume of blood in the veins without dramatically increasing resistance to blood flow. The resulting increase in the preload and increased sympathetic stimulation of the heart result in an elevated heart rate and stroke volume, which increases the cardiac output. As a consequence, blood pressure usually increases by 20–60 mm Hg, which also helps sustain the increased blood flow through skeletal muscle blood vessels.

In response to sympathetic stimulation, some decrease in blood flow through the skin can occur at the beginning of exercise. As body temperature increases in response to the increased muscular activity, however, temperature receptors in the hypothalamus are stimulated. As a result, action potentials in sympathetic nerve fibers causing vasoconstriction decrease, resulting in vasodilation of blood vessels in the skin. As a consequence, the skin turns a red or pinkish color, and a great deal of excess heat is lost as blood flows through the dilated blood vessels.

The overall effect of exercise on circulation is to greatly increase blood flow through exercising muscles and to keep blood flow through other organs at a value just adequate to supply their metabolic needs.

Carotid Sinus Syndrome

Occasionally, the application of pressure to the carotid arteries in the upper neck results in a dramatic decrease in blood pressure. This condition, called the **carotid sinus syndrome,** is most common in patients in whom arteriosclerosis of the carotid artery is advanced. In such patients, a tight collar can apply enough pressure to the region of the carotid sinuses to stimulate the baroreceptors. The increased action potentials from the baroreceptors initiate reflexes that result in a decrease in vasomotor tone and an increase in parasympathetic action potentials to the heart. As a result of the decreased peripheral resistance and heart rate, blood pressure decreases dramatically. As a consequence, blood flow to the brain decreases to such a low level that the person becomes dizzy or may even faint. People suffering from this condition must avoid applying external pressure to the neck region. If the carotid sinus becomes too sensitive, a treatment for this condition is surgical destruction of the innervation to the carotid sinuses.

Adrenal Medullary Mechanism

The adrenal medullary mechanism is activated when stimuli result in a substantial increase in sympathetic stimulation of the heart and blood vessels (figures 21.38 and 21.39). Large decreases in blood pressure, sudden and substantial increases in physical activity, and other stressful conditions are examples. The adrenal medullary mechanism results from stimulation of the adrenal

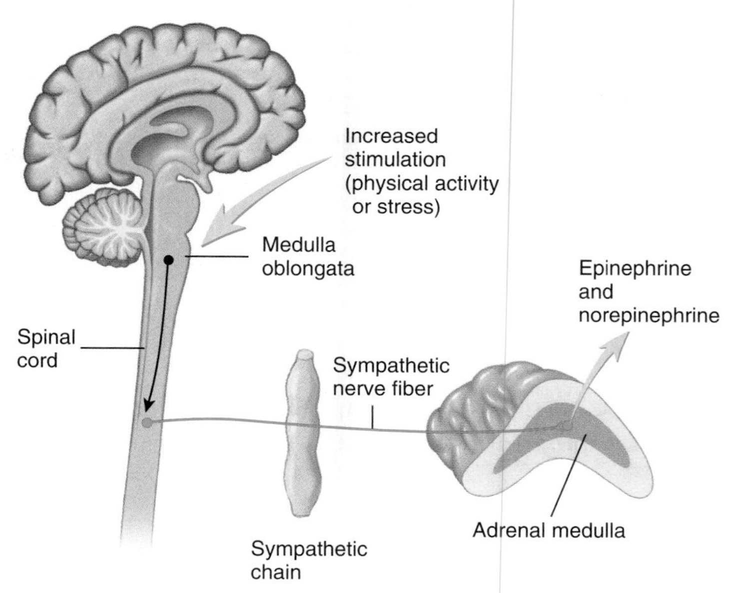

FIGURE 21.39 Adrenal Medullary Mechanism
Stimuli that increase sympathetic stimulation of the heart and blood vessels also result in increased sympathetic stimulation of the adrenal medulla and result in epinephrine and some norepinephrine secretion.

medulla by the sympathetic nerve fibers. The adrenal medulla releases epinephrine and smaller amounts of norepinephrine into the circulatory system (see figures 21.38 and 26.39). These hormones affect the cardiovascular system in a fashion similar to direct sympathetic stimulation, causing increased heart rate, increased stroke volume, and vasoconstriction in blood vessels to the skin and viscera. Epinephrine can indirectly cause vasodilation in blood vessels to the heart because of the increased rate of cardiac muscle metabolism (see chapter 20). The adrenal medullary mechanism is short-term and rapid-acting. It responds within seconds to minutes and is usually active for minutes to hours. Other hormonal mechanisms are long-term and slow-acting. They respond within minutes to hours and continue to function for many hours to days.

Chemoreceptor Reflexes

The **chemoreceptor** (kē′mō-rē-sep′tor) **reflexes** help maintain homeostasis when oxygen tension in the blood decreases or when carbon dioxide and hydrogen ion concentrations increase (figures 21.40 and 21.41).

Carotid bodies, small organs approximately 1–2 mm in diameter, lie near the carotid sinuses, and several **aortic bodies** lie adjacent to the aorta. Chemoreceptors are located in the carotid and aortic bodies. Afferent nerve fibers pass to the medulla oblongata through the glossopharyngeal nerve (IX) from the carotid bodies and through the vagus nerve (X) from the aortic bodies.

The chemoreceptors receive an abundant blood supply. When oxygen availability decreases in the chemoreceptor cells, the frequency of action potentials increases and stimulates the vasomotor center, resulting in increased vasomotor tone. The chemoreceptors act under emergency conditions and do not regulate the cardiovascular system under resting conditions. They normally do not respond strongly unless oxygen tension in the blood decreases markedly. The chemoreceptor cells are also stimulated by increased carbon dioxide and hydrogen ion concentrations to increase vasomotor tone. The increased vasomotor tone increases the mean arterial pressure. The increased mean arterial pressure increases blood flow through tissues in which blood vessels do not constrict. Blood vessels that are not constricted by the chemoreceptor reflex are blood vessels that deliver blood to the brain and cardiac muscle. Thus, the reflex helps provide an adequate oxygen supply to the brain and the heart when oxygen levels in the blood decrease.

Central Nervous System Ischemic Response

An elevation in blood pressure in response to a lack of blood flow to the medulla oblongata of the brain is called the **central nervous system (CNS) ischemic response.** The CNS ischemic response does not play an important role in regulating blood pressure under normal conditions. It functions primarily in response to emergency situations in which blood flow to the brain is severely restricted or when blood pressure falls below approximately 50 mm Hg.

Reduced blood flow results in reduced oxygen, increased carbon dioxide, and reduced pH within the medulla oblongata. Neurons of the vasomotor center are strongly stimulated. As a result, vasoconstriction is stimulated by the vasomotor center, and the systemic blood pressure rises dramatically.

The increase in blood pressure that occurs in response to CNS ischemia increases blood flow to the CNS, provided the blood

1. Chemoreceptors in the carotid and aortic bodies monitor blood O_2, CO_2, and pH.
2. Chemoreceptors in the medulla oblongata monitor blood CO_2 and pH.
3. Decreased blood O_2, increased CO_2, and decreased pH decrease parasympathetic stimulation of the heart, which increases the heart rate.
4. Decreased blood O_2, increased CO_2, and decreased pH increase sympathetic stimulation of the heart, which increases the heart rate and stroke volume.
5. Increased sympathetic stimulation of blood vessels increases vasoconstriction.

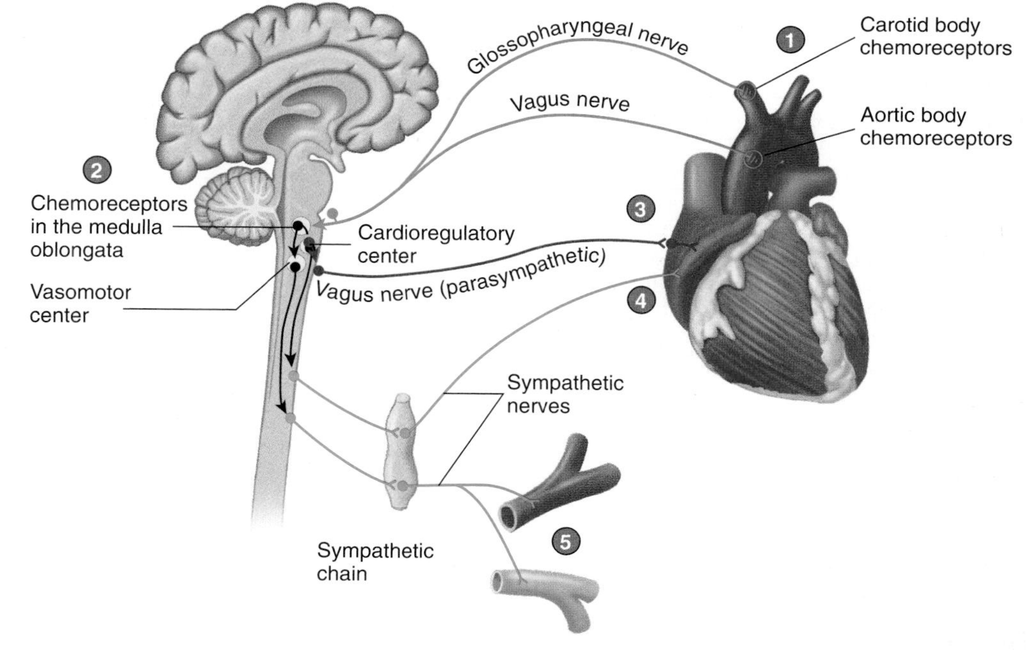

PROCESS FIGURE 21.40 Chemoreceptor Reflex Control of Blood Pressure

An increase in blood CO_2 and a decrease in pH and O_2 result in an increased heart rate and vasoconstriction. A decrease in blood CO_2 and an increase in blood pH result in a decreased heart rate and vasodilation.

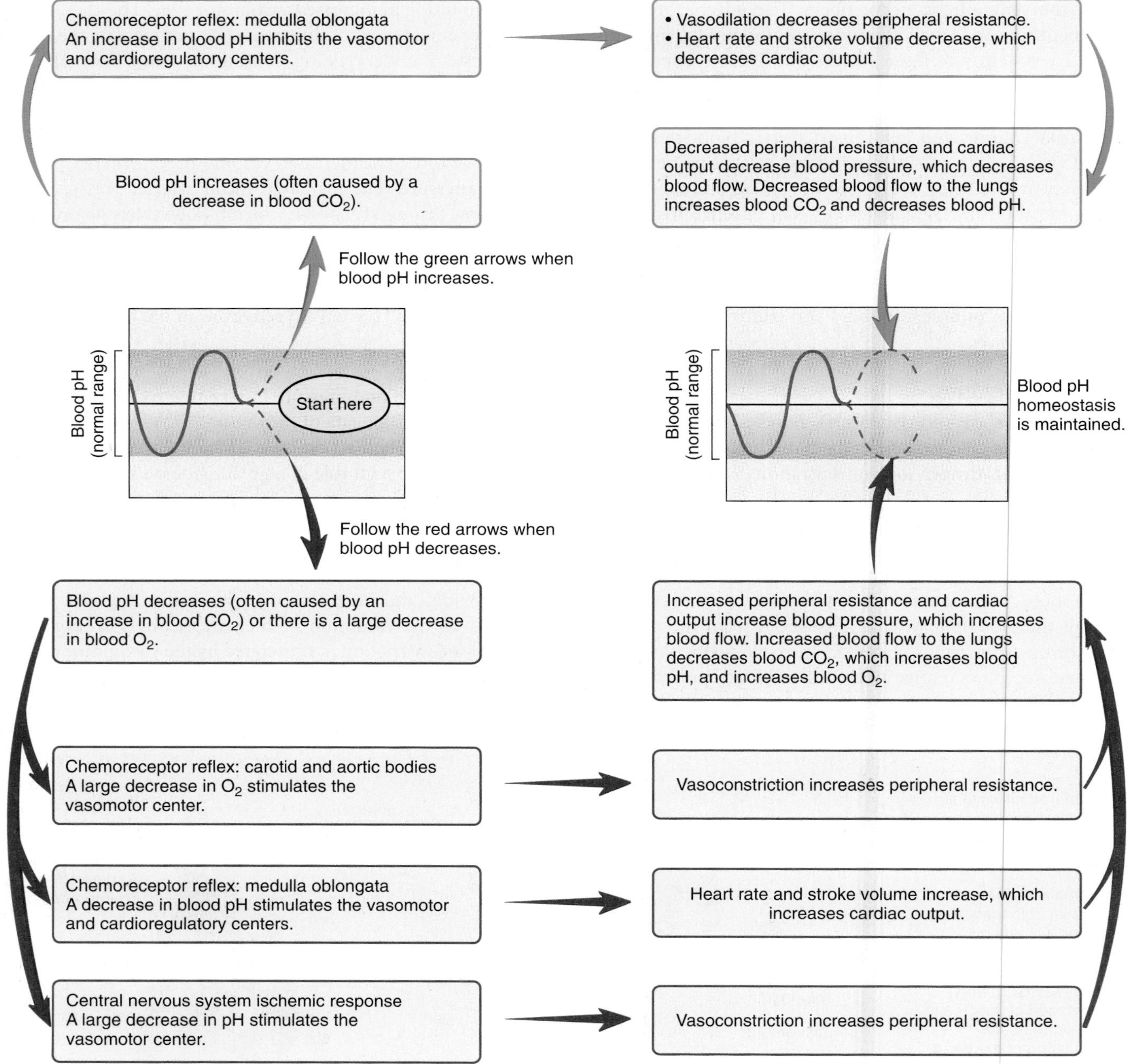

HOMEOSTASIS FIGURE 21.41 Summary of the Effects of pH and Gases on Blood Pressure

For more information on the chemoreceptor reflex, see figure 21.40 (p. 769); on the central nervous system ischemic response, see text p. 769.

vessels are intact. However, if severe ischemia lasts longer than a few minutes, metabolism in the brain fails because of the lack of oxygen. The vasomotor center becomes inactive, and extensive vasodilation occurs in the periphery as vasomotor tone decreases. Prolonged ischemia of the medulla oblongata leads to a massive decline in blood pressure and ultimately death.

In most circumstances throughout the day, the baroreceptor reflex mechanism is the most important short-term regulatory mechanism for maintaining blood pressure. The adrenal medullary mechanism plays a role during exercise and emergencies. The chemoreceptor mechanism is more important when oxygen tension in the blood is reduced, such as at high altitudes or when CO_2 is elevated or pH is reduced. Thus, it is more important in emergency situations. The CNS ischemic response is activated only in rare emergency conditions when the brain receives too little oxygen.

55 ***Define short-term regulation of blood pressure.***

56 ***Where are baroreceptors located? Describe the response of the baroreceptor reflex when blood pressure increases and decreases.***

57 ***Describe the adrenal medullary mechanism.***

58 ***Where are the chemoreceptors for carbon dioxide, pH, and oxygen located? Describe what happens when oxygen levels in the blood decrease.***

59 ***Describe the CNS ischemic response. Under what conditions does this mechanism operate?***

60 ***What mechanism is most important for short-term regulation of blood pressure under resting conditions?***

Long-Term Regulation of Blood Pressure

Regulation of the concentration and volume of blood by the kidneys, the movement of fluid across the wall of blood vessels, and alterations in the volume of the blood vessels all play a central role in the long-term regulation of blood pressure. Some of the long-term regulatory mechanisms begin to respond in minutes, but they continue to function for hours, days, or longer. They adjust the blood pressure precisely and keep it within a narrow range of values for years. Major regulatory mechanisms include the renin-angiotensin-aldosterone mechanism, vasopressin (ADH) mechanism, atrial natriuretic mechanism, fluid shift mechanism, and stress–relaxation response.

Renin-Angiotensin-Aldosterone Mechanism

The **renin-angiotensin-aldosterone mechanism** helps regulate kidney functions. This mechanism can also influence peripheral resistance by causing vasoconstriction. The kidneys increase urine output as the blood volume and arterial pressure increase, and they decrease urine output as the blood volume and arterial pressure decrease. Increased urine output reduces the blood volume and blood pressure, and decreased urine output resists a further decrease in the blood volume and blood pressure. The control of urine output is an important means by which blood pressure is regulated, and it continues to operate until the blood pressure is precisely within its normal range of values.

The kidneys release an enzyme called **renin** (rē′nin) into the circulatory system (figure 21.42) (see chapter 26) from specialized structures called the **juxtaglomerular** (jŭks′tă-glŏ-mer′ū-lăr) **apparatuses.** Renin acts on a plasma protein, synthesized by the liver, called **angiotensinogen** (an′jē-ō-ten-sin′ō-jen) to split a fragment off one end. The fragment, called **angiotensin** (an-jē-ō-ten′sin) **I,** contains 10 amino acids. Another enzyme, called **angiotensin-converting enzyme,** found primarily in small blood vessels of the lung, cleaves two additional amino acids from angiotensin I to produce a fragment consisting of 8 amino acids called **angiotensin II,** or **active angiotensin.**

Angiotensin II causes vasoconstriction in arterioles and to some degree in veins. As a result, it increases peripheral resistance and venous return to the heart, both of which raise blood pressure. Angiotensin II also stimulates aldosterone secretion from the adrenal cortex. **Aldosterone** (al-dos′ter-ōn) acts on the kidneys to increase the reabsorption of sodium and chloride ions from the filtrate into the extracellular fluid. If antidiuretic hormone (ADH; see chapter 18) is present, water moves by osmosis with the sodium and chloride ions. Consequently, aldosterone causes the kidney to retain solutes, such as sodium and chloride ions and water. The results are to decrease the production of urine and to conserve water to prevent a further reduction in blood volume caused by the formation of urine. If water intake is adequate, the effect of aldosterone is to increase blood volume (see chapter 26). Angiotensin II also increases the salt appetite, thirst, and ADH secretion.

Decreased blood pressure stimulates renin secretion and increased blood pressure decreases renin secretion. The renin-angiotensin-aldosterone mechanism is important in maintaining blood pressure on a daily basis. It also reacts strongly under conditions of circulatory shock, but it requires many hours to become maximally effective. Its onset of action is not as fast as nervous reflexes or the adrenal medullary response, but its duration of action is longer. Once renin is secreted, it remains active for approximately 1 hour and the effect of aldosterone is much longer (many hours).

Some stimuli can directly stimulate aldosterone secretion. For example, an increased plasma ion concentration of K^+ and a reduced plasma concentration of Na^+ directly stimulate aldosterone secretion from the adrenal cortex (see chapters 18 and 27). Aldosterone regulates the concentration of these ions in the plasma. A decreased blood pressure and an elevated K^+ concentration occur during plasma loss, during dehydration, and in response to tissue damage, such as burns and crushing injuries.

ACE Inhibitors and Hypertension

Angiotensin-converting enzyme (ACE) inhibitors are a class of drugs that inhibit angiotensin-converting enzyme, which converts angiotensin I to angiotensin II. These drugs were first identified as components of the venoms of pit vipers. Subsequently, several ACE inhibitors were synthesized. ACE inhibitors are effective in lowering blood pressure in many people who suffer from hypertension and have become one of the drugs commonly administered to people to combat hypertension.

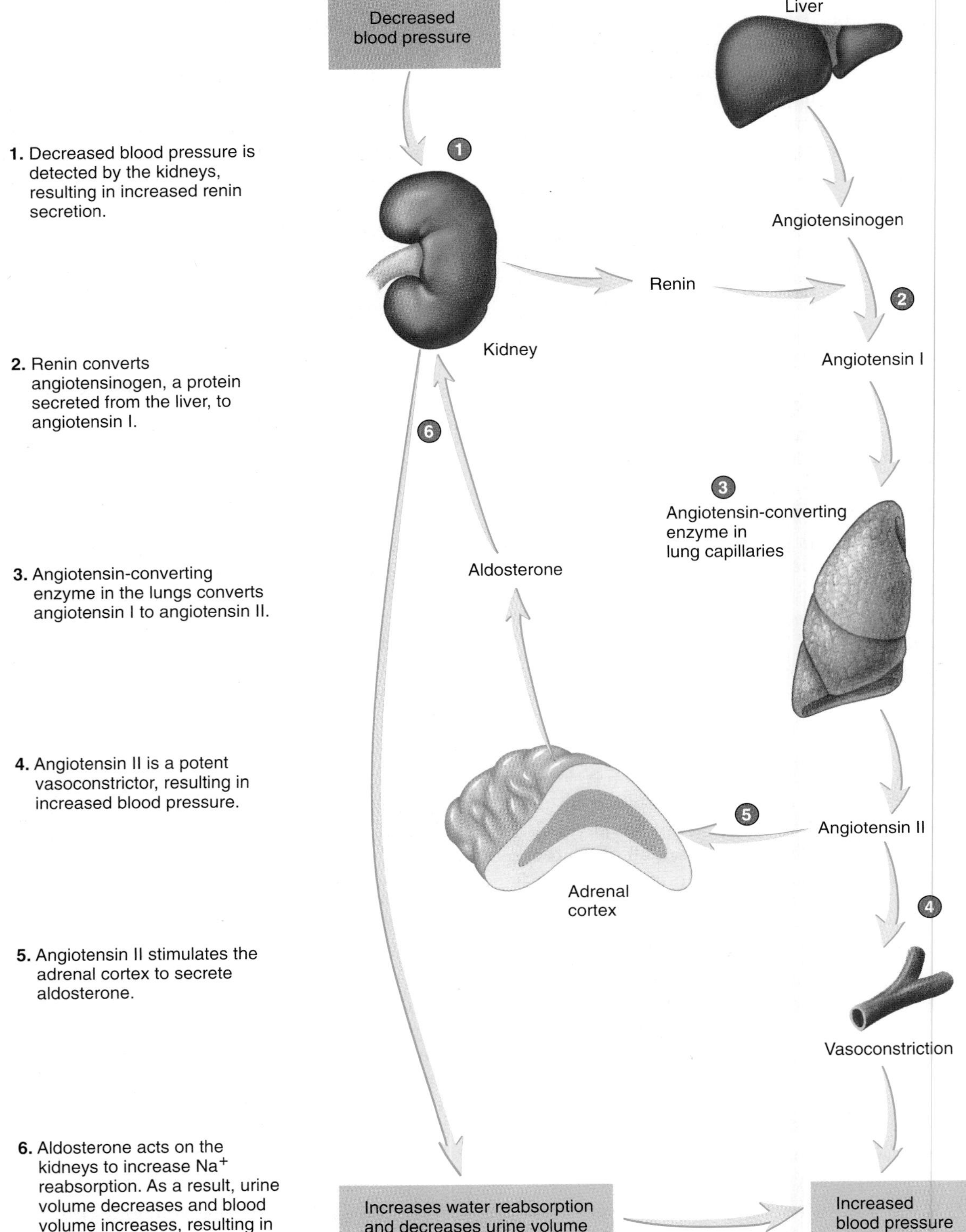

PROCESS FIGURE 21.42 Renin-Angiotensin-Aldosterone Mechanism

Decreased blood pressure is detected by the kidney, resulting in increased renin secretion. The result is vasoconstriction, increased water reabsorption, and decreased urine volume. These changes maintain blood pressure.

Vasopressin (ADH) Mechanism

The **vasopressin (ADH) mechanism** works in harmony with the renin-angiotensin-aldosterone mechanism in response to changes in blood pressure (figure 21.43). Baroreceptors are sensitive to changes in blood pressure, and decreases in blood pressure detected by the baroreceptors result in the release of vasopressin (vā-sō-pres′in, vas-ō-pres′in), or ADH, from the posterior pituitary, although the blood pressure must decrease substantially before the mechanism is activated.

ADH acts directly on blood vessels to cause vasoconstriction, although it is not as potent as other vasoconstrictor agents. Within minutes after a rapid and substantial decline in blood pressure, ADH is released in sufficient quantities to help reestablish normal blood pressure. ADH also decreases the rate of urine production by the kidneys, thereby helping maintain blood volume and blood pressure.

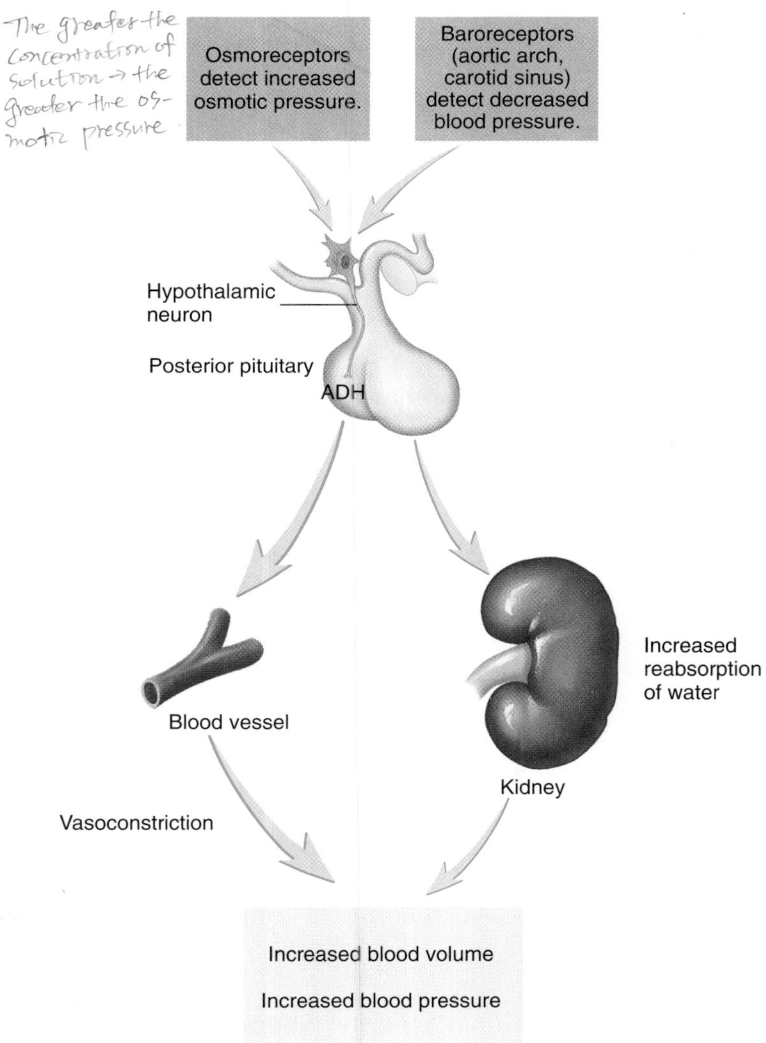

FIGURE 21.43 Vasopressin (ADH) Mechanism

Increases in osmolality of blood or decreases in blood pressure result in ADH secretion. ADH increases water reabsorption by the kidney, and large amounts of ADH result in vasoconstriction. These changes maintain blood pressure.

Neurons of the hypothalamus are sensitive to changes in the solute concentration of the plasma. Even small increases in the plasma concentration of solutes directly stimulate hypothalamic neurons that increase ADH secretion (see figure 21.43 and chapter 26). Increases in the concentration of the plasma, such as during dehydration, and decreases in blood pressure, such as in injuries involving plasma loss, such as extensive burns or crushing injuries, stimulate ADH secretion.

Atrial Natriuretic Mechanism

A polypeptide called **atrial natriuretic** (ā′trē-ăl nā′trē-ū-ret′ik) **hormone** is released from cells in the atria of the heart. A major stimulus for its release is increased venous return, which stretches atrial cardiac muscle cells. Atrial natriuretic hormone acts on the kidneys to increase the rate of urine production and Na^+ loss in the urine. It also dilates arteries and veins. Loss of water and Na^+ in the urine causes the blood volume to decrease, which decreases venous return, and vasodilation results in a decrease in peripheral resistance. These effects cause a decrease in blood pressure.

The renin-angiotensin-aldosterone, vasopressin (ADH), and atrial natriuretic mechanisms work simultaneously to help regulate blood pressure by controlling urine production by the kidneys. If blood pressure drops below 50 mm Hg, the volume of urine produced by the kidneys is reduced to nearly zero. If blood pressure is increased to 200 mm Hg, the urine volume produced is approximately six to eight times greater than normal. The mechanisms that regulate blood pressure in the long term are summarized in figure 21.44.

Fluid Shift Mechanism

The **fluid shift mechanism** begins to act within a few minutes but requires hours to achieve its full functional capacity. It plays a very important role when dehydration develops over several hours, or when a large volume of saline is administered over several hours. The fluid shift mechanism occurs in response to small changes in pressures across capillary walls. As blood pressure increases, some fluid is forced from the capillaries into the interstitial spaces. The movement of fluid into the interstitial spaces helps prevent the development of high blood pressure. As blood pressure falls, interstitial fluid moves into capillaries, which resists a further decline in blood pressure. The fluid shift mechanism is a powerful method through which blood pressure is maintained because the interstitial volume acts as a reservoir, and it is in equilibrium with the large volume of intercellular fluid.

Stress–Relaxation Response

A **stress–relaxation response** is characteristic of smooth muscle cells (see chapter 9). When blood volume suddenly declines, blood pressure also decreases, causing a reduction in the force applied to smooth muscle cells in blood vessel walls. As a result, during the next few minutes to an hour, the smooth muscle cells contract, reducing the volume of the blood vessels and thus resisting a further decline in blood pressure. Conversely, when blood volume increases rapidly, such as during a transfusion, blood pressure increases and smooth muscle cells of the blood vessel walls relax,

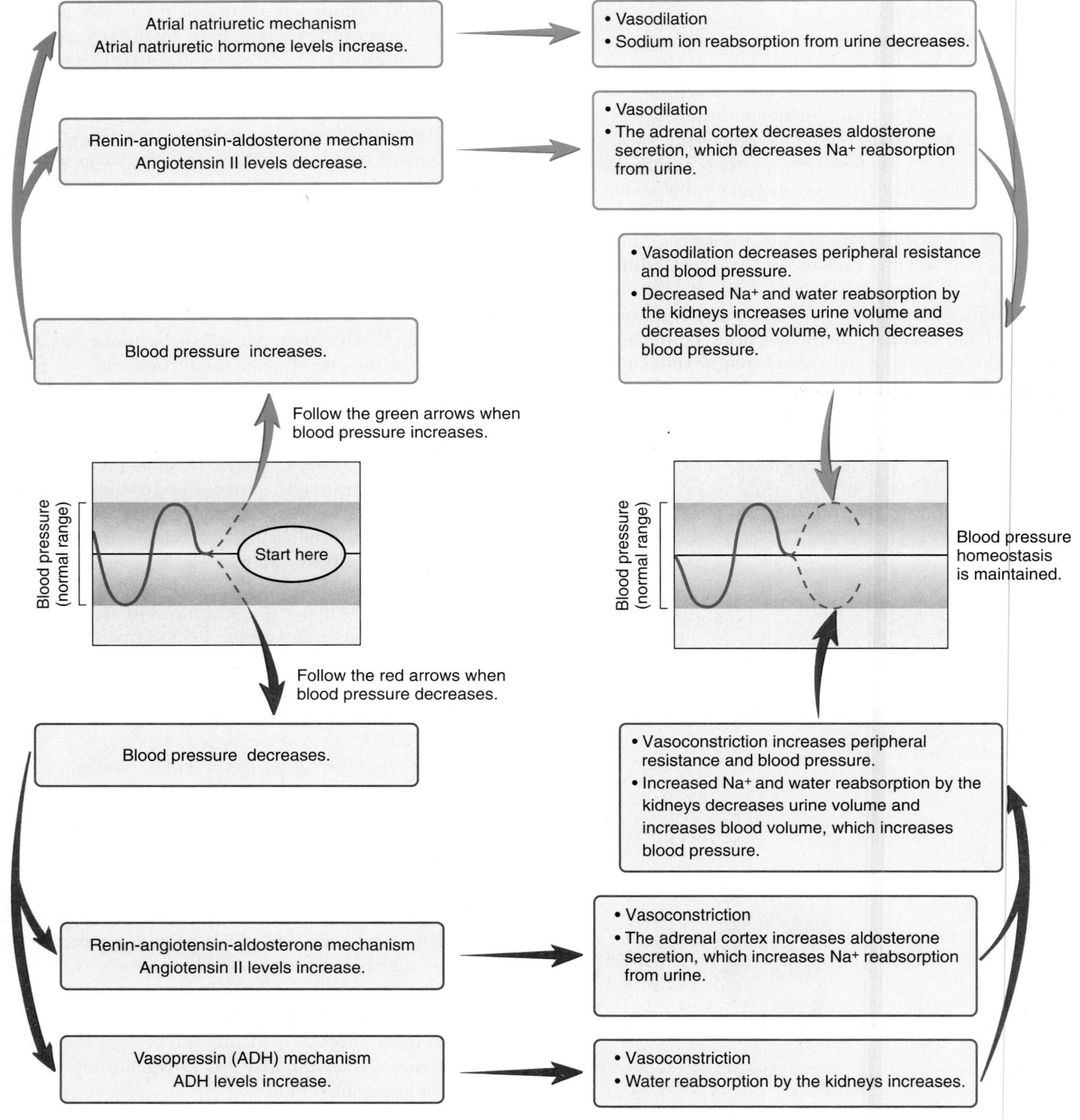

HOMEOSTASIS FIGURE 21.44 Summary of the Control of Blood Pressure Long-Term (Slow-Acting) Mechanisms

For more information on the rennin-angiotensin-aldosterone mechanism, see figure 21.42 (p. 773); on the vasopressin mechanism, see figure 21.43 (p. 774); on the atrial natriuretic mechanism, see figure 27.6 (p. 1010).

CLINICAL FOCUS

Shock

Circulatory shock is inadequate blood flow throughout the body. The failure of mechanisms that maintain blood pressure within a normal range of values results in dramatic decreases in blood pressure. As a consequence, tissues suffer damage as a result of too little delivery of oxygen to cells. Severe circulatory shock can damage vital body tissues to the extent that the individual dies.

Depending on its severity, circulatory shock can be divided into three stages: (1) compensated shock, (2) progressive shock, and (3) irreversible shock. All types of circulatory shock exhibit one or more of these stages, regardless of their cause. Several causes of circulatory shock exist, but hemorrhagic, or hypovolemic, shock is used to illustrate the characteristics of each stage.

In **compensated shock,** blood pressure decreases only a moderate amount, and the mechanisms that regulate blood pressure successfully reestablish normal blood pressure and blood flow. The baroreceptor reflexes, chemoreceptor reflexes, and ischemia within the medulla oblongata initiate strong sympathetic responses that result in intense vasoconstriction and increased heart rate. As blood volume decreases, the stress–relaxation response of blood vessels causes them to contract and helps sustain blood pressure. In response to reduced blood flow through the kidneys, increased amounts of renin are released. The elevated renin release results in a greater rate of angiotensin II formation, causing vasoconstriction and increased aldosterone release from the adrenal cortex. Aldosterone, in turn, promotes water and salt retention by the kidneys, thereby conserving water. In addition, ADH is released from the posterior pituitary gland and enhances the retention of water by the kidneys. Because of the fluid shift mechanism, water also moves from the interstitial spaces and the intestinal lumen to restore normal blood volume. An intense sensation of thirst increases water intake, also helping elevate normal blood volume.

In mild cases of compensated shock, baroreceptor reflexes can be adequate to compensate for blood loss until blood volume is restored, but, in more severe cases, all of the mechanisms described are required to compensate for the blood loss.

In **progressive shock,** the compensatory mechanisms are inadequate to compensate for the loss of blood volume. As a consequence, a positive-feedback cycle develops in which the blood pressure regulatory mechanisms are unable to compensate for circulatory shock. As circulatory shock worsens, regulatory mechanisms become even less able to compensate for the increasing severity of the circulatory shock. The cycle proceeds until the next stage of shock is reached or until medical treatment is applied that assists the regulatory mechanisms in reestablishing adequate blood flow to tissues.

During progressive shock, blood pressure declines to a very low level that is inadequate to maintain blood flow to cardiac muscle; thus, the heart begins to deteriorate. Substances that are toxic to the heart are released from tissues that suffer from severe ischemia. When blood pressure declines to a very low level, blood begins to clot in the small vessels. Eventually, blood vessel dilation begins as a result of decreased sympathetic activity and because of the lack of oxygen in capillary beds. Capillary permeability increases under ischemic conditions, allowing fluid to leave the blood vessels and enter the interstitial spaces, and finally intense tissue deterioration begins in response to inadequate blood flow.

Without medical intervention, progressive shock leads to **irreversible shock.** Irreversible shock leads to death, regardless of the amount or type of medical treatment applied. In this stage, the damage to tissues, including cardiac muscle, is so extensive that the patient is destined to die, even if adequate blood volume is reestablished and blood pressure is elevated to its normal value. Irreversible shock is characterized by decreasing heart function and progressive dilation of and increased permeability of peripheral blood vessels.

Patients suffering from shock are normally placed in a horizontal plane, usually with the head slightly lower than the feet, and oxygen is often supplied. **Replacement therapy** consists of transfusions of whole blood, plasma, artificial solutions called plasma substitutes, and physiologic saline solutions administered to increase blood volume. In some circumstances, drugs that enhance vasoconstriction are also administered. Occasionally, such as in patients in anaphylactic (an′ă-fī-lak′tik; allergic) shock, anti-inflammatory substances, such as glucocorticoids and antihistamines, are administered. The basic objective in treating shock is to reverse the condition so that progressive shock is arrested, to prevent it from progressing to the irreversible stage, and to reverse the condition so that normal blood flow through tissues is reestablished.

Several types of shock are classified by cause:

- **Hemorrhagic shock** is external or internal bleeding that causes a reduction in blood volume.
- **Plasma loss shock** is reduced blood volume that results from a loss of plasma into the interstitial spaces and greatly increased blood viscosity. Plasma loss shock includes **intestinal obstruction,** resulting in the movement of a large amount of plasma from the blood into the intestine, and **severe burns,** resulting in the loss of large amounts of plasma from the burned surface.
- **Dehydration** results from a severe and prolonged shortage of fluid intake.
- **Severe diarrhea** or **vomiting** cause a loss of plasma through the intestinal wall.
- **Neurogenic shock** is a rapid loss of vasomotor tone leading to vasodilation so extensive that a severe decrease in blood pressure results.
- **Anesthesia** includes deep general anesthesia or spinal anesthesia that decreases the activity of the medullary vasomotor center or the sympathetic nerve fibers.
- **Brain damage** leads to an ineffective medullary vasomotor function.
- **Emotional shock (vasovagal syncope)** results from emotions that cause strong parasympathetic stimulation of the heart and results in vasodilation in skeletal muscles and in the viscera.
- **Anaphylactic shock** results from an allergic response that causes the release of inflammatory substances that increase vasodilation and capillary permeability.
- **Septic shock,** or **blood poisoning,** results from peritoneal, systemic, and gangrenous infections that cause the release of toxic substances into the circulatory system, depressing the activity of the heart, leading to vasodilation, and increasing capillary permeability.
- **Cardiogenic shock** occurs when the heart stops pumping in response to conditions such as heart attack or electrocution.

resulting in a more gradual increase in blood pressure. The stress–relaxation mechanism is most effective when changes in blood pressure occur over a period of many minutes.

61 ***Define long-term regulation of blood pressure.***

62 ***For each of these hormones—renin, angiotensin, aldosterone, antidiuretic hormone, and atrial natriuretic hormone—state where the hormone is produced and what effects it has on the circulatory system.***

63 ***What is fluid shift, and what does it accomplish? Describe the stress–relaxation response of a blood vessel.***

PREDICT 11

Explain the differences in mechanisms that regulate blood pressure in response to hemorrhage that results in the rapid loss of a large volume of blood, compared with hemorrhage that results in the loss of the same volume of blood but over a period of several hours.

SUMMARY

Functions of the Peripheral Circulation (p. 722)

General Features of Blood Vessel Structure (p. 722)

1. Blood flows from the heart through elastic arteries, muscular arteries, and arterioles to the capillaries.
2. Blood returns to the heart from the capillaries through venules, small veins, and large veins.

Capillaries

1. The entire circulatory system is lined with simple squamous epithelium called endothelium. Capillaries consist only of endothelium.
2. Capillaries are surrounded by loose connective tissue, the adventitia, that contains pericapillary cells.
3. Three types of capillaries exist.
   - Fenestrated capillaries have pores, called fenestrae, that extend completely through the cell.
   - Sinusoidal capillaries are large-diameter capillaries with large fenestrae.
   - Continuous capillaries do not have fenestrae.
4. Materials pass through the capillaries in several ways: between the endothelial cells, through the fenestrae, and through the plasma membrane.
5. Blood flows from arterioles through metarterioles and then through the capillary network. Venules drain the capillary network.
   - Smooth muscle in the arterioles, metarterioles, and precapillary sphincters regulates blood flow into the capillaries.
   - Blood can pass rapidly through the thoroughfare channel.

Structure of Arteries and Veins

1. Except for capillaries and venules, blood vessels have three layers. The inner tunica intima consists of endothelium, basement membrane, and internal elastic lamina.
   - The tunica media, the middle layer, contains circular smooth muscle and elastic fibers.
   - The outer tunica adventitia is connective tissue.
2. The thickness and the composition of the layers vary with blood vessel type and diameter.
   - Large elastic arteries are thin-walled with large diameters. The tunica media has many elastic fibers and little smooth muscle.
   - Muscular arteries are thick-walled with small diameters. The tunica media has abundant smooth muscle and some elastic fibers.
   - Arterioles are the smallest arteries. The tunica media consists of smooth muscle cells and a few elastic fibers.
   - Venules are composed of endothelium surrounded by a few smooth muscle cells.
   - Small veins are venules covered with a layer of smooth muscle.
   - Medium-size veins and large veins contain less smooth muscle and fewer elastic fibers than arteries of the same size.
3. Valves prevent the backflow of blood in the veins.

Vasa Vasorum

1. Vasa vasorum are blood vessels that supply the tunica adventitia and tunica media.
2. Arteriovenous anastomoses allow blood to flow from arteries to veins without passing through the capillaries. They function in temperature regulation.

Nerves

Sympathetic nerve fibers supply the smooth muscle of the tunica media.

Aging of the Arteries

Arteriosclerosis results from a loss of elasticity in the aorta, large arteries, and coronary arteries.

Pulmonary Circulation (p. 728)

The pulmonary circulation moves blood to and from the lungs. The pulmonary trunk arises from the right ventricle and divides to form the pulmonary arteries, which project to the lungs. From the lungs, the pulmonary veins return to the left atrium.

Systemic Circulation: Arteries (p. 728)

Arteries carry blood from the left ventricle of the heart to all parts of the body.

Aorta

The aorta leaves the left ventricle to form the ascending aorta, aortic arch, and descending aorta (consisting of the thoracic and abdominal aortae).

Coronary Arteries

Coronary arteries supply the heart.

Arteries to the Head and the Neck

1. The brachiocephalic, left common carotid, and left subclavian arteries branch from the aortic arch to supply the head and the upper limbs. The brachiocephalic artery divides to form the right common carotid

and the right subclavian arteries. The vertebral arteries branch from the subclavian arteries.
2. The common carotid arteries and the vertebral arteries supply the head.
   - The common carotid arteries divide to form the external carotids, which supply the face and mouth, and the internal carotids, which supply the brain.
   - The vertebral arteries join within the cranial cavity to form the basilar artery, which supplies the brain.

Arteries of the Upper Limb

1. The subclavian artery continues (without branching) as the axillary artery and then as the brachial artery. The brachial artery divides into the radial and ulnar arteries.
2. The radial artery supplies the deep palmar arch, and the ulnar artery supplies the superficial palmar arch. Both arches give rise to the digital arteries.

Thoracic Aorta and Its Branches

The thoracic aorta has visceral branches that supply the thoracic organs and parietal branches that supply the thoracic wall.

Abdominal Aorta and Its Branches

1. The abdominal aorta has visceral branches that supply the abdominal organs and parietal branches that supply the abdominal wall.
2. The visceral branches are paired and unpaired. The paired arteries supply the kidneys, adrenal glands, and gonads. The unpaired arteries supply the stomach, spleen, and liver (celiac trunk); the small intestine and upper part of the large intestine (superior mesenteric); and the lower part of the large intestine (inferior mesenteric).

Arteries of the Pelvis

1. The common iliac arteries arise from the abdominal aorta, and the internal iliac arteries branch from the common iliac arteries.
2. The visceral branches of the internal iliac arteries supply the pelvic organs, and the parietal branches supply the pelvic wall and floor and the external genitalia.

Arteries of the Lower Limb

1. The external iliac arteries branch from the common iliac arteries.
2. The external iliac artery continues (without branching) as the femoral artery and then as the popliteal artery. The popliteal artery divides to form the anterior and posterior tibial arteries.
3. The posterior tibial artery gives rise to the fibular (peroneal) and plantar arteries. The plantar arteries form the plantar arch from which the digital arteries arise.

Systemic Circulation: Veins (p. 739)

1. The three major veins returning blood to the heart are the superior vena cava (head, neck, thorax, and upper limbs), the inferior vena cava (abdomen, pelvis, and lower limbs), and the coronary sinus (heart).
2. Veins are of three types: superficial, deep, and sinuses.

Veins Draining the Heart

Coronary veins enter the coronary sinus or the right atrium.

Veins of the Head and Neck

1. The internal jugular veins drain the venous sinuses of the anterior head and neck.
2. The external jugular veins and the vertebral veins drain the posterior head and neck.

Veins of the Upper Limb

1. The deep veins are the small ulnar and radial veins of the forearm, which join the brachial veins of the arm. The brachial veins drain into the axillary vein.
2. The superficial veins are the basilic, cephalic, and median cubital. The basilic vein becomes the axillary vein, which then becomes the subclavian vein. The cephalic vein drains into the axillary vein.

Veins of the Thorax

The left and right brachiocephalic veins and the azygos veins return blood to the superior vena cava.

Veins of the Abdomen and Pelvis

1. Ascending lumbar veins from the abdomen join the azygos and hemiazygos veins.
2. Vessels from the kidneys, adrenal gland, and gonads directly enter the inferior vena cava.
3. Vessels from the stomach, intestines, spleen, and pancreas connect with the hepatic portal vein. The hepatic portal vein transports blood to the liver for processing. Hepatic veins from the liver join the inferior vena cava.

Veins of the Lower Limb

1. The deep veins are the fibular (peroneal), anterior and posterior tibials, popliteal, femoral, and external iliac.
2. The superficial veins are the small and great saphenous veins.

Dynamics of Blood Circulation (p. 751)

The interrelationships among pressure, flow, resistance, and the central mechanisms that regulate blood pressure and blood flow play a critical role in the function of the circulatory system.

Laminar and Turbulent Flow in Vessels

Blood flow through vessels normally is streamlined, or laminar. Turbulent flow is disruption of laminar flow.

Blood Pressure

1. Blood pressure is a measure of the force exerted by blood against the blood vessel wall. Blood moves through vessels because of blood pressure.
2. Blood pressure can be measured by listening for Korotkoff sounds produced by turbulent flow in arteries as pressure is released from a blood pressure cuff.

Blood Flow and Poiseuille's Law

1. Blood flow is the amount of blood that moves through a vessel in a given period. Blood flow is directly proportional to pressure differences and is inversely proportional to resistance.
2. Resistance is the sum of all the factors that inhibit blood flow. Resistance increases when viscosity increases and when blood vessels become smaller in diameter or increase in length.
3. Viscosity is the resistance of a liquid to flow. Most of the viscosity of blood results from red blood cells. The viscosity of blood increases when the hematocrit increases.

Critical Closing Pressure and Laplace's Law

1. As pressure in a vessel decreases, the force holding it open decreases, and the vessel tends to collapse. The critical closing pressure is the pressure at which a blood vessel closes.

2. Laplace's law states that the force acting on the wall of a blood vessel is proportional to the diameter of the vessel times blood pressure.

Vascular Compliance

Vascular compliance is a measure of the change in volume of blood vessels produced by a change in pressure. The venous system has a large compliance and acts as a blood reservoir.

Physiology of Systemic Circulation (p. 755)

The greatest volume of blood is contained in the veins. The smallest volume is in the arterioles.

Cross-Sectional Area of Blood Vessels

As the diameter of vessels decreases, their total cross-sectional area increases, and the velocity of blood flow through them decreases.

Pressure and Resistance

Blood pressure averages 100 mm Hg in the aorta and drops to 0 mm Hg in the right atrium. The greatest drop occurs in the arterioles, which regulate blood flow through tissues.

Pulse Pressure

1. Pulse pressure is the difference between systolic and diastolic pressures. Pulse pressure increases when stroke volume increases or vascular compliance decreases.
2. Pulse pressure waves travel through the vascular system faster than the blood flows. Pulse pressure can be used to take the pulse.

Capillary Exchange and Regulation of Interstitial Fluid Volume

1. Blood pressure, capillary permeability, and osmosis affect movement of fluid from the capillaries.
2. A net movement of fluid occurs from the blood into the tissues. The fluid gained by the tissues is removed by the lymphatic system.

Functional Characteristics of Veins

Venous return to the heart increases because of an increase in blood volume, venous tone, and arteriole dilation.

Blood Pressure and the Effect of Gravity

In a standing person, hydrostatic pressure caused by gravity increases blood pressure below the heart and decreases pressure above the heart.

Control of Blood Flow in Tissues (p. 761)

Blood flow through tissues is highly controlled and matched closely to the metabolic needs of tissues.

Local Control of Blood Flow by the Tissues

1. Blood flow through a tissue is usually proportional to the metabolic needs of the tissue. Exceptions are tissues that perform functions that require additional blood.
2. Control of blood flow by the metarterioles and precapillary sphincters can be regulated by vasodilator substances or by lack of nutrients.
3. Only large changes in blood pressure have an effect on blood flow through tissues.
4. If the metabolic activity of a tissue increases, the number and the diameter of capillaries in the tissue increases over time.

Nervous and Hormonal Regulation of Local Circulation

1. The sympathetic nervous system (vasomotor center in the medulla) controls blood vessel diameter. Other brain areas can excite or inhibit the vasomotor center.
2. Vasomotor tone is a state of partial contraction of blood vessels.
3. The nervous system is responsible for routing the flow of blood and maintaining blood pressure.
4. Sympathetic action potentials stimulate epinephrine and norepinephrine release from the adrenal medulla, and these hormones cause vasoconstriction in most blood vessels.

Regulation of Mean Arterial Pressure (p. 765)

Mean arterial pressure (MAP) is proportional to cardiac output times peripheral resistance.

Short-Term Regulation of Blood Pressure

1. Baroreceptors are sensory receptors sensitive to stretch.
   - Baroreceptors are located in the carotid sinuses and the aortic arch.
   - The baroreceptor reflex changes peripheral resistance, heart rate, and stroke volume in response to changes in blood pressure.
2. Chemoreceptors are sensory receptors sensitive to oxygen, carbon dioxide, and pH levels in the blood.
3. Epinephrine and norepinephrine are released from the adrenal medulla as a result of sympathetic stimulation. They increase heart rate, stroke volume, and vasoconstriction.
4. The CNS ischemic response results from high carbon dioxide or low pH levels in the medulla and increases peripheral resistance.

Long-Term Regulation of Blood Pressure

1. Renin-angiotensin-aldosterone mechanism. Renin is released by the kidneys in response to low blood pressure. Renin promotes the production of angiotensin II, which causes vasoconstriction and an increase in aldosterone secretion.
2. The vasopressin (ADH) mechanism causes ADH release from the posterior pituitary in response to a substantial decrease in blood pressure causes vasoconstriction.
3. The atrial natriuretic mechanism causes atrial natriuretic hormone release from the cardiac muscle cells when atrial blood pressure increases. It stimulates an increase in urinary production, causing a decrease in blood volume and blood pressure.
4. The fluid shift mechanism causes fluid shift, which is a movement of fluid from the interstitial spaces into capillaries in response to a decrease in blood pressure to maintain blood volume.
5. The stress–relaxation response is an adjustment of the smooth muscles of blood vessels in response to a change in blood volume.

REVIEW AND COMPREHENSION

1. Given these blood vessels:
   1. arteriole
   2. capillary
   3. elastic artery
   4. muscular artery
   5. vein
   6. venule

   Choose the arrangement that lists the blood vessels in the order a red blood cell passes through them as it leaves the heart, travels to a tissue, and returns to the heart.
   a. 3,4,2,1,5,6
   b. 3,4,1,2,6,5
   c. 4,3,1,2,5,6
   d. 4,3,2,1,6,5
   e. 4,2,3,5,1,6
2. Given these structures:
   1. metarteriole
   2. precapillary sphincter
   3. thoroughfare channel

   Choose the arrangement that lists the structures in the order a red blood cell encounters them as it passes through a tissue.
   a. 1,3,2
   b. 2,1,3
   c. 2,3,1
   d. 3,1,2
   e. 3,2,1
3. In which of these blood vessels are elastic fibers present in the largest amounts?
   a. large arteries
   b. medium arteries
   c. arterioles
   d. venules
   e. large veins
4. Comparing and contrasting arteries and veins, veins have
   a. thicker walls.
   b. a greater amount of smooth muscle than arteries.
   c. a tunica media but arteries do not.
   d. valves but arteries do not.
   e. all of the above.
5. The structure that supplies the walls of blood vessels with blood is the
   a. venous shunt.
   b. tunic channel.
   c. arteriovenous anastomosis.
   d. vasa vasorum.
   e. coronary artery.
6. Given these blood vessels:
   1. aorta
   2. inferior vena cava
   3. pulmonary arteries
   4. pulmonary veins

   Which vessels carry oxygen-rich blood?
   a. 1,3
   b. 1,4
   c. 2,3
   d. 2,4
   e. 3,4
7. Given these arteries:
   1. basilar
   2. common carotid
   3. internal carotid
   4. vertebral

   Which of these arteries have *direct* connections with the cerebral arterial circle (circle of Willis)?
   a. 1,2
   b. 2,4
   c. 1,3
   d. 3,4
   e. 2,3
8. Given these blood vessels:
   1. axillary artery
   2. brachial artery
   3. brachiocephalic artery
   4. radial artery
   5. subclavian artery

   Choose the arrangement that lists the vessels in order, from the aorta to the right hand.
   a. 2,5,4,1
   b. 5,2,1,4
   c. 5,3,1,4,2
   d. 3,5,1,2,4
   e. 4,5,1,2,3
9. A branch of the aorta that supplies the liver, stomach, and spleen is the
   a. celiac trunk.
   b. common iliac.
   c. inferior mesenteric.
   d. superior mesenteric.
   e. renal.
10. Given these arteries:
    1. common iliac
    2. external iliac
    3. femoral
    4. popliteal

    Choose the arrangement that lists the arteries in order, from the aorta to the knee.
    a. 1,2,3,4
    b. 1,2,4,3
    c. 2,1,3,4
    d. 2,1,4,3
    e. 3,1,2,4
11. Given these veins:
    1. brachiocephalic
    2. internal jugular
    3. superior vena cava
    4. venous sinus

    Choose the arrangement that lists the veins in order, from the brain to the heart.
    a. 1,2,4,3
    b. 2,4,1,3
    c. 2,4,3,1
    d. 4,2,1,3
    e. 4,2,3,1
12. Blood returning from the arm to the subclavian vein passes through which of these veins?
    a. cephalic
    b. basilic
    c. brachial
    d. both a and b
    e. all of the above
13. Given these blood vessels:
    1. inferior mesenteric vein
    2. superior mesenteric vein
    3. hepatic portal vein
    4. hepatic vein

    Choose the arrangement that lists the vessels in order, from the small intestine to the inferior vena cava.
    a. 1,3,4
    b. 1,4,3
    c. 2,3,4
    d. 2,4,3
    e. 3,1,4
14. Given these veins:
    1. small saphenous
    2. great saphenous
    3. fibular (peroneal)
    4. posterior tibial

    Which are superficial veins?
    a. 1,2
    b. 1,3
    c. 2,3
    d. 2,4
    e. 3,4
15. If you could increase any of these factors that affect blood flow by twofold, which one would cause the greatest increase in blood flow?
    a. blood viscosity
    b. the pressure gradient
    c. vessel radius
    d. vessel length
16. Vascular compliance is
    a. greater in arteries than in veins.
    b. the increase in vessel volume divided by the increase in vessel pressure.
    c. the pressure at which blood vessels collapse.
    d. proportional to the diameter of the blood vessel times pressure.
    e. all of the above.
17. The resistance to blood flow is greatest in the
    a. aorta.
    b. arterioles.
    c. capillaries.
    d. venules.
    e. veins.

18. Pulse pressure
    a. is the difference between systolic and diastolic pressure.
    b. increases when stroke volume increases.
    c. increases as vascular compliance decreases.
    d. all of the above.
19. Veins
    a. increase their volume because of their large compliance.
    b. increase venous return to the heart when they vasodilate.
    c. vasodilate because of increased sympathetic stimulation.
    d. all of the above.
20. Local direct control of blood flow through a tissue
    a. maintains an adequate rate of flow despite large changes in arterial blood pressure.
    b. results from relaxation and contraction of precapillary sphincters.
    c. occurs in response to a buildup in carbon dioxide in the tissues.
    d. occurs in response to a decrease in oxygen in the tissues.
    e. all of the above.
21. An increase in mean arterial pressure can result from
    a. an increase in peripheral resistance.
    b. an increase in heart rate.
    c. an increase in stroke volume.
    d. all of the above.
22. In response to an increase in mean arterial pressure, the baroreceptor reflex causes
    a. an increase in sympathetic nervous system activity.
    b. a decrease in peripheral resistance.
    c. stimulation of the vasomotor center.
    d. vasoconstriction.
    e. an increase in cardiac output.
23. When blood oxygen levels markedly decrease, the chemoreceptor reflex causes
    a. peripheral resistance to decrease.
    b. mean arterial blood pressure to increase.
    c. vasomotor tone to decrease.
    d. vasodilation.
    e. all of the above.
24. When blood pressure is suddenly decreased a small amount (10 mm Hg), which of these mechanisms are activated to restore blood pressure to normal levels?
    a. chemoreceptor reflexes
    b. baroreceptor reflexes
    c. CNS ischemic response
    d. all of the above
25. A sudden release of epinephrine from the adrenal medulla
    a. increases heart rate.
    b. increases stroke volume.
    c. causes vasoconstriction in visceral blood vessels.
    d. all of the above.
26. When blood pressure decreases,
    a. renin secretion increases.
    b. angiotensin II formation decreases.
    c. aldosterone secretion decreases.
    d. all of the above.
27. In response to a decrease in blood pressure,
    a. ADH secretion increases.
    b. the kidneys decrease urine production.
    c. blood volume increases.
    d. all of the above.
28. In response to a decrease in blood pressure,
    a. more fluid than normal enters the tissues (fluid shift mechanism).
    b. smooth muscles in blood vessels relax (stress–relaxation response).
    c. the kidneys retain more salts and water than normal.
    d. all of the above.
29. A patient is found to have severe arteriosclerosis of his renal arteries, which has reduced renal blood pressure. Which of these is consistent with that condition?
    a. hypotension
    b. hypertension
    c. decreased vasomotor tone
    d. exaggerated sympathetic stimulation of the heart
    e. both a and c
30. During exercise, the blood flow through skeletal muscle may increase up to 20-fold. However, the cardiac output does not increase that much. This occurs because of
    a. vasoconstriction in the viscera.
    b. vasoconstriction in the skin (at least temporarily).
    c. vasodilation of skeletal muscle blood vessels.
    d. both a and b.
    e. all of the above.

Answers in Appendix E

CRITICAL THINKING

1. For each of the following destinations, name all the arteries that a red blood cell would encounter if it started its journey in the left ventricle.
    a. posterior interventricular groove of the heart
    b. anterior neck to the brain (give two ways)
    c. posterior neck to the brain (give two ways)
    d. external skull
    e. tip of the fingers of the left hand (what other blood vessel would be encountered if the trip were through the right upper limb?)
    f. anterior compartment of the leg
    g. liver
    h. small intestine
    i. urinary bladder
2. For each of the following starting places, name all the veins that a red blood cell would encounter on its way back to the right atrium.
    a. anterior interventricular groove of the heart (give two ways)
    b. venous sinus near the brain
    c. external posterior of skull
    d. hand (return deep and superficial)
    e. foot (return deep and superficial)
    f. stomach
    g. kidney
    h. left inferior wall of the thorax
3. In a study of heart valve functions, it is necessary to inject a dye into the right atrium of the heart by inserting a catheter into a blood vessel and moving the catheter into the right atrium. What route would

you suggest? If you wanted to do this procedure into the left atrium, what would you do differently?

4. In endurance-trained athletes, the hematocrit can be lower than normal because plasma volume increases more than red blood cell numbers increase. Explain why this condition would be beneficial.
5. All the blood that passes through the aorta, except the blood that flows into the coronary vessels, returns to the heart through the venae cavae. (*Hint:* The diameter of the aorta is 26 mm, and the diameter of a vena cava is 32 mm.) Explain why the resistance to blood flow in the aorta is greater than the resistance to blood flow in the venae cavae. Because the resistances are different, explain why blood flow can be the same.
6. As blood vessels increase in diameter, the amount of smooth muscle decreases and the amount of connective tissue increases. Explain why. (*Hint:* Remember Laplace's law.)
7. A patient is suffering from edema in the lower right limb. Explain why massage helps remove the excess fluid.
8. A very short nursing student is asked to measure the blood pressure of a very tall person. She decides to measure the blood pressure at the level of the tall person's foot while he is standing. What artery does she use? After taking the blood pressure, she decides that the tall person is suffering from hypertension because the systolic pressure is 200 mm Hg. Is her diagnosis correct? Why or why not?
9. Mr. D. was suffering from severe cirrhosis of the liver and hepatitis. He developed severe edema over a period of time. Explain how decreased liver function can result in edema.
10. During hyperventilation, carbon dioxide is "blown off," and carbon dioxide levels in the blood decrease. What effect does this decrease have on blood pressure? Explain. What symptoms do you expect to see as a result?
11. Epinephrine causes vasodilation of blood vessels in cardiac muscle but vasoconstriction of blood vessels in the skin. Explain why this is a beneficial arrangement.
12. One cool evening, Skinny Dip jumps into a hot Jacuzzi. Predict what will happen to Skinny's heart rate.

Answers in Appendix F

Visit this book's website at www.mhhe.com/seeley8 for chapter quizzes, interactive learning exercises, and other study tools.

22 Lymphatic System and Immunity

One of the basic themes of life is that many organisms consume or use other organisms to survive. For example, deer graze on grasses and wolves feed on deer. A parasite lives on or in another organism called the host. The host provides the parasite with the conditions and food necessary for survival. For example, hookworms can live in the sheltered environment of the human intestine, where they feed on blood. Humans are host to many kinds of organisms, including microorganisms, such as bacteria, viruses, fungi, and protozoans; insects; and worms. Often, parasites harm humans, causing disease and sometimes death. However, our bodies have ways to resist or destroy harmful microorganisms. The lymphatic system and immunity are the body's defense systems against threats arising from inside and outside the body. This chapter considers the *lymphatic system* (p. 783), *immunity* (p. 792), *innate immunity* (p. 792), *adaptive immunity* (p. 798), *immune interactions* (p. 814), *immunotherapy* (p. 814), *acquired immunity* (p. 816), and the *effects of aging on the lymphatic system and immunity* (p. 818).

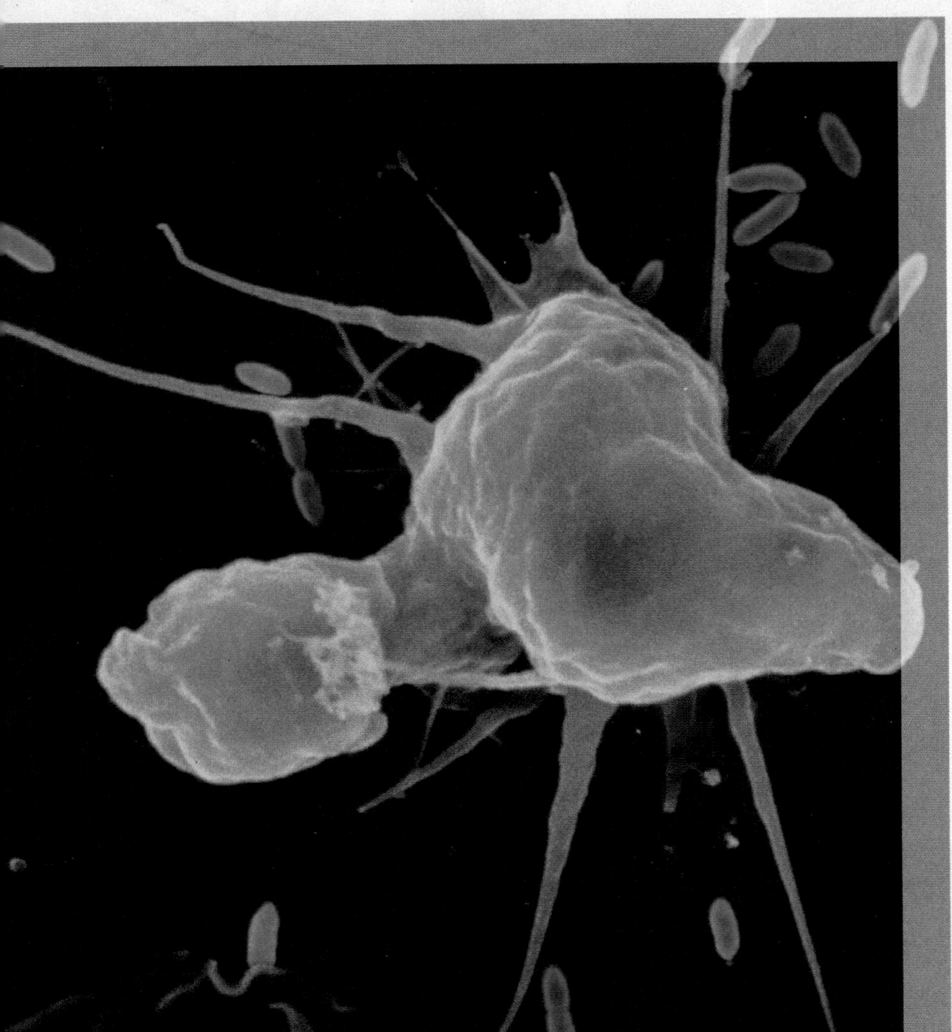

A macrophage (*large yellow cell*) about to phagocytize bacterial cells (*E. coli*) (*small blue cells*).

Lymphatic System

LYMPHATIC SYSTEM

The **lymphatic** (lim-fat′ik) **system** includes lymph, lymphatic vessels, lymphatic tissue, lymphatic nodules, lymph nodes, tonsils, the spleen, and the thymus (figure 22.1).

Functions of the Lymphatic System

The following are the three main functions of the lymphatic system:

1. *Fluid balance.* Approximately 30 L of fluid pass from the blood capillaries into the interstitial fluid each day, whereas only 27 L pass from the interstitial fluid back into the blood capillaries. If the extra 3 L of fluid were to remain in the interstitial fluid, edema would result, causing tissue damage and eventual death. Instead, the 3 L of fluid enters the lymphatic capillaries, where the fluid is called **lymph** (limf; clear spring water), and passes through the lymphatic vessels back to the blood (see chapter 21). In addition to water, lymph contains solutes derived from two sources: (1) Substances in plasma, such as ions, nutrients, gases, and some proteins, pass from blood capillaries into the interstitial fluid and become part of the lymph and (2) substances derived from cells, such as hormones, enzymes, and waste products, are also found in the lymph.
2. *Fat absorption.* The lymphatic system absorbs fats and other substances from the digestive tract (see chapter 24). Lymphatic vessels called **lacteals** (lak′tē-ălz) are located in the lining of the small intestine. Fats enter the lacteals and pass through the lymphatic vessels to the venous circulation. The lymph passing through these lymphatic vessels, called **chyle** (kīl), has a milky appearance because of its fat content.
3. *Defense.* Microorganisms and other foreign substances are filtered from lymph by lymph nodes and from blood by the spleen. In addition, lymphocytes and other cells are capable of destroying microorganisms and other foreign substances.

1 ***List the parts of the lymphatic system, and describe the three main functions of the lymphatic system.***

Lymphatic Vessels

The lymphatic vessels are essential for the maintenance of fluid balance. They begin as small, dead-end tubes called **lymphatic capillaries** (figure 22.2*a*). Fluids tend to move out of blood capillaries into tissue spaces (see "Capillary Exchange and Regulation of Interstitial Fluid Volume," chapter 21). Excess fluid passes through the tissue spaces and enters lymphatic capillaries to become lymph. Lymphatic capillaries are in most tissues of the body. Exceptions are the central nervous system, the bone marrow, and tissues without blood vessels, such as cartilage, epidermis, and the cornea. A superficial group of lymphatic capillaries is in the dermis of the skin and the hypodermis. A deep group of lymphatic capillaries drains the muscles, joints, viscera, and other deep structures.

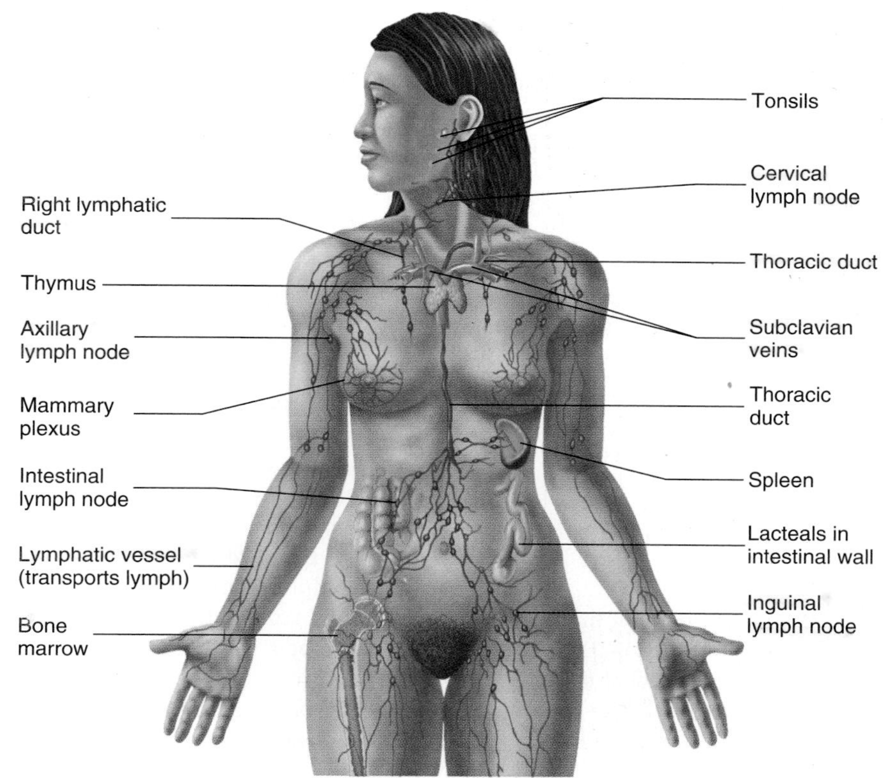

FIGURE 22.1 Lymphatic System

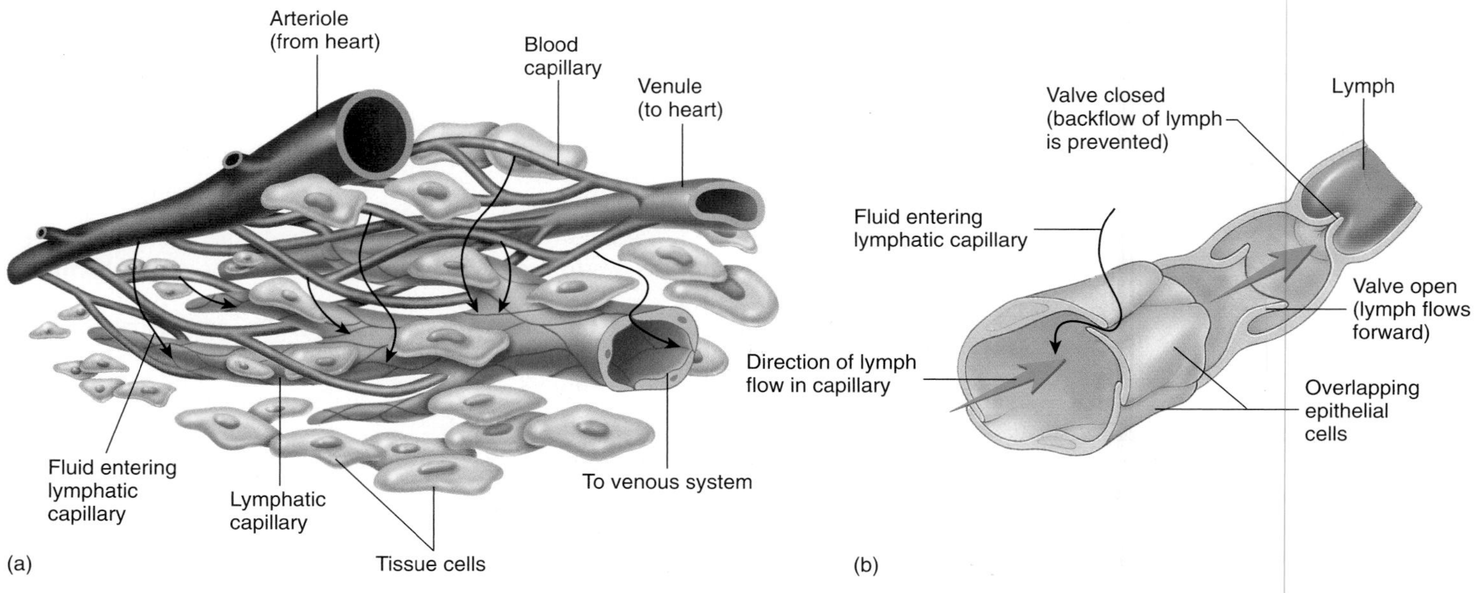

FIGURE 22.2 Lymph Formation and Movement

(*a*) Movement of fluid from blood capillaries into tissues and from tissues into lymphatic capillaries to form lymph. (*b*) The overlap of epithelial cells of the lymphatic capillary allows easy entry of fluid but prevents movement back into the tissue. Valves, located farther along in lymphatic vessels, also ensure one-way flow of lymph.

Lymphatic capillaries differ from blood capillaries in that they lack a basement membrane and the cells of the simple squamous epithelium slightly overlap and are attached loosely to one another (figure 22.2*b*). Two things occur as a result of this structure. First, the lymphatic capillaries are far more permeable than blood capillaries, and nothing in the interstitial fluid is excluded from the lymphatic capillaries. Second, the lymphatic capillary epithelium functions as a series of one-way valves that allow fluid to enter the capillary but prevent it from passing back into the interstitial spaces.

The lymphatic capillaries join to form larger **lymphatic vessels,** which resemble small veins. The inner layer of the lymphatic vessel consists of endothelium surrounded by an elastic membrane, the middle layer consists of smooth muscle cells and elastic fibers, and the outer layer is a thin layer of fibrous connective tissue.

Small lymphatic vessels have a beaded appearance because of the presence of one-way valves along their lengths that are similar to the valves of veins (see figure 22.2*b*). When a lymphatic vessel is compressed, backward movement of lymph is prevented by the valves; as a consequence, the lymph moves forward through the lymphatic vessel.

Three major mechanisms are responsible for the movement of lymph through lymphatic vessels:

1. *Contraction of lymphatic vessels.* In many parts of the body, lymphatic vessels pump lymph. The unidirectional valves divide lymphatic vessels into a series of chambers, which function as "primitive hearts." Lymph moves into a chamber, smooth muscle in the chamber wall contracts, and lymph moves into the next chamber. Some of the smooth muscle cells in the walls of lymphatic vessels are pacemaker cells (see "Electrical Properties of Smooth Muscle," chapter 21). The pacemaker cells spontaneously depolarize, resulting in periodic contractions of the lymphatic vessels.
2. *Contraction of skeletal muscles.* When surrounding muscle cells contract, lymphatic vessels are compressed, causing lymph movement.
3. *Thoracic pressure changes.* During inspiration, pressure in the thoracic cavity decreases, lymphatic vessels expand, and lymph flows into them. During expiration, pressure in the thoracic cavity increases and lymphatic vessels are compressed, causing lymph movement.

Lymph nodes are round, oval, or bean-shaped bodies distributed along the various lymphatic vessels (see "Lymph Nodes," p. 787). They filter lymph, which enters and exits the lymph nodes through the lymphatic vessels. The lymph nodes are connected together in a series, so that lymph leaving one lymph node is carried to another lymph node, and so on.

After passing through the lymph nodes, the lymphatic vessels converge to form larger vessels called **lymphatic trunks,** each of which drains a major portion of the body (figure 22.3*a* and *b*). The **jugular trunks** drain the head and neck; the **subclavian trunks** drain the upper limbs, superficial thoracic wall, and mammary glands; the **bronchomediastinal** (brong′kō-mē′dē-as-tī′năl) **trunks** drain thoracic organs and the deep thoracic wall; the **intestinal trunks** drain abdominal organs, such as the intestines, stomach, pancreas, spleen, and liver; and the **lumbar trunks** drain the lower limbs, pelvic and abdominal walls, pelvic organs, ovaries or testes, kidneys, and adrenal glands.

The lymphatic trunks connect to large veins in the thorax or join to yet larger vessels called **lymphatic ducts,** which then connect to the large veins. The connections of the lymphatic trunks and ducts to veins are quite variable. Many connect at the junction

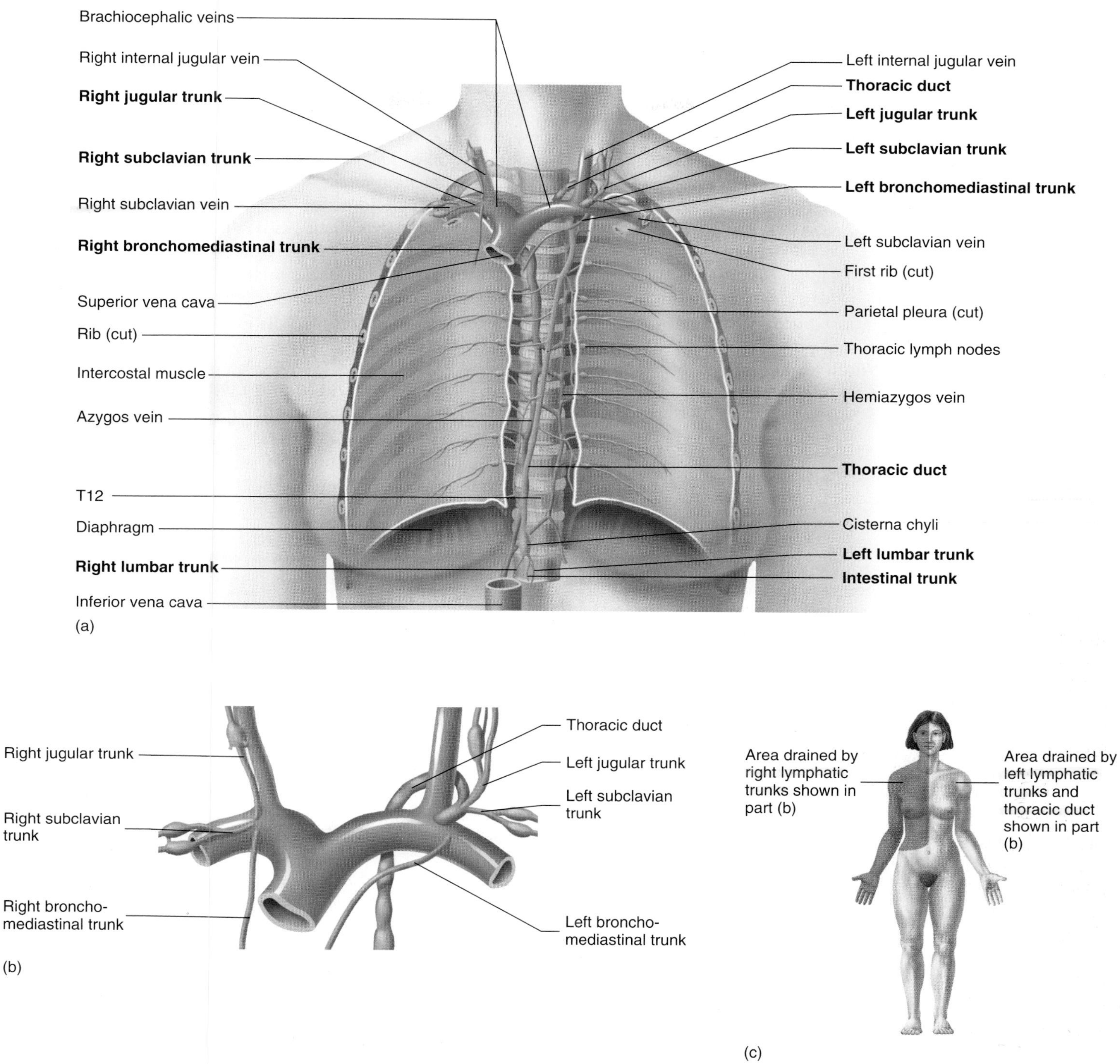

Brachiocephalic veins
Right internal jugular vein
Right jugular trunk
Right subclavian trunk
Right subclavian vein
Right bronchomediastinal trunk
Superior vena cava
Rib (cut)
Intercostal muscle
Azygos vein
T12
Diaphragm
Right lumbar trunk
Inferior vena cava
Left internal jugular vein
Thoracic duct
Left jugular trunk
Left subclavian trunk
Left bronchomediastinal trunk
Left subclavian vein
First rib (cut)
Parietal pleura (cut)
Thoracic lymph nodes
Hemiazygos vein
Thoracic duct
Cisterna chyli
Left lumbar trunk
Intestinal trunk
(a)

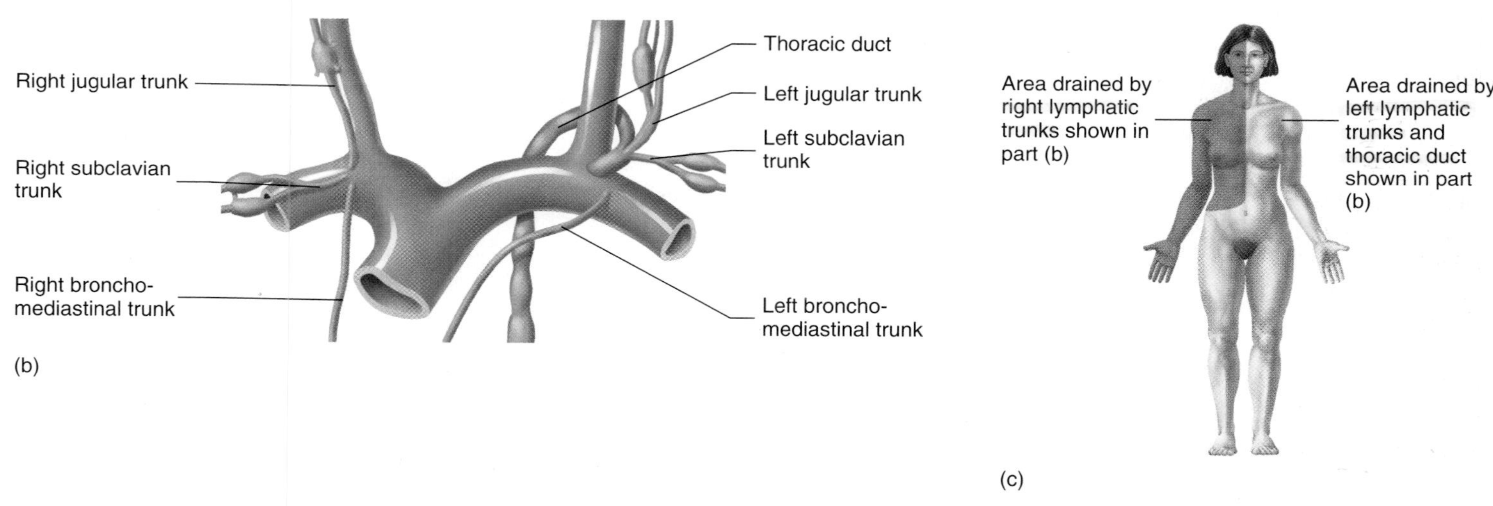

FIGURE 22.3 Lymph Drainage into Veins
(*a*) Anterior view of the major lymphatic vessels in the thorax and abdomen. (*b*) Close-up view of the lymphatic vessels from which lymph enters the blood. (*c*) Regions of the body drained by the right and left lymphatic vessels.

of the internal jugular and subclavian veins, but connections on the subclavian, jugular, and even brachiocephalic vein exist.

On the right side, the jugular, subclavian, and bronchomediastinal trunks typically join a right thoracic vein separately (see figure 22.3*b*). In about 20% of people, the three trunks join together to form a short duct 1 cm in length called the **right lymphatic duct** (see figure 22.1), which joins a right thoracic vein. These trunks drain the right side of the head, right-upper limb, and right thorax (figure 22.3*c*).

The **thoracic duct** (see figure 22.3*a* and *b*) drains lymph from the right side of the body inferior to the thorax and the entire left side of the body (see figure 22.3*c*). The thoracic duct is the largest lymphatic vessel. It is approximately 38–45 cm in length, extending from the twelfth thoracic vertebra to the base of the neck (see figure 22.3*c*). The jugular and subclavian trunks join the thoracic duct. The bronchomediastinal trunk sometimes connects to the thoracic duct but typically joins a vein. The intestinal and lumbar trunks, which drain lymph inferior to the diaphragm, supply the inferior end of the thoracic duct. They can directly join the thoracic duct or merge to form a network that connects to the thoracic duct. In a small proportion of cases, the lymphatic trunks form a sac called the **cisterna chyli** (sis-ter′nă kī′lī; a cistern or tank that contains juice).

In summary, lymph enters lymphatic capillaries, which converge to form larger lymphatic vessels. Lymph passes through the lymphatic vessels and through associated lymph nodes, where it is filtered. Lymphatic vessels converge to form larger lymphatic trunks, which drain lymph from major regions of the body. Lymphatic trunks empty directly into thoracic veins or combine to form larger lymphatic ducts, which empty into thoracic veins.

Metastasis of Cancer Cells

Cancer cells can separate from a tumor and metastasize, or spread, to other parts of the body by entering lymphatic vessels, blood capillaries, or body cavities. Cancer cells entering the lymphatic system eventually reach the blood, which carries them to other parts of the body. Cancer cells can induce the sprouting of new lymphatic vessels into the tumor. For many cancers, a greater number of lymphatic vessels within a tumor correlates with a greater risk of the cancer's spreading. The identification of lymphatic vessel growth factors and their receptors may provide new targets for drugs to help prevent the spread of cancer cells.

2 ***How is lymph formed?***

3 ***Describe the structure of a lymphatic capillary. Why is it easy for fluid and other substances to enter a lymphatic capillary?***

4 ***What is the function of the valves in lymphatic vessels? Name three mechanisms responsible for the movement of lymph through the lymphatic vessels.***

5 ***What are lymphatic trunks and ducts? Name the largest lymphatic vessel. What is the cisterna chyli?***

6 ***What areas of the body are drained by the right lymphatic trunks, left lymphatic trunks, and thoracic duct?***

Lymphatic Tissue and Organs

Lymphatic organs contain **lymphatic tissue,** which consists primarily of lymphocytes, but it also includes macrophages, dendritic cells, reticular cells, and other cell types. **Lymphocytes** are a type of white blood cell (see chapter 19). The two types of lymphocytes are **B cells** and **T cells.** They originate from red bone marrow and are carried by the blood to lymphatic organs and other tissues. When the body is exposed to microorganisms or other foreign substances, the lymphocytes divide, increase in number, and are part of the immune response that destroys microorganisms and foreign substances. Lymphatic tissue also has very fine collagen fibers, called **reticular fibers,** which are produced by **reticular cells.** Lymphocytes and other cells attach to these fibers. When lymph or blood filters through lymphatic organs, the fiber network traps microorganisms and other particles in the fluid.

Lymphatic tissue surrounded by a connective tissue capsule is said to be encapsulated, whereas lymphatic tissue without a capsule is called nonencapsulated. Lymphatic organs with a capsule include lymph nodes, the spleen, and the thymus. **Mucosa-associated lymphoid tissue (MALT)** consists of aggregates of nonencapsulated lymphatic tissue found in and beneath the mucous membranes lining the digestive, respiratory, urinary, and reproductive tracts. In these locations, the lymphatic tissue is well located to intercept microorganisms as they enter the body. Examples of MALT include diffuse lymphatic tissue, lymphatic nodules, and the tonsils.

Diffuse Lymphatic Tissue and Lymphatic Nodules

Diffuse lymphatic tissue contains dispersed lymphocytes, macrophages, and other cells; has no clear boundary; and blends with surrounding tissues (figure 22.4). It is located deep to mucous membranes, around lymphatic nodules, and within the lymph nodes and spleen.

Lymphatic nodules are denser arrangements of lymphoid tissue organized into compact, somewhat spherical structures, ranging in size from a few hundred microns to a few millimeters or more in diameter (see figure 22.4). Lymphatic nodules are numerous in the loose connective tissue of the digestive, respiratory, urinary, and reproductive systems. **Peyer's patches** are aggregations of lymphatic nodules found in the distal half of the small intestine and the appendix. In addition to MALT, lymphatic nodules are found within lymph nodes and the spleen, where they are usually referred to as **lymphatic follicles.**

Tonsils

Tonsils are large groups of lymphatic nodules and diffuse lymphatic tissue located deep to the mucous membranes within the pharynx

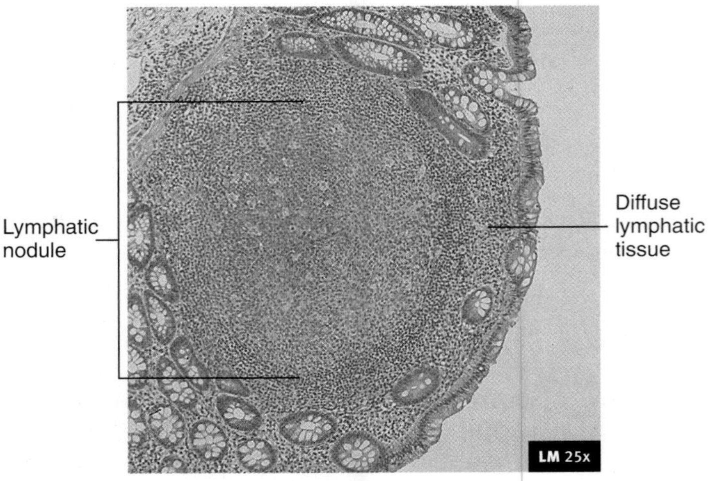

FIGURE 22.4 Diffuse Lymphatic Tissue and Lymphatic Nodule
Diffuse lymphatic tissue surrounding a lymphatic nodule in the small intestine.

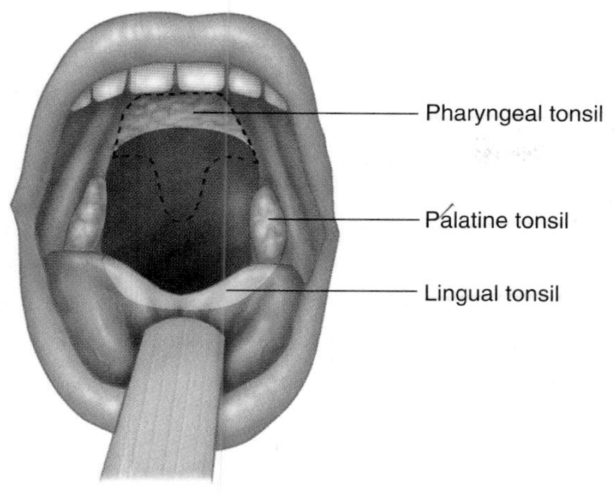

FIGURE 22.5 Tonsils

Anterior view of the oral cavity, showing the tonsils. Part of the palate is removed (*dotted line*) to show the pharyngeal tonsil.

(throat; figure 22.5). The tonsils provide protection against bacteria and other potentially harmful material entering the pharynx from the nasal or oral cavities. In adults, the tonsils decrease in size and eventually may disappear.

There are three groups of tonsils, but the **palatine tonsils** usually are referred to as "the tonsils." They are relatively large, oval lymphoid masses on each side of the junction between the oral cavity and the pharynx. The **pharyngeal** (fă-rin′jē-ăl) **tonsil,** is a collection of somewhat closely aggregated lymphatic nodules near the junction between the nasal cavity and the pharynx. When the pharyngeal tonsil is enlarged, it is commonly referred to as the **adenoid** (ad′ĕ-noyd; glandlike) or **adenoids** (ad′ĕ-noydz). An enlarged pharyngeal tonsil can interfere with normal breathing. The **lingual tonsil** is a loosely associated collection of lymphatic nodules on the posterior surface of the tongue.

Sometimes the palatine or pharyngeal tonsils become chronically infected and must be removed. The lingual tonsil becomes infected less often than the other tonsils and is more difficult to remove.

7 ***What are the functions of lymphocytes and reticular fibers in lymphatic tissue?***

8 ***What is mucosa-associated lymphoid tissue (MALT)? In what way is the location of MALT beneficial?***

9 ***Define diffuse lymphatic tissue, lymphatic nodule, Peyer's patches, and lymphatic follicle.***

10 ***Describe the structure, function, and location of the tonsils.***

Lymph Nodes

Lymph nodes are small, round, or bean-shaped structures, ranging in size from 1–25 mm long. They are distributed along the course of the lymphatic vessels (see figure 22.1; figure 22.6). They filter the lymph, removing bacteria and other materials. In addition, lymphocytes congregate, function, and proliferate within lymph nodes.

Lymph nodes are categorized as superficial or deep. **Superficial lymph nodes** are in the hypodermis beneath the skin, and **deep lymph nodes** are everywhere else. Most superficial and deep lymph nodes are located near or on blood vessels. Approximately 450 lymph nodes are found throughout the body. Cervical and head nodes (about 70) filter lymph from the head and neck, axillary nodes (about 30) filter lymph from the upper limbs and superficial thorax, thoracic nodes (about 100) filter lymph from the thoracic wall and organs, abdominopelvic nodes (about 230) filter lymph from the abdomen and pelvis, and inguinal and popliteal nodes (about 20) filter lymph from the lower limbs and superficial pelvis.

Femoral Hernia

Lymph from the lower limbs drains into the inguinal lymph nodes, which are located in the groin region. The **femoral canal** is a passageway through which lymphatic vessels from the inguinal nodes enter the abdominal cavity. The femoral canal is a potential weakness in the abdominal wall. A **femoral hernia** occurs when a loop of intestine pushes into, or even passes completely through, the femoral canal.

A dense connective tissue **capsule** surrounds each lymph node. Extensions of the capsule, called **trabeculae** (tră-bek′ū-lē), form a delicate internal skeleton in the lymph node. Reticular fibers extend from the capsule and trabeculae to form a fibrous network throughout the entire node. In some areas of the lymph node, lymphocytes and macrophages are packed around the reticular fibers to form lymphatic tissue; in other areas the reticular fibers extend across open spaces to form **lymphatic sinuses.** The lymphatic tissue and sinuses within the node are arranged into two somewhat indistinct layers, an outer cortex and an inner medulla. The **cortex** consists of a subcapsular sinus, beneath the capsule, and cortical sinuses, which are separated by diffuse lymphatic tissue, trabeculae, and lymphatic nodules. The inner **medulla** is organized into branching, irregular strands of diffuse lymphatic tissue, the **medullary cords,** separated by medullary sinuses.

Lymph nodes are the only structures to filter lymph. They have **afferent lymphatic vessels,** which carry lymph to the lymph nodes, where it is filtered, and **efferent lymphatic vessels,** which carry lymph away from the nodes. Lymph from afferent lymphatic vessels enters the subcapsular sinus, filters through the cortex and medulla, and exits the lymph node through efferent lymphatic vessels. The efferent vessels of one lymph node may become the afferent vessels of another node or may converge to form lymphatic trunks, which carry lymph to the blood at thoracic veins.

Macrophages lining the lymphatic sinuses remove bacteria and other foreign substances from the lymph as it slowly filters through the sinuses. Microorganisms and other foreign substances in the lymph can stimulate lymphocytes throughout the lymph node to undergo cell division, with proliferation especially evident in the lymphatic nodules of the cortex. These areas of rapid lymphocyte division are called **germinal centers.** The newly produced

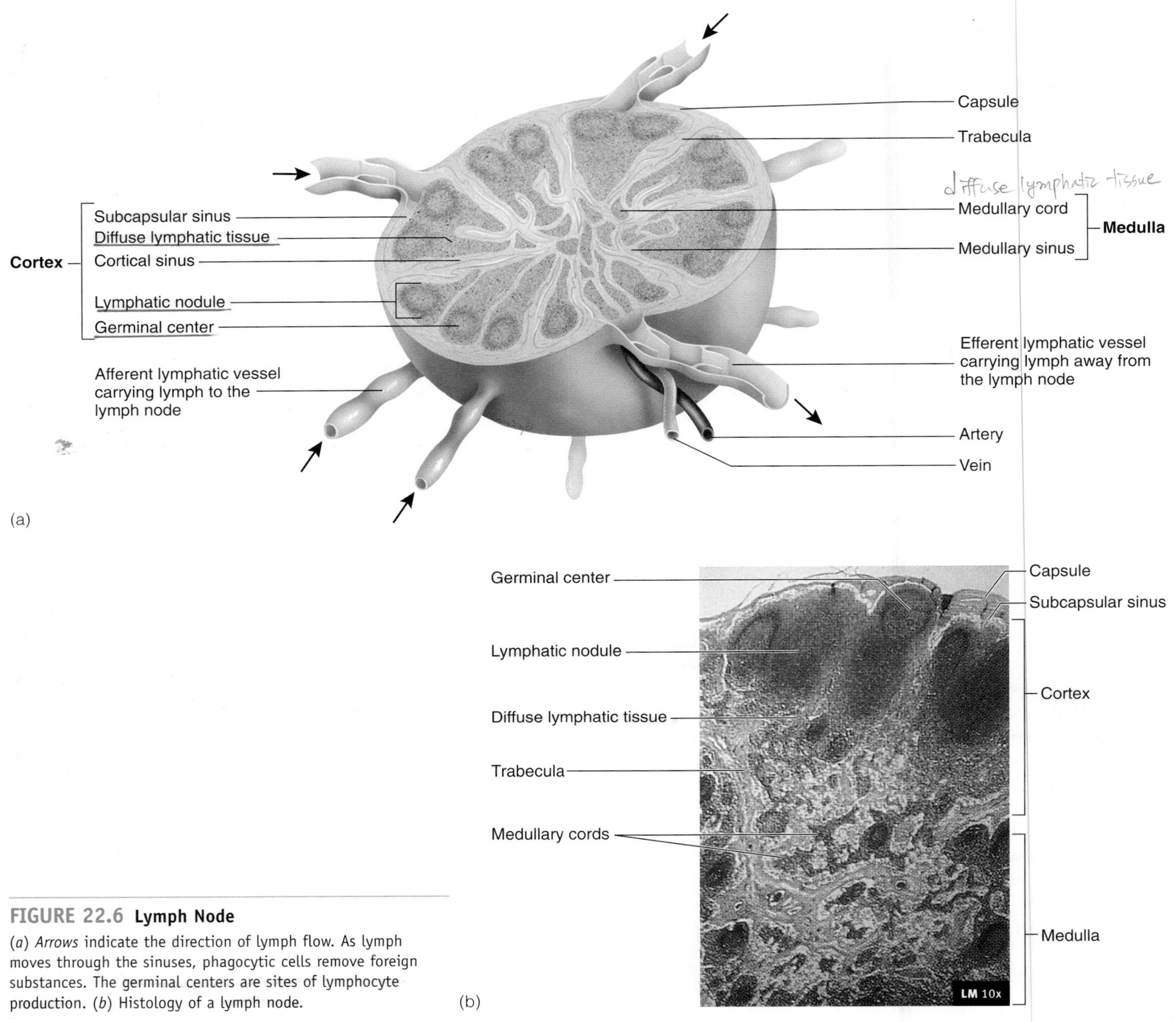

FIGURE 22.6 Lymph Node

(*a*) *Arrows* indicate the direction of lymph flow. As lymph moves through the sinuses, phagocytic cells remove foreign substances. The germinal centers are sites of lymphocyte production. (*b*) Histology of a lymph node.

lymphocytes are released into the lymph and eventually reach the bloodstream, where they circulate. Subsequently, the lymphocytes can leave the blood and enter other lymphatic tissues.

11 ***Where are lymph nodes found? Describe the parts of a lymph node, and explain how lymph flows through a lymph node.***

12 ***What are the functions of lymph nodes? How are they accomplished? What is a germinal center?***

Spleen

The **spleen,** which is roughly the size of a clenched fist, is located on the left side in the extreme, superior part of the abdominal cavity (figure 22.7). The average weight of the adult spleen is 180 g in males and 140 g in females. The size and weight of the spleen tends to decrease in older people, but in certain diseases the spleen can achieve weights of 2000 g or more.

The spleen has an outer **capsule** of dense irregular connective tissue and a small amount of smooth muscle. Bundles of connective tissue fibers from the capsule form **trabeculae,** which extend into the organ, subdividing it into small, interconnected compartments. Arteries, veins, and lymphatic vessels extend through the trabeculae to supply the compartments, which are filled with white and red pulp. **White pulp** is lymphatic tissue surrounding the arteries within the spleen. **Red pulp** is associated with the

CASE STUDY

Lymphedema

Cindy, a 40-year-old woman, had been diagnosed with breast cancer. Before removing the cancerous tumor from her left breast, her surgeon injected a dye and a radioactive tracer, called technetium-99, at the tumor site. The dye and tracer enabled the surgeon to find **sentinel lymph nodes,** which are the lymph nodes closest to the tumor. The sentinel lymph nodes were sampled for cancer. When cancer cells were found in all of them, the surgeon removed the axillary lymph nodes from under Cindy's left arm.

A few days after the surgery, Cindy noticed that the skin on her left arm felt tight and the arm felt heavy. In addition, her wedding ring was tighter than usual. Cindy had an abnormal accumulation of fluid in her upper limb, called **lymphedema** (limf′e-dē′mă), caused by the removal of her axillary lymph nodes. In the United States, the most common cause of lymphedema is the removal or damage of lymph nodes and vessels by cancer surgery or radiation treatment. Approximately 10%–20% of women who have their axillary lymph nodes removed develop lymphedema.

PREDICT 1

a. What is the rationale for testing sentinel lymph nodes for cancer?
b. What was the rationale for removing Cindy's axillary lymph nodes?
c. Why does removing the axillary lymph nodes result in lymphedema?
d. Exercise can help reduce lymphedema. Explain.
e. A compression bandage or garment can help reduce lymphedema. Explain.
f. In intermittent pneumatic pump compression therapy, the pressure of a garment enclosing a limb increases and decreases periodically. In addition, the pressure increases sequentially from the distal part to the proximal part of a limb. How does this therapy help reduce lymphedema?

veins. It consists of a fibrous network, filled with macrophages and red blood cells, and enlarged capillaries that connect to the veins. Approximately one-fourth of the volume of the spleen is white pulp and three-fourths is red pulp.

Branches of the **splenic** (splen′ik) **artery** enter the spleen at the **hilum,** and their branches follow the various trabeculae into the spleen (see figure 22.7*a* and *b*). From the trabeculae, arterial branches extend into the white pulp, which consists of the periarterial lymphatic sheath and lymphatic nodules (see figure 22.7*c*). The **periarterial lymphatic sheath** is diffuse lymphatic tissue surrounding arteries and arterioles extending to lymphatic nodules. Arterioles enter lymphatic nodules and give rise to capillaries supplying the red pulp, which consists of the splenic cords and venous sinuses. The **splenic cords** are a network of reticular cells that produce reticular fibers (see chapter 4). The spaces between the reticular cells are occupied by splenic macrophages and blood cells that have come from the capillaries. The **venous sinuses** are enlarged capillaries between the splenic cords. The venous sinuses typically connect to trabecular veins, which unite to form vessels that leave the spleen to form the **splenic vein.**

Blood flow through the spleen either takes a few seconds or takes minutes to an hour or more. Most blood flows through the spleen rapidly. The rapid flow results from the movement of blood from the ends of capillaries into the beginning of the venous sinuses. Although a few capillaries connect directly to venous sinuses, most capillaries are separated by a small gap (see figure 22.7*c*). The slower flow of blood occurs when blood leaves the ends of the capillaries, enters the splenic cords, percolates through them, and passes through the walls of the venous sinuses.

The spleen destroys defective red blood cells, detects and responds to foreign substances in the blood, and acts as a blood reservoir. As red blood cells age, they lose their ability to bend and fold. Consequently, the cells can rupture as they pass slowly through the meshwork of the splenic cords or the intercellular slits of the venous sinus walls. Splenic macrophages then phagocytize the cellular debris.

Foreign substances in the blood passing through the spleen can stimulate an immune response because of the presence in the white pulp of specialized lymphocytes (see "Adaptive Immunity, p. 798). There are high concentrations of T cells in the periarterial lymphatic sheath and B cells in the lymphatic nodules.

The splenic cords of the spleen are a limited reservoir for blood. For example, during exercise splenic volume can be reduced by 40%–50%. The resulting small increase in circulating red blood cells can promote better oxygen delivery to muscles during exercise or emergency situations. It is not presently known if this reduction results from the contraction of smooth muscle within the capsule, from the contraction of smooth muscle (myofibroblast) within the trabeculae, or from the reduced blood flow through the spleen caused by the constriction of blood vessels.

Ruptured Spleen

Although the spleen is protected by the ribs, it is often ruptured in traumatic abdominal injuries. A ruptured spleen can cause severe bleeding, shock, and death. Surgical intervention may stop the bleeding. Cracks in the spleen are repaired using sutures and blood-clotting agents. Mesh wrapped around the spleen can hold it together. A **splenectomy** (splē-nek′tō-mē), removal of the spleen, may be necessary if these techniques do not stop the bleeding. Other lymphatic organs and the liver compensate for the loss of the spleen's functions.

13 ***Where is the spleen located? Name the two components of white pulp and of red pulp.***

14 ***Explain the fast and slow flow of blood through the spleen.***

15 ***What are three functions of the spleen?***

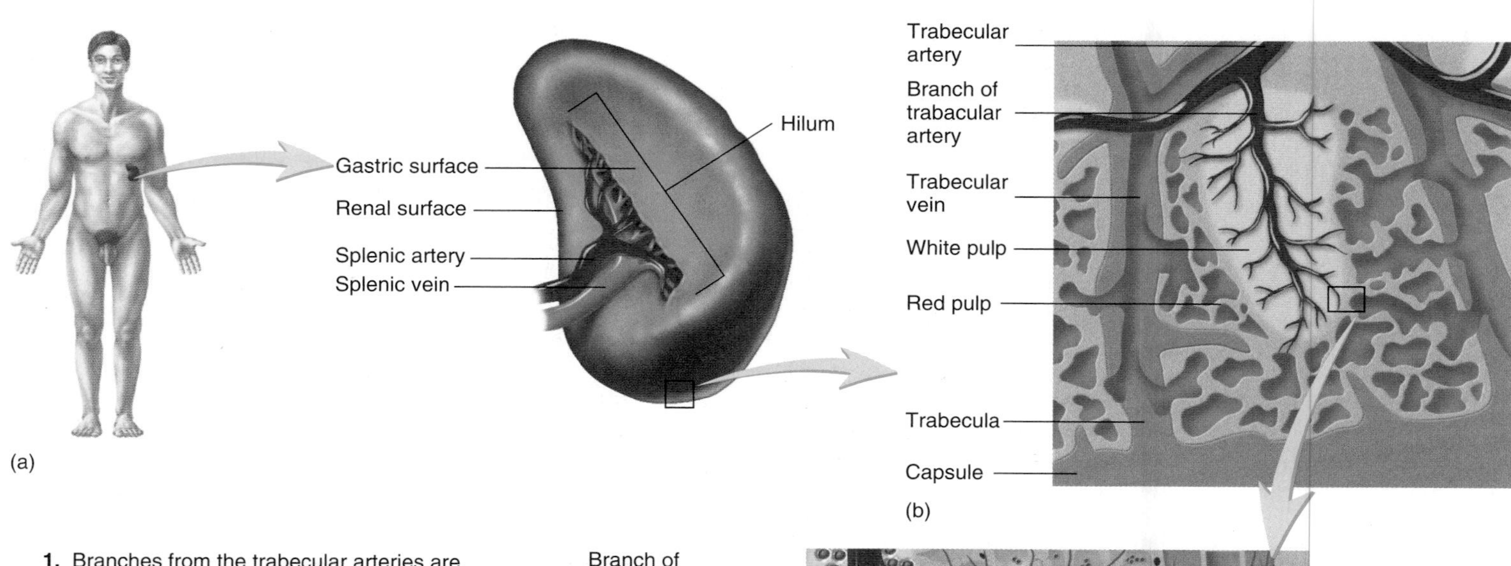

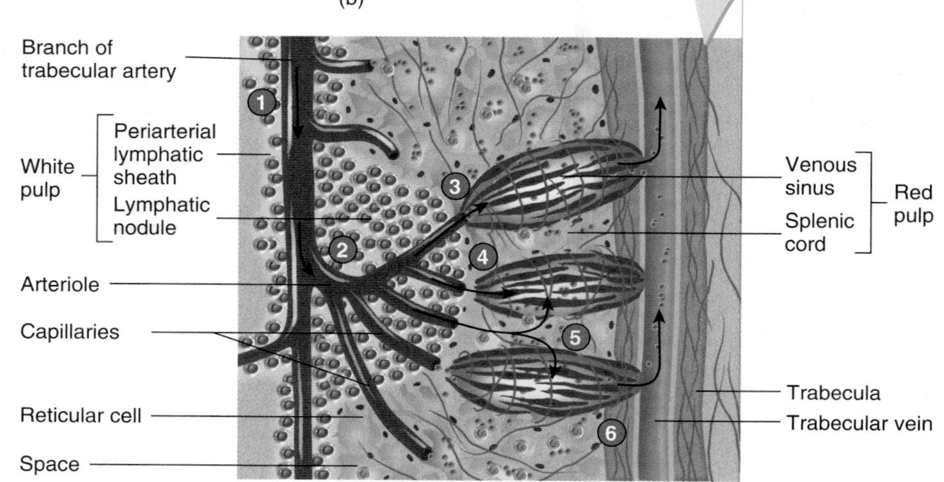

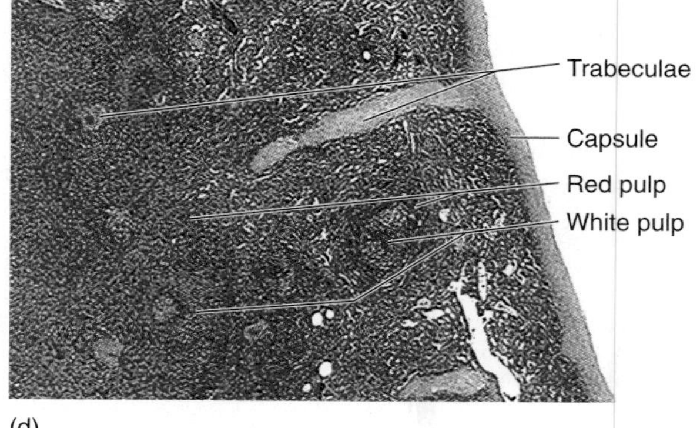

PROCESS FIGURE 22.7 Spleen

(*a*) Inferior view of the spleen. (*b*) Section showing the arrangement of arteries, veins, white pulp, and red pulp. White pulp is associated with arteries, and red pulp is associated with veins. (*c*) Blood flow through white and red pulp. (*d*) Histology of the spleen.

Thymus

The **thymus** (thī′mŭs) is a bilobed gland (figure 22.8) located in the superior mediastinum, the partition dividing the thoracic cavity into left and right parts. The thymus increases in size until the first year of life, after which it remains approximately the same size until 60 years of age, after which it decreases in size. Although the size of the thymus is fairly constant throughout much of life, by 40 years of age much of the thymic lymphatic tissue has been replaced with adipose tissue.

Each lobe of the thymus is surrounded by a thin connective tissue **capsule. Trabeculae** extend from the capsule into the substance of the gland, dividing it into **lobules.** Unlike other lymphatic tissue, which has a fibrous network of reticular fibers, the framework of thymic tissue consists of epithelial cells. The processes of the epithelial cells are joined by desmosomes, and the cells form small, irregularly shaped compartments filled with lymphocytes. Near the capsule and trabeculae, the lymphocytes are numerous and form dark staining areas of the lobules called

CLINICAL FOCUS

Disorders of the Lymphatic System

It is not surprising that many infectious diseases produce symptoms associated with the lymphatic system. The lymphatic system is involved with the production of lymphocytes, which fight infectious diseases, and the lymphatic system filters blood and lymph to remove microorganisms. **Lymphadenitis** (lim-fad'ĕ-nī'tis) is an inflammation of the lymph nodes, which causes them to become enlarged and tender. This inflammation is an indication that microorganisms are being trapped and destroyed within the lymph nodes. Sometimes the lymphatic vessels become inflamed to produce **lymphangitis** (lim-fan-jī'tis). This often results in visible red streaks in the skin that extend away from the site of infection. If microorganisms pass through the lymphatic vessels and nodes to reach the blood, **septicemia** (sep-ti-sē'mē-ă), or blood poisoning, can result (see chapter 19).

Bubonic (boo-bon'ik) **plague** is caused by bacteria (*Yersinia pestis*), which are transferred from rats to humans by the bite of the rat flea (*Xenopsylla*). The bacteria localize in the lymph nodes, causing them to enlarge. The term *bubonic* is derived from a Greek word referring to the groin because the disease often causes the inguinal lymph nodes of the groin to swell. Without treatment, the bacteria enter the blood, multiply, and infect tissues throughout the body, rapidly causing death in 70%–90% of those infected. In the sixth, fourteenth, and nineteenth centuries, the bubonic plague killed large numbers of people in Europe. Because of improved sanitation and the advent of antibiotics, relatively few cases occur today.

Lymphedema (limf'e-dē'mă) is the abnormal accumulation of lymph in tissues, often the upper or lower limbs due to disruption of lymph flow. **Primary lymphedema** is caused by a developmental defect of the lymphatic system that generally affects the lower limbs; it can be manifested anytime in life. For unknown reasons, it occurs primarily in women (70%–90% of cases). Some primary lymphedemas are inherited. **Hereditary lymphedema type I** (Milroy syndrome) has been mapped to mutations of the vascular endothelial growth factor receptor 3 gene on chromosome 5. Activation of this receptor on endothelial cells causes the proliferation of lymphatic vessels, but not blood vessels. **Secondary lymphedema** is caused by a disease or another pathologic condition that affects an otherwise normal lymphatic system. In the United States, secondary lymphedema is most commonly caused by certain cancer treatments. Worldwide, the major cause is a parasitic infection of the lymphatic system called **elephantiasis** (el-ĕ-fan-tī'ă-sis). Over 120 million people have elephantiasis, which is the most dramatic form of lymphedema. It is caused by long, slender roundworms (*Wuchereria bancrofti*). The adult worms lodge in the lymphatic vessels and block lymph flow. The resulting accumulation of fluid in the interstitial spaces and lymphatic vessels can cause permanent swelling and enlargement of a limb. The affected limb supposedly resembles an elephant's leg, the basis for its name. The offspring of the adult worms pass through the lymphatic system into the blood, from which they can be transferred to another human by mosquitoes.

A **lymphoma** (lim-fō'mă) is a neoplasm (tumor) of lymphatic tissue. Lymphomas are usually divided into two groups: (1) Hodgkin disease and (2) all other lymphomas, which are called non-Hodgkin lymphomas. Typically, lymphomas begin as an enlarged, painless mass of lymph nodes. The immune system is depressed, and the patient has an increased susceptibility to infections. Enlargement of the lymph nodes can also compress surrounding structures and produce complications. Fortunately, treatment with drugs and radiation is effective for many people who suffer from lymphoma.

the **cortex.** A lighter staining central portion of the lobules, called the **medulla,** has fewer lymphocytes. The medulla also contains rounded epithelial structures, called **thymic corpuscles** (Hassall corpuscles), whose function is unknown.

The thymus is the site of the maturation of T cells. Large numbers of lymphocytes are produced in the thymus, but most degenerate. The lymphocytes that survive the maturation process are capable of reacting to foreign substances, but they normally do not react to and destroy healthy body cells (see the "Origin and Development of Lymphocytes," p. 799). These surviving thymic lymphocytes migrate to the medulla, enter the blood, and travel to other lymphatic tissues.

16 ***Where is the thymus located? Describe its structure and function.***

Overview of the Lymphatic System

Figure 22.9 summarizes the parts of the lymphatic system and their functions. Lymphatic capillaries and vessels remove fluid from tissues and absorb fats from the small intestines. Lymph nodes filter lymph, and the spleen filters blood.

Figure 22.9 also illustrates two types of lymphocytes, the B and T cells. B cells originate and mature in red bone marrow. Pre-T cells are produced in red bone marrow and migrate to the thymus, where they mature to become T cells. B cells from red bone marrow and T cells from the thymus circulate to, and populate, other lymphatic tissues.

B and T cells are responsible for much of immunity. In response to infections, B and T cells increase in number and circulate to lymphatic and infected tissues. How B and T cells protect the body is discussed in the section "Adaptive Immunity," p. 798.

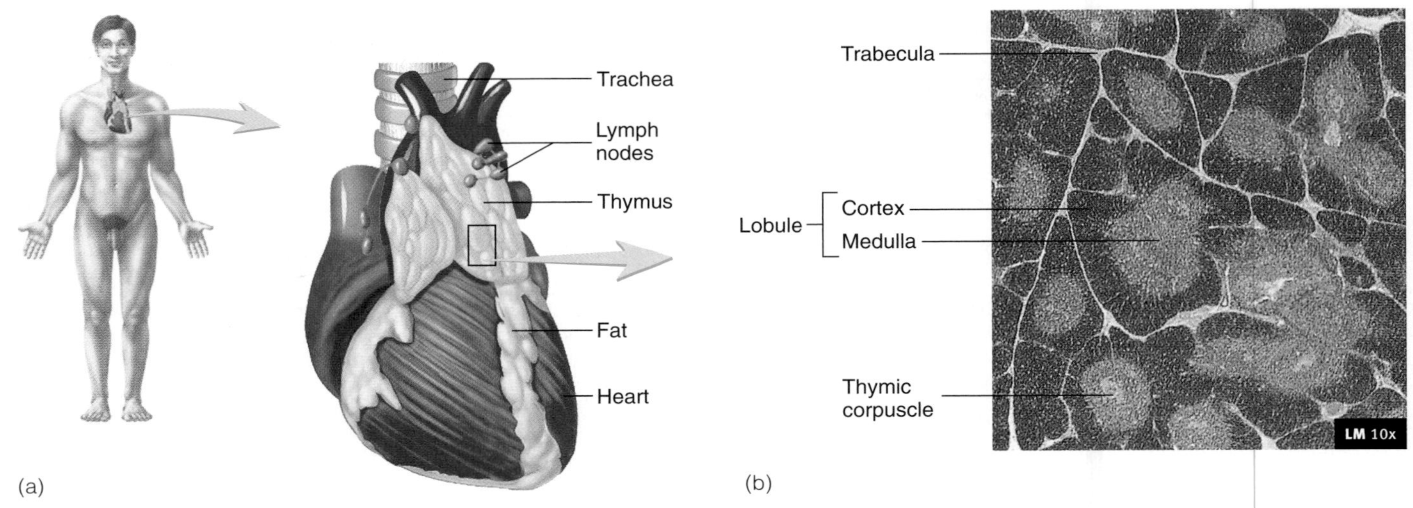

FIGURE 22.8 Thymus
(*a*) Location and shape of the thymus. (*b*) Histology of thymic lobules, showing outer cortex and inner medulla.

IMMUNITY

Immunity is the ability to resist damage from foreign substances, such as microorganisms; harmful chemicals; and internal threats, such as cancer cells. Immunity is categorized as **innate immunity** (also called nonspecific resistance) and **adaptive immunity** (also called specific immunity), although the two systems are fully integrated in the body. In innate immunity, the body recognizes and destroys certain foreign substances, but the response to them is the same each time the body is exposed to them. In adaptive immunity, the body recognizes and destroys foreign substances, but the response to them improves each time the foreign substance is encountered.

Specificity and memory are characteristics of adaptive immunity but not innate immunity. **Specificity** is the ability of adaptive immunity to recognize a particular substance. For example, innate immunity can act against bacteria in general, whereas adaptive immunity can distinguish among various kinds of bacteria. **Memory** is the ability of adaptive immunity to "remember" previous encounters with a particular substance. As a result, the response is faster and stronger and lasts longer.

In innate immunity, each time the body is exposed to a substance, the response is the same because specificity and memory of previous encounters are not present. For example, each time a bacterial cell is introduced into the body, it is phagocytized with the same speed and efficiency. Innate immunity is present to some degree in all multicellular organisms.

In adaptive immunity, the response during the second exposure is faster and stronger than the response to the first exposure because the immune system remembers the bacteria from the first exposure. For example, following the first exposure to the bacteria, the body can take many days to destroy them. During this time, the bacteria damage tissues and produce the symptoms of disease. Following the second exposure to the same bacteria, the response is rapid and effective. Bacteria are destroyed before any symptoms develop, and the person is said to be **immune.** Adaptive immunity is unique to vertebrates.

Innate and adaptive immunity are intimately linked. Innate immunity is required for the initiation and regulation of the adaptive immune response.

17 ***Define immunity, specificity, and memory.***

18 ***What are the differences between innate and adaptive immunity?***

INNATE IMMUNITY

The main components of innate immunity are (1) mechanical mechanisms that prevent the entry of microbes into the body or that physically remove them from body surfaces; (2) chemical mediators that act directly against microorganisms or that activate other mechanisms, leading to the destruction of the microorganisms; and (3) cells involved in phagocytosis and the production of chemicals that participate in the response of the immune system.

Mechanical Mechanisms

Mechanical mechanisms, such as the skin and mucous membranes, form barriers that prevent the entry of microorganisms and chemicals into the tissues of the body. They also remove microorganisms and other substances from the surface of the body in several ways. The substances are washed from the eyes by tears, from the mouth by saliva, and from the urinary tract by urine. In the respiratory tract, ciliated mucous membranes sweep microbes trapped in the mucus to the back of the throat, where they are swallowed. Coughing and sneezing also remove microorganisms from the respiratory tract.

(a) Lymphatic capillaries remove fluid from tissues. The fluid becomes lymph (see figure 22.2*a*).

(b) Lymph flows through lymphatic vessels, which have valves that prevent the backflow of lymph (see figure 22.2*b*).

(c) Lymph nodes filter lymph (see figure 22.6) and are sites where lymphocytes respond to infections, etc.

(d) Lymph enters lymphatic trunks and ducts (see figure 22.3*b*).

(e) Lymph enters the blood.

(f) Lacteals in the small intestine (see figure 24.16*c*) absorb fats, which enter the thoracic duct (see figure 22.3*a*).

(g) Chyle, which is lymph containing fats, enters the blood.

(h) The spleen (see figure 22.7) filters blood and is a site where lymphocytes respond to infections, etc.

(i) Lymphocytes (pre-B and pre-T cells) originate from stem cells in the red bone marrow (see figure 22.12). The pre-B cells become mature B cells in the red bone marrow and are released into the blood. The pre-T cells enter the blood and migrate to the thymus.

(j) The thymus (see figure 22.8) is where pre-T cells derived from red bone marrow increase in number and become mature T cells that are released into the blood (see figure 22.12).

(k) B and T cells from the blood enter and populate all lymphatic tissues. These lymphocytes can remain in the lymphatic tissues or pass through them and return to the blood. B and T cells can also respond to infections, etc., by dividing and increasing in number (see figures 22.18 and 22.22).

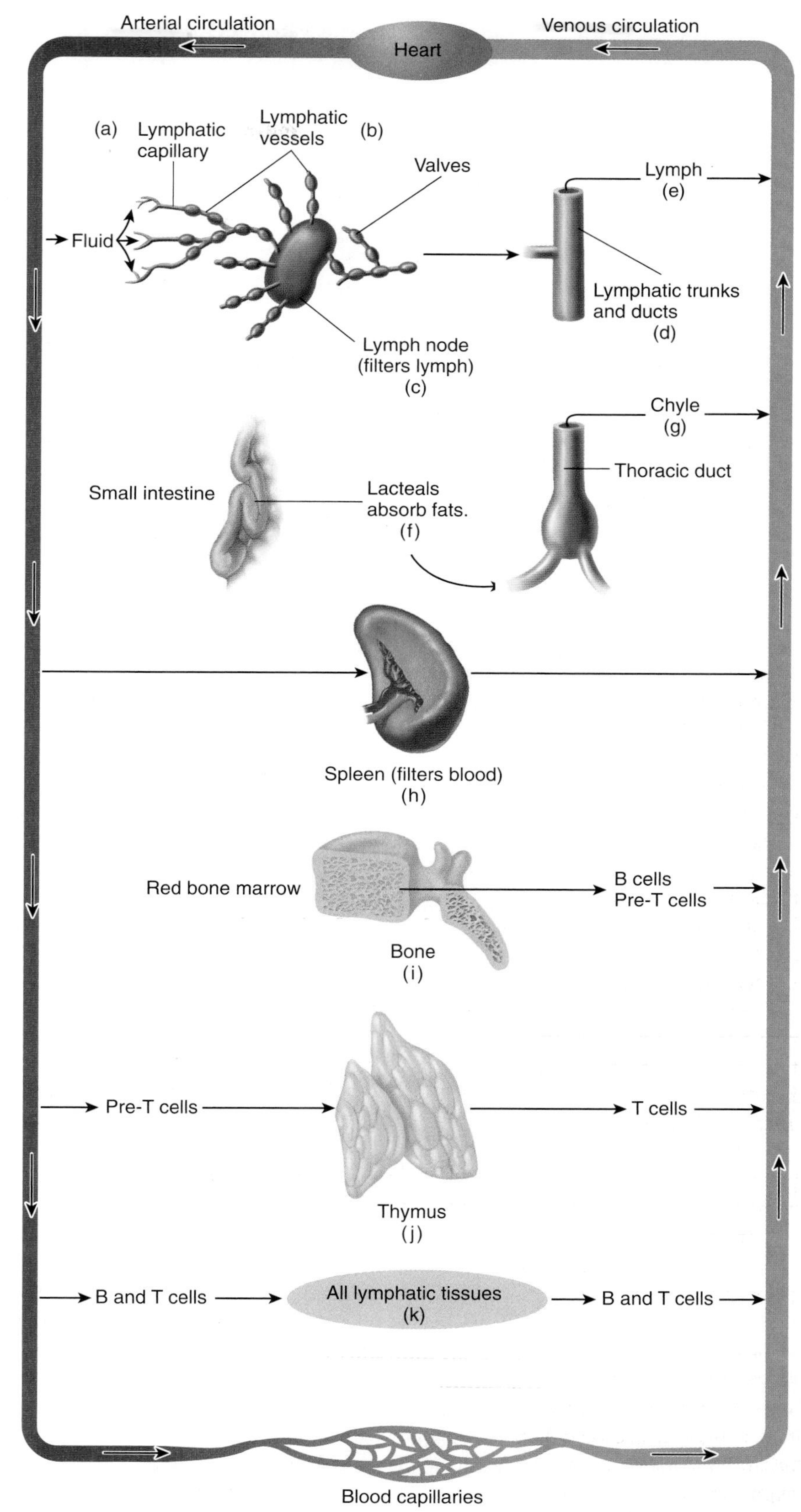

FIGURE 22.9 Overview of the Lymphatic System

TABLE 22.1 Chemical Mediators of Innate Immunity and Their Functions

Chemical	Description
Surface chemicals	Lysozymes (in tears, saliva, nasal secretions, and sweat) lyse cells; acid secretions (sebum in the skin and hydrochloric acid in the stomach) prevent microbial growth or kill microorganisms; mucus on the mucous membranes traps microorganisms until they can be destroyed
Histamine	Histamine is an amine released from mast cells, basophils, and platelets; histamine causes vasodilation, increases vascular permeability, stimulates gland secretions (especially mucus and tear production), causes smooth muscle contraction of airway passages (bronchioles) in the lungs, and attracts eosinophils
Kinins	Kinins are polypeptides derived from plasma proteins; kinins cause vasodilation, increase vascular permeability, stimulate pain receptors, and attract neutrophils
Interferons	Interferons are proteins, produced by most cells, that interfere with virus production and infection.
Complement	Complement is a group of plasma proteins that increase vascular permeability, stimulate the release of histamine, activate kinins, lyse cells, promote phagocytosis, and attract neutrophils, monocytes, macrophages, and eosinophils.
Prostaglandins	Prostaglandins are a group of lipids (PGEs, PGFs, thromboxanes, and prostacyclins), some of which cause smooth muscle relaxation and vasodilation, increase vascular permeability, and stimulate pain receptors.
Leukotrienes	Leukotrienes are a group of lipids, produced primarily by mast cells and basophils, that cause prolonged smooth muscle contraction (especially in the lung bronchioles), increase vascular permeability, and attract neutrophils and eosinophils.
Pyrogens	Pyrogens are chemicals, released by neutrophils, monocytes, and other cells, that stimulate fever production.

Abbreviations: PGE = prostaglandin E; PGF = prostaglandin F.

Chemical Mediators

Chemical mediators are molecules responsible for many aspects of innate immunity (table 22.1). Some chemical mediators found on the surface of cells, such as lysozyme, sebum, and mucus, kill microorganisms or prevent their entry into the cells. Other chemical mediators, such as histamine, complement, and eicosanoids (e.g., prostaglandins and leukotrienes), promote inflammation by causing vasodilation, increasing vascular permeability, attracting white blood cells, and stimulating phagocytosis. **Cytokines** (si′tō-kīnz) are proteins or peptides secreted by cells that bind to receptors on cell surfaces, stimulating a response. They usually bind to receptors on neighboring cells, but sometimes they bind to receptors on the secreting cell. Cytokines regulate the intensity and duration of immune responses and stimulate the proliferation and differentiation of cells. Examples of cytokines are interferons, interleukins, and lymphokines.

Complement

Complement is a group of about 20 proteins that make up approximately 10% of the globulin part of serum. They include proteins named C1–C9 and factors B, D, and P (properdin). Normally, complement proteins circulate in the blood in an inactive, nonfunctional form. They become activated in the **complement cascade,** a series of reactions in which each component of the series activates the next component (figure 22.10). The complement cascade begins through either the alternative pathway or the classical pathway. The **alternative pathway,** part of innate immunity, is initiated when the complement protein C3 becomes spontaneously active.

Activated C3 normally is quickly inactivated by proteins on the surface of the body's cells. Activated C3 can combine with some foreign substances, such as part of a bacterial cell or virus. It can become stabilized and cause activation of the complement cascade. The **classical pathway** is part of adaptive immunity, discussed on p. 798. The activation of complement by the alternative and classical pathways is one of several ways in which innate and adaptive responses are integrated.

Activated complement proteins provide protection in several ways (see figure 22.10). They can form a **membrane attack complex (MAC),** which produces a channel through the plasma membrane. MAC formation begins when activated C3 attaches to a plasma membrane, stimulating a series of reactions (C5–C9). The main component of a MAC is activated C9 molecules, which change shape, attach to each other, and form a channel through the membrane. When MACs form in the plasma membrane of a nucleated cell, Na^+ and water enter the cell through the channel and cause the cell to lysis. When MACs form in the outer membrane of certain bacteria (Gram negative), an enzyme called lysozyme passes through the channel and digests the bacterial cell wall. When the wall breaks apart, the bacterial cell undergoes lysis.

Complement proteins can also attach to the surface of bacterial cells and stimulate macrophages to phagocytize the bacteria. In addition, complement proteins attract immune system cells to sites of infection and promote inflammation.

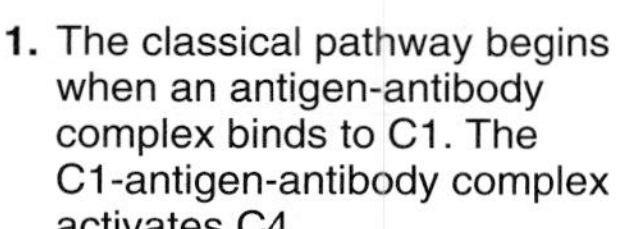

1. The classical pathway begins when an antigen-antibody complex binds to C1. The C1-antigen-antibody complex activates C4.
2. Activated C4 forms a complex with C2 that activates C3.
3. The alternate pathway begins when C3 is spontaneously activated.
4. Foreign substances and factors B, D, and P stabilize activated C3.
5. Once C3 is activated, the classical and alternate pathways are the same. C3 activates C5, C5 activates C6, C6 activates C7, C7 activates C8, and C8 activates C9.
6. Activated C3–C7 promote phagocytosis, inflammation, and chemotaxis (attracts cells).
7. Activated C5–C9 combine to form a membrane attack complex (MAC), which forms a channel through the plasma membrane (only C9 of MAC is shown).

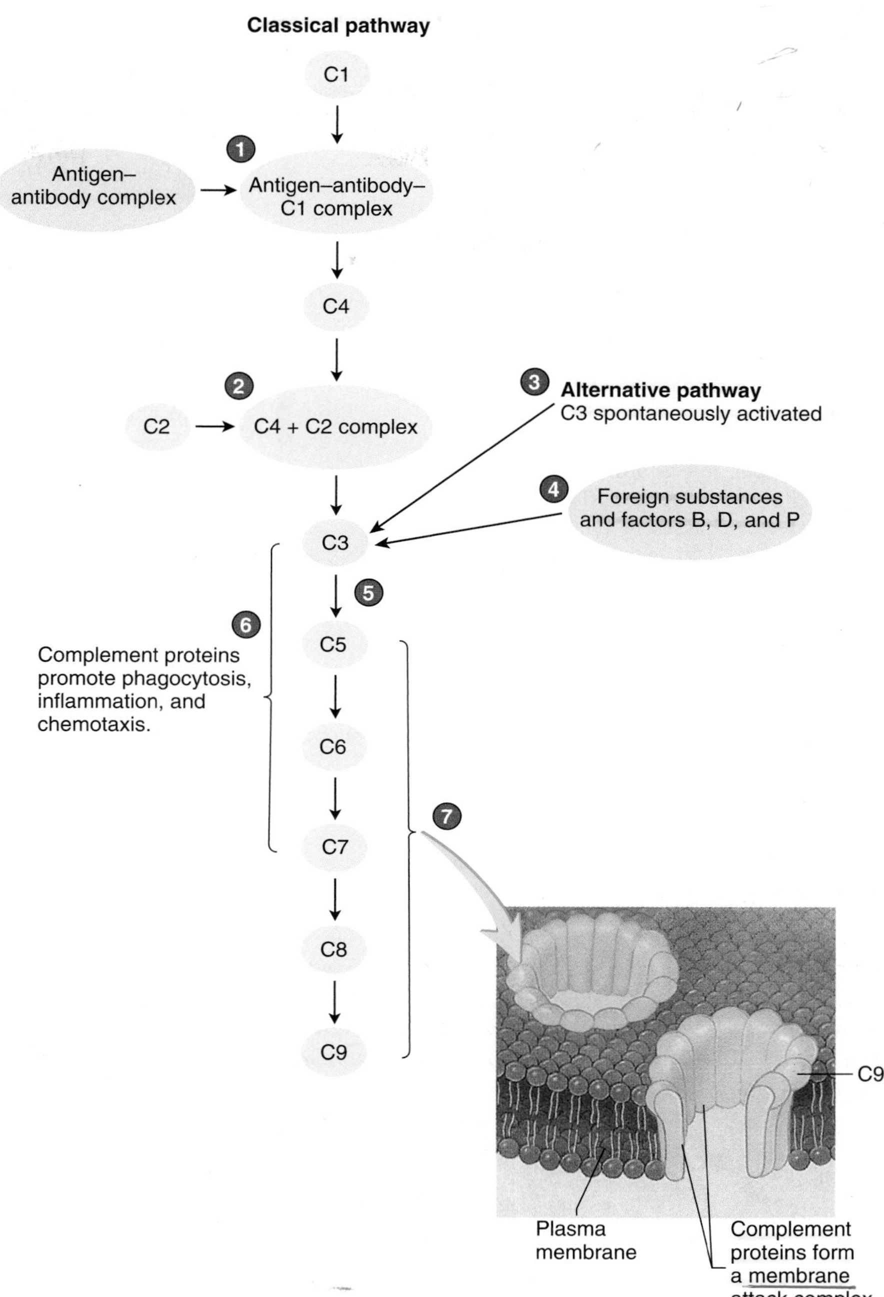

PROCESS FIGURE 22.10 Complement Cascade

Complement proteins circulate in the blood in an inactive from. Only the activated proteins are shown.

Interferons

Interferons (in-ter-fēr′onz) are proteins that protect the body against viral infection and perhaps some forms of cancer. After a virus infects a cell, viral replication can occur. Viral nucleic acids and proteins, which are produced using the cell's organelles, are assembled into new viruses. The new viruses are released from the infected cell to infect other cells. Because infected cells usually stop their normal functions or die during viral replication, viral infections are clearly harmful to the body. Fortunately, viruses and other substances can stimulate infected cells to produce interferons. Interferons neither protect the cell that produces them nor act directly against viruses. Instead, they bind to the surface of neighboring cells, where they stimulate them to produce antiviral proteins. These antiviral proteins stop viral reproduction in the neighboring cells by preventing the production of viral nucleic acids and proteins. Interferon viral resistance is innate rather than adaptive, and the same interferons act against many different viruses. Infection by one kind of virus actually can produce protection against infection by other kinds of viruses. Some interferons also play a role in the activation of immune cells, such as macrophages and natural killer cells.

Treating Viral Infections and Cancer with Interferons

Because some cancers are induced by viruses, interferons may play a role in controlling cancers. Interferons activate macrophages and natural killer cells (a type of lymphocyte), which attack tumor cells. Through genetic engineering, interferons are produced in sufficient quantities for clinical use and, along with other therapies, have been effective in treating certain viral infections and cancers. For example, interferons are used to treat hepatitis C, a viral disorder that can cause cirrhosis and cancer of the liver, and to treat genital warts, caused by the herpes virus. Interferons are also approved for the treatment of Kaposi sarcoma, a cancer that can develop in AIDS patients.

19 ***List the three main components of innate immunity.***

20 ***Name two mechanical mechanisms that prevent the entry of microorganisms into the body. In what ways are microorganisms removed from the surfaces of the body?***

21 ***What is complement? In what two ways is it activated? How does complement provide protection?***

22 ***What are interferons? How do they provide protection against viruses?***

Cells

White blood cells and the cells derived from them (see table 19.2) are the most important cellular components of the immune system (table 22.2). White blood cells are produced in red bone marrow and lymphatic tissue and are released into the blood, where they are transported throughout the body. To be effective, white blood cells must move into the tissues where they are needed. **Chemotactic** (kē-mō-tak′tik) **factors** are parts of microbes or chemicals released by tissue cells that act as chemical signals to attract white blood cells. Important chemotactic factors include complement, leukotrienes, kinins, and histamine. They diffuse from the area where they are released. White blood cells can detect small differences in chemotactic factor concentration and move from areas of lower chemotactic factor concentration to areas of higher concentration. Thus, they move toward the source of these substances, an ability called **chemotaxis.** White blood cells can move by ameboid movement over the surface of cells, can squeeze between cells, and sometimes pass directly through other cells.

Phagocytosis (fag-ō-sī-tō′sis) is the endocytosis and destruction of particles by cells called **phagocytes** (see figure 3.22). The particles can be microorganisms or their parts, foreign substances, or dead cells from the body. The most important phagocytic cells are neutrophils and macrophages.

Neutrophils

Neutrophils are small, phagocytic cells produced in large numbers in red bone marrow that are released into the blood, where they

TABLE 22.2 Immune System Cells and Their Primary Functions

Cell	Primary Function	Cell	Primary Function
Innate Immunity		**Adaptive Immunity**	
Neutrophil	Phagocytosis and inflammation; usually the first cell to leave the blood and enter infected tissues	B cell	After activation, differentiates to become plasma cell or memory B cell
Monocyte	Leaves the blood and enters tissues to become a macrophage	Plasma cell	Produces antibodies that are directly or indirectly responsible for the destruction of the antigen
Macrophage	Most effective phagocyte; important in later stages of infection and in tissue repair; located throughout the body to "intercept" foreign substances; processes antigens; involved in the activation of B and T cells	Memory B cell	Quick and effective response to an antigen against which the immune system has previously reacted; responsible for immunity
		Cytotoxic T cell	Responsible for the destruction of cells by lysis or by the production of cytokines
Basophil	Motile cell that leaves the blood, enters tissues, and releases chemicals that promote inflammation	Delayed hypersensitivity T cell	Produces cytokines that promote inflammation
		Helper T cell	Activates B and effector T cells
Mast cell	Nonmotile cell in connective tissues that promotes inflammation through the release of chemicals	Suppressor T cell	Inhibits B and effector T cells
		Memory T cell	Quick and effective response to an antigen against which the immune system has previously reacted; responsible for adaptive immunity
Eosinophil	Enters tissues from the blood and releases chemicals that inhibit inflammation	Dendritic cell	Processes antigen and is involved in the activation of B and T cells
Natural killer cell	Lyses tumor and virus-infected cells		

circulate for a few hours. Approximately 126 billion neutrophils per day leave the blood and pass through the wall of the gastrointestinal tract, where they provide phagocytic protection. The neutrophils are then eliminated as part of the feces. Neutrophils are usually the first cells to enter infected tissues in large numbers. They release chemical signals, such as cytokines and chemotactic factors, that increase the inflammatory response by recruiting and activating other immune cells. Neutrophils often die after a single phagocytic event.

Neutrophils also release lysosomal enzymes that kill microorganisms and cause tissue damage and inflammation. **Pus** is an accumulation of dead neutrophils, dead microorganisms, debris from dead tissue, and fluid.

Macrophages

Macrophages are monocytes that leave the blood, enter tissues, enlarge about fivefold, and increase their number of lysosomes and mitochondria. They are large phagocytic cells that outlive neutrophils, and they can ingest more and larger phagocytic particles than neutrophils. Macrophages usually accumulate in tissues after neutrophils and are responsible for most of the phagocytic activity in the late stages of an infection, including the cleanup of dead neutrophils and other cellular debris. In addition to their phagocytic role, macrophages produce a variety of chemicals, such as interferons, prostaglandins, and complement, that enhance the immune response.

Macrophages are beneath the free surfaces of the body, such as the skin (dermis), hypodermis, mucous membranes, and serous membranes, and around blood and lymphatic vessels. In these locations, macrophages provide protection by trapping and destroying microorganisms entering the tissues.

If microbes do gain entry to the blood or lymphatic system, macrophages are waiting within enlarged spaces, called **sinuses,** to phagocytize them. Blood vessels in the spleen, bone marrow, and liver have sinuses, as do lymph nodes. Within the sinuses, reticular cells produce a fine network of reticular fibers that slows the flow of blood or lymph and provides a large surface area for the attachment of macrophages. In addition, macrophages are on the endothelial lining of the sinuses.

Because macrophages on the reticular fibers and endothelial lining of the sinuses were among the first macrophages studied, these cells were referred to as the **reticuloendothelial system.** It is now recognized that macrophages are derived from monocytes and are in locations other than the sinuses. Because monocytes and macrophages have a single, unlobed nucleus, they are now called the **mononuclear phagocytic system.** Sometimes macrophages are given specific names—for instance, dust cells in the lungs, Kupffer cells in the liver, and microglia in the central nervous system.

Basophils, Mast Cells, and Eosinophils

Basophils, which are derived from red bone marrow, are motile white blood cells that can leave the blood and enter infected tissues. **Mast cells,** which are also derived from red bone marrow, are nonmotile cells in connective tissue, especially near capillaries. Like macrophages, mast cells are located at potential points of entry of microorganisms into the body, such as the skin, lungs, gastrointestinal tract, and urogenital tract.

Basophils and mast cells can be activated through innate immunity (e.g., by complement) or through adaptive immunity (see "Effects of Antibodies," p. 808). When activated, they release chemicals—for example, histamine and leukotrienes—that produce an inflammatory response or activate other mechanisms—for example, smooth muscle contraction in the lungs.

Eosinophils are produced in red bone marrow, enter the blood, and within a few minutes enter tissues. Enzymes released by eosinophils break down chemicals released by basophils and mast cells. Thus, as inflammation is initiated, mechanisms are activated that contain and reduce the inflammatory response. This process is similar to the blood clotting system, in which clot prevention and removal mechanisms are activated while the clot is being formed (see chapter 19). In patients with parasitic infections or allergic reactions with much inflammation, eosinophil numbers greatly increase. Eosinophils also secrete enzymes that effectively kill some parasites.

Natural Killer (NK) Cells

Natural killer (NK) cells are a type of lymphocyte produced in red bone marrow, and they account for up to 15% of lymphocytes. NK cells recognize classes of cells, such as tumor cells or virus-infected cells in general, rather than specific tumor cells or cells infected by a specific virus. For this reason and because NK cells do not exhibit a memory response, they are classified as part of innate immunity. NK cells use a variety of methods to kill their target cells, including the release of chemicals that damage plasma membranes, causing the cells to lyse.

23 ***Define chemotactic factor, chemotaxis, and phagocytosis.***

24 ***What are the functions of neutrophils and macrophages? What is pus?***

25 ***What effects are produced by the chemicals released from basophils, mast cells, and eosinophils?***

26 ***Describe the function of NK cells.***

Inflammatory Response

The **inflammatory response** is a complex sequence of events involving many of the chemical mediators and cells of innate immunity. Tissue injury, regardless of the type, can cause inflammation. Trauma, burns, chemicals, or infections can damage tissues, resulting in inflammation. A bacterial infection is used in this section to illustrate an inflammatory response (figure 22.11). The bacteria, or damage to tissues, cause the release or activation of chemical mediators, such as histamine, complement, kinins, and eicosanoids (e.g., prostaglandins and leukotrienes). The chemical mediators produce several effects: (1) vasodilation, which increases blood flow and takes phagocytes and other white blood cells to the area; (2) chemotactic attraction of phagocytes, which leave the blood and enter the tissue; and (3) increased vascular permeability, which allows fibrinogen and complement to enter the tissue from the blood. Fibrinogen is converted to fibrin, which prevents the

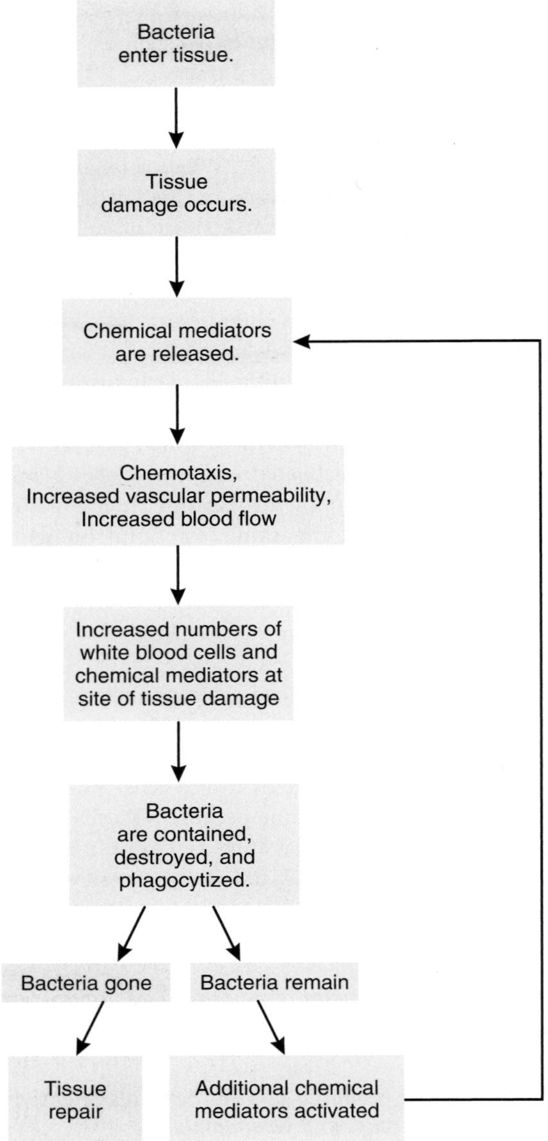

FIGURE 22.11 Inflammatory Response

Bacteria cause tissue damage and the release of chemical mediators, which initiate inflammation and phagocytosis, resulting in the destruction of the bacteria. If any bacteria remain, additional chemical mediators are activated. After all the bacteria have been destroyed, the tissue is repaired.

spread of infection by walling off the infected area. Complement further enhances the inflammatory response and attracts additional phagocytes. The process of releasing chemical mediators and attracting phagocytes and other white blood cells continues until the bacteria are destroyed. Phagocytes, such as neutrophils and macrophages, remove microorganisms and dead tissue, and the damaged tissues are repaired.

Inflammation can be localized or systemic. **Local inflammation** is an inflammatory response confined to a specific area of the body (see chapter 4). Symptoms of local inflammation include redness, heat, swelling, pain, and loss of function. Redness, heat, and swelling result from increased blood flow and increased vascular permeability. Pain is caused by swelling and by chemical mediators acting on pain receptors. Loss of function results from tissue destruction, swelling, and pain.

Systemic inflammation is an inflammatory response that occurs in many parts of the body. In addition to the local symptoms at the sites of inflammation, three additional features can be present. First, red bone marrow produces and releases large numbers of neutrophils, which promote phagocytosis. Second, **pyrogens** (pī′rō-jenz; fire-producing), chemicals released by microorganisms, macrophages, neutrophils, and other cells stimulate fever production. Pyrogens affect the body's temperature-regulating mechanism in the hypothalamus, heat is conserved, and body temperature increases. Fever promotes the activities of the immune system, such as phagocytosis, and inhibits the growth of some microorganisms. Third, in severe cases of systemic inflammation, increased vascular permeability is so widespread that large amounts of fluid are lost from the blood into the tissues. The decreased blood volume can cause shock and death.

27 ***Describe the events that take place during an inflammatory response.***

28 ***What are the symptoms of local and systemic inflammations?***

ADAPTIVE IMMUNITY

Adaptive immunity can recognize, respond to, and remember a particular substance. Substances that stimulate adaptive immunity are called **antigens** (an′ti-jenz). They usually are large molecules with a molecular weight of 10,000 or more. **Haptens** (hap′tenz) are small molecules (of low molecular weight) capable of combining with larger molecules, such as blood proteins, to stimulate an adaptive immune system response.

Antigens are divided into two groups: foreign antigens and self-antigens. **Foreign antigens** are not produced by the body but are introduced from outside it. Components of bacteria, viruses, and other microorganisms are examples of foreign antigens that cause disease. Pollen, animal dander (scaly, dried skin), feces of house dust mites, foods, and drugs are also foreign antigens and can trigger an overreaction of the immune system in some people, called an **allergic reaction.** Transplanted tissues and organs that contain foreign antigens result in the rejection of the transplant. **Self-antigens** are molecules produced by the body that stimulate an adaptive immune system response. The response to self-antigens can be beneficial or harmful. For example, the recognition of tumor antigens can result in tumor destruction, whereas **autoimmune disease** can result when self-antigens stimulate unwanted tissue destruction.

TABLE 22.3 Comparison of Innate and Adaptive Immunity

		Adaptive Immunity	
Characteristics	**Innate Immunity**	**Antibody-Mediated Immunity**	**Cell-Mediated Immunity**
Primary cells	Neutrophils, eosinophils, basophils, mast cells, monocytes, and macrophages	B cells	T cells
Origin of cells	Red bone marrow	Red bone marrow	Red bone marrow
Site of maturation	Red bone marrow (neutrophils, eosinophils, basophils, and monocytes) and tissues (mast cells and macrophages)	Red bone marrow	Thymus
Location of mature cells	Blood, connective tissue, and lymphatic tissue	Blood and lymphatic tissue	Blood and lymphatic tissue
Primary secretory products	Histamine, kinins, complement, prostaglandins, leukotrienes, and interferons	Antibodies	Cytokines
Primary actions	Inflammatory response and phagocytosis	Protection against extracellular antigens (bacteria, toxins, parasites, and viruses outside cells)	Protection against intracellular antigens (viruses, intracellular bacteria, and intracellular fungi) and tumors: regulates antibody-mediated immunity and cell-mediated immunity responses (helper T and suppressor T cells)
Hypersensitivity reactions	None	Immediate hypersensitivity (atopy, anaphylaxis, cytotoxic reactions, and immune complex disease)	Delayed hypersensitivity (allergy of infection and contact hypersensitivity)

Allergic Reactions to Penicillin

Penicillin is a hapten of clinical importance. It is a small molecule that does not evoke an immune system response. Penicillin can, however, break down and bind to serum proteins to form a combined molecule that can produce an allergic reaction. Most commonly, the reaction produces a rash and fever, but rarely a severe reaction causes death.

Adaptive immunity can be divided into antibody-mediated immunity and cell-mediated immunity. **Antibody-mediated immunity** involves proteins called **antibodies,** which are found in fluids outside of cells, such as blood, interstitial fluid, and lymph. **B cells** give rise to cells that produce antibodies. **Cell-mediated immunity** involves **T cells.** Several subpopulations of T cells exist, each of which is responsible for a particular aspect of cell-mediated immunity. For example, **effector T cells,** such as **cytotoxic T cells** and **delayed hypersensitivity T cells,** are responsible for producing the effects of cell-mediated immunity. **Regulatory T cells,** such as **helper T cells** and **suppressor T cells,** can promote or inhibit the activities of both antibody-mediated immunity and cell-mediated immunity.

Table 22.3 summarizes and contrasts the main features of innate and adaptive immunity.

29 ***Define antigen and hapten. Distinguish between a foreign antigen and a self-antigen.***

30 ***What are allergic reactions and autoimmune diseases?***

Origin and Development of Lymphocytes

All blood cells, including lymphocytes, are derived from stem cells in the red bone marrow (see chapter 19). The process of blood cell formation begins during embryonic development and continues throughout life. Some stem cells give rise to pre-T cells, which migrate through the blood to the thymus, where they divide and are processed into T cells (see figures 22.9; figure 22.12). The thymus produces hormones, such as thymosin, which stimulates T-cell maturation. Other stem cells produce pre-B cells, which are processed in the red bone marrow into B cells. A **positive selection** process results in the survival of pre-B and pre-T cells that are capable of an immune response. Cells that are incapable of an immune response die.

The B and T cells that can respond to antigens are composed of small groups of identical lymphocytes called **clones.** Although each clone can respond only to a particular antigen, such a large number of clones exist that the immune system can react to most molecules. Some of the clones can also respond to self-antigens. A **negative selection** process eliminates or suppresses clones acting against self-antigens, thereby preventing the destruction of one's own cells. Although the negative selection process occurs mostly during prenatal development, it continues throughout life (see "Inhibition of Lymphocytes," p. 805).

B cells are released from red bone marrow, T cells are released from the thymus, and both types of cells move through the blood to lymphatic tissue. There are approximately five T cells for every B cell in the blood. These lymphocytes live for a few months to many

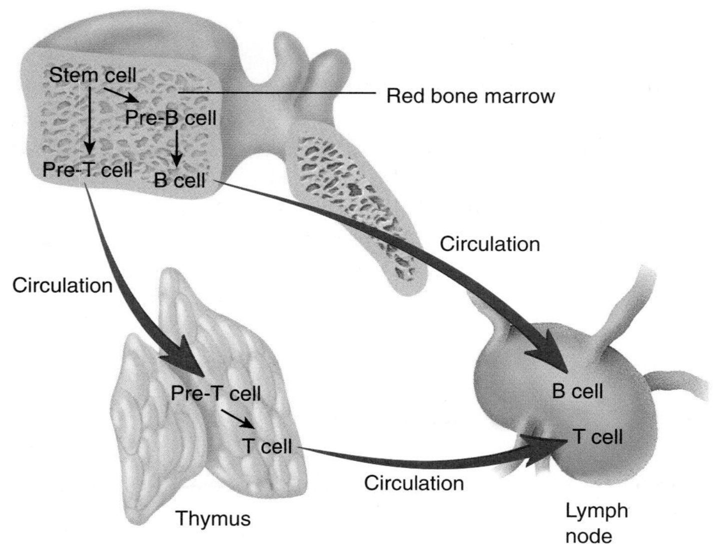

FIGURE 22.12 Origin and Processing of B and T Cells
Pre-B and pre-T cells originate from stem cells in red bone marrow. The pre-B cells remain in the red bone marrow and become B cells. The pre-T cells circulate to the thymus where they become T cells. Both B and T cells circulate to other lymphatic tissues, such as lymph nodes, where they can divide and increase in number in response to antigens.

years and continually circulate between the blood and the lymphatic tissues. Antigens can come into contact with and activate lymphocytes, resulting in cell divisions that increase the number of lymphocytes that can recognize the antigen. These lymphocytes can circulate in blood and lymph to reach antigens in tissues throughout the body.

In the **primary lymphatic organs,** lymphocytes mature into functional cells. These organs are the red bone marrow and thymus. In the **secondary lymphatic organs and tissues,** lymphocytes interact with each other, antigen-presenting cells, and antigens to produce an immune response. The secondary lymphatic organs and tissues include the diffuse lymphatic tissue, lymphatic nodules, tonsils, lymph nodes, and spleen.

31 ***Describe the origin and development of B and T cells.***

32 ***What are lymphocyte clones? Distinguish between positive and negative lymphocyte selection.***

33 ***What are primary and secondary lymphatic organs and tissues?***

Activation of Lymphocytes

Antigens activate lymphocytes in different ways, depending on the type of lymphocyte and the type of antigen involved. Despite these differences, however, two general principles of lymphocyte activation exist: (1) Lymphocytes must be able to recognize the antigen and, (2) after recognition, the lymphocytes must increase in number to destroy the antigen.

Antigenic Determinants and Antigen Receptors

If an adaptive immune system response is to occur, lymphocytes must recognize an antigen. Lymphocytes do not interact with an entire antigen, however. Instead, **antigenic determinants,** or **epitopes** (ep′i-tōps), are specific regions of a given antigen recognized by a lymphocyte, and each antigen has many different antigenic determinants (figure 22.13). All the lymphocytes of a given clone have, on their surfaces, identical proteins called **antigen receptors,** which combine with a specific antigenic determinant. The immune system response to an antigen with a particular antigenic determinant is similar to the lock-and-key model for enzymes (see chapter 2), and any given antigenic determinant can combine only with a specific antigen receptor. The **T-cell receptor** consists of two polypeptide chains, which are subdivided into a variable and a constant region (figure 22.14). The variable region can bind to an antigen. The many different types of T-cell

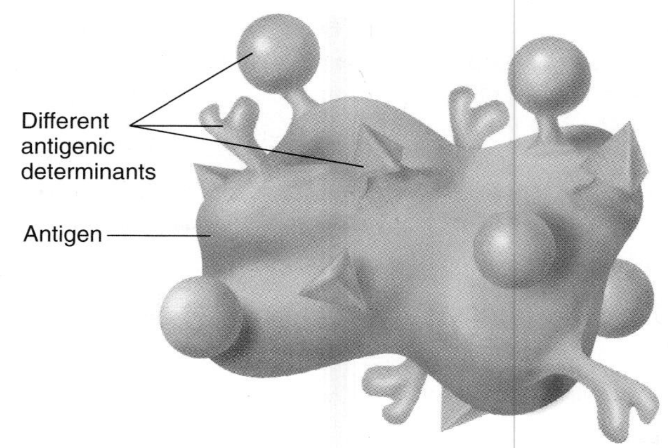

FIGURE 22.13 Antigenic Determinants
An antigen has many antigenic determinants to which lymphocytes can respond.

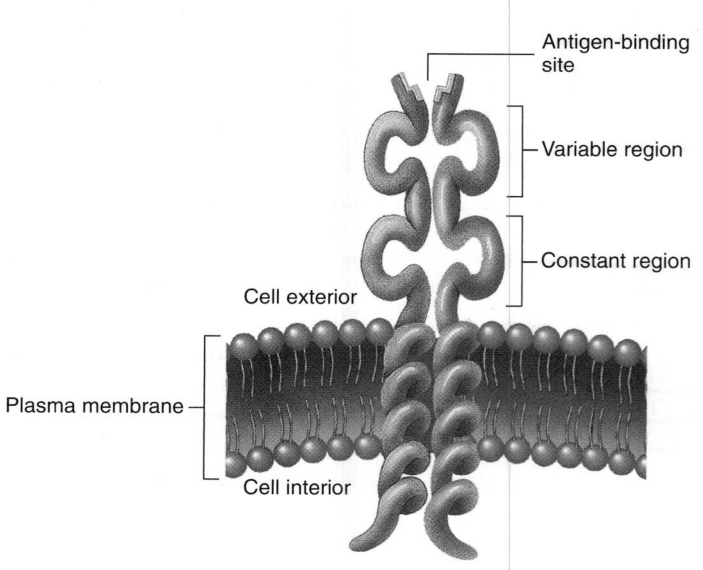

FIGURE 22.14 T-Cell Receptor
A T-cell receptor consists of two polypeptide chains. The variable region of each type of T-cell receptor is specific for a given antigen. The constant region attaches the T-cell receptor to the plasma membrane.

receptors respond to different antigens because they have different variable regions. The **B-cell receptor** consists of four polypeptide chains with two identical variable regions. It is a type of antibody (see p. 808).

Major Histocompatibility Complex Molecules

Although some antigens bind to their receptors and directly activate B cells and some T cells, most lymphocyte activation involves glycoproteins on the surfaces of cells called **major histocompatibility complex (MHC) molecules.** MHC molecules are attached to plasma membranes, and they have a variable region that can bind to foreign and self-antigens.

MHC class I molecules are found on nucleated cells; they display antigens produced inside the cells on their surfaces (figure 22.15*a*). This is necessary because the immune system cannot directly respond to an antigen inside a cell. For example, viruses reproduce inside cells, forming viral proteins that are foreign antigens. Some of these viral proteins are broken down in the cytoplasm. The protein fragments enter the rough endoplasmic reticulum and combine with MHC class I molecules to form

1. Foreign proteins or self-proteins within the cytosol are broken down into fragments that are antigens.
2. Antigens are transported into the rough endoplasmic reticulum.
3. Antigens combine with MHC class I molecules.
4. The MHC class I/antigen complex is transported to the Golgi apparatus, packaged into a vesicle, and transported to the plasma membrane.
5. Foreign antigens combined with MHC class I molecules stimulate cell destruction.
6. Self-antigens combined with MHC class I molecules do not stimulate cell destruction.

1. A foreign antigen is ingested by endocytosis and is within a vesicle.
2. The antigen is broken down into fragments to form processed foreign antigens.
3. The vesicle containing the processed foreign antigens fuses with vesicles produced by the Golgi apparatus that contain MHC class II molecules. Processed foreign antigens and MHC class II molecules combine.
4. The MHC class II/antigen complex is transported to the plasma membrane.
5. The displayed MHC class II/antigen complex can stimulate immune cells.

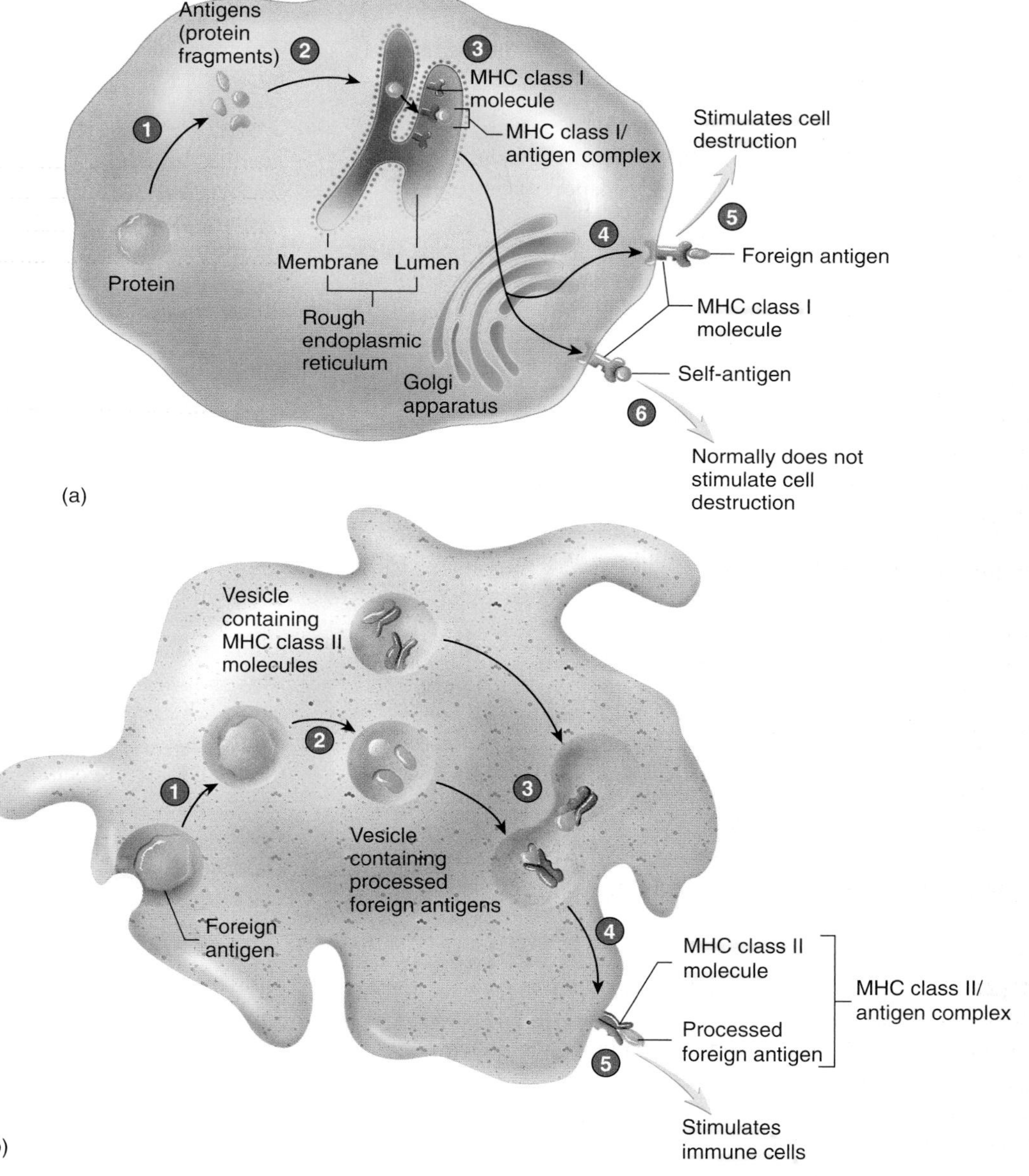

PROCESS FIGURE 22.15 Antigen Processing

(*a*) Foreign proteins, such as viral proteins, or self-proteins in the cytosol are processed and presented at the cell surface by MHC class I molecules. (*b*) Foreign antigens are taken into an antigen-presenting cell, processed, and presented at the cell surface by MHC class II molecules.

complexes that move through the Golgi apparatus to be distributed on the surface of the cell (see chapter 3).

MHC class I/antigen complexes on the surface of cells can bind to T-cell receptors on the surface of T cells. This combination is a signal that activates T cells. Activated T cells can destroy infected cells, which effectively stops viral replication (see "Cell-Mediated Immunity," p. 812). Thus, the MHC class I/antigen complex functions as a signal, or "red flag," that prompts the immune system to destroy the displaying cell. In essence, the cell is displaying a sign that says, "Kill me!" This process is said to be **MHC-restricted** because both the antigen and the individual organism's own MHC molecule are required.

PREDICT 2

In mouse A, T cells can respond to virus X. If these T cells are transferred to mouse B, which is infected with virus X, will the T cells respond to the virus? Explain.

The same process that moves foreign protein fragments to the surface of cells can also inadvertently transport self-protein fragments (see figure 22.15*a*). As part of normal protein metabolism, cells continually break down old proteins and synthesize new ones. Some self-protein fragments that result from protein breakdown can combine with MHC class I molecules and be displayed on the surface of the cell, thus becoming self-antigens. Normally, the immune system does not respond to self-antigens in combination with MHC molecules because the lymphocytes that could respond have been eliminated or inactivated (see "Inhibition of Lymphocytes," p. 805).

MHC class II molecules are found on **antigen-presenting cells,** which include B cells, macrophages, monocytes, and dendritic cells. **Dendritic** (den-drit′ik) **cells** are large, motile cells with long cytoplasmic extensions, and they are scattered throughout most tissues (except the brain), with their highest concentrations in lymphatic tissues and the skin. Dendritic cells in the skin are often called **Langerhans cells.**

Antigen-presenting cells can take in foreign antigens by endocytosis (figure 22.15*b*). Within the endocytotic vesicle, the antigen is broken down into fragments to form processed antigens. Vesicles from the Golgi apparatus containing MHC class II molecules combine with the endocytotic vesicles. The MHC class II molecules and processed antigens combine, and the MHC class II/antigen complexes are transported to the surface of the cell, where they are displayed to other immune cells.

MHC class II/antigen complexes on the surface of cells can bind to T-cell receptors on the surface of T cells. The presentation of antigen using MHC class II molecules is MHC-restricted because both the antigen and the individual's own MHC class II molecule are required. Unlike MHC class I molecules, however, this display does not result in the destruction of the antigen-presenting cell. Instead, the MHC class II/antigen complex is a "rally around the flag" signal that stimulates other immune system cells to respond to the antigen. The displaying cell is like Paul Revere, who spread the alarm for the militia to arm and organize. The militia then went out and killed the enemy. For example, when the lymphocytes of the B-cell clone that can recognize the antigen come into contact with the MHC class II/antigen complex, they are stimulated to divide. The activities of these lymphocytes, such as the production of antibodies, then result in the destruction of the antigen.

34 ***Define antigenic determinant and antigen receptor. How are they related to each other?***

35 ***What type of antigens are displayed by MHC class I and II molecules?***

36 ***What type of cells display MHC class I and II antigen complexes, and what happens as a result?***

37 ***Define MHC-restricted.***

PREDICT 3

How does elimination of an antigen stop the production of antibodies?

Costimulation

The combination of an MHC class II/antigen complex with an antigen receptor is usually only the first signal necessary to produce a response from a B or T cell. In many cases, **costimulation** by additional signals is also required. Costimulation is accomplished by cytokines released from cells and by molecules attached to the surface of cells (figure 22.16*a*). Cytokines produced by lymphocytes are often called **lymphokines** (lim′fō-kīnz). Table 22.4 lists important cytokines and their functions.

Certain pairs of surface molecules can also be involved in co stimulation (figure 22.16*b*). When the surface molecule on one cell combines with the surface molecule on another, the combination can act as a signal that stimulates a response from one of the cells, or the combination can hold the cells together. Typically, several kinds of surface molecules are necessary to produce a response. For example, a molecule called B7 on macrophages must bind with a molecule called CD28 on helper T cells before the helper T cells can respond to the antigen presented by the macrophage. In addition, helper T cells have a glycoprotein called CD4, which helps connect helper T cells to the macrophage by binding to MHC class II molecules. For this reason, helper T cells are sometimes referred to as **CD4,** or **T4, cells.** In a similar fashion, cytotoxic T cells are sometimes called **CD8,** or **T8, cells** because they have a glycoprotein called CD8, which helps connect cytotoxic T cells to cells displaying MHC class I molecules. The CD designation stands for "cluster of differentiation," which is a system used to classify many surface molecules.

38 ***What is costimulation? State two ways in which it can happen.***

39 ***Why are helper T cells sometimes called CD4 or T4 cells? Why are cytotoxic T cells sometimes called CD8 or T8 cells?***

Lymphocyte Proliferation

Before exposure to an antigen, the number of lymphocytes in a clone is too small to produce an effective response against the antigen. Exposure to an antigen results in an increase in lymphocyte number. First, there is an increase in the number of helper T cells.

(a) Helper T cells are activated by a first signal and by costimulation. The first signal is the binding of the MHC class II/antigen complex to the T-cell receptor. Costimulation is an additional signal, such as molecules released from another cell. For example, macrophages release cytokines that bind to receptors on helper T cells, resulting in costimulation.

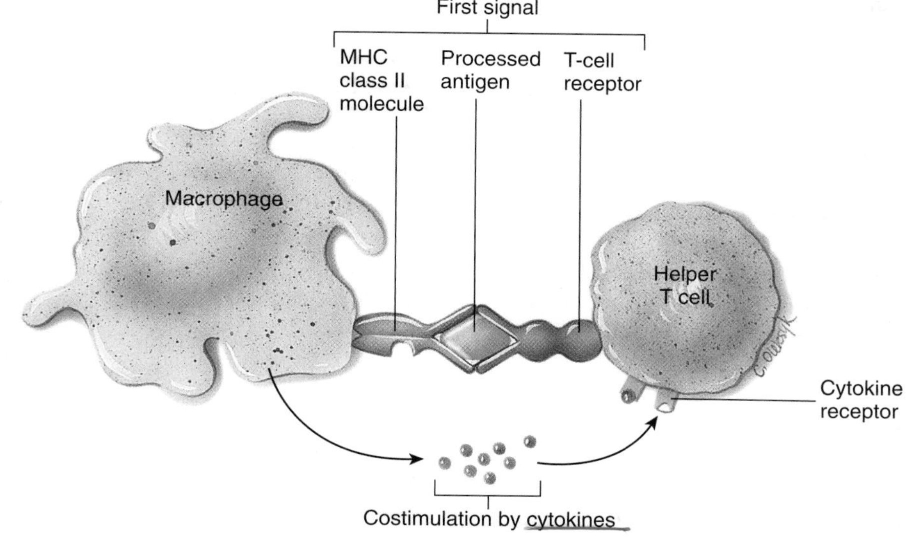

(b) Other costimulatory signals are the combining of surface molecules between cells, such as the binding of a B7 molecule of the macrophage with a CD28 molecule of the helper T cell. The CD4 molecule of the helper T cell binds to the macrophage's MHC class II molecule and helps hold the cells together.

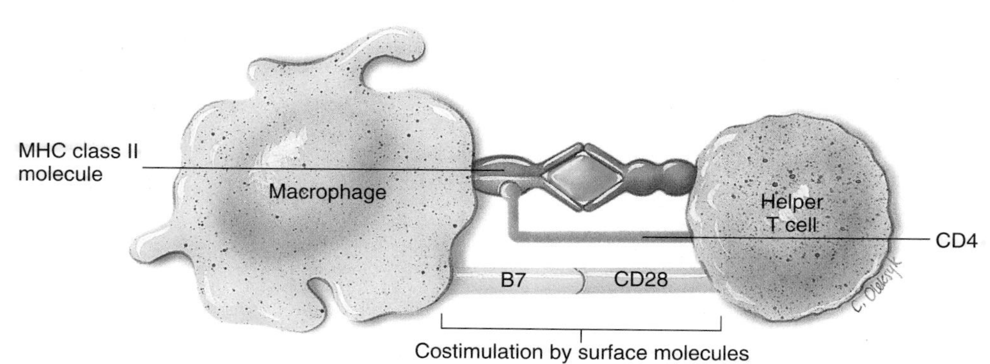

FIGURE 22.16 Costimulation

TABLE 22.4 Cytokines and Their Functions

Cytokine*	Description
Interferon alpha (IFNα)	Prevents viral replication and inhibits cell growth; secreted by virus-infected cells
Interferon beta (IFNβ)	Prevents viral replication, inhibits cell growth, and decreases the expression of major histocompatibility complex (MHC) class I and II molecules; secreted by virus-infected fibroblasts
Interferon gamma (IFNγ)	About 20 different proteins that activate macrophages and natural killer (NK) cells, stimulate adaptive immunity by increasing the expression of MHC class I and II molecules, and prevent viral replication; secreted by helper T, cytotoxic T, and NK cells
Interleukin-1 (IL-1)	Costimulation of B and T cells, promotes inflammation through prostaglandin production, and induces fever acting through the hypothalamus (pyrogen); secreted by macrophages, B cells, and fibroblasts
Interleukin-2 (IL-2)	Costimulation of B and T cells, activation of macrophages and NK cells; secreted by helper T cells
Interleukin-4 (IL-4)	Plays a role in allergic reactions by activation of B cells, resulting in the production of immunoglobulin E (IgE); secreted by helper T cells
Interleukin-5 (IL-5)	Part of the response against parasites by stimulating eosinophil production; secreted by helper T cells
Interleukin-8 (IL-8)	Chemotactic factor that promotes inflammation by attracting neutrophils and basophils; secreted by macrophages
Interleukin-10(IL-10)	Inhibits the secretion of interferon gamma and interleukins; secreted by suppressor T cells
Interleukin-15(IL-15)	Promotes inflammation, activates memory T cells, and natural killer cells
Lymphotoxin	Kills target cells; secreted by cytotoxic T cells
Perforin	Makes a hole in the membrane of target cells, resulting in lysis of the cell; secreted by cytotoxic T cells
Tumor necrosis factor α (TNFα)	Activates macrophages and promotes fever (pyrogen); secreted by macrophages

*Some cytokines were named according to the laboratory test first used to identify them; however, these names rarely are a good description of the actual function of the cytokine.

This is important because the increased number of helper T cells responding to the antigen can find and stimulate B or effector T cells. Second, the number of B or effector T cells increases. This is important because these cells are responsible for the immune response that destroys the antigen.

1. *Proliferation of helper T cells* (figure 22.17). Antigen-presenting cells use MHC class II molecules to present processed antigens to helper T cells. Only the helper T cells with the T-cell receptors that can bind to the antigen respond. These helper T cells respond to the MHC class II/antigen complex and costimulation by dividing. As a result, the number of helper T cells that recognize the antigen increases.
2. *Proliferation and activation of B or effector T cells.* Typically, the proliferation and activation of B or effector T cells involve helper T cells. This process is illustrated in figure 22.18 for

1. Antigen-presenting cells, such as macrophages, take in, process, and display antigens on the cell's surface.
2. The antigen is bound to an MHC class II molecule, which functions to present the processed antigen to a T-cell receptor of a helper T cell for recognition.
3. Costimulation occurs by a CD4 glycoprotein of the helper T cell or by cytokines. The macrophage secretes a cytokine called interleukin-1.
4. Interleukin-1 attaches to interleukin receptors and stimulates the helper T cell to secrete the cytokine interleukin-2 and to produce interleukin-2 receptors.
5. The helper T cell stimulates itself to divide when interleukin-2 binds to interleukin-2 receptors.
6. The "daughter" helper T cells resulting from this division can be stimulated to divide again if they are exposed to the same antigen that stimulated the "parent" helper T cell. This greatly increases the number of helper T cells.
7. The increased number of helper T cells can facilitate the activation of B cells or effector T cells.

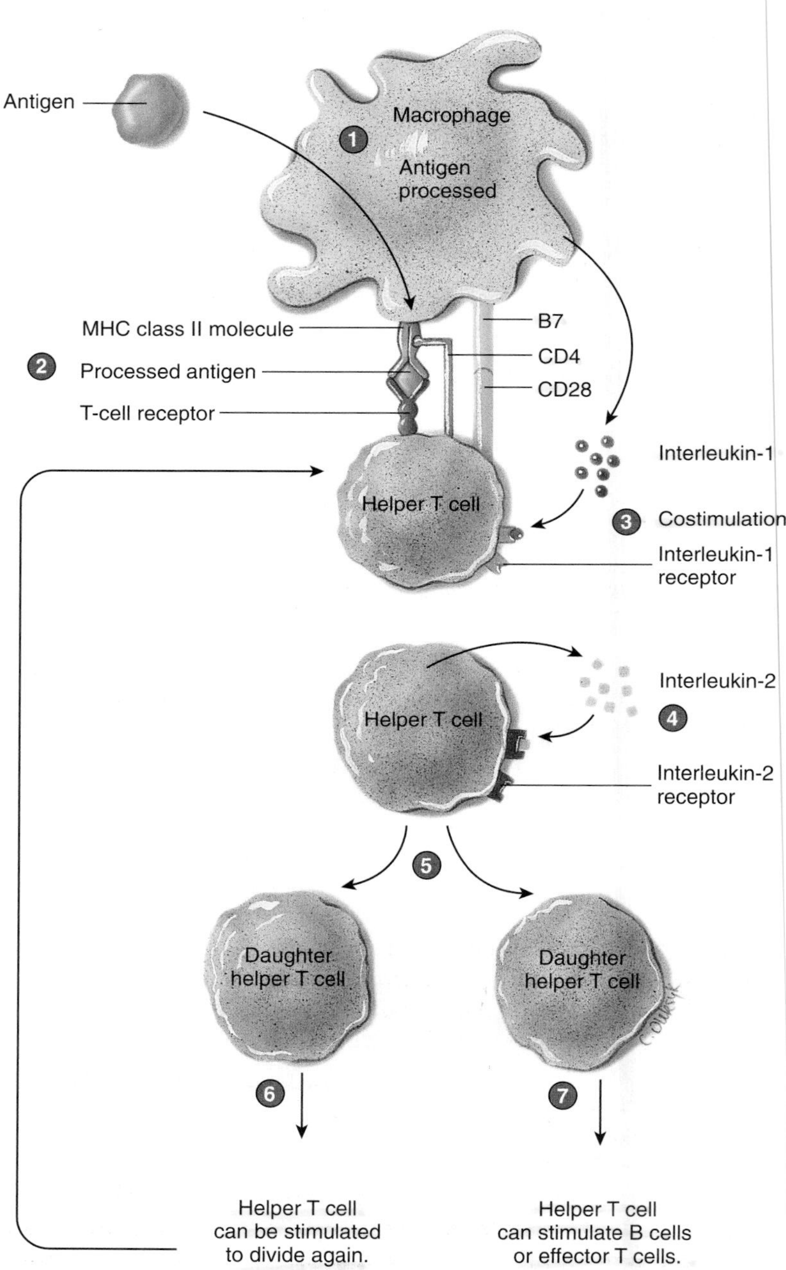

PROCESS FIGURE 22.17 Proliferation of Helper T Cells

An antigen-presenting cell (macrophage) stimulates helper T cells to divide and produce cytokines.

TABLE 22.5 Classes of Antibodies and Their Functions

Antibody	Total Serum Antibody(%)	Structure	Description
IgG	80–85	IgG	Activates complement and functions as an opsonin to increase phagocytosis; can cross the placenta and provide immune protection to the fetus and newborn; responsible for Rh reactions, such as hemolytic disease of the newborn
IgM	5–10	IgM	Activates complement and acts as an antigen-binding receptor on the surface of B cells; responsible for transfusion reactions in the ABO blood system; often the first antibody produced in response to an antigen
IgA	15	IgA	Secreted into saliva, into tears, and onto mucous membranes to provide protection on body surfaces; found in colostrum and milk to provide immune protection to newborns
IgE	0.002	IgE	Binds to mast cells and basophils and stimulates the inflammatory response
IgD	0.2	IgD	Functions as antigen-binding receptors on B cells

Heavy chain

Light chain

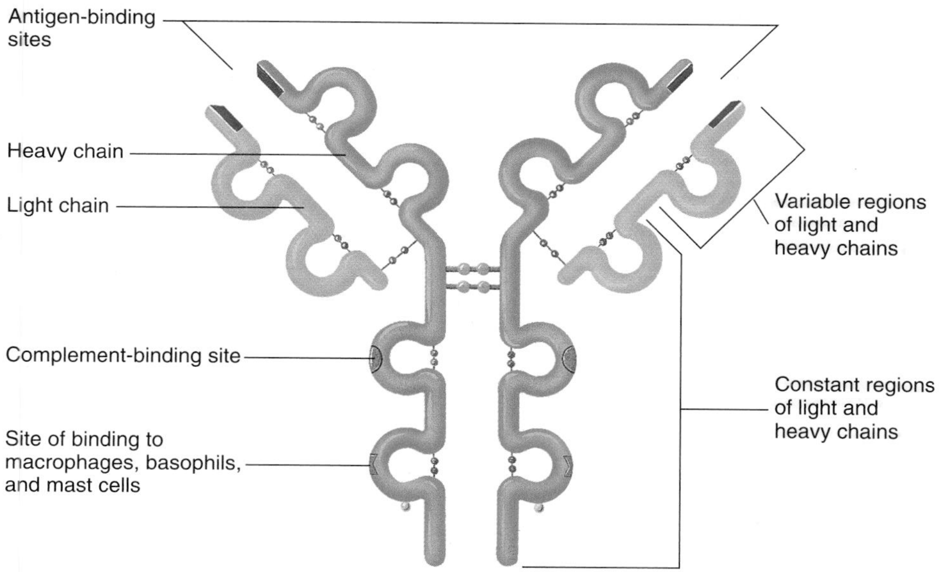

FIGURE 22.19 Structure of an Antibody

Antibodies consist of two heavy and two light polypeptide chains. The variable region of the antibody binds to the antigen. The constant region of the antibody can activate the classical pathway of the complement cascade. The constant region can also attach the antibody to the plasma membrane of cells such as macrophages, basophils, or mast cells.

antigens, rendering the antigens ineffective (figure 22.20*b*). The ability of antibodies to join antigens together is the basis for many clinical tests, such as blood typing, because, when enough antigens are bound together, they become visible as a clump or a precipitate.

Although antibodies can directly affect antigens, most of their effectiveness results from other mechanisms. When an antibody (IgG or IgM) combines with an antigen through the variable region, the constant region can activate the complement cascade through the **classical pathway** (see figure 22.10; figure 22.20*c*). Activated complement stimulates inflammation; attracts neutrophils, monocytes, macrophages, and eosinophils to sites of infection; and kills bacteria by lysis.

Antibodies (IgE) can initiate an inflammatory response (figure 22.20*d*). The antibodies attach to mast cells or basophils through their constant region. When antigens combine with the variable region of the antibodies, the mast cells or basophils release chemicals through exocytosis, and inflammation results.

Opsonins (op′sŏ-ninz) are substances that make an antigen more susceptible to phagocytosis. An antibody (IgG) acts as an opsonin by connecting to an antigen through the variable region of the antibody and to a macrophage through the constant region of the antibody. The macrophage then phagocytizes the antigen and the antibody (figure 22.20*e*).

(a) **Inactivates the antigen.** An antibody binds to an antigen and inactivates it.

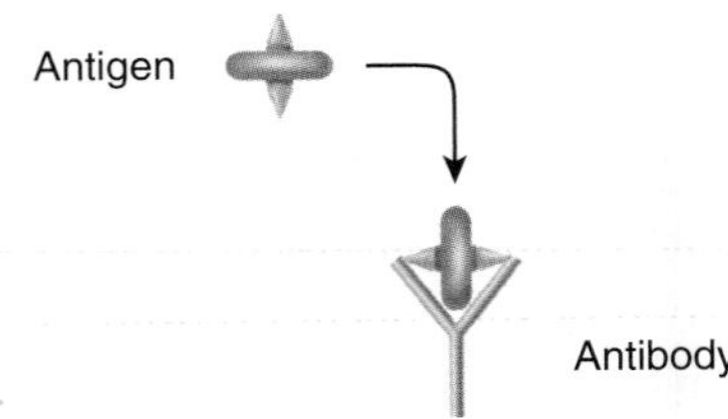

(b) **Binds antigens together.** Antibodies bind several antigens together.

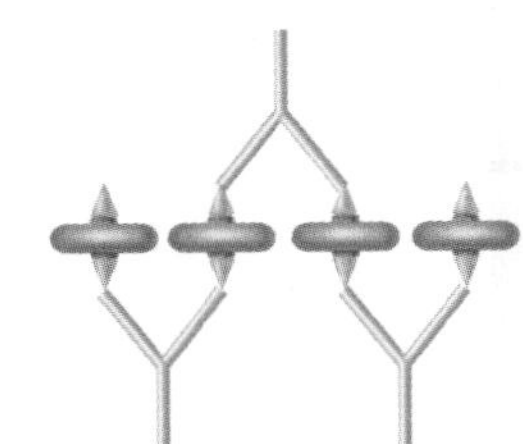

(c) **Activates the complement cascade.** An antigen binds to an antibody. As a result, the antibody can activate complement proteins, which can produce inflammation, chemotaxis, and lysis.

(d) **Initiates the release of inflammatory chemicals.** An antibody binds to a mast cell or basophil. When an antigen binds to the antibody, it triggers a release of chemicals that cause inflammation.

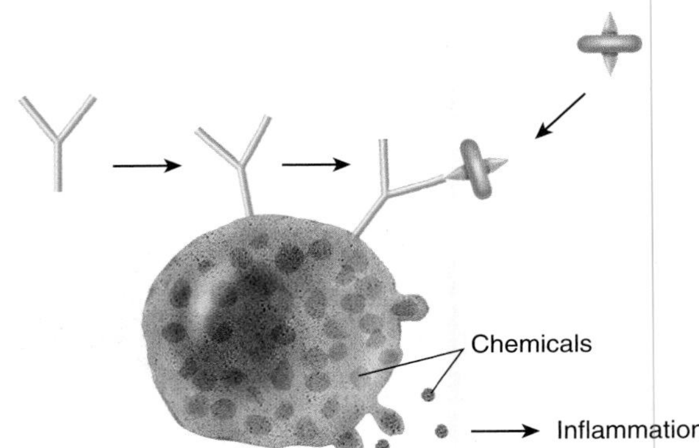

(e) **Facilitates phagocytosis.** An antibody binds to an antigen and then to a macrophage, which phagocytizes the antibody and antigen.

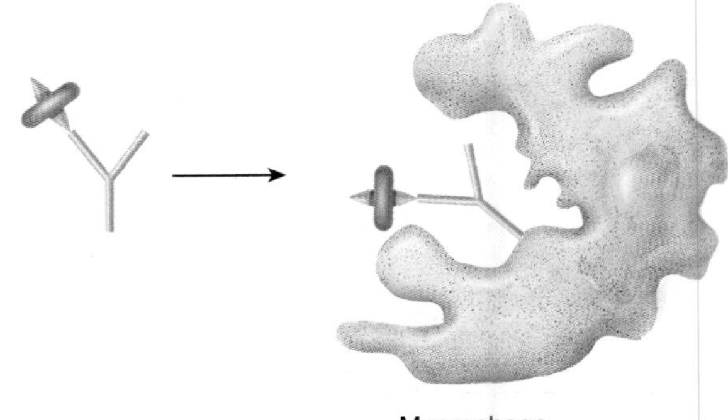

FIGURE 22.20 **Effects of Antibodies**

Antibodies directly affect antigens by inactivating the antigens or binding the antigens together. Antibodies indirectly affect antigens by activating other mechanisms through the constant region of the antibody. Indirect mechanisms include activation of complement, increased inflammation resulting from the release of inflammatory chemicals from mast cells or basophils, and increased phagocytosis resulting from antibody attachment to macrophages.

Antibody Production

The production of antibodies after the first exposure to an antigen is different from that after a second or subsequent exposure. The **primary response** results from the first exposure of a B cell to an antigen for which it is specific and includes a series of cell divisions, cell differentiation, and antibody production. The B-cell receptors on the surface of B cells are antibodies, usually IgM and IgD. The receptors have the same variable region as the antibodies that are eventually produced by the B cell. Before stimulation by an antigen, B cells are small lymphocytes. After activation, the B cells undergo a series of divisions to produce large lymphocytes. Some of these enlarged cells become **plasma cells,** which produce antibodies, and others revert back to small lymphocytes and become **memory B cells** (figure 22.21). Usually, IgM is the first antibody produced in response to an antigen, but later other classes of antibodies are produced as well. The primary response normally takes 3–14 days to produce enough antibodies to be effective against the antigen. In the meantime, the individual usually develops disease symptoms because the antigen has had time to cause tissue damage.

The **secondary,** or **memory, response** occurs when the immune system is exposed to an antigen against which it has already produced a primary response. The secondary response results from memory B cells, which rapidly divide to produce plasma cells and large amounts of antibody when exposed to the antigen. The secondary response provides better protection than the primary response for two reasons. First, the time required to

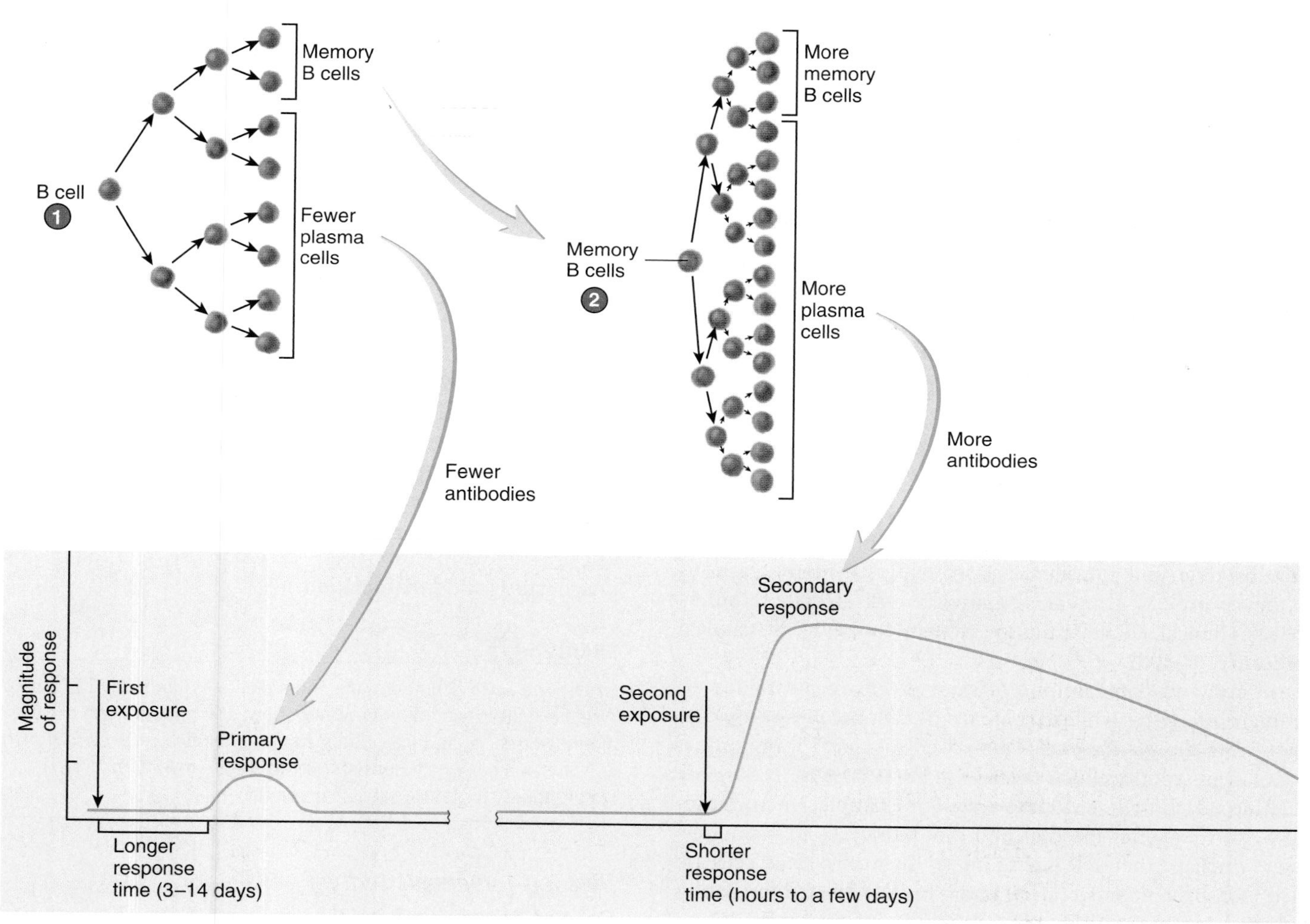

1. **Primary response.** The primary response occurs when a B cell is first activated by an antigen. The B cell proliferates to form plasma cells and memory cells. The plasma cells produce antibodies.

2. **Secondary response.** The secondary response occurs when another exposure to the same antigen causes the memory cells to rapidly form plasma cells and additional memory cells. The secondary response is faster and produces more antibodies than the primary response.

PROCESS FIGURE 22.21 Antibody Production

start producing antibodies is less (hours to a few days); and, second, the amount of antibody produced is much larger. As a consequence, the antigen is quickly destroyed, no disease symptoms develop, and the person is immune.

The memory response also includes the formation of new memory B cells, which provide protection against additional exposures to the antigen. Memory B cells are the basis for adaptive immunity. After destruction of the antigen, plasma cells die, the antibodies they released are degraded, and antibody levels decline to the point at which they can no longer provide adequate protection. Memory B cells may persist for many years and probably for life in some cases. If memory cell production is not stimulated, however, or if the memory B cells produced are short-lived, repeated infections of the same disease are possible. For example, the same cold virus can cause the common cold more than once in the same person.

45 ***Describe the different ways that antibodies participate in the destruction of antigens.***

46 ***What are plasma cells and memory cells, and what are their functions?***

47 ***What are the primary and secondary antibody responses? Why doesn't the primary response prevent illness but the secondary response does?***

PREDICT 4

One theory for long-lasting immunity assumes that humans are continually exposed to the disease-causing agent. Explain how this exposure can produce lifelong immunity.

Cell-Mediated Immunity

Cell-mediated immunity, a function of T cells, is most effective against intracellular microorganisms, such as viruses, fungi, intracellular bacteria, and parasites. Delayed hypersensitivity reactions and the control of tumors also involve cell-mediated immunity (see Clinical Focus "Immune System Problems of Clinical Significance," p. 806).

Antibody-mediated immunity is not effective against intracellular microorganisms while they are inside cells because antibodies cannot cross the plasma membrane. Cell-mediated immunity is effective against intracellular microorganisms because it destroys the cells in which they are sequestered. For example, when viruses infect cells, they enter the cells and direct the cells to make new viruses, which are then released to infect other cells. Thus, cells are turned into virus manufacturing plants. Cell-mediated immunity fights viral infections by destroying virally infected cells. When viruses infect cells, some viral proteins are broken down and become processed foreign antigens that are combined with MHC class I molecules and displayed on the surface of the infected cells (figure 22.22). T cells can distinguish between virally infected cells and noninfected cells because MHC class I/antigen complexes are on the surface of infected cells but not on the surface of uninfected cells. Binding of the T-cell receptor to the MHC class I/antigen complex is a signal for activating cytotoxic T cells. Costimulation by other surface molecules, such as CD8, also occurs. Helper T cells provide costimulation by releasing cytokines, such as interleukin-2, which stimulates activation and cell division of T cells. Unlike their interactions with macrophages and B cells, however, helper T cells do not connect to T cells through MHC class II/antigen complexes or other surface molecules.

When the number of helper T cells increases, it results in greater stimulation of cytotoxic T cells. In cell-mediated responses, helper T cells are activated and stimulated to divide in the same fashion as in antibody-mediated responses (see figure 22.17).

After T cells are activated by antigen on the surface of a target cell, they undergo a series of divisions to produce cytotoxic T cells and memory T cells (figure 22.23). The cytotoxic T cells are responsible for the cell-mediated immunity response. Memory T cells can provide a secondary response and long-lasting immunity in the same fashion as memory B cells.

Cytotoxic T Cells

Cytotoxic T cells have two main effects: They lyse cells and they produce cytokines (see figure 22.23). Cytotoxic T cells can come into contact with other cells and cause them to lyse. Virus-infected cells have viral antigens, tumor cells have tumor antigens, and tissue transplants have foreign antigens on their surfaces that can stimulate cytotoxic T-cell activity. A cytotoxic T cell binds to a target cell and releases chemicals that cause the target cell to lyse. The major method of lysis involves a protein called **perforin,** which is similar to the complement protein C9 (see figure 22.10). Perforin forms a channel in the plasma membrane of the target cell through which water enters the cell, causing lysis. The cytotoxic T cell then moves on to destroy additional target cells.

In addition to lysing cells, cytotoxic T cells release cytokines that activate additional components of the immune system. For example, one important function of cytokines is the recruitment of cells, such as macrophages. These cells are then responsible for phagocytosis and inflammation.

PREDICT 5

In patients with acquired immunodeficiency syndrome (AIDS), helper T cells are destroyed by a viral infection. The patients can die of pneumonia caused by an intracellular fungus (*Pneumocystis carinii*) or from Kaposi sarcoma, which consists of tumorous growths in the skin and lymph nodes. Explain what is happening.

Delayed Hypersensitivity T Cells

Delayed hypersensitivity T cells respond to antigens by releasing cytokines. Consequently, they promote phagocytosis and inflammation, especially in allergic reactions (see Clinical Focus "Immune System Problems of Clinical Significance," p. 806). For example, poison ivy antigens can be processed by Langerhans cells in the skin, which present the antigen to delayed hypersensitivity T cells, resulting in an intense inflammatory response with redness and itching.

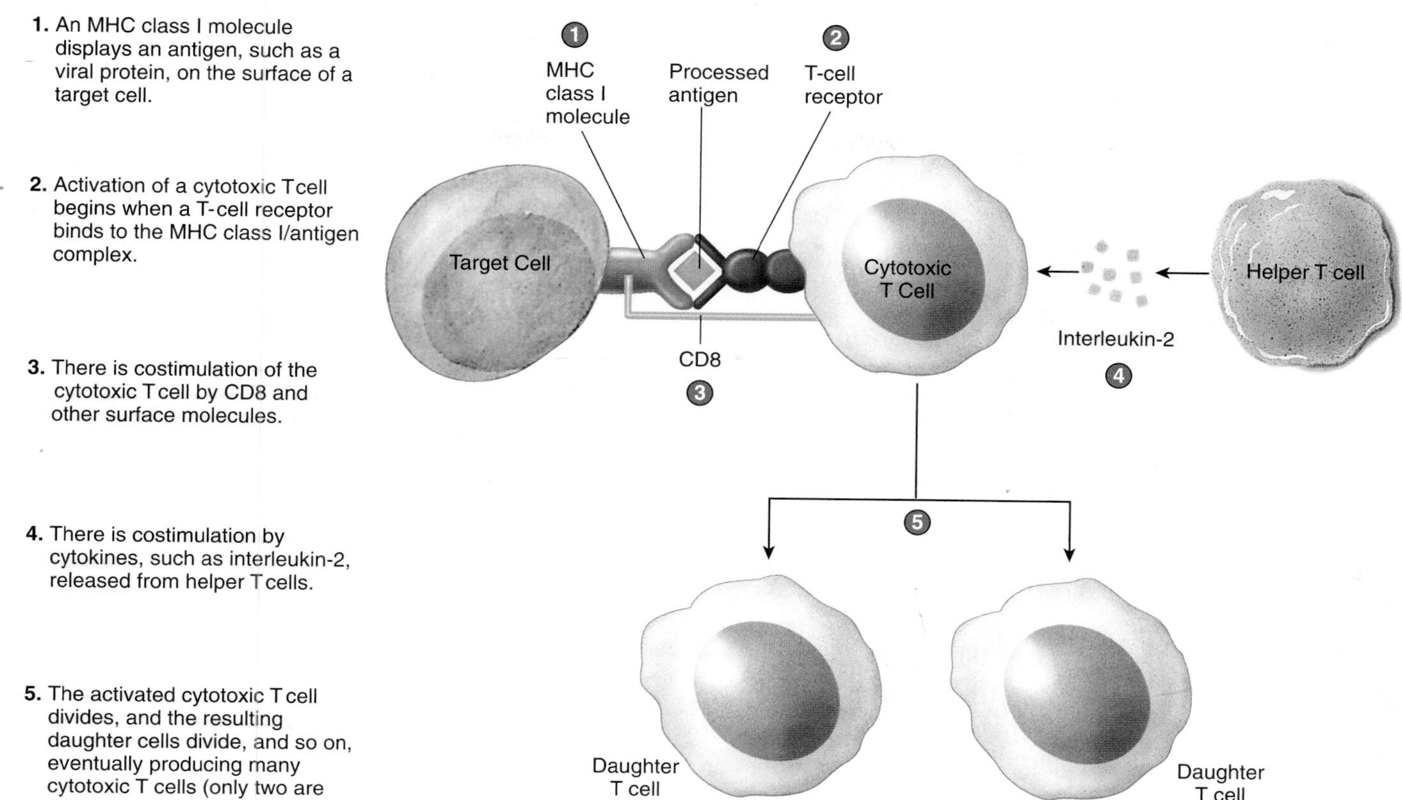

PROCESS FIGURE 22.22 Proliferation of Cytotoxic T Cells

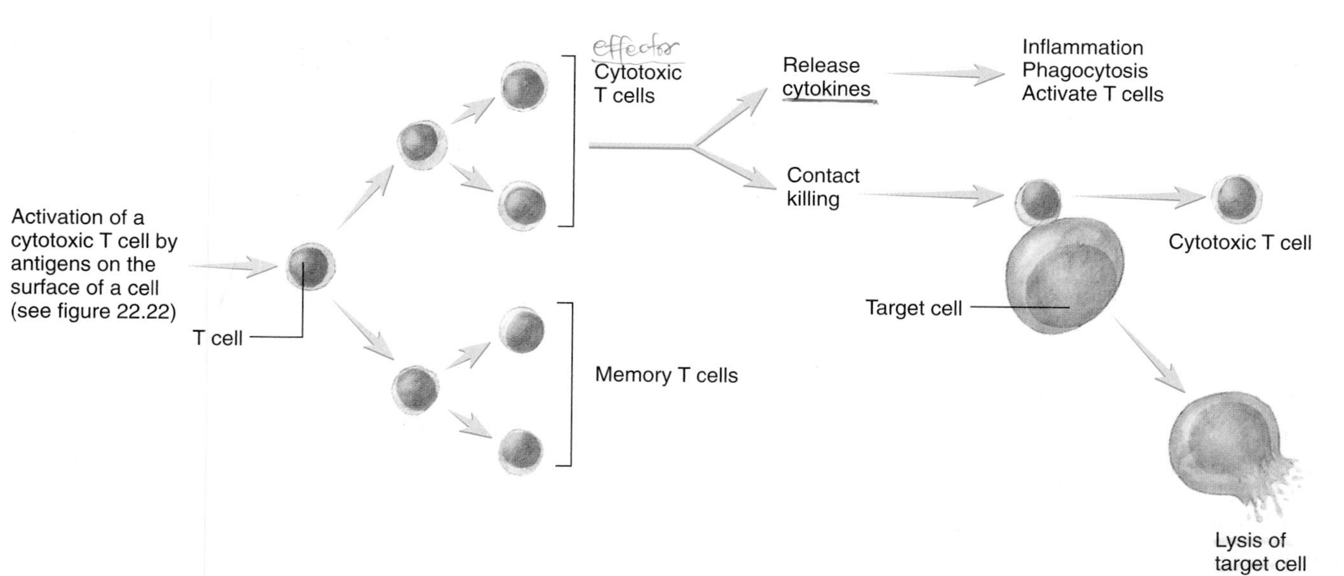

FIGURE 22.23 Stimulation and Effects of T Cells

When activated, cytotoxic T cells form many cytotoxic T cells and memory T cells. The cytotoxic T cells release cytokines that promote the destruction of the antigen or cause the lysis of target cells, such as virus-infected cells, tumor cells, or transplanted cells. The memory T cells are responsible for the secondary response.

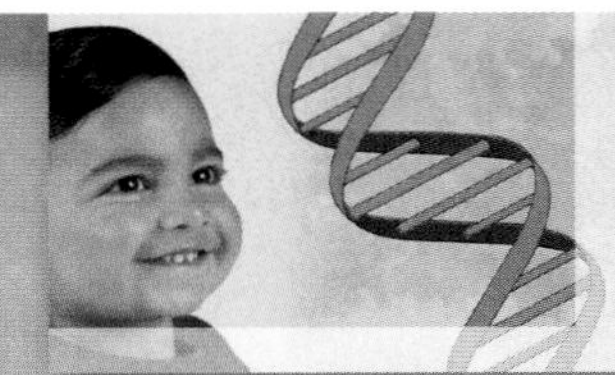

CLINICAL GENETICS

Celiac Disease

Celiac (sē′lē-ăk) **disease,** also called **gluten-sensitive enteropathy,** is a malabsorption disease, meaning nutrients are poorly absorbed. Celiac disease results from damage to the lining of the small intestine. The absorption of nutrients in the small intestine takes place at fingerlike projections, called villi, which increase the surface area for absorption (see figure 24.16). The absorptive surface of the villi is simple columnar epithelium. In a healthy intestine, the lining resembles a shag carpet. In celiac disease, the epithelium is damaged and the villi are flattened and inflamed.

The symptoms of celiac disease usually include gastrointestinal complications, such as diarrhea, painful abdominal cramping, bloating, and intestinal gas. Prolonged celiac disease leads to additional complications, including anemia, osteoporosis, and neurological problems, in part due to nutritional deficiencies. Celiac disease occurs at a frequency of about 1 in 133 people. The frequency may actually be even higher, however, because the widely varying symptoms and severity of the disease make diagnosis difficult.

Celiac disease is an autoimmune disease in which the mechanisms that normally protect the body can work against it. The damage is caused by an inappropriate immune response, which is triggered by the gluten proteins in wheat, barley, and rye. Gluten is not found in rice or corn, although it is often hidden in additives in prepared foods and sauces. The disease is often associated with other autoimmune disorders, such as systemic lupus erythematosus.

Celiac disease has a genetic component. Most patients have mutated variants of some of the MHC class II genes on chromosome 6. As a result, abnormal MHC class II molecules are produced and can bind to digested fragments of gluten. The genetics of celiac disease are complex, however, and not fully understood. For example, the variant MHC alleles alone are not sufficient to cause the disease because they are present in many people who do not develop celiac disease. Furthermore, the genetic expression of celiac disease is influenced by variable environmental factors because the onset and severity of the disease can be triggered by unknown factors at any time in life.

Celiac disease results from both adaptive and innate immune responses. On the surface of antigen-presenting cells, the MHC class II/gluten complex is presented to helper T cells to initiate an adaptive immune response (see figure 22.15*b*). The adaptive immune response includes antibody production and the activation of cytotoxic T cells. The innate immune response promotes inflammation through the activation of the alternative complement pathway and the release of cytokines, such as interleukin-15 (IL-15), from macrophages, dendritic cells, and other cells. Exposure to gluten can activate natural killer cells and possibly dendritic cells. The result of these immune responses is a deleterious attack on the epithelial lining of the intestine, leading to the damaged villi observed in celiac disease.

The only treatment for celiac disease is a strict, gluten-free diet. Before embarking on a life-long gluten-free diet, however, it is important to have a definitive diagnosis. Higher than normal levels of antibodies produced in celiac disease, such as anti-tissue transglutaminase, and a biopsy of the small intestine to examine the villi are recommended. In the future, early genetic diagnosis and manipulation of the immune response may be able to reduce the sensitivity to gluten.

48 ***What type of lymphocyte is responsible for cell-mediated immunity? What are the functions of cell-mediated immunity?***

49 ***How do intracellular microorganisms stimulate cytotoxic T cells? What role do helper T cells play in this process?***

50 ***State the two main responses of cytotoxic T cells.***

51 ***What kind of immune response is produce by delayed hypersensitivity T cells?***

52 ***How is long-lasting immunity achieved in cell-mediated immunity?***

IMMUNE INTERACTIONS

Although the immune system can be described in terms of innate, antibody-mediated, and cell-mediated immunity, only one immune system really exists. These categories are artificial divisions used to emphasize particular aspects of immunity. Actually, immune system responses often involve components of more than one type of immunity (figure 22.24). For example, although adaptive immunity can recognize and remember specific antigens, once recognition has occurred, many of the events that lead to the destruction of the antigen are innate immunity activities, such as inflammation and phagocytosis.

53 ***Describe how interactions among innate, antibody-mediated, and cell-mediated immunity can eliminate an antigen.***

IMMUNOTHERAPY

Knowledge of the basic ways that the immune system operates has produced two fundamental benefits: (1) an understanding of the cause and progression of many diseases and (2) the development or proposed development of effective methods to prevent, stop, or even reverse diseases.

Immunotherapy treats disease by altering immune system function or by directly attacking harmful cells. Some approaches

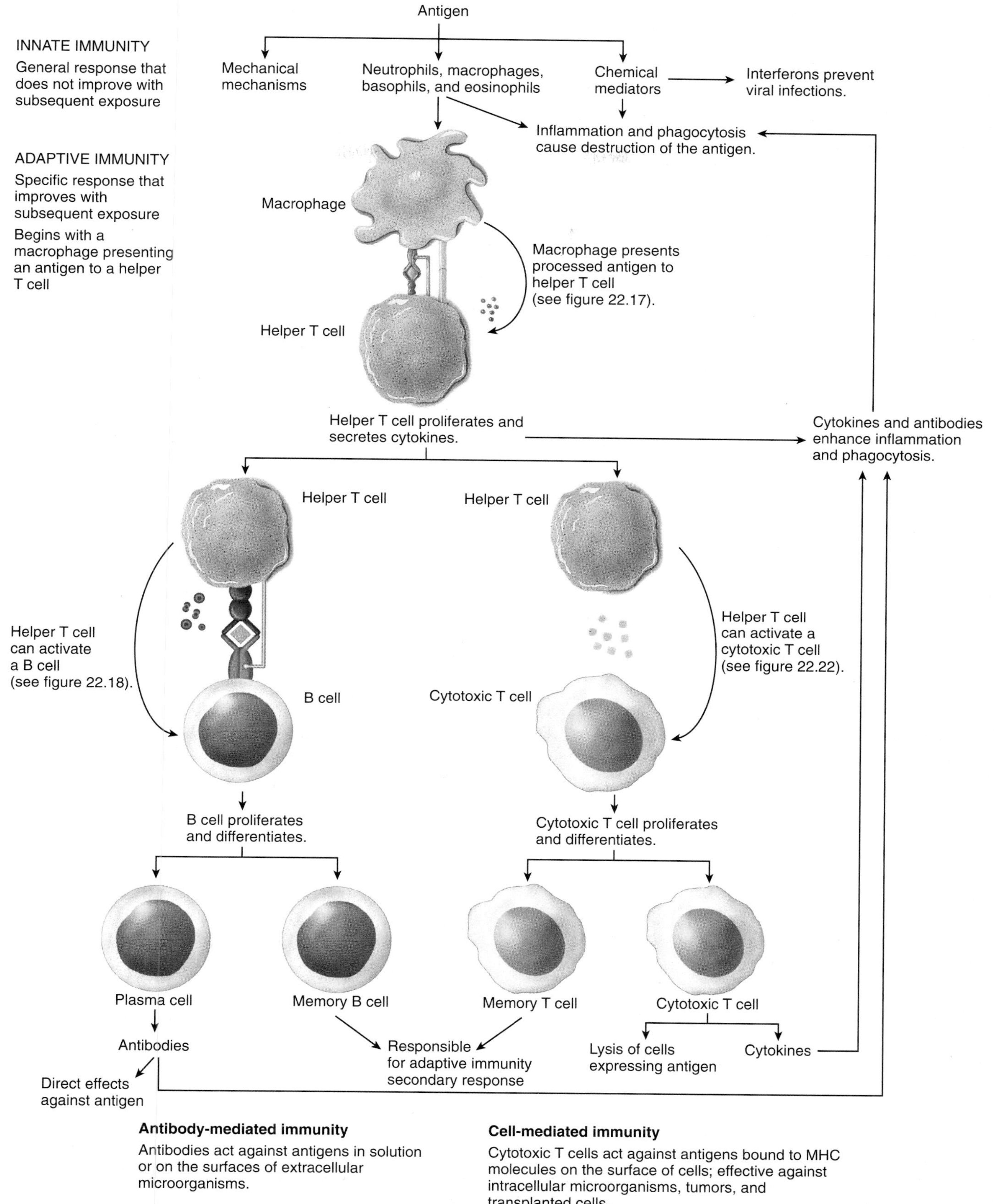

FIGURE 22.24 Immune Interactions

The major interactions and responses of innate and adaptive immunity to an antigen.

attempt to boost immune system function in general. For example, administering cytokines or other agents can promote inflammation and the activation of immune cells, which can help in the destruction of tumor cells. On the other hand, sometimes inhibiting the immune system is helpful. For example, multiple sclerosis is an autoimmune disease in which the immune system treats self-antigens as foreign antigens, thereby destroying the myelin that covers axons. Interferon beta (IFNβ) blocks the expression of MHC molecules that display self-antigens and is used to treat multiple sclerosis.

Some immunotherapy takes a more specific approach. For example, vaccination can prevent many diseases (see "Acquired Immunity," p. 816). The ability to produce monoclonal antibodies may result in therapies that are effective for treating tumors. If an antigen unique to tumor cells can be found, then monoclonal antibodies could be used to deliver radioactive isotopes, drugs, toxins, enzymes, or cytokines that can kill the tumor cell or can activate the immune system to kill the cell. Unfortunately, no antigen on tumor cells has been found that is not also found on normal cells. Nonetheless, this approach may be useful if damage to normal cells is minimal. For example, tumor cells may have more surface antigens of a particular type than normal cells, resulting in greater treatment delivery. Tumor cells may also be more susceptible to damage, or normal cells may be better able to recover from the treatment.

One problem with monoclonal antibody delivery systems is that the immune system recognizes the monoclonal antibody as a foreign antigen. After the first exposure, a memory response quickly destroys the monoclonal antibodies, rendering the treatment ineffective. In a process called **humanization,** the monoclonal antibodies are modified to resemble human antibodies. This approach has allowed monoclonal antibodies to sneak past the immune system.

The use of monoclonal antibodies to treat tumors is mostly in the research stage of development, but a few clinical trials are now yielding promising results. For example, monoclonal antibodies with radioactive iodine (^{131}I) have caused the regression of B-cell lymphomas and have produced few side effects. Herceptin is a monclonal antibody that binds to a growth factor that is overexpressed in 25%–30% of primary breast cancers. The antibodies serve to "tag" cancer cells, which are then lysed by natural killer cells. Herceptin slows disease progression and increases survival time, but it is not a cure for breast cancer.

Many other approaches for immunotherapy are being studied, and the development of treatments that use the immune system are certain to increase in the future.

54 ***What is immunotherapy? Give examples.***

Neuroendocrine Regulation of Immunity

An intriguing possibility for reducing the severity of diseases or even curing them is to use neuroendocrine regulation of immunity. The nervous system regulates the secretion of hormones, such as cortisol, epinephrine, endorphins, and enkephalins, for which lymphocytes have receptors. For example, cortisol released during times of stress inhibits the immune system. In addition, most lymphatic tissues, including some individual lymphocytes, receive sympathetic innervation. That a neuroendocrine connection exists with the immune system is clear. The question we need to answer is can we use this connection to control our own immunotherapy?

ACQUIRED IMMUNITY

It is possible to acquire adaptive immunity in four ways: active natural, active artificial, passive natural, and passive artificial immunity (figure 22.25). The terms *natural* and *artificial* refer to the method of exposure. Natural exposure implies that contact with an antigen or antibody occurs as part of everyday living and is not deliberate. Artificial exposure, also called **immunization,** is a deliberate introduction of an antigen or antibody into the body.

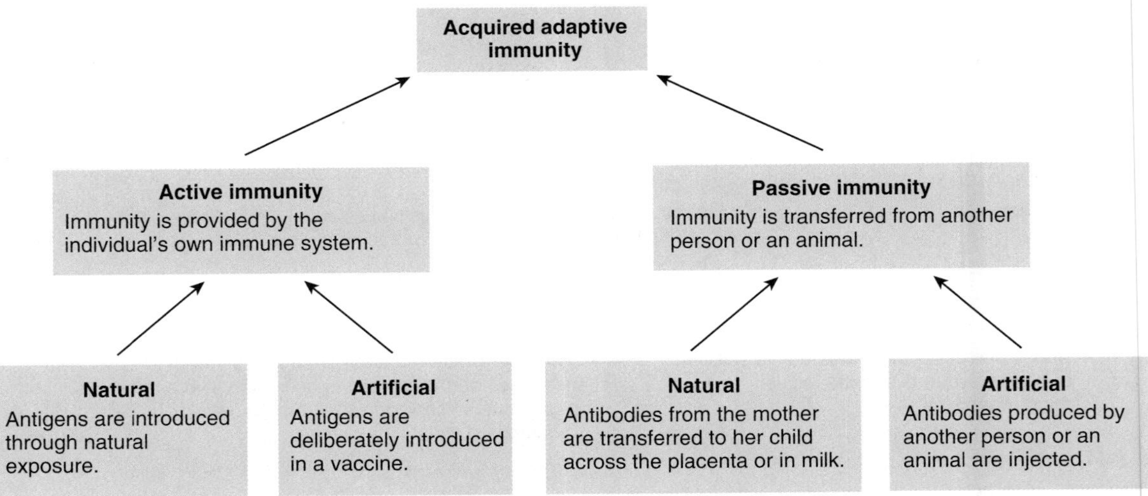

FIGURE 22.25 Ways to Acquire Adaptive Immunity

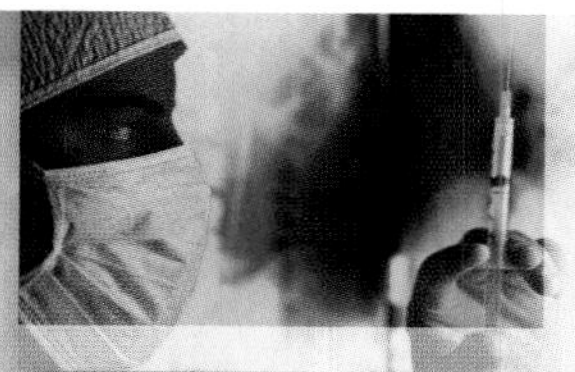

CLINICAL FOCUS

Acquired Immunodeficiency Syndrome

Acquired immunodeficiency syndrome (AIDS) is a life-threatening disease caused by the **human immunodeficiency virus (HIV).** HIV is transmitted from an infected to a noninfected person in body fluids, such as blood, semen, or vaginal secretions. The major methods of transmission are through unprotected intimate sexual contact, with contaminated needles used by intravenous drug users, through tainted blood products, and from a pregnant woman to her fetus. Present evidence indicates that household, school, and work contacts do not result in transmission.

HIV infection begins when a protein on the surface of the virus, called gp120, binds to a CD4 molecule on the surface of a cell. The CD4 molecule is found primarily on helper T cells, and it normally enables helper T cells to adhere to other lymphocytes—for example, during the process of antigen presentation. Certain monocytes, macrophages, neurons, and neuroglial cells also have CD4 molecules. Once attached to the CD4 molecules, the virus injects its genetic material (RNA) and enzymes into the cell and begins to replicate. Copies of the virus are manufactured using the organelles and materials within the cell. Replicated viruses escape from the cell and infect other cells.

Following infection by HIV, within 3 weeks to 3 months, many patients develop mononucleosis-like symptoms, such as fever, sweats, fatigue, muscle and joint aches, headache, sore throat, diarrhea, rash, and swollen lymph nodes. Within 1–3 weeks, these symptoms disappear as the immune system responds to the virus by producing antibodies and activating cytotoxic T cells that kill HIV-infected cells. The immune system is not able to eliminate HIV completely, however, and by about 6 months a kind of "set point" is achieved in which the virus continues to replicate at a low but steady rate. This chronic stage of infection lasts, on the average, 8–10 years, and the infected person feels good and exhibits few, if any, symptoms.

Although helper T cells are infected and destroyed during the chronic stage of HIV infection, the body responds by producing large numbers of helper T cells. Nonetheless, over a period of years the HIV numbers gradually increase and helper T cell numbers decrease. Normally, approximately 1200 helper T cells are present per cubic millimeter of blood. An HIV-infected person is considered to have AIDS when one or more of the following conditions appear: The helper T cell count falls below 200 cells/mm^3, an opportunistic infection occurs, or Kaposi sarcoma develops.

Opportunistic infections involve organisms that normally do not cause disease but can do so when the immune system is depressed. Without helper T cells, cytotoxic T- and B-cell activation is impaired, and adaptive resistance is suppressed. Examples of opportunistic infections include pneumocystis (noo-mō-sis′tis) pneumonia (caused by an intracellular fungus, *Pneumocystis carinii*), tuberculosis (caused by an intracellular bacterium, *Myocobacterium tuberculosis*), syphilis (caused by a sexually transmitted bacterium, *Treponema pallidum*), candidiasis (kan-di-dī′ă-sis; a yeast infection of the mouth or vagina caused by *Candida albicans*), and protozoans that cause severe, persistent diarrhea. Kaposi sarcoma is a type of cancer that produces lesions in the skin, lymph nodes, and visceral organs. Also associated with AIDS are symptoms resulting from the effects of HIV on the nervous system, including motor retardation, behavioral changes, progressive dementia, and possibly psychosis.

No cure for AIDS has yet been discovered. Management of AIDS can be divided into two categories: (1) management of secondary infections or malignancies associated with AIDS and (2) treatment of HIV. In order for HIV to replicate, the viral RNA is used to make viral DNA, which is inserted into the host cell's DNA. The inserted viral DNA directs the production of new viral RNA and proteins, which are assembled to form new HIV. Key steps in the replication of HIV require viral enzymes. **Reverse transcriptase** promotes the formation of viral DNA from viral RNA, and **integrase** (in′te-grās) inserts the viral DNA into the host cell's DNA. A viral **protease** (prō′tē-ās) breaks large viral proteins into smaller proteins, which are incorporated into the new HIV.

Blocking the activity of HIV enzymes can inhibit the replication of HIV. The first effective treatment of AIDS was the drug azidothymidine (AZT; az′i-dō-thī′mi-dēn), also called zidovudine (zī-dō′voo-dēn). AZT is a **reverse transcriptase inhibitor,** which prevents HIV RNA from producing viral DNA. AZT can delay the onset of AIDS but does not appear to increase the survival time of AIDS patients. However, the number of babies who contract AIDS from their HIV-infected mothers can be dramatically reduced by giving AZT to the mothers during pregnancy and to the babies following birth.

Protease inhibitors are drugs that interfere with viral proteases. The current treatment for suppressing HIV replication is **highly active antiretroviral therapy (HAART).** This therapy involves the use of drugs from at least two classes of antivirals. Treatment may involve the combination of three drugs, such as two reverse transcriptase inhibitors and one protease inhibitor. It is less likely that HIV will develop resistance to all three drugs. This strategy has proven very effective in reducing the death rate from AIDS and partially restoring health in some individuals.

Still in the research stage are **integrase inhibitors,** which prevent the insertion of viral DNA into the host cell's DNA. Perhaps someday integrase inhibitors will be part of a combination drug therapy for AIDS.

Another advance in AIDS treatment is a test for measuring **viral load,** which measures the number of viral RNA molecules in a milliliter of blood. The actual level of HIV is one-half the RNA count because each HIV has two RNA strands. Viral load is a good predictor of how soon a person will develop AIDS. If viral load is high, the onset of AIDS is much sooner than if it is low. It is also possible to detect developing viral resistance by an increase in viral load. In response, a change in drug dose or type may slow viral replication. Current treatment guidelines are to keep viral load below 500 RNA molecules per milliliter of blood.

An effective treatment for AIDS is not a cure. Even if viral load decreases to the point that the virus is undetected in the blood, the virus still remains in cells throughout the body. The virus may eventually mutate and escape drug suppression. The long-term goal for dealing with AIDS is to develop a vaccine that prevents HIV infection.

Because of improved treatment, people with HIV/AIDS can now live for many years. HIV/AIDS is therefore being viewed increasingly as a chronic disease, not as a death sentence. A multidisciplinary team of occupational therapists, physical therapists, nutritionists/dieticians, psychologists, infectious disease physicians, and others can work together to manage patients with HIV/AIDS to help them have a better quality of life.

The terms *active* and *passive* indicate whether or not an individual's immune system is directly responding to the antigen. When an individual is naturally or artificially exposed to an antigen, an adaptive immune system response can occur that produces antibodies. This is called **active immunity** because the individual's own immune system is the cause of the immunity. **Passive immunity** occurs when another person or an animal develops antibodies and the antibodies are transferred to a nonimmune individual. For example, infants can acquire antibodies from mother's milk. This is called passive immunity because the nonimmune individual did not produce the antibodies.

How long the immunity lasts differs for active and passive immunity. Active immunity can persist for a few weeks (e.g., common cold) to a lifetime (e.g., whooping cough and polio). Immunity can be long-lasting if enough B or T memory cells are produced and persist to respond to later antigen exposure. Passive immunity is not long-lasting because the individual does not produce his or her own memory cells. Because active immunity can last longer than passive immunity, it is the preferred method. Passive immunity is preferred, however, when immediate protection is needed.

Active Natural Immunity

Natural exposure to an antigen, such as a disease-causing microorganism, can cause an individual's immune system to mount an adaptive immune system response against the antigen and achieve **active natural immunity.** Because the individual is not immune during the first exposure, he or she usually develops the symptoms of the disease. Interestingly, exposure to an antigen does not always produce symptoms. Many people exposed to the poliomyelitis virus at an early age have an immune system response and produce poliomyelitis antibodies, yet they do not exhibit any disease symptoms.

Active Artificial Immunity

In **active artificial immunity,** an antigen is deliberately introduced into an individual to stimulate the immune system. This process is **vaccination,** and the introduced antigen is a **vaccine.** Injection of the vaccine is the usual mode of administration. Examples of injected vaccinations are the DTP injection against diphtheria, tetanus, and pertussis (whooping cough) and the MMR injection against mumps, measles, and rubella (German measles). Sometimes the vaccine is ingested, as in the oral poliomyelitis vaccine (OPV).

The vaccine usually consists of a part of a microorganism, a dead microorganism, or a live, altered microorganism. The antigen has been changed so that it will stimulate an immune response but will not cause the symptoms of disease. Because active artificial immunity produces long-lasting immunity without disease symptoms, it is the preferred method of acquiring adaptive immunity.

PREDICT 6

In some cases, a booster shot is used as part of a vaccination procedure. A booster shot is another dose of the original vaccine given some time after the original dose was administered. Why are booster shots given?

Passive Natural Immunity

Passive natural immunity results from the transfer of antibodies from a mother to her child across the placenta before birth. During her life, the mother has been exposed to many antigens, either naturally or artificially, and she has antibodies against many of these antigens. These antibodies protect the mother and the developing fetus against disease. Some of the antibodies (IgG) can cross the placenta and enter the fetal blood. Following birth, the antibodies provide protection for the first few months of the baby's life. Eventually, the antibodies are broken down, and the baby must rely on his or her own immune system. If the mother nurses her baby, antibodies (IgA) in the mother's milk may also provide some protection for the baby.

Passive Artificial Immunity

Achieving **passive artificial immunity** usually begins with vaccinating an animal, such as a horse. After the animal's immune system responds to the antigen, antibodies (sometimes T cells) are removed from the animal and injected into the individual requiring immunity. In some cases, a human who has developed immunity through natural exposure or vaccination is used as a source of antibodies. Passive artificial immunity provides immediate protection for the individual receiving the antibodies and is therefore preferred when time might not be available for the individual to develop his or her own immunity. This technique provides only temporary immunity, however, because the antibodies are used or eliminated by the recipient.

Antiserum is the general term used for serum, which is plasma minus the clotting factors, that contains antibodies responsible for passive artificial immunity. Antisera are available against microorganisms that cause diseases such as rabies, hepatitis, and measles; bacterial toxins such as tetanus, diphtheria, and botulism; and venoms from poisonous snakes and black widow spiders.

55 ***Distinguish between active and passive immunity.***

56 ***State four general ways of acquiring adaptive immunity. Which two provide the longest-lasting immunity?***

EFFECTS OF AGING ON THE LYMPHATIC SYSTEM AND IMMUNITY

Aging appears to have little effect on the ability of the lymphatic system to remove fluid from tissues, absorb fats from the digestive tract, or remove defective red blood cells from the blood.

Aging also seems to have little direct effect on the ability of B cells to respond to antigens, and the number of circulating B cells remains stable in most individuals. With age, thymic tissue is replaced with adipose tissue, and the ability to produce new, mature T cells in the thymus is eventually lost. Nonetheless, the number of T cells remains stable in most individuals due to the replication (not maturation) of T cells in secondary lymphatic tissues. In many individuals, however, there is a decreased ability of helper T cells to proliferate in response to antigens. Thus, antigen exposure produces fewer helper T cells, which results

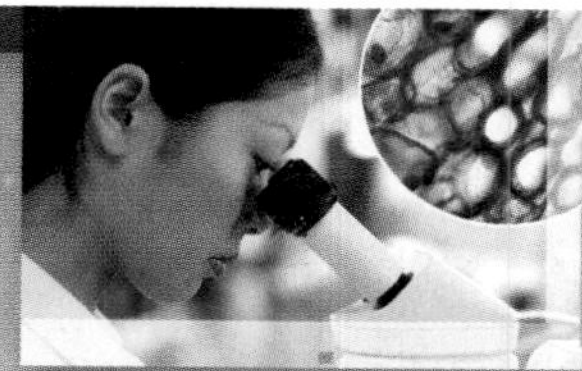

SYSTEMS PATHOLOGIES

Systemic Lupus Erythematosus

Lucy is a 30-year-old divorced woman with two children. Despite the fact that she has to work to support herself and the children, she entered college, determined to become a nurse and provide a better life for her family. Lucy was an excellent student, but her class attendance and her performance on tests were somewhat erratic. Sometimes she seemed very energetic and earned high grades, but other times she seemed depressed and did not do as well. Toward the end of the course, she developed a rash on her face (figure A), and a large, red lesion on her arm, and she was obviously not feeling well. Lucy asked her instructor if she could take an incomplete grade and take the last exam at a later time. She explained that she has had lupus since she was 25 years old. Normally, medication helps control her symptoms, but the stress of being a single parent combined with the challenges of school seemed to be making her condition worse. She further explained that the symptoms of lupus come and go, and bed rest is often helpful. Lucy finished the course requirements later that summer. She went on to complete her education and now has a full-time job as a nurse at a local hospital.

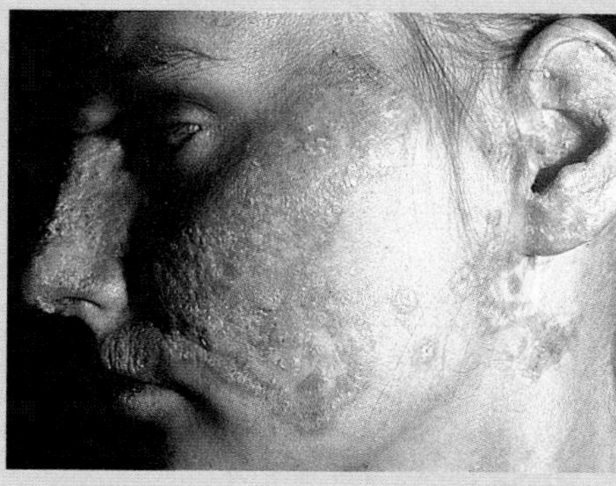

FIGURE A Systemic Lupus Erythematosus
The butterfly rash results from inflammation in the skin caused by systemic lupus erythematosus.

BACKGROUND INFORMATION

Systemic lupus erythematosus (SLE) is a disease of unknown cause in which tissues and cells are damaged by the immune system. The name describes some of the characteristics of the disease. The term *lupus* literally means wolf and was originally used to refer to eroded (as if gnawed by a wolf) lesions of the skin. *Erythematosus* refers to a redness of the skin resulting from inflammation. Unfortunately, as the term *systemic* implies, the disorder is not confined to the skin but can affect tissues and cells throughout the body. Another systemic effect is the presence of low-grade fever in most cases of active SLE.

SLE is an autoimmune disorder in which a large variety of antibodies are produced that recognize self-antigens, such as nucleic acids, phospholipids, coagulation factors, red blood cells, and platelets. The combination of the antibodies with self-antigens forms immune complexes that circulate throughout the body to be deposited in various tissues, in which they stimulate inflammation and tissue destruction. Thus, SLE is a disease that can affect many systems of the body. For example, the most common antibodies act against DNA that is released from damaged cells. Normally, the liver removes the DNA, but, when DNA and antibodies form immune complexes, they tend to be deposited in the kidneys and other tissues. Approximately 40%–50% of individuals with SLE develop renal disease. In some cases, the antibodies can bind to antigens on cells, resulting in lysis of the cells. For example, the binding of antibodies to red blood cells results in hemolysis and the development of anemia.

The cause of SLE is unknown. The most popular hypothesis is that a viral infection disrupts the function of suppressor T cells, resulting in loss of tolerance to self-antigens. The picture is probably more complicated, however, because not all SLE patients have reduced numbers of suppressor T cells. In addition, some patients have decreased numbers of the helper T cells that normally stimulate suppressor T-cell activity.

Genetic factors probably contribute to the development of the disease. The likelihood of developing SLE is much higher if a family member also has it. In addition, family members of SLE patients who do not have SLE are much more likely to have DNA antibodies than does the general population.

Approximately 1 of every 2000 individuals in the United States has SLE. The first symptoms usually appear between 15 and 25 years of age and affect women approximately nine times as often as men. The progress of the disease is unpredictable, with flare-ups of symptoms followed by periods of remission. The survival after diagnosis is greater than 90% after 10 years. The most frequent causes of death are kidney failure, central nervous system dysfunction, infections, and cardiovascular disease.

No cure for SLE exists, nor does one standard of treatment, because the course of the disease is highly variable and many differences can be found among patients. Treatment usually begins with mild medications and proceeds to more and more potent therapies as conditions warrant. Aspirin and nonsteroidal anti-inflammatory drugs are used to suppress inflammation. Antimalarial drugs are used to treat skin rash and arthritis in SLE, but the mechanism of action is unknown. Patients who do not respond to these drugs and those with severe SLE are helped by glucocorticoids. Although glucocorticoids effectively suppress inflammation, they can produce undesirable side effects, including suppression of normal adrenal gland functions. In patients with life-threatening SLE, very high doses of glucocorticoids are used.

PREDICT 7

The red lesion Lucy developed on her arm is called purpura (pŭr′poo-ră), which is caused by bleeding into the skin. The lesions gradually change color and disappear in 2–3 weeks. Explain how SLE produces purpura.

Continued

Continued

SYSTEM INTERACTIONS The Effect of Systemic Lupus Erythematosus on Other Systems

SYSTEM	INTERACTIONS
Integumentary	Skin lesions frequently occur and are made worse by exposure to the sun. There are three forms: (1) an inflammatory redness, which can take the form of a butterfly rash, which extends from the bridge of the nose to the cheeks; (2) small, localized, pimplelike eruptions accompanied by scaling of the skin; and (3) areas of atrophied, depigmented skin with borders of increased pigmentation. Diffuse thinning of the hair results from hair loss.
Skeletal	Arthritis, tendonitis, and death of bone tissue can occur.
Muscular	Destruction of muscle tissue and muscular weakness can occur.
Nervous	Memory loss, intellectual deterioration, disorientation, psychosis, reactive depression, headache, seizures, nausea, and loss of appetite can occur. Stroke is a major cause of dysfunction and death. Cranial nerve involvement results in facial muscle weakness, drooping of the eyelid, and double vision. Central nervous system lesion can cause paralysis.
Endocrine	Sex hormones may play a role in SLE because 90% of the cases occur in females, and females with SLE have reduced levels of androgens.
Cardiovascular	Inflammation of the pericardium (pericarditis) with chest pain can develop. Damage to heart valves, inflammation of cardiac tissue, tachycardia, arrhythmias, angina, and myocardial infarction can also occur. Hemolytic anemia and leukopenia can be present (see chapter 19). Antiphospholipid antibody syndrome, through an unknown mechanism, increases coagulation and thrombus formation, which increases the risk for stroke and heart attack.
Respiratory	Chest pain caused by inflammation of the pleural membranes; fever, shortness of breath, and hypoxemia caused by inflammation of the lungs; and alveolar hemorrhage can develop.
Digestive	Ulcers develop in the oral cavity and pharynx. Abdominal pain and vomiting are common, but no cause can be found. Inflammation of the pancreas and occasionally enlargement of the liver and minor abnormalities in liver function tests occur.
Urinary	Renal lesions and glomerulonephritis can result in progressive failure of kidney function. Excess proteins are lost in the urine, resulting in lower than normal blood proteins, which can produce edema.

in less stimulation of B cells and effector T cells. Consequently, both antibody-mediated immunity and cell-mediated immunity responses to antigens decrease.

Primary and secondary antibody responses decrease with age. More antigen is required to produce a response, the response is slower, less antibody is produced, and fewer memory cells result. Thus, the ability to resist infections and develop immunity decreases. It is recommended that vaccinations be given well before age 60 because these declines are most evident after age 60. Vaccinations, however, can be beneficial at any age, especially if the individual has reduced resistance to infection.

The ability of cell-mediated immunity to resist intracellular pathogens decreases with age. For example, the elderly are more susceptible to influenza (flu) and should be vaccinated every year. Some pathogens cause disease but are not eliminated from the body. With age, a decrease in immunity can result in reactivation of the pathogen. For example, the virus that causes chicken pox in children can remain latent within nerve cells, even though the disease seems to have disappeared. Later in life, the virus can leave the nerve cells and infect skin cells, causing painful lesions known as herpes zoster, or shingles.

Autoimmune disease occurs when immune responses destroy otherwise healthy tissue. There is very little increase in the number of new-onset autoimmune diseases in the elderly. However, the chronic inflammation and immune responses that begin earlier in life have a cumulative, damaging effect.

57 ***What effect does aging have on the major functions of the lymphatic system?***

58 ***Describe the effects of aging on B cells and T cells. Give examples of how they affect antibody-mediated immunity and cell-mediated immunity responses.***

SUMMARY

Lymphatic System (p. 783)

The lymphatic system consists of lymph, lymphatic vessels, lymphatic tissue, lymphatic nodules, lymph nodes, tonsils, the spleen, and the thymus.

Functions of the Lymphatic System

The lymphatic system maintains fluid balance in tissues, absorbs fats from the small intestine, and defends against microorganisms and foreign substances.

Lymphatic Vessels

1. Lymphatic vessels carry lymph away from tissues.
2. Lymphatic capillaries lack a basement membrane and have loosely overlapping epithelial cells. Fluids and other substances easily enter the lymphatic capillary.
3. Lymphatic capillaries join to form lymphatic vessels.
   - Lymphatic vessels have valves that ensure one-way flow of lymph.
   - Contraction of lymphatic vessel smooth muscle, skeletal muscle action, and thoracic pressure changes move the lymph.
4. Lymph nodes are along the lymphatic vessels. After passing through lymph nodes, lymphatic vessels form lymphatic trunks and lymphatic ducts.
5. Lymphatic trunks and ducts empty into the blood at thoracic veins (junctions of the internal jugular and subclavian veins).
   - Lymph from the right thorax, the upper-right limb, and the right side of the head and the neck enters right thoracic veins.
   - Lymph from the lower limbs, pelvis, and abdomen; the left thorax; the upper-left limb; and the left side of the head and the neck enters left thoracic veins.
6. The jugular, subclavian, and brochomediastinal trunks may unite to form the right lymphatic duct.
7. The thoracic duct is the largest lymphatic vessel.
8. The intestinal and lumbar trunks may converge on the cisterna chyli, a sac that joins the inferior end of the thoracic duct.

Lymphatic Tissue and Organs

1. Lymphatic tissue is reticular connective tissue that contains lymphocytes and other cells.
2. Lymphatic tissue can be surrounded by a capsule (lymph nodes, spleen, thymus).
3. Lymphatic tissue can be nonencapsulated (diffuse lymphatic tissue, lymphatic nodules, tonsils). Mucosa-associated lymphoid tissue is nonencapsulated lymphatic tissue located in and below the mucous membranes of the digestive, respiratory, urinary, and reproductive tracts.
4. Diffuse lymphatic tissue consists of dispersed lymphocytes and has no clear boundaries.
5. Lymphatic nodules are small aggregates of lymphatic tissue (e.g., Peyer's patches in the small intestines).
6. The tonsils
   - The tonsils are large groups of lymphatic nodules in the oral cavity and nasopharynx.
   - The three groups of tonsils are the palatine, pharyngeal, and lingual tonsils.
7. Lymph nodes
   - Lymphatic tissue in the node is organized into the cortex and the medulla. Lymphatic sinuses extend through the lymphatic tissue.
   - Substances in lymph are removed by phagocytosis, or they stimulate lymphocytes (or both).
   - Lymphocytes leave the lymph node and circulate to other tissues.
8. The spleen
   - The spleen is in the left superior side of the abdomen.
   - Foreign substances stimulate lymphocytes in the white pulp (periarterial lymphatic sheath and lymphatic nodules).
   - Foreign substances and defective red blood cells are removed from the blood by phagocytes in the red pulp (splenic cords and venous sinuses).
   - The spleen is a limited reservoir for blood.
9. The thymus
   - The thymus is a gland in the superior mediastinum and is divided into a cortex and a medulla.
   - Lymphocytes in the cortex are separated from the blood by reticular cells.
   - Lymphocytes produced in the cortex migrate through the medulla, enter the blood, and travel to other lymphatic tissues, where they can proliferate.

Overview of the Lymphatic System

See figure 22.9.

Immunity (p. 792)

Immunity is the ability to resist the harmful effects of microorganisms and other foreign substances.

Innate Immunity (p. 792)

Mechanical Mechanisms

Mechanical mechanisms prevent the entry of microbes (skin and mucous membranes) or remove them (tears, saliva, and mucus).

Chemical Mediators

1. Chemical mediators promote phagocytosis and inflammation.
2. Complement can be activated by either the alternative or the classical pathway. Complement lyses cells, increases phagocytosis, attracts immune system cells, and promotes inflammation.
3. Interferons prevent viral replication. Interferons are produced by virally infected cells and move to other cells, which are then protected.

Cells

1. Chemotactic factors are parts of microorganisms or chemicals that are released by damaged tissues. Chemotaxis is the ability of white blood cells to move to tissues that release chemotactic factors.
2. Phagocytosis is the ingestion and destruction of materials.
3. Neutrophils are small phagocytic cells.
4. Macrophages are large phagocytic cells.
   - Macrophages can engulf more than neutrophils can.
   - Macrophages in connective tissue protect the body at locations where microbes are likely to enter, and macrophages clean blood and lymph.
5. Basophils and mast cells release chemicals that promote inflammation.
6. Eosinophils release enzymes that reduce inflammation.
7. Natural killer cells lyse tumor cells and virus-infected cells.

Inflammatory Response

1. The inflammatory response can be initiated in many ways.
   - Chemical mediators cause vasodilation and increase vascular permeability, which allows the entry of other chemical mediators.
   - Chemical mediators attract phagocytes.

- The amount of chemical mediators and phagocytes increases until the cause of the inflammation is destroyed. Then the tissue undergoes repair.

2. Local inflammation produces the symptoms of redness, heat, swelling, pain, and loss of function. Symptoms of systemic inflammation include an increase in neutrophil numbers, fever, and shock.

Adaptive Immunity (p. 798)

1. Antigens are large molecules that stimulate an adaptive immune system response. Haptens are small molecules that combine with large molecules to stimulate an adaptive immune system response.
2. B cells are responsible for humoral, or antibody-mediated, immunity. T cells are involved with cell-mediated immunity.

Origin and Development of Lymphocytes

1. B cells and T cells originate in red bone marrow. T cells are processed in the thymus, and B cells are processed in bone marrow.
2. Positive selection ensures the survival of lymphocytes that can react against antigens, and negative selection eliminates lymphocytes that react against self-antigens.
3. A clone is a group of identical lymphocytes that can respond to a specific antigen.
4. B cells and T cells move to lymphatic tissue from their processing sites. They continually circulate from one lymphatic tissue to another.
5. Primary lymphatic organs (red bone marrow and thymus) are where lymphocytes mature into functional cells. Secondary lymphatic organs and tissues are where lymphocytes produce an immune response.

Activation of Lymphocytes

1. The antigenic determinant is the specific part of the antigen to which the lymphocyte responds. The antigen receptor (T-cell receptor or B-cell receptor) on the surface of lymphocytes combines with the antigenic determinant.
2. MHC class I molecules display antigens on the surface of nucleated cells, resulting in the destruction of the cells.
3. MHC class II molecules display antigens on the surface of antigen-presenting cells, resulting in the activation of immune cells.
4. MHC antigen complex and costimulation are usually necessary to activate lymphocytes. Costimulation involves cytokines and certain surface molecules.
5. Antigen-presenting cells stimulate the proliferation of helper T cells, which stimulate the proliferation of B or T effector cells.

Inhibition of Lymphocytes

1. Tolerance is suppression of the immune system's response to an antigen.
2. Tolerance is produced by the deletion of self-reactive cells, by the prevention of lymphocyte activation, and by suppressor T cells.

Antibody-Mediated Immunity

1. Antibodies are proteins.
   - The variable region of an antibody combines with the antigen. The constant region activates complement or binds to cells.
   - Five classes of antibodies exist: IgG, IgM, IgA, IgE, and IgD.
2. Antibodies affect the antigen in many ways.
   - Antibodies bind to the antigen and interfere with antigen activity or bind the antigens together.
   - Antibodies act as opsonins (substances that increase phagocytosis) by binding to the antigen and to macrophages.
   - Antibodies can activate complement through the classical pathway.
   - Antibodies attach to mast cells or basophils and cause the release of inflammatory chemicals when the antibody combines with the antigen.
3. The primary response results from the first exposure to an antigen. B cells form plasma cells, which produce antibodies and memory B cells.
4. The secondary response results from exposure to an antigen after a primary response, and memory B cells quickly form plasma cells and additional memory B cells.

Cell-Mediated Immunity

1. Cells infected with intracellular microorganisms process antigens that combine with MHC class I molecules.
2. Cytotoxic T cells are stimulated to divide, producing more cytotoxic T cells and memory T cells, when MHC class I/antigen complexes are presented to T-cell receptors. Cytokines released from helper T cells also stimulate cytotoxic T cells.
3. Cytotoxic T cells lyse virus-infected cells, tumor cells, and tissue transplants.
4. Cytotoxic T cells produce cytokines, which promote phagocytosis and inflammation.

Immune Interactions (p. 814)

Innate immunity, antibody-mediated immunity, and cell-mediated immunity can function together to eliminate an antigen.

Immunotherapy (p. 814)

Immunotherapy stimulates or inhibits the immune system to treat diseases.

Acquired Immunity (p. 816)

Active Natural Immunity

Active natural immunity results from natural exposure to an antigen.

Active Artificial Immunity

Active artificial immunity results from deliberate exposure to an antigen.

Passive Natural Immunity

Passive natural immunity results from the transfer of antibodies from a mother to her fetus or baby.

Passive Artificial Immunity

Passive artificial immunity results from transfer of antibodies (or cells) from an immune animal to a nonimmune animal.

Effects of Aging on the Lymphatic System and Immunity (p. 818)

1. Aging has little effect on the ability of the lymphatic system to remove fluid from tissues, absorb fats from the digestive tract, or remove defective red blood cells from the blood.
2. Decreased helper T cell proliferation results in decreased antibody-mediated immunity and cell-mediated immunity responses to antigens.
3. The primary and secondary antibody responses decrease with age.
4. The ability to resist intracellular pathogens decreases with age.

REVIEW AND COMPREHENSION

1. The lymphatic system
   a. removes excess fluid from tissues.
   b. absorbs fats from the digestive tract.
   c. defends the body against microorganisms and other foreign substances.
   d. all of the above.
2. Which of the following statements is true?
   a. Lymphatic vessels do not have valves.
   b. Lymphatic vessels empty into lymph nodes.
   c. Lymph from the right-lower limb passes into the right jugular or subclavian vein.
   d. Lymph from the jugular and subclavian trunks empties into the cisterna chyli.
   e. All of the above are true.
3. The tonsils
   a. consist of large groups of lymphatic nodules.
   b. protect against bacteria.
   c. can become chronically infected.
   d. decrease in size in adults.
   e. all of the above.
4. Lymph nodes
   a. filter lymph.
   b. are where lymphocytes divide and increase in number.
   c. contain a network of reticular fibers.
   d. contain lymphatic sinuses.
   e. all of the above.
5. Which of these statements about the spleen is *not* correct?
   a. The spleen has white pulp associated with the arteries.
   b. The spleen has red pulp associated with the veins.
   c. The spleen destroys defective red blood cells.
   d. The spleen is surrounded by trabeculae located outside the capsule.
   e. The spleen is a limited reservoir for blood.
6. The thymus
   a. increases in size in adults.
   b. produces lymphocytes that move to other lymphatic tissue.
   c. is located in the abdominal cavity.
   d. all of the above.
7. Which of these is an example of innate immunity?
   a. Tears and saliva wash away microorganisms.
   b. Basophils release histamine and leukotrienes.
   c. Neutrophils phagocytize a microorganism.
   d. The complement cascade is activated.
   e. All of the above are correct.
8. Neutrophils
   a. enlarge to become macrophages.
   b. account for most of the dead cells in pus.
   c. are usually the last cell type to enter infected tissues.
   d. are usually located in lymphatic and blood sinuses.
9. Macrophages
   a. are large, phagocytic cells that outlive neutrophils.
   b. develop from mast cells.
   c. often die after a single phagocytic event.
   d. have the same function as eosinophils.
   e. all of the above.
10. Which of these cells is the most important in the release of histamine, which promotes inflammation?
    a. monocyte
    b. macrophage
    c. eosinophil
    d. mast cell
    e. natural killer cell
11. Which of these conditions does *not* occur during the inflammatory response?
    a. the release of histamine and other chemical mediators
    b. the chemotaxis of phagocytes
    c. the entry of fibrinogen into tissues from the blood
    d. the vasoconstriction of blood vessels
    e. increased permeability of blood vessels
12. Antigens
    a. are foreign substances introduced into the body.
    b. are molecules produced by the body.
    c. stimulate an adaptive immune system response.
    d. all of the above.
13. B cells
    a. are processed in the thymus.
    b. originate in red bone marrow.
    c. once released into the blood remain in the blood.
    d. are responsible for cell-mediated immunity.
    e. all of the above.
14. MHC molecules
    a. are glycoproteins.
    b. attach to the plasma membrane.
    c. have a variable region that can bind to foreign and self-antigens.
    d. may form an MHC/antigen complex that activates T cells.
    e. all of the above.
15. Antigen-presenting cells can
    a. take in foreign antigens.
    b. process antigens.
    c. use MHC class II molecules to display the antigens.
    d. stimulate other immune system cells.
    e. all of the above.
16. Which of these participates in costimulation?
    a. cytokines
    b. complement
    c. antibodies
    d. histamine
    e. natural killer cells
17. Helper T cells
    a. respond to antigens from macrophages.
    b. respond to cytokines from macrophages.
    c. stimulate B cells with cytokines.
    d. all of the above.
18. The most important function of tolerance is to
    a. increase lymphocyte activity.
    b. increase complement activation.
    c. prevent the immune system from responding to self-antigens.
    d. prevent excessive immune system response to foreign antigens.
    e. process antigens.
19. Variable amino acid sequences on the arms of the antibody molecule
    a. make the antibody specific for a given antigen.
    b. enable the antibody to activate complement.
    c. enable the antibody to attach to basophils and mast cells.
    d. are part of the constant region.
    e. all of the above.
20. Antibodies
    a. prevent antigens from binding together.
    b. promote phagocytosis.
    c. inhibit inflammation.
    d. block complement activation.
    e. block the function of opsonins.

21. The secondary antibody response
    a. is slower than the primary response.
    b. produces fewer antibodies than the primary response.
    c. prevents disease symptoms from occurring.
    d. occurs because of cytotoxic T cells.
22. The type of lymphocyte that is responsible for the secondary antibody response is the
    a. memory B cell.
    b. B cell.
    c. T cell.
    d. helper T cell.
23. The largest percentage of antibodies in the blood are
    a. IgA.
    b. IgD.
    c. IgE.
    d. IgG.
    e. IgM.
24. Antibody-mediated immunity
    a. works best against intracellular antigens.
    b. regulates the activity of T cells.
    c. cannot be transferred from one person to another person.
    d. is responsible for immediate hypersensitivity reactions.
25. The activation of cytotoxic T cells can result in
    a. lysis of virus-infected cells.
    b. production of cytokines.
    c. production of memory T cells.
    d. all of the above.
26. Cytokines
    a. promote inflammation.
    b. activate macrophages.
    c. kill target cells by causing them to lyse.
    d. all of the above.

Answers in Appendix E

CRITICAL THINKING

1. A patient is suffering from edema in the right-lower limb. Explain why elevation of the limb and massage help remove the excess fluid.
2. If the thymus of an experimental animal is removed immediately after its birth, the animal exhibits the following characteristics: (a) It is more susceptible to infections, (b) it has decreased numbers of lymphocytes in lymphatic tissue, and (c) its ability to reject grafts is greatly decreased. Explain these observations.
3. If the thymus of an adult experimental animal is removed, the following observations can be made: (a) No immediate effect occurs and, (b) after 1 year, the number of lymphocytes in the blood decreases, the ability to reject grafts decreases, and the ability to produce antibodies decreases. Explain these observations.
4. Adjuvants are substances that slow but do not stop the release of an antigen from an injection site into the blood. Suppose injection A of a certain amount of antigen is given without an adjuvant and injection B of the same amount of antigen is given with an adjuvant that causes the release of antigen over a period of 2–3 weeks. Does injection A or B result in the greater amount of antibody production? Explain.
5. Tetanus is caused by bacteria that enter the body through wounds in the skin. The bacteria produce a toxin, which causes spastic muscle contractions. Death often results from the failure of the respiration muscles. A patient goes to the emergency room after stepping on a nail. If the patient has been vaccinated against tetanus, the patient is given a tetanus booster shot, which consists of the toxin altered so that it is harmless. If the patient has never been vaccinated against tetanus, the patient is given an antiserum shot against tetanus. Explain the rationale for this treatment strategy. Sometimes both a booster and an antiserum shot are given, but at different locations of the body. Explain why this is done and why the shots are given in different locations.
6. An infant appears to be healthy until about 9 months of age. Then he develops severe bacterial infections, one after another. Fortunately, the infections are successfully treated with antibiotics. When infected with the measles and other viral diseases, the infant recovers without unusual difficulty. Explain the different immune responses to these infections. Why did it take so long for this disorder to become apparent? (*Hint:* consider IgG.)
7. A baby is born with severe combined immunodeficiency disease (SCID). In an attempt to save her life, a bone marrow transplant is performed. Explain how this procedure might help. Unfortunately, there is a graft rejection, and the baby dies. Explain what happened.
8. A patient has many allergic reactions. As part of the treatment scheme, doctors decide to try to identify the allergen that stimulates the immune system's response. A series of solutions, each containing an allergen that commonly causes a reaction, is composed. Each solution is injected into the skin at different locations on the patient's back. The following results are obtained: (a) At one location, the injection site becomes red and swollen within a few minutes; (b) at another injection site, swelling and redness appear 2 days later; and (c) no redness or swelling develops at the other sites. Explain what happened for each observation by describing what part of the immune system was involved and what caused the redness and swelling.
9. Ivy Hurtt developed a poison ivy rash after a camping trip. Her doctor prescribed a cortisone ointment to relieve the inflammation. A few weeks later, Ivy scraped her elbow, which became inflamed. Because she had some of the cortisone ointment left over, she applied it to the scrape. Explain why the ointment was or was not a good idea for the poison ivy and for the scrape.
10. Suzy Withitt has just had her ears pierced. To her dismay, she finds that, when she wears inexpensive (but tasteful) jewelry, by the end of the day there is an inflammatory (allergic) reaction to the metal in the jewelry. Is this because of antibodies or cytokines?

Answers in Appendix F

Visit this book's website at www.mhhe.com/seeley8 for chapter quizzes, interactive learning exercises, and other study tools.

Respiratory System 23

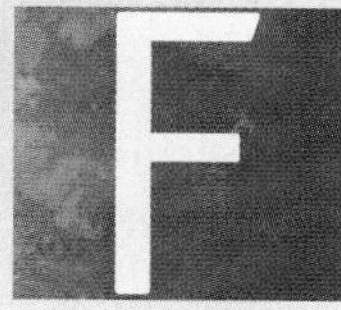

From our first breath at birth, the rate and depth of our breathing is unconsciously matched to our activities, whether studying, sleeping, talking, eating, or exercising. We can voluntarily stop breathing, but within a few seconds we must breathe again. Breathing is so characteristic of life that, along with the pulse, it is one of the first things we check for to determine if an unconscious person is alive.

Breathing is necessary because all living cells of the body require oxygen and produce carbon dioxide. The respiratory system allows the exchange of these gases between the air and the blood, and the cardiovascular system transports them between the lungs and the cells of the body. The capacity to carry out normal activity is reduced without healthy respiratory and cardiovascular systems.

Respiration includes (1) ventilation, the movement of air into and out of the lungs; (2) gas exchange between the air in the lungs and the blood, sometimes called external respiration; (3) the transport of oxygen and carbon dioxide in the blood; and (4) gas exchange between the blood and the tissues, sometimes called internal respiration. The term *respiration* is also used in reference to cell metabolism, which is discussed in chapter 25.

This chapter explains the *functions of the respiratory system* (p. 826), the *anatomy and histology of the respiratory system* (p. 826), *ventilation* (p. 841), *measurement of lung function* (p. 846), *physical principles of gas exchange* (p. 848), *oxygen and carbon dioxide transport in the blood* (p. 851), *regulation of ventilation* (p. 856), and *respiratory adaptations to exercise* (p. 863). The chapter concludes by looking at the *effects of aging on the respiratory system* (p. 863).

Colorized scanning electron micrograph of the lung, showing alveoli, which are small chambers where gas exchange takes place between the air and the blood.

Respiratory System

FUNCTIONS OF THE RESPIRATORY SYSTEM

Respiration is necessary because all living cells of the body require oxygen and produce carbon dioxide. The respiratory system assists in gas exchange and performs other functions as well:

1. *Gas exchange.* The respiratory system allows oxygen from the air to enter the blood and carbon dioxide to leave the blood and enter the air. The cardiovascular system transports oxygen from the lungs to the cells of the body and carbon dioxide from the cells of the body to the lungs. Thus, the respiratory and cardiovascular systems work together to supply oxygen to all cells and to remove carbon dioxide.
2. *Regulation of blood pH.* The respiratory system can alter blood pH by changing blood carbon dioxide levels.
3. *Voice production.* Air movement past the vocal folds makes sound and speech possible.
4. *Olfaction.* The sensation of smell occurs when airborne molecules are drawn into the nasal cavity.
5. *Protection.* The respiratory system provides protection against some microorganisms by preventing their entry into the body and by removing them from respiratory surfaces.

1 ***Explain the functions of the respiratory system.***

ANATOMY AND HISTOLOGY OF THE RESPIRATORY SYSTEM

The respiratory system consists of the external nose, the nasal cavity, the pharynx, the larynx, the trachea, the bronchi, and the lungs (figure 23.1). Although air frequently passes through the oral cavity, it is considered to be part of the digestive system instead of the respiratory system. The parts of the respiratory system are divided into the upper respiratory tract and the lower respiratory tract. The term **upper respiratory tract** refers to the external nose, nasal cavity, pharynx, and associated structures; the term **lower respiratory tract** refers to the larynx, trachea, bronchi, and lungs. These terms are not official anatomical terms, however, and there are several alternate definitions. For example, one alternate definition places the larynx in the upper respiratory tract. The diaphragm and the muscles of the thoracic and abdominal walls are responsible for respiratory movements.

2 ***Define upper and lower respiratory tract.***

Nose

The **nasus** (nā′sŭs), or **nose,** consists of the external nose and the nasal cavity. The **external nose** is the visible structure that forms a prominent feature of the face. The largest part of the external nose is composed of hyaline cartilage plates (see figure 7.9*b*). The bridge of the nose consists of the nasal bones plus extensions of the frontal and maxillary bones.

The **nasal cavity** extends from the nares to the choanae (figure 23.2). The **nares** (nā′res; sing. *naris*), or **nostrils,** are the external openings of the nasal cavity and the **choanae** (kō′an-ē) are the openings into the pharynx. The anterior part of the nasal cavity, just inside each naris, is the **vestibule** (ves′ti-bool; entry room). The vestibule is lined with stratified squamous epithelium, which is continuous with the stratified squamous epithelium of the skin. The **hard palate** (pal′ăt) is a bony plate covered by a mucous membrane that forms the floor of the nasal cavity. It separates the nasal cavity from the oral cavity. The **nasal septum**

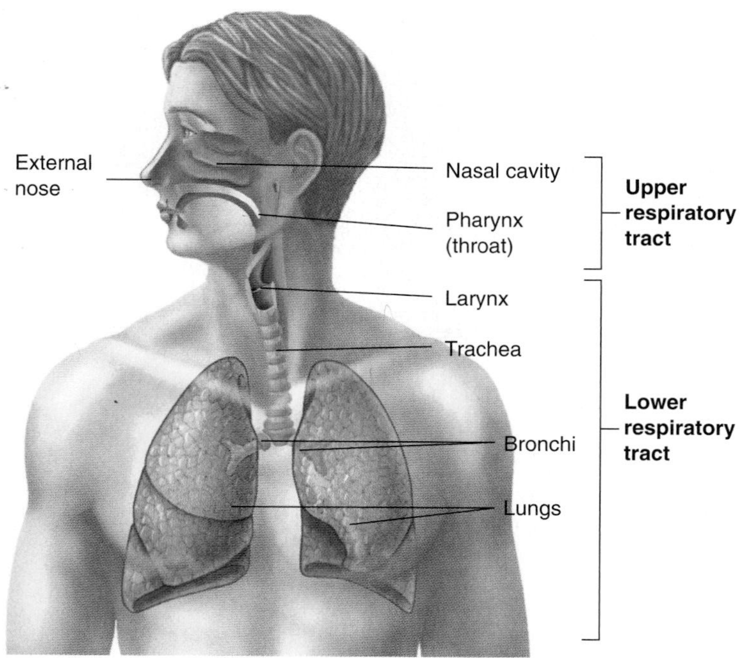

FIGURE 23.1 Respiratory System

The upper respiratory tract consists of the external nose, nasal cavity, and pharynx (throat). The lower respiratory tract consists of the larynx, trachea, bronchi, and lungs.

Cribriform plate
Nasal cavity
Superior concha
Middle concha
Inferior concha
Vestibule
Naris
Hard palate
Oral cavity
Tongue
Palatine tonsil
Lingual tonsil
Hyoid bone
Larynx
Epiglottis
Vestibular fold
Vocal fold
Thyroid cartilage
Cricoid cartilage
Trachea
Frontal sinus
Sphenoidal sinus
Paranasal sinuses
Superior meatus
Middle meatus
Inferior meatus
Choana
Nasal cavity
Pharyngeal tonsil
Opening of auditory tube
Nasopharynx
Soft palate
Uvula
Oropharynx
Laryngopharynx
Pharynx
Esophagus

(a) **Medial view**

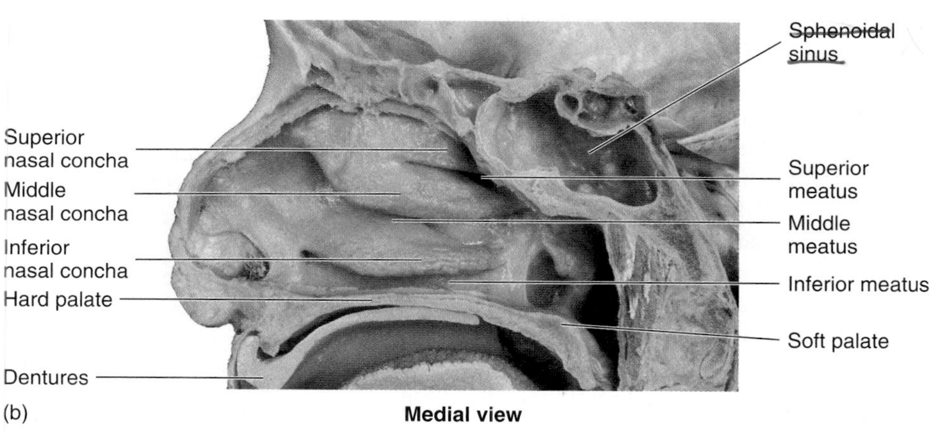

(b) **Medial view**

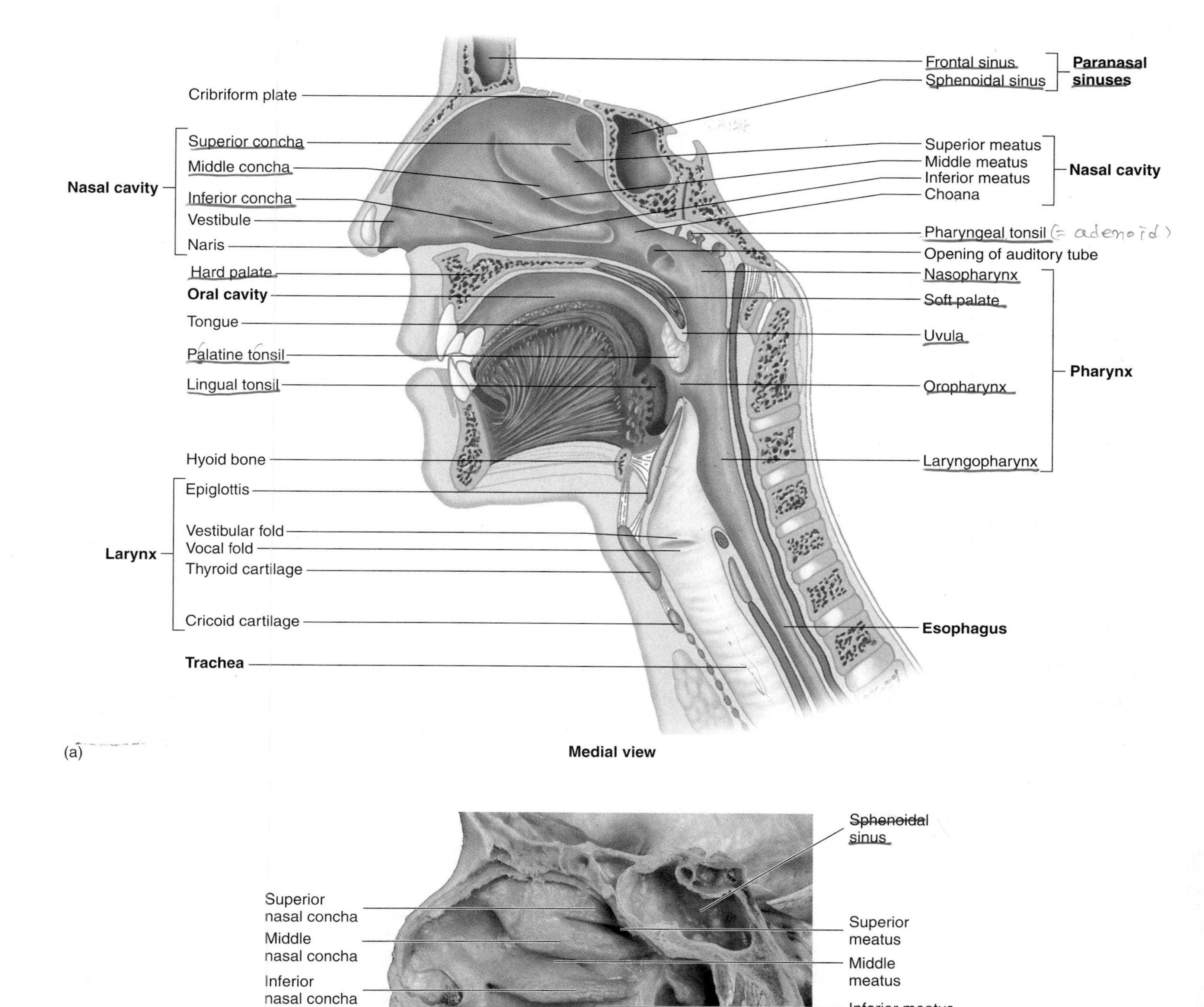

FIGURE 23.2 Nasal Cavity and Pharynx
(*a*) Sagittal section through the nasal cavity and pharynx. (*b*) Photograph of sagittal section of the head.

is a partition dividing the nasal cavity into right and left parts (see figure 7.9*a*). The anterior part of the nasal septum is cartilage, and the posterior part consists of the vomer bone and the perpendicular plate of the ethmoid bone. A deviated nasal septum occurs when the septum bulges to one side.

Three bony ridges, called **conchae** (kon′kē; resembling a conch shell), modify the lateral walls of the nasal cavity. Beneath each concha is a passageway called a **meatus** (mē-ā′tŭs; a tunnel or passageway). Within the superior and middle meatus are openings from the various **paranasal sinuses** (see figure 7.10), and the opening of a **nasolacrimal** (nā-zō-lak′ri-măl) **duct** is within each inferior meatus (see figure 15.9).

Sinusitis

Sinusitis (sī-nŭ-sī′tis) is an inflammation of the mucous membrane of any sinus, especially of one or more paranasal sinuses. Viral infections, such as the common cold, can cause mucous membranes to become inflamed, to swell, and to produce excess mucus. As a result, the sinus opening into the nasal cavity can be partially or completely blocked. Mucus accumulates within the sinus, which can promote the development of a bacterial infection. Treatments consist of taking antibiotics and promoting sinus drainage with decongestants, hydration, and steam inhalation. Sinusitis can also result from swelling caused by allergies or by polyps that obstruct a sinus opening into the nasal cavity.

The nasal cavity has five functions:

1. The nasal cavity is a passageway for air; it is open even when the mouth is full of food.
2. The nasal cavity cleans the air. The vestibule is lined with hairs, which trap some of the large particles of dust in the air. The nasal septum and nasal conchae increase the surface area of the nasal cavity and make airflow within the cavity more turbulent, thereby increasing the likelihood that air will come into contact with the mucous membrane lining the nasal cavity. This mucous membrane consists of pseudostratified ciliated columnar epithelium with goblet cells, which secrete a layer of mucus. The mucus traps debris in the air, and the cilia on the surface of the mucous membrane sweep the mucus posteriorly to the pharynx, where it is swallowed and eliminated by the digestive system.
3. The nasal cavity humidifies and warms the air. Moisture from the mucous epithelium and from excess tears that drain into the nasal cavity through the nasolacrimal duct is added to the air as it passes through the nasal cavity. Warm blood flowing through the mucous membrane warms the air within the nasal cavity before it passes into the pharynx, thus preventing damage to the rest of the respiratory passages caused by the cold air.
4. The nasal cavity contains the olfactory epithelium, which is the sensory organ for smell. The olfactory epithelium is located in the most superior part of the nasal cavity (see figure 15.1).
5. The nasal cavity and paranasal sinuses are resonating chambers for speech.

PREDICT 1

Explain what happens to your throat when you sleep with your mouth open, especially when your nasal passages are plugged as a result of having a cold. Explain what may happen to your lungs when you run a long way in very cold weather while breathing rapidly through your mouth.

3 ***Describe the structures of the nasal cavity and state their functions.***

Pharynx

The **pharynx** (far′ingks; throat) is the common opening of both the digestive and the respiratory systems. It receives air from the nasal cavity and receives air, food, and drink from the oral cavity. Inferiorly, the pharynx is connected to the respiratory system at the larynx and to the digestive system at the esophagus. The pharynx is divided into three regions: the nasopharynx, the oropharynx, and the laryngopharynx (see figure 23.2).

The **nasopharynx** (nā′zō-far′ingks) is located posterior to the choanae and superior to the **soft palate,** which is an incomplete muscle and connective tissue partition separating the nasopharynx from the oropharynx. The **uvula** (ū′vū-lă; a grape) is the posterior extension of the soft palate. The soft palate prevents swallowed materials from entering the nasopharynx and nasal cavity. The nasopharynx is lined with a mucous membrane containing pseudostratified ciliated columnar epithelium with goblet cells. Debris-laden mucus from the nasal cavity is moved through the nasopharynx and swallowed. Two auditory tubes from the middle ears open into the nasopharynx (see figures 15.23 and 23.2*a*). Air passes through them to equalize air pressure between the atmosphere and the middle ears. The posterior surface of the nasopharynx contains the pharyngeal tonsil, or adenoid (ad′ĕ-noyd), which helps defend the body against infection (see chapter 22). An enlarged pharyngeal tonsil can interfere with normal breathing and the passage of air through the auditory tubes.

The **oropharynx** (ōr′ō-far′ingks) extends from the soft palate to the epiglottis. The oral cavity opens into the oropharynx through the **fauces** (faw′sēz). Thus, air, food, and drink all pass through the oropharynx. Moist stratified squamous epithelium lines the oropharynx and protects it against abrasion. Two sets of tonsils, called the palatine tonsils and the lingual tonsils, are located near the fauces.

The **laryngopharynx** (lă-ring′gō-far-ingks) extends from the tip of the epiglottis to the esophagus and passes posterior to the larynx. Food and drink pass through the laryngopharynx to the esophagus. A small amount of air is usually swallowed with the food and drink. Swallowing too much air can cause excess gas in the stomach and belching. The laryngopharynx is lined with moist stratified squamous epithelium.

4 ***Name the three parts of the pharynx. With what structures does each part communicate?***

Larynx

The **larynx** (lar′ingks) is located in the anterior part of the throat and extends from the base of the tongue to the trachea (see figure 23.2*a*). It is a passageway for air between the pharynx and the trachea. The

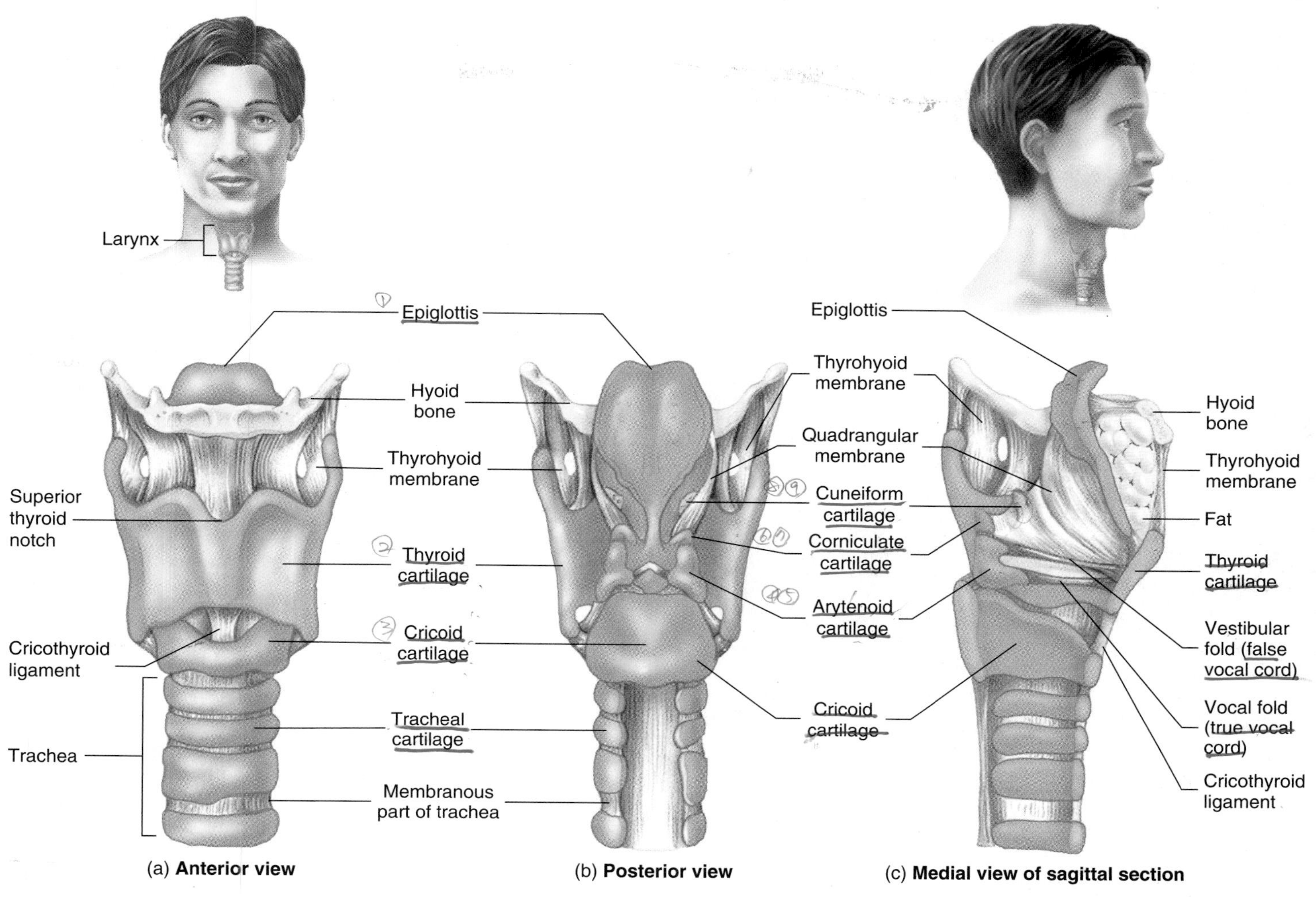

FIGURE 23.3 Anatomy of the Larynx

larynx is connected by membranes and/or muscles superiorly to the hyoid bone and consists of an outer casing of nine cartilages connected to one another by muscles and ligaments (figure 23.3). Six of the nine cartilages are paired, and three are unpaired. The largest of the cartilages is the unpaired **thyroid** (shield; refers to the shape of the cartilage) **cartilage,** or Adam's apple.

The most inferior cartilage of the larynx is the unpaired **cricoid** (krī′koyd; ring-shaped) **cartilage,** which forms the base of the larynx on which the other cartilages rest.

The third unpaired cartilage is the **epiglottis** (ep-i-glot′is; on the glottis). It is attached to the thyroid cartilage and projects superiorly as a free flap toward the tongue. The epiglottis differs from the other cartilages in that it consists of elastic rather than hyaline cartilage.

The paired **arytenoid** (ar-i-tē′noyd; ladle-shaped) **cartilages** articulate with the posterior, superior border of the cricoid cartilage, and the paired **corniculate** (kōr-nik′ū-lāt; horn-shaped) **cartilages** are attached to the superior tips of the arytenoid cartilages. The paired **cuneiform** (kū′nē-i-fōrm; wedge-shaped) **cartilages** are contained in a mucous membrane anterior to the corniculate cartilages (see figure 23.3*b*).

Two pairs of ligaments extend from the anterior surface of the arytenoid cartilages to the posterior surface of the thyroid cartilage. The superior ligaments are covered by a mucous membrane called the **vestibular folds,** or **false vocal cords** (see figure 23.3*c;* figure 23.4*a* and *b*).

The inferior ligaments are covered by a mucous membrane called the **vocal folds,** or **true vocal cords** (see figure 23.4). The vocal folds and the opening between them are called the **glottis** (glot′is; see figure 23.4). The vestibular folds and the vocal folds are lined with stratified squamous epithelium. The remainder of the larynx is lined with pseudostratified ciliated columnar epithelium. An inflammation of the mucosal epithelium of the vocal folds is called **laryngitis** (lar-in-jī′tis).

The larynx performs four important functions:

1. The thyroid and cricoid cartilages maintain an open passageway for air movement.
2. The larynx prevents the entry of swallowed materials into the lower respiratory tract and regulates the passage of air into and out of the lower respiratory tract. During swallowing, the epiglottis tips posteriorly until it lies below the horizontal plane and covers the opening into the larynx (see "Swallowing," chapter 24). Thus, food and liquid slide over the epiglottis toward the esophagus. The most important

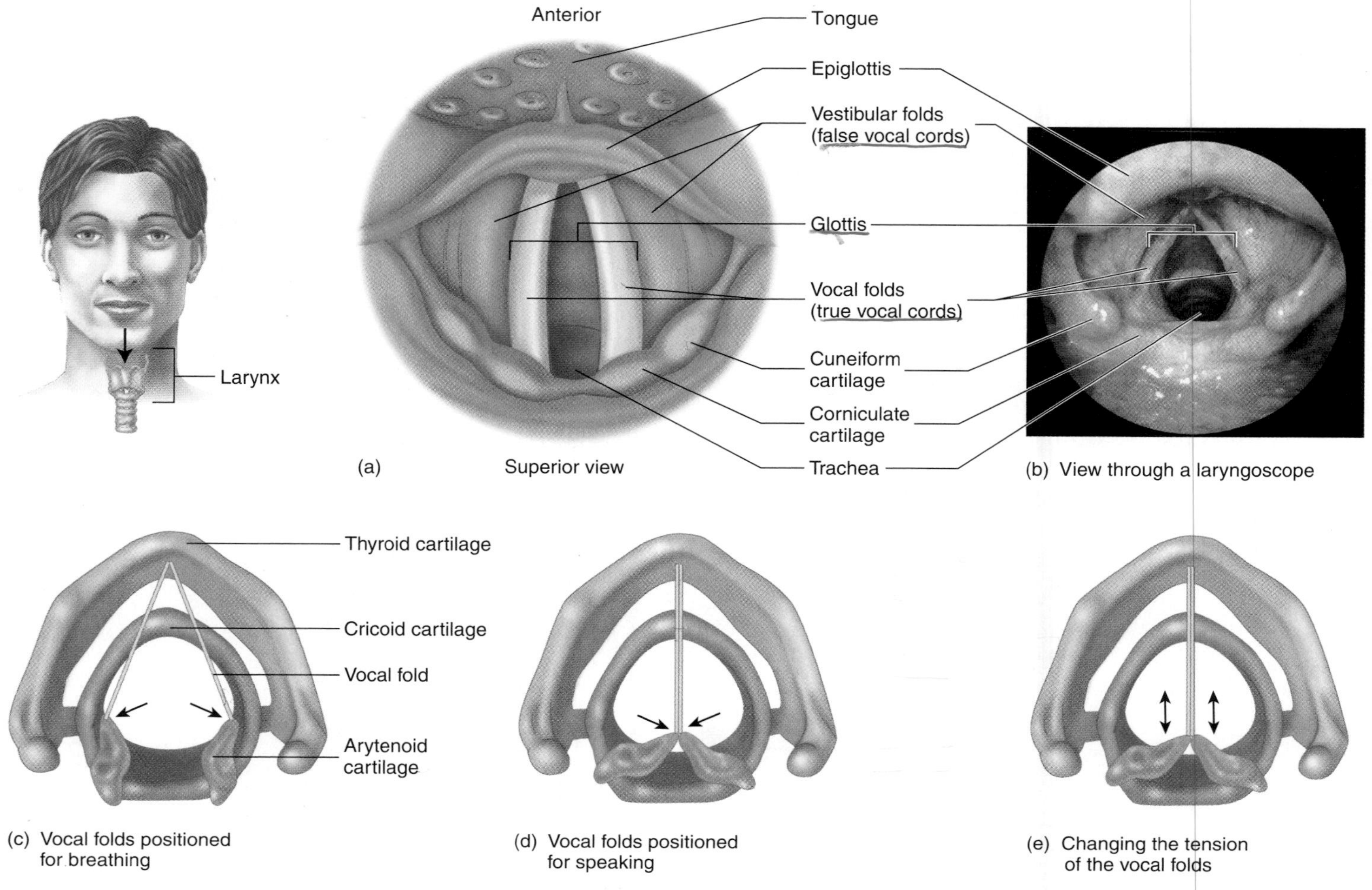

FIGURE 23.4 Vestibular Vocal Folds

Arrows show the direction of viewing the vestibular and vocal folds. (*a*) Relationship of the vestibular folds to the vocal folds and the laryngeal cartilages. (*b*) Laryngoscopic view of the vestibular and vocal folds. (*c*) Lateral rotation of the arytenoid cartilages moves the vocal folds laterally for breathing. (*d*) Medial rotation of the arytenoid cartilages moves the vocal folds medially for speaking. (*e*) Anterior/posterior movement of the arytenoid cartilages changes the length and tension of the vocal folds, changing the pitch of sounds (see text).

event for preventing the entry of materials into the larynx, however, is the closure of the vestibular and vocal folds. That is, the vestibular folds move medially and come together, as do the vocal folds. The closure of the vestibular and vocal folds can also prevent the passage of air, as when a person holds his or her breath or increases air pressure within the lungs prior to coughing or sneezing.

3. The vocal folds are the primary source of sound production. Air moving past the vocal folds causes them to vibrate and produce sound. The greater the amplitude of the vibration, the louder the sound. The force of air moving past the vocal folds determines the amplitude of vibration and the loudness of the sound. The frequency of vibrations determines pitch, with higher-frequency vibrations producing higher-pitched sounds and lower-frequency fibrations producing lower-pitched sounds. Variations in the length of the vibrating segments of the vocal folds affect the frequency of the vibrations. Higher-pitched tones are produced when only the anterior parts of the folds vibrate, and progressively lower tones result when longer sections of the folds vibrate. Because males usually have longer vocal folds than females, most males have lower-pitched voices. The sound produced by the vibrating vocal folds is modified by the tongue, lips, teeth, and other structures to form words. A person whose larynx has been removed because of carcinoma of the larynx can produce sound by swallowing air and causing the esophagus to vibrate.

Movement of the arytenoid and other cartilages is controlled by skeletal muscles, thereby changing the position and length of the vocal folds. When a person is only breathing, lateral rotation of the arytenoid cartilages abducts the vocal folds, which allows greater movement of air (see figure 23.4*c*). Medial rotation of the arytenoid cartilages adducts the vocal folds, places them in position for producing sounds, and changes the tension on them (see figure 23.4*d*). Anterior movement of the arytenoid cartilages decreases the length and tension of the vocal folds, lowering pitch. Posterior movement of the arytenoid cartilages increases the length and tension of the vocal folds, increasing pitch (figure 23.4*e*).

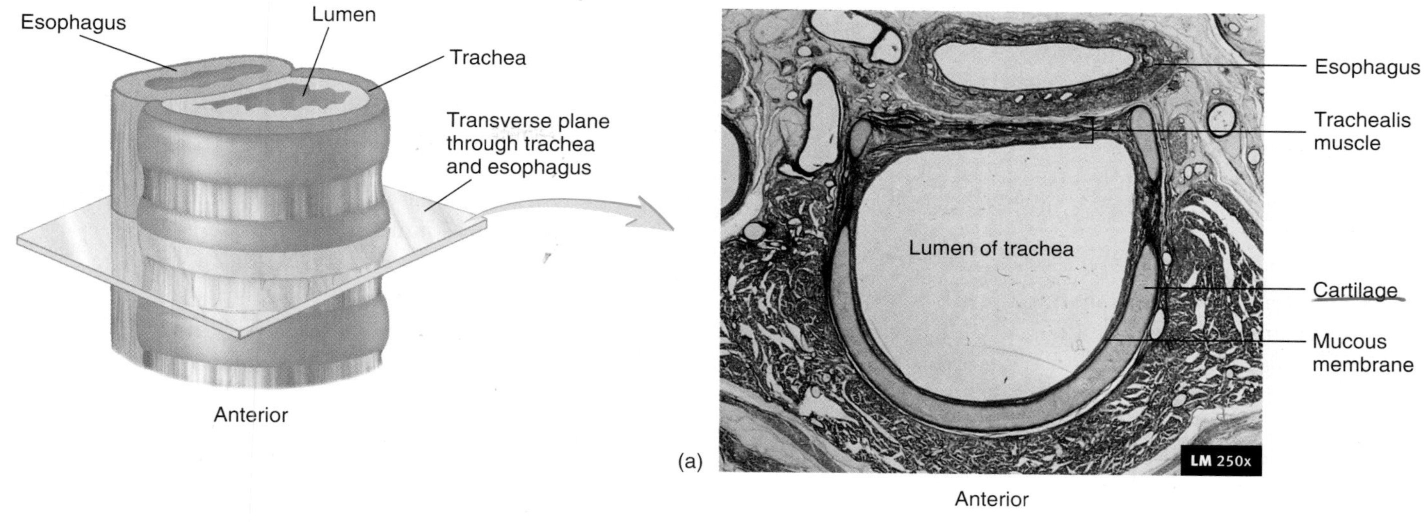

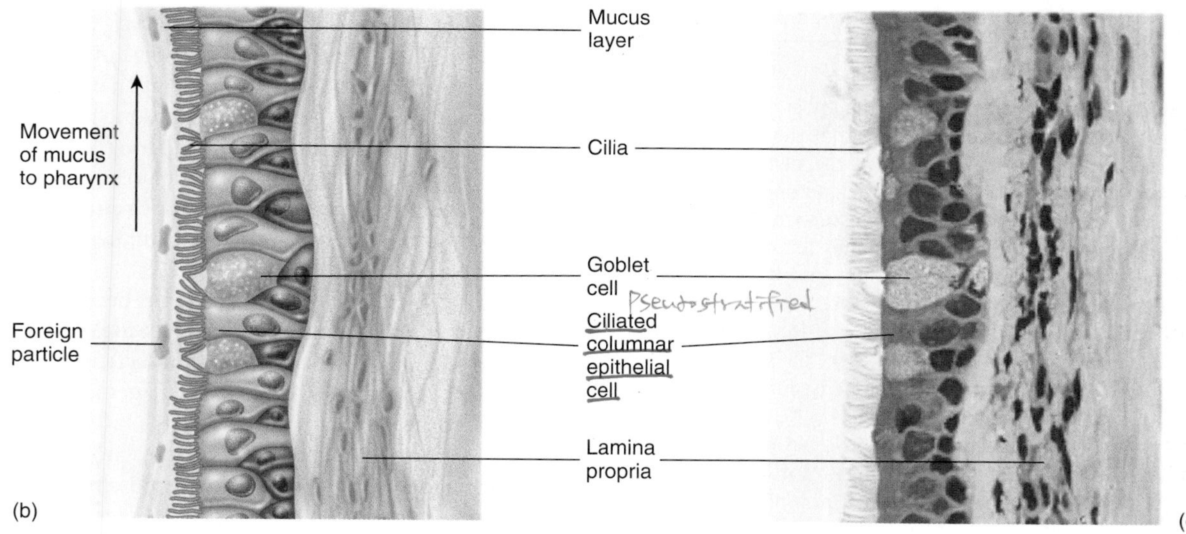

FIGURE 23.5 Trachea

(*a*) Light micrograph of a transverse section of the trachea. The esophagus is located posterior to the trachea, next to the smooth muscle connecting the ends of the C-shaped cartilages of the trachea. (*b*) Mucus, produced by the goblet cells, traps particles in the air. Movement of the cilia moves the mucus and particles to the laryngopharynx. (*c*) Light micrograph of the surface of the mucous membrane lining the trachea. Goblet cells are interspersed between ciliated cells.

4. The pseudostratified ciliated columnar epithelium lining the larynx produces mucus, which traps debris in air. The cilia move the mucus and debris into the pharynx.

5 ***Name and describe the three unpaired cartilages of the larynx. What are their functions?***

6 ***Distinguish between the vestibular and vocal folds. How are sounds of different loudness and pitch produced by the vocal folds?***

7 ***How does the position of the arytenoid cartilages change when a person is simply breathing versus making low-pitched and high-pitched sounds?***

Trachea

The **trachea** (trā′kē-ă), or windpipe, is a membranous tube attached to the larynx (see figure 23.3). It consists of dense regular connective tissue and smooth muscle reinforced with 15–20 C-shaped pieces of hyaline cartilage. The cartilages support the anterior and lateral sides of the trachea (figure 23.5*a*). They protect the trachea and maintain an open passageway for air. The posterior wall of the trachea is devoid of cartilage; it contains an elastic ligamentous membrane and bundles of smooth muscle called the **trachealis** (trā′kē-ā-lis) **muscle.** Contraction of the smooth muscle can narrow the diameter of the trachea. During coughing, this action causes air to move more rapidly through the trachea, which helps expel mucus and foreign objects. The esophagus lies immediately posterior to the cartilage-free posterior wall of the trachea.

PREDICT 2

Explain what happens to the shape of the trachea when a person swallows a large mouthful of food. Why is this change of shape advantageous?

The mucous membrane lining the trachea consists of pseudostratified ciliated columnar epithelium with numerous goblet cells (figure 23.5*b*). The goblet cells produce mucus, which traps inhaled foreign particles. The cilia move the mucus and foreign particles into the larynx, from which they enter the pharynx and are swallowed (figure 23.5*c*). Constant, long-term irritation to the trachea, such as occurs in smokers, can cause the tracheal epithelium to become moist stratified squamous epithelium that lacks cilia and goblet cells. Consequently, the normal function of the tracheal epithelium is lost.

Establishing Airflow

In cases of extreme emergency when the upper air passageway is blocked by a foreign object to the extent that the victim cannot breathe, quick reaction is required to save the person's life. The **Heimlich maneuver** is designed to force such an object out of the air passage by the sudden application of pressure to the abdomen. The person who performs the maneuver stands behind the victim, with his or her arms under the victim's arms and hands over the victim's abdomen between the navel and the rib cage. With one hand formed into a fist and the other hand over it, both hands are suddenly pulled toward the abdomen with an accompanying upward motion. This maneuver, if done properly, forces air up the trachea and dislodges most foreign objects.

There are other ways to establish air flow, but they should be performed only by trained medical personnel. **Intubation** is the insertion of a tube into an opening, a canal, or a hollow organ. A tube can be passed through the mouth or nose into the pharynx and then through the larynx to the trachea.

Sometimes it is necessary to make an opening through which to pass the tube. The preferred point of entry in emergency cases is through the membrane between the cricoid and thyroid cartilages, a procedure referred to as a **cricothyrotomy** (krī'kō-thī-rot'ō-mē).

A **tracheostomy** (trā'-kē-os'tō-mē, *tracheo-* + *stoma,* mouth) is an operation to make an opening into the trachea, usually between the second and third cartilage rings. Usually, the opening is intended to be permanent, and a tube is inserted into the trachea to allow airflow and provide a way to remove secretions. The term **tracheotomy** (trā-kē-ot'ō-mē, *tracheo-* + *tome,* incision) refers to the actual cutting into the trachea. Sometimes the terms *tracheostomy* and *tracheotomy* are used interchangeably. It is not advisable to enter the air passageway through the trachea in emergency cases because arteries, nerves, and the thyroid gland overlie the anterior surface of the trachea.

The trachea has an inside diameter of 12 mm and a length of 10–12 cm, descending from the larynx to the level of the fifth thoracic vertebra (figure 23.6). The trachea divides to form two smaller tubes called **main,** or **primary, bronchi** (brong'kī; sing. *bronchus,* brong'kŭs; windpipe), each of which extends to a lung. The most inferior tracheal cartilage forms a ridge called the **carina** (kă-rī'nă), which separates the openings into the main bronchi. The carina is an important radiologic landmark. In addition, the mucous membrane of the carina is very sensitive to mechanical stimulation, and materials reaching the carina stimulate a powerful cough reflex. Materials in the air passageways inferior to the carina do not usually stimulate a cough reflex.

Tracheobronchial Tree

Beginning with the trachea, all the respiratory passageways are called the **tracheobronchial** (trā'kē-ō-brong'kē-ăl) **tree** (see figure 23.6). The trachea divides to form main bronchi, which in turn divide to form smaller and smaller bronchi until, eventually, many microscopically small tubes and sacs are formed. The right main bronchus is larger in diameter and more directly in line with the trachea than the left main bronchus. As a result, swallowed objects that accidentally enter the lower respiratory tract are most likely to become lodged in the right main bronchus.

The main bronchi divide into **lobar,** or **secondary, bronchi** within each lung. Two lobar bronchi exist in the left lung, and three exist in the right lung. The lobar bronchi, in turn, give rise to **segmental,** or **tertiary, bronchi.** The bronchi continue to branch, finally giving rise to **bronchioles** (brong'kē-ōlz), which are less than 1 mm in diameter. The bronchioles also subdivide several times to become even smaller **terminal bronchioles.** Approximately 16 generations of branching occur from the trachea to the terminal bronchioles.

As the air passageways of the lungs become smaller, the structure of their walls changes. Like the trachea, the main bronchi are supported by C-shaped cartilage connected by smooth muscle. In the lobar bronchi, the C-shaped cartilages are replaced with cartilage plates, and smooth muscle forms a layer between the cartilage and the mucous membrane. As the bronchi become smaller, the cartilage becomes more sparse and smooth muscle becomes more abundant. The terminal bronchioles have no cartilage, and the smooth muscle layer is prominent. Relaxation and contraction of the smooth muscle within the bronchi and bronchioles can change the diameter of the air passageways and thereby change the volume of air moving through them. For example, during exercise the diameter can increase, which reduces the resistance to airflow and thereby increases the volume of air moved. During an **asthma attack,** however, contraction of the smooth muscle in the bronchi and bronchioles can decrease their diameter, resulting in increased resistance to airflow, which greatly reduces airflow. In severe cases, air movement can be so restricted that death results. Fortunately, medications, such as albuterol (al-bū'ter-ol), help counteract the effects of an asthma attack by promoting smooth muscle relaxation in the walls of terminal bronchioles so that air can flow more freely.

The bronchi are lined with a pseudostratified ciliated columnar epithelium. The larger bronchioles are lined with ciliated simple columnar epithelium, which changes to ciliated simple cuboidal epithelium in the terminal bronchioles. The ciliated epithelium functions as a mucus–cilia escalator, which traps debris in the air and moves it to the larynx.

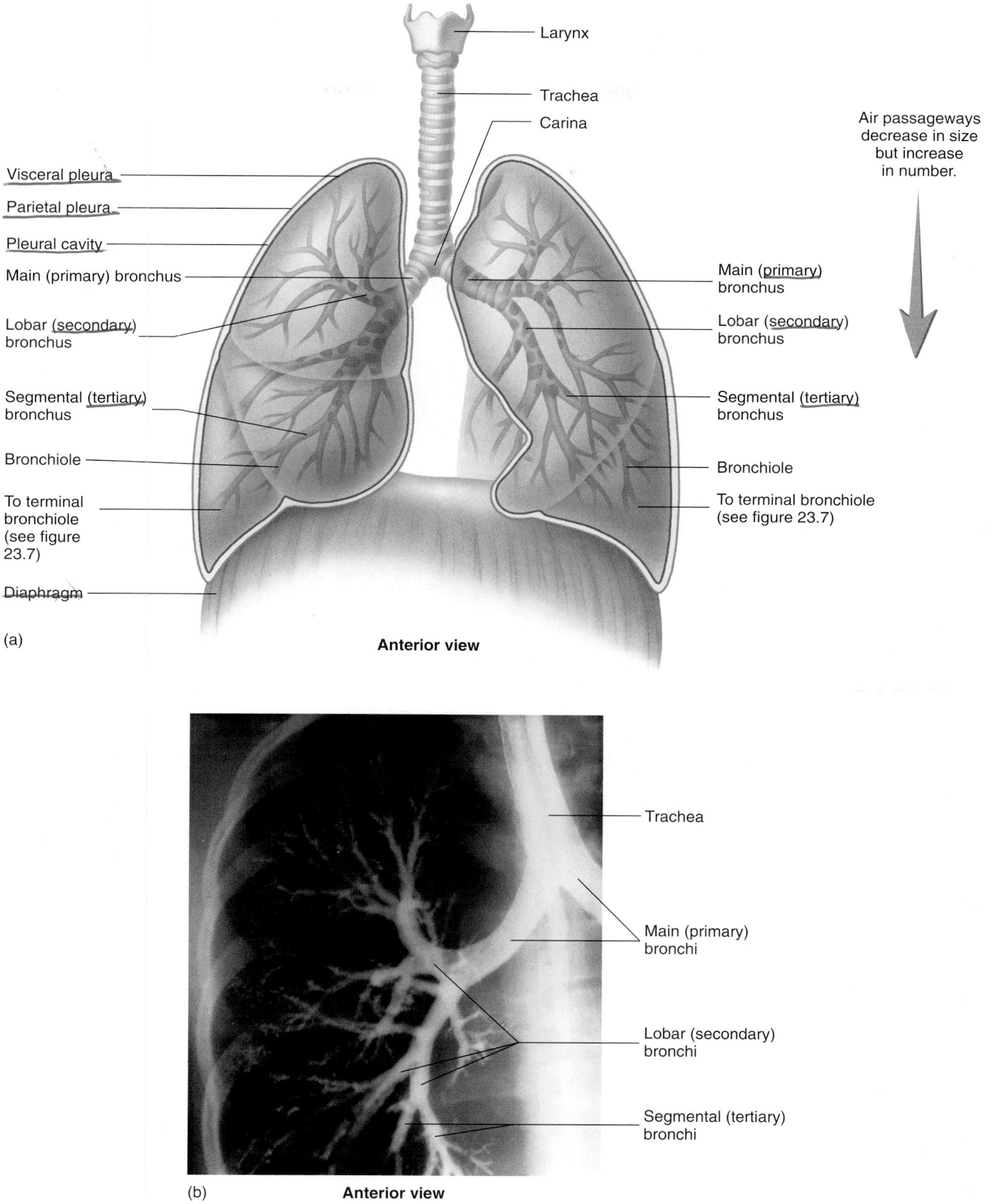

Larynx

Trachea

Carina

Air passageways decrease in size but increase in number.

Visceral pleura

Parietal pleura

Pleural cavity

Main (primary) bronchus

Lobar (secondary) bronchus

Segmental (tertiary) bronchus

Bronchiole

To terminal bronchiole (see figure 23.7)

Main (primary) bronchus

Lobar (secondary) bronchus

Segmental (tertiary) bronchus

Bronchiole

To terminal bronchiole (see figure 23.7)

Diaphragm

(a) Anterior view

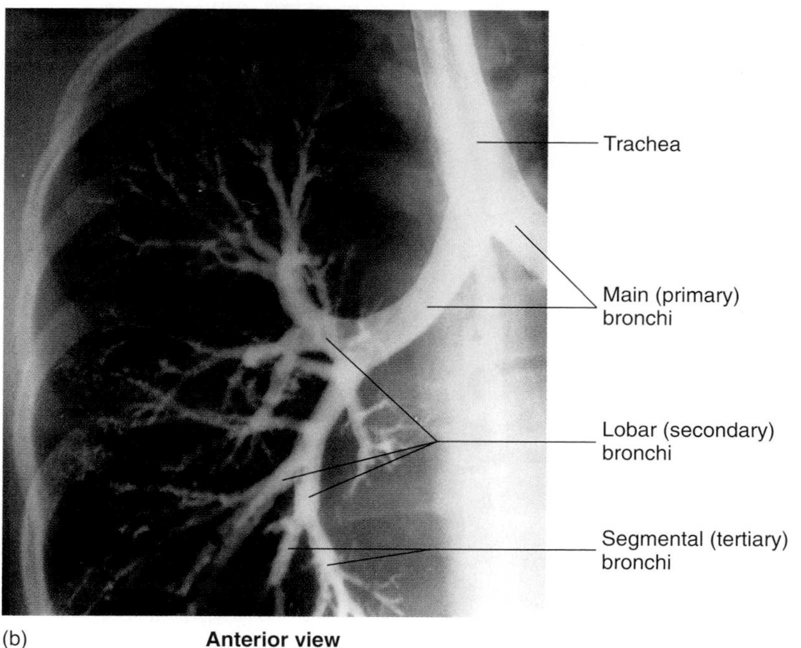

(b) Anterior view

FIGURE 23.6 Tracheobronchial Tree

(*a*) The conducting zone of the tracheobronchial tree begins at the trachea and ends at the terminal bronchioles. (*b*) A bronchogram is a radiograph of the tracheobronchial tree. A contrast medium, which makes the passageways visible, is injected through a catheter after a topical anesthetic is applied to the mucous membranes of the nose, pharynx, larynx, and trachea.

The terminal bronchioles divide to form **respiratory bronchioles** (figure 23.7), which have few attached alveoli. **Alveoli** (al-vē′ō-lī; hollow cavity) are small, air-filled chambers where gas exchange between the air and blood takes place. As the respiratory bronchioles divide to form smaller respiratory bronchioles, the number of attached alveoli increases. The respiratory bronchioles give rise to **alveolar** (al-vē′ō-lăr) **ducts,** which are like long, branching hallways with many open doorways. The "doorways" open into alveoli, which become so numerous that the alveolar duct wall is little more than a succession of alveoli. The alveolar ducts end as two or three **alveolar sacs,** which are chambers connected to two or more alveoli. Approximately seven

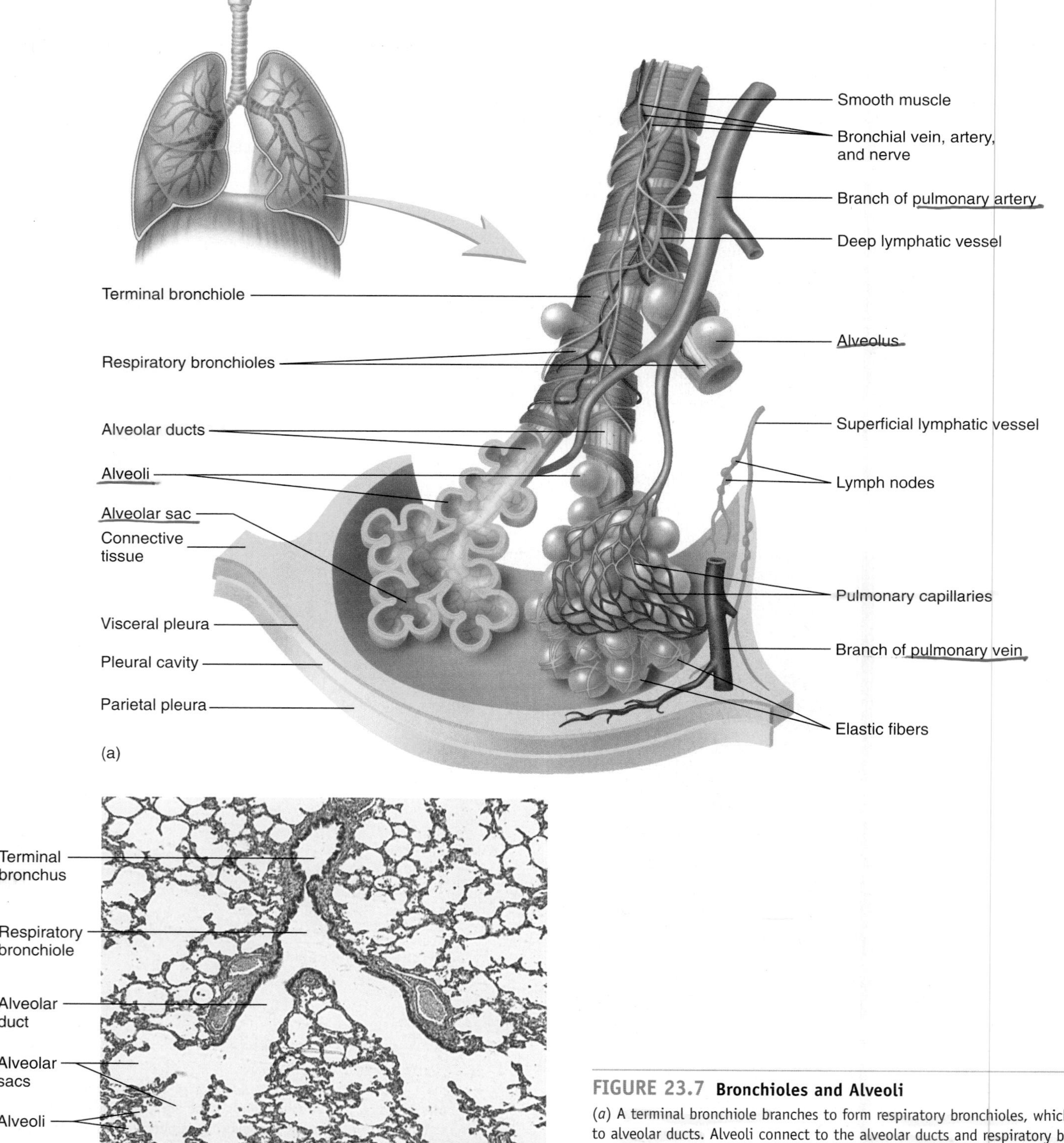

FIGURE 23.7 Bronchioles and Alveoli

(*a*) A terminal bronchiole branches to form respiratory bronchioles, which give rise to alveolar ducts. Alveoli connect to the alveolar ducts and respiratory bronchioles. The alveolar ducts end as two or three alveolar sacs. (*b*) Photomicrograph of lung tissue.

generations of branching occur from the terminal bronchioles to the alveolar ducts.

The tissue surrounding the alveoli contains elastic fibers, which allow the alveoli to expand during inspiration and recoil during expiration. The lungs are very elastic and, when inflated, are capable of expelling air and returning to their original, uninflated state. Even when not inflated, however, the lungs retain some air, which gives them a spongy quality.

The walls of respiratory bronchioles consist of collagenous and elastic connective tissue with bundles of smooth muscle. The epithelium in the respiratory bronchioles is a simple cuboidal epithelium. The alveolar ducts and alveoli consist of simple squamous epithelium. Although the epithelium of the respiratory zone is not ciliated, debris from the air can be removed by macrophages that move over the surfaces of the cells. The macrophages do not accumulate in the respiratory zone because they either move into nearby lymphatic vessels or enter terminal bronchioles, thereby becoming entrapped in mucus that is swept to the pharynx.

Approximately 300 million alveoli are in the two lungs. The average diameter of the alveoli is approximately 250 μm, and their walls are extremely thin. Two types of cells form the alveolar wall (figure 23.8*a*). **Type I pneumocytes** are thin squamous epithelial cells that form 90% of the alveolar surface. Most gas exchange between alveolar air and the blood takes place through these cells. **Type II pneumocytes** are round or cube-shaped secretory cells

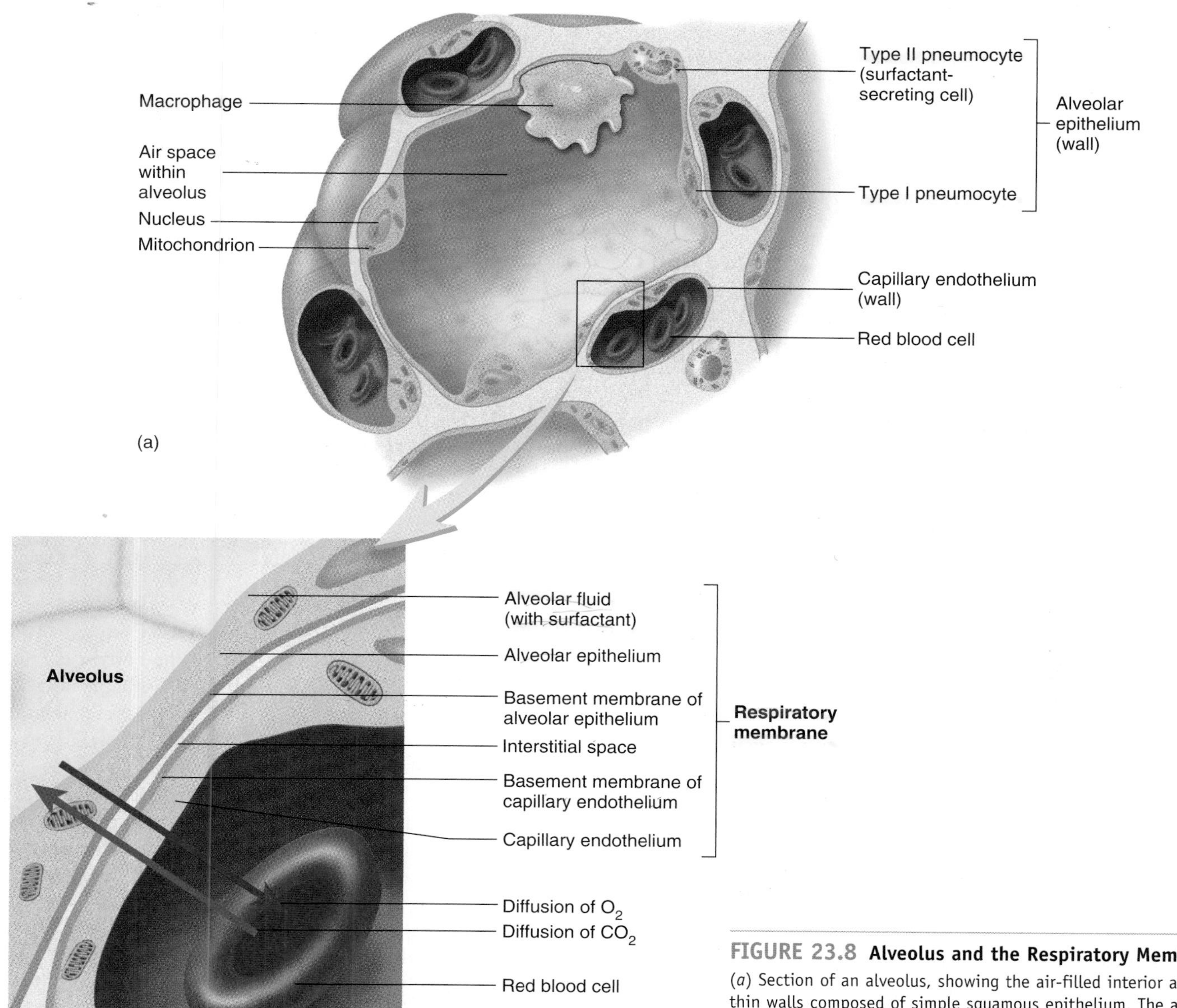

FIGURE 23.8 Alveolus and the Respiratory Membrane

(*a*) Section of an alveolus, showing the air-filled interior and thin walls composed of simple squamous epithelium. The alveolus is surrounded by elastic connective tissue and blood capillaries. (*b*) Diffusion of oxygen and carbon dioxide across the six thin layers of the respiratory membrane.

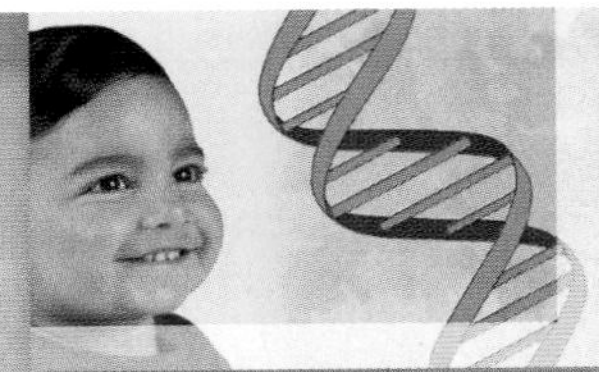

CLINICAL GENETICS

Alpha-1 Antitrypsin Deficiency

Emphysema (em-fi-zē′mă) is a condition in which lung alveoli become progressively destroyed and enlarged. Individuals suffering from emphysema experience shortness of breath and coughing. Chemicals in cigarette smoke damage lung tissues and stimulate inflammation. As part of the inflammatory response, neutrophils and macrophages release **proteases,** which are enzymes that break down proteins. Proteases in the lungs provide protection against some bacteria and foreign substances. Too much protease activity, however, can be harmful because it results in the breakdown of lung tissue proteins, especially elastin in elastic fibers. **Alpha-1 antitrypsin (AAT),** which is synthesized in the liver, is a **protease inhibitor (Pi).** Normally, AAT inhibits protease activity, preventing the destruction of lung tissue. Excess protease production stimulated by cigarette smoke, however, can cause lung damage, leading to emphysema.

Although cigarette smoking is the major risk factor for emphysema, approximately 1%–2% of emphysema cases are due to a deficiency of AAT caused by defects of the AAT gene located on chromosome 14. Multiple alleles for AAT have been identified. The normal allele is designated M. Individuals who are homozygous for the normal allele are designated PiMM, and they produce normal levels of AAT. That is, each M gene is responsible for 50% of the AAT produced. The most common abnormal allele is designated Z. Individuals with only one copy of Z (PiMZ) have about 60% of normal levels of AAT, which is sufficient to prevent protease damage. Individuals with two copies of the Z allele (PiZZ) produce only about 15%–20% of normal AAT levels. Smoking by these individuals accelerates the development of emphysema by 10–15 years. Other variant alleles cause different levels of AAT. The most severe form results in no AAT and the development of emphysema by age 30, even in nonsmokers.

Treatment of AAT deficiency follows the normal course of treatment for emphysema. Stopping smoking reduces the destruction of lung tissue by removing the stimulus for excess protease activity. Drugs, such as danazol and tamoxifen, can stimulate increased AAT production in the liver. In addition, individuals may receive intravenous infusions of AAT, a process called **alpha-1 antitrypsin augmentation.**

that produce surfactant, which makes it easier for the alveoli to expand during inspiration (see "Lung Recoil," p. 842).

8 ***Describe the arrangement of cartilage, smooth muscle, and epithelium in the tracheobronchial tree. Explain why breathing becomes more difficult during an asthma attack.***

9 ***How is debris removed from the tracheobronchial tree?***

10 ***Name the two types of cells in the alveolar wall, and state their functions.***

The Respiratory Membrane

The **respiratory membrane** of the lungs is where gas exchange between the air and blood takes place. It is formed mainly by the alveolar walls and surrounding pulmonary capillaries (figure 23.8*b*), but there is some contribution by the respiratory bronchioles and alveolar ducts. The respiratory membrane is very thin to facilitate the diffusion of gases. It consists of

1. a thin layer of fluid lining the alveolus
2. the alveolar epithelium composed of simple squamous epithelium
3. the basement membrane of the alveolar epithelium
4. a thin interstitial space
5. the basement membrane of the capillary endothelium
6. the capillary endothelium composed of simple squamous epithelium

11 ***List the parts of the respiratory membrane.***

PREDICT  3

Based on function, the respiratory passageways can be subdivided into the conducting and respiratory zones. The conducting zone functions as a passageway for the exchange of air between the outside of the body and the respiratory zone is where gas exchange between the air and blood takes place. Name in order the parts of the conducting and respiratory zones, starting with the nose and ending with the alveoli.

Lungs

The **lungs** are the principal organs of respiration, and on a volume basis they are among the largest organs of the body. Each lung is conical in shape, with its base resting on the diaphragm and its apex extending superiorly to a point approximately 2.5 cm superior to the clavicle. The right lung is larger than the left and weighs an average of 620 g, whereas the left lung weighs an average of 560 g.

The **hilum** (hī′lŭm) is a region on the medial surface of the lung where structures, such as the main bronchus, blood vessels, nerves, and lymphatic vessels, enter or exit the lung. All the structures passing through the hilum are referred to as the **root of the lung.**

The right lung has three **lobes,** and the left lung has two (figure 23.9). The lobes are separated by deep, prominent **fissures** on the surface of the lung, and each lobe is supplied by a lobar bronchus. The lobes are subdivided into **bronchopulmonary segments,** which are supplied by the segmental bronchi. Nine bronchopulmonary segments are present in the left lung, and 10 are present in the right lung. The bronchopulmonary segments are

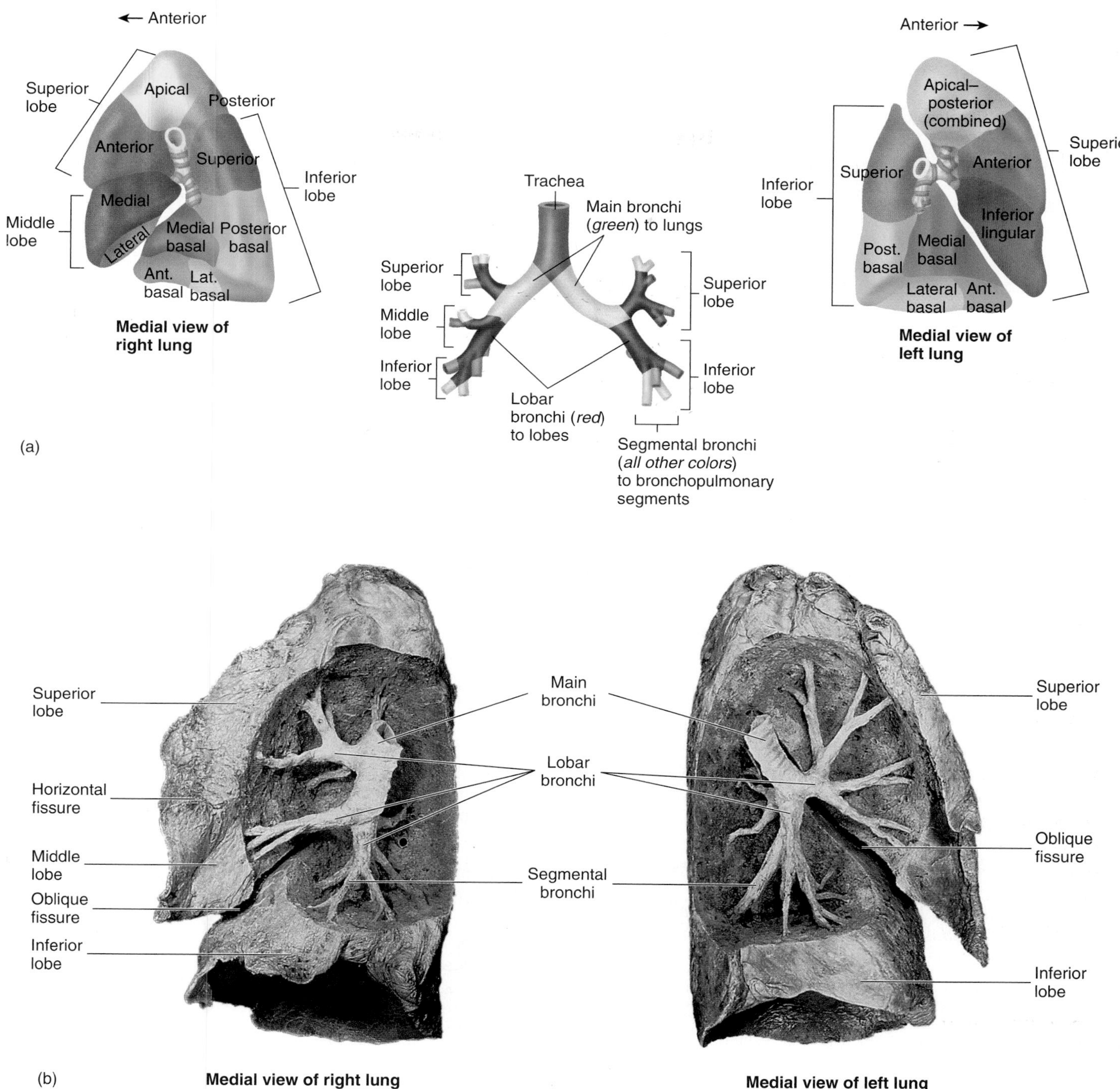

FIGURE 23.9 Lobes and Bronchopulmonary Segments of the Lungs

(*a*) The trachea (*blue*), main bronchi (*green*), lobar bronchi (*red*), and segmental bronchi (*all other colors*) are in the center of the figure, surrounded by a medial view of each lung, showing the bronchopulmonary segments. In general, each bronchopulmonary segment is supplied by a segmental bronchus (*color-coded to match the bronchopulmonary segment it supplies*). (*b*) Photograph of the lungs, showing the lung lobes and bronchi. The right lung is divided into three lobes by the horizontal and oblique fissures. The left lung is divided into two lobes by the oblique fissure. A main bronchus supplies each lung, a lobar bronchus supplies each lung lobe, and segmental bronchi supply the bronchopulmonary segments (*not visible*).

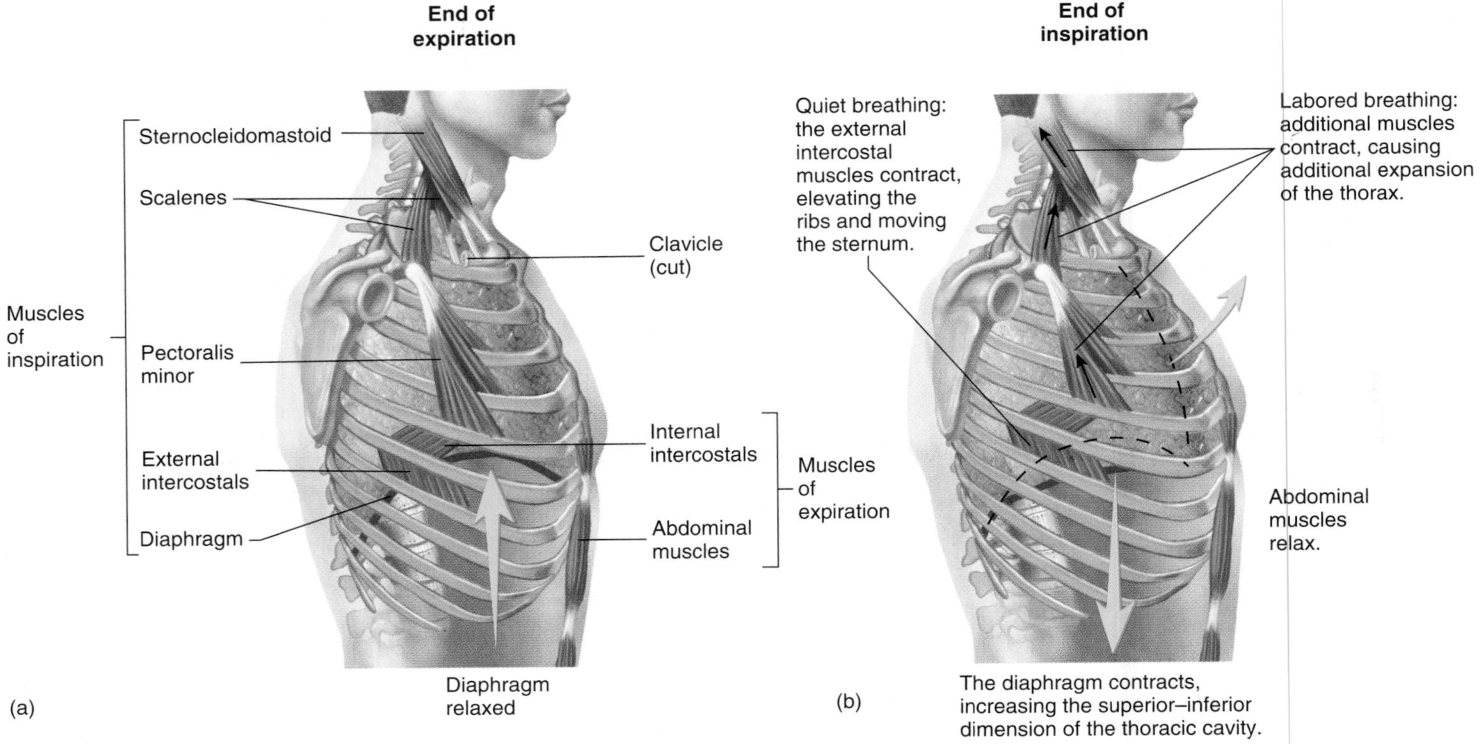

FIGURE 23.10 Effect of the Muscles of Respiration on Thoracic Volume
(*a*) Muscles of respiration at the end of expiration. (*b*) Muscles of respiration at the end of inspiration.

separated from each other by connective tissue partitions, which are not visible as surface fissures. Individual diseased bronchopulmonary segments can be surgically removed, leaving the rest of the lung relatively intact, because major blood vessels and bronchi do not cross the connective tissue partitions. The bronchopulmonary segments are subdivided into **lobules** by incomplete connective tissue walls. The lobules are supplied by the bronchioles.

12 ***Distinguish among a lung, a lung lobe, a bronchopulmonary segment, and a lobule. How are they related to the tracheobronchial tree?***

Thoracic Wall and Muscles of Respiration

The **thoracic wall** consists of the thoracic vertebrae, ribs, costal cartilages, sternum, and associated muscles (see chapters 7 and 10). The **thoracic cavity** is the space enclosed by the thoracic wall and the **diaphragm** (dī′ă-fram; partition), which separates the thoracic cavity from the abdominal cavity. The diaphragm and other skeletal muscles associated with the thoracic wall are responsible for respiration (figure 23.10). The **muscles of inspiration** include the diaphragm, external intercostals, pectoralis minor, and scalenes. Contraction of the diaphragm is responsible for approximately two-thirds of the increase in thoracic volume during inspiration. The external intercostals, pectoralis minor, and scalene muscles also increase thoracic volume by elevating the ribs. The **muscles of expiration** consist of the muscles that depress the ribs and sternum, such as the abdominal muscles and the internal intercostals. Although the internal intercostals are most active during expiration, and the external intercostals are most active during inspiration, the primary function of these muscles is to stiffen the thoracic wall by contracting at the same time. By doing so, they prevent inward collapse of the thoracic cage during inspiration.

Effect of Spinal Cord Injury on Ventilation

The diaphragm is supplied by the phrenic nerves, which arise from spinal nerves C3–C5 (see figure 12.16), descend along each side of the neck to enter the thorax, and pass to the diaphragm. The intercostal muscles are supplied by the intercostal nerves (see figure 12.15), which arise from spinal nerves T1–T11 and extend along the spaces between the ribs. Spinal cord injury superior to the origin of the phrenic nerves causes paralysis of the diaphragm and intercostal muscles and results in death unless artificial respiration is provided. A high spinal cord injury below the origin of the phrenic nerves causes paralysis of the intercostal muscles. Even though the diaphragm can function maximally, ventilation is drastically reduced because the intercostal muscles no longer prevent the thoracic wall from collapsing inward. Vital capacity is reduced to about 300 mL. With low spinal cord injury, below the origin of the intercostal nerves, both the diaphragm and the intercostal muscles function normally.

The diaphragm is dome-shaped, and the base of the dome attaches to the inner circumference of the inferior thoracic cage (see figure 10.18). The top of the dome is a flat sheet of connective tissue called the **central tendon.** In normal, quiet inspiration, contraction

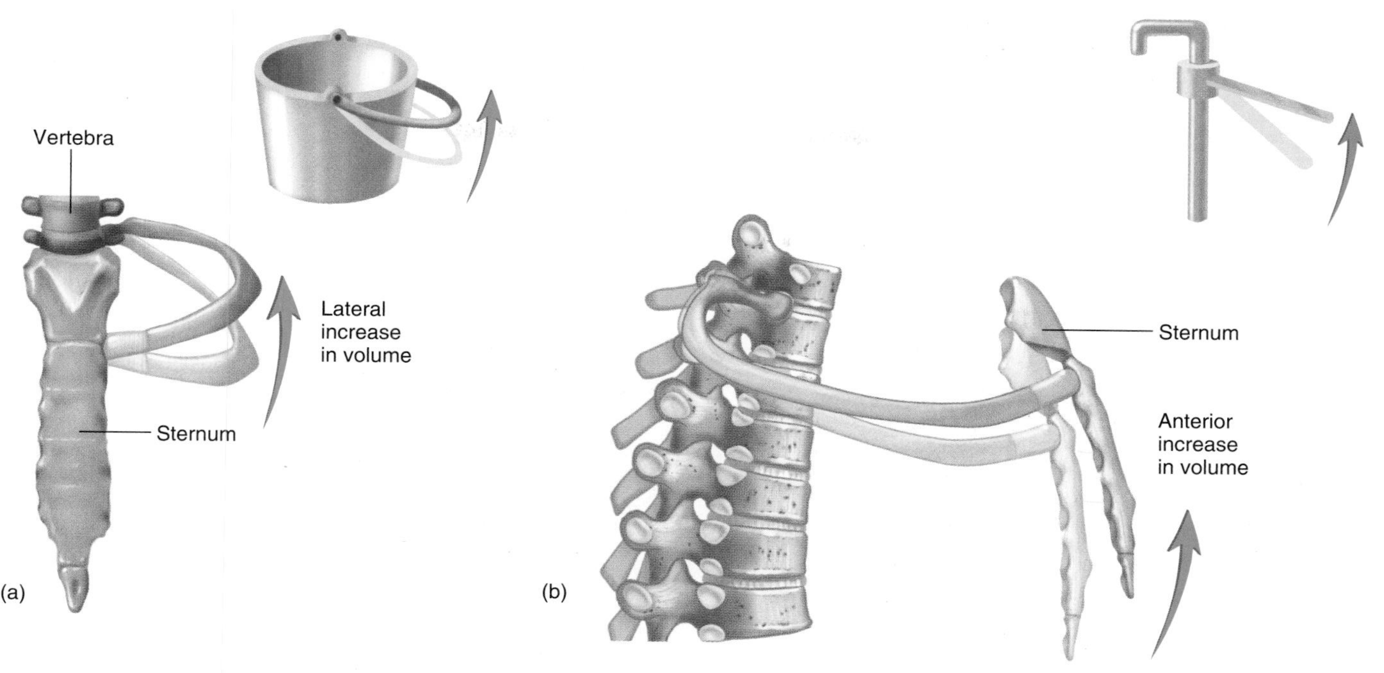

FIGURE 23.11 Effect of Rib and Sternum Movement on Thoracic Volume
(*a*) Elevation of the rib in the "bucket-handle" movement laterally increases thoracic volume. (*b*) As the rib is elevated, rotation of the rib in the "pump-handle" movement increases thoracic volume anteriorly.

of the diaphragm results in inferior movement of the central tendon with very little change in the overall shape of the dome. Inferior movement of the central tendon can occur because of relaxation of the abdominal muscles, which allows the abdominal organs to move out of the way of the diaphragm. As the depth of inspiration increases, inferior movement of the central tendon is prevented by the abdominal organs. Continued contraction of the diaphragm causes it to flatten as the lower ribs are elevated. In addition, other muscles of inspiration can elevate the ribs. As the ribs are elevated, the costal cartilages allow lateral rib movement and lateral expansion of the thoracic cavity (figure 23.11). The ribs slope inferiorly from the vertebrae to the sternum, and elevation of the ribs also increases the anterior–posterior dimension of the thoracic cavity.

Expiration during quiet breathing occurs when the diaphragm and external intercostals relax and the elastic properties of the thorax and lungs cause a passive decrease in thoracic volume. In addition, contraction of the abdominal muscles helps push abdominal organs and the diaphragm in a superior direction.

Role of Abdominal Muscles in Breathing

The importance of the abdominal muscles in breathing can be observed in a person with a spinal cord injury that causes flaccid paralysis of the abdominal muscles. In the upright position, the abdominal organs and diaphragm are not pushed superiorly and passive recoil of the thorax and lungs is inadequate for normal expiration. An elastic binder around the abdomen can help such patients. When a person is lying down, the weight of the abdominal organs can assist in expiration.

Several differences can be recognized between normal, quiet breathing and labored breathing. During labored breathing, all of the inspiratory muscles are active, and they contract more forcefully than during quiet breathing, causing a greater increase in thoracic volume (see figure 23.10*b*). During labored breathing, forceful contraction of the internal intercostals and the abdominal muscles produces a more rapid and greater decrease in thoracic volume than would be produced by the passive recoil of the thorax and lungs.

Pleura

The lungs are contained within the thoracic cavity, but each lung is surrounded by a separate **pleural** (ploor′ăl; relating to the ribs) **cavity** formed by the pleural serous membranes (figure 23.12). The **mediastinum** (mē′dē-as-tī′nŭm), a midline partition formed by the heart, trachea, esophagus, and associated structures, separates the pleural cavities. The **parietal pleura** covers the inner thoracic wall, the superior surface of the diaphragm, and the mediastinum. At the hilum, the parietal pleura is continuous with the **visceral pleura,** which covers the surface of the lung.

The pleural cavity is filled with pleural fluid, which is produced by the pleural membranes. The pleural fluid does two things: (1) It acts as a lubricant, allowing the parietal and visceral pleural membranes to slide past each other as the lungs and the thorax change shape during respiration, and (2) it helps hold the parietal and visceral pleural membranes together. When thoracic volume changes during respiration, lung volume changes because the parietal pleura is attached to the diaphragm and inner thoracic

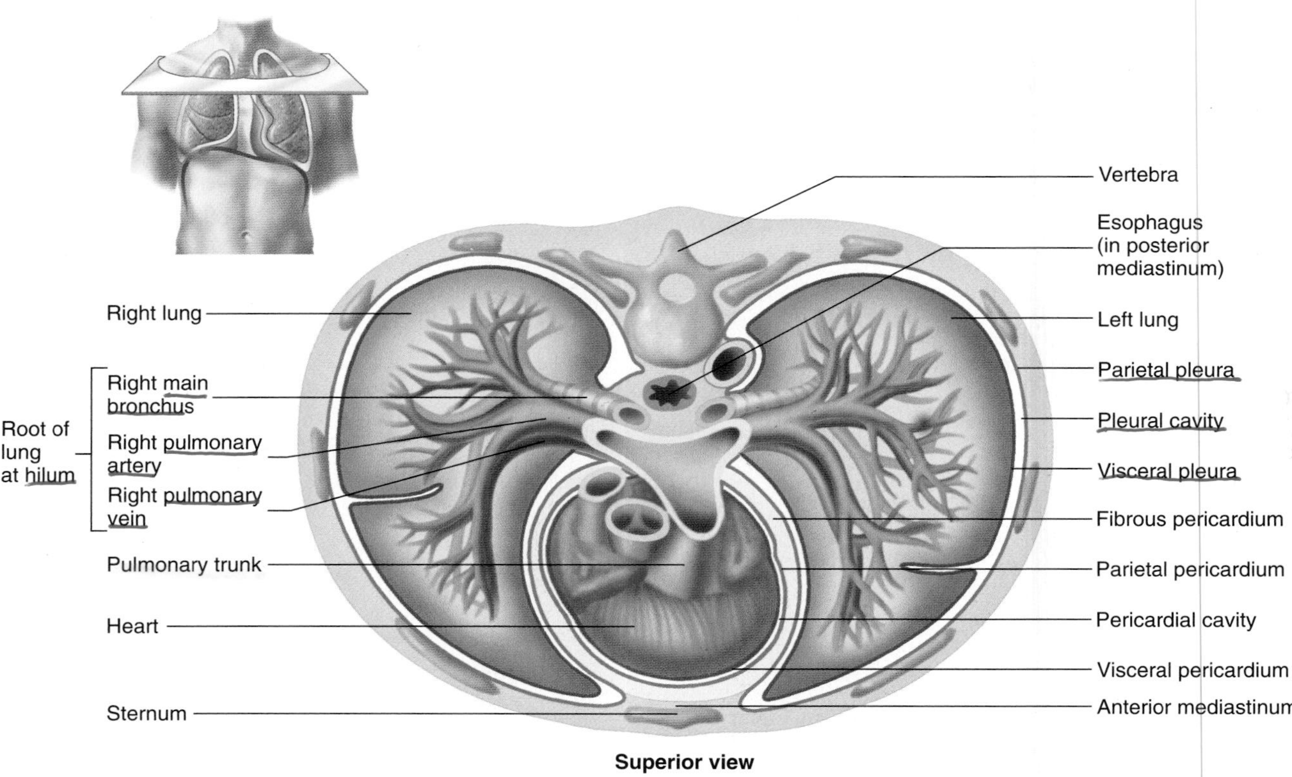

FIGURE 23.12 Pleural Cavities and Membranes

Transverse section of the thorax, showing the relationship of the pleural cavities to the thoracic organs. Each lung is surrounded by a pleural cavity. The parietal pleura lines the wall of each pleural cavity, and the visceral pleura covers the surface of the lungs. The space between the parietal and visceral pleurae is small and filled with pleural fluid.

wall, and the visceral pleura is attached to the lungs. The pleural fluid is analogous to a thin film of water between two sheets of glass (the visceral and parietal pleurae); the glass sheets can easily slide over each other, but it is difficult to separate them.

Blood Supply

Blood that has passed through the lungs and picked up oxygen is called **oxygenated blood,** and blood that has passed through the tissues and released some of its oxygen is called **deoxygenated blood.** Two blood flow routes to the lungs exist. The major route takes deoxygenated blood to the lungs, where it is oxygenated (see chapter 21 and figure 23.12). The deoxygenated blood flows through pulmonary arteries to pulmonary capillaries, becomes oxygenated, and returns to the heart through pulmonary veins. The other route takes oxygenated blood to the tissues of the bronchi down to the respiratory bronchioles. The oxygenated blood flows from the thoracic aorta through bronchial arteries to capillaries, where oxygen is released. Deoxygenated blood from the proximal part of the bronchi returns to the heart through the bronchial veins and the azygos venous system (see chapter 21). More distally, the venous drainage from the bronchi enters the pulmonary veins. Thus, the oxygenated blood returning from the alveoli in the pulmonary veins is mixed with a small amount of deoxygenated blood returning from the bronchi.

Lymphatic Supply

The lungs have two lymphatic supplies. The **superficial lymphatic vessels** are deep to the visceral pleura; they drain lymph from the superficial lung tissue and the visceral pleura. The **deep lymphatic vessels** follow the bronchi; they drain lymph from the bronchi and associated connective tissues. No lymphatic vessels are located in the walls of the alveoli. Both the superficial and deep lymphatic vessels exit the lung at the hilum.

Phagocytic cells within the lungs phagocytize carbon particles and other debris from inspired air and move them to the lymphatic vessels. In older people, the surface of the lungs can appear gray to black because of the accumulation of these particles, especially if the person smokes or has lived most of his or her life in a city with air pollution. Cancer cells from the lungs can spread to other parts of the body through the lymphatic vessels.

13 ***List the muscles of respiration and describe their role in quiet inspiration and expiration. How does this change during labored breathing?***

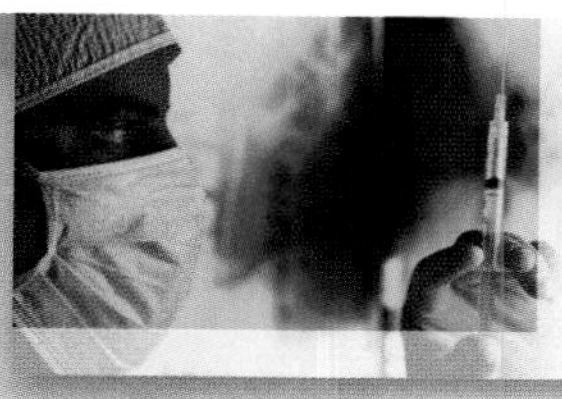

CLINICAL FOCUS

Cough and Sneeze Reflexes

The function of both the cough reflex and the sneeze reflex is to dislodge foreign matter or irritating material from the respiratory passages. The bronchi and trachea contain sensory receptors that are sensitive to foreign particles and irritating substances. The cough reflex is initiated when the sensory receptors detect such substances and initiate action potentials that pass along the vagus nerves to the medulla oblongata, where the cough reflex is triggered.

The movements resulting in a cough occur as follows: Approximately 2.5 L of air are inspired; the vestibular and vocal folds close tightly to trap the inspired air in the lungs; the abdominal muscles contract to force the abdominal contents up against the diaphragm; and the muscles of expiration contract forcefully. As a consequence, the pressure in the lungs increases to 100 mm Hg or more. Then the vestibular and vocal folds open suddenly, the soft palate is elevated, and the air rushes from the lungs and out the oral cavity at a high velocity, carrying foreign particles with it.

The sneeze reflex is similar to the cough reflex, but it differs in several ways. The source of irritation that initiates the sneeze reflex is in the nasal passages instead of in the trachea and bronchi, and the action potentials are conducted along the trigeminal nerves to the medulla oblongata, where the reflex is triggered. During the sneeze reflex, the soft palate is depressed so that air is directed primarily through the nasal passages, although a considerable amount passes through the oral cavity. The rapidly flowing air dislodges particulate matter from the nasal passages and can propel it a considerable distance from the nose.

About 17%–25% of people have a photic sneeze reflex, in which exposure to bright light, such as the sun, can stimulate a sneeze reflex. The pupillary reflex causes the pupils to constrict in response to bright light. It is speculated that the complicated "wiring" of the pupillary and sneeze reflexes are intermixed in some people so that, when bright light activates a pupillary reflex, it also activates a sneeze reflex. Sometimes the photic sneeze reflex is fancifully called ACHOO, which stands for *a*utosomal dominant *c*ompelling *h*elio-ophthalmic *o*utburst. As the name suggests, the reflex is inherited as an autosomal-dominant trait. A person needs to inherit only one copy of the gene to have a photic sneeze reflex.

14 ***Name the pleurae of the lungs. What is their function?***

15 ***What are the two major routes of blood flow to and from the lungs? What is the function of each route?***

16 ***Describe the lymphatic supply of the lungs.***

VENTILATION

Pressure Differences and Airflow

Ventilation is the process of moving air into and out of the lungs. The flow of air into the lungs requires a pressure gradient from the outside of the body to the alveoli, and airflow from the lungs requires a pressure gradient in the opposite direction. The physics of airflow in tubes, such as the ones that make up the respiratory passages, is the same as that of the flow of blood in blood vessels (see chapter 21). Thus, the following relationships hold:

$$F = \frac{P_1 - P_2}{R}$$

where F is airflow (milliliters per minute) in a tube, P_1 is pressure at point one, P_2 is pressure at point two, and R is resistance to airflow.

Air moves through tubes because of a pressure difference. When P_1 is greater than P_2, gas flows from P_1 to P_2 at a rate that is proportional to the pressure difference. For example, during inspiration, air pressure outside the body is greater than air pressure in the alveoli, and air flows through the trachea and bronchi to the alveoli.

Disorders That Decrease the Radius of Air Passageways

The flow of air decreases when the resistance to airflow is increased by conditions that reduce the radius of the respiratory passageways. According to Poiseuille's law (see chapter 21), the resistance to airflow is proportional to the radius (r) of a tube raised to the fourth power (r^4). Thus, a small change in radius results in a large change in resistance, which greatly decreases airflow. For example, asthma results in the release of inflammatory chemicals, such as leukotrienes, that cause severe constriction of the bronchioles. Emphysema produces increased airway resistance because the bronchioles are obstructed as a result of inflammation and because damaged bronchioles collapse during expiration, thus trapping air within the alveoli. Cancer can also occlude respiratory passages as the tumor replaces lung tissue. Increasing the pressure difference between alveoli and the atmosphere can help maintain airflow despite increased resistance. Within limits, this can be accomplished by increased contraction of the muscles of respiration.

Pressure and Volume

The pressure of a gas, such as air, in a container at a constant temperature is described according to **Boyle's law:**

$$P = k/V$$

where P is gas pressure, k is a constant for a given temperature, and V is the volume of the container. Body temperature in humans can be considered a constant. Thus, Boyle's law reveals that, in a

TABLE 23.1 Gas Laws

Description	Importance
Boyle's Law	
The pressure of a gas is inversely proportional to its volume at a given temperature.	Air flows from areas higher to lower in pressure. When alveolar volume increases, causing pleural pressure to decrease below atmosphereic pressure, air moves into the lungs. When alveolar volume decreases, causing pleural pressure to increase above atmospheric pressure, air moves out of the lungs.
Dalton's Law	
The partial pressure of a gas in a mixture of gases is the percentage of the gas in the mixture times the total pressure of the mixture of gases.	Gases move from areas of higher to areas of lower partial pressures. The greater the difference in partial pressure between two points, the greater the rate of gas movement. Maintaining partial pressure differences ensures gas movements.
Henry's Law	
The concentration of a gas dissolved in a liquid is equal to the partial pressure of the gas over the liquid times the solubility coefficient of the gas.	Only a small amount of the gases in air dissolves in the fluid lining the alveoli. Carbon dioxide, however, is 24 times more soluble than oxygen; therefore, carbon dioxide exits through the respiratory membrane more readily than oxygen enters.

container, such as the thoracic cavity or an alveolus, pressure is inversely proportional to volume. As volume increases, pressure decreases; as volume decreases, pressure increases (table 23.1).

17 ***Define ventilation.***

18 ***How do pressure differences and resistance affect airflow through a tube?***

19 ***What happens to the pressure within a container when the volume of the container increases?***

Airflow Into and Out of Alveoli

Respiratory physiologists use three conventions to help simplify the numbers used to express pressures:

1. **Barometric air pressure (P_B)**, which is atmospheric air pressure outside the body, is assigned a value of zero. Thus, whether at sea level with a pressure of 760 mm Hg or at 10,000 feet above sea level on a mountaintop with a pressure of 523 mm Hg, P_B is always assigned a value of zero.
2. The small pressures in respiratory physiology are usually expressed in centimeters of water (cm H_2O). A pressure of 1 cm H_2O is equal to 0.74 mm Hg.
3. Other pressures are measured in reference to barometric air pressure. For example, **alveolar pressure (P_{alv})** is the pressure inside an alveolus. An alveolar pressure of 1 cm H_2O is 1 cm H_2O greater pressure than barometric air pressure, and an alveolar pressure of -1 cm H_2O is 1 cm H_2O less pressure than barometric air pressure.

The movement of air into and out of the lungs results from changes in thoracic volume, which cause changes in alveolar volume. The changes in alveolar volume produce changes in alveolar pressure. The pressure difference between barometric air pressure and alveolar pressure ($P_B - P_{alv}$) results in air movement. The details of this process during quiet breathing are as follows:

1. *End of expiration* (figure 23.13, step 1). At the end of expiration, barometric air pressure and alveolar pressure are equal. Therefore, no movement of air into or out of the lungs takes place.
2. *During inspiration* (figure 23.13, step 2). As inspiration begins, contraction of inspiratory muscles increases thoracic volume, which results in expansion of the lungs and an increase in alveolar volume (see "Changing Alveolar Volume," on this page). The increased alveolar volume causes a decrease in alveolar pressure below barometric air pressure to approximately -1 cm H_2O. Air flows into the lungs because barometric air pressure is greater than alveolar pressure.
3. *End of inspiration* (figure 23.13, step 3). At the end of inspiration, the thorax stops expanding, the alveoli stop expanding, and alveolar pressure becomes equal to barometric air pressure because of airflow into the lungs. No movement of air occurs after alveolar pressure becomes equal to barometric pressure, but the volume of the lungs is larger than at the end of expiration.
4. *During expiration* (figure 23.13, step 4). During expiration, the volume of the thorax decreases as the diaphragm relaxes, and the thorax and lungs recoil. The decreased thoracic volume results in a decrease in alveolar volume and an increase in alveolar pressure over barometric air pressure to approximately 1 cm H_2O. Air flows out of the lungs because alveolar pressure is greater than barometric air pressure. As expiration ends, the decrease in thoracic volume stops and the alveoli stop changing size. The process repeats beginning at step 1.

20 ***Define barometric and alveolar pressures.***

21 ***Explain how changes in alveolar volume cause air to move into and out of the lungs.***

Changing Alveolar Volume

It is important to understand how alveolar volume is changed because these changes cause the pressure differences resulting in ventilation. In addition, many respiratory disorders affect how alveolar volume changes. Lung recoil and changes in pleural pressure cause changes in alveolar volume.

Lung Recoil

Lung recoil is a decrease in size of an expanded lung. A lung decreases in size as a result of a decrease in size (volume) of its alveoli. Expanded

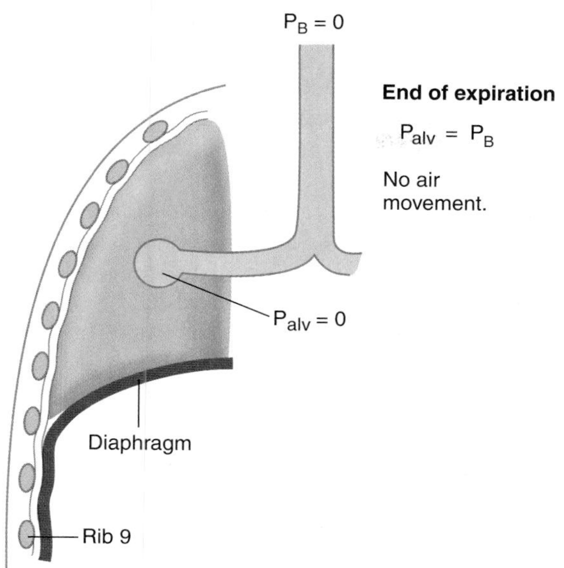

1. At the end of expiration, alveolar pressure (P_{alv}) is equal to barometric air pressure (P_B) and there is no air movement.

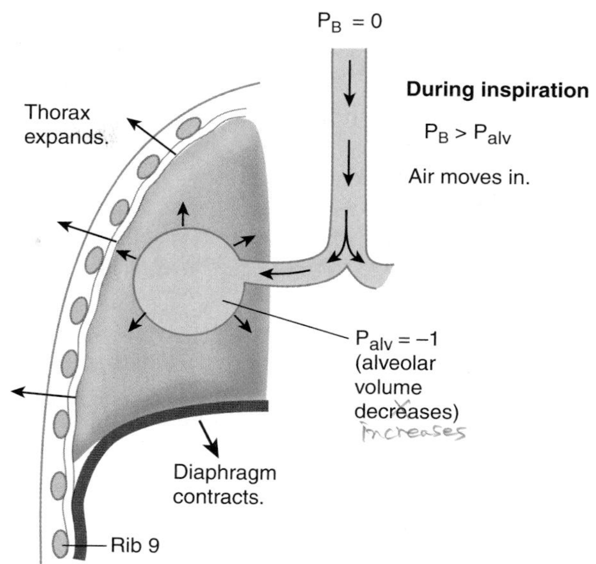

2. During inspiration, increased thoracic volume results in increased alveolar volume and decreased alveolar pressure. Barometric air presure is greater than alveolar pressure, and air moves into the lungs.

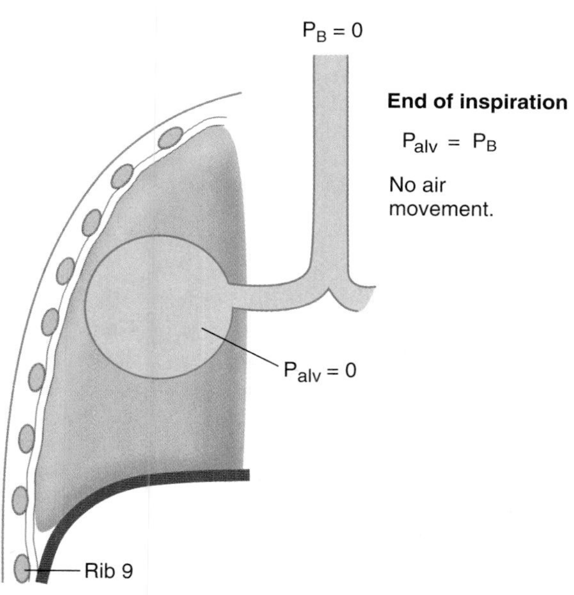

3. At the end of inspiration, alveolar pressure is equal to barometric air pressure and there is no air movement.

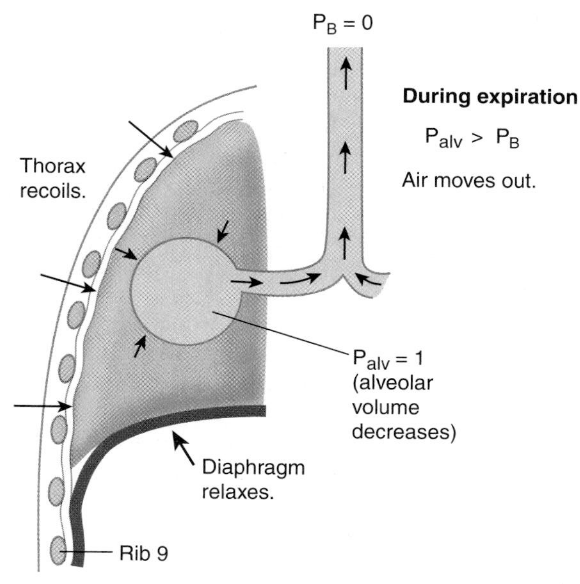

4. During expiration, decreased thoracic volume results in decreased alveolar volume and increased alveolar pressure. Alveolar pressure is greater than barometric air pressure, and air moves out of the lungs.

PROCESS FIGURE 23.13 Alveolar Pressure Changes During Inspiration and Expiration

The combined space of all the alveoli is represented by a large "bubble." The alveoli are actually microscopic and cannot be seen in the illustration.

alveoli decrease in size for two reasons: (1) elastic recoil caused by the elastic fibers in the alveolar walls and (2) surface tension of the film of fluid that lines the alveoli. Surface tension occurs at the boundary between water and air because the polar water molecules are attracted to one another more than they are attracted to the air molecules. Consequently, the water molecules are drawn together, tending to form a droplet. Because the water molecules of the alveolar fluid are also attracted to the surface of the alveoli, formation of a droplet causes the alveoli to collapse, thus producing fluid-filled alveoli with smaller volumes than air-filled alveoli.

Surfactant (ser-fak′tănt; *surf*ace *act*ing *agent*) is a mixture of lipoprotein molecules produced by the type II pneumocytes of the alveolar epithelium. The surfactant molecules form a monomolecular layer over the surface of the fluid within the alveoli to reduce the surface tension. With surfactant, the force produced by surface tension is approximately 4 cm H_2O; without surfactant, the force can be as high as 40 cm H_2O. Thus, surfactant greatly reduces the tendency of the lungs to collapse.

Infant Respiratory Distress Syndrome

In premature infants, **infant respiratory distress syndrome,** or **hyaline membrane disease,** is common, especially for infants with a gestation age of less than 7 months. This occurs because surfactant is not produced in adequate quantities until approximately 7 months of development. Thereafter, the amount produced increases as the fetus matures. Cortisol can be given to pregnant women who are likely to deliver prematurely because it crosses the placenta into the fetus and stimulates surfactant synthesis.

If a newborn produces insufficient surfactant, the lungs tend to collapse. Thus, a great deal of energy must be exerted by the muscles of respiration to keep the lungs inflated, and even then inadequate ventilation occurs. Without specialized treatment, most babies with this disease die soon after birth as a result of inadequate ventilation of the lungs and fatigue of the respiratory muscles. Positive end-expiratory pressure delivers oxygen-rich, pressurized air to the lungs through a tube passed through the respiratory passages. The pressure helps keep the alveoli inflated. In addition, surfactant administered with the pressurized air can reduce surface tension in the alveoli. Surfactant can be obtained from cow, pig, and human lungs; human amniotic fluid; and genetically modified bacteria. Synthetic surfactant is also available.

22 ***Name two things that cause the lungs to recoil. How does surfactant reduce lung recoil? What happens if there are inadequate amounts of surfactant in the alveoli?***

Pleural Pressure

Pleural pressure (P_{pl}) is the pressure in the pleural cavity. When pleural pressure is less than alveolar pressure, the alveoli tend to expand. This principle can be understood by considering a balloon. The balloon expands when the pressure outside the balloon is less than the pressure inside. This pressure difference is normally achieved by increasing the pressure inside the balloon by blowing forcefully into it. This pressure difference, however, can also be achieved by decreasing the pressure outside the balloon. For example, if the balloon is placed in a chamber from which air is removed, the pressure around the balloon becomes lower than atmospheric pressure, and the balloon expands. The lower the pressure outside the balloon, the greater the tendency for the higher pressure inside the balloon to cause it to expand. In a similar fashion, decreasing pleural pressure can result in expansion of the alveoli.

Normally, the alveoli are expanded because of a negative pleural pressure that is lower than alveolar pressure. At the end of a normal expiration, pleural pressure is −5 cm H_2O, and alveolar pressure is 0 cm H_2O. Pleural pressure is lower than alveolar pressure because of a suction effect caused by fluid removal by the lymphatic system and by lung recoil. As the lungs recoil, the visceral and parietal pleurae tend to be pulled apart. Normally, the lungs do not pull away from the thoracic wall because pleural fluid holds the visceral and parietal pleurae together. Nonetheless, this pull decreases pressure in the pleural cavity, an effect that can be appreciated by putting water on the palms of the hands and putting them together. A sensation of negative pressure is felt as the hands are gently pulled apart.

When pleural pressure is lower than alveolar pressure, the alveoli tend to expand. This expansion is opposed by the tendency of the lungs to recoil. Therefore, the alveoli expand when the pleural pressure is low enough that lung recoil is overcome. If the pleural pressure is not low enough to overcome lung recoil, the alveoli collapse.

Pneumothorax

A **pneumothorax** is the introduction of air into the pleural cavity through an opening in the thoracic wall or lung. A pneumothorax can result from penetrating trauma by a knife, a bullet, a broken rib, or another object; nonpenetrating trauma, such as a blow to the chest; medical procedures, such as inserting a catheter to withdraw pleural fluid; or disease, such as infections or emphysema or the cause can be unknown.

Pleural pressure increases and becomes equal to barometric air pressure when the pleural cavity is connected to the outside through an opening in the thoracic wall or the surface of the lung. The alveoli, therefore, do not tend to expand, lung recoil is unopposed, and the lung collapses and pulls away from the thoracic wall. A pneumothorax can occur in one lung while the other remains inflated because the two pleural cavities are separated by the mediastinum.

The most common symptoms of a pneumothorax are chest pain and shortness of breath. Treatment of a pneumothorax depends on its cause and severity. In patients with mild symptoms, the pneumothorax may resolve on its own. In other cases, a chest tube that aspirates the pleural cavity and restores a negative pressure can cause reexpansion of the lung. Surgery may also be necessary to close the opening into the pleural cavity.

In a **tension pneumothorax,** the pressure within the thoracic cavity is always higher than barometric air pressure. A tissue flap or an air passageway forms a flutter valve, which allows air to enter the pleural cavity during inspiration but not exit during expiration. The result is an increase in air and pressure within the pleural cavity, which can compress blood vessels returning blood to the heart, causing decreased venous return, low blood pressure, and inadequate delivery of oxygen to tissues. The insertion of a large bore needle into the pleural cavity allows air to escape and releases the pressure.

23 ***Define pleural pressure. What happens to alveolar volume when pleural pressure decreases? What causes pleural pressure to be lower than alveolar pressure?***

24 ***How does an opening in the chest wall cause the lung to collapse?***

25 ***During inspiration, what causes pleural pressure to decrease? What effect does this have on alveolar pressure and air movement?***

26 ***During expiration, what causes pleural pressure to increase? What effect does this have on alveolar pressure and air movement?***

Pressure Changes During Inspiration and Expiration

At the end of a normal expiration, pleural pressure is -5 cm H_2O, and alveolar pressure is equal to barometric pressure (0 cm H_2O) (figure 23.14). During normal, quiet inspiration, pleural pressure decreases to -8 cm H_2O. Consequently, the alveolar volume increases, alveolar pressure decreases below barometric air pressure, and air flows into the lungs. As air flows into the lungs, alveolar pressure increases and becomes equal to barometric pressure at the end of inspiration.

The decrease in pleural pressure during inspiration occurs for two reasons. First, because of the effect of changing volume on pressure (Boyle's law), when the volume of the thoracic cavity increases, pleural pressure decreases. Second, as the thoracic cavity expands, the lungs expand because they adhere to the inner thoracic wall through the pleurae. As the lungs expand, the tendency for the lungs to recoil increases, resulting in an increased suction effect and a lowering of pleural pressure. The tendency for the lungs to recoil increases as the lungs are stretched, similar to the increased force generated in a stretched rubber band.

Changes during inspiration

Changes during expiration

1. Pleural pressure decreases because thoracic volume increases.

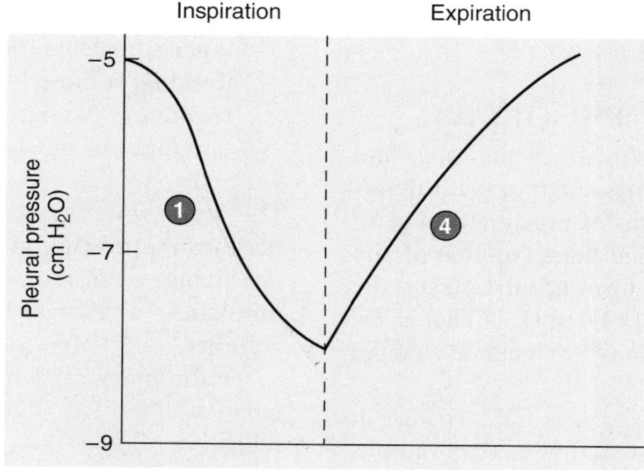

4. Pleural pressure increases because thoracic volume decreases.

2. As inspiration begins, alveolar pressure decreases below barometric air pressure (0 on the graph) because the decreased pleural pressure causes alveolar volume to increase. By the end of inspiration, alveolar and barometric air pressure are equal.

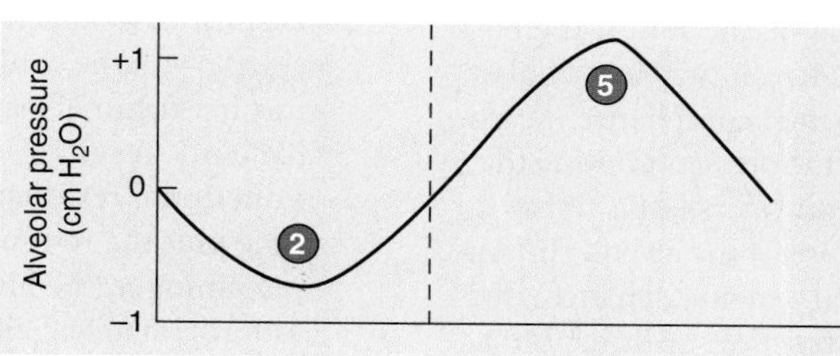

5. As expiration begins, alveolar pressure increases above barometric air pressure (0 on the graph) because the increased pleural pressure causes alveolar volume to decrease. By the end of expiration, alveolar and barometric air pressure are equal.

3. During inspiration, air flows into the lungs because alveolar pressure is lower than barometric air pressure.

Change in lung volume (L)

+0.5

0

0 1 2 3 4 5

Time (s)

6. During expiration, air flows out of the lungs because alveolar pressure is greater than barometric air pressure.

PROCESS FIGURE 23.14 Dynamics of a Normal Breathing Cycle

During expiration, pleural pressure increases because of decreased thoracic volume and decreased lung recoil (see figure 23.14). As pleural pressure increases, alveolar volume decreases, alveolar pressure increases above barometric air pressure, and air flows out of the lungs. As air flows out of the lungs, alveolar pressure decreases and becomes equal to barometric pressure at the end of expiration.

PREDICT 4

How does the pleural pressure at the end of expiration in a newborn with infant respiratory distress syndrome compare with that of a healthy newborn? How does the pleural pressure compare during inspiration? Explain.

MEASUREMENT OF LUNG FUNCTION

A variety of measurements can be used to assess lung function. Each of these tests compares a subject's measurements with a normal range. These measurements can be used to diagnose diseases, track the progress of diseases, and track recovery from diseases.

Compliance of the Lungs and Thorax

Compliance is a measure of the ease with which the lungs and thorax expand. The compliance of the lungs and thorax is the volume by which they increase for each unit of pressure change in alveolar pressure. It is usually expressed in liters (volume of air) per centimeter of water (pressure), and for a normal person the compliance of the lungs and thorax is 0.13 L/cm H_2O. That is, for every 1 cm H_2O change in alveolar pressure, the volume changes by 0.13 L.

The greater the compliance, the easier it is for a change in pressure to cause expansion of the lungs and thorax. For example, one possible result of emphysema is the destruction of elastic lung tissue. This reduces the elastic recoil force of the lungs, thereby making expansion of the lungs easier and resulting in a higher than normal compliance. A lower than normal compliance means that it is harder to expand the lungs and thorax. Conditions that decrease compliance include the deposition of inelastic fibers in lung tissue (pulmonary fibrosis), the collapse of the alveoli (infant respiratory distress syndrome and pulmonary edema), an increased resistance to airflow caused by airway obstruction (asthma, bronchitis, and lung cancer), and deformities of the thoracic wall that reduce the ability of the thoracic volume to increase (kyphosis and scoliosis).

Effects of Decreased Compliance

Pulmonary diseases can markedly affect the total amount of energy required for ventilation, as well as the percentage of the total amount of energy expended by the body. Diseases that decrease compliance can increase the energy required for breathing up to 30% of the total energy expended by the body.

27 ***Define compliance. What is the effect on lung expansion when compliance increases or decreases?***

Pulmonary Volumes and Capacities

Spirometry (spī-rom′ĕ-trē) is the process of measuring volumes of air that move into and out of the respiratory system, and a **spirometer** (spī-rom′ĕ-ter) is a device used to measure these pulmonary volumes. The four pulmonary volumes and representative values (figure 23.15) for a young adult male follow:

1. **Tidal volume** is the volume of air inspired or expired with each breath. At rest, quiet breathing results in a tidal volume of approximately 500 mL.
2. **Inspiratory reserve volume** is the amount of air that can be inspired forcefully after inspiration of the tidal volume (approximately 3000 mL at rest).
3. **Expiratory reserve volume** is the amount of air that can be forcefully expired after expiration of the tidal volume (approximately 1100 mL at rest).
4. **Residual volume** is the volume of air still remaining in the respiratory passages and lungs after the most forceful expiration (approximately 1200 mL).

The tidal volume increases when a person is more active. Because the maximum volume of the respiratory system does not change from moment to moment, an increase in tidal volume causes a decrease in the inspiratory and expiratory reserve volumes.

Pulmonary capacities are the sum of two or more pulmonary volumes (see figure 23.15). Some pulmonary capacities follow:

1. **Inspiratory capacity** is the tidal volume plus the inspiratory reserve volume, which is the amount of air that a person can inspire maximally after a normal expiration (approximately 3500 mL at rest).
2. **Functional residual capacity** is the expiratory reserve volume plus the residual volume, which is the amount of air remaining in the lungs at the end of a normal expiration (approximately 2300 mL at rest).
3. **Vital capacity** is the sum of the inspiratory reserve volume, the tidal volume, and the expiratory reserve volume, which is the maximum volume of air that a person can expel from the respiratory tract after a maximum inspiration (approximately 4600 mL).
4. **Total lung capacity** is the sum of the inspiratory and expiratory reserve volumes plus the tidal volume and the residual volume (approximately 5800 mL).

Factors such as sex, age, body size, and physical conditioning cause variations in respiratory volumes and capacities from one

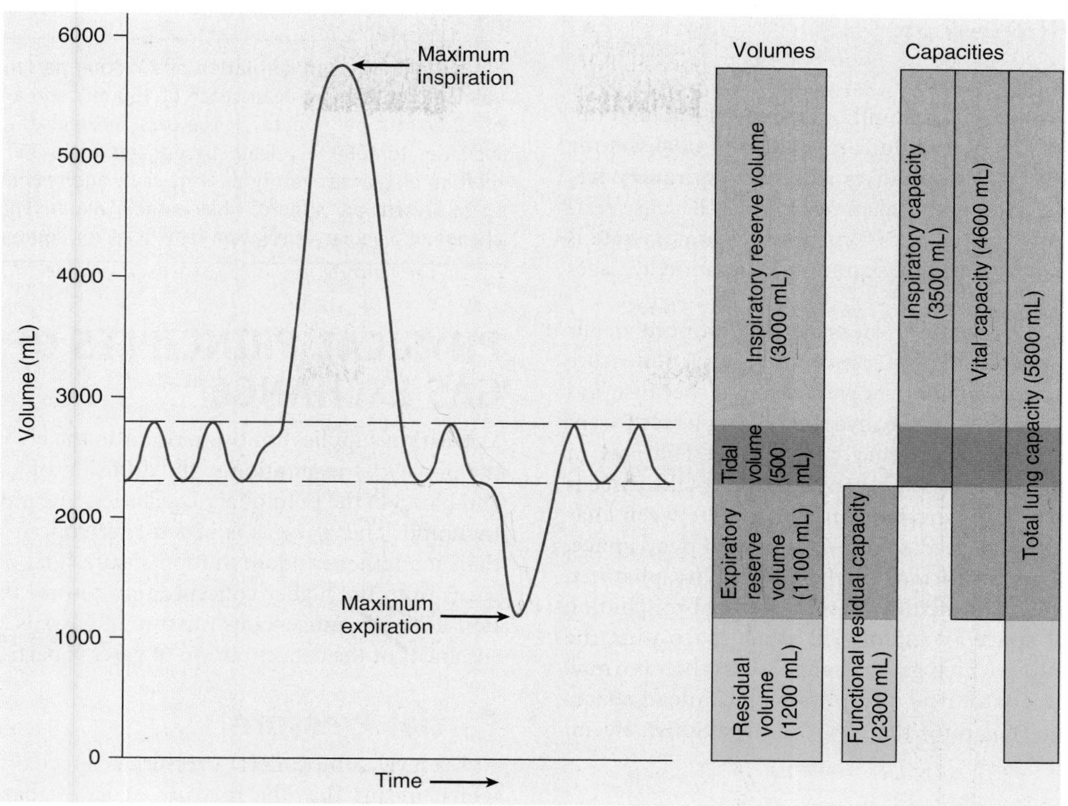

FIGURE 23.15 Lung Volumes and Lung Capacities

Lung volumes and capacities. The tidal volume in the figure is the tidal volume during resting conditions.

individual to another. For example, the vital capacity of adult females is usually 20%–25% less than that of adult males. The vital capacity reaches its maximum amount in young adults and gradually decreases in the elderly. Tall people usually have a greater vital capacity than short people, and thin people have a greater vital capacity than obese people. Well-trained athletes can have a vital capacity 30%–40% above that of untrained people. In patients whose respiratory muscles are paralyzed by spinal cord injury or diseases such as poliomyelitis or muscular dystrophy, vital capacity can be reduced to values not consistent with survival (less than 500–1000 mL). Factors that reduce compliance also reduce vital capacity.

The **forced expiratory vital capacity** is a simple and clinically important pulmonary test. The individual inspires maximally and then exhales maximally into a spirometer as rapidly as possible. The volume of air expired at the end of the test is the person's vital capacity. The spirometer also records the volume of air that enters it per second. The **forced expiratory volume in 1 second (FEV_1)** is the amount of air expired during the first second of the test. In some conditions, the vital capacity may not be dramatically affected, but how rapidly air is expired can be greatly decreased. Airway obstruction, caused by asthma, the collapse of bronchi in emphysema, or a tumor, and disorders that reduce the ability of the lungs or chest wall to deflate, such as pulmonary fibrosis, silicosis, kyphosis, and scoliosis, can cause a decreased FEV_1.

28 ***Define tidal volume, inspiratory reserve volume, expiratory reserve volume, and residual volume.***

29 ***Define inspiratory capacity, functional residual capacity, vital capacity, and total lung capacity.***

30 ***What is forced expiratory volume in 1 second, and why is it clinically important?***

Minute Ventilation and Alveolar Ventilation

Minute ventilation is the total amount of air moved into and out of the respiratory system each minute; it is equal to tidal volume times the respiratory rate. **Respiratory rate,** or **respiratory frequency,** is the number of breaths taken per minute. Because resting tidal volume is approximately 500 mL and respiratory rate is approximately 12 breaths per minute, minute ventilation averages approximately 6 L/min.

Although minute ventilation measures the amount of air moving into and out of the lungs per minute, it is not a measure of the amount of air available for gas exchange because gas exchange takes place mainly in the alveoli and to a lesser extent in the alveolar ducts and respiratory bronchioles. The part of the respiratory system where gas exchange does not take place is called the dead space. A distinction can be made between anatomical and physiological dead space. **Anatomical dead space,** which measures 150 mL, is formed by the nasal cavity, pharynx, larynx, trachea, bronchi, bronchioles, and terminal bronchioles. **Physiological dead space** is anatomical dead space plus the volume of any alveoli in which gas exchange is less than normal. In a healthy person, anatomical and physiological dead spaces are nearly the same, meaning that few nonfunctional alveoli exist.

Emphysema and Dead Space

In patients with **emphysema,** alveolar walls degenerate, and small alveoli combine to form larger alveoli. The result is fewer alveoli, but alveoli with an increased volume and decreased surface area. Although the enlarged alveoli are still ventilated, surface area is inadequate for complete gas exchange, and the physiological dead space increases.

During inspiration, much of the inspired air fills the dead space first before reaching the alveoli and, thus, is unavailable for gas exchange. The volume of air available for gas exchange per minute is called **alveolar ventilation ($\dot{V}_A$),** and it is calculated as follows:

$$\dot{V}_A = f(V_T - V_D)$$

where $\dot{V}_A$ is alveolar ventilation (milliliters per minute), *f* is respiratory rate (frequency; breaths per minute), V_T is tidal volume (milliliters per respiration), and V_D is dead space (milliliters per respiration).

31 ***Define minute ventilation and alveolar ventilation.***

32 ***What is dead space? What is the difference between anatomical and physiological dead space?***

PREDICT 5

What is the alveolar ventilation of a resting person with a tidal volume of 500 mL, a dead space of 150 mL, and a respiratory rate of 12 breaths per minute? If the person exercises and tidal volume increases to 4000 mL, dead space increases to 300 mL as a result of dilation of the respiratory passageways, and respiratory rate increases to 24 breaths per minute, what is the alveolar ventilation? How is the change in alveolar ventilation beneficial for doing exercise?

PHYSICAL PRINCIPLES OF GAS EXCHANGE

Ventilation supplies atmospheric air to the alveoli. The next step in the process of respiration is the diffusion of gases between alveoli and blood in the pulmonary capillaries. The molecules of gas move randomly, and, if a gas is in a higher concentration at one point than at another, random motion ensures that the net movement of gas is from the higher concentration toward the lower concentration until a homogeneous mixture of gases is achieved. One measurement of the concentration of gases is partial pressure.

Partial Pressure

At sea level, atmospheric pressure is approximately 760 mm Hg, which means that the mixture of gases that constitutes atmospheric air exerts a total pressure of 760 mm Hg. The major components of dry air are nitrogen (approximately 79%) and oxygen (approximately 21%). According to **Dalton's law,** in a mixture of gases, the part of the total pressure resulting from each type of gas is determined by the percentage of the total volume represented by each gas type (see table 23.1). The pressure exerted by each type of gas in a mixture is referred to as the **partial pressure** of that gas. Because nitrogen makes up 78.62% of the volume of atmospheric air, the partial pressure resulting from nitrogen is 0.7862 times 760 mm Hg, or 597.5 mm Hg. Because oxygen is 20.84% of the volume of atmospheric air, the partial pressure resulting from oxygen is 0.2084 times 760 mm Hg, or 158.4 mm Hg. It is traditional to designate the partial pressure of individual gases in a mixture with a capital P followed by the symbol for the gas. Thus, the partial pressure of nitrogen is denoted P_{N_2}, oxygen is P_{O_2}, and carbon dioxide is P_{CO_2}.

When air comes into contact with water, some of the water turns into a gas and evaporates into the air. Water molecules in gaseous form also exert a partial pressure. This partial pressure (P_{H_2O}) is sometimes referred to as **water vapor pressure.** The composition of dry, humidified, alveolar, and expired air is presented in table 23.2. The composition of alveolar air and of expired air is not identical to the composition of dry atmospheric air for three reasons. First, air entering the respiratory system during inspiration is humidified; second, oxygen diffuses from the alveoli into the blood, and carbon dioxide diffuses from the pulmonary capillaries into the alveoli; third, the air within the alveoli is only partially replaced with atmospheric air during each inspiration.

TABLE 23.2 Partial Pressures of Gases at Sea Level

	Dry Air		Humidified Air		Alveolar Air		Expired Air	
Gases	**mm Hg**	**%**	**mm Hg**	**%**	**mm Hg**	**%**	**mm Hg**	**%**
Nitrogen	597.5	78.62	563.4	74.09	569.0	74.9	566.0	74.5
Oxygen	158.4	20.84	149.3	19.67	104.0	13.6	120.0	15.7
Carbon dioxide	0.3	0.04	0.3	0.04	40.0	5.3	27.0	3.6
Water vapor	0.0	0.00	47.0	6.20	47.0	6.2	47.0	6.2

33 ***According to Dalton's law, what is the partial pressure of a gas? What is water vapor pressure?***

34 ***Why is the composition of inspired, alveolar, and expired air different?***

Diffusion of Gases Into and Out of Liquids

Gas molecules move from the air into a liquid, or from a liquid into the air, because of partial pressure gradients. If the partial pressure of a gas in the air is greater than in the liquid, there is net movement of gas molecules into the liquid. On the other hand, if the partial pressure of a gas in the air is less than in the liquid, there is net movement of the gas molecules out of the liquid. If the partial pressure of a gas in the air is equal to that in the liquid, the gas in the air is in equilibrium with the gas in the liquid, and there is no net movement of the gas into or out of the liquid. At a given temperature, **Henry's law** describes the concentration of a gas at equilibrium in a liquid (see table 23.1).

Concentration of dissolved gas = Pressure of gas × Solubility coefficient

The solubility coefficient is a measure of how easily the gas dissolves in the liquid. In water, the solubility coefficient for oxygen is 0.024; for carbon dioxide, it is 0.57. Thus, carbon dioxide is approximately 24 times more soluble in water than is oxygen.

35 ***According to Henry's law, how does the partial pressure and solubility of a gas affect its concentration in a liquid?***

PREDICT 6

As a SCUBA diver descends, the pressure of the water on the body prevents normal expansion of the lungs. To compensate, the diver breathes pressurized air, which has a greater pressure than air at sea level. What effect does the increased pressure have on the amount of gas dissolved in the diver's body fluids? A SCUBA diver who suddenly ascends to the surface from a great depth can develop decompression sickness (the bends), in which bubbles of nitrogen gas form. The expanding bubbles damage tissues or block blood flow through small blood vessels. Explain the development of the bubbles.

Diffusion of Gases Through the Respiratory Membrane

The factors that influence the rate of gas diffusion through the respiratory membrane are (1) the thickness of the membrane; (2) the diffusion coefficient of the gas in the substance of the membrane, which is approximately the same as the diffusion coefficient for the gas through water; (3) the surface area of the membrane; and (4) the partial pressure gradient of the gas across the membrane.

Respiratory Membrane Thickness

Increasing the thickness of the respiratory membrane decreases the rate of diffusion. The thickness of the respiratory membrane normally averages 0.6 μm, but diseases can cause an increase in the thickness. If the thickness of the respiratory membrane increases two or three times, the rate of gas exchange markedly decreases. Pulmonary edema caused by failure of the left side of the heart is the most common cause of an increase in the thickness of the respiratory membrane. Left side heart failure increases venous pressure in the pulmonary capillaries and results in the accumulation of fluid in the alveoli. Conditions that result in inflammation of the lung tissues, such as tuberculosis, pneumonia, or advanced silicosis, can also cause fluid accumulation within the alveoli.

Diffusion Coefficient

A **diffusion coefficient** is a measure of how easily a gas diffuses into and out of a liquid or tissue, taking into account the solubility of the gas in the liquid and the size of the gas molecule (molecular weight). If the diffusion coefficient of oxygen is assigned a value of 1, the relative diffusion coefficient of carbon dioxide is 20, which means carbon dioxide diffuses through the respiratory membrane about 20 times more readily than oxygen does.

When the respiratory membrane becomes progressively damaged as a result of disease, its capacity for allowing the movement of oxygen into the blood is often impaired enough to cause death from oxygen deprivation before the diffusion of carbon dioxide is dramatically reduced. If life is being maintained by extensive oxygen therapy, which increases the concentration of oxygen in the lung alveoli, the reduced capacity for the diffusion of carbon dioxide across the respiratory membrane can result in substantial increases in carbon dioxide in the blood.

Surface Area

In a healthy adult, the total surface area of the respiratory membrane is approximately 70 m^2 (approximately the floor area of a 25- by 30-foot room). Several respiratory diseases, including emphysema and lung cancer, cause a decrease in the surface area of the respiratory membrane. Even small decreases in this surface area adversely affect the respiratory exchange of gases during strenuous exercise. When the total surface area of the respiratory membrane is decreased to one-third or one-fourth of normal, the exchange of gases is significantly restricted, even under resting conditions.

A decreased surface area for gas exchange can also result from the surgical removal of lung tissue, the destruction of lung tissue by cancer, the degeneration of the alveolar walls by emphysema, or the replacement of lung tissue by connective tissue caused by tuberculosis. More acute conditions that cause the alveoli to fill with fluid also reduce the surface area for gas exchange because the increased thickness of the respiratory membrane caused by the fluid makes the alveoli nonfunctional. Examples include pneumonia and pulmonary edema resulting from failure of the left ventricle.

Partial Pressure Gradient

The partial pressure gradient of a gas across the respiratory membrane is the difference between the partial pressure of the gas in the alveoli and the partial pressure of the gas in the blood of the pulmonary capillaries. When the partial pressure of a gas is greater on one side of the respiratory membrane than on the other side, net diffusion occurs from the higher to the lower partial pressure (see figure 23.8*b*). Normally, the partial pressure of oxygen (Po_2) is greater in the alveoli than in the blood of the pulmonary capillaries, and the partial pressure of carbon dioxide (Pco_2) is greater in the blood than in the alveolar air.

By increasing alveolar ventilation, the partial pressure gradient for oxygen and carbon dioxide can be increased. The greater volume of atmospheric air exchanged with the residual volume raises alveolar Po_2, lowers alveolar Pco_2, and thus promotes gas exchange. Conversely, inadequate ventilation causes a lower than normal partial pressure gradient for oxygen and carbon dioxide, resulting in inadequate gas exchange.

36 ***Describe four factors that affect the diffusion of gases through the respiratory membrane. Give examples of diseases that decrease diffusion by altering these factors.***

37 ***Does oxygen or carbon dioxide diffuse more easily through the respiratory membrane?***

Relationship Between Alveolar Ventilation and Pulmonary Capillary Perfusion

Under normal conditions, alveolar ventilation and **pulmonary capillary perfusion,** which is blood flow through pulmonary capillaries, is such that effective gas exchange occurs between the air and the blood. During exercise, effective gas exchange is maintained because both ventilation and cardiac output increase.

The normal relationship between alveolar ventilation and pulmonary capillary perfusion can be disrupted in two ways. First, alveolar ventilation exceeds the blood's ability to pick up oxygen, which can happen because of inadequate cardiac output after a heart attack. Second, alveolar ventilation may not be great enough to provide the oxygen needed to oxygenate the blood flowing through the pulmonary capillaries. For example, constriction of the bronchioles in asthma can decrease air delivery to the alveoli.

Blood that is not completely oxygenated is called shunted blood. Two sources of shunted blood exist in the lungs. An **anatomical shunt** results when deoxygenated blood from the bronchi and bronchioles mixes with blood in the pulmonary veins (see "Blood Supply," p. 840). The other source of shunted blood is blood that passes through pulmonary capillaries but does not become fully oxygenated. The **physiological shunt** is the combination of deoxygenated blood from the anatomical shunt and deoxygenated blood that passes through the pulmonary capillaries. Normally, 1%–2% of cardiac output makes up the physiological shunt.

Disorders That Increase Shunted Blood

Any condition that decreases gas exchange between the alveoli and the blood can increase the amount of shunted blood. For example, obstruction of the bronchioles in conditions such as asthma can decrease ventilation beyond the obstructed areas. The result is a large increase in shunted blood because the blood flowing through the pulmonary capillaries in the obstructed area remains unoxygenated. In pneumonia or pulmonary edema, a buildup of fluid in the alveoli results in poor gas diffusion and less oxygenated blood.

When a person is standing, greater blood flow and ventilation occur in the base of the lung than in the top of the lung because of the effects of gravity. Arterial pressure at the base of the lung is 22 mm Hg greater than at the top of the lung because of hydrostatic pressure caused by gravity (see chapter 21). This greater pressure increases blood flow and distends blood vessels. The decreased pressure at the top of the lung results in less blood flow and vessels that are less distended, some of which are even collapsed during diastole.

During exercise, cardiac output and ventilation increase. The increased cardiac output increases pulmonary blood pressure throughout the lung, which increases blood flow. Blood flow increases most at the top of the lung, however, because the increased pressure expands the less distended vessels and opens the collapsed vessels. Thus, the effectiveness of gas exchange at the top of the lung increases because of greater blood flow.

Although gravity is the major factor affecting regional blood flow in the lung, under certain circumstances alveolar Po_2 can have an effect also. In most tissues, low Po_2 results in increased blood flow through the tissues (see chapter 21). In the lung, low Po_2 has the opposite effect, causing arterioles to constrict and reducing blood flow. This response reroutes blood away from areas of low oxygen toward parts of the lung that are better oxygenated. For example, if a bronchus becomes partially blocked, ventilation of alveoli past the blockage site decreases, which decreases gas exchange between the air and blood. The effect of this decreased gas exchange on overall gas exchange in the lungs is reduced by rerouting the blood to better-ventilated alveoli.

38 ***What effect do alveolar ventilation and pulmonary capillary perfusion have on gas exchange? What is the physiological shunt?***

39 ***What are the effects of gravity and alveolar Po_2 on blood flow in the lung?***

PREDICT 7

Even people in "good shape" can have trouble breathing at high altitudes. Explain how this can happen, even when ventilation of the lungs increases.

OXYGEN AND CARBON DIOXIDE TRANSPORT IN THE BLOOD

Once oxygen diffuses through the respiratory membrane into the blood, most of it combines reversibly with hemoglobin, and a smaller amount dissolves in the plasma. Hemoglobin transports oxygen from the pulmonary capillaries through the blood vessels to the tissue capillaries, where some of the oxygen is released. The oxygen diffuses from the blood to tissue cells, where it is used in aerobic respiration.

Cells produce carbon dioxide during aerobic metabolism. The carbon dioxide diffuses from the cells into the tissue capillaries. Once carbon dioxide enters the blood, it is transported in three ways: dissolved in the plasma, in combination with hemoglobin, and in the form of bicarbonate ions (HCO_3^-).

Oxygen Partial Pressure Gradients

The Po_2 within the alveoli averages approximately 104 mm Hg, whereas the Po_2 in blood flowing into the pulmonary capillaries is approximately 40 mm Hg (figure 23.16). Consequently, oxygen diffuses down its partial pressure gradient from the alveoli into the pulmonary capillary blood. By the time blood flows through the first third of the pulmonary capillary beds, an equilibrium has been achieved, and the Po_2 in the blood is 104 mm Hg, which is equivalent to the Po_2 in the alveoli. Even with the greater velocity of blood flow associated with exercise, by the time blood reaches the venous ends of the pulmonary capillaries, the Po_2 in the capillaries has achieved the same value as that in the alveoli.

Blood leaving the pulmonary capillaries has a Po_2 of 104 mm Hg, but blood leaving the lungs in the pulmonary veins has a Po_2 of approximately 95 mm Hg. This decrease in the Po_2 occurs because the blood from the pulmonary capillaries mixes with deoxygenated (shunted) blood from the bronchial veins.

The blood that enters the arterial end of the tissue capillaries has a Po_2 of approximately 95 mm Hg. The Po_2 of the interstitial fluid, in contrast, is close to 40 mm Hg and is probably near 20 mm Hg in the individual cells. Oxygen diffuses from the tissue capillaries to the interstitial fluid and from the interstitial fluid into the cells of the body, where it is used in aerobic metabolism. Because the cells use oxygen continuously, a constant partial pressure gradient exists for oxygen from the tissue capillaries to the cells.

Carbon Dioxide Partial Pressure Gradients

Carbon dioxide is continually produced as a by-product of cellular respiration, and a partial pressure gradient is established from tissue cells to the blood within the tissue capillaries. The intracellular Pco_2 is approximately 46 mm Hg, and the interstitial fluid Pco_2 is approximately 45 mm Hg. At the arterial end of the tissue capillaries, the Pco_2 is close to 40 mm Hg. As blood flows through the tissue capillaries, carbon dioxide diffuses from a higher Pco_2 to a lower Pco_2 until an equilibrium in Pco_2 is established. At the venous end of the capillaries, blood has a Pco_2 of 45 mm Hg (see figure 23.16).

After blood leaves the venous end of the capillaries, it is transported through the cardiovascular system to the lungs. At the arterial end of the pulmonary capillaries, the Pco_2 is 45 mm Hg. Because the Pco_2 is approximately 40 mm Hg in the alveoli, carbon dioxide diffuses from the pulmonary capillaries into the alveoli. At the venous end of the pulmonary capillaries, the Pco_2 has again decreased to 40 mm Hg.

40 ***Describe the partial pressures of oxygen and carbon dioxide in the alveoli, lung capillaries, tissue capillaries, and tissues. How do these partial pressures account for the movement of oxygen and carbon dioxide between air and blood and between blood and tissues?***

Hemoglobin and Oxygen Transport

Approximately 98.5% of the oxygen transported in the blood from the lungs to the tissues is transported in combination with hemoglobin in red blood cells, and the remaining 1.5% is dissolved in the water part of the plasma. The combination of oxygen with hemoglobin is reversible. In the pulmonary capillaries, oxygen binds to hemoglobin; in the tissue spaces, oxygen diffuses away from hemoglobin and enters the tissues.

1. Oxygen diffuses into the arterial ends of pulmonary capillaries and CO_2 diffuses into the alveoli because of differences in partial pressures.
2. As a result of diffusion at the venous ends of pulmonary capillaries, the P_{O_2} in the blood is equal to the P_{O_2} in the alveoli and the P_{CO_2} in the blood is equal to the P_{CO_2} in the alveoli.
3. The P_{O_2} of blood in the pulmonary veins is less than in the pulmonary capillaries because of mixing with deoxygenated blood from veins draining the bronchi and bronchioles.
4. Oxygen diffuses out of the arterial ends of tissue capillaries and CO_2 diffuses out of the tissue because of differences in partial pressures.
5. As a result of diffusion at the venous ends of tissue capillaries, the P_{O_2} in the blood is equal to the P_{O_2} in the tissue and the P_{CO_2} in the blood is equal to the P_{CO_2} in the tissue. Go back to step 1.

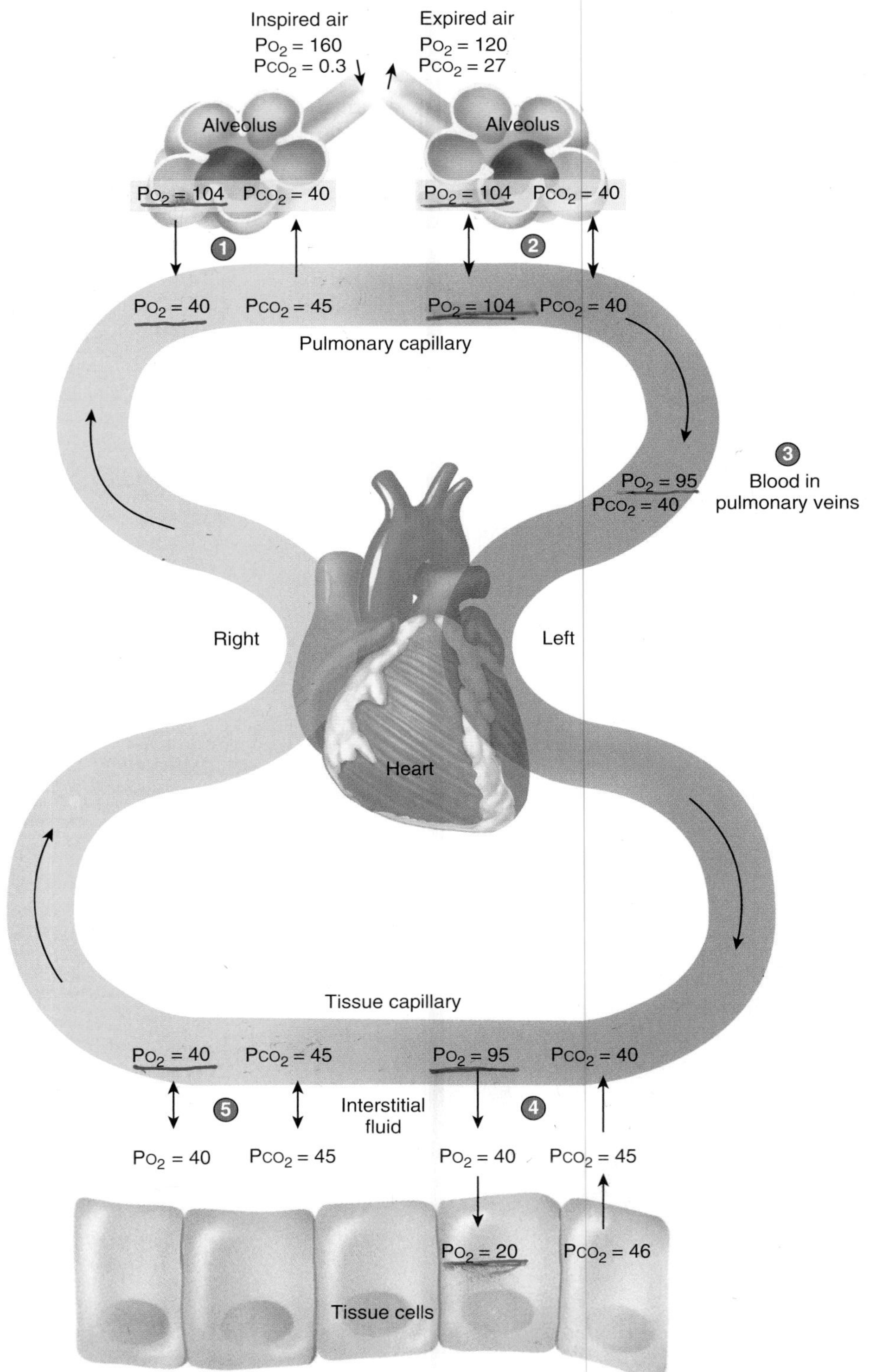

PROCESS FIGURE 23.16 Gas Exchange

Oxygen and carbon dioxide partial pressure gradients between the alveoli and the pulmonary capillaries and between the tissues and the tissue capillaries are responsible for gas exchange.

Effect of Po_2

The **oxygen–hemoglobin dissociation curve** describes the percent saturation of hemoglobin in the blood at different blood Po_2 values. Hemoglobin is 100% saturated with oxygen when four oxygen molecules are bound to each hemoglobin molecule in the blood. There are four heme groups in a hemoglobin molecule (see chapter 19), and an oxygen molecule is bound to each heme group. Hemoglobin is 50% saturated with oxygen when there is an average of two oxygen molecules bound to each hemoglobin molecule.

The Po_2 in the blood leaving the pulmonary capillaries is normally 104 mm Hg. At that partial pressure, hemoglobin is 98% saturated (figure 23.17*a*). Decreases in the Po_2 in the pulmonary capillaries have a relatively small effect on hemoglobin saturation, as shown by the fairly flat shape of the upper part of the oxygen–hemoglobin dissociation curve. Even if the blood Po_2 decreases from 104 mm Hg to 60 mm Hg, hemoglobin is still 90% saturated. Hemoglobin is very effective at picking up oxygen in the lungs, even if the Po_2 in the pulmonary capillaries decreases significantly.

In a resting person, the normal blood Po_2 leaving the tissue capillaries is 40 mm Hg and hemoglobin is 75% saturated. Thus, 23% (98% – 75%) of the oxygen picked up in the lungs is released from hemoglobin and diffuses into the tissues (figure 23.17*b*).

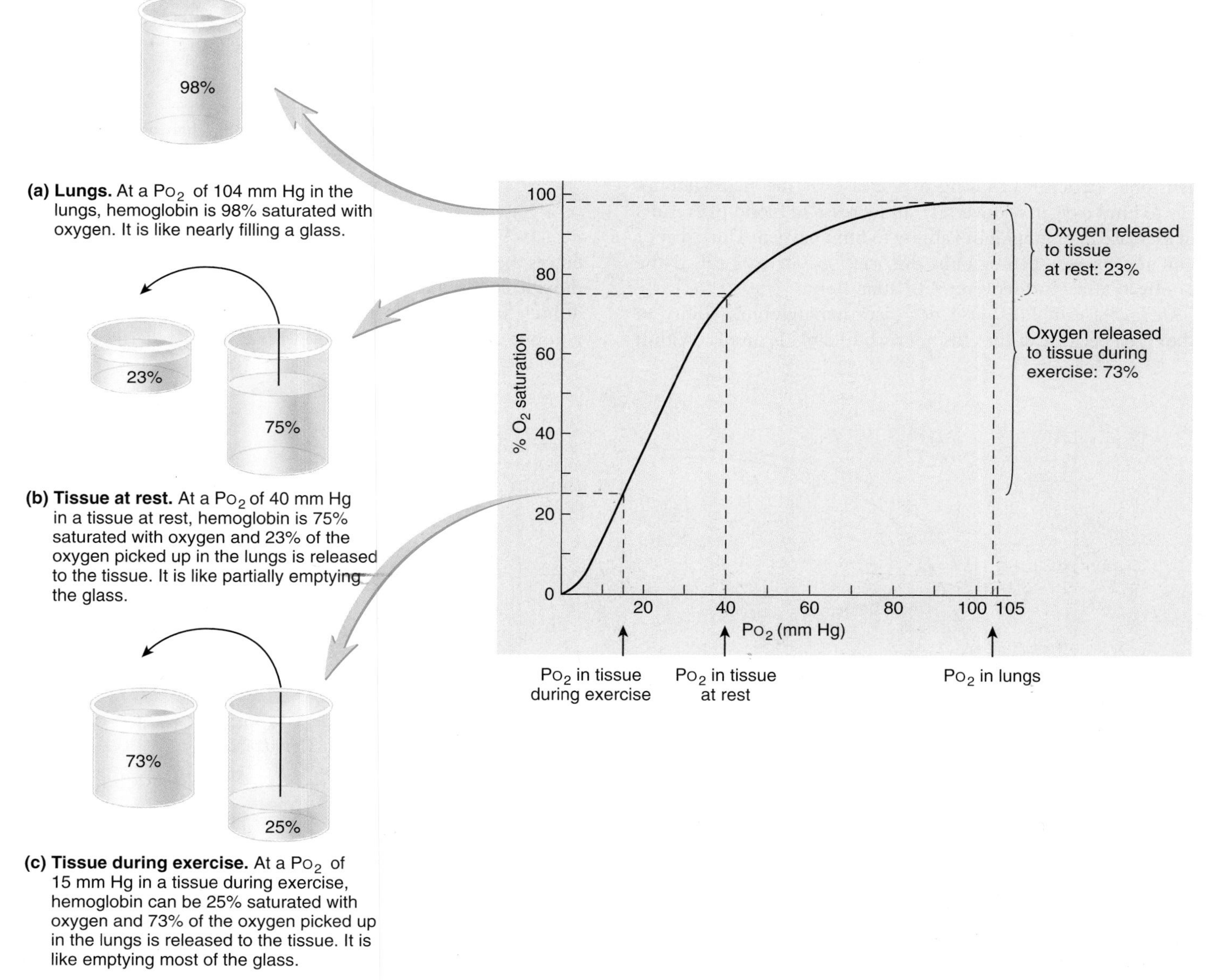

FIGURE 23.17 Oxygen–Hemoglobin Dissociation Curve

The oxygen–hemoglobin dissociation curve shows the percent saturation of hemoglobin as a function on Po_2. The ability of hemoglobin to pickup oxygen in the lungs and release it in the tissues is like a glass filling and emptying.

The 75% of oxygen still bound to the hemoglobin is an oxygen reserve, which can be released if blood P_{O_2} decreases further. In the tissues, a relatively small change in blood P_{O_2} results in a relatively large change in hemoglobin saturation, as shown by the steep slope of the oxygen–hemoglobin dissociation curve. For example, during vigorous exercise, the P_{O_2} in skeletal muscle capillaries can decline to levels as low as 15 mm Hg because of the increased use of oxygen during aerobic respiration in skeletal muscle cells (see chapter 9). At a P_{O_2} of 15 mm Hg, hemoglobin is only 25% saturated, resulting in the release of 73% (98% − 25%) of the oxygen picked up in the lungs (figure 23.17*c*). Thus, as tissues use more oxygen, hemoglobin releases more oxygen to those tissues.

Effect of pH, P_{CO_2}, and Temperature

In addition to P_{O_2}, blood pH, P_{CO_2}, and temperature influence the degree to which oxygen binds to hemoglobin (figure 23.18). As the pH of the blood declines, the amount of oxygen bound to hemoglobin at any given P_{O_2} also declines. This occurs because decreased pH results from an increase in H^+, and the H^+ combine with the protein part of the hemoglobin molecule and change its three-dimensional structure, causing a decrease in the hemoglobin's ability to bind oxygen. Conversely, an increase in blood pH results in an increase in hemoglobin's ability to bind oxygen. The effect of pH on the oxygen–hemoglobin dissociation curve is called the **Bohr effect,** after its discoverer, Christian Bohr.

An increase in P_{CO_2} also decreases hemoglobin's ability to bind oxygen because of the effect of carbon dioxide on pH. Within red blood cells, an enzyme called **carbonic anhydrase** catalyzes this reversible reaction.

$$\underset{\text{Carbon dioxide}}{CO_2} + \underset{\text{Water}}{H_2O} \overset{\text{Carbonic anhydrase}}{\rightleftarrows} \underset{\text{Carbonic acid}}{H_2CO_3} \rightleftarrows \underset{\text{Hydrogen ion}}{H^+} + \underset{\text{Bicarbonate ion}}{HCO_3^-}$$

As carbon dioxide levels increase, more H^+ are produced, and the pH declines. As carbon dioxide levels decline, the reaction proceeds in the opposite direction, resulting in a decrease in H^+ concentration and an increase in pH. Thus, changes in carbon dioxide levels indirectly produce a Bohr effect by altering pH. In addition, carbon dioxide can directly affect hemoglobin's ability to bind oxygen, to a small extent. When carbon dioxide binds to the α- and β-globin chains of hemoglobin (see chapter 19), hemoglobin's ability to bind with oxygen decreases.

PREDICT 8

Predict how the pH of the blood in a sprinter changes during a 400-meter race. How does the pH change affect O_2 delivery to skeletal muscles?

As blood passes through tissue capillaries, carbon dioxide enters the blood from the tissues. As a consequence, blood carbon dioxide levels increase, pH decreases, and hemoglobin has less affinity for oxygen in the tissue capillaries. Therefore, a greater amount of oxygen is released in the tissue capillaries than would

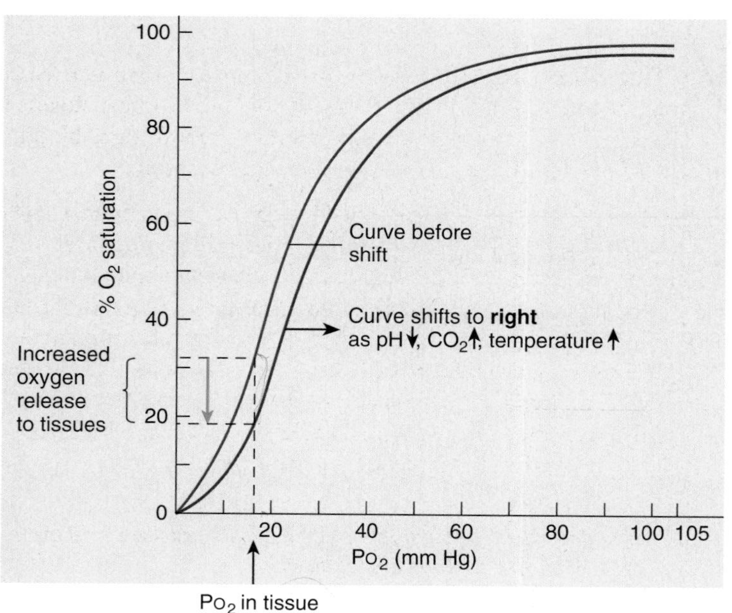

(a) **In the tissues, the oxygen–hemoglobin dissociation curve shifts to the right.** As pH decreases, P_{CO_2} increases, or temperature increases, the curve (*red*) shifts to the right (*blue*), resulting in an increased release of oxygen.

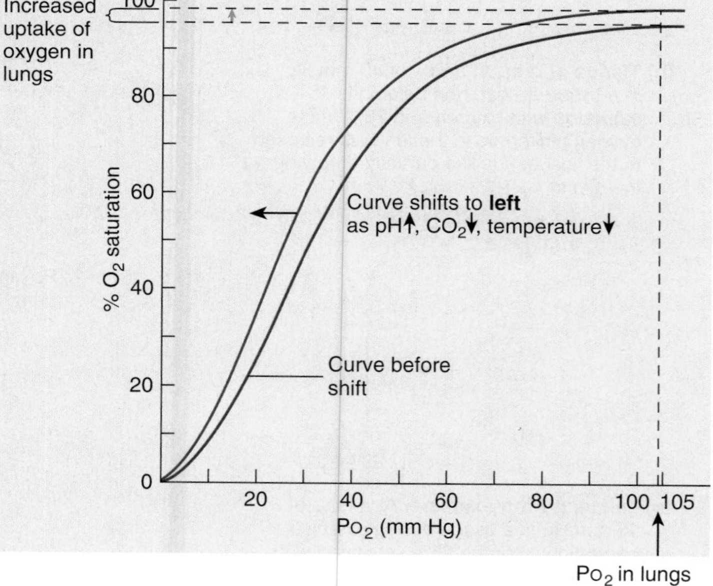

(b) **In the lungs, the oxygen–hemoglobin dissociation curve shifts to the left.** As pH increases, P_{CO_2} decreases, or temperature decreases, the curve (*blue*) shifts to the left (*red*), resulting in an increased ability of hemoglobin to pick up oxygen.

FIGURE 23.18 Effects of Shifting the Oxygen–Hemoglobin Dissociation Curve

be released if carbon dioxide were not present. When blood is returned to the lungs and passes through the pulmonary capillaries, carbon dioxide leaves the capillaries and enters the alveoli. As a result, carbon dioxide levels in the pulmonary capillaries are reduced, pH increases, and hemoglobin's affinity for oxygen increases.

An increase in temperature also decreases oxygen's tendency to remain bound to hemoglobin. Elevated temperatures resulting from increased metabolism, therefore, increase the amount of oxygen released into the tissues by hemoglobin. In less metabolically active tissues in which the temperature is lower, less oxygen is released from hemoglobin.

When hemoglobin's affinity for oxygen decreases, the oxygen–hemoglobin dissociation curve is shifted to the right, and hemoglobin releases more oxygen (see figure 23.18*a*). During exercise, when carbon dioxide and acidic substances, such as lactic acid, accumulate and the temperature increases in the tissue spaces, the oxygen–hemoglobin curve shifts to the right. Under these conditions, as much as 75%–85% of the oxygen is released from the hemoglobin. In the lungs, however, the curve shifts to the left because of the lower carbon dioxide levels, lower temperature, and lower lactic acid levels. Hemoglobin's affinity for oxygen, therefore, increases, and it becomes easily saturated (see figure 23.18*b*).

During resting conditions, approximately 5 mL of oxygen are transported to the tissues in each 100 mL of blood, and cardiac output is approximately 5000 mL/min. Consequently, 250 mL of oxygen are delivered to the tissues each minute. During exercise, this value can increase up to 15 times. Oxygen transport can be increased threefold because of a greater degree of oxygen release from hemoglobin in the tissue capillaries, and the rate of oxygen transport is increased another five times because of the increase in cardiac output. Consequently, the volume of oxygen delivered to the tissues can be as high as 3750 mL/min (15×250 mL/min). Highly trained athletes can increase this volume to as high as 5000 mL/min.

41 ***Name two ways that oxygen is transported in the blood, and state the percentage of total oxygen transport for which each is responsible.***

42 ***How does the oxygen–hemoglobin dissociation curve explain the uptake of oxygen in the lungs and the release of oxygen in tissues?***

43 ***What is the Bohr effect? How is it related to blood carbon dioxide?***

44 ***Why is it advantageous for the oxygen–hemoglobin dissociation curve to shift to the left in the lungs and to the right in tissues?***

PREDICT 9

In carbon monoxide (CO) poisoning, CO binds to hemoglobin, thereby decreasing the uptake of oxygen by hemoglobin. In addition, when CO binds to hemoglobin, the oxygen–hemoglobin dissociation curve shifts to the left. What are the consequences of this shift on the ability of tissues to get oxygen? Explain.

Effect of BPG

As red blood cells metabolize glucose for energy, they produce a by-product called **2,3-bisphosphoglycerate (BPG;** formerly called diphosphoglycerate). BPG binds to hemoglobin, reducing its affinity for oxygen, which increases its ability to release oxygen. A potent trigger for increased BPG production is low blood oxygen levels. For example, barometric pressure is lower at high altitudes than at sea level, causing the partial pressure of oxygen in the alveoli to be lower and the percent saturation of blood with oxygen in the pulmonary capillaries to be lower. Consequently, there is less oxygen in the blood to deliver to tissues. BPG helps increase oxygen delivery to tissues because increased levels of BPG increase the release of oxygen in tissues (the oxygen–hemoglobin dissociation curve shifts to the right). On the other hand, when blood is removed from the body and stored in a blood bank, the BPG levels in the stored blood decrease. As BPG levels decrease, the blood becomes unsuitable for transfusion purposes after approximately 6 weeks because the hemoglobin releases less oxygen to the tissues. Banked blood is, therefore, discarded after 6 weeks of storage.

45 ***How does BPG affect the release of oxygen from hemoglobin?***

PREDICT 10

If a person lacks the enzyme necessary for BPG synthesis, does he or she exhibit anemia (a lower than normal number of red blood cells) or erythrocytosis (a higher than normal number of red blood cells)? Explain.

Fetal Hemoglobin

As fetal blood circulates through the placenta, oxygen is released from the mother's blood into the fetal blood and carbon dioxide is released from fetal blood into the mother's blood. Fetal blood is very efficient at picking up oxygen for several reasons:

1. The concentration of fetal hemoglobin is approximately 50% greater than the concentration of maternal hemoglobin.
2. Fetal hemoglobin is different from maternal hemoglobin. It has an oxygen–hemoglobin dissociation curve that is to the left of the maternal oxygen–hemoglobin dissociation curve. Thus, for a given P_{O_2} fetal hemoglobin can hold more oxygen than maternal hemoglobin.
3. BPG has little effect on fetal hemoglobin. That is, BPG does not cause fetal hemoglobin to release oxygen.

46 ***How does fetal hemoglobin's affinity for oxygen compare with that of maternal hemoglobin?***

PREDICT 11

How does the movement of CO_2 from fetal blood into maternal blood increase the movement of oxygen from maternal blood into fetal blood? (*Hint:* Consider the shift of the oxygen–hemoglobin dissociation curve.)

Transport of Carbon Dioxide

Carbon dioxide is transported in the blood in three ways: approximately 7% as carbon dioxide dissolved in the plasma, approximately 70% in the form of HCO_3^- dissolved in plasma and in red blood cells, and approximately 23% bound to hemoglobin.

Carbon Dioxide Exchange in Tissues

Carbon dioxide diffuses from tissues into the plasma of blood (figure 23.19*a*). Most of the carbon dioxide diffuses into red blood cells, but approximately 7% of it is transported dissolved in the plasma. Carbon dioxide inside red blood cells reacts with water to form carbonic acid, a reaction catalyzed by carbonic anhydrase. Carbonic acid then dissociates to form HCO_3^- and H^+. Approximately 70% of blood carbon dioxide is transported in the form of HCO_3^-.

Removing HCO_3^- from inside the red blood cells promotes carbon dioxide transport because, as HCO_3^- concentration decreases, more carbon dioxide combines with water to form additional HCO_3^- and H^+ (see "Reversible Reactions," chapter 2). In a process called **chloride shift** (see figure 23.19*a*, step 4), antiporters exchange Cl^- for HCO_3^-. This exchange maintains electrical balance in the red blood cells and plasma as HCO_3^- diffuse out of, and Cl^- diffuse into, red blood cells.

Hydrogen ions bind to hemoglobin (see figure 23.19*a*, step 6). As a result, three effects are produced: (1) The transport of carbon dioxide increases because, as H^+ concentration decreases, more carbon dioxide combines with water to form additional HCO_3^- and H^+; (2) the pH inside the red blood cells does not decrease because hemoglobin is a buffer, preventing an increase in H^+ concentration; and (3) the affinity of hemoglobin for oxygen decreases. Hemoglobin releases oxygen in tissue capillaries because of decreased P_{O_2} (see figure 23.17). Hemoglobin's decreased affinity for oxygen results in a shift of the oxygen–hemoglobin curve to the right (the Bohr effect; see figure 23.18*a*) and an increase in the release of oxygen from hemoglobin.

Approximately 23% of blood carbon dioxide is transported bound to hemoglobin. Many carbon dioxide molecules bind in a reversible fashion to the α- and β-globin chains of hemoglobin molecules (figure 23.19*a*, step 7). Carbon dioxide's ability to bind to hemoglobin is affected by the amount of oxygen bound to hemoglobin. The smaller the amount of oxygen bound to hemoglobin, the greater the amount of carbon dioxide that can bind to it, and vice versa. This relationship is called the **Haldane effect.** In tissues, as hemoglobin releases oxygen, the hemoglobin has an increased ability to pick up carbon dioxide.

Carbon Dioxide Exchange in the Lungs

Carbon dioxide diffuses from red blood cells and plasma into the alveoli (figure 23.19*b*). As carbon dioxide levels in the red blood cells decrease, carbonic acid is converted to carbon dioxide and water. In response, HCO_3^- join with H^+ to form carbonic acid. As HCO_3^- and H^+ concentrations decrease because of this reaction, HCO_3^- enter red blood cells in exchange for Cl^- and H^+ are released from hemoglobin. Hemoglobin picks up oxygen in pulmonary capillaries because of increased P_{O_2} (see figure 23.17). The release of H^+ from hemoglobin increases hemoglobin's affinity for oxygen, resulting in a shift of the oxygen–hemoglobin curve to the left (Bohr effect; see figure 23.18*b*). Oxygen from the alveoli diffuses into the pulmonary capillaries and into the red blood cells, and it binds with hemoglobin. Carbon dioxide is released from hemoglobin and diffuses out of the red blood cells into the alveoli. As hemoglobin binds to oxygen, it more readily releases carbon dioxide (Haldane effect).

Carbon Dioxide and Blood pH

Blood pH refers to the pH in plasma, not inside red blood cells. In plasma, carbon dioxide can combine with water to form carbonic acid, a reaction that is catalyzed by carbonic anhydrase on the surface of capillary endothelial cells. The carbonic acid then dissociates to form HCO_3^- and H^+. Thus, as plasma carbon dioxide levels increase, H^+ levels increase, and blood pH decreases. An important function of the respiratory system is to regulate blood pH by changing plasma carbon dioxide levels (see chapter 27). Hyperventilation decreases plasma carbon dioxide, and hypoventilation increases it.

47 ***What is the effect of lowering HCO_3^- concentrations inside red blood cells on carbon dioxide transport? What is the chloride shift, and what does it accomplish?***

48 ***Name three effects produced by H^+ binding to hemoglobin.***

49 ***What is the Haldane effect?***

50 ***What effect does blood carbon dioxide level have on blood pH?***

PREDICT 12

What effect does hyperventilation and holding one's breath have on blood pH? Explain.

REGULATION OF VENTILATION

The generation of the basic rhythm of ventilation is controlled by neurons within the medulla oblongata that stimulate the muscles of respiration. The recruitment of muscle fibers and the increased frequency of stimulation of muscle fibers result in stronger contractions of the muscles and an increased depth of respiration. The rate of respiration is determined by how frequently the respiratory muscles are stimulated.

Respiratory Areas in the Brainstem

The classic view of respiratory areas held that distinct inspiratory and expiratory centers were located in the brainstem. This view is now known to be too simplistic. Although neurons involved with respiration are aggregated in certain parts of the brainstem, neurons that are active during inspiration are intermingled with neurons that are active during expiration.

The **medullary respiratory center** consists of two **dorsal respiratory groups,** each forming a longitudinal column of cells located bilaterally in the dorsal part of the medulla oblongata, and two **ventral respiratory groups,** each forming a longitudinal column of cells

1. In the tissues, carbon dioxide (CO_2) diffuses into the plasma and into red blood cells. Some of the carbon dioxide remains in the plasma.

2. In red blood cells, carbon dioxide reacts with water (H_2O) to form carbonic acid (H_2CO_3) in a reaction catalyzed by the enzyme carbonic anhydrase (CA).

3. Carbonic acid dissociates to form bicarbonate ions (HCO_3^-) and hydrogen ions (H^+).

4. In the chloride shift, as HCO_3^- diffuse out of the red blood cells, electrical neutrality is maintained by the diffusion of chloride ions (Cl^-) into them.

5. Oxygen (O_2) is released from hemoglobin (Hb). Oxygen diffuses out of red blood cells and plasma into the tissue.

6. Hydrogen ions combine with hemoglobin, which promotes the release of oxygen from hemoglobin (Bohr effect).

7. Carbon dioxide combines with hemoglobin. Hemoglobin that has released oxygen readily combines with carbon dioxide (Haldane effect).

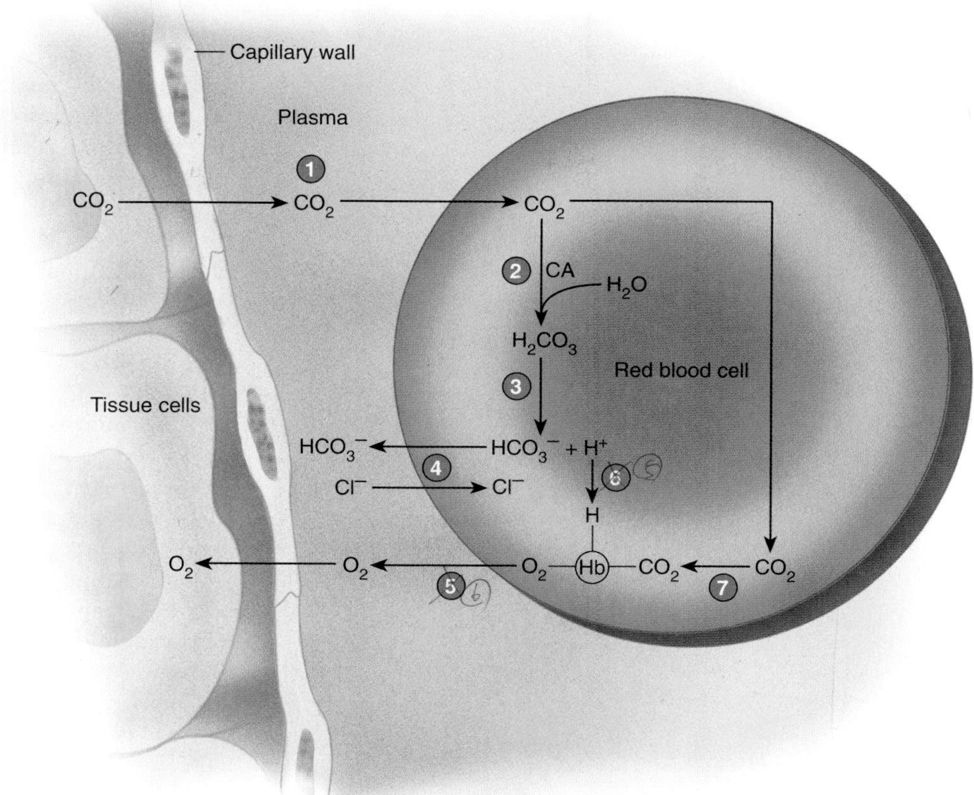

(a) **Gas exchange in the tissues**

1. In the lungs, carbon dioxide (CO_2) diffuses from red blood cells and plasma into the alveoli.

2. Carbonic anhydrase catalyzes the formation of CO_2 and H_2O from H_2CO_3.

3. Bicarbonate ions and H^+ combine to replace H_2CO_3.

4. In the chloride shift, as HCO_3^- diffuse into red blood cells, electrical neutrality is maintained by the diffusion of chloride ions (Cl^-) out of them.

5. Oxygen diffuses into the plasma and into red blood cells. Some of the oxygen remains in the plasma. Oxygen binds to hemoglobin.

6. Hydrogen ions are released from hemoglobin, which promotes the uptake of oxygen by hemoglobin (Bohr effect).

7. Carbon dioxide is released from hemoglobin. Hemoglobin that is bound to oxygen readily releases carbon dioxide (Haldane effect).

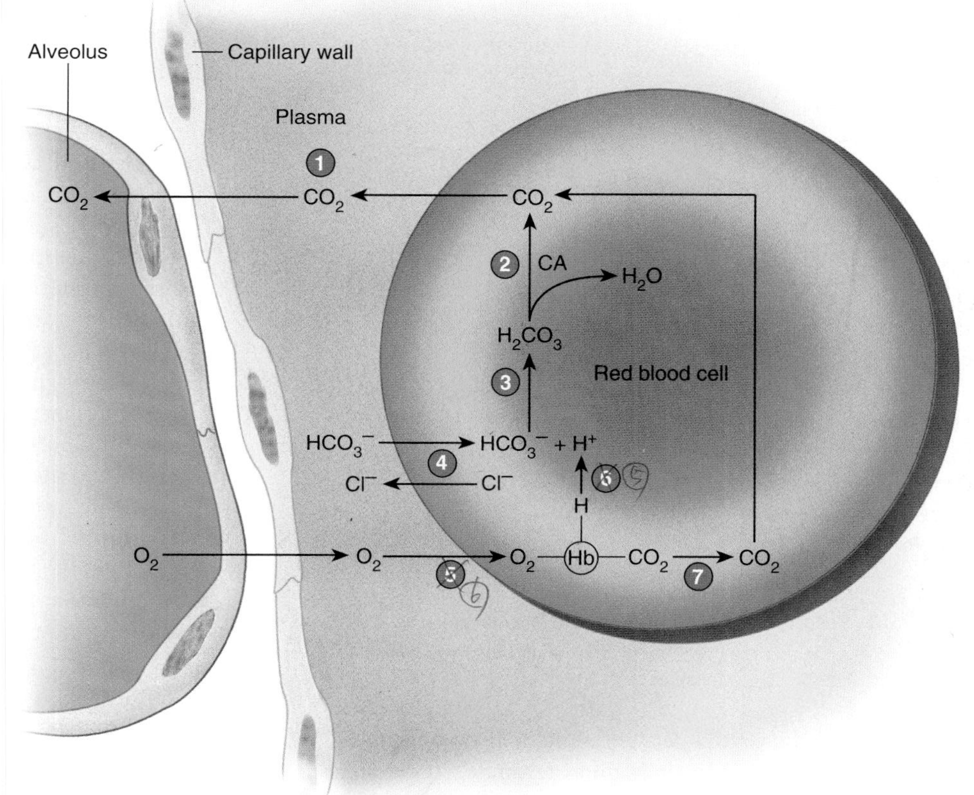

(b) **Gas exchange in the lungs**

PROCESS FIGURE 23.19 Gas Exchange

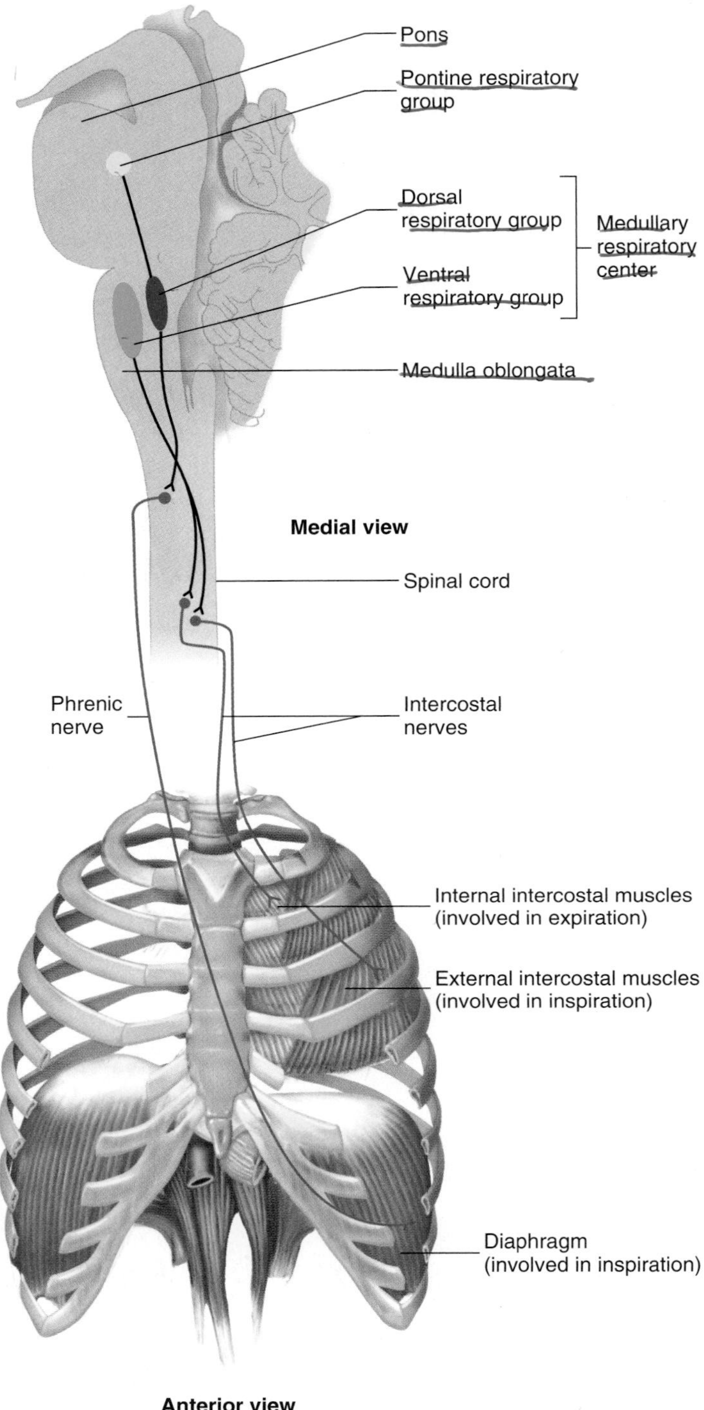

FIGURE 23.20 **Respiratory Structures in the Brainstem**
The relationship of respiratory structures to each other and to the nerves innervating the muscles of respiration.

located bilaterally in the ventral part of the medulla oblongata (figure 23.20). Although the dorsal and ventral respiratory groups are bilaterally paired, cross-communication exists between the pairs so that respiratory movements are symmetric. In addition, communication exists between the dorsal and ventral respiratory groups.

Each dorsal respiratory group is a collection of neurons that are most active during inspiration, but some are active during expiration. The dorsal respiratory groups are primarily responsible for stimulating contraction of the diaphragm. They receive input from other parts of the brain and peripheral receptors, which allows the modification of respiration.

Each ventral respiratory group is a collection of neurons that are active during inspiration and expiration. These neurons stimulate primarily the external intercostal, internal intercostal, and abdominal muscles. A part of the ventral respiratory group, the **pre-Bötzinger complex,** is believed to establish the basic rhythm of respiration.

The **pontine respiratory group,** formerly called the pneumotaxic center, is a collection of neurons in the pons (see figure 23.20). Some of the neurons are active only during inspiration, some only during expiration, and some during both inspiration and expiration. The precise function of the pontine respiratory group is unknown, but it has connections with the medullary respiratory center and appears to play a role in switching between inspiration and expiration, thus fine-tuning the breathing pattern. It is not considered to be essential for the generation of the respiratory rhythm.

Generation of Rhythmic Ventilation

One explanation for the generation of rhythmic ventilation involves the integration of stimuli that start and stop inspiration:

1. *Starting inspiration.* Neurons in the medullary respiratory center spontaneously establish the basic rhythm of respiration. The medullary respiratory center constantly receives stimulation from receptors that monitor blood gas levels, blood temperature, and the movements of muscles and joints. In addition, stimulation from the parts of the brain concerned with voluntary respiratory movements and emotions can occur. Inspiration starts when the combined input from all these sources causes the production of action potentials in the neurons that stimulate respiratory muscles.
2. *Increasing inspiration.* Once inspiration begins, more and more neurons are gradually activated. The result is progressively stronger stimulation of the respiratory muscles, which lasts for approximately 2 seconds.
3. *Stopping inspiration.* The neurons stimulating the muscles of respiration also stimulate the neurons in the medullary respiratory center that are responsible for stopping inspiration. The neurons responsible for stopping inspiration also receive input from the pontine respiratory group, stretch receptors in the lungs, and probably other sources. When these inhibitory neurons are activated, they cause the neurons stimulating respiratory muscles to be inhibited. Relaxation of respiratory muscles results in expiration, which lasts approximately 3 seconds. The next inspiration begins again at step 1.

Although the medullary neurons establish the basic rate and depth of breathing, their activities can be influenced by input from other parts of the brain and by input from peripherally located receptors.

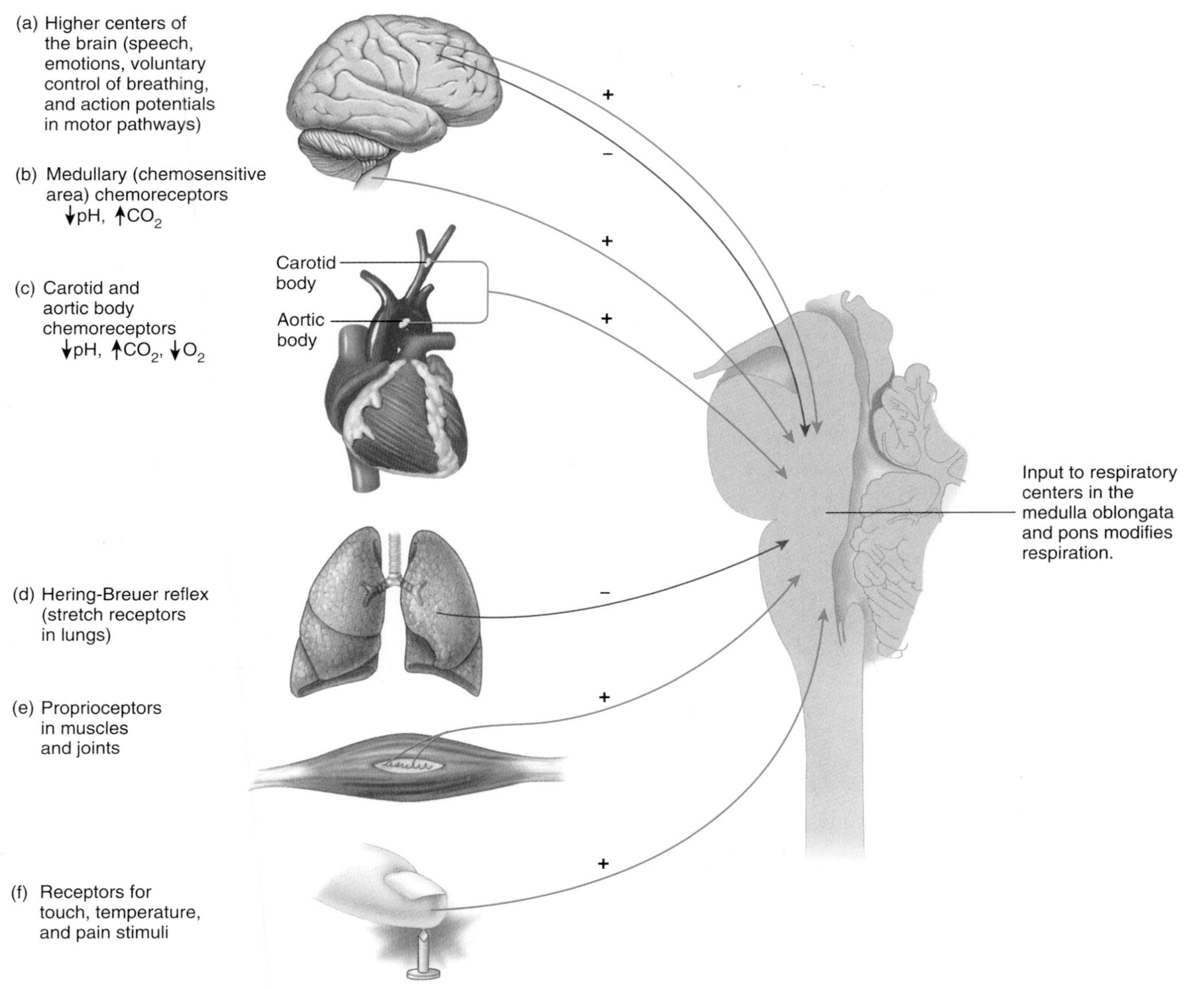

FIGURE 23.21 Major Regulatory Mechanisms of Ventilation
A plus sign indicates an increase in ventilation, and a minus sign indicates a decrease in ventilation.

51 ***Name the three respiratory groups and describe their main functions.***

52 ***How is rhythmic ventilation generated?***

Cerebral and Limbic System Control

Through the cerebral cortex, it is possible to consciously or unconsciously increase or decrease the rate and depth of the respiratory movements (figure 23.21). For example, during talking or singing, air movement is controlled to produce sounds, as well as to facilitate gas exchange.

Apnea (ap′nē-ă) is the absence of breathing. A person may stop breathing voluntarily. As the period of voluntary apnea increases, a greater and greater urge to breathe develops. That urge is primarily associated with increasing P_{CO_2} levels in the arterial blood. Finally, the P_{CO_2} reaches levels that cause the respiratory center to override the conscious influence from the cerebrum. Occasionally, people are able to hold their breath until the blood P_{O_2} declines to a level low enough that they lose consciousness. After consciousness is lost, the respiratory center resumes its normal function in automatically controlling respiration.

Voluntary hyperventilation can decrease blood P_{CO_2} levels sufficiently to cause vasodilation of the peripheral blood vessels and a decrease in blood pressure (see chapter 21). Dizziness or a giddy feeling can result because of decreased delivery of oxygen to the brain caused by the decreased rate of blood flow to the brain after blood pressure drops.

Emotions acting through the limbic system of the brain can also affect the respiratory center (see figure 23.21). For example,

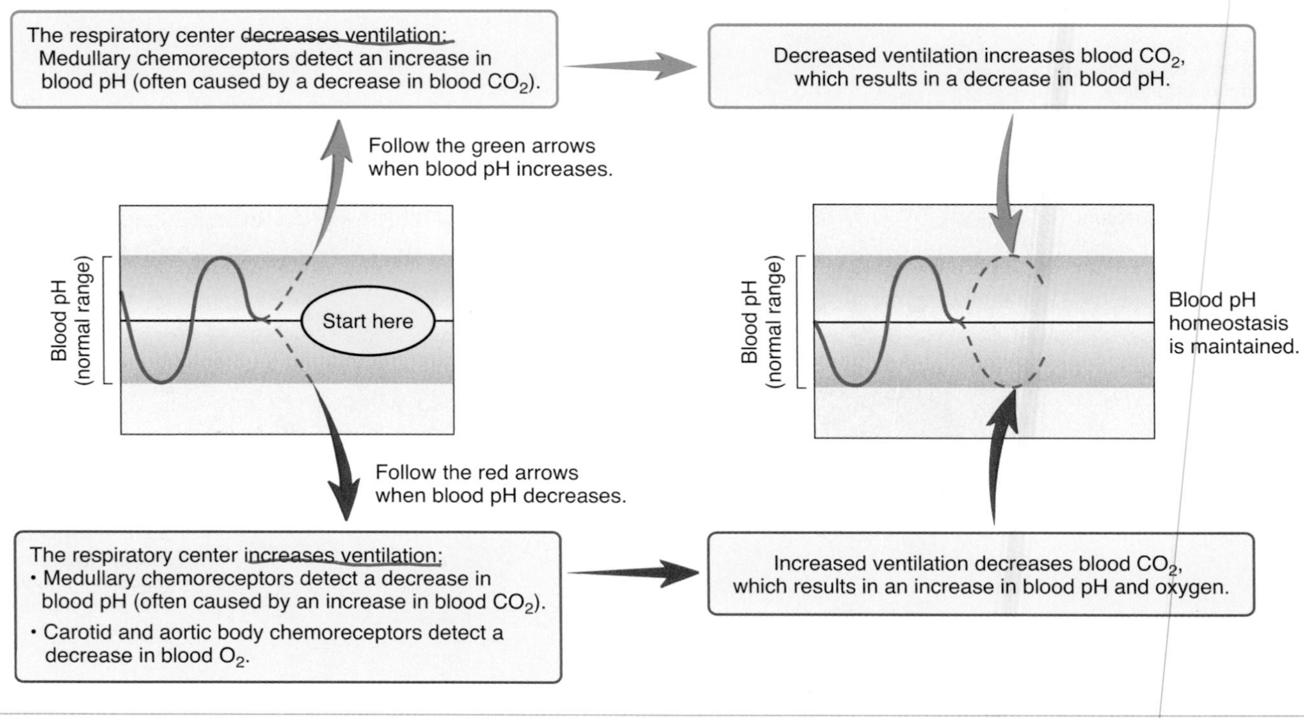

HOMEOSTASIS FIGURE 23.22 Summary of the Regulation of Blood pH and Gases

strong emotions can cause hyperventilation or produce the sobs and gasps of crying.

Chemical Control of Ventilation

The respiratory system maintains blood oxygen and carbon dioxide concentrations and blood pH within a normal range of values. A deviation in any of these parameters from their normal range has a marked influence on respiratory movements. The effect of changes in oxygen and carbon dioxide concentrations and in pH is superimposed on the neural mechanisms that establish rhythmic ventilation.

Chemoreceptors

Chemoreceptors are specialized neurons that respond to changes in chemicals in solution. The chemoreceptors involved in the regulation of respiration respond to changes in hydrogen ion concentrations, changes in P_{O_2}, or both (see figure 23.21; figure 23.22). **Central chemoreceptors** are located bilaterally and ventrally in the **chemosensitive area** of the medulla oblongata, and they are connected to the respiratory center. **Peripheral chemoreceptors** are found in the carotid and aortic bodies. These structures are small vascular sensory organs, which are encapsulated in connective tissue and located near the carotid sinuses and the aortic arch (see chapter 21). The respiratory center is connected to the carotid body chemoreceptors through the glossopharyngeal nerve (IX) and to the aortic body chemoreceptors by the vagus nerve (X).

Effect of pH

The chemosensitive area of the medulla oblongata and the carotid and aortic bodies respond to changes in blood pH. The chemosensitive area responds indirectly to changes in blood pH, whereas the carotid and aortic bodies respond directly. Hydrogen ions do not easily cross the blood–brain barrier and the blood–cerebrospinal fluid barrier (see chapter 11) to affect the chemosensitive area, but they do easily cross from the blood to the carotid and aortic bodies.

The chemosensitive area detects changes in blood pH through changes in blood carbon dioxide, which easily diffuses across the blood–brain barrier and the blood–cerebrospinal fluid barrier. For example, if blood carbon dioxide levels increase, carbon dioxide diffuses across the blood–brain barrier into the cerebrospinal fluid. The carbon dioxide combines with water to form carbonic acid, which dissociates into H^+ and HCO_3^-. The increased concentration of H^+ lowers the pH and stimulates the chemosensitive area, which then stimulates the respiratory center, resulting in a greater rate and depth of breathing. Consequently, carbon dioxide levels decrease as carbon dioxide is eliminated from the body and blood pH increases to normal levels.

Maintaining body pH levels within normal parameters is necessary for the proper functioning of cells. Because changes in

CASE STUDY

Asthma

Will is an 18-year-old track athlete in seemingly good health. Despite suffering from a slight cold, Will went jogging one morning with his running buddy, Al. After a few minutes of exercise, Will felt that he could hardly get enough air. Even though he stopped jogging, he continued to breathe rapidly and wheeze forcefully. Because his condition was not improving, Al took him to the emergency room of a nearby hospital.

The emergency room doctor used a stethoscope to listen to air movement in Will's lungs and noted that movement was poor. In addition, she ordered an arterial blood gas measurement for Will. He had a P_{O_2} of 60 mm Hg and a P_{CO_2} of 30 mm Hg. Although Will had no previous history of asthma, the emergency room doctor was convinced that he was having an asthma attack.

Asthma is a clinical condition characterized by airway inflammation, which episodically results in shortness of breath, coughing, and wheezing due to bronchoconstriction. Bronchoconstriction decreases compliance, which makes expansion of the lungs more difficult. An asthma attack can be provoked by viral infections, exercise, or exposure to environmental irritants, such as pollen or cigarette smoke (see Clinical Focus "Disorders of the Respiratory System," p. 864).

PREDICT 13

a. Are Will's arterial blood gas values above or below normal (see figure 23.16)?
b. Why did the asthma attack cause Will to breathe more rapidly (see figure 23.22)?
c. Why did the asthma attack cause Will to wheeze forcefully?
d. Did Will's rapid, forceful wheezing restore homeostasis? Explain.
e. Explain Will's blood P_{O_2} and P_{CO_2} values.
f. Is Will's blood pH lower or higher than normal? What effect does this blood pH normally have on respiration rate? Why didn't that happen?
g. Explain how β-adrenergic agents (see "The influence of Drugs on the Autonomic Nervous System," chapter 16) or inhaled glucocorticoids (see chapter 18) can help Will.

carbon dioxide levels can change pH, the respiratory system plays an important role in acid–base balance. For example, if blood pH decreases, the respiratory center is stimulated, resulting in the elimination of carbon dioxide and an increase in blood pH back to normal levels. Conversely, if blood pH increases, the respiratory rate decreases, and carbon dioxide levels increase, causing blood pH to decrease back to normal levels. The role of the respiratory system in maintaining pH is considered in greater detail in chapter 27.

Effect of Carbon Dioxide

Blood carbon dioxide levels are a major regulator of respiration during resting conditions and conditions when the carbon dioxide levels are elevated—for example, during intense exercise. Even a small increase in carbon dioxide in the circulatory system triggers a large increase in the rate and depth of respiration. An increase in P_{CO_2} of 5 mm Hg, for example, causes an increase in ventilation of 100%. A greater than normal amount of carbon dioxide in the blood is called **hypercapnia** (hī-per-kap′nē-ă). Conversely, lower than normal carbon dioxide levels, a condition called **hypocapnia** (hī-pō-kap′nē-ă), result in periods in which respiratory movements are reduced or do not occur.

PREDICT 14

Explain why a person who breathes rapidly and deeply (hyperventilates) for several seconds experiences a short period during which respiration does not occur (apnea) before normal breathing resumes.

The chemoreceptors in the chemosensitive area of the medulla oblongata and in the carotid and aortic bodies respond to changes in carbon dioxide because of the effects of carbon dioxide on blood pH (see figure 23.22). The chemosensitive area in the medulla oblongata is far more important for the regulation of P_{CO_2} and pH than are the carotid and aortic bodies. The carotid and aortic bodies are responsible for, at most, 15%–20% of the total response to changes in P_{CO_2} or pH. During intense exercise, however, the carotid bodies respond more rapidly to changes in blood pH than does the chemosensitive area of the medulla.

Effect of Oxygen

Changes in P_{O_2} can affect respiration (see figure 23.22), although P_{CO_2} levels detected by the chemosensitive area are responsible for most of the changes in respiration. A decrease in oxygen levels below normal values is called **hypoxia** (hī-pok′sē-ă). If P_{O_2} levels in the arterial blood are markedly reduced while the pH and P_{CO_2} are held constant, an increase in ventilation occurs. Within a normal range of P_{O_2} levels, however, the effect of oxygen on the regulation of respiration is small. Only after arterial P_{O_2} decreases to approximately 50% of its normal value does it begin to have a large stimulatory effect on respiratory movements.

At first, it is somewhat surprising that small changes in P_{O_2} do not cause changes in respiratory rate. Consideration of the oxygen–hemoglobin dissociation curve, however, provides an explanation. Because of the S shape of the curve, at any P_{O_2} above 80 mm Hg nearly all of the hemoglobin is saturated with oxygen. Consequently, until P_{O_2} levels change significantly, the oxygen-carrying capacity of the blood is unaffected.

The carotid and aortic body chemoreceptors respond to decreased P_{O_2} with increased stimulation of the respiratory center, which can keep it active despite decreasing oxygen levels. If P_{O_2} decreases sufficiently, however, the respiratory center can fail to function, resulting in death.

Importance of Reduced P_{O_2}

Carbon dioxide is much more important than oxygen as a regulator of normal alveolar ventilation, but under certain circumstances a reduced P_{O_2} in the arterial blood plays an important stimulatory role. During conditions of shock in which blood pressure is very low, the P_{O_2} in arterial blood can drop to levels sufficiently low to strongly stimulate carotid and aortic body sensory receptors. At high altitudes where barometric air pressure is low, the P_{O_2} in arterial blood can also drop to levels sufficiently low to stimulate carotid and aortic bodies. Although P_{O_2} levels in the blood are reduced, the respiratory system's ability to eliminate carbon dioxide is not greatly affected by low barometric air pressure. Thus, blood carbon dioxide levels become lower than normal because of the increased alveolar ventilation initiated in response to low P_{O_2}.

In people with emphysema, the destruction of the respiratory membrane results in decreased oxygen movement into the blood. The resulting low arterial P_{O_2} levels stimulate an increased rate and depth of respiration. At first, arterial P_{CO_2} levels may be unaffected by the reduced surface area of the respiratory membrane because carbon dioxide diffuses across the respiratory membrane 20 times more readily than does oxygen. However, if alveolar ventilation increases to the point that carbon dioxide exchange increases above normal, arterial carbon dioxide becomes lower than normal. More severe emphysema, in which the surface area of the respiratory membrane is reduced to a minimum, can decrease carbon dioxide exchange to the point that elevated arterial carbon dioxide occurs.

Hering-Breuer Reflex

The **Hering-Breuer** (her′ing-broy′er) **reflex** limits the degree to which inspiration proceeds and prevents overinflation of the lungs (see figure 23.21). This reflex depends on stretch receptors in the walls of the bronchi and bronchioles of the lung. Action potentials are initiated in these stretch receptors when the lungs are inflated and are passed along sensory neurons within the vagus nerves to the medulla oblongata. The action potentials have an inhibitory influence on the respiratory center and result in expiration. As expiration proceeds, the stretch receptors are no longer stimulated, and the decreased inhibitory effect on the respiratory center allows inspiration to begin again.

In infants, the Hering-Breuer reflex plays a role in regulating the basic rhythm of breathing and in preventing overinflation of the lungs. In adults, however, the reflex is important only when the tidal volume is large, such as during exercise.

Effect of Exercise on Ventilation

The mechanisms by which ventilation is regulated during exercise are controversial, and no one factor can account for all of the observed responses. Ventilation during exercise is divided into two phases:

1. *Ventilation increases abruptly.* At the onset of exercise, ventilation immediately increases. This initial increase can be as much as 50% of the total increase that occurs during exercise. The immediate increase in ventilation occurs too quickly to be explained by changes in metabolism or blood gases. As axons pass from the motor cortex of the cerebrum through the motor pathways, numerous collateral fibers project into the reticular formation of the brain. During exercise, action potentials in the motor pathways stimulate skeletal muscle contractions, and action potentials in the collateral fibers stimulate the respiratory center (see figure 23.21).

   Furthermore, during exercise, body movements stimulate proprioceptors in the joints of the limbs. Action potentials from the proprioceptors pass along sensory nerve fibers to the spinal cord and along ascending nerve tracts (the dorsal-column/medial-lemniscal system) of the spinal cord to the brain. Collateral fibers project from these ascending pathways to the respiratory center in the medulla oblongata. Movement of the limbs has a strong stimulatory influence on the respiratory center (see figure 23.21).

   A learned component may also exist in the ventilation response during exercise. After a period of training, the brain "learns" to match ventilation with the intensity of the exercise. Well-trained athletes match their respiratory movements more efficiently with their level of physical activity than do untrained individuals. Thus, the centers of the brain involved in learning have an indirect influence on the respiratory center, but the exact mechanism for this kind of regulation is unclear.
2. *Ventilation increases gradually.* After the immediate increase in ventilation, a gradual increase occurs and levels off within 4–6 minutes after the onset of exercise. Factors responsible for the immediate increase in ventilation may play a role in the gradual increase as well.

   Despite large changes in oxygen consumption and carbon dioxide production during exercise, the *average* arterial P_{O_2}, P_{CO_2}, and pH remain constant and close to resting levels as long as the exercise is aerobic (see chapter 9). This suggests that changes in blood gases and pH do not play an important role in regulating ventilation during aerobic exercise. During exercise, however, the values of arterial P_{O_2}, P_{CO_2}, and pH rise and fall more than at rest. Thus, even though their average values do not change, their oscillations may be a signal for helping control ventilation.

   The highest level of exercise that can be performed without causing a significant change in blood pH is called the **anaerobic threshold.** If the exercise intensity is high enough to exceed the anaerobic threshold, skeletal muscles produce and release lactic acid into the blood. The resulting change in blood pH stimulates the carotid bodies, resulting in increased ventilation. In fact, ventilation can increase so much that arterial P_{CO_2} decreases below resting levels and arterial P_{O_2} increases above resting levels.

Other Modifications of Ventilation

The activation of touch, thermal, and pain receptors can also affect the respiratory center (see figure 23.21). For example, irritants in the nasal cavity can initiate a sneeze reflex, and irritants in the lungs can stimulate a cough reflex. An increase in body temperature can stimulate increased ventilation.

53 ***Describe cerebral and limbic system control of ventilation.***

54 ***Define central and peripheral chemoreceptors. Which are most important for the regulation of blood pH and carbon dioxide?***

55 ***Define hypercapnia and hypocapnia.***

56 ***What effect does a decrease in blood pH or carbon dioxide have on respiratory rate?***

57 ***Describe the Hering-Breuer reflex and its function.***

58 ***Define hypoxia. Why must arterial Po_2 change significantly before it affects respiratory rate?***

59 ***What mechanisms regulate ventilation at the onset of exercise and during exercise? What is the anaerobic threshold?***

PREDICT 15

Describe the respiratory response when cold water is splashed onto a person. In the past, newborn babies were sometimes swatted on the buttocks. Explain the rationale for this procedure.

RESPIRATORY ADAPTATIONS TO EXERCISE

In response to training, athletic performance increases because the cardiovascular and respiratory systems become more efficient at delivering oxygen and picking up carbon dioxide. Ventilation in most individuals does not limit performance because ventilation can increase to a greater extent than does cardiovascular function.

After training, vital capacity increases slightly and residual volume decreases slightly. Tidal volume at rest and during submaximal exercise does not change. At maximal exercise, however, tidal volume increases. After training, the respiratory rate at rest or during submaximal exercise is slightly lower than in an untrained person, but at maximal exercise respiratory rate is generally increased.

Minute ventilation is affected by the changes in tidal volume and respiratory rate. After training, minute ventilation is essentially unchanged or slightly reduced at rest and is slightly reduced during submaximal exercise. Minute ventilation is greatly increased at maximal exercise. For example, an untrained person's minute ventilation of 120 L/min can increase to 150 L/min after training. Increases to 180 L/min are typical of highly trained athletes.

Gas exchange between the alveoli and blood increases at maximal exercise following training. The increased minute ventilation results in increased alveolar ventilation. In addition, increased cardiovascular efficiency results in greater blood flow through the lungs, especially in the superior parts of the lungs.

60 ***What effect does training have on resting, submaximal, and maximal tidal volumes and on minute ventilation?***

EFFECTS OF AGING ON THE RESPIRATORY SYSTEM

Most aspects of the respiratory system are affected by aging. Even though vital capacity, maximum ventilation rates, and gas exchange decrease with age, the elderly can engage in light to moderate exercise because the respiratory system has a large reserve capacity.

Vital capacity decreases with age because of a decreased ability to fill the lungs (decreased inspiratory reserve volume) and a decreased ability to empty the lungs (decreased expiratory reserve volume). As a result, maximum minute ventilation rates decrease, which in turn decreases the ability to perform intense exercise. These changes are related to the weakening of respiratory muscles and to decreased compliance of the thoracic cage caused by the stiffening of cartilage and ribs. Lung compliance actually increases with age, but this effect is offset by the decreased thoracic cage compliance. Lung compliance increases because parts of the alveolar walls are lost, which reduces lung recoil. There are no significant age-related changes in lung elastic fibers or surfactant.

Residual volume increases with age as the alveolar ducts and many of the larger bronchioles increase in diameter. This increases the dead space, which decreases the amount of air available for gas exchange (alveolar ventilation). In addition, gas exchange across the respiratory membrane is reduced because parts of the alveolar walls are lost, which decreases the surface area available for gas exchange, and the remaining walls thicken, which decreases the diffusion of gases. A gradual increase in resting tidal volume with age compensates for these changes.

With age, mucus accumulates within the respiratory passageways. The mucus–cilia escalator is less able to move the mucus because it becomes more viscous and because the number of cilia and their rate of movement decrease. As a consequence, the elderly are more susceptible to respiratory infections and bronchitis.

61 ***Why do vital capacity, alveolar ventilation, and the diffusion of gases across the respiratory membrane decrease with age?***

62 ***Why are the elderly more likely to develop respiratory infections and bronchitis?***

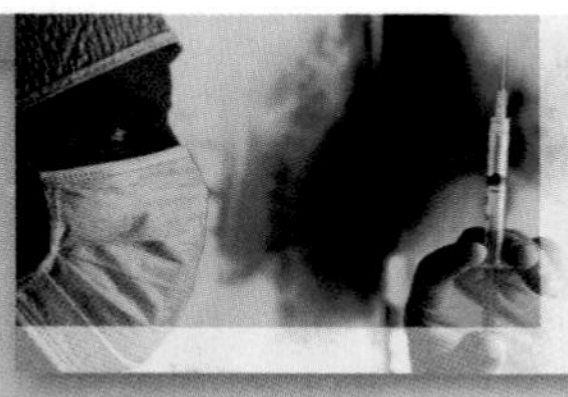

CLINICAL FOCUS

Disorders of the Respiratory System

BRONCHI AND LUNGS

Bronchitis (brong-kī′tis) is an inflammation of the bronchi caused by irritants, such as cigarette smoke, air pollution, or infections. The inflammation results in swelling of the mucous membrane lining the bronchi, increased mucus production, and decreased movement of mucus by cilia. Consequently, the diameter of the bronchi is decreased, and ventilation is impaired. Bronchitis can progress to emphysema.

Emphysema (em-fi-zē′mă) results in the destruction of the alveolar walls. Many smokers have both bronchitis and emphysema, which are often referred to as **chronic obstructive pulmonary disease (COPD).** Chronic inflammation of the bronchioles, usually caused by cigarette smoke or air pollution, probably initiates emphysema. Narrowing of the bronchioles restricts air movement, and air tends to be retained in the lungs. Coughing to remove accumulated mucus increases pressure in the alveoli, resulting in the rupture and destruction of alveolar walls. Loss of alveolar walls has two important consequences. The respiratory membrane has a decreased surface area, which decreases gas exchange, and loss of elastic fibers decreases the lung's ability to recoil and expel air. Symptoms of emphysema include shortness of breath and enlargement of the thoracic cavity. Treatment involves removing the sources of irritants (e.g., stopping smoking), promoting the removal of bronchial secretions, using bronchiodilators, retraining people to breathe so that expiration of air is maximized, and using antibiotics to prevent infections. The progress of emphysema can be slowed, but no cure exists.

Adult respiratory distress syndrome (ARDS) is caused by damage to the respiratory membrane. The damage stimulates an inflammatory response, which further damages the respiratory membrane. Water, ions, and proteins leave the blood and enter alveoli. Surfactant in the alveoli is reduced as surfactant-producing cells are damaged and surfactant present in the alveoli is diluted. The fluid-filled alveoli reduce gas exchange and make it more difficult for the lungs to expand. ARDS usually develops rapidly following an injurious event, such as an infection, inhalation of smoke from a fire, inhalation of toxic fumes, trauma, aspiration of gastric content associated with gastric reflux, or circulatory shock. Even with oxygen inhalation therapy, the mortality rate is high.

Asthma (az′mă) is a disease characterized by abnormally increased constriction of the bronchi and bronchioles in response to various stimuli, resulting in a narrowing of the air passageways and decreased ventilation efficiency. Symptoms include rapid, shallow breathing; wheezing; coughing; and shortness of breath. In contrast to many other respiratory disorders, however, the symptoms of asthma typically reverse either spontaneously or with therapy. The prevalence of asthma in the United States is from 3%–6% of the population. In approximately half the cases, the symptoms appear before age 10, and twice as many boys as girls develop asthma.

The exact cause or causes of asthma are unknown, but asthma and allergies are more common in some families. Multiple genes contribute to a person's susceptibility to asthma; genes on chromosomes 5, 6, 11, 12 and 14 have all been linked to asthma. Although no definitive pathologic feature or diagnostic test for asthma has yet been discovered, three important features of the disease are chronic airway inflammation, airway hyperreactivity, and airflow obstruction. The inflammatory response results in tissue damage, edema, and mucus buildup, which can block airflow through the bronchi and bronchioles. Airway hyperreactivity results in greatly increased contraction of the smooth muscle in the bronchi and bronchioles in response to a stimulus. As a result of airway hyperreactivity, the diameter of the airway decreases, and resistance to airflow increases. The effects of inflammation and airway hyperreactivity combine to cause airflow obstruction.

Many cases of asthma appear to be associated with a chronic inflammatory response. The number of immune cells in the bronchi and bronchioles, including mast cells, eosinophils, neutrophils, macrophages, and lymphocytes, increases. Inflammation appears to be linked to airway hyperreactivity by some chemical mediators released by immune cells (e.g., leukotrienes, prostaglandins, and interleukins), which increase the airway's sensitivity to stimulation and cause smooth muscle contraction.

The stimuli that prompt airflow obstruction in asthma vary from one individual to another. Some asthmatics have reactions to particular allergens, which are foreign substances that evoke an inappropriate immune system response (see chapter 22). Examples include inhaled pollen, animal dander, and dust mites. On the other hand, inhaled substances, such as chemicals in the workplace or cigarette smoke, can provoke an asthma attack in some people without stimulating an allergic reaction. Ingested substances, such as aspirin or nonsteroidal anti-inflammatory compounds, such as ibuprofen (ī-boo′-prō′-fen), can also stimulate an asthma attack. Acetaminophen (as-et-ă-mē′nō-fen; Tylenol), which does not stimulate an asthma attack, can be substituted for aspirin.

Other stimuli, such as strenuous exercise, especially in cold weather, can precipitate an asthma attack. Such episodes often can be avoided by using a bronchodilator drug prior to exercise. Viral infections, emotional upset, stress, and even reflux of stomach acid into the esophagus are also known to elicit an asthma attack.

Treatment of asthma involves avoiding the causative stimulus and administering drug therapy. Steroids and mast cell–stabilizing agents, which prevent the release of chemical mediators from mast cells, are used to reduce airway inflammation. Theophylline (thē-of′i-lēn, thē-of′i-lin) and β-adrenergic agents (see chapter 16) are commonly used to stimulate bronchiolar dilation. Treatment is generally effective in controlling asthma, although in rare cases death by asphyxiation occurs. Death can be prevented with early and intensive therapy.

Pulmonary fibrosis is the replacement of lung tissue with fibrous connective tissue, thereby making the lungs less elastic and breathing more difficult. Exposure to asbestos, silica (silicosis), or coal dust is the most common cause.

Lung, or **bronchiogenic, cancer** arises from the epithelium of the respiratory tract. Cancers arising from tissues other than respiratory epithelium are not called lung cancer, even though they occur in the lungs. Lung cancer is the most common cause of cancer death in males and females in the United States, and most cases occur in smokers or those exposed to secondhand smoke. Because of the rich lymph and blood supply in the lungs, cancer in the lung can readily spread to other parts of the lung or body. In addition, the disease is often advanced before symptoms become severe enough for the victim to seek medical aid. Typical symptoms include coughing, sputum production, and blockage of the airways. Treatments include removal of part or all of the lung, chemotherapy, and radiation. Promising new, early detection tests are being explored. These include blood tests for a lung cancer–specific protein and sputum DNA tests that detect genetic abnormalities.

NERVOUS SYSTEM

Sudden infant death syndrome (SIDS), or crib death, is the most frequent cause of the death of infants between 2 weeks and 1 year of age. Death results when the infant stops breathing during sleep. Although the cause of SIDS remains controversial, evidence exists that damage to the respiratory center during development is a factor. No treatment has yet been found, but at-risk babies can be placed on a monitor that sounds an alarm if the baby stops breathing.

Paralysis of the respiratory muscles can result from spinal cord damage in the cervical or thoracic region. The damage interrupts the nerve tracts that transmit action potentials to the muscles of respiration. Transection of the spinal cord can result from trauma, such as having an automobile accident or diving into water that is too shallow. Another cause of paralysis is poliomyelitis, a viral infection that damages neurons of the respiratory center or motor neurons that stimulate the muscles of respiration. Anesthetics and central nervous system depressants can also depress the function of the respiratory center if they are taken or administered in large enough doses.

INFECTIOUS DISEASES OF THE RESPIRATORY SYSTEM

Strep throat is caused by a streptococcal bacteria (*Streptococcus pyogenes*) and is characterized by inflammation of the pharynx and by fever. Frequently, inflammation of the tonsils and middle ear is involved. Without a throat analysis, the infection cannot be distinguished from viral causes of pharyngeal inflammation. Current techniques allow rapid diagnosis, within minutes to hours, and antibiotics are an effective treatment.

Laryngitis (lar-in-jī′tis) is an inflammation of the larynx, especially the vocal folds. Bacterial or viral infections can move from the upper respiratory tract to cause laryngitis.

The **common cold** is the result of a viral infection. Symptoms include sneezing, excessive nasal secretions, and congestion.

Flu (influenza) is a viral infection of the respiratory system and does not affect the digestive system, as is commonly assumed. Flu is characterized by chills, fever, headache, and muscular aches, in addition to coldlike symptoms.

Pneumonia (noo-mō′nē-ă) is a general term that refers to many infections of the lung. Most pneumonias are caused by bacteria, but some result from viral, fungal, or protozoan infections. Symptoms include fever, difficulty breathing, and chest pain. Inflammation of the lungs results in the accumulation of fluid within alveoli (pulmonary edema) and poor inflation of the lungs with air.

Other respiratory system infections include the bacterial infections **diphtheria** (dif-thē′rē-ă), **whooping cough** (pertussis; per-tŭs′is), and **tuberculosis** (tū-ber′kū-lō′sis) and the fungal infections **histoplasmosis** (his′tō-plaz-mō′sis) and **coccidioidomycosis** (kok-sid-ē-oy′dō-mī-kō′sis). Vaccines against diphtheria and whooping cough are part of the normal vaccination procedure for children in the United States.

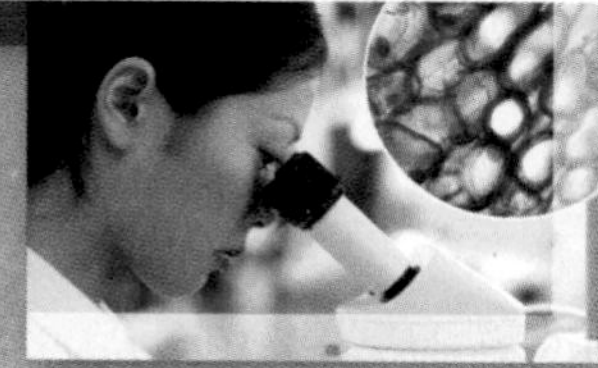

SYSTEMS PATHOLOGY

Cystic Fibrosis

Nicole is a 2-year-old who has experienced recurrent bouts of coughing and wheezing. Two months ago, she was diagnosed with bronchitis after suffering from an upper respiratory tract viral infection. Her condition worsened despite treatment with inhaled bronchodilators and steroids. Also, she has not been gaining weight, despite having a good appetite, and her stools are frequent, loose, foul-smelling, and greasy.

Nicole's parents have become worried about her persistent cough and lack of growth. Furthermore, Nicole's mother wonders whether a salty taste on Nicole's skin is due to a side effect of the medication she has been taking. The family medical history shows that Nicole's parents and 4-year-old sister are healthy, but her maternal grandmother has chronic bronchitis, seemingly due to having smoked cigarettes for 20 years.

Nicole's doctor decided to have an x-ray taken of Nicole's chest. The radiograph showed cloudy streaks in her right upper lung lobe, indicating **atelactasis** (at-ĕ-lek′tă-sis), which is the collapse of a portion of the lung. Atelactasis is sometimes caused by bronchioles being compacted with mucus. Subsequently, Nicole's doctor ordered a sweat-chloride test for cystic fibrosis and a cystic fibrosis genotype test. The sweat-chloride test was positive, and the genotype test revealed that Nicole has two copies of the cystic fibrosis gene; she has cystic fibrosis.

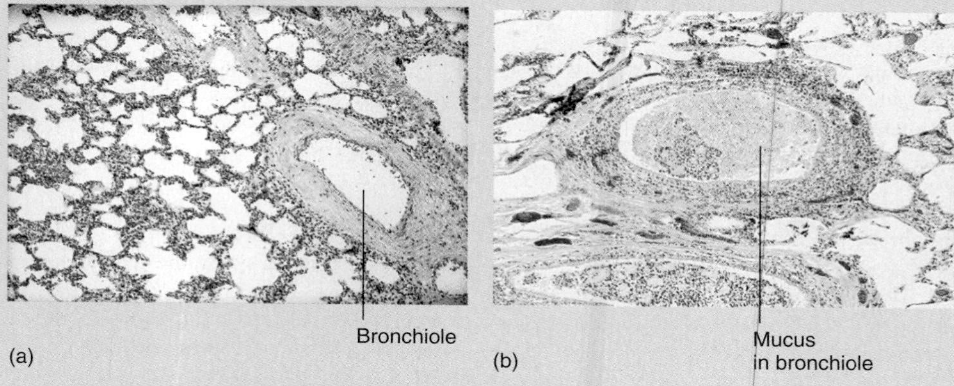

FIGURE A Bronchioles in Normal Lungs, Compared with Bronchioles in CF Lungs

(*a*) In normal lung tissue, bronchioles are the passageways for airflow. (*b*) In patients with CF, the bronchioles are obstructed with thickened mucus and airflow is restricted.

BACKGROUND INFORMATION

Cystic fibrosis (CF) is a disease characterized by frequent, serious respiratory infections; thick, sticky mucus in the lungs and digestive tract; and a salty taste to the skin. Often, the first sign of CF is the salty skin, which parents notice when they kiss their child. "A child that is salty to taste will die shortly after birth," says a 17th century English proverb.

Cystic fibrosis is the most common lethal genetic disorder in Caucasians. It occurs in 1 in 3000 Caucasians, 1 in 9200 Hispanics, 1 in 15,000 African-Americans, and 1 in 31,000 Asian-Americans. Cystic fibrosis is inherited as an autosomal-recessive gene on chromosome 7. Males and females are equally likely to have CF because the gene is on an autosomal chromosome. Individuals with CF are homozygous recessive for the CF gene, whereas carriers are heterozygous. A **carrier** is an individual who has one copy of a disease-causing gene but may have no symptoms or mild symptoms (see chapter 3). An estimated 12 million Americans are CF carriers.

In CF, a mutated gene, the **cystic fibrosis transmembrane conductance regulator (CFTR)** gene, results in the production of a defective chloride channel, also called CFTR. CFTR has two major functions: It is a chloride channel, which transports Cl^- across the plasma membrane of epithelial cells, and it is a regulator of epithelial sodium channels. Over 230 gene defects are known to be associated with CF. Over 70% of CF cases, however, result from the loss of the amino acid phenylalanine from the CFTR protein. As a result of this mutation, CFTR is not transported to the plasma membrane. Other mutations result in defective CFTR that are transported to the plasma membrane, but they either are nonfunctional or transport reduced amounts of Cl^-.

The air passageways of the tracheobronchial tree are lined with ciliated epithelium. The cilia are located in a layer of fluid called the **periciliary layer (PCL).** The PCL is approximately 7 μm thick, which is the length of a fully extended cilium. A layer of mucus that varies in height from 7 to 70 μm overlies the PCL. The mucus can trap microorganisms and inhaled debris. Beating of the cilia moves the PCL and the overlying mucus out of the respiratory system. This removal is part of innate immunity (see "Mechanical Mechanisms," chapter 22) and helps protect against respiratory tract infections.

The normal height of the PCL must be maintained for optimal movement of the mucus. Changes in PCL volume change the height of the PCL. Water movement between the mucus and PCL and across the layer of epithelial cells lining the respiratory tract alters PCL volume. The mucus acts as a reservoir to store and release water. For example, as PCL volume decreases, the mucus replenishes the volume by releasing water into the PCL.

Respiratory epithelial cells can regulate PCL volume by controlling the movement of Na^+ and Cl^-. As these ions move out of or into the PCL, water follows. Two opposing systems regulate the movement of Na^+ and Cl^-. When the Na^+ system is dominant, Na^+ are absorbed from the PCL, Cl^- and water follow, and PCL volume is reduced. When the Cl^- system is dominant, Cl^- are secreted into the PCL, Na^+ and water follow, and PCL volume is increased. The movement of Na^+ out of, and Cl^- into, the PCL occurs through Na^+ channels and CFTR channels, respectively. In addition to functioning as Cl^- channels, CFTR inhibits the activity of the Na^+ channels. Exactly how this inhibition is accomplished is not completely understood. CFTR usually acts as a "brake" on the Na^+ channels, thereby reducing the amount of Na^+ normally reabsorbed.

In cystic fibrosis, CFTR is either nonfunctional or marginally functional. Consequently, the normal inhibition of the Na^+ channels is lost or greatly diminished and Na^+ absorption increases dramatically, resulting in increased water loss from the PCL. As water is reabsorbed from the PCL, water is lost from the mucus. In addition, the normal response to decreasing PCL volume, the secretion of Cl^-, is lost or diminished. The net result of defective CFTR is the greatly reduced or absent transport of mucus because of the depletion of the PCL and a reduction in the water content of mucus, which causes the mucus to be thicker than normal (figure A).

A standard test for CF diagnosis is the **sweat-chloride test,** in which the chemical **pilocarpine** (pi-lō-kar′pēn) is swabbed onto the skin and a mild electric current is applied. Pilocarpine is a muscarinic agent that stimulates receptors in the sweat glands (see "The Influence of Drugs on the Autonomic Nervous System," chapter 16). The mild electric current drives the medication into the skin, producing localized sweating and avoiding systemic drug effects. The sweat that is produced is collected and tested for abnormally high levels of salt (NaCl). Normally, sweat glands produce a very dilute liquid, which cools the body without depleting salt from it. In CF, the malfunctioning CFTR results in a failure to absorb the normal amount of NaCl from sweat, resulting in high NaCl content in sweat.

Although CF tends to be primarily associated with respiratory malfunctions, the production of thickened mucus also has profound digestive tract effects. In fact, the original name of the disease was cystic fibrosis of the pancreas because, in 90% of CF patients, the pancreas is gradually destroyed and infiltrated by fibrous cysts. The pancreatic ducts of CF patients can become obstructed with sticky mucus, which prevents the secretion of adequate amounts of digestive enzymes, particularly fat-digesting enzymes. Children with CF can have severe nutritional deficiencies because of the decreased absorption of proteins and fat-soluble vitamins, such as vitamins A, D, E, and K. To aid food digestion and promote growth, children with CF may be given powdered digestive enzymes. Supplemental overnight feeding through a gastrostomy (gas-tros′tō-mē) tube (stomach tube, T-tube) may also be beneficial.

The main goal of CF treatment is to reduce lung infections, clear the lungs of mucus, improve airflow, and maintain sufficient calories and nutrition. People with CF must undergo **chest physical therapy,** also called **chest clapping** or **chest percussion.** This involves manually pounding the back and chest for 30 to 40 minutes three or four times daily to dislodge mucus trapped in the chest. Automated chest clappers are preferred by some CF patients. Antibiotics may be prescribed to help control lung infections. Mucus-thinning drugs, such as Pulmozyme, and bronchodilators can be inhaled to improve mucus clearance and open airways. Eventually, if breathing problems become too severe or the patient becomes resistant to antibiotics, a lung transplant may be necessary. The downside of a lung transplant is the need to take immunosuppressive drugs for life to prevent rejection of the transplanted lungs. These drugs produce side effects, such as increased susceptibility to infections, diabetes, tumors, and osteoporosis. The upside of lung transplantation is that it is a partial "cure" because the transplanted lung cells do not have the genetic defect. However, cells with the defective CFTR gene are still present elsewhere in the body. Scientists are also investigating the use of gene therapy, wherein a copy of the normal CFTR gene is inserted into epithelial cells by a harmless virus. So far, the effects of gene therapy have lasted for only a few days. With treatment, the current life expectancy for persons with cystic fibrosis is into the mid-30s. In 95% of CF cases, the patient dies due to complications from lung infections.

PREDICT 16

As cystic fibrosis becomes advanced, what happens to forced expiratory volume in 1 second (FEV1), residual volume, and physiological dead space?

SYSTEM INTERACTIONS — Effect of Cystic Fibrosis on Other Systems

SYSTEM	INTERACTIONS
Integumentary	Two to five times the normal amount of salt is secreted in sweat, which can cause rapid dehydration in hot conditions. Clubbing is an enlargement of the fingertips and toes due to a proliferation of connective tissue; the mechanism that produces clubbing is unclear, but it may be related to insufficient oxygen delivery, which stimulates an inflammatory response.
Skeletal	Low bone density is common because insufficient vitamin D is absorbed from the diet when the pancreatic ducts become blocked.
Cardiovascular	Lung disease may eventually cause the right ventricle of the heart to fail due to the increased force necessary to pump blood into damaged lungs.
Digestive	Mucus blockage of pancreatic ducts and liver bile ducts decreases fat digestion capabilities, resulting in bowel blockage; foul-smelling, greasy stools; and chronic diarrhea. Autodigestion of the pancreas by enzymes trapped in the pancreas can occur. Liver duct blockage may eventually lead to cirrhosis of the liver and gallstones.
Respiratory	Mucus buildup causes coughing, wheezing, and recurrent chest infections because bacteria are not effectively removed. Eventually, lung bleeding (hemoptysis) or collapsed lung (atelactasis) may result. There may also be polyps in the nasal cavity and paranasal sinuses due to thickening of the mucosa. Frequent instances of sinusitis are common.
Reproductive	Ninety-eight percent of men with CF are infertile because of a failure of the ductus deferens to develop. Up to 20% of women with CF may experience infertility related to mucus blockage of the uterine tubes or depression of the menstrual cycle because of malnutrition.
Immune	A decrease in innate immunity occurs because the thickened mucus in the respiratory tract impairs cilia movement. The beating of cilia in the respiratory tract is one of the important mechanical mechanisms that prevents the entry of microorganisms into the body.

SUMMARY

Respiration includes the movement of air into and out of the lungs, the exchange of gases between the air and the blood, the transport of gases in the blood, and the exchange of gases between the blood and tissues.

Functions of the Respiratory System (p. 826)

The major functions of the respiratory system are gas exchange, regulation of blood pH, voice production, olfaction, and protection against some microorganisms.

Anatomy and Histology of the Respiratory System (p. 826)

Nose

1. The nose consists of the external nose and the nasal cavity.
2. The bridge of the nose is bone, and most of the external nose is cartilage.
3. Openings of the nasal cavity
   - The nares open to the outside, and the choanae lead to the pharynx.
   - The paranasal sinuses and the nasolacrimal duct open into the nasal cavity.
4. Parts of the nasal cavity
   - The nasal cavity is divided by the nasal septum.
   - The anterior vestibule contains hairs that trap debris.
   - The nasal cavity is lined with pseudostratified ciliated columnar epithelium that traps debris and moves it to the pharynx.
   - The superior part of the nasal cavity contains the olfactory epithelium.
5. The nasal cavity is a passageway for air; it cleans and humidifies air; it is the location for the sense of smell; and, with the paranasal sinuses, it is a resonating chamber for speech.

Pharynx

1. The nasopharynx joins the nasal cavity through the internal nares and contains the openings to the auditory tube and the pharyngeal tonsils.
2. The oropharynx joins the oral cavity and contains the palatine and lingual tonsils.
3. The laryngopharynx opens into the larynx and the esophagus.

Larynx

1. Cartilage
   - Three unpaired cartilages exist. The thyroid cartilage and cricoid cartilage form most of the larynx. The epiglottis covers the opening of the larynx during swallowing.
   - Six paired cartilages exist. The vocal folds attach to the arytenoid cartilages.
2. The larynx maintains an open air passageway, regulates the passage of swallowed materials and air, produces sounds, and removes debris from the air.
3. Sounds are produced as the vocal folds vibrate when air passes through the larynx. Tightening the folds produces sounds of different pitches by controlling the length of the fold, which is allowed to vibrate.

Trachea

1. The trachea connects the larynx to the main bronchi.
2. The trachealis muscle regulates the diameter of the trachea.

Tracheobronchial Tree

1. The trachea divides to form two main bronchi, which go to the lungs. The main bronchi divide to form lobar bronchi, which divide to form segmental bronchi, which divide to form bronchioles, which divide to form terminal bronchioles.
2. The trachea to the terminal bronchioles is a passageway for air movement.
   - The area from the trachea to the terminal bronchioles is ciliated to facilitate the removal of inhaled debris.
   - Cartilage helps hold the tube system open (from the trachea to the bronchioles).
   - Smooth muscle controls the diameter of the tubes (terminal bronchioles).
3. Terminal bronchioles divide to form respiratory bronchioles, which give rise to alveolar ducts. Air-filled chambers called alveoli open into the respiratory bronchioles and alveolar ducts. The alveolar ducts end as alveolar sacs, which are chambers that connect to two or more alveoli.
4. Gas exchange occurs from the respiratory bronchioles to the alveoli.
5. The components of the respiratory membrane are a film of water, the walls of the alveolus and the capillary, and an interstitial space.

Lungs

1. The body contains two lungs.
2. The lungs are divided into lobes, bronchopulmonary segments, and lobules.

Thoracic Wall and Muscles of Respiration

1. The thoracic wall consists of vertebrae, ribs, sternum, and muscles that allow expansion of the thoracic cavity.
2. Contraction of the diaphragm increases thoracic volume.
3. Muscles can elevate the ribs and increase thoracic volume or can depress the ribs and decrease thoracic volume.

Pleura

The pleural membranes surround the lungs and provide protection against friction.

Blood Supply

1. Deoxygenated blood is transported to the lungs through the pulmonary arteries, and oxygenated blood leaves through the pulmonary veins.
2. Oxygenated blood is mixed with a small amount of deoxygenated blood from the bronchi.

Lymphatic Supply

The superficial and deep lymphatic vessels drain lymph from the lungs.

Ventilation (p. 841)

Pressure Differences and Airflow

1. Ventilation is the movement of air into and out of the lungs.
2. Air moves from an area of higher pressure to an area of lower pressure.

Pressure and Volume

Pressure is inversely related to volume.

Airflow Into and Out of Alveoli

1. Inspiration results when barometric air pressure is greater than alveolar pressure.
2. Expiration results when barometric air pressure is less than alveolar pressure.

Changing Alveolar Volume

1. Lung recoil causes alveoli to collapse.
   - Lung recoil results from elastic fibers and water surface tension.
   - Surfactant reduces water surface tension.
2. Pleural pressure is the pressure in the pleural cavity.
   - A negative pleural pressure can cause the alveoli to expand.
   - Pneumothorax is an opening between the pleural cavity and the air that causes a loss of pleural pressure.
3. Changes in thoracic volume cause changes in pleural pressure, resulting in changes in alveolar volume, alveolar pressure, and airflow.

Measurement of Lung Function (p. 846)

Compliance of the Lungs and Thorax

1. Compliance is a measure of lung expansion caused by alveolar pressure.
2. Reduced compliance means that it is more difficult than normal to expand the lungs.

Pulmonary Volumes and Capacities

1. Four pulmonary volumes exist: tidal volume, inspiratory reserve volume, expiratory reserve volume, and residual volume.
2. Pulmonary capacities are the sum of two or more pulmonary volumes and include inspiratory capacity, functional residual capacity, vital capacity, and total lung capacity.
3. The forced expiratory vital capacity measures vital capacity as the individual exhales as rapidly as possible.

Minute Ventilation and Alveolar Ventilation

1. The minute ventilation is the total amount of air moved in and out of the respiratory system per minute.
2. Dead space is the part of the respiratory system in which gas exchange does not take place.
3. Alveolar ventilation is how much air per minute enters the parts of the respiratory system in which gas exchange takes place.

Physical Principles of Gas Exchange (p. 848)

Partial Pressure

1. Partial pressure is the contribution of a gas to the total pressure of a mixture of gases (Dalton's law).
2. Water vapor pressure is the partial pressure produced by water.
3. Atmospheric air, alveolar air, and expired air have different compositions.

Diffusion of Gases Into and Out of Liquids

The concentration of a dissolved gas in a liquid is determined by its pressure and by its solubility coefficient (Henry's law).

Diffusion of Gases Through the Respiratory Membrane

1. The respiratory membrane is thin and has a large surface area that facilitates gas exchange.
2. The rate of diffusion of gases through the respiratory membrane depends on its thickness, the diffusion coefficient of the gas, the surface area of the membrane, and the partial pressure of the gases in the alveoli and the blood.

Relationship Between Alveolar Ventilation and Pulmonary Capillary Perfusion

1. Increased alveolar ventilation or increased pulmonary capillary perfusion increases gas exchange.
2. The physiological shunt is the deoxygenated blood returning from the lungs.

Oxygen and Carbon Dioxide Transport in the Blood (p. 851)

Oxygen Partial Pressure Gradients

1. Oxygen moves from the alveoli (Po_2 = 104 mm Hg) into the blood (Po_2 = 40 mm Hg). Blood is almost completely saturated with oxygen when it leaves the capillary.
2. The Po_2 in the blood decreases (Po_2 = 95 mm Hg) because of mixing with deoxygenated blood.
3. Oxygen moves from the tissue capillaries (Po_2 = 95 mm Hg) into the tissues (Po_2 = 40 mm Hg).

Carbon Dioxide Partial Pressure Gradients

1. Carbon dioxide moves from the tissues (Pco_2 = 45 mm Hg) into tissue capillaries (Pco_2 = 40 mm Hg).
2. Carbon dioxide moves from the pulmonary capillaries (Pco_2 = 45 mm Hg) into the alveoli (Pco_2 = 40 mm Hg).

Hemoglobin and Oxygen Transport

1. Oxygen is transported by hemoglobin (98.5%) and is dissolved in plasma (1.5%).
2. The oxygen–hemoglobin dissociation curve shows that hemoglobin is almost completely saturated when Po_2 is 80 mm Hg or above. At lower partial pressures, the hemoglobin releases oxygen.
3. A shift of the oxygen–hemoglobin dissociation curve to the right because of a decrease in pH (Bohr effect), an increase in carbon dioxide, or an increase in temperature results in a decrease in hemoglobin's ability to hold oxygen.
4. A shift of the oxygen–hemoglobin dissociation curve to the left because of an increase in pH (Bohr effect), a decrease in carbon dioxide, or a decrease in temperature results in an increase in hemoglobin's ability to hold oxygen.
5. The substance 2,3-bisphosphoglycerate increases hemoglobin's ability to release oxygen.
6. Fetal hemoglobin has a higher affinity for oxygen than does maternal hemoglobin.

Transport of Carbon Dioxide

1. Carbon dioxide is transported dissolved in plasma (7%), as HCO_3^- dissolved in plasma and in red blood cells (70%), and bound to hemoglobin (23%).
2. In tissue capillaries, the following occur.
   - Carbon dioxide combines with water inside red blood cells to form carbonic acid, which dissociates to form HCO_3^-. Decreasing HCO_3^- concentrations promotes carbon dioxide transport.
   - The chloride shift is the exchange of Cl^- for HCO_3^- between plasma and red blood cells.
   - Hydrogen ions binding to hemoglobin promote carbon dioxide transport, prevent a change in pH in red blood cells, and produce a Bohr effect.
   - In the Haldane effect, the smaller the amount of oxygen bound to hemoglobin, the greater the amount of carbon dioxide bound to it, and vice versa.
3. In pulmonary capillaries, the events occurring in the tissue capillaries are reversed.

Regulation of Ventilation (p. 856)

Respiratory Areas in the Brainstem

1. The medullary respiratory center consists of the dorsal and ventral respiratory groups.
   - The dorsal respiratory groups stimulate the diaphragm.
   - The ventral respiratory groups stimulate the intercostal and abdominal muscles.
2. The pontine respiratory group is involved with switching between inspiration and expiration.

Generation of Rhythmic Ventilation

1. Neurons in the medullary respiratory center establish the basic rhythm of respiration.
2. When stimuli from receptors or other parts of the brain exceed a threshold level, inspiration begins.
3. As respiratory muscles are stimulated, neurons that stop inspiration are stimulated. When the stimulation of these neurons exceeds a threshold level, inspiration is inhibited.

Cerebral and Limbic System Control

Respiration can be voluntarily controlled and can be modified by emotions.

Chemical Control of Ventilation

1. Carbon dioxide is the major regulator of respiration. An increase in carbon dioxide or a decrease in pH can stimulate the chemosensitive area, causing a greater rate and depth of respiration.
2. Oxygen levels in the blood affect respiration when a 50% or greater decrease from normal levels exists. Decreased oxygen is detected by receptors in the carotid and aortic bodies, which then stimulate the respiratory center.

Hering-Breuer Reflex

Stretch of the lungs during inspiration can inhibit the respiratory center and contribute to a cessation of inspiration.

Effect of Exercise on Ventilation

1. Collateral fibers from motor neurons and from proprioceptors stimulate the respiratory centers.
2. Chemosensitive mechanisms and learning fine-tune the effects produced through the motor neurons and proprioceptors.

Other Modifications of Ventilation

Touch, thermal, and pain sensations can modify ventilation.

Respiratory Adaptations to Exercise (p. 863)

Tidal volume, respiratory rate, minute ventilation, and gas exchange between the alveoli and blood remain unchanged or slightly lower at rest or during submaximal exercise but increase at maximal exercise.

Effects of Aging on the Respiratory System (p. 863)

1. Vital capacity and maximum minute ventilation decrease with age because of a weakening of respiratory muscles and decreased thoracic cage compliance.
2. Residual volume and dead space increase because of the increased diameter of respiratory passageways. As a result, alveolar ventilation decreases.
3. An increase in resting tidal volume compensates for decreased alveolar ventilation, loss of alveolar walls (surface area), and thickening of alveolar walls.
4. The ability to remove mucus from the respiratory passageways decreases with age.

REVIEW AND COMPREHENSION

1. The nasal cavity
   a. has openings for the paranasal sinuses.
   b. has a vestibule, which contains the olfactory epithelium.
   c. is connected to the pharynx by the nares.
   d. has passageways called conchae.
   e. is lined with squamous epithelium, except for the vestibule.
2. The nasopharynx
   a. is lined with moist stratified squamous epithelium.
   b. contains the pharyngeal tonsil.
   c. opens into the oral cavity through the fauces.
   d. extends to the tip of the epiglottis.
   e. is an area through which food, drink, and air pass.
3. The larynx
   a. connects the oropharynx to the trachea.
   b. has three unpaired and six paired cartilages.
   c. contains the vocal folds.
   d. contains the vestibular folds.
   e. all of the above.
4. Terminal bronchioles branch to form
   a. the alveolar duct.
   b. alveoli.
   c. bronchioles.
   d. respiratory bronchioles.

5. During an asthma attack, a person has difficulty breathing because of constriction of the
   a. trachea.
   b. bronchi.
   c. terminal bronchioles.
   d. alveoli.
   e. respiratory membrane.
6. During quiet expiration, the
   a. abdominal muscles relax.
   b. diaphragm moves inferiorly.
   c. external intercostal muscles contract.
   d. thorax and lungs passively recoil.
   e. all of the above.
7. The parietal pleura
   a. covers the surface of the lung.
   b. covers the inner surface of the thoracic cavity.
   c. is the connective tissue partition that divides the thoracic cavity into right and left pleural cavities.
   d. covers the inner surface of the alveoli.
   e. is the membrane across which gas exchange occurs.
8. Contraction of the bronchiolar smooth muscle has which of these effects?
   a. A smaller pressure gradient is required to get the same rate of airflow, compared with normal bronchioles.
   b. It increases airflow through the bronchioles.
   c. It increases resistance to airflow.
   d. It increases alveolar ventilation.
9. During expiration, the alveolar pressure is
   a. greater than the pleural pressure.
   b. greater than the barometric pressure.
   c. less than the barometric pressure.
   d. unchanged.
10. The lungs do not normally collapse because of
    a. surfactant.
    b. pleural pressure.
    c. elastic recoil.
    d. both a and b.
11. Immediately after the creation of an opening through the thorax into the pleural cavity,
    a. air flows through the hole and into the pleural cavity.
    b. air flows through the hole and out of the pleural cavity.
    c. air flows neither out nor in.
    d. the lung protrudes through the hole.
12. Compliance of the lungs and thorax
    a. is the volume by which the lungs and thorax change for each unit change of alveolar pressure.
    b. increases in emphysema.
    c. decreases because of lack of surfactant.
    d. all of the above.
13. Given these lung volumes:
    1. tidal volume = 500 mL
    2. residual volume = 1000 mL
    3. inspiratory reserve volume = 2500 mL
    4. expiratory reserve volume = 1000 mL
    5. dead space = 1000 mL

    The vital capacity is
    a. 3000 mL.
    b. 3500 mL.
    c. 4000 mL.
    d. 5000 mL.
    e. 6000 mL.
14. The alveolar ventilation is the
    a. tidal volume times the respiratory rate.
    b. minute ventilation plus the dead space.
    c. amount of air available for gas exchange in the lungs.
    d. vital capacity divided by the respiratory rate.
    e. inspiratory reserve volume times minute ventilation.
15. If the total pressure of a gas is 760 mm Hg and its composition is 20% oxygen, 0.04% carbon dioxide, 75% nitrogen, and 5% water vapor, the partial pressure of oxygen is
    a. 15.2 mm Hg.
    b. 20 mm Hg.
    c. 118 mm Hg.
    d. 152 mm Hg.
    e. 740 mm Hg.
16. The rate of diffusion of a gas across the respiratory membrane increases as the
    a. respiratory membrane becomes thicker.
    b. surface area of the respiratory membrane decreases.
    c. partial pressure gradient of the gas across the respiratory membrane increases.
    d. diffusion coefficient of the gas decreases.
    e. all of the above.
17. The partial pressure of carbon dioxide in the venous blood is
    a. greater than in the tissue spaces.
    b. less than in the tissue spaces.
    c. less than in the alveoli.
    d. less than in arterial blood.
18. Oxygen is mostly transported in the blood
    a. dissolved in plasma.
    b. bound to blood proteins.
    c. within HCO_3^-.
    d. bound to the heme portion of hemoglobin.
19. The oxygen–hemoglobin dissociation curve is adaptive because it
    a. shifts to the right in the pulmonary capillaries and to the left in the tissue capillaries.
    b. shifts to the left in the pulmonary capillaries and to the right in the tissue capillaries.
    c. does not shift.
20. Carbon dioxide is mostly transported in the blood
    a. dissolved in plasma.
    b. bound to blood proteins.
    c. within HCO_3^-.
    d. bound to the heme portion of hemoglobin.
    e. bound to the globin portion of hemoglobin.
21. When blood passes through the tissues, the hemoglobin in blood is better able to combine with carbon dioxide because of the
    a. Bohr effect.
    b. Haldane effect.
    c. chloride shift.
    d. Boyle effect.
    e. Dalton effect.
22. The chloride shift
    a. promotes the transport of carbon dioxide in the blood.
    b. occurs when Cl^- replace HCO_3^- within red blood cells.
    c. maintains electrical neutrality in red blood cells and the plasma.
    d. all of the above.
23. Which of these parts of the brainstem is correctly matched with its main function?
    a. ventral respiratory groups—stimulate the diaphragm
    b. dorsal respiratory groups—limit inflation of the lungs
    c. pontine respiratory group—switch between inspiration and expiration
    d. all of the above
24. The chemosensitive area
    a. stimulates the respiratory center when blood carbon dioxide levels increase.
    b. stimulates the respiratory center when blood pH increases.
    c. is located in the pons.
    d. stimulates the respiratory center when blood oxygen levels increase.
    e. all of the above.

25. Blood oxygen levels
    a. are more important than carbon dioxide levels in the regulation of respiration.
    b. need to change only slightly to cause a change in respiration.
    c. are detected by sensory receptors in the carotid and aortic bodies.
    d. all of the above.
26. At the onset of exercise, respiration rate and depth increase primarily because of
    a. increased blood carbon dioxide levels.
    b. decreased blood oxygen levels.
    c. decreased blood pH levels.
    d. input to the respiratory center from the cerebral motor cortex and proprioceptors.

Answers in Appendix E

CRITICAL THINKING

1. What effect does rapid (respiratory rate equals 24 breaths per minute), shallow (tidal volume equals 250 mL per breath) breathing have on minute ventilation, alveolar ventilation, and alveolar P_{O_2} and P_{CO_2}?
2. A person's vital capacity is measured while standing and while lying down. What difference, if any, in the measurement do you predict and why?
3. Ima Diver wanted to do some underwater exploration. She did not want to buy expensive SCUBA equipment, however. Instead, she bought a long hose and an inner tube. She attached one end of the hose to the inner tube so that the end was always out of the water, and she inserted the other end of the hose in her mouth and went diving. What happened to her alveolar ventilation and why? How can she compensate for this change? How does diving affect lung compliance and the work of ventilation?
4. The bacteria that cause gangrene (*Clostridium perfringens*) are anaerobic microorganisms that do not thrive in the presence of oxygen. Hyperbaric oxygenation (HBO) treatment places a person in a chamber that contains oxygen at three to four times normal atmospheric pressure. Explain how HBO helps in the treatment of gangrene.
5. Cardiopulmonary resuscitation (CPR) has replaced older, less efficient methods of sustaining respiration. The back-pressure/arm-lift method is one such technique that is no longer used. This procedure is performed with the victim lying face down. The rescuer presses firmly on the base of the scapulae for several seconds and then grasps the arms and lifts them. The sequence is then repeated. Explain why this procedure results in ventilation of the lungs.
6. A technique for artificial respiration is mouth-to-mouth resuscitation. The rescuer takes a deep breath, blows air into the victim's mouth, and then lets air flow out of the victim. The process is repeated. Explain the following: (1) Why do the victim's lungs expand? (2) Why does air move out of the victim's lungs? (3) What effect do the P_{O_2} and the P_{CO_2} of the rescuer's air have on the victim?
7. The left phrenic nerve supplies the left side of the diaphragm and the right phrenic nerve supplies the right side. Damage to the left phrenic nerve results in paralysis of the left side of the diaphragm. During inspiration, does the left side of the diaphragm move superiorly, move inferiorly, or stay in place?
8. Suppose that the thoracic wall is punctured at the end of a normal expiration, producing a pneumothorax. Does the thoracic wall move inward, outward, or not move?
9. During normal, quiet respiration, when does the maximum rate of diffusion of oxygen in the pulmonary capillaries occur? The maximum rate of diffusion of carbon dioxide?
10. There is experimental evidence that the overuse of erythropoietin (EPO; see chapter 19) reduces athletic performance. What side effects of EPO abuse reduce exercise stamina?
11. Predict what would happen to tidal volume if the vagus nerves were cut, the phrenic nerves were cut, or the intercostal nerves were cut.
12. You and your physiology instructor are trapped in an overturned ship. To escape, you must swim under water a long distance. You tell your instructor it would be a good idea to hyperventilate before making the escape attempt. Your instructor calmly replies, "What good would that do, since your pulmonary capillaries are already 100% saturated with oxygen?" What would you do and why?
13. Why does a can of soda eventually "lose its fizz" after it has been opened? How does the loss of carbon dioxide from a soda apply to gas exchange at the respiratory membrane?
14. Ima Anxious was hysterical and hyperventilating, so a doctor made her breathe into a paper bag. An especially astute student said to the doctor, "When Ima was hyperventilating, she was reducing blood carbon dioxide levels; when she breathed into the paper bag, carbon dioxide was trapped in the bag, and she was rebreathing it, thus causing blood carbon dioxide levels to increase. As Ima's blood carbon dioxide levels increased her urge to breathe should have increased. Instead, she began to breathe more slowly. Please explain." How do you think the doctor responded? (*Hint:* Recall that the effect of decreased blood carbon dioxide on the vasomotor center results in vasodilation and a sudden decrease in blood pressure.)

Answers in Appendix F

Visit this book's website at www.mhhe.com/seeley8 for chapter quizzes, interactive learning exercises, and other study tools.

Digestive System

Every cell of the body needs nourishment, yet most cells cannot leave their position in the body and travel to a food source, so the food must be converted to a usable form and delivered. The digestive system, with the help of the circulatory system, acts as a gigantic "meals on wheels," providing nourishment to over a hundred trillion "customer" cells in the body. It also has its own quality control and waste disposal system.

The digestive system provides the body with water, electrolytes, and other nutrients. To do this, the digestive system is specialized to ingest food, propel it through the digestive tract, digest it, and absorb water, electrolytes, and other nutrients from the lumen of the gastrointestinal tract. Once these useful substances are absorbed, they are transported through the circulatory system to cells, where they are used. The undigested portion of the food is moved through the digestive tract and eliminated through the anus.

This chapter presents the general *anatomy of the digestive system* (p. 874), followed by descriptions of the *functions of the digestive system* (p. 874), the *histology of the digestive tract* (p. 876), the *regulation of the digestive system* (p. 877), and the *peritoneum* (p. 878). The anatomy and physiology of each section of the digestive tract and its accessory structures are then presented—the *oral cavity* (p. 880), *pharynx* (p. 886), and *esophagus* (p. 886)—along with a section on *swallowing* (p. 886), the *stomach* (p. 888), *small intestine* (p. 896), *liver* (p. 899), *gallbladder* (p. 904), *pancreas* (p. 905), and *large intestine* (p. 907). *Digestion, absorption, and transport* (p. 912) of nutrients are then discussed, along with the *effects of aging on the digestive system* (p. 920).

* Essays :

1. Describe Carbohydrate digestion
2. What happens when blood CO_2 levels ↑ or ↓ ?
3. Describe the nephron

* 3 pr slides : . trachea (ciliated pseudo- transit epi — ?

- larynx : false vocal cords / true " "
- Kidney : renal corpuscle.

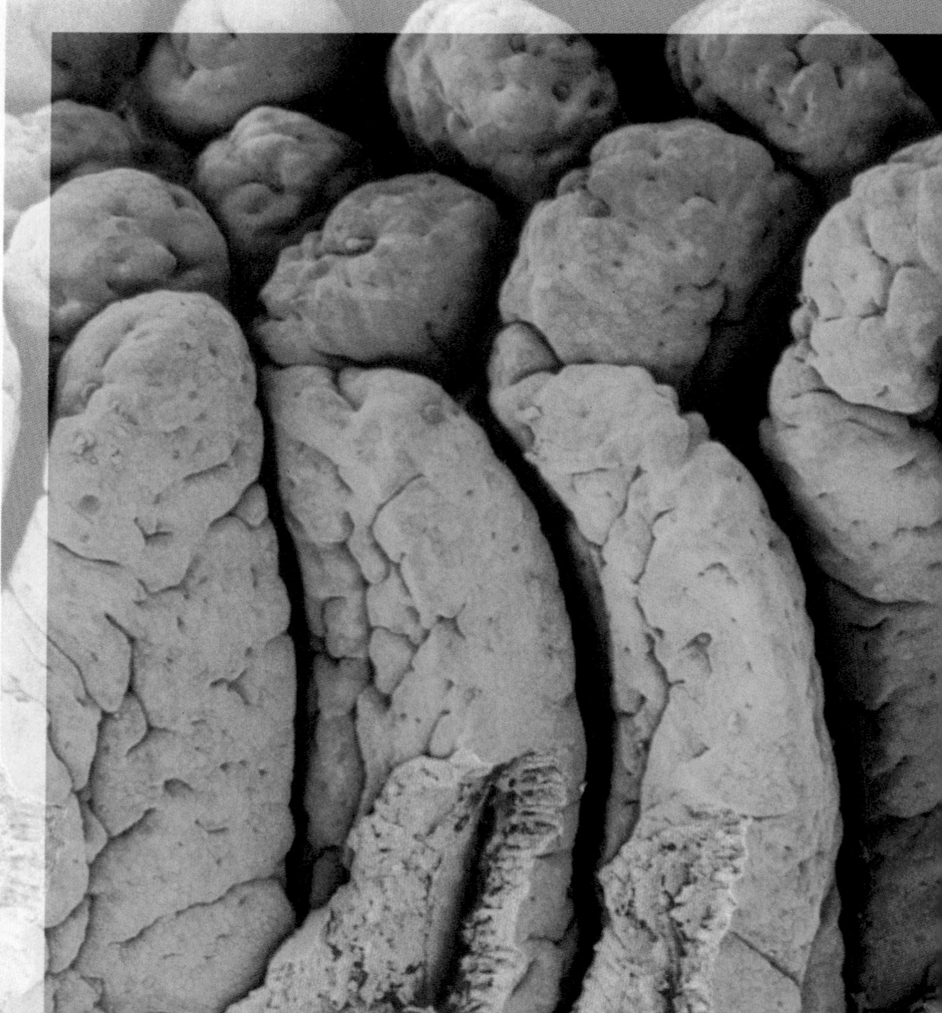

Colorized scanning electron micrograph of the interior surface of the small intestine, showing villi.

Digestive System

ANATOMY OF THE DIGESTIVE SYSTEM

The **digestive system** (figure 24.1) consists of the **digestive tract,** a tube extending from the mouth to the anus, and its associated **accessory organs**—primarily glands located outside the digestive tract, which secrete fluids into the digestive tract. The digestive tract is also called the **alimentary tract,** or **alimentary canal.** The term **gastrointestinal** (gas′trō-in-tes′tin-ăl; **GI**) **tract** technically refers only to the stomach and intestines but is often used as a synonym for the digestive tract.

The regions of the digestive tract include

1. the *oral cavity,* or *mouth,* which has salivary glands and tonsils as accessory organs
2. the *pharynx,* or throat, with tubular mucous glands
3. the *esophagus,* with tubular mucous glands
4. the *stomach,* which contains many tubelike glands, as well as mucous cells
5. the *small intestine,* consisting of the duodenum, jejunum, and ileum, with the liver, gallbladder, and pancreas as major accessory organs
6. the *large intestine,* including the cecum, colon, rectum, and anal canal, with mucous glands
7. the *anus.*

1 ***List the regions of the digestive tract.***

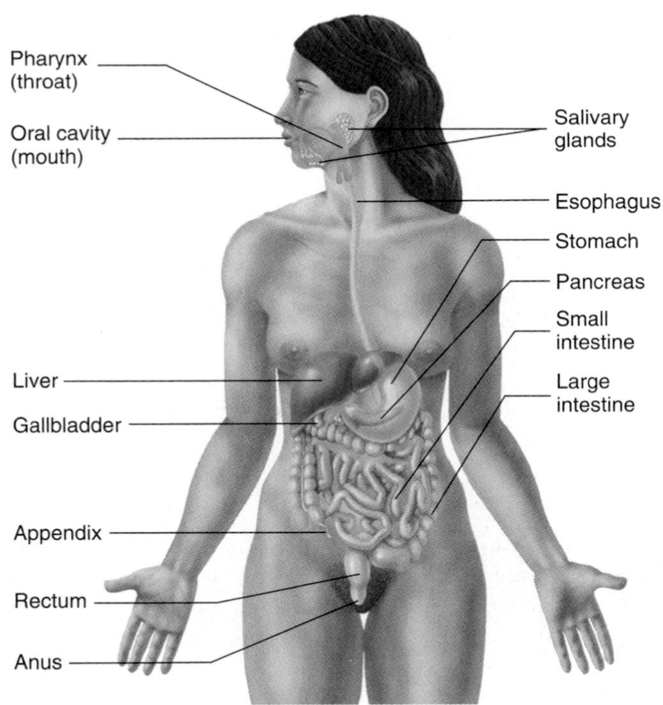

FIGURE 24.1 Digestive System

FUNCTIONS OF THE DIGESTIVE SYSTEM

The following are the major functions of the digestive system (table 24.1):

1. *Ingestion* is the introduction of solid or liquid food into the stomach. The normal route of ingestion is through the oral cavity, but food can be introduced directly into the stomach by a nasogastric, or stomach, tube.
2. *Mastication* is the process by which food taken into the mouth is chewed by the teeth. Digestive enzymes cannot easily penetrate solid food particles and can work effectively only on the surfaces of the particles. It is vital, therefore, that solid foods be mechanically broken down into small particles. Mastication breaks large food particles into many smaller particles, which have a much larger total surface area than do a few large particles.
3. *Propulsion* in the digestive tract is the movement of food from one end of the digestive tract to the other. The total time it takes food to travel the length of the digestive tract is usually about 24–36 hours. Each segment of the digestive tract is specialized to assist in moving its contents from the oral end to the anal end. **Deglutition** (dē′gloo-tish′ŭn), or swallowing, moves a mass of food or a liquid, called a **bolus** (bō′lŭs), from the oral cavity into the esophagus. **Peristalsis** (per-i-stal′sis; figure 24.2) is responsible for moving material through most of the digestive tract. Muscular contractions occur in **peristaltic** (per-i-stal′tik) **waves,** consisting of a wave of relaxation of the circular muscles in front of the bolus, followed by a wave of strong contraction of the circular muscles behind the bolus, which forces the bolus along the digestive tube. Each peristaltic wave travels the length of the esophagus in about 10 seconds. Peristaltic waves in the small intestine usually travel only for short distances. In some parts of the large intestine, material is moved by **mass movements,** which are contractions that extend over much larger parts of the digestive tract than peristaltic movements.
4. *Mixing.* Some contractions do not propel food from one end of the digestive tract to the other but, rather, move it back and forth within the digestive tract to *mix* it with digestive secretions and to help break it into smaller pieces. **Segmental contractions** (figure 24.3) are mixing contractions that occur in the small intestine.
5. *Secretion.* As food moves through the digestive tract, **secretions** are added to lubricate, liquefy, buffer, and digest the food. **Mucus,** secreted along the entire digestive tract, lubricates the food and the lining of the tract. The mucus coats and protects the epithelial cells of the digestive tract from mechanical abrasion, from the damaging effect of acid in the stomach, and from the digestive enzymes of the digestive tract. The secretions also contain large amounts of **water,** which liquefies the food, thereby making it easier to digest and absorb. Water also moves into the intestine by osmosis. Liver secretions break large fat droplets into much smaller droplets, which makes possible the digestion and absorption of fats. **Enzymes** secreted by the oral cavity, stomach,

TABLE 24.1 Functions of the Digestive Tract

Organ	Functions
Oral cavity	*Ingestion.* Solid food and fluids are taken into the digestive tract through the oral cavity. *Taste.* Tastants disolved in saliva stimulate taste buds in the tongue. *Mastication.* Movement of the mandible by the muscles of mastication cause the teeth to break food down into smaller pieces. The tongue and cheeks help place the food between the teeth. *Digestion.* Amylase in saliva begins carbohydrate (starch) digestion. *Swallowing.* The tongue forms food into a bolus and pushes the bolus into the pharynx. *Communication.* The lips, cheeks, teeth, and tongue are involved in speech. The lips change shape as part of facial expressions. *Protection.* Mucin and water in saliva provide lubrication, and lysozyme (an enzyme that lyses cells) kills microorganisms. Nonkeratinized stratified squamous epithelium prevents abrasion.
Pharynx	*Swallowing.* The involuntary phase of swallowing moves the bolus from the oral cavity to the esophagus. Materials are prevented from entering the nasal cavity by the soft palate and from entering the lower respiratory tract by the epiglottis and vestibular folds. *Breathing.* Air passes from the nasal or oral cavity through the pharynx to the lower respiratory tract. *Protection.* Mucus provides lubrication. Nonkeratinized stratified squamous epithelium prevents abrasion.
Esophagus	*Propulsion.* Peristaltic contractions move the bolus from the pharynx to the stomach. The lower esophageal sphincter limits reflux of the stomach contents into the esophagus. *Protection.* Glands produce mucus, which provides lubrication and protects the inferior esophagus from stomach acid.
Stomach	*Storage.* Rugae allow the stomach to expand and hold food until it can be digested. *Digestion.* Protein digestion begins as a result of the actions of hydrochloric acid and pepsin. *Absorption.* Except for a few substances (e.g., water, alcohol, aspirin) little absorption takes place in the stomach. *Mixing and propulsion.* Mixing waves churn ingested materials and stomach secretions into chyme. Peristaltic waves move the chyme into the small intestine. *Protection.* Mucus provides lubrication and prevents the digestion of the stomach wall. Stomach acid kills most microorganisms.
Small intestine	*Neutralization.* Bicarbonate ions from the pancreas and bile from the liver neutralize stomach acid to form a pH environment suitable for pancreatic and intestinal enzymes. *Digestion.* Enzymes from the pancreas and the lining of the small intestine complete the breakdown of food molecules. Bile salts from the liver emulsify fats. *Absorption.* The circular folds, villi, and microvilli increase surface area. Most nutrients are actively or passively absorbed. Most of the ingested water or the water in digestive tract secretions is absorbed. *Mixing and propulsion.* Segmental contractions mix the chyme, and peristaltic contractions move the chyme into the large intestine. *Excretion.* Bile from the liver contains bilirubin, cholestrol, fats, and fat-soluble hormones. *Protection.* Mucus provides lubrication, prevents the digestion of the intestinal wall, and protects the small intestine from stomach acid. Peyer's patches protect against microorganisms.
Large intestine	*Absorption.* The proximal half of the colon absorbs salts (e.g., sodium chloride), water, and the vitamins (e.g., K) produced by bacteria. *Storage.* The distal half of the colon holds feces until it is eliminated. *Mixing and propulsion.* Slight segmental mixing occurs. Mass movements propel feces toward the anus and defecation eliminates the feces. *Protection.* Mucus provides lubrication; mucus and bicarbonate ions protect against acids produced by bacteria.

intestine, and pancreas break large food molecules down into smaller molecules that can be absorbed by the intestinal wall.

6. *Digestion* is the breakdown of large organic molecules into their component parts: carbohydrates into monosaccharides, proteins into amino acids, and triglycerides into fatty acids and glycerol. Digestion consists of **mechanical digestion,** which involves the mastication and mixing of food, and **chemical digestion,** which is accomplished by digestive enzymes that are secreted along the digestive tract. Digestion of large organic molecules into their component parts must be accomplished before they can be absorbed by the digestive tract. Minerals and water are not broken down before being absorbed. Vitamins are also absorbed without digestion and lose their function if their structure is altered by digestion.
7. *Absorption* is the movement of molecules out of the digestive tract and into the circulation or into the lymphatic system.

1. A wave of circular smooth muscle relaxation moves ahead of the bolus of food, allowing the digestive tract to expand.

2. A wave of contraction of the circular smooth muscles behind the bolus of food propels it through the digestive tract.

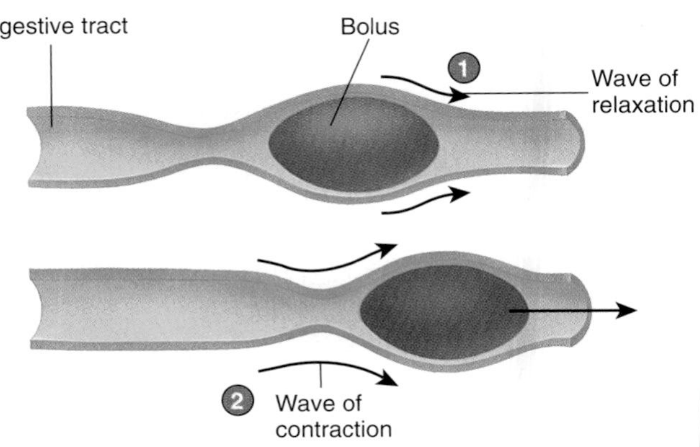

PROCESS FIGURE 24.2 Peristalsis

1. A secretion introduced into the digestive tract or food within the tract begins in one location.

2. Segments of the digestive tract alternate between contraction and relaxation.

3. Material (*brown*) in the intestine is spread out in both directions from the site of introduction.

4. The secretion or food is spread out in the digestive tract and becomes more diffuse (*lighter color*) through time.

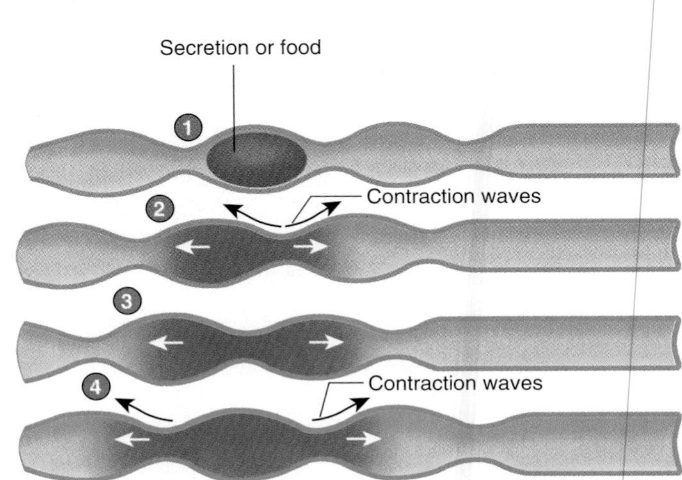

PROCESS FIGURE 24.3 Segmental Contractions

The mechanism by which absorption occurs depends on the type of molecule involved. Molecules pass out of the digestive tract by diffusion, facilitated diffusion, active transport, symport, or endocytosis (see chapter 3).

8. *Elimination* is the process by which the waste products of digestion are removed from the body. During this process, occurring primarily in the large intestine, water and salts are absorbed and change the material in the digestive tract from a liquefied state to a semi-solid state. These semi-solid waste products, called **feces,** are then eliminated from the digestive tract by the process of **defecation.**

2 ***Describe each of the processes involved in the normal functions of the digestive system.***

HISTOLOGY OF THE DIGESTIVE TRACT

Figure 24.4 depicts a generalized view of the digestive tract histology. The digestive tube consists of four major layers, or tunics: an internal mucosa and an external serosa, with a submucosa and muscularis in between. These four tunics are present in all areas of the digestive tract from the esophagus to the anus. Three major types of glands are associated with the intestinal tract: (1) unicellular mucous glands in the mucosa, (2) multicellular glands in the mucosa and submucosa, and (3) multicellular glands (accessory glands) outside the digestive tract.

Mucosa

The innermost tunic, the **mucosa** (mū-kō′să), or mucous membrane, consists of three layers: (1) the inner **mucous epithelium,** which is moist stratified squamous epithelium in the mouth, oropharynx, esophagus, and anal canal and simple columnar epithelium in the remainder of the digestive tract; (2) a loose connective tissue called the **lamina propria** (lam′i-nă prō′prē-ă); and (3) an outer, thin smooth muscle layer, the **muscularis mucosae.** The epithelium extends deep into the lamina propria in many places to form **intestinal glands,** or **crypts.** Specialized cells in the mucosa are mechanoreceptors involved in peristaltic reflexes.

Submucosa

The **submucosa** is a thick connective tissue layer containing nerves, blood vessels, lymphatic vessels, and small glands that lies beneath the mucosa. The plexus of nerve cells in the submucosa form the **submucosal plexus** (plek′sŭs; Meissner plexus), an enteric nerve plexus consisting of axons, many scattered neuron cell bodies, and neuroglial cells. Axons from the submucosal plexus extend to cells

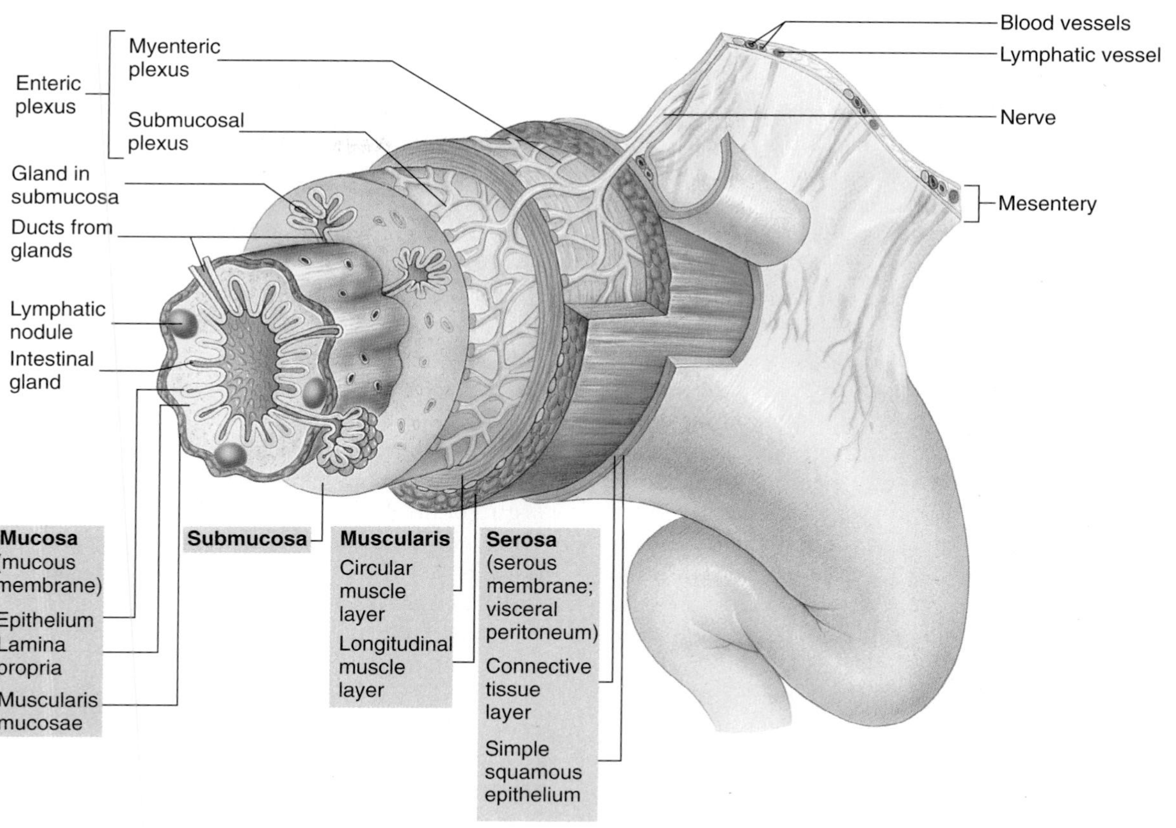

FIGURE 24.4 Digestive Tract Histology

The four tunics are the mucosa, submucosa, muscularis, and serosa or adventitia. Glands may exist along the digestive tract as part of the epithelium, as glands within the submucosa, or as large glands outside the digestive tract.

in epithelial intestinal glands, stimulating their secretion. The esophagus and stomach lack a submucosal plexus, but the plexus is extensive throughout the rest of the digestive tract.

Muscularis

The next tunic is the **muscularis,** which consists of an inner layer of circular smooth muscle and an outer layer of longitudinal smooth muscle. Two exceptions are the upper esophagus, where the muscles are striated, and the stomach, which has three layers of smooth muscle. Another nerve plexus, the **myenteric plexus** (mī-en-ter′ik; Auerbach plexus), which also consists of axons, many scattered neuron cell bodies, and neuroglial cells, is between these two muscle layers (see figure 24.4). The myenteric plexus is much more extensive than the submucosal plexus and controls the motility of the intestinal tract.

Within the myenteric plexus, specialized **interstitial cells** form a network of "pacemakers," which promote rhythmic contractions of smooth muscle along the GI tract. These cells also help transmit signals from neurons to muscle to regulate GI movement. Dysfunction of these pacemakers decreases motility in the GI tract.

Together, the submucosal and myenteric plexuses constitute the **enteric plexus** (en-tĕr′ik; relating to the intestine), or **intramural** (in′tră-mŭ′răl; within the walls) **plexus.** The enteric plexus is extremely important in the control of secretion and movement (see chapter 16).

Serosa or Adventitia

The fourth layer of the digestive tract is either the **serosa** or the **adventitia** (ad-ven-tish′ă; foreign or coming from outside), depending on the structure of the layer. Parts of the digestive tract that protrude into the peritoneal cavity have a serosa as the outermost layer. This serosa, or serous membrane, is called the visceral peritoneum. It consists of a thin layer of connective tissue and a simple squamous epithelium. When the outer layer of the digestive tract is derived from adjacent connective tissue, the tunic is called the adventitia and consists of a connective tissue covering that blends with the surrounding connective tissue. These areas include the esophagus and the retroperitoneal organs (see "Peritoneum" p. 878).

3 ***What are the major layers of the digestive tract? How do the serosa and adventitia differ?***

4 ***Describe the enteric plexus. In what layers of the digestive tract are the submucosal and myenteric plexuses found?***

REGULATION OF THE DIGESTIVE SYSTEM

Elaborate nervous and chemical mechanisms regulate the movement, secretion, absorption, and elimination processes.

Nervous Regulation of the Digestive System

The enteric plexus is extensive; it contains more neurons than the spinal cord. Most of the nervous control of the GI tract is local, occurring as a result of local reflexes within the enteric plexus, and some is more general, mediated largely by the parasympathetic division of the ANS through the vagus nerves and by sympathetic nerves.

Local neuronal control of the digestive tract occurs within the **enteric nervous system (ENS).** The ENS consists of the enteric plexus, made up of enteric neurons, and extensive associated neuroglial cells, within the wall of the digestive tract (see figure 24.4). There are three major types of enteric neurons: (1) Enteric sensory neurons detect changes in the chemical composition of the digestive tract contents or detect mechanical changes, such as stretch of the digestive tract wall; (2) enteric motor neurons stimulate or inhibit smooth muscle contraction and glandular secretion in the digestive system; (3) enteric interneurons connect enteric sensory and motor neurons. The ENS functions through **local reflexes** to control activities within specific, short regions of the digestive tract. The ENS is capable of controlling the complex peristaltic and mixing movements of the GI tract, as well as blood flow to the GI tract, without any outside influences. Although the ENS can control the activities of the digestive tract independent of the CNS, normally the two systems work together. For example, autonomic innervation from the CNS influences ENS activity.

General control of the digestive system by the CNS occurs when reflexes are activated by stimuli originating in the digestive tract. Action potentials are carried by sensory neurons in the vagus and sympathetic nerves to the CNS, where the reflexes are integrated. In addition, reflexes within the CNS can be activated by the sight, smell, or taste of food, which stimulate hunger. All of these reflexes influence activity in parasympathetic neurons of the CNS. Parasympathetic neurons extend to the digestive tract through the vagus nerves to control responses or alter the activity of the ENS and local reflexes. Some sympathetic neurons inhibit muscle contraction and secretion in the digestive system and decrease blood flow to the digestive system.

Hirschprung Disease

Hirschprung disease, also called **megacolon,** is a painful developmental disorder caused by the absence of enteric neurons in the distal large intestine. Mutations in the *RET* gene on chromosome 10 have been identified in Hirschprung patients. The *RET* gene encodes a receptor that is normally activated by the growth factors required for the survival and differentiation of the neural crest cells that become a subset of enteric neurons (see chapter 13). The mutations in *RET* that lead to a loss of receptor function result in a loss of enteric neurons, which results in poor intestinal motility and severe constipation. Conversely, a different set of mutations in the *RET* gene are linked to an inherited cancer called **multiple endocrine neoplasia type 2 (MEN2).** In contrast to the loss of function due to Hirschprung mutations, the MEN2 mutations cause a gain of RET receptor function so that it is active even in the absence of growth factors. Hence, two types of mutations in the same gene result in two very different syndromes. Rapid DNA tests are used to screen patients and family members for suspected Hirschprung and MEN2 mutations.

Chemical Regulation of the Digestive System

Over 30 neurotransmitters are associated with the ENS. Two major ENS neurotransmitters are acetylcholine and norepinephrine. In general, acetylcholine stimulates and norepinephrine inhibits GI tract motility and secretions. Another major ENS neurotransmitter is serotonin, which stimulates GI tract motility. Serotonin is also produced by endocrine cells within the digestive tract wall. Over 95% of the serotonin in the body is found in the GI tract, so drugs that increase serotonin levels and function, such as antidepressants (see chapter 11) and chemotherapeutics, can also affect GI tract activity.

The digestive tract produces a number of hormones, such as gastrin and secretin, which are secreted by endocrine cells in the digestive system and are carried through the circulation to target organs of the digestive system or to target tissues in other systems. These hormones help regulate many gastrointestinal tract functions, as well as the secretions of associated glands, such as the liver and pancreas.

In addition to the hormones produced by the digestive system, which enter the circulation, other paracrine chemicals, such as histamine, are released locally within the digestive tract and influence the activity of nearby cells. These localized chemical regulators help local reflexes within the ENS control local digestive tract environments, such as pH levels.

Serotonin and Cancer Treatment

An unintended consequence of many cancer therapies is nausea due to increased serotonin release from endocrine cells in the digestive tract. Serotonin binds to a subset of serotonin receptors, called 5HT3 receptors, on sensory neurons of the vagus nerve, which stimulate the vomiting center in the brain. This results in the nausea and vomiting associated with chemotherapy and radiotherapy. 5HT3 receptor blockers, such as ondansetron (ōn-dan′sē-tron), are commonly used to alleviate nausea.

5 ***What are the nervous and chemical mechanisms that regulate the digestive system?***

PERITONEUM

The body walls and organs of the abdominal cavity are lined with **serous membranes.** These membranes are very smooth and secrete a serous fluid, which provides a lubricating film between the layers of membranes. These membranes and fluid reduce friction as organs move within the abdomen. The serous membrane that covers the organs is the **visceral peritoneum** (per′i-tō-nē′ūm; to stretch over), and the one that covers the interior surface of the body wall is the **parietal peritoneum** (figure 24.5).

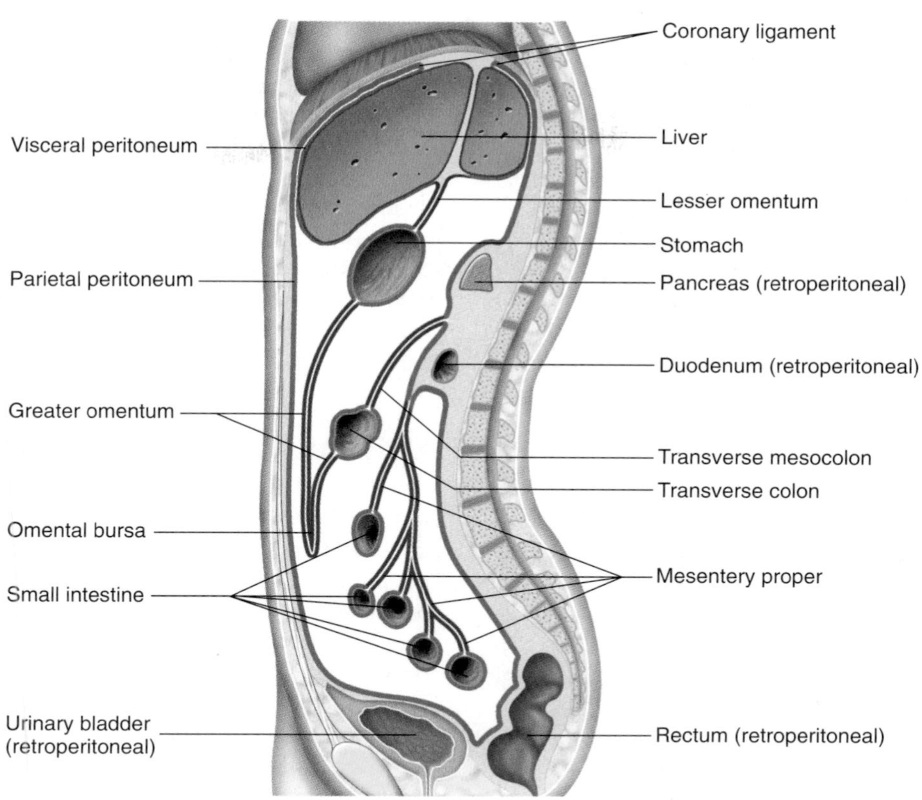

(a) **Medial view**

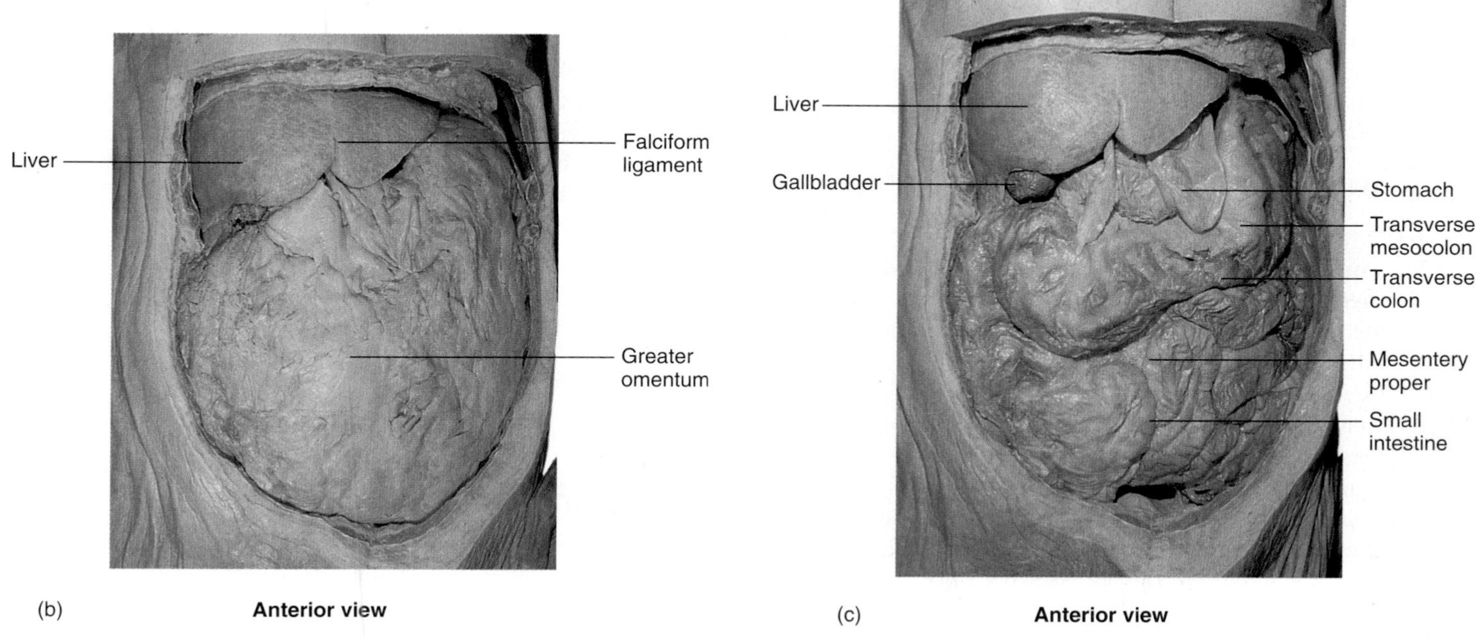

(b) **Anterior view**

(c) **Anterior view**

FIGURE 24.5 Peritoneum and Mesenteries

(*a*) Sagittal section through the trunk, showing the peritoneum and mesenteries associated with some abdominal organs. (*b*) Photograph of the abdomen of a cadaver, with the greater omentum in place. (*c*) Photograph of the abdomen of a cadaver, with the greater omentum removed to reveal the underlying viscera.

Peritonitis

Peritonitis is a potentially life-threatening inflammation of the peritoneal membranes. The inflammation can result from chemical irritation by substances, such as bile, that have escaped from a damaged digestive tract, or it can result from infection originating in the digestive tract, such as when the appendix ruptures. An accumulation of excess serous fluid in the peritoneal cavity, called **ascites** (ă-sī′tēz), can occur in peritonitis. Ascites may also accompany starvation, alcoholism, or liver cancer.

Connective tissue sheets called **mesenteries** (mes′en-ter′ēz; middle intestine) hold many of the organs in place within the abdominal cavity. The mesenteries consist of two layers of serous membranes with a thin layer of loose connective tissue between them. They provide a route by which vessels and nerves can pass from the body wall to the organs. Other abdominal organs lie against the abdominal wall, have no mesenteries, and are referred to as **retroperitoneal** (re′trō-per′i-tō-nē′ăl; behind the peritoneum; see chapter 1). The retroperitoneal organs include the duodenum, pancreas, ascending colon, descending colon, rectum, kidneys, adrenal glands, and urinary bladder.

The mesentery connecting the lesser curvature of the stomach and the proximal end of the duodenum to the liver and diaphragm is called the **lesser omentum** (ō-men′tŭm; membrane of the bowels), and the mesentery extending as a fold from the greater curvature and then to the transverse colon is called the **greater omentum** (see figure 24.5). The greater omentum forms a long, double fold of mesentery that extends inferiorly from the stomach over the surface of the small intestine. Because of this folding, a cavity called the **omental bursa** (ber′să; pocket) is formed between the two layers of mesentery. A large amount of fat accumulates in the greater omentum, and it is sometimes referred to as the "fatty apron." The greater omentum has considerable mobility in the abdomen.

PREDICT 1

If you placed a pin through the greater omentum, through how many layers of simple squamous epithelium would the pin pass?

The **coronary ligament** attaches the liver to the diaphragm. Unlike other mesenteries, the coronary ligament has a wide space in the center, the bare area of the liver, where no peritoneum exists. The **falciform ligament** attaches the liver to the anterior abdominal wall (see figure 24.5).

Although *mesentery* is a general term referring to the serous membranes attached to the abdominal organs, it is also used specifically to refer to the mesentery associated with the small intestine, sometimes called the **mesentery proper.** The mesenteries of parts of the colon are the **transverse mesocolon,** which extends from the transverse colon to the posterior body wall, and the **sigmoid mesocolon.** The vermiform appendix has its own little mesentery called the **mesoappendix.**

6 ***Where are visceral peritoneum and parietal peritoneum found? What is a retroperitoneal organ?***

7 ***Define mesentery. Name and describe the location of the mesenteries in the abdominal cavity.***

ORAL CAVITY

The **oral cavity** (figure 24.6), or mouth, is the part of the digestive tract bounded by the lips anteriorly, the **fauces** (faw′sēz; throat; opening into the pharynx) posteriorly, the cheeks laterally, the palate superiorly, and a muscular floor inferiorly. The oral cavity is divided into two regions: (1) the **vestibule** (ves′ti-bool; entry), which is the space between the lips or cheeks and the alveolar processes, which contain the teeth; and (2) the **oral cavity proper,** which lies medial to the alveolar processes. The oral cavity is lined with moist stratified squamous epithelium, which protects against abrasion.

Lips and Cheeks

The **lips,** or **labia** (lā′bē-ă; see figure 24.6), are muscular structures formed mostly by the **orbicularis oris** (ōr-bik′ū-lā′ris ōr′is) **muscle** (see figure 10.9) and connective tissue. The outer surfaces of the lips are covered by skin. The keratinized stratified epithelium of the skin is thin at the margin of the lips and is not as highly keratinized as the epithelium of the surrounding skin (see chapter 5); consequently, it is more transparent than the epithelium over the rest of the body. The color from the underlying blood vessels can be seen through the relatively transparent epithelium, giving the lips a reddish pink to dark red appearance, depending on the overlying pigment. At the internal margin of the lips, the epithelium is continuous with the moist stratified squamous epithelium of the mucosa in the oral cavity.

One or more **labial frenula** (fren′ū-lă; bridle), which are mucosal folds, extend from the alveolar processes of the maxilla to the upper lip and from the alveolar process of the mandible to the lower lip.

The **cheeks** form the lateral walls of the oral cavity. They consist of an interior lining of moist stratified squamous epithelium and an exterior covering of skin. The substance of the cheek includes the **buccinator muscle** (see chapter 10), which flattens the cheek against the teeth, and the **buccal fat pad,** which rounds out the profile on the side of the face.

The lips and cheeks are important in the processes of mastication and speech. They help manipulate food within the mouth and hold it in place while the teeth crush or tear it. They also help form words during the speech process. A large number of the muscles of facial expression are involved in movement of the lips. They are listed in chapter 10.

Palate and Palatine Tonsils

The **palate** (see figure 24.6) consists of two parts, an anterior bony part, the **hard palate** (see chapter 7), and a posterior, nonbony part, the **soft palate,** which consists of skeletal muscle and connective tissue. The **uvula** (ū′vū-lă; a grape) is the projection from the posterior edge of the soft palate. The palate is important in the swallowing process; it prevents food from passing into the nasal cavity.

The **palatine tonsils** are located in the lateral wall of the fauces (see chapter 22).

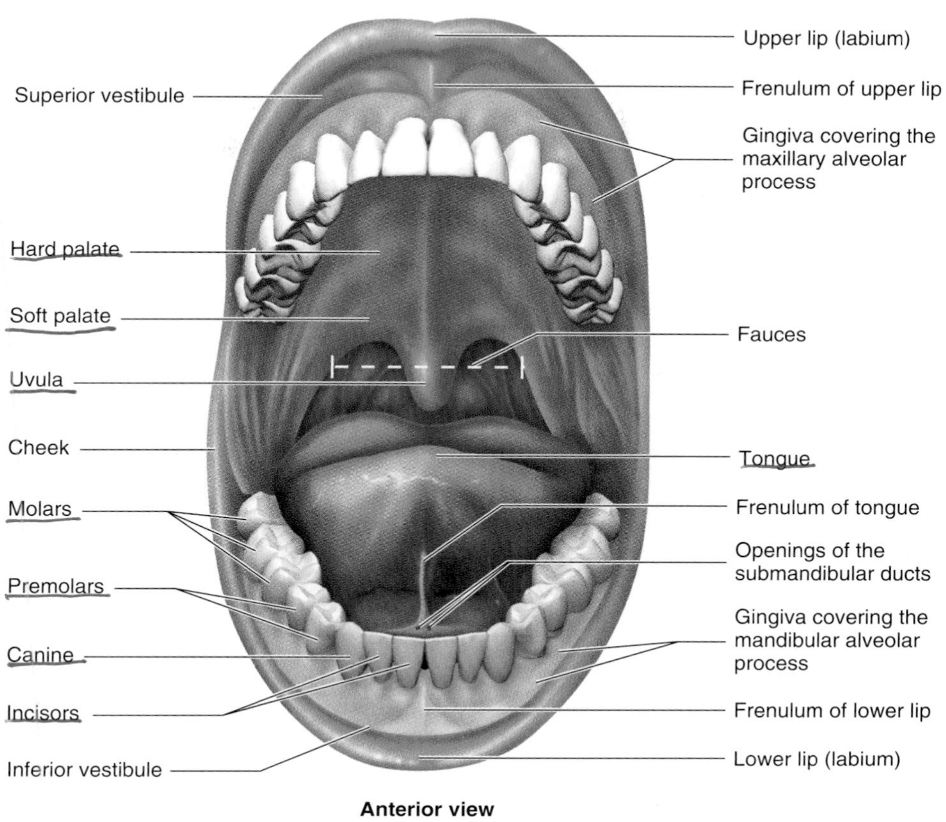

FIGURE 24.6 Oral Cavity

Tongue

The **tongue** is a large, muscular organ that occupies most of the oral cavity proper when the mouth is closed. Its major attachment in the oral cavity is through its posterior part. The anterior part of the tongue is relatively free and is attached to the floor of the mouth by a thin fold of tissue called the **lingual** (tongue) **frenulum.** The muscles associated with the tongue are divided into two categories: **intrinsic muscles,** which are within the tongue itself, and **extrinsic muscles,** which are outside the tongue but attached to it. The intrinsic muscles are largely responsible for changing the shape of the tongue, such as flattening and elevating it during drinking and swallowing. The extrinsic tongue muscles protrude and retract the tongue, move it from side to side, and change its shape (see chapter 10).

Tongue-Tied

A person is "tongue-tied" in a more literal sense if the frenulum extends too far toward the tip of the tongue, thereby inhibiting normal movement of the tongue and interfering with normal speech. Surgically cutting the frenulum can correct the condition.

A groove called the **terminal sulcus** divides the tongue into two parts. The part anterior to the terminal sulcus accounts for about two-thirds of the surface area and is covered by papillae, some of which contain taste buds (see chapter 15). The posterior one-third of the tongue is devoid of papillae and has only a few scattered taste buds. Instead, it has a few small glands and a large amount of lymphatic tissue, which form the **lingual tonsil** (see chapter 22). Moist stratified squamous epithelium covers the tongue.

Lipid-Soluble Drugs

Drugs that are lipid-soluble and can diffuse through the plasma membranes of the oral cavity can be quickly absorbed into the circulation. An example is nitroglycerin, which is a vasodilator used to treat angina pectoris. The drug is placed under the tongue, where, in less than 1 minute, it dissolves and passes through the very thin oral mucosa into the lingual veins.

The tongue moves food in the mouth and, in cooperation with the lips and gums, holds the food in place during mastication. It also plays a major role in the mechanism of swallowing (see p. 886). It is a major sensory organ for taste (see chapter 15) and one of the primary organs of speech.

Glossectomy and Speech

Patients who have undergone a glossectomy (tongue removal) as a result of glossal carcinoma can compensate for loss of the tongue's function in speech, and they can learn to speak fairly well. These patients, however, have substantial difficulty chewing and swallowing food.

Teeth

Normal adults have 32 **teeth,** which are distributed in two **dental arches.** One is called the maxillary arch and the other is called the mandibular arch. The teeth in the right and left halves of each dental arch are roughly mirror images of each other. As a result, the teeth are divided into four quadrants: right upper, left upper, right lower, and left lower. The teeth in each quadrant include one central and one lateral **incisor,** one **canine,** first and second **premolars,** and first, second, and third **molars** (figure 24.7*a*). The third molars are called **wisdom teeth** because they usually appear in a person's late teens or early twenties, when the person is old enough to have acquired some wisdom.

> **Impacted Wisdom Teeth**
>
> In some people with small dental arches, the third molars may not have room to erupt into the oral cavity and remain embedded within the jaw. Embedded wisdom teeth are referred to as impacted. They may cause pain or irritation and their surgical removal is often necessary.

The teeth of the adult mouth are **permanent,** or **secondary, teeth.** Most of them are replacements for **primary,** or **deciduous** (dē-sid'ū-ŭs; those that fall out; also called milk teeth), **teeth,** which are lost during childhood (figure 24.7*b*). The deciduous teeth erupt (the crowns appear within the oral cavity) between about 6 months and 24 months of age (see figure 24.7*b*). The permanent teeth begin replacing the deciduous teeth at about 5 years and the process is completed by about 11 years.

Each tooth consists of a **crown** with one or more **cusps** (points), a **neck,** and a **root** (figure 24.8). The **clinical crown** is the part of the tooth exposed in the oral cavity. The **anatomical crown** is the entire enamel-covered part of the tooth. The center of the tooth is a **pulp cavity,** which is filled with blood vessels, nerves, and connective tissue called **pulp.** The pulp cavity within the root is called the **root canal.** The nerves and blood vessels of the tooth enter and exit the pulp through a hole at the point of each root called the **apical foramen.** The pulp cavity is surrounded by a living, cellular, calcified tissue called **dentin.** The dentin of the tooth crown is covered by an extremely hard, nonliving, acellular substance called **enamel,** which protects the tooth against abrasion and acids produced by bacteria in the mouth. The surface of the dentin in the root is covered with a bonelike substance called **cementum,** which helps anchor the tooth in the jaw.

The teeth are set in **alveoli** (al-vē'ō-lī; sockets) along the alveolar processes of the mandible and maxilla. Dense fibrous connective tissue and stratified squamous epithelium, referred to as the **gingiva** (jin'ji-vă; gums), cover the alveolar processes (see figure 24.6). **Periodontal** (per'ē-ō-don'tăl; around a tooth) **ligaments** secure the teeth in the alveoli.

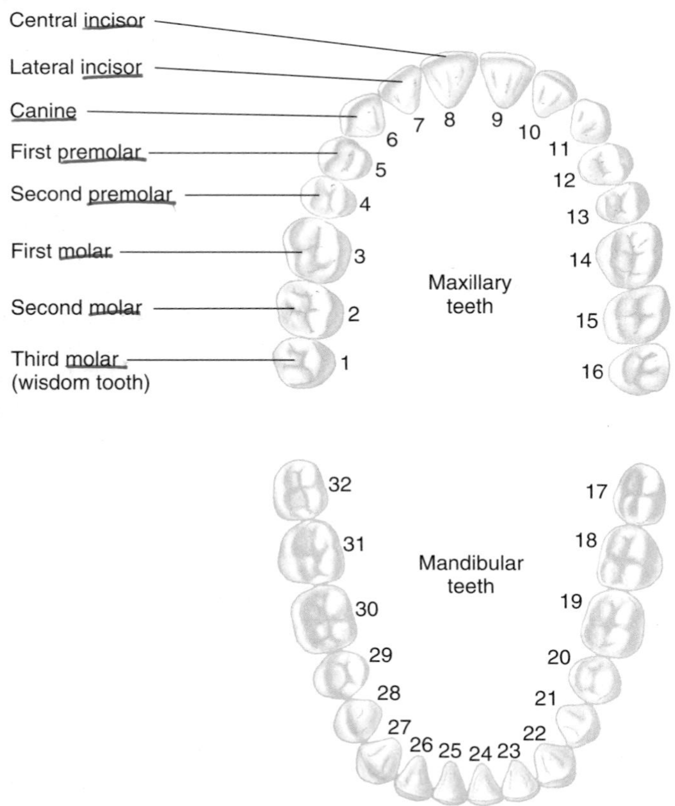

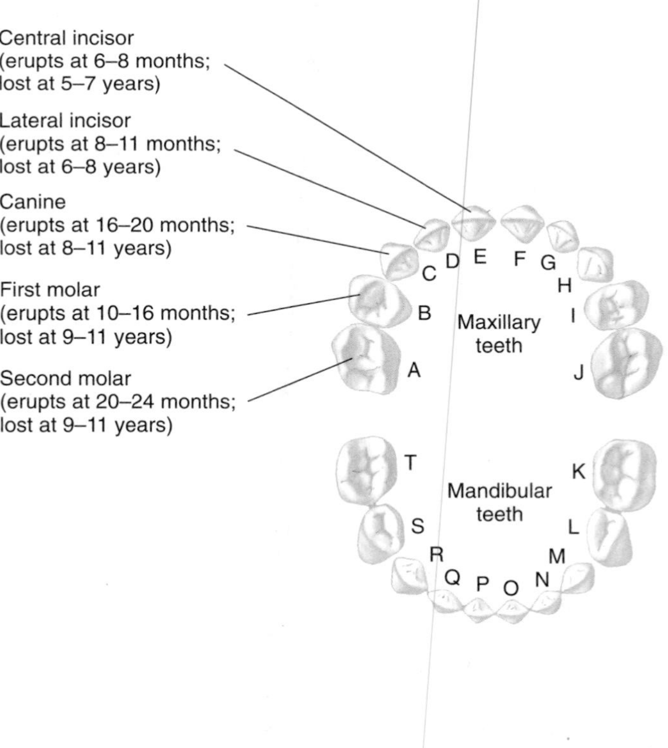

FIGURE 24.7 Teeth

(*a*) Permanent teeth. (*b*) Deciduous teeth. A "universal" numbering and lettering system has been developed by dental professionals for convenience in identifying individual teeth.

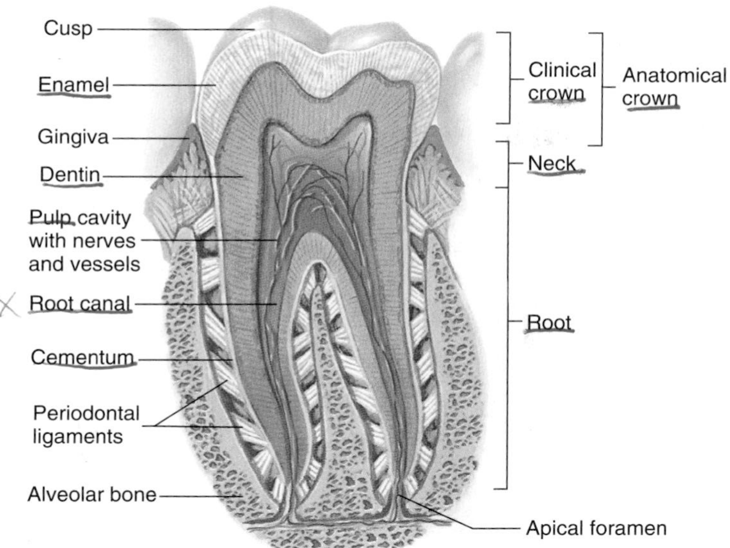

FIGURE 24.8 Molar Tooth in Place in the Alveolar Bone

The tooth consists of a crown and root. The root is covered with cementum, and the tooth is held in the socket by periodontal ligaments. Nerves and vessels enter and exit the tooth through the apical foramen.

The teeth play an important role in mastication and a role in speech.

8 ***Distinguish between the vestibule and the oral cavity proper.***

9 ***What are the functions of the lips and cheeks? What muscle forms the substance of the cheek?***

10 ***What are the hard and soft palates? Where is the uvula found?***

11 ***List the functions of the tongue. Distinguish between intrinsic and extrinsic tongue muscles.***

12 ***What are deciduous and permanent teeth? Name the kinds of teeth.***

13 ***Describe the parts of a tooth. What are dentin, enamel, cementum, and pulp?***

Dental Diseases

Dental caries, or tooth decay, is caused by a breakdown of enamel by acids produced by bacteria on the tooth surface. Because the enamel is nonliving and cannot repair itself, a dental filling is necessary to prevent further damage. If the decay reaches the pulp cavity, with its rich supply of nerves, toothache pain may result. In some cases in which decay has reached the pulp cavity, it is necessary to perform a dental procedure called a root canal, which consists of removing the pulp from the tooth.

Periodontal disease is the inflammation and degradation of the periodontal ligaments, gingiva, and alveolar bone. This disease is the most common cause of tooth loss in adults. **Gingivitis** (jin-ji-vī'tis) is an inflammation of the gingiva, often caused by food deposited in gingival crevices and not promptly removed by brushing and flossing. Gingivitis may eventually lead to periodontal disease. **Pyorrhea** (pī-ō-rē'ă) is a condition in which pus occurs with periodontal disease. **Halitosis** (hal-i-tō'sis), or bad breath, often occurs with periodontal disease and pyorrhea.

Mastication

Food taken into the mouth is **chewed,** or **masticated,** by the teeth. The anterior teeth, the incisors, and the canines primarily cut and tear food, whereas the premolars and molars primarily crush and grind it. Mastication breaks large food particles into smaller ones, which have a much larger total surface area. Because digestive enzymes digest food molecules only at the surface of the particles, mastication increases the efficiency of digestion.

Four pairs of muscles move the mandible during mastication: the **temporalis, masseter, medial pterygoid,** and **lateral pterygoid** (see chapter 10). The temporalis, masseter, and medial pterygoid muscles close the jaw; the lateral pterygoid muscle opens it. The medial and lateral pterygoids and the masseter muscles accomplish protraction and lateral and medial excursion of the jaw. The temporalis retracts the jaw. All these movements are involved in tearing, crushing, and grinding food.

The **mastication,** or **chewing, reflex,** which is integrated in the medulla oblongata, controls the basic movements involved in chewing. The presence of food in the mouth stimulates sensory receptors, which activate a reflex that causes the muscles of mastication to relax. The muscles are stretched as the mandible is lowered, and stretch of the muscles activates a reflex that causes contraction of the muscles of mastication. Once the mouth is closed, the food again stimulates the muscles of mastication to relax, and the cycle is repeated. Descending pathways from the cerebrum strongly influence the activity of the mastication reflex so that chewing can be initiated or stopped consciously. The rate and intensity of chewing movements can also be influenced by the cerebrum.

Salivary Glands

A considerable number of **salivary glands** are scattered throughout the oral cavity. Three pairs of large, multicellular salivary glands exist: the parotid, the submandibular, and the sublingual glands (figure 24.9). In addition to these large salivary glands, numerous small, coiled tubular salivary glands are located deep to the epithelium of the tongue (lingual glands), palate (palatine glands), cheeks (buccal glands), and lips (labial glands). The secretions from the salivary glands help keep the oral cavity moist and begin the process of digestion. They also suspend molecules in solution so they can be tasted.

All of the major large salivary glands are compound **alveolar glands,** which are branching glands with clusters of alveoli resembling grapes (see chapter 4). They produce thin serous secretions or thicker mucous secretions. Thus, saliva is a combination of serous and mucous secretions from the various salivary glands.

The largest salivary glands, the **parotid** (pă-rot'id; beside the ear) **glands,** are serous glands, which produce mostly watery saliva; they are located just anterior to the ear on each side of the head. Each **parotid duct** exits the gland on its anterior margin, crosses the lateral surface of the masseter muscle, pierces the buccinator muscle, and enters the oral cavity adjacent to the second upper molar (see figure 24.9).

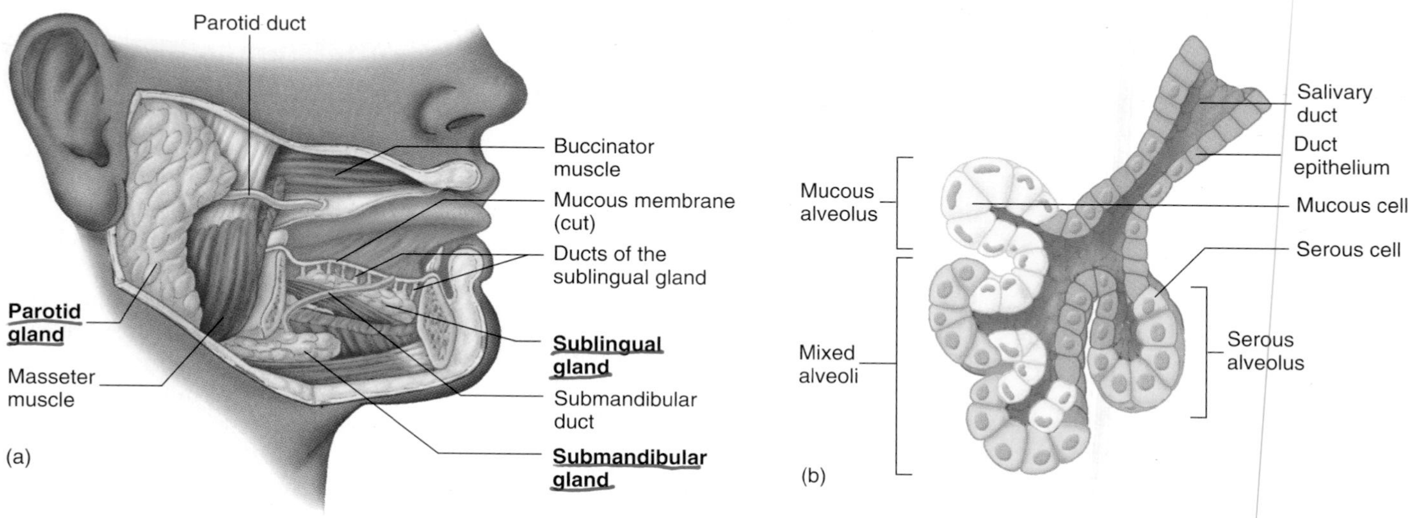

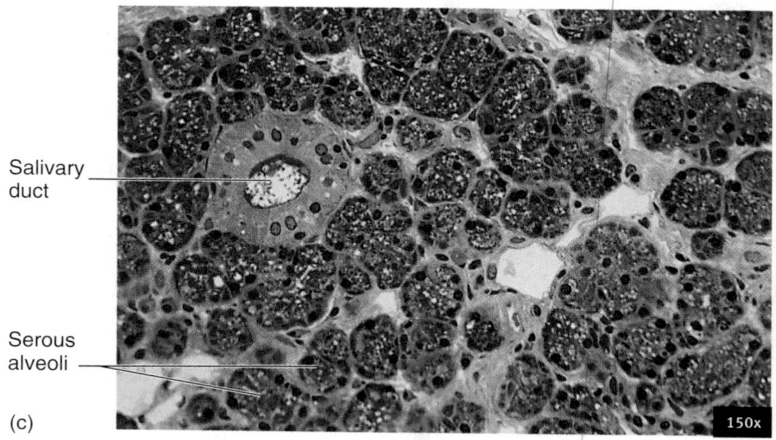

FIGURE 24.9 Salivary Glands

(*a*) The large salivary glands are the parotid glands, the submandibular glands, and the sublingual glands. The parotid duct extends anteriorly from the parotid gland. (*b*) An idealized schematic drawing of the histology of the large salivary glands. The figure is representative of all the glands and does not depict any one salivary gland. (*c*) Photomicrograph of the parotid gland.

Saliva and the Second Molar

Because the parotid secretions are released directly onto the surface of the second upper molar, it tends to have a considerable accumulation of mineral, secreted from the gland, on its surface.

Mumps

Inflammation of the parotid gland is called parotiditis. **Mumps,** which is caused by a virus, is the most common type of parotiditis.

The **submandibular** (below the mandible) **glands** are mixed glands with more serous than mucous alveoli. Each gland can be felt as a soft lump along the inferior border of the posterior half of the mandible. A submandibular duct exits each gland, passes anteriorly deep to the mucous membrane on the floor of the oral cavity, and opens into the oral cavity beside the frenulum of the tongue (see figure 24.6).

The **sublingual** (below the tongue) **glands,** the smallest of the three large, paired salivary glands, are mixed glands containing some serous alveoli but consisting primarily of mucous alveoli. They lie immediately below the mucous membrane in the floor of the mouth. These glands do not have single, well-defined ducts like those of the submandibular and parotid glands. Instead, each sublingual gland opens into the floor of the oral cavity through 10–12 small ducts.

Saliva is secreted at the rate of about 1–1.5 L/day. The serous part of saliva, produced mainly by the parotid and submandibular glands, contains a digestive enzyme called **salivary amylase** (am′il-ās; starch-splitting enzyme), which breaks the covalent bonds between glucose molecules in starch and other polysaccharides to produce the disaccharides maltose and isomaltose (table 24.2). The release of maltose and isomaltose gives starches a sweet taste in the mouth. Food spends very little time in the mouth, however; therefore, only about 3%–5% of the total carbohydrates are digested in the mouth. Most of the starches are covered by cellulose in plant tissues and are inaccessible to salivary amylase. Cooking and thorough chewing

TABLE 24.2 Functions of Major Digestive Secretions

Fluid or Enzyme	Function
Oral Cavity Secretions	
Serous (watery) saliva	Moistens food and mucous membrane; lysozyme kills bacteria
Salivary amylase	Digests starch (conversion to maltose and isomaltose)
Mucus	Lubricates food; protects gastrointestinal tract from digestion by enzymes
Lingual lipase	Digests a small amount (<10%) of lipids
Esophagus Secretions	
Mucus	Lubricates esophagus; protects esophagus lining from abrasion and allows food to move more smoothly through esophagus
Gastric Secretions	
Hydrochloric acid	Decreases stomach pH to activate pepsinogen to pepsin
Pepsin	Digests protein into smaller peptide chains; activates pepsinogen
Mucus	Protects stomach lining from digestion
Gastric lipase	Digests a minor amount of lipid
Liver Secretions	
Bile Sodium glycocholate (bile salt) Sodium taurocholate (bile salt) Cholesterol Biliverdin Bilirubin Mucus Fat Lecithin Cells and cell debris	Bile salts emulsify fats, making them available to intestinal lipases; help make end products soluble and available for absorption by the intestinal mucosa; aid peristalsis. Many of the other bile contents are waste products transported to the intestine for disposal.
Pancreas Secretions	
Trypsin	Digests proteins (breaks polypeptide chains at arginine or lysine residues); activates trypsinogen; activates other digestive enzymes
Chymotrypsin	Digests proteins (cleaves carboxyl links of hydrophobic amino acids)
Carboxypeptidase	Digests proteins (removes amino acids from the carboxyl end of peptide chains)
Pancreatic amylase	Digests carbohydrates (hydrolyzes starches and glycogen to form maltose and isomaltose)
Pancreatic lipase	Digests fat (breaks down triglycerides into monoglycerides and free fatty acids)
Ribonuclease	Digests ribonucleic acid (hydrolyzes phosphodiester bonds)
Deoxyribonuclease	Digests deoxyribonucleic acid (hydrolyzes phosphodiester bonds)
Cholesterol esterase	Hydrolyzes cholesterol esters to form cholesterol and free fatty acids
Bicarbonate ions	Provides appropriate pH for pancreatic enzymes
Small Intestine Secretions	
Mucus	Protects duodenum from stomach acid, gastric enzymes, and intestinal enzymes; provides adhesion for fecal matter; protects intestinal wall from bacterial action and acid produced in the feces
Aminopeptidase	Splits polypeptides into amino acids (from amino end of chain)
Peptidase	Splits amino acids from polypeptides
Enterokinase	Activates trypsin from trypsinogen
Amylase	Digests carbohydrates
Sucrase	Splits sucrose into glucose and fructose
Maltase	Splits maltose into two glucose molecules
Isomaltase	Splits isomaltose into two glucose molecules
Lactase	Splits lactose into glucose and galactose

of food destroy the cellulose covering and increase the efficiency of the digestive process.

Saliva prevents bacterial infection in the mouth by washing the oral cavity. Saliva also contains substances such as **lysozyme,** which has a weak antibacterial action, and immunoglobulin A, which helps prevent bacterial infection. Any lack of salivary gland secretion increases the risk for ulceration and infection of the oral mucosa and for caries in the teeth.

The mucous secretions of the submandibular and sublingual glands contain a large amount of **mucin** (mū′sin), a proteoglycan that gives a lubricating quality to the secretions of the salivary glands.

Salivary gland secretion is stimulated by the parasympathetic and sympathetic nervous systems, with the parasympathetic system being more important. Salivary nuclei in the brainstem increase salivary secretions by sending action potentials through parasympathetic fibers of the facial (VII) and glossopharyngeal (IX) cranial nerves in response to a variety of stimuli, such as tactile stimulation in the oral cavity or certain tastes, especially sour. Higher centers of the brain also affect the activity of the salivary glands. Odors that trigger thoughts of food or the sensation of hunger can increase salivary secretions.

14 ***List the muscles of mastication and the actions they produce. Describe the mastication reflex.***

15 ***Name and give the location of the three largest salivary glands. Name the other kinds of salivary glands.***

16 ***What substances are contained in saliva?***

17 ***What is the difference between serous and mucous saliva?***

PHARYNX

The **pharynx** was described in detail in chapter 23. The pharynx consists of three parts: the nasopharynx, oropharynx, and laryngopharynx. Normally, only the oropharynx and laryngopharynx transmit food. The **oropharynx** communicates with the nasopharynx superiorly, the larynx and **laryngopharynx** inferiorly, and the mouth anteriorly. The laryngopharynx extends from the oropharynx to the esophagus and is posterior to the larynx. The epiglottis covers the opening of the larynx and keeps food and drink from entering the larynx. The posterior walls of the oropharynx and laryngopharynx consist of three muscles: the superior, middle, and inferior **pharyngeal constrictors,** which are arranged like three stacked flowerpots, one inside the other. The oropharynx and the laryngopharynx are lined with moist stratified squamous epithelium, and the nasopharynx is lined with ciliated pseudostratified columnar epithelium.

18 ***Name the three parts of the pharynx. What are the pharyngeal constrictors?***

PREDICT 2

Explain the functional significance of the differences in epithelial types among the three pharyngeal regions.

ESOPHAGUS

The **esophagus** is the part of the digestive tube that extends between the pharynx and the stomach. It is about 25 cm long and lies in the mediastinum, anterior to the vertebrae and posterior to the trachea. It passes through the esophageal hiatus (opening) of the diaphragm and ends at the stomach. The esophagus transports food from the pharynx to the stomach.

Hiatal Hernia

A **hiatal hernia** is a widening of the esophageal hiatus. Hiatal hernias allow part of the stomach to extend through the opening into the thorax. They occur most commonly in adults. A hernia can decrease the resting pressure in the lower esophageal sphincter, allowing gastroesophageal reflux and subsequent esophagitis to occur. Hiatal herniation can also compress the blood vessels in the stomach mucosa, which can lead to gastritis or ulcer formation. Esophagitis, gastritis, and ulceration can be very painful.

The esophagus has thick walls consisting of the four tunics common to the digestive tract: mucosa, submucosa, muscularis, and adventitia. The muscular tunic has an outer longitudinal layer and an inner circular layer, as is true of most parts of the digestive tract, but it is different because it consists of skeletal muscle in the superior part of the esophagus and smooth muscle in the inferior part. An **upper esophageal sphincter** and a **lower esophageal sphincter,** at the upper and lower ends of the esophagus, respectively, regulate the movement of materials into and out of the esophagus. The mucosal lining of the esophagus is moist stratified squamous epithelium. Numerous mucous glands in the submucosal layer produce a thick, lubricating mucus, which passes through ducts to the surface of the esophageal mucosa.

19 ***Where is the esophagus located? Describe the layers of the esophageal wall and the esophageal sphincters.***

SWALLOWING

Swallowing, or **deglutition,** is divided into three phases: voluntary, pharyngeal, and esophageal. During the **voluntary phase** (figure 24.10, step 1), a bolus of food is formed in the mouth and pushed by the tongue against the hard palate, forcing the bolus toward the posterior part of the mouth and into the oropharynx.

The **pharyngeal phase** (see figure 24.10, steps 2–4) of swallowing is a reflex initiated by the stimulation of tactile receptors in the area of the oropharynx. Afferent action potentials travel through the trigeminal (V) and glossopharyngeal (IX) nerves to the **swallowing center** in the medulla oblongata. There, they initiate action potentials in motor neurons, which pass through the trigeminal (V), glossopharyngeal (IX), vagus (X), and accessory (XI) nerves to the soft palate and pharynx. This phase of swallowing begins with the elevation of the soft palate, which closes the passage between the nasopharynx and oropharynx. The pharynx elevates to receive the bolus of food from the mouth and moves the bolus down the pharynx into the esophagus. The superior, middle,

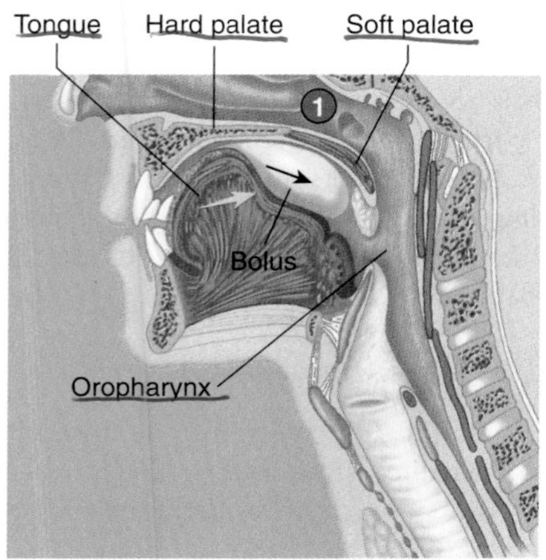

1. During the voluntary phase, a bolus of food (*yellow*) is pushed by the tongue against the hard and soft palates and posteriorly toward the oropharynx (*blue arrow* indicates tongue movement; *black arrow* indicates movement of the bolus). *Tan:* bone, *purple:* cartilage, *red:* muscle.

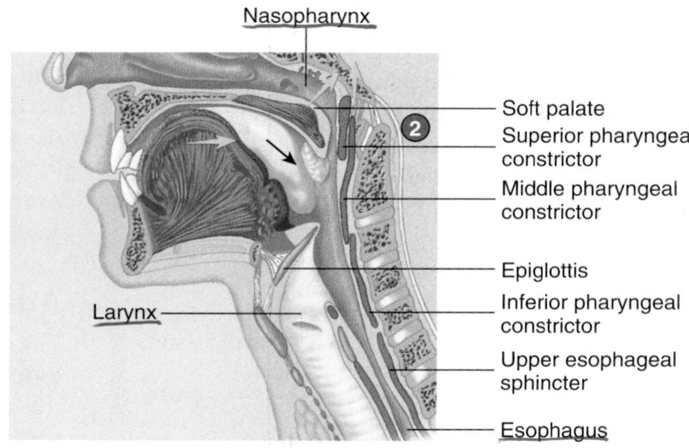

2. During the pharyngeal phase, the soft palate is elevated, closing off the nasopharynx. The pharynx and larynx are elevated (*blue arrows* indicate muscle movement).

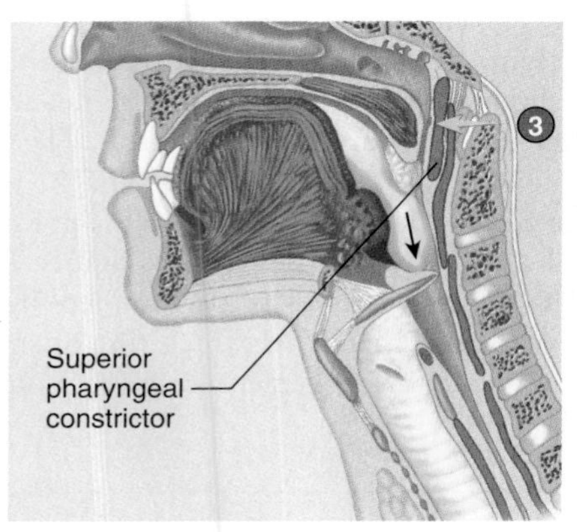

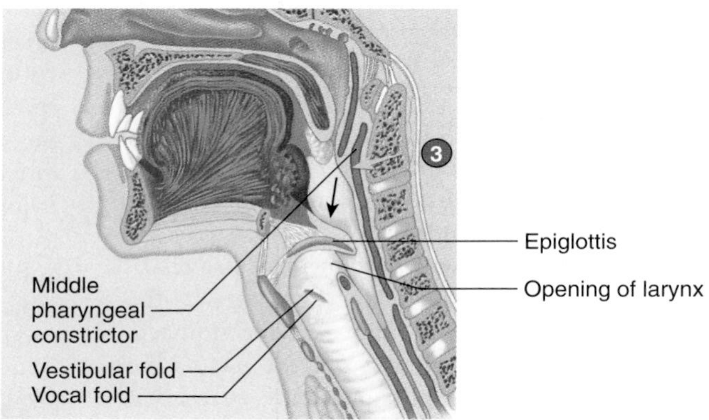

3. Successive constriction of the pharyngeal constrictors from superior to inferior (*blue arrows*) forces the bolus through the pharynx and into the esophagus. As this occurs, the vestibular and vocal folds expand medially to close the passage of the larynx. The epiglottis is bent down over the opening of the larynx largely by the force of the bolus pressing against it.

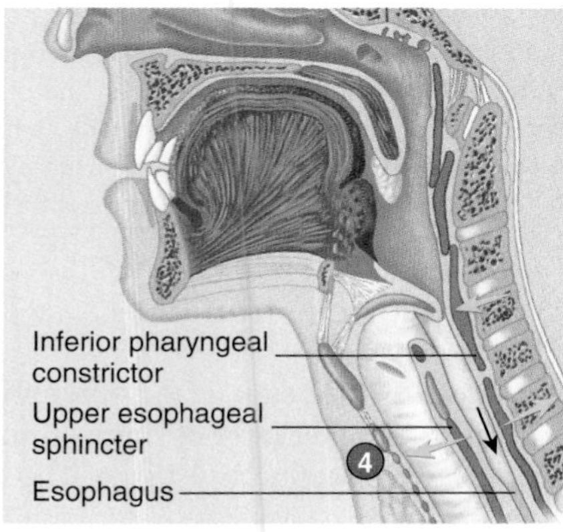

4. As the inferior pharyngeal constrictor contracts, the upper esophageal sphincter relaxes (outwardly directed *blue arrows*), allowing the bolus to enter the esophagus.

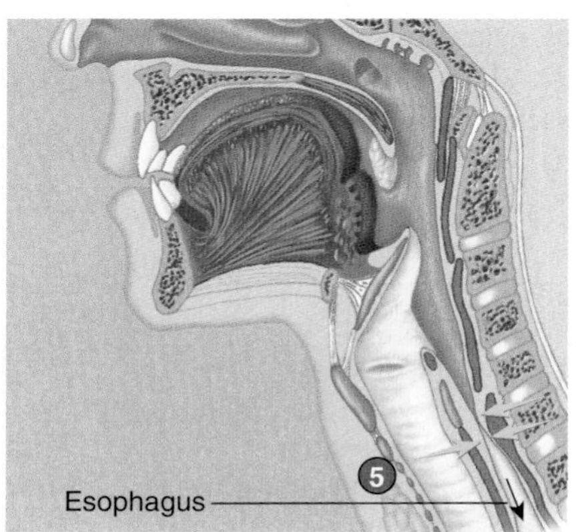

5. During the esophageal phase, the bolus is moved by peristaltic contractions of the esophagus toward the stomach (inwardly directed *blue arrows*).

PROCESS FIGURE 24.10 Three Phases of Swallowing (Deglutition)

and inferior **pharyngeal constrictor muscles** contract in succession, forcing the food through the pharynx. At the same time, the upper esophageal sphincter relaxes, the elevated pharynx opens the esophagus, and food is pushed into the esophagus. This phase of swallowing is unconscious and is controlled automatically, even though the muscles involved are skeletal. The pharyngeal phase of swallowing lasts about 1–2 seconds.

PREDICT 3

Why is it important to close the opening between the nasopharynx and oropharynx during swallowing? What may happen if a person has an explosive burst of laughter while trying to swallow a liquid?

During the pharyngeal phase, the vestibular folds and vocal cords close and the **epiglottis** (ep-i-glot′is; on the glottis) is tipped posteriorly so that the epiglottic cartilage covers the opening into the larynx, and the larynx is elevated. These movements of the larynx prevent food from passing through the opening into the larynx.

PREDICT 4

What can happen if you try to swallow and speak at the same time?

The **esophageal phase** (see figure 24.10, step 5) of swallowing, which takes about 5–8 seconds, is responsible for moving food from the pharynx to the stomach. Muscular contractions in the wall of the esophagus occur in peristaltic waves.

The peristaltic waves associated with swallowing cause relaxation of the lower esophageal sphincter in the esophagus as the peristaltic waves, and the bolus of food, approach the stomach. This sphincter is not anatomically distinct from the rest of the esophagus, but it can be identified physiologically because it remains tonically constricted to prevent the reflux of stomach contents into the lower part of the esophagus.

The presence of food in the esophagus stimulates the enteric plexus, which controls the peristaltic waves. The presence of food in the esophagus also stimulates tactile receptors, which send afferent impulses to the medulla oblongata through the vagus nerves. Motor impulses, in turn, pass along the vagal efferent fibers to the striated and smooth muscles within the esophagus, thereby stimulating their contractions and reinforcing the peristaltic contractions.

Swallowing and Gravity

Gravity assists the movement of material through the esophagus, especially when liquids are swallowed. The peristaltic contractions that move material through the esophagus are sufficiently forceful, however, to allow a person to swallow even while doing a headstand or floating in the zero-gravity environment of space.

20 ***What are the three phases of deglutition? List sequentially the processes involved in the last two phases, and describe how they are regulated.***

STOMACH

The **stomach** is an enlarged segment of the digestive tract in the left superior part of the abdomen (see figure 24.1). Its shape and size vary from person to person; even within the same individual, its size and shape change from time to time, depending on its food content and the posture of the body. Nonetheless, several general anatomical features can be described.

Anatomy of the Stomach

The opening from the esophagus into the stomach is the **gastroesophageal,** or **cardiac** (located near the heart), **opening,** and the region of the stomach around the cardiac opening is the **cardiac part** (figure 24.11). The lower esophageal sphincter, also called the **cardiac sphincter,** surrounds the cardiac opening. Recall that, although this is an important structure in the normal function of the stomach, it is a physiologic constrictor only and cannot be seen anatomically. A part of the stomach to the left of the cardiac part, the **fundus** (fŭn′dŭs; the bottom of a round-bottomed leather bottle), is actually superior to the cardiac opening. The largest part of the stomach is the **body,** which turns to the right, creating a **greater curvature** and a **lesser curvature.** The body narrows to form the funnel-shaped **pyloric** (pī-lōr′ik; gatekeeper) **part** of the stomach. The wider part of the funnel, toward the body of the stomach, is the **pyloric antrum.** The narrow part of the funnel is the **pyloric canal.** The pyloric canal opens through the **pyloric orifice** into the small intestine. The pyloric orifice is surrounded by the **pyloric sphincter,** or **pylorus,** a relatively thick ring of smooth muscle, which helps regulate the movement of gastric contents into the small intestine.

Hypertrophic Pyloric Stenosis

Hypertrophic pyloric stenosis is a common defect of the stomach in infants, occurring in 1 in 150 males and 1 in 750 females, in which the pyloric sphincter is greatly thickened, resulting in interference with normal stomach emptying. Infants with this defect develop symptoms 3 to 10 weeks after birth and often exhibit projectile vomiting. Because the pyloric sphincter is blocked, little food enters the intestine, and the infant fails to gain weight. Constipation is also a frequent complication. The defect can be corrected with surgery in which the pyloric orifice is enlarged by cutting through the mucosa and sphincter.

Histology of the Stomach

The **serosa,** or visceral peritoneum, is the outermost layer of the stomach. It consists of an outer layer of simple squamous epithelium and an inner layer of connective tissue. The **muscularis** of the stomach consists of three layers: an outer longitudinal layer, a middle circular layer, and an inner oblique layer (see figure 24.11*a*). In some areas of the stomach, such as in the fundus, the three layers blend with one another and cannot be separated. Deep to the muscular layer are the submucosa and the mucosa, which are thrown into large folds called **rugae** (roo′gē; wrinkles)

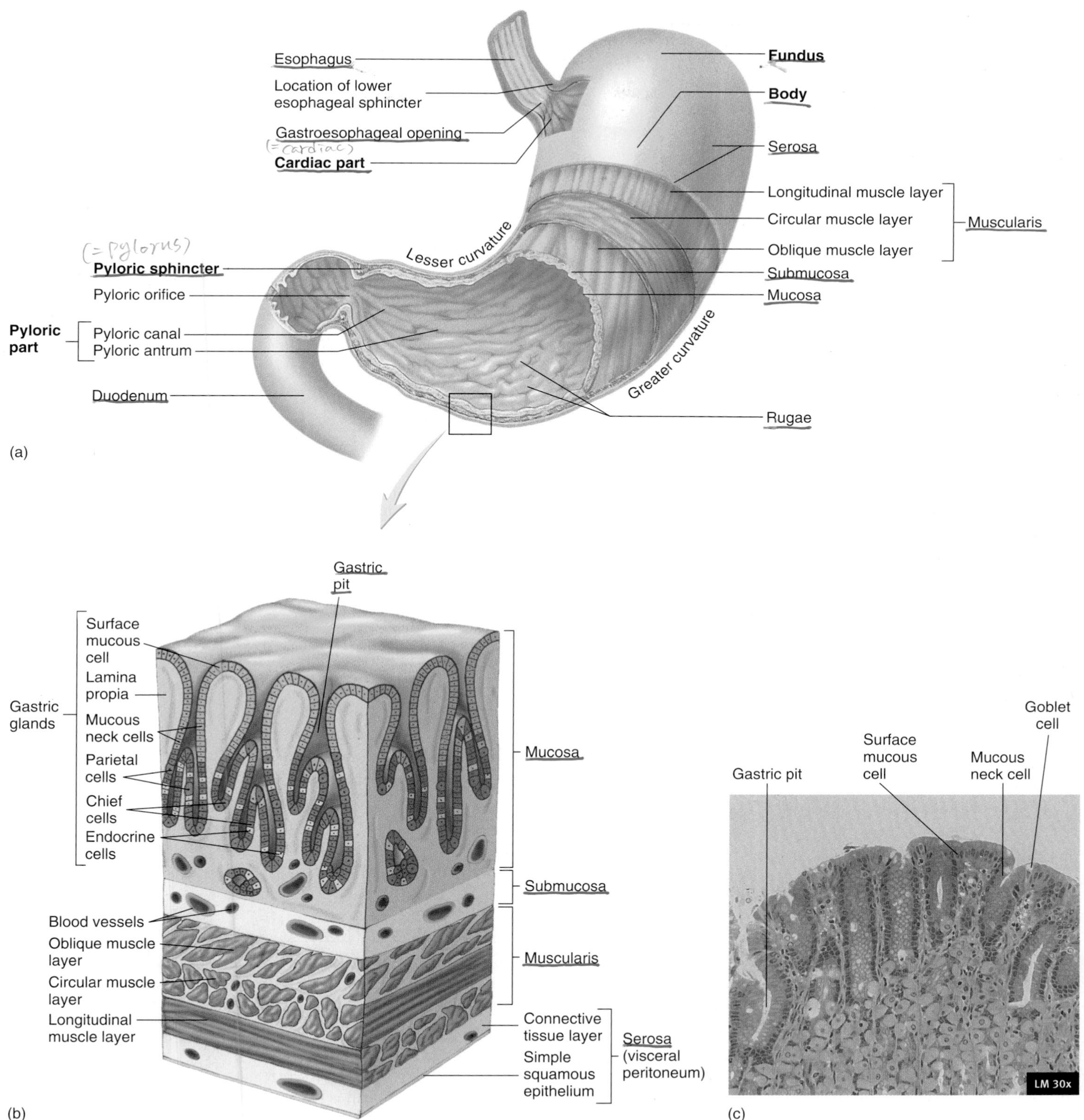

FIGURE 24.11 Anatomy and Histology of the Stomach

(*a*) Cutaway section reveals muscular layers and internal anatomy. (*b*) Section of the stomach wall that illustrates its histology, including several gastric pits and glands. (*c*) Photomicrograph of gastric glands.

when the stomach is empty. These folds allow the mucosa and submucosa to stretch, and the folds disappear as the stomach volume increases as it is filled.

The stomach is lined with simple columnar epithelium. The epithelium forms numerous, tubelike **gastric pits,** which are the openings for the **gastric glands** (see figure 24.11*b*). The epithelial cells of the stomach are of five types. The first type, **surface mucous cells,** which produce mucus, is on the surface and lines the gastric pit. The remaining four cell types are in the gastric glands. They are **mucous neck cells,** which produce mucus; **parietal** (oxyntic) **cells,** which produce hydrochloric acid and intrinsic factor; **chief** (zymogenic) **cells,** which produce pepsinogen; and **endocrine cells,** which produce regulatory hormones. The mucous neck cells are located near the openings of the glands, whereas the parietal, chief, and endocrine cells are interspersed in the deeper parts of the glands. There are several types of endocrine cells. Enterochromaffin-like cells produce histamine, which stimulates acid secretion by parietal cells. Gastrin-containing cells secrete gastrin, and somatostatin-containing cells secrete somatostatin, which inhibits gastrin and insulin secretion.

Surface mucous cells of the stomach protect the stomach wall from being damaged by acid and digestive enzymes. These cells produce an alkaline mucus on their surface that neutralizes the acid and is a barrier to the digestive enzymes. The surface mucous cells are connected by tight junctions, which provide an additional barrier that prevents acids and enzymes from reaching deeper tissues. In addition, when surface mucous cells are damaged, they are rapidly replaced.

21 ***Describe the parts of the stomach. List the layers of the stomach wall. How is the stomach different from the esophagus?***

22 ***What are gastric pits and gastric glands? Name the types of cells in the stomach and the secretions they produce.***

Secretions of the Stomach

Ingested food and stomach secretions, mixed together, form a semi-fluid material called **chyme** (kīm; juice). The stomach functions primarily as a storage and mixing chamber for the chyme. Although some digestion and absorption occur in the stomach, they are not its major functions.

Stomach secretions include mucus, hydrochloric acid, gastrin, histamine, intrinsic factor, and pepsinogen. Pepsinogen is the inactive form of the protein-digesting enzyme pepsin.

The surface mucous cells and mucous neck cells secrete a viscous and alkaline mucus that covers the surface of the epithelial cells, forming a layer 1–1.5 mm thick. The thick layer of mucus lubricates and protects the epithelial cells of the stomach wall from the damaging effect of the acidic chyme and pepsin. Irritation of the stomach mucosa results in stimulation of the secretion of a greater volume of mucus.

Parietal cells in the gastric glands of the pyloric region secrete intrinsic factor and a concentrated solution of hydrochloric acid. **Intrinsic factor** is a glycoprotein that binds with vitamin B_{12}, making the vitamin more readily absorbed in the ileum. Vitamin B_{12} is important in deoxyribonucleic acid (DNA) synthesis, which is required for red blood cell production. A lack of vitamin B_{12} absorption leads to pernicious anemia (see chapter 19).

Hydrochloric acid produces the low pH of the stomach's contents, which is normally between 1 and 3. Although the hydrochloric acid secreted into the stomach has a minor digestive effect on ingested food, one of its main functions is to kill bacteria that are ingested with essentially everything humans put into their mouths. Some pathogenic bacteria may avoid digestion in the stomach, however, because they have an outer coat that resists stomach acids.

The low pH of the stomach's contents has additional functions. It stops carbohydrate digestion by inactivating salivary amylase. Stomach acid also denatures many proteins so that proteolytic enzymes can reach internal peptide bonds. The acid environment provides the proper pH for the activation and function of pepsin.

Hydrogen ions are derived from carbon dioxide and water, which enter the parietal cell from its serosal surface, which is the side opposite the lumen of the gastric pit (figure 24.12). Once inside the cell, carbonic anhydrase catalyzes the reaction between carbon dioxide and water to form carbonic acid. Some of the carbonic acid molecules then dissociate to form H^+ and HCO_3^-. The H^+ are actively transported across the mucosal surface of the parietal cell into the lumen of the stomach by a H^+–K^+ exchange pump, often called a **proton pump.** Although H^+ are actively transported against a steep concentration gradient, Cl^- diffuse with the H^+ from the cell through the plasma membrane. Diffusion of Cl^- with the positively charged H^+ reduces the amount of energy needed to transport the H^+ against both a concentration gradient and an electrical gradient. Bicarbonate ions move down their concentration gradient from the parietal cell into the extracellular fluid. During this process, HCO_3^- are exchanged for Cl^- through an antiporter, which is located in the plasma membrane, and the Cl^- subsequently move into the cell.

PREDICT 5

Explain why a slight increase in blood pH may occur following a heavy meal. The elevated pH of blood, especially in the veins that carry blood away from the stomach, is called the postenteric alkaline tide.

Chief cells within the gastric glands secrete **pepsinogen** (pep-sin′ō-jen). Pepsinogen is packaged in **zymogen** (zī-mō-jen; related to enzymes) **granules,** which are released by exocytosis when pepsinogen secretion is stimulated. Once pepsinogen enters the lumen of the stomach, hydrochloric acid and previously formed pepsin molecules convert it to **pepsin.** Pepsin exhibits optimum enzymatic activity at a pH of 3 or less. Pepsin catalyzes the cleavage of some covalent bonds in proteins, thus breaking them into smaller peptide chains.

1. Carbon dioxide (CO_2) diffuses into the parietal cell.
2. Carbon dioxide combines with water (H_2O) in an enzymatic reaction that is catalyzed by carbonic anhydrase (CA) to form carbonic acid (H_2CO_3).
3. Carbonic acid dissociates into a bicarbonate ion (HCO_3^-) and a hydrogen ion (H^+).
4. Bicarbonate ions are transported back into the bloodstream. An antiporter in the plasma membrane exchanges HCO_3^- for a chloride ion (Cl^-).
5. A H^+–K^+ pump moves H^+ into the duct of the gastric gland and K^+ into the parietal cell.
6. Chloride ions diffuse with the charged H^+.

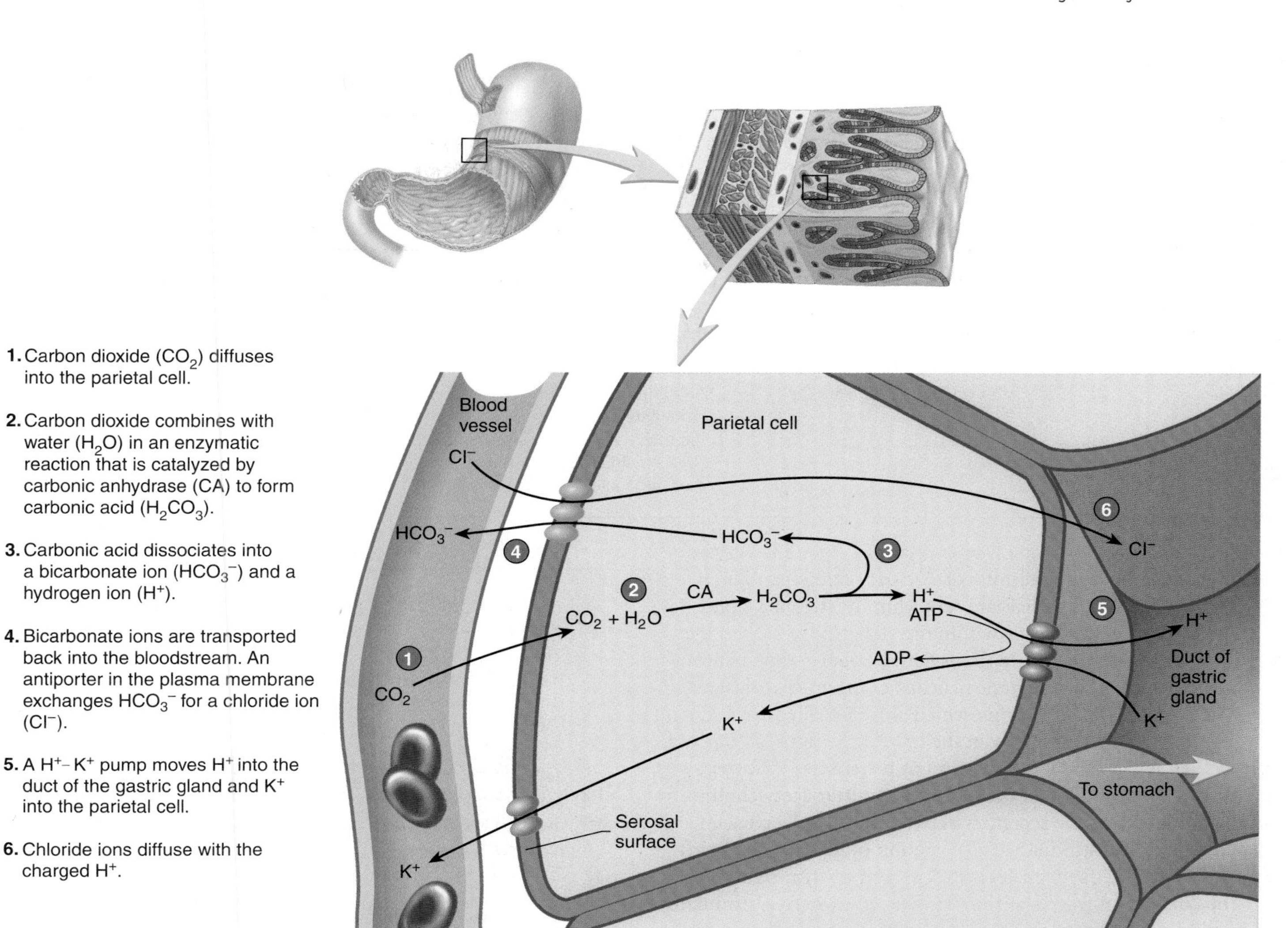

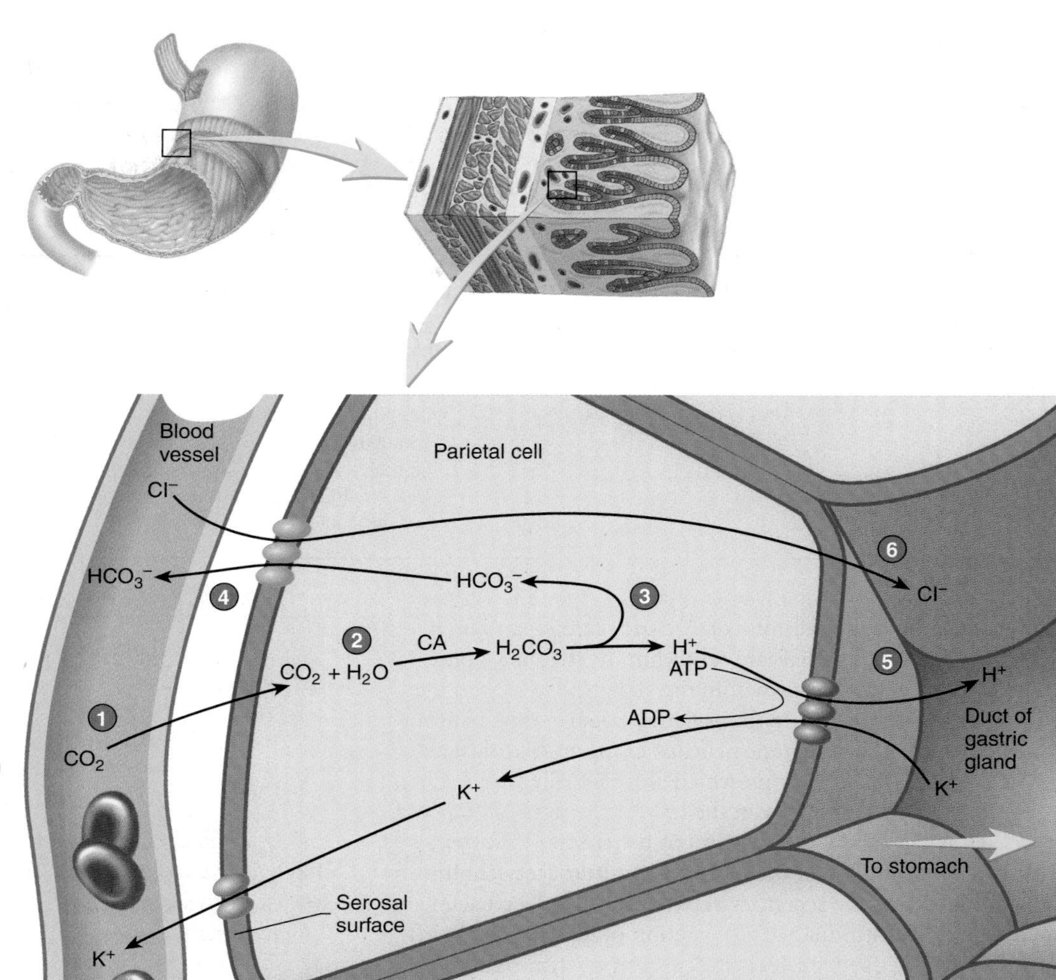

PROCESS FIGURE 24.12 **Hydrochloric Acid Production by Parietal Cells in the Gastric Glands of the Stomach**

Heartburn

Heartburn, or **pyrosis** (pī-rō′sis), is a painful or burning sensation in the chest usually associated with the reflux of acidic chyme into the esophagus. The pain is usually short-lived but may be confused with the pain of an ulcer or a heart attack. Overeating, eating fatty foods, lying down immediately after a meal, consuming too much alcohol or caffeine, smoking, and wearing extremely tight clothing can all cause heartburn. A hiatal hernia can also cause heartburn, especially in older people.

Regulation of Stomach Secretion

Approximately 2–3 L of gastric secretions (gastric juice) are produced each day. The amount and type of food entering the stomach and intestine dramatically affect the secretion amount, but up to 700 mL are secreted as a result of a typical meal. Both nervous and hormonal mechanisms regulate gastric secretions. The neural mechanisms involve reflexes integrated within the medulla oblongata and local reflexes integrated within the enteric plexus of the GI tract. In addition, higher brain centers influence the reflexes. The chemical signals that regulate stomach secretions include the hormones gastrin, secretin, and cholecystokinin (table 24.3), as well as the paracrine chemical signal histamine.

The regulation of stomach secretion is divided into three phases: cephalic, gastric, and gastrointestinal. The cephalic phase can be viewed as the "get started" phase when stomach secretions are increased in anticipation of incoming food. This is followed by the gastric "go for it" phase when most of the stimulation of secretion occurs. Finally, the gastrointestinal phase is the "slow down" phase, during which stomach secretion decreases.

1. **Cephalic phase.** In the cephalic phase of gastric regulation, the sensations of the taste and smell of food, the stimulation of tactile receptors during the process of chewing and swallowing, and pleasant thoughts of food stimulate centers within the

TABLE 24.3 Functions of the Gastrointestinal Hormones

Site of Production	Method of Stimulation	Secretory Effects	Motility Effects
Gastrin			
Stomach	Distention; partially digested proteins, autonomic stimulation, ingestion of alcohol or caffeine	Increases gastric secretion	Causes a minor increase in gastric motility
Secretin			
Duodenum	Acidity of chyme	Decreases gastric secretion; stimulates pancreatic and bile secretions high in bicarbonate ions	Decreases gastric motility
Cholecystokinin			
Duodenum	Fatty acids and peptides	Slightly decreases gastric secretion; stimulates pancreatic secretions high in digestive enzymes; causes contraction of the gallbladder and relaxation of the hepatopancreatic ampullar sphincter	Strongly decreases gastric motility

medulla oblongata that influence gastric secretions (figure 24.13*a*). Action potentials are sent from the medulla along parasympathetic neurons within the vagus (X) nerves to the stomach. Within the stomach wall, the preganglionic neurons stimulate the postganglionic neurons in the enteric plexus. The postganglionic neurons, which are primarily cholinergic, stimulate secretory activity in the cells of the stomach mucosa.

Parasympathetic stimulation of the stomach mucosa results in the release of the neurotransmitter acetylcholine, which increases the secretory activity of both the parietal and chief cells and stimulates the secretion of **gastrin** (gas′trin) and histamine from endocrine cells. The gastrin released into the circulation travels to the parietal cells, where it stimulates additional hydrochloric acid and pepsinogen secretion. In addition, gastrin stimulates enterochromaffin-like cells to release histamine, which stimulates parietal cells to secrete hydrochloric acid. Acetylcholine, histamine, and gastrin working together cause a greater secretion of hydrochloric acid than any of them does separately. Of the three, histamine has the greatest stimulatory effect.

2. **Gastric phase.** The greatest volume of gastric secretions is produced during the gastric phase of gastric regulation. The presence of food in the stomach initiates the gastric phase (figure 24.13*b*). The primary stimuli are distention of the stomach and the presence of amino acids and peptides in the stomach.

   Distention of the stomach wall, especially in the body or fundus, results in the stimulation of mechanoreceptors. Action potentials generated by these receptors initiate reflexes that involve both the CNS and the ENS. These reflexes result in acetylcholine release and the cascade of events in the cephalic phase. The presence of partially digested proteins or moderate amounts of alcohol or caffeine in the stomach also stimulates gastrin secretion.

   When the pH of the stomach contents falls below 2, increased gastric secretion produced by distention of the stomach is blocked. This negative-feedback mechanism limits the secretion of gastric juice.

Inhibitors of Gastric Acid Secretion

Histamine receptors on parietal cells are called H_2 receptors; they are different from the H_1 receptors involved in allergic reactions. Drugs that block allergic reactions do not affect histamine-mediated stomach acid secretion and vice versa. Cimetidine (Tagamet) and ranitidine (Zantac) are histamine receptor antagonists that prevent the binding of histamine to receptors on parietal cells. These chemicals are extremely effective inhibitors of gastric acid secretion. Cimetidine, one of the most commonly prescribed drugs, is used to treat gastric acid hypersecretion associated with gastritis and gastric ulcers.

3. **Gastrointestinal phase.** The entrance of acidic stomach contents into the duodenum of the small intestine controls the gastrointestinal phase of gastric regulation (figure 24.13*c*). The presence of chyme in the duodenum activates both neural and hormonal mechanisms. When the pH of the chyme entering the duodenum drops to 2 or below, or if the chyme contains fat digestion products, gastric secretions are inhibited.

   Acidic solutions in the duodenum cause the release of the hormone **secretin** (se-krē′tin) into the circulatory system. Secretin inhibits gastric secretion by inhibiting both parietal and chief cells.

   Fatty acids, other lipids, and, to a lesser degree, protein digestion products in the duodenum and the proximal jejunum initiate the release of the hormone **cholecystokinin** (kō′lē-sis-tō-kī′nin), which inhibits gastric secretion.

   The inhibition of gastric secretion is also under nervous control. Distention of the duodenal wall, the presence of irritating substances in the duodenum, reduced pH, and hypertonic or hypotonic solutions in the duodenum activate the enterogastric reflex. The **enterogastric reflex** consists of a local reflex and a reflex integrated within the medulla oblongata that reduce gastric secretion.

To summarize, gastric acid secretion is controlled by negative-feedback loops involving nerves and hormones. First, during the

Cephalic Phase

1. The taste, smell, or thought of food or tactile sensations of food in the mouth stimulate the medulla oblongata (*green arrows*).
2. Parasympathetic action potentials are carried by the vagus nerves to the stomach (*pink arrow*) where enteric plexus neurons are activated.
3. Postganglionic neurons stimulate secretion by parietal and chief cells and stimulate gastrin and histamine secretion by endocrine cells.
4. Gastrin is carried through the circulation back to the stomach (*purple arrow*), where, along with histamine, it stimulates secretion.

Gastric Phase

1. Distention of the stomach stimulates mechanoreceptors (stretch receptors) and activates a parasympathetic reflex. Action potentials generated by the mechanoreceptors are carried by the vagus nerves to the medulla oblongata (*green arrow*).
2. The medulla oblongata increases action potentials in the vagus nerves that stimulate secretions by parietal and chief cells and stimulate gastrin and histamine secretion by endocrine cells (*pink arrow*).
3. Distention of the stomach also activates local reflexes that increase stomach secretions (*orange arrow*).
4. Gastrin is carried through the circulation back to the stomach (*purple arrow*), where, along with histamine, it stimulates secretion.

Gastrointestinal Phase

1. Chyme in the duodenum with a pH less than 2 or containing fat digestion products (lipids) inhibits gastric secretions by three mechanisms (2–4).
2. Chemoreceptors in the duodenum are stimulated by H^+ (low pH) or lipids. Action potentials generated by the chemoreceptors are carried by the vagus nerves to the medulla oblongata (*green arrow*), where they inhibit parasympathetic action potentials (*pink arrow*), thereby decreasing gastric secretions.
3. Local reflexes activated by H^+ or lipids also inhibit gastric secretion (*orange arrows*).
4. Secretin and cholecystokinin produced by the duodenum (*brown arrows*) decrease gastric secretions in the stomach.

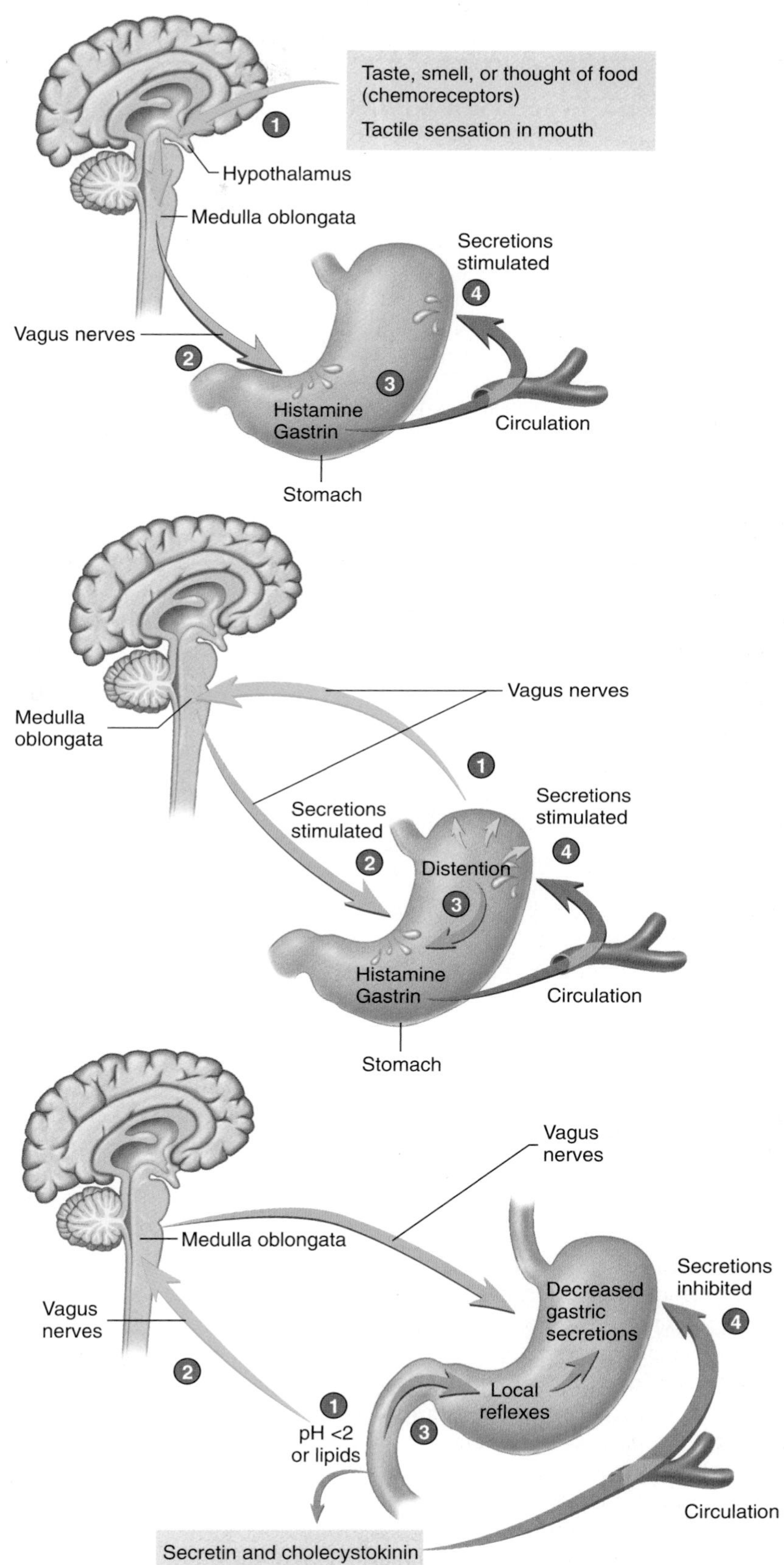

PROCESS FIGURE 24.13 Phases of Stomach Secretion

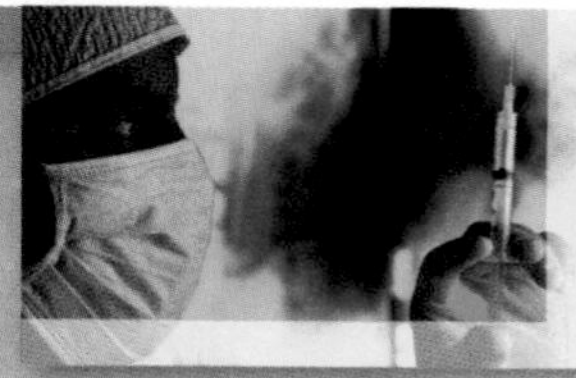

CLINICAL FOCUS

Peptic Ulcer

Approximately 10% of people in the United States will develop **peptic ulcers** during their lifetime. Peptic ulcers are caused when the gastric juices (acid and pepsin) digest the mucosal lining of the GI tract. Approximately 80% of peptic ulcers occur on the duodenal side of the pyloric sphincter, but peptic ulcers can also occur in the stomach (gastric ulcers) or esophagus (esophageal ulcers).

Nearly all peptic ulcers are due to infection by a specific bacterium, *Helicobacter pylori. H. pylori* is also linked to gastritis and gastric cancer. Because stress, diet, smoking, and alcohol cause excess acid secretion in the stomach, for years it was accepted that these lifestyle patterns were responsible for ulcers. Although they can contribute to ulcers, it is now clear that the root cause is *H. pylori.*

The presence of bacteria in the stomach mucosa was first discovered in 1892, but the finding was met with severe skepticism because it was believed that bacteria could not survive in the acidic stomach environment. In 1982, an Australian doctor, Barry Marshall, was finally able to culture an unusual bacterium, *H. pylori,* from stomach biopsies. To prove his belief that this bacterium could cause gastritis and ulcers, Marshall did something that no one should do at home. He drank a solution of *H. pylori* and subsequently developed gastric inflammation. Luckily, antibiotic treatment was able to cure him. In 2005, along with his colleague, Dr. Robin Warren, he received the Nobel Prize in Physiology or Medicine for his discovery.

Antibiotic treatment to eradicate *H. pylori* is the best therapy for ulcers. A combination of antibiotics and antacids cures 95% of gastric and 74% of duodenal ulcers within 2 months, with less than a 10% recurrence rate. In contrast, the previous conventional treatment with antacids yields only temporary relief, with about 90% recurrence rates within a year. Other treatments involve drugs that prevent histamine-stimulated acid secretion or that directly inhibit the proton pumps that secrete the acid. Such treatments are effective only for short-term relief, not for long-term treatment. These acid-blocking drugs, however, are effective agents in the treatment of **gastroesophageal reflux disease (GERD)** caused by acid reflux from the stomach. GERD is commonly known as heartburn.

Most bacteria cannot survive in the stomach. Hence, *H. pylori* is one of the most pervasive of human pathogens because it inhabits a niche without competition. It is estimated that well over half of the world's population is infected with *H. pylori.* The infection rate in the United States is about 1% per year of age—for example, 30% of all 30-year-olds are infected. In developing countries, nearly all people over age 25 are infected. This may contribute to the high rates of stomach cancer in some of those countries. Analyses of the *H. pylori* DNA sequences from various ethnic and geographic populations suggest that *H. pylori* infection has been present in humans for over 150,000 years, yet most people do not display the symptoms of *H. pylori* infection. Only about 15%–20% show gastric problems attributed to *H. pylori.* What triggers the bacterial infection to become symptomatic is a major unanswered question. It seems likely that both *H. pylori* infection and conditions that elevate acid secretion or damage the stomach wall, such as stress or the excessive ingestion of alcohol or aspirin, contribute to the development of an ulcer.

gastric phase, high acid levels in the stomach trigger a decrease in additional acid secretion. Second, during the gastrointestinal phase, acidic chyme entering the duodenum triggers a decrease in stomach acid secretion. These negative-feedback loops ensure that the acidic chyme entering the duodenum is neutralized, which is required for the digestion of food by pancreatic enzymes and for the prevention of peptic ulcer formation.

Movements of the Stomach

Stomach Filling

As food enters the stomach, the rugae flatten and the stomach volume increases. Despite the increase in volume, the pressure within the stomach does not increase until the volume nears maximum capacity because smooth muscle can stretch without an increase in tension (see chapter 9) and because of a reflex integrated within the medulla oblongata. This reflex inhibits muscle tone in the body of the stomach.

Mixing of Stomach Contents

Ingested food is thoroughly mixed with the secretions of the stomach glands to form chyme. This mixing is accomplished by gentle **mixing waves,** which are peristaltic-like contractions that occur about every 20 seconds, proceeding from the body of the stomach toward the pyloric sphincter to mix the ingested material with the stomach secretions. **Peristaltic waves** occur less frequently, are significantly more powerful than mixing waves, and force the chyme near the periphery of the stomach toward the pyloric sphincter. The more solid material near the center of the stomach is pushed superiorly toward the cardiac part for further digestion (figure 24.14). Roughly 80% of the contractions are mixing waves, and 20% are peristaltic waves. The back-and-forth movement of the chyme effectively mixes the ingested food with gastric juice.

Stomach Emptying

The amount of time food remains in the stomach depends on a number of factors, including the type and volume of food. Liquids exit the stomach within 1½–2½ hours after ingestion. After a typical meal, the stomach is usually empty within 3–4 hours. The pyloric sphincter usually remains partially closed because of mild tonic contraction. Each peristaltic contraction is sufficiently strong to force a small amount of chyme through the pyloric opening and into the duodenum. The peristaltic contractions responsible for the movement of chyme through the partially closed pyloric opening are called the **pyloric pump.**

1. A mixing wave initiated in the body of the stomach progresses toward the pyloric sphincter (*pink arrows directed inward*).

2. The more fluid part of the chyme is pushed toward the pyloric sphincter (*blue arrows*), whereas the more solid center of the chyme squeezes past the peristaltic constriction back toward the body of the stomach (*orange arrow*).

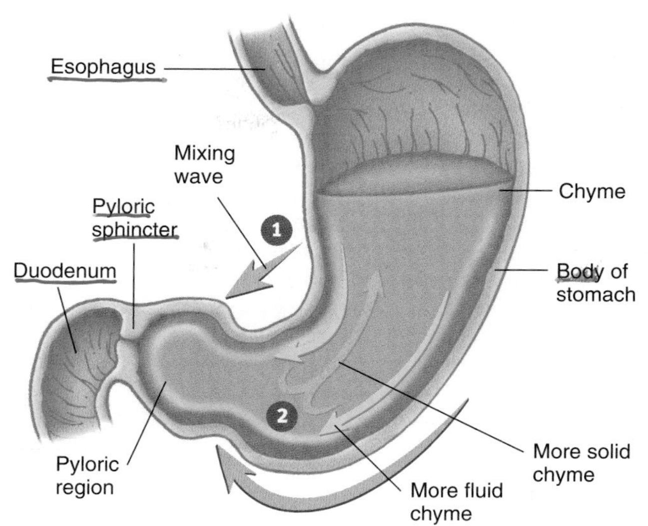

3. Peristaltic waves (*purple arrows*) move in the same direction and in the same way as the mixing waves but are stronger.

4. Again, the more fluid part of the chyme is pushed toward the pyloric region (*blue arrows*), whereas the more solid center of the chyme squeezes past the peristaltic constriction back toward the body of the stomach (*orange arrow*).

5. Peristaltic contractions force a few milliliters of the mostly fluid chyme through the pyloric opening into the duodenum (*small red arrows*). Most of the chyme, including the more solid portion, is forced back toward the body of the stomach for further mixing (*yellow arrow*).

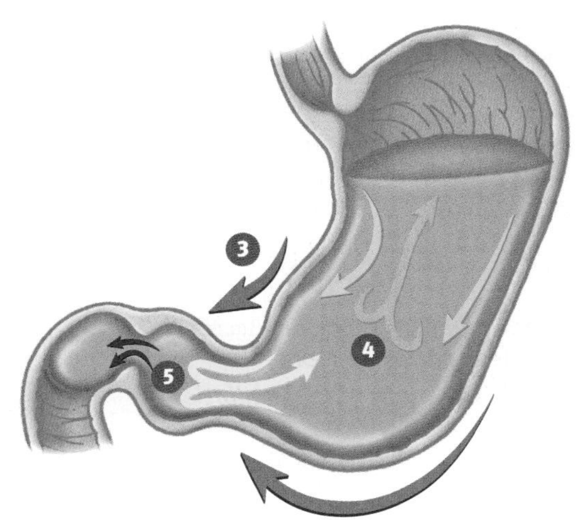

PROCESS FIGURE 24.14 Movements in the Stomach

Hunger Contractions

Hunger contractions are peristaltic contractions that approach tetanic contractions for periods of about 2–3 minutes. Low blood glucose levels cause the contractions to increase and become sufficiently strong to create uncomfortable sensations called hunger pangs. Hunger pangs usually begin 12–24 hours after a meal or in less time for some people. If nothing is ingested, they reach their maximum intensity within 3–4 days and then become progressively weaker.

Regulation of Stomach Emptying

If the stomach empties too fast, the efficiency of digestion and absorption is reduced, and acidic gastric contents dumped into the duodenum may damage its lining. If the rate of emptying is too slow, the highly acidic contents of the stomach may damage the stomach wall and reduce the rate at which nutrients are digested and absorbed. Stomach emptying is regulated to prevent these two extremes. The neural mechanisms that stimulate stomach secretions also are involved with increasing stomach motility. The major stimulus is distension of the stomach wall. Increased stomach motility increases stomach emptying. Conversely, the hormonal and neural mechanisms associated with the duodenum that decrease gastric secretions also inhibit gastric motility and increase constriction of the pyloric sphincter. The enterogastric reflex and release of the hormone cholecystokinin are major inhibitors of gastric motility. The result is a decrease in the rate of stomach emptying.

Vomiting

Vomiting can result from irritation (e.g., overdistension or overexcitation) anywhere along the GI tract. Action potentials travel through the vagus nerve and spinal visceral afferent nerves to the vomiting center in the medulla oblongata. Once the vomiting center is stimulated and the reflex is initiated, the following events occur: (1) A deep breath is taken; (2) the hyoid bone and larynx are elevated, opening the upper esophageal sphincter; (3) the opening of the larynx is closed; (4) the soft palate is elevated, closing the posterior nares; (5) the diaphragm and abdominal muscles are forcefully contracted, strongly compressing the stomach and increasing the intragastric pressure; (6) the lower esophageal sphincter is relaxed; and (7) the gastric contents are forced out of the stomach, through the esophagus and oral cavity, to the outside.

Pregame Meal

A meal of polysaccharide carbohydrates (starch and glycogen) is considered the best meal before engaging in a sporting activity. Polysaccharides help provide a steady source of glucose and have the fastest clearance time, typically 1 hour, from the stomach. For comparison, a meal containing both carbohydrates and proteins takes 3 hours to clear from the stomach, and a meal heavy with fats and proteins takes up to 6 hours. A major reason for the fast clearance of carbohydrates is that they do not increase cholecystokinin release, which is a major inhibitor of stomach emptying.

23 ***Describe the three phases of regulation of stomach secretion, and discuss the cause and result of each phase.***

24 ***How are gastric secretions inhibited? Why is this inhibition necessary?***

25 ***Why does pressure in the stomach not greatly increase as the stomach fills?***

26 ***What are two kinds of stomach movements? How are stomach movements regulated by hormones and nervous control?***

SMALL INTESTINE

The **small intestine** consists of three parts: the duodenum, the jejunum, and the ileum (figure 24.15). The entire small intestine is about 6 m long (range: 4.6–9 m). The duodenum is about 25 cm long (*duodenum* means 12, suggesting that it is 12 inches long). The jejunum, constituting about two-fifths of the total length of the small intestine, is about 2.5 m long; and the ileum, constituting three-fifths of the small intestine, is about 3.5 m long. Two major accessory glands, the liver and the pancreas, are associated with the duodenum.

The small intestine is where the greatest amount of digestion and absorption occur. Each day, about 9 L of water enters the digestive system. It comes from water that is ingested and from fluid secretions produced by glands along the length of the digestive tract. Most of the water, 8–8.5 L, moves by osmosis, with the absorbed solutes, out of the small intestine. A small part, 0.5–1 L, enters the colon.

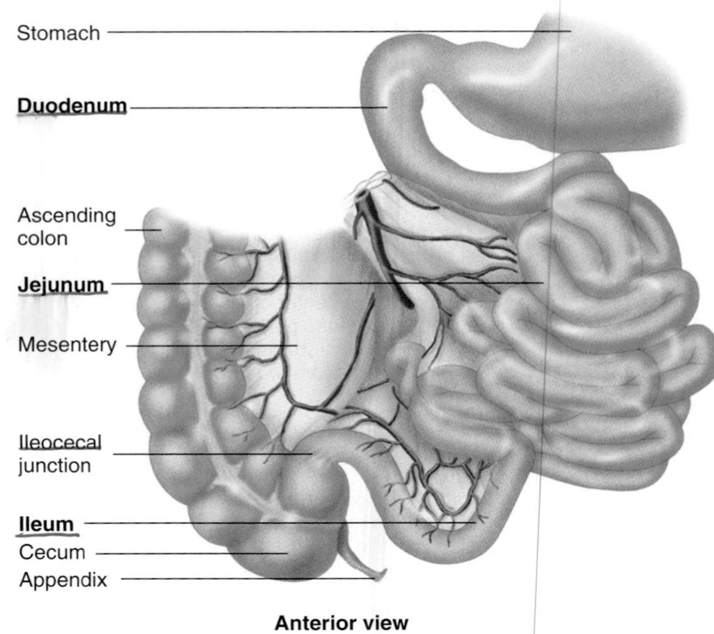

FIGURE 24.15 **Small Intestine**

Anatomy and Histology of the Small Intestine

Duodenum

The **duodenum** (doo-ō-dē′nŭm, doo-od′ĕ-nŭm) nearly completes a 180-degree arc as it curves within the abdominal cavity (figure 24.16), and the head of the pancreas lies within this arc. The duodenum begins with a short, superior part, which is where it exits the pylorus of the stomach, and ends in a sharp bend, which is where it joins the jejunum. Two small mounds are within the duodenum about two-thirds of the way down the descending part: the **major duodenal papilla** and the **minor duodenal papilla.** Ducts from the liver and/or pancreas open at these papillae (see p. 900 & p. 905).

The surface of the duodenum has several modifications that increase its surface area about 600-fold to allow for more efficient digestion and absorption of food. The mucosa and submucosa form a series of folds called the **circular folds,** or **plicae** (plī′sē; folds) **circulares** (see figure 24.16*a*), which run perpendicular to the long axis of the digestive tract. Tiny, fingerlike projections of the mucosa form numerous **villi** (vil′ī; shaggy hair), which are 0.5–1.5 mm in length (see figure 24.16*b*). Each villus is covered by simple columnar epithelium and contains a blood capillary network and a lymphatic capillary called a **lacteal** (lak′tē-ăl; see figure 24.16*c*). Most of the cells that make up the surface of the villi have numerous cytoplasmic extensions (about 1 μm long) called **microvilli,** which further increase the surface area (see figure 24.16*d*). The combined microvilli on the entire epithelial surface form the **brush border.** These various modifications greatly increase the surface area of the small intestine and, as a result, greatly enhance absorption.

The mucosa of the duodenum is simple columnar epithelium with four major cell types: (1) **Absorptive cells** are cells with microvilli, which produce digestive enzymes and absorb digested

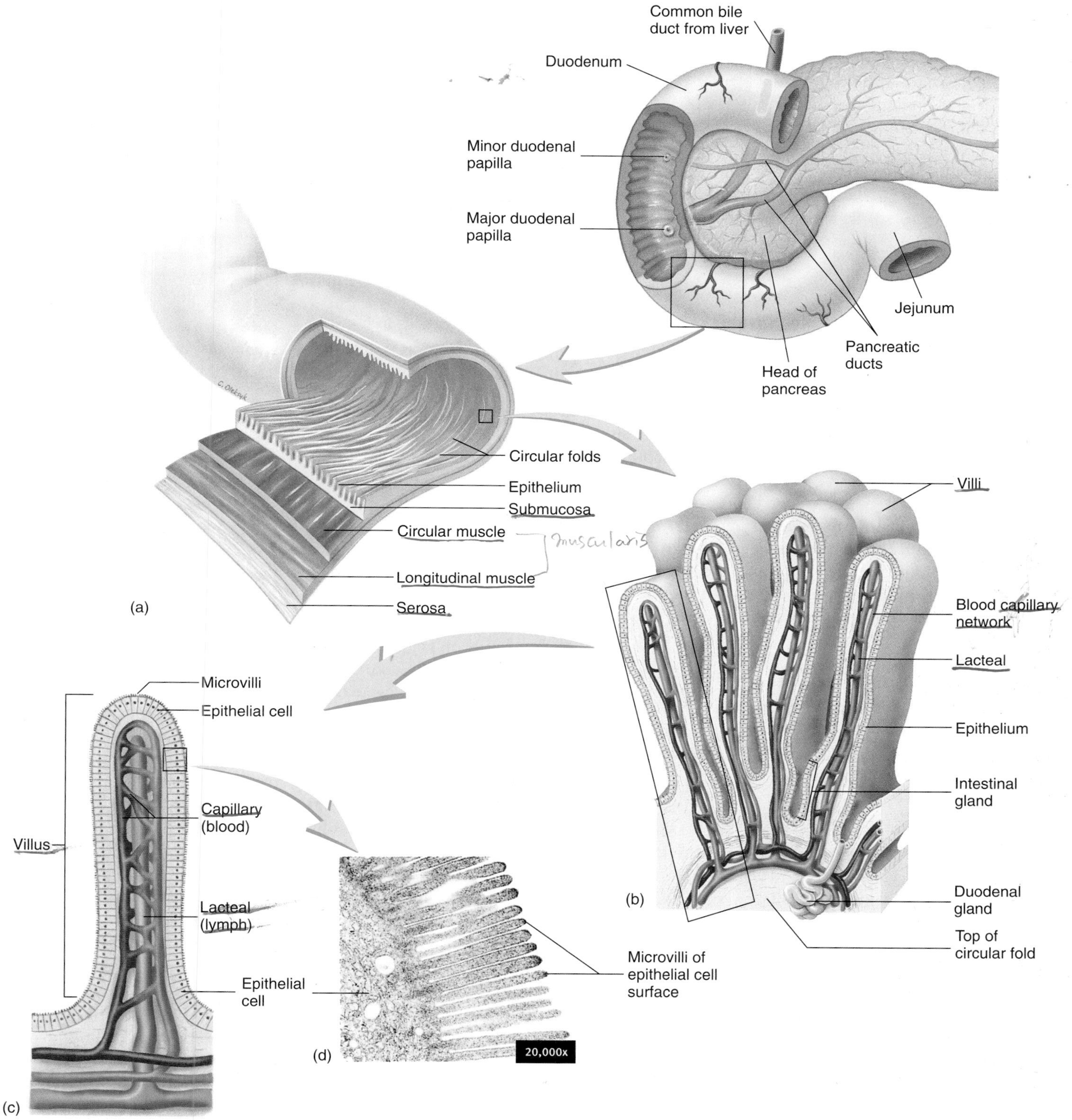

FIGURE 24.16 Anatomy and Histology of the Duodenum

(*a*) Wall of the duodenum, showing the circular folds. (*b*) The villi on a circular fold. (*c*) A single villus, showing the lacteal and capillary network. (*d*) Transmission electron micrograph of microvilli on the surface of a villus.

food; (2) **goblet cells** produce a protective mucus; (3) **granular cells** (Paneth cells) may help protect the intestinal epithelium from bacteria; and (4) **endocrine cells** produce regulatory hormones. The epithelial cells are produced within tubular invaginations of the mucosa, called **intestinal glands** (crypts of Lieberkühn), at the base of the villi. The absorptive and goblet cells migrate from the intestinal glands to cover the surface of the villi and eventually are shed from its tip. The granular and endocrine cells remain in the bottom of the glands. The submucosa of the duodenum contains coiled, tubular mucous glands called **duodenal glands** (Brunner glands), which open into the base of the intestinal glands.

Jejunum and Ileum

The **jejunum** (jĕ-joo′nŭm) and **ileum** (il′ē-ŭm) are similar in structure to the duodenum (see figure 24.15), except that a gradual decrease occurs in the diameter of the small intestine, the thickness of the intestinal wall, the number of circular folds, and the number of villi as one progresses through the small intestine. The duodenum and jejunum are the major sites of nutrient absorption, although some absorption occurs in the ileum. Lymphatic nodules called **Peyer's patches** are numerous in the mucosa and submucosa of the ileum. Peyer's patches and other mucosa-associated lymphoid tissue in the digestive tract initiate immune responses against microorganisms that enter the mucosa from injested food (see chapter 22).

The junction between the ileum and the large intestine is the **ileocecal junction.** It has a ring of smooth muscle, the **ileocecal sphincter,** and a one-way **ileocecal valve** (see figures 24.15 and 24.25).

27 ***Name and describe the three parts of the small intestine.***

28 ***What are the circular folds, villi, and microvilli in the small intestine? What are their functions?***

29 ***Name the four types of cells found in the duodenal mucosa, and state their functions.***

Secretions of the Small Intestine

The mucosa of the small intestine produces secretions that primarily contain mucus, electrolytes, and water. Intestinal secretions lubricate and protect the intestinal wall from the acidic chyme and the action of digestive enzymes. They also keep the chyme in the small intestine in a liquid form to facilitate the digestive process (see table 24.2). The intestinal mucosa produces most of the secretions that enter the small intestine, but the secretions of the liver and the pancreas also enter the small intestine and play essential roles in the process of digestion. Most of the digestive enzymes entering the small intestine are secreted by the pancreas. The intestinal mucosa also produces enzymes, but these remain associated with the intestinal epithelial surface.

The duodenal glands, intestinal glands, and goblet cells secrete large amounts of mucus. This mucus protects the wall of the intestine from the irritating effects of acidic chyme and from the digestive enzymes that enter the duodenum from the pancreas. Secretin and cholecystokinin are released from the intestinal mucosa and stimulate hepatic and pancreatic secretions (see figures 24.21 and 24.24).

The vagus nerve, secretin, and chemical or tactile irritation of the duodenal mucosa stimulate secretion from the duodenal glands. Goblet cells produce mucus in response to the tactile and chemical stimulation of the mucosa.

> **Duodenal Ulcer**
>
> Sympathetic nerve stimulation inhibits duodenal gland secretion, thus reducing the coating of mucus on the duodenal wall, which protects it against acid and gastric enzymes. If a person is highly stressed, elevated sympathetic activity may inhibit duodenal gland secretion and increase the person's susceptibility to **duodenal ulcers.**

Enzymes of the intestinal mucosa are bound to the membranes of the absorptive cell microvilli. These surface-bound enzymes include **disaccharidases,** which break disaccharides down to monosaccharides, and **peptidases,** which hydrolyze the peptide bonds between small amino acid chains (see table 24.2). Although these enzymes are not secreted into the intestine, they influence the digestive process significantly, and the large surface area of the intestinal epithelium brings these enzymes into contact with the intestinal contents. Small molecules, which are breakdown products of digestion, are absorbed through the microvilli and enter the circulatory or lymphatic systems.

Movement in the Small Intestine

Mixing and propulsion of chyme are the primary mechanical events that occur in the small intestine. These functions are the result of segmental or peristaltic contractions, which are accomplished by the smooth muscle in the wall of the small intestine and which are propagated only for short distances. Segmental contractions (see figure 24.3) mix the intestinal contents, and peristaltic contractions propel the intestinal contents along the digestive tract. A few peristaltic contractions may proceed the entire length of the intestine. Frequently, intestinal peristaltic contractions are continuations of peristaltic contractions that begin in the stomach. These contractions both mix and propel substances through the small intestine as the wave of contraction proceeds. The contractions move at a rate of about 1 cm/min. The movements are slightly faster at the proximal end of the small intestine and slightly slower at the distal end. It usually takes 3–5 hours for chyme to move from the pyloric region to the ileocecal junction.

Local mechanical and chemical stimuli are especially important in regulating the motility of the small intestine. Smooth muscle contraction increases in response to distension of the intestinal wall. Solutions that are either hypertonic or hypotonic, solutions with a low pH, and certain products of digestion, such as amino acids and peptides, also stimulate contractions of the small intestine. Local reflexes, which are integrated within the enteric plexus of the small intestine, mediate the response of the small intestine to these mechanical and chemical stimuli. Stimulation through parasympathetic nerve fibers may also increase the motility of the small intestine, but the parasympathetic influences in the small intestine are not as important as those in the stomach.

The ileocecal sphincter at the juncture between the ileum and the large intestine remains mildly contracted most of the time, but peristaltic waves reaching it from the small intestine cause it to relax and allow the movement of chyme from the small intestine into the cecum. Cecal distension, however, initiates a local reflex that causes more intense constriction of the ileocecal sphincter. Closure of the sphincter facilitates digestion and absorption in the small intestine by slowing the rate of chyme movement from the small intestine into the large intestine and prevents material from returning to the ileum from the cecum.

30 ***What are the functions of the intestinal glands and duodenal glands? State the factors that stimulate secretion from the duodenal glands and from goblet cells.***

31 ***List the enzymes of the small intestine wall and give their functions.***

32 ***What are two kinds of movement of the small intestine? How are they regulated?***

33 ***What is the function of the ileocecal sphincter?***

LIVER

Anatomy of the Liver

The **liver** is the largest internal organ of the body, weighing about 1.36 kg (3 pounds); it is in the right-upper quadrant of the abdomen, tucked against the inferior surface of the diaphragm (see figure 24.1; figure 24.17). The liver consists of two major **lobes, left** and **right,** and two minor lobes, **caudate** and **quadrate.**

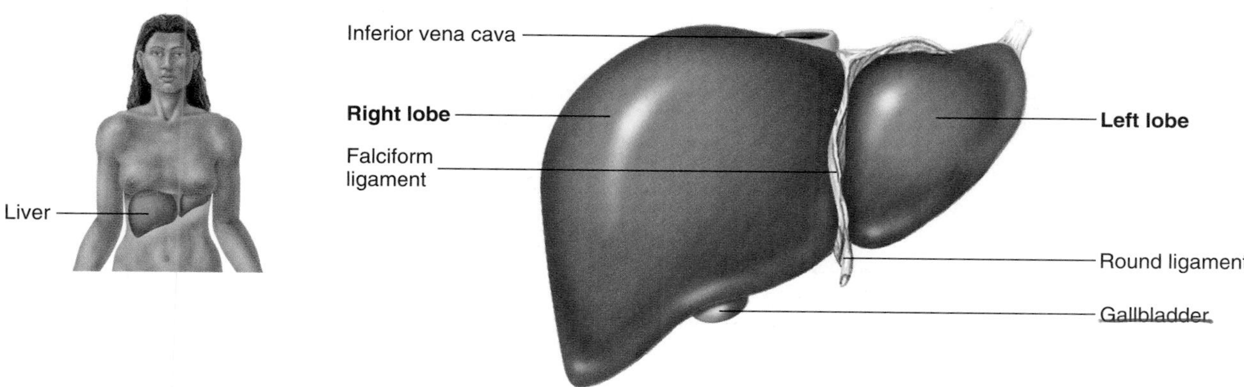

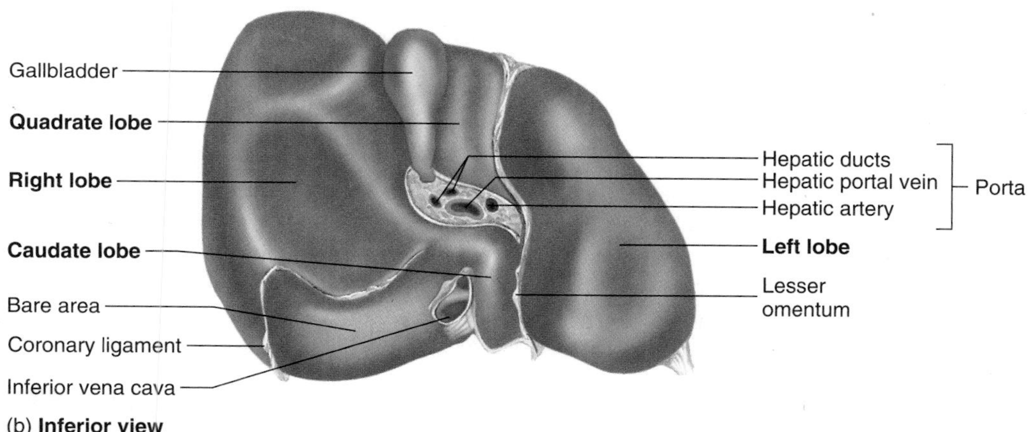

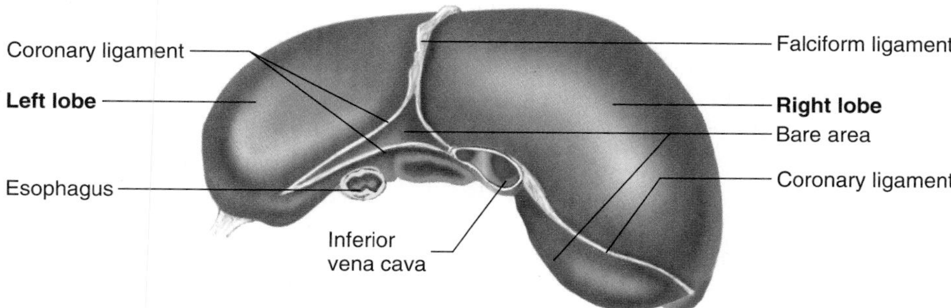

FIGURE 24.17 Liver

1. The hepatic ducts, which carry bile from the liver lobes, combine to form the common hepatic duct.
2. The common hepatic duct combines with the cystic duct from the gallbladder to form the common bile duct.
3. The common bile duct and the pancreatic duct combine to form the hepatopancreatic ampulla.
4. The hepatopancreatic ampulla empties bile and pancreatic secretions into the duodenum at the major duodenal papilla.
5. The accessory pancreatic duct empties pancreatic secretions into the duodenum at the minor duodenal papilla.

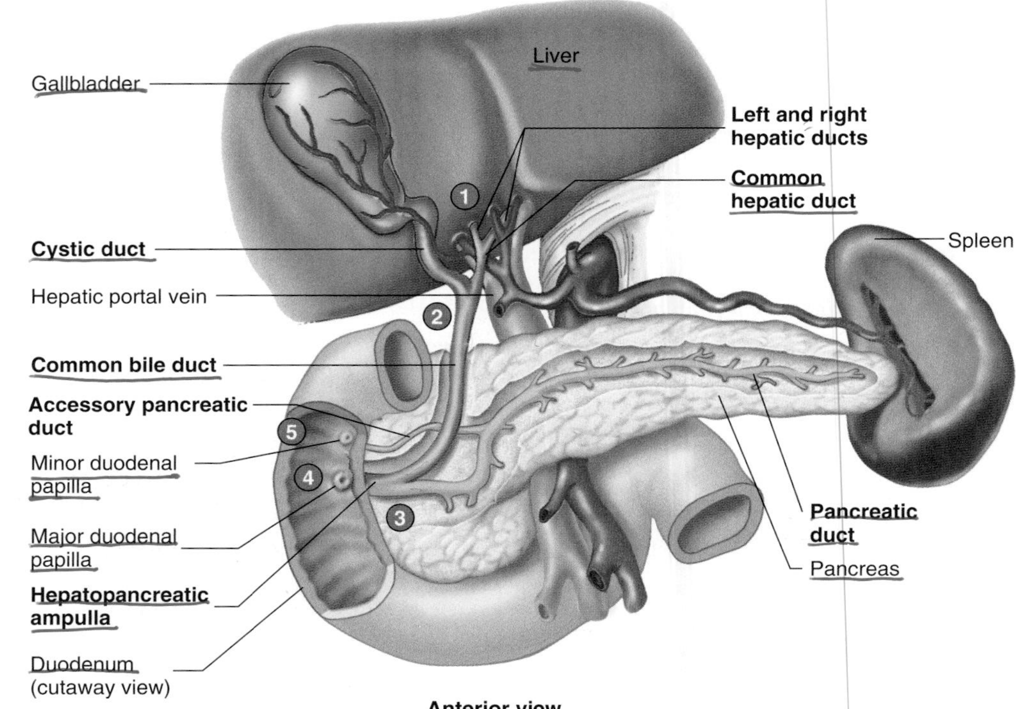

PROCESS FIGURE 24.18 **Liver, Gallbladder, Pancreas, and Duct System**

A **porta** (gate) is on the inferior surface of the liver, where the various vessels, ducts, and nerves enter and exit the liver. The **hepatic** (he-pat′ik; associated with the liver) **portal vein,** the **hepatic artery,** and a small hepatic nerve plexus enter the liver through the porta (see figure 24.17*b*). Lymphatic vessels and two hepatic ducts, one each from the right and left lobes, exit the liver at the porta. The hepatic ducts transport bile out of the liver. The right and left hepatic ducts unite to form a single **common hepatic duct** (figure 24.18). The **cystic duct** from the gallbladder joins the common hepatic duct to form the **common bile duct,** which joins the pancreatic duct at the **hepatopancreatic ampulla** (hĕ-pat′ō-pan-crē-at′ik am-pul′lă), an enlargement where the hepatic and pancreatic ducts come together. The hepatopancreatic ampulla empties into the duodenum at the major duodenal papilla (see figure 24.18). A smooth muscle sphincter surrounds the common bile duct where it enters the hepatopancreatic ampulla. The gallbladder is a small sac on the inferior surface of the liver that stores bile. Bile can flow from the gallbladder through the cystic duct into the common bile duct, or it can flow back up the cystic duct into the gallbladder.

Histology of the Liver

A connective tissue capsule and visceral peritoneum cover the liver, except for the **bare area,** which is a small area on the diaphragmatic surface that lacks a visceral peritoneum and is surrounded by the coronary ligament (see figure 24.17*c*). At the porta, the connective tissue capsule sends a branching network of septa (walls) into the substance of the liver to provide its main support. Vessels, nerves, and ducts follow the connective tissue branches throughout the liver.

The connective tissue septa divide the liver into hexagon-shaped **lobules** with a **portal triad** at each corner (figure 24.19). The triads are so-named because three vessels—the hepatic portal vein, hepatic artery, and hepatic duct—are located in them. Hepatic nerves and lymphatic vessels, often too small to be seen easily in light micrographs, are also located in these areas. A **central vein** is in the center of each lobule. Central veins unite to form **hepatic veins,** which exit the liver on its posterior and superior surfaces and empty into the inferior vena cava (see figure 24.19).

Hepatic cords radiate out from the central vein of each lobule like the spokes of a wheel. The hepatic cords are composed of **hepatocytes,** the functional cells of the liver. The spaces between the hepatic cords are blood channels called **hepatic sinusoids.** The sinusoids are lined with a very thin, irregular squamous endothelium consisting of two cell populations: (1) extremely thin, sparse **endothelial cells** and (2) **hepatic phagocytic cells** (Kupffer cells). A cleftlike lumen, the **bile canaliculus** (kan-ă-lik′ū-lŭs; little canal), lies between the cells within each cord (see figure 24.19).

Hepatocytes have six major functions: (1) bile production, (2) storage, (3) interconversion of nutrients, (4) detoxification, (5) phagocytosis, and (6) synthesis of blood components. Nutrient-rich, oxygen-poor blood from the viscera enters the hepatic sinusoids from branches of the hepatic portal vein and mixes with oxygen-rich, nutrient-depleted blood from the

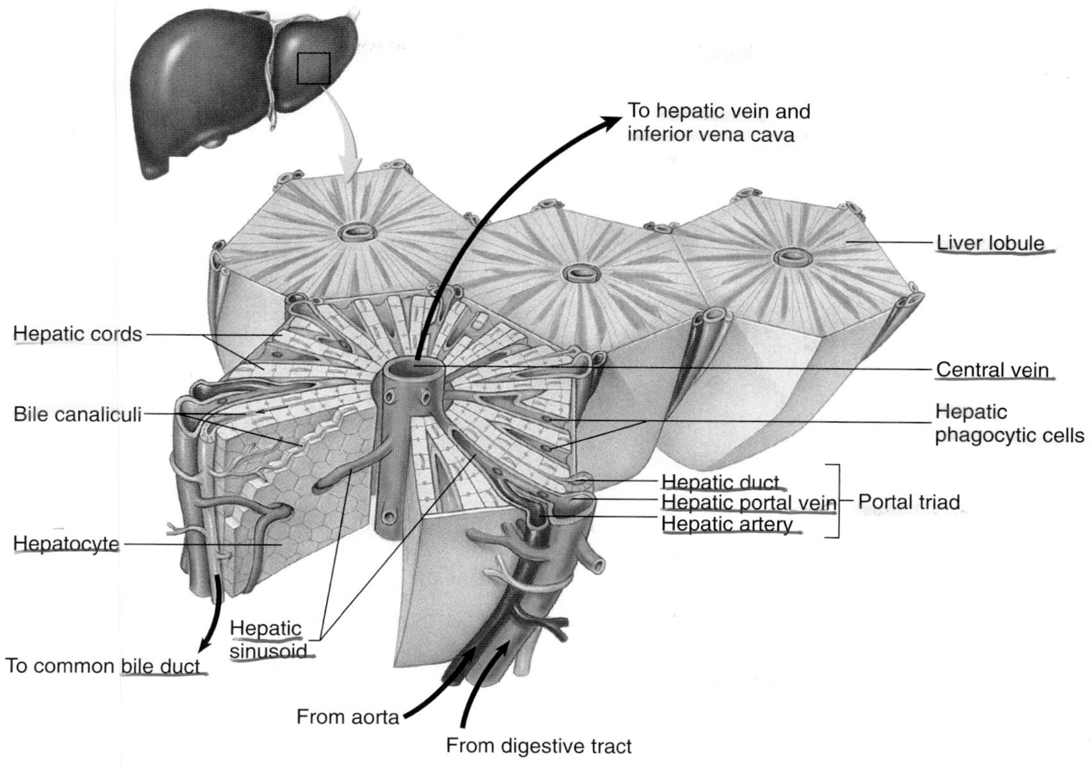

FIGURE 24.19 Histology of the Liver

hepatic arteries (figure 24.20). From the blood, the hepatocytes can take up the oxygen and nutrients, which are stored, detoxified, used for energy, or used to synthesize new molecules. Molecules produced by or modified in the hepatocytes are released into the hepatic sinusoids or into the bile canaliculi.

Mixed blood in the hepatic sinusoids flows to the central vein, where it exits the lobule and then exits the liver through the hepatic veins. **Bile,** produced by the hepatocytes and consisting primarily of metabolic by-products, flows through the bile canaliculi toward the hepatic triad and exits the liver through the hepatic ducts. Blood, therefore, flows from the triad toward the center of each lobule, whereas bile flows away from the center of the lobule toward the triad.

In the fetus, special blood vessels bypass the liver sinusoids. The remnants of fetal blood vessels can be seen in the adult as the round ligament (ligamentum teres) and the ligamentum venosum (see chapter 29).

Liver Rupture or Enlargement

The liver is easily ruptured because it is large, fixed in position, and fragile, or it can be lacerated by a broken rib. Liver rupture or laceration results in severe internal bleeding.

The liver may become enlarged as a result of heart failure, hepatic cancer, cirrhosis, or Hodgkin disease (a lymphatic cancer). However, as long as part of the liver contains normal tissue, the diseased portions can be removed and the normal tissue may restore the complete liver by regeneration. Likewise, an entire liver or a portion of a liver can be transplanted from one person to another.

34 ***Describe the lobes of the liver. What is the porta?***

35 ***Diagram the duct system from the liver, gallbladder, and pancreas that empties into the major duodenal papilla.***

36 ***What are the hepatic cords and the sinusoids?***

37 ***Describe the flow of blood to and through the liver. Describe the flow of bile away from the liver.***

Functions of the Liver

The liver performs important digestive and excretory functions, stores and processes nutrients, synthesizes new molecules, and detoxifies harmful chemicals.

Bile Production

The liver produces and secretes about 600–1000 mL of bile each day (see table 24.2). Bile contains no digestive enzymes, but it plays a role in digestion because it neutralizes and dilutes stomach acid and emulsifies fats. The pH of chyme as it leaves the stomach is too low for the normal function of pancreatic enzymes. Bile helps neutralize the acidic chyme and bring the pH up to a level at which pancreatic enzymes can function. **Bile salts** emulsify fats (see p. 913). Bile also contains excretory products, such as bile pigments; bilirubin is a bile pigment that results from the breakdown of hemoglobin. Bile also contains cholesterol, fats, fat-soluble hormones, and lecithin.

1. The hepatic artery carries oxygen-rich blood from the aorta through the porta of the liver. Hepatic artery branches become part of the portal triads. Blood from the hepatic artery branches enters the hepatic sinusoids and supplies hepatocytes in the hepatic cords with oxygen.
2. The hepatic portal vein carries nutrient-rich, oxygen-poor blood from the intestines through the porta of the liver. Hepatic portal vein branches become part of the portal triads. Blood from the hepatic portal vein branches enters the hepatic sinusoids and supplies hepatocytes in the hepatic cords with nutrients.
3. Blood in the hepatic sinusoids that comes from the hepatic artery and hepatic portal vein picks up plasma proteins, processed molecules, and waste products produced by the hepatocytes of the hepatic cords. The hepatic sinusoids empty into central veins. The central veins connect to hepatic veins, which connect to the inferior vena cava.
4. Bile produced by hepatocytes in the hepatic cords enters bile canaliculi, which connect to hepatic duct branches that are part of the portal triads.
5. The hepatic duct branches converge to form the left and right hepatic ducts, which carry bile out the porta of the liver.

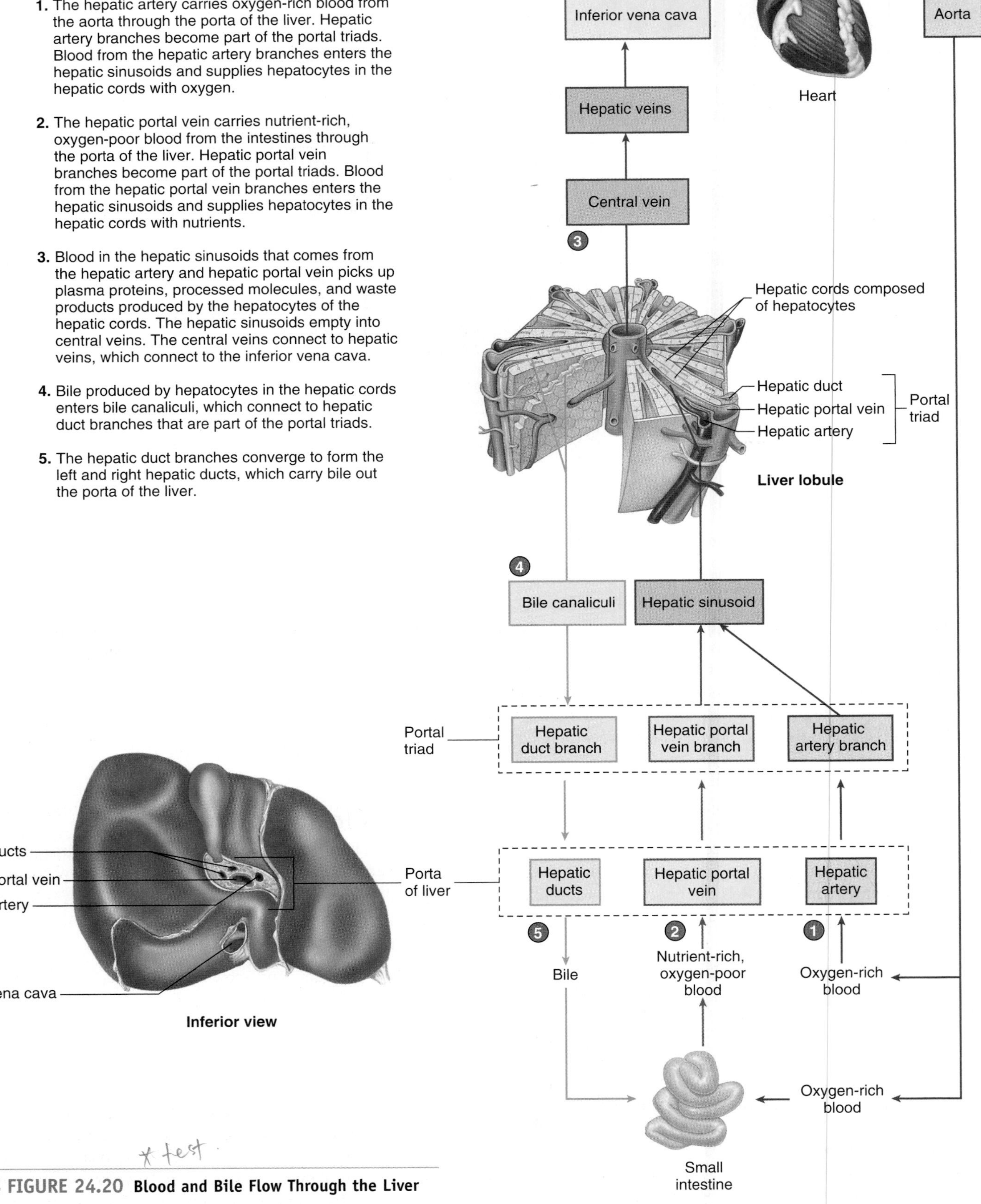

PROCESS FIGURE 24.20 Blood and Bile Flow Through the Liver

1. Secretin, produced by the duodenum (*purple arrows*) and carried through the circulation to the liver, stimulates bicarbonate secretion into bile (*green arrows inside the liver*).
2. Cholecystokinin, produced by the duodenum (*pink arrows*) and carried through the circulation to the gallbladder, stimulates the gallbladder to contract, thereby releasing bile into the duodenum (*green arrow outside the liver*).
3. Vagus nerve stimulation (*red arrow*) causes the gallbladder to contract, thereby releasing bile into the duodenum.
4. Bile salts stimulate bile secretion. Over 90% of bile salts are reabsorbed in the ileum and returned to the liver (*blue arrows*), where they stimulate additional secretion of bile salts.

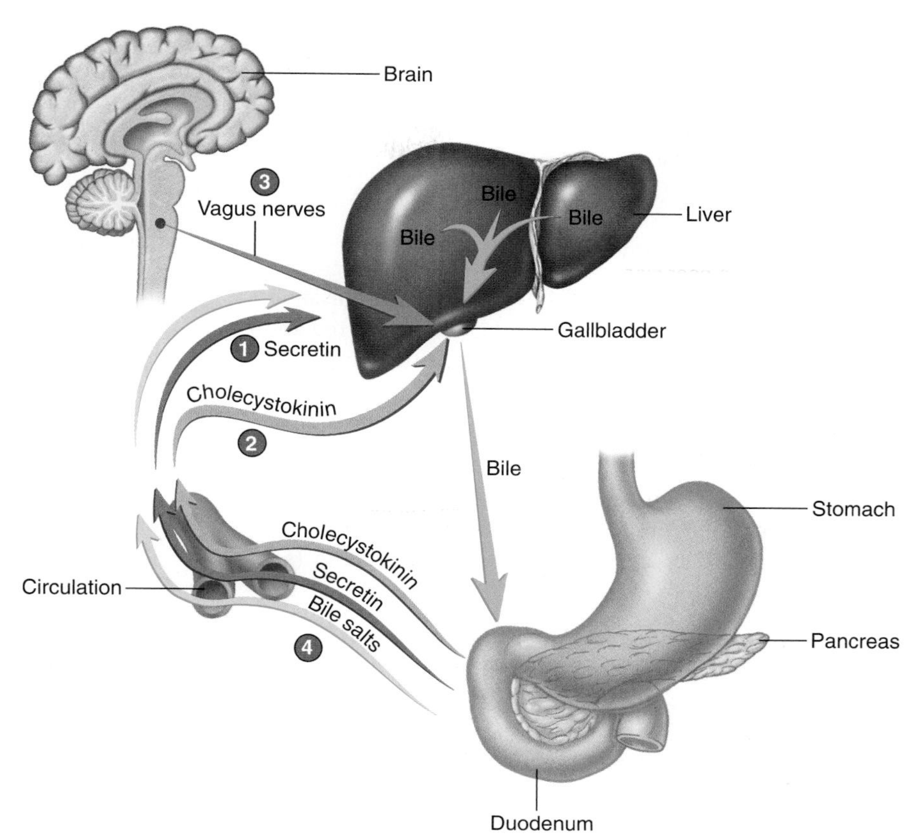

PROCESS FIGURE 24.21 Control of Bile Secretion and Release

Secretin stimulates bile secretion, primarily by increasing the water and bicarbonate ion content of bile (figure 24.21). Bile salts also increase bile secretion through a positive-feedback system. Over 90% of bile salts are reabsorbed in the ileum and carried in the blood back to the liver, where they contribute to further bile secretion. The loss of bile salts in the feces is reduced by this recycling process. Bile secretion into the duodenum continues until the duodenum empties.

Storage

Hepatocytes can remove sugar from the blood and store it in the form of **glycogen.** They can also store fat, vitamins (A, B_{12}, D, E, and K), copper, and iron. This storage function is usually short-term, and the amount of stored material in the hepatocytes and the cell size fluctuate during a given day.

Hepatocytes help control blood sugar levels within very narrow limits. If a large amount of sugar enters the general circulation after a meal, it will increase the osmolality of the blood and produce hyperglycemia. This is prevented because the blood from the intestine passes through the hepatic portal vein to the liver, where glucose and other substances are removed from the blood by hepatocytes, stored, and secreted back into the circulation when needed.

Nutrient Interconversion

The interconversion of nutrients is another important function of the liver. Ingested nutrients are not always in the proportion needed by the tissues. If this is the case, the liver can convert some nutrients into others. If, for example, a person is on a diet that is excessively high in protein, an oversupply of amino acids and an undersupply of lipids and carbohydrates may be delivered to the liver. The hepatocytes break down the amino acids and cycle many of them through metabolic pathways so they can be used to produce adenosine triphosphate, lipids, and glucose (see chapter 25).

Hepatocytes also transform substances that cannot be used by most cells into more readily usable substances. For example, ingested fats are combined with choline and phosphorus in the liver to produce phospholipids, which are essential components of plasma membranes. Vitamin D is hydroxylated in the liver. The hydroxylated form of vitamin D is the major circulating form of vitamin D, which is transported through the circulation to the kidney, where it is again hydroxylated. The double-hydroxylated vitamin D is the active form of the vitamin, which functions in calcium maintenance.

Detoxification

Many ingested substances are harmful to the cells of the body. In addition, the body itself produces many by-products of metabolism that, if accumulated, are toxic. The liver forms a major line of defense against many of these harmful substances. It detoxifies many substances by altering their structure to make them less toxic or to make their elimination easier. Ammonia, for example, a by-product of amino acid metabolism, is toxic and is not readily removed from the circulation by the kidneys. Hepatocytes remove ammonia from the circulation and convert it to urea, which is less toxic than ammonia, is secreted into the circulation, and is then eliminated by the kidneys in the urine. Other substances are removed from the circulation and excreted by the hepatocytes into the bile.

Phagocytosis

Hepatic phagocytic cells (Kupffer cells), which lie along the sinusoid walls of the liver, phagocytize "worn-out" and dying red and white blood cells, some bacteria, and other debris that enters the liver through the circulation.

Synthesis

The liver can produce its own new compounds. It produces many blood proteins, such as albumins, fibrinogen, globulins, heparin, and clotting factors, which are released into the circulation (see chapter 19).

Hepatitis, Cirrhosis, and Liver Damage

Strictly defined, **hepatitis** is an inflammation of the liver and does not necessarily result from an infection. Hepatitis can be caused by alcohol consumption or infection. Infectious hepatitis is caused by viral infections. Hepatitis A, also called infectious hepatitis, is responsible for about 30% of all hepatitis cases in the United States. Hepatitis B, also called serum hepatitis, is a more chronic infection and is responsible for half the hepatitis cases in the United States. Hepatitis C, also called non-A and non-B hepatitis, causes 20% of the U.S. hepatitis cases. It is caused by one or more virus types that cannot be identified in blood tests and it is spread by blood transfusions or sexual intercourse. If hepatitis is not treated, liver cells die and are replaced by scar tissue, resulting in loss of liver function. Death caused by liver failure can occur.

Cirrhosis (sir-rō′sis) of the liver involves the death of hepatocytes and their replacement by fibrous connective tissue. The liver becomes pale in color (the term *cirrhosis* means a tawny or orange condition) because of the presence of excess white connective tissue. It also becomes firmer, and the surface becomes nodular. The loss of hepatocytes eliminates the function of the liver, often resulting in jaundice. The buildup of connective tissue can impede blood flow through the liver. Cirrhosis frequently develops in alcoholics and may develop as a result of biliary obstruction, hepatitis, or nutritional deficiencies.

Under most conditions, mature hepatocytes can proliferate and replace lost parts of the liver. If the liver is severely damaged, however, the hepatocytes may not have enough regenerative power to replace the lost parts. In this case, a liver transplant may be necessary. The liver also maintains an undifferentiated stem cell population, called "oval" cells, which gives rise to two cell lines, one forming bile duct epithelium and the other producing hepatocytes. It is hoped that these stem cells can be used to reconstitute a severely damaged liver. It may even be possible in the future to remove stem cells from a person with hemophilia, genetically engineer the cells to produce the missing clotting factors, and then reintroduce the altered stem cells into the person's liver.

38 ***Explain and give examples of the major functions of the liver.***

39 ***What stimulates bile secretion from the liver?***

GALLBLADDER

The **gallbladder** is a saclike structure on the inferior surface of the liver that is about 8 cm long and 4 cm wide (see figure 24.18). Three tunics form the gallbladder wall: (1) an inner mucosa folded into rugae that allow the gallbladder to expand; (2) a muscularis, which is a layer of smooth muscle that allows the gallbladder to contract; and (3) an outer covering of serosa. The cystic duct connects the gallbladder to the common bile duct.

Bile is continually secreted by the liver and flows to the gallbladder, where 40–70 mL of bile can be stored. While the bile is in the gallbladder, water and electrolytes are absorbed, and bile salts and pigments become as much as 5–10 times more concentrated than they were when secreted by the liver. Shortly after a meal, the gallbladder contracts in response to stimulation by cholecystokinin and, to a lesser degree, in response to vagal stimulation, thereby dumping large amounts of concentrated bile into the small intestine (see figure 24.21).

Gallstones

Cholesterol, secreted by the liver, may precipitate in the gallbladder to produce **gallstones** (figure A). Cholesterol is not soluble in water and is ordinarily kept in solution by bile salts. Gallstones can form when there is excess cholesterol in the bile due to a high-cholesterol diet or when cholesterol is overly concentrated in the gallbladder. Occasionally, a gallstone passes out of the gallbladder and enters the cystic duct, blocking the release of bile. Such a condition interferes with normal digestion, and the gallstone often must be removed surgically. If the gallstone moves far enough down the duct, it can also block the pancreatic duct, resulting in pancreatitis.

FIGURE A Gallstones

40 ***Describe the three tunics of the gallbladder wall.***

41 ***What is the function of the gallbladder? What stimulates the release of bile from the gallbladder?***

PANCREAS

Anatomy of the Pancreas

The **pancreas** is a complex organ composed of both endocrine and exocrine tissues that perform several functions. The pancreas consists of a **head,** located within the curvature of the duodenum (figure 24.22*a*), a **body,** and a **tail,** which extends to the spleen.

The endocrine part of the pancreas consists of **pancreatic islets** (islets of Langerhans; figure 24.22*b*). The islet cells produce insulin and glucagon, which are very important in controlling the blood levels of nutrients, such as glucose and amino acids, and somatostatin, which regulates insulin and glucagon secretion and may inhibit growth hormone secretion (see chapter 18).

The exocrine part of the pancreas is a compound acinar gland (see chapter 4). The **acini** (as′i-nī; grapes; see figure 24.22*b*) produce digestive enzymes. Clusters of acini form lobules that are separated by thin septa. Lobules are connected by small **intercalated ducts** to **intralobular ducts,** which leave the lobules to join **interlobular ducts** between the lobules. The interlobular ducts attach to the main **pancreatic duct,** which joins the common bile duct at the hepatopancreatic ampulla (Vater's ampulla; see figures 24.18 and 24.22). The hepatopancreatic ampulla empties into the duodenum at the major duodenal papilla. A smooth muscle sphincter, the **hepatopancreatic ampullar sphincter** (sphincter of Oddi) regulates the opening of the ampulla. An accessory pancreatic duct, present in most people, opens at the minor duodenal papilla. The ducts are lined with simple cuboidal epithelium, and the epithelial cells of the acini are pyramid-shaped. A smooth muscle sphincter surrounds the pancreatic duct where it enters the hepatopancreatic ampulla.

42 ***Describe the parts of the pancreas responsible for endocrine and exocrine secretions. Diagram the duct system of the pancreas.***

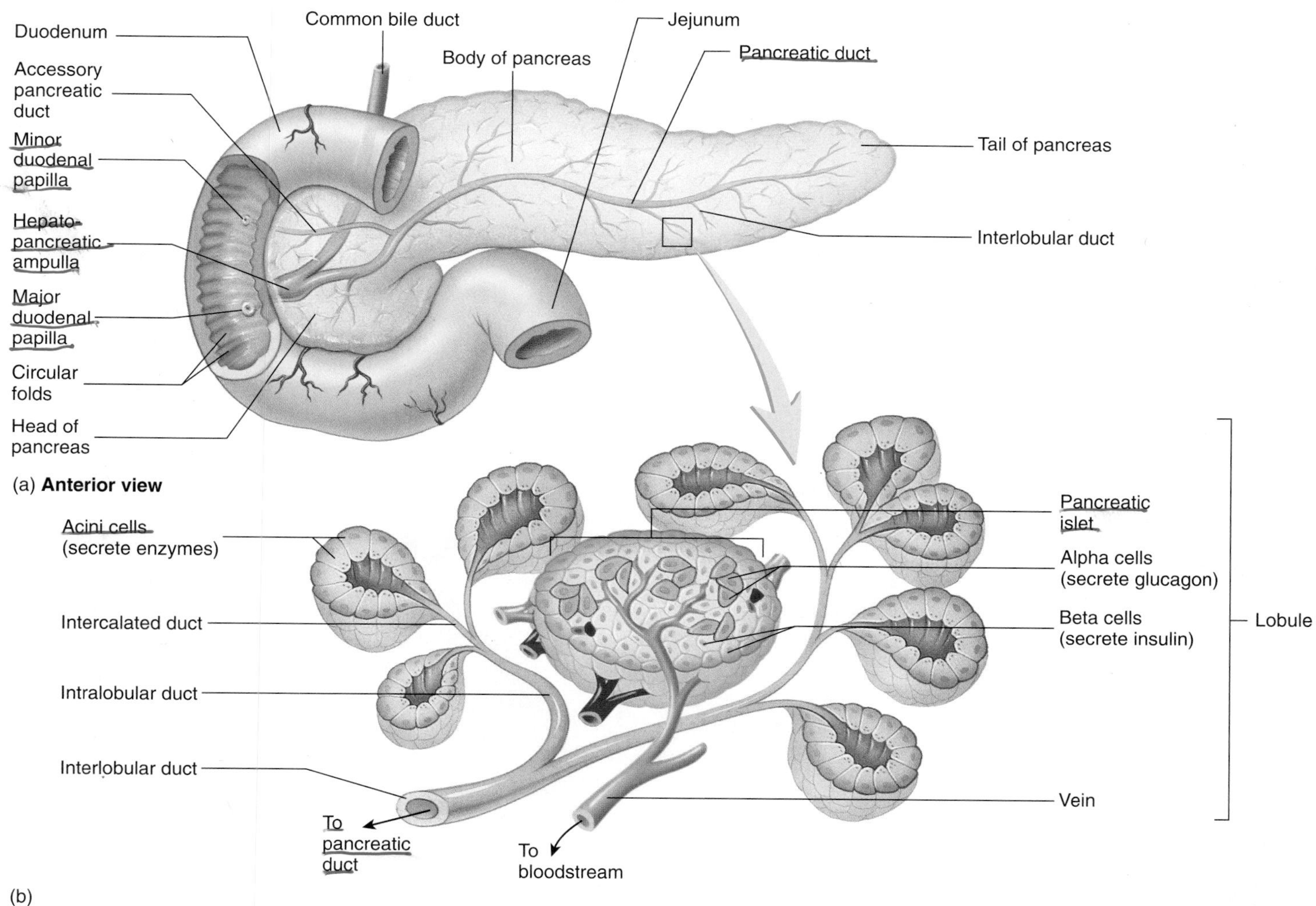

FIGURE 24.22 Anatomy and Histology of the Duodenum and Pancreas

(*a*) The head of the pancreas lies within the duodenal curvature, with the pancreatic duct emptying into the duodenum. (*b*) Histology of the pancreas, showing both the acini and the pancreatic duct system.

Pancreatic Secretions

The exocrine secretions of the pancreas are called **pancreatic juice** and have two major components: an aqueous component and an enzymatic component. Pancreatic juice is produced in the pancreas and is then delivered through the pancreatic ducts to the small intestine, where it functions in digestion. The **aqueous component** is produced principally by columnar epithelial cells that line the smaller ducts of the pancreas. It contains Na^+ and K^+ in about the same concentration found in extracellular fluid. Bicarbonate ions are a major part of the aqueous component, and they neutralize the acidic chyme that enters the small intestine from the stomach. The increased pH caused by pancreatic secretions in the duodenum stops pepsin digestion but provides the proper environment for the function of pancreatic enzymes. Bicarbonate ions are actively secreted by the duct epithelium, and water follows passively to make the pancreatic juice isotonic. The cellular mechanism responsible for the secretion of HCO_3^- is diagrammed in figure 24.23.

The enzymes of the pancreatic juice are produced by the acinar cells of the pancreas. They are important for the digestion of all major classes of food. Without the enzymes produced by the pancreas, lipids, proteins, and carbohydrates are not adequately digested (see tables 24.1 and 24.2).

The proteolytic pancreatic enzymes, which digest proteins, are secreted in inactive forms, whereas many of the other enzymes are secreted in active form. The major proteolytic enzymes are **trypsin, chymotrypsin,** and **carboxypeptidase.** They are secreted in their inactive forms as trypsinogen, chymotrypsinogen, and procarboxypeptidase and are activated by the removal of certain peptides from the larger precursor proteins. If these were produced in their active forms, they would digest the tissues producing them. The proteolytic enzyme **enterokinase** (en′tĕr-ō-kī′nās; intestinal enzyme), which is an enzyme attached to the brush border of the small intestine, activates trypsinogen. Trypsin then activates more trypsinogen, as well as chymotrypsinogen and procarboxypeptidase.

Pancreatic juice also contains **pancreatic amylase,** which continues the polysaccharide digestion initiated in the oral cavity. In addition, pancreatic juice contains a group of lipid-digesting enzymes called **pancreatic lipases,** which break down lipids into monoglycerides, free fatty acids, cholesterol, and other components.

Enzymes that reduce DNA and RNA to their component nucleotides, **deoxyribonucleases** and **ribonucleases,** respectively, are also present in pancreatic juice.

1. Water (H_2O) and carbon dioxide (CO_2) combine under the influence of carbonic anhydrase (CA) to form carbonic acid (H_2CO_3).
2. Carbonic acid dissociates to form hydrogen ions (H^+) and bicarbonate ions (HCO_3^-).
3. The H^+ are exchanged for sodium ions (Na^+) by an antiporter. Sodium ions are removed by the Na^+–K^+ pump.
4. The HCO_3^- are transported into the intercalated ducts in exchange for Cl^-, which return to the lumen by a channel. Sodium ions and H_2O follow the HCO_3^- into the ducts.
5. The ions and H_2O move through the intercalated duct toward the interlobular duct.

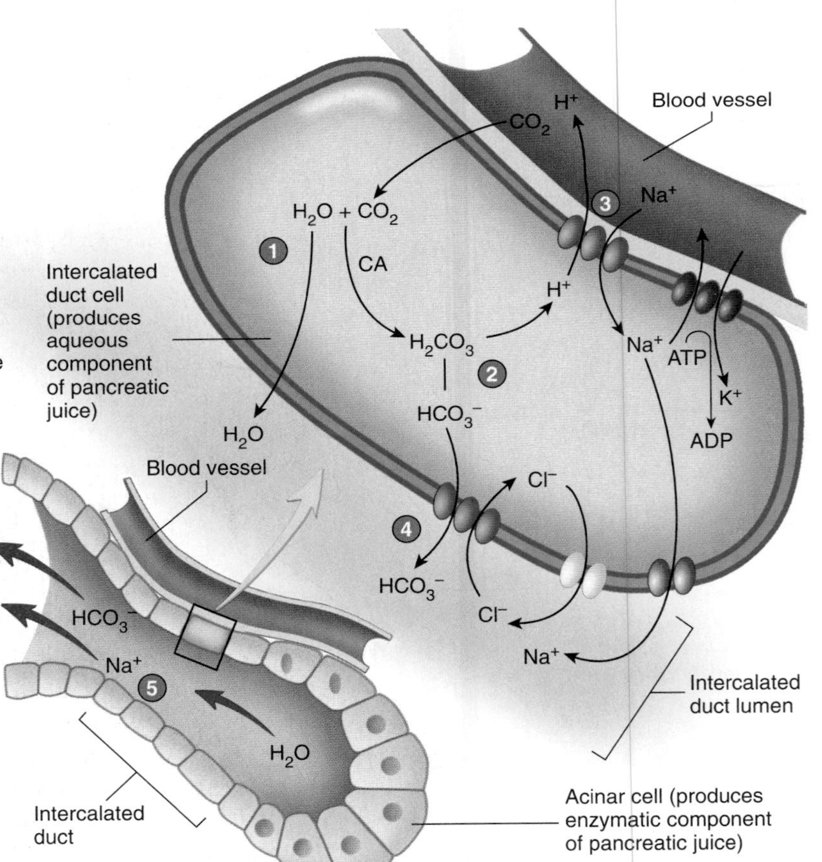

PROCESS FIGURE 24.23 Bicarbonate Ion Production in the Pancreas

1. Secretin (*purple arrows*) released from the duodenum stimulates the pancreas to release a watery secretion rich in HCO_3^-.
2. Cholecystokinin (*pink arrows*) released from the duodenum causes the pancreas to release a secretion rich in digestive enzymes.
3. Parasympathetic stimulation from the vagus nerve (*red arrow*) causes the pancreas to release a secretion rich in digestive enzymes.

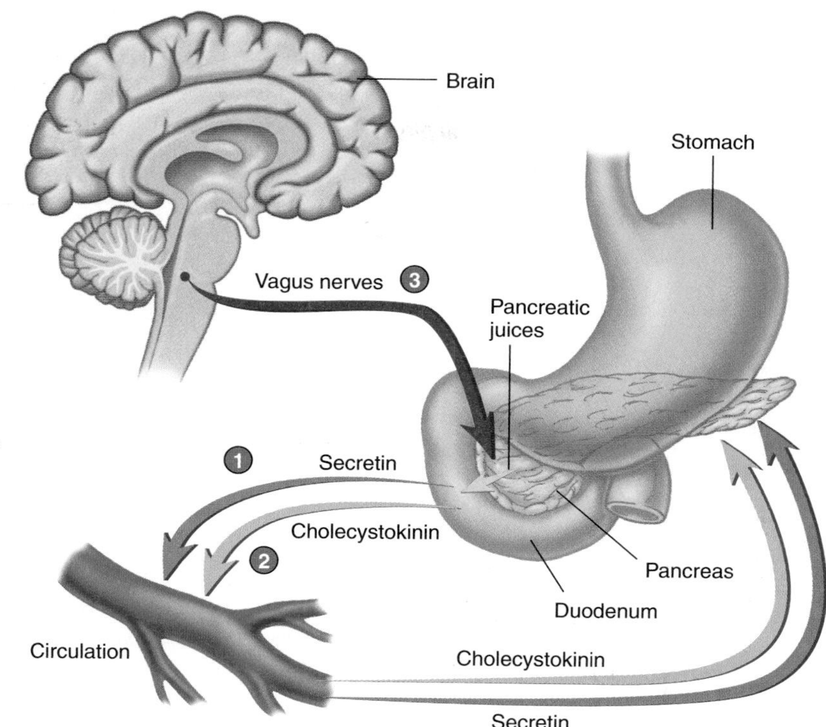

PROCESS FIGURE 24.24 **Control of Pancreatic Secretion**

> **Pancreatitis and Pancreatic Cancer**
>
> **Pancreatitis** is an inflammation of the pancreas that occurs quite commonly, Pancreatitis involves the release of pancreatic enzymes within the pancreas itself, which digest pancreatic tissue. It can result from alcoholism, the use of certain drugs, pancreatic duct blockage, cystic fibrosis, viral infection, or pancreatic cancer. Symptoms can range from mild abdominal pain to systemic shock and coma.
>
> Cancer of the pancreas can obstruct the pancreatic and common hepatic ducts, resulting in jaundice. Pancreatic cancer may not be detected until the tumor has become fairly large, and it can become so large as to block off the pyloric region of the stomach.

Regulation of Pancreatic Secretion

Both hormonal and neural mechanisms (figure 24.24) control the exocrine secretions of the pancreas. An acidic chyme in the duodenum stimulates the release of secretin. Secretin stimulates the secretion of a watery solution that contains a large amount of bicarbonate ions from the pancreas into the duodenum. Bicarbonate ions increase the pH of chyme in the duodenum so that the duodenum is not damaged by the low pH. In addition, pancreatic and brush border enzymes do not function at a low pH.

PREDICT 6

Explain why secretin production in response to acidic chyme and its stimulation of bicarbonate ion secretion constitute a negative-feedback mechanism.

Cholecystokinin stimulates the release of bile from the gallbladder and the secretion of pancreatic juice rich in digestive enzymes. The major stimulus for the release of cholecystokinin is the presence of fatty acids and other lipids in the duodenum.

Parasympathetic stimulation through the vagus (X) nerves also stimulates the secretion of pancreatic juices rich in pancreatic enzymes, and sympathetic impulses inhibit secretion. The effect of vagal stimulation on pancreatic juice secretion is greatest during the cephalic and gastric phases of stomach secretion.

43 ***Name the two kinds of exocrine secretions produced by the pancreas. What stimulates their production and what is their function?***

44 ***What are the enzymes present in pancreatic juice? Explain the function of each.***

LARGE INTESTINE

The **large intestine** is the portion of the digestive tract extending from the ileocecal junction to the anus. It consists of the cecum, colon, rectum, and anal canal. Normally, 18–24 hours are required for material to pass through the large intestine, in contrast to the 3–5 hours required for the movement of chyme through the small intestine. Thus, the movements of the colon are more sluggish than those of the small intestine. While in the colon, chyme is converted to feces. The absorption of water and salts, secretion of mucus, and extensive action of microorganisms are involved in the formation of feces, which the colon stores until the feces are

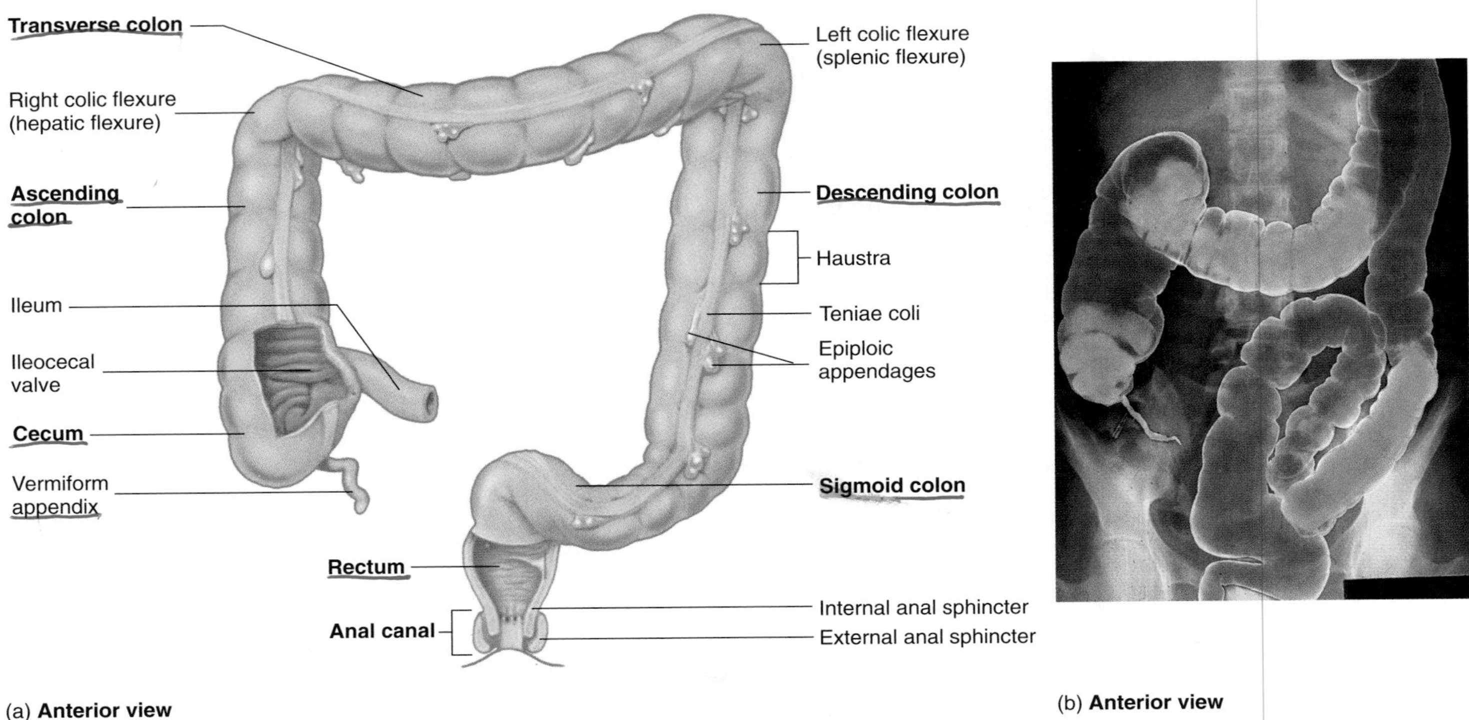

FIGURE 24.25 Large Intestine

(*a*) Large intestine (i.e., cecum, colon, and rectum) and anal canal. The teniae coli and epiploic appendages are along the length of the colon. (*b*) Radiograph of the large intestine following a barium enema.

eliminated by the process of defecation. About 1500 mL of chyme enters the cecum each day, but more than 90% of the volume is reabsorbed so that only 80–150 mL of feces are normally eliminated by defecation.

Anatomy of the Large Intestine

Cecum

The **cecum** (sē′kŭm; blind) is the proximal end of the large intestine. It is where the large and small intestines meet at the ileocecal junction. The cecum extends inferiorly about 6 cm past the ileocecal junction in the form of a blind sac (figure 24.25). Attached to the cecum is a small, blind tube about 9 cm long called the **vermiform** (ver′mi-fōrm; worm-shaped) **appendix.** The walls of the appendix contain many lymphatic nodules.

Colon

The **colon** (kō′lon), about 1.5–1.8 m long, consists of four parts: the ascending colon, transverse colon, descending colon, and sigmoid colon (see figure 24.25). The **ascending colon** extends superiorly from the cecum and ends at the right colic flexure (hepatic flexure) near the right inferior margin of the liver. The **transverse colon** extends from the right colic flexure to the left colic flexure (splenic flexure), and the **descending colon** extends from the left colic flexure to the superior opening of the true pelvis, where it becomes the sigmoid colon. The **sigmoid colon** forms an S-shaped tube that extends into the pelvis and ends at the rectum.

> **Appendicitis**
>
> **Appendicitis** is an inflammation of the vermiform appendix; it usually occurs because of an obstruction of the appendix. Secretions from the appendix cannot pass the obstruction and accumulate, causing enlargement and pain. Bacteria in the area can cause infection of the appendix. Symptoms include sudden abdominal pain, particularly in the right lower portion of the abdomen; a slight fever; loss of appetite; constipation or diarrhea; nausea; and vomiting. In the right-lower quadrant of the abdomen, about midway along a line between the umbilicus and the right anterior superior iliac spine, is an area on the body's surface called McBurney's point. This area becomes very tender in patients with acute appendicitis because of pain referred from the inflamed appendix to the body's surface. Each year, 500,000 people in the United States suffer from appendicitis. An appendectomy is removal of the appendix. If the appendix bursts, the infection can spread throughout the peritoneal cavity, causing peritonitis, with life-threatening results.

The circular muscle layer of the colon is complete, but the longitudinal muscle layer is incomplete. The longitudinal layer does not completely envelop the intestinal wall but forms three bands, called the **teniae coli** (tē′nē-ē kō′lī; a band or tape along the colon), that run the length of the colon (see figure 24.25; figure 24.26). Contractions of the teniae coli cause pouches called **haustra** (haw′stră; to draw up) to form along the length of the colon, giving it a puckered appearance. Small, fat-filled connective tissue pouches called **epiploic** (ep′i-plō′ik; related to the omentum)

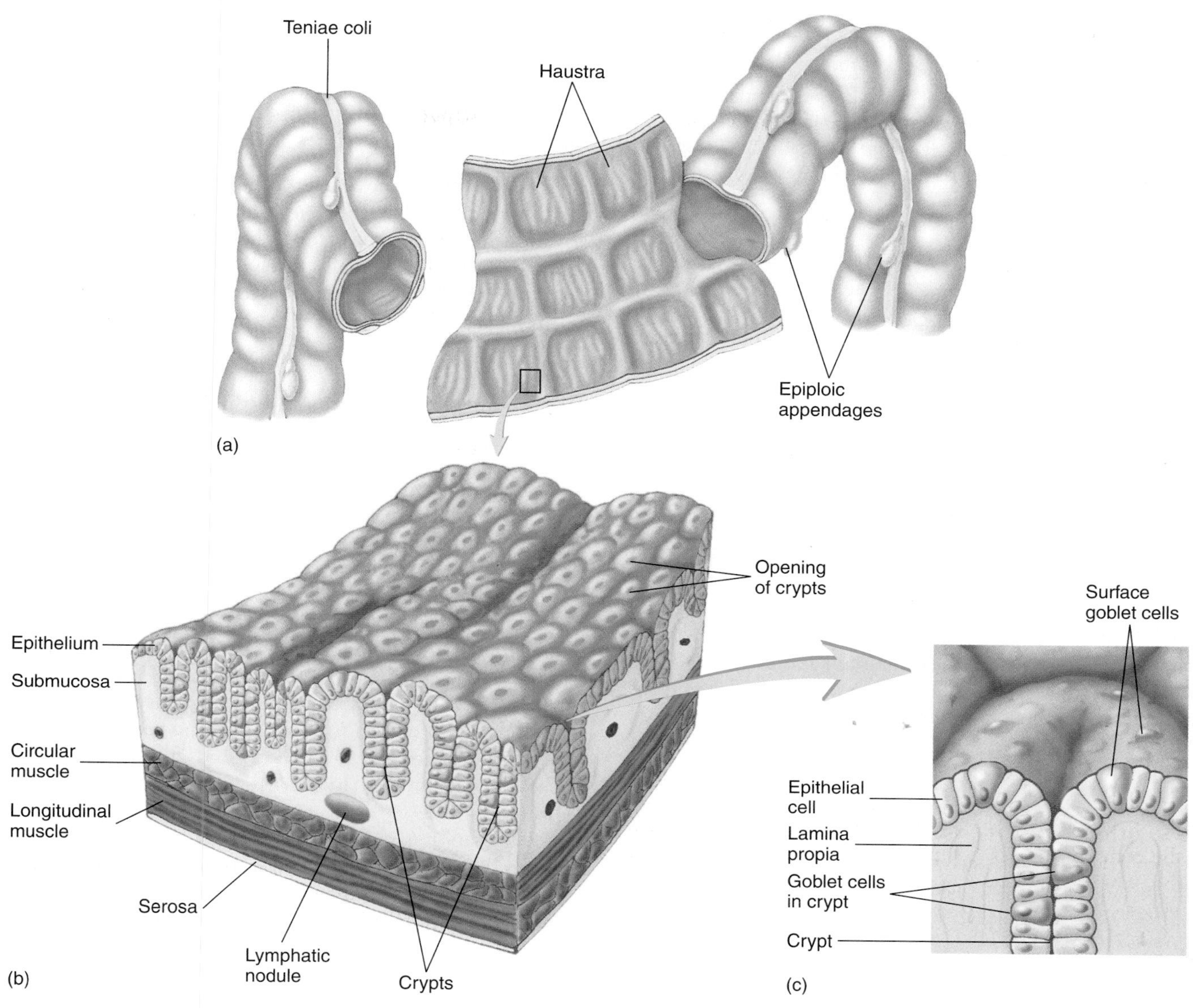

FIGURE 24.26 Histology of the Large Intestine

(*a*) Section of the transverse colon cut open to show the inner surface. (*b*) Enlargement of the inner surface, showing openings of the crypts. (*c*) Higher magnification of a single crypt.

appendages are attached to the outer surface of the colon along its length.

The mucosal lining of the large intestine consists of simple columnar epithelium. This epithelium is not formed into folds or villi like that of the small intestine but has numerous, straight, tubular glands called **crypts** (see figure 24.26). The crypts are somewhat similar to the intestinal glands of the small intestine, with three cell types—absorptive, goblet, and granular cells. The major difference is that, in the large intestine, goblet cells predominate and the other two cell types are greatly reduced in number.

Rectum

The **rectum** is a straight, muscular tube that begins at the termination of the sigmoid colon and ends at the anal canal (see figure 24.25). The mucosal lining of the rectum is simple columnar epithelium, and the muscular tunic is relatively thick, compared with the rest of the digestive tract.

Anal Canal

The last 2–3 cm of the digestive tract is the **anal canal** (see figure 24.25). It begins at the inferior end of the rectum and ends at the **anus** (external GI tract opening). The smooth muscle layer of the anal canal is even thicker than that of the rectum and forms the **internal anal sphincter** at the superior end of the anal canal. Skeletal muscle forms the **external anal sphincter** at the inferior end of the canal. The epithelium of the superior part of the anal canal is simple columnar and that of the inferior part is stratified squamous.

Hemorrhoids

Hemorrhoids are the enlargement, or inflammation, of the rectal veins, which supply the anal canal. Hemorrhoids, also called varicose rectal veins, cause pain, itching, and bleeding around the anus. They can be internal, involving the internal rectal veins, or external, involving the external rectal veins. Treatments include increasing the bulk (indigestible fiber) in the diet, taking sitz baths, and using hydrocortizone suppositories. Surgery may be necessary if the condition is extreme and does not respond to other treatments.

45 ***Describe the parts of the large intestine. What are teniae coli, haustra, and crypts?***

46 ***Explain the difference in structure between the internal anal sphincter and the external anal sphincter.***

Secretions of the Large Intestine

The mucosa of the colon has numerous goblet cells scattered along its length and numerous crypts lined almost entirely with goblet cells. Little enzymatic activity is associated with secretions of the colon because mucus is the major secretory product (see tables 24.1 and 24.2). Mucus lubricates the wall of the colon and helps the fecal matter stick together. Tactile stimuli and irritation of the wall of the colon trigger local enteric reflexes that increase mucous secretion. Parasympathetic stimulation also increases the secretory rate of the goblet cells.

Diarrhea

When the small or large intestine is irritated and inflamed, such as in patients with bacterial enteritis (inflamed intestine resulting from bacterial infection), the intestinal mucosa secretes large amounts of mucus and electrolytes, and water moves by osmosis into the colon. An abnormally frequent discharge of fluid feces is called **diarrhea.** Although such discharge increases fluid and electrolyte loss, it also moves the infected feces out of the intestine more rapidly and speeds recovery from the disease.

A molecular pump exchanges HCO_3^- for Cl^- in epithelial cells of the colon in response to acid produced by colic bacteria. Another pump exchanges Na^+ for H^+. Water crosses the wall of the colon through osmosis with the sodium chloride gradient.

The feces that leave the digestive tract consist of water, solid substances (e.g., undigested food), microorganisms, and sloughed-off epithelial cells.

Numerous microorganisms inhabit the colon. They reproduce rapidly and ultimately constitute about 30% of the dry weight of the feces. Some bacteria in the intestine synthesize vitamin K, which is passively absorbed in the colon, and break down a small amount of cellulose to glucose. However, the glucose cannot be absorbed in the large intestine.

Bacterial actions in the colon produce gases called **flatus** (flā′tŭs; blowing). The amount of flatus depends partly on the bacterial population in the colon and partly on the type of food consumed. For example, beans, which contain certain complex carbohydrates, are well known for their flatus-producing effect.

Movement in the Large Intestine

Segmental mixing movements occur in the colon much less often than in the small intestine. Peristaltic waves are largely responsible for moving chyme along the ascending colon. At widely spaced intervals (normally three or four times each day), large parts of the transverse and descending colon undergo several strong contractions, called **mass movements.** Each mass movement contraction extends over a much longer part of the digestive tract (≥20 cm) than does a peristaltic contraction and propels the colon contents a considerable distance toward the anus (figure 24.27). Mass movements are very common after meals because the presence of food in the stomach or duodenum initiates them. Mass movements are most common about 15 minutes after breakfast. They usually persist for 10–30 minutes and then stop for perhaps half a day. Local reflexes in the enteric plexus, which are called **gastrocolic reflexes** if initiated by the stomach or **duodenocolic reflexes** if initiated by the duodenum, integrate mass movements.

The gastrocolic and duodenocolic reflexes promote peristalsis of the small and large intestines, including mass movements. These reflexes are mediated by parasympathetic reflexes, local reflexes, and hormones, such as cholecystokinin and gastrin. The thought or smell of food, distension of the stomach, and the movement of chyme into the duodenum can stimulate them.

Defecation requires that the contractions that move feces toward the anus be coordinated with the relaxation of the internal and external anal sphincters. The movement of feces from the colon into the rectum distends the rectal wall, which acts as a stimulus initiating the defecation reflex. The **defecation reflex** consists of local and parasympathetic reflexes (see figure 24.27). Local reflexes cause weak contractions of the distal colon and rectum and relaxation of the internal anal sphincter. Parasympathetic reflexes are responsible for most of the defecation reflex. Action potentials produced in response to the distension travel along afferent nerve fibers to the defecation reflex center (S2–S4) in the conus medullaris of the spinal cord. Efferent action potentials are initiated that return through nerves to the colon and rectum, reinforcing peristaltic contractions and relaxation of the internal anal sphincter.

Action potentials from the sacral spinal cord ascend to the brain, where parts of the brainstem and hypothalamus inhibit or facilitate reflex activity in the spinal cord. In addition, action potentials ascend to the cerebrum, where awareness of the need to defecate is realized. The external anal sphincter is composed of skeletal muscle and is under conscious cerebral control. If this sphincter is relaxed voluntarily, feces are expelled. On the other hand, increased contraction of the external anal sphincter prevents defecation. The defecation reflex persists for only a few minutes and quickly declines. Generally, the reflex is reinitiated after a period that may be as long as several hours. Mass movements in

1. The thought or smell of food, distention of the stomach, and the movement of chyme into the duodenum can stimulate the gastrocolic and duodenocolic reflexes (*green arrows*).
2. The gastrocolic and duodenocolic reflexes stimulate mass movements in the colon, which propel the contents of the colon toward the rectum (*orange arrow*).
3. Distention of the rectum by feces stimulates local defecation reflexes. These reflexes cause contractions of the colon and rectum (*brown arrow*), which move feces toward the anus.
4. Local reflexes cause relaxation of the internal anal sphincter (*brown arrow*).
5. Distention of the rectum by feces stimulates parasympathetic reflexes. Action potentials are propagated to the defecation reflex center located in the spinal cord (*yellow arrow*).
6. Action potentials stimulate contraction of the colon and rectum and relaxation of the internal anal sphincter (*purple arrows*).
7. Action potentials are propagated through ascending nerve tracts to the brain (*blue arrow*).
8. Descending nerve tracts from the brain regulate the defecation reflex center (*pink arrow*).
9. Action potentials from the brain control the external anal sphincter (*purple arrow*).

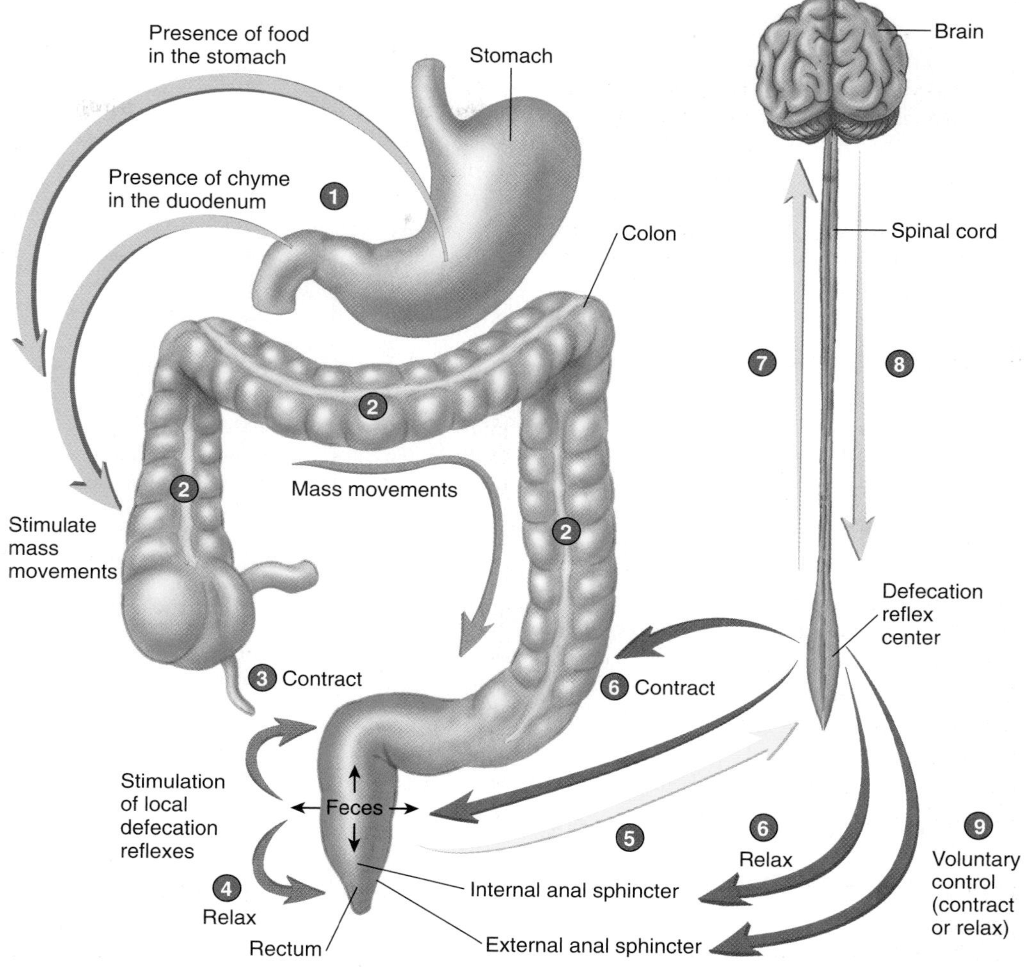

PROCESS FIGURE 24.27 Control of Defecation

the colon are usually the reason for the reinitiation of the defecation reflex.

Contraction of the internal and external anal sphincters prevents defecation. Resting sphincter pressure results from tonic muscle contractions, mostly of the internal anal sphincter. In response to increased abdominal pressure, reflexes mediated through the spinal cord cause contractions of the external anal sphincter. Thus, the untimely expulsion of feces during coughing or exertion is avoided.

Defecation can be initiated by voluntary actions that stimulate a defecation reflex. This "straining" includes a large inspiration of air, followed by closure of the larynx and forceful contraction of the abdominal muscles. As a consequence, the pressure in the abdominal cavity increases and forces feces into the rectum. Stretch of the rectum initiates a defecation reflex and input from the brain overrides the reflexive contraction of the external anal sphincter stimulated by increased abdominal pressure. The increased abdominal pressure also helps push feces through the rectum.

47 ***Name the substances secreted and absorbed by the large intestine. What is the role of microorganisms in the colon?***

48 ***What kind of movements occur in the colon? Describe the defecation reflex.***

On Being "Regular"

The importance of the regularity of defecation has been greatly overestimated. Many people have the misleading notion that a daily bowel movement is critical for good health. As with many other body functions, what is "normal" differs from person to person. Whereas many people defecate one or more times per day, some normal, healthy adults defecate on the average only every other day. A defecation rate of only twice per week, however, is usually described as constipation. Habitually postponing defecation when the defecation reflex occurs can lead to constipation and may eventually result in desensitization of the rectum so that the defecation reflex is greatly diminished.

CASE STUDY

Spinal Cord Injury and Defecation

Dan, a 17-year-old male, was driving home late at night from a ski trip when he missed a sharp curve and crashed. He suffered traumatic injury at the T11 level of the spinal cord, with complete paralysis of both lower limbs. Dan also has incontinence and is no longer able to control his bowel movements. A loss of the ability to control defecation commonly affects the quality of life of most spinal cord injury patients. About 10,000 new spinal cord injuries occur per year in the United States. About 80% of those injuries involve men, usually in their late teens or twenties. The most common cause is motor vehicle accidents, followed by violence, falls, and sports.

The spinal cord is required for a normal defecation reflex and voluntary control of the external anal sphincter (see figure 24.27). In regard to defecation, spinal cord injuries can be divided into two groups based on the level of injury: those that occur above the conus medullaris and those that damage the conus medullaris where the defecation reflex center is located. Immediately following a spinal cord injury, there is a loss of reflexes below the level of the injury, called **spinal shock.** The reflexes usually become functional again, however, and the defecation reflex is usually depressed for a few weeks but eventually returns.

PREDICT 7

Dan has injured his spinal cord above the conus medullaris and has recovered from spinal shock.

a. Explain how Dan's spinal cord injury results in the loss of voluntary control of defecation.

b. With Dan's spinal cord injury, he can induce defecation at a selected time using an enema. How does an enema cause a bowel movement in Dan?

c. Dan has found that an enema is usually most effective after breakfast. Why is this so?

d. Dan has found that straining to have a bowel movement actually makes it more difficult to do so. Explain.

e. If spinal cord injury damages the conus medullaris, it is still possible for defecation to occur. Explain.

DIGESTION, ABSORPTION, AND TRANSPORT

Digestion is the breakdown of food to molecules that are small enough to be absorbed into the circulation. **Mechanical digestion** breaks large food particles down into smaller ones. **Chemical digestion** is the breaking of covalent chemical bonds in organic molecules by digestive enzymes. Carbohydrates are broken down into monosaccharides, proteins are broken down into amino acids, and fats are broken down into fatty acids and glycerol. **Absorption** and **transport** are the means by which molecules are moved out of the digestive tract into the circulation for distribution throughout the body. Not all molecules (e.g., vitamins, minerals, and water) are broken down before being absorbed, however. Digestion begins in the oral cavity and continues in the stomach, but most digestion occurs in the proximal end of the small intestine, especially in the duodenum.

The absorption of certain molecules can occur all along the digestive tract. A few chemicals, such as nitroglycerin, can be absorbed through the thin mucosa of the oral cavity below the tongue. Some small molecules (e.g., alcohol and aspirin) can diffuse through the stomach epithelium into the circulation. Most absorption, however, occurs in the duodenum and jejunum, although some absorption occurs in the ileum.

Some molecules can diffuse, whereas others must be transported across the intestinal wall. Transport requires transport proteins, which work by facilitated diffusion, active-transport, or secondary-transport mechanisms, such as symport and antiport. The epithelial cells that form the intestinal wall have two distinct sides with different transport proteins on each side. The side that faces the digestive tract lumen is called the **apical membrane,** and the side that faces the circulation is called the **basolateral membrane.** The transport proteins in these membranes are responsible for the directional movement of molecules from the digestive tract to the rest of the body.

Once the digestive products have been absorbed, they are transported to other parts of the body by two routes. Water, ions, and water-soluble digestion products, such as glucose and amino acids, enter the hepatic portal system and are transported to the liver. The products of lipid metabolism are coated with proteins and transported into lymphatic capillaries called lacteals (see figure 24.16*c*). The lacteals are connected by lymphatic vessels to the thoracic duct (see chapter 22), which empties into the left subclavian vein. The protein-coated lipid products then travel in the circulation to adipose tissue or to the liver.

Carbohydrates

Ingested **carbohydrates** consist primarily of polysaccharides, such as starches; disaccharides, such as sucrose (table sugar) and lactose (milk sugar); and monosaccharides, such as glucose and fructose (the sugar found in many fruits). During the digestion process, polysaccharides are broken down into smaller chains and finally into disaccharides and monosaccharides. Disaccharides are broken down into monosaccharides. A minor amount of carbohydrate digestion begins in the oral cavity with the partial digestion of starches by **salivary amylase** (am′il-ās). Digestion continues in the stomach until the food is well mixed with acid, which inactivates salivary amylase. Carbohydrate digestion is resumed in the intestine by **pancreatic amylase** (table 24.4). A series of **disaccharidases** that are bound to the microvilli of the intestinal epithelium digest disaccharides into monosaccharides.

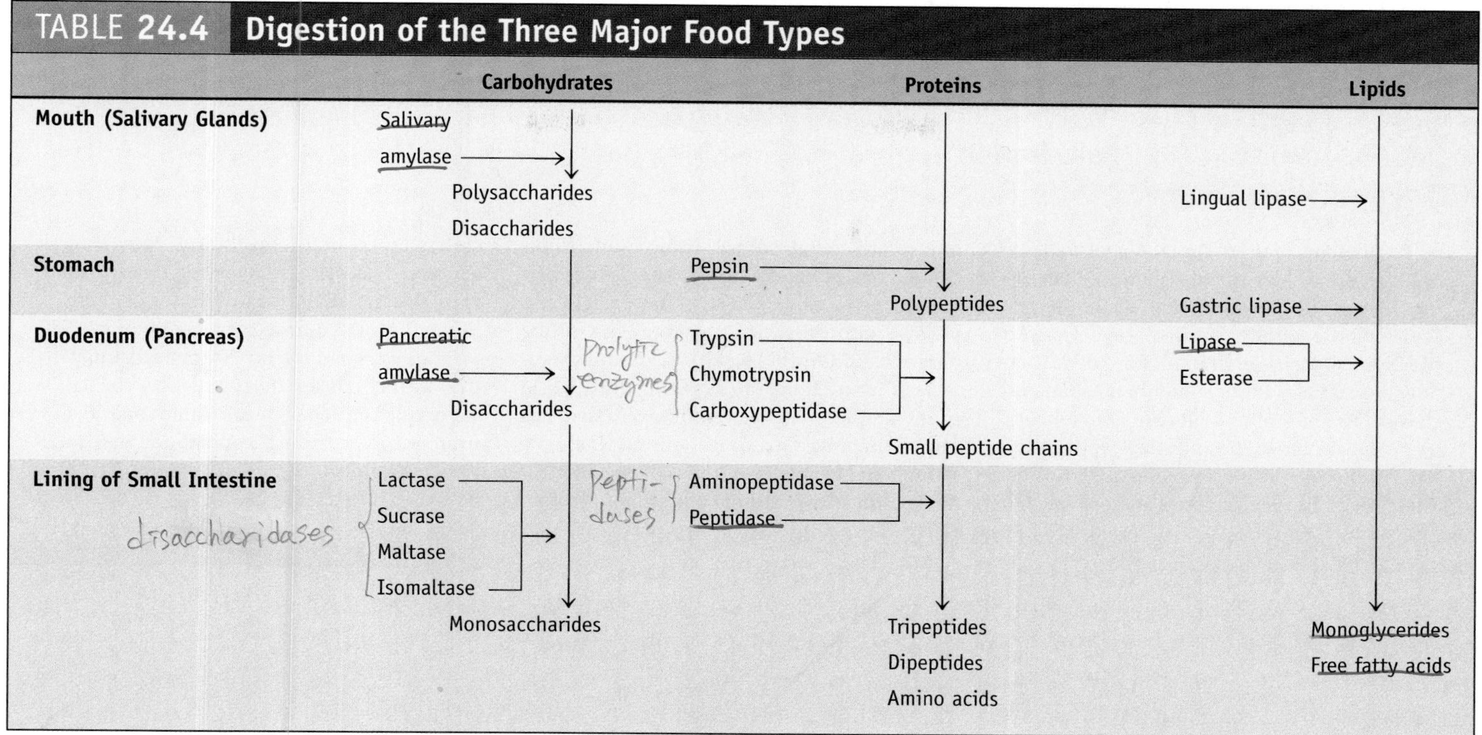

TABLE 24.4 Digestion of the Three Major Food Types

	Carbohydrates	Proteins	Lipids
Mouth (Salivary Glands)	Salivary amylase → Polysaccharides, Disaccharides		Lingual lipase →
Stomach		Pepsin → Polypeptides	Gastric lipase →
Duodenum (Pancreas)	Pancreatic amylase → Disaccharides	Trypsin, Chymotrypsin, Carboxypeptidase → Small peptide chains	Lipase, Esterase →
Lining of Small Intestine	Lactase, Sucrase, Maltase, Isomaltase → Monosaccharides	Aminopeptidase, Peptidase → Tripeptides, Dipeptides, Amino acids	Monoglycerides, Free fatty acids

Lactose Intolerance

Lactose intolerance is the inability to digest the lactose sugar found in milk and other dairy products. Adults in most of the world are lactose intolerant, although infants are not. Why can infants digest milk when their parents cannot? The reason is that many adults lack the enzyme lactase. Lactase is on the surface of absorptive cells in the intestinal mucosa; it digests the disaccharide lactose down to two monosaccharides. Lactase is made at birth; however, in 5%–15% of Europeans and 80%–90% of Africans and Asians, lactase is no longer synthesized after about age 6, so lactose can no longer be digested. The major exceptions are people of Northern European ancestry and some pastoral nomadic tribes in Africa and the Middle East. In these populations, a mutation in the promoter (see chapter 3) of the lactase gene on chromosome 2 permits the continued expression of lactase into adulthood. Lactase production normally stops because the promoter is "turned off" as the infant ages. The mutation allows the promoter to ignore this developmental switch.

It is believed that the dietary reliance on milk and milk products in some societies provided a selective advantage for lactase persistence. In the United States, most people are lactose tolerant, but it is still one of the most common GI disorders seen by primary care physicians. The consequence of lactose intolerance is diarrhea due to fluid loss as water follows lactose through the GI tract. In addition, a considerable amount of gas is generated from lactose metabolism by bacteria in the large intestine. Even though these colonic bacteria metabolize lactose to monosaccharides, it is too late for them to be absorbed. Gene therapy has proven successful in animal models of lactose intolerance, although at present the best treatment is simply to minimize lactose ingestion.

Monosaccharides, such as glucose and galactose, are taken up into intestinal epithelial cells by symport, powered by a Na^+ gradient (figure 24.28). The Na^+ gradient is generated by the **Na^+–K^+ pump** located on the basolateral membrane. Monosaccharides, such as fructose, are taken up by facilitated diffusion. The monosaccharides are transferred by facilitated diffusion to the capillaries of the intestinal villi and are carried by the hepatic portal system to the liver, where the nonglucose sugars are converted to glucose. Glucose enters the cells through facilitated diffusion. The rate of glucose transport into most types of cells is greatly influenced by **insulin** and may increase 10-fold in its presence (see chapter 18).

Lipids

Lipids are molecules that are insoluble or only slightly soluble in water. They include triglycerides, phospholipids, cholesterol, steroids, and fat-soluble vitamins. **Triglycerides** (trī-glis′er-īdz), also called **triacylglycerol** (trī-as′il-glis′er-ol), consist of three fatty acids and one glycerol molecule covalently bonded together. The first step in lipid digestion is **emulsification** (ē-mŭl′si-fi-kā′shŭn), which is the transformation of large lipid droplets into much smaller droplets. The enzymes that digest lipids (pancreatic lipase) are water-soluble and can digest the lipids only by acting at the surface of the droplets. The emulsification process increases the surface area of the lipid exposed to the digestive enzymes by decreasing the droplet size. Emulsification is accomplished by bile salts secreted by the liver and stored in the gallbladder.

Lipase (lip′ās) digests lipid molecules (see table 24.4). The vast majority of lipase is secreted by the pancreas. A minor amount of lingual lipase is secreted in the oral cavity, is swallowed with

CLINICAL FOCUS

Rehydration

What does an endurance sport athlete and a person with severe diarrhea have in common? Both share the risk of dehydration from excessive water loss. An effective rehydration strategy for both situations is to drink water containing sodium and glucose. As sodium and glucose are absorbed by symport across the intestinal epithelium, water follows by osmosis. As an added value, this strategy also replaces ions and provides an immediate energy source. Most sport drinks contain sodium and glucose, which efficiently rehydrate the athlete. The same principle is used in oral rehydration therapy for severe diarrhea. The World Health Organization estimates that millions of people in third world countries die every year from severe diarrhea caused by intestinal infections. Drinking a sodium and glucose solution is often sufficient to prevent dehydration until the infection clears. Since oral rehydration therapy was adopted as a main strategy for treating diarrhea, the annual number of deaths of children under 5 dropped from over 4 million in 1980 to about 1.5 million in 2000. Unfortunately, this simple, cheap treatment and the clean water that it requires are often unavailable in areas of exploding populations and poor sanitary conditions.

Monosaccharide transport

1. Monosaccharides are absorbed by symport with Na^+ into intestinal epithelial cells.
2. Symport is driven by a sodium gradient established by a Na^+–K^+ pump.
3. Monosaccharides move out of the intestinal epithelial cells by facilitated diffusion.
4. Monosaccharides enter the capillaries of the intestinal villi and are carried through the hepatic portal vein to the liver.

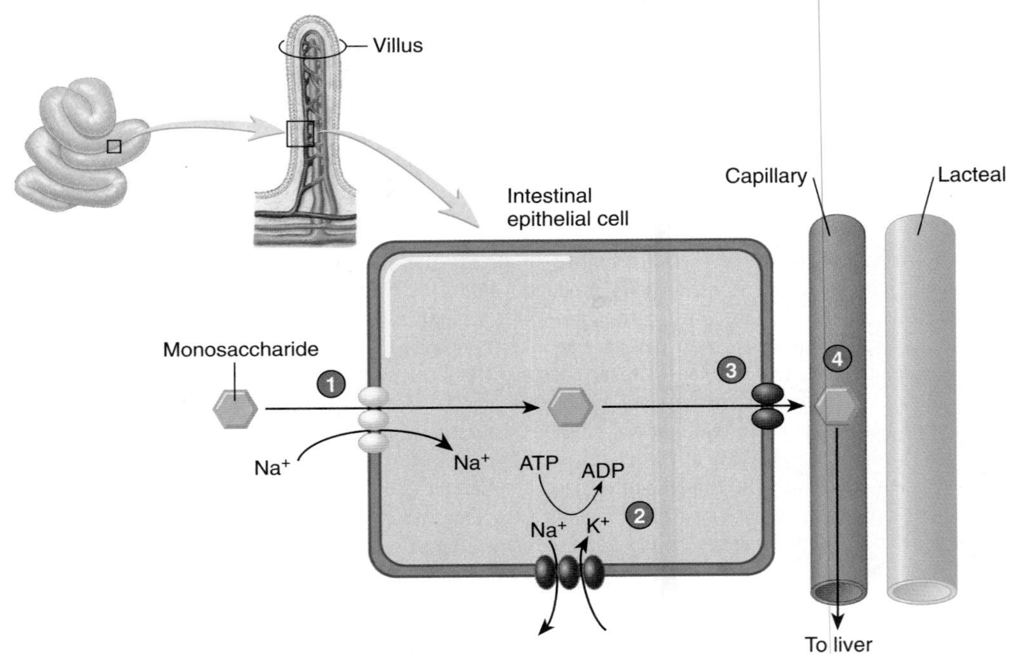

PROCESS FIGURE 24.28 Transport of Monosaccharides Across the Intestinal Epithelium

food, and digests a small amount (<10%) of lipid in the stomach. The stomach also produces very small amounts of gastric lipase. The primary products of lipase digestion are free fatty acids and monoglycerides. Cholesterol and phospholipids are digested by esterase and phospholipase, respectively.

Once lipids are digested in the intestine, bile salts aggregate around the small droplets to form **micelles** (mi-selz′, mī-selz′; a small morsel; figure 24.29). The hydrophobic ends of the bile salts are directed toward the free fatty acids, cholesterol, and monoglycerides at the center of the micelle; the hydrophilic ends are directed outward toward the water environment. When a micelle comes into contact with the epithelial cells of the small intestine, the lipid contents of the micelle pass by means of simple diffusion through the plasma membrane of the epithelial cells. The bile salts are not absorbed until they reach the epithelium of the distal ileum.

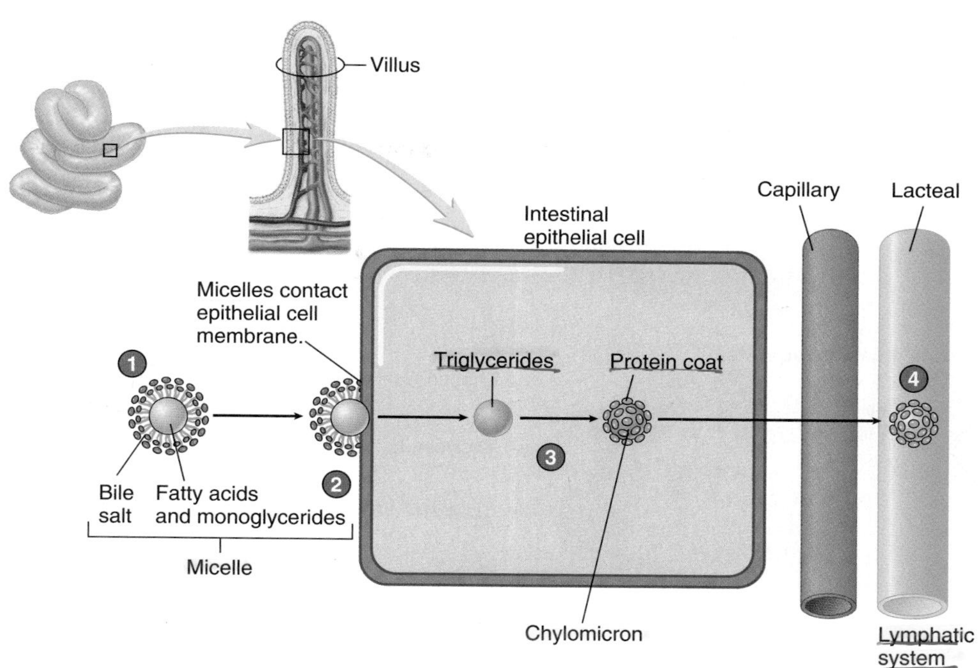
Lipid transport
1. Bile salts surround fatty acids and monoglycerides to form micelles.
2. Micelles attach to the cell membranes of intestinal epithelial cells, and the fatty acids and monoglycerides pass by simple diffusion into the intestinal epithelial cells.
3. Within the intestinal epithelial cell, the fatty acids and monoglycerides are converted to triglycerides; proteins coat the triglycerides to form chylomicrons, which move out of the intestinal epithelial cells by exocytosis.
4. The chylomicrons enter the lacteals of the intestinal villi and are carried through the lymphatic system to the general circulation.
Villus
Capillary
Lacteal
Intestinal epithelial cell
Micelles contact epithelial cell membrane.
Triglycerides
Protein coat
Bile salt
Fatty acids and monoglycerides
Micelle
Chylomicron
Lymphatic system

PROCESS FIGURE 24.29 Transport of Lipids Across the Intestinal Epithelium

> **Cystic Fibrosis**
>
> **Cystic fibrosis** is a hereditary disorder that occurs in 1 of every 2000 births and affects 33,000 people in the United States; it is the most common lethal genetic disorder among Caucasians. The most critical effects of the disease, accounting for 90% of the deaths, are on the respiratory system. Several other problems occur, however, in affected people. Because the disease is a disorder in a Cl^- transport channel protein—which affects chloride transport and, as a result the movement of water—all exocrine glands are affected. The buildup of thick mucus in the pancreatic and hepatic ducts causes blockage of the ducts so that bile salts and pancreatic digestive enzymes are prevented from reaching the duodenum. As a result, fats and fat-soluble vitamins, which require bile salts to form micelles and which cannot be adequately digested without pancreatic enzymes, are not well digested and absorbed. The person suffers from vitamin A, D, E, and K deficiencies, which result in conditions such as night blindness, skin disorders, rickets, and excessive bleeding. Therapy includes administering the missing vitamins to the person and reducing dietary fat intake.

Lipid Transport

Within the smooth endoplasmic reticulum of the intestinal epithelial cells, free fatty acids are combined with monoglyceride molecules to form triglycerides. Proteins synthesized in the epithelial cells attach to droplets of triglycerides, phospholipids, and cholesterol to form **chylomicrons** (kī-lō-mi′kronz; small particles in the chyle, or fat-filled lymph). The chylomicrons leave the epithelial cells and enter the lacteals of the lymphatic system within the villi.

Chylomicrons enter the lymphatic capillaries rather than the blood capillaries because the lymphatic capillaries lack a basement membrane and are more permeable to large particles, such as chylomicrons (about 0.3 mm in diameter). Chylomicrons are about 90% triglyceride, 5% cholesterol, 4% phospholipid, and 1% protein (figure 24.30). They are carried through the lymphatic system to the bloodstream and then by the blood to adipose tissue. Before entering the adipose cells, triglyceride is broken back down into fatty acids and glycerol, which enter the fat cells and are once more converted to triglyceride. Triglycerides are stored in adipose tissue until an energy source is needed elsewhere in the body. In the liver, the chylomicron lipids are stored, converted into other molecules, or used as energy. The chylomicron remnant, minus the triglyceride, is conveyed through the circulation to the liver, where it is broken up.

Because lipids are either insoluble or only slightly soluble in water, they are transported through the blood in combination with proteins, which are water-soluble. Lipids combined with proteins are called **lipoproteins.** Chylomicrons are one type of lipoprotein. Other lipoproteins are referred to as high- or low-density lipoproteins. Density describes the compactness of a substance and is the ratio of mass to volume. Lipids are less dense than water and tend to float in water. Proteins, which are denser than water, tend to sink in water. A lipoprotein with a high lipid content has a very low density, whereas a lipoprotein with a high protein content has a relatively high density. Chylomicrons, which are made up of 99% lipid and only 1% protein, have an extremely low density. The other major transport lipoproteins are **very low-density lipoprotein (VLDL),** which is 92% lipid and 8% protein, **low-density lipoprotein (LDL),** which is 75% lipid and

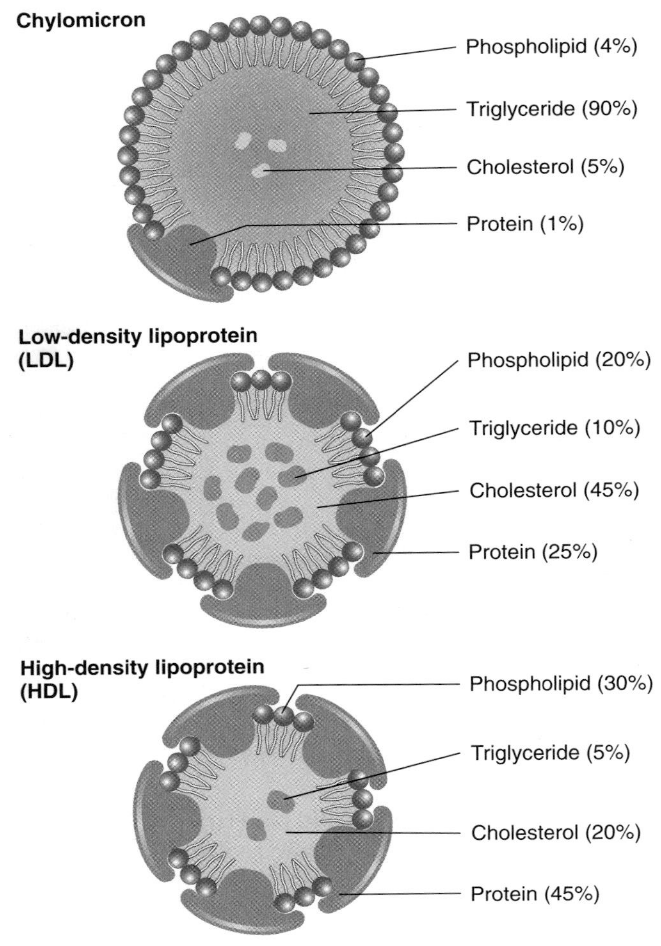

FIGURE 24.30 **Lipoproteins**

25% protein, and **high-density lipoprotein (HDL),** which is 55% lipid and 45% protein (see figure 24.30).

About 15% of the cholesterol in the body is ingested in the food we eat, and the remaining 85% is manufactured in the cells of the body, mostly in the liver and intestinal mucosa. Most of the lipid taken into or manufactured in the liver leaves the liver in the form of VLDL. Most of the triglycerides are removed from the VLDL to be stored in adipose tissue; as a result, VLDL becomes LDL.

The cholesterol in LDL is critical for the production of steroid hormones in the adrenal cortex and the production of bile acids in the liver. It is also an important component of plasma membranes. LDL is delivered to cells of various tissues through the circulation. Cells have **LDL receptors** in "pits" on their surfaces, which bind the LDL. Once LDL is bound to the receptors, the pits on the cell surface become endocytotic vesicles, and the cell takes in LDL by receptor-mediated endocytosis (figure 24.31). Each fibroblast, as an example of a tissue cell, has 20,000–50,000 LDL receptors on the surface. Those receptors are confined to cell surface pits, however, which occupy only 2% of the cell surface. Once inside the cell, the endocytotic vesicle combines with a lysosome, and the LDL components are separated for use in the cell.

Cells not only take in cholesterol and other lipids from LDLs, but they also make their own cholesterol. When the combined intake and manufacture of cholesterol exceeds a cell's needs, a negative-feedback system reduces the amount of LDL receptors and cholesterol manufactured by the cell. Excess lipids are also packaged into HDLs by the cells. These are transported back to the liver for recycling or excretion in bile.

LDL is commonly considered to be "bad" because, when in excess, it deposits cholesterol in arterial walls. On the other hand, HDL is considered to be "good" because it transports cholesterol from the tissues via blood to the liver for removal from the body in the bile. A high HDL/LDL ratio in the bloodstream is related to a lower risk for heart disease.

Cholesterol and Coronary Heart Disease

Cholesterol is a major component of atherosclerotic plaques. The level of plasma cholesterol is positively linked to coronary heart disease (CHD). Cholesterol levels of over 200 mg/100 mL increase the risk for CHD. Other risk factors, which are additive to high cholesterol levels, are hypertension, diabetes mellitus, cigarette smoking, and low plasma HDL levels. Low HDL levels are linked to obesity, and weight reduction increases HDL levels. Aerobic exercise can decrease LDL levels and increase HDL levels. The ingestion of saturated fatty acids raises plasma cholesterol levels by stimulating LDL production and inhibiting LDL receptor production. The ingestion of unsaturated fatty acids lowers plasma cholesterol levels. Replacing fats by carbohydrates in the diet can also reduce blood cholesterol levels. The American Heart Association recommends that no more than 30% of an adult's total caloric intake should be from fats and that only 10% be from saturated fats. Total cholesterol intake should be no more than 300 mg/day. People should eat no more than 7 ounces of meat per day, and that should be chicken, fish, or lean meat. People should eat less than one egg a day and drink milk with 1% or less butter fat. Young children, however, require more fats in their diet to stimulate normal brain development, and whole milk is recommended in their diets. Cholesterol is required for normal membrane structure in cells. Abnormally low cholesterol levels may lead to weakened blood vessel walls and an increased risk for cerebral hemorrhage.

Proteins

Proteins are taken into the body from a number of dietary sources. **Pepsin** secreted by the stomach (see table 24.3) catalyzes the cleavage of covalent bonds in proteins to produce smaller polypeptide chains. Gastric pepsin digests as much as 10%–20% of the total ingested protein. Once the proteins and polypeptide chains leave the stomach, proteolytic enzymes produced in the pancreas continue the digestive process and produce small peptide chains (see table 24.4). These are broken down into tripeptides, dipeptides, and amino acids by **peptidases** bound to the microvilli of the small intestine. Each peptidase is specific for a certain peptide chain length or for a certain amino acid sequence.

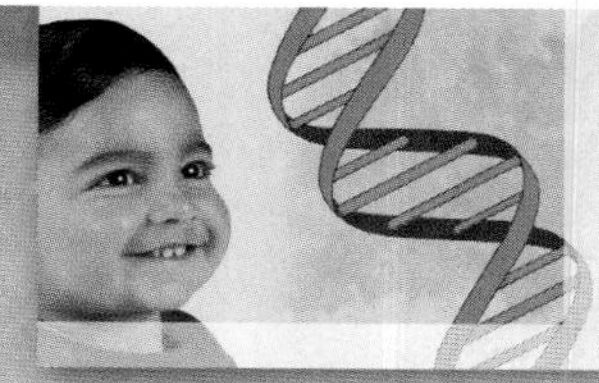

CLINICAL GENETICS

Familial Hypercholesterolemia

Familial hypercholesterolemia (FH) is a common genetic disorder in Europe and North America that affects 1 out of 500 people. The clinical sign of the disease is increased blood levels of LDL cholesterol. The elevated cholesterol levels accelerate the development of atherosclerosis, which often leads to coronary artery disease and heart attacks among people in their forties and fifties. Another common feature of FH is the presence of **xanthomas** (zan-thō′măs), which are nodules of cholesterol and other lipids just under the skin, especially at joints.

FH is caused by mutations in the LDL receptor gene on chromosome 19, which result in defective LDL receptors. The LDL receptor normally removes cholesterol from the blood by transporting LDL cholesterol into cells. Once inside the cell, LDL cholesterol is metabolized and there is a negative-feedback inhibition of cholesterol synthesis. A reduction in LDL cholesterol transport into cells results in elevated blood LDL cholesterol for two reasons: (1) The normal removal of LDL cholesterol does not occur and (2) there is less inhibition of cholesterol synthesis. The usual treatment for FH is statin drugs, which lower blood LDL levels by inhibiting the synthesis of cholesterol.

FH is an example of an incomplete dominance disorder, in which the dominant allele does not completely mask the effects of the recessive allele in the heterozygote (see chapter 3). The severity of FH depends on whether a person is homozygous dominant, heterozygous, or homozygous recessive. Homozygous-dominant individuals have two mutant alleles and completely lack LDL receptors. These patients have the most severe form of FH. Their blood LDL cholesterol levels are very elevated and they often have heart attacks in their teens. Most FH patients are heterozygous, with one normal and one mutant allele. These patients have half the number of normal LDL receptors and are at increased risk for heart attacks at midlife. Homozygous-recessive individuals have two normal LDL receptor alleles and have a normal number of receptors. Although genetic testing is not yet standard, there are automated assays for the more common mutations in the LDL receptor, which can allow early diagnosis and treatment before the development of life-threatening symptoms. In the future, FH will be an excellent candidate for gene therapy because it is a well-studied, severe, single-gene defect.

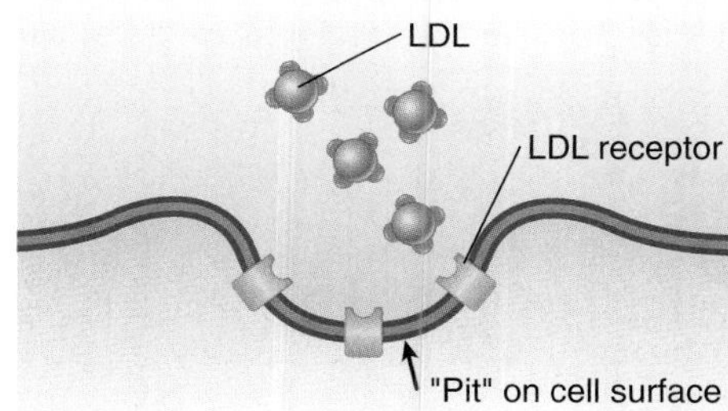

1. Cells have pits on the surface, which contain LDL receptors.

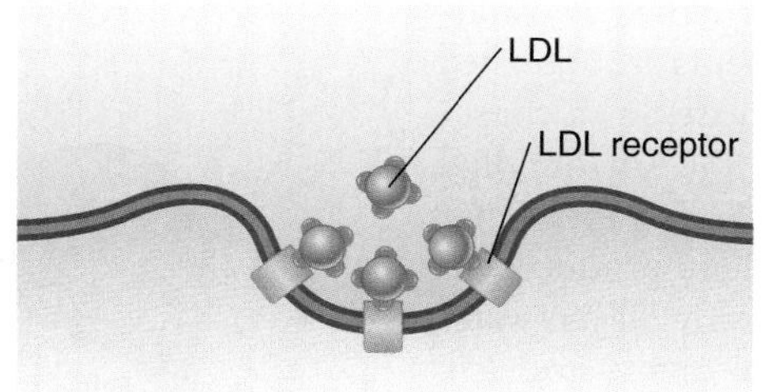

2. LDL binds to the LDL receptors in the pits.

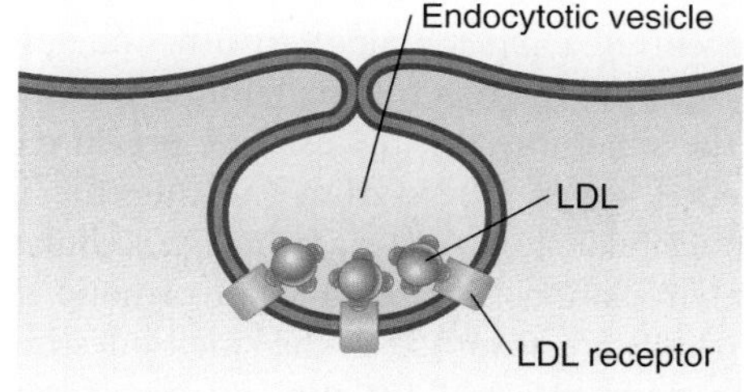

3. The LDL, bound to LDL receptors, is taken into the cell by endocytosis.

PROCESS FIGURE 24.31 **Transport of LDL into Cells**

Dipeptides and tripeptides enter intestinal epithelial cells through a group of related carrier molecules, by a symport mechanism powered by a Na^+ concentration gradient similar to that for glucose. Separate carrier molecules transport basic, acidic, and neutral amino acids into the epithelial cells. Acidic and most neutral amino acids enter by symport with a Na^+ gradient, whereas basic amino acids enter the epithelial cells by facilitated diffusion. The total amount of each amino acid that enters the intestinal epithelial cells as dipeptides or tripeptides is considerably more than the amount that enters as single amino acids. Once inside the cells, dipeptidases and tripeptidases split the dipeptides and tripeptides into their component amino acids. Individual amino acids then leave the epithelial cells and enter the hepatic portal system, which transports them to the liver (figure 24.32). The amino acids may be modified in the liver or released into the bloodstream and distributed throughout the body.

Amino acids are actively transported into the various cells of the body. This transport is stimulated by growth hormone and insulin. Most amino acids are used as building blocks to form new proteins (see chapter 2), but some amino acids may be used for energy.

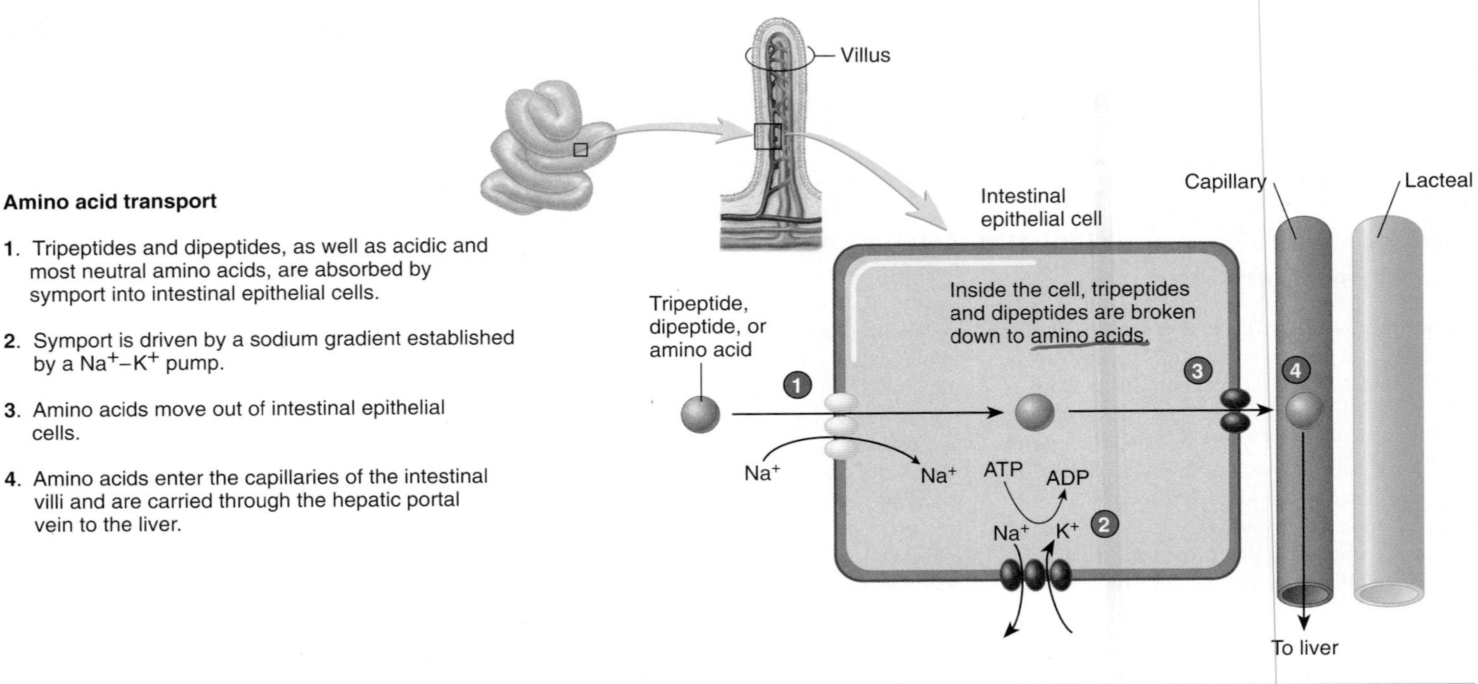

PROCESS FIGURE 24.32 **Amino Acid Transport Across the Intestinal Epithelium**

Water

About 9 L of **water** enter the digestive tract each day, of which about 92% is absorbed in the small intestine, and another 6%–7% is absorbed in the large intestine (figure 24.33). Water can move in either direction across the wall of the small intestine. Osmotic gradients across the epithelium determine the direction of its diffusion. When the chyme is dilute, water is absorbed by osmosis across the intestinal wall into the blood. When the chyme is very concentrated and contains very little water, water moves by osmosis into the lumen of the small intestine. As nutrients are absorbed in the small intestine, its osmotic pressure decreases; as a consequence, water moves from the intestine into the surrounding extracellular fluid. Water in the extracellular fluid can then enter the circulation. Because of the osmotic gradient produced as nutrients are absorbed in the small intestine, nearly all the water that enters the small intestine by way of the oral cavity, stomach, or intestinal secretions is reabsorbed.

Ions

Active transport mechanisms for **sodium** ions are present within the epithelial cells of the small intestine. **Potassium, calcium, magnesium,** and **phosphate** are also actively transported. **Chloride** ions move passively through the intestinal wall of the duodenum and the jejunum following the positively charged sodium ions, but chloride ions are actively transported from the ileum. Although calcium ions are actively transported along the entire length of the small intestine, vitamin D is required for that transport process. The absorption of calcium is under hormonal control, as is its excretion and storage. Parathyroid hormones, calcitonin, and vitamin D all play a role in regulating blood levels of calcium in the circulatory system (see chapters 6, 18, and 27).

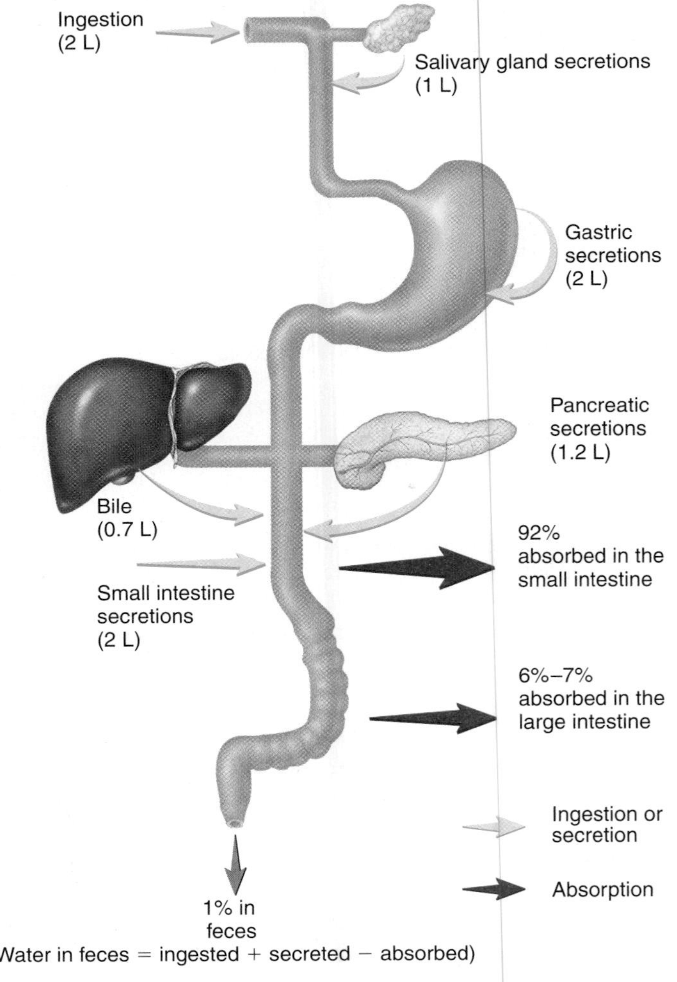

FIGURE 24.33 **Fluid Volumes in the Digestive Tract**

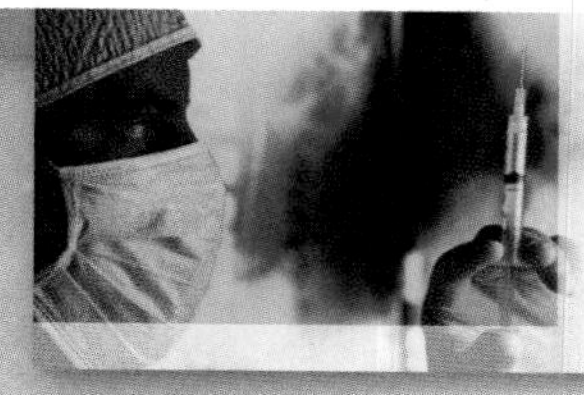

CLINICAL FOCUS

Intestinal Disorders

INFLAMMATORY BOWEL DISEASE

Enteritis is inflammation of the intestine, and **colitis** is inflammation of the colon. Both can result from an infection, chemical irritation, or unknown cause. **Inflammatory bowel disease (IBD)** is the general name given to either Crohn disease (regional enteritis) or ulcerative colitis. IBD occurs at a rate in Europe and North America of approximately 4 to 8 new cases per 100,000 people per year, which is much higher than in Asia and Africa. Males and females are affected about equally. IBD is of unknown cause, but infectious, autoimmune, and hereditary factors have been implicated. **Crohn disease** involves localized inflammatory degeneration, which may occur anywhere along the digestive tract but most commonly involves the distal ileum and proximal colon. The degeneration involves the entire thickness of the digestive tract wall. The intestinal wall often becomes thickened, constricting the lumen, with ulcerations and fissures in the damaged areas. The disease causes diarrhea, abdominal pain, fever, and weight loss. Treatment centers around anti-inflammatory drugs, but other treatments, including the avoidance of foods that increase symptoms and even surgery, are used. **Ulcerative colitis** is limited to the mucosa of the large intestine and rectum. The involved mucosa exhibits inflammation, including edema, vascular congestion, hemorrhage, and the accumulation of plasma cells, lymphocytes, neutrophils, and eosinophils. Patients may experience abdominal pain, fever, malaise, fatigue, and weight loss, as well as diarrhea and hemorrhage. In rare cases, severe diarrhea and hemorrhage require transfusions. Treatment includes the use of anti-inflammatory drugs and, in some cases, the avoidance of foods that increase symptoms.

IRRITABLE BOWEL SYNDROME

Irritable bowel syndrome (IBS), also called spastic colon, is a disorder of unknown cause in which intestinal mobility is abnormal. The disorder accounts for over half of all referrals to gastroenterologists. Male and female children are affected equally, but women are affected twice as often as men. IBS patients experience abdominal pain mainly in the left-lower quadrant, especially after eating. They also have alternating bouts of constipation and diarrhea. There is no specific histopathology in the digestive tracts of IBS patients. There are no anatomical abnormalities, no indication of infection, and no sign of metabolic causes. Patients with IBS appear to exhibit greater than normal levels of psychologic stress or depression and show increased contractions of the esophagus and small intestine during times of stress. There is a high familial incidence. Some patients present with a history of traumatic events, such as physical or sexual abuse. Treatments include psychiatric counseling and stress management, diets with increased fiber and limited gas-producing foods, and loose clothing. In some patients, drugs that reduce parasympathetic stimulation of the digestive system are useful.

MALABSORPTION SYNDROME

Malabsorption syndrome (sprue) is a spectrum of disorders of the small intestine that results in abnormal nutrient absorption. One type of malabsorption, called **celiac disease,** or **gluten-sensitive enteropathy,** results from an immune response to gluten, which is present in certain types of grains and involves the destruction of newly formed epithelial cells in the intestinal glands. These cells fail to migrate to the villi surface, the villi become blunted, and the surface area decreases. As a result, the intestinal epithelium is less capable of absorbing nutrients. Another type of malabsorption, tropical malabsorption, is apparently caused by bacteria, although no specific bacterium has been identified.

COLON CANCER

Colon cancer is the second leading cause of cancer-related deaths in the United States; it accounts for 55,000 deaths a year. Susceptibility to colon cancer can be familial; however, a correlation exists between colon cancer and lifestyle factors, including diets low in fiber and high in fat and red or processed meats.

A gene for colon cancer may be present in as many as 1 in 200 people, making colon cancer one of the most common inherited diseases. Eleven different genes have been associated with colon cancer. One group of those genes is involved in cell regulation—that is, keeping cell growth in check. The second group is required for genetic stability. As a result of mutations in the second group, there are wholesale errors and mutations throughout the genome. Such genetic instability has been identified in at least 15% of sporadic (not occurring in families) colon cancer. Screening for colon cancer includes testing the stool for blood content and performing a colonoscopy, which allows physicians to see into the colon.

CONSTIPATION

Constipation is the slow movement of feces through the large intestine. The feces often become dry and hard because of increased fluid absorption during the extended time they are retained in the large intestine. In the United States, 2.5 million doctor visits occur each year from people complaining of constipation, and $400 million is spent each year on laxatives.

Constipation often results after a prolonged time of inhibiting normal defecation reflexes. A change in habits, such as travel, dehydration, depression, disease, metabolic disturbances, certain medications, pregnancy, or dependency on laxatives, can cause constipation. Irritable bowel syndrome can also cause constipation. Constipation can also occur with diabetes, kidney failure, colon nerve damage, or spinal cord injuries or as the result of an obstructed bowel. Of greatest concern, the obstruction could be caused by colon cancer. Chronic constipation can result from the slow movement of feces through the entire colon, in just the distal part (descending colon and rectum), or in just the rectum. Interestingly, in one large study of people who claimed to be suffering from chronic constipation, one-third were found to have normal movement of feces through the large intestine. Defecation frequency was often normal. Many of those people were suffering from psychologic distress, anxiety, or depression and just thought they had abnormal defecation frequencies.

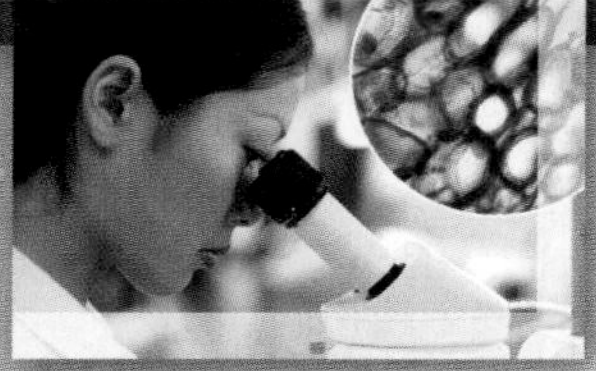

SYSTEMS PATHOLOGY

Diarrhea

While on vacation in Mexico, Mr. T was shopping with his wife when he started to experience sharp pains in his abdominal region. He also began to feel hot and sweaty and felt an extreme urge to defecate. His wife quickly looked up the word *toilet* in their handy Spanish–English pocket travel dictionary, and Mr. T anxiously inquired of a local resident where the nearest facility could be found. Once the immediate need was taken care of, Mr. and Mrs. T went back to their hotel room, where they remained while Mr. T recovered. Over the next 2 days, his stools were frequent and watery. He also vomited a couple of times. Because they were in a foreign country, Mr. T did not consult a physician. He rested, took plenty of fluids, and was feeling much better, although a little weak, in a couple of days.

BACKGROUND INFORMATION

Diarrhea is one of the most common complaints in clinical medicine and affects more than half of the tourists in developing countries. Diarrhea is not itself a disease but is a symptom of a wide variety of disorders. Normally, about 600 mL of fluid enters the colon each day and all but 150 mL is reabsorbed. The loss of more than 200 mL of stool per day is considered abnormal.

Mucous secretion by the colon increases dramatically in response to diarrhea. This mucus contains large quantities of HCO_3^-, which comes from the dissociation of carbonic acid into HCO_3^- and H^+ within the blood supply to the colon. The HCO_3^- enter the mucus secreted by the colon, whereas the H^+ remain in the circulation; as a result, the blood pH decreases. Thus, a condition called metabolic acidosis can develop (see chapter 27).

Diarrhea in tourists usually results from the ingestion of food or water contaminated with bacteria or bacterial toxins. Acute diarrhea is defined as lasting less than 2–3 weeks, and diarrhea lasting longer than that is considered chronic. Acute diarrhea is usually self-limiting, but some forms of diarrhea can be fatal if not treated. Diarrhea results from either a decrease in fluid absorption in the gut or an increase in fluid secretion. Some bacterial toxins and other chemicals can also cause an increase in bowel motor activity. As a result, chyme is moved more rapidly through the digestive tract, fewer nutrients and water are absorbed out of the small intestine, and more water enters the colon. Symptoms can occur in as little as 1–2 hours after bacterial toxins are ingested to as long as 24 hours or more for some strains of bacteria.

In cases of short-term acute diarrhea, the infectious agent is seldom identified. Nearly any bacterial species is capable of causing diarrhea. Some types of bacterial diarrhea are associated with severe vomiting, whereas others are not. Some bacterial toxins also induce fever. Some viruses and amebic parasites can also cause diarrhea. In most cases, a laboratory analysis of the food or stool is necessary to identify the causal organism. In cases of mild diarrhea away from home, laboratory evaluation is not practical, and empiric therapy is usually applied. Fluids and electrolytes must be replaced, and the consumption of fluids with electrolytes is important. The diet should be limited to clear fluids during at least the first day or so. Bismuth subsalicylate (Pepto-Bismol) or loperamide (Imodium; except in cases of fever) may also be used to help combat secretory diarrhea. Milk and milk products should be avoided. Breads, toast, rice, and baked fish or chicken can be added to the diet with improvement. A normal diet can be resumed after 2–3 days.

PREDICT 8

Predict the effects of prolonged diarrhea.

49 ***Describe the mechanism of absorption and the route of transport for water-soluble and lipid-soluble molecules.***

50 ***Describe the enzymatic digestion of carbohydrates, lipids, and proteins, and list the breakdown products of each.***

51 ***Explain how fats are emulsified. Describe the role of micelles, chylomicrons, VLDLs, LDLs, and HDLs in the absorption and transport of lipids in the body.***

52 ***Explain how tripeptides, dipeptides, and amino acids enter intestinal epithelial cells.***

53 ***Describe the movement of water through the intestinal wall.***

54 ***When and where are various ions absorbed?***

EFFECTS OF AGING ON THE DIGESTIVE SYSTEM

As a person ages, gradual changes occur throughout the entire digestive tract. The connective tissue layers of the digestive tract, the submucosa and serosa, tend to thin. The blood supply to the digestive tract decreases. There is also a decrease in the number of smooth muscle cells in the muscularis, resulting in decreased motility in the digestive tract. In addition, goblet cells within the mucosa secrete less mucus. Glands along the digestive tract, such as the gastric glands, the liver, and the pancreas, also tend to secrete less with age. These changes by themselves do not appreciably decrease the function of the digestive system.

Through the years, the digestive tract, like the skin and lungs, is directly exposed to materials from the outside environment. Some of those substances can cause mechanical damage to the digestive tract and others are toxic to the tissues. Because the connective tissue of the digestive tract becomes thin with age and because the protective mucous covering is reduced, the digestive tract of elderly people becomes less and less protected from these outside influences. In addition, the mucosa of elderly people tends to heal more slowly following injury. The liver's ability to detoxify certain chemicals tends to decline, the hepatic phagocytic cells' ability to remove particulate contaminants decreases, and the liver's ability to store glycogen decreases. These problems are increased in people who smoke.

SYSTEM INTERACTIONS Effects of Diarrhea on Other Systems

SYSTEM	INTERACTIONS
Integumentary	Pallor occurs due to vasoconstriction of blood vessels in the skin, resulting from a decrease in blood fluid levels. Pallor and sweating increase in response to abdominal pain and anxiety.
Muscular	Muscular weakness may result due to electrolyte loss, metabolic acidosis, fever, and general malaise. The involuntary stimulus to defecate may become so strong as to overcome the voluntary control mechanisms.
Nervous	Local reflexes in the colon respond to increased colon fluid volume by stimulating mass movements and the defecation reflex. Abdominal pain, much of which is felt as referred pain, can occur as the result of inflammation and distension of the colon. Decreased function is due to electrolyte loss. Reduced blood fluid levels stimulate a sensation of thirst in the CNS.
Endocrine	A decrease in extracellular fluid volume, due to the loss of fluid in the feces, stimulates the release of hormones (antidiuretic hormone from the posterior pituitary and aldosterone from the adrenal cortex) that increase water retention and electrolyte reabsorption in the kidney. In addition, decreased extracellular fluid volume and anxiety result in an increased release of epinephrine and norepinephrine from the adrenal medulla.
Cardiovascular	The movement of extracellular fluid into the colon results in a decreased blood volume. The reduced blood volume activates the baroreceptor reflex, antidiuretic hormone release, the renin-angiotensin-aldosterone mechanism, and atrial naturitic hormone release, which all elevate blood volume or increase blood pressure.
Lymphatic and Immune	White blood cells migrate to the colon in response to infection and inflammation. In the case of bacterial diarrhea, the immune response is initiated to begin the production of antibodies against bacteria and bacterial toxins.
Respiratory	As a result of reduced blood pH, respiration rate increases to eliminate carbon dioxide, which helps eliminate excess H^+.
Urinary	A decrease in urine volume and an increase in urine concentration result from activation of the baroreceptor reflex, which decreases blood flow to the kidney; antidiuretic hormone secretion, which increases water reabsorption in the kidney; and aldosterone secretion, which increases electrolyte and water reabsorption in the kidney. After approximately 24 hours, the kidney is activated to compensate for metabolic acidosis by increasing H^+ secretion and HCO_3^- reabsorption.

This overall decline in the defenses of the digestive tract with advancing age leaves elderly people more susceptible to infections and to the effects of toxic agents. Elderly people are more likely to develop ulcerations and cancers of the digestive tract. Colorectal cancers, for example, are the second leading cause of cancer deaths in the United States, with an estimated 135,000 new cases and 57,000 deaths each year.

Gastroesophageal reflux disorder (GERD) increases with advancing age. It is probably the main reason that elderly people take antacids, H_2 antagonists, and proton pump inhibitors. Disorders that are not necessarily age-induced, such as hiatal hernia and irregular or inadequate esophageal motility, can be worsened by the effects of aging, because of a general decreased motility in the digestive tract.

The enamel on the surface of elderly people's teeth becomes thinner with age and may expose the underlying dentin. In addition, the gingiva covering the tooth root recedes, exposing additional dentin. Exposed dentin may become painful and change the person's eating habits. Many elderly people also lose teeth, which can have a marked effect on eating habits unless artificial teeth are provided. The muscles of mastication tend to become weaker and, as a result, older people tend to chew their food less before swallowing.

Another complication of the age-related changes in the digestive system is the way medications and other chemicals are absorbed from the digestive tract. The decreased mucous covering and the thinned connective tissue layers allow chemicals to pass more readily from the digestive tract into the circulatory system. However, a decline in the blood supply to the digestive tract hinders the absorption of such chemicals. Drugs administered to treat cancer, which occurs in many elderly people, may irritate the mucosa of the digestive tract, resulting in nausea and loss of appetite.

55 ***What is the general effect of aging on digestive tract secretions?***

56 ***What are the effects of the overall decline in the defenses of the digestive tract with advancing age?***

SUMMARY

Anatomy of the Digestive System (p. 874)

1. The digestive system consists of a digestive tube and its associated accessory organs.
2. The digestive system consists of the oral cavity, pharynx, esophagus, stomach, small intestine, large intestine, and anus.
3. Accessory organs, such as the salivary glands, liver, gallbladder, and pancreas, are located along the digestive tract.

Functions of the Digestive System (p. 874)

The functions of the digestive system are ingestion, mastication, propulsion, mixing, secretion, digestion, absorption, and elimination.

Histology of the Digestive Tract (p. 876)

The digestive tract is composed of four tunics: mucosa, submucosa, muscularis, and serosa or adventitia.

Mucosa

1. The mucosa consists of a mucous epithelium, a lamina propria, and a muscularis mucosae.
2. The epithelium extends into the lamina propria to form intestinal glands.

Submucosa

The submucosa is a connective tissue layer containing the submucosal plexus (part of the enteric plexus), blood vessels, and small glands.

Muscularis

1. The muscularis consists of an inner layer of circular smooth muscle and an outer layer of longitudinal smooth muscle.
2. The myenteric plexus is between the two muscle layers.
3. Interstitial pacemaker cells are located throughout the myenteric plexus.

Serosa or Adventitia

The serosa or adventitia forms the outermost layer of the digestive tract.

Regulation of the Digestive System (p. 877)

Nervous, hormonal, and local chemical mechanisms regulate digestion.

Nervous Regulation of the Digestive System

Nervous regulation involves the ENS and CNS reflexes.

Chemical Regulation of the Digestive System

1. Over 30 neurotransmitters are associated with the ENS.
2. The digestive tract produces hormones that regulate digestion.
3. Other chemicals produced by the digestive tract exercise local control of digestion.

Peritoneum (p. 878)

1. The peritoneum is a serous membrane that lines the abdominal cavity and organs.
2. Mesenteries are peritoneum that extends from the body wall to many of the abdominal organs.
3. Retroperitoneal organs are located behind the peritoneum.

Oral Cavity (p. 880)

The oral cavity includes the vestibule and oral cavity proper.

Lips and Cheeks

The lips and cheeks are involved in facial expression, mastication, and speech.

Palate and Palatine Tonsils

1. The roof of the oral cavity is divided into the hard and soft palates.
2. The palatine tonsils are located in the lateral wall of the fauces.

Tongue

1. The tongue is involved in speech, taste, mastication, and swallowing.
2. The intrinsic tongue muscles change the shape of the tongue, and the extrinsic tongue muscles move the tongue.
3. The anterior two-thirds of the tongue is covered with papillae; the posterior one-third is devoid of papillae.

Teeth

1. Twenty deciduous teeth are replaced by 32 permanent teeth.
2. The types of teeth are incisors, canines, premolars, and molars.
3. A tooth consists of a crown, a neck, and a root.
4. The root is composed of dentin. Within the dentin of the root is the pulp cavity, which is filled with pulp, blood vessels, and nerves. The crown is dentin covered by enamel.
5. Periodontal ligaments hold the teeth in the alveoli.

Mastication

The muscles of mastication are the masseter, temporalis, medial pterygoid, and lateral pterygoid.

Salivary Glands

1. Salivary glands produce serous and mucous secretions.
2. The three pairs of large salivary glands are the parotid, submandibular, and sublingual.

Pharynx (p. 886)

The pharynx consists of the nasopharynx, oropharynx, and laryngopharynx.

Esophagus (p. 886)

1. The esophagus connects the pharynx to the stomach. The upper and lower esophageal sphincters regulate movement.
2. The esophagus consists of an outer adventitia, a muscular layer (longitudinal and circular), a submucosal layer (with mucous glands), and a stratified squamous epithelium.

Swallowing (p. 886)

1. During the voluntary phase of deglutition, a bolus of food is moved by the tongue from the oral cavity to the pharynx.
2. The pharyngeal phase is a reflex caused by the stimulation of stretch receptors in the pharynx.
   - The soft palate closes the nasopharynx, and the epiglottis and vestibular folds close the opening into the larynx.
   - Pharyngeal muscles move the bolus to the esophagus.

3. The esophageal phase is a reflex initiated by the stimulation of stretch receptors in the esophagus. A wave of contraction (peristalsis) moves the food to the stomach.

Stomach (p. 888)

Anatomy of the Stomach

The openings of the stomach are the gastroesophageal (to the esophagus) and the pyloric (to the duodenum).

Histology of the Stomach

1. The wall of the stomach consists of an external serosa, a muscle layer (longitudinal, circular, and oblique), a submucosa, and simple columnar epithelium (surface mucous cells).
2. Rugae are the folds in the stomach when it is empty.
3. Gastric pits are the openings to the gastric glands, which contain mucous neck cells, parietal cells, chief cells, and endocrine cells.

Secretions of the Stomach

1. Mucus protects the stomach lining.
2. Pepsinogen is converted to pepsin, which digests proteins.
3. Hydrochloric acid promotes pepsin activity and kills microorganisms.
4. Intrinsic factor is necessary for vitamin B_{12} absorption.
5. The sight, smell, taste, or thought of food initiates the cephalic phase. Nerve impulses from the medulla stimulate hydrochloric acid, pepsinogen, gastrin, and histamine secretion.
6. Distension of the stomach, which stimulates gastrin secretion and activates CNS and local reflexes that promote secretion, initiates the gastric phase.
7. Acidic chyme, which enters the duodenum and stimulates neuronal reflexes and the secretion of hormones that inhibit gastric secretions, initiates the gastrointestinal phase.

Movements of the Stomach

1. The stomach stretches and relaxes to increase volume.
2. Mixing waves mix the stomach contents with stomach secretions to form chyme.
3. Peristaltic waves move the chyme into the duodenum.
4. Gastrin and stretching of the stomach stimulate stomach emptying.
5. Chyme entering the duodenum inhibits movement through neuronal reflexes and the release of hormones.

Small Intestine (p. 896)

The small intestine is divided into the duodenum, jejunum, and ileum.

Anatomy and Histology of the Small Intestine

1. The wall of the small intestine consists of an external serosa, muscles (longitudinal and circular), submucosa, and simple columnar epithelium.
2. Circular folds, villi, and microvilli greatly increase the surface area of the intestinal lining.
3. Absorptive, goblet, and endocrine cells are in intestinal glands. Duodenal glands produce mucus.

Secretions of the Small Intestine

1. Mucus protects against digestive enzymes and stomach acids.
2. Digestive enzymes (disaccharidases and peptidases) are bound to the intestinal wall.
3. Chemical or tactile irritation, vagal stimulation, and secretin stimulate intestinal secretion.

Movement in the Small Intestine

1. Segmental contractions mix intestinal contents. Peristaltic contractions move materials distally.
2. Stretch of smooth muscles, local reflexes, and the parasympathetic nervous system stimulate contractions. Distension of the cecum initiates a reflex that inhibits peristalsis.

Liver (p. 899)

Anatomy of the Liver

1. The liver has four lobes: right, left, caudate, and quadrate.
2. The liver is divided into lobules.
   - The hepatic cords are composed of columns of hepatocytes separated by the bile canaliculi.
   - The sinusoids are enlarged spaces filled with blood and lined with endothelium and hepatic phagocytic cells.

Histology of the Liver

1. The portal triads supply the lobules.
   - The hepatic arteries and the hepatic portal veins take blood to the lobules and empty into the sinusoids.
   - The sinusoids empty into central veins, which join to form the hepatic veins, which leave the liver.
   - Bile canaliculi converge to form hepatic ducts, which leave the liver.
2. Bile leaves the liver through the hepatic duct system.
   - The hepatic ducts receive bile from the lobules.
   - The cystic duct from the gallbladder joins the hepatic duct to form the common bile duct.
   - The common bile duct joins the pancreatic duct at the point at which it empties into the duodenum.

Functions of the Liver

1. The liver produces bile, which contains bile salts that emulsify fats.
2. The liver stores and processes nutrients, produces new molecules, and detoxifies molecules.
3. Hepatic phagocytic cells phagocytize red blood cells, bacteria, and other debris.
4. The liver produces blood components.

Gallbladder (p. 904)

1. The gallbladder is a small sac on the inferior surface of the liver.
2. The gallbladder stores and concentrates bile.
3. Cholecystokinin stimulates gallbladder contraction.

Pancreas (p. 905)

Anatomy of the Pancreas

1. The pancreas is an endocrine and an exocrine gland. Its exocrine function is the production of digestive enzymes.
2. The pancreas is divided into lobules that contain acini. The acini connect to a duct system that eventually forms the pancreatic duct, which empties into the duodenum.

Pancreatic Secretions

Secretin stimulates the release of a watery bicarbonate solution that neutralizes acidic chyme.

Regulation of Pancreatic Secretion

Cholecystokinin and the vagus nerve stimulate the release of digestive enzymes.

Large Intestine (p. 907)

Anatomy of the Large Intestine

1. The cecum forms a blind sac at the junction of the small and large intestines. The vermiform appendix is a blind tube off the cecum.
2. The ascending colon extends from the cecum superiorly to the right colic flexure. The transverse colon extends from the right to the left colic flexure. The descending colon extends inferiorly to join the sigmoid colon.
3. The sigmoid colon is an S-shaped tube that ends at the rectum.
4. Longitudinal smooth muscles of the large intestine wall are arranged into bands, called teniae coli, that contract to produce pouches called haustra.
5. The mucosal lining of the large intestine is simple columnar epithelium with mucus-producing crypts.
6. The rectum is a straight tube that ends at the anus.
7. An internal anal sphincter (smooth muscle) and an external anal sphincter (skeletal muscle) surround the anal canal.

Secretions of the Large Intestine

1. Mucus protects the intestinal lining.
2. Epithelial cells secrete HCO_3^-. Sodium is absorbed by active transport, and water is absorbed by osmosis.
3. Microorganisms are responsible for vitamin K production, gas production, and much of the bulk of feces.

Movement in the Large Intestine

1. Segmental movements mix the colon's contents.
2. Mass movements are strong peristaltic contractions that occur three or four times a day.
3. Defecation is the elimination of feces. Reflex activity moves feces through the internal anal sphincter. Voluntary activity regulates movement through the external anal sphincter.

Digestion, Absorption, and Transport (p. 912)

1. Digestion is the breakdown of organic molecules into their component parts.
2. Absorption and transport are the means by which molecules are moved out of the digestive tract and are distributed throughout the body.
3. Transportation occurs by two routes.
   - Water, ions, and water-soluble products of digestion are transported to the liver through the hepatic portal system.
   - The products of lipid digestion are transported through the lymphatic system to the circulatory system.

Carbohydrates

1. Carbohydrates consist of starches, glycogen, sucrose, lactose, glucose, and fructose.
2. Polysaccharides are broken down into monosaccharides by a number of different enzymes.
3. Monosaccharides are taken up by intestinal epithelial cells by symport that is powered by a Na^+ gradient or by facilitated diffusion.
4. The monosaccharides are carried to the liver, where the nonglucose sugars are converted to glucose.
5. Glucose is transported to the cells that require energy.
6. Glucose enters the cells through facilitated diffusion.
7. Insulin influences the rate of glucose transport.

Lipids

1. Lipids include triglycerides, phospholipids, steroids, and fat-soluble vitamins.
2. Emulsification is the transformation of large lipid droplets into smaller droplets and is accomplished by bile salts.
3. Lipase digests lipid molecules to form free fatty acids and glycerol.
4. Micelles form around lipid digestion products and move to epithelial cells of the small intestine, where the products pass into the cells by simple diffusion.
5. Within the epithelial cells, free fatty acids are combined with glycerol to form triglyceride.
6. Proteins coat triglycerides, phospholipids, and cholesterol to form chylomicrons.
7. Chylomicrons enter lacteals within intestinal villi and are carried through the lymphatic system to the bloodstream.
8. Triglyceride is stored in adipose tissue, converted into other molecules, or used as energy.
9. Lipoproteins include chylomicrons, VLDL, LDL, and HDL.
10. LDL transports cholesterol to cells, and HDL transports it from cells to the liver.
11. LDLs are taken into cells by receptor-mediated endocytosis, which is controlled by a negative-feedback mechanism.

Proteins

1. Pepsin in the stomach breaks proteins into smaller polypeptide chains.
2. Proteolytic enzymes from the pancreas produce small peptide chains.
3. Peptidases, bound to the microvilli of the small intestine, break down peptides.
4. Amino acids are absorbed by symport that is powered by a Na^+ gradient.
5. Amino acids are transported to the liver, where the amino acids can be modified or released into the bloodstream.
6. Amino acids are actively transported into cells under the stimulation of growth hormone and insulin.
7. Amino acids are used as building blocks or for energy.

Water

Water can move in either direction across the wall of the small intestine, depending on the osmotic gradients across the epithelium.

Ions

1. Sodium, potassium, calcium, magnesium, and phosphate are actively transported.
2. Chloride ions move passively through the wall of the duodenum and jejunum but are actively transported from the ileum.
3. Calcium ions are actively transported, but vitamin D is required for transport, and the transport is under hormonal control.

Effects of Aging on the Digestive System (p. 920)

The mucous layer, the connective tissue, the muscles, and the secretions all tend to decrease as a person ages. These changes make an older person more open to infections and toxic agents.

REVIEW AND COMPREHENSION

1. Which layer of the digestive tract is in direct contact with the food that is consumed?
   a. mucosa
   b. muscularis
   c. serosa
   d. submucosa
2. The enteric plexus is found in the
   a. submucosa layer.
   b. muscularis layer.
   c. serosa layer.
   d. both a and b.
   e. all of the above.
3. The tongue
   a. holds food in place during mastication.
   b. plays a major role in swallowing.
   c. helps form words during speech.
   d. is a major sense organ for taste.
   e. all of the above.
4. Dentin
   a. forms the surface of the crown of the teeth.
   b. holds the teeth to the periodontal ligaments.
   c. is found in the pulp cavity.
   d. makes up most of the structure of the teeth.
   e. is harder than enamel.
5. The number of premolar deciduous teeth is
   a. 0.
   b. 2.
   c. 4.
   d. 8.
   e. 12.
6. Which of these glands does *not* secrete saliva into the oral cavity?
   a. submandibular glands
   b. pancreas
   c. sublingual glands
   d. parotid glands
7. The portion of the digestive tract in which digestion begins is the
   a. oral cavity.
   b. esophagus.
   c. stomach.
   d. duodenum.
   e. jejunum.
8. During deglutition (swallowing),
   a. the movement of food results primarily from gravity.
   b. the swallowing center in the medulla oblongata is activated.
   c. food is pushed into the oropharynx during the pharyngeal phase.
   d. the soft palate closes off the opening into the larynx.
9. The stomach
   a. has large folds in the submucosa and mucosa called rugae.
   b. has two layers of smooth muscle in the muscularis layer.
   c. opening from the esophagus is the pyloric opening.
   d. has an area closest to the duodenum called the fundus.
   e. all of the above.
10. Which of these stomach cell types is *not* correctly matched with its function?
    a. surface mucous cells—produce mucus
    b. parietal cells—produce hydrochloric acid
    c. chief cells—produce intrinsic factor
    d. endocrine cells—produce regulatory hormones
11. HCl
    a. is an enzyme.
    b. creates the acid condition necessary for pepsin to work.
    c. is secreted by the small intestine.
    d. activates salivary amylase.
    e. all of the above.
12. Why doesn't the stomach digest itself?
    a. The stomach wall is not composed of protein, so it is not affected by proteolytic enzymes.
    b. The digestive enzymes of the stomach are not strong enough to digest the stomach wall.
    c. The lining of the stomach wall has a protective layer of epithelial cells.
    d. The stomach wall is protected by large amounts of mucus.
13. Which of these hormones stimulates stomach secretions?
    a. cholecystokinin
    b. insulin
    c. gastrin
    d. secretin
14. Which of these phases of stomach secretion is correctly matched?
    a. cephalic phase—the largest volume of secretion is produced
    b. gastric phase—gastrin secretion is inhibited by distension of the stomach
    c. gastric phase—initiated by chewing, swallowing, or thinking of food
    d. gastrointestinal phase—stomach secretions are inhibited
15. Which of these structures increase the mucosal surface of the small intestine?
    a. circular folds
    b. villi
    c. microvilli
    d. length of the small intestine
    e. all of the above
16. Given these parts of the small intestine:
    1. duodenum
    2. ileum
    3. jejunum

    Choose the arrangement that lists the parts in the order food encounters them as it passes from the stomach through the small intestine.
    a. 1,2,3
    b. 1,3,2
    c. 2,1,3
    d. 2,3,1
    e. 3,1,2
17. Which cells in the small intestine have digestive enzymes attached to their surfaces?
    a. mucous cells
    b. goblet cells
    c. endocrine cells
    d. absorptive cells
18. The hepatic sinusoids
    a. receive blood from the hepatic artery.
    b. receive blood from the hepatic portal vein.
    c. empty into the central veins.
    d. all of the above.
19. Given these ducts:
    1. common bile duct
    2. common hepatic duct
    3. cystic duct
    4. hepatic ducts

    Choose the arrangement that lists the ducts in the order bile passes through them when moving from the bile canaliculi of the liver to the small intestine.
    a. 3,4,2
    b. 3,2,1
    c. 3,4,1
    d. 4,1,2
    e. 4,2,1
20. Which of these might occur if a person suffers from a severe case of hepatitis that impairs liver function?
    a. Fat digestion is difficult.
    b. By-products of hemoglobin breakdown accumulate in the blood.
    c. Plasma proteins decrease in concentration.
    d. Toxins in the blood increase.
    e. All of the above occur.
21. The gallbladder
    a. produces bile.
    b. stores bile.
    c. contracts and releases bile in response to secretin.
    d. contracts and releases bile in response to sympathetic stimulation.
    e. both b and c.

22. The aqueous component of pancreatic secretions
    a. is secreted by the pancreatic islets.
    b. contains HCO_3^-.
    c. is released primarily in response to cholecystokinin.
    d. passes directly into the blood.
    e. all of the above.
23. Given these structures:
    1. ascending colon
    2. descending colon
    3. sigmoid colon
    4. transverse colon

    Choose the arrangement that lists the structures in the order that food encounters them as it passes between the small intestine and the rectum.
    a. 1,2,3,4
    b. 1,4,2,3
    c. 2,3,1,4
    d. 2,4,1,3
    e. 3,4,1,2
24. Which of these is *not* a function of the large intestine?
    a. absorption of fats
    b. absorption of certain vitamins
    c. absorption of water and salts
    d. production of mucus
    e. all of the above
25. Defecation
    a. can be initiated by stretch of the rectum.
    b. can occur as a result of mass movements.
    c. involves local reflexes.
    d. involves parasympathetic reflexes mediated by the spinal cord.
    e. all of the above.
26. Which of these structures produces enzymes that digest carbohydrates?
    a. salivary glands
    b. pancreas
    c. lining of the small intestine
    d. both a and b
    e. all of the above
27. Bile
    a. is an important enzyme for the digestion of fats.
    b. is made by the gallbladder.
    c. contains breakdown products from hemoglobin.
    d. emulsifies fats.
    e. both c and d.
28. Micelles are
    a. lipids surrounded by bile salts.
    b. produced by the pancreas.
    c. released into lacteals.
    d. stored in the gallbladder.
    e. reabsorbed in the colon.
29. If the thoracic duct were tied off, which of these classes of nutrients would *not* enter the circulatory system at their normal rate?
    a. amino acids
    b. glucose
    c. lipids
    d. fructose
    e. nucleotides
30. Which of these lipoprotein molecules transports excess lipids from cells back to the liver?
    a. high-density lipoprotein (HDL)
    b. low-density lipoprotein (LDL)
    c. very low-density lipoprotein (VLDL)

Answer in Appendix E

CRITICAL THINKING

1. While anesthetized, patients sometimes vomit. Given that the anesthetic eliminates the swallowing reflex, explain why it is dangerous for an anesthetized patient to vomit.
2. Achlorhydria is a condition in which the stomach stops producing hydrochloric acid and other secretions. What effect would achlorhydria have on the digestive process? On red blood cell count?
3. Victor Worrystudent experienced the pain of a duodenal ulcer during final examination week. Describe the possible reasons. Explain what habits could have caused the ulcer, and recommend a reasonable remedy.
4. Gallstones sometimes obstruct the common bile duct. What are the consequences of such a blockage?
5. A patient has a spinal cord injury at level L2 of the spinal cord. How will this injury affect the patient's ability to defecate? What components of the defecation response are still present, and which are lost?
6. The bowel (colon) occasionally can become impacted. Given what you know about the functions of the colon and the factors that determine the movement of substances across the colon wall, predict the effect of the impaction on the contents of the colon above the point of impaction.
7. The bacterium *Vibrio cholerae* produces cholera toxin, which activates a chloride channel in the intestinal epithelium. In contrast, mutations that inactivate the same channel cause cystic fibrosis. Explain how increased chloride channel activity causes severe diarrhea, whereas decreased activity causes cystic fibrosis.
8. Discuss why the most effective oral rehydration therapy is water with sodium and glucose instead of water alone or water with fructose.
9. Would a patient with FH benefit from dietary changes?

Answers in Appendix F

Visit this book's website at www.mhhe.com/seeley8 for chapter quizzes, interactive learning exercises, and other study tools.

Nutrition, Metabolism, and Temperature Regulation

25

We are often more concerned with the taste of food than with its nutritional value when choosing from a menu or selecting food to prepare. Knowing about nutrition is important, however, because food provides us with energy and the building blocks necessary to synthesize new molecules. What happens if we do not obtain enough vitamins, or if we eat too much sugar and fats? Health claims about foods and food supplements bombard us every day. Which ones are ridiculous, and which ones have merit? A basic understanding of nutrition can answer these and other questions so that we can develop a healthy diet. It also allows us to know which questions currently do not have good answers.

This chapter explains *nutrition* (p. 928), *metabolism* (p. 937), *carbohydrate metabolism* (p. 938), *lipid metabolism* (p. 946), *protein metabolism* (p. 948), the *interconversion of nutrient molecules* (p. 950), *metabolic states* (p. 951), *metabolic rate* (p. 953), and *body temperature regulation* (p. 954).

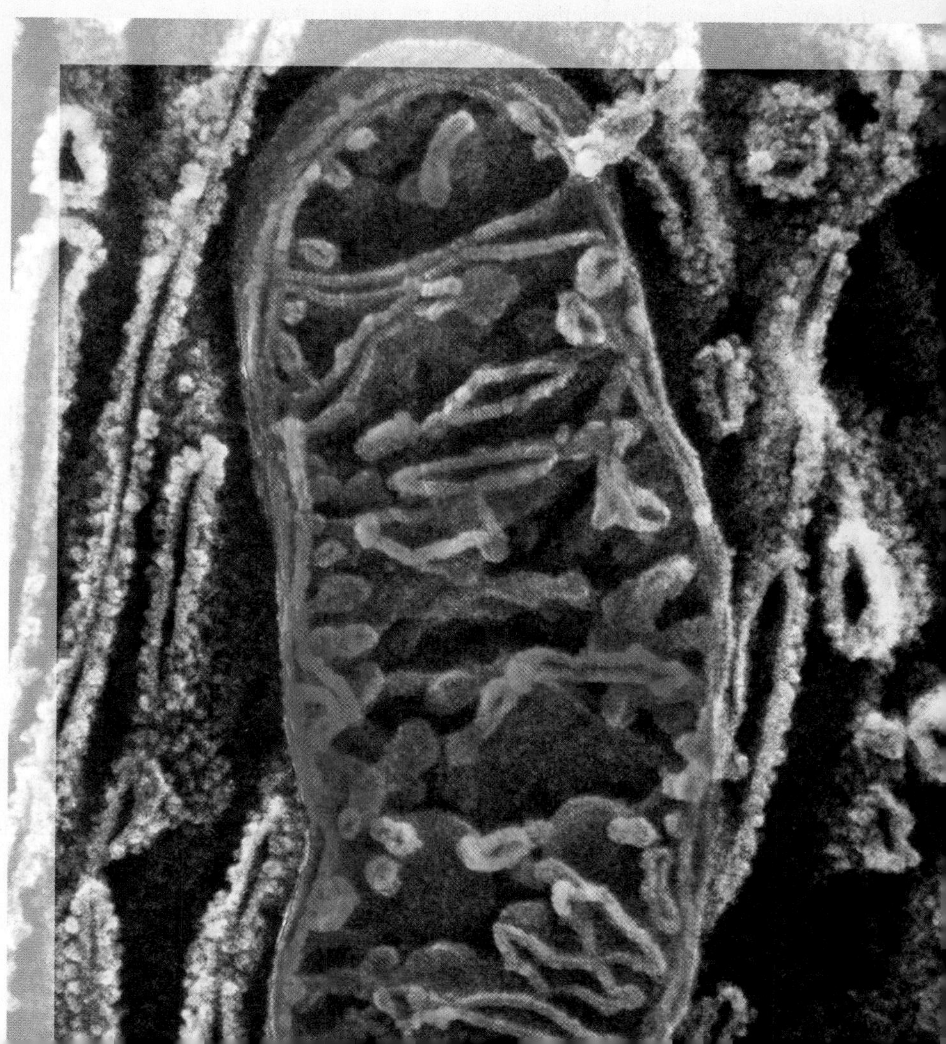

Colorized scanning electron micrograph, showing a mitochondrion (*pink*) in the cytoplasm of an intestinal epithelial cell. The mitochondrion has an outer and inner membrane. The inner membrane has numerous folds, which project into the interior of the mitochondrion. Enzymes, necessary for producing ATP, are located in these folds.

Digestive System

NUTRITION

Nutrition is the process by which the body obtains and uses certain components of food. The process includes digestion, absorption, transportation, and cell metabolism. Nutrition is also the evaluation of food and drink requirements for normal body function.

Nutrients

Nutrients are the chemicals taken into the body that are used to produce energy, to provide building blocks for new molecules, and to function in other chemical reactions. Some important substances in food, such as nondigestible plant fibers, are not nutrients. Nutrients are divided into six major classes: carbohydrates, proteins, lipids, vitamins, minerals, and water. Carbohydrates, lipids, and proteins are the major organic nutrients and they are broken down by enzymes into their individual components during digestion. Many of these subunits are broken down further to supply energy, whereas others are used as building blocks for making new carbohydrates, lipids, and proteins. Carbohydrates, lipids, proteins, and water are required in fairly substantial quantities, whereas vitamins and minerals are required in only small amounts. Vitamins, minerals, and water are taken into the body without being digested.

Essential nutrients are nutrients that must be ingested because the body cannot manufacture them or adequate amounts of them. The essential nutrients include certain amino acids, certain fatty acids, most vitamins, minerals, water, and a minimum amount of carbohydrates. The term *essential* does not mean, however, that the body requires only the essential nutrients. Other nutrients are necessary, but, if they are not part of the diet, they can be synthesized from the essential nutrients. Most of this synthesis takes place in the liver, which has a remarkable ability to transform and manufacture molecules.

1 ***Define nutrient and essential nutrient, and list the six major classes of nutrients.***

Kilocalories

The body uses the energy stored within the chemical bonds of certain nutrients. A **calorie** (kal′ō-rē; **cal**) is the amount of energy (heat) necessary to raise the temperature of 1 g of water 1°C. A **kilocalorie** (kil′ō-kal-ō-rē; **kcal**) is 1000 calories and is used to express the larger amounts of energy supplied by foods and released through metabolism.

A kilocalorie is often called a Calorie (with a capital *C*). Unfortunately, this usage has resulted in confusion between the terms *calorie* (with a lowercase *c*) and *Calorie* (with a capital *C*). It is common practice on food labels and in nutrition books to use *calorie* when *Calorie* (*kilocalorie*) is the proper term.

Most of the kilocalories supplied by food come from carbohydrates, proteins, or fats. For each gram of carbohydrate or protein that the body metabolizes, about 4 kcal of energy are released. Fats contain more energy per unit of weight than carbohydrates and proteins and yield about 9 kcal/g. Table 25.1 lists the kilocalories supplied by some typical foods. A typical diet in the United States consists of 50%–60% carbohydrates, 35%–45% fats, and 10%–15% protein. Table 25.1 also lists the carbohydrate, fat, and protein composition of some foods.

2 ***Define kilocalorie, and state the number of kilocalories supplied by a gram of carbohydrate, lipid, and protein.***

MyPyramid

Every 5 years, the U.S. Department of Health and Human Services (HHS) and the Department of Agriculture (USDA) jointly make its recommendations on what Americans should eat to be healthy. The latest recommendations, "The Dietary Guidelines for Americans 2005," were published in January 2005. Unlike the previous, single food guide pyramid, there are now 12 pyramids, which take into account a person's age, sex, and activity level. Thus, you can pick the pyramid, called MyPyramid, that best describes you (www.mypyramid.gov). All of the new pyramids have the same form (figure 25.1). Six colored bands represent the approximate, recommended proportions of grains (orange), vegetables (green), fruits (red), fats and oils (yellow), milk and milk products (blue), and meat and beans (purple). A balanced diet includes a variety of foods from each of the major food groups. Variety is necessary because no one food contains all of the nutrients necessary for health. Moderation is indicated by the narrowing of each food group from bottom to top. The wider base represents foods with little or no solid fats, added sugars, or caloric sweeteners. The climbing stick figure stresses the importance of daily exercise.

Recommendations and Criticisms of MyPyramid

The 2005 MyPyramid guidelines make several important recommendations: weight control and exercise are emphasized. Moderation, how much we eat, is at least as important as what we eat. Reducing intake by 50–100 kilocalories per day may prevent weight gain in many people. Thirty minutes a day of exercise is good for the heart, but even more is needed for the waist. There are good and bad fats. We should limit the intake of saturated fats and *trans* fats, while using unsaturated fats. Carbohydrates are good, especially whole grains, but sugars and highly refined grains should be avoided. We should consume foods rich in vitamins, minerals, and fiber. Nine servings a day of fruits and vegetables and three cups a day of low fat or fat free milk or yogurt, or three servings of other dairy products, are recommended.

Although the MyPyramid guidelines are considered a positive step forward, there are some criticisms. The guidelines permit half of the carbohydrates in the diet to be refined starch, such as white bread, white rice, and chips. The body's response to these carbohydrates is similar to its response to sugar. The guidelines recommend protein sources that are "lean, low-fat, or fat-free." Thus, the guidelines do not distinguish between lean red meat and other sources of protein, such as poultry, fish, and beans, which have less saturated fats and more unsaturated fats. The guidelines' recommendation for dairy products is intended to increase calcium intake to help prevent osteoporosis. Supplements of calcium and vitamin D, along with exercise, are effective preventive measures that do not increase daily intake of kilocalories. Also, many people are lactose intolerant and cannot consume dairy products (see chapter 24).

TABLE 25.1 Food Consumption

Food	Quantity	Food Energy (kcal)	Carbohydrate (g)	Fat (g)	Protein (g)
Dairy Products					
Whole milk (3.3% fat)	1 cup	150	11	8	8
Low-fat milk (2% fat)	1 cup	120	12	5	8
Butter	1 tablespoon	100	—	12	—
Grain					
Bread, white enriched	1 slice	75	24	1	2
Bread, whole-wheat	1 slice	65	14	1	3
Fruit					
Apple	1	80	20	1	—
Banana	1	100	26	—	1
Orange	1	65	16	—	1
Vegetables					
Corn, canned	1 cup	140	33	1	4
Peas, canned	1 cup	150	29	1	8
Lettuce	1 cup	5	2	—	—
Celery	1 cup	20	5	—	1
Potato, baked	1 large	145	33	—	4
Meat, Fish, and Poultry					
Lean ground beef (10% fat)	3 ounces	185	—	10	23
Shrimp, french fried	3 ounces	190	9	9	17
Tuna, canned	3 ounces	170	—	7	24
Chicken breast, fried	3 ounces	160	1	5	26
Bacon	2 slices	85	—	8	4
Hot dog	1	170	1	15	7
Fast Foods					
McDonald's Egg McMuffin	1	327	31	15	19
McDonald's Big Mac	1	563	41	33	26
Taco Bell's beef burrito	1	466	37	21	30
Arby's roast beef	1	350	32	15	22
Pizza Hut Super Supreme	1 slice	260	23	13	15
Long John Silver's fish	2 pieces	366	21	22	22
Dairy Queen malt, large	1	840	125	28	22
Desserts					
Chocolate chip cookie	4	200	29	9	2
Apple pie	1 piece	345	51	15	3
Dairy Queen cone, large	1	340	52	10	10
Beverage					
Cola soft drink	12 ounces	145	37	—	—
Beer	12 ounces	144	13	—	1
Wine	3-1/2 ounces	73	2	—	—
Hard liquor (86 proof)	1-1/2 ounces	105	—	—	—
Miscellaneous					
Egg	1	80	1	6	6
Mayonnaise	1 tablespoon	100	—	11	—
Sugar	1 tablespoon	45	12	—	—

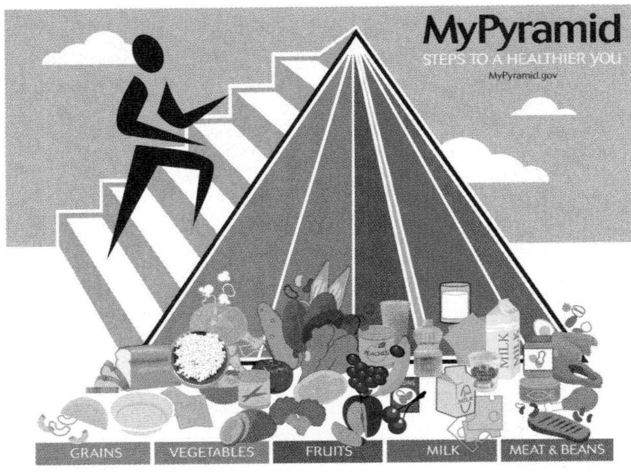

FIGURE 25.1 My Pyramid

The pyramid suggests the approaches to a healthy diet: Eat various foods (different-colored bands), eat different amounts of each food type (band width), eat in moderation (bands narrow from bottom to top), use fats and sugars sparingly (the wide base stands for foods with little or no solid fats or added sugars), and exercise (stick figure climbing stairs).

Source: U.S. Department of Agriculture.

3 ***List the six food groups shown in MyPyramid. How is the moderation of eating and exercise indicated in MyPyramid?***

Carbohydrates

Sources in the Diet

Carbohydrates include monosaccharides, disaccharides, and polysaccharides (see chapter 2). Most of the carbohydrates humans ingest come from plants. An exception is lactose (milk sugar), which is found in animal and human milk.

The most common monosaccharides in the diet are glucose and fructose. Plants capture the energy in sunlight and use the energy to produce glucose which can be found in vegetables. Fructose (fruit sugar) and galactose are isomers of glucose (see figure 2.14). Glucose is found in vegetables and fructose is found in fruits, berries, honey, and high-fructose corn syrup, which is used to sweeten soft drinks and desserts. Galactose is usually found in milk.

The disaccharide sucrose (table sugar) is what most people think of when they use the term *sugar.* Sucrose is a glucose and a fructose molecule joined together, and its principal sources are sugarcane, sugar beets, maple sugar, and honey. Maltose (malt sugar), derived from germinating cereals, is a combination of two glucose molecules, and lactose (in milk) consists of a glucose and a galactose molecule (see figure 2.14).

The **complex carbohydrates** are the polysaccharides: starch, glycogen, and cellulose. These polysaccharides consist of many glucose molecules bound together to form long chains. Starch is an energy storage molecule in plants found primarily in vegetables, fruits, and grains. Glycogen is an energy storage molecule in animals and is located primarily in muscle and in the liver. By the time meats have been processed, they contain little, if any, glycogen because it is used up by the dying muscle cells (see "Anaerobic Respiration," p. 941). Cellulose forms cell walls, which surround plant cells.

Uses in the Body

During digestion, polysaccharides and disaccharides are split into monosaccharides, which are absorbed into the blood (see chapter 24). Humans can digest starch and glycogen because they can break the bonds between the glucose molecules of starch and glycogen. Humans are unable to digest cellulose because they cannot break the bonds between its glucose molecules. Instead, cellulose provides fiber, or "roughage," thereby increasing the bulk of feces, making it easier to defecate.

The liver converts fructose, galactose, and other monosaccharides absorbed by the blood into glucose. Glucose, whether absorbed from the digestive tract or synthesized in the liver, is an energy source used to produce **adenosine triphosphate (ATP)** molecules (see "Anaerobic Respiration," p. 941 and "Aerobic Respiration," p. 942). Because the brain relies almost entirely on glucose for its energy, blood glucose levels are carefully regulated (see chapter 18).

If excess amounts of glucose are present, the glucose is converted into glycogen, which is stored in muscle and in the liver. Because cells can store only a limited amount of glycogen, any additional glucose is converted into fat, which is stored in adipose tissue. Glycogen can be rapidly converted back to glucose when energy is needed. For example, during exercise muscles convert glycogen to glucose and in-between meals the liver helps maintain blood sugar levels by converting glycogen to glucose, which is released into the blood.

In addition to being used as a source of energy, sugars have other functions. They form part of deoxyribonucleic acid (DNA), ribonucleic acid (RNA), and ATP molecules (see chapter 2); they also combine with proteins to form glycoproteins, such as the glycoprotein receptor molecules on the outer surface of the plasma membrane (see chapter 3).

Recommended Amounts

According to the Dietary Guidelines Advisory Committee, the **Acceptable Macronutrient Distribution Range (AMDR)** for carbohydrates is 45%–65% of total kilocalories. Although a minimum level of carbohydrates is not known, it is assumed that amounts of 100 g or less per day result in overuse of the body's proteins and fats for energy sources. Because muscles are primarily protein, the use of proteins for energy can result in the breakdown of muscle tissue. The extensive use of fats as an energy source can result in acidosis (see chapter 18).

Complex carbohydrates are recommended in the diet because starchy foods often contain other valuable nutrients, such as vitamins and minerals, and because the slower rate of digestion and absorption of complex carbohydrates does not result in large increases and decreases in blood glucose levels, as the consumption of large amounts of simple sugars does. Foods containing large amounts of simple sugars, such as soft drinks and candy, are rich in carbohydrates, but they have few other nutrients. For example, a typical soft drink is mostly sucrose, containing 9 teaspoons of sugar per 12 oz container. In excess, the consumption of these kinds of foods usually results in obesity and tooth decay.

4 ***What are the most common monosaccharides in the diet? What are sucrose, maltose, and lactose?***

5 ***Give three examples of complex carbohydrates. How does the body use them?***

6 ***How does the body use glucose and other monosaccharides?***

7 ***What quantities of carbohydrate should be ingested daily?***

Lipids

Sources in the Diet

About 95% of the lipids in the human diet are **triglycerides** (trī-glis′er-īdz). Triglycerides, which are sometimes called **triacylglycerols** (trī-as′il-glis′er-olz), consist of three fatty acids attached to a glycerol molecule (see chapter 2). Triglycerides are often referred to as fats or oils. Fats are solid at room temperature, whereas oils are liquid. Fats and oils can be categorized as saturated or unsaturated. **Saturated fats and oils** have only single covalent bonds between the carbon atoms of their fatty acids (see figure 2.17). They are found in the fats of meats (e.g., beef, pork), dairy products (e.g., whole milk, cheese, butter), eggs, coconut oil, and palm oil. **Unsaturated fats and oils** have one or more double covalent bonds between the carbon atoms of their fatty acids (see figure 2.17). **Monounsaturated** fats have one double bond and **polyunsaturated** fats have two or more double bonds. Monounsaturated fats include olive and peanut oils; polyunsaturated fats are in fish, safflower, sunflower, and corn oil.

Unsaturated fatty acids can be classified according to the location of their first double bond from the omega (methyl) end of the fatty acid. The first double bond of an omega-3 fatty acid starts three carbon atoms after the omega end, an omega-6 fatty acid after six carbons, and an omega-9 fatty acid after nine carbons (see figure 2.17).

Saturating Fats

Solid fats, such as shortening and margarine, work better than liquid oils do for preparing some foods, such as pastries. Polyunsaturated vegetable oils can be changed from a liquid to a solid by making them more saturated—that is, by decreasing the number of double covalent bonds in their polyunsaturated fatty acids. To saturate an unsaturated oil, hydrogen gas is bubbled through it. As hydrogen binds to the fatty acids, double covalent bonds are converted to single covalent bonds to produce a change in molecular shape that solidifies the oil. The more saturated the product, the harder it becomes at room temperature.

Unprocessed polyunsaturated fats are found mostly in the ***cis* form,** which means the hydrogen atoms are on the same side of the carbon-carbon double bond in their fatty acids (see figure 2.17). During hydrogenation, some of the hydrogen atoms are transferred to the opposite side of the double bond to make the ***trans* form,** in which one hydrogen atom is on one side of the double bond and another is on the opposite side. Processed foods and oils account for most of the *trans* fats in the American diet, although some *trans* fats occur naturally in food from animal sources. *Trans* fatty acids raise the concentration of low-density lipoproteins and lower the concentration of high-density lipoproteins in the blood (see chapter 24). These changes are associated with a greater risk for cardiovascular disease.

The remaining 5% of lipids include cholesterol and phospholipids, such as **lecithin** (les′i-thin). Cholesterol is a steroid (see chapter 2) found in high concentrations in liver and egg yolks, but it is also present in whole milk, cheese, butter, and meats. Cholesterol is not found in plants. Phospholipids are major components of plasma membranes, and they are found in a variety of foods, such as egg yolks.

Uses in the body

Triglycerides are important sources of energy that are used to produce ATP molecules. A gram of triglyceride delivers more than twice as many kilocalories as a gram of carbohydrate. Some cells, such as skeletal muscle cells, derive most of their energy from triglycerides.

After a meal, excess triglycerides that are not immediately used are stored in adipose tissue or the liver. Later, when energy is required, the triglycerides are broken down, and their fatty acids are released into the blood, where they are taken up and used by various tissues. In addition to storing energy, adipose tissue surrounds and pads organs, and under the skin adipose tissue is an insulator, which prevents heat loss.

Cholesterol is an important molecule with many functions in the body. It can either be obtained in food or manufactured by the liver and most other tissues. Cholesterol is a component of the plasma membrane, and it can be modified to form other useful molecules, such as bile salts and steroid hormones. Bile salts are necessary for fat digestion and absorption. Steroid hormones include the sex hormones estrogen, progesterone, and testosterone, which regulate the reproductive system.

The eicosanoids (ī′kō-să-noydz), which include prostaglandins and leukotrienes, are derived from fatty acids. The molecules are involved in activities such as inflammation, blood clotting, tissue repair, and smooth muscle contraction. Phospholipids, such as lecithin, are part of the plasma membrane and are used to construct the myelin sheath around the axons of nerve cells. Lecithin is also found in bile and helps emulsify fats.

Recommended Amounts

The AMDR for fats is 20%–35% for adults, 25%–35% for children and adolescents 4 to 18 years of age, and 30%–35% for children 2 to 3 years of age. Saturated fats should be 10% of total kilocalories, or as low as possible. Most dietary fat should come from sources of polyunsaturated and monounsaturated fats. Cholesterol should be limited to 300 mg (the amount in one egg yolk) or less per day and *trans* fat consumption should be as low as possible. These guidelines reflect the belief that excess amounts of fats, especially saturated fats, *trans* fats, and cholesterol, contribute to cardiovascular disease. The typical American diet derives 35%–45% of its kilocalories from fats, indicating that most Americans need to reduce fat consumption.

Most of the lecithin consumed in the diet is broken down in the digestive tract. The liver can manufacture all of the lecithin necessary to meet the body's needs, so it is not necessary to consume lecithin supplements.

Alpha-linolenic (lin-ō-len′ik) **acid** is an omega-3 fatty acid, and **linoleic** (lin-ō-lē′ik) **acid** is an omega-6 fatty acid. They are

essential fatty acids, which must be ingested because humans lack the enzymes necessary to synthesize them. Other fatty acids, such as omega-9 fatty acids, can be synthesized from the essential fatty acids. Seeds, nuts, and legumes are good sources of alpha-linolenic and linoleic acids. Alpha-linolenic acid is in the green leaves of plants, and linoleic acid is found in grains.

Fatty Acids and Blood Clotting

The essential fatty acids are used to synthesize prostaglandins that affect blood clotting. Linoleic acid can be converted to **arachidonic** (ă-rak-i-don′ik) **acid,** which is an omega-6 fatty acid. The arachidonic acid is used to produce thromboxanes, which increase blood clotting. Alpha-linolenic acid can be converted to **eicosapentaenoic** (ī′kō-să-pen-tă-nō′ik) **acid (EPA)** and **docasahexaenoic** (dō′kō-să-heks-ă-nō′ik) **acid (DHA),** which are omega-3 fatty acids. They can be used to synthesize prostaglandins that decrease blood clotting. Individuals who consume foods rich in EPA and DHA, such as herring, salmon, tuna, and sardines, increase the synthesis of prostaglandins from EPA and DHA. Individuals who eat these fish two or more times per week have a lower risk for heart attack than those who do not, possibly because of reduced blood clotting. EPA and DHA are also known to reduce blood triglyceride levels. Those who do not like to eat fish can take fish oil supplements as a source of EPA and DHA. Flaxseed is a source of alpha-linolenic acid, from which EPA and DHA can be synthesized. Whether or not flaxseed can provide the same benefits as EPA and DHA is under investigation. Individuals who have bleeding disorders, take anticoagulants, or anticipate surgery should follow their physician's advice regarding the use of these supplements, however, because they can increase the risk for bleeding and hemorrhagic stroke.

8 ***What is the major source of lipids in the diet? What are other sources?***

9 ***Define saturated and unsaturated fats.***

10 ***How does the body use triglycerides, cholesterol, prostaglandins, and lecithin?***

11 ***Describe the recommended dietary intake of lipids. List the essential fatty acids.***

Proteins

Sources in the Diet

Proteins are chains of amino acids (see chapter 2). Proteins in the body are constructed of 20 kinds of amino acids, which are divided into two groups: essential and nonessential. The body cannot synthesize **essential amino acids,** so they must be obtained in the diet. The nine essential amino acids are histidine, isoleucine, leucine, lysine, methionine, phenylalanine, threonine, tryptophan, and valine. Although the **nonessential amino acids** are necessary to construct our proteins, they are nonessential in the sense that it is not necessary to ingest them because they can be synthesized from the essential amino acids.

A **complete protein** food contains adequate amounts of all nine essential amino acids, whereas an **incomplete protein** food does not. Examples of complete proteins are meat, fish, poultry, milk, cheese, and eggs; incomplete proteins include leafy green vegetables, grains, and legumes (peas and beans).

Uses in the Body

The body uses essential and nonessential amino acids to synthesize proteins. Proteins perform numerous functions, as the following examples illustrate. Collagen provides structural strength in connective tissue, as does keratin in the skin, and the combination of actin and myosin makes muscle contraction possible. Enzymes regulate the rate of chemical reactions, and protein hormones regulate many physiologic processes (see chapter 18). Proteins in the blood prevent changes in pH (buffers), promote blood clotting (coagulation factors), and transport oxygen and carbon dioxide in the blood (hemoglobin). Transport proteins (see chapter 3) move materials across plasma membranes, and other proteins in the plasma membrane function as receptor molecules. Antibodies, lymphokines, and complement are part of the immune system response that protects against microorganisms and other foreign substances.

The body also uses proteins as a source of energy, yielding the same amount of energy as carbohydrates. If excess proteins are ingested, the energy in the proteins can be stored by converting their amino acids into glycogen or fats.

Recommended Amounts

The AMDR for protein is 10%–35% of total kilocalories. If two incomplete proteins, such as rice and beans are ingested, each can provide the amino acids lacking in the other. Thus, a correctly balanced vegetarian diet can provide all of the essential amino acids.

When protein intake is adequate, the synthesis and breakdown of proteins in a healthy adult occur at the same rate. The amino acids of proteins contain nitrogen, so saying that a person is in **nitrogen balance** means that the nitrogen content of ingested protein is equal to the nitrogen excreted in urine and feces. A starving person is in negative nitrogen balance because the nitrogen gained in the diet is less than that lost by excretion. In other words, when proteins are broken down for energy, more nitrogen is lost than is replaced in the diet. A growing child or a healthy pregnant woman, on the other hand, is in positive nitrogen balance because more nitrogen is going into the body to produce new tissues than is lost by excretion.

12 ***Distinguish between essential and nonessential amino acids and between complete and incomplete protein foods.***

13 ***Describe some of the functions performed by proteins in the body.***

14 ***What is the AMDR of proteins? Define nitrogen balance.***

Vitamins

Vitamins (vīt′ă-minz; life-giving chemicals) are organic molecules that exist in minute quantities in food and are essential to normal metabolism (table 25.2). **Essential vitamins** cannot be produced by the body and must be obtained through the diet.

TABLE 25.2 **Principal Vitamins**

Vitamin	Fat (F)- or Water (W)-Soluble	Source	Function	Symptoms of Deficiency	Reference Daily Intake (RDI)s*
A (retinol)	F	From provitamin beta carotene found in yellow and green vegetables: preformed in liver, egg yolk, butter, and milk	Necessary for rhodopsin synthesis, normal health of epithelial cells, and bone and tooth growth	Rhodopsin deficiency, night blindness, retarded growth, skin disorders, and increase in infection risk	900 RE[†]
B_1 (thiamine)	W	Yeast, grains, and milk	Involved in carbohydrate and amino acid metabolism, necessary for growth	Beriberi—muscle weakness (including cardiac muscle), neuritis, and paralysis	1.2 mg
B_2 (riboflavin)	W	Green vegetables, liver, wheat germ, milk, and eggs	Component of flavin adenine dinucleotide; involved in citric acid cycle	Eye disorders and skin cracking, especially at corners of the mouth	1.3 mg
B_3 (niacin)	W	Fish, liver, red meat, yeast, grains, peas, beans, and nuts	Component of nicotinamide adenine dinucleotide; involved in glycolysis and citric acid cycle	Pellagra—diarrhea, dermatitis, and nervous system disorder	16 mg
Pantothenic acid	W	Liver, yeast, green vegetables, grains, and intestinal bacteria	Constituent of coenzyme-A, glucose production from lipids and amino acids, and steroid hormone synthesis	Neuromuscular dysfunction and fatigue	5 mg
Biotin	W	Liver, yeast, eggs, and intestinal bacteria	Fatty acid and nucleic acid synthesis, movement of pyruvic acid into citric acid cycle	Mental and muscle dysfunction, fatigue, and nausea	30 µg
B_6 (pyridoxine)	W	Fish, liver, yeast, tomatoes, and intestinal bacteria	Involved in amino acid metabolism	Dermatitis, retarded growth, and nausea	1.7 mg
Folate	W	Liver, leafy green vegetables, and intestinal bacteria	Nucleic acid synthesis, hematopoiesis, prevent birth defects	Macrocytic anemia (enlarged red blood cells) and neural tube defects	0.4 mg
B_{12} (cobalamins)	W	Liver, red meat, milk, and eggs	Necessary for red blood cell production, some nucleic acid and amino acid metabolism	Pernicious anemia and nervous system disorders	2.4 µg
C (ascorbic acid)	W	Citrus fruit, tomatoes, and green vegetables	Collagen synthesis, general protein metabolism	Scurvy—defective bone formation and poor wound healing	90 mg
D (cholecalciferol, ergosterol)	F	Fish liver oil, enriched milk, and eggs; provitamin D converted by sunlight to cholecalciferol in the skin	Promotes calcium and phosphorus use, normal growth and bone and teeth formation	Rickets—poorly developed, weak bones, osteomalacia; and bone reabsorption	10 µg[‡]
E (tocopherol, tocotrienols)	F	Wheat germ, cottonseed, palm, and rice oils; grain; liver; and lettuce	Prevents the oxidation of cell membranes and DNA	Hemolysis of red blood cells	15 mg
K (phylloquinone)	F	Alfalfa, liver, spinach, vegetable oils, cabbage, and intestinal bacteria	Required for synthesis of a number of clotting factors	Excessive bleeding due to retarded blood clotting	120 µg

*RDIs for people over 4 years of age; IU = international units.

[†]Retinol equivalents (RE). 1 retinol equivalent = 1 µg retinol or 6 µg beta carotene.

[‡]As cholecalciferol, 1 µg cholecalciferol = 40 IU vitamin D.

Because no single food item or nutrient class provides all the essential vitamins, it is necessary to maintain a balanced diet by eating a variety of foods. The absence of an essential vitamin in the diet can result in a specific deficiency disease. A few vitamins, such as vitamin K, are produced by intestinal bacteria, and a few can be formed by the body from substances called provitamins. A **provitamin** is a part of a vitamin that can be assembled or modified by the body into a functional vitamin. Beta carotene is an example of a provitamin that can be modified by the body to form vitamin A. The other provitamins are **7-dehydrocholesterol** (dē-hī′dro-kō-les′ter-ol), which can be converted to vitamin D, and **tryptophan** (trip′tō-fan), which can be converted to niacin.

Vitamins are not broken down by catabolism but are used by the body in their original or slightly modified forms. After the chemical structure of a vitamin is destroyed, its function is usually lost. The chemical structure of many vitamins is destroyed by heat, such as when food is overcooked.

Many vitamins function as **coenzymes,** which combine with enzymes to make the enzymes functional (see chapter 2). Without coenzymes and their enzymes, many chemical reactions would occur too slowly to support good health and even life. For example, vitamins B_2 and B_3, biotin (bī′ō-tin), and pantothenic (pan-tō-then′ik) acid are critical for some of the chemical reactions involved in the production of ATP. Folate (fō′lāt) and vitamin B_{12} are involved in nucleic acid synthesis. Vitamins A, B_1, B_6, B_{12}, C, and D are necessary for growth. Vitamin K is necessary for the synthesis of proteins involved in blood clotting (see table 25.2).

Vitamins are either fat-soluble or water-soluble. **Fat-soluble vitamins,** such as vitamins A, D, E, and K, dissolve in lipids. They are absorbed from the intestine along with lipids. Some of them can be stored in the body for a long time. Because they can be stored, these vitamins can accumulate in the body to the point of toxicity. **Water-soluble vitamins,** such as the B vitamins and vitamin C, dissolve in water. They are absorbed from the water in the intestinal tract and typically remain in the body only a short time before being excreted in the urine.

Vitamins were discovered at the beginning of the twentieth century. They were found to be associated with certain foods known to protect people from diseases such as rickets and beriberi. In 1941, the first Food and Nutrition Board established the **Recommended Dietary Allowances (RDAs),** which are the nutrient intakes sufficient to meet the needs of nearly all people in certain age and gender groups. RDAs were established for different-aged males and females, starting with infants and continuing on to adults. RDAs are also set for pregnant and lactating women. The RDAs have been reevaluated every 4–5 years and updated when necessary on the basis of new information.

The RDAs establish a minimum intake of vitamins and minerals that should protect almost everyone (97%) in a given group from diseases caused by vitamin or mineral deficiencies. Although personal requirements can vary, the RDAs are a good benchmark. The further dietary intake is below the RDAs, the more likely a nutritional deficiency is to occur. On the other hand, the consumption of too large a quantity of some vitamins and minerals can have harmful effects. For example, the long-term ingestion of 3–10 times the RDA for vitamin A can cause bone and muscle pain, skin disorders, hair loss, and increased liver size. The long-term consumption of 5–10 times the RDA of vitamin D can result in the deposition of calcium in the kidneys, heart, and blood vessels, and the regular consumption of more than 2 g of vitamin C daily can cause stomach inflammation and diarrhea.

Free Radicals and Antioxidants

Free radicals are molecules, produced as part of normal metabolism, that are missing an electron. Free radicals can replace the missing electron by taking an electron from cell molecules, such as fats, proteins, or DNA, resulting in damage to the cell. Damage from free radicals may contribute to aging and certain diseases, such as atherosclerosis and cancer. The loss of an electron from a molecule is called oxidation. **Antioxidants** are substances that prevent the oxidation of cell components by donating an electron to free radicals. Examples of antioxidants include beta carotene (provitamin A), vitamin C, and vitamin E.

Many studies have been done to determine whether or not taking large doses of antioxidants is beneficial. Although future research may suggest otherwise, the consensus among scientists establishing the RDAs is that the best evidence presently available does not support claims that taking large doses of antioxidants prevents chronic disease or otherwise improves health. On the other hand, the amount of antioxidants normally found in a balanced diet that includes fruits and vegetables rich in antioxidants, combined with the complex mix of other chemicals found in food, can be beneficial.

15 ***What are vitamins, essential vitamins, and provitamins? Name the water-soluble vitamins and the fat-soluble vitamins.***

16 ***List some of the functions of vitamins.***

17 ***What are Recommended Dietary Allowances (RDAs)? Why are they useful?***

PREDICT 1

What would happen if vitamins were broken down during the process of digestion rather than being absorbed intact into the circulation?

Minerals

Minerals (min′er-ălz) are inorganic nutrients that are necessary for normal metabolic functions. Based on the amount of the mineral required in the diet for good health, the minerals are divided into two groups. The daily requirement for **major minerals** is 100 mg or more daily, whereas for **trace minerals** it is less than 100 mg daily. The requirement for some trace minerals is unknown. Minerals constitute about 4%–5% of total body weight and are components of coenzymes, a few vitamins, hemoglobin, and other organic molecules. Minerals are involved in a number of important functions, such as establishing resting membrane potentials and generating action potentials, adding

TABLE 25.3 Important Minerals

Mineral	Function	Symptoms of Deficiency	Reference Daily Intake (RDIs)*
Calcium	Bone and teeth formation, blood clotting, muscle activity, and nerve function	Spontaneous action potential generation in neurons and tetany	1300 mg
Chlorine	Blood acid–base balance, hydrochloric acid production in stomach	Acid–base imbalance	2.3 g†
Chromium	Associated with enzymes in glucose metabolism	Unknown	35 μg
Cobalt	Component of vitamin B_{12}, red blood cell production	Anemia	Unknown
Copper	Hemoglobin and melanin production, electron-transport system	Anemia and loss of energy	0.9 mg
Fluorine	Provides extra strength in teeth, prevents dental caries	No real pathology	4 mg
Iodine	Thyroid hormone production, maintenance of normal metabolic rate	Goiter and decrease in normal metabolism	150 μg
Iron	Component of hemoglobin, ATP production in electron-transport system	Anemia, decreased oxygen transport, and energy loss	18 mg
Magnesium	Coenzyme constituent, bone formation, and muscle and nerve function	Increased nervous system irritability, vasodilation, and arrhythmias	420 mg
Manganese	Hemoglobin synthesis, growth, and activation of several enzymes	Tremors and convulsions	2.3 mg
Molybdenum	Enzyme component	Unknown	45 μg
Phosphorus	Bone and teeth formation, energy transfer (ATP), and component of nucleic acids	Loss of energy and cellular function	1250 mg
Potassium	Muscle and nerve function	Muscle weakness, abnormal electrocardiogram, and alkaline urine	4.7 g
Selenium	Component of many enzymes	Unknown	55 μg
Sodium	Osmotic pressure regulation and nerve and muscle function	Nausea, vomiting, exhaustion, and dizziness	1.5 g†
Sulfur	Component of hormones, several vitamins, and proteins	Unknown	Unknown
Zinc	Component of several enzymes, carbon dioxide transport and metabolism, and protein metabolism	Deficient carbon dioxide transport and deficient protein metabolism	11 mg

*RDIs for people over 4 years of age, except for sodium.
†3.8 g sodium chloride (table salt).

mechanical strength to bones and teeth, combining with organic molecules, and acting as coenzymes, buffers, and regulators of osmotic pressure. Table 25.3 lists important minerals and their functions.

Minerals are ingested by themselves or in combination with organic molecules, and they are obtained from animal and plant sources. Mineral absorption from plants, however, can be limited because the minerals tend to bind to plant fibers. Refined breads and cereals have hardly any minerals or vitamins because they are lost in the processing of the seeds used to make them. The seeds are crushed and the outer parts of the seeds, which contain most of their minerals and vitamins, are removed. The inner part of the seeds, which has few minerals and vitamins, is used to make the refined breads and cereals. Minerals and vitamins are often added to refined breads and cereals to compensate for their loss during the refinement process.

A balanced diet can provide all the vitamins and minerals required for good health for most people. Some nutritionists, however, recommend taking a once-a-day multiple vitamin and mineral supplement as insurance because many people do not have a balanced diet.

18 ***What are minerals? List some of the important functions of minerals.***

Vegetarian Diet

Plants alone can provide all of the protein required for good health. In order to get adequate amounts of the essential amino acids, a variety of protein sources, such as grains and legumes, should be consumed.

The Dietary Guidelines for Americans recommend that vegan diets be supplemented with vitamin B_{12}, vitamin D, calcium, iron, and zinc. This is especially important for children and pregnant and lactating women. Plant sources do not supply vitamin B_{12} or sufficient amounts of vitamin D, although the body can produce vitamin D with adequate exposure to sunlight (see chapter 5). Calcium is found in leafy green vegetables and nuts. Iron and zinc are in whole grains, nuts, and legumes. However, these minerals are either in low amounts or not easily absorbed.

Daily Values

Daily Values are dietary reference values that appear on food labels to help consumers plan a healthy diet. However, not all possible Daily Values are required to be listed on food labels. Daily Values are based on two other sets of reference values: Reference Daily Intakes and Daily Reference Values. The **Reference Daily Intakes (RDIs)** are based on the 1968 RDAs for certain vitamins and minerals. RDIs have been set for four categories of people: infants, toddlers, people over 4 years of age, and pregnant or lactating women. Generally, the RDIs are set to the highest 1968 RDA value of an age category. For example, the highest RDA for iron in males over 4 years of age is 10 mg/day and for females over 4 years of age is 18 mg/day. Thus, the RDI for iron is set at 18 mg/day. The **Daily Reference Values (DRVs)** are set for total fat, saturated fat, cholesterol, total carbohydrate, dietary fiber, sodium, potassium, and protein.

Having two standards on food labels, RDIs for vitamins and minerals and DRVs for other nutrients, was thought to be more confusing for consumers than having one standard. Therefore, the RDIs and DRVs were combined to form the Daily Values.

The Daily Values appearing on food labels are based on a 2000 kcal reference diet, which approximates the weight maintenance requirements of postmenopausal women, women who exercise moderately, teenage girls, and sedentary men (figure 25.2). On large food labels, additional information is listed based on a daily intake of 2500 kcal, which is adequate for young men.

The Daily Values for energy-producing nutrients are determined as a percentage of daily kilocaloric intake: 60% for carbohydrates, 30% for total fats, 10% for saturated fats, and 10% for proteins. The Daily Value for fiber is 11.5 g for each 1000 kcal of intake. The Daily Values for a nutrient in a 2000 kcal/day diet can be calculated on the basis of the recommended daily percentage of the nutrient and the kilocalories in a gram of the nutrient. For example, carbohydrates should be 60% of a 2000 kcal/day diet, or 1200 kcal/day (0.60×2000). Since there are 4 kcal in a gram of carbohydrate, the Daily Value for carbohydrate is 300 g/day (1200/4).

The Daily Values for some nutrients are the uppermost limits considered desirable because of the link between these nutrients and certain diseases. Thus, the Daily Values for total fats are less than 65 g; saturated fats, less than 20 g; and cholesterol, less than 300 mg because of their association with increased risk for heart disease. The Daily Value for sodium is less than 2400 mg because of its association with high blood pressure in some people.

Nutrition Facts

Serving Size 1 oz. (28g/About 32 chips)
Servings Per Container 2.5

Amount Per Serving	
Calories 160	Calories from Fat 90
	% Daily Value*
Total Fat 10g	**16%**
Saturated Fat 1.5g	**7%**
Cholesterol 0mg	**0%**
Sodium 170mg	**7%**
Total Carbohydrate 15g	**5%**
Dietary Fiber 1g	**4%**
Sugars less than 1g	
Protein 2g	
Vitamin A 0% •	Vitamin C 0%
Calcium 2% •	Iron 0%

* Percent Daily Values are based on a 2,000 calorie diet. Your daily values may be higher or lower depending on your calorie needs:

	Calories:	2,000	2,500
Total Fat	Less than	65g	80g
Sat Fat	Less than	20g	25g
Cholesterol	Less than	300mg	300mg
Sodium	Less than	2,400mg	2,400mg
Total Carbohydrate		300g	375g
Dietary Fiber		25g	30g

Calories per gram:
Fat 9 • Carbohydrate 4 • Protein 4

FIGURE 25.2 Food Label

For a particular food, the Daily Value is used to calculate the **Percent Daily Value (% Daily Value)** for some of the nutrients in one serving of the food (see figure 25.2). For example, if a serving of food has 3 g of fat and the Daily Value for total fat is 65 g, the % Daily Value is 5% ($3/65 = 0.05$, or 5%). The Food and Drug Administration (FDA) requires % Daily Values to be on food labels so that the public has useful and accurate dietary information.

PREDICT 2

One serving of a food has 30 g of carbohydrate. What % Daily Value for carbohydrate is on the food label for this food?

The % Daily Values for nutrients related to energy consumption are based on a 2000 kcal/day diet. For people who maintain their weight on a 2000 kcal/day diet, the total of the % Daily Values for each of these nutrients should add up to no more than 100%. For individuals consuming more or fewer kilocalories per day than 2000 cal, however, the total of the % Daily Values can be more or fewer than 100%. For example, for a person consuming 2200 kcal/day, the total of the % Daily Values for each of these nutrients should add up to no more than 110% because $2200/2000 = 1.10$, or 110%.

PREDICT 3

Suppose a person consumes 1800 kcal/day. What total % Daily Values for energy-producing nutrients is recommended?

When using the % Daily Values of a food to determine how the amounts of certain nutrients in the food fit into the overall diet, the number of servings in a container or package needs to be considered. For example, suppose a small (2.25-ounce) bag of corn chips has a % Daily Value of 16% for total fat. One might suppose that eating the bag of chips accounts for 16% of total fat for the day. The bag, however, contains 2.5 servings. Therefore, if all the chips in the bag are consumed, they account for 40% (16% × 2.5) of the maximum recommended total fat.

19 ***What are the Reference Daily Intakes and the Daily Reference Values? When combined, what reference set of values is established?***

20 ***Define % Daily Values. The % Daily Values appearing on food labels is based on how many kilocalories per day?***

METABOLISM

Metabolism (mĕ-tab′ō-lizm; change) is the total of all the chemical reactions that occur in the body. It consists of **catabolism** (kă-tab′ō-lizm), the energy-releasing process by which large molecules are broken down into smaller molecules, and **anabolism** (ă-nab′ō-lizm), the energy-requiring process by which small molecules are joined to form larger molecules.

Catabolism begins during the process of digestion and is concluded within individual cells. The energy derived from catabolism is used to drive anabolic reactions and processes such as active transport and muscle contraction. Anabolism occurs in all the cells of the body as they divide to form new cells, maintain their own intracellular structure, and produce molecules such as hormones, neurotransmitters, and extracellular matrix molecules for export.

Large nutrient molecules, such as carbohydrates, lipids, and proteins, are broken down by digestion into smaller molecules, such as glucose, amino acids, and fatty acids, which are absorbed from the digestive tract into the blood (see chapter 24). These smaller molecules are taken into cells, they are catabolized, and the energy from them is used to combine adenosine diphosphate (ADP) and an inorganic phosphate group (P_i) to form ATP (figure 25.3):

$$ADP + P_i + Energy \rightarrow ATP$$

The energy in small nutrient molecules is used to produce many ATP molecules, each of which stores a small amount of energy. The smaller amount of energy in each of the many ATP molecules is more readily available for use in cells than is the larger amount of energy stored in nutrient molecules. ATP is often called the energy currency of the cell because, when it is spent, or broken down to ADP, energy becomes available for use by the cell. If a quarter represents an ATP molecule, then a $20 bill is analogous to a small nutrient molecule. The quarter (ATP) can be used in various vending machines (chemical reactions), but the $20 bill (nutrient molecule) cannot.

The chemical reactions responsible for the transfer of energy from the chemical bonds of nutrient molecules to ATP molecules involve oxidation–reduction reactions (see chapter 2). A molecule is reduced when it gains electrons and is oxidized when it loses electrons. A nutrient molecule has many hydrogen atoms covalently bonded to the carbon atoms that form the "backbone" of the molecule. Because a hydrogen atom is a H^+ (proton) and an electron, the nutrient molecule has many electrons and is, therefore, highly reduced. When a H^+ and an associated electron are lost from the

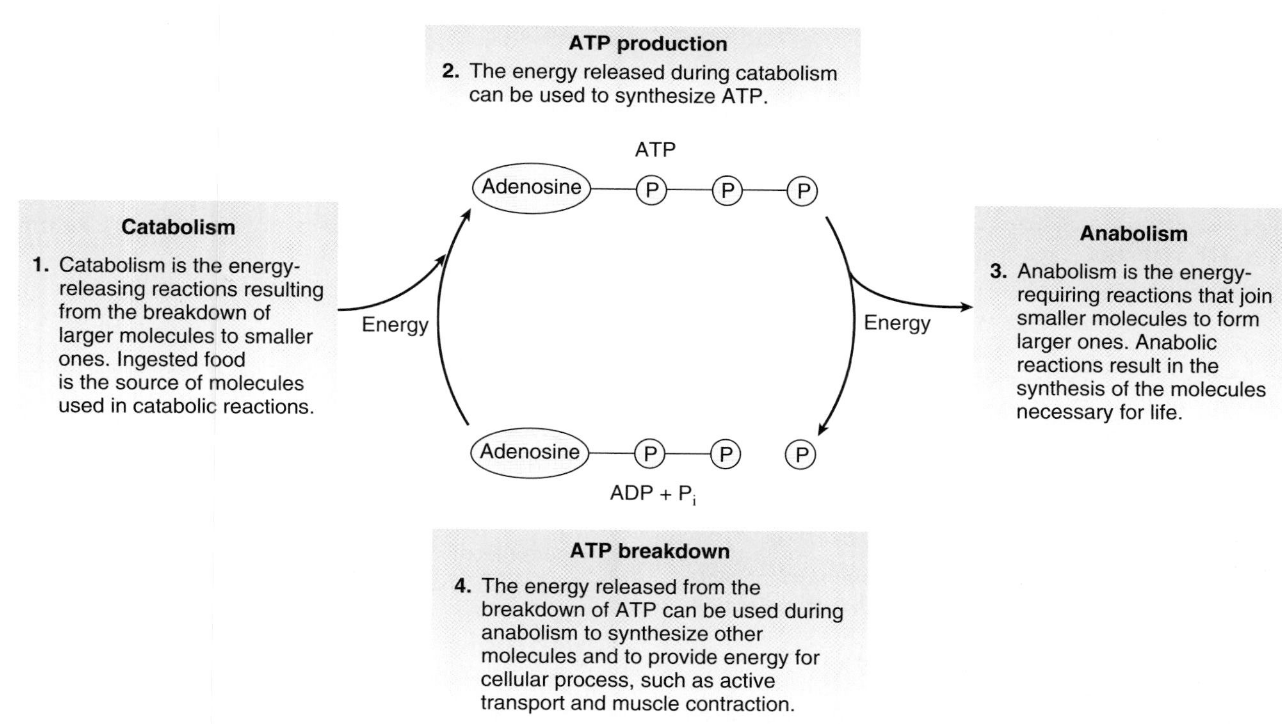

PROCESS FIGURE 25.3 ATP Coupling of Catabolism and Anabolism

Energy from catabolism is required to form ATP from ADP and phosphate (P). Energy and a phosphate are given off when ATP is converted back to ADP.

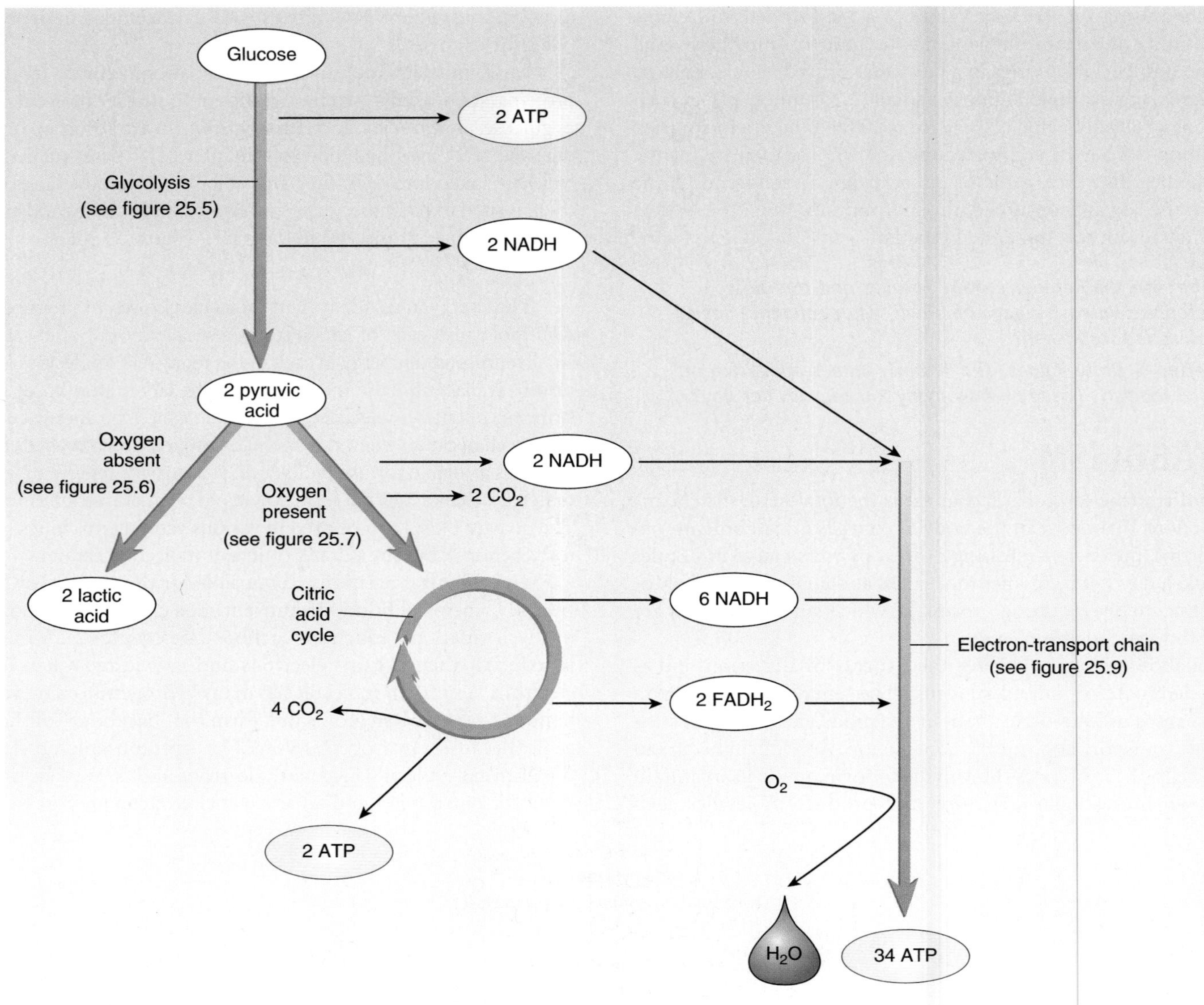

FIGURE 25.4 ATP Synthesis
Glycolysis, the citric acid cycle, and the electron-transport chain.

nutrient molecule, the molecule loses energy and becomes oxidized. The energy in the electron is used to synthesize ATP. The major events of ATP synthesis are summarized in figure 25.4.

21 ***Define metabolism, catabolism, and anabolism. How is the energy derived from catabolism used to drive anabolic reactions?***

22 ***How does the removal of hydrogen atoms from nutrient molecules result in a loss of energy from the nutrient molecule?***

CARBOHYDRATE METABOLISM

Monosaccharides are the breakdown products of carbohydrate digestion. Of these, glucose is the most important as far as cellular metabolism is concerned. Glucose is transported in the circulation to all the tissues of the body, where it is used as a source of energy. Any excess glucose in the blood following a meal can be used to form **glycogen** (glī′kō-jen; *glyks*, sweet), or it can be partially broken down and the components used to form fat. Glycogen is a short-term energy storage molecule, which can only be stored by the body in limited amounts, whereas fat is a long-term energy-storage molecule that can be stored in the body in large amounts. Most of the body's glycogen is in skeletal muscle and in the liver.

Glycolysis

Carbohydrate metabolism begins with **glycolysis** (glī-kol′i-sis), which is a series of chemical reactions in the cytosol that results in the breakdown of glucose into two **pyruvic** (pī-roo′vik) **acid** molecules (figure 25.5).

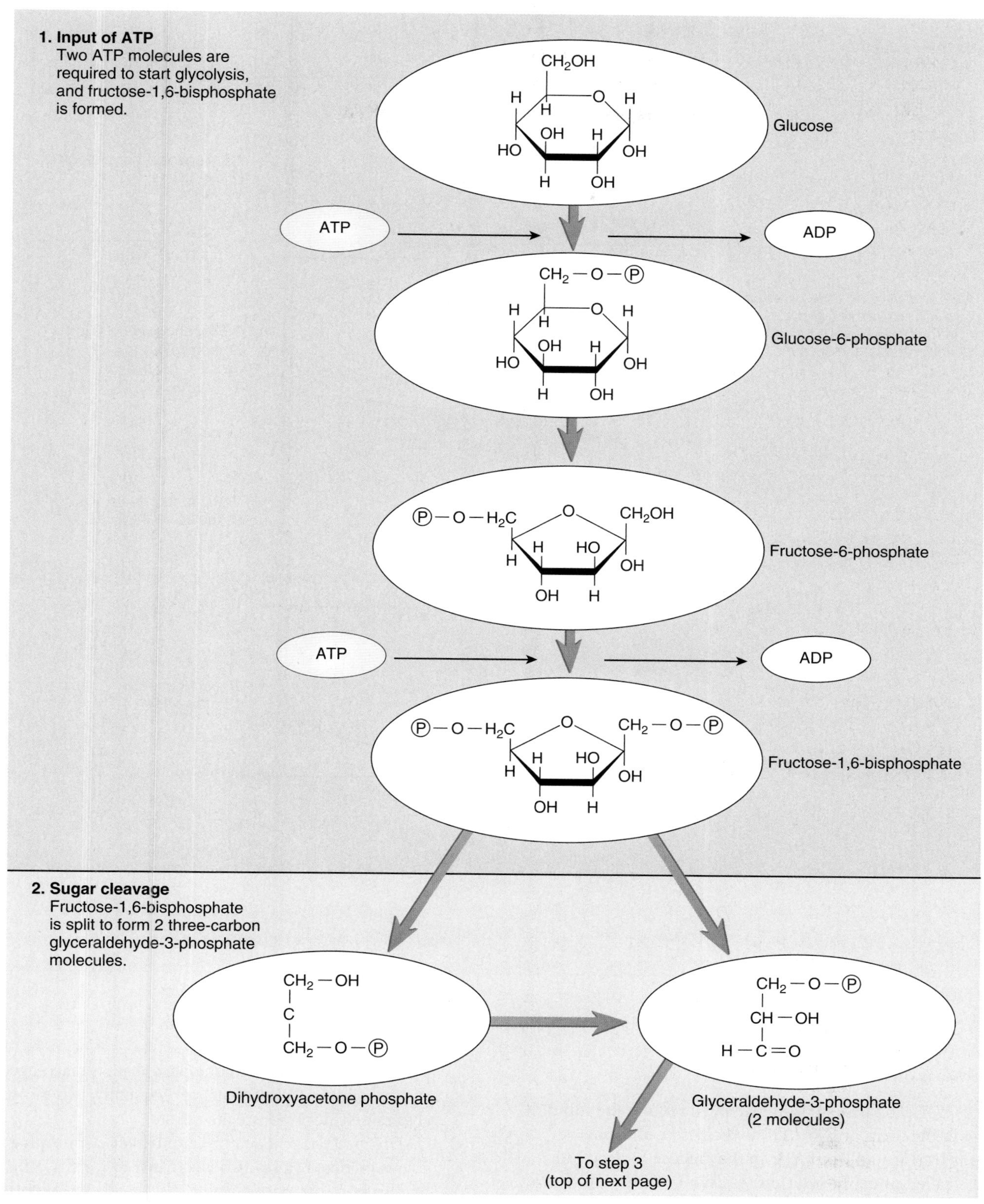

PROCESS FIGURE 25.5 Glycolysis

The chemical reactions of glycolysis take place in the cytosol.

(Figure continues)

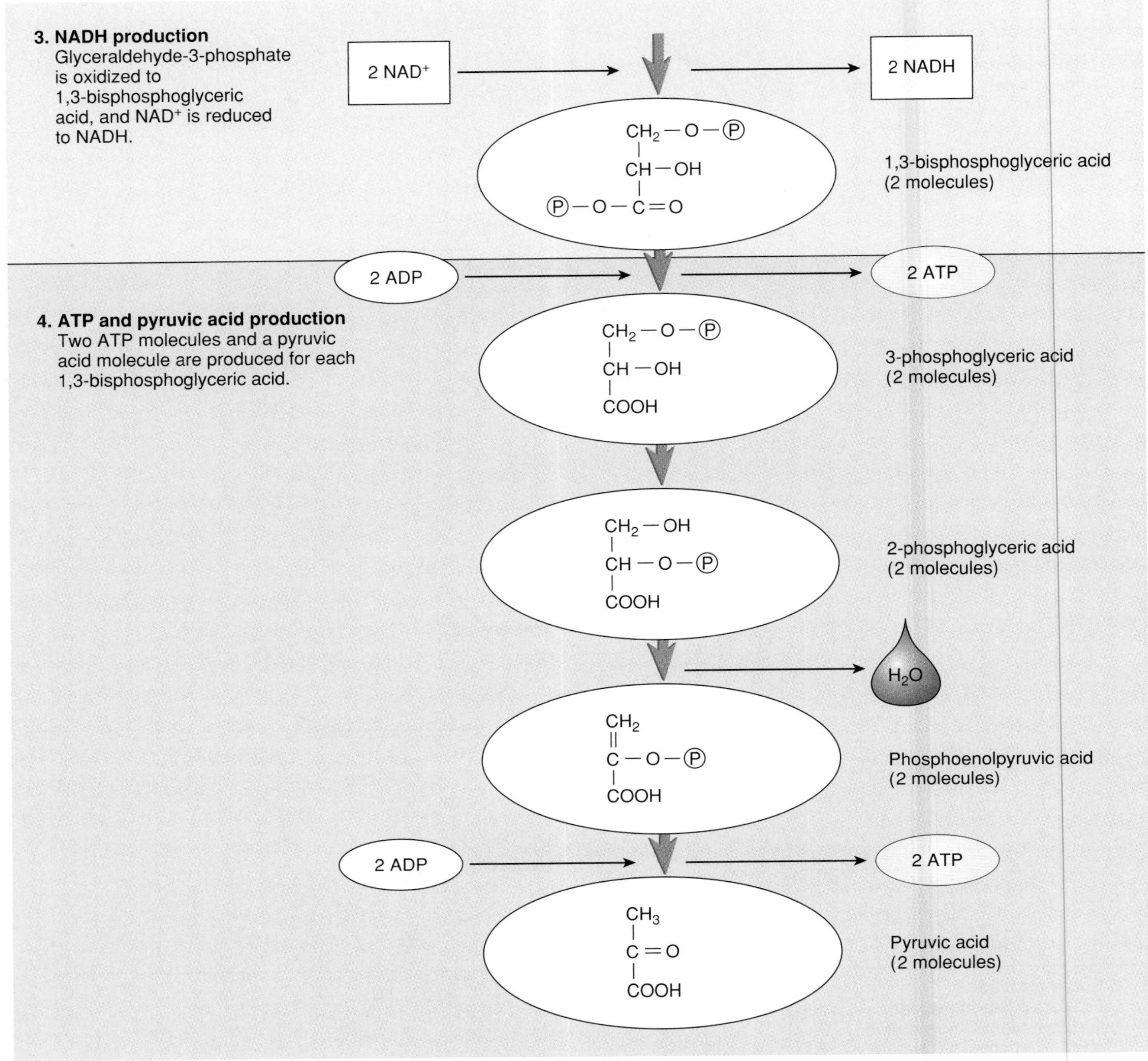

PROCESS FIGURE 25.5 ***(continued)***

Glycolysis is divided into four phases:

1. *Input of ATP.* The first steps in glycolysis require the input of energy in the form of two ATP molecules. A phosphate group is transferred from ATP to the glucose molecule, a process called **phosphorylation** (fos′fōr-i-lā′shŭn), to form glucose-6-phosphate. The glucose-6-phosphate atoms are rearranged to form fructose-6-phosphate, which is then converted to fructose-1,6-bisphosphate by the addition of another phosphate group from another ATP.
2. *Sugar cleavage.* Fructose-1,6-bisphosphate is cleaved into two three-carbon molecules, glyceraldehyde (glis-er-al′dĕ-hīd)-3-phosphate and dihydroxyacetone (dī′hī-drok-sē-as′e-tōn) phosphate. Dihydroxyacetone phosphate is rearranged to form glyceraldehyde-3-phosphate; consequently, two molecules of glyceraldehyde-3-phosphate result.
3. *NADH production.* Each glyceraldehyde-3-phosphate molecule is oxidized (loses two electrons) to form 1,3-bisphosphoglyceric (biz′phos-fo-gli′sĕr′ik) acid, and **nicotinamide adenine** (nik-ō-tin′ă-mīd ad′ĕ-nēn) **dinucleotide (NAD^+)** is reduced (gains two electrons) to **NADH.** Glyceraldehyde-3-phosphate also loses two H^+, one of which binds to NAD^+.

$$NAD^+ + 2\,e^- + 2\,H^+ \rightarrow NADH + H^+$$

NAD^+ is the oxidized form of nicotinamide adenine dinucleotide, and NADH is the reduced form. NADH is a carrier molecule with two high-energy electrons (e^-) that can be used to produce ATP molecules through the electron-transport chain (see "Electron-Transport Chain," p. 944).

4. *ATP and pyruvic acid production.* The last four steps of glycolysis produce two ATP molecules and one pyruvic acid molecule from each 1,3-bisphosphoglyceric acid molecule.

The events of glycolysis are summarized in table 25.4. Each glucose molecule that enters glycolysis forms two glyceraldehyde-3-phosphate molecules at the sugar cleavage phase. Each glyceraldehyde-3-phosphate molecule produces two ATP molecules, one NADH molecule, and one pyruvic acid molecule. Each glucose molecule, therefore, forms four ATP, two NADH, and two pyruvic acid molecules. Because the start of glycolysis requires the input of two ATP molecules, however, the final yield of each glucose molecule is two ATP, two NADH, and two pyruvic acid molecules (see figure 25.4).

If the cell has adequate amounts of oxygen, the NADH and pyruvic acid molecules are used in aerobic respiration to produce ATP. In the absence of sufficient oxygen, they are used in anaerobic respiration.

23 ***Describe the four phases of glycolysis. What are the products of glycolysis?***

24 ***What determines whether the pyruvic acid produced in glycolysis is used in aerobic or anaerobic respiration?***

Anaerobic Respiration

Anaerobic (an-ăr-ō′bik) **respiration** is the breakdown of glucose in the absence of oxygen to produce two molecules of **lactic** (lak′tik) **acid** and two molecules of ATP (figure 25.6). The ATP thus produced is a source of energy during activities such as intense exercise, when insufficient oxygen is delivered to tissues. Anaerobic respiration can be divided into two phases:

1. *Glycolysis.* Glucose undergoes several reactions to produce two pyruvic acid molecules and two NADH. There is also a net gain of two ATP molecules.

TABLE 25.4 ATP Production from One Glucose Molecule

Process	Product	Total ATP Produced*
Glycolysis	4 ATP	2 ATP (4 ATP produced minus 2 ATP to start)
	2 NADH	6 ATP (or 4 ATP; see text)
Acetyl-CoA production	2 NADH	6 ATP
Citric acid cycle	2 ATP	2 ATP
	6 NADH	18 ATP
	2 $FADH_2$	4 ATP
Total		38 ATP (or 36 ATP; see text)

*NADH and $FADH_2$ are used in the production of ATP in the electron-transport chain.
Abbreviations: ATP = adenosine triphosphate, NADH = reduced nicotinamide adenine dinucleotide, $FADH_2$ = reduced flavin adenine diphosphate, acetyl-CoA = acetyl coenzyme A.

1. Glycolysis converts glucose to two pyruvic acid molecules. The many reactions within the pathway are not shown. There is a net gain of two ATP and two NADH from glycolysis.

2. Anaerobic respiration, which does not require oxygen, includes glycolysis and converts the two pyruvic acid molecules produced by glycolysis to two lactic acid molecules. This conversion requires energy, which is derived from the NADH generated in glycolysis.

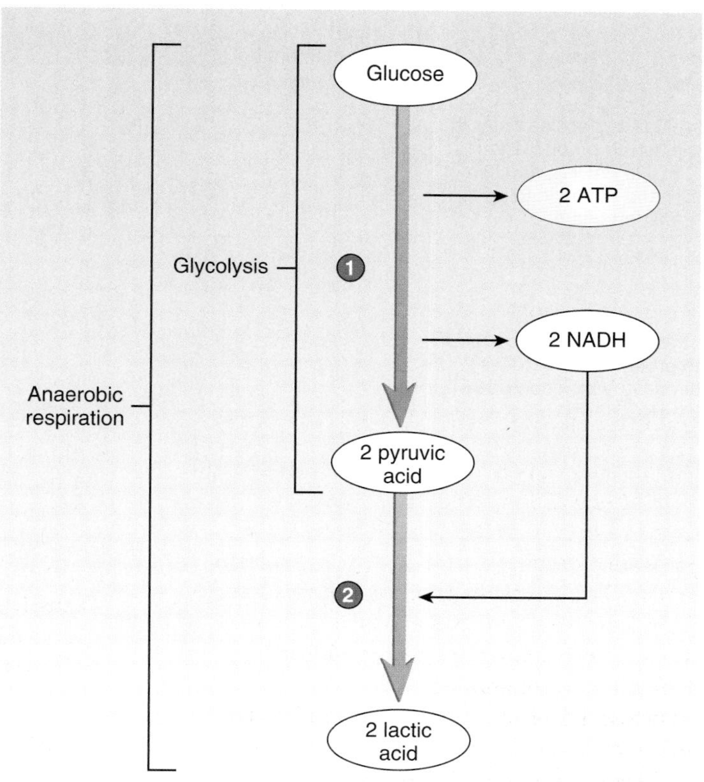

PROCESS FIGURE 25.6 **Glycolysis and Anaerobic Respiration**

2. *Lactic acid formation.* Pyruvic acid is converted to lactic acid, a reaction that requires the input of energy from the NADH produced in glycolysis.

Lactic acid is released from the cells that produce it and is transported by the blood to the liver. When oxygen becomes available, the lactic acid in the liver can be converted through a series of chemical reactions into glucose. The glucose then can be released from the liver and transported in the blood to cells that use glucose as an energy source. This process of converting lactic acid to glucose is called the **Cori cycle.** Some of the reactions involved in converting lactic acid into glucose require the input of energy derived from ATP that is produced by aerobic respiration. The oxygen necessary for the synthesis of the ATP is part of the **oxygen deficit** (see chapter 9).

25 ***Describe the two phases of anaerobic respiration. How many ATP molecules are produced by anaerobic respiration?***

26 ***What happens to the lactic acid produced in anaerobic respiration when oxygen becomes available?***

Aerobic Respiration

Aerobic (ār-ō′bik) **respiration** is the breakdown of glucose in the presence of oxygen to produce carbon dioxide, water, and 38 ATP molecules. Most of the ATP molecules required to sustain life are produced through aerobic respiration, which can be considered in four phases. The first phase of aerobic respiration, as in anaerobic respiration, is glycolysis. The remaining phases are acetyl-CoA formation, the citric acid cycle, and the electron-transport chain (figure 25.7).

1. Glycolysis in the cytosol converts glucose to two pyruvic acid molecules and produces two ATP and two NADH. The NADH can go to the electron-transport chain in the inner mitochondrial membrane.
2. The two pyruvic acid molecules produced in glycolysis are converted to two acetyl-CoA molecules, producing two CO_2 and two NADH. The NADH can go to the electron-transport chain.
3. The two acetyl-CoA molecules enter the citric acid cycle, which produces four CO_2, six NADH, two $FADH_2$, and two ATP. The NADH and $FADH_2$ can go to the electron-transport chain.
4. The electron-transport chain uses NADH and $FADH_2$ to produce 34 ATP. This process requires O_2, which combines with H^+ to form H_2O.

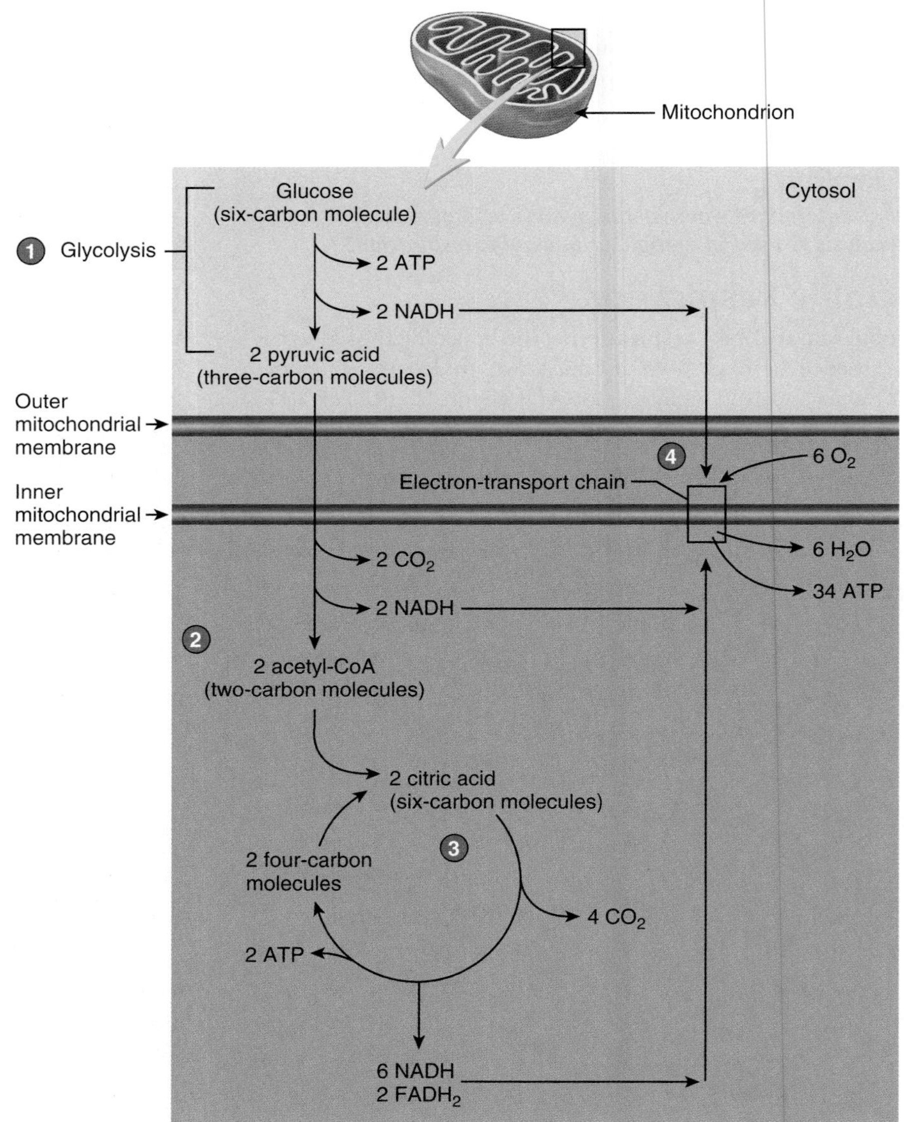

PROCESS FIGURE 25.7 Aerobic Respiration

Aerobic respiration involves four phases: (1) glycolysis, (2) acetyl-CoA formation, (3) the citric acid cycle, and (4) the electron-transport chain. The number of carbon atoms in a molecule is indicated after the molecule's name. As glucose is broken down, the carbon atoms from glucose are incorporated into carbon dioxide.

Acetyl-CoA Formation

In the second phase of aerobic respiration, pyruvic acid moves from the cytosol into a mitochondrion, which is separated into inner and outer compartments by the inner mitochondrial membrane. Within the inner compartment, enzymes remove a carbon and two oxygen atoms from the three-carbon pyruvic acid molecule to form carbon dioxide and a two-carbon acetyl (as′e-til) group (figure 25.8). Energy is released in the reaction and is used to reduce NAD^+ to NADH. The acetyl group combines with coenzyme A (CoA) to form acetyl-CoA. For each two pyruvic acid molecules from glycolysis, two acetyl-CoA molecules, two carbon dioxide molecules, and two NADH are formed (see figure 25.4).

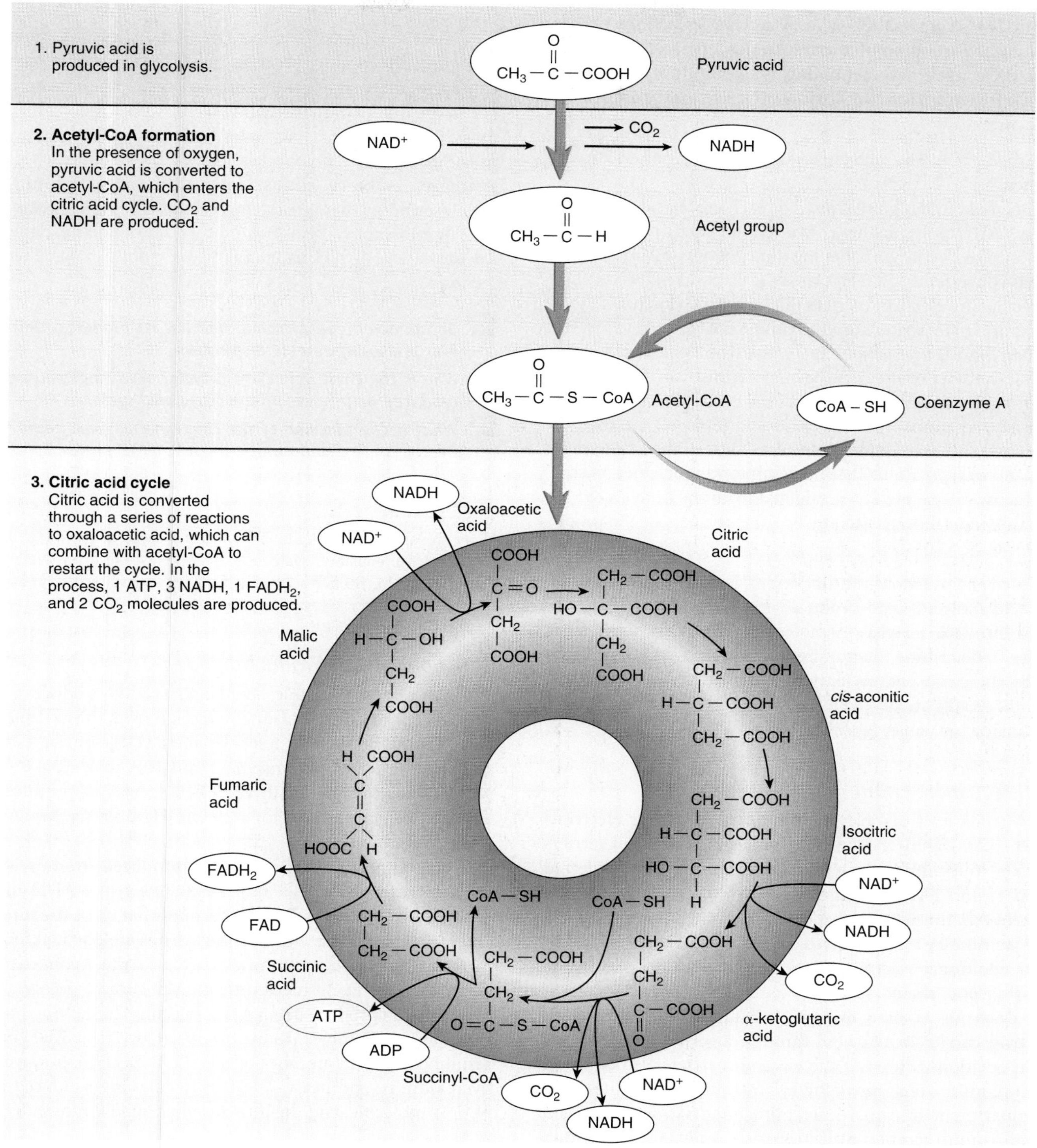

PROCESS FIGURE 25.8 Acetyl-CoA and the Citric Acid Cycle
Pyruvic acid from the cytosol is converted to acetyl-CoA in mitochondria. The acetyl-CoA enters the citric acid cycle.

Citric Acid Cycle

The third phase of aerobic respiration is the **citric acid cycle,** which is named after the six-carbon citric acid molecule formed in the first step of the cycle (see figure 25.8). It is also called the Krebs cycle after its discoverer, British biochemist Sir Hans Krebs. The citric acid cycle begins with the production of citric acid from the combination of acetyl-CoA and a four-carbon molecule called oxaloacetic (ok′să-lō-ă-sē′tik) acid. A series of reactions occurs, resulting in the formation of another oxaloacetic acid, which can start the cycle again by combining with another acetyl-CoA. During the reactions of the citric acid cycle, three important events occur:

1. *ATP production.* For each citric acid molecule, one ATP is formed.
2. *NADH and $FADH_2$ production.* For each citric acid molecule, three NAD^+ molecules are converted to NADH molecules, and one flavin (flā′vin) adenine dinucleotide (FAD) molecule is converted to $FADH_2$. The NADH and $FADH_2$ molecules are electron carriers that enter the electron-transport chain and are used to produce ATP.
3. *Carbon dioxide production.* Each six-carbon citric acid molecule at the start of the cycle becomes a four-carbon oxaloacetic acid molecule at the end of the cycle. Two carbon and four oxygen atoms from the citric acid molecule are used to form two carbon dioxide molecules. Thus, some of the carbon and oxygen atoms that make up food molecules, such as glucose are eventually eliminated from the body as carbon dioxide. Humans literally breathe out part of the food they eat.

For each glucose molecule that begins aerobic respiration, two pyruvic acid molecules are produced in glycolysis, and they are converted into two acetyl-CoA molecules, which enter the citric acid cycle. To determine the number of molecules produced from glucose by the citric acid cycle, two "turns" of the cycle must be counted; the results are two ATP, six NADH, two $FADH_2$, and four carbon dioxide molecules (see figure 25.4).

Electron-Transport Chain

The fourth phase of aerobic respiration involves the **electron-transport chain** (figure 25.9), which is a series of electron carriers in the inner mitochondrial membrane. Electrons are transferred from NADH and $FADH_2$ to the electron-transport carriers, and H^+ are released from NADH and $FADH_2$. After the loss of the electrons and the H^+, the oxidized NAD^+ and FAD are reused to transport additional electrons from the citric acid cycle to the electron-transport chain.

The electrons released from NADH and $FADH_2$ pass from one electron carrier to the next through a series of oxidation–reduction reactions. Three of the electron carriers also function as proton pumps, which move the H^+ from the inner mitochondrial compartment into the outer mitochondrial compartment. Each proton pump accepts an electron, uses some of the electron's energy to export a H^+, and passes the electron to the next electron carrier. The last electron carrier in the series collects the electrons and combines them with oxygen and H^+ to form water.

$$1/2\ O_2 + 2\ H^+ + 2\ e^- \rightarrow H_2O$$

Without oxygen to accept the electrons, the reactions of the electron-transport chain cease, effectively stopping aerobic respiration.

The H^+ released from NADH and $FADH_2$ are moved from the inner mitochondrial compartment to the outer mitochondrial compartment by active transport. As a result, the concentration of H^+ in the outer compartment exceeds that of the inner compartment, and H^+ diffuse back into the inner compartment. The H^+ pass through certain channels formed by an enzyme called **ATP synthase.** As the H^+ diffuse down their concentration gradient, they lose energy that is used to produce ATP. This process is called the **chemiosmotic** (kem-ē-os-mot′ik) **model** because the chemical formation of ATP is coupled to a diffusion force similar to osmosis.

27 ***Define aerobic respiration, and list its products. Describe the four phases of aerobic respiration.***

28 ***Why is the citric acid cycle a cycle? What molecules are produced as a result of the citric acid cycle?***

29 ***What is the function of the electron-transport chain? Describe the chemiosmotic model of ATP production.***

PREDICT 4

Many poisons function by blocking certain steps in the metabolic pathways. For example, cyanide blocks the last step in the electron-transport chain. Explain why this blockage causes death.

Summary of ATP Production

For each glucose molecule, aerobic respiration produces a net gain of 38 ATP molecules: 2 from glycolysis, 2 from the citric acid cycle, and 34 from the NADH molecules and $FADH_2$ molecules that pass through the electron-transport chain (see table 25.4). For each NADH molecule formed, three ATP molecules are produced by the electron-transport chain; for each $FADH_2$ molecule, two ATP molecules are produced.

The number of ATP molecules produced from each glucose molecule can be 36 ATP molecules. The two NADH molecules produced by glycolysis in the cytosol cannot cross the inner mitochondrial membrane; thus, their electrons are donated to a shuttle molecule, which carries the electrons to the electron-transport chain. Depending on the shuttle molecule, each glycolytic NADH molecule can produce 2 or 3 ATP molecules. In skeletal muscle and the brain, 2 ATP molecules are produced for each NADH molecule, resulting in a total number of 36 ATP molecules; however, in the liver, kidneys, and heart, 3 ATP molecules are produced for each NADH molecule, and the total number of ATP molecules formed is 38.

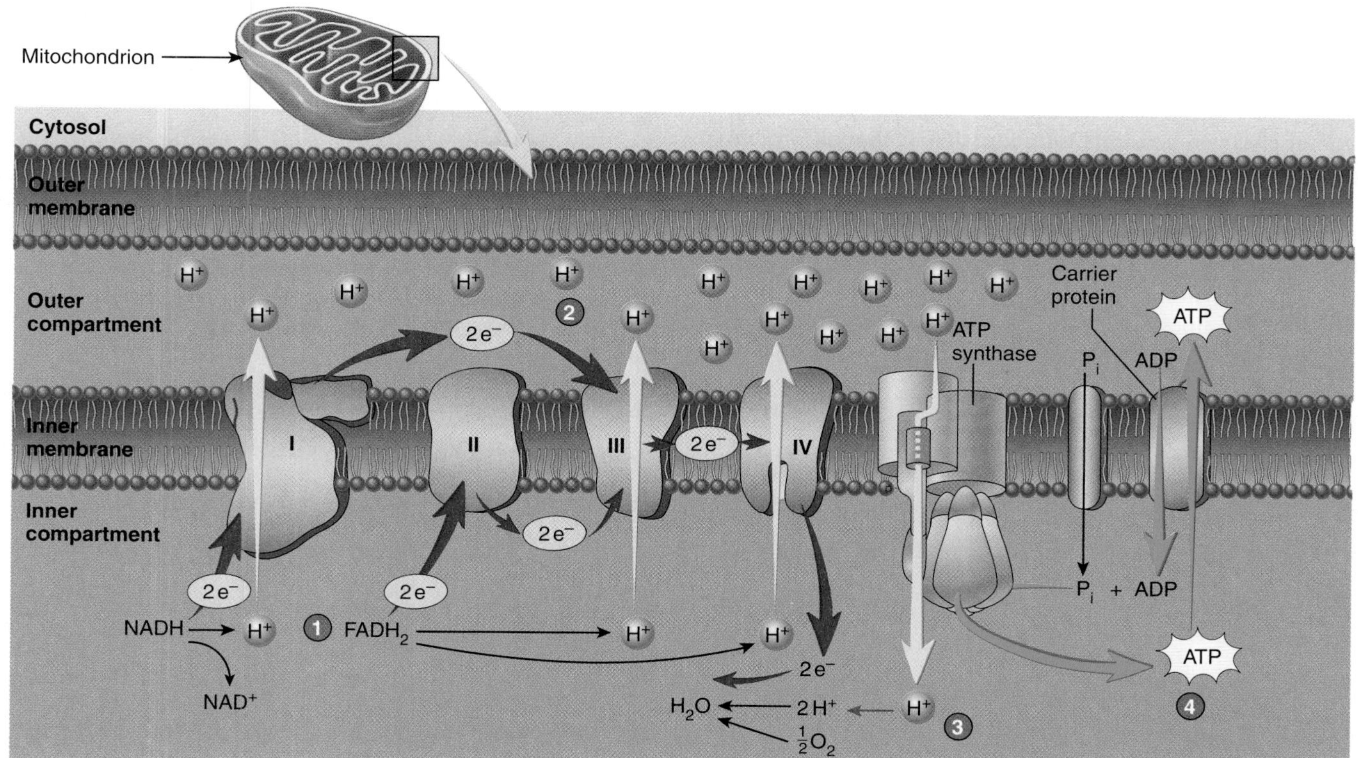

1. NADH or $FADH_2$ transfer their electrons to the electron-transport chain.
2. As the electrons move through the electron-transport chain, some of their energy is used to pump H^+ into the outer compartment, resulting in a higher concentration of H^+ in the outer than in the inner compartment.
3. The H^+ diffuse back into the inner compartment through special channels (ATP synthase) that couple the H^+ movement with the production of ATP. The electrons, H^+, and O_2 combine to form H_2O.
4. ATP is transported out of the inner compartment by a carrier protein that exchanges ATP for ADP. A different carrier protein moves phosphate into the inner compartment.

PROCESS FIGURE 25.9 Electron-Transport Chain

The electron-transport chain in the mitochondrial inner membrane consists of four protein complexes (*purple;* numbered I to IV) with carrier proteins. As electrons are transferred from one carrier protein to another, they lose energy that is used to move H^+ out of the inner compartment. Hydrogen ions move back into the inner compartment through special channels (ATP synthase; *green*), which produce ATP. Carrier proteins (*brown*) move ATP out of and ADP and P_i into the inner compartment.

Six carbon dioxide molecules are produced in aerobic respiration. Water molecules are reactants in some of the chemical reactions of aerobic respiration and products in others. Six water molecules are used, but 12 are formed, for a net gain of 6 water molecules. Thus, aerobic respiration can be summarized as follows:

$$C_6H_{12}O_6 + 6\ O_2 + 6\ H_2O + 38\ ADP + 38\ P_i \rightarrow 6\ CO_2 + 12\ H_2O + 38\ ATP$$

30 ***In aerobic respiration, how many ATP molecules are produced from one molecule of glucose through glycolysis, the citric acid cycle, and the electron-transport chain?***

31 ***Why is the total number of ATP produced in aerobic respiration listed as 38 or 36?***

Quantity of ATP Produced from Glucose

The number of ATP molecules produced per glucose molecule is a theoretical number that assumes that two H^+ are necessary for the formation of each ATP. If more than two are required, the efficiency of aerobic respiration decreases. In addition, it costs energy to get ADP and phosphates into the mitochondria and to get ATP out. Considering all these factors, each glucose molecule yields about 25 ATP molecules instead of 38 ATP molecules.

LIPID METABOLISM

Lipids are the body's main energy-storage molecules. In a healthy person, lipids are responsible for about 99% of the body's energy storage, and glycogen accounts for about 1%. Although proteins are used as an energy source, they are not considered storage molecules because the breakdown of proteins normally involves the loss of molecules that perform other functions.

Lipids are stored primarily as triglycerides in adipose tissue. There is constant synthesis and breakdown of triglycerides; thus, the fat present in adipose tissue today is not the same fat that was there a few weeks ago. Between meals, when triglycerides are broken down in adipose tissue, some of the fatty acids produced are released into the blood, where they are called **free fatty acids.** Other tissues, especially skeletal muscle and the liver, use the free fatty acids as a source of energy.

The metabolism of fatty acids occurs by **beta-oxidation,** a series of reactions in which two carbon atoms at a time are removed from the end of a fatty acid chain to form acetyl-CoA. The process of beta-oxidation continues to remove two carbon atoms at a time until the entire fatty acid chain is converted into acetyl-CoA molecules. Acetyl-CoA can enter the citric acid cycle and be used to generate ATP (figure 25.10).

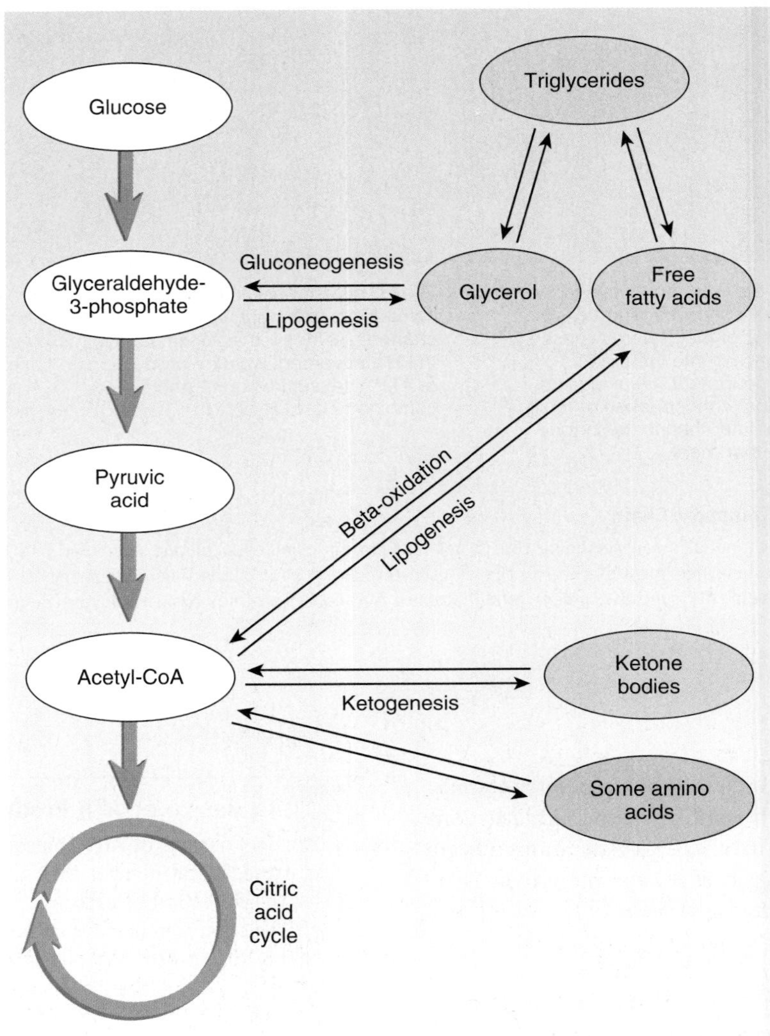

FIGURE 25.10 Lipid Metabolism

Triglyceride is broken down into glycerol and fatty acids. Glycerol enters glycolysis to produce ATP. The fatty acids are broken down by beta-oxidation into acetyl-CoA, which enters the citric acid cycle to produce ATP. Acetyl-CoA can also be used to produce ketone bodies (ketogenesis). Lipogenesis is the production of lipids. Glucose is converted to glycerol, and amino acids are converted to acetyl-CoA molecules. Acetyl-CoA molecules can combine to form fatty acids. Glycerol and fatty acids join to form triglycerides.

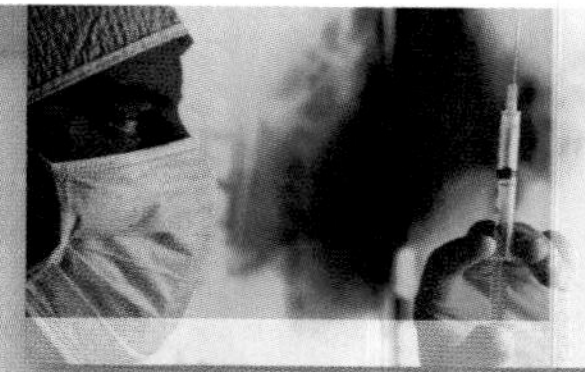

CLINICAL FOCUS

Starvation

Starvation results from the inadequate intake of nutrients or the inability to metabolize or absorb nutrients. It has a number of causes, such as prolonged fasting, anorexia, deprivation, and disease. No matter what the cause, starvation follows the same course and consists of three phases. The events of the first two phases occur even during relatively short periods of fasting or dieting, but the third phase occurs only in prolonged starvation and can end in death.

During the first phase of starvation, blood glucose levels are maintained through the production of glucose from glycogen, proteins, and fats. At first, glycogen is broken down into glucose; however, only enough glycogen is stored in the liver to last a few hours. Thereafter, blood glucose levels are maintained by the breakdown of proteins and fats. Fats are decomposed into fatty acids and glycerol. Fatty acids can be used as a source of energy, especially by skeletal muscle, thus decreasing the use of glucose by tissues other than the brain. Glycerol can be used to make a small amount of glucose, but most of the glucose is formed from the amino acids of proteins. In addition, some amino acids can be used directly for energy.

In the second phase, which can last for several weeks, fats are the primary energy source. The liver metabolizes fatty acids into ketone bodies, which can be used as a source of energy. After about a week of fasting, the brain begins to use ketone bodies, as well as glucose, for energy. This usage decreases the demand for glucose, and the rate of protein breakdown diminishes but does not stop. In addition, the proteins not essential for survival are used first.

The third phase of starvation begins when the fat reserves are depleted and a switch to proteins as the major energy source takes place. Muscles, the largest source of protein in the body, are rapidly depleted. At the end of this phase, proteins essential for cellular functions are broken down, and cell function degenerates.

In addition to weight loss, the symptoms of starvation include apathy, listlessness, withdrawal, and increased susceptibility to infectious disease. Few people die directly from starvation because they usually die of an infectious disease first. Other signs of starvation include changes in hair color, flaky skin, and massive edema in the abdomen and lower limbs, causing the abdomen to appear bloated.

During starvation, the body's ability to consume normal volumes of food also decreases. Foods high in bulk but low in protein content often cannot reverse the process of starvation. Intervention involves feeding the starving person low-bulk food that provides ample proteins and kilocalories and is fortified with vitamins and minerals. Starvation also results in dehydration, and rehydration is an important part of intervention. Even with intervention, a victim may be so affected by disease or weakness that he or she cannot recover.

Acetyl-CoA is also used in **ketogenesis** (kē-tō-jen′ĕ-sis), the formation of ketone bodies. In the liver, when large amounts of acetyl-CoA are produced, not all of the acetyl-CoA enters the citric acid cycle. Instead, two acetyl-CoA molecules combine to form a molecule of acetoacetic (as′e-tō-a-sē′tik) acid, which is converted mainly into β-hydroxybutyric (hī-drŏk′sē-bū-tir′ik) acid and a smaller amount of acetone (as′e-tōn). Acetoacetic acid, β-hydroxybutyric acid, and acetone are called **ketone** (kē′tōn) **bodies;** they are released into the blood, where they travel to other tissues, especially skeletal muscle. In these tissues, the ketone bodies are converted back into acetyl-CoA, which enters the citric acid cycle to produce ATP.

32 ***Define beta-oxidation, and explain how it results in ATP production.***

33 ***What are ketone bodies, how are they produced, and for what are they used?***

The Danger of Excessive Amounts of Ketones

Small amounts of ketone bodies in the blood are normal and beneficial. An excessive production of ketone bodies is called **ketosis** (kē-tō′sis). If the increased number of acidic ketone bodies exceeds the capacity of the body's buffering systems, acidosis, a decrease in blood pH, can occur (see chapter 27). Conditions that increase fat metabolism can increase the rate of ketone body formation. Examples are starvation (see Clinical Focus "Starvation," above), diets consisting of proteins and fats with few carbohydrates, and untreated diabetes mellitus (see Systems Pathology "Insulin-Dependent Diabetes Mellitus," chapter 18). Ketone bodies are excreted by the kidneys and diffuse into the alveoli of the lungs. Ketone bodies in the urine and "acetone breath" are characteristic of untreated diabetes mellitus.

Glucose
Gycolysis
Pyruvic acid
Alanine, cysteine glycine, serine, threonine
Isoleucine, leucine, tryptophan
Acetyl-CoA
Acetoacetyl-CoA
Phenylalanine, tyrosine, leucine, lysine, tryptophan
Aspartate, asparagine
Oxaloacetic acid
Tyrosine, phenylalanine
Fumaric acid
Citric acid cycle
Citric acid
Succinyl-CoA
α-ketoglutaric acid
Arginine, histidine, glutamine, proline
Isoleucine, methionine, valine

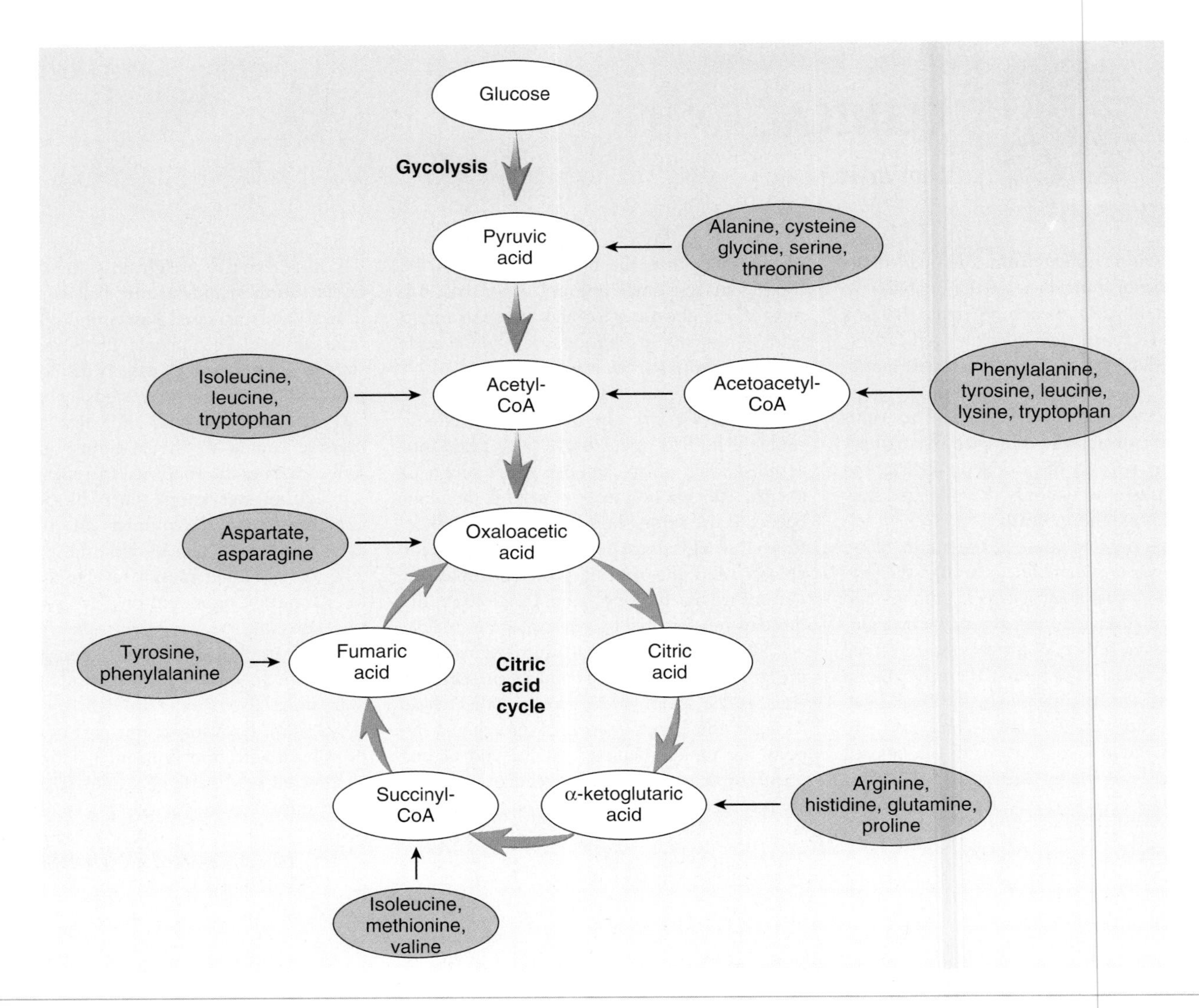

FIGURE 25.11 Amino Acid Metabolism
Various entry points for amino acids (*tan ovals*) into carbohydrate metabolism.

PROTEIN METABOLISM

Once absorbed into the body, amino acids are quickly taken up by cells, especially in the liver. Amino acids are used to synthesize needed proteins (see chapter 3) or as a source of energy (figure 25.11). Unlike glycogen and triglycerides, amino acids are not stored in the body.

The synthesis of nonessential amino acids usually begins with keto acids (figure 25.12). A keto acid can be converted into an amino acid by replacing its oxygen with an amine group. Usually, this conversion is accomplished by transferring an amine group from an amino acid to the keto acid, a reaction called **transamination** (trans-am′i-nā′shŭn). For example, α-ketoglutaric acid (a keto acid)

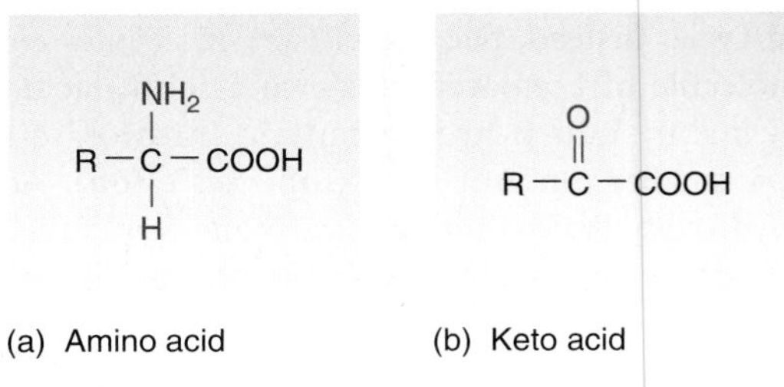

FIGURE 25.12 General Formulas of an Amino Acid and a Keto Acid
(*a*) Amino acid with a carboxyl group (—COOH), an amine group (NH_2), a hydrogen atom (H), and a group called "R," which represents the rest of the molecule. (*b*) Keto acid with a double-bonded oxygen replacing the amine group and the hydrogen atom of the amino acid.

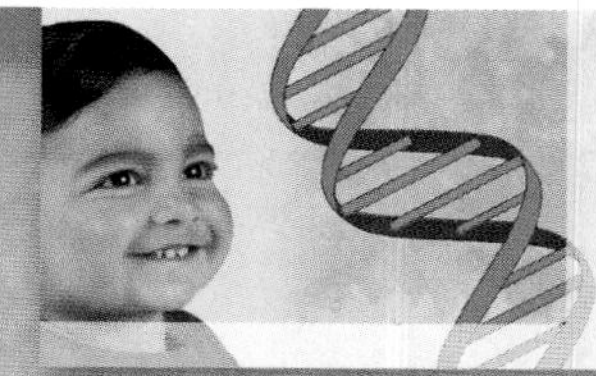

CLINICAL GENETICS

Newborn Screening

Metabolic disorders, or inborn errors of metabolism, are genetic disorders resulting in biochemical defects of metabolism. Metabolic disorders affect the body's ability to break down or use the nutrients needed for energy, growth, and repair. Too little or too much of certain substances can cause significant health problems. In general, metabolic disorders can hinder an infant's mental and physical development. For many metabolic disorders, a specific treatment can prevent or limit harm if it is started early, so early detection through newborn screening is vital.

In the United States, most states require the mandatory screening of newborns. The specific disorders for which tests are performed, however, vary from state to state; there is no national standard. Although over 300 genetic disorders are known, most are so rare that it is not cost-effective to test for them. Table 25.A lists the most common tests performed for metabolic disorders using a blood sample. All of them are autosomal-recessive. Other genetic disorders tested for include sickle-cell anemia (see chapter 19), cystic fibrosis (see chapter 23), and Duchenne muscular dystrophy (see chapter 9). Additional tests screen for hormonal disorders, such as congenital adrenal hyperplasia and congenital hypothyroidism (see chapter 18), and diseases, such as toxoplasmosis (a parasitic infection transmitted across the placenta) and HIV infection (see chapter 22).

Table 25.A Metabolic Disorders

Disorder	Description	Effect	Treatment
Phenylketonuria (PKU) (chromosome 12)	Inability to metabolize the amino acid phenylalanine (see chapter 3)	Mental retardation	Restrict dietary phenylalanine.
Galactosemia (chromosome 9)	Inability to convert the sugar galactose to glucose, resulting in a buildup of galactose	Mental retardation, growth deficiency, cataracts, severe infections, and death	Eliminate milk and other dairy products from the diet. Galactose is one of two sugars in lactose (milk sugar).
Biotinidase deficiency (chromosome 3)	Inability to separate the vitamin biotin from other chemicals, resulting in a biotin deficiency	Seizures, hearing loss, optic atrophy, mental retardation, and poor muscle control	Take oral biotin supplements.
Maple syrup urine disease (four types; chromosome 1, 6, 7, or 19)	Deficiency in an enzyme complex, resulting in an inability to metabolize the amino acids leucine, isoleucine, and valine	Mental retardation in those surviving past 3 months of age	Restrict dietary intake of the affected amino acids.
Homocystinuria (chromosome 21)	Defect in methionine metabolism, leading to an accumulation of homocysteine	Dislocated lenses of the eyes, mental retardation, skeletal abnormalities, and abnormal blood clotting	Take high doses of vitamin B_6; eat methionine-restricted diet supplemented with cystine.
Tyrosinemia (three types; chromosome 12, 15, or 16)	Deficiency in a series of enzymes that break down the amino acid tyrosine	Mild retardation, language skill difficulties, and liver and kidney failure	Restrict dietary tyrosine and phenylalanine.

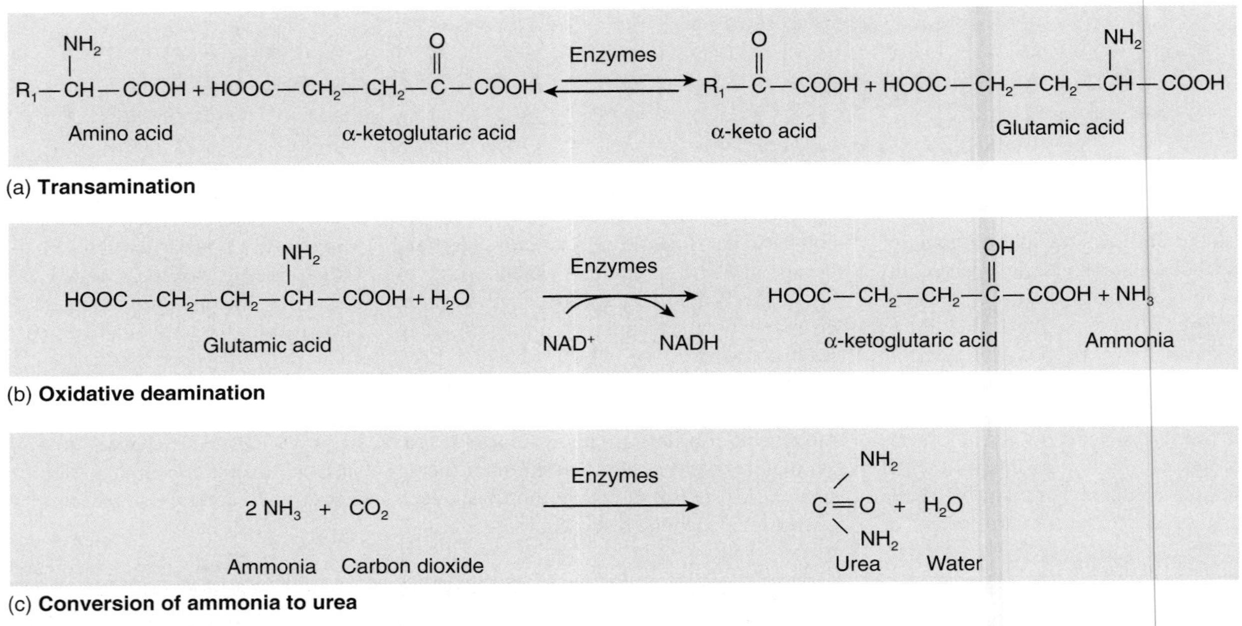

$$R_1{-}\overset{NH_2}{\overset{|}{C}H}{-}COOH + HOOC{-}CH_2{-}CH_2{-}\overset{O}{\overset{\|}{C}}{-}COOH \xrightleftharpoons{\text{Enzymes}} R_1{-}\overset{O}{\overset{\|}{C}}{-}COOH + HOOC{-}CH_2{-}CH_2{-}\overset{NH_2}{\overset{|}{C}H}{-}COOH$$

Amino acid α-ketoglutaric acid α-keto acid Glutamic acid

(a) **Transamination**

$$HOOC{-}CH_2{-}CH_2{-}\overset{NH_2}{\overset{|}{C}H}{-}COOH + H_2O \xrightarrow[NAD^+ \rightarrow NADH]{\text{Enzymes}} HOOC{-}CH_2{-}CH_2{-}\overset{OH}{\overset{\|}{C}}{-}COOH + NH_3$$

Glutamic acid NAD⁺ NADH α-ketoglutaric acid Ammonia

(b) **Oxidative deamination**

$$2\,NH_3 + CO_2 \xrightarrow{\text{Enzymes}} C(=O)(NH_2)_2 + H_2O$$

Ammonia Carbon dioxide Urea Water

(c) **Conversion of ammonia to urea**

FIGURE 25.13 Amino Acid Reactions

(*a*) Transamination reaction in which an amine group is transferred from an amino acid to a keto acid to form a different amino acid. (*b*) Oxidative deamination reaction in which an amino acid loses an amine group to become a keto acid and to form ammonia. In the process, NADH, which can be used to generate ATP, is formed. (*c*) Ammonia is converted to urea in the liver. The actual conversion of ammonia to urea is more complex, involving a number of intermediate reactions that constitute the urea cycle.

reacts with an amino acid to form glutamic acid (an amino acid; figure 25.13*a*). Most amino acids can undergo transamination to produce glutamic acid. The glutamic acid is used as a source of an amine group to construct most of the nonessential amino acids. A few nonessential amino acids are formed by other chemical reactions from the essential amino acids.

Amino acids can be used as a source of energy. In **oxidative deamination** (dē-am-i-nā′shŭn) or **deaminization** (dē-am′i-nizā′shŭn), an amine group is removed from an amino acid (usually glutamic acid), leaving ammonia and a keto acid (figure 25.13*b*). In the process, NAD^+ is reduced to NADH, which can enter the electron-transport chain to produce ATP. Ammonia is toxic to cells. An accumulation of ammonia to toxic levels is prevented because the liver converts it into urea, which is carried by the blood to the kidneys, where the urea is eliminated (figure 25.13*c*; see chapter 26).

Amino acids are also used as a source of energy by converting them into the intermediate molecules of carbohydrate metabolism (see figure 25.11). These molecules are then metabolized to yield ATP. The conversion of an amino acid often begins with a transamination or oxidative deamination reaction, in which the amino acid is converted into a keto acid (see figure 25.12). The keto acid enters the citric acid cycle or is converted into pyruvic acid or acetyl-CoA.

34 ***What is accomplished by transamination and oxidative deamination?***

35 ***How are proteins (amino acids) used to produce energy?***

INTERCONVERSION OF NUTRIENT MOLECULES

Blood glucose enters most cells by facilitated diffusion and is immediately converted to glucose-6-phosphate, which cannot recross the plasma membrane (figure 25.14*a*). Glucose-6-phosphate then continues through glycolysis to produce ATP. If, however, excess glucose is present (e.g., after a meal), it is used to form glycogen through a process called **glycogenesis** (glī-kō-jen′ĕ-sis). Most of the body's glycogen is contained in skeletal muscle and the liver.

Once glycogen stores, which are quite limited, are filled, glucose and amino acids are used to synthesize lipids, a process called **lipogenesis** (lip-ō-jen′ĕ-sis; see figure 25.10). Glucose molecules can be used to form glyceraldehyde-3-phosphate and acetyl-CoA. Amino acids can also be converted to acetyl-CoA. Glyceraldehyde-3-phosphate is converted to glycerol, and the two-carbon acetyl-CoA molecules are joined together to form fatty acid chains. Glycerol and three fatty acids then combine to form triglycerides.

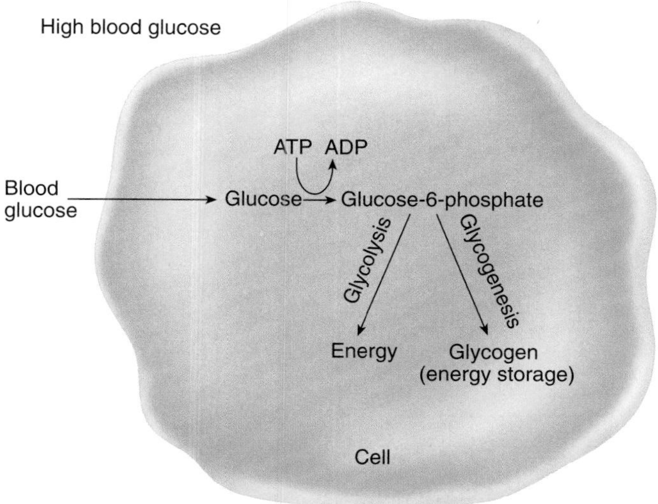

(a) When blood glucose levels are high, glucose enters the cell and is phosphorylated to form glucose-6-phosphate, which can enter glycolysis or glycogenesis.

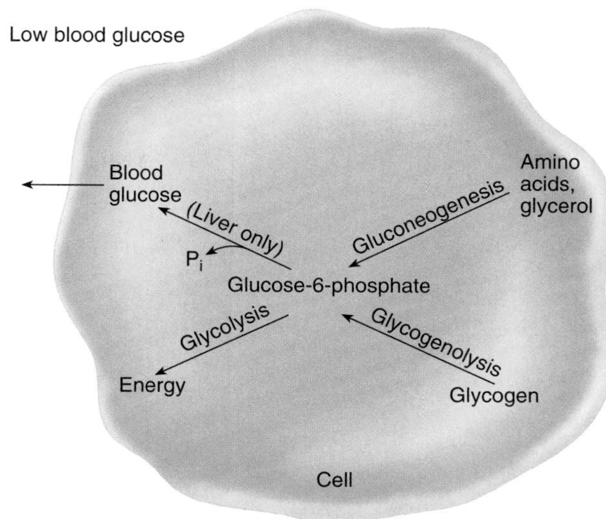

(b) When blood glucose levels drop, glucose-6-phosphate can be produced through glycogenolysis or gluconeogenesis. Glucose-6-phosphate can enter glycolysis, or the phosphate group can be removed in liver tissue and glucose released into the blood.

FIGURE 25.14 Interconversion of Nutrient Molecules

> **Alcoholism and Cirrhosis of the Liver**
>
> Enzymes in the liver convert ethanol (beverage alcohol) into acetyl-CoA; in the process, two NADH molecules are produced. The NADH molecules enter the electron-transport chain and are used to produce ATP molecules. Each gram of ethanol provides 7 kcal of energy. Because of the high level of NADH in the cell that results from the metabolism of ethanol, the production of NADH by glycolysis and by the citric acid cycle is inhibited. Consequently, sugars and amino acids are not broken down but are converted into fats, which accumulate in the liver. Chronic alcohol abuse can, therefore, result in **cirrhosis** (sir-rō′sis) **of the liver,** which involves fat deposition, cell death, inflammation, and scar tissue formation. Death can occur because the liver is unable to carry out its normal functions.

When glucose is needed, glycogen can be broken down into glucose-6-phosphate through a set of reactions called **glycogenolysis** (glī′kō-jĕ-nol′i-sis; figure 25.14*b*). In skeletal muscle, glucose-6-phosphate continues through glycolysis to produce ATP. The liver can use glucose-6-phosphate for energy or can convert it to glucose, which diffuses into the blood. The liver can release glucose, but skeletal muscle cannot because it lacks the necessary enzymes to convert glucose-6-phosphate into glucose.

The release of glucose from the liver is necessary to maintain blood glucose levels between meals. Maintaining these levels is especially important to the brain, which normally uses only glucose for an energy source and consumes about two-thirds of the total glucose used each day. When liver glycogen levels are inadequate to supply glucose, it is synthesized from molecules other than carbohydrates, such as amino acids from proteins and glycerol from triglycerides, in a process called **gluconeogenesis** (gloo′kō-nē-ō-jen′ĕ-sis). Most amino acids can be converted into citric acid cycle molecules, acetyl-CoA, or pyruvic acid (see figure 25.11). Through a series of chemical reactions, these molecules are converted into glucose. Glycerol enters glycolysis by becoming glyceraldehyde-3-phosphate.

 ***Define glycogenesis, lipogenesis, glycogenolysis, and gluconeogenesis.***

METABOLIC STATES

The absorptive and postabsorptive states are the two major metabolic states of the body. The regulation of these states is discussed in chapter 18 (see "Hormonal Regulation of Nutrients"). The **absorptive state** is the period immediately after a meal when nutrients are being absorbed through the intestinal wall into the circulatory and lymphatic systems (figure 25.15). The absorptive state usually lasts about 4 hours after each meal, and the cells use most of the glucose that enters the circulation for the energy they require. The remainder of the glucose is converted into glycogen or fats. Most of the absorbed fats are deposited in adipose tissue. Many of the absorbed amino acids are used by cells in protein synthesis, some are used for energy, and others enter the liver and are converted into fats or carbohydrates.

The **postabsorptive state** occurs late in the morning, late in the afternoon, or during the night after each absorptive state is concluded (figure 25.16). It is vital to the body's homeostasis that normal blood glucose levels be maintained, especially for normal functioning of the brain. During the postabsorptive state, blood

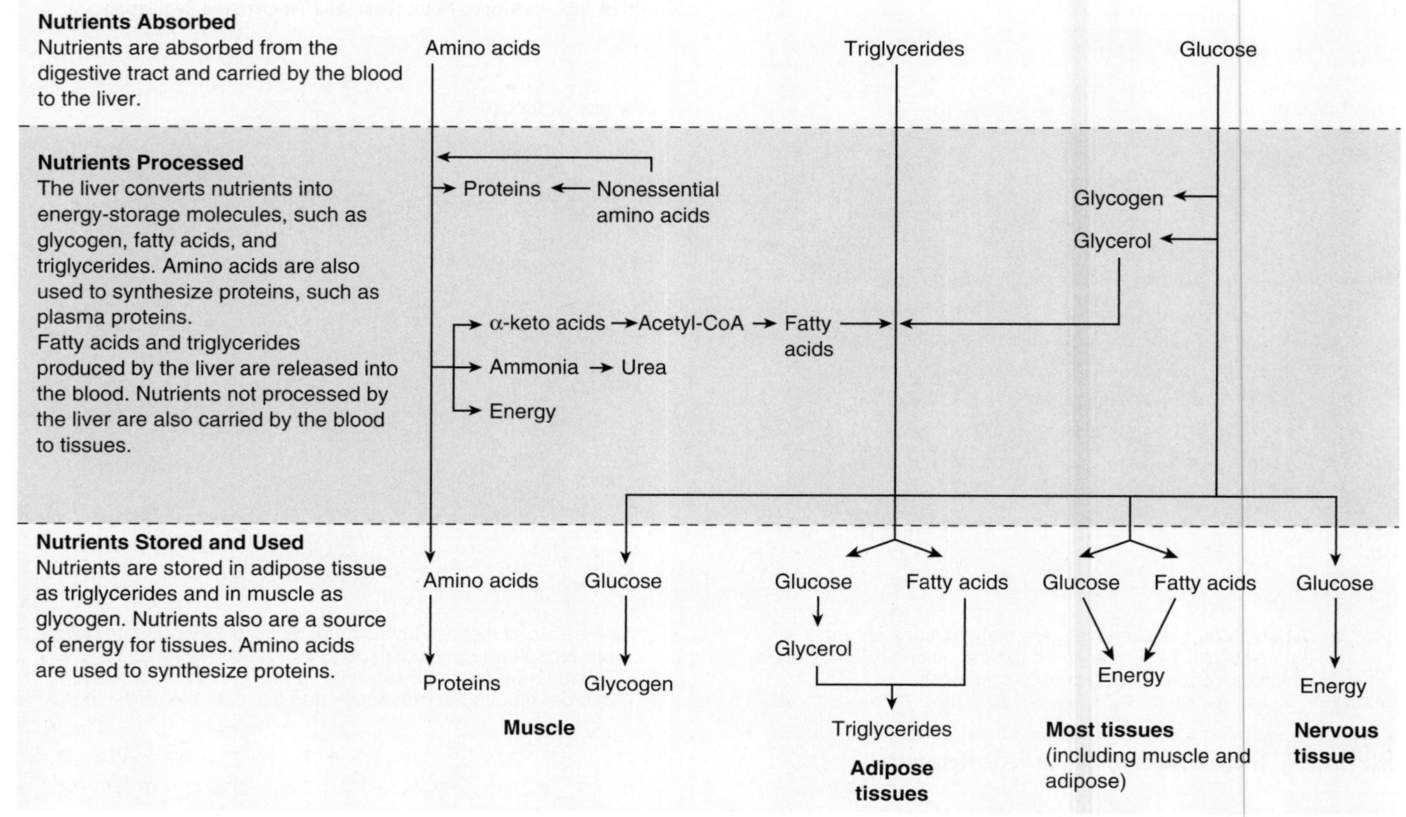

FIGURE 25.15 Events of the Absorptive State

Absorbed molecules, especially glucose, are used as sources of energy. Molecules not immediately needed for energy are stored: Glucose is converted to glycogen or triglycerides, triglycerides are deposited in adipose tissue, and amino acids are converted to triglycerides or carbohydrates.

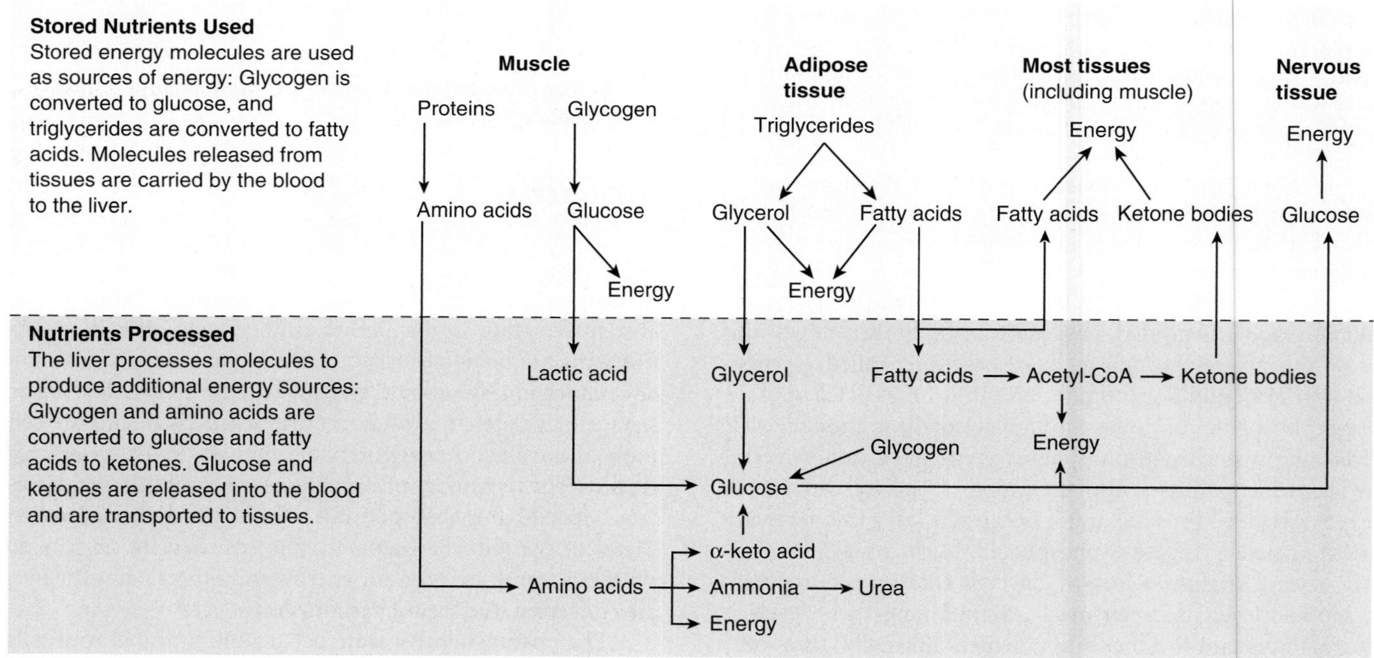

FIGURE 25.16 Events of the Postabsorptive State

Stored energy molecules are used as sources of energy: Glycogen is converted to glucose; triglycerides are broken down to fatty acids, some of which are converted to ketones; and proteins are converted to glucose.

glucose levels are maintained by the conversion of other molecules to glucose. The first source of blood glucose during the postabsorptive state is the glycogen stored in the liver. This glycogen supply, however, can provide glucose for only about 4 hours. The glycogen stored in skeletal muscles can also be used during times of vigorous exercise. As glycogen stores are depleted, fats are used as an energy source. The glycerol from triglycerides can be converted to glucose. The fatty acids from fat can be converted to acetyl-CoA, moved into the citric acid cycle, and used as a source of energy to produce ATP. In the liver, acetyl-CoA is used to produce ketone bodies, which other tissues use for energy. The use of fatty acids as an energy source partly eliminates the need to use glucose for energy, resulting in reduced glucose removal from the blood and the maintenance of blood glucose levels at homeostatic levels. Proteins can also be used as a source of glucose or can be used for energy production, again sparing the use of blood glucose.

37 ***What happens to glucose, fats, and amino acids during the absorptive state?***

38 ***Why is it important to maintain blood glucose levels during the postabsorptive state? Name three sources of this glucose.***

METABOLIC RATE

Metabolic rate is the total amount of energy produced and used by the body per unit of time. A molecule of ATP exists for less than 1 minute before it is degraded back to ADP and inorganic phosphate. For this reason, ATP is produced in cells at about the same rate as it is used. Thus, in examining metabolic rate, ATP production and use can be roughly equated. Metabolic rate is usually estimated by measuring the amount of oxygen used per minute because most ATP production involves the use of oxygen. One liter of oxygen consumed by the body is assumed to produce 4.825 kcal of energy.

The daily input of energy should equal the metabolic expenditure of energy; otherwise, a person will gain or lose weight. For a typical 23-year-old, 70 kg (154-pound) male to maintain his weight, the daily input should be 2700 kcal/day; for a typical 58 kg (128-pound) female of the same age, 2000 kcal/day is necessary. A pound of body fat provides about 3500 kcal. Reducing kilocaloric intake by 500 kcal/day can result in the loss of 1 pound of fat per week. Clearly, adjusting kilocaloric input is an important way to control body weight.

Proportion of Fat in the Diet and Body Weight

Not only the number of kilocalories ingested but also the proportion of fat in the diet has an effect on body weight. To convert dietary fat into body fat, 3% of the energy in the dietary fat is used, leaving 97% for storage as fat deposits. On the other hand, the conversion of dietary carbohydrate to fat requires 23% of the energy in the carbohydrate, leaving just 77% as body fat. If two people have the same kilocaloric intake, the one with the higher proportion of fat in his or her diet is more likely to gain weight because fewer kilocalories are used to convert the dietary fat into body fat.

Metabolic energy is used in three ways: for basal metabolism, for the thermic effect of food, and for muscular activity.

Basal Metabolic Rate

The **basal metabolic rate (BMR)** is the energy needed to keep the resting body functional. It is the metabolic rate calculated in expended kilocalories per square meter of body surface area per hour. BMR is determined by measuring the oxygen consumption of a person who is awake but restful and has not eaten for 12 hours. The liters of oxygen consumed are then multiplied by 4.825 because each liter of oxygen used results in the production of 4.825 kcal of energy. A typical BMR for a 70 kg (154-pound) male is 38 $kcal/m^2/h$.

In the average person, basal metabolism accounts for about 60% of energy expenditure. Basal metabolism supports active-transport mechanisms, muscle tone, maintenance of body temperature, beating of the heart, and other activities. A number of factors can affect the BMR. Muscle tissue is metabolically more active than adipose tissue, even at rest. Younger people have a higher BMR than older people because of increased cell activity, especially during growth. Fever can increase BMR 7% for each degree Fahrenheit increase in body temperature. During dieting or fasting, greatly reduced kilocaloric input can depress BMR, which apparently is a protective mechanism to prevent weight loss. Thyroid hormones can increase BMR on a long-term basis, and epinephrine can increase BMR on a short-term basis (see chapter 18). Males have a greater BMR than females because men have proportionately more muscle tissue and less adipose tissue than women do. During pregnancy, a woman's BMR can increase 20% because of the metabolic activity of the fetus.

Thermic Effect of Food

The second component of metabolic energy is the assimilation of food. When food is ingested, the accessory digestive organs and the intestinal lining produce secretions, the motility of the digestive tract increases, active transport increases, and the liver is involved in the synthesis of new molecules. The energy cost of these events is called the **thermic effect of food,** and it accounts for about 10% of the body's energy expenditure.

Muscular Activity

Muscular activity consumes about 30% of the body's energy. Physical activity resulting from skeletal muscle movement requires the expenditure of energy. In addition, energy must be provided for the increased contraction of the heart and muscles of respiration. The number of kilocalories used in an activity depends almost entirely on the amount of muscular work performed and on the duration of the activity. Despite the fact that studying can make a person feel tired, intense mental concentration produces little change in BMR.

Energy loss through muscular activity is the only component of energy expenditure that a person can reasonably control. A comparison of the number of kilocalories gained from food versus the number of kilocalories lost in exercise reveals why losing weight can be difficult. For example, if brisk walking uses 225 kcal/h, it takes 20 minutes of brisk walking to burn off the 75 kcal in one slice of bread (75/225 = 0.33 h). Research suggests that a combination of appropriate physical activity and appropriate kilocaloric intake is the best approach to maintaining a healthy body composition and weight.

PREDICT 5

If watching TV uses 95 kcal/h, how long does it take to burn off the kilocalories in one cola or beer (see table 25.1)? If jogging at a pace of 6 mph uses 580 kcal/h, how long does it take to use the kilocalories in one cola or beer?

39 ***Define metabolic rate.***

40 ***What is BMR? What factors can alter BMR?***

41 ***What is the thermic effect of food?***

42 ***BMR, the thermic effect of food, and muscular activity each accounts for what percent of total energy expenditure?***

43 ***How are kilocaloric input and output adjusted to maintain body weight?***

BODY TEMPERATURE REGULATION

Humans can maintain a relatively constant internal body temperature despite changes in the temperature of the surrounding environment. The maintenance of a constant body temperature is very important to homeostasis. Most enzymes are very temperature-sensitive, functioning only in narrow temperature ranges. Environmental temperatures are too low for normal enzyme function, and the heat produced by metabolism helps maintain body temperature at a steady, elevated level that is high enough for normal enzyme function.

Free energy is the total amount of energy liberated by the complete catabolism of food. It is usually expressed in terms of kilocalories (kcal) per mole of food consumed. For example, the complete catabolism of 1 mole of glucose (168 g; see chapter 2) releases 686 kcal of free energy. About 43% of the total energy released by catabolism is used to produce ATP and to accomplish biologic work, such as anabolism, muscular contraction, and other cellular activities. The remaining energy is lost as **heat.**

PREDICT 6

Explain why we become warm during exercise and why we shiver when it is cold.

The average normal body temperature usually is considered 37°C (98.6°F) when measured orally and 37.6°C (99.7°F) when measured rectally. Rectal temperature comes closer to the true core body temperature, but an oral temperature is more easily obtained in older children and adults and therefore is the preferred measure.

Heat can be exchanged with the environment in a number of ways (figure 25.17). **Radiation** is the loss of heat as infrared radiation, a type of electromagnetic radiation. For example, the coals in a fire give off radiant heat, which can be felt some distance away from the fire. **Conduction** is the exchange of heat between objects in direct contact with each other, such as the bottoms of the feet and the floor. **Convection** is a transfer of heat between the body and the air or water. A cool breeze results in the movement of air over the body and loss of heat from the body. **Evaporation** is the conversion of water from a liquid to a gas, a process that requires heat. The evaporation of 1 g of water from the body's surface results in the loss of 580 cal of heat.

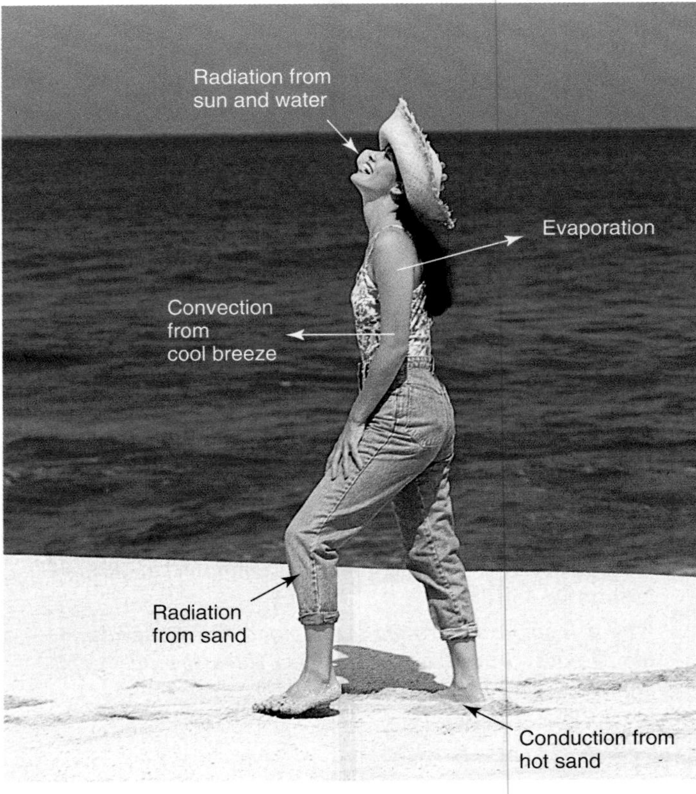

FIGURE 25.17 Heat Exchange
Heat exchange between a person and the environment occurs by convection, radiation, evaporation, and conduction. *Arrows* show the direction of net heat gain or loss in this environment.

Body temperature is maintained by balancing heat gain with heat loss. When heat gain equals heat loss, body temperature is maintained. If heat gain exceeds heat loss, body temperature increases; if heat loss exceeds heat gain, body temperature decreases. Heat gain occurs through metabolism and the muscular contractions of shivering, whereas heat loss occurs through evaporation. Heat gain or loss can occur by radiation, conduction, or convection, depending on skin temperature and the environmental temperature. If the skin temperature is lower than the environmental temperature, heat is gained; however, if skin temperature is higher than the environmental temperature, heat is lost.

The difference in temperature between the body and the environment determines the amount of heat exchanged between the environment and the body. The greater the temperature difference,

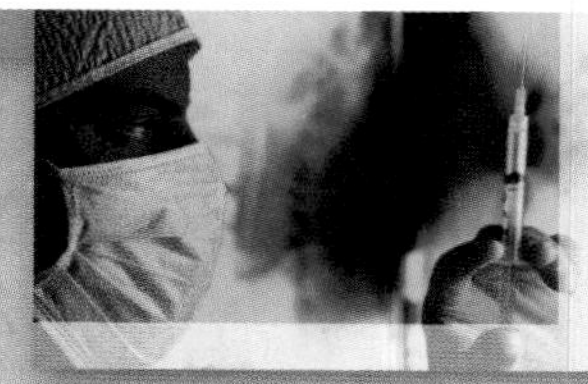

CLINICAL FOCUS

Obesity

Obesity is the presence of excess body fat, resulting from the ingestion of more food than is necessary for the body's energy needs. Obesity can be defined on the basis of body weight, body mass index, or percent body fat. "Desirable body weight" is listed in the Metropolitan Life Insurance Table (1983) and indicates, for any height, the weight that is associated with a maximum life span. Being overweight is defined as weighing 10% to 20% more than the "desirable weight," and being obese is defined as weighing 20% or more than the "desirable weight." Body mass index (BMI) can be calculated by dividing a person's weight (Wt) in kilograms by the square of his or her height (Ht) in meters: $BMI = Wt/Ht^2$. A BMI greater than 25–27 is overweight, and a value greater than 30 is defined as obese. About 10% of people in the United States have a BMI of 30 or greater. In terms of the percent of the total body weight contributed by fat, 15% body fat or less in men and 25% body fat or less in women is associated with reduced health risks. Obesity is defined to be more than 25% body fat in men and 30%–35% in women.

Obesity is classified according to the number and size of fat cells. The greater the amount of lipids stored in the fat cells, the larger their size. In **hyperplastic obesity,** a greater than normal number of fat cells occurs, and they are larger than normal. This type of obesity, which is associated with massive obesity, begins at an early age. In nonobese children, the number of fat cells triples or quadruples between birth and 2 years of age and then remains relatively stable until puberty, when a further increase in the number occurs. In obese children, however, between 2 years of age and puberty, an increase also occurs in the number of fat cells. **Hypertrophic obesity** results from a normal number of fat cells that have increased in size. This type of obesity is more common, is associated with moderate obesity or being "overweight," and typically develops in adults. People who were thin or of average weight and quite active when they were young become less active as they become older. They begin to gain weight between ages 20 and 40, and, although they no longer use as many kilocalories, they still take in the same amount of food as when they were younger. The unused kilocalories are turned into fat, causing fat cells to increase in size. At one time, it was believed that the number of fat cells did not increase after adulthood. It is now known that the number of fat cells can increase in adults. Apparently, if all the existing fat cells are filled to capacity with lipids, new fat cells are formed to store the excess lipids. Once fat cells are formed, however, dieting and weight loss do not result in a decrease in the number of fat cells—instead, they become smaller in size as their lipid content decreases.

The distribution of fat in obese individuals varies. Fat can be found mainly in the upper body, such as in the abdominal region, or it can be associated with the hips and buttocks. These distribution differences are clinically significant because upper body obesity is associated with an increased likelihood of diabetes mellitus, cardiovascular disease, stroke, and death.

In some cases, a specific cause of obesity can be identified. For example, a tumor in the hypothalamus can stimulate overeating. In most cases, however, no specific cause is apparent. In fact, obesity occurs for many reasons, and obesity in an individual can have more than one cause. Obesity seems to have a genetic component and, if one or both parents are obese, their children are more likely to be obese also. Environmental factors, such as eating habits, however, can also play an important role. For example, adopted children can exhibit the obesity of their adoptive parents. In addition, psychologic factors, such as overeating as a means for dealing with stress, can contribute to obesity.

The regulation of body weight is actually a matter of regulating body fat because most changes in body weight reflect changes in the amount of fat in the body. According to the "set point" theory of weight control, the body maintains a certain amount of body fat. If the amount decreases below or increases above this level, mechanisms are activated to return the amount of body fat to its normal value.

It is a common belief that the main cause of obesity is overeating. Certainly, for obesity to occur, at sometime energy intake must have exceeded energy expenditure. A comparison of the kilocaloric intake of obese and that of lean individuals at their usual weights, however, reveals that, on a per kilogram basis, obese people consume fewer kilocalories than lean people.

When people lose a large amount of weight, their feeding behavior changes. They become hyperresponsive to external food cues, think of food often, and cannot get enough to eat without gaining weight. This behavior is typical of both lean and obese individuals who are below their relative set point for weight. Other changes, such as a decrease in basal metabolic rate, take place in a person who has lost a large amount of weight. Most of this decrease in BMR probably results from a decrease in muscle mass associated with weight loss. In addition, some evidence exists that energy lost through exercise and the thermic effect of food are also reduced.

Thus, a person who has lost a large amount of weight has an increased appetite and a decreased ability to expend energy. It is no surprise that only a small percentage of obese people maintain weight loss over the long term. Instead, the typical pattern is one of repeated cycles of weight loss followed by a rapid regain of the lost weight.

The message emerging from current research is that body weight results from many complicated genetic and metabolic factors that go awry in many ways. Obesity is being regarded as a chronic condition that may respond to medication in much the same way that diabetes does. Nonetheless, medication is only part of the story. Drugs can help, but eating less and exercising more is still necessary for optimal health.

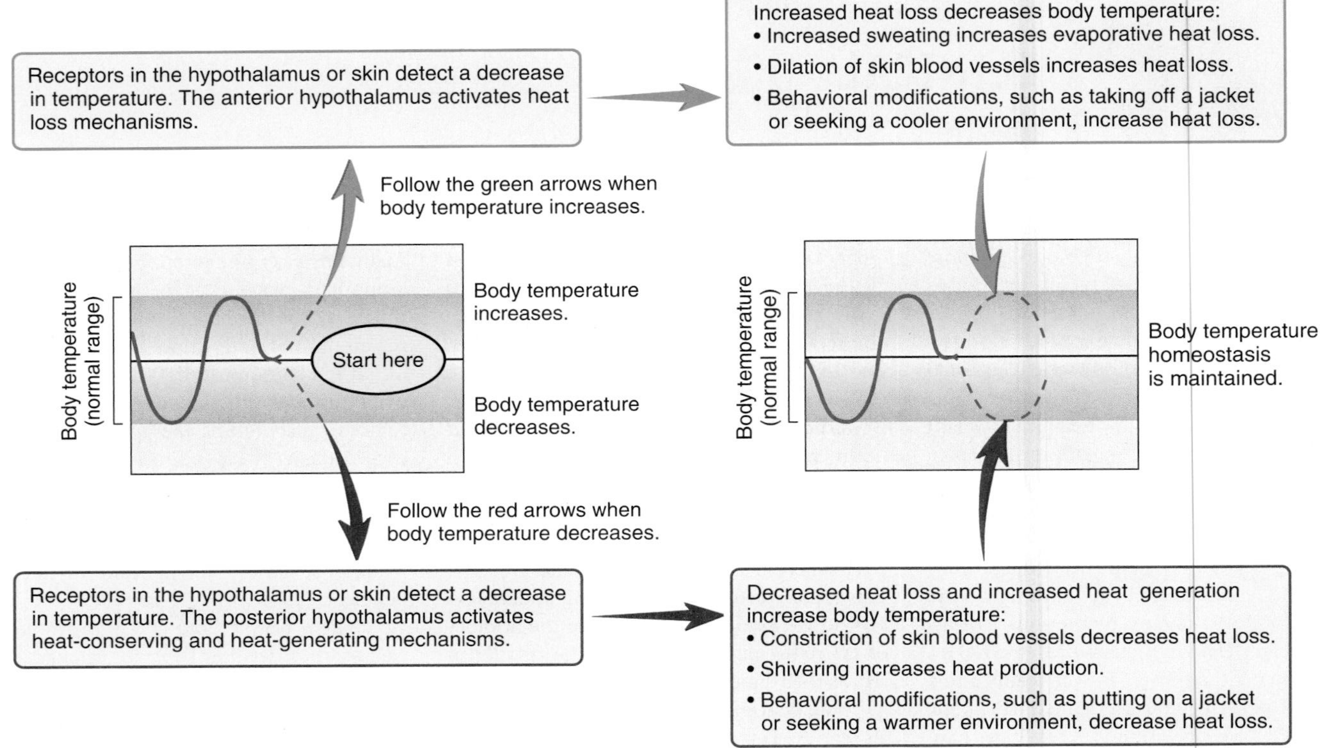

HOMEOSTASIS FIGURE 25.18 **Summary of Temperature Regulation**

the greater the rate of heat exchange. Control of the temperature difference is used to regulate body temperature. For example, if environmental temperature is very cold, as on a cold winter day, a large temperature difference exists between the body and the environment, and a large loss of heat occurs. The heat loss can be decreased by behaviorally selecting a warmer environment—for example, by going inside a heated house. Heat loss can also be decreased by insulating the exchange surface, such as by putting on extra clothes. Physiologically, temperature difference can be controlled through the dilation and constriction of blood vessels in the skin. When these blood vessels dilate, they bring warm blood to the surface of the body, raising skin temperature; conversely, vasoconstriction decreases blood flow and lowers skin temperature.

PREDICT 7

Explain why vasoconstriction of the skin's blood vessels on a cool day is beneficial.

When the environmental temperature is greater than body temperature, vasodilation brings warm blood to the skin, causing an increase in skin temperature, which decreases heat gain from the environment. At the same time, evaporation carries away excess heat to prevent heat gain and overheating.

Body temperature regulation is an example of a negative-feedback system that is controlled by a set point. A small area in the anterior part of the hypothalamus detects slight increases in body temperature through changes in blood temperature (figure 25.18). As a result, mechanisms are activated that cause heat loss, such as vasodilation and sweating, and body temperature decreases. A small area in the posterior hypothalamus can detect slight decreases in body temperature and can initiate heat gain by increasing muscular activity (shivering) and vasoconstriction.

Under some conditions, the set point of the hypothalamus is actually changed. For example, during a fever, the set point is raised, heat-conserving and heat-producing mechanisms are stimulated, and body temperature increases. In recovery from a fever, the set point is lowered to normal, heat loss mechanisms are initiated, and body temperature decreases.

44 ***Define homeotherm and free energy. How much of the free energy is lost as heat from the body?***

45 ***What are four ways that heat is exchanged between the body and the environment?***

46 ***How is body temperature behaviorally and physiologically maintained in a cold and in a hot environment?***

47 ***How does the hypothalamus regulate body temperature?***

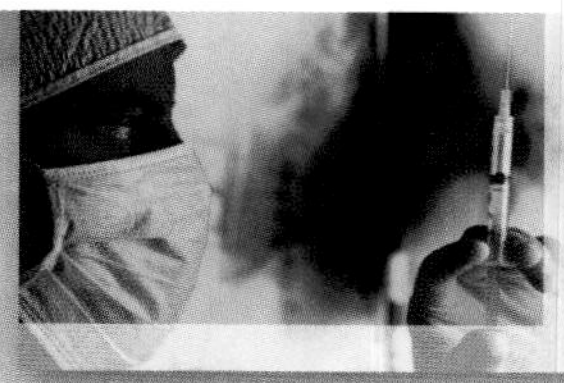

CLINICAL FOCUS

Hyperthermia and Hypothermia

HYPERTHERMIA

Hyperthermia, an elevated body temperature, develops when heat gain exceeds the body's ability to lose heat. Hyperthermia can result from exercise, exposure to hot environments, fever, or anesthesia.

Exercise increases body temperature because of the heat produced as a by-product of muscle activity (see chapter 9). Normally, vasodilation and increased sweating prevent harmful body temperature increases. In a hot, humid environment, the evaporation of sweat is decreased, and exercise levels have to be reduced to prevent overheating.

Exposure to a hot environment normally results in the activation of heat loss mechanisms, and body temperature is maintained at normal levels, a negative-feedback mechanism. Prolonged exposure to a hot environment, however, can result in **heat exhaustion.** The normal negative-feedback mechanisms for controlling body temperature are operating, but they are unable to prevent an increase in body temperature above normal levels. Heavy sweating results in dehydration, decreased blood volume, decreased blood pressure, and increased heart rate. Individuals suffering from heat exhaustion have a wet, cool skin because of the heavy sweating. They usually feel weak, dizzy, and nauseated. Treatment includes reducing heat gain by moving to a cooler environment, ceasing activity to reduce the heat produced by muscle metabolism, and restoring blood volume by drinking fluids.

Heat stroke is more severe than heat exhaustion because it results from a breakdown in the normal negative-feedback mechanisms of temperature regulation. If the temperature of the hypothalamus becomes too high, it no longer functions appropriately. Sweating stops, and the skin becomes dry and flushed. The person becomes confused, irritable, or even comatose. In addition to the treatment for heat exhaustion, heat loss from the skin should be increased. This can be accomplished by increasing evaporation from the skin by applying wet cloths or by increasing conductive heat loss by immersing the person in a cool bath.

Fever is the development of a higher than normal body temperature following the invasion of the body by microorganisms or other foreign substances. Lymphocytes, neutrophils, and macrophages release chemicals called **pyrogens** (pī′rō-jenz). Examples of pyrogens include certain interleukins, interferons, and tissue necrosis factor. Pyrogens increase the synthesis of prostaglandins, which stimulate a rise in the temperature set point of the hypothalamus. Consequently, body temperature and metabolic rate increase. Fever is believed to be beneficial because it speeds up the chemical reactions of the immune system (see chapter 22) and inhibits the growth of some microorganisms. Although beneficial, body temperatures greater than 41°C (106°F) can be harmful. Aspirin, nonsteroidal anti-inflammatory drugs (NSAIDs), and acetaminophen lower body temperature by inhibiting the synthesis of prostaglandins.

Malignant hyperthermia is an inherited muscle disorder. Certain drugs used to induce general anesthesia for surgery cause sustained, uncoordinated muscle contractions in individuals with this disorder. Consequently, body temperature increases.

Therapeutic hyperthermia is an induced local or general body increase in temperature. It is a treatment sometimes used on tumors and infections.

HYPOTHERMIA

If heat loss exceeds the body's ability to produce heat, body temperature decreases below normal levels. **Hypothermia** is a decrease in body temperature to 35°C (95°F) or below. Hypothermia usually results from prolonged exposure to cold environments. At first, normal negative-feedback mechanisms maintain body temperature. Heat loss is decreased by constricting blood vessels in the skin, and heat production is increased by shivering. If body temperature decreases despite these mechanisms, hypothermia develops. The individual's thinking becomes sluggish, and movements are uncoordinated. Heart, respiratory, and metabolic rates decline, and death results unless body temperature is restored to normal. Rewarming should occur at a rate of a few degrees per hour.

Frostbite is damage to the skin and deeper tissues, resulting from prolonged exposure to the cold. Damage results from direct cold injury to cells, injury from ice crystal formation, and reduced blood flow to affected tissues. The fingers, toes, ears, nose, and cheeks are most commonly affected. Damage from frostbite can range from redness and discomfort to loss of the affected part. The best treatment is immersion in a warm-water bath. Rubbing the affected area and local, dry heat should be avoided.

Therapeutic hypothermia is sometimes used to slow metabolic rate during surgical procedures, such as heart surgery. Because the metabolic rate is decreased, the tissues do not require as much oxygen as normal and are less likely to be damaged.

SUMMARY

Nutrition (p. 928)

Nutrition is the taking in and use of food.

Nutrients

1. Nutrients are the chemicals used by the body. They are carbohydrates, lipids, proteins, vitamins, minerals, and water.
2. Essential nutrients are nutrients that must be ingested because the body cannot manufacture them or is unable to manufacture adequate amounts of them.

Kilocalories

1. A calorie (cal) is the heat (energy) necessary to raise the temperature of 1 g of water 1°C. A kilocalorie (kcal), or Calorie (Cal), is 1000 calories.
2. A gram of carbohydrate or protein yields 4 kcal, and a gram of fat yields 9 kcal.

MyPyramid

The MyPyramid guide recommends the amounts of different food types and fiber necessary for good health, based on a person's age, sex, and physical activity.

Carbohydrates

1. Carbohydrates are ingested as monosaccharides (glucose, fructose), disaccharides (sucrose, maltose, lactose), and polysaccharides (starch, glycogen, cellulose).
2. Monounsaturated fats and oils have one double bond, and polyunsaturated fats and oils have two or more double bonds.
3. Most unprocessed polyunsaturated oils occur in the *cis* form, whereas hydrogenated polyunsaturated oils are in the *trans* form.
4. Polysaccharides and disaccharides are converted to glucose. Glucose can be used for energy or stored as glycogen or fats.
5. The Acceptable Macronutrient Distribution Range (AMDR) for carbohydrates is 45%–65% of total kilocalories.

Lipids

1. Lipids are ingested as triglycerides (95%) or cholesterol and phospholipids (5%).
2. Triglycerides are used for energy or are stored in adipose tissue. Cholesterol forms other molecules, such as steroid hormones. Cholesterol and phospholipids are part of the plasma membrane.
3. The AMDR for lipids is 20%–35%.

Proteins

1. Proteins are ingested and broken down into amino acids.
2. Proteins perform many functions: protection (antibodies), regulation (enzymes, hormones), structure (collagen), muscle contraction (actin and myosin), transportation (hemoglobin, transport proteins) and receptors.
3. The AMDR for protein is 10%–35% of total kilocalories.

Vitamins

1. Many vitamins function as coenzymes or as parts of coenzymes.
2. Most vitamins are not produced by the body and must be obtained in the diet. Some vitamins can be formed from provitamins.
3. Vitamins are classified as either fat-soluble or water-soluble.
4. Recommended Dietary Allowances (RDAs) are a guide for estimating the nutritional needs of groups of people based on their age, sex, and other factors.

Minerals

1. Minerals are necessary for normal metabolism, add mechanical strength to bones and teeth, function as buffers, and are involved in osmotic balance.
2. The daily requirement for major minerals is 100 mg or more daily, whereas for trace minerals it is less than 100 mg daily.

Daily Values

1. Daily Values are dietary references that can be used to help plan a healthy diet.
2. Daily Values for vitamins and minerals are based on Reference Daily Intakes, which are generally the highest 1968 RDA values of age categories.
3. Daily Values are based on Daily Reference Values.
   - The Daily Reference Values for energy-producing nutrients (carbohydrates, total fat, saturated fat, and proteins) and dietary fiber are recommended percentages of the total kilocalories ingested daily for each nutrient.
   - The Daily Reference Values for total fats, saturated fats, cholesterol, and sodium are the uppermost limit considered desirable because of their link to diseases.
4. The % Daily Value is the percent of the recommended Daily Value of a nutrient found in one serving of a particular food.

Metabolism (p. 937)

1. Metabolism consists of catabolism and anabolism. Catabolism is the breaking down of molecules and gives off energy. Anabolism is the building up of molecules and requires energy.
2. The energy in carbohydrates, lipids, and proteins is used to produce ATP through oxidation–reduction reactions.

Carbohydrate Metabolism (p. 938)

Glycolysis

Glycolysis is the breakdown of glucose into two pyruvic acid molecules. Also produced are two NADH molecules and two ATP molecules.

Anaerobic Respiration

1. Anaerobic respiration is the breakdown of glucose in the absence of oxygen into two lactic acid and two ATP molecules.
2. Lactic acid can be converted to glucose (Cori cycle) using aerobically produced ATP (oxygen debt).

Aerobic Respiration

1. Aerobic respiration is the breakdown of glucose in the presence of oxygen to produce carbon dioxide, water, and 38 (or 36) ATP molecules.
2. The first phase is glycolysis, which produces two ATP, two NADH, and two pyruvic acid molecules.
3. The second phase is the conversion of the two pyruvic acid molecules into two molecules of acetyl-CoA. These reactions also produce two NADH and two carbon dioxide molecules.

4. The third phase is the citric acid cycle, which produces two ATP, six NADH, two FADH2, and four carbon dioxide molecules.
5. The fourth phase is the electron-transport chain. The high-energy electrons in NADH and FADH2 enter the electron-transport chain and are used in the synthesis of ATP and water.

Lipid Metabolism (p. 946)

1. Adipose triglycerides are broken down and released as free fatty acids.
2. Free fatty acids are taken up by cells and broken down by beta-oxidation into acetyl-CoA.
   - Acetyl-CoA can enter the citric acid cycle.
   - Acetyl-CoA can be converted into ketone bodies.

Protein Metabolism (p. 948)

1. New amino acids are formed by transamination, the transfer of an amine group to a keto acid.
2. Amino acids are used to synthesize proteins. If used for energy, ammonia is produced as a by-product of oxidative deamination. Ammonia is converted to urea and is excreted.

Interconversion of Nutrient Molecules (p. 950)

1. Glycogenesis is the formation of glycogen from glucose.
2. Lipogenesis is the formation of lipids from glucose and amino acids.
3. Glycogenolysis is the breakdown of glycogen to glucose.
4. Gluconeogenesis is the formation of glucose from amino acids and glycerol.

Metabolic States (p. 951)

1. In the absorptive state, nutrients are used as energy or stored.
2. In the postabsorptive state, stored nutrients are used for energy.

Metabolic Rate (p. 953)

Metabolic rate is the total energy expenditure per unit of time, and it has three components.

Basal Metabolic Rate

Basal metabolic rate is the energy used at rest. It is 60% of the metabolic rate.

Thermic Effect of Food

The thermic effect of food is the energy used to digest and absorb food. It is 10% of the metabolic rate.

Muscular Activity

Muscular energy is used for muscle contraction. It is 30% of the metabolic rate.

Body Temperature Regulation (p. 954)

1. Body temperature is a balance between heat gain and heat loss.
   - Heat is produced through metabolism.
   - Heat is exchanged through radiation, conduction, convection, and evaporation.
2. The greater the temperature difference between the body and the environment, the greater the rate of heat exchange.
3. Body temperature is regulated by a set point in the hypothalamus.

REVIEW AND COMPREHENSION

1. Which of these statements concerning kilocalories is true?
   a. A kilocalorie is the amount of energy required to raise the temperature of 1 g of water 1°C.
   b. There are 9 kcal in a gram of protein.
   c. There are 4 kcal in a gram of fat.
   d. A pound of body fat contains 3500 kcal.
2. Complex carbohydrates include
   a. sucrose.
   b. milk sugar (lactose).
   c. starch, an energy storage molecule in plants.
   d. all of the above.
3. What type of nutrient is recommended as the primary energy source in the diet?
   a. carbohydrates
   b. fats
   c. proteins
   d. cellulose
4. A source of monounsaturated fats is
   a. fat associated with meat.
   b. egg yolks.
   c. whole milk.
   d. fish oil.
   e. olive oil.
5. A complete protein food
   a. provides the daily amount (grams) of protein recommended for a healthy diet.
   b. can be used to synthesize the nonessential amino acids.
   c. contains all 20 amino acids.
   d. includes beans, peas, and leafy green vegetables.
6. Concerning vitamins,
   a. most can be synthesized by the body.
   b. they are normally broken down before they can be used by the body.
   c. A, D, E, and K are water-soluble vitamins.
   d. many function as coenzymes.
7. Minerals
   a. are inorganic nutrients.
   b. compose about 4%–5% of total body weight.
   c. act as buffers and osmotic regulators.
   d. are components of enzymes.
   e. all of the above.
8. Glycolysis
   a. is the breakdown of glucose to two pyruvic acid molecules.
   b. requires the input of two ATP molecules.
   c. produces two NADH molecules.
   d. does not require oxygen.
   e. all of the above.
9. Anaerobic respiration ____________ oxygen and produces ____________ energy (ATP) for the cell than aerobic respiration.
   a. does not require, more
   b. does not require, less
   c. requires, more
   d. requires, less
10. Which of these reactions takes place in both anaerobic and aerobic respiration?
    a. glycolysis
    b. citric acid cycle
    c. electron-transport chain
    d. acetyl-CoA formation

11. The molecule that moves electrons from the citric acid cycle to the electron-transport chain is
    a. tRNA. b. mRNA. c. ADP. d. NADH. e. pyruvic acid.
12. The production of ATP molecules by the electron-transport chain is accompanied by the synthesis of
    a. alcohol. b. water. c. oxygen. d. lactic acid. e. glucose.
13. The carbon dioxide you breathe out comes from
    a. glycolysis. b. the electron-transport chain. c. anaerobic respiration. d. the food you eat.
14. Lipids are
    a. stored primarily as triglycerides.
    b. synthesized by beta-oxidation.
    c. broken down by oxidative deamination.
    d. all of the above.
15. Amino acids
    a. are classified as essential or nonessential.
    b. can be synthesized in a transamination reaction.
    c. can be used as a source of energy.
    d. can be converted to keto acids.
    e. all of the above.
16. Ammonia is
    a. a by-product of lipid metabolism.
    b. formed during ketogenesis.
    c. converted into urea in the liver.
    d. produced during lipogenesis.
    e. converted to keto acids.
17. The conversion of amino acids and glycerol into glucose is called
    a. gluconeogenesis. b. glycogenolysis. c. glycogenesis. d. ketogenesis.
18. Which of these events takes place during the absorptive state?
    a. Glycogen is converted into glucose.
    b. Glucose is converted into fats.
    c. Ketones are produced.
    d. Proteins are converted into glucose.
19. The major use of energy by the body is in
    a. basal metabolism.
    b. physical activity.
    c. the thermic effect of food.
20. The loss of heat resulting from the loss of water from the body's surface is
    a. radiation. b. evaporation. c. conduction. d. convection.

Answers in Appendix E

CRITICAL THINKING

1. One serving of a food has 2 g of saturated fat. What % Daily Value for saturated fat would appear on a food label for this food? (See the bottom of figure 25.2 for information needed to answer this question.)
2. An active teenage boy has a daily intake of 3000 kcal/day. What is the maximum amount (weight) of total fats he should consume, according to the Daily Values?
3. If the teenager in question 2 eats a serving of food that has a total fat content of 10 g/serving, what is his % Daily Value for total fat?
4. Suppose the food in question 3 is in a package that lists a serving size of 1/2 cup, with 4 servings in the package. If the teenager eats half of the contents of the package (1 cup), how much of his % Daily Value does he consume?
5. Why does a vegetarian usually have to be more careful about his or her diet than a person who includes meat in the diet?
6. Explain why a person suffering from copper deficiency feels tired all the time.
7. Some people claim that occasionally fasting for short periods can be beneficial. How can fasts be damaging?
8. Why can some people lose weight on a 1200 kcal/day diet but others cannot?
9. Lotta Bulk, a muscle builder, wanted to increase her muscle mass. Knowing that proteins are the main components of muscle, she began a high-protein diet in which most of her daily kilocalories were supplied by proteins. She also exercised regularly with heavy weights. After 3 months of this diet and exercise program, Lotta increased her muscle mass, but not any more than her friend, who did the same exercises but did not have a high-protein diet. Explain what happened. Was Lotta in positive or negative nitrogen balance?
10. On learning that sweat evaporation results in the loss of calories, an anatomy and physiology student enters a sauna in an attempt to lose weight. He reasons that a liter (about a quart) of water weighs 1000 g, which is equivalent to 580,000 cal, or 580 kcal, of heat when lost as sweat. Instead of reducing his diet by 580 kcal/day, if he loses a liter of sweat every day in the sauna, he believes he will lose about a pound of fat a week. Will this approach work? Explain.
11. Thyroid hormone is known to increase the activity of the Na^+–K^+ pump. If a person produced excess amounts of thyroid hormone, what effect would this have on basal metabolic rate, body weight, and body temperature? How would the body attempt to compensate for the changes in body weight and temperature?
12. In some diseases, an infection causes a high fever, resulting in a crisis state. The person is on the way to recovery when body temperature begins to return to normal. If you were looking for symptoms in a person who had just passed through the crisis state, would you look for a dry, pale skin or a wet, flushed skin? Explain.

Answers in Appendix F

Visit this book's website at www.mhhe.com/seeley8 for chapter quizzes, interactive learning exercises, and other study tools.

Urinary System 26

The kidneys make up the body's main purification system. They control the composition of blood by removing waste products, many of which are toxic, and conserving useful substances. The kidneys help control blood volume and consequently play a role in regulating blood pressure. The kidneys also play an essential role in regulating blood pH. Approximately one-third of one kidney is all that is needed to maintain homeostasis. Even after extensive damage, the kidneys can still perform their life-sustaining function. If the kidneys are damaged further, however, death results unless specialized medical treatment is administered.

The **urinary system** consists of two kidneys; a single, midline urinary bladder; two ureters, which carry urine from the kidneys to the urinary bladder; and a single urethra, which carries urine from the bladder to the outside of the body (figure 26.1).

This chapter explains the *functions of the urinary system* (p. 962), *kidney anatomy and histology* (p. 962), *urine production* (p. 970), *regulation of urine concentration and volume* (p. 983), *plasma clearance and tubular maximum* (p. 991), and *urine movement* (p. 992). The chapter concludes with a look at the *effects of aging on the kidneys* (p. 996).

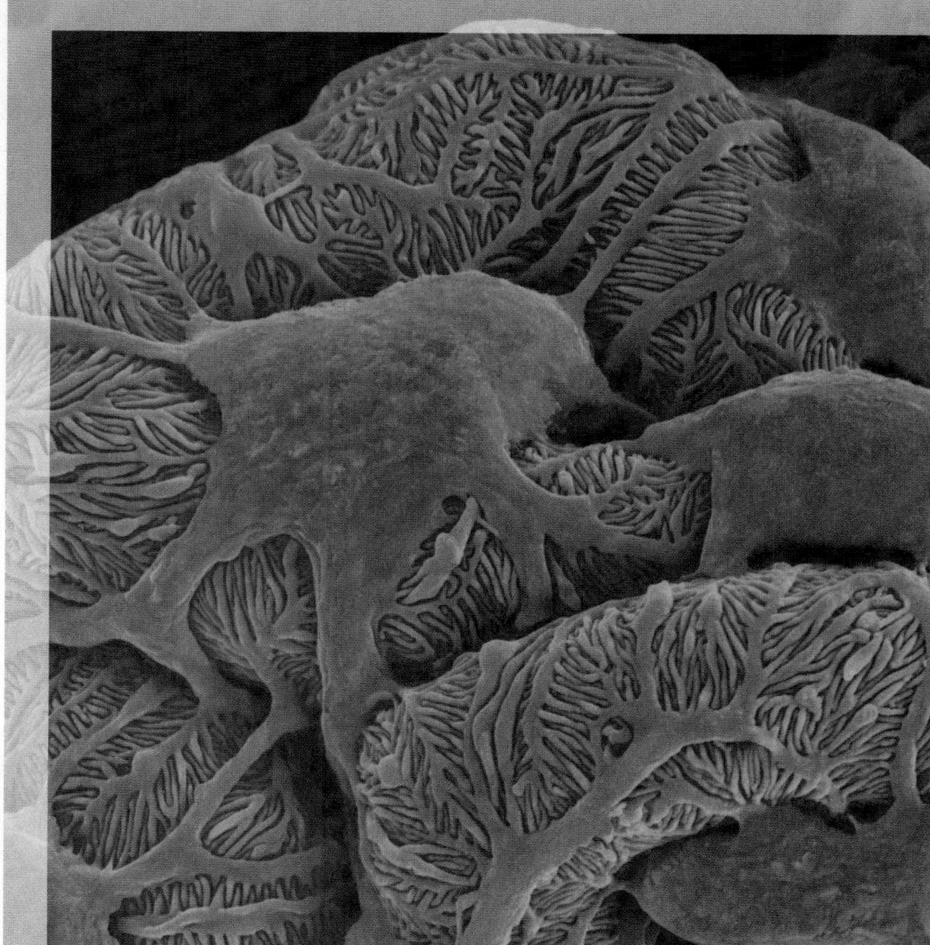

Color-enhanced scanning electron micrograph of podocytes wrapped around the glomerular capillaries.

Urinary System

FUNCTIONS OF THE URINARY SYSTEM

The kidneys are the major excretory organs of the body. The skin, liver, lungs, and intestines eliminate some waste products, but, if the kidneys fail to function, these other excretory organs cannot adequately compensate. The following functions are performed by the kidneys:

1. *Excretion.* The kidneys filter blood, and a large volume of filtrate is produced. Large molecules, such as proteins and blood cells, are retained in the blood, whereas smaller molecules and ions enter the filtrate. Most of the filtrate volume is reabsorbed back into the blood, along with useful molecules and ions. Metabolic wastes, toxic molecules, and excess ions remain in a small volume of filtrate. Additional waste products are secreted into the filtrate and the result is urine formation.
2. *Regulation of blood volume and pressure.* The kidneys play a major role in controlling the extracellular fluid volume in the body by producing either a large volume of dilute urine or a small volume of concentrated urine. Consequently, the kidneys regulate blood volume and blood pressure.
3. *Regulation of the concentration of solutes in the blood.* The kidneys help regulate the concentration of the major ions, such as Na^+, Cl^-, K^+, Ca^{2+}, HCO_3^-, and HPO_4^{2-}.
4. *Regulation of the pH of the extracellular fluid.* The kidneys secrete variable amounts of H^+ to help regulate the extracellular fluid pH.

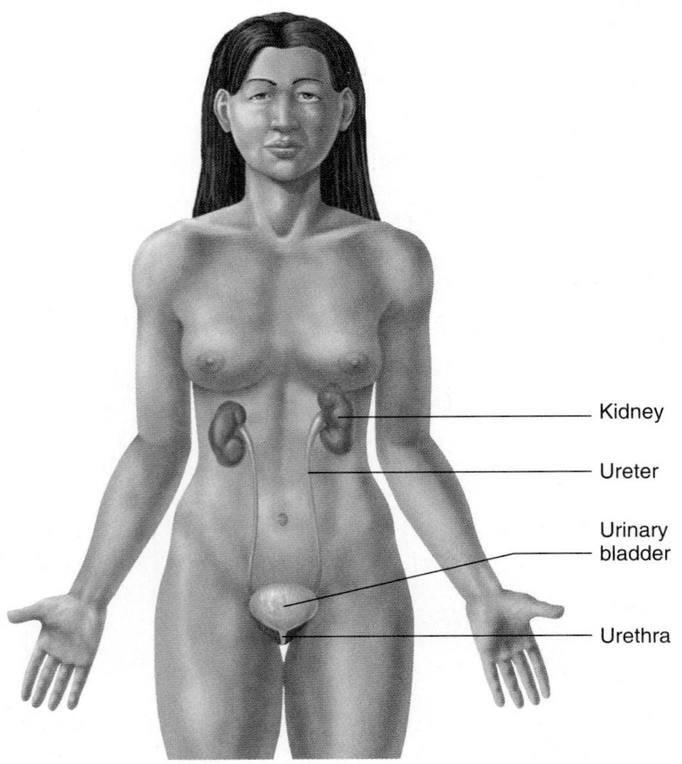

FIGURE 26.1 Urinary System

The urinary system consists of two kidneys, two ureters, a urinary bladder, and a urethra.

5. *Regulation of red blood cell synthesis.* The kidneys secrete a hormone, erythropoietin, which regulates the synthesis of red blood cells in bone marrow (see chapter 19).
6. *Vitamin D synthesis.* The kidneys play an important role in controlling blood levels of Ca^{2+} by regulating the synthesis of vitamin D (see chapter 6).

1 ***List the functions performed by the kidneys, and briefly describe each.***

KIDNEY ANATOMY AND HISTOLOGY

Location and External Anatomy of the Kidneys

The **kidneys** are bean-shaped and each is about the size of a tightly clenched fist. They lie behind the peritoneum on the posterior abdominal wall on each side of the vertebral column near the lateral borders of the psoas major muscles (figure 26.2). The kidneys extend from the level of the last thoracic (T_{12}) to the third lumbar (L_3) vertebrae and the rib cage partially protects them. The liver is superior to the right kidney, causing the right kidney to be slightly lower than the left. Each kidney measures about 11 cm long, 5 cm wide, and 3 cm thick and weighs about 130 g. A **renal capsule,** a layer of fibrous connective tissue, surrounds each kidney. **Perirenal fat,** a dense layer of adipose tissue, in turn, engulfs the renal capsule. This perirenal fat acts as a shock absorber, cushioning the kidneys against mechanical shock. A thin layer of connective tissue, the **renal fascia,** surrounds the perirenal fat and helps anchor the kidneys and surrounding adipose tissue to the abdominal wall. The fat around the renal fascia is between the renal fascia and the parietal peritoneum.

The **hilum** (hī′lŭm) is a small area where the renal artery and nerves enter, and the renal vein and ureter exit, the kidney. It is located on the concave, medial side of the kidney. The hilum opens into the **renal sinus,** a cavity filled with fat and connective tissue. Structures that enter and leave the kidney pass through the renal sinus (figure 26.3).

2 ***Describe the location, size, and shape of the kidneys.***

3 ***Describe the renal capsule and the structures that surround the kidney.***

4 ***List the structures found at the hilum and in the renal sinus of a kidney.***

Internal Anatomy and Histology of the Kidneys

A frontal section of the kidney reveals that the kidney is divided into an outer **cortex** and an inner **medulla,** which surrounds the renal sinus (see figure 26.3). **Renal pyramids** are cone-shaped structures that make up the medulla. **Medullary rays** extend from the renal pyramids into the cortex. **Renal columns** consist of the same tissue as the cortex that projects between the renal pyramids. The bases of the pyramids form the boundary between the cortex and the medulla. The tips of the pyramids, the **renal papillae,** point toward the renal sinus. **Minor calyces** (kal′i-sēz; cup of flower) are funnel-shaped chambers into which the renal papillae extend.

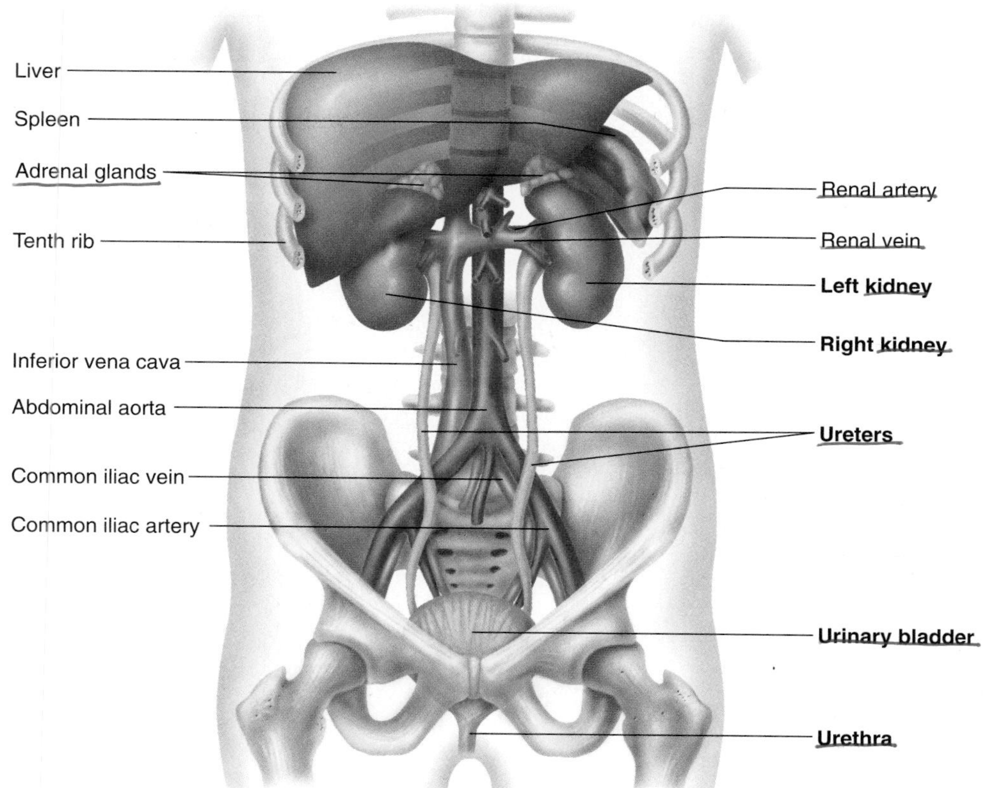

Anterior view

(a) The kidneys are located in the abdominal cavity, with the right kidney just below the liver and the left kidney below the spleen. A ureter extends from each kidney to the urinary bladder within the pelvic cavity. An adrenal gland is located at the superior pole of each kidney.

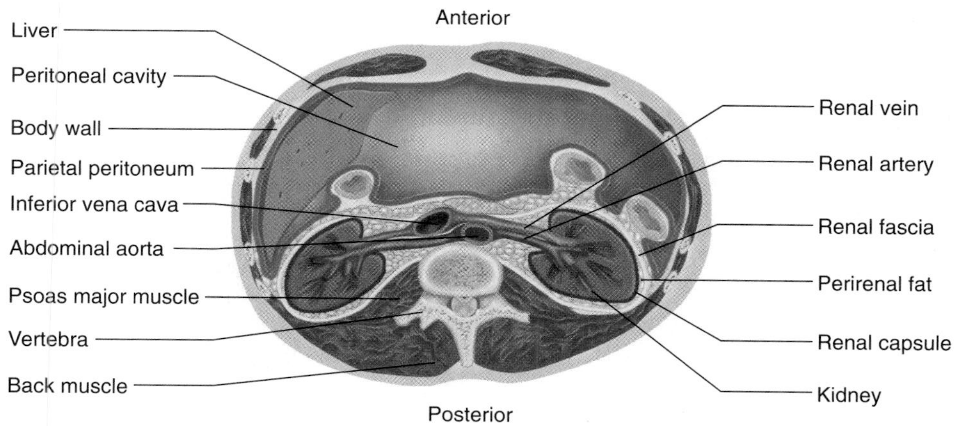

Inferior view

(b) The kidneys are located behind the parietal peritoneum. Surrounding each kidney is the perirenal fat. A connective tissue layer, the renal fascia, anchors the kidney to the abdominal wall. Additional fat is found between the renal fascia and the parietal peritoneum. The renal arteries extend from the abdominal aorta to each kidney, and the renal veins extend from the kidneys to the inferior vena cava.

FIGURE 26.2 Anatomy of the Urinary System

The minor calyces of several pyramids merge to form larger funnels, the **major calyces.** Each kidney contains 8–20 minor calyces and 2 or 3 major calyces. The major calyces converge to form an enlarged chamber called the **renal pelvis,** which is surrounded by the renal sinus. The renal pelvis narrows into a small-diameter tube, the **ureter,** which exits the kidney at the hilum and connects to the urinary bladder. Urine formed within the kidney flows from the renal papillae into the minor calyces. From the minor calyces, urine flows into the major calyces, collects in the renal pelvis, and then leaves the kidney through the ureter.

The **nephron** (nef′ron) is the histological and functional unit of the kidney (figure 26.4). Each nephron is a tubelike structure

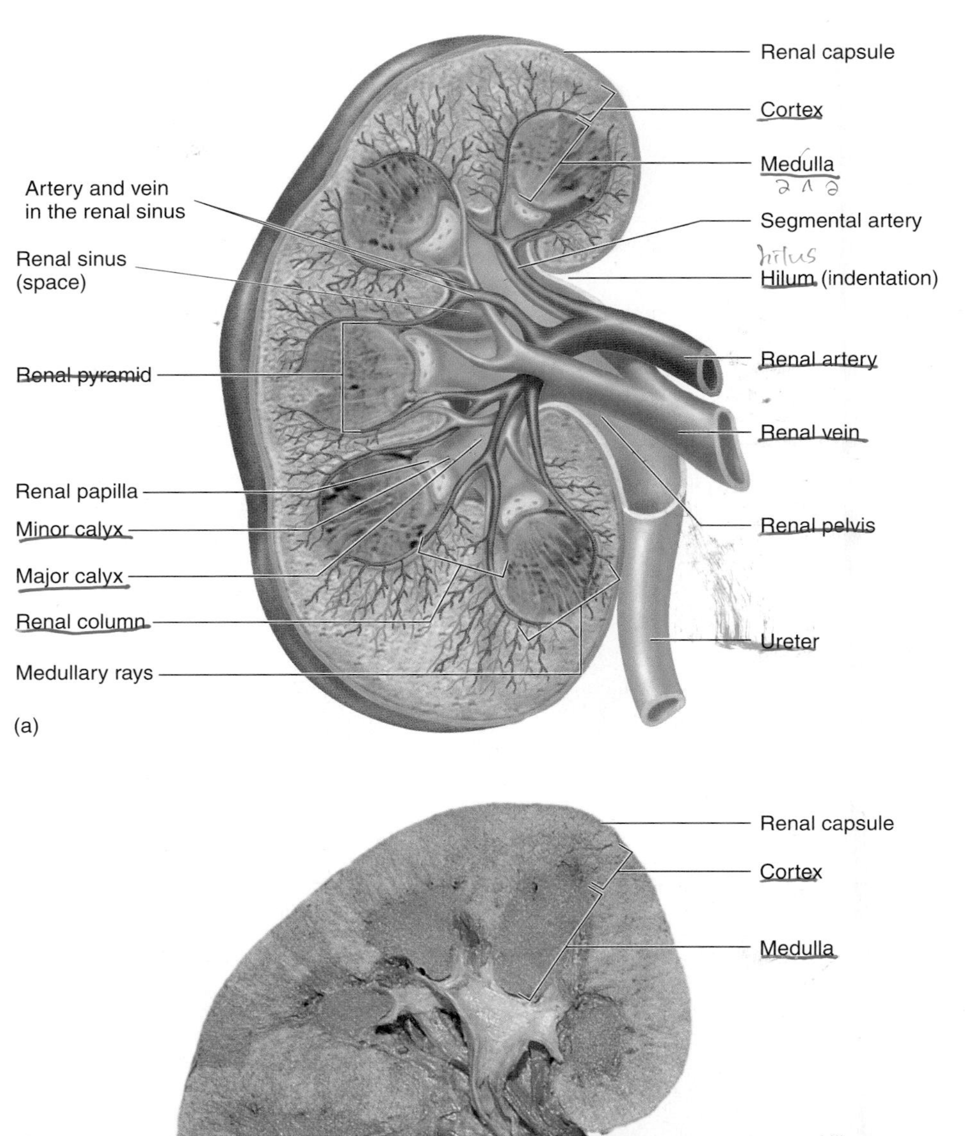

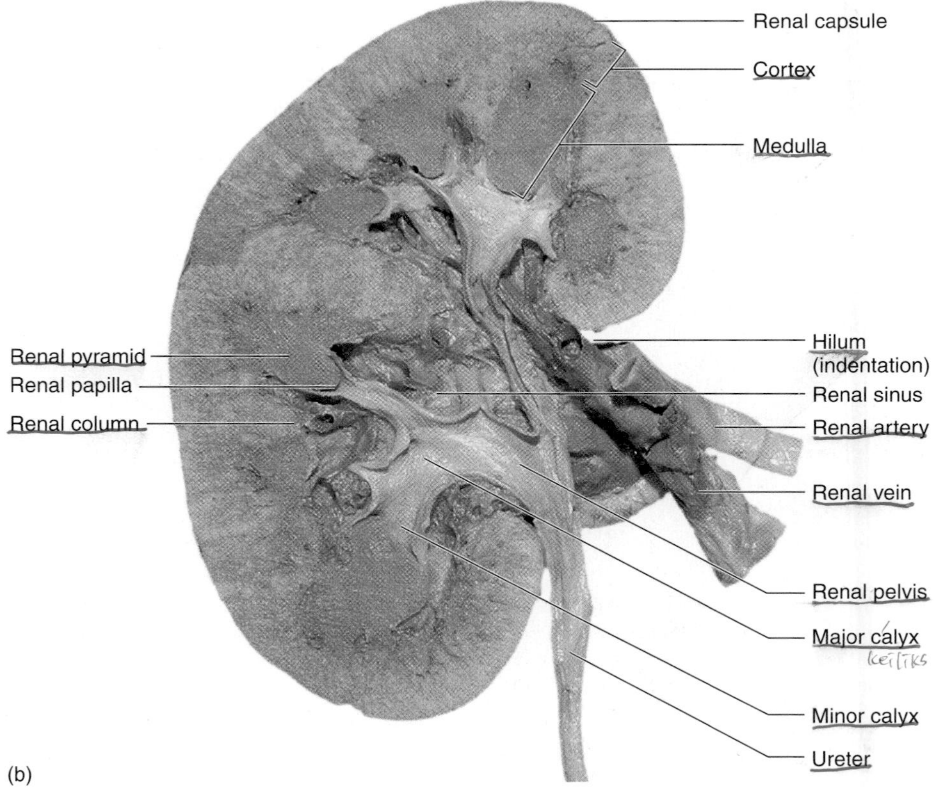

FIGURE 26.3 Longitudinal Section of the Kidney and Ureter

(*a*) The cortex forms the outer part of the kidney, and the medulla forms the inner part. A central cavity called the renal sinus contains the renal pelvis. The renal columns of the kidney project from the cortex into the medulla and separate the pyramids. (*b*) Photograph of a longitudinal section of a human kidney and ureter.

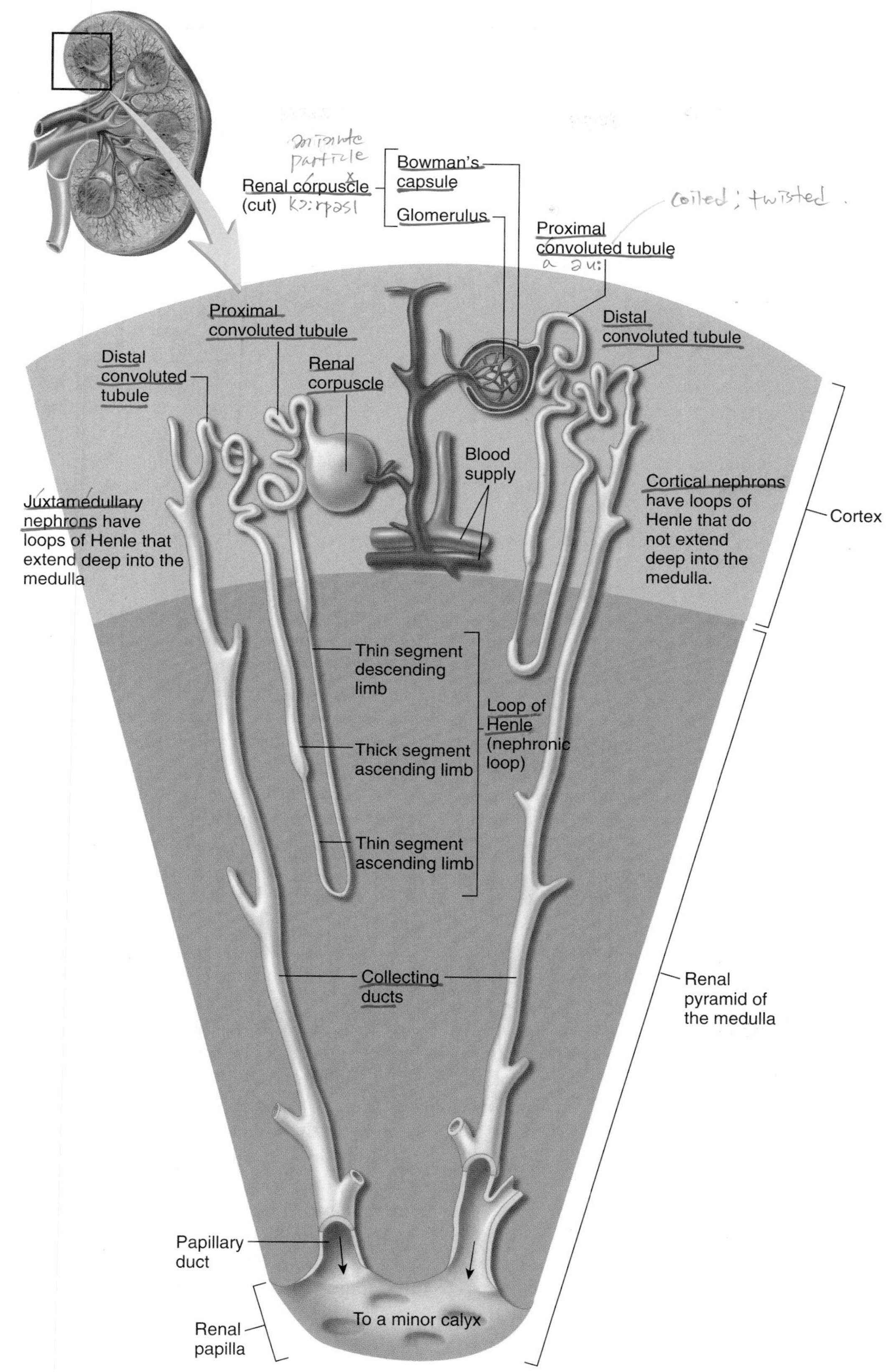

FIGURE 26.4 Functional Unit of the Kidney—the Nephron

A nephron consists of a renal corpuscle, proximal convoluted tubule, loop of Henle, and distal convoluted tubule. The distal convoluted tubule empties into a collecting duct. Juxtamedullary nephrons (those near the medulla of the kidney) have loops of Henle that extend deep into the medulla of the kidney, whereas other nephrons do not. Collecting ducts undergo a transition to larger-diameter papillary ducts near the tip of the renal papilla. The papillary ducts empty into a minor calyx.

with an enlarged end called **Bowman's capsule** (glomerular capsule), a proximal convoluted tubule, a loop of Henle (nephronic loop), and a distal convoluted tubule. The distal convoluted tubule empties into a collecting duct, which carries urine from the cortex of the kidney toward the renal papilla. Near the tip of the renal papilla, several collecting ducts merge into a larger-diameter tubule called a **papillary duct,** which empties into a minor calyx. The Bowman's capsules, proximal convoluted tubules, and distal convoluted tubules are located in the renal cortex, but the collecting ducts, parts of the loops of Henle, and the papillary ducts are in the renal medulla.

There are approximately 1.3 million nephrons in each kidney. Most nephrons measure 50–55 mm in length. Nephrons whose Bowman's capsules lie near the medulla are called **juxtamedullary** (juks′ta-med′ŭ-lār-ē; *juxta* is Latin, meaning next to) **nephrons.** They have long loops of Henle, which extend deep into the medulla. Only about 15% of the nephrons are juxtamedullary nephrons. The remainder of the nephrons are called **cortical nephrons,** and their loops of Henle do not extend deep into the medulla (see figure 26.4).

Each **renal corpuscle** consists of a Bowman's capsule, and a network of capillaries called the **glomerulus** (glō-măr′ū-lŭs; figure 26.5*a* and *b*). The wall of Bowman's capsule is indented to form a double-walled chamber. The glomerulus, which looks like a wad of yarn, fills the indentation. Fluid is filtered from the glomerulus into Bowman's capsule. The filtered fluid then flows into the proximal convoluted tubule, which carries it away from Bowman's capsule.

Bowman's capsule has an outer layer, called the **parietal layer,** and an inner layer, called the **visceral layer** (see figure 26.5*b*). The parietal layer is constructed of simple squamous epithelium, which becomes cube-shaped at the beginning of the proximal convoluted tubule. The visceral layer is constructed of specialized **podocyte cells,** which wrap around the glomerular capillaries.

Numerous, windowlike openings, called **fenestrae** (fe-nes′trē), are in the endothelial cells of the glomerular capillaries. Gaps, called **filtration slits,** are between cell processes of the podocytes that make up the visceral layer of Bowman's capsule (figure 26.5*c*). A basement membrane lies sandwiched between the endothelial cells of the glomerular capillaries and the podocytes of Bowman's capsule. Together, the capillary endothelium, the basement membrane, and the podocytes of Bowman's capsule form the kidney's **filtration membrane** (figure 26.5*d*), which performs the first major step in urine formation. Urine formation begins when fluid from the glomerular capillaries is filtered across the filtration membrane into the lumen, or space, inside Bowman's capsule.

An **afferent** (af′er-ent) **arteriole** supplies blood to the glomerulus and an **efferent** (ef′er-ent) **arteriole** drains it (see figure 26.5*a*). A layer of smooth muscle lines both the afferent and efferent arterioles. At the point where the afferent arteriole enters the renal corpuscle, the smooth muscle cells form a cufflike arrangement around the arteriole. These cells are called **juxtaglomerular cells.** Lying between the afferent and efferent arterioles adjacent to the renal corpuscle is part of the distal convoluted tubule of the nephron. Specialized tubule cells in this section are collectively called the **macula** (mak′ū-lă) **densa.** The juxtaglomerular cells of the afferent arteriole and the macula densa cells intimately contact each other. Coupled together, they are called the **juxtaglomerular apparatus** (see figure 26.5*b*). The juxtaglomerular apparatus secretes the enzyme renin and plays an important role in the autoregulation of filtrate formation and blood pressure regulation.

The **proximal convoluted tubule** measures approximately 14 mm long and 60 μm in diameter. Simple cuboidal epithelium makes up its wall. The cells rest on a basement membrane, which forms the outer surface of the tubule. Many microvilli project from the luminal surface of the cells (figure 26.6*a* and *b*).

The **loops of Henle** (nephronic loops) are continuations of the proximal convoluted tubules. Each loop has two limbs, one **descending** and one **ascending.** The first part of the descending limb is similar in structure to the proximal convoluted tubules. The loops of Henle that extend into the medulla become very thin near the end of the loop (see figure 26.6*a;* figure 26.6*c*). The lumen in the thin part narrows, and an abrupt transition occurs from simple cuboidal epithelium to simple squamous epithelium. Like the descending limb, the first part of the ascending limb is thin and is made of simple squamous epithelium. Soon, however, it becomes thicker, and simple cuboidal epithelium replaces the simple squamous epithelium. The thick part of the ascending limb returns toward the renal corpuscle and ends by giving rise to the distal convoluted tubule near the macula densa.

The **distal convoluted tubules** are not as long as the proximal convoluted tubules. The epithelium is simple cuboidal, but the cells are smaller than the epithelial cells in the proximal convoluted tubules and do not possess a large number of microvilli (figure 26.6*d*). The distal convoluted tubules of many nephrons connect to the **collecting ducts,** which are composed of simple cuboidal epithelium (see figure 26.6*c*). The collecting ducts, which are larger in diameter than other segments of the nephron, form much of the medullary rays and extend through the medulla toward the tips of the renal pyramids.

Approximately 90% of people who have **polycystic kidney disease** inherited the condition as an autosomal-dominant trait. Consequently, if one parent carries an allele for this disorder, each child has a 50% chance of also having the disorder (see chapter 3). The kidneys in affected people are enlarged and often contain large, fluid-filled cysts varying in size from a few millimeters to centimeters. The cysts increase in number and enlarge as the affected person ages. Development of the cysts results from abnormal cell–cell interactions and results in the excess proliferation of the epithelial cells that make up the kidney nephrons and collecting ducts. The gene for this condition is located on chromosome 16 and codes for a protein that may regulate cell–cell interactions. Polycystic kidney disease is usually diagnosed when patients are between 30 and 50 years of age, and it is the third leading cause of renal failure after diabetes mellitus and high blood pressure. Approximately 50% of people with the condition require renal dialysis (see Systems Pathology "Acute Renal Failure," p. 998) by 70 years of age. Polycystic kidney disease is often detected by using ultrasound techniques.

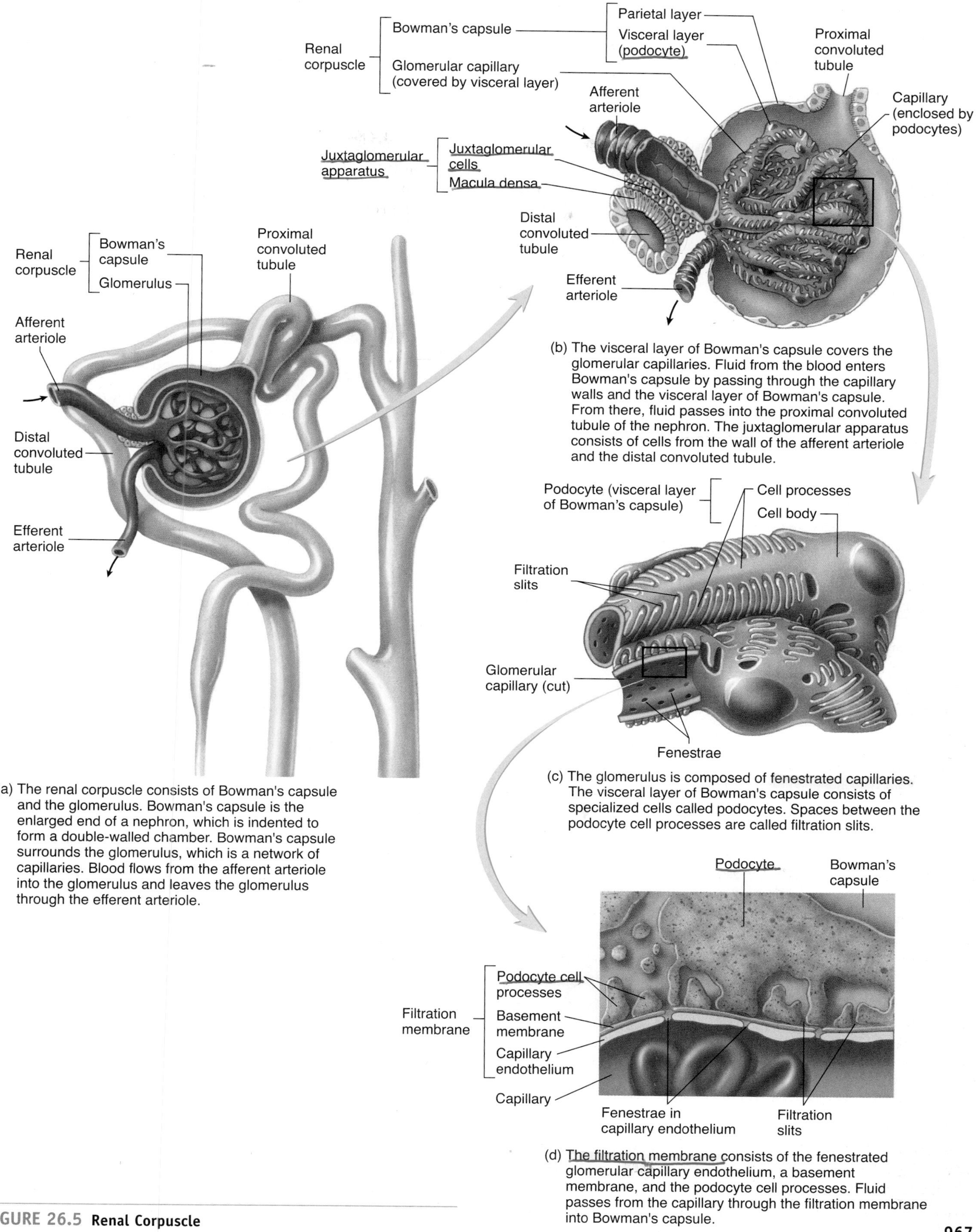

(a) The renal corpuscle consists of Bowman's capsule and the glomerulus. Bowman's capsule is the enlarged end of a nephron, which is indented to form a double-walled chamber. Bowman's capsule surrounds the glomerulus, which is a network of capillaries. Blood flows from the afferent arteriole into the glomerulus and leaves the glomerulus through the efferent arteriole.

(b) The visceral layer of Bowman's capsule covers the glomerular capillaries. Fluid from the blood enters Bowman's capsule by passing through the capillary walls and the visceral layer of Bowman's capsule. From there, fluid passes into the proximal convoluted tubule of the nephron. The juxtaglomerular apparatus consists of cells from the wall of the afferent arteriole and the distal convoluted tubule.

(c) The glomerulus is composed of fenestrated capillaries. The visceral layer of Bowman's capsule consists of specialized cells called podocytes. Spaces between the podocyte cell processes are called filtration slits.

(d) The filtration membrane consists of the fenestrated glomerular capillary endothelium, a basement membrane, and the podocyte cell processes. Fluid passes from the capillary through the filtration membrane into Bowman's capsule.

FIGURE 26.5 Renal Corpuscle

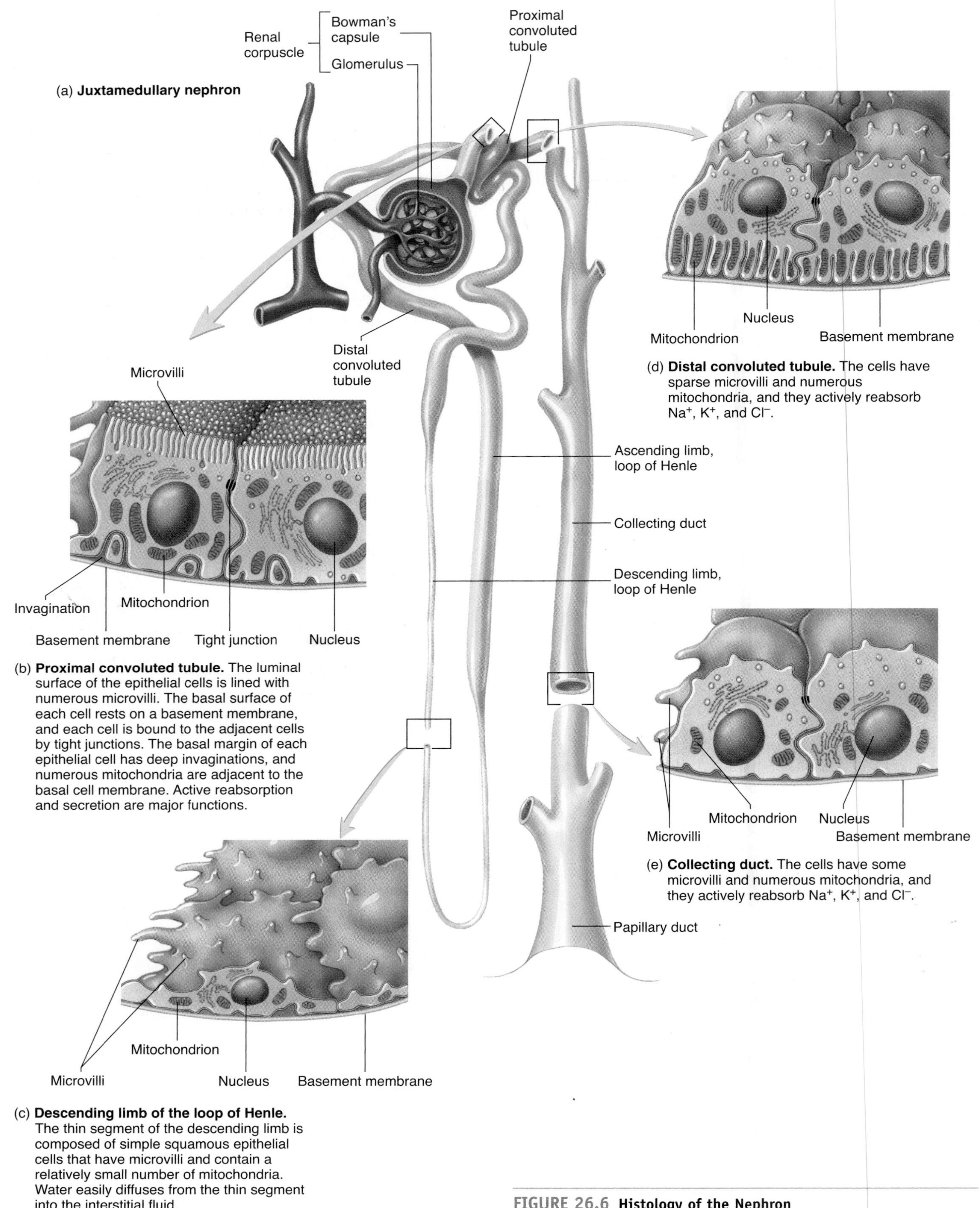

(b) **Proximal convoluted tubule.** The luminal surface of the epithelial cells is lined with numerous microvilli. The basal surface of each cell rests on a basement membrane, and each cell is bound to the adjacent cells by tight junctions. The basal margin of each epithelial cell has deep invaginations, and numerous mitochondria are adjacent to the basal cell membrane. Active reabsorption and secretion are major functions.

(c) **Descending limb of the loop of Henle.** The thin segment of the descending limb is composed of simple squamous epithelial cells that have microvilli and contain a relatively small number of mitochondria. Water easily diffuses from the thin segment into the interstitial fluid.

(d) **Distal convoluted tubule.** The cells have sparse microvilli and numerous mitochondria, and they actively reabsorb Na^+, K^+, and Cl^-.

(e) **Collecting duct.** The cells have some microvilli and numerous mitochondria, and they actively reabsorb Na^+, K^+, and Cl^-.

FIGURE 26.6 Histology of the Nephron

Arteries and Veins of the Kidneys

A **renal artery** branches off the abdominal aorta and enters the renal sinus of each kidney (figure 26.7*a*). **Segmental arteries** diverge from the renal artery to form **interlobar arteries,** which ascend within the renal columns toward the renal cortex. Branches from the interlobar arteries diverge near the base of each pyramid and arch over the bases of the pyramids to form the **arcuate** (ar′kū-āt) **arteries. Interlobular arteries** project from the arcuate arteries into the cortex, and afferent arterioles are derived from the interlobular arteries or their branches. The afferent arterioles supply blood to the glomerular capillaries of the renal corpuscles. Efferent arterioles arise from the glomerular capillaries and carry blood away from the glomeruli. After each efferent arteriole exits the glomerulus, it gives rise to a plexus of capillaries, called the **peritubular capillaries,** around the proximal and distal tubules. Specialized parts of the peritubular capillaries, called **vasa recta** (vā′să rek′tă), course into the medulla along with loops of Henle of juxtamedullary nephrons (figure 26.7*b*) and then back toward the cortex. The peritubular capillaries drain into **interlobular veins,** which in turn drain into the **arcuate veins.** The arcuate veins empty into the **interlobar veins,** which drain into the **renal vein.** The renal vein exits the kidney and connects to the inferior vena cava.

5 ***What is the functional unit of the kidney? Name its parts.***

6 ***Distinguish between cortical and juxtamedullary nephrons.***

7 ***List the components of a renal corpuscle.***

8 ***Describe the structure of Bowman's capsule, the glomerulus, and the filtration membrane.***

9 ***Describe the structure of the afferent and efferent arterioles and the juxtaglomerular apparati.***

10 ***Describe the structure and location of the proximal convoluted tubules, loops of Henle, distal convoluted tubules, collecting ducts, and papillary ducts.***

11 ***Describe the blood supply for the kidney.***

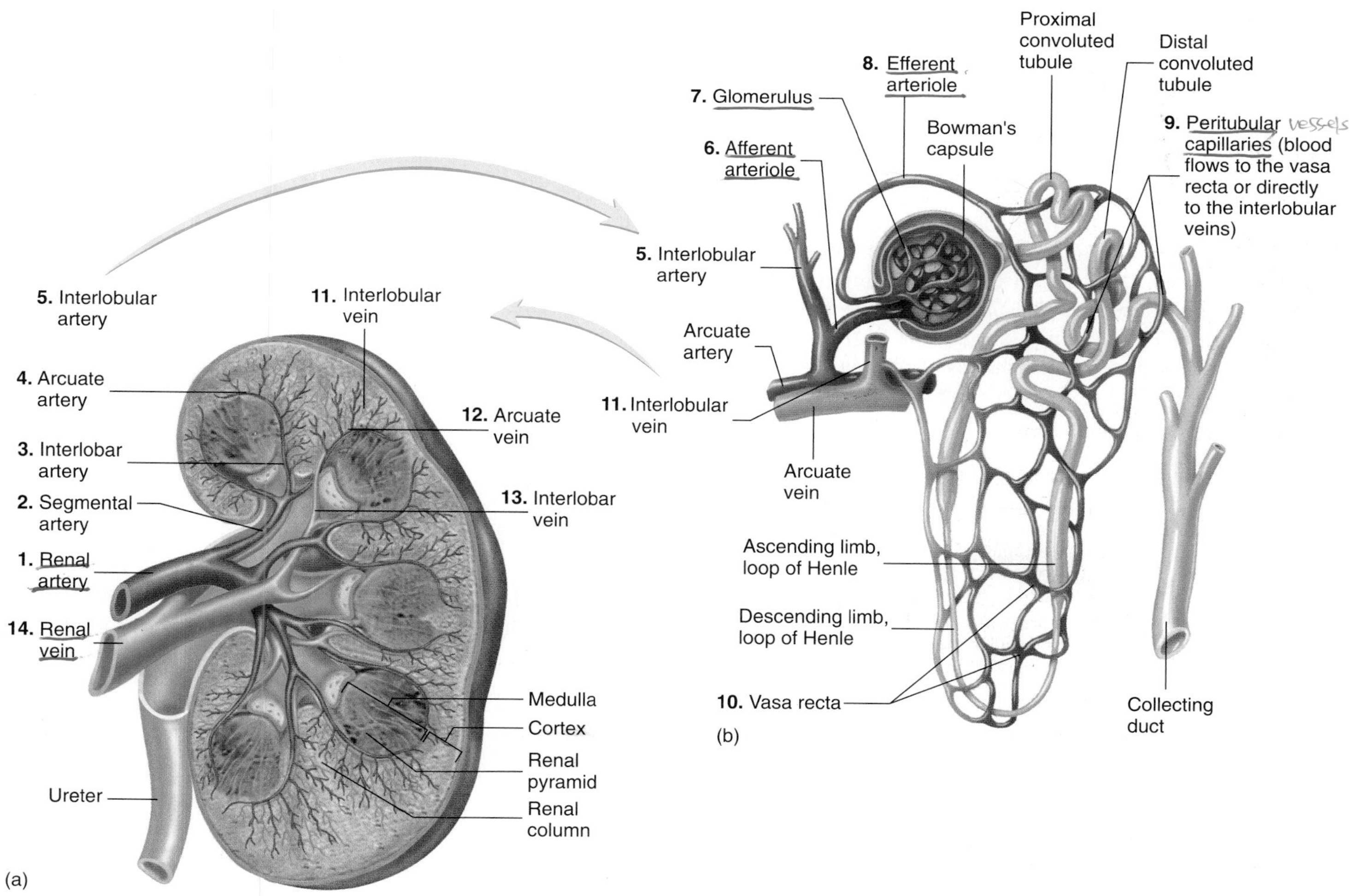

FIGURE 26.7 Blood Flow Through the Kidney

(*a*) Blood flow through the larger arteries and veins of the kidney. (*b*) Blood flow through the arteries, capillaries, and veins that provide circulation to the nephrons.

URINE PRODUCTION

Nephrons are called the **functional units of the kidney** because they are the smallest structural components capable of producing urine. Three major processes are essential for urine formation: filtration, tubular reabsorption, and tubular secretion (figure 26.8). All three are essential for the regulation of body fluid composition. **Filtration** is the movement of water and small solutes across the filtration membrane as a result of a pressure difference. The fluid entering the nephron is called the **filtrate. Tubular reabsorption** is the movement of water and solutes from the filtrate back into the blood. In general, most of the water and useful solutes are reabsorbed, whereas waste products, excess solutes, and a small amount of water are not (table 26.1).

Tubular secretion is the active transport of solutes across the walls of the nephron into the filtrate. Consequently, urine consists of substances that are filtered across the filtration membrane and those that are secreted from the peritubular capillaries into the nephron, minus the substances that are reabsorbed.

Filtration

Filtration is the movement of water and small solutes, derived from blood flowing through the glomerulus, across the porous filtration membrane. The filtrate includes water and small molecules and ions that easily pass through the filtration membrane. The filtration membrane prevents larger molecules, such as proteins and blood cells, from passing through it. A pressure difference forces the filtrate across the filtration membrane.

The part of the total cardiac output that passes through the kidneys is called the **renal fraction.** Table 26.2 presents the calculation of renal blood flow rate and other kidney flow rates. The renal fraction varies from 12%–30% of the cardiac output in healthy, resting adults, but it averages 21%. This means the average **renal blood flow rate** is 1176 mL/min. The plasma that flows through the kidneys each minute, called the **renal plasma flow rate,** is equal to the renal blood flow rate multiplied by the portion of the blood that is made up of plasma, which is approximately 55% (1176 mL/min $\times$ 0.55 = 646.8 mL plasma/min, or approximately 650 mL plasma/min).

The part of the plasma flowing through the kidney that is filtered through the filtration membranes into the lumen of Bowman's capsules to become filtrate is called the **filtration fraction.** The filtration fraction averages 19% of the plasma flowing through the kidney (650 mL plasma/min $\times$ 0.19 = 123.5 mL plasma/min). Thus, approximately 125 mL of filtrate is produced each minute. The amount of filtrate produced each minute is called the **glomerular filtration rate (GFR),** which is equivalent to approximately 180 L of filtrate produced daily. Because only 1–2 L of urine are produced each day by a healthy person, it is obvious that not all of the filtrate becomes urine. Approximately 99% of the filtrate volume is reabsorbed as it passes through the nephron, and less than 1% becomes urine.

PREDICT 1

If the filtration fraction increases from 19% to 22% and if 99.2% of the filtrate is reabsorbed, how much urine is produced in a normal person with a cardiac output of 5600 mL/min? (*Hint:* See table 26.2.)

Filtration Membrane

The **filtration membrane** is a filtration barrier, which prevents the entry of blood cells and proteins into the lumen of Bowman's capsule but allows other blood components to enter. The filtration membrane is many times more permeable than a typical capillary. Water and solutes of a small molecular diameter readily pass from the glomerular capillaries through the filtration membrane into Bowman's capsule, but larger molecules do not. The fenestrae of the glomerular capillary, the basement membrane, and the podocyte cells (see figure 26.5*d*) prevent molecules larger than 7 nm in diameter or with a molecular mass of

Urine formation results from the following three processes:

1. Filtration. Filtration (*blue arrow*) is the movement of materials across the filtration membrane into Bowman's capsule to form filtrate.

2. Tubular reabsorption. Solutes are reabsorbed (*purple arrow*) across the wall of the nephron into the interstitial fluid by transport processes, such as active transport and cotransport.

Water is reabsorbed (*green arrow*) across the wall of the nephron by osmosis. Water and solutes pass from the interstitial fluid into the peritubular capillaries.

3. Tubular secretion. Solutes are secreted (*orange arrow*) across the wall of the nephron into the filtrate.

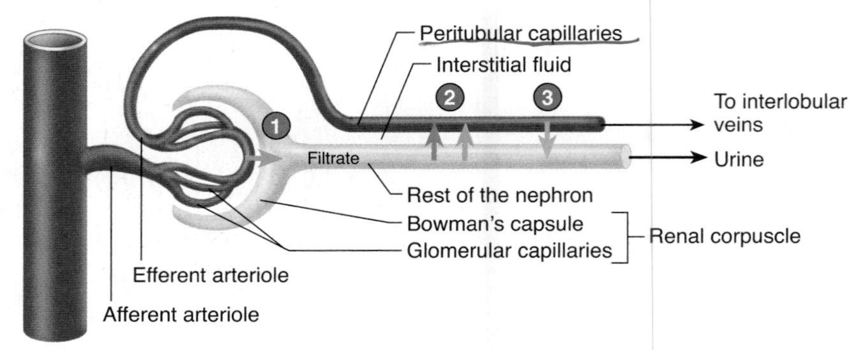

PROCESS FIGURE 26.8 Urine Formation

TABLE 26.1 Concentrations of Major Solutes

Substance	Plasma	Filtrate	Net Movement of Solute*	Urine	Urine Concentration / Plasma Concentration
Water (L)	180	180	178.6	1.4	———
Organic molecules (mg/100mL)					
Protein	3900–5000	6–11	−100.0	0†	0
Glucose	100	100	−100.0	0	0
Urea	26	26	−11.4	1820	70
Uric acid	3	3	−2.7	42	14
Creatinine	1.1	1.1	0.5	196	140
Ions (mEq/L)					
Na^+	142	142	−141.0	128	0.9
K^+	5	5	−4.5	60	12.0
Cl^-	103	103	−101.9	134	1.3
HCO_3^-	28	28	−27.9	14	0.5

*In many cases, solute moves into and out of the nephron. Numbers indicate net movement. Negative numbers are net movement out of the filtrate, and positive numbers are net movement into the filtrate.

†Trace amounts of protein can be found in the urine. A value of zero is assumed here.

TABLE 26.2 Calculation of Renal Flow Rates

	Amount per Minute (mL)	Calculation
Renal Blood Flow	1176	Amount of blood flowing through the kidneys per minute; equals cardiac output (5600 mL blood/min) times the percent (21%; renal fraction) of cardiac output that enters the kidneys 5600 mL blood/min × 0.21 = 1176 mL blood/min
Renal Plasma Flow	650	Amount of plasma flowing through the kidneys per minute; equals renal blood flow times percent of the blood that is plasma. Because the hematocrit is the percent of the blood that consists of formed elements, the percent of the blood that is plasma is 100 minus the hematocrit. Assuming a hematocrit of 45, the percent of the blood that is plasma is 55% (100 − 45). Renal plasma flow is therefore 55% of renal blood flow. 1176 mL blood/min × 0.55 ≈ 650 mL plasma/min
Glomerular Filtration Rate (GFR)	125	Amount of plasma (filtrate) that enters Bowman's capsule per minute; equals renal plasma flow times the percent (19%; filtration fraction) of the plasma that enters the renal capsule 650 mL plasma/min × 0.19 ≈ 125 mL filtrate/min
Urine	1	Nonreabsorbed filtrate that leaves the kidneys per minute; equals glomerular filtration rate times the percent (0.8%) of the filtrate that is not reabsorbed into the blood 125 mL filtrate/min × 0.008 = 1 mL urine/min Milliliters of urine per minute can be converted to liters of urine per day by multiplying by 1.44. 1 mL urine/min × 1.44 = 1.4 L/day

40,000 daltons from passing through. Most plasma proteins are slightly larger than 7 nm in diameter and are retained in the glomerular capillaries. Albumin, which has a diameter just slightly less than 7 nm, enters the filtrate in small amounts so that the filtrate contains about 0.03% protein. Protein hormones are also small enough to pass through the filtration barrier. Proteins that do pass through the filtration membrane are actively reabsorbed by endocytosis and metabolized by the cells in the proximal convoluted tubule. Consequently, little protein is found in the urine of healthy people.

PREDICT 2

Hemoglobin has a smaller diameter than albumin, but very little hemoglobin passes from the blood into the filtrate. Explain why. Under what circumstances would large amounts of hemoglobin enter the filtrate?

Hematuria and Glomerulonephritis

Hematuria (hē-mă-too′rē-ă, hem-ă-too′rē-ă) occurs when red blood cells are found in the urine. The source of red blood cells in the urine can be the outside of the kidney or the kidney itself. Conditions outside the kidney that result in hematuria include kidney stones or tumors in the renal pelvis, ureter, urinary bladder, prostate, or urethra. Infections of the urinary tract, resulting in inflammation of the urinary bladder (cystitis), of the prostate gland (prostatitis, pros-tă-tī′tis), and of the urethra (urethritis, ū-rē-thrī′tis) can also cause hematuria. Conditions inside the kidney include those that affect the filtration membrane or other areas of the kidney. Inflammation of the glomeruli, called **glomerulonephritis** (glō-mār′ū-lō-nef-rī′tis), can increase the permeability of the filtration membrane and allow blood cells to cross. Other areas of the kidney can be a source of blood in response to inflammation of the nephrons due to infections, tumors in the kidney tissue, and infarcted areas of the kidney, where an artery is blocked, resulting in necrosis of part of the kidney tissue.

Filtration Pressure

The formation of filtrate depends on a pressure gradient, called the **filtration pressure,** which forces fluid from the glomerular capillary across the filtration membrane into the lumen of Bowman's capsule. The filtration pressure results from the sum of the forces that move fluid out of the glomerular capillary into the lumen of Bowman's capsule and those that move fluid out of the lumen of Bowman's capsule into the glomerular capillary (figure 26.9). The **glomerular capillary pressure, (GCP),** the blood pressure inside the capillary, moves fluid out of the capillary into Bowman's capsule. The glomerular capillary pressure averages approximately 50 mm Hg, which is much higher than that in most capillaries.

The high glomerular capillary pressure results from a low resistance to blood flow in the afferent arterioles and glomerular capillaries and from a higher resistance to blood flow in the efferent arterioles. As the diameter of a vessel decreases, resistance to flow through the vessel increases (see chapter 21), and pressure upstream from the point of decreased vessel diameter is higher than the pressure downstream from the point of decreased diameter. For example, in the extreme case of a completely closed vessel, pressure is higher upstream from the constriction and it falls to zero downstream from the constriction. The efferent arterioles have a small diameter, and the blood pressure is higher within the glomerular capillaries because of the low resistance to blood flow in the afferent arterioles and glomerular capillaries and because of the higher resistance to blood flow in the efferent arterioles. Also, the pressure is lower in the peritubular capillaries which are downstream from the efferent arterioles. Consequently, filtrate is forced across the filtration membrane into the lumen of Bowman's capsule. The low pressure in the peritubular capillaries allows fluid to move into them from the interstitial fluid.

The smooth muscle cells in the walls of the afferent and efferent arterioles can alter the vessel diameter and the glomerular filtration pressure. For example, dilation of the afferent arterioles or constriction of the efferent arterioles increases glomerular capillary pressure, increasing filtration pressure and glomerular filtration.

Opposing the movement of fluid into the lumen of Bowman's capsule is the **capsular pressure, (CP),** which is approximately 10 mm Hg, caused by the pressure of filtrate already inside Bowman's capsule. The **blood colloid osmotic pressure (BCOP)** within the glomerular capillary exists because plasma proteins do

1. Glomerular capillary pressure (GCP), the blood pressure (50 mm Hg) within the glomerulus, moves fluid from the blood into Bowman's capsule.
2. Capsular pressure (CP), the pressure inside Bowman's capsule (10 mm Hg), moves fluid from Bowman's capsule into the blood.
3. Blood colloid osmotic pressure (BCOP), produced by the concentration of blood proteins (30 mm Hg), moves fluid from Bowman's capsule into the blood by osmosis.
4. Filtration pressure is equal to the glomerular capillary pressure minus the capsular and blood colloid osmotic pressures.

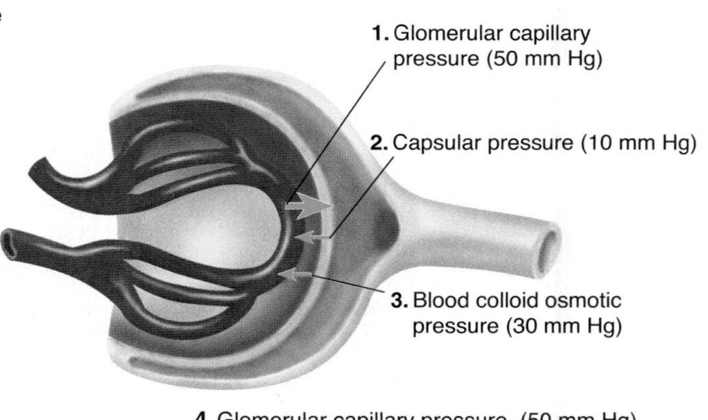

PROCESS FIGURE 26.9 Filtration Pressure

Filtration pressure across the filtration membrane is equal to the glomerular capillary pressure (GCP) minus the blood colloid osmotic pressure (BCOP) in the glomerular capillary minus the pressure in Bowman's capsule (CP).

not pass through the filtration membrane. Instead, they remain within the glomerular capillary and produce an osmotic force of about 30 mm Hg, which favors fluid movement into the glomerular capillary from Bowman's capsule. The BCOP is greater at the end of the glomerular capillary than at its beginning because, as fluid leaves the capillary and enters Bowman's capsule, there is a higher protein concentration in the glomerular capillary. The average BCOP is approximately 30 mm Hg. The filtration pressure is therefore approximately 10 mm Hg.

Filtration pressure		Glomerular capillary pressure		Capsular pressure		Blood colloid osmotic pressure
(10 mm Hg)	=	(50 mm Hg)	−	(10 mm Hg)	−	(30 mm Hg)

The blood colloid osmotic pressure in Bowman's capsule is normally very close to zero because few proteins cross the filtration membrane. However, if the permeability of the filtration membrane increases, as it does in **glomerular nephritis,** the blood colloid osmotic pressure in Bowman's capsule increases. This results in an elevated filtration pressure and an increase in the filtrate volume.

12 ***Name the three general processes involved in the production of urine.***

13 ***Define renal blood flow, renal plasma flow, and glomerular filtration rate. How do they affect urine production?***

14 ***Describe the filtration membrane. What substances do not pass through it?***

15 ***What is filtration pressure? How does glomerular capillary pressure affect filtration pressure and the amount of urine produced?***

16 ***How do systemic blood pressure and afferent arteriole diameter affect glomerular capillary pressure?***

PREDICT 3

Karl was pouring gasoline from a small can into his lawnmower when it ignited. He experienced third-degree burns on his chest, neck, arms, and hands. In the hospital emergency room, his blood pressure was found to be in the low normal range, his heart rate was very high, and his pulse was weak. His skin appeared very pale. There was a delay of a couple of hours from the time of the accident until an IV was started. For several hours after the accident, he produced almost no urine. Explain that response.

Regulation of Glomerular Filtration Rate

The mean arterial blood pressure can vary from 90 to 180 mm Hg without dramatically affecting the glomerular filtration rate (GFR). Two important mechanisms that regulate renal blood flow and the glomerular filtration rate are autoregulation and sympathetic stimulation. Severe hemorrhage and dehydration can cause the mean arterial pressure to decrease below 90 mm Hg. As a consequence, both renal blood flow and the glomerular filtration rate decrease dramatically.

Autoregulation

Autoregulation is the maintenance, within the kidneys, of a relatively stable GFR over a wide range of systemic blood pressures. For example, GFR is relatively constant as the systemic blood pressure changes between 90 and 180 mm Hg.

Autoregulation involves changes in the degree of constriction in the afferent arterioles. The precise mechanism by which autoregulation is achieved is unclear but, as systemic blood pressure increases, the afferent arterioles constrict and prevent an increase in renal blood flow and filtration pressure across the filtration membrane of the renal corpuscle. Conversely, a decrease in systemic blood pressure results in dilation of the afferent arterioles, thus preventing a decrease in renal blood flow and filtration pressure across the filtration membrane.

Autoregulation is also influenced by the rate of flow of filtrate past cells of the macula densa. The macula densa detects an increased flow rate and sends a signal to the juxtaglomerular apparatus to constrict the afferent arteriole. The result is a decrease in the filtration pressure across the filtration membrane of the renal corpuscle.

17 ***Describe autoregulation.***

Sympathetic Stimulation

Sympathetic neurons with norepinephrine as their neurotransmitter innervate the blood vessels of the kidneys. Sympathetic stimulation constricts the small arteries and afferent arterioles, thereby decreasing renal blood flow and filtrate formation. Intense sympathetic stimulation, such as during shock or intense exercise, decreases the rate of filtrate formation to only a few milliliters per minute. Small changes in sympathetic stimulation have a minimal effect on renal blood flow and filtrate formation. Autoregulation maintains renal blood flow and filtrate formation at a relatively constant rate unless sympathetic stimulation is intense.

In response to severe stress or circulatory shock, renal blood flow can decrease to such low levels that the blood supply to the kidney is inadequate to maintain normal kidney metabolism. As a consequence, kidney tissues can be damaged and thus be unable to perform their normal functions if blood flow is not reestablished. Consequently, shock should be treated quickly. Reduced blood flow to the kidneys during stress or shock, however, is consistent with homeostasis. Intense vasoconstriction, which maintains blood pressure at levels adequate to sustain blood flow to organs such as the heart and brain, is essential to homeostasis. A short-term reduction in blood flow to organs such as the kidneys is only harmful if the lack of blood flow is prolonged.

18 ***Explain the effect that sympathetic stimulation has on the kidney and GFR during rest, exercise, and shock.***

Tubular Reabsorption

The filtrate leaves the lumen of Bowman's capsule and flows through the proximal convoluted tubule, the loop of Henle, and the distal convoluted tubule and then into the collecting ducts. As it passes through these structures, many of the substances in the filtrate undergo **tubular reabsorption.** Tubular reabsorption results from processes such as diffusion, facilitated diffusion, active transport, symport, and osmosis. Inorganic salts, organic molecules, and about 99% of the filtrate volume leave the nephron and enter the interstitial fluid. Because the pressure is low in the peritubular capillaries, these substances enter the peritubular

TABLE 26.3 Reabsorption of Major Solutes from the Nephron

Apical Membrane	Basal Membrane
Proximal Convoluted Tubule	
Substances symported with Na^+	Active transport
	Na^+ (exchanged for K^+)
K^+	Facilitated diffusion
Cl^-	K^+
Ca^{2+}	Cl^-
Mg^{2+}	Ca^{2+}
HCO_3^-	HCO_3^-
PO_4^{3-}	PO_4^{3-}
Amino acids	Amino acids
Glucose	Glucose
Fructose	Fructose
Galactose	Galactose
Lactate	Lactate
Succinate	Succinate
Citrate	Citrate
Diffusion between nephron cells	
K^+	
Ca^{2+}	
Mg^{2+}	
Thick Ascending Limb of the Loop of Henle	
Substances symported with Na^+	Active transport
	Na^+ (exchanged for K^+)
K^+	Facilitated diffusion
Cl^-	K^+
	Cl^-
Diffusion between nephron cells	
K^+	
Ca^{2+}	
Mg^{2+}	
Distal Convoluted Tubule and Collecting Duct	
Substances symported with Na^+	Active transport
	Na^+ (exchanged for K^+)
Cl^-	Facilitated diffusion
K^+	K^+
	Cl^-

capillaries and flow through the renal veins to enter the general circulation (see figure 26.8).

Solutes reabsorbed from the lumen of the nephron to the interstitial fluid include amino acids, glucose, and fructose, as well as Na^+, K^+, Ca^{2+}, HCO_3^-, and Cl^-. A more complete list is provided in table 26.3 for each part of the nephron.

Water follows the solutes that are reabsorbed across the wall of the nephron. As the solutes in the nephron are reabsorbed, water follows the solutes by the process of osmosis. The transport processes and the permeability characteristics of each portion of the nephron are responsible for reabsorption of the filtrate. The small volume of the filtrate (approximately 1% of the filtrate volume) that forms urine contains a relatively high concentration of urea, uric acid, creatinine, K^+, and other substances that are toxic in high concentrations. The regulation of solute reabsorption and the permeability characteristics of portions of the nephron allow for the production of a small volume of very concentrated urine or a large volume of very dilute urine. The mechanisms in the wall of the nephron that are responsible for reabsorption are described in the following sections. The mechanisms that regulate urine concentration are described in the section "Urine Concentration Mechanism," p. 979).

Reabsorption in the Proximal Convoluted Tubule

Most reabsorbed substances must pass through the cells that make up the wall of the nephrons. These cells have an **apical surface,** which makes up the inside surface of the nephron; a **basal surface,** which forms the outer wall of the nephron; and **lateral surfaces,** which are bound to the lateral surfaces of other cells of the nephron. Reabsorption of most solute molecules from the proximal convoluted tubule is linked to the active transport of Na^+ across the **basal membrane** of the nephron epithelial cells from the cytoplasm into the interstitial fluid, thereby creating a low concentration of Na^+ inside the cells (figure 26.10). At the basal cell membrane, ATP provides the energy for the antiport mechanism, which moves Na^+ out of the cell and K^+ into the cell. Because the concentration of Na^+ in the lumen of the tubule is high, a large concentration gradient is present from the lumen of the nephron to the intracellular fluid of the cells lining the nephron. This concentration gradient for Na^+ is responsible for the symport of many solute molecules from the lumen of the nephron into the nephron cells.

Carrier proteins that transport amino acids, glucose, and other solutes are located within the apical membrane, which separates the lumen of the nephron from the cytoplasm of epithelial cells. Each of these carrier proteins binds specifically to one of the substances to be transported and to Na^+. The concentration gradient for Na^+ provides the source of energy that moves both the Na^+ and the other molecules or ions bound to the carrier protein from the lumen into the cell of the nephron. Once the symported molecules are inside the cell, they cross the basal membrane of the cell by facilitated diffusion. The number of carrier proteins limits the rate at which a substance can be transported. For example, the concentration of glucose in the filtrate exceeds the rate at which glucose can be transported in people suffering from untreated diabetes mellitus. The excess glucose remains in the filtrate and becomes part of the urine (see "Plasma Clearance and Tubular Maximum," p. 991).

Some solutes also diffuse between the cells from the lumen of the nephron into the interstitial fluid. The concentration of these solutes increases as other solutes are symported and water moves by osmosis from the lumen into interstitial fluid. As the concentration gradient for these solutes increases above the concentration in the interstitial fluid, they diffuse between the epithelial cells. Some

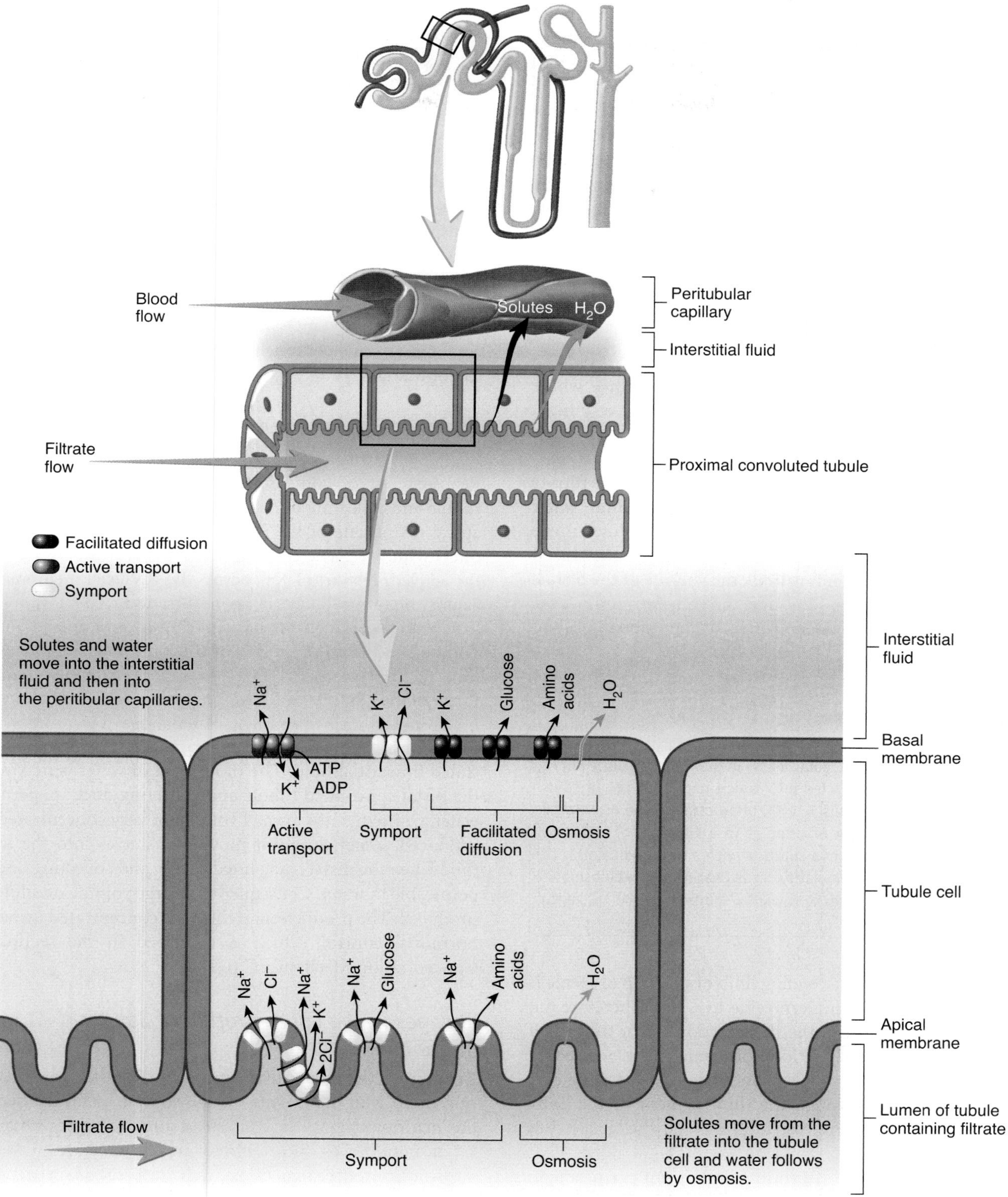

FIGURE 26.10 Reabsorption of Solutes in the Proximal Convoluted Tubule

The symport of molecules and ions across the epithelial lining of the nephron depends on the active transport of Na^+, in exchange for K^+, across the basal membrane. Symport is the process by which carrier proteins move molecules or ions with Na^+ across the apical membrane. The Na^+ concentration gradient provides the energy for symport. Amino acids, glucose, K^+, Cl^-, and most other solutes are transported into the cells of the nephron with Na^+. Water enters and leaves the cell by osmosis. Glucose, amino acids, Na^+, Cl^-, and many other solutes leave the cells across the basal membrane by facilitated diffusion.

K^+, Ca^{2+}, and Mg^{2+} diffuse between the cells of the proximal convoluted tubule wall from the lumen of the tubule to the interstitial fluid. Reabsorption of these solutes by diffusion occurs, even though the same ions are also reabsorbed by symport processes.

Reabsorption of solutes in the proximal convoluted tubule is extensive, and the tubule is permeable to water. As solute molecules are transported from the nephron to the interstitial fluid, water moves by osmosis in the same direction. By the time the filtrate has reached the end of the proximal convoluted tubule, its volume has been reduced by approximately 65%. The concentration of the filtrate in the proximal convoluted tubule remains about the same as that of the interstitial fluid (300 mOsm/kg) because the wall of the nephron is permeable to water.

Reabsorption in the Loop of Henle

The loop of Henle descends into the medulla of the kidney, where the concentration of solutes in the interstitial fluid is very high. The thin segment of the descending limb of the loop of Henle (figure 26.11) is highly permeable to water and moderately permeable to urea, sodium, and most other ions. As the filtrate passes through the thin segment of the descending limb, water moves out of the nephron by osmosis, and some solutes move into the nephron. By the time the filtrate has reached the end of the thin segment of the descending limb, the volume of the filtrate has been reduced by another 15%, and the concentration of the filtrate is equal to the high concentration of the interstitial fluid (1200 mOsm/L).

Osmoles, Osmolality, and Osmosis

An **osmole** is a measure of the number of particles in a solution. One osmole is the molecular mass, in grams, of a solute times the number of ions or particles into which it dissociates in solution. A milliosmole (mOsm) is 1/1000 of an osmole. The osmolality of a solution is the number of osmoles in a kilogram of solution. Water moves by osmosis from a solution with a lower osmolality to a solution with a higher osmolality. Thus, water moves by osmosis from a solution of 100 mOsm/kg toward a solution of 300 mOsm/kg (see Appendix C).

The thin portion of the ascending limb of the loop of Henle is permeable to solutes but is impermeable to water. Therefore, no additional water diffuses from the thin portion of the ascending limb. The ascending limb of the loop of Henle is surrounded by interstitial fluid, which becomes less concentrated toward the cortex. As the filtrate flows through the thin segment of the limb, solutes diffuse into the interstitial fluid, making the filtrate less concentrated.

The thick segment of the ascending limb is not permeable to either water or solutes. Solutes such as Na^+, K^+, and Cl^-, however, are transported from the thick segment of the ascending limb of the loop of Henle into the interstitial fluid. Symport is responsible for the movement of K^+ and Cl^- with Na^+ across the apical membrane of the ascending limb of the loop of Henle (figure 26.12). Once inside the cells of the ascending limb, Cl^- and K^+ cross the basal cell membrane into the interstitial fluid from a higher concentration inside the cells to a lower concentration outside the cells by facilitated diffusion.

The concentration of Na^+in the lumen of the nephron is high and the concentration inside the nephron cells is low. This concentration gradient is created by the active transport of the Na^+ out of the cell in exchange for K^+ across the basal membrane (see figure 26.12).

Because the thin and thick segments of the ascending limb of the loop of Henle are impermeable to water, and because ions are transported out of the thick portion of the ascending limb, the concentration of solutes in the nephron is reduced to about 100 mOsm/kg by the time the fluid reaches the distal convoluted tubule. In contrast, the concentration of the interstitial fluid in the cortex is about 300 mOsm/kg. Thus, the filtrate entering the distal convoluted tubule is much more dilute than the interstitial fluid surrounding it.

Reabsorption in the Distal Convoluted Tubule and Collecting Duct

Reabsorption of some solutes occurs in the distal convoluted tubule and collecting duct. Chloride ions are symported across the apical membrane of the distal convoluted tubules and collecting ducts with Na^+. The concentration gradient for Na^+ depends on the active transport of Na^+ across the basal cell membrane, which causes the intracellular Na^+ concentration to be low, in comparision with the lumen of the nephron. In addition, the collecting ducts extend from the cortex of the kidney, where the concentration of the interstitial fluid is approximately 300 mOsm/kg, through the medulla of the kidney, where the concentration of the interstitial fluid is very high. Water moves by osmosis from the distal convoluted tubule and collecting duct into the more concentrated interstitial fluid. The movement of water is maximal when the distal convoluted tubule and collecting duct are permeable to water. Consequently, a small volume of very concentrated urine is produced. Water does not move by osmosis into the interstitial fluid when the distal convoluted tubule and collecting duct are not permeable to water. Consequently, a large volume of dilute urine is produced. The production of dilute or concentrated urine is under hormonal control, which is described in the section "Urine Concentration Mechanism," p. 979.

Changes in the Concentration of Solutes in the Nephron

Urea enters the glomerular filtrate and is in the same concentration there as in the plasma. As the volume of filtrate decreases in the nephron, the concentration of urea increases because renal tubules are not as permeable to urea as they are to water. Only 40%–60% of the urea is passively reabsorbed in the nephron, although about 99% of the water is reabsorbed. In addition to urea, urate ions, creatinine, sulfates, phosphates, and nitrates are reabsorbed, but not to the same extent as water. They therefore become more concentrated in the filtrate as the volume of the filtrate becomes smaller. These substances are toxic if they accumulate in the body, so their accumulation in the filtrate and elimination in urine help maintain homeostasis.

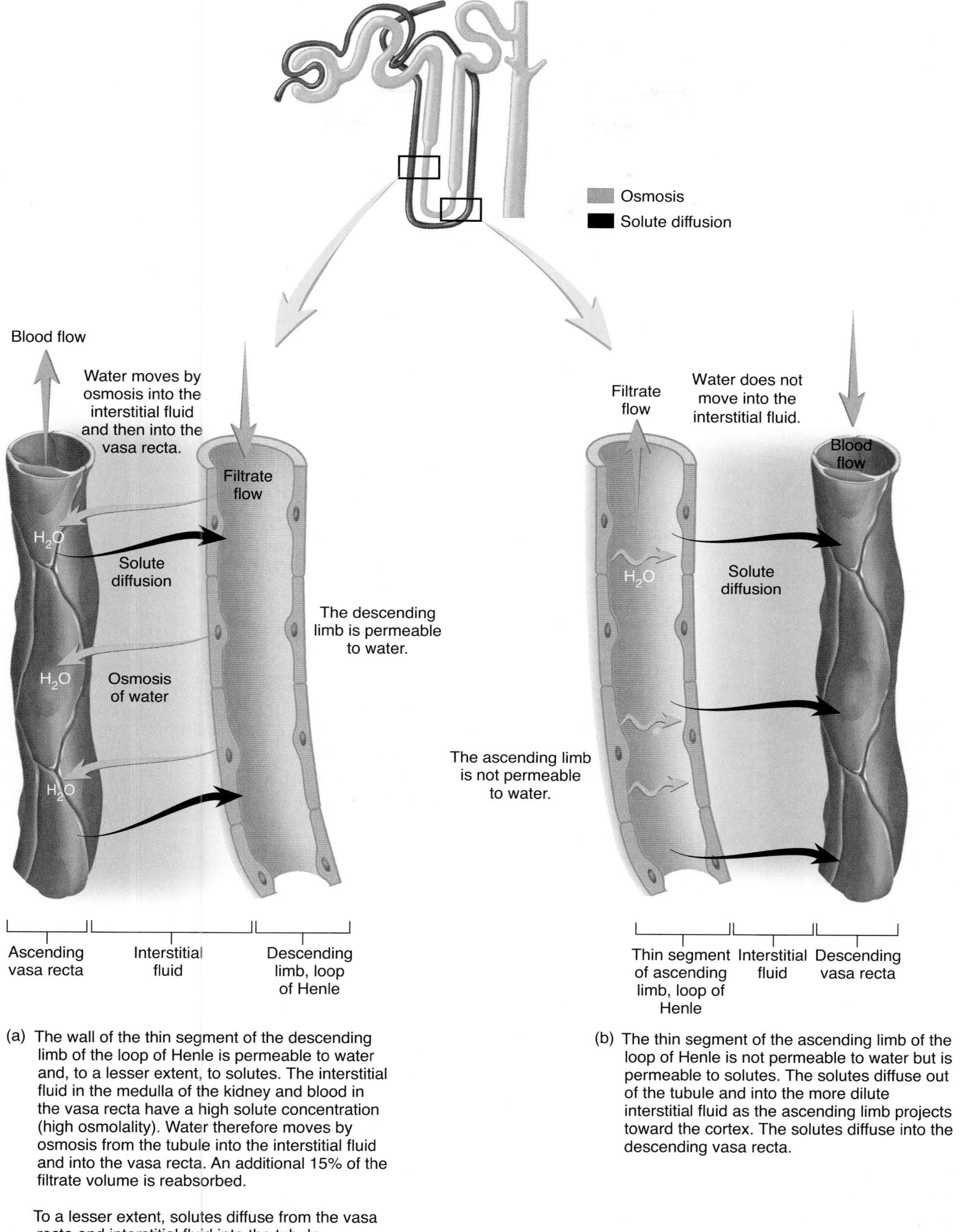

(a) The wall of the thin segment of the descending limb of the loop of Henle is permeable to water and, to a lesser extent, to solutes. The interstitial fluid in the medulla of the kidney and blood in the vasa recta have a high solute concentration (high osmolality). Water therefore moves by osmosis from the tubule into the interstitial fluid and into the vasa recta. An additional 15% of the filtrate volume is reabsorbed.

To a lesser extent, solutes diffuse from the vasa recta and interstitial fluid into the tubule.

(b) The thin segment of the ascending limb of the loop of Henle is not permeable to water but is permeable to solutes. The solutes diffuse out of the tubule and into the more dilute interstitial fluid as the ascending limb projects toward the cortex. The solutes diffuse into the descending vasa recta.

FIGURE 26.11 Reabsorption in the Loop of Henle: The Descending Limb and the Thin Segment of the Ascending Limb

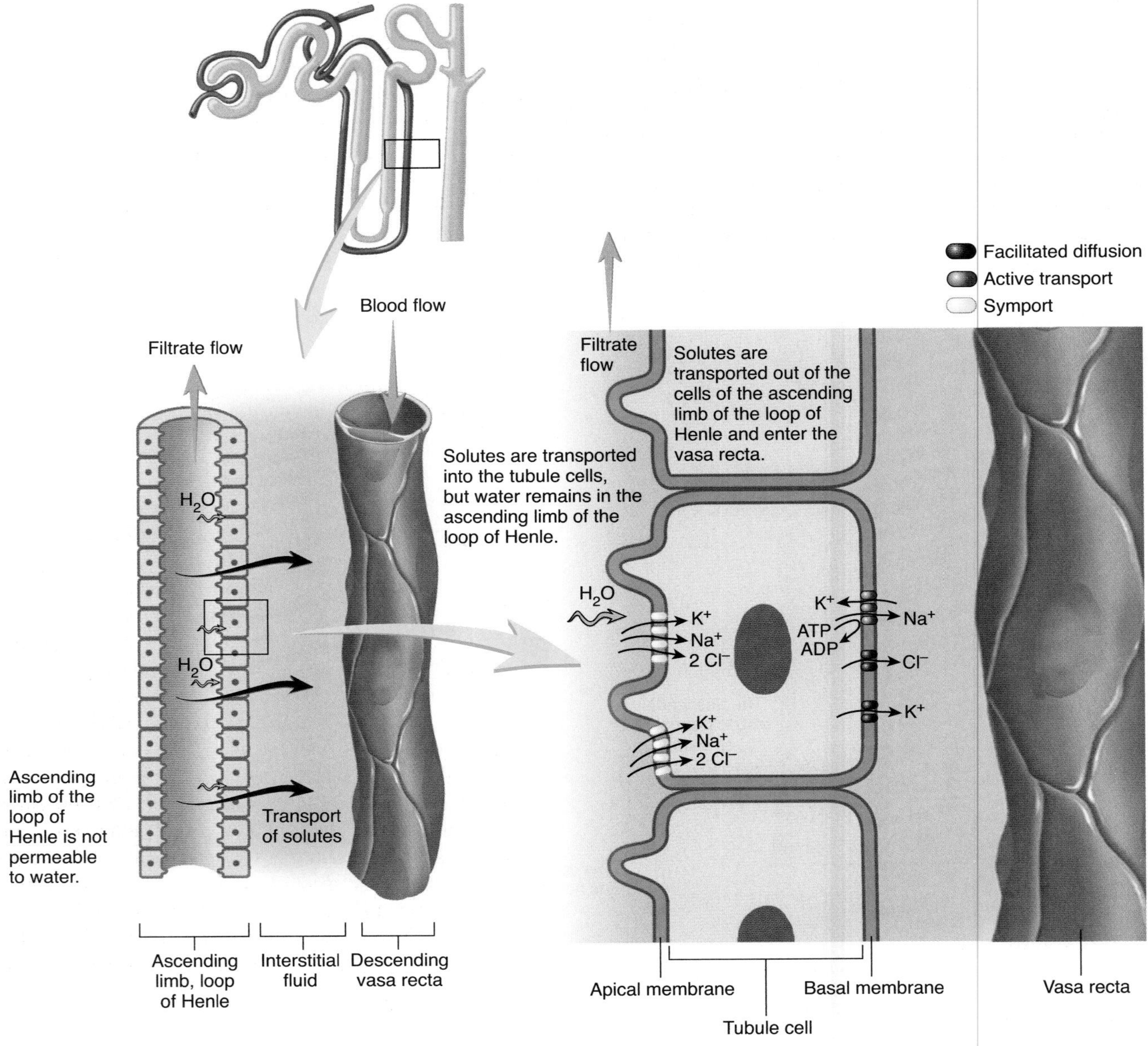

The wall of the ascending limb of the loop of Henle is not permeable to water. Sodium ions move across the wall of the basal membrane by active transport, establishing a concentration gradient for Na^+. Potassium ions and Cl^- are symported with Na^+ across the apical membrane and ions pass by facilitated diffusion across the basal membrane of the tubule cells.

FIGURE 26.12 Reabsorption in the Thick Segment of the Ascending Limb of the Loop of Henle

19 ***Describe what happens to most of the filtrate that enters the nephron.***

20 ***On what side of the nephron tubule cell does active transport take place during reabsorption of materials?***

21 ***Describe how symport works in the nephron.***

22 ***Name the substances that are moved by active and passive transport. In what part of the nephron does this movement take place?***

Tubular Secretion

Tubular secretion involves the movement of some substances, such as by-products of metabolism that become toxic in high concentrations and drugs or molecules not normally produced by the body, into the nephron (table 26.4). As with tubular reabsorption, tubular secretion can be either active or passive. Ammonia is produced in the epithelial cells of the nephron as amino groups are removed from amino acids and diffuse into the lumen of the

Conjugation and Excretion

Some drugs, environmental pollutants, and other foreign substances that gain access to the circulatory system are reabsorbed from the nephron. These substances are usually lipid-soluble, nonpolar compounds. They enter the glomerular filtrate and are reabsorbed passively by a process similar to that by which urea is reabsorbed. Because these substances are passively reabsorbed within the nephron, they are not rapidly excreted. The liver cells attach other molecules to them by a process called **conjugation** (kon-jū-gā'shŭn), which converts them to more water-soluble molecules. These more water-soluble substances enter the filtrate but do not pass as readily through the wall of the nephron, are not reabsorbed from the renal tubules, and consequently are more rapidly excreted in the urine. One of the important functions of the liver is to convert nonpolar toxic substances to more water-soluble forms, thus increasing the rate at which they are excreted in the urine.

nephron. Hydrogen ions, K^+, penicillin, and ***para*-aminohippuric acid** (par'ă-a-mē'nō-hi-pūr'ik; *p*-aminohippuric acid; **PAH**), among others, are actively secreted by either active transport or antiport processes into the nephron.

For example, an antiport process moves H^+ from cells of the nephron into the nephron's lumen. Hydrogen ions bind to carrier proteins on the inside of the plasma membrane, and Na^+ bind to the carrier proteins on the outside of the plasma membrane. As Na^+ move into the cell, H^+ move out of the cell (figure 26.13). The secreted H^+ are produced as a result of carbon dioxide and water reacting to form H^+ and HCO_3^-. The antiporters secrete H^+ into the nephron's lumen, and Na^+ enter the nephron cell. Sodium ions and HCO_3^- are symported across the basal membrane of the cell and enter the peritubular capillaries. Hydrogen ions are secreted into the proximal and distal convoluted tubules, and K^+ are actively secreted in the distal convoluted tubule. Secretion of H^+ by the nephron plays a major role in regulating body fluid pH (see chapter 27). Penicillin and *para*-aminohippuric acid are examples of substances not normally produced by the body that are actively secreted into the proximal convoluted tubules.

TABLE 26.4 Secretion of Substances into the Nephron

Transport Process	Substance Transported
Proximal Convoluted Tubule	
Antiport	H^+
Active transport	Hydroxybenzoates
	Para-aminohippuric acid
	Neurotransmitters
	Dopamine
	Acetylcholine
	Epinephrine
	Bile pigments
	Uric acid
	Drugs and toxins
	Penicillin
	Atropine
	Morphine
	Saccharin
Diffusion	Ammonia
Distal Convoluted Tubule	
Active transport	K^+
Antiport	K^+
	H^+

23 ***Where does tubular secretion take place? What substances are secreted? List the mechanisms by which these substances are transported.***

Urine Concentration Mechanism

When a large volume of water is consumed, it is necessary to eliminate the excess without losing large amounts of electrolytes or other substances essential for the maintenance of homeostasis. The kidneys' response is to produce a large volume of dilute urine. On the other hand, when drinking water is not available, producing a large volume of dilute urine would lead to rapid dehydration. When water intake is restricted, the kidneys produce a small volume of concentrated urine that conserves water and contains sufficient waste products to prevent their accumulation in the circulatory system. The kidneys are able to produce urine with concentrations that range from a minimum of 65 to a maximum of 1200 mOsm/kg while maintaining the extracellular fluid concentration very close to 300 mOsm/kg. The maintenance of a high concentration of solutes in the medulla, the countercurrent functions of the loops of Henle, and the mechanism that controls the permeability of the distal convoluted tubules and collecting ducts to water are essential for the kidneys to control the volume and concentration of the urine they produce.

Medullary Concentration Gradient

The kidneys' ability to concentrate urine depends on maintaining a high concentration of solutes in the medulla. The interstitial fluid concentration is about 300 mOsm/kg in the cortical region of the kidney. Moving from the cortex toward the medulla, the interstitial fluid becomes progressively more concentrated until it achieves a maximum concentration of 1200 mOsm/kg at the tip of the renal pyramid. This is more concentrated than seawater (figure 26.14)! Maintenance of the high solute concentration in the kidney medulla depends on the functions of the loops of Henle, the vasa recta, and on the distribution of urea. The major mechanisms that create and maintain the high solute concentration in the renal medulla are the following:

- The active transport of Na^+ and the symport of K^+, and Cl^-, and other ions out of the thick portion of the ascending limb of the loop of Henle into the interstitial fluid of the medulla.
- The impermeability of the thin and thick segments of the ascending limb of the loop of Henle to water.

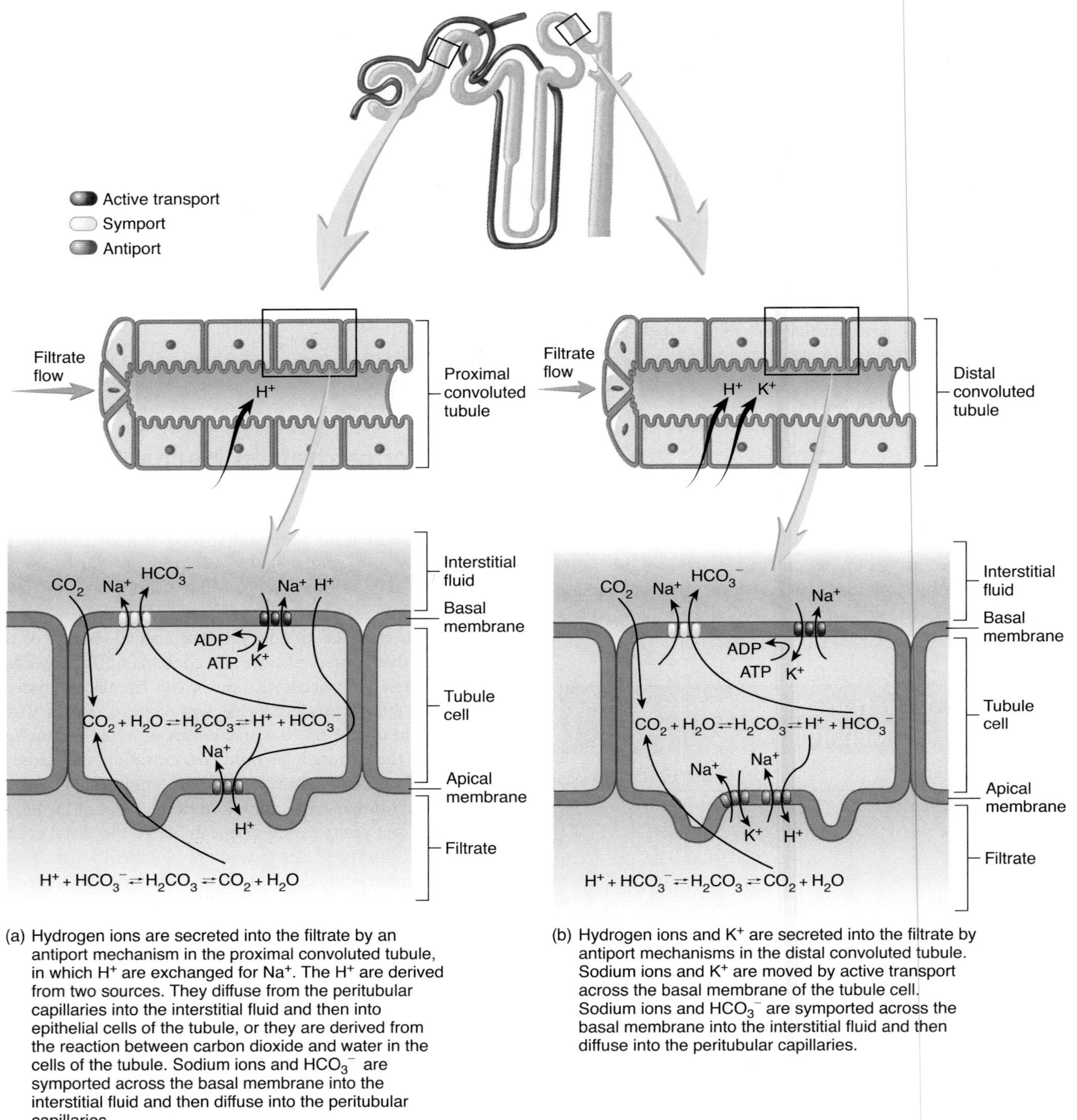

(a) Hydrogen ions are secreted into the filtrate by an antiport mechanism in the proximal convoluted tubule, in which H^+ are exchanged for Na^+. The H^+ are derived from two sources. They diffuse from the peritubular capillaries into the interstitial fluid and then into epithelial cells of the tubule, or they are derived from the reaction between carbon dioxide and water in the cells of the tubule. Sodium ions and HCO_3^- are symported across the basal membrane into the interstitial fluid and then diffuse into the peritubular capillaries.

(b) Hydrogen ions and K^+ are secreted into the filtrate by antiport mechanisms in the distal convoluted tubule. Sodium ions and K^+ are moved by active transport across the basal membrane of the tubule cell. Sodium ions and HCO_3^- are symported across the basal membrane into the interstitial fluid and then diffuse into the peritubular capillaries.

FIGURE 26.13 Secretion of H^+ and K^+ into the Nephron

- The vasa recta supply of blood to the medulla of the kidney and the removal of excess water and solutes that enter the medulla without destroying the high concentration of solutes in the interstitial fluid of the medulla.
- The active transport of ions from the collecting ducts into the interstitial fluid of the medulla.
- Urea cycling due to the diffusion of urea from the collecting duct into the interstitial fluid of the medulla into the descending limb of the loop of Henle.

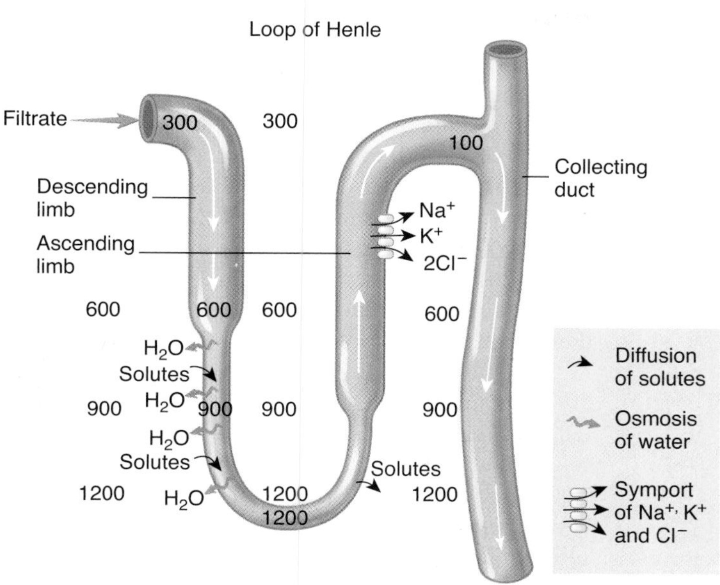

(a) Water diffuses out of the thin segment of the descending limb of the loop of Henle, and a small amount of solute diffuses into the descending limb. At the tip of the loop of Henle, the osmolality of the filtrate is 1200 mOsm. The ascending limb of the loop of Henle is not permeable to water. Solutes diffuse out of the thin segment of the ascending limb, and solutes, such as Na^+, K^+, and Cl^-, are symported out of the thick segment of the ascending limb of the loop of Henle. At the end of the thick segment of the ascending limb, the filtrate osmolality is 100 mOsm.

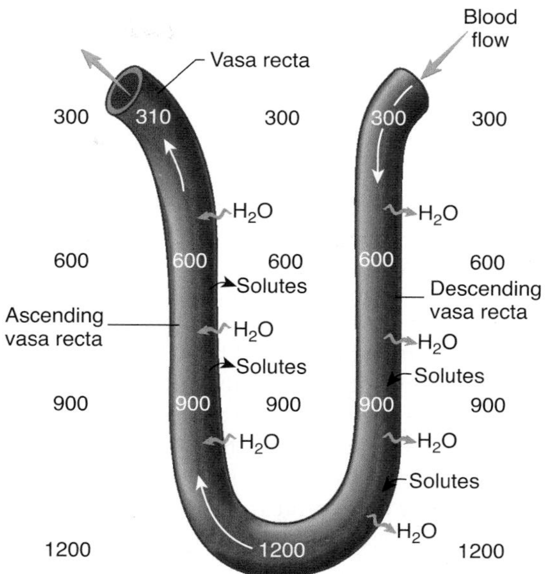

(b) Water diffuses out and solutes diffuse into the descending portion of the vasa recta. At the tip of the vasa recta, the osmolality of blood is 1200 mOsm. Water diffuses in and solutes diffuse out of the ascending limb of the vasa recta. At the end of the vasa recta, the blood osmolality is 310 mOsm, a concentration only slightly greater than the osmolality of the blood that enters the vasa recta.

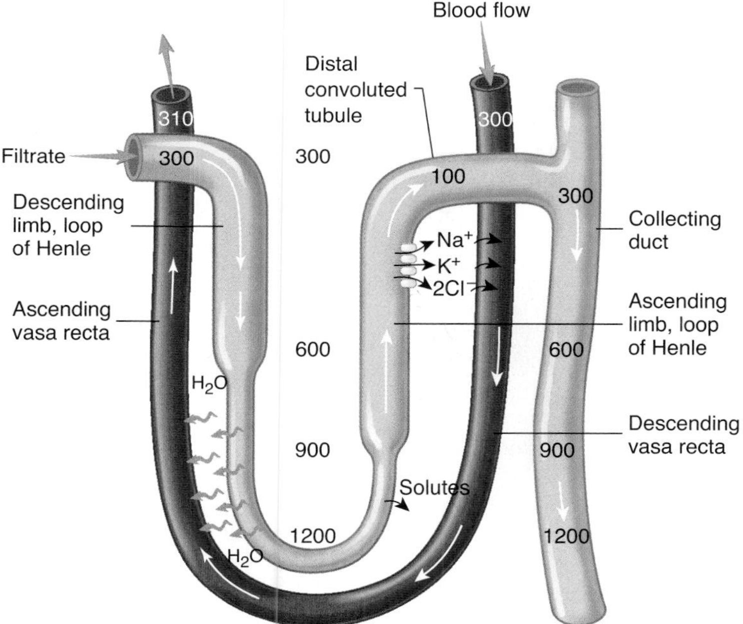

(c) The loop of Henle and the vasa recta function together. Water moves out of the descending limb of the loop of Henle and enters the ascending vasa recta. Solutes diffuse out of the ascending thin segment of the loop of Henle and enter the vasa recta, but water does not. Solutes transported out of the thick segment of the ascending limb of the loop of Henle enter the descending vasa recta. The vasa recta carry excess water and solutes away from the medulla without reducing the high concentration of solutes. The concentration of the filtrate is reduced to 100 mOsm/kg by the time it reaches the distal convoluted tubule.

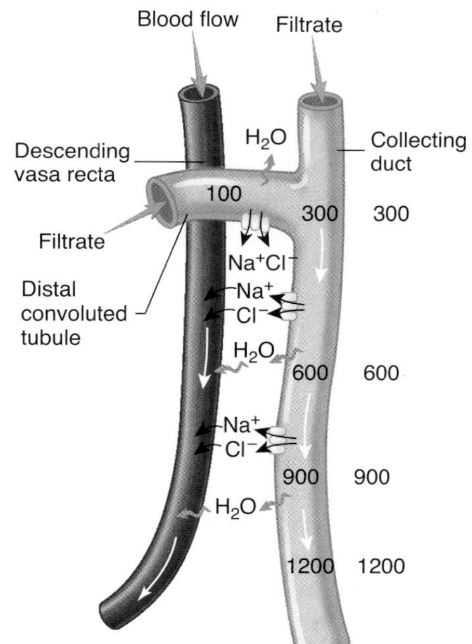

(d) Solutes, such as Na^+ and Cl^-, are transported and water moves out of the distal convoluted tubule and the collecting duct into the descending vasa recta.

FIGURE 26.14 Filtrate Concentration and the Medullary Concentration Gradient

The loop of Henle and the vasa recta function together to maintain a high concentration of solutes in the medulla of the kidney.

The roles of these mechanisms listed above in the maintenance of the high concentration of solutes in the medulla of the kidney are described in the following three sections:

1. *Transport of solutes and diffusion of water across the wall of the loop of Henle.* The long loops of Henle of the juxtamedullary nephrons are essential for the maintenance of a high medullary solute concentration. The walls of the descending limbs of the loops of Henle are permeable to water. As filtrate flows into the medulla of the kidney through the descending limbs, water diffuses out of the nephrons into the more concentrated interstitial fluid. The excess water that enters the interstitial fluid enters the vasa recta and is removed from the medulla. See section 2. Some solutes diffuse into the descending limb, but in relatively small amounts. By the time the filtrate reaches the tip of the loop of Henle, it is very concentrated. The walls of the thick and thin segments of the ascending limbs of the loops of Henle are impermeable to water. Consequently, solutes diffuse out of the thin segment of the ascending limb as it passes through progressively less concentrated interstitial fluid on its way back to the cortex of the kidney. Also, Na^+, K^+, and Cl^- are symported out of the thick segment of the ascending limb into the interstitial fluid. Thus, water enters the interstitial fluid from the descending limbs and solutes enter the interstitial fluid from the ascending limbs of the loops of Henle (see figure 26.14*a*). The solutes that diffuse from the thin segments, and those that are symported from the thick segments, add solutes to the medulla.
2. *Diffusion of water and solutes across the wall of the vasa recta.* The vasa recta supply blood to the medulla of the kidney, and they are countercurrent mechanisms, which remove excess water and solutes from the medulla of the kidney without changing the high concentration of solutes in the medullary interstitial fluid. A **countercurrent** (kown′ter-ker′ent) **mechanism** consists of parallel tubes in which fluid flows, but in opposite directions, and heat or substance, such as water or solute, diffuse from the tubes carrying fluid in one direction to the tubes carrying fluid in the opposite direction so that the fluid in both sets of tubes has nearly the same composition. The vasa recta are a countercurrent mechanism because blood flows through them to the kidney medulla and, after the vessels turn near the tip of the renal pyramid, the blood is carried in the opposite direction, back toward the cortex. The walls of the vasa recta are permeable to water and to solutes. As blood flows toward the medulla, water moves out of the vasa recta, and some solutes diffuse into them. As blood flows back toward the cortex, water moves into the vasa recta, and some solutes diffuse out of them (figure 26.14*b*). The directions of diffusion are such that slightly more water and slightly more solute are carried from the medulla by the vasa recta than enter it. Thus, the composition of the blood at both ends of the vasa recta is nearly the same, with the volume and osmolality being slightly greater as the blood once again reaches the cortex.

   The loops of Henle and the vasa recta are in parallel with one another, and their functions are closely related. The water and solutes that leave the loops of Henle enter the vasa recta. The vasa recta carry away the water and solutes without diminishing the high concentration of solutes in the medulla of the kidney (figure 26.14*c*). Water passes by osmosis from the collecting ducts, and solutes, such as Na^+ and Cl^-, are actively transported out of the collecting ducts and pass into the interstitial fluid of the medulla of the kidney (figure 26.14*d*). The water and solutes that leave the collecting ducts also enter the vasa recta and are carried away from the medulla.

   The loops of Henle and vasa recta function together to (1) maintain a high concentration of solutes in the interstitial fluids of the medulla and (2) carry away the water and solutes that enter the medulla from the loops of Henle and collecting ducts.
3. *Urea cycling.* **Urea** (ū-rē′ă) molecules are responsible for a substantial part of the high osmolality in the medulla of the kidney (figure 26.15). The walls of the descending limbs of the loops of Henle are permeable to urea, and urea

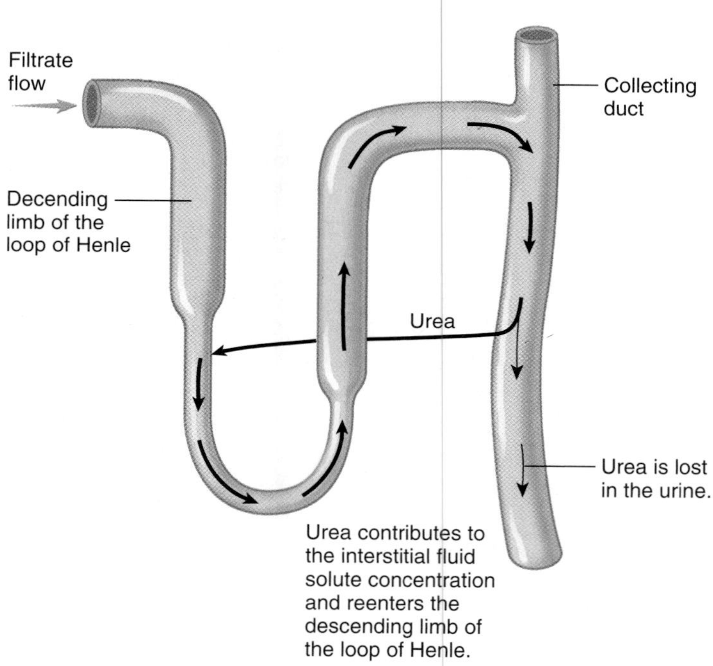

FIGURE 26.15 Medullary Concentration Gradient and Urea Cycling
The concentration of urea in the medulla of the kidney is high and contributes to the overall high concentration of solutes there. The wall of the collecting duct is permeable to urea. Urea diffuses out of the collecting duct into the interstitial fluid of the medulla. The wall of the descending limb of the loop of Henle is also permeable to urea. Urea diffuses from the interstitial fluid into the descending limb. Thus, a cycle is produced in which urea flows into the descending limb, through the ascending limb, through the distal convoluted tubule, through the collecting duct, out of the collecting duct, and back into the descending limb.

diffuses into the descending limbs from the interstitial fluid. The ascending limbs of the loops of Henle and the distal convoluted tubules are impermeable to urea. The collecting ducts are permeable to urea and some urea diffuses out of the collecting ducts into the interstitial fluid of the medulla. Thus, urea is recycled from the interstitial fluid into the descending limbs of the loops of Henle, through the ascending limbs, through the distal convoluted tubules, and into the collecting ducts. Urea then diffuses from the collecting ducts back into the interstitial fluid of the medulla. Consequently, a high urea concentration is maintained in the medulla of the kidney.

24 ***List the major mechanisms that create and maintain the high solute concentration in the renal medulla.***

25 ***Describe the role of the loops of Henle, vasa recta, and urea in maintaining a high interstitial solute concentration in the kidney medulla.***

Summary of Changes in Filtrate Volume and Concentration

In the average person, about 180 L of filtrate enter the proximal convoluted tubules daily. Glucose, amino acids, Na^+, Ca^{2+}, K^+, Cl^-, and other substances (see table 26.3) are transported, and water moves by osmosis from the lumens of the proximal convoluted tubules into the interstitial fluid. The excess solutes and water then enter the peritubular capillaries. Consequently, approximately 65% of the filtrate is reabsorbed as water, and solutes move from the proximal convoluted tubules into the interstitial fluid. The osmolality of both the interstitial fluid and the filtrate is maintained at about 300 mOsm/kg.

The filtrate then passes into the descending limbs of the loops of Henle, which are highly permeable to water and solutes. As the descending limbs penetrate deep into the medulla of the kidney, the surrounding interstitial fluid has a progressively greater osmolality. Water diffuses out of the nephrons as solutes slowly diffuse into them. By the time the filtrate reaches the deepest part of the loops of Henle, its volume has been reduced by an additional 15% of the original volume and its osmolality has increased to about 1200 mOsm/kg (figure 26.16). By the time the filtrate has reached the tips of the loops of Henle, at least 80% of the filtrate volume has been reabsorbed.

After passing through the descending limbs of the loops of Henle, the filtrate enters the ascending limbs. Both the thin and thick segments are impermeable to water, but solutes diffuse out of the thin segment and Na^+, Cl^-, and K^+ are symported from the filtrate into the interstitial fluid in the thick segments (see figure 26.16). The movement of solutes, but not water, across the wall of the ascending limbs causes the osmolality of the filtrate to decrease from 1200 to about 100 mOsm/kg by the time the filtrate again reaches the cortex of the kidney. The volume of the filtrate does not change as it passes through the ascending limbs. As a result, the filtrate entering the distal convoluted tubules is dilute, compared with the concentration of the surrounding interstitial fluid, which has an osmolality of about 300 mOsm/kg.

The changes just described are obligatory; that is, they occur regardless of the concentration and the volume of the urine that is finally produced by the kidney. The mechanisms by which concentrated and dilute urine are formed by the kidney are described in the following sections.

26 ***Describe how the filtrate volume and concentration change as they flows through the nephrons and collecting ducts.***

REGULATION OF URINE CONCENTRATION AND VOLUME

Urine can be diluted or very concentrated, and it can be produced in large or small amounts. Urine concentration and volume are regulated by mechanisms that maintain the extracellular fluid osmolality and volume within narrow limits.

Filtrate reabsorption in the proximal convoluted tubules and the descending limbs of the loops of Henle is obligatory and therefore remains relatively constant. Filtrate reabsorption in the distal convoluted tubules and collecting ducts is regulated, however, and can change dramatically, depending on the conditions to which the body is exposed. If homeostasis requires the elimination of a large volume of dilute urine, the dilute filtrate in the distal convoluted tubules and collecting ducts can pass through them with little change in concentration. On the other hand, if water must be conserved to maintain homeostasis, water is reabsorbed from the filtrate as it passes through the distal convoluted tubules and collecting ducts. This results in the production of a small volume of very concentrated urine. The regulation of urine volume and concentration involves hormonal mechanisms, described below, as well as autoregulation, and the sympathetic nervous system.

Hormonal Mechanisms

Two major hormonal mechanisms are involved in regulating urine concentration and volume: the renin-angiotensin-aldosterone mechanism and the antidiuretic hormone (ADH) mechanism. Each mechanism is activated by different stimuli, but they work together to achieve homeostasis. The renin-angiotensin-aldosterone mechanism is more sensitive to changes in blood pressure, and the ADH mechanism is more sensitive to changes in blood osmolality.

Renin-Angiotensin-Aldosterone

Renin is an enzyme secreted by cells of the juxtaglomerular apparatus. The rate of renin secretion increases if blood pressure in the afferent arteriole decreases, or if the Na^+ concentration of the filtrate decreases as it passes by the macula densa cells of the juxtaglomerular apparatuses. Renin enters the general circulation, acts on **angiotensinogen,** a plasma protein produced by the liver, and converts it to **angiotensin I.** Subsequently, a proteolytic enzyme called **angiotensin-converting enzyme (ACE),** found in capillary beds in organs such as the lungs, converts angiotensin I to

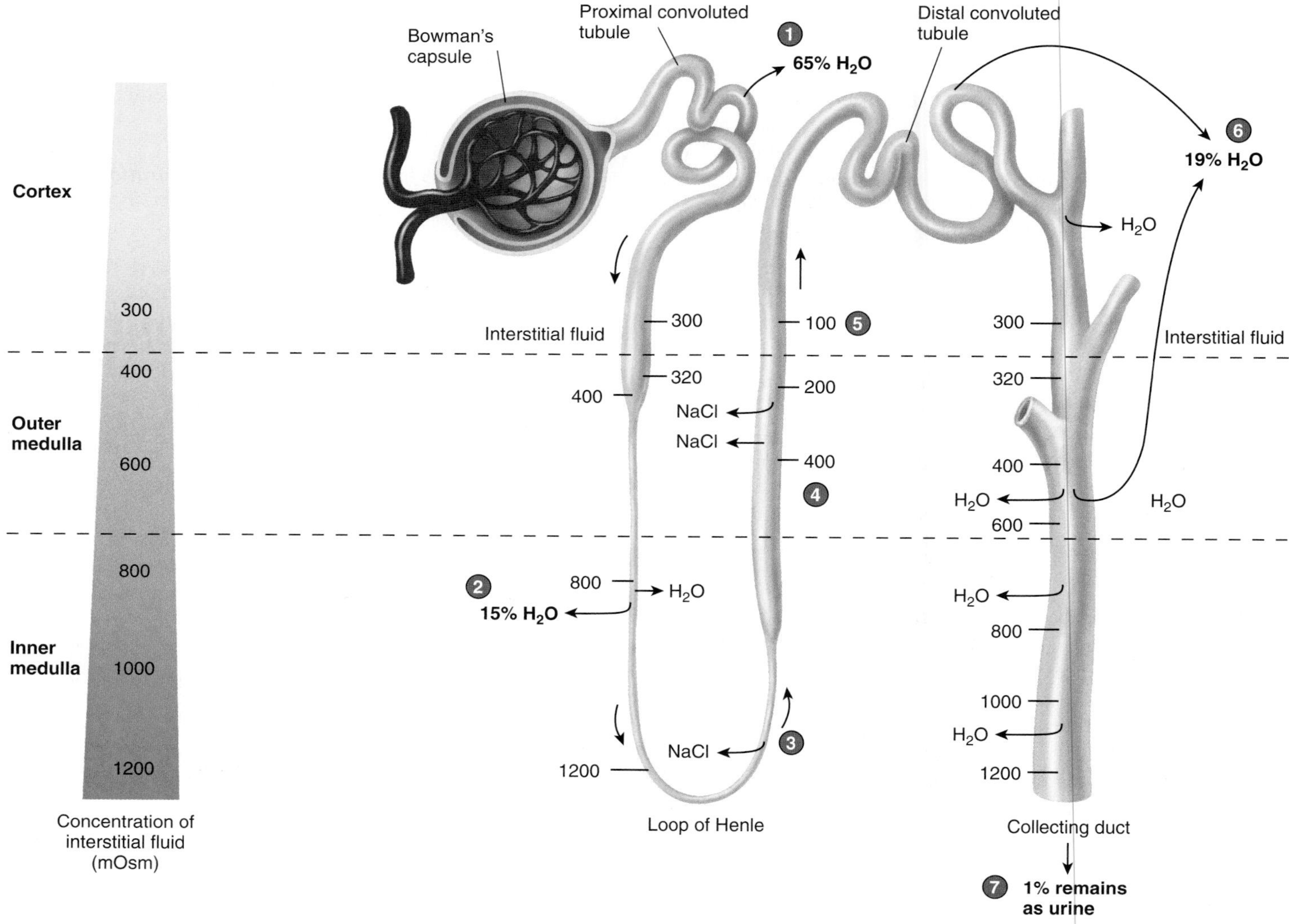

1. Approximately 180 L of filtrate enters the nephrons each day; of that volume, 65% is reabsorbed in the proximal convoluted tubule. In the proximal convoluted tubule, solute molecules move by active transport and symport from the lumen of the tubule into the interstitial fluid. Water moves by osmosis because the cells of the tubule wall are permeable to water (see figure 26.10).
2. Approximately 15% of the filtrate volume is reabsorbed in the thin segment of the descending limb of the loop of Henle. The descending limb passes through the concentrated interstitial fluid of the medulla. Because the wall of the descending limb is permeable to water, water moves by osmosis from the tubule into the more concentrated interstitial fluid (see figure 26.11*a*). By the time the filtrate reaches the tip of the renal pyramid, the concentration of the filtrate is equal to the concentration of the interstitial fluid.
3. The ascending limb of the loop of Henle is not permeable to water. Solutes diffuse out of the thin segment (see figure 26.11*b*).
4. A Na^+- K^+- $2Cl^-$ symporter is in the apical membrane of the thick segment. Sodium ions are actively transported and K^+ and Cl^- diffuse across the basal membrane of epithelial cells of the thick segment into the interstitial fluid (see figure 26.12).
5. The volume of the filtrate does not change as it passes through the ascending limb, but the concentration is greatly reduced (see figure 26.12). By the time the filtrate reaches the cortex of the kidney, the concentration is approximately 100 mOsm/L, which is less concentrated than the interstitial fluid of the cortex (300 mOsm/L).
6. The distal convoluted tubule and collecting duct are permeable to water if ADH is present. If ADH is present, water moves by osmosis from the less concentrated filtrate into the more concentrated interstitial fluid. By the time the filtrate reaches the tip of the renal pyramid, an additional 19% of the filtrate is reabsorbed.
7. One percent or less of the filtrate remains as urine when ADH is present.

PROCESS FIGURE 26.16 Urine Concentrating Mechanism

* test

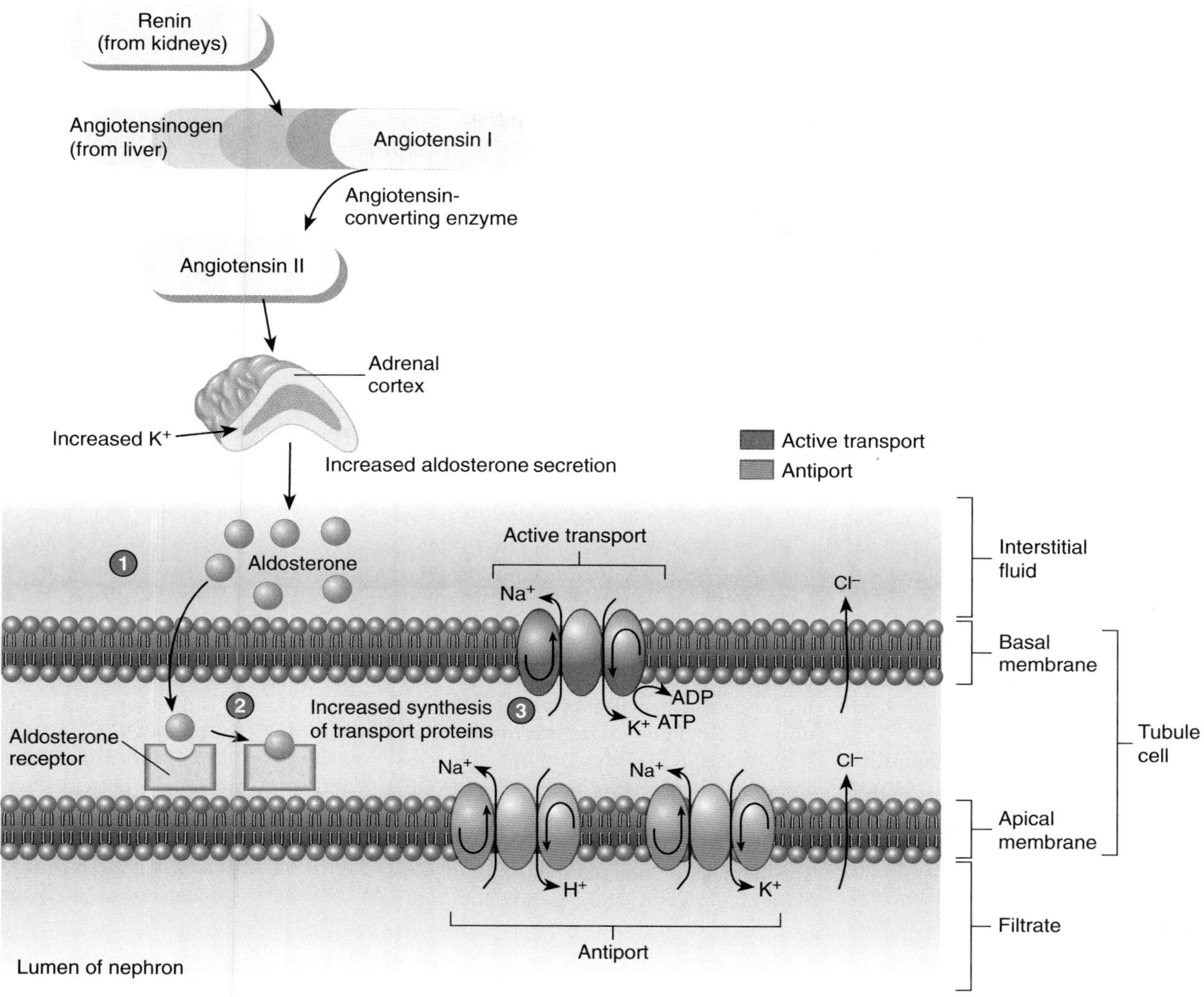

1. Aldosterone secreted from the adrenal cortex enters cells of the distal convoluted tubule.
2. Aldosterone binds to intracellular receptors and increases the synthesis of transport proteins of the apical and basal membranes.
3. Newly synthesized transport proteins increase the rate at which Na^+ are absorbed and K^+ and H^+ are secreted. Chloride ions move with the Na^+ because they are attracted to the positive charge of Na^+.

PROCESS FIGURE 26.17 **Effect of Aldosterone on the Distal Convoluted Tubule**

angiotensin II (figure 26.17). Angiotensin II is a potent vasoconstricting substance that increases peripheral resistance, causing blood pressure to increase. However, angiotensin II is rapidly broken down, so its effect lasts for only a short time. Angiotensin II also increases the rate of aldosterone secretion, the sensation of thirst, salt appetite, and ADH secretion.

The rate of renin secretion decreases if blood pressure in the afferent arteriole increases, or if the Na^+ concentration of the filtrate increases as it passes by the macula densa of the juxtaglomerular apparatuses.

A large decrease in the concentration of Na^+ in the interstitial fluids acts directly on the aldosterone-secreting cells of the adrenal cortex to increase the rate of aldosterone secretion. Angiotensin II is much more important than the blood level of Na^+, however, in regulating aldosterone secretion.

Aldosterone, a steroid hormone secreted by the cortical cells of the adrenal glands (see chapter 18), passes through the circulatory system from the adrenal glands to the cells in the distal convoluted tubules and the collecting ducts. Aldosterone molecules diffuse through the plasma membranes and bind to receptor molecules within the cells. The combination of aldosterone molecules with their receptor molecules increases synthesis of the transport proteins that increase the transport of Na^+ across the basal and apical membranes of nephron cells. As a result, the rate of Na^+ transport out of the filtrate back into the blood increases (see figure 26.17).

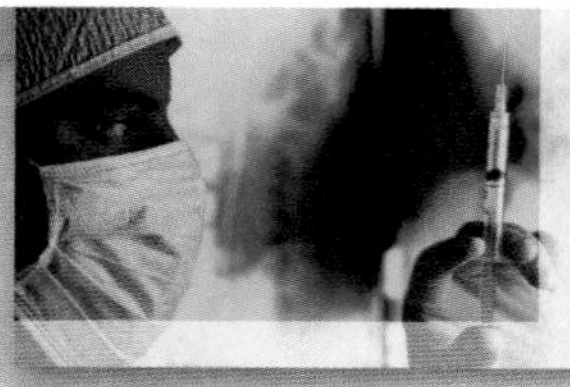

CLINICAL FOCUS

Type 2 Diabetes Mellitus, Diabetic Nephropathy, and Renal Failure

Diabetic nephropathy (ne-frop′ă-thē) is a disease of the kidney associated with diabetes mellitus, and it is the principal cause of chronic renal failure. It damages renal glomeruli and ultimately results in the destruction of functional nephrons through progressive scar tissue formation, which is mediated in part by an inflammatory response. The damaged glomeruli no longer filter the blood effectively and proteins are able to pass through the filtration membrane and are lost in the urine. The presence of protein in the urine of people who have type 2 diabetes strongly suggests significant diabetic nephropathy, which can lead to end-stage renal failure. About one in five Americans over age 30 have some degree of type 2 diabetes mellitus, and most of the people who receive renal dialysis have type 2 diabetes mellitus.

The development of diabetic nephropathy is complex. Although the mechanism is not completely understood, angiotensin II, a powerful vasoconstrictor (see "Renin-Angiotensin-Aldosterone Mechanism," in chapters 18 and 21), is elevated in diabetes mellitus. This causes exaggerated efferent arteriole vasoconstriction and, consequently, increased glomerular capillary pressure. The increased glomerular capillary pressure damages the glomerular basement membrane, causing it to thicken and become more permeable. The glomerular basement membrane is also damaged by the production of glycoproteins called **advanced glycosylation end products (AGEs).** AGEs are produced when glucose forms irreversible cross-links with kidney and plasma proteins. The AGEs stimulate the secretion of growth factors from glomerular cells, which promote glomerular basement membrane thickening.

Because the permeability of the glomerular basement membrane is increased in patients with diabetes mellitus, plasma proteins cross the filtration membrane and enter the urine. Initially, a small amount of protein enters the urine, a condition called microalbuminuria (mī′krō-al-boo-min-ū′rē-ă). Eventually, microalbuminuria progresses to overt proteinuria (prō-tē-noo′rē-ă), which is defined as the secretion of more than 300 mg albumin/day. The number of functional nephrons in the kidney is decreased. This may take 10–15 years to develop. By the time overt proteinuria has developed, the number of functional nephrons has decreased so that the kidneys no longer have the ability to excrete adequate amounts of waste products and is called **end-stage renal disease (ESRD).** In ESRD, renal failure has worsened to the point that kidney function is less than 10% of normal. End-stage renal disease must be treated by dialysis or kidney transplantation, or death results.

The use of angiotensin converting enzyme (ACE) inhibitors slows or, in some cases, even halts the progression of proteinuria and end-stage renal disease. ACE inhibitors prevent the formation of angiotensin II; consequently, arterial blood pressure and glomerular capillary pressure remain within their normal range of values. When ACE inhibitors are used in combination with drugs called **angiotensin receptor blockers (ARBs),** which prevent angiotensin II molecules from binding to their receptors, proteinuria decreases up to 45%, when compared with the use of no drugs. People with type 2 diabetes who maintain their blood glucose within normal levels have a much lower incidence of diabetic nephropathy and ESRD.

Reduced secretion of aldosterone decreases the rate of Na^+ transport. As a consequence, the concentration of Na^+ in the distal convoluted tubules and the collecting ducts remains high. Because the concentration of filtrate passing through the distal convoluted tubules and the collecting ducts has a greater than normal concentration of solutes, the capacity for water to move by osmosis from the distal convoluted tubules and the collecting ducts is diminished, urine volume increases, and the urine has a greater concentration of Na^+.

Increases in blood K^+ levels act directly on the adrenal cortex to stimulate aldosterone secretion, whereas decreases in blood K^+ levels decrease aldosterone secretion (see chapter 27).

PREDICT 4

Drugs that increase urine volume are called diuretics. Some diuretics inhibit the active transport of Na^+ in the nephron. Explain how these diuretic drugs could cause increased urine volume.

27 ***What are the effects of aldosterone on Na^+ and Cl^- transport? How does aldosterone affect urine concentration, urine volume, and blood pressure?***

28 ***Where is aldosterone produced? What factors stimulate aldosterone secretion?***

29 ***How is angiotensin II activated? What effects does it produce?***

30 ***What factors cause an increase in renin production?***

Antidiuretic Hormone

renal tubule Collecting tubule

The distal convoluted tubules and collecting ducts remain relatively impermeable to water in the absence of ADH (figure 26.18). When little ADH is secreted, a large part of the 19% of the filtrate that is normally reabsorbed in the distal convoluted tubules and the collecting ducts becomes part of the urine. People who do not secrete sufficient ADH often produce 10–20 L of urine per day and develop major problems, such as dehydration and ion imbalances. Insufficient ADH secretion results in a condition called **diabetes insipidus** (dī-ă-bē′tēz in-sip′i-dŭs); the word *diabetes* implies the production of a large volume of urine, and the word *insipidus* implies the production of a clear, tasteless, dilute urine. This condition is in contrast to **diabetes mellitus** (me-lī′tŭs), which implies the production of a large volume of urine that contains a

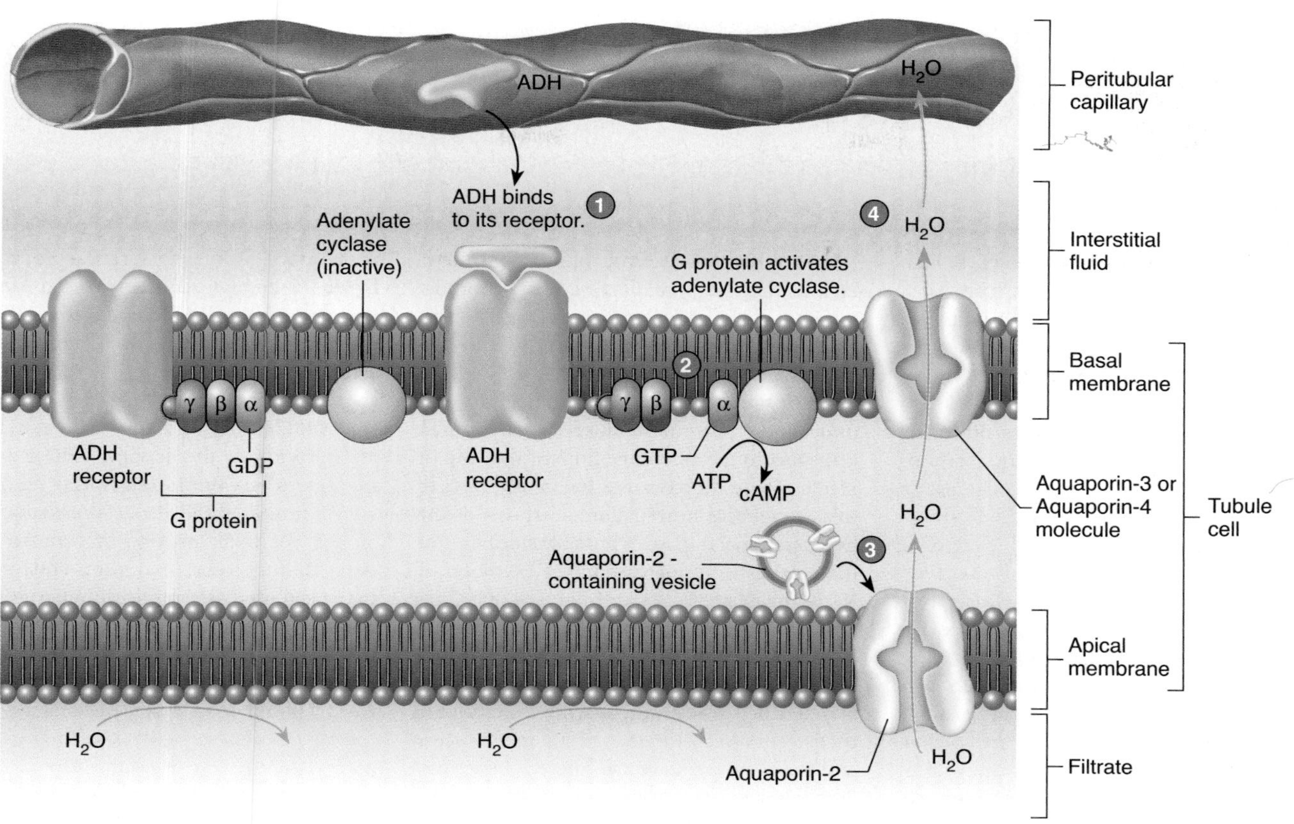

1. ADH moves from the peritubular capillaries and binds to ADH receptors in the plasma membranes of the distal convoluted tubule cells and the collecting duct cells.
2. When ADH binds to its receptor, a G protein mechanism is activated, which in turn activates adenylate cyclase.
3. Adenylate cyclase increases the rate of cAMP synthesis. Cyclic AMP promotes the insertion of aquaporin-2 containing cytoplasmic vesicles into the apical membranes of the distal convoluted tubules and collecting ducts, thereby increasing their permeability to water. Water then moves by osmosis out of the distal convoluted tubules and collecting ducts into the tubule cells through the aquaporin-2 water channels.
4. Water exits the tubule cells and enters the interstitial fluid through aquaporin-3 and aquaporin-4 water channels in the basal membranes.

PROCESS FIGURE 26.18 **Effect of Antidiuretic Hormone (ADH) on Nephron Water Movement**

high concentration of glucose. The word *mellitus* means honeyed, or sweet.

The posterior pituitary secretes ADH. Neurons with cell bodies primarily in the supraoptic nuclei of the hypothalamus have axons that course to the posterior pituitary gland. ADH is released into the circulatory system from these neuron terminals. Cells called **osmoreceptor cells** in the supraoptic nuclei are very sensitive to even slight changes in the osmolality of the interstitial fluid. If the osmolality of the blood and interstitial fluid increases, these cells stimulate the ADH-secreting neurons. Action potentials are then propagated along the axons of the ADH-secreting neurons to the posterior pituitary gland, where the axons release ADH from their ends. Reduced osmolality of the interstitial fluid within the supraoptic nuclei inhibits ADH secretion from the posterior pituitary gland (see figure 18.5).

Baroreceptors that monitor blood pressure in the atria of the heart, large veins, carotid sinuses, and aortic arch also influence ADH secretion when the blood pressure increases or decreases in excess of 5%–10%. Decreases in blood pressure are detected by the baroreceptors, which consequently decrease the frequency of action potentials sent along the afferent pathways that ultimately extend to the supraoptic region of the hypothalamus. The result is an increase in ADH secretion.

When blood osmolality increases or when blood pressure declines significantly, ADH secretion increases and acts on the kidneys to increase the reabsorption of water. The reabsorption of water decreases blood osmolality. It also increases blood volume, which increases blood pressure. Conversely, when blood osmolality decreases or when blood pressure increases, ADH secretion

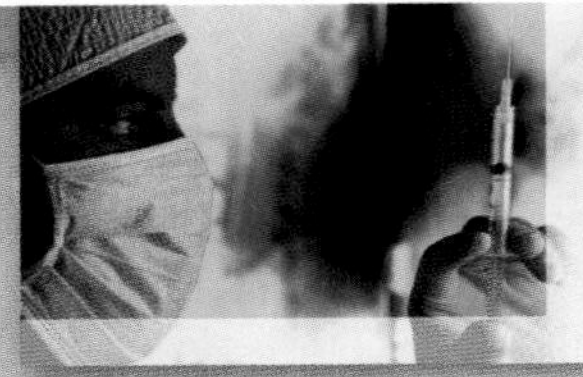

CLINICAL FOCUS

Diuretics

Diuretics (dī-ū-ret′iks) are agents that increase the rate of urine formation. Although the definition is simple, a number of physiological mechanisms are involved.

Diuretics are used to treat disorders such as hypertension and several types of edema caused by congestive heart failure, cirrhosis of the liver, and other anomalies. The use of diuretics can lead to complications, however, including dehydration and electrolyte imbalances.

The action of **carbonic anhydrase** (kar-bon′ik an-hī′drās) **inhibitors** reduces the rate of H^+ secretion and the reabsorption of HCO_3^-. As H^+ are secreted into the nephron, they combine with HCO_3^- to form carbonic acid. Carbonic acid forms water and carbon dioxide, which can diffuse across the wall of the nephron. Reduced H^+ secretion causes HCO_3^- to remain in the nephron. The HCO_3^- increase tubular osmotic pressure, causing osmotic diuresis. With the long-term use of carbonic anhydrase inhibitors, the diuretic effect tends to be lost. The diuretic effect is useful in treating conditions such as glaucoma and altitude sickness.

Sodium ion reabsorption inhibitors include thiazide-type diuretics. They promote the loss of Na^+, Cl^-, and water in the urine. These diuretics are given to some people who have hypertension. Other inhibitors of Na^+ reabsorption, such as bumetanide, furosemide, and ethacrynic acid, specifically inhibit transport in the ascending limb of the loop of Henle. These diuretics are frequently used to treat congestive heart failure, cirrhosis of the liver, and renal disease. A possible side effect of these drugs is an increase in the excretion of K^+ in the urine.

Certain **potassium-sparing diuretics** act on the distal convoluted tubules and the collecting ducts to reduce the exchange between Na^+ and K^+. Potassium-sparing diuretics are used to diminish the loss of K^+ in the urine and therefore preserve, or "spare," these ions. Some potassium-sparing diuretic drugs act by competitive inhibition of aldosterone, whereas others inhibit the symporters for Na^+ in apical membranes of cells in the distal convoluted tubules and collecting ducts. Both types result in Na^+ diuresis and K^+ retention. Because, K^+ depletion is one of the side effects of prolonged treatment with inhibitors of Na^+–Cl^- symporters in the ascending limb of the loop of Henle that are frequently used as diuretics, potassium-sparing diuretics are commonly used in combination with these Na^+–Cl^- symport inhibitors.

Osmotic diuretics freely pass by filtration into the filtrate, and they undergo limited reabsorption by the nephron. These diuretics increase urine volume by elevating the osmotic concentration of the filtrate, thus reducing the amount of water moving by osmosis out of the nephron. Urea, mannitol, and glycerine have been used as osmotic diuretics. Although they are not commonly used, they are effective in treating people who are suffering from cerebral edema and edema in acute renal failure (see Clinical Focus "Renal Pathologies").

Xanthines (zan′thēnz), including caffeine and related substances, act as diuretics partly because they increase renal blood flow and the rate of glomerular filtrate formation. They also influence the nephron by decreasing Na^+ and Cl^- reabsorption.

Alcohol acts as a diuretic, although it is not used clinically for that purpose. It inhibits ADH secretion from the posterior pituitary and results in increased urine volume.

declines. The reduced ADH levels cause the kidneys to reabsorb less water and to produce a larger volume of dilute urine. The increased loss of water in the urine increases blood osmolality and decreases blood pressure. ADH secretion occurs in response to small changes in osmolality, whereas a substantial change in blood pressure is required to alter ADH secretion. Thus, ADH is more important in the regulation of blood osmolality than in the regulation of blood pressure.

PREDICT 5

Ethyl alcohol inhibits ADH secretion. Given this information, describe the mechanism by which alcoholic beverages affect urine production.

31 ***Where is ADH produced? What factors stimulate an increase in ADH secretion?***

32 ***What effect does ADH have on urine volume and concentration?***

Formation of Concentrated Urine

Filtrate enters the distal convoluted tubules after passing through the loops of Henle and then passes through the collecting ducts. Near the ends of the distal convoluted tubules and in the collecting ducts, the wall of the tubules become very permeable to water, provided antidiuretic hormone (ADH) is present. Water then diffuses from the lumens of the nephrons into the more concentrated interstitial fluid.

ADH increases the permeability of the apical plasma membranes of the distal convoluted tubules and collecting ducts to water by binding to membrane-bound receptors. This activates a G protein mechanism that increases cAMP synthesis inside of these cells. Cyclic AMP promotes the insertion of aquaporins into the apical plasma membrane (see figure 26.18). **Aquaporins** are water channel proteins that increase the apical plasma membrane's permeability to water. There are multiple forms of aquaporins. In cells of the distal convoluted tubules and collecting ducts, the basal plasma membranes contain aquaporin molecules—aquaporin-3 and aquaporin-4—that are insensitive to ADH. These aquaporin molecules provide channels for water to exit from the collecting duct cells into the interstitial fluid. Aquaporin-2 molecules regulate water movement into the cells. In cells that have not been exposed to ADH, the aquaporin-2 molecules are found in the membranes of vesicles in the cytoplasm (see figure 3.3*b*). In response to ADH, the increased cAMP initiates the incorporation of the membranes of vesicles containing aquaporin-2 channels into the apical plasma membrane. Thus, when ADH is present, water moves by osmosis out of the distal convoluted tubules and collecting ducts, whereas, in the absence of ADH, water remains in the distal convoluted tubules and collecting ducts to become urine (figure 26.19). Abnormal aquaporin-2 genes can result in excessive urine production because these genes code for abnormal aqaporin-2 molecules that do not function normally. Thus, there is a decrease in functional aquaporins and water remains in the nephron.

The filtrate flows into the distal convoluted tubules and collecting ducts that pass through the medulla of the kidney with its high concentration of solutes. If ADH is present, water moves by osmosis from the distal convoluted tubules and the collecting ducts into the interstitial fluid. By the time the filtrate has reached the end of the collecting ducts, another 19% of the filtrate has been reabsorbed. Thus, 1% of the filtrate remains as urine, and 99% of the filtrate has been reabsorbed. The osmolality of the filtrate at the ends of the collecting ducts is approximately 1200 mOsm/kg (see figure 26.16).

In addition to the dramatic decrease in filtrate volume and the increase in filtrate osmolality, a marked alteration occurs in the filtrate composition. Waste products, such as creatinine and urea and excess ions such as K^+, H^+, phosphate, and sulfate, are at a much higher concentration in urine than in the original filtrate because water has been removed from the filtrate. Many substances are selectively reabsorbed from the nephron and others are secreted into the nephron so that beneficial substances are retained in the body and toxic substances are eliminated.

33 ***Describe how the presence of ADH causes the formation of a small volume of concentrated urine.***

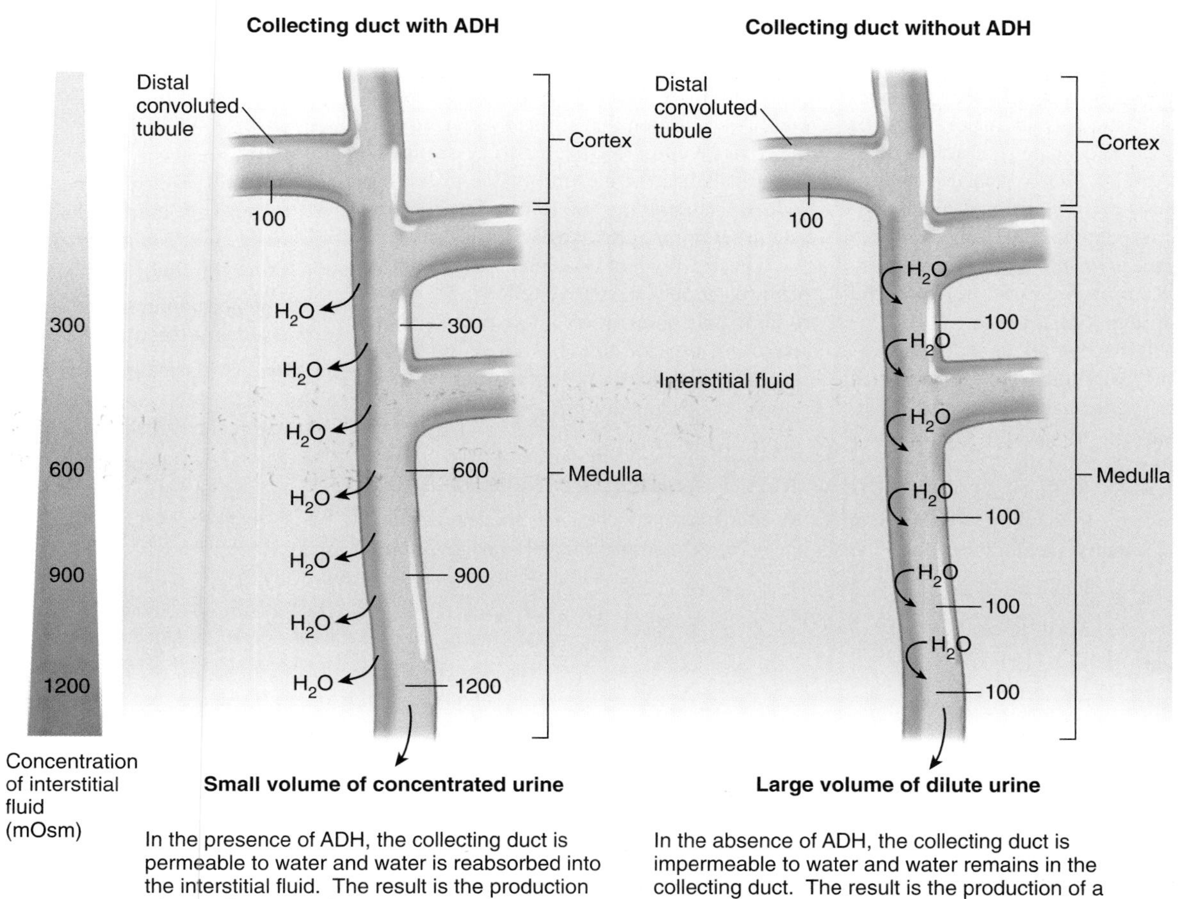

PROCESS FIGURE 26.19 **Effect of ADH on Urine Concentration and Volume**

CASE STUDY

Diabetes Insipidus

Not long after two infants arrived home from the hospital, their parents noticed that their diapers were excessively wet hour after hour throughout the day and night. After the first week, both sets of parents took their infants to the doctor because they both had slight fevers and had vomited, even though they had not eaten for several feedings. The infants were irritable, but their parents thought this was due to a virus causing the vomiting. The doctor took a blood sample, which indicated that both of the infants had high blood Na^+ levels. Subsequently, the doctor ordered a water deprivation test, during which plasma levels of ADH were monitored. Based on the test results, the doctor diagnosed the infants' condition to be **nephrogenic diabetes insipidus (NDI).** One infant was diagnosed with an ADH receptor abnormality, whereas the other was diagnosed with an aquaporin-2 abnormality.

The term *diabetes* refers to a disease state characterized by polyuria, an excess production of urine. There are two major causes of diabetes: (1) an inadequate production of or response to insulin, called diabetes mellitus (see chapter 18) and (2) an inadequate production of or response to ADH, called diabetes insipidus. Diabetes insipidus is a relatively rare disease that occurs in two varieties: (1) **central diabetes insipidus (CDI),** which is caused by a failure of ADH secretion, and (2) **nephrogenic diabetes insipidus,** which results when ADH secretion is normal but the ADH receptor, or the response to ADH, in the kidney is abnormal. Consequently, the G protein mechanism, which normally functions in the insertion of the aquaporin-2 water channel protein in the apical plasma membranes, does not occur. In most cases, NDI results from an inherited condition that affects the function of the ADH receptor. NDI can also be acquired, but usually later in life, and is due to several factors, including the use of certain prescription drugs or an underlying systemic disease.

The treatment of NDI includes ensuring a plentiful supply of water, following a low-sodium and sometimes a low-protein diet, and using thiazide diuretics (Na^+ ion reabsorption inhibitors) in combination with a potassium-sparing diuretic.

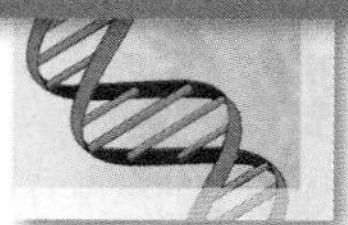

CLINICAL GENETICS

Nephrogenic Diabetes Insipidus (NDI)

There are three types of inherited NDI. X-linked NDI is the most common form and affects males more often than females. X-linked NDI is caused by a mutation in the V_2 ADH receptor gene on the X chromosome. Mutations in this gene result in defective ADH receptors, which prevent a normal response to ADH in the kidney.

Autosomal-recessive NDI is more rare than the X-linked form, and males and females are affected equally. This form of NDI requires both parents to be a carrier for an abnormal aquaporin-2 gene on chromosome 12. For children to have autosomal-recessive NDI, they must inherit a recessive allele from each parent. In autosomal-recessive NDI, there is a 25% chance that each child of heterozygous parents will have NDI.

Autosomal-dominant NDI is the most rare form of NDI and also affects males and females equally. With this form, only one parent must have a dominant allele for an abnormal aquaporin-2 gene, but that parent will also have symptoms of NDI. There is a 50% chance that each child will have NDI.

PREDICT 6

Use your knowledge of kidney physiology and figure 26.18 to answer the following questions.

a. Why do the infants have high blood Na^+ levels and dilute urine?
b. Predict how the infants' plasma levels of ADH will change during the water deprivation test, given a diagnosis of NDI. (*Hint:* See "Formation of Concentrated Urine," p. 988.)
c. Why does an abnormal aquaporin-2 gene result in excessive urine production?
d. Predict plasma levels of ADH following a water deprivation test in an individual with central diabetes insipidus.
e. Why is treatment with a thiazide diuretic helpful to patients with NDI? (*Hint:* See Clinical Focus "Diuretics," p. 988)?

Formation of Dilute Urine

If ADH is not present or if its concentration is reduced, the distal convoluted tubules and collecting ducts have a low permeability to water. The amount of water moving by osmosis from the distal convoluted tubule and collecting duct, therefore, is low. The concentration of the urine produced is less than 1200 mOsm/kg, and the volume is increased. The volume of this more dilute urine can be much larger than 1% of the filtrate formed each day. If no ADH is secreted, the osmolality of the urine may be close to the osmolality of the filtrate in the distal convoluted tubule, and the volume of urine may approach 20–30 L/day, which is the same volume as 10–15 2-liter soda bottles per day (see figure 26.19).

In a healthy person, even when the kidney produces dilute urine, the concentration of waste products in the urine is large enough to maintain homeostasis in the body. Consequently, beneficial substances are retained, and both toxic substances and excess water are eliminated.

PREDICT 7

Amanda, an inexperienced runner, competed in her first marathon last spring in Phoenix, Arizona. During the run, the temperature reached 35°C (95°F) with 30% humidity. Amanda drank very little water during the race. When she finished the race 4½ hours later, she was dizzy and disoriented and had an increased heart rate. She was also very pale. She was taken to a hospital, where the doctor diagnosed severe dehydration and prescribed the administration of IV fluids. Amanda did not urinate until nearly 12 hours later. Explain the physiological responses that resulted in her reduced urine production. (*Hint:* See blood pressure regulation in chapter 21.)

34 ***Describe how the absence of ADH causes the formation of a large volume of dilute urine.***

Urine Concentration and Juxtamedullary Nephrons

Only the juxtamedullary nephrons have loops of Henle that descend deep into the medulla, but enough of them exist to maintain a high interstitial concentration of solutes in the interstitial fluid of the medulla. Not all of the nephrons need to have loops of Henle that descend into the medulla to concentrate urine effectively. The cortical nephrons function as the juxtamedullary nephrons do, with the exception that their loops of Henle are not as efficient at concentrating urine. Because the filtrate from the cortical nephrons passes through the collecting ducts, however, water can diffuse out of the collecting ducts into the interstitial fluid. Thus, the filtrate becomes concentrated. Animals that concentrate urine more effectively than humans have a greater percentage of nephrons that descend into the medulla of the kidney. For example, desert mammals have many nephrons that descend into the medulla of the kidney, and the renal pyramids are longer than in humans and most other mammals.

Other Hormones

A polypeptide hormone, called **atrial natriuretic** (nā′trē-ū-ret′ik) **hormone (ANH),** is secreted from cardiac muscle cells in the right atrium of the heart when blood volume in the right atrium increases and stretches the cardiac muscle cells (see chapter 21). Atrial natriuretic hormone inhibits Na^+ reabsorption in the kidney tubules. ANH also inhibits ADH secretion from the posterior pituitary gland. Consequently, increased ANH secretion increases the volume of urine produced and lowers blood volume and blood pressure. Atrial natriuretic hormone also dilates arteries and veins, which reduces peripheral resistance and lowers blood pressure. Thus, a decrease occurs in venous return and blood volume in the right atrium.

Two other hormone-like substances, prostaglandins and kinins, are formed in the kidneys and affect kidney function. Their roles are unclear, but both substances influence the rate of filtrate formation and Na^+ reabsorption. Prostaglandins probably moderate the sensitivity of the renal blood vessels to neural stimuli and to angiotensin II.

35 ***Where is atrial natriuretic hormone produced, and what effect does it have on urine production?***

PLASMA CLEARANCE AND TUBULAR MAXIMUM

Plasma clearance is a calculated value representing the volume of plasma that is cleared of a specific substance each minute. For example, if the clearance value is 100 mL/min for a substance, the substance is completely removed from 100 mL of plasma each minute.

Clearance

The plasma clearance can be calculated for any substance that enters the circulatory system according to the following formula:

$$\text{Plasma clearance (mL/min)} = \text{Quantity of urine (mL/min)} \times \frac{\text{Concentration of substance in urine}}{\text{Concentration of substance in plasma}}$$

Plasma clearance can be used to estimate GFR if the appropriate substance is monitored (see table 26.2). Such a substance must have the following characteristics: (1) It must pass through the filtration membrane of the renal corpuscle as freely as water or other small molecules, (2) it must not be reabsorbed, (3) it must not be secreted into the nephron, and (4) it must not be either metabolized or produced in the kidney. **Inulin** (in′ū-lin) is a nonphysiological polysaccharide that has these characteristics. As filtrate is formed, it has the same concentration of inulin as plasma; however, as the filtrate flows through the nephron, all the inulin remains in the nephron to enter the urine. As a consequence, all the volume of plasma that becomes filtrate is cleared of inulin, and the plasma clearance for inulin is equal to the rate of glomerular filtrate formation.

GFR is reduced when the kidney fails. Measurement of the GFR indicates the degree to which kidney damage has occurred. The clearance value for urea and creatinine can be used clinically. Because these substances are naturally occuring metabolites, foreign substances do not have to be injected. A high plasma concentration and a lower than normal clearance value for them indicates a reduced GFR and kidney failure. Creatinine clearance can be used to monitor the progress of changes in GFR in people who suffer from renal failure.

PREDICT 8

During surgery, a patient's blood pressure drops to very low levels, and ischemia of the kidney develops. Within 1 day following the surgery, the GFR and the urine volume decrease to very low levels. Given that the structure of the glomeruli did not dramatically change, but that the epithelium of nephrons suffered from ischemia and sloughed into the nephron to form casts of epithelial cells in the nephron, explain why the GFR was reduced.

Plasma clearance can also be used to calculate renal plasma flow (see table 26.2). Substances with the following characteristics, however, must be used: (1) The substance must pass through the filtration membrane of the renal corpuscle and (2) it must be secreted into the nephron at a sufficient rate so that very little of it remains in the blood as the blood leaves the kidney. A substance that meets these requirements is *para*-aminohippuric acid (PAH). As blood flows through the kidney, essentially all the PAH is either filtered or secreted into the nephron. The clearance calculation for PAH is therefore a good estimate of the volume of plasma flowing through the kidney each minute. Also, if the hematocrit is known, the total volume of blood flowing through the kidney each minute can be calculated easily.

The concept of plasma clearance can be used to make the measurements described previously, or it can be used to determine the means by which drugs or other substances are excreted by the kidney. A plasma clearance value greater than the inulin clearance value suggests that the substance is secreted by the nephron into the filtrate.

PREDICT 9

A person is suspected of suffering from chronic renal failure. To assess kidney function, urea clearance is measured and found to be very low. Explain what a very low urea clearance indicates in a person suffering from chronic renal failure.

The **tubular load** of a substance is the total amount of the substance that passes through the filtration membrane into the nephrons each minute. Normally, glucose is almost completely reabsorbed from the nephron by the process of active transport. The nephron's capacity to actively transport glucose across the epithelium of the nephron is limited, however. If the tubular load is greater than the nephron's capacity to reabsorb it, the excess glucose remains in the urine.

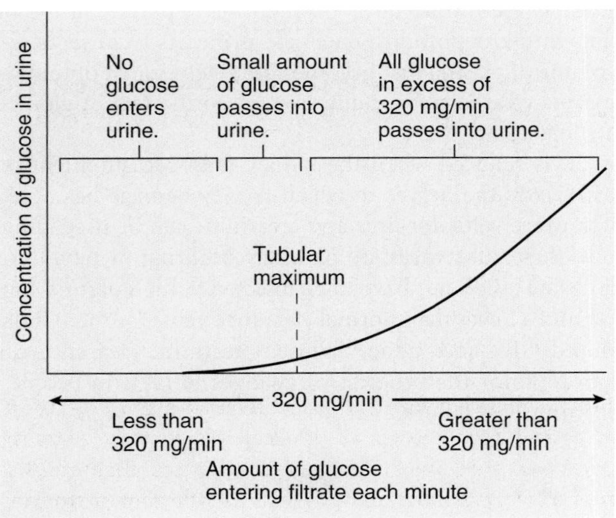

FIGURE 26.20 Tubular Maximum for Glucose

As the concentration of glucose increases in the filtrate, it reaches a point that exceeds the nephron's ability to actively reabsorb it. That concentration is called the tubular maximum. Beyond that concentration, the excess glucose enters the urine.

The maximum rate at which a substance can be actively reabsorbed is called the **tubular maximum** (figure 26.20). Each substance that is reabsorbed has its own tubular maximum, determined by the number of active transport carrier proteins and the rate at which they are able to transport molecules of the substance. For example, in people suffering from diabetes mellitus, the tubular load for glucose can exceed the tubular maximum by a substantial amount, thus allowing glucose to appear in the urine. Urine volume is also greater than normal because the glucose molecules in the filtrate increase the osmolality of the filtrate in the nephron and reduce the effectiveness of water reabsorption by osmosis.

36 ***Define and explain the significance of plasma clearance, tubular load, and tubular maximum.***

URINE MOVEMENT

Anatomy and Histology of the Ureters and Urinary Bladder

The **ureters** are tubes through which urine flows from the kidneys to the urinary bladder. The ureters extend inferiorly and medially from the renal pelvis at the renal hilum of each kidney to reach the urinary bladder (see figures 26.1 and 26.2; figure 26.21), which stores urine. The **urinary bladder** is a hollow, muscular container that lies in the pelvic cavity just posterior to the symphysis pubis. The ureters enter on its posterolateral surface. In males, the urinary bladder is just anterior to the rectum; in females, it is just anterior to the vagina and inferior and anterior to the uterus. Its volume increases and decreases depending on how much or how little urine is stored in it.

The urethra, which transports urine to the outside of the body, exits the urinary bladder inferiorly and anteriorly (see figure 26.21*a*). The triangular area of its wall between the two ureters posteriorly and the urethra anteriorly is called the **trigone** (trī′gōn). This region differs histologically from the rest of the urinary bladder wall and expands minimally during filling.

Transitional epithelium lines both the ureters and the urinary bladder. The rest of the walls of these structures consists of a lamina propria, a muscular coat, and a fibrous adventitia (see figure 26.21*b* and *c*). The wall of the urinary bladder is much thicker than the wall of a ureter. This thickness is caused by layers composed primarily of smooth muscle, sometimes called the **detrusor** (dē-troo′ser) **muscle,** external to the epithelium. Contraction of the urinary bladder smooth muscle forces urine out of the urinary bladder. The epithelium itself ranges from four or five cells thick when the urinary bladder is empty to two or

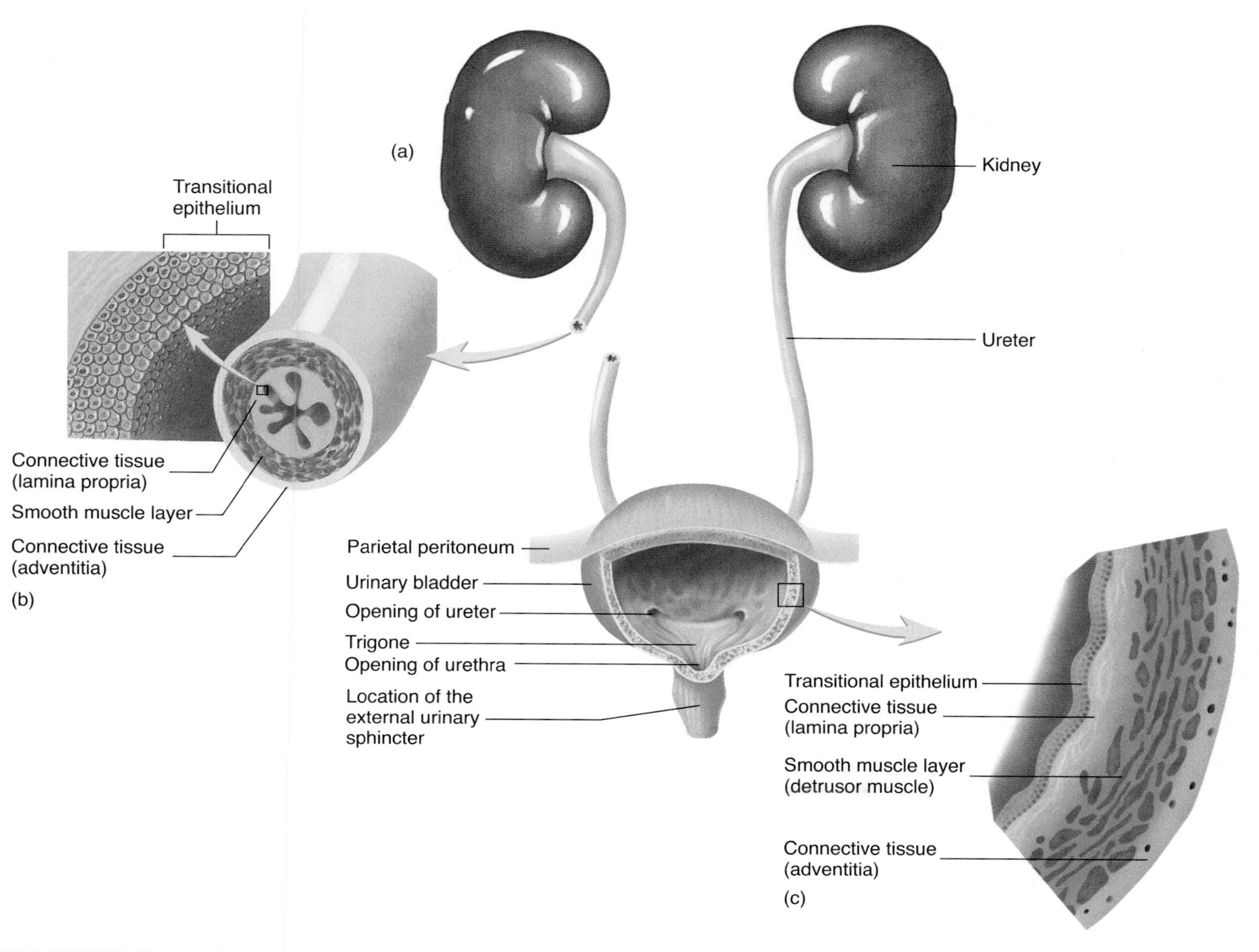

FIGURE 26.21 Ureters and Urinary Bladder

(*a*) Ureters extend from the pelvis of the kidney to the urinary bladder. (*b*) The walls of the ureters and the urinary bladder are lined with transitional epithelium, which is surrounded by a connective tissue layer (lamina propria), smooth muscle layers, and a fibrous adventitia. (*c*) Section through the wall of the urinary bladder.

three cells thick when it is distended. Transitional epithelium is specialized so that the cells slide past one another, and the number of cell layers decreases as the volume of the urinary bladder increases. The epithelium of the urethra is stratified or pseudostratified columnar epithelium.

Where the urethra exits the urinary bladder, elastic connective tissue and smooth muscle keep urine from flowing out of the urinary bladder until the pressure in the urinary bladder is great enough to force urine to flow from it. The elastic tissue and smooth muscle forms an **internal urinary sphincter** in males. There is no functional internal urinary sphincter in females. In males, the internal urinary sphincter contracts to keep semen from entering the urinary bladder during sexual intercourse (see chapter 28). The **external urinary sphincter** is skeletal muscle that surrounds the urethra as the urethra extends through the pelvic floor. It acts as a valve that controls the flow of urine through the urethra.

In males, the urethra extends to the end of the penis, where it opens to the outside (see chapter 28). The urethra is much shorter in females than in males, and it opens into the vestibule anterior to the vaginal opening.

Urinary Bladder Cancer

In the United States, urinary bladder cancer affects more than 60,000 new patients each year and is among the 10 most common cancers in men and women. However, with early detection (when the cancer is confined to the urinary bladder) the chance for survival is 94%, compared with 6% if the cancer has spread to other areas of the body. Unfortunately, early detection of urinary bladder cancer is especially challenging due to its rapid growth rate. Frequently, blood in the urine is a symptom but, because this symptom is also associated with other, less serious problems, it tends to be ignored.

Half the diagnosed cases of urinary bladder cancer can be attributed to cigarette smoking, even 10 years or more after cessation of smoking. Other risk factors include exposure to industrial aniline dyes and two prescription drugs, phenacetin (an analgesic) and chlornaphazine (a drug used to treat polycythemia, see chapter 19).

Scientists are investigating ways to detect urinary bladder cancer early and noninvasively. Currently, urinary bladder cancer tests screen for abnormal cells; and the bladder is visually examined with a catheter in a process called **cystoscopy** (sis-tos'kŏ-pē). New tests that hold promise include a urine test that measures the levels of an enzyme called telomerase, present in nearly all human cancer cells. Measurements of telomerase levels are especially promising for the detection of urinary bladder cancer because telomerase levels are measurable earlier in urinary bladder cancer than in many other cancers.

37 ***What are the functions of the ureters, urethra, and urinary bladder? Describe their structure, including the epithelial lining of their inner surfaces. What is the trigone?***

PREDICT 10

Cystitis (sis-tī'tis) is inflammation of the urinary bladder. It typically results from infections that often occur when bacteria from outside the body enter the bladder. Are males or females more prone to cystitis? Explain.

Urine Flow Through the Nephron and Ureters

Hydrostatic pressure averages 10 mm Hg in Bowman's capsule and nearly 0 mm Hg in the renal pelvis. This pressure gradient forces the filtrate to flow from Bowman's capsule through the nephron into the renal pelvis. Because the pressure is 0 mm Hg in the renal pelvis, no pressure gradient exists to force urine to flow to the urinary bladder through the ureters. The circular smooth muscle in the walls of the ureters exhibits peristaltic contractions, which force urine to flow through the ureters. The peristaltic waves progress from the region of the renal pelvis to the urinary bladder. They occur from once every few seconds to once every 2–3 minutes. Parasympathetic stimulation increases their frequency, and sympathetic stimulation decreases it.

The peristaltic contractions of each ureter proceed at a velocity of approximately 3 cm/s and can generate pressures in excess of 50 mm Hg. Where the ureters penetrate the urinary bladder, they course obliquely through the trigone. Pressure inside the urinary bladder compresses that part of the ureter to prevent the backflow of urine.

When no urine is present in the urinary bladder, internal pressure is about 0 mm Hg. When the volume is 100 mL of urine, pressure is elevated to only 10 mm Hg. Pressure increases slowly as its volume increases to approximately 300 mL, but above urinary bladder volumes of 400 mL the pressure rises rapidly.

Kidney Stones

Kidney stones are hard objects usually found in the pelvis of the kidney. They are normally 2–3 mm in diameter, with either a smooth or a jagged surface, but occasionally a large, branching kidney stone, called a **staghorn stone,** forms in the renal pelvis. About 1% of all autopsies reveal kidney stones, and many of the stones occur without causing symptoms. The symptoms associated with kidney stones occur when a stone passes into the ureter, resulting in referred pain down the back, side, and groin area. The ureter contracts around the stone, causing the stone to irritate the epithelium and produce bleeding, which appears as blood in the urine, a condition called **hematuria.** In addition to causing intense pain, kidney stones can block the ureter, cause ulceration in the ureter, and increase the probability of bacterial infections.

About 65% of all kidney stones are composed of calcium oxylate mixed with calcium phosphate, 15% are magnesium ammonium phosphate, and 10% are uric acid or cystine; approximately 2.5% of each kidney stone is composed of mucoprotein.

The cause of kidney stones is usually obscure. Predisposing conditions include a concentrated urine and an abnormally high calcium concentration in the urine, although the cause of the high calcium concentration is usually unknown. Magnesium ammonium phosphate stones are often found in people with recurrent kidney infections, and uric acid stones often occur in people suffering from gout. Severe kidney stones must be removed surgically. Lithotripsy (lith'ō-trip-sē), the use of instruments that pulverize kidney stones with ultrasound or lasers, however, has replaced most traditional surgical procedures.

Micturition Reflex

The flow of urine from the kidney to the urinary bladder through the ureter is relatively continuous. The urinary bladder acts as a reservoir for urine until it can be eliminated relatively quickly at an appropriate time and place. The urinary bladder can distend to accommodate a large volume of fluid; at its maximum volume, the urinary bladder can contain 1 L (about 1 quart) of urine, but discomfort becomes noticeable when urine volume exceeds approximately 500 mL. The urinary bladder's capacity to distend is due to three factors. First, the wall of the urinary bladder contains large folds, similar to those of the stomach, which unfold to enlarge the lumen of the urinary bladder. Second, the lining of the urinary bladder is transitional epithelium, which stretches. Third, the smooth muscle wall of the urinary bladder, with the exception of the trigone, also stretches to accommodate fluid. As urine enters

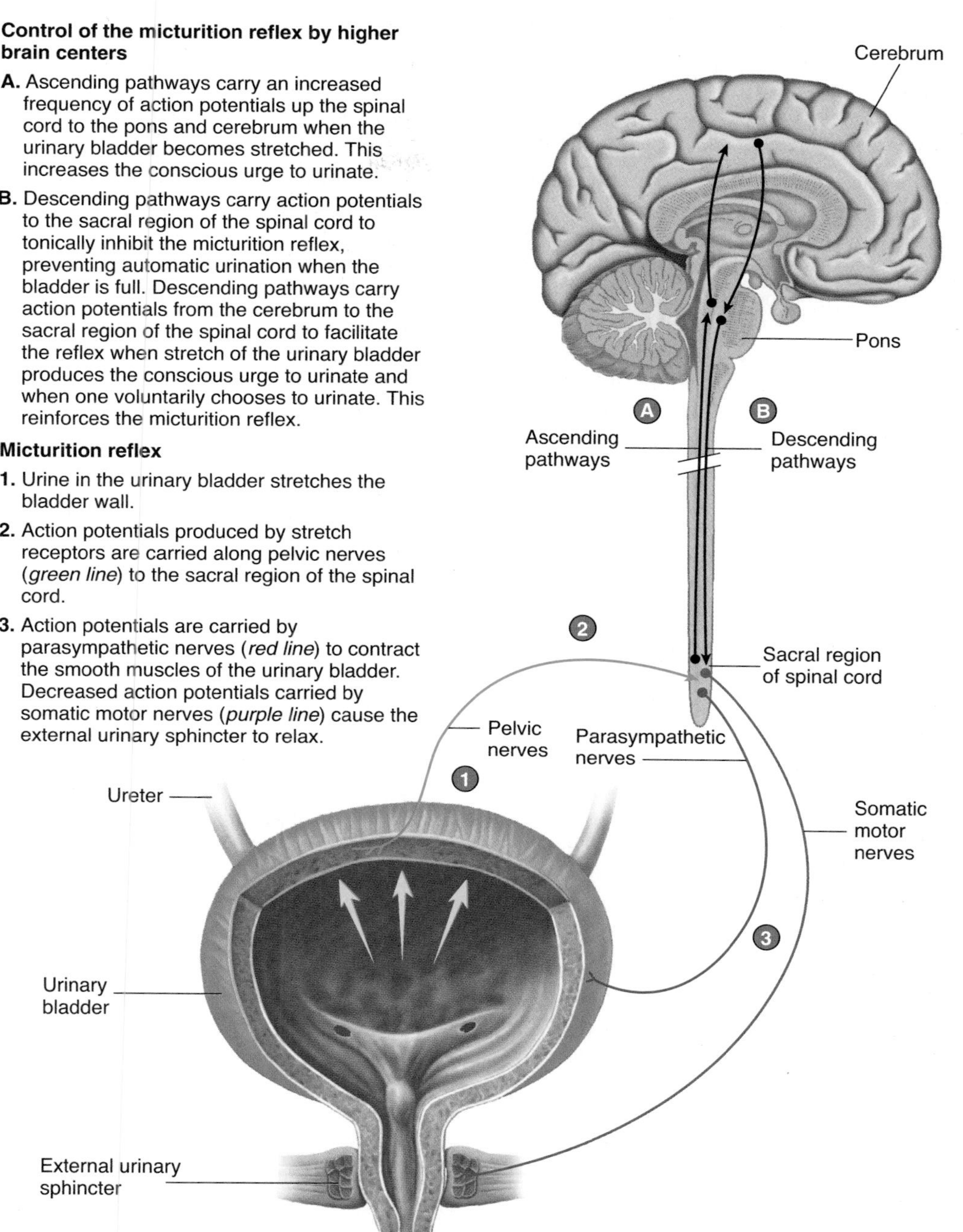

PROCESS FIGURE 26.22 Micturition Reflex

the urinary bladder, the urinary bladder lifts and expands superiorly to accommodate the fluid. Because of the changes that occur in the transitional epithelium, the smooth muscle, and connective tissue of the urinary bladder wall, the internal pressure of the urinary bladder does not increase at the same rate as fluid filling. Pressure in the urinary bladder remains low until its volume exceeds approximately 300 mL.

The **micturition** (mik-choo-rish′un) **reflex** is activated when the urinary bladder wall is stretched, resulting in **micturition,** the elimination of urine from the urinary bladder. Integration of the micturition reflex occurs in the sacral region of the spinal cord and is modified by centers in the pons and cerebrum.

Urine filling the urinary bladder stimulates stretch receptors, which produce action potentials. The action potentials are carried by sensory neurons to the sacral segments of the spinal cord through the pelvic nerves. In response, action potentials are carried to the urinary bladder through parasympathetic fibers in the pelvic nerves (figure 26.22). The parasympathetic action potentials cause the smooth muscle of the urinary bladder (the detrusor muscle) to contract. In addition, decreased somatic motor action potentials cause the external urinary sphincter, which consists of skeletal muscle, to relax. Urine flows from the urinary bladder when the pressure there is great enough to force urine to flow through the urethra while the external urinary sphincter is relaxed. The micturition reflex normally produces a series of contractions of the urinary bladder.

Action potentials carried by sensory neurons from stretch receptors in the urinary bladder wall also ascend the spinal cord to a micturition center in the pons and to the cerebrum. Descending action potentials are sent from these areas of the brain to the sacral region of the spinal cord, where they modify the activity of the micturition reflex in the spinal cord. The micturition reflex, integrated in the spinal cord, predominates in infants. The ability to inhibit micturition voluntarily develops at the age of 2–3 years; subsequently, the influence of the pons and cerebrum on the spinal micturition reflex predominates. The micturition reflex integrated in the spinal cord is automatic, but it is either stimulated or inhibited by descending action potentials. Higher brain centers prevent micturition by sending action potentials from the cerebrum and pons through spinal pathways to inhibit the spinal micturition reflex. Consequently, parasympathetic stimulation of the urinary bladder is inhibited and somatic motor neurons that keep the external urinary sphincter contracted are stimulated.

The slow increase in pressure helps explain why there is little urge to urinate when the urinary bladder contains less than 300 mL. The pressure in the urinary bladder increases rapidly once its volume exceeds approximately 400 mL, and there is an increase in the frequency of action potentials carried by sensory neurons. The increased frequency of action potentials conducted by the ascending spinal pathways to the pons and cerebrum results in an increased urge to urinate.

The voluntary initiation of micturition involves an increase in action potentials sent from the cerebrum to facilitate the micturition reflex and to voluntarily relax the external urinary sphincter. In addition to facilitating the micturition reflex, there is an increased voluntary contraction of abdominal muscles, which causes an increase in abdominal pressure. This enhances the micturition reflex by increasing the pressure applied to the urinary bladder wall.

The urge to urinate normally results from stretch of the urinary bladder wall, but irritation of the urinary bladder or the urethra by bacterial infections or other conditions can also initiate the urge to urinate, even if the urinary bladder is nearly empty.

38 ***What is responsible for the movement of urine through the nephron and ureters?***

39 ***Describe the micturition reflex. How is voluntary control of micturition accomplished?***

Automatic, Noncontracting, and Hyperexcitable Urinary Bladder

If the spinal cord is damaged above the sacral region, no micturition reflex exists for a time; however, if the urinary bladder is emptied frequently, the micturition reflex eventually becomes adequate to cause it to empty. Some time is generally required for the micturition reflex integrated within the spinal cord to begin to operate. A typical micturition reflex can exist, but there is no conscious control over its onset or duration. This condition is called the **automatic bladder.**

Damage to the sacral region of the spinal cord or to the nerves that carry action potentials between the spinal cord and the urinary bladder can result in failure of the urinary bladder to contract although the external urinary sphincter is relaxed. As a result, the micturition reflex cannot occur. The bladder fills to capacity, and urine is forced in a slow dribble through the external urinary sphincter. In elderly people or in patients with damage to the brainstem or spinal cord, a loss of inhibitory action potentials to the sacral region of the spinal cord can occur. Without inhibition, the sacral centers are hyperexcitable, and even a small amount of urine in the bladder can elicit an uncontrollable micturition reflex.

EFFECTS OF AGING ON THE KIDNEYS

Aging causes a gradual decrease in the size of the kidneys, beginning as early as age 20, becoming obvious by age 50, and continuing until death. The loss of kidney size appears to be related to changes in the blood vessels of the kidney. The amount of blood flowing through the kidneys gradually decreases. Starting at age 20, there appears to be an approximately 10% decrease every 10 years. Small arteries, including the afferent and efferent arterioles, become irregular and twisted. Functional glomeruli are lost. By age 80, 40% of the glomeruli are not functioning. About 30% of the glomeruli that stop functioning no longer have a lumen through which blood flows. Other glomeruli thicken and assume a structure similar to that of arterioles. Some nephrons and collecting ducts become thicker, shorter, and more irregular in structure. The capacity to secrete and absorb declines, and whole nephrons stop functioning. The kidney's ability to concentrate urine gradually declines. Eventually, changes in the kidney increase the risk for dehydration because of the kidney's reduced ability to produce a concentrated urine. There is also a decreased ability to eliminate uric acid, urea, creatine, and toxins from the blood.

An age-related loss of responsiveness to ADH and to aldosterone occurs. The kidney decreases renin secretion. A reduced

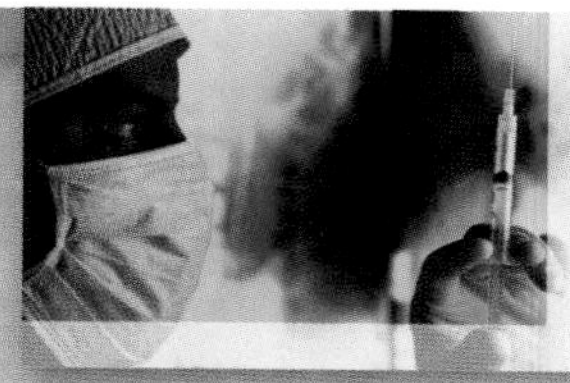

CLINICAL FOCUS

Renal Pathologies

Glomerular nephritis (glō-mār′ū-lăr ne-frī′tis) results from inflammation of the filtration membrane within the renal corpuscle. It is characterized by an increased permeability of the filtration membrane and the accumulation of numerous white blood cells in the area. As a consequence, a high concentration of plasma proteins enters the filtrate, along with numerous white blood cells. A greater than normal urine volume accompanies the increase in plasma proteins in the urine.

Acute glomerular nephritis often occurs 1–3 weeks after a severe bacterial infection, such as streptococcal sore throat or scarlet fever. Antigen–antibody complexes associated with the disease become deposited in the filtration membrane and cause its inflammation. This acute inflammation normally subsides after several days.

Chronic glomerular nephritis is long-term and usually progressive. The filtration membrane thickens and eventually is replaced by connective tissue. Although in the early stages chronic glomerular nephritis resembles the acute form, in the advanced stages many of the renal corpuscles have been replaced by fibrous connective tissue, and the kidney eventually ceases to function.

Pyelonephritis (pī′ĕ-lō-ne-frī′tis) is inflammation of the renal pelvis, medulla, and cortex. It often begins as a bacterial infection of the renal pelvis and then extends into the kidney itself. It can result from several types of bacteria, including *Escherichia coli.* Pyelonephritis can destroy nephrons and renal corpuscles, but, because the infection starts in the pelvis of the kidney, it affects the medulla more than the cortex. As a consequence, the kidney's ability to concentrate urine is dramatically affected.

Renal failure can result from any condition that interferes with kidney function. **Acute renal failure** occurs when kidney damage is extensive and leads to the accumulation of urea in the blood and to acidosis (see chapter 27). In complete renal failure, death can occur in 1–2 weeks. Acute renal failure can result from acute glomerular nephritis, or it can be caused by damage to or blockage of the renal tubules. Some poisons, such as mercuric ions or carbon tetrachloride, which are common to certain industrial processes, cause necrosis of the nephron epithelium. If the damage does not interrupt the basement membrane surrounding the nephrons, extensive regeneration can occur within 2–3 weeks. Severe ischemia associated with circulatory shock resulting from sympathetic vasoconstriction of the renal blood vessels can cause necrosis of the epithelial cells of the nephron.

Chronic renal failure results when so many nephrons are permanently damaged that the nephrons that remain functional cannot adequately compensate. Chronic renal failure can result from chronic glomerular nephritis, trauma to the kidneys, the absence of kidney tissue caused by congenital abnormalities, or tumors. Urinary tract obstruction by kidney stones, damage resulting from pyelonephritis, and severe arteriosclerosis of the renal arteries also cause degeneration of the kidney.

In chronic renal failure, the GFR is dramatically reduced, and the kidney is unable to excrete excess excretory products, including electrolytes and metabolic waste products. The accumulation of solutes in the body fluids causes water retention and edema. Potassium levels in the extracellular fluid are elevated, and acidosis occurs because the distal convoluted tubules and collecting ducts cannot excrete sufficient quantities of K^+ and H^+. Acidosis, elevated potassium levels in the body fluids, and the toxic effects of metabolic waste products cause mental confusion, coma, and finally death when chronic renal failure is severe.

ability to participate in vitamin D synthesis occurs, which contributes to Ca^{2+} deficiency, osteoporosis, and bone fractures.

Recall that one-third of one kidney is required to maintain homeostasis. The additional kidney tissue beyond this constitutes a reserve capacity. The age-related changes in the kidney cause a reduction in the kidney's reserve capacity. As the functional kidney mass is reduced substantially in older people, high blood pressure, atherosclerosis, and diabetes have greater adverse effects.

40 ***Describe the effect of aging on the kidneys. Why do the kidneys gradually decrease in size?***

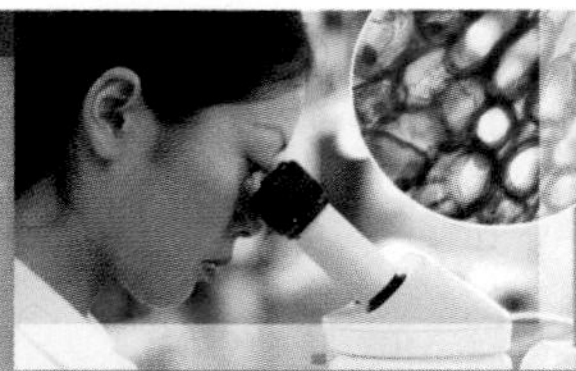

SYSTEMS PATHOLOGY

Acute Renal Failure

A large piece of machinery overturned at the construction site where Hugh worked, trapping him beneath it. His legs were severely crushed, although they healed after several months. Hugh nearly lost his life, however, because of the acute renal failure that developed as a result of his injury. Hugh was trapped for several hours in a very difficult place to reach. During that time, his blood pressure decreased to very low levels because of the blood loss, the edema in the inflamed tissues, and emotional shock. After he was rescued, he received fluid replacement in the form of both intravenous saline solutions and blood transfusions, and his blood pressure returned to its normal range. Twenty-four hours after the accident, however, his urine volume began to decrease. His urinary Na^+ concentration increased, his urine osmolality decreased, his urine specific gravity decreased, and cellular debris was evident in his urine.

For approximately 7 days, Hugh exhibited oliguria (reduced urine production). During this period, renal dialysis was required to maintain Hugh's blood volume and electrolyte concentrations within normal ranges. After approximately 7 days, his kidneys gradually began to produce large quantities of urine (a diuretic phase). Careful observation was required to keep his blood pressure and electrolyte concentrations within normal ranges. Substantial water, Na^+, and K^+ had to be administered to him. After about 3 weeks, the function of his kidneys slowly began to improve, although many months passed before his kidney function returned to normal.

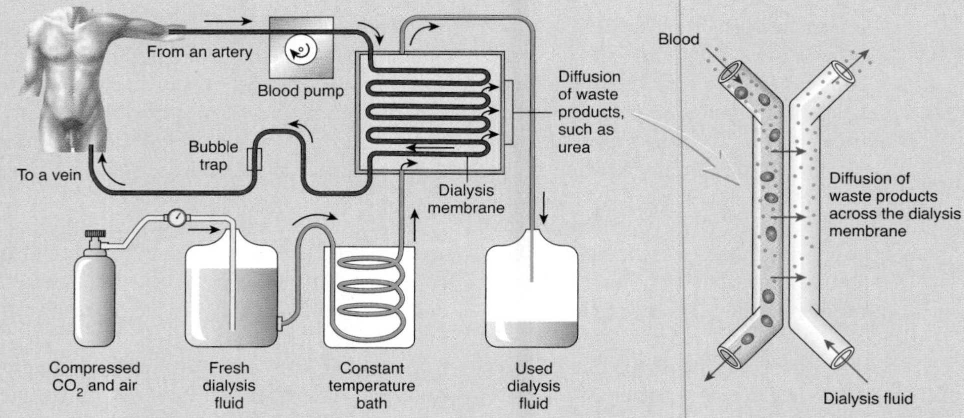

FIGURE A Kidney Dialysis

During kidney dialysis, blood flows through a system of tubes composed of a selectively permeable membrane. Dialysis fluid, the composition of which is similar to that of blood, except that the concentration of waste products is very low, flows in the opposite direction on the outside of the dialysis tubes. Consequently, waste products, such as urea, diffuse from the blood into the dialysis fluid. Other substances, such as sodium, potassium, and glucose, do not rapidly diffuse from the blood into the dialysis fluid because there is no concentration gradient for these substances between the blood and the dialysis fluid.

BACKGROUND INFORMATION

The events after 24 hours were consistent with acute renal failure caused by prolonged hypotension and ischemia of the kidney. While Hugh was suffering from hypotension, blood flow to his kidneys was very low. The reduced blood flow was severe enough to result in damage to the epithelial lining of the kidney tubules (tubular necrosis). The period of reduced urine volume resulted from tubular damage. Renal ischemia results in the necrosis of tubular cells, which then slough off into the tubules and block them so that filtrate cannot flow through the tubules. In addition, the filtrate leaks from the blocked or partially blocked tubules back into the interstitial spaces and therefore back into the circulatory system. As a result, the amount of filtrate that becomes urine is markedly reduced. Blood levels of urea and of creatine increase because of the reduction in filtrate formation and reduced function of the tubular epithelium. The kidneys' ability to eliminate metabolic waste products is therefore reduced. The small amount of urine produced has a high Na^+ concentration but an osmolality that is close to the concentration of the body fluids because the kidneys are not able to reabsorb Na^+ and because the kidneys' urine-concentrating ability is severely damaged.

Renal dialysis allows blood to flow though tubes made of a selectively permeable membrane. On the outside of the dialysis tubes is a fluid that contains the same concentration of solutes as the plasma, except for the metabolic waste products. As a consequence, a diffusion gradient exists for the metabolic waste products from the blood to the dialysis fluid. The dialysis membrane has pores that are too small to allow plasma proteins to pass through them. For smaller solutes, the dialysis fluid contains the same beneficial solutes as the plasma, so the net movement of these substances is zero. In contrast, the dialysis fluid contains no metabolic waste products, so metabolic waste products diffuse rapidly from the blood into the dialysis fluid.

Blood usually is taken from an artery, passed through tubes of the dialysis machine, and then returned to a vein. The rate of blood flow is normally several hundred milliliters per minute, and the total surface area of exchange in the machine is close to 10,000–20,000 cm^2 (figure A). Kidney dialysis is not convenient for those suffering from kidney failure, and it can be emotionally difficult. Clearly, kidney dialysis is not a good substitute for healthy kidneys.

During the diuretic phase, Hugh began to produce large quantities of urine because the nephrons were partially healed and could produce urine, but the nephrons' ability to concentrate urine was not yet normal. Large volumes of urine that contained significant amounts of Na^+ and K^+ were therefore produced. The kidneys were able to produce urine that was more concentrated than the body fluids, but the kidneys' concentrating ability was still below normal. As time passed, the kidneys' concentrating ability improved and eventually became normal once again.

PREDICT 11

Nine days after the accident, Hugh began to appear pale, and he became dizzy and lethargic. His hematocrit was elevated and his heart was arrhythmic. He was very weak. Explain these manifestations.

SYSTEM INTERACTIONS Effect of Acute Renal Failure on Other Systems

SYSTEM	INTERACTIONS
Integumentary	Pallor results from anemia, and bruising results from reduced clotting proteins in the blood because they are lost in the urine. A waxy, yellow cast develops to the skin of light-skinned people, an ashen gray cast in black-skinned people, and a yellowish brown cast in brown-skinned people due to an accumulation of urinary pigments. When the urea concentration in the blood is very high, white crystals of urea, called uremic frost, may appear on areas of the skin where heavy perspiration occurs.
Skeletal	Changes in the skeletal system are not marked unless kidney damage results in chronic kidney failure. Bone resorption may result during a prolonged diuretic phase because of excessive loss of Ca^{2+} in the urine. Also, vitamin D levels may be reduced during both the oliguric and diuretic phases.
Muscular	Neuromuscular irritability results from the toxic effect of metabolic wastes on the central nervous system and ionic imbalances, such as hyperkalemia. Involuntary jerking and twitching may occur as neuromuscular irritability develops. Tremors of the hands are an indication of the toxic effects of metabolic wastes on the cerebrum.
Nervous	Elevated blood K^+ levels and the toxic effects of metabolic wastes result in the depolarization of neurons. Slowing of nerve conduction, burning sensations, pain, numbness, or tingling results. Also, decreased mental acuity, reduced ability to concentrate, apathy, and lethargy result. Periods of lethargy may alternate with restlessness and insomnia. In severe cases, the patient may become confused and comatose.
Endocrine	Major predictable hormone deficiencies include vitamin D deficiency. In addition, a decrease occurs in the secretion of reproductive hormones due to the effects of metabolic wastes and ionic imbalances on the hypothalamus.
Cardiovascular	Water and Na^+ retention may result in edema in peripheral tissues and in the lung. Also, hypertension and congestive heart failure may result. Hyperkalemia results in dysrhythmias and may cause cardiac arrest. Anemia due to decreased erythropoietin production by the damaged kidney and decreased half-life of red blood cells may result. Anemia is more likely because of the blood lost as a result of the crushing injury. Nosebleeds and bruising occur due to a reduced concentration of clottting factors because they are lost in the urine.
Lymphatic	No major direct effects occur to the lymphatic system, with the exception that increased lymph flow happens as a result of edema.
Respiratory	Early during acute renal failure, the depth of breathing increases, and it becomes labored as acidosis develops because the kidney is not able to secrete H^+. Pulmonary edema often develops because of water and Na^+ retention. The likelihood of pulmonary infection increases secondary to pulmonary edema.
Digestive	Anorexia, nausea, and vomiting result from altered gastrointestinal functions due to the effects of ionic imbalances on the nervous system. An odor of ammonia may occur on the breath and a metallic taste in the mouth. These effects are the result of the accumulation of metabolic waste products in the gastrointestinal tract and the action of the normal gastrointestinal microorganisms on the waste products, which convert urea to ammonia. The ammonia and other metabolic waste products predispose the mouth to inflammation and infection.

SUMMARY

The urinary system consists of the kidneys, ureters, urinary bladder, and urethra.

Functions of the Urinary System (p. 962)

The urinary system eliminates wastes; regulates blood volume, ion concentration, and pH; and is involved with red blood cell and vitamin D production.

Kidney Anatomy and Histology (p. 962)

The kidney is surrounded by a renal capsule and perirenal fat and is held in place by the renal fascia.

Location and External Anatomy of the Kidneys

1. A kidney lies behind the peritoneum on the posterior abdominal wall on each side of the vertebral column.
2. The renal capsule surrounds each kidney, and the perirenal fat and the renal fascia surround each kidney and anchor it to the abdominal wall.
3. The hilum, on the medial side of each kidney, where blood vessels and nerves enter and exit the kidney, opens into the renal sinus, containing fat and connective tissue.

Internal Anatomy and Histology of the Kidneys

1. The two layers of the kidney are the cortex and the medulla.
   - The renal columns extend toward the medulla between the renal pyramids.
   - The renal pyramids of the medulla project to the minor calyces.
2. The minor calyces open into the major calyces, which open into the renal pelvis. The renal pelvis leads to the ureter.
3. The functional unit of the kidney is the nephron. The parts of a nephron are the renal corpuscle, the proximal convoluted tubule, the loop of Henle, and the distal convoluted tubule.
   - The renal corpuscle is Bowman's capsule and the glomerulus. Materials leave the blood in the glomerulus and enter Bowman's capsule through the filtration membrane.
   - The nephron empties through the distal convoluted tubule into a collecting duct.
4. The juxtaglomerular apparatus consists of the macula densa (part of the distal convoluted tubule) and the juxtaglomerular cells of the afferent arteriole.

Arteries and Veins of the Kidneys

1. Arteries branch as follows: renal artery to segmental artery to interlobar artery to arcuate artery to interlobular artery to afferent arteriole.
2. Afferent arterioles supply the glomeruli.
3. Efferent arteries from the glomeruli supply the peritubular capillaries and vasa recta.
4. Veins form from the peritubular capillaries as follows: interlobular vein to arcuate vein to interlobar vein to renal vein.

Urine Production (p. 970)

Urine is produced by the processes of filtration, tubular reabsorption, and tubular secretion.

Filtration

1. The renal filtrate is plasma minus blood cells and blood proteins. Most (99%) of the filtrate is reabsorbed.
2. The filtration membrane is fenestrated endothelium, basement membrane, and the slitlike pores formed by podocytes.
3. Filtration pressure is responsible for filtrate formation.
   - Filtration pressure is glomerular capillary pressure minus capsule pressure minus blood colloid osmotic pressure.
   - Filtration pressure changes are primarily caused by changes in glomerular capillary pressure.

Regulation of Glomerular Filtration Rate

Two important mechanisms regulating GFR are autoregulation and sympathetic stimulation.

Autoregulation

Autoregulation dampens systemic blood pressure changes by altering afferent arteriole diameter.

Sympathetic Stimulation

Sympathetic stimulation decreases afferent arteriole diameter.

Tubular Reabsorption

1. Filtrate is reabsorbed by passive transport, including simple diffusion, facilitated diffusion, active transport, and symport from the nephron into the peritubular capillaries.
2. Specialization of tubule segments
   - The thin segment of the loop of Henle is specialized for passive transport.
   - The rest of the nephron and collecting ducts perform active transport, symport, and passive transport.
3. Substances transported
   - Active transport moves mainly Na^+ across the wall of the nephron. Other ions and molecules are moved primarily by symport.
   - Passive transport moves water, urea, and lipid-soluble, nonpolar compounds.

Tubular Secretion

1. Substances enter the proximal or distal convoluted tubules and the collecting ducts.
2. Hydrogen ions, K^+, and some substances not produced in the body are secreted by antiport mechanisms.

Urine Concentration Mechanism

1. The vasa recta, the loop of Henle, and the distribution of urea are responsible for the concentration gradient in the medulla. The concentration gradient is necessary for the production of concentrated urine.
2. Production of urine
   - In the proximal convoluted tubule, Na^+ and other substances are removed by active transport. Water follows passively, filtrate volume is reduced 65%, and the filtrate concentration is 300 mOsm/L.
   - In the descending limb of the loop of Henle, water exits passively and solute enters. The filtrate volume is reduced 15%, and the osmolality of the filtrate concentration is 1200 mOsm/kg.

- In the ascending limb of the loop of Henle, Na^+, Cl^-, and K^+ are transported out of the filtrate, but water remains because this segment of the nephron is impermeable to water. The osmolality of the filtrate concentration is 100 mOsm/kg.

Regulation of Urine Concentration and Volume (p. 983)

Hormonal Mechanisms

1. Aldosterone is produced in the adrenal cortex and affects Na^+ and Cl^- transport in the nephron and collecting ducts.
   - A decrease in aldosterone results in less Na^+ reabsorption and an increase in urine concentration and volume. An increase in aldosterone results in greater Na^+ reabsorption and a decrease in urine concentration and volume.
   - Aldosterone production is stimulated by angiotensin II, increased blood K^+ concentration, and decreased blood Na^+ concentration.
2. Renin, produced by the kidneys, causes the production of angiotensin II.
   - Angiotensin II acts as a vasoconstrictor and stimulates aldosterone secretion, causing a decrease in urine production and an increase in blood volume.
   - Decreased blood pressure or decreased Na^+ concentration stimulates renin production.
3. ADH is secreted by the posterior pituitary and increases water permeability in the distal convoluted tubules and collecting ducts.
   - ADH decreases urine volume, increases blood volume, and thus increases blood pressure.
   - ADH release is stimulated by increased blood osmolality or a decrease in blood pressure.
   - Water movement out of the distal convoluted tubules and collecting ducts is regulated by ADH. If ADH is absent, water is not reabsorbed and a dilute urine is produced. If ADH is present, water moves out and a concentrated urine is produced.
4. Atrial natriuretic hormone, produced by the heart when blood pressure increases, inhibits ADH production and reduces the kidney's ability to concentrate urine.

Plasma Clearance and Tubular Maximum (p. 991)

1. Plasma clearance is the volume of plasma that is cleared of a specific substance each minute.
2. Tubular load is the total amount of substance that enters the nephron each minute.
3. Tubular maximum is the fastest rate at which a substance is reabsorbed from the nephron.

Urine Movement (p. 992)

Anatomy and Histology of the Ureters and Urinary Bladder

1. Structure
   - The walls of the ureter and urinary bladder consist of the epithelium, the lamina propria, a muscular coat, and a fibrous adventitia.
   - The transitional epithelium permits changes in size.
2. Function
   - The ureters transport urine from the kidney to the urinary bladder.
   - The urinary bladder stores urine.

Urine Flow Through the Nephron and Ureters

1. Hydrostatic pressure forces urine through the nephron.
2. Peristalsis moves urine through the ureters.

Micturition Reflex

1. Stretch of the urinary bladder stimulates a reflex that causes the urinary bladder to contract and inhibits the urinary sphincters.
2. Higher brain centers can stimulate or inhibit the micturition reflex.

Effects of Aging on the Kidneys (p. 996)

1. There is a gradual decrease in the size of the kidney.
2. The decrease in kidney size is associated with a decrease in renal blood flow.
3. The number of functional nephrons decreases.
4. Renin secretion and vitamin D synthesis decrease.
5. The nephron's ability to secrete and absorb declines.

REVIEW AND COMPREHENSION

1. Which of these is *not* a general function of the kidneys?
   a. regulation of blood volume
   b. regulation of solute concentration in the blood
   c. regulation of the pH of the extracellular fluid
   d. regulation of vitamin A synthesis
   e. regulation of red blood cell synthesis
2. The cortex of the kidney contains the
   a. hilus.
   b. glomeruli.
   c. perirenal fat.
   d. renal pyramids.
   e. renal pelvis.
3. The layer of fibrous connective tissue that surrounds each kidney is the
   a. hilum.
   b. renal pelvis.
   c. renal sinus.
   d. renal capsule.
   e. perirenal fat.
4. Given these structures:
   1. major calyx
   2. minor calyx
   3. renal papilla
   4. renal pelvis

   Choose the arrangement that lists the structures in order as urine leaves the collecting duct and travels to the ureter.
   a. 1,4,2,3
   b. 2,3,1,4
   c. 3,2,1,4
   d. 4,1,3,2
   e. 4,3,2,1
5. Which of these structures contains blood?
   a. glomerulus
   b. vasa recta
   c. distal convoluted tubule
   d. Bowman's capsule
   e. both a and b

6. Given these vessels:
   1. arcuate artery
   2. interlobar artery
   3. segmental artery

   A red blood cell has just passed through the renal artery. Choose the path the red blood cell must take to reach the interlobular artery.
   a. 1,2,3
   b. 2,1,3
   c. 2,3,1
   d. 3,1,2
   e. 3,2,1
7. The juxtaglomerular cells of the ____________ and the macula densa cells of the ____________ form the juxtaglomerular apparatus.
   a. afferent arteriole, proximal convoluted tubule
   b. afferent arteriole, distal convoluted tubule
   c. efferent arteriole, proximal convoluted tubule
   d. efferent arteriole, distal convoluted tubule
8. Given these blood vessels:
   1. afferent arteriole
   2. efferent arteriole
   3. glomerulus
   4. peritubular capillaries

   Choose the correct order as blood passes from an interlobular artery to an interlobular vein.
   a. 1,2,3,4
   b. 1,3,2,4
   c. 2,1,4,3
   d. 3,2,4,1
   e. 4,3,1,2
9. Kidney function is accomplished by which of these means?
   a. filtration
   b. secretion
   c. reabsorption
   d. both a and b
   e. all of the above
10. The amount of plasma that enters Bowman's capsule per minute is the
    a. GFR.
    b. renal plasma flow.
    c. renal fraction.
    d. renal blood flow.
11. Given these structures:
    1. basement membrane
    2. fenestra
    3. filtration slit

    Choose the arrangement that lists the structures in the order a molecule of glucose encounters them as the glucose passes through the filtration membrane to enter Bowman's capsule.
    a. 1,2,3
    b. 2,1,3
    c. 2,3,1
    d. 3,1,2
    e. 3,2,1
12. If the glomerular capillary pressure is 40 mm Hg, the capsule pressure is 10 mm Hg, and the blood colloid osmotic pressure within the glomerulus is 30 mm Hg, the filtration pressure is
    a. -20 mm Hg.
    b. 0 mm Hg.
    c. 20 mm Hg.
    d. 60 mm Hg.
    e. 80 mm Hg.
13. Which of these conditions reduces filtration pressure in the glomerulus?
    a. elevated blood pressure
    b. constriction of the afferent arterioles
    c. decreased plasma protein in the glomerulus
    d. dilation of the afferent arterioles
    e. decreased capsule pressure
14. If blood pressure increases by 50 mm Hg,
    a. the afferent arterioles constrict.
    b. glomerular capillary pressure increases by 50 mm Hg.
    c. GFR increases dramatically.
    d. efferent arterioles constrict.
    e. all of the above.
15. Glucose usually is completely reabsorbed from the filtrate by the time the filtrate has reached
    a. the end of the proximal convoluted tubule.
    b. the tip of the loop of Henle.
    c. the end of the distal convoluted tubule.
    d. the end of the collecting duct.
    e. Bowman's capsule.
16. The greatest volume of water is reabsorbed from the nephron by the
    a. proximal convoluted tubule.
    b. loop of Henle.
    c. distal convoluted tubule.
    d. collecting duct.
17. Water leaves the nephron by
    a. active transport.
    b. filtration into the capillary network.
    c. osmosis.
    d. facilitated diffusion.
    e. symport.
18. Potassium ions enter the ____________ by ____________.
    a. proximal convoluted tubule, diffusion
    b. proximal convoluted tubule, active transport
    c. distal convoluted tubule, diffusion
    d. distal convoluted tubule, antiport
19. Reabsorption of most solute molecules from the proximal convoluted tubule is linked to the active transport of Na^+
    a. across the apical membrane and out of the cell.
    b. across the apical membrane and into the cell.
    c. across the basal membrane and out of the cell.
    d. across the basal membrane and into the cell.
20. Which of these ions is used to symport amino acids, glucose, and other solutes through the apical membrane of nephron epithelial cells?
    a. K^+
    b. Na^+
    c. C^-
    d. Ca^{2+}
    e. Mg^{2+}
21. Which of the following contribute to the formation of a hyperosmotic environment in the medulla of the kidney?
    a. the effects of ADH on water permeability of the ascending limb of the loop of Henle
    b. the impermeability of the ascending limb of the loop of Henle to water
    c. the symport of Na^+, K^+, and Cl^- out of the ascending limb of the loop of Henle
    d. both a and c
    e. both b and c
22. At which of these sites is the osmolality lowest (lowest concentration)?
    a. glomerular capillary
    b. proximal convoluted tubule
    c. bottom of the loop of Henle
    d. initial section of the distal convoluted tubule
    e. collecting duct
23. Increased aldosterone causes
    a. increased reabsorption of Na^+.
    b. decreased blood volume.
    c. decreased reabsorption of Cl^-.
    d. increased permeability of the distal convoluted tubule to water.
    e. increased volume of urine.
24. Juxtaglomerular cells are involved in the secretion of
    a. ADH.
    b. angiotensin.
    c. aldosterone.
    d. renin.

25. Angiotensin II
    a. causes vasoconstriction.
    b. stimulates aldosterone secretion.
    c. stimulates ADH secretion.
    d. increases the sensation of thirst.
    e. all of the above.
26. ADH governs the
    a. Na^+ pump of the proximal convoluted tubules.
    b. water permeability of the loop of Henle.
    c. Na^+ pump of the vasa recta.
    d. water permeability of the distal convoluted tubules and collecting ducts.
    e. Na^+ reabsorption in the proximal convoluted tubule.
27. A decrease in blood osmolality results in
    a. increased ADH secretion.
    b. increased permeability of the collecting ducts to water.
    c. decreased urine osmolality.
    d. decreased urine output.
    e. all of the above.
28. The amount of a substance that passes through the filtration membrane into the nephrons per minute is the
    a. renal plasma flow.
    b. tubular load.
    c. plasma clearance.
    d. tubular maximum.
29. The urinary bladder
    a. is made up of skeletal muscle.
    b. is lined by simple columnar epithelium.
    c. is connected to the outside of the body by the ureter.
    d. is located in the pelvic cavity.
    e. has two urethras and one ureter attached to it.
30. Given these events:
    1. loss of voluntary control of urination
    2. loss of the sensation or desire to urinate
    3. loss of reflex emptying of the urinary bladder

    Which of these events occurs following transection of the spinal cord at level L5?
    a. 1
    b. 2
    c. 1,2
    d. 2,3
    e. 1,2,3

Answers in Appendix E

CRITICAL THINKING

1. To relax after an anatomy and physiology examination, Mucho Gusto goes to a local bistro and drinks 2 quarts of low-sodium beer. What effect does this beer have on urine concentration and volume? Explain the mechanisms involved.
2. A male eats a full bag of salty potato chips. What effect does this have on urine concentration and volume? Explain the mechanisms involved.
3. During severe exertion in a hot environment, a person can lose up to 4 L of hypoosmotic (less concentrated than plasma) sweat per hour. What effect does this loss have on urine concentration and volume? Explain the mechanisms involved.
4. Harry Macho is doing yard work one hot summer day and refuses to drink anything until he is finished. He then drinks glass after glass of plain water. Assume that he drinks enough water to replace all the water he lost as sweat. How does this much water affect urine concentration and volume? Explain the mechanisms involved.
5. A patient has the following symptoms: slight increase in extracellular fluid volume, a large decrease in plasma sodium concentration, very concentrated urine, and cardiac fibrillation. An imbalance of what hormone is responsible for these symptoms? Are the symptoms caused by oversecretion or undersecretion of the hormone?
6. Propose as many ways as you can to decrease the GFR.
7. Design a kidney that can produce hypoosmotic urine, which is less concentrated than plasma, or hyperosmotic urine, which is more concentrated than plasma, by the active transport of water instead of Na^+. Assume that the anatomical structure of the kidney is the same as that in humans. Feel free to change anything else you choose.
8. If only a very small amount of urea were present in the interstitial fluid of the kidney instead of its normal concentration, how would it affect the kidney's ability to concentrate urine?
9. Some patients with hypertension are kept on a low-salt (low-sodium) diet. Propose an explanation for this therapy.
10. Research has shown that mammals with kidneys having relatively thicker medullas have the ability to produce more concentrated urine than humans. Explain why this is so.
11. Prednisone is a steroid that reduces inflammation and is prescribed to treat conditions such as autoimmune diseases. It has an aldosterone-like effect on the kidney. Predict the effect of this treatment on blood pressure, blood volume, and edema. Explain.
12. Addison disease is a consequence of degeneration of the adrenal cortex. Predict the effect of this condition on blood pressure, blood volume, and urine volume. Explain.
13. Renin-secreting tumors are usually found in the kidneys but rarely are found in other organs, such as the liver, lungs, pancreas, and ovaries. Predict the effects of renin-secreting tumors on blood K^+ levels and explain the effects on action potential conduction in nerves and muscle tissues.
14. Even though mutations of aquaporin-3 and aquaporin-4 in the collecting duct have not been described in the literature, if mutations occurred that resulted in a reduced number of these aquaporins in the cells of the collecting ducts, what effect would it have on urine volume and concentration? Would ADH be an effective treatment?

Answers in Appendix F

Visit this book's website at www.mhhe.com/seeley8 for chapter quizzes, interactive learning exercises, and other study tools.

27 Water, Electrolytes, and Acid–Base Balance

Life depends on many complex and highly regulated chemical reactions, all of which occur in water. Many of these reactions are catalyzed by enzymes that can function only within a narrow range of conditions. Changes in the total amount of water, the pH, or the concentration of specific electrolytes can alter the chemical reactions on which life depends. Homeostasis requires the maintenance of these parameters within a narrow range of values, and the failure to maintain homeostasis can result in illness or death.

The kidneys, along with the respiratory, integumentary, and gastrointestinal systems, regulate water volume, electrolyte concentrations, and pH. The nervous and endocrine systems coordinate the activities of these systems.

This chapter covers *body fluids* (p. 1005), the *regulation of body fluid concentration and volume* (p. 1006), the *regulation of intracellular fluid composition* (p. 1011), the *regulation of specific electrolytes in the extracellular fluid* (p. 1012), and the *regulation of acid–base balance* (p. 1020).

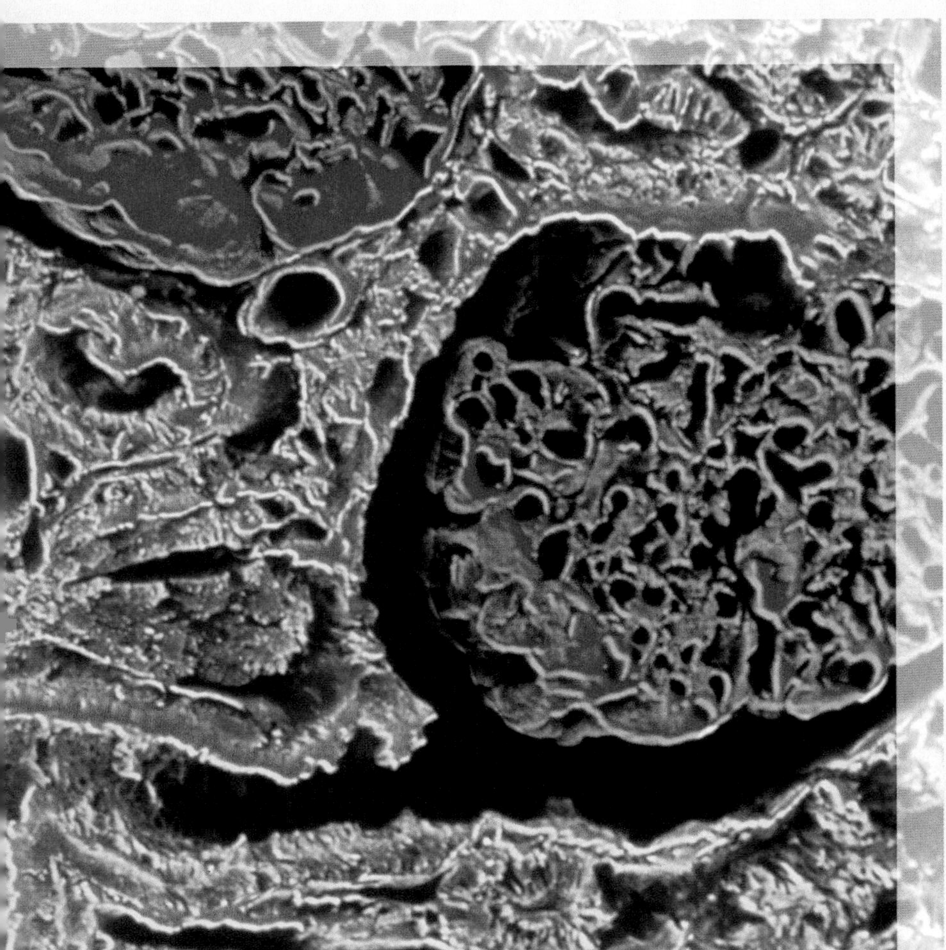

Color-enhanced scanning electron micrograph of a cross section of renal corpuscles in the renal cortex.

Urinary System

BODY FLUIDS

The proportion of body weight composed of water decreases from birth to old age, with the greatest decrease occurring during the first 10 years of life (table 27.1). Because the water content of adipose tissue is relatively low, the fraction of the body's weight composed of water decreases as the amount of adipose tissue increases. The relatively lower water content of adult females compared with that of adult males reflects the greater development of subcutaneous adipose tissue characteristic of women.

For people of all ages and body compositions, the two major fluid compartments are the intracellular and extracellular fluid compartments. The **intracellular** (in-tră-sel'ū-lăr) **fluid compartment** includes all the fluid in the several trillion cells of the body. The intracellular fluid from all cells has a similar composition, and it accounts for approximately 40% of total body weight.

The **extracellular** (eks-tră-sel'ū-lăr) **fluid compartment** includes all the fluid outside the cells, constituting nearly 20% of total body weight. The extracellular fluid compartment can be divided into several subcompartments. The major ones are interstitial fluid and plasma; some of the others are lymph, cerebrospinal fluid, and synovial fluid. **Interstitial** (in-ter-stish'ăl) **fluid** occupies the extracellular spaces outside the blood vessels, and **plasma** (plaz'mă) occupies the extracellular space within blood vessels. All the other subcompartments of the extracellular compartment constitute relatively small volumes.

Although the fluid contained in each subcompartment differs somewhat in composition from that in the others, continuous and extensive exchange occurs between the subcompartments. Water diffuses from one subcompartment to another, and small molecules and ions are either transported or diffuse freely between them. Large molecules, such as proteins are much more restricted in their movement because of the permeability characteristics of the membranes that separate the fluid subcompartments (table 27.2).

The osmotic pressure of most fluid subcompartments is approximately equal. For example, the osmotic pressure of the hyaluronic acid in synovial joints is roughly equal to the osmotic pressure of the proteins in intraocular fluid.

TABLE 27.1 Approximate Volumes of Body Fluid Compartments*

			Extracellular Fluid		
	Total Body Water	Intracellular Fluid	Plasma	Interstitial Fluid	Total
Infants	75	45	4	26	30
Adult Males	60	40	5	15	20
Adult Females	50	35	5	10	15

*Expressed as percentage of body weight.

TABLE 27.2 Approximate Concentration of Major Solutes in Body Fluid Subcompartments*

Solute	Plasma	Interstitial Fluid	Intracellular Fluid†
Cations			
Sodium (Na^+)	153.2	145.1	12.0
Potassium (K^+)	4.3	4.1	150.0
Calcium (Ca^{2+})	3.8	3.4	4.0
Magnesium (Mg^{2+})	1.4	1.3	34.0
TOTAL	162.7	153.9	200.0
Anions			
Chloride (Cl^-)	111.5	118.0	4.0
Bicarbonate (HCO_3^-)	25.7	27.0	12.0
Phosphate (HPO_4^{2-} plus HPO_4^-)	2.2	2.3	40.0
Protein	17.0	0.0	54.0
Other	6.3	6.6	90.0
TOTAL	162.7	153.9	200.0

*Expressed as milliequivalents per liter (mEq/L).
†Data are from skeletal muscle.

> **Edema**
>
> **Edema** is an example of a fluid shift from one extracellular fluid subcompartment, the plasma, to another, the interstitial fluid. Edema commonly results from an increase in the permeability of the capillary walls due to inflammation, which allows proteins to diffuse from the plasma into the interstitial fluid. Water moves in the same direction by osmosis. Edema can also result from a change in the hydrostatic pressure across capillary walls. An increased hydrostatic pressure in capillaries due to the blockage of veins or heart failure forces fluid from plasma into the interstitial spaces (see chapter 21).

1. ***Define intracellular fluid, extracellular fluid, interstitial fluid, and plasma.***
2. ***How do age and percent body fat affect the proportion of body weight composed of water?***
3. ***Compare the osmotic concentration among most fluid subcompartments.***

REGULATION OF BODY FLUID CONCENTRATION AND VOLUME

Regulation of Water Content

The body's water content is regulated so that the total volume of water in the body remains constant. Thus, the volume of water taken into the body is equal to the volume lost each day. Changes in the water volume in body fluids alter the osmolality of body fluids, blood pressure, and interstitial fluid pressure. The total volume of water entering the body each day is 1500–3000 mL. Most of that volume (90%) comes from ingested fluids, some comes from food, and a smaller amount, approximately 10%, is derived from the water produced during cellular metabolism (table 27.3 and see figure 24.33).

The movement of water across the wall of the gastrointestinal tract depends on osmosis, and the volume of water entering the body depends, to a large degree, on the volume of water consumed. If a large volume of dilute liquid is consumed, the rate at which water enters the body fluids increases. If a small volume of concentrated liquid is consumed, the rate decreases.

TABLE 27.3 Summary of Water Intake and Loss

Sources of Water	Routes by Which Water Is Lost
Ingestion (90%)	Urine (61%)
Cellular metabolism (10%)	Evaporation (35%)
	Perspiration
	Insensible
	Sensible
	Respiratory passages
	Feces (4%)

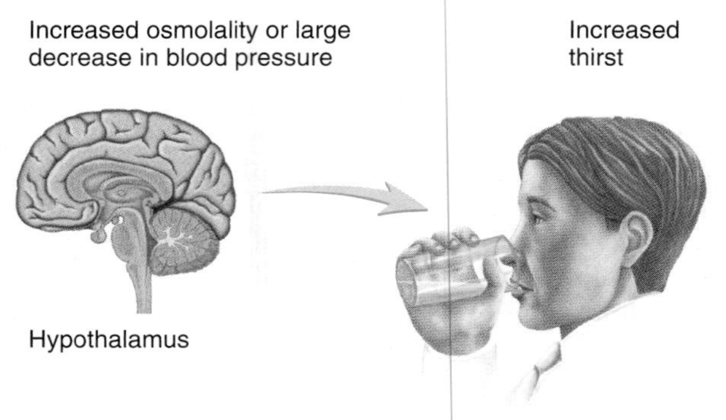

FIGURE 27.1 Effect of Blood Osmolality and Pressure on Water Intake

Increased blood osmolality affects hypothalamic neurons, and large decreases in blood pressure affect baroreceptors in the aortic arch, carotid sinuses, and atrium. As a result of these stimuli, an increase in thirst results, which increases water intake. Increased water intake reduces blood osmolality.

Although fluid consumption is heavily influenced by habit and by social settings, water ingestion does depend, at least in part, on regulatory mechanisms. The sensation of thirst results primarily from an increase in the osmolality of the extracellular fluids and from a reduction in plasma volume. Cells of the supraoptic nucleus within the hypothalamus can detect an increased extracellular fluid osmolality and initiate activity in neural circuits, resulting in a conscious sensation of thirst (figure 27.1).

Baroreceptors can also influence the sensation of thirst. When they detect a substantial decrease in blood pressure, action potentials are conducted to the brain along sensory neurons to influence the sensation of thirst. Low blood pressure associated with hemorrhagic shock, for example, is correlated with an intense sensation of thirst.

Renin is released from the juxtaglomerular apparatuses of the kidneys in response to reduced blood flow to the kidney. Renin increases the formation of angiotensin II in the circulatory system (see chapters 21 and 26). Angiotensin II opposes a decrease in blood pressure by acting on the brain to stimulate the sensation of thirst, by acting on the adrenal cortex to increase aldosterone secretion, and by acting on blood vessel smooth muscle cells to increase vasoconstriction.

When people who are dehydrated drink water, they eventually consume a quantity sufficient to reduce the osmolality of the extracellular fluid to its normal value. They do not normally consume the water all at once. Instead, they drink intermittently until the proper osmolality of the extracellular fluid is established. The thirst sensation is temporarily reduced after the ingestion of small amounts of liquid. At least two factors are responsible for this temporary interruption of the thirst sensation. First, when the oral mucosa becomes wet after it has been dry, sensory neurons conduct action potentials to the thirst center of the hypothalamus and temporarily decrease the sensation of thirst. Second, consumed fluid increases the gastrointestinal tract volume, and stretch of the gastrointestinal wall initiates sensory action potentials in stretch receptors. The sensory neurons conduct action potentials to the thirst center of the hypothalamus, where they temporarily suppress the sensation of thirst. Because the absorption of water from the gastrointestinal tract requires time, mechanisms that temporarily suppress the sensation of thirst prevent the consumption of extreme volumes of fluid that would exceed the amount required to reduce blood osmolality. A longer-term suppression of the thirst sensation results when the extracellular fluid osmolality and blood pressure are within their normal ranges.

Learned behavior can be very important in avoiding periodic dehydration through the consumption of fluids either with or without food, even though blood osmolality is not reduced. The volume of fluid ingested by a healthy person usually exceeds the minimum volume required to maintain homeostasis, and the kidneys eliminate the excess water in urine.

Water loss from the body occurs through three routes (see table 27.3). The greatest amount of water, approximately 61%, is lost through the urine. Approximately 35% of water loss occurs through evaporation from respiratory passages, evaporation of water that diffuses through the skin, and perspiration. Approximately 4% is lost in the feces.

The volume of water lost through the respiratory system depends on the temperature and humidity of the air, body temperature, and the volume of air expired. Water lost through the diffusion and evaporation of water from the skin is called **insensible perspiration** (see chapter 25) and it plays a role in heat loss. For each degree that the body temperature rises above normal, an increased volume of 100–150 mL of water is lost each day in the form of insensible perspiration.

Sweat, or **sensible perspiration,** is secreted by the sweat glands (see chapters 5 and 25); in contrast to insensible perspiration, it contains solutes. Sweat resembles extracellular fluid in its composition, with sodium chloride as the major component, but it also contains some potassium, ammonia, and urea (table 27.4). The volume of fluid lost as sweat is negligible for a person at rest in a cool environment. The volume of sweat produced is determined primarily by neural mechanisms that regulate body temperature, although some sweat is produced as a result of sympathetic stimulation in response to stress. During exercise, elevated environmental temperature, or fever, the volume increases substantially and plays an important role in heat loss. Sweat losses of 8–10 L/day have been measured in outdoor workers in the summer.

Adequate fluid replacement during extensive sweating is important. Sweat is hyposmotic to plasma. The loss of a large volume of hyposmotic sweat causes a decrease in body fluid volume and an increase in body fluid concentration. Fluid volume is lost primarily from the extracellular space, which leads to an increase in extracellular fluid osmolality, a reduction in plasma volume, and an increase in hematocrit. During severe dehydration, the change can be great enough to cause blood viscosity to increase substantially. The increased workload created for the heart by that increase in viscosity can result in heart failure.

TABLE 27.4 Composition of Sweat

Solute	Concentration (mM*)
Sodium	9.8–77.2
Potassium	3.9–9.2
Chloride	5.5–65.1
Ammonia	1.7–5.6
Urea	6.5–12.1

*1 mM is 1/1000 of a mole of solute in 1 liter of solution (see appendix C).

Relatively little water is lost by way of the digestive tract. Although the total volume of fluid secreted into the gastrointestinal tract is large, nearly all the fluid is reabsorbed under normal conditions (see chapter 24). Severe vomiting and diarrhea, however, are exceptions and can result in a large volume of fluid loss.

The kidneys are the primary organs that regulate the composition and volume of body fluids by controlling the volume and concentration of water excreted in the form of urine (see chapter 26). Urine production varies greatly, ranging from a small volume of concentrated urine to a large volume of dilute urine in response to the mechanisms that regulate the body's water content. The mechanisms that respond to changes in extracellular fluid osmolality and extracellular fluid volume keep the total body water levels within a narrow range of values.

4 ***List three factors that increase thirst. Name two factors that inhibit the sense of thirst.***

5 ***Describe three routes for the loss of water from the body. Contrast insensible and sensible perspiration.***

6 ***What are the primary organs that regulate the composition and volume of body fluids?***

Regulation of Extracellular Fluid Osmolality

Adding water to a solution, or removing water from it, changes the osmolality of the solution. Consider a solution contained in a pan on a stove. Adding water to the solution decreases its osmolality, or dilutes it. Boiling the solution in the pan removes water by evaporation, increasing its osmolality and making it more concentrated. Adding water to, or removing water from, body fluids maintains their osmolality between 285 and 300 mOsm/kg.

An increase in the osmolality of the extracellular fluid triggers thirst and antidiuretic hormone (ADH) secretion. Water that is consumed is absorbed from the intestine and enters the extracellular fluid. ADH acts on the distal convoluted tubules and collecting ducts of the kidneys to increase the reabsorption of water from the filtrate. The increase in the amount of water entering the extracellular fluid causes a decrease in osmolality (figures 27.2 and 27.3). The ADH and thirst mechanisms are sensitive to even small changes in extracellular fluid osmolality and the response is fast (from minutes to a few hours). Larger increases in extracellular fluid osmolality, such as during dehydration, result in an even greater increase in thirst and ADH secretion.

A decrease in extracellular fluid osmolality inhibits thirst and ADH secretion. Less water is consumed and reabsorbed from the filtrate in the kidneys. Consequently, more water is lost as a large volume of dilute urine. The result is an increase in the osmolality of the extracellular fluid (see figure 27.3). For example, consumption of a large volume of water in a beverage results in reduced

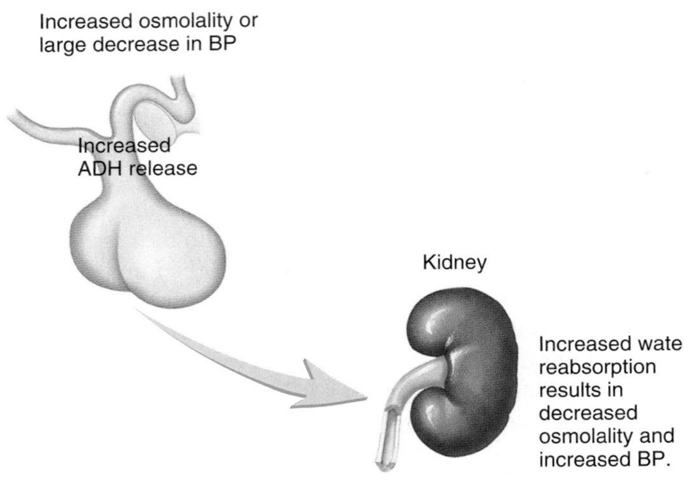

FIGURE 27.2 Effect of Blood Osmolality and Blood Pressure on Water Reabsorption in the Kidneys

Increased blood osmolality affects hypothalamic neurons, and decreased blood pressure affects baroreceptors in the aortic arch, carotid sinuses, and atrium. As a result of these stimuli, the rate of antidiuretic hormone (ADH) secretion from the posterior pituitary increases, which increases water reabsorption by the kidney.

extracellular fluid osmolality. This results in reduced ADH secretion, less reabsorption of water from the filtrate in the kidneys, and the production of a large volume of dilute urine. This response occurs quickly enough that the osmolality of the extracellular fluid is maintained within a normal range of values.

7 ***What two mechanisms are triggered by an increase in the osmolality of the extracellular fluid?***

8 ***Describe how the osmolality of the extracellular fluid is affected when these mechanisms are activated.***

Regulation of Extracellular Fluid Volume

The volume of extracellular fluid can increase or decrease even if the osmolality of the extracellular fluid is maintained within a narrow range of values. Sensory receptors that detect changes in blood pressure are important in the regulation of extracellular fluid volume. Carotid sinus and aortic arch baroreceptors monitor blood pressure in large arteries, receptors in the juxtaglomerular apparatuses monitor pressure changes in the afferent arterioles of the kidneys, and receptors in the walls of the atria of the heart and large veins are sensitive to the smaller changes in blood pressure that occur within them. These receptors activate neural mechanisms and three hormonal mechanisms that regulate extracellular fluid volume (figure 27.4):

1. *Neural mechanisms.* Neural mechanisms change the frequency of action potentials carried by sympathetic neurons to the afferent arterioles of the kidney in response to changes in blood pressure. When baroreceptors detect an increase in arterial and venous blood pressure, the frequency of action potentials carried by sympathetic neurons to the afferent arterioles decreases. Consequently, the afferent arterioles dilate. This increases glomerular capillary pressure, resulting

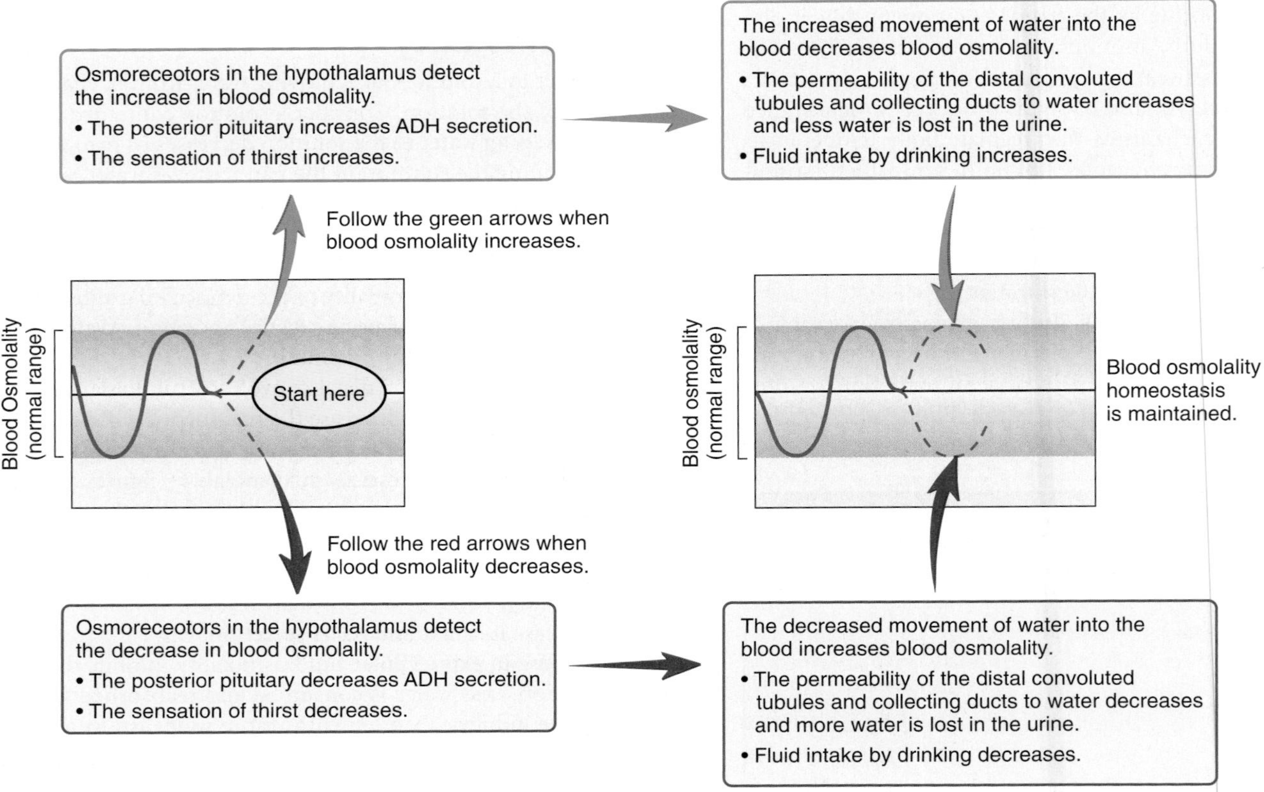

HOMEOSTASIS FIGURE 27.3 Summary of Blood Osmolality Regulation by Hormones

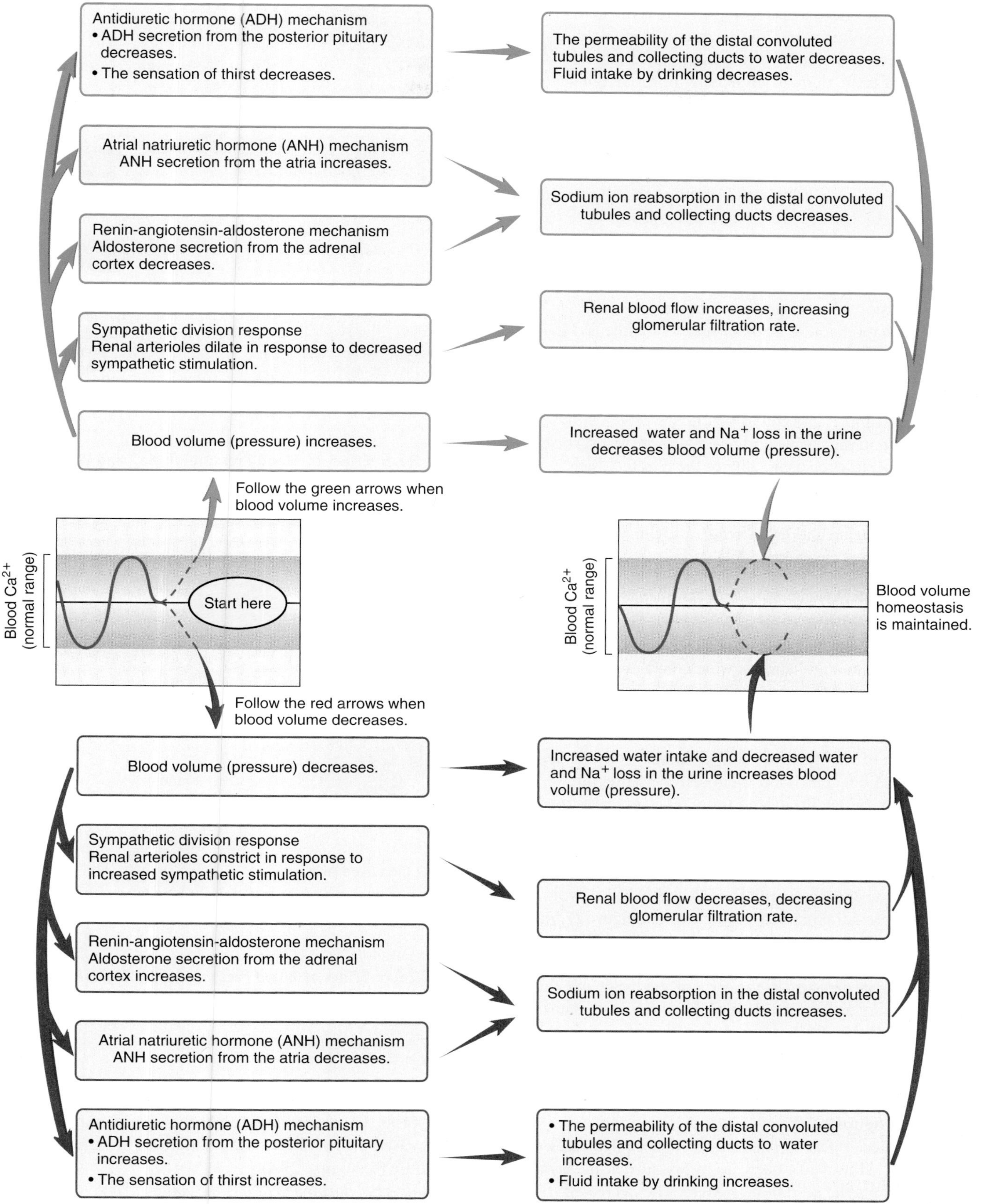

HOMEOSTASIS FIGURE 27.4 Summary of Blood Volume Regulation

For more information on the sympathetic division response, see p. 1008; on the renin-angiotensin-aldosterone mechanism, see figure 27.5; on the atrial natriuretic hormone mechanism, see figure 27.6; on the antidiuretic hormone mechanism, see figure 27.2; on the thirst mechanism, see figure 27.1.

in an increase in the glomerular filtration rate (GFR), an increase in filtrate volume, and an increase in urine volume.

When baroreceptors detect a decrease in arterial and venous blood pressure, there is an increase in the frequency of action potentials carried by sympathetic neurons to the afferent arterioles. Consequently, the afferent arterioles constrict. This decreases GFR, filtrate volume, and urine volume.

2. *Renin-angiotensin-aldosterone mechanism.* The renin-angiotensin-aldosterone mechanism responds to small changes in blood volume. Increased blood pressure results from increased blood volume. Juxtaglomerular cells detect increases in blood pressure in the afferent arterioles and decrease the rate of renin secretion. The decrease in renin secretion results in a decreased conversion of angiotensinogen to angiotensin I, which slows the conversion of angiotensin I to angiotensin II. Reduced angiotensin II causes a decrease in the rate of aldosterone secretion from the adrenal cortex. Decreased aldosterone levels reduce the rate of Na^+ reabsorption, primarily from the distal convoluted tubules and collecting ducts. Consequently, more Na^+ remain in the filtrate and fewer Na^+ are reabsorbed. The effect is to increase the osmolality of the filtrate, which reduces the kidney's ability to reabsorb water. The water remains, with the excess Na^+, in the filtrate. Thus, the volume of urine produced by the kidney increases and the extracellular fluid volume decreases (see figure 27.4).

A decrease in blood volume causes a decrease in blood pressure in the afferent arterioles, which results in an increased rate of renin secretion by the juxtaglomerular cells. The increase in renin secretion results in an increased conversion of angiotensinogen to angiotensin I, which increases the conversion of angiotensin I to angiotensin II. The increased angiotensin II causes an increase in the rate of aldosterone secretion from the adrenal cortex. Increased aldosterone increases the rate of Na^+ reabsorption, primarily from the distal convoluted tubules and collecting ducts. Consequently, fewer Na^+ remain in the filtrate and more Na^+ are reabsorbed. The effect is to decrease the osmolality of the filtrate. This increases the kidney's ability to reabsorb water and to increase extracellular fluid volume. Thus, the volume of urine produced by the kidney decreases and the extracellular fluid volume and blood pressure increase (see figure 27.4; figure 27.5).

3. *Atrial natriuretic hormone (ANH) mechanism.* The ANH mechanism responds to increases in extracellular fluid volume. An increase in pressure in the atria of the heart, which usually results from an increase in blood volume, stimulates the secretion of ANH, which decreases Na^+ reabsorption in the distal convoluted tubules and collecting ducts. This increases the rate of Na^+ and water loss in the urine. Thus, increased ANH secretion decreases extracellular fluid volume (see figure 27.4; figure 27.6).

ANH does not appear to respond strongly to decreases in blood volume. However, a decrease in pressure in the atria of the heart inhibits the secretion of ANH. The decreased ANH decreases the inhibition of Na^+ reabsorption in the distal convoluted tubules and collecting ducts. Therefore, the rate of Na^+ reabsorption increases and water reabsorption increases. Thus, decreased ANH secretion is consistent with a decreased urine volume and an increase in extracellular fluid volume (see figure 27.4).

4. *Antidiuretic hormone (ADH) mechanism.* The ADH mechanism plays an important role in regulating extracellular fluid volume in response to large changes in blood pressure (of 5%–10%). An increase in blood pressure results in a decrease

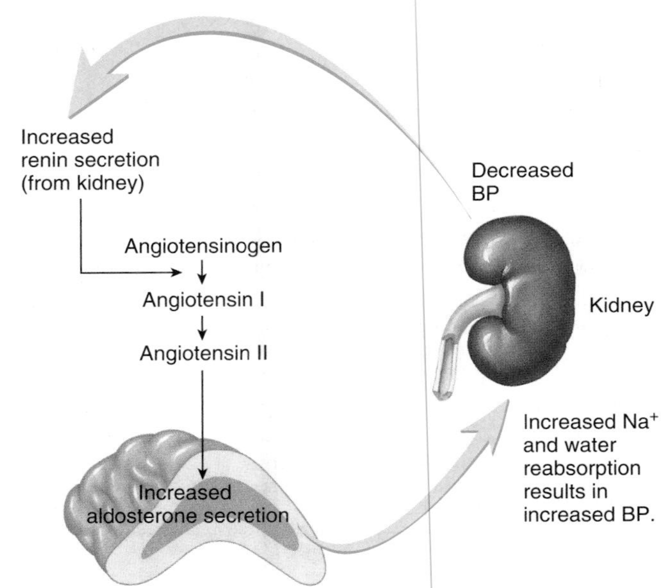

FIGURE 27.5 Effect of Blood Pressure on Na^+ and Water Reabsorption in the Kidneys

Low blood pressure (BP) stimulates renin secretion from the kidney. Renin converts angiotensinogen to angiotensin I, which is converted to angiotensin II. Angiotensin II stimulates aldosterone secretion from the adrenal cortex. Aldosterone increases Na^+ and water reabsorption in the kidney.

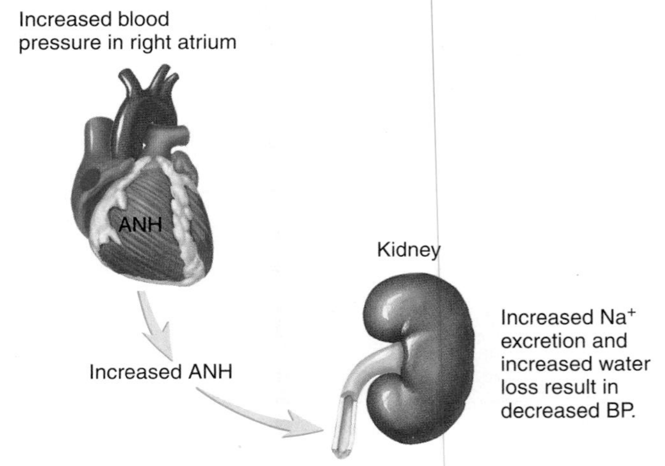

FIGURE 27.6 Effect of Blood Pressure in the Right Atrium on Na^+ and Water Excretion

Increased blood pressure in the right atrium of the heart causes increased secretion of atrial natriuretic hormone (ANH), which increases Na^+ excretion and water loss in the form of urine.

in ADH secretion. As a result, the reabsorption of water from the lumen of the distal convoluted tubules and collecting ducts decreases, resulting in a larger volume of dilute urine. This response helps decrease extracellular fluid volume and blood pressure (see figures 27.2 and 27.4).

A decrease in blood pressure results in an increase in ADH secretion. Consequently, the reabsorption of water from the lumen of the distal convoluted tubules and collecting ducts increases, resulting in a smaller volume of concentrated urine. This response helps increase extracellular fluid volume and blood pressure (see figures 27.2 and 27.4).

The mechanisms that maintain extracellular fluid concentration and volume function together. However, when mechanisms that maintain fluid volume do not function normally, extracellular fluid volume may increase even though the extracellular concentration of fluids is maintained within a normal range of values. For example, increased aldosterone secretion from an enlarged adrenal cortex increases Na^+ reabsorption by the kidney and the total volume of extracellular fluid increases. Mechanisms, such as the regulation of ADH secretion, keep the concentration of the body fluids constant. The blood pressure can be elevated and edema can result, but the osmolality of the extracellular fluid is maintained between 285 and 300 mOsm/kg. Similarly, in people suffering from heart failure, the resulting reduced blood pressure activates mechanisms that increase blood pressure to its normal range of values. Those mechanisms include the release of renin from the kidneys. Consequently, aldosterone secretion increases, and the result is an increase in the extracellular fluid volume and edema in the periphery, including edema in the lungs.

9 ***What sensory receptors are responsible for activating neural and hormonal mechanisms that regulate extracellular fluid volume?***

10 ***What is the effect on sympathetic stimulation, afferent arterioles, GFR, filtrate volume, urine volume, and extracellular fluid volume when baroreceptors detect an increase in arterial and venous blood pressure?***

11 ***Describe the response of the renin-angiotensin-aldosterone mechanism to a decrease in blood pressure. How is extracellular fluid volume and urine volume affected?***

12 ***What effect does ANH have on extracellular fluid volume?***

13 ***How does an increase in blood pressure affect the secretion of ADH? How does ADH affect extracellular fluid volume?***

REGULATION OF INTRACELLULAR FLUID COMPOSITION

The composition of intracellular fluid is substantially different from that of extracellular fluid. Plasma membranes, which separate the two compartments, are selectively permeable—they are relatively impermeable to proteins and other large molecules and have limited permeability to smaller molecules and ions. Consequently, most large molecules synthesized within cells, such as proteins, remain within the intracellular fluid. Some substances, such as electrolytes, are actively transported across the plasma membrane, and their concentrations in the intracellular fluid are determined by the transport processes and by the electric charge difference across the plasma membrane (figure 27.7).

1. Large organic molecules, such as proteins, which cannot cross the plasma membrane, are synthesized inside cells and influence the concentration of solutes inside the cells.
2. The transport of ions across the plasma membrane, such as Na^+, K^+, and Ca^{2+}, influences the concentration of ions inside and outside the cell.
3. An electric charge difference across the plasma membrane influences the distribution of ions inside and outside the cell.
4. The distribution of water inside and outside the cell is determined by osmosis.

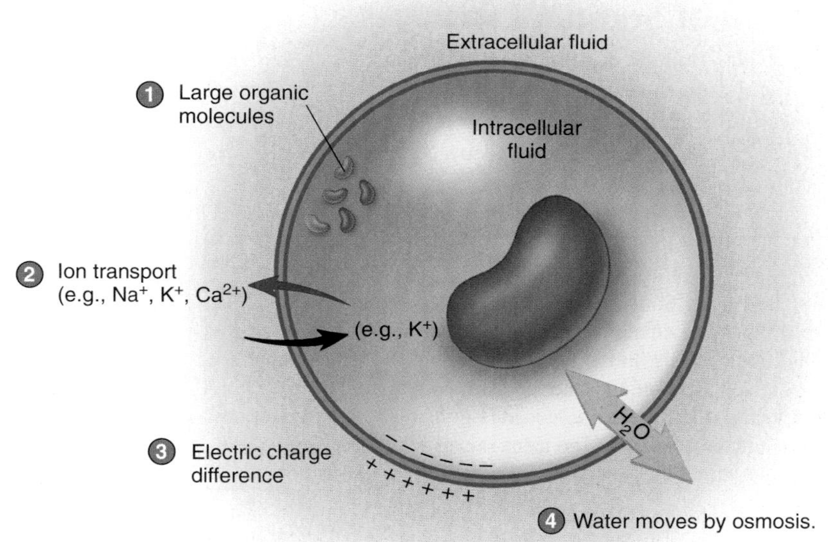

PROCESS FIGURE 27.7 Regulation of Intracellular and Extracellular Distribution of Water and Solutes

Water movement across the plasma membrane is controlled by osmosis. Thus, the net movement of water is affected by changes in the concentration of solutes in the extracellular and intracellular fluids. For example, as dehydration develops, the concentration of solutes in the extracellular fluid increases, resulting in the movement of water by osmosis from the intracellular fluid into the extracellular fluid. If dehydration is severe, enough water moves from the intracellular fluid to cause the cells to function abnormally. If water intake increases after a period of dehydration, the concentration of solutes in the extracellular fluids decreases, which results in the movement of water back into the cells.

14 ***What factors determine the composition of intracellular fluid? What characteristic of plasma membranes is responsible for maintaining the differences between intracellular and extracellular fluid?***

REGULATION OF SPECIFIC ELECTROLYTES IN THE EXTRACELLULAR FLUID

Electrolytes (ē-lek′trō-līts) are molecules or ions with an electric charge. The major extracellular ions are Na^+, Cl^-, K^+, Ca^{2+}, Mg^{2+}, and phosphate ions. The ingestion of electrolytes in food and water adds them to the body, whereas organs—such as the kidneys and, to a lesser degree, the liver, skin, and lungs—remove them from the body. The concentrations of electrolytes in the extracellular fluid are regulated so that they do not change unless the individual is growing, gaining weight, or losing weight. The regulation of each electrolyte involves the coordinated participation of several organ systems.

Regulation of Sodium Ions

Sodium ions (Na^+) are the dominant extracellular cations. Because of their abundance in the extracellular fluids, they exert substantial osmotic pressure. Approximately 90%–95% of the osmotic pressure of the extracellular fluid is caused by Na^+ and the negative ions associated with them.

Diet and Na^+ Homeostasis

In the United States, the quantity of Na^+ ingested each day is 20–30 times the amount needed. Less than 0.5 g is required to maintain homeostasis, but the average individual ingests approximately 10–15 g of sodium chloride daily. Regulation of the Na^+ content in the body, therefore, depends primarily on the excretion of excess quantities of Na^+. The mechanisms for conserving Na^+ in the body are effective, however, when the Na^+ intake is very low.

The kidneys are the major route by which Na^+ are excreted. Sodium ions readily pass from the glomerulus into the lumen of Bowman's capsule and is present in the same concentration in the filtrate as in the plasma. The concentration of Na^+ excreted in the urine is determined by the amount of Na^+ and water reabsorbed from filtrate in the nephron. If Na^+ reabsorption from the nephron decreases, large quantities are lost in the urine. If Na^+ reabsorption from the nephron increases, only small quantities are lost in the urine.

The rate of Na^+ transport in the proximal convoluted tubule is relatively constant, but the Na^+ transport mechanisms of the distal convoluted tubule and the collecting duct are under hormonal control. **Aldosterone** increases Na^+ reabsorption from the distal convoluted tubule and collecting duct. As little as 0.1 g of sodium is excreted in the urine each day in the presence of high blood levels of aldosterone. When aldosterone is absent, Na^+ reabsorption in the nephron is greatly reduced, and as much as 30–40 g of sodium can be lost in the urine daily.

Sodium ions are also excreted from the body in **sweat.** Normally, only a small quantity of Na^+ is lost each day in the form of sweat, but the amount increases during heavy exercise in a warm environment. The mechanisms that regulate sweating control the quantity of Na^+ excreted through the skin. As the body temperature increases, thermoreceptor neurons within the hypothalamus respond by increasing the rate of sweat production. As the rate of sweat production increases, the quantity of Na^+ lost in the urine decreases to keep the extracellular concentration of Na^+ constant. The loss of Na^+ in sweat is rarely physiologically significant.

The primary mechanisms that regulate Na^+ concentrations in the extracellular fluid do not directly monitor Na^+ levels but are sensitive to changes in extracellular fluid osmolality or changes in blood pressure (see figure 27.4 and table 27.5). The quantity of Na^+ in the body has a dramatic effect on extracellular osmotic pressure and extracellular fluid volume. For example, if the quantity of Na^+ increases, the osmolality of the extracellular fluid increases. An increase in the osmolality of the extracellular fluids stimulates ADH secretion, which increases the reabsorption of water by the kidney and causes a small volume of concentrated urine to be produced. An increase in extracellular fluid osmolality also increases the sensation of thirst. Consequently, extracellular fluid volume increases. A decrease in the quantity of Na^+ in the body decreases the osmolality of the extracellular fluid. This inhibits ADH secretion, which stimulates a large volume of dilute urine to be produced and decreases the sensation of thirst. Thus, extracellular osmolality increases. By regulating extracellular fluid osmolality and extracellular fluid volume, the concentration of Na^+ in the body fluids is maintained within a narrow range of values.

Elevated blood pressure under resting conditions increases Na^+ and water excretion (see figure 27.4 and table 27.5). If blood pressure is low, the total Na^+ content of the body is usually also low. In response to low blood pressure, mechanisms such as the renin-angiotensin-aldosterone mechanism are activated, increasing Na^+ concentration and water volume in the extracellular fluid (see figure 27.4 and table 27.5).

PREDICT 1

In response to hemorrhagic shock, the kidneys produce a small volume of very concentrated urine. Explain how the rate of filtrate formation changes and how Na^+ transport changes in the distal part of the nephron in response to hemorrhagic shock.

TABLE 27.5 Homeostasis: Mechanisms Regulating Blood Sodium

Mechanism	Stimulus	Response to Stimulus	Effect of Response	Result
Response to Changes in Blood Osmolality				
Antidiuretic hormone (ADH); the most important regulator of blood osmolality	Increased blood osmolality (e.g., increased Na^+ concentration)	Increased ADH secretion from the posterior pituitary; mediated through cells in the hypothalamus	Increased water reabsorption in the kidney; production of a small volume of concentrated urine	Decreased blood osmolality as reabsorbed water dilutes the blood
	Decreased blood osmolality (e.g., decreased Na^+ concentration)	Decreased ADH secretion from the posterior pituitary; mediated through cells in the hypothalamus	Decreased water reabsorption in the kidney; production of a large volume of dilute urine	Increased blood osmolality as water is lost from the blood into the urine
Response to Changes in Blood Pressure				
Renin-angiotensin-aldosterone	Decreased blood pressure in the kidney's afferent arterioles	Increased renin release from the juxtaglomerular apparatuses; renin initiates the conversion of angiotensinogen to angiotensin; angiotensin I is converted to angiotensin II, which increases aldosterone secretion from the adrenal cortex	Increased Na^+ reabsorption in the kidney (because of increased aldosterone); increased water reabsorption as water follows the Na^+; decreased urine volume	Increased blood pressure as blood volume increases because of increased water reabsorption; blood osmolality is maintained because both Na^+ and water are reabsorbed*
	Increased blood pressure in the kidney's afferent arterioles	Decreased renin release from the juxtaglomerular apparatuses, resulting in reduced formation of angiotensin I; reduced angiotensin I leads to reduced angiotensin II, which causes a decrease in aldosterone secretion from the adrenal cortex	Decreased Na^+ reabsorption in the kidney (because of decreased aldosterone); decreased water reabsorption as fewer Na^+ are reabsorbed; increased urine volume	Decreased blood pressure as blood volume decreases because water is lost in the urine; blood osmolality is maintained because both Na^+ and water are lost in the urine*
Atrial natriuretic hormone (ANH)	Decreased blood pressure in the atria of the heart	Decreased ANH released from the atria	Increased Na^+ reabsorption in the kidney; increased water reabsorption as water follows the Na^+; decreased urinary volume	Increased blood pressure as blood volume increases because of increased water reabsorption; blood osmolality is maintained because both Na^+ and water are reabsorbed*
	Increased blood pressure in the atria of the heart	Increased ANH released from the atria	Decreased Na^+ reabsorption in the kidney; decreased water reabsorption as water is lost with Na^+ in the urine; increased urinary volume	Decreased blood osmolality as blood volume decreases because water is lost in the urine; blood osmolality is maintained because both Na^+ and water are lost in the urine*
ADH—activated by significant decreases in blood pressure; normally regulates blood osmolality (see above)	Decreased arterial blood pressure	Increased ADH secretion from the posterior pituitary; mediated through baroreceptors	Increased water reabsorption in the kidney; production of a small volume of concentrated urine	Increased blood pressure resulting from increased blood volume; decreased blood osmolality
	Increased arterial blood pressure	Decreased ADH secretion from the posterior pituitary; mediated through baroreceptors	Decreased water reabsorption in the kidney; production of a large volume of dilute urine	Decreased blood pressure resulting from decreased blood volume; increased blood osmolality

Abbreviation: ADH = antidiuretic hormone.

*Assumes normal levels of ADH.

TABLE 27.6 Consequences of Abnormal Plasma Levels of Sodium Ions

Major Causes	Symptoms
Hypernatremia	
High dietary sodium rarely causes symptoms.	Thirst, fever, dry mucous membranes, restlessness; most serious symptoms—convulsions and pulmonary edema
Administration of hypertonic saline solutions (e.g., sodium bicarbonate treatment for acidosis)	
Oversecretion of aldosterone (i.e., aldosteronism)	When occurring with an increased water volume—weight gain, edema, elevated blood pressure, and bounding pulse
Water loss (e.g., because of fever, respiratory infections, diabetes insipidus, diabetes mellitus, and diarrhea)	
Hyponatremia	
Inadequate dietary intake of sodium rarely causes symptoms—can occur in those on low-sodium diets and those taking diuretics.	Lethargy, confusion, apprehension, seizures, and coma
Extrarenal losses—vomiting, prolonged diarrhea, gastrointestinal suctioning, and burns	When accompanied by reduced blood volume—reduced blood pressure, tachycardia, and decreased urine output
Dilution—intake of large water volume after excessive sweating	When accompanied by increased blood volume—weight gain, edema, and distension of veins
Hyperglycemia, which attracts water into the circulatory system but reduces the concentration of Na^+	

Cells in the walls of the atria synthesize ANH, which is secreted in response to an elevation in blood pressure within the right atrium. ANH acts on the kidneys to increase urine production by inhibiting the reabsorption of Na^+ (see figure 27.6 and table 27.5). It also inhibits the effect of ADH on the distal convoluted tubules and collecting ducts and inhibits ADH secretion (see chapter 26, p. 991).

Deviations from the normal concentration range for Na^+ in body fluids result in significant symptoms. Some major causes of **hypernatremia** (hī′per-nă-trē′mē-ă), or an elevated plasma Na^+ concentration, and **hyponatremia** (hī′pō-nă-trē′mē-ă), or a reduced plasma Na^+ concentration, and the major symptoms of each are listed in table 27.6.

15 ***Name the substance responsible for most of the osmotic pressure of the extracellular fluid.***

16 ***How does aldosterone affect the concentration of Na^+ in the urine?***

17 ***What role does sweating play in Na^+ balance?***

18 ***How does increased blood pressure result in a loss of water and salt? What happens when blood pressure decreases?***

19 ***What effect does ANH have on Na^+ and water loss in urine?***

PREDICT 2

If a person consumes an excess amount of Na^+ and water, predict the effect on (a) blood pressure, (b) urine volume, and (c) urine concentration.

Regulation of Chloride Ions

The predominant anions in the extracellular fluid are **chloride ions (Cl^-).** The electrical attraction of anions and cations makes it difficult to separate these charged particles. Consequently, the regulatory mechanisms that influence the concentration of cations in the extracellular fluid also influence the concentration of anions. The mechanisms that regulate Na^+, K^+, and Ca^{2+} levels in the body are important in influencing Cl^- levels. Because Na^+ predominate, the mechanisms that regulate extracellular Na^+ concentration are those that are most important in regulating the extracellular Cl^- concentration.

20 ***What mechanisms regulate Cl^- concentrations?***

Regulation of Potassium Ions

The extracellular concentration of **potassium ions (K^+)** must be maintained within a narrow range. The concentration gradient of K^+ across the plasma membrane has a major influence on the resting membrane potential, and cells that are electrically excitable are highly sensitive to slight changes in that concentration gradient. An increase in extracellular K^+ concentration leads to depolarization, and a decrease in extracellular K^+ concentration leads to hyperpolarization of the resting membrane potential. **Hyperkalemia** (hī′per-kă-lē′mē-ă) is an abnormally high level of K^+ in the extracellular fluid, and **hypokalemia** (hī′pō-ka-lē′mē-ă) is an abnormally low level of K^+ in the extracellular fluid. Major causes of hyperkalemia and hypokalemia and their symptoms are listed in table 27.7.

Potassium ions pass freely through the filtration membrane of the renal corpuscle. They are actively reabsorbed in the proximal convoluted tubules and actively secreted in the distal convoluted tubules and collecting ducts. Potassium ion secretion into the distal convoluted tubule and collecting duct is highly regulated and primarily responsible for controlling the extracellular concentration of K^+.

Aldosterone plays a major role in regulating the concentration of K^+ in the extracellular fluid by increasing the rate of K^+ secretion in the distal convoluted tubule and collecting duct. Aldosterone secretion from the adrenal cortex is stimulated by elevated K^+ blood levels (figure 27.8; see chapter 26). Aldosterone secretion is

TABLE 27.7 Consequences of Abnormal Concentrations of Potassium Ions

Major Causes	Symptoms
Hyperkalemia	
Movement of K^+ from intracellular to extracellular fluid resulting from cell trauma (e.g., burns or crushing injuries) and alterations in plasma membrane permeability (e.g., acidosis, insulin deficiency, and cell hypoxia)	Mild hyperkalemia (caused mainly by partial depolarization of plasma membranes):
Decreased renal excretion of K^+ (e.g., from decreased secretion of aldosterone in persons with Addison disease)	Increased neuromuscular irritability and restlessness
	Intestinal cramping and diarrhea
	Electrocardiogram—alterations, including rapid repolarization with narrower and taller T waves and shortened QT intervals
	Severe hyperkalemia (caused mainly by partial depolarization of plasma membranes severe enough to hamper action potential conduction): muscle weakness, loss of muscle tone, and paralysis
	Electrocardiogram—alterations, including changes caused by reduced rate of action potential conduction (e.g., depressed ST segment, prolonged PR interval, wide QRS complex, arrhythmias, and cardiac arrest)
Hypokalemia	
Alkalosis (K^+ shift into cell in exchange for H^+)	Symptoms are mainly due to hyperpolarization of membranes.
Insulin administration (promotes cellular uptake of K^+)	Decreased neuromuscular excitability—skeletal muscle weakness
Reduced K^+ intake (especially with anorexia nervosa and alcoholism)	Decreased tone in smooth muscle
Increased renal loss (excessive aldosterone secretion, improper use of diuretics, and kidney diseases that result in reduced ability to reabsorb Na^+)	Cardiac muscle—delayed ventricular repolarization, bradycardia, and atrioventricular block

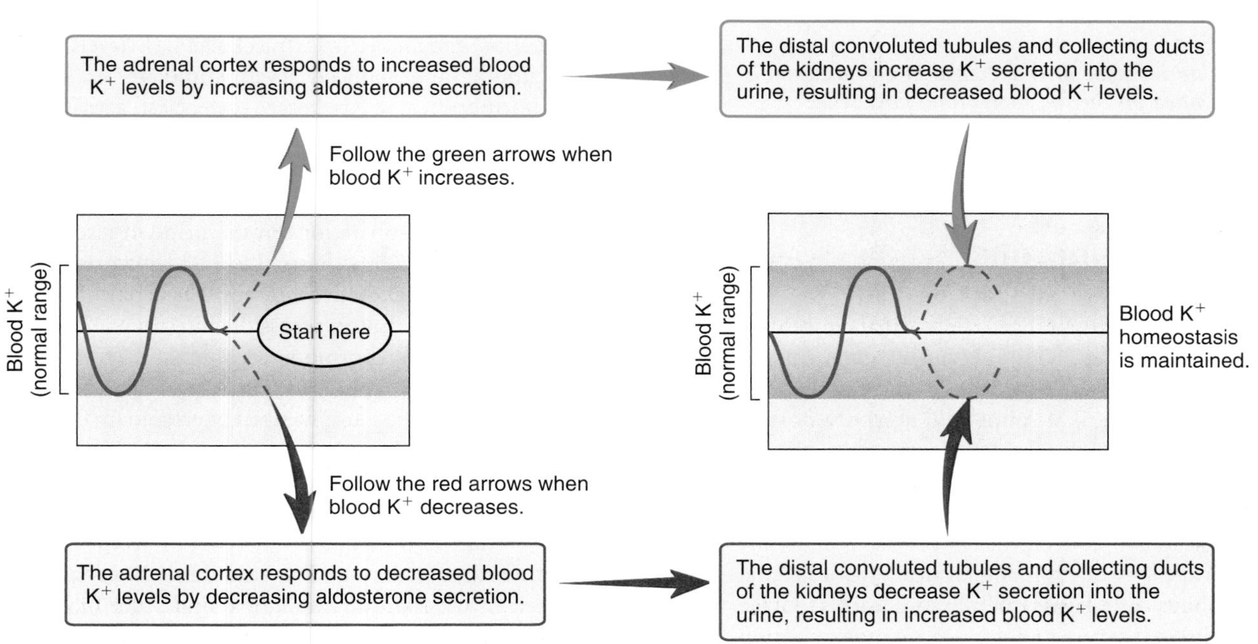

HOMEOSTASIS FIGURE 27.8 **Summary of Blood K^+ Regulation**

TABLE 27.8 Consequences of Abnormal Concentrations of Calcium

Major Causes	Symptoms
Hypocalcemia	
Nutritional deficiencies	Symptoms due mainly to increased permeability of plasma membranes to Na^+
Vitamin D deficiency	Increase in neuromuscular excitability—confusion, muscle spasms, hyperreflexia, and intestinal cramping
Decreased parathyroid hormone secretion	
Malabsorption of fats (reduced vitamin D absorption)	Severe neuromuscular excitability—convulsions, tetany, and inadequate respiratory movements
Bone tumors that increase Ca^{2+} deposition	Electrocardiogram—prolonged QT interval (prolonged ventricular depolarization)
	Reduced absorption of phosphate from the intestine
Hypercalcemia	
Excessive parathyroid hormone secretion	Symptoms due mainly to decreased permeability of plasma membranes to Na^+
Excess vitamin D	Loss of membrane excitability—fatigue, weakness, lethargy, anorexia, nausea, and constipation
	Electrocardiogram—shortened QT segment and depressed T waves
	Kidney stones

also stimulated in response to increased angiotensin II. Elevated aldosterone concentrations in the circulatory system increase K^+ secretion into the nephron, thereby lowering blood levels of K^+.

Circulatory system shock can result from plasma loss, dehydration, and tissue damage, such as occurs in burn patients. This shock causes the extracellular K^+ to be more concentrated than normal, which stimulates aldosterone secretion from the adrenal cortex. Aldosterone secretion also occurs in response to decreased blood pressure, which stimulates the renin-angiotensin-aldosterone mechanism. Homeostasis is reestablished as K^+ excretion increases. Also, increased Na^+ and water reabsorption stimulated by aldosterone results in an increase in extracellular fluid volume, which dilutes the K^+ in the body fluids. Blood pressure increases toward normal as water reabsorption increases and when vasoconstriction is stimulated by angiotensin II.

21 ***What effect does an increase or a decrease in extracellular K^+ concentration have on resting membrane potential?***

22 ***Where are K^+ secreted in the nephron?***

23 ***How is the secretion of K^+ regulated?***

Regulation of Calcium Ions

The extracellular concentration of **calcium ions (Ca^{2+}),** like that of K^+, is regulated within a narrow range. The normal concentration of Ca^{2+} in plasma is 9.4 mg/100 mL. **Hypocalcemia** (hī′pō-kal-sē′mē-ă) is a below normal level of Ca^{2+} in the extracellular fluid, and **hypercalcemia** (hī′per-kal-sē′mē-ă) is an above normal level of Ca^{2+} in the extracellular fluid. Major symptoms develop when the extracellular concentration of Ca^{2+} declines below 6 mg/100 mL or increases above 12 mg/100 mL. Decreases and increases in the extracellular concentration of Ca^{2+} markedly affect the electrical properties of excitable tissues. Hypocalcemia increases the permeability of plasma membranes to Na^+. As a result, nerve and muscle tissues undergo spontaneous action potential generation. Hypercalcemia decreases the permeability of the plasma membrane to Na^+, thus preventing normal depolarization of nerve and muscle cells. High extracellular Ca^{2+} levels cause the deposition of calcium carbonate salts in soft tissues, resulting in irritation and inflammation of those tissues. Table 27.8 lists the major causes and symptoms of hypocalcemia and hypercalcemia.

The kidneys, intestinal tract, and bones are important in maintaining extracellular Ca^{2+} levels (figure 27.9). Almost 99% of total body calcium is contained in bone. Part of the extracellular Ca^{2+} regulation involves the regulation of Ca^{2+} deposition into and resorption from bone (see chapter 6). Long-term regulation of Ca^{2+} levels, however, depends on maintaining a balance between Ca^{2+} absorption across the wall of the intestinal tract and Ca^{2+} excretion by the kidneys.

Parathyroid (par-ă-thī′royd) **hormone,** secreted by the parathyroid glands, increases extracellular Ca^{2+} levels and reduces extracellular phosphate levels (see figure 27.9). The rate of parathyroid hormone secretion is regulated by extracellular Ca^{2+} levels. Elevated Ca^{2+} levels inhibit and reduced levels stimulate its secretion. Parathyroid hormone causes increased osteoclast activity, which results in the degradation of bone and the release of Ca^{2+} and phosphate ions into body fluids. Parathyroid hormone increases the rate of Ca^{2+} reabsorption from nephrons in the kidneys and increases the concentration of phosphate ions in the urine. It also increases the rate at which vitamin D is converted to 1,25-dihydroxycholecalciferol, or **active vitamin D.** Active vitamin D acts on the intestinal tract to increase Ca^{2+} absorption across the intestinal mucosa.

A lack of parathyroid hormone secretion results in a rapid decline in extracellular Ca^{2+} concentration. A reduction in the rate of absorption of Ca^{2+} from the intestinal tract, increased Ca^{2+} excretion by the kidneys, and reduced bone resorption cause this decline. A lack of parathyroid hormone secretion can result in death because of tetany of the respiratory muscles caused by hypocalcemia.

Vitamin D can be obtained from food or from vitamin D biosynthesis. Normally, vitamin D biosynthesis is adequate, but prolonged lack of exposure to sunlight reduces the biosynthesis because ultraviolet light is required for one step in the process (see chapter 5). Dietary vitamin D can involve the ingestion of active vitamin D or one of its precursors.

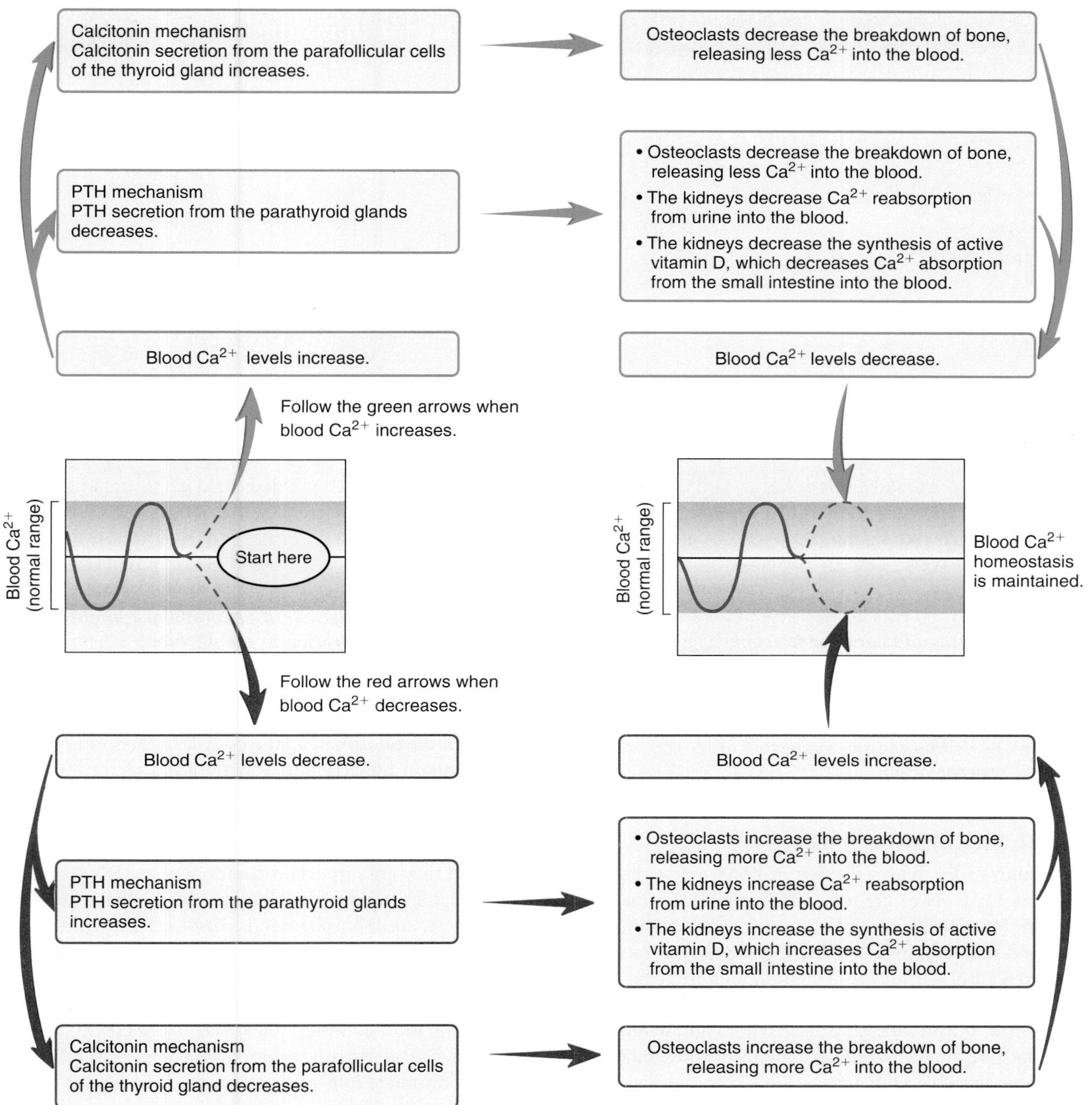

HOMEOSTASIS FIGURE 27.9 Summary of Blood Ca^{2+} Regulation
For more information on the PTH and calcitonin mechanisms, see figure 6.20.

Without vitamin D, the transport of Ca^{2+} across the wall of the intestinal tract is negligible. This leads to inadequate Ca^{2+} absorption, even though large amounts of these ions may be present in the diet. Thus, Ca^{2+} absorption depends on both the consumption of an adequate amount of calcium in food and the presence of an adequate amount of vitamin D.

Calcitonin (kal-si-tō′nin), which is secreted by the parafollicular cells of the thyroid gland, reduces extracellular Ca^{2+} levels. It is most effective when Ca^{2+} levels are elevated, although greater than normal calcitonin levels in the blood are not consistently effective in causing blood levels of Ca^{2+} to decline below normal values. The major effect of calcitonin is on bone. It inhibits osteoclast activity and prolongs the activity of osteoblasts. Thus, it decreases bone demineralization and increases bone mineralization (see chapter 6).

Elevated Ca^{2+} levels stimulate calcitonin secretion, whereas reduced Ca^{2+} levels inhibit it. Increased calcitonin secretion reduces blood levels of Ca^{2+}, but large doses of calcitonin do not consistently reduce blood levels of Ca^{2+} below normal levels.

TABLE 27.9 Consequences of Abnormal Plasma Levels of Magnesium Ions

Major Causes	Symptoms
Hypomagnesemia (Rare)	
Malnutrition Alcoholism Reduced absorption of magnesium in the intestine Renal tubular dysfunction Some diuretics	Symptoms result from increased neuromuscular excitability and include irritability, increased reflexes, muscle weakness, tetany, and convulsions.
Hypermagnesemia (Rare)	
Renal failure Magnesium-containing antacids	Symptoms result from depressed skeletal muscle contractions and nerve functions and include nausea, vomiting, muscle weakness, hypotension, bradycardia, and reduced respiration.

Although calcitonin reduces the blood levels of Ca^{2+} when they are elevated, it is not as important as parathyroid hormone in the regulation of blood Ca^{2+} levels (see figure 27.9).

24 ***Describe the effect of hypocalcemia and hypercalcemia on membrane potentials.***

25 ***Explain the role of parathyroid hormone in regulating the extracellular Ca^{2+} concentration.***

26 ***Describe the role of vitamin D in regulating the extracellular Ca^{2+} concentration.***

27 ***Describe the role of calcitonin in the regulation of extracellular Ca^{2+} concentration.***

Regulation of Magnesium Ions

Most of the magnesium in the body is stored in bones or in the intracellular fluid. Less than 1% of the total are ions found in the extracellular fluid. Approximately one-half of those ions are bound to plasma proteins and one-half are free. The free magnesium ion (Mg^{2+}) concentration is 1.8–2.4 mEq/L. Magnesium ions are cofactors for intracellular enzymes, such as the Na^+–K^+ ATPase involved in actively transporting Na^+ out of and K^+ into cells. **Hypomagnesemia** is a below normal blood level of magnesium, and **hypermagnesemia** is a higher than normal blood level of magnesium. Low and high levels of plasma magnesium produce symptoms (table 27.9) associated with the effect of magnesium on Na^+–K^+ active transport.

Free Mg^{2+} pass through the filtration membranes of the kidney into the filtrate. About 85%–90% of those ions are reabsorbed from the filtrate, and only about 10%–15% enter the urine. Of the Mg^{2+} reabsorbed, most are reabsorbed by the loop of Henle. The remainder are reabsorbed by the proximal convoluted tubule, distal convoluted tubule, and collecting duct.

The kidney's capacity to reabsorb Mg^{2+} is limited. If the level of free Mg^{2+} increases in the extracellular fluid, the excess Mg^{2+} remain in the filtrate, and there is an increase in the rate of Mg^{2+} loss in the urine. If the level of free Mg^{2+} decreases in the extracellular fluid, nearly all of the Mg^{2+} are reabsorbed, and there is a decrease in the rate of Mg^{2+} loss in the urine. The control of Mg^{2+} reabsorption is not clear, but a decreased extracellular concentration of Mg^{2+} causes an increase in the rate of reabsorption in the nephron (figure 27.10).

28 ***In what part of the nephron are most Mg^{2+} reabsorbed?***

29 ***What effect does a decreased extracellular concentration of Mg^{2+} have on reabsorption in the nephron?***

Regulation of Phosphate Ions

About 85% of the phosphate in the body is in the form of calcium phosphate salts in bone (hydroxyapatite) and teeth. Most of the remaining phosphate is inside cells. Many of the phosphate ions are covalently bound to other organic molecules. Phosphate ions are bound to lipids (to form phospholipids), proteins, and carbohydrates, and they are important components of DNA, RNA, and ATP. Phosphates also play important roles in the regulation of enzyme activity, and phosphate ions dissolved in the intracellular fluid act as buffers (see "Regulation of Acid–Base Balance," p. 1020). The extracellular concentration of phosphate ions is between 1.7 and 2.6 mEq/L. Phosphate ions are in the form of $H_2PO_4^-$, HPO_4^{2-}, and PO_4^{3-}. The most common phosphate ion is HPO_4^{2-}.

The kidneys' capacity to reabsorb phosphate ions is limited. If the level of phosphate ions increases in the extracellular fluid, the excess phosphate remains in the filtrate, and there is an increase in the rate of phosphate loss in the urine. If the level of phosphate ions decreases in the extracellular fluid, nearly all of the phosphate ions are reabsorbed, and there is a decrease in the rate of phosphate ion loss in the urine (figure 27.11).

Over time, a diet low in phosphate can increase the rate of phosphate reabsorption. Consequently, most of the phosphate that enters the filtrate is reabsorbed to maintain the extracellular phosphate concentration. Parathyroid hormone (PTH) can play a significant role in regulating extracellular phosphate levels. PTH promotes bone resorption. Thus, large amounts of both Ca^{2+} and phosphate ions are released into the extracellular space. PTH decreases the reabsorption of phosphate ions from renal tubules so that a greater proportion of the tubular phosphate is lost in the

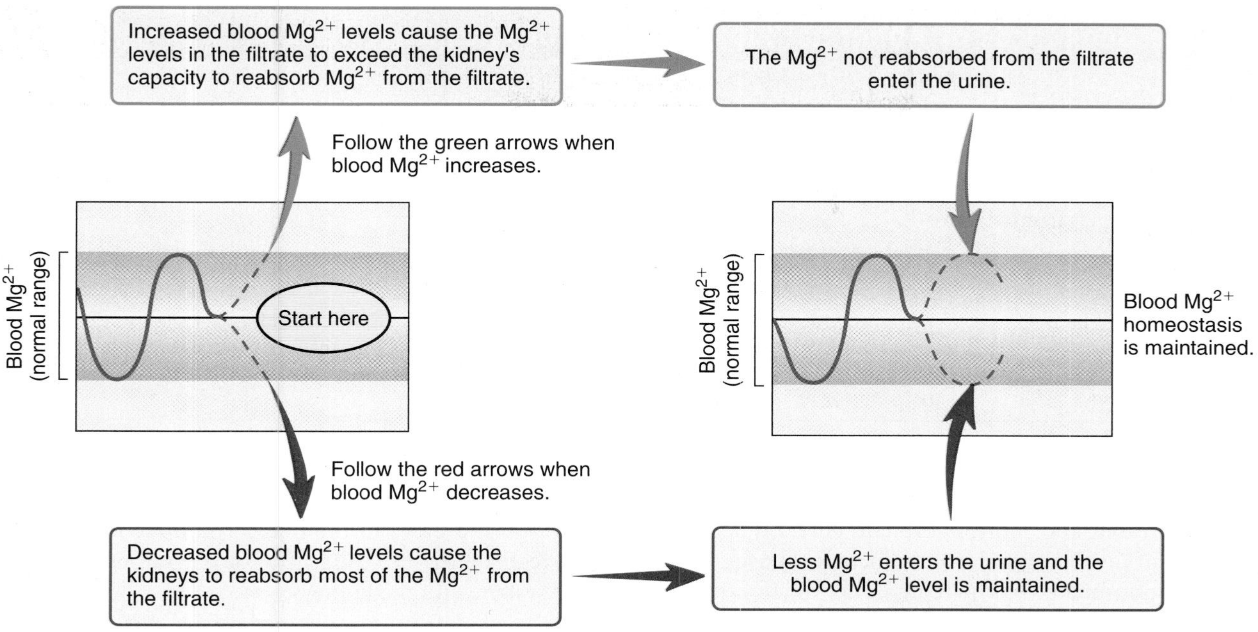

HOMEOSTASIS FIGURE 27.10 **Summary of Blood Mg^{2+} Regulation**

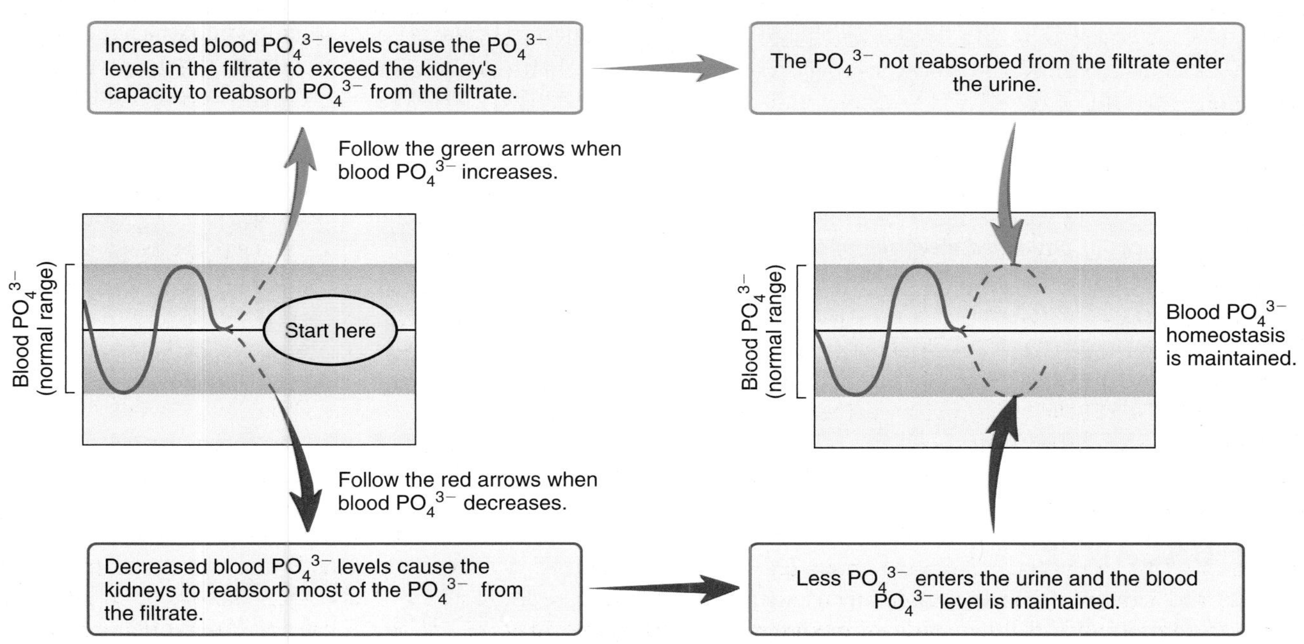

HOMEOSTASIS FIGURE 27.11 **Summary of Blood Phosphate Ion Regulation**

TABLE 27.10 **Consequences of Abnormal Plasma Levels of Phosphate Ions**

Major Causes	Symptoms
Hypophosphatemia	
Reduced absorption from the intestine associated with vitamin D deficiency and alcohol abuse	Reduced rate of metabolism, reduced transport of oxygen, reduced white blood cell functions, and blood clotting
Increased renal excretion with hyperparathyroidism	
Hyperphosphatemia	
Renal failure	
Tissue destruction with chemotherapy used to treat metastatic tumors	Symptoms related to reduced plasma Ca^{2+} concentrations due to calcium phosphate deposited in tissues such as the lungs, kidneys, and joints
Hyperparathyroidism initially leading to elevated plasma Ca^{2+} combined with reduced excretion of phosphate by the kidney	

urine. Thus, whenever plasma PTH is increased, tubular phosphate reabsorption is decreased and more phosphate enters the urine. If phosphate levels in the extracellular fluid increase above normal levels, Ca^{2+} and phosphate ions precipitate as calcium phosphate salts in soft tissues.

Elevated blood levels of phosphate may occur with acute or chronic renal failure as a result of a very reduced rate of filtrate formation by the kidney. The rate of phosphate excretion is consequently reduced. Also, the chronic use of laxatives containing phosphates may result in elevated blood levels of phosphate. Symptoms of elevated phosphate levels are related to reduced blood Ca^{2+} levels because phosphate ions and Ca^{2+} precipitate out of solution and are deposited in the soft tissues of the body. Prolonged elevation of blood levels of phosphate can result in calcium phosphate deposits in the joints and other tissues such as the lungs and kidneys. **Hypophosphatemia** (hī′pō-fos-fă-tē′mē-ă) is a below normal blood level of phosphate, and **hyperphosphatemia** (hī′per-fos-fă-tē′mē-ă) is an above normal blood level of phosphate. The consequences of increased or reduced plasma levels of phosphates are presented in table 27.10.

30 ***Explain how the kidneys control plasma levels of phosphate ions.***

31 ***How does an increased level of PTH affect tubular phosphate reabsorption?***

32 ***What are the consequences of prolonged elevation of blood levels of phosphate ions?***

PREDICT 3

Mary Thon runs several miles each day. List the mechanisms through which water loss changes during her run.

REGULATION OF ACID–BASE BALANCE

Hydrogen ions affect the activity of enzymes and interact with many electrically charged molecules. Consequently, most chemical reactions within the body are highly sensitive to the H^+ concentration of the fluid in which they occur. The maintenance of the H^+ concentration within a narrow range of values is essential for normal metabolic reactions. The H^+ concentration is determined by acids and bases in the body. The three major mechanisms that regulate the H^+ concentration are buffer systems, the respiratory system, and the kidneys.

Acids and Bases

Acids are substances that release H^+ into a solution; **bases** bind to H^+ and remove them from solution. Many bases release hydroxide ions (OH^-), which react with H^+ to form water (H_2O). Acids and bases are categorized as either strong or weak. Strong acids and strong bases completely dissociate to form ions in solution. Hydrochloric acid is a strong acid that dissociates to form H^+ and Cl^- (figure 27.12), and sodium hydroxide is a strong base, which dissociates to form Na^+ and OH^-. In contrast to strong acids, weak acids dissociate, but most molecules remain intact. Many of the weak acid molecules do not dissociate to release H^+ into the solution. For each type of weak acid an equilibrium is established (see figure 27.12). The proportion of weak acid molecules that release H^+ into solution is very predictable and is influenced by the pH of the solution into which the weak acid is placed. Weak acids are common in living systems, and they play important roles in preventing large changes in body fluid pH.

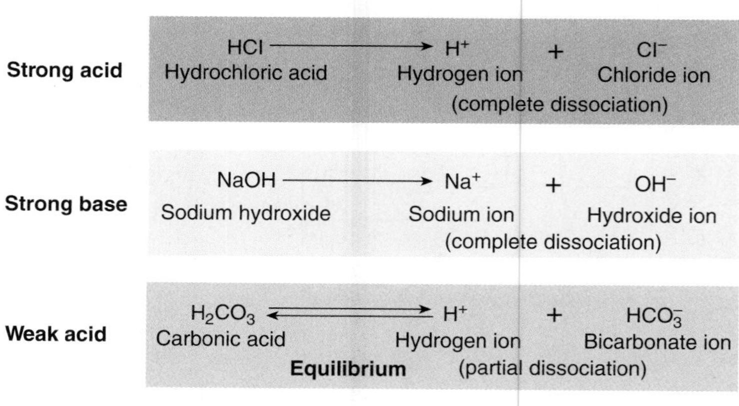

FIGURE 27.12 **Comparison of Strong Acids and Bases with Weak Acids**

Strong acids and bases completely dissociate when dissolved in water. Weak acids do not completely dissociate. Weak acids partially dissociate so that an equilibrium is established between the acid and the ions that are formed when the dissociation occurs.

TABLE 27.11 Buffer Systems

	Characteristics of the Buffer Systems
Protein Buffer System	Intracellular proteins and plasma proteins form a large pool of protein molecules that can act as buffer molecules. Because of their high concentration, they provide approximately three-fourths of the buffer capacity of the body. Hemoglobin in red blood cells is an important intracellular protein. Other intracellular molecules, such as histone proteins and nucleic acids, also act as buffers.
Bicarbonate Buffer System	Components of the bicarbonate buffer system are not present in high enough concentrations in the extracellular fluid to constitute a powerful buffer system. However, the concentrations of the components of the buffer system are regulated. Therefore it plays an exceptionally important role in controlling the pH of extracellular fluid.
Phosphate Buffer System	Components of the phosphate buffer system are low in the extracellular fluids, compared with the other buffer systems, but it is an important intracellular buffer system.

33 ***Define acid and base.***

34 ***Describe weak acids. Why are weak acids important in living systems?***

Buffer Systems

Buffers (bŭf′erz; see chapter 2) resist changes in the pH of a solution. Buffers within body fluids stabilize the pH by chemically binding to excess H^+ when H^+ are added to a solution or by releasing H^+ when the H^+ concentration in a solution begins to fall.

Several important buffer systems function together to resist changes in the pH of body fluids (table 27.11). The carbonic acid/bicarbonate buffer system; protein molecules, such as hemoglobin and plasma proteins; and phosphate compounds all act as buffers.

Carbonic Acid/Bicarbonate Buffer System

Carbonic acid (H_2CO_3) is a weak acid. When it is dissolved in water, the following equilibrium is established:

$$H_2CO_3 \rightleftarrows HCO_3^- + H^+$$

The carbonic acid/bicarbonate buffer system depends on the equilibrium that is quickly established between H_2CO_3 and the H^+ and bicarbonate (HCO_3^-). When H^+ are added to this solution, by adding a small amount of a strong acid a large proportion of the H^+ binds to HCO_3^- to form H_2CO_3, and only a small percentage remain as free H^+. Thus, a large decrease in pH is resisted by the carbonic acid/bicarbonate buffer system when acidic substances are added to a solution containing H_2CO_3.

When H^+ are removed from a solution containing H_2CO_3, by adding a small amount of a strong base many of the H_2CO_3 form HCO_3^- and H^+. Thus, a large increase in pH is resisted when basic substances are added to a solution containing H_2CO_3.

The carbonic acid/bicarbonate buffer system plays an important role in regulating the extracellular pH. It quickly responds to the addition of substances such as CO_2 or lactic acid produced by increased metabolism during exercise (see chapter 23) and increased fatty acid and ketone body production during periods of elevated fat metabolism (see chapter 25). It also responds to the addition of basic substances, such as large amounts of $NaHCO_3$ consumed as an antacid. The carbonic acid/bicarbonate buffer system has a limited capacity to resist changes in pH, but it is very important because it plays an essential role in the control of pH by both the respiratory system and the kidneys (see "Mechanisms of Acid–Base Balance Regulation," below).

Protein Buffer System

Intracellular proteins and plasma proteins form a large pool of protein molecules that act as buffer molecules. They provide approximately three-fourths of the buffer capacity of the body because of their high concentration. Hemoglobin in red blood cells is one of the most important intracellular proteins. Other intracellular molecules, such as histone proteins, associated with nucleic acids also act as buffers. The capacity of proteins to function as buffers is due to the functional groups of amino acids, such as carboxyl (–COOH) or amino (–NH_2) groups. These functional groups can act as weak acids and bases. Consequently, as the H^+ concentration increases, more H^+ bind to the functional groups, and, when the H^+ concentration decreases, H^+ are released from the functional groups (see table 27.11).

Phosphate Buffer System

The concentration of phosphate and phosphate-containing molecules is low in the extracellular fluid, compared with that of the other buffer systems, but it is an important intracellular buffer system. Phosphate-containing molecules, such as DNA, RNA, ATP, and phosphate ions (such as HPO_4^{2-}), in solution act as buffers. Phosphate ions act as weak acids and therefore can bind to H^+, to form $H_2PO_4^-$, when the pH decreases and ions, such as $H_2PO_4^-$, release H^+ into the solution when the pH begins to increase (see table 27.11).

35 ***Define buffer. Describe how a buffer works when H^+ are added to a solution or when they are removed from a solution.***

36 ***Name the three buffer systems of the body. Which of these systems provides the largest proportion of buffer capacity in the body?***

Mechanisms of Acid–Base Balance Regulation

Buffers and the mechanisms of acid–base balance regulation work together and play essential roles in the regulation of acid–base balance (figure 27.13). Buffers almost instantaneously resist changes in the pH of body fluids. The mechanisms of acid–base regulation

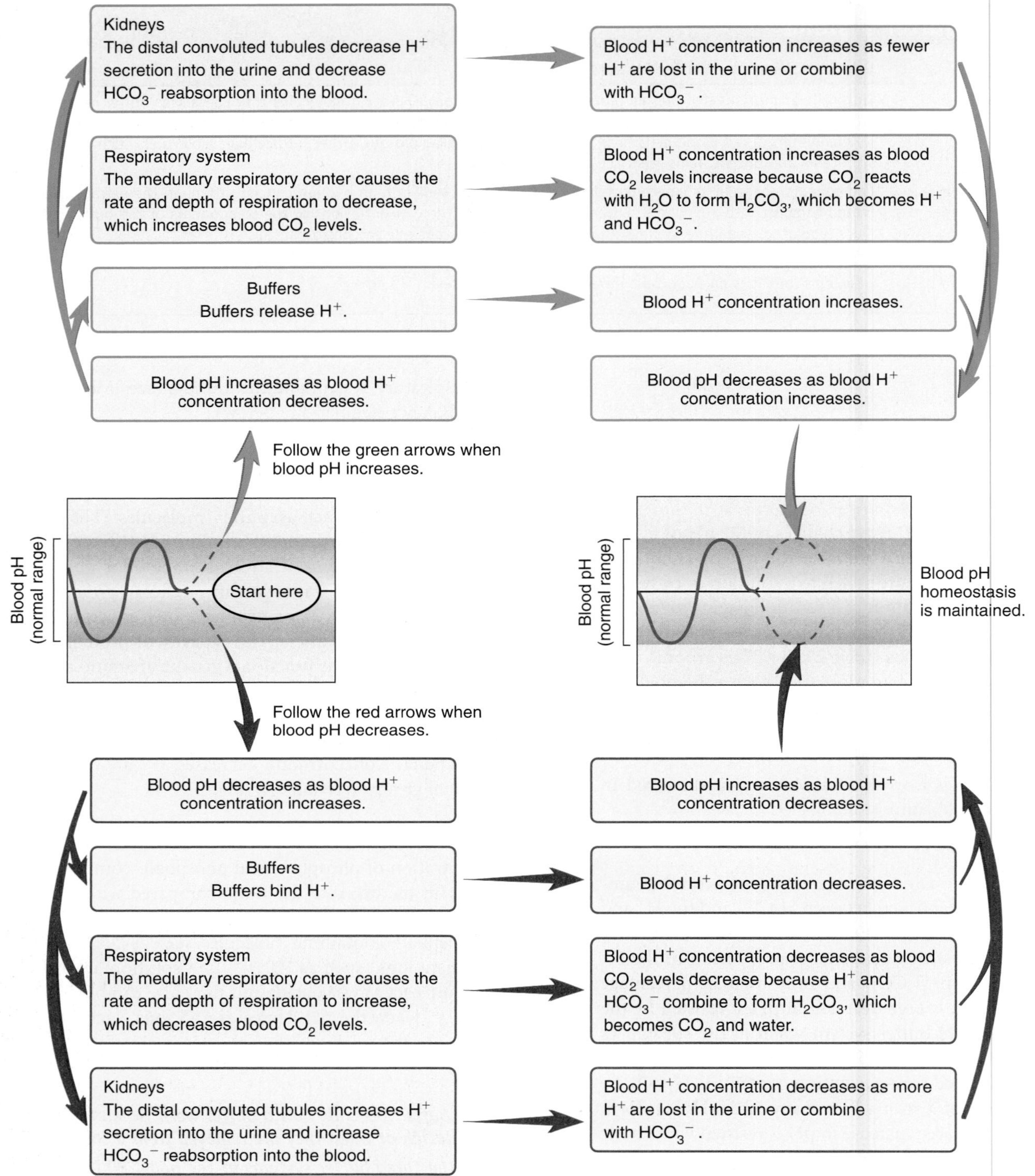

HOMEOSTASIS FIGURE 27.13 Summary of Acid–Base Balance Regulation
For more information on buffers, see "Buffer Systems" (p. 1021); on the respiratory system, see figure 27.14; on the kidneys, see figure 27.15.

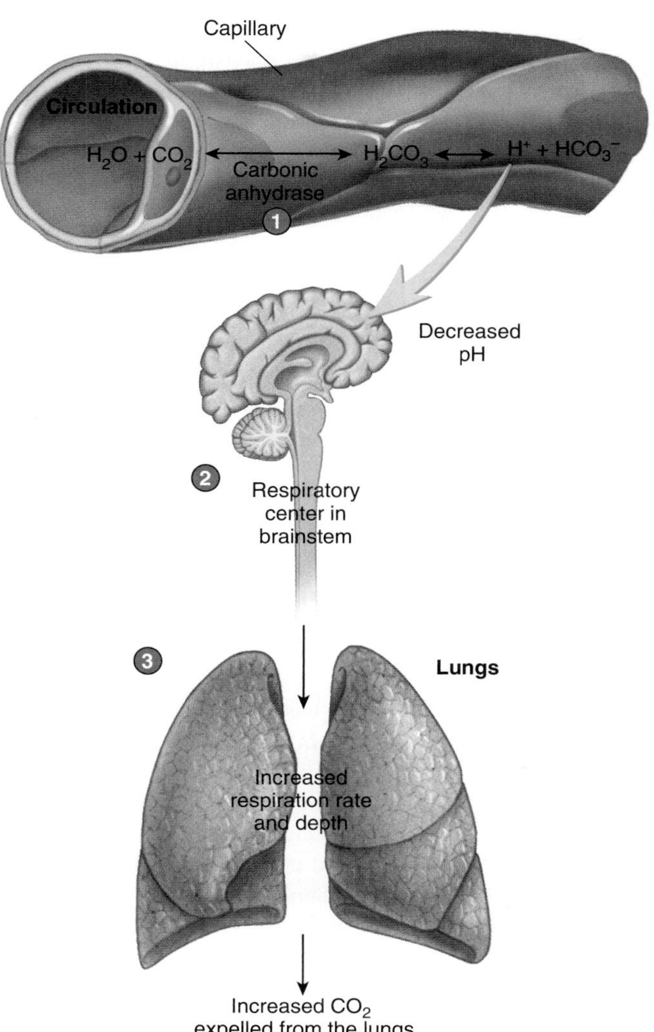

1. Carbon dioxide reacts with H_2O to form H_2CO_3. An enzyme, carbonic anhydrase, found in red blood cells and on the surface of blood vessel epithelium, catalyzed the reaction. Bicarbonate ion dissociates to form H^+ and HCO_3^-. An equilibrium is quickly established.

2. A decreased pH of the extracellular fluid stimulates the respiratory center and causes an increased rate and depth of respiration.

3. An increased rate and depth of respiration causes CO_2 to be expelled from the lungs, thus reducing the extracellular CO_2 levels. As CO_2 levels decrease,the extracellular concentration of H^+ decreases, and the extracellular fluid pH increases.

PROCESS FIGURE 27.14 Respiratory Regulation of Body Fluid Acid–Base Balance

depend on the regulation of respiration and on kidney function. The respiratory system responds within a few minutes to changes in pH to bring the pH of body fluids back toward its normal range. The respiratory system's capacity to regulate pH, however, is not as great as that of the kidneys, nor does the respiratory system have the same ability to return the pH to its precise range of normal values. In contrast, the kidneys respond more slowly, within hours to days, to alterations of body fluid pH, and their capacity to respond is substantial.

37 ***Compare the capacity and rate at which the respiratory system and kidneys control body fluid pH.***

Respiratory Regulation of Acid–Base Balance

The respiratory system regulates acid–base balance by influencing the carbonic acid/bicarbonate buffer system. Carbon dioxide (CO_2) reacts with water (H_2O) to form carbonic acid (H_2CO_3), which dissociates to form H^+ and HCO_3^- as follows:

$$H_2O + CO_2 \leftrightarrows H_2CO_3 \leftrightarrows H^+ + HCO_3^-$$

This reaction is in equilibrium. As CO_2 increases, CO_2 combines with H_2O. The higher the concentration of CO_2, the greater the amount of H_2CO_3 formed. Many H_2CO_3 molecules then dissociate to form H^+ and HCO_3^-. If CO_2 levels decline, however, the equilibrium shifts in the opposite direction so that many H^+ and HCO_3^- combine to form H_2CO_3, which then forms CO_2 and H_2O. Thus, H^+ and HCO_3^- decrease in the solution.

The reaction between CO_2 and H_2O is catalyzed by an enzyme, **carbonic anhydrase,** which is found in a relatively high concentration in red blood cells and on the surface of capillary epithelial cells (figure 27.14). This enzyme does not influence equilibrium but accelerates the rate at which the reaction proceeds in either direction so that equilibrium is achieved quickly.

Decreases in body fluid pH, regardless of the cause, stimulate neurons in the respiratory center in the brainstem and cause the rate and depth of ventilation to increase. The increased rate and depth of ventilation cause CO_2 to be eliminated from the body through the lungs at a greater rate, and the concentration of CO_2 in the body fluids decreases. As CO_2 levels decline, the carbonic acid/bicarbonate buffer system reacts. Hydrogen ions combine with HCO_3^- to form H_2CO_3, which then form CO_2 and H_2O. Consequently, the concentration of H^+ decreases toward its normal range as CO_2 exits through the respiratory system (see figure 27.14).

Increases in body fluid pH, regardless of the cause, inhibit neurons in the respiratory center in the brainstem and cause the rate and depth of ventilation to decrease. The decreased rate and depth of ventilation cause less CO_2 to be eliminated from the body through the lungs. The concentration of CO_2 in the body fluids increases because CO_2 is continually produced as a by-product of metabolism in all tissues. Thus, the body fluid concentration of H_2CO_3 increases also. As H_2CO_3 increases, many H_2CO_3 molecules dissociate to form H^+ and HCO_3^-. This results in an increase in the H^+ concentration, and the pH decreases toward its normal range.

PREDICT 4

Under stressful conditions, some people hyperventilate. What effect does the rapid rate of ventilation have on blood pH? Explain why a person who is hyperventilating may benefit from breathing into a paper bag.

Renal Regulation of Acid–Base Balance

Cells of the kidney tubules directly regulate acid–base balance by increasing or decreasing the rate of H^+ secretion into the filtrate (figure 27.15) and the rate of HCO_3^- reabsorption. Carbonic anhydrase present within nephron cells catalyzes the formation of H_2CO_3 from CO_2 and H_2O. The carbonic acid molecules dissociate to form H^+ and HCO_3^-. An antiport system then exchanges H^+ for Na^+ across the apical membrane of the cells. Thus, cells of the kidney tubules secrete H^+ into the filtrate and reabsorb Na^+. The Na^+ and HCO_3^- are symported across the basal membrane. After the Na^+ and HCO_3^- are symported from the kidney tubule cells, they diffuse into the peritubular capillaries. As a result, H^+ are secreted into the lumen of the kidney tubules, and HCO_3^- pass into the extracellular fluid.

The reabsorbed HCO_3^- combine with excess H^+ in the extracellular fluid to form H_2CO_3. This combination removes H^+ from the extracellular fluid and increases extracellular pH. The rate of H^+ secretion and HCO_3^- reabsorption increases when the pH of the body fluids decreases, and this process slows when the pH of the body fluids increases (see figure 27.15).

Some of the H^+ secreted by cells of the kidney tubules into the filtrate combine with HCO_3^-, which enter the filtrate through the filtration membrane, in the form of sodium bicarbonate ($NaHCO_3$). Within the kidney tubules, H^+ combines with HCO_3^- to form H_2CO_3, which then dissociate to form CO_2 and H_2O. The CO_2 diffuses from the filtrate into the cells of the kidney tubules, where it can react with H_2O to form H_2CO_3, which subsequently dissociates to form H^+ and HCO_3^- (see figure 27.15). Once again, H^+ are antiported into the lumen of the kidney tubules in exchange for Na^+, whereas HCO_3^- enter the extracellular fluid. As a result, many of the HCO_3^- entering the filtrate reenter the extracellular fluid.

Hydrogen ions secreted into the nephron normally exceed the amount of HCO_3^- that enters the kidney tubules through the filtration membrane. Because the H^+ combine with HCO_3^-, almost all the HCO_3^- are reabsorbed from the kidney tubules (see figure 27.15). Few HCO_3^- are lost in the urine unless the pH of the body fluids becomes elevated.

The rate of H^+ secretion into the filtrate and the rate of HCO_3^- reabsorption into the extracellular fluid decrease if the pH of the body fluids increases. As a result, the amount of bicarbonate filtered into the kidney tubules exceeds the amount of secreted H^+,

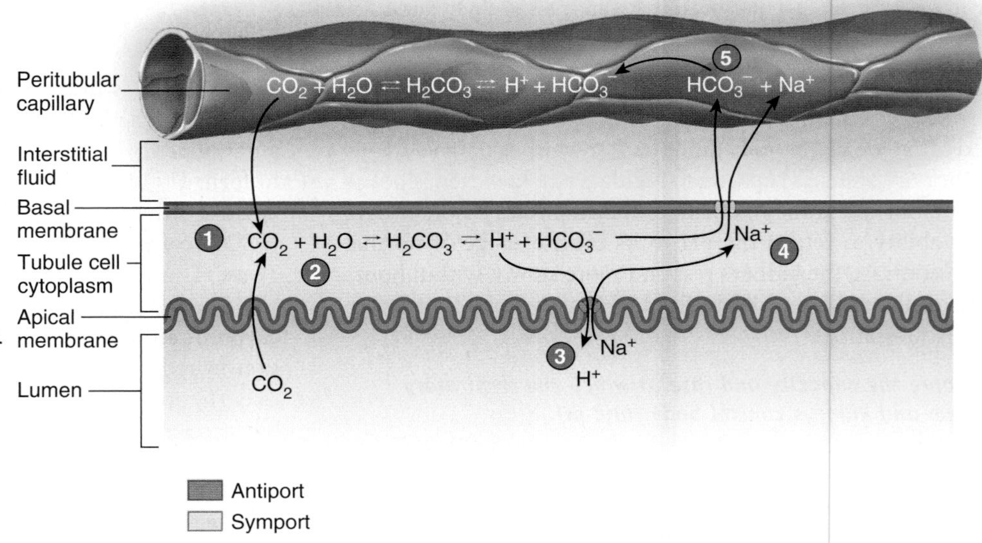
1. When the filtrate or blood pH decreases, H+ combine with HCO3− to form carbonic acid that is converted into CO2 and H2O. The CO2 diffuses into tubule cells.
2. In the tubule cells, CO2 combines with H2O to form H2CO3 that dissociates to form H+ and HCO3−.
3. An antiport mechanism secretes H+ into the filtrate in exchange for Na+ from the filtrate. As a result, filtrate pH decreases.
4. Bicarbonate ions are symported with Na+ into the interstitial fluid. They then diffuse into capillaries.
5. In capillaries, HCO3− combine with H+. This decreases the H+ concentration and increases blood pH.
Peritubular capillary
Interstitial fluid
Basal membrane
Tubule cell cytoplasm
Apical membrane
Lumen
CO2 + H2O ⇌ H2CO3 ⇌ H+ + HCO3−
HCO3− + Na+
CO2 + H2O ⇌ H2CO3 ⇌ H+ + HCO3−
Na+
Na+
H+
CO2
1
2
3
4
5
Antiport
Symport

PROCESS FIGURE 27.15 Kidney Regulation of Body Fluid Acid–Base Balance

As the extracellular pH decreases, the rate of H^+ secretion by tubule cells of the kidney and HCO_3^- reabsorption increase.

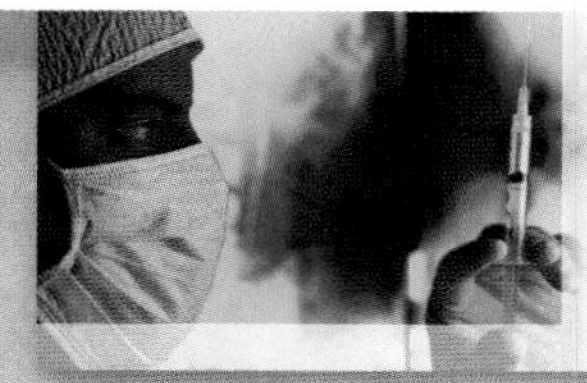

CLINICAL FOCUS

Acidosis and Alkalosis

The normal pH value of the body fluids is between 7.35 and 7.45. When the pH value of body fluids is below 7.35, the condition is called **acidosis** (as-i-dō′sis), and, when the pH is above 7.45, it is called **alkalosis** (al′kă-lō′sis).

Metabolism produces acidic products that lower the pH of the body fluids. For example, CO_2 is a by-product of metabolism, and CO_2 combines with water to form H_2CO_3. Also, lactic acid is a product of anaerobic metabolism, protein metabolism produces phosphoric and sulfuric acids, and lipid metabolism produces fatty acids. These acidic substances must continuously be eliminated from the body to maintain pH homeostasis. The failure to eliminate the acidic products of metabolism results in acidosis. The excess elimination of acidic products of metabolism results in alkalosis.

The major effect of acidosis is depression of the central nervous system. When blood pH falls below 7.35, the central nervous system malfunctions, and the individual becomes disoriented and possibly comatose as the condition worsens.

A major effect of alkalosis is hyperexcitability of the nervous system. Peripheral nerves are affected first, resulting in spontaneous nervous stimulation of muscles. Spasms, tetanic contractions, and possibly extreme nervousness or convulsions result. Severe alkalosis can cause death as a result of tetany of the respiratory muscles.

Although buffers help resist changes in the pH of body fluids, the respiratory system and the kidneys regulate the pH of the body fluids. Malfunctions in either the respiratory system or the kidneys can result in acidosis or alkalosis.

Acidosis and alkalosis are categorized by the cause of the condition. **Respiratory acidosis** or **respiratory alkalosis** results from abnormalities in the respiratory system. **Metabolic acidosis** or **metabolic alkalosis** results from all causes other than abnormal respiratory functions.

Inadequate ventilation of the lungs causes respiratory acidosis (table A). The rate at which CO_2 is eliminated from the body fluids through the lungs falls. This increases the concentration of CO_2 in the body fluids. As CO_2 levels increase, CO_2 reacts with water to form H_2CO_3. Carbonic acid forms H^+ and HCO_3^-. The increase in H^+ concentration causes the pH of the body fluids to decrease. If the pH of the body fluids falls below 7.35, the symptoms of respiratory acidosis become apparent.

Buffers help resist a decrease in pH, and the kidneys help compensate for the failure of the lungs to prevent respiratory acidosis by increasing the rate at which they secrete H^+ into the filtrate and reabsorb HCO_3^-. The capacity of buffers to resist changes in pH can be exceeded, however, and a period of 1–2 days is required for the kidneys to become maximally

Continued

Table A	Acidosis and Alkalosis
Acidosis	
Respiratory Acidosis	
Reduced elimination of CO_2 from the body fluids	
Asphyxia	
Hypoventilation (e.g., impaired respiratory center function due to trauma, tumor, shock, or renal failure)	
Advanced asthma	
Severe emphysema	
Metabolic Acidosis	
Elimination of large amounts of HCO_3^- resulting from mucous secretion (e.g., severe diarrhea and vomiting of lower intestinal contents)	
Direct reduction of body fluid pH as acid is absorbed (e.g., ingestion of acidic drugs, such as aspirin)	
Production of large amounts of fatty acids and other acidic metabolites, such as ketone bodies (e.g., untreated diabetes mellitus)	
Inadequate oxygen delivery to tissue, resulting in anaerobic respiration and lactic acid buildup (e.g., exercise, heart failure, or shock)	
Alkalosis	
Respiratory Alkalosis	
Reduced CO_2 levels in the extracellular fluid (e.g., hyperventilation due to emotions)	
Decreased atmospheric pressure reduces oxygen levels, which stimulates the chemoreceptor reflex (e.g., causes hyperventilation at high altitudes)	
Metabolic Alkalosis	
Elimination of H^+ and reabsorption of HCO_3^- in the stomach or kidney (e.g., severe vomiting or formation of acidic urine in response to excess aldosterone)	
Ingestion of alkaline substances (e.g., large amounts of sodium bicarbonate)	

Continued

functional. Thus, the kidneys are not effective if respiratory acidosis develops quickly. The kidneys are very effective if respiratory acidosis develops slowly or if it lasts long enough for the kidneys to respond. For example, the kidneys cannot compensate for respiratory acidosis occurring in response to a severe asthma attack that begins quickly and subsides within hours. If, however, respiratory acidosis results from emphysema, which develops over a long time, the kidneys play a significant role in helping compensate.

Respiratory alkalosis results from hyperventilation of the lungs (see table A). This increases the rate at which CO_2 is eliminated from the body fluids and results in a decrease in the concentration of CO_2 in the body fluids. As CO_2 levels decrease, H^+ react with HCO_3^- to form H_2CO_3. The H_2CO_3 forms H_2O and CO_2. The resulting decrease in the concentration of H^+ causes the pH of the body fluids to increase. If the pH of body fluids increases above 7.45, the symptoms of respiratory alkalosis become apparent.

The kidneys help compensate for respiratory alkalosis by decreasing the rate of H^+ secretion into the filtrate and the rate of HCO_3^- reabsorption. If an increase in pH occurs, the kidneys need 1–2 days to compensate. Thus, the kidneys are not effective if respiratory alkalosis develops quickly. They are very effective, however, if respiratory alkalosis develops slowly. For example, the kidneys are not effective in compensating for respiratory alkalosis that occurs in response to hyperventilation triggered by emotions, which usually begins quickly and subsides within minutes or hours. If alkalosis results, however, from staying at a high altitude over a 2- or 3-day period, the kidneys play a significant role in helping compensate.

Metabolic acidosis results from all conditions that decrease the pH of the body fluids below 7.35, with the exception of conditions resulting from altered function of the respiratory system (see table A). As H^+ accumulate in the body fluids, buffers first resist a decline in pH. If the buffers cannot compensate for the increase in H^+, the respiratory center helps regulate body fluid pH. The reduced pH stimulates the respiratory center, which causes hyperventilation. During hyperventilation, CO_2 is eliminated at a greater rate. The elimination of CO_2 also eliminates excess H^+ and helps maintain the pH of the body fluids within a normal range.

If metabolic acidosis persists for many hours and if the kidneys are functional, the kidneys can also help compensate for metabolic acidosis by secreting H^+ at a greater rate and increasing the rate of HCO_3^- reabsorption. The symptoms of metabolic acidosis appear if the respiratory and renal systems are not able to maintain the pH of the body fluids within its normal range.

Metabolic alkalosis results from all conditions that increase the pH of the body fluids above 7.45, with the exception of conditions resulting from altered function of the respiratory system. As H^+ decrease in the body fluids, buffers first resist an increase in pH. If the buffers cannot compensate for the decrease in H^+, the respiratory center helps regulate body fluid pH. The increased pH inhibits respiration. Reduced respiration allows CO_2 to accumulate in the body fluids. Carbon dioxide reacts with water to produce H_2CO_3. If metabolic alkalosis persists for several hours and if the kidneys are functional, the kidneys reduce the rate of H^+ secretion to help reverse alkalosis (see table A).

and the excess HCO_3^- pass into the urine. The excretion of excess HCO_3^- in the urine diminishes the amount of HCO_3^- in the extracellular fluid. This allows the amount of extracellular H^+ to increase; as a consequence, the pH of the body fluids decreases toward its normal range.

Cushing Syndrome and Alkalosis

Aldosterone increases the rate of Na^+ reabsorption and K^+ secretion by the kidneys, but in high concentrations aldosterone also stimulates H^+ secretion. Elevated aldosterone levels, such as occur in patients with Cushing syndrome, can therefore elevate body fluid pH above normal (alkalosis). The major factor that influences the rate of H^+ secretion, however, is the pH of the body fluids.

PREDICT 5

Predict the effect of aldosterone hyposecretion on body fluid pH.

The secretion of H^+ into the urine can decrease the filtrate pH to approximately 4.5. A filtrate pH below 4.5 inhibits the secretion of additional H^+. The H^+ that pass into the filtrate are greater than the quantity required to decrease the pH of an unbuffered solution below 4.5. Buffers in the filtrate combine with many of the secreted H^+. Bicarbonate ions, phosphate ions (HPO_4^{2-}), and **ammonia** (NH_3) in the filtrate act as buffers. Both HCO_3^- and HPO_4^{2-} enter the kidney tubules through the filtration membrane along with the rest of the filtrate, and NH_3 diffuses across the wall of nephron cells to enter the filtrate. These ions combine with H^+ secreted by the nephron (figure 27.16).

NH_3 is produced in the cells of the nephron when amino acids, such as glycine, are deaminated. Subsequently, NH_3 diffuses from the nephron cells into the filtrate and combines with H^+ in the filtrate to form **ammonium ions** (NH_4^+; see figure 27.14). The rate of NH_3 production increases when the pH of the body fluids has been depressed for 2–3 days, such as during prolonged respiratory or metabolic acidosis. The elevated ammonia production increases the buffering capacity of the filtrate, allowing the secretion of additional H^+ into the urine.

HCO_3^-, HPO_4^{2-}, and NH_3 constitute major buffers within the filtrate, but other weak acids, such as lactic acid in the filtrate, also combine with H^+ and increase the amount of H^+ that can be secreted into the filtrate.

PREDICT 6

Mr. Puffer suffers from severe emphysema. Gas exchange in his lungs is not adequate, and he must have a supply of oxygen. His blood carbon dioxide level is elevated. Nevertheless, his blood pH is close to normal. Explain. Also explain how his blood bicarbonate ion (HCO_3^-) level is likely to differ from normal values.

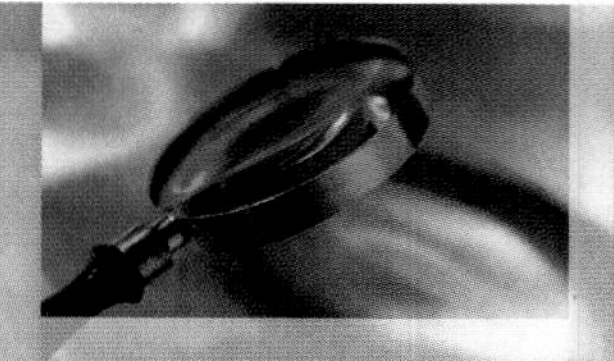

CASE STUDY
Gastroenteritis

Dan is a college student who works at the campus library most evenings. After work on a Friday night, Dan and some friends had a late dinner at a fast-food restaurant. Shortly after dinner, Dan was not feeling very well, so he went to his dorm room. By midnight, he was feeling very nauseated. He vomited repeatedly over the next several hours. He was still nauseated and was vomiting at least once each hour the next morning, so he went to the student health center at about 10:00 A.M. At the student health center, about 12 hours after Dan had begun to feel nauseated, the following observations were made (see table). Each time Dan vomited, a significant volume of acidic gastric fluid was expelled. Dan continued to vomit at least once each hour. The next day, about 36 hours after Dan had begun to feel nauseated, the following observations were made.

	12 Hours	36 Hours	Normal
Body Weight	71 kg	68 kg	Not known
Blood Pressure	115/75 mm Hg	90/60 mmHg	120/80 mmHg
Heart Rate	77 beats/min	105 beats/min	72 beats/min
Plasma pH	7.48	7.5	7.35 to 7.45
Plasma HCO_3^-	32 mEq/L	36 mEq/L	22 to 28 mEq/L
Plasma P_{CO_2}	44 mmHg	48 mmHg	35 to 43 mmHg
Skin Color	Pallor	Pallor	

Dan began to feel better by the end of the second day (about 48 hours after he had begun to be nauseated), and the parameters slowly returned to normal. At 72 hours, he felt fairly normal again. After the nauseous feeling disappeared, Dan was very thirsty and he drank a substantial amount of water and juice over several hours. It was assumed that Dan had had gastroenteritis caused by an unknown microorganism. The medication given to him for the infection appeared to be effective.

PREDICT 7

a. Explain the weight change, blood pressure change, heart rate change, and skin color change in Dan that occurred between 12 and 36 hours.

b. Explain the changes in plasma pH and plasma HCO_3^- between 12 and 36 hours.

c. Explain the plasma P_{CO_2} at 12 and 36 hours.

d. Explain why all of the parameters had returned to normal by 72 hours.

1. Hydrogen ions secreted into the filtrate are buffered.
2. Hydrogen ions can react with HCO_3^- that enters the filtrate to form H_2CO_3, which is in equilibrium with H_2O and CO_2.
3. Hydrogen ions can react with HPO_4^{2-} that enters the filtrate to form $H_2PO_4^-$.
4. Hydrogen ions can react with NH_3 formed by amino acid deamination and secreted into the nephron to form NH_4^+.

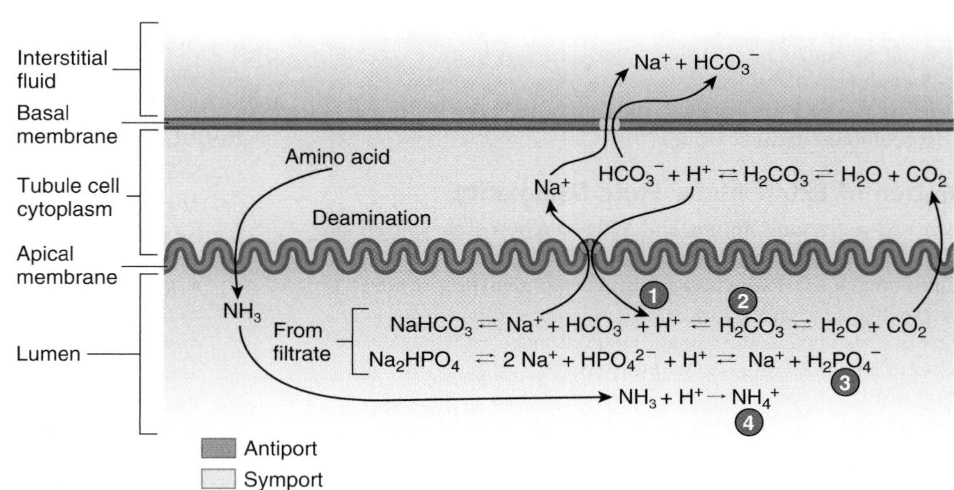

PROCESS FIGURE 27.16 Hydrogen Ion Buffering in the Filtrate

The secretion of H^+ into the filtrate decreases filtrate pH. As the concentration of H^+ increases in the filtrate, the ability of tubule cells to secrete additional H^+ becomes limited. Buffering the H^+ in the filtrate decreases their concentration and enables tubule cells to secrete additional H^+.

38 ***Name the three mechanisms that play essential roles in the regulation of acid–base balance.***

39 ***What happens to blood pH when blood CO_2 levels go up or down? What causes this change?***

40 ***What effect do increased CO_2 levels or decreased pH have on respiration? How does this change in respiration affect blood pH?***

41 ***Describe the process by which nephron cells move H^+ into the kidney tubule lumen and HCO_3^- into the extracellular fluid.***

42 ***Describe the process by which HCO_3^- are reabsorbed from the kidney tubule lumen.***

43 ***Name the factors that cause an increase and a decrease in H^+ secretion.***

44 ***What is the purpose of buffers in the urine? Describe how the ammonia buffer system operates.***

SUMMARY

Water, acid, base, and electrolyte levels are maintained within a narrow range of concentrations. The urinary, respiratory, gastrointestinal, integumentary, nervous, and endocrine systems play a role in maintaining fluid, electrolyte, and pH balance.

Body Fluids (p. 1005)

1. Intracellular fluid is inside cells.
2. Extracellular fluid is outside cells and includes interstitial fluid and plasma.

Regulation of Body Fluid Concentration and Volume (p. 1006)

Regulation of Water Content

1. Water crosses the gastrointestinal tract through osmosis.
2. An increase in extracellular osmolality or a decrease in blood pressure stimulates the sense of thirst.
3. Wetting of the oral mucosa or stretch of the gastrointestinal tract inhibits thirst.
4. Learned behavior plays a role in the amount of fluid ingested.
5. The routes of water loss are listed below.
   - Water is lost through evaporation from the respiratory system and the skin (insensible perspiration and sweat).
   - Water loss into the gastrointestinal tract normally is small. Vomiting or diarrhea can significantly increase this loss.
   - The kidneys are the primary regulator of water excretion. Urine output can vary from a small amount of concentrated urine to a large amount of dilute urine.

Regulation of Extracellular Fluid Osmolality

1. Increased water consumption and ADH secretion occur in response to increases in extracellular fluid osmolality. Decreased water consumption and ADH secretion occur in response to decreases in extracellular fluid osmolality.
2. Increased water consumption and ADH decrease extracellular fluid osmolality by increasing water absorption from the intestine and water reabsorption from the nephrons. Decreased water consumption and ADH increase extracellular fluid osmolality by decreasing absorption from the intestine and water reabsorption from the nephrons.

Regulation of Extracellular Fluid Volume

1. Increased extracellular fluid volume results in decreased aldosterone secretion, increased ANH secretion, decreased ADH secretion, and decreased sympathetic stimulation of afferent arterioles. The effects of these changes are to decrease Na^+ reabsorption and to increase urine volume to decrease extracellular fluid volume.
2. Decreased extracellular fluid volume results in increased aldosterone secretion, decreased ANH secretion, increased ADH secretion, and increased sympathetic stimulation of the afferent arterioles. The effects of these changes are to increase Na^+ reabsorption and to decrease urine volume so as to increase extracellular fluid volume.

Regulation of Intracellular Fluid Composition (p. 1011)

1. Substances used or produced inside the cell and substances exchanged with the extracellular fluid determine the composition of intracellular fluid.
2. Intracellular fluid is different from extracellular fluid because the plasma membrane regulates the movement of materials.
3. The difference between intracellular and extracellular fluid concentrations determines water movement.

Regulation of Specific Electrolytes in the Extracellular Fluid (p. 1012)

The intake and elimination of substances from the body and the exchange of substances between the extracellular and intracellular fluids determine extracellular fluid composition.

Regulation of Sodium Ions

1. Sodium is responsible for 90%–95% of extracellular osmotic pressure.
2. The amount of Na^+ excreted in the kidneys is the difference between the amount of Na^+ that enters the nephron and the amount that is reabsorbed from the nephron.
   - Glomerular filtration rate determines the amount of Na^+ entering the nephron.
   - Aldosterone determines the amount of Na^+ reabsorbed.
3. Small quantities of Na^+ are lost in sweat.
4. Increased blood osmolality leads to the production of a small volume of concentrated urine and to thirst. Decreased blood osmolality leads to the production of a large volume of dilute urine and to decreased thirst.
5. Increased blood pressure increases water and salt loss.
   - Baroreceptor reflexes reduce ADH secretion.
   - Renin secretion is inhibited, leading to reduced aldosterone production.

Regulation of Chloride Ions

Chloride ions are the dominant negatively charged ions in extracellular fluid.

Regulation of Potassium Ions

1. The extracellular concentration of K^+ affects resting membrane potentials.
2. The amount of K^+ excreted depends on the amount that enters with the glomerular filtrate, the amount actively reabsorbed by the nephron, and the amount secreted into the distal convoluted tubule.
3. Aldosterone increases the amount of K^+ secreted.

Regulation of Calcium Ions

1. Elevated extracellular calcium levels prevent membrane depolarization. Decreased levels lead to spontaneous action potential generation.
2. Parathyroid hormone increases extracellular Ca^{2+} levels and decreases extracellular phosphate levels. It stimulates osteoclast activity, increases calcium reabsorption from the kidneys, and stimulates active vitamin D production.
3. Vitamin D stimulates Ca^{2+} uptake in the intestines.
4. Calcitonin decreases extracellular Ca^{2+} levels.

Regulation of Magnesium Ions

The kidney's capacity to reabsorb magnesium is limited so that excess magnesium is lost in the urine and decreased extracellular magnesium results in a greater degree of magnesium reabsorption.

Regulation of Phosphate Ions

1. Under normal conditions, reabsorption of phosphate occurs at a maximum rate in the nephron.
2. An increase in plasma phosphate increases the amount of phosphate in the nephron beyond that which can be reabsorbed, and the excess is lost in the urine.

Regulation of Acid–Base Balance (p. 1020)

Acids and Bases

Acids release H^+ into solution, and bases remove them.

Buffer Systems

1. A buffer resists changes in pH.
   - When H^+ are added to a solution, the buffer removes them.
   - When H^+ are removed from a solution, the buffer replaces them.
2. Carbonic acid/bicarbonate, proteins, and phosphate compounds are important buffers.

Mechanisms of Acid–Base Balance Regulation

Buffers, the respiratory system, and the kidneys regulate acid–base balance.

1. Respiratory regulation of pH is achieved through the carbonic acid/bicarbonate buffer system.
   - As carbon dioxide levels increase, pH decreases.
   - As carbon dioxide levels decrease, pH increases.
   - Carbon dioxide levels and pH affect the respiratory centers. Hypoventilation increases blood carbon dioxide levels, and hyperventilation decreases blood carbon dioxide levels.
2. The secretion of H^+ into the filtrate and the reabsorption of HCO_3^- into extracellular fluid cause extracellular pH to increase.
   - Carbonic acid dissociates to form H^+ and HCO_3^- in nephron cells.
   - An antiport mechanism moves H^+ into the nephron lumen and Na^+ into the nephron cell.
   - Sodium ions and HCO_3^- diffuse into the extracellular fluid.
3. Bicarbonate ions in the filtrate are reabsorbed.
   - Bicarbonate ions combine with H^+ to form carbonic acid, which dissociates to form carbon dioxide and water.
   - Carbon dioxide diffuses into nephron cells and forms carbonic acid, which dissociates to form HCO_3^- and H^+.
   - Bicarbonate ions diffuse into the extracellular fluid, and H^+ are secreted into the filtrate.
4. The rate of H^+ secretion increases as body fluid pH decreases or as aldosterone levels increase.
5. Secretion of H^+ is inhibited when urine pH falls below 4.5.
   - Carbonic acid/bicarbonate, ammonia, and phosphate buffers in the urine resist a drop in pH.
   - As the buffers absorb H^+, more H^+ are pumped into the urine.

REVIEW AND COMPREHENSION

1. The sensation of thirst increases when
   a. the levels of angiotensin II increase.
   b. the osmolality of the blood decreases.
   c. blood pressure increases.
   d. renin secretion decreases.
2. Insensible perspiration
   a. is lost through sweat glands.
   b. results in heat loss from the body.
   c. increases when ADH secretion increases.
   d. results in the loss of solutes, such as Na^+ and Cl^-.
3. The composition and volume of body fluid are regulated primarily by the
   a. skin. b. lungs. c. kidneys. d. heart. e. spleen.
4. Which of these conditions *decreases* extracellular fluid volume?
   a. constriction of afferent arterioles
   b. increased ADH secretion
   c. decreased ANH secretion
   d. decreased aldosterone secretion
   e. stimulation of sympathetic nerves to the kidneys
5. Which of these results in an increased blood Na^+ concentration?
   a. decrease in ADH secretion
   b. decrease in aldosterone secretion
   c. increase in ANH
   d. decrease in renin secretion
6. Which of these mechanisms is the most important for regulating blood osmolality?
   a. ADH c. ANH
   b. renin-angiotensin-aldosterone d. parathyroid hormone
7. A decrease in extracellular K^+
   a. produces depolarization of the plasma membrane.
   b. results when aldosterone levels increase.
   c. occurs when tissues are damaged (e.g., in burn patients).
   d. increases ANH secretion.
   e. increases PTH secretion.
8. Calcium ion concentration in the blood decreases when
   a. vitamin D levels are lower than normal.
   b. calcitonin secretion decreases.
   c. parathyroid hormone secretion increases.
   d. all of the above.

9. An acid
   a. solution has a pH greater than 7.
   b. is a substance that releases H^+ into a solution.
   c. is considered weak if it completely dissociates in water.
   d. all of the above.
10. Buffers
    a. release H^+ when pH increases.
    b. resist changes in the pH of a solution.
    c. include the proteins of the blood.
    d. all of the above.
11. Which of these is *not* a buffer system in the body?
    a. sodium chloride buffer system
    b. carbonic acid/bicarbonate buffer system
    c. phosphate buffer system
    d. protein buffer system
12. Which of these systems regulating blood pH is the fastest-acting?
    a. respiratory
    b. kidney
13. An increase in blood carbon dioxide levels is followed by a (an) ____________ in H^+ and a (an) ____________ in blood pH.
    a. increase, increase
    b. increase, decrease
    c. decrease, increase
    d. decrease, decrease
14. High levels of bicarbonate ions in the urine indicate
    a. a low level of H^+ secretion into the urine.
    b. that the kidneys are causing blood pH to increase.
    c. that urine pH is decreasing.
    d. all of the above.
15. High levels of ammonium ions in the urine indicate
    a. a high level of H^+ secretion into the urine.
    b. that the kidneys are causing blood pH to decrease.
    c. that urine pH is too alkaline.
    d. all of the above.
16. Blood plasma pH is normally
    a. slightly acidic.
    b. strongly acidic.
    c. slightly alkaline.
    d. strongly alkaline.
    e. neutral.
17. Acidosis
    a. increases neuron excitability.
    b. can produce tetany by affecting the peripheral nervous system.
    c. may lead to coma.
    d. may produce convulsions through the central nervous system.
18. Respiratory alkalosis is caused by ____________ and can be compensated for by the production of a more ____________ urine.
    a. hypoventilation, alkaline
    b. hypoventilation, acidic
    c. hyperventilation, acidic
    d. hyperventilation, alkaline

Answers in Appendix E

CRITICAL THINKING

1. In patients with diabetes mellitus, not enough insulin is produced; as a consequence, blood glucose levels increase. If blood glucose levels rise high enough, the kidneys are unable to absorb the glucose from the glomerular filtrate, and glucose "spills over" into the urine. What effect does this glucose have on urine concentration and volume? How does the body adjust to the excess glucose in the urine?
2. A patient suffering from a tumor in the hypothalamus produces excessive amounts of ADH, a condition called syndrome of inappropriate ADH (SIADH) production. For this patient, the excessive ADH production is chronic and has persisted for many months. A student nurse keeps a fluid intake–output record on the patient. She is surprised to find that fluid intake and urinary output are normal. What effect was she expecting? Can you explain why urinary output is normal?
3. A patient exhibits elevated urine ammonia and increased rate of respiration. Does she have metabolic acidosis or metabolic alkalosis?
4. Swifty Trotts has an enteropathogenic *Escherichia coli* infection that produces severe diarrhea. What does this diarrhea do to his blood pH, urine pH, and respiratory rate?
5. Acetazolamide is a diuretic that blocks the activity of the enzyme carbonic anhydrase inside kidney tubule cells. This blockage prevents the formation of carbonic acid from carbon dioxide and water. Normally, carbonic acid dissociates to form H^+ and HCO_3^-, and the H^+ are exchanged for Na^+ from the urine. Blocking the formation of H^+ in the cells of the nephron tubule blocks sodium reabsorption, thus inhibiting water reabsorption and producing the diuretic effect. With this information in mind, what effect does acetazolamide have on blood pH, urine pH, and respiratory rate?
6. As part of a physiology experiment, Hardy Breath, an anatomy and physiology student, is asked to breathe through a 3-foot-long glass tube. What effect does this action have on his blood pH, urine pH, and respiratory rate?
7. A young girl is suspected of having epilepsy and therefore is prone to having convulsions. Based on your knowledge of acid–base balance and respiration, propose a hypothetical experiment that might suggest that the girl is susceptible to convulsions.
8. Hardy Explorer climbed to the top of a mountain. To celebrate, she drank a glass of whiskey. Alcohol stimulates hydrochloric acid secretion in the stomach. What do you expect to happen to Hardy's respiratory rate and the pH of her urine?

Answers in Appendix F

PART 5

REPRODUCTION AND DEVELOPMENT

Reproductive System 28

The reproductive system controls the development of the structural and functional differences between males and females, and it has profound effects on human behavior. Not only are male and female reproductive structures different, but also the differences that exist among other systems, such as the integumentary, muscular, and skeletal systems, result from the effects of the reproductive system on them. Reproduction is an essential characteristic of living organisms, and functional male and female reproductive systems are necessary for individuals to reproduce. Even though some individuals do not reproduce, their reproductive systems play important roles. The male reproductive system produces sperm cells and can transfer them to the female. The female reproductive system produces oocytes and can receive sperm cells, one of which may unite with an oocyte. The female reproductive system is then intimately involved with nurturing the development of a new individual until birth and usually for some considerable time after birth.

Although the male and female reproductive systems show such striking differences, they also have a number of similarities. Many reproductive organs of males and females are derived from the same embryologic structures (see chapter 29). In addition, some hormones are the same in males and females, even though they act in very different ways (table 28.1).

This chapter discusses the *anatomy of the male reproductive system* (p. 1033), the *physiology of male reproduction* (p. 1045), the *anatomy of the female reproductive system* (p. 1049), the *physiology of female reproduction* (p. 1059), and the *effects of aging on the reproductive system* (p. 1071).

Color-enhanced scanning electron micrograph of the ciliated epithelial surface of a uterine tubule.

Reproductive System

TABLE 28.1 Major Reproductive Hormones

Hormone	Source	Target Tissue	Response
Males			
Gonadotropin-releasing hormone (GnRH)	Hypothalamus	Anterior pituitary	Stimulates secretion of LH and FSH
Luteinizing hormone (LH) (also called interstitial cell-stimulating hormone [ICSH] in males)	Anterior pituitary	Interstitial cells in the testes	Stimulates synthesis and secretion of testosterone
Follicle-stimulating hormone (FSH)	Anterior pituitary	Seminiferous tubules (sustentacular cells)	Supports spermatogenesis
Testosterone	Leydig cells in the testes	Testes and body tissues	Supports spermatogenesis, stimulates development and maintenance of reproductive organs, causes development of secondary sexual characteristics
		Anterior pituitary and hypothalamus	Inhibits GnRH, LH, and FSH secretion through negative feedback
Females			
Gonadotropin-releasing hormone (GnRH)	Hypothalamus	Anterior pituitary	Stimulates production of LH and FSH
Luteinizing hormone (LH)	Anterior pituitary	Ovaries	Causes follicles to complete maturation and undergo ovulation; causes ovulated follicle to become the corpus luteum
Follicle-stimulating hormone (FSH)	Anterior pituitary	Ovaries	Causes follicles to begin development
Prolactin	Anterior pituitary	Mammary glands	Stimulates milk secretion following parturition
Estrogen	Follicles of Ovaries	Uterus	Causes proliferation of endometrial cells
		Mammary glands	Causes development of the mammary glands (especially duct systems)
		Anterior pituitary and hypothalamus	Has a Positive feedback effect before ovulation, resulting in increased LH and FSH secretion: has a negative feedback effect with progesterone on the hypothalamus and anterior pituitary after ovulation, resulting in decreased LH and FSH secretion
		Other tissues	Causes development of secondary sexual characteristics
Progesterone	Corpus luteum of ovaries	Uterus	Causes hypertrophy of endometrial cells and secretion of fluid from uterine glands
		Mammary glands	Causes development of the mammary glands (especially alveoli)
		Anterior pituitary	Has a negative feedback effect with estrogen, on the hypothalamus and anterior pituitary after ovulation, resulting in decreased LH and FSH secretion
		Other tissues	Causes development of secondary sexual characteristics
Oxytocin*	Posterior pituitary	Uterus and mammary glands	Causes contraction of uterine smooth muscle during intercourse and childbirth; causes contraction of myoepithelial cells in the breast resulting in milk letdown in lactating women
Human chorionic gonadotropin (HCG)	Placenta	Corpus luteum of ovaries	Maintains the corpus luteum and increases its rate of progesterone secretion during the first one-third (first trimester) of pregnancy

*Covered in chapter 29.

ANATOMY OF THE MALE REPRODUCTIVE SYSTEM

The male reproductive system consists of the testes (sing. *testis*), a series of ducts, accessory glands, and supporting structures. The ducts include the epididymides (sing. *epididymis*), ductus deferentia (sing. *deferens;* also *vas deferens*), and urethra. Accessory glands include the seminal vesicles, prostate gland, and bulbourethral glands. Supporting structures include the scrotum and penis (figure 28.1).

Sperm cells are very temperature-sensitive and do not develop normally at usual body temperatures. The testes and epididymides, in which the sperm cells develop, are located outside the body cavity in the scrotum, where the temperature is lower. The ductus deferentia lead from the testes into the pelvis, where they join the ducts of the seminal vesicles to form the ampullae. Extensions of the ampullae, called the ejaculatory ducts, pass through the prostate and empty into the urethra within the prostate. The urethra, in turn, exits from the pelvis and passes through the penis to the outside of the body.

1 ***List the structures that make up the male reproductive system.***

Scrotum

The **scrotum** (skrō′tŭm) contains the testes and is divided into two internal compartments by an incomplete connective tissue septum. Externally, the scrotum is marked in the midline by an irregular ridge, the **raphe** (rā′fē; a seam), which continues posteriorly to the anus and anteriorly onto the inferior surface of the penis. The outer layer of the scrotum includes the skin, a layer of superficial fascia consisting of loose connective tissue, and a layer of smooth muscle called the **dartos** (dar′tōs; to skin) **muscle.**

When the scrotum is exposed to cool temperatures, the dartos muscle contracts, causing the skin of the scrotum to become firm and wrinkled and reducing its overall size. At the same time, the

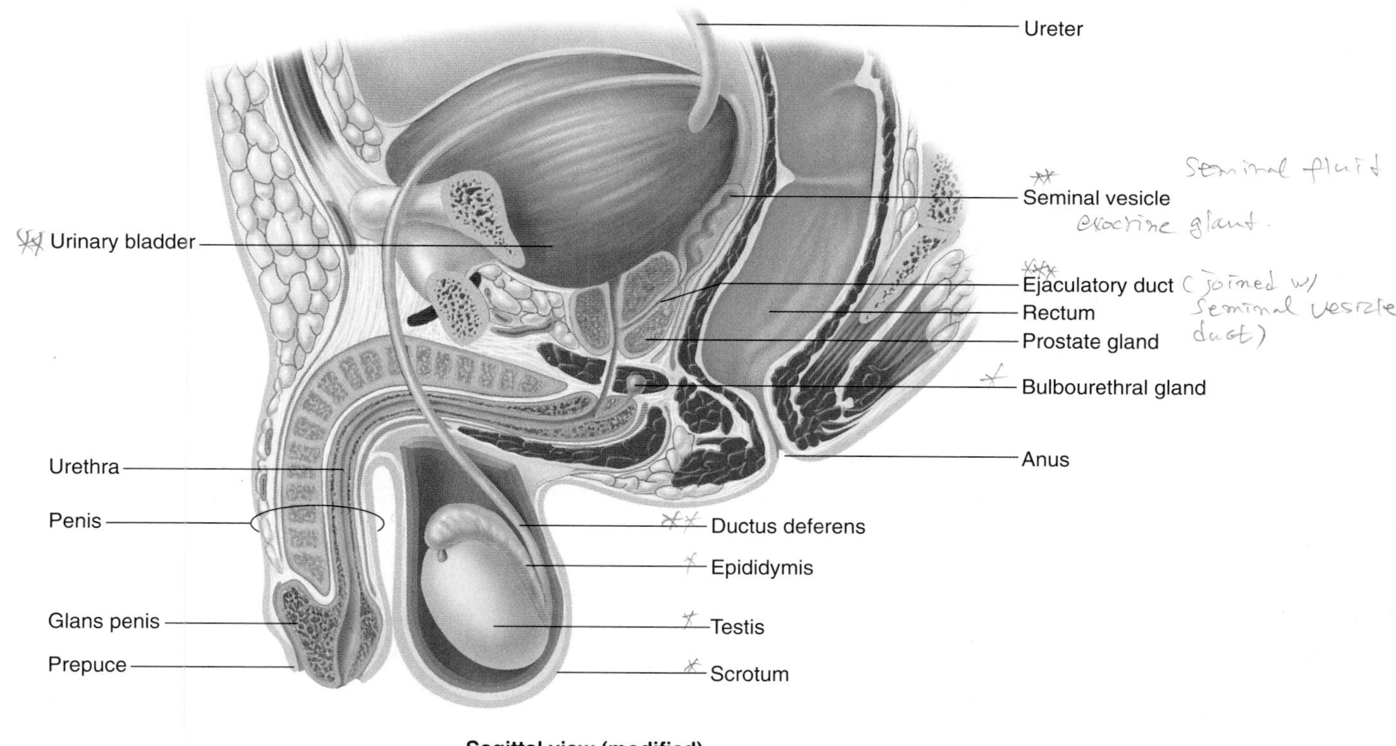

FIGURE 28.1 Male Reproductive Structures
Sagittal section of the male pelvis showing the male reproductive structures. Some structures are shown as a modified sagittal section so that the structure of the testis, seminal vesicles can be shown, and to show the relationship of the ductus deferens to the ureter and urinary bladder.

cremaster (krē-mas′ter) **muscles** (see figure 28.6), which are extensions of abdominal muscles into the scrotum, contract and help pull the testes nearer the body, which helps keep the testes warm. When the scrotum is exposed to warm temperatures or becomes warm because of exercise or fever, the dartos and cremaster muscles relax, and the skin of the scrotum becomes loose and thin, allowing the testes to descend away from the body, which helps keep the testes cool. The response of the dartos and cremaster muscles is important in the regulation of temperature in the testes. If the testes become too warm or too cold, normal sperm cell formation does not occur.

2 ***Describe the structure of the scrotum.***

3 ***Describe the role of the dartos and cremaster muscles in temperature regulation of the testes.***

Perineum

The area between the thighs, which is bounded by the symphysis pubis anteriorly, the coccyx posteriorly, and the ischial tuberosities laterally, is called the **perineum** (per′i-nē′ŭm). The perineum is divided into two triangles by a set of muscles, the superficial transverse and deep transverse perineal muscles, which run transversely between the two ischial tuberosities (see figure 10.21). The anterior, or **urogenital** (ū′rō-jen′i-tăl), **triangle,** contains the base of the penis and the scrotum. The smaller posterior, or **anal, triangle,** contains the anal opening (figure 28.2).

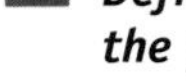

4 ***Define perineum. Describe the two triangles that make up the perineum.***

Testes

Testicular Histology

The **testes** (tes′tēz) are small, ovoid organs, each about 4–5 cm long, within the scrotum (see figure 28.1). They are both exocrine and endocrine glands. Sperm cells form a major part of the exocrine secretions of the testes, and testosterone is the major endocrine secretion of the testes.

The outer part of each testis is a thick, white capsule consisting of mostly fibrous connective tissue called the **tunica albuginea** (al-bū-jin′ē-ă; white). Connective tissue of the tunica albuginea enters the testis and forms incomplete **septa** (sep′tă; figure 28.3). The septa divide each testis into about 300–400 cone-shaped **lobules.** The substance of the testis between the septa includes **seminiferous** (sem′i-nif′er-ŭs; seed carriers) **tubules,** in which sperm cells develop, and a loose connective tissue stroma that surrounds the tubules and contains clusters of endocrine cells called **interstitial cells,** or **Leydig cells,** which secrete testosterone.

The combined length of the seminiferous tubules in both testes is nearly half a mile. The seminiferous tubules empty into a set of short, straight tubules, the **tubuli recti,** which in turn empty into a tubular network called the **rete** (rē′tē; net) **testis.** The rete testis empties into 15–20 tubules called **efferent ductules** (dŭk′tools). They have a ciliated pseudostratified columnar epithelium, which helps move sperm cells out of the testis. The efferent ductules pierce the tunica albuginea to exit the testis.

5 ***Describe the covering and connective tissue of the testis.***

6 ***Where are the seminiferous tubules and interstitial cells located? What are their functions?***

7 ***Describe the structures of the tubuli recti, rete testis, and efferent ductules.***

Descent of the Testes

The testes have developed as retroperitoneal organs in the abdominopelvic cavity by approximately 5 weeks following fertilization. By approximately 14 weeks, each testis has connected to a labioscrotal swelling, which becomes the scrotum, by a **gubernaculum**

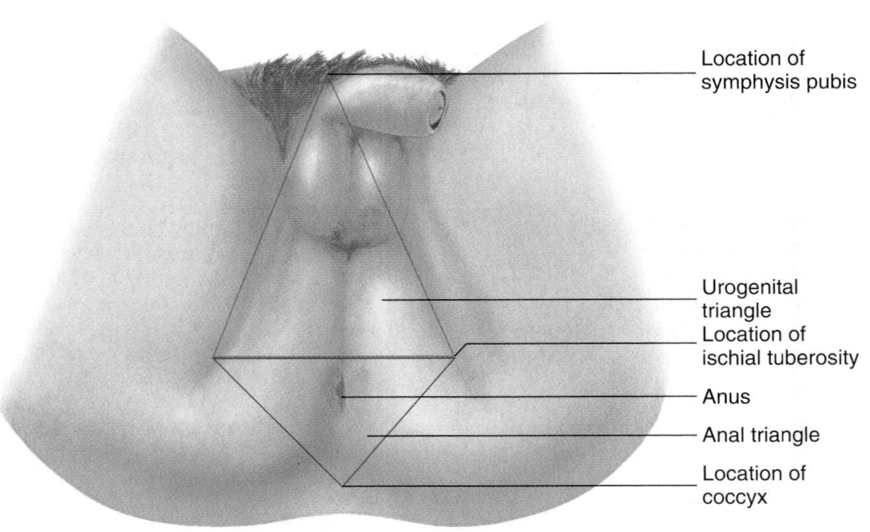

FIGURE 28.2 Male Perineum
Inferior view of the male perineum.

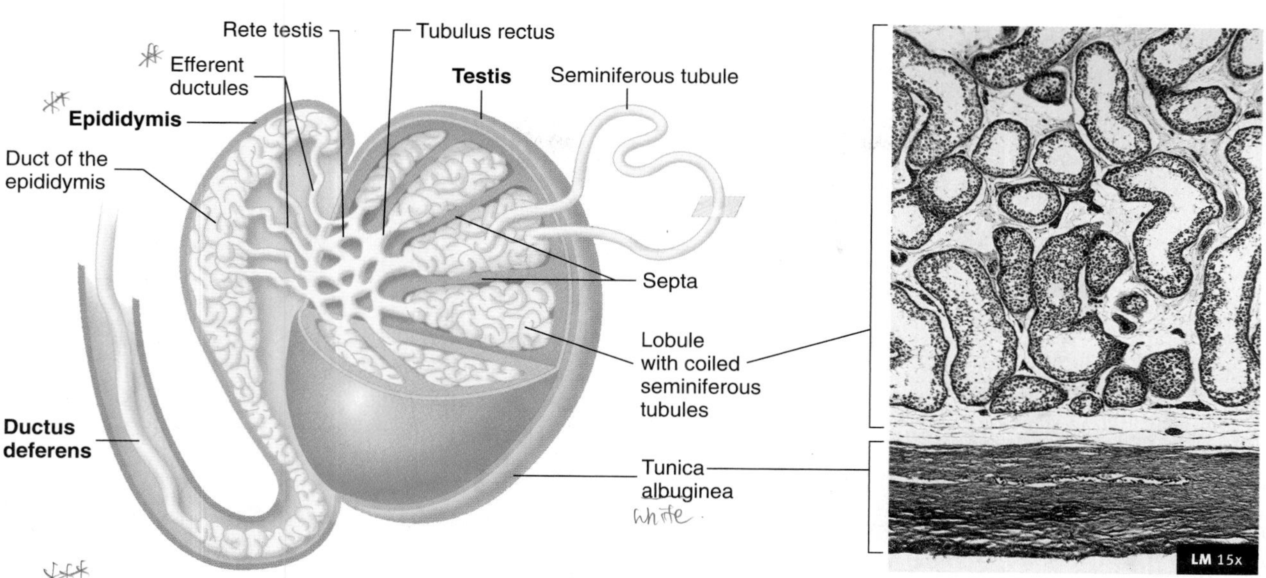

(a) Gross anatomy of the testis, with a section cut away to reveal the internal structures and histology.

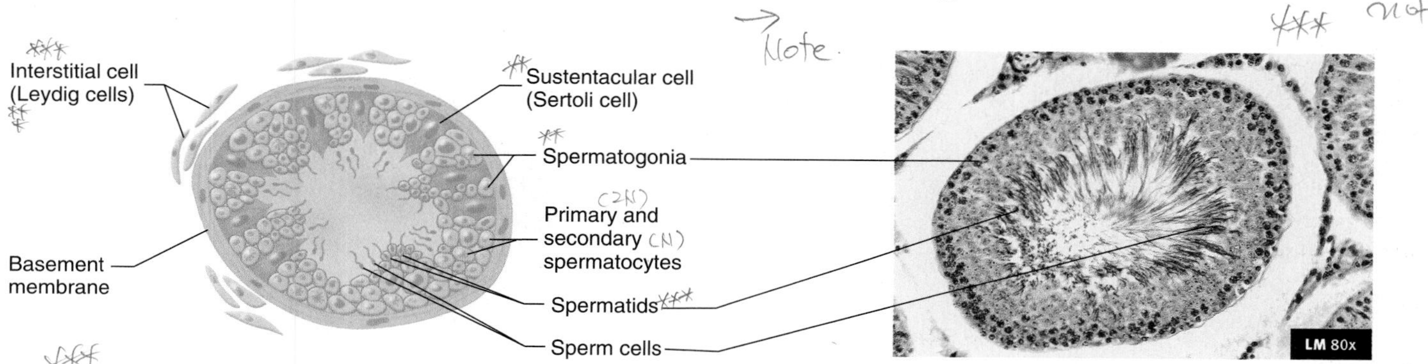

(b) Cross section of a seminiferous tubule. Spermatogonia are near the periphery, and mature sperm cells are near the lumen of the seminiferous tubules.

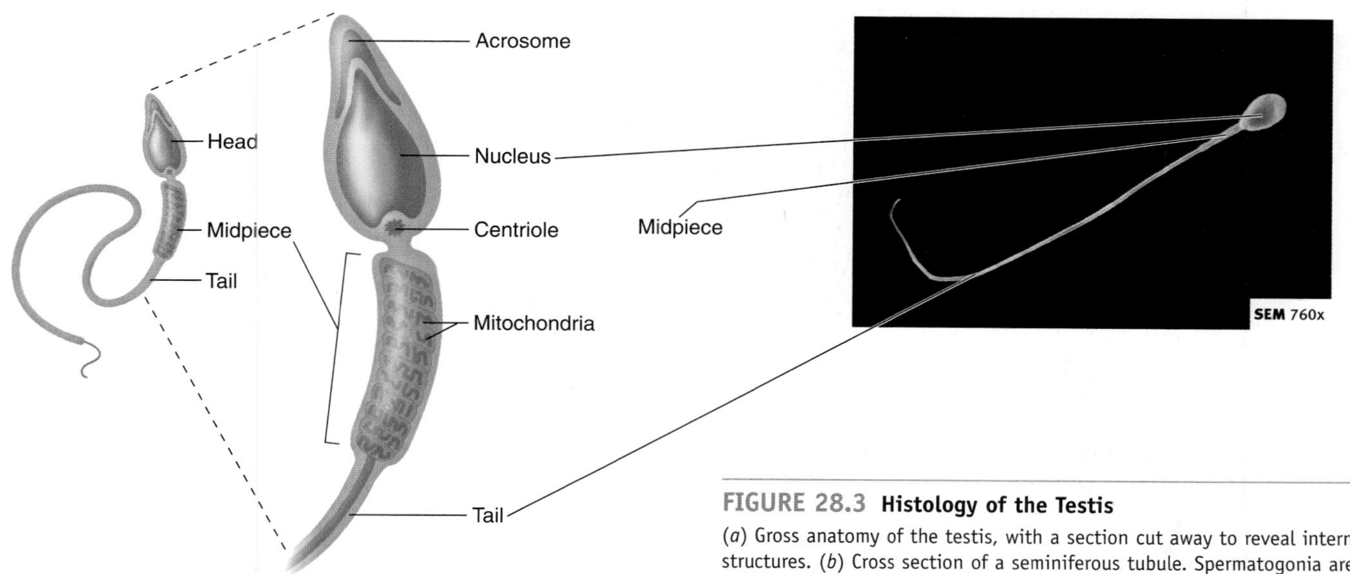

(c) Head, midpiece, and tail of a mature sperm cell.

FIGURE 28.3 Histology of the Testis

(*a*) Gross anatomy of the testis, with a section cut away to reveal internal structures. (*b*) Cross section of a seminiferous tubule. Spermatogonia are near the periphery, and mature sperm cells are near the lumen of the seminiferous tubule. (*c*) Mature sperm cell. (*d*) Head of a mature sperm cell.

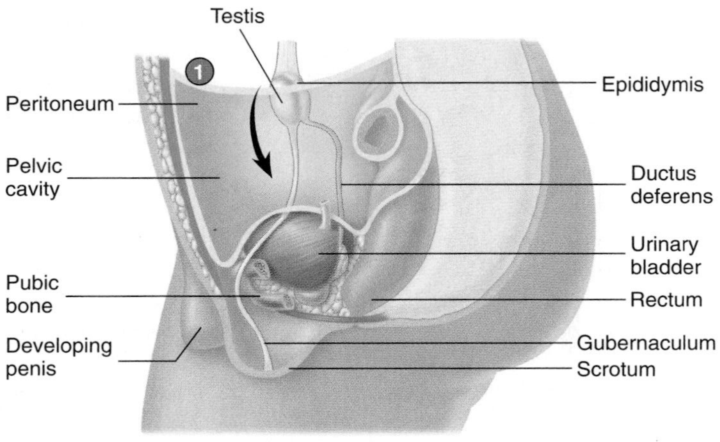

(1) Each testis forms as a retroperitoneal structure near the level of the kidney and begins to descend.

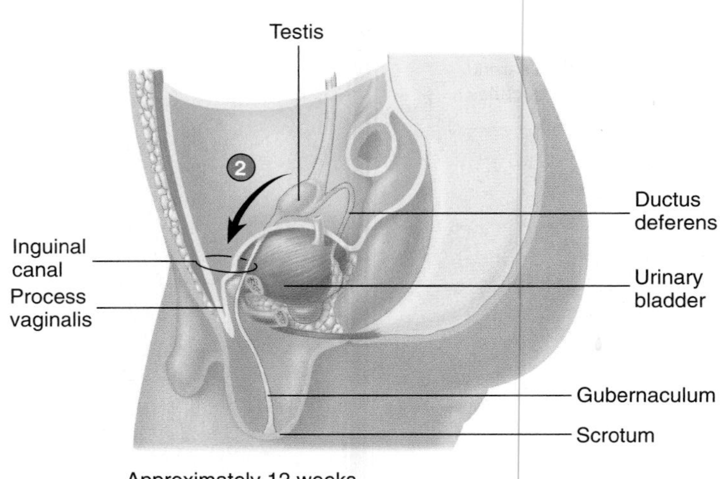

(2) The testis beneath the parietal peritoneum descends toward the inguinal canal.

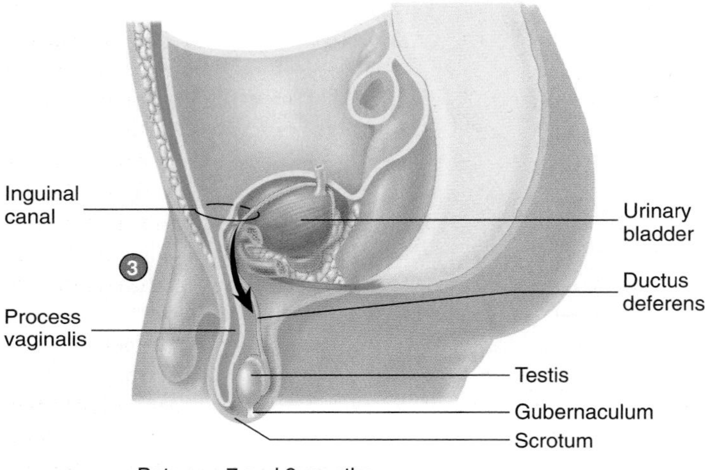

(3) The testis descends through the inguinal canal into the scrotum.

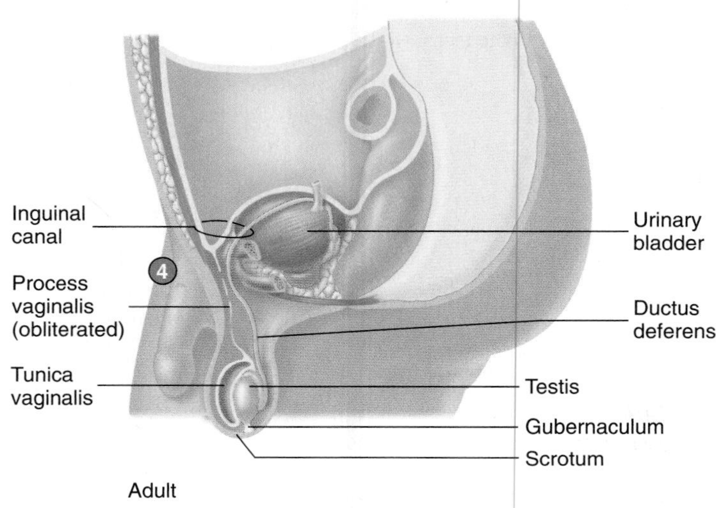

(4) Between birth and adulthood, the process vaginalis is obliterated and its inferior portion becomes the tunica vaginalis.

PROCESS FIGURE 28.4 **Descent of the Testes**

(goo′ber-nak′ū-lŭm), a fibromuscular cord (figure 28.4, step 1; see chapter 29). Between 14 and 28 weeks, the **inguinal** (ing′gwi-năl) **canals** (figure 28.4, step 2) form and the testes move through them into the scrotum (figure 28.4, step 3). As it moves into the scrotum, each testis is preceded by an outpocketing of the peritoneum called the **process vaginalis** (vaj′i-nă-lis). The superior part of each process vaginalis usually becomes obliterated, and the inferior part remains as a small, closed sac, the **tunica** (too′ni-kă) **vaginalis** (figure 28.4, step 4). The tunica vaginalis surrounds most of the testis in much the same way that the pericardium surrounds the heart, and the small amount of fluid in it allows the testes to move with little friction. The visceral layer of the tunica vaginalis covers the anterior surface of the testis, and the parietal layer lines the scrotum. The tunica vaginalis is a serous membrane consisting of a layer of simple squamous epithelium resting on a basement membrane. The testes have descended into the scrotum of approximately 79% of fetuses delivered prior to 28 weeks of development. By 28 weeks of development, the testes have descended into the scrotum of greater than 97% of fetuses. At 9 months of age, the testes have descended into the scrotum of 98.2% of male infants.

The inguinal canals are bilateral oblique passageways in the anterior abdominal wall. They originate at the **deep inguinal rings**, which open through the aponeuroses of the transversus abdominis muscles. The canals extend inferiorly and obliquely and end at the **superficial inguinal rings**, openings in the aponeuroses of the external abdominal oblique muscles. In females, the inguinal canals do develop, but they are much smaller than in males and the ovaries do not normally descend through them.

Cryptorchidism (krip-tōr'ki-dizm) is the failure of one or both of the testes to descend into the scrotum. The higher temperature of the abdominal cavity prevents normal sperm cell development and, if it involves both testes, results in no sperm cell production (see chapter 29, page 1100).

Inguinal Hernia

Normally, the inguinal canals are closed, but they represent weak spots in the abdominal wall. **Inguinal hernias** (her'nē-ăz) are abnormal openings in the abdominal wall in the inguinal region through which structures such as a portion of the small intestine can protrude. If the deep inguinal ring remains open, or if it is weak and enlarges later in life, a loop of intestine can protrude into or even pass through the inguinal canal, resulting in indirect inguinal hernia. A direct inguinal hernia results from a tear, or rupture, in a weakened area of the anterior abdominal wall near the inguinal canal, but not through the inguinal canal. These hernias can be quite painful and even very dangerous, especially if a portion of the small intestine is compressed so its blood supply is cut off. Fortunately, inguinal hernias can be repaired surgically. Males are much more prone to inguinal hernias than are females because a male's inguinal canals are larger and weakened because the testes pass through them on their way into the scrotum.

8 ***When and how do the testes descend into the scrotum?***

9 ***Describe the tunica vaginalis.***

10 ***Define cryptorchidism and its effect on sperm cell production.***

Sperm Cell Development

Before puberty, the testes remain relatively simple and unchanged from the time of their initial development. The interstitial cells are not particularly prominent during this period, and the seminiferous tubules lack a lumen and are not yet functional. At 12–14 years of age, the interstitial cells increase in number and size, a lumen develops in each seminiferous tubule, and sperm cell production begins. It takes approximately 74 days for sperm cells to be produced. About 50 of those days are spent in the seminiferous tubules.

Sperm cell development, or **spermatogenesis** (sper'mă'-tō-jen'ĕ-sis; see figure 28.3*b*; figure 28.5*a*), occurs in the seminiferous tubules and is the process by which **germ cells** divide and differentiate to form sperm cells. The seminiferous tubules contain two types of cells, germ cells and **sustentacular** (sŭs-ten-tak'ū-lăr), or **Sertoli** (sĕr-tō'lē; named for an Italian histologist), **cells.** The sustentacular cells are also sometimes referred to as **nurse cells.**

Sustentacular cells are large cells that extend from the periphery to the lumen of the seminiferous tubule (see figure 28.3*b*; figure 28.5*b*). They nourish the germ cells and probably produce, together with the interstitial cells, a number of hormones, such as androgens, estrogens, and inhibins. In addition, tight junctions between the sustentacular cells form a **blood–testis barrier** between spermatogonia and sperm cells. It isolates the sperm cells from the immune system (see figure 28.5*b*). This barrier is necessary because, as the sperm cells develop, they form surface antigens that could stimulate an immune response, resulting in their destruction (see figure 28.5*b*).

Testosterone, produced by the interstitial cells, passes into the sustentacular cells and binds to receptors. The combination of testosterone with the receptors is required for the sustentacular cells to function normally. In addition, testosterone is converted to two other steroids in the sustentacular cells: **dihydrotestosterone** (dī-hī'drō-tes-tos'ter-ōn) and **estrogen** (es'trō-jen). The sustentacular cells also secrete a protein called **androgen-binding** (an'drō-jen) **protein** into the seminiferous tubules. Testosterone and dihydrotestosterone bind to androgen-binding protein and are carried along with other secretions of the seminiferous tubules to the epididymis. Estradiol and dihydrotestosterone may be the active hormones that promote sperm cell formation.

Scattered between the sustentacular cells are smaller germ cells from which sperm cells are derived. The germ cells are arranged so that the most immature cells are at the periphery and the most mature cells are near the lumen of the seminiferous tubules. The most peripheral cells, those adjacent to the basement membrane of the seminiferous tubules, are **spermatogonia** (sper'mă-tō-gō'nē-ă, which divide by mitosis (see figure 28.5*a*). Some of the daughter cells produced from these mitotic divisions remain spermatogonia and continue to produce additional spermatogonia. The others divide through mitosis and differentiate to form **primary spermatocytes** (sper'mă-tō-sītz).

Meiosis (see Clinical Focus "Meiosis") begins when the primary spermatocytes divide. Each primary spermatocyte passes through the first meiotic division to become two **secondary spermatocytes.** Each secondary spermatocyte undergoes a second meiotic division to produce two even smaller cells called **spermatids** (sper'mă-tidz). Each spermatid contains one of each of the homologous pairs of chromosomes. Therefore, each sperm cell contains 22 autosomes and either an X or a Y chromosome. Each spermatid undergoes the last phase of spermatogenesis called **spermiogenesis** (sper'mē-ō-jen'ĕ-sis) to form a mature **sperm cell,** or **spermatozoon** (sper'mă-tō-zō'on; pl. *spermatozoa*, sper'mă-tō-zō'ă; see figures 28.3*c* and *d* and 28.5). Each spermatid develops a head, midpiece, and tail, or flagellum. The head contains chromosomes, and at the leading end it has a cap, the **acrosome** (ak'rō-sōm), which contains enzymes necessary for the sperm cell to penetrate the female sex cell. The flagellum is similar to a cilium (see chapter 3), and movement of microtubules past one another within the tail causes the tail to move and propel the sperm cell forward. The midpiece has large numbers of mitochondria, which produce the ATP necessary for microtubule movement.

At the end of spermatogenesis, the developing sperm cells gather around the lumen of the seminiferous tubules, with their heads directed toward the surrounding sustentacular cells and their tails directed toward the center of the lumen (see figures 28.3*b* and 28.5). Finally, sperm cells are released into the lumen of the seminiferous tubules.

CLINICAL FOCUS

Meiosis

Sperm cell development and oocyte development involve meiosis (mī-ō′sis). This kind of cell division occurs only in the testes and ovaries. It consists of two consecutive nuclear divisions without a second replication of the genetic material between the divisions. Four daughter cells are produced, and each has half as many chromosomes as the parent cell (figure A).

The normal chromosome number in human cells is 46. This number is called a **diploid** (dip′loyd), or a **2*n*, number** of chromosomes. The chromosomes consist of 23 pairs, each of which is called a **homologous** (hŏ-mol′ō-gŭs) **pair.** The homologous pairs consist of 22 autosomal pairs and 1 pair of sex chromosomes. The sex chromosomes are an X and a Y chromosome in males and two X chromosomes in females. One chromosome of each homologous pair is from the male parent, and the other is from the female parent. The chromosomes of each homologous pair look alike, and they contain genes for the same traits.

In sperm cells and oocytes, the number of chromosomes is 23. This number is called a **haploid** (hap′loyd), or ***n*, number** of chromosomes. Each gamete contains one chromosome from each of the homologous pairs. Reduction in the number of chromosomes in sperm cells or oocytes to an n number is important. When a sperm cell and an oocyte fuse to form a fertilized egg, each provides an n number of chromosomes, which reestablishes a $2n$ number of chromosomes. If meiosis did not take place, each time fertilization occurred the number of chromosomes in the fertilized oocyte would double. The extra chromosomal material would be lethal to the developing offspring.

Sperm cells have 22 autosomal chromosomes and either an X or Y chromosome. Oocytes contain 1 autosomal chromosome from each of the 22 homologous pairs and an X chromosome. During fertilization, when a sperm cell fuses with an oocyte, the normal number of 46 chromosomes, consisting of 23 pairs of chromosomes, is reestablished. The sex of the baby is determined by the sperm cell that fertilizes the oocyte. The sex is male if the sperm cell that fertilizes the oocyte carries a Y chromosome, female if the sperm cell carries an X chromosome.

The two divisions of meiosis are called **meiosis I** and **meiosis II.** The stages of meiosis have the same names as the stages in mitosis—that is, prophase, metaphase, anaphase, and telophase—but distinct differences exist between mitosis and meiosis.

Before meiosis begins, all the DNA in the chromosomes is duplicated. At the beginning of meiosis, each of the 46 chromosomes consists of two sister **chromatids** (krō′mă-tid) connected by a **centromere** (sen′trō-mēr; see figure A). In prophase of meiosis I, the chromosomes become visible, and the homologous pairs of chromosomes come together. This process is called **synapsis** (si-nap′sis). Because each chromosome consists of two chromatids, the pairing of the homologous chromosomes brings two chromatids of each chromosome close together, an arrangement called a **tetrad.** Occasionally, part of a chromatid of one homologous chromosome breaks off and is exchanged with part of another chromatid from the other homologous chromosome of the tetrad. This exchange of genetic material is called **crossing-over.** Crossing-over allows the exchange of genetic material between maternal and paternal chromosomes.

During metaphase, homologous pairs of chromosomes line up near the center of the cell. For each pair of homologous chromosomes, however, the side of the cell on which the maternal or paternal chromosome is located is random. The way the chromosomes align during synapsis results in the random assortment of maternal and paternal chromosomes in the daughter cells during meiosis. Crossing-over and the random assortment of maternal and paternal chromosomes are responsible for the large degree of diversity in the genetic composition of sperm cells and oocytes produced by each individual.

During anaphase I, the homologous pairs are separated to each side of the cell. During telophase I, new nuclei form and the cell completes division of the cytoplasm to form two cells. As a consequence, when meiosis I is complete, each daughter cell has 1 chromosome from each of the homologous pairs. Each of the 23 chromosomes in each daughter cell consists of two chromatids joined by a centromere.

It is during the first meiotic division that the chromosome number is reduced from a $2n$ number (46 chromosomes, or 23 pairs) to an n number (23 chromosomes, or 1 from each homologous pair). The first meiotic division is therefore called a **reduction division.**

The second meiotic division is similar to mitosis. The chromosomes, each consisting of two chromatids, line up near the middle of the cell (see figure A). Then the chromatids separate at the centromere, and each daughter cell receives one of the chromatids from each chromosome. When the centromere separates, each of the chromatids is called a chromosome. Consequently, each of the four daughter cells produced by meiosis contains 23 chromosomes.

First Meiotic Division (Meiosis I)

1. Early prophase I
The duplicated chromosomes become visible chromatids (shown separated for emphasis; they actually are so close together that they appear as a single strand).

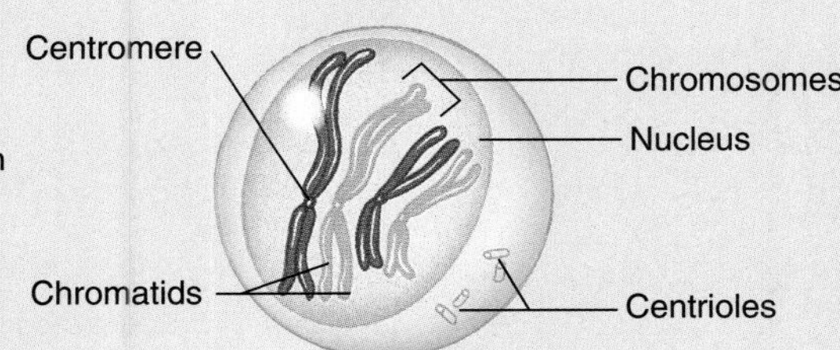

2. Middle prophase I
Homologous chromosomes synapse to form tetrads. Crossing over may occur at this stage.

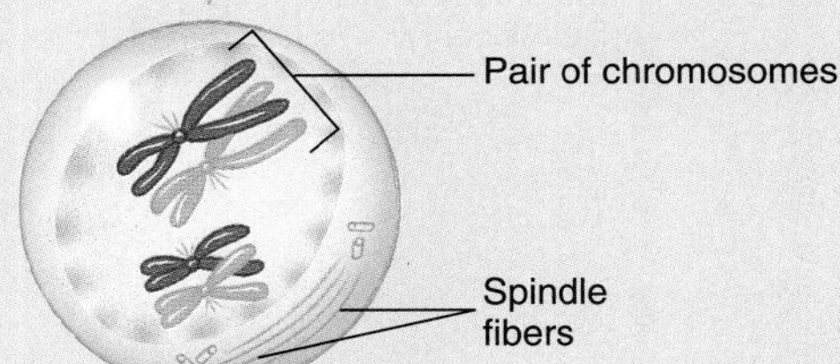

3. Metaphase I
Tetrads align at the center of the cell. Random assortment of homologous chromosomes occurs.

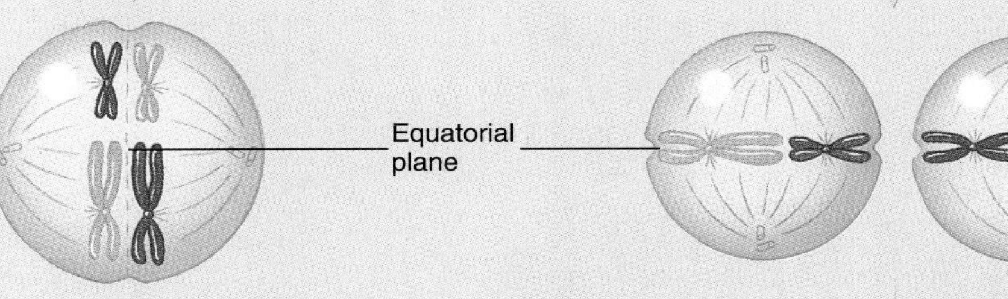

4. Anaphase I
Homologous chromosomes move apart to opposite sides of the cell.

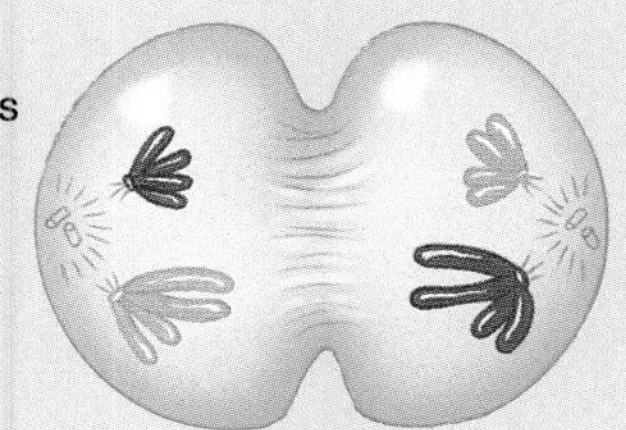

5. Telophase I
New nuclei form, and the cell divides.

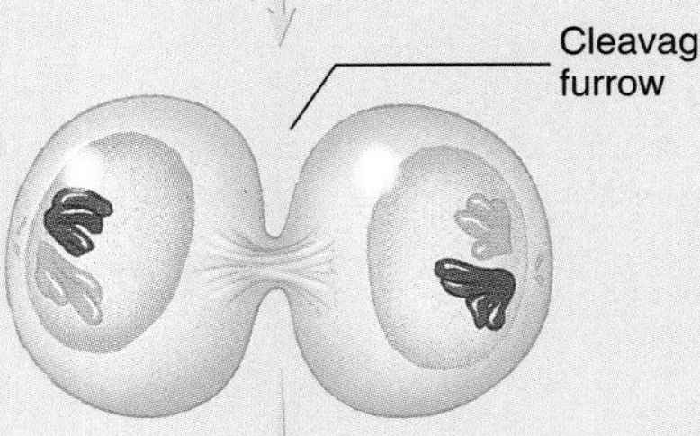

Prophase II (top of next column)

Second Meiotic Division (Meiosis II)
(continued from the bottom of previous column)

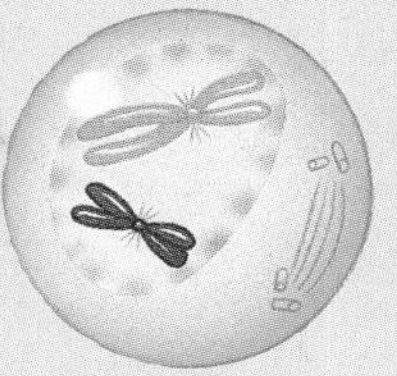
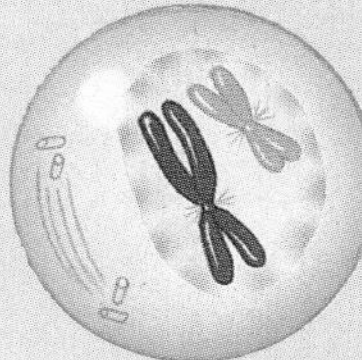

6. Prophase II
Each chromosome consists of two chromatids.

7. Metaphase II
Chromosomes align along the center of the cell.

8. Anaphase II
Chromatids separate and each is now called a chromosome.

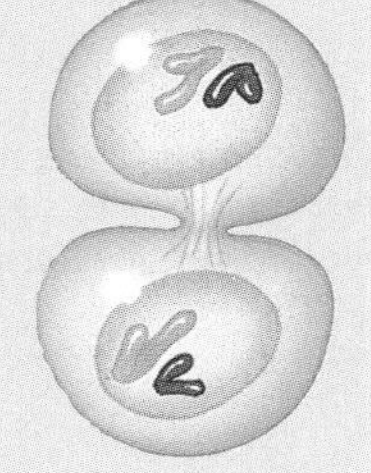
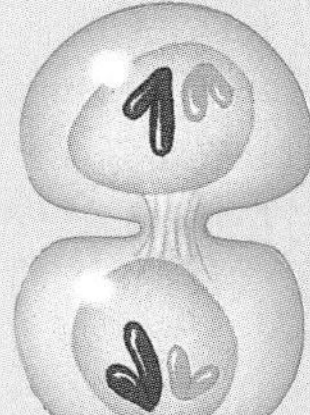

9. Telophase II
New nuclei form around the chromosomes. The cells divide to form four daughter cells with a haploid (*n*) number of chromosomes.

PROCESS FIGURE A **Meiosis**

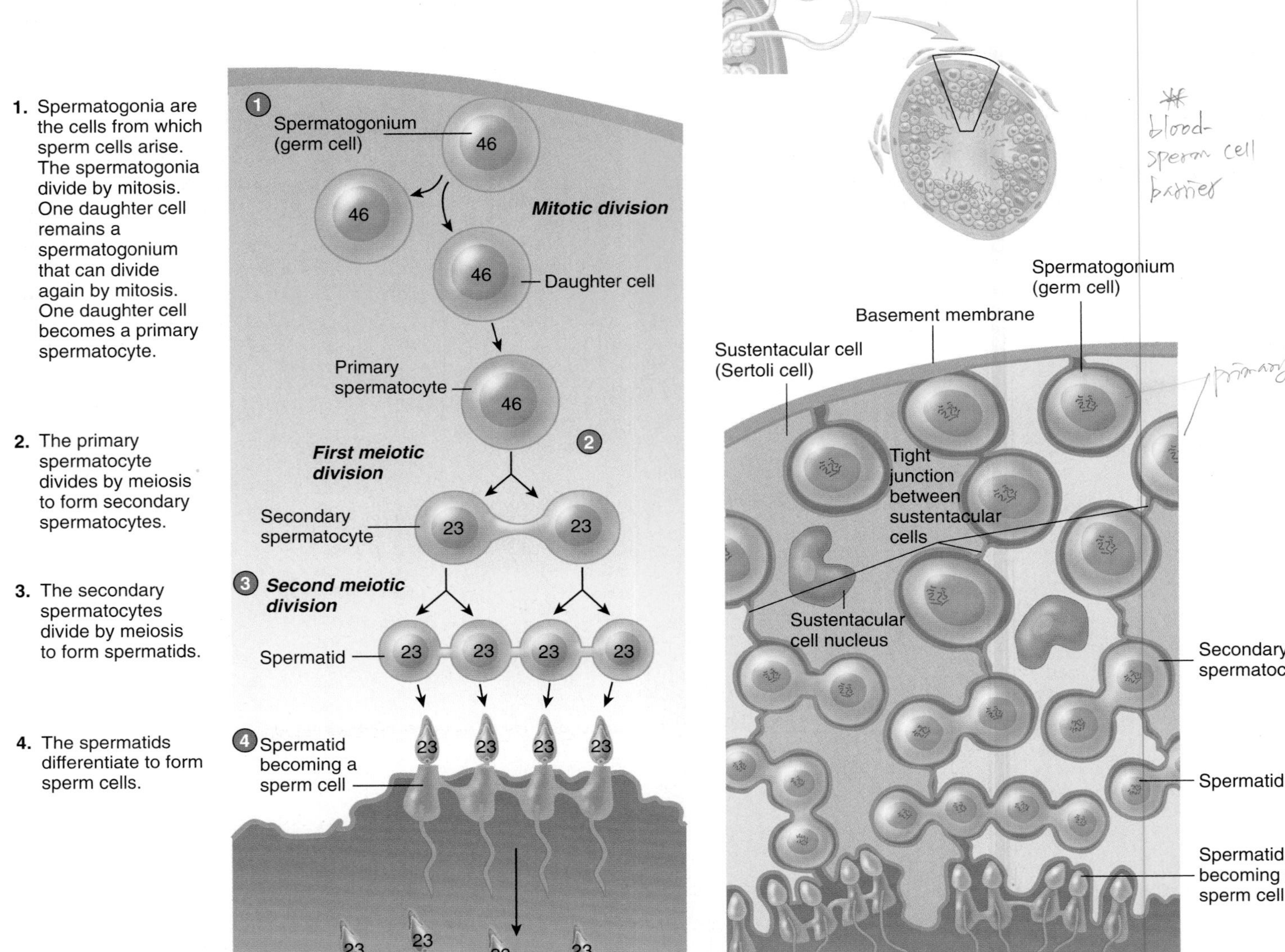

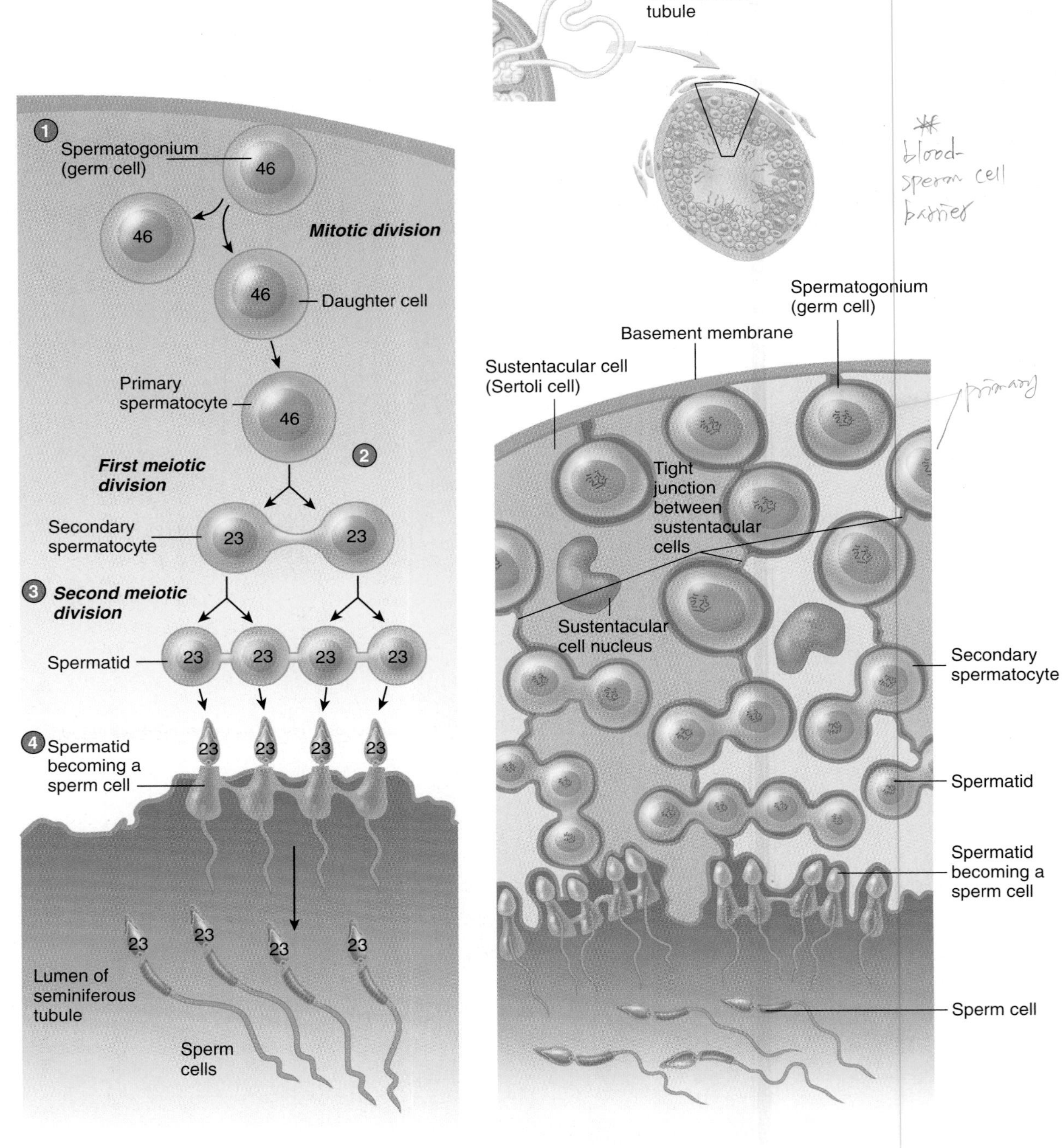

PROCESS FIGURE 28.5 Spermatogenesis

(*a*) Meiosis during spermatogenesis. A section of the seminiferous tubule illustrating the process of meiosis and sperm cell formation. (*b*) The tight junctions that form between adjacent sustentacular cells form the blood-testis barrier. The sustentacular cells extend from the periphery to the lumen of the seminiferous tubules. Spermatogonia are peripheral to the blood-testis barrier, and spermatocytes are central to it.

11 ***Describe the role of germ cells, sustentacular cells, and the blood–testis barrier in the structure of the seminiferous tubules.***

12 ***Describe the conversion of testosterone to other hormones in the sustentacular cells.***

13 ***Where specifically are sperm cells produced in the testis?***

14 ***Describe the role of meiosis in sperm cell formation.***

15 ***Describe the function of the parts of a mature sperm cell and explain the events that change a spermatid to a sperm cell.***

Ducts

After their release into the seminiferous tubules, the sperm cells pass through the tubuli recti to the rete testis. From the rete testis, they pass through the efferent ductules, which leave the testis and enter the epididymis to join the duct of the epididymis. The sperm cells then leave the epididymis, passing through the ductus epididymis, ductus deferens, ejaculatory duct, and urethra to reach the exterior of the body.

Epididymis

The **efferent ductules** from each testis become extremely convoluted and form a comma-shaped structure on the posterior side of the testis called the **epididymis** (ep-i-did′i-mis; pl. *epididymides,* ep-i-di-dim′i-dēz; on the twin; "twin" refers to the paired, or twin, testes; figure 28.6).

The final maturation of the sperm cells occurs within the epididymis. It takes 12–16 days to travel through the epididymis and appear in the ejaculate. Changes that occur in sperm cells as

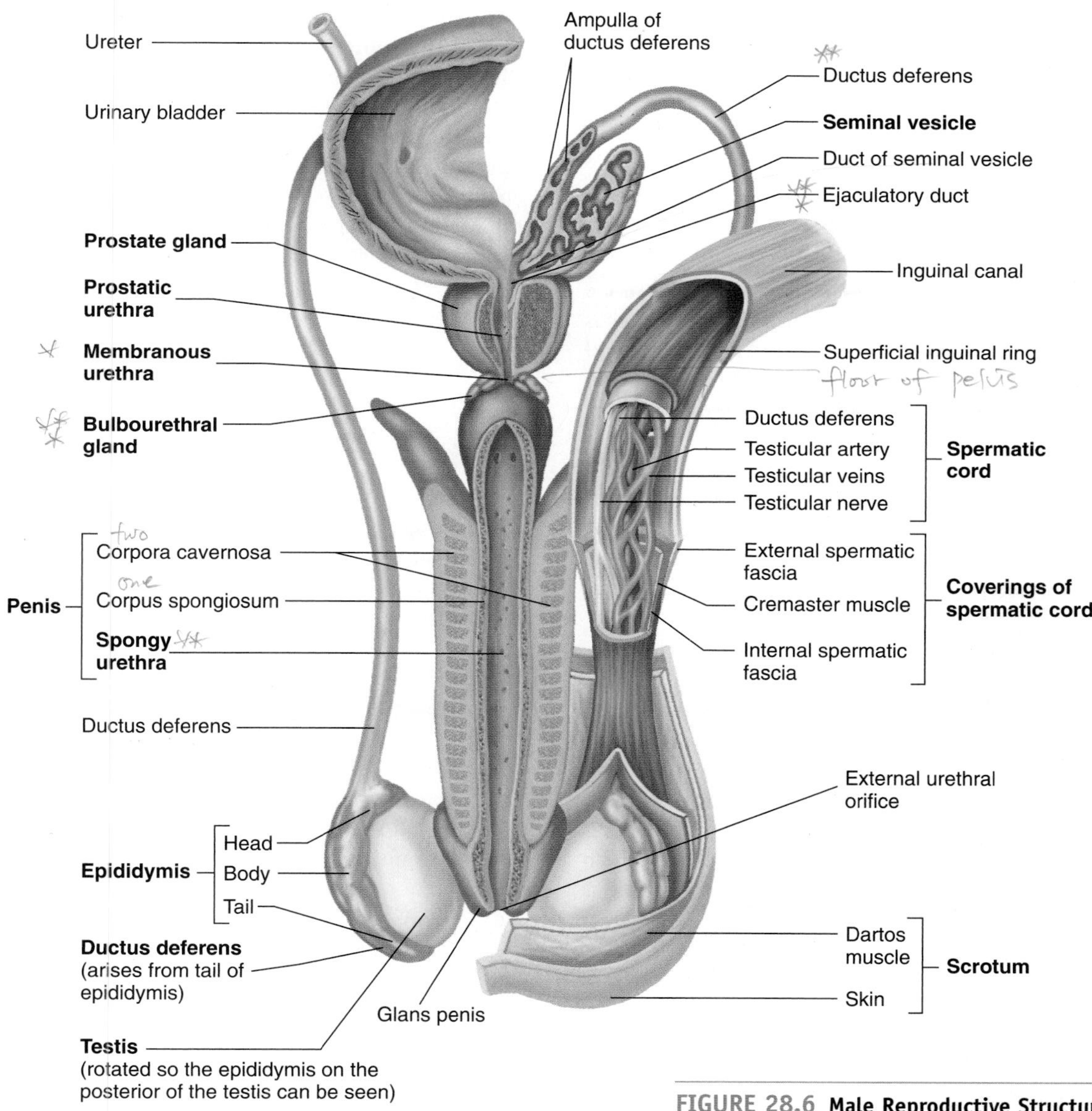

FIGURE 28.6 Male Reproductive Structures

Testes, epididymis, ductus deferens, and glands of the male reproductive system. The urethra is cut open along its dorsal side.

they pass through the epididymis include a further reduction in cytoplasm, maturation of the acrosome, and the ability to bind to the zona pellucida of the secondary oocyte (see "Oocyte Development and Fertilization," p. 1051). Sperm cells taken from the head of the epididymis are unable to fertilize secondary oocytes, and they are not yet able to become motile. Sperm cells taken from the tail of the epididymis are able to fertilize secondary oocytes and are capable of becoming motile.

Each epididymis consists of a head, a body, and a long tail (see figures 28.3*a* and 28.6). The head contains the convoluted efferent ductules, which empty into a single convoluted tube, the **duct of the epididymis,** located primarily within the body of the epididymis (see figure 28.3*a*). This duct alone, if unraveled, would extend for several meters. The duct of the epididymis has a pseudostratified columnar epithelium with elongated microvilli called **stereocilia** (ster′ē-ō-sil′ē-ă). The stereocilia increase the surface area of epithelial cells that absorb fluid from the lumen of the duct of the epididymis. The duct of the epididymis ends at the tail of the epididymis, which is located at the inferior border of the testis.

16 ***Describe how sperm cells move from the epididymis.***

17 ***List the changes that occur in sperm cells while in the epididymis and the functions of the epididymis.***

Ductus Deferens and Ejaculatory Duct

The **ductus deferens,** or **vas deferens** emerges from the tail of the epididymis and ascends along the posterior side of the testis medial to the epididymis and becomes associated with the blood vessels and nerves that supply the testis (see figures 28.1, 28.3*a*, and 28.6). These structures constitute the **spermatic cord.** The spermatic cord consists of (1) the ductus deferens, (2) the testicular artery and venous plexus, (3) lymphatic vessels, (4) nerves, and (5) fibrous remnants of the process vaginalis. The coverings of the spermatic cord include the **external spermatic fascia** (fash′ē-ă); the **cremaster muscle**, an extension of the muscle fibers of the internal abdominal oblique muscle of the abdomen; and the **internal spermatic fascia** (see figure 28.6).

The ductus deferens and the rest of the spermatic cord structures ascend and pass through the inguinal canal to enter the pelvic cavity (see figures 28.1 and 28.6). The ductus deferens crosses the lateral and posterior walls of the pelvic cavity, travels over the ureter, and loops over the posterior surface of the urinary bladder to approach the prostate gland. The end of the ductus deferens enlarges to form an **ampulla** (am-pul′lă). The lumen of the ductus deferens is lined by pseudostratified columnar epithelium, which is surrounded by smooth muscle. Peristaltic contractions of this smooth muscle tissue help propel sperm cells through the ductus deferens.

Adjacent to the ampulla of each ductus deferens is a sac-shaped gland called the seminal vesicle. A short duct from the seminal vesicle joins the ampulla of the ductus deferens to form the **ejaculatory** (ē-jak′ū-lă-tōr-ē) **duct.** The ejaculatory ducts are approximately 2.5 cm long. These ducts project into the prostate gland and end by opening into the urethra (see figures 28.1 and 28.6).

18 ***Describe the structure and functions of the ductus deferens.***

19 ***List the components and coverings of the spermatic cord.***

20 ***Describe the route by which the ductus deferens extends from the testis to the prostate gland.***

Urethra

The **male urethra** (ū-rē′thră) is about 20 cm long and extends from the urinary bladder to the distal end of the penis (see figures 28.1, and 28.6; figure 28.7). The urethra is a passageway for both urine and male reproductive fluids. The urethra is divided into three parts: the prostatic part, the membranous part, and the spongy part. The **prostatic** (pros-tat′ik) **urethra** is connected to the bladder and passes through the prostate gland. Fifteen to 30 small ducts from the prostate gland and the two ejaculatory ducts empty into the prostatic urethra. The **membranous urethra** is the shortest part of the urethra, extending from the prostate gland through the perineum, which is part of the muscular floor of the pelvis. The **spongy urethra,** also called the **penile** (pē′nīl) **urethra,** is by far the longest part of the urethra; it extends from the membranous urethra through the length of the penis. In rare cases, the penis does not develop completely and the urethra may open to the exterior along the inferior surface of the penis (see chapter 29, p. 1100). Stratified columnar epithelium lines most of the urethra, but transitional epithelium is in the prostatic urethra near the bladder, and stratified squamous epithelium is near the opening of the spongy urethra to the outside. Several minute, mucus-secreting **urethral glands** empty into the urethra.

21 ***Distinguish among the ejaculatory duct and the prostatic, membranous, and spongy parts of the male urethra.***

Penis

The **penis** contains three columns of erectile tissue (see figure 28.7); engorgement of this erectile tissue with blood causes the penis to enlarge and become firm, a process called **erection.** The penis is the male organ of copulation, through which sperm cells are transferred from the male to the female. Two of the erectile columns form the dorsum and sides of the penis and are called the **corpora cavernosa** (kōr′pōr-ă kav-er-nos′ă). The third column, the **corpus spongiosum** (kōr′pŭs spŭn′jē-ō′sŭm), forms the ventral portion of the penis; it expands to form a cap, the **glans penis,** over the distal end of the penis. The spongy urethra passes through the corpus spongiosum, penetrates the glans penis, and opens as the **external urethral orifice.** At the base of the penis, the corpus spongiosum expands to form the **bulb of the penis,** and each corpus cavernosum expands to form **crus** (kroos; pl. *crura,* kroo′ră) **of the penis.** Together, these structures constitute the **root of the penis,** and the crura attach the penis to the coxae.

Skin is loosely attached to the connective tissue that surrounds the erectile columns in the shaft of the penis. The skin is firmly attached at the base of the glans penis, and a thinner layer of skin tightly covers the glans penis. The skin of the penis, especially the glans penis, is well supplied with sensory receptors. A loose fold of skin called the **prepuce** (prē′poos), or **foreskin,** covers the glans penis (see figure 28.1).

FIGURE 28.7 Penis

(*a*) Section through the spongy, or penile, urethra laid open and viewed from above. The prostate is also cut open to show the prostatic urethra. (*b*) Cross section of the penis, showing principal nerves, arteries, and veins along the dorsum of the penis. The *line* and *arrows* depict the manner in which (*a*) is cut and laid open.

> ## Circumcision
>
> **Circumcision** (ser-kŭm-sizh′ŭn) is the surgical removal of the prepuce, usually near the time of birth. Compelling medical reasons for circumcision do not appear to exist. Uncircumcised males have a higher incidence of penile cancer, but the underlying cause appears to be related to chronic infections and poor hygiene. In the few cases in which the prepuce is "too tight" to be moved over the glans penis, circumcision can be necessary to avoid chronic infections and maintain normal circulation.

The primary nerves, arteries, and veins of the penis pass along its dorsal surface (see figure 28.7*b*). Dorsal arteries, with dorsal nerves lateral to them, exist on each side of a single, midline dorsal vein. Additional deep arteries lie within the corpora cavernosa.

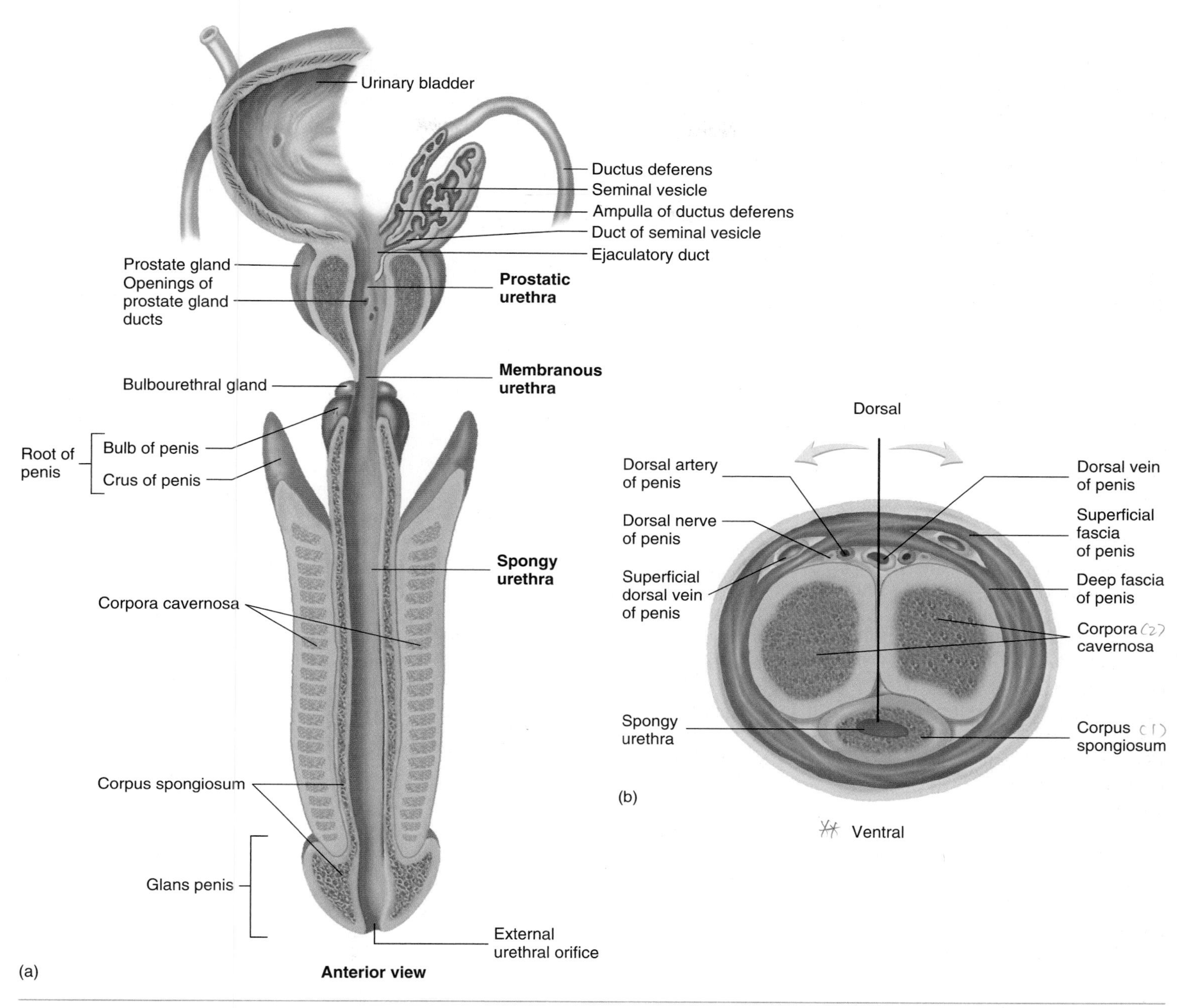

22 ***Describe the erectile tissue of the penis.***

23 ***Define glans penis, crus, bulb, and prepuce, and describe their structure.***

CASE STUDY

Prostate Cancer

Sixty-five-year-old Vern has a physical examination every year. Eight years ago, a **prostate specific antigen (PSA)** test indicated that Vern's blood prostate specific antigen levels were higher than the results from his previous tests. His physician reported moderate enlargement of the prostate gland, but no obvious tumorlike structures could be detected by a digital examination. Because of the increasing PSA levels, Vern's physician recommended a biopsy, so needle biopsies of his prostate gland were taken through his rectum. The pathology report described suspicious cells consistent with prostate cancer in one of the tissue samples. Vern's physician had the biopsy samples examined by another pathology laboratory, which did not confirm the first pathology report. As a consequence, Vern's physician explained that one option was to do nothing and continue having regular checkups because prostate cancer typically develops slowly. Eight years later, a PSA test showed another substantial increase in Vern's PSA levels, although no tumor could be detected by a digital exam, and Vern had no complaints, such as difficulty urinating. Needle biopsies of the prostate gland indicated cancer cells were present in two of the six biopsy samples. His physician explained that Vern's chances of surviving are high because it appeared that the cancer was discovered before it had metastasized outside the prostate gland. Vern could choose to do nothing, have his prostate gland surgically removed, or treat the cancer with radiation therapy, hormonal therapy, or chemotherapy. Vern elected to have radiation therapy, which focuses radiation on the prostate gland to kill the cancer cells. Statistics indicate that surgery and radiation therapy have similar success rates for small, localized tumors like Vern's. The trauma of surgery and the higher probability of erectile dysfunction following surgery convinced Vern that radiation therapy was preferable. Vern's physician indicated that doing nothing is a reasonable option for men who are significantly older than Vern because older men diagnosed with prostate cancer often die of other conditions before they succumb to prostate cancer. Vern's physician explained that, for patients like him, approximately 85% of patients are cancer free 5 years after radiation treatments. Vern was grateful that he had annual physical examinations. Prostate cancer represents 29% of cancers in males in the United States and 14% of the deaths due to cancer. Only lung cancer results in more cancer deaths in men.

PREDICT 1

a. Given the elevated PSA levels that were detected and the suspicious cells appearing in the first biopsy, was it an error or was it reasonable to conclude that cancer cells were not present in Vern's prostate gland? (*Hint:* See Chapter 4, p. 142.)
b. Explain why, for the first biopsy, the second pathology report did not confirm the results of the first pathology report.
c. Explain why cancer cells were present in only two of the six needle biopsy samples.
d. Since increased blood levels of PSA are a good diagnostic tool for prostate cancer, explain why Vern's PSA levels were elevated prior to the first biopsy.

Accessory Glands

Seminal Vesicles

The **seminal vesicles** (sem′i-năl ves′i-klz) are sac-shaped glands located next to the ampullae of the ductus deferentia (see figure 28.6). Each gland is about 5 cm long and tapers into a short excretory duct that joins the ampulla of the ductus deferens to form the ejaculatory duct. The seminal vesicles have a capsule containing fibrous connective tissue and smooth muscle cells.

Prostate Gland

The **prostate** (pros′tāt; one standing before) **gland,** consisting of both glandular and muscular tissue, is about the size and shape of a walnut, about 4 cm long and 2 cm wide. It is dorsal to the symphysis pubis at the base of the urinary bladder, where it surrounds the prostatic urethra and the two ejaculatory ducts (see figure 28.1). The prostate gland is composed of a fibrous connective tissue capsule containing distinct smooth muscle cells and numerous fibrous partitions, also containing smooth muscle, that radiate inward toward the urethra. Covering these muscular partitions is a layer of columnar epithelial cells that form saccular dilations into which the columnar cells secrete prostatic fluid. Fifteen to 30 small prostatic ducts carry these secretions into the prostatic urethra.

PREDICT 2

The prostate gland can enlarge for several reasons, including infections, tumor, and old age. The detection of enlargement or changes in the prostate is an important way to detect prostatic cancer. Suggest a way other than surgery that the prostate gland can be examined by palpation for any abnormal changes.

Bulbourethral Glands

The **bulbourethral** (bŭl′bō-ū-rē′thrăl) **glands** are a pair of small glands located near the membranous part of the urethra (see figures 28.1 and 28.6). In young males, each is about the size of a pea, but they decrease in size with age and are almost impossible to see in old men. Each bulbourethral gland is a compound mucous gland (see chapter 4). The small ducts of each gland unite to form a single duct. The single duct from each bulbourethral gland then enters the spongy urethra at the base of the penis.

Semen

Semen (sē′men) is a composite of sperm cells and secretions from the male reproductive glands. The seminal vesicles produce about 60% of the fluid, the prostate gland contributes about 30%, the testes contribute 5%, and the bulbourethral glands contribute 5%. **Emission** is the discharge of the secretions of the seminal vesicles, prostate gland, and bulbourethral glands with sperm cells from the epididymides into the urethra to form semen. **Ejaculation** is the forceful expulsion of semen from the urethra caused by the contraction of the urethra, the skeletal muscles in the floor of the pelvis, and the muscles at the base of the penis.

The bulbourethral glands and urethral mucous glands produce a mucous secretion just before ejaculation. This mucus lubricates the urethra, neutralizes the contents of the normally acidic spongy urethra, provides a small amount of lubrication during intercourse, and helps reduce vaginal acidity.

Testicular secretions include sperm cells, a small amount of fluid, and metabolic by-products. The thick, mucuslike secretions of the seminal vesicles contain large amounts of fructose, citric acid, and other nutrients that nourish the sperm cells. The seminal vesicle secretions also contain fibrinogen, which is involved in a weak coagulation reaction of the semen immediately after ejaculation, and prostaglandins, which can cause uterine contractions.

The thin, milky secretions of the prostate have a rather high pH; with secretions of the seminal vesicles, bulborurethral glands, and urethral mucous glands help neutralize the acidic urethra. The secretions of the prostate and seminal vesicles also help neutralize the acidic secretions of the testes and of the vagina. The prostatic secretions are also important in the transient coagulation of semen because they contain clotting factors that convert fibrinogen from the seminal vesicles to fibrin, resulting in coagulation. The coagulated material keeps the semen as a single, sticky mass for a few minutes after ejaculation, and then fibrinolysin from the prostate causes the coagulum to dissolve, thereby releasing the sperm cells to make their way up the female reproductive tract.

PREDICT 3

Explain a possible reason for the coagulation reaction.

Before ejaculation, the ductus deferens begins to contract rhythmically to propel sperm cells and testicular and epididymal secretions from the tail of the epididymis to the ampulla of the ductus deferens. Contractions of the ampullae, seminal vesicles, and ejaculatory ducts cause the sperm cells, along with testicular and epididymal secretions, to move into the prostatic urethra with the prostatic secretions. Secretions of the seminal vesicles then enter the prostatic urethra, where they mix with the other secretions.

Normal sperm cell counts in the semen range from 75 to 400 million sperm cells per milliliter of semen, and a normal ejaculation usually consists of about 2–5 mL of semen. The semen with the highest sperm count is expelled from the penis first because it contains the greater percentage of sperm-containing fluid from the epididymis. Sperm cells become motile after ejaculation once they are mixed with secretions of the male accessory glands and with secretions of the female reproductive tract. The alkaline pH (an average of 7.5), nutrients, and removal of inhibitory substances from the surface of sperm cells appear to increase sperm cell motility. Most of the sperm cells (millions) are expended in moving the general group of sperm cells through the female reproductive system. Enzymes carried in the acrosomal cap of each sperm cell help digest a path through the mucoid fluids of the female reproductive tract and through materials surrounding the oocyte. Once the acrosomal fluid is depleted from a sperm cell, the sperm cell is no longer capable of fertilization.

24 ***State where the seminal vesicles, prostate gland, and bulbourethral glands empty into the male reproductive duct system.***

25 ***Define emission and ejaculation.***

26 ***Define semen. Describe the contribution to semen of the accessory sex glands.***

27 ***What is the function of the secretions of each of the accessory glands?***

PHYSIOLOGY OF MALE REPRODUCTION

Normal function of the male reproductive system depends on both hormonal and neural mechanisms. Hormones are primarily responsible for the development of reproductive structures and maintenance of their functional capacities, the development of secondary sexual characteristics, the control of sperm cell formation, and influence over sexual behavior. Neural mechanisms are primarily involved in sexual behavior and control of the sexual act.

Regulation of Sex Hormone Secretion

Hormonal mechanisms that influence the male reproductive system involve the hypothalamus, the pituitary gland, and the testes (figure 28.8). A small peptide hormone called **gonadotropin-releasing hormone (GnRH),** or **luteinizing hormone–releasing hormone (LHRH),** is released from neurons in the hypothalamus. GnRH passes through the hypothalamohypophyseal portal system to the anterior pituitary gland (see chapter 18). In response to GnRH, cells within the anterior pituitary gland secrete two hormones, referred to as **gonadotropins** (gō′nad-ō-trō′pinz, gon′ă-dō-trō′pinz) because they influence the function of the **gonads** (gō′nadz; testes or ovaries).

The two gonadotropins are **luteinizing hormone** (**LH**) and **follicle-stimulating hormone** (**FSH**). They are named for their

1. GnRH from the hypothalamus stimulates the secretion of LH and FSH from the anterior pituitary.
2. LH stimulates testosterone secretion from the interstitial cells.
3. FSH stimulates sustentacular cells of the seminiferous tubules to increase spermatogenesis and to secrete inhibin.
4. Testosterone has a stimulatory effect on the sustentacular cells of the seminiferous tubules, and it has a stimulatory effect on the development of sex organs and secondary sex characteristics.
5. Testosterone has a negative-feedback effect on the hypothalamus and pituitary to reduce LH and FSH secretion.
6. Inhibin has a negative-feedback effect on the anterior pituitary to reduce FSH secretion.

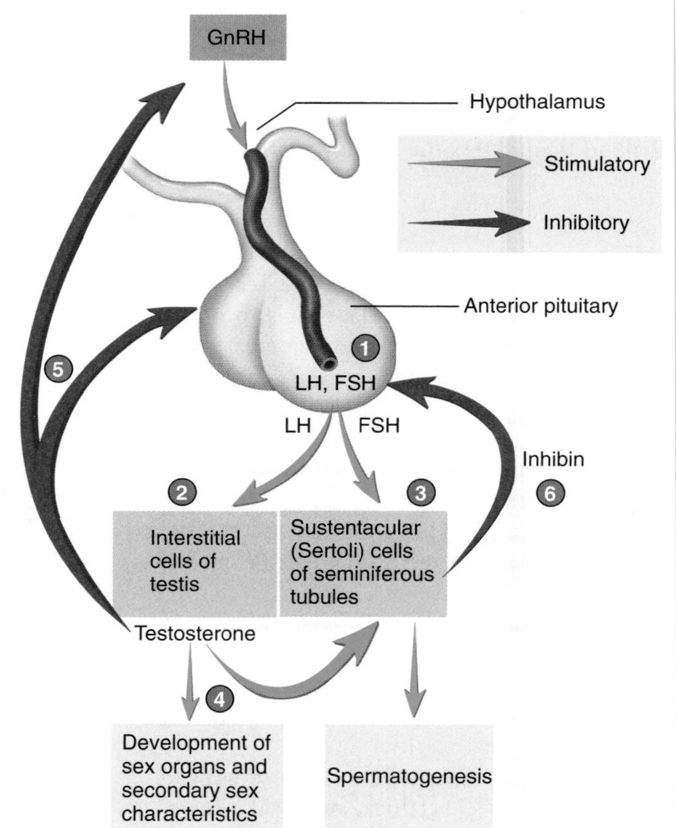

PROCESS FIGURE 28.8 Regulation of Reproductive Hormone Secretion in Males

functions in females, but they also have important functions in males. LH in males is sometimes called **interstitial cell–stimulating hormone (ICSH).** LH binds to the interstitial cells in the testes and causes them to increase their rate of testosterone synthesis and secretion. FSH binds primarily to sustentacular cells in the seminiferous tubules and promotes sperm cell development. Both gonadotropins bind to specific receptor molecules on the membranes of the cells they influence, and cyclic adenosine monophosphate (cAMP) is an important intracellular mediator in those cells.

For GnRH to stimulate the secretion of large quantities of LH and FSH, the anterior pituitary must be exposed to a series of brief increases and decreases in GnRH. Chronically elevated GnRH levels in the blood cause the anterior pituitary cells to become insensitive to stimulation by GnRH molecules, and little LH or FSH is secreted.

GnRH Pulses and Infertility

GnRH can be produced synthetically, and, if administered in small amounts in frequent pulses or surges, it can be useful in treating males who are infertile. GnRH can also inhibit reproduction because the chronic administration of it can sufficiently reduce LH and FSH levels to prevent sperm cell production in males or ovulation in females.

Testosterone is the major male hormone secreted by the testes. It is classified as an **androgen** (*andros* is Greek for male human being) because it stimulates the development of male reproductive structures (see chapter 29) and male secondary sexual characteristics. The testes secrete other androgens, but they are produced in smaller concentrations and are less potent than testosterone. In addition, the testes secrete small amounts of estrogen and progesterone.

Testosterone has a major influence on many tissues. It plays an essential role in the embryonic development of reproductive structures, their further development during puberty, the development of secondary sexual characteristics during puberty, the maintenance of sperm cell production, and the regulation of gonadotropin secretion. It also influences behavior.

Inside some target tissue cells such as cells of the scrotum and penis, an enzyme converts testosterone to dihydrotestosterone. In these cells, dihydrotestosterone is the active hormone. If the enzyme is not active, these structures do not fully develop normally. In other target tissue cells, an enzyme converts testosterone to estrogen and estrogen is the active hormone. Some brain cells convert testosterone to estrogen; in these cells, estrogen may be the active hormone responsible for some aspects of male sexual behavior.

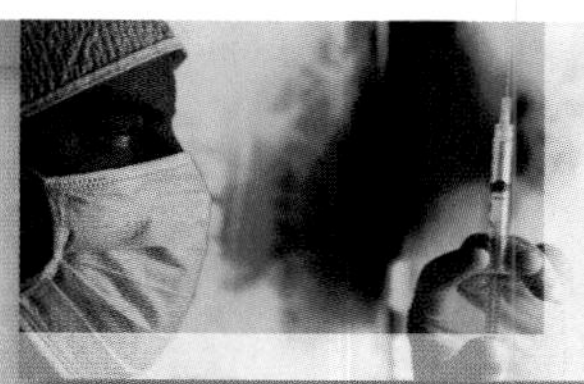

CLINICAL FOCUS

Male Infertility

Infertility (in-fer-til′i-tē) is reduced or diminished fertility. The most common cause of infertility in males is a low sperm cell count. If the sperm cell count drops to below 20 million sperm cells per milliliter, the male is usually infertile.

A decreased sperm cell count can occur because of damage to the testes as a result of trauma, radiation, cryptorchidism, or infections, such as mumps. Varicocele (var′i-kō-sēl) is an abnormal dilation of a vein in the spermatic cord. It can result from incompetent or absent valves in spermatic veins, from thrombi, or from tumors. There is a decrease in testicular blood flow and decreased spermatogenesis. Reduced sperm cell counts can also result from inadequate secretion of luteinizing hormone and follicle-stimulating hormone, which can be caused by hypothyroidism, trauma to the hypothalamus, infarctions of the hypothalamus or anterior pituitary gland, and tumors. Decreased testosterone secretion also reduces the sperm cell count. Some reports suggest that the average sperm cell count has decreased substantially since the end of World War II (1945), although there is some controversy about the accuracy of these reports. It is speculated that certain synthetic chemicals are responsible.

Fertility is reduced if the sperm cell count is normal but sperm cell structure is abnormal. Abnormal sperm cell structure can be due to chromosomal abnormalities or abnormal genes. Reduced sperm cell motility also results in infertility. A major cause of reduced sperm cell motility is antisperm antibodies produced by the immune system, which bind to sperm cells.

Fertility can sometimes be achieved by collecting several ejaculations and concentrating the sperm cells. The more concentrated sperm cells can then be introduced into the female's reproductive tract, a process called **artificial insemination** (in-sem-i-nā′shŭn).

The sustentacular cells of the testes secrete a polypeptide hormone called **inhibin** (in-hib′in), which inhibits FSH secretion from the anterior pituitary.

28 ***Where are GnRH, FSH, LH, and inhibin produced? What effects do they produce?***

29 ***Where is testosterone produced?***

30 ***Explain the regulation of testosterone secretion.***

Puberty

A gonadotropin-like hormone called **human chorionic** (kō-rē-on′ik) **gonadotropin (HCG),** which the placenta secretes, stimulates the synthesis and secretion of testosterone by the fetal testes before birth. After birth, however, no source of stimulation is present, and the testes of the newborn baby atrophy slightly and secrete only small amounts of testosterone until puberty, which normally begins when a boy is 12–14 years old.

Puberty (pū′ber-tē) is the age at which individuals become capable of sexual reproduction. Before puberty, small amounts of testosterone and other androgens in males inhibit GnRH release from the hypothalamus. At puberty, the hypothalamus becomes much less sensitive to the inhibitory effect of androgens, and the rate of GnRH secretion increases, leading to increased LH and FSH release. Elevated FSH levels promote sperm cell formation, and elevated LH levels cause the interstitial cells to secrete larger amounts of testosterone. Testosterone still has a negative-feedback effect on GnRH secretion after puberty but is not capable of completely suppressing it.

31 ***What changes in hormone production occur at puberty?***

Effects of Testosterone

Testosterone is by far the major androgen in males. Nearly all of the androgens, including testosterone, are produced by the interstitial cells, with small amounts produced by the adrenal cortex and possibly by the sustentacular cells. Testosterone causes the enlargement and differentiation of the male genitals and reproductive duct system, is necessary for sperm cell formation, and is required for the descent of the testes near the end of fetal development. Testosterone stimulates hair growth in the following regions: (1) the pubic area and extending up the linea alba, (2) the legs, (3) the chest, (4) the axillary region, (5) the face, and (6) the back. It causes vellus hair to be converted to terminal hair, which is more pigmented and coarser.

Male Pattern Baldness

Some men have a genetic tendency for **male pattern baldness,** which develops in response to testosterone and other androgens. When testosterone levels increase at puberty, the density of the hair on the top of the head begins to decrease. Baldness usually reaches its maximum rate of development when the individual is in the third or fourth decade of life.

Testosterone also causes the texture of the skin to become rougher or coarser. The quantity of melanin in the skin also increases, making the skin darker. Testosterone increases the rate of secretion from the sebaceous glands, especially on the face, frequently near puberty, with the development of acne. Beginning near puberty, testosterone also causes hypertrophy of the larynx and reduced tension on the vocal folds. The structural changes can first result in a voice that is difficult to control, but ultimately the voice reaches its normal masculine quality.

Testosterone has a general stimulatory effect on metabolism so that males have a slightly higher metabolic rate than females. The red blood cell count is increased by nearly 20% as a result of effects of testosterone on erythropoietin production. Testosterone also has a minor mineralocorticoid-like effect, causing the retention of sodium in the body and, consequently, an increase in the volume of body fluids. Testosterone promotes protein synthesis in most tissues; as a result, skeletal muscle mass increases at puberty. The average percentage of the body weight composed of skeletal muscle is greater for males than for females because of the effect of androgens.

Synthetic Androgens and Muscle Mass

Some athletes, especially weightlifters, ingest synthetic androgens in an attempt to increase muscle mass. The side effects of the large doses of androgens are often substantial and include testicular atrophy, kidney damage, liver damage, heart attacks, and strokes. The use of synthetic androgens is highly discouraged by the medical profession and is a violation of the rules of most athletic organizations.

Testosterone causes rapid bone growth and increases the deposition of calcium in bone, resulting in an increase in height. The growth in height is limited, however, because testosterone also stimulates ossification of the epiphyseal plates of long bones (see chapter 6). Males who mature sexually at an earlier age grow rapidly but reach their maximum height earlier. Males who mature sexually at a later age do not exhibit a rapid period of growth, but they grow for a longer period and can become taller than those who mature sexually at an earlier age.

32 ***Describe the effects of testosterone on male reproductive structures and secondary sexual characteristics.***

Male Sexual Behavior and the Male Sex Act

Testosterone is required to initiate and maintain male sexual behavior. Testosterone enters cells within the hypothalamus and the surrounding areas of the brain and influences their function, resulting in sexual behavior. Male sexual behavior may depend, in part, however, on the conversion of testosterone to other steroids, such as estrogen, in the cells of the brain.

The blood levels of testosterone remain relatively constant throughout the lifetime of a male from puberty until about 40 years of age. Thereafter, the levels slowly decline to about 20% of this value by 80 years of age, causing a slow decrease in sex drive and fertility.

The male sex act is a complex series of reflexes that result in erection of the penis, secretion of mucus into the urethra, emission, and ejaculation. Sensations that are normally interpreted as pleasurable occur during the male sexual act and result in a climactic sensation, **orgasm** (ōr′gazm), associated with ejaculation. After ejaculation, a phase called **resolution** occurs, in which the penis becomes flaccid, an overall feeling of satisfaction exists, and the male is unable to achieve erection and a second ejaculation for many minutes to many hours or longer.

Sensory Action Potentials and Integration

Action potentials are conducted by sensory neurons from the genitals through the pudendal nerve to the sacral region of the spinal cord, where reflexes that result in the male sexual act are integrated. Action potentials travel from the spinal cord to the cerebrum to produce conscious sexual sensations.

Rhythmic massage of the penis, especially the glans penis, provides an extremely important source of sensory action potentials that initiate erection and ejaculation. Sensory action potentials, produced in surrounding tissues, such as the scrotum and the anal, perineal, and pubic regions, reinforce sexual sensations. Engorgement of the prostate and seminal vesicles with their secretions also cause sexual sensations. In some cases, mild irritation of the urethra, such as from an infection, can cause sexual sensations.

Psychic stimuli, such as sight, sound, odor, or thoughts, have a major effect on sexual reflexes. Thinking sexual thoughts or dreaming about erotic events tends to reinforce stimuli that trigger sexual reflexes, such as erection and ejaculation. Ejaculation while sleeping is a relatively common event in young males and is thought to be triggered by psychic stimuli associated with dreaming. Psychic stimuli can also inhibit the sexual act, and thoughts that are not sexual tend to decrease the effectiveness of the male sexual act. The inability to concentrate on sexual sensations is one of the causes of **impotence** (im′pŏ-tens), or **erectile dysfunction (ED),** the inability to achieve or maintain an erection and to accomplish the male sexual act. Impotence can also be caused by nerve damage, such as can result from prostate surgery, or restricted circulation.

Action potentials from the cerebrum that reinforce the sacral reflexes are not absolutely required for the culmination of the male sexual act, and the male sexual act can be accomplished by males who have suffered spinal cord injuries superior to the sacral region.

Erection, Emission, and Ejaculation

When **erection** (ē-rek′shŭn) occurs, the penis becomes enlarged and rigid. Erection is the first major component of the male sexual act. Action potentials travel from the spinal cord through the pudendal nerve to the arteries that supply blood to the erectile tissues. The nerve fibers release acetylcholine as well as **nitric oxide (NO)** as neurotransmitter substances. Acetylcholine binds to muscarinic receptors and activates a G protein mechanism that causes smooth muscle relaxation. Nitric oxide diffuses into the smooth muscle cells of blood vessels and combines with the enzyme guanylate cyclase, which converts guanosine triphosphate (GTP) to cyclic guanosine monophosphate (cGMP). The cGMP causes smooth muscle cells to relax and blood vessels to dilate (figure 28.9). At the same time, other arteries of the penis constrict to shunt blood to the erectile tissues. As a consequence, blood fills the sinusoids of the erectile tissue and compresses the veins. Because venous outflow is partially occluded, the blood

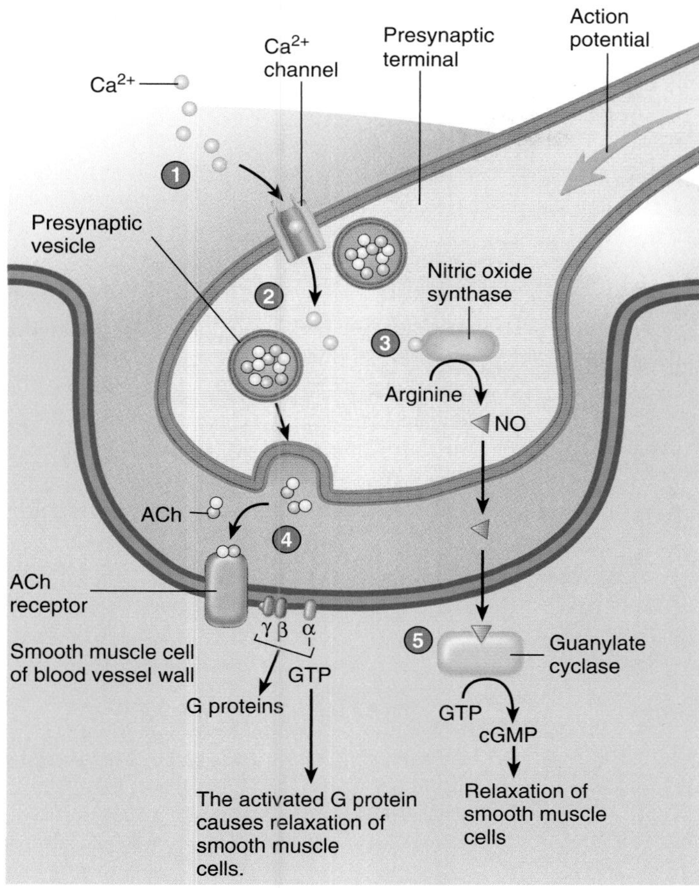

1. Action potentials in parasympathetic neurons cause voltage-gated Ca^{2+} channels to open and Ca^{2+} diffuse into the presynaptic terminals.
2. Calcium ions initiate the release of acetylcholine (ACh) from presynaptic vesicles.
3. Calcium ions also activate nitric oxide synthase, which promotes the synthesis of nitric oxide (NO) from arginine.
4. ACh binds to ACh receptors on the smooth muscle cells and activates a G protein mechanism. The activated G protein causes the relaxation of smooth muscle cells and erection of the penis.
5. NO binds to guanylate cyclase enzymes and activates them. The activated enzymes convert GTP to cGMP, which causes relaxation of the smooth muscle cells and erection of the penis.

PROCESS FIGURE 28.9 Neural Control of Erection

pressure in the sinusoids causes inflation and rigidity of the erectile tissue. Nerve action potentials that result in erection come from parasympathetic centers (S2–S4) or sympathetic centers (T2–L1) in the spinal cord. Normally, the parasympathetic centers are more important, but in cases of damage to the sacral region of the spinal cord, erection can occur through the sympathetic system.

Parasympathetic action potentials also cause the mucous glands within the penile urethra and the bulbourethral glands at the base of the penis to secrete mucus.

Impotence, or Erectile Dysfunction

Failure to achieve erections—impotence, or erectile dysfunction (ED)—can be a major source of frustration for some men and can contribute to disharmony in relationships. ED can be due to reduced testosterone secretion, which can result from hypothalamic, pituitary, or testicular complications. In other cases, ED is caused by defective stimulation of the erectile tissue by nerve fibers or reduced response of the blood vessels to neural stimulation. Erection can be achieved in some people by oral medication, such as sildenafil (Viagra), tadafil (Cialis), or verdenafil (Livitra), or by the injection of specific drugs into the base of the penis, which increase blood flow into the sinusoids of the erectile tissue of the penis, resulting in erection for many minutes. Sildenafil blocks the activity of the enzyme that converts cGMP to GMP. Consequently, it allows cGMP to accumulate in smooth muscle cells in the arteries of erectile tissues and causes them to relax. This response is effective in enhancing erection in males. Sildenafil's action is not specific to the erectile tissue of the penis, however. It causes vasodilation in other tissues and can increase the workload of the heart.

Emission (ē-mish′ŭn) is the accumulation of sperm cells and secretions of the prostate gland and seminal vesicles in the urethra. Sympathetic centers (T12–L1) in the spinal cord, which are stimulated as the level of sexual tension increases, control emission. Sympathetic action potentials cause peristaltic contractions of the reproductive ducts and stimulate the seminal vesicles and the prostate gland to release their secretions. Consequently, semen accumulates in the prostatic urethra and produces sensory action potentials that pass through the pudendal nerves to the spinal cord. Integration of these action potentials results in both sympathetic and somatic motor output. Sympathetic action potentials cause constriction of the internal sphincter of the urinary bladder so that semen and urine are not mixed. Somatic motor action potentials are sent to the skeletal muscles of the urogenital diaphragm and the base of the penis, causing several rhythmic contractions that force the semen out of the urethra. The movement of semen out of the urethra is called ejaculation (ē-jak-ū-lā′shŭn). In addition, an increase in muscle tension occurs throughout the body.

33 ***What effects do psychic, tactile, parasympathetic, and sympathetic stimulation have on the male sex act?***

34 ***Describe the processes of erection, emission, and ejaculation.***

ANATOMY OF THE FEMALE REPRODUCTIVE SYSTEM

The female reproductive organs consist of the ovaries, uterine tubes, uterus, vagina, external genital organs, and mammary glands. The internal reproductive organs of the female are within the pelvis between the urinary bladder and the rectum (figure 28.10). The uterus and the vagina are in the midline, with the ovaries to each side of the uterus. A group of ligaments holds the internal reproductive organs in place. The most conspicuous is the **broad ligament,** an extension of the peritoneum that spreads out on both sides of the uterus and to which the ovaries and uterine tubes are attached (figure 28.11).

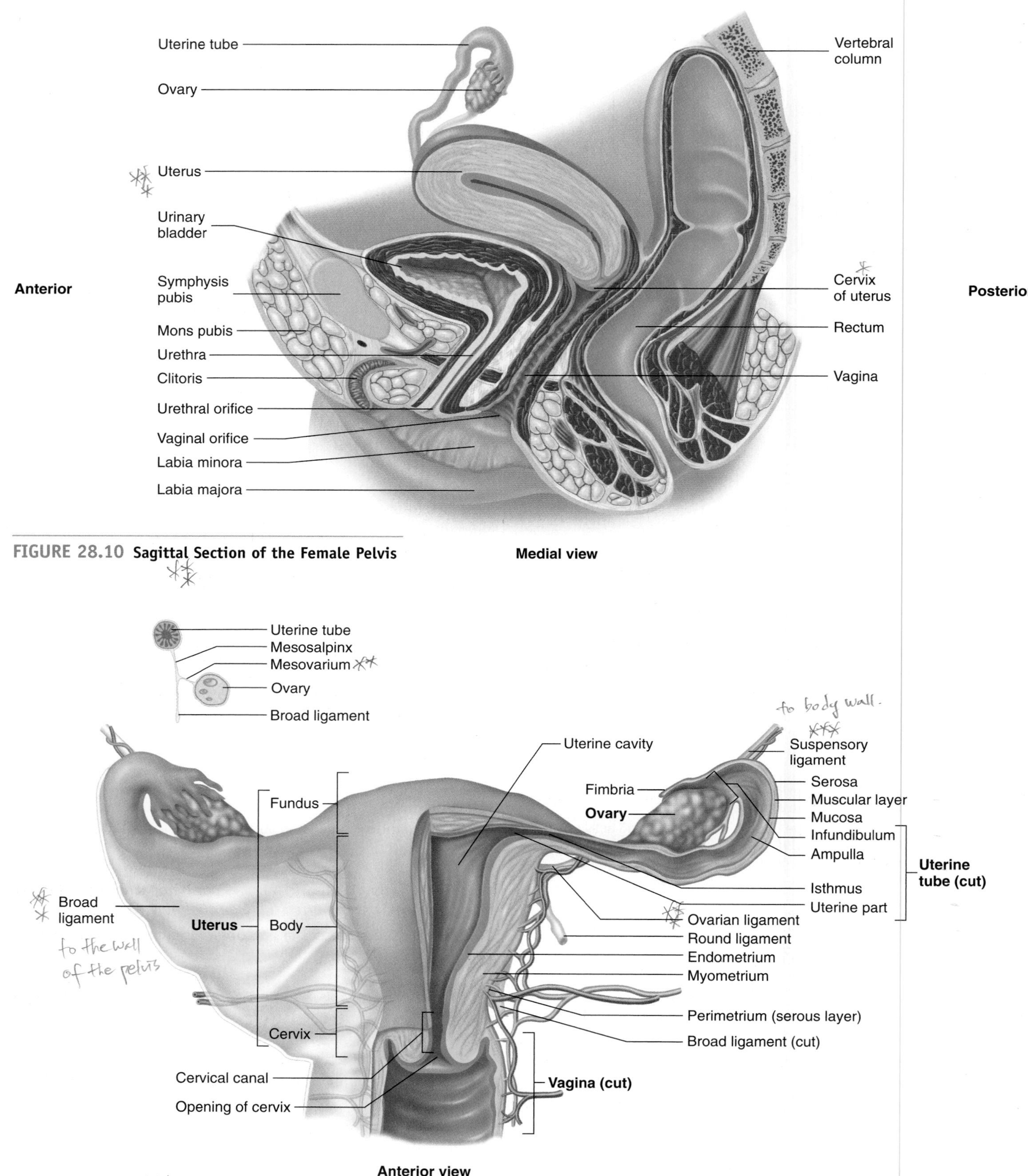

FIGURE 28.10 Sagittal Section of the Female Pelvis

FIGURE 28.11 Uterus, Vagina, Uterine Tubes, Ovaries, and Supporting Ligaments

The uterus and uterine tubes are cut in section (on the left side), and the vagina is cut to show the internal anatomy. The inset shows the relationships among the ovary, the uterine tube, and the ligaments that suspend them in the pelvic cavity.

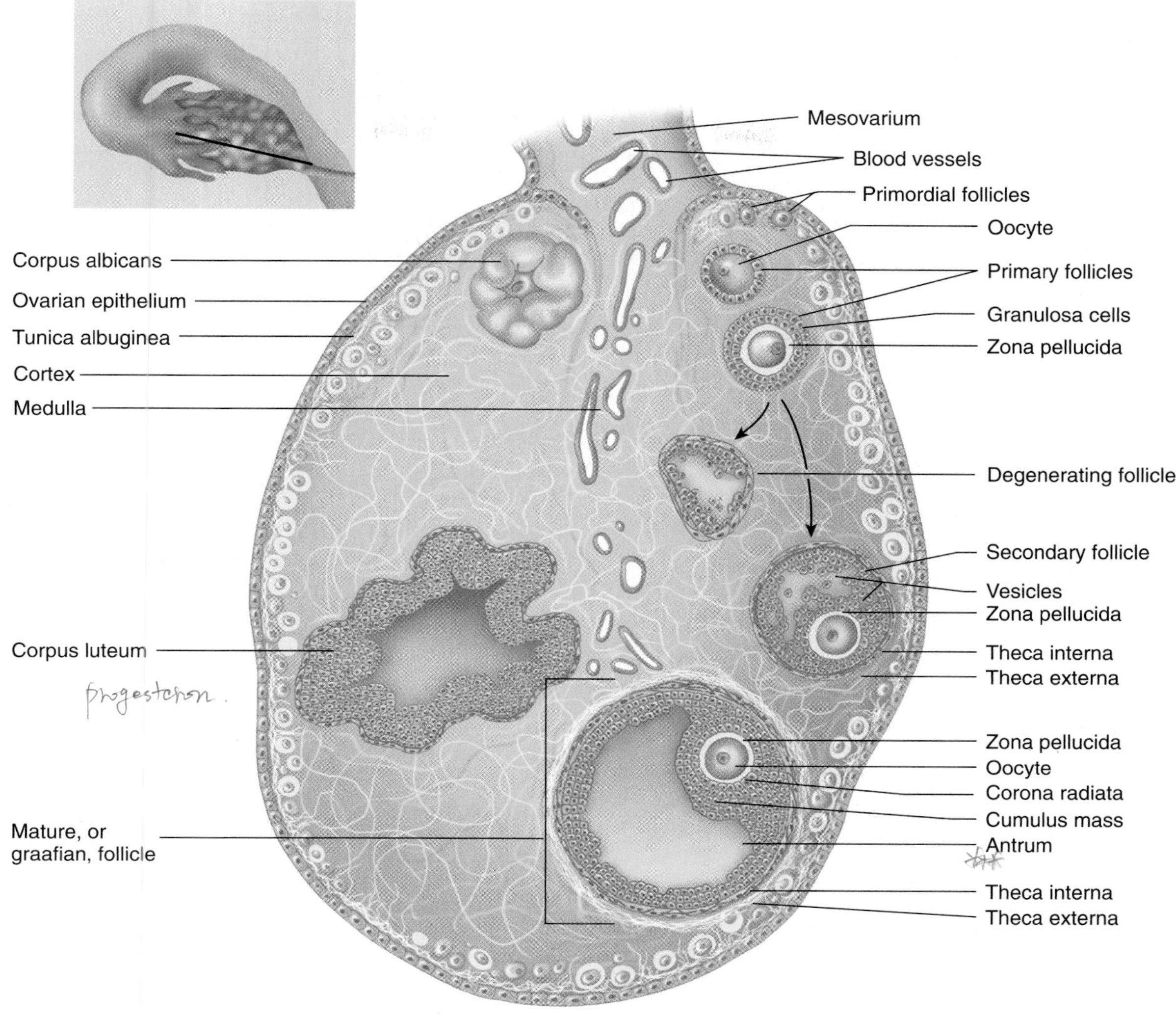

FIGURE 28.12 Histology of the Ovary

The ovary is sectioned to illustrate its internal structure (the inset shows plane of section). Ovarian follicles from each major stage of development are shown.

Ovaries

The two **ovaries** (ō′var-ēz) are small organs about 2–3.5 cm long and 1–1.5 cm wide (see figure 28.11). A peritoneal fold called the **mesovarium** (mez′ō-vā′rē-ŭm; mesentery of the ovary) attaches each ovary to the posterior surface of the broad ligament. Two other ligaments are associated with the ovary: the **suspensory ligament,** which extends from the mesovarium to the body wall, and the **ovarian ligament,** which attaches the ovary to the superior margin of the uterus. The ovarian arteries, veins, and nerves traverse the suspensory ligament and enter the ovary through the mesovarium.

Ovarian Histology

The visceral peritoneum, which is made up of simple cuboidal epithelium, where it covers the surface of the ovary, is called the **ovarian,** or **germinal, epithelium.** The germinal epithelium is so-named because the cuboidal epithelial cells were once thought to produce oocytes. Immediately below the epithelium is a capsule of dense fibrous connective tissue, the **tunica albuginea** (al-bū-jin′ē-ă). The denser, outer part of the ovary is called the **cortex** and the looser, inner part is called the **medulla** (figure 28.12). The connective tissue of the ovary is called the **stroma.** Blood vessels, lymphatic vessels, and nerves from the mesovarium enter the medulla. Numerous small vesicles called ovarian follicles, each of which contains an **oocyte** (ō′ō-sīt), are distributed throughout the stroma of the cortex.

35 ***Name and describe the ligaments that hold the ovaries in place.***

36 ***Describe the coverings and structure of the ovary.***

Oocyte Development and Fertilization

The formation of female sex cells begins in the fetus. By the fourth month of development, the ovaries contain 5 million **oogonia** (ō-ō-gō′nē-ă; *oon,* egg + *gone,* generation), the cells from which

1. Oogonia are the cells from which oocytes arise. The oogonia divide by mitosis to produce other oogonia and primary oocytes.

2. Five million oocytes may be produced by the fourth month of prenatal life. Primary oocytes begin the first meiotic division but stop at prophase I. All of the primary oocytes remain in this state until puberty.

3. The first meiotic division is completed in a single mature follicle just before ovulation during each menstrual cycle. A secondary oocyte and the first polar body result from the unequal division of the cytoplasm.

4. The secondary oocyte begins the second meiotic division but stops at metaphase II.

5. The second meiotic division is completed after ovulation and after a sperm cell unites with the secondary oocyte. A secondary oocyte and a second polar body are formed.

6. Fertilization is completed after the nuclei of the secondary oocyte and the sperm cell unite. The resulting cell is called a zygote.

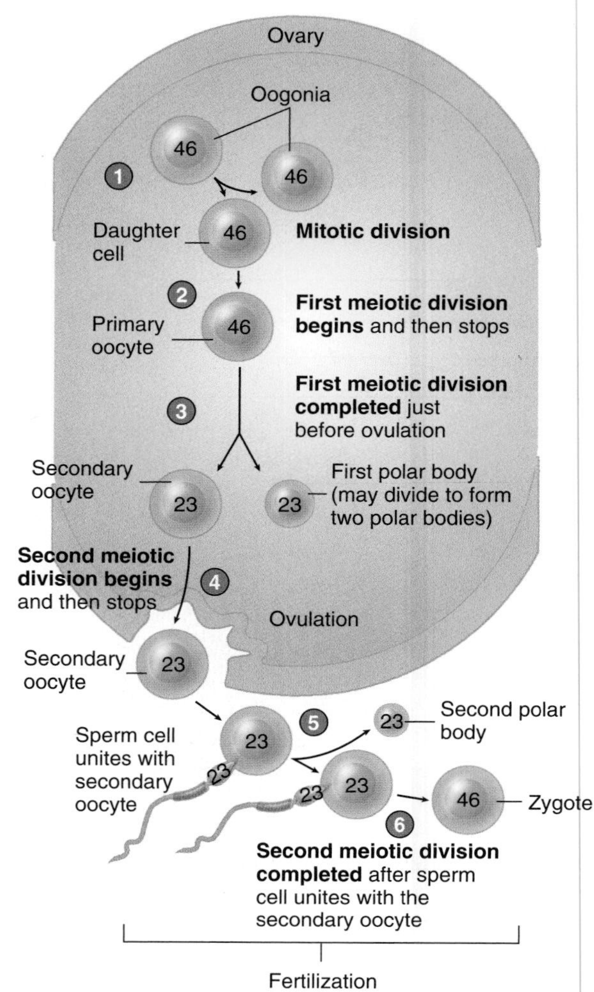

PROCESS FIGURE 28.13 Maturation and Fertilization of the Oocyte

The primary oocyte undergoes meiosis and gives off the first polar body to become a secondary oocyte just before ovulation. Sperm cell penetration initiates the completion of the second meiotic division and the expulsion of a second polar body. The nuclei of the oocyte and the sperm cell unite. Fertilization results in the formation of a zygote.

oocytes develop (figure 28.13). By the time of birth, many of the oogonia have degenerated, and the remaining have begun meiosis. Also, oogonia can form after birth from stem cells, but the extent to which this occurs, and how long it occurs, is not clear. Meiosis stops, however, during the first meiotic division at a stage called prophase I. The cell at this stage is called a **primary oocyte,** and at birth there are about 2 million of them. From birth to puberty, the number of primary oocytes decreases to around 300,000 to 400,000. Only about 400 primary oocytes will complete development and give rise to the secondary oocytes that are released from the ovaries during ovulation.

Ovulation (ov′ū-lā′shŭn, ō′vū-lā′shŭn) is the release of a **secondary oocyte** from an ovary (see figure 28.13). Just before ovulation, the primary oocyte completes the first meiotic division to produce a secondary oocyte and a **polar body.** Unlike meiosis in males, cytoplasm is not split evenly between the two cells. Most of the cytoplasm of the primary oocyte remains with the secondary oocyte. The cytoplasm contains organelles, such as mitochondria, and nutrients that increase the viability of the secondary oocyte. The polar body either degenerates or divides to form two polar bodies. Eventually, the polar bodies degenerate. The secondary oocyte begins the second meiotic division, but it stops in metaphase II.

After ovulation, the secondary oocyte may be fertilized by a sperm cell (see figure 28.13). **Fertilization** (fer′til-i-zā-shŭn) begins when a sperm cell binds to the plasma membrane and penetrates into the cytoplasm of a secondary oocyte. Subsequently, the secondary oocyte completes the second meiotic division to form two cells, each containing 23 chromosomes. One of these cells has very little cytoplasm and is another polar body that degenerates. In the other, larger cell, the 23 chromosomes from the sperm cell nucleus joins with the 23 chromosomes from the female to form a

zygote (zī'gōt; *zygotos,* yolked) and complete fertilization. The zygote has 23 pairs (46) of chromosomes. All cells of the human body contain 23 pairs of chromosomes, except for the male and female sex cells. The zygote divides by mitosis to form two cells, which divide to form four cells, and so on. The mass of cells formed, after about 7 days after ovulation, may implant or attach to, the wall of the uterus. The implanted mass of cells continues to develop to form, after approximately 9 months, a new individual (see chapter 29).

37 ***Starting with oogonia, describe the formation of secondary oocytes by meiosis. What are polar bodies?***

38 ***Define ovulation and fertilization.***

39 ***Describe the formation of a zygote. How many pairs of chromosomes are in a zygote, and where do they come from? What can a zygote become?***

Follicle Development

A **primordial follicle** is a primary oocyte surrounded by a single layer of flat cells, called **granulosa cells** (figure 28.14). Beginning with puberty, some of the primordial follicles are converted to **primary follicles** when the oocyte enlarges and the single layer of granulosa cells becomes enlarged and cuboidal. Subsequently, several layers of granulosa cells form and a layer of clear material is deposited around the primary oocyte called the **zona pellucida** (zō'nă pellū'sid-dă; *zone,* girdle + *pellucidus,* passage of light).

Approximately every 28 days, hormonal changes stimulate some of the primary follicles to continue to develop (see figure 28.14). The primary follicle becomes a **secondary follicle** as fluid-filled spaces called **vesicles** form among the granulosa cells, and a capsule called the **theca** (thē'kă; a box) forms around the follicle. Cells of the **theca interna** surround the granulosa cells and participate with the granulosa cells in the synthesis of ovarian hormones (see p. 1062). The **theca externa** is primarily connective tissue that merges with the stroma of the ovary.

The secondary follicle continues to enlarge; when the fluid-filled vesicles fuse to form a single, fluid-filled chamber called the **antrum** (an'trŭm), the follicle is called a **mature,** or **Graafian** (graf'ē-ăn; named for Dutch histologist Reijnier de Graaf [1641]), **follicle.** The oocyte is pushed off to one side and lies in a mass of granulosa cells called the **cumulus cells,** or **cumulus oophorus** (kū'mū-lŭs ō-of'ōr-ŭs). The innermost cells of this mass resemble a crown radiating from the oocyte and are thus called the **corona radiata.**

The mature follicle forms a lump on the surface of the ovary. During ovulation, the mature follicle ruptures, forcing a small amount of blood, follicular fluid, and the oocyte, surrounded by the cumulus cells, into the peritoneal cavity.

Usually, only one Graafian follicle reaches the most advanced stages of development and is ovulated. The other follicles degenerate, a process called **atresia.**

40 ***Distinguish among primordial, primary, secondary, and mature follicles.***

41 ***Describe the process of ovulation.***

42 ***Starting with the oogonia, describe the development and production of a mature follicle that contains a secondary oocyte.***

Fate of the Follicle

After ovulation, the follicle still has an important function. It is transformed into a glandular structure called the **corpus luteum** (kōr'pŭs loo'tē-ŭm; yellow body), which has a convoluted appearance as a result of its collapse after ovulation (see figure 28.14). The granulosa cells and the theca interna, now called **luteal cells,** enlarge and begin to secrete hormones—progesterone and smaller amounts of estrogen.

If pregnancy occurs, the corpus luteum enlarges and remains throughout pregnancy as the **corpus luteum of pregnancy.** If pregnancy does not occur, the corpus luteum remains functional for about 10–12 days and then begins to degenerate. Progesterone and estrogen secretion decreases, and connective tissue cells become enlarged and clear, giving the whole structure a whitish color; it is therefore called the **corpus albicans** (al'bī-kanz; white body). The corpus albicans continues to shrink and eventually disappears after several months or even years.

43 ***What is the corpus luteum? What happens to the corpus luteum if fertilization occurs? If fertilization does not occur?***

Uterine Tubes

Two **uterine tubes,** also called **fallopian** (fa-lō'pē-an) **tubes** or **oviducts** (ō'vi-dŭkts), are present. There is a uterine tube on each side of the uterus associated with an ovary (see figure 28.11). Each tube is located along the superior margin of the broad ligament. The part of the broad ligament most directly associated with the uterine tube is called the **mesosalpinx** (mez'ō-sal'pinks; mesothelium of the trumpet-shaped uterine tube).

The uterine tube opens directly into the peritoneal cavity to receive the oocyte from the ovary. It expands to form the **infundibulum** (in-fŭn-dib'ū-lŭm; funnel), and long, thin processes called **fimbriae** (fim'brē-ē; fringe) surround the opening of the infundibulum. The inner surfaces of the fimbriae consist of a ciliated mucous membrane.

The part of the uterine tube that is nearest the infundibulum is called the **ampulla.** It is the widest and longest part of the tube and accounts for about 7.5–8 cm of the total 10 cm length of the tube. The part of the uterine tube nearest the uterus, the **isthmus,** is much narrower and has thicker walls than does the ampulla. The **uterine,** or **intramural, part** of the tube passes through the uterine wall and ends in a very small uterine opening.

The wall of each uterine tube consists of three layers (see figure 28.11). The outer **serosa** is formed by the peritoneum, the middle **muscular layer** consists of longitudinal and circular smooth muscle cells, and the inner **mucosa** consists of a mucous membrane of simple ciliated columnar epithelium. The mucosa is arranged into numerous longitudinal folds.

The mucosa of the uterine tubes provides nutrients for the oocyte or, if fertilization has occurred, for the developing embryonic

1. The primordial follicle consists of a primary oocyte surrounded by a single layer of squamous granulosa cells.

2. A primordial follicle becomes a primary follicle as the granulosa cells become enlarged and cuboidal.

3. The primary follicle enlarges. Granulosa cells form more than one layer of cells. A zona pellucida forms around the primary oocyte.

4. A secondary follicle forms when fluid-filled vesicles (spaces) develop among the granulosa cells and a well-developed theca becomes apparent around the granulosa cells.

5. A mature follicle forms when the fluid-filled vesicles form a single antrum. When a follicle becomes fully mature, it is enlarged to its maximum size, a large antrum is present, and the primary oocyte is located in the cumulus mass. Cells of the cumulus mass called the corona radiata surround the oocyte. Just prior to ovulation, the first meiotic division is completed to produce a secondary oocyte and a polar body (see figure 28.13).

6. During ovulation the secondary oocyte is released from the follicle, along with cells of the corona radiata.

7. Following ovulation, the granulosa cells divide rapidly and enlarge to form the corpus luteum.

8. When the corpus luteum degenerates, it forms the corpus albicans.

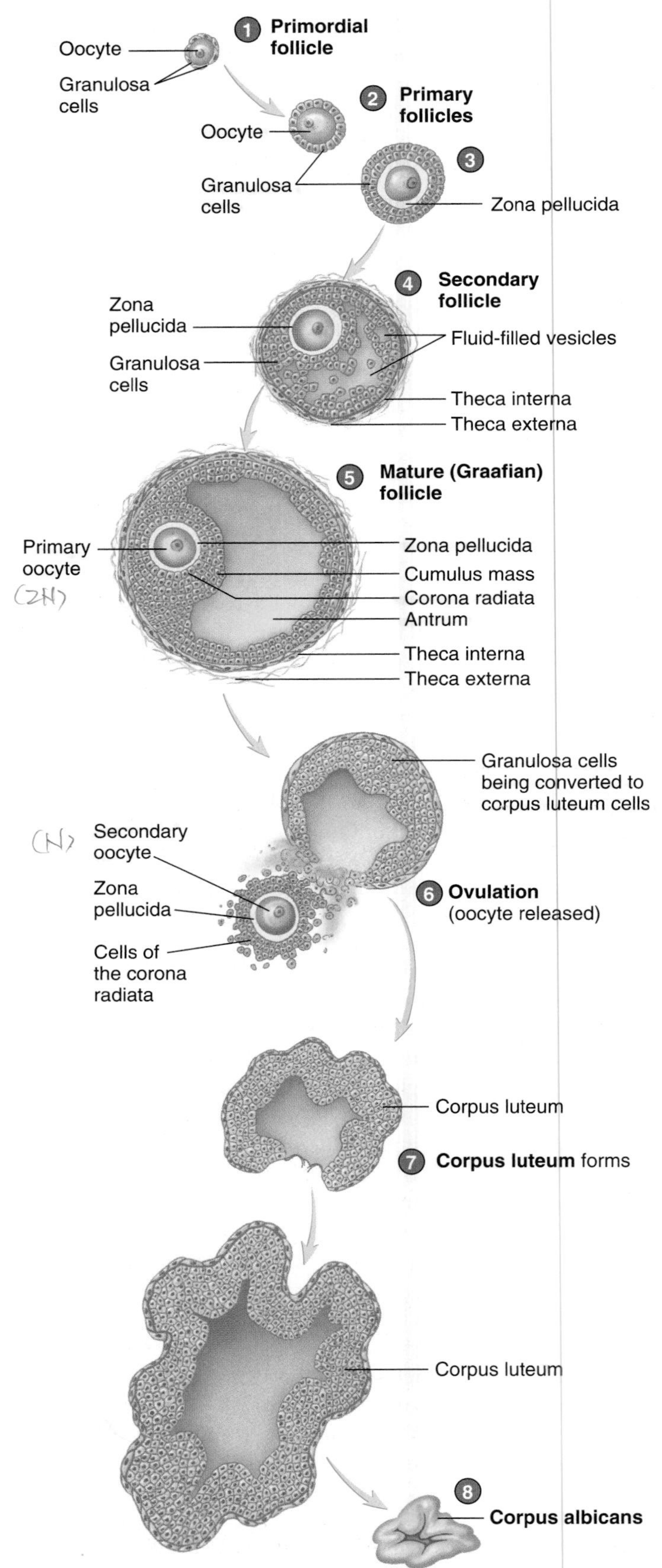

PROCESS FIGURE 28.14 Maturation of the Follicle and Oocyte

mass (see chapter 29) as it passes through the uterine tube. The ciliated epithelium helps move the small amount of fluid and the oocyte, or the developing embryonic mass, through the uterine tubes.

44 ***Describe the structures of the uterine tube.***

45 ***How are the uterine tubes involved in moving the oocyte or the zygote?***

Uterus

The **uterus** (ū′ter-ŭs) is the size and shape of a medium-sized pear—about 7.5 cm long and 5 cm wide (see figures 28.10 and 28.11). It is slightly flattened anteroposteriorly and is oriented in the pelvic cavity with the larger, rounded part, the **fundus** (fŭn′dŭs; bottom of a rounded flask), directed superiorly and the narrower part, the **cervix** (ser′viks; neck), directed inferiorly. The main part of the uterus, the **body,** is between the fundus and the cervix. A slight constriction called the **isthmus** marks the junction of the cervix and the body. Internally, the uterine cavity continues as the **cervical canal,** which opens through the **ostium** into the vagina.

Cancer of the Cervix

Cancer of the cervix is a relatively common type of cancer of the reproductive organs in females; it can be detected and treated successfully. Early in the development of cervical cancer, the cells of the cervix change in a characteristic way. This change can be observed by examining a cell sample microscopically. The most common technique is to obtain a **Papanicolaou (Pap) smear,** which is named after a U.S. physician who developed the technique. Pap smears have a reliability of 90% for detecting cervical cancer.

The major ligaments holding the uterus in place are the **broad ligament,** the **round ligaments** (see figure 28.11), and the **uterosacral ligaments.** The broad ligament is a peritoneal fold extending from the lateral margin of the uterus to the wall of the pelvis on either side. It also ensheaths the ovaries and the uterine tubes. The round ligaments extend from the uterus through the inguinal canals to the labia majora of the external genitalia, and the uterosacral ligaments attach the lateral wall of the uterus to the sacrum. Normally, the uterus is anteverted, meaning that the body of the uterus is tipped slightly anteriorly. In some women, the uterus is retroverted, or tipped posteriorly. In addition to the ligaments, skeletal muscles of the pelvic floor provide much support inferiorly to the uterus. If these muscles are weakened (e.g., in childbirth), the uterus can extend inferiorly into the vagina, a condition called a **prolapsed uterus.**

The uterine wall is composed of three layers: perimetrium (serous), myometrium, and endometrium (see figure 28.11). The **perimetrium** (per-i-mē′trē-ŭm), or **serous layer,** of the uterus is the peritoneum that covers the uterus. The next layer, just deep to the perimetrium, is the **myometrium** (mī′ō-mē′trē-ŭm), or **muscular layer,** which consists of a thick layer of smooth muscle. The myometrium accounts for the bulk of the uterine wall and is the thickest layer of smooth muscle in the body. In the cervix, the muscular layer contains less muscle and more dense connective tissue. The cervix is therefore more rigid and less contractile than the rest of the uterus. The innermost layer of the uterus is the **endometrium** (en′dō-mē′trē-ŭm), or **mucous membrane.** The endometrium consists of a simple columnar epithelial lining and a connective tissue, the lamina propria. Simple tubular glands are scattered about the lamina propria and open through the epithelium into the uterine cavity. The endometrium consists of two layers: a thin, deep **basal layer,** which is the deepest part of the lamina propria and is continuous with the myometrium, and a thicker, superficial **functional layer,** which consists of most of the lamina propria and the endothelium, lining the cavity itself. The functional layer is so-named because it undergoes changes and sloughing during the female menstrual cycle. Small **spiral arteries** of the lamina propria provide the blood supply to the functional layer of the endometrium. These blood vessels play an important role in cyclic changes of the endometrium.

Columnar epithelial cells line the cervical canal, which contains **cervical mucous glands.** The mucus fills the cervical canal and acts as a barrier to substances that could pass from the vagina into the uterus. Near ovulation, the consistency of the mucus changes, making the passage of sperm cells from the vagina into the uterus easier.

46 ***Name the parts of the uterus.***

47 ***Describe the major ligaments holding the uterus in place.***

48 ***Describe the layers of the uterine wall.***

Vagina

The **vagina** (vă-jī′nă) is a tube about 10 cm long that extends from the uterus to the outside of the body (see figure 28.11). The vagina is the female organ of copulation, functioning to receive the penis during intercourse, and it allows menstrual flow and childbirth. Longitudinal ridges called **columns** extend the length of the anterior and posterior vaginal walls, and several transverse ridges called **rugae** (roo′gē) extend between the anterior and posterior columns. The superior, domed part of the vagina, the **fornix** (fōr′niks; domed), is attached to the sides of the cervix so that a part of the cervix extends into the vagina.

The wall of the vagina consists of an outer muscular layer and an inner mucous membrane. The muscular layer is smooth muscle that allows the vagina to increase in size to accommodate the penis during intercourse and to stretch greatly during childbirth. The mucous membrane is moist stratified squamous epithelium that forms a protective surface layer. The vaginal mucous membrane releases most of the lubricating secretions produced by the female during intercourse.

A thin mucous membrane called the **hymen** (hī′men) covers the **vaginal opening,** or **orifice.** Sometimes, the hymen completely closes the vaginal opening (a condition called imperforate hymen), and it must be removed to allow menstrual flow. More commonly, the hymen is perforated by one or several holes. The openings in the hymen are usually greatly enlarged during the first sexual intercourse. In addition, the hymen can be perforated at some earlier time in a young woman's life, such as during strenuous physical exercise. Thus, the absence of an intact hymen does not necessarily indicate that a woman has had sexual intercourse, as was once thought.

49 ***Where is the vagina located?***

50 ***Describe the layers of the vaginal wall. What are rugae and columns?***

51 ***What is the hymen?***

External Genitalia

The external female genitalia, also referred to as the **vulva** (vŭl′vă), or **pudendum** (pū-den′dŭm), consist of the vestibule and its surrounding structures (figure 28.15). The **vestibule** (ves′ti-bool) is the space into which the vagina opens posteriorly and the urethra opens anteriorly. A pair of thin, longitudinal skin folds called the **labia** (lā′bē-ă; lips) **minora** (sing. *labium minus*) form a border on each side of the vestibule. A small erectile structure called the **clitoris** (klit′ō-ris) is located in the anterior margin of the vestibule. Anteriorly, the two labia minora unite over the clitoris to form a fold of skin called the **prepuce.**

The clitoris is usually less than 2 cm in length and consists of a shaft and a distal glans. Well supplied with sensory receptors, it initiates and intensifies levels of sexual tension. The clitoris contains two erectile structures, the **corpora cavernosa,** each of which expands at the base end of the clitoris to form the **crus of the clitoris** and attaches the clitoris to the coxal bones. The corpora cavernosa of the clitoris are comparable to the corpora cavernosa of the penis, and they become engorged with blood as a result of sexual excitement. In most women, this engorgement results in an increase in the diameter, but not the length, of the clitoris. With increased diameter, the clitoris makes better contact with the prepuce and surrounding tissues and is more easily stimulated.

Erectile tissue that corresponds to the corpus spongiosum of the male lies deep to and on the lateral margins of the vestibular floor on each side of the vaginal orifice. Each erectile body is called a **bulb of the vestibule.** Like other erectile tissue, it becomes engorged with blood and is more sensitive during sexual arousal. Expansion of the bulbs causes narrowing of the vaginal orifice and produces better contact of the vagina with the penis during intercourse.

On each side of the vestibule, between the vaginal opening and the labia minora, is an opening of the duct of the **greater vestibular gland.** Additional small mucous glands, the **lesser vestibular glands,** or **paraurethral glands,** are located near the clitoris and urethral opening. They produce a lubricating fluid that helps maintain the moistness of the vestibule.

Lateral to the labia minora are two prominent, rounded folds of skin called the **labia majora.** Subcutaneous fat is primarily responsible for the prominence of the labia majora. The two labia majora unite anteriorly in an elevation over the symphysis pubis called the **mons pubis** (monz; mound; pū′bis). The lateral surfaces of the labia majora and the surface of the mons pubis are covered with coarse hair. The medial surfaces are covered with numerous sebaceous and sweat glands. The space between the labia majora is called the **pudendal** (pū-den′dăl) **cleft.** Most of the time, the labia majora are in contact with each other across the midline, closing the pudendal cleft and concealing the deeper structures within the vestibule.

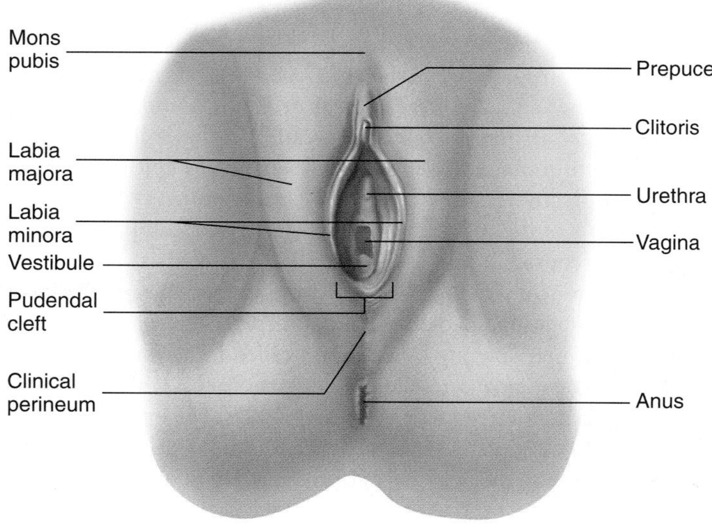

FIGURE 28.15 Female External Genitalia

52 ***What are the vulva, pudendum, and vestibule?***

53 ***What erectile tissue is in the clitoris and bulb of the vestibule?***

54 ***What is the function of the clitoris and bulb of the vestibule?***

55 ***Describe the labia minora, the prepuce, the labia majora, the pudendal cleft, and the mons pubis.***

56 ***Where are the greater and lesser vestibular glands located? What is their function?***

Perineum

The **perineum,** is divided into two triangles by the superficial and deep transverse perineal muscles (figure 28.16). The anterior, urogenital triangle contains the external genitalia, and the posterior, anal triangle contains the anal opening. The region between the vagina and the anus is the **clinical perineum.** The skin and muscle of this region can tear during childbirth. To prevent such tearing, an incision called an **episiotomy** (e-piz-ē-ot′ō-mē, e-pis-ē-ot′ō-mē) is sometimes made in the clinical perineum. This clean, straight incision is easier to repair than a tear would be. Alternatively, allowing the perineum to stretch slowly during delivery may prevent tearing, thereby making an episiotomy unnecessary.

57 ***Define perineum.***

58 ***What is the clinical perineum?***

59 ***Define and give the purpose of an episiotomy.***

Mammary Glands

The **mammary glands** the organs of milk production, are located within the **mammae** (mam′ē), or **breasts** (figure 28.17). The mammary glands are modified sweat glands. Externally, the

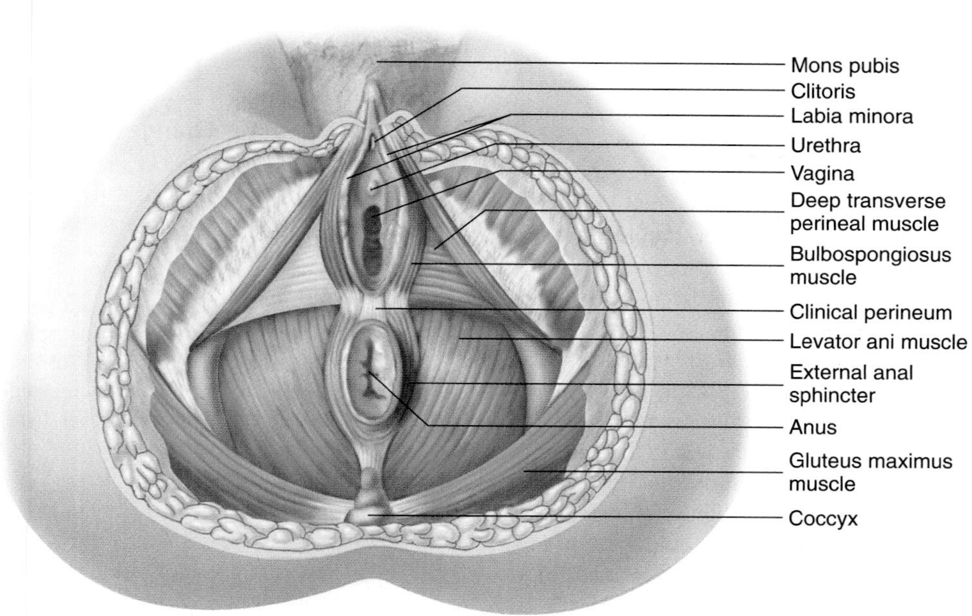

FIGURE 28.16 Inferior View of the Female Perineum

Lactiferous duct

Lobule

Lobe

(a)

Fascia

Rib

Pectoralis major

Duct

Epithelium

Nonlactating

Fat

Suspensory ligaments

Lactiferous sinus

Lactiferous ducts

(b) **Nonlactating breast**

(c) **Lactating breast**

Nipple

Bumps caused by areola glands

Areola

Lobule

Lobe

Myoepithelial cell

Epithelium

Alveoli

Lactating

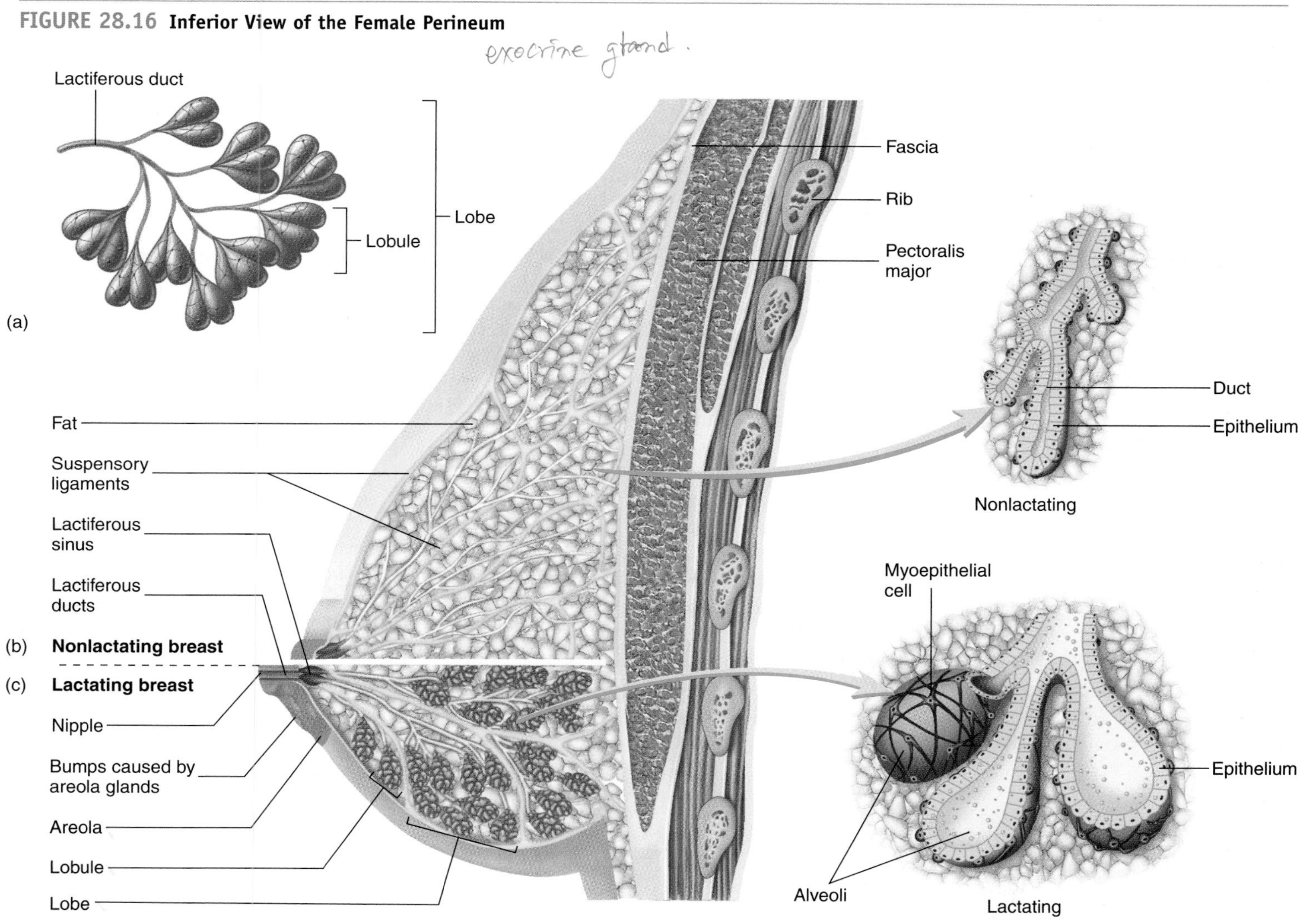

FIGURE 28.17 Anatomy of the Breast

(*a*) Lactiferous ducts divide to supply lobules, which form lobes. (*b*) In the nonlactating breast, only the duct system is present. (*c*) In the milk-producing breast, the ends of the ducts of the mammary glands have secretory sacs, called alveoli, which produce milk. Surrounding the alveoli are myoepithelial cells, which can contract, causing the movement of milk out of the alveoli.

breasts of both males and females have a raised **nipple** surrounded by a circular, pigmented **areola** (ă-rē′ō-lă). The areolae normally have a slightly bumpy surface caused by the presence of rudimentary mammary glands, called **areolar glands,** just below the surface. Secretions from these glands protect the nipple and the areola from chafing during nursing.

The general structure of the breasts is similar in both males and females before puberty. The breasts possess a rudimentary glandular system, which consists mainly of ducts with sparse alveoli. The female breasts begin to enlarge during puberty, primarily under the influence of estrogen and progesterone. Increased sensitivity or pain in the breasts often accompanies this enlargement. Males often experience these same sensations during early puberty, and their breasts can even develop slight swellings; however, these symptoms usually disappear fairly quickly. On rare occasions, the breasts of a male become enlarged, a condition called **gynecomastia** (gī′nĕ-kō-mas′tē-ă).

Each adult female mammary gland usually consists of 15–20 glandular **lobes** covered by a considerable amount of adipose tissue. It is primarily this superficial fat that gives the breast its form. The lobes of each mammary gland form a conical mass, with the nipple located at the apex. Each lobe has a single **lactiferous** (lak-tif′er-ŭs; milk-producing) **duct,** which opens independently of other lactiferous ducts on the surface of the nipple. Just deep to the surface, each lactiferous duct enlarges to form a small, spindle-shaped **lactiferous sinus,** in which accumulates milk during milk ejection. The lactiferous duct supplying a lobe subdivides to form smaller ducts, each of which supplies a **lobule.** Within a lobule, the ducts branch and become even smaller. In the milk-producing breast, the ends of these small ducts expand to form secretory sacs called **alveoli.** Myoepithelial cells surround the alveoli. These cells can contract to expell milk from the alveoli in the milk-producing breast (see chapter 29). In non-milk producing breasts, only the duct system is present.

A group of **suspensory,** or **Cooper's, ligaments** support and hold the breasts in place. These ligaments extend from the fascia over the pectoralis major muscles to the skin over the mammary glands and prevent the breasts from excessive sagging. In older adults, however, these ligaments weaken and elongate, allowing the breasts to sag to a greater extent than when the person was younger.

The nipples are very sensitive to tactile stimulation and contain smooth muscle cells that contract, causing the nipple to become erect in response to stimulation. These smooth muscle cells respond, like other erectile tissues, during sexual arousal.

60 ***Describe the anatomy of the mammary glands.***

61 ***Describe the route taken by a drop of milk from its site of production to the outside of the body.***

62 ***What are Cooper's ligaments?***

Fibrocystic Changes in the Breast

Fibrocystic changes in the breast are benign changes. They include the formation of fluid-filled cysts, hyperplasia of the duct system of the breast, and the deposition of fibrous connective tissue. These changes occur in approximately 10% of women who are less than 21 years of age, 25% of women in their reproductive years, and 50% of women who are postmenopausal. The cause of the condition is not known. Major manifestations are breast pain, especially during the luteal phase of the menstrual cycle and continuing until menstruation. Some evidence suggest that some women with certain types of duct hyperplasia, when associated with a family history of breast cancer, have an increased likelihood of developing breast cancer.

Breast Cancer

Breast cancer is a serious, often fatal, disease in women. It is the most common cancer in North American women. Greater than 75% of the breast cancer cases occur in women older than 50 years of age, and 85% involve cancer of the epithelium of the ducts of the mammary glands (ductal carcinoma).

The younger a woman is age when her first child is born, the lower her risk of developing breast cancer. However, the risk of breast cancer is not lowered in young women who become pregnant but who's pregnancies do not go full term. Also, women who have never given birth are at a greater risk than those who have given birth. It appears that differentiation of the breast epithelium caused by the first pregnancy decreases the risk for cancer because it reduces the amount of undifferentiated epithelium in the breast. Abnormal cells in fibrocystic disease of the breast also increase the risk of breast cancer.

The effect of reproductive hormones on the differentiation of epithelium of the breast during the first pregnancy reduces the incidence of breast cancer. A reduced cancer risk because of early menopause, or removal of the ovaries, is probably due to the decreased exposure of the breast epithelium to estrogen and progesterone. An increased duration of the use of postmenopausal estrogens and progesterone is correlated with an increased incidence of cancer. There is a decrease in breast cancer incidence with the use of drugs that block the effect of estrogen receptors.

A number of environmental factors have been associated with the development of breast cancer. Exposure to ionizing radiation, especially during adolescence and during pregnancy when epithelial cells are dividing more rapidly than at other times, is correlated with breast cancer. High fat intake and obesity are correlated with breast cancer, but the fish oil–derived omega-3 fatty acids may help prevent cancer. Breast cancer rates are low in Japanese women, but Japanese women who immigrate to the United States and who adopt a Western diet have breast cancer rates close to those found in women from the United States. There are a number of environmental factors that mimic estrogen, such as components of plastics, fuels, pharmaceuticals, and chlorine-based chemicals such as DDT, PCBs, and chlorofluorocarbons, and these may increase the incidence of breast cancer.

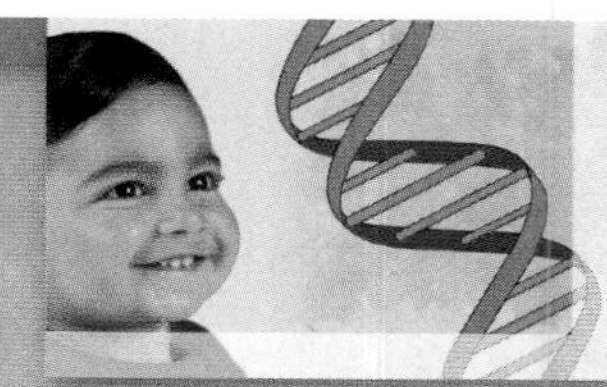

CLINICAL GENETICS

Breast Cancer

Women who have inherited specific gene mutations have an increased risk for breast cancer. A history of breast cancer in a close relative, such as a mother or sister, increases the risk for cancer. The risk increases further if there is more than one close relative who has had breast cancer, especially if the cancer affected both breasts and if it occurred before menopause. Between 5% and 10% of breast cancers may be hereditary. The presence of genes that make breast cancer likely does not guarantee that cancer will develop. Genetic tests can identify genes that increase the risk of breast cancer. Some women who are known to be at high risk because of their genetic makeup have frequent breast examinations while others elect to have their breasts surgically removed before cancer develops.

Mutations in the *BRCA1* gene on chromosome 17 and the *BRCA2* gene on chromosome 13 are responsible for about 30%–40% of inherited breast cancers. The *BRCA* genes are tumor suppressor genes, which normally suppress cell division (see "Cancer," chapter 4). Normal *BRCA* alleles are dominant over mutated alleles. Thus, people with two normal alleles and people with one normal and one mutated allele have normal tumor suppressor activity, whereas people who have two mutated alleles do not have normal tumor suppressor activity, and they are much more likely to develop breast cancer. If people have two normal alleles, both alleles must mutate before there is a loss of tumor suppressor function. However, if people have one normal allele and one mutated allele, only one mutation in the normal allele is necessary to eliminate tumor suppressor activity. Consequently, people who have inherited two normal alleles are much less likely to develop breast cancer than people who have inherited one normal and one mutated allele.

Another tumor suppressor gene, called p53, is on chromosome 17. The cells of almost 50% of all cancers have a mutated p53 gene, and these cancers are more aggressive and more fatal than cancers without a mutated p53 gene. Approximately 20%–40% of individuals with hereditary breast cancer have a mutated p53 gene. P53 is a regulatory gene that increases DNA repair, and, if DNA damage is extensive, p53 causes cell death. Thus, the p53 gene normally helps repair or eliminate cells that may become cancer cells. When p53 is mutated, its regulatory properties are reduced, conferring a loss of tumor suppressor activity, and it may increase tumor formation.

PHYSIOLOGY OF FEMALE REPRODUCTION

Female reproduction is under the control of hormonal and nervous regulation. The development of the female reproductive organs and their normal function depend on a number of hormones in the body.

Puberty

During puberty, females experience their first episode of menstrual bleeding, which is called **menarche** (me-nar′kē), between the ages of 11 and 16. The vagina, uterus, uterine tubes, and external genitalia begin to enlarge. Fat is deposited in the breasts and around the hips, causing them to enlarge and assume an adult form. The ducts of the breasts develop, pubic and axillary hair grows, and the voice changes, although this is a more subtle change than in males. The development of sexual drive is also associated with puberty.

Elevated rates of estrogen and progesterone secretion by the ovaries are primarily responsible for the changes associated with puberty. Before puberty, estrogen and progesterone are secreted in very small amounts. LH and FSH levels also remain very low. The low secretory rates are due to a lack of GnRH released from the hypothalamus. At puberty, not only are GnRH, LH, and FSH secreted in greater quantities than before puberty, but also the adult pattern is established in which a cyclic pattern of FSH and LH secretion occurs. The cyclic secretion of LH and FSH, ovulation, the monthly changes in secretion of estrogen and progesterone, and the resultant changes in the uterus characterize the menstrual cycle.

63 ***Define puberty and describe the changes that occur during puberty.***

64 ***Define menarche. What changes occur in LH, FSH, estrogen, and progesterone secretion during puberty?***

Menstrual Cycle

The term **menstrual** (men′stroo-ăl) **cycle** technically refers to the cyclic changes that occur in sexually mature, nonpregnant females and culminate in menses. Typically, the menstrual cycle is about 28 days long, although it can be as short as 18 days in some women and as long as 40 days in others (figure 28.18 and table 28.2). The term **menses** (men′sēz) is derived from a Latin word meaning month. It is a period of mild hemorrhage, which occurs approximately once each month, during which the uterine epithelium is sloughed and expelled from the uterus. **Menstruation** (men-stroo-ā′shŭn) is the discharge of the blood and elements of the uterine mucous membrane. Although the term *menstrual cycle* refers specifically to changes that occur in the uterus, several other cyclic changes are associated with it, and the term is often used to

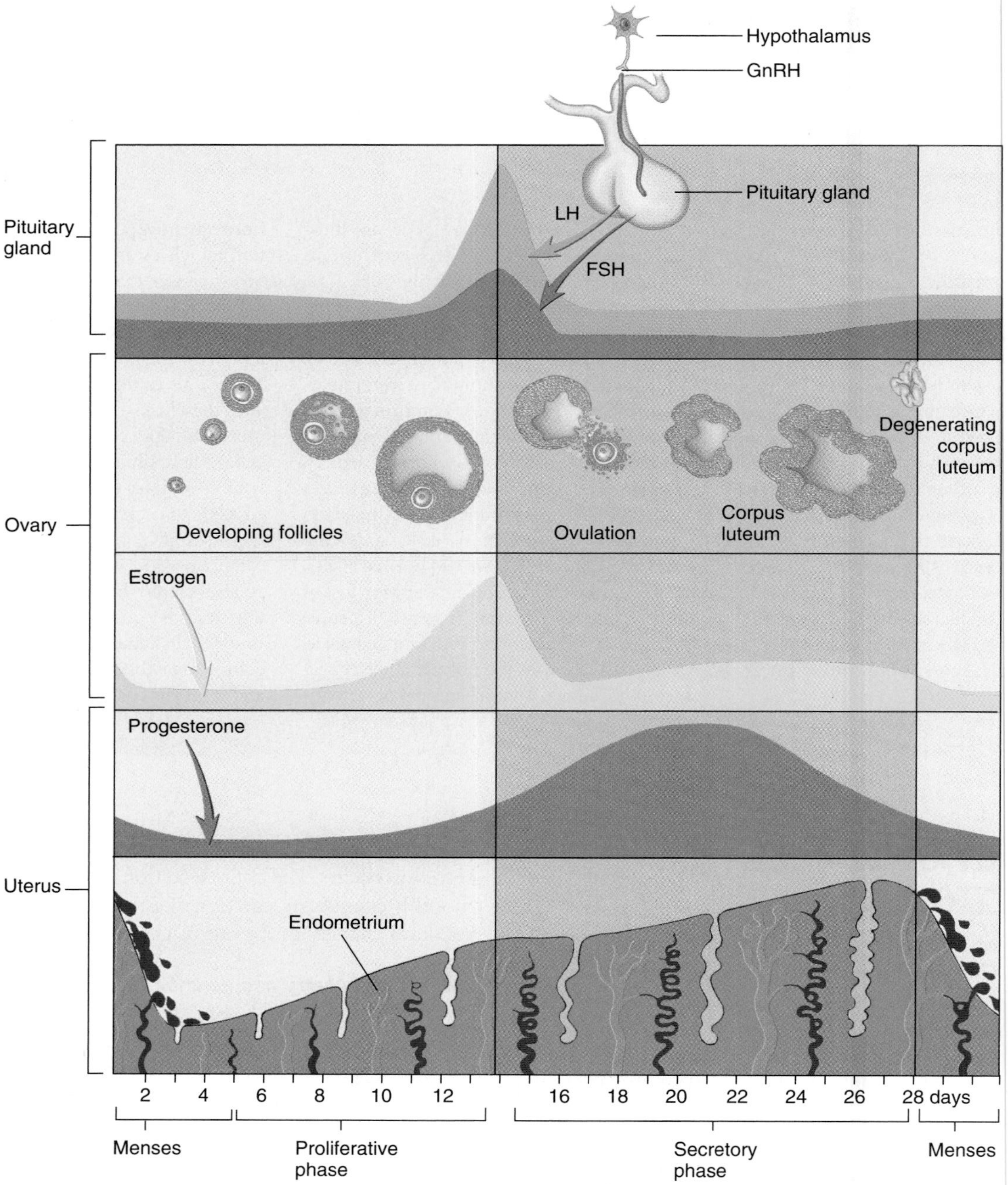

FIGURE 28.18 Menstrual Cycle

The various lines depict the changes in blood hormone levels, the development of the follicles, and the changes in the endometrium during the cycle.

refer to all of the cyclic events that occur in the female reproductive system. These changes include the cyclic changes in hormone secretion, in the ovary, and in the uterus.

The first day of menses is day 1 of the menstrual cycle, and menses typically lasts 4–5 days. Ovulation occurs on about day 14 of a 28–day menstrual cycle, although the timing of ovulation varies from individual to individual and varies within a single individual from one menstrual cycle to the next. The time between ovulation, on day 14, and the next menses is typically 14 days. The time between the first day of menses and the day of ovulation is more variable than the time between ovulation and the next menses. The time between the ending of menses and ovulation is called the **follicular phase** because of the rapid development of ovarian follicles or the **proliferative phase** because of the rapid proliferation of the uterine mucosa. The period after ovulation and before the next menses is called the **luteal phase** because of the existence of the corpus luteum or the **secretory phase** because of maturation of and secretion by uterine glands.

TABLE 28.2 Menstrual Cycle

Menses	Proliferative Phase	Ovulation	Secretory Phase
Pituitary Hormones			
LH levels are low and remain low; FSH increases somewhat.	LH and FSH levels begin to increase rapidly in response to increases in estrogen near the end of the proliferative phase.	Increasing levels of LH trigger ovulation. Ovulation generally occurs after LH levels have reached their peak. FSH reaches a peak about the time of ovulation and initiates the development of follicles that may complete maturation during a later cycle.	LH and FSH levels decline to low levels following ovulation and remain at low levels during the secretory phase in response to increases in estrogen and progesterone.
Developing Follicles			
FSH secreted during menses causes several follicles to begin to enlarge.	Several follicles continue to enlarge. As they enlarge, they begin to secrete estrogen. In addition, many of them degenerate. Only one of the follicles becomes a mature follicle that is capable of ovulating by the end of the proliferative phase.	Normally, a single follicle reaches maturity and ovulates in response to LH. The oocyte and some cumulus cells are released during ovulation.	Following ovulation, the granulosa cells of the ovulated follicle change to luteal cells and begin secreting large amounts of progesterone and some estrogen.
Estrogen			
The ovarian follicles secrete very little estrogen.	Near the end of the proliferative phase, the enlarging follicles begin to secrete increasing amounts of estrogen. The estrogen causes the pituitary gland to secrete increasing quantities of LH and smaller quantities of FSH. The positive-feedback relationship between estrogen and LH results in rapidly increasing LH and estrogen levels several days prior to ovulation. The rapid increase in LH triggers ovulation.	Estrogen, secreted by developing follicles, reaches a peak at ovulation.	Following ovulation, estrogen levels decline. After the luteal cells have been established, smaller amounts of estrogen are secreted by the corpus luteum.
Progesterone			
The ovarian follicles secrete very little progesterone.	Progesterone levels are low during the proliferative phase.	Progesterone levels are low.	Following ovulation, progesterone levels increase due to the secretion of progesterone by the corpus luteum. Progesterone levels remain high throughout the secretory phase and fall rapidly just before menses unless pregnancy occurs.
Uterine Endometrium			
The endometrium of the uterus undergoes necrosis and is eliminated in the menstrual fluid during menses. The necrosis is a result of decreasing progesterone concentrations near the end of the proliferative phase.	In response to estrogen, endometrial cells of the uterus undergo rapid cell division and proliferate rapidly. In addition, the number of progesterone receptors in the endometrial cells increases in response to estrogen.	Ovulation occurs over a short time, and it signals the end of the proliferative phase, as estrogen levels decline, and the onset of the secretory phase, as the progesterone levels begin to increase.	Progesterone causes the endometrial cells to enlarge and secrete a small amount of fluid. The endometrium continues to thicken throughout the secretory phase. Near the end of the secretory phase, declining progesterone levels allow the spiral arteries of the endometrium to constrict, causing ischemia, and the endometrium becomes necrotic unless pregnancy occurs.

65 ***Define the menstrual cycle.***

66 ***What is the length of a typical menstrual cycle and list the major events that occur.***

67 ***Which day is ovulation, usually?***

68 ***Define the follicular and proliferative phases and the secretory and luteal phases.***

Ovarian Cycle

The **ovarian cycle** specifically refers to the events that occur in a regular fashion in the ovaries of sexually mature, nonpregnant women during the menstrual cycle. The hypothalamus and anterior pituitary release hormones that control these events. FSH from the anterior pituitary is primarily responsible for initiating the development of primary follicles, and as many as 25 begin to mature during each menstrual cycle. The follicles that start to develop in response to FSH may not ovulate during the same menstrual cycle in which they begin to mature, but they may ovulate one or two cycles later.

Although several follicles begin to mature during each cycle, normally only one is ovulated. The remaining follicles degenerate. Larger and more mature follicles appear to secrete estrogen and other substances that have an inhibitory effect on other, less mature follicles.

Early in the menstrual cycle, the release of GnRH from the hypothalamus increases, and sensitivity of the anterior pituitary to GnRH increases. These changes stimulate the production and release of a small amount of FSH and LH by the anterior pituitary.

FSH and LH stimulate follicular growth and maturation. They also cause an increase in estradiol secretion by the developing follicles. FSH exerts its main effect on the granulosa cells. LH exerts its initial effect on the cells of the theca interna and later on the granulosa cells.

LH stimulates the theca interna cells to produce androgens, which diffuse from these cells to the granulosa cells. FSH stimulates the granulosa cells to convert androgens to estrogen. In addition, FSH gradually increases LH receptors in the granulosa cells, and estrogen produced by the granulosa cells increases LH receptors in the theca interna cells. Consequently, theca interna cells and granulosa cells cooperate to produce estrogen. Estrogen, in turn, increases receptors for LH in both theca interna cells and granulosa cells.

After LH receptors in the granulosa cells have increased, LH stimulates the granulosa cells to produce some progesterone, which diffuses from the granulosa cells to the theca interna cells, where it is converted to androgens. These androgens are also converted to estrogen by the granulosa cells. Thus, the production of androgens by the theca interna cells increases, and the conversion of androgens to estrogen by the granulosa cells is responsible for a gradual increase in estrogen secretion by these cells throughout the follicular phase, even though only a small increase in LH secretion occurs. FSH levels actually decrease during the follicular phase because developing follicles produce inhibin, and inhibin has a negative-feedback effect on FSH secretion.

As estrogen levels increase during the follicular phase, they have a negative-feedback effect on the secretion of LH and FSH by the anterior pituitary. However, the gradual increase in estrogen levels, especially late in the follicular phase, begins to have a positive-feedback effect on LH and FSH release from the anterior pituitary. Consequently, as the estrogen level in the blood increases, it stimulates greater LH and FSH secretion. The sustained increase in estrogen is necessary for the development of this positive-feedback effect. In response to this positive-feedback effect, LH and FSH secretion increases rapidly and in large amounts just before ovulation (figure 28.19). The increase in blood levels of both LH and FSH is called the **LH surge,** and the increase in FSH is called the **FSH surge.** The LH surge occurs several hours earlier and to a greater degree than the FSH surge, and the LH surge can last up to 24 hours.

The LH surge initiates ovulation and causes the ovulated follicle to become the corpus luteum. FSH can make the follicle more sensitive to the influence of LH by stimulating the synthesis of additional LH receptors in the follicles and by stimulating the development of follicles that may ovulate during later ovarian cycles.

The LH surge causes the primary oocyte to complete the first meiotic division just before or during the process of ovulation. Also, the LH surge triggers several events that are very much like inflammation in a mature follicle. These events result in ovulation. The follicle becomes edematous, proteolytic enzymes cause the degeneration of the ovarian tissue around the follicle, the follicle ruptures, and the oocyte and some surrounding follicle cells are slowly extruded from the ovary.

Shortly after ovulation, the follicle's production of estrogen decreases. The remaining granulosa cells of the ovulated follicle are converted to corpus luteum cells and begin to secrete progesterone. After the corpus luteum forms, progesterone levels become much higher than before ovulation, and some estrogen is produced. The increased progesterone and estrogen have a negative-feedback effect on GnRH release from the hypothalamus. As a result, LH and FSH release from the anterior pituitary decreases. Estrogen and progesterone also cause the down-regulation of GnRH receptors in the anterior pituitary, and the anterior pituitary cells become less sensitive to GnRH. Because of the decreased secretion of GnRH and decreased sensitivity of the anterior pituitary to GnRH, the rate of LH and FSH secretion declines to very low levels after ovulation (see figures 28.18 and 28.19).

If fertilization of the ovulated oocyte does take place, the outer layer of the developing embryo begins to secrete the LH-like substance **human chorionic gonadotropin (HCG)**, which keeps the corpus luteum from degenerating. As a result, blood levels of estrogen and progesterone do not decrease, and menses does not occur. If fertilization does not occur, HCG is not produced. The cells of the corpus luteum begin to atrophy after day 25 or 26, and the blood levels of estrogen and progesterone decrease rapidly, which results in menses.

69 ***Describe the events of the ovarian cycle. What role do FSH and LH play in the ovarian cycle?***

70 ***Describe how the cyclic increase and decrease in FSH and LH is produced.***

71 ***Where is HCG produced, and what effect does it have on the ovary?***

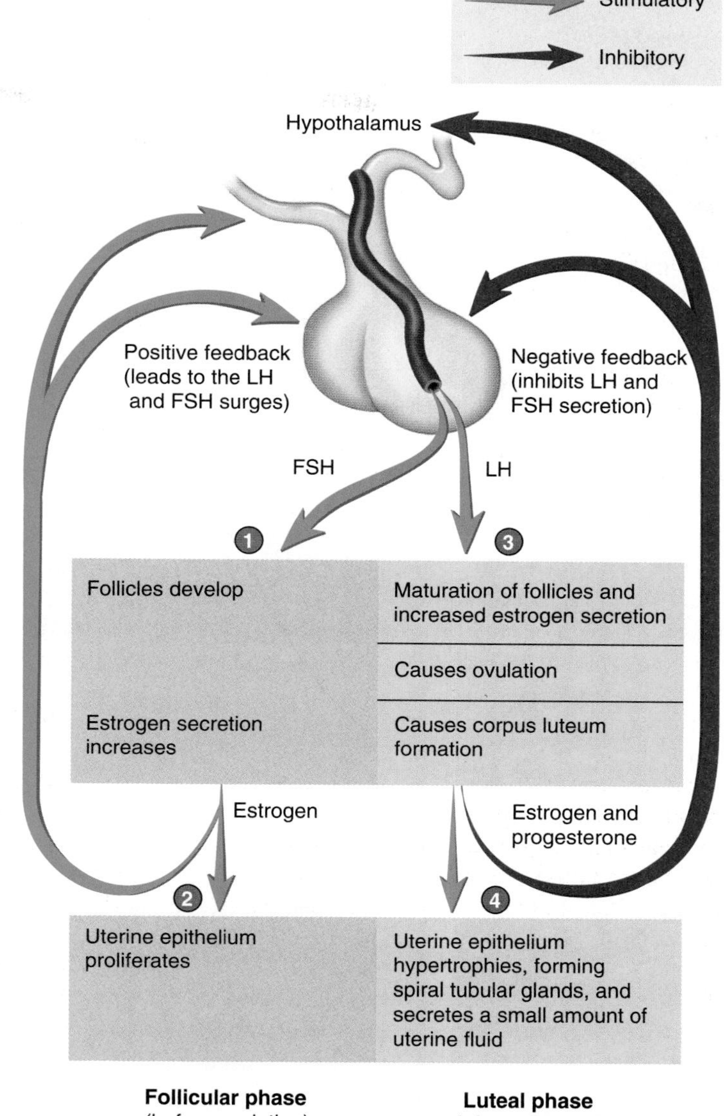

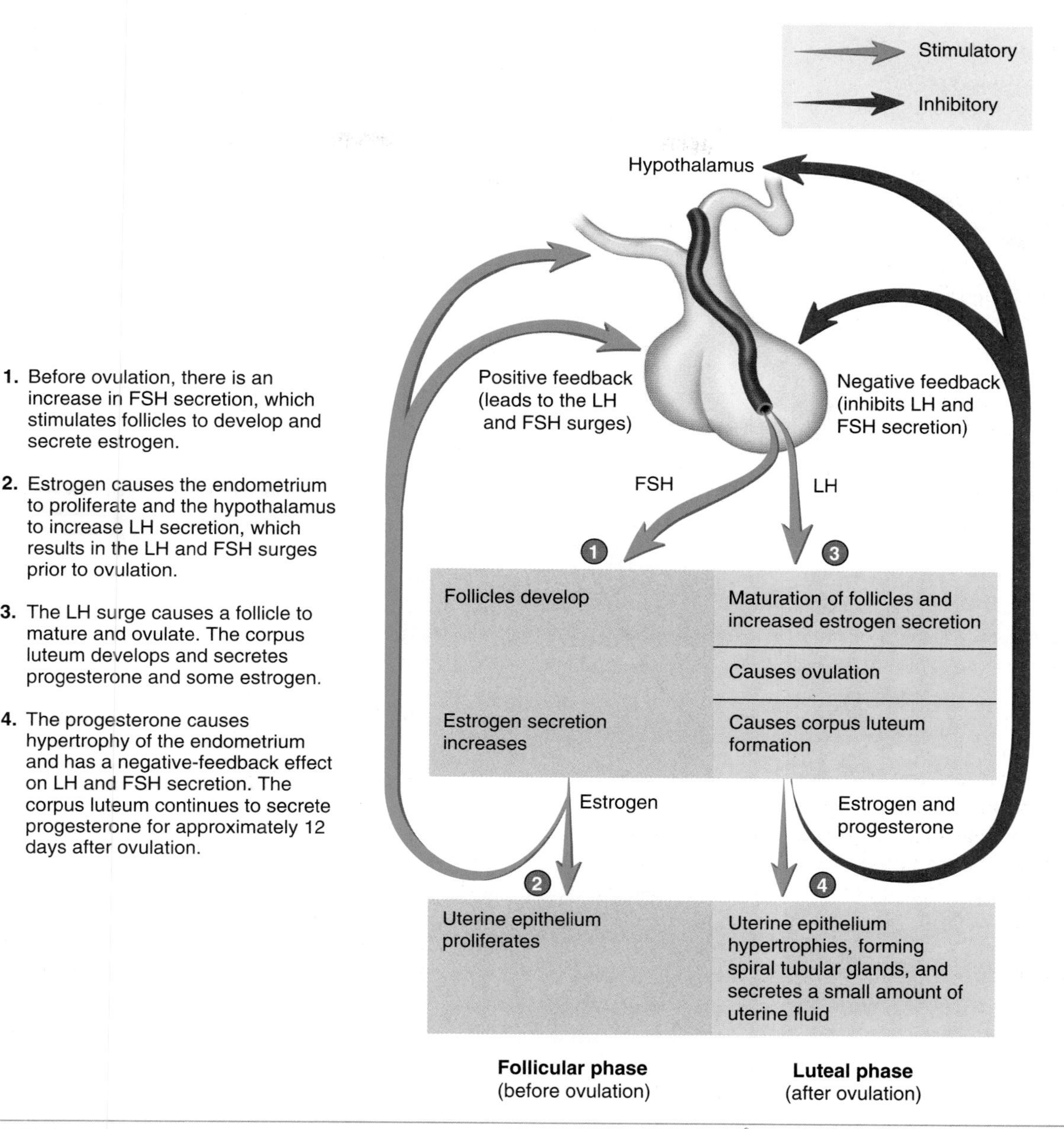

PROCESS FIGURE 28.19 Regulation of Hormone Secretion During the Menstrual Cycle
The regulation of hormone secretion from the anterior pituitary and the ovary before and after ovulation is shown.

PREDICT 4

Predict the effect on the ovarian cycle of administering a relatively large amount of estrogen and progesterone just before the preovulatory LH surge. Also predict the consequences of continually administering high concentrations of GnRH.

Uterine Cycle

The term **uterine cycle** refers to changes that occur primarily in the endometrium of the uterus during the menstrual cycle (see figure 28.18; figure 28.20). Other, more subtle changes also occur in the vagina and other structures during the menstrual cycle. Cyclic secretions of estrogen and progesterone are the primary cause of these changes.

The endometrium of the uterus begins to proliferate after menses. The remaining epithelial cells rapidly divide and replace the cells of the functional layer that were sloughed during the last menses. A relatively uniform layer of low cuboidal endometrial cells is produced. It later becomes columnar and is folded to form tubular **spiral glands.** Blood vessels called **spiral arteries** project through the delicate connective tissue that separates the individual spiral glands to supply nutrients to the endometrial cells. After ovulation, the endometrium becomes thicker, and the spiral glands develop to a greater extent and begin to secrete small

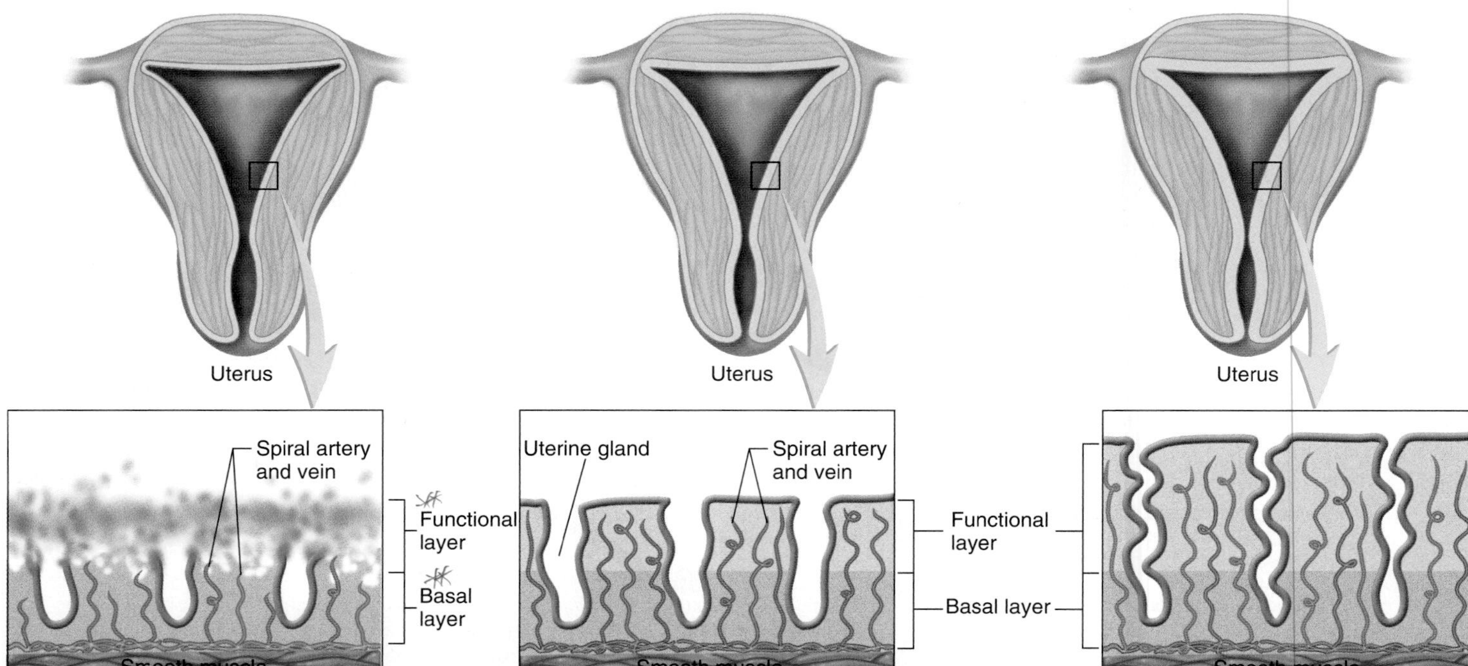

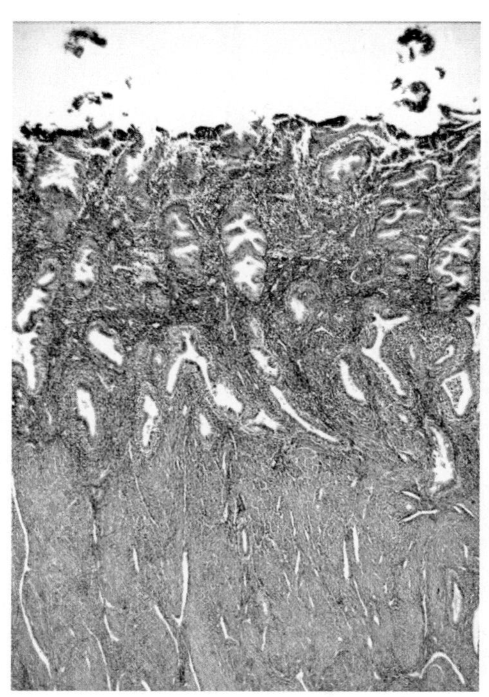

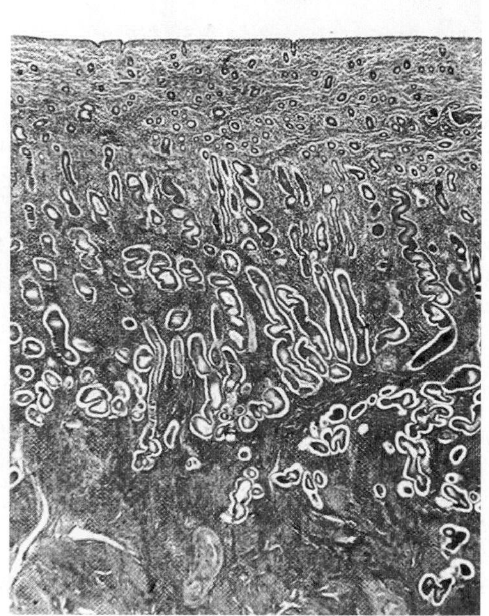

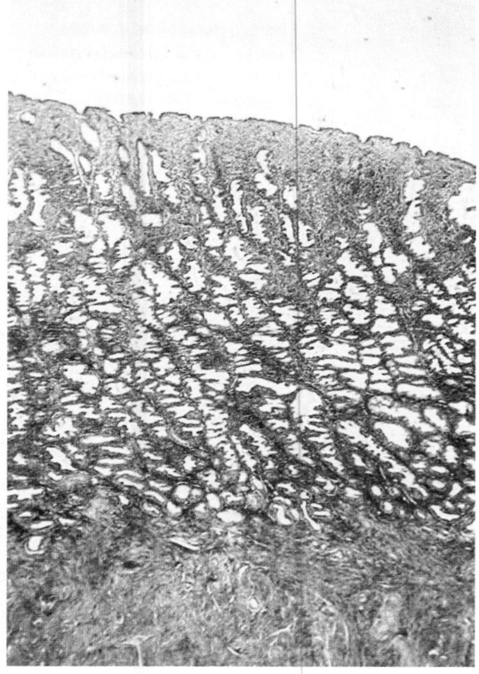

1. **Menses**
   The functional layer of the endometrium sloughs off as the spiral arteries remain in a constricted state in response to low levels of progesterone, depriving the functional layer of an adequate blood supply. Functional layer tissue and some blood make up most of the menstrual fluid. The basal layer remains.

2. **Proliferative phase**
   The more basal portion of the endometrium is called the basal layer. Epithelial cells of the basal layer of the endometrium proliferate in response to estrogen. As a result, the epithelial cells and loose connective tissue on which they rest form the tubular, spiral-shaped uterine glands. The apical portion of the thickening endometrium is the functional layer. The spiral arteries found in the loose connective tissue between the uterine glands nourish the functional layer.

3. **Secretory phase**
   Epithelial cells of the basal and functional layers undergo hypertrophy in response to progesterone. As a result, the uterine glands become more elongated and more spiral. Consequently, the endometrial layer reaches its greatest thickness. The spiral arteries can be seen in the loose connective tissue between the uterine glands. The spiral arteries remain dilated due to the presence of progesterone.

PROCESS FIGURE 28.20 **Uterine Cycle**

amounts of a fluid rich in glycogen. Approximately 7 days after ovulation, or about day 21 of the menstrual cycle, the endometrium is prepared to receive the developing embryonic mass, if fertilization has occurred. If the developing embryonic mass arrives in the uterus too early or too late, the endometrium does not provide a suitable environment for it.

Estrogen causes the endometrial cells and, to a lesser degree, the myometrial cells to proliferate. It also makes the uterine tissue more sensitive to progesterone by stimulating the synthesis of progesterone receptor molecules within the uterine cells. After ovulation, progesterone from the corpus luteum binds to the progesterone receptors, resulting in cellular hypertrophy in the endometrium and myometrium and causing the endometrial cells to become secretory. Estrogen increases the tendency of the smooth muscle cells of the uterus to contract in response to stimuli, but progesterone inhibits smooth muscle contractions. When progesterone levels increase while estrogen levels are low, contractions of the uterine smooth muscle are reduced.

PREDICT 5

Predict the effect on the endometrium of maintaining high progesterone levels in the circulatory system, including the period of time during which estrogen normally increases following menstruation.

In menstrual cycles in which pregnancy does not occur, progesterone and estrogen levels decline to low levels as the corpus luteum degenerates. As a consequence, the uterine lining also begins to degenerate. The spiral arteries constrict in a rhythmic pattern for longer and longer periods as progesterone levels fall. As a result, all but the basal parts of the spiral glands become ischemic and then necrotic. As the cells become necrotic, they slough into the uterine lumen. The necrotic endometrium, mucous secretions, and a small amount of blood released from the spiral arteries make up the menstrual fluid. Decreases in progesterone levels and increases in inflammatory substances that stimulate myometrial smooth muscle cells cause uterine contractions, which expel the menstrual fluid from the uterus through the cervix and into the vagina.

72 ***Name the stages of the uterine cycle, and describe the events that take place in each stage.***

73 ***What are the effects of estrogen and progesterone on the uterus?***

Female Sexual Behavior and the Female Sex Act

Sexual drive in females, like sexual drive in males, depends on hormones. The adrenal gland and other tissues, such as the liver, convert steroids, such as progesterone, to androgens. Androgens and possibly estrogens affect cells in the brain, especially in the hypothalamus, to influence sexual behavior. Androgens and estrogen alone do not control sexual drive, however. For example, sexual drive cannot be predictably increased simply by injecting these hormones into healthy women or men. Psychologic factors also affect sexual behavior. For example, after removal of the ovaries or after menopause, many women report having an increased sex drive because they no longer fear pregnancy.

The neural pathways, both sensory and motor, involved in controlling sexual responses are the same for males and females. Sensory action potentials are conducted from the genitals to the sacral region of the spinal cord, where reflexes that govern sexual responses are integrated. Ascending pathways, primarily the spinothalamic tracts (see chapter 14), conduct sensory information through the spinal cord to the brain, and descending pathways conduct action potentials back to the sacrum. As a result, cerebral influences modulate the sacral reflexes. Motor action potentials are conducted from the spinal cord to the reproductive organs by both parasympathetic and sympathetic nerve fibers and to skeletal muscles by the somatic motor nerve fibers.

During sexual excitement, erectile tissue within the clitoris and around the vaginal opening becomes engorged with blood as a result of parasympathetic stimulation. The nipples of the breast often become erect as well. The mucous glands within the vestibule, especially the vestibular glands, secrete small amounts of mucus. Large amounts of mucuslike fluid are also extruded into the vagina through its wall, although no well-developed mucous glands are within the vaginal wall. These secretions provide lubrication that allows for easy entry of the penis into the vagina and easy movement of the penis during intercourse. Tactile stimulation of the female's genitals that occurs during sexual intercourse, along with psychologic stimuli, normally triggers an orgasm. The vaginal, uterine, and perineal muscles contract rhythmically, and muscle tension increases throughout much of the body. After the sexual act, a period of resolution characterized by an overall sense of satisfaction and relaxation occurs. In contrast to males, females can be receptive to further stimulation and can experience successive orgasms. Although orgasm is a pleasurable component of sexual intercourse, it is not necessary for females to experience an orgasm for fertilization to occur.

74 ***Compare the female sexual act with the male sexual act.***

75 ***Is an orgasm required for fertilization to occur?***

Female Fertility and Pregnancy

After sperm cells are ejaculated into the vagina during sexual intercourse, they are transported through the cervix, the body of the uterus, and the uterine tubes to the ampulla (figure 28.21). The forces responsible for the movement of sperm cells through the female reproductive tract involve the swimming ability of the sperm cells and possibly the muscular contractions of the uterus and the uterine tubes. During sexual intercourse, oxytocin is released from the posterior pituitary of the female, and the semen introduced into the vagina contains prostaglandins. Both of these hormones stimulate smooth muscle contractions in the uterus and uterine tubes.

Menstrual Cramps, PMS, and Amenorrhea

Menstrual cramps are the result of strong myometrial contractions that occur before and during menstruation. The cramps can result from excessive prostaglandin secretion. Sloughing of the endometrium of the uterus results in an inflammation in the endometrial layer of the uterus, and prostaglandins are produced as part of the inflammatory process. Progesterone inhibits sloughing of the endometrium and contractions of uterine smooth muscle. Estrogen stimulates sloughing of the endometrium because it increases uterine smooth muscle contractions. In some women, menstrual cramps are extremely uncomfortable. Many women can alleviate painful menstruation by taking nonsteroidal anti-inflammatory drugs (NSAIDs), such as aspirin or ibuprofen, which inhibit prostaglandin biosynthesis, just before the onset of menstruation. These treatments, however, are not effective in treating all painful menstruation, especially when the causes of pain are more complicated than inflammation associated with normal menstruation, such as from tumors of the myometrium or obstruction of the cervical canal.

A topic of continuing research is **premenstrual syndrome (PMS).** Some women suffer from severe changes in mood that often result in aggression and other socially unacceptable behaviors prior to menses. It has been hypothesized that hormonal changes associated with the menstrual cycle trigger these mood changes. Some women are treated successfully with steroid hormones; however, this treatment does not appear to be effective for everyone with the condition. Similarly, reducing caffeine, alcohol, sugar, and animal fat consumption helps some women. It is unclear how many women are affected by PMS. A precise definition of the premenstrual period is not well established, symptoms vary among individuals and are not easily monitored, and its precise cause and physiologic mechanisms are unknown. Whether all women diagnosed as having PMS are suffering from the same condition is unclear. Improvements in defining the symptoms have resulted in improved treatment.

The absence of a menstrual cycle is called **amenorrhea** (ă-men-ō-rē′ă). If the pituitary gland does not function properly because of abnormal development, the female will not begin to menstruate at puberty. This condition is called **primary amenorrhea.** In contrast, if a female has had normal menstrual cycles and later stops menstruating, the condition is called **secondary amenorrhea.** One cause of secondary amenorrhea is anorexia, a condition in which lack of food causes the hypothalamus to decrease GnRH secretion to levels so low that the menstrual cycle cannot occur. Female athletes or ballet dancers who have rigorous training schedules have a high frequency of secondary amenorrhea. The physical stress that can be coupled with an inadequate food intake also results in very low GnRH secretion. Increased food intake, for anorexic women, and reduced training, for women who exercise intensely, generally restores normal hormone secretion and normal menstrual cycles.

Secondary amenorrhea can result from pituitary tumors that decrease FSH and LH secretion, or from a lack of GnRH secretion from the hypothalamus. Head trauma and tumors that affect the hypothalamus can result in lack of GnRH secretion.

Secondary amenorrhea can result from a lack of normal hormone secretion from the ovaries, which can be caused by autoimmune diseases that attack the ovary or by polycystic ovarian disease, in which cysts in the ovary produce large amounts of androgen that are converted to estrogens by other tissues in the body. The increased estrogen prevents the normal cycle of FSH and LH secretion required for ovulation to occur. Other hormone-secreting tumors of the ovary can also disrupt the normal menstrual cycle and result in amenorrhea.

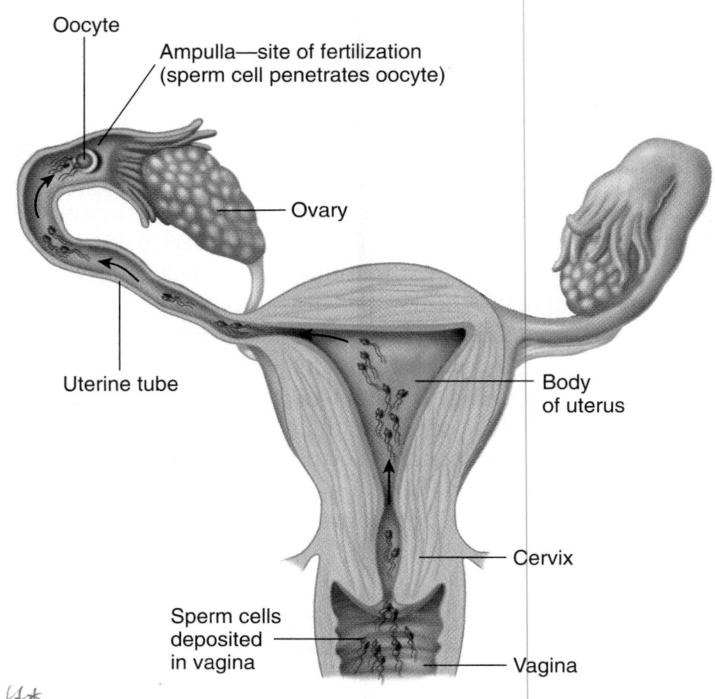

FIGURE 28.21 Sperm Cell Movement

Sperm cells are deposited into the vagina as part of the semen when the male ejaculates. Sperm cells pass through the cervix, the body of the uterus, and the uterine tube. Fertilization normally occurs when the oocyte is in the upper one-third of the uterine tube (the ampulla).

While passing through the vagina, uterus, and uterine tubes, the sperm cells undergo **capacitation** (kă-pas′i-tā′shŭn), the removal of proteins and the modification of glycoproteins of the sperm cell membranes. Following capacitation, acrosomal enzymes of some sperm cells are released as the sperm cells move through the uterus and oviducts. The acrosomal enzymes allow penetration of the cervical mucus, cumulus mass, zona pellucida, and oocyte plasma membrane.

The oocyte can be fertilized for up to 24 hours after ovulation, and some sperm cells remain viable in the female reproductive tract for up to 6 days, although most of them have degenerated after 24 hours. For fertilization to occur successfully, sexual intercourse must therefore occur between 5 days before and 1 day after ovulation.

One sperm cell enters the secondary oocyte, and fertilization occurs (see chapter 29). For the next several days, a sequence of cell divisions occurs while the developing embryo passes through the uterine tube to the uterus. By 7 or 8 days after ovulation, which is day 21 or 22 of the average menstrual cycle, the endometrium of the uterus has been prepared for implantation. Estrogen and progesterone have caused it to reach its maximum thickness and secretory activity, and the developing embryo begins to implant. The outer layer of the developing embryo, the **trophoblast** (trof′ō-blast, trō′fō-blast), secretes proteolytic enzymes that digest the cells of the thickened endometrium (see chapter 29), and the developing embryo digests its way into the endometrium.

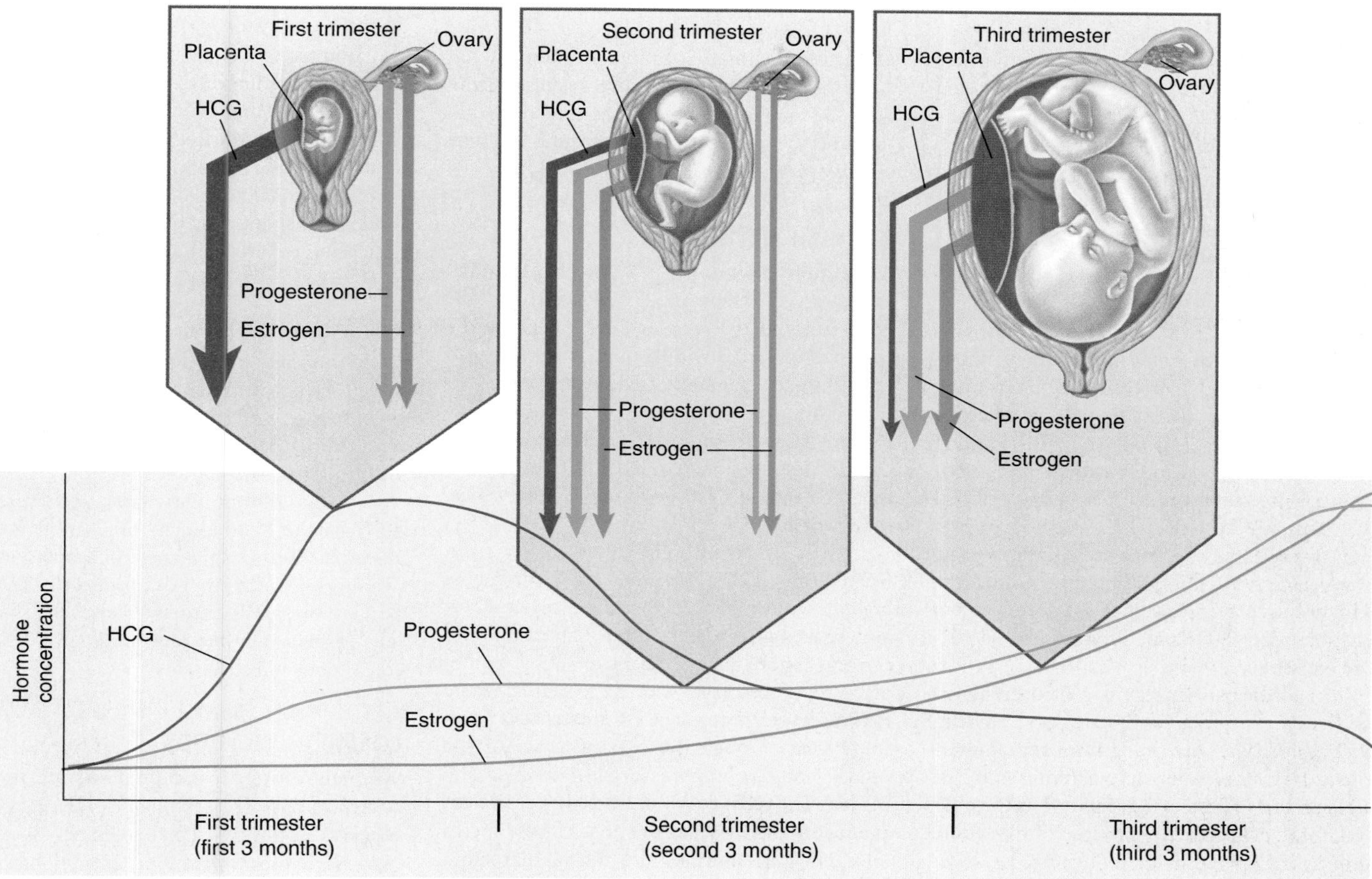

PROCESS FIGURE 28.22 Changes in Hormone Concentrations and Changes in Hormone Secretion During Pregnancy

Ectopic Pregnancy

An ectopic pregnancy results if implantation occurs anywhere other than in the uterine cavity. The most common site of ectopic pregnancy is the uterine tube. Implantation in the uterine tube eventually is fatal to the fetus and can cause the tube to rupture. The possibility of hemorrhage makes ectopic pregnancy dangerous to the mother. In rare cases, implantation occurs in the mesenteries of the abdominal cavity, and the fetus can develop normally but must be delivered by cesarean section.

The trophoblast secretes HCG, which is transported in the blood to the ovary and causes the corpus luteum to remain functional. As a consequence, both estrogen and progesterone levels continue to increase rather than decrease. The secretion of HCG increases rapidly and reaches a peak about 8–9 weeks after fertilization. Subsequently, HCG levels in the circulatory system have declined to a lower level by 16 weeks and remain at a relatively constant level throughout the remainder of pregnancy. The detection of HCG excreted in the urine is the basis for some pregnancy tests.

The estrogen and progesterone secreted by the corpus luteum are essential for the maintenance of pregnancy. After the **placenta** (plă-sen′tă) forms from the trophoblast and uterine tissue, however, it also begins to secrete estrogen and progesterone. After the first 3 months of pregnancy, the corpus luteum is no longer needed to maintain pregnancy; the placenta has become an endocrine gland that secretes sufficient quantities of estrogen and progesterone to maintain pregnancy. Estrogen and progesterone levels increase in the woman's blood throughout pregnancy (figure 28.22).

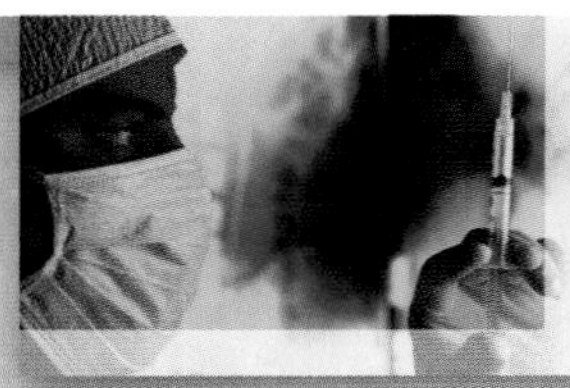

CLINICAL FOCUS

Control of Pregnancy

Many methods are used to prevent or terminate pregnancy (figure B), including methods that prevent fertilization (contraception), prevent implantation of the developing embryo (IUDs), or remove the implanted embryo or fetus (abortion). Many of these techniques are quite effective when done properly and used consistently (table A).

BEHAVORIAL METHODS

Abstinence, or refraining from sexual intercourse, is a sure way to prevent pregnancy when practiced consistently. It is not an effective method when used only occasionally.

Coitus (kō'i-tŭs) **interruptus** is removal of the penis from the vagina just before ejaculation. This is a very unreliable method of preventing pregnancy because it requires perfect awareness and willingness to withdraw the penis at the correct time. It also ignores the fact that some sperm cells are found in preejaculatory emissions.

Periodic abstinence, also called the **natural family planning method** or the **rhythm method,** requires abstaining from sexual intercourse near the time of ovulation. A major factor in the success of this method is the ability to predict accurately the time of ovulation. Although the rhythm method provides some protection against becoming pregnant, it has a relatively high rate of failure, resulting from both the inability to predict the time of ovulation and the failure to abstain around the time of ovulation.

BARRIER METHODS

A **condom** (kon'dom) is a sheath of animal membrane, rubber, or latex. Placed over the erect penis, it is a barrier device because the semen is collected within the condom instead of within the vagina. Condoms also provide protection against sexually transmitted diseases. A **vaginal condom** also acts as a barrier device. The vaginal condom can be placed into the vagina by the woman before sexual intercourse.

Methods to prevent sperm cells from reaching the oocyte once they are in the vagina include the use of a diaphragm, a cervical cap, and spermicidal agents. The diaphragm and cervical cap are flexible plastic or rubber domes placed over the cervix within the vagina, where they prevent the passage of sperm cells from the vagina through the cervical canal of the uterus. The diaphragm is larger than the cervical cap. The most commonly used **spermicidal agents** are foams or creams that kill the sperm cells. They are inserted into the vagina before sexual intercourse. When used in combination, a condom and foam or cream are much more effective than when they are used alone. A **spermicidal douche** (doosh) is a stream of fluid containing a chemical toxic to sperm cells that is injected into the vagina. The stream of fluid removes and kills sperm cells. Spermicidal douches used alone are not very effective.

LACTATION

Lactation (lak-tā'shŭn) prevents the menstrual cycle for a few months after childbirth. Action potentials sent to the hypothalamus in response to suckling that cause the release of oxytocin and prolactin also inhibit FSH and LH release from the anterior pituitary. Lactation, therefore, prevents the development of ovarian follicles and ovulation. Despite continual lactation, the ovarian and uterine cycles eventually resume. Because ovulation occurs before menstruation, relying on lactation to prevent pregnancy is not consistently effective.

CHEMICAL METHODS

Synthetic estrogen and progesterone in **oral contraceptives** (birth-control pills) effectively suppress fertility in females. These substances may have more than one action, but they reduce LH and FSH release from the anterior pituitary. Estrogen and progesterone are present in high enough concentrations to have a negative-feedback effect on the pituitary, which prevents the large increase in LH and FSH secretion that triggers ovulation. Over the years, the dose of estrogen and progesterone in birth-control pills has been reduced. The current lower-dose birth-control pills have fewer side effects than earlier dosages. An increased risk for heart attack or stroke exists in women using oral contraceptives who smoke or have a history of hypertension or coagulation disorders. For most women, the pill is effective and has a minimum frequency of complications, until at least age 35. The **patch** is an adhesive skin patch containing synthetic estrogen and progesterone. It is worn on the lower abdomen, buttocks, or upper body. The **vaginal contraceptive ring** is inserted into the vagina and releases synthetic estrogen and progesterone.

Table A Effectiveness of Various Methods for Preventing Pregnancy

Technique	Optimal Effectiveness (%)
Abortion	100
Sterilization	100
Combination (estrogens and progesterones) pill	99.9
Intrauterine device	98
Minipill (low dose of estrogens and progesterones)	99
Condom plus spermicide	99
Condom alone	97
Diaphragm plus spermicide	97
Foam	97
Rhythm	97

The **mini-pill** is an oral contraceptive that contains only synthetic progesterone. It reduces and thickens the mucus of the cervix, which prevents sperm cells from reaching the egg. It also prevents blastocytes from implanting in the uterus.

Progesterone-like chemicals, such as medroxy progesterone (Depo-Provera), which are injected intramuscularly and slowly released into the circulatory system, can act as effective contraceptives. Injected progesterone-like chemicals can provide protection from pregnancy for up to 3 months, depending on the amount injected. Thin silastic tubes containing these chemicals can be implanted beneath the skin, usually in the upper arm, from which they are slowly released into the circulatory system.

Continued

(a) (b) (c)

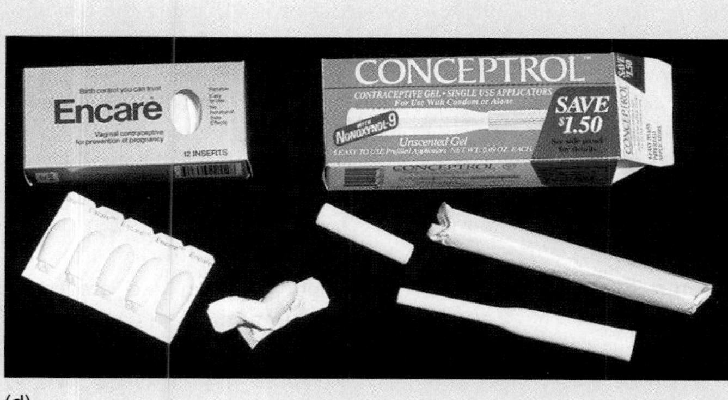

(d)

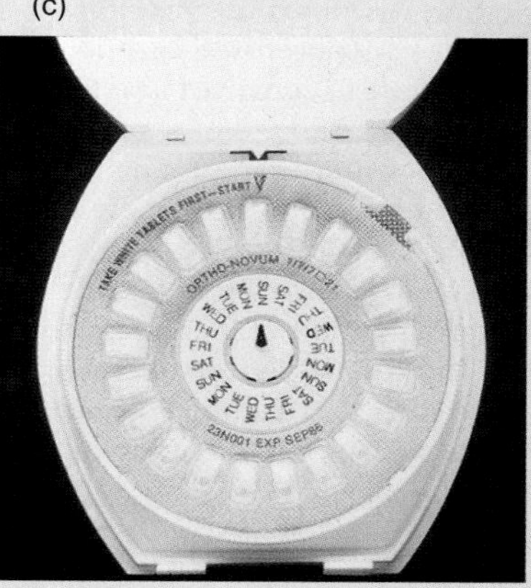

(e)

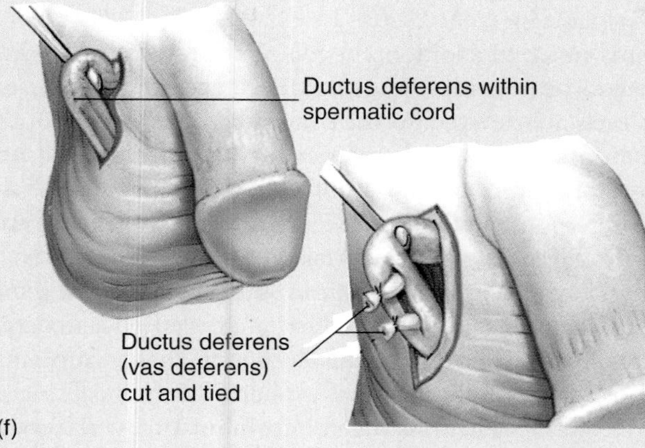

(f)

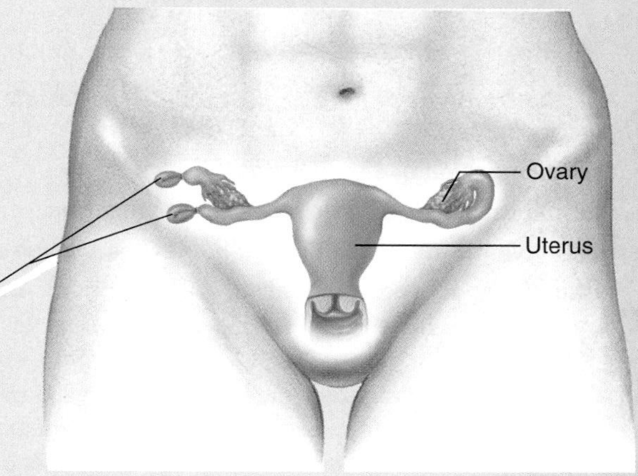

(g)

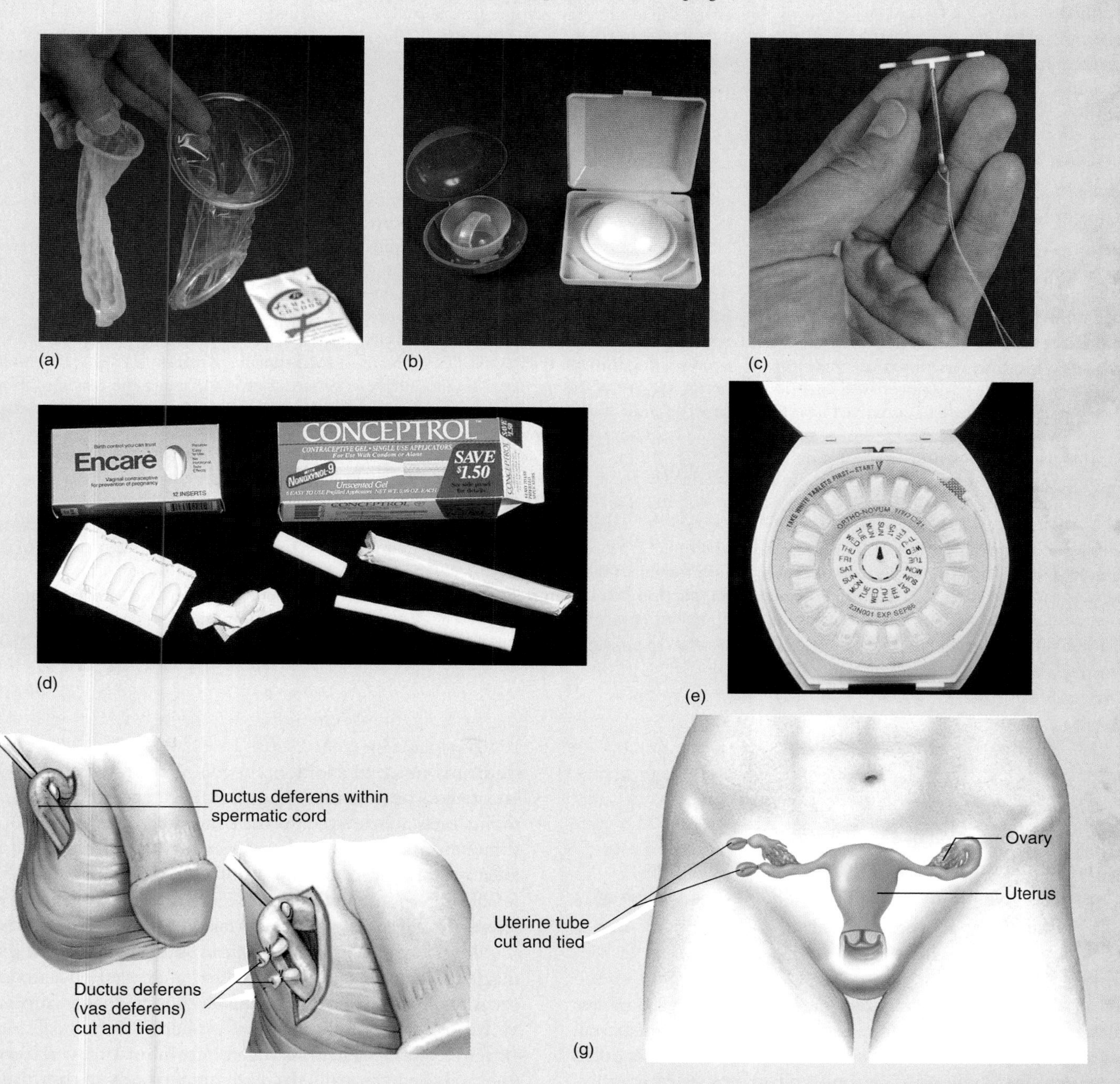

FIGURE B Contraceptive Devices and Techniques

(*a*) Condom. (*b*) Cervical cap and diaphragm used with spermicidal jelly. (*c*) Intrauterine device (IUD). (*d*) Spermicidal foam. (*e*) Oral contraceptives. (*f*) Vasectomy. (*g*) Tubal ligation.

Continued

The implants can be effective for up to 5 years. Menstruation does not normally occur in women using these techniques while the progesterone levels are elevated.

The advantages of injected and implanted progesterone-like contraceptives over other chemical methods of birth control are that they do not require taking pills on a daily basis. The long-term effects of injected and implanted progesterone-like chemicals have not been as thoroughly studied as the long-term effects of birth-control pills have, and they are still being evaluated.

A drug called **mifepristone** (mif′pris-tōn), or **RU486,** blocks the action of progesterone. It causes the endometrium of the uterus to slough off and to be expelled from the uterus, as it does at the time of menstruation. Because it blocks progesterone receptors, the endometrium undergoes changes similar to those caused by decreasing progesterone levels. It is therefore used to induce menstruation and reduce the possibility of implantation when sexual intercourse has occurred near the time of ovulation. It can also be used to terminate pregnancies. **Morning-after pills,** similar in composition to birth-control pills, are available. Doubling the number of birth-control pills after sexual intercourse within 3 days and again after 12 more hours is sometimes recommended. This or similar techniques can be used after intercourse but are only about 75% effective. The elevated estrogen and progesterone levels may inhibit the preovulatory LH surge in some cases, it may alter the rate of transport of the fertilized ovum from the uterine tube to the uterus, or it may inhibit implantation. The precise effect of the elevated estrogen like and progesterone-like substances depends on the stage of menstrual cycle when they are taken.

SURGICAL METHODS

Vasectomy (va-sek′tō-mē) is a common method used to render males incapable of fertilization without affecting the performance of the sex act. Vasectomy is a surgical procedure used to cut and tie the ductus deferens from each testis within the scrotal sac. In the majority of cases the procedure is permanent, but in some cases vasectomy can be surgically reversed. This procedure prevents sperm cells from passing through the ductus deferens and becoming part of the ejaculate. Because such a small volume of ejaculate comes from the testis and epididymis, vasectomy has little effect on the volume of the ejaculated semen. The sperm cells are reabsorbed in the epididymis.

A common method of permanent birth control in females is **tubal ligation** (lī-gā′shŭn), a procedure in which the uterine tubes are tied and cut or clamped through an incision made through the wall of the abdomen. This procedure closes off the pathway between the sperm cells and the oocyte. **Laparoscopy** (lap-ă-ros′kŏ-pē), a procedure in which an instrument is inserted into the abdomen through a small incision, is commonly used so that only small openings are required to perform the operation.

In some cases, pregnancies are terminated by surgical procedures called **abortions.** The most common method for performing abortions is the insertion of an instrument through the cervix into the uterus. The instrument scrapes the endometrial surface while a strong suction is applied. The endometrium and the embedded embryo are disrupted and sucked out of the uterus. This technique is normally used only in pregnancies that have progressed less than 3 months.

PREVENTION OF IMPLANTATION

Intrauterine devices (IUDs) are inserted into the uterus through the cervix to prevent the normal implantation of the developing embryonic mass within the endometrium. Some early IUD designs produced serious side effects, such as perforation of the uterus; as a result, many IUDs were removed from the market. Data indicate, however, that IUDs are effective in preventing pregnancy. The effective IUDs include those containing copper and progesterone.

76 ***Describe the transport of sperm cells through the female reproductive system.***

77 ***What is capacitation of sperm cells?***

78 ***Describe the events that occur following fertilization.***

79 ***Describe implantation of the embryo and the formation of the placenta.***

Menopause

When a woman is 40–50 years old, menstrual cycles become less regular, and ovulation often does not occur. Eventually, menstrual cycles stop completely. The cessation of menstrual cycles is called **menopause** (men′ō-pawz). The time from the onset of irregular cycles to their complete cessation, which is often 3 to 5 years, is called the **female climacteric** (klī-mak′ter-ik, klī-mak-ter′ik), or **perimenopause.**

Menopause is associated with changes in the ovaries. The number of follicles remaining in the ovaries of menopausal women is small. In addition, the follicles that remain become less sensitive to stimulation by LH and FSH, even though these hormone levels are elevated. As the ovaries become less responsive to stimulation by FSH and LH, fewer mature follicles and corpora lutea are produced. Gradual morphologic changes occur in the female in response to the reduced amount of estrogen and progesterone produced by the ovaries (table 28.3).

A variety of symptoms occur in some women during the climacteric, including "hot flashes," irritability, fatigue, anxiety, and occasionally severe emotional disturbances. These symptoms can be treated successfully in many women by administering small amounts of estrogen, or estrogen in combination with progesterone, and then gradually decreasing the treatment over time or by providing psychological counseling. It appears that administering estrogen following menopause also helps prevent osteoporosis and may reduce colorectal cancer. Although estrogen therapy has been successful, in many women it prolongs the symptoms associated with menopause. Some potential side effects of hormone

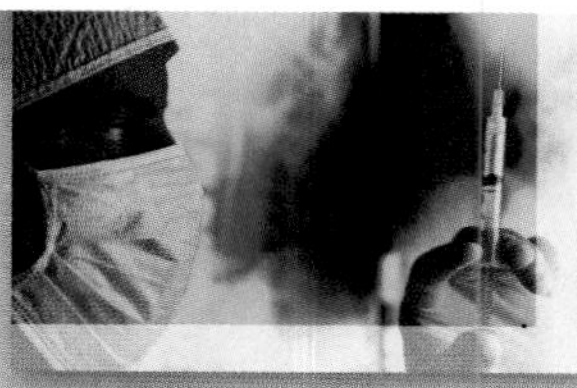

CLINICAL FOCUS

Causes of Female Infertility

The causes of infertility in females include malfunctions of the uterine tubes, reduced hormone secretion from the pituitary or ovary, and interruption of implantation.

Adhesions from pelvic inflammatory conditions caused by a variety of infections can cause the blockage of one or both uterine tubes and is a relatively common cause of infertility in women.

Reduced ovulation can result from inadequate secretion of LH and FSH, which can be caused by hypothyroidism, trauma to the hypothalamus, infarctions of the hypothalamus or anterior pituitary gland, and tumors.

Interruption of implantation may result from uterine tumors or conditions causing abnormal ovarian hormone secretion.

Endometriosis (en'dō-mē-trē-ō'sis), a condition in which endometrial tissue is found in abnormal locations, reduces fertility. Generally, endometriosis is thought to result from some endometrial cells passing from the uterus through the uterine tubes into the pelvic cavity. The endometrial cells invade the peritoneum of the pelvic cavity. Because the endometrium is sensitive to estrogen and progesterone, periodic inflammation of the areas where the endometrial cells have invaded occurs. Endometriosis is a cause of abdominal pain associated with menstruation and it can reduce fertility.

TABLE 28.3 Possible Changes Caused by Decreased Ovarian Hormone Secretion in Postmenopausal Women

Affected Structures and Functions	Changes
Menstrual cycle	Five to 7 years before menopause, the cycle becomes more irregular; finally, the number of cycles in which ovulation does not occur increases, and corpora lutea do not develop.
Uterine tubes	Little change occurs.
Uterus	Irregular menstruation is gradually followed by no menstruation; the chance of cystic glandular hypertrophy of the endometrium increases; the endometrium finally atrophies, and the uterus becomes smaller.
Vagina and external genitalia	The dermis and epithelial lining become thinner; the vulva becomes thinner and less elastic; the labia majora become smaller; public hair decreases; the vaginal epithelium produces less glycogen; vaginal pH increases; reduced secretion leads to dryness; the vagina is more easily inflamed and infected.
Skin	The epidermis becomes thinner; melanin synthesis increases.
Cardiovascular system	Hypertension and atherosclerosis occur more frequently.
Vasomotor instability	Hot flashes and increased sweating are correlated with the vasodilation of cutaneous blood vessels; hot flashes are not caused by abnormal FSH and LH secretion but are related to decreased estrogen levels.
Sex drive	Temporary changes, such as either decreases or increases in sex drive, are often associated with the onset of menopause.
Fertility	Fertility begins to decline approximately 10 years before the onset of menopause; by age 50, almost all oocytes and follicles have been lost; the loss is gradual, and no increased follicular degeneration is associated with the onset of menopause.

therapy are of concern, such as an increased risk for breast, ovarian, and uterine cancer. In addition, the increased risk for heart disease for the first few years after the beginning of menopause is not reduced by estrogen or estrogen plus progesterone. Some data indicate that the risk for heart attacks, strokes, and blood clots is also increased.

80 ***Define menopause and female climacteric.***

81 ***What causes the changes that lead to menopause.***

EFFECTS OF AGING ON THE REPRODUCTIVE SYSTEM

Age-Related Changes in Males

Several age-related changes occur in the male reproductive system. In some but not all men, there is an age-related decrease in the size and the weight of the testes, with an associated decrease in the number of interstitial cells and a thinning of the wall of the

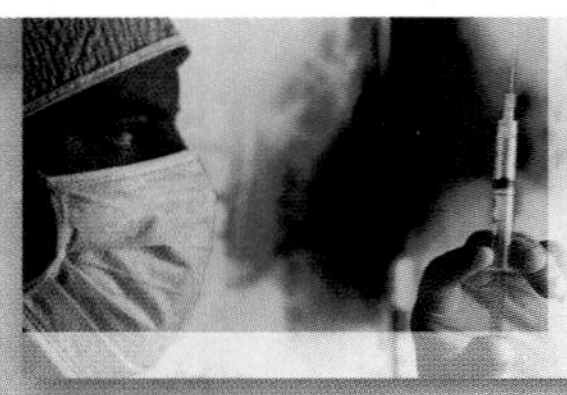

CLINICAL FOCUS

Infectious Diseases

SEXUALLY TRANSMITTED DISEASES

Sexually transmitted diseases (STDs) are a class of infectious diseases spread by intimate sexual contact between individuals. These diseases include the major venereal diseases nongonococcal urethritis, trichomoniasis, gonorrhea, genital herpes, genital warts, syphilis, and acquired immunodeficiency syndrome.

Nongonococcal urethritis is any inflammation of the urethra that is not caused by gonorrhea. Factors such as trauma or the passage of a nonsterile catheter through the urethra can cause this condition, but many cases are acquired through sexual contact. In most cases, the bacterium *Chlamydia trachomatis* is responsible, but other bacteria may be involved. This infection often is unrecognized in people who have it and is responsible for many cases of pelvic inflammatory disease. Left untreated, it can also result in sterility, but antibiotics are usually effective in curing the condition.

Trichomonas vaginalis is a protozoan commonly found in the vagina of females and the urethra of males. If the normal acidity of the vagina is disturbed, *Trichomonas* can grow rapidly. *Trichomonas* infection is more common in females than in males because the vagina provides a suitable environment in which these organisms can survive. The rapid growth of these organisms results in inflammation and a greenish yellow discharge characterized by a foul odor.

Gonorrhea (gon-ō-rē'ă) is caused by the bacterium *Neisseria gonorrhoeae.* The organisms attach to the epithelial cells of the vagina or to the male urethra. The invasion of bacteria establishes an inflammatory response in which pus is formed. Males become aware of a gonorrheal infection by painful urination and the discharge of pus-containing material from the urethra. Symptoms appear within a few days to a week. Recovery may eventually occur without complication but, when complications do occur, they can be serious. The urethra can become partially blocked, or sterility can result from the blockage of the reproductive ducts with scar tissue. In some cases, other organ systems, such as the heart, meninges of the brain, or joints, may become infected. In females, the early stages of infection may pass unnoticed, but the infection can lead to pelvic inflammatory disease. Gonorrheal eye infections may occur in the newborn children of females with gonorrheal infections. Antibiotics are usually effective in treating gonorrheal infections, and the immune system often successfully combats gonorrheal infections in untreated individuals.

Genital herpes (her'pēz) is a viral infection usually caused by herpes simplex type 2. Lesions appear after an incubation period of about 1 week and cause a burning sensation. After this, blisterlike areas of inflammation appear. In males and females, urination can be painful, and walking or sitting can be unpleasant, depending on the location of the lesions. The blisterlike areas heal in about 2 weeks. The lesions may reoccur. The viruses exist in a latent condition in nerve cells and may produce inflamed lesions on the genitals in response to factors such as menstruation, emotional stress, or illness. If active lesions

seminiferous tubules. These changes may be secondary to a decrease in blood flow to the testes or to a gradual decrease in sex hormone production. There is a decrease in the rate of sperm cell production and an increase in the number of abnormal sperm cells. However, sperm cell production does not stop, and it remains adequate for fertility for most men.

Age-related changes have become obvious in the prostate gland by 40 years of age. By 60 years of age, there is a clear decrease in blood flow, an increased thickness in the epithelial cell lining of the prostate gland, and a decrease in functional smooth muscle cells in the wall of the prostate. The changes in the prostate gland do not decrease fertility. There is a substantial increase in the incidence of benign prostatic hypertrophy, which can create difficulty in urination because it compresses the prostatic urethra. A significant number of men older than 60 (approximately 15%) require medical treatment for benign prostatic hypertrophy. By the age of 80, this number is 50%. Before 50 years of age, prostatic cancer is rare. After 55, it is the third leading cause of death from cancer in men. Enlargement of the prostate because of prostatic cancer can also create difficulty in urination.

Impotence increases in men with age. By 60 years of age, approximately 15% of men have difficulty with impotence; by 80, it is 50%. There is an increase in fibrous connective tissue in the erectile tissue of the penis, which, by the time males are 60 years of age, generally decreases the speed of erection.

Although there is great variation among males, many exhibit a decrease in the frequency of sexual activity and a decrease in sexual performance. Psychological changes, age-related changes in the nervous system, and decreased blood flow explain some of the decrease. Many exhibit decreased sexual activity because of the side effects of medications taken for other conditions.

82 ***List the major age-related changes that occur in males.***

83 ***What are the age-related changes that occur in the prostate gland?***

are present in the mother's vagina or external genitalia, a cesarean delivery is performed to prevent newborns from becoming infected with the herpes virus. Because genital herpes is caused by a virus, no effective antibiotic cure for it is available.

Genital warts also result from a viral infection (human papillomavirus) and are quite contagious. This disease is common, and its frequency is increasing. Genital warts can also be transmitted from infected mothers to their infants. Genital warts vary from separate, small, warty growths to large, cauliflower-like clusters. The lesions are usually not painful, but they can cause painful intercourse, and they bleed easily. Women who have genital warts have an increased risk for cervical cancer. Treatments for genital warts include topical agents, cryosurgery, and other surgical methods.

Syphilis (sif′i-lis) is caused by the bacterium *Treponema pallidum,* which can be spread by sexual contact of all kinds. Syphilis exhibits an incubation period from 2 weeks to several months. The disease progresses through several recognized stages. In the primary stage, the initial symptom is a small, hard-based **chancre** (shan′ker), or sore, that usually appears at the site of infection. Several weeks after the primary stage, the disease enters the secondary stage, characterized mainly by skin rashes and mild fever. The symptoms of secondary syphilis usually subside after a few weeks, and the disease enters a latent period, in which no symptoms are present. In less than half the cases, a tertiary stage develops after many years. In the tertiary stage, many lesions develop that can cause extensive tissue damage and can lead to paralysis, insanity, and even death. Syphilis can be passed on to newborns by an infected mother. Damage to mental development and other neurologic symptoms are among the more serious consequences. Females who have syphilis in the latent phase are most likely to have babies who are infected. Antibiotics are used to treat syphilis, although some strains are very resistant to certain antibiotics.

Acquired immunodeficiency syndrome (AIDS) is caused by infection with the human immunodeficiency virus (HIV), which appears to result in the destruction of the immune system (see chapter 22). The most common mechanisms of viral transmission are through sexual contact with a person infected with HIV and through the sharing of needles with an infected person during the administration of illicit drugs. Although transmission did occur during the early 1980s through tainted blood products, screening techniques now make the transmission of HIV through blood transfusions very rare. Some rare cases of transmission of HIV through accidental needle sticks in hospitals and other health-care facilities have been documented. No evidence exists that casual contact with a person who has AIDS or who is infected with HIV results in transmission of the disease. Transmission appears to require exposure to the body fluids of an infected person in a way that allows HIV to enter the interior of another person. Normal casual contact, including touching an HIV-infected person, does not increase the risk for infection.

PELVIC INFLAMMATORY DISEASE

Pelvic inflammatory disease (PID) is a bacterial infection of the female pelvic organs. It usually involves the uterus, uterine tubes, or ovaries. A vaginal or uterine infection may spread throughout the pelvis. Gonorrhea and chlamydia are the most common causes of PID; however, other bacteria can be involved. Early symptoms of PID include increased vaginal discharge and pelvic pain. Early treatment with antibiotics can stop the spread of PID, but lack of treatment results in a life-threatening infection. PID can also lead to sterility.

Age-Related Changes in Females

The most significant age-related change in females is menopause. By age 50, few viable follicles remain in the ovaries. As a result, there is a decrease in the estrogen and progesterone produced by the ovaries.

The uterus decreases in size and the endometrium decreases in thickness. The time between menstruations becomes irregular and longer. Finally, menstruations stop. As the uterus decreases in size, it tips posteriorly and assumes a lower position in the pelvic cavity. Occasionally, uterine prolapse, in which the ligaments of the uterus allow it to descend and protrude into the vagina, occurs. Within 15 years after menopause, the uterus is 50% of its original size.

The vaginal wall becomes thinner and less elastic. There is less lubrication of the vagina and the epithelial lining is more fragile. An increased tendency occurs for vaginal infections. Vaginal contractions during intercourse decrease and the vagina narrows with age. In healthy females, sexual excitement requires greater time to develop, the peak levels of sexual activity are lower, and return to the resting state occurs more quickly.

Approximately 10% of all women will have breast cancer. The increase is most rapid between 45 and 65 years of age, and the incidence is greater for those who have a history of breast cancer in their families than for those who do not. Cancer of the endometrium and cancer of the uterine cervix increases between 50 and 65 years of age. Ovarian cancer increases in frequency in older women, and it is the second most common cancer of the reproductive system in older women.

84 ***List the major age-related changes that occur in females.***

85 ***What are some of the long-term consequences of menopause?***

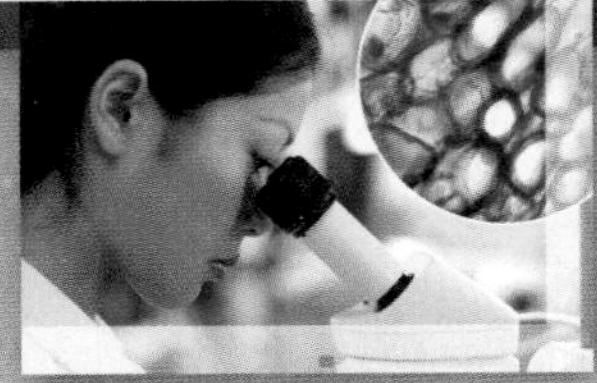

SYSTEMS PATHOLOGY

Benign Uterine Tumors

Mrs. M had four children and was 43 years old. She noticed that menstruation was becoming gradually more severe and lasting up to several days longer each time it started. After she menstruated almost continuously for 2 months, she made an appointment with her physician, who performed a pelvic examination, including tests for conditions such as cervical cancer and uterine cancer. Palpation of the uterus indicated the presence of enlarged masses in Mrs. M's uterus. The results of a dilation and curettage (D&C—dilation of the cervix and scraping [curettage] of the endometrium to remove growths or other abnormal tissues) indicated that Mrs. M suffered from leiomyomas, or fibroid tumors of the uterus.

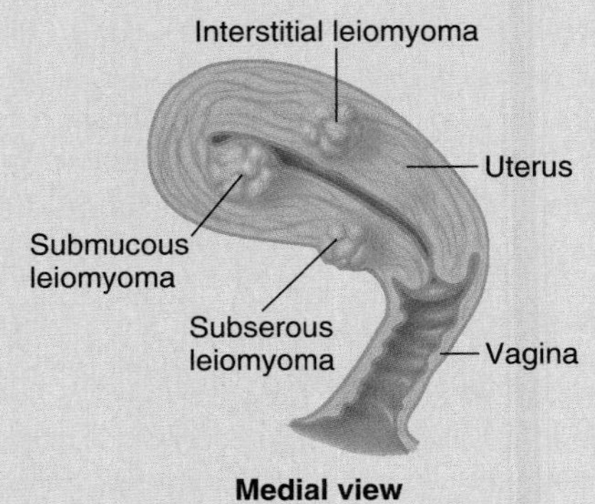

FIGURE C Leiomyomas or Fibroid Tumors

Leiomyomas, or fibroid tumors, are enlarged masses of smooth muscle. They are located near the mucosa (submucous), within the myometrium (interstitial), or near the serosa (subserous).

BACKGROUND INFORMATION

Uterine **leiomyomas** (lī′-ō-mī′măs; figure C), also called uterine fibroids, are enlarged masses of smooth muscle in the myometrium, and they are one of the most common disorders of the uterus. They are the most frequent tumor in women, affecting one of every four. Three-fourths of the women with this condition, however, experience no symptoms. The enlarged mass compresses the uterine lining (endometrium), resulting in ischemia and inflammation and in frequent and severe menstruations. Abdominal cramping because of strong uterine contractions can be present. Constant menstruation is a frequent manifestation of these tumors, and it is one of the most common reasons that women elect to have the uterus removed, a procedure called a **hysterectomy** (his-ter-ek′tō-mē).

PREDICT 6

When discussing her condition with her mother, Mrs. M discovered that her mother recalled frequent menstruations that were irregular and prolonged when she was in her late forties. Her mother did not have a hysterectomy, and in a few years the frequency of menstruation began to subside gradually. Explain.

SYSTEM INTERACTIONS — Effect of Leiomyomas on Other Systems

SYSTEM	INTERACTION
Integumentary	If anemia does not develop, skin appearance is normal, but, if anemia does develop, the skin can appear pale because of the reduced hemoglobin in red blood cells. The continual loss of blood often results in iron-deficiency anemia. The hemoglobin concentration of blood and the hematocrit are therefore reduced.
Muscular	If anemia develops and is severe, muscle weakness may result because of the reduced ability of the cardiovascular system to deliver adequate oxygen to muscles.
Skeletal	The rate of red blood cell synthesis in the red bone marrow increases.
Digestive	An enlarged tumor can put pressure on the rectum or sigmoid colon, resulting in constipation.
Cardiovascular	A chronic loss of blood as in prolonged menstruation over many months to years frequently results in iron-deficiency anemia. Manifestations of anemia include reduced hematocrit, reduced hemoglobin concentration, smaller than normal red blood cells (microcytic anemia), and increased heart rate.
Respiratory	Because of anemia, the oxygen-carrying capacity of the blood is reduced. Increased respiration during physical exertion and rapid fatigue are likely to occur if anemia develops.
Urinary	The kidneys increase erythropoietin secretion in response to the loss of red blood cells. The erythropoietin increases red blood cell synthesis in red bone marrow. An enlarged tumor can put pressure on the urinary bladder, resulting in increased frequency of and painful urination.

SUMMARY

The male reproductive system produces sperm cells and transfers them to the female. The female reproductive system produces the oocyte and nurtures the developing child.

Anatomy of the Male Reproductive System (p. 1033)

The male reproductive system includes the testes, ducts, accessory glands, and supporting structures.

Scrotum

1. The scrotum is a two-chambered sac that contains the testes.
2. The dartos and cremaster muscles help regulate testicular temperature.

Perineum

The perineum, the diamond-shaped area between the thighs, consists of a urogenital triangle and an anal triangle.

Testes

1. The tunica albuginea is the outer connective tissue capsule of the testes.
2. The testes are divided by septa into compartments that contain the seminiferous tubules and the interstitial cells.
3. The seminiferous tubules become straight to form the tubuli recti, which lead to the rete testis. The rete testis opens into the efferent ductules of the epididymis.
4. During development, the testes pass from the abdominal cavity through the inguinal canal to the scrotum.

Sperm Cell Development

1. Sperm cells (spermatozoa) are produced in the seminiferous tubules.
2. Spermatogonia divide (mitosis) to form primary spermatocytes.
3. Primary spermatocytes divide (first division of meiosis) to form secondary spermatocytes, which divide (second division of meiosis) to form spermatids.
4. Spermatids develop an acrosome and a flagellum to become sperm cells.
5. Sertoli cells nourish the sperm cells, form a blood–testes barrier, and produce hormones.

Ducts

1. Efferent ductules extend from the testes to the head of the epididymis.
2. The epididymis is a coiled tube system located on the testis that is the site of sperm cell maturation. It consists of a head, body, and tail.
3. The ductus deferens passes from the epididymis into the abdominal cavity.
4. The end of the ductus deferens, called the ampulla, and the seminal vesicle join to form the ejaculatory duct.
5. The prostatic urethra extends from the urinary bladder to join with the ejaculatory ducts to form the membranous urethra.
6. The membranous urethra extends through the urogenital diaphragm and becomes the spongy urethra, which continues through the penis.
7. The spermatic cord consists of the ductus deferens, blood and lymphatic vessels, nerves, and remnants of the process vaginalis. Coverings of the spermatic cord consist of the external spermatic fascia, cremaster muscle, and internal spermatic fascia.
8. The spermatic cord passes through the inguinal canal into the abdominal cavity.

Penis

1. The penis consists of erectile tissue.
   - The two corpora cavernosa form the dorsum and the sides of the penis.
   - The corpus spongiosum forms the ventral part and the glans penis.
2. The bulb of the penis and the crura form the root of the penis and the crura attaches the penis to the coxae.
3. The prepuce covers the glans penis.

Accessory Glands

1. The seminal vesicles empty into the ejaculatory ducts.
2. The prostate gland consists of glandular and muscular tissue and empties into the prostatic urethra.
3. The bulbourethral glands are compound mucous glands that empty into the spongy urethra.
4. Semen
   - Semen is a mixture of gland secretions and sperm cells.
   - The bulbourethral glands and the urethral mucous glands produce mucus, which neutralizes the acidic pH of the urethra.
   - The testicular secretions contain sperm cells.
   - The seminal vesicle fluid contains fructose and fibrinogen.
   - The prostate secretions make the seminal fluid more pH-neutral. Clotting factors activate fibrinogen, and fibrinolysin breaks down fibrin.

Physiology of Male Reproduction (p. 1045)

Normal function of the male reproductive system depends on hormonal and neural mechanisms.

Regulation of Sex Hormone Secretion

1. GnRH is produced in the hypothalamus and released in surges.
2. GnRH stimulates LH and FSH release from the anterior pituitary.
   - LH stimulates the interstitial cells to produce testosterone.
   - FSH stimulates sperm cell formation.
3. Inhibin, produced by sustentacular cells, inhibits FSH secretion.

Puberty

1. Before puberty, small amounts of testosterone inhibit GnRH release.
2. During puberty, testosterone does not completely suppress GnRH release, resulting in increased production of FSH, LH, and testosterone.

Effects of Testosterone

1. Interstitial cells, the adrenal cortex, and possibly the sustentacular cells produce testosterone.
2. Testosterone causes the development of male sex organs in the embryo and stimulates the descent of the testes.
3. Testosterone causes enlargement of the genitals and is necessary for sperm cell formation.
4. Other effects of testosterone occur.
   - Hair growth stimulation (pubic area, axilla, and beard) and inhibition (male pattern baldness) occur.
   - Enlargement of the larynx and deepening of the voice occur.
   - Increased skin thickness and melanin and sebum production occur.
   - Increased protein synthesis (muscle), bone growth, blood cell synthesis, and blood volume occur.
   - Metabolic rate increases.

Male Sexual Behavior and the Male Sex Act

1. Testosterone is required for normal sex drive.
2. Stimulation of the sexual act can be tactile or psychologic.
3. Afferent action potentials pass through the pudendal nerve to the sacral region of the spinal cord.
4. Parasympathetic stimulation
   - Erection is due to vasodilation of the blood vessels that supply the erectile tissue.
   - The glands of the urethra and the bulbourethral glands produce mucus.
5. Sympathetic stimulation causes erection, emission, and ejaculation.

Anatomy of the Female Reproductive System (p. 1049)

The female reproductive system includes the ovaries, uterine tubes, uterus, vagina, external genitals, and summary glands.

Ovaries

1. The broad ligament, the mesovarium, the suspensory ligaments, and the ovarian ligaments hold the ovaries in place.
2. The peritoneum (ovarian epithelium) covers the surface of the ovaries.
3. The ovary has an outer capsule (tunica albuginea) and is divided internally into a cortex (contains follicles) and a medulla (receives blood and lymph vessels and nerves).
4. Oocyte development and fertilization
   - Oogonia proliferate and become primary oocytes that are in prophase I of meiosis.
   - Ovulation is the release of an oocyte from an ovary.
   - Prior to ovulation, a primary oocyte continues meiosis and produces a secondary oocyte, which in metaphase II of meiosis, and a polar body, which degenerates or divides to form two polar bodies.
   - Fertilization is the joining of a sperm cell and a secondary oocyte to form a zygote. A sperm cell enters a secondary oocyte, which then completes the second meiotic division and produces a polar body. A zygote is formed when the nuclei of the sperm cell and oocyte fuse to form a diploid nucleus.
   - The haploid nuclei then fuse to form a diploid nucleus.
5. Follicle development
   - Primordial follicles are surrounded by a single layer of flat granulosa cells.
   - Primary follicles are primary oocytes surrounded by cuboidal granulosa cells.
   - The primary follicles become secondary follicles as granulosa cells increase in number and fluid begins to accumulate in the vesicles. The granulosa cells increase in number, and theca cells form around the secondary follicles.
   - Mature follicles are enlarged secondary follicles at the surface of the ovary.
6. Ovulation occurs when the follicle swells and ruptures and the secondary oocyte is released from the ovary.
7. Fate of the follicle
   - The mature follicle becomes the corpus luteum.
   - If pregnancy occurs, the corpus luteum persists. If no pregnancy occurs, it becomes the corpus albicans.

Uterine Tubes

1. The mesosalpinx holds the uterine tubes.
2. The uterine tubes transport the oocyte or zygote from the ovary to the uterus.
3. Structures
   - The ovarian end of the uterine tube is expanded as the infundibulum. The opening of the infundibulum is the ostium, which is surrounded by fimbriae.
   - The infundibulum connects to the ampulla, which narrows to become the isthmus. The isthmus becomes the uterine part of the uterine tube and passes through the uterus.
4. The uterine tube consists of an outer serosa, a middle muscular layer, and an inner mucosa with simple ciliated columnar epithelium.
5. Movement of the oocyte
   - Cilia move the oocyte over the fimbriae surface into the infundibulum.
   - Peristaltic contractions and cilia move the oocyte within the uterine tube.
   - Fertilization occurs in the ampulla, where the zygote remains for several days.

Uterus

1. The uterus consists of the body, the isthmus, and the cervix. The uterine cavity and the cervical canal are the spaces formed by the uterus.
2. The uterus is held in place by the broad, round, and uterosacral ligaments.
3. The wall of the uterus consists of the perimetrium (serous membrane), the myometrium (smooth muscle), and the endometrium (mucous membrane).

Vagina

1. The vagina connects the uterus (cervix) to the vestibule.
2. The vagina consists of a layer of smooth muscle and an inner lining of moist stratified squamous epithelium.
3. The vagina is folded into rugae and longitudinal folds.
4. The hymen covers the vestibular opening of the vagina.

External Genitalia

1. The vulva, or pudendum, comprises the external genitalia.
2. The vestibule is the space into which the vagina and the urethra open.
3. Erectile tissue
   - The two corpora cavernosa form the clitoris.
   - The corpora spongiosa form the bulbs of the vestibule.
4. The labia minora are folds that cover the vestibule and form the prepuce.

5. The greater and lesser vestibular glands produce a mucous fluid.
6. When closed, the labia majora cover the labia minora.
   - The pudendal cleft is a space between the labia majora.
   - The mons pubis is an elevated fat deposit superior to the labia majora.

Perineum

The clinical perineum is the region between the vagina and the anus.

Mammary Glands

1. The mammary glands are modified sweat glands located in the breasts.
   - The mammary glands consist of glandular lobes and adipose tissue.
   - The lobes consist of lobules that are divided into alveoli.
   - The lobes connect to the nipple through the lactiferous ducts.
   - The areola surround the nipple.
2. Cooper's ligaments support the breasts.

Physiology of Female Reproduction (p. 1059)

Puberty

1. Puberty begins with the first menstrual bleeding (menarche).
2. Puberty begins when GnRH levels increase.

Menstrual Cycle

1. Ovarian cycle
   - FSH initiates the development of the primary follicles.
   - The follicles secrete a substance that inhibits the development of other follicles.
   - LH stimulates ovulation and completion of the first meiotic division by the primary oocyte.
   - The LH surge stimulates the formation of the corpus luteum. If fertilization occurs, HCG stimulates the corpus luteum to persist. If fertilization does not occur, the corpus luteum becomes the corpus albicans.
2. A positive-feedback mechanism causes FSH and LH levels to increase near the time of ovulation.
   - Estrogen produced by the theca cells of the follicle stimulates GnRH secretion.
   - GnRH stimulates FSH and LH, which stimulate more estrogen secretion, and so on.
   - Inhibition of GnRH levels causes FSH and LH levels to decrease after ovulation. Inhibition is due to the high levels of estrogen and progesterone produced by the corpus luteum.
3. Uterine cycle
   - Menses (day 1 to day 4 or 5). The spiral arteries constrict, and endometrial cells die. The menstrual fluid is composed of sloughed cells, secretions, and blood.
   - Proliferation phase (day 5 to day 14). Epithelial cells multiply and form glands, and the spiral arteries supply the glands.
   - Secretory phase (day 15 to day 28). The endometrium becomes thicker, and the endometrial glands secrete.
   - Estrogen stimulates proliferation of the endometrium and synthesis of progesterone receptors.
   - Increased progesterone levels cause hypertrophy of the endometrium, stimulate gland secretion, and inhibit uterine contractions. Decreased progesterone levels cause the spiral arteries to constrict and start menses.

Female Sexual Behavior and the Female Sex Act

1. Female sex drive is partially influenced by androgens (produced by the adrenal gland) and steroids (produced by the ovaries).
2. Parasympathetic effects
   - The erectile tissue of the clitoris and the bulbs of the vestibule become filled with blood.
   - The vestibular glands secrete mucus, and the vagina extrudes a mucuslike substance.

Female Fertility and Pregnancy

1. Intercourse must take place 5 days before to 1 day after ovulation if fertilization is to occur.
2. Sperm cell transport to the ampulla depends on the ability of the sperm cells to swim and possibly on contractions of the uterus and the uterine tubes.
3. Implantation of the developing embryo into the uterine wall occurs when the uterus is most receptive.
4. Estrogen and progesterone secreted first by the corpus luteum and later by the placenta are essential for the maintenance of pregnancy.

Menopause

The female climacteric begins with irregular menstrual cycles and ends with menopause, the cessation of the menstrual cycle.

Effects of Aging on the Reproductive System (p. 1071)

Several age-related changes occur in the male and female reproductive systems.

Age-Related Changes in Males

1. There is an age-related decrease in the size and weight of the testes, a decrease in the number of interstitial cells, a thinning of the seminiferous tubule wall, and a decrease in sperm production. Sperm cell production is still adequate for fertilization.
2. The prostate gland enlarges, and there is an age-related increase in prostatic cancer.
3. Impotence is age-related and there is a gradual decrease in sexual activity.

Age-Related Changes in Females

1. The most significant age-related change in females is menopause.
2. The uterus decreases in size and the vaginal wall thins.
3. There is an age-related increase in breast, uterine, and ovarian cancer.

REVIEW AND COMPREHENSION

1. If an adult male walked into a swimming pool of cold water, which of these muscles would be expected to contract?
   a. cremaster muscle c. gubernaculum e. both a and b
   b. dartos muscle d. prepuce muscle
2. Testosterone is produced in the
   a. interstitial cells.
   b. seminiferous tubules of the testes.
   c. anterior lobe of the pituitary.
   d. sperm cells.
3. Early in development (4 months after fertilization), the testes
   a. are found in the peritoneal cavity.
   b. move through the inguinal canal.
   c. produce a membrane that becomes the scrotum.
   d. produce sperm cells.
   e. all of the above.
4. The site of spermatogenesis in the male is the
   a. ductus deferens. d. rete testis.
   b. seminiferous tubules. e. efferent ductule.
   c. epididymis.
5. The location of final maturation and storage of sperm cells before their ejaculation is the
   a. seminal vesicles. d. epididymis.
   b. seminiferous tubules. e. sperm bank.
   c. glans penis.
6. Given these structures:
   1. ductus deferens
   2. efferent ductule
   3. epididymis
   4. ejaculatory duct
   5. rete testis

   Choose the arrangement that lists the structures in the order a sperm cell passes through them from the seminiferous tubules to the urethra.
   a. 2,3,5,4,1 c. 3,2,4,1,5 e. 5,2,3,1,4
   b. 2,5,3,4,1 d. 3,4,2,1,5
7. Concerning the penis,
   a. the membranous urethra passes through the corpora cavernosa.
   b. the glans penis is formed by the corpus spongiosum.
   c. the penis contains four columns of erectile tissue.
   d. the crus of the penis is part of the corpus spongiosum.
   e. the bulb of the penis is covered by the prepuce.
8. Given these glands:
   1. prostate gland
   2. bulbourethral gland
   3. seminal vesicle

   Choose the arrangement that is in the order the glands contribute their secretions to the formation of semen.
   a. 1,2,3 c. 2,3,1 e. 3,2,1
   b. 2,1,3 d. 3,1,2
9. Which of these glands is correctly matched with the function of its secretions?
   a. bulbourethral gland—neutralizes acidic contents of the urethra
   b. seminal vesicles—contain large amounts of fructose, which nourishes the sperm cells
   c. prostate gland—contains clotting factors that cause cogulation of the semen
   d. all of the above
10. LH in the male stimulates
    a. the development of the seminiferous tubules.
    b. spermatogenesis.
    c. testosterone production.
    d. both a and b.
    e. all of the above.
11. Which of these factors causes a decrease in GnRH release?
    a. decreased inhibin c. decreased FSH
    b. increased testosterone d. decreased LH
12. In the male, before puberty
    a. FSH levels are higher than after puberty.
    b. LH levels are higher than after puberty.
    c. GnRH release is inhibited by testosterone.
    d. all of the above.
13. Testosterone
    a. stimulates the development of terminal hairs.
    b. decreases red blood cell count.
    c. prevents closure of the epiphyseal plate.
    d. decreases blood volume.
    e. all of the above.
14. Which of these is consistent with erection of the penis?
    a. parasympathetic stimulation
    b. dilation of arterioles
    c. engorgement of sinusoids with blood
    d. occlusion of veins
    e. all of the above
15. The first polar body
    a. is normally formed before fertilization.
    b. is normally formed after fertilization.
    c. normally receives most of the cytoplasm.
    d. is larger than an oocyte.
    e. both a and b.
16. After ovulation, the mature follicle collapses, taking on a yellowish appearance to become the
    a. degenerating follicle. d. tunica albuginea.
    b. corpus luteum. e. cumulus mass.
    c. corpus albicans.
17. The ampulla of the uterine tube
    a. is the opening of the uterine tube into the uterus.
    b. has long, thin projections called the ostium.
    c. is connected to the isthmus of the uterine tube.
    d. is lined with ciliated columnar epithelium.
18. The layer of the uterus that undergoes the greatest change during the menstrual cycle is the
    a. perimetrium. c. endometrium. e. broad ligament.
    b. hymen. d. myometrium.
19. The vagina
    a. consists of skeletal muscle.
    b. has ridges called rugae.
    c. is lined with simple squamous epithelium.
    d. all of the above.
20. During sexual excitement, which of these structures fills with blood and causes the vaginal opening to narrow?
    a. bulb of the vestibule d. labia majora
    b. clitoris e. prepuce
    c. mons pubis

21. Given these vestibular-perineal structures:
    1. vaginal opening
    2. clitoris
    3. urethral opening
    4. anus

    Choose the arrangement that lists the structures in their proper order from the anterior to the posterior aspect.
    a. 2,3,1,4
    b. 2,4,3,1
    c. 3,1,2,4
    d. 3,1,4,2
    e. 4,2,3,1
22. Concerning the breasts,
    a. lactiferous ducts open on the areola.
    b. each lactiferous duct supplies an alveolus.
    c. they are attached to the pectoralis major muscles by mammary ligaments.
    d. even before puberty, the female breast is quite different from the male breast.
23. The major secretory product of the mature follicle is
    a. estrogen.
    b. progesterone.
    c. LH.
    d. FSH.
    e. relaxin.
24. In the average adult female, ovulation occurs at day ___________ of the menstrual cycle.
    a. 1
    b. 7
    c. 14
    d. 21
    e. 28
25. Which of these processes or phases in the monthly reproductive cycle of the human female occur at the same time?
    a. maximal LH secretion and menstruation
    b. early follicular development and the secretory phase of the uterus
    c. regression of the corpus luteum and an increase in ovarian progesterone secretion
    d. ovulation and menstruation
    e. proliferation stage of the uterus and increased estrogen production
26. During the proliferative phase of the menstrual cycle, one would normally expect
    a. the highest levels of estrogen that occur during the menstrual cycle.
    b. the mature follicle to be present in the ovary.
    c. an increase in the thickness of the endometrium.
    d. both a and b.
    e. all of the above.
27. The cause of menses in the menstrual cycle appears to be
    a. increased progesterone secretion from the ovary, which produces blood clotting.
    b. increased estrogen secretion from the ovary, which stimulates the muscles of the uterus to contract.
    c. decreased progesterone secretion by the ovary.
    d. decreased production of oxytocin, causing the muscles of the uterus to relax.
28. After fertilization, the successful development of a mature, full-term fetus depends on the
    a. release of human chorionic gonadotropin (HCG) by the developing placenta.
    b. production of estrogen and progesterone by the placental tissues.
    c. maintenance of the corpus luteum for all 9 months.
    d. both a and b.
    e. all of the above.
29. A woman with a 28-day menstrual cycle is most likely to become pregnant as a result of coitus on days
    a. 1–3.
    b. 5–8.
    c. 9–14.
    d. 15–20.
    e. 21–28.
30. Menopause
    a. develops when follicles become less responsive to FSH and LH.
    b. results from elevated estrogen levels in 40- to 50-year-old women.
    c. occurs because too many follicles develop during each cycle.
    d. results when follicles develop but contain no oocytes.
    e. occurs because FSH and LH levels decline.

Answers in Appendix E

CRITICAL THINKING

1. If an adult male were castrated, what would happen to the levels of GnRH, FSH, LH, and testosterone in his blood? What effect would these hormonal changes have on his sexual characteristics and behavior?
2. If a 9-year-old boy were castrated, what would happen to the levels of GnRH, FSH, LH, and testosterone in his blood? What effect would these hormonal changes have on his sexual characteristics and behavior?
3. Suppose you want to produce a birth-control pill for men. On the basis of what you know about the male hormone system, what do you want the pill to do? Discuss any possible side effects of the pill.
4. If the ovaries are removed from a postmenopausal woman, what happens to the levels of GnRH, FSH, LH, estrogen, and progesterone in her blood? What symptoms do you expect to observe?
5. During the secretory phase of the menstrual cycle, you normally expect
    a. the highest levels of progesterone that occur during the menstrual cycle.
    b. a follicle present in the ovary that is ready to undergo ovulation.
    c. that the endometrium reaches its greatest degree of development.
    d. both a and b.
    e. both a and c.
6. If the ovaries are removed from a 20-year-old woman, what happens to the levels of GnRH, FSH, LH, estrogen, and progesterone in her blood? What side effects do these hormonal changes have on her sexual characteristics and behavior?
7. A study divides healthy women into two groups (A and B). Both groups are composed of women who have been sexually active for

at least 2 years and are not pregnant at the beginning of the experiment. The subjects weigh about the same amount, and none smokes cigarettes, although some drink alcohol occasionally. Group A women receive a placebo in the form of a sugar pill each morning during their menstrual cycles. Group B women receive a pill containing estrogen and progesterone each morning of their menstrual cycles. Then plasma LH levels are measured before, during, and after ovulation. The results are as follows:

Group	4 Days Before Ovulation	The Day of Ovulation	4 Days After Ovulation
A	18 mg/100 mL	300 mg/100 mL	17 mg/100 mL
B	21 mg/100 mL	157 mg/100 mL	15 mg/100 mL

The number of pregnancies in group A is 37/100 women/year. The number of pregnancies in group B is 1.5/100 women/year. What conclusion can you reach on the basis of these data? Explain the mechanism involved.

8. A woman who is taking birth-control pills that consist of only progesterone experiences the hot flash symptoms of menopause. Explain why.
9. GnRH can be used to treat some women who want to have children but have not been able to get pregnant. Explain why it is critical to administer the correct concentration of GnRH at the right time during the menstrual cycle.

Answers in Appendix F

Visit this book's website at www.mhhe.com/seeley8 for chapter quizzes, interactive learning exercises, and other study tools.

Development, Growth, and Aging

29

The stages of life and associated activities are issues of great interest in today's society. We tend to view life stages very differently today than in the past. For example, in 1960, 20% of males and 12% of females graduating from high school attended college. Today, over half of all people 25 and older have attended some college. In addition, there are many more nontraditional college students than there were twenty years ago.

In 1900, only 5% of the U.S. population was over age 65. Today, about 13% of the population is over age 65, and by 2030 more than 20% will be older than 65. The average life expectancy in 1900 was about 47 years, in 1940 it was about 63 years, and today it is about 78 years. In 1900, nearly 70% of all males over age 65 were still working; today, only about 20% are still working past age 65. Older people are healthier and more active than they have ever been.

The life span is usually considered the period between birth and death; however, the 9 months before birth are a critical part of a person's existence. What happens in these 9 months profoundly affects the rest of a person's life. Although most people develop normally and are born without defects, approximately 3 out of every 100 people are born with a birth defect so severe that it requires medical attention during the first year of life. Later in life, many more people discover previously unknown problems, such as the tendency to develop asthma, certain brain disorders, or cancer. This chapter discusses *prenatal development* (p. 1082), *parturition* (p. 1104), *the newborn* (p. 1106), *lactation* (p. 1110), the *first year after birth* (p. 1111), *life stages* (p. 1111), *aging* (p. 1112), and *death* (p. 1113).

Colorized scanning electron micrograph of an oocyte with sperm cells on the surface.

Reproductive System

PRENATAL DEVELOPMENT

The **prenatal period** is the period from conception until birth, and it is divided into three parts: (1) approximately the first 2 weeks of development, during which the primitive germ layers are formed; (2) from about the second to the end of the eighth week of development, during which the major organ systems come into existence; and (3) the last 30 weeks of the prenatal period, during which the organ systems grow and become more mature.

The developing human between the time of fertilization and 8 weeks of development is called an **embryo** (em′brē-ō). From 8 weeks to birth, the developing human is called a **fetus** (fē′tus).

The medical community in general uses the mother's **last menstrual period (LMP)** to calculate the **clinical age** of the unborn child. Most embryologists, on the other hand, use **postovulatory age** to describe the timing of developmental events. Postovulatory age is used in this book. Because ovulation occurs about 14 days after LMP and fertilization occurs near the time of ovulation, it is assumed that postovulatory age is 14 days less than clinical age.

1. ***Describe the three parts of the prenatal period. Give the length of time for each part.***
2. ***Define clinical age and postovulatory age, and distinguish between the two.***

Fertilization

Fertilization is the process by which a sperm cell attaches to a secondary oocyte, the contents of the sperm head enter the oocyte cytoplasm and join the oocyte pronucleus to form a new nucleus (figure 29.1). Only a few dozen sperm cells, of the several hundred million deposited in the vagina during sexual intercourse, reach the vicinity of the secondary oocyte in the ampulla of the uterine tube. The **corona radiata,** which are cells of the cumulus mass expelled from the follicle with the oocyte (see figure 28.12), is a barrier to the sperm cells reaching the oocyte. The sperm cells are propelled through the loose matrix between the follicular cells of the corona radiata by the action of their flagella. The **zona pellucida** is an extracellular membrane, comprised mostly of glycoproteins, between the corona radiata and the oocyte. One particular zona pellucida

1. Many sperm cells attach to the corona radiata of a secondary oocyte.

Corona radiata
Sperm cell
Oocyte nucleus
First polar body
Zona pellucida

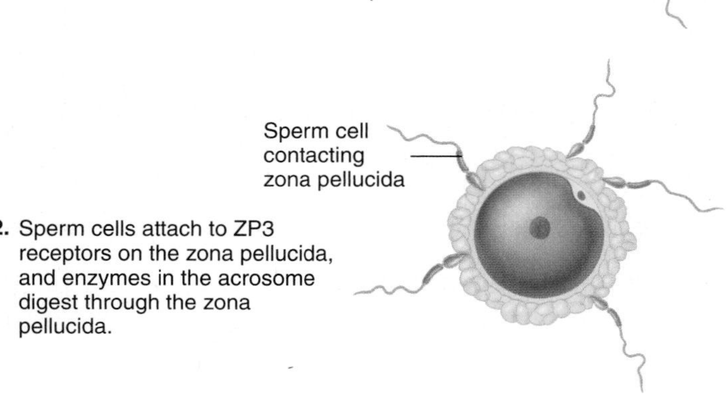

2. Sperm cells attach to ZP3 receptors on the zona pellucida, and enzymes in the acrosome digest through the zona pellucida.

3. The head of one sperm cell penetrates the zona pellucida, attaches to α6β1 integrins on the oocyte plasma membrane, and penetrates the oocyte plasma membrane to enter the oocyte cytoplasm. Changes in the zona pellucida (moving away from the oocyte) form a perivitelline space and prevent additional sperm cells from entering the oocyte.

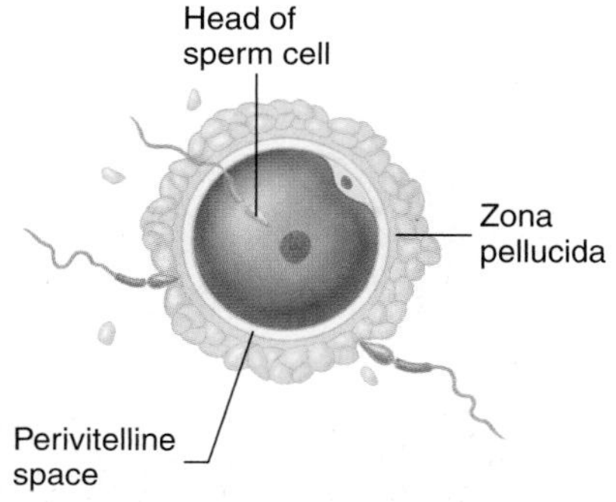

4. In response to the entry of the contents of the sperm head into the oocyte, the oocyte nucleus moves to one side of the oocyte, where it completes the second meiotic division and gives off a second polar body. When the second meiosis is complete, the oocyte nucleus, now called the female pronucleus, moves back toward the center of the oocyte.

5. The contents of the sperm head enlarge and become the male pronucleus.

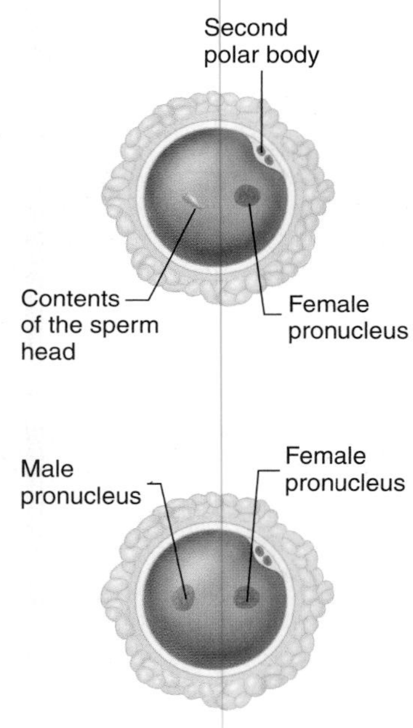

6. The two pronuclei fuse to form a single nucleus. Fertilization is complete and a zygote results.

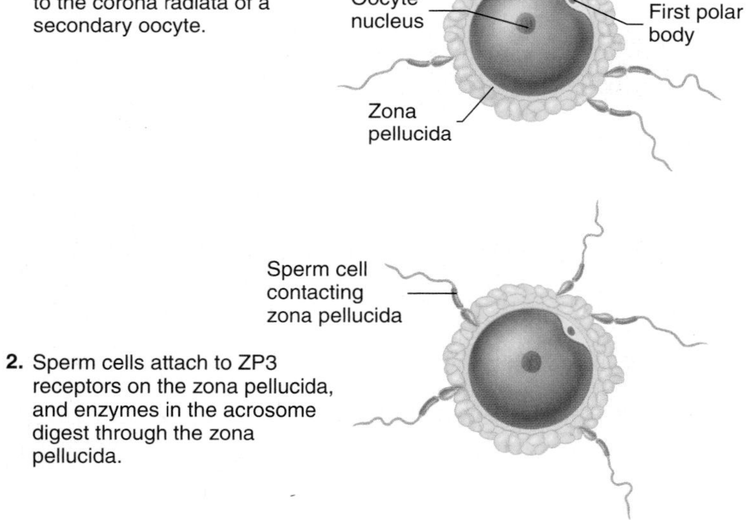

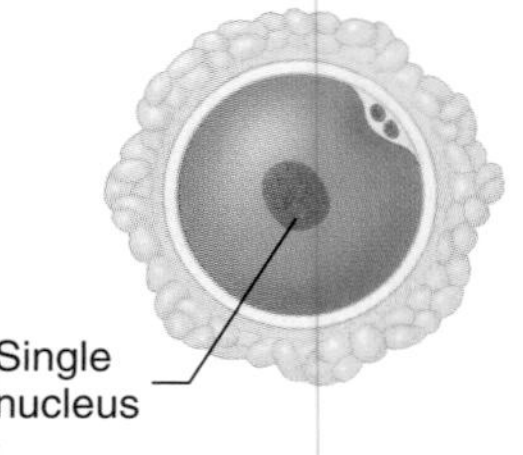

PROCESS FIGURE 29.1 Fertilization

glycoprotein, called **ZP3,** is a species-specific sperm cell receptor, to which molecules on the acrosomal cap of the sperm cell bind. This binding initiates the **acrosomal reaction,** which results in the activation of digestive enzymes, primarily hyaluronidase, in the acrosome.

The first sperm cell through the zona pellucida attaches to the integrin α6β1 on the surface of the oocyte plasma membrane. This attachment causes depolarization of the oocyte plasma membrane within 2–3 seconds of sperm cell attachment. This depolarization, called the **fast block to polyspermy,** prevents additional sperm from attaching to the oocyte plasma membrane. Depolarization also causes the intracellular release of Ca^{2+}, which in turn causes the exocytosis of water and other molecules from secretory vesicles, referred to as cortical granules, on the inner surface of the oocyte plasma membrane. The released fluid causes the oocyte to shrink and the zona pellucida to denature and expand away from the oocyte. As the result of denaturation of the zona pellucida, ZP3 is inactivated, and no additional sperm cells can attach. This reaction is referred to as the **slow block to polyspermy.** The fast block and slow block to polyspermy ensure that the oocyte is fertilized by only one sperm cell. The fluid-filled space between the oocyte plasma membrane and the zona pellucida is called the **perivitelline space.**

The entrance of a sperm cell into the oocyte stimulates the female nucleus to undergo the second meiotic division, and the second polar body is formed. The nucleus that remains after the second meiotic division, called the **female pronucleus,** moves to the center of the oocyte, where it meets the **male pronucleus** of the sperm cell. Both the male and female pronuclei are haploid, each having one-half of each chromosome pair (see chapter 28). Fusion of the pronuclei completes the process of fertilization and restores the diploid number of chromosomes. The product of fertilization is a single cell, the **zygote** (zī′gōt; figure 29.2*a*).

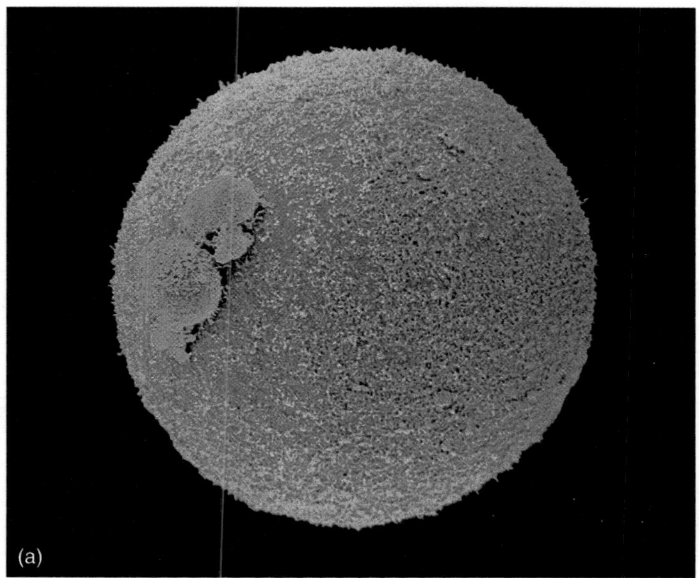

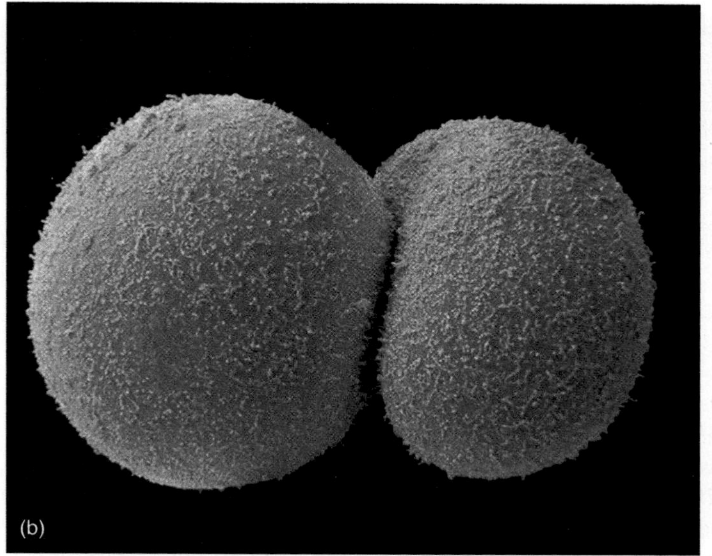

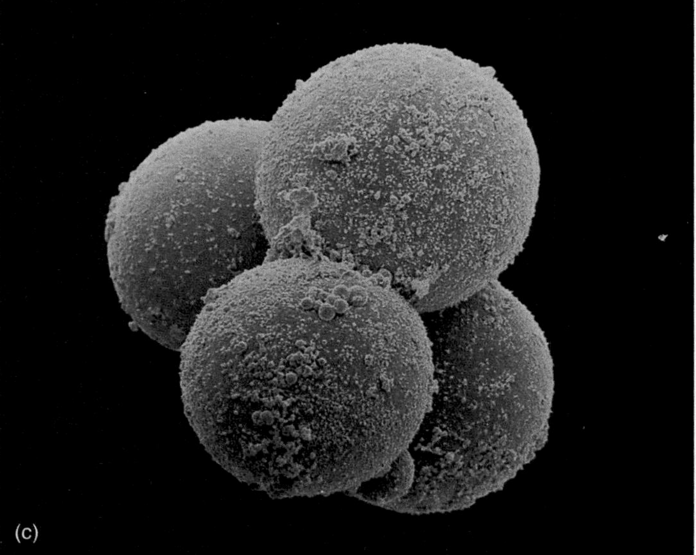

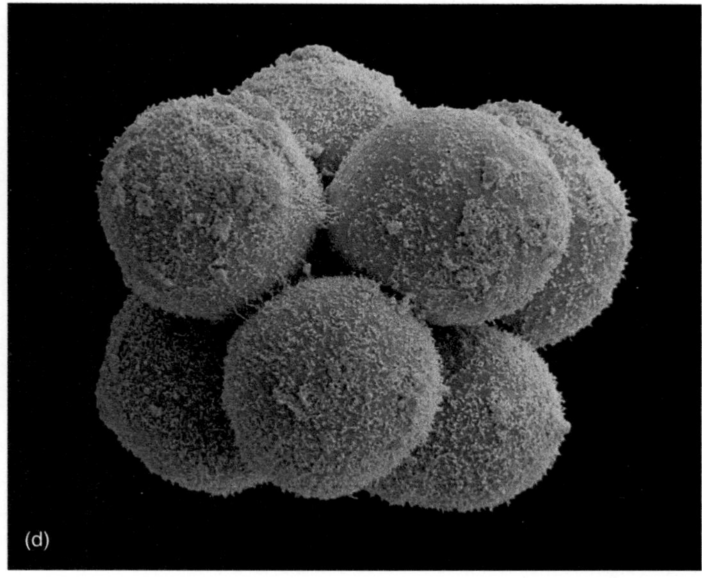

FIGURE 29.2 Early Stages of Human Development

(*a*) Zygote (120 μm in diameter) with two polar bodies attached. (*b*)–(*d*) During the early cell divisions, the embryo divides into more and more cells, but the total size of the embryo remains relatively constant. (*b*) Two cells. (*c*) Four cells. (*d*) Eight cells.

3 ***Describe the events of fertilization. Where does fertilization occur?***

Cloning

Fusion of the male and female pronuclei forms a new nucleus in the zygote. Alternatively, the nucleus can be removed from an oocyte and the nucleus from another cell can be introduced. This complete, diploid nucleus forms a zygote. This process of introducing a nucleus from a cell into an oocyte, whose own nucleus has been removed, is called **cloning.** The first cloned mammal was a sheep named Dolly in Scotland in 1996.

Early Cell Division

About 18–36 hours after fertilization, the zygote divides to form two cells. Those two cells divide to form four cells, which divide to form eight cells, and so on (figure 29.2*b–d*). In the very early stages of development (days 1–4), the cells are said to be **totipotent** (tō-tīp′ō-tent; whole-powered), meaning each cell has the potential to give rise to any tissue type necessary for development. At this point, if a cell separates from the embryo, the totipotent cell could give rise to another individual, as in identical twins. However, the cells of the developing embryo soon undergo **differentiation,** or specialization. Once differentiation occurs the dividing cells of the embryo are referred to as **pluripotent** (ploo-rip′ō-tent; multiple-powered), which means that any cell has the ability to develop into a wide range of tissues, but not all the tissues necessary for development, as is the case with totipotent cells. As a result, the total number of embryonic cells can decrease, increase, or reorganize without affecting the normal development of the embryo.

Twins

In rare cases, following early cell divisions, the cells separate and develop to form two individuals, called "identical," or **monozygotic, twins.** Identical twins have identical genetic information in their cells. Other mechanisms that occur a little later in development can also cause identical twins. Occasionally, a woman ovulates two or more secondary oocytes at the same time. Fertilization of two oocytes by different sperm cells results in "fraternal," or **dizygotic, twins.** Multiple ovulations can occur naturally or be stimulated by the injection of drugs that stimulate gonadotropin release. These drugs are sometimes used to treat certain forms of infertility. This drug treatment may result in multiple births in women undergoing the treatment.

Morula and Blastocyst

Once the dividing embryo is a solid ball of 12 or more cells, it is a sphere composed of numerous, smaller spheres and is therefore called a **morula** (mōr′oo-lă, mōr′ū-lă; mulberry; figure 29.3). Four or 5 days after ovulation, the morula consists of about

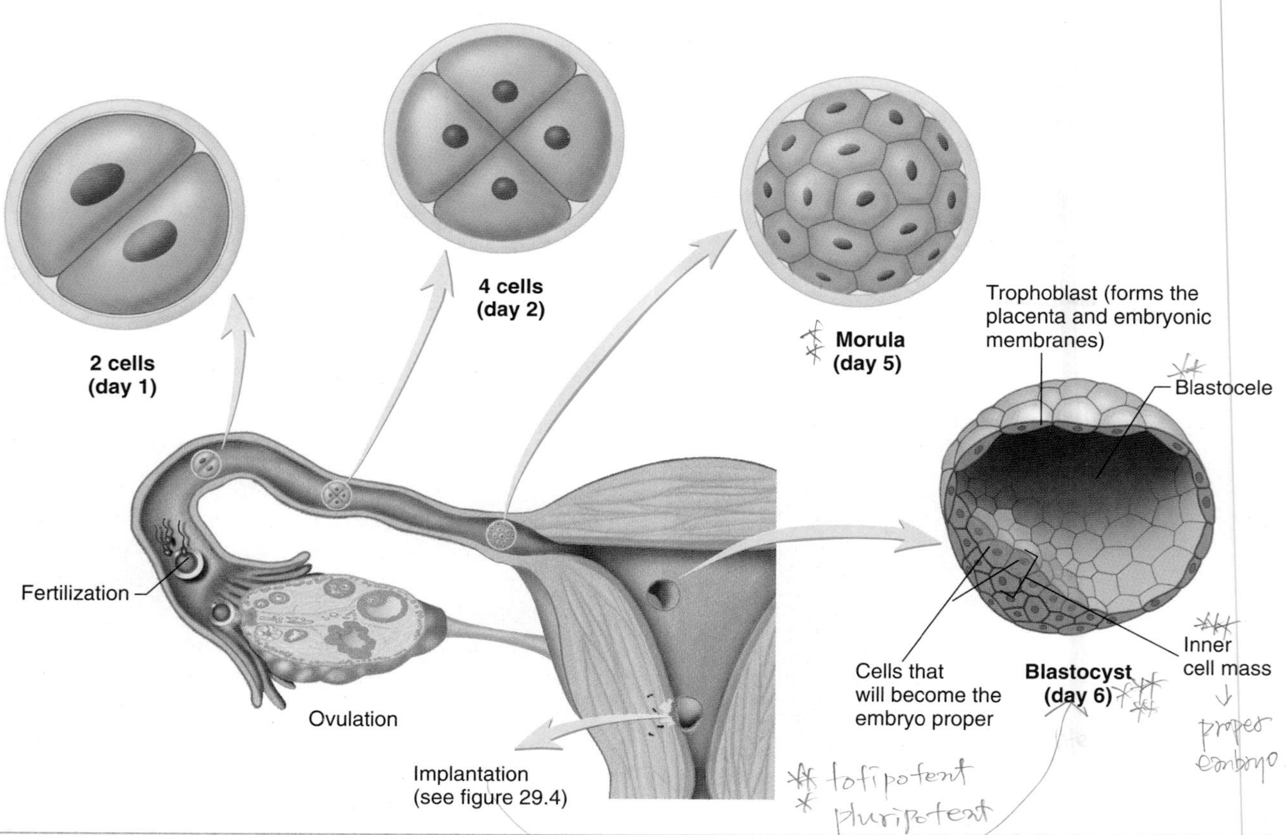

FIGURE 29.3 Blastocyst

Green cells are trophoblastic, and *orange* cells will form the embryo proper.

32 cells. Near this time, a fluid-filled cavity called the **blastocele** (blas′tō-sēl) begins to appear approximately in the center of the cellular mass. The hollow sphere that results is called a **blastocyst** (blas′tō-sist; see figure 29.3). A single layer of cells, the **trophoblast** (trof′ō-blast, trō′fō-blast; feeding layer), surrounds most of the blastocele, but at one end of the blastocyst the cells are several layers thick. The thickened area is the **inner cell mass** and is the tissue from which the embryo proper develops. The pluripotent cells of the inner cell mass do not form all the tissues necessary for normal development. The trophoblast forms the placenta and the membranes (chorion and amnion) surrounding the embryo that are essential for development.

PREDICT 1

What is the expected fate of a cell separated from the inner cell mass? How is this different from the fate of a totipotent cell separated from an embryo in the very early stage of development?

Stem Cell Research

During growth and development, many cells differentiate for a particular function and many lose the ability to divide. However, some cells do not become fully specialized and retain the ability to undergo mitosis and differentiation. These cells are called **stem cells.** Recently, stem cells have become a hot topic for discussion, not only in the scientific world but also in the political arena. Stem cells have the potential to treat many diseases by replacing dysfunctional cells with normal cells. Bone marrow transplants have become a common treatment for certain types of leukemia. Blood stem cells, found in the red bone marrow, are harvested from a donor and introduced into a leukemia patient. The stem cells provide a new source of normal blood cells for that individual. The question for many other diseases, however, is whether or not a source of stem cells can be isolated and used to treat them.

Stem cells can be obtained from different sources, including adult tissues. Adult stem cells, however, are usually limited as to the type of cell they can produce. For example, adult liver stem cells can give rise to many cell types found in the liver, but they cannot give rise to other tissue cell types, such as blood cells. The most versatile stem cell is the totipotent zygote because it can give rise to all cell types. However, the zygote also gives rise to placental tissue. The pluripotent cells of the inner cell mass are called **embryonic stem cells.** These cells are also very versatile because theoretically they can give rise to all tissues of a new individual, but they cannot give rise to the tissues necessary for implantation and development of the placenta. Embryonic stem cells were first isolated in 1998; since then, many people have suggested the great potential of these cells for treating many diseases. One problem, however, lies in the ability to artificially stimulate embryonic stem cells to differentiate into the right type of cell. Scientists are still in the early stages of determining the biochemical processes of how embryonic stem cells communicate to allow for the proper development of all tissues in multicellular organisms. Embryonic stem cells differentiate into several cell types. Ideally, scientists would prefer to limit the number of cell types a particular stem cell can give rise to; otherwise, when these stem cells are grown in the lab, further work has to be done to isolate the correct cell type for a given treatment procedure.

Adult stem cells have had limited success in treating certain diseases. Stem cells obtained from adult blood may be used to address the complications associated with coronary heart disease, particularly the growth of new vessels around areas of blockage. It appears that adult stem cells are not effective for treatments of other diseases, such as type I diabetes mellitus and Parkinson disease. For these cases, research suggests that embryonic stem cells have a greater potential to provide a longer-lasting cure.

Important ethical issues are involved in the use of embryonic stem cells. To harvest the embryonic cells, an embryo must be destroyed. Opponents of embryonic stem cell use argue that the zygote is a living human and the destruction of the zygote or cells derived from the zygote is unacceptable. Their argument is that the sanctity of human life extends well into the prenatal period to include the earliest stages of human development. Proponents of embryonic stem cell use argue that the definition of a human does not extend to the zygote and the sanctity and quality of human life for those suffering from debilitating, often fatal, and currently incurable diseases are worth the price. Furthermore, proponents argue that the zygotes providing the embryonic stem cells were produced for some other reason, such as infertility treatment, and will more than likely be destroyed. The debate probably will not end soon. Current research is attempting to obtain embryonic stem cells without the need to destroy embryos.

Implantation of the Blastocyst and Development of the Placenta

All of the events of the early germinal phase, including the first cell division through formation of the blastocele and the inner cell mass, occur as the embryo moves from the site of fertilization in the ampulla of the uterine tube to the site of implantation in the uterus. About 7 days after fertilization, the blastocyst attaches itself to the uterine wall, usually in the area of the uterine fundus, and begins the process of **implantation,** which is the burrowing of the blastocyst into the uterine wall.

As the blastocyst invades the uterine wall, two populations of trophoblast cells develop and form the embryonic portion of the **placenta** (figure 29.4), the organ of nutrient and waste product exchange between the embryo and the mother. The first is a proliferating population of individual trophoblast cells called the **cytotrophoblast** (sī-tō-trof′ō-blast). The other is a nondividing syncytium, or multinucleated cell, called the **syncytiotrophoblast** (sin-sish′ē-ō-trō′fō-blast). The cytotrophoblast remains nearer the other embryonic tissues, and the syncytiotrophoblast invades the endometrium of the uterus. The syncytiotrophoblast is nonantigenic, which means that, as it invades the maternal tissue, no immune reaction is triggered.

As the syncytiotrophoblast encounters maternal blood vessels, it surrounds them and digests the vessel wall, forming pools of maternal blood within cavities called **lacunae** (lă-koo′nē; see

1. Frontal section of the uterus and uterine tube, showing development 7 days after fertilization.

Ovary
Uterus
Endometrium
Myometrium
Implantation of blastocyst

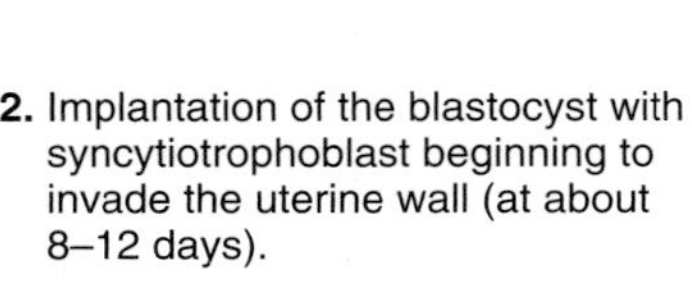

2. Implantation of the blastocyst with syncytiotrophoblast beginning to invade the uterine wall (at about 8–12 days).

Uterine endometrium
Uterine gland
Uterine epithelium
Maternal arteriole
Syncytiotrophoblast invading uterine wall
Cytotrophoblast
Implanted blastocyst
Cells that will become the embryo proper

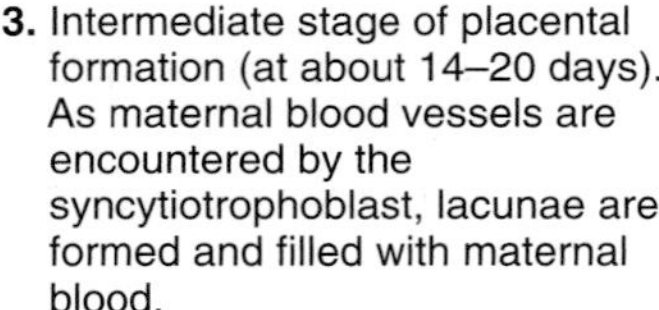

3. Intermediate stage of placental formation (at about 14–20 days). As maternal blood vessels are encountered by the syncytiotrophoblast, lacunae are formed and filled with maternal blood.

Maternal arteriole
Syncytiotrophoblast
Cytotrophoblast
Connecting stalk to developing embryo
Maternal arteriole wall
Arteriole wall digested by syncytiotrophoblast
Lacuna filled with maternal blood

4. Cytotrophoblast cords surround the syncytiotrophoblast and lacunae, and embryonic blood vessels enter the cord (at about 1 month).

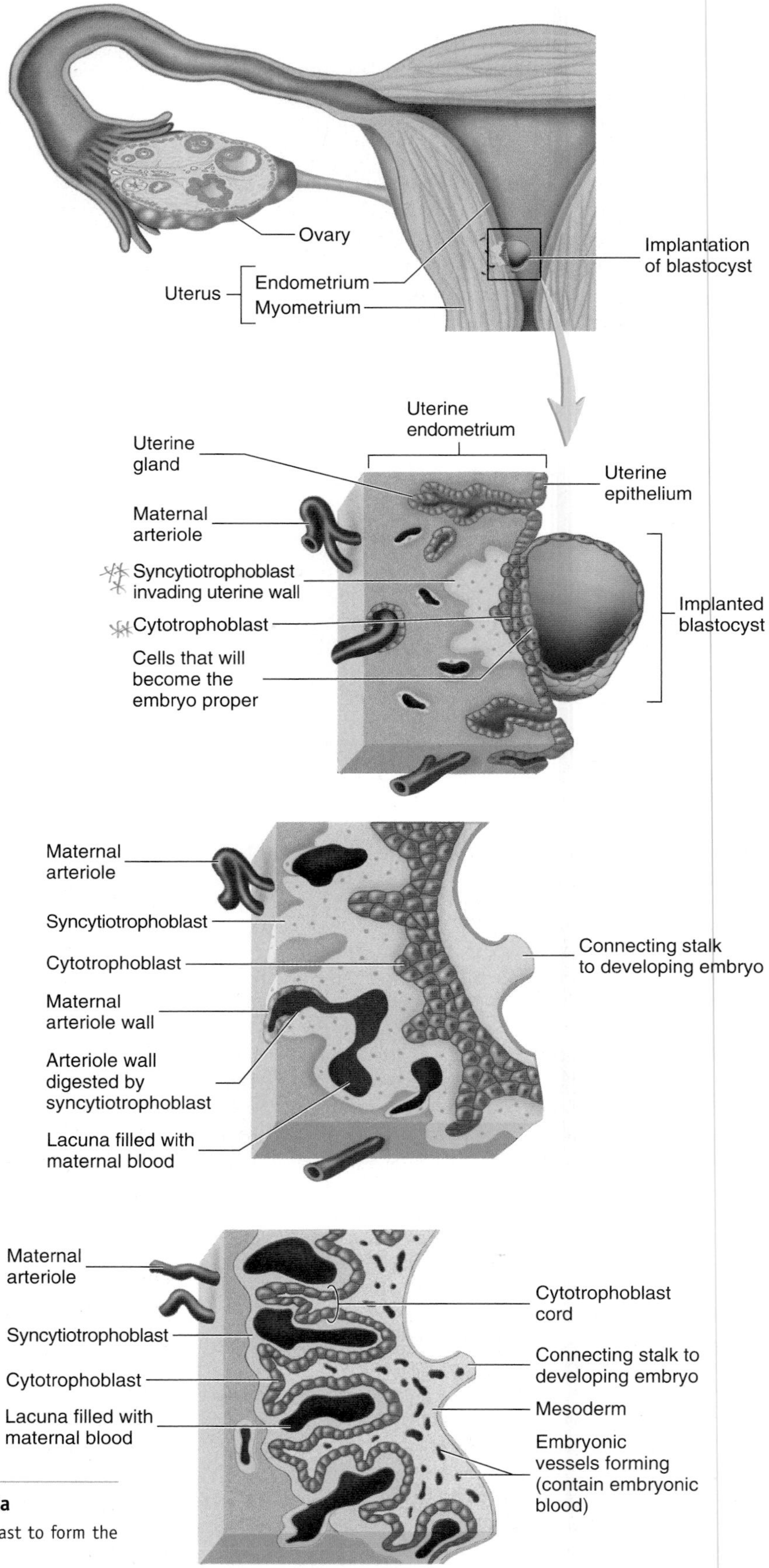

Maternal arteriole
Cytotrophoblast cord
Syncytiotrophoblast
Connecting stalk to developing embryo
Cytotrophoblast
Mesoderm
Lacuna filled with maternal blood
Embryonic vessels forming (contain embryonic blood)

PROCESS FIGURE 29.4 Formation of the Placenta

Implantation of the blastocyst and invasion of the trophoblast to form the placenta.

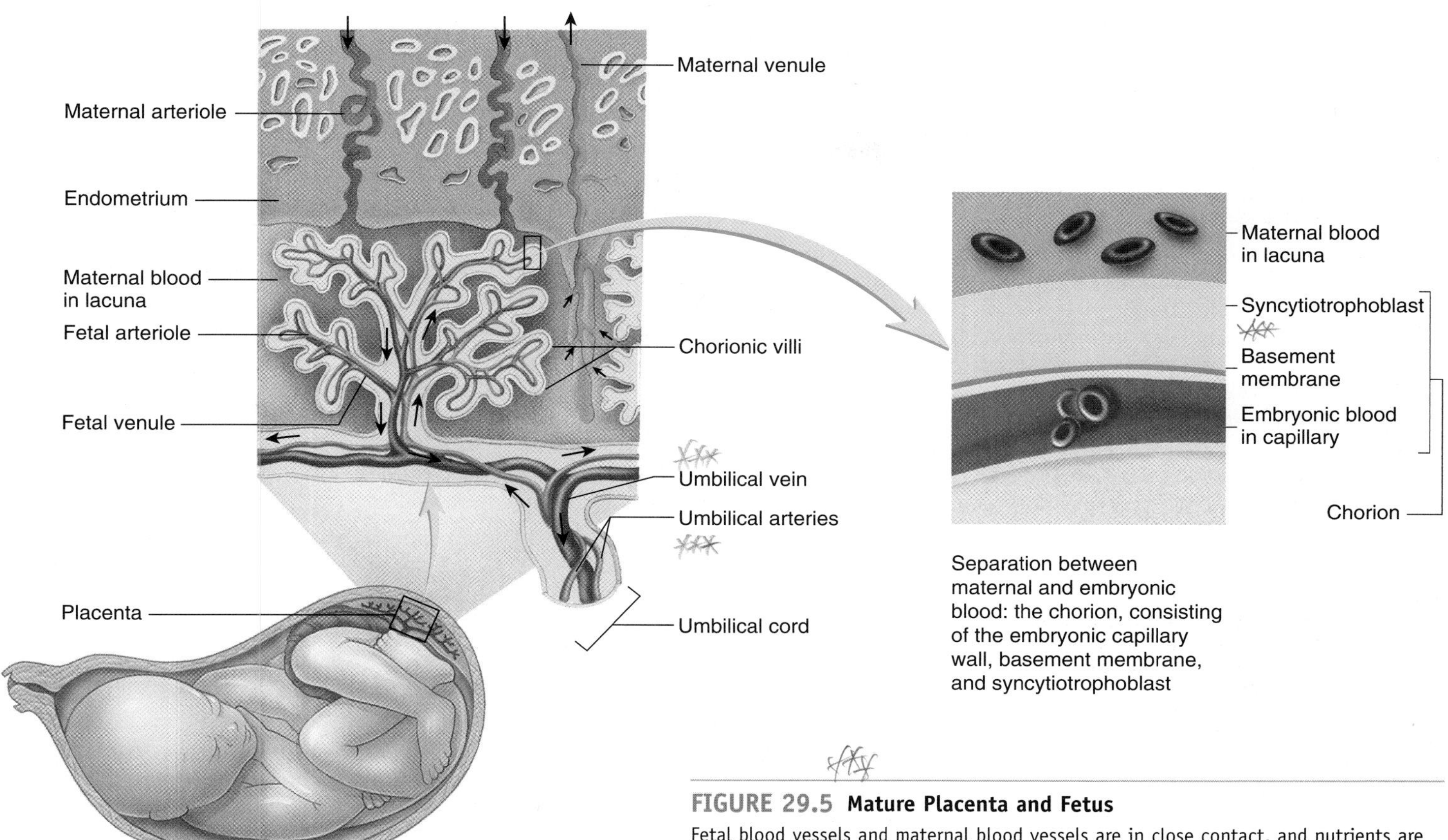

FIGURE 29.5 Mature Placenta and Fetus

Fetal blood vessels and maternal blood vessels are in close contact, and nutrients are exchanged between fetal and maternal blood, but fetal and maternal blood do not mix.

figure 29.4*c*). The lacunae are still connected to intact maternal vessels so that blood circulates from the maternal vessels through the lacunae. Cords of cytotrophoblast surround the syncytiotrophoblast and lacunae (see figure 29.4*d*). Branches, called **chorionic** (kō-rē-on′ik) **villi,** sprout from these cords and protrude into the lacunae, like fingers. The entire embryonic structure facing the maternal tissues is called the **chorion** (kō′rē-on). Embryonic mesoderm and blood vessels grow into the cords and villi as they protrude into the lacunae. In the mature placenta (figure 29.5), the cytotrophoblast disappears so that the embryonic blood supply is separated from the maternal blood supply by only the embryonic capillary wall, a basement membrane, and a thin layer of syncytiotrophoblast.

Placental Problems

If the blastocyst implants near the cervix, a condition called **placenta previa** (prē′vē-ă) occurs. In this condition, as the placenta grows, it may extend partially or completely across the internal cervical opening. As the fetus and placenta continue to grow and the uterus stretches, the region of the placenta over the cervical opening may tear, and hemorrhaging may occur. **Abruptio** (ab-rŭp′shē-ō) **placentae** is a tearing away of a normally positioned placenta from the uterine wall accompanied by hemorrhaging. Both of these conditions can result in miscarriage and can be life-threatening to the mother.

4 ***What are the events during the first week after fertilization? Define zygote, morula, blastocyst, blastocele, and pluripotent.***

5 ***Describe the trophoblast and inner cell mass, and explain what develops from each.***

6 ***Describe the implantation of the blastocyst and development of the placenta.***

Formation of the Germ Layers

After implantation, a new cavity called the **amniotic** (am-nē-ot′ik) **cavity** forms inside the inner cell mass and is surrounded by a layer of cells called the **amnion** (am′nē-on), or **amniotic sac.** Formation of the amniotic cavity causes part of the inner cell mass nearest the blastocele to separate as a flat disk of tissue called the **embryonic disk** (figure 29.6). This embryonic disk is composed of two layers of cells: an **ectoderm** (ek′tō-derm; outside layer) adjacent to the amniotic cavity and an **endoderm** (en′dō-derm; inside layer) on the side of the disk opposite the amnion. A third cavity, the **yolk sac,** forms inside the blastocele from the endoderm. The amniotic sac, yolk sac, and intervening double-layered embryonic disk can be thought of as resembling two balloons pushed together. One balloon represents the amniotic sac, and the other represents the yolk sac. The circular double layer of balloon wall where the two balloons are pressed together

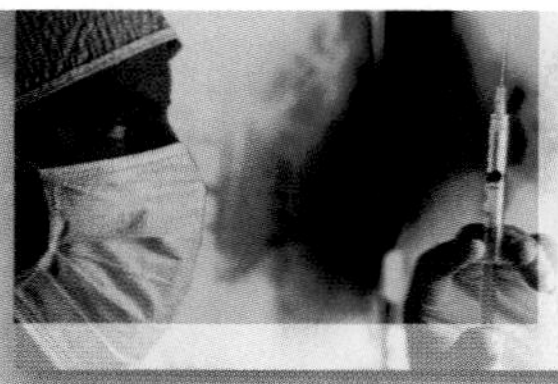

CLINICAL FOCUS

In Vitro Fertilization and Embryo Transfer

In a small number of women, normal pregnancy is not possible because of an anatomic or physiologic condition. In 87% of these cases, the uterine tubes are incapable of transporting the zygote to the uterus or of allowing sperm cells to reach the oocyte. In vitro fertilization and embryo transfer have made pregnancy possible in hundreds of such women since 1978. **In vitro fertilization (IVF)** involves the removal of secondary oocytes from a woman, placing the oocytes into a petri dish, and adding sperm cells to the dish to allow fertilization and early development to occur in vitro, which means in glass. **Embryo transfer** involves the removal of the developing embryo from the petri dish and its introduction into the uterus of a recipient female.

For IVF and embryo transfer to be accomplished, a woman is first injected with an LH-like substance, which causes more than one follicle to mature at one time. Just before the follicles rupture, the secondary oocytes are surgically removed from the ovary. The oocytes are then incubated in a dish and maintained at body temperature for 6 hours. Then sperm cells are added to the dish.

After 24–48 hours, when the zygotes have divided to form two- to eight-cell masses, several of the embryos are transferred to the uterus. Several embryos are transferred because only a small percentage of them survive. Implantation and subsequent development then proceed in the uterus as they would for natural implantation.

Typically, three embryos are transferred at a time. The rate of complications, such as multiple pregnancies, miscarriage, and prematurity, however, also increases with the greater numbers of embryos per transfer.

About one-third of transfers of three embryos end in multiple pregnancies. Of triplets born as a result of IVF, 64% require intensive care after birth, and 75% of quadruplets require intensive care, often for several weeks.

Prematurity from IVF pregnancies in the United Kingdom results in newborn mortality in 2.7% of cases, a rate three times that of natural pregnancies. As a result of the possible complications, no more than two to three embryos are transferred per IVF in the United Kingdom.

The success rate for IVF has dramatically increased through time. The success of IVF is dependent on many factors, including the reason for infertility, the age of the mother, and the expertise and experience of the physician. The current success rate for achieving pregnancy following IVF treatment is 31%. Of these pregnancies, 83% result in live births. This success rate has exceeded natural limits, because only 50% or less of natural fertilizations result in a successful delivery. The increased success rate probably results from the fact that embryos that appear abnormal are not transferred and that more than one embryo is transferred during IVF.

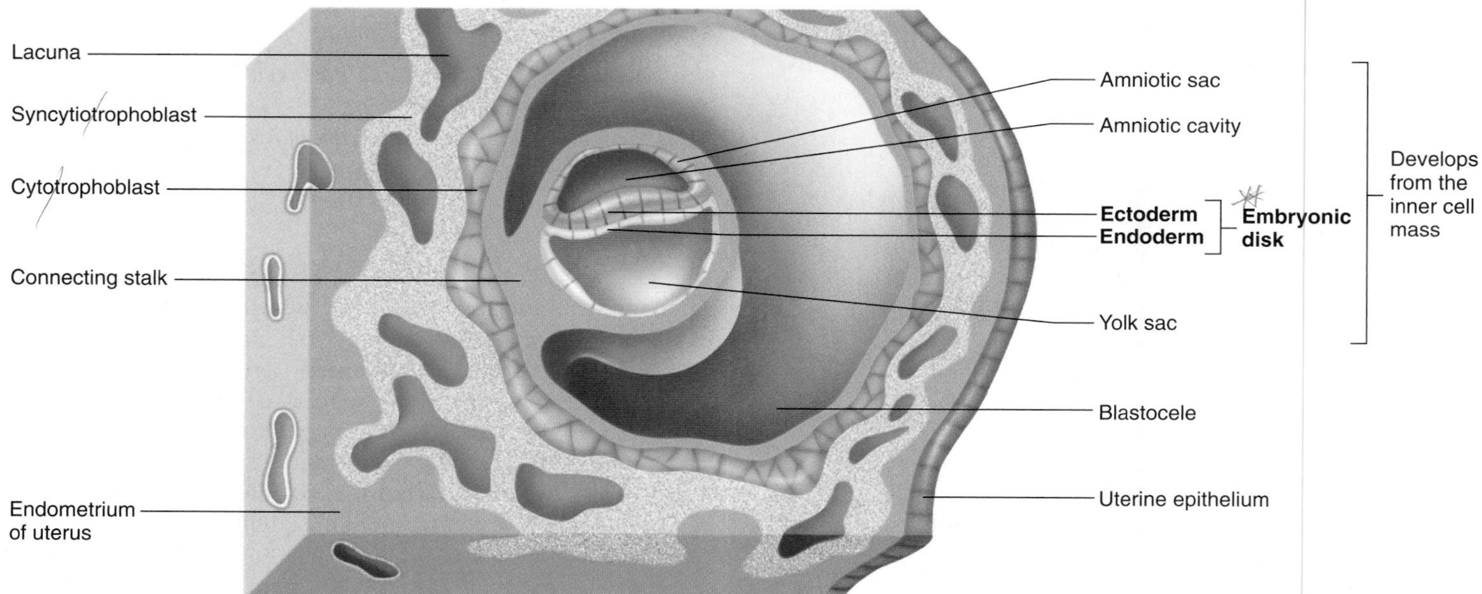

FIGURE 29.6 Embryonic Disk
Embryonic disk consisting of ectoderm (*blue*) and endoderm (*yellow*), with the amniotic cavity and yolk sac.

1. Cells in the surface ectoderm move toward the primitive streak and migrate through the streak (*blue arrow tails*).
2. Cells of the ectoderm that migrate through the primitive streak become mesodermal cells (*red arrows*).
3. The mesoderm (*red*) lies between the ectoderm (*blue*) and endoderm (*yellow*).

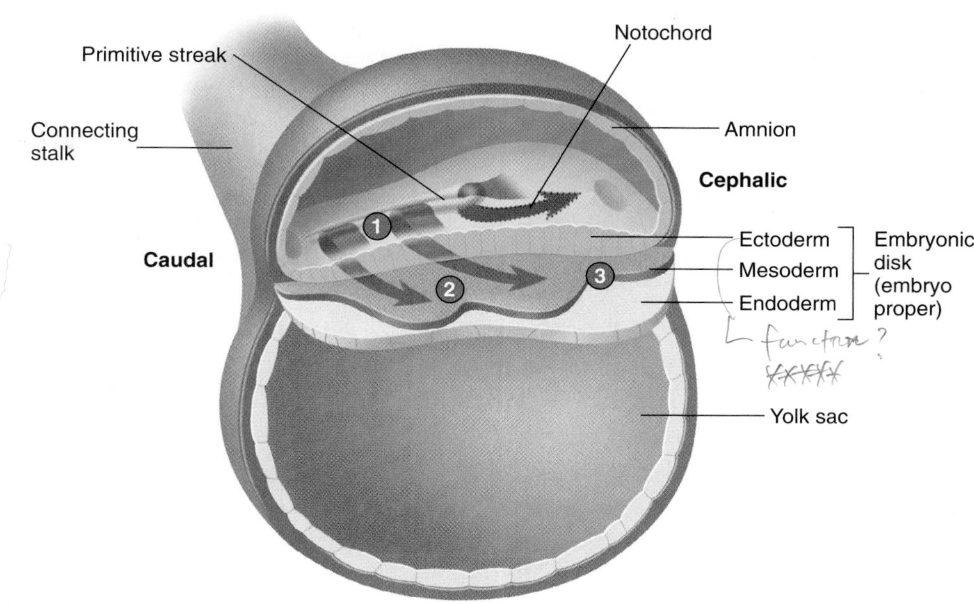

PROCESS FIGURE 29.7 Embryonic Disk with a Primitive Streak
The head of the embryo will develop over the notochord.

represents the embryonic disk. The amniotic sac eventually enlarges to surround the developing embryo, providing it with a protective fluid environment, the "bag of waters," where the embryo forms.

About 13 or 14 days after fertilization, the embryonic disk becomes a slightly elongated oval structure. This phase of development, known as **gastrulation,** involves the movement of cells, resulting in the formation of three distinct germ layers that will give rise to the many structures of the body. Proliferating cells of the ectoderm migrate toward the center and the caudal end of the disk, forming a thickened line called the **primitive streak.** Some ectoderm cells leave the ectoderm, migrate through the primitive streak, and emerge between the ectoderm and endoderm as a new germ layer, the **mesoderm** (mez′o-derm; middle layer; figure 29.7). These three germ layers, the ectoderm, mesoderm, and endoderm, are the beginning of the embryo proper. All tissues of the adult can be traced to them (table 29.1). A cordlike structure called the **notochord** extends from the cephalic end of the primitive streak.

The development of the germ layers and the subsequent development of the organ systems is heavily dependent on cell communication. Some of the cell communication depends on direct cell–cell contact, whereas other communication depends on diffusable molecules, such as **growth factors.** Two important families of growth factors are epidermal growth factors (EGF) and fibroblast growth factors (FGF).

7 ***Describe the formation of the germ layers and the role of the primitive streak.***

PREDICT 2

Predict the results of two primitive streaks forming in one embryonic disk. What if the two primitive streaks are touching each other?

Neural Tube and Neural Crest Formation

The ectoderm near the cephalic end of the primitive streak is stimulated about 18 days after fertilization to form a thickened **neural plate.** The lateral edges of the plate begin to rise, like two ocean waves coming together. These edges are called the **neural folds,** and a **neural groove** lies between them (figure 29.8). The underlying notochord stimulates the folding of the neural plate at the neural groove. The crests of the neural folds begin to meet in

TABLE 29.1 Germ Layer Derivatives

Ectoderm	Mesoderm
Epidermis of skin	Dermis of skin
Tooth enamel	Circulatory system
Lens and cornea of eye	Parenchyma of glands
Outer ear	Muscle
Nasal cavity	Bones (except facial)
Anterior pituitary	Microglia
Neuroectoderm	Kidneys
Brain and spinal cord	**Endoderm**
Somatic motor neurons	Lining of gastrointestinal tract
Preganglionic autonomic neurons	Lining of lungs
Neuroglia cells (except microglia)	Lining of hepatic, pancreatic, and other exocrine ducts
Posterior pituitary	Urinary bladder
Neural crest cells	Thymus
Melanocytes	Thyroid
Sensory neurons	Parathyroid
Postganglionic autonomic neurons	Tonsils
Adrenal medulla	
Facial bones	
Teeth (dentin, pulp, and cementum) and gingiva	
A few skeletal muscles in head	

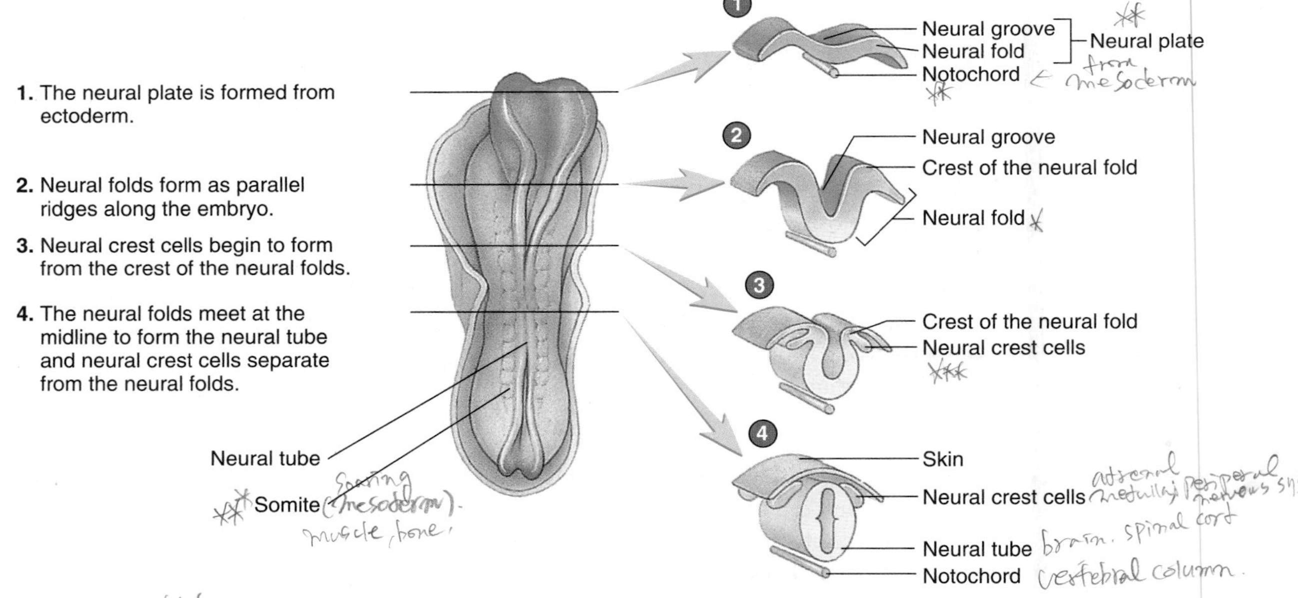

PROCESS FIGURE 29.8 Formation of the Neural Tube

The neural folds come together in the midline and fuse to form a neural tube. This fusion begins in the center and moves both cranially and caudally. The embryo shown is about 21 days after fertilization. The insets to the *right* show progressive closure of the neural tube. The *lines* indicate locations of cross sections.

the midline and fuse into a **neural tube,** which has completely closed by 26 days. The neural tube becomes the brain and the spinal cord, and the cells of the neural tube are called **neuroectoderm** (see table 29.1).

As the neural folds come together and fuse, a population of cells breaks away from the neuroectoderm all along the crests of the folds. In the body of the embryo, **neural crest cells** migrate along one of three distinct routes as they leave the neural folds. Those that migrate down along the side of the developing neural tube become autonomic ganglia neurons, adrenal medullary cells, or enteric nervous system neurons. Neural crest cells that migrate into the somites (see "Somite Formation" below) become sensory ganglia neurons. Those that migrate laterally between the somites and ectoderm become melanocytes. In the head, neural crest cells perform additional functions; they contribute to the skull, the dentin of teeth, a few small skeletal muscles, and general connective tissue. Because neural crest cells in the head give rise to many of the same tissues as the mesoderm in the head and trunk, the general term **mesenchyme** (mez′en-kīm) is sometimes applied to cells of either neural crest or mesoderm origin.

Malformations

During the first 2 weeks of development, the embryo is quite resistant to outside influences that may cause malformations. Factors that adversely affect the embryo at this age are more likely to kill it. Between 2 weeks and the next 4–7 weeks (depending on the structure considered), the embryo is more sensitive to outside influences that cause malformations than at any other time.

8 ***Describe the formation of the neural tube and neural crest.***

Somite Formation

As the neural tube develops, the mesoderm immediately adjacent to the tube forms distinct segments called **somites** (sō′mītz; see figure 29.8). In the head, the first few somites never become clearly divided but develop into indistinct, segmented structures called **somitomeres.** The somites and somitomeres eventually give rise to a part of the skull, the vertebral column, and skeletal muscle. Most of the head muscles are derived from the somitomeres.

9 ***What is a somite?***

Formation of the Gut and Body Cavities

At the same time its neural tube is forming, the embryo itself is becoming a tube along the upper part of the yolk sac. The **foregut** and **hindgut** develop as the cephalic and caudal ends of the yolk sac are separated from the main yolk sac. This is the beginning of the digestive tract (figure 29.9*a*). The developing digestive tract pinches off from the yolk sac as a tube but remains attached in the center to the yolk sac by a yolk stalk.

The ends of the foregut and hindgut (figure 29.9*b*) are in close relationship to the overlying ectoderm and form membranes called the oropharyngeal membrane and the cloacal membrane, respectively. The **oropharyngeal membrane** opens to form the mouth, and the **cloacal membrane** opens to form the urethra and anus. Thus, the digestive tract becomes a tube that opens to the outside at both ends.

A considerable number of **evaginations** (ē-vaj-i-nā′shŭnz; outpocketings) occur along the early digestive tract (figure 29.9*c*). The first to form is the allantois (see figure 29.9*b*), part of which will form the urinary bladder. Other evaginations develop into structures such as the anterior pituitary, the thyroid gland, the lungs, the liver, the pancreas, and the urinary bladder. At the same time, solid bars of tissue known as **branchial arches** (see figure 29.9*c;* figure 29.10) form along the lateral sides of the head, and the sides of the foregut expand as pockets between the branchial arches. The central expanded foregut is called the **pharynx,** and the pockets along both sides of the pharynx are called **pharyngeal pouches.** Adult derivatives of the pharyngeal pouches include the auditory tubes, tonsils, thymus, and parathyroids.

At about the same time the gut is developing, a series of isolated cavities starts to form within the embryo, thus beginning development of the **celom** (sē′lom; see figure 29.9), or body cavities. The most cranial group of cavities enlarges and fuses to form the **pericardial cavity.** Shortly thereafter, the celomic cavity extends toward the caudal end of the embryo as the **pleural** and **peritoneal cavities.** Initially, all three of these cavities are continuous, but they eventually separate into three distinct adult cavities (see chapter 1).

10 ***Describe the formation of the gut and body cavities.***

Limb Bud Development

Arms and legs first appear at about 28 days as limb buds (see figure 29.10). The **apical ectodermal ridge,** a specialized thickening of the ectoderm, develops on the lateral margin of each limb bud and stimulates its outgrowth. As the buds elongate, limb tissues are laid down in a proximal-to-distal sequence. For example, in the upper limb, the arm is formed before the forearm, which is formed before the hand.

11 ***Describe limb formation. What does a proximal-to-distal growth sequence mean?***

Development of the Face

The face develops by fusion of five embryonic structures: the **frontonasal process,** which forms the forehead, nose, and midportion of the upper jaw and lip; two **maxillary processes,** which form the

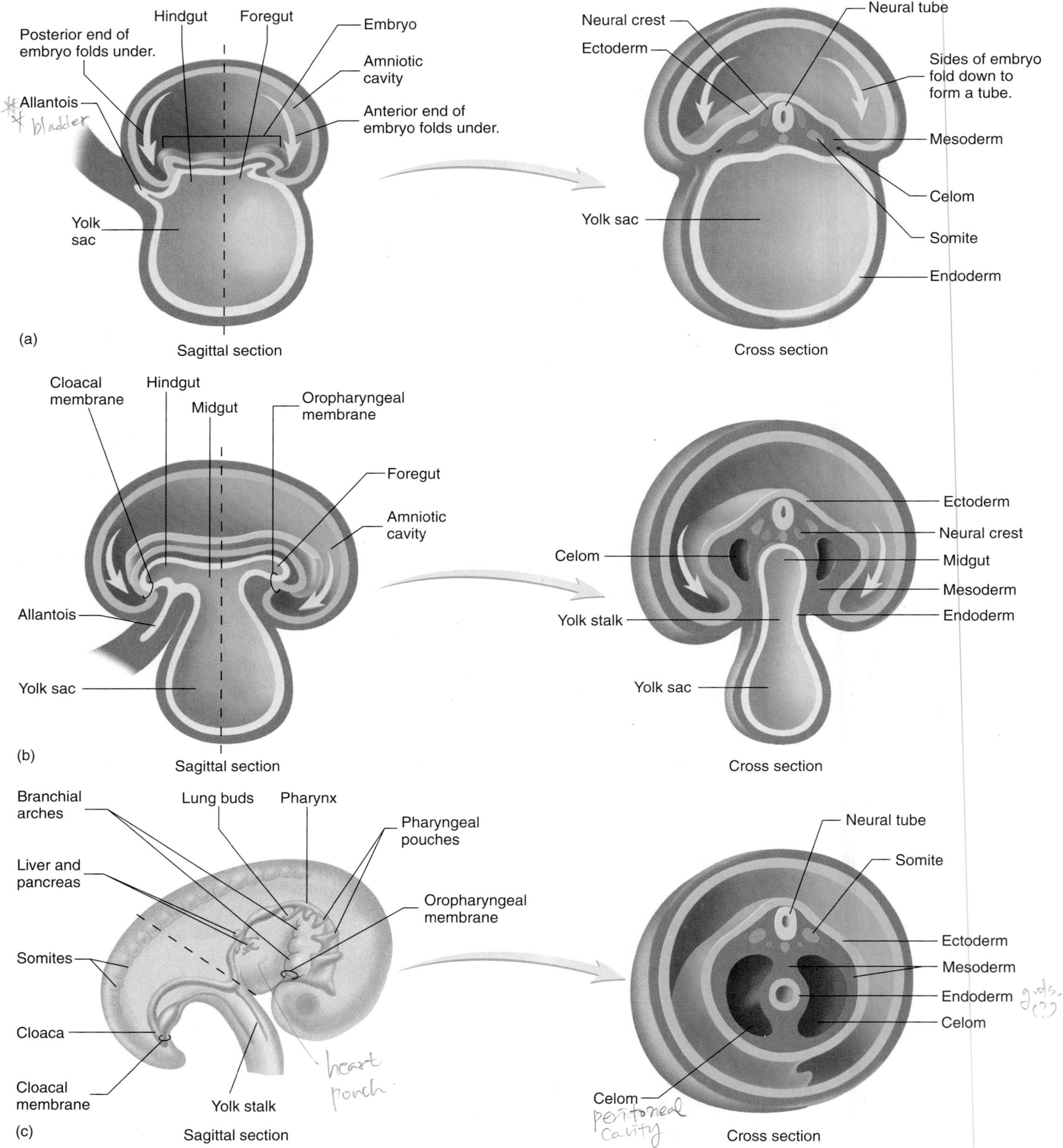

FIGURE 29.9 Formation of the Digestive Tract

Blue arrows show the folding of the digestive tract into a tube. *Dotted lines* show the plane of the section from which cross sections were taken. (*a*) Twenty days after fertilization. (*b*) Twenty-five days after fertilization. (*c*) Thirty days after fertilization. Evaginations are identified along the pharynx and digestive tract.

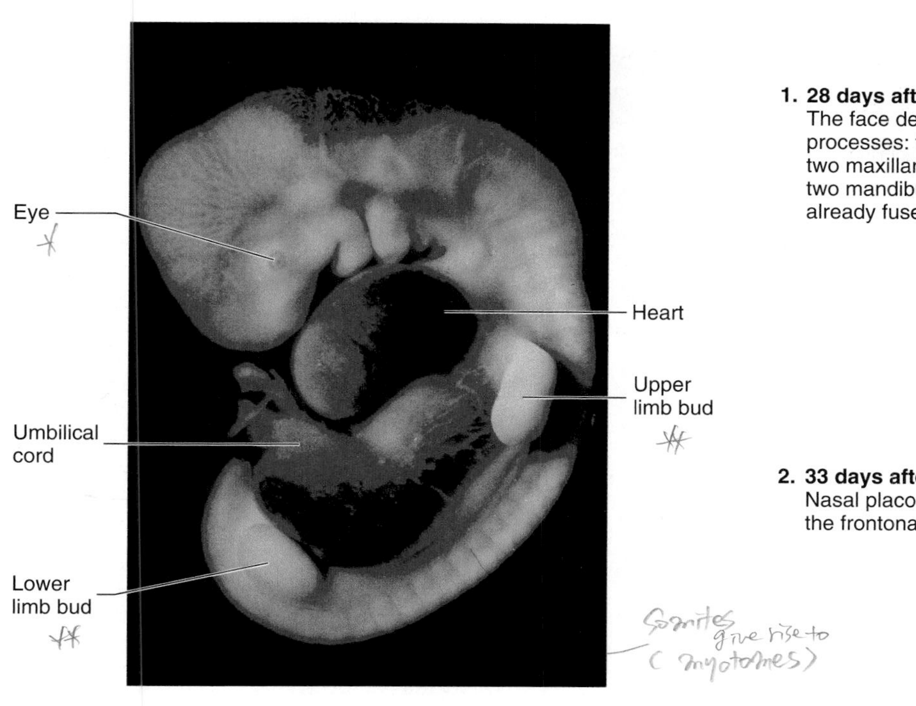

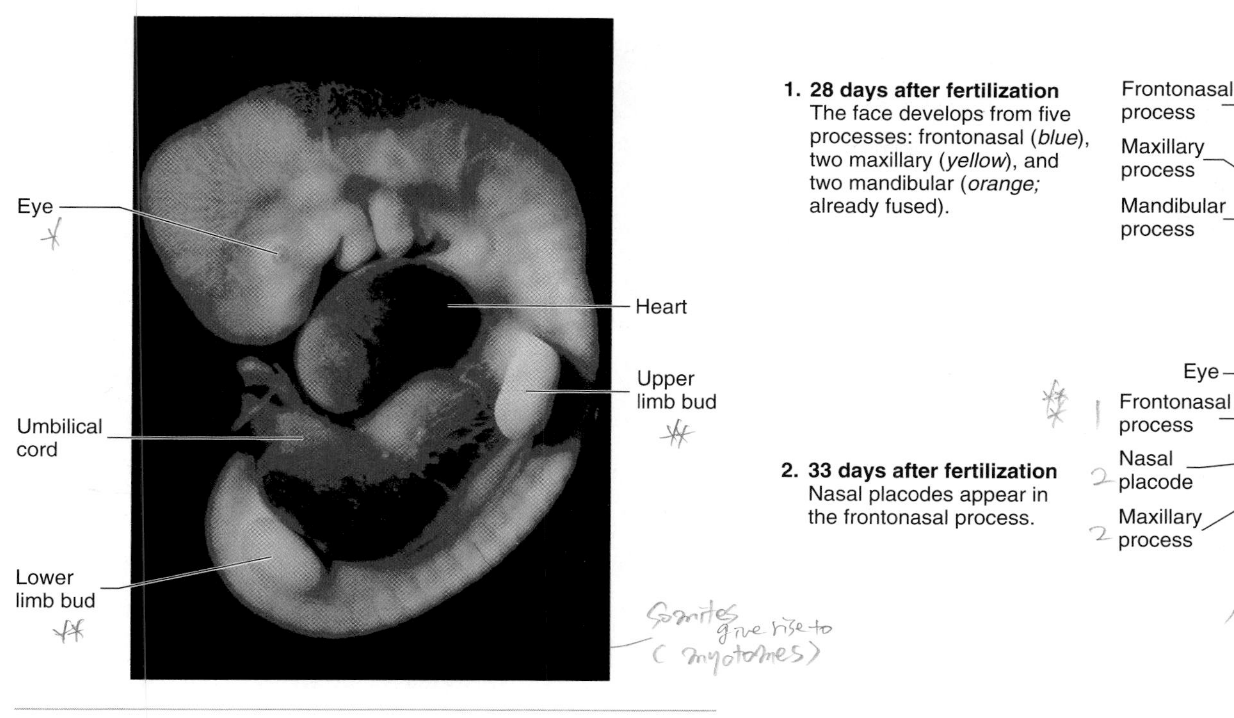

FIGURE 29.10 Human Embryo 35 Days After Fertilization

lateral parts of the upper jaw and lip; and two **mandibular processes,** which form the lower jaw and lip (figure 29.11, step 1). **Nasal placodes** (plak′ōdz), which develop at the lateral margins of the frontonasal process, develop into the nose and the center of the upper jaw and lip (figure 29.11, step 2).

As the brain enlarges and the face matures, the nasal placodes approach each other in the midline. The medial edges of the placodes fuse to form the midportion of the upper jaw and lip (figure 29.11, steps 3 and 4). This part of the frontonasal process is between the two maxillary processes, which are expanding toward the midline, and fuses with them to form the upper jaw and lip, known as the **primary palate** (see figure 29.11, step 4).

Cleft Lip

A **cleft lip** results from failure of the frontonasal and one or both maxillary processes to fuse (see figure 29.11). Because the frontonasal process is a midline structure that normally fuses with the two lateral maxillary processes during formation of the primary palate, cleft lips usually do not occur in the midline but to one side (or both sides). The cleft can vary in severity from a slight indentation in the lip to a fissure that extends from the mouth to the nares (nostril).

At about the same time that the primary palate is forming, the lateral edges of the nasal placodes fuse with the maxillary processes to close off the groove extending from the mouth to the eye (figure 29.11, steps 4 and 5). On rare occasions, these structures fail to meet, resulting in a facial cleft extending from the mouth to the eye. The inferior margins of the maxillary processes fuse with the superior margins of the mandibular processes to decrease the size of the mouth. All of the previously described fusions and the growth of the brain give the face a recognizably "human" appearance by about 50 days.

1. 28 days after fertilization
The face develops from five processes: frontonasal (*blue*), two maxillary (*yellow*), and two mandibular (*orange; already fused*).

Frontonasal process
Maxillary process
Mandibular process

2. 33 days after fertilization
Nasal placodes appear in the frontonasal process.

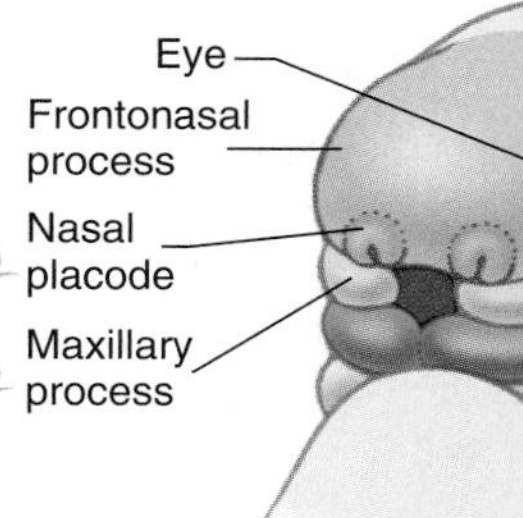

3. 40 days after fertilization
Maxillary processes extend toward the midline. The nasal placodes also move toward the midline and fuse with the maxillary processes to form the jaw and lip.

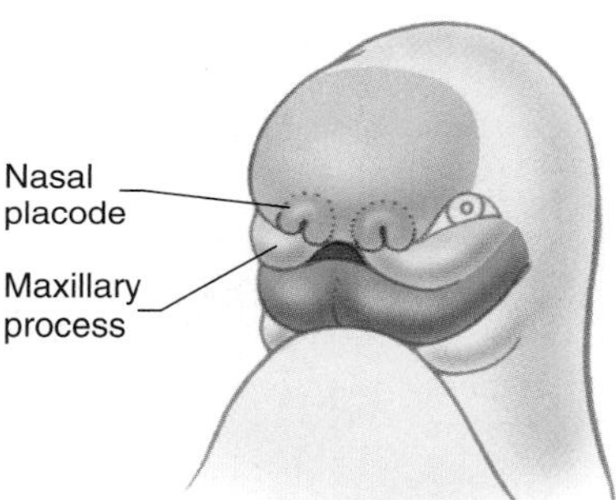

4. 48 days after fertilization
Continued growth brings structures more toward the midline.

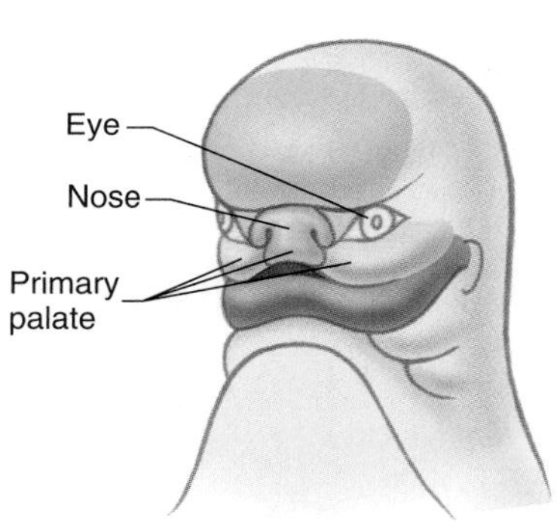

5. 14 weeks after fertilization
Colors show the contributions of each process to the adult face.

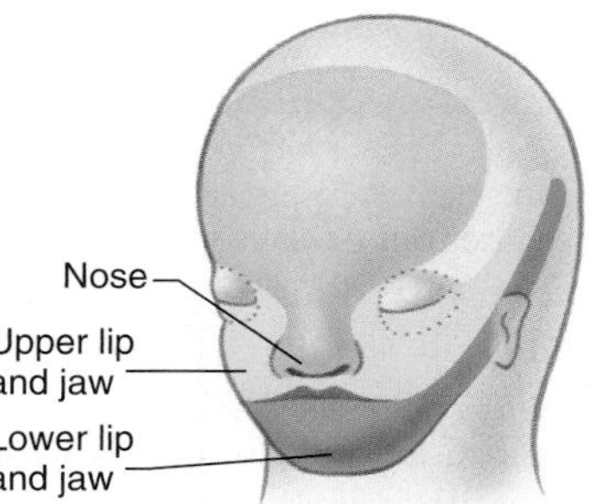

PROCESS FIGURE 29.11 Development of the Face

TABLE 29.2 Development of the Organ Systems

	Age (Days Since Fertilization)					
	1–5	6–10	11–15	16–20	21–25	26–30
General Features	Fertilization Morula Blastocyst	Blastocyst implants	Primitive streak Three germ layers	Neural plate	Neural tube closed	Limb buds and other "buds" appear.
Integumentary System			Ectoderm Mesoderm			Melanocytes from neural crest
Skeletal System			Mesoderm		Neural crest (will form facial bones)	Limb buds
Muscular System			Mesoderm	Somites begin to form.		Somites all present
Nervous System			Ectoderm	Neural plate	Neural tube complete Neural crest Eyes and ears begin.	Lens begins to form.
Endocrine System			Ectoderm Mesoderm Endoderm	Thyroid begins to develop.		Parathyroids appear.
Cardiovascular System			Mesoderm	Blood islands form. Two heart tubes	Single-tubed heart begins to beat.	Interatrial septum begins to form.
Lymphatic System			Mesoderm			Thymus appears.
Respiratory System			Mesoderm Endoderm		Diaphragm begins to form.	Trachea forms as single bud. Lung buds (primary bronchi)
Digestive System			Mesoderm Endoderm		Neural crest (will form tooth dentin) Foregut and hindgut form.	Liver and pancreas appear as buds. Tongue bud appears.
Urinary System			Mesoderm Endoderm		Pronephros develops. Allantois appears.	Mesonephros appears.
Reproductive System			Mesoderm Endoderm		Primordial germ cells on yolk sac	Mesonephros appears. Genital tubercle forms.

The roof of the mouth, known as the **secondary palate,** begins to form as vertical shelves, which swing to a horizontal position and begin to fuse with each other at about 56 days of development. Fusion of the entire palate is not completed until about 90 days. If the secondary palate does not fuse, a midline fissure in the roof of the mouth, called a **cleft palate,** results. A cleft palate can range in severity from a slight cleft of the uvula (see figure 24.6) to a fissure extending the entire length of the palate. A cleft lip and cleft palate can occur together, forming a continuous fissure.

12 ***Describe the processes involved in the formation of the face. What clefts may result if these processes fail to fuse?***

Development of the Organ Systems

The major organ systems appear and begin to develop during the embryonic period. The period between 14 and 60 days is therefore called the period of organogenesis (table 29.2).

Age (Days Since Fertilization)					
31–35	36–40	41–45	46–50	51–55	56–60
Hand and foot plates on limbs	Fingers and toes appear. Lips formed Embryo 15 mm	External ear forming Embryo 20 mm	Embryo 25 mm	Limbs elongate to a more adult relationship. Embryo 35 mm	Face is distinctly human in appearance.
Sensory receptors appear in skin.		Collagen fibers clearly present in skin		Extensive sensory endings in skin	
Mesoderm condensation in areas of future bone	Cartilage in site of future humerus	Cartilage in site of future ulna and radius	Cartilage in site of hand and fingers		Ossification begins in clavicle and then in other bones.
Muscle precursor cells enter limb buds.			Functional muscle		Nearly all muscles appear in adult form.
Nerve processes enter limb buds.		External ear forming Olfactory nerves begin to form.		Semicircular canals in inner ear complete	Eyelids form. Cochlea in inner ear complete
Pituitary appears as evaginations from brain and mouth.	Gonadal ridges form. Adrenal glands forming		Pineal body appears.	Thyroid gland in adult position and attachment to tongue lost	Anterior pituitary loses its connection to the mouth.
Interventricular septum begins to form.		Interventricular septum complete	Interatrial septum complete but still has opening until birth		
Large lymphatic vessels form in neck.	Spleen appears.			Adult lymph pattern forms.	
Secondary bronchi to lobes form.	Tertiary bronchi to bronchopulmonary segments form.		Tracheal cartilage begins to form.		
Oropharyngeal membrane ruptures.		Secondary palate begins to form. Tooth buds begin to form.			Secondary palate begins to fuse (fusion complete by 90 days).
Metanephros begins to develop.				Mesonephros degenerates.	Anal portion of cloacal membrane ruptures.
	Gonadal ridges form.	Primordial germ cells enter gonadal ridges.	Paramesonephric ducts appear.		Uterus forming Beginning of differentiation of external genitalia in male and female

Skin

The **epidermis** of the skin is derived from ectoderm, and the **dermis** is derived from mesoderm, or from neural crest cells in the case of the face. Nails, hair, and glands develop from the epidermis (see chapter 5). Melanocytes and sensory receptors in the skin are derived from neural crest cells.

Skeleton

The skeleton develops from either mesoderm or the neural crest cells through intramembranous or endochondral bone formation (see chapter 6). The bones of the face develop from neural crest cells, whereas the rest of the skull, the vertebral column, and the ribs develop from somite- or somitomere-derived mesoderm. The appendicular skeleton develops from limb bud mesoderm.

Muscle

Myoblasts (mī′ō-blastz) are the early, embryonic cells that give rise to skeletal muscle fibers. Myoblasts migrate from somites or somitomeres to sites of future muscle development, where they continue to divide and begin to fuse and form multinuclear cells called **myotubes,** which enlarge to become the muscle fibers of the skeletal muscles. Shortly after myotubes form, nerves grow into the area and innervate the developing muscle fibers. After the basic form of each muscle is established, an increase in the number of muscle fibers causes continued muscle growth. The total number of muscle fibers is established before birth and remains relatively constant thereafter. Muscle enlargement after birth results from an increase in the size of individual fibers.

Nervous System

The nervous system is derived from the neural tube and neural crest cells. Neural tube closure begins at about 21 days of development in the upper cervical region and proceeds into the head and down the spinal cord. Soon after the neural tube has closed at about 25 days of development, the part of the neural tube that becomes the brain begins to expand and develops a series of pouches (see figure 13.14). The central cavity of the neural tube becomes the ventricles of the brain and the central canal of the spinal cord.

Neural Tube Defects

Anencephaly (an′en-sef′ă-lē; no brain) is a birth defect in which much of the brain fails to form. It results when the neural tube fails to close in the region of the head. A baby born with anencephaly cannot survive. **Spina bifida** (spī′nă bi′fi-dă; split spine) is a general term describing defects of the spinal cord, vertebral column or both (figure A). Spina bifida can range from a simple defect with no clinical manifestations and with one or more vertebral spinous processes split or missing to a more severe defect that results in paralysis of the limbs or the bowels and bladder, depending on where the defect occurs.

The inclusion of **folic acid,** the B vitamin folate, in the diet of a woman during the early stages of her pregnancy significantly reduces the risk for neural tube defects in her developing embryo.

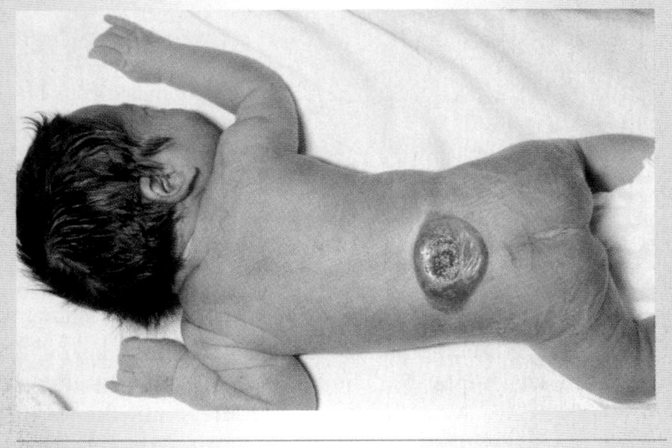

FIGURE A **Spina Bifida**

The neuron cell bodies of somatic motor neurons and preganglionic neurons of the autonomic nervous system, which provide axons to the peripheral nervous system, are located within the neural tube. Sensory neurons and postganglionic neurons of the autonomic nervous system are derived from neural crest cells.

Alcohol and Cigarette Smoke

A number of drugs and other chemicals are known to affect the embryo and fetus during development. The two most common are alcohol and cigarette smoke. Alcohol consumption can result in **fetal alcohol syndrome,** which includes decreased mental function. Though excessive alcohol consumption, such as alcoholism and binge drinking, are known to cause fetal alcohol syndrome, research results are inconsistent about the effects of lower levels of consumption. Most physicians recommend that alcohol not be consumed at all during pregnancy. Exposure of the fetus to **cigarette smoke** throughout pregnancy can stunt the physical growth and mental development of the fetus.

Special Senses

The **olfactory bulbs** and **nerves** develop as an evagination from the telencephalon (see figure 13.14). The eyes develop as evaginations from the diencephalon. Each evagination elongates to form an **optic stalk,** and a bulb called the **optic vesicle** develops at its terminal end. The optic vesicle reaches the side of the head and stimulates the overlying ectoderm to thicken into a **lens.** The sensory part of the ear appears as an ectodermal thickening, or placode, that invaginates and pinches off from the overlying ectoderm.

Endocrine System

An evagination from the floor of the diencephalon forms the **posterior pituitary gland.** The **anterior pituitary gland** develops from an evagination of ectoderm in the roof of the embryonic oral cavity and grows toward the floor of the brain. It eventually loses its connection with the oral cavity and becomes attached to the posterior pituitary gland (see chapter 18).

The **thyroid gland** originates as an evagination from the floor of the pharynx in the region of the developing tongue and moves into the lower neck, eventually losing its connection with the pharynx. The **parathyroid glands,** which are derived from the third and fourth pharyngeal pouches, migrate inferiorly and become associated with the thyroid gland.

The **adrenal medulla** arises from neural crest cells and consists of specialized postganglionic neurons of the sympathetic division of the autonomic nervous system (see chapter 16). The **adrenal cortex** is derived from mesoderm.

The **pancreas** originates as two evaginations from the duodenum, which come together to form a single gland (see figure 29.9*c*).

Circulatory System

The heart develops from two endothelial tubes (figure 29.12, step 1), which fuse into a single, midline heart tube (figure 29.12, step 2). Blood vessels and blood cells form from blood islands on the surface of the yolk sac and inside the embryo. **Blood islands** are small masses of mesoderm that become blood vessels on the outside and blood cells on the inside. These islands expand and

1. 20 days after fertilization
At this age, the heart consists of two parallel tubes.

2. 22 days after fertilization
The two parallel tubes have fused to form one tube. This tube bends as it elongates (*blue arrows* suggest the direction of bending) within the confined space of the pericardium.

3. 31 days after fertilization
The interatrial septum (septum primum, *green*) and the interventricular septum grow toward the center of the heart.

4. 35 days after fertilization
The interventricular septum is nearly complete. A foramen opens in the septum primum (*green*) as the septum secundum begins to form (*blue*).

5. The final embryonic condition of the interatrial septum
Blood from the right atrium can flow through the foramen ovale into the left atrium. After birth, as blood begins to flow in the other direction, the left side of the interatrial septum is forced against the right side, closing the foramen ovale.

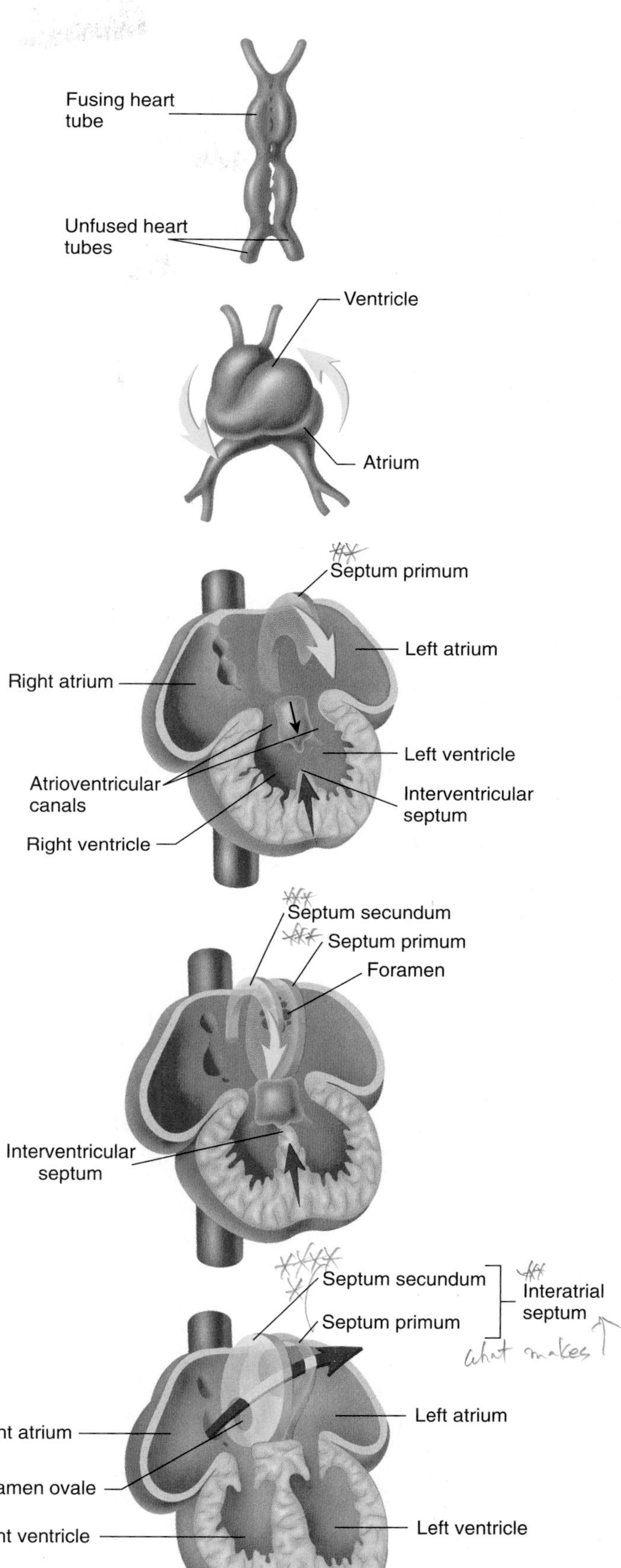

PROCESS FIGURE 29.12 Development of the Heart

fuse to form the circulatory system. A series of dilations appears along the length of the primitive heart tube, and four major regions can be identified: the **sinus venosus,** the site where blood enters the heart; a single **atrium;** a single **ventricle;** and the **bulbus cordis,** where blood exits the heart (see figure 29.12, step 2).

The elongating heart, confined within the pericardium, becomes bent into a loop, the apex of which is the ventricle (see figure 29.12, step 2). The major chambers of the heart, the atrium and the ventricle, expand rapidly. The right part of the sinus venosus becomes absorbed into the atrium, and the bulbus cordis is absorbed into the ventricle. The embryonic sinus venosus initiates contraction at one end of the tubular heart. Later in development, part of the sinus venosus becomes the sinoatrial node, which is the adult pacemaker.

PREDICT 3

What would happen if the sinus venosus did not contract before other areas of the primitive heart?

The **interatrial septum,** which separates the two atria in the adult heart, is formed from two parts: the **septum primum** (primary septum) and the **septum secundum** (secondary septum). An opening in the interatrial septum called the **foramen ovale** (ō-val′ē) connects the two atria and allows blood to flow from the right to the left atrium in the embryo and fetus. An **interventricular septum** (figure 29.12, steps 3–5) develops, which divides the single ventricle into two chambers.

Heart Defects

If the septum secundum fails to grow far enough or if the foramen secundum becomes too large, an **atrial septal defect (ASD)** occurs, allowing blood to flow from the left atrium to the right atrium in the newborn. If the interventricular septum does not grow enough to completely separate the ventricles, a **ventricular septal defect (VSD)** results. VSDs are more common than ASDs. Both ASDs and VSDs result in abnormal heart sounds called **heart murmurs.** Blood passes through the ASD and VSD from the left to right side of the heart. The right side of the heart usually hypertrophies. In many cases, septal defects are not serious. In severe cases of VSD (Eisenmenger syndrome), the increased pressure in the pulmonary blood vessels and a decreased blood flow through the systemic blood vessels result in pulmonary edema, cyanosis (a bluish color due to deficient blood oxygenation), or heart failure.

Respiratory System

The lungs begin to develop as a single midline evagination from the foregut in the region of the future esophagus. This evagination branches to form two **lung buds** (figure 29.13, step 1). The lung buds elongate and branch, first forming the bronchi that project to the lobes of the lungs (figure 29.13, step 2) and then the bronchi that project to the bronchopulmonary segments of the lungs (figure 29.13, step 3). This branching continues (figure 29.13, step 4) until, by the end of the sixth month, about 17 generations of branching have occurred. Even after birth, some branching continues as the lungs grow larger, and in the adult about 24 generations of branches have been established.

1. 28 days after fertilization
A single bud forms and divides into two buds, which will become the lungs and primary bronchi.

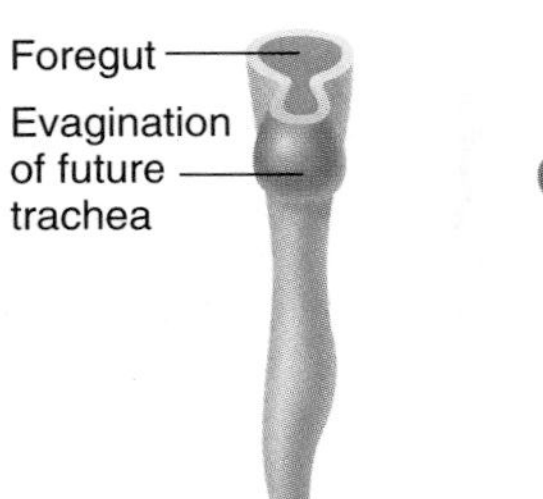

2. 32 days after fertilization
Primary bronchi branch to form secondary bronchi, which supply the lung lobes.

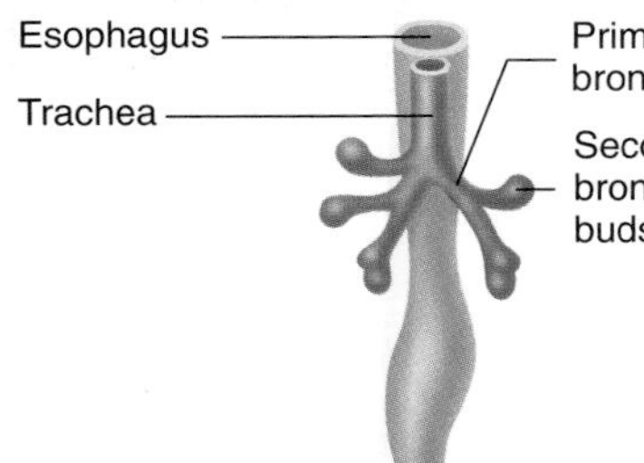

3. 35 days after fertilization
Secondary bronchi branch to form tertiary bronchi, which supply the bronchopulmonary segments.

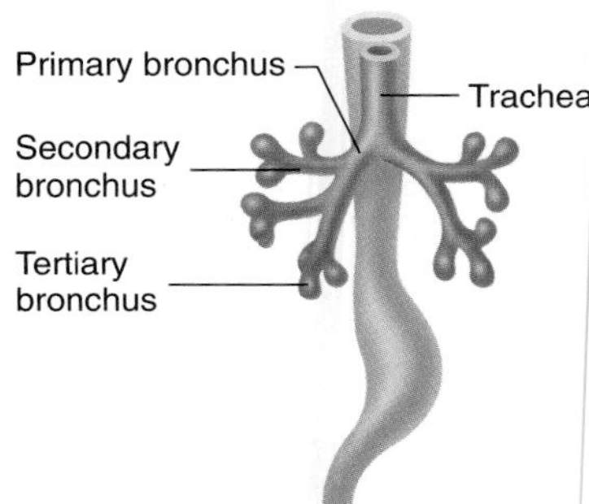

4. 50 days after fertilization
The branching will continue to eventually form the extensive respiratory passages in the lungs.

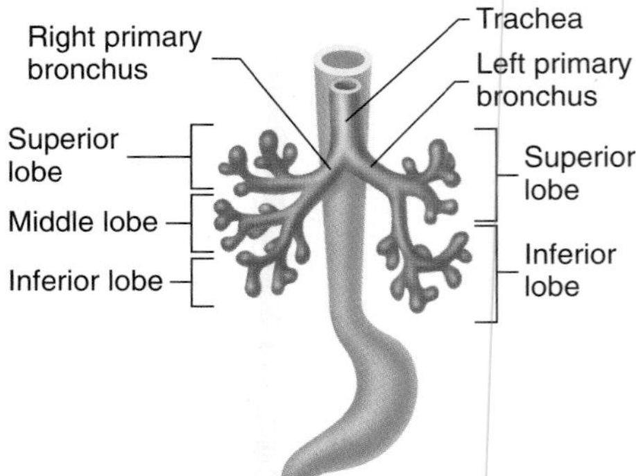

PROCESS FIGURE 29.13 Development of the Lung

Urinary System

The kidneys develop from mesoderm located between the somites and the lateral part of the embryo. About 21 days after fertilization, the mesoderm in the cervical region differentiates into a structure called the **pronephros** (the most forward or earliest kidney; figure 29.14*a*), which consists of a duct and simple tubules connecting the duct to the open celomic cavity. This type of kidney is the functional adult kidney in some lower chordates, but it is probably not functional in the human embryo and soon disappears.

The **mesonephros** (middle kidney; see figure 29.14*a*) is a functional organ in the embryo. It consists of a duct, which is a caudal extension of the pronephric duct, and a number of minute tubules, which are smaller and more complex than those of the pronephros. One end of each tubule opens into the mesonephric duct, and the other end forms a glomerulus (see chapter 26).

As the mesonephros is developing, the caudal end of the hindgut begins to enlarge to form the **cloaca** (klō-ā′kă; sewer), the common junction of the digestive, urinary, and genital systems (figure 29.14*b*). A **urorectal septum** divides the cloaca into two parts: a digestive part called the **rectum** and a urogenital part called the **urethra** (figure 29.14*c*). The cloaca has two tubes associated with it: the hindgut and the **allantois** (ă-lan′tō-is; sausage), which is a blind tube extending into the umbilical cord (see figures 29.9 and 29.14). The part of the allantois nearest the cloaca enlarges to form the urinary bladder, and the remainder, which is from the bladder to the umbilicus, forms the median umbilical ligament (figure 29.14*d*).

The mesonephric duct extends caudally as it develops and eventually joins the cloaca. At the point of junction, another tube, the **ureter,** begins to form. Its distal end enlarges and branches to form the duct system of the adult kidney, called the **metanephros** (last kidney), which takes over the function of the degenerating mesonephros.

Reproductive System

The male and female gonads appear as **gonadal ridges** along the ventral border of each mesonephros (figure 29.15*a*). **Primordial germ**

(a) The three parts of the developing kidney: pronephros, mesonephros, metanephros

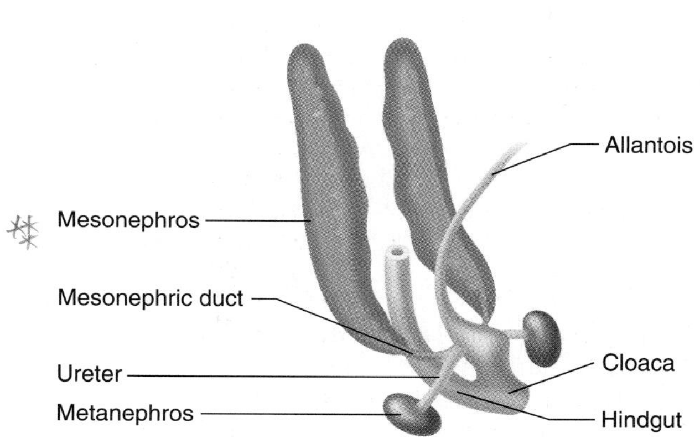

(b) The metanephros (adult kidney) enlarges as the mesonephros degenerates.

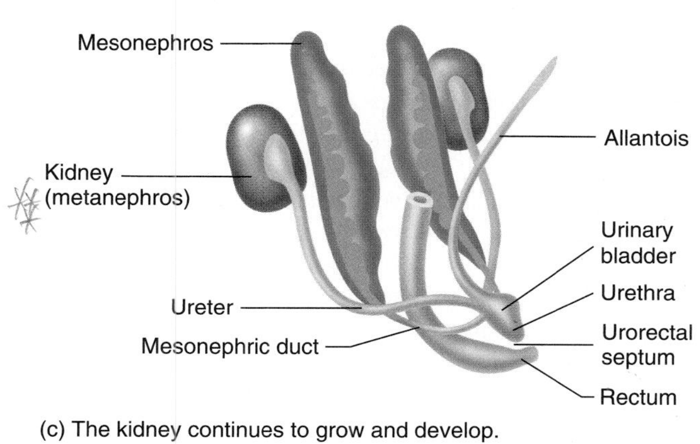

(c) The kidney continues to grow and develop.

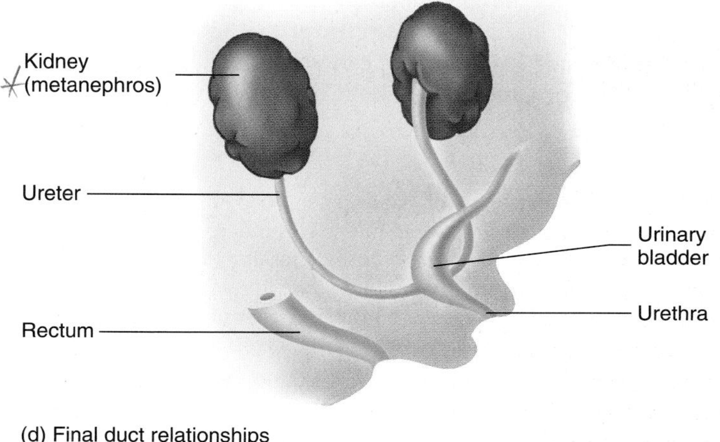

(d) Final duct relationships

FIGURE 29.14 Development of the Kidney and Urinary Bladder

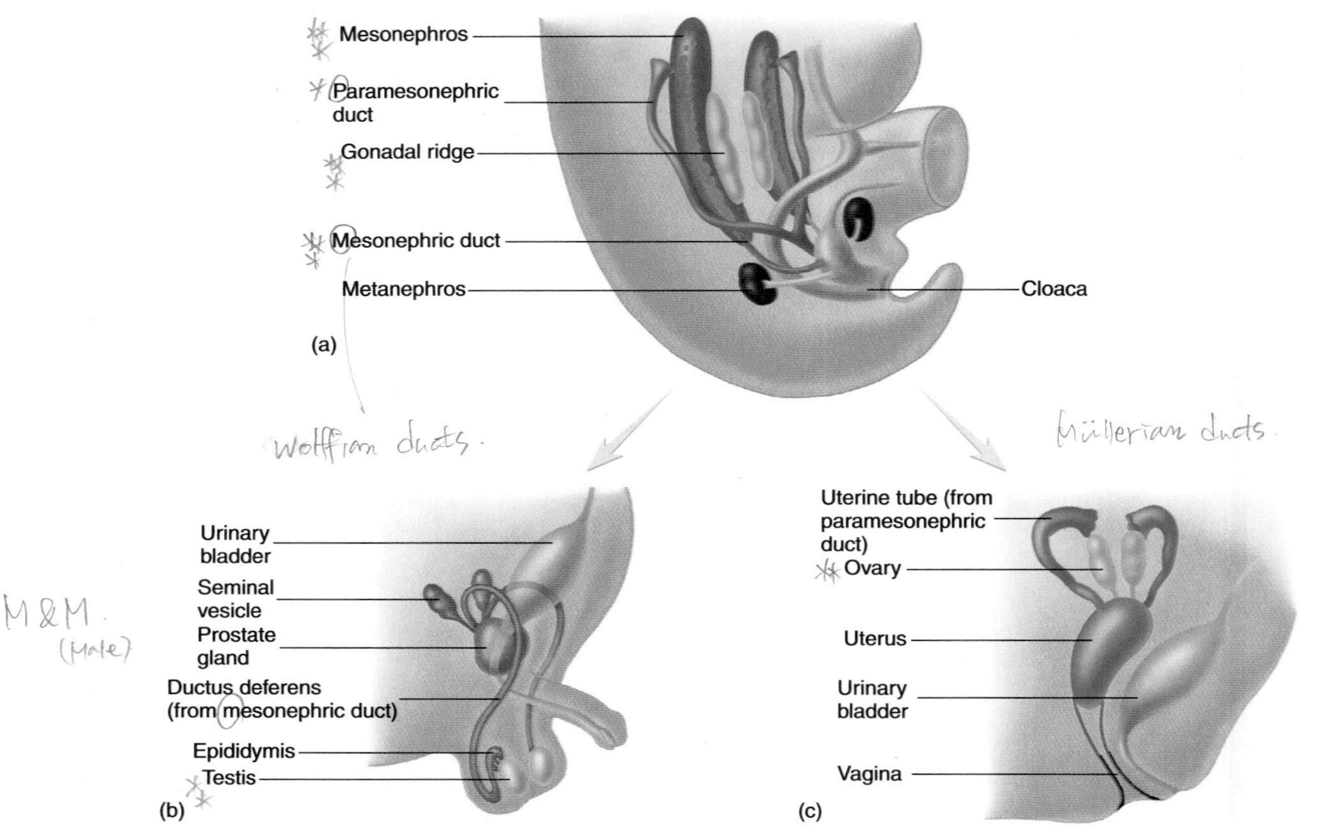

FIGURE 29.15 Development of the Reproductive System
(*a*) Indifferent stage. (*b*) The male, under the influence of male hormones, develops a ductus deferens from the mesonephric duct, and the paramesonephric duct degenerates. (*c*) The female, without male hormones, develops a uterus and uterine tubes from the paramesonephric duct, and the mesonephros disappears.

cells, destined to become oocytes or sperm cells, form on the surface of the yolk sac, migrate into the embryo, and enter the gonadal ridge.

In the female, the ovaries descend from their original position high in the abdomen to a position within the pelvis. In the male, the testes descend even farther. As the testes reach the anteroinferior abdominal wall, a pair of tunnels called the **inguinal canals** form through the abdominal musculature. The testes pass through these canals, leaving the abdominal cavity and coming to lie within the **scrotum** (see figure 28.4). Descent of the testes through the canals begins about 7 months after conception, and the testes enter the scrotum about 1 month before the infant is born.

Cryptorchidism

In approximately 3% of male children, one or both testes fail to enter the scrotum. This condition is called undescended testes, or **cryptorchidism** (krip-tōr′ki-dizm). Because testosterone is required for the testes to descend into the scrotum, cryptorchidism is often the result of inadequate testosterone secreted by the fetal testes. If neither testis descends and the defect is not corrected, the male will be infertile because the slightly higher temperature of the body cavity, compared with that of the scrotal sac, causes the spermatogonia to degenerate. Cryptorchidism is an important risk factor for testicular cancer. Cryptorchidism is treated with hormone therapy, or may be surgically corrected.

Paramesonephric ducts (also called müllerian duct) begin to develop just lateral to the **mesonephric ducts** (also called wolffian ducts) and grow inferiorly to meet one another, where they enter the cloaca as a single, midline tube.

In male embryos, testosterone is secreted by the testes. It causes the mesonephric duct system to enlarge and differentiate to form the epididymis, ductus deferens, seminal vesicles, and prostate gland (figure 29.15*b*). Müllerian-inhibiting hormone, which the testes also secrete, causes the paramesonephric müllerian ducts to degenerate. In female embryos, neither testosterone nor müllerian-inhibiting hormone is secreted. As a result, the mesonephric duct system atrophies, and the paramesonephric duct system develops to form the uterine tubes, the uterus, and part of the vagina in females (figure 29.15*c*).

Like the other sexual organs, the external genitalia begin as the same structures in the male and female and then diverge (figure 29.16). An enlargement called the **genital tubercle** develops in the groin of the embryo. **Genital folds** develop on each side of a **urethral groove,** and **labioscrotal swellings** develop lateral to the folds.

In the male, under the influence of dihydrotestosterone, which is derived from testosterone, the genital tubercle and the genital folds close over the urethral groove to form the penis. If this closure does not proceed all the way to the end of the penis, a defect known as **hypospadias** (hī′pō-spā′dē-ăs) results. The

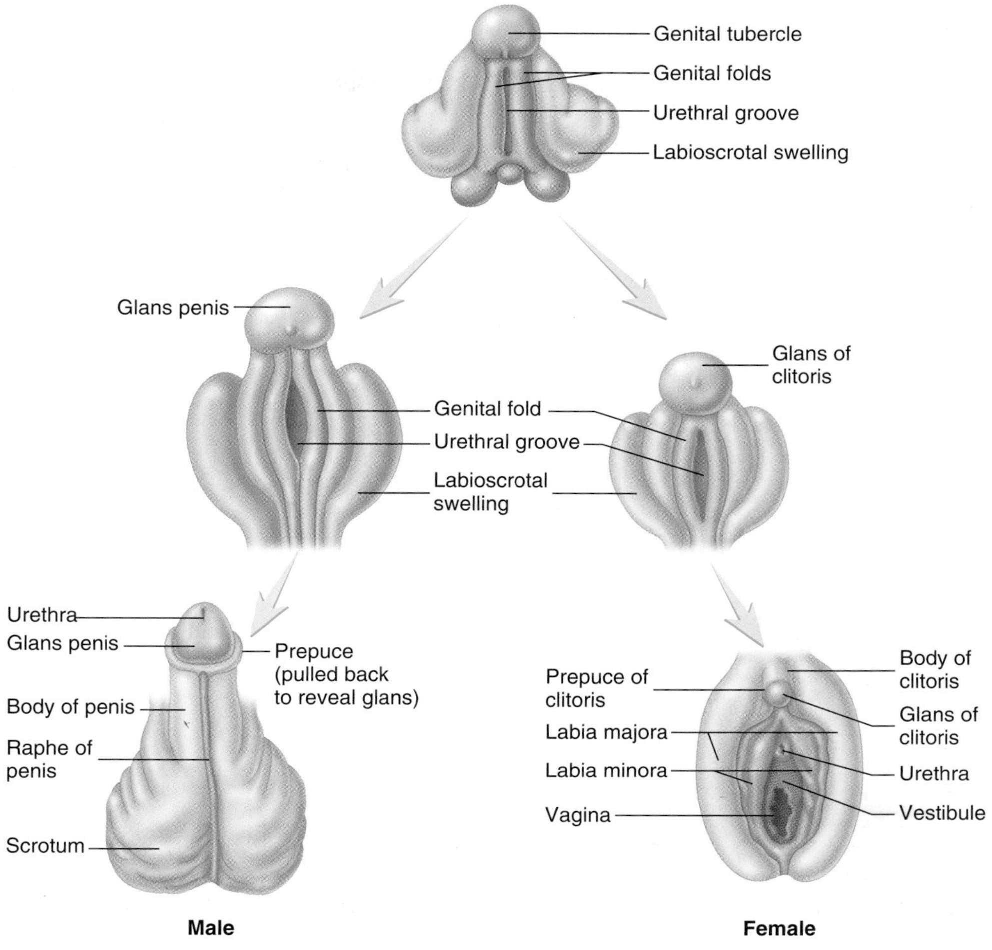
1. At approximately 5 weeks after fertilization, the genital tubercle, genital folds, and labioscrotal swellings are the same for the male and female.
Genital tubercle
Genital folds
Urethral groove
Labioscrotal swelling
2. By 10 weeks of development, the male penile structures are somewhat larger than those of the female clitoris. The urethral groove is still open in the male and female.
Glans penis
Genital fold
Urethral groove
Labioscrotal swelling
Glans of clitoris
3. Near term, the general adult condition is achieved. The labioscrotal swellings become the scrotum in the male, the labia majora in the female. The genital folds fuse in the male to form the body of the penis, whereas they form the labia minora in the female. The urethral groove in the male is closed by the raphe of the penis. It remains open in the female as the vestibule.
Urethra
Glans penis
Prepuce (pulled back to reveal glans)
Body of penis
Raphe of penis
Scrotum
Male
Prepuce of clitoris
Labia majora
Labia minora
Vagina
Body of clitoris
Glans of clitoris
Urethra
Vestibule
Female

PROCESS FIGURE 29.16 **Development of the External Genitalia**

testes move into the labioscrotal swellings, which become the scrotum of the male.

In the female, in the absence of testosterone, the genital tubercle becomes the clitoris. The urethral groove disappears; genital folds do not fuse. As a result, the urethra opens somewhat posterior to the clitoris but anterior to the vaginal opening. The unfused genital folds become the labia minora, and the labioscrotal folds become the labia majora.

13 ***Describe the formation of these major organs and systems: skin, bones, skeletal muscles, nervous system, eyes, and respiratory system.***

14 ***Explain the formation of the following endocrine glands: anterior pituitary, posterior pituitary, thyroid, parathyroid, adrenal medulla, adrenal cortex, and pancreas.***

15 ***Explain the process whereby a one-chambered heart becomes a four-chambered heart.***

16 ***Describe the development of the pronephros, mesonephros, and metanephros in the development of the kidneys.***

17 ***Describe the effect of hormones on the development of the male and female reproductive systems.***

18 ***Compare the male and female structures formed from each of the following: genital tubercle, genital folds, and labioscrotal swellings.***

PREDICT 4

How would the failure to produce müllerian-inhibiting hormone affect the development of the internal reproductive system and external genitalia in a male embryo?

Growth of the Fetus

The embryo becomes a **fetus** approximately 60 days after fertilization (a 50-day-old embryo is shown in figure 29.17*a*). The major difference between the embryo and the fetus is that in the embryo most of the organ systems are developing, whereas in the fetus the organs are present. Most morphologic changes occur in the embryonic phase of development, whereas the fetal period is primarily a "growing phase."

The fetus grows within the uterus from about 3 cm and 2.5 g at 60 days to 50 cm and 3300 g at term—more than a 15-fold

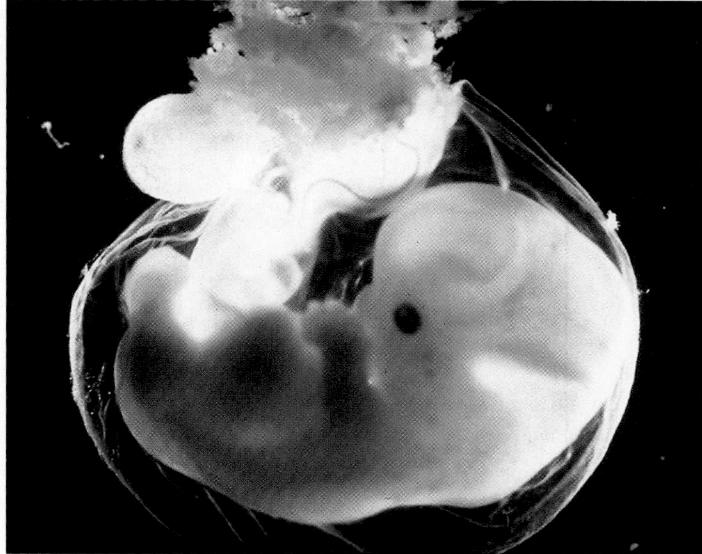
(a)

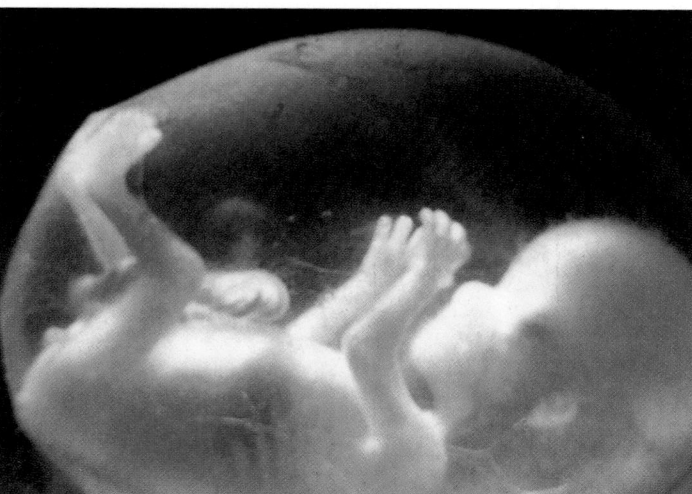
(b)

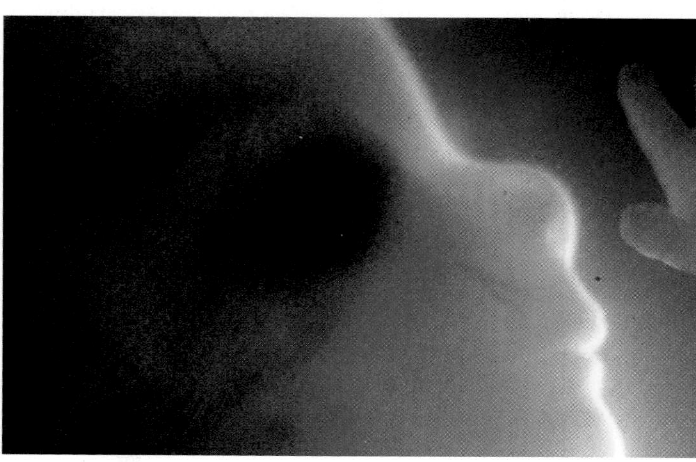
(c)

FIGURE 29.17 Embryos and Fetuses at Different Ages
(*a*) Fifty days after fertilization. (*b*) Three months after fertilization. (*c*) Four months after fertilization.

increase in length and a 1300-fold increase in weight. These changes cause a concomitant increase in the size of the uterus (figure 29.18). Although growth is certainly a major feature of the fetal period, it is not the only feature. The major organ systems still continue to develop during the fetal period.

Fine, soft hair called **lanugo** (la-noo′gō) covers the fetus, and a waxy coat of sloughed epithelial cells called **vernix caseosa** (ver′niks kā-se-ō′să) protects the fetus from the somewhat toxic nature of the amniotic fluid formed by the accumulation of waste products from the fetus.

Subcutaneous fat that accumulates in the older fetus and newborn provides a nutrient reserve, helps insulate the baby, and aids the baby in sucking by strengthening and supporting the cheeks so that negative pressure can be developed in the oral cavity.

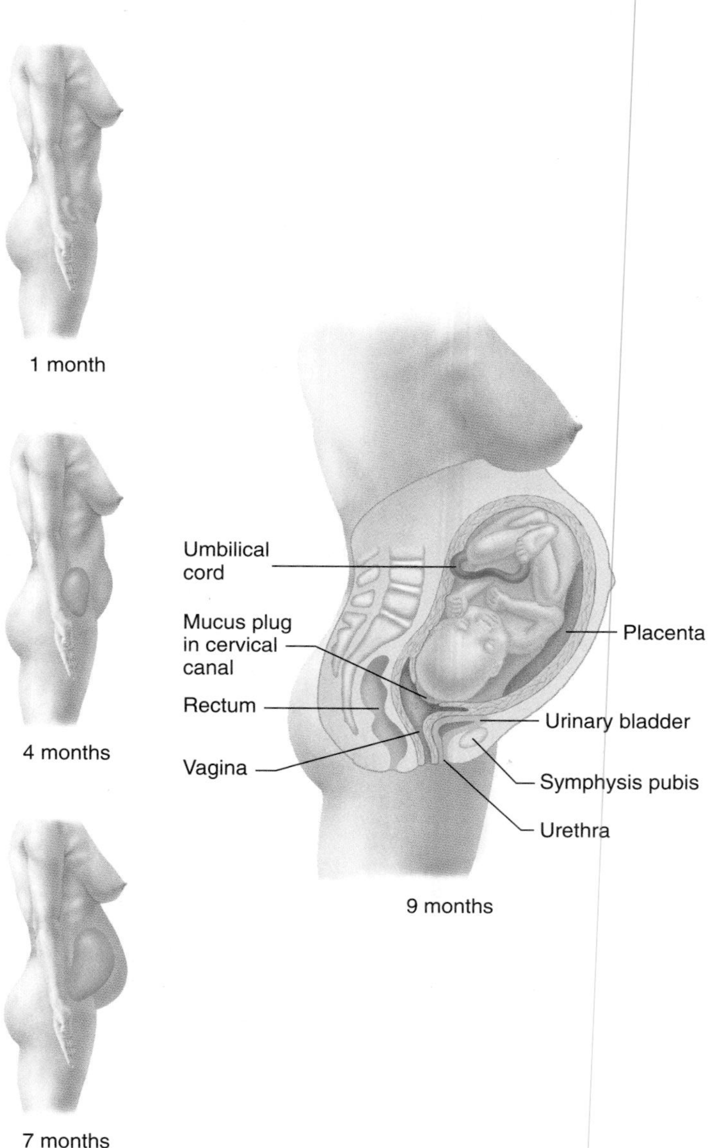

FIGURE 29.18 Enlargement of the Uterus During Fetal Development

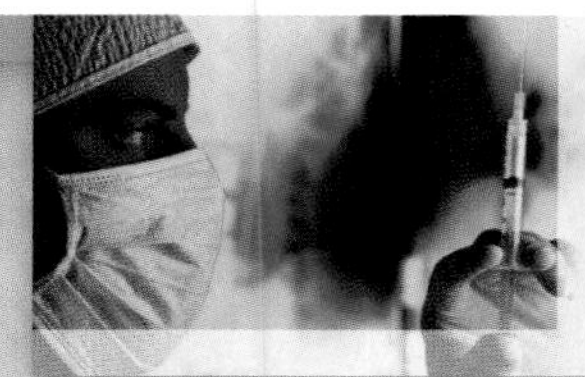

CLINICAL FOCUS

Fetal Monitoring

Amniocentesis (am'nē-ō-sen-tē'sis) is the removal of amniotic fluid from the amniotic cavity (figure B). As the fetus develops, it expels molecules of various types, as well as living cells, into the amniotic fluid. These molecules and cells can be collected and analyzed. A number of normal conditions can be evaluated, and a number of metabolic disorders can be detected by analyzing the types of molecules that the fetus expels. The cells collected by amniocentesis can be grown in culture, and additional metabolic disorders can be evaluated. Chromosome analysis, called a **karyotype** (see p. 96), can also be performed on the cultured cells. Amniocentesis has been done as early as 10 weeks after fertilization, but the success rate at that time is quite low. It is most commonly performed at 13–16 weeks after fertilization.

Fetal tissue samples can also be obtained by **chorionic villus sampling,** in which a probe is introduced into the uterine cavity through the cervix and a small piece of chorion removed. This technique has an advantage over amniocentesis in that it can be used earlier in development, as early as 8–10 weeks after fertilization. Furthermore, cells can be used directly for analyses such as karyotyping, rather than having to culture cells, as in amniocentesis.

One of the molecules normally produced by the fetus and released into the amniotic fluid is **α-fetoprotein.** If the fetus has tissues exposed to the amniotic fluid that are normally covered by skin, such as nervous tissue, resulting from failure of the neural tube to close, or abdominal tissues, resulting from failure of the abdominal wall to fully form, an excessive amount of α-fetoprotein is lost into the amniotic fluid.

Some of the metabolic by-products from the fetus, such as α-fetoprotein and estriol, a weak form of estrogen produced in the placenta after 20 weeks of gestation, can enter the maternal blood. In some cases, the by-products are processed and passed to the maternal urine. The levels of these fetal products can then be measured in the mother's blood or urine.

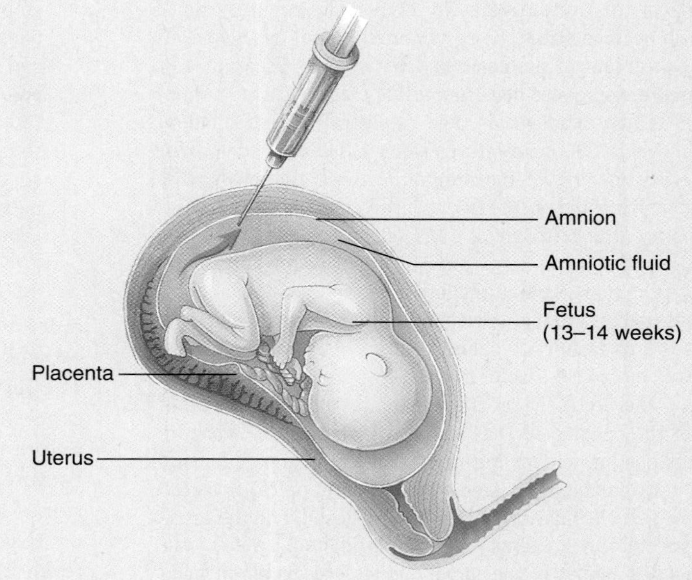

FIGURE B Removal of Amniotic Fluid for Amniocentesis

The fetus can be seen within the uterus by **ultrasound,** which uses sound waves that are bounced off the fetus like sonar and then analyzed and enhanced by computer, or by **fetoscopy,** in which a fiberoptic probe is introduced into the amniotic cavity. Because of the constantly increasing resolution in ultrasound and because it is noninvasive compared with fetoscopy, it is usually the preferred technique. However, some fetal defects cannot be adequately a ssessed by ultrasound and require fetoscopy. Ultrasound is used to guide the fetoscope. Ultrasound has not been found to pose any risk to the fetus or mother. It is accomplished by placing a transducer on the abdominal wall (transabdominal) or by inserting the transducer into the woman's vagina (transvaginal). The latter technique produces much higher resolution because fewer layers of tissue exist between the transducer and the uterine cavity. Transvaginal ultrasound can be used to identify the yolk sac of a developing embryo as early as 17 days after fertilization, and the embryo can be visualized at 25 days. Transabdominal ultrasound allows for fetal monitoring by 6–8 weeks after fertilization.

Fetal heart rate can be detected with an ultrasound stethoscope by the tenth week after fertilization and with a conventional stethoscope by 20 weeks. Fetal heart rate is now more commonly monitored electronically, either indirectly by transducers on the mother's abdomen or by a probe attached to the skin of the fetus. The normal fetal heart rate is 140 bpm (normal range is 110–160).

Peak body growth occurs late in gestation, but, as placental size and blood supply limits are approached, the growth rate slows. Growth of the placenta essentially stops at about 35 weeks, thus restricting further intrauterine growth.

Fetal Surgery

Fetal surgery performed while the fetus is still in the uterus was first done in the United States in 1979 to drain excess fluid associated with hydrocephalus. These surgeries did not usually solve the underlying neurologic problems and have been discontinued. Since 1981, in utero surgeries have successfully removed excess fluid from enlarged urinary bladders of male fetuses. The fluid buildup occurs in 1 in every 2000 male fetuses when a flap of tissue grows over the internal opening of the urethra. Without treatment, the amount of amniotic fluid is greatly reduced, and most of those babies die shortly after birth. Since 1989, more advanced surgeries have repaired diaphragmatic hernia, in which part of the abdominal organs push up through a hole in the left side of the diaphragm into the left pleural space so that the left lung fails to develop fully. The defect occurs in 1 of every 2000 babies, and, without surgery, babies with this defect run a 75% risk of dying before or soon after birth. During surgery, the uterus is cut open, and the fetus is pulled far enough out of the opening so that a small incision can be made in its side. The abdominal organs are moved back into the abdomen, the hole in the diaphragm is covered with a surgical patching material called Gore-Tex, the incision in the fetus is closed, and the fetus is tucked back into the uterus. The amniotic fluid, which was removed and saved earlier in the surgery, is replaced, and the incision in the uterus and mother's skin is repaired.

At about 38 weeks of development, the fetus has progressed to the point at which it can survive outside the mother. The average weight at this point is 3250 g for a female fetus and 3300 g for a male fetus.

Exercise During Pregnancy

Healthy women with normal pregnancies are encouraged to engage in moderate exercise during pregnancy. Women who are active before becoming pregnant are able to continue most of their activities during early pregnancy. Later in pregnancy, changes in weight, balance, and joint stability can affect exercise. Exercise during pregnancy is beneficial to the woman and helps her recover more quickly after delivery.

19 ***What major events distinguish between embryonic and fetal development?***

PARTURITION

Parturition (par-toor-ish'ŭn) is the process by which a baby is born. Physicians usually calculate the gestation period, or length of the pregnancy, as 280 days (40 weeks) from the last menstrual period (LMP) to the date of delivery of the infant.

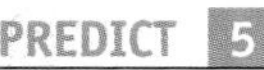
PREDICT 5

Compare and contrast clinical age and developmental age for fertilization, implantation, beginning of the fetal period, and parturition.

Prematurity

Occasionally, the fetus is delivered before it has sufficiently matured. It is then considered to be **premature.** Prematurity is one of the most significant problems in pediatrics, the branch of medical science dealing with children, because of all the associated complications. The most significant of these complications is **infant respiratory distress syndrome,** which occurs because very young premature infants cannot produce **surfactant,** a mixture of phospholipids and protein that lines the inner surface of the lungs and allows the lungs to expand as we breathe. Each year, 65,000 premature infants suffer from respiratory distress syndrome in the United States. Until recently, 10% of those infants died. Now, surfactant substitutes are being developed, and glucocorticoid administration can stimulate surfactant production. These therapies have cut the death rate in half, and more effective replacements are being investigated.

Near the end of pregnancy, the uterus becomes progressively more excitable and usually exhibits occasional contractions, which become stronger and more frequent until parturition is initiated. The cervix gradually dilates, and strong uterine contractions help expel the fetus from the uterus through the vagina (figure 29.19). Before expulsion of the fetus from the uterus, the amniotic sac ruptures, and amniotic fluid flows through the vagina to the exterior of the woman's body.

Labor is the period during which the contractions occur that result in expulsion of the fetus from the uterus. It occurs as three stages:

1. *First stage.* The first stage, often called the dilation stage, begins with the onset of regular uterine contractions and extends until the cervix dilates to a diameter about the size of the fetus's head. This stage of labor commonly lasts 8–24 hours, but it may be as short as a few minutes in some women who have had more than one child. Normally (95% of the time), the head of the fetus is in an inferior position within the woman's pelvis during labor. The head acts as a wedge, forcing the cervix and vagina to open as the uterine contractions push against the fetus. During this stage of labor, the amniotic sac ruptures, releasing the amniotic fluid. This event is commonly referred to as when the "water breaks."

The Perineum in Pregnancy

The **central tendon of the perineum** (see figure 10.21) is very important in supporting the uterus and vagina. Tearing or stretching of the tendon during childbirth may weaken the inferior support of these organs, and prolapse of the uterus may occur. **Prolapse** is a "sinking" of the uterus so that the uterine cervix moves down into the vagina (first degree), moves down near the vaginal orifice (second degree), or protrudes through the vaginal orifice (third degree). During childbirth, an **episiotomy,** a cut in the clinical perineum, prevents tearing of the skin and muscle of this region. The cut relieves the pressure and heals better than would a tear; however, there appears to be evidence that many episiotomies may be unnecessary.

2. *Second stage.* The second stage of labor, often called the expulsion stage, lasts from the time of maximum cervical

dilation until the baby exits the vagina. This stage may last from a minute to an hour or more. During this stage, contractions of the abdominal muscles assist the uterine contractions. The contractions generate enough pressure to compress blood vessels in the placenta so that blood flow to the fetus is stopped. During periods of relaxation, blood flow to the placenta resumes.

> **Oxytocin**
>
> Occasionally, synthetic oxytocin (pitocin) is administered to women during labor to increase the force of the uterine contractions. Caution must be exercised in the use of this drug, however, so that tetanic-like contractions, which would drastically reduce the blood flow through the placenta, do not occur.

3. *Third stage.* The third stage of labor, often called the placental stage, involves the expulsion of the placenta from the uterus. Contractions of the uterus cause the placenta to tear away from the wall of the uterus. Some bleeding occurs because of the intimate contact between the placenta and the uterus; however, bleeding normally is restricted because uterine smooth muscle contractions compress the blood vessels to the placenta.

Blood levels of estrogen and progesterone fall dramatically after parturition. Once the placenta has been dislodged from the uterus, the source of these hormones is gone. In addition, during the 4 or 5 weeks after parturition, the uterus becomes much smaller, but it remains somewhat larger than it was before pregnancy. The cells of the uterus become smaller, and many of them degenerate. A vaginal discharge composed of small amounts of blood and degenerating endometrium persists for 1 week or more after parturition.

The precise signal that triggers parturition is unknown, but many of the factors that support it have been identified (figure 29.20). Before parturition, the progesterone concentration in the maternal circulation is at its highest level. Progesterone has an inhibitory effect on uterine smooth muscle cells. Near the end of pregnancy, however, estrogen levels rapidly increase in the maternal circulation, and the excitatory influence of estrogens on uterine smooth muscle cells overcomes the inhibitory influence of progesterone.

The adrenal glands of the fetus are greatly enlarged before parturition. The stress of the confined space of the uterus and the limited oxygen supply resulting from a more rapid increase in the size of the fetus than in the size of the placenta increase the rate of adrenocorticotropic hormone (ACTH) secretion by the fetus's anterior pituitary gland. ACTH causes the fetal adrenal cortex to produce glucocorticoids, which travel to the placenta, where they decrease the rate of progesterone secretion and increase the rate of estrogen synthesis. In addition, prostaglandin synthesis is initiated. Prostaglandins strongly stimulate uterine contractions.

During parturition, stretch of the uterine cervix initiates nervous reflexes that cause oxytocin to be released from the woman's posterior pituitary gland. Oxytocin stimulates uterine contractions, which move the fetus farther into the cervix, causing further stretch.

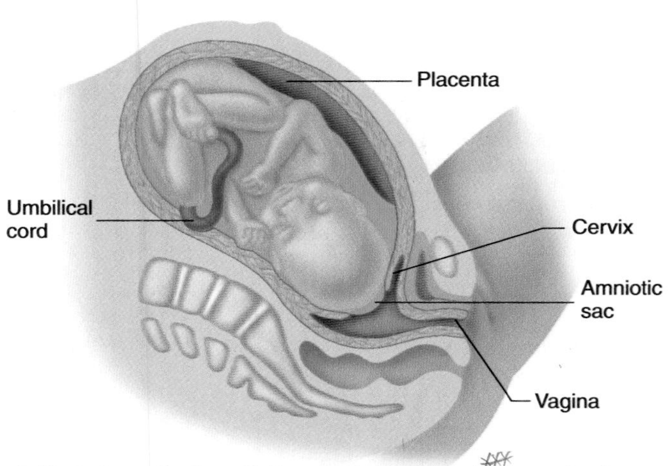

1. **First stage:** Begins with the onset of regular uterine contractions. The cervix begins to dilate. The first stage commonly lasts 8–24 hours.

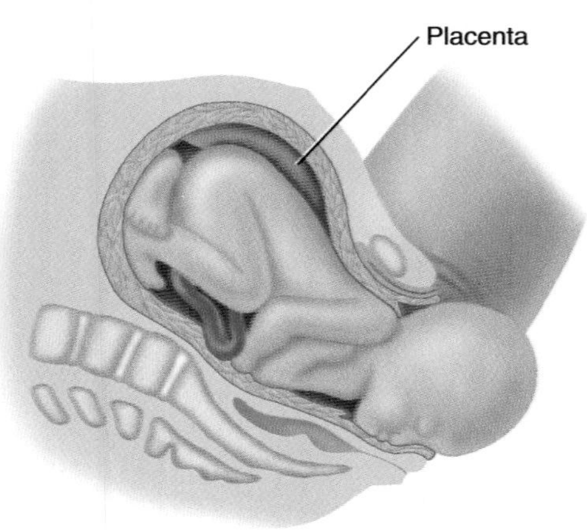

2. **Second stage:** Extends from maximal cervical dilation until the baby exits the vagina. It lasts from a few minutes to an hour.

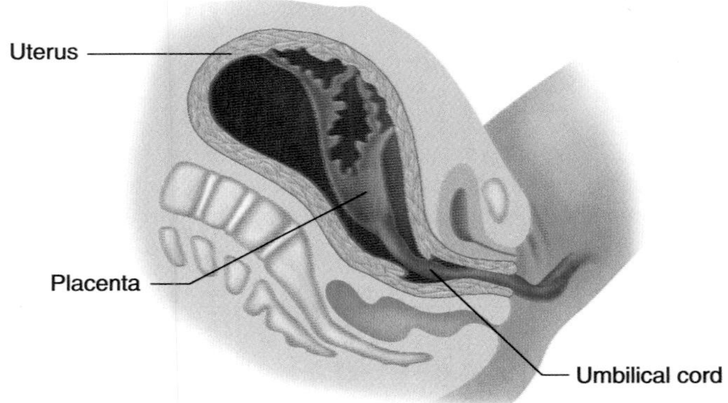

3. **Third stage:** The placenta is expelled.

PROCESS FIGURE 29.19 Parturition

1. The fetal hypothalamus secretes a releasing hormone, corticotropin-releasing hormone (CRH), which stimulates adrenocorticotropic hormone (ACTH) secretion from the pituitary. The fetal pituitary secretes ACTH in greater amounts near parturition.

2. ACTH causes the fetal adrenal gland to secrete greater quantities of adrenal cortical steroids.

3. Adrenal cortical steroids travel in the umbilical blood to the placenta.

4. In the placenta, the adrenal cortical steroids cause progesterone synthesis to level off and estrogen and prostaglandin synthesis to increase, making the uterus more excitable.

5. The stretching of the uterus produces action potentials that are transmitted to the brain through ascending pathways.

6. Action potentials stimulate the secretion of oxytocin from the posterior pituitary.

7. Oxytocin causes the uterine smooth muscle to contract.

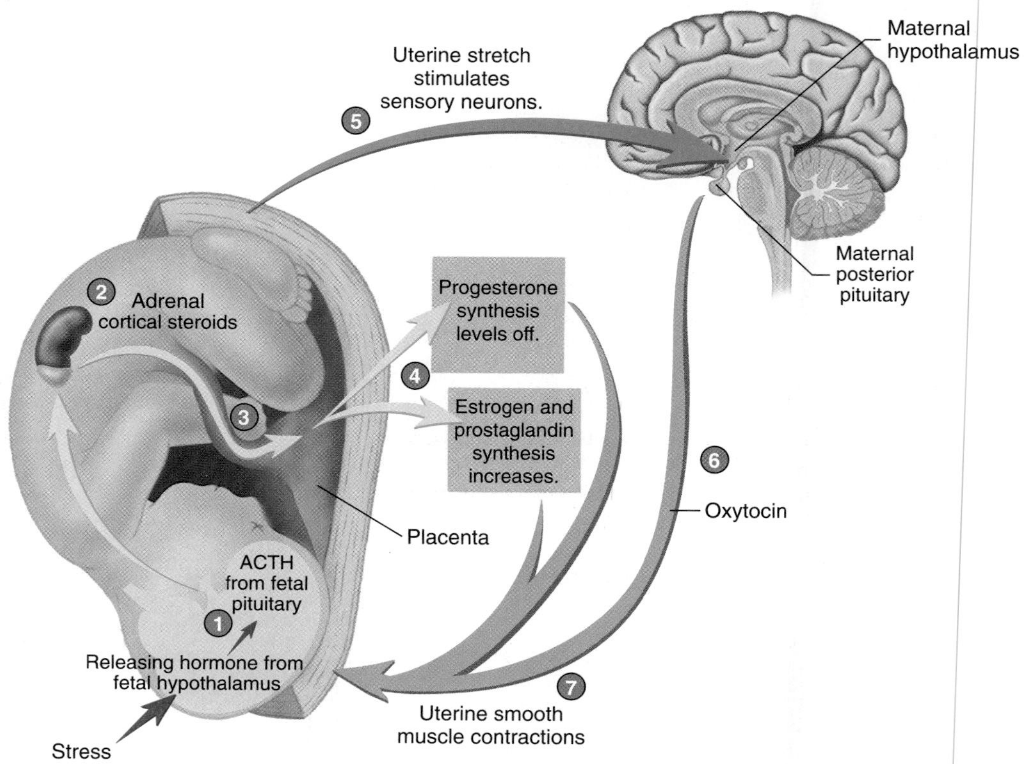

PROCESS FIGURE 29.20 Factors That Influence the Process of Parturition
Although the precise control of parturition is unknown, these changes appear to play a role.

Thus, a positive-feedback mechanism is established in which stretch stimulates oxytocin release and oxytocin causes further stretch. This positive-feedback system stops after delivery, when the cervix is no longer stretched.

Progesterone inhibits oxytocin release and decreases the number of oxytocin receptors. Decreased progesterone levels in the maternal circulation results in increased oxytocin secretion and an increase in the number of oxytocin receptors in the uterus. In addition, estrogens make the uterus more sensitive to oxytocin stimulation by increasing the synthesis of receptor sites for oxytocin. Estrogen may also increase the formation of gap junctions between myometrial cells, thereby increasing the contractility of the uterus. Some evidence suggests that oxytocin also stimulates prostaglandin synthesis in the uterus. All these events support the development of strong uterine contractions.

20 ***List the stages of labor, and indicate when each stage begins and its approximate length of time.***

21 ***Describe the hormonal changes that take place before and during delivery. How is stretch of the cervix involved in delivery?***

PREDICT 6

A woman is having an extremely prolonged labor. From her anatomy and physiology course, she remembers the role of calcium in muscle contraction and asks the doctor to give her a calcium injection to speed the delivery. Explain why the doctor would or would not do as she requests.

THE NEWBORN

The newborn baby, or **neonate,** experiences several dramatic changes at the time of birth. The major and earliest changes in the infant are separation from the maternal circulation and transfer from a fluid to a gaseous environment. The large, forced gasps of air that occur when the infant cries at the time of delivery help inflate the lungs.

Respiratory and Circulatory Changes

The initial inflation of the lungs causes important changes in the circulatory system (figure 29.21). Before birth, very little blood flows through the pulmonary arteries to the lungs and back to the heart through the pulmonary veins. As a result, the left atrium has little blood and very low blood pressure. Therefore, blood flows from the right atrium, through the foramen ovale, and into the left atrium.

Expansion of the lungs reduces the resistance to blood flow through the lungs, resulting in increased blood flow through the pulmonary arteries. Consequently, more blood flows from the right atrium to the right ventricle and into the pulmonary arteries, and less blood flows from the right atrium through the foramen ovale to the left atrium. In addition, an increased volume of blood

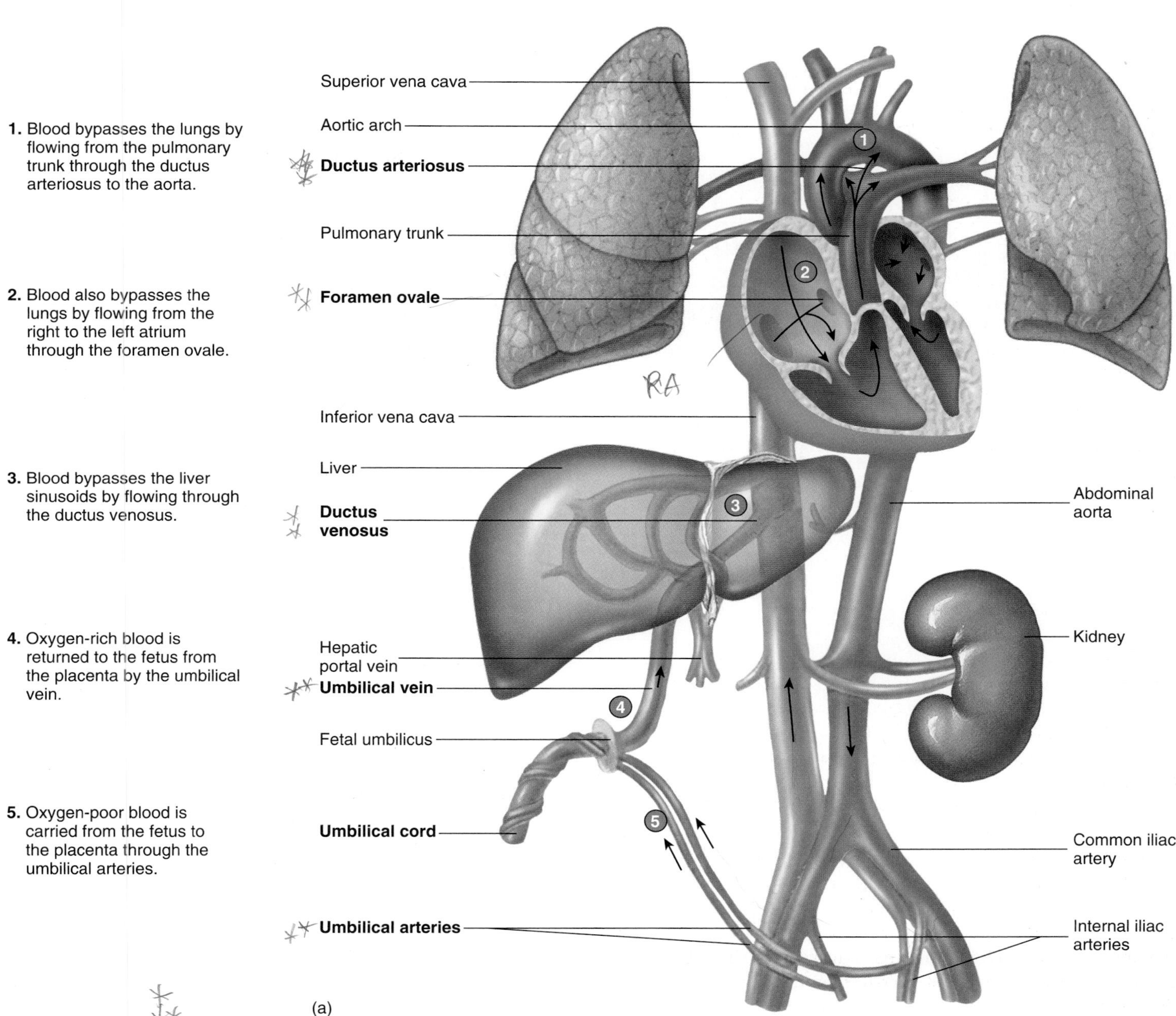

PROCESS FIGURE 29.21 Circulatory Changes at Birth
(*a*) Circulatory conditions in the fetus.

1. When air enters the lungs, blood is forced through the pulmonary arteries to the lungs. The ductus arteriosus closes and becomes the ligamentum arteriosum (*gray*).

2. The foramen ovale closes and becomes the fossa ovalis. Blood can no longer flow from the right to the left atrium.

3. The ductus venosus degenerates and becomes the ligamentum venosum (*gray*).

4. The umbilical arteries and vein are cut. The umbilical vein becomes the round ligament of the liver (*gray*).

5. The umbilical arteries also degenerate and become the cords of the umbilical arteries (*gray*).

Superior vena cava

Aortic arch

Ligamentum arteriosum (closed ductus arteriosus)

Pulmonary trunk

Fossa ovalis (foramen ovale closed)

Inferior vena cava

Liver

Ligamentum venosum (degenerated ductus venosus)

Hepatic portal vein

Round ligament of liver (degenerated umbilical vein)

Umbilicus

Cords of the umbilical arteries (degenerated umbilical arteries)

Abdominal aorta

Kidney

Common iliac artery

Internal iliac arteries

(b)

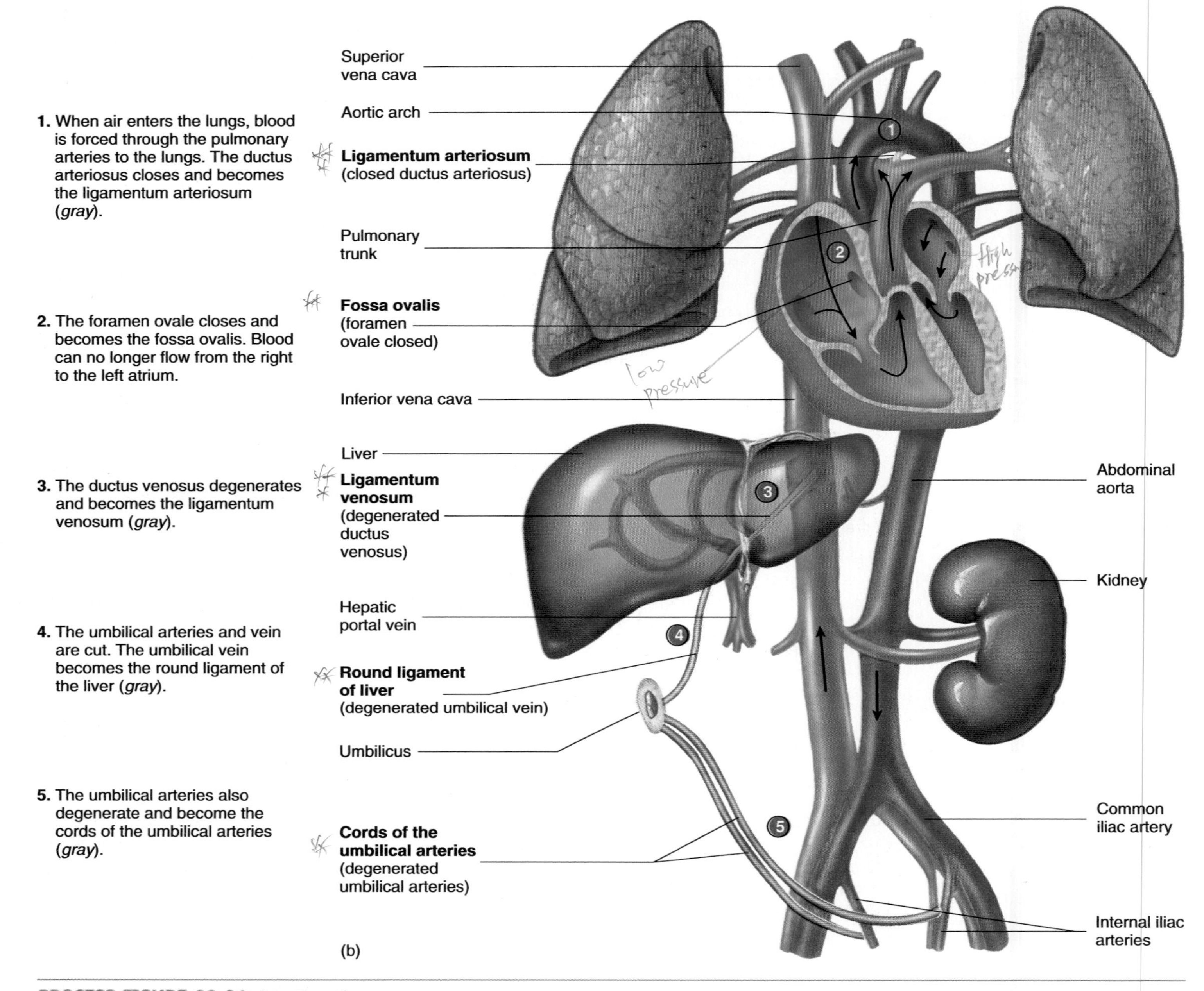

PROCESS FIGURE 29.21 ***(continued)***
(*b*) Circulatory changes that occur at birth.

returns from the lungs through the pulmonary veins to the left atrium, which increases the pressure in the left atrium. The increased left atrial pressure and decreased right atrial pressure, resulting from decreased pulmonary resistance, forces blood against the septum primum, causing the foramen ovale to close. This action functionally completes the separation of the heart into two pumps: the right side of the heart and the left side of the heart. The closed foramen ovale becomes the **fossa ovalis.**

The **ductus arteriosus,** which connects the pulmonary trunk to the aorta and allows blood to flow from the pulmonary trunk to the systemic circulation, closes off within 1 or 2 days after birth. This closure occurs because of the sphincterlike constriction of the artery and is probably stimulated by local changes in blood pressure and blood oxygen content. Once closed, the ductus arteriosus is replaced by connective tissue and is known as the **ligamentum arteriosum.**

> **Patent Ductus Arteriosus**
>
> If the ductus arteriosus does not close completely, it is said to be **patent.** This is a serious birth defect, resulting in marked elevation in pulmonary blood pressure because blood flows from the left ventricle to the aorta, through the ductus arteriosus to the pulmonary arteries. If not corrected, it can lead to irreversible degenerative changes in the heart and lungs.

The fetal blood supply passes to the placenta through umbilical arteries from the internal iliac arteries and returns through an umbilical vein. The blood passes through the liver via the ductus venosus, which joins the inferior vena cava. When the umbilical cord is tied and cut, no more blood flows through the umbilical vein and arteries, and they degenerate. The remnant of the umbilical vein becomes the **ligamentum teres,** or **round ligament,** of the liver, and the ductus venosus becomes the **ligamentum venosum.** The remnants of the umbilical arteries becomes the medial umbilical ligaments.

Digestive Changes

When a baby is born, it is suddenly separated from its source of nutrients provided by the maternal circulation. Because of this separation and the stress of birth and new life, the neonate usually loses 5%–10% of its total body weight during the first few days of life. Although the digestive system of the fetus becomes somewhat functional late in development, it is still very immature, compared with that of an adult, and can digest only a limited number of food types.

Late in gestation, the fetus swallows amniotic fluid from time to time. Shortly after birth, this swallowed fluid plus cells sloughed from the mucosal lining, mucus produced by intestinal mucous glands, and bile from the liver pass from the digestive tract as a greenish anal discharge called **meconium** (mē-kō′nē-ŭm).

The pH of the stomach at birth is nearly neutral because of the presence of swallowed alkaline amniotic fluid. Within the first 8 hours of life, a striking increase in gastric acid secretion occurs, causing the stomach pH to decrease. Maximum acidity is reached at 4–10 days, and the pH gradually increases for the next 10–30 days.

The neonatal liver is also functionally immature. It lacks adequate amounts of the enzyme required in the production of bilirubin. This enzyme system usually develops within 2 weeks after birth in a healthy neonate, but, because this enzyme system is not fully developed at birth, some full-term babies temporarily develop jaundice, with elevated blood levels of bilirubin. Jaundice often occurs in premature babies.

The newborn digestive system is capable of digesting lactose (milk sugar) from the time of birth. The pancreatic secretions are sufficiently mature for a milk diet, but the digestive system only gradually develops the ability to digest more solid foods over the first year or two; therefore, new foods should be introduced gradually during the first 2 years. It is also advised that only one new food be introduced at a time into the infant's diet so that, if an allergic reaction occurs, the cause is more easily determined.

Amylase secretion by the salivary glands and the pancreas remains low until after the first year. Lactase activity in the small intestine is high at birth but declines during infancy, although the levels still exceed those in adults. Lactase activity is lost in many adults (see chapter 24).

Apgar Scores

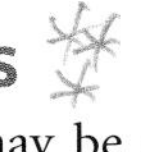

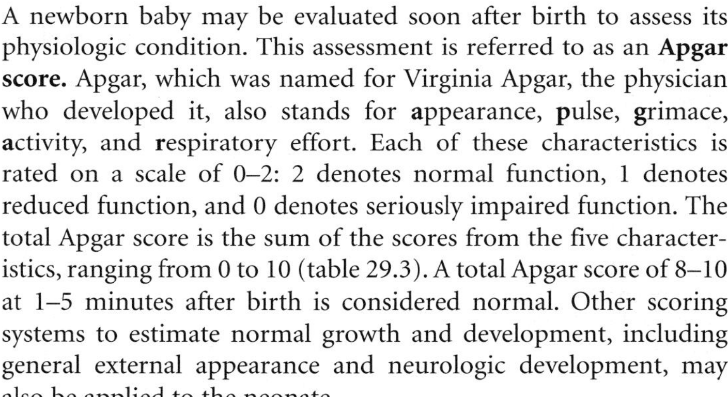

A newborn baby may be evaluated soon after birth to assess its physiologic condition. This assessment is referred to as an **Apgar score.** Apgar, which was named for Virginia Apgar, the physician who developed it, also stands for **a**ppearance, **p**ulse, **g**rimace, **a**ctivity, and **r**espiratory effort. Each of these characteristics is rated on a scale of 0–2: 2 denotes normal function, 1 denotes reduced function, and 0 denotes seriously impaired function. The total Apgar score is the sum of the scores from the five characteristics, ranging from 0 to 10 (table 29.3). A total Apgar score of 8–10 at 1–5 minutes after birth is considered normal. Other scoring systems to estimate normal growth and development, including general external appearance and neurologic development, may also be applied to the neonate.

TABLE 29.3 Examples of Apgar Rating Scales

Physiologic Conditions	0	1	2
Appearance (skin color)	White or blue	Limbs blue, body pink	Pink
Pulse (rate)	No pulse	100 bpm	>100 bpm
Grimace (reflexive grimace initiated by stimulating the plantar surface of the foot)	No response	Facial grimaces, slight body movement	Facial grimaces, extensive body movement
Activity (muscle tone)	No movement, muscles flaccid	Limbs partially flexed, little movement, poor muscle tone	Active movement, good muscle tone
Respiratory effort (amount of respiratory activity)	No respiration	Slow, irregular respiration	Good, regular respiration; strong cry

Congenital Disorders

The term *congenital* means "present at birth" and **congenital disorders** are abnormalities commonly referred to as birth defects. Approximately 15% of all congenital disorders have a known genetic cause, and approximately 70% of all birth defects are of unknown cause. The remaining 15% are the result of environmental causes or a combination of environmental and genetic causes. In the case of environmental causes, the birth defect results from damage to the fetus during development. Agents that cause birth defects are called **teratogens** (ter′ă-tō-jenz). For example, fetal alcohol syndrome results when a pregnant woman drinks alcohol, which crosses the placenta and damages the fetus. The baby is born with a smaller than normal head and mental retardation and may exhibit other birth defects. Researchers are working to identify various teratogens. With this information, women can avoid known teratogens and reduce the risk for birth defects.

22 ***What changes take place in the newborn's circulatory system shortly after birth? What do each of the following become: foramen ovale, ductus arteriosus, umbilical vein, and ductus venosus?***

23 ***What changes take place in the newborn's digestive system shortly after birth?***

24 ***Define meconium. Why does jaundice often develop after birth?***

25 ***What is an Apgar score?***

26 ***What are congenital disorders? What causes these disorders?***

27 ***What is a teratogen? Give an example.***

LACTATION

Lactation is the production of milk by the mother's breasts (mammary glands; figure 29.22). It normally occurs in females after parturition and may continue for 2 or 3 years, or even longer, provided suckling occurs often and regularly.

During pregnancy, the high concentration and continuous presence of estrogens and progesterone cause expansion of the duct system and secretory units of the breasts. The ducts grow and branch repeatedly to form an extensive network. Additional adipose tissue is deposited also; thus, the size of the breasts increases substantially throughout pregnancy. Estrogen is primarily responsible for breast growth during pregnancy, but normal development of the breast does not occur without the presence of several other hormones. Progesterone causes development of the breasts' secretory alveoli, which enlarge but do not usually secrete milk during pregnancy. The other hormones are growth hormone, prolactin, thyroid hormones, glucocorticoids, and insulin. The placenta secretes a growth hormonelike substance (human somatotropin) and a prolactin-like substance (human placental lactogen), and these substances help support the development of the breasts.

1. Stimulation of the nipple by the baby's suckling initiates action potentials in sensory neurons that connect with the hypothalamus.
2. In response, the hypothalamus stimulates the posterior pituitary to release oxytocin and the anterior pituitary to release prolactin.
3. Oxytocin stimulates milk release from the breast. Prolactin stimulates additional milk production.

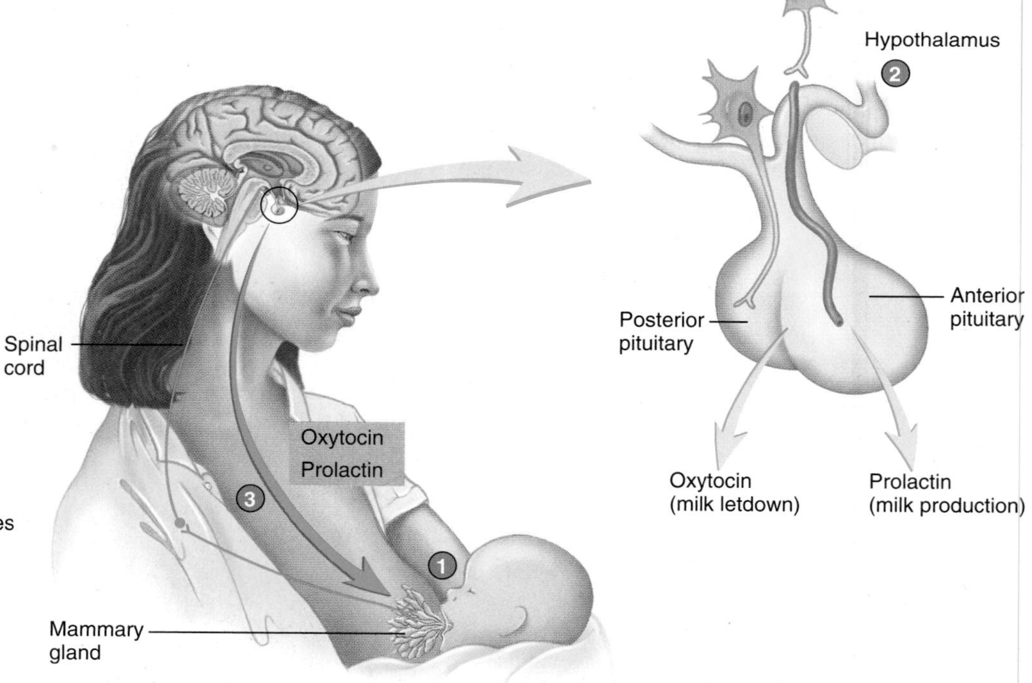

PROCESS FIGURE 29.22 Hormonal Control of Lactation

Prolactin, which is produced by the anterior pituitary gland, is the hormone responsible for milk production. Before parturition, high levels of estrogen stimulate an increase in prolactin production. Milk production is inhibited during pregnancy, however, because high levels of estrogen and progesterone inhibit the effect of prolactin on the mammary gland. After parturition, estrogen, progesterone, and prolactin levels decrease, and, with lower estrogen and progesterone levels, prolactin stimulates milk production. Despite a decrease in the basal levels of prolactin, a reflex response produces surges of prolactin release. During suckling, mechanical stimulation of the breasts initiates nerve impulses that reach the hypothalamus, causing the secretion of **prolactin-releasing factor (PRF)** and inhibiting the release of **prolactin-inhibiting factor (PIF).** Consequently, prolactin levels temporarily increase and stimulate milk production.

For the first few days after birth, the mammary glands secrete **colostrum** (kō-los′trŭm), a very nutritious substance that contains little fat and less lactose than milk. The breasts may secrete some colostrum during pregnancy. Eventually, milk with a higher fat and lactose content is produced. Colostrum and milk not only provide nutrition but also contain antibodies (see chapter 22) that help protect the nursing baby from infections.

HIV and the Newborn

The human immunodeficiency virus (HIV) can be transmitted from a mother to her child in utero, during parturition, or during breast-feeding. HIV has been isolated from human breast milk and colostrum. In developed countries, HIV patients are advised not to breast-feed but to use formula and bottle-feeding. In underdeveloped countries, bottle-feeding is not yet a viable option. Newborn babies given a single dose of an AIDS drug at birth have half the risk for HIV infection. Most of those gains are lost, however, when babies are breast-fed. The current proposal for underdeveloped countries is that daily AIDS drug treatment for newborns of HIV patients continue as long as the child is breast-feeding. In one study, treated infants had a risk of 1% for HIV infection, compared with a 15% rate in untreated infants.

Repeated stimulation of prolactin release, by suckling, makes nursing (breast-feeding) possible for several years. If nursing stops, however, within a few days the ability to produce prolactin ceases, and milk production stops.

Because it takes time to produce milk, an increase in prolactin results in the production of milk to be used in the next nursing period. At the time of nursing, stored milk is released as a result of a reflex response. Mechanical stimulation of the breasts produces nerve impulses that cause the release of oxytocin from the posterior pituitary, which stimulates cells surrounding the alveoli to contract. Milk is then released from the breasts, a process called **milk letdown** or milk ejection. In addition, higher brain centers can stimulate oxytocin release, and such things as hearing an infant cry can result in milk letdown.

28 ***What hormones are involved in preparing the breast for lactation? Describe the events involved in milk production and milk letdown. What is colostrum?***

PREDICT 7

While nursing her baby, a woman notices that she develops "uterine cramps." Explain what is happening.

FIRST YEAR AFTER BIRTH

Many changes occur in the life of the newborn from birth until 1 year of age. The time of these changes vary considerably from child to child, and the dates are only rough estimates. The brain is still developing, and much of what the neonate can do depends on how much brain development has occurred. It is estimated that the total adult number of neurons is present in the CNS at birth, but subsequent growth and maturation of the brain involve the addition of new neuroglial cells, some of which form new myelin sheaths, and the addition of new connections between neurons, which may continue throughout life.

By 6 weeks, the infant usually can hold up its head when placed in a prone position and begins to smile in response to people or objects. At 3 months of age, the infant's limbs move apparently aimlessly. The infant has enough control of the arms and hands, however, that voluntary thumb sucking can occur. The infant can follow a moving person with its eyes. At 4 months, the infant begins to raise itself by its arms. It can begin to grasp objects placed in its hand, coo and gurgle, roll from its back to its side, listen quietly when hearing a person's voice or music, hold its head erect, and play with its hands. At 5 months, the infant can usually laugh, reach for objects, turn its head to follow an object, lift its head and shoulders, sit with support, and roll over. At 8 months, the infant recognizes familiar people, sits up without support, and reaches for specific objects. At 12 months, the infant may pull itself to a standing position and may be able to walk without support. It can pick up objects in its hands and examine them carefully. It can understand much of what is said to it and may say several words of its own.

29 ***List the major changes that occur during the first year of life. What do most of these activities depend on?***

LIFE STAGES

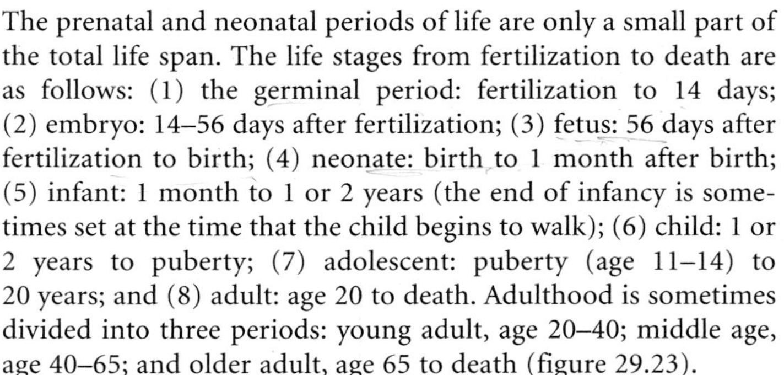

The prenatal and neonatal periods of life are only a small part of the total life span. The life stages from fertilization to death are as follows: (1) the germinal period: fertilization to 14 days; (2) embryo: 14–56 days after fertilization; (3) fetus: 56 days after fertilization to birth; (4) neonate: birth to 1 month after birth; (5) infant: 1 month to 1 or 2 years (the end of infancy is sometimes set at the time that the child begins to walk); (6) child: 1 or 2 years to puberty; (7) adolescent: puberty (age 11–14) to 20 years; and (8) adult: age 20 to death. Adulthood is sometimes divided into three periods: young adult, age 20–40; middle age, age 40–65; and older adult, age 65 to death (figure 29.23).

During childhood, the individual develops considerably. Many of the emotional characteristics that a person possesses throughout life are formed during early childhood.

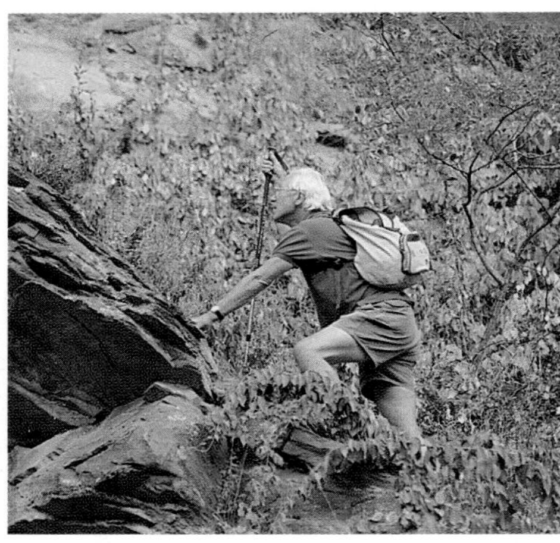

FIGURE 29.23 **Active Older Adults**
A conspicuous feature of the population of older adults is the range of variability. In some adults over 70 years, many systems are beginning to fail. Others can look forward to at least 10 more years of healthy living.

Major physical and physiologic changes occur during adolescence that also affect the emotions and behavior of the individual. Other emotional changes occur as the adolescent attempts to fit into an adult world. Puberty usually occurs somewhat earlier in females (at about 11–13 years) than in males (at about 12–14 years). A period of rapid growth usually accompanies the onset of puberty. This rapid growth period is followed by a period of slower growth. Full adult stature is usually achieved by age 17 or 18 in females and 19 or 20 years in males.

30 ***Define the different life stages, starting with the germinal stage and ending with the adult.***

AGING

The development of a new human being begins at fertilization, as does the process of aging. Cells proliferate at an extremely rapid rate during early development and then the process begins to slow as various cells become committed to specific functions within the body.

Many cells of the body, such as liver and skin cells, continue to proliferate throughout life, replacing dead or damaged tissue. Many other cells, however, such as the neurons in the central nervous system, cease to proliferate once they have reached a certain number, and dead cells are not replaced. After the number of neurons reaches a peak, at about the time of birth, their numbers begin to decline. Neuronal loss is most rapid early in life and later decreases to a slower, steadier rate.

A natural, but as yet unexplained, decline occurs in mitochondrial DNA function with age. If this decline reaches a threshold, which apparently differs from tissue to tissue, the normal function of the mitochondrion is lost, and the tissue or organ may exhibit disease symptoms. In a small number of people, this mitochondrial degeneration occurs very early in life, resulting in premature aging.

The physical plasticity (i.e., the state of being soft and pliable) of young embryonic tissues results largely from the presence of large amounts of hyaluronate and relatively small amounts of collagen; furthermore, the collagen and other related proteins that are present are not highly cross-linked; thus, the tissues are very flexible and elastic. Many of these proteins produced during development are permanent components of the individual, however, and, as the individual ages, more and more cross-links form between these protein molecules, thereby rendering the tissues more rigid and less elastic.

The tissues with the highest content of collagen and other related proteins are those most severely affected by the collagen cross-linking and tissue rigidity associated with aging. The lens of the eye is one of the first structures to exhibit pathologic changes as a result of this increased rigidity. Vision of close objects becomes more difficult with advancing age until most middle-aged people require reading glasses (see chapter 15). Loss of elasticity also affects other tissues, including the joints, blood vessels, kidneys, lungs, and heart, and greatly reduces their functional ability.

Like nervous tissue, mature muscle cells do not normally proliferate after terminal differentiation occurs before birth. As a result, the total number of skeletal and cardiac muscle fibers declines with age. The strength of skeletal muscle reaches a peak between 20 and 30 years of life and declines steadily thereafter. Furthermore, like the collagen of connective tissue, the macromolecules of muscle undergo biochemical changes during aging and render the muscle tissue less functional. A good exercise program, however, can slow or even reverse this process.

The decline in muscular function also contributes to a decline in cardiac function with advancing age. The heart loses elastic recoil ability and muscular contractility. As a result, total cardiac output declines, and less oxygen and fewer nutrients reach cells, such as neurons in the brain and cartilage cells in the joints, thereby contributing to the decline in these tissues. Reduced cardiac function also may result in decreased blood flow to the kidneys, thus contributing to decreases in their filtration ability. Degeneration of the connective tissues as a result of collagen cross-linking and other factors also decreases the filtration efficiency of the glomerular basement membrane.

Atherosclerosis (ath′er-ō-skler-ō′sis) is the deposit and subsequent hardening of a soft, gruel-like material, containing lipids, cholesterol, calcium, and other materials, in lesions of the intima of large and medium-sized arteries. These deposits then become fibrotic and calcified, resulting in **arteriosclerosis** (ar-tēr′ē-ō-skler-ō′sis; hardening of the arteries). Arteriosclerosis interferes with normal blood flow and can result in a **thrombus,** which is a clot or plaque formed inside a vessel. A piece of the plaque, called an **embolus** (em′bō-lŭs) can break loose, float through the circulation, and lodge in smaller arteries to cause myocardial infarctions or strokes. Atherosclerosis occurs to some extent in all middle-aged and older people and even may occur in young people. People with high blood cholesterol, however, appear to be at increased risk of developing atherosclerosis. In addition to dietary influences, this condition seems to have a heritable component, and blood tests are available to screen people for high blood cholesterol levels.

Many other organs, such as the liver, pancreas, stomach, and colon, undergo degenerative changes with age. The ingestion of harmful agents may accelerate such changes. Examples include the degenerative changes induced in the lungs, aside from lung cancer, by cigarette smoke and sclerotic changes in the liver as a result of alcohol consumption.

In addition to the previously described changes associated with aging, cellular wear and tear, or cytologic aging, contributes to aging. Progressive damage from many sources, such as radiation and toxic substances, may result in irreversible cellular insults and may be one of the major factors leading to aging. It has been speculated that ingestion of the antioxidant vitamins C and E in combination may help slow this part of aging by stimulating cell repair. Vitamin C also stimulates collagen production and may slow the loss of tissue plasticity associated with aging collagen.

According to the **free radical theory of aging,** free radicals, which are atoms or molecules with an unpaired electron, can react with and alter the structure of molecules that are critical for normal cell function. Alteration of these molecules can result in cell dysfunction, cancer, or other types of cellular damage. Free radicals are produced as a normal part of metabolism and are introduced into the body from the environment through the air we breathe and the food we eat. The damage caused by free radicals may accumulate with age. Antioxidants, such as beta carotene (provitamin A), vitamin C, and vitamin E, can donate electrons to free radicals, without themselves becoming harmful. Thus, antioxidants may prevent the damage caused by the free radicals and may ward off age-related disorders, ranging from wrinkles to cancer. Experiments designed to test this hypothesis, however, have not consistently produced positive results.

As a result of poor diet, many people over age 50 do not get the minimum daily allotment of several vitamins and minerals. Feeling "bad" is not necessarily a part of aging but is mostly a result of poor nutrition and lack of exercise. Engaging in moderate exercise and avoiding overeating can prolong life. Moderate exercise can reduce the risk for heart attack by as much as 20%. It can also reduce the risk for stroke, high blood pressure, and some forms of cancer. Exercise can also increase a person's ability to reason and remember. Walking 30 minutes a day is recommended.

One characteristic of aging is an overall decrease in ATP production. This decrease is associated with a decline in oxidative phosphorylation, which has been shown in many cases to be associated with **mitochondrial DNA mutations.** Mitochondria lack the DNA repair mechanisms that exist in the nucleus for nuclear DNA. Such mutations are also often associated with Alzheimer disease. There are also genes in nuclear DNA associated with Alzheimer disease.

Immune system changes may also be a major factor contributing to aging. The aging immune system loses its ability to respond to outside antigens but becomes more sensitive to the body's own antigens. These autoimmune changes add to the degeneration of the tissues and may be responsible for such conditions as arthritic joint disorders, chronic glomerular nephritis, and hyperthyroidism. In addition, T lymphocytes tend to lose their functional capacity with aging and cannot destroy abnormal cells as efficiently. This change may be one reason that certain types of cancer occur more readily in older people.

Genetic traits may also cause many of the changes associated with aging. As a general rule, animals with a very high metabolic rate have a shorter life span than those with a lower metabolic rate. In humans, a very small number of exceptional people have a slightly reduced normal body temperature, suggesting a lower metabolic rate. The same people often have an unusually long life span. This tendency appears to run in families and probably has some genetic basis. Studies of the general population suggest that, if your parents and grandparents have lived long, so will you.

Another piece of evidence suggesting a strong genetic component to aging comes from a disorder called **progeria** (prō-jēr′ē-ă; premature aging). This genetic trait causes the degenerative changes of aging to occur shortly after the first year of life, and the child may look like a very old person by age 7.

One of the greatest disadvantages of aging is the increasing lack of ability to adjust to stress. Older people have a far more precarious homeostatic balance than younger people, and eventually some stress is encountered that is so great that the body's ability to recover is surpassed, and death results.

31 ***How does the loss of cells that are not replaced affect the aging process? Give examples.***

32 ***How does loss of tissue plasticity affect the aging process? Give examples.***

33 ***Explain the free radical theory of aging.***

34 ***How does aging affect the immune system?***

35 ***What role does genetics play in aging?***

DEATH

Death is usually not attributed to old age. Another problem, such as heart failure, renal failure, or stroke, is usually the cause of death.

Death was once defined as the loss of heartbeat and respiration. In recent years, however, more precise definitions of death have been developed because both the heart and the lungs can be kept working artificially, and the heart can even be replaced by an artificial device. Modern definitions of death are based on the permanent cessation of life functions and the cessation of integrated tissue and organ function. The most widely accepted indication of death in humans is **brain death,** which is defined as irreparable brain damage manifested clinically by the absence of response to stimulation, the absence of spontaneous respiration and heart beat, and an isoelectric ("flat") electroencephalogram for at least 30 minutes, in the absence of known CNS poisoning or hypothermia.

36 ***Define death.***

SUMMARY

Prenatal development is an important part of an individual's life.

Prenatal Development (p. 1082)

1. Prenatal development is divided into three parts: formation of the germ layers, formation of the organ systems, and growth and maturation.
2. Postovulatory age is 14 days less than clinical age.

Fertilization

Fertilization, the union of the oocyte and sperm, results in a zygote.

Early Cell Division

The cells of the early embryo are pluripotent (capable of making any cell of the body). In the very early stages of development the cells are totipotent, meaning each cell can give rise to any tissue necessary for development.

Morula and Blastocyst

The product of fertilization undergoes divisions until it becomes a mass, called a morula, and then a hollow ball of cells, called a blastocyst.

Implantation of the Blastocyst and Development of the Placenta

The blastocyst implants into the uterus about 7 days after fertilization. The placenta is derived from the trophoblast of the blastocyst.

Formation of the Germ Layers

All tissues of the body are derived from three primary germ layers: ectoderm, mesoderm, and endoderm.

Neural Tube and Neural Crest Formation

The nervous system develops from a neural tube that forms in the ectodermal surface of the embryo and from neural crest cells derived from the developing neural tube.

Somite Formation

Segments, called somites, that develop along the neural tube give rise to the musculature, vertebral column, and ribs.

Formation of the Gut and Body Cavities

1. The gastrointestinal tract develops as the developing embryo closes off part of the yolk sac.
2. The celom develops from small cavities that fuse within the embryo.

Limb Bud Development

The limbs develop from proximal to distal as outgrowths called limb buds.

Development of the Face

The face develops from the fusion of five major tissue processes.

Development of the Organ Systems

1. The skin develops from the ectoderm (epithelium), mesoderm and neural crest (dermis), and neural crest (melanocytes).
2. The skeletal system develops from mesoderm or neural crest cells.
3. Muscle develops from myoblasts, which migrate from somites.
4. The brain and spinal cord develop from the neural tube, and the peripheral nervous system develops from the neural tube and the neural crest cells.
5. The special senses develop mainly as neural tube or neural crest cell derivatives.
6. Many endocrine organs develop mainly as evaginations of the brain or digestive tract.
7. The heart develops as two tubes fuse into a single tube, which bends and develops septa to form four chambers.
8. The peripheral circulation develops from mesoderm as blood islands become hollow and fuse to form a network.
9. The lungs form as evaginations of the digestive tract. These evaginations undergo repeated branching.
10. The urinary system develops in three stages—pronephros, mesonephros, and metanephros—from the head to the tail of the embryo. The ducts join the allantois, part of which becomes the urinary bladder.
11. The reproductive system develops in conjunction with the urinary system. The presence or absence of certain hormones is very important to sexual development.

Growth of the Fetus

1. The embryo becomes a fetus at 60 days.
2. The fetal period is from day 60 to birth. It is a time of rapid growth.

Parturition (p. 1104)

1. The total length of gestation is 280 days (clinical age).
2. Uterine contractions force the baby out of the uterus during labor.
3. Increased estrogen levels and decreased progesterone levels help initiate parturition.
4. Fetal glucocorticoids act on the placenta to decrease progesterone synthesis and to increase estrogen and prostaglandin synthesis.
5. Stretch of the uterus and decreased progesterone levels stimulate oxytocin secretion, which stimulates uterine contraction.

The Newborn (p. 1106)

The newborn baby experiences several dramatic changes when it is separated from the maternal circulation that provided nutrients, oxygen and waste removal before parturition.

Respiratory and Circulatory Changes

1. The foramen ovale closes, separating the two atria.
2. The ductus arteriosus closes, and blood no longer flows between the pulmonary trunk and the aorta.
3. The umbilical vein and arteries degenerate.

Digestive Changes

1. Meconium is a mixture of cells from the digestive tract, amniotic fluid, bile, and mucus excreted by the newborn.
2. The stomach begins to secrete acid.
3. The liver does not form adult bilirubin for the first 2 weeks.
4. Lactose can be digested, but other foods must be gradually introduced.

Apgar Scores

1. *Apgar* represents appearance, pulse, grimace, activity, and respiratory effort.
2. Apgar and other methods are used to assess the physiologic condition of the newborn.

Congenital Disorders

1. Congenital disorders are abnormalities present at birth.
2. Teratogens are environmental agents that cause some congenital disorders.

Lactation (p. 1110)

1. Estrogen, progesterone, and other hormones stimulate the growth of the breasts during pregnancy.
2. Suckling stimulates prolactin and oxytocin synthesis. Prolactin stimulates milk production, and oxytocin stimulates milk letdown.

First Year After Birth (p. 1111)

1. The number of neuron connections and glial cells increases.
2. Motor skills gradually develop, especially head, eye, and hand movements.

Life Stages (p. 1111)

The life stages include the following: germinal, embryo, fetus, neonate, infant, child, adolescent, and adult.

Aging (p. 1112)

1. Loss of cells that are not replaced contributes to aging.
   - A loss of neurons occurs.
   - Loss of muscle cells can affect skeletal and cardiac muscle function.
2. Loss of tissue plasticity results from cross-link formation between collagen molecules. The lens of the eye loses the ability to accommodate. Other organs, such as the joints, kidneys, lungs, and heart, also have reduced efficiency with advancing age.
3. The immune system loses the ability to act against foreign antigens and may attack self-antigens.
4. Many aging changes are probably genetic.

Death (p. 1113)

Death is the loss of brain functions.

REVIEW AND COMPREHENSION

1. The major development of organ systems takes place in the
   a. first 2 weeks of development.
   b. third to eighth week of development.
   c. eighth to twentieth week of development.
   d. last 30 weeks of development.
2. Given these structures:
   1. blastocyst
   2. morula
   3. zygote

   Choose the arrangement that lists the structures in the order in which they are formed during development.
   a. 1,2,3 c. 2,3,1 e. 3,2,1
   b. 1,3,2 d. 3,1,2
3. The embryo proper develops from the
   a. inner cell mass. c. blastocele.
   b. trophoblast. d. yolk sac.
4. The placenta
   a. develops from the trophoblast.
   b. allows maternal blood to mix with embryonic blood.
   c. invades the lacunae of the embryo.
   d. all of the above.
5. The embryonic disk
   a. forms between the amniotic cavity and the yolk sac.
   b. contains the primitive streak.
   c. becomes a three-layered structure.
   d. all of the above.
6. The brain develops from
   a. ectoderm. b. endoderm. c. mesoderm.
7. Most of the skeletal system develops from
   a. ectoderm. b. endoderm. c. mesoderm.
8. Given these structures:
   1. neural crest 2. neural plate 3. neural tube

   Choose the arrangement that lists the structure in the order in which they form during development.
   a. 1,2,3 c. 2,1,3 e. 3,2,1
   b. 1,3,2 d. 2,3,1
9. The somites give rise to the
   a. circulatory system. c. lungs. e. brain.
   b. skeletal muscle. d. kidneys.
10. The pericardial cavity forms from
    a. evagination of the early gastrointestinal tract.
    b. the neural tube.
    c. the celom.
    d. the branchial arches.
    e. pharyngeal pouches.
11. The parts of the limbs develop
    a. in a proximal-to-distal sequence.
    b. in a distal-to-proximal sequence.
    c. at approximately the same time.
    d. before the primitive streak is formed.
12. Concerning development of the face,
    a. the face develops by the fusion of five embryonic structures.
    b. the maxillary processes normally meet at the midline to form the lip.
    c. the primary palate forms the roof of the mouth.
    d. clefts of the secondary palate normally occur to one side of the midline.
13. Concerning the development of the heart,
    a. the heart develops from a single tube, which results from the fusion of two tubes.
    b. the SA node develops in the wall of the sinus venosus.
    c. the foramen ovale lets blood flow from the right atrium to the left atrium.
    d. the bulbis cordis is absorbed into the ventricle.
    e. all of the above.
14. Given these structures:
    1. mesonephros
    2. metanephros
    3. pronephros

    Choose the arrangement that lists the structures in the order in which they form during development.
    a. 1,2,3 c. 2,3,1 e. 3,2,1
    b. 1,3,2 d. 3,1,2

15. A study of the early embryo indicates that the glans penis of the male develops from the same embryonic structure as which of these female structures?
    a. labia majora
    b. uterus
    c. clitoris
    d. vagina
    e. urinary bladder
16. Which hormone causes the differentiation of sex organs in the developing male fetus?
    a. FSH and LH
    b. LH and testosterone
    c. testosterone and dihydrotestosterone
    d. estrogen and progesterone
    e. GnRH and FSH
17. The onset of labor may be a result of
    a. increased estrogen secretion by the placenta.
    b. increased glucocorticoid secretion by the fetus.
    c. increased secretion of oxytocin.
    d. stretch of the uterus.
    e. all of the above.
18. Following birth,
    a. the ductus arteriosus closes.
    b. the pH of the stomach increases.
    c. the fossa ovalis becomes the foramen ovale.
    d. blood flow through the pulmonary arteries decreases.
    e. all of the above.
19. Which of these can result in a congenital disorder?
    a. a parent with the disorder
    b. a teratogen
    c. a mutation
    d. all of the above
20. The hormone involved in milk production is
    a. oxytocin.
    b. prolactin.
    c. estrogen.
    d. progesterone.
    e. ACTH.
21. Which of these most appropriately predicts the consequences of removing the sensory neurons from the areola of a lactating rat (or human)?
    a. Blood levels of oxytocin decrease.
    b. Blood levels of prolactin decrease.
    c. Milk production and letdown decrease.
    d. All of the above occur.
22. Which of these life stages is correctly matched with the time that the stage occurs?
    a. neonate—birth to 1 month after birth
    b. infant—1 month to 6 months
    c. child—6 months to 5 years
    d. puberty—10–12 years
    e. middle age—20–40 years
23. Which of these occurs as we get older?
    a. Neurons replicate to replace lost neurons.
    b. Skeletal muscle cells replicate to replace lost muscle cells.
    c. Cross-links between collagen molecules increase.
    d. The immune system become less sensitive to the body's own antigens.
    e. Free radicals help prevent cancer.

Answers in Appendix E

CRITICAL THINKING

1. Triploidy is the presence of three sets of chromosomes in a cell. Though rare, triploidy does occur in humans. What failures during fertilization could lead to triploidy?
2. A woman is told by her physician that she is pregnant and that she is 44 days past her LMP. Approximately how many days has the embryo been developing, and what developmental events are occurring?
3. A high fever can prevent neural tube closure. If a woman has a high fever approximately 35–45 days after her LMP, what kinds of birth defects may be seen in the developing embryo?
4. If the apical ectodermal ridge is damaged during embryonic development when the limb bud is about one-half grown, what kinds of birth defects might be expected? Describe the anatomy of the affected structure.
5. What are the results of exposing a female embryo to high levels of testosterone while she is developing?
6. A woman goes into labor during the thirtieth week of her pregnancy. What would be the effect of administering progesterone at this stage?
7. Three minutes after birth, a newborn has an Apgar score of 5 as follows: A, 0; P, 1; G, 1; A, 1; and R, 2. What are some of the possible causes for this low score? What might be done for this neonate?
8. When a woman nurses, milk letdown can occur in the breast that is not being suckled. Explain how this response happens.

Answers in Appendix F

Visit this book's website at www.mhhe.com/seeley8 for chapter quizzes, interactive learning exercises, and other study tools.

APPENDIX A

Periodic Table of the Elements

9 **F** Fluorine 19.00	
9	Atomic number
19.00	Atomic mass

1 1A	2 2A	3 3B	4 4B	5 5B	6 6B	7 7B	8	9 8B	10	11 1B	12 2B	13 3A	14 4A	15 5A	16 6A	17 7A	18 8A
1 **H** Hydrogen 1.008																	2 **He** Helium 4.003
3 **Li** Lithium 6.941	4 **Be** Beryllium 9.012											5 **B** Boron 10.81	6 **C** Carbon 12.01	7 **N** Nitrogen 14.01	8 **O** Oxygen 16.00	9 **F** Fluorine 19.00	10 **Ne** Neon 20.18
11 **Na** Sodium 22.99	12 **Mg** Magnesium 24.31											13 **Al** Aluminum 26.98	14 **Si** Silicon 28.09	15 **P** Phosphorus 30.97	16 **S** Sulfur 32.07	17 **Cl** Chlorine 35.45	18 **Ar** Argon 39.95
19 **K** Potassium 39.10	20 **Ca** Calcium 40.08	21 **Sc** Scandium 44.96	22 **Ti** Titanium 47.88	23 **V** Vanadium 50.94	24 **Cr** Chromium 52.00	25 **Mn** Manganese 54.94	26 **Fe** Iron 55.85	27 **Co** Cobalt 58.93	28 **Ni** Nickel 58.69	29 **Cu** Copper 63.55	30 **Zn** Zinc 65.39	31 **Ga** Gallium 69.72	32 **Ge** Germanium 72.59	33 **As** Arsenic 74.92	34 **Se** Selenium 78.96	35 **Br** Bromine 79.90	36 **Kr** Krypton 83.80
37 **Rb** Rubidium 85.47	38 **Sr** Strontium 87.62	39 **Y** Yttrium 88.91	40 **Zr** Zirconium 91.22	41 **Nb** Niobium 92.91	42 **Mo** Molybdenum 95.94	43 **Tc** Technetium (98)	44 **Ru** Ruthenium 101.1	45 **Rh** Rhodium 102.9	46 **Pd** Palladium 106.4	47 **Ag** Silver 107.9	48 **Cd** Cadmium 112.4	49 **In** Indium 114.8	50 **Sn** Tin 118.7	51 **Sb** Antimony 121.8	52 **Te** Tellurium 127.6	53 **I** Iodine 126.9	54 **Xe** Xenon 131.3
55 **Cs** Cesium 132.9	56 **Ba** Barium 137.3	57 **La** Lanthanum 138.9	72 **Hf** Hafnium 178.5	73 **Ta** Tantalum 180.9	74 **W** Tungsten 183.9	75 **Re** Rhenium 186.2	76 **Os** Osmium 190.2	77 **Ir** Iridium 192.2	78 **Pt** Platinum 195.1	79 **Au** Gold 197.0	80 **Hg** Mercury 200.6	81 **Tl** Thallium 204.4	82 **Pb** Lead 207.2	83 **Bi** Bismuth 209.0	84 **Po** Polonium (210)	85 **At** Astatine (210)	86 **Rn** Radon (222)
87 **Fr** Francium (223)	88 **Ra** Radium (226)	89 **Ac** Actinium (227)	104 **Rf** Rutherfordium (257)	105 **Db** Dubnium (260)	106 **Sg** Seaborgium (263)	107 **Bh** Bohrium (262)	108 **Hs** Hassium (265)	109 **Mt** Meitnerium (266)	110 **Ds** Darmstadtium (269)	111 **Rg** Roentgenium (272)	112	(113)	114	(115)	116	(117)	(118)

58 **Ce** Cerium 140.1	59 **Pr** Praseodymium 140.9	60 **Nd** Neodymium 144.2	61 **Pm** Promethium (147)	62 **Sm** Samarium 150.4	63 **Eu** Europium 152.0	64 **Gd** Gadolinium 157.3	65 **Tb** Terbium 158.9	66 **Dy** Dysprosium 162.5	67 **Ho** Holmium 164.9	68 **Er** Erbium 167.3	69 **Tm** Thulium 168.9	70 **Yb** Ytterbium 173.0	71 **Lu** Lutetium 175.0
90 **Th** Thorium 232.0	91 **Pa** Protactinium (231)	92 **U** Uranium 238.0	93 **Np** Neptunium (237)	94 **Pu** Plutonium (242)	95 **Am** Americium (243)	96 **Cm** Curium (247)	97 **Bk** Berkelium (247)	98 **Cf** Californium (249)	99 **Es** Einsteinium (254)	100 **Fm** Fermium (253)	101 **Md** Mendelevium (256)	102 **No** Nobelium (254)	103 **Lr** Lawrencium (257)

Metals

Metalloids

Nonmetals

The 1–18 group designation has been recommended by the International Union of Pure and Applied Chemistry (IUPAC) but is not yet in wide use.

The modern periodic table of the elements lists the known elements in order of their atomic masses. Each element has a box that contains the name of the element and its chemical symbol, atomic number, and atomic mass. The boxes are organized into a grid of horizontal rows, called periods, and vertical columns, called groups. Within a period, the elements are listed in order of increasing atomic number from left to right. Elements in a period have different chemical properties, whereas elements in a group have similar chemical properties.

The atomic number is the number of protons in an element. Each element has a unique number of protons and therefore a unique atomic number. There are 90 naturally occurring elements. Scientists have been able to create new elements by changing the number of protons in the nuclei of existing elements. Protons, neutrons, or electrons from one atom are accelerated to very high speeds and then smashed into the nucleus of another atom. The resulting changes in the nucleus produce a new element with a new atomic number. These artificially produced elements are usually unstable, and they quickly convert back to more stable elements. The synthetic elements are technetium (Tc, atomic number 43), promethium (Pm, atomic number 61), and all the elements with an atomic number of 93 or higher. An element with an atomic number of 112 has the highest number officially recognized by the International Union of Pure and Applied Chemistry, but elements with higher atomic numbers have reportedly been made.

APPENDIX B

Scientific Notation

Very large numbers with many zeros, such as 1,000,000,000,000,000, or very small numbers, such as 0.0000000000000001, are very cumbersome to work with. Consequently, the numbers are expressed in a kind of mathematical shorthand known as scientific notation. Scientific notation has the following form:

$$M \times 10^n$$

where n specifies how many times the number M is raised to the power of 10. The exponent n has two meanings, depending on its sign. If n is positive, M is multiplied by 10 n times. For example, if $n = 2$ and $M = 1.2$, then

$$1.2 \times 10^2 = 1.2 \times 10 \times 10 = 120$$

In other words, if n is positive, the decimal point of M is moved to the right n times. In this case the decimal point of 1.2 is moved two places to the right.

$$1.20.$$

If n is negative, M is divided by 10 n times.

$$1.2 \times 10^{-2} = \frac{1.2}{(10 \times 10)} = \frac{1.2}{100} = 0.012$$

In other words, if n is negative, the decimal point of M is moved to the left n times. In this case, the decimal point of 1.2 is moved two places to the left.

$$0.01.2$$

If M is the number 1.0, it often is not expressed in scientific notation. For example, 1.0×10^2 is the same thing as 10^2, and 1.0×10^{-2} is the same thing as 10^{-2}.

Two common examples of the use of scientific notation in chemistry are Avogadro's number and pH. Avogadro's number, 6.023×10^{23}, is the number of atoms in 1 molar mass of an element. Thus

$$6.023 \times 10^{23} = 602{,}300{,}000{,}000{,}000{,}000{,}000{,}000$$

which is a very large number of atoms.

The pH scale is a measure of the concentration of hydrogen ions in a solution. A neutral solution has 10^{-7} moles of hydrogen ions per liter. In other words

$$10^{-7} = 0.0000001$$

which is a very small amount (1 ten-millionth of a gram) of hydrogen ions.

APPENDIX C

Solution Concentrations

Physiologists often express solution concentration in terms of percent, molarity, molality, and equivalents.

Percent

The weight-volume method of expressing percent concentrations states the weight of a solute in a given volume of solvent. For example, to prepare a 10% solution of sodium chloride, 10 g of sodium chloride is dissolved in a small amount of water (solvent) to form a salt solution. Then additional water is added to the salt solution to form 100 mL of salt solution. Note that the sodium chloride is dissolved in water and then diluted to the required volume. The sodium chloride is not dissolved directly in 100 mL of water.

Molarity

Molarity determines the number of moles of solute dissolved in a given volume of solvent. A 1 molar (1 M) solution is made by dissolving 1 mole (mol) of a substance in enough water to make 1 L of solution. For example, 1 mol of sodium chloride solution is made by dissolving 58.44 g of sodium chloride in enough water to make 1 L of solution. One mol of glucose solution is made by dissolving 180.2 g of glucose in enough water to make 1 L of solution. Both solutions have the same number (Avogadro's number) of formula units (NaCl) and molecules (glucose) in solution.

Molality

Although 1 M solutions have the same number of solute molecules, they do not have the same number of solvent (water) molecules. Because 58.5 g of sodium chloride occupies less volume than 180 g of glucose, the sodium chloride solution has more water molecules. **Molality** is a method of calculating concentrations that takes into account the number of solute and solvent molecules. A 1 molal solution (1 *m*) is 1 mol of a substance dissolved in 1 kg of water. Thus, all 1-molal solutions have the same number of solvent molecules.

When sodium chloride, which is an ionic compound, is dissolved in water, it dissociates to form two ions, a sodium cation (Na^+) and a chloride anion (Cl^-). Glucose does not dissociate when dissolved in water, however, because it is a molecule. Thus, the sodium chloride solution contains twice as many particles as the glucose solution (one Na^+ and one Cl^- for each glucose molecule). To report the concentration of these substances in a way that reflects the number of particles in a given mass of solvent, the concept of **osmolality** is used. The osmolality of a solution is the molality of the solution times the number of particles into which the solute dissociates in 1 kg of solvent. Thus, 1 mol of sodium chloride in 1 kg of water is a 2-osmolal (osm) solution because sodium chloride to form two ions.

The osmolality of a solution is a reflection of the number, not the type, of particles in a solution. Thus, a 1 osm solution contains 1 osm of particles per kilogram of solvent, but the particles may be all one type or a complex mixture of different types.

The concentration of particles in body fluids is so low that the measurement milliosmole (mOsm), 1/1000 of an osmole, is used. Most body fluids have an osmotic concentration of approximately 300 mOsm and consist of many different ions and molecules. The osmotic concentration of body fluids is important because it influences the movement of water into or out of cells (see chapter 3).

Equivalents

Equivalents are a measure of the concentrations of ionized substances. One equivalent (Eq) is 1 mol of an ionized substance multiplied by the absolute value of its charge. For example, 1 mol of NaCl dissociates into 1 mol of Na^+ and 1 mol of Cl^-. Thus, there is 1 Eq of Na^+ (1 mol × 1) and 1 Eq of Cl^- (1 mol × 1). One mole of $CaCl_2$ dissociates into 1 mol of Ca^{2+} and 2 mol of Cl^-. Thus, there are 2 Eq of Ca^{2+} (1 mol × 2) and 2 Eq of Cl^- (2 mol × 1). In an electrically neutral solution, the equivalent concentration of positively charged ions is equal to the equivalent concentration of the negatively charged ions. One milliequivalent (mEq) is 1/1000 of an equivalent.

APPENDIX D

pH

Pure water weakly dissociates to form small numbers of hydrogen and hydroxide ions:

$$H_2O \leftrightarrow H^+ + OH^-$$

At 25°C, the concentration of both hydrogen ions and hydroxide ions is 10^{-7} mol/L. Any solution that has equal concentrations of hydrogen and hydroxide ions is considered **neutral.** A solution is an **acid** if it has a higher concentration of hydrogen ions than hydroxide ions, and a solution is a **base** if it has a lower concentration of hydrogen ions than hydroxide ions. In any aqueous solution (at 25°C), the hydrogen ion concentration $[H^+]$ times the hydroxide ion concentration $[OH^-]$ is a constant that is equal to 10^{-14}.

$$[H^+] \times [OH^-] = 10^{-14}$$

Consequently, as the hydrogen ion concentration decreases, the hydroxide ion concentration increases, and vice versa—for example,

	$[H^+]$	$[OH^-]$
Acidic solution	10^{-3}	10^{-11}
Neutral solution	10^{-7}	10^{-7}
Basic solution	10^{-12}	10^{-2}

Although the acidity or basicity of a solution could be expressed in terms of either hydrogen or hydroxide ion concentration, it is customary to use hydrogen ion concentration. The pH of a solution is defined as

$$pH = -\log_{10}(H^+)$$

Thus, a neutral solution with 10^{-7} mol of hydrogen ions per liter has a pH of 7.

$$\begin{aligned} pH &= -\log_{10}(H^+) \\ &= -\log_{10}(10^{-7}) \\ &= -(-7) \\ &= 7 \end{aligned}$$

In simple terms, to convert the hydrogen ion concentration to the pH scale, the exponent of the concentration (e.g., −7) is used, and it is changed from a negative to a positive number. Thus, an acidic solution with 10^{-3} mol of hydrogen ions/L has a pH of 3, whereas a basic solution with 10^{-12} hydrogen ions/L has a pH of 12.

APPENDIX E

Answers to Review and Comprehension Questions

Chapter One
1. a; 2. b; 3. a; 4. c; 5. d; 6. e; 7. a; 8. b; 9. c; 10. d; 11. d; 12. c; 13. d; 14. d; 15. a; 16. b; 17. b; 18. a; 19. e; 20. c; 21. a; 22. b; 23. a; 24. b; 25. e

Chapter Two
1. e; 2. a; 3. b; 4. b; 5. a; 6. d; 7. b; 8. e; 9. c; 10. e; 11. c; 12. d; 13. a; 14. b; 15. c; 16. c; 17. d; 18. c; 19. d; 20. e; 21. b; 22. a; 23. b; 24. d; 25. b; 26. a; 27. c; 28. d; 29. a; 30. e

Chapter Three
1. a; 2. e; 3. c; 4. d; 5. e; 6. c; 7. e; 8. d; 9. b; 10. b; 11. a; 12. b; 13. b; 14. a; 15. d; 16. d; 17. b; 18. b; 19. c; 20. c; 21. c; 22. c; 23. b; 24. e; 25. c; 26. a; 27. d; 28. c; 29. d; 30. c; 31. b; 32. c; 33. c

Chapter Four
1. e; 2. c; 3. a; 4. a; 5. b; 6. d; 7. c; 8. d; 9. d; 10. b; 11. a; 12. b; 13. b; 14. e; 15. a; 16. d; 17. b; 18. b; 19. d; 20. d; 21. a; 22. b; 23. e; 24. b; 25. c; 26. c; 27. b; 28. c; 29. b; 30. d

Chapter Five
1. b; 2. d; 3. e; 4. b; 5. a; 6. c; 7. d; 8. e; 9. c; 10. d; 11. a; 12. b; 13. a; 14. b; 15. b; 16. c; 17. c; 18. d; 19. b; 20. c; 21. b; 22. a; 23. b; 24. a.; 25. d; 26. c; 27. d; 28. c; 29. c; 30. c

Chapter Six
1. e; 2. e; 3. d; 4. a; 5. b; 6. d; 7. c; 8. e; 9. a; 10. a; 11. b; 12. e; 13. c; 14. d; 15. d; 16. c; 17. b; 18. c; 19. e; 20. d; 21. a; 22. b; 23. c; 24. e; 25. a; 26. e; 27. c; 28. e; 29. d; 30. a

Chapter Seven
1. c; 2. c; 3. e; 4. c; 5. d; 6. c; 7. a; 8. d; 9. a; 10. b; 11. a; 12. c; 13. a; 14. b; 15. a; 16. d; 17. e; 18. b; 19. a; 20. b; 21. c; 22. b; 23. c; 24. a; 25. a

Chapter Eight
1. b; 2. d; 3. d; 4. e; 5. c; 6. d; 7. b; 8. a; 9. e; 10. a; 11. d; 12. c; 13. d; 14. e; 15. a; 16. c; 17. b; 18. c; 19. d; 20. b; 21. b; 22. a; 23. c; 24. c

Chapter Nine
1. c; 2. e; 3. a; 4. d; 5. c; 6. e; 7. b; 8. a; 9. b; 10. b; 11. d; 12. b; 13. d; 14. c *or* f; 15. e; 16. e; 17. d; 18. c; 19. c; 20. a; 21. d; 22. b; 23. d; 24. a; 25. d; 26. c; 27. c; 28. c; 29. b; 30. a

Chapter Ten
1. d; 2. b; 3. c; 4. c; 5. c; 6. c; 7. d; 8. d; 9. b; 10. c; 11. a; 12. a; 13. c; 14. d; 15. a; 16. b; 17. d; 18. b; 19. a; 20. d; 21. a; 22. c; 23. a; 24. a; 25. e; 26. c; 27. b; 28. b; 29. d

Chapter Eleven
1. a; 2. b; 3. b; 4. a; 5. c; 6. c; 7. b; 8. a; 9. c; 10. a; 11. b; 12. d; 13. a; 14. c; 15. b; 16. a; 17. a; 18. b; 19. d; 20. c; 21. e; 22. b; 23. d; 24. d; 25. b; 26. b; 27. e; 28. e; 29. e; 30. a

Chapter Twelve
1. d; 2. c; 3. c; 4. d; 5. b; 6. b; 7. d; 8. c; 9. c; 10. b; 11. e; 12. c; 13. d; 14. d; 15. c; 16. e; 17. c; 18. d; 19. c; 20. d

Chapter Thirteen
1. a; 2. c; 3. b; 4. e; 5. c; 6. d; 7. b; 8. b; 9. b; 10. c; 11. a; 12. e; 13. d; 14. d; 15. a; 16. c; 17. d; 18. b; 19. b; 20. a; 21. a; 22. d; 23. b; 24. b; 25. e; 26. b; 27. c; 28. e; 29. c; 30. b; 31. b

Chapter Fourteen
1. c; 2. d; 3. e; 4. c; 5. a; 6. d; 7. b; 8. e; 9. a; 10. b; 11. b; 12. e; 13. c; 14. c; 15. b; 16. d; 17. a; 18. b; 19. b; 20. d; 21. b; 22. b; 23. d; 24. e; 25. d; 26. b; 27. b; 28. d; 29. e; 30. c

Chapter Fifteen
1. e; 2. e; 3. c; 4. a; 5. b; 6. e; 7. b; 8. a; 9. c; 10. d; 11. b; 12. b; 13. c; 14. d; 15. c; 16. b; 17. a; 18. d; 19. c; 20. d; 21. e; 22. a; 23. d; 24. c; 25. a; 26. a; 27. b; 28. d; 29. a; 30. c

Chapter Sixteen
1. e; 2. d; 3. d; 4. a; 5. e; 6. b; 7. d; 8. e; 9. d; 10. d; 11. d; 12. a; 13. d; 14. c; 15. d; 16. c; 17. d; 18. a; 19. c

Chapter Seventeen
1. c; 2. c; 3. b; 4. e; 5. d; 6. b; 7. a; 8. e; 9. b; 10. a; 11. c; 12. e; 13. d; 14. e; 15. d; 16. c; 17. d; 18. a; 19. c; 20. c

Chapter Eighteen
1. e; 2. d; 3. e; 4. b; 5. a; 6. c; 7. b; 8. b; 9. e; 10. c; 11. d; 12. e; 13. a; 14. d; 15. b; 16. c; 17. a; 18. a; 19. d; 20. c; 21. d; 22. e; 23. e; 24. d; 25. b; 26. d; 27. a; 28. d; 29. a; 30. b; 31. e; 32. c; 33. c; 34. e; 35. d

Chapter Nineteen
1. e; 2. c; 3. a; 4. d; 5. a; 6. e; 7. d; 8. b; 9. c; 10. d; 11. b; 12. b; 13. a; 14. c; 15. b; 16. b; 17. e; 18. e; 19. b; 20. a; 21. a; 22. d; 23. d; 24. c; 25. c; 26. c

Chapter Twenty
1. d; 2. e; 3. c; 4. a; 5. b; 6. d; 7. c; 8. c; 9. e; 10. a; 11. c; 12. d; 13. a; 14. b; 15. e; 16. a; 17. a; 18. b; 19. e; 20. b; 21. c; 22. d; 23. a; 24. a; 25. d; 26. c; 27. c; 28. c; 29. e; 30. c

Chapter Twenty-One
1. b; 2. a; 3. a; 4. d; 5. d; 6. b; 7. c; 8. d; 9. a; 10. a; 11. d; 12. e; 13. c; 14. a; 15. c; 16. b; 17. b; 18. d; 19. a; 20. e; 21. d; 22. b; 23. b; 24. b; 25. d; 26. a; 27. d; 28. c; 29. b; 30. e

Chapter Twenty-Two
1. d; 2. b; 3. e; 4. e; 5. d; 6. b; 7. e; 8. b; 9. a; 10. d; 11. d; 12. d; 13. b; 14. e; 15. e; 16. a; 17. d; 18. c; 19. a; 20. b; 21. c; 22. a; 23. d; 24. d; 25. d; 26. d

Chapter Twenty-Three
1. a; 2. b; 3. e; 4. d; 5. c; 6. d; 7. b; 8. c; 9. b; 10. d; 11. a; 12. d; 13. c; 14. c; 15. d; 16. c; 17. b; 18. d; 19. b; 20. c; 21. b; 22. d; 23. c; 24. a; 25. c; 26. d

Chapter Twenty-Four
1. a; 2. d; 3. e; 4. d; 5. a; 6. b; 7. a; 8. b; 9. a; 10. c; 11. b; 12. d; 13. c; 14. d; 15. e; 16. b; 17. d; 18. d; 19. e; 20. e; 21. b; 22. b; 23. b; 24. a; 25. e; 26. e; 27. e; 28. a; 29. c; 30. a

Chapter Twenty-Five
1. d; 2. c; 3. a; 4. e; 5. b; 6. d; 7. e; 8. e; 9. b; 10. a; 11. d; 12. b; 13. d; 14. a; 15. e; 16. c; 17. a; 18. b; 19. a; 20. b

Chapter Twenty-Six
1. d; 2. b; 3. d; 4. c; 5. e; 6. e; 7. b; 8. b; 9. e; 10. a; 11. b; 12. b; 13. b; 14. a; 15. a; 16. a; 17. c; 18. d; 19. c; 20. b; 21. e; 22. d; 23. a; 24. d; 25. e; 26. d; 27. c; 28. b; 29. d; 30. c

Chapter Twenty-Seven
1. a; 2. b; 3. c; 4. d; 5. a; 6. a; 7. b; 8. a; 9. b; 10. d; 11. a; 12. a; 13. b; 14. a; 15. a; 16. c; 17. c; 18. d

Chapter Twenty-Eight
1. e; 2. a; 3. a; 4. b; 5. d; 6. e; 7. b; 8. d; 9. d; 10. c; 11. b; 12. c; 13. a; 14. e; 15. a; 16. b; 17. c *or* d; 18. c; 19. b; 20. a; 21. a; 22. c; 23. a; 24. c; 25. e; 26. e; 27. c; 28. d; 29. c; 30. a

Chapter Twenty-Nine
1. b; 2. e; 3. a; 4. a; 5. d; 6. a; 7. c; 8. c; 9. b; 10. c; 11. a; 12. a; 13. e; 14. d; 15. c; 16. c; 17. e; 18. a; 19. d; 20. b; 21. d; 22. a; 23. c

APPENDIX F

Answers to Critical Thinking Questions

Chapter 1

1. Student B is correct. Body temperature begins to rise as a result of exposure to the hot environment. Sweating eliminates heat from the body and lowers body temperature. Body temperature returning to its ideal normal value is an example of negative feedback. Student A probably thought that it was positive feedback because sweating continued to increase. Sweating, however, is the response. The variable being regulated by sweating is body temperature.
2. Answer *e* is correct. Positive-feedback mechanisms result in movement away from homeostasis and are usually harmful. The continually decreasing blood pressure is an example. Negative-feedback mechanisms result in a return to homeostasis. The elevated heart rate is a negative-feedback mechanism that attempts to return blood pressure to a normal value. In this case, the negative-feedback mechanism was inadequate to restore homeostasis, and medical intervention (a transfusion) was necessary.
3. When a boy is standing on his head, his nose is superior to his mouth. Directional terms refer to a person in the anatomical position, not to the body's current position.
4. a. Inferior or caudal
   b. Anterior, ventral, or superficial
   c. Proximal, superior, or cephalic
   d. Medial
5. The pancreas is located in the right-upper and left-upper quadrants; it is located in the epigastric and left hypochondriac regions. The top of the urinary bladder is located in the right-lower and left-lower quadrants; it is located in the hypogastric region. The rest of the urinary bladder is located in the pelvic cavity.
6. Answer *a* is correct. The best way to reach the anterior surface of the heart begins with the patient lying on his or her back so that the anterior surfaces of the thorax and heart are facing the surgeon. The heart is located in the anterior portion of the thoracic cavity within the mediastinum and is surrounded by the pericardial cavity. The pericardial cavity is lined with the pericardial serous membranes, through which a cut can be made to reach the heart.
7. The uterus is located in the pelvic cavity. The pelvic cavity, however, is surrounded by the bones of the pelvis and does not increase in size during pregnancy. Instead, as the fetus grows the expanding uterus must move into the abdominal cavity, thereby crowding abdominal organs and dramatically increasing the size of the abdominal cavity.
8. After passing through the left thoracic wall, the first membrane encountered is the parietal pleura. Continuing through the pleural cavity, the visceral pleura of the left lung and then the left lung are pierced. Leaving the lung, the bullet penetrates the visceral pleura, the pleural cavity, and the parietal pleura (the lung is surrounded by a double-membrane sac). Next, the parietal pericardium, the pericardial cavity, the visceral pericardium, and the heart are penetrated.
9. After the pole passes through the abdominal wall, it pierces the parietal peritoneum. In passing through the stomach, it penetrates the visceral peritoneum, the stomach itself, and the visceral peritoneum on the other side of the stomach. Because the diaphragm is lined inferiorly by parietal peritoneum and superiorly by parietal pleura, these are the next two membranes pierced. The pole then passes through the pleural space and visceral pleura to enter the lung.

Chapter 2

1. An atom of iron has 26 protons (the atomic number), 30 neutrons (the mass number minus the atomic number), and 26 electrons (because the number of electrons is equal to the number of protons). If an atom of iron loses 3 electrons, it has 3 more protons (positive charges) than electrons (negative charges). Therefore, the iron ion has an overall charge of +3, which is represented symbolically as Fe^{3+}.
2. The formation of free fatty acids and glycerol from a triglyceride is a decomposition reaction because a larger molecule breaks down into smaller molecules. All of the decomposition reactions in the body are collectively referred to as catabolism. This reaction can also be classified as a hydrolysis reaction because as part of the reaction a water molecule is split into hydrogen, which becomes part of the glycerol molecule, and hydroxide, which becomes part of a fatty acid molecule.
3. The slight amount of heat functions as activation energy and starts a chemical reaction. The reaction releases a large amount of heat, thus causing the solution to become hot.
4. Heating (boiling) has destroyed the ability of the molecules in one or both of the solutions to function in the chemical reaction. This is called denaturation. There are two possibilities as to what is denatured. It might be the reactants themselves or an enzyme that catalyzes the reaction.
5. Muscle contains proteins. To increase muscle mass, proteins must be synthesized from amino acids. The synthesis of molecules in living organisms requires the input of energy. That energy comes from the potential energy stored in the chemical bonds of food molecules, which is released during the decomposition of food molecules.
6. pH is a measure of H^+ concentration. If equal amounts of solutions A and B are mixed, the resulting H^+ concentration is the average value of the two solutions—that is, the pH is $(8 + 2) / 2 = 5$. A pH of 5 is acidic. This question illustrates an important point: The pH of a solution can be changed by adding a more acidic or more basic solution to it.
7. The $NaHCO_3$ dissociates, thereby increasing the amount of HCO_3^- in the solution. The HCO_3^- combines with H^+ to form H_2CO_3, which becomes CO_2 and H_2O. The decrease in H^+ causes the pH of the solution to increase (become more alkaline).
8. As A and B are added to the solution, the enzyme E catalyzes the formation of C. However, when C binds to the active site of E, the ability of E to catalyze the formation of C is blocked. As more and more C is produced, the rate of formation of C is slowed. Because the reaction of C with E is reversible, there will always be some E that has a functional (not blocked) active site, and some A will therefore always combine with B.
9. One might try heating the substances because proteins can be denatured and can coagulate (as in frying an egg). Another possibility is to try dissolving the substances in water. Most lipids are insoluble in water, whereas many proteins either are soluble in water or form colloids with water.
10. Most proteins (i.e., typical proteins) contain sulfur, which is not found in phospholipids.

Typical phospholipids and proteins contain carbon, hydrogen, oxygen, nitrogen, and phosphate.

Chapter 3

1. The cells within the wound swell with water and lyse by the introduction of a hypotonic solution. This kills potentially metastatic cells that may still be present in the wound.
2. Water moves by osmosis from solution B into solution A. Because solution A is hyperosmotic to solution B, solution A has more solutes and less water than does solution B. Water therefore moves from solution B (with more water) to solution A (with less water).
3. Answer *b* is correct. At point A on the graph, the extracellular concentration is equal to the intracellular concentration. If movement were by simple diffusion or by facilitated diffusion, at this point the rate of movement would be zero. Because it is not zero, it is reasonable to conclude that the mechanism involved is active transport.
4. Answer *b* is correct. Because the solution is isotonic, there is no exchange of water. Because the solution contains the same concentration of all substances except that it has no urea, only a net movement of urea occurs across the membrane.
5. A reduced intracellular K^+ concentration reduces the concentration difference for K^+ across the plasma membrane. Thus, the rate at which K^+ diffuses out of the cell is reduced, and a smaller charge difference is required across the plasma membrane to oppose the diffusion of the K^+ out of the cell. The potential difference across the plasma membrane is therefore reduced.
6. Because the drug inhibits mRNA synthesis, protein synthesis is stopped. If the cell releases proteins as they are synthesized, the rate of protein secretion dramatically decreases following the administration of the drug. On the other hand, if the cell releases stored proteins, the rate of secretion at first is normal and then gradually declines.
7. The well-developed rough endoplasmic reticulum is indicative of protein synthesis, and a well-developed Golgi apparatus is indicative of secretion. It is likely that this cell synthesizes and secretes proteins.
8. Primary nondisjunction. In secondary nondisjunction where one chromosome fails to separate, one gamete receives an extra chromosome and one receives one chromosome too few. Therefore two of four gametes are affected. In primary nondisjunction two replicated chromosomes fail to separate during first meiosis. Therefore one nucleus undergoing second meiosis starts with one extra chromosome ant the other nucleus starts with one too few. As a result of second meiosis, each gamete receives an extra chromosome from the nucleus with the extra replicated chromosome and each gamete receives one too few chromosomes from the nucleus that is missing a chromosome. Therefore all four gametes are affected.
9. Yes. If D represents the dominant gene for dimpled cheeks and d is the recessive gene for non-dimpled cheeks, the parents could both be Dd. As a result, 1/4 of their children should be DD, 1/2 Dd, and 1/4 dd. Therefore, 3/4 of the children should have dimples and 1/4 should not.
10. Not from this information alone. If tongue rolling is designated T and the inability to roll the tongue is designated t, then a Tt woman married to a tt man will have half Tt children and half tt children. Therefore a man who cannot roll his tongue and a woman who can roll her tongue can have a child who can roll his tongue. This connection alone, however, is not sufficient to establish paternity.
11. A person who is AB blood type has the genotype I^AI^B but has no i allele. A person of blood type O has a genotype of ii. As the AB parent has no i allele, an AB individual cannot be the parent of a child with blood type O.

Chapter 4

1. The tissue is epithelial tissue, as it is lining a free surface, and the epithelium is stratified because it consists of more than one layer. The types of stratified epithelium are stratified squamous, stratified cuboidal, stratified columnar, and transitional epithelium. The structure of the cells in the surface layers enables the determination of a specific tissue type. Flat cells in the surface layer indicate stratified squamous epithelium. Cuboidal cells in the surface layer indicate stratified cuboidal epithelium, and columnar cells in the surface layer point to stratified columnar epithelium. The surface cells of transitional epithelium are roughly cuboidal with cubelike or columnar cells beneath them. When transitional epithelium is stretched, the surface cells are still roughly cuboidal, but underlying layers can be somewhat flattened.
2. In general, epithelial cells undergo cell division (mitosis) in response to injury, and the newly produced cells replace the damaged cells. If the basement membrane is destroyed, however, nothing is present to provide scaffolding for the newly formed epithelial cells. Without the basement membrane, there is no effective way for the newly formed epithelial cells to repair a structure, such as a kidney tubule. Since the basement membranes appear to be mostly present, the person is likely to survive and the kidney will regain most of its ability to function.
3. Epithelium that resists abrasion is stratified squamous epithelium. The nonkeratinized stratified squamous epithelium lining the mouth and the keratinized stratified squamous epithelium of the skin are examples. The cells at the surface are flattened, and when scraped away due to abrasion they are replaced by the cells beneath them. In contrast, epithelial cells that carry out absorption are either simple cuboidal or simple columnar. Because they are one layer thick, they are more susceptible to damage and are not resistant to abrasion. In addition, these cells are large in volume, which allows them to contain the organelles involved in transport, such as mitochondria to produce ATP in the case of active transport. The surface of the cells that absorb are likely to contain microvilli, which increase the surface area for absorption. The flat cells that resist abrasion have no microvilli.
4. Glands producing merocrine secretions do so with no loss of actual cellular material, whereas glands producing holocrine secretions shed entire cells. The cells rupture and die, and the entire cell becomes part of the secretion. You could chemically analyze the secretions for the types of molecules found in cellular organelles. For example, if phospholipids and proteins normally found in membranes are in the secretion, the secretion is a holocrine secretion. If the secretion is watery or contains products that are not found in membranes or organelles, it is a merocrine secretion.
5. The tissue described is dense it is regularly arranged collagenic connective tissue. Injury to this type of tissue affects structures made up of this type of connective tissue, which includes tendons. It does not include ligaments that connect neck vertebrae. These ligaments contain abundant dense regularly arranged elastic connective tissue. The majority of the intervertebral disk consists of dense irregularly arranged collagenic connective tissue.
6. The statement is not appropriate. A tissue capable of contracting is muscle. Both cardiac muscle and smooth muscle cells are mononucleated, although some cardiac muscle cells can have two nuclei, and they are both under involuntary control.

Cardiac muscle is striated, but smooth muscle is not.

7. Histamine is one of the mediators of inflammation released in response to tissue damage. Several other mediators of inflammation, however, are released during inflammation in addition to histamine. Antihistamines might reduce the inflammatory response somewhat, but they are not likely to have a major effect because of the other mediators of inflammation released at the same time. In certain types of inflammatory responses, such as allergic responses, histamines are released in large amounts. Under these conditions, antihistamines do reduce the inflammatory response.
8. The secretions produced by this gland are consistent with a merocrine secretion process. A merocrine secretion involves the transport of solutes into the lumen of the gland by the epithelial cells that line the gland. If the epithelial cells of the gland are permeable to water, water will follow the solutes because of osmosis and the secrection will be isotonic. Aprocrine and holocrine secretions have many organic molecules in them because portions of the cell become part of the secretion.
9. Because this tissue has a free surface and consists of a single layer of cells, it is simple epithelium. Because the cells are narrow and tall, it is simple columnar epithelium. The microvilli increase the surface area of the free surface of the epithelial cells and the mictochondria synthesize ATP. Therefore, this cell type is probably involved in active transport of solutes. The goblet cells indicate that it also secretes mucus. These characteristics are consistent with the epithelium lining the small intestine.
10. Cancer results from mutations in somatic cells. The BRCA1 and BRCA2 genes are susceptible to mutations in breast tissue cells. If a mutation occurs, the result is an increased mitosis of the affected cells. The develfopment of cancer depends on mutations in these genes and additional mutations of genes in some of the cells that result from the increased rate of mitosis. The mutations necessary for the development of cancer do not occur in all people who have inherited the genes that make them susceptible to cancer. Therefore, people with these genes do not always develop cancer. Also, a person who is heterozygous for one of these genes (e.g., a persons who has one BRCA1 and one normal allele) has less of a chance of developing cancer than a person who has two BRCA1 alleles and therefore is homozygous for the BRCA1 gene. The person who has two BRCA1 alleles has a much greater chance of a mutation in that gene than a person who has one BRCA1 allele.

Chapter 5

1. The stratum corneum, the outermost layer of the skin, consists of many rows of flat, dead epithelial cells. The many rows of cells, which are continuously shed and replaced, are responsible for the protective function of the integument. In infants, there are fewer rows of cells, resulting in skin that is more easily damaged than that in adults.
2. Melanocytes produce melanin, which protects underlying tissue from ultraviolet radiation. Therefore, we expect melanocytes to be as superficial as possible. Also, melanin production varies, depending on exposure to the sun. Response to stimulation is a characteristic of living cells. Thus, melanocytes should be found in the most superficial living layer of the epidermis, the stratum basale.
3. Alcohol is a solvent that dissolves lipids (see chapter 2). It removes the lipids from the skin, especially in the stratum corneum. The rate of water loss increases after soaking the hand in alcohol because of the removal of the lipids that normally prevent water loss.
4. Carotene, a yellow pigment from ingested plants, accumulates in lipids. The stratum corneum of a callus has more layers of cells than other noncallused parts of the skin, and the cells in each layer are surrounded by lipids. The carotene in the lipids makes the callus appear yellow.
5. When first exposed to the cold temperature just before starting the run, the blood vessels in the skin constrict to conserve heat. This produces a pale skin color. After a person has been running for awhile, as a result of the excess heat generated by the exercise, the blood vessels in the skin dilate. This results in heat loss and helps prevent overheating. Increased blood flow through the skin causes it to turn red. After the run, the body still has excess heat to eliminate, so the skin remains red for some time.
6. Yes, the skin (dermis) can be overstretched due to obesity or rapid growth.
7. The vermillion border is covered by keratinized epithelium that is transitional between the nonkeratinized stratified epithelium of the mucous membrane and the keratinized stratified epithelium of the facial skin. The mucous membrane has mucous glands, which secrete mucus (see chapter 4). The mucus helps keep the inner surface of the lips moist. In addition, the inner surface of the lips is "sealed off" from the outside air most of the time and is moistened by saliva. The keratinized stratified epithelium of the skin has a stratum corneum, with lipids that reduce water loss. The skin also has sebaceous glands, which produce sebum that reduces water loss. The skin of the vermillion border is not moistened by mucus or saliva and it does not have sebaceous glands. Without sebaceous glands, the surface of the vermillion border is not protected against drying by sebum. Also, the vermillion border is not as heavily keratinized as facial skin—that is, there are fewer cell layers with surrounding lipids. For these reasons, the vermillion border dries out more easily than the mucous membrane of the oral cavity or the skin of the face.
8. Eyelashes have a short growth stage (30 days) and are therefore short. Fingernails grow continuously but are short because they are cut, broken off, or worn down.
9. The hair follicle, but not the hair, is surrounded with nerve endings that can detect movement or pulling of the hair. The hair is dead, keratinized epithelium, so cutting the hair is not painful.
10. Several methods have some degree of success in treating acne. (a) Kill the bacteria. One effective agent is benzoyl peroxide, found in some acne medications. (b) Prevent blockage of the hair follicle. A vitamin A derivative (tretinoin; Retin-A) has proven effective in keeping the follicular epithelial cells and sebum from building up and closing off the hair follicle. (c) Unplug the follicle. Some sulfur compounds (e.g., Acnederm) speed up peeling of the skin and thus unplug the follicle.
11. Probably not, because, following removal of the nail from the nail fold, it may grow back into the nail fold and the ingrown toenail may reoccur. One solution is to remove the small part of the nail responsible for the ingrown toenail. Prior to this drastic approach, sterile gauze can be placed between the nail and the nail fold to force the nail away from the nail fold. After the nail fold is healed, the gauze can be removed.
12. Rickets is a disease of children resulting from inadequate vitamin D intake. With inadequate vitamin D, there is insufficient absorption of calcium from the intestine, resulting in soft bones. If adequate vitamin D is ingested, rickets is prevented, whether one is dark- or fair-skinned. However, if dietary vitamin D is inadequate, when the skin is exposed to ultraviolet light, 7-dehydrocholesterol is converted into cholecalciferol, which can be converted to vitamin D. Dark-skinned children are more

susceptible to rickets because the additional melanin in their skin screens out the ultraviolet light and they produce less vitamin D.

Chapter 6

1. Normally, bone matrix and bone trabeculae are organized to be strongest along lines of stress. Random organization of the collagen fibers of bone matrix results in weaker bones. In addition, the reduced amount of trabecular bone makes the bone weaker. Fractures of the bone can occur when the weakened bone is subjected to stress.
2. Replacement of cartilage of the epiphyseal plate by bone normally occurs on the diaphyseal side of the plate. As growth ceases, the cartilage cells stop dividing and producing new cartilage. Replacement of cartilage with bone continues from the diaphyseal side, and eventually all of the cartilage of the plate is bone.
3. Mechanical stress applied to bone stimulates osteoblast activity, so the patient with a walking cast should heal faster.
4. Osteoporosis is depletion of bone matrix that results when more bone is destroyed than is formed. Because mechanical stress stimulates bone formation (osteoblast activity), running helps prevent osteoporosis in the bones being stressed. This includes the bones of the lower limbs and the spine.
5. The loss of bone density results because the bones are not bearing weight in the weightless environment. Therefore, osteoblasts are not sufficiently stimulated and bone resorption exceeds bone building. Bone loss can be slowed by stressing the bones using exercises against resistance, such as cycling.
6. The kidneys are the site of production of active vitamin D (see chapter 5), which is needed for Ca^{2+} absorption in the small intestine. Kidney failure can result in inadequate vitamin D production, too little uptake of Ca^{2+}, and therefore osteomalacia.
7. Testosterone normally causes a growth spurt at puberty, followed by slower growth and closure of the epiphyseal plate. Without testosterone, growth is slower but proceeds longer, resulting in a taller than normal person.
8. Blood vessels in central canals run parallel to the long axis of the bone, and perforating canals run at approximately a right angle to the central canals. Thus, perforating canals connect to central canals, which allows blood vessels in the perforating canals to connect with blood vessels in the central canals. After a fracture, blood flow through the central canals stops back to the point where the blood vessels in the central canals connect to the blood vessels in the perforating canals. The regions of bone on both sides of the fracture associated with this lack of blood delivery die.
9. Hyperparathyroidism stimulates increased bone breakdown and could cause osteitis fibrosa cystica, a condition in which the bone is eaten away as Ca^{2+} are released from the bone. The result can be a deformed bone that is likely to fracture. Vitamin D therapy might help because vitamin D promotes an increase in blood Ca^{2+} levels and therefore increased deposition of Ca^{2+} in bone.

Chapter 7

1. An infection in the nasal cavity could spread to adjacent cavities and fossae, including the paranasal sinuses: (1) frontal, (2) maxillary, (3) ethmoidal, and (4) sphenoidal; (5) the orbit (through the nasolacrimal duct); (6) the cranial cavity (through the cribriform plate); and (7) the throat (through the posterior opening of the nasal cavity).
2. Falling on the top of the head could drive the occipital condyles into the superior articulating processes of the atlas, causing a fracture. An uppercut to the jaw would slightly lift the occipital condyles away from the superior articulating processes of the atlas and usually does not result in a fractured atlas. Such a blow to the jaw can, however, fracture the temporal bone where it articulates with the mandible.
3. Forceful rotation of the vertebral column is most likely to damage the articular processes, especially in the lumbar region, where the articular processes tend to prevent excessive rotation (the superior articular processes face medially and the inferior articular processes face laterally).
4. Weaker back muscles on one side can cause the vertebral column to bend laterally (scoliosis) toward the opposite side. Lordosis can result from pregnancy. As the fetus causes the abdomen to move anteriorly, the thorax and head tend to pull posteriorly, to restore the center of gravity. This posture increases the lumbar curvature. The same effect can be seen in people who are "pot-bellied."
5. If the ulna and radius become fused, the radius can no longer rotate relative to the ulna. As a result, most of the rotation of the forearm and hand is lost.
6. Measure from the anterior superior iliac spine (a "stationary" point relative to the limb, which can be easily found as a surface landmark) to the lateral malleolus. The inferior side of the foot can also be used, if the person is standing on a flat surface. A defect of the foot or ankle may occur, however, in which the ankle on one side is elevated. If the length of the thigh is the only part to be measured, measure to the lateral epicondyle.
7. The ischial tuberosity is the bony protuberance.
8. Women's hips are wider than men's. As the knees are positioned toward the midline, the slope of the femur from its proximal end toward its distal end is greater in women. As a result, more women than men tend to be knock-kneed.
9. The lateral malleolus extends further distally than does the medial malleolus, thus making it more difficult to turn the foot laterally than to turn it medially. The styloid process of the radius extends farther distally than the styloid process of the ulna, thus making it more difficult to cock the wrist toward the thumb (laterally) than toward the little finger (medially).
10. Landing on the heels could fracture the calcaneus. Heavy objects, such as the firefighters, landing on the dorsal surface of the foot could fracture the metatarsals or even the tarsals.

Chapter 8

1. If the sternocostal synchondrosis were to ossify, becoming a synostosis, there would no longer be any stretch through the costal cartilage, the thorax could not expand, and, as a result, respiration would be severely hampered.
2. a. suture, little or no movement
   b. syndesmosis, some movement
   c. complex synovial joints: the humeroulnoradial joint is a hinge joint, the radioulnar joint is a pivot joint; all have considerable movement
3. a. flexion and supination
   b. flexion of the hip and extension of the knee
   c. abduction of the arm at the shoulder
   d. flexion of the knee and plantar flexion of the foot
4. The anterior drawer test determines the integrity of the anterior cruciate ligament, and the posterior drawer test determines the integrity of the posterior cruciate ligament. Unusual movement during the posterior drawer test indicates damage to the posterior cruciate ligament.

Chapter 9

1. Botulism poisoning results from the consumption of botulism toxin produced by the bacterium *Clostridium botulinum.* The toxin binds to presynaptic nerve terminals and prevents the release of acetylcholine. Thus, action potentials in nerves cannot produce action potentials in skeletal muscles, and the result is paralysis of skeletal muscles, which explains the difficulty in breathing and swallowing. Other reasonable

explanations are that the toxin binds to and blocks the receptors for acetylcholine, that the toxin blocks the entry of Ca^{2+} into the presynaptic terminal and thus prevents acetylcholine release, and that the toxin specifically prevents the entry of ions through Na^+ channels of skeletal muscle cells.

2. Muscular dystrophy results from gradual atrophy of skeletal muscle fibers and their replacement with connective tissue. Myasthenia gravis results from the degeneration of the receptors for acetylcholine on the postsynaptic membranes of skeletal muscle cells. If an inhibitor of acetylcholinesterase is administered, the result should be an increase in the concentration of acetylcholine in the nerve muscle synapse. Thus, more acetylcholine is available to bind to acetylcholine receptors. In people suffering from myasthenia gravis, the increased concentration of acetylcholine in the synapse allows acetylcholine to bind a greater percentage of the acetylcholine receptors present and causes the muscle contractions to increase in strength. In people who have muscular dystrophy, the muscle contractions do not increase in strength because muscle atrophy is the cause of the weakness. The additional acetylcholine in the neuromuscular synapse has no effect on the weakened muscle fibers.
3. Placing sarcoplasmic reticulum from skeletal muscle cells into the beaker would remove calcium from the solution because sarcoplasmic reticulum transports Ca^{2+} from the solution into the Sarcoplasmic reticulum. In addition, ATP would have to be added for two reasons: (1) The sarcoplasmic reticulum actively transports calcium and therefore requires ATP and (2) ATP must bind to the heads of the myosin molecules before the myosin heads can release from the active sites on the actin molecules.
4. A lower than normal temperature decreases the rate of all of the processes that occur in the lag phase of muscular contraction because a lower temperature decreases the rate of all chemical reactions and the rate of ion diffusion. As a consequence, the lag phase requires a longer time.
5. Start with a subthreshold stimulus and increase the stimulus strength by very small increments. Apply the stimulus to the nerve of muscle A and muscle B. If the number of motor units is the same for both preparations, each time the stimulus strength is increased the degree of tension produced by the muscles will also increase to the same degree in each muscle. If one muscle has more motor units than the other, the muscle with the greater number of motor units will exhibit a greater number of separate increases in tension, and the magnitude of the increases in tension will be smaller than those seen in the muscle with fewer motor units.
6. While the weight is being held steady, the cross-bridges are pulling the z-disks closer together, but the external load (the weight) is pulling the sarcomeres apart with equal force. Because the internal and external forces are equal, the cross-bridges are producing enough force to hold the weight steady, but not enough to shorten the muscle, so the sarcomeres remain the same length. When the individual lowers the weight, the cross-bridges are producing less force than the weight. Thus, each time a cross-bridge detaches from actin, the thin filaments "slip" and the sarcomeres lengthen. When the individual raises the weight, the cross-bridges are producing more force than the external load. Thus, the cross-bridges collectively are able to produce enough force to pull the Z lines closer together and the sarcomeres get shorter.
7. The shape of an active tension curve for skeletal muscle can be seen in figure 9.21. In contrast, an active tension curve is much flatter for smooth muscle. That is, for each increase in the length of a muscle fiber, there is little change in the active tension produced by the smooth muscle fiber. Smooth muscle has the ability to increase in length without much increase in the tension produced by the smooth muscle cells.
8. Both the 100 m run and weight lifting involve rapid and intense contractions of skeletal muscles that are completed quickly. These contractions depend on anaerobic respiration for a significant amount of the ATP produced. In contrast, the 10,000 m run involves sustained muscular contractions that are not as rapid, but the slower contractions are repeated many times during the run. Aerobic respiration produces the majority of the ATP for the 10,000 m run. Anaerobic respiration is associated with a decrease in creatine phosphate, an increase in creatine, an increase in lactic acid, and a decrease in glycogen, and the enzymes responsible for anaerobic respiration function more rapidly. Aerobic respiration is associated with increased enzyme activity in the mitochondria and an increase in carbon dioxide production. Oxygen is used more rapidly during aerobic respiration.
9. During intense exercise, it is possible to experience physiologic contracture. Being unable to either contract or relax the muscles for a short time while exercising suggests the existence of physiologic contracture.
10. Smooth muscle contractions in the stomach result from the autorhythmic contraction of visceral smooth muscle tissue. Because stretching this type of muscle does not dramatically affect the force of the rhythmic contractions, there is little difference between the contractions 5 minutes and 1 hour after a meal.
11. During the 100 *m* race, Shorty depended on ATP produced by anaerobic metabolism. That produced an oxygen deficit at the end of the run, which resulted in an elevated rate of respiration for a time. During the longer and slower run, most of the ATP for muscle contractions was produced by aerobic respiration, and very little oxygen deficit developed. Prolonged aerobic respiration is required to "pay back" the oxygen deficit. Shorty's rate of respiration was, therefore, prolonged after the 100 m race but not after the longer but slower run.
12. High blood K^+ concentration also results in depolarization of smooth muscle plasma membranes. Depolarization of the smooth muscle plasma membrane results in increased muscle contractions and increased permeability of the plasma membrane to both Na^+ and Ca^{2+}, which results in further depolarization and an increase in the intracellular concentration of Ca^{2+}. These changes result in the production of action potentials and muscle contractions.
13. The muscles would contract. ATP would be available to bind to the myosin heads, thus allowing myosin molecules to be released from actin molecules. The cross-bridges would immediately re-form, and complete cross-bridge cycling would result in contraction of the muscle fibers. As long as Ca^{2+} were present at high concentrations in the sarcoplasm, contraction of the muscles would occur. If the sarcoplasmic reticulum were intact, ATP would be available to drive the active transport of Ca^{2+} into the sarcoplasmic reticulum. As the Ca^{2+} decreased in the sarcoplasm, relaxation would result. If the sarcoplasmic reticulum were not intact, however, and could not transport Ca^{2+} into the sarcoplasmic reticulum as fast as they leaked out, the muscle would remain contracted until it fatigued.
14. Hormones can bind to ligand-gated Ca^{2+} channels, and the channels, in response, open. Calcium ions diffuse into the cell and cause contraction to occur. Only a small amount of depolarization results as Ca^{2+} diffuse into the cell, and, since Na^+ channels do not open, a large change in the resting membrane potential does not occur.
15. In experiment A, the students used anaerobic respiration as they started to run in place, but aerobic respiration also increased to meet most of their energy needs. When they stopped, their respiration rate was increased over resting levels because of repayment of

the oxygen deficit due to anaerobic respiration. In experiment B, almost all of the students' respiration came from anaerobic respiration because the students held their breath while in running place. Consequently, the students had a much larger oxygen deficit. The students' respiratory rate and depth would be greater than in experiment A, or their respiration rates would be elevated for a longer period of time than in experiment A.

16. The color of the meat depends on the number of capillaries (blood is red) within the muscle and is based on its myoglobin content (myoglobin is also red). After cooking, the tissues look darker, not red, because the blood and myoglobin have been broken down by the heat. Thus, the dark meat is darker because it contains more capillaries and more myoglobin. This is consistent with slow-twitch muscle fibers, which are used for the maintenance of posture and slow movements, such as walking. The white muscle, with fewer capillaries and lower myoglobin content, is consistent with fast-twitch muscle fibers, which are used for quick movements, such as running or flying.

Chapter 10

1.

Muscle	Action	Synergist	Antagonist
Longus capitis	Flexes neck	Rectus capitis anterior Longis coli	Most of the posterior neck muscles
Erector spinae	Extends vertebral column	Interspinales Multifidus Semispinalis thoracis	Most anterior abdominal muscles
Coraco-brachialis	Adducts arm	Latissimus dorsi Pectoralis major Teres major Teres minor	Deltoid Supra-spinatus
	Flexes arm	Deltoid (anterior) Pectoralis major Biceps brachii	Deltoid (posterior) Latissimus dorsi Teres major Teres minor Infra-spinatus Sub-scapularis Triceps brachii

2. Biceps brachii: pull-ups with hands supinated
Triceps brachii: push-ups
Deltoid: abduction of the arms to shoulder height, with weights in the hands (abduction past shoulder height involves mostly scapular rotation by the trapezius)
Rectus abdominis: sit-ups to 45 degrees (sit-ups past 45 degrees involve mostly the psoas major)
Quadriceps femoris: extending the legs against a force
Gastrocnemius: plantar flexion of the feet against a force, such as toe raises with a weight on the shoulders
3. The brachioradialis originates on the humerus and inserts onto the distal end of the radius. The fulcrum of this lever system is the elbow joint. With a weight held in the hand, the pull, applied between the weight and the fulcrum, is a class III lever system. With the weight on the forearm, the weight is between the pull and the fulcrum and is a class II lever system. A greater weight can be lifted if placed on the forearm rather than in the hand, but weights placed on the forearm cannot be lifted as far.
4. The muscles that flex the head also oppose extension of the neck. In an accident causing hyperextension of the neck, these muscles could be stretched and torn. The muscles involved could include the sternocleidomastoid, longus capitis, rectus capitis anterior, and longus coli. Automobile headrests are designed so that, if adjusted correctly, the back of the head hits the headrest during a rear-end accident, thereby preventing hyperextension of the neck.
5. The only muscle that elevates the lower eyelid is the orbicularis oculi, which "closes the eye." With this muscle not functioning, the lower eyelid would droop. The levator anguli oris, which elevates the angle of the mouth, was also apparently affected, allowing the corner of the mouth to droop. The zygomaticus major may also have been affected, as it inserts onto the corner of the mouth (see figure 10.9).
6. The genioglossus muscle protrudes the tongue. If it becomes relaxed, or paralyzed, the tongue may fall back and obstruct the airway. This can be prevented or reversed by pulling forward and down on the mandible, thus opening the mouth. The genioglossus originates on the genu of the mandible. As the mandible is pulled down and forward, the genioglossus is pulled forward with the mandible, thus pulling the tongue forward also.
7. The rotator cuff muscles are the primary muscles holding the head of the humerus in the glenoid fossa, especially the supraspinatus. In fact, a torn rotator cuff, which usually involves a tear of the supraspinatus muscle, often results in dislocation of the shoulder.
8. With the quadriceps femoris paralyzed, the leg could not be extended, and the lower limb could not bear weight unless the knee were passively extended, such as by pushing back on the distal end of the thigh with the hand. Walking would be almost impossible, except by taking very small steps and by pushing back on the knee with each step, or by bracing the knee in an extended position.
9. Speedy has ruptured the calcaneal tendon, and the gastrocnemius and soleus muscles have retracted, thereby causing the abnormal bulging of the calf muscles. Because the major plantar flexors are no longer connected to the calcaneus, the runner cannot plantar flex the foot, and the foot is abnormally dorsiflexed because the antagonists have been disconnected.
10. If your are sitting at a desk with the book open on the desk in front of you, then very few muscles are required to turn the page. First the arm is extended slightly to push the forearm and hand along the edge of the book. This can be accomplished with the anterior portion of the deltoid muscle. The hand is then slightly supinated by means of the supinator muscle. The thumb and index finger are then flexed to grasp the page to be turned. This movement involves the flexor pollicis brevis and flexor digitorum superficialis to the index finger (the flexor digitorum profundus may also be involved). The page is then turned by extending the fingers (extensor digitorum), pronating the hand (pronator quadratus and pronator teres), and medially rotating the arm (pectoralis major, teres major, latisimus dorsi).

Chapter 11

1. A reduced intracellular concentration of K^+ causes depolarization of the resting membrane potential. Because the intracellular concentration of K^+ is reduced, the concentration gradient for K^+ from the inside to the outside of the plasma membrane is also reduced. Thus, the rate at which K^+ diffuse out of the cell is reduced, and a smaller charge difference across the plasma membrane is required to oppose the diffusion of the K^+ out of the cell. Therefore, the potential difference across the plasma membrane is reduced, and the cell is depolarized.
2. Because the plasma membrane is much less permeable to Na^+ than to K^+, changes in the extracellular concentration of Na^+ affect the resting membrane potential less than do changes in the extracellular concentration of K^+. Therefore, increases in extracellular Na^+ have a minimal effect on the resting membrane potential. Because the membrane is much more permeable to

Na^+ during the action potential, the elevated concentration of Na^+ in the extracellular fluid results in Na^+ diffusing into the cell at a more rapid rate during the action potential, resulting in a greater degree of depolarization during the depolarization phase of the action potential.

3. Because lithium ions reduce the permeability of plasma membranes to Na^+, the Na^+ channels in the plasma membrane tend to remain closed. A normal stimulus causes Na^+ channels to open, allowing Na^+ to diffuse into the cell, thereby resulting in depolarization. The cell is less sensitive to stimuli because the membrane is less permeable to Na^+.
4. Smooth muscle cells contract spontaneously in response to spontaneous depolarizations that produce action potentials. One way action potentials can be produced spontaneously is if membrane permeability to Na^+ spontaneously increases. As a result, a few Na^+ enter the smooth muscle cells and cause a small depolarizing graded potential. The small depolarization can cause voltage-gated Na^+ channels to open, which results in further depolarization, thereby stimulating additional Na^+ voltage-gated ion channels to open. This positive-feedback cycle can continue until the plasma membrane is depolarized to its threshold level and an action potential is produced.
5. Action potential conduction along a myelinated nerve fiber is more energy-efficient because the action potential is propagated by saltatory conduction, which produces action potentials at the nodes of Ranvier. Compared with an unmyelinated nerve fiber, only a small portion of the myelinated neuron's membrane has action potentials. Thus, there is less flow of Na^+ into the neuron (depolarization) and less flow of K^+ out of the neuron (repolarization). Consequently, the Na^+–K^+ pump has to move fewer ions in order to restore ion concentrations. Because the Na^+–K^+ pump requires ATP, myelinated axons use less ATP than unmyelinated axons.
6. The inhibitory neuromodulator causes the postsynaptic neuron to become less sensitive to excitatory stimuli, probably by causing hyperpolarization of the postsynaptic neuron. As a result, the excitatory neurotransmitter released from the excitatory neuron is less likely to produce postsynaptic action potentials.
7. With aging, there is a decrease in the amount of myelin surrounding axons, which decreases the speed of action potential propagation. At synapses, there is also an increase in the time it takes for action potentials in the presynaptic terminal to cause the production of action potentials in the postsynaptic membrane. It is believed this results from a reduced release of neurotransmitter by the presynaptic terminal and a reduced number of receptors in the postsynaptic membrane.
8. Organophosphates inhibit acetylcholinesterase, thereby causing an increase in acetylcholine in the synaptic cleft, leading to overproduction of action potentials, tetanus of muscles, and possible death resulting from respiratory failure (see chapter 9). Curare is the best antidote because it blocks the effect of acetylcholine and counteracts the organophosphate. Too much curare, however, can cause flaccid paralysis of the respiratory muscles. Injecting acetylcholine would make the effect of the organophosphate worse. Potassium chloride causes depolarization of muscle cell membranes, thereby making them more sensitive to acetylcholine.
9. If the motor neurons supplying skeletal muscle are innervated by both excitatory and inhibitory neurons, blocking the activity of the inhibitory neurons with strychnine results in overstimulation of the motor neurons by the excitatory neurons.
10. When the neurotoxin binds to ligand-gated Na^+ channels in the postsynaptic membrane of a skeletal muscle fiber, they open and Na^+ enter the cell, producing graded potentials. When the graded potentials reach threshold, an action potential is produced, stimulating the muscle fiber to contract. The neurotoxin tends to remain bound to the ligand-gated Na^+ channels, however, which prevents ACh from binding. Thus, the nervous system's ability to stimulate the muscle fiber decreases as more and more neurotoxin binds to ligand-gated Na^+ channels. Because the ligand-gated Na^+ channels with bound neurotoxin remain open, Na^+ continue to enter the muscle fiber, causing its resting membrane potential to depolarize. Eventually, the membrane becomes so depolarized that it is unresponsive to stimulation. Death from a cobra bite usually occurs because of paralysis of respiratory muscles.
11. A Na^+ channelopathy in which Na^+ channels open more readily that normal or stay open longer than normal could cause an increased production of action potentials. A Ca^{2+} channelopathy in which more Ca^{2+} than normal enter the presynaptic terminal could result in increased release of neurotransmitter from the presynaptic terminal and thus an increased production of action potentials.

Chapter 12

1. If the neuron with its cell body in the cerebrum is an inhibitory neuron and if it also synapses with the motor neuron of a reflex arc, stimulation of the cerebral neuron can inhibit the reflex.
2. The phrenic nerve is cut in the thorax, and the surgery is performed while the lung is being removed.
3. The ulnar nerve supplies the medial third of the hand, little finger, and medial half of the ring finger. The median nerve supplies the lateral two-thirds of the palm and thumb and the surface of the index, middle, and lateral half of the ring finger. The radial nerve supplies the lateral two-thirds of the dorsum of the hand.
4. Pulling on the upper limb when it is raised over the head can damage the lower brachial plexus—in this case, the origin of the ulnar nerve. The ulnar nerve innervates muscles that abduct/adduct the fingers and flex the wrist.
5. The sciatic nerve has rootlets from L4 to S3. Depending on the rootlet compressed, pain can be felt in different locations.
6. a. obturator nerve
   b. femoral nerve
   c. sciatic (tibial) nerve
   d. obturator nerve
   e. obturator nerve, some from femoral nerve
7. The plaster cast is pressing against the common fibular (peroneal) nerve at the neck of the fibula. Tingling is expected along the lateral and anterior leg and the dorsum of the foot.

Chapter 13

1. A condition in which a patient loses all sense of feeling in the left side of the back, below the upper limb, and extending in a band around to the chest, also below the upper limb, but all sensation on the right is normal, suggests that the patient's dorsal roots have been damaged on the left side adjacent to the part of the spinal cord supplying that part of the body. (The basis of this condition is explained more fully in chapter 14.)
2. If CSF does not drain properly, the fluid accumulates and exerts pressure on the brain (hydrocephalus). In the developing fetus, the ventricles enlarge because of the excess fluid pressure. The head also enlarges because the skull bones have not fused. The expansion of the head is not sufficient,

however, to relieve all the pressure exerted on the developing brain by the expanding ventricles. As a result, the cerebral cortex becomes proportionately thinner as it is compressed between the ventricles and skull. In many cases, less gyri form in the cerebral cortex. Brain damage may or may not result, depending on the amount of excess CSF, the ventricular pressure generated, and the areas of the brain damaged by the pressure.

3. Enlargement of the lateral and third ventricles, without enlargement of the fourth ventricle, suggests a blockage between the third and fourth ventricles in the cerebral aqueduct. This defect, called aqueductal stenosis, is a common congenital problem.
4. Blood in the CSF taken through a spinal tap indicates the presence of blood in the subarachnoid space and suggests that the patient has a damaged blood vessel in the subarachnoid space.
5. I: Test vision; II: Have the person describe the smell of something placed under their nose; III: Test eye movement; IV: Test the ability to move the eye down and out; V: Test the sense of feeling in the face; VI: Test the ability to move the eye to the side; VII: Test for the ability to taste an item on the front of the tongue. Also check for facial expression; VIII: Test the person's ability to hear; IX: Test the ability of the person to swallow; X: Test the person's ability to swallow. Also check the uvula when the mouth is opened. The uvula will "point" away from the side where X is not working; XI: Test the ability of the person to turn their head; XII: Have the person protrude their tongue. If XII is not functioning on one side, the tongue will "point" toward the damaged side.
6. The cerebellum was damaged in this patient. The cerebellum is in charge of controlling coordinated muscle movement and maintaining muscle tone.
7. The olfactory nerves (CN I) travel through the olfactory foramina within the cribriform plate of the ethmoid bone as they pass from the nasal epithelium to the olfactory bulbs (see table 13.5). In this case, the cribriform plate of the ethmoid bone was probably fractured, severing the connections of the olfactory nerves to the olfactory bulb and resulting in a loss of the sense of smell.

Chapter 14

1. The first sensations that occur when a woman picks up an apple and bites into it are visual (special), tactile (general), and proprioceptive (general). The woman holds the apple in her hand and looks at it. The tactile sensations from mechanoreceptors in the hand tell her that the apple is firm and smooth. The proprioceptive sensations originating in the joints of the hand tell the woman the size and shape of the apple. Visual input also tells her the size of the apple and that it has a smooth surface, as well as its color. As the woman bites into the apple and begins to chew, proprioceptive sensations from the teeth and jaw provide information on how widely the jaws must be opened to accommodate the bite and how hard to bite down. Tactile sensations originating in the tongue and cheeks tell her the location of the bite of apple and its texture as it is moved about in the mouth. Taste sensations (special, chemoreceptor) from the tongue provide information that the apple has characteristics of being both sweet and sour. Olfactory sensations (special) provide more specific information that the "fruity taste" is that of an apple.
2. a. The most likely explanation is that the olfactory neurons accommodate and no longer respond to odor stimulus.
   b. The fact that one can hear the sound when one tries indicates that the hair cells in the spiral organ have not accommodated and are still able to detect the sound stimulus. Many action potentials arriving in the brain are prevented from causing conscious perception, until we consciously "pay attention" to the stimulus. For example, you may not be paying attention to general conversations in a crowded room or hall, until someone says your name. The sound of your name leaps out of the surrounding babble, and you are suddenly interested in what was being said by the person who spoke your name.
3. It is possible that the dorsal-column/medial-lemniscal system within the right side of the spinal cord is damaged. However, it is also possible that this system is damaged within the medulla oblongata, where neurons synapse and cross over to the left side of the brain, or within the tracts on the left side that ascend from the medulla oblongata to the thalamus. Another possibility is damage to the cerebral cortex on the left side. Additional information is needed to determine exactly where the injury is located.
4. The fibers of the dorsal-column/medial-lemniscal system carry two-point discrimination and proprioceptive information. Primary neurons from the right side of the body ascend the spinal cord in the dorsal column and synapse with secondary neurons in the medulla oblongata. The secondary neurons cross over in the upper medulla and ascend through the left side of the pons to the thalamus. A patient suffering from a loss of two-point discrimination and proprioception on the right side of the body as a result of a lesion in the medial-lemniscal system in the pons has a lesion in the left side of the pons.
5. The fibers of the lateral spinothalamic tract carry impulses for pain and temperature. A lesion in the area where these fibers decussate results in the bilateral loss of pain and temperature sensations only at the level of the lesion, and there is no loss of sensation below the lesion. This occurs because fibers decussating above or below the lesion, as well as tracts that pass lateral to the lesion, are unaffected.
6. The damaged tracts are the lateral corticospinal tract, controlling motor functions on the right side of the body, and the lateral spinothalamic tract for pain and temperature sensations from the left side of the body. Damage to these tracts in the right side of the spinal cord produces the observed symptoms, because, in the cord the lateral spinothalamic tract crosses over at the level of entry and is therefore located on the opposite side of the cord from its peripheral nerve endings, whereas the corticospinal tract lies on the same side of the cord as its target muscles.
7. Complete unilateral transection of the right side of the spinal cord results in loss of motor function (lateral corticospinal tract), proprioception, and two-point discrimination (dorsal-column/medial-lemniscal system) on the same side of the body as the lesion, below the level of the lesion. Pain and temperature sensations (lateral spinothalamic tract) are lost on the opposite side of the body from below the level of the lesion. These symptoms describe the Brown-Séquard syndrome. Light touch is not greatly affected on either side because of the large number of collateral branches in the anterior spinothalamic tract.
8. The right cerebral cortex controls the left side of the body. The motor cortex has a topographic representation of the opposite side of the body, with the hand, forearm, arm, and shoulder located approximately in the center of the precentral gyrus. The lesion is therefore in the center of the right precentral gyrus of the cerebrum. Some grosser control of the left-upper limb may still exist because of the indirect pathways, but there is spastic paralysis.
9. Damage to the cerebellum can result in decreased muscle tone, balance impairment, a tendency to overshoot when reaching for

or touching something, and an intention tremor. These symptoms are opposite to those seen with basal ganglia dysfunction. Cerebellar dysfunction exhibits symptoms very similar to those seen in an inebriated person, and the same tests can be applied, such as having the person touch his or her nose or walk a straight line.

10. Memory storage for the 10 minutes prior to the accident was in short-term memory and was disrupted before it could be transferred to long-term memory. Anytime a person suffers a concussion, there is a possibility that he or she will later develop postconcussion syndrome. Symptoms include muscle tension or migraine headaches, reduced alcohol tolerance, difficulty learning new things, reduction in creativity and motivation, fatigue, and personality changes, and the syndrome may last a month to a year. Postconcussion syndrome may be the result of a slowly occurring subdural hematoma, which may be missed by an early examination.
11. The descending pathway affected is most likely the corticospinal tract. The lesions in the cerebral cortex would have to be contralateral to the affected side. This is because descending pathways arising from the cerebral cortex cross over at the medulla oblongata (lateral corticospinal tract) and spinal cord (anterior corticospinal tract) before they synapse with lower motor neurons.

Chapter 15

1. The lens of the eye is biconvex and causes light rays to converge. If the lens is removed, the replacement lens should also cause light rays to converge. A biconvex lens or a lens with a single convex surface would work. Bifocals or trifocals can also be recommended because of the loss of accommodation.
2. The light reflected by the tapetum lucidum can stimulate photoreceptors and increase the sensitivity of the eye to light, which can be an advantage when light levels are low. Because the same light image can stimulate different photoreceptors, however, there is a loss of visual acuity and a blurring of vision.
3. Carrots contain vitamin A (retinoic acid), which can be used to form retinal. Retinal and opsin combine to form rhodopsin, which is found in rods. Rhodopsin is necessary for rods to respond to low levels of light. Lack of vitamin A can result in lack of rhodopsin and night blindness.
4. If the person looks a few inches to the side, the image of the needle and thread is projected to the periphery of the retina, where there is the highest concentration of rods. The rods function better than cones in low-light intensities. If Jean looks directly at the needle and thread, their image falls on the macula, which has few rods and mostly cones, which do not function well in dim light. By looking to the side, however, she is using a part of the retina where the photoreceptor cells are not as densely packed as in the macula, and the image is fuzzy rather than sharp.
5. This phenomenon is called a negative afterimage. While a man is staring at the clock, the darkest portion of the image (the black clock) causes dark adaptation in part of the retina. That is, part of the retina becomes more sensitive to light. At the same time, the lightest part of the image (the white wall) causes light adaptation in the rest of the retina, and that part of the retina becomes less sensitive to light. When the man looks at a black wall, the dark adapted portion of the retina, which is more sensitive to light, produces more action potentials than does the light adapted part of the retina. Consequently, he perceives a light clock against a darker background.
6. A lesion of the optic chiasm results in visual loss in both the right and left temporal fields, a condition called bitemporal hemianopsia, or tunnel vision. Tunnel vision can cause problems for normal functions, such as when driving a car, because the peripheral vision is severely limited. The occurrence of this condition can also suggest a much more serious problem, such as a pituitary tumor, which sits just posterior to the optic chiasm.
7. The most likely area damaged is the spiral organ, where waves result in the production of action potentials. The action is much like ocean waves breaking on the shore during a violent storm, as compared with those breaking in from a calm ocean. Specifically, damage likely occurs in the part of the spiral organ near the oval window because it is this part of the basilar membrane that vibrates the most in response to high-frequency sounds.
8. Normally, as pressure changes, the auditory tubes open to allow an equalization of pressure between the middle ear and the external environment. If this does not occur, the buildup of pressure in the middle ear can rupture the tympanic membrane, or the pressure can be transmitted to the inner ear and cause sensoneural damage.
9. Normally, airborne sounds cause the tympanic membrane to vibrate, resulting in the movement of the middle ear ossicles and the production of waves in the perilymph of the scala vestibuli. Vibration of the skull bones can also cause vibration of the perilymph in the scala vestibuli.

Chapter 16

1. The sympathetic division of the ANS is responsible for dilation of the pupil. Preganglionic fibers from the upper thoracic region of the spinal cord pass through spinal nerves (T1 and T2), into the white rami communicantes, and into the sympathetic chain ganglia. The preganglionic fibers ascend the sympathetic chain and synapse with postganglionic neurons in the superior cervical sympathetic chain ganglia. The axons of the postganglionic neurons leave the sympathetic chain ganglia as small nerves that project to the iris of the eye.
2. Reduced salivary and lacrimal gland secretions can indicate damage to the facial nerves, which innervate the submandibular, sublingual, and lacrimal glands. The glossopharyngeal nerves innervate the parotid glands but not the lacrimal glands.
3. Cutting the preganglionic fibers in the white rami of T2–T3 is the best way to eliminate innervation of the blood vessels in the skin. Cutting the gray rami at levels T2–T3 is inappropriate because the postganglionic fibers that innervate the hand blood vessels exit from the first thoracic and inferior cervical sympathetic chain ganglia. Cutting the spinal nerves is inappropriate because it eliminates all sensory and motor functions to the area supplied.
4. a. pelvic splanchic nerves
   b. gray rami
   c. vagus nerves
   d. cranial nerves
   e. pelvic splanchic nerves
5. The parasympathetic division innervates the heart through the vagus nerves. The postganglionic nerve fibers of the vagus nerves release acetylcholine, which reduces heart rate. Methacholine can bind to the same receptors as acetylcholine and reduce heart rate. Side effects result from stimulating other parasympathetic effector organs. For example, stimulating the salivary glands results in increased salivation. Dilation of the pupils and sweating are effects expected from sympathetic stimulation. The muscles of respiration are not regulated by the ANS, but they do respond to acetylcholine through somatic neurons. Methacholine would be expected to make contractions of respiratory muscles more likely.
6. Inactivation of acetylcholinesterase results in a buildup of acetylcholine in synapses and overstimulation of muscarinic receptors.

One would expect mostly parasympathetic effects because the effects of acetylcholine are enhanced: blurring of vision as a result of contraction of ciliary muscles, excess tear formation because of overstimulation of the lacrimal glands, and frequent or involuntary urination because of overstimulation of the urinary bladder. Pallor resulting from vasoconstriction in the skin is a sympathetic effect that would not be expected because skin blood vessels respond to norepinephrine. Muscle twitching or cramps of skeletal muscles might occur because they normally respond to acetylcholine. Atropine, a muscarinic blocking agent, can be used to treat exposure to malathion.

7. Epinephrine causes vasoconstriction and confines the drug to the site of administration. This increases the drug's duration of action locally and decreases its systemic effects. Vasoconstriction also reduces bleeding if a dry field (an area clear of blood on its surface) is required.
8. Because normal action potentials are produced, the drug does not act at the synapse between the preganglionic and postganglionic neurons. Because injected norepinephrine works, sympathetic receptors in the heart are functioning and are not affected by the drug. Therefore, the drug must somehow affect the postganglionic neurons. Possibly it inhibits neurotransmitter production or release from the postganglionic neurons.
9. Because cutting the white rami of T1–T4 does not affect the drug's action, sympathetic preganglionic neurons in the spinal cord and sympathetic centers in the brain can be ruled out as a site of action. Because cutting the vagus nerves eliminates the drug's effect, the drug cannot be acting at the synapse between the preganglionic neurons and the postganglionic neurons, or between the synapse of the postganglionic neuron and the effector organ of either division of the ANS. The drug must therefore excite parasympathetic centers in the brainstem, resulting in a decrease in heart rate.
10. a. Responses in a person who is extremely angry are primarily controlled by the sympathetic division of the ANS. These responses include increased heart rate and blood pressure, decreased blood flow to the internal organs, increased blood flow to skeletal muscles, decreased contractions of the intestinal smooth muscle, flushed skin in the face and neck region, and dilation of the pupils of the eyes.
    b. For a person who has just finished eating and is now relaxing, the parasympathetic reflexes are more important than sympathetic reflexes. The blood pressure and heart rate are at normal resting levels, the blood flow to the internal organs is greater, contractions of smooth muscle in the intestines are greater, and secretions that achieve digestion are more active. If the urinary bladder or the colon becomes distended, autonomic reflexes that result in urination or defecation can result. Blood flow to the skeletal muscles is reduced.

Chapter 17

1. Liver disease and kidney disease increase the concentration of this hormone in the blood, and the concentration remains high for a longer time. The liver modifies the hormone to cause it to be excreted by the kidney more rapidly. In the case of liver disease, the hormone is not modified and excreted rapidly. Therefore, the concentration becomes higher than normal and the concentration of the hormone remains high for longer than normal. A similar result is seen if the kidney is diseased and the hormone cannot be excreted rapidly.
2. Secretion of hormones is usually controlled by a negative-feedback mechanism. If a hormone controls the concentration of a substance in the circulatory system, the hormone is secreted in smaller amounts if the substance increases in the circulatory system. If a tumor begins to secrete the substance in large amounts, the presence of the substance has a negative-feedback effect on the secretion of the hormone and the concentration of the hormone in the circulatory system is very low.
3. Usually, intracellular mediator mechanisms respond quickly, and the hormone's effect is brief. Intracellular receptor mechanisms usually take a long time (several hours) to respond, and their effects last much longer. If the hormone is large and water-soluble, it is probably functioning through an intracellular mediator mechanism; if the hormone is lipid-soluble, it is probably an intracellular receptor mechanism. If you have the ability to monitor the concentration of a suspected intracellular mediator and it increases in response to the hormone, or if you can inhibit the synthesis of an intracellular mediator and it prevents the target cells' response to the hormone, it is an intracellular mediator mechanism. If you can inhibit the synthesis of mRNA and this inhibits the action of the hormone, or if you can measure an increase in mRNA synthesis in response to the hormone, then the mechanism is an intracellular receptor mechanism.
4. When the hormone binds to its receptor, the α subunit of the G protein is released. GTP must bind to the α subunit, however, before it can have its normal effect. If the α subunit cannot bind GTP, the hormone has no effect on the target tissue.
5. Inhibitors of prostaglandin synthesis reduce prostaglandin synthesis in all tissues, not just in the tissues in which prostaglandins produce undesirable effects. Symptoms such as inflammation, vomiting, and fever are reduced. Because prostaglandins also play a role in producing beneficial effects in some tissues, however, these benefits would not occur normally. Inhibitors of prostaglandin synthesis may cause labor to be delayed or produce other undesirable responses due to their inhibitory effects on the synthesis of prostaglandins.
6. Phosphodiesterase causes the conversion of cAMP to AMP, thus reducing the concentration of cAMP. A drug that inhibits phosphodiesterase, therefore, increases the amount of cAMP in cells where cAMP is produced. Therefore, an inhibitor of phosphodiesterase increases the response of a tissue to a hormone that has cAMP as an intracellular mediator.
7. A short half-life for epinephrine allows epinephrine to produce a short-lived response. The response to a potentially harmful or dangerous situation is terminated shortly after the harmful or dangerous situation passes. If epinephrine had a long half-life, heart rate and blood glucose level would be elevated for a long time, even if the harmful or dangerous situation was very brief.
8. Because thyroid hormones are important in regulating the basal metabolic rate, a long half-life is an advantage. Thyroid hormones are secreted and have a prolonged effect without large fluctuations in the basal metabolic rate. If thyroid hormones had a short half-life, the basal metabolic rate could fluctuate with changes in the rate of secretion of thyroid hormones. Certainly, the rate of secretion of thyroid hormones would have to be controlled within narrow limits if it did have a short half-life.
9. If liver disease results in a decrease of plasma proteins to which thyroid hormones bind, higher than normal concentrations of free (unbound) thyroid hormones occur in the circulatory system. Because of the higher than normal concentration of thyroid hormones that are unbound, the responses to thyroid hormones increase. In addition, the half-life of the thyroid hormones is shortened. Thus, as thyroid hormone secretion increases, the concentration of thyroid hormone also increases. As the thyroid hormone secretion decreases, the concentration of thyroid hormone also decreases. Thyroid

hormones fluctuate in concentration in the circulatory system more than normal.

10. Elevated GnRH levels in the blood as a result of the GnRH-secreting tumor causes the down-regulation of GnRH receptors in the anterior pituitary. This decreases the ability of GnRH to stimulate the anterior pituitary, and the rate of luteinizing hormone and follicle-stimulating hormone secretion by the anterior pituitary decreases and remains decreased as long as the GnRH levels are chronically elevated. Therefore, the functions of the reproductive system controlled by luteinizing hormone and follicle-stimulating hormone decrease.
11. Insulin levels normally change in order to maintain normal blood sugar levels, despite periodic fluctuations in sugar intake. A constant supply of insulin from a skin patch might result in insulin levels that are too low when blood sugar levels are high (after a meal) and might be too high when blood sugar levels are low (between meals). In addition, insulin is a protein hormone that would not readily diffuse through the lipid barrier of the skin (see chapter 5). Estrogen is a lipid-soluble steroid hormone.

Chapter 18

1. The hypothalamohypophyseal portal system allows neurohormones that function as releasing and inhibiting hormones, which are secreted by neurons in the hypothalamus, to be carried directly from the hypothalamus to the anterior pituitary gland. Consequently, the releasing and inhibiting hormones are not diluted or destroyed by the enzymes, which are abundant in the kidneys, liver, lungs, and general circulation, before they reach the anterior pituitary. Also, the time it takes for releasing and inhibiting hormones to reach the anterior pituitary is less than if they were secreted into the general circulation.
2. A hot environment increases ADH secretion. Because the amount of water lost in the form of sweat can be quite large, and because sweat is more dilute than the body fluids, sweating gradually increases the osmolality of the body fluids. The increasing osmolality of body fluids stimulates an increase in ADH secretion. Thus, a hot environment can result in increased ADH secretion because of an increasing osmolality of the body fluids. Increased ADH secretion in a hot environment reduces the amount of water lost in the form of urine. Therefore, water is conserved.
3. Polydipsia and polyuria are consistent with both diabetes mellitus or diabetes insipidus. Diabetes mellitus, however, is consistent with an increased urine osmolality because of the large amount of glucose lost in the urine. Diabetes insipidus is consistent with urine with a low specific gravity because little water is reabsorbed by the kidney. Thus, urine has an osmolality close to that of the body fluids, and the rapid loss of dilute urine results in a decrease in blood pressure. Thus, polyuria with a low specific gravity is not consistent with diabetes mellitus but is consistent with diabetes insipidus. The administration of ADH reverses the symptoms of diabetes insipidus. Neither polydipsia nor polyuria results from a lack of glucagon or aldosterone.
4. The symptoms are consistent with acromegaly, which is a consequence of elevated GH secretion after the epiphyses have closed. Increased GH causes enlarged finger bones, the growth of bony ridges over the eyes, and increased growth of the jaw. The anterior pituitary tumor increases pressure at the base of the brain near the optic nerves as it enlarges. The pituitary rests in the sella turcica of the sphenoid bone; as it enlarges, pressure increases because the pituitary is nearly surrounded by rigid bone and the brain is located just superior to the pituitary. As the anterior pituitary enlarges because of a tumor, it pushes superiorly and pressure is applied to the ventral portion of the brain. The GH also causes bone deposition on the inner surface of skull bones, which also increases the pressure inside the skull.
5. If hyperthyroidism results from a pituitary abnormality, laboratory tests should show elevated TSH levels in the circulatory system in addition to elevated T_3 and T_4 levels. If hyperthyroidism results from the production of a nonpituitary thyroid-stimulating substance, laboratory tests should also show elevated T_3 and T_4 levels, but TSH levels will be low because of the negative-feedback effects of T_3 and T_4 on the hypothalamus and pituitary gland.
6. The second student is correct. Low levels of vitamin D reduce calcium uptake in the gastrointestinal tract, which results in a decreased blood level of calcium ions. As blood calcium levels decrease, the rate of PTH secretion increase. Parathyroid hormone increases bone breakdown, which maintains blood calcium levels, even if vitamin D deficiency exists for a prolonged time. Osteomalacia results because of the increased bone reabsorption necessary to maintain normal blood calcium levels.
7. A glucose tolerance test can distinguish between these conditions. The person would consume glucose after a period of fasting. Over the next few hours, the blood glucose levels in a healthy person would increase and then return to fasting levels. The blood glucose levels would always remain within the normal range, however. In a person with diabetes, the blood glucose levels would increase to above normal levels and would remain elevated for several hours. In a person who secreted large amounts of insulin, blood glucose levels would increase, and then they would decrease to below normal levels within a relatively short time.
8. Because the person is a diabetic and probably is taking insulin, the condition is more likely to be insulin shock than a diabetic coma. To confirm the condition, however, a blood sample should be taken. If the condition is due to a diabetic coma, the blood glucose levels will be elevated. If the condition is due to insulin shock, the blood glucose levels will be below normal. In the case of insulin shock, glucose can be administered intravenously. In the case of diabetic coma, insulin should be administered. An isotonic solution containing insulin can be administered to reduce the osmolality of the extracellular fluid.
9. Adrenal diabetes results from the elevated and uncontrolled secretion of glucocorticoid hormones, such as cortisol, from the adrenal gland. Because glucocorticoid hormones increase blood glucose levels, elevated secretion of these hormones results in elevated blood glucose levels and symptoms similar to those of diabetes mellitus. Pituitary diabetes results from elevated secretion of GH from the anterior pituitary. Elevated GH levels cause an increase in blood glucose levels and, therefore, produce symptoms similar to diabetes mellitus. Prolonged elevation of both glucocorticoids and growth hormone secretion can lead to the development of diabetes mellitus if the insulin-secreting cells of the pancreatic islets degenerate because of the prolonged need to secrete insulin in response to the elevated blood glucose levels.
10. Elevated epinephrine from the adrenal medulla promotes elevated blood pressure and increases the work load on the heart, increases the rate of metabolism, and results in increased sweating and nervousness. The risks for heart attack and stroke are increased. Elevated cortisol causes hyperglycemia and can lead to diabetes mellitus, a depressed immune system with increased susceptibility to infections, and the destruction of proteins, leading to tissue wasting.

Chapter 19

1. Because of the rapid destruction of the red blood cells, we would expect erythropoiesis to increase in an attempt to replace the

lost red blood cells. The reticulocyte count would therefore be above normal. Jaundice is a symptom of hereditary hemolytic anemia because the destroyed red blood cells release hemoglobin, which is converted into bilirubin. Removal of the spleen cures the disease because the spleen is the major site of red blood cell destruction.

2. Blood doping increases the number of red blood cells in the blood, thereby increasing its oxygen-carrying capacity. The increased number of red blood cells also makes it more difficult for the blood to flow through the blood vessels, increasing the heart's workload.
3. Symptoms resulting from decreased red blood cells are associated with a decreased ability of the blood to carry oxygen: shortness of breath, weakness, fatigue, and pallor. Symptoms resulting from decreased platelets are associated with a decreased ability to form platelet plugs and clots: small areas of hemorrhage in the skin (petechiae), bruises, and decreased ability to stop bleeding. Symptoms resulting from decreased white blood cells include an increased susceptibility to infections.
4. Hypoventilation results in decreased blood oxygen levels, which stimulate erythropoiesis. Therefore, the number of red blood cells increases and produces secondary polycythemia.
5. Removal of the stomach removes intrinsic factor, which is necessary for vitamin B_{12} absorption. Therefore, the patient develops pernicious anemia. Lack of stomach acid can decrease iron absorption in the small intestine and results in iron-deficiency anemia.
6. Vitamin B_{12} and folic acid are necessary for blood cell division. Lack of these vitamins results in pernicious anemia. Iron is necessary for the production of hemoglobin. Lack of iron results in iron-deficiency anemia. Vitamin K is necessary for the production of many blood clotting factors. Lack of vitamin K can greatly increase blood clotting time, resulting in excessive bleeding.
7. The anemia results from too little hemoglobin. Because there is less hemoglobin, less hemoglobin is broken down into bilirubin. Consequently, less bilirubin is excreted as part of the bile into the small intestine. With decreased bilirubin in the intestine, bacteria produce fewer of the pigments that normally color the feces.
8. Reddie Popper has hemolytic anemia. The RBC is lower than normal because the red blood cells are being destroyed faster than they are being replaced. With fewer red blood cells, hemoglobin and hematocrit is lower than normal. Bilirubin levels are above normal because of the breakdown of the hemoglobin released from the ruptured red blood cells.

Chapter 20

1. The walls of the ventricles are thicker than the walls of the atria because the ventricles must produce a greater pressure to pump blood into the arteries. Only a small pressure is required to pump blood from the atria into the ventricles during diastole. The wall of the left ventricle is thicker than the wall of the right ventricle because the left ventricle produces a much greater pressure to force blood through the aorta than the right ventricle produces to move blood through the pulmonary trunk and pulmonary arteries.
2. During systole, the cardiac muscle in the right and left ventricles contracts, which compresses the coronary arteries. During diastole, the cardiac muscle of the ventricles relaxes and blood flow through the coronary arteries increases. The diastolic pressure is sufficient to cause blood to flow through coronary arteries during diastole.
3. A drug that prolongs the plateau of cardiac muscle cell action potentials prolongs the time each action potential exists and increases the refractory period. Therefore, the drug slows the heart. A drug that shortens the plateau shortens the length of time each action potential exists and shortens the refractory period. Therefore, the drug can allow the heart rate to increase further.
4. Endurance-trained athletes have decreased heart rates because their cardiac muscle undergoes hypertrophy in response to exercise. The hypertrophied cardiac muscle causes the stroke volume to increase substantially. The increased stroke volume is sufficient to maintain an adequate cardiac output and blood pressure even though the heart rate is slower.
5. The two heartbeats occurring close together can be heard through the stethoscope, because the heart valves open and close normally during each of the heartbeats even if they are close together. The second heartbeat, however, produces a greatly reduced stroke volume because there is not enough time for the ventricles to fill with blood between the first and second contractions of the heart. Thus, the preload is reduced. Because the preload is reduced, the second heartbeat has a greatly reduced stroke volume. The reduced stroke volume fails to produce a normal pulse. The pulse deficit, therefore, results from the reduced stroke volume of the second of the two beats that are very close together.
6. Aerobic training causes hypertrophy of the cardiac muscle in the heart and causes the heart to produce a greater stroke volume as a consequence. The heart rate can decrease while the cardiac output remains the same because cardiac output is equal to the stroke volume times the heart rate. If the stroke volume increases, the heart rate can decrease and the cardiac output can remain the same.
7. Atrial contractions complete ventricular filling, but atrial contractions are not primarily responsible for ventricular filling. Therefore, even if the atria are fibrillating, blood can still flow into the ventricles and ventricular contractions can occur. As long as the ventricles contract rhythmically, the heart can pump an adequate amount of blood, even though the atria are not effective pumps. If the ventricles fail to contract forcefully and rhythmically, however, they cannot function as pumps. Thus, the stroke volume will become too low to maintain adequate blood flow to tissues. Therefore, atrial transplants are not essential but ventricle transplants are.
8. The results depend on Cee Saw's response to the conditions of the laboratory. First, as Cee Saw's head is lowered, gravity causes blood pressure in the carotid sinuses and aortic arch to increase. The increased blood pressure stimulates baroreceptors, which detect the increased blood pressure and send action potentials indicating that blood pressure increased to the cardioregulatory center in the medulla oblongata along sensory nerve fibers. The cardioregulatory center increases parasympathetic stimulation and reduces sympathetic stimulation of the heart. Thus, the heart rate decreases. Second, if as her head is lowered Cee Saw becomes excited, the sympathetic division of the ANS becomes more active. The resulting increase in sympathetic stimulation of the heart causes the heart rate to increase.
9. After Cee Saw is tilted so that her head is higher than her feet for a few minutes, the regulatory mechanisms that control blood pressure adjust so that the heart pumps sufficient blood to supply the needs of her tissues. If she is then tilted so that her head is higher than her feet, gravity causes blood to flow toward her feet, and the blood pressure in the carotid sinus and aortic arch decreases. The decrease in blood pressure would be detected by the baroreceptors in these vessels and would activate baroreceptor reflexes. The result would be increased sympathetic and decreased parasympathetic stimulation

of the heart and an increase in the heart rate. The increased heart rate would increase the blood pressure to its normal value.

10. An ECG measures the electrical activity of the heart and does not indicate a slight heart murmur. Heart murmurs are detected by listening to the heart sounds. The boy may have a heart murmur, but the mother does not understand the basis for making such a diagnosis.
11. An incompetent aortic semilunar valve does not close completely at the end of systole. Consequently, blood flows from the aorta back into the left ventricle at the end of systole and during diastole. This causes the diastolic blood pressure to be lower than normal, and at the end of diastole the volume of blood in the left ventricle is greater than normal. Because of the increased volume of blood in the left ventricle at the end of diastole, the stroke volume during systole is greater than normal. As a result, the systolic blood pressure is greater than normal. The heart sounds are affected by an incompetent aortic semilunar valve. Turbulent blood flow and, consequently, an abnormal heart sound is heard at the end of systole as blood flows backward through the incompetent aortic semilunar valve. This heart sound follows the second heart sound, which results from closure of the semilunar valves.
12. When both common carotid arteries are clamped, the blood presure within the internal carotid arteries drops dramatically. The decreased blood pressure is detected, and the baroreceptor reflex increases heart rate and stroke volume. The resulting increase in cardiac output causes the increase in blood pressure.
13. Venous return declines markedly in hemorrhagic shock because of the loss of blood volume. With decreased venous return, stroke volume decreases (Starling's law of the heart). The decreased stroke volume results in a decreased cardiac output, which produces a decreased blood pressure. In response to the decreased blood pressure, the baroreceptor reflex causes an increase in heart rate in an attempt to restore normal blood pressure. However, with inadequate venous return the increased heart rate is not able to restore normal blood pressure.

Chapter 21

1. a. aorta, left coronary artery, circumflex artery, posterior interventricular artery or aorta, right coronary artery, posterior interventricular artery
   b. aorta, brachiocephalic artery, right common carotid artery, right internal carotid artery or aorta, left common carotid artery, left internal carotid artery
   c. aorta, brachiocephalic artery, right subclavian artery, right vertebral artery, basilar artery or aorta, left subclavian artery, left vertebral artery, basilar artery
   d. aorta, left or right common carotid artery, left or right external carotid artery
   e. aorta, left subclavian artery, axillary artery, brachial artery, radial or ulnar artery, deep or superficial palmar arch, digital artery (on the right: the brachiocephalic artery would be included)
   f. aorta, common iliac artery, external iliac artery, femoral artery, popliteal artery, anterior tibial artery
   g. aorta, celiac artery, common hepatic artery
   h. aorta, superior mesenteric artery, intestinal branches
   i. aorta, left or right internal iliac artery
2. a. great cardiac vein, coronary sinus or anterior cardiac vein
   b. transverse sinus, sigmoid sinus, internal jugular vein, brachiocephalic vein, superior vena cava
   c. retromandibular vein, external jugular vein, subclavian vein, brachiocephalic vein, superior vena cava
   d. deep: vein of hand, radial or ulnar vein, brachial vein, axillary vein, subclavian vein, brachiocephalic vein, superior vena cava
   superficial: vein of hand, radial or ulnar vein, cephalic or basilic vein, axillary vein, subclavian vein, brachiocephalic vein, superior vena cava
   e. deep: vein of foot, dorsalis veins of foot, anterior tibial vein, popliteal vein, femoral vein, external iliac vein, common iliac vein, inferior vena cava
   superficial: vein of foot, great saphenous vein, external iliac vein, common iliac vein, inferior vena cava; or vein of foot, small saphenous vein, popliteal vein, femoral vein, external iliac vein, common iliac vein, inferior vena cava
   f. gastric vein or gastroepiploic vein, hepatic portal vein, hepatic sinusoids, hepatic vein, inferior vena cava
   g. renal vein, inferior vena cava
   h. hemiazygous vein or accessory hemiazygous vein, azygous vein, superior vena cava
3. A superficial vessel is easiest, such as the right cephalic or basilic vein. The catheter is passed through the cephalic (or brachial) vein and the superior vena cava to the right atrium. Because the pulmonary veins are not readily accessible, dye is not normally placed directly into them. Instead, the dye is placed in the right atrium using the procedure just described. The dye passes from the right atrium into the right ventricle, the pulmonary arteries, the lungs, the pulmonary veins, and into the left atrium. If the catheter has to be placed in the left atrium, it can be inserted through an artery, such as the femoral artery, and passed via the aorta to the left ventricle and then into the left atrium.
4. The viscosity of the blood is affected primarily by the hematocrit. As hematocrit increases, the viscosity of the blood increases logarithmically so that even a small increase in hematocrit results in a large increase in viscosity. Greater force is therefore not needed to cause blood to flow through the blood vessels. With the increased blood volume, blood flow through vessels is adequate without an increase in viscosity.
5. The resistance to blood flow is less in the vena cavae for two reasons: First, the diameter of one vena cava is greater than the diameter of the aorta; second, an increased diameter of a blood vessel reduces resistance to flow (Poiseuille's law). In addition, there are two venae cavae, the superior vena cava and the inferior vena cava, but only one aorta. The blood flow through the aorta and the venae cavae is about equal, but the velocity of blood flow is much higher in the aorta than it is in the venae cavae.
6. According to Laplace's law, as the diameter of a blood vessel increases the force applied to the vessel wall increases, even if the pressure remains constant. The increased connective tissue in the walls of the large blood vessels therefore makes the wall of those vessels stronger and more capable of resisting the force applied to the wall.
7. Veins and lymphatic vessels have one-way valves in them. Massage creates a cycle of increasing and decreasing pressure to the veins, which rhythmically compresses them. The compression of the veins forces fluid to move out of the limb through both veins and lymphatic vessels. The movement of fluid through the veins lowers the pressure within the venous end of the capillary. Thus, the forces that move fluid into the capillaries at their venous ends are greater and they move more interstitial fluid into the capillaries. Compression of the lymphatic capillaries also causes more lymphatic fluid to move into the lymphatic vessels. Because there is less fluid in the limb, the edema decreases.
8. The nursing student's diagnosis was incorrect. Blood pressure measurements are normally made in either the right or the left arm, both of which are close to the level of the heart. Blood pressure taken in the leg is influenced by pressure created by the

pumping action of the heart, but the effect of gravity on the blood, as it flows into the leg, also influences the blood pressure in a substantial way. In this case, gravity increases blood pressure from about 120 mm Hg for the systolic pressure to 200 mm Hg.

9. Decreased liver function includes a decrease in the synthesis of plasma proteins. Consequently, the concentration of plasma proteins decreases, and the colloid osmotic pressure of the blood decreases. Therefore, less water moves by osmosis into the capillaries at the venous ends and the result is edema.
10. Chemoreceptors in the medulla oblongata detect carbon dioxide and the pH of the blood. The normal blood levels of CO_2 and pH stimulate these chemoreceptors, which in turn stimulate the vasomotor center. The vasomotor center keeps blood vessels partially constricted under resting conditions. This basal level of activity is called the vasomotor tone. Blowing off CO_2 reduces the blood levels of carbon dioxide and increases the pH of the body fluids. These changes reduce vasomotor tone and result in vasodilation. If a person hyperventilates and blows off CO_2, the stimulus to the vasomotor center decreases, which results in a decrease in vasomotor tone. The decrease in vasomotor tone results in a decrease in systemic blood pressure. If the blood pressure decreases enough, the blood flow to the brain decreases and can cause a sensation of dizziness or even loss of consciousness.
11. Epinephrine is secreted from the adrenal medulla in response to stressful stimuli and the epinephrine stimulates responses that are consistent with increased physical activity. Vasoconstriction of the blood vessels in the skin shunts blood away from the skin to skeletal muscles. Vasodilation occurs in blood vessels of exercising skeletal muscles. Blood flow through the exercising skeletal muscles therefore increases. Because epinephrine causes vasodilation of the blood vessels of cardiac muscle, blood flow through the cardiac muscle increases. This response is consistent with the increased work performed by the heart under conditions of increased physical activity.
12. The hot Jacuzzi increases Skinny's skin and body temperature. As a result, the blood vessels of the skin dilate. Because the blood vessels dilate, peripheral resistance decreases, causing the blood pressure to decrease. The baroreceptors of the carotid sinus and aortic arch detect the decrease in blood pressure and send action potentials to the cardioregulatory center in the medulla oblongata. As a result, the sympathetic stimulation to the heart increases and the heart rate, in response, increases. The increased heart rate elevates the blood pressure back to within its normal range of values.

Chapter 22

1. Elevation of the limb reduces blood pressure in the limb, resulting in less fluid movement from the blood into the tissues (see chapter 21). Thus, the edema is reduced as the lymphatic system removes fluid from the tissues faster than it enters them. Massage moves lymph through the lymphatic vessels in the same fashion as contraction of skeletal muscle. The application of pressure periodically to lymphatic vessels forces lymph to flow toward the trunk of the body, but valves prevent the flow of lymph in the reverse direction. The removal of lymph from the tissue helps relieve edema.
2. Normally, T cells are processed in the thymus and then migrate to other lymphatic tissues. Without the thymus, this processing is prevented. Because there are normally five T cells for every one B cell, the number of lymphocytes is greatly reduced. The loss of T cells results in an increased susceptibility to infection and an inability to reject grafts because of the loss of cell-mediated immunity. In addition, since helper T cells are involved with the activation of B cells, antibody-mediated immunity is also depressed.
3. That there is no immediate effect indicates there is a reservoir of T cells in the lymphatic tissue. As the reservoir is depleted through time, the number of lymphocytes decreases and cell-mediated immunity is depressed, the animal is more susceptible to infections, and the ability to reject grafts decreases. The ability to produce antibodies decreases because of the loss of helper T cells that are normally involved with the activation of B cells.
4. Injection B results in the greatest amount of antibody production. At first, the antigen causes a primary response. A few weeks later, the slowly released antigen causes a secondary response, resulting in a greatly increased production of antibodies. Injection A does not cause a secondary response because all of the antigen is eliminated by the primary response.
5. If the patient has already been vaccinated, the booster shot stimulates a memory (secondary) response and rapid production of antibodies against the toxin. If the patient has never been vaccinated, vaccinating now is not effective because there is not enough time for the patient to develop his or her own primary response. Therefore, antiserum is given to provide immediate, but temporary, protection. Sometimes both are given: The antiserum provides short-term protection and the tetanus vaccine stimulates the patient's immune system to provide long-term protection. If the shots are given at the same location in the body, the antiserum (antibodies against the tetanus toxin) can cancel the effects of the tetanus vaccine (tetanus toxin altered to be nonharmful).
6. The infant's antibody-mediated immunity is not functioning properly, whereas his cell-mediated immunity is working properly. This explains the susceptibility to extracellular bacterial infections and the resistance to intracellular viral infections. It took so long to become apparent because IgG from the mother crossed the placenta and provided the infant with protection. The infant began to get sick after these antibodies degraded.
7. Bone marrow is the source of the lymphocytes responsible for adaptive immunity. If successful, the transplanted bone marrow starts producing lymphocytes and the baby has a functioning immune response. In this case, there is a graft versus host rejection in which the lymphocytes in the transplanted red marrow mount an immune attack against the baby's tissues, resulting in death.
8. At the first location, an antibody-mediated response results in an immediate hypersensitivity reaction, which produces inflammation. Most likely, the response resulted from IgE antibodies. At the second location, a cell-mediated response results in a delayed hypersensitivity reaction, which produces inflammation. This probably involves the release of cytokines and the lysis of cells. At the other locations, there is neither an antibody-mediated nor a cell-mediated response.
9. The ointment is a good idea for the poison ivy, which causes a delayed hypersensitivity reaction—for example, too much inflammation. For the scrape, it is a bad idea because a normal amount of inflammation is beneficial and helps fight infection in the scrape.
10. Because both antibodies and cytokines produce inflammation, the fact that the metal in the jewelry results in inflammation is not enough information to answer the question. However, the fact that it took most of the day (many hours) to develop the reaction indicates a delayed hypersensitivity reaction and therefore cytokines.

Chapter 23

1. Minute respiratory volume is equal to the respiration rate times the tidal volume. With a respiration rate of 12 breaths per minute and a tidal volume of 500 mL per

breath, normal minute ventilation is 6000 mL/min (12×500). Rapid (24 breaths per minute), shallow (250 mL per breath) breathing results in the same minute ventilation—that is, 6000 mL/min (24×250). Alveolar ventilation rate (V_A) is the respiratory rate (frequency; f) times the difference between the tidal volume (V_T) and dead space (V_D).

$$V_A = f(V_T - V_D)$$

$$\text{Normal resting } V_A = 12 \times (500 - 150) = 4200 \text{ mL/min}$$

In this case of rapid shallow breathing,

$$V_A = 24 \times (250 - 150) = 2400 \text{ mL/min}$$

Thus, even though the minute ventilation is the same in both cases, the alveolar ventilation rate is less during rapid, shallow breathing because there is a less effective exchange of gases between the atmosphere and the dead space. Because there is less exchange of gases, the partial pressures of alveolar gases become closer to the partial pressure of blood gases. Consequently, the alveolar partial pressure of O_2 decreases and the alveolar partial pressure of CO_2 increases. This decreases the concentration gradients for gases, resulting in less gas exchange between alveolar air and blood.

2. We expect vital capacity to be greatest when standing because the abdominal organs move inferiorly, thereby allowing greater depression of the diaphragm and a greater inspiratory reserve volume.
3. The hose increases dead space and therefore decreases alveolar ventilation. Ima Diver has to compensate by increasing respiratory rate or tidal volume. If the hose is too long, she will not be able to compensate. Furthermore, with a long hose, air is simply moved back and forth in the hose with little exchange of air between the atmosphere and the lungs taking place. Another consideration is the effect of water pressure on the thorax, which decreases compliance and increases the work of ventilation. In fact, a few feet underwater there is enough pressure on the thorax to prevent the intake of air, even through a short hose connected to the atmosphere.
4. The increase in atmospheric pressure increases the partial pressure of oxygen. According to Henry's law, as the partial pressure of oxygen increases, the amount of oxygen dissolved in the body fluids increases. The increase in dissolved oxygen is detrimental to the gangrene bacteria. Because hemoglobin is already saturated with oxygen, the HBO treatment does not increase hemoglobin's ability to pick up oxygen in the lungs.
5. Compression causes a decrease in thoracic volume and therefore lung volume. Consequently, pressure in the lungs increases over atmospheric pressure and air moves out of the lungs. Raising the arms expands the thorax and lungs. This results in a lower than atmospheric pressure in the lungs, and air moves into the lungs.
6. The victim's lungs expand because of the pressure generated by the rescuer's muscles of expiration. This fills the lungs with air that has a greater pressure than atmospheric pressure. Air flows out of the victim's lungs as a result of this pressure difference and because of the recoil of the thorax and lungs. Although the partial pressure of oxygen of the rescuer's expired air is less than atmospheric, enough oxygen can be provided to sustain the victim. The lower partial pressure of oxygen can also activate the chemoreceptor reflex and stimulate the victim to breathe. In addition, the rescuer's partial pressure of carbon dioxide is higher than atmospheric and this can activate the chemosensitive area in the medulla.
7. The left side of the diaphragm moves superiorly. During inspiration, thoracic volume increases as the right side of the diaphragm moves inferiorly and the intercostal muscles move the ribs outward. Increased thoracic volume causes a decrease in pressure in the thoracic cavity. As a result, the pressure on the superior surface of the diaphragm is less than on the inferior surface. The paralyzed left side of the diaphragm moves superiorly because of this pressure difference.
8. At the end of expiration, the thoracic wall is not moving. Therefore, the forces causing the thoracic wall to move inward and outward must be equal. At the end of expiration, the lungs are adhering to the thoracic wall through the pleurae, and recoil of the lungs is pulling the thoracic wall inward. As a result of the pneumothorax, air enters the pleural cavity and the visceral and parietal pleurae separate from each other. The recoil of the lungs causes them to collapse. Without the inward force produced by lung recoil, the thoracic wall expands outward.
9. All else being equal (i.e., the thickness of the respiratory membrane, the diffusion coefficient of the gas, and the surface area of the respiratory membrane), diffusion is a function of the partial pressure difference of the gas across the respiratory membrane. The greater the difference in partial pressure, the greater the rate of diffusion. The greatest rate of oxygen diffusion should therefore occur at the end of inspiration when the partial pressure of oxygen in the alveoli is at its highest. The greatest rate of carbon dioxide diffusion should occur at the end of inspiration when the partial pressure of carbon dioxide in the alveoli is at its lowest.
10. EPO increases the number of red blood cells and consequently the blood viscosity. Thus, even though the blood has a higher oxygen-carrying capacity, it increases the workload on the heart and can cause a heart attack or stroke.
11. Cutting the vagus nerves eliminates the Hering-Breuer reflex and results in a greater than normal inspiration. This increases tidal volume. Cutting the phrenic nerves eliminates contraction of the diaphragm. Tidal volume decreases drastically, and death probably results. Cutting the intercostal nerves eliminates raising of the ribs and sternum and decreases tidal volume, unless the diaphragm compensates.
12. While hyperventilating and making ready to leave your instructor behind, you might make the following arguments:
    - Hyperventilation increases the oxygen content of the air in the lungs; therefore, you would have more oxygen to use when holding your breath.
    - It is hemoglobin that is saturated. Hyperventilation increases the amount of oxygen dissolved in the blood plasma.
    - Hyperventilation decreases the amount of carbon dioxide in the blood. This makes it possible to hold one's breath for a longer time because of a decreased urge to take a breath.
    - Hyperventilation activates alveoli not in use because increasing alveolar oxygen and decreasing alveolar carbon dioxide causes lung arterioles to relax, thereby increasing blood flow through the lungs.
13. A can of soda is a pressurized container holding a liquid (mostly water) into which the gas carbon dioxide is dissolved. When the tab on the soda can is popped, the liquid is exposed to air that has a lower partial pressure of carbon dioxide than that in the liquid. According to Henry's law, a gas diffuses out of the liquid, down its partial pressure gradient until it reaches equilibrium. Thus, because carbon dioxide diffuses out of the soda, it eventually tastes "flat" after it has been opened. At the respiratory membrane, a partial pressure gradient exists between the blood and alveolar air for both oxygen and carbon dioxide. The partial pressure gradient for oxygen causes it to diffuse from the alveolar air into the blood, whereas the partial pressure gradient for carbon dioxide causes it to diffuse from the blood into the alveolar air. Solubility of the gases is also an important consideration.

Because carbon dioxide is 24 times more soluble in water than is oxygen, a lower partial pressure gradient for carbon dioxide than for oxygen does not diminish the ability to exhale carbon dioxide.

14. When Ima is hyperventilating, the stimulus for the hyperventilation is anxiousness, and the anxiousness is more important than the carbon dioxide in controlling respiratory movements. As the blood levels of carbon dioxide decrease during hyperventilation, vasodilation occurs in the periphery. As a result, the systemic blood pressure decreases. The systemic blood pressure can decrease enough that the blood flow to the brain decreases. Decreased blood flow to the brain results in a reduced oxygen level in the brain tissue, causing dizziness. Breathing into a paper bag increases Ima's blood levels of carbon dioxide toward normal. Because the carbon dioxide does not increase above normal, it does not increase the urge to breathe. The more normal level of carbon dioxide prevents the peripheral vasodilation. As Ima breathes into the paper bag, the anxiousness is likely to subside. As the anxiousness subsides, the normal regulation of respiration resumes.

Chapter 24

1. With the loss of the swallowing reflex, the vocal folds no longer occlude the glottis. Consequently, vomit can enter the larynx and block the respiratory tract.
2. Without adequate amounts of hydrochloric acid, the pH in the stomach is not low enough for the activation of pepsin. This loss of pepsin function results in inadequate protein digestion. If the food is well chewed, however, proteolytic enzymes in the small intestine (e.g., trypsin, chymotrypsin) can still digest the protein. If the stomach secretion of intrinsic factor decreases, the absorption of vitamin B_{12} is hindered. Inadequate amounts of vitamin B_{12} can result in decreased red blood cell production (pernicious anemia).
3. Even though ulcers are apparently ultimately caused by bacteria, overproduction of hydrochloric acid due to stress is a possible contributing factor. Reducing hydrochloric acid production is recommended. In addition to antibiotic therapy, commonly recommended solutions include relaxation, drugs that reduce stomach acid secretion, and antacids to neutralize the hydrochloric acid. Smaller meals are also advised because distension of the stomach stimulates acid production. Proper diet is also important. The patient is also advised to avoid alcohol, caffeine, and large amounts of protein because they stimulate acid production. Ingestion of fatty acids is recommended because they inhibit acid production by causing the release of gastric inhibitory polypeptide and cholecystokinin. Stress also stimulates the sympathetic nervous system, which inhibits duodenal gland secretion. As a result, the duodenum has less of a mucous coating and is more susceptible to gastric acid and enzymes. Relaxing after a meal helps decrease sympathetic activities and increase parasympathetic activities.
4. Lack of bile due to blockage of the common bile duct can result in jaundice (due to an accumulation of bile pigments in the blood) and clay-colored stools (due to lack of bile pigments in the feces). Blockage of the bile duct causes abdominal pain, nausea, and vomiting. Fat absorption is impaired because of the absence of bile salts in the duodenum and a loose, bulky stool results. Lack of fat absorption reduces the absorption of fat-soluble vitamins, such as vitamin K, resulting in a lack of normal clotting function.
5. The patient would still be able to defecate. Following a meal, the gastrocolic and duodenocolic reflexes could initiate mass movement of the feces into the rectum. In the rectum, local reflexes and the defecation reflex (integrated in the sacral level of the cord and not requiring connections to high brain centers) would cause defecation. Awareness of the need to defecate would be lost (due to loss of sensory input to the brain) and the ability to prevent defecation voluntarily via the external anal sphincter would also be lost.
6. The accumulation of materials above the site of impaction and the action of bacteria on the material would result in an increase in osmotic pressure in the area. Water would move by osmosis into the colon above the site of impaction. Bowel impaction is very dangerous and must be treated quickly. The increased volume and distension of the digestive tract above the site of impaction causes compression of the mucosa. This compression can occlude blood vessels in the mucosa and lead to necrosis. Necrosis of the mucosa results in increased permeability of the mucosa, allowing toxic organisms and substances in the digestive tract to enter the circulation, resulting in septic shock.
7. Cholera toxin irreversibly activates a G protein that causes persistent activation of the chloride channel. The activated channel allows excessive movements of chloride from the cells into the digestive tract. Water follows the osmotic gradient generated by the chloride, which leads to diarrhea. Conversely, mutations in the channel reduce the movement of chloride into the digestive tract. A thick mucus builds up on the surface of the epithelial cells, leading to some symptoms of cystic fibrosis.
8. Oral rehydration therapy relies on the principle of osmosis. Water follows solutes as they are absorbed across the intestinal epithelium. The combination of sodium and glucose is optimal, since the two molecules are cotransported by a symporter that is driven by a sodium gradient established by the Na^+-K^+ pump. Hence, the presence of sodium aids glucose absorption. Fructose is absorbed by a facilitated diffusion transporter that is not coupled to a sodium gradient.
9. Dietary changes are beneficial but alone will not likely be sufficient for FH patients. Only about 15% of the cholesterol in the body comes from diet; the remainder is synthesized, primarily in the liver. A diet with greatly reduced total fat and cholesterol, along with exercise, would help lower serum LDL cholesterol levels. However, since FH patients no longer have the normal LDL receptor-mediated feedback loop, there is still an increased cellular synthesis of cholesterol, which often requires statin drug therapy to restore cholesterol levels to normal.

Chapter 25

1. In figure 25.2, the Daily Value for saturated fat is listed as less than 20 g for a 2000 kcal/day diet. The % Daily Values appearing on food labels are based on a 2000 kcal/day diet. Therefore, the % Daily Value for saturated fat for one serving of this food is 10% (2/20 = .10, or 10%).
2. According to the Daily Value guidelines, total fats should be no more than 30% of total kilocaloric intake. For someone consuming 3000 kcal/day, this is 900 kcal (3000 kcal × 0.30). There are 9 kcal in a gram of fat. Therefore, the maximum amount (weight) of fats the active teenage boy should consume is 100 g (900/9).
3. The % Daily Value is the amount of the nutrient in one serving divided by its Daily Value. Therefore, the % Daily Value is 10% (10/100 = .10, or 10%).
4. The % Daily Value for one serving of the food is 10% (see answer to question 3). Since there are four servings in the package, if the teenager eats half of the food in the package, he consumes two servings. Thus, he eats 20% (10% × 2) of the recommended maximum total fat.

5. The protein in meat contains all of the essential amino acids and is a complete protein food. Although plants contain proteins, a variety of plants must be consumed to ensure that all the essential amino acids are included in adequate amounts. Also, plants contain less protein per unit weight than meat, so a larger quantity of plants must be consumed to get the same amount of protein.
6. Copper is necessary for the proper functioning of the electron-transport chain. Inadequate copper in the diet results in reduced ATP production—that is, not enough energy.
7. Fasting can be damaging because proteins are used to produce glucose. The glucose enters the blood and provides an energy source for the brain. This breakdown of proteins can damage tissues, such as muscle, and disrupt chemical reactions regulated by enzyme systems. A single day without food, however, is unlikely to cause permanent harm.
8. Weight is lost when kilocalories used per day exceeds kilocalories ingested per day. About 60% of the kilocalories used per day is due to basal metabolic rate. A person with a high basal metabolic rate loses weight faster than a person with a low basal metabolic rate, all else being equal. Another factor to consider is the amount of physical activity, which accounts for about 30% of kilocalories used per day. An active person loses more weight than a sedentary person does.
9. Amino acids, derived from ingested proteins, are necessary to build muscles. As Lotta and her friend discovered, excess proteins do not accelerate this process. Excess proteins can be used as an energy source in oxidative deamination, for the formation of the intermediate molecules of carbohydrate metabolism, or in gluconeogenesis. Excess proteins are also converted into storage molecules through glycogenesis or lipogenesis. Lotta is in positive nitrogen balance because the amount of nitrogen she gains from her diet is greater than the amount she loses by excretion. Some of the nitrogen in the amino acids she ingests is incorporated into the proteins of her muscles as they enlarge.
10. No, this approach does not work because he is not losing stored energy from adipose tissue. In the sauna, he gains heat, primarily by convection from the hot air and by radiation from the hot walls. The evaporating sweat is removing heat gained from the sauna. The loss of water will make him thirsty, and he will regain the lost weight from fluids he drinks and food he eats.
11. As ATP breakdown increases, more ATP is produced to replace that used. Over an extended time, the ATP must be produced through aerobic respiration. Therefore, oxygen consumption and basal metabolic rate increase. The production of ATP requires the metabolism of carbohydrates, lipids, or proteins. As these molecules are used at a faster than normal rate, body weight decreases. Increased appetite and increased food consumption resist the loss in body weight. As ATP is produced and used, heat is released as a by-product. The heat raises body temperature, which is resisted by the dilation of blood vessels in the skin and by sweating.
12. During fever production, the body produces heat by shivering. The body also conserves heat by the constriction of blood vessels in the skin (producing pale skin) and by a reduction in sweat loss (producing dry skin). When the fever breaks—that is, "the crisis is over"—heat is lost from the body to lower body temperature to normal. This is accomplished by the dilation of blood vessels in the skin (producing flushed skin) and increased sweat loss (producing wet skin).

Chapter 26

1. The large volume of hypoosmotic fluid ingested increases blood volume and causes blood osmolality to decrease. The increased blood volume is detected by baroreceptors, and the decreased blood osmolality is detected by osmoreceptors in the hypothalamus. The response to these stimuli is an inhibition of ADH secretion. The alcohol in the beer also inhibits ADH secretion. The increased volume inhibits the renin-angiotensin-aldosterone mechanism, which in turn inhibits aldosterone secretion. The changes in aldosterone, however, take much longer to influence kidney function than changes in ADH. As a result of these changes, a large volume of dilute urine is produced until the blood osmolality and blood volume return to normal.
2. Once the salt is absorbed, the osmolality of the blood increases. The increased osmolality of blood is detected by osmoreceptor neurons in the hypothalamus, thereby stimulating ADH secretion and inhibiting aldosterone secretion. A small volume of concentrate urine is produced as a result, until the excess salt is eliminated and the blood osmolality returns to its normal value.
3. The hypoosmotic sweat loss results in more loss of water than electrolytes. This simultaneously decreases plasma volume and increases blood osmolality, thereby stimulating increased ADH secretion. In addition, the decreased plasma volume stimulates the renin-angiotensin-aldosterone mechanism, resulting in a decreased glomerular filtration rate and increased aldosterone secretion. The effect of the changes is to produce a small amount of concentrated urine.
4. The loss of sweat results in a loss of water and electrolytes. Replacing just the water restores blood volume and decreases blood osmolality. At first, the decreased osmolality inhibits ADH secretion, and dilute urine is produced. As blood volume decreases as a result of urine production, however, ADH secretion and the renin-angiotensin-aldosterone mechanisms are stimulated. Consequently, urine concentration increases, and only a small amount of urine is produced.
5. As aldosterone levels decrease, sodium reabsorption in the nephron decreases and, consequently, plasma sodium levels decrease. The sodium is lost in the urine, and water follows the sodium by osmosis. Thus, a large amount of urine that has a high concentration of sodium is produced. The loss of water reduces blood volume, which causes the low blood pressure. As aldosterone levels decrease, potassium secretion into the nephron decreases, resulting in an increase in plasma potassium levels. The increased extracellular potassium causes the depolarization of nerve and muscle membranes, leading to tremors of skeletal muscles and cardiac arrhythmias, including fibrillation.
6. There are several ways to decrease glomerular filtration rate:
   a. Decrease hydrostatic pressure in the glomerulus.
      1. Decrease systemic arterial blood pressure.
         a. Decrease extracellular fluid volume.
         b. Decrease peripheral resistance.
         c. Decrease cardiac output.
      2. Constrict or occlude the afferent arteriole.
      3. Relax the efferent arteriole.
   b. Increase glomerular capsule pressure.
   c. Increase the colloid osmotic pressure of the plasma.
   d. Decrease the permeability of the filtration barrier.
   e. Decrease the total area of the glomeruli available for filtration.
7. Assume that the ascending limb of the loop of Henle and the distal convoluted tubules are impermeable to sodium and other ions but actively pump out water. Other characteristics of the kidney are assumed to be unchanged. As the urine moves up the ascending limb, it becomes hyperosmotic, because sodium remains behind as water is pumped out. Assuming that the collecting ducts are impermeable to sodium, on

reaching the collecting ducts the presence or absence of ADH determines the final concentration of the urine. If ADH is absent, there is little or no exchange of water as the urine passes down the collecting ducts and a hyperosmotic urine will be produced. On the other hand, if ADH is present, water moves from the interstitial fluid into the collecting ducts, thus diluting the urine and producing a hypoosmotic urine.

8. Urea is partially responsible for the high osmolality of the interstitial fluid in the medulla of the kidney. Since a high osmolality of the interstitial fluid must exist for the kidney to produce a concentrated urine, a small amount of urea in the kidney results in the production of dilute urine by the kidney.
9. A low-salt diet tends to reduce the osmolality of the blood. Consequently, ADH secretion is inhibited, producing dilute urine and thus eliminating water. This in turn reduces blood volume and blood pressure.
10. As the loops of Henle become longer, the mechanisms that increase the concentration of the interstitial fluid of the medulla become more efficient, thus raising the concentration of the interstitial fluid. The maximum concentration for urine is determined by the concentration of the interstitial fluid deep in the medulla of the kidneys. The higher the concentration of interstitial fluid in the medulla of the kidney, the greater the concentration of the urine the kidney is able to produce.
11. Prednisone treatment, because of its aldosterone-like effects, increases the blood volume and it can increase blood pressure. It can also lead to the development of edema as the blood volume increases. The effect of prednisone is to increase Na^+ reabsorption. This increases the Na^+ concentration of the blood and interstitial fluid, and water is retained.
12. Hormones of the adrenal cortex are required to maintain the normal volume of blood and blood pressure. Degeneration of the adrenal cortex results in reduced blood volume and blood pressure. Reduced aldosterone causes less Na^+ to be reabsorbed and leads to an increased urine volume.
13. High blood renin levels would result in an increase in blood aldosterone levels (see figure 26.17). The aldosterone increases the number of Na^+-K^+ transport proteins in the basal membranes of cells in the distal convoluted tubules and collecting ducts. More Na^+ would be reabsorbed, and more K^+ would be excreted into the urine. Consequently, blood K^+ levels would be lower than normal. The low blood potassium levels result in the hyperpolarization of membrane potentials. As a result, stimuli of a greater strength than normal are required to produce action potentials in nerve and muscle tissue.
14. Water that moves into the collecting duct cells would diffuse into the interstitial fluid at a reduced rate because of the reduced number of aquaporins. Urine volume would increase because of water retention in the distal convoluted tubule and collecting ducts. Urine concentration would decrease because of dilution by the water. ADH can cause an increase in the number of aquaporin-2 water channel proteins in the apical membrane of the distal convoluted tubules and collecting ducts, but not in aquaporin-3 or aquaporin-4 channels in the basal membranes. These channels determine the permeability of the basal membranes to water. Therefore, ADH would not be an effective treatment. The net effect would be polyuria and is similar to nephrogenic diabetes insipidus caused by an abnormal aquaporin-2 water channel protein.

Chapter 27

1. When excess glucose is not reabsorbed, it osmotically obligates water to remain in the nephron. This results in a large production of urine, called polyuria, with a consequent loss of water, salts, and glucose. The loss of water can be compensated for by increasing fluid intake. The intense thirst that stimulates increased fluid intake is called polydipsia. The loss of salts can be compensated for by increasing the salt intake. The high glucose levels in the blood would increase the blood osmolality, thus stimulating the secretion of ADH. This increases the permeability of the distal convoluted tubule and collecting duct to water. Normally, this would allow reabsorption of water from the collecting ducts and thus conserve water. If glucose levels in the urine are high enough, however, water loss increases even with high levels of ADH being present.
2. When ADH levels first increase, the reabsorption of water increases and urinary output is reduced. This also causes an increase in blood volume and therefore an increase in blood pressure. The increased blood pressure increases glomerular filtration rate, which increases urinary output to normal levels. In addition, the increased blood volume inhibits the renin-angiotensin-aldosterone mechanism, inhibits aldosterone secretion, and stimulates natriuretic hormone secretion. These responses also increase urinary output.
3. Elevated ammonia ions in the urine results from an increased secretion of H^+. Increased secretion of H^+ occurs in response to either metabolic or respiratory acidosis. Because an elevated respiratory rate increases blood pH, the most logical conclusion is that the condition is metabolic acidosis, and the observed increase in respiration rate compensates for the metabolic acidosis by lowering H^+ levels.
4. Diarrhea is one of the most common causes of metabolic acidosis, resulting from the loss of bicarbonate ions. Increasing the respiration rate and producing an acidic urine help increase the blood pH.
5. Blocking H^+ secretion produces acidosis. Because H^+ are exchanged for Na^+, the Na^+ remain in the urine as sodium bicarbonate. This effectively prevents the reabsorption of HCO_3^- and produces an alkaline urine. The blood pH is reduced because H^+ are not being secreted as rapidly by the nephron. The respiration rate increases because of the stimulatory effect of decreased blood pH on the respiratory center.
6. Breathing through the glass tube increases the dead air space and decreases the efficiency of gas exchange. Consequently, blood carbon dioxide levels increase and produce a decrease in blood pH. Compensatory responses include an increased respiration rate and the production of acidic urine.
7. A major effect of alkalosis is hyperexcitability of the nervous system. If the girl is prone to having convulsions, inducing alkalosis might result in a seizure. This can be accomplished by having the girl hyperventilate. The resulting loss of carbon dioxide from the blood causes an increase in blood pH.
8. At high altitudes, we expect stimulation of the chemoreceptor reflex and an increase in respiration rate. This can result in hyperventilation, a decrease in blood carbon dioxide, and respiratory alkalosis. The increased secretion of hydrochloric acid into the stomach can also increase blood pH and contribute to the problem. The kidney produces a more alkaline urine.

Chapter 28

1. Removing the testes would eliminate the major source of testosterone. Blood levels of testosterone would therefore decrease. Because testosterone has a negative-feedback effect on the hypothalamus and pituitary gland, GnRH, FSH, and LH secretion would increase and the blood levels of these hormones would increase.
2. Prior to puberty, the levels of GnRH are very low because the hypothalamus is very sensitive to the inhibitory effects of testosterone. Since GnRH levels are low, so are

FSH and LH levels. Loss of the testes and testosterone production would result in an increase in GnRH, FSH, and LH levels. Because little testosterone is produced, the boy would not develop sexually and would have no sex drive. Small amounts of androgens would be produced because the adrenal cortex produces some androgens. He would be taller than normal as an adult, with thin bones and weak musculature. His voice would not deepen and the normal masculine distribution of hair would not develop.

3. Ideally, the pill would inhibit spermatogenesis. Using the same approach as in females the inhibition of FSH and LH secretion should work. It is known that chronic administration of GnRH suppresses FSH and LH levels enough to cause infertility, through down-regulation. Lack of LH can also result in reduced testosterone levels and a loss of sex drive, however. Some evidence indicates that administration of testosterone in the proper amounts reduces FSH and LH secretion, thus leading to a reduced sperm cell production. The testosterone, however, maintains normal sex drive. The technique appears to work for a large percentage of males, resulting in a sperm concentration in the semen that is too low to result in fertilization. The technique is not sufficiently precise, however, to be used as a standard birth-control technique.
4. In a postmenopausal woman, the ovaries have stopped producing estrogen and progesterone. Without the negative-feedback effect of these hormones the levels of GnRH, FSH, and LH increase. Removal of the nonfunctioning ovaries in a postmenopausal woman does not change the level of any of these hormones or produce any symptoms not already occurring due to the lack of ovarian function.
5. Answer *e* is correct. The secretory phase of the menstrual cycle occurs after ovulation. It is following ovulation that the corpus luteum forms and produces progesterone. In addition, the progesterone acts on the endometrium of the uterus to cause its maximum development. Progesterone secretion therefore reaches its maximum levels and the endometrium reaches its greatest degree of development during the secretory phase of the menstrual cycle.
6. The removal of the ovaries from a 20-year-old woman eliminates the major site of estrogen and progesterone production, thereby causing an increase in GnRH, FSH, and LH levels due to lack of negative feedback. One expects to see the symptoms of menopause, such as cessation of menstruation and reduction in the size of the uterus, vagina, and breasts. There may also be a temporary reduction in sex drive.
7. It is clear that estrogen and progesterone administration resulted in a large decrease in the amount of LH in the plasma the day of ovulation. The differences in plasma LH levels between the groups at other times are very small. The incidence of pregnancies suggests that the reduced plasma LH levels may result in no ovulation.
8. The progesterone inhibits GnRH in the hypothalamus. Consequently, the anterior pituitary is not stimulated to produce LH and FSH. Lack of LH prevents ovulation and lack of FSH prevents development of the follicles. LH also is required for the maturation of follicles prior to ovulation. Without follicle development, there is inadequate estrogen production, which causes the hot flash symptoms.
9. GnRH administered either before or after the normal time of ovulation does not result in ovulation because the anterior pituitary is less sensitive to the effect of GnRH during those times. Also, follicles in the ovary are not adequately developed. The concentration of GnRH must be controlled carefully because too little results in inadequate FSH and LH being released from the anterior pituitary. Too little FSH and LH fails to cause ovulation. Too much GnRH given at the proper time results in the maturation of more than one follicle and the release of an oocyte from more than one of the follicles. If the oocytes are fertilized, multiple pregnancies can result.

Chapter 29

1. Polyspermy, the fertilization of one oocyte with two sperm cells, can result in triploidy. Polyspermy is usually prevented by the fast and slow block to polyspermy, both of which depend on the depolarization of the oocyte membrane. If depolarization of the oocyte membrane does not occur, the zona pellucida will not degenerate and other sperm cells can attach to the oocyte membrane, leading to polyspermy.
2. Postovulatory age, the approximate length of time the embryo has been developing, is 14 days less than the time since the last menstrual period (LMP). In this case, the postovulatory age is 30 days (44 − 14). By this time, the neural tube has closed, the somites have formed, the digestive tract is developing, the limb buds have appeared, a tubular beating heart is present, and the lungs are developing. Based on reproductive structures, which are just forming, male and female embryos are indistinguishable at this age.
3. The fever would have occurred on days 21–31 of development, which is during part of the time of neural tube closure (days 18–25). If the fever prevented neural tube closure, the child might be born with anencephalus or spina bifida.
4. The limb buds develop in a proximal-to-distal sequence. If the apical ectodermal ridge is damaged during embryonic development when the limb bud is about one-half grown, the proximal structures, the arm and forearm, develop normally but the distal structures, the wrist and hand, do not form normally. Depending on the degree of damage, the wrist and hand might be completely absent or underdeveloped.
5. The mesonephric duct system develops, because of testosterone, to form portions of the male reproductive duct system. Without the production of müllerian-inhibiting hormone, the paramesonephric duct system also develops to form the uterus and uterine tubes. Although ovaries are present, the clitoris may be enlarged because of testosterone to produce somewhat the appearance of male external genitalia. The amount of masculinization depends on the levels of testosterone and how long it was administered. High levels of testosterone over an extended period completely masculinize the external genitalia.
6. Progesterone reduces the release of oxytocin, the hormone that stimulates uterine contractions. Also, progesterone reduces the number of oxytocin receptors in the uterus. Both of these effects are expected to reduce labor contractions and can prevent preterm delivery.
7. This total Apgar score of 5 indicates appearance (A, 0) white or blue; pulse (P, 1) low; grimace (G, 1) slight; activity (A, 1) little movement and poor muscle tone; and respiration (R, 2) normal. The white or blue appearance (A, 0) is consistent with a poor circulation indicated by a reduced pulse (P, 1). The reduced heart rate, resulting in the low pulse, may indicate a circulatory system problem. The reduced reflexes and motor activity (G, 1; A, 1) can result from the lack of oxygen in the muscles resulting from poor circulation. Because the infant has poor circulation despite a normal respiration, clearing the airway (if obstructed) and administering oxygen are in order. This Apgar score can have several causes, and additional information is necessary to determine the actual cause.
8. Suckling the breast stimulates the release of oxytocin from the neurohypophysis (posterior pituitary). Once the oxytocin is in the blood, it travels to both breasts and causes milk letdown.

APPENDIX G

Answers to Predict Questions

Chapter 1

1. The chemical level is the level at which correction is currently being accomplished. Insulin can be purchased and injected into the circulation to replace the insulin normally produced by the pancreas. Therapy at the cellular level includes drugs that stimulate pancreatic cells to increase insulin production or make cells more responsive to insulin. Transplantation of pancreatic cells, pancreatic tissue, and even the entire pancreas are also possibilities.
2. a. The normal response to a decrease in blood pressure is for receptors to detect the decrease and for the brain to stimulate the heart to contract faster, which increases blood pressure and maintains homeostasis.
   b. Just before Molly fainted, her heart rate increased. However, it did not increase enough to accommodate her change in position, and her blood pressure dropped below normal. The decreased blood pressure dropped below normal. The decreased blood pressure resulted in the delivery of less blood to the brain, homeostasis of brain tissue was disrupted, and Molly fainted. Note that the regulation of blood pressure involves more than changes in heart rate and is discussed more thoroughly in chapters 20 and 21.
   c. When Molly fell to the floor, she returned to a lying down position. This eliminated the pooling effect of the blood in the veins below the heart because of gravity. Therefore, blood return to the heart increased, blood pressure increased, blood flow to the brain increased, homeostasis of brain tissue was restored, and she regained consciousness.
3. Negative-feedback mechanisms work to control respiratory rates so that body cells have adequate oxygen and are able to eliminate carbon dioxide. The greater the respiratory rate, the greater the exchange of gases between the body and the air. When a person is at rest, there is less of a demand for oxygen, and less carbon dioxide is produced than during exercise. At rest, homeostasis can be maintained with a low respiration rate. During exercise, there is a greater demand for oxygen, and more carbon dioxide must be eliminated. Consequently, to maintain homeostasis during exercise, the respiratory rate increases.
4. The sensation of thirst is involved in a negative-feedback mechanism that maintains body fluids. The sensation of thirst increases with a decrease in body fluids. The thirst mechanism causes a person to drink fluids, which returns body fluid levels to normal, thereby maintaining homeostasis.
5. In the cat, cephalic and anterior are toward the head; dorsal and superior are toward the back. In humans, cephalic and superior are toward the head; dorsal and posterior are toward the back.
6. Your kneecap is both proximal and superior to the heel. It is also anterior to the heel because it is on the anterior side of the lower limb, whereas the heel is on the posterior side.
7. The spleen is in the left-upper quadrant, the gallbladder is in the right-upper quadrant, the left kidney is in the left-upper quadrant, the right kidney is in the right-upper quadrant, the stomach is mostly in the left-upper quadrant, and the liver is mostly in the right-upper quadrant.
8. There are two ways in which an organ can be located within the abdominopelvic cavity but not be within the peritoneal cavity. First, the visceral peritoneum wraps around organs. Thus, the peritoneal cavity surrounds the organ, but the organ is not inside the peritoneal cavity. The peritoneal cavity contains only peritoneal fluid. Second, retroperitoneal organs are in the abdominopelvic cavity, but they are between the wall of the abdominopelvic cavity and the parietal peritoneal membrane.

Chapter 2

1. The mass (amount of matter) of the astronaut on the surface of the earth and in outer space does not change. In outer space, where the force of gravity from the earth is very small, the astronaut is "weightless," compared with his or her weight on the earth's surface.
2. Potassium has 19 protons (the atomic number), 20 neutrons (the mass number minus the atomic number), and 19 electrons (because the number of electrons equals the number of protons).
3. There are more atoms in 12.01 g of carbon than in 12.01 g of magnesium. There is Avogadro's number or 1 mole (mol) of carbon atoms in 12.01 g of carbon. There is Avogadro's number or 1 mole (mol) of magnesium atoms in 24.305 g of magnesium. Therefore, there is approximately half a mol ($0.5 = 12.01/24.305$) of magnesium atoms in 12.01 g of magnesium.
4. The molecular formula for glucose is $C_6H_{12}O_6$. The atomic mass of carbon is 12.01, hydrogen is 1.008, and oxygen is 16.00. The molecular mass of glucose is therefore $(6 \times 12.01) + (12 \times 1.008) + (6 \times 16.00)$, or 180.2.
5. A decrease in blood CO_2 decreases the amount of H_2CO_3 and therefore the blood H^+ level. Because CO_2 and H_2O are in equilibrium with H^+ and HCO_3^-, with H_2CO_3 as an intermediate, a decrease in CO_2 causes some H^+ and HCO_3^- to join together to form H_2CO_3, which then forms CO_2 and H_2O. Consequently, the H^+ concentration decreases.
6. When two hydrogen atoms combine with an oxygen atom to form water, a polar covalent bond forms between each hydrogen atom and the oxygen atom. Unequal sharing of electrons occurs, and the electrons are associated with the oxygen atom more than with the hydrogen atoms. In this sense, the hydrogen atoms lose their electrons, and the oxygen atom gains electrons. The hydrogen atoms are therefore oxidized, and the oxygen atom is reduced.
7. During exercise, muscle contractions increase, which requires energy. This energy is obtained from the energy in the chemical bonds of ATP. As ATP is broken down, energy is released. Some of the energy is used to drive muscle contractions, and some becomes heat. Because the rate of these reactions increases during exercise, more heat is produced than when at rest, and body temperature increases.
8. Monohydrogen phosphate ion (HPO_4^{2-}) is the conjugate base formed when the conjugate acid, dihydrogen phosphate ion ($H_2PO_4^-$), loses a H^+. If H^+ are added to the solution, they combine with the conjugate base, HPO_4^{2-}, to form $H_2PO_4^-$, which helps prevent an increase in H^+ concentration. If OH^- are added to the solution, they combine with H^+ to form water. Then

the conjugate acid, $H_2PO_4^-$, dissociate to replace the H^+, which helps prevent a decrease in H^+ concentration.

Chapter 3

1. Urea is continually produced by metabolizing cells and diffuses from the cells into the interstitial spaces and from the interstitial spaces into the blood. If the kidneys stop eliminating urea, it begins to accumulate in the blood. Because the concentration of urea increases in the blood, urea cannot diffuse from the interstitial spaces. As urea accumulates in the interstitial spaces, the rate of diffusion from cells into the interstitial spaces slows because the urea must pass from a higher to a lower concentration by the process of diffusion. The urea finally reaches concentrations high enough to be toxic to cells, thereby causing cell damage followed by cell death.
2. If the membrane is freely permeable, the solutes in the tube diffuse from the tube (higher concentration of solutes) into the beaker (lower concentration of solutes) until equal amounts of solutes exist inside the tube and beaker (i.e., equilibrium). In a similar fashion, water in the beaker diffuses from the beaker (higher concentration of water) into the tube (lower concentration of water) until equal amounts of water are inside the tube and beaker. Consequently, the solution concentrations inside the tube and beaker are the same because they both contain the same amount of solutes and water. Under these conditions, no net movement of water into the tube occurs. This simple experiment demonstrates that osmosis and osmotic pressure require a membrane that is selectively permeable.
3. Glucose transported by facilitated diffusion across the plasma membrane moves from a higher to a lower concentration. If glucose molecules are quickly converted to some other molecule as they enter the cell, a steep concentration gradient is maintained. The rate of glucose transport into the cell is directly proportional to the magnitude of the concentration gradient.
4. Digitalis should increase the force of heart contractions. By interfering with Na^+ transport, digitalis decreases the concentration gradient for Na^+ because fewer Na^+ are pumped out of cells by active transport. Consequently, fewer Na^+ diffuse into cells, and fewer Ca^{2+} move out of the cells by antiport. The higher intracellular levels of Ca^{2+} promote more forceful contractions.
5. a. Cells highly specialized to synthesize and secrete proteins have large amounts of rough endoplasmic reticulum (ribosomes attached to endoplasmic reticulum) because these organelles are important for protein synthesis. Golgi apparatuses are well developed because they package materials for release in secretory vesicles. Also, numerous secretory vesicles exist in the cytoplasm.
   b. Cells highly specialized to actively transport substances into the cell have a large surface area exposed to the fluid from which substances are actively transported, and numerous mitochondria are present near the membrane across which active transport occurs.
   c. Cells highly specialized to synthesize lipids have large amounts of smooth endoplasmic reticulum. Depending on the kind of lipid produced, lipid droplets may accumulate in the cytoplasm.
   d. Cells highly specialized to phagocytize foreign substances have numerous lysosomes in their cytoplasm and evidence of phagocytic vesicles.
6. The mRNA sequence is GCAUGCGGCUCUGCAGUUG. The complementary strand sequence is GCATGCGGCTCTGCAGTTG. The difference between the complementary strand and the mRNA sequence is the presence of thymine (T) in the complementary strand and the presence of uracil (U) in the mRNA sequence.
7. Because adenine pairs with thymine (no uracil exists in DNA) and cytosine pairs with guanine, the sequence of DNA replicated from strand 1 is TACGAT. This sequence is also the sequence of DNA in the original strand 2. A replicate of strand 2 is therefore ATGCTA, which is the same as the original strand 1.
8. Genotype *DD* (homozygous dominant) would have the polydactyly phenotype, genotype *Dd* (heterozygous) would have the polydactyly phenotype, and genotype *dd* (homozygous recessive) would have the normal phenotype.
9. The number of chromosomes in a gamete is half the number in a somatic cell because two gametes combine at fertilization to form a new diploid zygote. If the gametes had the same number of chromosomes as somatic cells, the chromosome number would double with each generation.
10. None. One in two will be homozygous normal, and one in two will be normal heterozygous carriers.

Chapter 4

1. a. The secretion of mucus and digestive enzymes and the absorption of nutrients normally occur in the digestive tract. Simple columnar epithelial cells contain organelles that are specialized to carry out nutrient absorption and secretion of mucus and digestive enzymes. Stratified squamous epithelium is not specialized to either absorb or secrete, and the layers of epithelial cells reduce the ability of nutrient molecules to pass through them and therefore to be absorbed. The ability of digestive enzymes to pass through the layers of epithelial cells, and therefore be secreted, is also reduced.
   b. Keratinized stratified epithelium forms a tough layer that is a barrier to the movement of water. Replacing the epithelium of skin with moist stratified squamous epithelium increases the loss of water across the skin because water can diffuse through nonkeratinized stratified squamous epithelium, and it is more delicate and provides less protection than keratinized stratified squamous epithelium.
   c. The stratified squamous epithelium that lines the mouth provides protection. Replacement of it with simple columnar epithelium makes the lining of the mouth much more susceptible to damage because the single layer of epithelial cells is easier to damage.
2. The zonula occludens prevents substances from passing between the epithelial cells. NaCl must pass across the epithelial layer. NaCl could pass by facilitated diffusion across one membrane of the epithelial cells and by active transport across the other epithelial membrane. If the membranes have a high permeability to water, water will move in the same direction as NaCl because of osmosis.
3. Elastic ligaments attached to the vertebrae help the vertebral column return to its normal, upright position after it is flexed. The elastic ligaments act much as elastic bands do. Tendons attach muscles to bones. When muscles contract, they pull on the tendons, which in turn pull on bones. Because they are not elastic, when the muscle pulls on the tendon, all of the force is applied to the bone, causing it to move. If tendons were elastic, when the muscle contracted the tendon would stretch, and not all of the tension would be applied to the bone.
4. Collagen synthesis is required for scar formation. If collagen synthesis does not occur because of a lack of vitamin C or if collagen synthesis is slowed, wound healing does not occur or is slower than normal. One might expect that the density of collagen fibers in a scar is reduced and the scar is not as durable as a normal scar.

5. Hyaline cartilage provides a smooth surface so that bones in joints can move easily. When the smooth surface provided by hyaline cartilage is replaced by dense fibrous connective tissue, the smooth surface is replaced by a less smooth surface, and the movement of bones in joints is much more difficult. The increased friction helps increase the inflammation and pain that occurs in the joints of people who have rheumatoid arthritis.
6. In severely damaged tissue in which cells are killed and blood vessels are destroyed, the usual symptoms of inflammation cannot occur. Surrounding these areas of severe tissue damage, however, where blood vessels are still intact and cells are still living, the classic signs of inflammation do develop. The signs of inflammation therefore appear around the periphery of severely injured tissues.

Chapter 5

1. Because the permeability barrier is composed mainly of lipids surrounding the epidermal cells, substances that are lipid-soluble easily pass through, whereas water-soluble substances have difficulty.
2. a. The lips are pinker or redder than the palms of the hand. Several explanations for this are possible: There might be more blood vessels in the lips, increased blood flow might occur in the lips, or the blood vessels might be easier to see through the epidermis of the lips. The last possibility explains most of the difference in color between the lips and the palms. The epidermis of the lips is thinner and not as heavily keratinized as that of the palms. In addition, the papillae containing the blood vessels in the lips are "high" and closer to the surface.
   b. A person who does manual labor has a thicker stratum corneum on the palms (and possibly calluses) than a person who does not perform manual labor. The thicker epidermis masks the underlying blood vessels, and the palms do not appear as pink. In addition, carotene accumulating in the lipids of the stratum corneum might impart a yellowish cast to the palms.
   c. The posterior surface of the forearm appears darker because of the tanning effect of ultraviolet light from the sun.
   d. The genitals normally have more melanin and appear darker than the soles of the feet.
3. The story is not true. Hair color results from the transfer of melanin from melanocytes to keratinocytes in the hair matrix as the hair grows. The hair itself is dead. To turn white, the hair must grow out without the addition of melanin, a process that takes weeks.
4. a. Billy's ears and nose turned pale because of decreased blood flow. As ambient temperature decreases, body temperature decreases. Constriction of blood vessels in the skin reduces heat loss and helps maintain body temperature.
   b. Billy's ears and nose turned red because of increased blood flow. As skin temperature decreases and blood vessels constrict, tissues can be damaged by the lack of blood flow and ice crystal formation. Cold-induced vasodilation periodically increases blood flow and prevents or slows the rate of ice crystal formation. This strategy for maintaining tissue homeostasis is beneficial as long as body temperature can be maintained.
   c. At colder temperatures, cold-induced vasodilation stops and tissues are no longer protected by an infusion of warm blood. Damaged tissues are white because blood no longer flows through them.
   d. One function of the skin is to prevent the entry of microorganisms. If frostbite results in the destruction of skin cells, this function can be compromised, resulting in infections.
5. Reducing water loss is one of the normal functions of the skin. Loss of skin, or damage to the skin, can greatly increase water loss. In addition, burning large areas of the skin results in increased capillary permeability and additional loss of fluid from the burn and into tissue spaces. The loss of fluid reduces blood volume, which results in reduced blood flow to the kidneys. Consequently, urine output by the kidneys decreases, which reduces fluid loss and thereby helps compensate for the fluid loss caused by the burn. The reduced blood flow to the kidneys can cause tissue damage, however. To counteract this effect, during the first 24 hours following the injury, part of the treatment for burn victims is the administration of large volumes of fluid. The question, however, is how much fluid should be given. The amount of fluid given should be sufficient to match that lost plus enough to prevent kidney damage and allow the kidneys to function. Urine output is therefore monitored. If it is too low, more fluid is administered; if it is too high, less fluid is given. An adult receiving intravenous fluids should produce 30–50 mL of urine/hour, and children should produce 1 mL/kg of body weight/hour.

Chapter 6

1. In the absence of a good blood supply, nutrients, chemicals, and cells involved in tissue repair enter cartilage tissue very slowly. As a result, the ability of cartilage to undergo repair is poor. Within a joint, the articular cartilage of one bone presses against and moves against the articular cartilage of another bone. If the articular cartilages were covered by perichondrium, or contained blood vessels and nerves, the resulting pressure and friction could damage these structures.
2. In the elderly, the bone matrix contains proportionately less collagen than hydroxyapatite, compared with the bones of younger people. Collagen provides bone with flexible strength, and a reduction in collagen results in brittle bones. In addition, the elderly have less dense bones with less matrix. The combination of reduced matrix that is more brittle results in a greater likelihood of bones breaking.
3. Cancellous bone consists of trabeculae with spaces between them. Blood vessels can pass through these spaces. In compact bone, the blood vessels pass through the perforating and central canals. The trabeculae in cancellous bone are thin enough that nutrients and gases can diffuse from blood vessels around the trabeculae to the osteocytes through the canaliculi.
4. Chondroblasts are surrounded by cartilage matrix and receive oxygen and nutrients by diffusion through the matrix. When the matrix becomes calcified, diffusion is reduced to the point that the cells die. When osteoblasts form bone matrix, they connect to one another by their cell processes. Thus, when the matrix is laid down, canaliculi are formed. Even though the ossified bone matrix is dense and prevents significant diffusion, the osteocytes can receive gases and nutrients through the canaliculi or by movement from one osteocyte to another.
5. Interstitial growth of cartilage results from the division of chondrocytes within the cartilage, followed by the addition of new cartilage matrix between the chondrocytes. The resulting expansion of the cartilage matrix is possible because cartilage matrix is not too rigid. Bones cannot undergo interstitial growth because bone matrix is rigid and cannot expand from within. New bone must therefore be added to the surface by apposition.

6. Damage to the epiphyseal plate interferes with bone elongation; as a result, the bone, and therefore the thigh, will be shorter than normal. Recovery is difficult because cartilage repairs very slowly.
7. Growth of articular cartilage results in an increase in the size of epiphyses. This is only one of the functions of articular cartilage; it also forms a smooth, resilient covering over the ends of the epiphyses within joints. Ossified articular cartilage could not perform that function.
8. Her growth for the next few months increases, and she may be taller than a typical 12-year-old female. Because the epiphyseal plates ossify earlier than normal, however, her height at age 18 will be less than otherwise expected.
9. Taking in adequate Ca^{2+} and vitamin D through the digestive system during adulthood increases Ca^{2+} absorption from the small intestine. The increased Ca^{2+} is used to increase bone mass. The greater the bone mass before the onset of osteoporosis, the greater the tolerance for bone loss later in life. For this reason, it is important for adults, especially women in their twenties and thirties, to ingest adequate amounts of calcium. Exercising the muscular system places stress on bone, which also increases bone density. The granddaughter should not smoke because this reduces estrogen levels.
10. a. Henry's bone density is less than normal for a man his age. Less dense bone is more likely to break.
    b. Henry's eating habits have resulted in inadequate dietary intake of Ca^{2+} and vitamin D. Therefore, the absorption of Ca^{2+} from his intestine into his blood has been inadequate.
    c. One might expect that Henry's blood Ca^{2+} would be low because of his diet. Low blood Ca^{2+} levels, however, stimulate increased PTH secretion. An increase in PTH maintains normal blood Ca^{2+} levels by increasing the number of osteoclasts, which break down bone and release Ca^{2+} into the blood. Thus, Henry's blood Ca^{2+} levels are maintained at the expense of his bones, which become less dense as more matrix than usual is broken down. Increased PTH levels also promote increased Ca^{2+} reabsorption from the urine.
    d. Normally, exposure to ultraviolet light, especially in sunlight, activates a precursor molecule in the skin that eventually becomes activated vitamin D in the kidneys (see chapter 5). Henry produces few, if any, precursor molecules because of his nocturnal lifestyle. Even though PTH normally increases the formation of active vitamin D in the kidneys, Henry's elevated PTH levels have not resulted in adequate vitamin D levels because of his lack of precursor molecules. Low levels of vitamin D result in less absorption of Ca^{2+} from the small intestine, contributing to Henry's low bone density.
    e. Exercise increases the mechanical stress on bones, which increases osteoblast activity. Lack of exercise has resulted in decreased osteoblast activity. Bone density has decreased because less bone matrix has been produced by osteoblasts at the same time that more bone matrix has been broken down by osteoclasts under the influence of PTH.

Chapter 7

1. The sagittal suture is so-named because it is in line with the midsagittal plane of the head. The coronal suture is so-named because it is in line with the coronal plane (see chapter 1).
2. The bones most often broken in a "broken nose" are the nasals, ethmoid, vomer, and maxillae.
3. The lumbar vertebrae support a greater weight than the other vertebrae. The vertebrae are more massive because of the greater weight they support.
4. The anterior support of the scapula is lost with a broken clavicle, and the shoulder is located more inferiorly and anteriorly than normal. In addition, since the clavicle normally holds the upper limb away from the body, the upper limb moves medially and rests against the side of the body.
5. The olecranon processes moves into the olecranon fossa as the elbow is straightened. The coronoid process moves into the coronoid fossa as the elbow is bent.
6. The dried skeleton seems to have longer "fingers" than the hand with soft tissue intact because the soft tissue fills in the space between the metacarpals. With the soft tissue gone, the metacarpals seem to be an extension of the fingers, which appear to extend from the most distal phalanx to the carpals.
7. The femoral neck is commonly injured in elderly people because it is the smallest portion of the femur, which supports the weight of the body. It also forms an angle between the pelvis and the shaft of the femur, so the downward force of gravity on the body places an enormous amount of pressure on this part of the femur. That pressure is usually resisted in younger people with strong bones. The bone matrix begins to deteriorate in the elderly (osteoporosis), however. Because of hormonal differences between men and women, osteoporosis is much more common among elderly women.
8. The top of modern ski boots is placed high up the leg to protect the weakest point of the fibula to make it less susceptible to great strain during a fall. Modern ski boots are also designed to reduce ankle mobility, which increases comfort and performance.
9. Decubitus ulcers form over bony prominences where the bone is close to the overlying skin and where the body contacts the bed when lying down. Such sites are the back and front of the skull and the cheeks (over zygomatic bones), acromion process, scapula, olecranon process, coccyx, greater trochanter, lateral epicondyle of the femur, patella, and lateral malleolus.

Chapter 8

1. The joint between the metacarpals and the phalanges is the metacarpophalangeal joint.
2. Premature sutural synostosis can result in abnormal skull shape, interfere with normal brain growth, and result in brain damage if not corrected. Such an abnormality is usually corrected surgically by removing some of the bone around the suture and creating an artificial fontanel, which then undergoes normal synostosis.
3. The synovial membrane is very thin and delicate. A considerable amount of pressure is exerted on the articular cartilages within a joint, and the articualr cartilage is very tough, yet flexible, to withstand the pressure. If the synovial membrane covered the articualr cartilage, it would be easily damaged during movement.
4. The movements required are abduction of the upper limb and flexion of the elbow. Flexion of the shoulder and elbow also works.
5. A shoulder separation involves stretching or tearing of the acromioclavicular ligament and may involve tearing of the coracoclavicular ligament as well. Because the only bony attachment of the upper limb to the body is from the scapula through the clavicle to the sternum, separation of the acromioclavicular joint greatly reduces the stability of the shoulder. The scapula and humerus tend to be displaced inferiorly, and the proximal pivot point for the upper limb is destabilized.

Chapter 9

1. When a muscle changes length, the I bands and the H zones change in width, but the A band does not. When a muscle is stretched, the I bands and the H zones increase in width as the length of the

sarcomere increases. When a muscle contracts, cross-bridges form and cause the actin myofilaments to slide over the myosin myofilaments. The result is that the I bands and H zones decrease in width as the sarcomeres shorten. When a muscle relaxes, cross-bridges release, and actin myofilaments slide past myosin myofilaments as the sarcomeres lengthen. The I bands and H zones increase in width.

2. When gated K^+ channels open, K^+ diffuse from an area of higher concentration inside of the cell to an area of lower concentration outside of the cell. This leaves more negative charges inside the cell, which causes the resting membrane potential to be more negative.
3. If insufficient acetylcholine is released from the presynaptic terminal of an axon, an action potential is not produced in the muscle fiber and the muscle cannot contract. An action potential must be produced in the muscle fiber for contraction to occur. If inadequate acetylcholine is released from the presynaptic terminal of an axon, several action potentials in the axons must occur to cause the presynaptic terminal neurons to release enough acetylcholine to produce an action potential in the muscle fibers. Each action potential releases some acetylcholine. In response, a local potential may be produced in the postsynaptic membrane. If the local potentials were produced over a short period, they could summate (see chapter 11) and reach threshold. If threshold is reached, an action potential is produced.
4. a. If Na^+ cannot enter the muscle fiber, no action potentials are produced in the muscle fiber because the influx of Na^+ causes the depolarization phase of the action potential. Without action potentials, the muscle fiber cannot contract at all. The result is flaccid paralysis.
   b. If ATP levels are low in a muscle fiber before stimulation, the following events occur. Energy from the breakdown of ATP already is stored in the heads of the myosin molecules. After stimulation, cross-bridges form. If not enough additional ATP molecules are in the muscle cells to bind to the myosin molecules to allow for cross-bridge release, however, the muscle becomes stiff without contracting or relaxing.
   c. If ATP levels in the muscle fiber are adequate but the action potential frequency is so high that Ca^{2+} accumulate around the myofilaments, the muscle contracts continuously without relaxing. As long as the ions are numerous within the sarcoplasm in the area of the myofilaments, cross-bridge formation is possible. If ATP levels are adequate, cross-bridge formation, release, and formation can proceed again, resulting in a continuously contracting muscle.
5. A decrease occurs in muscle control when reinnervation of muscle fibers occurs after poliomyelitis because the number of motor units in the muscle is decreased. Reinnervation results in a greater number of muscle fibers per motor unit. Control is reduced because the number of motor units that can be recruited is decreased. The greater the number of motor units in a muscle, the greater the ability to have fine gradations of muscle contraction as motor units are recruited. A smaller number of motor units means that gradations of muscle contraction are not as fine.
6. a. Organophosphate poisons inhibit the activity of acetylcholinesterase, which breaks down acetylcholine at the neuromuscular junction and limits the length of time the acetylcholine stimulates the postsynaptic terminal of the muscle fiber. Consequently, acetylcholine accumulates in the synaptic cleft and continuously stimulates the muscle fiber. As a result, the muscle remains contracted until it fatigues. Death is caused by the victim's inability to breathe. Either the respiratory muscles are in spastic paralysis or they are so depleted of ATP that they cannot contract at all.
   b. Curare binds to acetylcholine receptors and thus prevents acetylcholine from binding to them. Because curare does not activate the receptors, the muscles do not respond to nervous stimulation. The person suffers from flaccid paralysis and dies from suffocation because the respiratory muscles are not able to contract.
7. As a weight is lifted, the muscle contractions are concentric contractions. When a weight lifter lifts a heavy weight above the head, most of the muscle groups contract with a force while the muscle is shortening. Concentric contractions are a category of isotonic contractions in which tension in the muscle increases or remains about the same while the muscle shortens. While the weight is held above the head, the contractions are isometric contractions, because the length of the muscles does not change. While the weight is lowered, unless the weight lifter simply drops the weight, the length of the muscles increases as the weight is lowered for most of the muscle groups. Eccentric contractions are contractions in which tension is maintained in a muscle while the muscle increases in length. The major muscle groups are therefore contracting eccentrically while the weight is lowered.
8. The increase in breathing after running up the stairs is recovery oxygen consumption and is partially due to the oxygen deficit that occurs at the beginning of exercise. Because the duration and intensity of the exercise were similar in both individuals, the only explanation for the difference in recovery oxygen consumption is the general condition of the individual. Because of conditioning, Eric produced more of his ATP aerobically, whereas John relied more on anaerobic respiration. Therefore, John had a greater oxygen deficit.
9. Susan was unable to keep pace during the finishing sprint because her muscles were not trained for quick movements, such as sprinting. She had trained them only for aerobic, steady-state activity, such as the activity occurring during most of the race. As her coach, you would advise Susan to incorporate high-intensity sprinting activities into her training. Such activities will cause her fast-twitch muscle fibers to increase their performance capacity, which will increase her ability to sprint at the end of the race.
10. A ligand that binds to its receptor and results in a sustained increase in the permeability of the plasma membrane to Ca^{2+} results in a sustained contraction without a large increase in ATP breakdown. The increased intracellular concentration of Ca^{2+} increases the number of phosphate groups removed from the myosin molecules while cross-bridges are attached. Because these cross-bridges release slowly, the result is a sustained contraction.
11. A boy with advanced DMD would have muscles that contain fewer muscle fibers and more fibrous connective tissue and fat. His posture would be much worse due to abnormal curvatures of the spinal column (such as kyphoscoliosis) and more deformity of the thoracic cage. In general, his physical condition would be worse because of weakened heart muscle and difficulty breathing due to thoracic cage and vertebral column deformities. A boy with advanced DMD would much more likely be wheelchair-bound due to severe weakness of his skeletal muscles.

Chapter 10

1. Shortening the right sternocleidomastoid muscle rotates the head to the left. It also slightly elevates the chin.
2. Raising eyebrows—occipitofrontalis; winking—orbicularis oculi and then levator

palpebrae superioris; whistling—orbicularis oris and buccinator; smiling—levator anguli oris, risorius, zygomaticus major, and zygomaticus minor; frowning—corrugator supercilii and procerus; flaring nostrils—levator labii superioris alaeque nasi and nasalis.

3. Weakness of the lateral rectus allows the eye to deviate medially.
4. As the arm is maximally abducted, the supraspinatus tendon rubs against the acromion process of the scapula. This can result in damage to, and pain in, the supraspinatus muscle, commonly referred to as a torn rotator cuff.
5. Two arm muscles are involved in flexion of the elbow: the brachialis and the biceps brachii. The brachialis only flexes, whereas the biceps brachii both flexes the elbow and supinates the forearm. With the forearm supinated, both muscles can flex the elbow optimally; when pronated, the biceps brachii does less to flex the elbow. Chin-ups with the elbow supinated are therefore easier because both muscles flex the forearm optimally in this position. Bodybuilders who wish to build up the brachialis muscle perform chin-ups with the forearms pronated.

Chapter 11

1. When the axon of a neuron is severed, the proximal portion of the axon remains attached to the neuron cell body. The distal portion is detached, however, and has no way to replenish the enzymes and other proteins essential to its survival. Because the DNA in the nucleus provides the information that determines the structure of proteins by directing mRNA synthesis, the distal portion of the axon has no source of new proteins. Consequently, it degenerates and dies. On the other hand, the proximal portion of the axon is still attached to the nucleus and therefore has a source of new proteins. It remains alive and, in many cases, grows to replace the severed distal axon.
2. Tissue A has the larger resting membrane potential. There is a greater tendency for K^+ to diffuse out of the cell because it has significantly more K^+ leak channels. As a result, a greater negative charge develops on the inside of the plasma membrane, resulting in a larger resting membrane potential.
3. If the intracellular concentration of K^+ is increased, the concentration gradient from the inside to the outside of the plasma membrane increases. This situation is similar to decreasing the extracellular concentration of K^+. The greater concentration gradient for K^+ increases their tendency to diffuse out of the cell across the plasma membrane. A greater negative charge then develops inside the cell (hyperpolarization).
4. Calcium ions bind to gating proteins that regulate the voltage-gated Na^+ channels. Low concentrations of Ca^{2+} cause the voltage-gated Na^+ channels to open, and high concentrations of Ca^{2+} cause the voltage-gated Na^+ channels to close. If the extracellular concentration of Ca^{2+} decreases, the resting membrane potential depolarizes because voltage-gated Na^+ channels open and Na^+ diffuse into the cell.
5. When the cells are stimulated, there is an increase in the permeability of their plasma membranes to Na^+. These ions diffuse into the cells down their concentration gradients and cause depolarization of the plasma membranes. If the concentration gradient for Na^+ is reduced, the tendency for Na^+ to diffuse into the cell decreases. In cell A, with the reduced Na^+ concentration gradient, the depolarization is of a smaller magnitude than in cell B because fewer ions are able to diffuse into the cell in response to the stimulus.
6. If the extracellular concentration of Na^+ decreases, the magnitude of the action potential is reduced. The smaller extracellular concentration of Na^+ reduces the tendency for Na^+ to diffuse into the cell when the Na^+ channels are open during an action potential. Consequently, the inside of the plasma membrane does not become as positive as it does in cells with a high extracellular concentration of Na^+. Even though the magnitude of action potentials is reduced when the extracellular Na^+ concentration is reduced, all of the action potentials are the same magnitude (all-or-none principle).
7. A prolonged stronger-than-threshold stimulus produces more action potentials than a prolonged threshold stimulus of the same duration. A prolonged stronger-than-threshold stimulus can stimulate more action potentials because the permeability of the membrane to Na^+ is increased. A very strong stimulus can even stimulate action potentials during the relative refractory period, whereas a prolonged threshold stimulus stimulates a low frequency of action potentials. Thus, when a prolonged stronger-than-threshold stimulus is applied, less time elapses between the production of one action potential and the next, resulting in the production of a greater number of action potentials.
8. If the duration of the absolute refractory period is 1 ms, that means action potentials can be generated no faster than every millisecond. The maximal frequency of action potentials is 1000 per second because there are 1000 milliseconds in a second.
9. Action potentials are transmitted fastest by electrical synapses because the ionic currents can quickly flow through the connexons. In contrast, chemical synapses are slower because the synaptic vesicles must be stimulated to release neurotransmitter, which diffuses across the synaptic cleft. The neurotransmitter must stimulate ligand-gated Na^+ channels to open. The resulting movement of Na^+ into the cell can produce an action potential. All of these events take time.
10. Temporal summation resulting from stimulation by neuron B produces more action potentials in the postsynaptic neuron than temporal summation resulting from stimulation by neuron A. The neuromodulator from neuron B produces EPSPs, which depolarize the membrane potential of neuron C, bringing the membrane potential closer to threshold. A smaller amount of neurotransmitter is therefore required to produce an action potential. Although neurons A and B release the same amount and type of neurotransmitter, the neuromodulator makes the neurotransmitter from neuron B more effective, resulting in more action potentials.

Chapter 12

1. The cord is enlarged in the cervical and lumbosacral regions because of the large numbers of motor nerve fibers exiting from the cord to the limbs and sensory nerve fibers entering the cord from the limbs. Also, more neuron cell bodies in the spinal cord regions are associated with the increased numbers of sensory and motor fibers.
2. The common site of injection into the vertebral column is between L3 and L4 or between L4 and L5 (see chapter 7).
3. Dorsal root ganglia contain neuron cell bodies, which are larger in diameter than the axons of the dorsal roots. Action potentials are propogated in both directions in spinal nerves, toward the spiral cord in the dorsal root and away from the spinal cord in the ventral root.
4. Nerves C5–T1, which innervate the left arm, forearm, and hand, are damaged.
5. Damage to the right phrenic nerve results in the absence of muscular contraction in the right half of the diaphragm. Because the phrenic nerves originate from C3–C5, damage to the upper cervical region of the

spinal cord eliminates their functions; damage in the lower cord below the point where the spinal nerves originate does not affect the nerves to the diaphragm. Breathing is affected, however, because the intercostal nerves to the intercostal muscles, which move the ribs, are paralyzed.

6. The radial nerve lies along the shaft of the humerus about midway along its length. If the humerus is fractured, the radial nerve can be lacerated by bone fragments or, more commonly, pinched between two fragments of bone, decreasing or eliminating the function of the nerve.
7. Pain, tingling, and numbness radiating from the elbow, down the posteromedial portion of the forearm, and to the medial side of the hand and fingers (4 and 5) is consistent with damage to the C8–T1 dermatome (see figure 12.14). Ulnar nerve damage results in the same symptoms in the medial hand and fingers, but the symptoms do not radiate into the forearm and elbow (see figure 12.21). Careful examination of Sarah's right-upper limb to map the extent of the pain and numbness allows for a tentative differential diagnosis between ulnar nerve damage and damage to the C8–T1 brachial plexus roots. An x-ray of the neck can indicate the presence of an extra rib, which most commonly affects roots C8–T1. As the ulnar nerve arises from the C8–T1 roots of the brachial plexus, all muscles innervated by that nerve can be affected. Because the radial and median nerves also obtain some axons from the C8–T1 roots, some of their more distal-lateral muscles may also be affected.

Chapter 13

1. a. Often when one part of the head suffers a heavy blow, the brain moves within the cranial cavity and hits the opposite side of the cranial cavity. In this case, the blow to the back of the head forced the brain anteriorly and the frontal lobes struck the frontal bones with enough force to tear blood vessels between the brain and the dura. Subsequent bleeding from these vessels into the subdural space created the subdural hematoma. This type of injury is a contrecoup brain injury because it occurred on the opposite side of the brain from the point of traumatic impact.
   b. The subdural hematoma forcing the medulla oblongata to herniate into the vertebral canal caused the tissues of the medulla to become compressed and fail to function. The medulla oblongata contains the centers for the control of respiration and heart rate. Thus, damage to these areas of the brain interrupted the woman's ability to regulate her breathing and heart rate, which was most likely a major contributing factor in her death.
2. The oculomotor nerve innervates four eye muscles and the levator palpebrae superioris muscle. One cause of ptosis, a drooping upper eyelid, can be oculomotor nerve damage and subsequent paralysis of the levator palpebrae superioris muscle. The four eye muscles innervated by the oculomotor nerve move the eyeball so that the gaze is directed superiorly, inferiorly, medially, or superolaterally. Damage to this nerve can be tested by having the patient look in these directions. The abducens nerve directs the gaze laterally, and the trochlear nerve directs the gaze inferolaterally. If the patient can move the eyes in these directions, the associated nerves are intact.
3. The sternocleidomastoid muscle pulls the mastoid process (located behind the ear) toward the sternum, thus turning the face to the opposite side. If the innervation to one sternocleidomastoid muscle is eliminated (accessory nerve injury), the opposite muscle is unopposed and turns the face toward the side of injury. A person with wry neck whose head is turned to the left most likely has an injured left accessory nerve.
4. The tongue is protruded by contraction of the geniohyoid muscle, which pulls the back of the tongue forward, thereby pushing the muscle mass of the tongue forward. With one side pushed forward and unopposed by muscles of the opposite side, the tongue deviates toward the nonfunctional side. In the example, therefore, the right hypoglossal nerve is damaged.

Chapter 14

1. Because hot and cold objects may not be perceived any differently for temperatures of 0° to 12°C or above 47°C (both temperature ranges stimulate pain fibers), the nervous system may not be able to discriminate between the two temperatures. At low temperatures, both cold and pain receptors are stimulated; thus, after the object has been in the hand for a very short time, it is possible to discriminate between cold and pain. If, however, the CNS has been preprogrammed to think that the object to be placed in the hand is hot, a cold object can elicit a rapid withdrawal reflex.
2. Lesions on one side of the spinal cord that interrupt the anterolateral system eliminate pain and temperature sensation below that level on the opposite side of the body. This occurs because the pathway is contralateral (ascending on the opposite side of the spinal cord) to the area of skin innervated by the receptors. Light touch and the like show few, if any, clinical changes because some fibers cross and ascend contralaterally, whereas others ascend ipsilaterally for some distance before crossing.
3. The damage to Bill's spinal cord would be on the left side. The fasciculus gracilis conveys sensations of proprioception, fine touch, and vibration through the spinal cord on the same side of the body as the sensory receptors. The damage to Mary's brainstem would be on the right side if the damage occurred above the medulla oblongata or on the left if it occurred in the inferior part of the medulla oblongata. The secondary neurons in the nucleus gracilis cross over in the medulla through the decussations of the medial lemniscus and, once crossed, are on the opposite side of the body from the nerve endings where the sensations would be initiated.
4. Most proprioception from the lower limbs is unconscious, whereas that from the upper limbs is mostly conscious. This difference is valuable because walking and standing (balance) are not activities on which we want to focus our attention, whereas proprioceptive activities of the arms and hands are essential for gaining information about the environment.
5. In the visual cortex, the brain "sees" an object. Without a functional visual cortex, a person is blind. The visual association areas allow us to relate objects seen to previous experiences and to interpret what has been seen. Similarly, other association areas allow us to relate the sensory information integrated in the primary sensory areas with previous experiences and to make judgments about the information.
6. Constipation, with painful distension and cramping of the colon, results in the sensation of diffuse pain. Deep, visceral pain is not highly localized because few mechanoreceptors are present in deeper structures, such as the colon. The pain is perceived as occurring in the skin over the lower central portion of the abdomen (in the hypogastric region) because it is referred to that location because of converging CNS pathways.
7. Stimulating the reticular activating system (RAS) promotes consciousness. Acoustic stimuli arrive at the RAS via collateral branches of cranial nerve VIII. Therefore, continuous acoustic stimuli coming from a dripping faucet stimulates the RAS, thereby preventing sleep.
8. The vagus nerve supplies the muscles of the larynx that aid in voice production; therefore, minor injury to this nerve in the

neck area can lead to hoarseness. The vagus nerve also supplies the pharyngeal muscles involved in swallowing and coughing, so these reflexes may also be affected due to minor injury to the vagus nerve.

9. If a person holds an object in her right hand, tactile sensations of various types travel up the spinal cord to the brain, where they reach the somatic sensory cortex of the left hemisphere and the object is recognized. Action potentials then travel to Wernicke's area (probably on both sides of the cerebrum), where the object is given a name. From there, action potentials travel to Broca's area, where the spoken word is mitiated. Action potentials from Broca's area travel to the premotor area and primary motor cortex, where action potentials are initiated that stimulate the muscles necessary to form the word.

10. a. In myelinated neurons, action potentials are propagated at great speed but, when the myelin is destroyed, as in multiple sclerosis, the speed of impulse conduction is slowed. Demyelination of the optic nerve slows down the conduction of sensory impulses to the cerebral cortex, as shown by the results of the VEP test. This explains the blurry vision.
    b. It is possible that other neurons than the optic nerve are also affected and this explains Betty's motor (weakness) and sensory (tingling) symptoms. The muscle weakness may have been caused by damage to CNS neurons of the descending motor pathway, and the tingling may have been caused by damage to neurons of the ascending spinal pathway.

11. The stroke was on the left side of the brainstem. Both the motor and sensory neurons to the right side of the body are located in the left cerebral cortex. At the level of the upper medulla oblongata, neither the motor nor sensory pathways to the limbs have yet crossed over to the left side of the CNS. Most of the motor fibers cross at the inferior end of the medulla oblongata, whereas sensory pain and temperature fibers cross over at the level where they enter the CNS. Loss of pain and temperature on the left side of the face indicates that the lesion occurred at a level where the nerve fibers from the face had entered the CNS but had not yet crossed.

Chapter 15

1. Inhaling slowly and deeply allows a large amount of air to be drawn into the olfactory region, whereas not as much air enters during normal breaths. Sniffing (rapid, repeated air intake) is effective for the same reason.
2. Adaptation can occur at several levels in the olfactory system. First, adaptation can occur at the receptor cell membrane, where receptor sites are filled or become less sensitive to a specific odor. Second, olfactory bulb neurons within the olfactory bulb can modify sensitivity to an odor by inhibiting mitral cells or tufted cells. Third, neurons from the intermediate olfactory area of the cerebrum can send action potentials to the olfactory bulb neurons to inhibit further sensory action potentials.
3. Eyedrops placed into the eye tend to drain through the nasolacrimal duct into the nasal cavity. Recall that much of what is considered taste is actually smell. The medication is detected by the olfactory neurons and is interpreted by the brain as taste sensation. Crying produces extra tears, which are conducted to the nasal cavity, causing a "runny" nose.
4. Inflammation of the cornea involves edema, the accumulation of fluid. Fluid accumulation in the cornea increases its water content and, because water causes the proteoglycans to expand, the transparency of the lens decreases, interfering with normal vision.
5. Eyestrain, or eye fatigue, occurs primarily in the ciliary muscles. It occurs because close vision requires accommodation. Accommodation occurs as the ciliary muscles contract, releasing the tension of the suspensory ligaments and allowing the lens to become more rounded. Continued close vision requires the maintenance of accommodation, which requires that the ciliary muscles remain contracted for a long time, resulting in their fatigue.
6. Rhodopsin breakdown is associated with adaptation to bright light and occurs rapidly, whereas rhodopsin production is associated with adaptation to conditions of little light and occurs slowly. Eye adapt rather quickly to bright light but quite slowly to very dim light.
7. Rod cells distributed over most of the retina are involved in both peripheral vision (out of the corner of the eye) and vision under conditions of very dim light. When attempting to focus directly on an object, however, a person relies on the cones within the macula lutea; although the cones are involved in visual acuity, they do not function well in dim light; thus, the object may not be seen at all.
8. A lesion in the right optic nerve at *B* results in loss of vision in the right visual field (see the following illustration).

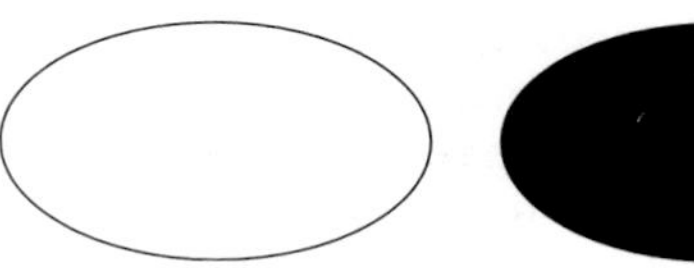

9. The stapedius muscle, attached to the stapes, is innervated by the facial nerve (VII). Loss of facial nerve function eliminates part of the sound attenuation reflex, although not all of it because the tensor tympani muscle, innervated by the trigeminal nerve, is still functional. A reduction in the sound attenuation reflex results in sounds being excessively loud in the affected ear. A reduced reflex can also leave the ear more susceptible to damage by prolonged loud sounds.
10. Perfect pitch is the ability to reproduce a pitch precisely just by being told its name or reading it on a sheet of music, with no other musical support, such as from piano accompaniment. This remarkable talent, as well as conditions such as tone deafness (the complete inability to recognize or reproduce musical pitches) or a decreased ability to perceive tone differences, can occur at a number of locations. The structure of the basilar membrane may be such that tones are not adequately spaced along the cochlear duct in some people to facilitate clear separation of tones. The reflex from the superior olive to the spiral organ may have a very narrow "window of function" for people with perfect pitch but may not be functioning in some other people. The auditory cortex may not be able to translate as accurately in some people to distinguish differences in tones.
11. It is much easier to perceive subtle musical tones when music is played somewhat softly as opposed to very loudly because loud sounds have sound waves with a greater amplitude, which causes the basilar membrane to vibrate more violently over a wider range. The spreading of the wave in the basilar membrane to some extent counteracts the reflex from the superior olive that is responsible for enabling a person to hear subtle tone differences.
12. The brain compares sensory input from the semicircular canals, eyes, and proprioceptors in the lower limbs, and perceived differences in the input may result in motion sickness. Closing the eyes eliminates one of these inputs so that the brain has less input to compare. Less input may decrease the perceived differences and thus reduce the symptoms of motion sickness. Likewise, we can perceive more motion in close objects than in distant objects. Therefore, looking at distant objects, where there is less

perceived motion, reduces the visual input of perceived motion to the brain. This reduced input may decrease the perceived differences and reduce the symptoms of motion sickness.

Chapter 16

1. Terminal ganglia are found near or embedded within the wall of organs supplied by the parasympathetic division and contribute to the enteric nervous system. Postganglionic parasympathetic axons from the terminal ganglia also contribute to the enteric nervous system. Chain ganglia and collateral ganglia contain the cell bodies of sympathetic neurons. They are not embedded within the walls of organs supplied by the sympathetic division. Instead, postganglionic neurons extend from them to organs. Thus, postganglionic sympathetic axons are found in the enteric nervous system.
2. For a sensory axon running alongside sympathetic axons, the sensory axon leaves the wall of the small intestine, joins the superior mesenteric plexus, and passes through the superior mesenteric ganglion and from there through a splanchnic nerve to a sympathetic chain ganglion. From the sympathetic chain ganglion, the sensory axon passes through a white ramus communicans, the ventral rami of a spinal nerve, a spinal nerve, the dorsal root of a spinal nerve, to a dorsal root ganglion. For a sensory axon running alongside parasympathetic axons, the sensory axon leaves the wall of the small intestine, joins the superior mesenteric plexus, and passes to the esophageal plexus. From there, the sensory axon passes through a vagus nerve to its sensory ganglion.
3. Nicotinic receptors are located within the autonomic ganglia as components of the membranes of the postganglionic neurons of the sympathetic and parasympathetic divisions. Nicotine binds to the nicotinic receptors of the postganglionic neurons, resulting in action potentials. Consequently, the postganglionic neurons stimulate their effector organs. After the consumption of nicotine, structures innervated by both the sympathetic and the parasympathetic divisions are affected. After the consumption of muscarine, only the effector organs that respond to acetylcholine are affected. This includes all the effector organs innervated by the parasympathetic division and the sweat glands, which are innervated by the sympathetic division.
4. a. The radial muscles of the iris cause the pupil to dilate, and the circular muscles cause the pupil to constrict.
   b. The radial muscles are controlled by the sympathetic division, and the circular muscles are controlled by the parasympathetic division.
   c. An adrenergic drug could activate α_1 receptors on the radial muscles of the iris, causing dilation. A muscarinic blocking agent could block muscarinic receptors on the circular muscles, causing less constriction—that is, causing dilation.
   d. An inability to see close-up objects indicates that the ciliary muscles of the eye are affected by the drug. A muscarinic blocking agent could produce this effect by blocking the contraction of the ciliary muscles. Usually, pupil-dilating eyedrops contain a muscarinic blocking agent and an adrenergic agent.
   e. Contraction of the radial muscles of the iris causes dilation of the pupil. An adrenergic blocking agent could block α_1 receptors on the radial muscles, causing less dilation. A muscarinic agent could stimulate muscarinic receptors on the circular muscles of the iris, causing pupil constriction.
   f. Sympathetic stimulation of blood vessels normally keeps them in a state of partial contraction. Decreased sympathetic stimulation of blood vessels in the conjunctiva of the eye can result in dilation of the blood vessels and blood-shot eyes. The pupil-dilating eyedrops contain an adrenergic blocking agent, which dilates the pupils and, as a side effect, produces blood-shot eyes.
5. The frequency of action potentials in sympathetic neurons to the sweat glands increases as the body temperature increases. The increasing body temperature is detected by the hypothalamus, which activates the sympathetic neurons. Sweating cools the body by evaporation. As the body temperature declines, the frequency of action potentials in sympathetic neurons to the sweat glands decreases. A lack of sweating helps prevent heat loss from the body.
6. In response to an increase in blood pressure, information is transmitted in the form of action potentials along sensory neurons to the medulla oblongata. From the medulla oblongata, the frequency of action potentials delivered along sympathetic nerve fibers to blood vessels decreases. As a result, blood vessels dilate, causing the blood pressure to decrease. In response to a decrease in blood pressure, fewer action potentials are transmitted along sensory neurons to the medulla oblongata, which responds by increasing the frequency of action potentials delivered along sympathetic nerves to blood vessels. As a result, blood vessels constrict, causing blood pressure to increase.
7. The parasympathetic division releases acetylcholine, which binds to muscarinic receptors on organs. Bethanechol chloride produces effects similar to the stimulation of organs by the parasympathetic division. Thus, this drug should stimulate the urinary bladder to contract. Side effects can be produced by the stimulation of muscarinic receptors elsewhere in the body. The stimulation of smooth muscle in the digestive tract can produce abdominal cramps. The stimulation of air passageways can cause an asthmatic attack. Decreased tear production, salivation, and dilation of the pupils are not expected side effects because parasympathetic stimulation causes increased tear production, salivation, and constriction of the pupils. Sweat glands are innervated by the sympathetic division but have muscarinic receptors. Bethanechol chloride can increase sweating.

Chapter 17

1. Because the abnormal substance acts like TSH, it acts on the thyroid gland to increase the rate of secretion of T_3 and T_4, which increase in concentration in the circulatory system. The thyroid hormones have a negative-feedback effect on the secretion of TSH from the anterior pituitary gland, thereby decreasing the concentration of TSH in the circulatory system to low levels. Because the abnormal substance is not regulated, it can cause T_3 and T_4 levels to become very elevated.
2. Josie's blood T_3 and T_4 levels were very high. The T_3 and T_4 have a negative feedback effect on TSH secretion from the anterior pituitary. Therefore, the blood TSH levels were very low. Josie's condition was not due to a TSH-secreting tumor of the anterior pituitary because the TSH levels were very low. LATS is an immunoglobulin that binds to receptors for TSH and activates the receptors. The high blood levels of LATS is a result of their abnormal immune response. After a large portion of the adrenal gland was destroyed by ^{131}I, the ability of the thyroid gland to secrete T_3 and T_4 was decreased dramatically so that blood levels of T_3 and T_4 would fall below their normal range of values unless Josie took T_3 and T_4 in the form of a pill.
3. A major function of plasma proteins, to which hormones bind, is to increase the half-life of the hormone. If the

concentration of the plasma protein decreases, the half-life and consequently, the concentration of the hormone in the circulatory system decrease. The half-life of the hormone decreases because the rate at which the hormone leaves the circulatory system increases. If the secretion rate for the hormone does not increase, its concentration in the blood declines.

4. If too little estrogen is secreted, the up-regulation of receptors in the uterus for progesterone cannot occur. As a result, progesterone cannot prepare the uterus for the embryo to attach to its wall following ovulation, and pregnancy cannot occur. Because of the lack of up-regulation, the uterus cannot respond adequately to progesterone, regardless of how much is secreted. If some progesterone receptors are present, the uterus will require a much larger amount of progesterone to produce its normal response.
5. A drug can increase the cAMP concentration in a cell by stimulating its synthesis or by inhibiting its breakdown. Drugs that bind to a receptor that increases adenylate cyclase activity will increase cAMP synthesis. Because phosphodiesterase normally causes the breakdown of cAMP, an inhibitor of phosphodiesterase decreases the rate of cAMP breakdown and causes cAMP to increase in the smooth muscle cells of the airway and produces relaxation.
6. Intracellular receptor mechanisms result in the synthesis of new proteins that exist within the cell for a considerable amount of time. Intracellular receptors are therefore better adapted for mediating responses that last a relatively long time (i.e., for many minutes, hours, or longer). On the other hand, membrane-bound receptors that increase the synthesis of intracellular mediators, such as cAMP, normally activate enzymes already existing in the cytoplasm of the cell for shorter periods. The synthesis of cAMP occurs quickly, but the duration is short because cAMP is broken down quickly, and the activated enzymes are then deactivated. Membrane-bound receptor mechanisms are therefore better adapted to short-term and rapid responses.

Chapter 18

1. The cell bodies of the neurosecretory cells that produce ADH are in the hypothalamus, and their axons extend into the posterior pituitary, where ADH is stored and secreted. Removing the posterior pituitary severs the axons, resulting in a temporary reduction in secretion. The cell bodies still produce ADH, however; as the ADH accumulates at the ends of severed axons, ADH secretion resumes.
2. If GH is administered to young people before the growth of their long bones is complete, it causes their long bones to grow and they will grow taller. To accomplish this, however, GH has to be administered over a considerable length of time. It is likely that some symptoms of acromegaly will develop. In addition to undesirable changes in the skeleton, nerves frequently are compressed as a result of the proliferation of connective tissue. Because GH spares glucose usage, chronic hyperglycemia results, frequently leading to diabetes mellitus and the development of severe atherosclerosis. Mr. Hoops's doctor would therefore not prescribe GH.
3. Surgical removal of the thyroid gland causes T_3 and T_4 levels to decline in the blood. TRH and TSH levels in the blood increase because, as T_3 and T_4 levels in the blood decrease, the negative-feedback effect of T_3 and T_4 on TRH and TSH is removed. The oral administration of T_3 and T_4 causes blood levels of T_3 and T_4 to increase and, because of negative feedback, TRH and TSH levels decline.
4. In response to a reduced dietary intake of calcium, the blood levels of calcium begin to decline. In response to the decline in blood levels of calcium, an increase of PTH secretion from the parathyroid glands occurs. The PTH increases calcium resorption from bone. Consequently, blood levels of calcium are maintained within the normal range but, at the same time, bones are being decalcified. Severe dietary calcium deficiency results in bones that become soft and eaten away because of the decrease in calcium content.
5. Removal of the thyroid gland means that the tissue responsible for thyroid hormone (T_3 and T_4) secretion from thyroid follicles, and calcitonin from parafollicular cells, no longer remain. However, blood Ca^{2+} remains within its normal range. Calcitonin is not essential for the maintenance of normal blood Ca^{2+} levels. Removal of the parathyroid gland eliminates PTH secretion. Without PTH, blood levels of calcium fall. When the blood levels of calcium fall below normal, the permeability of nerve and muscle cells to Na^+ increases. As a consequence, spontaneous action potentials are produced that cause tetanus of muscles. Death can result from tetany of respiratory muscles.
6. High aldosterone levels in the blood lead to elevated Na^+ levels in the circulatory system and low blood levels of K^+. The effect of low blood levels of K^+ is hyperpolarization of muscle and neurons. The hyperpolarization results from the lower levels of K^+ in the extracellular fluid and a greater tendency for K^+ to diffuse from the cell. As a result, a greater than normal stimulus is required to cause the cells to depolarize to threshold and generate an action potential. Symptoms of low serum K^+ levels therefore include lethargy and muscle weakness. Elevated Na^+ concentrations result in a greater than normal amount of water retention in the circulatory system, which can result in elevated blood pressure. The major effect of a low rate of aldosterone secretion is elevated blood K^+ levels. As a result, nerve and muscle cells partially depolarize. Because of their partial depolarization, they produce action potentials spontaneously or in response to very small stimuli. The result is muscle spasms, or tetanus.
7. Large doses of cortisone can damage the adrenal cortex because cortisone inhibits ACTH secretion from the anterior pituitary. ACTH is required to keep the adrenal cortex from undergoing atrophy. Prolonged use of large doses of cortisone can cause the adrenal gland to atrophy to the point at which it cannot recover if ACTH secretion does increase again.
8. a. Fred's symptoms are consistent with those commonly found in people suffering from Cushing syndrome. Elevated blood glucose levels and reduced blood K^+ levels are also consistent with Cushing syndrome. Cortisol increases gluconeogenesis and decreases the use of glucose in metabolism. Therefore, blood glucose levels increase. Increased blood glucose levels stimulate insulin secretion and insulin causes glucose to be taken up by fat cells, where they convert glucose to fat. The decrease in muscle mass occurs because cortisol decreases protein synthesis and increases protein breakdown. Reduced blood levels of K^+ result because of an increased rate of K^+ transport by the renal tubules and their elimination in the urine.
   b. Fred's blood cortisol levels were very high and blood levels of ACTH were very low. Blood CRH levels are not normally be measured. ACTH from the anterior pituitary normally stimulates cortisol secretion from the adrenal glands. Because cortisol has a negative-feedback effect on the rate of ACTH secretion, and because the blood ACTH levels were low, the high blood levels of cortisol could not be due to elevated ACTH secretion from the anterior pituitary. A possible cause of

increased cortisol secretion is an adrenal gland tumor, which was confirmed by imaging techniques.

c. After the surgical removal of Fred's left adrenal gland, one would expect the blood levels of cortisol to decrease because the tumor was the source of the high blood cortisol levels. As the blood cortisol levels decreased to normal levels, the rate of ACTH secretion from the anterior pituitary gland would increase. Blood cortisol levels can be within the normal range of values, if production of cortisol by the right adrenal gland compensates for the missing left adreal gland. If cortisol production by the remaining adrenal gland is not adequate, additional cortisol will have to be administered.

d. If Fred's high blood level of cortisol had been due to a hormone-secreting tumor of the anterior pituitary gland, blood levels of both ACTH and cortisol levels would have been elevated. There was no evidence that cancer was involved, but, if cancer tissue was the source of ACTH, blood ACTH and cortisol levels would have been increased.

9. An increase in insulin secretion in response to parasympathetic stimulation and gastrointestinal hormones is consistent with the maintenance of homeostasis because parasympathetic stimulation and increased gastrointestinal hormones result from conditions such as eating a meal. Insulin levels therefore increase just before large amounts of glucose and amino acids enter the circulatory system. The elevated insulin levels prevent a large increase in blood glucose and the loss of glucose in the urine.

10. In response to a meal high in carbohydrates, insulin secretion is increased and glucagon secretion is reduced. The stimulus for the insulin secretion comes from parasympathetic stimulation and, more importantly, from elevated blood levels of glucose. Target tissues take up glucose and blood glucose levels remain within a normal range. In response to a meal high in protein but low in carbohydrates, insulin secretion is increased slightly and glucagon secretion is increased. Insulin secretion is stimulated by the parasympathetic system and an increase in blood amino acid levels. Glucagon is stimulated by low blood glucose levels and by some amino acids. The lower insulin secretion causes some increase in the rate of glucose uptake and amino acid uptake, but the rate of uptake is not great enough to cause blood glucose levels to fall below normal values. Glucagon also causes glucose to be released from the liver. During exercise, sympathetic stimulation inhibits insulin secretion. As blood glucose levels decline, an increase of glucagon secretion occurs. The lower rate of insulin secretion decreases the rate at which tissues, such as skeletal muscle, take up glucose. Muscle depends on intracellular glycogen and fatty acids for energy. Blood glucose levels are maintained within normal range of values. Glucagon prevents glucose levels from decreasing too much.

11. Sympathetic stimulation during exercise inhibits insulin secretion. Blood glucose levels are not high because skeletal muscle tissue continues to take up some glucose and metabolizes it. Muscle contraction depends on glucose stored in the form of glycogen in muscles and fatty acid metabolism. During a long run, glycogen levels are depleted. The "kick" at the end of the race results from increased energy production through anaerobic respiration, which uses glucose or glycogen as an energy source. Because blood glucose levels and glycogen levels are low, the source of energy is insufficient for greatly increased muscle activity.

12. Increased sugar intake will result in elevated blood glucose levels. The elevated blood glucose levels can lead to polyuria and to increased osmolality of the body fluids. That results in the dehydration of neurons. As a result, some of the neural symptoms of untreated diabetes, such as irritability and a general sensation of not feeling well, occur. Billy may also experience a sudden increase in weight gain because of increased sugar intake and insulin administration. In addition, he may have an increased chance of getting infections, such as urinary tract infections. Many of the long-term consequences of diabetes, such as nephropathies, neuropathies, and atherosclerosis, develop much more rapidly.

Chapter 19

1. Fetal hemoglobin must be more effective at binding oxygen than adult hemoglobin so that the fetal circulation can draw the needed oxygen away from the maternal circulation. If maternal blood had an equal or greater oxygen affinity, the fetal blood would not be able to draw away the required oxygen and the fetus would die.
2. An elevated reticulocyte count indicates that erythropoiesis and the demand for red blood cells are increased and that immature red blood cells (reticulocytes) are entering the circulation in large numbers. An elevated reticulocyte count can occur for a number of reasons, including loss of blood; therefore, after a person donates a unit of blood, the reticulocyte count increases.
3. Carbon monoxide binds to the iron of hemoglobin and prevents the transport of oxygen. The decreased oxygen stimulates the release of erythropoietin from the kidneys. Erythropoietin increases red blood cell production in red bone marrow, thereby causing the number of red blood cells in the blood to increase.
4. The white blood cells shown in figure 19.8 are (a) lymphocyte, (b) basophil, (c) monocyte, (d) neutrophil, and (e) eosinophil.
5. Platelets become activated at sites of tissue damage, where it is advantageous to form a clot to stop bleeding.
6. People with type AB blood were called universal recipients because they could receive type A, B, AB, or O blood with little likelihood of a transfusion reaction. Type AB blood does not have antibodies against type A or B antigens; therefore, the transfusion of these antigens in type A, B, or AB blood does not cause a transfusion reaction in a person with type AB blood. The term is misleading, however, for two reasons. First, other blood groups can cause a transfusion reaction. Second, antibodies in the donor's blood can cause a transfusion reaction. For example, type O blood contains anti-A and anti-B antibodies that can react against the A and B antigens in type AB blood.
7. a. HDN causes a decrease in the number of red blood cells. A transfusion treats this anemia by increasing the number of red blood cells, which promotes oxygen and carbon dioxide transport.
   b. An exchange transfusion not only increases the number of red blood cells but also decreases bilirubin and anti-Rh antibody levels by removing them in the withdrawn blood. Fewer anti-Rh antibodies decreases the agglutination and lysis of red blood cells.
   c. Erythropoietin stimulates red blood cell production. Thus, Billy has two sources of red blood cells: the donor blood and himself.
   d. The destruction of red blood cells in a fetus with HDN results in lower than normal numbers of red blood cells. The resulting decrease in oxygen transport stimulates increased erythropoietin production.
   e. After birth, Billy breathes on his own. The ability to oxygenate the blood using the lungs is greater than the ability to oxygenate the blood across the placenta. Billy's blood oxygen levels increase and erythropoietin levels decrease. Thus, his

production of red blood cells decreases and his anemia gets worse.

f. The donor blood used in exchange transfusions for the treatment of HDN should be Rh-negative, even though the newborn is Rh-positive. Rh-negative red blood cells do not have Rh antigens. Therefore, any anti-Rh antibodies in the newborn's blood do not react with the transfused Rh-negative red blood cells. Note, however, that Rh-positive blood is often used because supplies of Rh-negative blood are limited. A judgment is made on the severity of the HDN. In severe cases, Rh-negative blood is used.

g. Giving Rh-negative blood to an Rh-positive newborn does not change the blood type of the newborn because blood type is genetically determined. Eventually, all of the Rh-negative red blood cells die and the newborn produces only Rh-positive red blood cells.

8. A white blood count (WBC) should be done. An elevated WBC, leukocytosis, can be an indication of bacterial infections. A differential WBC should also be done. An increase in the number of neutrophils supports the diagnosis of a bacterial infection. Coupled with other symptoms, this can mean appendicitis. If these tests are normal, appendicitis is still a possibility and the physician must rely on other clinical signs. Diagnostic accuracy for appendicitis is approximately 75%–85% for experienced physicians.

Chapter 20

1. The heart tissues supplied by the artery lose their oxygen and nutrient supply and die. This part of the heart (and possibly the entire heart) stops functioning. Restricted blood flow through the anterior interventricular artery reduces blood flow to the anterior portion of the ventricles. If this condition develops rapidly, it is called a heart attack, or myocardial infarction.
2. The heart must continue to function under all conditions and requires energy in the form of ATP. During heavy exercise, lactic acid is produced in skeletal muscle as a by-product of anaerobic respiration. The ability to use lactic acid provides the heart with an additional energy source.
3. Contraction of the ventricles, beginning at the apex and moving toward the base of the heart, forces blood out of the ventricles and toward their outflow vessels—the aorta and pulmonary trunks. The aorta and pulmonary trunks are located at the base of the heart.
4. Ectopic foci cause various regions of the heart to contract at different times. As a result, pumping effectiveness is reduced. Cardiac muscle contraction is not coordinated, which interrupts the cyclic filling and emptying of the ventricles.
5. If cardiac muscle could undergo tetanic contraction, it would contract for a long time without relaxing. Its pumping action then would stop because that action requires alternating contraction and relaxation.
6. During isovolumic contraction, the volume of the ventricles does not change because no blood leaves the ventricle. Therefore, the pressure increases but the length of the cardiac muscle does not change significantly. Therefore, the contraction is isometric (see chapter 9).
7. The left ventricle has the thickest wall. The pressure produced by the left ventricle is much higher than the pressure produced by the right ventricle, when the ventricles contract. It is important for each ventricle to pump the same amount of blood because, with two connected circulation loops, the blood flowing into one must equal the blood flowing into the other so that one does not become overfilled with blood at the expense of the other. For example, if the right ventricle pumps less blood than the left ventricle, blood must accumulate in the systemic blood vessels. If the left ventricle pumps less blood than the right ventricle, blood accumulates in the pulmonary blood vessels.
8. Fibrillation makes cardiac muscle an ineffective pump. The pumping action of the heart depends on coordinated contractions of the cardiac muscle. Fibrillation destroys the coordinated contractions and the cardiac muscle loses its function as a pump. The ventricles are the primary pumping chambers of the heart. Ventricular fibrillation results in death because of the heart's inability to pump blood. The atria function primarily as reservoirs. Their pumping action is most important during exercise. Therefore, atrial fibrillation does not destroy the ventricles' ability to pump blood.
9. An incompetent bicuspid valve allows blood to flow from the left ventricle to the left atrium when the left ventricle contracts. As a result, the stroke volume (the volume of blood ejected into the aorta) is reduced. At the end of ventricular systole, the volume of the left atrium is increased because blood flowed back into the left atrium during ventricular systole. At the end of ventricular diastole, the volume of the left ventricle is greater than normal because the blood that flowed back into the left atrium during the previous ventricular systole now flows into the ventricle, along with the blood that normally returns from the lungs through the pulmonary veins.
10. Sympathetic stimulation increases heart rate. If venous return remains constant, stroke volume decreases as the number of beats per minute increases. Dilation of the coronary arteries is important because, as the heart does more work, the cardiac tissue requires more energy and therefore a greater blood supply to carry more oxygen.
11. a. The stenosed (narrowed) valve makes it more difficult for contractions of the left ventricle to force blood to flow into the aorta because of the narrowed opening. When the resistance is sufficiently high, the stroke volume decreases because the left ventricle is unable to overcome the increased resistance to blood flow through the stenosed valve.

b. Because of the increased resistance caused by the stenosed aortic semilunar valve, the left ventricle must generate a greater force during ventricular systole to cause blood to flow into the aorta. Consequently, the left ventricular pressure is very high during ventricular systole, even though the systolic pressure is low. Just as skeletal muscle hypertrophies in response to the resistance of moving heavy weights, cardiac muscle hypertrophies in response to the resistance to blood flow through the stenosed valve.

c. The mean arterial pressure is equal to cardiac output (heart rate × stroke volume) × peripheral resistance. The mean arterial pressure decreases when stroke volume decreases, unless there is a compensating increase in heart rate or peripheral resistance.

d. When Norma rises to a standing position, gravity affects blood in arteries and veins, and the blood tends to settle in blood vessels below the heart. Consequently, there is decreased venous return to the heart, which results in decreased cardiac output (Starling's law of the heart). The decreased cardiac output results in decreased blood pressure, decreased blood flow to the brain, and dizziness.

e. When Norma stands, her blood pressure decreases, activating the baroreceptor reflex. Her heart rate increases. Although Norma has a normal baroreceptor reflex, the heart is not able to pump enough blood past the stenosed valve to maintain blood pressure.

f. Angina results from inadequate blood delivery to heart muscle through the coronary circulation. Norma's stenosed valve has reduced blood pressure in the aorta, which reduces blood delivery to the body and to the coronary blood vessels. At rest, there is sufficient blood delivery to cardiac muscle, and there is no pain; however, when Norma starts exercising, the cardiac muscle requires greater delivery of oxygen by the blood, which is not possible with the stenosed valve. This results in an increase in anaerobic respiration and a decrease in pH, which is responsible for the pain of angina pectoris.

12. Rupture of the left ventricle, as experienced by Mr. P, is more likely several days after a myocardial infarction. As the necrotic tissues are removed by macrophages, the wall of the ventricle becomes thinner and may bulge during systole. If the wall of the ventricle becomes very thin before new connective tissue is deposited, it may rupture. If the left ventricle ruptures, blood flows from the left ventricle into the pericardial sac. As blood fills the pericardial sac, it compresses the ventricle from the outside. This is called cardiac tamponade. Thus, the ventricle is not able to fill with blood and its pumping ability is eliminated. Death occurs quickly in response to a ruptured wall of the left ventricle.

Chapter 21

1. Arteriosclerosis slowly reduces blood flow through the carotid arteries and therefore the amount of blood that flows to the brain. As the resistance to flow increases in the carotid arteries during the late stages of arteriosclerosis, the blood flow to the brain is compromised, resulting in reduced oxygen delivery. Confusion, loss of memory, and loss of the ability to perform other normal brain functions occur.
2. a. Vasoconstriction of blood vessels in the skin in response to exposure to cold results in a decreased flow of blood through the skin and in a dramatic increase in resistance (Poiseuille's law). Vasoconstriction makes the skin appear pale.
   b. Vasodilation of blood vessels in the skin results in increased blood flow through the skin. Vasodilation makes the skin appear flushed or red in color.
   c. In a patient with erythrocytosis, the hematocrit increases dramatically. As a result, the viscosity of the blood increases, which increases resistance to flow. Consequently, flow decreases or a greater pressure is needed to maintain the same flow.
3. An aneurysm in the aorta is a major problem because the tension applied to the aneurysm becomes greater as its size increases (Laplace's law). The aneurysm usually develops because of a weakness in the wall of the aorta. Arteriosclerosis complicates the matter by making the wall of the artery less elastic and by increasing the systolic blood pressure. The decreased elasticity and the increased blood pressure increase the probability that the aneurysm will rupture.
4. Premature beats of the heart and ectopic beats result in contraction of the heart muscle before the heart has had time to fill to its normal capacity. Consequently, a reduced stroke volume occurs, which results in a weak pulse in response to that premature contraction. Other contractions and the resulting pulses are normal. Strong bounding pulses in a person who received too much saline solution in an intravenous transfusion result from an increase in venous return to the heart because of the increased volume of fluid in the circulatory system. Because of the increased venous return (increased preload), the heart contracts with greater force and produces a larger stroke volume. The strong bounding pulse results from the increased stroke volume. Weak pulses occur in response to hemorrhagic shock because of a decreased venous return. The heart does not fill with blood between contractions (decreased preload); the stroke volume is therefore reduced and the pulse is weak.
5. Arterial blood pressure can increase substantially without resulting in edema. As arterial blood pressure increases the precapillary sphincters constrict to match capillary blood flow with the metabolic needs of tissues. Thus, the capillary blood pressure does not change substantially, even though the blood pressure may increase to high levels. The blood pressure must increase above approximately 175 mm Hg before edema results. In contrast, a small increase in venous pressure leads to edema because there is no sphincter muscle that protects the capillary from an increase in pressure. Thus, small increases in venous pressure can lead to edema. Blockage of veins by blood clots or increases in venous pressure due to heart failure result in edema.
6. Keeping the legs elevated reduces the blood pressure in the capillaries of the legs because of the effect of gravity on blood flow. A major effect is that the force that moves fluid out of the capillary is decreased. As a result, the net movement of fluid out of the arterial ends decreases and the net movement into the venous ends of the capillaries increases. Therefore, excess interstitial fluid is carried away from the legs. In addition, the effect of gravity increases lymph flow into the lymphatic capillaries, which also increases the rate that interstitial fluid is drained from the legs.
7. a. The blocked vein in Harry's right leg caused edema and led to tissue ischemia. Edema developed inferior to the blocked vein. Blockage of the vein increased the capillary hydrostatic pressure in the capillary beds drained by the blocked vein. The increased capillary hydrostatic pressure increased the amount of fluid that flowed from the capillaries into the tissue spaces and reduced the amount of fluid that returned to the capillaries. Consequently, fluid accumulated in the tissue spaces and caused edema (see figure 21.34). The ischemia resulted in pain, much the way ischemia of the heart causes pain during myocardial infarctions (see chapter 20).
   b. Emboli that originate in the posterior tibial vein pass through the following parts of the circulatory system before they lodge in the pulmonary arteries of the lungs: posterior tibial vein, popliteal vein, femoral vein, external iliac vein, common iliac vein, inferior vena cava, right atrium, right ventricle, pulmonary trunk, right or left pulmonary artery. Emboli will lodge in branches of the pulmonary arteries. Emboli are most likely to lodge in the lungs because the pulmonary arteries branch many times before they deliver blood to the pulmonary capillaries and, as they branch, their diameters decrease. Even small emboli will eventually lodge in the smaller branches of the pulmonary arteries. The other parts of the circulatory system through which the emboli pass have much larger diameters and emboli can pass readily through them.
   c. When emboli are large enough or numerous enough to block blood flow through a significant part of the lungs, resistance to blood flow through the lungs increases. The increased resistance to flow increases the pulmonary venous pressure, which increases the afterload for the right ventricle of the heart. If the right ventricle is unable to overcome the increased afterload, failure of the right side of the heart can occur, and blood flow through the lungs is reduced.
   d. Pulmonary emboli large enough to significantly reduce blood flow through the lungs reduces the lungs' ability to carry out gas exchange with blood, and blood oxygen levels decrease. If blood flow through the lungs to the left side of the heart is reduced significantly,

hypotension can develop. Reduced blood flow to the left side of the heart will result in a reduced cardiac output because of decreased venous return (Starling's law of the heart). As a result, blood pressure falls and the mechanisms that regulate blood pressure are activated (see "Regulation of Mean Arterial Pressure"). Manifestations of hypotension, such as an increased heart rate, weak pulse, and pallor, may be present.

e. Heparin and coumadin are anticoagulants. They are prescribed because they decrease the rate at which coagulation proceeds (see chapter 19). Heparin is administered intravenously, whereas coumadin can be taken orally, which makes home use possible. Prothrombin time is checked periodically because enough anticoagulant must be administered to prevent enlargement of the thrombus in the deep vein of Harry's leg and to prevent additional emboli from forming. Enzymes continually break down coagulated blood. Clots are removed because the slower rate of coagulation allows them to be broken down faster than they can be formed. It is also important to monitor Harry's prothrombin time because excess blockage of coagulation can result in bleeding (see chapter 19). Coumadin is prescribed for a substantial amount of time after a venous thrombosis or a pulmonary embolus has formed to prevent them from reoccurring.

8. Reactive hyperemia can be explained on the basis of any of the theories for the local control of blood pressure. When a blood vessel is occluded, nutrients are depleted and waste products accumulate in tissue that is suffering from a lack of adequate blood supply. Both of these effects cause vasodilation and a greatly increased blood flow through the area after the occlusion has been removed.
9. While this athlete is relaxing, the sympathetic stimulation of the arteries in her skeletal muscles, the arteries in her digestive system, and the large veins decreases. As a result, vasoconstriction increases in the arteries of her muscles, and vasodilation occurs in the blood vessels of her digestive system and in the large veins. Blood flow decreases to her skeletal muscles, and blood flow increases to her digestive system. In addition, more blood accumulates in the large veins. Consequently, venous return to the heart decreases, which is consistent with the reduced cardiac output.
10. During a headstand, gravity acting on the blood causes the blood pressure in the area of the aortic arch and carotid sinus baroreceptors to increase. The increased pressure activates the baroreceptor reflexes, increasing parasympathetic stimulation of the heart and decreasing sympathetic stimulation. Thus, the heart rate decreases. Because standing on one's head also causes blood from the periphery to run downhill to the heart, the venous return increases, causing the stroke volume to increase because of Starling's law of the heart. Some peripheral vasodilation also can occur because the elevated baroreceptor pressure causes a decrease in vasomotor tone.
11. The baroreceptor reflex, ADH, and renin-angiotensin-aldosterone mechanisms function similarly in both cases. The fluid shift mechanism, however, is important when the loss of blood occurs over several hours, but it does not operate within a short period. The fluid shift mechanism plays a very important role in the maintenance of blood volume when blood loss or dehydration develops over several hours. When the blood pressure decreases, interstitial fluids pass into the capillaries, which prevents a further decline in blood pressure. The fluid shift mechanism is a powerful method through which blood pressure is maintained because the interstitial fluid acts as a fluid reservoir. The stress–relaxation mechanism responsive to changes in blood pressure, but it is most responsive to sudden changes in blood volume and it responds within minutes to hours.

Chapter 22

1. a. The lymphatic vessles remove fluid from tissues. Cancer cells that break free from a tumor can spread, or metastasize, to other parts of the body by entering lymphatic or blood capillaries (see chapter 4). If they enter lymphatic capillaries, they are carried by lymph to lymph nodes, which filter lymph. The first lymph nodes in which cancer cells are likely to become trapped are the sentinel lymph nodes.
   b. All of Cindy's sentinel lymph nodes contained cancer cells, indicating that the cancer cells have spread into her lymphatic system. In order to minimize the risk of further metastasis, her axillary lymph nodes were removed because they drain the superficial thorax and upper limb. If none of the sentinel lymph nodes had contained cancer cells, however, an option would have been to not remove the axillary lymph nodes.
   c. When lymph nodes and their attached lymphatic vessels are removed, it disrupts the normal removal of fluid from tissues. Recall from chapter 21 that approximately one-tenth of the fluid entering tissues is normally removed by the lymphatic system. A reduction in this fluid removal can result in an accumulation of fluid in the tissues.
   d. Exercise increases skeletal muscle contractions and increases breathing, both of which increase lymph flow. Consequently, more fluid can enter lymphatic capillaries, reducing edema.
   e. Compression bandages or garments produce an external pressure on tissues and reduce the movement of fluid into tissues from blood capillaries (see chapter 21). It is recommended that compression bandages or garments be worn during daily activities, especially during exercise.
   f. Lymphatic vessels periodically contract and relax, moving lymph in one direction because of the valves within lymphatic vessels. The compression pump mimics this effect through external pressure changes. Increased pressure compresses the lymphatic vessels, moving lymph. Decreased pressure allows the lymphatic vessels to expand and fill with lymph. The sequential pressure changes help move lymph in a distal-to-proximal direction.
2. The T cells transferred to mouse B do not respond to the antigen. The T cells are MHC-restricted and must have the MHC proteins of mouse A as well as antigen X to respond.
3. When the antigen is eliminated, it is no longer available for processing and combining with MHC class II molecules. Consequently, no signal takes place to cause lymphocytes to proliferate and produce antibodies.
4. The first exposure to the disease-causing agent (antigen) evokes a primary response. Gradually, however, antibodies degrade and memory cells die. If, before all the memory cells are eliminated, a second exposure to the antigen occurs, a secondary response results. The memory cells produced then can provide immunity until the next exposure to the antigen.
5. With the depression of helper T-cell activity, the ability of antigens to activate effector T cells is greatly decreased. The depression of cell-mediated immunity results in an inability to resist intracellular microorganisms and cancer.
6. The booster shot stimulates a secondary (memory) response, resulting in the formation of large amounts of antibodies and memory cells. Consequently, there is better, longer-lasting immunity.

7. SLE is an autoimmune disorder in which self-antigens activate immune responses. Often, this results in the formation of immune complexes and inflammation, but sometimes antibodies bind to antigens on cells, resulting in the lysis of the cells. Purpura results from bleeding into the skin, which means that platelet plug formation, the normal mechanism for repairing small breaks in blood vessels, is not working. In this case of SLE, antibodies are causing the destruction of platelets, and the decreased number of platelets results in decreased platelet plug formation and coagulation (see chapter 19). The condition is called thrombocytopenia.

Chapter 23

1. When you sleep with your mouth open, less air passes through the nasal passages. This is especially true when the nasal passages are plugged because you have a cold. As a consequence, inspired air is not humidified and warmed. The dry air dries and irritates the throat. Air breathed through the mouth while running in very cold weather is not warmed and humidified, which can irritate the throat, larynx, and trachea.
2. When food moves down the esophagus, the normally collapsed esophagus expands. If the cartilage rings were solid, expansion of the esophagus, and therefore swallowing, would be more difficult.
3. The conducting zone consists of the nose (i.e., external nose, nostrils, nasal cavity, and choanae), pharynx (i.e., nasopharynx, oropharynx, and laryngopharynx), larynx, and most of the tracheobronchial tree (i.e., trachea, main bronchi, lobar bronchi, segmental bronchi, bronchioles, and terminal bronchioles). The respiratory zone consists of the parts of the tracheobronchial tree distal to the terminal bronchioles—the respiratory bronchioles (with a few alveoli), alveolar ducts (with many alveoli), and alveolar sacs.
4. Respiratory distress syndrome results from inadequate surfactant, which results in increased water surface tension. Consequently, lung recoil is increased. At the end of expiration, pleural pressure is lower than normal because of the increased lung recoil. Although the decreased pleural pressure increases the tendency for the alveoli to expand, the alveoli do not expand because the increased force of expansion is only counteracting the increased lung recoil. The alveoli collapse if the lung recoil becomes larger than the force of expansion caused by the difference between alveolar and pleural pressure. During inspiration, pleural pressure has to be lower than normal to overcome the effect of the larger than normal lung recoil. A larger than normal increase in thoracic volume can cause a greater than normal decrease in pleural pressure. The effort of overcoming the increased lung recoil, however, can cause muscular fatigue and death.
5. The alveolar ventilation is 4200 mL/min (12 $\times$ [500 $-$ 150]). During exercise, the alveolar ventilation is 88,800 mL/min (24 $\times$ [4000 $-$ 300]), a 21-fold increase. The increased air exchange increases P_{O_2} and decreases P_{CO_2} in the alveoli, thus increasing gas exchange between the alveoli and the blood.
6. The air the diver is breathing has a greater total pressure than atmospheric pressure at sea level. Consequently, the partial pressure of each gas in the air increases. According to Henry's law, as the partial pressure of a gas increases, the amount (concentration) of gas dissolved in the liquid (e.g., body fluids) with which the gas is in contact increases. When the diver suddenly ascends, the partial pressure of gases in the body returns toward sea level barometric pressure. As a result, the amount (concentration) of gas that can be dissolved in body fluids suddenly decreases. When the fluids can no longer hold all the gas, gas bubbles form.
7. At high altitudes, the atmospheric P_{O_2} decreases because of a decrease in atmospheric pressure. The decreased atmospheric P_{O_2} results in a decrease in alveolar P_{O_2} and less oxygen diffusion into lung tissue. If the person's arterioles are especially sensitive to the decreased oxygen levels, constriction of the arterioles reduces blood flow through the lungs and the ability to oxygenate blood decreases. Such generalized hypoxemia can also be caused by certain respiratory diseases, such as emphysema and cystic fibrosis.
8. Blood pH decreases because of increased CO_2 production by the actively working skeletal muscles. As carbon dioxide levels increase, carbonic anhydrase in red blood cells catalyzes the conversion of CO_2 to carbonic acid, which in turn dissociates into H^+ and HCO_3^-. Therefore, pH declines. In addition, pH declines because some of the skeletal muscle fibers produce lactic acid as a result of anaerobic respiration (see chapter 9). Oxygen delivery to the skeletal muscles increases during the race. Under acidic conditions, hemoglobin's affinity for oxygen decreases and more oxygen is released into the tissues (the oxygen–hemoglobin dissociation curve shifts to the right).
9. Remember that oxygen–hemoglobin dissociation curve normally shifts to the right in tissues. The shift of the curve to the left caused by CO reduces hemoglobin's ability to release oxygen to tissues, which contributes to the detrimental effects of CO poisoning. In the lungs, the shift to the left may slightly increase the hemoglobin's ability to pick up oxygen, but this effect is offset by its decreased ability to release oxygen to tissues.
10. A person who cannot synthesize BPG has mild erythrocytosis. Her hemoglobin releases less oxygen to tissues. Consequently, one would expect increased erythropoietin release from the kidneys and increased red blood cell production in red bone marrow.
11. The movement of CO_2 from fetal blood into maternal blood increases CO_2 levels inside maternal red blood cells. As a result, pH inside maternal red blood cells decreases, the affinity of maternal hemoglobin for oxygen decreases, and more oxygen is released (the maternal oxygen–hemoglobin curve shifts to the right). The movement of CO_2 from fetal blood into maternal blood decreases CO_2 levels inside fetal red blood cells. Consequently, pH inside fetal red blood cells increases, the affinity of fetal hemoglobin for oxygen increases, and more oxygen binds to fetal hemoglobin (the fetal oxygen–hemoglobin dissociation curve shifts to the left). This shifting of the maternal and fetal oxygen–hemoglobin curves is called the double Bohr effect. The double Bohr effect increases the delivery of oxygen from maternal blood to fetal blood because maternal hemoglobin releases more oxygen and fetal hemoglobin is more effective at picking up that oxygen.
12. Hyperventilation decreases blood carbon dioxide levels, causing an increase in blood pH. Holding one's breath increases blood carbon dioxide levels and decreases blood pH.
13. a. Normal arterial P_{O_2} is 95 mm Hg and P_{CO_2} is 40 mm Hg. A P_{O_2} of 60 mm Hg and a P_{CO_2} of 30 mm Hg are below normal.
    b. Asthma causes constriction of the terminal bronchioles, which reduces alveolar ventilation and alveolar P_{O_2}. Consequently, oxygen exchange across the respiratory membrane decreases and blood P_{O_2} decreases. The carotid and aortic body chemoreceptors detect the decrease in blood P_{O_2}, resulting in increased stimulation of the respiratory centers and an increased respiratory rate.

c. Constriction of the terminal bronchioles increases resistance to air flow. Will attempted to compensate for the increased resistance by breathing more forcefully.
d. No, Will's blood P_{O_2} and P_{O_2} are abnormally low. Despite the forceful breathing, Will's tidal volume was so low that his increased respiratory rate did not increase alveolar ventilation enough to restore homeostasis.
e. Will's blood P_{O_2} was lower than normal because of low alveolar P_{O_2}. The low blood P_{O_2} stimulated him to hyperventilate. Despite the hyperventilation, alveolar ventilation was insufficient to restore normal blood P_{O_2}. The hyperventilation, however, effectively removed carbon dioxide from the blood because the diffusion coefficient for carbon dioxide is 20 times greater than that of oxygen. As a result, blood P_{CO_2} is lower than normal.
f. As Will's blood P_{CO_2} decreased, his blood pH would increase (become more alkaline). Normally, if just blood pH increased, that would be detected by the medullary chemoreceptors, leading to a decrease in ventilation (see figure 23.22). However, respiration results from the integration of many stimuli (see figure 23.21). During Will's asthma attack, the effect of the low blood P_{O_2} on the respiratory centers was greater than the effect of the decreased blood pH.
g. Inhaled β-adrenergic agents cause relaxation of the smooth muscle in the bronchi and bronchioles. Bronchodilation improves air flow and facilitates the return of blood gases to homeostatic levels. Inhaled glucocorticoids help reduce the inflammatory responses and mucus buildup associated with asthma.

14. When a person hyperventilates, P_{CO_2} in the blood decreases. Consequently, carbon dioxide moves out of cerebrospinal fluid into the blood. As carbon dioxide levels in cerebrospinal fluid decrease, H^+ and HCO_3^- combine to form carbonic acid, which forms carbon dioxide. The result is a decrease in H^+ concentration in cerebrospinal fluid and decreased stimulation of the respiratory center by the chemosensitive area. Until blood P_{CO_2} levels increase, the chemosensitive area is not stimulated and apnea results.
15. Through touch, thermal, or pain receptors, the respiratory center can be stimulated to cause a sudden inspiration of air.
16. A buildup of mucus and inflammatory damage to the air passageways increases resistance to airflow. It becomes difficult to expel air, so FEV1 decreases. As air is trapped in the alveoli, residual volume gradually increases. Because air exchange in the alveoli is less than normal, physiologic dead space increases.

Chapter 24

1. A pin placed through the greater omentum passes through four layers of simple squamous epithelium. The greater omentum is actually a folded mesentery, with each part consisting of two layers of serous squamous epithelium.
2. The moist stratified squamous epithelium of the oropharynx and the laryngopharynx protects these regions from abrasive food when it is first swallowed. The ciliated pseudostratified epithelium of the nasopharynx helps move mucus produced in the nasal cavity and the nasopharynx into the oropharynx and esophagus. It is not as necessary to protect the nasopharynx from abrasion because food does not normally pass through this cavity.
3. It is important for the nasopharynx to be closed during swallowing so that food does not reflux into it or the nasal cavity. An explosive burst of laughter can relax the soft palate, open the nasopharynx, and cause the liquid to enter the nasal cavity.
4. Usually, if a person tries to swallow and speak at the same time, the epiglottis is elevated, the laryngeal muscles closing the opening to the larynx are mostly relaxed, and food or liquid can enter the larynx, causing the person to choke.
5. After a heavy meal, blood pH may increase because, as bicarbonate ions pass from the cells of the stomach into the extracellular fluid, the pH of the extracellular fluid increases. As the extracellular fluid exchanges ions with the blood, the blood pH also increases.
6. Secretin production and its stimulation of bicarbonate ion secretion constitute a negative-feedback mechanism because, as the pH of the chyme in the duodenum decreases as a result of the presence of acid, secretin causes an increase in bicarbonate ion secretion, which increases the pH and restores the proper pH balance in the duodenum.
7. a. The spinal cord injury has disrupted ascending and descending nerve tracts. Dan has no awareness of the need to defecate, and he has lost the ability to relax or contract the external anal sphincter.
   b. The defecation reflex center in the conus medullaris is still functional. Distension caused by an enema can activate the defecation reflex.
   c. After breakfast, the gastrocolic and duodenocolic reflexes can cause mass movement of feces into the rectum, causing distension and activation of the defecation reflex. After breakfast may be better than after other meals because there is more time for feces to have moved into the colon.
   d. Straining increases abdominal pressure, which stimulates reflexive contraction of the external anal sphincter. This reflex is still functional in Dan. During straining in a noninjured person, the brain is able to inhibit this reflex through a descending nerve tract. Because of the disruption of Dan's descending nerve tracts, however, his brain no longer can inhibit the reflex, so, when Dan strains, the external anal sphincter contracts.
   e. Even if the defecation reflex center is damaged, the local reflexes of the enteric nervous system are still functional. Although this local component of the defecation reflex is weak and lacks voluntary control, defecation can still occur with assistance from large enemas and strong cathartics. Cathartics function by increasing fecal volume, softening feces, or stimulating intestinal muscle contractions.
8. The major effect of prolonged diarrhea is on the cardiovascular system and is much like massive blood loss. Hypovolemia continues to increase. Blood pressure declines in a positive-feedback cycle and, without intervention, can lead to heart failure.

Chapter 25

1. If vitamins are broken down during the process of digestion, their structures are destroyed and, as a result,their ability to function is lost.
2. The Daily Value for carbohydrate is 300 g/day. One serving of food with 30 g of carbohydrate has a % Daily Value of 10% (30/300 = .10, or 10%).
3. On a 1800 kcal/day diet, the total percentage of Daily Values for energy-producing nutrients should add up to no more than 90%, because 1800/2000 = 0.9, or 90%.
4. If the electron of the electron-transport chain cannot be donated to oxygen, the entire electron-transport chain and the citric acid cycle stop, no ATP can be produced aerobically, and death results because too little energy is available for the body to perform vital functions. Anaerobic respiration is not adequate to provide all the energy needed to maintain human life, except for a short time.
5. The kilocalories in a beer or cola is about 145 kcal. It takes about 1.5 h to burn off these kilocalories while watching TV

(145/95 = 1.5 h) and about 15 min while jogging (145/580 = 0.25 h). Although it may be difficult to burn off kilocalories through exercise, it is clear that exercise can significantly increase kilocalorie use.

6. When muscles contract, they use ATP. As a result of the chemical reactions necessary to synthesize ATP, heat is also produced. During exercise, the large amounts of heat can raise body temperature, and we feel warm. Shivering consists of small, rapid muscle contractions that produce heat in an effort to prevent a decrease in body temperature in the cold.
7. Vasoconstriction reduces blood flow to the skin, which reduces skin temperature because less warm blood from the deeper parts of the body reaches the skin. As the difference in temperature between the skin and the environment decreases, less loss of heat occurs. If the skin temperature decreases too much, however, dilation of blood vessels to the skin occurs, which prevents the skin from becoming so cold that it is damaged.

Chapter 26

1. If the cardiac output is 5600 mL of blood per minute and the hematocrit is 45, renal plasma flow is 650 mL of plasma per minute (see table 26.2). If the filtration fraction increased from 19% to 22%, the GFR would be 143 mL of filtrate per minute (650 mL of plasma × 0.22). If 99.2% of the filtrate is reabsorbed, 0.8% becomes urine. Thus, the urine produced is 1.14 mL of urine per minute (143 mL of filtrate × 0.008). Compared with the rate of urine production when the filtrate fraction was 19% (i.e., 1 mL per minute), the 3% increase in filtration fraction has caused a 14% increase in urine production. Converting 1.14 mL of urine per minute to liters of urine produced per day yields 1.64 L/day (1.14 mL/min × 1 L/1000 mL × 1440 min/day).
2. Even though hemoglobin is a smaller molecule than albumin, it does not normally enter the filtrate because hemoglobin is contained within red blood cells and these cells cannot pass through the filtration membrane. If red blood cells rupture, however, a process called hemolysis, the hemoglobin is released into the plasma, and large amounts of hemoglobin enter the filtrate. Conditions that cause red blood cells to rupture in the circulatory system result in large amounts of hemoglobin entering the urine.
3. Karl was suffering from plasma loss due to the extensive burns, which results in intense vasoconstriction, including vasoconstriction of arteries and arterioles that supply the kidneys. Constriction of the renal arteries and the renal afferent arterioles decreases the blood pressure in the glomeruli. As a consequence, the filtration pressure decreases. After an IV is administered and blood volume increases, the blood pressure returns to normal and renal arteries and arterioles once again dilate; thus, blood pressure in the glomeruli increases to normal levels.
4. Without the normal active transport of Na^+, the concentration of Na^+ and ions symported with them remains elevated in the nephron. Movement of water by osmosis out of the nephron into the interstitial spaces is decreased, resulting in an increased volume of urine.
5. Inhibition of ADH secretion is one of the numerous effects alcohol has on the body. The lack of ADH secretion causes the distal convoluted tubules and the collecting ducts to be relatively impermeable to water. The water cannot therefore move by osmosis from the distal nephrons and collecting ducts and remains in the nephrons to become urine. In addition, because other fluids are normally consumed with the alcohol, the increased water intake also results in an increase in dilute urine production.
6. a. Decreased water reabsorption in the distal convoluted tubules and collecting ducts results in increased water loss in urine and a large urine volume. Additionally, because the infants' kidneys are unable to reabsorb water from the filtrate, the infants' blood osmolality will increase.
   b. The infants' plasma levels of ADH will increase during the water deprivation test. The hallmark of NDI is a normal secretion of ADH from the pituitary followed by an abnormal response in the kidney.
   c. Normally, following exposure to ADH, the cells of the distal convoluted tubules and collecting ducts insert aquaporin-2 water channels into the apical membranes, which increases their permeability to water. However, in autosomal-recessive and autosomal-dominant NDI, the aquaporin-2 is dysfunctional and water stays in the collecting duct to exit the body as part of urine.
   d. Plasma levels of ADH would be abnormally low to nondetectable because central diabetes insipidus results from a deficiency in ADH secretion.
   e. Although seemingly counterintuitive to treat a patient already suffering from polyuria with a drug that increases urine output, treatment with a Na^+ reabsorption inhibitor in conjunction with a low-sodium diet ultimately promotes homeostatic water balance in the patient's cell fluids. By preventing plasma levels of sodium from becoming abnormally elevated, water is not prone to move by osmosis out of cells' cytoplasm into the extracellular fluid toward the blood plasma.
7. During the race, Amanda's blood volume decreased because of increased water loss through respiration and sweating. Consequently, because of Amanda's severe dehydration, her body immediately initiated the mechanisms that maintain blood pressure. The vasoconstriction and reduced blood volume explain her paleness. Increased sympathetic stimulation caused the increased heart rate. The renal arteries also became vasoconstricted. Subsequently, the GFR decreased and urine production was reduced. In addition, Amanda's ADH levels increased, which increased water reabsorption from the distal convoluted tubules and collecting ducts.
8. Anything that reduces the formation of filtrate reduces the GFR. If the epithelium of the nephrons sloughs off and forms casts in the nephrons, normal flow of filtrate through them is blocked. Consequently, the blocked flow of filtrate in the nephron causes the pressure in Bowman's capsule to increase enough that the pressure inside Bowman's capsule is close to the pressure in the glomerulus. Unless the pressure in the glomerulus is higher than the pressure in Bowman's capsule, no filtrate forms and the GFR is very low. If very little filtrate forms, the volume of urine produced is reduced. Also, filtrate that enters a nephron that is blocked, with the epithelial lining of the nephron disrupted, can flow directly into the peritubular capillaries.
9. Low urea clearance indicates that the amount of blood cleared of urea, a metabolic waste product, per minute is lower than normal. It is consistent with the reduction in the number of functional nephrons that occurs in advanced cases of renal failure. In addition, a low urea clearance is an indication that the GFR is reduced and that the blood levels of urea are increasing.
10. The urethra of females is much shorter than the urethra of males. In addition, the opening of the urethra in females is closer to the anus, which is a potential source of bacteria. The female urinary bladder is therefore more accessible to bacteria from the exterior. This accessibility is a major reason that urinary bladder infections are more common in females than in males.
11. After 7 days, Mr. H's kidneys began to produce a large volume of urine with

larger than normal Na^+ and K^+ concentrations. The observations are consistent with Mr. H becoming dehydrated by day 10. Dehydration results in reduced blood volume. The pale skin was the result of vasoconstriction, which was triggered by the reduced blood pressure. Dizziness resulted from reduced blood flow to the brain when Mr. H tried to stand and walk. He was lethargic in part because of reduced blood volume but also because of low blood levels of K^+ and Na^+. The arrythmia of his heart was due to low blood levels of K^+ and increased sympathetic stimulation, which also was triggered by low blood pressure.

Chapter 27

1. During hemorrhagic shock, blood pressure decreases and visceral blood vessels constrict (see chapter 21). As a consequence, blood flow to the kidneys and the blood pressure in the glomeruli decrease dramatically. The total filtration pressure decreases, and the amount of filtrate formed each minute decreases. The rate at which Na^+ enter the nephron therefore decreases. In addition, renin is secreted from the kidneys in large amounts. Renin causes the formation of angiotensin I from angiotensinogen. Angiotensin I converts to angiotensin II, which stimulates aldosterone secretion. Aldosterone increases the rate at which Na^+ are reabsorbed from the filtrate in the distal convoluted tubule and collecting ducts.
2. a. If the amount of Na^+ and water ingested in food exceeds that needed to maintain a constant extracellular fluid composition, it increases total blood volume and increases blood pressure.
   b. Excessive Na^+ and water intake causes an increase in total blood volume and blood pressure. The elevated blood pressure causes a reflex response that results in decreased ADH secretion. The elevated pressure also causes reduced renin secretion from the kidneys, resulting in a reduction in the rate at which angiotensin II is formed. The reduction in angiotensin II reduces the rate of aldosterone secretion. Together these changes cause increased loss of Na^+ in the urine and an increase in the volume of urine produced. Increased Na^+ and increased blood pressure also cause the secretion of ANH, which inhibits ADH secretion and Na^+ reabsorption in the nephron.
   c. If the amount of water ingested is large, urine concentration is reduced, urine volume is increased, and the concentration of Na^+ in the urine is low. If the amount of salt ingested is great, the concentration of salt in the urine can high, and the urine volume is larger and contains a substantial concentration of salt.
3. During exercise, the amount of water lost is increased because of increased evaporation from the respiratory system, increased insensible perspiration, and increased sweat. The amount of water lost in the form of sweat can increase substantially. The amount of urine formed decreases during exercise.
4. Hyperventilation results in a greater than normal rate of carbon dioxide loss from the circulatory system. Because carbon dioxide is lost from the circulatory system, H^+ concentration decreases and the pH of body fluids increases. Breathing into a paper bag corrects for the effects of hyperventilation because the person rebreathes air that has a higher concentration of carbon dioxide. The result is an increase in carbon dioxide in the body. Consequently, the H^+ concentration increases and pH decreases toward normal levels.
5. Aldosterone hyposecretion results in acidosis. Aldosterone increases the rate at which Na^+ are reabsorbed from nephrons, but it also increases the rate at which K^+ and H^+ are secreted. Hyposecretion of aldosterone decreases the rate at which H^+ are secreted by the nephrons and therefore can result in acidosis.
6. Elevated blood carbon dioxide levels cause an increase in H^+ and a decrease in blood pH due to the following reaction:

   $$CO_2 + H_2O \rightleftarrows H_2CO_3 \rightleftarrows H^+ + HCO_3^-$$

   However, the kidney plays an important role in the regulation of blood pH. The kidney's rate of H^+ secretion into the urine and reabsorption of HCO_3^- increase. This helps prevent high blood H^+ levels and low blood pH in Mr. Puffer. Because of increased resorption of HCO_3^-, the blood levels of HCO_3^- would be greater than normal.
7. a. Each time Dan vomited, he lost a substantial amount of fluid. Because he was very nauseated, he did not consume any food or liquid. Water crosses the wall of the stomach by osmosis. Secretion of acid by the gastric glands of Dan's stomach produced the acidic gastric fluid and vomiting eliminated the fluid from the stomach. As a result of the lost fluid volume, Dan's body weight decreased. His body weight before he became nauseated, was not known. The 71 kg body weight at 12 hours is probably less than his normal body weight. By 36 hours, Dan had lost an additional 3 kg. Because of the decrease in the plasma volume, Dan's blood pressure began to fall. The mechanisms that regulate blood pressure and blood volume were activated (see figure 27.4 and "Regulation of Mean Arterial Pressure"). The decrease in blood pressure and the increase in heart rate at 36 hours were a result of the continued fluid loss as vomiting continued. Because of the baroreceptor reflex (see figure 21.39), there is increased sympathetic stimulation of the heart and peripheral blood vessels. As a result, there is an increase in heart rate and pallor due to vasoconstriction of the blood vessels of the skin. It is likely that his heart rate was increased somewhat by 12 hours, but it was clearly increased by 36 hours. Although not reported, the ADH mechanism caused his urine volume to decrease dramatically. Thirst would have increased if Dan had not been so nauseated. At 12 hours, the mechanisms that control blood pressure were adequate to maintain blood pressure within its normal range. By 36 hours, Dan's blood pressure had decreased in spite of the homeostatic mechanisms.
   b. The gastric glands of the stomach secrete H^+. The rate of H^+ secretion had to increase in response to nausea and vomiting because Dan expelled a significant volume of acidic gastric fluid each time he vomited. The increased H^+ secretion caused metabolic acidosis (see Clinical Focus "Acidosis and Alkalosis"). At the same time H^+ were secreted, HCO_3^- were absorbed into the circulatory system (see figure 24.12). Therefore, the plasma HCO_3^- levels and plasma pH were increased at both 12 and 36 hours. The secretion of H^+ into the stomach and the absorption of HCO_3^- into the blood caused the plasma pH to increase. The mechanisms that control blood pH were activated (see figure 27.13). However, they were not adequate to keep the plasma pH within its normal range of values. The plasma pH was increased at 12 hours and was further increased at 36 hours.
   c. The plasma CO_2 levels were increased at both 12 and 36 hours. As the pH of the body fluids increased, stimulation of the respiratory centers decreased (see figure 23.23). Consequently, CO_2 elimination was slowed and plasma CO_2 levels increased.
   d. After Dan stopped vomiting, the rate of fluid loss was decreased. The rate of H^+ loss and HCO_3^- absorption slowed dramatically. He was also able

to consume liquids. These changes and the mechanisms that regulate body fluid volume and plasma pH continued to function until homeostasis was reestablished.

Chapter 28

1. a. Generally, cancer results from several mutations that occur in cells. The accumulation of mutations occurs over many generations of cells and may require several years to develop. Once a mutation alters a gene in a cell, that altered gene can be passed to the daughter cells when that cell undergoes mitosis. For cells that survive and undergo cell division, altered genes are passed to all of the offspring of the original cell in which the mutation occurred. For example, a single mutation may cause a cell to undergo mitosis at an increased rate. This mutation is passed to the offspring of that cell. Another mutation may occur in one of these cells and it, along with the original mutation, is passed to its offspring. In this fashion, mutations responsible for the development of cancer accumulate in cells. Therefore, cells that had accumulated some mutations leading to cancer might have existed in Vern's prostate at the time of the first biopsy, but cancer cells might not have been present. However, a small number of cancer cells might have been present and they could have been missed in the first biopsy.

   b. Cancer cells are identified because of their microscopic anatomy. Pathologists examine biopsies and make conclusions based on their observations. Cells do not always fit neatly into the categories of either cancer cells or normal cells. Cells that may become cancer cells in the future are likely to have accumulated some mutations consistent with cancer and therefore some abnormal structural characteristics, but they may not appear as unambiguous cancer cells.

   c. Obviously, the cancer cells had not yet formed a large tumor in Vern's prostate gland by the time of his second biopsy. He was fortunate that the cancer was detected in its early stages. It is possible therefore that most of the needle biopsies did not sample cancer cells and that only two of the needles passed through the developing tumor and therefore sampled cancer cells.

   d. The first biopsy could have missed cancer cells that were present at the time of the first biopsy. It also likely that mutations that cause cells to secrete increased amounts of PSA could have accumulated in some cells of the prostate, but all of the mutations that cause cells to cancerous may not have been present.

2. The prostate gland is anterior to the wall of the rectum. A finger inserted into the rectum can palpate the prostate gland through the rectal wall.
3. Coagulation may help keep the sperm cells within the female reproductive tract, thereby increasing the likelihood of fertilization.
4. If administered before the preovulatory LH surge, estrogen stimulates the hypothalamus to secrete GnRH. Estrogen and progesterone, in large amounts, inhibit GnRH and LH releases. A large amount of estrogen and progesterone administered at this time should therefore reduce the surge of LH. Continual administration of high levels of GnRH causes anterior pituitary cells to become insensitive to GnRH. Thus, LH and FSH levels remain low and the ovarian cycle stops.
5. High progesterone levels after menses inhibit GnRH secretion from the hypothalamus and therefore FSH and LH secretion from the anterior pituitary. Without FSH and LH, the events of the ovarian cycle, including estrogen production, are inhibited. Because estrogen causes proliferation of the endometrium, thickening of the endometrium is not expected. Also, estrogen increases the synthesis of uterine progesterone receptors, and without estrogen the secretory response of the endometrium to the elevated progesterone is inhibited.
6. Mrs. M's mother could have had leiomyomas also, although, without direct data from medical examinations, one cannot be certain. If that was the cause of her irregular menstruations, they may have become less frequent as Mrs. M's mother experienced menopause. During menopause, the uterus gradually becomes smaller, and eventually the cyclic changes in the endometrial lining cease. If the leiomyomas were relatively mild, the onset of menopause could explain the gradual disappearance of the irregular and prolonged menstruations. (*Note:* If the tumors are large, constant and severe menstruations are likely, even if regular menstrual cycles stop due to menopause.)

Chapter 29

1. The cells of the inner cell mass are pluripotent, meaning they can give rise to the different cells of the developing embryo; however, pluripotent cells do not give rise to the placental tissues necessary for implantation and normal development. If a cell of the inner cell mass is separated from the blastocyst, theoretically it will not develop normally. Totipotent cells, however, can give rise to all the cells necessary for normal development, both embryonic cells and placental tissues. If a totipotent cell is separated from the embryo, the development of an identical twin is possible.
2. Two primitive streaks on one embryonic disk result in the development of two embryos. If the two primitive streaks are touching or are very close to each other, the embryos may be joined. This condition is called conjoined (Siamese) twins.
3. Because the early embryonic heart is a simple tube, blood must be forced through the heart in almost a peristaltic fashion, and the contraction begins in the sinus venosus. If the sinus venosus did not contract first, blood could flow in the opposite direction.
4. The lack of müllerian-inhibiting hormone allows the paramesonephric duct system to develop into the internal female reproductive structures, particularly the uterus and the uterine tube. Assuming that testosterone secretion is normal, the internal male reproductive system will also develop. In the presence of dihydrotestosterone, derived from testosterone, the external male genitalia also develop.
5.

	Clinical Age	Developmental Age
Fertilization	14 days	0 days
Implantation	21 days	7 days
Fetal period	70 days	56 days
Parturition	280 days	266 days

6. Elevation of calcium levels might cause the uterine muscles to contract tetanically. This tetanic contraction might compress blood vessels and cut the blood supply to the fetus. Hypercalcemia can also result in arrhythmias and muscle weakness (see chapter 27). The doctor would therefore not administer calcium to the woman in labor but could give oxytocin, which strengthens contractions but is less likely to produce tetany.
7. Nursing stimulates the release of oxytocin from the mother's posterior pituitary gland, which is responsible for milk letdown. Oxytocin can also cause uterine contractions and cramps.

GLOSSARY

Many of the words in this glossary and the text are followed by a simplified phonetic spelling showing pronunciation. The pronunciation key reflects standard clinical usage as presented in *Stedman's Medical Dictionary* (27th edition), a leading reference volume in the health sciences.

The phonetic system used is a basic one and has only a few conventions:

- Two diacritical marks are used; the macron (¯) for long vowels; and the breve (˘) for short vowels.
- Principal stressed syllables are followed by a prime (′); monosyllables do not have a stress mark.
- Other syllables are separated by hyphens.

The following pronunciation key provides examples and consonant sounds encountered in the phonetic system. No attempt has been made to accommodate the slurred sounds common in speech or regional variations in speech sounds. Note that a vowel with a breve (˘) is used for the indefinite vowel sound of the schwa (ə). Native pronunciation of foreign words is approximated as closely as possible.

Pronunciation Key

Vowels

ā	day, care, gauge
a	mat, damage
ă	about, para
ah	father
aw	fall, cause, raw
ē	be, equal, ear
ĕ	taken, genesis
e	term, learn
ī	pie
ĭ	pit, sieve, build
ō	note, for so,
o	not, oncology, ought
oo	food
ow	cow, out
oy	troy, void
ū	unit, curable
ŭ	cut

Consonants

b	bad
ch	child
d	dog
dh	this, smooth
f	fit
g	got
h	hit
j	jade
k	kept
ks	tax
kw	quit
l	law
m	me
n	no
ng	ring
p	pan
r	rot
s	so, miss
sh	should
t	ten
th	thin, with
v	very
w	we
y	yes
z	zero
zh	azure, measure

In some words the initial sound is not that of the initial letter(s), or the initial letter(s) is not sounded or has a different sound, as in the following examples:

aerobe (ar′ob)	phthalein (thal′e-in)
eimuria (ime′re-a)	pneumonia (nu-mo′ne-a)
gnathic (nath′ik)	psychology (si-kol′o-je)
knuckle (nuk-l)	ptosis (to′sis)
oedipism (ed′i-pizm)	xanthoma (zan-tho′ma)

A

A band Length of the myosin myofilament in a sarcomere.

abdomen (ab-dō′men, ab′dō-men) Belly, between the thorax and the pelvis.

abduction (ab-dŭk′shŭn) [L., *abductio,* take away] Movement away from the midline.

absolute refractory period (ab′sō-loot rē-frak′tōr-ē) Portion of the action potential during which the membrane is insensitive to all stimuli, regardless of their strength.

absorptive cell (ab-sōrp′tiv) Cell on the surface of villi of the small intestines and the luminal surface of the large intestine that is characterized by having microvilli; secretes digestive enzymes and absorbs digested materials on its free surface.

absorptive state Immediately after a meal when nutrients are being absorbed from the intestine into the circulatory system.

accommodation (ă-kom′ŏ-dā′shŭn) [L., *ac* + *commodo,* to adapt] Ability of electrically excitable tissues, such as nerve or muscle cells, to adjust to a constant stimulus so that the magnitude of the local potential decreases through time; also called adaptation.

acetabulum (as-ĕ-tab′ū-lŭm) [L., shallow vinegar vessel or cup] Cup-shaped depression on the external surface of the coxa.

acetylcholine (ACh) (as-e-til-kō′lēn) Neurotransmitter substance released from motor neurons, all preganglionic neurons of the parasympathetic and sympathetic divisions, all postganglionic neurons of the parasympathetic division, some postganglionic neurons of the sympathetic division, and some central nervous system neurons.

acetylcholinesterase (as′ē-til-kō-lin-es′ter-ās) Enzyme found in the synaptic cleft that causes the breakdown of acetylcholine to acetic acid and choline, thus limiting the stimulatory effect of acetylcholine.

Achilles tendon *See* calcaneal tendon.

acid (as′id) Molecule that is a proton donor; any substance that releases hydrogen ions (H^+).

acidic Solution containing more than 10^{-27} mol of hydrogen ions per liter; has a pH less than 7.

acinus; pl. **acini** (as′i-nŭs, as′ĭ-nī) [L., berry, grape] Grape-shaped secretory portion of a gland. The terms *acinus* and *alveolus* are sometimes used interchangeably. Some authorities differentiate the terms: Acini have a constricted opening into the excretory duct, whereas alveoli have an enlarged opening.

acromion (ă-krō′mē-on) [Gr., *akron,* extremity + omos, shoulder] Bone comprising the tip of the shoulder.

acrosome (ak′rō-sōm) [Gr., *akron,* extremity + *soma,* body] Cap on the head of the spermatozoon, with hydrolytic enzymes that help the spermatozoon penetrate the ovum.

actin myofilament (ak′tin) Thin myofilament within the sarcomere; composed of two F actin molecules, tropomyosin, and troponin molecules.

action potential [L., *potentia,* power, potency] Change in membrane potential in an excitable tissue that acts as an electric signal and is propagated in an all-or-none fashion.

activation energy (ak-ti-vā′shŭn) Energy that must be added to molecules to initiate a reaction.

active site Portion of an enzyme in which reactants are brought into close proximity and that plays a role in reducing activation energy of the reaction.

active tension Tension produced by the contraction of a muscle.

active transport Carrier-mediated process that requires ATP and can move substances against a concentration gradient.

adaptive immunity Immune status in which there is an ability to recognize, remember, and destroy a specific antigen.

adenohypophysis (ad′ĕ-nō-hī-pof′ĭ-sis) Portion of the hypophysis derived from the oral ectoderm; also called anterior pituitary.

adenosine diphosphate (ADP) (ă-den′ō-sēn) Adenosine, an organic base, with two phosphate groups attached to it. Adenosine diphosphate combines with a phosphate group to form adenosine triphosphate.

adenosine triphosphate (ATP) Adenosine, an organic base, with three phosphate groups attached to it. Energy stored in ATP is used in nearly all of the endergonic reactions in cells.

adenylate cyclase (ad′e-nil sīklās) An enzyme acting on ATP to form 3′,5′-cyclic AMP plus pyrophosphate (two phosphate groups). A crucial step in the regulation and formation of the intracellular chemical signal 3′,5′-cyclic AMP.

adipocyte (ad′i-pō-sīt) Fat cell.

adipose (ad′i-pōs) [L., *adeps,* fat] Fat.

adrenal gland (ă-drē′năl) [L., *ad,* to + *ren,* kidney] Located near the superior pole of each kidney, it is composed of a cortex and a medulla. The adrenal medulla is a highly modified sympathetic ganglion that secretes the hormones epinephrine and norepinephrine; the cortex secretes aldosterone and cortisol as its major secretory products. Also called suprarenal gland.

adrenergic neuron (ad-rĕ-ner′jik) Nerve fiber that secretes norepinephrine (or epinephrine) as a neurotransmitter substance.

adrenergic receptor (ad-rĕ-ner′jik) Receptor molecule that binds to adrenergic agents, such as epinephrine and norepinephrine.

adrenocorticotropic hormone (ACTH) (ă-drē′nō-kōr′ti-kō-trō′pik) Hormone of the adenohypophysis that governs the nutrition and growth of the adrenal cortex, stimulates it to functional activity, and causes it to secrete cortisol.

adventitia (ad-ven-tish′ă) [L., *adventicius,* coming from abroad, foreign] Outermost covering of any organ or structure that is properly derived from outside the organ and does not form an integral part of the organ.

aerobic respiration (ār-ō′bik) Breakdown of glucose in the presence of oxygen to produce carbon dioxide, water, and approximately 38 ATPs; includes glycolysis, the citric acid cycle, and the electron-transport chain.

afferent arteriole (af′er-ent) Branch of an interlobular artery of the kidney that conveys blood to the glomerulus.

afferent division Nerve fibers that send impulses from the periphery to the central nervous system.

agglutination (ă-gloo-ti-nā-shŭn) [L., *ad,* to + *gluten,* glue] Process by which blood cells, bacteria, or other particles are caused to adhere to one another and form clumps.

agglutinin (ă-gloo′ti-nin) Antibody that binds to an antigen and causes agglutination.

agglutinogen (ă-gloo-tin′ō-jen) Antigen on surface of red blood cells that can stimulate the production of antibodies (agglutinins) that combine with the antigen and cause agglutination.

agranulocyte (ă-gran′ū-lō-sīt) Nongranular leukocyte (monocyte or lymphocyte).

ala; pl. **alae** (ā′lă, ā′lē) [L., a wing] Wing-shaped structure.

aldosterone (al-dos′ter-ōn) Steroid hormone produced by the zona glomerulosa of the adrenal cortex that facilitates potassium exchange for sodium in the distal renal tubule, causing sodium reabsorption and potassium and hydrogen secretion.

alkaline (al′kă-līn) Solution containing less than 10^{-7} mol of hydrogen ions per liter; has a pH greater than 7.0.

alkalosis (al-kă-lō′sis) Condition characterized by blood pH of 7.45 or above.

allantois (ă-lan′tō-is) Tube extending from the embryonic hindgut into the umbilical cord; forms the urinary bladder.

allele (ă-lēl′) [Gr., *allelon,* reciprocally] Any one of a series of two or more different genes that may occupy the same position or locus on a specific chromosome.

all-or-none principle When a stimulus is applied to a cell, an action potential is either produced or not. In muscle cells, the cell either contracts to the maximum extent possible (for a given condition) or does not contract.

alternative pathway Part of the nonspecific immune system for activation of complement.

alveolar duct (al-vē′ō-lăr) Part of the respiratory passages beyond a respiratory bronchiole; from it arise alveolar sacs and alveoli.

alveolar gland Gland in which the secretory unit has a saclike form and an obvious lumen.

alveolar sac Two or more alveoli that share an opening.

alveolus; pl. **alveoli** (al-vē′ō-lŭs, al-vē′ō-lī) Cavity. Examples include the sockets into which teeth fit, the endings of the respiratory system, and the terminal endings of secretory glands.

amino acid (ă-mē′nō) Class of organic acids that constitute the building blocks for proteins.

amplitude-modulated signal (am′pli-tood) Signal that varies in magnitude or intensity, such as with large versus small concentrations of hormones.

ampulla (am-pul′lă, am-pul′lē) [L., two-handled bottle] Saclike dilatation of a semicircular canal; contains the crista ampullaris. Wide portion of the uterine tube between the infundibulum and the isthmus.

amygdala (a-mig′da-la, -lē) [L. fr. Gr., *amygydale,* almond] Nucleus in the temporal lobe of the brain, amygdaloid nucleus; also called amydaloid nuclear complex.

amylase (am′il-ās) One of a group of starch-splitting enzymes that cleave starch, glycogen, and related polysaccharides.

anabolism (ă-nab′ō-lizm) [Gr., *anabole,* a raising up] All of the synthesis reactions that occur within the body; requires energy.

anaerobic respiration (an-ār-ō′bik) Breakdown of glucose in the absence of oxygen to produce lactic acid and two ATPs; consists of glycolysis and the reduction of pyruvic acid to lactic acid.

anal canal Terminal portion of the digestive tract.

anal triangle Posterior portion of the perineal region through which the anal canal opens.

analgesic (an-al-jē′zik) Compound capable of producing analgesia, without producing anesthesia or loss of consciousness, characterized by reduced response to painful stimuli.

anaphase (an′ă-fāz) Time during cell division when chromatids divide (or, in the case of first meiosis, when the chromosome pairs divide).

anastomoses (ă-nas′tō-mō′sez) Natural communication, direct or indirect, between two blood vessels or other tubular structures. An opening created by surgery, trauma, or disease between two or more normally separate spaces or organs.

anatomical dead air space Volume of the conducting airways from the external environment down to the terminal bronchioles.

androstenedione (an-drō-stēn-dī′ōn) Androgenic steroid of weaker potency than testosterone; secreted by the testis, ovary, and adrenal cortex.

anencephaly (an′en-sef′ă-lē) [Gr., *an* + *enkephalos,* no brain] Defective development of the brain and absence of the bones of the cranium. Only a rudimentary brainstem and a trace of basal ganglia are present.

aneurysm (an′ū-rizm) [Gr., *eurys,* wide] Dilated portion of an artery.

angiotensin I (an-jē-ō-ten′sin) Peptide derived when renin acts on angiotensinogen.

angiotensin II Peptide derived from angiotensin I; stimulates vasoconstriction and aldosterone secretion; also called active angiotensin.

anion (an′ī-on) Ion carrying a negative charge.

antagonist (an-tag′ŏ-nist) Muscle that works in opposition to another muscle.

anterior chamber Chamber of the eye between the cornea and the iris.

anterior interventricular sulcus Groove on the anterior surface of the heart, marking the location of the septum between the two ventricles.

anterior pituitary *See* adenohypophysis.

antibody (an′tē-bod-ē) Protein found in the plasma that is responsible for humoral immunity; binds specifically to antigen.

antibody-mediated immunity Immunity due to B cells and the production of antibodies.

anticoagulant (an'tē-kō-ag'ū-lant) Agent that prevents coagulation.

antidiuretic hormone (ADH) (an'tē-dī-ū-ret'ik) Hormone secreted from the neurohypophysis that acts on the kidney to reduce the output of urine; also called vasopressin because it causes vasoconstriction.

antigen (an'ti-jen) [anti(body) + Gr., *gen,* producing] Substance that induces a state of sensitivity or resistance to infection or toxic substances after a latent period; substance that stimulates the specific immune system; also called epitope.

antigenic determinant (an-ti-jen'ik) Specific part of an antigen that stimulates an immune system response by binding to receptors on the surface of lymphocytes; also called epitope.

antithrombin (an-tē-throm'bin) Substance that inhibits or prevents the effects of thrombin so that blood does not coagulate.

antrum (an'trŭm) [Gr., *antron,* a cave] Cavity of an ovarian follicle filled with fluid containing estrogen.

anulus fibrosus (an'ū-lŭs fī-brō'sus) [L., fibrous ring] Fibrous material forming the outer portion of an intervertebral disk.

anus (ā'nŭs) Lower opening of the digestive tract through which fecal matter is extruded.

aorta (ā-ōr'tă) [Gr., *aorte* from *aeiro,* to lift up] Large, elastic artery that is the main trunk of the systemic arterial system; carries blood from the left ventricle of the heart and passes through the thorax and abdomen.

aortic arch [L., bow] Curve between the ascending and descending portions of the aorta.

aortic body One of the smallest bilateral structures, similar to the carotid bodies, attached to a small branch of the aorta near its arch; contains chemoreceptors that respond primarily to decreases in blood oxygen; less sensitive to decreases in blood pH or increases in carbon dioxide.

apex (ā'peks) [L., summit or tip] Extremity of a conical or pyramidal structure. The apex of the heart is the rounded tip directed anteriorly and slightly inferiorly.

Apgar score Named for U.S. anesthesiologist Virginia Apgar (1909–1974). Evaluation of a newborn infant's physical status by assigning numerical values to each of five criteria: appearance (skin color), pulse (heart rate), grimace (response to stimulation), activity (muscle tone), and respiratory effort; a score of 10 indicates the best possible condition.

apical ectodermal ridge Layer of surface ectodermal cells at the lateral margin of the embryonic limb bud; stimulates growth of the limb.

apical foramen [L., aperture] Opening at the apex of the root of a tooth, gives passage to the nerve and blood vessels.

apocrine gland (ap'ō-krin) [Gr., *apo,* away from + *krino,* to separate] Gland whose cells contribute cytoplasm to its secretion (e.g., mammary glands). Sweat glands that produce organic secretions traditionally are called apocrine. These sweat glands, however, are actually merocrine glands.

appendicular skeleton (ap'en-dik'ū-lăr) The portion of the skeleton consisting of the upper limbs and the lower limbs and their girdles.

appositional growth (ap-ō-zish'ŭn-al) [L., *ap* + *pono,* to put or place] To place one layer of bone, cartilage, or other connective tissue against an existing layer.

aqueous humor (ak'wē-ŭs, ā'kwē-ŭs) Watery, clear solution that fills the anterior and posterior chambers of the eye.

arachnoid (ă-rak'noyd) [Gr., *arachne,* spider, cobweb] Thin, cobweb-appearing meningeal layer surrounding the brain; the middle of the three layers.

arcuate artery (ar'kū-āt) Artery that originates from the interlobar arteries of the kidney and forms an arch between the cortex and medulla of the kidney.

areola (ă-rē'ō-lă, -lē) [L., area] Circular, pigmented area surrounding the nipple; its surface is dotted with little projections caused by the presence of the areolar glands beneath.

areolar gland (ă-rē'ō-lăr) Gland forming small, rounded projections from the surface of the areola of the mamma.

arrectores pilorum; pl. **arrector pili** (ă-rek-tō'rez pī-lōr'um, ă-rek'tōr pī'lī) [L., that which raises; hair] Smooth muscle attached to the hair follicle and dermis that raises the hair when it contracts.

arterial capillary (ar-tē'rē-ăl) Capillary opening from an arteriole or a metarteriole.

arteriole (ar-tēr'ē-ōl) Minute artery with all three tunics that transports blood to a capillary.

arteriosclerosis (ar-tēr'ē-ō-skler-ō'sis) [L., *arterio* + Gr., *sklerosis,* hardness] Hardening of the arteries.

arteriovenous anastomosis (ar-tēr'ē-ō-vē'nŭs ă-nas'tō-mō'sis) Vessel through which blood is shunted from an arteriole to a venule without passing through the capillaries.

artery (ar'ter-ē) Blood vessel that carries blood away from the heart.

articular cartilage (ar-tik'ū-lăr kar'ti-lij) Hyaline cartilage covering the ends of bones within a synovial joint.

articulation (ar-tik-ū-lā'shŭn) Place where two bones come together; also called joint.

arytenoid cartilages (ar-i-tē'noyd) Small, pyramidal laryngeal cartilages that articulate with the cricoid cartilage.

ascending aorta Part of the aorta from which the coronary arteries arise.

ascending colon (kō'lon) Portion of the colon between the small intestine and the right colic flexure.

asthma (az'mă) Condition of the lungs in which widespread narrowing of airways occurs, caused by contraction of smooth muscle, edema of the mucosa, and mucus in the lumen of the bronchi and bronchioles.

astrocyte (as'trō-sīt) [Gr., *astron,* star + *kytos,* a hollow, a cell] Star-shaped neuroglia cell involved with forming the blood–brain barrier.

atherosclerosis (ath'er-ō-skler-ō'sis) Arteriosclerosis characterized by irregularly distributed lipid deposits in the intima of large and medium-sized arteries.

atomic number (ă-tom'ik) Number of protons in each type of atom.

atrial diastole (ā'trē-ăl dī-as'tō-lē) Dilation of the heart's atria.

atrial natriuretic hormone (ANH) (ā'trē-ăl nā'trē-ū-ret'ik) Peptide released from the atria when atrial blood pressure is increased; lowers blood pressure by increasing the rate of urinary production, thus reducing blood volume.

atrial systole (ā'trē-ăl sis'tō-lē) Contraction of the atria.

atrioventricular (AV) bundle (ā'trē-ō-ven-trik'ū-lar) Bundle of modified cardiac muscle fibers that projects from the AV node through the interventricular septum.

atrioventricular (AV) node Small node of specialized cardiac muscle fibers that gives rise to the atrioventricular bundle of the conduction system of the heart.

atrioventricular valve One of two valves closing the openings between the atria and ventricles.

atrium; pl. **atria** (ā'trē-ŭm, ā'trē-ă) [L., entrance hall] One of two chambers of the heart into which veins carry blood.

auditory cortex (aw'di-tōr-ē kōr'teks) Portion of the cerebral cortex that is responsible for the conscious sensation of sound; in the dorsal portion of the temporal lobe within the lateral fissure and on the superolateral surface of the temporal lobe.

auditory ossicle (os'i-kl) Bone of the middle ear; includes the malleus, incus, and stapes.

auricle (aw'ri-kl) [L., *auris,* ear] Part of the external ear that protrudes from the side of the head; also called pinna. Small pouch projecting from the superior, anterior portion of each atrium of the heart.

auscultatory (aws-kŭltă-tō-rē) Relating to auscultation, listening to the sounds made by the various body structures as a diagnostic method.

autoimmune disease (aw-tō-i-mūn' di-zēz') Disease resulting from a specific immune system reaction against self-antigens.

autonomic ganglia (aw-tō-nom′ik gang′glē-ă) Ganglia containing the nerve cell bodies of the autoimmune division of the nervous system.

autonomic nervous system (ANS) Nervous system composed of nerve fibers that send impulses from the central nervous system to smooth muscle, cardiac muscle, and glands.

autophagia (aw-tō-fā′jē-ă) [Gr., *auto,* self + *phagein,* to eat] Segregation and disposal of organelles within a cell.

autoregulation (aw′tō-reg-ŭ-lā′shŭn) Maintenance of a relatively constant blood flow through a tissue despite relatively large changes in blood pressure; maintenance of a relatively constant glomerular filtration rate despite relatively large changes in blood pressure.

autorhythmic Spontaneous and periodic—for example, in smooth muscle, spontaneous (without nervous or hormonal stimulation) and periodic contractions.

autosome (aw′tō-sōm) [Gr., *auto,* self + *soma,* body] Any chromosome other than a sex chromosome; normally exist in pairs in somatic cells and singly in gametes.

axial skeleton (ak′sē-ăl) Skull, vertebral column, and rib cage.

axillary (ak′sil-ār-ē) Relating to the axilla; the space below the shoulder joint, bounded by the pectoralis major anteriorly, the latissimus dorsi posteriorly, the serratus anterior medially, and the humerus laterally.

axolemma (ak′sō-lem′ă) [Gr., *axo* + *lemma,* husk] Plasma membrane of the axon.

axon (ak′son) [Gr., axis] Main central process of a neuron that normally conducts action potentials away from the neuron cell body.

axon hillock Area of origin of the axon from the nerve cell body.

axoplasm (ak′sō-plazm) Neuroplasm or cytoplasm of the axon.

B

baroreceptor (bar′ō-rē-sep′ter, bar′ō-rē-sep′tōr) Sensory nerve ending in the walls of the atria of the heart, venae cavae, aortic arch, and carotid sinuses; sensitive to stretching of the wall caused by increased blood pressure; also called pressoreceptor.

baroreceptor reflex Detects changes in blood pressure and produces changes in heart rate, heart force of contraction, and blood vessel diameter that return blood pressure to homeostatic levels.

basal ganglia (bā′săl gang′glē-ă) Nuclei at the base of the cerebrum involved in controlling motor functions.

base (bās) Molecule that is a proton acceptor; any substance that binds to hydrogen ions. Lower part or bottom of a structure; the base of the heart is the flat portion directed posteriorly and superiorly; veins and arteries project into and out of the base, respectively.

basement membrane (bās′ment mem′brān) Specialized extracellular material located at the base of epithelial cells and separating them from the underlying connective tissues.

basilar membrane (bas′i-lăr mem′brān) Wall of the membranous labyrinth bordering the scala tympani; supports the organ of Corti.

basophil (bā′sō-fil) [Gr., *basis,* baso + *phileo,* to love] White blood cell with granules that stain specifically with basic dyes; promotes inflammation.

B cell Type of lymphocyte responsible for antibody-mediated immunity.

belly (bel′ē) Largest portion of muscle between the origin and insertion.

beta-oxidation (bā′tă ok-si-dā′shŭn) Metabolism of fatty acids by removing a series of two-carbon units to form acetyl-CoA.

bicarbonate ion (bī-kar′bon-āt) Anion (HCO_3^-) remaining after the dissociation of carbonic acid.

bicuspid valve (bī-kŭs′pid) Valve closing the orifice between left atrium and left ventricle of the heart; also called mitral valve.

bile (bīl) Fluid secreted from the liver into the duodenum; consists of bile salts, bile pigments, bicarbonate ions, cholesterol, fats, fat-soluble hormones, and lecithin.

bile canaliculus (bīl kan′ă-lik′ū-lŭs) One of the intercellular channels approximately 1 μm or less in diameter that occurs between liver cells into which bile is secreted; empties into the hepatic ducts.

bile salt Organic salt secreted by the liver that functions as an emulsifying agent.

bilirubin (bil-i-roo′bin) [L., *bili* + *ruber,* red] Bile pigment derived from hemoglobin during the destruction of red blood cells.

biliverdin (bil-i-ver′din) Green bile pigment formed from the oxidation of bilirubin.

binocular vision (bin-ok′ū-lăr) [L., *bini,* paired + *oculus,* eye] Vision using two eyes at the same time; responsible for depth perception when the visual fields of the eyes overlap.

bipolar neuron (bī-pō′ler) One of the three categories of neurons consisting of a neuron with two processes—one dendrite and one axon—arising from opposite poles of the cell body.

blastocele (blas′tō-sēl) [Gr., *blastos,* germ + *koilos,* hollow] Cavity in the blastocyst.

blastocyst (blas′tō-sist) [Gr., *blastos,* germ + *kystis,* bladder] Stage of mammalian embryos that consists of the inner cell mass and a thin trophoblast layer enclosing the blastocele.

bleaching In response to light, retinal separates from opsin.

blind spot (blīnd) Point in the retina where the optic nerve penetrates the fibrous tunic; contains no rods or cones and therefore does not respond to light.

blood clot Coagulated phase of blood.

blood colloid osmotic pressure Osmotic pressure due to the concentration difference of proteins across a membrane that does not allow passage of the proteins.

blood groups Classification of blood based on the type of antigen found on the surface of red blood cells.

blood island Aggregation of mesodermal cells in the embryonic yolk sac that forms vascular endothelium and primitive blood cells.

blood pressure (BP) [L., *pressus,* to press] Tension of the blood within the blood vessels; commonly expressed in units of millimeters of mercury (mm Hg).

blood–brain barrier Permeability barrier controlling the passage of most large-molecular compounds from the blood to the cerebrospinal fluid and brain tissue; consists of capillary endothelium and may include the astrocytes.

blood–thymic barrier Layer of reticular cells that separates capillaries from thymic tissue in the cortex of the thymus gland; prevents large molecules from leaving the blood and entering the cortex.

Bohr effect Named for Danish physiologist Christian Bohr (1855–1911). Shift of the oxygen–hemoglobin dissociation curve to the right or left because of changes in blood pH. The definition sometimes is extended to include shifts caused by changes in blood carbon dioxide levels.

bony labyrinth (lab′i-rinth) Part of the inner ear; contains the membranous labyrinth that forms the cochlea, vestibule, and semicircular canals.

Boyle's law The pressure of a gas is equal to the volume of a container times the constant, K, for a given temperature. Assuming a constant temperature, the pressure of a gas is inversely proportional to its volume.

brachial (brā′kē-āl) [L., *brachium,* arm] Relating to the arm.

branchial arch Typically, six arches in vertebrates; in the lower vertebrates, they bear gills, but they appear transiently in the higher vertebrates and give rise to structures in the head and neck.

broad ligament Peritoneal fold passing from the lateral margin of the uterus to the wall of the pelvis on each side.

bronchiole (brong′kē-ōl) One of the finer subdivisions of the bronchial tubes, less than 1 mm in diameter; has no cartilage in its wall but does have relatively more smooth muscle and elastic fibers.

brush border Epithelial surface consisting of microvilli.

buffer (bŭf′er) Mixture of an acid and a base that reduces any changes in pH that would otherwise occur in a solution when acid or base is added to the solution.

bulb of the penis Expanded posterior part of the corpus spongiosum of the penis.

bulb of the vestibule Mass of erectile tissue on each side of the vagina.

bulbar conjunctiva (bŭl′bar kon-jŭnk-tī′vă) Conjunctiva that covers the surface of the eyeball.

bulbourethral gland (bŭl′bō-ū-rē′thrăl) One of two small compound glands that produce a mucoid secretion; it discharges through a small duct into the spongy urethra.

bulbus cordis (bŭl′bŭs) [L., plant bulb] End of the embryonic cardiac tube where blood leaves the heart; becomes part of the ventricle.

bursa; pl. bursae (ber′să, ber′sē) [L., purse] Closed sac or pocket containing synovial fluid, usually found in areas where friction occurs.

bursitis (ber-sī′tis) [L., *purse* + Gr., *ites,* inflammation] Inflammation of a bursa.

C

calcaneal tendon (kal-kā′nē-al) Common tendon of the gastrocnemius, soleus, and plantaris muscle that attaches to the calcaneus; also called Achilles tendon.

calcitonin (kal-si-tō′nin) Hormone released from parafollicular cells that acts on tissues to cause a decrease in blood levels of calcium ions.

calmodulin (kal-mod′ū-lin) [*calcium* + *modulate*] Protein receptor for Ca^{2+} that plays a role in many Ca^{2+}-regulated processes, such as smooth muscle contraction.

calorie (cal) (kal′ō-rē) [L., *calor,* heat] Unit of heat content or energy. The quantity of energy required to raise the temperature of 1 g of water 1°C.

calpain (kal′pān) Enzyme involved in changing the shape of dendrites; involved with long-term memory.

calyx; pl. calyces (kā′liks, kal′i-sēz) [Gr., cup of a flower] Flower-shaped or funnel-shaped structure; specifically, one of the branches or recesses of a renal pelvis into which the tips of the renal pyramids project.

cancellous bone (kan′sē-lŭs) [L., grating or lattice] Bone with a latticelike appearance; also called spongy bone.

cancer (kan′ser) Any of various types of malignant neoplasms, most of which invade surrounding tissues, may metastasize to several sites, and are likely to recur after attempted removal and to cause death of the patient unless adequately treated.

canine (kā′nīn) Referring to the cuspid tooth.

cannula (kan′ū-lă) [L., *canna,* reed] Tube; often inserted into an artery or a vein.

capacitation (kă-pas′i-tā′shŭn) [L., *capax,* capable of] Process whereby spermatozoa acquire the ability to fertilize ova; occurs in the female genital tract.

capitulum (kă-pit′ū-lŭm) [L., *caput,* head] Head-shaped structure.

carbaminohemoglobin (kar-bam′i-nō-hē-mō-glō′bin) Carbon dioxide bound to hemoglobin by means of a reactive amino group on the hemoglobin.

carbohydrate (kar-bō-hī′drāt) Monosaccharide (simple sugar) or the organic molecules composed of monosaccharides bound together by chemical bonds—for example, glycogen. For each carbon atom in the molecule, there are typically one oxygen molecule and two hydrogen molecules.

carbonic acid/bicarbonate buffer system One of the major buffer systems in the body; major components are carbonic acid and bicarbonate ions.

carbonic anhydrase Enzyme that catalyzes the reaction between carbon dioxide and water to form carbonic acid.

carcinoma (kar-si-nō′mă) Malignant neoplasm derived from epithelial tissue.

cardiac [Gr., *kardia,* heart] Related to the heart.

cardiac cycle [Gr., *kyklos,* circle] Complete round of cardiac systole and diastole.

cardiac nerve Nerve that extends from the sympathetic chain ganglia to the heart.

cardiac output (CO) Volume of blood pumped by the heart per minute; also called minute volume.

cardiac part Region of the stomach near the opening of the esophagus.

cardiac reserve [L., *re* + *servo,* to keep back, reserve] Work that the heart is able to perform beyond that required during ordinary circumstances of daily life.

carotid body (ka-rot′id) One of the small organs near the carotid sinuses; contains chemoreceptors that respond primarily to decreases in blood oxygen; less sensitive to decreases in blood pH or increases in carbon dioxide.

carotid sinus Enlargement of the internal carotid artery near the point where the internal carotid artery branches from the common carotid artery; contains baroreceptors.

carpal (kar′păl) [Gr., *karpos,* wrist] Bone of the wrist.

carrier Person in apparent health whose chromosomes contain a pathologic mutant gene, which may be transmitted to his or her children.

cartilage (kar′ti-lij) [L., *cartilage,* gristle] Firm, smooth, resilient, nonvascular connective tissue.

cartilaginous joint (kar-ti-laj′i-nŭs) Bones connected by cartilage; includes synchondroses and symphyses.

catabolism (kă-tab′ō-lizm) [Gr., *katabole,* a casting down] All of the decomposition reactions that occur in the body; releases energy.

catalyst (kat′ă-list) Substance that increases the rate at which a chemical reaction proceeds without being changed permanently.

cataract (kat′a-rakt) Complete or partial opacity of the lens of the eye.

cations (kat′ī-on) [Gr., *kation,* going down] Ions carrying a positive charge.

caveola; pl. caveolae (kav-ē-ō′lă, kav-ē-ō′lē) [L., small pocket] Shallow invagination in the membranes of smooth muscle cells that may perform a function similar to that of both the T tubules and sarcoplasmic reticulum of skeletal muscle.

cecum (sē′kŭm, sē′kă) [L., *caecus,* blind] Cul-de-sac forming the first part of the large intestine.

cell-mediated immunity Immunity due to the actions of T cells and null cells.

celom (sē′lom, sē-lō′mă) [Gr., *koilo* + *amma,* a hollow] Principal cavities of the trunk—for example, the pericardial, pleural, and peritoneal cavities. Separate in the adult, they are continuous in the embryo.

cementum (se-men′tŭm) [L., *caementum,* rough quarry stone] Layer of modified bone covering the dentin of the root and neck of a tooth; blends with the fibers of the periodontal membrane.

central nervous system (CNS) Major subdivision of the nervous system, consisting of the brain and spinal cord.

central vein Terminal branches of the hepatic veins that lie centrally in the hepatic lobules and receive blood from the liver sinusoids.

centrosome (sen′trō-sōm) Specialized zone of cytoplasm close to the nucleus and containing two centrioles.

cerebellum (ser-e-bel′ŭm) [L., little brain] Separate portion of the brain attached to the brainstem at the pons; important in maintaining muscle tone, balance, and coordination of movement.

cerebrospinal fluid (CSF) (ser′ĕ-brō-spī-năl) Fluid filling the ventricles and surrounding the brain and spinal cord.

ceruminous glands (sĕ-roo′mi-nŭs) Modified sebaceous glands in the external acoustic meatus that produce cerumen (earwax).

cervical canal (ser′vĭ-kal) Canal extending from the isthmus of the uterus to the opening of the uterus into the vagina.

cervix; pl. cervices (ser′viks, ser-vī′sēz) [L., neck] Lower part of the uterus extending from the isthmus of the uterus into the vagina.

chalazion (ka-lā′zē-on) Chronic inflammation of a meibomian gland; also called meibomian cyst.

cheek (chēk) Side of the face forming the lateral wall of the mouth.

chemical signal Molecule that binds to a macromolecule, such as receptors or enzymes, and alters their function; also called ligand.

chemoreceptor (kē′mō-rē-sep′tor) Sensory cell that is stimulated by a change in the concentration of chemicals to produce action potentials. Examples include taste receptors, olfactory receptors, and carotid bodies.

chemoreceptor reflex Chemoreceptors detect a decrease in blood oxygen, an increase in carbon dioxide, or a decrease in pH and produce an increased rate and depth of respiration and, by means of the vasomotor center, vasoconstriction.

chemosensitive area (kem-ō-sen′si-tiv, kē-mō-sen′si-tiv) Chemosensitive neurons in the medulla oblongata detect changes in blood, carbon dioxide, and pH.

chemotactic factor (kē-mō-tak′tik) Part of a microorganism or chemical released by tissues and cells that act as chemical signals to attract leukocytes.

chemotaxis (kē-mo-tak′sis) [Gr., *chemo* + *taxis,* orderly arrangement] Attraction of living protoplasm (cells) to chemical stimuli.

chief cell Cell of the parathyroid gland that secretes parathyroid hormone. Cell of a gastric gland that secretes pepsinogen.

chloride (klōr′īd) Compound containing chlorine—for example, salts of hydrochloric acid.

chloride shift Diffusion of chloride ions into red blood cells as bicarbonate ions diffuse out; maintains electrical neutrality inside and outside the red blood cells.

choana; pl. **choanae** (kō′an-ă, kō-ā′nē) *See* internal naris.

cholecystokinin (kō′lē-sis-tō-kī′nin) Hormone liberated by the upper intestinal mucosa on contact with gastric contents; stimulates the contraction of the gallbladder and the secretion of pancreatic juice high in digestive enzymes.

cholinergic neuron (kol-in-er′jik) Nerve fiber that secretes acetylcholine as a neurotransmitter substance.

chondroblast (kon′drō-blast) [Gr., *chondros,* gristle, cartilage + *blastos,* germ] Cartilage-producing cell.

chondrocyte (kon′drō-sīt) [Gr., *chondros,* gristle, cartilage + *kytos,* a cell] Mature cartilage cell.

chorda tympani; pl. **chordae** (kōr′dă tim′pan-ē, kōr′dē) Branch of the facial nerve that conveys taste sensation from the front two-thirds of the tongue.

chordae tendineae (kōr′dă ten′di-nē-ē) [L., cord] Tendinous strands running from the papillary muscles to the atrioventricular valves.

choroid (ko′royd) Portion of the vascular tunic associated with the sclera of the eye.

choroid plexus [Gr., *chorioeides,* membrane-like] Specialized plexus located within the ventricles of the brain that secretes cerebrospinal fluid.

chromatid (krō′mă-tid) One-half of a chromosome; separates from its partner during cell division.

chromatin (krō′ma-tin) Colored material; the genetic material in the nucleus.

chromosome (krō′mō-sōm) Colored body in the nucleus, composed of DNA and proteins and containing the primary genetic information of the cell; 23 pairs in humans.

chronic pain Prolonged pain.

chylomicron (kī-lō-mī′kron) [Gr., *chylos,* juice + *micros,* small] Microscopic particle of lipid surrounded by protein; in chyle and blood.

chymotrypsin (kī-mō-trip′sin) Proteolytic enzyme formed in the small intestine from the pancreatic precursor chymotrypsinogen.

ciliary body (sil′ē-ar-ē) Structure continuous with the choroid layer at its anterior margin that contains smooth muscle cells; functions in accommodation.

ciliary gland Modified sweat gland that opens into the follicle of an eyelash, keeping it lubricated.

ciliary muscle Smooth muscle in the ciliary body of the eye.

ciliary process Portion of the ciliary body of the eye that attaches by suspensory ligaments to the lens.

ciliary ring Portion of the ciliary body of the eye that contains smooth muscle cells.

circumduction (ser-kŭm-dŭk′shŭn) [L., around + *ductus,* to draw] Movement in a circular motion.

circumferential lamellae (ser-kŭm-fer-en′shē-al lă-mel′ē) Lamellae covering the surface of and extending around compact bone inside the periosteum.

circumvallate papilla (ser-kŭm-val′āt pă-pil′ă) Type of papilla on the surface of the tongue surrounded by a groove.

cisterna; pl. **cisternae** (sis-ter′nă, sis-ter′nē) Interior space of the endoplasmic reticulum.

cisterna chyli (kīl′ē) [L., tank + Gr., *chylos,* juice] Enlarged inferior end of the thoracic duct that receives chyle from the intestine.

citric acid cycle (sit′rik) Series of chemical reactions in which citric acid is converted into oxaloacetic acid, carbon dioxide is formed, and energy is released. The oxaloacetic acid can combine with acetyl-CoA to form citric acid and restart the cycle. The energy released is used to form NADH, FADH, and ATP.

classical pathway Part of the specific immune system for activation of complement.

clavicle (klav′i-kl) The collarbone, between the sternum and scapula.

cleavage furrow (klēv′ij) Inward pinching of the plasma membrane that divides a cell into two halves, which separate from each other to form two new cells.

cleft palate (kleft) Failure of the embryonic palate to fuse along the midline, resulting in an opening through the roof of the mouth.

clinical age (klin′i-kl) Age of the developing fetus from the time of the mother's last menstrual period before pregnancy.

clinical perineum (klin′i-kl per′i-nē′ŭm) Portion of the perineum between the vaginal and anal openings.

clitoris (klit′ō-ris) Small, cylindrical, erectile body, rarely exceeding 2 cm in length, situated at the most anterior portion of the vulva and projecting beneath the prepuce.

cloaca (klō-ā′kă) [L., sewer] In early embryos, the endodermally lined chamber into which the hindgut and allantois empty.

cloning (klōn′ing) Growing a colony of genetically identical cells or organisms.

clot retraction Condensation of a clot into a denser, compact structure; caused by the elastic nature of fibrin.

coagulation (kō-ag-ū-lā′shŭn) Process of changing from liquid to solid, especially of blood; formation of a blood clot.

cochlear duct (kok′lē-ăr) Interior of the membranous labyrinth of the cochlea; also called cochlear canal or scala media.

cochlear nerve Nerve that carries sensory impulses from the organ of Corti to the vestibulocochlear nerve.

cochlear nucleus Neurons from the cochlear nerve synapse within the dorsal or ventral cochlear nucleus in the superior medulla oblongata.

codon (kō′don) Sequence of three nucleotides in mRNA or DNA that codes for a specific amino acid in a protein.

cofactor (kō′fak′ter, kō-fak′ tōr) Nonprotein component of an enzyme, such as coenzymes and inorganic ions essential for enzyme action.

collagen fibers (kol′lă-jen) [Gr., *koila,* glue + *gen,* producing] Ropelike protein of the extracellular matrix.

collateral ganglia (ko-lat′er-ăl gang′glē-ă) Sympathetic ganglia at the origin of large abdominal arteries; include the celiac, superior, and inferior mesenteric arteries; also called prevertebral ganglia.

collecting duct Straight tubule that extends from the cortex of the kidney to the tip of the renal pyramid. Filtrate from the distal convoluted tubes enters the collecting duct and is carried to the calyces.

colloid (kol′oyd) [Gr., *kolla,* glue + *eidos,* appearance] Atoms or molecules dispersed in a gaseous, liquid, or solid medium that resist separation from the liquid, gas, or solid.

colloidal solution (ko-loyd′ăl) Fine particles suspended in a liquid; resistant to sedimentation or filtration.

colon (kō′lon) Division of the large intestine that extends from the cecum to the rectum.

colostrum (kō-los′trŭm) Thin, white fluid; the first milk secreted by the breast at the termination of pregnancy; contains less fat and lactose than the milk secreted later.

columnar Shaped like a column.

commissure (kom′i-shŭr) [L., *commissura,* a joining together] Connection of nerve fibers between the cerebral hemispheres or from one side of the spinal cord to the other.

common bile duct Duct formed by the union of the common hepatic and cystic ducts; it empties into the small intestine.

common hepatic duct Part of the biliary duct system formed by the joining of the right and left hepatic ducts.

compact bone Bone that is denser and has fewer spaces than cancellous bone.

competition Similar molecules binding to the same carrier molecule or receptor site.

complement (kom′plĕ-ment) Group of serum proteins that stimulates phagocytosis and inflammation.

complement cascade Series of reactions in which each component activates the next component, resulting in activation of complement proteins.

compliance (kom-plī′ans) Change in volume (e.g., in lungs or blood vessels) caused by a given change in pressure.

compound (kom′pownd) Substance composed of two or more different types of atoms that are chemically combined.

concha; pl. **conchae** (kon′kă, kon′kē) [L., shell] Structure comparable to a shell in shape—for example, the three bony ridges on the lateral wall of the nasal cavity.

conduction (kon-dŭk′shŭn) [L., *con* + *ductus,* to lead, conduct] Transfer of energy, such as heat, from one point to another without evident movement in the conducting body.

cone (kōn) Photoreceptor in the retina of the eye; responsible for color vision.

congenital (kon-jen′i-tăl) [L., *congenitus,* born with] Occurring at birth; may be genetic or due to some influence (e.g., drugs) during development.

conjunctiva (kon-jŭnk-tī′vă) [L., *conjungo,* to bind together] Mucous membrane covering the anterior surface of the eyeball and lining the lids.

conjunctival fornix (kon-jŭnk-tī′văl fōr′niks) Area in which the palpebral and bulbar conjunctiva meet.

constant region Portion of an antibody that does not combine with an antigen and is the same in different antibodies.

continuous capillary [L., *capillaris,* relating to hair] Capillary in which pores are absent; is less permeable to large molecules than are other types of capillaries.

contraction phase (kon-trak′shŭn) One of the three phases of muscle contraction; the time during which tension is produced by the contraction of muscle.

convection (kon-vek′shŭn) [L., *con* + *vectus,* to carry or bring together] Transfer of heat in liquids or gases by movement of the heated particles.

coracoid (kōr′ă-koyd) [Gr., *korakodes,* crow's beak] Resembling a crow's beak—for example, a process on the scapula.

Cori cycle Named for Czech-U.S. biochemist and Nobel laureate Carl F. Cori (1896–1984). Lactic acid, produced by skeletal muscle, is carried in the blood to the liver, where it is aerobically converted into glucose. The glucose may return through the blood to skeletal muscle or may be stored as glycogen in the liver.

cornea (kōr′nē-ă) Transparent portion of the fibrous tunic that makes up the outer wall of the anterior portion of the eye.

corniculate cartilage (kōr-nik′ū-lāt) Conical nodule of elastic cartilage surmounting the apex of each arytenoid cartilage.

corona radiata Single layer of columnar cells derived from the cumulus mass, which anchor on the zona pellucida of the oocyte in a secondary follicle.

coronary (kōr′o-nār-ē) [L., *coronarius,* a crown] Resembling a crown; encircling.

coronary artery One of two arteries that arise from the base of the aorta and carry blood to the muscle of the heart.

coronary ligament Peritoneal reflection from the liver to the diaphragm at the margins of the bare area of the liver.

coronary sinus Short trunk that receives most of the veins of the heart and empties into the right atrium.

coronoid (kōr′ŏ-noyd) [Gr., *korone,* a crow] Shaped like a crow's beak—for example, a process on the mandible.

corpus; pl. **corpora** (kōr′pŭs, -pōr-ă) [L., *body*] Any body or mass; the main part of an organ.

corpus albicans (al′bi-kanz) Atrophied corpus luteum, leaving a connective tissue scar in the ovary.

corpus callosum (kăl-lō′sŭm) [L., *body* + *callous*] Largest commissure of the brain, connecting the cerebral hemispheres.

corpus cavernosum; pl. **corpora cavernosa** One of two parallel columns of erectile tissue forming the dorsal part of the body of the penis or the body of the clitoris.

corpus luteum (lū′tē-ŭm) Yellow endocrine body formed in the ovary in the site of a ruptured vesicular follicle immediately after ovulation; secretes progesterone and estrogen.

corpus luteum of pregnancy Large corpus luteum in the ovary of a pregnant female; secretes large amounts of progesterone and estrogen.

corpus spongiosum (spŭn′jē-ō′sŭm) Median column of erectile tissue located between and ventral to the two corpora cavernosa in the penis; posteriorly it forms the bulb of the penis, and anteriorly it terminates as the glans penis; it is traversed by the urethra. In the female, it forms the bulb of the vestibule.

corpus striatum (strī-ā′tŭm) [L., *corpus,* body + *striatus,* striated or furrowed] Caudate nucleus, putamen, and globus pallidus; so-named because of the striations caused by intermixing of gray and white matter, which result from the number of tracts crossing the anterior portion of the corpus striatum.

cortex; pl. **cortices** (kōr′teks, kōr′ti-sēz) [L., bark] Outer portion of an organ (e.g., adrenal cortex or cortex of the kidney).

corticotropin-releasing hormone (kōr′ti-kō-trō′pin) Hormone from the hypothalamus that stimulates the anterior pituitary gland to release adrenocorticotropic hormone.

cortisol (kōr′ti-sol) Steroid hormone released by the zona fasciculata of the adrenal cortex; increases blood glucose and inhibits inflammation.

covalent bond (kō-vāl′ent) Chemical bond characterized by the sharing of electrons.

coxal bone (kok′să) Hipbone.

cranial nerve (krā′nē-ăl) Nerve that originates from a nucleus within the brain; there are 12 pairs of cranial nerves.

cranial vault Eight skull bones that surround and protect the brain; braincase.

craniosacral division (krā′nē-ō-sā′krăl) Parasympathetic division of the autonomic nervous system.

cranium (krā′nē-ŭm) [Gr., *kranion,* skull] Skull; in a more limited sense, the braincase.

cremaster muscle (krē-mas′ter) Extension of abdominal muscles originating from the internal oblique muscles; in the male, raises the testicles; in the female, envelops the round ligament of the uterus.

crenation (krē-nā′shŭn) [L., *crena,* notched] Denoting the outline of a shrunken cell.

cricoid cartilage (krī′koyd) Most inferior laryngeal cartilage.

cricothyrotomy (krī′kō-thī-rot′ō-mē) Incision through the skin and cricothyroid membrane for relief of respiratory obstruction.

crista, cristae (kris′tă, kris′tē) [L., crest] Shelflike infolding of the inner membrane of a mitochondrion.

crista ampullaris (kris′tă am-pul′ăr-ĭs) [L., crest] Elevation on the inner surface of the ampulla of each semicircular duct for dynamic or kinetic equilibrium.

critical closing pressure Pressure in a blood vessel below which the vessel collapses, occluding the lumen and preventing blood flow.

crown Part of a tooth that is covered with enamel.

cruciate (kroo′shē-āt) [L., *cruciatus,* cross] Resembling or shaped like a cross.

crus of the penis (krūs) Posterior portion of the corpus cavernosum of the penis attached to the ischiopubic ramus.

crypt (kript) Pitlike depression or tubular recess.

cryptorchidism (krip-tōr′ki-dizm) Failure of the testis to descend.

crystalline (kris′tă-lēn) Protein that fills the epithelial cells of the lens in the eye.

cuboidal Resembling a cube.

cumulus oophorus (ō-of′ōr-ŭs) [L., a heap] Mass of epithelial cells surrounding the oocyte; also called cumulus mass.

cuneiform cartilage (kū′nē-i-fōrm) Small rod of elastic cartilage above each corniculate cartilage in the larynx.

cupula; pl. **cupulae** (koo′poo-lă, kū′pū-lă, koo′poo-lē) [L., *cupa,* tub] Gelatinous mass that overlies the hair cells of the cristae ampullares of the semicircular ducts.

cuticle (kū′ti-kl) [L., *cutis,* skin] Outer, thin layer, usually horny—for example, the outer covering of hair or the growth of the stratum corneum onto the nail.

cystic duct (sis′tik) Duct leading from the gallbladder; joins the common hepatic duct to form the common bile duct.

cytokine (sī′tō-kīn) Protein or peptide secreted by a cell that regulates the activity of neighboring cells.

cytokinesis (sī′tō-ki-nē′sis) [Gr., *cyto,* cell + *kinsis,* movement] Division of the cytoplasm during cell division.

cytology (sī-tol′ō-jē) [Gr., *kytos,* a hollow (cell) + *logos,* study] Study of anatomy, physiology, pathology, and chemistry of the cell.

cytoplasm (sī′tō-plazm) Protoplasm of the cell surrounding the nucleus.

cytoplasmic inclusion (sī-tō-plaz′mik) Any foreign or other substance contained in the cytoplasm of a cell.

cytotoxic reaction (sī′tō-tok′sik) [Gr., *cyto,* cell + L., *toxic,* poison] Antibodies (IgG or IgM) combine with cells and activate complement, and cell lysis occurs.

cytotrophoblast (sī′tō-trof′ō-blast) Inner layer of the trophoblast, composed of individual cells.

D

Daily Values Dietary reference values useful for planing a healthy diet. The Daily Values are taken from the Reference Daily Intakes (RDIs) and the Daily Reference Values.

Daily Reference Values (DRVs) Recommended amounts in the diet for total fat, saturated fat, cholesterol, total carbohydrate, dietary fiber, sodium, potassium, and protein. The values for total fat, saturated fat, cholesterol, and sodium are the uppermost limits considered desirable because of their link to certain diseases.

Dalton's law Named for English chemist John Dalton (1766–1844). In a mixture of gases, the portion of the total pressure resulting from each type of gas is determined by the percentage of the total volume represented by each gas type.

dartos muscle (dar′tōs) Layer of smooth muscle in the skin of the scrotum; contracts in response to lower temperature and relaxes in response to higher temperature; raises and lowers testes in the scrotum.

deciduous tooth (dē-sid′ū-ŭs) Tooth of the first set of teeth; also called primary tooth.

decussate (dē′kŭ-sāt, dē-kŭs′āt) [L., *decusso,* X-shaped, from *decussis,* ten (X)] To cross.

deep inguinal ring (ing′gwi-năl) Opening in the transverse fascia through which the spermatic cord (or round ligament in the female) enters the inguinal canal.

defecation (def-ĕ-kā′shŭn) [L., *defaeco,* to remove the dregs, purify] Discharge of feces from the rectum.

defecation reflex Combination of local and central nervous system reflexes initiated by distension of the rectum and resulting in the movement of feces out of the lower colon.

deglutition (dē-gloo-tish′ŭn) [L., *de* + *glutio,* to swallow] Act of swallowing.

dendrite (den′drīt) [Gr., *dendrites,* tree] Branching processes of a neuron; receives stimuli and conducts potentials toward the cell body.

dendritic cell (den-drit′ik) Large cells with long, cytoplasmic extensions that are capable of taking up and concentrating antigens, leading to the activation of B or T lymphocytes.

dendritic spine Extension of nerve cell dendrites where axons form synapses with the dendrites; also called gemmule.

dental arch (den′tăl) [L., *arcus,* bow] Curved maxillary or mandibular arch in which the teeth are located.

dentin (den′tin) Bony material forming the mass of the tooth.

deoxyhemoglobin (dē-oks′ē-hē-mō-glō′bin) Hemoglobin without oxygen bound to it.

deoxyribonuclease (de-oks′ē-rī-bō-noo′klē-ās) Enzyme that splits DNA into its component nucleotides.

deoxyribonucleic acid (DNA) (dē-oks′ē-rī′bō-noo-klē′ic) Type of nucleic acid containing deoxyribose as the sugar component, found principally in the nuclei of cells; constitutes the genetic material of cells.

depolarization (dē-pō′lăr-i-zā′shŭn) Change in the electric charge difference across the plasma membrane that causes the difference to be smaller or closer to 0 mV; phase of the action potential in which the membrane potential moves toward zero, or becomes positive.

depression (dē-presh′ŭn) Movement of a structure in an inferior direction.

depth perception (per-sep′shun) Ability to distinguish between near and far objects and to judge their distance.

dermatome (der′mă-tōm) Area of skin supplied by a spinal nerve.

dermis (der′mis) [Gr., *derma,* skin] Dense irregular connective tissue that forms the deep layer of the skin.

descending aorta Part of the aorta, further divided into the thoracic aorta and abdominal aorta.

descending colon Part of the colon extending from the left colonic flexure to the sigmoid colon.

desmosome (dez′mō-sōm) [Gr., *desmos,* a band + *soma,* body] Point of adhesion between cells. Each contains a dense plate at the point of adhesion and a cementing extracellular material between the cells.

desquamate (des′kwă-māt) [L., *desquamo,* to scale off] Peeling or scaling off of the superficial cells of the stratum corneum.

diabetes insipidus (dī-ă-bē′tēz in-sip′ĭ-dŭs) Chronic excretion of large amounts of urine of low specific gravity accompanied by extreme thirst; results from inadequate output of antidiuretic hormone.

diabetes mellitus (me-lī′tŭs) Metabolic disease in which carbohydrate use is reduced and that of lipid and protein enhanced; caused by a deficiency of insulin or an inability to respond to insulin and is characterized, in more severe cases, by hyperglycemia, glycosuria, water and electrolyte loss, ketoacidosis, and coma.

diapedesis (dī′ă-pĕ-dē′sis) [Gr., *dia,* through + *pedesis,* a leaping] Passage of blood or any of its formed elements through the intact walls of blood vessels.

diaphragm (dī′ă-fram) Musculomembranous partition between the abdominal and thoracic cavities.

diaphysis (dī-af′i-sis) [Gr., growing between] Shaft of a long bone.

diastole (dī-as′tō-lē) [Gr., *diastole,* dilation] Relaxation of the heart chambers, during which they fill with blood; usually refers to ventricular relaxation.

diencephalon (dī-en-sef′ă-lon) [Gr., *dia,* through + *enkephalos,* brain] Second portion of the embryonic brain; in the inferior core of the adult cerebrum.

diffuse lymphatic tissue Dispersed lymphocytes and other cells with no clear boundary; found beneath mucous membranes, around lymph nodules, and within lymph nodes and spleen.

diffusion (di-fū′zhŭn) [L., *diffundo,* to pour in different directions] Tendency for solute molecules to move from an area of high concentration to an area of low concentration in solution; the product of the constant random motion of all atoms, molecules, or ions in a solution.

diffusion coefficient Measure of how easily a gas diffuses through a liquid or tissue.

digestive tract (di-jes′tiv, dī-jes′tiv) Mouth, oropharynx, esophagus, stomach, small intestine, and large intestine.

digit (dij′it) Finger, thumb, or toe.

dilator pupillae (dī′lā-tĕr pū-pil′ē) Radial smooth muscle cells of the iris diaphragm that cause the pupil of the eye to dilate.

diploid (2*n*) number (dip′loyd) Normal number of chromosomes (in humans, 46 chromosomes) in somatic cells.

disaccharide (dī-sak′ă-rīd) Condensation product of two monosaccharides by the elimination of water.

dissociate (di-sō′sē-āt) [L., *dis* + *socio,* to disjoin, separate] Ionization in which ions are dissolved in water and the cations and anions are surrounded by water molecules.

distal convoluted tubule Convoluted tubule of the nephron that extends from the ascending limb of the loop of Henle and ends in a collecting duct.

distributing artery Medium-sized artery with a tunica media composed principally of smooth muscle; regulates blood flow to different regions of the body.

dominant (dom′i-nant) [L., *dominus,* a master] Gene that is expressed phenotypically to the exclusion of a contrasting recessive gene.

dorsal root (dōr′săl) Sensory (afferent) root of a spinal nerve.

dorsal root ganglion (gang′glē-on) Collection of sensory neuron cell bodies within the dorsal root of a spinal nerve; also called spinal ganglion.

ductus arteriosus (dŭk′tŭs ar-tēr′ē-ō-sŭs) Fetal vessel connecting the left pulmonary artery with the descending aorta.

ductus deferens (def′er-enz) Duct of the testicle, running from the epididymis to the ejaculatory duct; also called vas deferens.

duodenal gland (doo′ō-dē′năl, doo-od′ĕ-năl) Small gland that opens into the base of intestinal glands; secretes a mucoid alkaline substance.

duodenocolic reflex (doo-ō-dē′nō-kō-lik) Local reflex resulting in a mass movement of the contents of the colon; produced by stimuli in the duodenum.

duodenum (doo-ō-dē′nŭm, doo-od′ĕ-nŭm) [L., *duodeni*] First division of the small intestine; connects to the stomach.

dura mater (doo′ră mā′ter) [L., hard mother] Tough, fibrous membrane forming the outer covering of the brain and spinal cord.

E

eardrum (ēr′drŭm) Cellular membrane that separates the external from the middle ear; vibrates in response to sound waves; also called tympanic membrane.

ectoderm (ek′tō-derm) Outermost of the three germ layers of an embryo.

ectopic focus; pl. **foci** (ek-top′ik fōkŭs, fō′sī) Any pacemaker other than the sinus node of the heart; abnormal pacemaker; an ectopic pacemaker.

edema (e-dē′mă) [Gr., *oidema,* a swelling] Excessive accumulation of fluid within or around calls, usually causing swelling.

effector T cell (ē-fek′tŏr, ē-fek′tōr) Subset of T lymphocytes that is responsible for cell-mediated immunity.

efferent arteriole (ef′er-ent ar-tēr′ē-ōl) Vessel that carries blood from the glomerulus to the peritubular capillaries.

efferent division Nerve fibers that send impulses from the central nervous system to the periphery.

efferent ductule (ef′er-ent dŭk′tool) [L., *ductus,* duct] One of a number of small ducts leading from the testis to the head of the epididymis.

ejaculation (ē-jak-ū-lā′shŭn) Reflexive expulsion of semen from the penis.

ejaculatory duct (ē-jak′ū-lă-tōr-ē) Duct formed by the union of the ductus deferens and the excretory duct of the seminal vesicle; opens into the prostatic urethra.

ejection period (ē-jek′shŭn) Time in the cardiac cycle when the semilunar valves are open and blood is being ejected from the ventricles into the arterial system.

elastin (ĕ-las′tin) Yellow, elastic, fibrous mucoprotein that is the major connective tissue protein of elastic structures (e.g., large blood vessels and elastic ligaments).

electrocardiogram (ECG, EKG) (ē-lek-trō-kar′dē-ō-gram) [Gr., *elektron,* amber + *kardia,* heart + *gramma,* a drawing] Graphic record of the heart's electric currents obtained with an electrocardiograph.

electrolyte (ē-lek′trō-līt) [Gr., *electro* + *lytos,* soluble] Cation or anion in solution that conducts an electric current.

electron (ē-lek′tron) Negatively charged subatomic particle in an atom.

electron-transport chain Series of electron carriers in the inner mitochondrial membrane; they receive electrons from NADH and $FADH_2$, using the electrons in the formation of ATP and water.

element (el′ĕ-ment) [L., *elementum,* rudiment, beginning] Substance composed of atoms of only one kind.

elevation (el-ĕ-vā′shŭn) Movement of a structure in a superior direction.

embolism (em′bō-lizm) [Gr., *embolisma,* a piece of patch, literally something thrust in] Obstruction or occlusion of a vessel by a transported clot, a mass of bacteria, or other foreign material.

embolus; pl. **emboli** (em′bō-lŭs, em′bō-lī) [Gr., *embolos,* plug, wedge, or stopper] Plug, composed of a detached clot, a mass of bacteria, or another foreign body, occluding a blood vessel.

embryo (em′brē-ō) Developing human from the first to the eighth week of development.

embryonic disk (em-brē-on′ik) Point in the inner cell mass at which the embryo begins to be formed.

embryonic period From approximately the second to the eighth week of development, during which the major organ systems are organized.

emission (ē-mish′ŭn) [L., *emissio,* to send out] Discharge; accumulation of semen in the urethra prior to ejaculation. A nocturnal emission is a discharge of semen while asleep.

emmetropia (em-ĕ-trō′pē-ă) [Gr., *emmetros,* according to measure + *ops,* eye] In the eye, the state of refraction in which parallel rays are focused exactly on the retina; no accommodation is necessary.

emulsify (ē-mŭl′si-fī) To form an emulsion.

enamel (ē-nam′ĕl) Hard substance covering the exposed portion of the tooth.

endocardium; pl. **endocardia** (en-dō-kar′dē-ŭm, en-dō-kar′dē-ă) Innermost layer of the heart, including endothelium and connective tissue.

endocrine gland (en′dō-krin) [Gr., *endon,* inside + *krino,* to separate] Ductless gland that secretes a hormone internally, usually into the circulation.

endocytosis (en′dō-sī-tō′sis) Bulk uptake of material through the cell membrane.

endoderm (en′dō-derm) Innermost of the three germ layers of an embryo.

endolymph (en′dō-limf) [Gr., *endo* + L., *lympha,* clear fluid] Fluid found within the membranous labyrinth of the inner ear.

endometrium; pl. **endometria** (en′dō-mē′trē-ŭm, en′dō-mē′trē-ă) Mucous membrane composing the inner layer of the uterine wall; consists of a simple columnar epithelium and a lamina propria that contains simple tubular uterine glands.

endomysium (en′dō-miz′ē-ŭm, en′dō-mis′ē-ŭm) [Gr., *endo,* within + *mys,* muscle] Fine connective tissue sheath surrounding a muscle fiber.

endoneurium (en-dō-noo′rē-ŭm) [Gr., *endo,* within + *neuron,* nerve] Delicate connective tissue surrounding individual nerve fibers within a peripheral nerve.

endoplasmic reticulum; pl. **reticula** (en′dō-plas′mik re-tik′ū-lŭm, re-tik′ū-lă) Double-walled membranous network inside the cytoplasm; rough has ribosomes attached to the surface; smooth does not have ribosomes attached.

endorphin (en-dōr′fin) Opiate-like polypeptide found in the brain and other parts of the body; binds in the brain to the same receptors that bind exogenous opiates.

endosteum (en-dos′tē-ŭm) [Gr., *endo,* within + *osteon,* bone] Membranous lining of the medullary cavity and the cavities of spongy bone.

endothelium; pl. **endothelia** (en-dō-thē′lē-ŭm, en-dō-thē′lē-ă) [Gr., *endo,* within + *thele,* nipple] Layer of flat cells lining blood and lymphatic vessels and the chambers of the heart.

enkephalin (en-kef′ă-lin) Pentapeptide found in the brain; binds to specific receptor sites, some of which may be pain-related opiate receptors.

enteric nervous system Complex network of neuron cell bodies and axons within the wall of the digestive tract; capable of controlling the digestive tract independently of the central nervous system through local reflexes.

enterokinase (en′tĕr-ō-kī′nās) Intestinal proteolytic enzyme that converts trypsinogen into trypsin.

enzyme (en′zīm) [Gr., *en,* in + *zyme,* leaven] Protein that acts as a catalyst.

eosinophil (ē-ō-sin′ō-fil) [Gr., *eos,* dawn + *philos,* fond] White blood cell that stains with acidic dyes; inhibits inflammation.

epicardium (ep-i-kar′dē-ŭm) [Gr., *epi,* on + *kardia,* heart] Serous membrane covering the surface of the heart; also called visceral pericardium.

epidermis (ep-i-derm′is) [Gr., *epi,* on + *derma,* skin] Outer portion of the skin formed of epithelial tissue that rests on or covers the dermis.

epididymis; pl. **epididymides** (ep-i-did′i-mis, -di-dim′i-dēz) [Gr., *epi,* on + *didymos,* twin] Elongated structure connected to the posterior surface of the testis, which consists of the head, body, and tail; site of storage and maturation of the spermatozoa.

epiglottis (ep-i-glot′is) [Gr., *epi,* on + *glottis,* mouth of the windpipe] Plate of elastic cartilage covered with mucous membrane; serves as a valve over the glottis of the larynx during swallowing.

epimysium (ep-i-mis′ē-ŭm) [Gr., *epi,* on + *mys,* muscle] Fibrous envelope surrounding a skeletal muscle.

epinephrine (ep′i-nef′rin) Hormone (amino acid derivative) similar in structure to the neurotransmitter norepinephrine; major hormone released from the adrenal medulla; increases cardiac output and blood glucose levels; also called adrenaline.

epineurium (ep-i-noo′rē-ŭm) [Gr., *epi,* on + *neuron,* nerve] Connective tissue sheath surrounding a nerve.

epiphyseal line (ep-i-fiz′ē-ăl) Dense plate of bone in a bone that is no longer growing, indicating the former site of the epiphyseal plate.

epiphyseal plate Site at which bone growth in length occurs; located between the epiphysis and diaphysis of a long bone; area of hyaline cartilage where cartilage growth is followed by endochondral ossification; also called metaphysis or growth plate.

epiphysis; pl. **epiphyses** (e-pif′i-sis, e-pif′i-sēz) [Gr., *epi,* on + *physis,* growth] Portion of a bone developed from a secondary ossification center and separated from the remainder of the bone by the epiphyseal plate.

epiploic appendage (ep′i-plō′ik) One of a number of little processes of peritoneum projecting from the serous coat of the large intestine except the rectum; they are generally distended with fat.

epithelium (ep-i-thē′lē-ŭm) [Gr., *epi,* on + *thele,* nipple] One of the four primary tissue types. *Nipple* refers to the tiny capillary-containing connective tissue in the lips, which is where the term was first used. The use of the term was later expanded to include all covering and lining surfaces of the body.

epitope (ep′i-tōp) [Gr., *epi,* on + *top,* place] *See* antigenic determinant.

eponychium (ep-ō-nik′ē-ŭm) [Gr., *epi,* on + *onyx,* nail] Outgrowth of the skin that covers the proximal and lateral borders of the nail; also called cuticle.

erection (ē-rek′shŭn) [L., *erectio,* to set up] Condition of erectile tissue when filled with blood; tissue becomes hard and unyielding; especially refers to this state of the penis.

erythroblastosis fetalis (ē-rith′rō-blas-tō′sis fē-tă′lis) [*erythroblast* + *osis,* condition] Destruction of erythrocytes in the fetus or newborn caused by antibodies produced in an Rh-negative mother acting on the Rh-positive blood of the fetus or newborn.

erythrocyte (ĕ-rith′rō-sīt) [Gr., *erythros,* red + *kytos,* cell] Red blood cell; biconcave disk containing hemoglobin.

erythropoiesis (ĕ-rith′rō-poy-ē′sis) [*erythrocyte* + Gr., *poiesis,* a making] Production of erythrocytes.

erythropoietin (ĕ-rith′rō-poy′ĕ-tin) Protein that enhances erythropoiesis by stimulating the formation of proerythroblasts and the release of reticulocytes from bone marrow.

esophagus; pl. **esophagi** (ē-sof′ă-gŭs, ē-sof′ă-gī, ē-sof′ă-jī) [Gr., *oisophagos,* gullet] Portion of the digestive tract between the pharynx and stomach.

essential amino acid Amino acid, required by animals, that must be supplied in the diet.

estrogen (es′trō-jen) Substance that exerts biologic effects characteristic of estrogen hormone, such as stimulating female secondary sexual characteristics, growth, and maturation of long bones, and help controlling the menstrual cycle.

eustachian tube (ū-stā′shŭn, ū-stā′kē-ăn) Named for Italian anatomist Bartolommeo Eustachio (1524–1574). Auditory canal; extends from the middle ear to the nasopharynx.

evagination (ē-vaj-i-nā′shŭn) [L., *e,* out + *vagina,* sheath] Protrusion of some part or organ from its normal position.

evaporation (ē-vap-ŏ-ra′shŭn) [L., *e,* out + *vaporare,* to emit vapor] Change from liquid to vapor form.

eversion (ē-ver′zhŭn) [L., *everto,* to overturn] Turning outward.

excitation–contraction coupling (ek-sī-tā′shŭn kon-trak′shŭn kŭp′ling) Stimulation of a muscle fiber produces an action potential that results in contraction of the muscle fiber.

excitatory postsynaptic potential (EPSP) (ek-sī′tă-tō-rē pōst-si-nap′tik pō-ten′shăl) Depolarization in the postsynaptic membrane that brings the membrane potential close to threshold.

exocrine gland (ek′sō-krin) [Gr., *exo,* outside + *krino,* to separate] Gland that secretes to a surface or outward through a duct.

exocytosis (ek′sō-sī-to′sis) Elimination of material from a cell through the formation of vacuoles.

expiratory reserve volume Maximum volume of air that can be expelled from the lungs after a normal expiration.

extension (eks-ten′shŭn) [L., *extensio,* to stretch out] To stretch out.

external acoustic meatus (mē-ā′tŭs) Short canal that opens to the exterior environment and terminates at the eardrum; part of the external ear.

external anal sphincter Ring of striated muscular fibers surrounding the anus.

external ear Portion of the ear that includes the auricle and external acoustic meatus; terminates at the eardrum.

external nose Nostril; anterior or external opening of the nasal cavity.

external spermatic fascia Outer fascial covering of the spermatic cord.

external urethral orifice Slitlike opening of the urethra in the glans penis.

external urinary sphincter Sphincter skeletal muscle around the base of the urethra external to the internal urinary sphincter.

exteroceptor (eks′ter-ō-sep′ter, eks′ter-ō-sep′tōr) [L., *exterus,* external + *receptor,* receiver] Sensory receptor in the skin or mucous membranes that responds to stimulation by external agents or forces.

extracellular (eks-tră-sel′ū-lăr) Outside the cell.

extracellular matrix; pl. **matrices** (eks-tră-sel′ū-lăr mā′triks, mā′tri-sēz) Nonliving chemical substances located between connective tissue cells.

extrinsic clotting pathway (eks-trin′sik) Series of chemical reactions resulting in clot formation; begins with chemicals (e.g., tissue thromboplastin) found outside the blood.

extrinsic muscle Muscle located outside the structure being moved.

eyebrow Short hairs on the bony ridge above the eyes.

eyelash Hair at the margins of the eyelids.

eyelid Movable fold of skin in front of the eyeball; also called palpebra.

F

F actin (ak′tin) Fibrous actin molecule that is composed of a series of globular actin molecules (G actin).

facilitated diffusion Carrier-mediated process that does not require ATP and moves substances into or out of cells from a high to a low concentration.

falciform ligament (fal′si-fōrm lig′ă-ment) Fold of peritoneum extending to the surface of the liver from the diaphragm and anterior abdominal wall.

fallopian tube (fa-lō′pē-an) *See* uterine tube.

false pelvis Portion of the pelvis superior to the pelvic brim; composed of the bone on the posterior and lateral sides and by muscle on the anterior side; also called greater pelvis.

falx cerebelli (falks ser-ĕ-bel′ī) Dural fold between the two cerebellar hemispheres.

falx cerebri (falx ser′ĕ-brī) Dural fold between the two cerebral hemispheres.

far point of vision Distance from the eye where accommodation is not needed to have the image focused on the retina.

fascia; pl. **fasciae** (fash′ē-ă, fash′ē-ē) [L., band or fillet] Loose areolar connective tissue found beneath the skin (hypodermis) or dense connective tissue that encloses and separates muscles.

fasciculus (fă-sik′ū-lŭs) [L., *fascis,* bundle] Band or bundle of nerve or muscle fibers bound together by connective tissue.

fat [A.S., *faet*] Greasy, soft-solid material found in animal tissues and many plants; composed of two types of molecules: glycerol and fatty acids.

fatigue (fă-tēg′) [L., *fatigo,* to tire] Period characterized by a reduced capacity to do work.

fat-soluble vitamin Vitamin, such as A, D, E, and K, that is soluble in lipids and absorbed from the intestine along with lipids.

fauces (faw′sēz) [L., throat] Space between the cavity of the mouth and the pharynx.

female climacteric Period of life occurring in women, encompassing termination of the reproductive period and characterized by endocrine, somatic, and transitory psychologic changes and ultimately menopause; also called perimenopause.

female pronucleus Nuclear material of the ovum after the ovum has been penetrated by the spermatozoon. Each pronucleus carries the haploid number of chromosomes.

fertilization (fer′til-i-zā′shŭn) Process that begins with the penetration of the secondary oocyte by the spermatozoon and is completed with the fusion of the male and female pronuclei.

fetus (fē′tus) Developing human following the embryonic period (after 8 weeks of development).

fibrin (fī′brin) Elastic filamentous protein derived from fibrinogen by the action of thrombin, which releases peptides from fibrinogen in coagulation of the blood.

fibroblast (fī′brō-blast) [L., *fibra,* fiber + Gr., *blastos,* germ] Spindle-shaped or stellate cells that form connective tissue.

fibrocyte (fī′brō-sīt) Mature cell of fibrous connective tissue.

fibrous joint (fī′brŭs) Bones connected by fibrous tissue with no joint cavity; includes sutures, syndesmoses, and gomphoses.

fibrous layer Outer layer of the eye; composed of the sclera and the cornea.

filiform (fil′i-fōrm) Filament-shaped.

filtrate (fil′trāt) Liquid that has passed through a filter—for example, fluid that enters the nephron through the filtration membrane of the glomerulus.

filtration (fil-trā′shŭn) Movement, due to a pressure difference, of a liquid through a filter that prevents some or all of the substances in the liquid from passing through.

filtration fraction Fraction of the plasma entering the kidney that filters into Bowman's capsule. Normally, it is around 19%.

filtration membrane Membrane formed by the glomerular capillary endothelium, the basement membrane, and the podocytes of Bowman's capsule.

filtration pressure Pressure gradient that forces fluid from the glomerular capillary through the filtration membrane into Bowman's capsule; glomerular capillary pressure minus glomerular capsule pressure minus colloid osmotic pressure.

fimbria; pl. **fimbriae** (fim′brē-ă, fim′brē-ē) [L., fringe] Fringelike structure located at the ostium of the uterine tube.

first messenger *See* intercellular chemical signal.

fixator (fik-sā′ter) Muscle that stabilizes the origin of a prime mover.

flagellum; pl. **flagella** (flă-jel′ŭm, flă-jel′ă) [L., whip] Whiplike locomotory organelle of constant structural arrangement consisting of double peripheral microtubules and two single central microtubules.

flatus (flā′tŭs) [L., a blowing] Gas or air in the gastrointestinal tract that may be expelled through the anus.

flexion (flek′shŭn) [L., *flectus*] Bending.

focal point Point at which light rays cross after passing through a concave lens, such as the lens of the eye.

foliate (fō′lē-āt) Leaf-shaped.

follicle-stimulating hormone (FSH) (fol′i-kl) Hormone of the adenohypophysis that, in females, stimulates the Graafian follicles of the ovary and assists in follicular maturation and the secretion of estrogen; in males, FSH stimulates the epithelium of the seminiferous tubules and is partially responsible for inducing spermatogenesis.

follicular phase (fō-lik′ū-lăr) Time between the end of menses and ovulation, characterized by rapid division of endometrial cells and development of follicles in the ovary; also called proliferative phase.

foramen; pl. **foramina** (fō-rā′men, fō-ram′i-nă) Hole.

foramen ovale (o-val′ē) In the fetal heart, the oval opening in the septum secundum; the persistent part of septum primum acts as a valve for this interatrial communication during fetal life; postnatally, the septum primum becomes fused to the septum secundum to close the foramen ovale, forming the fossa ovale.

force That which produces a motion in the body; pull.

foregut Cephalic portion of the primitive digestive tube in the embryo.

foreskin *See* prepuce.

formed elements Cells (i.e., red and white blood cells) and cell fragments (i.e., platelets) of blood.

formula unit Relative number of cations and ions in an ionic compound.

fornix (fōr′niks) [L., arch, vault] Recess at the cervical end of the vagina. Recess deep to each eyelid where the palpebral and bulbar conjunctivae meet.

fovea centralis (fō′vē-ă) Depression in the middle of the macula where there are only cones and no blood vessels.

free energy Total amount of energy that can be liberated by the complete catabolism of food.

frenulum (fren′ū-lŭm) [L., *frenum,* bridle] Fold extending from the floor of the mouth to the midline of the undersurface of the tongue.

frequency-modulated signals Signals, all of which are identical in amplitude, that differ in their frequency—for example, strong stimuli may initiate a high frequency of action potentials and weak stimuli may initiate a low frequency of action potentials.

FSH surge Increase in plasma follicle-stimulating hormone (FSH) levels before ovulation.

fulcrum (ful′krŭm) Pivot point.

fundus (fŭn′dŭs) [L., bottom] Bottom, or rounded end, of a hollow organ—for example, the fundus of the stomach or uterus.

fungiform (fŭn′ji-fōrm) Mushroom-shaped.

G

G actin (jē ak′tin) Globular protein molecules that, when bound together, form fibrous actin (F actin).

gallbladder (gawl′blad-er) Pear-shaped receptacle on the inferior surface of the liver; serves as a storage reservoir for bile.

gamete (gam′ēt) Ovum or spermatozoon.

gamma globulin (gam′ă glob′ū-lin) [L., *globulus,* globule] Plasma proteins that include the antibodies.

ganglion; pl. **ganglia** (gang′glē-on, gang′glē-ă) [Gr., swelling, or knot] Any group of nerve cell bodies in the peripheral nervous system.

gap junction Small channel between cells that allows the passage of ions and small molecules between cells; provides means of intercellular communication.

gastric gland (gas′trik) Gland located in the mucosa of the fundus and body of the stomach.

gastric inhibitory polypeptide Hormone secreted by the duodenum that inhibits gastric acid secretion.

gastric pit Small pit in the mucous membrane of the stomach, at the bottom of which are the mouths of the gastric glands that secrete mucus, hydrochloric acid, intrinsic factor, pepsinogen, and hormones.

gastrin (gas′trin) Hormone secreted in the mucosa of the stomach and duodenum that stimulates the secretion of hydrochloric acid by the parietal cells of the gastric glands.

gastrocolic reflex (gas′trō-kol′ik) Local reflex resulting in mass movement of the contents of the colon, which occurs after the entrance of food into the stomach.

gastroesophageal opening (gas′trō-ē-sof′ā-jē′āl) Opening of the esophagus into the stomach; also called cardiac opening.

gene (jēn) [Gr., *genos,* birth, descent] Functional unit of heredity. Each gene occupies a specific place, or locus, on a chromosome; it is capable of reproducing itself exactly at each cell division; it often is capable of directing the formation of an enzyme or another protein.

genetics (jĕ-net′iks) [Gr., *genesis,* origin or production] Branch of science that deals with heredity.

genital fold (jen′i-tăl) Paired longitudinal ridges developing in the embryo on each side of the urogenital orifice. In the male, they form part of the penis; in the female, they form the labia minora.

genital tubercle (jen′i-tăl) Median elevation just cephalic to the urogenital orifice of an embryo; gives rise to the penis of the male or the clitoris of the female.

genotype (jen′ō-tīp) [Gr., *genos,* birth, descent + *typos,* type] Genetic makeup of an individual.

germ cell (jerm) Spermatozoon or ovum.

germ layer One of three layers in the embryo (ectoderm, endoderm, or mesoderm) from which the four primary tissue types arise.

germinal center Lighter-staining center of a lymphatic nodule; area of rapid lymphocyte division.

gingiva (jin′ji-vă) Dense fibrous tissue, covered by mucous membrane, that covers the alveolar processes of the upper and lower jaws and surrounds the necks of the teeth.

girdle (ger′dl) Belt or zone; the bony region where the limbs attach to the body.

gland [L., *glans,* acorn] Secretory organ from which secretions may be released into the blood, into a cavity, or onto a surface.

glans penis [L., *glans,* acorn] Conical expansion of the corpus spongiosum that forms the head of the penis.

globin (glō′bin) Protein portion of hemoglobin.

glomerular capillary pressure (glō-măr′ū-lăr) Blood pressure within the glomerulus.

glomerular filtration rate (GFR) Amount of plasma (filtrate) that filters into Bowman's capsules per minute.

glomerulus (glō-măr′ū-lŭs) [L., *glomus,* ball of yarn] Mass of capillary loops at the beginning of each nephron, nearly surrounded by Bowman's capsule.

glottis (glot′is) [Gr., aperture of the larynx] Vocal apparatus; includes vocal folds and the cleft between them.

glucocorticoid (gloo-kō-kōr′ti-koyd) Steroid hormone (e.g., cortisol) released by zonula fasciculata of the adrenal cortex; increases blood glucose and inhibits inflammation.

gluconeogenesis (gloo′kō-nē-ō-jen′ē-sis) [Gr., *glykys,* sweet + *neos,* new + *genesis,* production] Formation of glucose from noncarbohydrates, such as proteins (amino acids) or lipids (glycerol).

glycogenesis (glī′kō-jĕ-nō′-sis) Formation of glycogen from glucose molecules.

glycolysis (glī-kol′i-sis) [Gr., *glykys,* sweet + *lysis,* a loosening] Anaerobic process during which glucose is converted to pyruvic acid; net of two ATP molecules is produced during glycolysis.

goblet cell Mucus-producing epithelial cell that has its apical end distended with mucin.

Golgi apparatus (gol′jē) Named for Camillo Golgi, Italian histologist and Nobel laureate (1843–1926). Specialized endoplasmic reticulum that concentrates and packages materials for secretion from the cell.

Golgi tendon organ Proprioceptive nerve ending in a tendon.

gomphosis (gom-fō′sis) [Gr., *gomphos,* bolt, nail + *osis,* condition] Fibrous joint in which a peglike process fits into a hole.

gonad (gō′nad) [Gr., *gone,* seed] Organ that produces sex cells; testis of a male or ovary of a female.

gonadal ridge (gō-nad′āl) Elevation on the embryonic mesonephros; primordial germ cells become embedded in it, establishing it as a testis or an ovary.

gonadotropin (gō′nad-ō-trō′pin) Hormone capable of promoting gonadal growth and function. Two major gonadotropins are luteinizing hormone (LH) and follicle-stimulating hormone (FSH).

gonadotropin-releasing hormone (GnRH) Hypothalamic-releasing hormone that stimulates the secretion of gonadotropins (LH and FSH) from the adenohypophysis; also called luteinizing hormone–releasing hormone (LHRH).

granulocyte (gran′ū-lō-sīt) Mature granular white blood cell (neutrophil, basophil, or eosinophil).

granulosa cell (gran-ū-lō′să) Cell in the layer surrounding the primary follicle.

gray matter Collection of nerve cell bodies, their dendritic processes, and associated neuroglial cells within the central nervous system.

gray ramus communicans; pl. **rami communicantes** (rā′mŭs kō-mū′nī-kans, rā′mī kō-mū-nī-kan′tēz) Connection between a spinal nerve and a sympathetic chain ganglion through which unmyelinated postganglionic axons project.

greater omentum Peritoneal fold passing from the greater curvature of the stomach to the transverse colon, hanging like an apron in front of the intestines.

greater vestibular gland One of two mucus-secreting glands on each side of the lower part of the vagina. The equivalent of the bulbourethral glands in the male.

growth hormone Hormone that stimulates general growth of the individual; stimulates cellular amino acid uptake and protein synthesis; also called somatotropin.

gubernaculum (goo′ber-nak′ū-lŭm) [L., helm] Column of tissue that connects the fetal testis to the developing scrotum; involved in testicular descent.

gustatory (gŭs′tă-tōr-ē) Associated with the sense of taste.

gustatory hair Microvillus of gustatory cell in a taste bud.

gynecomastia (gī′nē-kō-mas′tē-ă) [Gr., *gyne,* woman + *mastos,* breast] Excessive development of the male mammary glands, which sometimes secrete milk.

H

H zone Area in the center of the A band in which there are no actin myofilaments; contains only myosin.

hair [A.S., hear] Columns of dead keratinized epithelial cells.

hair follicle Invagination of the epidermis into the dermis; contains the root of the hair and receives the ducts of sebaceous and apocrine glands.

Haldane effect Named for Scottish physiologist John S. Haldane (1860–1936). Hemoglobin that is not bound to carbon dioxide binds more readily to oxygen than hemoglobin that is bound to carbon dioxide.

half-life Time it takes for one-half of an administered substance to be lost through biologic processes.

haploid (*n*) number (hap′loyd) One set of chromosomes, in contrast to diploid; characteristic of gametes.

hapten (hap′ten) [Gr., *hapto,* to fasten] Small molecule that binds to a large molecule; together they stimulate the specific immune system.

hard palate Floor of the nasal cavity that separates the nasal cavity from the oral cavity; composed of the palatine processes of the maxillary bones and the horizontal plates of the palatine bones; also called bony palate.

haustra (haw′stră) [L., machine for drawing water] Sacs of the colon, caused by contraction of the taeniae coli, which are slightly shorter than the gut so that the latter is thrown into pouches.

haversian canal (ha-ver′shan) Named for seventeenth-century English anatomist Clopton Havers (1650–1702). Canal containing blood vessels, nerves, and loose connective tissue and running parallel to the long axis of the bone.

heart skeleton Fibrous connective tissue that provides a point of attachment for cardiac muscle cells, electrically insulates the atria from the ventricles, and forms the fibrous rings around the valves.

heat energy Energy that results from the random movement of atoms, ions, or molecules; the greater the amount of heat energy in an object, the higher the object's temperature.

helicotrema (hel′i-kō-trē′mă) [Gr., *helix,* spiral + *traema,* hole] Opening at the apex of the cochlea through which the scala vestibuli and the scala tympani of the cochlea connect.

helper T cell Subset of T lymphocytes that increases the activity of B cells and T cells.

hematocrit (hē′mă-tō-krit) [Gr., *hemato,* blood + *krin,* to separate] Percentage of blood volume occupied by erythrocytes.

hematoma (he-ma-to′ma) Localized mass of blood released from blood vessels but confined within an organ or a space; the blood is usually clotted.

heme (hēm) Oxygen-carrying, color-furnishing part of hemoglobin.

hemidesmosome (hem-ē-des′mō-sōm) Similar to half a desmosome, attaching epithelial cells to the basement membrane.

hemocytoblast (he′mo-si′to-blast) Blood stem cell derived from mesenchyme that can give rise to red and white blood cells and platelets.

hemoglobin (hē-mō-glō′bin) Red, respiratory protein of red blood cells; consists of 6% heme and 94% globin; transports oxygen and carbon dioxide.

hemolysis (hē-mol′i-sis) [Gr., *haima* + *lysis,* destruction] Destruction of red blood cells in such a manner that hemoglobin is released.

hemopoiesis (hē′mō-poy-ē′sis) [Gr., *haima,* blood + *poiesis,* a making] Formation of the formed elements of blood—that is, red blood cells, white blood cells, and platelets; also called hematopoiesis.

hemopoietic tissue (hē′mō-poy-et′ik) [Gr., *haima,* blood + *poiesis,* to make] Blood-forming tissue.

hemostasis (hē′mō-stā-sis) Arrest of bleeding.

Henry's law Named for English chemist William Henry (1775–1837). The concentration of a gas dissolved in a liquid is equal to the partial pressure of the gas over the liquid times the solubility coefficient of the gas.

heparin (hep′ă-rin) Anticoagulant that prevents platelet agglutination and thus prevents thrombus formation.

hepatic artery (he-pa′tik) Branch of the aorta that delivers blood to the liver.

hepatic cord Plate of liver cells that radiates away from the central vein of a liver lobule.

hepatic portal system System of portal veins that carries blood from the intestines, stomach, spleen, and pancreas to the liver.

hepatic portal vein Portal vein formed by the superior mesenteric and splenic veins and entering the liver.

hepatic sinusoid (si′nŭ-soyd) Terminal blood vessel having an irregular and larger caliber than an ordinary capillary within the liver lobule.

hepatic vein Vein that drains the liver into the inferior vena cava.

hepatocyte (hep′ă-tō-sīt) Liver cell.

hepatopancreatic ampulla Dilation within the major duodenal papilla that normally receives both the common bile duct and the main pancreatic duct.

hepatopancreatic ampullar sphincter Smooth muscle sphincter of the hepatopancreatic ampulla; also called sphincter of Oddi.

Hering-Breuer reflex (her′ing broy′er) Named for German physiologist Heinrich Ewald Hering (1866–1948) and Austrian internist Josef Breuer (1842–1925). Sensory impulses from stretch receptors in the lungs arrest inspiration; expiration then occurs.

heterozygous (het′er-ō-zī′gŭs) [Gr., *heteros,* other + *zygon,* yoke] Having different allelic genes at one or more paired loci in homologous chromosomes.

hiatus (hī-ā′tŭs) [L., aperture, to yawn] Opening.

hilum (hī′lŭm) [L., small bit or trifle] Indented surface on many organs, serving as a point where nerves and vessels enter or leave.

hindgut Caudal or terminal part of the embryonic gut.

histamine (his′tă-mēn) Amine released by mast cells and basophils that promotes inflammation.

histology (his-tol′ō-jē) [Gr., *histo,* web (tissue) + *logos,* study] Science that deals with the microscopic structure of cells, tissues, and organs in relation to their function.

holocrine gland (hol′ō-krin) [Gr., *holos,* complete + *krino,* to separate] Gland whose secretion is formed by the disintegration of entire cells (e.g., sebaceous gland).

homeostasis (hō′mē-ō-stā′sis) [Gr., *homoio,* like + *stasis,* a standing] State of equilibrium in the body with respect to functions and composition of fluids and tissues.

homologous (hŏ-mol′ō-gŭs) [Gr., ratio or relation] Alike in structure or origin.

homozygous (hō-mō-zī′gŭs) [Gr., *homos,* the same + *zygon,* yoke] State of having identical allelic genes at one or more paired loci in homologous chromosomes.

hormone (hōr′mōn) [Gr., *hormon,* to set into motion] Substance secreted by endocrine tissues into the blood that acts on a target tissue to produce a specific response.

hormone receptor Protein or glycoprotein molecule of cells that specifically binds to hormones and produces a response.

horn Subdivision of gray matter in the spinal cord. The axons of sensory neurons synapse with neurons in the posterior horn, the cell bodies of motor neurons are in the anterior horn, and the cell bodies of autonomic neurons are in the lateral horn.

human chorionic gonadotropin (HCG) Hormone produced by the placenta; stimulates the secretion of testosterone by the fetus; during the first trimester, it stimulates ovarian secretion from the corpus luteum of the estrogen and progesterone required for the maintenance of the placenta. In a male fetus, it stimulates the secretion of testosterone by the fetal testis.

humoral immunity (hū'mōr-ăl) [L., *humor,* a fluid] Immunity due to antibodies in serum.

hyaline cartilage (hī'ă-lin) [Gr., *hyalos,* glass] Gelatinous, glossy cartilage tissue consisting of cartilage cells and their matrix; contains collagen, proteoglycans, and water.

hyaluronic acid (hī'ă-loo-ron'ik) Mucopolysaccharide made up of alternating β-(1,4)-linked residues of hyalobiuronic acid, forming a gelatinous material in the tissue spaces and acting as a lubricant and shock absorbant generally throughout the body.

hydrochloric acid (HCl) (hī-drō-klōr'ik) Acid of gastric juice.

hydrogen bond (hī'drō-jen) Hydrogen atoms bound covalently to either N or O atoms have a small positive charge that is weakly attracted to the small negative charge of other atoms, such as O or N; it can occur within a molecule or between different molecules.

hydrophilic (hi-dro-fil'ik) [Gr., *hydro,* water + *philos,* love] Denoting the property of attracting or associating with water molecules, possessed by polar molecules and ions; the opposite of hydrophobic.

hydrophobic (hi-dro-fob'ik) [Gr., *hydro,* water + *phobos,* fear] Lacking an attraction to water, possessed by nonpolar molecules; the opposite of hydrophilic.

hydroxyapatite (hī-drok'sē-ap'ă-tīt) Mineral with the empiric formula 3 $Ca_3(PO_4)_2 \cdot Ca(OH)_2$; the main mineral of bone and teeth.

hymen (hī'men) [Gr., membrane] Thin, membranous fold partly occluding the vaginal external orifice; normally disrupted by sexual intercourse or other mechanical phenomena.

hyoid bone (hī'oyd)(Gr., *hyoeides,* shaped like the Greek letter epsilon [ϵ]) U-shaped bone between the mandible and larynx.

hypercalcemia (hī'per-kal-sē'mē-ā) Abnormally high levels of calcium in the blood.

hypercapnia (hī'per-kap'nē-ā) Higher than normal levels of carbon dioxide in the blood or tissues.

hyperkalemia (hī'per-kă-lē'mē-ā) Greater than normal concentration of potassium ions in the circulating blood.

hypernatremia (hī'per-nă-trē'mē-ă) Abnormally high plasma concentration of sodium ions.

hyperosmotic (hī'per-oz-mot'ĭk) [Gr., *hyper,* above + *osmos,* an impulsion] Having a greater osmotic concentration or pressure than a reference solution.

hyperpolarization (hī'per-pō'lăr-i-zā'shŭn) Increase in the charge difference across the plasma membrane; causes the charge difference to move away from 0 mV.

hypertonic (hī-per-ton'ik) [Gr., *hyper,* above + *tonos,* tension] Solution that causes cells to shrink.

hypertrophy (hī-per'trō-fē) [Gr., *hyper,* above + *trophe,* nourishment] Increase in bulk or size; not due to an increase in number of individual elements.

hypocalcemia (hī-pō-kal-sē'mē-ă) Abnormally low levels of calcium in the blood.

hypocapnia (hī'pō-kap'nē-ă) Lower than normal levels of carbon dioxide in the blood or tissues.

hypodermis (hī'pō-der'mis) [Gr., *hypo,* under + *dermis,* skin] Loose areolar connective tissue found deep to the dermis that connects the skin to muscle or bone.

hypokalemia (hī'pō-ka-lē'mē-ă) Abnormally small concentration of potassium ions in the blood.

hyponatremia (hī'pō-nā-trē'mē-ă) Abnormally low plasma concentration of sodium ions.

hyponychium (hī-pō-nik'ē-ūm) [Gr., *hypo,* under + *onyx,* nail] Thickened portion of the stratum corneum under the free edge of the nail.

hypopolarization Change in the electric charge difference across the plasma membrane that causes the charge difference to be smaller or move closer to 0 mV.

hyposmotic (hī'pos-mot'ik) [Gr., *hypo,* under + *osmos,* an impulsion] Having a lower osmotic concentration or pressure than a reference solution.

hypospadias (hī'pō-spā'dē-ăs) [Gr., one having the orifice of the penis too low; *hypospao,* to draw away from under] Developmental anomaly in the wall of the urethra so that the canal is open for a greater or lesser distance on the undersurface of the penis; a similar defect in the female in which the urethra opens into the vagina.

hypothalamohypophysial portal system (hī'pō-thal'ă-mō-hī'pō-fiz'ē-ăl) Series of blood vessels that carries blood from the area of the hypothalamus to the anterior pituitary gland; originates from capillary beds in the hypothalamus and terminates as a capillary bed in the anterior pituitary gland.

hypothalamohypophyseal tract Nerve tract consisting of the axons of neurosecretory cells and extending from the hypothalamus into the posterior pituitary gland. Hormones produced in the neurosecretory cell bodies in the hypothalamus are transported through the hypothalamohypophyseal tract to the posterior pituitary gland, where they are stored for later release.

hypothalamus (hī'pō-thal'ă-mūs) [Gr., *hypo,* under + *thalamus,* bedroom] Important autonomic and neuroendocrine control center beneath the thalamus.

hypothenar (hī-pō-thē'nar) [Gr., *hypo,* under + *thenar,* palm of the hand] Fleshy mass of tissue on the medial side of the palm; contains muscles responsible for moving the little finger.

hypotonic (hī-pō-ton'ik) [Gr., *hypo,* under + *tonos,* tension] Solution that causes cells to swell.

I

I band Area between the ends of two adjacent myosin myofilaments within a myofibril; Z disk divides the I band into two equal parts.

ileocecal sphincter (il'ē-ō-sē'kăl) Thickening of circular smooth muscle between the ileum and the cecum, forming the ileocecal valve.

ileocecal valve Valve formed by the ileocecal sphincter between the ileum and the cecum.

ileum (il'ē-ŭm) [Gr., *eileo,* to roll up, twist] Third portion of the small intestine, extending from the jejunum to the ileocecal opening into the large intestine; the posterior inferior bone of the coxal bone.

immunity (i-mū'ni-tē) [L., ***immunis,*** free from service] Resistance to infectious disease and harmful substances.

immunization (im-mū'ni-zā'shun) Process by which a subject is rendered immune by deliberately introducing an antigen or antibody into the subject.

immunoglobulin (IG) (im'ū-nō-glob'ū-lin) Antibody found in the gamma globulin portion of plasma.

implantation (im-plan-tā'shŭn) Attachment of the blastocyst to the endometrium of the uterus, occurring 6 or 7 days after fertilization of the ovum.

impotence (im'pŏ-tens) Inability to accomplish the male sexual act; caused by psychic or physical factors; also called erectile dysfunction (ED).

incisor (in-sī'zŏr) [L., *incido,* to cut into] One of the anterior, cutting teeth.

incisura (in′sī-soo′ră) [L., a cutting into] Notch or indentation at the edge of any structure.

incus (ing′kus) [L., anvil] Middle of the three ossicles in the middle ear.

inferior colliculus (ko-lik′ū-lŭs) [L., *collis,* hill] One of two rounded eminences of the midbrain; involved with hearing.

inferior vena cava Vein that returns blood from the lower limbs and the greater part of the pelvic and abdominal organs to the right atrium.

inflammatory response (in-flam′ă-tōr-ē) Complex sequence of events involving chemicals and immune cells that results in the isolation and destruction of antigens and tissues near the antigens.

infundibulum (in-fŭn-dib′ū-lŭm) [L., funnel] Funnel-shaped structure or passage—for example, the infundibulum that attaches the hypophysis to the hypothalamus or the funnel-like expansion of the uterine tube near the ovary.

Infusion Introduction of a fluid other than blood, such as a saline or glucose solution, into the blood.

inguinal canal (ing′gwi-năl) Passage through the lower abdominal wall that transmits the spermatic cord in the male and the round ligament in the female.

inhibin (in-hib′in) Polypeptide secreted from the testes that inhibits FSH secretion.

inhibitory neuron (in-hib′i-tōr-ē) Neuron that produces IPSPs and has an inhibitory influence.

inhibitory postsynaptic potential (IPSP) Hyperpolarization in the postsynaptic membrane, which causes the membrane potential to move away from threshold.

innate immunity (i′nāt, i-nāt′) Immune system response that is the same with each exposure to an antigen; there is no ability for the system to remember a previous exposure to the antigen.

inner cell mass Group of cells at one end of the blastocyst, part of which forms the body of the embryo.

inner ear Part of the ear that contains the sensory organs for hearing and balance; contains the bony and membranous labyrinth.

insensible perspiration [L., *per,* through + *spiro,* to breathe everywhere] Perspiration that evaporates before it is perceived as moisture on the skin; the term sometimes includes evaporation from the lungs.

insertion (in-ser′shŭn) More movable attachment point of a muscle; usually, the lateral or distal end of a muscle associated with the limbs; also called mobile end.

inspiratory capacity (in-spī′ră-tō-rē) Volume of air that can be inspired after a normal expiration; the sum of the tidal volume and the inspiratory reserve volume.

inspiratory reserve volume Maximum volume of air that can be inspired after a normal inspiration.

insulin (in′sŭ-lin) Protein hormone secreted from the pancreas that increases the uptake of glucose and amino acids by most tissues.

interatrial septum (in-ter-ā′trē-ăl) [L., *saeptum,* partition] Wall between the atria of the heart.

intercalated disk (in-ter′kă-lā-ted) Cell-to-cell attachment with gap junctions between cardiac muscle cells.

intercalated duct Minute duct of glands, such as the salivary gland, and the pancreas; leads from the acini to the interlobular ducts.

intercellular (in-ter-sel′ū-lăr) Between cells.

intercellular chemical signal Chemical that is released from cells and passes to other cells; acts as a signal that allows cells to communicate with each other; also called first messenger.

interferons (in-ter-fēr′onz) Proteins that prevent viral replication.

interlobar artery (in-ter-lō′bar) Branch of the segmental arteries of the kidney; runs between the renal pyramids and gives rise to the arcuate arteries.

interlobular artery (in-ter-lob′ū-lăr) Artery that passes between lobules of an organ; branches of the interlobar arteries of the kidney pass outward through the cortex from the arcuate arteries and supply the afferent arterioles.

interlobular duct Any duct leading from a lobule of a gland and formed by the junction of the intercalated ducts draining the acini.

interlobular vein Vein that parallels the interlobular arteries; in the kidney, it drains the peritubular capillary plexus, emptying into arcuate veins.

intermediate olfactory area Part of the olfactory cortex responsible for the modulation of olfactory sensations.

internal anal sphincter [Gr., *sphinkter,* band or lace] Smooth muscle ring at the upper end of the anal canal.

internal naris; pl. nares (nā′ris, nā′rēs) Opening from the nasal cavity into the nasopharynx.

internal spermatic fascia Inner connective tissue covering of the spermatic cord.

internal urinary sphincter Traditionally recognized as a sphincter composed of a thickening of the middle smooth muscle layer of the bladder around the urethral opening.

interphase (in′ter-fāz) Period between active cell divisions when DNA replication occurs.

interstitial (in-ter-stish′ăl) [L., *inter,* between + *sisto,* to stand] Space within tissue. Interstitial growth is growth from within.

interstitial cell Cell between the seminiferous tubules of the testes; secretes testosterone; also called Leydig cell.

interventricular septum (in-ter-ven-trik′ū-lăr) Wall between the ventricles of the heart.

intestinal gland (in-tes′ti-năl) Tubular gland in the mucous membrane of the small and large intestines; also called crypt.

intracellular (in-tră-sel′ū-lăr) Inside a cell.

intracellular mediator Molecule produced within a cell that binds to a macromolecule, such as receptors or enzymes inside that cell, that regulates their activities; also called second messenger.

intramural plexus (in′tră-mū′răl plek′sus) Combined submucosal and myenteric plexuses.

intrinsic clotting pathway (in-trin′sik) Series of chemical reactions resulting in clot formation that begins with chemicals (e.g., plasma factor XII) within the blood.

intrinsic factor Factor secreted by the parietal cells of gastric glands and required for adequate absorption of vitamin B_{12}.

intrinsic muscles Muscles located within the structure being moved.

intubation (in-too-ba′shun) Insertion of a tube into an opening, a canal, or a hollow organ.

inversion (in-ver′zhŭn) [L., *inverto,* to turn about] Turning inward.

ion (ī′on) [Gr., *ion,* going] Atom or group of atoms carrying a charge of electricity by virtue of having gained or lost one or more electrons.

ion channel Pore in the plasma membrane through which ions, such as sodium and potassium, move.

ionic bonding (ī-on′ik) Chemical bond that is formed when one atom loses an electron and another accepts that electron.

iris (ī′ris) Specialized portion of the vascular tunic; the "colored" portion of the eye that can be seen through the cornea.

ischemia (is-kē′mē-ă) [Gr., *ischo,* to keep back + *haima,* blood] Reduced blood supply to an area of the body.

ischium (is′kē-ŭm) Superior bone of the coxal bone.

isomers (ī′sō-merz) [Gr., *isos,* equal + *meros,* part] Molecules having the same number and types of atoms but differing in their three-dimensional arrangement.

isometric contraction (ī-sō-met′rik) [Gr., *isos,* equal + *metron,* measure] Muscle contraction in which the length of the muscle does not change but the tension produced increases.

isosmotic (ī′sō-os-mot′ik) [Gr., *isos,* equal + *osmos,* an impulsion] Having the same osmotic concentration or pressure as a reference solution.

isotonic (ī′sō-ton′ik) [Gr., *isos,* equal + *tonos,* tension] Type of solution that causes cells to neither shrink nor swell.

isotope (ī′sō-tōp) [Gr., *isos,* equal + *topos,* part, place] Either of two or more atoms that have the same atomic number but a different number of neutrons.

isthmus (is′mŭs) Constriction connecting two larger parts of an organ, such as the constriction between the body and the cervix of the uterus or the portion of the uterine tube between the ampulla and the uterus.

J

jaundice (jawn′dis) [Fr., *jaune,* yellow] Yellowish staining of the integument, the sclerae, and the other tissues with bile pigments.

jejunum (jĕ-joo′nŭm) [L., *jejunus,* empty] Second portion of the small intestine; located between the duodenum and the ileum.

juxtaglomerular apparatus (jŭks′tă-glŏ-mer′ū-lăr) Complex consisting of juxtaglomerular cells of the afferent arteriole and macular densa cells of the distal convoluted tubule near the renal corpuscle; secretes renin.

juxtaglomerular cell Modified smooth muscle cell of the afferent arteriole located at the renal corpuscle; a component of the juxtaglomerular apparatus.

juxtamedullary nephron (jŭks′tă-med′ŭ-lăr-ē) Nephron located near the junction of the renal cortex and medulla.

K

karyotype (kar′ē-ō-tīp) Display of chromosomes arranged by pairs.

keratinization (ker′ă-tin-i-zā′shŭn) Production of keratin and changes in the chemical and structural character of epithelial cells as they move to the skin surface.

keratinized (ker′ă-ti-nīzd) [Gr., *keras,* horn] Having become a structure that contains keratin, a protein found in skin, hair, nails, and horns.

keratinocyte (ke-rat′i-nō-sīt) [Gr., *keras,* horn + *kytos,* cell] Epidermal cell that produces keratin.

keratohyalin (ker′ă-tō-hī′ă-lin) Nonmembrane-bound protein granule in the cytoplasm of stratum granulosum cells of the epidermis.

ketogenesis (kē-tō-jen′ĕ-sis) Production of ketone bodies, such as from acetyl-CoA.

ketone body (kē′tōn) One of a group of ketones, including acetoacetic acid, β-hydrobutyric acid, and acetone.

ketosis (kē-tō′sis) [*ketone* + *osis,* condition] Condition characterized by the enhanced production of ketone bodies, as in diabetes mellitus or starvation.

kidney (kid′nē) [A.S., *cwith,* womb, belly + *neere,* kidney] One of the two organs that excrete urine. The kidneys are bean-shaped organs approximately 11 cm long, 5 cm wide, and 3 cm thick lying on each side of the spinal column, posterior to the peritoneum, approximately opposite the twelfth thoracic and first three lumbar vertebrae.

kilocalorie (Kcal) (kil′ō-kal-ō-rē) Quantity of energy required to raise the temperature of 1 kg of water 1°C; 1000 calories. Equal to one dietary calorie.

kinetic energy (ki-net′ik) Motion energy or energy that can do work.

kinetic labyrinth (lab′i-rinth) Part of the membranous labyrinth composed of the semicircular canals; detects dynamic or kinetic equilibrium, such as movement of the head.

kinetochore (ki-nē′tō-kōr, ki-net′o-) [Gr., *kinēto,* moving + Gr., *chōra,* space] Structural portion of the chromosome to which microtubules attach.

Korotkoff sounds (kō-rot′kof) Named for Russian physician Nikolai S. Korotkoff (1874–1920). Sounds heard over an artery when blood pressure is determined by the auscultatory method; caused by turbulent flow of blood.

L

labium majus; pl. **labia majora** (lā′bē-ŭm, lā′bē-ă) One of two rounded folds of skin surrounding the labia minora and vestibule; homolog of the scrotum in males.

labium minus; pl. **labia minora** One of two narrow longitudinal folds of mucous membrane enclosed by the labia majora and bounding the vestibule; anteriorly they unite to form the prepuce.

lacrimal apparatus (lak′ri-măl) Lacrimal, or tear, gland in the superolateral corner of the orbit of the eye and a duct system that extends from the eye to the nasal cavity.

lacrimal canaliculus Canal that carries excess tears away from the eye; located in the medial canthus and opening on the lacrimal papilla.

lacrimal gland Tear gland located in the superolateral corner of the orbit.

lacrimal papilla Small lump of tissue in the medial canthus or corner of the eye; the lacrimal canal opens within the lacrimal papilla.

lacrimal sac Enlargement in the lacrimal canal that leads into the nasolacrimal duct.

lactation (lak-tā′shŭn) [L., *lactatio,* suckle] Period after childbirth during which milk is formed in the breasts.

lacteal (lak′tē-ăl) Lymphatic vessel in the wall of the small intestine that carries chyle from the intestine and absorbs fat.

lactiferous duct (lak-tif′er-ŭs) One of 15–20 ducts that drain the lobes of the mammary gland and open onto the surface of the nipple.

lactiferous sinus Dilation of the lactiferous duct just before it enters the nipple.

lacuna; pl. **lacunae** (lă-koo′nă, -koo′nē) [L., *lacus,* a hollow, a lake] Small space or cavity; potential space within the matrix of bone or cartilage normally occupied by a cell that can be visualized only when the cell shrinks away from the matrix during fixation; space containing maternal blood within the placenta.

lag phase One of the three phases of muscle contraction; time between the application of the stimulus and the beginning of muscular contraction. Also called latent phase.

lamella; pl. **lamellae** (lă-mel′ă, lă-mel′ē) Thin sheet or layer of bone.

lamellated corpuscle (lam′ĕ-lāt-ed) Oval receptor found in the deep dermis or hypodermis (responsible for deep cutaneous pressure and vibration) and in tendons (responsible for proprioception); also called Pacinian corpuscle.

lamina; pl. **laminae** (lam′i-nă, lam′i-nē) [L., *lamina,* plate, leaf] Thin plate—for example, the thinner portion of the vertebral arch.

lamina propria (prō′prē-ă) Layer of connective tissue underlying the epithelium of a mucous membrane.

laminar flow (lam′i-nar) Relative motion of layers of a fluid along smooth, concentric, parallel paths.

Langerhans cell Named for German anatomist Paul Langerhans (1847–1888); dendritic cell found in the skin.

lanugo (lă-noo′gō) [L., *lana,* wool] Fine, soft, unpigmented fetal hair.

Laplace's law Named for French mathematician Pierre S. de Laplace (1749–1827); the force that stretches the wall of a blood vessel is proportional to the radius of the vessel times the blood pressure.

large intestine Portion of the digestive tract extending from the small intestine to the anus.

laryngitis (lar-in-jī′tis) Inflammation of the mucous membrane of the larynx.

laryngopharynx (lă-ring′gō-far-ingks) Part of the pharynx lying posterior to the larynx.

larynx; pl. **larynges** (lar′ingks, lă-rin′jēz) Organ of voice production located between the pharynx and the trachea; it consists of a framework of cartilages and elastic membranes housing the vocal folds and the muscles that control the position and tension of these elements.

last menstrual period (LMP) Beginning of the last menstruation before pregnancy; used clinically to time events during pregnancy.

lateral geniculate nucleus (je-nik′ū-lāt) Nucleus of the thalamus where fibers from the optic tract terminate.

lateral olfactory area (ol-fak′tŏ-rē) Part of the olfactory cortex involved in the conscious perception of olfactory stimuli.

lens Transparent biconvex structure lying between the iris and the vitreous humor.

lens fiber Epithelial cell that makes up the lens of the eye.

lesser omentum (ō-men′tŭm) [L., membrane that encloses the bowels] Peritoneal fold passing from the liver to the lesser curvature of the stomach and to the upper border of the duodenum for a distance of approximately 2 cm beyond the pylorus.

lesser vestibular gland (ves-tib′ū-lăr) Number of minute mucous glands opening on the surface of the vestibule between the openings of the vagina and urethra; also called paraurethral gland.

leukocyte (loo′kō-sīt) White blood cell.

leukocytosis (loo′kō-sī-tō′sis) Abnormally large number of white blood cells in the blood.

leukopenia (loo-kō-pē′nē-ă) Lower than normal number of white blood cells in the blood.

leukotriene (loo-kō-trī′ēn) Specific class of physiologically active fatty acid derivatives present in many tissues.

lever Rigid shaft capable of turning about a fulcrum or pivot point.

LH surge Increase in plasma luteinizing hormone (LH) levels before ovulation and responsible for initiating it.

ligamentum arteriosum (lig′ă-men′tŭm) Remains of the ductus arteriosus.

ligamentum venosum Remnant of the ductus venosus.

ligand *See* chemical signal.

ligand-gated ion channel Ion channel in a plasma membrane caused to either open or close by a ligand binding to a receptor.

limbic system (lim′bik) [L., *limbus,* border] Part of the brain involved with emotions and olfaction; includes the cingulate gyrus, hippocampus, habenular nuclei, parts of the basal ganglia, the hypothalamus (especially the mammillary bodies, the olfactory cortex, and various nerve tracts).

lingual tonsil (ling′gwăl) Collection of lymphoid tissue on the posterior portion of the dorsum of the tongue.

lipase (lip′ās) Any fat-splitting enzyme.

lipid (li′pid) [Gr., *lipos,* fat] Substance composed principally of carbon, oxygen, and hydrogen; contains a lower ratio of oxygen to carbon and is less polar than carbohydrates; generally soluble in nonpolar solvents.

lipid bilayer Double layer of lipid molecules forming the plasma membrane and other cellular membranes.

lipochrome (lip′ō-krōm) Lipid-containing pigment that is metabolically inert.

lipotropin (li-pō-trō′pin) One of the peptide hormones released from the adenohypophysis; increases lipolysis in fat cells.

liver (liv′er) Largest gland of the body, lying in the upper-right quadrant of the abdomen just inferior to the diaphragm; secretes bile and is of great importance in carbohydrate and protein metabolism and in detoxifying chemicals.

lobar bronchus (brong′kus) Branch from a primary bronchus that conducts air to each lobe of the lungs. There are two branches in the left lung and three branches from the primary bronchus in the right lung. Also called secondary bronchus.

lobe (lōb) Rounded, projecting part, such as the lobe of a lung, the liver, or a gland.

lobule (lob′ūl) Small lobe or subdivision of a lobe, such as a lobule of the lung or a gland.

local inflammation Inflammation confined to a specific area of the body. Symptoms include redness, heat, swelling, pain, and loss of function.

local potential Depolarization that is not propagated and that is graded or proportional to the strength of the stimulus.

local reflex Reflex of the intramural plexus of the digestive tract that does not involve the brain or spinal cord.

locus; pl. **loci** (lō′kŭs, lō′sī) Place; usually a specific site.

loop of Henle Named for German anatomist Friedrich G. J. Henle (1809–1885). U-shaped part of the nephron extending from the proximal to the distal convoluted tubule and consisting of descending and ascending limbs. Some of the loops of Henle extend into the renal pyramids.

lower respiratory tract Larynx, trachea, and lungs.

lung recoil Decrease in the size of an expanded lung as a result of a decrease in the size (volume) of its alveoli; due to elastic recoil of elastic fibers surrounding alveoli and water surface tension of a thin film of water within alveoli.

lunula; pl. **lunulae** (loo′noo-lă, loo′noo-lē) [L., *luna,* moon] White, crescent-shaped portion of the nail matrix visible through the proximal end of the nail.

luteal phase (loo′tē-ăl) Portion of the menstrual cycle extending from the time of formation of the corpus luteum after ovulation to the time when menstrual flow begins; usually 14 days in length; also called secretory phase.

luteinizing hormone (LH) (loo′tē-ĭ-nīz-ing) In females, hormone stimulating the final maturation of the follicles and the secretion of progesterone by them, with their rupture releasing the ovum, and the conversion of the ruptured follicle into the corpus luteum; in males, stimulates the secretion of testosterone in the testes.

lymph (limf) [L., *lympha,* clear spring water] Clear or yellowish fluid derived from interstitial fluid and found in lymph vessels.

lymph node Encapsulated mass of lymph tissue found among lymph vessels.

lymphatic capillary Beginning of the lymphatic system of vessels; lined with flattened endothelium lacking a basement membrane.

lymphatic nodule Small accumulation of lymph tissue lacking a distinct boundary.

lymphatic sinus Channels in a lymph node crossed by a reticulum of cells and fibers.

lymphatic vessel One of the system of vessels carrying lymph from the lymph capillaries to the veins.

lymphedema (limf′e-dē′mă) Swelling of tissues resulting from the excessive accumulation of fluid caused by the removal, damage, or blockage of lymphatic vessels or lymph nodes; usually results in swelling of the arm or leg.

lymphoblast (lim′fō-blast) Cell that matures into a lymphocyte.

lymphocyte (lim′fō-sīt) Nongranulocytic white blood cell formed in lymphoid tissue.

lymphokine (lim′fō-kīn) Chemical produced by lymphocytes that activates macrophages, attracts neutrophils, and promotes inflammation.

lysis (lī′sis) [Gr., *lysis,* a loosening] Process by which a cell swells and ruptures.

lysosome (lī′sō-sōm) [Gr., *lysis,* loosening + *soma,* body] Membrane-bounded vesicle containing hydrolytic enzymes that function as intracellular digestive enzymes.

lysozyme (lī′sō-zīm) Enzyme that is destructive to the cell walls of certain bacteria; present in tears and some other fluids of the body.

M

M line Line in the center of the H zone made of delicate filaments that holds the myosin myofilaments in place in the sarcomere of muscle fibers.

macrophage (mak′rō-fāj) [Gr., *makros,* large + *phagein,* to eat] Any large, mononuclear phagocytic cell.

macula; pl. **maculae** (mak′ū-lă, mak′ū-lē) [L., a spot] Sensory structure in the utricle and saccule, consisting of hair cells and a gelatinous mass embedded with otoliths.

macula densa Cells of the distal convoluted tubule located at the renal corpuscle and forming part of the juxtaglomerular apparatus.

main bronchus; pl. **bronchi** (brong′kŭs, brong′kī) One of two tubes arising at the inferior end of the trachea; each primary bronchus extends into one of the lungs; also called primary bronchus.

major duodenal papilla Point of opening of the common bile duct and pancreatic duct into the duodenum.

major histocompatibility complex (MHC) molecules Genes that control the production of major histocompatibility complex proteins, which are glycoproteins found on the surfaces of cells. The major histocompatibility proteins serve as self-markers for the immune system and are used by antigen-presenting cells to present antigens to lymphocytes.

male pronucleus Nuclear material of the sperm cell after the ovum has been penetrated by the sperm cell.

malignant (mă-lig′nănt) Resistant to treatment; occurring in severe form and frequently fatal; having locally invasive and destructive growth and metastasis.

malleus; pl. **mallei** (mal′ē-ŭs, mal′ē-ī) [L., hammer] Largest of the three auditory ossicles; attached to the tympanic membrane.

mamillary bodies (mam′i-lār-ē) [L., breast- or nipple-shaped] Nipple-shaped structures at the base of the hypothalamus.

mamma; pl. **mammae** (mam′ă, mam′ē) Breast; the organ of milk secretion; one of two hemispheric projections of variable size situated in the subcutaneous layer over the pectoralis major muscle on each side of the chest; it is rudimentary in the male.

mammary ligaments (mam′ă-rē) Well-developed ligaments that extend from the overlying skin to the fibrous stroma of mammary gland; also called Cooper's ligaments.

manubrium; pl. **manubria** (mă-noo′brē-ŭm, mă-noo′bre-ă) [L., handle] Part of a bone representing the handle, such as the manubrium of the sternum representing the handle of a sword.

mass movement Forcible peristaltic movement of short duration, occurring only three or four times a day, which moves the contents of the large intestine.

mass number Number of protons plus the number of neutrons in each atom.

mastication (mas-ti-kā′shŭn) [L., *mastico,* to chew] Process of chewing.

mastication reflex Repetitive cycle of relaxation and contraction of the muscles of mastication.

mastoid (mas′toyd) [Gr., *mastos,* breast] Resembling a breast.

mastoid air cells Spaces within the mastoid process of the temporal bone connected to the middle ear by ducts.

mature follicle Ovarian follicle in which the oocyte attains its full size. The follicle contains a fluid-filled antrum and is surrounded by the theca interna and externa. Also called Graafian follicle.

maximal stimulus Stimulus resulting in a local potential just large enough to produce the maximum frequency of action potentials.

meatus (mē-ā′tŭs) [L., to go, pass] Passageway or tunnel.

mechanoreceptor (mek′ă-nō-rē-sep′tŏr) Sensory receptor that responds to mechanical pressures—for example, pressure receptors in the carotid sinus or touch receptors in the skin.

meconium (mē-kō′nē-ŭm) [Gr., *mekon,* poppy] First intestinal discharges of the newborn infant, greenish in color and consisting of epithelial cells, mucus, and bile.

medial olfactory area Part of the olfactory cortex responsible for the visceral and emotional reactions to odors.

medulla oblongata (me-dool′ă ob-long-gah′tă) Inferior portion of the brainstem that connects the spinal cord to the brain and contains autonomic centers controlling functions such as heart rate, respiration, and swallowing.

medullary cavity (med′ul-er-ē, med′oo-lār-ē) Large, marrow-filled cavity in the diaphysis of a long bone.

medullary ray Extension of the kidney medulla into the cortex, consisting of collecting ducts and loops of Henle.

megakaryoblast (meg-ă-kar′ē-ō-blast) [Gr., *mega* + *karyon,* nut, nucleus + *blastos,* germ] Cell that gives rise to platelets or thrombocytes.

meibomian cyst (mī-bō′mē-an) Named for German anatomist Hendrik Meibom (1638–1700). A chronic inflammation of a meibomian gland.

meibomian gland Sebaceous gland near the inner margins of the eyelid; secretes sebum that lubricates the eyelid and retains tears.

meiosis (mī-ō′sis) [Gr., a lessening] Cell division that results in the formation of gametes. Consists of two divisions, which result in one (female) or four (male) gametes, each of which contains one-half the number of chromosomes in the parent cell.

Meissner corpuscle (mīs′nerz kōr′pŭs-l) Named for German histologist Georg Meissner (1829–1905). *See* tactile corpuscle.

melanin (mel′ă-nin) [Gr., *melas,* black] Group of related molecules responsible for skin, hair, and eye color. Most melanins are brown to black pigments; some are yellowish or reddish.

melanocyte (mel′ă-nō-sīt) [Gr., *melas,* black + *kytos,* cell] Cell found mainly in the stratum basale; produces the brown or black pigment melanin.

melanocyte-stimulating hormone (MSH) Peptide hormone secreted by the anterior pituitary; increases melanin production by melanocytes, making the skin darker in color.

melanosome (mel′ă-nō-sōm) [Gr., *melas,* black + *soma,* body] Membranous organelle containing the pigment melanin.

melatonin (mel-ă-tōn′in) Hormone (amino acid derivative) secreted by the pineal body; inhibits the secretion of gonadotropin-releasing hormone from the hypothalamus.

membrane attack complex (MAC) Channel through a plasma membrane produced by activated complement proteins, primarily complement protein C9; in nucleated cells, water enters the channel, causing lysis of cells.

membrane-bound receptor Receptor molecule, such as a hormone receptor, that is bound to the plasma membrane of the target cell.

membranous labyrinth (mem′bră-nŭs lab′i-rinth) Membranous structure within the inner ear consisting of the cochlea, vestibule, and semicircular canals.

membranous urethra (ū-rē′thră) Portion of the male urethra, approximately 1 cm in length, extending from the prostate gland to the beginning of the penile urethra.

memory cell Small lymphocyte that is derived from a B cell or T cell and that rapidly responds to a subsequent exposure to the same antigen.

menarche (me-nar′kē) [Gr., *mensis,* month + *arche,* beginning] Establishment of menstrual function; the time of the first menstrual period or flow.

meninx; pl. **meninges** (mē′ninks, mĕ-nin′jes) [Gr., membrane] Connective tissue membrane surrounding the brain.

meniscus; pl. **menisci** (me-nis′kus, me-nis′sī) Crescent-shaped intraarticular fibrocartilage found in certain joints, such as the crescent-shaped fibrocartilaginous structure of the knee.

menopause (men′ō-pawz) [Gr., *mensis,* month + *pausis,* cessation] Permanent cessation of the menstrual cycle.

menses (men′sēz) [L., *mensis,* month] Periodic hemorrhage from the uterine mucous membrane, occurring at approximately 28-day intervals.

menstrual cycle (men′stroo-ăl) Series of changes that occur in sexually mature, nonpregnant women and result in menses. Specifically refers to the uterine cycle but is often used to include both the uterine and ovarian cycles.

Merkel (tactile) disk (mer′kelz) Named for German anatomist Friedrich Merkel (1845–1919). *See* tactile disk.

merocrine gland (mer′ō-krin) [Gr., *meros,* part + *krino,* to separate] Gland that secretes products with no loss of cellular material—for example, water-producing sweat glands.

mesencephalon (mez-en-sef′ă-lon) [Gr., *mesos,* middle + *enkephalos,* brain] Midbrain in both the embryo and adult; consists of the cerebral peduncle and the corpora quadrigemini.

mesentery (mes′en-ter-ē) [Gr., *mesos,* middle + *enteron,* intestine] Double layer of peritoneum extending from the abdominal wall to the abdominal viscera, conveying to it its vessels and nerves.

mesoderm (mez′ō-derm) Middle of the three germ layers of an embryo.

mesonephros (mez′ō-nef′ros) One of three excretory organs appearing during embryonic development; forms caudal to the pronephros as the pronephros disappears. It is well developed and is functional for a time before the establishment of the metanephros, which gives rise to the kidney. It undergoes regression as an excretory organ, but its duct system is retained in the male as the efferent ductule and epididymis.

mesosalpinx (mez′ō-sal′pinks) [Gr., *mesos,* middle + *salpinx,* trumpet] Part of the broad ligament supporting the uterine tube.

mesothelium (mez-ō-thē′lē-ŭm) Single layer of flattened cells forming an epithelium that lines serous cavities, such as peritoneum, pleura, pericardium.

mesovarium (mez′ō-vā′rē-ŭm) Short peritoneal fold connecting the ovary with the broad ligament of the uterus.

messenger ribonucleic acid (mRNA) Type of RNA that moves out of the nucleus and into the cytoplasm, where it is used as a template to determine the structure of proteins.

metabolism (mĕ-tab′ō-lizm) [Gr., *metabole,* change] Sum of all the chemical reactions that take place in the body, consisting of anabolism and catabolism. *Cellular metabolism* refers specifically to the chemical reactions within cells.

metacarpal (met′ă-kar′păl) Relating to the fine bones of the hand between the carpus (wrist) and the phalanges.

metanephros (met-ă-nef′ros) Most caudally located of the three excretory organs appearing during embryonic development; becomes the permanent kidney of mammals. In mammalian embryos, it is formed caudal to the mesonephros and develops later as the mesonephros undergoes regression.

metaphase (met′ă-fās) Time during cell division when the chromosomes line up along the equator of the cell.

metarteriole (met′ar-tēr′ē-ōl) One of the small peripheral blood vessels that contain scattered groups of smooth muscle fibers in their walls; located between the arterioles and the true capillaries.

metastasis (mĕ-tas′tă-sis) Shifting of a disease or its local manifestations or the spread of a disease from one part of the body to another, as in a malignant neoplasm.

metatarsal (met′ă-tar′sal) [Gr., *meta,* after + *tarsos,* sole of the foot] Distal bone of the foot.

metencephalon (met′en-sef′ă-lon) [Gr., *meta,* after + *enkephalos,* brain] Second most posterior division of the embryonic brain; becomes the pons and cerebellum in the adult.

micelle (mi-sel′, mī-sel′) [L., *micella,* small morsel] Droplets of lipid surrounded by bile salts in the small intestine.

microfilament (mī-krō-fil′ă-ment) Small fibril forming bundles, sheets, or networks in the cytoplasm of cells; provides structure to the cytoplasm and mechanical support for microvilli and stereocilia.

microglia (mī-krog′lē-ă) [Gr., *micro* + *glia,* glue] Small neuroglial cells that become phagocytic and mobile in response to inflammation; considered to be macrophages within the central nervous system.

microtubule (mī-krō-too′būl) Hollow tube composed of tubulin, measuring approximately 25 nm in diameter and usually several micrometers long. Helps provide support to the cytoplasm of the cell and is a component of certain cell organelles, such as centrioles, spindle fibers, cilia, and flagella.

microvillus; pl. **microvilli** (mī′krō-vil′ŭs, mī′-krō-vil′ī) Minute projection of the cell membrane that greatly increases the surface area.

micturition reflex (mik-choo-rish′ŭn) Contraction of the urinary bladder stimulated by stretching of the bladder wall; results in emptying of the bladder.

middle ear Air-filled space within the temporal bone; contains auditory ossicles; between the external and internal ear.

milk letdown Expulsion of milk from the alveoli of the mammary glands; stimulated by oxytocin.

mineral Inorganic nutrient necessary for normal metabolic functions.

mineralocorticoid (min′er-al-ō-kōr′ti-koyd) Steroid hormone (e.g., aldosterone) produced by the zona glomerulosa of the adrenal cortex; facilitates exchange of potassium for sodium in the distal renal tubule, causing sodium reabsorption and potassium and hydrogen ion secretion.

minor duodenal papilla Site of the opening of the accessory pancreatic duct into the duodenum.

minute ventilation Product of tidal volume times the respiratory rate.

mitochondrion; pl. **mitochondria** (mī-tō-kon′drē-on, mī-tō-kon′drē-ă) [Gr., *mitos,* thread + *chandros,* granule] Small, spherical, rod-shaped or thin filamentous structure in the cytoplasm of cells that is a site of ATP production.

mitosis (mī-tō′sis) [Gr., thread] Cell division resulting in two daughter cells with exactly the same number and type of chromosomes as the mother cell.

modiolus (mō-dī′ō′lŭs) [L., nave of a wheel] Central core of spongy bone about which turns the spiral canal of the cochlea.

molar (mō′lăr) Tricuspid tooth; the three posterior teeth of each dental arch.

molecule (mol′ĕ-kūl) Substance composed of two or more atoms chemically combined to form a structure that behaves as an independent unit.

monoblast (mon′ō-blast) Cell that matures into a monocyte.

monocyte (mon′o-sīt) Type of white blood cell; large phagocytic white blood cell that moves from the blood into tissues and becomes a macrophage.

mononuclear phagocytic system (mon-ō-noo′klē-ăr fag-ō-sit′ik) Phagocytic cells, each with a single nucleus; derived from monocytes; also called reticuloendothelial system.

monosaccharide (mon-ō-sak′ă-rīd) Simple sugar carbohydrate that cannot form any simpler sugar by hydrolysis.

mons pubis (monz pū′bis) [L., mountain; the genitals] Prominence caused by a pad of fatty tissue over the symphysis pubis in the female.

morula (mōr′oo-lă, mōr′ū-lă) [L., *morus,* mulberry] Mass of 12 or more cells resulting from the early cleavage divisions of the zygote.

motor neuron Neuron that innervates skeletal, smooth, or cardiac muscle fibers.

motor unit Single neuron and the muscle fibers it innervates.

mucosa (mū-kō′să) [L., *mucosus* mucous] Mucous membrane consisting of epithelium and lamina propria. In the digestive tract, there is also a layer of smooth muscle.

mucous membrane (mū′kŭs) Thin sheet consisting of epithelium and connective tissue (lamina propria) that lines cavities that open to the outside of the body; many contain mucous glands that secrete mucus.

mucous neck cell One of the mucous-secreting cells in the neck of a gastric gland.

mucus (mū′kŭs) Viscous secretion produced by and covering mucous membranes; lubricates mucous membranes and traps foreign substances.

multiple motor unit summation Increased force of contraction of a muscle due to recruitment of motor units.

multiple-wave summation Increased force of contraction of a muscle due to increased frequency of stimulation.

multipolar neuron One of three categories of neurons consisting of a neuron cell body, an axon, and two or more dendrites.

muscarinic receptor (mŭs′kă-rin′ik) Class of cholinergic receptor that is specifically activated by muscarine in addition to acetylcholine.

muscle fiber Muscle cell.

muscle spindle Three to 10 specialized muscle fibers supplied by gamma motor neurons and wrapped in sensory nerve endings; detects stretch of the muscle and is involved in maintaining muscle tone.

muscle tone Relatively constant tension produced by a muscle for long periods as a result of asynchronous contraction of motor units.

muscle twitch Contraction of a whole muscle in response to a stimulus that causes an action potential in one or more muscle fibers.

muscular fatigue Fatigue due to a depletion of ATP within the muscle fibers.

muscularis (mŭs-kū-lā′ris) [Modern L., muscular] Muscular coat of a hollow organ or tubular structure.

muscularis mucosa Thin layer of smooth muscle found in most parts of the digestive tube; located outside the lamina propria and adjacent to the submucosa.

musculi pectinati (mŭs-kū-lī pek′tĭ-nă′tē) Prominent ridges of atrial myocardium located on the inner surface of much of the right atrium and both auricles.

mutation (mū-tā′shŭn) Change in the number or kinds of nucleotides in the DNA of a gene.

myelencephalon (mī′el-en-sef′ă-lon) [Gr., *myelos,* medulla, marrow + *enkephalos* brain] Most caudal portion of the embryonic brain; also called medulla oblongata.

myelin sheath (mī′ĕ-lin) Envelope surrounding most axons; formed by Schwann cell membranes being wrapped around the axon.

myelinated axon (mī′ĕ-li-nāt-ed ak′son) Nerve fiber having a myelin sheath.

myeloblast (mī′ĕ-lō-blast) Immature cell from which the different granulocytes develop.

myenteric plexus (mī′en-ter′ik) Plexus of unmelinated fibers and postganglionic autonomic cell bodies lying in the muscular coat of the esophagus, stomach, and intestines; communicates with the submucosal plexuses.

myoblast (mī′ō-blast) [Gr., *mys,* muscle + *blastos,* germ] Primitive, multinucleated cell with the potential to develop into a muscle fiber.

myofilament (mī-ō-fil′ă-ment) Extremely fine molecular thread helping form the myofibrils of muscle; thick myofilaments are formed of myosin, and thin myofilaments are formed of actin.

myometrium (mī′ō-mē′trē-ŭm) Muscular wall of the uterus; composed of smooth muscle; also called muscular layer.

myosin myofilament (mī′ō-sin mī-ō-fil′ă-ment) Thick myofilament of muscle fibrils; composed of myosin molecules.

N

nail (nāl) [A.S., naegel] Several layers of dead epithelial cells containing hard keratin on the ends of the digits.

nail bed Epithelial tissue resting on dermis under the nail between the nail matrix and hyponychium; contributes to the formation of the nail.

nail matrix Epithelial tissue resting on dermis under the proximal end of a nail; produces most of the nail.

nasal cavity (nā′zăl) Cavity between the external nares and the pharynx. It is divided into two chambers by the nasal septum and is bounded inferiorly by the hard and soft palates.

nasal septum Bony partition that separates the nasal cavity into left and right parts; composed of the vomer, the perpendicular plate of the ethmoid, and hyaline cartilage.

nasolacrimal duct (nā-zō-lak′ri-măl) Duct that leads from the lacrimal sac to the nasal cavity.

nasopharynx (nā-zō-far′ingks) Part of the pharynx that lies above the soft palate; anteriorly it opens into the nasal cavity.

near point of vision Closest point from the eye at which an object can be held without appearing blurred.

neck Slightly constricted part of a tooth, between the crown and the root.

neoplasm (nē′ō-plazm) Abnormal tissue that grows by cellular proliferation more rapidly than normal and continues to grow after the stimuli that initiated the new growth ceases.

nephron (nef′ron) [Gr., *nephros,* kidney] Functional unit of the kidney, consisting of the renal corpuscle, the proximal convoluted tubule, the loop of Henle, and the distal convoluted tubule.

nerve tract Bundles of parallel axons with their associated sheaths in the central nervous system.

neural crest (noor′ăl) Edge of the neural plate as it rises to meet at the midline to form the neural tube.

neural crest cells Cells derived from the crests of the forming neural tube in the embryo; together with the mesoderm, they form the mesenchyme of the embryo; they give rise to part of the skull, the teeth, melanocytes, sensory neurons, and autonomic neurons.

neural layer Portion of the retina containing rods and cones.

neural plate Region of the dorsal surface of the embryo that is transformed into the neural tube and neural crest.

neural tube Tube formed from the neuroectoderm by the closure of the neural groove; develops into the spinal cord and brain.

neuroectoderm (noor-ō-ek′tō-derm) Part of the ectoderm of an embryo giving rise to the brain and spinal cord.

neurohormone (noor′ō-hōr′mōn) Hormone secreted by a neuron.

neuromodulator Substance that influences the sensitivity of neurons to neurotransmitters but neither strongly stimulates nor strongly inhibits neurons by itself.

neuromuscular junction (noor-ō-mŭs′kū-lăr) Specialized synapse between a motor neuron and a muscle fiber.

neuron (noor′on) [Gr., nerve] Morphologic and functional unit of the nervous system, consisting of the nerve cell body, the dendrites, and the axon; also called nerve cell.

neuron cell body Enlarged portion of the neuron containing the nucleus and other organelles; also called nerve cell body.

neurotransmitter (noor′ō-trans-mit′er) [Gr., *neuro,* nerve + L., *transmitto,* to send across] Any specific chemical agent released by a presynaptic cell on excitation that crosses the synaptic cleft and stimulates or inhibits the postsynaptic cell.

neutral solution (noo′trăl) Solution, such as pure water, that has 10^{-7} mol of hydrogen ions per liter and an equal concentration of hydroxide ions; has a pH of 7.

neutron (noo′tron) [L., *neuter,* neither] Electrically neutral particle in the nuclei of atoms (except hydrogen).

neutrophil (noo′trō-fil) [L., *neuter,* neither + Gr., *philos,* fond] Type of white blood cell; small phagocytic white blood cell with a lobed nucleus and small granules in the cytoplasm.

nicotinic receptor (nik-ō-tin′ik) Class of cholinergic receptor molecule that is specifically activated by nicotine and by acetylcholine.

nipple (nip′l) Projection at the apex of the mamma, on the surface of which the lactiferous ducts open; surrounded by a circular pigmented area, the areola.

Nissl substance (nis′l) Named after German neurologist Franz Nissl (1860–1919). Areas in the neuron cell body containing rough endoplasmic reticulum.

nociceptor (nō-si-sep′ter) [L., *noceo,* to injure + *capio,* to take] Sensory receptor that detects painful or injurious stimuli; also called pain receptor.

nonelectrolyte (non-ē-lek′trō-līt) [Gr., *electro* + *lytos,* soluble] Molecules that do not dissociate and do not conduct electricity.

norepinephrine (nōr′ep-i-nef′rin) Neurotransmitter substance released from most of the postganglionic neurons of the sympathetic division; hormone released from the adrenal cortex that increases cardiac output and blood glucose levels; also called noradrenaline.

nose, or **nasus** (nōz, nā′sŭs) Visible structure that forms a prominent feature of the face; nasal cavities.

notochord (nō′tō-kōrd) [Gr., *notor,* back + *chords,* cord] Small rod of tissue lying ventral to the neural tube. A characteristic of all vertebrates, in humans it becomes the nucleus pulposus of the intervertebral disks.

nuchal (noo'kăl) Back of the neck.

nuclear envelope (noo'klē-er) Double membrane structure surrounding and enclosing the nucleus.

nuclear pores Porelike openings in the nuclear envelope where the inner and outer membranes fuse.

nucleic acid (noo-klē'ik, noo-klā'ik) Polymer of nucleotides, consisting of DNA and RNA, forms a family of substances that comprise the genetic material of cells and control protein synthesis.

nucleolus; pl. **nucleoli** (noo-klē'ō-lŭs, noo-klē'ō-lī) Somewhat rounded, dense, well-defined nuclear body with no surrounding membrane; contains ribosomal RNA and protein.

nucleotide (noo'klē-ō-tīd) Basic building block of nucleic acids consisting of a sugar (either ribose or deoxyribose) and one of several types of organic bases.

nucleus; pl. **nuclei** (noo'klē-ŭs, noo'klē-ī) [L., inside of a thing] Cell organelle containing most of the genetic material of the cell; collection of nerve cell bodies within the central nervous system; center of an atom consisting of protons and neutrons.

nucleus pulposus (pŭl-pō'sŭs) [L., central pulp] Soft central portion of the intervertebral disk.

nutrient (noo'trē-ent) [L., *nutriens,* to nourish] Chemicals taken into the body that are used to produce energy, provide building blocks for new molecules, or function in other chemical reactions.

O

olecranon process (ō-lek'ră-non, ō'lē-krā'non) Process on the distal end of the ulna, forming the point of the elbow.

olfaction (ol-fak'shŭn) [L., *olfactus,* smell] Sense of smell.

olfactory area Extreme superior region of the nasal cavity.

olfactory bulb (ol-fak'tŏ-rē) Ganglion-like enlargement at the rostral end of the olfactory tract that lies over the cribriform plate; receives the olfactory nerves from the nasal cavity.

olfactory cortex Termination of the olfactory tract in the cerebral cortex within the lateral fissure of the cerebrum.

olfactory epithelium Epithelium of the olfactory recess containing olfactory receptors.

olfactory tract Nerve tract that projects from the olfactory bulb to the olfactory cortex.

oligodendrocyte (ol'i-gō-den'drō-sīt) Neuroglial cell that has cytoplasmic extensions that form myelin sheaths around axons in the central nervous system.

oncogene (ong'kō-jēn) Gene that can change or be activated to cause cancer.

oncology (ong-kol'ō-jē) Study of neoplasms.

oocyte (ō'ō-sīt) [Gr., *oon,* egg + *kytos,* a hollow (cell)] Immature ovum.

oogenesis (ō-ō-jen'ĕ-sis) Formation and development of a secondary oocyte or ovum.

oogonium (ō-ō-gō'nē-ŭm) [Gr., *oon,* egg + *gone,* generation] Primitive cell from which oocytes are derived by meiosis.

opposition Movement of the thumb and little finger toward each other; movement of the thumb toward any of the fingers.

opsin (op'sin) Protein portion of the rhodopsin molecule; a class of proteins that bind to retinal to form the visual pigments of the rods and cones of the eye.

opsonin (op'sŏ-nin) [Gr., *opsonein* to prepare food] Substance, such as antibody or complement, that enhances phagocytosis.

optic chiasma (op'tik kī'az'mă) [Gr., two crossing lines; *chi* the letter χ] Point of crossing of the optic tracts.

optic disc Point at which axons of ganglion cells of the retina converge to form the optic nerve, which then penetrates through the fibrous tunic of the eye.

optic nerve Nerve carrying visual signals from the eye to the optic chiasm.

optic stalk Constricted proximal portion of the optic vesicle in the embryo; develops into the optic nerve.

optic tract Tract that extends from the optic chiasma to the lateral geniculate nucleus of the thalamus.

optic vesicle One of the paired evaginations from the walls of the embryonic forebrain from which the retina develops.

oral cavity (ōr'ăl) The mouth; consists of the space surrounded by the lips, cheeks, teeth, and palate; limited posteriorly by the fauces.

orbit (ōr'bit) Eye socket; formed by seven skull bones that surround and protect the eye.

organ of Corti (ōr'găn) Named for Italian anatomist Marquis Alfonso Corti (1822–1888). Spiral organ; rests on the basilar membrane and supports the hair cells that detect sounds.

organelle (or'gă-nel) [Gr., *organon,* tool] Specialized part of a cell with one or more specific individual functions.

orgasm (ōr'gazm) [Gr., *orgao,* to swell, be excited] Climax of the sexual act, associated with a pleasurable sensation.

origin (ōr'i-jin) Less movable attachment point of a muscle; usually the medial or proximal end of a muscle associated with the limbs; also called fixed end.

oropharynx (ōr'ō-far'ingks) Portion of the pharynx that lies posterior to the oral cavity; it is continuous above with the nasopharynx and below with the laryngopharynx.

oscillating circuit Neuronal circuit arranged in a circular fashion that allows action potentials produced in the circuit to keep stimulating the neurons of the circuit.

osmolality (os-mō-lal'i-tē) Osmotic concentration of a solution; the number of moles of solute in 1 kg of water times the number of particles into which the solute dissociates.

osmoreceptor cell (os-mō-rē-sep'ter, os'mō-rē-sep'tōr) [Gr., *osmos,* impulsion] Receptor in the central nervous system that responds to changes in the osmotic pressure of the blood.

osmosis (os-mō'sis) [Gr., *osmos,* thrusting or an impulsion] Diffusion of solvent (water) through a membrane from a less concentrated solution to a more concentrated solution.

osmotic pressure (os-mot'ik) Force required to prevent the movement of water across a selectively permeable membrane.

ossification (os'i-fi-kā'shŭn) [L., *os,* bone + *facio,* to make] Bone formation; also called osteogenesis.

osteoblast (os'tē-ō-blast) [Gr., *osteon,* bone + *blastos,* germ] Bone-forming cell.

osteoclast (os'tē-ō-klast) [Gr., *osteon,* bone + *klastos,* broken] Large, multinucleated cell that absorbs bone.

osteocyte (os'tē-ō-sīt) [Gr., *osteon,* bone + *kytos,* cell] Mature bone cell surrounded by bone matrix.

osteomalacia (os'tē-ō-mă-lā'shē-ă) Softening of bones due to calcium depletion; adult rickets.

osteon (os'tē-on) Central canal containing blood capillaries and the concentric lamellae around it; occurs in compact bone; also called haversian system.

osteoporosis (os'tē-ō-pō-rō'sis) [Gr., *osteon,* bone + *poros,* pore + *osis,* condition] Reduction in quantity of bone, resulting in porous bone.

ostium (os'tē-ŭm) [L., door, entrance, mouth] Small opening—for example, the opening of the uterine tube near the ovary or the opening of the uterus into the vagina.

otolith (ō'tō-lith) Crystalline particles of calcium carbonate and protein embedded in the maculae.

oval window (ō'văl) Membranous structure to which the stapes attaches; transmits vibrations to the inner ear.

ovarian cycle (ō-var'ē-an) Series of events that occur in a regular fashion in the ovaries of sexually mature, nonpregnant females; results in ovulation and the production of the hormones estrogen and progesterone.

ovarian epithelium Peritoneal covering of the ovary; also called germinal epithelium.

ovarian ligament Bundle of fibers passing to the uterus from the ovary.

ovary (ō'vă-rē) One of two female reproductive glands located in the pelvic cavity;

produces the secondary oocyte, estrogen, and progesterone.

oviduct (ō′vi-dŭkt) *See* uterine tube.

ovulation (ov′ū-lā′shun) Release of an ovum, or secondary oocyte, from the vesicular follicle.

oxidation (ok-si-dā′shŭn) Loss of one or more electrons from a molecule.

oxidation–reduction reaction Reaction in which one molecule is oxidized and another is reduced.

oxidative deamination (ok-si-dā′tiv) Removal of the amine group of an amino acid to form a keto acid, ammonia, and NADH.

oxygen deficit (ok′sē-jen) Oxygen necessary for the synthesis of the ATP required to remove lactic acid produced by anaerobic respiration.

oxygen–hemoglobin dissociation curve Graph describing the relationship between the percentage of hemoglobin saturated with oxygen and a range of oxygen partial pressures.

oxyhemoglobin (ox′sē-hē-mō-glō′bin) Oxygenated hemoglobin.

P

P wave First complex of the electrocardiogram representing depolarization of the atria.

Pacinian corpuscle (pa-sin′ē-an) Named for Italian anatomist Filippo Pacini (1812–1883). *See* lamellated corpuscle.

palate (pal′ăt) [L., *palatum,* palate] Roof of the mouth.

palatine tonsil (pal′ă-tīn) One of two large, oval masses of lymphoid tissue embedded in the lateral wall of the oral pharynx.

palpebra; pl. **palpebrae** (pal-pē′bră, pal-pē′brē) [L., eyelid] Eyelid.

palpebral conjunctiva (pal-pē′brăl kon-jŭnk-tī′vă) Conjunctiva that covers the inner surface of the eyelids.

palpebral fissure Space between the upper and lower eyelids.

pancreas (pan′krē-as) [Gr., *pankreas,* sweetbread] Abdominal gland that secretes pancreatic juice into the intestine and insulin and glucagon from the pancreatic islets into the bloodstream.

pancreatic duct (pan-krē-at′ik) Excretory duct of the pancreas that extends through the gland from tail to head, where it empties into the duodenum at the greater duodenal papilla.

pancreatic islet Cellular mass varying from a few to hundreds of cells lying in the interstitial tissue of the pancreas; composed of different cell types that make up the endocrine portion of the pancreas and are the source of insulin and glucagon; also called islets of Langerhans.

pancreatic juice [L., *jus,* broth] External secretion of the pancreas; clear, alkaline fluid containing several enzymes.

papilla (pă-pil′ă) [L., nipple] Small, nipplelike process; projection of the dermis, containing blood vessels and nerves, into the epidermis; projections on the surface of the tongue.

papillary muscle (pap′i-lăr′ē) Nipplelike, conical projection of myocardium within the ventricle; the chordae tendineae are attached to the apex of the papillary muscle.

parafollicular cell (par-ă-fo-lik′ū-lăr) Endocrine cell in the thyroid gland; secretes the hormone calcitonin.

paramesonephric duct (par-ă-mes-ō-nef′rik) One of two embryonic tubes extending along the mesonephros and emptying into the cloaca; in the female, the duct forms the uterine tube, the uterus, and part of the vagina; in the male, it degenerates.

paranasal sinus (par-ă-nā′săl) Air-filled cavities within certain skull bones that connect to the nasal cavity; located in the frontal, maxillary, sphenoid, and ethmoid bones.

parasympathetic division (par-ă-sim-pa-thet′ik) Subdivision of the autonomic nervous system; characterized by having the cell bodies of its preganglionic neurons located in the brainstem and the sacral region of the spinal cord (craniosacral division); usually involved in activating vegetative functions, such as digestion, defecation, and urination.

parathyroid gland (par-ă-thī′royd) One of four glandular masses embedded in the posterior surface of the thyroid gland; secretes parathyroid hormone.

parathyroid hormone (PTH) Peptide hormone produced by the parathyroid gland; increases bone breakdown and blood calcium levels.

parietal (pă-rī′ĕ-tăl) [L., *paries,* wall] Relating to the wall of any cavity.

parietal cell Gastric gland cell that secretes hydrochloric acid.

parietal pericardium Serous membrane lining the fibrous portion of the pericardial sac.

parietal peritoneum Layer of peritoneum lining the abdominal walls.

parietal pleura Serous membrane that lines the different parts of the wall of the pleural cavity.

parotid gland (pă-rot′id) Largest of the salivary glands; situated anterior to each ear.

partial pressure Pressure exerted by a single gas in a mixture of gases.

passive tension Tension applied to a load by a muscle without contracting; produced when an external force stretches the muscle.

patella (pa-tel′ă) [L., *patina,* shallow disk] Kneecap.

pectoral girdle (pek′tŏ-răl) Site of attachment of the upper limb to the trunk; consists of the scapula and the clavicle; also called shoulder girdle.

pedicle (ped′ĭ-kl) [L., *pes,* feet] Stalk or base of a structure, such as the pedicle of the vertebral arch.

pelvic brim (pel′vik) Imaginary plane passing from the sacral promontory to the pubic crest.

pelvic girdle Site of attachment of the lower limb to the trunk; ring of bone formed by the sacrum and the coxal bones.

pelvic inlet Superior opening of the true pelvis.

pelvic outlet Inferior opening of the true pelvis.

pelvis; pl. **pelves** (pel′vis, pel′vēz) [L., basin] Any basin-shaped structure; cup-shaped ring of bone at the lower end of the trunk, formed from the ossa coxal bones, sacrum, and coccyx.

pennate (bipennate) (pen′āt) [L., *penna,* feather] Muscles with fasciculi arranged like the barbs of a feather along a common tendon.

pepsin (pep′sin) [Gr., *pepsis,* digestion] Principal digestive enzyme of the gastric juice, formed from pepsinogen; digests proteins into smaller peptide chains.

pepsinogen (pep-sin′ō-jen) [*pepsin* + Gr. *gen,* producing] Proenzyme formed and secreted by the chief cells of the gastric mucosa; the acidity of the gastric juice and pepsin itself converts pepsinogen into pepsin.

peptidase (pep′ti-dās) Enzyme capable of hydrolyzing one of the peptide links of a peptide.

peptide bond (pep′tīd) Chemical bond between amino acids.

Percent Daily Value (% Daily Value) Percent of the recommended daily value of a nutrient found in one serving of a particular food.

perforating canal Canal containing blood vessels and nerves and running through bone perpendicular to the haversian canals; also called Volkmann's canal.

periarterial sheath (per′ē-ar-tē′rē-ăl) Dense accumulation of lymphocytes (white pulp) surrounding arteries within the spleen.

pericapillary cell One of the slender connective tissue cells in close relationship to the outside of the capillary wall; it is relatively undifferentiated and may become a fibroblast, macrophage, or smooth muscle cell.

pericardial cavity (per-i-kar′dē-ăl) Space within the mediastinum in which the heart is located.

pericardial fluid Viscous fluid contained within the pericardial cavity between the visceral and parietal pericardium; functions as a lubricant.

pericardium (per-i-kar′dē-ŭm) [Gr., *pericardion,* membrane around the heart] Membrane covering the heart; also called pericardial sac.

perichondrium (per-i-kon′drē-ŭm) [Gr., *peri,* around + *chondros,* cartilage] Double-layered connective tissue sheath surrounding cartilage.

perilymph (per′i-limf) [Gr., *peri,* around + L., *lympha,* clear fluid (lymph)] Fluid contained within the bony labyrinth of the inner ear.

perimetrium (per-i-mē′trē-ŭm) Outer serous coat of the uterus; also called serous layer.

perimysium (per-i-mis′ē-ŭm, per-i-miz′ē-ŭm) [Gr., *peri,* around + *mys,* muscle] Fibrous sheath enveloping a bundle of skeletal muscle fibers (muscle fascicle).

perineum (per′i-nē′ŭm) Area inferior to the pelvic diaphragm between the thighs; extends from the coccyx to the pubis.

perineurium (per-i-noo′rē-ŭm) [L., *peri,* around + Gr., *neuron,* nerve] Connective tissue sheath surrounding a nerve fascicle.

periodontal ligament (per′ē-ō-don′tăl) Connective tissue that surrounds the tooth root and attaches it to its bony socket.

periosteum (per-ē-os′tē-ŭm) [Gr., *peri,* around + *osteon,* bone] Thick, double-layered connective tissue sheath covering the entire surface of a bone, except the articular surface, which is covered with cartilage.

peripheral nervous system (PNS) (pē-rif′ē-răl) Major subdivision of the nervous system consisting of nerves and ganglia.

peripheral resistance Resistance to blood flow in all the blood vessels.

peristaltic wave (per-i-stal′tik) Contraction in a tube, such as the intestine, characterized by a wave of contraction in smooth muscle preceded by a wave of relaxation that moves along the tube.

peritubular capillary Capillary network located in the cortex of the kidney; associated with the distal and proximal convoluted tubules.

permanent tooth One of the 32 teeth belonging to the permanent dentition; also called secondary tooth.

peroneal (per-ō-nē′ăl) [Gr., *perone,* fibula] Associated with the fibula.

peroxisome (per-ok′si-sōm) Membrane-bounded body similar to a lysosome in appearance but often smaller and irregular in shape; contains enzymes that either decompose or synthesize hydrogen peroxide.

Peyer's patch Named for Swiss anatomist Johann K. Peyer (1653–1712). Lymph nodule found in the lower half of the small intestine and the appendix.

phagocyte (fag′ō-sīt) Cell that ingests bacteria, foreign particles, and other cells.

phagocytosis (fag′ō-sī-tō′sis) [Gr., *phagein,* to eat + *kytos,* cell + *osis,* condition] Cells' ingestion of solid substances, such as other cells, bacteria, bits of necrosed tissue, and foreign particles.

phalange; pl. **phalanges** (fă-lanj′, fă-lan′jēz) [Gr., *phalanx,* line of soldiers] Bone of a finger or toe.

pharyngeal pouch (fă-rin′jē-ăl) Paired evagination of embryonic pharyngeal endoderm between the brachial arches that gives rise to the thymus, thyroid gland, tonsils, and parathyroid glands.

pharyngeal tonsil (fă-rin′jē-ăl) One of two collections of aggregated lymphoid nodules on the posterior wall of the nasopharynx.

pharynx (far′ingks) [Gr., *pharynx,* throat, the joint opening of the gullet and windpipe] Upper expanded portion of the digestive tube between the esophagus below and the oral and nasal cavities above and in front.

phenotype (fē′nō-tīp) [Gr., *phaino,* to display, show forth + *typos,* model] Characteristic observed in an individual due to the expression of his or her genotype.

phosphodiesterase (fos′fō-dī-es′ter-ās) Enzyme that splits phosphodiester bonds—that is, that breaks down cyclic AMP to AMP.

phospholipid (fos-fō-lip′id) Lipid with phosphorus, resulting in a molecule with a polar end and a nonpolar end; main component of the lipid bilayer.

phosphorylation (fos′fōr-i-lā′shŭn) Addition of phosphate to an organic compound.

photoreceptor (fō′tō-rē-sep′ter, fō′tō-rē-sep′tōr) [L., *photo,* light + *ceptus,* to receive] Sensory receptor that is sensitive to light—for example, rods and cones of the retina.

phrenic nerve (fren′ik) Nerve derived from spinal nerves C3–C5; supplies the diaphragm.

physiologic contracture (fiz-ē-ō-loj′ik kon-trak′chūr) Temporary inability of a muscle to either contract or relax because of a depletion of ATP so that active transport of calcium ions into the sarcoplasmic reticulum cannot occur.

physiologic dead space Sum of anatomical dead air space plus the volume of any nonfunctional alveoli.

physiologic shunt Deoxygenated blood from the alveoli plus deoxygenated blood from the bronchi and bronchioles.

pia mater (pī′ă mā′ter, pē′a mah′ter) [L., tender mother] Delicate membrane forming the inner covering of the brain and spinal cord.

pigmented layer Pigmented portion of the retina.

pineal body (pin′ē-ăl) [L., *pineus,* relating to pine trees] A small, pine cone–shaped structure that projects from the epiphysis of the diencephalon; produces melatonin; also called pineal gland.

pinna (pin′ă) [L., *pinna* or *penna,* feather] *See* auricle.

pinocytosis (pin′ō-sī-tō′sis, pī′no-sī-tō′sis) [Gr., *pineo,* to drink + *kytos,* cell + *osis,* condition] Cell drinking; uptake of liquid by a cell.

pituitary gland Endocrine gland attached to the hypothalamus by the infundibulum; also called hypophysis.

plane (plān) [L., *planus,* flat] Flat surface; an imaginary surface formed by extension through any axis or two points—for example, a midsagittal plane, a coronal plane, and a transverse plane.

plasma (plaz′mă) [Gr., something formed] Fluid portion of blood.

plasma cell Cell derived from B cells; produces antibodies.

plasma clearance Volume of plasma per minute from which a substance can be completely removed by the kidneys.

plasmin (plaz′min) Enzyme derived from plasminogen; dissolves clots by converting fibrin into soluble products.

plateau phase of action potential Prolongation of the depolarization phase of a cardiac muscle cell membrane; results in a prolonged refractory period.

platelet (plāt′let) Irregularly shaped disk found in blood; contains granules in the central part and clear protoplasm peripherally but has no definite nucleus; also called thrombocyte.

platelet plug Accumulation of platelets that stick to each other and to connective tissue; prevents blood loss from damaged blood vessels.

pleural cavity (ploor′ăl) Potential space between the parietal and visceral layers of the pleura.

plexus; pl. **plexuses** (plek′sŭs, plek′sŭs-ez) [L., a braid] Intertwining of nerves or blood vessels.

plicae circulares (plī′kă, plī′sē) Numerous folds of the mucous membrane of the small intestine.

pluripotent (ploo-rip′ō-tent) [L., *pluris,* more + *potentia,* power] In development, a cell or group of cells that have not yet become fixed or determined as to what specific tissues they are going to become.

podocyte cells (pod′ō-sīt) [Gr., *pous, podos,* foot + *kytos,* a hollow (cell)] Epithelial cell of Bowman's capsule attached to the outer surface of the glomerular capillary basement membrane by cytoplasmic foot processes.

Poiseuille's law (pwah-zuh′yez) Named for French physiologist and physicist Jean Léonard Marie Poiseuille (1797–1869). The volume of a fluid passing per unit of time through a tube is directly proportional to the pressure difference between its ends and to the fourth power of the internal radius of the tube and inversely proportional to the tube's length and the viscosity of the fluid.

polar body (pō′lăr) One of the two small cells formed during oogenesis because of unequal division of the cytoplasm.

polar covalent bond Covalent bond in which atoms do not share their electrons equally.

polarization Development of differences in potential between two points in living tissues, as between the inside and outside of the plasma membrane.

polycythemia (pol′ē-sī-thē′mē-ă) Increase in red blood cell number above the normal.

polygenic (pol-ē-jen′ik) Relating to a hereditary disease or normal characteristic controlled by the interaction of genes at more than one locus.

polysaccharide (pol-ē-sak′ă-rīd) Carbohydrate containing a large number of monosaccharide molecules.

polyunsaturated fat Fatty acid that contains two or more double covalent bonds between its carbon atoms.

pons (ponz) [L., bridge] Portion of the brainstem between the medulla and midbrain.

popliteal (pop-lit′ē-ăl, pop-li-tē′ăl) [L., ham] Posterior region of the knee.

porta (pōr′tă) [L., gate] Fissure on the inferior surface of the liver where the portal vein, hepatic artery, hepatic nerve plexus, hepatic ducts, and lymphatic vessels enter or exit the liver.

portal system (pōr′tal) System of vessels in which blood, after passing through one capillary bed, is conveyed through a second capillary network.

portal triad Branches of the portal vein, hepatic artery, and hepatic duct bound together in the connective tissue that divides the liver into lobules.

postabsorptive state State following the absorptive state; blood glucose levels are maintained because of the conversion of other molecules to glucose.

posterior chamber (pos-tēr′ē-ŏr) Chamber of the eye between the iris and the lens.

posterior interventricular sulcus Groove on the diaphragmatic surface of the heart, marking the location of the septum between the two ventricles.

posterior pituitary Portion of the hypophysis derived from the brain. Major secretions include antidiuretic hormone and oxytocin. Also called neurohypophysis.

postganglionic neuron (pōst′gang-glē-on′ik) Autonomic neuron that has its cell body located within an autonomic ganglion and sends its axon to an effector organ.

postovulatory age Age of the developing fetus based on the assumption that fertilization occurs 14 days after the last menstrual period before the pregnancy.

postsynaptic (pōst-si-nap′tik) Relating to the membrane of a nerve, muscle, or gland that is in close association with a presynaptic terminal. The postsynaptic membrane has receptor molecules within it that bind to neurotransmitter molecules.

potential difference (pō-ten′shăl) Difference in electrical potential, measured as the charge difference across the plasma membrane.

potential energy [Gr., *en,* in + *ergon,* work] Energy in a chemical bond that is not being exerted or used to do work.

PQ interval Time elapsing between the beginning of the P wave and the beginning of the QRS complex in the electrocardiogram; also called PR interval.

precapillary sphincter (prē-kap′i-lār-ē sfingk′ter) Smooth muscle sphincter that regulates blood flow through a capillary.

preganglionic neuron (prē′gang-glē-on′ik) Autonomic neuron that has its cell body located within the central nervous system and sends its axon through a nerve to an autonomic ganglion, where it synapses with postganglionic neurons.

premolar (prē-mō′lăr) Bicuspid tooth.

prenatal (prē-nā′tal) [*pre* + L., *natus,* born] Preceding birth.

prepuce (prē′poos) In males, the free fold of skin that more or less completely covers the glans penis; the foreskin. In females, the external fold of the labia minora that covers the clitoris.

pressoreceptor (pres′ō-rē-sep′ter, pres′ō-rē-sep′tōr) *See* baroreceptor.

presynaptic terminal (prē′si-nap′tik) Enlarged axon terminal or terminal bouton.

primary palate In the early embryo, the structure that gives rise to the upper jaw and lips.

primary response Immune response that occurs as a result of the first exposure to an antigen.

primary spermatocyte (sper′mă-tō-sīt) Spermatocyte arising by a growth phase from a spermatogonium; gives rise to secondary spermatocytes after the first meiotic division.

prime mover Muscle that plays a major role in accomplishing a movement.

primitive streak (prim′i-tiv) Ectodermal ridge in the midline of the embryonic disk, from which arises the mesoderm by the inward and then lateral migration of cells.

primordial germ cell (prī-mōr′dē-ăl) Most primitive undifferentiated sex cell, found initially outside the gonad on the surface of the yolk sac.

PR interval *See* PQ interval.

process (pros′es, prō′ses) Projection on a bone.

processus vaginalis (prō-ses′ŭs vaj′i-năl-ŭs) Peritoneal outpocketing in the embryonic lower anterior abdominal wall that traverses the inguinal canal; in the male, it forms the tunica vaginalis testis and normally loses its connection with the peritoneal cavity.

proerythroblast Cell that matures into an erythrocyte.

progeria (prō-jēr′ē-ă) [Gr., *pro,* before + *ge* + *amras,* old age] Severe retardation of growth after the first year accompanied by a senile appearance and death at an early age.

progesterone (prō-jes′tĕ-rōn) Steroid hormone secreted by the corpus luteum and one of the hormones secreted by the placenta.

prolactin (prō-lak′tin) Hormone of the adenohypophysis that stimulates the production of milk.

prolactin-inhibiting hormone (PIH) Neurohormone released from the hypothalamus that inhibits prolactin release from the adenohypophysis.

prolactin-releasing hormone (PRH) Neurohormone released from the hypothalamus that stimulates prolactin release from the adenohypophysis.

proliferative phase (prō-lif′er-ă-tiv) *See* follicular phase.

pronation (prō-nā′shŭn) [L., *pronare,* to bend forward] Rotation of the forearm so that the anterior surface is down (prone).

pronephros (prō-nef′ros) In the embryos of higher vertebrates, a series of tubules emptying into the celomic cavity. It is a temporary structure in the human embryo, followed by the mesonephros and still later by the metanephros, which gives rise to the kidney.

prophase (prō′fāz) First stage in cell division when chromatin strands condense to form chromosomes.

proprioception (prō-prē-ō-sep′shun) [L., *proprius,* one's own + *capio,* to take] Information about the position of the body and its various parts.

proprioceptor (prō′prē-ō-sep′ter) Sensory receptor associated with joints and tendons.

prostaglandin (pros′tă-glan′din) Class of physiologically active substances present in many tissues; among its effects are vasodilation, stimulation and contraction of uterine smooth muscle and the promotion of inflammation and pain.

prostate gland (pros′tāt) [Gr., *prostates,* one standing before] Gland that surrounds the beginning of the urethra in the male. The secretion of the gland is a milky fluid that is discharged by 20–30 excretory ducts into the prostatic urethra as part of the semen.

prostatic urethra (pros-tat′ik) Part of the male urethra, approximately 2.5 cm in length, that passes through the prostate gland.

protease (prō′tē-ās) Enzyme that breaks down proteins.

protein (prō′tēn, prō′tē-ĭn) [Gr., *proteios,* primary] Macromolecule consisting of long sequences of amino acids linked together by peptide bonds.

protein kinase (kin-āz) Class of enzymes that phosphorylates other proteins. Many of these kinases are responsive to other chemical signals (e.g., cAMP, cGMP, insulin, epidermal growth factor, calcium and calmodulin).

proteoglycan (prō′tē-ō-glī′kan) Macromolecule consisting of numerous polysaccharides attached to a common protein core.

prothrombin (prō-throm′bin) Glycoprotein present in blood that, in the presence of prothrombin activator, is converted to thrombin.

proton (prō′ton) [Gr., *protos,* first] Positively charged particle in the nuclei of atoms.

protraction (prō-trak′shŭn) [L., *protractus,* to draw forth] Movement forward or in the anterior direction.

provitamin (prō-vī′tă-min) Substance that may be converted into a vitamin.

proximal convoluted tubule (prok′si-măl) Part of the nephron that extends from the glomerulus to the descending limb of the loop of Henle.

pseudostratified epithelium Epithelium consisting of a single layer of cells but having the appearance of multiple layers.

psychologic fatigue (sī-kō-loj′-ik) Fatigue caused by the central nervous system.

ptosis (tō′sis) [G., *ptosis,* a falling] Falling down of an organ—for example, the drooping of the upper eyelid.

puberty (pū′ber-tē) [L., *pubertas,* grown up] Series of events that transform a child into a sexually mature adult; involves an increase in the secretion of GnRH.

pubis (pū′bis) Anterior inferior bone of the coxal bone.

pudendal cleft (pū-den′dăl) Cleft between the labia majora.

pudendum (pū-den′dŭm) *See* vulva.

pulmonary artery (pŭl′mō-nār-ē) One of the arteries that extend from the pulmonary trunk to the right or left lungs.

pulmonary capacity Sum of two or more pulmonary volumes.

pulmonary trunk Large, elastic artery that carries blood from the right ventricle of the heart to the right and left pulmonary arteries.

pulmonary vein One of the veins that carry blood from the lungs to the left atrium of the heart.

pulp (pŭlp) [L., *pulpa,* flesh] Soft tissue within the pulp cavity of the tooth, consisting of connective tissue containing blood vessels, nerves, and lymphatics.

pulse pressure (pŭls) Difference between systolic and diastolic pressure.

pupil (pū′pĭl) Circular opening in the iris through which light enters the eye.

Purkinje fiber (pŭr-kĭn′jē) Named for Bohemian anatomist Johannes E. von Purkinje (1787–1869). Modified cardiac muscle cells found beneath the endocardium of the ventricles. Specialized to conduct action potentials.

pus (pŭs) Fluid product of inflammation; contains white blood cells, the debris of dead cells, and tissue elements liquefied by enzymes.

pyloric opening (pī-lōr′ik) Opening between the stomach and the superior part of the duodenum.

pyloric sphincter Thickening of the circular layer of the gastric musculature encircling the junction between the stomach and duodenum; also called pylorus.

pyrogen (pī′rō-jen) Chemical released by microorganisms, neutrophils, monocytes, and other cells that stimulates fever production by acting on the hypothalamus.

Q

QRS complex Principal deflection in the electrocardiogram, representing ventricular depolarization.

QT interval Time elapsing from the beginning of the QRS complex to the end of the T wave, representing the total duration of electrical activity of the ventricles.

R

radial pulse (rā′dē-ăl) Pulse detected in the radial artery.

radiation (rā′dē-ā′shŭn) [L., *radius,* ray, beam] Sending forth of light, short radiowaves, ultraviolet or x-rays, or any other rays for treatment or diagnosis or for other reasons; radiant heat.

radioactive isotope (rā′dē-ō-ak′tiv) Isotope with a nuclear composition that is unstable from which subatomic particles and electromagnetic waves are emitted.

ramus; pl. rami (rā′mŭs, rā′mī) [L., branch] One of the primary subdivisions of a nerve or blood vessel; the part of a bone that forms an angle with the main body of the bone.

raphe (rā′fē) [Gr., *rhaphe,* suture, seam] Central line running over the scrotum from the anus to the root of the penis.

receptor Structural protein or glycoprotein molecule on the cell surface or within the cytoplasm that binds to a specific factor (chemical signal).

recessive (rē-ses′iv) Gene that may not be expressed because of suppression by a contrasting dominant gene.

Recommended Dietary Allowances (RDAs) Guide for estimating the nutritional needs of groups of people based on their age, sex, and other factors; first established in 1941.

Recommended Daily Intake (RDI) Generally, the highest RDA value in each of four categories: infants, toddlers, people over 4 years of age, and pregnant or lactating women. The RDIs used to determine the Daily Values are based on the 1968 RDAs.

rectum (rek′tŭm) [L., *rectus,* straight] Portion of the digestive tract that extends from the sigmoid colon to the anal canal.

red marrow (mar′o) Soft, pulpy connective tissue filling the cavities of bones; consists of reticular fibers and the development stages of blood cells and platelets; gradually replaced by yellow marrow in long bones and the skull.

red pulp [L., *pulpa,* flesh] Reddish brown substance of the spleen consisting of venous sinuses and the tissues intervening between them, called pulp cords.

reduction (rē-dŭk′shŭn) Gain of one or more electrons by a molecule.

refraction (rē-frak-shŭn) Bending of a light ray when it passes from one medium into another of different density.

refractory period (rē-frak′tōr-ē) [Gr., *periodos,* a way around, a cycle] Period following effective stimulation during which excitable tissue, such as heart muscle, fails to respond to a stimulus of threshold intensity.

regeneration (rē′jen-er-ā′shŭn) Reproduction or reconstruction of a lost or injured part.

regulatory gene Gene involved with controlling the activity of structural genes.

relative refractory period Portion of the action potential following the absolute refractory period during which another action potential can be produced with a greater-than-threshold stimulus strength.

relaxation phase (rē-lak-sā′shŭn) Phase of muscle contraction following the contraction phase; the time from maximal tension production until tension decreases to its resting level.

relaxin Polypeptide hormone secreted by the corpus luteum and placenta during pregnancy; facilitates the birth process by causing a softening and lengthening of the pubic symphysis and cervix.

renal artery (rē′năl) Artery that originates from the aorta and delivers blood to the kidney.

renal blood flow rate Volume at which blood flows through the kidneys per minute; an average of approximately 1200 mL/min.

renal column Cortical substance separating the renal pyramids.

renal corpuscle Glomerulus and Bowman's capsule that encloses it.

renal fascia (făsh′i-ā) Connective tissue surrounding the kidney that forms a sheath or capsule for the organ.

renal fat pad Fat layer that surrounds the kidney and functions as a shock-absorbing material.

renal fraction Portion of the cardiac output that flows through the kidneys; averages 21%.

renal pelvis Funnel-shaped expansion of the upper end of the ureter that receives the calyces.

renal pyramid One of a number of pyramidal masses seen on longitudinal section of the kidney; they contain part of the loops of Henle and the collecting tubules.

renin (rē′nin) Enzyme secreted by the juxtaglomerular apparatus that converts angiotensinogen to angiotensin I.

renin-angiotensin-aldosterone mechanism Renin, released from the kidneys in response to low blood pressure, converts angiotensinogen to angiotensin I. Angiotensin I is converted by angiotensin-converting enzyme to angiotensin II, which causes vasoconstriction, resulting in increased blood pressure. Angiotensin II also increases aldosterone secretion, which increases blood pressure by increasing blood volume.

repolarization (rē′pō-lăr-i-zā′shŭn) Phase of the action potential in which the membrane potential moves from its maximum degree of depolarization toward the value of the resting membrane potential.

reposition Return of a structure to its original position.

residual volume (rē-zid′ū-ăl) Volume of air remaining in the lungs after a maximum expiratory effort.

resolution (rez-ō-loo′shŭn) [L., *resolutio,* a slackening] Phase of the male sexual act after ejaculation during which the penis becomes flaccid; feeling of satisfaction; inability to achieve erection and second ejaculation. Last phase of the female sexual act, characterized by an overall sense of satisfaction and relaxation.

respiration (res-pi-ră′shŭn) [L., *respiratio,* to exhale, breathe] Process of life in which oxygen is used to oxidize organic fuel molecules, providing a source of energy, carbon dioxide, and water; movement of air into and out of the lungs, the exchange of gases with blood, the transportation of gases in the blood, and gas exchange between the blood and the tissues.

respiratory bronchiole (res′pi-ră-tōr-ē, rĕ-spīr′ă-tōr-ē) Smallest bronchiole (0.5 mm in diameter) that connects the terminal bronchiole to the alveolar duct.

respiratory membrane Membrane in the lungs across which gas exchange occurs with blood.

resting membrane potential (RMP) Electric charge difference inside a plasma membrane, measured relative to just outside the plasma membrane.

reticular (re-tik′ū-lăr) [L., *rete,* net] Relating to a fine network of cells or collagen fibers.

reticular cell Cell with processes making contact with those of other similar cells to form a cellular network; along with the network of reticular fibers, the reticular cells form the framework of bone marrow and lymphatic tissues.

reticulocyte (re-tik′ū-lō-sīt) Young red blood cell with a network of basophilic endoplasmic reticulum occurring in larger numbers during the process of active red blood cell synthesis.

reticuloendothelial system (re-tik′ū-lō-en-dō-thē′lē-ăl) *See* mononuclear phagocytic system.

retina (ret′i-nă) Nervous tunic of the eyeball.

retinaculum (ret-i-nak′ū-lŭm) [L., band, halter, to hold back] Dense regular connective tissue sheath holding down the tendons at the wrist, ankle, or other sites.

retraction (rē-trak′shŭn) [L., *retractio,* a drawing back] Movement in the posterior direction.

retroperitoneal (re′trō-per′i-tō-nē′ăl) Behind the peritoneum.

rhodopsin (rō-dop′sin) Light-sensitive substance found in the rods of the retina; composed of opsin loosely bound to retinal.

ribonuclease (rī-bō-nū′klē-ās) Enzyme that splits RNA into its component nucleotides.

ribonucleic acid (RNA) (rī′bō-noo-klē′ik) Nucleic acid containing ribose as the sugar component; found in all cells in both nuclei and cytoplasm; helps direct protein synthesis.

ribosomal RNA (rRNA) (rī′bō-sōm-ăl) RNA that is associated with certain proteins to form ribosomes.

ribosome (rī′bō-sōm) Small, spherical, cytoplasmic organelle where protein synthesis occurs.

right lymphatic duct Lymphatic duct that empties into the right subclavian vein; drains the right side of the head and neck, the right-upper thorax, and the right-upper limb.

rigor mortis (rig′er mōr′tĭs) Increased rigidity of muscle after death due to cross-bridge formation between actin and myosin as calcium ions leak from the sarcoplasmic reticulum.

rod Photoreceptor in the retina of the eye; responsible for noncolor vision in low-intensity light.

root Part below the neck of a tooth covered by cementum rather than enamel and attached by the periodontal ligament to the alveolar bone.

root of the penis Proximal attached part of the penis, including the two crura and the bulb.

rotation (rō-tā′shun) Movement of a structure about its axis.

rotator cuff (rō-tā′ter, rō-tā′tor) Four deep muscles that attach the humerus to the scapula.

round ligament Fibromuscular band that is attached to the uterus on each side in front of and below the opening of the uterine tube; it passes through the inguinal canal to the labium majus.

round ligament of the liver Remains of the umbilical vein.

round window Membranous structure separating the scala tympani of the inner ear from the middle ear.

Ruffini end organ (roo-fē′nēz) Named for Italian histologist Angelo Ruffini (1864–1929); receptor located deep in the dermis and responding to continuous touch or pressure.

ruga; pl. rugae (roo′gă, roo′gē) [L., a wrinkle] Fold or ridge; fold of the mucous membrane of the stomach when the organ is contracted; transverse ridge in the mucous membrane of the vagina.

S

saccule (sak′yūl) Part of the membranous labyrinth; contains a sensory structure, the macula, that detects static equilibrium.

salivary amylase (sal′i-vār-ē am′il-ās) Enzyme secreted in the saliva that breaks down starch to maltose and isomaltose.

salivary gland Gland that produces and secretes saliva into the oral cavity. The three major pairs of salivary glands are the parotid, submandibular, and sublingual glands.

salt Molecule consisting of a cation other than hydrogen and an anion other than hydroxide.

sarcolemma (sar′kō-lem′ă) [Gr., *sarco,* muscle + *lemma,* husk] Plasma membrane of a muscle fiber.

sarcoma (sar-kō′mă) Malignant neoplasm derived from connective tissue.

sarcomere (sar′kō-mēr) [Gr., *sarco,* muscle + *meros,* part] Part of a myofibril between adjacent Z disks.

sarcoplasm (sar′kō-plazm) [Gr., *sarco,* muscle + *plasma,* a thing formed] Cytoplasm of a muscle fiber, excluding the myofilaments.

sarcoplasmic reticulum (sar′kō-plaz′mik) [Gr., *sarco,* muscle + *plasma,* a thing formed + *reticulum,* net] Endoplasmic reticulum of muscle.

satellite cell (sat′ĕ-līt) Specialized cell that surrounds the cell bodies of neurons within ganglia.

saturated (satch′ŭ-rāt-ĕd) Fatty acid in which the carbon chain contains only single bonds between carbon atoms.

saturation (satch-ŭ-rā′shun) Point when all carrier molecules or enzymes are attached to substrate molecules and no more molecules can be transported or reacted.

scala tympani (skā′lă tim′pă-nī) [L., stairway] Division of the spiral canal of the cochlea lying below the spiral lamina and basilar membrane.

scala vestibuli (skā′lă ves-tib′ū-lī) Division of the cochlea lying above the spiral lamina and vestibular membrane.

scapula (skap′ū-lă) Bone forming the shoulder blade.

scar (skar) [Gr., *eschara,* scab] Fibrous tissue replacing normal tissue; also called cicatrix.

sciatic nerve (sī-at′ik) Tibial and common peroneal nerves bound together; also called ischiadic nerve.

sclera (sklĕr′ă) White of the eye; white, opaque portion of the fibrous tunic of the eye.

scleral venus sinus Series of veins at the base of the cornea that drain excess aqueous humor from the eye.

scrotum; pl. **scrota, scrotums** (skrō′tŭm, skrō′tă, skrō′tŭmz) Musculocutaneous sac containing the testes.

sebaceous gland (sē-bā′shŭs) [L., *sebum,* tallow] Gland of the skin, usually associated with a hair follicle, that produces sebum.

second messenger *See* intracellular mediator.

secondary follicle Follicle in which the secondary oocyte is surrounded by granulosa cells at the periphery; contains fluid-filled antral spaces.

secondary oocyte (ō′ō-sīt) Oocyte in which the second meiotic division stops at metaphase II unless fertilization occurs.

secondary palate Roof of the mouth in the early embryo that gives rise to the hard and the soft palates.

secondary response Immune response that occurs when the immune system is exposed to an antigen against which it has already produced a primary response; also called memory response.

secondary spermatocyte (sper′mă-tō-sīt) Spermatocyte derived from a primary spermatocyte by the first meiotic division; each secondary spermatocyte gives rise by the second meiotic division to two spermatids.

secretin (se-krē′tin) Hormone formed by the epithelial cells of the duodenum; stimulates secretion of pancreatic juice high in bicarbonate ions.

secretion (se-krē′shŭn) Substance produced inside a cell and released from the cell.

secretory phase (se-krēt′ĕ-rē, sē′krĕ-tōr-ē) *See* luteal phase.

segmental artery One of five branches of the renal artery, each supplying a segment of the kidney.

segmental bronchus (brong′kŭs) Extends from the secondary bronchus and conducts air to each lobule of the lungs.

self-antigen Antigen produced by the body that is capable of initiating an immune response against the body.

semen (sē′men) [L., seed (of plants, men, animals)] Penile ejaculate; thick, yellowish white, viscous fluid containing spermatozoa and secretions of the testes, seminal vesicles, prostate, and bulbourethral glands.

semicircular canal (sem′ē-sir′kū-lăr) Canal in the petrous portion of the temporal bone that contains sensory organs that detect kinetic or dynamic equilibrium. Three semicircular canals are within each inner ear.

seminal vesicle (sem′i-năl) One of two glandular structures that empty into the ejaculatory ducts; its secretion is one of the components of semen.

seminiferous tubule (sem′i-nif′er-ŭs) Tubule in the testis in which spermatozoa develop.

sensible perspiration (sen′si-bl pers-pi-rā′shŭn) Perspiration excreted by the sweat glands that appears as moisture on the skin; produced in large quantity when there is much humidity in the atmosphere.

septum primum (sep′tŭm prī′mŭm) First septum in the embryonic heart that arises on the wall of the originally single atrium of the heart and separates it into right and left chambers.

septum secundum (sek′ŭn-dŭm) Second of two major septal structures involved in the partitioning of the atrium, arising later than the septum primum and located to the right of it; it remains an incomplete partition until after birth, with its unclosed area constituting the foramen ovale.

serosa (se-rō′să) [L., *serosus,* serous] Outermost covering of an organ or a structure that lies in a body cavity.

serous fluid (ser′ŭs) Fluid similar to lymph that is produced by and covers serous membrane; it lubricates the serous membrane.

serous membrane Thin sheet composed of epithelial and connective tissues; it lines cavities that do not open to the outside of the body or contain glands but do secrete serous fluid.

serous pericardium Lining of the pericardial sac composed of a serous membrane.

Sertoli cell (ser-tō′lē) Named for Italian histologist Enrico Sertoli (1842–1910). Elongated cell in the wall of the seminiferous tubules to which spermatids are attached during spermatogenesis.

serum (sēr′ŭm) [L., whey] Fluid portion of blood after the removal of fibrin and blood cells.

sesamoid bone (ses′ă-moyd) [Gr., *sesamoceies,* like a sesame seed] Bone found within a tendon, such as the patella.

sex chromosomes Pair of chromosomes responsible for sex determination, XX in female and XY in male.

sex-linked trait Characteristic resulting from the expression of a gene on a sex chromosome.

sigmoid colon (sig′moyd) Part of the colon between the descending colon and the rectum.

sigmoid mesocolon Fold of peritoneum attaching the sigmoid colon to the posterior abdominal wall.

simple epithelium Epithelium consisting of a single layer of cells.

sinoatrial (SA) node (si′nō-ā′trē-ăl) Mass of specialized cardiac muscle fibers; acts as the "pacemaker" of the cardiac conduction system.

sinus (sī′nŭs) [L., cavity] Hollow in a bone or other tissue; enlarged channel for blood or lymph.

sinus venosus End of the embryonic cardiac tube where blood enters the heart; becomes a portion of the right atrium, including the SA node.

sinusoid (si′nŭ-soyd) [L., sinus + Gr., *eidos,* resemblance] Terminal blood vessel having a larger diameter than an ordinary capillary.

sinusoidal capillary (si-nŭ-soy′dăl) Capillary with caliber of 10–20 μm or more; lined with a fenestrated type of endothelium.

small intestine [L., *intestinus,* entrails] Portion of the digestive tube between the stomach and the cecum; consists of the duodenum, jejunum, and ileum.

sodium–potassium (Na^+–K^+) pump Biochemical mechanism that uses energy derived from ATP to achieve the active transport of potassium ions opposite to that of sodium ions; also called sodium potassium ATP-ase.

soft palate Posterior muscular portion of the palate, forming an incomplete septum between the mouth and the oropharynx and between the oropharynx and the nasopharynx.

solute (sol′ūt, sō′loot) [L., *solutus,* dissolved] Dissolved substance in a solution.

solution (sō-loo′shŭn) [L., *solutio*] Homogenous mixture formed when a solute is dissolved in a solvent.

solvent (sol′vent) [L., *solvens,* to dissolve] Liquid that holds another substance in solution.

soma (sō′mă) [Gr., body] Neuron cell body or the enlarged portion of the neuron containing the nucleus and other organelles.

somatic (sō-mat′ik) [Gr., *somatikos,* bodily] Relating to the body; the cells of the body except the reproductive cells.

somatic nervous system Composed of nerve fibers that send impulses from the central nervous system to skeletal muscle.

somatomedin (sō′mă-tō-mē′din) Peptide synthesized in the liver capable of stimulating certain anabolic processes in bone and cartilage, such as synthesis of DNA, RNA, and protein.

somatotropin (sō′mă-tō-trō′pin) Protein hormone of the anterior pituitary gland; it promotes body growth, fat mobilization, and inhibition of glucose utilization.

somite (sō′mīt) [Gr., *soma,* body + *ite*] One of the paired segments consisting of cell masses formed in the early embryonic mesoderm on each side of the neural tube.

somitomere (sō′mīt-ō-mēr) Indistinct somite in the head region of the embryo.

spatial summation Summation of the local potentials in which two or more action potentials arrive simultaneously at two or more presynaptic terminals that synapse with a single neuron.

specific heat Heat required to raise the temperature of any substance 1°C compared with the heat required to raise the same volume of water 1°C.

speech Use of the voice in conveying ideas.

spermatic cord (sper-mat′ik) Cord formed by the ductus deferens and its associated structures; extends through the inguinal canal into the scrotum.

spermatid (sper′mă-tid) [Gr., *sperma,* seed + *id*] Cell derived from the secondary spermatocyte; gives rise to a spermatozoon.

spermatogenesis (sper′mă-tō-jen′ĕ-sis) Formation and development of the spermatozoon.

spermatogonium (sper′mă-tō-gō′nē-ŭm) [Gr., *sperma,* seed + *gone,* generation] Cell that divides by mitosis to form primary spermatocytes.

spermatozoon; pl. **spermatozoa** (sper′mă-tō-zō′on, sper′mă-to-zō′ă) [Gr., *sperma,* seed + *zoon,* animal] Male gamete or sex cell, composed of a head and a tail. The spermatozoon contains the genetic information transmitted by the male. Also called sperm cell.

sphenoid (sfē′noyd) [Gr., *shen,* wedge] Wedge-shaped.

sphincter pupillae (sfingk′ter pū-pil′ē) Circular smooth muscle fibers of the iris's diaphragm that constrict the pupil of the eye.

sphygmomanometer (sfig′mō-mă-nom′ĕ-ter) [Gr., *sphygmos,* pulse + *manos,* thin, scanty + *metron,* measure] Instrument for measuring blood pressure.

spinal nerve (spī′năl) One of 31 pairs of nerves formed by the joining of the dorsal and ventral roots that arise from the spinal cord.

spindle fiber (spin′dl) Specialized microtubule that develops from each centrosome and extends toward the chromosomes during cell division.

spiral artery (spī′răl) One of the corkscrew-like arteries in premenstrual endometrium; most obvious during the secretory phase of the uterine cycle.

spiral ganglion Cell bodies of sensory neurons that innervate hair cells of the organ of Corti are located in the spiral ganglion.

spiral lamina Attached to the modiolus and supports the basilar and vestibular membranes.

spiral ligament Attachment of the basilar membrane to the lateral wall of the bony labyrinth.

spiral organ Organ of Corti; rests on the basilar membrane and consists of the hair cells that detect sound; also called organ of Corti.

spiral tubular gland Well-developed simple or compound tubular gland; spiral in shape; within the endometrium of the uterus; prevalent in the secretory phase of the uterine cycle.

spirometer (spī-rom′ĕ-ter) [L., *spiro,* to breathe + Gr., *metron,* measure] Gasometer used for measuring the volume of respiratory gases; usually understood to consist of a counterbalanced cylindrical bell sealed by dipping into a circular trough of water.

spirometry (spī-rom′ĕ-trē) Making pulmonary measurements with a spirometer.

spleen (splēn) Large lymphatic organ in the upper part of the abdominal cavity on the left side between the stomach and diaphragm, composed of white and red pulp. It responds to foreign substances in the blood, destroys worn-out red blood cells, and is a storage site for blood cells.

spongy urethra Portion of the male urethra, approximately 15 cm in length, that traverses the corpus spongiosum of the penis.

squamous (skwā′mŭs) [L., *squama,* a scale] Scalelike, flat.

stapedius (stā-pē′dē-ŭs) Small skeletal muscles attached to the stapes.

stapes (stā′pēz) [L., stirrup] Smallest of the three auditory ossicles; attached to the oval window.

Starling's law of the heart Named for English physiologist Ernest H. Starling (1866–1927). Force of contraction of cardiac muscle is a function of the length of its muscle fibers at the end of diastole; the greater the ventricular filling, the greater the stroke volume produced by the heart.

sternum (ster′nŭm) [L., *sternon,* chest] Breastbone.

steroid (stēr′oyd, ster′oyd) Large family of lipids, including some reproductive hormones, vitamins, and cholesterol.

stomach (stŭm′ŭk) Large sac between the esophagus and the small intestine, lying just beneath the diaphragm.

stratified epithelium (strat′i-fīd ep-i-thē′lē-ŭm) Epithelium consisting of more than one layer of cells.

stratum basale (strat-ŭm băh-săl′ē) [L., layer; basal] Basal, or deepest, layer of the epidermis; also called stratum germinativum.

stratum corneum (kōr′nē-ŭm) [L., layer + *corneus,* horny] Most superficial layer of the epidermis consisting of flat, keratinized, dead cells.

stratum germinativum *See* stratum basale.

stratum granulosum (gran′ū-lō′sŭm) [L., layer; granulum, a small grain] Layer of cells in the epidermis filled with granules of keratohyalin.

stratum lucidum (lū′sid-ŭm) [L., layer + *lucidus,* clear] Clear layer of the epidermis found in thick skin between the stratum granulosum and the stratum corneum.

stratum spinosum (spī′nōs-ŭm) [L., layer + *spina,* spine] Layer of many-sided cells in the epidermis with intercellular connections (desmosomes) that give the cells a spiny appearance.

stria; pl. **striae** (strī′ă, strī′ē) [L., channel] Line or streak in the skin that is a different texture or color from the surrounding skin; also called stretch mark.

striated (strī′āt-ēd) [L., *striatus,* furrowed] Striped; marked by stripes or bands.

stroke volume (SV) [L., *volumen,* something rolled up, scroll, from *volvo,* to roll] Volume of blood pumped out of one ventricle of the heart in a single beat.

structural gene Gene that determines the structure of a specific protein or peptide.

sty (stī) Inflamed ciliary gland of the eye.

subcutaneous (sŭb′-koo-tā′nē-ŭs) [L., *sub,* under + *cutis,* skin] Under the skin; same tissue as the hypodermis.

sublingual gland (sŭb-ling′gwăl) One of two salivary glands in the floor of the mouth beneath the tongue.

submandibular gland (sŭb-man-dib′ū-lăr) One of two salivary glands in the neck, located in the space bounded by the two bellies of the digastric muscle and the angle of the mandible.

submucosa (sŭb-moo-kō′să) Layer of tissue beneath a mucous membrane.

submucosal plexus (sŭb-mū-kō′săl) [L., a braid] Gangliated plexus of unmyelinated nerve fibers in the intestinal submucosa.

substantia nigra (sŭb-stan′shē-ă nī′gră) [L., substance; black] Black nuclear mass in the midbrain; involved in coordinating movement and maintaining muscle tone.

subthreshold stimulus Stimulus resulting in a local potential so small that it does not reach threshold and produce an action potential.

sucrose (soo′krōs) Disaccharide composed of glucose and fructose; table sugar.

sulcus; pl. **sulci** (sool′kūs, sŭl′sī) [L., furrow or ditch] Furrow or groove on the surface of the brain between the gyri; may also refer to a fissure.

superficial inguinal ring (ing′gwi-năl) Slitlike opening in the aponeurosis of the external oblique muscle of the abdominal wall through which the spermatic cord (round ligament in the female) emerges from the inguinal canal.

superior colliculus (ko-lik′ū-lūs) [L., *collis,* hill] One of two rounded eminences of the midbrain; aids in coordination of eye movements.

superior vena cava (vē′nă cā′vă) Vein that returns blood from the head and neck, upper limbs, and thorax to the right atrium.

supination (soo′pi-nā′shŭn) [L., *supino,* to bend backward, place on back] Rotation of the forearm (when the forearm is parallel to the ground) so that the anterior surface is up (supine).

supramaximal stimulus Stimulus of greater magnitude than a maximal stimulus; however, the frequency of action potentials is not increased above that produced by a maximal stimulus.

suppressor T cell Subset of T lymphocytes that decreases the activity of B cells and T cells.

surfactant (ser-fak′tănt) Lipoproteins forming a monomolecular layer over pulmonary alveolar surfaces; stabilizes alveolar volume by reducing surface tension and the tendency for the alveoli to collapse.

suspension (sŭs-pen′shŭn) Liquid through which a solid is dispersed and from which the solid separates unless the liquid is kept in motion.

suspensory ligament (sŭs-pen′sŏ-rē) Band of peritoneum that extends from the ovary to the body wall; contains the ovarian vessels and nerves. Small ligament attached to the margin of the lens in the eye and the ciliary body to hold the lens in place.

suture (soo′choor) [L., *sutura,* a seam] Junction between flat bones of the skull.

sweat (swet) [A.S., *swat*] Perspiration; secretions produced by the sweat glands of the skin; also called sensible perspiration.

sweat gland Usually means structure that produces a watery secretion called sweat. Some sweat glands, however, produce viscous organic secretions. Also called sudoriferous gland.

sympathetic chain ganglion (sim-pă-thet′ik) Collection of sympathetic postganglionic neurons that are connected to each other to form a chain along both sides of the spinal cord; also called paravertebral ganglion.

sympathetic division Subdivision of the autonomic division of the nervous system characterized by having the cell bodies of its preganglionic neurons located in the thoracic and upper lumbar regions of the spinal cord (thoracolumbar division); usually involved in preparing the body for physical activity; also called thoracolumbar division.

symphysis; pl. **symphyses** (sim′fi-sis, sim′fă-sēz) [Gr., a growing together] Fibrocartilage joint between two bones.

synapse; pl. **synapses** (sin′aps, sĭ-naps′, sĭ-nap′sēz) [Gr., *syn,* together + *haptein,* to clasp] Functional membrane-to-membrane contact of a nerve cell with another nerve cell, muscle cell, gland cell, or sensory receptor; functions in the transmission of action potentials from one cell to another; also called neuromuscular junction.

synaptic cleft (si-nap′tik) Space between the presynaptic and the postsynaptic membranes.

synaptic fatigue Fatigue due to depletion of neurotransmitter vesicles in the presynaptic terminals.

synaptic vesicle Secretory vesicle in the presynaptic terminal containing neurotransmitter substances.

synchondrosis; pl. **synchondroses** (sin′kon-drō′sis, -sēz) [Gr., *syn,* together + *chondros,* cartilage + *osis,* condition] Union between two bones formed by hyaline cartilage.

syncytiotrophoblast (sin-sish′ē-ō-trō′fō-blast) Outer layer of the trophoblast composed of multinucleated cells.

syndesmosis; pl. **syndesmoses** (sin′dez-mō′sis, sin′dez-mō′sēz) [Gr., *syndeo,* to bind + *osis,* condition] Form of fibrous joint in which opposing surfaces that are some distance apart are united by ligaments.

synergist (sin′er-jist) Muscle that works with other muscles to cause a movement.

synovial (si-nō′vē-ăl) [Gr., *syn,* together + *oon,* egg] Relating to or containing synovia (a substance that serves as a lubricant in a joint, tendon sheath, or bursa).

synovial fluid Slippery fluid found inside synovial joints and bursae; produced by the synovial membranes.

systemic inflammation (sis-tem′ik) Inflammation that occurs in many areas of the body. In addition to symptoms of local inflammation, increased neutrophil numbers in the blood, fever, and shock can occur.

systole (sis′tō-lē) [Gr., *systole,* a contracting] Contraction of the heart chambers during which blood leaves the chambers; usually refers to ventricular contraction.

T

T cell Thymus-derived lymphocyte of immunologic importance; it is of long life and is responsible for cell-mediated immunity.

T tubule (tū′bul) Tubelike invagination of the sarcolemma that conducts action potentials toward the center of the cylindrical muscle fibers.

T wave Deflection in an electrocardiogram following the QRS complex, representing ventricular repolarization.

tactile corpuscle (tak′til) Oval receptor found in the papillae of the dermis; responsible for fine, discriminative touch; also called Meissner corpuscle.

tactile disk Cuplike receptor found in the epidermis; responsible for light touch and superficial pressure; also called Merkel's disk.

talus (tā′lŭs) [L., ankle bone, heel] Tarsal bone contributing to the ankle.

target tissue Tissue on which a hormone acts.

tarsal bone (tar′săl) [Gr., *tarsos,* sole of foot] One of seven ankle bones.

tarsal plate (tar′săl) Crescent-shaped layer of connective tissue that helps maintain the shape of the eyelid.

taste (tāst) Sensations created when a chemical stimulus is applied to the taste receptors in the tongue.

taste bud Sensory structure, mostly on the tongue, that functions as a taste receptor.

tectum (tek′tŭm) Roof of the midbrain.

tegmentum (teg-men′tŭm) Floor of the midbrain.

telencephalon (tel-en-sef′ă-lon) [Gr., *telos,* end + *enkephalos,* brain] Anterior division of the embryonic brain from which the cerebral hemispheres develop.

telophase (tel′ō-fāz) Time during cell division when the chromosomes are pulled by spindle fibers away from the cell equator and into the two halves of the dividing cell.

temporal summation (tem′pŏ-răl) Summation of the local potential that results when two or more action potentials arrive at a single synapse in rapid succession.

tendon (ten′dŏn) Band or cord of dense connective tissue that connects a muscle to a bone or another structure.

tensor tympani (ten′sōr tim′pa-nī) Small skeletal muscle attached to the malleus.

tentorium cerebelli (ten-tō′rē-ŭm ser′ĕ-bel′ī) Dural folds between the cerebrum and the cerebellum.

terminal bouton (bū-ton′) [Fr., button] Enlarged axon terminal or presynaptic terminal.

terminal cisterna (sis-ter′nă) [L., *terminus,* limit + *cista,* box] Enlarged end of the sarcoplasmic reticulum in the area of the T tubules.

terminal hair [L., *terminus,* a boundary, limit] Long, coarse, usually pigmented hair found in the scalp, eyebrows, and eyelids and replacing vellus hair.

terminal sulcus (sŭl′kŭs) [L., furrow or ditch] V-shaped groove on the surface of the tongue at the posterior margin.

testis; pl. **testes** (tes′tis, tes′tēz) One of two male reproductive glands located in the scrotum; produces spermatozoa, testosterone, and inhibin.

testosterone (tes-tos′tĕ-rōn) Steroid hormone secreted primarily by the testes; aids in spermatogenesis, maintenance and development of male reproductive organs, secondary sexual characteristics, and sexual behavior.

tetraiodothyronine (T_4) (tet′ră-ī-ō′dō-thī′rō-nēn) One of the iodine-containing thyroid hormones; also called thyroxine.

thalamus (thal′ă-mŭs) [Gr., *thalamos,* bed, bedroom] Large mass of gray matter that forms the larger dorsal subdivision of the diencephalon.

theca (thē′kă) [Gr., *theke,* box] Sheath or capsule.

theca externa External fibrous layer of the theca of a vesicular follicle.

theca interna Inner vascular layer of the theca of the secondary and mature follicle; produces estrogen and contributes to the formation of the corpus luteum after ovulation.

thenar eminence (thē′nar) [Gr., palm of the hand] Fleshy mass of tissue at the base of the thumb; contains muscles responsible for thumb movements.

thick skin Skin in the palms, soles, and tips of the digits; has all five epidermal strata.

thin skin Skin over most of the body, usually without a stratum lucidum; has fewer layers of cells than thick skin.

thoracic cavity (thō-ras′ik) Space within the thoracic walls, bounded below by the diaphragm and above by the neck.

thoracic duct Largest lymph vessel in the body, beginning at the cisterna chyli and emptying into the left subclavian vein; drains the left side of the head and neck, the left-upper thorax, the left-upper limb, and the inferior half of the body.

thoracolumbar division (thōr′ă-kō-lŭm′bar) *See* sympathetic division.

thoroughfare channel Channel for blood through a capillary bed from an arteriole to a venule.

threshold potential (thresh′ōld) Value of the membrane potential at which an action potential is produced as a result of depolarization in response to a stimulus.

threshold stimulus Stimulus resulting in a local potential just large enough to reach threshold and produce an action potential.

thrombocyte (throm′bō-sīt) Platelet.

thrombocytopenia (throm′bō-sī′tō-pē′nē-ă) [*thrombocyte* + Gr., *penia,* poverty] Condition in which there is an abnormally small number of platelets in the blood.

thromboxane (throm-bok′sān) Specific class of physiologically active fatty acid derivatives present in many tissues.

thrombus; pl. **thrombi** (throm′bŭs, throm′bī) [Gr., *thrombos,* a clot] Clot in the cardiovascular system formed from constituents of blood; may be occlusive or attached to the vessel or heart wall without obstructing the lumen.

thymus (thī′mŭs) [Gr., *thymos,* sweetbread] Bilobed lymph organ located in the inferior neck and superior mediastinum; secretes the hormone thymosin.

thyroid cartilage (thī′royd) Largest laryngeal cartilage. It forms the laryngeal prominence, or Adam's apple.

thyroid gland [Gr., *thyreoeides,* shield] Endocrine gland located inferior to the larynx and consisting of two lobes connected by the isthmus; secretes the thyroid hormones triiodothyronine (T_3) and tetraiodothyronine (T_4).

thyroid-stimulating hormone (TSH) Glycoprotein hormone released from the hypothalamus; stimulates thyroid hormone secretion from the thyroid gland; also called thyrotropin.

thyrotropin (thī-rot′rō-pin, thī-rō-trō′pin) *See* thyroid-stimulating hormone (TSH).

thyroxine (thi-rok′sēn, thi-rok′sin) *See* tetraiodothyronine.

tidal volume (tī′dăl) Volume of air that is inspired or expired in a single breath during regular, quiet breathing.

tissue repair (tish′ū) Substitution of viable cells for damaged or dead cells by regeneration or replacement.

tolerance (tol′er-ăns) Failure of the specific immune system to respond to an antigen.

tongue (tŭng) Muscular organ occupying most of the oral cavity when the mouth is closed; major attachment is through its posterior portion.

tonicity (tō-nis′ī-tē) [Gr., *tonos,* tone] Osmotic pressure or tension of a solution, usually relative to that of blood; a state of continuous activity or tension caused by muscle contraction beyond the tension related to physical properties of muscle.

tonsil; pl. **tonsils** (ton′sil, ton′silz) [L., *tonsilla,* stake] Collection of lymphoid tissue; usually refers to a large collection of lymphatic tissue beneath the mucous membrane of the oral cavity and pharynx; lingual, pharyngeal, and palatine tonsils.

total lung capacity Volume of air contained in the lungs at the end of a maximum inspiration; equals vital capacity plus residual volume.

total tension Sum of active and passive tension.

trabecula; pl. **trabeculae** (tră-bek′ū-lă, tră-bek′ū-lē) [L., *trabs,* beam] One of the supporting bundles of fibers traversing the substance of a structure, usually derived from the capsule or one of the fibrous septa, such as trabeculae of lymph nodes, testes; a beam or plate of cancellous bone.

trachea (trā′kē-ă) [Gr., *tracheia arteria,* rough artery] Air tube extending from the larynx into the thorax, where it divides to form the bronchi; composed of 16–20 rings of hyaline cartilage.

tracheostomy (tra′ke-os′to-me) [*tracheo, trachea* + Gr., *stoma,* mouth] Operation to make an opening into the trachea; usually the opening is intended to be permanent, and a tube is inserted into the trachea to allow airflow.

tracheotomy (tra′ke-ot′o-me) [*tracheo, trachea* + Gr., *tome,* incision] Act of cutting into the trachea.

transcription (tran-skrip′shŭn) Process of forming RNA from a DNA template.

transfer RNA (tRNA) RNA that attaches to individual amino acids and transports them to the ribosomes, where they are connected to form a protein polypeptide chain.

transfusion (trans-fū′zhŭn) [L., *trans,* across + *fundo,* to pour from one vessel to another] Transfer of blood from one person to another.

transitional epithelium (tran-sish′ŭn-ăl) Stratified epithelium that may be either cuboidal or squamouslike, depending on the presence or absence of fluid in the organ (as in the urinary bladder).

translation (trans-lā′shŭn) Synthesis of polypeptide chains at the ribosome in response to information contained in mRNA molecules.

transverse colon (trans-vers′ kō′lon) Part of the colon between the right and left colic flexures.

transverse mesocolon (mez′ō-kō′lon) Fold of peritoneum attaching the transverse colon to the posterior abdominal wall.

transverse (T) tubule [L., *tubus,* tube] Tubule that extends from the sarcolemma to a myofibril of striated muscles.

treppe (trep′eh) [Ger., staircase] Series of successively stronger contractions that occur when a rested muscle fiber receives closely spaced stimuli of the same strength but with a sufficient stimulus interval to allow complete relaxation of the fiber between stimuli.

triacylglycerol (trī-as′il-glis′er-ol) *See* triglyceride.

triad (trī′ad) Two terminal cisternae and a T tubule between them.

tricuspid valve (trī-kŭs′pid) Valve closing the orifice between the right atrium and the right ventricle of the heart.

triglyceride (tri-glis′er-id) Three-carbon glycerol molecule with a fatty acid attached to each carbon; constitute approximately 95% of the fats in the human body. Also called triacylglycerol.

trigone (trī′gōn) [Gr., *trigonon,* triangle] Triangular, smooth area at the base of the bladder between the openings of the two ureters and that of the urethra.

triiodothyronine (T_3) (trī-ī′ō-dō-thī′rō-nēn) One of the iodine-containing thyroid hormones.

trochlea (trok′lē-ă) [L., pulley] Structure shaped like or serving as a pulley or spool.

trochlear nerve (trok′lē-ar) [L., *trochlea,* pulley] Cranial nerve IV, to the muscle (superior oblique) turning around a pulley.

trophoblast (trof′ō-blast) [Gr., *trophe,* nourishment + *blastos,* germ] Cell layer forming the outer layer of the blastocyst, which erodes the uterine mucosa during implantation; the trophoblast does not become part of the embryo but contributes to the formation of the placenta.

tropomyosin (trō-pō-mī′ō-sin) Fibrous protein found as a component of the actin myofilament.

troponin (trō′pō-nin) Globular protein component of the actin myofilament.

true pelvis Portion of the pelvis inferior to the pelvic brim.

true rib (ver-tĕ′brō-ster′năl) Rib that attaches by an independent costal cartilage directly to the sternum; also called vertebrosternal rib.

trypsin (trip′sin) Proteolytic enzyme formed in the small intestine from the inactive pancreatic precursor trypsinogen.

tubercle (too′ber-kl) Lump on a bone.

tubular load (too′bū-lăr) Amount of a substance per minute that crosses the filtration membrane into Bowman's capsule.

tubular maximum Maximum rate of secretion or reabsorption of a substance by the renal tubules.

tubular reabsorption Movement of materials, by means of diffusion, active transport, or symport, from the filtrate within a nephron to the blood.

tubular secretion Movement of materials, by means of active transport, from the blood into the filtrate of a nephron.

tumor (too′mŏr) Any swelling or growth; a neoplasm.

tunic (too′nik) [L., coat] One of the enveloping layers of a part; one of the coats of a blood vessel; one of the coats of the eye; one of the coats of the digestive tract.

tunica adventitia (too′ni-kă ad-ven-tish′ă) Outermost fibrous coat of a vessel or an organ that is derived from the surrounding connective tissue.

tunica albuginea (al-bū-jin′ē-ă) Dense, white, collagenous tunic surrounding a structure, such as the capsule around the testis.

tunica intima (in′ti-mă) Innermost coat of a blood vessel; consists of endothelium, a lamina propria, and an inner elastic membrane.

tunica media Middle, usually muscular, coat of an artery or another tubular structure.

turbulent flow Flow characterized by eddy currents exhibiting nonparallel blood flow.

tympanic membrane (tim-pan′ik) Eardrum; cellular membrane that separates the external from the middle ear; vibrates in response to sound waves.

U

unipolar neuron (oo-ni-pō′lar) One of the three categories of neurons consisting of a nerve cell body with a single axon projecting from it; also called a pseudounipolar neuron.

unmyelinated axon (ŭn-mī′ĕ-li-nā-ted) Nerve fibers lacking a myelin sheath.

unsaturated (ŭn-sach′ŭr-āt-ed) Carbon chain of a fatty acid that possesses one or more double or triple bonds.

upper respiratory tract Nasal cavity, pharynx, and associated structures.

up-regulation Increase in the concentration of receptors in response to a signal.

ureter (ū-rē′ter, ū′rē-ter) [Gr., *oureter,* urinary canal] Tube conducting urine from the kidney to the urinary bladder.

urethral gland (ū-rē′thrăl) One of numerous mucous glands in the wall of the spongy urethra in the male.

urogenital triangle Anterior portion of the perineal region containing the openings of the urethra and vagina in the female and the urethra and root structures of the penis in the male.

uterine cycle (ū′ter-in, ū′ter-īn) Series of events that occur in a regular fashion in the uterus of sexually mature, nonpregnant females; prepares the uterine lining for implantation of the embryo.

uterine part Portion of the uterine tube that passes through the wall of the uterus.

uterine tube One of the tubes leading on each side from the uterus to the ovary; consists of the infundibulum, ampulla, isthmus, and uterine parts; also called fallopian tube or oviduct.

uterus (ū′ter-ŭs) Hollow, muscular organ in which the fertilized ovum develops into a fetus.

utricle (oo′tri-kl) Part of the membranous labyrinth; contains a sensory structure, the macula, that detects static equilibrium.

uvula (ū′vū-lă) [L., *uva,* grape] Small, grape-like appendage at the posterior margin of the soft palate.

V

vaccination (vak′si-nā′shŭn) Deliberate introduction of an antigen into a subject to stimulate the immune system and produce immunity to the antigen.

vaccine (vak′sēn, vak-sēn′) [L., *vaccinus,* relating to a cow] Preparation of killed microbes, altered microbes, or derivatives of microbes or microbial products intended to produce immunity. The method of administration is usually inoculation, but ingestion is preferred in some instances, and nasal spray is used occasionally.

vagina (vă-jī′nă) [L., sheath] Genital canal in the female, extending from the uterus to the vulva.

vapor pressure Partial pressure exerted by water vapor.

variable region Part of an antibody that combines with an antigen.

vas deferens (vas def′er-enz) *See* ductus deferens.

vasa recta (vā′să rek′tă) Specialized capillary that extends from the cortex of the kidney into the medulla and then back to the cortex.

vasa vasorum (vā′sor-ŭm) [L., vessel, dish] Small vessels distributed to the outer and middle coats of larger blood vessels.

vascular layer (vas′kū-lăr) Middle layer of the eye; contains many blood vessels.

vasoconstriction (vā′sō-kon-strik′shŭn, vas′ō-kon-strik′shŭn) Decreased diameter of blood vessels.

vasodilation (vā′sō-dī-lā′shŭn) Increased diameter of blood vessels.

vasomotion (vā-sō-mō′shŭn) Periodic contraction and relaxation of the precapillary sphincter, resulting in cyclic blood flow through capillaries.

vasomotor center (vā-sō-mō′ter, vas-ō-mō′ter) Area within the medulla oblongata that regulates the diameter of blood vessels by way of the sympathetic nervous system.

vasomotor tone Relatively constant frequency of sympathetic impulses that keep blood vessels partially constricted in the periphery.

vasopressin (vā-sō-pres′in, vas-ō-pres′in) *See* antidiuretic hormone.

vellus hair (vel′ŭs) [L., fleece] Short, fine, usually unpigmented hair that covers the body except for the scalp, eyebrows, and eyelids. Much of the vellus is replaced at puberty by terminal hairs.

venous capillary (vē′nŭs) Capillary opening into a venule.

venous return Volume of blood returning to the heart.

venous sinus Endothelium-lined venous channel in the dura mater that receives cerebrospinal fluid from the arachnoid granulations.

ventilation (ven-ti-lā′shŭn) [L., *ventus,* the wind] Movement of gases into and out of the lungs.

ventral root (ven′trăl) Motor (efferent) root of a spinal nerve.

ventricle (ven′tri-kl) [L., *venter,* belly] Chamber of the heart that pumps blood into arteries (i.e., the left and right ventricles); in the brain, a fluid-filled cavity.

ventricular diastole (ven-trik′ū-lăr) Dilation of the heart ventricles.

ventricular systole Contraction of the ventricles.

venule (ven′ool, vē′nool) Minute vein, consisting of endothelium and a few scattered smooth muscles, that carries blood away from capillaries.

vermiform appendix (ver′mi-fōrm) [L., *vermis,* worm + *forma,* form; appendage] Wormlike sac extending from the blind end of the cecum.

vesicle (ves′i-kl) [L., *vesica,* bladder] Small sac containing a liquid or gas, such as a blister in the skin or an intracellular, membrane-bounded sac.

vestibular fold (ves-tib′ū-lăr) One of two folds of mucous membrane stretching across the laryngeal cavity from the angle of the thyroid cartilage to the arytenoid cartilage superior to the vocal cords; helps close the glottis; also called false vocal cord.

vestibular membrane Membrane separating the cochlear duct and the scala vestibuli.

vestibule (ves′ti-bool) [L., antechamber, entrance court] Anterior part of the nasal cavity just inside the external nares that is enclosed by cartilage; space between the lips and the alveolar processes and teeth; middle region of the inner ear containing the utricle and saccule; space behind the labia minora containing the openings of the vagina, urethra, and vestibular glands.

vestibulocochlear nerve (ves-tib′ū-lō-kok′lē-ăr) Nerve formed by the cochlear and vestibular nerves; extends to the brain.

villus; pl. villi (vil′ŭs, vil′ī) [L., shaggy hair (of beasts)] Projection of the mucous membrane of the intestine; leaf-shaped in the duodenum; becomes shorter, more finger-shaped, and sparser in the ileum.

visceral (vis′er-ăl) Relating to the internal organs.

visceral pericardium (per′i-kar′dē-ŭm) Serous membrane covering the surface of the heart; also called epicardium.

visceral peritoneum (per′i-tō-nē′ŭm) [Gr., *periteino,* to stretch over] Layer of peritoneum covering the abdominal organs.

visceral pleura (vis′er-ăl plūr′ă) Serous membrane investing the lungs and dipping into the fissures between the several lobes.

visceroreceptor (vis′er-ō-rē-sep′tŏr) Sensory receptor associated with the organs.

viscosity (vis-kos′i-tē) [L., *viscosus,* viscous] Resistance to flow or alteration of shape by any substance as a result of molecular cohesion.

visual cortex (vizh′oo-ăl) Area in the occipital lobe of the cerebral cortex that integrates visual information and produces the sensation of vision.

visual field Area of vision for each eye.

vital capacity (vī-tăl) Greatest volume of air that can be exhaled from the lungs after a maximum inspiration.

vitamin (vīt′ă-min) [L., *vita,* life + amine] One of a group of organic substances present in minute amounts in natural foodstuffs that are essential to normal metabolism; insufficient amounts in the diet may cause deficiency diseases.

vitamin D Fat-soluble vitamin produced from precursor molecules in skin exposed to ultraviolet light; increases calcium and phosphate uptake from the intestines.

vitreous humor (vit′rē-ŭs) Transparent, jelly-like material that fills the space between the lens and the retina.

Volkmann's canal Named for German surgeon Richard Volkmann (1830–1889). Canal in bone containing blood vessels; not surrounded by lamellae; runs perpendicular to the long axis of the bone and the haversian canals, interconnecting the latter with each other and the exterior circulation.

vulva (vŭl′vă) [L., wrapper or covering, seed covering, womb] External genitalia of the female, composed of the mons pubis, the labia majora and minora, the clitoris, the vestibule of the vagina and its glands, and the opening of the urethra and of the vagina. Also called pudendum.

W

water-soluble vitamin Vitamin, such as B complex and C, that is absorbed with water from the intestinal tract.

white matter Bundles of parallel axons with their associated sheath in the central nervous system.

white pulp Part of the spleen consisting of lymphatic nodules and diffuse lymphatic tissue; associated with arteries.

white ramus communicans; pl. rami communicantes (rā′mŭs kō-mū′nĭ-kans, rā′mī kō-mū-nī-kan′tēz) Connection between a spinal nerve and a sympathetic chain ganglion through which myelinated preganglionic axons project.

wisdom tooth Third molar tooth on each side in each jaw.

X

xiphoid (zi′foyd) [Gr., *xiphos,* sword] Sword-shaped, with special reference to the sword tip; the inferior part of the sternum.

X-linked Gene located on an X chromosome.

Y

yellow marrow (mar′o) Connective tissue filling the cavities of bones; consists primarily of reticular fibers and fat cells; replaces red marrow in long bones and the skull.

Y-linked Gene located on a Y chromosome.

yolk sac (yōk, yōlk) Highly vascular layer surrounding the yolk of an embryo.

Z

Z disk Delicate, membranelike structure found at each end of a sarcomere to which actin myofilaments attach.

zona fasciculata (zō′nă fa-sik′ū-lă′tă) [L., *zone,* a girdle, one of the zones of the sphere] Middle layer of the adrenal cortex that secretes cortisol.

zona glomerulosa (glō-măr-ū-lōs-ă) Outer layer of the adrenal cortex that secretes aldosterone.

zona pellucida (pe-lū′sĭ-dă) Layer of viscous fluid surrounding the oocyte.

zona reticularis (rē-tik′ū-lar′is) Inner layer of the adrenal cortex that secretes androgens and estrogens.

zonula adherens (zō′nū-lă ad-her′enz) [L., a small zone; adhering] Small zone holding or adhering cells together.

zonula occludens (ō-klūd′enz) [L., occluding] Junction between cells in which the plasma membranes may be fused; occludes or blocks off the space between the cells.

zygomatic (zī-gō-mat′ik) [Gr., *zygon,* yoke] Yoking or joining; bony arch created by the junction of the zygomatic and temporal bones.

zygote (zī′gōt) [Gr., *zygotos,* yoked] Diploid cell resulting from the union of a sperm cell and an oocyte.

PHOTO CREDITS

Chapter 1

Opener: © Quest/Science Photo Library/Photo Researchers, Inc.; **1.1:** © Bart Harris/CORBIS; **Figure A:** © Omikron/Photo Researchers, Inc.; **Figure B:** © Bernard Benoit/Science Photo Library/Photo Researchers, Inc.; **Figure C:** © Scott Camazine & Sue Trainor; **Figure D:** © D. P. M. Ribotsky/Custom Medical Stock Photo; **Figure E:** © ISM/Phototake NYC; **Figure F:** © Photodisc; **Figure G:** © McGill University/CNRI/Phototake; **1.10 both, 1.11a&b, 1.13a:** © McGraw-Hill Higher Education, Inc./Eric Wise, photographer; **1.13b–d:** © R. T Hutchings.

Chapter 2

Opener: © Prof. P. Motta/Dept. of Anatomy/University "La Sapienza", Rome/Science Photo Library/Custom Medical Stock Photo; **2.4c:** © Trent Stephens; **2.15b:** © Barry King/Tom Stack & Assocs.

Chapter 3

Opener: © Dennis Kunkel Microscopy, Inc.; **3.2b:** © Don W. Fawcett/Photo Researchers, Inc.; **3.23b:** © Manfred Kage/Peter Arnold, Inc.; **3.26b:** © Courtesy of Dr. Birgit H. Satir; **3.27b:** © Don Fawcett/Photo Researchers, Inc.; **3.28b:** © Don Fawcett/Photo Researchers, Inc.; **3.28c:** © Bernard Gilula/Photo Researchers, Inc.; **3.29:** Photo © and courtesy of Dr. Victoria Foe, from *Molecular Biology of the Cell* by Bruce Alberts et al, Garland Publishing, 1994. **3.31b:** © J. David Robertson, from Charles Flickinger, Medical Cell Biology, Philadelphia; **3.32b:** © Robert Bollender/Don Fawcett/Visuals Unlimited; **3.35b:** © Don Fawcett/Visuals Unlimited; **3.36b:** © Biology Media/Photo Researchers, Inc.; **3.37b:** © E. de Harven/Photo Researchers, Inc.; **3.37c:** © Dr. Gopal Murti/Science Photo Library/Photo Researchers, Inc.; **3.39b:** © Courtesy Susumi Ito, Ph.D..; **3.47:** © Ed Reschke; **3.48:** © Norman Lightfoot/Photo Researchers, Inc. **3.49:** © Dr. M.A. Ansary/Photo Researchers, Inc.; **3.50:** © CNRI/Science Photo Library/Photo Researchers, Inc.

Chapter 4

Opener: © Prof. P. Motta & E. Vizza/Science Photo Library/Photo Researchers, Inc.; **4.1b:** © Victor Eroschenko; **4.1c:** © Ed Reschke; **Table 4.2a:** © Dr. Fred Hossler/Visuals Unlimited; **Table 4.2b&c, Table 4.3a&b:** © Victor Eroschenko; **Table 4.3c:** © R. Kessel/Visuals Unlimited; **Table 4.4a&b(both), Table 4.7a&b:** © Victor Eroschenko; **Table 4.8:** © Ed Reschke; **Table 4.9a1:** © Victor Eroschenko; **Table 4.9a2:** © Ed Reschke/Peter Arnold; **Table 4.9b1&b2, Table 4.9c1:** © Victor Eroschenko; **Table 4.9c2:** © Ed Reschke/Peter Arnold; **Table 4.9d1&d2:** © Ed Reschke; **Table 4.10a:** © Carolina Biological Supply/Phototake; **Table 4.10b:** © Victor Eroschenko; **Table 4.11a:** © Carolina Biological Supply/Phototake; **Table 4.11b&c, Table 4.11c, Table 4.12a:** © Victor Eroschenko; **Table 4.12b:** © Trent Stephens; **Table 4.13a&b, Table 4.15a&b:** © Ed Reschke; **Table 4.15c:** © Victor Eroschenko; **Table 4.16a&b, 4.6:** © Trent Stephens.

Chapter 5

Opener: © Dennis Kunkel Microscopy, Inc.; **5.2a:** © Alan Stevens, University of Nottingham, Med School, UK; **5.2b:** © The Bergman Collection; **Figure Aa:** © Dr. P. Marazzi/Photo Researchers, Inc.; **Figure Ab:** © Caliendo/Custom Medical Stock Photo; **Figure Ac:** © Thomas B. Habif; **5.10a:** Courtesy of A. M. Kligman, Professor of Dermatology, University of Pennsylvania School of Medicine, Philadelphia, PA; **5.10b:** Edward Curtis photo from Library of Congress; Courtesy of A. M. Kligman, Professor of Dermatology, University of Pennsylvania School of Medicine, Philadelphia, PA. **Figure Da:** © James Stevenson/Science Photo Library/Photo Researchers, Inc.; **Figure Db:** © Stan Levy/Photo Researchers, Inc.

Chapter 6

Opener: © Prof. P. Motta/Dept. of Anatomy/University "La Sapienza", Rome/Science Photo Library/Photo Researchers, Inc.; **6.1:** © Ed Reschke; **6.2a–c:** © Trent Stephens; **6.3c:** © Bio-Photo Assocs/Photo Researchers, Inc.; **Figure A:** © Hilt & Cogburn, Manual of Orthopedics; **6.5:** © Robert Caladine/Visuals Unlimited; **6.6b:** © Trent Stephens; **6.10(1):** © Victor Eroschenko; **6.10(2):** © R. Kessel/Visuals Unlimited; **6.10(3):** © Victor Eroschenko; **6.11:** © Visuals Unlimited; **6.13a:** © Ed Reschke/Peter Arnold, Inc.; **6.13b:** © Bio-Photo Assocs/Photo Researchers, Inc.; **6.15:** © J. M. Booher; **Figure B:** © Ewing Galloway, Inc./Index Stock; **Figure Da:** © Princess Margaret Rose Orthopaedic Hospital/Science Photo Library/Photo Researchers, Inc.; **Figure Db:** © Dr. Michael Klein/Peter Arnold, Inc.

Chapter 7

Opener: © Fred Hossler/Visuals Unlimited; **7.5, 7.7:** © McGraw-Hill Higher Education, Inc./Eric Wise, photographer; **7.10c&d:** © Jupiter Media/Alamy; **Figure A:** © Brashear & Beverly, Handbook of Orthopedic Surgery; **Table 7.9c&d:** © Trent Stephens; **7.14:** © McGraw-Hill Higher Education, Inc./Eric Wise, photographer; **7.16g, 7.17c, 7.18c, 7.20c:** © Trent Stephens; **7.21:** © McGraw-Hill Higher Education, Inc./Eric Wise, photographer; **7.23d:** © Trent Stephens; **7.27, 7.31, 7.36:** © McGraw-Hill Higher Education, Inc./Eric Wise, photographer.

Chapter 8

Opener: © CNRI/Phototake; **8.14, 8.15a–c, 8.16, 8.17:** © McGraw-Hill Higher Education, Inc./Eric Wise, photographer; **8.18a:** © McGraw-Hill Companies/Jill Braaten, Photographer; **8.18b, 8.19, 8.20, 8.21, 8.22, 8.23, 8.24, 8.25, 8.26:** © McGraw-Hill Higher Education, Inc./Eric Wise, photographer; **8.31e:** © R. T. Hutchings; **Figure Ca:** © James Stevenson/Science Photo Library/Photo Researchers, Inc.; **Figure Cb:** © CNRI/Science Photo Library/Photo Researchers, Inc.

Chapter 9

Opener: © Professors P. M. Motta, P. M. Andrews, K. R. Porter & J. Vial/Science Photo Library/Photo Researchers, Inc.; **9.1:** © Ed Reschke; **9.5a:** © Richard Rodewald; **9.11c:** © Fred Hossler/Visuals Unlimited; **9.17b:** © Don Fawcett/Photo Researchers, Inc.; **9.22:** © Victor Eroschenko; **Figure A:** © Andrew J. Kornberg; **Figure B:** © Dr. Richard Kessel/Visuals Unlimited; **Figure C:** © Roberta Seidman.

Chapter 10

Opener: © Eye of Science/Science Photo Library/Photo Researchers, Inc.; **10.6b, 10.7b:** © McGraw-Hill Companies/Jill Braaten, Photographer; **10.10a–d:** © McGraw-Hill Higher Education, Inc./Eric Wise, photographer; **10.19b:** © McGraw-Hill Companies/Jill Braaten, Photographer; **10.23b:** © Christine Eckel; **10.23c:** © McGraw-Hill Companies/Jill Braaten, Photographer; **10.24b:** © Christine Eckel; **10.24c:** © McGraw-Hill Companies/Jill Braaten, Photographer; **10.26b:** © McGraw-Hill Higher Education, Inc./Rebecca Gray, photographer/Don Kincaid, dissections; **10.26c, 10.28d:** © McGraw-Hill Companies/Jill Braaten, Photographer; **10.29c:** © McGraw-Hill Higher Education, Inc./Rebecca Gray, photographer/Don Kincaid, dissections; **10.29d, 10.30c:** © McGraw-Hill Companies/Jill Braaten, Photographer; **10.32c** © Trent Stephens; **10.33b:** © Christine Eckel; **10.33c:** © McGraw-Hill Companies/Jill Braaten, Photographer; **10.34:** © McGraw-Hill Higher Education, Inc./Rebecca Gray, photographer/Don Kincaid, dissections; **10.36:** © McGraw-Hill Companies/Jill Braaten, Photographer; **10.38c:** © Christine Eckel; **10.38d, 10.39d** © McGraw-Hill Companies/Jill Braaten, Photographer; **Figure A:** © Duomo/ Paul J. Sutton.

Chapter 11

Opener: © Dennis Kunkel Microscopy, Inc.

Chapter 12

Opener: © Loyola University Medical Education Network (LUMEN). (http://www.meddean.luc.edu/lumen/); **12.3:** © Ed Reschke/Peter Arnold, Inc.; **12.15b:** © McGraw-Hill Higher Education, Inc./Rebecca Gray, photographer/Don Kincaid, dissections.

Chapter 13

Opener: © Quest/Science Photo Library/Photo Researchers, Inc.; **13.1:** © Courtesy of Branislav Vidic; **13.6a–c:** © McGraw-Hill Higher Education, Inc./Rebecca Gray, photographer/Don Kincaid, dissections; **13.8a:** © R. T. Hutchings; **13.8b, 13.9b:** © McGraw-Hill Higher Education, Inc./Rebecca Gray, photographer/Don Kincaid, dissections; **13.10b&c:** © Trent Stephens.

Chapter 14

Opener: © Omikron/Photo Researchers, Inc.; **14.3:** © McGraw-Hill Higher Education, Inc./Eric Wise, photographer; **14.20b:** © Marcus Raichle, MD, Washington University School of Medicine; **Figure Ba:** © Howard J. Radzyner/Phototake; **Figure Bb:** © CNRI/Science Photo Library/Photo Researchers, Inc.

Chapter 15

Opener: © Mireille Lavigne-Rebillard, INSERM Unit 254, Montpellier ("from the site Promenade around the cochlea (http://www.iurc.montp.inserm.fr/cric/audition/fran%E7ais/cochlea/fcochlea.htm)" by R. Pujol, S Blatrix and T. Pujol, CRIC, University of Montpellier/INSERM; **15.4h:** © Omikran/Photo Researchers, Inc.; **15.7:** © McGraw-Hill Higher Education, Inc./Eric Wise, photographer; **15.11:** © McGraw-Hill Higher Education, Inc./Rebecca Gray, photographer/Don Kincaid, dissections; **15.14a:** © A. L. Blum/Visuals Unlimited; **15.17a:** © Steve Gschmeissner/Photo Researchers, Inc.; **15.22d:** © McGraw-Hill Higher Education, Inc./Rebecca Gray, photographer/Don Kincaid, dissections; **Figure A:** © National Eye Institute, Bethedsa, MD; **Figure C:** Reproduced from Isihara's Tests for Color Blindness published by Kanehara & Co., Ltd., Tokyo, Japan, but tests for color blindness cannot be conducted with this material. For accurate testing, the original plates should be used.; **Figure Da–c:** © National Eye Institute, Bethedsa, MD; **15.24:** © McGraw-Hill Higher Education, Inc./Eric Wise, photographer; **15.28a:** Courtesy of R. A. Jacobs and A. J. Hudspeth; **15.28b:** © Fred Hossler/Visuals Unlimited; **15.29b:** Courtesy of A. J. Hudspeth; **15.29c:** © David Corey, Harvard Medical School; **15.35d** © Susumu Nishinag/Photo Researchers, Inc.; **15.36a&b:** © Trent Stephens; **15.37d:** © Dr. Richard Kessel & Dr. Randy Kardon/Tissues and Organs/Visuals Unlimited; **15.38b:** © Jerry Wachter/Photo Researchers, Inc.

Chapter 16

Opener: © Innserspace Imaging/Science Photo Library/Photo Researchers, Inc.

Chapter 17

Opener: © Quest/Science Photo Library/Photo Researchers, Inc.

Chapter 18

Opener: © Dr. L. Orci, University of Geneva/Science Photo Library/Photo Researchers, Inc.; **18.8c. 18.11b, 18.13b:** © Victor Eroschenko; **Figure A, 18.16:** © Bio-Photo Assocs/Photo Researchers, Inc.; **Figure B (Boy):** © Mark Clark/Science Photo Library/Photo Researchers, Inc.; **Figure B (Kit):** © Visuals Unlimited.

Chapter 19

Opener: © Dennis Kunkel Microscopy, Inc.; **19.1:** © McGraw-Hill Higher Education, Inc./Eric Wise, photographer; **19.3a:** © National Cancer Institute/Science Photo Library/Photo Researchers, Inc.; **Figure A:** © Dr. Stanley Flegler/Visuals Unlimited; **19.7:** © Ed Reschke; **19.8a–e:** © Victor Eroschenko; **19.10:** © Oliver Meckes/Photo Researchers, Inc.

Chapter 20

Opener: © David M. Phillips/Photo Researchers, Inc.; **20.5b:** © R. T. Hutchings; **20.8a & b:** © McMinn & Hutchings, Color Atlas of Human Anatomy/Mosby; **20.12b:** © Ed Reschke; **20.20:** © Terry Cockerham/Cynthia Alexander/Synapse Media Productions; **Figure A:** © Hank Morgan/Science Source/Photo Researchers, Inc.

Chapter 21

Opener: © Prof. P. Motta/Dept. of Anatomy/University "La Sapienza", Rome/Science Photo Library/Photo Researchers, Inc. **21.5:** © Visuals Unlimited.

Chapter 22

Opener: © Dennis Kunkel Microscopy, Inc.; **22.4:** © Victor Eroschenko; **22.6b:** © Trent Stephens; **22.7d:** © McGraw-Hill Higher Education, Inc./APR; **22.8b:** © Trent Stephens; **Figure A:** © Kenneth Greer/Visuals Unlimited.

Chapter 23

Opener: © David Phillips/Photo Researchers, Inc.; **23.2b:** © The McGraw-Hill Companies, Inc./Photo and dissection by Christine Eckel; **23.4b:** © CNRI/Phototake; **23.5a:** © John Cunningham/Visuals Unlimited; **23.5c:** © Ed Reschke/Peter Arnold, Inc.; **23.6b:** © Walter Reiter/Phototake; **23.7b:** © Carolina Biological Supply Company/Phototake; **23.9b:** © Branislav Vidic; **Figure Aa&b:** © Custom Medical Stock Photo.

Chapter 24

Opener: © Dennis Kunkel Microscopy, Inc.; **24.5b & c:** © McGraw-Hill Higher Education, Inc./Rebecca Gray, photographer/Don Kincaid, dissections; **24.9c:** © Ed Reschke; **24.11c:** © Victor Eroschenko; **24.16d:** © David M. Phillips/Visuals Unlimited; **Figure A:** © SIU Biomedical/Photo Researchers, Inc.; **24.25b:** © CNRI/Science Photo Library/Photo Researchers, Inc.

Chapter 25

Opener: © Professors P. Motta & T. Naguro/Science Photo Library/Photo Researchers, Inc; **25.17:** © Kyle Rothenborg/Pacific Stock.

Chapter 26

Opener: © Dennis Kunkel Microscopy, Inc.; **26.3b:** © McGraw-Hill Higher Education, Inc./Rebecca Gray, photographer/Don Kincaid, dissections.

Chapter 27

Opener: © Custom Medical Stock Photo.

Chapter 28

Opener: © Prof. P. Motta & S. Makabe/Science Photo Library/Photo Researchers, Inc.; **28.3a:** © Biodisc/Visuals Unlimited; **28.3b:** © Ken Wagner/Phototake NYC; **28.3c:** © Dennis Kunkel/Phototake NYC; **28.20(1):** © Educational Images/Custom Medical Stock Photo; **28.20(2) & (3):** © Biophoto Associates/Photo Researchers, Inc.; **Figure Ba:** © M. Long/Visuals Unlimited; **Figure Bb:** © Ray Ellis/Photo Researchers, Inc.; **Figure Bc:** © Hank Morgan/Photo Researchers, Inc.; **Figure Bd:** © McGraw-Hill Higher Education; **Figure Be:** © McGraw-Hill Higher Education Group, Inc./Bob Coyle photographer.

Chapter 29

Opener & 29.2a–d: © Dr. Yorgus Nikas/Jason Burns/Phototake; **29.10:** © John Giannicchi/Photo Researchers, Inc.; **Figure A:** © NMSB/Custom Medical Stock Photo; **29.17a–c:** © Petit Format/Nestle/Photo Researchers, Inc.; **29.23:** © Will & Deni McIntyre/Photo Researchers, Inc.

INDEX

Note: Page numbers followed by *f* or *t* indicate figures and tables, respectively.

Selected Abbreviations

α alpha
ACE angiotensin-converting enzyme
acetyl-CoA acetyl coenzyme A
ACh acetylcholine
ADH antidiuretic hormone
ADP adenosine diphosphate
ANH atrial natriuretic hormone
ANS autonomic nervous system
apo E apolipoprotein E
ATP adenosine triphosphate
AV atrioventricular beta
B̄COP blood colloid osmotic pressure
BMI body mass index
BMR basal metabolic rate
BP blood pressure
BPG 2,3-bisphosphoglycerate
bpm beats per minute
BUN blood urea nitrogen
$C_6H_{12}O_6$ glucose
$Ca_{10}(PO_4)_6(OH)_2$ hydroxyapatite
Ca^{2+} calcium ion
cal calorie
cAMP cyclic adenosine monophosphate
CBC complete blood count
cGMP cyclic guanosine monophosphate
CH_3COOH acetic acid
Cl^- chloride ion
CNS central nervous system
CO cardiac output
CO carbon monoxide
CO_2 carbon dioxide
—COOH carboxyl group
COX-1 cyclooxygenase-1
CP capsule pressure
CRH corticotrophin-releasing hormone
CSF cerebrospinal fluid
DAG diacylglycerol
DNA deoxyribonucleic acid
ECG or EKG electrocardiogram
EEG electroencephalogram
EGF epidermal growth factor
ENS enteric nervous system
EPSP excitatory postsynaptic potential
FAD flavin adenine dinucleotide
$FADH_2$ reduced flavin adenine diphosphate
Fe^{2+} iron ion
FEV_1 forced expiratory volume in one second
FGF fibroblast growth factor
FSH follicle-stimulating hormone gamma
ḡ gram
GABA gamma-aminobutyric acid
GCP glomerular capillary pressure
GDP guanosine diphosphate
GFR glomerular filtration rate
GH growth hormone
GHIH growth hormone-inhibiting hormone
GHRH growth hormone-releasing hormone
GnRH gonadotropin-releasing hormone
GTP guanosine triphosphate
H^+ hydrogen ion
H_2CO_3 carbonic acid
H_2O water
H_2O_2 hydrogen peroxide
$H_2PO_4^-$ dihydrogen phosphate ion
HCG human chorionic gonadotropin
HCl hydrochloric acid
HCO_3^- bicarbonate ion
HDL high-density lipoprotein
Hg mercury
HIV human immunodeficiency virus
HLA human leukocyte antigen
HPO_4^{2-} monohydrogen phosphate ion
HR heart rate
I^- iodide ion
ICSH interstitial cell-stimulating hormone
IFP interstitial fluid pressure
Ig immunoglobulin
IP_3 inositol triphosphate
IPSP inhibitory postsynaptic potential
IU international units
K^+ potassium ion
kcal kilocalorie
kg kilogram
L liter
LDL low-density lipoprotein
LH luteinizing hormone
LHRH luteinizing hormone-releasing hormone
MAC membrane attack complex
MALT mucosa-associated lymphoid tissue
MAO monoamine oxidase
MAP mean arterial pressure
mEq milliequivalent
Mg^{2+} magnesium ion
MHC major histocompatibility complex
mOsm milliosmole
mRNA messenger ribonucleic acid
mV millivolt
Na^+ sodium ion
NaCl sodium chloride
NAD^+ nicotinamide adenine dinucleotide
NADH reduced nicotinamide adenine dinucleotide
$NaHCO_3$ sodium bicarbonate
NaOH sodium hydroxide
NFP net filtration pressure
$—NH_2$ amine group
NH_3 ammonia
NH_4^+ ammonium ion
NK cells natural killer cells
NO nitric oxide
O_2 oxygen
OH^- hydroxide ion
PAH *para*-aminohippuric acid
P_{ALV} alveolar pressure
P_B barometric air pressure
PGE prostaglandin E
PGF prostaglandin F
P_i inorganic phosphate
PIF prolactin-inhibiting factor
PIH prolactin-inhibiting hormone
PIP_2 phosphoinositol
PMNs polymorphonuclear neutrophils
PNS peripheral nervous system
PO_4^{3-} phosphate ion
P_{PL} pleural pressure
PR peripheral resistance
PRF prolactin-releasing factor
PRH prolactin-releasing hormone
PTH parathyroid hormone
RANKL receptor for activation of nuclear factor kappa B ligand
RAS reticular activating system
RBC red blood count
RDA recommended daily allowances
RDI reference daily intake
RhoGAM Rh_o (D) immune globulin
RMP resting membrane potential
RNA ribonucleic acid
SA sinoatrial
SV stroke volume
T_3 triiodothyronine
T_4 tĕtraiodothyronine
TF tissue factor
TGF-_ transforming growth factor beta
TMJ temporomandibular joint
TNF tumor necrosis factor
TRH thyroid-releasing hormone
TSH thyroid-stimulating hormone
V_A alveolar ventilation
VLDL very low-density lipoprotein
vWF von Willebrand factor
WBC white blood count

Prefixes, Suffixes, and Combining Forms

The ability to break down medical terms into separate components or to recognize a complete word depends on mastery of the combining forms (roots or stems) and the prefixes and suffixes that alter or modify their meanings. Common prefixes, suffixes, and combining forms are listed below in boldface type, followed by the meaning of each form and an example illustrating its use.

a-, an- without, lack of: *a*phasia (lack of speech), *an*aerobic (without oxygen)
ab- away from: *ab*ductor (leading away from)
-able capable: vi*able* (capable of living)
acou- hearing: *acou*stics (science of sound)
acr- extremity: *acro*megaly (large extremities)
ad- to, toward, near to: *ad*renal (near the kidney)
adeno- gland: *adeno*ma (glandular tumor)
-al expressing relationship: neur*al* (referring to nerves)
-algia pain: gastr*algia* (stomach pain)
angio- vessel: *angio*graphy (radiography of blood vessels)
ante- before, forward: *ante*cubital (before elbow)
anti- against, reversed: *anti*peristalsis (reversed peristalsis)
arthr- joint: *arthr*itis (inflammation of a joint)
-ary associated with: urin*ary* (associated with urine)
-asis condition, state of: homeost*asis* (state of staying the same)
auto- self: *auto*lysis (self breakdown)
bi- twice, double: *bi*cuspid (two cusps)
bio- live: *bio*logy (study of living)
-blast bud, germ: fibro*blast* (fiber-producing cell)
brady- slow: *brady*cardia (slow heart rate)
-c expressing relationship: cardia*c* (referring to heart)
carcin- cancer: *carcin*ogenic (causing cancer)
cardio- heart: *cardio*pathy (heart disease)
cata- down, according to: *cata*bolism (breaking down)
cephal- head: *cephal*ic (toward the head)
-cele hollow: blasto*cele* (hollow cavity inside a blastocyst)
cerebro- brain: *cerebro*spinal (referring to brain and spinal cord)
chol- bile: a*chol*ic (without bile)
cholecyst- gallbladder: *cholecyst*okinin (hormone causing the gallbladder to contract)
chondr- cartilage: *chondr*ocyte (cartilage cell)
-cide kill: bacteri*cide* (agent that kills bacteria)
circum- around, about: *circum*duction (circular movement)
-clast smash, break: osteo*clast* (cell that breaks down bone)
co-, com-, con- with, together: *co*enzyme (molecule that functions with an enzyme), *com*misure (coming together), *con*vergence (to incline together)
contra- against, opposite: *contra*lateral (opposite side)
crypto- hidden: *crypt*orchidism (undescended or hidden testes)
cysto- bladder, sac: *cysto*cele (hernia of a bladder)
-cyte-, cyto- cell: erythro*cyte* (red blood cell), *cyto*skeleton (supportive fibers inside a cell)
de- away from: *de*hydrate (remove water)
derm- skin: *derm*atology (study of the skin)
di- two: *di*ploid (two sets of chromosomes)
dia- through, apart, across: *dia*pedesis (ooze through)
dis- reversal, apart from: *dis*sect (cut apart)
-duct- leading, drawing: ab*duct* (lead away from)
-dynia pain: masto*dynia* (breast pain)
dys- difficult, bad: *dys*mentia (bad mind)
e- out, away from: *e*viscerate (take out viscera)
ec- out from: *ec*topic (out of place)
ecto- on outer side: *ecto*derm (outer skin)
-ectomy cut out: append*ectomy* (cut out the appendix)
-edem- swell: myo*edem*a (swelling of a muscle)
em-, en- in: *em*pyema (pus in), *en*cephalon (in the brain)
-emia blood: an*emia* (deficiency of blood)
endo- within: *endo*metrium (within the uterus)
entero- intestine: *enter*itis (inflammation of the intestine)
epi- upon, on: *epi*dermis (on the skin)
erythro- red: *erythro*cyte (red blood cell)
eu- well, good: *eu*phoria (well-being)
ex- out, away from: *ex*halation (breathe out)
exo- outside, on outer side: *exo*genous (originating outside)
extra- outside: *extra*cellular (outside the cell)
-ferent carry: af*ferent* (carrying to the central nervous system)
-form expressing resemblance: fusi*form* (resembling a fusion)
gastro- stomach: *gastro*dynia (stomach ache)
-genesis produce, origin: patho*genesis* (origin of disease)
gloss- tongue: hypo*gloss*al (under the tongue)
glyco- sugar, sweet: *glyco*lysis (breakdown of sugar)
-gram a drawing: myo*gram* (drawing of a muscle contraction)
-graph instrument that records: myo*graph* (instrument for measuring muscle contraction)
hem- blood: *hem*opoiesis (formation of blood)
hemi- half: *hemi*plegia (paralysis of half of the body)
hepato- liver: *hepa*titis (inflammation of the liver)
hetero- different, other: *hetero*zygous (different genes for a trait)
hist- tissue: *hist*ology (study of tissues)
homeo-, homo- same: *homeo*stasis (state of staying the same), *homo*logous (alike in structure or origin)
hydro- wet, water: *hydro*cephalus (fluid within the head)
hyper- over, above, excessive: *hyper*trophy (overgrowth)
hypo- under, below, deficient: *hypo*tension (low blood pressure)
-ia, -id expressing condition: neural*gia* (pain in nerve), flacc*id* (state of being weak)
-iatr- treat, cure: ped*iatr*ics (treatment of children)
-im not: *im*permeable (not permeable)
in- in, into: *in*jection (forcing fluid into)
infra- below, beneath: *infra*orbital (below the eye)
inter- between: *inter*costal (between the ribs)
intra- within: *intra*ocular (within the eye)
-ism condition, state of: dimorph*ism* (condition of two forms)